Immature Insects

Volume 2

Immature Insects

Volume 2

Edited by

Frederick W. Stehr
Department of Entomology
Michigan State University

KENDALL/HUNT PUBLISHING COMPANY
4050 Westmark Drive Dubuque, Iowa 52002

Credits

Figure 33.1 reproduced by permission of the Commonwealth Agricultural Bureaux, Farmham Royal, England.

Figures 34.1, 34.2, 34.4, 34.5, 34.6, 34.7, and 34.69a & b from he Insects of Australia (Melbourne University Press).

Figures 34.136-34.140 from Dybas, H.D. 1976. The Larval Characters of featherwing and limulodid beetles and their family relationships in the Staphylinoidea (Coleoptera: Ptiliidae and Limulodidae). Fieldiana Zoology, 70:29-78.

Figures 34.698 and 34.709. Reprinted with permission from the Annals of the Entomological Society of America; copyright 1969 and 1964 by the Entomological Society of America.

Figures 36.1a-c. Reprinted by permission of the editor from the Journal of Medical Entomology, 1976, Volume 13:107-111.

Many Figures used in Chapter 37 are taken from the Manual of Nearctic Diptera Volumes 1 and 2 (McAlpine, et al. 1981, 1987), and are reproduced with permission from the Minister of Supply and Services, Canada.

Figures 37.78, 37.79, 37.150, 37.151, 37.153c-h; Key Figures 37 numbers 32, 35, 37 and 51 from Merrit-Cummins: An Introduction to the Aquatic Insects in North America, 2nd Edition. © 1984 Kendall/Hunt Publishing Company, Dubuque, IA. All rights reserved.

Cover: Larvae of tortoise beetle, *Physonota alutacea* Boheman (Chrysomelidae: Cassidinae), on wild olive, *Cordia boissieri*, near Weslaco, Hidalgo Co., Texas, October 1983. Photograph courtesy of Charles W. Melton.

ISBN 978-0-7575-5611-1

Printed in the United States of America
10 9 8 7 6 5

Dedication

To the memory of Alvah Peterson (1888–1972) who first introduced immature insects to me, and whose pioneering books, *Larvae of Insects,* have been of immeasurable value in stimulating and facilitating their study.

Contents

Preface

The identification of immature insects and the location of information about them is a common and continuing problem for many. The literature available to accomplish this is widely scattered, limited to certain groups (e.g., aquatics), outdated, difficult to use, or non-existent. This book has been conceived to address all of these needs.

Specifically, it has been designed to serve as: 1) a textbook for courses on immature insects; 2) an introduction to and partial source of the literature on immature insects; 3) a means to identify the larvae of all orders to family; 4) a means to identify a number of common, economically important, or unusual species (largely North American) via the abundant illustrations, descriptions and/or selected keys; and 5) a source of information on the biology and ecology of the families and selected important species. The emphasis is on larvae, with minimal coverage of eggs and pupae since knowledge of the latter is *relatively* meager and the need to identify them is not great compared with larvae or adults.

All orders are covered, not just the Holometabola. Some non-holometabolous immatures can be identified using existing keys for adults, but others, especially groups where wing characters are used, cannot be identified satisfactorily. In addition, in existing keys to families for adults, the user is never informed whether the key will work for immatures.

Volume 1 contains:

Chapter 1. Introduction, including the use of "larvae" to include "nymphs," the use of "stemmata" and "ocelli," and the use of terms relating to kinds of metamorphosis and various life stages and forms.

Chapter 2. Techniques, including proper killing, fixation and preservation, permanent storage, study methods, and shipping procedures.

Chapter 3. Key to orders, covering immature insects, brachypterous and wingless adult insects, and other terrestrial and freshwater invertebrates that are likely to be collected with or possibly confused with immature insects. Selected illustrations of basic life forms and stages, and two plates showing some of the great diversity of form found in parasitic insect larvae are also provided.

The Following orders are covered in Volume 1.

Chapter

4. Protura	12. Isoptera	20. Plecoptera
5. Collembola	13. Mantodea	21. Psocoptera
6. Diplura	14. Grylloblattodea	22. Mallophaga
7. Microcoryphia	15. Phasmatodea	23. Anoplura
8. Thysanura	16. Orthoptera	24. Mecoptera
9. Ephemeroptera	17. Dermaptera	25. Trichoptera
10. Odonata	18. Embiidina	26. Lepidoptera
11. Blattodea	19. Zoraptera	27. Hymenoptera

Volume 2 covers the remaining orders and includes a glossary largely independent of Volume 1, as well as a comprehensive index, so Volume 2 can be used relatively independently of Volume 1, providing the user is sure of the order to which the immature belongs, and knows how to kill and preserve it properly.

A few abbreviations have been used. The terms prothorax, mesothorax, and metathorax, and various combinations and numbers of the abdominal segments are repeatedly used, especially in the descriptive sections and keys. The following abbreviations for these terms have been used in many cases whenever it did not appear that confusion would result: prothorax (**T1**), mesothorax (**T2**), metathorax (**T3**), abdominal segments 1 through 10 (**A1–10**), etc. A mature or full-grown larva is abbreviated (**f.g.l.**). Other abbreviations are defined in the sections where used.

The area emphasized is America north of Mexico, although some of the coverage (e.g., Coleoptera and Strepsiptera) is worldwide. Although the keys may work reasonably well for many specimens from other parts of the world (especially Eurasia), there are taxa that do not occur or are minimally represented in America north of Mexico and that will not key satisfactorily.

In the keys, direct character comparisons between couplet halves are used, but supplementary information is added at times. Illustrations have been abundantly used in the keys, and in the longer ones figures are inserted where referred to, and usually duplicated if used again in the key.

Quick recognition characters (diagnoses) for distinguishing between closely related or superficially similar groups are supplied for all taxa.

Family coverage begins with a presentation of relationships and the diagnosis. This is followed by a section on biology and ecology, a description of the family, and comments on a variety of topics such as economic importance, size, distribution and state of knowledge. An author-year index to selected literature for the family follows, and illustrations are provided for nearly all families, ranging from a single figure for small families to multiple figures for the larger ones.

The descriptions for each family are relatively comprehensive. These descriptions, along with the diagnoses and illustrations, are important tools for confirming identifications made with the keys. They are particularly important for immatures since many users will not have access to an adequate reference collection as they more often do for adults. In addition, virtually all entomology textbooks emphasize adults and offer little help in recognition of immatures below the order level, although some contain general descriptive statements, and some information on biology and common species is often presented.

The literature for each order is extensive but not complete; however, it is believed that sufficient publications have been included to enable the user to find the additional information necessary for further study.

Acknowledgments

Frederick W. Stehr

On behalf of the contributing authors of Volume 2, grateful acknowledgment is made to all of our colleagues, students and friends not specifically acknowledged below who made special efforts to collect specimens, loaned specimens, tested keys, reviewed manuscripts, answered questions and in many other ways made completion easier. Their help has made possible a more complete and more accurate publication. We deeply appreciate it.

Administrative support is essential, and I wish to thank James E. Bath, former Chairman, Department of Entomology, Michigan State University, for his support in the past, and Mark Scriber, present Chairman, for his continuing support. In addition, the Michigan Agricultural Experiment Station and the College of Natural Science have supported this project.

As in Volume 1, the illustrations come from many sources. Some are reproduced from other publications and are acknowledged where used. A considerable number of Alvah Peterson's original drawings have been used (some with slight modifications) by agreement with his widow, Helen Peterson; they are marked with Ⓟ and/or acknowledged in the legends. Contributing authors have been responsible for providing most of their own illustrations, either via originals or by other arrangements as acknowledged below or in the legends. I would like to thank Lana Tackett LT, formerly Michigan State University, for making a few drawings of Diptera, and Peter Carrington, Michigan State University, for drawing or redrawing a number of figures, mostly Diptera, sometimes on short notice. Other artists are acknowledged under the appropriate chapters below.

Many thanks to R. W. Merritt, Michigan State University, who reviewed parts of Vol. 1 and this volume, and who has made other helpful suggestions; to L. V. Knutson, Systematic Entomology Laboratory, USDA, who identified on short notice reared adults of *Villa lateralis* (Say) (Bombyliidae) so the larva could be illustrated; to G. W. Ulrich, University of California, Berkeley, who provided badly needed specimens of Phengodidae (Coleoptera) for the Key to Orders in Vol. 1; and to J. Capizzi, Oregon State University, for the loan of many *Delia* root maggots (Anthomyiidae).

Lois Connington Caprio has continued the fine proofreading she did on Vol. 1 by reviewing the entire set of galley proofs for Vol. 2. This has again made the final product significantly more polished than if only the contributing authors and editor had checked them. We deeply appreciate her dedication and exceptional work.

Acknowledgments by Chapter

Order Thysanoptera: B. S. Heming

I thank Brian Pitkin and Laurence Mound of the British Museum (Natural History) and Charles Gerdes of the Illinois Natural History Survey for supplying mounted immatures of many species. The manuscript was constructively criticised by George Ball and Douglas Craig of the University of Alberta; I thank them for their assistance.

Order Hemiptera: Thomas R. Yonke

I wish to express appreciation to John T. Polhemus, Englewood, Colorado, Lauren D. Anderson, University of California, Riverside, and A. G. Wheeler, Jr., Pennsylvania Department of Agriculture, Bureau of Plant Industry, who provided specimens and loans of selected references, and to Robert L. Blinn and Dennis D. Kopp for assisting in many ways during the preparation of the manuscript.

A special thank you is extended to my son, Thaddeus T. Yonke, who prepared 28 excellent illustrations of various families. Additional illustrations were prepared by several of my former students, including Robert L. Blinn, North Carolina State University, Dennis D. Kopp, North Dakota State University, Ronald D. Oetting, Georgia Experiment Station, Experiment, Georgia, Ronald D. Sheeley, and Stephen O. Swadener. Their contributions are gratefully recognized.

The manuscript was reviewed by Jay E. McPherson of Southern Illinois University. His critique has been most helpful, is greatly appreciated, and sincerely acknowledged.

Order Homoptera: Lois B. O'Brien

I wish to thank Steve Wilson, Central Missouri State University for the illustrations of the Fulgoroidea, and Dennis Kopp, North Dakota State University for the illustrations of the rest of the Auchenorrhyncha. I also wish to thank them and the following for critical reading of this chapter or parts thereof: Saul Frommer, University of California, Riverside; Murray Hanna, Michigan Department of Agriculture, Division of Plant Industry; Frank Mead, Florida Department of Agriculture; Thomas Moore, University of Michigan; Paul Oman, Oregon State University; and Thomas Wood, University of Delaware. Without the contributions of Manya Stoetzel and Douglass Miller this chapter could not have been written.

Order Megaloptera: H. H. Neunzig and J. R. Baker

Our special thanks to Elwin Evans, Michigan Department of Natural Resources, who generously gave us many excellent western North American Megaloptera larvae, thereby making possible a much more comprehensive treatment than would have otherwise been possible.

The figures for the key were prepared by Shuling Tung, and the habitus illustrations by Paul Kooaroon.

Orders Raphidioptera and Neuroptera: Catherine A. Tauber

I thank P. A. Adams, California State University, Fullerton, H. Aspöck, Hygiene-Institut der Universität Wien, L. A. Stange, Florida Department of Agriculture and Consumer Services, and M. J. Tauber, Cornell University for their help.

Order Coleoptera: John F. Lawrence

The following artists have drawn the figures indicated. All are from CSIRO, Division of Entomology unless otherwise indicated. Drawings for other contributors may be credited by means of logos or in the legends (the Alvah Peterson drawings for example).

A. A. Atkins: *Introduction:* figures 34.9–34.11; 34.13–34.17; 34.19–34.21; 34.23. *Key to Families:* figures 1–216. *Family Coverage:* figures 34.68h, i; 34.70b; 34.400c; 34.402a; 34.403; 34.407; 34.409c, e; 34.457d; 34.458b; 34.513; 34.520a; 34.527; 34.531–34.533; 34.537; 34.538a, b, c; 34.541; 34.551; 34.554c; 34.558; 34.561b; 34.564; 34.566; 34.626; 34.632a; 34.639a, b; 34.640a, b; 34.642a, b; 34.643a; 34.645; 34.648; 34.649; 34.653; 34.658; 34.662a; 34.664; 34.669a; 34.671; 34.710a; 34.712.

A. Hastings: *Introduction:* figures 34.12; 34.18; 34.22; 34.24–34.26. *Key to Families:* figure modifications. *Family Coverage:* figure 34.325.

S. P. Kim: *Family Coverage:* figures 34.323b, c, d, e; 34.387; 34.388; 34.392; 34.404a, b, c; 34.435a, b; 34.516a, b; 34.518a, b; 34.522; 34.523a, b; 34.524; 34.544; 34.548; 34.557a, b; 34.560; 34.562a, b; 34.565; 34.567; 34.627a, b; 34.628a, b; 34.636; 34.644a; 34.659.

S. Poulakis, Cambridge, Massachusetts: *Introduction:* figure 34.3. *Family Coverage:* figures 34.67a; 34.68a; 34.141; 34.168; 34.170a; 34.316a; 34.318; 34.323a; 34.386; 34.400a; 34.409a; 34.424; 34.457a; 34.458a; 34.465a; 34.468a; 34.483a; 34.484; 34.511a; 34.512; 34.517; 34.525a; 34.526; 34.534; 34.539; 34.542; 34.543; 34.545; 34.546; 34.549; 34.550; 34.554a; 34.556; 34.559; 34.561a; 34.563; 34.569a; 34.625; 34.630; 34.633; 34.634a; 34.641; 34.660; 34.668; 34.689; 34.695a, b; 34.696a, b; 34.713a, b, e, f, g; 34.714a, b, c, d; 34.726; 34.731; 34.734.

M. K. Ryan, Systematic Entomology Laboratory, USDA: *Family Coverage:* figures 34.672–34.676; 34.678–34.681; 34.683–34.685; 34.686c, d; 34.688; 34.690; 34.693.

L. Tackett, formerly Michigan State University: *Family Coverage:* figures 34.452x, y, z.

The following scanning electron micrographs were taken at the Museum of Comparative Zoology, Harvard University, by E. Seling, and were funded in part by NSF facilities grant BMS 7412494 and NSF research grant BMS 7502606. *Introduction:* figures 34.27–34.66. *Family Coverage:* figures 34.67b, c, d, e; 34.68b, c, d, e, f, g; 34.316b; 34.317; 34.319; 34.323f, g; 34.390; 34.391; 34.400b; 34.402b; 34.405a, b; 34.406a, b; 34.409b, d; 34.525a, b, c, d; 34.457b, c; 34.468b; 34.469; 34.470b, c; 34.483b; 34.492; 34.493; 34.503b; 34.511b, c; 34.520a, b; 34.521a, b; 34.525b; 34.554b; 34.569d; 34.632c; 34.634b, c, d; 34.635b; 34.637b; 34.638a, b; 34.640c; 34.643b; 34.644b; 34.652; 34.661a, b; 34.662b; 34.666a, b; 34.667d, e; 34.668b; 34.669b; 34.696c, d; 34.710b; 34.711b; 34.713c, d; 34.715c, d; 34.717b, c; 34.730a; 34.732a, b; 34.734c, d; 34.735; 34.737; 34.740; 34.741; 34.730a; 34.732a, b; 34.734c, d; 34.735; 34.737; 34.740; 34.741; 34.743; 34.744.

Photography and graphics were carried out by E. Alyta, R. Clegg, A. G. Edwards, J. Green and S. Smith, CSIRO, Division of Entomology, and by A. Coleman and P. Chandoha, Museum of Comparative Zoology, Harvard University.

The following persons donated specimens, made special collecting efforts or loaned specimens from the collections under their care:

A. Allen, Boise, Idaho.
D. M. Anderson, Systematic Entomology Laboratory, USDA.
F. Andrews, California Department of Agriculture, Sacramento.
J. Chemsak, University of California, Berkeley.
K. W. Cooper, University of California, Davis.
R. A. Crowson, University of Glasgow.
E. C. Dahms, Queensland Museum, Brisbane, Australia.
J. Donaldson, Queensland Department of Primary Industries, Indooroopilly, Australia.
J. T. Doyen, University of California, Berkeley.
H. S. Dybas (deceased), Field Museum of Natural History, Chicago.
H. Elliot, Tasmanian Forestry Commission, Hobart, Tasmania.
T. Erwin, Smithsonian Institution.
R. L. Fischer, Michigan State University.
N. Hayashi, Yokohama, Japan.
T. F. Hlavac, San Francisco, California.
K. Houston, Queensland Department of Primary Industries, Indooroopilly, Australia.
H. F. Howden, Carleton University, Ottawa.
P. Johnson, University of Idaho.
D. Kavanaugh, California Academy of Sciences.
J. Kingsolver, Systematic Entomology Laboratory, USDA.
C. Kitayama, University of California, Berkeley.
G. Kuschel, New Zealand Arthropod Collection, DSIR, Auckland, NZ.
H. B. Leach, California Academy of Sciences.
B. Mamaev, Department of Forest Protection, Moscow, USSR.
J. E. Marshall, British Museum (Natural History).
E. G. Matthews, South Australian Museum, Adelaide.
R. P. McMillan, Western Australian Museum, Perth.

G. Montieth, Queensland Museum, Brisbane, Australia.
B. P. Moore, CSIRO, Division of Entomology.
A. Neboiss, Museum of Victoria, Melbourne, Australia.
A. F. Newton, Jr., Field Museum of Natural History, Chicago.
C. W. O'Brien, Florida A & M University, Tallahassee.
T. K. Pal, Zoological Survey of India, Calcutta.
G. Parsons, Oregon State University.
S. B. Peck and J. Peck, Carleton University, Ottawa.
J. D. Pinto, University of California, Riverside.
F. Plaumann, Nova Teutonia, Brazil.
J. Powell, University of California, Berkeley.
E. Schlinger, University of California, Berkeley.
J. Scott, Museum of Comparative Zoology, Harvard University.
T. Sen Gupta, Zoological Survey of India, Calcutta.
D. Simberloff, Florida State University, Tallahassee.
T. Spilman, Systematic Entomology Laboratory, USDA.
F. W. Stehr, Michigan State University.
K. Stephan, Red Oak, Oklahoma.
M. K. Thayer, Field Museum of Natural History, Chicago.
G. Ulrich, University of California, Berkeley.
J. C. Watt, DSIR, Entomology Division, Auckland, NZ.
F. Werner, University of Arizona.
D. K. Young, University of Wisconsin.

The following provided miscellaneous assistance as indicated; all are from CSIRO, Division of Entomology unless otherwise indicated:

M. J. Dallwitz (DELTA system, computer advice).
E. Parker (advice on typesetting system).
W. Allen, T. Weir, M. L. Johnson, J. Pyke (technical assistance).
L. E. Watrous, Field Museum of Natural History, Chicago (advice).
J. S. Stribling, Ohio State University (advice).
P. B. Carne (read MS).
B. P. Moore (advice, read MS).
E. S. Nielsen (read MS).

Harvard University provided research support prior to September, 1977, and CSIRO since then. The National Science Foundation provided support through Facilities Grant BMS 7412494 and Research Grant 7502606.

Coleoptera Acknowledgments by Contributors

D. M. Anderson

The following figures in the family coverages were drawn by artists of the Systematic Entomology Laboratory, USDA:

M. K. Ryan: figures 34.851a, b, c; 34.852a, b, c; 34.853a, b, c; 34.854a, b, c; 34.855a, b; 34.857a, b, c, d; 34.858a, b; 34.859a, b, c, d; 34.860a, b, c, d; 34.861a, b, d, e; 34.862a, b; 34.863a, b, d; 34.864a, b, c, d; 34.865a, b, c, d; 34.866; 34.867a, b; 34.869a; 34.870a, b; 34.871a, b, c, d; 34.874a, b; 34.875a, b, c, d; 34.876a, b, c; 34.877a, b, c, d, e; 34.878a, b, c, d; 34.879b, d; 34.880a, b, c, d; 34.881a, b; 34.882a, b, c, d; 34.883a, b, c, d, e.

I. C. Feller: figures 34.869b, c, d; 34.873; 34.879a; 34.884.

T. B. Griswold: figures 34.861c; 34.863c; 34.868a; 34.872a, b, c, d.

D. D. Carlson

Much of the information in the family coverage for the Lucanidae, Passalidae, and Scarabaeidae is the work of P. O. Ritcher, Emeritus Professor, Oregon State University, who prepared the initial drafts.

K. W. Cooper

D. M. Anderson, Systematic Entomology Laboratory, USDA (loans).
L. E. Anderson, Duke University (identification of host mosses).
T. F. Hlavac, San Francisco (loans, advice).
J. F. Lawrence, CSIRO, Division of Entomology (loans, advice, and much help).
J. W. Stubblefield, formerly Harvard University (collection of fresh specimens in New Hampshire).

F. A. Lawson

I am most grateful to the following:

F. W. Stehr, Michigan State University (loans).
C. A. Triplehorn, Ohio State University (loans).
J. A. Wilcox, New York State Museum (loans).
K. Johnston, University of Wyoming (technical assistance).

A. F. Newton, Jr

The following donated specimens, made special collections, or loaned specimens from collections under their care:

D. M. Anderson, Systematic Entomology Laboratory, USDA.
R. S. Anderson, Texas A & M University.
J. S. Ashe, Field Museum of Natural History, Chicago.
C. Besuchet, Museum d'Histoire Naturelle, Geneva.
R. Fogel, Forest Service, USDA, Portland, Oregon.
P. M. Hammond, British Museum (Natural History).
H. E. Hinton (deceased), Bristol University.
P. J. Johnson, University of Wisconsin.
D. H. Kavanaugh, California Academy of Sciences.
J. F. Lawrence, CSIRO, Division of Entomology.
L. LeSage, Canadian National Collection, Ottawa.
B. M. Mamaev, Department of Forest Protection, Moscow, USSR.
S. B. Peck, Carleton University, Ottawa.
M. K. Thayer, Field Museum of Natural History, Chicago.
J. C. Watt, DSIR, Entomology Division, Auckland, NZ.
R. L. Wenzel, Field Museum of Natural History, Chicago.

G. S. Pfaffenberger

I would like to thank the following individuals for the loan of specimens:

D. H. Habeck, University of Florida.
C. A. Johansen, Washington State University.
C. D. Johnson, Northern Arizona University.
J. M. Kingsolver, Systematic Entomology Laboratory, USDA.
R. E. Woodruff, Florida Department of Agriculture.

My utmost appreciation is extended to C. D. Johnson for many thought-provoking and informative conversations and for reviewing this manuscript. I extend my gratitude to W. F. Barr and M. A. Brusven for their tutelage in immatures.

T. J. Spilman

The following figures in the Tenebrionidae were drawn by Molly K. Ryan, artist, Systematic Entomology Laboratory, USDA: 34.672a-d; 34.673a-d; 34.674a,b; 34.675a,b; 34.676a,b; 34.678a,b; 34.679a,b; 34.680a,b; 34.681a,b; 34.683a,b; 34.684a,b; 34.685a.b; 34.686c,d; 34.688a,b; 34.690a,b; 34.693a,b.

Order Strepsiptera: Marcos Kogan

I thank John Sherrod for most of the illustrations in this chapter and Jenny Kogan for keeping the Strepsiptera file updated. Both are staff of the Illinois Natural History Survey.

Order Siphonaptera: Robert E. Elbel

For the loan of larvae appreciation is expressed to: A. M. Barnes, B. W. Hudson, present members, and W. L. Jellison, Leo Kartmen, H. B. Morlan and H. E. Stark, former members of the U.S. Public Health Service; A. H. Benton, State University College, Fredonia New York; W. W. Cudmore, Chemeketa Community College, Salem, Oregon; D. Gettinger, University of Oklahoma; D. E. Johnson, formerly University of Utah; W. L. Krinsky, Yale University; R. E. Lewis, Iowa State University; R. G. Robbins, U.S. National Museum of Natural History; R. C. H. Shepherd, formerly Keith Turnbull Research Institute, Frankston, Victoria, Australia; M.-L. A. Tracy, formerly Immaculata Preparatory School, Washington, DC; and N. Wilson, University of Northern Iowa.

Special thanks are extended to L. T. Nielsen, Department of Biology, University of Utah, for providing space and facilities, and to R. L. C. Pilgrim, University of Canterbury, Christchurch, New Zealand for loan of larvae, for helpful advice and criticism, and for characterization of the larvae of certain species in his collection.

Order Diptera: B. A. Foote (and F. W. Stehr)

We thank Don Webb, Illinois Natural History Survey, for reviewing most of the family writeups of the lower Brachycera and some of the Nematocera. We also thank the dipterists at the United States National Museum for reviewing nearly the entire Diptera manuscript, according to their special areas of responsibility. These include from the Systematic Entomology Laboratory, USDA, R. J. Gagné, L. V. Knutson, A. L. Norrbom, R. V. Peterson, C. W. Sabrosky, G. C. Steyskal, F. C. Thompson, and N. E. Woodley; and from the Department of Entomology, Smithsonian Institution, W. N. Mathis. Their suggestions, modifications and additions have been most helpful, as has the loan or gift of a few specimens.

We (B. A. Foote and F. W. Stehr) owe a special acknowledgment to H. J. Teskey (retired) from the Biosystematics Research Institute, Agriculture Canada, who obtained permission for use of the fine illustrations from the Manual of Nearctic Diptera, Volumes 1 and 2, most of which were drawn by Mr. Idema. They are reproduced by permission of the Minister of Supply and Services Canada. Their use has made the illustrations for the Diptera immeasurably better and more comprehensive than would otherwise have been possible. We deeply appreciate that.

Gregory A. Dahlem

I deeply appreciate the comments and suggestions provided by William L. Downes, Jr., Michigan State University, on the larvae of the Sarcophagidae.

List of Contributors

D. M. ANDERSON, Systematic Entomology Laboratory, USDA, U.S. National Museum, Washington, D.C. 20560

J. R. BAKER, Department of Entomology, North Carolina State University, Raleigh, NC 27650

R. S. BEAL, Jr., Department of Biological Sciences, Northern Arizona University, Flagstaff, AZ 86011

E. C. BECKER, Biosystematics Research Institute, Agriculture Canada, Ottawa, Ontario, Canada K1A 0C6

R. T. BELL, Department of Zoology, University of Vermont, Burlington, VT 05405-0083

Y. BOUSQUET, Biosystematics Research Institute, Agriculture Canada, Ottawa, Ontario, Canada K1A 0C6

D. E. BRIGHT, Biosystematics Research Institute, Agriculture Canada, Ottawa, Ontario, Canada K1A 0C6

H. P. BROWN, Department of Zoology, University of Oklahoma, Norman, OK 73019

D. C. CARLSON, Environmental Specialist, 5229 Butterwood Circle, Orangevale, CA 91702

E. LUNA DE CARVALHO, R. do Mercado 28, 2725 Algueirao, Portugal

K. W. COOPER, Professor Emeritus, Department of Biology, University of California, Riverside, CA 92521

G. A. DAHLEM, Department of Entomology, Michigan State University, East Lansing, Michigan 48824

D. S. DENNIS, 5875 E. Weaver Circle, Englewood, CO 80111

J. R. DOGGER, Systematic Entomology Laboratory, USDA, U.S. National Museum, Washington, D.C. 20560

H. S. DYBAS (deceased), Field Museum of Natural History, Roosevelt Rd. at Lakeshore Dr., Chicago, IL 60605

R. E. ELBEL, Department of Biology, University of Utah, Salt Lake City, UT 84112

B. A. FOOTE, Department of Biological Sciences, Kent State University, Kent, OH 44240

D. E. FOSTER, Department of Entomology, Iowa State University, Ames, IA 50011

J. H. FRANK, Department of Entomology, University of Florida, Gainesville, FL 32611

B. S. HEMING, Department of Entomology, University of Alberta, Edmonton, Alberta, Canada T6G 2E3

D. H. KAVANAUGH, Department of Entomology, California Academy of Sciences, Golden Gate Park, San Francisco, CA 94118

M. KOGAN, Section of Economic Entomology, Illinois Natural History Survey, Champaign, IL 61820

D. M. LABELLA, Department of Entomology and Nematology, University of Florida, Gainesville, FL 32611

J. F. LAWRENCE, Division of Entomology, CSIRO-Box 1700, Canberra City, A.C.T. 2601, Australia

F. A. LAWSON, 4210 Grays Gable Rd., Laramie, WY 82070

L. LESAGE, Biosystematics Research Institute, Agriculture Canada, Ottawa, Ontario, Canada K1A 0C6

J. E. LLOYD, Department of Entomology and Nematology, University of Florida, Gainesville, FL 32611

D. R. MILLER, Systematic Entomology Laboratory, 11B111, ARS USDA, Beltsville, MD 20705

H. H. NEUNZIG, Department of Entomology, North Carolina State University, Raleigh, North Carolina 27650

A. F. NEWTON, Jr., Field Museum of Natural History, Roosevelt Rd. at Lakeshore Dr., Chicago, IL 60605

L. B. O'BRIEN, Laboratory of Aquatic Entomology, Florida A & M University, Tallahassee, FL 32307

G. S. PFAFFENBERGER, Department of Life Sciences, Eastern New Mexico University, Portales, NM 88130

H. REICHARDT (deceased), Museo de Zoologia, Universidade de São Paulo, São Paulo, Brazil

R. B. SELANDER, Department of Entomology, University of Illinois, Urbana, IL 61801

P. J. SPANGLER, Department of Entomology, Smithsonian Institution, Washington, D.C. 20560

T. J. SPILMAN, Systematic Entomology Laboratory, USDA, U.S. National Museum, Washington, D.C. 20560

T. A. STASNY, Department of Entomology, West Virginia University, Morgantown, WV 26505

F. W. STEHR, Department of Entomology, Michigan State University, East Lansing, MI 48824

M. B. STOETZEL, Systematic Entomology Laboratory, 11B111, ARS, USDA, Beltsville, MD 20705

C. A. TAUBER, Department of Entomology, Cornell University, Ithaca, NY 14853

H. J. TESKEY, Biosystematics Research Institute, Agriculture Canada, Ottawa, Ontario, Canada K1A 0C6

F. C. THOMPSON, Systematic Entomology Laboratory, USDA, U.S. National Museum, Washington, D.C. 20560

MANUEL G. DE VIEDMA, (deceased) Universidad Politecnica de Madrid, Escuela Tecnica Superior de, Ingenieros de Montes, Madrid, Spain

Q. D. WHEELER, Department of Entomology, Cornell University, Ithaca, NY 14853

T. R. YONKE, Department of Entomology, University of Missouri, Columbia, MO 65211

D. K. YOUNG, Department of Entomology, University of Wisconsin, Madison, WI 53706

Introduction

This introduction is intended to supplement that of Volume 1. Volume 2 is being published about three years after Volume 1 (1987). The original idea for putting these books together was generated about 20 years ago and evolved along the way. It may seem unusually long, but 15–20 years seems to be a rather "normal" gestation period for a large work on larvae in North America, since Böving and Craighead (1931) indicate that they began working together on their publication on Coleoptera larvae in 1915 when they both realized the other was working independently on a similar project, and Alvah Peterson (1948 and 1951) began working on his two volumes in the 1930s.

The importance of immatures cannot be overemphasized, since they are found just about everywhere, cause much more economic damage overall than adults, are key indicators of soil and water quality, are important decomposers, are key items in the diets of many vertebrates, especially young birds and fish, and are essential components of terrestrial and fresh water ecosystems. In addition, they have unusual uses such as forensic entomology (Smith 1986), and the data from immatures have made and will continue to make major contributions toward better classifications and better phylogenies of insect taxa.

The original classification of any group of insects has almost always been based exclusively on adults, but classifications and phylogenies will always be better when all possible data, including that from immatures, are used in their construction. A summary of the taxonomic significance of the characters of immature insects which is still useful was provided by Emden (1957). If we believe that a classification is a framework within which all of the data about organisms should be filed, then we must use all the available data in building that framework, if it is to be useful to all who seek to file or retrieve data.

It is obvious that most larvae of the Holometabola are adapted for living in very different habitats than the adults, and have evolved substantially different structures and behaviors than the adults. Despite this fact, there is relatively little discordance between suites of adult and larval characters, and many times the immatures have pointed the way toward the correct placement of a puzzling species or group, whether as a part of an existing taxon or as a new one.

At higher levels, the classification of the major suborders of the Hymenoptera, Diptera and Coleoptera works well for adults and larvae, and characters of immatures and adults have been used in developing the suborders. However, in the Lepidoptera, the two major suborders, Monotrysia and Ditrysia, do not separate as cleanly, perhaps because of the extreme specialization of many monotrysian larvae for a diversity of stem-boring and leaf-mining habits. Occasionally, as in the Siphonaptera in this book, there is some disagreement about classifications based on adult or larval characters, but there can be only one classification and it must include all life stages, so any differences must eventually be resolved.

A review of the contributions of the use of immature stages to classification and phylogeny would be a long chapter in itself, but a small sampling follows, including some contributions by contributors to these volumes. Edmunds and Allen (1966) have discussed the significance of larvae in the study of Ephemeroptera, where larvae are more readily available and where identification of larvae is needed much more often than adults. Immature Coccoidea (Homoptera), especially crawlers and second instar males, have been of great value in the study of their classification and evolution (Miller and Kosztarab 1979; Howell 1980), especially since the adult female scales are heavily sclerotized and have relatively few characters. For the Trichoptera, Wiggins (1981) has presented a strong review of the relevance of immatures to their systematics. Lepidoptera larvae, especially first instars, have been widely used in the establishment of classifications and phylogenetic relationships (Hinton 1946), and the placement of some groups has only been resolved after the larvae have become known (Kristensen and Nielsen 1983). In the Coleoptera, Crowson (1955) used larval data and all other available data in arriving at his classification, and he reiterated the theory that the larval and life cycle data were the best evidence for believing that the Coleoptera are most closely related to the Neuroptera, particularly the Megaloptera. Hymenoptera larvae have long been used, and Yuasa's larval work (1923) made an early contribution to symphytan systematics and evolution. Among the Apocrita, Michener (1953), Evans (1964), Bohart and Menke (1976), McGinley (1981) and Carpenter (1982) have all made extensive use of larvae in arriving at better classifications.

Individuals, both amateur and professional, have made important contributions to our immature collections by their long and dedicated work in rearing many species. The association of immatures with adults is without doubt one of the most important contributions that can be made by amateurs—it can be done with minimal equipment and some ingenuity, and without the aid of a good microscope as long as the immatures are properly killed, labeled and preserved. It must be emphasized that immatures of any life stage should not be placed with other immature stages or adults unless the association is known to be correct. All key data must be with all life stages, as well as an entry or a special label listing all the associated life stages. Log books are very useful, but tend to become misplaced, lost or unavailable, so as much data as is reasonable must be on the label(s). Ideally, all stages would be stored together, but this may not be practical if the adults are best preserved pinned and the immatures are best preserved in alcohol. A list of useful publications on techniques is given in Chapter 2 in Volume 1, to which can be added Steyskal et al. (1986).

We have a long way to go before the immatures are as well known as the adults. Those who study or are interested in immatures are urged to collect and rear larvae, or to obtain eggs from females when feasible and rear larvae (this is the best technique since the association is certain and larvae of all instars can be preserved). All life stages (including the cast larval and pupal skins) and the associated adults should be deposited in major museums for future study, and publications should require that voucher specimens of newly described life stages must be deposited in major museums. An increased knowledge of immatures will strengthen the science of entomology in many ways, but especially in the economic, ecological, systematic, and evolutionary arenas.

Literature cited:

Bohart, R. M., and A. S. Menke. 1976. Sphecid wasps of the world. A generic revision. Berkeley and Los Angeles: Univ. Calif. Press. 695 pp.

Böving, A. G., and F. C. Craighead. 1931. An illustrated synopsis of the principal larval forms of the order Coleoptera. Brooklyn Entomol. Soc. 351 pp.

Carpenter, J. M. 1982. The phylogenetic relationships and natural classification of the Vespoidea (Hymenoptera). Systematic Entomology 7:11–38.

Crowson, R. A. 1955. The natural classification of the families of Coleoptera. N. Lloyd, London. 187 pp.

Edmunds, G. F., Jr. and R. K. Allen. 1966. The significance of nymphal stages in the study of Ephemeroptera. Annals Ent. Soc. Amer. 59:300–303.

Emden, F. I. van. 1957. The taxonomic significance of the characters of immature insects. Annu. Rev. Ent. 2:91–106.

Evans, H. E. 1964. The classification and evolution of digger wasps as suggested by larval characters (Hymenoptera: Sphecoidea). Ent. News 75:225–237.

Hinton, H. E. 1946. On the homology and nomenclature of setae of lepidopterous larvae, with some notes on the phylogeny of the Lepidoptera. Trans. Roy Ent. Soc. Lond. 97:1–37.

Howell, J. O. 1980. The value of second-stage males in armoured scale insects (Diaspididae) phyletics. Israel J. Ent. 14:87–96.

Kristensen, N. P. and E. S. Nielsen. 1983. The *Heterobathmia* life history elucidated: immature stages contradict assignment to suborder Zeugloptera (Insecta, Lepidoptera). Sonderdruck aus Zeitschrift für Zool. Systematik u Evolutionsforschung 21:101–24.

McGinley, R. J. 1981. Systematics of the Colletidae based on mature larvae with phenetic analysis of apoid larvae (Hymenoptera: Apoidea) J. Kan. Ent. Soc. 53:539–552.

Michener, C. D. 1953. Comparative morphological and systematic studies of bee larvae with a key to the families of hymenopterous larvae. Univ. Kan. Sci Bull. 35:987–1102.

Miller, D. R. and M. Kosztarab. 1979. Recent advances in the study of scale insects. Annu. Rev. Ent. 24:1–27.

Peterson, A. 1948, 1951. Larvae of insects. An introduction to Nearctic species. Printed for the author by Edwards Bros., Ann Arbor, Mich. Pt. 1, 315 pp.; Pt. 2, 416 pp.

Smith, K. G. B. 1986. A manual of forensic entomology. Cornell Univ. Press, Ithaca, N.Y. 205 pp.

Steyskal, G. C., W. L. Murphy and E. M. Hoover (eds.) 1986. Insects and mites: techniques for collection and preservation. U.S. Dept. Agr., Agr. Res. Ser., Misc. Publ. 1443. 103 pp.

Wiggins, G. B. 1981. Considerations on the relevance of immature stages to the systematics of Trichoptera. Proc. 3rd Symp. on Trichoptera, G. P. Moretti (ed.). Series Entomologica 20:395–407.

Yuasa, H. 1923 (1922). A classification of the larvae of the Tenthredinoidea. Ill. Biol. Monogr. 7. 172 pp.

Order Thysanoptera

28

B. S. Heming
University of Alberta

THRIPS

Thrips are among the smallest insects, adults of most species rarely exceeding 2 mm. They appear to have evolved from litter-dwelling psocopteroid ancestors sometime in the upper Carboniferous or lower Permian and to constitute the sister group of the Hemiptera (Hennig, 1981).

DIAGNOSIS

Larvae of Thysanoptera (figs. 28.3–28.5; 28.8–28.10) are distinguished from the young of all other insects by their asymmetric, "punch and suck", mouthparts located in a mouthcone below the head (fig. 28.2) (see Heming, 1978 for details), by their lack of cerci, and by the protrusible bladders (aroliae) at the tips of their legs (Heming, 1972).

They are usually small (≂2 mm) and slender, from white to purple or scarlet, and are often gregarious. When viewed with a hand lens or the naked eye, both larvae I and II can be easily confused with larvae of cecidomyiid flies, and larvae I with larval or imaginal eriophyid mites.

The quiescent instars are not easily confused with the young of other insects except for the propupae and pupae of some male Coccoidea. They bear unsegmented antennae of various shapes (figs. 28.11–28.17). Long slender wing pads, of various extent depending on instar, are usually present but are absent from propupal phlaeothripids (fig. 28.13) and in propupae and pupae of all thrips that are wingless as adults.

BIOLOGY AND ECOLOGY

Although thrips are varied in feeding habits (predation, phytophagy, fungivory, pollen- and spore-feeding) and often occur in prodigious numbers in practically all habitats of the world, their ecological significance remains obscure (Lewis, 1973).

Larval thrips live on flowers and leaves, on or under bark of recently-dead branches, in leaf litter and associated with grass. They damage crops and ornamental plants by direct feeding, soiling with faeces, forming galls and, in a few cases, by transmitting tomato spotted wilt virus (Mound, 1973). Temperate grasslands and leaf litter of temperate and subtropical woodlands are particularly rich in species. Phytophagous and pollen-feeding thrips can be quite host-specific but, more commonly, are catholic in their feeding preferences.

Thrips usually have haplo-diploid reproduction (= arrhenotoky, in which males develop from unfertilized, haploid eggs and females from fertilized diploid eggs), and a type of metamorphosis intermediate between holometabolous and hemimetabolous (Lewis, 1973). Two active, feeding instars, the larvae I (figs. 28.3–28.5) and II (figs. 28.8–28.10), are followed by two or three quiescent, non-feeding instars—the propupae (figs. 28.11–28.13) and pupae (figs. 28.14–28.17). In Phlaeothripidae, there are two pupal instars, designated pupae I (fig. 28.14) and II (fig. 28.17). In some families, notably the Aeolothripidae, these non-feeding stages are contained within a cocoon spun by the larva II.

DESCRIPTION

Larvae of many Palearctic and some Holarctic and introduced species are known and can be keyed in Priesner (1928, 1964, 1964a), but few have been illustrated or adequately described and too much emphasis has been placed on color. Full and accurate descriptions of the structure of all instars are provided by Priesner (1964), and Ananthakrishnan and Sen (1980) presented keys to larvae II of Indian thrips to the subfamily level that will work for North American forms. The most important papers on larval thrips are those of Speyer and Parr (1941) on some English representatives of the suborder Terebrantia, of Miyazaki and Kudo (1986) on larvae of thrips occurring on solanaceous and cucurbitaceous crops, of Kirk (1987) on larvae of common flower thrips in Australia, and of Vance (1974) on some larval thripids of Illinois. Speyer and Parr (1941) presented a detailed description of integumentary structure and chaetotaxy of larvae, noted the common plan of the latter throughout the Terebrantia and developed a system of notation that, in modified form, I use here (figs. 28.1, 28.2). They also discovered a way for distinguishing between larvae I and II. In Vance (1974) can be found a detailed analysis of larval characteristics, keys to instar, family and some genera, and detailed descriptions and accurate illustrations of representatives of common terebrantian genera and species.

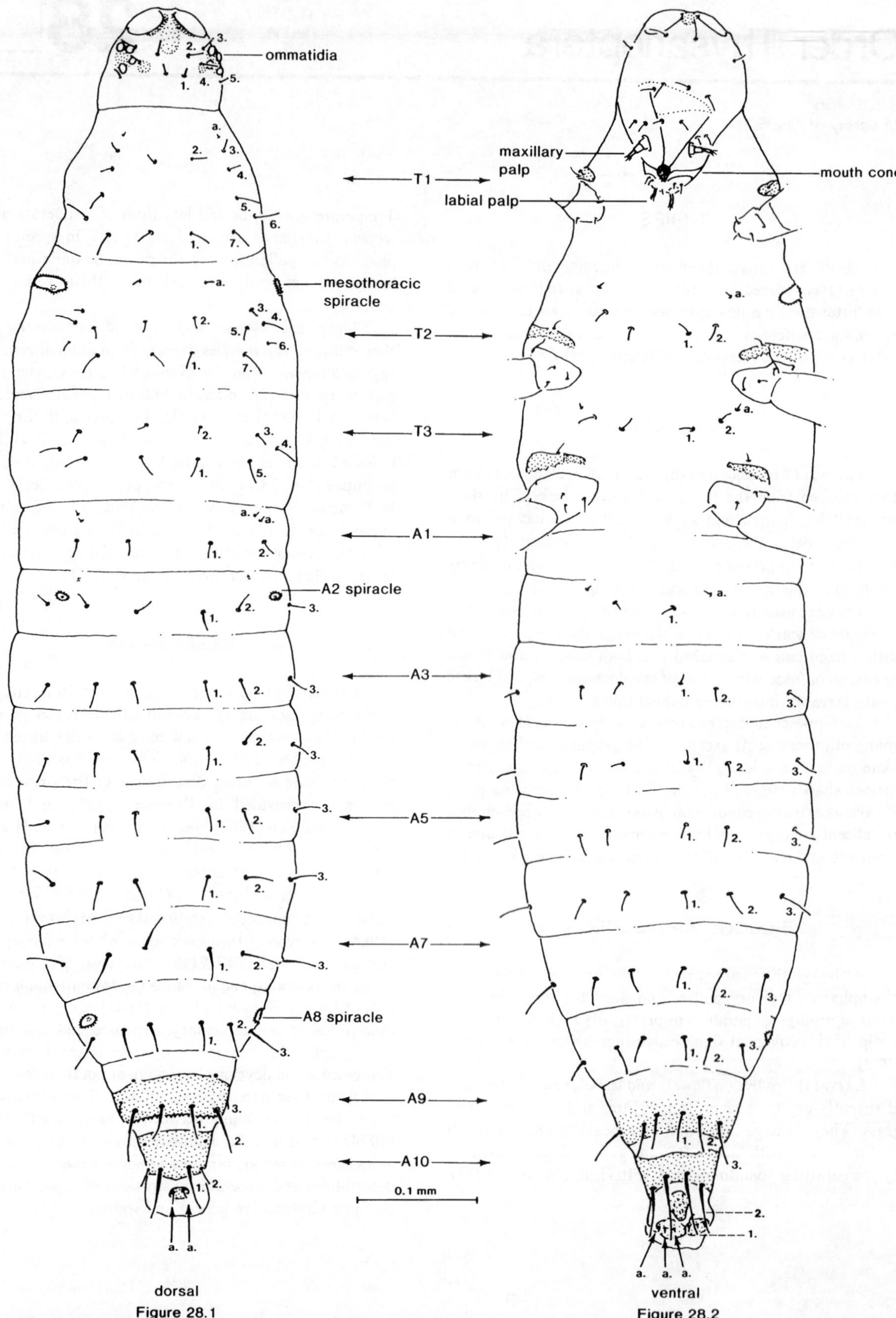

Figures 28.1–28.2. Thripidae. Setal Map of larva II male of the Tobacco thrips, *Frankliniella fusca* (Hinds) **28.1** Dorsal; **28.2** Ventral.

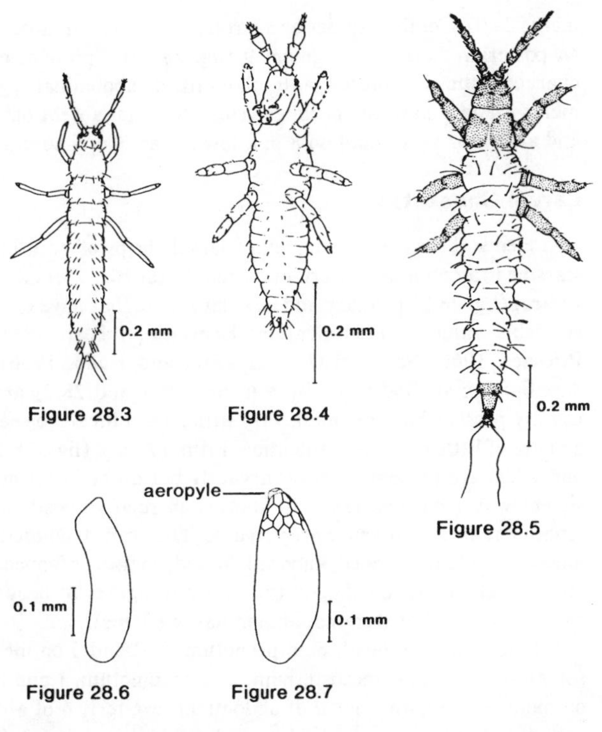

Figure 28.3 Figure 28.4 Figure 28.5 Figure 28.6 Figure 28.7

Figures 28.3–28.7. Larvae I and eggs.

Figure 28.3. Aeolothripidae. *Aeolothrips intermedius* Bagnall, dorsal. (redrawn from Derbeneva, 1967)

Figure 28.4. Thripidae. The Onion thrips, *Thrips tabaci* Lindeman, ventral. (redrawn from Lange and Razvyazkina, 1953)

Figure 28.5. Phlaeothripidae. The Mullein thrips, *Haplothrips verbasci* (Osborn), dorsal.

Figure 28.6 Egg of *A. intermedius.* (redrawn from Derbeneva, 1967)

Figure 28.7. Egg of *H. verbasci.*

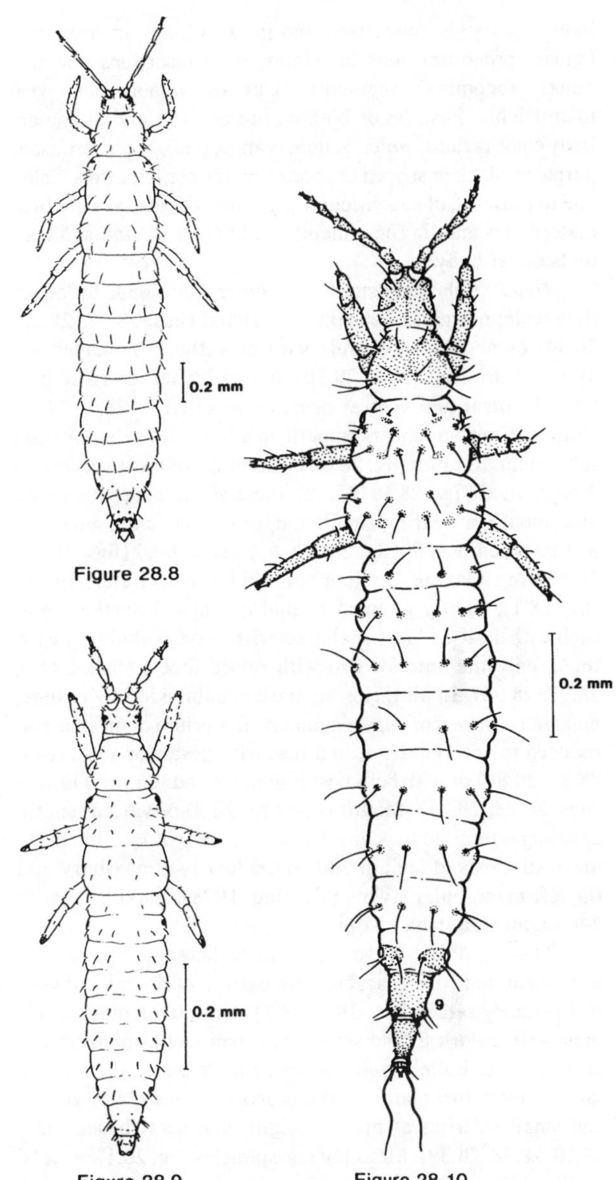

Figure 28.8 Figure 28.9 Figure 28.10

Figures 28.8–28.10. Larvae II.

Figure 28.8. Aeolothripidae. *A. intermedius.* (redrawn from Derbeneva, 1967)

Figure 28.9. Thripidae. *T. tabaci.* (redrawn from Lange and Razvyazkina, 1953)

Figure 28.10. Phlaeothripidae. *H. verbasci.*

Eggs

Figures 28.6, 28.7

Eggs minute, 0.2–0.5 × 0.1–0.25 mm. In Terebrantia, oval to reniform but, of some species, with operculum, or cut off obliquely at anterior end (fig. 28.6); chorion smooth and thin to reticulate or thick and pale to brown. In Terebrantia, eggs of most species deposited singly into leaves, flower parts or cotyledons. Eggs in Tubulifera, oval, with chorion strong, white to pink or grey, with large pentagonal or hexagonal reticulations; aeropyle well-developed at anterior end in eggs of *Haplothrips* spp. (fig. 28.7); micropyle single, situated posteroventrally. In Tubulifera, eggs deposited singly on surface of leaves or flower parts, or under fungus, inside galls, in bark crevices or in galleries made by other insects; in most instances glued to substrate with long axis parallel to surface, occasionally erect and in groups. Hatching of eggs of four species described in detail by Kirk (1985).

Larvae

Figures 28.1–28.5; 28.8–28.10

Mature larvae II minute to small, 0.5–5.0 mm long in North America but up to 14 mm elsewhere. Body slender, elongate, subcylindrical to depressed, more or less tapered posteriorly and with characteristic array of dorsal and ventral primary setae (figs. 28.1, 28.2). Cuticle usually colorless, in a few yellow or orange; mostly unsclerotized and flexible, but light to strongly sclerotized and grey to black in head, antennae, pronotum, muscle origins, setal insertions and terminal abdominal segments. Cuticle smooth or with microtrichia, pustules or both on pterothorax and abdomen. Body color various: white, yellow, orange, pink, green, crimson, purple or black or striped or spotted in any combination. Color due to presence of subcuticular pigment in epidermis or chromatophores and, in some members, to color of stomach contents or fat body.

Head: Relatively small and wider than long, to longer than wide; margins of cheeks constricted (figs. 28.29, 28.32, 28.34) or not. Head capsule with or without Y-shaped ecdysial suture (figs. 28.5, 28.10) and with characteristic pattern of dorsal and ventral primary setae (figs. 28.1, 28.2). Cuticle in most members smooth; in a few, with tubercles and sub-antennal spines (fig. 28.86). Antennae of most specimens 7-segmented (figs. 28.18–28.22), these of characteristic shape and smooth or bearing annuli and/or microtrichia; antennae of a few members 5- (fig. 28.96) or 6-segmented (figs. 28.21, 28.98). Most larvae with four ommatidia on each side of head (fig. 28.1), tightly grouped behind antennal insertions, and each with a raised facet of characteristic size. Tubulifera with three, only the anterior two with raised facets (figs. 28.90, 28.92, 28.94). In all thrips ommatidia subtended by orange, pink, red, crimson or purple pigment. Tentorium of most larvae reduced to anterior arms, in a few, with posterior arms (figs. 28.82, 28.84) or with both posterior arms and tentorial bridge (figs. 28.23, 28.25). Mouth-cone (fig. 28.2) of varied length, opisthognathous to hypognathous, consisting of labrum, maxillary stipites and labium and concealing two maxillary and the left mandibular stylets (Heming, 1978); maxillary palpi 2–4 segmented; labial palpi 1–3 segmented.

Thorax: Pro-, meso- and metathoraces broad and prominent and with characteristic pattern of dorsal and ventral primary setae (figs. 28.1, 28.2). Cuticle of most specimens soft, colorless, and smooth; in some, with microtrichia or pustules or both. In some larvae, cuticle is sclerotized dorsally to form two plates on the pronotum (figs. 28.5, 28.10) and small sclerites at muscle origins and setal bases (figs. 28.10, 28.38, 28.39). Mesothoracic spiracles (fig. 28.1) present in all larvae and of characteristic size. Legs slender, of various length, and consisting of coxa, femur, tibiotarsus and pretarsus, the latter with claws and eversible arolium (Heming, 1972); leg cuticle lightly sclerotized, yellow to light or dark brown, in most larvae smooth, but in a few, with whorls of microtrichia.

Abdomen: Slender, 11-segmented, and with characteristic pattern of dorsal and ventral primary setae (figs. 28.1, 28.2). Cuticle in most members soft, colorless, flexible and smooth or with characteristic pattern of microtrichia, pustules, or both (figs. 28.24, 28.26, 28.28, 28.45–28.60, 28.83, 28.85); in many, lightly sclerotized and yellow to black in segments 8–10. Comb of spines or microtrichia present or absent on posterior margin of segment 9 (fig. 28.59). Spiracles of characteristic size, present anterolaterally on abdominal segments 2 and 8 in most specimens (fig. 28.1) but absent on 2 and absent or very small on 8 in a few larvae. Cerci absent.

Larval Chaetotaxy

Head, thorax and abdomen of larval thrips of both instars bear a common pattern of paired dorsal (fig. 28.1) and ventral (fig. 28.2) primary and auxiliary setae that have considerable value in identification (Speyer and Parr, 1941; Priesner, 1960; Vance, 1974; Miyazaki and Kudo, 1986). Auxiliary setae (indicated by *a* in figs. 28.1 and 28.2) are usually present but are small, vary little, are difficult to see, and are of little use in identification. Primary setae (figs. 28.1 and 28.2) are present in most larvae II but differ in a consistent way from one taxon to another in relative position, length and form and hence are valuable. They can be pointed, lanceolate, blunt, rounded, knobbed, forked, fringed or fanned, and funnel—or spoon-shaped (as shown in figures of heads and pronota and of terminal abdominal segments).

Seta pair 5 on head, 6 on pronotum, 2, 3 and 5 on mesonotum, 1 and 2 on mesosternum, 2 on metanotum, 1 and 2 on metasternum, anterior *a* of abdominal tergite 1, *a* of abdominal sternite 2, and 2 and 3 of abdominal sternites 3–8 are usually absent in larvae I (figs. 28.3–28.5, 28.25, 28.84). Seta pair 4 on pronotum and 2 and 3 of meso- and metanotum are frequently absent in larvae II of panchaetothripine Thripidae (figs. 28.30–28.32). Seta pair 3 on metanotum is absent in larvae II of *Sericothrips* spp. Seta pair 4 on head, 5 on mesonotum and 2 on abdominal sternites 3–8 are absent in larvae II of Phlaeothripidae and an additional pair of setae, 6, is present on the metanotum (figs. 28.10, 28.86–28.94). Seta pair 1 and/or 2 of abdominal tergite 9 are short, stout spines in some larvae II (figs. 28.8, 28.24, 28.28, 28.55, 28.83).

In most larval thrips, the number of setae on abdominal segment 9 differs between males and females (figs. 28.76, 28.77, 28.100, 28.101) as can be corroborated by referring to the rudiments of the external genitalia or gonads in particularly good preparations (this was first noted by Priesner, 1964). The number of setae involved varies according to instar and family.

Setae and sense cones of antennae and setae of legs also have a common plan throughout the order (Speyer and Parr, 1941; Heming, 1972, 1975; Miyazaki and Kudo, 1986) but usually do not vary enough to have much use in identification. However, the number of annuli and whorls of microtrichia on antennal segments 2–7 are useful in larval analysis at the family and generic levels (figs. 28.18–28.22).

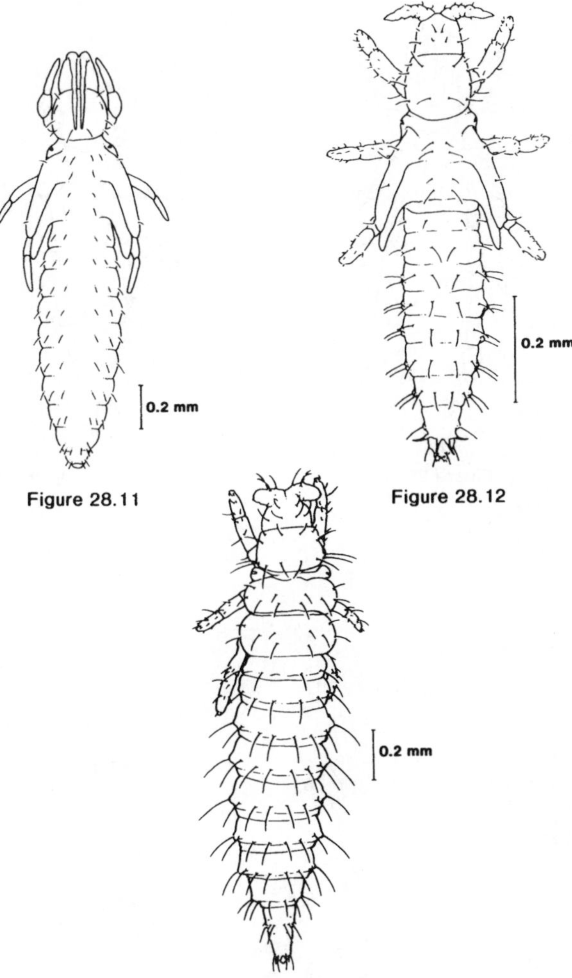

Figure 28.11

Figure 28.12

Figure 28.13

Figures 28.11–28.13. Propupae.

Figure 28.11. Aeolothripidae. *Aeolothrips intermedius.* (redrawn from Derbeneva, 1967)

Figure 28.12. Thripidae. *Thrips tabaci.* (redrawn from Lange and Razvyazkina, 1953)

Figure 28.13. Phlaeothripidae. *Haplothrips verbasci.*

Propupae

Figures 28.11–28.13

Propupae exarate, quiescent, and non-feeding; in some species they are found beneath the soil surface within a cocoon spun by the larva II. Body similar in size, shape and color to that of mature larvae II and with a similar pattern of long to very long dorsal and ventral setae which are inserted on raised bases. Cuticle delicate, colorless, unsclerotized, flexible and smooth throughout; but a few specimens may have small pustules on sides of pronotum and terminal abdominal segments.

Antennae without distinct sementation. Gnathal stylets absent and palpi reduced to short stubs. Legs with coxa, femur, and tibiotarsus, and with pretarsus usually reduced (Heming, 1973). Abdomen 9-segmented, with spiracles absent or greatly reduced and with dorsal seta pair 1 and/or 2 of segment 9 sometimes modified into spines. Lobes of developing ovipositor on abdominal sternites 8 and 9 of most terebrants (fig. 28.78). Other characters mentioned in key and family descriptions.

Pupae

Figures 28.14–28.17

Body similar to that of propupae but with longer setae and additional differences noted below. Head relatively larger and antennae longer. Developing compound eyes, dorsal ocelli and gnathal stylets of adult visible through pupal cuticle in pharate adults. Wing pads, if present, relatively longer. Legs relatively longer and expanded apically. Spiracles usually absent but well-developed on abdominal segment 2 in some members. Lobes of developing ovipositor longer in terebrants (fig. 28.80). Sternum 9 of male more (figs. 28.103, 28.105) or less (fig. 28.81) prolonged caudad. Other characters are mentioned in the key and in family descriptions.

Below the subfamily level, propupae and pupae have few identifying characteristics.

COMMENTS

Some 788 species in 150 genera and 6 families are so far known for North America—slightly less than 1/6 of the presently known world fauna.

Thrips are of relatively minor economic importance, with about 30 species listed as injurious in the United States, but only 7 of major significance (Bailey, 1940).

Immatures of the vast majority of North American thrips remain undescribed, and for many pest species, inadequately so (descriptions given in most life history studies). Most pests are in the subtribe Thripina of the family Thripidae and their larvae lack easily-seen characteristics by which they can be distinguished.

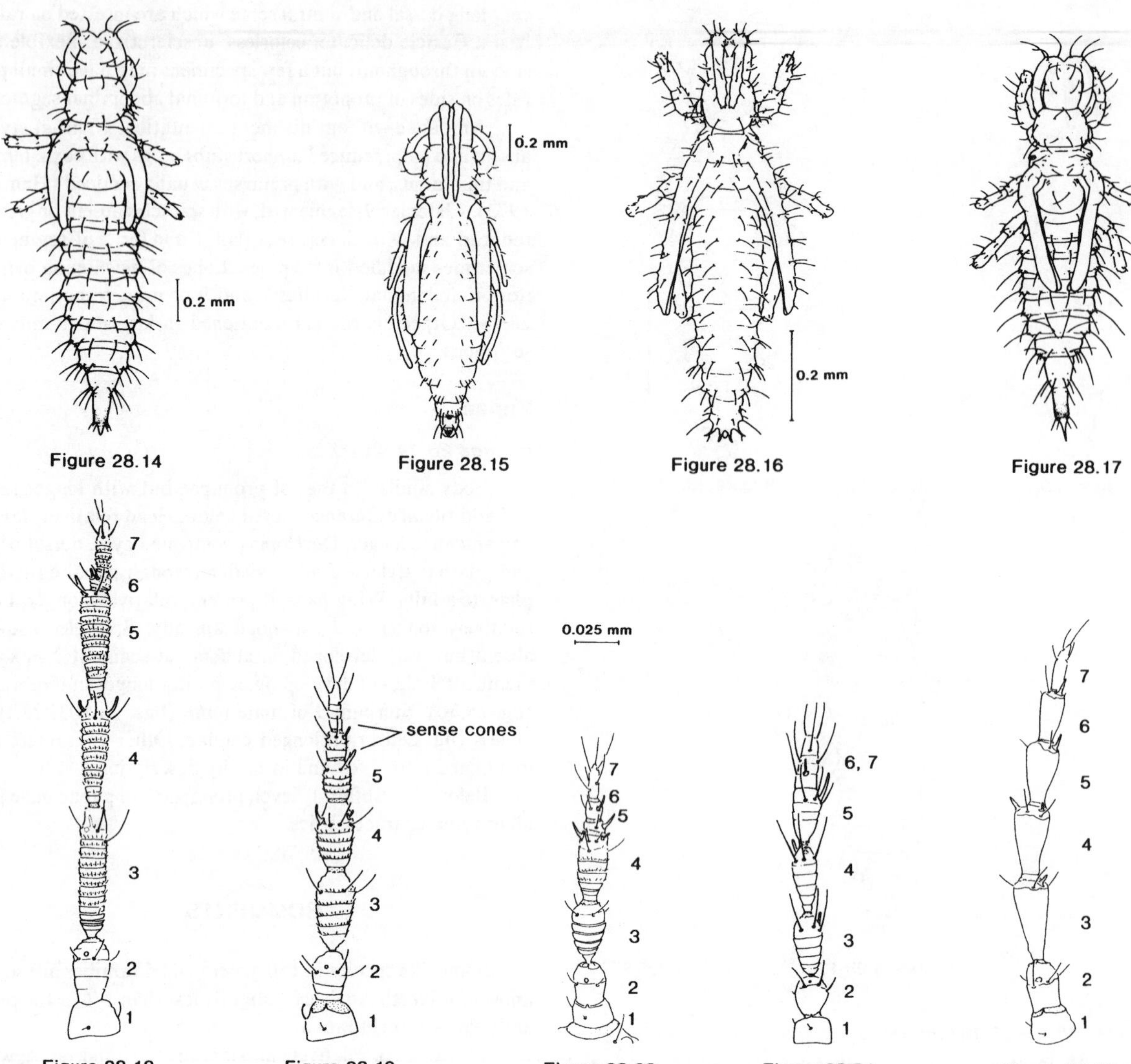

Figures 28.14–28.17. Pupae.

Figure 28.14. Phlaeothripidae. Pupa I of *H. verbasci.*

Figure 28.15. Aeolothripidae. *Aeolothrips intermedius.* (redrawn from Derbeneva, 1967)

Figure 28.16. Thripidae. *Thrips tabaci.* (redrawn from Lange and Razvyazkina, 1973)

Figure 28.17. Phlaeothripidae. Pupa II of *Haplothrips verbasci.*

Figures 28.18–28.22. Antennae of larvae II.

Figure 28.18. Aeolothripidae. *Aeolothrips* sp.—left.

Figure 28.19. Heterothripidae. *Heterothrips arisaemae* Hood—right. (redrawn from Vance, 1974)

Figure 28.20. Thripidae. The Tobacco thrips, *Frankliniella fusca* (Hinds)—right. (redrawn from Heming, 1975)

Figure 28.21. Merothripidae. *Merothrips morgani* Hood—right. (redrawn from Vance, 1974)

Figure 28.22. Phlaeothripidae. *H. verbasci*—right. (redrawn from Heming, 1975)

TECHNIQUES

Because of their small size and fragility, immature thrips require special care to correctly prepare specimens.

Collecting

Thrips can be swept from vegetation with a finely-woven sweep net, but better results are obtained by beating or jarring specimens into a large, pastel-tinted, plastic wash basin with its sides cut down to leave a small lip (Mound and Pitkin, 1972). The "bladder feet" of the larvae tend to stick to the plastic and the thrips contrast with the color of the basin. Individuals are picked up on a dampened, camel's hair brush and placed in 10% ethanol to which a wetting agent has been added (Heming, 1969). A.G.A. (60% ethanol—10 parts; Glycerine—1 part; acetic acid—1 part), a good collecting fluid for adult thrips (Mound and Pitkin, 1972), is inferior to 10% ethanol for immatures.

Larvae of leaf-litter species are best collected with a Tullgren Funnel using 2 or more 40 watt bulbs as a heat source. Other sampling techniques are described in Lewis (1973).

Rearing: Most larval thrips are difficult to keep in cages because they are small and easily damaged and are able to squeeze through incredibly small crevices. Therefore, cages must be thoroughly sealed with no gaps larger than about 0.065 mm. Such cages restrict air movement and may encourage condensation on the cage walls to which thrips stick and drown. Thus, large, cloth-covered windows in the walls of the cage, sub-surface watering of growing plants and a constant rearing temperature should be used (Lewis, 1973). Moist sand or earth at least 3 cm deep and fine enough to pass through a 2 mm sieve should be provided beneath the host plant for those thrips pupating in soil. Methods discussed under "Plaster of Paris" in the general "techniques section" can be used for rearing litter- and wood-dwelling forms (vol. 1).

Killing, Fixing and Preserving

To be identified, immature thrips must be correctly mounted on glass slides and examined with a compound microscope with both conventional optics and phase contrast. Satisfactory killing, preserving and mounting methods should retain natural color, not distort shape, make all cuticular structures including setae and other sense organs visible and preserve features for an indefinite period when the slides are stored in the dark (Lewis, 1973).

After two days to a week or more in 10% ethanol thrips distend and some of their internal organs disintegrate. Distension is facilitated by higher temperatures (37° C) and by placing grass seeds in the collecting solution. If they cannot be mounted within two or three months, they should be taken up gradually to 60% ethanol (with 1/2 h each in 30% and 50% ethanol) to which a drop of glycerine has been added, and stored in the dark. Heavily-pigmented specimens should be macerated in *cold* 5% NaOH for 1/2 h to 4 h to reveal greater cuticular detail (Mound and Pitkin, 1972).

Mounting: Immature thrips should be mounted individually on glass slides in neutral Canada balsam. Use of water-based mountants such as Berlese or Hoyer's is faster but should be avoided because such media are not permanent and cause distortion (Mound and Pitkin, 1972).

Thrips are passed successively through 35%, 50%, 70%, 80%, 90%, 95% and absolute ethanol, leaving them in each concentration about 1/2 h. They are then transferred individually with a camel's hair brush into a mixture of clove or cedarwood oil and absolute ethanol (Heming, 1969). Because the oil is heavier than the alcohol, the alcohol floats on top and a sharp interface is present. Thrips sink quickly through the alcohol to the surface of the oil and, in an hour or so, they become infiltrated with oil and sink to the bottom—becoming clear as they do. With this gentle infiltration, little collapsing occurs. The entire collection is then transferred via medicine dropper to a syracuse watch glass filled with pure oil.

Each specimen is mounted *venter upward* in a *small* drop of balsam on a 13 mm cover slip together with two small fragments of crushed cover slip positioned in front of and behind the specimen. A slide is then lowered *carefully* onto the coverslip and is then flipped right side up. Only one specimen should be mounted per slide, each with its head toward the preparator so its image will be erect under the compound microscope. Finished slides must be dried horizontally at 40° C on a hot plate for up to 6 weeks if the specimens are not to slip when the slides are stored vertically (but they should be stored horizontally). Further details of preparation can be found in Mound and Pitkin (1972).

CLASSIFICATION

Order **THYSANOPTERA**[1]
Suborder Terebrantia
- Family Uzelothripidae—(not represented in North America)
- Merothripidae—large-legged thrips (6; 15 World)
- Aeolothripidae—broad-winged or banded thrips (61; 220 World)
- Adiheterothripidae (1; 3 World)
- Fauriellidae—(not represented in North America)
- Heterothripidae (25; 55 World)
- Thripidae (Thripididae)—common thrips (322; >1500 World)

Suborder Tubulifera
- Family Phlaeothripidae (365; >2700 World)

1. I use the classification of Jacot-Guillarmod (1970–1975) and Mound *et al.* (1980) for the suborder Terebrantia and that of Mound *et al.* (1976) for the suborder Tubulifera. The number of species given is for North America (including Mexico).

KEY TO KNOWN INSTARS AND FAMILIES OF WORLD THYSANOPTERA[2]

1. Antennae with distinct segmentation; wing pads absent; most specimens active (figs. 28.3–28.5; 28.8–28.10) .... larvae 6

1′. Antennae usually without distinct segmentation; wing pads present or absent; most specimens inactive and hidden in crevices or in soil, a few within a silken cocoon .... quiescent instars 2

2(1′). Antennae projecting laterally as short stubs (fig. 28.13), anteriorly (fig. 28.12), or reflexed dorsally over the head (fig. 28.11); if wing pads present, these extend caudad at most to middle of abdominal segment 2 (figs. 28.11, 28.12) .... propupae 3

2′. Antennae reflexed laterally beside (figs. 28.14, 28.17) or dorsally over the head (figs. 28.15, 28.16); if wing pads present, these extend caudad beyond middle of abdominal segment 2 (figs. 28.14–28.17) .... pupae 4

3(2). Antennae relatively long, projecting forward and laterally (fig. 28.12) or reflexed dorsally over the head (fig. 28.11); wing pads present or absent .... propupae, suborder TEREBRANTIA

3′. Antennae projecting laterally as short stubs; wing pads absent (fig. 28.13) .... propupae, ***(Phlaeothripidae)*** suborder TUBULIFERA

4(2′). Antennae reflexed dorsally over the head (figs. 28.15, 28.16) .... pupae, suborder TEREBRANTIA

4′. Antennae reflexed laterally beside the head (figs. 28.14, 28.17) .... pupae, ***(Phlaeothripidae)*** suborder TUBULIFERA 5

5(4′). Antennae extending caudad to pronotum; wing pads, if present, reaching caudad at most to posterior margin of abdominal segment 2 (fig. 28.14) .... (pupae I) ***Phlaeothripidae*** (p. 17)

5′. Antennae extending caudad underneath anterior margin of pronotum; wing pads, if present, reaching caudad to abdominal segment 5 (fig. 28.17) .... (pupae II) ***Phlaeothripidae*** (p. 17)

6(1). Abdominal segment 10 never elongate and tubular, in most specimens broader than long (figs. 28.1–28.4; 28.8, 28.9); maxillary stylets subequal to length of mouthcone; middle antennal segments with annuli which bear microtrichia in most specimens (figs. 28.18–28.21) .... suborder TEREBRANTIA 7

6′. Abdominal segment 10 elongate and tubular (figs. 28.5, 28.10); maxillary stylets usually longer than mouthcone; middle antennal segments without annulations or microtrichia (fig. 28.22) .... ***(Phlaeothripidae)*** suborder TUBULIFERA 12

7(6). Pronotum usually with 6 pairs of setae (seta pair 6 absent; figs. 28.3, 28.5, 28.84); abdominal sternites 3–8 with one pair of setae each (fig. 28.4) .... (larvae I) suborder TEREBRANTIA

7′. Pronotum usually with 7 pairs of setae (figs. 28.1, 28.8, 28.9); abdominal sternites 3–8 with 3 pairs of setae each (fig. 28.2) .... (larvae II) suborder TEREBRANTIA 8

8(7′). Antennal segment 5 from half as long to same length or longer than segment 4 (figs. 28.18, 28.19, 28.21) 9

8′. Antennal segment 5 much less than half as long as segment 4 (fig. 28.20) .... ***Thripidae*** (p. 11)

9(8). Length of antennal segment 5 half that of segment 4; segments 3–6 with microtrichia-bearing annuli (fig. 28.19) .... (genus *Heterothrips*) ***Heterothripidae*** (p. 10)

9′. Length of antennal segment 5 equal to or longer than that of segment 4; segments 3–6 with annuli and with or without microtrichia (figs. 28.18, 28.21) 10

10(9′). Antennal segments 6 and 7 fused; segments 3–5 relatively short and devoid of microtrichia (fig. 28.21) .... (genus *Merothrips*) ***Merothripidae*** (p. 17)

10′. Antennal segments 6 and 7 separate 11

11(10′). Length of antennal segment 5 equal or subequal to that of segment 4; segments 3–5 elongate with numerous whorls of microtrichia-bearing annuli (fig. 28.18) .... ***Aeolothripidae*** (p. 9)

11′. Length of antennal segment 5 almost double that of 4; segments 5–7 elongate with annuli devoid of microtrichia .... (*Uzelothrips scabrosus* Hood, E. Brazil & Singapore) .... ***Uzelothripidae***

12(6′). Pronotum with 6 pairs of setae (seta pair 6 absent, fig. 28.5); abdominal segment 9 sclerotized only in posterior half (figs. 28.5, 28.100, 28.101) .... (larvae I) ***Phlaeothripidae*** (p. 17)

12′. Pronotum with 7 pairs of setae in most specimens (fig. 28.10): abdominal segment 9 sclerotized throughout (figs. 28.10, 28.89, 28.91, 28.93) .... (larvae II) ***Phlaeothripidae*** (p. 17)

2. NOTE: Larvae of Fauriellidae and Adiheterothripidae are unknown but probably similar to those of Heterothripidae. The family Fauriellidae is distributed in South Africa and southern Europe.

Selected Bibliography

General
Ananthakrishnan 1984.
Ananthakrishnan and Sen 1980.
Bailey 1940.
Hennig 1981.
Jacot-Guillarmod 1970–1979.
Kirk 1985.
Lewis 1973.
Mound 1973.
Mound et al. 1976, 1980.
Priesner 1928, 1964, 1964a.
Speyer and Parr 1941.
Vance 1974.

Techniques
Heming 1969.
Lewis 1973.
Mound and Pitkin 1972.

Structure
Derbeneva 1967.
Heming 1972, 1973, 1975, 1978.
Lange and Razvyazkina 1953.
Miyazaki and Kudo 1986.
Priesner 1964.
Speyer and Parr 1941.

Suborder TEREBRANTIA

AEOLOTHRIPIDAE

The Broad-Winged or Banded Thrips

Figures 28.3, 28.6, 28.8, 28.11, 28.15, 28.18, 28.23–28.26.

Relationships and Diagnosis: The aeolothripids are a well-defined family apparently constituting the sister group of the Merothripidae (Mound, *et al.*, 1980). Larvae of both instars are distinguished from those of all other thrips by their long, slender legs (figs. 28.3, 28.8) and by their antennae which have segments 3–5 elongate and widened at their tips, segments 3–7 with numerous whorls of microtrichia and segment 5 as long as or longer than 4 (fig. 28.18). In addition, a fully-developed tentorium is present within their heads (figs. 28.23, 28.25).

Biology and Ecology: Larvae of known species are very active and occur on flowers or leaves of grasses, trees, or herbs. They are pollen feeders or (at least in larvae II) facultative predators on thrips, mites and other small arthropods (Lewis, 1973); most are quite host specific.

Larvae II fall to the ground after completing development, enter the soil and spin a loosely-woven cocoon of material emanating from their anus. The larva remains quiescent within this cocoon until it pupates in the fall. Both quiescent instars have unsegmented antennae reflexed dorsally over their heads and pronota, and wing pads (figs. 28.11, 28.15) and are not capable of moving even if provoked (Derbeneva, 1967). Adults of most species emerge in spring.

Description: Mature larvae of most species not exceeding 2 mm in length. Body elongate, slender, cylindrical and pale to crimson (figs. 28.3, 28.8). Cuticle soft and colorless except for head, antennae, spots or sclerites on pronotum, legs and terminal abdominal segments which are lightly sclerotized and grey to yellow or light brown. Cuticle of pterothorax and abdomen densely stippled throughout with fine microtrichia (figs. 28.24, 28.26). Setae long, moderately stout and pointed or knobbed.

Head: Head capsule short, broad, and rounded anteriorly below antennal insertions; cheeks parallel or divergent caudad. Cuticle of head smooth, lightly sclerotized between antennae, on inner margins of antennal sockets, in immediate vicinity of ommatidia and on caudal margin of vertex. Cephalic setae with arrangement shown in figs. 28.23 and 28.25. Antennae 7-segmented, segments 3–5 elongate, 3–7 with numerous whorls of microtrichia, and 5 as long as or longer than 4; sense cones relatively long and slender (fig. 28.18). Ommatidia, 4, with raised facets, on either side of head behind antennal insertions. Tentorium complete (figs. 28.23, 28.25). Mouthcone short, hypognathous; maxillary palpi 4-segmented (indistinct in larvae II), labial palpi 3-segmented.

Thorax: Pronotum with 2 lightly sclerotized plates in larvae I (fig. 28.25), colorless except for a few sclerotized spots caudad in larvae II (fig. 28.23). Mesothoracic spiracles small. Legs long, slender, lightly sclerotized and with numerous whorls of microtrichia (Heming, 1972).

Abdomen: Long, slender and cylindrical in mature larvae II. Spiracles very small, situated antero-laterally on segments 2 and 8. Seta pairs 1 and 2 on abdominal tergite 9 of larvae II usually modified into stout spines (fig. 28.24), but normal in larvae I (fig. 28.26) and in both larval instars of mymarothripines and franklinothripines. Chaetotaxy of segment 9 sexually dimorphic in both larval instars: in larvae I, 3 pairs in females, 4 pairs in males; in larvae II, 5 pairs in females (including spines) and 6 pairs in males. Setae of segment 10, 3 pairs and very long (fig. 28.26) in larvae I.

Comments: Sixty-one species of aeolothripids in 9 genera are so far recognized in North America (including Mexico), 33 being in the genus *Aeolothrips* (Jacot-Guillarmod, 1970).

No species are of economic importance, but individuals of the multivoltine species, *Aeolothrips fasciatus* (L.), are often cited as being predators of mites and other thrips in papers on economic entomology.

In the eastern and southeastern U.S., *Aeolothrips albicinctus* Haliday, a brachypterous species, and *A. bicolor* Hinds are common on grasses and *A. melaleucus* Haliday on trees and shrubs. Desert species are *A. nitidus* Moulton, *A. aureus* Moulton and *A. duvali* Moulton. *A. hartleyi* Moulton and *A. vittipennis* Hood occur in the north and at high altitudes (Bailey, 1951).

Vance (1974) described larvae I and II of *A. vittipennis;* otherwise immatures of North American aeolothripids are unknown. Detailed descriptions of all aeolothripid instars are in Priesner (1928, 1964) and keys to known larvae II of European *Franklinothrips, Rhipidothrips, Aeolothrips, Melanthrips* and *Ankothrips* species in Priesner (1964, 1964a). Adults of most North American species are keyed in Bailey (1951, 1957).

Selected Bibliography

Bailey 1951, 1957.
Derbeneva 1967.
Heming 1972.
Jacot-Guillarmod 1970.
Lewis 1973.
Mound et al. 1980.
Priesner 1928, 1964, 1964a.
Vance 1974.

ADIHETEROTHRIPIDAE (Immatures unknown)

The Adiheterothripids

Relationships and Diagnosis: Adiheterothripids resemble heterothripids and formerly (Jacot-Guillarmod, 1970) were considered a tribe within that family. However, they are less specialized than heterothripids and probably constitute the sister group of the (Fauriellidae (old world) + Heterothripidae) + Thripidae (Mound *et al.* 1980). Their immatures are unknown.

Biology and Ecology: *Oligothrips oreios* Moulton, the only North American species, occurs in California and Oregon in blossoms of madrone and manzanita. There is one generation a year, the active feeding stages being found only in early spring (Bailey, 1957).

Description: Immatures of *Oligothrips oreios* have yet to be described but larvae II will probably be found to resemble those of *Heterothrips arisaemae* (figs. 28.19, 28.27, 28.28) except for the presence of a tentorial bridge.

Comments: Two species of *Holarthrothrips* occur: one in India and the other in France, Italy, Corsica and Greece (Jacot-Guillarmod, 1970).

Selected Bibliography

Bailey 1957.
Jacot-Guillarmod 1970.
Mound et al. 1980.

HETEROTHRIPIDAE

The Heterothripids

Figures 28.19, 28.27, 28.28

Relationships and Diagnosis: Heterothripids most closely resemble fauriellids (old world) and thripids and share with them many of their derived characteristics (Mound, *et al.* 1980). In other respects, they are intermediate in structure to them and adiheterothripids. I thus consider them and fauriellids to constitute the sister group of the Thripidae, all three families probably sharing a common ancestor with the stem group of the Adiheterothripidae.

Larvae of *Heterothrips arisaemae* Hood, the only species for which larvae are known, are distinguished from others by their antennae which have segments 3–6 with 2–5 whorls of microtrichia and segment 5 half as long as 4 (fig. 28.19).

Biology and Ecology: Little is known about heterothripid life history. Bailey and Cott (1954) considered most of them to be univoltine because adults of most species were collected only in spring and early summer. They occur on flowers of various trees, shrubs and herbs and most appear to be quite host specific. Specimens of *Heterothrips vitifloridus* Bailey and Cott have been reared from flowers of wild grape

Figure 28.23 Figure 28.25

0.05 mm

Figure 28.24 Figure 28.26

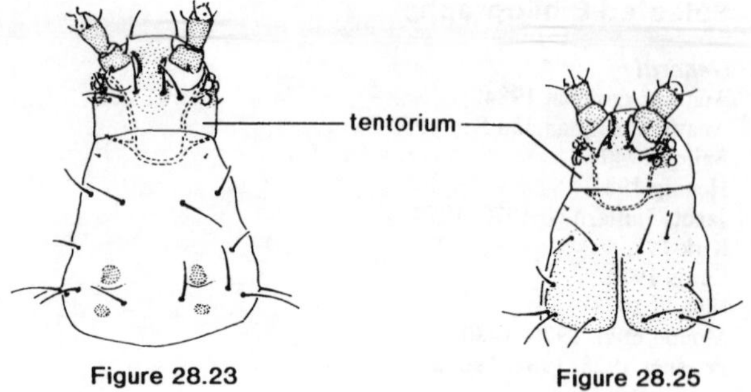

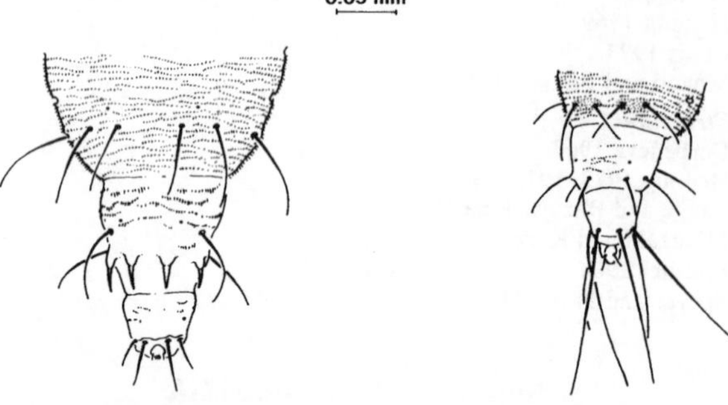

Figures 28.23–28.26. Aeolothripidae. Larvae I and II of *Aeolothrips* sp.

Figure 28.23. Head and pronotum of larva II male.

Figure 28.24. Abdominal segments 8–11 of larva II male.

Figure 28.25. Head and pronotum of larva I female.

Figure 28.26. Abdominal segments 8–11 of larva I female.

0.05 mm

Figure 28.27 Figure 28.28

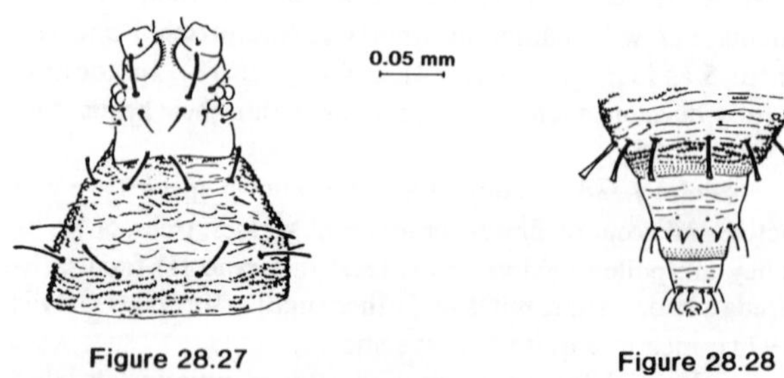

Figures 28.27, 28.28. Heterothripidae. Larva II, *Heterothrips arisaemae.*

Figure 28.27. Head and pronotum. (redrawn from Vance, 1974)

Figure 28.28. Abdominal segments 8–11. (redrawn from Vance, 1974)

(Bailey, 1940a) and larvae II pupate in debris beneath the host plant within a cocoon similar to that of known aeolothripids.

Larvae (except for those of *H. arisaemae*) and quiescent instars are unknown.

Description: (***Heterothrips arisaemae*** Hood) Mature larva less than 1.5 mm long. Body elongate but somewhat depressed and less slender than that of aeolothripids. Cuticle soft and colorless except for head, antennae, legs and terminal abdominal segments which are lightly sclerotized and grey; thorax and abdomen covered with numerous, small, microtrichia-bearing pustules (figs. 28.27, 28.28). Setae short to long and lanceolate, and sharp to blunt or terminally-funneled.

Head: Longer than broad; cheeks slightly divergent caudad. Cuticle of head capsule smooth and lightly sclerotized on inner margins of antennal sockets. Seta pair 2 on vertex long, lanceolate and blunt to funneled (fig. 28.27). Antennae 7-segmented, segments 3–6 with 2–5 whorls of microtrichia; segment 5 about half as long as 4; sense cones relatively short (fig. 28.19). Ommatidia, 4, with bulging facets, on either side of head behind antennal insertions and subtended by red pigment. Tentorium consisting only of anterior arms. Mouthcone relatively short and hypognathous; maxillary palpi 3-segmented, labial palpi 2-segmented.

Thorax: Of unsclerotized cuticle throughout and covered with numerous, small, microtrichia-bearing pustules. Seta pair 4 on pronotum small and sharp; 6 short and blunt; all others long, lanceolate and funneled. Mesothoracic spiracles small. Legs relatively short and stout, lightly sclerotized throughout and having annuli, each with numerous short, blunt microtrichia (Heming, 1972).

Abdomen: Thripid-like. Spiracles small and situated antero-laterally on segments 2 and 8. Seta pair 1 and 2 on abdominal tergum 9 of larvae II modified into stout spines similar to those of *Aeolothrips* spp. (fig. 28.28). (According to Vance (1974), these are also present in larvae I, but this observation probably in error). Chaetotaxy of segment 9 sexually dimorphic at least in larvae II: 5 pairs in females (including spines) and 6 pairs in males.

Comments: Twenty-five species of *Heterothrips* are so far recorded from North America (including Mexico). *H. arisaemae* is common throughout the eastern U.S.A. in spring on flowers of green dragon (*Arisaema dracontium* (L.) Schott) and jack-in-the-pulpit (*A. triphyllum* (L.) Schott), and *H. salicis* Shull on catkins of *Salix interior* Rowlee. Most others are less common with little known about them. None are of economic importance. Adults of North American species can be keyed in Bailey (1957) and Bailey and Cott (1954).

Selected Bibliography

Bailey 1940a, 1957.
Bailey and Cott 1954.
Heming 1972.
Mound et al. 1980.

THRIPIDAE

The Common Thrips

Figures 28.1, 28.2, 28.4, 28.9, 28.12, 28.16, 28.20, 28.29–28.81

Relationships and Diagnosis: This family is the largest in the suborder Terebrantia, and its species the most highly derived. It appears to constitute the sister group of the Heterothripidae and Fauriellidae (old world) (Mound, *et al.* 1980).

Thripid larvae are distinguished from those of all other families by their antennae in which segment 5 is much less than half the length of segment 4 (fig. 28.20).

Biology and Ecology: This family consists of two subfamilies: the Panchaetothripinae (= Heliothripinae) and Thripinae which is further subdivided into four tribes: the Dendrothripini, Sericothripini, Chirothripini and Thripini. The tribe Thripini, in turn, contains two subtribes: the Aptinothripina and Thripina, the latter subtribe containing about two-thirds of all species in the family. Larvae and adults of most panchaetothripines, dendrothripines, and sericothripines and of many aptinothripines feed almost exclusively on the leaves of dicotyledonous plants. Most thripines are flower-dwelling pollen-feeders and most chirothripines feed on leaves and flowers of grasses (Mound *et al.* 1976). The family is present throughout the world at practically all latitudes and altitudes, three species even occuring at 81°49′N on Canada's Ellesmere Is. (Chiasson, 1986). Larvae of some species are quite host-specific while those of others, for example the onion thrips, *Thrips tabaci* Lindeman, will feed on practically any plant.

Larvae (figs. 28.1, 28.2, 28.4, 28.9) vary considerably in color and structure, depending on species. They usually move relatively rapidly over the substrate. Pupation can occur on the host (most panchaetothripines) or in the soil in a silk-lined chamber formed by the larva II (Lewis, 1973). Propupae have short, unsegmented antennae extending forward from the head (fig. 28.12), while those of pupae are reflexed dorsally over the head and pronotum (fig. 28.16). In those species having fully winged adults, the quiescent instars have well-developed wing pads. Neither propupae nor pupae move unless disturbed and neither instar has functional mouthparts.

Description: Mature larva II 0.5–2.0 mm long, in most species 1.5 mm. Body slender, elongate and subcylindrical to depressed. Cuticle in most specimens colorless, unsclerotized and flexible, but more or less sclerotized and grey to yellow or light brown in head, antennae, pronotum, muscle origins, setal insertions, legs and terminal abdominal segments. Cuticle of most larvae with characteristic pattern of microtrichia, pustules or both, particularly on pterothorax and abdomen. Primary setae of characteristic color, shape, size and position in larvae of different species (figs. 28.1, 28.2, 28.29–28.60). Body color variable; white to crimson or purple.

Head: Wider than long to longer than wide; cheeks constricted behind eyes (figs. 28.29–28.35) or not. Head capsule (figs. 28.36, 28.39) with or without Y-shaped ecdysial suture. Antennae 7-segmented with segment 5 much less than half length of 4 (fig. 28.20) and of 2 basic types: in Panchaetothripinae, length of segment 7, 7–8 times its greatest width, segments 3–7 with annulations usually free of microtrichia (figs. 28.61–28.64); in Thripinae, length of segment 7 at most 2–3 times its greatest width, only segments 3–4 with annuli, these with or without microtrichia (figs. 28.20, 28.65–28.75); sense cones short to long and slender. Ommatidia, 4, on each side of head behind antennal insertions; with or without bulging facets. Tentorium reduced to anterior arms. Mouthcone (figs. 28.2, 28.4) hypognathous to opisthognathous, of variable length; maxillary palpi 2–4 segmented, labial palpi 2-segmented.

Thorax: Pronotum colorless (figs. 28.1, 28.9, 28.29–28.35, 28.37, 28.40–28.44), with brown, sclerotized spots over muscle origins (figs. 28.38, 28.39) or with 2 pronotal plates (fig. 28.36). Mesothoracic spiracles small to large and characteristic for a particular species (figs. 28.1, 28.4, 28.9). Legs short and stout to long and slender, lightly sclerotized throughout and smooth (Heming 1972).

Abdomen: Short and broad and somewhat depressed to long, slender and subcylindrical. Segments 9 and 10 lightly sclerotized or not (figs. 28.45–28.60). Segment 9 (figs. 28.59, 28.60) with or without posterior comb of microtrichia. Spiracles of characteristic size, usually present antero-laterally on segments 2 and 8; those on 2 absent in some members. Seta pair 1 and/or 2 on tergum 9 and 1 on tergum 10 modified into short spines in some larvae (fig. 28.55); seta pair 3 of segment 9 very small in most. Chaetotaxy of segment 9 sexually dimorphic in larvae of both instars: in larvae I, 3 pairs in females, 4 pairs in males; in larvae II, 5 pairs in females (fig. 28.76), 6 pairs in males (fig. 28.77).

Detailed descriptions of subfamily and generic characteristics are in Vance (1974).

Comments: As of this writing, 322 species of Thripidae in 64 genera are known from North America (including Mexico), with many additional species probably remaining to be described.

Twenty-five of the 32 injurious thrips listed by Bailey (1940) for the U.S. are thripids, including the onion thrips, *Thrips tabaci* Lindeman (figs. 28.4, 28.9, 28.12, 28.16, 28.44, 28.60, 28.75, 28.78–28.80), which has the dubious distinction of being the most destructive species in the Thysanoptera, and the tobacco thrips, *Frankliniella fusca* (Hinds) (figs. 28.1, 28.2, 28.20, 28.73, 28.76, 28.77, 28.81). *Scolothrips pallidus* (Beach) (figs. 28.42, 28.58), a species long confused with the 6-spotted thrips, *S. sexmaculatus* (Pergande), is common in the eastern U.S. on soybeans and other crops where it feeds on spider mites. Other important species are illustrated (figs. 28.29–28.74) and Lewis (1973) lists others in Appendix 6. Larvae of species in the genera *Thrips, Taeniothrips* and *Frankliniella* are very conservative in structure, as are their adults, and it is almost impossible to make positive identifications. *See* Miyazaki and Kudo (1986) for differences between larvae of the genus *Thrips*. However, Vance (1974) has shown that the equally-conservative larvae of *Sericothrips* can be distinguished from each other. More detailed information on the importance and life history of all of these thrips is in Stannard (1968) and Lewis (1973).

The most important paper on thripid larvae of North America is that of Vance (1974). In this are keys, descriptions, phylogenetic analyses, and illustrations of representative larvae of species of many thripid genera. Important European references are Priesner (1928—descriptions of all instars of many species; 1964—detailed descriptions of all instars with variations, and of larvae and pupae of Mediterranean species; 1964a—brief descriptions of larvae of some European species) and Speyer and Parr (1941—the best discussion of integumentary sculpture and chaetotaxy available together with illustrations and descriptions of larvae I and II of many common thripids). Miyazaki and Kudo (1986) describe, key and illustrate larvae II of 12 species of thrips on solanaceous and cucurbitaceous crops in Japan, some of which occur here.

Adults are discussed and keyed in Bailey (1957) and Stannard (1968).

Selected Bibliography

Bailey 1940, 1957.
Chiasson 1986.
Heming 1972.
Lewis 1973.
Miyazaki and Kudo 1986.
Mound et al. 1976, 1980.
Priesner 1928, 1964, 1964a.
Speyer and Parr 1941.
Stannard 1968.
Vance 1974.

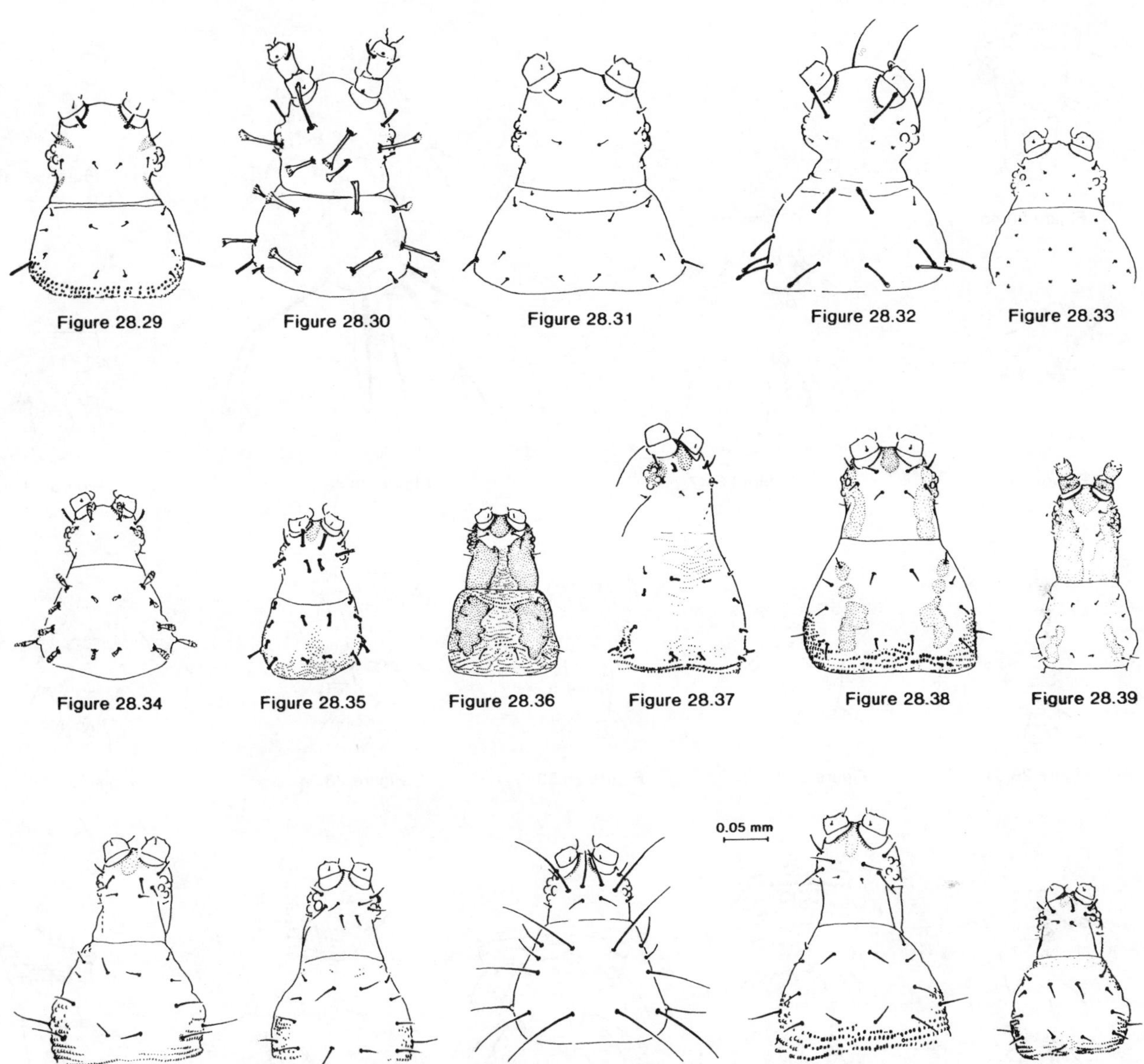

Figure 28.29 Figure 28.30 Figure 28.31 Figure 28.32 Figure 28.33

Figure 28.34 Figure 28.35 Figure 28.36 Figure 28.37 Figure 28.38 Figure 28.39

Figure 28.40 Figure 28.41 Figure 28.42 Figure 28.43 Figure 28.44

Figures 28.29–28.44. Thripidae. Heads and pronota of larvae II.

Figure 28.29. The Greenhouse thrips, *Heliothrips haemorrhoidalis* (Bouché). (redrawn from Vance, 1974).

Figure 28.30. The Palm thrips, *Parthenothrips dracaenae* (Heeger).

Figure 28.31. The Grapevine thrips, *Rhipiphorothrips cruentatus* Hood.

Figure 28.32. The Red-banded or Cacao thrips, *Selenothrips rubrocinctus* (Giard).

Figure 28.33. The Privet thrips, *Dendrothrips ornatus* (Jablonowski).

Figure 28.34. The Soybean thrips, *Sericothrips variabilis* (Beach).

Figure 28.35. *Scirtothrips* sp.

Figure 28.36. The Grain thrips, *Limothrips cerealium* (Haliday). (redrawn from Vance, 1974).

Figure 28.37. The Grass thrips, *Anaphothrips obscurus* (Müller).

Figure 28.38. *Apterothrips secticornis* (Trybom).

Figure 28.39. *Aptinothrips stylifer* Trybom.

Figure 28.40. *Chilothrips pini* Hood.

Figure 28.41. The Cotton bud thrips, *Frankliniella schultzei* (Trybom).

Figure 28.42. *Scolothrips pallidus* (Beach) (redrawn from Vance, 1974).

Figure 28.43. The Gladiolus thrips, *Thrips simplex* (Morison).

Figure 28.44. The Onion thrips, *Thrips tabaci* Lindeman.

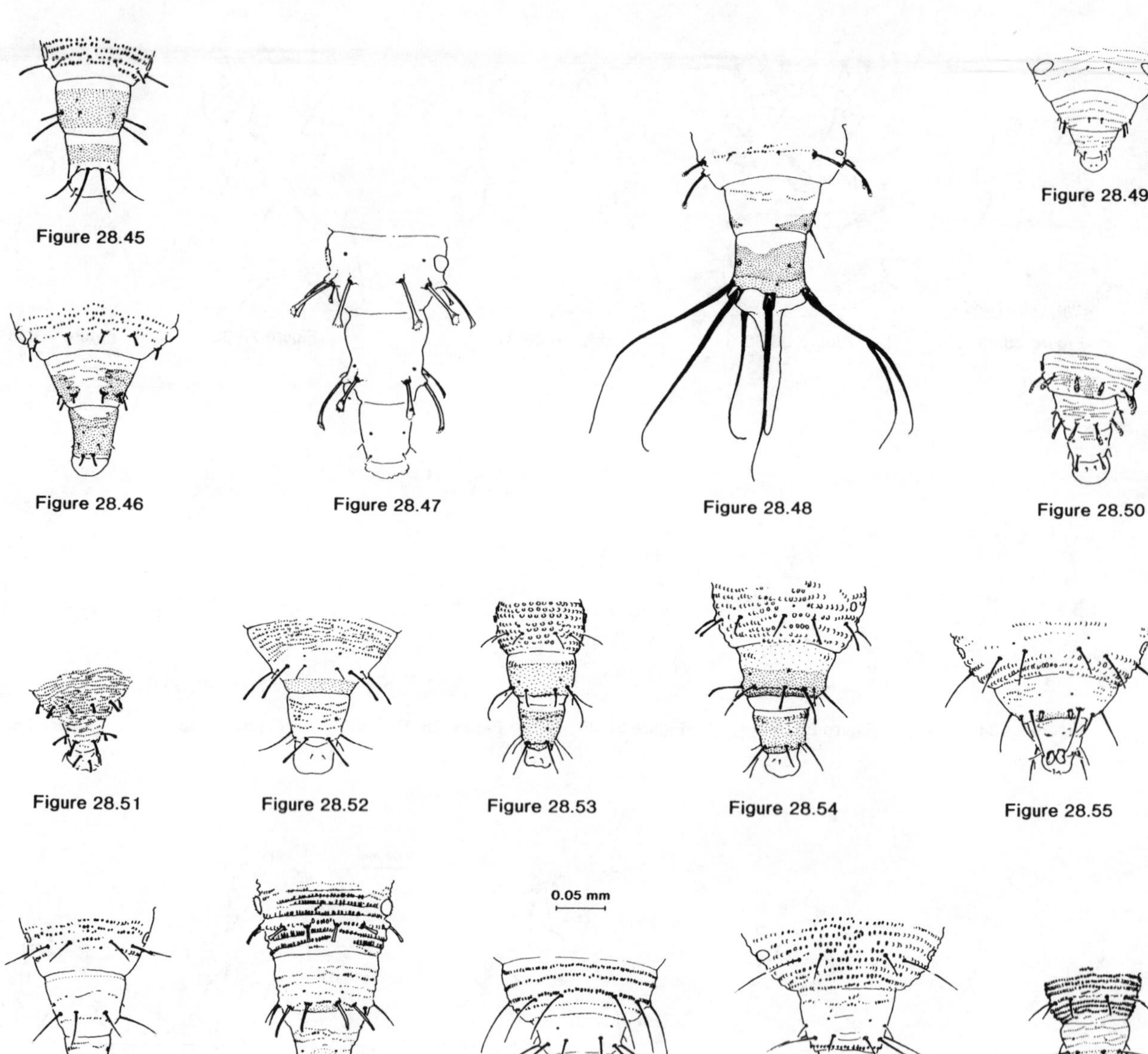

Figures 28.45–28.60. Thripidae. Abdominal segments 8-11 of larvae II.

Figure 28.45. *Heliothrips haemorrhoidalis* (redrawn from Vance, 1974).

Figure 28.46. The Banded Greenhouse thrips, *Hercinothrips femoralis* (O. M. Reuter) (redrawn from Vance, 1974).

Figure 28.47. *Parthenothrips dracaenae* female.

Figure 28.48. *Selenothrips rubrocinctus* female.

Figure 28.49. *Dendrothrips ornatus* female.

Figure 28.50. *Sericothrips variabilis* female.

Figure 28.51. *Scirtothrips* sp.

Figure 28.52. *Limothrips cerealium* (redrawn from Vance, 1974).

Figure 28.53. *Aptinothrips stylifer* female.

Figure 28.54. *Apterothrips secticornis* female.

Figure 28.55. *Chilothrips pini* female.

Figure 28.56. *Frankliniella schultzei* female.

Figure 28.57. *Anaphothrips obscurus* female.

Figure 28.58. *Scolothrips pallidus.* (redrawn from Vance, 1975).

Figure 28.59. *Thrips simplex* female.

Figure 28.60. *Thrips tabaci* female.

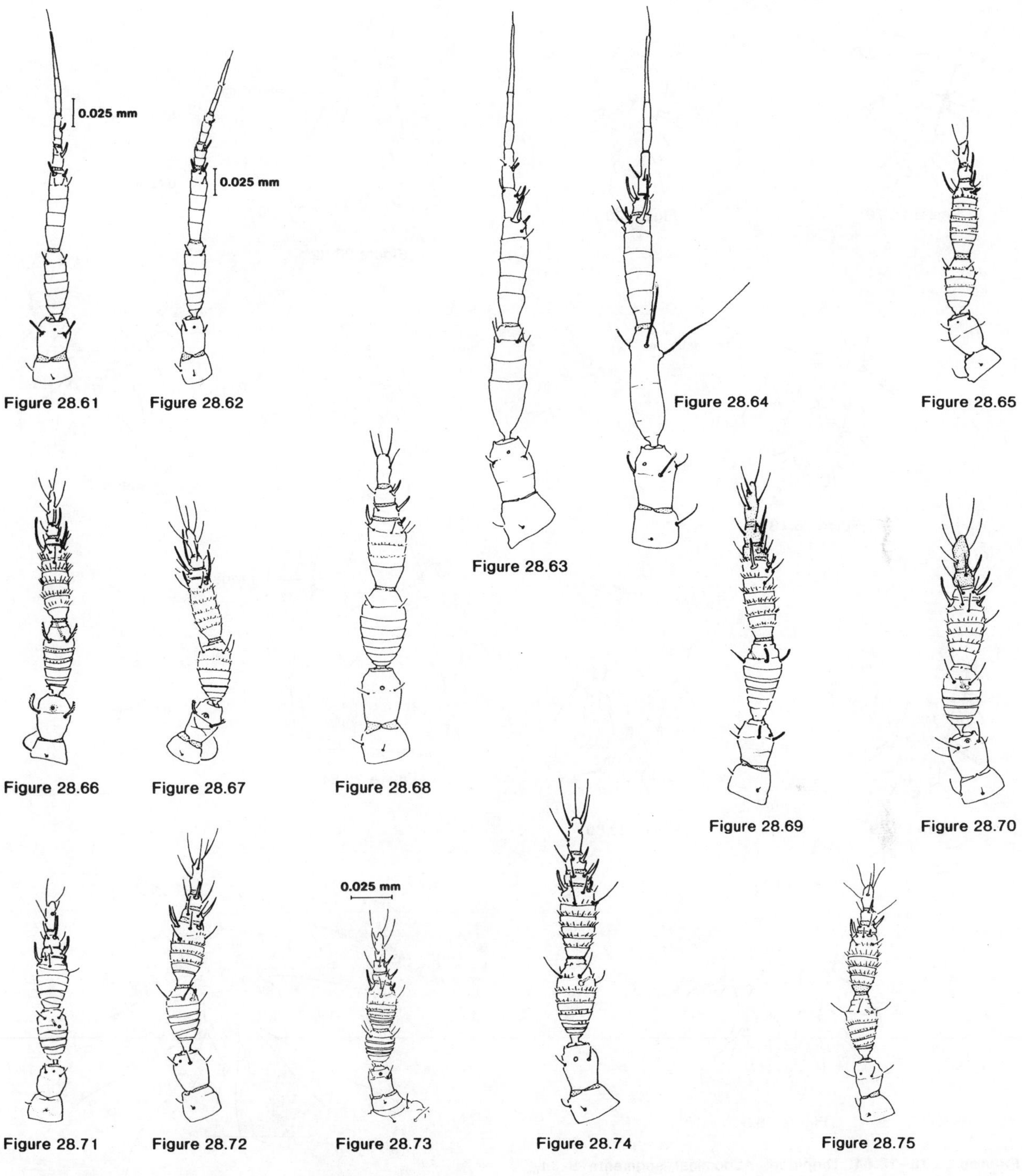

Figures 28.61–28.75. Thripidae. Antennae of larvae II (right except where indicated).

Figure 28.61. *Heliothrips haemorrhoidalis* (redrawn from Vance, 1974).

Figure 28.62. *Hercinothrips femoralis* (redrawn from Vance, 1974).

Figure 28.63. *Rhipiphorothrips cruentatus.*

Figure 28.64. *Selenothrips rubrocinctus* (left).

Figure 28.65. *Dendrothrips ornatus.*

Figure 28.66. *Sericothrips variabilis.*

Figure 28.67. *Scirtothrips* sp. (left).

Figure 28.68. *Limothrips cerealium* (redrawn from Vance, 1974).

Figure 28.69. *Anaphothrips obscurus.*

Figure 28.70. *Apterothrips secticornis.*

Figure 28.71. *Aptinothrips stylifer.*

Figure 28.72. *Chilothrips pini.*

Figure 28.73. The Tobacco thrips, *Frankliniella fusca* (Hinds) (larva I) (redrawn from Heming, 1975).

Figure 28.74. *Thrips simplex.*

Figure 28.75. *Thrips tabaci.*

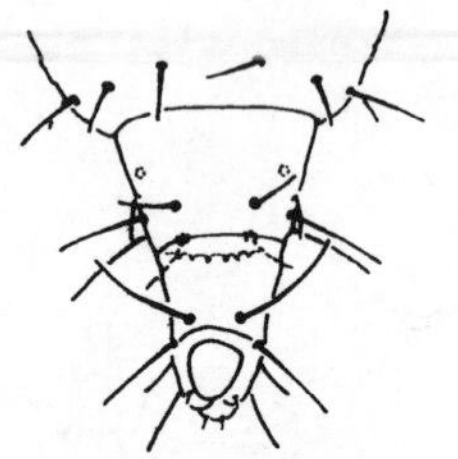

Figure 28.76

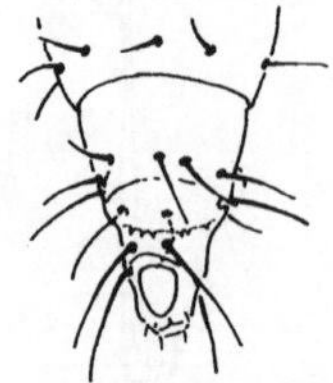

Figure 28.77

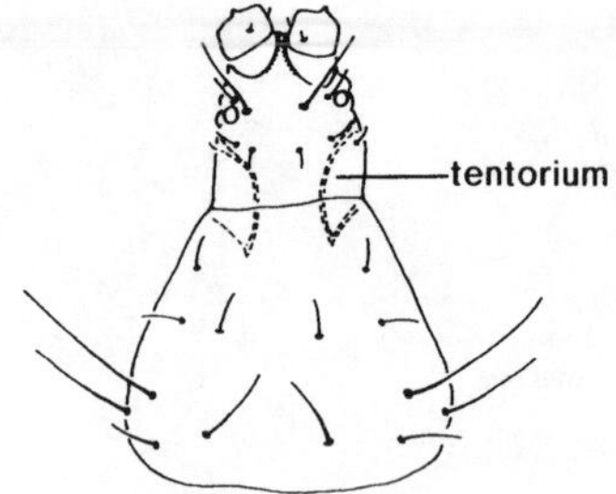

Figure 28.82

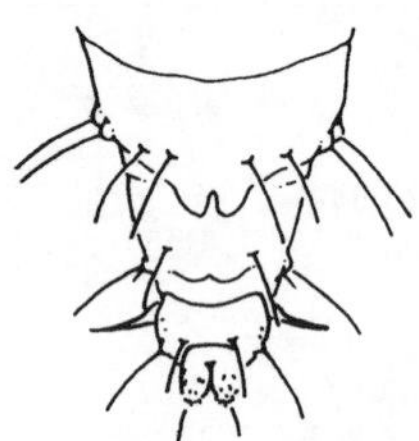

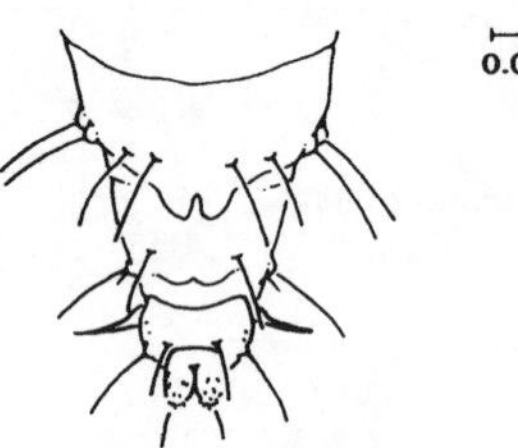

Figure 28.78

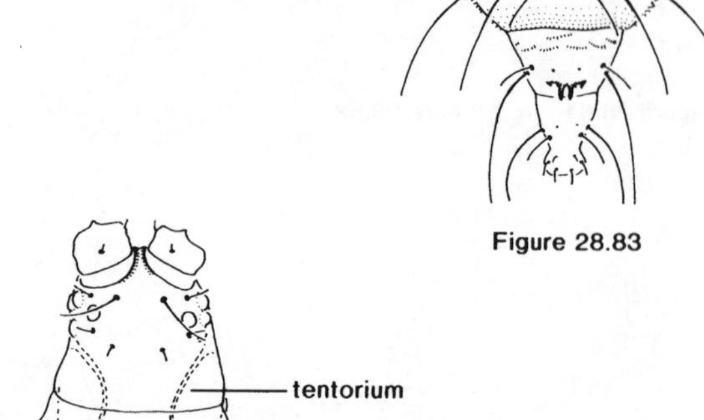

Figure 28.83

Figure 28.84

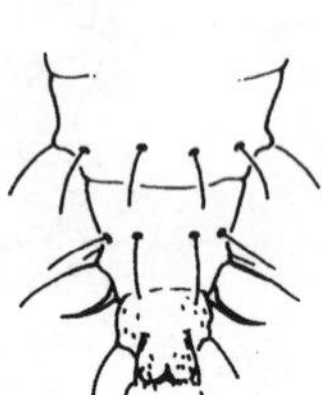

Figure 28.79

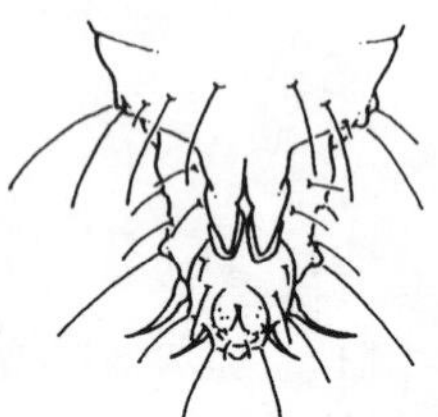

Figure 28.80

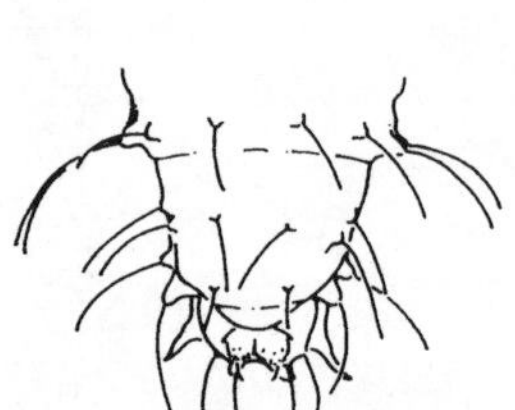

Figure 28.81

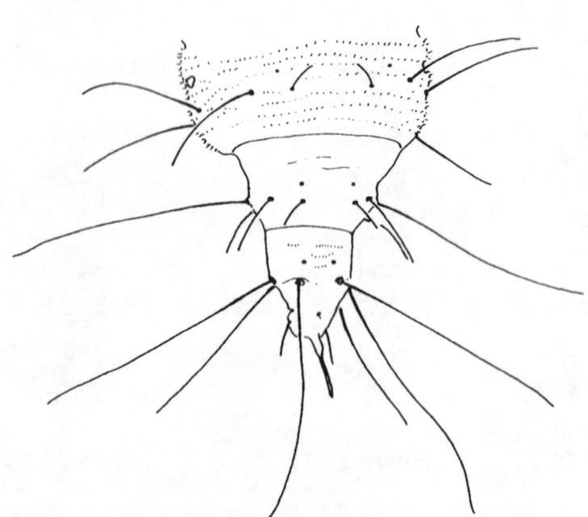

Figure 28.85

Figures 28.76–28.81. Thripidae. Abdominal segments 8–11, ventral, showing sex differences.

Figure 28.76. *Frankliniella fusca,* larva II female.

Figure 28.77. *F. fusca,* larva II male.

Figure 28.78. *Thrips tabaci,* propupa female (redrawn from Lange and Razvyazkina, 1953).

Figure 28.79. *T. tabaci,* propupa male (redrawn from Lange and Razvyazkina, 1953).

Figure 28.80. *T. tabaci,* pupa female (redrawn from Lange and Razvyazkina, 1953).

Figure 28.81. *F. fusca,* pupa male.

Figures 28.82–28.85. Merothripidae. Larvae I and II of *Merothrips morgani* Hood.

Figure 28.82. Head and pronotum, larva II.

Figure 28.83. Abdominal segments 8–11, larva II.

Figure 28.84. Head and pronotum, larva I.

Figure 28.85. Abdominal segments 8–11, larva I.

MEROTHRIPIDAE

The Large-Legged Thrips

Figures 28.21, 28.82–28.85

Relationships and Diagnosis: This small family, believed by Mound and O'Neill (1974) and Mound *et al.* (1980) to be the most primitive extant family in the order, constitutes the sister group of the Aeolothripidae. These two families probably share a common ancestor with the stem group of the monobasic Uzelothripidae, a highly specialized relic taxon known from Brazil and Singapore.

Larvae of *Merothrips morgani* Hood, the only North American species for which larvae are known, are distinguished from those of other thrips by their antennae (fig. 28.21). These have segments 6 and 7 fused, annuli only weakly developed and devoid of microtrichia, and segment 5 as long as 4. In addition, larvae of both instars have well-developed anterior and posterior tentorial arms within their head capsules (figs. 28.82, 28.84).

Biology and Ecology: Collecting records indicate that most species live in the upper layer of leaf litter or on freshly dead wood where they feed on fungal mycelia or on the products of fungal action (Mound and O'Neill, 1974).

Description: (*Merothrips morgani* (Hood)) Mature larva II < 1.0 mm long. Body elongate, slender and pale to yellow. Cuticle delicate, soft and colorless except for head, antennae, legs and portions of terminal abdominal segments which are lightly sclerotized and yellow to light grey; cuticle of pterothorax and abdominal segments 1–8 stippled with very fine microtrichia (figs. 28.83, 28.85). Setae long, delicate and pointed.

Head: Longer than broad, cheeks slightly divergent caudad. Cuticle of head capsule smooth, lightly sclerotized and slightly pigmented on inner margins of antennal sockets. Seta pair 2 on vertex much longer than 1, or 3–5 (figs. 28.82, 28.84). Antennae 6-segmented, with segments 6 and 7 fused, segments 3–6 bearing 3–5 weakly-developed annuli devoid of microtrichia and segment 5 about the same length as 4, (fig. 28.21); sense cones and setae long and slender. Ommatidia, 4, with slightly bulging facets, on either side of head behind antennal insertions and subtended by orange to red pigment. Tentorium consisting of anterior and posterior arms joined at each side of head (figs. 28.82, 28.84). Mouthcone short, blunt, hypognathous, and almost symmetrical, with left mandibular stylet very small as in larvae of *Melanthrips* spp. (Aeolothripidae); maxillary palpi 3-segmented (segments 2 and 3 fused in some specimens); labial palpi 2-segmented.

Thorax: Pronotum relatively long and slender. Cuticle unsclerotized; that of pterothorax stippled with fine microtrichia. Seta pair 2 on pronotum far caudad of 3 and in line with 4 (figs. 28.82, 28.84); seta pair 5 very long and slender. Mesothoracic spiracles very large and resembling those of some larval phlaeothripids. Legs relatively short and stout with fore and hind femora relatively broad (Heming, 1972).

Abdomen: Long and slender. Spiracles very small and situated antero-laterally on segments 2 and 8. Seta pair 1 and 2 on tergite 9 of larvae I, normal (fig. 28.85), those of larvae II modified into short spines, with 2 smaller than 1 (fig. 28.83). Chaetotaxy of segment 9 sexually dimorphic at least in larvae II: 5 pairs in females (including spines), 6 pairs in males. Terminal abdominal setae of larvae I relatively much longer than those of larvae II (fig. 28.85). Tip of abdomen in larvae I produced caudad to encompass tips of developing terminal setae of larvae II (fig. 28.85).

Comments: Six species of Merothripidae, all in the genus *Merothrips,* occur in North America (including Mexico). None are of economic importance. Only *M. morgani* and *M. floridensis* Watson are relatively common, the former in the southern and eastern U.S., the latter in the south only.

Mound and O'Neill (1974) treated the fauna of the world and provided keys to genera and species, illustrations of adults, and discussions of life history, phylogeny and distribution.

Selected Bibliography

Heming 1972.
Mound and O'Neill 1974.
Mound et al. 1980.

Suborder TUBULIFERA

PHLAEOTHRIPIDAE

The Phlaeothripids

Figures 28.5, 28.7, 28.10, 28.13, 28.14, 28.17, 28.22, 28.86–28.105

Relationships and Diagnosis: This family, considered by most authors as the only one in the suborder Tubulifera, is the largest in the order, and its member species are the most highly derived. It probably is the sister group of all other Thysanoptera, although some authors consider it to be the sister group of the subfamily Panchaetothripinae of the Thripidae (Mound et al., 1980: figures 42, 43).

Larvae are distinguished from those of all other thrips by the following combination of characteristics: usually sluggish in their movements; abdominal segment 10 strongly sclerotized, elongate and tubular (figs. 28.5, 28.10, 28.89, 28.91, 28.93, 28.100, 28.101); body cuticle smooth; maxillary stylets in most specimens longer than mouthcone and invaginated into head (figs. 28.87, 28.90); and antennae with middle segments (3–6) devoid of annuli or microtrichia (figs. 28.22, 28.95–28.99).

Biology and Ecology: Phlaeothripids are classified into two subfamilies, the Phlaeothripinae and Idolothripinae, based on the structure of their maxillary stylets (Mound et al. 1976; Mound and Palmer, 1983). The vast majority of species are in the Phlaeothripinae and are characterized in both larvae and adults by their relatively slender stylets (e.g. fig. 28.87).

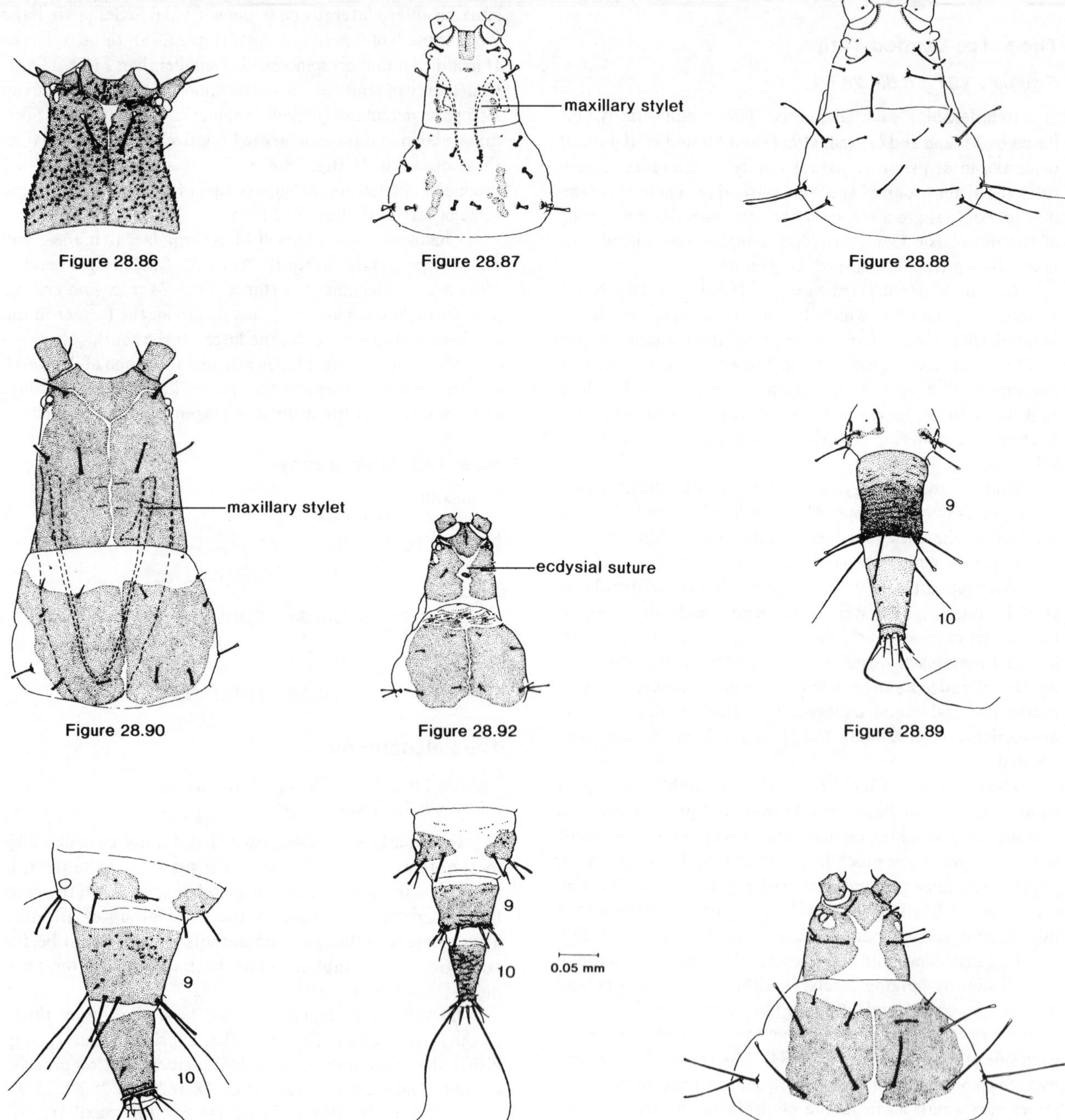

Figures 28.86–28.94. Phlaeothripidae. Heads, pronota, and abdominal segments 8–11 of larvae II.

Figure 28.86. Head of *Acanthothrips nodicornis* (O. M. Reuter).

Figure 28.87. *Cephalothrips monilicornis* (O. M. Reuter).

Figure 28.88. The Cuban laurel thrips, *Gynaikothrips ficorum* (Marchal).

Figure 28.89. *G. ficorum*.

Figure 28.90. *Compsothrips* sp.

Figure 28.91. *Compsothrips* sp.

Figure 28.92. *Haplothrips halophilus* Hood.

Figure 28.93. *H. halophilus*.

Figure 28.94. *Liothrips* sp.

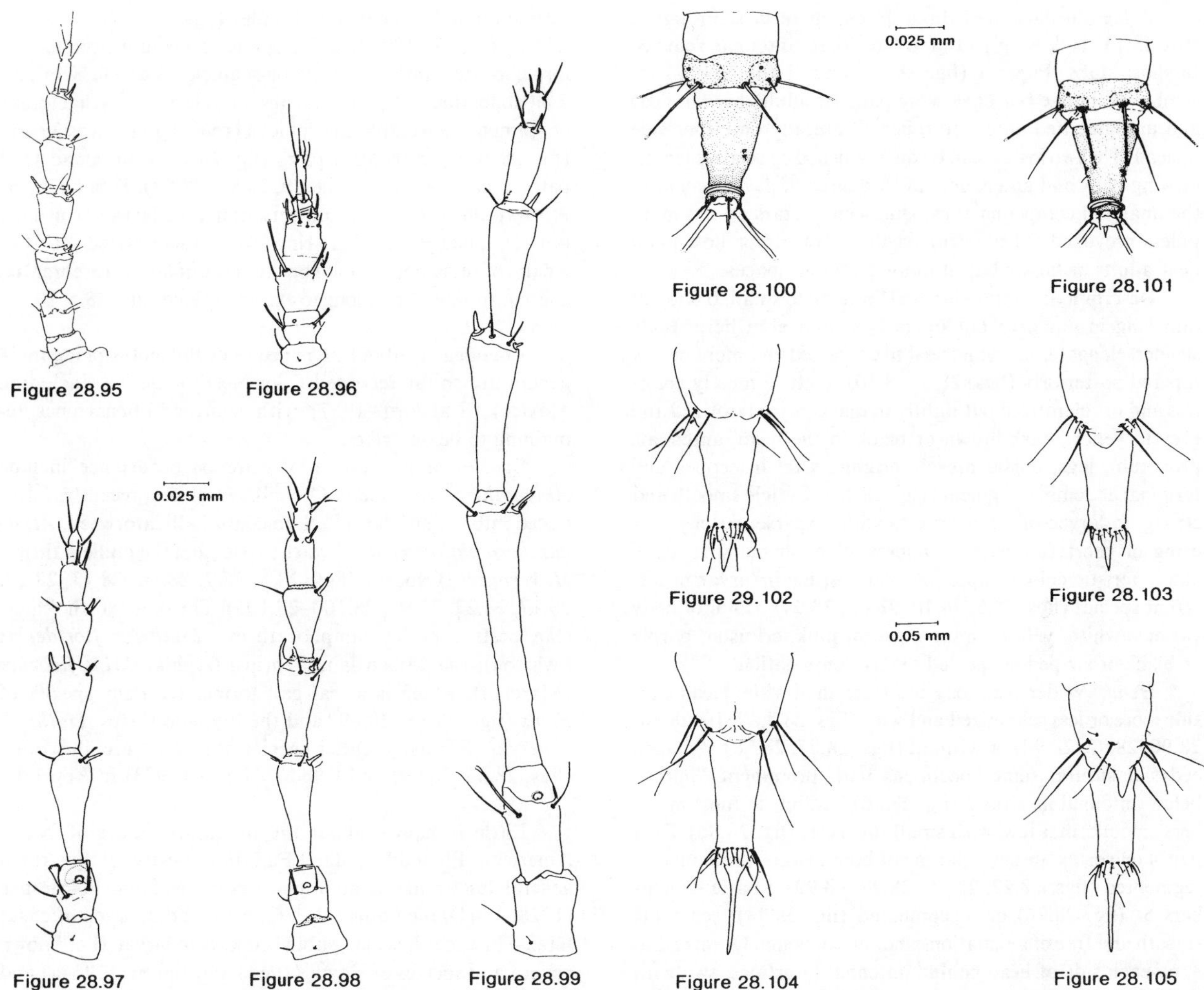

Figures 28.95–28.99. Phlaeothripidae. Right antennae of larvae I and II.

Figure 28.95. The Mullein thrips, *Haplothrips verbasci* (Osborn)—larva I (redrawn from Heming, 1975).

Figure 28.96. *Stephanothrips* sp. larva I.

Figure 28.97. *Cephalothrips monilicornis*—larva II.

Figure 28.98. The Virginia Creeper thrips, *Liothrips russelli* (Hood)—larva II.

Figure 28.99. *Compsothrips* sp.—larva II.

Figures 28.100–28.105. Phlaeothripidae. Abdominal segments 9-11 of *Haplothrips verbasci* showing sex differences.

Figure 28.100. Larva I female, dorsal aspect.

Figure 28.101. Larva I male, dorsal aspect.

Figure 28.102. Pupa I female, ventral aspect.

Figure 28.103. Pupa I male, ventral aspect.

Figure 28.104. Pupa II female, ventral aspect.

Figure 28.105. Pupa II male, ventral aspect.

Most phlaeothripines live in leaf litter or on recently dead wood and eat fungal hyphae or the products of fungal action. In the tropics and subtropics, many are leaf feeders and some cause galls on leaves (Ananthakrishnan, 1978) e.g. *Gynaikothrips ficorum* (Marchal) on *Ficus* spp. (figs. 28.88, 28.89) and *Liothrips* spp. (fig. 28.94). In North America, members of the tribe Haplothripini are particularly common and occur in large numbers on flowers of Asteraceae, Poaceae and Cyperaceae (e.g. *Haplothrips halophilus* Hood, figs. 28.92, 28.93). A few are known predators on other small arthropods.

The Idolothripinae is a small taxon containing our largest thrips (Mound and Palmer, 1983). Both larvae and adults are recognized by their long, band-like maxillary stylets (fig. 28.90), which they use to suck up fungal spores in their leaf litter or dead wood habitat. A species in Panama, *Anactinothrips gustaviae,* has advanced, parasocial behavior and lives in colonies on smooth barked trees, reproduces communally, cooperatively broods young, and forages in a highly coordinated fashion (Keister and Strates, 1984). All phlaeothripids are more or less specialized for a cryptophilous existence.

Phlaeothripids have three quiescent instars. Propupae (fig. 28.13) lack wing pads and have their antennae reduced to short stubs. Pupae I (fig. 28.14) and II (fig. 28.17) resemble propupae but have wing pads (if adults are winged) and unsegmented antennae reflexed laterally on either side of head. The two instars can be distinguished by relative length of wing pads and antennae and by degree of development of the imaginal compound eyes. Quiescent instars do not move unless provoked. They occur on the substrate or host plant near adults or larvae but in more confining spaces.

Description: Mature larvae II minute to small, 0.5 to 5.0 mm long in our area but up to 14.0 mm elsewhere. Body slender, elongate, subcylindrical to depressed and more or less tapered posteriorly (figs. 28.5, 28.10). Cuticle mostly colorless and unsclerotized but lightly to markedly sclerotized and grey to yellow, dark brown or black in the head, antennae, pronotum, legs, dorsal muscle origins, setal insertions and terminal abdominal segments (fig. 28.10). Cuticle smooth and strong, in larvae of a few species with a sparse, patchy covering of short tubercles or microtrichia. Primary setae of characteristic color, shape, size and position in larvae of different species (figs. 28.5, 28.10, 28.86–28.94). Color of body various: white, yellow, orange, green, pink, crimson, purple or black or striped or spotted in any combination.

Head: Wider than long to longer than wide. Head capsule more or less sclerotized and with (figs. 28.5, 28.10, 28.86, 28.90, 28.92, 28.94) or without (figs. 28.87, 28.88) y-shaped ecdysial suture; some specimens with prominent "horns" below antennal insertions (fig. 28.86). Cuticle in most members smooth, in a few, with small tubercles (fig. 28.86). Seta pair 4 on vertex absent in larvae of both instars. Antennae 7-segmented (figs. 28.22, 28.95, 28.97, 28.99), in a few members 5- (fig. 28.96) or 6-segmented (fig. 28.98); segments smooth and free of annulations and microtrichia. Ommatidia, 3, on each side of head behind antennal insertions, two with bulging facets, the posterior one marked by an area of flat, transparent cuticle (figs. 28.86–28.88, 28.90, 28.92, 28.94). Tentorium reduced to anterior arms. Mouthcone opisthognathous and of characteristic length and shape; maxillary stylets slender (fig. 28.87) or broad and band-like (fig. 28.90), in most larvae long and extended cephalad into head capsule; maxillary palpi 2-segmented, labial palpi 1-segmented.

Thorax: Pronotum colorless (fig. 28.88), with brown sclerotized spots over muscle insertions (fig. 28.87), or with 2 pronotal plates (figs. 28.5, 28.10, 28.90, 28.92, 28.94). Mesothoracic spiracles large and protuberant in most larvae (figs. 28.5, 28.10). Seta pair 5 of mesonotum absent in larvae II; an extra pair of setae (6) on metanotum (fig. 28.10). Legs short and stout to long and slender, usually brown to dark brown or black (figs. 28.5, 28.10) and smooth (Heming, 1972).

Abdomen: Slender, elongate, subcylindrical to depressed and more or less tapered at posterior end. Segments 9 and 10 and, in larvae II of some species, segment 8, more or less sclerotized and light to dark brown or black; segment 10 and often 9 elongate and tubular (figs. 28.5, 28.10, 28.89, 28.91, 28.93, 28.100, 28.101). Spiracles medium-sized to very large, located anterolaterally on segments 2 and 8. Seta pair 2 on abdominal sternites 3–8 absent in larvae II. Chaetotaxy of segment 9 sexually-dimorphic in larvae I: in females 3 pairs (fig. 28.100), in males 4 pairs (fig. 28.101), in larvae II, 5 pairs in both sexes (figs. 28.89, 28.91, 28.93). Seta pair 2 on 9th tergum in larvae I and seta pair 3 in larvae II usually minute. Last segment (11) ring-like and sclerotized with 4 or 5 pairs of setae not homologous to those of larval Terebrantia; one ventro-lateral pair long to extremely long (figs. 28.5, 28.10, 28.89, 28.93).

Comments: About 365 species of Phlaeothripidae in 74 genera are so far recorded from North America (including Mexico) (Stannard, 1957) with many additional ones remaining to be described.

Species of interest or of common occurrence include *Amynothrips andersoni* O'Neill, a species recently introduced into the southern U.S. to control alligatorweed, *Alternanthera philoxeroides* (Mart.) Gris.; and the mullein thrips, *H. verbasci* (Osborn) (figs. 28.5, 28.7, 28.10, 28.13, 28.14, 28.17, 28.22, 28.95, 28.100–28.105). The only North American pests are the camphor thrips, *Liothrips floridensis* (Watson); the Cuban laurel thrips, *Gynaikothrips ficorum* (Marchal), which is a leaf-gall former on many species of *Ficus* (figs. 28.88, 28.89) and the lily bulb thrips, *Liothrips vaneecki* (Priesner). Other species of interest are illustrated (figs. 28.86–28.99) and listed by Lewis (1973) in Appendix 6.

Little is known about the immature stages of North American Phlaeothripidae. Full but mostly unillustrated general descriptions of all immatures can be found in Priesner (1928, 1964), including a brief, accurate discussion of chaetotaxy. Priesner has also published keys to larvae II of known genera and species of Egypt (1964) and Europe (1964a) and has provided detailed descriptions of some of these (1928, 1964).

Keys to and descriptions of adults of North American phlaeothripids can be found in Cott (1956—California species), Stannard (1957—North America; 1968—Illinois and eastern U.S.) and Thomasson and Post (1966—North Dakota), together with information on habitat and life history.

Selected Bibliography

Ananthakrishnan 1978.
Cott 1956.
Heming 1972.
Jacot-Guillarmod 1978.
Mound and Palmer 1983.
Mound et al. 1980.
Priesner 1928, 1964, 1964a.
Stannard 1957, 1968.
Thomasson and Post 1966.

BIBLIOGRAPHY

Ananthakrishnan, T. N. 1978. Thrips galls and gall thrips. Zool. Surv. India. Tech. Monogr. No. 1, 95 pp.

Ananthakrishnan, T. N. 1984. Bioecology of thrips. Indira Publ. House, Oak Park. 233 pp.

Ananthakrishnan, T. N. and S. Sen. 1980—Taxonomy of Indian Thysanoptera. Zool. Surv. India, Handbook Series No. 1., 258 pp.

Bailey, S. F. 1940. The distribution of injurious thrips in the United States. J. Econ. Ent. 33: 133–136.

Bailey, S. F. 1940a. Cocoon-spinning Thysanoptera. Pan.-Pacif. Ent. 16: 77–79.

Bailey, S. F. 1951. The genus *Aeolothrips* Haliday in North America. Hilgardia 21(2): 43–80.

Bailey, S. F. 1957. The thrips of California. Part I: suborder Terebrantia. Bull. Calif. Ins. Surv. 4(5): 139–220.

Bailey, S. F. and H. E. Cott. 1954. A review of the genus *Heterothrips* Hood (Thysanoptera: Heterothripidae) in North America, with descriptions of two new species. Ann. Ent. Soc. Amer. 47: 614–35.

Chiasson, H. 1986. A synopsis of the Thysanoptera (thrips) of Canada. Macdonald Coll., Lyman Ent. Mus. Res. Lab., Mem No. 17. 153 pp.

Cott, H. E. 1956. Systematics of the suborder Tubulifera (Thysanoptera) in California. Univ. Calif. Publs. Ent. 13: 1–216.

Derbeneva, N. N. 1967. New data on the biology and structure of the preimaginal phases and stages of the predatory thrips, *Aeolothrips intermedius* Bagnall (Thysanoptera, Aeolothripidae) (In Russian). Ent. Obozr. 46: 629–644.

Heming, B. S. 1969. A modified technique for mounting Thysanoptera in Canada Balsam. Ent. News. 80: 323–328.

Heming, B. S. 1972. Functional morphology of the pretarsus in larval Thysanoptera. Can. J. Zool. 50: 751–766.

Heming, B. S. 1973. Metamorphosis of the pretarsus in *Frankliniella fusca* (Hinds) (Thripidae) and *Haplothrips verbasci* (Osborn) (Phlaeothripidae) (Thysanoptera). Can. J. Zool. 51: 1211–1234.

Heming, B. S. 1975. Antennal structure and metamorphosis in *Frankliniella fusca* (Hinds) (Thripidae) and *Haplothrips verbasci* (Osborn) (Phlaeothripidae) (Thysanoptera). Quaest. ent. 11: 25–68.

Heming, B. S. 1978. Structure and function of the mouthparts in larvae of *Haplothrips verbasci* (Osborn) (Thysanoptera, Tubulifera, Phlaeothripidae). J. Morph. 156: 1–38.

Hennig, W. 1981. Insect phylogeny. John Wiley and Sons. 514 pp.

Jacot-Guillarmod, C. F. Catalogue of the Thysanoptera of the world. Ann. Cape Prov. Mus. (Nat. Hist.). 7 Part I: 1–216 (1970) (Aeolothripidae, Merothripidae, Heterothripidae, Uzelothripidae), Part 2: 217–515 (1971) (Thripidae), Part 3: 517–976 (1974) (Thripidae), Part 4: 977–1255 (1975) (Thripidae) Part 5: 1257–1556 (1978) (Phlaeothripidae), Part 6: 1557–1724 (1979) (Phlaeothripidae), 17 Part 1: 1–93 (1986) (Phlaeothripidae) (Continuing).

Keister, A. R. and E. Strates. 1984. Social Behavior in a thrips from Panama. J. Nat. Hist. 18:303–314.

Kirk, W. D. J. 1985. Egg-hatching in thrips (Insecta: Thysanoptera). J. Zool. Lond. (A) 207:181–190.

Kirk, W. D. J. 1987. A key to the larvae of some common Australian flower thrips (Insecta: Thysanoptera), with a host-plant survey. Austral. J. Zool. 35:173–185.

Kono, T. and C. S. Papp. 1977. Handbook of Agricultural pests. Calif. Dept. Food and Agr., Div. Plt. Ind., Lab. Ser.—Entomology. Thrips, p. 87–134.

Lange, A. B. and G. M. Razvyazkina. 1953. Morphology and development of the Tobacco thrips. (In Russian). Zool. Zhurn. 32: 576–593.

Lewis, T. R. 1973. Thrips: their Biology, Ecology and Economic Importance. Academic Press, New York and London. 349 p.

Miyazaki, M. and I. Kudo. 1986. Descriptions of thrips larvae which are noteworthy on cultivated plants (Thysanoptera: Thripidae). I. Species occurring on solanaceous and curcurbitaceous crops. Akitu. N.S. No. 79. 25 pp.

Mound, L. A., 1973. Thrips and whitefly, p. 229–242. *In* Gibbs, A. J. (ed.), Viruses and Invertebrates. North Holland Publishing Co., Amsterdam. 673 p.

Mound, L. A., B. S. Heming and J. M. Palmer. 1980. Phylogenetic relationships between the families of recent Thysanoptera (Insecta). Zool. J. Linn. Soc. 69:111–141.

Mound, L. A., G. D. Morison, B. R. Pitkin and J. M. Palmer. 1976. Thysanoptera. Handbooks for the Identification of British Insects. Roy. Ent. Soc. London. Volume 1 Part II, 79 p.

Mound, L. A. and K. O'Neill. 1974. Taxonomy of the Merothripidae, with ecological and phylogenetic considerations (Thysanoptera). J. Nat. Hist. 8: 481–509.

Mound, L. A. and J. M. Palmer. 1983. The generic and tribal classification of the spore-feeding Thysanoptera (Phlaeothripidae: Idolothripinae). Bull. Brit. Mus. (Nat. Hist.) 46(1): 1–174.

Mound, L. A. and B. R. Pitkin. 1972. Microscopic wholemounts of thrips (Thysanoptera). Ent. Gaz. 23: 121–125.

Priesner, H. 1928. Die Thysanopteren Europas. Verlag F. Wagner, Wien. 755 p.

Priesner, H. 1964. A monograph of the Thysanoptera of the Egyptian deserts. Publs. Inst. Desert, Cairo No. 13 pp. 1–549.

Priesner, H. 1964a. Ordnung Thysanoptera. Bestimmungsbucher zur Bodenfauna Europas. Liefrung 2. Akademie Verlag, Berlin. 242 p.

Speyer, E. R. and W. J. Parr. 1941. The external structure of some thysanopterous larvae. Trans. Roy. Ent. Soc. Lond. 91:559–635.

Stannard, L. J. Jr. 1957. The phylogeny and classification of the North American genera of the suborder Tubulifera (Thysanoptera). Ill. Bio. Monogr. No. 25. 200 p.

Stannard, L. J. Jr. 1968. The thrips or Thysanoptera of Illinois. Ill. Nat. Hist. Surv. Bull. 29(4): 215–552.

Thomasson, G. L. and R. L. Post. 1966. North Dakota Tubulifera (Thysanoptera). N. D. Ins. Publ. No. 6 Dept. Ent., Agric. Expt. Sta. N. D. State Univ. 58 p.

Vance, T. C. 1974. Larvae of the Sericothripini (Thysanoptera: Thripidae), with reference to other larvae of the Terebrantia of Illinois. Ill. Nat. Hist. Surv. Bull. 31(5): 145–208.

Order Hemiptera

29

Thomas R. Yonke
University of Missouri–Columbia

THE TRUE BUGS

The Hemiptera is composed of those insects usually referred to as true bugs. This is in contrast to the generic term, "bug", used for any insect-like form. The Hemiptera is herein treated in the strict sense (*s. str.*). The Homoptera, sometimes treated as a suborder of the Hemiptera, is in this book given ordinal rank.

DIAGNOSIS

Larvae of true bugs are minute, ca. 1 mm, to very large, 25 mm; oval to elongate oval, robust, sometimes flattened; often ornamented with spines and setae; often brownish and with bright color markings; antennae usually composed of 4 elongate, filiform segments; antennal and leg segments often modified with triangular, bulbular and leaf-like dilations; mouthparts modified into an elongate, thick labium which holds mandibular and maxillary stylets in a deep groove; compound eyes well developed. Meso- and metathoracic wing pads well developed in 4th and 5th instars. Abdominal scent gland openings often present on abdominal tergites.

BIOLOGY AND ECOLOGY

Hemiptera are predominantly phytophagous, but many are predators, some are ectoparasites, and others feed on fungi and phytoplankton.

True bugs can be found in a wide variety of aquatic and terrestrial habitats. Most species are relatively host specific, albeit that we do not clearly understand their host and habitat relationships and behaviors. Some may utilize, if not require, a succession of feeding hosts during development. A "critical host" for development may be one that is required by the second and third instars, while the fifth instar and adults may feed on a wider variety of hosts. This is especially true for plant-feeding species. Consequently, the collector should pay special attention to recording feeding observations on hosts for each instar as well as for adults.

Most hemipterans are either uni- or bivoltine with 5 instars, and overwintering usually occurs in the adult stage; however, some species pass the winter as eggs or larvae.

The eggs of the true bugs are represented by a wide variety of forms from barrel-shaped, to oval, to cylindrical (figs. 29.1a,b), to columnar, to elongate. Those laid in the open on substrates or dropped into the litter have a thick chorion and are often ornately sculptured, while eggs laid in plant tissue or in tight crevices such as under bracts are usually soft and elongate. In some species single eggs are laid separately, but in many species eggs are laid in batches which are noted for the number of eggs and their pattern.

DESCRIPTION

Head (fig. 29.2a,b) triangular to quadrate with antennae arising on lateral aspect, often from distinct antenniferous tubercles; anterior margin usually 3-lobed, median tylus and paired lateral juga. Labium usually 3- or 4-segmented, arising from anterior region of head, and extending caudad along ventral groove, segment 1 in part or totally positioned between shallow to deep ventrally extending plates, the bucculae; labrum elongate, triangular, tongue-like, positioned over basal aspect of labium.

Pronotum quadrate, nearly flat to slightly rounded dorsally; anterior margin often extending over caudal region of head; median, paired, oval areas, the calli, sometimes present on anterior third of pronotum; meso- and metanotum bearing distinct wing pads in 4th and 5th instars of alate forms (figs. 29.lf,g), those of the 5th extending well onto the 3rd abdominal segment; 3rd instar with some posteriolateral bud-like development of wings (fig. 29.le); thoracic nota of 1st and 2nd instars undistinguished (figs. 29.lc,d).

Abdomen oval to elongate, robust, terminating in ring-like segment bearing anal opening; tergites usually flat to slightly raised, often sculptured with tubercles, spines, scoli and differentially pigmented areas; caudal margins of tergites 3, 4, 5 and/or 6, and often anterior margins of respective next caudal segment, tuberculate and bearing abdominal scent gland opening(s) (figs. 29.lc–g); abdomen bearing specialized setae, trichobothria, on sternites 3–7 (Schaefer 1975).

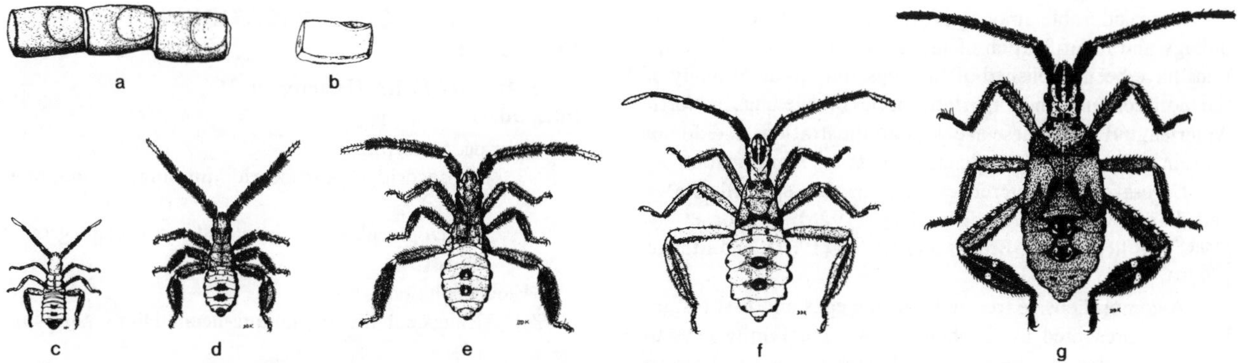

Figure 29.1a-g. Coreidae. *Leptoglossus corculus* (Say), a. dorsal, and b. lateral view of eggs; c. 1st, d. 2nd, e. 3rd, f. 4th, g. 5th instars, respectively.

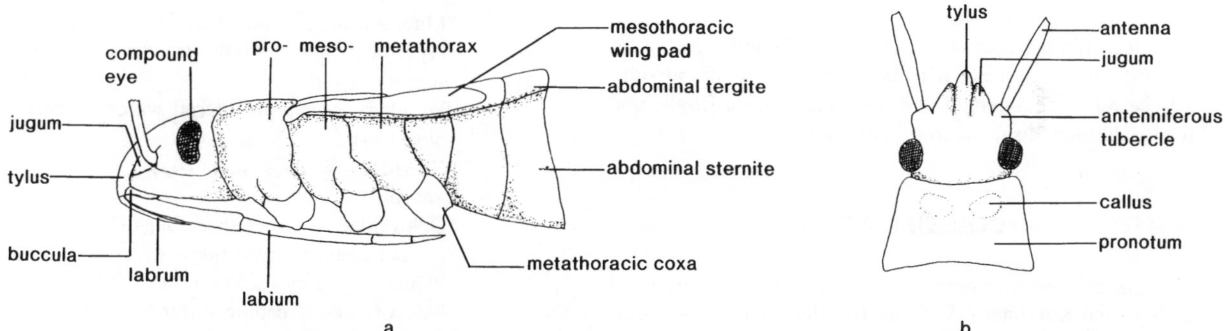

Figure 29.2a-b. a. Lateral view of true bug, b. dorsal view of head and pronotum.

A simple progression of development can generally be seen, such as in the leaffooted pine seed bug, *Leptoglossus corculus* (Say) (figs. 29.lc–g). However, some species undergo rather distinct changes in both color and certain structures between instars, making recognition difficult. It is usually necessary to either rear them to adults or collect eggs and larvae in direct association with adults to obtain accurate species identification, unless one has known specimens for comparison or adequate literature is available.

Aptery and brachyptery are frequent occurrences among the Hemiptera, however, these adults can usually be recognized by the development of the external genitalia. In contrast, the caudal abdominal segments of larvae are ring-like and usually bear only the anus. Fifth-instars of some species can be sexed, but their genitalia are not well developed. Metathoracic scent gland openings are present immediately in front of, sometimes slightly above, the metacoxae in adults of most families of Hemiptera. Immatures have no such metathoracic scent glands. Ocelli are present in some wingless adults; but are usually absent, or if present are poorly developed in larvae of Hemiptera. By carefully assessing the combination of the above mentioned features one should be able to successfully separate larvae from wingless adults.

Measurement data are given for 5th instars and the following terms apply: minute, less than or equal to 1 mm; very small, 2–4 mm; small, 5–6 mm; medium, 7–15 mm; large, 16–25 mm; and very large, 26+ mm.

COMMENTS

Many species are of considerable direct importance to humans as pests of agricultural crops and health. The former group includes such notable true bugs as the tarnished plant bug, *Lygus lineolaris* (Palisot de Beauvois), the chinch bug, *Blissus leucopterus leucopterus* (Say), and the squash bug, *Anasa tristis* (De Geer). The bed bug, *Cimex lectularius* L., and the kissing bugs, *Triatoma* spp., are examples of noxious hemipterans that bite humans and other animals.

Some families are included for which immatures are extremely rare or unknown in collections. They are indicated in the classification by an asterisk. Their treatment is presented with the hope that more deliberate attempts will be made by collectors to obtain these specimens and make them available for comparative studies. Eggs and larvae of many species, even of the more common families, remain unknown.

A considerable amount of work is yet to be done on the biology and identification of immature Hemiptera. Descriptions have been published of the eggs and larvae of many of the more common and important species of true bugs in North America, but often these are without illustrations and do not contain sufficient diagnostic statements for identification.

Several general references will be useful, especially Slater and Baranowski (1978), Blatchley (1926), Borror et al. (1981), Miller (1971), Brooks and Kelton (1967), and Menke (1979).

A comprehensive treatment of the eggs and first instars has been presented by Cobben (1968), and family keys to larvae provided by Leston and Scudder (1956), Lawson (1959), DeCoursey (1971), Herring and Ashlock (1971), Slater and Baranowski (1978), and Menke (1979).

TECHNIQUES

Eggs and larvae should be stored in 75–80% alcohol for general study. Larvae of Miridae should be stored individually because the antennal and leg segments are fragile and tend to become dismembered in alcohol.

CLASSIFICATION

The classification employed is, in general, that proposed by Stys and Kerzhner (1975) for the Heteroptera (= Hemiptera s. str.). The inclusion of 7 infraorders more appropriately reflects the taxonomic relationships of hemipteran families than does the use of the taxa Hydrocorizae, Amphibicorizae and Geocorizae of China and Miller (1959). The latter classification has been generally used (ex. Borror et al. 1981) in textbooks which serve to introduce students to the classification of Hemiptera.

Stys and Kerzhner (1975) summarized the historical treatment of heteropteran higher taxa names. The "morpha" infraordinal classification will serve to better orient the student to current literature on the systematics of Hemiptera, such as Cobben (1968, 1978), Polhemus (1977, 1981), and Anderson (1979, 1982). The current intense interest in the higher classification of Hemiptera will likely suggest changes in some family relationships. Here they are discussed only in the context of the North American fauna. The student wishing to know a group's relationships within Hemiptera on a world scale is directed to more comprehensive references such as Cobben (1978), Stys and Kerzhner (1975), and those suggested for each family.

Different views of the origin of Heteroptera and Homoptera and the behaviors of ancestral families are discussed by Cobben (1968, 1978, 1979), Schaefer (1975), Polhemus (1977), and Sweet (1979), among others. The debate essentially focuses on whether the ancestral hemipterans were carnivorous and hydrophilous or phytophagous and terrestrial.

CLASSIFICATION

True Bugs

Order **HEMIPTERA** (Heteroptera)[1]

Infraorders

- Dipsocoromorpha
  - *Dipsocoridae—dipsocorid jumping ground bugs (4)
  - *Schizopteridae—schizopterid jumping ground bugs (5)
- Enicocephalomorpha
  - *Enicocephalidae—unique-headed bugs, gnat bugs (7)
- Leptopodomorpha
  - *Leptopodidae—spiny shore bugs (1)
  - Saldidae—shore bugs (75)
- Gerromorpha
  - Mesoveliidae—water treaders (4)
  - Hebridae—velvet water bugs (18)
  - *Macroveliidae—macroveliid shore bugs (2)
  - Hydrometridae—marsh treaders, water measurers (7)
  - Veliidae—broad-shouldered water striders, riffle bugs (34)
  - Gerridae—water striders (45)
- Nepomorpha
  - *Ochteridae—velvety shore bugs (5)
  - Gelastocoridae—toad bugs (6)
  - Pleidae—pygmy backswimmers (5)
  - Naucoridae—creeping water bugs (18)
  - Belostomatidae—giant water bugs (19)
  - Nepidae—water scorpions (15)
  - Notonectidae—backswimmers (35)
  - Corixidae—water boatmen (120)
- Cimicomorpha
  - *Thaumastocoridae—royal palm bugs (1)
  - Tingidae—lace bugs (150)
  - *Microphysidae—microphysid bugs (1)
  - Miridae—plant bugs (1800)
  - Nabidae—damsel bugs (30)
  - Anthocoridae—minute pirate bugs, flower bugs (70)
  - Cimicidae—bed bugs (15)
  - *Polyctenidae—bat bugs (2)
  - Phymatidae—ambush bugs (25)
  - Reduviidae—assassin bugs (140)
- Pentatomomorpha
  - Aradidae—flat bugs, fungus bugs (100)
  - *Piesmatidae—ash-gray leaf bugs (10)
  - Berytidae—stilt bugs (11)
  - Lygaeidae—seed bugs (250)
  - Largidae—largid bugs (12)
  - Pyrrhocoridae—red bugs, cotton stainers (7)
  - Alydidae—broad-headed bugs (20)
  - Rhopalidae—scentless plant bugs (35)
  - Coreidae—leaf footed bugs (75)
  - Cydnidae—burrower bugs (30)
  - Corimelaenidae—negro bugs (30)
  - Scutelleridae—shield-backed bugs (34)
  - Pentatomidae—stink bugs (180)

*Immatures are rare or unknown in collections.

1. The number of species given is for America north of Mexico.

KEY TO THE FAMILIES OF HEMIPTERA LARVAE[1]

Fred A. Lawson and Thomas R. Yonke

1. Compound eyes absent; antennal or genal ctenidia present; no dorsal abdominal scent glands; ectoparasites of bats .... ***Polyctenidae*** p. 48

1′. Compound eyes present; antennal and genal ctenidia absent; dorsal abdominal scent glands usually present 2

2(1′). Antennae shorter than head, inserted beneath eyes (figs. 1, 2), not visible from above except in Ochteridae (fig. 3), true aquatics and a few shore bugs 3

2′. Antennae longer than head, inserted in front of eyes and visible from above (may be tucked under body in preservation) (figs. 4, 6) 10

3(2). Labium very short, broad, scarcely distinguishable from apex of head, not distinctly segmented (fig. 5); protarsi spatulate; 2 or 3 dorsal abdominal scent gland openings .... ***Corixidae*** p. 42

3′. Labium longer, cylindrical or coneshaped, segmented (fig. 6); protarsi not spatulate; scent gland openings lacking or on only 1 segment 4

4(3′). Posterior abdominal siphon tube at least ¼ length of body (fig. 29.17), shorter in *Nepa* spp. (fig. 29.18) .... ***Nepidae*** p. 40

4′. Siphon tube inconspicuous or absent 5

5(4′). Meso- and/or metathoracic legs with more or less extensive fringes of long swimming hairs (figs. 7, 29.16); water bugs 6

5′. Meso- and/or metathoracic legs without swimming hairs (fig. 8) 9

6(5). Prothoracic legs distinctly raptorial, femora enlarged, tibiae articulating against femora (fig. 9); body with subflattened dorsum 7

6′. Prothoracic legs not raptorial (fig. 10); body with convex dorsum 8

7(6). Eyes protruding laterally beyond head margin (fig. 11); protarsi with claws; dorsal abdominal scent glands absent .... ***Belostomatidae*** p. 40

7′. Eyes not protruding (fig. 12); protarsi without claws; 1 pair minute dorsal scent gland openings between 3rd and 4th abdominal tergites .... ***Naucoridae*** p. 39

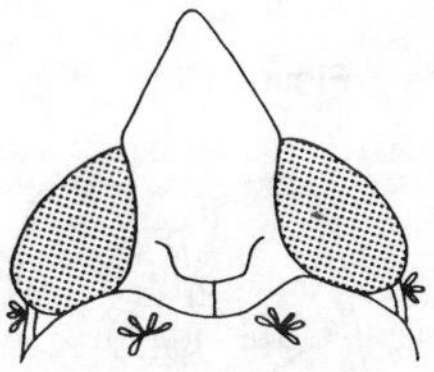

Figure 1

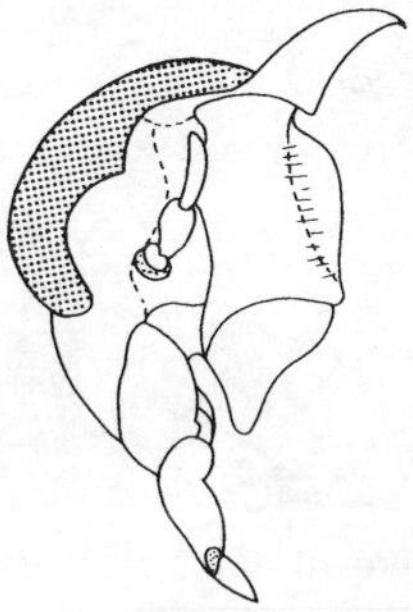

Figure 2

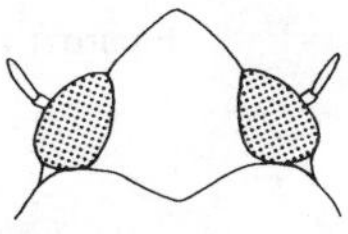

Figure 3

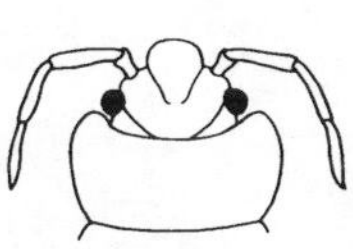

Figure 4

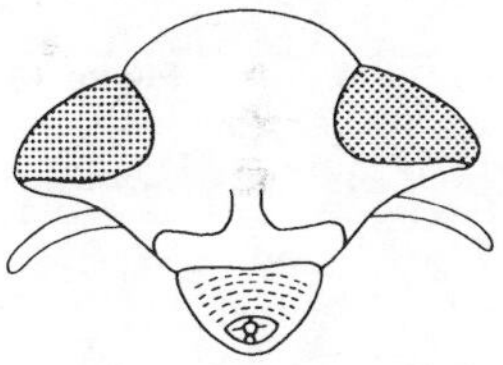

Figure 5

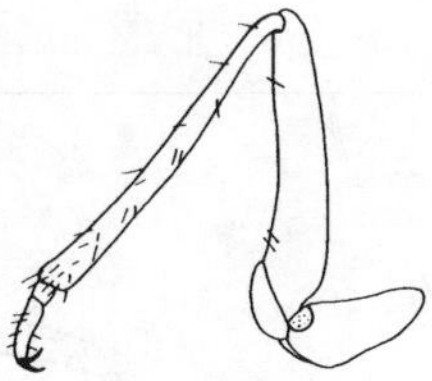

Figure 8

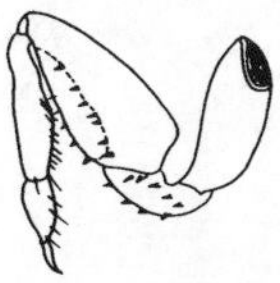

Figure 9

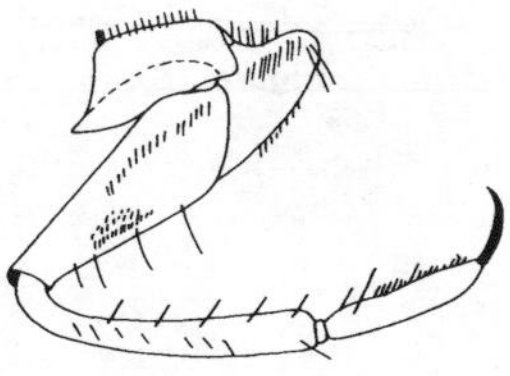

Figure 10

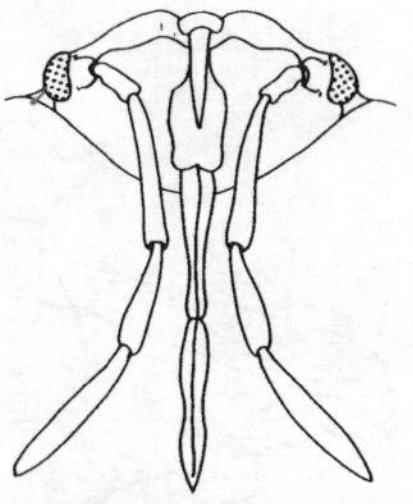

Figure 6

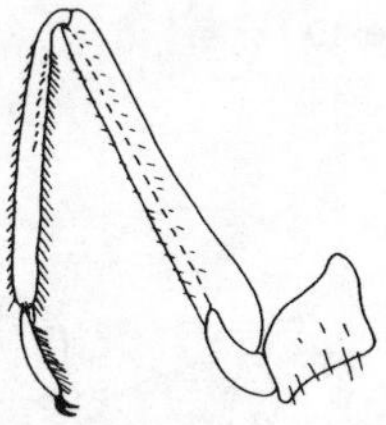

Figure 7

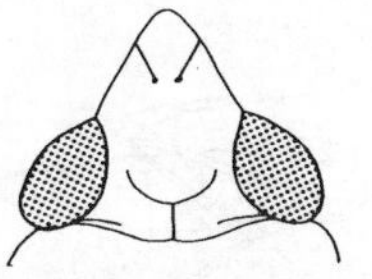

Figure 11

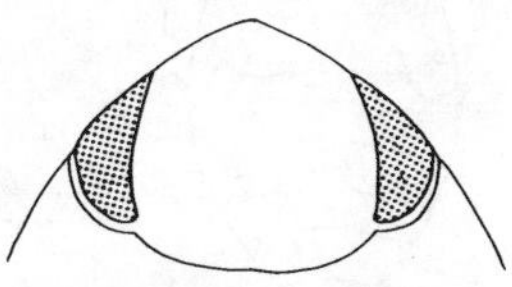

Figure 12

1. The key, although based primarily on 5th (last) instars, will identify earlier instars of most species.

8(6′). Very small species with 2 distinct claws on metatarsi (fig. 13); form oval; metathoracic legs not long and oarlike .... ***Pleidae*** p. 38

8′. Larger species, without distinct claws on metatarsi (fig. 14); form elongate-oval; metathoracic legs long and oarlike .... ***Notonectidae*** p. 41

9(5′). Prothoracic legs raptorial, femora very broad, grooved along inner edges near curved tibiae; antennae concealed in grooves beneath strongly protruding eyes (fig. 15) .... ***Gelastocoridae*** p. 38

9′. Prothoracic legs similar to mesothoracic legs, fitted for running; antennae exposed (figs. 3, 29.12) .... ***Ochteridae*** p. 37

10(2′). Head as long as entire thorax and very slender; eyes small, over halfway back on sides of head (fig. 16); dorsal abdominal scent glands absent .... ***Hydrometridae*** p. 35

10′. Head shorter, stouter, not exceeding length of thorax (fig. 17), (except *Myodocha* spp., Lygaeidae, fig. 56); eyes larger, located forward on head; body heavier 11

11(10′). Tarsal claws of at least the prothoracic legs attached anteapically (fig. 18) 12

11′. Tarsal claws all attached apically (fig. 19) 13

12(11). Metafemora surpassing tip of abdomen; meso- and metacoxae adjacent, both far removed from procoxae (fig. 20); median longitudinal groove on head absent .... ***Gerridae*** p. 36

12′. Metafemora scarcely, if at all, surpassing tip of abdomen; coxae spaced more equally on venter (fig. 21); median longitudinal groove on head present .... ***Veliidae*** p. 36

13(11′). Legs much longer than body and thread-like; body very long and slender (stick-like) (fig. 22); last antennal segment short and swollen (stick-like Emesinae, Reduviidae, with 4th segment slender) .... ***Berytidae*** p. 52

13′. Legs, body, and antennae with different proportions 14

14(13′). Profemora ca. 2X as long or less when compared to height in side view (figs. 23, 24), almost triangular 15

14′. Profemora over 3X as long as high, not triangular (figs. 25, 28) 16

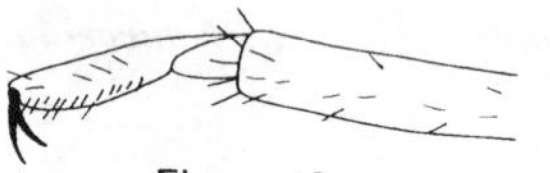

Figure 13

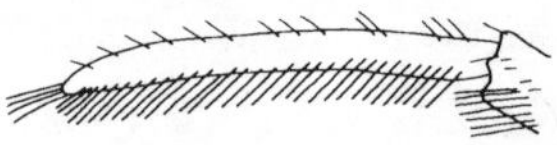

Figure 14

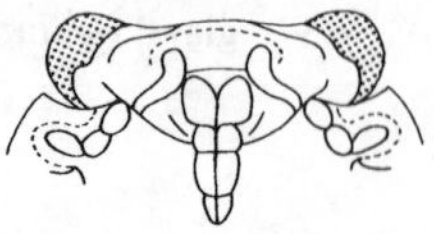

Figure 15

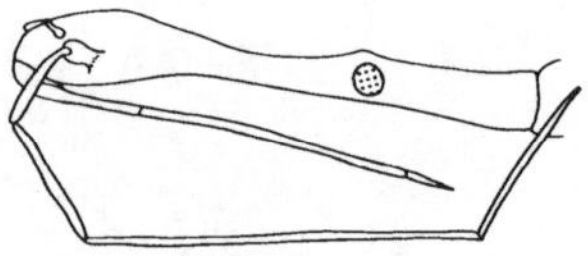

Figure 16

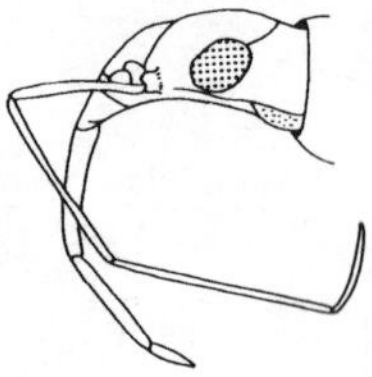

Figure 17

Figure 18

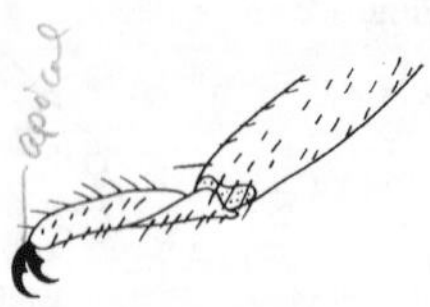

Figure 19

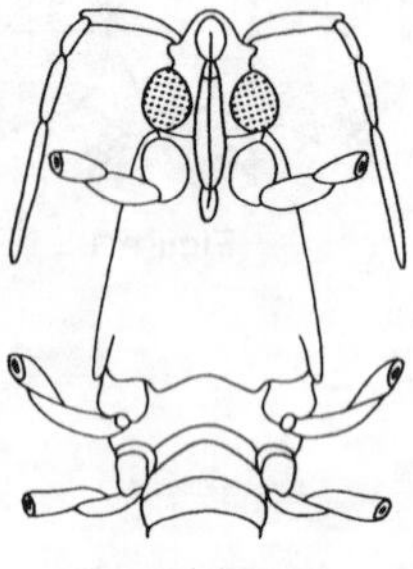

Figure 20

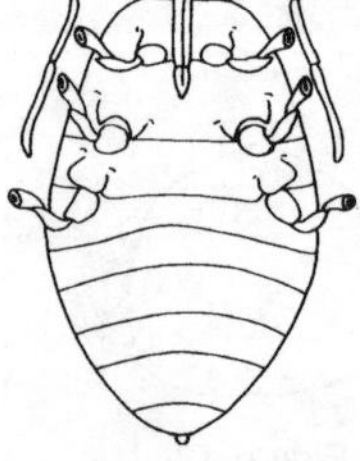
Figure 21

Figure 22

Figure 23

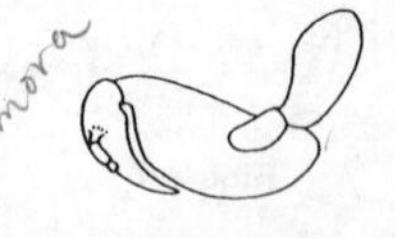
Figure 24

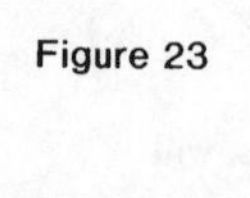

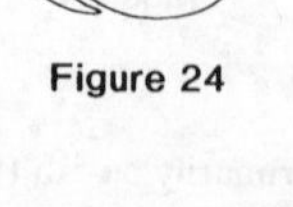

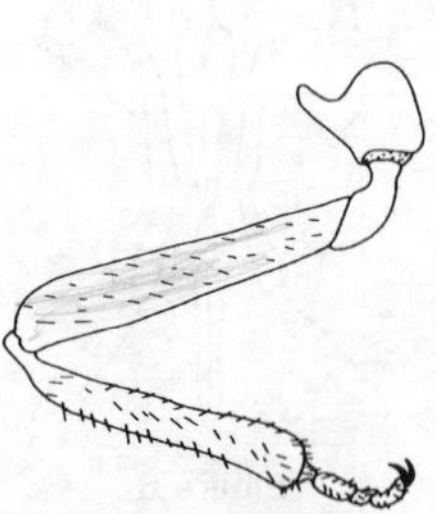
Figure 25

15(14). Body slender, minute; head and labium extended forward (fig. 26); head 2-lobed .... ***Enicocephalidae*** p. 32

15′. Body stout, deep, small to medium size; head directed ventrad, labium directed caudad (fig. 27) .... ***Phymatidae*** p. 48

16(14′). Profemora long, "thigh" shaped, with dense fine setae and 4 or 5 evenly spaced long, usually black setae on ventral surface (fig. 28); protibiae with 2 long rows of tiny black spines on posterior surface; 4-segmented labium; 3 pairs of dorsal abdominal scent glands .... ***Nabidae*** p. 46

16′. Profemora with different characteristics; if similar, then labium 3-segmented or with 1 dorsal abdominal scent gland opening 17

17(16′). Labium short, often stout, may extend between but not beyond procoxae (fig. 29) 18

17′. Labium long, usually slender and extending beyond procoxae to base of mesothoracic legs or beyond (fig. 30) 27

18(17). Body very wide and flat, little depth compared to width (fig. 31) 19

18′. Body not notably flat, nearly as deep as wide or deeper (fig. 32) 23

19(18). Bucculae well developed, forming deep groove for labium and extending to or nearly to caudal margin of head (*Atheas* spp.) .... (part) ***Tingidae*** p. 43

19′. Bucculae short, at most extending to mid-venter of head 20

20(19′). Labium 4-segmented; host royal palm .... ***Thaumastocoridae*** p. 43

20′. Labium 3-segmented 21

21(20′). Head elongate; prosternum with stridulatory (ridged) groove (figs. 29, 34) .... (part) ***Reduviidae*** p. 49

21′. Head triangular to quadrangular; prosternal groove absent 22

22(21′). Coxae widely spaced on venter (fig. 33); surface finely tuberculate or segments with 1 or parts of 2 regular rows of warty bumps, or both; under dead bark .... ***Aradidae*** p. 50

22′. Coxae close together on venter; body without warty bumps; covered with long, curved, recumbent hairs; ectoparasites of birds and mammals .... (part) ***Cimicidae*** p. 47

23(18′). Head elongate, tubular; tip of labium usually fitting into median stridulatory (ridged) groove in prosternum (fig. 34) .... (part) ***Reduviidae*** p. 49

23′. Head transverse, flattened or at most moderately elongate; labium not fitting into prosternal groove 24

24(23′). Tarsi 1-segmented; labium 3-segmented; minute bugs, ca. 1 mm .... ***Microphysidae*** p. 44

24′. Tarsi 2-segmented; labium 3 or 4-segmented; very small to small bugs, 2 mm+ 25

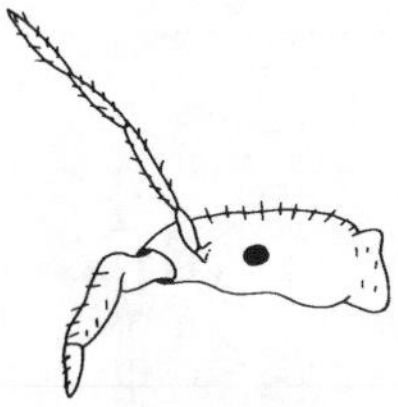

Figure 26

Figure 27

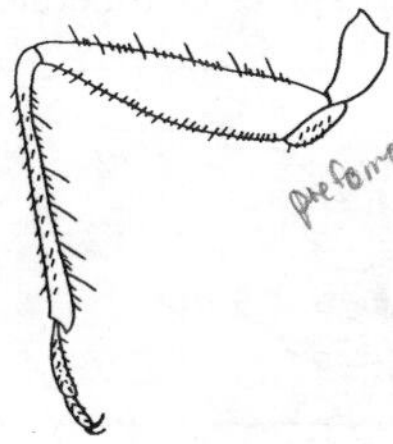

Figure 28

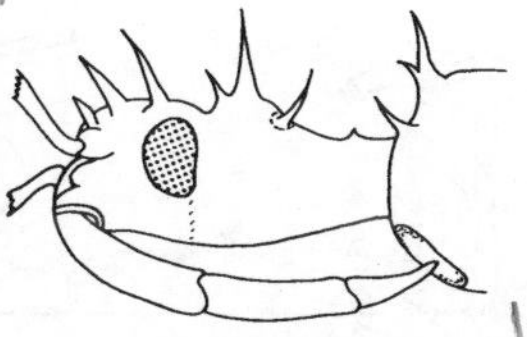

Figure 29

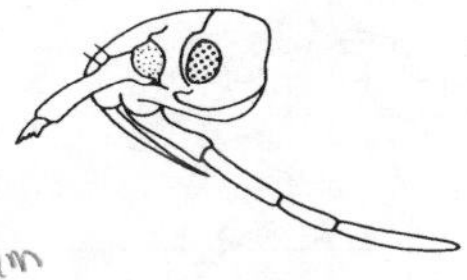

Figure 30

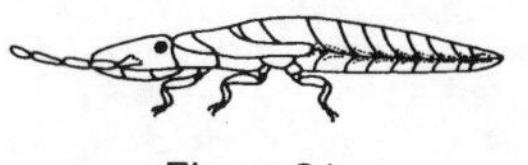
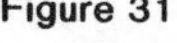

Figure 31

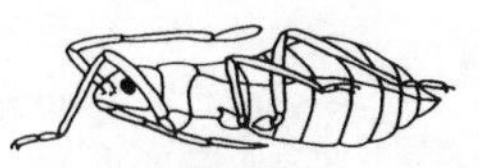

Figure 32

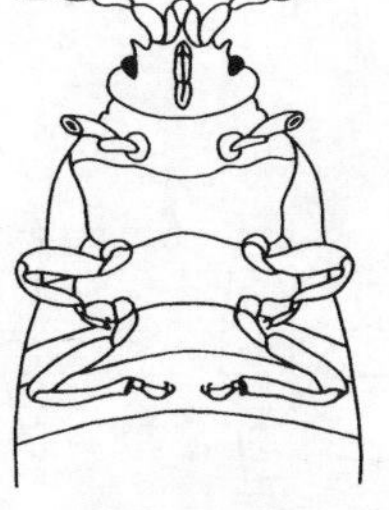

Figure 33

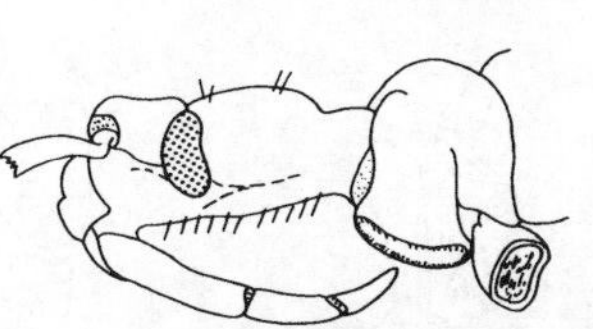

Figure 34

25(24′). Mesocoxae spaced farther apart than metacoxae (fig. 35); labium 3-segmented, short or long, often projecting forward; 3 pairs of abdominal scent gland openings, minute, widely separated, on 3rd, 4th, and 5th segments .... ***Anthocoridae*** p. 46

25′. Coxae spaced differently; labium longer than head; with either 1 small or 3 well developed abdominal scent glands 26

26(25′). Coxae spaced evenly (fig. 36); labium 4-segmented .... (part) ***Miridae*** p. 45

26′. Metacoxae farther apart than the pro- or mesocoxae (fig. 37); labium 3-segmented; may be inflated due to gorging or preservation; ectoparasites of birds or mammals .... (part) ***Cimicidae*** p. 47

27(17′). Labium very slender but if thicker, then bowed, with 3 segments, 1st very small, ring-like, 2nd very long, 3rd—⅓ to ½ as long as 2nd (fig. 38); mostly semiaquatic bugs 28

27′. Labium thicker, 4-segmented, with different proportions among segments (fig. 39); terrestrial bugs 34

28(27). Ventral surface of head deeply grooved to form a labial sulcus (fig. 40) .... ***Hebridae*** p. 34

28′. Ventral surface of head not grooved, base of labium visible (fig. 38) 29

29(28′). Three or 4 pairs of dorsal abdominal scent gland openings; minute bugs (1 mm or smaller) .... ***Dipsocoridae*** p. 31

29′. A single or 1 pair of abdominal scent gland opening(s) (figs. 41, 42) 30

30(29′). One pair of widely separated scent gland openings on 4th abdominal tergite; minute bugs (<2 mm) .... ***Schizopteridae*** p. 31

30′. Single scent gland opening on 3rd or 4th abdominal tergite; if paired, bugs larger 31

31(30′). Scent gland opening on caudal margin of 3rd abdominal tergite (fig. 41) 32

31′. Scent gland opening in center of 3rd or 4th visible abdominal tergite (fig. 42) 33

32(31). Profemora and labium with long, black, stout bristles .... ***Leptopodidae*** p. 32

32′. Profemora and labium with a few light-colored setae .... ***Saldidae*** p. 32

33(31′). Legs with scattered stiff black bristles (fig. 43) .... ***Mesoveliidae*** p. 33

33′. Legs without scattered stiff black bristles (fig. 44) .... ***Macroveliidae*** p. 34

34(27′). Pronotum with wide lateral projections (fig. 45); usually highly spinose, spines very long, single or with 2 or more shorter spines arising from common base (scolus), very small to small bugs (2–5 mm) .... (part) ***Tingidae*** p. 43

34′. Pronotum without wide lateral projections; head and body not spinose above or on margins; if spinose, bugs much larger; if short spines or tubercles with minute spines occur, spines never branched 35

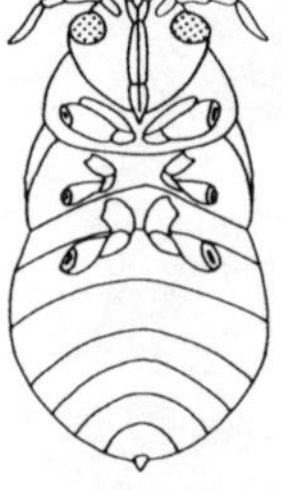

Figure 35

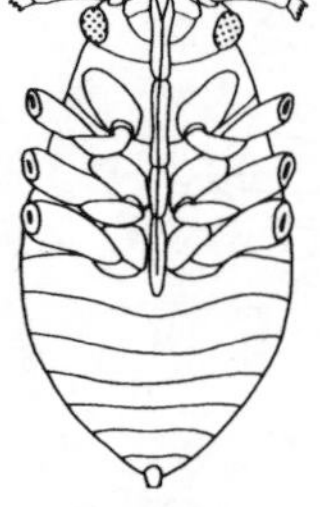

Figure 36

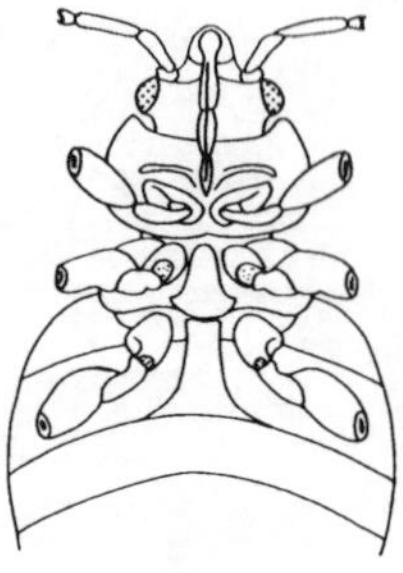

Figure 37

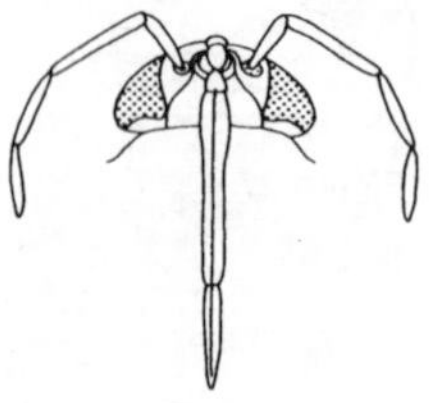

Figure 38

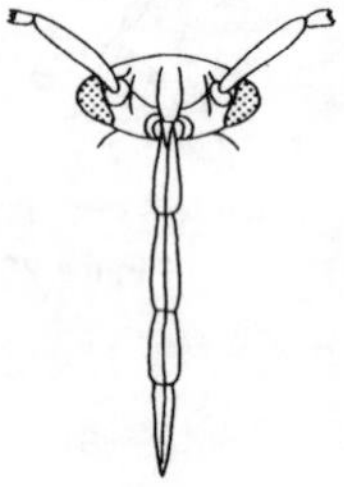

Figure 39

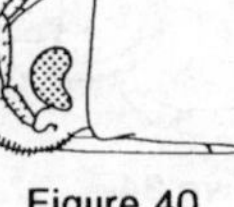
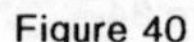

Figure 40

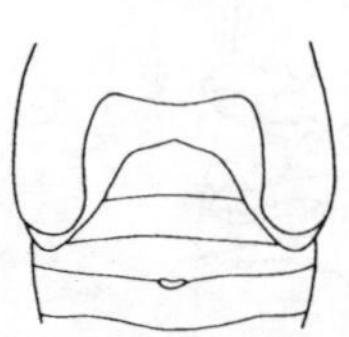

Figure 41

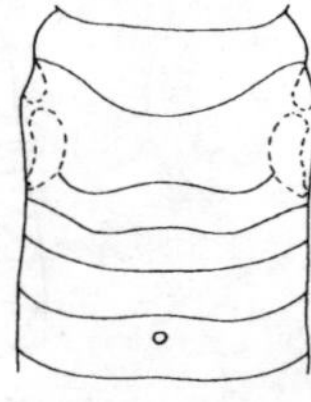

Figure 42

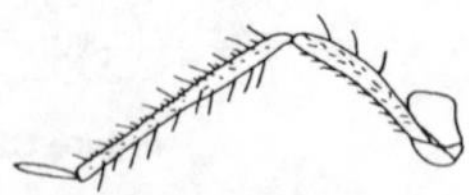

Figure 43

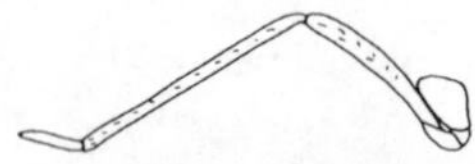

Figure 44

35(34′). Juga extending well beyond tylus (fig. 46); very small, (2–3 mm), somewhat flattened bugs .... ***Piesmatidae*** p. 51

35′. Juga shorter than or about same length as tylus (figs. 50, 61 62); juga exceeds tylus in some larger bugs; very small (2–3 mm) to large bugs 36

36(35′). One dorsal abdominal scent gland opening or none, minute when present, located medially on posterior margin of 3rd tergum and often in a small pigmented area (figs. 47, 48) .... (part) ***Miridae*** p. 45

36′. Two or 3 dorsal abdominal scent gland openings, paired or single; circular, transverse, or slit-like (figs. 58, 66) 37

37(36′). Dorsal abdominal scent gland openings on 3 segments (figs. 53, 54, 55) 38

37′. Dorsal abdominal scent gland openings on 2 segments only (figs. 63, 64, 65, 66) 45

38(37). Head distinctly wider than long, transverse, somewhat retracted into prothorax (fig. 49) 39

38′. Head nearly as long as wide or longer, truncate or pointed (fig. 50) 40

39(38). Dorsum strongly arched (fig. 51) .... ***Corimelaenidae*** p. 59

39′. Dorsum nearly flat (fig. 52) .... ***Cydnidae*** p. 58

40(38′). First, and sometimes 2nd and 3rd abdominal terga much reduced or divided in center by enlarged mesonotum; 1st dorsal abdominal scent gland openings far apart (fig. 53) .... ***Scutelleridae*** p. 59

40′. First 3 abdominal terga continuous; 1st dorsal scent gland openings as close together as other pairs (figs. 54, 55) 41

41(40′). Sclerite around 1st dorsal scent gland openings narrow, transverse, not over ½ the size of those around 2nd and 3rd gland openings (figs. 54, 55) .... ***Pentatomidae*** p. 60

41′. Pigmented areas or sclerites around all scent gland openings about same size or absent 42

42(41′). Head slender, black, elongate (neck-like); body relatively slender (fig. 56, *Myodocha* spp.) .... (part) ***Lygaeidae*** p. 52

42′. Head without such neck-like modification (fig. 57); body usually stouter 43

43(42′). Dorsal abdominal scent glands minute, paired, openings laterad of mid-dorsal line; posterior margins of tergites 4 and 5 angled sharply caudad from sides of body (fig. 58) .... (part) ***Lygaeidae*** p. 52

43′. Dorsal abdominal scent glands with single or double opening on mid-dorsal line; posterior margins of tergites 4 and 5 straight or with short median caudal extension (figs. 59, 60) 44

Figure 45

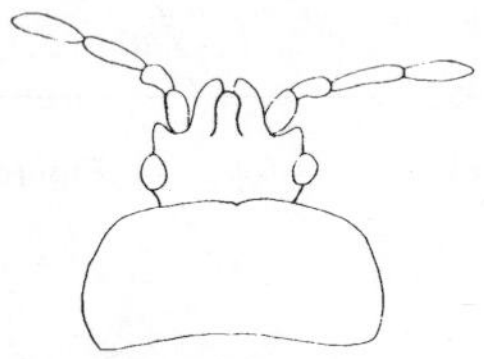

Figure 46

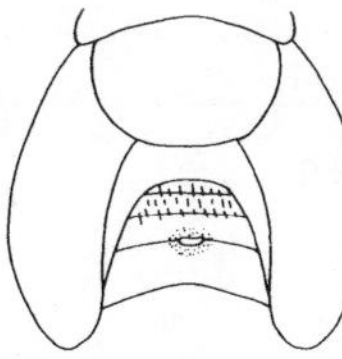

Figure 47

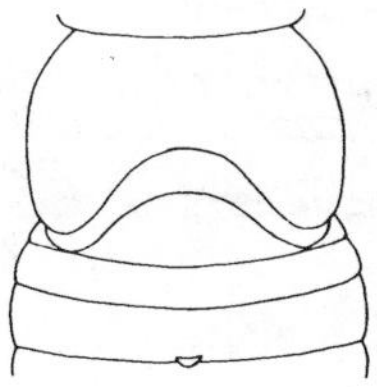

Figure 48

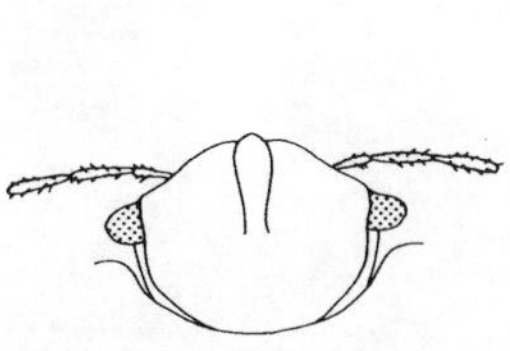

Figure 49

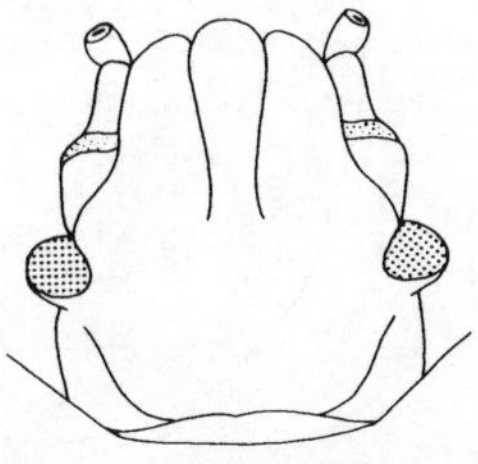

Figure 50

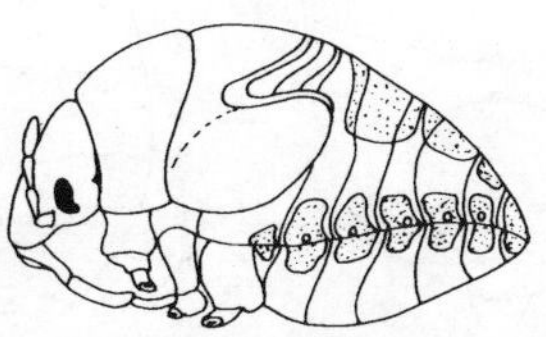

Figure 51

Corimelaenidae

Figure 52

Cydnidae

Figs. 53–66 ⟶

44(43′). Body elongate-oblong; yellow to dark red (non-metallic); abdominal scent gland openings large, transverse (fig. 59) .... ***Pyrrhocoridae*** p. 54

44′. Body ovate-rounded; dark to metallic; some ant-like, with round head (*Arhaphe* spp.); abdominal scent gland openings minute, circular or slit-like (fig. 60) .... ***Largidae*** p. 54

45(37′). Head, including eyes, as wide as or wider than prothorax (fig. 61); base of buccula not extending proximally to base of antenna; slender or ant-like .... ***Alydidae*** p. 55

45′. Head, including eyes, not as wide as prothorax (fig. 62); base of buccula extending proximally to base of antenna; usually thick-bodied 46

46(45′). Dorsal scent gland openings large, on transverse, often pigmented and/or spinose swellings or extensions of median caudal areas of segments (fig. 63) .... ***Coreidae*** p. 57

46′. Dorsal scent gland openings small to minute; truncated swellings absent from median caudal area of segments, but flat pigmented areas may surround openings 47

47(46′). Dorsal scent gland openings circular, minute, a single opening on caudal margins of tergites 4 and 5; tergite 5 constricted in middle (fig. 64) .... ***Rhopalidae*** p. 56

47′. Dorsal scent gland openings transverse, oval or slit-like; openings single or paired, usually in a pigmented area, tergite 5 not constricted (figs. 65, 66) .... (part) ***Lygaeidae*** p. 52

Figure 53 Figure 54 Figure 55 Figure 56 Figure 57

Figure 58 Figure 59 Figure 60 Figure 61 Figure 62

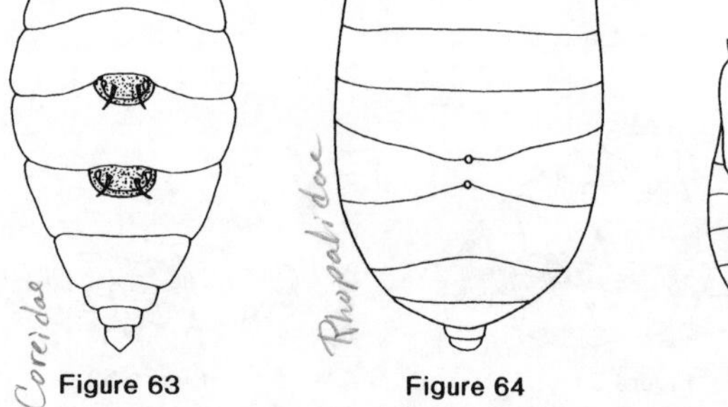

Figure 63 Figure 64

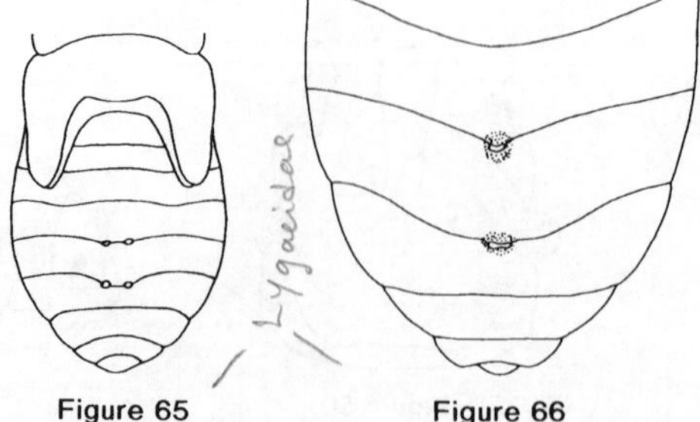

Figure 65 Figure 66

DIPSOCORIDAE

Dipsocorid Jumping Ground Bugs

Figure 29.3

Relationships and Diagnosis: These nearly microscopic bugs are most closely related to a rare family in our fauna, the Schizopteridae. Their relationship to other families is less clear; however, the Dipsocoridae forms the nominal taxon for the Dipsocoromorpha. Immatures from both families may be recognized by very long dark brown setae over the body, elongate coxae and a very thick labium tapering to a fine point. Dipsocorids have three or four pairs of dorsal abdominal scent gland openings while schizopterids have only one pair.

Biology and Ecology: Virtually nothing is known about the biology of this family except that its species occur in ground litter, in aged wood piles, and under stones. They are thought to be predaceous. A 5th instar has been taken from a Berlese sample of material under rotting white oak in March, 3rd and 4th instars were collected from rotting logs in a wood pile in early June, and 5th instars and adults have been taken by the Berlese method in September in Missouri.

Description: Minute, ca. 1 mm as mature larva; elongate, brown. Dorsum of head with several long setae; thorax and abdomen bearing 1–2 pairs of large setae (as long as 2nd antennal segment) per segment. Head elongate with tylus raised and extending well beyond antennal socket. Well developed compound eyes laterally positioned. Antennae 4-segmented, filiform to setaceous from base to apex; 1st short, cylindrical; 2nd, 2–3X length of 1st; 3rd, ca. 2X length of 2nd; 4th setaceous, ca. 1½X length of 3rd. Labium 3- or 4-segmented, joints may be indistinct, arising from extreme front of head, very thick at base, tapering to needlelike apex, very long, extending to mid-abdominal venter. Prothorax slightly arched, transverse. Coxae very long, thick, subequal in length to femora and tibiae. Tibiae slender, tarsi 1-segmented. Legs bearing many long and medium length setae. Abdomen oval and dorsoventrally rounded, bearing at least 3 scent glands opening on caudal margins of tergites 3, 4 and 5. Tergites darkly pigmented, venter light.

Comments: This group is poorly collected because of its minute size and cryptic habits. However, the diligent collector can obtain specimens by using a Berlese funnel to process litter samples from moist forest floors, by searching in and under debris from rotting logs, and by looking on and under stones along pond and stream banks. Four species in 2 genera are known from New York to Florida to California. *Ceratocombus vagans* McAtee and Malloch is apparently the most common species.

Selected Bibliography

Emsley, M. G. 1969
McAtee, W. L. and J. R. Malloch 1925

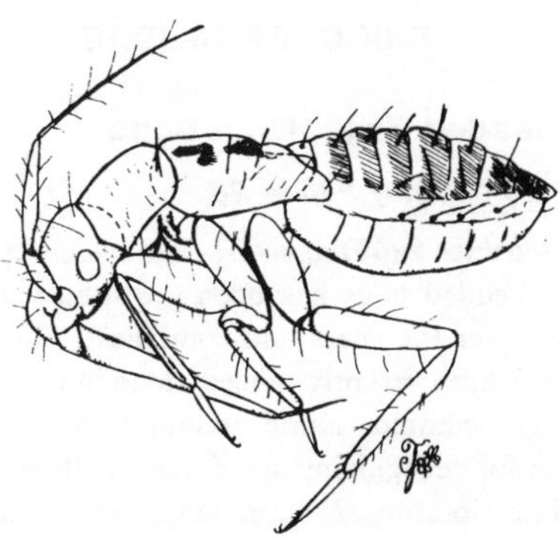

Figure 29.3. Dipsocoridae. Jumping ground bug, *Ceratocombus* sp.

SCHIZOPTERIDAE

Schizopterid Jumping Ground Bugs

Relationships and Diagnosis: The schizopterids are closely related to the dipsocorids and placed in the Dipsocoromorpha. Because of the small size, microphysid bugs also might be confused with this group; however, schizopterids have the 3rd and 4th antennal segments, taken together, over twice as long as the 1st plus the 2nd segments. Schizopterids have eyes projecting outward and backward onto the anterolateral aspects of the pronotum and do not possess long setae on the head or legs as do the dipsocorids.

Biology and Ecology: The habits of these small creatures are largely unknown. Apparently forest litter and a moist soil habitat are preferred by some species. Alate forms are attracted to lights while others are known to be inquilines with ants. Eggs of at least some species are known to be attached to soil substrate. There are 5 instars. These bugs jump when disturbed instead of running away.

Description: Minute, less than 2 mm; compact rotund bugs; soft or hard bodied, usually without pigment, sometimes reddish; compound eyes red. Head sub-triangular and deflexed, generally with a few paired large primary setae. Compound eyes with large facets, close to and often overlapping anterior margin of pronotum. Antennae 4-segmented, 1 and 2 subequal, short; 3 and 4 subequal, long, with long setae. Labium 3- or 4-segmented, truncated (*Corixidea*) or pointed (*Schizoptera*) at tip. Legs generally devoid of long setae. Abdomen 10-segmented with 1 pair of widely separated scent gland openings caudally on tergite 6. Large setae with bifid, trifid or capitate tips.

Comments: Three species in 3 genera are known from the United States. *Glyptocombus saltator* Heidemann is known from Maryland, Tennessee and Michigan, *Corixidea major* McAtee and Malloch has been found in Tennessee, and *Schizoptera bispina* McAtee and Malloch is known from Florida. Schizopterids will undoubtedly be found in additional states.

Selected Bibliography

Emsley, 1969
Slater and Baranowski, 1978

ENICOCEPHALIDAE

Unique-Headed Bugs, Gnat Bugs

Figures 29.4 and Key Figure 26

Relationships and Diagnosis: The systematic position of the unique-headed bugs has been the subject of discussion and dispute over the years and continues to be highly speculative. They are currently placed alone in the Enicocephalomorpha. The common name, unique-headed bugs, is most appropriate for recognizing specimens of this family. These very small hemipterans, 2–4 mm long, should not be confused with any others. The "2-part" head resulting from a constriction just caudad to the compound eyes sets off a posterior bulbular region and an anterior, narrow, porrect nose-like "segment." The modified raptorial prothoracic legs and protarsi each with 2 large claws are also distinctive.

Biology and Ecology: As with many other tiny true bugs, little is known of the biology. They live secluded among leaves and litter and under bark of fallen trees. They are predacious upon other small arthropods. Adults, looking like gnats, can be found in dense mating swarms at dusk in open areas, especially during September and October. Fifth instars have been collected from Berlese samples in September and March in Missouri. This, along with the swarming adults, suggests that they may overwinter as both mature larvae and adults.

Description: Very small, 2–4 mm, slender. Yellowish brown, eyes red. Body with many long white setae. Head elongate, divided into posterior bulbular region and anterior cylindrical region with compound eyes, well developed tylus-jugal "nose", bearing cone-shaped thick 3-segmented labium extending antero-ventrally. Four antennal segments subequal, together about as long as head plus labium; bearing numerous long clear setae. Anterior region of prothorax elongate, posterior region with slightly transverse pronotum and bearing the legs. Coxae well developed; femora thickened, especially profemora; protibiae thickened, modified raptorial, apically bearing large brown spurs; tarsi 1-segmented, each bearing 2 large claws; meso- and metalegs arising close together and distant from prothoracic legs. Abdomen cylindrical, elongate; median scent gland opening on tergite 4.

Comments: Only a few species in 3 genera are known from N. America. Collection records are poor with some genera and species being known only from restricted localities. Species most likely to be encountered belong to the genus *Systelloderes*. Very little is known of the group and no detailed study has been conducted other than on its taxonomy (Usinger 1945).

Selected Bibliography

Usinger, 1945

LEPTOPODIDAE

Spiny Shore Bugs

Relationships and Diagnosis: The Leptopodidae constitutes the nominal taxon for the Leptopodomorpha to which it and the Saldidae belong. A single immigrant species, *Patapius spinosus* (Rossi), reported from California (Usinger 1941) has been collected over a range of several hundred miles. This small bug of about 3 mm length can be recognized primarily by the presence of long, black spines on the profemora, the short, 3-segmented labium, and a single, dorsal abdominal scent-gland opening located on the caudal margin of tergite 3. Care may need to be taken to separate them from some Miridae which have a 4-segmented labium.

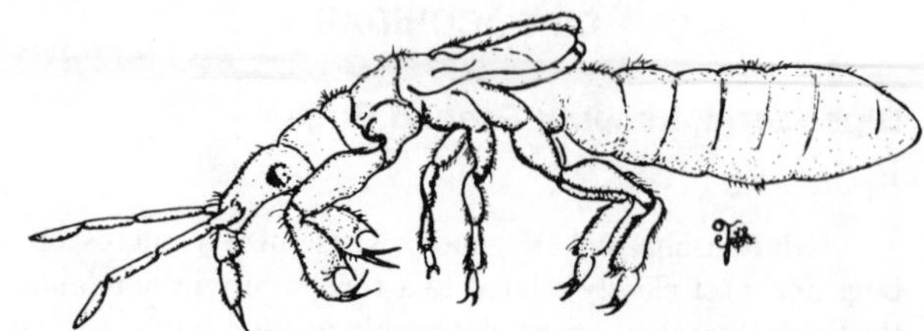

Figure 29.4. Enicocephalidae. Unique-headed bug, *Systelloderes* sp.

Biology and Ecology: Almost nothing is known of the species except that it often can be found on the ground away from water. It is thought to be predacious.

Description: As given above under diagnosis.

Comments: Immatures of the species are scarce in collections. Their paucity precludes a more detailed description. *P. spinosus* is known to occur from Los Angeles to north of Sacramento. Collectors in this area are encouraged to collect and observe ground-running larvae in an attempt to learn more.

Selected Bibliography

Usinger, 1941

SALDIDAE

Shore Bugs

Figure 29.5

Relationships and Diagnosis: In recent years, considerable taxonomic activity has focused on the Saldidae and its relationship to other taxa of Hemiptera. Studies by Cobben (1968), Stys and Kerzhner (1975), Polhemus (1977) and others have satisfactorily demonstrated its placement in the Leptopodomorpha, apart from the other semiaquatic bugs of the Gerromorpha. In our fauna, it is most closely related to the Leptopodidae.

Many terrestrial true bugs resemble the shore bugs. The saldids can be distinguished by rugulose areas on the dorsum and the presence of an abdominal scent gland on tergite 3 which possesses 1 pair of openings, often with distinct lateral canals.

Biology and Ecology: The shore bugs in N. America most often occupy habitats depicted by the common name, but some can be found far from water, especially in drier periods of late summer. Other species are present in intertidal zones. Saldids are both predators and scavengers.

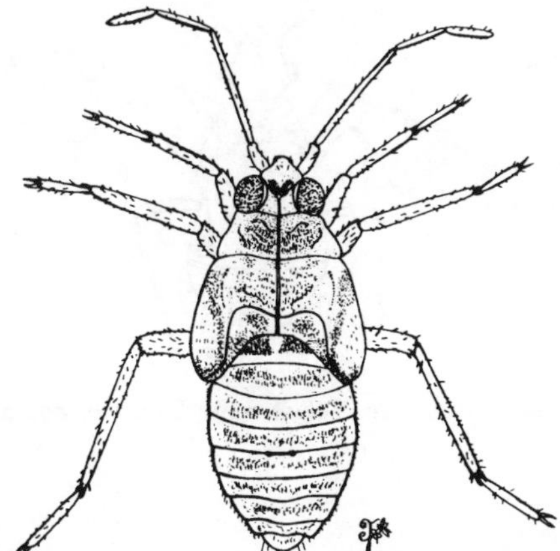

Figure 29.5. Saldidae. Shore bug, *Pentacora signoreti* (Guerin).

Saldids copulate in a side by side position. Eggs are generally deposited in plant tissue, less frequently glued to a substrate. Few detailed life history studies have been conducted. Wiley's (1922) observations showed that incubation took 7–9 days for *Saldula pallipes* (Fabricius) and 12 days for *Salda lugubris* (Say) while larval development required 16 and 17 days, respectively. Different species of shore bugs apparently overwinter as eggs or adults.

Description: 3–7 mm long, oval, somewhat robust, light yellowish-brown to mottled grey-black. Small to large rugulose areas on dorsum, including wing pads; several short and long dark setae on body and appendages. Head transverse with large eyes occupying almost entire lateral margin; tylus deflexed, set off by posterior suture, rounded; pair of short mid-lateral grooves and pits on frons; 4-segmented antennae long, arising from below and in front of eyes; labium apparently 3-segmented, extending to metacoxae, 2nd very long; labrum short, broad, triangular; bucculae extending to about mid-point of eyes. Pronotum nearly trapezoidal, reflexed along lateral and posterior margins, lateral margins shelf-like; proepisternum large, plate-like, fused in middle, forming recessed coxal cavities. Coxae elongate; femora moderately thickened; tibiae elongate, especially on metalegs; pro- and mesotarsi 2-segmented, metatarsi 3. Abdomen about as wide as long, moderately tapering toward end; 1 pair dorsal abdominal scent glands opening on caudal margin of tergite 3.

Comments: The N. American fauna includes over 70 species, mostly in *Saldula*. *Pentacora ligata* (Say) is one of the largest shore bugs commonly found over much of the United States. Saldids are noted for exhibiting a unique jumping behavior which should entertain as well as challenge the collector. At least some saldids are attracted to lights and can thus be collected near lights at night. A special point of interest is made by Polhemus (1967) that some sibling *Salda* spp. may be identified on the basis of the larvae.

Selected Bibliography

Cobben, 1968
Polhemus, 1967, 1976, 1977
Polhemus and Chapman (*In* Menke 1979)
Stys and Kerzhner, 1975
Wiley, 1922

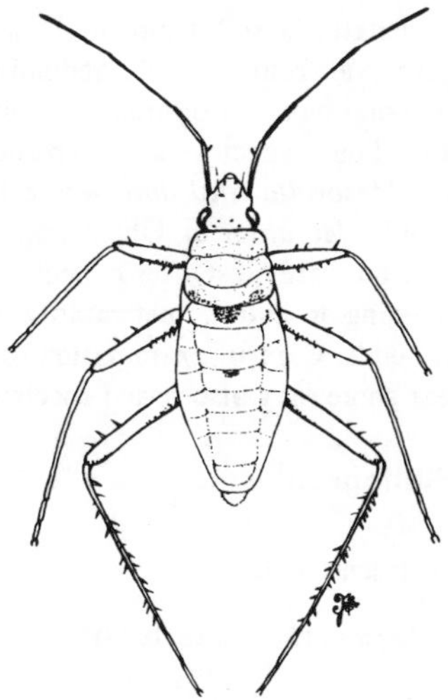

Figure 29.6. Mesoveliidae. Water treader, *Mesovelia* sp.

MESOVELIIDAE

Water Treaders

Figure 29.6

Relationships and Diagnosis: These bugs are usually placed in the Gerromorpha, but family relationships remain obscure. Anderson (1979) has proposed that the Mesoveliidae is close to the ground plan of the Gerromorpha as a whole and, therefore, should be placed as a sister-group of all other gerromorphan bugs.

The water treaders can be recognized by their very small size and light colored, elongate-fusiform body. Apical rather than anteapical claws and the more slender fusiform body distinguish water treaders from gerrids or veliids in the N. American fauna. Numerous long dark setae separate them from the macroveliids. Mesoveliids are not nearly as long or slender as the hydrometrids, both of which are found in similar habitats.

Biology and Ecology: Water treaders are sometimes called pondweed bugs because of their preference for ponds that bear thick, surface or emergent vegetation. Also, they may be found on open water across which they can easily move.

Mesoveliids are known to be predacious, feeding on insects and other arthropods at the water or vegetation surfaces of ponds. Their prey includes dead or disabled organisms. Eggs are laid in plant tissue near the water surface. Some species may overwinter as eggs. Development of the egg and larvae requires about 30 days, allowing, potentially, several generations per year.

Description: Very small, 2 mm, to small, 4.5 mm; elongate, tapering at both ends, greenish-yellow to light brown; many short light brown setae over body. Head elongate, almost conical near labium; eyes medium size; a few longer dark setae on vertex and frons; 4-segmented, elongate antennae arising far in front of head and below level of eyes; labium 4-segmented, elongate, especially 3rd, extending to beyond mesocoxae. Thoracic segments about even in size, tergites rectangular. Legs with simple moderately elongate segments; femora with a few long, dark apical setae; tibiae with several setae over the length; tarsi 1-segmented, claws apical. Abdomen elongate oval, round; single abdominal scent gland opening at midpoint on anterior area of tergite 4.

Comments: Four species are recorded for North America, with *Mesovelia mulsanti* White having a broad distribution while *M. amoena* Uhler apparently is more southerly. These bugs are most often collected singly or in low numbers by passing an aquatic net rapidly through surface vegetation on ponds. Careful examination of the water and vegetation near shore may also reveal specimens.

Selected Bibliography

Anderson, 1979
Anderson and Polhemus, 1980
Hoffmann, 1932
Polhemus and Chapman (*In* Menke, 1979)

HEBRIDAE

Velvet Water Bugs

Figure 29.7

Relationships and Diagnosis: The hebrids form a rather compact group and are most closely related to the Mesoveliidae. They might be confused with very small mesoveliids except that hebrids have a distinct labial groove on the head venter, and their bodies are covered with a dense hydrofuge pile.

Biology and Ecology: These tiny semiaquatic bugs occupy both moist shorelines and floating vegetation, even being found on the undersides of floating leaves. The hydrofuge pile allows them to go beneath the surface and remain there for some time. However, its more common benefit may be the protection from drowning as they are easily dislodged and blown about.

Hebrids are predacious, feeding on collembolans and other small arthropods. Apparently, relatively few eggs are laid. They are laid singly, over a 2-month period, glued to substrates including moss, algae and stones, and hatch in 8 to 12 days. The 5 instars complete development in 20–36 days (Porter 1950). Adults overwinter.

Description: Very small, 1–2 mm long, oval, robust, light yellowish brown to light brown. Head slightly elongate, with broad, raised tylus; small lateral eyes; antennae with 4 or 5 subequal segments arising in front of eyes; labium residing in distinct ventral groove on head, appearing 3-segmented, 2nd very long, 3rd with needle-like tip extending to beyond mesocoxae. Thorax box-like with many small fine setae on dorsum. Legs evenly developed with trochanters about as large as coxae; coxae widely separated on each sternite; tarsi 1-segmented. Abdomen almost round, thick; tergites narrow and wide; abdominal scent gland opening between tergites 3 and 4.

Comments: Two genera occur in North America, *Hebrus* with 15 species and *Merragata* with 3. *M. hebroides* White is widely distributed over the continental United States and is commonly encountered. Because of their small size the collector will find it helpful to use an alcohol-moistened camel's hair brush to pick up hebrids and place them in alcohol.

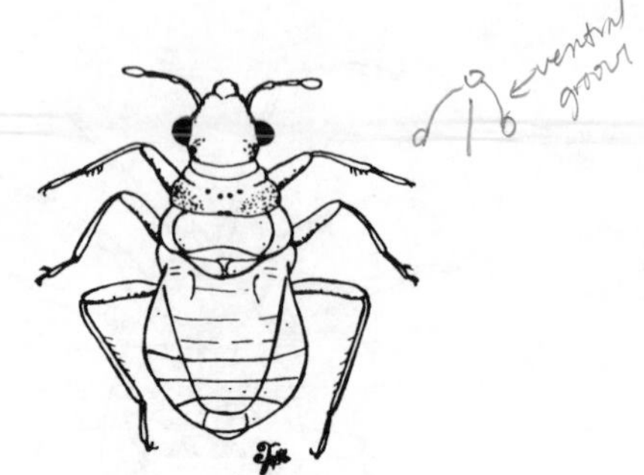

Figure 29.7. Hebridae. Velvet water bug, *Merragata hebroides* White.

Selected Bibliography

Anderson, 1979
Polhemus and Chapman (*In* Menke, 1979)
Porter, 1950

MACROVELIIDAE

Macroveliid Shore Bugs

Figure 29.8

Relationships and Diagnosis: Macroveliids have been linked with both Mesoveliidae and Veliidae. They have been treated with the Mesoveliidae in some American literature. Whatever their closest family relationship, macroveliids generally are accepted for inclusion with the Gerromorpha.

Macroveliids are recognized by their small size, elongate body, and legs that lack the long, dark setae found in mesoveliids. Their claws are apical, not anteapical as in the veliids. Macroveliids, which bear elongate prothoracic legs, should not be confused with gerrids where the prothoracic legs are disproportionally short as compared to the meso- and metathoracic legs.

Biology and Ecology: What little is known about these uncommon bugs is summarized by Polhemus and Chapman (*in* Menke 1979). They report that macroveliids "inhabit spring or seep areas, usually where there is abundant vegetation, especially watercress and mosses . . . ," and that they are negatively phototrophic. Although they are called shore bugs, macroveliids may be found on the water surface. Also, macroveliids are apparently active in winter during warm periods. Much more needs to be known about the biology of these species to better understand the family relationships. Adults may be brachypterous or macropterous in *Macrovelia hornii* Uhler, but are only known to be apterous in *Oravelia pege* Drake & Chapman.

Description: Ca. 6.5 mm long, elongate oval, generally brown on dorsum, whitish laterally on abdomen, lighter brown on venter; numerous short, dense brown setae over body. Head moderately elongate, narrowing somewhat to tylus; tylus appearing "nose-like", set off by distinct sutures on all sides; large bulbular eyes on caudal half of head, not near prothorax, facets distinctly convex; bucculae large; labium 4-segmented, extending to beyond metacoxae, 1 and 2 very short, 4 short, ca. 2X 1 and 2 together, 3 very long, ca. 3X

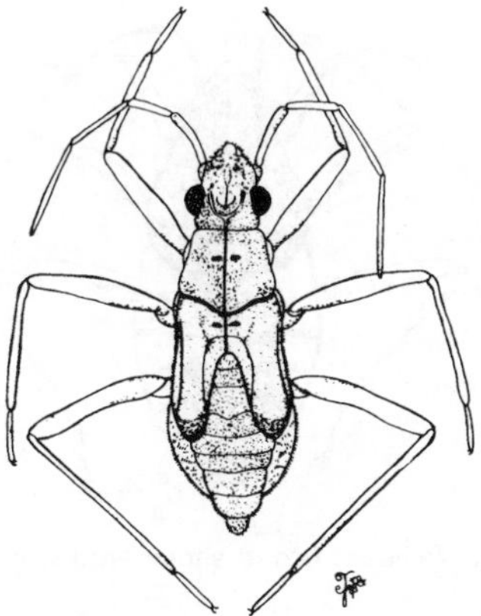

Figure 29.8. Macroveliidae. Macroveliid shore bug, *Macrovelia hornii* Uhler.

segment 4. Thorax moderately elongate; respective epimerum and episternum fused, deeply cleft ventrally over coxa; pronotum ca. as long as wide with 2 pits near midline on anterior 1/3; wind pads very long, extending to near 5th abdominal tergite. Legs similarly developed, meso- and metacoxae more widely spaced than procoxae; femora, tibiae and tarsi long; tarsi 1-segmented, claws apical, metathoracic legs longest. Abdomen much longer than wide; single scent gland opening at midpoint on anterior 3rd of tergite 4; each tergite with ca. middle 2/3 sclerotized and darkly pigmented, lateral margins white; each sternite, 3–5, more or less divided into 3 sclerotized plates.

Comments: Both species known are from the western United States. *Macrovelia hornii* is widely distributed, having been recorded from N. Dakota to southern California. *Oravelia pege* is known only from Fresno Co., California.

Selected Bibliography

Polhemus and Chapman (*In* Menke 1979)

HYDROMETRIDAE

Marsh Treaders, Water Measurers

Figures 29.9 and Key Figure 16

Relationships and Diagnosis: The marsh treaders form an easily recognizable group without any apparent "near" relatives. They were associated with the mesoveliid-gerrid-veliid lineage by China and Miller (1959), but have been identified as a sister-group of the Macroveliidae by Anderson (1979).

Because of the extremely slender body of N. American marsh treaders, one is not likely to confuse them with other true bugs in an aquatic habitat. A few terrestrial Hemiptera are also long and very slender such as some of the stilt bugs (Berytidae) and emesine assassin bugs (Reduviidae: Emesinae); however, the hydrometrids can be recognized by the following combination of characters: head about as long as thorax, 3-segmented labium, and pro-, meso-, and metathoracic legs thread-like.

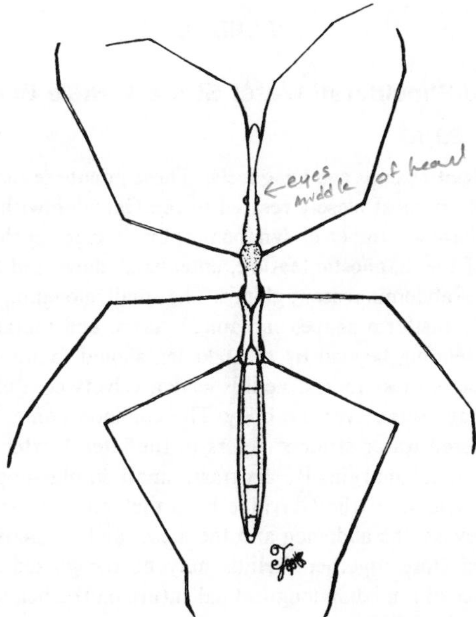

Figure 29.9. Hydrometridae. Marshtreader, *Hydrometra martini* Kirkaldy.

Biology and Ecology: Marsh treaders are found on small ponds and other quiet water pools with lush surface vegetation. They reside predominantly on or near vegetation and are predacious on other small arthropods. Sprague (1956) has studied the biology of *Hydrometra martini* Kirkaldy. Eggs are glued singly to objects just above the water surface with eclosion in about 12 days. Larvae require from 21 to 35 days to complete 5 instars. From 3 to 5 or more generations may be passed depending on climatic conditions, with adults overwintering.

Description: Ca. 8 to 10 mm long, slender cylindrical head, thorax and abdomen; light brownish yellow. Head longer than thorax, widened cephalad; eyes protruding laterad, located just over half way back from apex; distinct antenniferous tubercules; antennae 4-segmented, 2nd–4th very long; labium 3-segmented, 2nd very long, extending caudad from eyes. Pronotum long, extending over mesonotum, prosternum short. Legs about evenly developed with short coxae and trochanters; femora and tibiae, especially of metathoracic legs, very long; tarsi 1-segmented. Abdomen about as long as head and thorax together; no abdominal scent gland openings.

Comments: Seven species of *Hydrometra* occur in N. America with *H. martini* being found over much of the United States. Small, relatively shallow, undisturbed ponds with much emergent vegetation are the best places to collect marsh treaders. Careful observation will allow one to see them on the surface.

Selected Bibliography

Anderson, 1979
China and Miller, 1959
Hungerford, 1920
Polhemus and Chapman (*In* Menke 1979)
Sprague, 1956

VELIIDAE

Broad-Shouldered Water Striders, Riffle Bugs

Figure 29.10

Relationships and Diagnosis: These minute semiaquatic insects are most closely related to the Gerridae with which they share a number of synapomorphies including the presence of the diagnostic feature, anteapical claws and the absence of abdominal scent glands. The small, elongate, oval to broadly fusiform shape, anteapical claws, and metafemora not extending beyond tip of abdomen should facilitate recognition of most veliids. Veliids have a velvety coat of dense hydrofuge setae over the body. The common name, broad-shouldered water striders, refers to the lateral extension of the pronotal margin. By contrast, small similar-appearing water striders of the Gerridae have metafemora extending well beyond the abdomen and the meso- and metacoxae positioned close together. Veliids may be recognized by the presence of a median longitudinal suture on the head.

Biology and Ecology: Veliids inhabit ponds, the quiet water margins of streams, and riffle or ripple areas of streams. The term riffle bugs refers especially to species of *Rhagovelia* which occur in large numbers on the surface of fast moving water. Veliids are predacious on other small arthropods in contact with the water surface or adjacent shore. Eggs are laid singly or in clusters on various objects near the water surface. Some species have 4 instars rather than 5, and development to adult appears to vary considerably depending on conditions, resulting in overlapping generations. Many veliids overwinter as adults while others may do so in the egg stage. In the N. American species there is a predominance of apterous morphs.

Description: Very small to small (1.5 to 5 mm long) as mature larvae; somewhat elongate to oval; generally dark brown to black (*Microvelia* spp. and *Rhagovelia* spp.), light brown in *Paravelia* spp. Species variously covered with small setae and dark longer setae. Head transverse with median longitudinal groove or suture and large oval eyes in *Rhagovelia* or broadly triangular with large eyes in *Microvelia;* antennae 4-segmented, long, arising from below eyes; tylus distinctly raised; 4-segmented labium short, thick, extending to beyond procoxae. Thoracic dorsum consisting of simple transverse segments or convoluted. Legs moderately long and cylindrical; meso- and metafemora not generally extending beyond tip of abdomen; tarsi 1-segmented, mesotarsi deeply cleft (appearing as "2 fingers" in *Rhagovelia*); pretarsus and claws anteapical. Abdomen elongate, moderately oval, tapering to apex or "compacted" and triangular; longitudinal suture mid-laterally causing recess of tergites; scent gland openings absent.

Comments: These small water striders can be seen skimming the surface in large numbers. In this behavior they are likened to the gerrids. Species of *Microvelia* are known to feed on mosquito eggs and larvae.

Among the semiaquatic bugs the Veliidae is an extremely successful group. It contains about 420 species in some 30 genera, with species in the tropics well adapted to land in the moist rain forest.

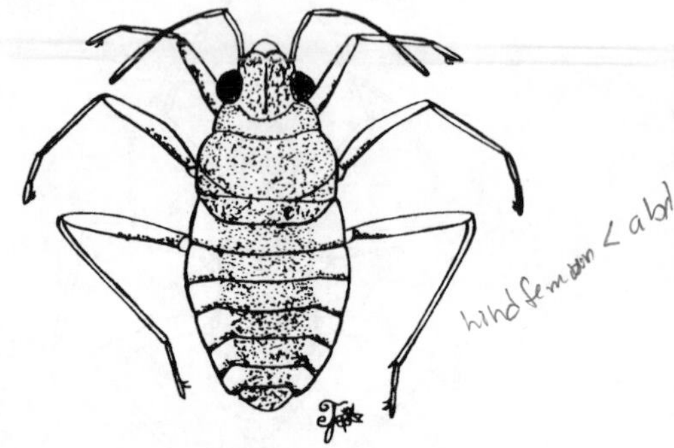

Figure 29.10. Veliidae. Broad-shouldered water strider, *Microvelia* sp.

Our fauna includes 19 species of *Microvelia,* 9 of *Rhagovelia,* 4 of *Paravelia,* and very uncommon marine species of *Trochopus* and *Husseyella* occurring among mangroves in Florida.

Selected Bibliography

Cheng and Fernando, 1971
Polhemus and Chapman (*In* Menke 1979)

GERRIDAE

Water Striders

Figure 29.11 and Key Figures 18, 20

Relationships and Diagnosis: This group constitutes the nominal family of the Gerromorpha within which it is thought to be highly specialized and most closely related to the Veliidae (Anderson 1979). The taxonomic limits of the Gerridae are not uniformly agreed upon; however, this has relatively little effect on the N. American fauna.

The large size of most species relative to those of other families found on the water surface allows easy recognition of most specimens. The meso- and metathoracic legs of water striders are very long, with the metafemora extending beyond the abdomen, while the prothoracic legs are shorter and arise relatively far-removed from the mesothoracic legs. Also, the pretarsi and claws are anteapical and the body is covered with a velvety coat of hydrofuge setae. Both apterous and alate morphs occur. Species are brown to black or greyish with black markings. Smaller specimens must be examined more carefully for the above features. In addition to leg placement, the absence of a median longitudinal suture or groove on the head will help distinguish gerrids from veliids.

Biology and Ecology: Water striders are commonly found in a variety of aquatic habitats from small ponds to rivers and estuaries. Species of *Halobates* are oceanic. As with other semiaquatic species, gerrids are predacious at the water surface on a variety of small arthropods. Eggs are laid on vegetation and debris just under the water surface, hatch in a few days, and develop through 5 instars in from 20 to about 40 days. The number of generations per year is apparently dependent upon the climate.

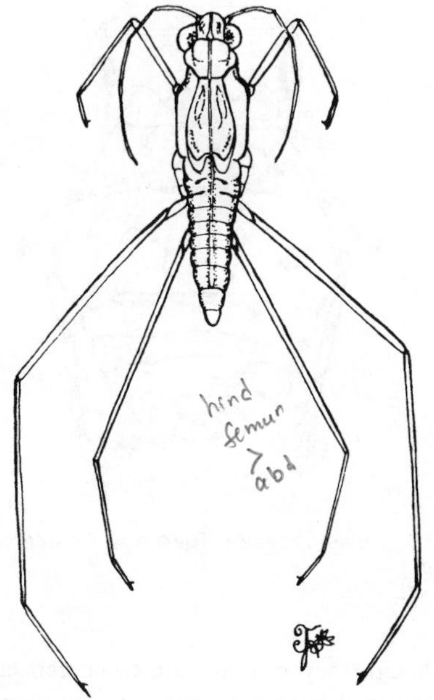

Figure 29.11. Gerridae. Water strider, *Gerris* sp.

Description: Very small, ca. 2 mm in *Rheumatobates,* to large, ca. 18 mm in some *Limnoporus;* short-oval to elongate-oval; light yellowish brown to black; appendages and body covered with dense short setae. Head without longitudinal groove or suture; eyes large, ball-like, laterally protruding; tylus raised, extending forward; antennae arising from socket in front of eyes, 4-segmented, 1 and 2 short, nearly fused, 3 and 4 elongate; labium 4-segmented, extending to or beyond procoxae. Prothorax small; mesothorax large, longitudinally grooved and variously raised on dorsum; metathorax small. Prothoracic legs relatively short; meso- and metathoracic legs very long; meso- and metafemora extending well beyond apex of abdomen; mesothoracic legs arising caudally near metathoracic legs; tarsal claws anteapical. Abdomen short (*Metrobates, Trepobates* and *Rheumatobates*) or long (*Gerris* and *Limnogonus*), dorsum often with distinct mid-lateral longitudinal sutures; abdominal scent gland opening absent.

Comments: The gerrids form the most successful group of Gerromorpha with more than 56 genera and 450 species worldwide. The N. American fauna includes some 40 species, primarily in 6 genera. A number of species range over much of the United States and are commonly collected in large numbers. Males ride on the backs of females during copulation. Wing polymorphism is notable.

Selected Bibliography

Anderson, 1979
Calabrese, 1974
Matsuda, 1960
Polhemus and Chapman (*In* Menke 1979)

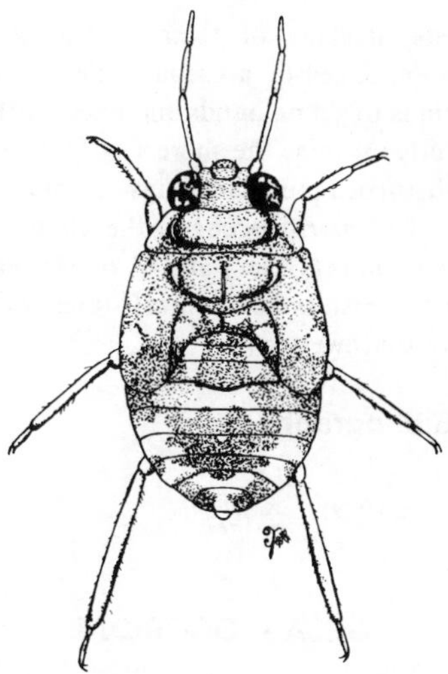

Figure 29.12. Ochteridae. Velvety shore bug, *Ochterus americanus* (Uhler).

OCHTERIDAE

Velvety Shore Bugs

Figure 29.12

Relationships and Diagnosis: The velvety shore bugs are most often associated with the toad bugs as a sister-group. Together they have been placed in the more truly aquatic Nepomorpha. Ochterids resemble some of the small oval-shaped saldids from which they are distinguished by the presence of a 4-, rather than 3-segmented labium. In contrast to the toad bugs, the velvety shore bugs have long, cylindrical prothoracic legs and clearly visible antennae.

Biology and Ecology: The velvety shore bugs, so called because of their shiny, dark integument, are found on dark or sandy soil adjacent to ponds and larger bodies of water. Bobb (1951) reported that *Ochterus banksi* Barber laid eggs singly along the shore, overall development required 316 days and the 4th instar, which overwintered, took 214 days. Ochterids are predacious on fly larvae, collembolans, and other small arthropods.

Description: Small, ca. 2 to 3 mm long, oval, somewhat dorsoventrally flattened; mottled dark brown and black, cryptic; a few short dark setae present. Head transverse from above, partly retracted into prothorax, strongly deflexed frons and tylus; tylus set off by distinct lateral and posterior sutures; compound eyes large, round; elongate, 4-segmented antennae arising beneath eyes, about as long as head and thorax combined; labrum broadly triangular; bucullae short but high at base; 4-segmented labium extending to metacoxae. Pronotum transverse; lateral margins of pronotum and mesonotum reflexed; large, vertical, plate-like covers located anterior to procoxae in *Octerus americanus* (Uhler), fused at midline. Legs adapted for running; femora somewhat elongate and thickened; tibiae subequal to femora in length. Abdomen rounded at caudal end, somewhat flattened, especially at lateral margins; slit-like opening of scent gland at caudal margin of tergite 3.

Comments: Because of their cryptic coloration ochterids can go unnoticed by a casual collector. The best way to collect them is to get on hands and knees with aspirator or jar and carefully examine the shore's surface to detect movement. Once disturbed these tiny bugs can move rapidly. Five or so species of *Ochterus* occur in the United States. Both *Ochterus americanus* (Uhler) and *O. banksi* are widely distributed over the eastern and central states while *O. barberi* Schell is more southwestern.

Selected Bibliography

Bobb, 1951
Menke (*In* Menke 1979)

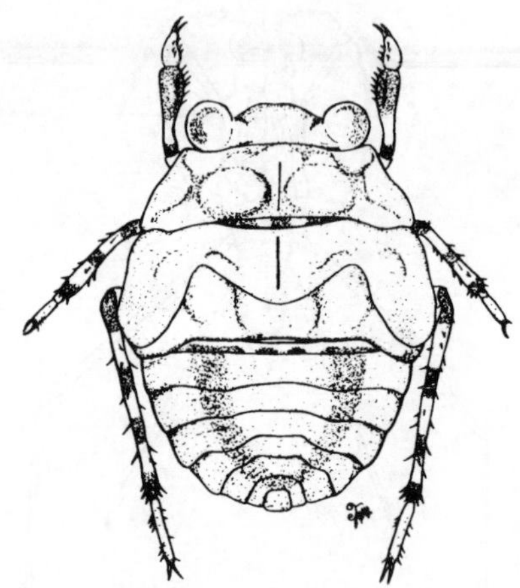

Figure 29.13. Gelastocoridae. Toad bug, *Gelastocoris oculatus* (Fabricius).

GELASTOCORIDAE

Toad Bugs

Figure 29.13 and Key Figure 15

Relationships and Diagnosis: Gelastocorids and ochterids are often placed in the same superfamily, Gelastocoroidea, in the Nepomorpha. The broadly oval shape, dorsoventrally flattened body, and the wart-like elevations of the toad bug's body make it distinct from other true bugs.

Biology and Ecology: Toad bugs live on and in the mud and sand of ponds and streams. At times, both adults and immatures of *Gelastocoris oculatus* (Fabricius) can be found in large numbers on mud flats or previously shallow littoral areas that have been exposed from the lowering of a lake or stream level. These cryptic bugs commonly burrow in mud or other substrates. They also occur in water. Hungerford (1922) reported *G. oculatus* lays about 200 eggs and develops in from 60 to 100 days, indicating a single generation per year. Species are actively predacious on other insects and mites, jumping on and grasping prey with their raptorial prothoracic legs. *Nerthra* spp. are more commonly found under stones near water.

Description: Medium, ca. 7 to 8 mm; broadly oval, dorsoventrally compressed; yellowish brown to darkly mottled. Head transverse, compressed anteroposteriorly, appearing "bar-belled" from above due to protruding, bulbular compound eyes; frons broadly triangular; antennae short, 3 bulbular segments, arising below eyes from caudal side; labium 4-segmented, short, thickly cone-shaped. Thoracic segments transverse, somewhat sinuate, lateral edges greatly flattened; prothoracic legs short and thick, especially femora; meso- and metathoracic legs elongate, especially metatibiae; protarsi and tibiae fused (*Nerthra*) or protarsi 1-segmented (*Gelastocoris*); mesotarsi 1- or 2-segmented, metatarsi 2-segmented; venter of leg segments with short thick spines and setae. Abdomen short, wide, flat, broadly rounded caudally; without scent gland opening.

Comments: Only 2 species of *Gelastocoris* and 4 of *Nerthra* occur in the United States. *G. oculatus* is widely distributed while *N. martini* Todd is common only in the western states. Although they can sometimes be collected in large numbers, few toad bugs have been closely observed. Much more study of the field biology is needed on this group.

Selected Bibliography

Hungerford, 1922
Menke (*In* Menke 1979)

PLEIDAE

Pygmy Backswimmers

Figure 29.14 and Key Figure 13

Relationships and Diagnosis: Taxonomically, the pleids have been historically associated with the backswimmers. Some authors have treated them as belonging to the Notonectidae, but they are properly treated as a unique family in the Nepomorpha.

The humpbacked shape of these miniature backswimmers allows for easy recognition and distinguishes them from any other hemipterans.

Biology and Ecology: Very little is known about the biology of the pygmy backswimmers. As the name implies they swim upside down. They are most commonly found in quiet water covered with much plant growth. On occasions pleids can be seen in large numbers swimming about in an erratic pattern beneath the water's surface. The air trapped around the body by the dense hydrofuge pile gives them a silvery appearance. They are predacious upon microscopic Crustacea belonging to the Entomostraca (Blatchley 1926).

Description: Very small, ca. 1.5 mm; elongate-oval, convex dorsally; yellowish brown, silvery appearance in water. Head transverse from above, strongly deflexed and tapering to short, cone-shaped 3-segmented labium; antennae minute, nearly indistinguishable; compound eyes moderate in size, laterally positioned. Thorax convex dorsally. Legs about

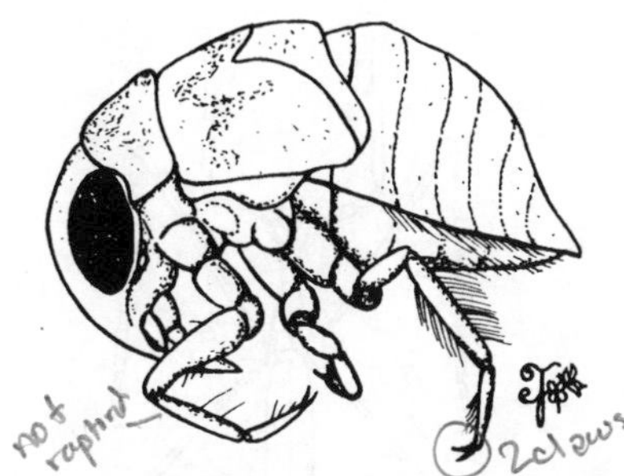

Figure 29.14. Pleidae. Pygmy backswimmer, *Neoplea striola* (Fieber).

equally developed; coxae elongate and thick; femora and tibiae subequal; tarsi 1-segmented, elongate; metatibiae and tarsi with moderate fringes of light-colored setae. Abdominal dorsum strongly convex with fringe of long setae along lateroventral margin to tapering apex; single scent gland opening on caudal area of tergite 3.

Comments: Only 5 species are recorded for N. America. *Neoplea striola* (Fieber) is a widely distributed species occurring over much of the eastern and central United States.

Selected Bibliography

Drake and Chapman, 1953

NAUCORIDAE

Creeping Water Bugs

Figure 29.15 and Key Figure 12

Relationships and Diagnosis: Close family relationships of these truly aquatic bugs are unresolved; however, they are included in the Nepomorpha. Superficially they look like small belostomatids, to which they are only distantly related. The creeping water bugs can be recognized by their small size, ca. 8 mm, raptorial prothoracic legs with claw-like tarsi, eyes and lateral margins of head entire, and single, paired scent gland openings on abdominal tergite 3. Giant waterbugs are larger, with laterally protruding eyes, have 2 claws on protarsus, and lack a scent gland opening. The creeping water bugs are about the same size and general shape as toad bugs, but the latter have bulbular, laterally protruding eyes and no scent gland opening.

Biology and Ecology: Creeping water bugs occupy both well vegetated ponds and streams (*Pelocoris* spp.), and clearer streams or the more open water of lakes (*Ambrysus* spp.). They swim and move among aquatic plants where they prey upon a variety of small organisms. Eggs are glued to aquatic plants. Apparently, naucorids require longer development periods in the egg (3–8 weeks) and larval (7–10 weeks) stages

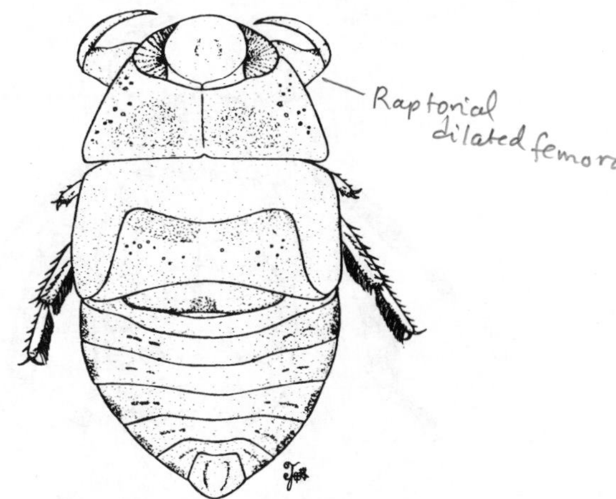

Figure 29.15. Naucoridae. Creeping water bug, *Pelocoris femoratus* Palisot de Beauvois.

than do most other aquatic and semiaquatic bugs. They overwinter as adults in soft oozing mud among aquatic vegetation.

Description: Small, ca. 5–10 mm, oval, tapering caudally, dorsum moderately convex, dorsoventrally compressed; light to dark brown. Head partly retracted into prothorax, transversely oval with compound eyes somewhat disc-shaped, anterior margin and eyes entire, vertex smooth; labrum broadly triangular; labium short, 3-segmented, broadly cone-shaped; antennae short, 3-segmented, arising on venter from behind eyes, segments 1 and 2 short, 3 long. Pronotum transverse, anterior margin concave; pro-, meso-, and metanota shelf-like laterally. Prothoracic legs strongly raptorial with large thick femora, elongate cylindrical tibiae and claw-like tarsi; meso- and metathoracic legs with elongate femora, tibiae and tarsi spined, and metatibiae and tarsi with long fringes of setae on inner margins. Abdominal dorsum and venter moderately convex, about as wide as long, moderately tapering to apex; one pair of widely separated, small, oval, dorsal abdominal scent gland openings on posterior suture of tergite 3.

Comments: About 18 species of creeping water bugs in 5 genera, *Ambrysus* and *Pelocoris* spp. being most common, are known from the United States and Canada. Polhemus (*in* Menke 1979) has provided a useful key to larvae of the 5 genera based on 5th instars. Some specimens are known only from restricted locations in the South and West. As with a few other aquatic bugs, creeping water bugs can inflict a painful bite so they should be handled with care.

Selected Bibliography

Hungerford, 1927
Polhemus (*In* Menke 1979)
Usinger, 1946

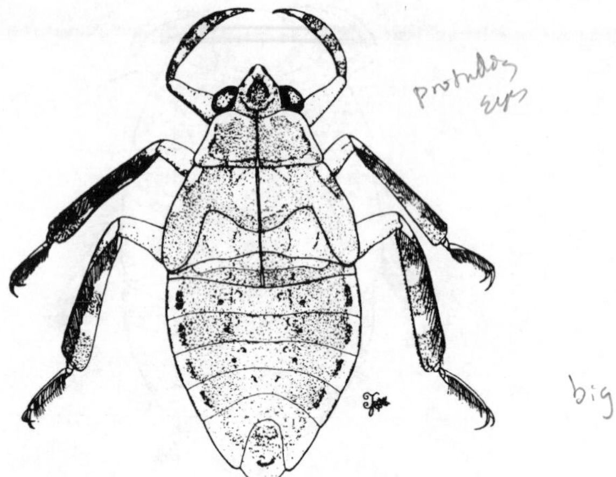

Figure 29.16. Belostomatidae. Giant water bug, *Belostoma* sp.

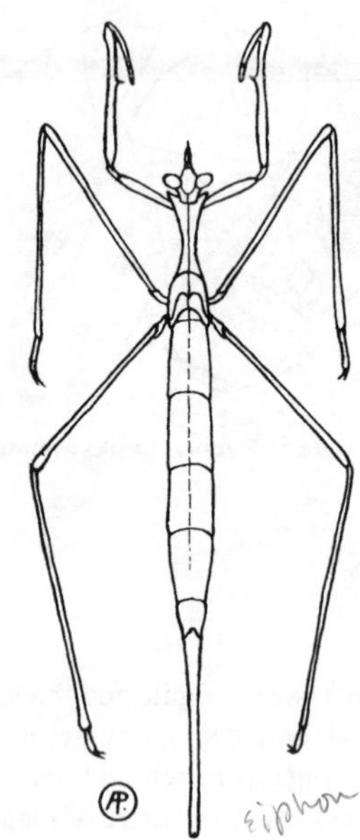

Figure 29.17. Nepidae. Water scorpion, *Ranatra* sp. (from Peterson, 1948).

BELOSTOMATIDAE

Giant Water Bugs

Figure 29.16 and Key Figure 11

Relationships and Diagnosis: This group of large to very large aquatic bugs is probably most closely related to the water scorpions (Nepidae) and are included in the Nepomorpha. Their size and shape allow for easy recognition of most giant water bugs. Water scorpions of the genus *Nepa* are elongate-oval and might be confused with immatures of smaller *Lethocerus* or with *Belostoma* spp. Also, small *Belostoma* spp. or early instars of larger ones may be mistaken for creeping water bugs (Naucoridae). However, the giant water bugs do not have a long respiratory tube as do *Nepa* spp., nor do they have dorsal abdominal scent gland openings as found in the naucorids.

Biology and Ecology: These cryptic, secluded predators feed on small fish and other aquatic animals. They usually occupy ponds, small lakes and quiet pools of streams (*Lethocerus* and *Belostoma*) or flowing water of streams (*Abedus* spp.). Egg deposition by *Lethocerus* females is in the usual manner of gluing them to plants or other surfaces. Their eggs are laid in linear batches. In contrast, an interesting behavior is exhibited by females of both *Belostoma* and *Abedus*. They lay eggs in large batches on the backs of their respective males which carry the eggs until hatching. Smith (1976a,b) has described the complexity of this brooding behavior in *Abedus herberti* Hidalgo and *Belostoma flumineum* Say, respectively. Incubation generally takes less than 2 weeks under optimum conditions and larval development 5 to 9 weeks. Adults overwinter in mud or sheltered places.

Description: Medium to very large, 14 to ca. 45 mm; elongate to broadly oval; somewhat dorsoventrally flattened; light to dark brown. Head broadly triangular, ca. twice as wide as long, extending to rounded tylus, modified at apex into tubular socket holding modified base of labium; labium thick, 3-segmented, extending to or just beyond procoxae; antennae 4-segmented, last 3 segments pectinate, hidden beneath and behind head. Thorax large, flattened; prothorax nearly trapezoidal, moderately convex. Legs elongate; coxae close together; prothoracic legs large, raptorial, with thickened elongate femora, profemora, tibiae and tarsi (1-segmented) with juxtapositioned short, dense mats of stiff setae; meso- and metatibiae and tarsi with fringe of long swimming setae; large metasternal plate present, bearing fringed setae. Abdomen as wide or about as wide as long, or longer than wide, tapering somewhat caudally, dorsoventrally flattened; scent gland openings absent; venter with dense clear setae.

Comments: The 3 N. American genera contain 19 species in the United States. *Lethocerus americanus* Leidy and *B. flumineum* are common and distributed over the entire country. Giant water bugs are also known to be attracted to lights and are often found in swimming pools. As with some other water bugs, belostomatids are known to inflict painful bites. Some *Lethocerus* spp. are pests in hatcheries where they feed on fish fry.

Menke (1979) provided a key to immatures of the 3 genera.

Selected Bibliography

Menke (*In* Menke 1979)
Rankin, 1935
Smith, 1976a,b

NEPIDAE

Water Scorpions

Figures 29.17, 29.18

Relationships and Diagnosis: Water scorpions are most closely related to the giant water bugs and form the nominal taxon for the Nepomorpha. The slender *Ranatra* spp. are easily recognized by their raptorial prothoracic legs, sticklike meso- and metathoracic legs, and long respiratory-siphon

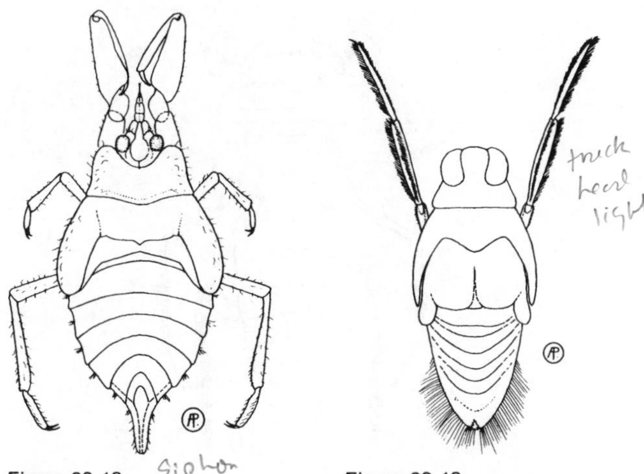

Figure 29.18 Figure 29.19

Figure 29.18. Nepidae. Water scorpion, *Nepa* sp. (from Peterson, 1948).

Figure 29.19. Notonectidae. Backswimmer, *Notonecta* sp., (from Peterson, 1948).

projecting from the abdomen. The siphon also serves to distinguish the water scorpions from the giant water bugs and terrestrial bugs that are slender.

Biology and Ecology: Water scorpions are crawling insects that are most often found among dense vegetation in ponds. They may also be found under the overhanging banks of slow streams, sometimes in mud in shallow water (*Nepa*) and in streams on aquatic plants where they remain motionless in wait for prey, including tadpoles and odonate and mosquito larvae, among others. Egg laying behaviors are varied, eggs sometimes being deposited in plant tissue, other times in muddy banks. Several eggs are laid during a short period of time. The eggs are equipped with long respiratory siphons opening to the air, but are also capable of being submerged in water. There is apparently only 1 or 2 generations per year, larvae requiring about 2 months for development, with overwintering in the adult stage.

Description: *Ranatra* spp., very long, ca. 18 to 35 mm, slender tubular body with long slender legs, and long posterior siphon tube extending from tip of abdomen; *Nepa* spp. elongate, ca. 14 mm long, broad body and thick legs; color light to dark brown. Head porrect, with large bulbular compound eyes extending laterad, otherwise longer than wide; tylus and juga rounded and separated; antennae short, thick, tucked in below eyes, 3-segmented, slightly pectinate on 2nd and 3rd; 3-segmented, short cone-shaped labium projecting cephalad from head in *Ranatra* spp. Prothorax elongate with coxae arising from sockets which open anteriorly and ventrally; wing pads extending along side of body; prothoracic legs raptorial, with coxae and femora very long and slender in *Ranatra* spp., thicker femora in *Nepa* spp., and tibiae and tarsi slightly sickle shaped, without protarsal claws in *Ranatra,* but with single claw in *Nepa;* distal 2/5 of femora from the broad triangular spine with a dense short setal mat juxtaposed to tibiae and tarsi; meso- and metathoracic legs elongate, slender, arising close together, distant from prothoracic legs; all tarsi 1-segmented. Abdomen tube-like with elongate segments; no dorsal abdominal scent gland openings; spiracles on venter; posterior paired siphons ca. 1/2 length of abdomen, with ventral grooves extending from siphons to 2 continuous broad grooves mid-laterally on venter formed from lateral and midventral dense rows of hydrofuge setae.

Comments: In the United States 11 species of *Ranatra,* 3 of *Curicta* and 1 of *Nepa* are known. *Ranatra fusca* P. de B. is common over most of the country while *Nepa apiculata* Uhler is widely distributed, but uncommon, and *Curicta* spp. are southwestern. McPherson and Packauskas (1987) confirmed only four larval instars for *N. apiculata,* one of the few cases where four rather than five instars has been found among the true bugs.

The collector of either *N. apiculata* or *Curicta* spp. should pay special attention to recording details of habitat and biology so as to increase our knowledge of these species.

Menke (1979) provided a key to the immatures of the 3 genera.

Selected Bibliography

McPherson and Packauskas, 1987.
Menke (*In* Menke 1979)
Radinovsky, 1964

NOTONECTIDAE

Backswimmers

Figure 29.19 and Key Figure 14

Relationships and Diagnosis: The backswimmers have been often linked to the pygmy backswimmers (Pleidae) and both are presently included in the Nepomorpha. However, notonectids lack the dorsal abdominal scent gland openings present in the pleids. A close relationship to other nepomorphan families remains obscure. Backswimmers are superficially similar to the water boatmen with which they may be confused since they both are elongate, about the same size, and occupy the same waters. However, notonectids can be recognized by their more rounded body with a keeled dorsum and flat venter, lighter color, elongate prothoracic legs, and long oar-like metathoracic legs which are often held in an anterolateral position. The water boatmen are more dorsoventrally flattened, usually darker, and have short, flattened prothoracic legs. Backswimmers are much larger than pygmy backswimmers.

Biology and Ecology: The group is actively predacious upon a variety of other aquatic organisms including small fish. Gittelman (1974) has outlined different strategies of predation for *Buenoa, Martarega* and *Notonecta* spp., respectively. Backswimmers usually live in ponds and lakes, but may also be found in other quiet water habitats. Some species are more commonly collected in slow moving water. Apparently, different species overwinter as adults, as eggs, or as eggs or adults. Eggs are laid in plant tissue or on the surface of aquatic plants. Acoustical communication has been noted and various stridulatory structures have been identified by several investigators including Wilcox (1969, 1975).

Description: Small, ca. 4 mm, to medium, ca. 15 mm, elongate, dorsally convex; generally light or whitish with black or reddish orange markings. Head transverse (*Notonecta* spp.) to semi-spherical (*Buenoa* spp.); compound eyes very large, occupying most of head, nearly contiguous along midline in some species (*Buenoa*); labium short, appearing 3-segmented, cone-shaped; antennae short, thick, arising ventrally behind eyes, 2-segmented in *Buenoa,* 3 in *Notonecta* and *Martarega.* Thorax appearing cylindrical, dorsum strongly convex; pronotum transverse, meso- and meta-wingpads positioned laterally rather than dorsally; venter with dense dark setal pattern. Legs elongate, covered with many very long light to dark, thick setae, and short dark spines; pro- and mesothoracic legs arising close together, distant from metathoracic legs, and modified for grasping prey; pro- and mesofemora moderately thick, mesofemora of *Notonecta* spp. with large light-colored spine near apex; pro- and mesotibiae elongate, slightly scalloped on inner sides; pro- and mesotarsi 1-segmented, about as long as tibiae, each bearing 2 long claws; metafemora elongate with rows of short, stout spines, metatibiae and tarsi with dense fringes of long setae for swimming, tarsi with one claw. Abdominal dorsum strongly convex, tapering to apex with some clear setae on surface; scent gland openings absent; venter flat to slightly concave with median and midlateral fringes of dense, dark, long setae.

Comments: Several of the 35 or so species in the 3 genera of our fauna are widely distributed and locally common. Backswimmers are commonly collected at lights and will bite if handled carelessly.

Truxal (*In* Menke 1979) has provided a key to 5th instars of 7 California species.

Selected Bibliography

Gittelman, 1974
Hungerford, 1920
McPherson, 1965, 1967
Truxal (*In* Menke 1979)
Wilcox, 1969, 1975

CORIXIDAE

Water Boatmen

Figure 29.20 and Key Figure 5

Relationships and Diagnosis: The Corixidae forms a rather homogeneous and unique group without clear, close relationships to other families in the Nepomorpha where it continues to be placed. The presence of a 1-segmented labium, capability of ingesting particulate as well as liquid food, highly modified scoop-like protarsi, and uniformly structured mesowings set the Corixidae apart from other taxa.

Immatures of water boatmen can be readily identified by their dark color and elongate-oval shape, along with the labial and protarsal features just mentioned.

Biology and Ecology: Water boatmen inhabit nearly all quiet water habitats. A few species live in streams and rivers. Eggs are laid in large batches attached to various surfaces including the backs of crayfish (Griffith 1945). Eggs hatch in a week or so and development is completed in 5 to 7 weeks.

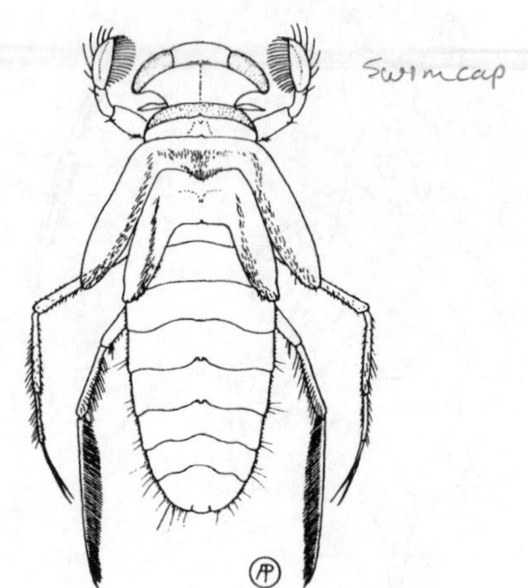

Figure 29.20. Corixidae. Water boatman, *Corixa* sp., (from Peterson, 1948).

Corixids go through 1 to 2 generations per year with different species overwintering as eggs, larvae and/or adults. Hilsenhoff (1970) reported large numbers of water boatmen of 15 or more species aggregating in late fall in the Wisconsin River. Another collection in the St. Croix River (Wisconsin-Minnesota border) produced 23 species. He concluded that they were congregating in the rivers to pass the winter.

Description: Small, ca. 3 mm, to medium, ca. 10 mm; elongate-oval, often dorsoventrally flattened; mottled light yellowish brown to dark brown. Body often covered with setae, especially fringed along margins. Head transversely triangular, somewhat disc-shaped from above, with frons and apex of head strongly reflexed; compound eyes, large, located at posteriolateral angles of head; head greatly scalloped along caudal surface; labium short, unsegmented, broadly triangular, almost indistinguishable from head; antennae arising from caudoventral aspect of head, 2-segmented, 1st short, 2nd long and cylindrical, extending to near outer margin of eye. Pronotum narrowly transverse and reflexed anteriorly; meso- and metanota together nearly square except for posteriorly directed wing pads. Coxae nearly touching; prothoracic legs short, "appearing" raptorial, femora and tibia thickened, highly modified protarsus (called palae) 1-segmented, long and spatulate, spines or pegs present on inner surface, fringed with setae; mesothoracic legs long, with very long femora, tibiae shorter, tarsi long with 2 very long claws; metathoracic legs long, femora and tibiae often subequal, tarsi long, somewhat spatulate, with dense fringe of setae on inner margin, claws short. Abdomen somewhat flattened dorsoventrally with alternating color bands, dorsum moderately convex. Three pairs of abdominal scent gland openings (sometimes 3rd indistinct) on caudal margins of tergites 3, 4 and 5, respectively.

Comments: This is a large taxon of some 120 species in 17 genera for N. America. *Sigara* contains about 50 species and *Hesperocorixa* 18, including some of the larger species. Corixids are known to feed on a variety of food materials including plant as well as animal. They sometimes generate large local populations and are attracted to lights. Water boatmen serve as food for many other animals including fish. The eggs, larvae, and adults are eaten by humans in Mexico.

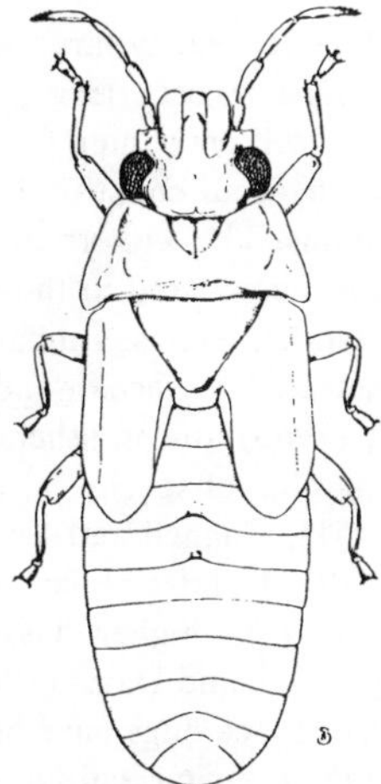

Figure 29.21. Thaumastocoridae. Royal palm bug, *Xylastodoris luteolus* Barber, (from Schaefer, 1962).

Selected Bibliography

Griffith, 1945
Hilsenhoff, 1970
Hungerford, 1948
Lauck (*In* Menke 1979)

THAUMASTOCORIDAE

Royal Palm Bugs

Figure 29.21

Relationships and Diagnosis: The phytogenetic position of this little known family remains somewhat obscure. Schaefer (1969) and Stys and Kerzhner (1975) treated it as an isolated family of the Cimicomorpha. Cobben (1968) considered it as possibly annectant between the Cimicomorpha and the Pentatomomorpha.

Because of the unique host plant relationship it is not likely that the collector will confuse the single United States species with other hemipterans. Superficially they resemble the unornamented tingid, *Atheas austroriparius* Heidemann, which are the same size. However, the royal palm bug has antennal segments 2, 3 and 4 subequal and the juga are well developed and parallel-sided, whereas the tingid has the 2nd segment very long and the juga are poorly developed.

Biology and Ecology: A single species, *Xylastodoris luteolus* Barber, occurs in royal palms, *Roystonea regia,* of southern Florida. Some biological information is available (Baranowski 1958, Moznette 1921). The bug feeds on newly opened fronds, causing damage. They seem to prefer larger, older trees. Only a few eggs, deposited singly, are laid. Egg and immature development appears to be in sequence with new frond production on palms throughout the season.

Description: Very small, ca. 2 mm, elongate-oval to parallel-sided, dorsoventrally compressed; pale yellowish-brown. Head squared with broad juga and tylus; antenniferous tubercle pointed, distinct; compound eyes large, laterally positioned near thorax; antennae 4-segmented, ca. as long as head and pronotum together; labium 4-segmented, short, thick, extending to near procoxae. Prothorax much wider than long, lateral margins compressed, shelf-like. Legs short, femora thickened, tibiae short. Abdomen oval, dorsoventrally compressed, 2 pairs of scent gland openings at caudal margins of tergites 3 and 4.

Comments: In the United States the family consists of a single species, *X. luteolus* which occurs on royal palms in southern Florida. It will only be taken by the most diligent collectors as it is found under the fronds of certain royal palms. Three references will be of special interest, including Drake and Slater (1957), Baranowski (1958) and Schaefer (1969).

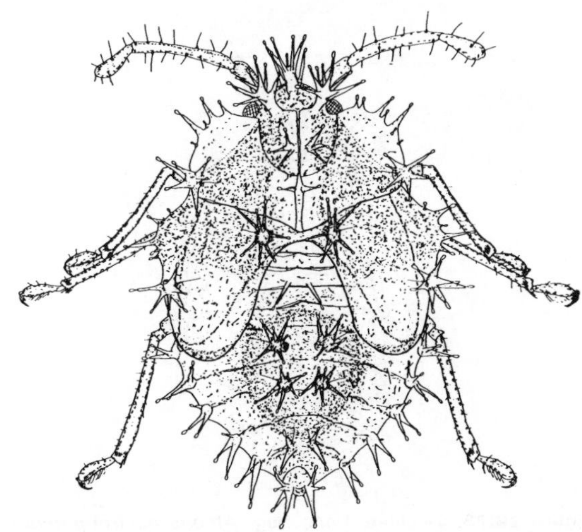

Figure 29.22. Tingidae. Lace bug, *Corythucha ulmi* Osborn & Drake.

Selected Bibliography

Baranowski, 1958
Cobben, 1968
Drake and Slater, 1957
Schaefer, 1969

TINGIDAE

Lace Bugs

Figures 29.22, 29.23 and Key Figure 45

Relationships and Diagnosis: The common name, lace bugs, comes from the lace-like appearance of the adults. It is an interesting and unique family. Relationships of the Tingidae to other true bugs remain obscure except that tingids are generally accepted as belonging to the Cimicomorpha to which the Reduvioidea and Cimicidae also belong. Piesmatids resemble the tingids in size and general appearance, but are not phylogenetically close.

Most lace bugs are very small and ornamented with large spines, highly modified setae, and scoli. This and their secluded position in nature on the underside of leaves provide for easy recognition.

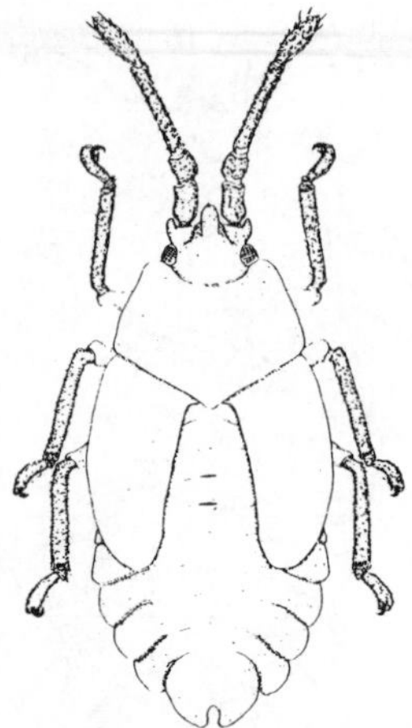

Figure 29.23. Tingidae. Lace bug, *Atheas austroriparius* Heidemann, (from Sheeley and Yonke, 1977a).

Biology and Ecology: Tingids are exclusively phytophagous, primarily feeding on the lower epidermis of leaves on herbs, shrubs, and trees. A few species are most frequently picked up on the ground (*Hesperotingis* spp.) and others are associated with mosses (*Acalypta* spp.) (Slater and Baranowski 1978). Adults overwinter and most species go through 2 or 3 generations per year. A prolonged oviposition period results in overlapping generations. Eggs of many species are inserted in leaf tissue to the egg cap, but in some species eggs are glued to the leaf surface and covered with a protective secretion. Larvae display an aggregation behavior, are sessile, and feed in small groups of 10 or fewer on the underside of leaves. They remain closely appressed to the plant surface and generally do not move even when disturbed. Most species are specific to one or a few closely related hosts such as *Atheas austroriparius* Heidemann on *Schrankia uncinata* Willd. (sensitive brier) (Sheeley and Yonke 1977a) and *Corythucha bulbosa* Osborn and Drake on *Staphylea trifilia*, bladdernut (Sheeley and Yonke 1977b).

Description: Very small, generally less than 2.5 mm; oval and dorsoventrally flattened, most often bearing numerous pronounced setae and scoli, some elaborately ornamented; generally dull grey to dark brown in patches, antennae and legs light yellowish brown. Head with long setae arising from scoli; antennae 4-segmented, 1st and 2nd short, 3rd very long, 4th ca. as long or somewhat longer than 1 and 2 combined, 3 and 4 with several medium to long setae; bucculae usually very well developed, deeply grooved and extending to caudal margin of head; labium 4-segmented, thickened, extending to or beyond mesocoxae, short in *Atheas*. Pronotum broadly expanded laterally, often reflexed and with raised median pronotal hump (esp. *Corythucha* spp.); margins of pronotum, wing pads, and hump bearing stout scoli and chalazae. Abdomen with chalazae and scoli on dorsum and along lateral margins; single minute scent gland opening at midline on caudal margins of tergites 3 and 4. Note: immatures of *Atheas* spp. (fig. 29.23) are devoid of chalazae and scoli, but recognized by well developed bucculae, by 3rd antennal segment being much longer than 1, 2, and 4 combined, and by prothorax flattened and expanded somewhat laterally.

Comments: This is a fairly large family with some 16 genera and over 150 species occurring north of Mexico. However, most collectors will encounter *Corythucha* spp. which are widely distributed, occur on a variety of plants, and often form large populations. The western and southwestern species are poorly known compared to those from the Midwest and northeastern United States. Immatures have been described and/or illustrated by Sheeley and Yonke (1977a) and Drake and Ruhoff (1965) among others. Biological and life history data on a number of species have been presented including Bailey (1951), Connell and Beacher (1947), Dickerson and Weiss (1917) and Sheeley and Yonke (1977b). Significant research on the higher classification of the Tingidae was done by Drake and Davis (1960).

Several species of lace bugs have been cited as causing considerable damage on various cultivated shrubs and trees, but few are considered of economic importance.

Selected Bibliography

Bailey, 1951
Connell and Beacher, 1947
Dickerson and Weiss, 1917
Drake and Davis, 1960
Drake and Ruhoff, 1965
Horn et al. 1979
Sheeley and Yonke, 1977a,b
Slater and Baranowski, 1978

MICROPHYSIDAE

Microphysid Bugs

Relationships and Diagnosis: This small group of very small bugs belongs to the Cimicomorpha in which it is most clearly related to Miridae, and less closely to Anthocoridae. The microphysids can be distinguished from the above 2 families to which they are similar by their minute size of about 1 mm, possession of an apparently 3-segmented, (actually 4) labium, and 1-segmented tarsi. Mirids have a distinct 4-segmented labium and 2 tarsal segments. Anthocorids have an apparent 3-segmented labium and usually 2 tarsal segments.

Biology and Ecology: Almost nothing is known of the field biology of microphysids, but they are thought to be predacious, especially on psocids. They have been reported as inhabiting old and dying trees, and living under bark and among lichen and moss.

Description: Minute, ca. 1 mm, head porrect, labium apparently 3-segmented; antennae 4-segmented, filiform, with 3rd and 4th segments together less than 2X length of 1st and 2nd together.

Comments: A single species has been found in the United States, *Mallochiola gagates* (McAtee and Malloch 1925). To enhance chances of collecting this and perhaps other species of Microphysidae look carefully on old and fallen trees where many other small insects abound. Also, be alert to the accidental find, and record all possible details if you do collect these rare bugs.

Selected Bibliography

McAtee and Malloch, 1925

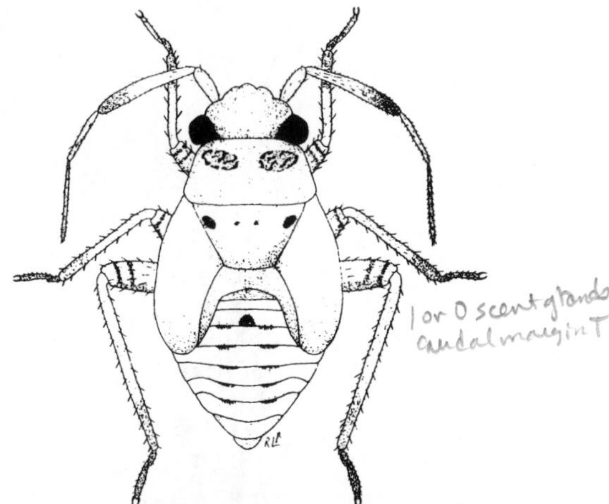

Figure 29.24. Miridae. Plant bug, *Capsus ater* (L.).

MIRIDAE

Plant Bugs

Figures 29.24 and Key Figures 36, 47, 48

Relationships and Diagnosis: The plant bugs belong to the Cimicomorpha, but do not have an obvious "closest" family other than Microphysidae. Most species are soft bodied and delicate as immatures. They resemble some lygaeids and rhopalids in general appearance as they are oval or elongate, cursorial, and have a 4-segmented labium. However, plant bugs have no dorsal abdominal scent gland openings or only 1 opening located on the caudal margin of tergite 3.

Biology and Ecology: In this large and diverse group most species are plant feeders. Plant feeding species may be either broadly phytophagous, such as *Lygus lineolaris* (P. de B.) and *Poecilocapsus lineatus* (Fabricius) which feed on a number of species in many plant families, or relatively host specific. There are also many predacious plant bugs such as species of *Phytocoris, Deraeocoris* and *Ceratocapsus*. They feed upon mites, aphids, psyllids and lace bugs, among others. Many predacious species seem to be associated with particular plants on which they pursue prey. Other plant bugs, such as species of *Lygus*, are known to be both predatory and plant feeders. For a group as large, diverse and economically important as this one, relatively little is definitively known about the biology of its species. Most species of plant bugs go through only 1 generation per year. Thus, immatures can be collected only for a few weeks, usually in spring and early summer. However, many pest species complete a life cycle from egg to egg in about 35 days, producing overlapping generations throughout the growing season as long as environmental conditions favor development. Unlike most other Hemiptera many mirids overwinter as eggs inserted in plant tissue. A large number of species are cryptic. Brachyptery occurs in several mirids.

Description: Most 5th instars very small to small, 2–5 mm. Immatures basically of 3 types: small-oval, elongate, or ant-like. Some species dark; most are light yellowish-green to reddish-brown; body often with color patterns and covered with setae. Head transverse to triangular with large compound eyes; antennae with 4 elongate segments, especially 2 and 3, often covered with dense long dark setae, 3 and 4 often threadlike; labium 4-segmented, usually extending to mesocoxae or beyond, or short, extending only to procoxae. Pronotum moderately transverse; calli often evident on anterior region of pronotum. Legs cursorial with long femora and tibiae, coxae about evenly spaced. Abdomen elongate to oval and bulbular; none or 1 abdominal scent gland opening at midline on caudal margin of tergite 3.

Comments: There are more species of plant bugs than any other family of Hemiptera. Slater and Baranowski (1978) treated over 200 genera in their key to adults. While several genera are known only from 1 or a few species, most genera are moderate in size (15–30 species) or large. *Pilophorus* has 45 ant-like species, *Ceratocapsus* 56 species, *Parthenicus* 68 mostly western species, and *Phytocoris* has about 200 species. Many species are widely distributed and easily collected. Others require careful searching of potential host plants.

Some authors have treated the isometopines as a separate family; however, they are included here with the Miridae (Wheeler and Henry 1977b).

The Miridae includes several species that are significant as agronomic pests. Most notable are the so-called lygus-bugs (*Lygus* spp.), the meadow plant bug (*Leptopterna dolabrata* (L.)), rapid plant bug (*Adelphocoris rapidus* (Say)), alfalfa plant bug (*A. lineolatus* (Goeze)) and the cotton fleahopper (*Pseudatomoscelis seriatus* (Reuter)), to name a few.

Some important selected references for more detailed information on the biology and descriptions of immatures are Wheeler and Henry (1976, 1977a,b, 1978), Akingbohungbe (1974), Akingbohungbe et al. (1973), Arrand and McMahon (1974), and Wheeler et al. (1975).

Special handling is advised for plant bugs. Only a few larvae should be placed in the same vial since antennae and legs will easily separate from the body. It is especially important in this family to associate larvae and adults of the same species at the time of collecting, since larvae of different species are very similar.

Selected Bibliography

Akingbohungbe, 1974
Akingbohungbe et al. 1973
Arrand and McMahon, 1974
Knight, 1941
Slater and Baranowski, 1978
Wheeler, A. G., Jr. and Henry 1976, 1977a,b, 1978
Wheeler, A. G., Jr. et al. 1975

NABIDAE

Damsel Bugs

Figures 29.25 and Key Figure 28

Relationships and Diagnosis: The Nabidae are most closely related to the Anthocoridae and the Cimicidae. Immatures can be recognized by their small to medium size, slender head and thorax, long, bowed, 4-segmented labium, thickened profemora, and protibiae and profemora with numerous long hairs.

Damsel bugs can be confused with some slender reduviids and anthocorids. Examination will reveal a 3-segmented, or apparently 3-segmented, labium in both the assassin bugs and anthocorids rather than 4 as in the damsel bugs. Damsel bugs do not have a ridged prosternal groove as do assassin bugs. Mature larvae of damsel bugs are almost always much larger (5 mm+) than those of anthocorids (4 mm). Also, most anthocorid larvae have a short labium extending only to the procoxae.

Biology and Ecology: Most species are aggressively predacious and while some species occupy more secretive habitats such as galls, many are found in open fields on both vegetation and the ground. Overwintering for most species apparently occurs in the adult stage. *Nabicula subcoleopteratus* (Kirby) overwinters as an egg. The more common species such as *Nabis americoferus* (Carayon) and *N. alternatus* (Parshley) go through 2 or 3 generations per year depending upon location. Eggs are laid in plant tissue. In some species, brachypterous adults are more common than macropterous forms.

Description: Small to medium, ca. 5–10 mm; slender, somewhat dorso-ventrally flattened; larvae of some species ant-mimics; yellowish-red to light brown to brown in most genera or black in *Pagasa* spp. and *N. subcoleopteratus,* bright pigment spots on some immatures. Head elongate, almost appearing cylindrical in profile; compound eyes large, extending laterad; tylus very long, juga less so and deflexed, reaching large, short bucculae which form prominent "nose-like" structure giving rise to thick, short 1st labial segment; labium 4-segmented, bowed, long, usually extending to mesothoracic legs, only reaching prothoracic legs in *Pagasa* spp. Antennae usually 4-segmented, elongate, 5-segmented in *Pagasa* spp., 1 and 2 short, 3–5 long. Thorax elongate, not deep. Legs cursorial with elongate coxae, femora, tibiae and tarsi; prothoracic legs and to some extent mesothoracic legs modified raptorial with thickened femora and numerous long and short setae on underside of femur, and long setae and rows of black spines on underside of tibiae; metafemora longer and more slender than pro- or meso-femora. Abdomen elongate, somewhat flattened, with 3 pairs of slit-like scent gland openings on posterior midline of tergites 3, 4 and 5.

Comments: This is a small group of only about 8 genera and 30 species in our fauna. Some species are common in various habitats and are widespread. The group is considered to be beneficial since active predation by a few species occurs

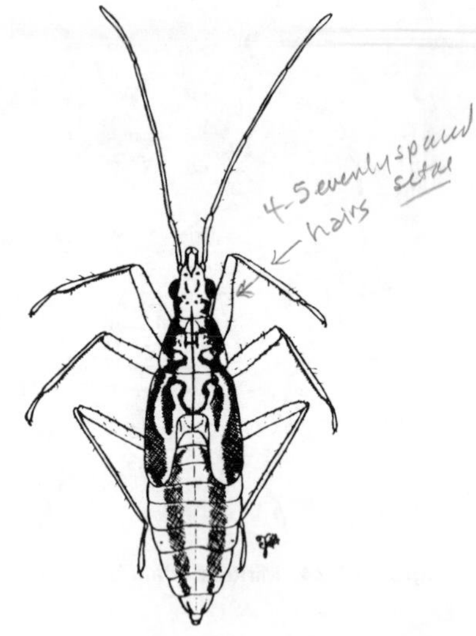

Figure 29.25. Nabidae. Damsel bug, *Nabis* sp.

in cultivated crops on pests such as aphids and leafhoppers. Some species will bite when handled; however, bites are usually not severe.

Little is known of the immatures and only a few references are available on the family including Readio (1928), Taylor (1949), and Hormchan et al. (1976).

Selected Bibliography

Harris, 1928
Hormchan et al. 1976
Readio, 1928
Taylor, 1949

ANTHOCORIDAE

Minute Pirate Bugs, Flower Bugs

Figures 29.26 and Key Figure 35

Relationships and Diagnosis: The Anthocoridae belong to the Cimicomorpha and are probably most closely related to the Cimicidae. Larvae are not likely to be confused with most other Hemiptera and are recognized by the well developed and protruding jugal-tylus "nose" and the apparently 3-segmented labium. The labium has 4 segments, but the 1st is usually obscured by the labrum. Minute pirate bugs somewhat resemble certain nabids; however, 5th instar anthocorids are smaller and do not have long well developed profemora bearing long setae, nor protibiae with 2 rows of tiny black spines on the posterior surface as do the nabids.

Biology and Ecology: Most species are predatory, feeding on other small arthropods. A few species such as *Orius insidiosus* (Say) are widely distributed and commonly collected in diverse habitats and in the flowers of a wide variety of plants where they search for prey. They are known to feed on aphids, caterpillar eggs, leafhopper larvae, mites, thrips,

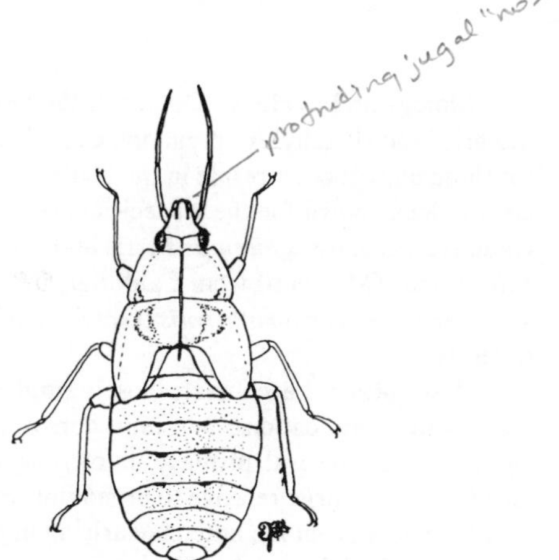

Figure 29.26. Anthocoridae. Minute pirate bug.

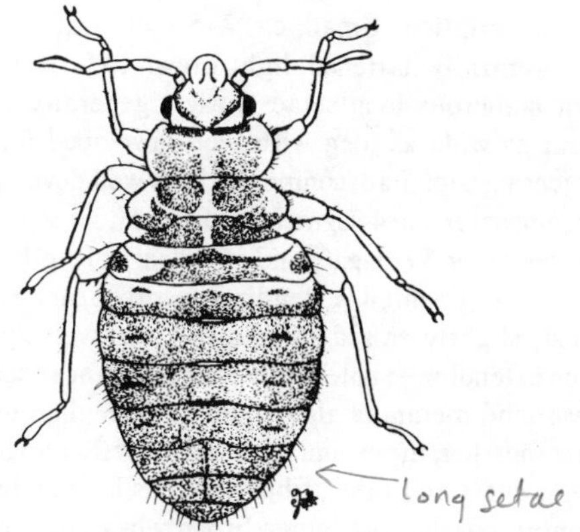

Figure 29.27. Cimicidae. Bed bug, *Oeciacus vicarius* Horvath.

and whiteflies. For most other species little or nothing is known about their biologies. They may be collected on both deciduous and coniferous trees, in ground litter, in aphid-formed galls and other selected, secretive habitats.

Description: Small, 2.5–4 mm long, elongate-oval with rounded abdomen tapering to triangular head; somewhat dorsoventrally flattened; cursorial; yellowish brown to dark brown or red; body punctate. Head triangular, tapering to broad, well developed tylus and juga; rounded laterally; antennae 4-segmented, filiform, 1st segment not reaching tip of tylus, 2nd twice length of 1st; labium appearing 3-segmented, arising from broad bucculae-tylus snout, short, extending to procoxae, or long and extending to metacoxae, distinctly bowed in repose, sometimes projecting anteriorly. Pronotum transverse, twice as wide as long, rounded at anterolateral angles. Femora short, somewhat thickened, tibiae cylindrical with few short setae; 1st tarsal segment short, 2nd long. Abdomen broadly oval, somewhat flattened dorsoventrally, with 3 pairs of scent gland openings easily visible in midlateral position on caudal margins of tergites 3, 4 and 5, 4th pair closer to midline and nearly indistinct.

Comments: Specimens may be taken while sweeping; however, larvae will more likely be collected by placing flower heads and litter samples in Berlese funnels. The family is of moderate size with some 20 genera and 70 species occurring in the continental United States and Canada. The general collector is likely to encounter only a few species since most, other than *O. insidiosus,* are not found in large numbers. Little is known about the immatures, but they are generally considered to be beneficial, and *O. insidiosus* has been studied as a biological control agent against Lepidoptera eggs, thrips and mites (Isenhour and Marston, 1981).

Selected Bibliography

Anderson, 1962
Askari and Stern, 1972
Bacheler and Baranowski, 1975
Isenhour and Marston, 1981
Sands, 1957

CIMICIDAE

Bed Bugs

Figure 29.27 and Key Figure 37

Relationships and Diagnosis: Bed bugs are related most closely to the minute pirate bugs (Anthocoridae) and the bat bugs (Polyctenidae), and form the nominal taxon for the Cimicomorpha. They can be separated from most minute pirate bugs by habitat (nests of birds, roosts of bats and in human dwellings) and by their generally larger size and oval to pear shape. The bat bugs are elongate, very small ectoparasites that have no eyes, while cimicids have compound eyes.

Biology and Ecology: This family is composed of blood-sucking ectoparasites of bats, birds, and, often humans. They are found on the host only when feeding. Otherwise they remain secluded in the nest or on the walls and in cracks and crevices of the host dwelling. The human bed bugs, *Cimex lectularius* L. and *C. hemipterus* (Fabricius), and other cimicids appear to have a broad range of temperature tolerance that closely regulates development, frequency of feeding, egg laying, and longevity. Consequently, depending upon conditions *C. lectularius* may go through only 1 generation per year or as many as 12. At the other extreme adults may survive for a year or more without feeding. The number of eggs laid is proportional to the volume of the blood meal. Under favorable conditions bed bugs will feed about once per week. Immatures begin feeding within 24 hours after hatching and continue feeding more frequently throughout development than adults. Species are relatively host specific. The more commonly encountered cimicids include *C. lectularius* associated with humans, *C. adjunctus* Barber with bats, *Oeciacus vicarius* Horvath on cliff swallows, and *Hesperocimex* spp. with purple martins.

The method of insemination in the bed bugs is unique. It is extragenital and traumatic in that the male punctures the abdomen of the female, causing an integumental wound, and deposits sperm into the cavity outside the normal reproductive system.

Description: Small, ca. 3–5 mm; oval to elongate oval, dorsoventrally flattened; light brown to brown; body usually with numerous long setae; cuticle generally smooth. Head about as wide as long with well developed jugal-tylus lobe projecting cephalad; compound eyes well developed on lateral margin; antennae 4-segmented, filiform, 1st segment short and thicker, other 3 elongate and subequal in length; labium short, thick, 3-segmented, extending to procoxae. Pronotum flattened, slightly raised to midline, transversely rectangular, often extending at anterolateral angles to near compound eyes; meso- and metanota also much broader than long. Procoxae close together, meso- and metacoxae farther apart; coxa large, femur short and thick, tibia about as long as femur and trochanter together. Abdomen broadly oval, dorsoventrally flattened; scent glands opening on caudal margins of tergites 3, 4, 5 and sometimes 6.

Comments: This small group of 8 genera and about 15 species contains a few species such as *C. lectularius* that are widely distributed. Others are much more restricted. The high degree of host and habitat specificity requires the collector to search the nests and domiciles of the host for specimens. Although cimicids can be taken from the hosts this is not common. Humans may be bitten by either the human bed bugs or by bat bugs. If populations of bats and their associated bat bugs are large or the hosts are suddenly not available bat bugs will search for other hosts. Likewise, *C. lectularius* is known to be an occasional pest of chickens and other livestock. Although cimicids have been repeatedly implicated in disease transmission little conclusive evidence is available to substantiate the claims.

Numerous illustrations and descriptions of the eggs and larvae of several species are provided by Usinger (1966). He also provides partial keys to the eggs, 1st instars, and late instars. Anyone interested in the group must consult this comprehensive treatment of the family.

Selected Bibliography

Usinger, 1966

POLYCTENIDAE

Bat Bugs

Relationships and Diagnosis: These bugs, ectoparasites of bats, are placed in the Cimicomorpha where they are most often linked to the Cimicidae. However, their true lineages remain obscure, awaiting considerably more information on biology and comparative morphology of both immature and adult stages.

The bat bugs will not be confused with any other true bugs. Since they are found on bats, as are species of the dipteran bat fly families Streblidae and Nycteribiidae, the inexperienced collector should look carefully for the presence of antennae longer than the head which will separate the polyctenids. Also, bat bugs do not have eyes while bed bugs occurring in association with bats do.

Biology and Ecology: Details of the biology of bat bugs are brief and sketchy. A "pseudoplacento" viviparity is noted for these bugs. Young reside in the bodies of the female. Host associations known for the 2 species in the United States include *Hesperoctenes eumops* Ferris and Usinger on *Eumops californicus* (Merriam) from California and *H. hermsi* Ferris and Usinger on *Tarida macrotis* (Gray) from Texas, both free-tailed bats.

Description: Late instars closely similar to adults in ectoparasitic modifications: 3–4 mm, dorsoventrally flattened, elongate oval; several primary, numerous secondary setae. Head box-like, large relative to pronotum, with antennal and genal ctenidia (combs); antennae arising in pocket on venter, 4-segmented, filiform; labium arising on venter removed from cephalad margin, "said to be 4-segmented"; compound eyes and ocelli absent; 1 or more sets of setal-combs present on venter. Prothorax flattened, broad laterally. Prothoracic legs short, thick; meso- and meta-thoracic legs elongate, all with paired claws. Abdomen with equally differentiated segments, somewhat compacted, no dorsal abdominal scent glands.

Comments: The polyctenids, when first described, were placed in the Nycteribiidae, later as a family in the Anoplura, and finally in the Hemiptera.

Ferris and Usinger (1939) presented as comprehensive a piece of work as was possible on the Polyctenidae in hope that others would meet the call to study bat bugs in greater detail. Their appeal has gone unanswered. One might say that in this group most of the research is yet to be done. Bat bugs mainly occur in tropical and subtropical regions.

Selected Bibliography

Ferris and Usinger, 1939

PHYMATIDAE

Ambush Bugs

Figure 29.28 and Key Figure 27

Relationships and Diagnosis: The ambush bugs are most closely related to the assassin bugs and are sometimes treated as a subfamily of the Reduviidae. However, they are not likely to be confused with reduviids or other true bugs. Their stout, yellow, green and black bodies along with the large, thickened profemora and sickle-shaped protibiae make them distinct.

Biology and Ecology: Ambush bugs are most noted for their cryptic color and motionless behavior as they lie in wait to ambush prey, such as small flies and bees. Host plant and places on the host occupied by the ambush bug are relatively specific, and, in part, determined by the size, shape, and cryptic color of the bug. Almost nothing other than casual observations is recorded about the biology of these interesting Hemiptera.

Figure 29.28. Phymatidae. Ambush bug, *Phymata americana* Melin.

Description: Medium, ca. 8–10 mm; some species smaller, ca. 5 mm; elongate oval and robust; body granulate especially along margins; yellow, some species greenish yellow. Head somewhat box-like, about as high (markedly deflexed ventrad and caudad) as long; vertex heavily spined in some species of *Phymata* and with 2 horn-like anterior projections; compound eyes positioned laterally; tylus and juga vertical, bucculae well developed; antennae arising in front of eyes under vertex shelf, 4-segmented, cylindrical, moderately long (longer than head); labium short, thick, cone-shaped, 3-segmented, extending in repose to prosternal cleft. Thorax elongate, with lateral margins and wing pads projected laterad forming extended "shelf," reflexed dorsally; wing pads greatly expanded posterolaterally. Prothoracic legs raptorial, with greatly thickened coxae, trochanters and femora, tibiae thick and sickle-shaped with small antagonistic spines on femora and tibiae; meso- and metathoracic legs cursorial, femora with large spines. Abdomen round to oval from above, with greatly expanded and upturned margin, paired dorsal abdominal scent gland openings on caudal margins of tergites 4 and 5.

Comments: Larvae and adults can be collected from flower heads and by sweeping low vegetation. The small family contains only 2 genera and about 25 species in our fauna. A few species are common and widely distributed such as *Phymata pennsylvania* Handlirsch and *P. americana* Melin, while most are less common. Little is known about this group other than the taxonomy, which is based on adults. Because ambush bugs are predators they may be thought of as beneficial; however, some of their prey includes small pollinating bees and they are not known to be significant predators on any pest insects. Rather, interest in this group lies in the unique structural modifications, cryptic conditions, behaviors, and their role in the natural environment. Kormilev's (1960) revision of the Phymatinae is worth noting. As mentioned previously, the ambush bugs are often grouped in the Reduviidae; however, their unique form and easy recognition by the general collector as well as some taxonomic considerations warrant their treatment here as a family.

Selected Bibliography

Kormilev, 1960

REDUVIIDAE

Assassin Bugs

Figures 29.29, 29.30 and Key Figure 34

Relationships and Diagnosis: The assassin bugs are placed in the Cimicomorpha and are closely related to, or include, the Phymatidae. However, they are not likely to be confused with ambush bugs even though both possess a median, ridged prosternal groove. The short, greatly thickened raptorial prothoracic legs characterize the phymatids. Reduviids are quite variable in size and form so the beginner should expect to find considerably different-looking species in this group. Other thin, "stick-like" hemipterans such as the marsh treaders (Hydrometridae) and stilt-legged bugs (Berytidae) are easily separated from similar slender assassin bugs. In marsh treaders the head is very long, as long as the thorax, and the prothoracic legs are in no way raptorial, while the head is shorter in the reduviids and some genera (*Brace* and *Ploiaria*) have raptorial prothoracic legs. The stilt bugs have a very long, slender 4-segmented labium while that of assassin bugs is 3-segmented, short and thickly cone-shaped. Some other moderately slender broad-headed bugs and damsel bugs may also look superficially like assassin bugs and vice versa; however, characters of the elongate, transversely grooved head, short labium extending to the prosternal groove, and sometimes moderately raptorial prothoracic legs will serve to separate the reduviids.

Biology and Ecology: This large and interesting group of predators has received considerable study of its biology and immatures compared with most true bugs. Most species occupy rather selected habitats, some being ground dwelling while most live up on plants. Some species live in wooded areas, others in open fields. Eggs are usually laid in batches, but some species place them singly.

Most larvae and adults are cryptic and often are motionless hunters. Most are predacious upon other insects; but some, such as *Triatoma* spp., are notorious because of their blood-sucking habits on mammals, including humans. Overwintering may occur in any life stage, depending upon the species. This feature is somewhat unusual among hemipteran families which tend to have more uniformity than is realized here. Assassin bugs are usually univoltine. The common and widely distributed *Sinea diadema* (Fabricius) and another species, *S. complexa* Caudell, are bivoltine. Polymorphism, with respect to wings, occurs in adults of a number of species. There are a number of apterous species usually occupying secluded, compressed habitats such as under loose bark, logs, and stones. Also, many species are attracted to lights where they feed on other insects. Reduviids can be easily reared in the laboratory.

Description: Extremely variable group. Mostly medium ca. 10 mm, to very large, 25 mm; some species small, ca. 5 mm; stick-like to elongate and robust; light yellowish-brown or green to black, some species with red and orange markings; often spined. Head elongate, cylindrical, porrect, with or

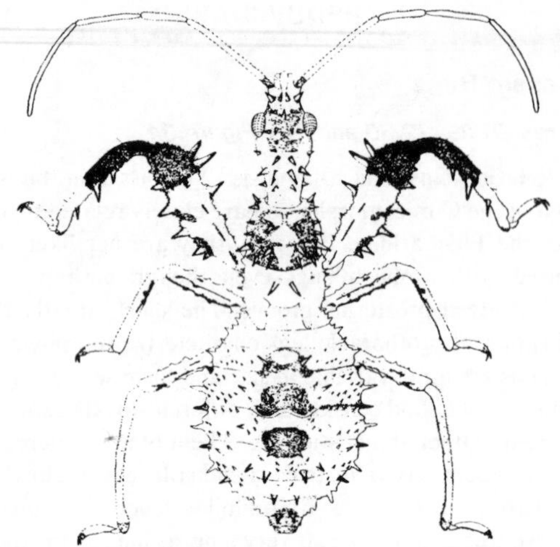

Figure 29.29. Reduviidae. Assassin bug, *Sinea complexa* Caudell, (from Swadener and Yonke, 1973b).

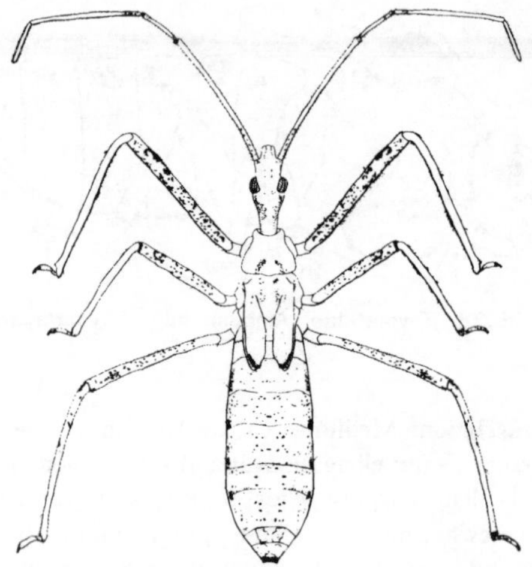

Figure 29.30. Reduviidae. Assassin bug, *Zelus socius* Uhler, (from Swadener and Yonke, 1973c).

without distinct transverse groove just behind eyes, often bearing large spines, large compound eyes, tylus well developed, juga less so. Antennae 4-segmented, filiform, frequently as long as or longer than body, sometimes bearing long setae; labium 3-segmented, usually bowed, short, thick, cone-shaped extending to distinct prosternal groove. Thorax more elongate than in most Hemiptera; prothorax distinct and well developed, placing prothoracic legs in anterior position, pronotum often with collar region, prosternal groove with microscopic transverse ridges for stridulation by labium, present from 1st instar through adult stage. Legs generally long, cursorial; prothoracic legs often distinctly raptorial or moderately thickened and bearing numerous long setae and/or spines on tibiae and femora; procoxae nearly contiguous, meso- and metacoxae separated; meso- and metathoracic legs separated from each other and by greater distance from prothoracic legs. Abdomen slender, elongate-oval or oval, robust when fed; paired dorsal abdominal scent glands variable, none, or opening on caudal margins of tergites 3, 4 and/or 5.

Comments: The family is of moderate size in our fauna with about 35 genera and 140 species, most genera having only 1 or a few species. Several species are quite common and widely distributed. Immatures are most easily collected in late summer. Careful searching in wood piles and litter or in mammal nests will be necessary to discover some species. A number of species will inflict a bite if given the chance. Bites of the "wheel-bug", *Arilus cristatus* (L.), and some *Triatoma* spp. are especially painful. Reduviids are known for their sound-producing prosternal groove, which is rasped by the thick cone-shaped labium.

Assassin bugs are generally considered to be very beneficial because of their highly voracious predatory appetites and the common occurrence of some species. *Sinea diadema* (Fabricius) is frequently found in agricultural crops feeding on pest insects, and *Zelus exsanguis* (Stål) consumes a variety of larval defoliators on trees. However, the blood-feeding, home-inhabiting reduviid species are considered, at best, to be a nuisance, except for those that transmit Chagas disease to humans, which is a serious threat to health.

A few publications are of note to one interested in the biology and immatures of reduviids including Readio (1926, 1927), Underhill (1954), Ryckman (1962), Swadener and Yonke (1973a, b, c, 1975), and Yonke and Medler (1970).

Selected Bibliography

Readio, 1926, 1927
Ryckman, 1962
Swadener and Yonke, 1973a,b,c, 1975
Underhill, 1954
Wygodzinsky, 1966
Yonke and Medler, 1970

ARADIDAE

Flat Bugs

Figures 29.31 and Key Figure 33

Relationships and Diagnosis: The flat bugs are highly specialized in both structural development and biology. Because of this, their phylogenetic relationship to other families of true bugs remains somewhat obscure. They are often placed by themselves somewhere in or near the Pentatomomorpha. Immatures can be recognized easily by their dark color and very flattened-oval bodies which are covered with small tubercles. They are not likely to be confused with other Hemiptera.

Biology and Ecology: As the common name implies, flat bugs are generally found in shallow spaces particularly under loose bark. They may also be found on fungi or occasionally on the surface of bark. A few species reside on living trees while most species are found on dead trees or under bark of logs and fallen trees. A most interesting relationship exists between this group and the development of fungi. Species of this mycetophagous family are probably closely associated with particular fungi that are present under bark. The presence of a particular flat bug may be linked to the stage of succession and deterioration of the log or tree that provides a suitable microhabitat for specific fungal development. Tree

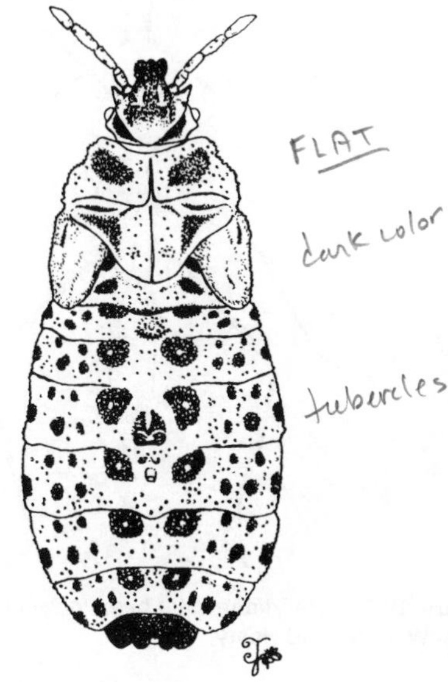

Figure 29.31. Aradidae. Flat bug.

host specificity is quite well known for many flat bugs such as *Aradus cinnamomeus* Panzer on *Pinus* spp., but in most cases it may be the specificity to the species of fungus that determines the distribution and biology of the flat bug. Larvae and adults are often found together in large numbers. Small oval eggs are laid in late spring and early summer, sometimes free and sometimes attached to the substrate by a sticky material. One or 2 generations are passed depending on the species and overwintering is usually in the adult stage.

Description: Most species small, ca. 3–8 mm, oval, dorsoventrally very flat; brown to dark brown or black. Head wider than long, granulate, with tylus and juga projecting cephalad, snout-like; compound eyes small, on lateral margin; antenniferous tubercles prominent, projecting at least half length of segment 1; antennae 4-segmented, short, granulate, segments 3 and 4 generally not over 2X length of segment 1 and 2 combined; labium arising ventrally, caudad from tylus, appearing 2- or 3-segmented, short, thick, extending just beyond head. Pronotum transverse, flat. Legs short, coxae widely separated. Abdomen broadly oval, lateral margins often scalloped and granulate; paired dorsal scent gland openings variously on caudal margins of tergites 3, 4 and 5 and differing in structure. Thorax and abdomen granulate or tuberculate and often with distinct patterns of color and fine structure reticulations.

Comments: Specimens are most easily collected by peeling bark from fallen trees and logs or loose bark near the base of trees where flat bugs will often be found clinging to the underside of the bark. Some species are attracted to lights and 1 species in California is found associated with termites. Usinger and Matsuda (1959) listed 99 species in 9 genera for the Nearctic region. Of these, 4 genera (*Aradus, Aneurus, Neuroctenus* and *Mezira*) are cosmopolitan and contain commonly collected species. *Aradus* includes ca. 75 species in our region.

Little is known of the biology of this group and virtually no literature is available on the immatures. The group is not known to be of economic importance, but presents fascinating

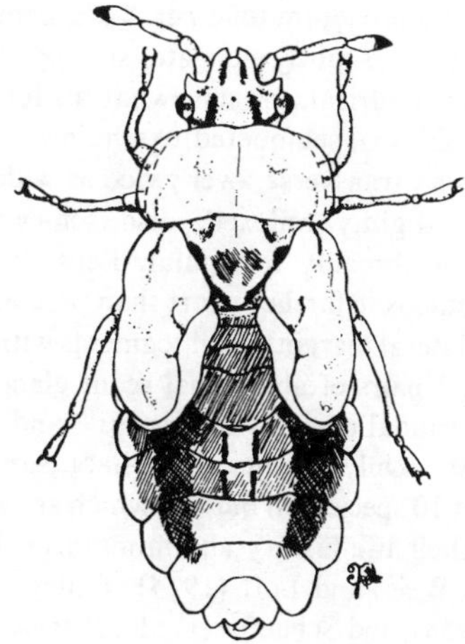

Figure 29.32. Piesmatidae. Ash-grey leaf bug, *Piesma cinerea* (Say).

potential for study by ecologists, especially someone trained in mycology and in aradid taxonomy. The above mentioned reference contains some useful information for the beginner as well as the specialist.

Selected Bibliography

Usinger and Matsuda, 1959

PIESMATIDAE

Ash-Gray Leaf Bugs

Figure 29.32

Relationships and Diagnosis: The Piesmatidae is probably most closely related to the lygaeid tribe Cymini (Drake and Davis 1958) and is placed in the Pentatomomorpha. This family is small both in number and size of species. Immatures may be confused with some of the lace bugs of the unornamented genus *Atheas*. The ash-gray leaf bugs are very small, ca. 2 mm long, and can be distinguished by the juga projecting noticeably beyond the tylus while in the tingids the tylus exceeds the juga. It is also conceivable that these bugs could be mistaken for flat-bugs; however, piesmatids are smaller, lighter in color, and are collected on low herbaceous vegetation, not under bark. Also piesmatids lack an aereolate (network-like) cuticle which most aradids have.

Biology and Ecology: Little is known about the ash-gray leaf bugs except that they show a preference for plants of the Chenopodiaceae. The most common of the N. American species, *Piesma cinerea* (Say), has also been taken from *Amaranthus* spp. (pigweeds). Eggs are deposited singly on the under surfaces of leaves where larvae hatch and feed. One generation per year is likely and overwintering occurs in the adult stage.

Description: Very small, ca. 2.5 mm, elongate oval, somewhat flattened, light yellowish-green to brown. Head transverse with large laterally projecting eyes; prominent bifurcated antenniferous tubercles with projections above and to side of antennae; juga exceeding tylus. Antennae about as long as head and pronotum together, 4-segmented; 1st short, bulbular; 2nd short, slightly elongate; 3rd and 4th longest and subequal; 3rd cylindrical, 4th somewhat fusiform with apical half darker. Labium 4-segmented, extending just beyond procoxae. Prothorax transverse, over twice as wide as long, lateral margins slightly reflexed, disc somewhat elevated; scutellum large, broadly triangular. Legs short, cursorial, meso- and metacoxae farther apart than procoxae. Abdomen broadly oval, lateral margins of all segments with large scallop on each side; 2 pairs of abdominal scent glands with transverse slits on caudal margins of tergites 3 and 4.

Comments: Only 1 genus occurs in N. America. *Piesma* contains about 10 species, all but 1 of which are western. Little is known of their life history and immatures. References of some use are Weiss and Loft (1924), Bailey (1951), Drake and Davis (1958), and Schaefer (1981). *Piesma cinerea* (Say) has been found to be a vector of sugar beet savoy, a virus disease. *Piesma* spp. also have both brachypterous and macropterous forms about which nothing is known.

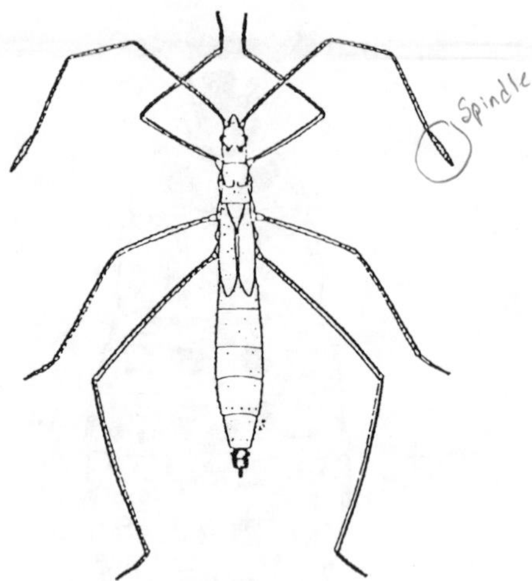

Figure 29.33. Berytidae. Stilt bug, *Jalysis wickhami* Van Duzee, (from Wheeler and Henry, 1981).

Selected Bibliography

Bailey, 1951
Drake and Davis, 1958
McAtee, 1919
Schaefer, 1981
Weiss and Loft, 1924

BERYTIDAE

Stilt Bugs

Figures 29.33 and Key Figure 22

Relationships and Diagnosis: The stilt bugs are placed in the Pentatomomorpha and are probably most closely related to the lygaeids of our fauna. However, the long thin body, antennae, and legs serve to quickly separate them from other immature Hemiptera. Superficially they resemble the marsh treaders (Hydrometridae) and the thread-legged assassin bugs (Emesinae). Marsh treaders have a very long head, longer than the thorax, and they are found in a different habitat, on the surface of water. In the emesines the front legs are raptorial and arise at the anterior end of a very long prothorax.

Biology and Ecology: Species may be found in a variety of habitats from roadsides and old fields to agricultural crops. Wheeler and Henry (1981) have summarized information and corrected previous errors in identification of berytid species and their host plant relationships. *Jalysis wickhami* Van Duzee is primarily associated with plant species that are glandular and hairy such as the Malvaceae, Onagraceae, Oxalidaceae, Scrophulariaceae, and Solanaceae. *J. spinosus* (Say) appears to be primarily associated with grasses in the genus *Panicum*. Another common species of more northern states, *Neides muticus* (Say), is found on common mullein. An unusual feature of the biology is that species seem to require animal food, insect eggs and other small soft-bodied insects, to complete development (Elsey and Stinner 1971). Various aspects of the biology of *J. wickhami* have been determined by these researchers, especially as it relates to their utility as predators of eggs. One prolonged generation, perhaps 2 overlapping generations, occurs, with immatures first appearing in late May to late June. Adults overwinter.

Description: Small to medium, 4–7 mm, very thin; yellowish to light brown body, antennae and legs darker, eyes red. Head moderately elongate, distinct tylus; antennae arising above an imaginary line running from midpoint of compound eye to tylus, 4-segmented, very long, 4th segment much shorter than 1–3; labium 4-segmented, elongate, extending to metacoxae or beyond. Thorax slightly elongate. Legs very long with cylindrical femora, tibiae, and tarsi. Abdomen very long and cylindrical; small middorsal scent gland openings on caudal margins of tergites 3 and 4 or only 3.

Comments: This small family contains only about 11 species in the United States, of which *N. muticus, J. spinosus* and *J. wickhami* are the most likely to be encountered. Several larvae will often be found at one time on a host. *J. wickhami* (identified as *J. spinosus*) has often been noted as a pest of a variety of crops; but, since it is an active predator on Lepidoptera eggs and aphids, benefits probably outweigh any detrimental effects from plant feeding.

Wheeler and Henry (1981) have provided brief descriptions and a key to separate 5th instars of *J. spinosus* and *J. wickhami*.

Selected Bibliography

Elsey and Stinner, 1971
Wheeler, A. G., Jr. and Henry, 1981
Wheeler, A. G., Jr. and Schaefer, 1982

LYGAEIDAE

Seed Bugs

Figures 29.34, 29.35 and Key Figures 56, 58, 65, 66

Relationships and Diagnosis: Lygaeids, which are in the Pentatomomorpha, are related most closely to the Berytidae in our fauna; and more remotely to Largidae, Pyrrhocoridae, Coreidae and Rhopalidae, among others. Immatures of the

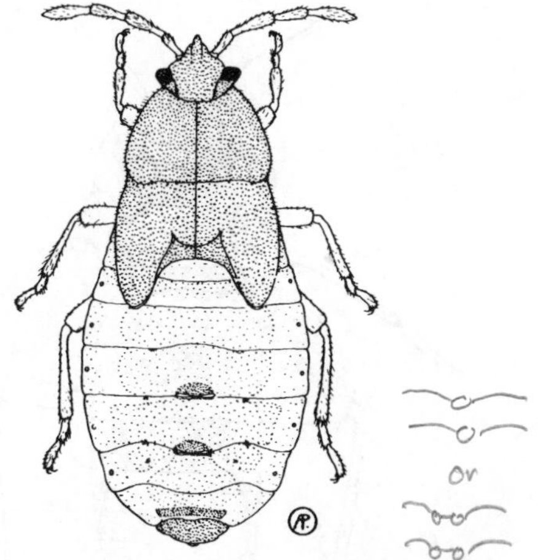

Figure 29.34. Lygaeidae. Seed bug, (from Peterson, 1948).

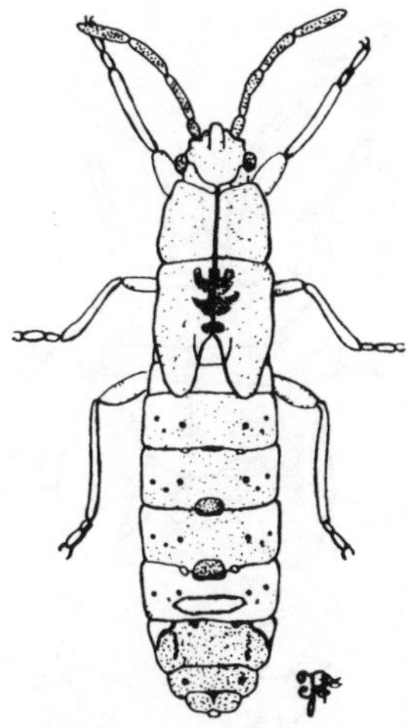

Figure 29.35. Lygaeidae. Seed bug, *Ischnodemus falicus* (Say).

Lygaeidae are fairly diverse in appearance, some small and oval, pear-shaped, or elongate; and others of medium size, 8–12 mm, and oval to pear-shaped. Smaller larvae may be confused with the plant bugs (Miridae) and rhopalids, but they can be separated by the abdominal scent glands. Lygaeids have 2 or 3 paired scent glands on the caudal margins of abdominal tergites 3 and 4, 4 and 5, or 3, 4 and 5, while plant bugs have only 1. The scent gland openings of lygaeids are paired or large and transverse. In rhopalids there are 2 glands with single very small openings lacking sclerites on the midlines caudally on tergites 4 and 5. Larger lygaeids may be confused with largids and pyrrhocorids. They may also be separated on the basis of the abdominal scent gland characters. Both largids and pyrrhocorids have 3 scent glands, largids with very small openings on the midline and pyrrhocorids with larger single transverse openings on the caudal margins of tergites 3, 4 and 5. Some Rhyparochrominae lygaeids have the prothorax and head projected and rounded laterally and may be confused with some nabids. However, the nabids have a 3-segmented, and lygaeids a 4-segmented, labium.

Biology and Ecology: Lygaeids occur in a variety of habitats from ground dwelling species in grasslands and open fields on low and emergent vegetation to forest inhabiting species which are also stratified on the vegetation. Overwintering occurs as adults with 1 or 2 generations per year. Although most species are phytophagous, several are predacious or omnivorous. The most significant treatment of the biology of this group was by Sweet (1964a,b), which included 39 species of Rhyparochrominae from New England. Tamaki and Weeks (1972) also presented a detailed biology of the big-eyed bugs, *Geocoris pallens* Stål and *G. bullatus* (Say). Most species of lygaeids are seed feeders, thus the common name of the family. Species such as the milkweed bug, *Oncopeltus fasciatus* (Dallas), are restricted to milkweed species while others such as the chinch bug, *Blissus leucopterus* (Say), feed on a wide variety of graminaceous hosts and migrate from host to host. Some species are sap feeders and a few are predacious.

Description: Most species small to medium, 3–7 mm, oval to elongate; dull brown to black; body with few prominent setae; larger species 8–12 mm long, often more brightly colored with red and orange patterns. Head about as long as wide, shorter in some species, tylus pronounced, projecting beyond juga, in other species head very long with cylindrical "neck"; antennae 4-segmented, cylindrical; labium 4-segmented, short in some species, extending to between procoxae, longer in other species, extending to beyond metacoxae. Prothorax trapezoidal, disc slightly arched, some species with greatly rounded and somewhat elongate prothorax. Profemora of some species slightly to greatly swollen. Abdomen elongate to oval, generally somewhat thickened; scent glands on abdominal tergites generally well developed, 2 or 3 pairs with single, double or transverse openings on caudal margins of tergites 3 and 4, 4 and 5, or 3, 4 and 5; posterior suture of tergites 4 and 5 angled caudally from lateral margins.

Comments: This is the second largest family of Hemiptera in N. America with over 50 genera and about 250 species occurring north of Mexico. It is an especially well developed group in the tropics. A few *Geocoris* species are considered to be beneficial in biological control of pests because of their predatory habits and occurrence in agricultural crops. The chinch bug has been considered to be a serious pest of grasses for several decades, especially of wheat, grain sorghum and corn, and has received much attention in the economic literature. A few other species are occasionally noted as pests of garden plants and fruits. The milkweed bug has served as a model laboratory animal for various physiological and behavioral experiments. A key to the larvae of 53 Nearctic genera was presented by Sweet and Slater (1961). Additional descriptions of immatures were provided for Cyminae (Slater 1952), Cyminae and Ischnorhynchinae (Slater 1963), *Ischnodemus* (Harrington 1972), and for the chinch bug (Packard and Benton 1937). Slater's (1964) catalog of the Lygaeidae of the world serves as an invaluable reference.

Selected Bibliography

Harrington, 1972
Packard and Benton, 1937
Slater, 1952, 1963, 1964
Sweet, 1964a,b
Sweet and Slater, 1961
Tamaki and Weeks, 1972

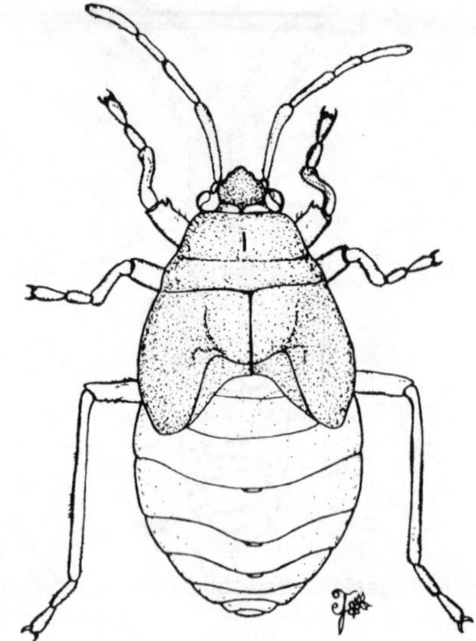

Figure 29.36. Largidae. Largid bug, *Largus* sp.

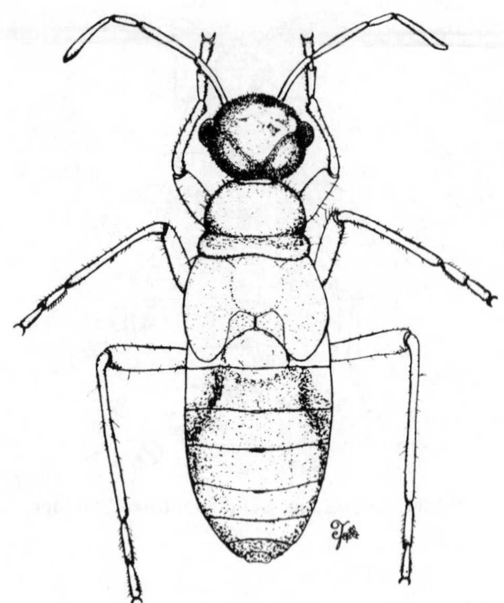

Figure 29.37. Largidae. Largid bug, *Arhaphe* sp.

LARGIDAE

Largid Bugs

Figures 29.36, 29.37 and Key Figure 60

Relationships and Diagnosis: The largids are generally grouped taxonomically close to the Pyrrhocoridae and less closely to Lygaeidae in the Pentatomomorpha. Largid larvae should not be confused with others although they are of 2 basic types: the most common type, *Largus* spp. (fig. 29.36), are large, glabrous, thick, metallic-dark-blue bodied; the second, *Arhaphe* spp. (fig. 29.37) consists of medium-sized, ant-like, black larvae. The latter species are very uncommon, but could be confused with alydid ant-mimicking species from which they can be separated by possession of 3 single, dorsal abdominal scent gland openings, rather than 2 pairs as in alydids.

Biology and Ecology: Almost nothing is known on the biology of this group. They are apparently phytophagous, living close to the ground and feeding on low vegetation. They overwinter as adults and lay eggs singly in the litter. Population levels never seem to be very high. Much work needs to be done on this family to better understand its biology and phylogenetic relationships.

Description: *Largus* spp. large, 10–13 mm, robust, with smooth, metallic, dark blue cuticle and with red to yellowish-brown horseshoe pattern on venter and on tergite 1; antennae 4-segmented, filiform, arising from large, cylindrical antenniferous base; head triangular, retracted to level of large compound eyes into prothorax; labium 4-segmented, extending to mesothoracic coxae. Profemur with 1 very large and 1 smaller, but prominent, spine near apex; meso- and metafemora devoid of spines. Abdomen oval and robust with 3 characteristic single middorsal openings of scent glands on caudal margins of tergites 3, 4 and 5.

Larvae of *Arhaphe* somewhat ant-like with nearly round head; generally black with some white areas. Antennae and labium each 4-segmented. Anterior 2/3 of pronotum distinctly rounded, posterior 1/3 transverse and broadly rounded. Abdomen elongate and thickly rounded.

Comments: The 2 most common genera, *Largus* and *Arhaphe,* have 8 and 4 species, respectively, in the continental United States. Larvae and adults will be most often collected singly on the ground in relatively open woods as they move about the litter. The family has no economically important species.

Selected Bibliography

Halstead, 1972a,b

PYRRHOCORIDAE

Red Bugs, Cotton Stainers

Figure 29.38 and Key Figure 59

Relationships and Diagnosis: Pyrrhocorids are most closely related to largids and lygaeids and belong in the Pentatomomorpha. They closely resemble some of the larger milkweed bug type lygaeids that are bright red and orange with black markings. Pyrrhocorids can be separated by the location of the scent glands which open middorsally rather than laterad of the middorsal line. Also, in lygaeids the posterior margins of tergites 4 and 5 are often angled sharply caudad from the lateral abdominal margins while in red bugs the caudal margins are transverse with only a slight caudal extension at the midpoint.

Biology and Ecology: Little is known of the field biology of the group except for those species (*Dysdercus, Pyrrhocoris*) that have become pests of cotton. Natural hosts of most species are in the Malvaceae where the red bugs feed on seeds and fruits.

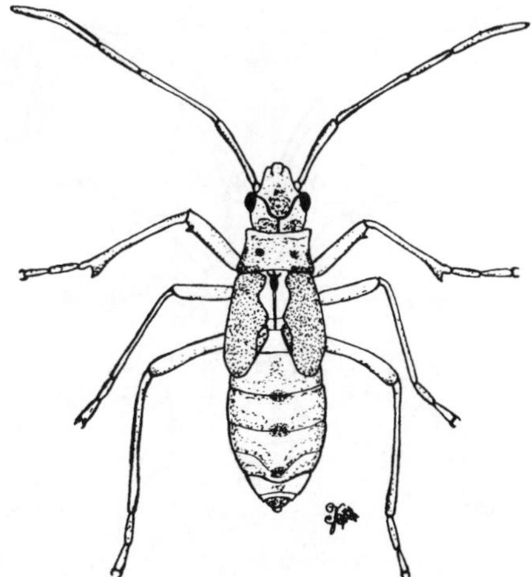

Figure 29.38. Pyrrhocoridae. Red bug, *Dysdercus mimulus* Hussey.

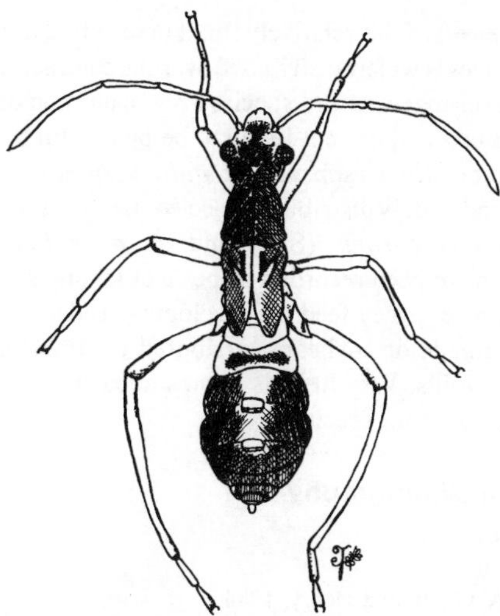

Figure 29.39. Alydidae. Broad-headed bug, *Alydus eurinus* (Say).

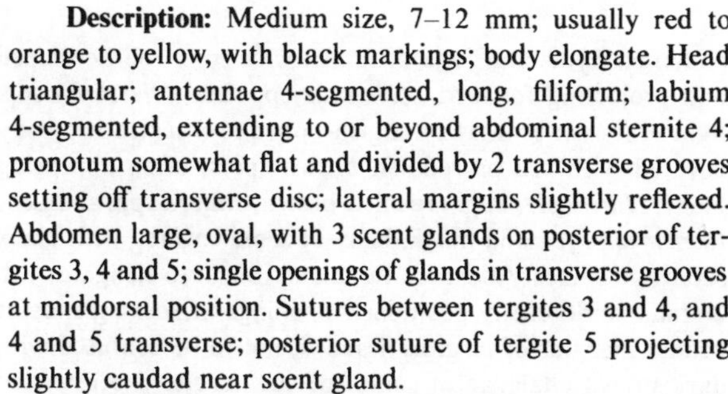

Description: Medium size, 7–12 mm; usually red to orange to yellow, with black markings; body elongate. Head triangular; antennae 4-segmented, long, filiform; labium 4-segmented, extending to or beyond abdominal sternite 4; pronotum somewhat flat and divided by 2 transverse grooves setting off transverse disc; lateral margins slightly reflexed. Abdomen large, oval, with 3 scent glands on posterior of tergites 3, 4 and 5; single openings of glands in transverse grooves at middorsal position. Sutures between tergites 3 and 4, and 4 and 5 transverse; posterior suture of tergite 5 projecting slightly caudad near scent gland.

Comments: The family is primarily tropical with our fauna being limited to the southern regions of the United States where 7 species of *Dysdercus* are found. Several pyrrhocorids have been recorded as pests of cotton and are known as cotton stainers and red bugs.

Selected Bibliography

Van Doesburg, 1968

ALYDIDAE

Broad-Headed Bugs

Figure 29.39 and Key Figure 61

Relationships and Diagnosis: Alydid immatures are of 2 basic types, the more common being ant mimics and the less common (except in Florida) being very thin and elongate. The Alydidae are most closely related to Coreidae and Rhopalidae of the Pentatomomorpha and belong to the Coreoidea. In our fauna one will have little difficulty separating alydids from coreids or rhopalids. However, other groups, including the Miridae, contain ant-mimicking species, and some ground dwelling Lygaeidae have larvae which superficially resemble ants. Alydids have 2 dorsal scent gland openings on the abdomen; mirids have only 1 opening or none. Alydids are separated from those lygaeids that have 2 scent gland openings by examining the heads. In alydids the antennae arise above an imaginary line extending between the midpoint of the compound eye and the top of the tylus, while in the lygaeids the antennae arise on or below the line. The elongate grass-feeding alydid larvae (*Protenor* spp. and *Leptocorisa* spp.) may be confused with elongate mirids, lygaeids, reduviids, and berytids. They can be separated from the former 2 by the same features as indicated above, from the berytids which have a distinctly swollen 4th antennal segment as compared with the cylindrical one of alydids, and from the reduviids which have a 3-segmented labium as compared to the 4-segmented one of alydids.

Biology and Ecology: The biologies of 3 species of *Alydus* and of *Megalotomus quinquespinosus* (Say) have been studied (Yonke and Medler 1965, 1968). These ant-like larvae feed primarily upon developing seed heads and on fallen seeds of legumes. *Esperanza texana* Barber, *Protenor* spp. and *Leptocorisa* spp. feed primarily upon grasses and sedges. Species where the biology is known undergo 2 generations per year. Eggs, which are round or triangular (*Protenor belfragei* Haglund), are dropped free in the litter where they overwinter. Little is known about the biology of most species, especially those inhabiting grasses and sedges. Wheeler and Henry (1984) reported Bermuda grass as the host of *E. texana.* An assessment of feeding strategies of the family was presented by Schaefer (1972).

Description: Mature ant-like larvae of *Alydus, Megalotomus, Tollius,* and *Stachyocnemus* medium, ca. 8–14 mm; elongate oval larvae of *Esperanza* and *Darmistus* small to medium, ca. 5–8 mm, and *Protenor* and *Leptocorisa* spp. medium, ca. 10–15 mm long and slender; ant-like species dark reddish brown to black, slender grass-feeding species light yellowish-brown to green. Head more or less triangular, as wide as or wider than pronotum; antennae 4-segmented, filiform, especially long in the grass-feeding species; labium 4-segmented, extending to or beyond metacoxae. Thorax relatively narrow and long. Abdomen either bulbular in ant-like forms or slender and elongate in grass-feeding groups. Two pairs of dorsal abdominal scent gland openings on caudal margins of tergites 4 and 5.

Comments: This relatively small taxon of 10 or so genera and 20 species is widely distributed over the continental United States. Several genera and species are uncommon or occur in select habitats and are not likely to be picked up by the general collector. *Alydus* spp. and *M. quinquespinosus,* the most common and widely distributed species, are often found in old fields. *Alydus eurinus* (Say) and *A. pilosulus* Herrich-Schaeffer have been recorded as pests of legumes, especially soybeans, where they feed on developing seeds. Some of the feeding damage on soybeans attributed to stink bugs is the result of alydids. Very little is known about the biology and immatures of other species.

Selected Bibliography

Fracker, 1918
Schaefer, 1972
Wheeler, A. G., Jr. and Henry, 1984
Yonke and Medler, 1965, 1968

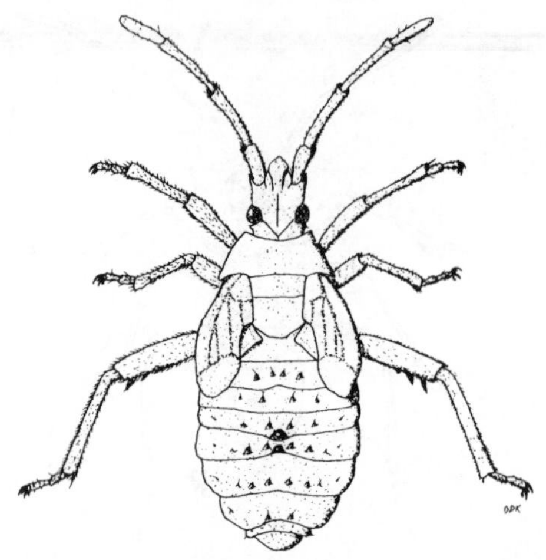

Figure 29.40. Rhopalidae. Scentless plant bug, *Harmostes reflexulus* (Say), (from Yonke and Walker, 1970b).

RHOPALIDAE

Scentless Plant Bugs

Figures 29.40 and Key Figure 64

Relationships and Diagnosis: The Rhopalidae belongs to the superfamily Coreoidea of the Pentatomomorpha. Its immatures are generally small (less than 5 mm long), robust, and very lygaeid-like. *Jadera* and *Leptocoris* spp. are larger, ca. 10 mm, and resemble the milkweed-bug types of the Lygaeidae. Rhopalids are separated from some lygaeids by having 2 small, dorsal, abdominal scent gland openings rather than 3. From the lygaeids with only 2 transverse scent gland openings, rhopalids may be distinguished by their single minute scent gland openings on the caudal margins of tergites 3 and 4.

Biology and Ecology: Little definitive information is available on the biology of the scentless plant bugs. All are phytophagous. Individual species of *Arhyssus* (Chopra 1968) have been recorded from a wide variety of plants from legumes to composites and from pines to cotton. Notes on the biology of *Arhyssus lateralis* (Say), *Liorhyssus hyalinus* (Fabricius) and *Niesthrea sidae* (Fabricius) were provided by Readio (1928), *Stictopleurus punctiventris* (Dallas) by Yonke and Medler (1967), *Harmostes reflexulus* (Say) by Yonke and Walker (1970a) and *H. fraterculus* (Say) by Wheeler and Miller (1983). Paskewitz and McPherson (1983) gave detailed information on the life history of *A. lateralis.* Most species appear to be bivoltine, overwintering as adults. Eggs are usually laid singly in flower heads (*H. reflexulus*), in small clusters (*N. sidae* and *L. hyalinus*) on plant surfaces, or deposited or dropped free in the ground litter (*J. haematoloma* (H.-S.) and *Leptocoris trivittatus* (Say)). *L. trivittatus* is mostly associated with the boxelder tree and *J. haematoloma* is found on the ground under and on soapberry.

Description: Very small to medium, 3–10 mm, robust. Head more or less triangular; antennae 4-segmented, filiform; labium 4-segmented, extending to metacoxae or onto abdomen; tylus and juga of *Harmostes* spp. well developed and projecting forward. *Arhyssus* spp. and *Niesthrea* spp. light yellowish-brown to orangish-brown to orangish-red, body covered with red or reddish brown spots, often appearing hairy; *Harmostes* spp. greenish-brown; *Jadera* spp. and *Leptocoris* spp. red and black, body of some covered with black setae and spines. Abdomen thick and robust to elongate oval; 2 small middorsally positioned openings of scent glands at caudal margins of tergites 4 and 5, not to be confused with dark brown chalazae of some species; tergite 5 constricted middorsally (fig. 29.40).

Comments: The immatures of only a few species have been described, including *H. reflexulus* (Yonke and Walker 1970b), *H. fraterculus* (Say) (Wheeler and Miller 1983), *A. lateralis, L. hyalinus,* and *N. sidae* (Readio 1928) and *A. lateralis* (Paskewitz and McPherson 1983). The continental United States fauna is represented by 9 genera and about 35 species. Many of the species are limited geographically. A few are widespread and common including *A. lateralis, H. reflexulus,* and *L. trivittatus.* The latter species, the boxelder bug, is probably the most commonly encountered and recognized of the rhopalids. It is associated with the boxelder tree, *Acer negundo,* builds up large populations, and is often considered a pest in houses where it seeks shelter for the winter.

Selected Bibliography

Aldrich et al., 1972
Chopra, 1968
Paskewitz and McPherson, 1983
Readio, 1928
Schaefer and Chopra, 1982
Wheeler and Miller, 1983
Yonke and Medler, 1967
Yonke and Walker, 1970a,b

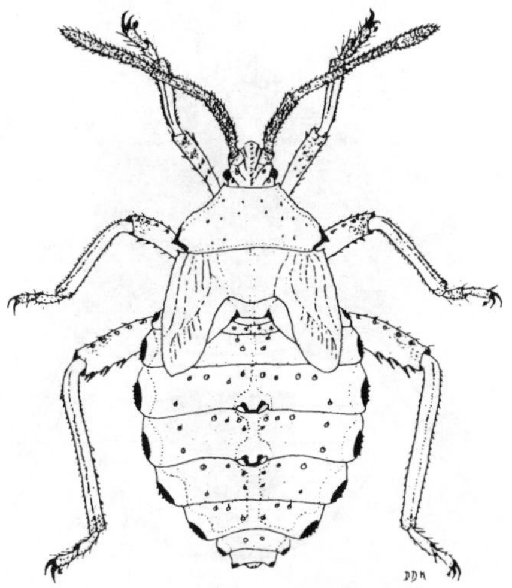

Figure 29.41. Coreidae. Coreid bug, *Mozena obesa* Montandon.

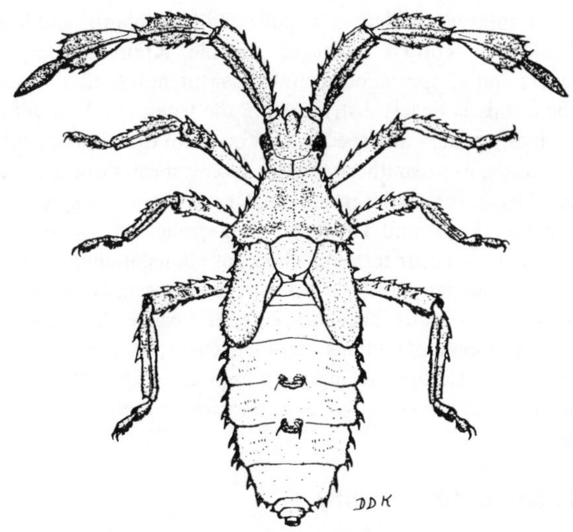

Figure 29.42. Coreidae. Coreid bug, *Chariesternus antennator* (Fab.).

COREIDAE

Squash Bugs, Leaf-Footed Plant Bugs

Figures 29.41, 29.42 and Key Figure 63

Relationships and Diagnosis: The Coreidae, Alydidae, and Rhopalidae are usually combined in the Coreoidea. They belong in the Pentatomomorpha. The coreids of the continental United States will not likely be confused with alydids or most rhopalids, although small species (ca. 5 mm in 5th instar) will have to be examined carefully to separate them from similar looking rhopalids or lygaeids. Scent gland characters of the abdomen will separate the 2 groups as given in the key (couplets 37 and 45–47). Some coreids bear superficial resemblance to pentatomids, pyrrhocorids, and lygaeids but are easily separated using the key.

Immatures mostly medium to large and with body elongate to pear-shaped and covered with some ornamentation such as brightly colored areas, large spines, chalazae, and scoli, especially along lateral edges of pronotum and abdomen.

Biology and Ecology: Coreids are exclusively plant feeders. Most species within a genus, and some genera and tribes, show preferences for closely related hosts. For example, *Anasa* spp. generally are collected from cucurbits. Although immatures will often be found feeding upon fruiting bodies, one is more likely to find them feeding on stems and petioles where they will cause a wilting of the terminal shoots. Coreids are generally univoltine, overwintering as adults. Eggs are laid singly in some species (*Archimerus alternatus* (Say)) or clustered in a rather defined pattern (*Anasa tristis* (De Geer)). Egg shape varies considerably from rectangular (*Catorhintha mendica* Stål) to barrel (*Leptoglossus* spp.) to oval (*Anasa* spp.).

Biologies of some species have been determined including *Archimerus alternatus* (Say), *Acanthocephala terminalis* (Dallas), and *Euthochtha galeator* (Fabricius) (Yonke and Medler (1969a); *Cimolus obscurus* Stål (Jones 1924); *Catorhintha mendica* Stål (Balduf 1942, 1957); *Anasa tristis* (De Geer) (Beard 1940); and *Chelinidea, Narnia,* and *Leptoglossus* spp. on cactus (Mann 1969).

Description: Mostly medium, ca. 7–10 mm, to large, ca. 11–16 mm; elongate to oval; many thick bodied; *Nisoscolopocerus apiculatus* Barber, dorsoventrally flattened; dark to mottled brown, light green; early instars of some species white, bright red or black, later instars brown. Antennae 4-segmented, usually long, filiform; antennae of many species modified, *E. galeator* and *Thasus acutangulus* Stål 3rd antennal segment flattened and dilated, *Chariesterus antennator* (Fabricius) 2nd and 3rd dilated (fig. 29.42), and *Chelinidea* spp. segments triangular in cross section. Head, thorax and abdomen generally smooth, with few setae; small spines often scattered over dorsum; some species, *E. galeator* and others, with well developed chalazae and scoli and numerous spines over body. Labium 4-segmented, extending to or beyond metacoxae. Legs cylindrical, elongate; femora, especially of metathoracic legs often thickened and bearing stout spines; tibiae cylindrical in most species, flattened and dilated in *Acanthocephala, Chariesterus, Leptoglossus,* and *Narnia,* and triangular in cross section in *Chelinidea.* Two pairs of raised abdominal scent glands prominent along caudal margin of abdominal tergites 4 and 5.

Comments: Detailed descriptions of a few species are available in the literature including those for *A. tristis* (Beard 1940), *E. galeator* (Yonke and Medler 1969b), *A. terminalis* (Yonke and Medler 1969c), *A. alternatus* (Yonke and Medler 1969d), *C. obscurus* (Jones 1924).

Immatures will often be collected from plants, and hosts should be recorded whenever possible. Approximately 30 genera and 75 species occur in the continental United States. The family is widely distributed in the tropics and, therefore has many genera and species that occur in the United States only in the most southern regions. Some species are only collected by searching the ground, under rocks, and low spreading plants on the ground. However, most species are found up on the vegetation near terminal shoots of plants, including trees.

The squash bugs, *Anasa* spp., are especially important as pests of various Cucurbitaceae, especially the squashes. The leaf-footed plant bugs, *Leptoglossus* spp., are occasional pests of fruit crops, having been implicated in cat-facing of peaches, and are known to feed on various nut crops including pecans.

Selected Bibliography

Balduf, 1942, 1957
Beard, 1940
Jones, 1924
Mann, 1969
Schaefer and Mitchell, 1983
Yonke and Medler, 1969a,b,c,d

CYDNIDAE

Burrower Bugs

Figure 29.43 and Key Figure 52

Relationships and Diagnosis: The cydnids are probably most closely related to the Corimelaenidae. They are also related to the stink bugs and shield-backed bugs and are included in the Pentatomomorpha. Most of the burrower bugs are similar in size, shape and coloration. They can be distinguished most easily by the large number of dark, well developed setae and spurs on the tibiae, and the flattened, almost spatulate, condition of the protibiae. Although other true bugs may have setae and some spines developed on the tibiae they are not markedly noticeable as with the cydnids, except for the negro bugs (Corimelaenidae). The latter group has a strongly arched dorsum compared with the relatively flattened dorsum of the cydnids.

Biology and Ecology: The common name of these true bugs gives a useful clue to their habitat and ecology. Many species appear to prefer sandy soil. The group is mostly found in soil at the bases of plants where they feed on roots. Some species such as the common *Sehirus cinctus* (P. de B.) are collected frequently on herbaceous vegetation and on cultivated crops including alfalfa. As with the European species of *Sehirus*, American species are known to utilize members of the mint family (Lamiaceae) as hosts. The life history of *S. cinctus cinctus* has been determined on *Lamium purpureum*, dead nettle, (Sites and McPherson 1982). Unlike other families of Pentatomoidea, the burrowing bugs lay their eggs in clumps in the ground. Also, the females remain with and appear to guard the eggs, but the extent and effectiveness of this maternalism is unknown for most species. Cydnids are univoltine and overwinter as adults. Very little is known about species other than *S. cinctus*.

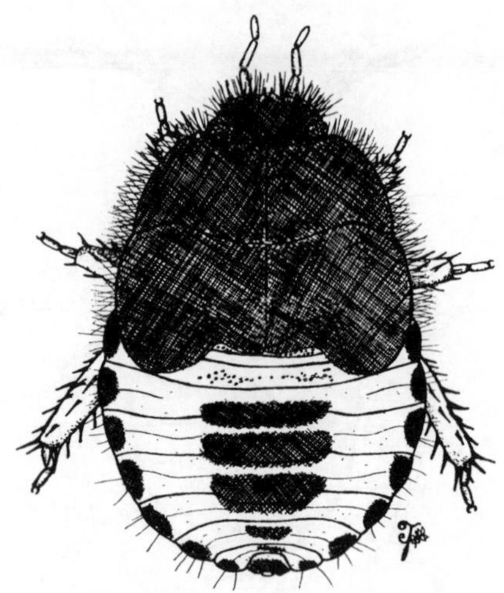

Figure 29.43. Cydnidae. Burrower bug.

Description: Small, 2 mm (*Amnestus pusillus* Uhler), to medium, 8mm; oval; body somewhat dorsoventrally flattened, brown to black. Margins of head, thorax, and abdomen of species in some genera (e.g., *Microporus* and *Tominotus*) with long brown setae, other species generally devoid of very long setae. Head flattened on dorsum, distinctly wider than long; jugal lobes very broad and rounded, long rectangular tylus; margin of juga and tylus of *Microporus* and *Tominotus* spp. with long setae and short, round-tipped, peg-like setae, other species without peg-like setae; head retracted to the compound eyes within scalloped or overhanging anterior pronotal margin; antennae 4-segmented, elongate, arising from below lateral carina in front and below compound eyes, 1st often shorter than 2–4; bucculae well developed, elongate, grooved for 1st labial segment; labium 4-segmented, somewhat thick, extending to mesocoxae. Thorax well developed, nota transverse, pro- and mesonota deflexed under sharp lateral carina; scutellum projecting moderately caudad, not exceeding metanotum. Legs short; coxae fixed in position, bearing fringe of apical setae; femora somewhat thickened bearing some dark setae; metatibiae longer than pro- or mesotibiae, tibiae bearing numerous dark thick spurs over full length, protibiae modified for fossorial habits, somewhat flattened to spatulate; tarsi 2-segmented. Abdomen broadly oval with median and lateral darkly pigmented, sclerotized plates on dorsum and venter; transverse scent gland slits with small openings at each end of slit on caudal margins of tergites 3, 4, and 5.

Comments: This is a small group of some 8 genera and about 30 species, of which about 6 are common and widely distributed. To collect most species, especially immatures, one must look in sandy soil near the bases of plants. Some specimens may be collected with the use of a Berlese funnel. Adults of several species, including *Amnestus pusillus* Uhler and *Cyrtomenus mirabilis* (Perty), will commonly be taken at black light.

Little has been published on this group. Froeschner's (1960) treatment of the family in the Western Hemisphere is a useful source of information on adults. Sites and McPherson (1982) have described the eggs and immatures of *S. cinctus cinctus* and reported on its life history.

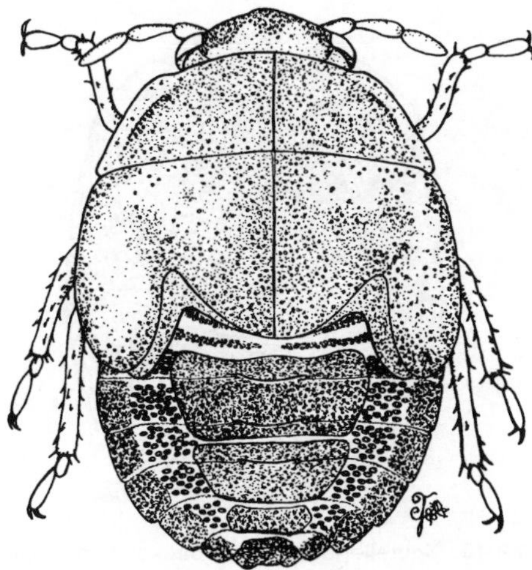

Figure 29.44. Corimelaenidae. Negro bug.

Selected Bibliography

Froeschner, 1960
Sites and McPherson, 1982

CORIMELAENIDAE

Negro Bugs

Figure 29.44 and Key Figure 51

Relationships and Diagnosis: The negro bugs are most closely related to the burrower bugs. They are included in the Pentatomomorpha. Their small size and oval shape help to distinguish them from most other true bugs. They are smaller than most stink bugs and shield-backed bugs, and can be separated from the former by a well developed scutellum, more transverse head, and body being strongly arched; and from the latter by the 1st pair of scent gland openings being closer together than the last 2 pairs in the negro bugs compared to the scutellerids where the 1st pair of openings is much wider than the last 2 pairs. Although some cydnids are as small as the larger negro bugs, most negro bugs are more distinctly arched on the dorsum and they have fewer, less well developed spurs and setae on their legs than the burrower bugs.

Biology and Ecology: The negro bugs can most often be found on low herbaceous vegetation in open fields, especially old fields, and along railroad embankments and other disturbed sites which foster a succession of opportunistic hosts. Several species are very common in different regions of the country, and numerous hosts (perhaps more accurately called collection records) have been recorded. The life histories of only a few species have received any attention. *Galgupha ovalis* Hussey is known to feed on bracted plantain (*Plantago aristata*). Biehler and McPherson (1982) determined this species to have 1 generation per year, overwintering as adults, as is true for the following species. *Corimelaena lateralis* (Fabricius) is commonly collected from wild carrot, *Daucus carota,* which may be considered one of its hosts. Adults of *C. pulicaria* (Germar) have been taken repeatedly from alfalfa and red clover. Likewise some *Galgupha* species have been collected frequently from alfalfa, red clover, and wheat fields. However, again these are not representative of true host records, those required for development of immatures. Rather, the adults may be only opportunistic, utilizing what is most commonly available in a succulent state.

Description: Very small to small, 1.5–5 mm, oval, robust, dorsum notably arched; dark brown to black, often with some clear areas on abdomen; surface densely punctate. Head transverse, wider than long, somewhat deflexed, broadly triangular, and retracted partly into prothorax; tylus and juga well developed, lateral ridges extending from compound eyes; antennae 4-segmented, subequal, elongate, arising in front of and below eyes, 4th segment somewhat fusiform; bucculae well developed; labium elongate, 4-segmented, extending to metacoxae. Thorax well developed, transversely arched, pronotum reflexed mesally below sharp lateral carina, scutellum well developed, projecting caudad to about half length of wing pads. Legs with relatively short femora and tibiae, cursorial; coxae fixed; femora moderately thickened; tibiae cylindrical, bearing several dark spines or setae. Abdomen oval to round and dorsoventrally thickened; tergites 1 and 2 often narrow and interrupted by clear area at midline; segments 3–7 broadly oval with median and lateral darkly colored sclerotized plates on dorsum and venter; transverse dorsal abdominal scent gland slits with minute openings at each end of slit on tergites 3, 4, and 5.

Comments: The family has only 3 genera and about 30 species in our fauna. *Cydnoides ciliatus* (Uhler) and several species of *Galgupha* and *Corimelaena* are very common and widely distributed. The small size and somewhat secretive habits have resulted in little being known about the group. Useful references include the comprehensive taxonomic treatment of the adults (McAtee and Malloch 1933), a paper on the life history and immatures of *Galgupha ovalis* Hussey (Biehler and McPherson 1982), and McPherson's (1972) study on *Corimelaena lateralis lateralis* (Fabricius).

Selected Bibliography

Biehler and McPherson, 1982
McAtee and Malloch, 1933
McPherson, 1972

SCUTELLERIDAE

Shield-Backed Bugs

Figure 29.45 and Key Figure 53

Relationships and Diagnosis: The shield-backed bugs are often included in the Pentatomidae as a subfamily thus reflecting the close relationship between the 2 taxa. They are members of the Pentatomomorpha. Other related families are the Corimelaenidae and the Cydnidae. Shield-backed bugs can be separated from the latter 2 families by the head being nearly as long as wide or longer in the scutellerids; they can be separated from the stink bugs by the 1st abdominal tergite being interrupted by the scutellum projecting caudad. In a few species this may be difficult to determine in preserved, "distended" specimens. A more reliable character is the 1st pair of dorsal abdominal scent gland openings which open farther apart than those of the 2nd and 3rd pairs on scutellerids, whereas the 1st pair in the other 3 families is closer together or about the same distance apart as the 2nd and 3rd pairs.

Biology and Ecology: Not a great deal is known about the biology. Adults apparently overwinter, emerging in the spring to mate and lay eggs in batches or rows adhering to vegetation. Most species appear to be univoltine. All scutellerids are phytophagous, but definitive host associations are known for only a few species including *Stethaulax marmoratus* (Say) on *Rhus glabra*, *Acantholomidea denticulata* Stål on *Ceanothus pubescens* and *C. ovatus*, and *Tetyra bipunctata* (Herrich-Schaeffer) on red and jack pine.

Description: Small, 3–5 mm, to medium, 6–9 mm, nearly oval, robust; light to dark brown, usually with distinct color patterns on dorsum; body often densely punctate. Head often strongly deflexed and somewhat retracted into prothorax, usually as long as wide or longer; juga elongate and triangular to tylus; tylus long and broad; bucculae well developed and long; antennae elongate, 4-segmented, subequal, arising under vertex-jugal shelf, below and in front of compound eyes; labium somewhat thickened, 4-segmented, elongate, extending to metacoxae. Thorax rather strongly arched, pronotum transverse, sometimes projecting cephalad at anterolateral angles, lateral pronotal margins forming shallow shelf; scutellum well developed, extending in broadly rounded triangle onto abdomen, nearly or completely bisecting 1st abdominal tergite. Legs with femora and tibiae short, cursorial. Abdomen oval, segments somewhat narrow, dense dark punctations dorsally and ventrally; tergites bearing paired scent gland openings usually darkly pigmented and elevated as a bump, openings on 3rd tergite farther from each other than those on 4th or 5th tergites.

Comments: This is a small family of 15 genera and 34 species. Most species have to be deliberately searched for, as they are somewhat cryptic as both immatures and adults, and we do not know much about their biologies. One is most likely to collect immatures and adults in late August through September, especially in the northern states and Canada. Fifth instars of an unknown species have been collected from grass in December at the Flamingo area of Everglades National Park, Florida.

Some useful references on the biology and descriptions of the immatures include those of *A. denticulata* (Harris and Andre 1934), *T. bipunctata* (Gilbert et al. 1967), and *S. marmorata* (Walt and McPherson 1972, 1973).

Selected Bibliography

Gilbert et al. 1967
Harris and Andre, 1934
Walt and McPherson, 1972, 1973

Figure 29.45. Scutelleridae. Shield-backed bug, *Homaemus* sp.

PENTATOMIDAE

Stink Bugs

Figure 29.46 and Key Figures 54, 55

Relationships and Diagnosis: This family forms the nominal taxon of the Pentatomomorpha. Its closest families include the Corimelaenidae, Cydnidae and Scutelleridae. The elongate-oval to oval shape and stout body of the stink bugs and related families serve to easily distinguish them from most other true bugs. In the stink bugs the head is about as long as or longer than wide, while the heads of the corimelaenids and cydnids are distinctly wider than long. Scutellerids are sometimes treated as a subfamily of the Pentatomidae, but in any case stink bug (*sensu st.*) larvae can be distinguished by their anterior abdominal tergites being continuous segments, not divided by the mesonotum. Also, their 3 pairs of scent glands open in similar places on the respective tergites.

Biology and Ecology: Most species of stink bugs are phytophagous, feeding on petioles, succulent stems, and/or fruiting bodies of their hosts. As such, life histories of species are often synchronized to the development and rapid growth of these parts of the hosts. Species of the subfamily Asopinae are predacious. The vast majority of stink bugs overwinter as adults and pass through 1 or 2 generations per year. Eggs are laid in batches of a few to many, usually on the surface of vegetation. The barrel-shaped, sometimes highly sculptured, eggs of stink bugs make them easily recognizable. First instars tend to stay aggregated around the egg mass, but will later disperse. A broad range of plant hosts is utilized and some species are known to feed on plants from several different families including cultivated crops. Others are more restricted in their host affiliations, such as the cryptic species of *Brochymena* which feed on oak, elm and willow. Predacious species such as *Podisus maculiventris* (Say) tend to be generalist in their feeding habits and often occur in different habitats.

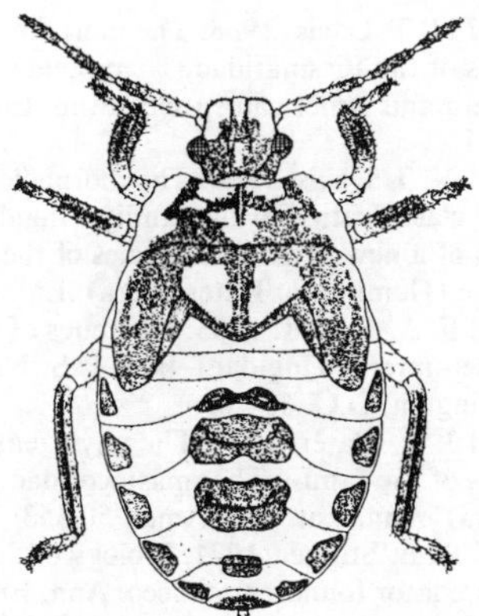

Figure 29.46. Pentatomidae. Stink bug, *Podisus placidus* Uhler, (from Oetting and Yonke, 1971a).

Description: Some species small, 4–6 mm, most species larger, 7–14 mm; mostly elongate-oval, several oval, usually thick bodied, some dorsoventrally compressed, a few elongate; color variable, often brown, but also green, black, sometimes red or orange or combination of above in patterns; body often sculptured with marginal spines and pitted over surface. Head as long as or longer than wide, somewhat triangular in profile; large compound eyes on lateral margin near prothorax; vertex usually flat to slightly rounded with long, wide juga; antennae 4-segmented, filiform, 1st segment short, other 3 long, arising in front of eyes on venter; bucculae well developed, in plant feeding forms restricting 1st labial segment; labium 4-segmented, extending beyond mesocoxae; labium of predacious species thicker, not restricted by bucculae, extending to mesocoxae in repose, extending forward while feeding. Pro- and mesothorax well developed; pronotum usually broadly triangular to transverse, lateral margins often compressed, humeral angles widely separated; median region of mesonotum well developed, broadly triangular, not completely intercepting abdominal tergite 1. Legs cursorial; femora and tibiae subequal, tibiae nearly triangular in cross section. Abdomen oval to elongate, robust, with median and lateral sclerotized, darkly pigmented plates; median plates of adjacent tergites 3–4, 4–5 and 5–6 bearing paired scent gland openings in similar positions on caudal margins of tergites 3, 4, and 5.

Comments: The stink bugs comprise a large group of very common and widely distributed species. Over 50 genera and 180 species are known from the continental United States and Canada. Specimens are easily collected by sweeping vegetation.

The biology and immature stages of this group are probably better known than any other family of true bugs. Some useful references include general treatments on biology (Esselbaugh 1947), egg structure (Esselbaugh 1946), and immatures (DeCoursey and Esselbaugh 1962); Jones and Coppel (1963) on *Apateticus cynicus* (Say); Mukerji and LeRoux (1965) on *P. maculiventris* (Say); and Oetting and Yonke on *P. placidus* (1971a), *Hymenarcys* spp. (1971b, 1972), and *Euthyrhynchus floridanus* (L.) (1975). DeCoursey and Allen (1968) provided a key to larvae of 25 genera of stink bugs.

Many species are of economic importance including the harlequin bug, *Murgantia histrionica* (Hahn), on vegetables (Canerday 1965); the southern green stink bug, *Nezara viridula* (L.), the green stink bug, *Acrosternum hilare* (Say), and *Euschistus* spp. on soybeans and other legumes (Miner 1966). A treatment by McPherson (1982) on the Pentatomoidea provides additional summary and reference information on species in northeastern N. America. The Podopinae, sometimes treated as a separate family (Borror et al. 1981), are here included with the Pentatomidae. Likewise, the acanthosomatids (Rolston and McDonald 1979) are included in the Pentatomidae.

Selected Bibliography

Canerday, 1965
DeCoursey and Allen, 1968
DeCoursey and Esselbaugh, 1962
Esselbaugh, 1946, 1947
Jones and Coppel, 1963
McPherson, 1982
Miner, 1966
Oetting and Yonke, 1971a,b, 1972, 1975
Rolston and McDonald, 1979

BIBLIOGRAPHY

Akingbohungbe, A. E. 1974. Nymphal characters and higher classification analysis in the Miridae, with a subfamily key based on the nymphs. Can. Ent. 106:687–694.

Akingbohungbe, A. E., J. L. Libby and R. D. Shenefelt. 1973. Nymphs of Wisconsin Miridae (Hemiptera: Heteroptera). Res. Bull. Univ. of Wisconsin–Madison R2561:1–25.

Aldrich, J. R., T. R. Yonke and R. D. Oetting. 1972. Histology and morphology of the abdominal scent apparatus in three alydids. J. Kansas Ent. Soc. 45:162–171.

Anderson, N. H. 1962. Anthocoridae of the Pacific Northwest with notes on distribution, life-histories, and habits (Heteroptera). Can. Ent. 94:1325–1334.

Anderson, N. M. 1979. Phylogenetic inference as applied to the evolutionary diversification of semiaquatic bugs (Hemiptera: Gerromorpha). Syst. Zool. 28:554–578.

Anderson, N. M. 1982. The semiaquatic bugs (Hemiptera, Gerromorpha). Phylogeny, adaptations, biogeography and classification. Entomonograph Vol. 3. Scandinavian Science Press, Ltd., Hesselbakke, Ganlose, DK-2760 Molov., Denmark. 454 p.

Anderson, N. M. and J. T. Polhemus. 1980. Four new genera of Mesoveliidae (Hemiptera, Gerromorpha) and the phylogeny and classification of the family. Ent. Scand. 11:369–392.

Arrand, J. C. and H. McMahon. 1974. *Plagiognathus medicagus* (Hemiptera: Miridae): Description of egg and five nymphal instars. Can. Ent. 106:433–435.

Askari, A. and V. M. Stern. 1972. Biology and feeding habits of *Orius tristicolor* (Hemiptera: Anthocoridae). Ann. Ent. Soc. Amer. 65:96–100.

Bacheler, J. and R. M. Baranowski. 1975. *Paratriphleps laeviusculus*, a phytophagous anthocorid new to the United States (Hemiptera: Anthocoridae). Fla. Ent. 58:157–163.

Bailey, N. S. 1951. The Tingoidea of New England and their biology. Ent. Amer. 31:1–140.

Balduf, W. V. 1941. Bionomics of *Catorhintha mendica* Stål (Coreidae, Hemiptera). Bull. Brooklyn Ent. Soc. 37:158–166.

Balduf, W. V. 1957. The spread of *Catorhintha mendica* Stål (Coreidae, Hemiptera). Proc. Ent. Soc. Wash. 59:176–185.

Baranowski, R. 1958. Notes on the biology of the royal palm bug, *Xylastodoris luteolus* Barker (Hemiptera, Thaumastocoridae). Ann. Ent. Soc. Amer. 51:547–551.

Beard, R. L. 1940. The biology of *Anasa tristis* De Geer with particular reference to the tachinid parasite, *Trichopoda pennipes* Fabr. Conn. (New Haven) Agric. Exp. Sta. Bull. 440:594–679.

Biehler, J. A. and J. E. McPherson. 1982. Life history and laboratory rearing of *Galgupha ovalis* (Hemiptera: Corimelaenidae), with descriptions of immature stages. Ann. Ent. Soc. Amer. 75:465–470.

Blatchley, W. S. 1926. Heteroptera or true bugs of Eastern North America. Nature Publ. Co., Indianapolis, Indiana. 1116 p.

Bobb, M. L. 1951. Life history of *Ochterus banksi* Barb. (Hemiptera: Ochteridae). Bull. Brooklyn Ent. Soc. 46:92–100.

Borror, Donald J., Dwight M. DeLong and Charles A. Triplehorn. 1981. An introduction to the study of insects. Fifth Edition. Saunders College Publishing, Philadelphia. 827 p.

Brooks, A. R. and L. A. Kelton. 1967. Aquatic and semiaquatic Heteroptera of Alberta, Saskatchewan and Manitoba (Hemiptera). Mem. Ent. Soc. Can. 51. 92 p.

Calabrese, D. M. 1974. Keys to the adults and nymphs of the species of *Gerris* Fabricius occurring in Connecticut. Mem. Conn. Ent. Soc. 1974:227–266.

Canerday, T. D. 1965. On the biology of the harlequin bug, *Murgantia histrionica* (Hemiptera: Pentatomidae). Ann. Ent. Soc. Amer. 58:931–932.

Cheng, L. and C. H. Fernando. 1971. Life history and biology of the riffle bug *Rhagovelia obesa* Uhler (Heteroptera: Veliidae) in southern Ontario. Can. J. Zool. 49:435–442.

China, W. E. and N. C. E. Miller. 1959. Checklist and keys to the families and subfamilies of the Hemiptera-Heteroptera. Bull. Brit. Museum (Nat. Hist.) Ent. Series 8:(1):1–45.

Chopra, N. P. 1968. A revision of the genus *Arhyssus* Stål. Ann. Ent. Soc. Amer. 61:629–655.

Cobben, R. H. 1968. Evolutionary trends in Heteroptera. Part I. Eggs, architecture of the shell, gross embryology and eclosion. H. Veenman and Zonen, B. V. Wageningen 151:1–475.

Cobben, R. H. 1978. Evolutionary trends in Heteroptera Part II. Mouth-part-structures and feeding strategies. Laboratorium voor Entomologie. H. Veenman and Zonen, B. V. Wageningen 289:1–407.

Cobben, H. R. 1979. On the original feeding habits of the Hemiptera (Insecta): a reply to Merrill Sweet. Ann. Ent. Soc. Amer. 72:711–715.

Connell, W. A. and J. H. Beacher. 1947. Life history and control of the oak lace bug. Univ. Delaware Agric. Exp. Sta. Bull. No. 265, Tech. No. 37:1–28.

DeCoursey, R. M. 1971. Keys to the families and subfamilies of the nymphs of North American Hemiptera-Heteroptera. Prot. Ent. Soc. Wash. 73:413–428.

DeCoursey, R. M. and R. C. Allen. 1968. A generic key to the nymphs of the Pentatomidae of the eastern United States (Hemiptera: Heteroptera). Univ. Connecticut Occasional Papers Biol. Sci. Ser. 1:141–151.

DeCoursey, R. M. and C. O. Esselbaugh. 1962. Descriptions of the nymphal stages of some North American Pentatomidae (Hemiptera-Heteroptera). Ann. Ent. Soc. Amer. 55:323–342.

Dickerson, E. L. and H. B. Weiss. 1917. The azalea lace bug, *Stephanitis pyrioides* Scott (Tingitidae, Hemiptera). Ent. News 28:101–105.

Drake, C. J. and H. C. Chapman. 1953. Preliminary report on the Pleidae (Hemiptera) of the Americas. Proc. Biol. Soc. Wash. 66:53–59.

Drake, C. J. and N. T. Davis. 1958. The morphology and systematics of the Piesmatidae (Hemiptera), with keys to world genera and American species. Ann. Ent. Soc. Amer. 51:567–581.

Drake, C. J. and N. T. Davis. 1960. The morphology, phylogeny, and higher classification of the family Tingidae, including the description of a new genus and species of the subfamily Vianaidinae (Hemiptera: Heteroptera). Ent. Amer. 39:1–100.

Drake, C. J. and F. A. Ruhoff. 1965. Lacebugs of the world: a catalog (Hemiptera: Tingidae). Bull. U.S. National Museum, 243, Washington, D.C. 634 p.

Drake, C. J. and J. A. Slater. 1957. The phylogeny and systematics of the family Thaumastocoridae (Hemiptera: Heteroptera). Ann. Ent. Soc. Amer. 50:353–370.

Elsey, K. D. and R. E. Stinner. 1971. Biology of *Julysus spinosus,* an insect predator found on tobacco. Ann. Ent. Soc. Amer. 64:779–783.

Emsley, M. G. 1969. The Schizopteridae (Hemiptera: Heteroptera) with the description of new species from Trinidad. Mem. Amer. Ent. Soc. 25:1–154.

Esselbaugh, C. O. 1946. A study of the eggs of the Pentatomidae (Hemiptera). Ann. Ent. Soc. Amer. 39:667–691.

Esselbaugh, C. O. 1947. Some notes on the biology of *Hymenarcys aequalis* Say (Pentatomidae). Bull. Brooklyn Ent. Soc. 42:25–30.

Ferris, G. F. and R. L. Usinger. 1939. The family Polyctenidae (Hemiptera: Heteroptera). Microentomology 4:(1):1–50.

Fracker, S. B. 1918. The Alydinae of the United States. Ann. Ent. Soc. Amer. 11:255–280.

Froeschner, R. C. 1960. Cydnidae of the Western Hemisphere. Proc. U.S. National Museum 111:337–680.

Gilbert, B. L., S. J. Barras and D. M. Morris. 1967. Bionomics of *Tetyra bipunctata* (Hemiptera: Pentatomidae: Scutellerinae) as associated with *Pinus banksiana* in Wisconsin. Ann. Ent. Soc. Amer. 60:698–701.

Griffith, M. E. 1945. The environment, life history and structure of the water boatman, *Ramphocorixa acuminata* (Uhler). Univ. Kansas Sci. Bull. 30:241–365.

Gittelman, S. H. 1974. Locomotion and predatory strategy in backswimmers, Amer. Midl. Nat. 92:496–500.

Halstead, T. F. 1972a. A review of the genus *Arhaphe* Herrich-Schaeffer (Hemiptera: Largidae). Pan-Pac. Ent. 48:1–7.

Halstead, T. F. 1972b. Notes and synonymy in *Largus* Hahn with a key to United States species (Hemiptera: Largidae). Pan-Pac. Ent. 48:246–248.

Harrington, B. J. 1972. Notes on the biology of *Ischnodemus* species of America north of Mexico (Hemiptera: Lygaeidae: Blissinae). Univ. Conn. Occas. Papers Biol. Sci. Ser. 2(6):47–56.

Harris, H. M. 1928. A monographic study of the hemipterous family Nabidae as it occurs in North America. Ent. Amer. IX (1 and 2):1–97.

Harris, H. M. and F. Andre. 1934. Notes on the biology of *Acantholoma denticulata* Stål (Hemiptera: Scutelleridae). Ann. Ent. Soc. Amer. 27:5–15.

Herring, J. L. and P. D. Ashlock. 1971. A key to the nymphs of the families of Hemiptera (Heteroptera) of America north of Mexico. Florida Ent. 54:207–212.

Hilsenhoff, W. L. 1970. Corixidae (water boatmen) of Wisconsin. Wis. Acad. Sci., Arts and Letters 58:203–235.

Hoffman, C. H. 1932. The biology of the three North American species of *Mesovelia.* Can. Ent. 64:88–94, 113–120, 126–133.

Hormchan, P., L. W. Hepner and M. F. Schuster. 1976. Predacious damsel bugs: identification and distribution of the subfamily Nabinae in Mississippi. Miss. Ag. and Forestry Exp. Sta. Tech. Bull. 76. 4 p.

Horn, K. F. *et al.* 1979. Identification of the fifth nymphal stage of ten species of *Corythuca.* J. Georgia Ent. Soc. 14(2):131–136.

Hungerford, H. B. 1917. The life history of the backswimmer, *Notonecta undulata* Say (Hem., Het.). Ent. News 28:267–278.

Hungerford, H. B. 1920. The biology and ecology of aquatic and semi-aquatic Hemiptera. Univ. Kansas Sci. Bull. 11:1–328.

Hungerford, H. B. 1922. The life history of the toadbug *Gelastocoris oculatus* Fabr. (Gelastocoridae). Univ. Kansas Sci. Bull. 14:145–171.

Hungerford, H. B. 1927. Life history of the creeping water bug, *Pelocoris carolinensis* Bueno. Bull. Brooklyn Ent. Soc. 22:77–82.

Hungerford, H. B. 1948. The Corixidae of the Western Hemisphere. Univ. Kansas Sci. Bull. 32:1–827.

Isenhour, D. J. and N. L. Marston. 1981. Seasonal cycles of *Orius insidiosus* (Hemiptera: Anthocoridae) in Missouri soybeans. J. Kan. Ent. Soc. 54:129–142.

Jones, T. H. 1924. The life history and stages of *Cimolus obscurus* Stål (Hemiptera). Proc. Ent. Soc. Wash. 26(8):197–200.

Jones, P. A. and H. C. Coppel. 1963. Immature stages and biology of *Apateticus cynicus* (Say) (Hemiptera: Pentatomidae). Can. Ent. 95:770–779.

Knight, H. H. 1941. The plant bugs, or Miridae, of Illinois. Bull. Ill. Nat. Hist. Surv. 22(1):1–234.

Kormilev, N. 1960. Revision of Phymatinae (Hemiptera, Phymatidae). Philippine J. Sci. 89(3–4):287–486.

Lawson, F. A. 1959. Identification of the nymphs of common families of Hemiptera. J. Kansas Ent. Soc. 32:88–92.

Leston, D., J. G. Pendergrast and T. R. E. Southwood. 1954. Classification of the terrestrial Heteroptera (Geocorisae). Nature 174:91–92.

Mann, J. 1969. Cactus-feeding insects and mites. Bull. U.S. National Museum 256. 158 p.

Matsuda, R. 1960. Morphology, evolution, and a classification of the Gerridae (Hemiptera-Heteroptera). Univ. Kansas Sci. Bull. 41:25–632.

McAtee, W. L. 1919. Key to the Nearctic species of Piesmatidae (Heteroptera). Bull. Brooklyn Ent. Soc. 14:80–93.

McAtee, W. L. and J. R. Malloch. 1924. Some annectant bugs of the super-family Cimicoidea. Bull. Brooklyn Ent. Soc. 19:69–82.

McAtee, W. L. and J. R. Malloch. 1925. Revision of bugs of the family Cryptostemmatidae in the collection of the U.S. National Museum. Proc. U.S. National Museum 67:1–42.

McAtee, W. L. and J. R. Malloch. 1933. Revision of the subfamily Thyreocorinae of the Pentatomidae (Hemiptera-Heteroptera). Ann. Carnegie Museum 21:191–411.

McPherson, J. E. 1965. Notes on the life history of *Notonecta hoffmanni*. Pan-Pac. Ent. 41:86–89.

McPherson, J. E. 1967. Brief descriptions of the external anatomy of the various stages of *Notonecta hoffmanni*. Pan-Pac. Ent. 43:117–121.

McPherson, J. E. 1972. Life history of *Corimelaena lateralis lateralis* (Hemiptera: Thyreocoridae) with descriptions of immature stages and list of other species of Scutelleroidea found with it on wild carrot. Ann. Ent. Soc. Amer. 65:985–987.

McPherson, J. E. 1982. The Pentatomoidea (Hemiptera) of Northeastern North America with emphasis on the fauna of Illinois. Southern Illinois Univ. Press, Carbondale and Edwardsville. 240 p.

McPherson, J. E. and R. J. Packauskas. 1987. Life history and laboratory rearing of *Nepa apiculata* (Heteroptera: Nepidae), with descriptions of immature stages. Ann. Ent. Soc. Amer. 80:680–85.

Menke, A. S. 1979. The semiaquatic and aquatic Hemiptera of California (Heteroptera: Hemiptera). Bull. California Insect Surv. 21:1–166.

Miller, N. C. E. 1971. The Biology of the Heteroptera. Second Edition. E. W. Classey Ltd, Hampton, Middlesex, England. 206 p.

Miner, F. D. 1966. Biology and control of stink bugs on soybeans. Arkansas Agric. Exp. Sta. Bull. 708:1–40.

Moznette, G. F. 1921. Notes on the royal palm bug. Quar. Bull. 86. U.S. Bur. Ent. 110 p.

Mukerji, M. K. and E. J. LeRoux. 1965. Laboratory rearing of a Quebec strain of the pentatomid predator, *Podisus maculiventris* (Say) (Hemiptera: Pentatomidae). Phytoprotection. 46:40–60.

Oetting, R. D. and T. R. Yonke. 1971a. Immature stages and biology of *Podisus placidus* and *Stiretrus fimbriatus* (Hemiptera: Pentatomidae). Can. Ent. 103:1505–1516.

Oetting, R. D. and T. R. Yonke. 1971b. Immature stages and biology of *Hymenarcys nervosa* and *H. aequalis* (Hemiptera: Pentatomidae). Ann. Ent. Soc. Amer. 64:1289–1296.

Oetting, R. D. and T. R. Yonke. 1972. Immature stages and notes on the biology of *Hymenarcys crassa* (Hemiptera: Pentatomidae). Ann. Ent. Soc. Amer. 65:474–478.

Oetting, R. D. and T. R. Yonke. 1975. Immature stages and notes in biology of *Euthyrhynchus floridanus* (L.) (Hemiptera: Pentatomidae). Ann. Ent. Soc. Amer. 68:659–662.

Packard, C. M. and C. Benton. 1937. How to fight the chinch bug. USDA Farmers' Bull. 1780. 21 p.

Paskewitz, S. M. and J. E. McPherson. 1983. Life history and laboratory rearing of *Arhyssus lateralis* (Hemiptera: Rhopalidae) with descriptions of immature stages. Ann. Ent. Soc. Amer. 76:477–482.

Peterson, A. 1948. Larvae of insects. Part 1. Published by the author, Columbus, Ohio, 315 p.

Polhemus, J. T. 1967. Notes on North American Saldidae. Proc. Ent. Soc. Wash. 69:24–30.

Polhemus, J. T. 1976. Shore bugs (Hemiptera: Saldidae, etc.). *In* Cheng, L. (ed.), Marine Insects. North-Holland Publ. Co. Ch. 9:225–262.

Polhemus, J. T. 1977. The biology and systematics of the Saldidae of Mexico and Middle America. Ph.D. Diss., Univ. of Colorado, Boulder, 606 p.

Polhemus, J. T. 1981. The Phylogeny of the Leptopodomorpha and Relationship to other Heteroptera. Rostria 33 Suppl. 17–27.

Porter, T. W. 1950. Taxonomy of the American Hebridae and the natural history of selected species. Ph.D. Diss., Univ. of Kansas, Lawrence, 185 p. + 10 plates.

Rankin, K. P. 1935. Life history of *Lethocerus americanus* (Leidy) (Hemiptera, Belostomatidae). Univ. Kansas Sci. Bull. 22:479–491.

Radinovsky, S. 1964. Cannibal of the pond. Natural History 73:16–25.

Readio, P. A. 1926. Studies of the eggs of some Reduviidae. Univ. Kansas Sci. Bull. 16:157–179.

Readio, P. A. 1927. Studies on the biology of the Reduviidae of America north of Mexico. Univ. Kansas Sci. Bull. 17(1):5–291.

Readio, P. A. 1928. Studies on the biology of the genus *Corizus* (Coreidae, Hemiptera). Ann. Ent. Soc. Amer. 21:189–201.

Rolston, L. H. and F. J. D. McDonald. 1979. Keys and diagnosis for the families of Western Hemisphere Pentatomoidea, subfamilies of Pentatomidae and tribes of Pentatominae (Hemiptera). J. N. Y. Ent. Soc. 87(3):189–207.

Ryckman, R. E. 1962. Biosystematics and hosts of the *Triatoma protracta* complex in North America (Hemiptera: Reduviidae) (Rodentia: Cricetidae). Univ. Calif. Publ. Ent., Univ. Calif. Press 27(2):93–240.

Sands, W. A. 1957. The immature stages of some British Anthocoridae (Hemiptera). Trans. R. Ent. Soc. London 109(10):295–310.

Schaefer, C. W. 1969. Morphological and phylogenetic notes on the Thaumastocoridae (Hemiptera-Heteroptera). J. Kansas Ent. Soc. 42:251–256.

Schaefer, C. W. 1972. Clades and grades in the Alydidae. J. Kansas Ent. Soc. 45:135–141.

Schaefer, C. W. 1975. Heteropteran trichobothria (Hemiptera: Heteroptera). Int. J. Insect Morph. Embryol. 4(3):193–264.

Schaefer, C. W. 1981. Improved cladistic analysis of the Piesmatidae and consideration of known host plants. Ann. Ent. Soc. Amer. 74:536–539.

Schaefer, C. W. and N. P. Chopra. 1982. Cladistic analysis of the Rhopalidae, with a list of food plants. Ann. Ent. Soc. Amer. 75:224–233.

Schaefer, C. W. and P. L. Mitchell. 1983. Food plants of the Coreoidea (Hemiptera: Heteroptera). Ann. Ent. Soc. Amer. 76:591–615.

Sheeley, R. D. and T. R. Yonke. 1977a. Immature stages and biology of *Atheas austroriparius* and *Leptoypha costata* (Hemiptera: Tingidae). Ann. Ent. Soc. Amer. 70:603–614.

Sheeley, R. D. and T. R. Yonke. 1977b. Biological notes on seven species of Missouri tingids (Hemiptera: Tingidae). J. Kansas Ent. Soc. 50:342–356.

Sites, R. W. and J. E. McPherson. 1982. Life history and laboratory rearing of *Sehirus cinctus cinctus* (Hemiptera: Cydnidae), with descriptions of immature stages. Ann. Ent. Soc. Amer. 75:210–215.

Slater, J. A. 1952. A contribution to the biology of the subfamily Cyminae (Heteroptera: Lygaeidae). Ann. Ent. Soc. Amer. 45:315–325.

Slater, J. A. 1963. Immature stages of the subfamily Cyminae and Ischnorhynchinae (Hemiptera: Lygaeidae). J. Kansas Ent. Soc. 36:84–93.

Slater, J. A. 1964. A catalogue of the Lygaeidae of the world, 2 vols. University of Connecticut, Storrs. 1668 p.

Slater, J. A. 1982. Hemiptera, p. 417–447. *In* Parker, S. P. (ed.), Synopsis and Classification of Living Organisms. McGraw-Hill. N.Y.

Slater, J. A. and R. M. Baranowski. 1978. How to know the true bugs (Hemiptera-Heteroptera). Wm. C. Brown Company Publishers, Dubuque, Iowa. 256 p.

Smith, R. L. 1976a. Male brooding behavior of the water bug *Adeus herberti* (Hemiptera: Belostomatidae). Ann. Ent. Soc. Amer. 69:740–747.

Smith, R. L. 1976b. Brooding behavior of a male water bug *Belostoma flumineum* (Hemiptera: Belostomatidae). J. Kansas Ent. Soc. 49:333–343.

Sprague, I. B. 1956. The biology and morphology of *Hydrometra martini* Kirkaldy. Univ. Kansas Sci. Bull. 38:579–693.

Stys, P. and I. Kerzhner. 1975. The rank and nomenclature of higher taxa in recent Heteroptera. Acta Ent. Bohemosolv 72:65–79.

Swadener, S. O. and T. R. Yonke. 1973a. Immature stages and biology of *Apiomerus crassipes* (Hemiptera: Reduviidae). Ann. Ent. Soc. Amer. 66:188–196.

Swadener, S. O. and T. R. Yonke. 1973b. Immature stages and biology of *Sinea complexa* with notes on four additional reduviids. J. Kansas Ent. Soc. 46:123–136.

Swadener, S. O. and T. R. Yonke. 1973c. Immature stages and biology of *Zelus socius* (Hemiptera: Reduviidae). Can. Ent. 105:231–238.

Swadener, S. O. and T. R. Yonke. 1975. Immature stages and biology of *Pselliopus cinctus* and *P. barberi* (Hemiptera: Reduviidae). J. Kansas Ent. Soc. 48:477–492.

Sweet, M. H. 1964a. The biology and ecology of the Rhyparochrominae of New England (Heteroptera: Lygaeidae). Pt. I. Ent. Amer. 43:1–124.

Sweet, M. H. 1964b. The biology and ecology of the Rhyparochrominae of New England (Heteroptera: Lygaeidae). Pt. II. Ent. Amer. 44:1–201.

Sweet, M. H. 1979. On the original feeding habits of the Hemiptera (Insecta). Ann. Ent. Soc. Amer. 72:575–579.

Sweet, M. H. and J. A. Slater. 1961. A generic key to the nymphs of North American Lygaeidae (Hemiptera-Heteroptera). Ann. Ent. Soc. Amer. 54:333–340.

Tamaki, G. and R. E. Weeks. 1972. Biology and ecology of two predators, *Geocoris pallens* Stål and *G. bullatus* (Say). USDA Tech. Bull. 1446. 46 p.

Tashiro, H. 1987. Turfgrass insects of the United States and Canada. Cornell Univ. Press, Ithaca, N.Y., 371 pp.

Taylor, E. J. 1949. A life history study of *Nabis alternatus*. J. Econ. Ent. 42:991.

Underhill, R. A. 1954. Habits and life history of *Arilus cristatus* (Hemiptera: Reduviidae). Walla Walla College Publ. Dept. Biol. Sci. 11:1–15.

Usinger, R. L. 1941. A remarkable immigrant leptopolid in California. Bull. Brooklyn Ent. Soc. 36:164–165.

Usinger, R. L. 1945. Classification of the Enicocephalidae (Hemiptera, Reduvioidea). Ann. Ent. Soc. Amer. 38(3):321–342.

Usinger, R. L. 1946. Notes and descriptions of *Ambrysus* Stal with an account of the life history of *Ambrysus mormon* Montd. Univ. Kansas Sci. Bull. 31:185–210.

Usinger, R. L. 1966. Monograph of Cimicidae (Hemiptera-Heteroptera). Thomas Say Foundation 7. 585 p.

Usinger, R. L. and R. Matsuda. 1959. Classification of the Aradidae (Hemiptera-Heteroptera). British Museum (Nat. Hist.) London. 410 p.

Van Doesburg, P. H.. Jr. 1968. A revision of the New World species of *Dysdercus* Guérin-Méneville (Heteroptera, Pyrrhocoridae). Zool. Verhand. (Leiden) 97:1–215.

Walt, J. F. and J. E. McPherson. 1972. Laboratory rearing of *Stethaulax marmoratus* (Hemiptera: Scutelleridae). Ann. Ent. Soc. Amer. 65:1242–1243.

Walt, J. F. and J. E. McPherson. 1973. Descriptions of immature stages of *Stethaulax marmorata* (Hemiptera: Scutelleridae) with notes on its life history. Ann. Ent. Soc. Amer. 66:1103–1107.

Weiss, H. B. and R. F. Loft. 1924. Notes on *Piesma cinerea* Say in New Jersey (Hemiptera). Psyche 31:233–235.

Wheeler, A. G., Jr. and T. J. Henry. 1976. Biology of the honey locust plant bug, *Diaphnocoris chlorionis,* and other mirids associated with ornamental honey locust. Ann. Ent. Soc. Amer. 69:1095–1104.

Wheeler, A. G., Jr. and T. J. Henry. 1977a. Miridae associated with Pennsylvania conifers 1. Species on Arborvitae, false cypress, and juniper. Trans. Amer. Ent. Soc. 13:623–656.

Wheeler, A. G., Jr. and T. J. Henry. 1977b. Isometopinae (Hemiptera: Miridae) in Pennsylvania: Biology and descriptions of fifth instars with observation on predation on obscure scale. Ann. Ent. Soc. Amer. 70:607–614.

Wheeler, A. G., Jr. and T. J. Henry. 1978. *Ceratocapsus modestus,* a predator of grape phylloxera: seasonal history and description of fifth instar. Melsheimer Ent. Ser. 25:6–10.

Wheeler, A. G., Jr. and T. J. Henry. 1981. *Jalysus spinosus* and *J. wickhami:* Taxonomic clarification, review of host plants and distribution, and keys to adults and 5th instars. Ann. Ent. Soc. Amer. 74:606–615.

Wheeler, A. G., Jr. and T. J. Henry. 1984. Host plants, distribution, and description of fifth-instar *Arhyssus hirtus* (Rhopalidae) and *Esperanza texana* (Alydidae). Fla. Ent. 67:521–529.

Wheeler, A. G., Jr. and G. L. Miller. 1983. *Harmostes fraterculus* (Hemiptera: Rhopalidae): field history, laboratory rearing, and descriptions of immature stages. Proc. Ent. Soc. Wash. 85(3):426–434.

Wheeler, A. G., Jr. and C. W. Schaefer. 1982. Review of stilt bug (Hemiptera: Berytidae) host plants. Ann. Ent. Soc. Amer. 75:498–506.

Wheeler, A. G., Jr., B. R. Stinner and T. J. Henry. 1975. Biology and nymphal stages of *Deraeocoris nebulosus* (Hemiptera: Miridae), a predator of arthropod pests on ornamentals. Ann. Ent. Soc. Amer. 68:1063–1068.

Wilcox, R. S. 1969. Acoustical behavior, sound-producing structures and biology of *Buenoa.* Ph.D. Diss., Dept. Zool., Univ. Michigan, Ann Arbor, Michigan.

Wilcox, R. S. 1975. Sound-producing mechanisms of *Buenoa macrotibialis* Hungerford. Int. J. Insect Morphol. Embryol. 4:169–182.

Wiley, G. O. 1922. Life history notes on two species of Saldidae found in Kansas. Univ. Kansas Sci. Bull. 14:301–311.

Wygodzinsky, P. W. 1966. A monograph of the Emesinae (Reduviidae, Hemiptera). Bull. Amer. Mus. Natural History 133:1–614.

Yonke, T. R. and J. T. Medler. 1965. Biology of *Megalotomus quinquespinosus* (Hemiptera: Alydidae). Ann. Ent. Soc. Amer. 58:222–224.

Yonke, T. R. and J. T. Medler. 1967. Observations on some Rhopalidae (Hemiptera). Proc. N. Central States Branch Ent. Soc. Amer. 22:74–75.

Yonke, T. R. and J. T. Medler. 1968. Biologies of three species of *Alydus* in Wisconsin. Ann. Ent. Soc. Amer. 61:526–531.

Yonke, T. R. and J. T. Medler. 1969a. Biology of the Coreidae in Wisconsin. Trans. Wisconsin Acad. Science, Arts and Letters 57:163–188.

Yonke, T. R. and J. T. Medler. 1969b. Description of immature stages of Coreidae. 1. *Euthochtha galeator*. Ann. Ent. Soc. Amer. 62:469–473.

Yonke, T. R. and J. T. Medler. 1969c. Description of immature stages of Coreidae. 2. *Acanthocephala terminalis*. Ann. Ent. Soc. Amer. 62:474–476.

Yonke, T. R. and J. T. Medler. 1969d. Description of immature stages of Coreidae. 3. *Archimerus alternatus*. Ann. Ent. Soc. Amer. 62:477–480.

Yonke, T. R. and J. T. Medler. 1970. New records of parasites from *Zelus exsanguis* and *Pselliopus cinctus*. J. Kansas Ent. Soc. 43:441–443.

Yonke, T. R. and D. L. Walker. 1970a. Field history, parasites, and biology of *Harmostes reflexulus* (Say) (Hemiptera: Rhopalidae). J. Kansas Ent. Soc. 43:444–450.

Yonke, T. R. and D. L. Walker. 1970b. Description of the egg and nymphs of *Harmostes reflexulus* (Hemiptera: Rhopalidae). Ann. Ent. Soc. Amer. 63:1749–1754.

Order Homoptera

30

Lois B. O'Brien, Coordinator[1]
Florida A & M University

Manya B. Stoetzel[2] and Douglass R. Miller[3]
Systematic Entomology Laboratory, ARS, USDA

CICADAS, HOPPERS, PSYLLIDS, APHIDS, WHITEFLIES, AND SCALES

The Homoptera are a large and very diverse group of insects closely related to and sometimes regarded as a suborder of the Hemiptera. They vary in length from 1 to 110 mm and their structural features range from very complicated to very degenerate, with some adult female scale insects lacking marked body regions, compound eyes, wings, and legs.

DIAGNOSIS

The most characteristic feature of the order as a whole is the structure of the mouthparts. The beak consists, as in the Hemiptera, of 2 pairs of stylets formed by the maxillae and the mandibles and sheathed in the labium. The stylets form a double tube, one channel for saliva, the other for food. The beak apparently arises from the caudal margin of the head or from the meson between the thoracic legs, and this characteristic separates the Homoptera from all Hemiptera but the Peloridiidae, an unusual intermediate family of circum-subantarctic Hemiptera, and some aquatic Hemiptera. No homopteran is considered aquatic, although some cicada and cercopid immatures live in water-filled burrows and some scales and delphacids live on plants in the intertidal zone. Immatures of the suborder Auchenorrhyncha have compound eyes and wing pads as in Hemiptera; some immatures of the suborder Sternorrhyncha lack compound eyes and wing pads. The antennae of Auchenorrhyncha are usually short and bristlelike; those of Sternorrhyncha are usually long and filiform. As adults, Auchenorrhyncha have three tarsal segments; but as immatures they may have 1, 2, or 3, often adding more as they molt to a later instar. Adults and immatures in the Sternorrhyncha have 1 or 2 tarsal segments.

BIOLOGY AND ECOLOGY

All Homoptera feed on plant juices and most feed above ground, although some feed on roots or under bark as immatures and above ground as adults, and some are subterranean in all stages. Some are important pests of crops, either because of feeding damage or because they are vectors of plant diseases caused by viruses, rickettsiae, bacteria, or mycoplasma-like organisms. Some are beneficial through the production of dyes, waxes, shellac, the manna referred to in the Bible, and other products.

Their life histories may be simple or complex. Sometimes hermaphroditism or parthenogenesis is involved, sometimes alternation of parthenogenetic and bisexual generations or winged and wingless generations, sometimes alternation of host plants. The majority are oviparous, but Coccoidea are oviparous or ovoviviparous, and many Aphididae are viviparous during some or all of their life cycle.

Eggs may be inserted into plant tissue, glued to the surface of plants, covered by mud cases, deposited in waxy egg sacs, protected by the body of the adult female, or simply dropped. Immatures have 2 to 7 instars and are frequently gregarious. Aleyrodidae have a more or less inactive last instar which is commonly called a "pupa", and male coccoids have "prepupal" and "pupal" instars.

Immatures may be protected by secreted wax, hard shell-like coverings, protective coloration, or by association with their mother, ants, or stingless bees. They may produce galls in plants. "Spittle", a foam produced by trapping air in anal secretions mixed with epidermal secretions, protects most cercopid immatures from desiccation and also from predators and parasitoids. (One family, Machaerotidae, not found in the U.S., lives in calcareous tubes).

Although the report of Fulgoridae or lanternflies being luminescent has existed for almost 200 years, it seems to be in error. Ridout (1983) tested all of the enzyme systems known to produce luminescence in animals against frozen *Fulgora* and found no trace of luminescence. He found no new witnesses and reported that Heyde recanted his letter cited in "Bioluminescence" (Harvey, 1952).

TECHNIQUES

Auchenorrhyncha may be preserved in alcohol or mounted on pins or points as adults are. The methods of preservation for the Sternorrhyncha are given in the discussion of the superfamilies or families.

1. Introduction and Auchenorrhyncha.
2. Aphidoidea, Aleyrodidae, and Psyllidae.
3. Coccoidea.

CLASSIFICATION[4]

Order **HOMOPTERA** (7863)
- Suborder Auchenorrhyncha (3865)
  - Superfamily Fulgoroidea—planthoppers
    - Cixiidae—cixiid planthoppers (172)
    - Delphacidae—delphacid planthoppers (300)
    - Derbidae—derbid planthoppers (62)
    - Kinnaridae—kinnarid planthoppers (immatures not yet found) (6)
    - Dictyopharidae—dictyopharid planthoppers (76)
    - Fulgoridae—lantern flies, fulgorid planthoppers (17)
    - Achilidae—achilid planthoppers (56)
    - Tropiduchidae—tropiduchid planthoppers (2)
    - Flatidae—flatid planthoppers (30)
    - Issidae—issid planthoppers (128)
    - Acanaloniidae—acanaloniid planthoppers (18)
  - Superfamily Cicadoidea
    - Cicadidae—cicadas (180)
  - Superfamily Cercopoidea
    - Cercopidae—froghoppers, spittlebugs (60)
  - Superfamily Cicadelloidea
    - Aetalionidae—aetalionids (2)
    - Membracidae—treehoppers (255)
    - Cicadellidae—leafhoppers (2,500)
- Suborder Sternorrhyncha (3998)
  - Superfamily Aphidoidea
    - Aphididae [includes Pemphigidae (Eriosomatidae)]—aphids (1500)
    - Adelgidae (Chermidae suppressed)—adelgids (20)
    - Phylloxeridae—phylloxera, phylloxerans, phylloxerids (50)
  - Superfamily Aleyrodoidea
    - Aleyrodidae—whiteflies (500)
  - Superfamily Psylloidea
    - Psyllidae—psyllids (1,000)
  - Superfamily Coccoidea
    - Margarodidae—giant scales, ground pearls (42)
    - Ortheziidae—ensign scales (31)
    - Putoidae—giant mealybugs (20)
    - Pseudococcidae—mealybugs (290)
    - Coccidae—soft scales (93)
    - Tachardiidae (=Lacciferidae)—lac scales (7)
    - Aclerdidae—flat grass scales (15)
    - Dactylopiidae—cochineal scales (3)
    - Kermesidae (=Kermidae)—gall-like scales (34)
    - Eriococcidae—felt scales (54)
    - Lecanodiaspididae—false pit scales (5)
    - Cerococcidae—ornate pit scales (5)
    - Asterolecaniidae—pit scales (30)
    - Phoenicococcidae—palm scales (1)
    - Conchaspididae—false armored scales (1)
    - Diaspididae—armored scales (297)

KEY TO THE FAMILIES OF IMMATURE HOMOPTERA IN AMERICA NORTH OF MEXICO[5]

(Except Kinnaridae)

1. Mouthparts apparently arising from caudal portion of head capsule (fig. 1); antennae usually very short and bristlelike; compound eyes and wing pads present; tarsi usually 2- or 3-segmented .... **Suborder AUCHENORRHYNCHA** 2

1′. Mouthparts apparently arising between front coxae (fig. 2); antennae usually long and filiform; most early instars lack compound eyes and wing pads except Psyllidae; tarsi 1- or 2-segmented .... **Suborder STERNORRHYNCHA** 16

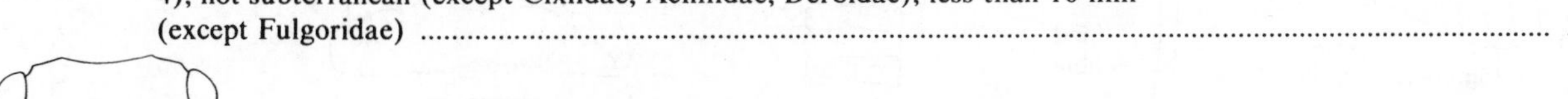

2(1). **AUCHENORRHYNCHA: Note: In couplets 2 through 15, key late instars only. The characters are not always reliable for first and second instars.**
Foreleg enlarged and fitted for digging, foretarsus attached before apex of tibia (fig. 3); subterranean; usually 10 mm or larger .... ***Cicadidae*** p. 83

2′. Foreleg not modified for digging, or, if so, foretarsus attached to apex of tibia (fig. 4); not subterranean (except Cixiidae, Achilidae, Derbidae); less than 10 mm (except Fulgoridae) 3

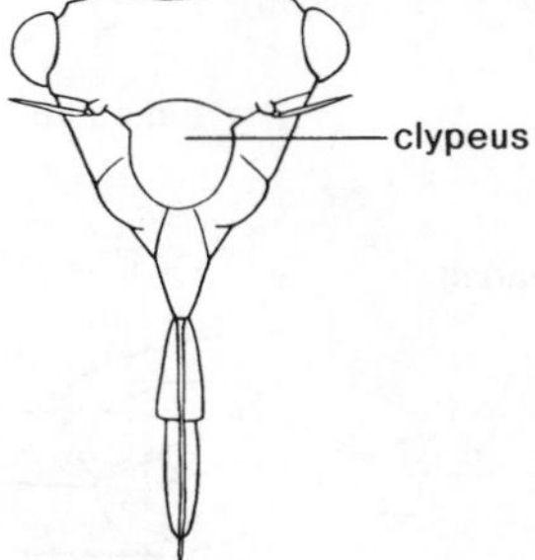

Figure 1

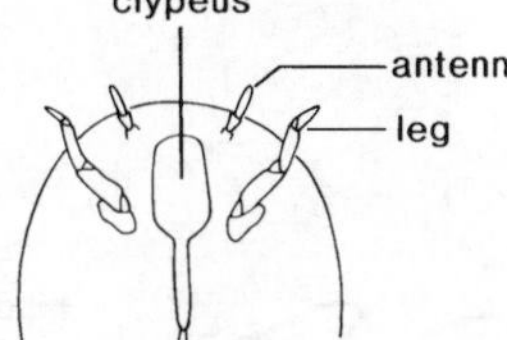

Figure 2

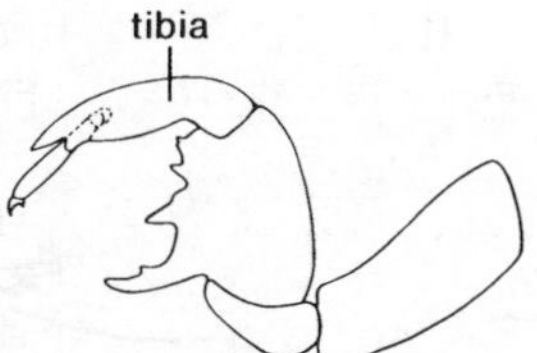

Figure 3

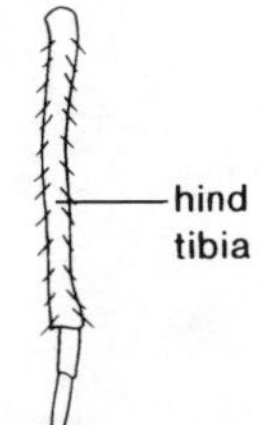

Figure 4

4. The number of species given is for America north of Mexico.
5. For the Auchenorrhyncha the key is for last instars (it may not work for early instars); for the Sternorrhyncha all instars are keyed.

3(2′). Abdominal terga formed into ventral channel for forming "spittle" (fig. 5); frons prominent, globose (fig. 6) .... ***Cercopidae*** p. 83

3′. Abdominal terga not modified for forming "spittle" although terga may reach ventral surface; frons not globose (fig. 7) 4

4(3′). Antennae and eyes separated from front of head by lateral carinae (figs. 7 and 8); antennae usually inserted on side of head beneath eyes, 3rd segment threadlike; frons usually with sensory pits; usually producing wax (figs. 9 and 10) .... (Fulgoroidea) 7

4′. Antennae and eyes not separated from front of head by carinae; antennae inserted in front of eyes (fig. 1) 5

5(4′). Ninth (apparent eighth) abdominal segment elongate (figs. 11 and 12); hind tibiae without rows of spines (fig. 13), or, if so, then hairs or spines approximately same size on all legs; thorax and abdomen often with tubercles or spines 6

5′. Abdominal segments nearly equal in length (fig. 14); hind tibiae with 1 or more, usually 2, rows of small spines (fig. 4), these spines larger than those of other legs ; thorax and abdomen without prominent tubercles or spines .... ***Cicadellidae*** p. 85

6(5). Body dorso-ventrally compressed; head as seen from above short and wide and shaped as in fig. 15 .... ***Aetalionidae*** p. 84

6′. Body triangular in cross section, usually with spines or projections; head not as above .... ***Membracidae*** p. 84

7(4). Posterior tibia with a movable apical spur (fig. 16) .... ***Delphacidae*** p. 78

7′. Posterior tibia without a movable spur (figs. 17, 18, 19) 8

8(7′). Second posterior tarsal segment armed with a row of spines at apex (fig. 17); apex truncate or emarginate 9

8′. Second posterior tarsal segment with a single or no spine at each side of apex; apical margin usually rounded or subconical (figs. 18 and 19) 12

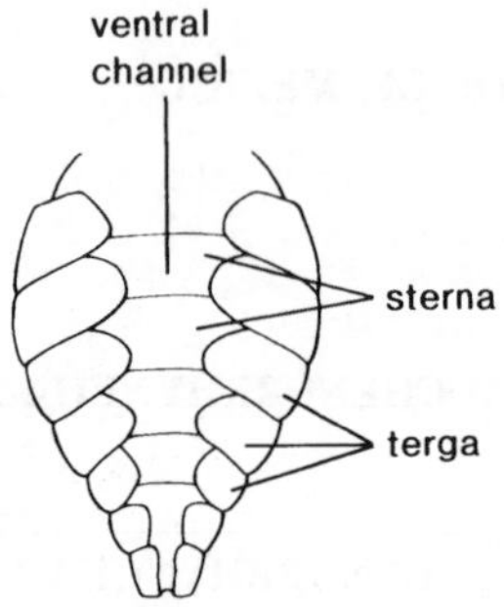

Figure 5 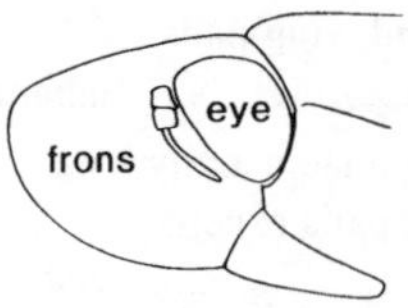 Figure 6 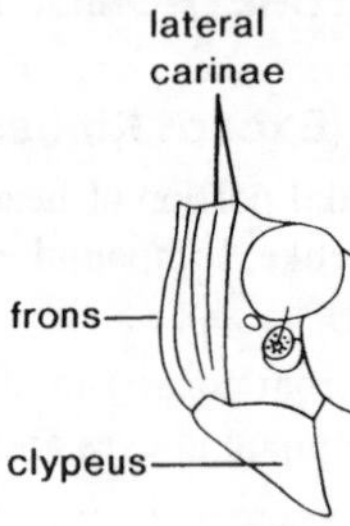 Figure 7 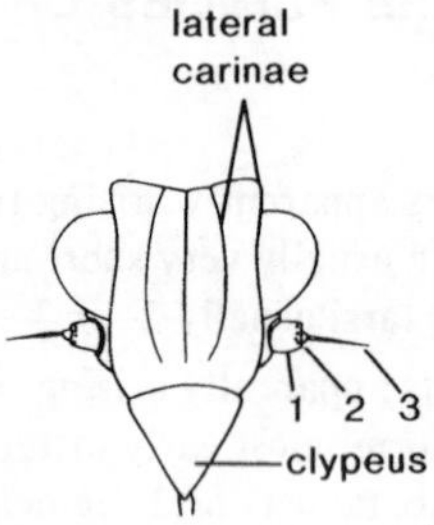 Figure 8 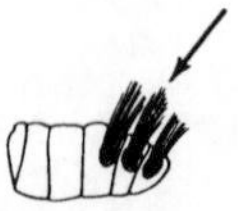 Figure 9 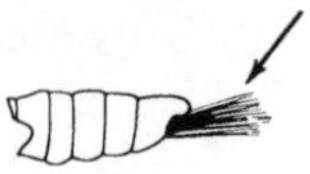 Figure 10

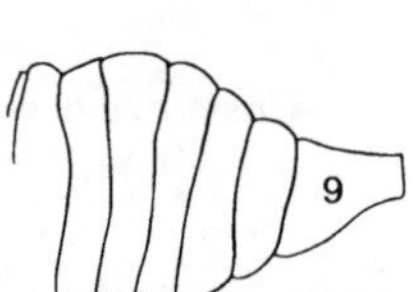

Figure 11 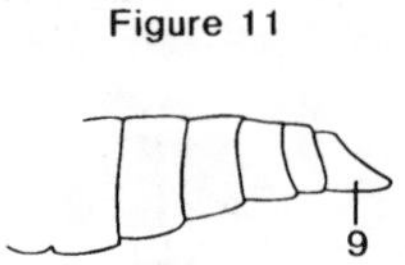 Figure 12 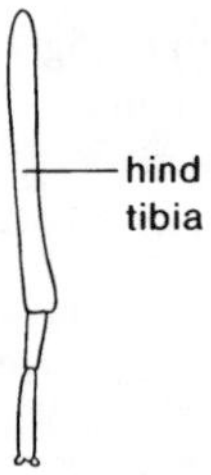 Figure 13 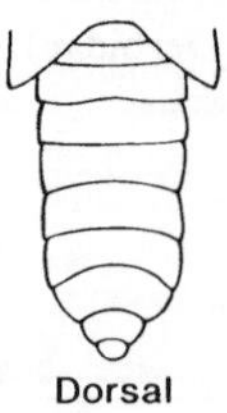 Figure 14 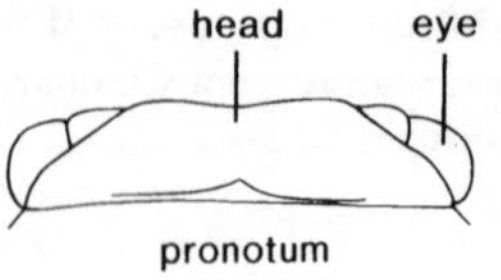 Figure 15 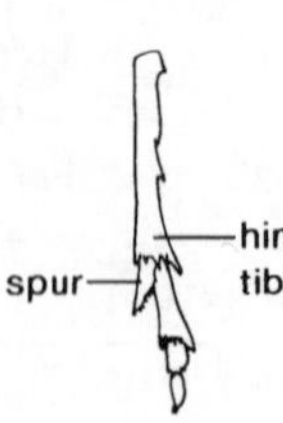 Figure 16

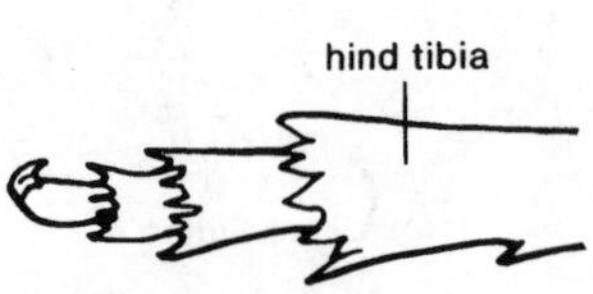

Figure 17 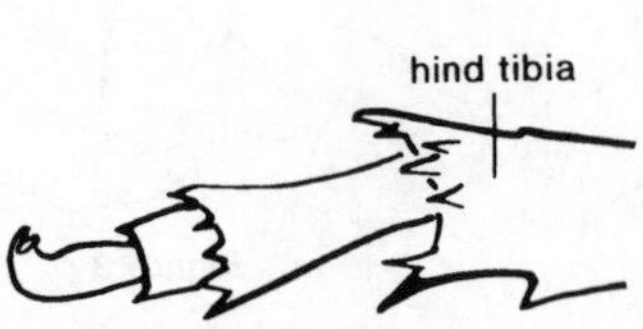 Figure 18 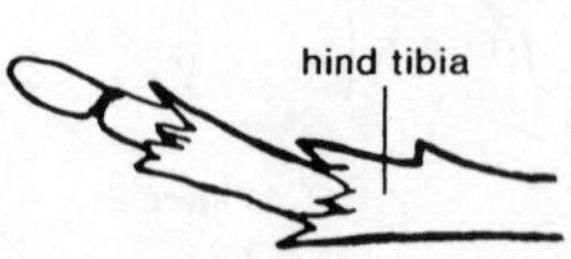 Figure 19 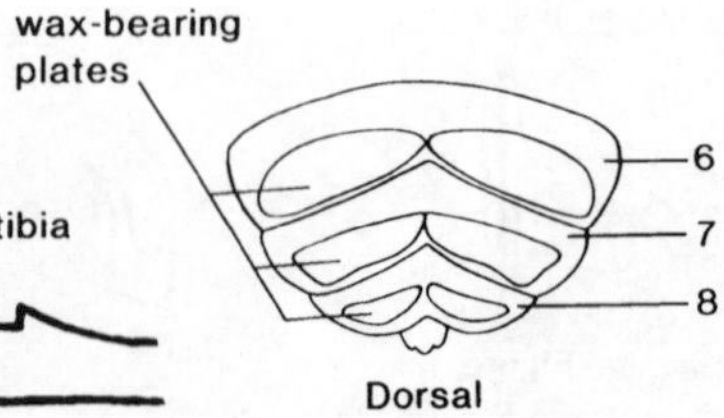 Figure 20

9(8). Eyes usually reduced, diameter about 2x diameter of first antennal segment; living beneath bark, underground, or in litter; lateral carinae of frons not continued on the clypeus (fig. 7); wax-bearing plates directed dorsad (figs. 9, 20, 21) 10

9′. Eyes normal, at least 3x diameter of first antennal segment; found above ground and in the open; lateral carinae of frons continued strongly on to the clypeus (fig. 24); wax-bearing plates directed caudad (figs. 10, 22, 23, 26) 11

10(9). Wax-bearing plates covering more than ¾ width of terga 6, 7, and 8, almost meeting medially (fig. 20); living under stones or underground .... ***Cixiidae*** p. 78

10′. Wax-bearing plates covering about ½ width of terga 6, 7, and 8 (fig. 21); living beneath bark or in litter .... ***Achilidae*** p. 81

11(9′). Two pairs of transversely oriented (long axis horizontal), oval, wax-bearing plates on caudal aspect of 7th and 8th abdominal terga (fig. 22); southern states north to southern Illinois; large, adults usually 10 mm long .... ***Fulgoridae*** p. 80

11′. No wax-bearing plates on caudal aspect of abdominal terga (tribe Orgeriini, California and Southwest) or 2 or 3 pairs of wax-bearing plates on tergum 6 and/or 7 and 8, at least 1 pair round or, if oval, at least 1 with long axis vertical (fig. 23); not restricted to the South; adults usually smaller than 10 mm (our most common genus, *Scolops*, with head produced (fig. 24) .... ***Dictyopharidae*** p. 79

12(8′). Clypeus laterally carinate (as in fig. 24); first posterior tarsal segment usually lacking apical series of spines; living in cavities in rotten logs .... ***Derbidae*** p. 79

12′. Clypeus not laterally carinate (fig. 7); hind tibiae and first tarsal segment with apical series of spines (fig. 17); living on plants 13

13(12′). Two pairs of wax-bearing plates visible on 7th and 8th terga in caudal view (fig. 26), 6th tergum usually bilobed in dorsal view, lobes usually projecting caudally to cover wax-bearing plates (fig. 25); legs with a few hairs which are not longer than spines on 2nd posterior tarsal segment .... ***Flatidae*** p. 81

13′. Wax-bearing plates absent or variable, not as above; 6th tergum chevron-shaped (fig. 27) or rectangular; legs hairy, usually hairs longer than spines of 2nd posterior tarsal segment 14

14(13′). Rostrum (mouthparts) not reaching hind trochanters; 2nd and 3rd posterior tarsal segments equal in width (fig. 19); recorded only from Florida, Mississippi, and North Carolina in the U.S. .... ***Tropiduchidae*** p. 81

14′. Rostrum (mouthparts) usually reaching hind trochanters (fig. 24); 2nd posterior tarsal segment wider than third (fig. 18) 15

15(14′). Very strongly humpbacked (fig. 28), often dorsum concave behind metathorax .... ***Acanaloniidae*** p. 82

15′. At most moderately to weakly humpbacked (fig. 29) .... ***Issidae*** p. 82

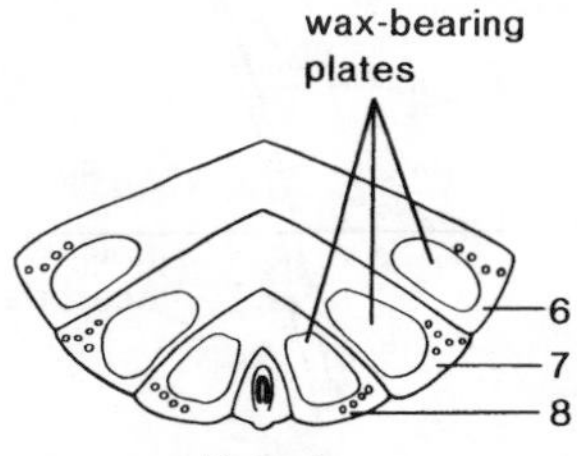

wax-bearing plates
6
7
8
Dorsal

Figure 21

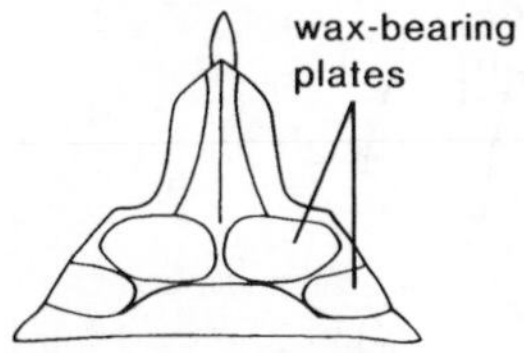

wax-bearing plates
tip of abdomen (caudal)

Figure 22

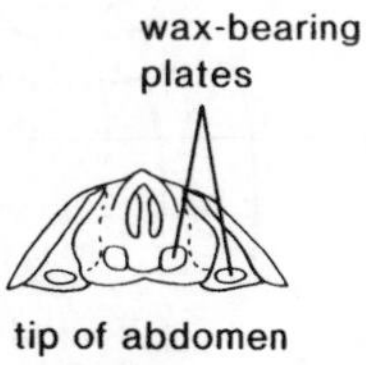

wax-bearing plates
tip of abdomen (caudal)

Figure 23

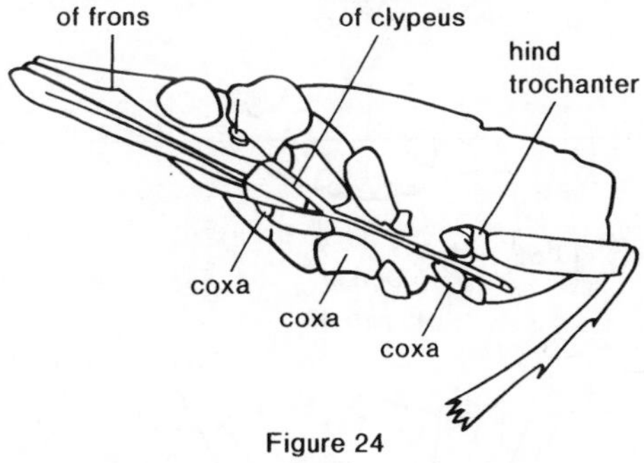

lateral carina of frons
lateral carina of clypeus
hind trochanter
coxa
coxa
coxa

Figure 24

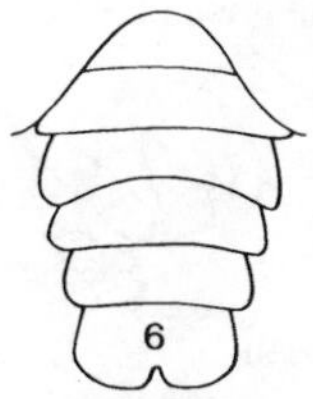

6
Dorsal

Figure 25

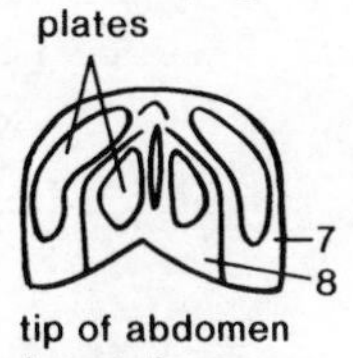

wax-bearing plates
7
8
tip of abdomen (caudal)

Figure 26

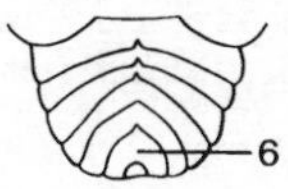

6
Dorsal

Figure 27

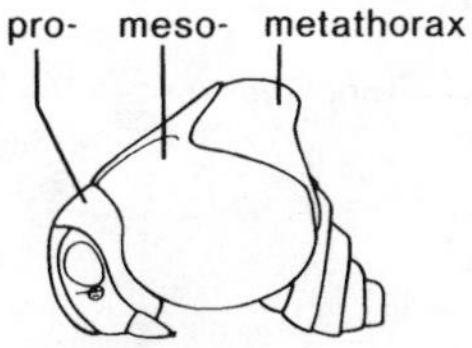

pro- meso- metathorax

Figure 28

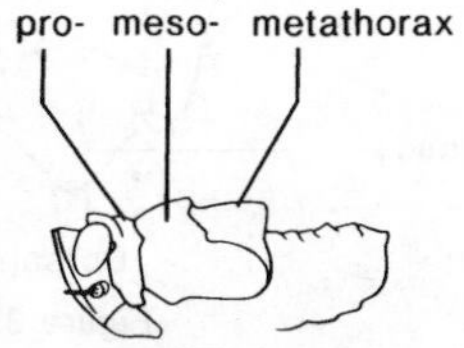

pro- meso- metathorax

Figure 29

STERNORRHYNCHA: Split drawings: right side = ventral (V); left side = dorsal (D)

16(1′). Legs usually present and with 2 claws, tarsi 1- or 2-segmented, rostrum usually 4- to 5-segmented; immatures of winged forms with 2 pairs of wing pads 17

16′. Legs present or absent, when present with a single claw, tarsi usually 1-segmented; immatures of winged forms with 1 pair of wing pads .... (Coccoidea) 21

17(16). Anal area with a vasiform orifice, lingula, and operculum (fig. 30) .... ***Aleyrodidae*** p. 88

17′. Anal area without a vasiform orifice, lingula, or operculum 18

18(17′). Anal area with a circumanal ring (fig. 31) .... ***Psyllidae*** p. 89

18′. Anal area without a circumanal ring .... (Aphidoidea) 19

19(18′). Antenna 3–6 segmented with one primary sensorium located subapically on the penultimate segment and another on the last segment at the division between the base and the unguis (fig. 32); cauda present but poorly developed in immature stages (fig. 33); cornicles usually present dorsally on the 5th or 6th abdominal segment (fig. 33) .... ***Aphididae*** p. 86

19′. Antenna 3-segmented with one primary sensorium located subapically on last segment; cauda and cornicles absent 20

20(19′). Dorsum with wax glands distributed over body (fig. 34), tubercles absent. Living on conifers .... ***Adelgidae*** p. 87

20′. Dorsum without wax glands, tubercles sometimes present (fig. 35). Living on deciduous plants .... ***Phylloxeridae*** p. 87

21(16′). Abdominal spiracles present (very small in immatures, fig. 36) 22

21′. Abdominal spiracles absent 23

22(21). Apex of each antenna with stout seta (fig. 37); trochanter and femur fused (fig. 38); anal ring at dermal surface with setae and pores (fig. 39); sessile pores predominantly of quadrilocular type (fig. 39) .... (ensign scales) ***Ortheziidae*** p. 92

22′. Apex of each antenna without stout seta (fig. 40); trochanter and femur separate (fig. 41); anal ring absent or orifice with thin sclerotized ring without setae and pores or ring located internally on invaginated tube without setae, occasionally with pores; sessile pores (fig. 36) rarely of quadrilocular type .... (giant scales) ***Margarodidae*** p. 91

- a. Antennae usually 5- or 6-segmented (fig. 40) .... first instar
- a′. Antennae usually with 7 or more segments, occasionally with fewer than 5 (fig. 42) .... remaining immature instars

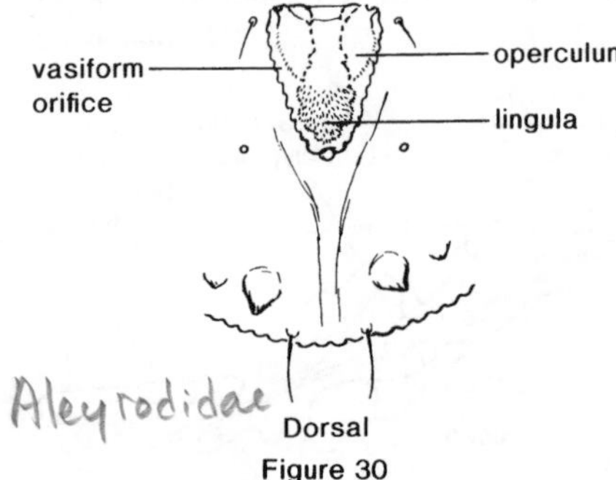

Dorsal
Figure 30

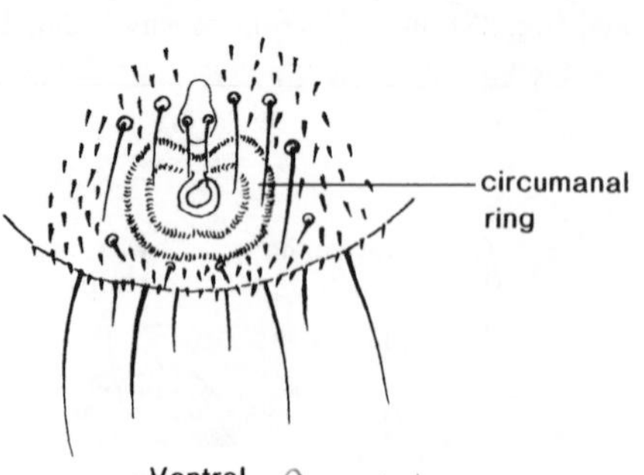

Ventral
Figure 31

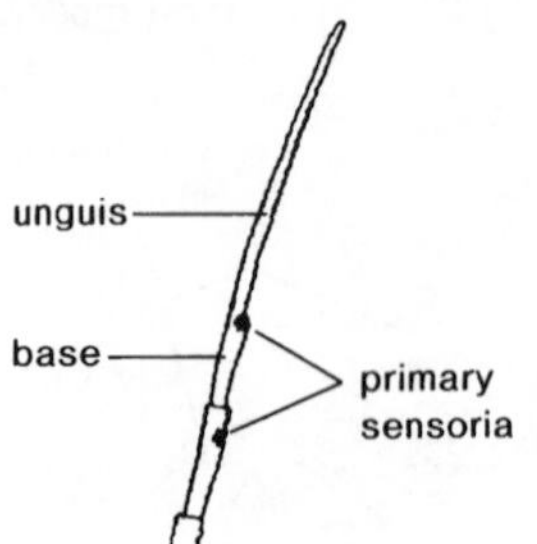

Figure 32

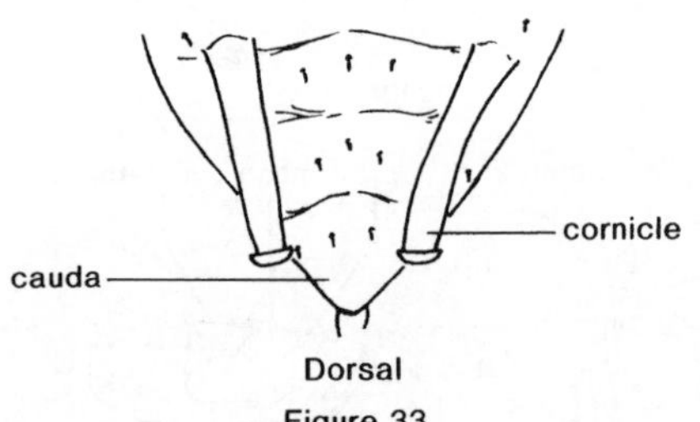

Dorsal
Figure 33

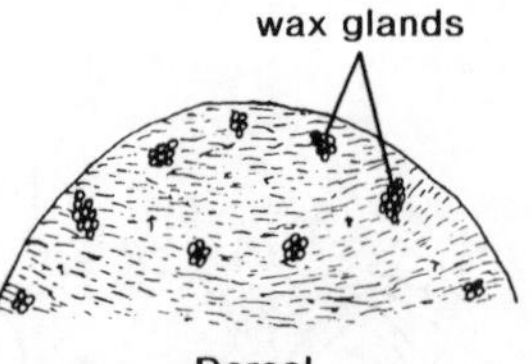

Dorsal
Figure 34

tubercles

Dorsal
Figure 35

Phylloxeridae

23(21′). Posterior abdominal segments coalesced into pygidium (fig. 43) 24
23′. Posterior abdominal segments separate, not coalesced into pygidium 25
24(23). At least one pair of lobes on pygidium (fig. 43); ocellar spot absent; legs, if present, without denticle on claw; without setae on femur (fig. 44) .... (armored scales) ***Diaspididae*** p. 106
- a. With legs; antennae 5- or 6-segmented (fig. 43) .... first instar
- a′. Without legs; antennae 1-segmented (fig. 45) .... second instar

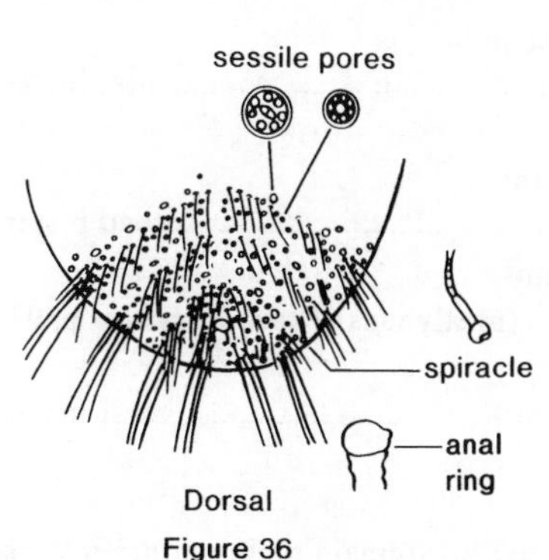

Figure 36

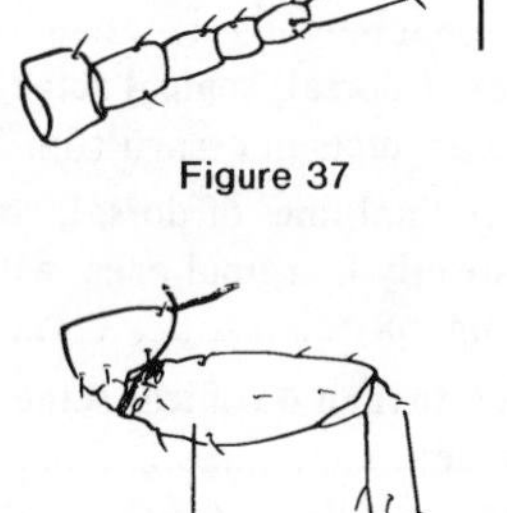
Figure 37

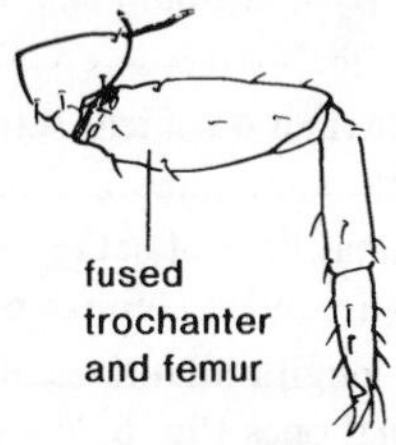

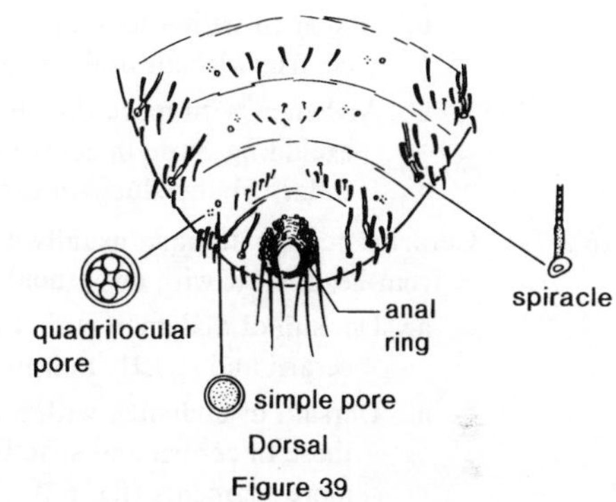

Figure 39

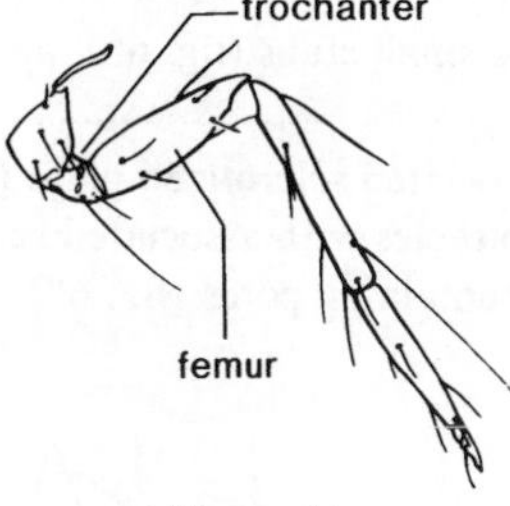

Figure 38

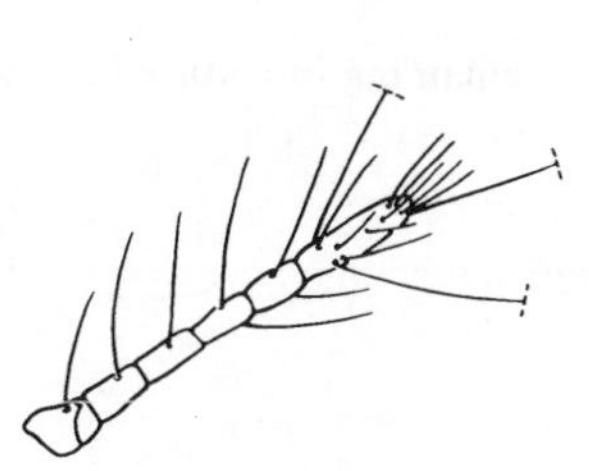
Figure 40

Figure 41

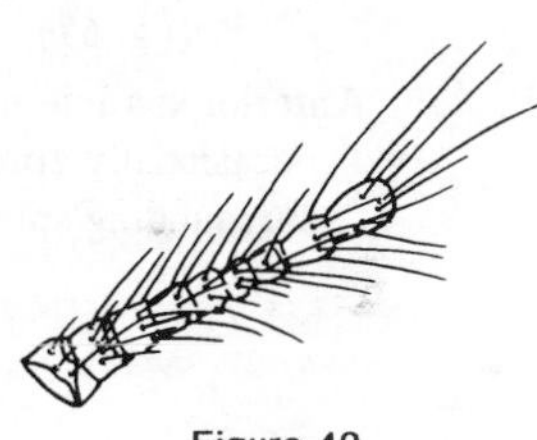
Figure 42

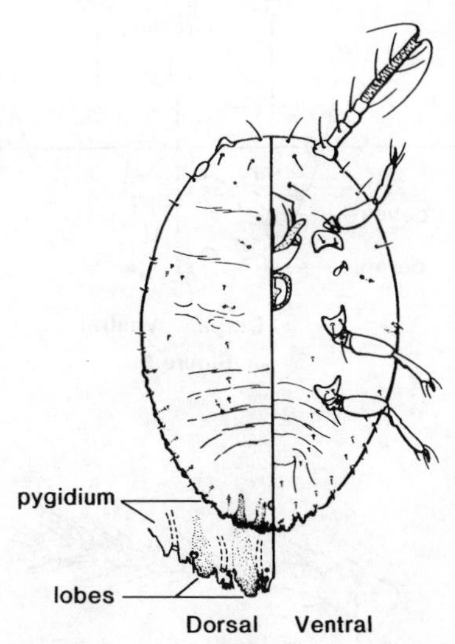

Figure 43

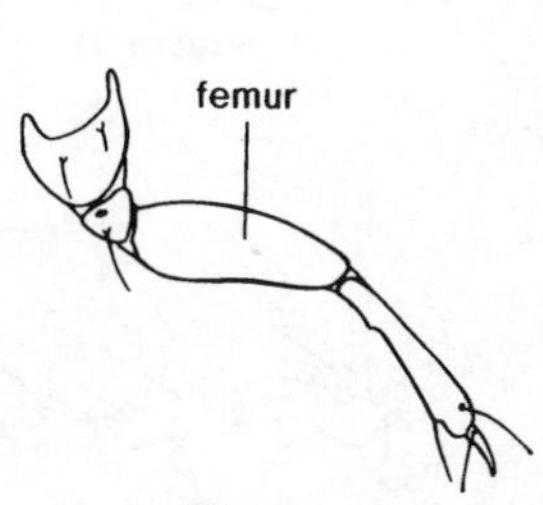

Figure 44

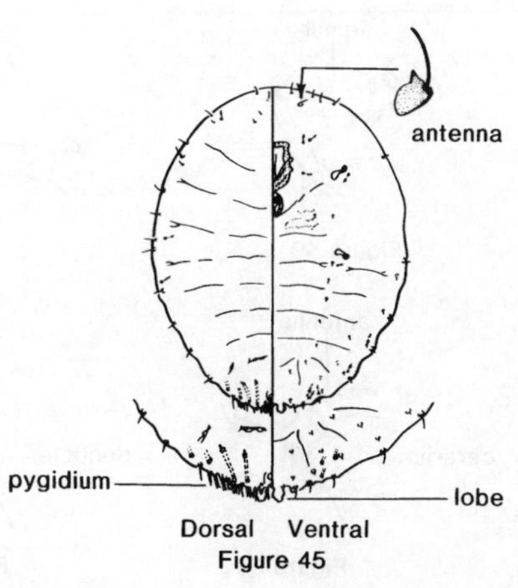

Figure 45

24′. Without lobes on pygidium (fig. 46); ocellar spot present (fig. 47); legs with small denticle on claw (fig. 48); with setae on femur (fig. 49) .... (false armored scales) ***Conchaspididae*** p. 105

- a. Antennae slender, 6-segmented (fig. 50); without scale cover .... first instar
- a′. Antennae broad, with fewer than 6 segments (fig. 51); with scale cover .... remaining immature instars

25(23′). Ostioles present; trilocular pores, cerarii, and circulus usually present (fig. 52) 26

25′. Ostioles, cerarii, and circulus absent; trilocular pores usually absent 27

26(25). Cerarius nearest antenna with 4 or more conical setae (fig. 53); claw with denticle (fig. 54) .... (giant mealybugs) ***Putoidae*** p. 92

- a. Antennae 7-segmented (fig. 55) .... b
- a′. Antennae 8-segmented (fig. 56) .... third instar female
- b. Abdomen with 4 longitudinal lines of dorsal, conical setae excluding those in cerarii; without oral-collar tubular ducts in cerarii (fig. 57) .... first instar
- b′. Abdomen with more than 6 longitudinal lines of dorsal, conical setae excluding those in cerarii and strictly marginal ones; with or without oral-collar tubular ducts in cerarii (fig. 58) .... second instar

26′. Cerarius nearest antenna usually with fewer than 4 conical setae or cerarii absent from head; claw with or without denticle .... (mealybugs) ***Pseudococcidae*** p. 93

- a. Dorsum of abdomen with 4 longitudinal lines of setae excluding those in cerarii and strictly marginal ones (fig. 59); antennae 6-segmented (fig. 60) .... first instar
- a′. Dorsum of abdomen with 6 or more longitudinal lines of setae excluding those in cerarii and strictly marginal ones (fig. 61); antennae with 6 or more segments (fig. 62) .... remaining immature instars

27(25′). Anterior spiracle with associated sclerotized plate (brachial plate) containing several pores (fig. 63) .... (lac scales) ***Tachardiidae*** p. 96

- a. Legs (fig. 64) and antennae (fig. 65) well developed .... first instar
- a′. Legs usually absent or reduced to small stubs (fig. 66), antennae reduced (fig. 67) .... remaining immature instars

27′. Anterior spiracle usually without an associated sclerotized plate (brachial plate); occasionally anterior and posterior spiracles with associated sclerotization surrounding spiracular opening and containing pores (fig. 68) 28

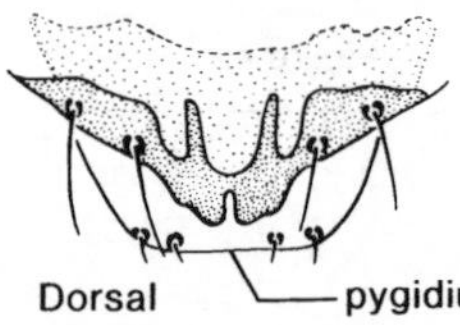

Dorsal — pygidium

Figure 46

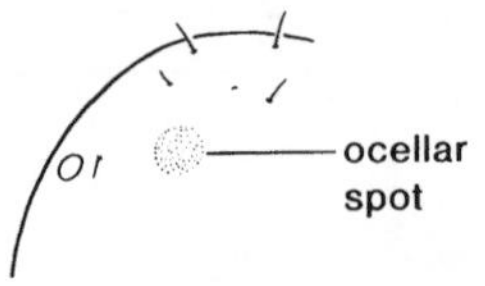

ocellar spot

Figure 47

denticle

Figure 48

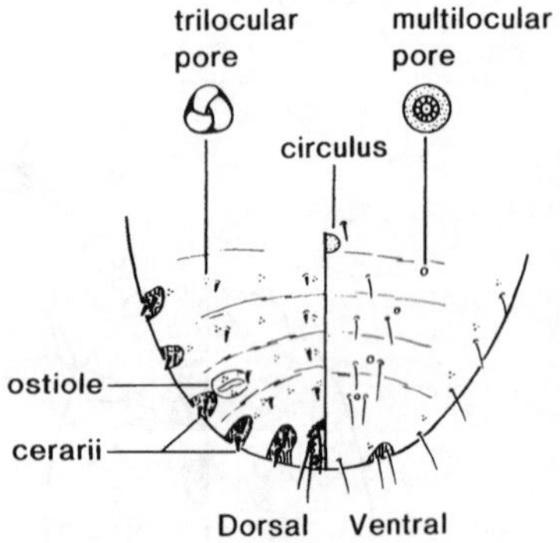

femur

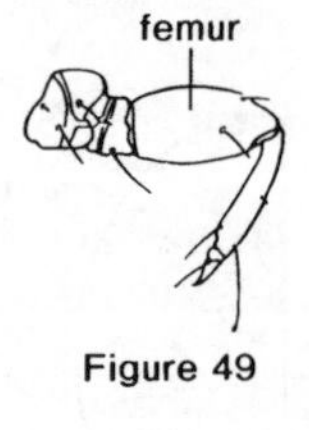

Figure 49

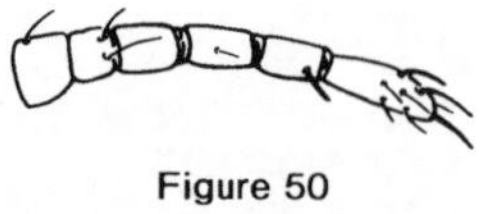

Figure 50

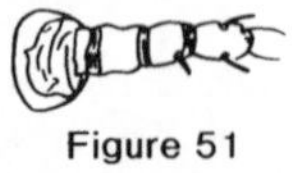

Figure 51

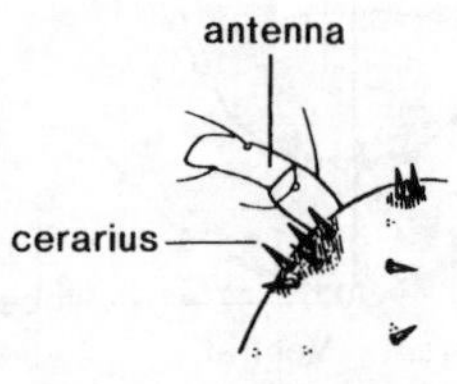

trilocular pore — multilocular pore — circulus — ostiole — cerarii — Dorsal — Ventral

Figure 52

antenna — cerarius

Figure 53

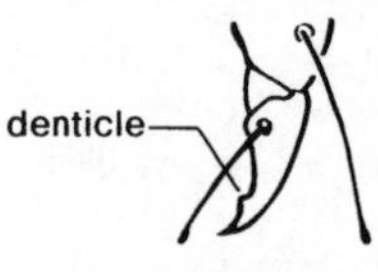

denticle

Figure 54

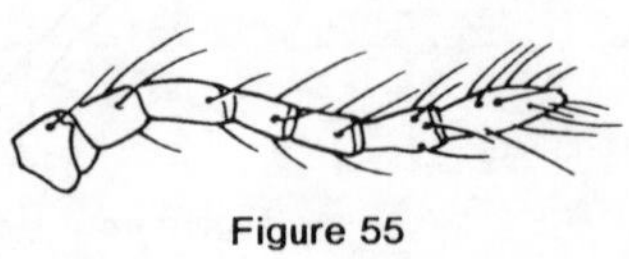

Figure 55

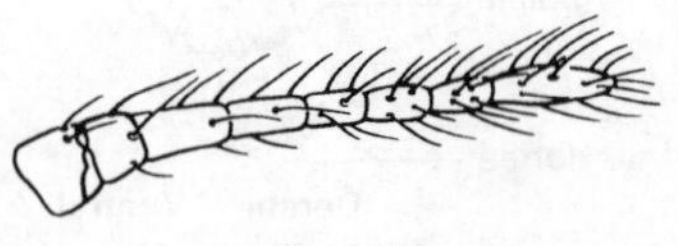

Figure 56

28(27′). Body with clusters of quinquelocular pores (fig. 69); enlarged setae truncate (fig. 69); body bright red, turning preservative red .... (cochineal scales) ***Dactylopiidae*** p. 98

a. Quinquelocular pores on dorsal surface of abdomen, and invaginated tubular ducts absent (fig. 70) .... first instar

a′. Quinquelocular pores scattered over dorsal surface, and invaginated tubular ducts present (fig. 71) .... b

b. Dorsal abdominal segments, excluding posterior segment, each with more than 10 truncate setae (fig. 71); invaginated tubular ducts associated with clusters of quinquelocular pores (fig. 71); antennae 6-segmented (fig. 72) .... second instar female

b′. Dorsal abdominal segments, excluding posterior segment, each with 7 to 9 truncate setae; invaginated tubular ducts scattered over dorsal surface, not associated with clusters of quinquelocular pores (fig. 73); antennae 7-segmented (fig. 74) .... second instar male

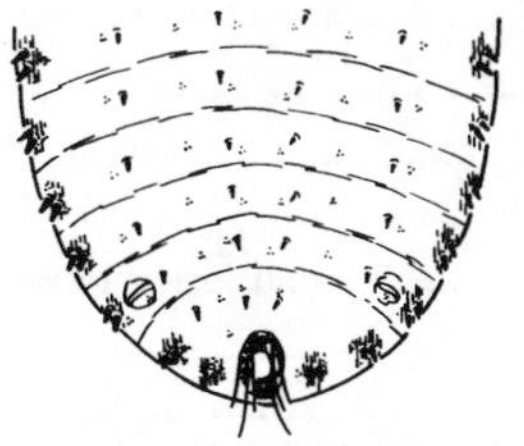

Dorsal
Figure 57

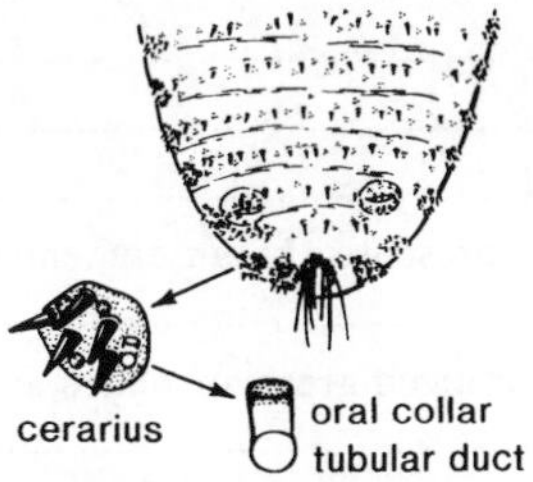

Dorsal
Figure 58

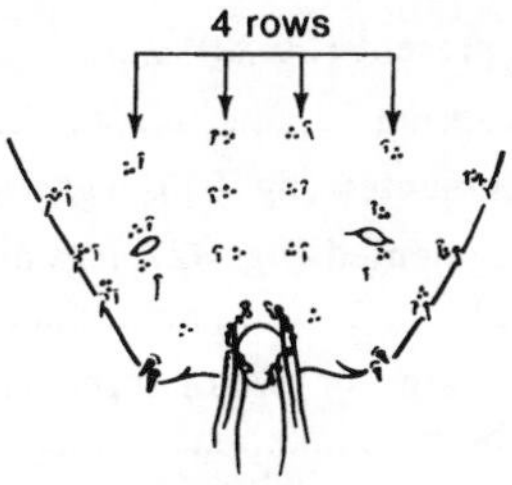

Dorsal
Figure 59

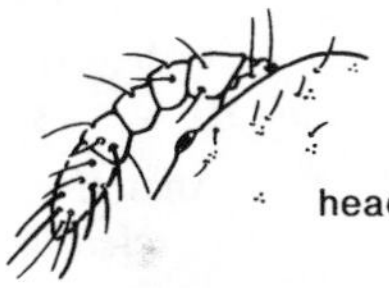

Dorsal
Figure 60

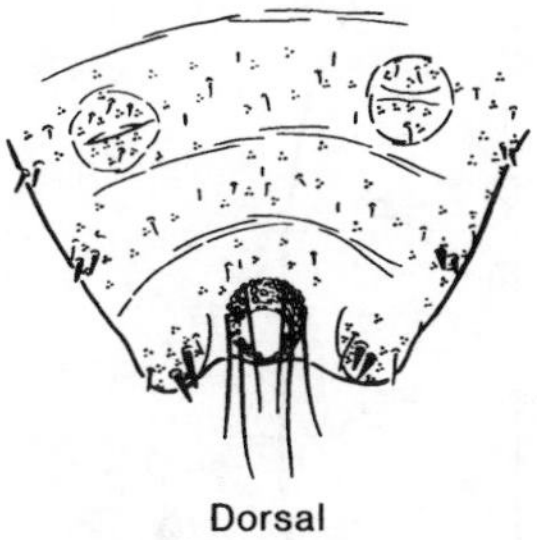

Dorsal
Figure 61

Figure 62

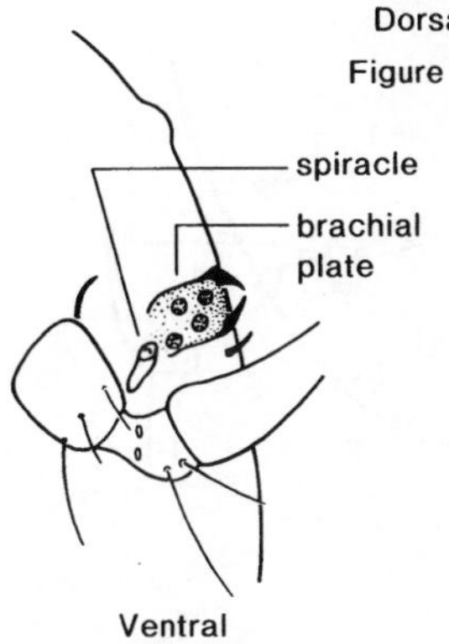

Ventral
Figure 63

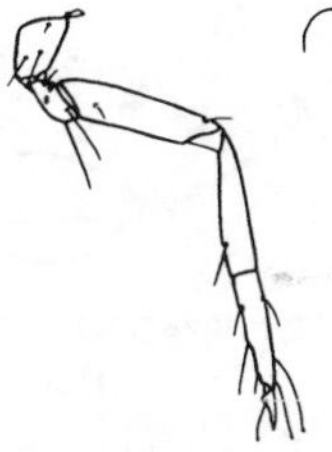

Figure 64

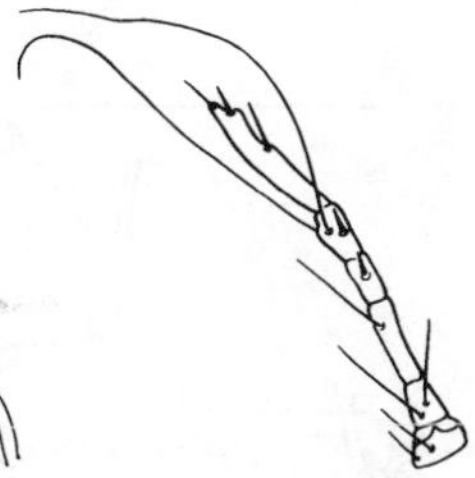

Figure 65

Figure 66

Figure 67

Figure 68

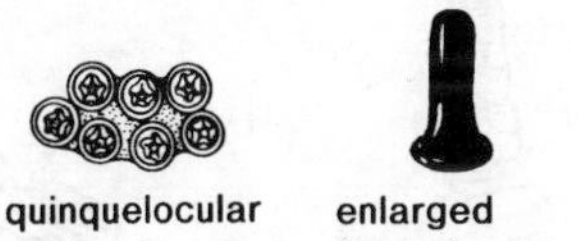

Figure 69

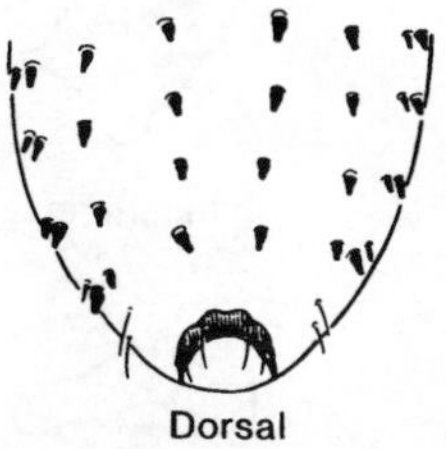

Dorsal
Figure 70

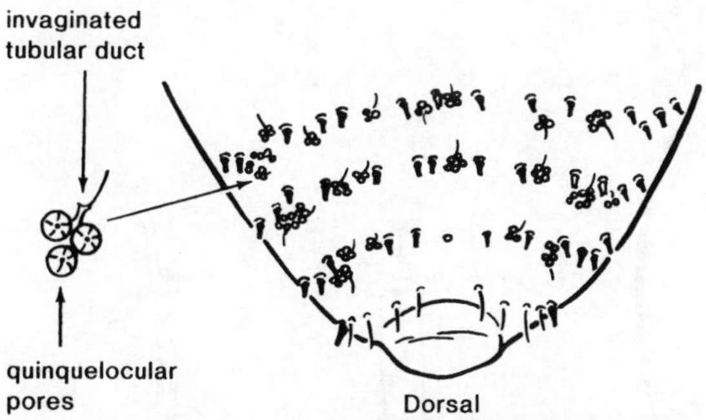

Dorsal
Figure 71

Figure 72

28′. Body without clusters of quinquelocular pores; enlarged setae, when present, rarely truncate; body rarely red, not turning preservative red 29

29(28′). Spiracles with at least one pore in atrium (fig. 76); enlarged setae present and restricted to marginal areas of body (fig. 75) .... (flat grass scales) ***Aclerdidae*** p. 97

- a. Legs and antennae segmented; spiracles each with one or two pores in atrium (fig. 76) .... first instar
- a′. Legs and antennae absent or represented by unsegmented stubs (fig. 77); spiracles each with 5 or more pores in atrium (fig. 68) .... remaining immature instars

29′. Spiracles without pores in atrium; enlarged setae usually absent except in some Eriococcidae; when present, generally not restricted to marginal areas of body 30

30(29′). With 8-shaped pores (fig. 78) 31

30′. Without 8-shaped pores; occasionally with 8-shaped tubular ducts (fig. 79) 33

31(30). Anal area with single plate between anal lobes (fig. 78) .... (ornate pit scales) ***Cerococcidae*** p. 102

- a. Tubular ducts present (fig. 80) .... second instar male
- a′. Tubular ducts absent .... b
- b. Antennae 6-segmented (fig. 81); legs segmented .... first instar
- b′. Antennae unsegmented (fig. 82); legs abortive, represented by unsegmented stubs (fig. 83) .... second instar female

31′. Anal area with three plates, or two or three narrow sclerotized areas, or plates and sclerotized areas absent 32

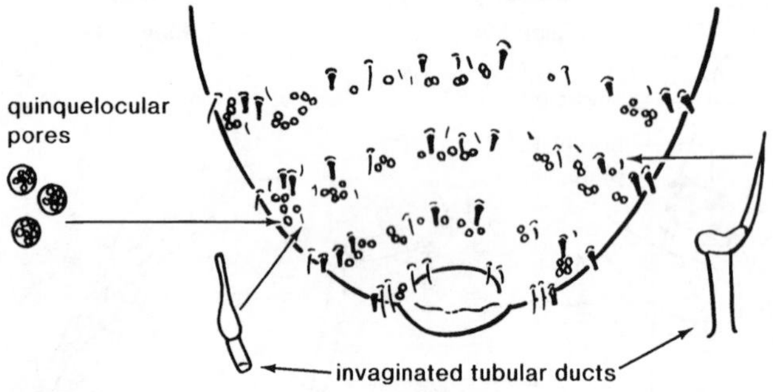

Figure 73

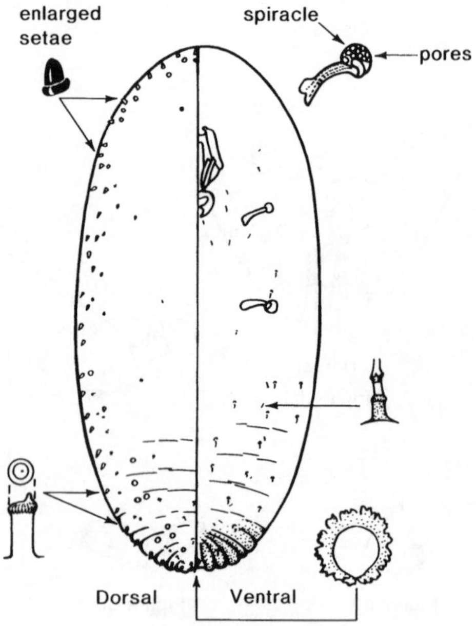

Figure 74

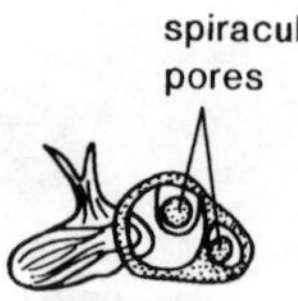

Figure 76

Figure 77

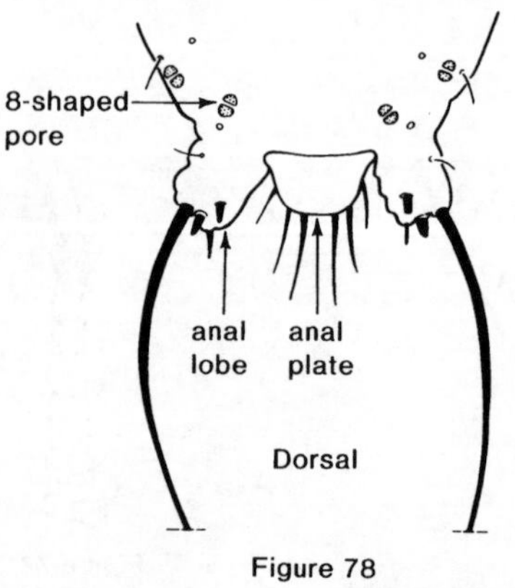

Figure 75

Figure 78

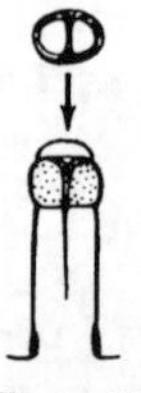

Figure 79

Figure 80

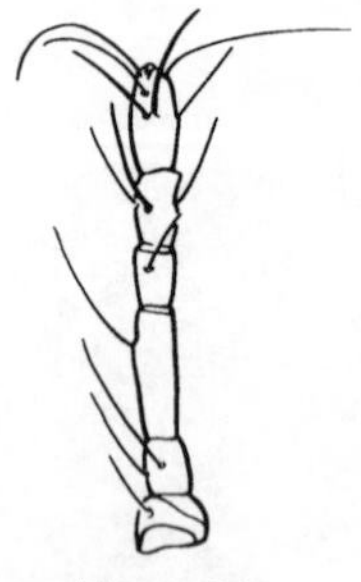

Figure 81

Figure 82

Figure 83

32(31′). Anal area with three sclerotized plates, anterior plate represented by single narrow arch plate, posterior plates represented by two broad sclerotizations (fig. 84); spiracular setae normally near body margin laterad of anterior spiracle (fig. 85) .... (false pit scales) ***Lecanodiaspididae*** p. 101
- a. Tubular ducts present (fig. 86) .... second instar male
- a′. Tubular ducts absent .... b
- b. Legs segmented .... first instar
- b′. Legs represented by small, unsegmented stubs (fig. 87) .... second instar female

32′. Anal area without sclerotized plates, occasionally with two or three narrow sclerotized areas (fig. 88); without modified spiracular setae .... (pit scales) ***Asterolecaniidae*** p. 103
- a. Legs well developed, segmented .... first instar
- a′. Legs absent or represented by unsegmented stubs .... b
- b. Tubular ducts present (fig. 89) .... second instar male
- b′. Tubular ducts absent .... second instar female

33(30′). With anal operculum (fig. 90); often with differentiated spiracular setae and row of pores from spiracle to body margin (fig. 90) .... (soft scales) ***Coccidae*** p. 94
- a. Anal opercula each with apical seta at least as long as length of operculum (fig. 91) .... first instar
- a′. Anal opercula with setae shorter than length of operculum (fig. 92) .... b
- b. Line of invaginated tubular ducts usually present near body margin (fig. 93) .... second instar male
- b′. Invaginated tubular ducts absent near body margin .... second and third instar females

33′. Without anal opercula; without differentiated spiracular setae; usually without row of pores from spiracle to body margin 34

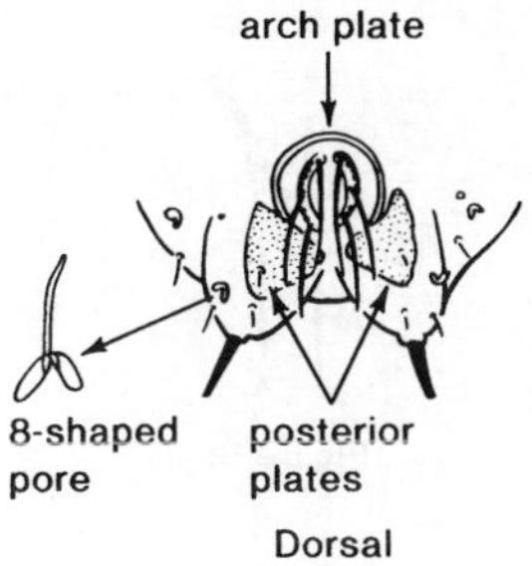

Figure 84

Figure 85

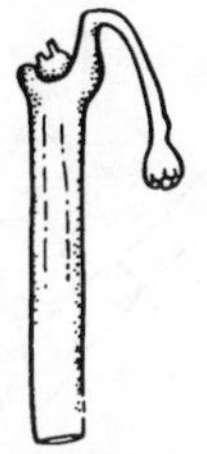

Figure 86

Figure 87

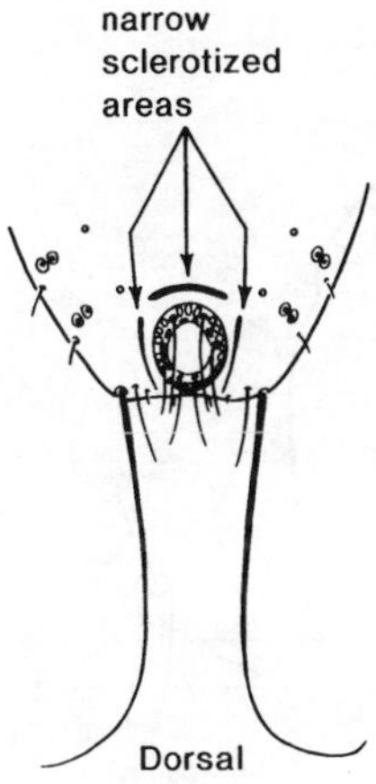

Figure 88

Figure 89

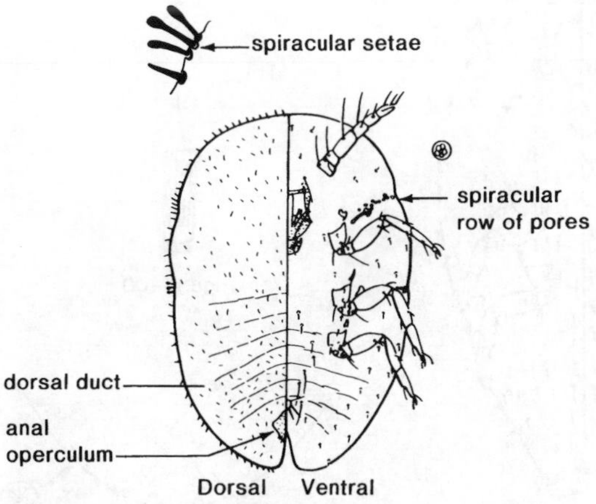

Figure 90

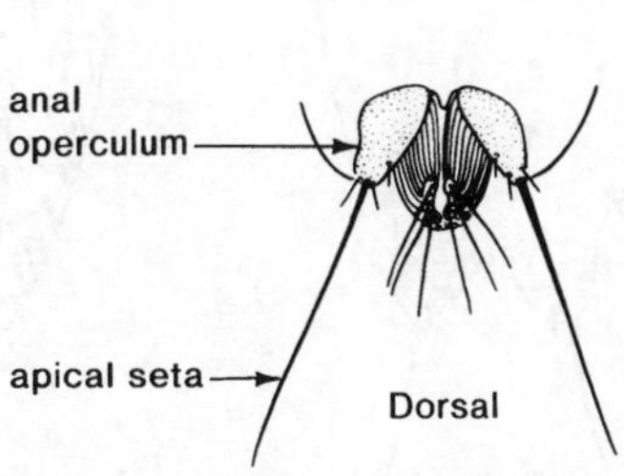

Figure 91

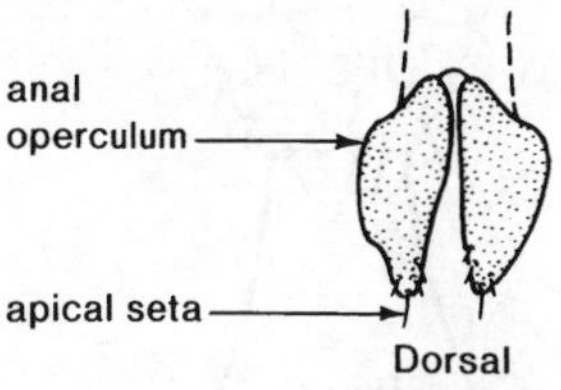

Figure 92

34(33′). With 8-shaped tubular ducts (fig. 79) .... (palm scales) ***Phoenicococcidae*** p. 104
- a. Legs well developed (fig. 94) .... first instar
- a′. Legs absent or represented by unsegmented stubs (fig. 95) .... b
- b. Legs absent .... second instar females
- b′. Legs represented by small stubs (fig. 95) .... second instar males

34′. Without 8-shaped tubular ducts .... (felt scales) ***Eriococcidae*** and (gall-like scales) ***Kermesidae*** p. 99
- a. Tubular ducts absent (excluding microtubular ducts) .... b
- a′. Tubular ducts present (fig. 96) .... d
- b. Four dorsal setae on each abdominal segment excluding posterior segment (fig. 97) .... first instar, ***Kermesidae*** p. 99
- b′. Six or more dorsal setae on each abdominal segment excluding posterior segment (fig. 98) c
- c. Six dorsal setae on each abdominal segment (fig. 98) .... first instar, ***Eriococcidae*** p. 101
- c′. More than 6 dorsal setae on some abdominal segments (fig. 99) .. second instar female, ***Eriococcidae*** p. 101
- d. Microtubular ducts present (fig. 100) .... second instar male, ***Eriococcidae*** p. 101
- d′. Microtubular ducts absent .... e
- e. Without protruding anal lobes (fig. 101) .... third instar female, ***Kermesidae*** p. 99
- e′. With protruding anal lobes (fig. 102) .... f
- f. Antennae 7-segmented (fig. 103); some tubular ducts on dorsum same size as those on venter (fig. 104) .... second instar males, ***Kermesidae*** p. 99
- f′. Antennae 6-segmented (fig. 105); dorsal tubular ducts absent, larger, or shaped differently than tubular ducts on venter (fig. 106) .... second instar females, ***Kermesidae*** p. 99.

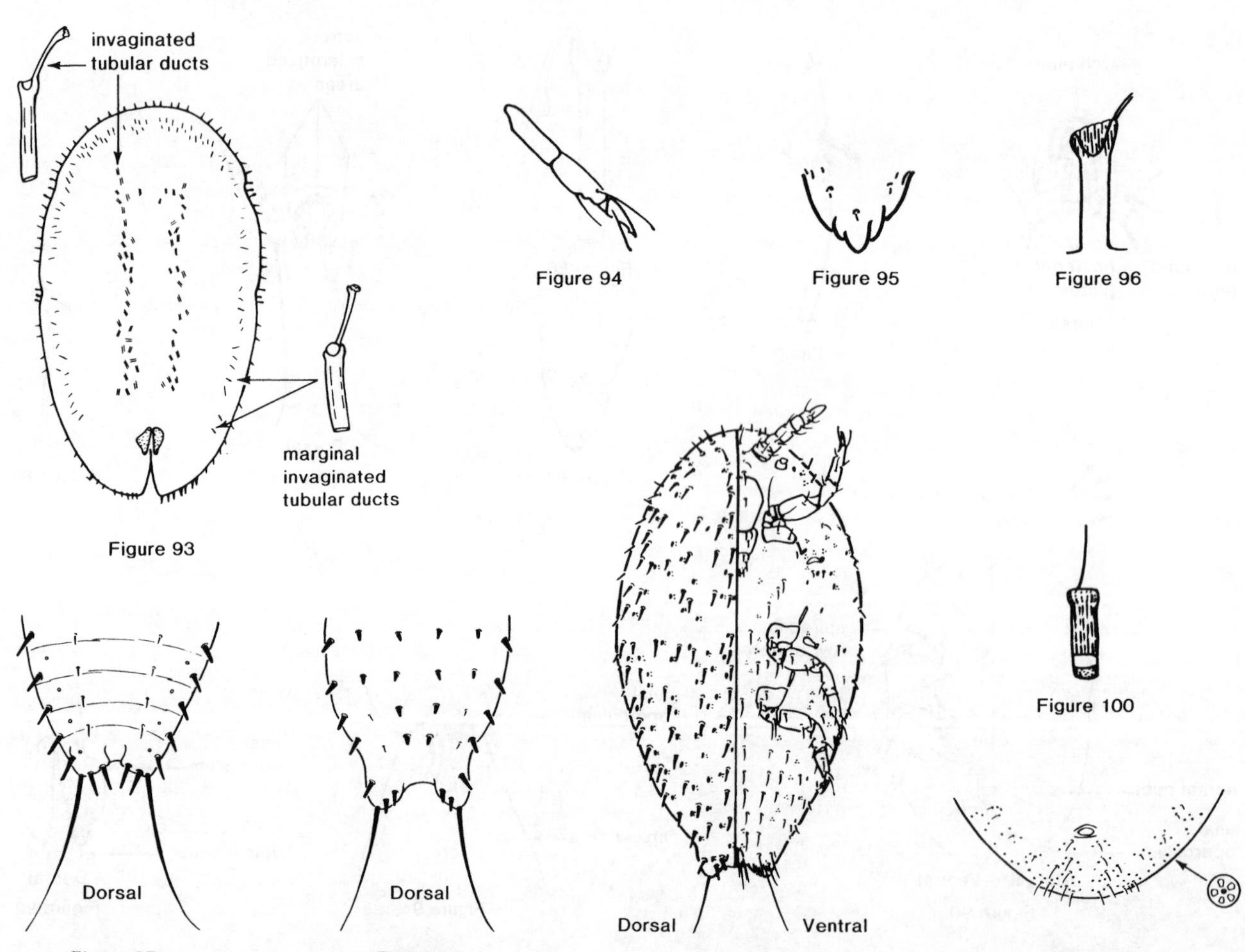

Figure 93

Figure 94

Figure 95

Figure 96

Figure 97

Figure 98

Figure 99

Figure 100

Figure 101

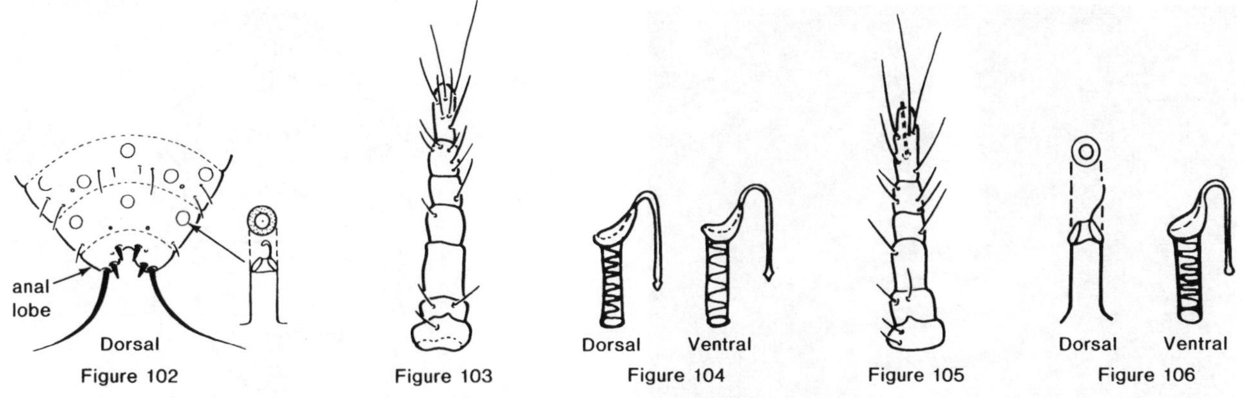

Figure 102 Figure 103 Figure 104 Figure 105 Figure 106

SUBORDER AUCHENORRHYNCHA

Lois B. O'Brien, *Florida A & M University*

In general, little work has been done on the immatures of this suborder except for economically important species. This is the first attempt to key the immatures of these families of the Fulgoroidea, although Vilbaste (1968) keys the 5 northern European families (Tettigometridae, Cixiidae, Delphacidae, Achilidae, and Issidae). The higher classification of the Auchenorrhyncha is also unstable; current usage favors the superfamilies as given here.

Auchenorrhyncha may be separated from Sternorrhyncha as discussed earlier.

Metcalf has published a bibliography (1942) and an excellent catalogue (1932–1968) of the Auchenorrhyncha in which all known references are listed and characterized by the following terms: listed, key, described, illustrated, ecology, economics, food plants, parasites, disease vector.

SUPERFAMILY FULGOROIDEA

The Planthoppers, Lanternflies or Fulgoroids

The Fulgoroidea are a diverse group of 20 families which are grouped together because of the shape and location of the antennae and ocelli and the spination of the hind tibiae and tarsi. They feed on plant juices and the immatures and many adults secrete wax filaments. Fifteen families occur in the New World, 11 in the U.S. and 7 or 8 of these in Canada.

Most temporate species are not economically important, but some tropical species are important pests of monocots, especially rice, wheat, corn, and palms. Several Delphacidae transmit virus diseases and some Cixiidae are vectors of MLO (mycoplasma-like organism) diseases, such as *Myndus crudus* Van Duzee, the vector of lethal yellowing of palms. Brewbaker (1979) suggests sustained corn crop failures caused by a virus transmitted by *Peregrinus maidis* Ashmead may have been the cause of the fall of the Mayan empire in the ninth century, but Nault (1983) believes *P. maidis* is of African origin on sorghum and was introduced to the New World in post-Columbian times. Immatures have rarely been studied, except for a few of these economic species. Usually there are five instars. Metcalfe (1968) lists five characters that change in the Delphacidae in progressive instars; these changes also occur in other U.S. fulgoroids (Tropiduchidae, Fulgoridae, Achilidae, and Derbidae have not been available for this sort of study). These characters are increasing size, differentiating wing buds (much less pronounced in brachypterous species), increasing number and size of spines of the metatibiae and tarsi, changes in pigmentation, and increasing number of sensory receptors on the antennae.

Most fulgoroid immatures have wax secreting plates from which tufts several millimeters long are extruded (fig. 30.1). The rest of the body also produces smaller amounts of wax. All of this may be brushed off easily and usually it is not present in preserved specimens. Tsai and Kirsch (1978) say the underground immatures of *Myndus crudus* Van Duzee [Cixiidae] "wave their abdomens back and forth around the nest area until all of the cottony material is brushed off the tips of their abdomens. They then pack the material down on the grass stolons and roots to form a "nest", remaining in one "nest", usually in groups of 2–10 individuals". Cumber (1952) says this material waterproofs the nest. Often eggs of Fulgoroidea are covered with the wax from females.

A few genera are worldwide in distribution (*Oliarus* and *Cixius* in the Cixiidae, *Delphacodes* in the Delphacidae), but these are rare. Cixiids and delphacids are good colonizers (*e.g.*, in the Hawaiian Islands, and they and other families in the West Indies), and pest species have been widely spread. Derbidae, Fulgoridae, and Tropiduchidae are primarily tropical. Issidae are widely distributed in, but not restricted to, deserts and areas of low summer rainfall (Mediterranean climates). Kinnaridae are not common, but are found in the Southwest, Mexico, and the West Indies. The other families are well distributed in temperate as well as tropical areas.

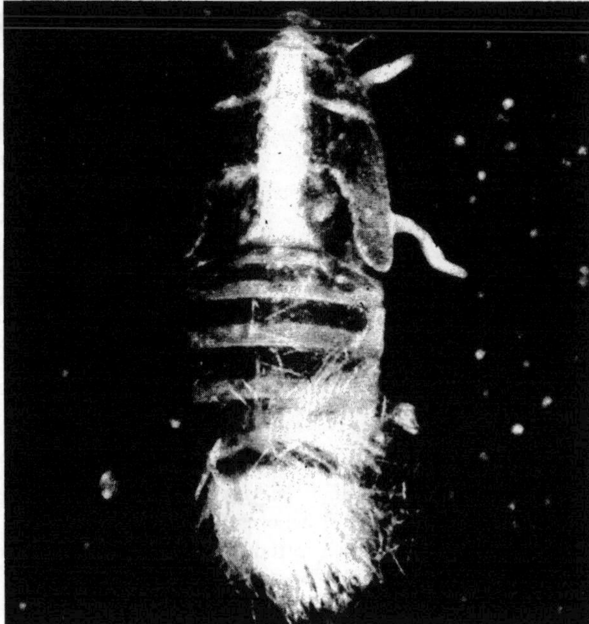

Figure 30.1 Fulgoroid immature, Cixiidae, fifth instar, *Myndus crudus* (Van Duzee) with wax. Courtesy of J. Tsai, University of Florida.

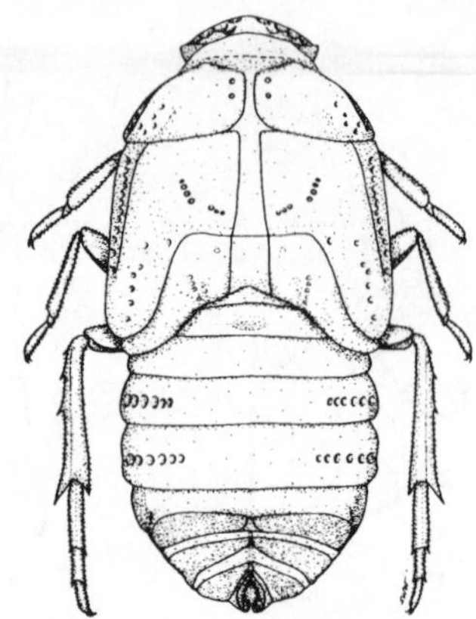

Figure 30.2. Cixiidae. *Oliarus placidus* Van Duzee

CIXIIDAE

The Cixiids

Figures 30.1, 30.2, Key Figures 9, 20

Relationships and Diagnosis: The cixiids are considered the most primitive Fulgoroidea, in part because adults often have all 3 ocelli. The immatures may be distinguished from other Fulgoroidea by the large wax-producing plates on the 6th, 7th, and 8th abdominal terga. All 3 families of fulgoroids (Cixiidae, Achilidae, and Derbidae) with immatures that live in litter, under bark, or underground, have eyes reduced in size. It is possible that when kinnarid immatures are known they will be similar to the cixiids, as are the adults. Fennah (pers. comm.) says they feed on roots underground in Trinidad. He did not preserve any. Myers (1929b) described an immature cixiid, *Bothriocera signoreti* Stål, with fossorial front legs, but the tarsi are attached to the tip of the tibiae, not before the tip as in cicadas.

Biology and Ecology: The immatures live in cracks underground (Hacker, 1925) or in galleries among dead leaf litter (Cumber, 1952), feeding on roots. The wax they exude is highly water repellent and lines the areas where they live. The wax also covers them, so that, when flooded, they are surrounded by an air bubble. Cumber (1953) reports that perhaps 50% of *Oliarus atkinsoni* Myers in New Zealand can withstand flooding for 3 days, although 14 days submergence will kill them. At least one species has a 2 year life cycle (Cumber, 1953). Some species deposit eggs in the soil, but I have seen eggs of *Oecleus nolinus* Ball and Klingenberg inserted near the leaf tips of *Agave lechuguilla* in Texas. In either case, they are covered with wax. Immatures are also found in ant nests (Myers, 1929b). There are blind cave-dwelling species in Hawaii, Mexico, and New Zealand (Fennah, 1973).

Comments: Metcalf (1936) lists 84 genera and 786 species in the world. Including recent revisions (Kramer 1983), more than 13 genera and 172 species are now recorded for the U.S. and 22 of these species are found in Canada. Three species have been found to be MLO (mycoplasma-like organism) vectors (O'Brien and Wilson 1985). None has been considered to be economically important here until *Myndus crudus* was investigated as a probable vector of "lethal yellowing" of the coconut palm in Florida and the West Indies (Tsai et al., 1976).

Selected Bibliography

Cumber 1952, 1953.
Fennah 1973.
Hacker 1925.
Ishihara 1969.
Kramer 1983.
Metcalf 1936.
Myers 1929b.
Tsai and Kirsch 1978.
Tsai et al. 1976.

DELPHACIDAE

The Delphacids

Figure 30.3, Key Figure 16

Relationships and Diagnosis: The Delphacidae are a well-defined family of small Fulgoroidea distinguished by the movable spur at the apex of the hind tibia in both immatures and adults. Subfamilies have been designated based on the shape of the spur.

Biology and Ecology: All immatures are exposed feeders, primarily on grasses and herbaceous plants. Eggs are laid in slits in leaves and stems, 1–12 being placed in each chamber. The scar is covered with wax. Up to six generations a year have been recorded. Delphacidae are excellent colonizers, as shown by their presence on oceanic islands and the wide distribution of pest species.

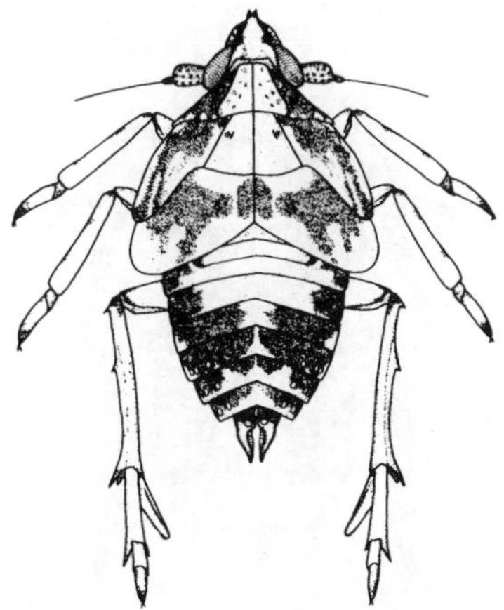

Figure 30.3. Delphacidae. *Megamelus davisi* Van Duzee

Comments: The family is the largest of the Fulgoroidea in number of species and contains most of the economically important ones, some of which carry virus diseases. Metcalf (1943) gives 137 genera and 1114 species for the world. About 42 genera and 300 species occur in the U.S. A modern key to genera and species of adults is needed.

The sugarcane leafhopper, *Perkinsiella saccharicida* Kirkaldy, was very destructive in Hawaii in 1903–4. Its subsequent control there is one of the classic examples of biological control. It also transmits Fiji disease to sugarcane. The corn planthopper, *Peregrinus maidis* Ashmead, is the vector of maize stripe virus and maize mozaic virus in the Americas (Nault, 1983). The rice delphacid, *Sogatodes orizicola* (Muir), transmits Hoja Blanca virus. Florida, Mississippi, and Louisiana were formerly infested, but have been considered free of the pest since 1959 (Crane and Paddock, 1974).

Selected Bibliography

Brewbaker 1979.
Crane and Paddock 1974.
Ishihara 1969.
Metcalf 1943.
Van Dine 1911.

DERBIDAE

The Derbids

Figure 30.4

Relationships and Diagnosis: The Derbidae are grouped with the Cixiidae, Delphacidae, Kinnaridae, Dictyopharidae, Fulgoridae, and Achilidae, which all have a row of spines on the second posterior tarsal segment in the adults. Some derbids lack these spines as immatures. The derbids, as do fulgorids and dictyopharids, have lateral carinae on the clypeus. Last instar derbids have a very marked ecdysial membrane on the dorsal midline of the thorax and abdomen dividing each tergum into 2 tergites. The other families have this in varying but much lesser degree.

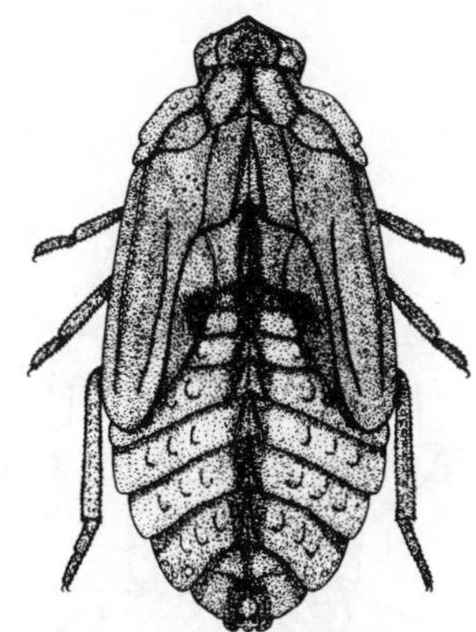

Figure 30.4. Derbidae. *Apache degeerii* Kirby

Biology and Ecology: Immatures are found in chambers in rotting wood or under bark. They are believed to feed on fungi. Adults of many species are found on monocotyledonous plants from grasses to palms, but other plants are utilized as well.

Comments: Metcalf (1945) lists 111 genera and 733 species. Fourteen genera and 62 species are found in the U.S., mostly in the South and Southwest. None is considered economically important.

Selected Bibliography

Metcalf 1945.

KINNARIDAE

The Kinnarids

Relationships and Diagnosis: The immatures have not been described. Fennah (pers. comm.) found them living underground in Trinidad, feeding on roots. No specimens were preserved.

Comments: Metcalf (1945) lists 8 genera and 42 species. The largest numbers in the New World occur in the West Indies. One genus, *Oeclidius,* with six species, is found in the southwestern U.S. They have not been economically important.

Selected Bibliography

Metcalf 1945.

DICTYOPHARIDAE

The Dictyopharids

Figure 30.5, Key Figures 17, 23, 24

Relationships and Diagnosis: One tribe of Dictyopharidae, the Orgeriini, contains many species in the western U.S. Wings are reduced and the adults look very much like flattened, round, brown immatures. This tribe does not have wax-bearing plates on the 7th and 8th abdominal terga in the immatures. The other species found in the U.S. have either round

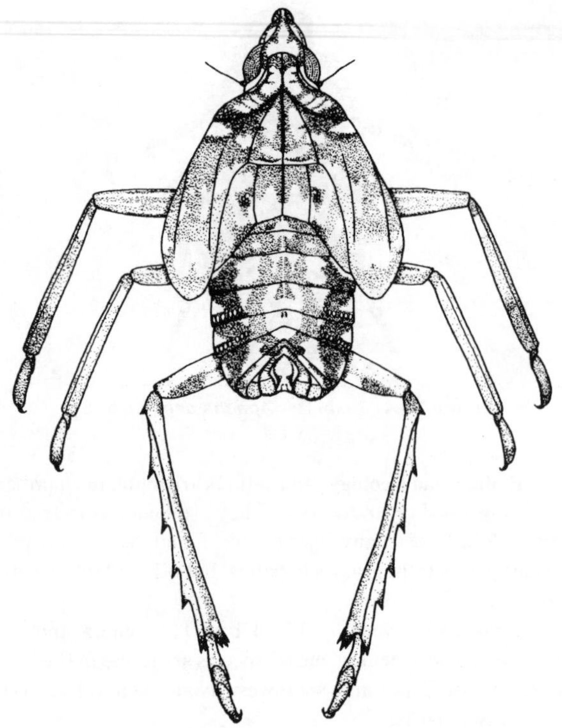

Figure 30.5. Dictyopharidae. *Nersia florens* Stäl

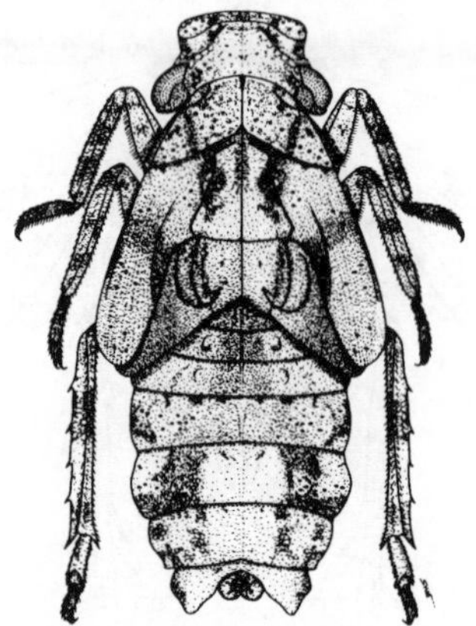

Figure 30.6. Fulgoridae.

or oval wax-bearing plates. Fulgoridae also have oval wax-bearing plates, but the long axis of one oval is vertical in dictyopharids; both are horizontal in fulgorids (key fig. 22). All species have lateral carinae on the clypeus, as do fulgorids and derbids.

Biology and Ecology: Immatures and adults feed primarily on herbs.

Comments: Metcalf (1946) records 119 genera and 489 species. Fifteen genera and 76 species occur in the U.S. The genus most widely distributed in the U.S., *Scolops,* is easily identified by the projection of the head (key fig. 24). Most species are not considered to be economically important.

Selected Bibliography

Metcalf 1946.

FULGORIDAE

The Planthoppers, Lanternflies, Fulgorids

Figure 30.6, Key Figure 22

Relationships and Diagnosis: The Fulgoridae contains species as large as the cicadas and as varied as the membracids. Often they have long, bizarrely-shaped projections of their heads which were formerly thought to be luminescent (*see* the introduction to Homoptera). Dictyopharids may also have head projections, as may delphacids, cixiids, and cicadellids in other parts of the world. Immatures may be separated from other Fulgoroidea by a combination of 3 characters, the many spines on the second segment of the posterior tarsi, the lateral carinae on the clypeus, and the 2 pairs of wax-producing plates on the 7th and 8th abdominal terga directed caudad, with the plates oval and the long axis horizontal. Dictyopharids share the first two characters, but either have no wax plates or three pairs, or if only two, then one pair is round or if both pairs are oval, then one is oriented with the long axis vertical.

Biology and Ecology: Immatures and adults both prefer to feed through the thick bark of the trunks or branches of trees or shrubs, according to Kershaw and Kirkaldy's studies of *Pyrops* (1910). Eggs are glued to the tree trunk or branch in rows, covered with colleterial fluid to form an oötheca, and rubbed with wax. The immatures sit gregariously in long and fairly ordered rows on the bark, although they tend to disperse as they grow older. *Pyrops* spp. have one generation a year and overwinter as immatures in south China.

Comments: The family is tropical, with 108 genera and 543 species in the world (Metcalf, 1947a). Most species are not economically important. However, the peanut bug, *Fulgora* spp., which is found from Mexico to Argentina, was feared by the Indians in the Amazon jungles. It was rumored to fly a zig-zag path through the forest, killing all it touched. If one was caught and proven to be harmless, the natives said that it, of course, was not the insect they were talking about (Branner, 1885). Eight genera and 17 species occur in the U.S., mostly in the Southwest.

Selected Bibliography

Branner 1885.
Kershaw and Kirkaldy 1910.
Metcalf 1947a.
Ridout 1983.

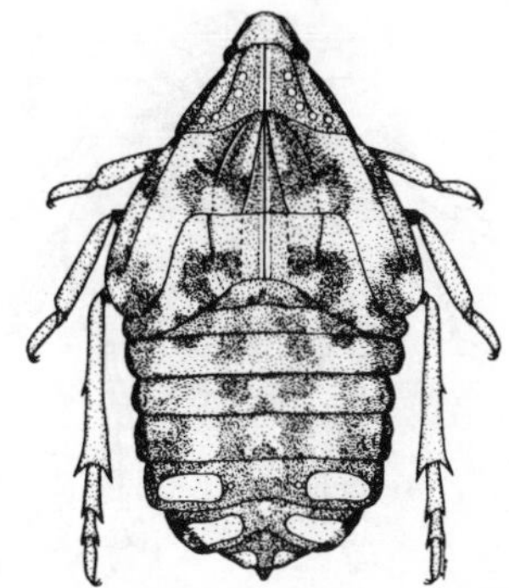

Figure 30.7. Achilidae. *Epiptera opaca* Say

ACHILIDAE

The Achilids

Figure 30.7, Key Figure 21

Relationships and Diagnosis: The Achilidae are the last of the five families with the second posterior tarsal segment armed with a row of spines at the apex. The eyes of the immatures are reduced and the wax-bearing plates cover about half the width of the tergites.

Biology and Ecology: Most species are found in wooded areas. Immatures are reported to feed on fungi, either under bark or in leaf litter. Fletcher (1979) got Australian achilids to lay eggs when they were provided with bits of bark. Each egg was coated with bark bits, then dropped into the leaf litter.

Comments: Metcalf (1947b) lists 77 genera and 224 species. Eight genera and 56 species occur in the U.S. The immatures most frequently seen in the U.S. and Canada are those of *Epiptera,* which are large, up to 7 mm, and found under bark of downed trees. *Epiptera* is a northern genus usually associated with conifers. Ten of the 14 species are found in Canada.

Selected Bibliography

Fletcher 1979.
Metcalf 1947b.

TROPIDUCHIDAE

The Tropiduchid Planthoppers

Figure 30.8, Key Figure 19

Relationships and Diagnosis: The Tropiduchidae are one of the group of families (including Flatidae, Issidae, and Acanaloniidae in the U.S.) with a single spine on each side of the apex of the 2nd posterior tarsal segment. Wax-bearing plates are absent, and the 2nd posterior tarsal segment is narrow, not flared apically. The rostrum usually does not reach the hind trochanters.

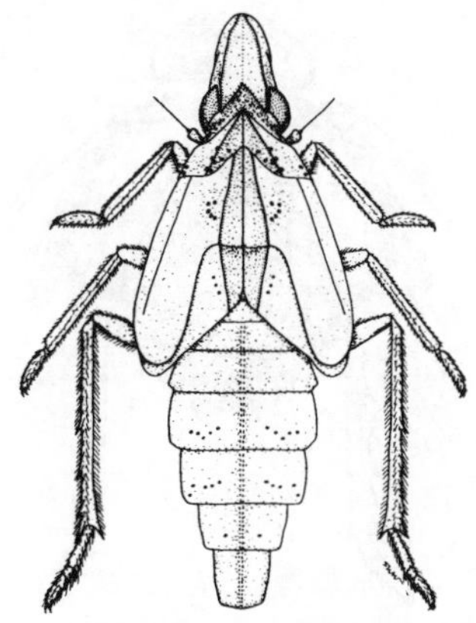

Figure 30.8. Tropiduchidae.

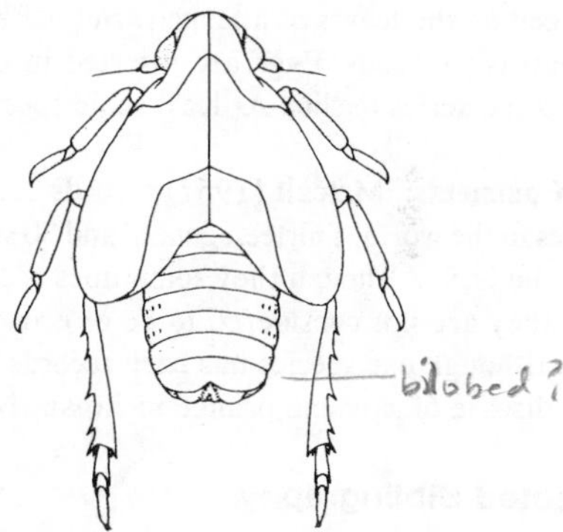

Figure 30.9. Flatidae. *Metcalfa pruinosa* Say

Biology and Ecology: Immatures can be beaten from the same bushes or trees on which adults are found, including *Coccoloba,* sea grape.

Comments: Metcalf (1954a) lists 106 genera and 280 species in the world. Two genera and two species have been reported from Florida; one of these is found in Mississippi and North Carolina also. Many genera and species occur in the West Indies. They are not economically important.

Selected Bibliography

Metcalf 1954a.

FLATIDAE

The Flatid Planthoppers

Figure 30.9, Key Figures 10, 25, 26

Relationships and Diagnosis: The Flatidae may be distinguished by the bilobed 6th tergum extending over the wax-bearing plates, concealing them from dorsal view (key figs. 25, 26). In some species, the indentation is shallow, and they might be confused with issids, whose 6th tergum is chevron-shaped or rectangular.

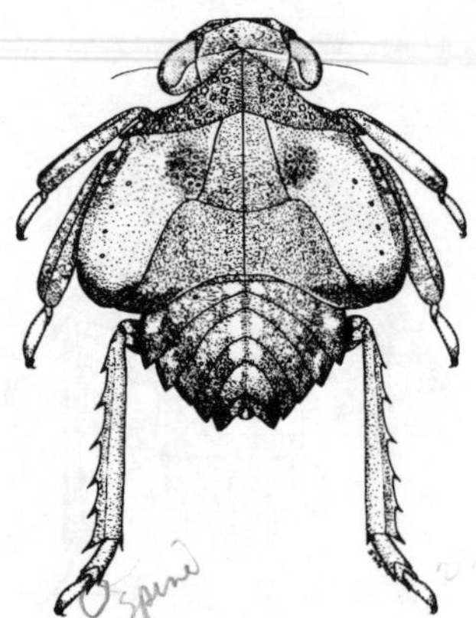

Figure 30.10. Issidae. *Thionia,* probably *bullata* Say

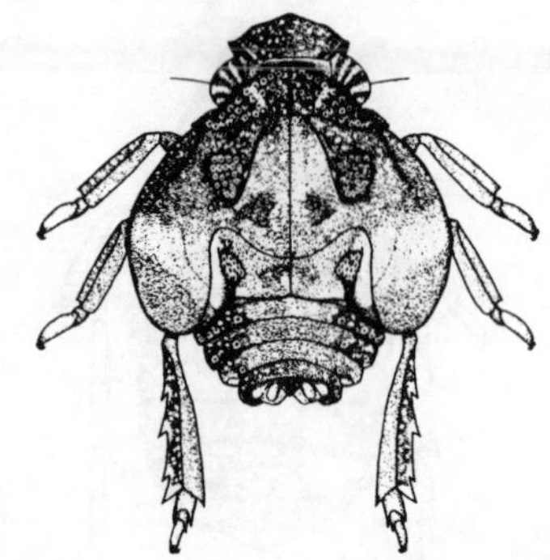

Figure 30.11. Acanaloniidae. *Acanalonia conica* Say

Biology and Ecology: Many species are not host specific and feed on the leaves of a large variety of hosts from palms and citrus to weeds. Eggs are inserted in bark singly or in end-to-end series in slits. At least some species overwinter as eggs.

Comments: Metcalf (1957) records 212 genera and 918 species in the world. Thirteen genera and 30 species are known from the U.S. Although they sometimes occur in large numbers, they are not considered to be of economic importance here, although one species has been recorded as a vector of a virus disease of satsuma orange in Japan (Mead, 1969).

Selected Bibliography

Mead 1965, 1969.
Metcalf 1957.
Wilson and McPherson 1981b.

ISSIDAE

The Issid Planthoppers

Figure 30.10, Key Figures 7, 8, 18, 27, 29

Relationships and Diagnosis: Fennah (1954) combined the Acanaloniidae with the Issidae and transferred most of the Californian issids to the family Nogodinidae (1984), but these changes have not been generally accepted as yet. The Issidae contains one subfamily, Caliscelinae, composed of many small species with reduced wings that appear to be immatures or weevils until examined carefully (*Bruchomorpha* etc. in the U.S.). The immatures may be identified to family by a combination of characters. They have a single spine on each side of the 2nd posterior tarsal segment, which is flared apically, as in flatids and acanaloniids. The mouthparts reach the hind trochanters, which usually does not occur in flatids. The 6th tergum is either chevron-shaped (key fig. 27) or rectangular, not bilobed as in flatids. Acanaloniids may usually be separated by their very strongly humpbacked appearance, but some issids have this to a lesser degree. Errors may occur in using this subjective character, so mildly humpbacked specimens should be reared for positive identification.

Biology and Ecology: Immatures and adults feed on the stems and leaves of grasses, weeds and shrubs. One genus, *Hysteropterum,* makes mud egg cases attached to branches (Schlinger, 1958).

Comments: Metcalf (1958) lists 206 genera and 981 species in the world. Twenty-three genera and 128 species are known from the U.S., mostly from the arid Southwest. Two of these species have been introduced from the Old World, *Caliscelis bonellii* (Latreille) (O'Brien, 1967) and *Asarcopus palmarum* Horvath, the datebug, which feeds in the growing tips and bases of fruit stalks of the date palm. The datebug was economically important because of the large amount of honeydew it produced (Essig, 1929).

Selected Bibliography

Essig 1929.
Fennah 1954, 1984.
Metcalf 1958.
O'Brien 1967.
Schlinger 1958.
Wilson and McPherson 1981c.

ACANALONIIDAE

The Acanaloniid Planthoppers

Figure 30.11, Key Figure 28

Relationships and Diagnosis: The Acanaloniidae are closely related to the Issidae, and Fennah has combined them (1954). They are treated separately here to follow American tradition. The immatures have a very strongly humpbacked appearance (key fig. 28). Some have a concave area behind the dorsally visible part of the metathorax, especially in dried specimens, which issids do not have, but this is not present in all specimens. This is a very subjective character, so rearing is recommended when determinations are in doubt.

Biology and Ecology: Immatures and adults feed on shrubs. In some species, the eggs are inserted into woody tissue in which they overwinter (Wilson and McPherson 1981a).

Comments: Metcalf (1954c) lists 13 genera and 81 species in the world. Eighteen species of *Acanalonia* occur in the U.S. They are not economically important.

Selected Bibliography

Fennah 1954.
Metcalf 1954c.
Wilson and McPherson 1981a.

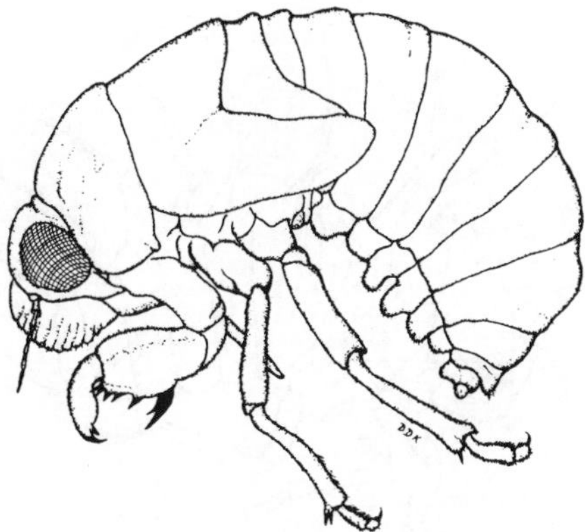

Figure 30.12. Cicadidae, lateral

CICADIDAE (CICADOIDEA)

The Cicadas

Figure 30.12, Key Figure 3

Relationships and Diagnosis: Immature Cicadidae can be separated from other Homoptera by their fossorial front legs. These have elongate coxae, thickened enlarged tibiae with dentate projections, and the tarsi attached subapically to the dentate tibiae (key fig. 3). Their large size, 12–50 mm, bulbous frons, and tubular body shape are helpful preliminary characters.

Biology and Ecology: All immatures live underground feeding on roots for from 4 to 17 years, depending upon the species. Most probably require about 10 years to mature. Eggs are laid in series in slits in grass or twigs cut with the sawlike ovipositor. The row may be 25–100 mm long. Although some overwinter as eggs, most species hatch the season they are laid, drop to the ground, and enter the soil with the aid of their burrowing forelegs. A short time before they are ready to become adults, they burrow toward the surface, making a tunnel near the surface and sometimes even a tube above surface level. They leave the soil and climb on some vertical surface for the molt to adult. One Cameroon species has been reported living in water-filled tunnels in sugarcane fields (Boulard, 1969).

Comments: In the U.S. and Canada we have 16 genera and 180 species. Metcalf's catalogue (1962c, 1963a & b) lists 265 genera and 1940 species in the world. The 17-year periodical cicadas, *Magicicada septendeci* (L.), *M. cassini* (Fisher) and *M. septendecula* Alexander and Moore, occur together in many broods, are well known in the eastern U.S., and sometimes live in water-filled burrows. There are also three species of 13-year periodical cicadas, *M. tredecula* and *M. tredecassini* Alexander and Moore and *M. tredecim* (Walsh and Riley). The damage done by root feeding is usually not assessed, but damage done by oviposition can be marked as the twig tips usually die and this can be important in fruit trees and ornamentals (Metcalf et al., 1951). A generic key to the immatures is being prepared by Moore (pers. comm.).

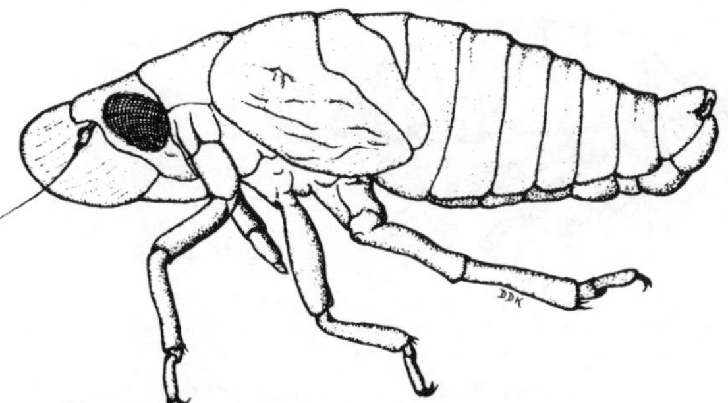

Figure 30.13. Cercopidae. *Philaenus spumarius* (L.)

Selected Bibliography

Beamer 1928.
Boulard 1969.
Metcalf *et al.* 1951.
Metcalf 1962c, 1963a, 1963b.
Moore 1961.
Myers 1929a.
Snodgrass 1919.

CERCOPIDAE (CERCOPOIDEA)

The Spittlebugs, Froghoppers, Cercopids

Figure 30.13, Key Figures 5, 6

Relationships and Diagnosis: Cercopidae are here treated as a family and superfamily, but some authorities consider them a superfamily composed of Cercopidae, Aphrophoridae, Machaerotidae, and Clastopteridae, or any combination of the above. They are closely related to the Cicadidae (Moore, 1961) as shown by small tymbal structures behind the metathorax, by both immature and adult cicadas and spittlebugs being known to feed on xylem rather than phloem tissue, and by none being tended by ants as are many honeydew-producing phloem feeders such as aphids and membracids. Also, both cicadas and spittlebugs are known to discharge anal fluids toward approaching animals in disturbance situations (Moore, pers. comm.).

Immatures may be distinguished from other Homoptera by the bulbous frons and the spittle-making ventral abdominal channel (key fig. 5) which can be closed by bringing the terga together ventrally, or is permanently closed basally in *Clastoptera* immatures. Cicada immatures flood their bodies with anal fluids too, but cannot close their broader ventral abdominal channels. (In Machaerotinae, the tube-dwelling group from the Orient, Africa, and Australia, a membrane closes this area). Last instar larvae range in size from 2–11 mm.

Biology and Ecology: These insects feed mostly on grasses and herbaceous plants, but many feed on trees and shrubs as immatures or adults. Eggs are inserted into twigs, leaves, or stems, and some are covered by a protective material (Barber and Ellis, 1922). Either eggs or adults overwinter. "Egg-bursters" have been reported in *Clastoptera* and *Philaenus* by Hanna (1967, 1969). Several species practice subterranean feeding (Doering, 1942). Most species are univoltine, but more generations per year can occur.

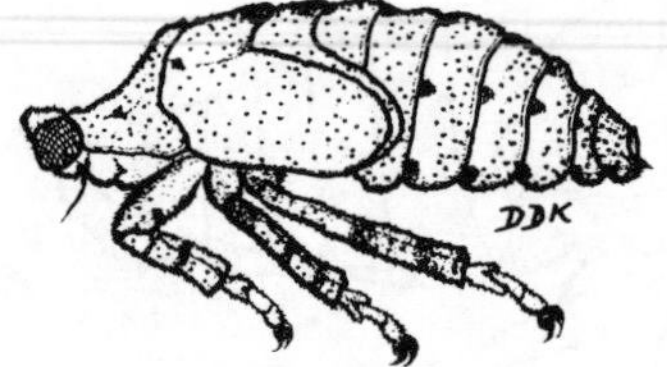

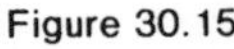

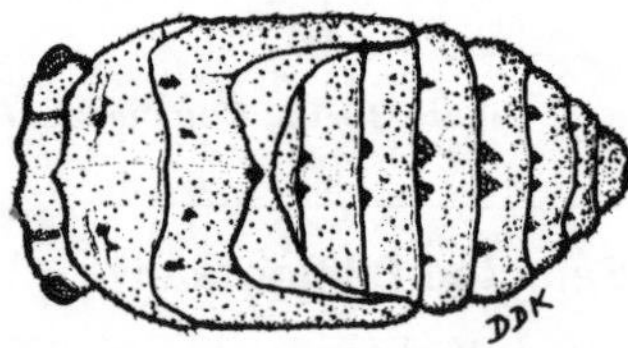

Figure 30.14. Aetalionidae, lateral

Figure 30.15. Aetalionidae, dorsal

Comments: Hannah and Moore (1966) key the immature and adult U.S. cercopids to genus. Metcalf's catalogue (1960a,b, 1961, 1962a,b) lists 426 genera and 2368 species for the world. Five genera and about 60 species are found in the U.S. and Canada, and many of their hosts are listed by Doering (1942). Four species have been found to be vectors of peach yellows and lucerne dwarf virus diseases (Ishihara, 1969). *Philaenus spumarius* (L.), the meadow spittlebug, *Clastoptera achatina* Germar, the pecan spittlebug, and *Aphrophora parallela* (Say), the pine spittlebug are among the economically important species.

Selected Bibliography

Barber and Ellis 1922.
Doering 1942.
Hanna 1967, 1969.
Hanna and Moore 1966.
Ishihara 1969.
Metcalf 1960, 1961, 1962a, 1962b.
Moore 1961.

SUPERFAMILY CICADELLOIDEA

Three families, Aetalionidae, Membracidae, and Cicadellidae are present in our fauna.

AETALIONIDAE

The Aetalionids

Figures 30.14, 30.15, Key Figures 12, 13, 15

Relationships and Diagnosis: The aetalionids are a primitive family of Cicadelloidea. The genera and species have often been placed in the membracids in the past. They share the tubular 9th abdominal segment of the membracids, but are dorsoventrally flattened and have a characteristically shaped short broad head (key fig. 15).

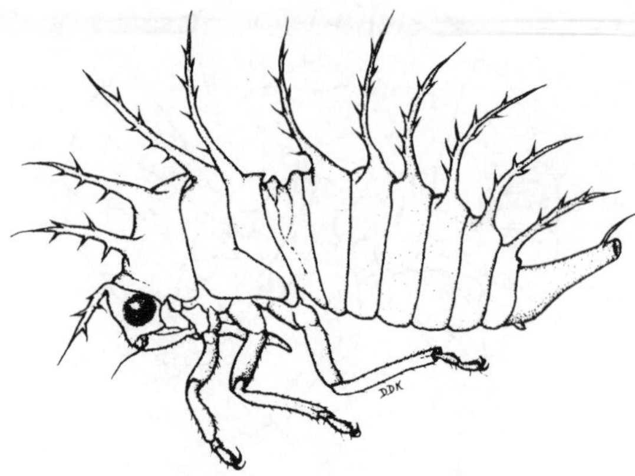

Figure 30.16. Membracidae. *Stictocephala* sp.

Biology and Ecology: Approximately 25 eggs are found in a round or oval egg mass flattened against a branch of trees or bushes of a number of plant families. The female sits on or below the egg mass, guarding it. Often several families oviposit on the same branch, and the resulting aggregation may have all instars plus the female parents. They may be tended by ants and melliponid bees as well.

Comments: Hamilton (1971) places 21 genera in the family. Metcalf and Wade (1965) included 8 genera and 47 species in the world. Two genera, *Aetalion* in Florida and southern Arizona and California, and *Microcentrus,* occur in the United States.

Selected Bibliography

Hamilton 1971.
Marques 1928.
Metcalf and Wade 1965.

MEMBRACIDAE

The Treehoppers

Figure 30.16, Key Figure 11

Relationships and Diagnosis: The membracids are an easily recognized family, often triangular in cross-section and covered with projections and/or elongate ornate spines. Immature Membracidae and Aetalionidae may be separated from other Auchenorrhyncha by their long tubular 9th abdominal tergite (the apparent 8th segment since the first is not visible) which enfolds the extrusible 10th segment. Membracids are triangular in cross-section and usually have spines and projections on their thorax. The pronotum is usually as long or longer medially than the meso- and metanotum combined; in aetalionids usually it is not longer.

Biology and Ecology: Three types of life histories have been reported (Kopp and Yonke, 1973–4). In the first group, adults overwinter in grass clumps or litter. In spring they find herbaceous hosts, mate, and oviposit. They are usually bivoltine. Winter climatic conditions destroy the herbaceous hosts and thus would destroy the eggs.

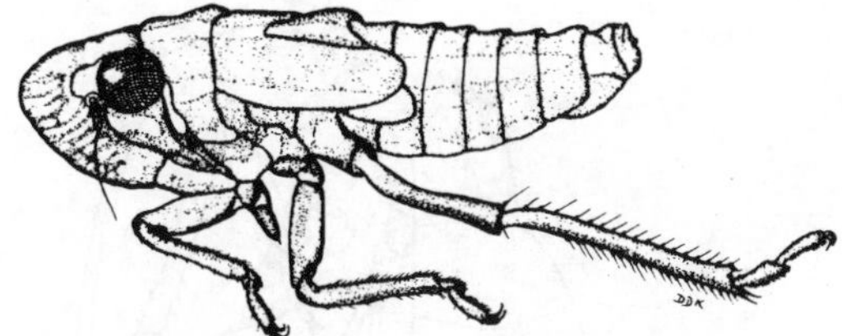

Figure 30.17. Cicadellidae. *Cuerna* sp.

In the second group, eggs overwinter in deciduous hosts, but the immatures drop to the ground in spring and feed on herbaceous hosts. The females of the single yearly generation again oviposit in deciduous hosts.

In the third group, eggs are again the overwintering stage, but all development takes place on a deciduous host.

In oviposition, the female cuts a single slit in a twig and forces 6–12 eggs to the left and right into the inner bark. Twigs usually have a double row or single rows of scars. Scars of the buffalo treehopper, *Stictocephala bisonia* Kopp and Yonke, later separate to form a characteristic double crescent. In some species, froth is deposited over the egg slit, presumably to protect the overwintering eggs (Wood and Patton, 1971).

Immatures are often gregarious. Wood (1976a, 1976b, 1977) documented presocial behavior. He found the presence of the female protects the immatures; in some species, they are attended by both the female and ants.

Comments: Quisenberry et al. (1978) keys 24 genera of immature membracids of Missouri. Metcalf (1965) lists 299 genera and 2251 species in the world. The buffalo treehopper, one of the four species given an accepted common name in the U.S., damages fruit and shade trees by laying its eggs in the small branches, which either die or grow poorly. The three-cornered alfalfa hopper, *Spissistilus festinus* (Say), damages alfalfa and other legumes by feeding around the stems, essentially girdling them (Kopp & Yonke, 1973–4). *Micrutalis malleifera* Fowler is a vector of pseudo-curlytop virus, a minor disease of tomatoes in Florida (Simons and Coe, 1958).

Selected Bibliography

Kopp and Yonke 1973–4.
Metcalf and Wade 1963–1965.
Quisenberry et al. 1978.
Simons and Coe 1958.
Wood 1976a, 1976b, 1977.
Wood and Patton, 1971.

CICADELLIDAE

The Leafhoppers, Cicadellids

Figure 30.17, Key Figures 4, 14

Relationships and Diagnosis: The Cicadellidae are placed in the Cicadelloidea with the Membracidae and Aetalionidae. Some Homopterists raise subfamilies of either Cicadellidae or Membracidae to family. Cicadellid immatures and adults may be separated from other Homoptera by the row or double row of spines on the hind tibiae.

Biology and Ecology: Leafhoppers feed on almost all types of plants, usually feeding on the leaves and stems. Most of our species have at least two generations per year, spending the winter in the egg or adult stage, depending upon the species. Eggs are inserted into the stem, midrib, or large veins of leaves. Several generations a year are sometimes found in the tropics and the southern U.S. Some species are attended by ants feeding on their honeydew.

Comments: Metcalf's catalogue (1964–1968) lists 1074 genera and 9078 species for the world, making this the largest family of Auchenorrhyncha. They also are the most important economically. One hundred and fifty species of 8 subfamilies have been implicated in the transmission of virus diseases of such important crops as sugar beets, rice, potatoes, clover, grapes, peaches, sugarcane, corn, and wheat (Nielson, 1985). In some cases, viruses are transmitted to immatures by transovarial passage. Borror et al. (1981) give other ways leafhoppers can damage plants with examples of each. They are by removing sap, plugging the phloem and xylem vessels, by ovipositing in twigs, by inhibiting growth on the underside of leaves where they feed, causing stunting and leaf curling, and by transmitting other kinds of plant pathogens.

Selected Bibliography

Borror, DeLong, and Triplehorn 1981.
Metcalf 1964–1968.
Nielson 1968, 1985.

SUBORDER STERNORRHYNCHA

Manya B. Stoetzel, Douglass R. Miller, *Systematic Entomology Laboratory, ARS, USDA*

Members of this suborder have the beak appearing to arise from between the front coxae, antennae that are usually long and filiform, and tarsi that are 1- or 2-segmented. Many groups within the suborder are relatively sedentary.

APHIDOIDEA

Manya B. Stoetzel, *Systematic Entomology Laboratory, ARS, USDA*

In a conservative classification, three families of Aphidoidea are recognized: Aphididae, Phylloxeridae, and Adelgidae. The biologies of aphids, phylloxerans, and adelgids are complex, with host alternation common in some groups but unknown in others. Species in the Aphidoidea are all plant feeders, with some being serious pests of agricultural crops and ornamental plants, and others efficient virus vectors. The many and varied life forms and the intricate life cycles make the Aphidoidea one of the most interesting groups of insects to study and at the same time one of the most difficult groups to decifer.

Aphids, phylloxerans, and adelgids are soft-bodied insects and should never be preserved dry on a point. They should be collected in ethyl alcohol and then properly prepared and mounted in Canada balsam on microscope slides. Illustrative plant damage such as galls, dwarfed and twisted leaves, and waxy secretions can be freeze-dried and kept in Riker mounts.

APHIDIDAE (APHIDOIDEA)

The Aphids

Figure 30.18, Key Figures 32, 33

Relationships and Diagnosis: Aphids or plant lice belong to the superfamily Aphidoidea, along with the Adelgidae and Phylloxeridae. The antennae are 3–6 segmented with 2 primary sensoria (rhinaria), one located subapically on the penultimate antennal segment and the other on the last segment at the division between the base and the unguis (processus terminalis). Immatures do not have secondary sensoria on the antennal segments, while some adults do. Aphids have a characteristic rostrum or beak (= labium) which is 4-segmented, but which may be vestigal. When present, tarsi are 2-segmented with a pair of claws. There are 2 pairs of thoracic spiracles and one pair on each of the first 7 abdominal segments. The 9th abdominal tergum of aphids is modified into a cauda (tail) that varies in shape and is located above the anal opening. Aphids are unique in having a cauda; and most, but not all, aphids also have a pair of cornicles (siphunculi) dorsally on the 5th or 6th abdominal segment. There are 4 immature instars.

Biology and Ecology: Most aphids live on only one species of host plant, but some exhibit an alternation of hosts (Stoetzel, 1987). The primary host is the plant on which an aphids overwinters, while the secondary host is that on which it spends the summer months. The overwintering egg hatches on the primary host in the spring, and the young of the resultant female (stem mother or fundatrix) develop into more females which may or may not be winged. If winged, they are called spring migrants, and they move to a secondary host. Several generations of females may be produced on the secondary host before autumn migrants and males are produced and migrate back to the primary host. Egg-laying females (oviparae) are produced on the overwintering host where they mate, lay eggs, and die. Aphids without sexual forms do not produce overwintering eggs but reproduce parthenogenetically throughout the year. Sexual females and males are not larviform and most have a rostrum. Parthenogenetic females are viviparous and do not have ovipositors.

Some aphids are host specific while others are polyphagous. They can be flower, fruit, leaf, root, or bark feeders. Each species has its own developmental cycle, and often there are deviations within a species from one geographical area to another or from one host to another. See Dixon (1973, 1985) and Blackman (1974) for a more complete discussion on the biology of aphids.

Aphids are injurious to fruit, nut, and shade trees, greenhouse and ornamental plants, and a variety of cultivated plants (Blackman and Eastop, 1984). Aphids are efficient virus vectors (Kennedy et al., 1962; Harris and Maramorosch, 1977); and more than 200 species have been reported as virus vectors, more than any other arthropod group. Aphids produce honeydew which serves as a medium for sooty mold and which can be a problem in and of itself.

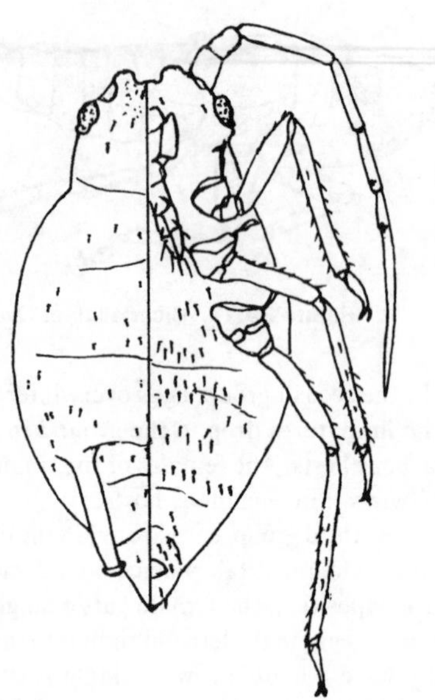

Figure 30.18. Aphididae. *Myzus persicae* (Sulzer), green peach aphid immature, dorsal on left, ventral on right.

Comments: There are about 3,500 species of aphids worldwide. In North America there are approximately 1,500 species of which about 100 are of economic importance. See the bibliography in Dixon (1985) for additional information on Aphididae. Some of the economically important species are: *Acyrthosiphon pisum* (Harris), pea aphid; *Aphis craccivora* Koch, cowpea aphid; *Aphis fabae* Scopoli, bean aphid; *Aphis gossypii* Glover, melon or cotton aphid; *Brevicoryne brassicae* (L.), cabbage aphid; *Diuraphis noxia* (Mordvilko), Russian wheat aphid; *Macrosiphum euphorbiae* (Thomas), potato aphid; *Myzus persicae* (Sulzer), green peach aphid; *Rhopalosiphum maidis* (Fitch), corn leaf aphid; *Schizaphis graminum* (Rondani), greenbug; and *Therioaphis trifolii* (Monell), yellow clover aphid.

Collection and Preservation: Whenever possible representatives of all stages present in a population should be collected in 95% ethyl alcohol. Information on the feeding site(s) and identification of the host is very important. For preservation and identification, aphids should be properly prepared and mounted in Canada balsam on microscope slides. Aphids should never be preserved dry on a point.

Selected Bibliography

Blackman 1974.
Blackman and Eastop 1984.
Bowers et al. 1972.
Dixon 1973, 1985.
Eastop and Hille Ris Lambers 1976.
Harris and Maramorosch (eds.) 1977.
Kennedy et al. 1962.
Palmer 1952.
Smith 1972.
Smith and Parron 1978.
Stoetzel 1987.

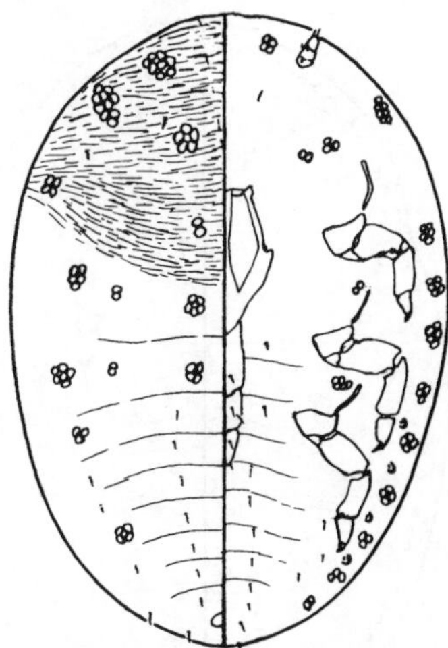

Figure 30.19. Adelgidae. *Pineus strobi* (Hartig), pine bark adelgid immature, dorsal on left, ventral on right.

ADELGIDAE (APHIDOIDEA)

The Adelgids

Figure 30.19, Key Figure 34

Relationships and Diagnosis: Adelgids belong to the Aphidoidea, and while they are similar to the aphids, they should be called adelgids and not aphids or chermids. The International Commission on Zoological Nomenclature (1965) approved the generic name *Adelges* Vallot and suppressed the generic name *Chermes* L. The Commission also approved the subfamily name Adelginae which Annand (1928) had placed in the Phylloxeridae but which most workers treat as a separate family, the Adelgidae.

Within the Adelgidae, determination at the specific level is difficult. Many authors (Annand, 1928; Pschorn-Walcher and Zwolfer, 1956; Boerner and Heinze, 1957; Carter, 1971) have reported that only first instars have reliable, constant morphological characters.

The immatures have 3-segmented antennae with one sensorium on the terminal segment. The rostrum is 3-segmented and appears to arise between the front coxae. The tarsi are 2-segmented and end in a pair of claws. There are 2 pairs of thoracic spiracles. Species of *Pineus* have 4 distinct pairs of abdominal spiracles, while species of *Adelges* have 5 distinct pairs. These abdominal spiracles are difficult to see in the first instars. The anal opening is dorsal on the 9th abdominal segment. A cauda and cornicles are absent. Wax glands are often distributed over the body. There are 4 immature instars.

Biology and Ecology: Adelgids feed exclusively on conifers (Pinaceae). The primary host is spruce (*Picea*), and the secondary or intermediate host is another conifer such as hemlock (*Tsuga*), pine (*Pinus*), larch (*Larix*), Douglas-fir (*Pseudotsuga*), or silver fir (*Abies*). The life cycle is very complex and usually requires two years. The overwintering stage is usually the first instar. Detailed discussions of these complex life cycles can be found in Annand (1928) and Carter (1971).

Sexual females and males are larviform and have a rostrum. Parthenogenetic females are oviparous and have a short ovipositor.

Comments: There are about 50 species of adelgids worldwide with 20 occurring in North America where the economically important species are *Adelges abietis* (L.), eastern spruce gall adelgid; *Adelges picea* (Ratzeburg), balsam woolly adelgid; *Pineus pinifoliae* (Fitch), pine leaf adelgid; and *Pineus strobi* (Hartig), pine bark adelgid. For additional information on the Adelgidae, see the bibliography in Carter (1971).

Collection and Preservation: Adelgids can be found in white, waxy material on the needles and/or bark of conifers. They should be collected in 95% ethyl alcohol. Like aphids, adelgids should be properly prepared and mounted in Canada balsam on microscope slides. To illustrate plant damage, preserve adelgids and their galls and/or waxy secretion in Riker mounts.

Selected Bibliography

Annand 1928.
Boerner and Heinze 1957.
Carter 1971.
International Commission on Zoological Nomenclature 1965.
Pschorn-Walcher and Zwolfer 1956.

PHYLLOXERIDAE (APHIDOIDEA)

The Phylloxera, Phylloxerans, or Phylloxerids

Figure 30.20, Key Figure 35

Relationships and Diagnosis: These insects belong to the Aphidoidea and are closely related to the Adelgidae and Aphididae. They are commonly called phylloxera, phylloxerans, or phylloxerids.

Immatures have 3-segmented antennae with only one sensorium at the end of the last segment. When present, the rostrum is 3–5-segmented and appears to arise between the front coxae. The tarsi are 2-segmented and end in a pair of claws. There are 2 pairs of thoracic spiracles and five or six pairs of abdominal spiracles. A cauda and cornicles are absent. Some of the immature stages on *Quercus* and *Castanea* have dorsal tubercles distributed over the body. There are four immature instars.

Biology and Ecology: Phylloxerans feed exclusively on dicotyledonous plants; oaks, pecans, and hickories are common hosts. Grapevines, elms and pears are also attacked.

Phylloxerans hatch from overwintering eggs in early spring when the buds are bursting and the young twigs are beginning to grow. Galls produced by phylloxerans differ in shape and location on the host.

A basic life cycle takes but two to three months in the spring, and consists of at least four adult stages: stem-mothers, sexuparae, and sexual females and males. In some species adult females continue to produce feeding first instars that begin new galls and additional feeding generations until leaf drop in the fall. These parthenogenetic females are oviparous but do not have an ovipositor.

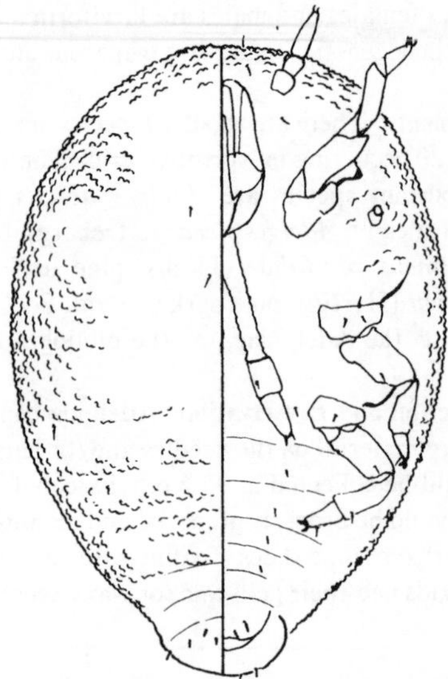

Figure 30.20. Phylloxeridae. *Daktulosphaira vitifoliae* (Fitch), grape phylloxera immature, dorsal on left, ventral on right.

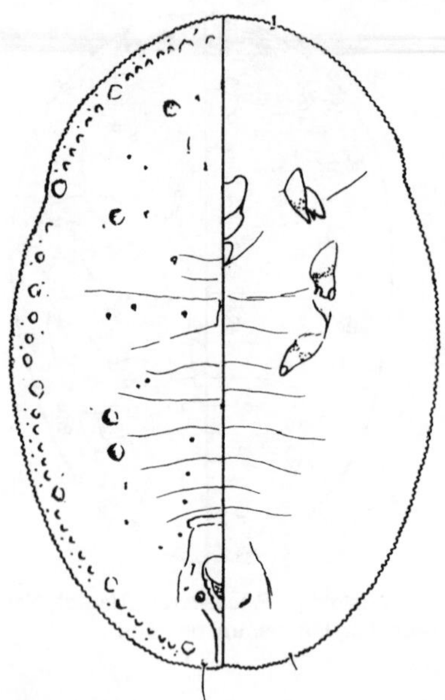

Figure 30.21. Aleyrodidae. *Trialeurodes vaporariorum* (Westwood), greenhouse whitefly immature, dorsal on left, ventral on right.

Stoetzel (1985a) has reported the existence of host alternation in the Phylloxeridae from one plant family (Juglandaceae) to another (Fagaceae). *Phylloxera texana* Stoetzel and *Phylloxera castanea* Pergande have been shown to have hickory (*Carya*) as their primary host with the secondary host being a species of oak (*Quercus*) or chestnut (*Castanea*).

The sexual females and males (sexuales) are larviform, and the female lays but one overwintering egg. Contrary to some reports (Whitehead and Eastop, 1937; Caldwell and Schuder, 1979) sexuales do not hatch directly from the egg. Instead there is a "pupiform" larva (Stoetzel, 1985b) which hatches from the egg and develops through four molts into the sexuales. Sexual males and females lack a rostrum. Stoetzel (1981) and Stoetzel and Tedders (1981) give detailed information on the morphology and biology of several species of *Phylloxera* on pecan.

The grape phylloxera forms small galls on the leaves and gall-like swellings on the roots of grape. For additional information on the grape phylloxera, see Russell (1974). For additional information on phylloxerans of hickories and oaks, see Pergande (1904) and Duncan (1922).

Comments: There are about 70 species worldwide with 50 occurring in North America. The economically important species in North America are: *Daktulosphaira vitifoliae* (Fitch), the grape phylloxera; *Phylloxera caryaecaulis* (Fitch), hickory gall phylloxera; *Phylloxera devastratix* Pergande, pecan phylloxera; and *Phylloxera notabilis* Pergande, pecan leaf phylloxera.

Collection and Preservation: Phylloxerans should be collected in 95% ethyl alcohol. Like aphids and adelgids, they should be properly prepared and mounted in Canada balsam on microscope slides. The galls are characteristic in shape and color, but they are difficult to preserve.

Selected Bibliography

Caldwell and Schuder 1979.
Duncan 1922.
Pergande 1904.
Russell 1974.
Stoetzel 1981, 1985a, 1985b.
Stoetzel and Tedders 1981.
Whitehead and Eastep 1937.

ALEYRODIDAE (ALEYRODOIDEA)

Manya B. Stoetzel, *Systematic Entomology Laboratory, ARS, USDA*

The Whiteflies or Aleyrodids

Figure 30.21, Key Figure 30

Relationships and Diagnosis: The whiteflies or aleyrodids are alone in the Aleyrodoidea. In structure, they seem to be related to the psyllids, but they share biological similarities with the aphids.

Immatures are scalelike and are usually found on the underside of leaves of the host plant. Identification of most species is based on the structure of the "pupal case" of the fourth instar.

Antennae are usually 1-segmented, and the rostrum is reduced. The legs are reduced or vestigal. There are two pairs of thoracic spiracles and two pairs of abdominal spiracles. A unique character of aleyrodids is the structure of their anal apparatus which is dorsal and the opening of which is called the vasiform orifice. Within the vasiform orifice is an operculum (lid) and a lingula (tonguelike process). There are four immature instars.

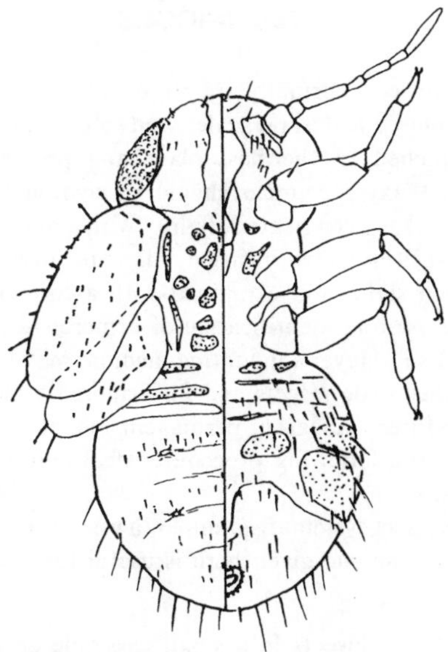

Figure 30.22. Psyllidae. *Psylla pyricola* Foerster, pear psylla immature, dorsal on left, ventral on right.

Biology and Ecology: Aleyrodids have a world-wide distribution, but they are most abundant in the tropics and subtropics. Large populations can develop, and crops can be severely damaged within a few weeks.

In the tropics development is continuous, while in colder areas the puparia hibernate. Aleyrodid eggs have a short pedicel or stalk by which they are attached to the leaf tissue of the host. The first instars are active crawlers, but they usually do not leave the leaf on which they hatched. The second, third, and fourth instars are sessile and are often covered with a white powdery wax or various waxy secretions. Adult females and males have two pairs of wings which are about equal in size and which have reduced venation.

Aleyrodids are pests of crops and ornamental plants, and several are known to be virus vectors (Mound, 1973). They also produce honeydew which serves as a medium for sooty mold.

Comments: There are about 1,200 species in the world with 500 occurring in North America. Some of the economically important species in North America are: *Aleurocanthus spiniferus* (Quaintance), orange spiny whitefly; *Aleurocanthus woglumi* Ashby, citrus blackfly; *Aleurothrixus floccosus* (Maskell), woolly whitefly; *Bemisia tabaci* (Gennadius), sweetpotato whitefly; *Dialeurodes citri* (Ashmead), citrus whitefly; *Dialeurodes citrifolii* (Morgan), cloudywinged whitefly; *Trialeurodes abutilonea* (Haldeman), bandedwinged whitefly; *Trialeurodes floridensis* (Quaintance), avocado whitefly; and *Trialeurodes vaporariorum* (Westwood), greenhouse whitefly.

Collection and Preservation: When kept dry on leaves in envelopes or boxes, the waxy secretions of the puparia are preserved. Aleyrodids can also be collected in 95% ethyl alcohol. For identification and study, they must be mounted on slides. If dark specimens fail to clear or lighten in KOH, put them in peroxide solution (one drop of ammonia for every six drops of hydrogen peroxide) until they are the desired color.

Selected Bibliography

Mound 1973.
Mound and Halsey 1978.
Quaintance and Baker 1913–1915, 1917.
Russell 1947, 1948.

PSYLLIDAE (PSYLLOIDEA)

Manya B. Stoetzel, *Systematic Entomology Laboratory, ARS, USDA*

The Psyllids

Figure 30.22, Key Figure 31

Relationships and Diagnosis: The psyllids or jumping plant lice are alone in the Psylloidea.

The number of antennal segments is variable and increases as development continues through five instars. The rostrum is 2-segmented and appears to arise between the front coxae. The tarsi are 2-segmented and end in a pair of claws. There are two pairs of thoracic spiracles and six pairs of abdominal spiracles. Unlike the Aphidoidea and Aleyrodoidea, all of the immature stages of psyllids possess compound eyes, wing pads, and a ventral, circumanal ring. Sexual dimorphism has been reported in the immatures of several species (Carter, 1961; Ball and Jensen, 1966).

Biology and Ecology: Psyllids are plant feeders and the majority live on trees and shrubs. A few species live on herbaceous plants, and a few are injurious to vegetable crops. Most psyllids overwinter as adults or immatures, and there is one or more generations through the summer. The immatures produce honeydew and some secrete wax; most are active and free living, while others cause galls or pseudogalls on leaves or twigs. Species of *Pachypsylla* produce galls on the leaves, twigs, or branches of *Celtis* spp.

Reproduction is bisexual, and the adults resemble very small cicadas.

Several psyllid species have been proven to be virus vectors.

Comments: There are about 2,000 species in the world with 1,000 occurring in North America. Some of the economically important species in North America are *Pachypsylla* spp., hackberry gall makers; *Paratrioza cockerelli* (Sulc), potato or tomato psyllid; *Psylla pyricola* Foerster, pear psylla; and *Trioza diospyri* (Ashmead), persimmon psylla.

Collection and Preservation: The immatures are dorsoventrally flattened and are best preserved mounted on slides. The galls can be preserved dry in Riker mounts.

Selected Bibliography

Ball and Jensen 1966.
Carter 1961.
Crawford 1914.
Tuthill, 1943.
White and Hodkinson 1985.

SUPERFAMILY COCCOIDEA

Douglass R. Miller, *Systematic Entomology Laboratory, ARS, USDA*

The Scales, Scale Insects, or Coccoids

Scale insects are some of the more specialized Homoptera and probably are most closely related to the Aphidoidea. Adult females lack wings, may or may not have legs, and are sacklike with no definite head, thorax, or abdomen. Adult males are more insectlike in appearance than adult females and have one pair of wings, one pair of halterelike structures, well-developed legs, and a definite head, thorax, and abdomen. Adult males are rarely collected because they are small and normally live for only a day or two. Most scale insects produce a waxy secretion that covers the body either as a domicilelike structure (scale cover) detached from the body or as a substance on the integumental surface. This waxy secretion may vary from a thin translucent sheet to a thick, wet mass or to a powdery, bloomlike substance.

Scale insects are serious plant pests and as small, often cryptic components of the phytophagous ecosystem, they frequently are not detected until they have caused obvious damage. They are most important as pests of perennial plants and can cause especially serious damage to nut and fruit trees, woody ornamentals, forest vegetation, and house plants. Damage is usually caused by removal of large quantities of plant sap but also may be caused by plant pathogens, toxins, and the production of large quantities of honeydew with resultant growth of sooty mold fungi that cover leaf surfaces and reduce photosynthesis.

On the other hand scales may be beneficial. They have been used as sources of dyes (cochineal scales, gall-like scales, giant scales, and lac scales), of shellac and lacquerlike substances (lac insects and giant scales), of candle wax (soft scales), of the "manna" of the Israelites (mealybugs), of "pearls" for necklaces (ground pearls or giant scales), and even of chewing gum (ornate pit scales) (Miller and Kosztarab, 1979). Cochineal scales and mealybugs have also been used to control certain noxious weeds (Miller and Kosztarab, 1979).

Life histories within the Coccoidea are varied. Females usually have two or three immature instars and males usually have four. Males normally have two pupalike instars called the prepupa and pupa, respectively, that develop in a waxy enclosure produced by the second instar. The pupalike instars and the adult males do not feed. The neotenic female lays eggs or first instars either in a cavity under her body or in a waxy cover that may or may not be attached to her body. First instars are the principle agents of feeding site location and dispersal; other immature instars generally are sessile. Scales possess a diversity of reproductive systems including hermaphroditism, seven kinds of parthenogenesis, and six major types of sexual chromosome systems (Miller and Kosztarab, 1979).

Scale insects occur in nearly all available botanical habitats from the tundra to the tropics. They are found on nearly all parts of the host including the leaves, branches, trunks, fruits, and roots. They sometimes occur under bark and may cause various kinds of plant deformities including chlorotic spots, pits, and galls.

TECHNIQUES

Scale insects normally are preserved in 70% alcohol, although armored scales, pit scales, and soft scales are best collected attached to the host plant and preserved dry in envelopes. Waxy specimens should be vigorously shaken in alcohol that has been heated slightly with a match or lighter.

For accurate identification, all scale insects must be mounted on slides and examined with a compound microscope. For routine identification of armored scales, temporary mounts in Hoyer's mounting medium may be adequate, but for other scale insects and for armored scales to be retained for future research, permanent mounts are required. Therefore, the following procedures deal only with permanent mounts.

Scale insect systematists employ a vast array of mounting techniques. The one given here is useful for most scale insects.

1. Place scale insects in a small casserole or watch glass containing a 10% solution of potassium hydroxide (KOH). When specimens are pliable, make a small incision on one side. Specimens may stand in cold KOH solution for 24 hours or until the body contents are clear, or they may be heated gently at a low simmer (not a boil) on a hot plate until the specimens are translucent or clear. Be careful not to "overcook" the specimens.
2. With a spatula, carefully press down on the specimens so that the body contents are forced out. Repeat this action until the body contents are removed.
3. Transfer the specimens to distilled water and wash off excess KOH. Wash again in water one or two more times.
4. Place specimens in stain. Acid fuchsin (5% aqueous solution of: Acid Fuchsin 0.5 gm; 10% HCL 25 ml; distilled water 300 ml) frequently is used although it tends to fade with time. Allow to stand until sclerotized areas are clearly stained. This may require a few minutes or several hours depending on the specimens. If the specimens are overstained, transferring them to 70% ethyl alcohol frequently will remove some of the excess.
5. Wash specimens in 95% alcohol. Remove remaining body contents. Transfer to absolute alcohol for about one minute.
6. Place in clove oil or carboxylol (1 part carbolic acid: 2 parts xylene) for a few minutes or until clear.
7. With a glass rod, spread a small drop of balsam in the center of a 2.5 × 7.5 cm microscope slide and place each specimen dorsal side up in the balsam. With a pair of fine forceps, carefully place a cover glass over the specimens. It is critical that the proper amount of balsam be placed on the slide. If too much is used, the specimens may be washed out from under the cover glass, making remounting necessary. Excess balsam also makes specimens difficult to view with the compound microscope. If too little balsam is used, it may be necessary to add additional balsam later.
8. Place each slide on a hot plate. This will force air bubbles out from under the cover glass. Be careful not to allow the balsam to boil.

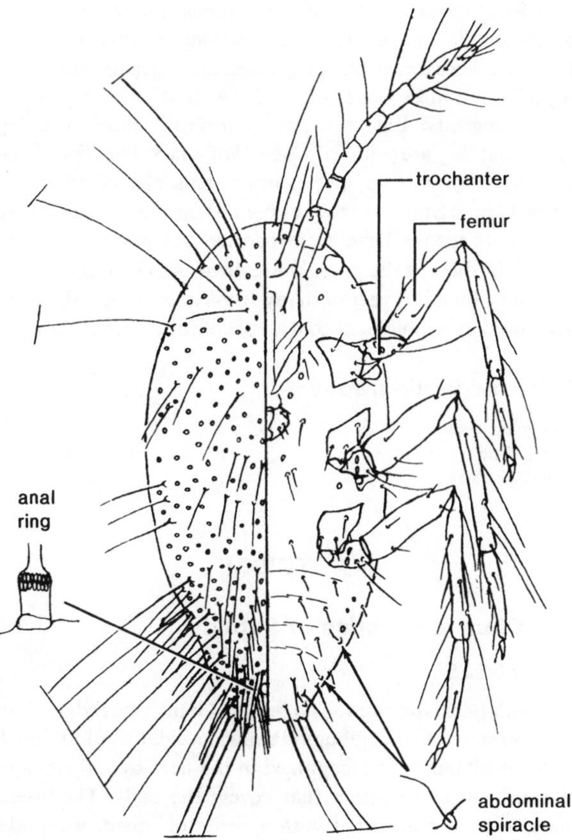

Figure 30.23. Margarodidae. First instar—dorsal on left, ventral on right.

9. Place the specimens in a warm oven (40° C.) for 2–4 weeks. If an oven is not available, the slides should be allowed to cure at room temperature for a month.
10. Place a label on the right side of the slide giving the following data: locality, collector, host, date of collection, and identifying number. Store slides horizontally.

MARGARODIDAE

The Giant Scales, Ground Pearls, or Margarodids

Figure 30.23

Relationships and Diagnosis: The giant scales apparently are closely related to the ensign scales (Ortheziidae) and giant mealybugs (Putoidae). Because of the diversity within the family Margarodidae, it is sometimes divided into several families. Adult females in the field take a variety of body forms; many are elliptical, but some are oval and rotund or flattened and elongate. Margarodids tend to be larger than other scale insects, up to 35 mm in length; few are as small as 1.5 mm. Wax coverings vary from a conspicuous, ornate wax in *Icerya* species to a simple powder in some species of *Llaveia.* In many groups the body is brightly colored with orange or red and contrasting black markings, while some are pale yellow, brown, or white. Most species have large black legs and antennae.

First instar giant scales can be separated from first instars of other families by having small abdominal spiracles, a separate trochanter and femur, and an anal opening usually with a thin, sclerotized ring that lacks pores and setae. Other immature instars have the same aggregation of diagnostic features.

Determination of the instar of giant-scales must be done on a species-by-species basis because there is considerable variation among taxa. With a few exceptions, the crawler has 5-, 6-, or rarely 7-segmented antennae, whereas older instars have progressively more segments. The numbers of setae and pores usually increase as the margarodid passes through each life stage. The first instar often has one or more long pairs of setae along the body margin that are absent from the other instars.

Giant scales are similar to ensign scales (for a comparison see the description of the Ortheziidae) and to giant mealybugs. Giant scales differ from giant mealybugs by having abdominal spiracles and by lacking trilocular pores, cerarii, ostioles, and a sclerotized, dermal anal opening without setae and pores. Giant mealybugs lack abdominal spiracles and usually have trilocular pores, cerarii, ostioles, and a sclerotized, dermal anal opening that has setae and pores.

Biology and Ecology: Giant scales may have two or three or perhaps more immature instars in the female and four or five in the male. Giant scales have two general types of life-history patterns—i.e., those species that form a cyst and those that do not. The cyst formers have one or more intermediate instars that lack legs and have reduced antennae. The cyst stage is apparently highly resistant to unfavorable environmental conditions; in *Margarodes vitis* (Phillipi) the cyst has been reported to remain viable for as long as 17 years without water or food. The cyst stage of various species of the fossorial genus *Margarodes* and its relatives is usually nearly round, lightly irridescent, and is referred to as a ground pearl. In the development of groups that do not form cysts each consecutive instar is similar to the preceding instar. Unlike other scale insects, the male margarodid passes through only one "pupal" stage. Eggs may be laid in ovisacs, in an internal sac called a marsupium, or in loose masses of filamentous wax. In some species the eggs hatch inside the body before being laid. Margarodids may be found on the roots and be fossorial; they may occur under bark; they may occur on the bark or leaves; or they may be found in needle sheaths of conifers. Sexuality varies from hermaphroditic in some species of *Icerya,* to parthenogenetic, to functionally bisexual. In most species males are known. Species are known on all parts of the host. Woody shrubs are the most common hosts.

Comments: Giant scales occur in all six major zoogeographic areas. Although they seem to be most abundant in tropical regions, there is a diversity of taxa in temperate areas also. There are approximately 250 species in 65 genera; in the United States there are about 42 species in 16 genera.

Ten species are considered to be of economic importance in the United States; the most notorious are the centipedegrass ground pearl, *Dimargarodes meridionalis* (Morrison), the cottony cushion scale, *Icerya purchasi* Maskell, the red pine scale, *Matusucoccus resinosae* Bean and Godwin, the Prescott pine scale, *M. vexillorum* Morrison, and the sycamore scale, *Stomacoccus platani* Ferris.

Selected Bibliography

Boratynski 1952.
Jakubski 1965.
Morrison 1928.

ORTHEZIIDAE

The Ensign Scales or Ortheziids

Figure 30.24

Relationships and Diagnosis: Ensign scales apparently are closely related to giant scales (Margarodidae). Both families possess abdominal spiracles. Adult females in the field are easily recognized by having a thick, waxy ovisac that is attached to the body of the scale but not to the host. This attachment allows the female to change sites and transport the eggs. The body of the female usually is adorned with thick patches of wax, which give this scale insect an unusual, ornate appearance.

First-instar ensign scales can be separated from first instars of other families by having small abdominal spiracles, a fused trochanter and femur, an anal ring with pores and setae, a stout seta at the apex of each antenna, predominantly quadrilocular pores, and characteristically shaped setae. Second and third instars have the same aggregation of diagnostic features.

First instar ensign scales must be separated from later instars on a species-by-species basis since there is considerable variation among taxa. First instars have fewer setae and pores than older instars and have 6-segmented antennae in species that have 7- or 8-segmented antennae as adults. Second instars also have 6-segmented antennae whereas third instars have seven. Generally, the number of setae and pores increases in successive instars.

Ensign scales are similar to giant scales but can be separated by having an anal ring with setae and pores on the dermal surface, a stout seta at the apex of each antenna, predominantly quadrilocular pores, and characteristically shaped setae. Giant scales generally lack a distinct anal ring or have a ring at the apex of an invaginated tube that may have pores but lacks setae; giant scales normally lack quadrilocular pores and a stout seta at the apex of the antenna.

Biology and Ecology: Ensign scales apparently have three immature instars in the female; the number of male instars is unknown. Biological information on this group is scant. As many as three generations may be produced each year. Mature adult females lay eggs in an ovisac that is attached to the body and is transported by the adult. Males of most species are uncommon.

Most species are found on the stems or leaves of the host, although a few are subterranean. Woody shrubs are common hosts, but herbaceous plants, grasses, and perhaps even fungus mycelia and moss sometimes serve as hosts.

Comments: Ensign scales occur on all continents except Australia; they are primarily New World in distribution. There are approximately 75 species of ensign scales in six genera; in the United States there are about 31 species in four genera. Five or six species have been reported as pests; the most notorious of these is the greenhouse orthezia, *Orthezia insignis* Browne which is a pest of many tropical ornamentals. It has been used as a biological control agent for control of lantana.

Selected Bibliography

Beingolea 1969.
Ezzat 1956.
Morrison 1952.

PUTOIDAE

The Giant Mealybugs or Putoids

Figure 30.25

Relationships and Diagnosis: Giant mealybugs are closely related to mealybugs (Pseudococcidae). Adult female giant mealybugs are recognized in the field by a thick layer of mealy waxy secretion that covers the body. The lateral margins are ornamented with a series of broad, waxy filaments. Filamentous ovisacs that are common in mealybugs apparently are not produced by giant mealybugs. Infestations may occur on all parts of the host including the roots.

First-instar giant mealybugs can be separated from first instars of other families by having trilocular pores, multilocular pores, 7-segmented antennae, cerarii each with basal sclerotization, the cerarius near the eye with four or more conical setae, a denticle on the claw, and by usually having a circulus. The remaining immature instars have not been studied in detail. They can be distinguished from immature instars of other families by having the above combination of characters except for the antennal segmentation, which is variable, and the cerarii, nearly all of which have at least four conical setae.

First instar giant mealybugs can be separated from later instars by having 7-segmented antennae, most cerarii with two or three conical setae, and by lacking tubular ducts. Second instar females have most cerarii with three or more conical setae, tubular ducts scattered over the dorsum, and may have tubular ducts in the cerarii. Third instar females are similar to adult females except that the adults have a vulva and usually have 9-segmented antennae.

Giant mealybugs are similar to mealybugs and giant scales; for a comparison refer to the descriptions of the Pseudococcidae and Margarodidae.

Biology and Ecology: Giant mealybugs apparently have three immature instars in the female and four in the male. Unfortunately, *Puto sandini* Washburn is the only putoid whose life history has been studied. Because it occurs at high

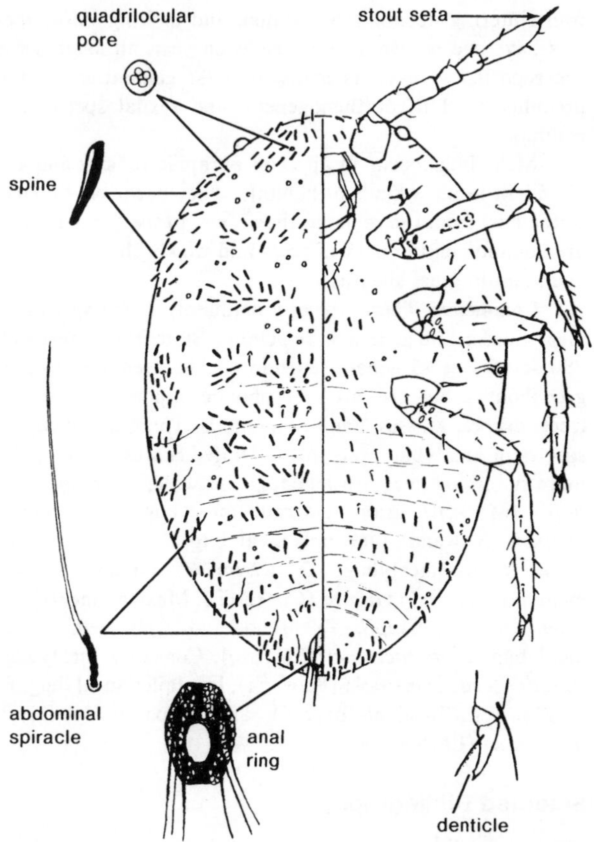

Figure 30.24. Ortheziidae. First instar—dorsal on left, ventral on right.

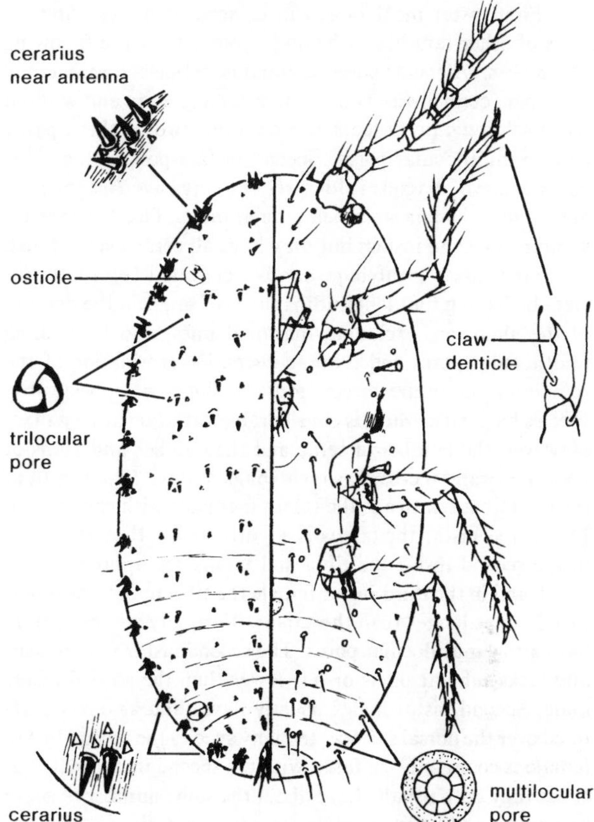

Figure 30.25. Putoidae. First instar—dorsal on left, ventral on right.

elevations, it probably has a life history that is not representative of the family. The species has one generation every four years, gives birth to first instars rather than laying eggs, and requires mating to reproduce. Giant mealybugs overwinter in the duff beneath the tree as first instars, second instars, third instar females, and mated adult females. Feeding takes place on the foliage and molting occurs under the bark of the bole. Most species of *Puto* apparently have one generation each year, give birth to first instars rather than laying eggs, and require mating to reproduce. They occur on all parts of the host, including the roots.

Comments: Giant mealybugs occur in the Nearctic, Neotropical, and Palearctic regions. They are most abundant in the western United States. There are approximately 35 species in three genera; in the United States there are 20 species in one genus, *Puto*. Some workers consider the family to include only the genus *Puto* with the genera *Ceroputo* and *Macrocerococcus* treated as junior synonyms. Giant mealybugs are occasional pests; *Puto cupressi* (Coleman) and *P. sandini* may cause considerable damage to conifers under some circumstances in the United States; *Puto ulter* Ferris often infests orchids in Mexico, Central and South America; *Puto barberi* (Cockerell) may be a pest in some South American countries.

Selected Bibliography

McKenzie 1967.
Washburn 1965.

PSEUDOCOCCIDAE

The Mealybugs or Pseudococcids

Figure 30.26

Relationships and Diagnosis: Mealybugs are most closely related to giant mealybugs (Putoidae). Adult females normally are characterized in the field by a white, mealy secretion that covers the body. Species that occur in concealed habitats such as grass sheaths or under bark either lack mealy secretions or have only small amounts of wax. Frequently marginal areas of the body have a series of protruding lateral filaments. These filaments may be absent, confined to the posterior one or two visible abdominal segments, or occur around the entire body margin. A filamentous secretion often is produced that encloses the eggs and at least part of the body of the adult female.

First instar mealybugs can be separated from first instars of other families by having a majority of the following characters: trilocular pores, a circulus, ostioles, 6-segmented antennae, cerarii with two or three conical setae and without basal sclerotization except the posterior two or three pairs, and no multilocular pores. Second instars possess the same aggregation of diagnostic features but may have 7-segmented antennae and noninvaginated tubular ducts. Third instars are similar to second instars but may have all cerarii sclerotized.

First instar mealybugs can be separated from later instars by having four longitudinal lines of setae on the dorsum of the abdomen, excluding marginal lines, and by lacking multilocular pores and tubular ducts. Determination of the remaining immature insects must be done on a species-by-species basis since there is considerable variation among taxa. However, the number of setae and the number and kinds of pores generally increase as mealybugs mature. The length of the hind tibia relative to the tarsus increases with each instar. In the first instar the tarsus is slightly longer than the tibia; in the second instar the tibia and tarsus are approximately equal; and in the third-instar female the tibia may be as much as 1.5 times longer than the tarsus. Normally second instars have a few multilocular pores. The second instar female usually lacks tubular ducts or has fewer than the second instar male. Second instar males usually have tubular ducts scattered over the dorsal surface. In many species the second instar female is commonly confused with the second instar male because they both usually have about the same number of setae and pores, and have the same tibia/tarsus ratio.

Mealybugs are similar to giant mealybugs. Mealybug first instars differ from giant mealybugs by having 6-segmented antennae, cerarii each with two or three conical setae, and cerarii without basal sclerotization except on the posterior two or three pairs, and by lacking multilocular pores. Giant mealybug first instars have multilocular pores, 7-segmented antennae, the cerarius nearest the eye with four or five conical setae, and most cerarii with basal sclerotization. Other immature instars of mealybugs differ from giant mealybug immatures by lacking a denticle on the claw, but in the mealybug tribe Phenacoccini a large denticle is usually present. Immature instars of this tribe differ from giant mealybugs by usually having quinquelocular pores, by usually having cerarii with less than four conical setae, and by lacking tubular ducts in the cerarii. Immature instars of giant mealybugs usually lack quinquelocular pores, have several cerarii with more than four conical setae, and often have tubular ducts in the cerarii.

Biology and Ecology: Mealybugs have three immature instars in the female and four in the male. Because the family Pseudococcidae is large and diverse, a generalized life history will have many exceptions. Many mealybugs overwinter as second instars, although adults, first instars, and eggs occasionally overwinter. Eggs or first instars may be produced by the female. Eggs are normally laid in an ovisac that may enclose all or part of the body of the adult female. Even though the majority of species have legs in all instars, most species remain relatively stationary throughout their life; a few species are reported to move to different areas of the host for overwintering, feeding, oviposition, and molting. Most species have one or two generations each year, although some are reported to have as many as eight generations in the greenhouse. Both parthenogenetic and sexual species are common.

Mealybugs occur in all zoogeographic regions and are abundant in most kinds of habitats. They occur on all parts of the host from the roots, to the crown, to the foliage. They are common on the bark of trees and woody shrubs and are abundant in grass sheaths.

Comments: There are approximately 1,000 species of mealybugs in 190 genera; in the United States there are about 290 species in 45 genera. Pseudococcids often are pests in greenhouses, but also are of economic importance on fruit trees, grapes, grasses, and ornamentals. There are about 40 species of mealybug pests in the United States; some of the more notorious ones are Rhodesgrass scale, *Antonina graminis* (Maskell), gray sugarcane mealybug, *Dysmicoccus boninsis* (Kuwana), pineapple mealybug, *D. brevipes* (Cockerell), taxus mealybug, *D. wistariae* (Green), striped mealybug, *Ferrisia virgata* (Cockerell), Mexican mealybug, *Phenacoccus gossypii* Townsend and Cockerell, citrus mealybug, *Planococcus citri* (Risso), Comstock mealybug, *Pseudococcus comstocki* (Kuwana), longtailed mealybug, *P. longispinus* (Targioni-Tozzetti), and grape mealybug, *P. maritimus* (Ehrhorn).

Selected Bibliography

Ferris 1950, 1953.
McKenzie 1967.
Miller 1975.
Yang and Kosztarab 1967.

COCCIDAE

The Soft Scales or Coccids

Figure 30.27

Relationships and Diagnosis: Soft scales are considered to be closely related to flat grass scales (Aclerdidae). Adult females in the field are diverse in body form. In most species the body is flat ventrally and slightly to highly convex dorsally. The flat ventral part of the body adheres tightly to the host substrate. When the adult female is removed from the substrate, a pair of white waxy bands can usually be seen on the host and on the underside of the insect. These bands are formed by the wax pores in the spiracular furrows. The body usually lacks an obvious wax covering, although in the genus *Ceroplastes* the wax is usually very ornate, and in a few other groups, a powdery wax is produced. *Pulvinaria* produces a conspicuous, white ovisac that may be several times the length of the body. Most other genera produce small, inconspicuous ovisacs or protect the eggs with the body of the adult female. The test formed by the second instar male is characteristic of soft scales; it is semitransparent, glassy in appearance, and composed of a series of platelike waxy structures. Newly molted adult females often have color patterns that are distinctive and sometimes quite striking.

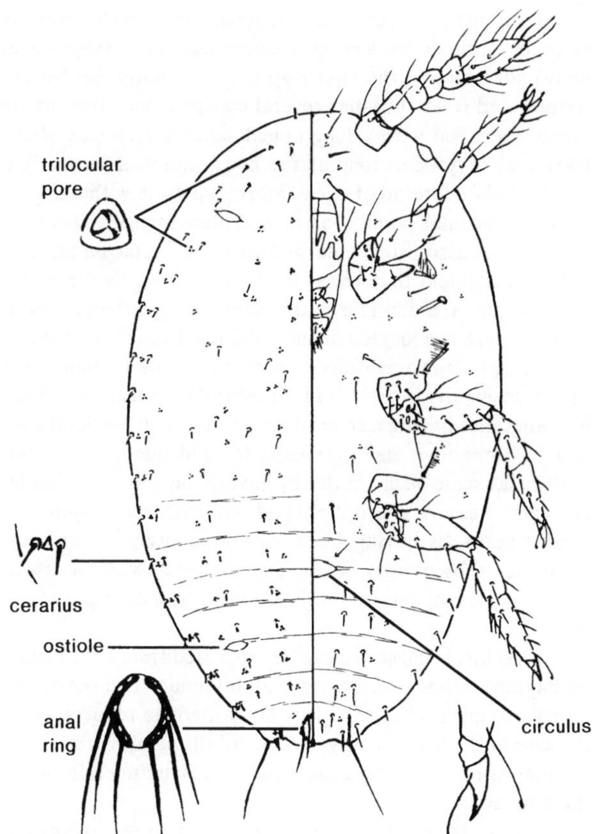

Figure 30.26. Pseudococcidae. First instar—dorsal on left, ventral on right.

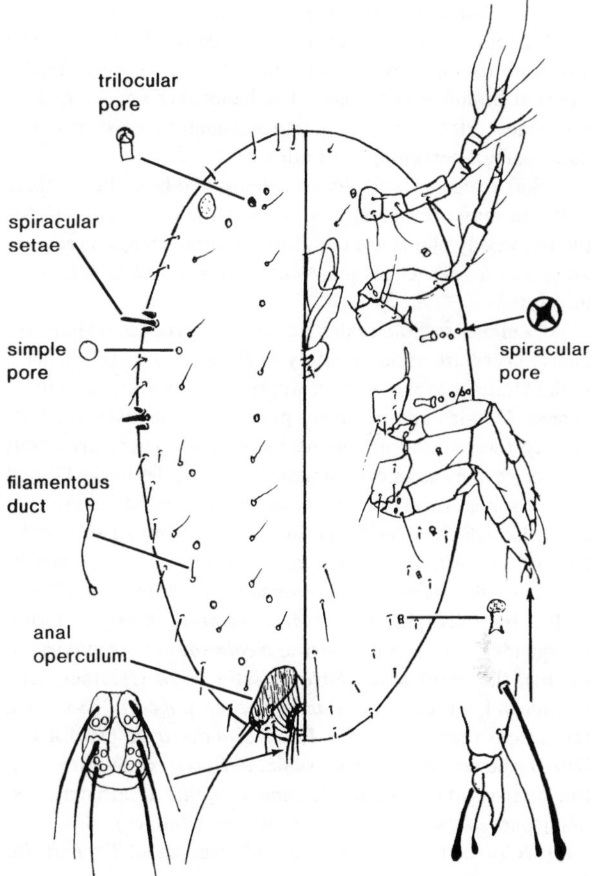

Figure 30.27. Coccidae. First instar—dorsal on left, ventral on right.

First instar soft scales can be separated from first instars of other families by having two sclerotized anal opercula, by usually having differentiated spiracular setae and a row of spiracular pores, and by lacking protruding anal lobes and pores in the spiracular atrium. Second instars have the same aggregation of features and usually have a well-developed anal cleft. Second instar males usually have invaginated tubular ducts. Third instar females are similar to second instar females but have more setae and pores.

First instar soft scales can be separated from later soft scale instars by having two elongate or triangular sclerotized anal opercula that have a long apical seta. Second instars have anal opercula that lack a long apical seta and usually have a conspicuous anal cleft. Second instar males have many tubular ducts that are usually absent from females. Third instar females, when present, usually are similar to the adult but lack multilocular pores, a vulva, and tubular ducts.

Soft scales are similar to flat grass scales. Soft scale first instars differ by having distinct anal opercula and an anal ring with setae and pores, by usually having a spiracular band of pores and differentiated spiracular setae, but lack protruding lobes and pores in the spiracular atria. Flat grass scale first instars lack anal opercula and pores and setae on the anal ring, spiracular pore bands, and differentiated spiracular setae, but have protruding anal lobes and at least one pore in each spiracular atrium. Second instar soft scales usually have a distinct spiracular pore band, differentiated spiracular setae, well-developed legs and antennae and two anal opercula, and lack large clusters of pores in the spiracular atria and a crenulate posterior body margin. Second instar flat grass scales lack spiracular setae, legs, and anal opercula, and have large clusters of pores in the spiracular atria and a crenulate posterior body margin. Third instar female aclerids apparently are quite similar to second instars.

Biology and Ecology: Soft scales have two or three immature instars in the female and four in the male. Life history data are highly variable throughout the soft scales. In the United States many species have only one generation each year, although two generations per year is not uncommon. Several species have a single generation in northern areas but have two or more in southern areas. Greenhouse pests may have as many as six generations each year. Some tropical species develop continuously throughout the year but slow their growth rate during cool periods. Overwintering may occur in any stage except the third, fourth, or fifth instar males. Most species overwinter in the same stage; second instars and mated adult females are the most common overwintering stages in the United States. Eggs or first instars may be laid by the adult. Most species produce large numbers of offspring; some species of *Ceroplastes* are reported to lay 2,000 or more eggs.

In many instances the first instars settle on the leaves. Immature females usually continue to feed on the leaves until late summer or early fall, at which time they move to the stems or branches of the host. Parthenogenesis is common in soft scales, although some parthenogenetic forms may produce a small percentage of males.

Soft scales occur on leaves, stems, and branches of their host and may occasionally be found on fruits and roots. Although woody plants are the most important hosts, perennial grasses and herbaceous plants are not immune to soft scale infestation.

Comments: Soft scales occur in all zoogeographic regions. There are approximately 1,000 species in 100 genera; in the United States there are approximately 93 species in 22 genera. Coccids are important pests in greenhouses, fruit orchards, forests, and ornamental situations. There are about 40 species of economically important soft scales in the United States; some of the more notorious pests are the Indian wax scale, *Ceroplastes cerifer* (Fabricius), the brown soft scale, *Coccus hesperidum* Linneaus, the green scale, *C. viridis* (Green), the calico scale, *Eulecanium cerasorum* (Cockerell), the terrapin scale, *Mesolecanium nigrofasciatum* (Pergande), the magnolia scale, *Neolecanium cornuparvum* (Thro), the nigra scale, *Parasaissetia nigra* (Nietner), the European fruit lecanium, *Parthenolecanium corni* (Bouché), the cottony maple leaf scale, *Pulvinaria acericola* (Walsh and Riley), the cottony camellia scale, *P. floccifera* (Westwood), the cottony maple scale, *P. innumerabilis* (Rathvon), the hemispherical scale, *Saissetia coffeae* (Walker), the black scale, *S. oleae* (Olivier), and the tuliptree scale, *Toumeyella liriodendri* (Gmelin).

Selected Bibliography

Borchsenius 1957.
Gill et al. 1977.
Hamon and Williams 1984.
Phillips 1962.
Steinweden 1929.
Williams and Kosztarab 1972.

TACHARDIIDAE

The Lac Scales or Tachardiids

Figure 30.28

Relationships and Diagnosis: Lac scales have become so specialized that affinities are difficult to determine. They fall within the lecanoid group of families but relationships within the group are unclear. Structural features of the first instars and adult males suggest affinities with soft scales (Coccidae). Adult females in the field are easily recognized by having a thick resinous test that has three openings. Each opening has a series of white, waxy filaments which originate from the area of the anterior spiracles and the anal opening.

First instar lac scales can be separated from first instars of other families by having a sclerotized area (supra-anal plate) surrounding the anal ring that is usually divided anteriorly and frequently has several conspicuous, often ornate protrusions, and with a long apical seta on each side of the body that may be as long as the body, and pseudocerarii, a brachial plate laterad of each anterior spiracle with a cluster of quinquelocular pores, the brachial plate sclerotized or with two or three associated, enlarged setae, two setae on the fifth antennal segment that are at least half as long as the length of the antenna, a denticle on the claw, two tarsal digitules on each leg with one located distally and one located subdistally, by usually having a modified type of invaginated tubular duct that occasionally is replaced by a bilocular or trilocular duct, by sometimes having a sclerotized process near the hind legs, and by sometimes having spermatozooid ducts. Other immature lac scale instars differ by having the anterior spiracles enlarged, heavily sclerotized, and with associated quinquelocular pores, by having a brachial plate laterad of one pair of spiracles, by usually having the anal area with an associated anal fringe, and by lacking well-developed legs and antennae.

First instar lac scales can be separated from later instars by having well-developed legs and antennae. The remaining immature instars have the legs and antennae reduced or, in the case of the legs, usually absent. Adult females are usually distinguished by having a long spinelike structure anterior of the anal area.

Lac scales probably are most similar to soft scales but they may be separated in all instars by having a brachial plate and a different anal structure (*see* figures 30.42, 30.43, 30.73).

Biology and Ecology: Lac scales have two (Chamberlin, 1923) or three (Glover, 1937) immature instars in the female and four in the male. In the true lac scale, *Kerria lacca* (Kerr), in India there normally are two generations each year, although a third generation occurs in the Mysore area on *Shorea*. Eggs are laid inside the lac test. A brood chamber is formed as the adult female shrivels during the egg-laying process. As many as 1,000 eggs may be laid by a female. The eggs hatch and first instars emerge from the test through the hole formed by the anal area. The first instars settle on the new stem growth and produce a broad, lac test. The female molts three times and enlarges the test until it coalesces with the secretions of other females. Males also produce a test, but the male test is narrow and has a large opercular opening. The adult male emerges through the operculum and may be winged or wingless. A population may contain as many as 30% males, but they apparently are unnecessary for reproduction. The biology of *Tachardiella larrae* Comstock in the United States also has been examined. There is one generation each year. First instars are laid inside the brood chamber and appear in early spring. Mating apparently occurs in July, and eggs begin developing in females in late summer. Lac scales normally infest stems and branches of their woody hosts. Some species form large aggregations completely coating the infested stems.

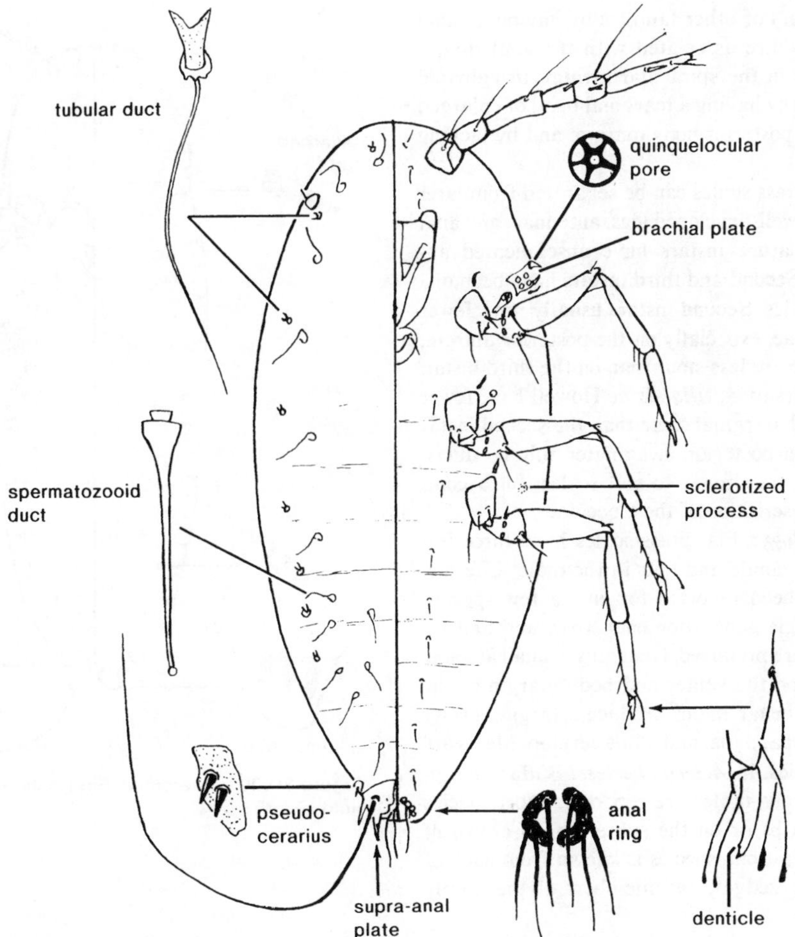

Figure 30.28. Tachardiidae. First instar—dorsal on left, ventral on right.

Comments: Lac scales occur throughout the world except in Europe. They apparently prefer warm dry areas. There are approximately 80 species of lac scales in eight genera; in the United States there are seven species in one genus which occur primarily in the desert areas of the Southwest. Lac scales are rarely reported as pests. The true lac scale is of importance as a source of lac which is used in shellac and other resinous products. At one time this insect was also used as a source of certain red and yellow dyes, but the development of synthetic products and the discovery of the cochineal dyes essentially have eliminated lac dyes from the market.

Selected Bibliography

Chamberlin 1923.
Colton 1943, 1944.
Glover 1937.
Varshney 1970.
Varshney and Ganguly 1971.

ACLERDIDAE

The Flat Grass Scales or Aclerdids

Figure 30.29

Relationships and Diagnosis: Flat grass scales are considered to be most closely related to soft scales (Coccidae). Adult females usually can be recognized in the field by occurring in grass leaf sheaths and root crowns, and by having an elongate body, no wax, and a sclerotized posterior apex. The body is often pink, red, or brown.

First instars of flat grass scales can be distinguished from first instars of all other scale families by having protruding anal lobes with two or three enlarged setae, at least one pore in each spiracular atrium, a row of enlarged setae around the body margin, lacking an anal plate, dorsal body setae on the abdomen, and setae and pores on the anal ring. Second and third instars have not been studied in detail. They differ from second and third instars of other families by having an anal cleft, a platelike structure associated with the anal ring, a large cluster of pores in the spiracular atrium, invaginated tubular ducts, by usually having a marginal band of enlarged setae and a crenulate posterior body margin, and by lacking legs.

First instar flat grass scales can be separated from later immatures by having well-developed legs, antennae, and anal lobes; the other immature instars have unsegmented antennae and lack legs. Second and third instars have been distinguished in two species. Second instars usually have fewer marginal enlarged setae, especially on the posterior margin, and the enlarged setae are less stout than on the third instar. The second instar males of *A. tillandsiae* Howell have fewer and differently-shaped marginal setae than the second instar females and have fewer posterior, invaginated tubular ducts.

Flat grass scales are similar to soft scales; for a comparison refer to the description of the Coccidae.

Biology and Ecology: Flat grass scales have three immature instars in the female and four in the male. Life history information has been reported for only a few species. There usually is a single generation each year, and first instars rather than eggs are produced. Generally a small amount of mealy wax occurs on the venter and body margin of the adult female. Second instar males produce a fragile, glassy test in which the prepupa, pupa, and adult develop. Males are produced in some species. In *Aclerda berlesei* Buffa macropterous and brachypterous males are reported. Most species occur in the leaf sheaths or on the subterranean crown of grasses, although at least one species is known from each of Spanish moss, orchids, sedges, and members of the family Combretaceae.

Comments: Flat grass scales occur in all zoogeographic regions, with the greatest diversity of species occurring in the United States. There are approximately 50 species of flat grass scales in three genera; in the United States there are approximately 15 species in one genus, *Aclerda*. Flat grass scales are occasionally pests of sugar cane. The genus *Rhodesaclerda* is quite different from *Aclerda* and *Nipponaclerda* in that the species occur on exposed portions of the host and produce a thick waxy test.

Selected Bibliography

Borchsenius 1960.
Gomez-Menor Ortega 1937.
Howell 1973.
LaFace 1916.
McConnell 1954.

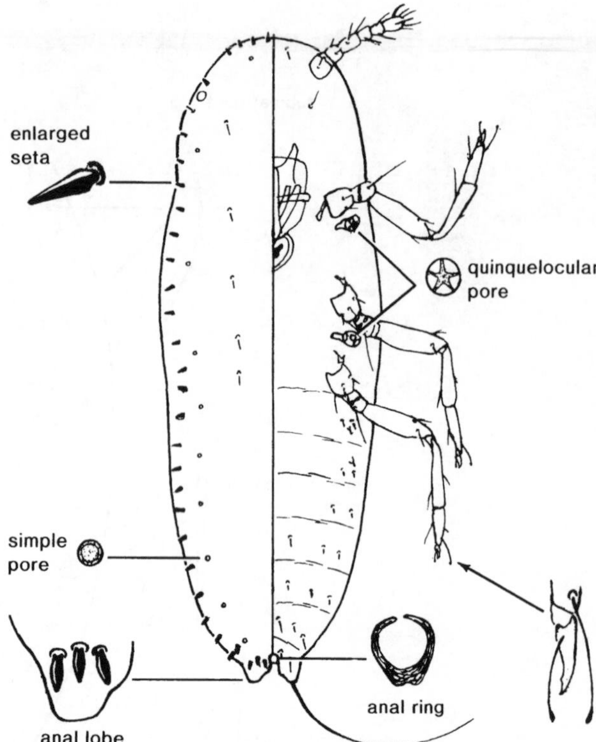

Figure 30.29. Aclerdidae. First instar—dorsal on left, ventral on right.

DACTYLOPIIDAE

The Cochineal Scales or Dactylopiids

Figure 30.30

Relationships and Diagnosis: Cochineal scales seem to be related to the felt scales (Eriococcidae) and gall-like scales (Kermesidae), but this relationship needs study. Some coccidologists believe that cochineal scales are more closely related to mealybugs. Adult female cochineal scales are easily recognized in the field by the sticky, weblike strands of waxy secretions that cover the body, the bright red color of the body contents, and the occurrence of this insect on cactus of the genera *Opuntia* and *Nopalea*. *Dactylopius coccus* Costa lacks a filamentous secretion. Infestations usually occur in protected and shaded parts of the cactus.

First instar cochineal scales can be separated from first instars of other families by having truncate, enlarged setae on the dorsum, clusters of quinquelocular pores near the body margins and near the spiracles, 6-segmented antennae, an anal ring with no pores and with two pairs of setae, only the anterior portion of the ring sclerotized, and by lacking a denticle on the claw. Second instar females have the same features as the first instars and also have invaginated tubular ducts associated with the clusters of quinquelocular pores. Second instar males are similar to second instar females but have 7-segmented antennae, invaginated tubular ducts that are not associated with the clusters of quinquelocular pores, and have fewer truncate enlarged setae.

First instar cochineal scales can be separated from later instars by having the quinquelocular pores on the abdomen restricted to the marginal areas, six longitudinal lines of enlarged setae, and by lacking tubular ducts. Second instar females can be separated by having 6-segmented antennae, quinquelocular pores scattered over the dorsal surface of the abdomen, more than six longitudinal lines of enlarged setae, and invaginated tubular ducts associated with the quinquelocular pore clusters. Second instar males have 7-segmented

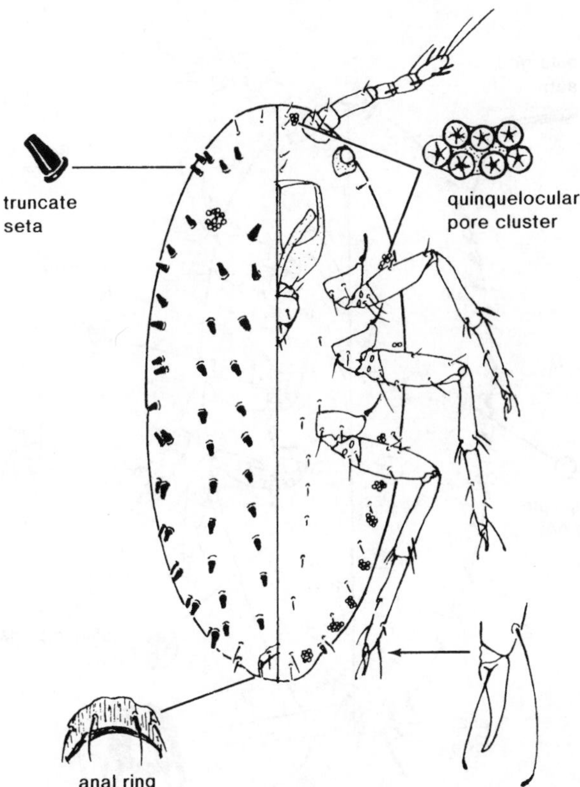

Figure 30.30. Dactylopiidae. First instar—dorsal on left, ventral on right.

antennae, quinquelocular pores scattered over the dorsal surface of the abdomen, six longitudinal lines of enlarged setae, and invaginated tubular ducts scattered over the body surface that are not associated with the quinquelocular pore clusters.

Cochineal scales are similar to felt scales and mealybugs. Cochineal scale first instars differ by having truncate enlarged setae, clusters of quinquelocular pores along the body margin, an anal ring that lacks pores and is sclerotized anteriorly only, by lacking microtubular ducts and a denticle on the claw. Felt scale first instars rarely have truncate enlarged setae, have an anal ring that usually has pores and is sclerotized completely, rarely have clusters of quinquelocular pores and when present they are not restricted to the marginal areas, and have a denticle on the claw. Mealybugs usually have trilocular pores and ostioles; these structures are absent from cochineal scales. Second instar cochineal insects have the same aggregation of differences as the first instars except the quinquelocular pores are more generally distributed.

Biology and Ecology: Cochineal scales have two immature instars in the female and four in the male. There are three to six generations each year, and development is continuous. Eggs hatch from a few minutes to several hours after being laid; it is likely that in some instances first instars hatch inside the body of the female before being laid. The first instars produce long filaments from some of the truncate setae. The filaments produced from the head are long on the female and short on the male. The filaments apparently cause increased buoyancy and aid wind dispersal. The first instar produces small amounts of weblike secretion. The second and third instar females produce large amounts of weblike secretions that enclose the entire body. Males are common in most species. Cochineal scales occur in protected, often shaded areas of the nonsubterranean portions of the plant and usually are in large aggregations.

Comments: Cochineal scales have been transported to all parts of the world but are endemic to the New World. Most occur in desert areas of the southwestern United States, Mexico, and South America. There are nine species in one genus, *Dactylopius;* in the United States there are three species. The cochineal scale, *Dactylopius coccus* Costa has been used as a source of red dyes for the past several centuries. Even today a cochineal industry persists in southern Mexico, Peru, and the Canary Islands. Cochineal insects also have been used successfully in the control of opuntia cactus. In South Africa *Dactylopius ceylonicus* (Green) has nearly decimated *Opuntia monacantha* (Gunn 1978). On two of the California Channel Islands the once large, widespread populations of *Opuntia* have been restricted to small, isolated populations by *Dactylopius opuntiae* (Cockerell). (Goeden, Fleschner, and Ricker 1967).

Selected Bibliography

De Lotto 1974.
Ferris 1955.
Goeden et al. 1967.
Gunn 1978.
Karny 1972.
Mann 1969.

KERMESIDAE

The Gall-Like Scales or Kermesids

Figure 30.31

Relationships and Diagnosis: Gall-like scales are considered to be closely related to felt scales (Eriococcidae) based on the remarkable similarity of the first and second instars of these families. Adult females and males demonstrate little of the relationship exemplified by the first two instars. Adult female characteristics in the field are distinctive. The body of fully mature adult females swells to form a rotund structure similar in appearance to a gall. Newly molted adult females often are ornately colored with stripes, spots, or stippling, but as the body enlarges these patterns disappear. Some species produce wax over the body but the secretion wears with age.

First instar gall-like scales can be separated from first instars of other families by having protruding and completely sclerotized anal lobes, usually three dorsal setae on each anal lobe, usually with two submedial setae on each abdominal segment excluding posterior one, a denticle on the claw, quinquelocular pores, bilocular pores (probably the same as cruciform pores on felt scales), simple pores on the dorsum, by sometimes having enlarged dorsal setae, and by lacking microtubular ducts. Second instar females have the same features as the first instar including 6-segmented antennae except simple and bilocular pores are sometimes absent, small invaginated tubular ducts are present on the venter, and large invaginated ducts are sometimes present on the dorsum. Second instar males are similar to females but have 7-segmented antennae, small invaginated ducts on the dorsum and venter and sometimes lack simple pores. Second instars may have two partial, longitudinal lines of dorsal setae between the submedial lines that usually are absent from the first instar. Third instar females possess many of the features of the second instars. They have a submarginal band of invaginated tubular ducts on the venter, abortive antennae, usually have abortive legs, and lack protruding anal lobes. Third instars are most similar to adult females but lack a vulva.

First instar gall-like scales can be separated from later gall-like scale instars by lacking invaginated tubular ducts and by having protruding anal lobes, and well-developed legs and antennae. Second instars can be distinguished by having invaginated tubular ducts, protruding anal lobes, and well-developed legs and antennae. Second instar males have 7-segmented antennae and some tubular ducts on the dorsum of the same size as those on the venter. Second instar females have 6-segmented antennae and no dorsal tubular ducts or larger tubular ducts on the dorsum than on the venter. Third instar females lack protruding anal lobes, but have a submarginal cluster of tubular ducts on the venter, and abortive antennae.

Gall-like scales are similar to felt scales. Gall-like scale first instars differ by usually having two submedial dorsal setae on each abdominal segment excluding the posterior segment, by having anal lobes that are completely sclerotized, simple pores, and by lacking microtubular ducts. Felt scale first instars have four submedial dorsal setae on each abdominal segment excluding the posterior one, anal lobes that are partially sclerotized, microtubular ducts, and lack simple pores. Second instar gall-like scales differ in the same ways as the first instars. Felt scales do not have an immature third instar female.

Biology and Ecology: Gall-like scales have three immature instars in the female and four in the male. Life history data have been gathered for only a limited number of species. There is one generation each year. Overwintering stages occur on the large stems or bole of the tree in cracks and crevices. In early spring those species that overwinter as first instars molt and second instar females migrate to the new growth or leaves and feed; second instar males either remain in the overwintering site or migrate to duff beneath the tree and develop into prepupae, pupae, and adults. Those species that overwinter as second instars remain and feed at the overwintering site as females or migrate to large branches on the bole of the tree as males. In late spring or early summer adults occur and mating takes place. Females enlarge rapidly, assuming the appearance of a gall or bud. Egg laying occurs in summer inside of a special brood chamber formed by lobelike extensions that are developed from the body wall

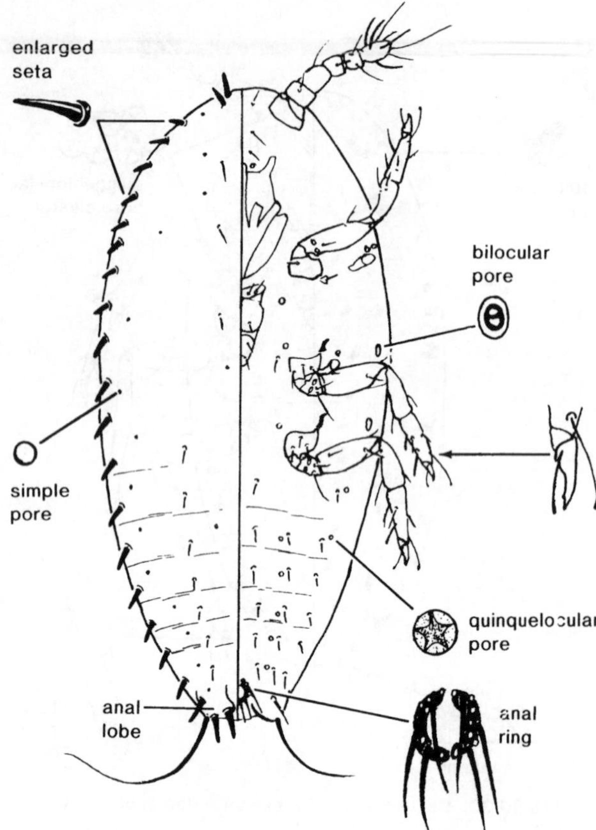

Figure 30.31. Kermesidae. First instar—dorsal on left, ventral on right.

(Bullington and Kosztarab, 1985). As many as 3,000 eggs may be laid by a single female. The eggs hatch, and the first instars leave the brood chamber through a small opening in the posterior end of the female body. Some gall-like scales cause branch and leaf deformation.

Gall-like scales primarily occur in the Palearctic and Nearctic regions and are restricted to *Quercus* (oaks), and rarely to *Lithocarpus* (tanoak) and *Chrysolepsis* (chinquapins). They occur on the stems, branches, boles, and leaves of the host.

Comments: There are approximately 65 species of gall-like scales in nine genera; in North America there are 34 species in five genera. The European kermesid, *Kermes roboris* (Fourcroy) was used as a source of crimson dye before the introduction of New World dyes produced by the cochineal scale. Gall-like scales occasionally cause damage to ornamental oaks. They may cause flagging, especially during times of stress. Species implicated as causing damage in the United States are *Kermes kingi* Cockerell and *K. pubescens* Bogue.

Selected Bibliography

Baer 1980.
Baer and Kosztarab 1985.
Borchsenius 1960.
Bullington and Kosztarab 1985.
Hamon et al. 1976.

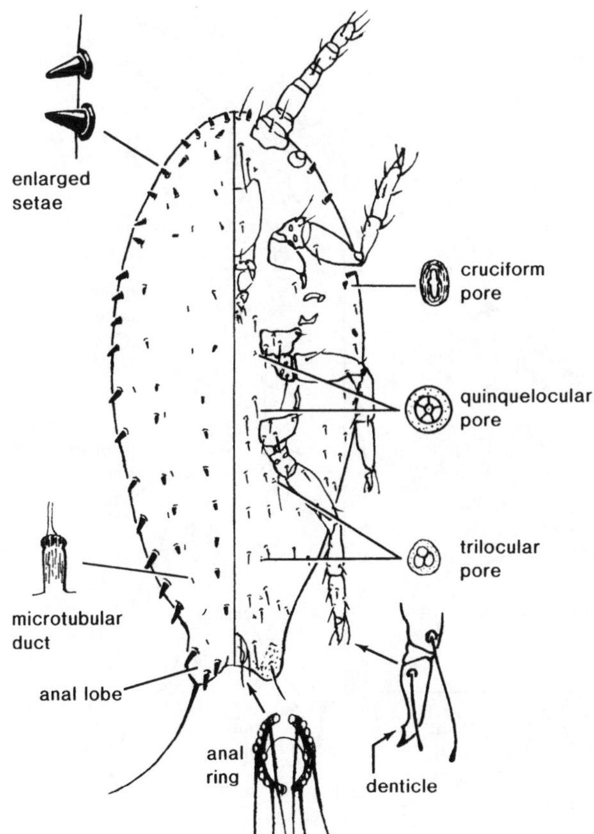

Figure 30.32. Eriococcidae. First instar—dorsal on left, ventral on right.

ERIOCOCCIDAE

The Felt Scales or Eriococcids

Figure 30.32

Relationships and Diagnosis: Felt scales apparently are most closely related to the gall-like scales (Kermesidae); the first instars are nearly inseparable. Adult females of most felt scales are characterized in the field by producing a white, felt ovisac that encloses the pyriform body. Body color varies from bright pink or red to purple or brown. Infestation may occur on all parts of the host including the roots.

First instar felt scales can be separated from first instars of other families by having enlarged setae, cruciform pores, microtubular ducts, six dorsal setae on each abdominal segment except the last segment, quinquelocular pores, a denticle on the claw, protruding anal lobes that are incompletely sclerotized and which possess one to three enlarged setae, and by lacking simple pores. Second instar females have the same aggregation of characters but have more setae, pores, and microtubular ducts. Second instar males usually have invaginated tubular ducts and 7-segmented antennae.

First instar felt scales can be separated from later felt scale instars by having six dorsal setae on each abdominal segment except the posterior one, 6-segmented antennae, and by lacking invaginated tubular ducts. Second instars have at least eight dorsal setae on each abdominal segment except the posterior one. Second instar females usually have 6-segmented antennae and normally lack invaginated tubular ducts. Second instar males usually have 7-segmented antennae and invaginated tubular ducts.

Felt scales are similar to gall-like scales and cochineal scales; for comparisons refer to the descriptions of the Kermesidae and Dactylopiidae.

Biology and Ecology: Felt scales have two immature instars in the female and four in the male. Most species have one or two generations each year. The overwintering stage often is the adult female or egg in the ovisac. First instars appear in early spring by escaping from the ovisac through a small hole in the posterior end. Settling usually occurs within hours of emergence from the ovisac. Second instar males feed for a short period then produce a narrow, felt sack that encloses the body. Development of the prepupa, pupa, and adult male occurs within the sack. Soon after molting to the adult, females mate and produce the ovisac several days later. Usually 50 to 100 eggs are laid. A few groups of felt scales do not produce ovisacs. Males are common. Felt scales generally occur on stems, branches or roots.

Comments: Felt scales occur in all zoogeographic regions, have very poor representation in the Ethiopian and Oriental regions, and are most abundant in New Zealand and Australia. There are approximately 450 species in 60 genera; in the United States there are 54 species in 10 genera. Several species are pests of ornamentals. Eight species are considered of economic importance in the United States; some of the more notorious ones are beech scale, *Cryptococcus fagisuga* Lindinger, azalea bark scale, *Eriococcus azaleae* Comstock, cactus felt scale, *E. coccineus* Cockerell, and European elm scale, *Gossyparia spuria* (Modeer). The beech scale, in association with a fungus, has caused the loss of large stands of beech in the New England states and Canada. The azalea bark scale may cause dieback and an unsightly appearance in azaleas. The cactus felt scale can kill ornamental cactus if left unchecked. European elm scale may cause stunting, dieback, and even death of elm.

Selected Bibliography

Ferris 1955.
Hoy 1963.
Miller 1969.
Miller and McKenzie 1967.

LECANODIASPIDIDAE

The False Pit Scales or Lecanodiaspidids

Figure 30.33

Relationships and Diagnosis: False pit scales are most closely related to ornate pit scales (Cerococcidae); the lecanodiaspid-cerococcid lineage is thought to be closely related to the pit scales. Adult females are characterized in the field by producing a waxy test that encloses the body. The test usually is papery in texture, giving a corrugated appearance. Color varies from yellow to reddish brown. Tests are produced by adult females and second instar males. The male test is similar to the adult female test but is smaller and narrower. Mature females produce pit malformation on the stems of some hosts.

First instar false pit scales can be separated from first instars of other families by having two anal plates, an arched plate, strongly protruding anal lobes that are membranous, 8-shaped pores, quinquelocular pores near the spiracles, three pairs of setae on the anal ring, and by usually having spiracular setae and a denticle on the claw. Second instar females possess the same aggregation of differentiating features as the first instars except for the legs which are represented by small stubs or are absent. Also, the pores near the spiracles often form a band from the spiracles to the base of the spiracular setae. Second instar males are similar to second instar females but have invaginated tubular ducts and longer legs.

First instar false pit scales can be separated from later instars by having 6-segmented antennae with constrictions at each intersegmental area, well-developed legs, and no tubular ducts. Second instars have antennae that lack noticeable constrictions between segments and usually have more than six segments. Second instar females have abortive legs and lack tubular ducts. Second instar males have tubular ducts and weakly segmented legs.

False pit scales are similar to ornate pit scales and pit scales; for comparisons refer to the descriptions of the Cerococcidae and Asterolecaniidae.

Biology and Ecology: False pit scales have two immature instars in the female and 4 in the male. Life history data are nearly non-existent. Howell and Kosztarab (1972) published a brief description of *Lecanodiaspis prosopidis* (Maskell); it very likely is typical of many temperate species. There is 1 generation each year and overwintering takes place in the egg stage inside of the adult female test. Eggs hatch in early spring, and first instars leave the test through a small hole at the posterior end. Second instars appear in early to mid summer. Adults are present in mid to late summer, and eggs are laid in the fall. Males probably occur in most species. Species in other genera are frequently tended by ants and are found in carton tents or hollow stems. Two species form stem galls on their host. False pit scales normally occur on stems and branches of the host, but they occasionally are collected on the leaves and trunks.

Comments: False pit scales occur in all zoogeographic regions. *Lecanodiaspis* is worldwide, but the remaining 10 genera have relatively local distributions. The greatest diversity of genera occurs in the southern Palearctic region and in the Oriental region. There are 72 species in 10 genera; in the United States there are five species in the genus *Lecanodiaspis*. In the United States the common pit scale, *L. prosopidis* (Maskell) is an occasional pest of ornamentals. Damage includes severe pitting of the stems and branches. Howell and Kosztarab (1972) mention eight economic species of *Lecanodiaspis* in the world. Lambdin and Kosztarab (1973) state that several false pit scale genera include species of economic importance.

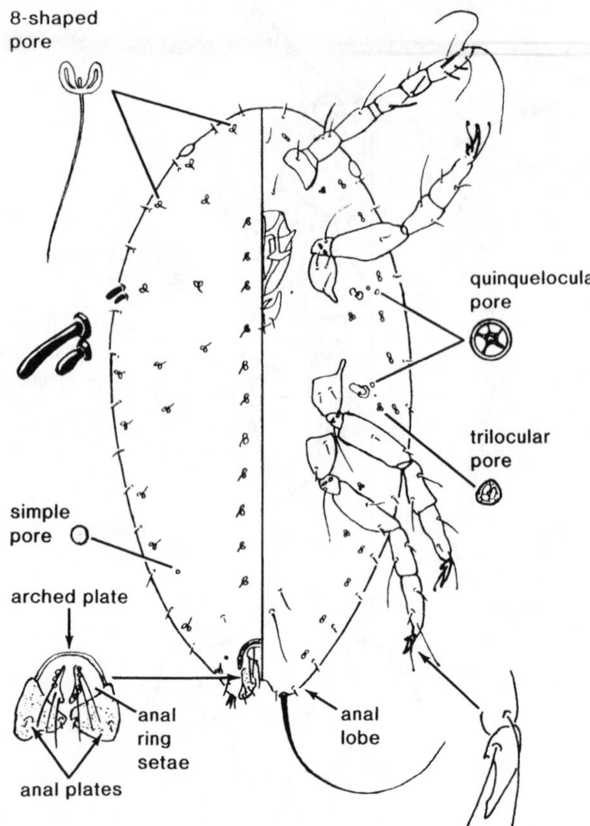

Figure 30.33. Lecanodiaspididae. First instar—dorsal on left, ventral on right.

Selected Bibliography

Borchsenius 1960.
Howell and Kosztarab 1972.
Lambdin and Kosztarab 1973.
Williams and Kosztarab 1970.

CEROCOCCIDAE

The Ornate Pit Scales or Cerococcids

Figure 30.34

Relationships and Diagnosis: Ornate pit scales are most closely related to the false pit scales (Lecanodiaspididae); the cerococcid-lecanodiaspidid lineage is most closely related to the pit scales (Lambdin and Kosztarab, 1977). Adult females are characterized in the field by producing a waxy test that encloses the body. The test may be smooth, corrugated, stellate, checkered, or woollike. In most species it is cream to dark brown, but a few species produce tests that are orange, yellow, pink, red, or white. Tests are produced by adult females and second instar males. The male test is smaller and

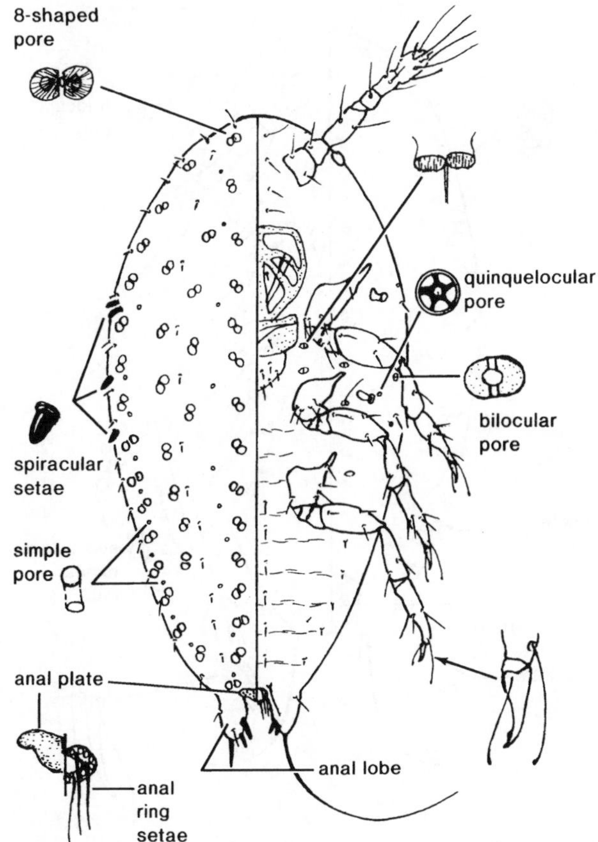

Figure 30.34. Cerococcidae. First instar—dorsal on left and ventral on right.

narrower than the adult female test. The tests of *Asterococcus* and *Solenophora* incorporate the shed skin of the first instar and probably have a test in the second instar female as well as the second instar male. Mature females usually occur in a pit on the stem of the host.

First instar ornate pit scales can be separated from first instars of other families by having a single plate near the anal ring, strongly protruding anal lobes that are heavily sclerotized, 8-shaped pores, quinquelocular pores near the spiracles, a denticle on the claw, three or four pairs of setae on the anal ring, by usually having spiracular setae, and by lacking an arched plate. Second instar females possess the diagnostic features of the first instar except for unsegmented antennae and legs, which are absent or vestigial. Also, the quinquelocular pores near the spiracles often form a band from the spiracles to the base of the spiracular setae. Second instar males are similar to second instar females but usually have invaginated tubular ducts.

First instar ornate pit scales may be separated from later instars by having 6-segmented antennae and well developed legs. Second instars can be separated by lacking a vulva and by having 1-segmented antennae and legs that are abortive or absent.

Ornate pit scales are similar to pit scales; for a comparison refer to the description of the Asterolecaniidae. Ornate pit scales are most closely related to false pit scales. Both families produce similar tests, have 8-shaped pores, and anal plates. Cerococcids differ by having a single anal plate and heavily sclerotized anal lobes and by lacking an arched plate. False pit scales have two anal plates, membranous anal lobes, and an arched plate.

Biology and Ecology: Ornate pit scales have two immature instars in the female and four in the male. Life history data are scant. Based on information from a few United States species of *Cerococcus* the following life history apparently is representative. There is one generation each year and overwintering takes place in the egg stage inside the test. Eggs hatch in the spring and first instars escape from the test through a small hole at the posterior end. Second instars appear in early summer and adults occur in mid to late summer. Eggs are laid in the test in the fall. Males are known in most species. Ornate pit scales feed on stems and branches, apparently preferring 1- or 2-year old stems.

Comments: Ornate pit scales occur in all zoogeographic regions. They seem to be most abundant in temperate and subtropical areas; species diversity is greatest in the Palearctic region. There are about 60 species in three genera; in the United States all five species are in the genus *Cerococcus*. Some species may be of economic importance on ornamental trees and shrubs. Damage includes premature leaf or fruit drop, dieback, loss of plant vigor, or unsightly appearance. United States species reported to cause damage are *C. deklei* Kosztarab and Vest, *C. parrotti* (Hunter), and *C. quercus* Comstock. Lambdin and Kosztarab (1977) list 15 species as economic pests in the world.

Selected Bibliography

Borchsenius 1960.
Hamon and Kosztarab 1979.
Howell et al. 1971.
Lambdin and Kosztarab 1976, 1977.

ASTEROLECANIIDAE

The Pit Scales or Asterolecaniids

Figure 30.35

Relationships and Diagnosis: Pit scales appear to be a phylogenetically diverse group and probably are representatives of at least two different lineages. To maintain a reasonable degree of consistency, comments will be restricted for the most part to the group of genera similar to *Asterolecanium*. Asterolecaniids appear to be at the most advanced end of an aggregation of families grouped in the lecanoid assemblage. Although pit scales have features typical of lecanoid families, they also demonstrate some affinities with the more advanced diaspidoid assemblage. Adult females often are characterized in the field by the presence of a pitlike depression in the host under the body of the insect. The body of the insect is enclosed in a translucent, waxy test that acts like a cover similar to that formed by the armored scales. The cover does not include the skins of immatures and is separate from the body.

First instar pit scales can be separated from first instars of other families by usually having 8-shaped pores around the body margin, by having quinquelocular or trilocular pores near the spiracles, membranous anal lobes that are weakly protruding, two or three narrow sclerotizations occasionally present in the anal area, and by lacking spiracular setae, pygidial lobes, anal plates, and a denticle on the claw. Second instar females lack legs and have 1-segmented antennae but otherwise possess the characteristics of the first instar. Second instar males are similar to second instar females but usually differ by having invaginated tubular ducts.

First instar pit scales can be separated from later instars by having 5- or 6-segmented antennae and well-developed legs. Second instars can be separated by lacking legs and a vulva and by having unsegmented antennae.

Pit scales are similar to palm scales; for a comparison refer to the descriptions of the Phoenicococcidae. Pit scales also are similar to ornate pit scales and to false pit scales. Pit scale first instars differ by having small anal lobes, and by lacking spiracular setae and a denticle on the claw. The anal lobes on ornate pit scales are large and heavily sclerotized, whereas on false pit scales they are large but not as sclerotized. Ornate pit scale and false pit scale first instars have spiracular setae in most species and a denticle on the claw. Second instar pit scales have the same aggregation of differences as first instars except for the legs which are absent. The antennae on pit scale and ornate pit scale second instars are unsegmented, whereas on false pit scales there usually are more than six segments.

Biology and Ecology: Pit scales have two immature instars in the female and four in the male. Life histories of only a few species have been studied. Most species have a single generation per year and overwinter as adult females. During egg laying, the female shrivels into the anterior end of the test as the eggs fill the posterior end. Eggs hatch in late spring and early summer. First instars form a small pit soon after they feed. Adults appear in mid to late summer. The test is produced by the adult female. Males of most asterolecaniid species are unknown; many species are believed to be parthenogenetic. They generally occur on stems and branches.

Comments: Pit scales occur in all zoogeographic regions. There are over 250 species; in the United States there are about 30 species in four genera. Several species are important pests of ornamentals and forest trees including *Asterolecanium minus* Lindinger, *A. puteanum* Russell, and *A. variolosum* (Ratzeburg). *Pollinia pollini* (Costa) is a pest of olive in Europe. Damage may involve severe distortion of twigs and branches, dieback of stems, and delayed leaf production in spring.

Selected Bibliography

Boratynski 1961.
Borchsenius 1960.
Ferris 1955.
Russell 1941.

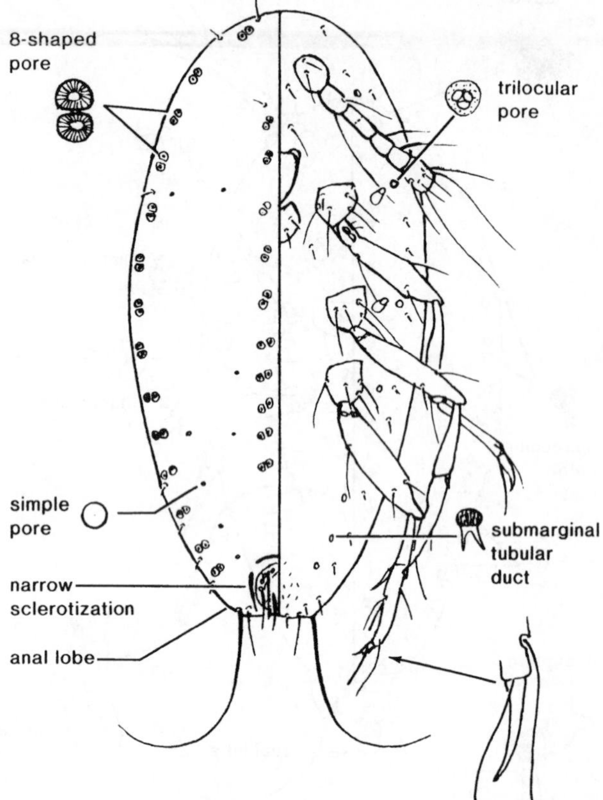

Figure 30.35. Asterolecaniidae. First instar—dorsal on left, ventral on right.

PHOENICOCOCCIDAE

The Palm Scales or Phoenicococcids

Figure 30.36

Relationships and Diagnosis: Palm scales seem to be an annectant group sharing similarities of armored scales (Diaspididae) and pit scales (Asterolecaniidae). Adult females are characterized in the field by producing a white, filamentous, waxy covering that encloses the body. The cover frequently is brushed off on adults, exposing the conspicuous red body. First and second instars also produce a filamentous, waxy cover.

First instar palm scales can be separated from first instars of other families by having 8-shaped tubular ducts on the abdomen, quinquelocular pores near the spiracles, three pairs of setae associated with the anal ring, a denticle on the claw, and by lacking a pygidium and pygidial lobes. Second instars lack well-developed legs and have reduced antennae but otherwise possess the characteristics of the first instar.

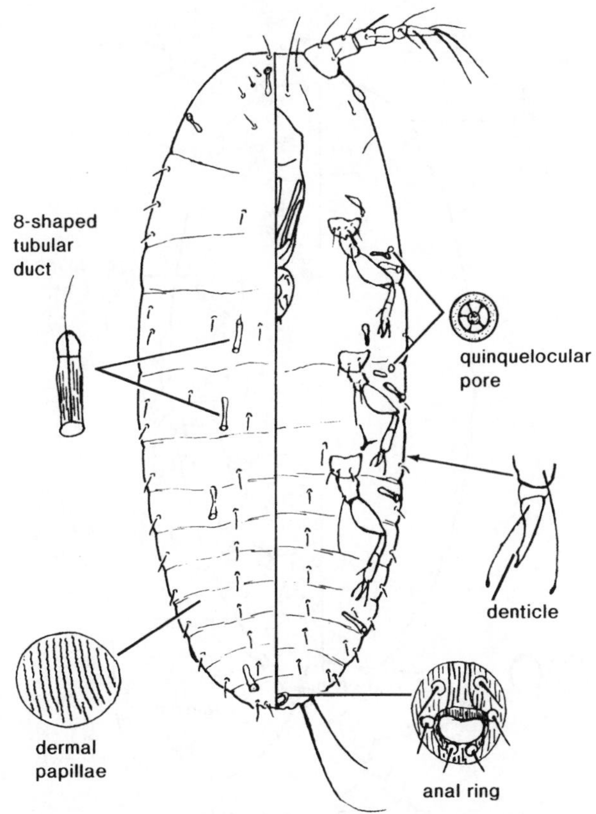

Figure 30.36. Phoenicococcidae. First instar—dorsal on left, ventral on right.

First instar palm scales can be separated from later instars by having 6-segmented antennae and well-developed legs. Second instar females can be separated by lacking leg stubs and a vulva. Second instar males have leg stubs and lack a vulva.

Palm scales are similar to pit scales. Palm scale first instars differ by having quinquelocular pores near the spiracles and a denticle on the claw and by lacking 8-shaped pores. Pit scales have trilocular or quinquelocular pores near the spiracles, lack a denticle on the claw, and often have characteristic 8-shaped pores. Second instar female palm scales have 8-shaped tubular ducts and dermal papillae and lack 8-shaped pores. Pit scales lack dermal papillae, usually lack 8-shaped tubular ducts, and have 8-shaped pores. Second instar male palm scales, in addition to the aforementioned characters of the female, differ from second instar male pit scales by having small leg stubs and no invaginated tubular ducts. Pit scales lack legs and have many invaginated tubular ducts.

Biology and Ecology: Palm scales have two immature instars in the female and four in the male. Many overlapping generations occur with all instars present at any time during the year. In warm months development requires about 60 days but is considerably slower during fall and winter. Both eggs and first instars are laid by adult females. Some eggs hatch within the body of the female, some hatch immediately after being laid, and some hatch an hour or more after oviposition. All instars except third, fourth, and fifth instar males produce a white, filamentous sac that encloses the body. Population highs are in the spring.

The red date palm scale has been reported in Arizona, California, and Texas. In the Old World it occurs in Egypt, Iraq, Israel, Pakistan, and Saudi Arabia. It infests *Pandanus,* and palms including date palm, *Phoenix* spp., *Calamus,* and *Daemonorops.* Preferred feeding sites are on white tissue below the fiber. Heavily invested hosts may have specimens on exposed roots and young leaves.

Comments: Phoenicococcidae includes only one species, the red date palm scale, *Phoenicococcus marlatti* Cockerell. Although several peculiar scales have been included in the family, it seems best to include only the type species. The red date palm scale apparently causes premature drying of fruit, early death of older leaves, and general unthriftiness of date palms.

Selected Bibliography

Borden 1921.
Stickney 1934.
Stickney et al. 1950.

CONCHASPIDIDAE

The False Armored Scales or Conchaspidids

Figure 30.37

Relationships and Diagnosis: False armored scales are similar in appearance to armored scales (Diaspididae), but their phylogenetic affinities have not been carefully analyzed. Adult females are characterized in the field by having a domicile-type cover that is not attached to the body and consists of waxlike secretions but does not incorporate the exuviae of immatures. First instars do not produce a cover. Second and third instar females make a cover similar to that of the adult.

First instar false armored scales can be separated from first instars of other families by having a sclerotized pygidium, a weakly developed dorsal ocellar spot, a denticle on the claw, a trilocular or quinquelocular pore associated with the anterior spiracle, femur with setae, and by lacking sensoria on the trochanter and a separate tibia and tarsus. Other immature instars have the same aggregation of characters.

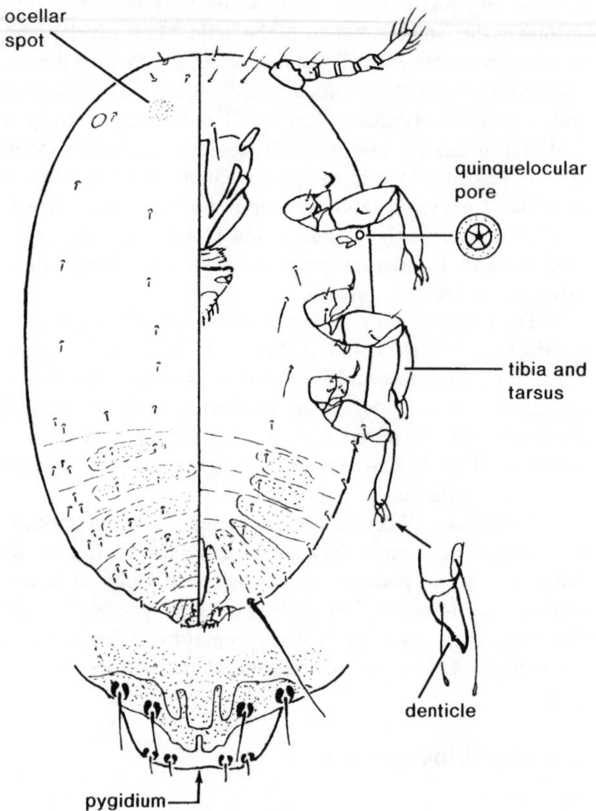

Figure 30.37. Conchaspididae. First instar—dorsal on left, ventral on right.

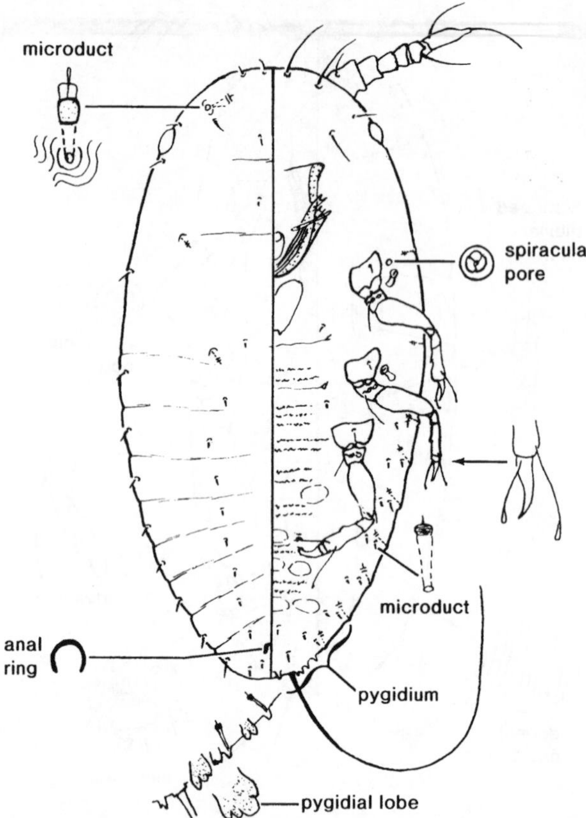

Figure 30.38. Diaspididae. First instar—dorsal on left, ventral on right.

First instar false armored scales can be separated from later instars by having slender 6-segmented antennae, a small body, and no scale cover. Second instars have broad antennae with less than 6 segments and lack a vulva. The second instar male is similar to the second instar female but differs by having more tubular ducts.

False armored scales are similar to armored scales; for a comparison refer to the description of the Diaspididae.

Biology and Ecology: False armored scales have three immature instars in the female and four in the male. Males are known in some species. Feeding occurs on leaves or branches.

Comments: The family Conchaspididae occurs in tropical and subtropical areas and is endemic in the Ethiopian, Neotropical, Oriental, and southern Nearctic Regions. The greatest diversity of species occurs in Madagascar. There are approximately 23 species in two genera. A third genus, *Fagisuga,* is sometimes considered to be a member, but its familial placement is questionable. Only one species, *Conchaspis angraeci* Cockerell, occurs in the United States. In Florida it may be a pest of sea grape, *Coccoloba* spp., hibiscus, orchids, and pittosporum.

Selected Bibliography

Ben-Dov 1974, 1981.
D'Ascoli and Kosztarab 1969.
Hamon 1979.
Mamet 1954.

DIASPIDIDAE

The Armored Scales or Diaspidids

Figure 30.38

Relationships and Diagnosis: Armored scales are considered to be the most advanced scale insects. Adult females are characterized in the field by a domicile-type cover that is not attached to the body and which consists of waxlike secretions and exuviae of immatures. First instars produce two general types of covers, those that are composed of a filamentous secretion or a simple transparent layer, and those that are composed of two parts, a cap that is formed of fine waxy filaments, and a series of concentric rings that are formed of thick secretory material. The rings are attached to the edge of the cap as the first instar enlarges. The second instar female incorporates the cover and shed skin of the first instar and adds secretory material as the body enlarges. The cover of the second instar male is usually smaller and more elongate than the cover of the second instar female and sometimes is feltlike with one to three longitudinal ridges; these characters are absent from female covers. The prepupal and pupal covers are the same as those of second instar males.

First instar armored scales can be separated from first instars of other families by having a posterior pygidium and at least one pair of pygidial lobes, and by lacking setae on the femur. Second instars also possess these diagnostic features except the legs are absent or vestigial.

First instar armored scales can be separated from later instars by having 5- or 6-segmented antennae and well developed legs. Second instars have 1-segmented antennae and lack legs. Second instars differ from adult females by having fewer ducts and by lacking a vulva.

Armored scales are similar in appearance to false armored scales; both families produce a domicile-type scale cover and have a pygidium. Diaspidid first instars differ by having pygidial lobes and a scale cover, and by lacking a dorsal ocellar spot, and a denticle on the claw. Conchaspidid first instars lack pygidial lobes and a scale cover and have a dorsal ocellar spot, and a denticle on the claw. Second instar armored scales have unsegmented antennae, lack legs or have small stubs; second instar false armored scales have segmented legs and antennae.

Biology and Ecology: Armored scales have two immature instars in the female and four in the male. Life histories are diverse; there may be from one to six generations each year and overwintering may occur in any stage except the third, fourth, or fifth instar male. In many species the number of generations and overwintering stage may vary considerably depending upon climatic conditions. Some species develop continuously in some areas but are slower in the cooler months. Eggs or first instars are laid under the scale cover. A small slit is present at the posterior end of the cover through which the first instars escape. Natural dispersal is undertaken by the first instar either passively by wind or actively by crawling. All other instars are apodous except the third, fourth, and fifth instar males. Transportation of infested plant materials is also an important mechanism of dispersal. Armored scales primarily infest leaves and branches of the host. They sometimes occur under the epidermis of the host and are exceedingly cryptic. They rarely are found on subterranean parts of the plant.

Comments: Armored scales are found nearly everywhere that perennial plants occur, with the greatest diversity of taxa found in the tropics and subtropics. They are the largest scale-insect family, there being about 1,700 species in 338 genera in the world and 297 species in 85 genera in the United States. Many are pests, particularly of ornamental trees and shrubs. Some of the more notorious species are California red scale, *Aonidiella aurantii* (Maskell), pine needle scale, *Chionaspis pinifoliae* (Fitch), tea scale, *Fiorinia theae* Green, white peach scale, *Pseudaulacaspis pentagona* (Targioni-Tozzetti), San José scale, *Quadraspidiotus perniciosus* (Comstock), and arrowhead scale, *Unaspis yanonensis* (Kuwana). Schmutterer et al. (1957) lists about 135 species of economic armored scales in the world; Beardsley and Gonzalez (1975) give 43 species in their world list of principal diaspidid pests; there are approximately 100 species of some economic importance in the United States.

Selected Bibliography

Beardsley and Gonzalez 1975.
Borchsenius 1966.
Ferris 1937–1942.
Schmutterer 1952.
Stoetzel 1975, 1976.
Stoetzel and Davidson 1974.

BIBLIOGRAPHY

Annand, P. N. 1928. A contribution toward a monograph of the Adelginae (Phylloxeridae) of North America. Stanford Univ. Publ. Univ. Ser. Biol. Sci. 6(1):1–146.

Baer, R. G. 1980. A new species of gall-like coccid from southeastern United States. J. Ga. Ent. Soc. 15:20–25. (Kermesidae)

Baer, R. G. and M. Kosztarab. 1985. A morphological and systematic study of the first and second instars of the family Kermesidae in the Nearctic Region. Va. Polytech. Inst. State Univ. Agr. Expt. Sta. Bull. 85–11, pp. 119–261.

Ball, J. C. and D. D. Jensen. 1966. Sexual dimorphism in nymphs of *Psylla pryicola* (Hemiptera: Psyllidae). Ann. Ent. Soc. Am. 59:1292–1294.

Barber, G. W. and W. O. Ellis. 1922. Eggs of three Cercopidae. Psyche 29:1–3.

Beamer, R. H. 1928. Studies on the biology of Kansas Cicadidae. Univ. Kansas Sci. Bull., 18(2):155–263.

Beardsley, J. W., Jr. and Gonzalez, R. H. 1975. The biology and ecology of armored scales. Ann. Rev. Ent. 20:47–73.

Beingolea, G. O. 1969. Notas sobre *Orthezia pseudoinsignis* Morrison, plaga de algunas plantas cultivadas y ornamentales en el Peru. Rev. Peru. Ent. 12:96–118. (Ortheziidae)

Ben-Dov, Y. 1974. On the species of Conchaspididae from Africa and Madagascar with description of a new species. Rev. Zool. Afr. 88:363–373.

Ben-Dov, Y. 1981. A catalog of the Conchaspididae of the world. Ann. Ent. Soc. France (N.S.) 17:143–156.

Blackman, R. L. 1974. Aphids. Ginn & Company Ltd., London. 175 pp.

Blackman, R. L. and V. F. Eastop. 1984. Aphids on the world's crops. John Wiley and Sons, New York. 474 pp.

Boerner, C. and K. Heinze. 1957. Adelgidae (Chermesidae), Tannen-gallause. Handb. Pfl-Krankh. (Ed. Sorauer). 5:323–354.

Boratynski, J. 1952. *Matsucoccus pini* (Green 1925): Bionomics and external anatomy with reference to the variability of some taxonomic characters. Trans. R. Ent. Soc. Lond. 103:285–326. (Margarodidae)

Boratynski, K. 1961. A note on the species of *Asterolecanium* Targioni-Tozzetti, 1869 on oak in Britain. Proc. R. Ent. Soc. Lond. (B) 30:4–14. (Asterolecaniidae)

Borchsenius, N. S. 1957. Fauna of USSR, Coccidae. Akad. Nauk Zool. Inst. (n.s. 66) 9, 493 pp. (in Russian)

Borchsenius, N. S. 1960. Fauna of USSR, Kermococcidae, Asterolecaniidae, Lecaniodiaspididae, Aclerdidae. Akad. Nauk SSSR Zool. Inst. (n.s. 77) 8, 282 pp. (in Russian)

Borchsenius, N. S. 1966. A catalog of the armored scale insects of the world. Akad. Nauk SSSR Zool. Inst., Leningrad. 449 pp. (in Russian)

Borden, A. D. 1921. A biological study of the red date-palm scale, *Phoenicococcus marlatti*. J. Agr. Res. 21:659–668. (Phoenicococcidae)

Borror, D. J., D. M. DeLong, and C. A. Triplehorn. 1981. An introduction to the study of insects. Fifth edition. Holt, Rinehart and Winston, New York. 827 pp.

Boulard, M. 1969. L'adaptation à la vie aquatique chez les larves de *Muansa clypealis* (Homoptera, Cicadellidae). C. R. Acad. Sci. Paris (D) 268:2602–2604.

Bowers, W. S., L. R. Nault, R. E. Webb, and S. R. Dutky. 1972. Aphid alarm pheromone: Isolation, identification, synthesis. Science. 177:1121–1122.

Branner, John C. 1885. The reputation of the lanternfly. Amer. Nat. 19:835–838.

Brewbaker, J. L. 1979. Diseases of maize in the wet lowland tropics and the collapse of the classic Maya civilization. Econ. Bot. 33:101–118.

Bullington, S. W. and M. Kosztarab. 1985. Revision of the family Kermesidae in the Nearctic Region based on adult and third instar females. Va. Polytech. Inst. State Univ. Agr. Expt. Sta. Bull. 85–11, pp. 1–118.

Caldwell, D. L. and D. L. Schuder. 1979. The life history and description of *Phylloxera caryaecaulis* on shagbark hickory. Ann. Ent. Soc. Am. 72(3):384–390.

Carter, C. I. 1971. Conifer woody aphids (Adelgidae) in Britain. For. Comm. Bull. 42. 51 pp.

Carter, R. D. 1961. Distinguishing sexes in nymphs of the tomato psyllid, *Paratrioza cockerelli*. Ann. Ent. Soc. Am. 54:464–465.

Chamberlin, J. C. 1923. A systematic monograph of the Tachardiinae or lac insects. Bull. Ent. Res. 14:147–212.

Colton, H. S. 1943. Life history and economic possibilities of the American lac insect, *Tachardiella larrea*. Plateau 16:21–32. (Tachardiidae)

Colton, H. S. 1944. The anatomy of the female American lac insect, *Tachardiella larrea*. Mus. North. Ariz. Bull. 21, 24 pp. (Tachardiidae)

Crane, P. S., and E. L. Paddock. 1974. Rice Delphacid. State of California Dept. Food & Agr., Div. Plant Ind. Detection Manual (Add.) DT 3:75.

Crawford, D. L. 1914. A monograph of the jumping plant-lice or Psyllidae of the New World. U.S. Nat. Mus. Bull. 85. 186 pp.

Cumber, R. A. 1952. Studies on *Oliarus atkinsoni* Meyers (Hem. Cixiidae), vector of the "yellow-leaf" disease of *Phormium tenax* Forst. II. The nymphal instars and seasonal changes in the composition of nymphal populations. New Zealand J. Sci. & Tech., B. 34:160–165.

Cumber, R. A. 1953. Studies on *Oliarus atkinsoni* Meyers (Hem. Cixiidae), vector of the "yellow-leaf" disease of *Phormium tenax* Forst. III. Resistance of nymphal forms to submergence-control by inundation. New Zealand J. Sci. & Tech., B. 34:260–266.

D'Ascoli, A. and Kosztarab, M. 1969. Morphological studies on the three nymphal instars of *Conchaspis lata* Hempel. Va. Polytech. Inst. Res. Div. Bull. 36, pp. 39–51. (Conchaspididae)

DeLotto, G. 1974. On the status and identity of the cochineal insects. J. Ent. Soc. S. Afr. 37:167–193. (Dactylopiidae)

Dixon, A. F. G. 1973. Biology of Aphids. The Institute of Biology's Studies in Biology No. 44. The Camelot Press, Ltd., London. 58 pp.

Dixon, A. F. G. 1985. Aphid ecology. Methuen, Inc., New York. 157 pp.

Doering, Kathleen C. 1942. Host plant records of Cercopidae in North America, north of Mexico (Homoptera). J. Kansas Ent. Soc. 15:65–92.

Duncan, C. D. 1922. The North American species of *Phylloxera* infesting oak and chestnut (Hemiptera: Phylloxeridae). Can. Ent. 54(12):267–276.

Eastop, V. F. and D. Hille Ris Lambers. 1976. Survey of the World's Aphids. Dr. W. Junk b.v., The Hague. 573 pp.

Essig, E. O. 1929. Insects of Western North America. MacMillan Co., New York. 1035 pp.

Ezzat, Y. M. 1956. Studies of the "Kew bug", *Orthezia insignis* Browne. Bull. Soc. Ent. Egypte 40:415–431. (Ortheziidae)

Fennah, R. G. 1954. The higher classification of the family Issidae (Homoptera: Fulgoroidea) with descriptions of new species. Trans. R. Ent. Soc. Lond. 105:455–474.

Fennah, R. G. 1973. The cavernicolous fauna of Hawaiian lava tubes: IV. Two new blind *Oliarus* (Fulgoroidea: Cixiidae). Pacific Insects 15:181–184.

Fennah, R. G. 1984. Revisionary notes on the classification of the Nogodinidae (Homoptera, Fulgoroidea) with descriptions of a new genus and a new species. Ent. Monthly Mag. 120:81–86.

Ferris, G. F. 1937–1942. Atlas of the scale insects of North America. Ser. I–IV. The family Diaspididae. Stanf. Univ. Press, Stanford, Calif.

Ferris, G. F. 1950 and 1953. Atlas of the scale insects of North America. Vols. V and VI. The family Pseudococcidae. Stanf. Univ. Press, Stanford, Calif. 506 pp.

Ferris, G. F. 1955. Atlas of the scale insects of North America. Vol. VII. The families Aclerdidae, Asterolecaniidae, Conchaspididae, Dactylopiidae, and Lacciferidae. Stanf. Univ. Press, Stanford, Calif. 233 pp.

Fletcher, M. J. 1979. Egg types and oviposition behaviour in some fulgoroid leafhoppers (Homoptera, Fulgoroidea). Aust. Ent. Mag. 6:13–18.

Gill, R. J., Nakahara, S., and Williams, M. L. 1977. A review of the genus *Coccus* Linnaeus in America north of Panama. Occas. Papers Ent., State Calif., Dept. Food Agr. 24, 44 pp. (Coccidae)

Glover, P. M. 1937. Lac cultivation in India. Indian Lac Res. Inst., Ranchi, 147 pp. (Tachardiidae)

Goeden, R. D., Fleschner, C. A., and Ricker, D. W. 1967. Biological control of prickly pear cacti on Santa Cruz Island, California. Hilgardia 38:579–606. (Dactylopiidae)

Gomez-Menor Ortega, J. 1937. Coccidos de Espana. Madrid Inst. Invest. Agron. Estac. Fitopat. Agr. Almeria, 432 pp.

Gunn, B. H. 1978. Sexual dimorphism in the first instar of the cochineal insect *Dactylopius austrinus* De Lotto. J. Ent. Soc. S. Afr. 41:333–338. (Dactylopiidae)

Hacker, H. 1925. The life history of *Oliarus felis* Kirk. (Homoptera). Mem. Queensland Mus. 8(II):113–114.

Hamilton, K. G. A. 1971. Placement of the genus *Microcentrus* in the Aetalionidae (Homoptera: Cicadelloidea) with a redefinition of the family. J. Georgia Ent. Soc. 9:229–236.

Hamon, A. B. 1979. Angraecum scale, *Conchaspis angraeci* Cockerell. Fla. Dept. Agr. Consumer Serv. Div. Plant Industry Ent. Circ. No. 201, 2 pp. (Conchaspididae)

Hamon, A. B. and Kosztarab, M. 1979. Morphology and systematics of the first instars of the genus *Cerococcus*. Va. Polytech. Inst. State Univ. Res. Div. Bull. 146, 122 pp. (Cerococcidae)

Hamon, A. B., Lambdin, P. L., and Kosztarab, M. 1976. Life history and morphology of *Kermes kingi* in Virginia. Va. Polytech. Inst. State Univ. Res. Div. Bull. 111, pp. 1–32. (Kermesidae)

Hamon, A. B. and M. L. Williams. 1984. The soft scale insects of Florida. Arthropods Fla. and neighboring land areas. Vol. II. Fl. Dept. Ag. & Consumer Ser., Div. Plt. Ind., Gainesville. 194 pp.

Hanna, Murray. 1967. The life history of *Philaenus abjectus* Uhler in Michigan (Homoptera: Cercopidae). Pap. Michigan Acad. Science, Arts, Letters. 52:69–76.

Hanna, Murray. 1969. The life history of *Clastoptera hyperici* McAtee in Michigan (Homoptera: Cercopidae). Michigan Academician 1:141–147.

Hanna, Murray and Thomas E. Moore. 1966. The spittlebugs of Michigan (Homoptera: Cercopidae). Pap. Michigan Acad. Science, Arts, Letters 51:39–73.

Harris, K. F. and K. Maramorosch (eds.). 1977. Aphids as Virus Vectors. Academic Press, New York. 559 pp.

Harvey, E. Newton 1952. Bioluminescence. Academic Press, New York. 649 pp.

Howell, J. O. 1973. The immature stages of *Aclerda tillandsiae*. Ann. Ent. Soc. Amer. 66:1335–1342. (Aclerdidae)

Howell, J. O. and Kosztarab, M. 1972. Morphology and systematics of the adult females of the genus *Lecanodiaspis*. Va. Polytech. Inst. State Univ. Res. Div. Bull. 70, 248 pp. (Lecanodiaspididae)

Howell, J. O., Williams, M. L., and Kosztarab, M. 1971. Morphology and systematics of *Cerococcus parrotti* (Hunter) with notes on its biology. Va. Polytech. Inst. State Univ. Res. Div. Bull. 64, 23 pp. (Cerococcidae)

Hoy, J. M. 1963. A catalogue of the Eriococcidae of the world. N. Zeal. Dept. Sci. Indust. Res. Bull. 150, 260 pp.

International Commission on Zoological Nomenclature. 1965. Opinion 731. Bull. Zool. Nomen. 22(2):86–87. (Adelgidae)

Ishihara, Tamotsu. 1969. Families and genera of leafhopper vectors. 235–254. *In* Maramorosch, Karl (ed.), Viruses, vectors, and vegetation. John Wiley & Sons, New York, London, Sidney, Toronto.

Jakubski, A. W. 1965. A critical revision of the families Margarodidae and Termitococcidae. Trustees Brit. Mus. (Nat. Hist.), London, 187 pp.

Karny, M. 1972. Comparative studies on three *Dactylopius* species attacking introduced opuntias in South Africa. Ent. Mem. Dept. Agr. Tech. Serv. Rep. S. Afr. Bull. no. 26, 19 pp. (Dactylopiidae)

Kennedy, J. S., M. F. Day, and V. F. Eastop. 1962. A conspectus of aphids as vectors of plant viruses. Commonwealth Institute of Entomology, London. 114 pp.

Kershaw, J. C. W. and G. W. Kirkaldy. 1910. A memoir on the anatomy and life-history of the Homopterous insect *Pyrops candelaria* (or "Candle-fly"). Zool. Jahrb. Syst. 29:105–124.

Kono, T. and C. S. Papp. 1977. Handbook of agricultural pests. Calif. Dept. Food & Agr., Div. Plt. Ind., Lab. Ser.-Entomology. Aphids, pp. 9–86.

Kopp, Dennis D. and Thomas R. Yonke. 1973–4. Treehoppers of Missouri, Parts 1–4. J. Kansas Ent. Soc. 46:42–64; 46(3):375–421; 47(1):80–130.

Kramer, J. P. 1983. Taxonomic study of the planthopper family Cixiidae in the United States (Homoptera: Fulgoroidea). Trans. Amer. Ent. Soc. 109:1–58.

LaFace, L. 1916. La metamorfosi dell' *Aclerda berlesii* Buffa. Portici R. Scuola Super. Agr. Lab. zool. Gen. Agr. Bol. 11:235–249. (Aclerdidae)

Lambdin, P. L. and Kosztarab, M. 1973. A revision of the seven genera related to *Lecanodiaspis*. Va. Polytech. Inst. State Univ. Res. Div. Bull. 83, 110 pp. (Lecanodiaspididae)

Lambdin, P. L. and Kosztarab, M. 1976. The genus *Solenophora* and its type species. Va. Polytech. Inst. State Univ. Res. Div. Bull. 111, pp. 33–41. (Cerococcidae)

Lambdin, P. L. and Kosztarab, M. 1977. Morphology and systematics of the adult females of the genus *Cerococcus*. Va. Polytech. Inst. State Univ. Res. Div. Bull. 128, 252 pp. (Cerococcidae)

Mamet, R. 1954. A monograph of the Conchaspididae Green. Trans. R. Ent. Soc. London. 105:189–239.

Mann, J. 1969. Cactus-feeding insects and mites. U.S. Nat. Mus. Bull. 256, 158 pp.

Marques, Luiz A. de Azevedo. 1928. Cigarrinha nociva a varias especies de vegetaes. Biologia do membracideo *Aethalion reticulatum* (L.) Bol. Inst. Biol. Defesa Agr., 6:1–25.

McConnell, H. S. 1954 (1953). A classification of the coccid family Aclerdidae. Univ. Md. Agric. Exp. Sta. Bull. A-75, 121 pp.

McKenzie, H. L. 1967. Mealybugs of California. Univ. Calif. Press, Berkeley, 525 pp.

Mead, Frank W. 1965. *Ormenaria rufifascia* (Walker), a planthopper pest of palms (Homoptera: Flatidae). Florida Dept. Agr., Div. Plant Industry, Ent. Circ. 37:1–2.

Mead, Frank W. 1969. Citrus flatid planthopper, *Metcalfa pruinosa* (Say). Florida Dept. Agr., Div. Plant Industry, Ent. Circ. 85:1–2.

Metcalf, C. L., W. P. Flint and R. L. Metcalf. 1951. Destructive and useful insects. Third Edition. McGraw-Hill Book Company, Inc. New York, Toronto, London. 1071 pp.

Metcalf, Z. P. 1942. A bibliography of the Homoptera (Auchenorhyncha). Vol. I. Authors' list A–Z. 886 pp. Vol. II. List of journals and topical index. North Carolina State College Agric. and Engr., Raleigh.

Metcalf, Z. P. 1932–1947. General catalogue of the Hemiptera. Fasc. IV. Fulgoroidea: Part 1. Tettigometridae, 74 pp. (1932); Part 2. Cixiidae, 274 pp. (1936); Part 3. Araeopidae (Delphacidae), 556 pp. (1943); Parts 4. Derbidae, 5. Achilixiidae, 6. Meenoplidae, 7. Kinnaridae, 256 pp. (1945); Part 8. Dictyopharidae, 250 pp. (1946); Part 9. Fulgoridae, 280 pp. (1947a); Part 10. Achilidae, 85 pp. (1947b). Smith College, Northhampton, Mass.

Metcalf, Z. P. 1954–1956. General catalogue of the Homoptera. Fasc. IV. Fulgoroidea: Part 11. Tropiduchidae, 176 pp. (1954a); Part 12. Nogodinidae, 84 pp. (1954b); Part 13. Flatidae and Hypochthonellidae, 574 pp. (1957); Part 14. Acanaloniidae, 64 pp. (1954c); Part 15. Issidae, 580 pp. (1958); Part 16. Ricaniidae, 208 pp. (1955a); Part 17. Lophopidae, 84 pp. (1955b); Part 18. Eurybrachidae and Gengidae, 90 pp. (1956); [see V. Wade, 1960, for Species Index]. North Carolina State University, Raleigh.

Metcalf, Z. P. 1960–1962. General catalogue of the Homoptera. Fasc. VII. Cercopoidea: A Bibliography of the Cercopoidea, 266 pp. (1960a); Part 1. Machaerotidae, 56 pp. (1960b); Part 2. Cercopidae, 616 pp. (1961); Part 3. Aphrophoridae, 608 pp. (1962a); Part 4. Clastopteridae, 66 pp. (1962b); [*see* V. Wade, 1963, for Species Index]. North Carolina State University, Raleigh.

Metcalf, Z. P. 1962–1968. General catalogue of the Homoptera. Fasc. VI. Cicadelloidea: A Bibliography of the Cicadelloidea, 349 pp. (1964); Part 1. Tettigellidae, 730 pp. (1965); Part 2. Hylicidae, 18 pp. (1962d); Part 3. Gyponidae, 229 pp. (1962e); Part 4. Ledridae, 147 pp. (1962f); Part 5. Ulopidae, 101 pp. (1962g); Part 6. Evacanthidae, 63 pp. (1963d); Part 7. Nirvanidae, 35 pp. (1963e); Part 8. Aphrodidae 268 pp. (1963f); Part 9. Hecalidae, 123 pp. (1963g); Part 10. Euscelidae, 2695 pp. (1967); Part 11. Coelidiidae, 182 pp. (1964a); Part 12. Eurymelidae, 44 pp. (1964b); Part 13. Macropsidae, 261 pp. (1966a); Part 14. Agalliidae, 173 pp. (1966b); Part 15. Iassidae, 229 pp. (1966c); Part 16. Idioceridae, 237 pp. (1966d); Part 17. Cicadellidae, 1513 pp. (1968) [*see* V. Wade, 1971, for Index . . . to Parts 1–17]. USDA, Agr. Research Service, Washington, D.C.
Macropsidae, 261 pp. (1966a); Part 14. Agalliidae, 173 pp. (1966b); Part 15. Issidae, 229 pp. (1966c); Part 16. Idioceridae, 237 pp. (1966d); Part 17. Cicadellidae, 1513 pp. (1968) [*see* V. Wade, 1971, for Index . . . to Parts 1–17]. USDA, Agr. Research Service, Washington, D.C.

Metcalf, Z. P. and Virginia Wade. 1963–1965. General catalogue of the Homoptera. Fasc. I. Supplement. A Bibliography of the Membracoidea and Fossil Homoptera (Homoptera: Auchenorrhyncha), 204 pp. (1963); Membracoidea, 1552 pp. (1965). North Carolina State Univ., Raleigh.

Metcalfe, J. R. 1968. Studies on the biology of the sugar-cane pest *Saccharosydne saccharivora* (Westw.) (Hom., Delphacidae). Bull. Ent. Res. 59:393–408.

Miller, D. R. 1969. A systematic study of *Eriococcus* Targioni-Tozzetti, with a discussion of the zoogeography of the Eriococcidae. Ph.D. dissertation, Univ. Calif. Davis, 253 pp.

Miller, D. R. 1975. A revision of the genus *Heterococcus* Ferris with a diagnosis of *Brevennia* Goux. U.S. Dept. Agr. Tech. Bull. 1497, 61 pp. (Pseudococcidae)

Miller, D. R. and Kosztarab, M. 1979. Recent advances in the study of scale insects. Ann. Rev. Ent. 24:1–27.

Miller, D. R. and McKenzie, H. L. 1967. A systematic study of *Ovaticoccus* Kloet and its relatives, with a key to North American genera of Eriococcidae. Hilgardia 38:471–539.

Moore, Thomas E. 1961. Audiospectrographic analysis of sounds of Hemiptera and Homoptera. Ann. Ent. Soc. America 54:273–291.

Morrison, H. 1928. A classification of the higher groups and genera of the coccid family Margarodidae. U.S. Dept. Agr. Tech. Bull. No. 52, 240 pp.

Morrison, H. 1952. Classification of the Ortheziidae. Supplement to classification of scale insects of the subfamily Ortheziinae. U.S. Dept. Agr. Tech. Bull. 1052, 80 pp.

Mound, L. A. 1973. Chapter 13. Thrips and Whitefly. *In* Gibbs, A. J. (ed.). Viruses and Invertebrates. North Holland Publ. Co., Amsterdam. pp. 229–242.

Mound, L. A. and S. M. Halsey. 1978. Whitefly of the World. A systematic catalogue of the Aleyrodidae (Homoptera) with host plant and natural enemy data. British Museum (Natural History), London and John Wiley and Sons, Chichester. 340 pp.

Myers, J. G. 1929a. Insect singers, a natural history of the cicadas. George Routledge and Sons, Ltd. London, xix + 304 pp.

Myers, J. G. 1929b. Observations on the biology of two remarkable Cixiid plant-hoppers (Homoptera) from Cuba. Psyche 36:283–292.

Nault, L. R. 1983. Origins of leafhopper vectors of maize pathogens in Mesoamerica. pp. 75–82 *in* Gordon, D. J., K. Knoke, L. R. Nault and R. M. Ritter (eds.). Proc. Int. maize virus disease Colloq. and workshop, 2–6 Aug. 1982. The Ohio State Univ., Ohio Agric. and Develop. Center, Wooster. 266 pp.

Nielson, M. W. 1968. The leafhopper vectors of phytopathogenic viruses (Homoptera, Cicadellidae). Taxonomy, biology and virus transmission. United States Dept. Agr., Agr Res. Serv. Tech. Bull. 1382. 386 pp.

Nielson, M W. 1985. Leafhopper Systematics. pp. 11–39. *In* Nault, L. R. and J. G. Rodriquez (eds.). The leafhoppers and planthoppers. John Wiley and Sons, New York. 500 pp.

O'Brien, Lois B. 1967. *Caliscelis bonellii* (Latreille), a European genus of Issidae new to the United States. Pan-Pacific Ent. 43:130–133.

O'Brien, L. B. and S. W. Wilson. 1985. Planthopper systematics and external morphology. pp. 61–102. *In* Nault, L. R. and J. G. Rodriquez (eds.). The leafhoppers and planthoppers. John Wiley and Sons, New York.

O'Brien, L. B. and S. W. Wilson. 198–. A systematic catalog of the Fulgoroidea (Homoptera) of the United States and Canada. (in press) (larval descriptions listed under each genus)

Palmer, M. A. 1952. Aphids of the Rocky Mountain Region. The Thomas Say Foundation. Vol. 5. 452 pp.

Pergande, T. 1904. North American Phylloxerinae affecting *Hicoria* (*Carya*) and other trees. Proc. Davenport Acad. Sci. 9:185–273.

Phillips, J. H. H. 1962. Description of the immature stages of *Pulvinaria vitis* (L.) and *P. innumerabilis* (Rathvon), with notes on the habits of these species in Ontario, Canada. Canad. Ent. 94:497–502. (Coccidae)

Pschorn-Walcher, H. and H. Zwolfer. 1956. Neuere Untersuchunger über die Weisstannenlaüse der Gattung *Dreyfusia* C. B. und ihren Vertilgerkreis. Anz. Schaedlingskd. 29:116–122. (Adelgidae)

Quaintance, A. L. and A. C. Baker. 1913–1915. Classification of the Aleyrodidae, Part I, Part II, and Contents and Index, U.S. Dept. Agric. Bur. Ent. Tech. Ser. No. 27. 114 pp.

Quaintance, A. L. and A. C. Baker. 1917. A contribution to our knowledge of the whiteflies of the subfamily Aleyrodinae (Aleyrodidae). Proc. U.S. Nat. Mus. 51:335–445.

Quisenberry, S. S., T. R. Yonke and D. D. Kopp. 1978. Key to the genera of certain immature treehoppers of Missouri with notes on their host plants. J. Kansas Ent. Soc. 51:109–122.

Ridout, B. V. 1983. Structure, form and function of the lanternfly head process (*Fulgora laternaria* L.). Unpublished Ph.D. thesis, University of London.

Russell, L. M. 1941. A classification of the scale insect genus *Asterolecanium*. U.S. Dept. Agr. Misc. Pub. No. 424, 322 pp. (Asterolecaniidae)

Russell, L. M. 1947. A classification of the whiteflies of the new tribe Trialeurodini (Homoptera: Aleyrodidae). Rev. Ent. 18(1–2):1–44.

Russell, L. M. 1948. The North American species of whiteflies of the genus *Trialeurodes*. U.S. Dept. Agric. Misc. Publ. No. 653. 85 pp.

Russell, L. M. 1974. *Daktulosphaira vitifoliae* (Fitch), the correct name of the grape phylloxeran (Hemiptera: Homoptera: Phylloxeridae). J. Wash. Acad. Sci. 64(4):303–308.

Schlinger, Evert I. 1958. Notes on the biology of a mud egg-case making Fulgorid, *Hysteropterum beameri* Doering (Homoptera: Fulgoridae). J. Kansas Ent. Soc. 31:104–106.

Schmutterer, H. 1952. Die Okologie der Cocciden Frankens. Ztschr. Angew. Ent. 33:369–420, 544–584; 34:65–100.

Schmutterer, H., Kloft, W. and Ludicke, M. 1957. Coccoidea, Schildause, scale insects, cochenilles. pp. 403–520. *In* Sorauer, P. (ed.) Handbuch der Pflanzenkrankh. Paul Parey, Berlin.

Simons, J. N. and D. M. Coe. 1958. Transmission of pseudo-curly top virus in Florida by a treehopper. Virology 6:43–48.

Smith, C. F. 1972. Bibliography of the Aphididae of the World. N. C. Agric. Exp. Stn. Tech. Bull. 216. 717 pp.

Smith, C. F. and C. S. Parron. 1978. An annotated list of Aphididae (Homoptera) of North America. N. Carolina Agr. Expt. Stn. Tech. Bull. 255. 428 pp.

Snodgrass, R. E. 1919. The seventeen-year locust. Ann. Report, Smithsonian Inst., 1919:381–409.

Steinweden, J. B. 1929. Bases for the generic classification of the coccoid family Coccidae. Ann. Ent. Soc. Amer. 22:197–243.

Stickney, F. S. 1934. The external anatomy of the red date scale, *Phoenicococcus marlatti* Cockerell, and its allies. U.S. Dept. Agr. Tech. Bull. no. 404, 163 pp. (Phoenicococcidae)

Stickney, F. S., Barnes, D. F. and Simmons, P. 1950. Date palm insects in the United States. U.S. Dept. Agr. Circ. No. 846, 57 pp.

Stoetzel, M. B. 1975. Seasonal history of seven species of armored scale insects of the Aspidiotini. Ann. Ent. Soc. Amer. 68:489–492. (Diaspididae)

Stoetzel, M. B. 1976. Scale-cover formation in the Diaspididae. Proc. Ent. Soc. Wash. 78:323–332.

Stoetzel, M. B. 1981. Two new species of *Phylloxera* (Homoptera: Phylloxeridae) on pecan. J. Ga. Ent. Soc. 16(2):127–144.

Stoetzel, M. B. 1985a. Host alternation: a newly discovered attribute of the Phylloxeridae (Homoptera: Aphidoidea). Proc. Ent. Soc. Washington. 87:265–68.

Stoetzel, M. B. 1985b. Pupiform larvae in the Phylloxeridae (Homoptera: Aphidoidea). Proc. Ent. Soc. Washington 87:535–37.

Stoetzel, M. B. 1987. Host alternation in the Aphidoidea (Homoptera). pp. 204–208 *in* Proc. Intl. Symposium on population structure, genetics and taxonomy of aphids and Thysanoptera at Smolenice, Czechoslovakia, 9–14 Sept. 1985. SPB Academic Publishing, The Hague, Netherlands.

Stoetzel, M. B. and Davidson, J. A. 1974. Biology, morphology, and taxonomy of immature stages of 9 species in the Aspidiotini. Ann. Ent. Soc. Amer. 67:475–509. (Diaspididae)

Stoetzel, M. B. and W. L. Tedders. 1981. Investigation of two species of *Phylloxera* on pecan in Georgia. J. Ga. Ent. Soc. 16(2):144–150.

Sutherland, D. W. S. 1978. Common names of insects and related organisms. Ent. Soc. Amer. Spec. Publ. 78–1, 132 pp.

Tsai, James H., N. L. Woodiel and O. H. Kirsch. 1976. Rearing techniques for *Haplaxius crudus* (Homoptera: Cixiidae). Florida Ent. 59:41–43.

Tsai, James H. and Oscar H. Kirsch. 1978. Bionomics of *Haplaxius crudus* (Van Duzee) (Homoptera: Cixiidae). Environ. Ent. 7(2):305–308.

Tuthill, L. D. 1943. The psyllids of America north of Mexico (Psyllidae: Homoptera). Iowa State Coll. J. Sci. 17(4):443–667.

Van Dine, D. L. 1911. The sugar-cane insects of Hawaii. United States Dept. Agr. Bur. Ent. Bull. 93:1–34.

Varshney, R. K. 1970. Lac literature. A bibliography of lac insects and shellac. Shellac Export Promotion Council, Calcutta, India, 216 pp. (Tachardiidae)

Varshney, R. K. and Ganguly, G. 1971 (1969). On some lac insects of India and neighbouring countries in the collection of the U.S. National Museum. Patna Univ. J. 24:17–22. (Tachardiidae)

Vilbaste, J. 1968. Preliminary key for the identification of the nymphs of North European Homoptera Cicadina. Ann. Ent. Fenn. 34:65–74.

Wade, Virginia. 1960–1966. *In* Metcalf, General catalogue of the Homoptera. Fasc. IV. Fulgoroidea, Species Index. 80 pp. (1960). Fasc. VII. Cercopoidea, Species Index. 36 pp. (1963). Fasc. VIII. Cicadoidea, Species Index. 28 pp. (1964). Fasc. I. Membracoidea and Fossil Homoptera, Species Index. 40 pp. (1966) North Carolina State Univ., Raleigh.

Wade (Burnside), Virginia. 1971. *In* Metcalf, General catalogue of the Homoptera. Fasc. VI. Cicadelloidea, Index to Genera And Species with Addenda and Corrigenda to Parts 1–17. 269 pp. USDA. Agr. Research Service, Washington, D.C.

Washburn, R. I. 1965. Description and bionomics of a new species of *Puto* from Utah. Ann. Ent. Soc. Amer. 58:293–297. (Putoidae)

White, J. M. and I. D. Hodkinson. 1985. Nymphal taxonomy and systematics of the Psylloidea (Homoptera). Bull. Brit. Mus. (Nat. Hist.) (Ent.) 50:153–301.

Whitehead, F. E. and O. Eastep. 1937. The seasonal cycle of *Phylloxera notabilis* Pergande. Ann. Ent. Soc. Am. 30:71–74.

Williams, M. L. and Kosztarab, M. 1970. A morphological and systematic study of the first instar nymphs of the genus *Lecanodiaspis*. Va. Polytech. Inst. State Univ. Res. Div. Bull. 52, 96 pp. (Lecanodiaspididae)

Williams, M. L. and Kosztarab, M. 1972. Morphology and systematics of the Coccidae of Virginia with notes on their biology. Va. Polytech. Inst. State Univ. Res. Div. Bull. 74, 215 pp.

Wilson, S. W. and J. E. McPherson. 1981a. Life histories of *Acanalonia bivittata* and *A. conica* with descriptions of immature stages. Ann. Ent. Soc. Amer. 74(3):289–298. (Acanaloniidae)

Wilson, S. W. and J. E. McPherson. 1981b. Life histories of *Anormenis septentrionalis, Metcalfa pruinosa,* and *Ormenoides venusta* with descriptions of immature stages. Ann. Ent. Soc. Amer. 74(3):299–311. (Flatidae)

Wilson, S. W. and J. E. McPherson. 1981c. Descriptions of the immature stages of *Bruchomorpha oculata* with notes on laboratory rearing. Ann. Ent. Soc. Amer. 74(4):341–344. (Issidae)

Wilson, S. W. and L. B. O'Brien. 1987. A survey of planthopper pests of economically important plants (Homoptera: Fulgoroidea). pp. 343–360 *in* Wilson, M. R. and L. R. Nault (eds.). Proc. 2nd international workshop on leafhoppers and planthoppers of economic importance, 28 July–1 Aug. 1986. Provo, Utah. Commonwealth Inst. Ent.

Wood, Thomas K. 1976a. Alarm behavior of brooding female *Umbonia crassicornis* (Homoptera: Membracidae). Ann. Ent. Soc. Amer. 69:340–344.

Wood, Thomas K. 1976b. Biology and presocial behavior of *Platycotis vittata* (Homoptera: Membracidae). Ann. Ent. Soc. Amer. 69:807–811.

Wood, Thomas K. 1977. Role of parent females and attendant ants in the maturation of the treehopper, *Entylia bactriana* (Homoptera: Membracidae). Sociobiology 2:257–272.

Wood, Thomas K. and R. L. Patton. 1971. Egg froth distribution and deposition by *Enchenopa binotata* (Homoptera: Membracidae). Ann. Ent. Soc. Amer. 64(5):1190–1191.

Yang, S. P. and Kosztarab, M. 1967. A morphological and taxonomical study on the immature stages of *Antonina* and of the related genera. Va. Polytech. Inst. Res. Div. Bull. 3, 73 pp. (Pseudococcidae)

Order Megaloptera

31

H. H. Neunzig and J. R. Baker
North Carolina State University

ALDERFLIES, DOBSONFLIES AND FISHFLIES

The Megaloptera are sometimes included as a suborder (Sialodea) within the order Neuroptera. Most recent authors, however, on the basis of distinct differences in the morphology of the immatures, lesser, but nevertheless evident, morphological differences in the adults, and major differences in biology, follow the classifications of Handlirsch (1903, 1906) and recognize the Megaloptera as a separate order.

Common names for the larger larvae include dobsons, hellgramites or arnly.

DIAGNOSIS

The following combination of characters will separate larvae of Megaloptera from larvae of other insect orders: labrum present (the presence of a labrum will separate Megaloptera from similar Coleoptera larvae which lack a labrum); mandibles simple, with several teeth along the anteromesal margin (without a blood groove or tube); abdominal segments with lateral filaments (also called lateral gills, appendages or styli) on segments 1–7 or 1–8 and 10; caudal abdominal segment either elongate and tapering to a median pointed filament (Sialidae), or not elongate and with a pair of anal prolegs, each with 2 hooks (Corydalidae).

BIOLOGY AND ECOLOGY

Larvae of all species whose biology is known are aquatic or semi-aquatic. They can be found in spring seeps, streams, rivers, lakes, ponds, swamps, and in temporarily-dry stream beds. Small animals, mostly other insect larvae, are used as food. Larvae pass through many instars and take 1–5 years to complete development. Last-stage larvae are usually most abundant in late winter or early spring. Pupation occurs in cells in the soil adjacent to a body of water, in logs in or near the water, or in dry stream-beds.

The most significant contributions to our knowledge of the morphology and biology of the immature stages of the North American Megaloptera are the following: Azam (1969), Azam and Anderson (1969), Baker and Neunzig (1968), Canterbury (1978), Canterbury and Neff (1980), Chandler (1956), Cuyler (1956, 1958, 1965), Evans (1972), Evans and Neunzig (1984), Gurney and Parfin (1959), Kelsey (1954, 1957), Leischner and Pritchard (1973), Needham and Betten (1901), Neunzig (1966), Pritchard and Leischner (1973), Riley (1873, 1876, 1877, 1878), Ross (1937), Tarter et al. (1975), Woodrum and Tarter (1973) and Tarter et al. (1979). A more complete listing is given in the bibliography.

DESCRIPTION

Larvae of Megaloptera are slightly flattened, possess a distinct head with conspicuous mouthparts, a 3-segmented thorax with 3 pairs of segmented legs, and a distinct 10-segmented abdomen.

Head: Heavily sclerotized; labrum distinct; mandibles generalized chewing-type; prominent maxillae with 2–3 appendages; labium also with several appendages. Antennae present, relatively short; stemmata (ocelli) on each side of the head.

Thorax: Prothorax large, more strongly developed than the meso- or metathorax; each segment with a distinct, heavily-sclerotized shield; most of the remainder of the thorax relatively soft; legs well-developed, 5-segmented, terminating in two claws.

Abdomen: Elongate; relatively soft; segments 1–7, or 1–8 and 10, each with tapering lateral filaments; 10th segment either elongate and tapering or short with pair of anal prolegs with hooks.

COMMENTS

In America north of Mexico, there are presently recognized two families, and 43 species. The study of Evans (1972) on western species has added considerably to our knowledge of the immatures and helped clarify relationships within the order. However, the immatures of several North American species are still undescribed and the biology of a number of species is unknown.

TECHNIQUES

Larvae of many aquatic insects are killed and preserved in 70–80% alcohol. However, with larvae of the Megaloptera, *alcohol by itself should not be used for killing and fixing.* Severe shriveling and distortion of specimens almost always occurs. Live specimens should be placed in KAAD or a similar fixative. After a day or so they can be transferred to 75–80% ETOH for preservation. Evans (1972) suggests orally injecting KAAD into larger Megaloptera. This insures adequate distension with rapid fixation and maximum color retention. After injection, larvae are placed in KAAD for a day or so and then transferred to alcohol.

CLASSIFICATION[1]

Superorder Neuropteroidea
- Order Megaloptera
  - Sialidae—sialids, alderflies (23)
  - Corydalidae—dobsonflies, fishflies, hellgrammites (20)
- Order Raphidioptera
- Order Neuroptera (= Planipennia)

1. The number of species given is for America north of Mexico.

KEY TO THE FAMILIES AND GENERA OF LAST-STAGE LARVAE OF NORTH AMERICAN MEGALOPTERA

1. Caudal abdominal segment with median setiferous terminal filament; without anal prolegs (fig. 31.1) .... (*Sialis*) ***Sialidae*** (p. 115)

   Caudal abdominal segment without median terminal filament; with 2 anal prolegs with large hooks (fig. 31.2) .... ***Corydalidae*** (p. 116) 2

2(1). Lateral abdominal filaments with ventral gill tufts at base (fig. 31.3); patches of hydrofuge pile on venter of abdominal segments 9 and 10 .... *Corydalus*

   Lateral abdominal filaments without ventral gill tufts at base; hydrofuge pile absent on venter of abdominal segments 9 and 10 3

3(2). Head with posterolateral aspect of gena strongly developed (fig. 31.4); western North America 4

   Head with gena more weakly developed posterolaterally (figs. 31.5, 31.6); eastern and western North America 5

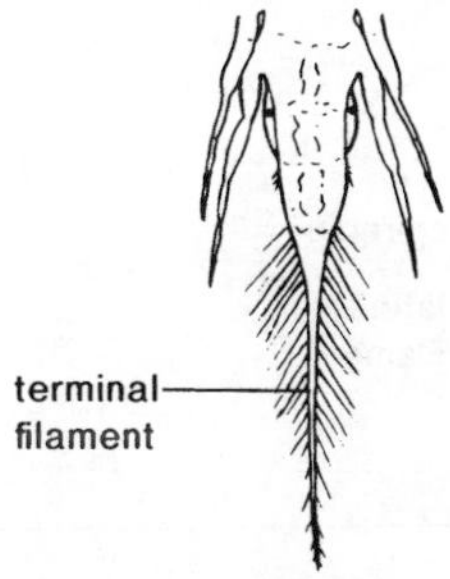

terminal filament

Figure 31.1

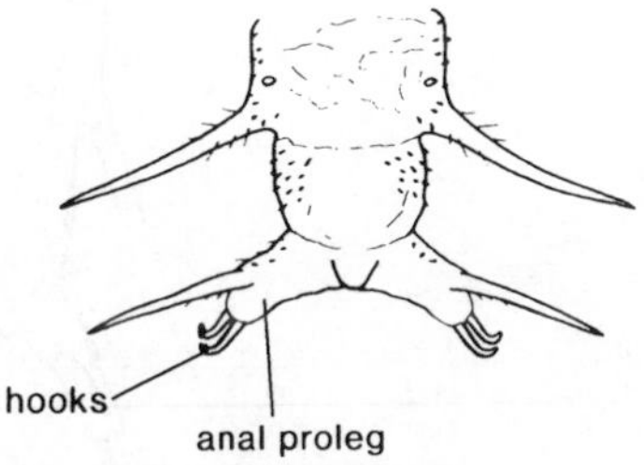

hooks

anal proleg

Figure 31.2

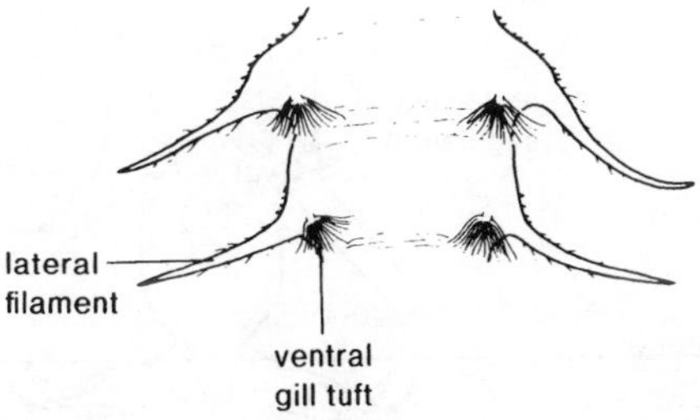

lateral filament

ventral gill tuft

Figure 31.3

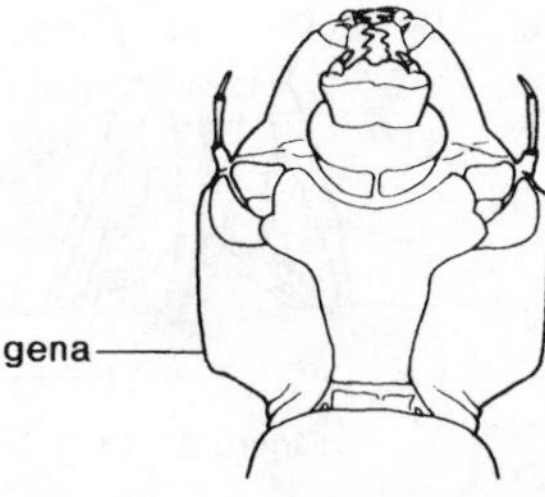

Figure 31.4

Figure 31.5

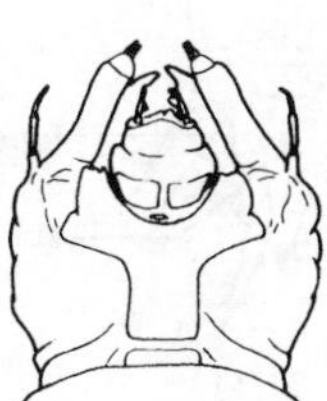

Figure 31.6

4(3). Spiracles of abdominal segment 8 small, located dorsoanterior to base of lateral filaments (fig. 31.7); lateral filaments of mid-abdominal segments almost as long as width of mid-abdominal segments .... *Orohermes*

Spiracles of abdominal segment 8 large, on subconical protuberances of integument, located mesally near posterior margin of segment (fig. 31.8); lateral filaments of mid-abdominal segments distinctly shorter than width of mid-abdominal segments .... *Dysmicohermes*

5(3). Spiracles of abdominal segment 8 sessile or only slightly elevated (figs. 31.9, 31.10); dorsum of head usually with distinct pale markings 6

Spiracles of abdominal segment 8 on distinct short to long tubes (figs. 31.11, 31.12, or 31.13); dorsum of head, anterior to occiput, usually without distinct pale markings 7

6(5). Spiracles on abdominal segment 8 distinct, associated with slightly raised areas of the integument (fig. 31.9) .... *Neohermes*

Spiracles on abdominal segment 8 less conspicuous, integument not raised around spiracles (fig. 31.10) .... *Protochauliodes*

7(5). Spiracles of abdominal segment 8 on long contractile tubes (fig. 31.13) .... *Chauliodes*

Spiracles of abdominal segment 8 on shorter tubes (figs. 31.11, 31.12) .... *Nigronia*

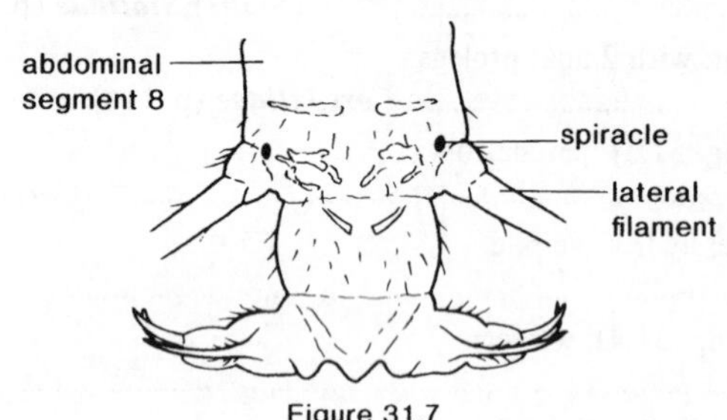

Figure 31.7

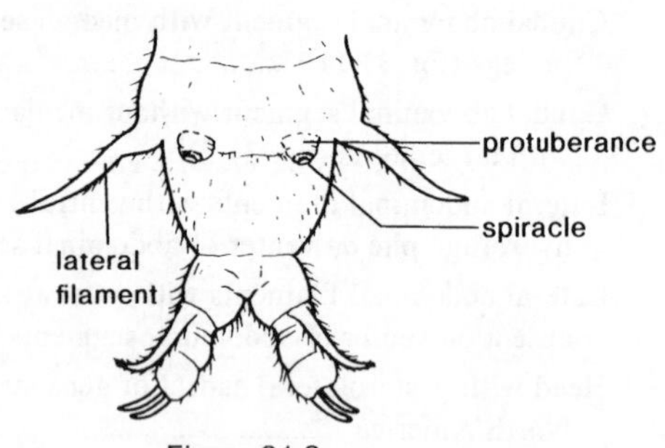

Figure 31.8

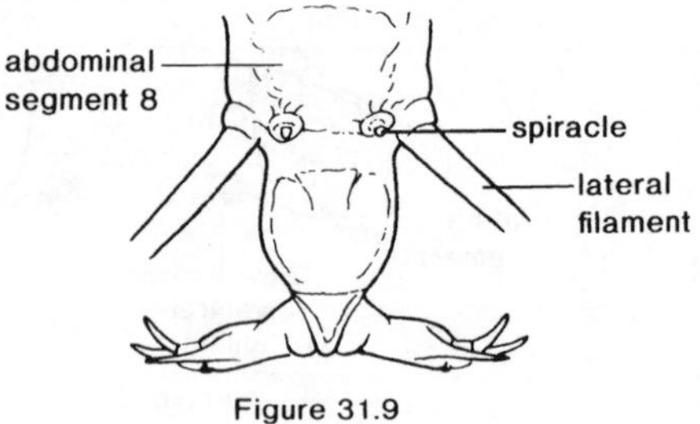

Figure 31.9

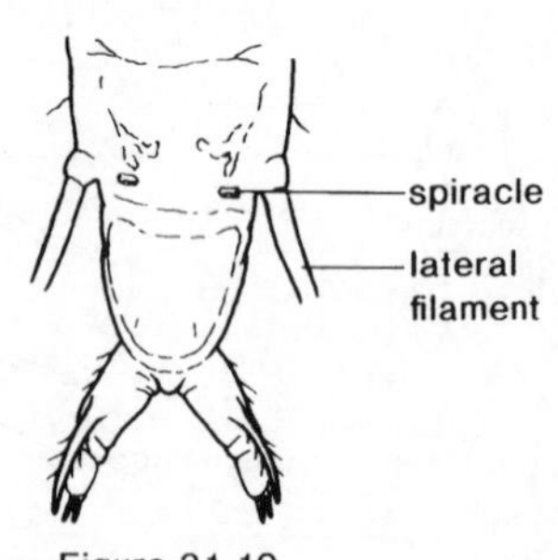

Figure 31.10

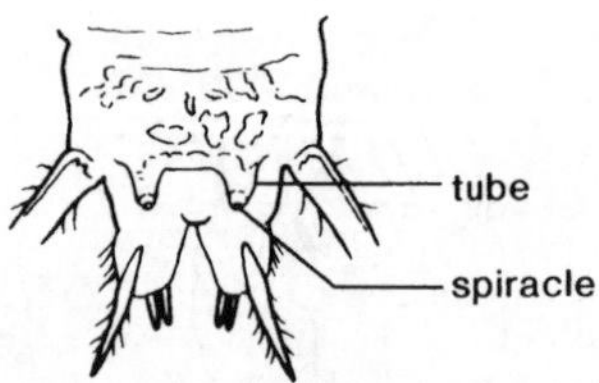

Figure 31.11

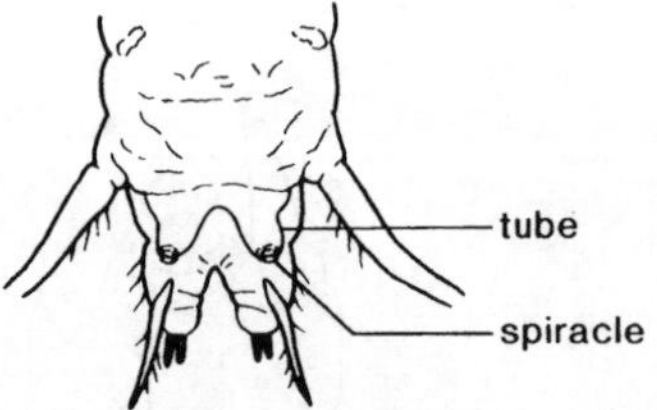

Figure 31.12

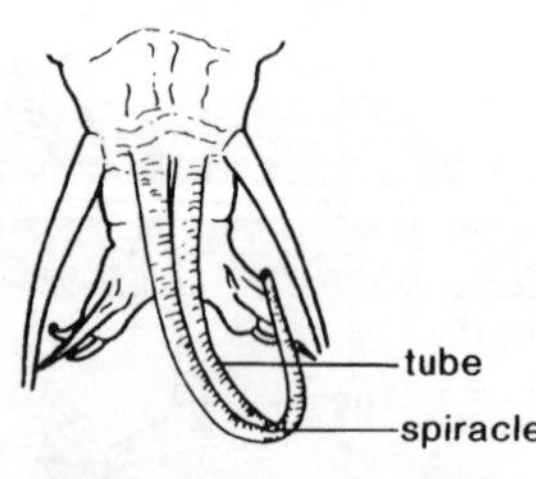

Figure 31.13

SIALIDAE

Sialids, Alderflies

Figure 31.14

Relationships and Diagnosis: Sialid larvae can be separated from Corydalidae in that they possess a setiferous, terminal filament (fig. 31.1), lack anal prolegs, have lateral filaments only on abdominal segments 1–7, and have distinct maxillary laciniae.

Biology and Ecology: Sialid larvae are found in many aquatic situations, ranging from small springs to large rivers, and from ponds to large lakes. Most species are abundant where the substrate is soft and where dead leaves or other plant material has accumulated, and most occur in a variety of environments, but some species are restricted to, or more abundant in, certain habitats. *Sialis rotunda* Banks, and *Sialis cornuta* Ross, for example, occur primarily in lentic situations, such as ponds and lakes. *S. rotunda* occasionally occurs in streams, but usually in a backwater microhabitat. *Sialis californica* Banks, on the other hand, prefers streams and rivers. Larvae dig into the substrate, sometimes to a depth of 12 inches (Davis, 1903). Usually a J- or U-shaped hole with 1 or 2 openings at the substrate and water interface is formed (Pritchard and Leischner, 1973). This burrow provides protection from predators and close proximity to many prey organisms.

One or two years are required to complete the life cycle. Larvae seem to be indiscriminate feeders, consuming whatever small animals are present, and cannibalism has been reported. Larvae leave the water to pupate, usually in an unlined chamber made in the soil, under moss, or among other low-growing vegetation or debris.

Adults usually stay in the general area where the larvae and pupae occur. They spend most of the time passively clinging to vegetation near, or overhanging, the water. Flight and other activities are most pronounced toward mid-day. Adult life is brief and, apparently, no feeding occurs.

The unprotected egg masses are laid under leaves of plants growing in, or immediately adjacent to, the larval habitat, or on the under surface of substrates such as uprooted trees, branches, bridges, and large rocks overhanging or projecting from the water. Eggs hatch at night in from 8 days to 2 weeks.

Description: Last-stage larvae 10–28 mm long (including the caudal filament), somewhat elongate and slightly flattened, with head and most of thorax heavily sclerotized and abdomen relatively soft.

Head: Prominent, yellowish brown or reddish brown with paler markings. Ecdysial line distinct. Anterior margin of frons formed by distinct epistomal suture which completely separates frons from transverse postclypeus and pale narrow anteclypeus. Attached to anteclypeus is a distinct labrum which is attenuated distally and has a crenulate margin.

Genae rounded posteriorly. Six stemmata on each side of head. Antennae 4-segmented. Mandibles with long slender distal tooth and several additional much shorter teeth along mesal margin. Maxillae consisting of basal cardo and truncate stipes, the latter bearing an inner anterior falciform lacinia, a single-segmented galea, and an outer 4-segmented palpus.

Ventral surface of head with median gulamentum, with mentum and prementum attached anteriorly. Prementum with 3-segmented labial palpi and a single median lobe.

Thorax: Of 3 distinct segments. Legs 5-segmented, with two unequal claws. Prothorax larger than meso- or metathorax and with large, dorsal, heavily-sclerotized, quadrate shield. A narrow curved sclerite (the basisternum?) occurs anterior to each coxa on venter of the prothorax. Dorsal shields also on the meso- and metathorax, but smaller and less heavily sclerotized than prothoracic shield. Venter of meso- and metathorax without distinct sclerotized areas.

Sclerotized shield of prothorax similar in color to the head, being brownish and usually having lighter markings. Smaller shields of meso- and metathorax also brownish and maculated but suffused with purple. Softer parts of thorax white or purplish white. Legs usually pale brown.

A pair of annular spiracles just anterior to mesothoracic coxae. A vestigial pair also frequently evident anterior to metathoracic coxae. Several rows or groups of distinct setae found on legs. Less obvious slender, moderately long or very short setae associated with the sclerotized shields and other softer parts of thorax. Most softer regions of thorax appearing glabrous, having only a few inconspicuous setae.

Abdomen: Abdomen with 10 distinct, relatively soft segments, the first 7 possessing a pair of 4- to 5-segmented lateral filaments. 10th segment very elongate and strongly tapered distally. Dorsum brownish or purplish with lighter whitish markings, at times with pale longitudinal median stripe. Venter similar to dorsum but usually much paler. Lateral filaments and 10th abdominal segment usually mostly white.

A lateral pair of small annular spiracles on segments 1–8. Filaments and 9th and modified 10th segments fringed with numerous slender setae along lateral margins. Remainder of abdomen appearing glabrous, having only a few inconspicuous setae.

Comments: Sialidae occur throughout North America. Species such as *Sialis velata* Ross and *Sialis mohri* Ross are widespread, being found from the Atlantic states to the Rocky Mountains. On the other hand, *Sialis iola* Ross is restricted to the Appalachian region. Several species are distinctly western, occurring in the Rocky Mountain region or westward, as for example *S. californica, S. hamata* and *S. rotunda*. There are presently 23 Nearctic species, all in the genus *Sialis*.

Canterbury (1978) has prepared a key to the last stage larvae of 10 species of eastern North American *Sialis*. There is also some information on western species, including *S. californica, S. rotunda, S. cornuta, S. arvalis, S. hamata* and *S. nevadensis* (Azam and Anderson, 1969; Evans, 1972; Leischner and Pritchard, 1973). Information on the morphology of early-stage larvae is given by Azam and Anderson (1969), Evans (1972), and Leischner and Pritchard (1973).

The pupa of *S. cornuta* has been described rather thoroughly by Leischner and Pritchard (1973). No other detailed study of the pupae has been made.

Canterbury (1978) and Canterbury and Neff (1980) have provided detailed descriptions of, and a key to, the eggs and/or egg masses of 10 eastern North American species of *Sialis*. The micropylar process appears to be particularly diagnostic for species.

Selected Bibliography

Combined with Corydalidae for the order.

CORYDALIDAE

Dobsonflies, Fishflies, Hellgramites

Figures 31.15–31.28

Relationships and Diagnosis: Corydalid larvae can be distinguished from Sialidae in that they possess a relatively short caudal segment with anal prolegs (fig. 31.2), have lateral filaments on abdominal segments 1–8 and 10, and each maxilla has only two appendages, the palpus and galea.

Biology and Ecology: Corydalid larvae occur in almost all kinds of permanent freshwater habitats, and, at times, are found in dry or semi-dry beds of intermittent streams.

Species of *Corydalus* usually abound in fast flowing, rocky-bottomed streams or rivers, where they most frequently occur under rocks in the more rapidly flowing, well-aerated parts. There are some reports of *Corydalus* larvae occurring in intermittent streams and near hot springs (32° C). Species of *Dysmicohermes* and *Orohermes* prefer small, cold (water temperatures seldom exceeding 20° C.) streams. They usually are found in areas where detritus collects.

In western North America, larvae of *Neohermes* and *Protochauliodes* frequently coexist in rocky intermittent streams or relatively warm permanent streams (over 20° C.) for most of the year (Evans, 1972). *Neohermes californicus* (Walker) prefers streams that are permanent or mostly permanent with relatively deep water and rocky substrates. *Neohermes filicornis* (Banks) occurs in similar streams but in those with a clay substrate. *Protochauliodes spenceri* Munroe can be found among leaves and debris in the currents of small intermittent streams having a soft substrate. *Protochauliodes aridus* Maddux occurs in intermittent streams also, but in those with a hard, mostly rocky substrate. During the drier months of the year (usually June to October) *P. aridus* and larvae of several other species of western corydalids inhabiting intermittent streams, burrow into the dry stream beds. In eastern North America, larvae of *Neohermes concolor* (Davis) are associated with detritus in spring seeps (Tarter, et al., 1979).

According to Cuyler (1956), species of *Chauliodes* are restricted to lentic situations where water movement is nil or very slight. *Chauliodes pectinicornis* (L.) is most abundant in shaded swamps or small woodland pools, and only occasionally is found in open ponds or lakes. *Chauliodes rastricornis* Rambur, on the other hand, appears to prefer a more sunny situation and occurs most frequently in exposed or mostly exposed ponds or lakes. Both species are most abundant in bodies of water having deep layers of debris and decaying logs in, or at the edge of the water. *Chauliodes* has occasionally been reported from streams or rivers (Parfin, 1952), but probably where water movement was negligible and an essentially lentic habitat prevailed. It has not been definitely established, but apparently *Chauliodes* larvae burrow into the debris and mud. Food is usually abundant in this sediment and they are also protected from predators. However, oxygen is very scarce in the mud at the bottom of lentic environments, particularly where large quantities of decaying organic matter accumulate. Apparently the long, contractile respiratory tubes of *Chauliodes* can reach to the mud-water or debris-water interface, or perhaps at times to the surface of the water, enabling the larvae to respire and develop in this habitat.

Nigronia are generally found in rather rapidly flowing rivers or streams (Cuyler, 1956, 1965; Neunzig 1966; Tarter et al., 1975). However, species such as *Nigronia serricornis* are much more common in the shallow parts of slow-moving small to medium-sized streams. Another species, *N. fasciatus,* prefers even less water movement, and mostly occurs in very small, cool, shaded woodland streams. Frequently, these streams are very shallow and during the summer months they may partially dry up. When this occurs, *N. fasciatus* larvae live for several months under stones in the moist stream bed.

Like sialid larvae, corydalid larvae feed on all kinds of small animals in their environment (insect larvae, annelids, crustaceans and molluscs). They are strong, active predators and in captivity are cannibalistic. However, corydalid larvae (and sialid larvae, too) are protected against each other, and small predators in general, by their tactile filaments which radiate from the relatively soft abdomen. Frequently, in an encounter with another predator of about equal size, only a small portion of a lateral filament is lost, which apparently has no effect on the development of the individual.

The life cycle varies from 2–5 years. The following information has been established for each genus: *Chauliodes*—2–3 years; *Nigronia*—2–3 years; *Corydalus*—at least 3 years and probably longer under unfavorable conditions; *Neohermes*—2–5 years, depending on habitat; *Protochauliodes*—2–5 years; *Dysmicohermes* and *Orohermes*—about 4–5 years. The low temperatures prevailing in some mountainous habitats account for the prolonged life cycle in some genera. Also, larvae developing in intermittent streams usually take longer than those occurring in permanent streams or rivers. According to Evans (1972) "the selection of intermittent streams by an insect with a larval stage of several years duration is unique among aquatic insects."

The only information available on the number of instars in corydalid larvae is that of Evans (1972), who reports probably 12 instars for *Neohermes californicus* (Walker).

Corydalid larvae pupate in a variety of places. *Corydalus* larvae crawl under stones or debris such as logs or boards, usually adjacent to the larval habitat. They usually pupate farther from the water than other corydalid larvae and sometimes travel as much as 30 meters from the water's edge.

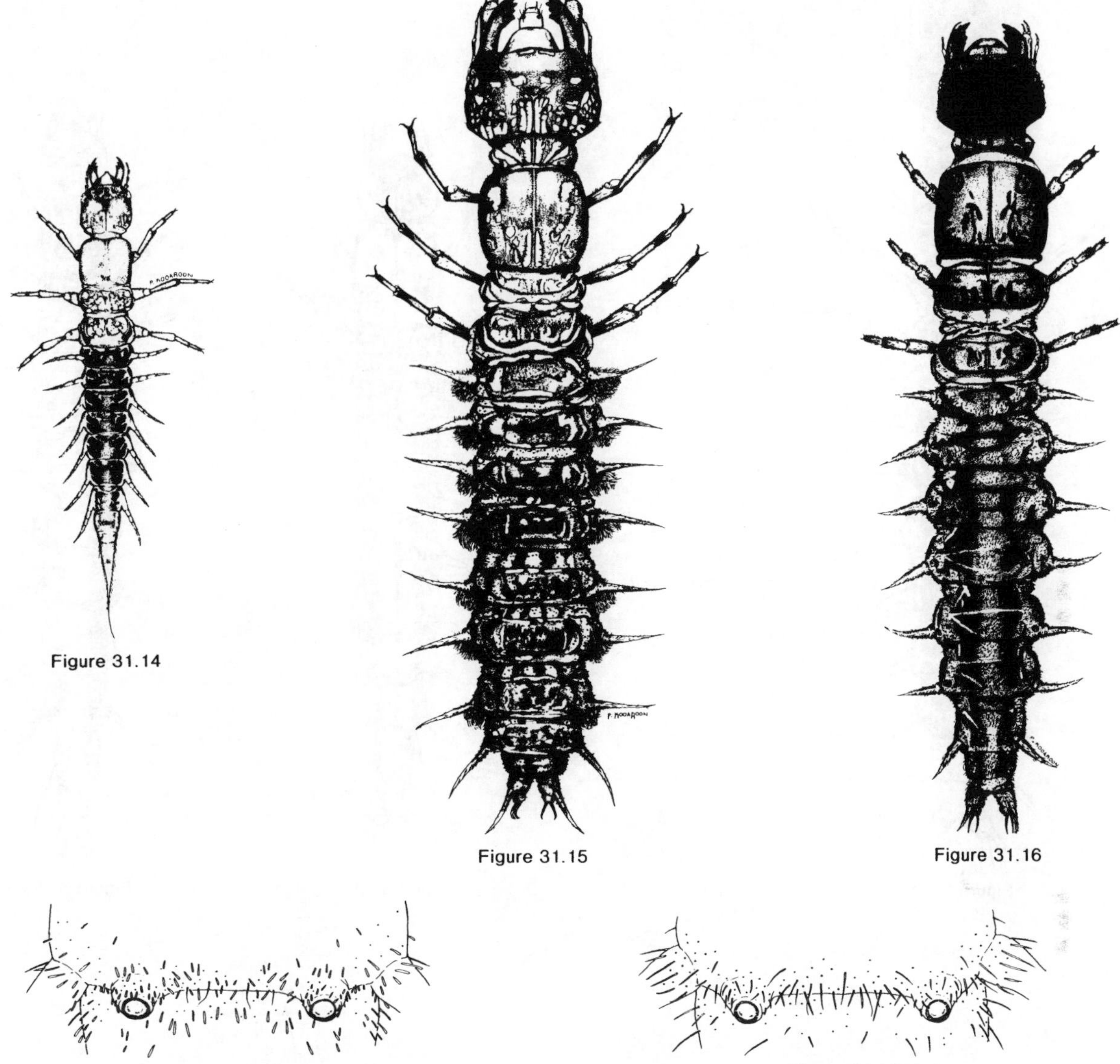

Figure 31.14

Figure 31.15

Figure 31.16

Figure 31.17

Figure 31.18

Figure 31.14. Sialidae. *Sialis aequalis* Banks. F.g.l. 10–17 mm. Head and prothorax yellowish brown to brown dorsally without distinct markings. Meso- and metathorax with some brown or purplish-brown markings. Abdomen purple to dark purple dorsally, with small white markings. Lateral filaments whitish. Collected from small, slow-moving stream. Wake Co., North Carolina, by J. R. Baker and T. R. Weaver.

Figure 31.15. Corydalidae. *Corydalus cornutus* (L.). F.g.l. 75–90 mm. Head and thoracic shields brown to dark reddish brown, distinctly maculated. Softer parts of thorax and abdomen pale grayish brown to purplish brown, darkened by a covering of dark brown carinate microspines. Muscle attachment sites paler than surrounding integument. Abdomen with ventral gill tufts at base of lateral filaments. Collected under stone in rapid stream, Ashe Co., North Carolina, by H. H. Neunzig.

Figure 31.16. Corydalidae. *Dysmicohermes disjunctus* (Walker). F.g.l. 55–65 mm. Head and thoracic shields usually dark reddish brown. Head usually without distinct pale markings. Softer parts of thorax and abdomen purplish brown without distinct pale markings. Spiracles on abdominal segment 8 large, associated with subconical protuberances of integument, and located on dorsum near posterior margin of segment. Lateral filaments of mid-abdominal segments distinctly shorter than width of mid-abdominal segments. Collected from falls in a creek in Benton Co., Oregon by C. Kerst.

Figure 31.17. Corydalidae. *Dysmicohermes disjunctus* (Walker). Enlarged view of spiracles of 8th abdominal segment, and surrounding integument. Note that most setae are blunt and shorter than the height of the protuberances bearing the spiracles.

Figure 31.18. Corydalidae. *Dysmicohermes ingens* Chandler. Enlarged view of spiracles of 8th abdominal segment, and surrounding integument. Note that all of the setae are pointed and some are as long as the height of the protuberances bearing the spiracles. Collected from small cold stream in Tehama Co., California by E. D. Evans.

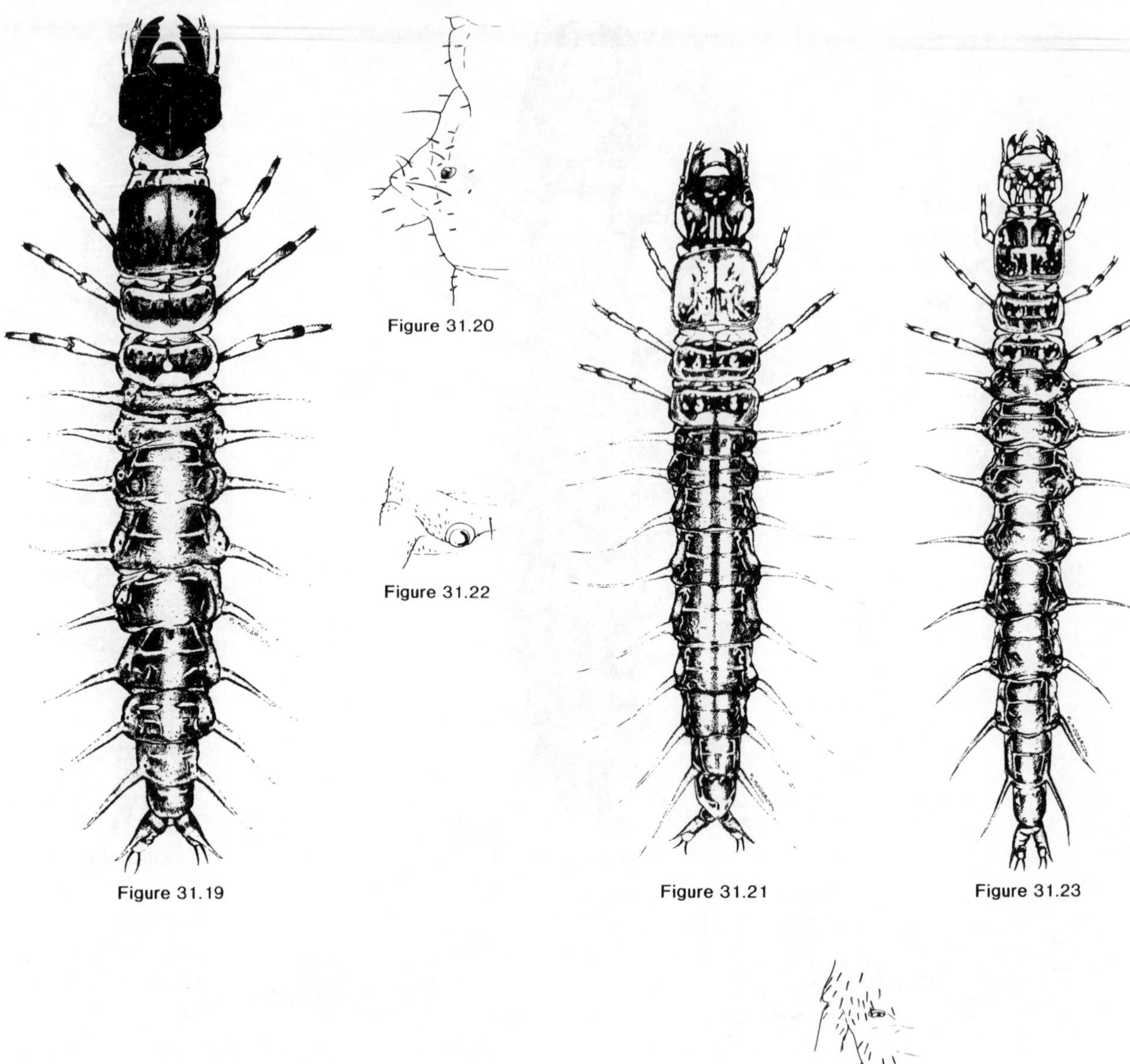

Figure 31.19 Figure 31.20 Figure 31.21 Figure 31.22 Figure 31.23 Figure 31.24

Figure 31.19. Corydalidae. *Orohermes crepusculus* (Chandler). F.g.l. 55–66 mm. Head and thoracic shields reddish brown to dark reddish brown, usually without distinct pale markings. Softer parts of thorax and abdomen grayish brown to purplish brown, without distinct pale markings. Spiracles on abdominal segment 8 small, not distinctly raised above integument, located dorsoanterior to base of lateral filaments. Lateral filaments of mid-abdominal segments almost as long as width of mid-abdominal segments. Collected from a creek in Sierra Co., California by E. D. Evans and A. Knight.

Figure 31.20. Corydalidae. *Orohermes crepusculus* (Chandler). Enlarged view of left spiracle of 8th abdominal segment, and surrounding integument.

Figure 31.21. Corydalidae. *Neohermes filicornis* (Banks). F.g.l. 40–62 mm. Head reddish brown to dark brown with pale markings. Thoracic shields paler than head, purplish brown with pale maculation. Softer parts of thorax and abdomen, including lateral filaments, brownish purple. Spiracles on abdominal segment 8 on dorsum, large, and associated with slightly raised areas of the integument. Lateral filaments of mid-abdominal segments longer than width of mid-abdominal segments. Collected from a small stream in San Bernadino Co., California by E. D. Evans.

Figure 31.22. Corydalidae. *Neohermes filicornis* (Banks). Enlarged view of left spiracle of 8th abdominal segment, and surrounding integument.

Figure 31.23. Corydalidae. *Protochauliodes aridus* Maddux. F.g.l. 35–50 mm. Head and thoracic shields brown to dark brown with paler markings. Softer parts of thorax and abdomen mostly brownish pruple with a pale lateral stripe and pale ventral patches. Spiracles on abdominal segment 8 on dorsum, small. Integument not raised around spiracles (spiracles on segment 8 comparable in elevation to spiracles elsewhere on abdomen). Lateral filaments of mid-abdominal segments slender, as long as, or shorter than, width of mid-abdominal segments. Collected from a creek in Butte Co., California by E. D. Evans.

Figure 31.24. Corydalidae. *Protochauliodes aridus* Maddux. Enlarged view of left spiracle of 8th abdominal segment, and surrounding integument.

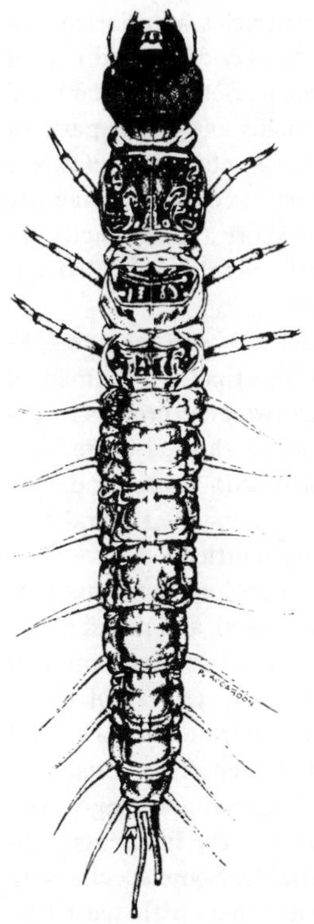
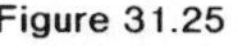

Figure 31.25

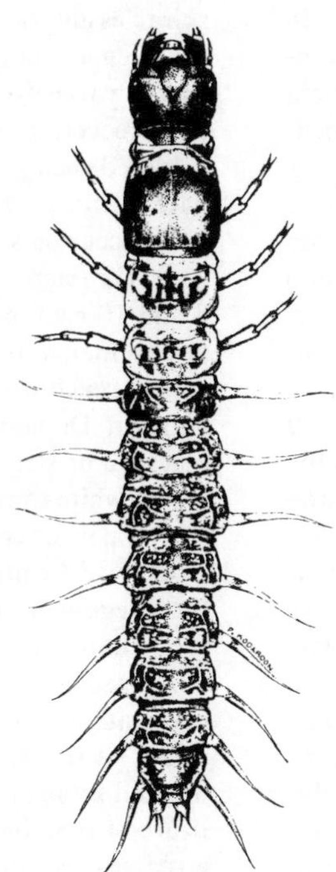

Figure 31.26

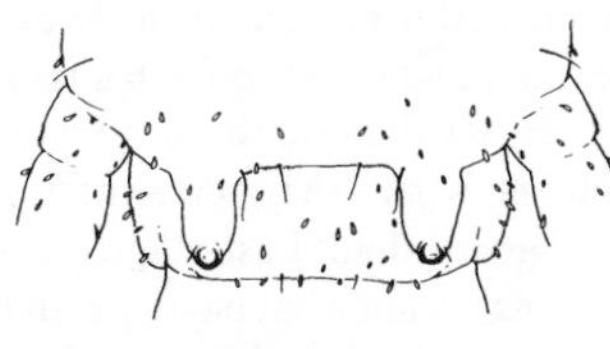

Figure 31.27

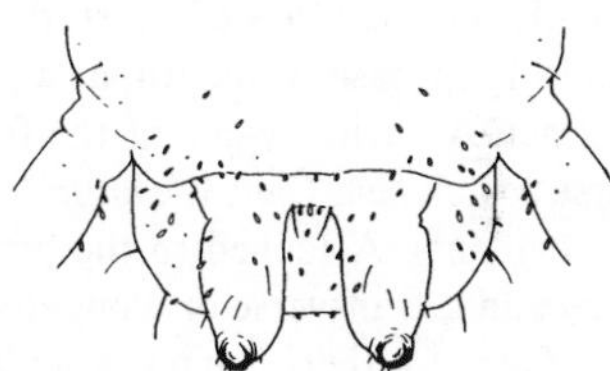

Figure 31.28

Figure 31.25. Corydalidae. *Chauliodes pectinicornis* (L). F.g.l. 38–42 mm. Head and thoracic shields brown to dark brown with pale markings (more obscure on the head). Softer parts of thorax and abdomen brown to brownish purple with small pale spots and middorsal pale abdominal stripe. Spiracles on abdominal segment 8 at ends of long, unequal, contractile tubes originating dorsally near the posterior margin of the segment. Lateral filaments of mid-abdominal segments slightly shorter than width of mid-abdominal segments. Collected from wooded, swampy area in Johnston Co., North Carolina by H. H. Neunzig and J. R. Baker. *Chauliodes rastricornis* Rambur is similar to *C. pectinicornis* but has a dark brown to black middorsal stripe on the thorax and abdomen.

Figure 31.26. Corydalidae. *Nigronia serricornis* (Say). F.g.l. 25–40 mm. Head and prothoracic shield brown to dark reddish brown, usually with obscure markings. Meso- and metathoracic shields light brown with brown markings. Softer parts of thorax and abdomen mostly purple (anterior lateral filaments purple at base, mostly whitish distally). Spiracles on abdominal segment 8 at end of tubes that are dorsad of the lateral filaments, widely separated at their bases, and about 1.5 times as long as wide. Lateral filaments of mid-abdominal segments longer than width of mid-abdominal segments. Collected from a stream in Wake Co., North Carolina by H. H. Neunzig.

Figure 31.27. Corydalidae. *Nigronia serricornis* (Say). Enlarged view of respiratory tubes and spiracles of 8th abdominal segment, and surrounding integument. Note bases of tubes are widely separated.

Figure 31.28. Corydalidae. *Nigronia fasciatus* (Walker). Enlarged view of respiratory tubes and spiracles of 8th abdominal segment, and surrounding integument. Note bases of tubes are close together and tubes are longer than in *N. serricornis*. Collected from a slow-moving, small woodland stream in Wake Co., North Carolina by H. H. Neunzig.

Pupae of *Neohermes* and *Protochauliodes* have been found beneath rocks in the beds of dried-up or partially dried-up rivers and arroyos above larger streams. *Chauliodes* larvae usually construct pupal cells under the bark of moist, soft rotting logs lying in or along the edge of the water. If the logs are in the water the cell is made above the water level. *Nigronia* larvae crawl from several centimeters to about a meter from the stream and pupate in shallow cells in the soil under moss or under debris along the stream banks. No information is available on the pupation sites of *Dysmicohermes* and *Orohermes*.

The pupal stage lasts 7–14 days; most authors report 11–12 days. Adult corydalids are diurnal or nocturnal. Species active during the day include *Nigronia serricornis, Nigronia fasciatus* and *Protochauliodes minimus* (Davis). These species are much darker in color than other corydalid adults, and they inhabit shaded woodlands.

Most corydalid adults are nocturnal and apparently travel farther from their place of development than diurnal species. Light traps frequently catch these species, although Evans (1972) reports that *O. crepusculus* is not attracted to light, and that *Dysmicohermes* spp. only infrequently enter a trap. Adult longevity is short, and food intake nil or negligible. If anything is ingested, it is probably liquids.

Eggs are laid above the larval habitat on leaves, branches, rocks, bridges, and similar substrates. Usually several thousand eggs are deposited in a single- or multilayered mass. In the genus *Corydalus* the egg mass is covered with a very conspicuous white secretion, but the egg masses of other genera are bare or only lightly and indistinctly covered. The incubation period is usually 7–14 days. Like sialid larvae, hatching occurs almost simultaneously for most of the individuals of a particular egg mass, and usually at night. When egg masses are located over permanent bodies of water, larvae fall into the water and crawl beneath a rock or debris. Over dry stream beds larvae dig into the substrate to a depth of 10 to 30 centimeters, where they remain until rains renew the water.

Description: Last-stage larvae medium to large (25–90 mm long), somewhat elongate and slightly flattened, with the head and a considerable part of the cervical region and thorax heavily sclerotized and the abdomen relatively soft.

Head: Prominent, brown or reddish brown, at times with paler markings. Setae sparse and indistinct dorsally. Venter usually with patches of short, distinct secondary setae, particularly at base of mouthparts and on genae. Ecdysial line distinct. Anterior region of the frons confluent with a transverse postclypeus (only the lateral parts of the epistomal suture are evident). Attached to the postclypeus is a pale anteclypeus and a transverse or elongate labrum.

Genae either sharply angulate or weakly angulate to rounded posteriorly. Six stemmata on each side of head. Antennae 4–5 segmented. Mandibles strongly sclerotized with several symmetrical or asymmetrical teeth on anteromesal margin. Most distal tooth of each mandible slightly larger than the others. Maxillae consisting of short basal cardo and long stipes, the latter bearing an outer maxillary palpus and an inner galea. Lacinia absent.

Ventral surfaces of the head with large, median T-shaped gulamentum. Apices of lateral arms of gulamentum either entire and usually pointed, or emarginate. Anterior to gulamentum is the mentum, divided mesally, and more distally the prementum, bearing labial palpi and paraglossae. Glossae absent.

Cervical Region and Thorax: Between the head and the prothorax is the jugulare, consisting of two inconspicuous, thinly sclerotized dorsal plates, and a large brownish, heavily sclerotized ventral plate.

Thorax of three distinct segments. Legs 5-segmented, with two unequal or subequal claws. Prothorax larger than meso- or metathorax and with large, dorsal, heavily sclerotized quadrate shield. A ventral shield, formed by the basisternum and episternum, and about one-half the size of the dorsal shield, also present. Smaller, less heavily sclerotized dorsal shields on the meso- and metathorax. No distinct sclerotized regions present on venter of the meso- and metathorax.

Sclerotized dorsal and ventral shields of prothorax similar in color to the head, being brown or reddish brown. In some species, shields are relatively uniform in color, whereas in others distinct pale markings occur. Dorsal shields of the meso- and metathorax also brownish and usually with maculation. Softer parts of thorax whitish to dark purplish. Legs various shades of brown.

A pair of distinct, cribriform spiracles anterior to mesothoracic coxae. A much smaller, weakly developed pair of spiracles anterior to metathoracic coxae. Setae abundant and conspicuous on legs. Vestiture of shields and softer parts of thorax varies from a few inconspicuous setae to a relatively dense covering. In a few species, some secondary setae are modified, being short and blunt, or short and distinctly inflated. In *Corydalus,* numerous dark, closely appressed spinules occur on softer parts of thorax.

Abdomen: With 10 distinct, relatively soft segments, 1–8 and 10 each possessing a pair of indistinctly 2-segmented lateral filaments. 10th segment with two anal prolegs or pygopods, each with two stout hooks and a short, 1-segmented filament. Dorsum brownish or purplish with lighter or darker maculae or stripes. Venter paler than dorsum. Lateral filaments white to purple and frequently mottled.

A pair of cribriform spiracles on each of segments 1–8. Those on 8 frequently more dorsal, enlarged, and raised above the integument. In some genera, the 8th abdominal spiracles are situated at the ends of distinct tubes. Larvae of *Corydalus* possess dense tufts of gills on venter at base of lateral filaments. Lateral filaments sparsely to densely covered with setae. At times, the setae occur as longitudinal fringes with inflated setae and spinules at the base of the filaments. Vestiture of remainder of abdomen variable. Some species with a sparse covering of unmodified setae; others with dense coverings of mixed modified and unmodified setae, and, at times, spinules covering the segments. Larvae of *Corydalus* also with patches of hydrofuge pile on venter of caudal segments.

Comments: *Chauliodes* and *Nigronia* are widely distributed in the eastern United States. Confined to the West are *Dysmicohermes, Orohermes* and *Protochauliodes.* Two genera have representatives in both regions; *Corydalus cornutus* L. in the East apparently is replaced by *Corydalus* possibly *cognatus* Hagen in the southwestern United States, and there are three eastern and two western species of *Neohermes.* Approximately 20 species of Corydalidae are presently recognized in America north of Mexico.

The last-stage larvae of the following species are unknown: *Neohermes angusticollis* (Hagen), *Neohermes matheri* Flint, *Protochauliodes simplus* Chandler, *Protochauliodes cascadius* Evans and *Protochauliodes minimus.* The first-stage larvae of several eastern corydalids have been treated by Baker and Neunzig (1968); however, little is known of the early stages of western species.

The egg masses and eggs of a number of corydalids have been examined (Baker and Neunzig, 1968; Evans, 1972); however, more detailed diagnostic information is needed, particularly for western species.

The pupa of *Corydalus cornutus* has been studied in detail by Kelsey (1954, 1957). No other detailed study of the pupae of this family has been attempted.

Selected Bibliography

Combined with Sialidae for the order.

BIBLIOGRAPHY

Azam, K. M. 1969. Life history and production studies of *Sialis californica* Banks and *Sialis rotunda* Banks (Megaloptera: Sialidae). Ph.D. thesis. Oregon State University. Corvallis, 111 p.

Azam, K. M. and N. H. Anderson. 1969. Life history and habitats of *Sialis rotunda* and *S. californica* in western Oregon. Ann. Ent. Soc. Amer. 62:549–558.

Baker, J. R. and H. H. Neunzig. 1968. The egg masses, eggs, and first-instar larvae of eastern North American Corydalidae. Ann. Ent. Soc. Amer. 61:1181–1187.

Balduf, W. V. 1939. The bionomics of entomophagous insects. Part II. John S. Swift Co. St. Louis. 384 p.

Brigham, W. V. 1982. Megaloptera. pp. 7.1–7.12. *In* Brigham, A. K., W. V. Brigham and A. Gnilke (Eds.). Aquatic insects and oligochaetes of North and South Carolina. Midwest Aquatic Ent., Mahomet, Illinois.

Brimley, C. S. 1908. Notes on some neuropteroids from Raleigh, N.C. Ent. News 19:133.

Canterbury, L. E. 1978. Studies of the genus *Sialis* (Sialidae: Megaloptera) in eastern North America. Ph.D. thesis Univ. Louisville. 93 p.

Canterbury, L. E. and S. E. Neff. 1980. Eggs of *Sialis* (Sialidae: Megaloptera) in eastern North America. Can. Ent. 112:409–419.

Chandler, H. P. 1956. Megaloptera, p. 229–233. *In* Usinger, R. L. (ed.). Aquatic insects of California. Univ. Calif. Press. Berkeley, 508 p.

Cuyler, R. D. 1956. Taxonomy and ecology of larvae of sialoid Megaloptera of east-central North Carolina with key to and descriptions of larvae of genera known to occur in the United States. M.S. thesis. North Carolina State College. Raleigh. 150 p.

Cuyler, R. D. 1958. The larvae of *Chauliodes* (Megaloptera: Corydalidae). Ann. Ent. Soc. Amer. 51:582–586.

Cuyler, R. D. 1965. The larva of *Nigronia fasciatus* Walker (Megaloptera: Corydalidae). Ent. News 76:192–195.

Davis, K. C. 1903. Sialididae of North and South America, p. 442–486. *In* Felt, E. P. Aquatic insects in New York State. N.Y. State Mus. Bull. 68, 517 p.

Dolin, P. S. and D. C. Tarter, 1982. Life history and ecology of *Chauliodes rastricornis* Rambur and *C. pectinicornis* (L.) in Greenbottom Swamp, Cabell Co., W. Virginia. Brimleyana 7:111–120.

Essig, E. O. 1947. College Entomology. MacMillan Co., New York, 900 p.

Evans, E. D. 1972. A study of the Megaloptera of the Pacific coastal region of the United States. Ph.D. thesis. Oregon State University. Corvallis, 210 p.

Evans, E. D. and H. H. Neunzig. 1984. Megaloptera and Aquatic Neuroptera, p. 261–270. *In* Merritt, R. W. and Cummins, K. W. (Eds.) An introduction to the aquatic insects of North America. Second Edition. Kendall/Hunt, Dubuque. 722 pp.

Gurney, A. B. and S. Parfin. 1959. Neuroptera, p. 937–980. *In* Edmonson, W. T. (Ed.) Fresh-water biology. Second Edition. Wiley, New York. 1248 pp.

Haldeman, A. M. 1849. History and transformations of *Corydalus cornutus*. Mem. Amer. Acad. Arts and Science 4:157–161.

Handlirsch, A. 1903. Zur Phylogenie der Hexapoden. Sitzber. Akad. Wissensch. Wien. Math.-nat. Kl. 107:716–738.

Handlirsch, A. 1906. Uber Phylogenie der Arthropoden. Verh. Zool-bot. Ges. Wien, 56:88–103.

Howard, L. O. 1896. A coleopterous enemy of *Corydalus cornutus*. Proc. Ent. Soc. Wash. 3:310–313.

Kelsey, L. P. 1954. The skeleto-motor mechanisms of the dobsonfly, *Corydalus cornutus*. Part 1. Head and prothorax. Cornell Univ. Memoir 334, 51 p.

Kelsey, L. P. 1957. The skeleto-motor mechanisms of the dobsonfly, *Corydalus cornutus*. Part II. Pterothorax. Cornell Univ. Memoir 346, 31 p.

Leischner, T. G. and G. Pritchard. 1973. The immature stages of the alderfly, *Sialis cornuta* (Megaloptera: Sialidae). Can. Ent. 105:411–418.

Lintner, J. A. 1893. 8th Rpt. on the injurious and other insects of the State of N.Y. for 1891, 300 p.

Maddux, D. C. 1952. A study of the dobsonflies (Megaloptera) of the Chico, California area. M.S. thesis. Chico State College. Chico. 44 p.

Maddux, D. C. 1954. A new species of dobsonfly from California (Megaloptera: Corydalidae). Pan-Pacific Ent. 30:70–71.

McCafferty, W. P. 1981. Fishflies, dobsonflies, and alderflies (order Megaloptera). p. 189–196. *In* Aquatic Entomology. Science Books Int. Boston, 448 p.

Minshall, G. W. 1965. Community dynamics and economics of a woodland spring brook. Ph.D. thesis. Univ. Louisville, Louisville. 261 p.

Minshall, G. W. 1968. Community dynamics of the benthic fauna in a woodland spring brook. Hydrobiol. 32:305–339.

Moody, H. L. 1877. The larva of *Chauliodes*. Psyche 2:52–53.

Needham, J. G. and C. Betten. 1901. Aquatic Insects in the Adirondacks. N.Y. State Mus. Bull. 47, 612 p.

Neunzig, H. H. 1966. Larvae of the genus *Nigronia* Banks. Proc. Ent. Soc. Wash. 68:11–16.

Parfin, S. I. 1952. The Megaloptera and Neuroptera of Minnesota. Amer. Midland Nat. 47:421–434.

Penland, D. R. 1953. A detailed study of the life cycle and respiratory system of a new species of western dobsonfly, *Neohermes aridus*. M.S. thesis. Chico State College. Chico. 34 p.

Peterson, Alvah. 1951. Larvae of insects. Part 2. Printed for the author by Edwards Bros. Ann. Arbor. 416 p.

Peterson, R. C. 1974. Life history and bionomics of *Nigronia serricornis* (Say) (Megaloptera: Corydalidae). Ph.D. thesis. Mich. State Univ., East Lansing, 210 p.

Pritchard, G. and T. G. Leischner. 1973. The life history and feeding habits of *Sialis cornuta* Ross in a series of abandoned beaver ponds. (Insecta: Megaloptera). Can. J. Zool. 51:121–131.

Riley, C. V. 1873. The hellgrammite fly—*Corydalus cornutus* (Linn.). *In* 5th Ann. Rpt. noxious, beneficial, and other insects of the State of Missouri, p. 142–145.

Riley, C. V. 1876. On the curious egg mass of *Corydalus cornutus* (Linn.) and on the eggs that have hitherto been referred to that species. Proc. Amer. Assoc. Adv. Sci. 25:275–279.

Riley, C. V. 1877. The hellgrammite, *Corydalus cornutus* (Linn.). *In* 9th Ann. Rpt. Nox., beneficial, and other insects of Missouri, p. 125–129.

Riley, C. V. 1878. On the larval characteristics of *Corydalus* and *Chauliodes* and on the development of *Corydalus cornutus*. Proc. Amer. Assoc. Adv. Sci. 27:285–287.

Riley, C. V. 1879. On the larval characteristics of *Corydalus* and *Chauliodes* and on the development of *Corydalus cornutus*. Can. Ent. 11:96–98.

Ross, H. H. 1937. Nearctic alder flies of the genus *Sialis*, p. 57–78. *In* Ross, H. H. and T. H. Frison. Studies of Nearctic Aquatic Insects. Bull. Ill. Nat. Hist. Survey 21, 99 p.

Smith, E. L. 1970. Biology and structure of the dobsonfly, *Neohermes californicus* (Walker) (Megaloptera: Corydalidae). Pan-Pacific Ent. 46:142–150.

Smith, R. C. 1920. The process of hatching in *Corydalus cornuta* Linn. Ann. Ent. Soc. Amer. 13:70–74.

Smith, R. C. 1922. Hatching in three species of Neuroptera. Ann. Ent. Soc. Amer. 15:169–176.

Stewart, K. W., G. P. Friday, and R. E. Rhame. 1973. Food habits of hellgrammite larvae, *Corydalus cornutus* (Megaloptera: Corydalidae) in the Brazos River, Texas. Ann. Ent. Soc. Amer. 66:959–963.

Tarter, D. C. 1976. Limnology in West Virginia: A lecture and laboratory manual. Marshall Univ. Book Store, Huntington, West Va. 249 p.

Tarter, D. C., W. D. Watkins and D. A. Etnier. 1979. Larval description and habitat notes of the fishfly *Neohermes concolor* (Davis) (Megaloptera: Corydalidae). Ent. News 90:29–32.

Tarter, D. C., W. D. Watkins, and M. L. Little. 1975. Life history of the fishfly *Nigronia fasciatus* (Megaloptera: Corydalidae). Psyche 82:81–88.

Townsend, L. H. 1935. Key to larvae of certain families and genera of Nearctic Neuroptera. Proc. Ent. Soc. Wash. 37:25–29.

Walsh, B. D. 1863. Observations on certain N. A. Neuroptera, by H. Hagen, M.D., of Koenigsberg, Prussia; translated from the original French MS, and published by permission of the author, with notes and descriptions of about twenty new N. A. species of Pseudoneuroptera. Proc. Ent. Soc. Phila. 2:167–182.

Walsh, B. D. and C. V. Riley. 1868. The hellgrammite fly (*Corydalis cornutus* Linn.). Amer. Ent. 1:61–62.

Weed, C. M. 1889. Studies in pond life. Article II. Ohio Agr. Exp. Sta. Tech. Bull. 1, 46 p.

Woodrum, J. E. and D. C. Tarter. 1973. The life history of the alderfly, *Sialis aequalis* Banks, in an acid mine stream. Amer. Midl. Nat. 89(2):360–368.

Order Raphidioptera

32

Catherine A. Tauber
Cornell University

SNAKEFLIES

Although included by some within the Megaloptera (*sensu lato*), the Raphidioptera [= Raphidiodea in Vol. 1] are here given equal rank with the other two neuropteroid orders (see Achtelig 1967; Aspöck *et al.*; 1980; Gepp 1984). Aspöck *et al.* (1980) consider the Raphidioptera a sister group of the Megaloptera.

Diagnosis: Raphidioptera larvae are distinguished from Coleoptera larvae by their 10-segmented abdomens and 3-segmented labial palpi and from the Megaloptera by their terrestrial habits and the absence of abdominal processes or appendages. Chewing mouthparts (with mandibles independent of maxillae and with both maxillary and labial palpi present), a large, well-sclerotized head and prothorax, and an intestinal tract that is open during the entire larval period distinguish raphidiopteran from neuropteran larvae.

BIOLOGY AND ECOLOGY

Clusters of elongate eggs are laid in crevices of bark and in other hidden areas. The larvae are robust and active; their short legs and pygidial ambulatory organ subserve rapid movement in both forward and backward directions. Larvae are known to live under loose bark of trees and in the soil (Aspöck & Aspöck 1971; Aspöck *et al.* 1975). No studies have established precisely the natural diet of raphidiopteran larvae, but all instars are believed to prey on a variety of soft-bodied arthropods.

The life cycle is usually semivoltine with the number of larval stadia varying around 10 or 11 (Woglum & McGregor 1958, 1959; Aspöck, Rausch and Aspöck 1974). Overwintering generally occurs in the larval stage, occasionally in the pupal. Pupation occurs in cells under bark or in crevices of other organic material. Mating behavior has been described for some species (Acker 1966).

DESCRIPTION

Mature larva robust, depressed, relatively large, around 25 mm long (fig. 32.1). Integument smooth with few short, straight setae. Head large, dorsoventrally flattened, somewhat rectangular; antennae short, 4-segmented; eyes small, semiglobular, arranged in two groups on sides of head just behind antennae, with 4, 6, or 7 stemmata. Mouthparts (fig. 32.2) consisting of labrum, maxillae (with galeae and 5-segmented palpi), stout mandibles, labium (with ligula and 3-segmented palpi).

Prothorax flat, rectangular, almost as long as meso- and metathoraces combined, relatively darkly sclerotized dorsally. Meso- and metathoraces also sclerotized. Legs short, outstretched from body; tarsus unsegmented, bearing two claws apically. Abdomen with 10 visible segments; pygidial area with rosette of soft pads.

COMMENTS

The Raphidioptera is a small order with only two families. Its geographical distribution is limited to the Holarctic region, primarily Europe, Central Asia, and western North America (west of the eastern slopes of the Rocky Mountains). Aspöck *et al.* (1980) and Aspöck (1986) estimated that there are approximately 200 species, with about 170 valid, described species; more than 25, from both families, are from western North America (Carpenter 1936; Aspöck 1975, 1986). The Raphidioptera in America are clearly distinct from those in other parts of the world. Aspöck (1986) proposed that the American raphidiopteran fauna had its origins in immigrations from Asia during the Tertiary.

Adult and larval snakeflies are infrequently collected. Although they are known to accept a variety of arthropod pests (e.g., aphids, Lepidoptera larvae, scales) as prey, their economic value is unknown. Unsuccessful attempts were made to introduce Raphidioptera (probably mixed species) into New Zealand for codling moth control; no major effort has been made to introduce them into eastern North America.

CLASSIFICATION

Superorder Neuropteroidea
- Order Megaloptera
- Order Raphidioptera (snakeflies)[1]
  - Inocelliidae (2)
  - Raphidiidae (19)
- Order Neuroptera (= Planipennia)

1. The number of species given is for America north of Mexico.

KEY TO FAMILIES OF RAPHIDIOPTERA LARVAE OF NORTH AMERICA[2]

1. Eyes with 4 stemmata; caudal region of abdominal terga 1 to 8 with small, rectangular or trapezoidal pale area (fig. 32.3) .... ***Inocelliidae*** p. 124

   Eyes with 6 or 7 stemmata; abdominal terga 1 to 8 without pale area in caudal region (fig. 32.4) .... ***Raphidiidae*** p. 124

Selected Bibliography

Acker 1966.
Achtelig 1967.
Aspöck 1975, 1986.
Aspöck & Aspöck 1971.
Aspöck, Aspöck and Hölzel 1980.
Aspöck, Aspöck and Rausch 1974, 1975.
Aspöck, Rausch and Aspöck 1974.
Carpenter 1936.
Woglum & McGregor 1958, 1959.

INOCELLIIDAE

Snakeflies

Figure 32.3

Relationships and Diagnosis: The 2 raphidiopteran families clearly represent distinct lineages (Aspöck *et al.* 1980). Both families occur in North America.

Unlike raphidiid larvae, the larvae of Inocelliidae (*Inocellia*) have 4 stemmata on each side of the head and a small, rectangular or trapezoidal, pale area in the caudal region of abdominal segments 1 to 8 (Fig. 32.3) (Aspöck & Aspöck 1971; Aspöck, Aspöck and Rausch 1974; Aspöck *et al.* 1980).

Biology and Ecology: As in Raphidioptera. Few biological features have been described specifically for the inocelliids. Unlike the raphidiids, the larvae of inocelliids are only known to occur under the loose bark of trees, not in the soil (Aspöck *et al.* 1975). Some courtship and mating patterns may also characterize the inocelliids (Acker 1966).

Description: As in Raphidioptera. The only unique characters that are ascribed to inocelliid larvae are four stemmata and the markings of abdominal segments 1 to 8 (*see* above).

Comments: Two species are described from western North America (Aspöck 1975). The larval taxonomy is in need of work.

Selected Bibliography

Acker 1966.
Aspöck 1975.
Aspöck & Aspöck 1971.
Aspöck, Aspöck and Rausch 1974.
Aspöck, Aspöck and Rausch 1975.
Aspöck, Rausch & Aspöck 1974.
Aspöck, Aspöck & Hölzel 1980.

RAPHIDIIDAE

Snakeflies

Figures 32.1, 32.2, 32.4

Relationships and Diagnosis: Raphidiid larvae are distinguished from inocelliids by the presence of 6 to 7 stemmata on each side of the head and the absence of a small, rectangular or trapezoidal pale area in the caudal region of abdominal segments 1 to 8 (Figure 32.4) (Aspöck & Aspöck 1971; Aspöck, Aspöck and Rausch 1974; Aspöck *et al.* 1980).

Biology and Ecology: As in Raphidioptera. Few unique features have been described for the raphidiids. The larvae tend to occur both under the loose bark of trees and in the soil; inocelliid larvae are not known from the soil (Aspöck *et al.* 1975). Some courtship and mating patterns may also be characteristic of the family (Acker 1966).

Description: As in Raphidioptera. The only unique characters that are ascribed to raphidiid larvae are eyes with 6 to 7 stemmata and the absence of pale areas on abdominal segments 1 to 8 (*see* above).

Comments: The Raphidiidae contains only 2 genera in N. America, *Agulla* and *Alena;* 19 species are known from western regions in America north of Mexico (Aspöck 1975, 1986). Larvae of 2 species have been described (Woglum & McGregor 1958, 1959).

Selected Bibliography

Acker 1966.
Aspöck 1975, 1986
Aspöck & Aspöck 1971.
Aspöck, Aspöck and Rausch 1974, 1975.
Aspöck, Rausch & Aspöck 1974.
Aspöck, Aspöck & Hölzel 1980.
Woglum and McGregor 1958, 1959.

BIBLIOGRAPHY

Achtelig, M. 1967. Über die Anatomic des Kopfes von *Raphidia flavipes* Stein und die Verwandtschaftsbeziehungen der Raphidiidae zu den Megaloptera. Zool. Jb. Anat. 84:249–312.

Acker, T. S. 1966. Courtship and mating behavior in *Agulla* species (Neuroptera: Raphidiidae). Ann. Ent. Soc. Amer. 59:1–6.

2. After Aspöck et al. 1980.

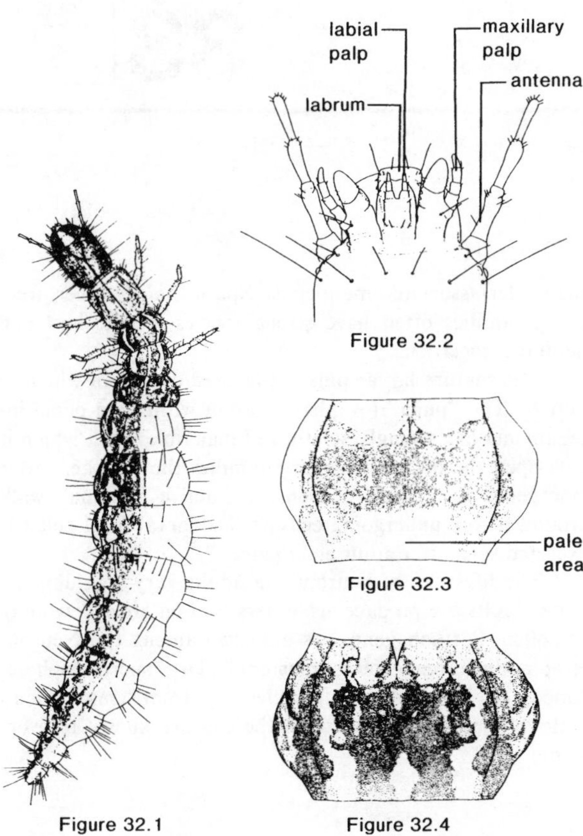

Figure 32.1 Raphidiidae, *Agulla*. Mature larva (from Woglum & McGregor 1958).

Figure 32.2. Raphidiidae, *Agulla*. Mouthparts, ventral view (from Woglum & McGregor 1958).

Figure 32.3. Inocelliidae. Abdominal tergum (from Aspöck, Aspöck & Rausch 1974).

Figure 32.4. Raphidiidae. Abdominal tergum (from Aspöck, Aspöck & Rausch 1974).

Aspöck, U. 1975. The present state of knowledge on the Raphidioptera of America (Insecta, Neuropteroidea). Polskie Pismo Ent. 45:537–46.

Aspöck, H. 1986. The Raphidioptera of the world: a review of present knowledge. pp. 15–29 *in* Gepp, J., H. Aspöck and H. Hölzel (eds.). Recent research in neuropterology. Proc. Int. Sym. neuropterology, Hamburg. 176 pp.

Aspöck, H. and Aspöck, U. 1971. Raphidioptera (Kamelhalsfliegen). In Handbuch der Zoologie. Band 4, Arthropoda; Hälfe 2, Insecta (J. G. Helmcke, D. Stack & H. Wermuth, eds.). pp. 1–48.

Aspöck, H., Aspöck, U. and Hölzel, H. 1980. Die Neuropteren Europas. Goecke & Evers, Krefeld. Vol. I, 495 pp.; Vol. II, 355 pp.

Aspöck, H., Aspöck, U. and Rausch, H. 1974. Bestimmungsschlüssel der Larven der Raphidiopteren Mitteleuropas (Insecta: Neuropteroidea). Zeit. Ang. Zool. 61:45–62.

Aspöck, H., Aspöck, U. and H. Rausch. 1975. Raphidiopteren-Larven als Bodenbewohner (Insecta, Neuropteroidea). Zeit. Angew. Zool. 62:361–75.

Aspöck, H., Rausch, H. and Aspöck, U. 1974. Untersuchungen über die Ökologie der Raphidiopteren Mitteleuropas (Insecta, Neuropteroidea). Zeit. Angew. Ent. 61:1–30.

Carpenter, F. M. 1936. Revision of the Nearctic Raphidiodea (recent and fossil). Proc. Amer. Acad. Arts Sci. 71:89–157, 2 pl.

Gepp, J. 1984. Erforschungsstand der Neuropteren-Larven der Erde. pp. 183–239. *In* Progress in World's Neuropterology (J. Gepp, H. Aspöck, & H. Hölzel, eds.). Proc. 1st Int. Symp. Neuropterology, Graz. 265 pp.

Wöglum, R. S. and McGregor, E. A. 1958. Observations on the life history and morphology of *Agulla bractea* Carpenter (Neuroptera: Raphidiodea: Raphidiidae). Ann. Ent. Soc. Amer. 51:129–41.

Wöglum, R. S. and E. A. McGregor. 1959. Observations on the life history and morphology of *Agulla astuta* (Banks) (Neuroptera: Raphidiodea: Raphidiidae). Ann. Ent. Soc. Amer. 52:489–502.

Order Neuroptera

33

Catherine A. Tauber
Cornell University

LACEWINGS, ANTLIONS, OWLFLIES, ETC.

The Neuroptera (*sensu stricto,* = Planipennia of many authors) comprise a relatively well-defined, presumably monophyletic, but diverse group of holometabolous insects (see Aspöck *et al.* 1980; Henry 1982; Gepp 1984). Together with the Megaloptera and Raphidioptera, they comprise the Neuropteroidea (Neuroptera, *sensu lato* of many authors), an ancient group that is perhaps ancestral to the Coleoptera.

DIAGNOSIS

The Neuroptera (*s.s.*) are characterized by a number of synapomorphies, the primary one being suctorial larval mouthparts consisting of sickle-shaped, grooved mandibles and maxillae (laciniae) that form feeding tubes. The mouth is reduced to a small, narrow cleft and is closed off by the modified labrum and labium. Other synapomorphies include the absence of maxillary palpi, the presence of 1-segmented tarsi, and the closure of the mid- and hindgut. Undigested food is stored during the entire larval period in the caudal part of the ventriculus and is excreted immediately after emergence of the adult. There are 8 Malpighian tubules (6 in the Coniopterygidae), of which 6 are attached to the hindgut and modified as silk-producing organs in the last instar.

BIOLOGY AND ECOLOGY

Oviposition behavior varies among the Neuroptera. In many, but not all families, the eggs are laid singly or in clusters at the end of silken threads; in other families the eggs are deposited onto soil or attached to a substrate.

All neuropteran families except the Ithonidae and the Dilaridae have 3 larval instars. With major exceptions (the aquatic Sisyridae and Neurorthidae and the semiaquatic Osmylidae), the larvae are terrestrial, and without exception they are carnivorous. Most prey on soft-bodied arthropods; however, the sisyrids feed on freshwater sponges (*Spongilla*), and some mantispids are parasitoids of wasps and bees. Highly specialized lifestyles characterize many families, e.g., the subterranean ithonids, the termite-associated berothids, and the spider-associated mantispids. Species in the other, free-living families often have strong species-specific host and habitat associations.

The mature larva spins a 2-layered cocoon in which to pupate. The "pupa" (= pharate adult within the pupal integument) has strongly sclerotized mandibles with which it cuts open the cocoon. In many families the "pupae," after emergence from their cocoons, are active and may walk around before undergoing ecdysis. The larval fecal pellet is excreted soon after adult emergence.

The lifestyles of neuropteran adults vary considerably. Some adults are predaceous; others feed on honeydew and/or pollen. Parthenogenesis is unknown among the Neuroptera; mating often involves transfer of a large spermatophore. Among the Neuroptera, life cycles vary from semivoltine to multivoltine. All stages except the egg are known to overwinter.

DESCRIPTION

Mature larva from ~3 to ~20 mm long; body campodaeiform, fusiform, or scarabaeiform. Spiracles present on mesothorax and abdominal segments 1 to 8 (absent from Sisyridae).

Head variable in shape, from rounded (Polystoechotidae and Ithonidae) to flattened (other families). Eyes with 0 to 7 stemmata; antennae from short and stub-like to long and filiform. Mouthparts composed of suctorial jaws (grooved mandibles and elongate, grooved laciniae, which contain a feeding tube) and a labium; jaws usually sickle-shaped, long, exserted; mouth closed by labrum and modified labium; maxillary palpi absent; labial palpi present (absent in Sisyridae), with 2 to 5 segments.

Thorax well developed; prothorax usually relatively large, bearing pair of dorsal sclerites; meso- and metathoraces with smaller, less pronounced sclerites. Legs usually well developed (reduced in Ithonidae, second and third instar Mantispidae); tarsi 1-segmented, usually not fused with tibiae [except in Ithonidae (all legs), Myrmeleontidae and Ascalaphidae (hindlegs)].

Abdomen with 10 segments, no cerci; terminal segment often with adhesive organ to aid in walking or holding.

COMMENTS

Among the Insecta, the Neuroptera (*s.s.*) is a relatively small order. There are approximately 4000 species in 17 families, all of which except 6 [Rapismatidae (Asia), Osmylidae (worldwide except North America), Neurorthidae (Europe, Asia, Australia), Psychopsidae (South Africa, Asia, Australia), Nemopteridae (Mediterranean area, Asia, Africa, Australia, South America), and Nymphidae (Australian region)] occur in America north of Mexico. The families roughly fall into 6 suborders (*see* classification below), all of which, except the Myrmeleontoidea, are poorly defined. The Ithonoidea and Osmyloidea may be polyphyletic.

Larval characters are of primary importance in defining the neuropteran families and their phylogenetic relationships (Withycombe 1925; MacLeod 1970; Gepp 1984). Despite this and despite the importance of the larvae of several families as natural enemies of important agricultural and forest insect pests, the larval taxonomy of most neuropteran families is in need of considerable work, especially at the generic level.

CLASSIFICATION[1]

- Superorder Neuropteroidea
  - Order Megaloptera
  - Order Raphidioptera
  - Order Neuroptera (= Planipennia)
    - Suborder Ithonoidea
      - Ithonidae—moth lacewings (1)
      - Polystoechotidae—giant lacewings (2)
      - *Rapismatidae
    - Suborder Coniopterygoidea
      - Coniopterygidae—dustywings (50)
    - Suborder Osmyloidea
      - *Osmylidae
      - Sisyridae—spongillaflies (6)
      - *Neurorthidae
    - Suborder Mantispoidea
      - Dilaridae—pleasing lacewings (2)
      - Mantispidae—mantidflies (10)
      - Berothidae—beaded lacewings (10)
    - Suborder Hemerobioidea
      - Chrysopidae—green lacewings (100)
      - Hemerobiidae—brown lacewings (60)
    - Suborder Myrmeleontoidea
      - Myrmeleontidae—antlions (100)
      - Ascalaphidae—owlflies (5)
      - *Psychopsidae—silky lacewings
      - *Nemopteridae
      - *Nymphidae

KEY TO FAMILIES OF NEUROPTERA LARVAE OF NORTH AMERICA[2]

1. Head robust, rounded behind; jaws short, stout, almost straight; mature larva large (~20 mm long); subterranean, in litter, under stones 2

   Head dorsoventrally flattened throughout; jaws long and curved or straight and tapering distally; if short and stout, larva parasitic in wasp or bee nests or confined to spider egg sac; size variable 3

2(1). Scarabaeiform; stemmata absent (figs. 33.1, 33.2) ..... ***Ithonidae*** p. 128

   Carabaeiform or slender; stemmata present (fig. 33.3) ..... ***Polystoechotidae*** p. 129

3(1). Body small (3.5 mm maximum); thorax wider than and as long as abdomen (fig. 33.4); labrum present, at least partially covering jaws (figs. 33.5–33.8); antennae and labial palpi each with 2 segments ..... ***Coniopterygidae*** p. 130

   Size variable; thorax not wider than abdomen, generally shorter; labrum absent; antennae and labial palpi (when present) with more than 2 segments 4

4(3). Spiracles and labial palpi absent; mouthparts comprised of long, flexible, needle-like mandibles and maxillary stylets (figs. 33.9, 33.10); aquatic; abdomen of second and third instars with tracheal gills on venter (fig. 33.11) ..... ***Sisyridae*** p. 130

   Spiracles and labial palpi present; mouthparts not as above; terrestrial; abdomen without gills 5

1. The number of species given is for America north of Mexico.
*absent from North America

2. Key characters for the Ithonidae and Polystoechotidae (couplets 1, 2) are tentative; no larvae of the North American ithonid species are known, and only the first instar polystoechotid has been described.

5(4). Jaws straight, or even slightly bent outward, narrow distally; if bent inward slightly, then stout and with short, stout labial palpi (Mantispidae, first instar *Plega*) 6

Jaws bending inward conspicuously; mandibles smooth, sickle-shaped or toothed, curving inward distally 8

6(5). Abdomen more than 3 times length of thorax (fig. 33.12); 1 pair of stemmata; labial palpi with 5–8 segments (fig. 33.13); under bark .... ***Dilaridae*** p. 132

Abdomen less than 3 times length of thorax; 2 to 3 pairs of stemmata; labial palpi with 2 or 3 segments; usually associated with ant or termite nests, spider egg sacs, or bee or wasp larval cells 7

7(6). First instar: campodeiform (fig. 33.14), with large head, long antennae, stout mandibles, labial palpi with stout midsegment (fig. 33.15), free-living or phoretic on spiders or adult Hymenoptera; second and third instars: scarabaeiform (fig. 33.16), with small, pale head, short antennae and labial palpi, small mouthparts (fig. 33.17), small stub-like legs, confined to single bee or wasp cell or to single spider egg sac .... ***Mantispidae*** p. 133

First and third instars: elongate, fusiform (fig. 33.18) with dark head; all instars: head elongate, with elongate mouthparts (fig. 33.19); antennae and labial palpi longer than jaws, legs well developed, associated with termite and ant nests .... ***Berothidae*** p. 135

8(5). Antennae and labial palpi longer than jaws; mandibles without teeth; body fusiform, globular or cucujid-like; setae straight, curved, hooked or serrated, not dolichasterine (trumpet-shaped) 9

Antennae and labial palpi much smaller than jaws; stout ovoid body; at least some parts of body with setae that are dolichasterine (trumpet-shaped), scale-like, stellate, or peg-like (fig. 33.35) 10

9(8). Thorax and abdomen with setigerous tubercles of various lengths (figs. 33.20, 33.21); tarsi with trumpet-shaped empodia between claws (fig. 33.23); first abdominal segment approximately 2/3 length of second segment .... ***Chrysopidae*** p. 135

Thorax and abdomen without setigerous tubercles (figs. 33.24, 33.25); empodia absent from tarsi of second and third instars (fig. 33.26), present on first; first abdominal segment usually nearly same length as second .... ***Hemerobiidae*** p. 137

10(8). Hind margin of head not bilobed (figs. 33.30, 33.31); thorax and abdomen with lateral setigerous projections (scoli) small or absent (fig. 33.29) .... ***Myrmeleontidae*** p. 138

Hind margin of head bilobed; thorax and abdomen with large lateral setigerous projections (scoli) (fig. 33.32) .... ***Ascalaphidae*** p. 140

Selected Bibliography

Aspöck et al. 1980.
Gepp 1984.
Henry 1982.
MacLeod 1970.
Withycombe 1925.

ITHONIDAE

Moth Lacewings

Figures 33.1, 33.2

Relationships and Diagnosis: The ithonids share some important characteristics with the Megaloptera (e.g., large, stout body, more than three larval instars), but the larval mouthparts are clearly neuropteran. Together with the Polystoechotidae, the Ithonidae is considered an archaic family among the Neuroptera.

Larvae of only one species, *Ithone fusca* Newman, from Australia, have been described (Tillyard 1922; Withycombe 1925). They are diagnosed by their scarabaeiform character, lack of eyes, distinctive mouthparts, and fused tibia and tarsus on all legs.

Biology and Ecology: Large eggs are deposited in soil. The female rolls each egg on the substrate, and soil or sand particles adhere to a sticky substance on the egg and form a protective covering. The larvae are subterranean; large groups may occur near a single tree. There are at least five instars, and the complete life cycle probably takes two years. Larvae will feed on scarabaeid larvae (Tillyard 1922), but there is considerable question as to what they feed upon in nature (Smithers 1979). On warm evenings, males form large swarms around females, and mating takes place on the sides of trees, etc. These swarms may suffer heavy predation (Tillyard 1922).

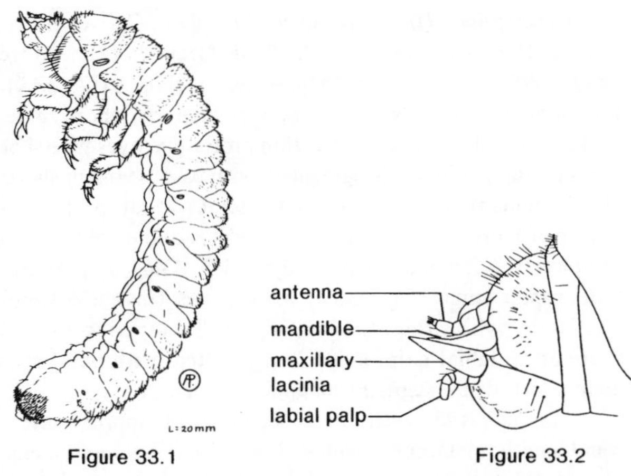

Figure 33.1 Figure 33.2

Figure 33.1. Ithonidae. Mature larva (*Ithone fusca*) (from Tillyard 1922. Reproduced by permission of the Commonwealth Agricultural Bureaux, Farnham Royal, England) (Redrawn by Peterson, 1951).

Figure 33.2. Ithonidae. Lateral view of head and mouthparts (*Ithone fusca*).

Description: (largely from Tillyard 1922; Withycombe 1925). Mature larva scarabaeiform (fig. 33.1), large, about 20 mm long, white. Integument with macro- and microtrichiae. Head fairly large, without eyes (fig. 33.2). Antennae short, with five segments. Mouthparts small; mandibles short, straight, pointed, curved upwards, smaller than maxillae; maxillae blunt-ended, with large bases, devoid of setae at tips; labial palpi short, with five segments.

Thorax and abdomen soft, white. Legs short, with tibiae and tarsi fused; each leg terminating in two claws. Last abdominal segment large, rounded, with small anal papillae.

Comments: The family is very small, largely restricted to Australia, with 1 species, *Oliarces clara* Banks, recorded from North America (southern California) (Adams 1950; Carpenter 1951; Riek 1974). The Ithonidae and the rare Rapismatidae (known only from the highlands of India and Nepal, southeastern Asia, and Indonesia) appear to be very closely related (Barnard 1981). Discovery and comparative study of additional ithonid larvae and rapismatid larvae would be of special interest in determining the origins of the primitive neuropteran families.

Selected Bibliography

Adams 1950.
Barnard 1981.
Carpenter 1951.
Riek 1974.
Smithers 1979.
Tillyard 1922.
Withycombe 1925.

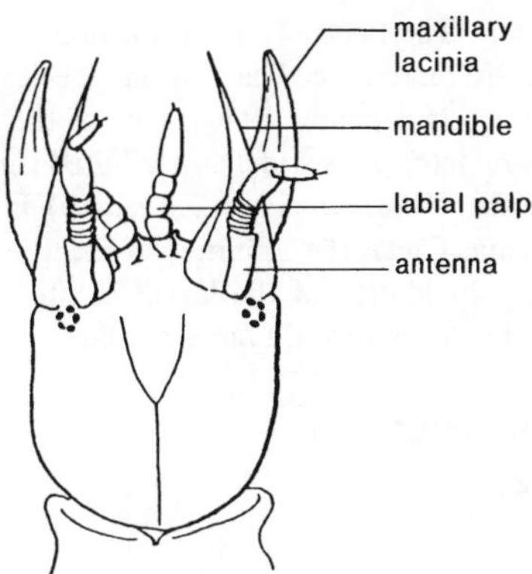

Figure 33.3. Polystoechotidae. Dorsal view of head (*Polystoechotes punctatus*) (from Withycombe 1925).

POLYSTOECHOTIDAE

Giant Lacewings

Figure 33.3

Relationships and Diagnosis: The Polystoechotidae is a small, poorly known Nearctic family. Only the first instar larva of one species, *Polystoechotes punctatus* Fabricius, is described. It shares several characteristics with megalopteran larvae, i.e., a rounded, well-chitinized head, short and almost straight jaws, and large thorax with heavily sclerotized thoracic terga and a large prothorax; however, the mandibular and maxillary structure is clearly neuropteran.

Among the neuropteran families, the polystoechotids appear to be most closely related to the ithonids (Henry 1982). In addition to their affinities with the Megaloptera, both families have larvae with short, relatively straight jaws and a labrum, but unlike ithonids, polystoechotid larvae are carabaeiform.

Biology and Ecology: In captivity, females lay single, unstalked eggs. First instar larvae are active; later instar larvae, as well as their prey preference, are unknown. Presumably they are ground-dwelling predators, with specialized food habits.

Description: (largely from Welch 1914; Withycombe 1925). First instar only. Body somewhat hemerobiid-like, about 1.5 mm long. Head (fig. 33.3) broad, rounded behind, somewhat flattened dorsoventrally. Eyes probably with five stemmata; antennae short, stout, with four segments. Labrum present, with two papillae at tip. Jaws short, stout, slightly curved inward; mandibles slender, pointed apically; maxillae stout, blunt-ended; labial palpi 5-segmented, with terminal segment long.

Thorax large; prothorax about as long as meso- and metathoraces together; each segment with complete dorsal tergum. Legs well developed; tibio-tarsal joint articulated; each tarsus with pair of simple, straight claws and long, knobbed empodium. Abdomen less sclerotized than thorax, with numerous, long macrotrichae.

Comments: The Polystoechotidae contains three genera, of which two are represented, each by one species, in America north of Mexico (Carpenter 1940). One of the species (*Polystoechotes punctatus*) is widely distributed throughout the region; the other (*Platystoechotes lineatus*) is known only from California. Given the affinities of the family with the Megaloptera, elucidation of the larval habits and the morphology of later instars would be of considerable interest.

Selected Bibliography

Carpenter 1940.
Henry 1982.
Welch 1914.
Withycombe 1925.

CONIOPTERYGIDAE

The Dustywings

Figures 33.4–33.8

Relationships and Diagnosis: The Coniopterygidae is a homogeneous, monophyletic family, characterized by apomorphic characters in the larvae and adults (Meinander 1972); it is the only family in the Coniopterygoidea. Like the Ithonoidea, the Coniopterygoidea represents a sideline divergent from the main phylogenetic line of Neuroptera. Both groups share some characters with the Megaloptera. However, the Coniopterygidae does not have any synapomorphic characters with either of the ithonoid families, and it is therefore considered an independent line.

Coniopterygid larvae are distinguished from other neuropteran larvae by their small size, a protruding labrum (a character occurring in the Megaloptera and shared only with the Polystoechotidae and Psychopsidae); straight, forward-directed jaws (shared by some other neuropteran families); 6 Malpighian tubules (as opposed to 8 in other families, but the same as in Megaloptera), with unique histological properties (Withycombe 1925); three abdominal ganglia; no cerci or apical rosette of sensory setae; testes functioning only during larval stage, absent from adults; no spermatophore.

Biology and Ecology: Eggs are cemented singly, or occasionally in groups of two or three, to leaves or bark. Both larvae and adults are most frequently found on bushes or trees, although some species appear to be confined to low vegetation. Larvae feed on a variety of small, inactive prey, such as aphids, coccids, mites, and whiteflies. Many species are associated with certain types of vegetation, or even a single species of plant, indicating preferences for certain kinds of food or even monophagy. There are 3 larval instars, although one report describes 4 (Badgley *et al.* 1955).

Mature larvae spin flat, circular cocoons, consisting of two layers of white silk. Upon emergence the adult deposits the larval excrement in several small, black viscous masses (unlike the other Neuroptera, which excrete a single pellet). Adults are predaceous, but their feeding habits are poorly known. Flight is with a darting, fluttering motion; adults may feign death.

Description: (from Withycombe 1923; Collyer 1951; Rousset 1966; Meinander 1972, 1974b; Greve 1974). Mature larva small, under 3.5 mm long, widest at thorax (fig. 33.4). Integument with very small microtrichiae and short setae. Head small, often concealed within prothorax; eyes with 4 or 5 stemmata; antennae 2-segmented, with basal segment short, distal segment long, bearing long setae. Mouthparts consisting of labrum that projects beyond head and covers jaws completely (Coniopteryginae, figs. 33.5, 33.6) or partially (Aleuropteryginae, figs. 33.7, 33.8); jaws (mandibles and maxillary styli) straight, enlarged basally, tapering toward acute apex; labial palpi large, 2-segmented, with basal segment short, distal segment long, densely haired.

Thorax large, with three segments of approximately equal length; metathorax widest. Legs long (*Coniopteryx* and *Conwentzia*) or short (*Semidalis, Aleuropteryx,* and *Heteroconis*); tibia and tarsus not well articulated; tarsus with 2 terminal claws, pad-like empodium between them. Abdomen tapering toward apex, with alimentary canal visible through abdominal integument.

Comments: The Coniopterygidae are not commonly collected and were long considered to be rare, but recent work (e.g., Meinander 1972) illustrates that the group is widespread (worldwide), diverse (~300 described species in 2 subfamilies), and relatively abundant in certain areas. Although they are important natural enemies of many soft-bodied insect and mite pests, their potential as biological control agents is far from realized. There are 8 genera, containing about 50 species, recorded from America north of Mexico; most species occur in western North America (Meinander 1972, 1974a, 1975, 1986; Johnson 1981a, b). Larvae of some species have been described (e.g., see Meinander 1972, 1974b).

Selected Bibliography

Badgley *et al.* 1955.
Collyer 1951.
Gepp 1984.
Greve 1974.
Johnson 1981a, 1981b.
Meinander 1972, 1974a, 1974b, 1975, 1986.
Rousset 1966.
Withycombe 1923, 1925.

SISYRIDAE

Spongillaflies

Figures 33.9–33.11

Relationships and Diagnosis: Primarily on the basis of larval characters the Osmylidae and Neurorthidae are considered sister-groups; together, they constitute a sister-group of the Sisyridae (Zwick 1967; Aspöck *et al.* 1980). Larvae of all 3 families are aquatic or semiaquatic; of the 3, only the sisyrids occur in America north of Mexico.

Sisyrid larvae are distinguished from other neuropteran larvae by their long, multi-segmented (5–16 joints) antennae; long, needle-like jaws; lack of labial palpi; tarsi that bear a single apical claw and lack an empodium; and tracheal gills on the first 7 abdominal segments (second and third instars).

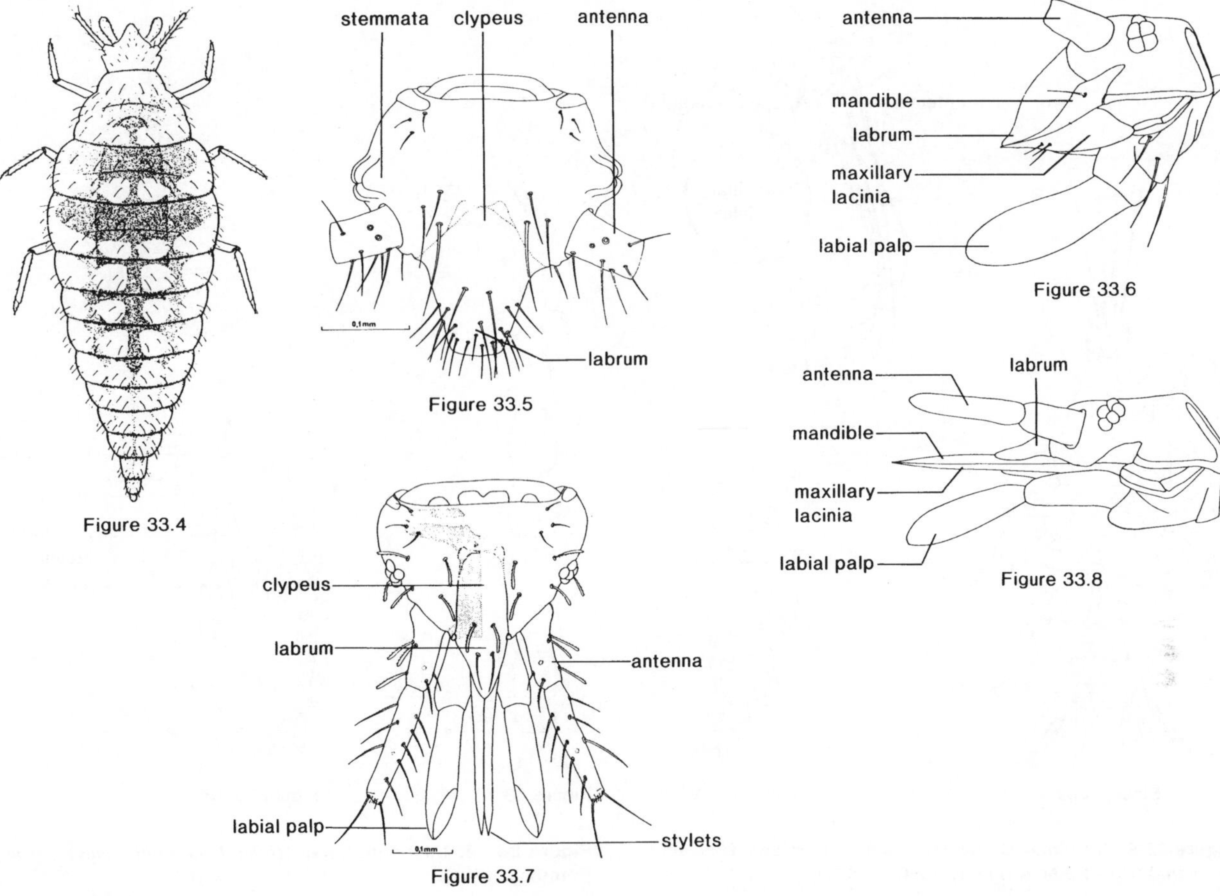

Figure 33.4

Figure 33.5

Figure 33.6

Figure 33.7

Figure 33.8

Figure 33.4. Coniopterygidae. Third instar larva (*Conwentzia pineticola*) (from Collyer 1951).

Figure 33.5. Coniopterygidae. Dorsal view of head (Coniopteryginae) (from Rousset 1966).

Figure 33.6. Coniopterygidae. Lateral view of head and mouthparts (Coniopteryginae) (from Rousset 1966).

Figure 33.7. Coniopterygidae. Dorsal view of head (Aleuropteryginae) (Stylets = mandibles and maxillary laciniae) (from Rousset 1966).

Figure 33.8. Coniopterygidae. Lateral view of head and mouthparts (Aleuropteryginae) (from Rousset 1966).

Biology and Ecology: Clusters of eggs are laid on objects above fresh water and are covered with a silken web of 3 to 4 layers. Newly hatched larvae drop to the water and penetrate the surface film. During the first instar, respiration apparently is cuticular; during the second and third larval stages it is by means of jointed tracheal gills on the abdomen.

Larval nourishment is derived from freshwater sponges (*Spongilla*). Larvae live on the outside of the sponge or they may enter the canals. Long setae covering the body entangle debris which serves to conceal the larvae (Withycombe 1923, 1925; Brown 1952).

Mature third-instar larvae leave the water and spin cocoons composed of a coarse outer layer and a fine, compact inner layer. Adults apparently feed on pollen or small insects. Among species in the family, voltinism is variable—some species being univoltine, others multivoltine.

Description: (largely from Withycombe 1923, 1925; Parfin & Gurney 1956). Mature larvae greenish to brownish, somewhat oval in shape, about 5 mm long (fig. 33.9). Integument with long setae. Head small, round; eyes composed of 6 stemmata; antennae longer than jaws, with 5 (*Climacia*) to 15 or 16 (*Sisyra*) segments. Jaws composed of elongate, needle-like mandibles and maxillary styli that are flexible and can be curled; labial palpi absent (fig. 33.10).

Thoracic segments each with pair of dark dorsal sclerites bearing 3 (meso- and metathoraces) to 5 or 6 (prothorax) setae; meso- and metathoraces with pair of large dorsolateral protuberances each bearing three setae. Legs with only 1 apical claw, no empodium.

Abdominal segments 1 to 7 each with pair of enlarged, dorsal plates bearing three long setae, each with pair of large lateral tubercles bearing three long setae. Abdominal segment 8 with two pairs of large, lateral setae-bearing tubercles. Second and third instars with venter of each abdominal segment 1 through 7 bearing pair of folded, tracheal gills; first pair of gills largest, but with only 2 joints; succeeding gills 3-jointed and decreasing in size posteriorly (fig. 33.11).

First instars are very different from later instars; their jaws are short and stout, antennae are 5-segmented, and tracheal gills are absent.

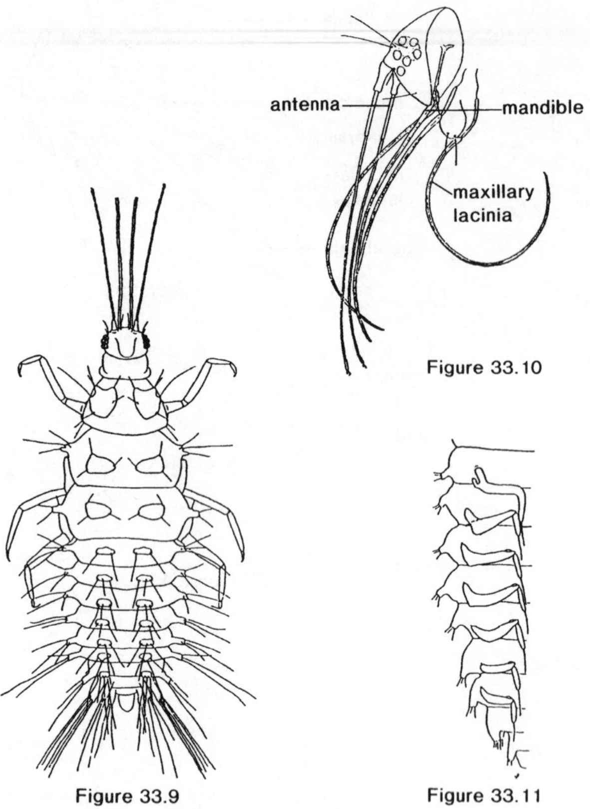

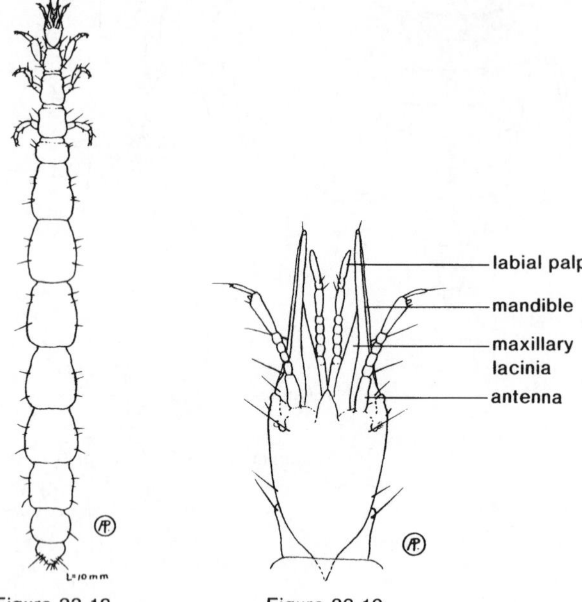

Figure 33.9. Sisyridae. Dorsum of second instar larva (*Climacia areolaris*) (from Parfin & Gurney 1956).

Figure 33.10. Sisyridae. Lateral view of head and internal bases of mouthparts (*Climacia areolaris*) (from Parfin & Gurney 1956).

Figure 33.11. Sisyridae. Tracheal gills on venter of abdominal segments 1-7 (*Sisyra vicaria*) (from Parfin & Gurney 1956).

Figure 33.12. Dilaridae. Larva (*Nallachius americanus*). (From Peterson, 1951).

Figure 33.13. Dilaridae. Head (*Nallachius americanus*). (From Peterson, 1951).

Comments: The Sisyridae occurs worldwide and contains three well-defined genera that are not designated to subfamilies (Parfin & Gurney 1956; Monserrat 1977). About 6 species in two genera (*Climacia* and *Sisyra*) are known from America north of Mexico, both genera occurring in eastern and western regions (Parfin & Gurney 1956; Poirrier 1969; Poirrier & Arceneaux 1972; Grigarick 1975; Merritt & Cummins 1984). Larvae of the 2 genera can be differentiated on the basis of thoracic and abdominal morphology and setation (Poirrier & Arceneaux 1972; Pupedis 1980). Larvae of some species have been described (Parfin & Gurney 1956; Gepp 1984).

Selected Bibliography

Aspöck *et al.* 1980.
Brown 1952.
Gepp 1984.
Grigarick 1975.
Merritt and Cummins 1984.
Monserrat 1977.
Parfin & Gurney 1956.
Poirrier 1969.
Poirrier & Arceneaux 1972.
Pupedis 1980.
Withycombe 1923, 1925.
Zwick 1967.

DILARIDAE

Pleasing Lacewings

Figures 33.12, 33.13

Relationships and Diagnosis: The systematic position of the dilarids in relation to other members of the Neuroptera is unclear; they appear to be most closely related to the berothids and mantispids.

Larvae are recognized by their very long, slender bodies, soft unpigmented cuticle, abdominal length of three or more times that of the thorax, long straight jaws, single pair of stemmata, 5- to 8-segmented antennae with enlarged penultimate segments, and 5-segmented labial palpi.

Biology and Ecology: Eggs of the North American species, *Nallachius americanus,* are laid in groups in the cracks and crevices of bark on recently killed or rotting trees, primarily oaks (*Quercus*) and tuliptrees (*Liriodendron*) (MacLeod & Spiegler 1961). Larvae occur between the bark and outer wood and presumably feed on wood-inhabiting beetle larvae. Unlike the larvae of all other neuropteran families except the Ithonidae, the dilarids have a variable number of larval instars; in addition, their method of ecdysis apparently is unique (MacLeod & Spiegler 1961). Mature larvae spin tightly woven, silk cocoons; fragments of wood, frass and other small particles are incorporated into the outer layer.

Description: (largely from Gurney 1947; Gilyarov 1962). Mature larva up to 12 mm long; body elongate, thin, not flattened (fig. 33.12). Integument unpigmented, with setae sparse, long; spiracles well developed, but difficult to see without magnification. Head yellowish; single pair of stemmata; antennae long, with 5 to 8 segments, with penultimate segment enlarged, bearing an apical peg in *Nallachius;* jaws (fig. 33.13) elongate, forward-pointing, broad at base, tapering apically; labial palpi almost as long as jaws, with 5 to 8 segments.

Thorax pale; mesothoracic spiracles in fold between prothorax and mesothorax; forelegs somewhat more robust and elongate than mid- or hindlegs; tarsi distinct but capable of little or no movement on tibiae; each leg with two claws, pair of pulvilli at base of claws, slender empodium. Abdominal segments slightly wider than thoracic segments; each segment with two lobes, each with pair of strong setae.

Comments: The Dilaridae is a small family, with approximately 50 described species from southern Europe, northern Africa, Asia, North America, and South America. There are two subfamilies—Dilarinae restricted to the Old World and Nallachiinae in the New World (Carpenter 1940; Adams 1970). Two species are known from America north of Mexico—*Nallachius americanus* from eastern deciduous forests and *Nallachius pulchellus* from Arizona (Adams 1970). The larvae of only two species are described [*Dilar turcicus* Hag., a soil-inhabiting species from the Soviet Union (Gilyarov 1962) and *N. americanus* from the United States (Gurney 1947)].

Selected Bibliography

Adams 1970.
Carpenter 1940.
Gilyarov 1962.
Gurney 1947.
MacLeod & Spiegler 1961.

MANTISPIDAE

Mantidflies

Figures 33.14–33.17

Relationships and Diagnosis: The Mantispidae is considered a sister-group of the Berothidae (MacLeod & Adams 1967; Aspöck *et al.* 1980), but the relationship of the two families to other Neuroptera is not clear. They appear to be most closely allied to the dilarids. Larvae of all three families are characterized by small heads, relatively straight jaws, and antennae placed far forward. The mantispids and berothids both undergo hypermetamorphosis, but only in the Mantispidae are the first-instar larvae different in form and habits from both the second and third instars.

Mantispid larvae are distinguished morphologically from berothid and dilarid larvae by their short jaws, which do not exceed half the length of the head, two pairs of stemmata (one pair in Dilaridae, one or two in Berothidae), and the length of the thorax, which is at least one-third the length of the abdomen (less in the Dilaridae). Third instar mantispids are easily recognized by their C-shaped body, enlarged abdomen, reduced head and mouthparts, 1-segmented antennae, 2-segmented labial palpi, and small, stub-like legs.

Biology and Ecology: Clusters of several hundred to several thousand eggs are laid at the end of short silken stalks. The eggs are small, but each female is capable of producing large numbers during her lifetime (Lucchese 1955, 1956; Redborg and MacLeod 1985). The first instar is free-living and must find the food upon which the entire larval stage will develop. The food source differs between the two subfamilies: mantispine larvae prey exclusively on spider eggs (within one egg sac), whereas the platymantispines are parasitoids of vespoid, sphecoid, and solitary apoid Hymenoptera or they feed on subterranean insects (Parfin 1958; Parker & Stange 1965; Werner & Butler 1965; MacLeod & Redborg 1982; Gilbert & Rayor 1983). Host finding by first-instar mantispines occurs by two methods. Either the larva burrows directly into a previously constructed spider egg sac or it crawls onto a female spider and enters her egg sac as she produces it (Redborg 1982). First-instar larvae of some mantispine species may feed on spider haemolymph while awaiting oviposition of an egg sac. Mantispids that feed on larval Hymenoptera may also attach themselves to adult bees and wasps for transport to the nest (*see* Batra 1972).

After food is found, mantispid larvae develop exclusively within either a single spider egg sac or one cell of the hymenopteran host. The mature larva spins a silken cocoon within the egg sac or the hymenopteran cell; pupation occurs within the larval skin. Adults are active during the day or night, and they feed on a variety of insect prey. Some mimic wasps (Batra 1972; Opler 1981). Overwintering usually occurs during the first larval instar (McKeown & Mincham 1948); there can be one, two, or three generations per year.

Description: (largely from Lucchese 1955, 1956; Parker & Stange 1965). First instar campodeiform, with body tapering slightly at posterior (fig. 33.14). Head roughly quadrate (fig. 33.15); eyes with three stemmata; antennae 4-segmented, longer than jaws, with long bristle distally; jaws broad at base, projecting anteriorly and tapering to narrow, toothed or smooth blades distally (*Mantispa*) or curved and tapering only slightly (*Plega*); labial palpi 3-segmented, reaching nearly to tip of jaws (*Plega*) or beyond (*Mantispa*). Thorax membranous with sclerotization on first (*Plega*) or all (*Mantispa*) segments. Legs well developed, each with 2 tarsal claws, empodium apically. Abdomen membranous, with some sclerites bearing setae.

Second and third instars scarabaeiform (fig. 33.16); mature larva about 10 mm long, without pigmentation except on jaws and small abdominal tergites. Head (fig. 33.17) small relative to body; antennae 1-segmented, with short apical seta; labial palpi 2-segmented, without setae, arising from median paligeral swelling; jaws straight or nearly straight; mandibles very slightly tapered; maxillae broad at base, tapering medially. Thorax membranous; legs reduced to small 2-segmented (*Plega*) or 4-segmented (*Mantispa*) stubs. Abdomen large, with patterns of setation; segments 2 to 8 with small, paired swellings ventrally.

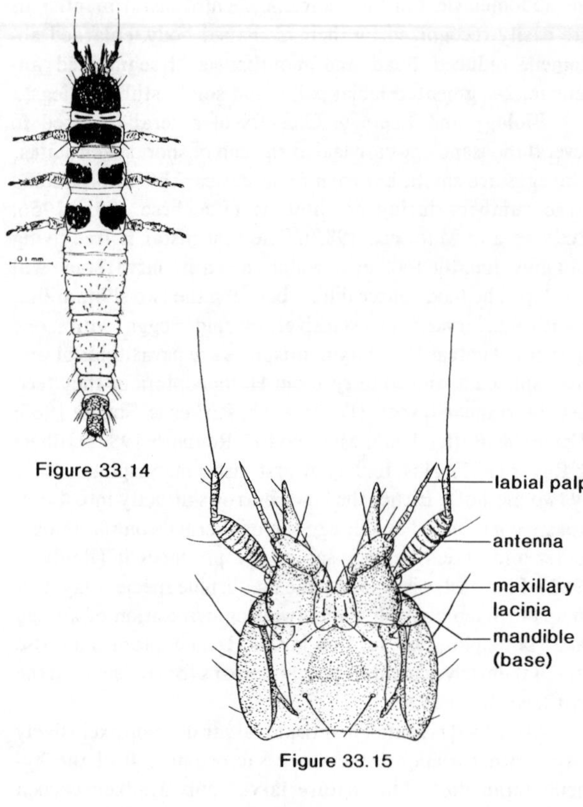

Figure 33.14

Figure 33.15

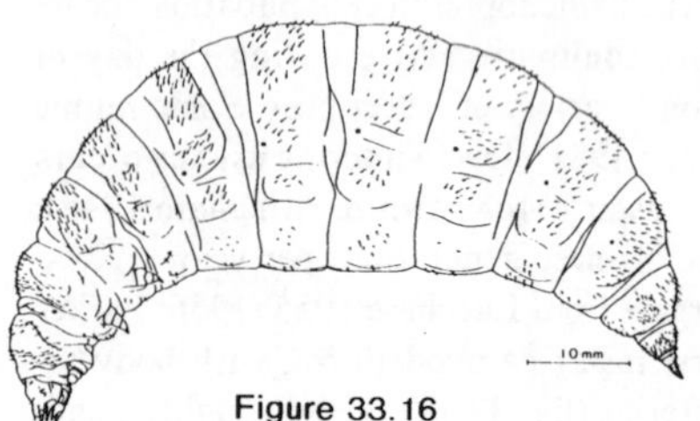

Figure 33.16

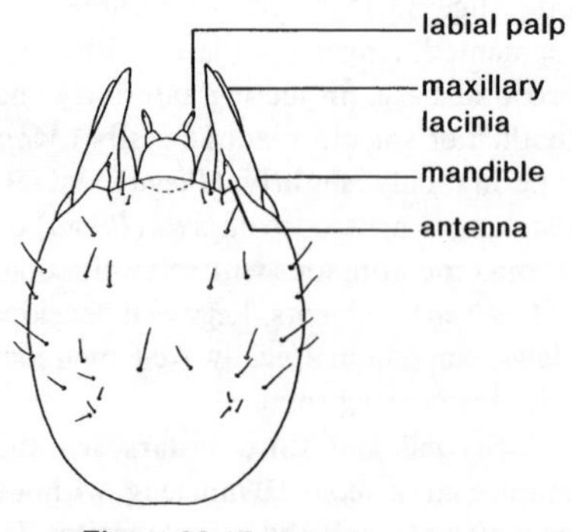

Figure 33.17

Figure 33.14. Mantispidae. First instar larva (*Mantispa uhleri*) (from Redborg 1982).

Figure 33.15. Mantispidae. Ventral view of first instar head (*Mantispa perla*) (from Lucchese 1955).

Figure 33.16. Mantispidae. Mature third instar larva (*Mantispa uhleri*) (from Redborg 1982).

Figure 33.17. Mantispidae. Dorsum of third instar head (*Plega yucatanea*) (from Parker & Stange 1965).

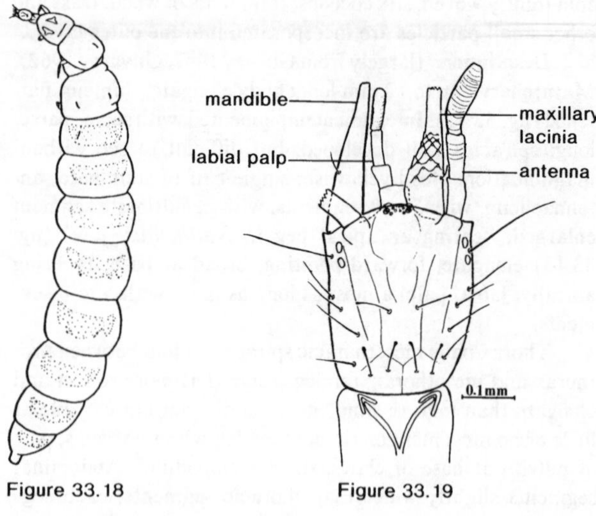

Figure 33.18

Figure 33.19

Figure 33.18. Berothidae. Mature third instar larva (*Lomamyia*) (from Gurney 1947).

Figure 33.19. Berothidae. Head of third instar larva (*Lomamyia latipennis*) (from Tauber & Tauber 1968).

Comments: The Mantispidae occurs worldwide, but is most prominent in the tropics and subtropics. There are about 350 described species, and the classification is in need of work (Aspöck *et al.* 1980). The two subfamilies (Mantispinae and Platymantispinae) are both represented in America north of Mexico. *Mantispa,* which is known from eastern and western U.S., is the most common mantispine genus; *Climaciella* (eastern and western U.S.) and *Plega* (western U.S.) are the most common platymantispines. Larvae of *Mantispa* and *Plega* are described (Rehn 1939; Kuroko 1961; Parker & Stange 1965; Gepp 1984; Redborg and MacLeod 1985).

Selected Bibliography

Aspöck *et al.* 1980.
Batra 1972.
Gepp 1984.
Gilbert & Rayor 1983.
Kuroko 1961.
Lucchese 1955, 1956.
MacLeod & Adams 1967.
MacLeod & Redborg 1982.
McKeown & Mincham 1948.
Opler 1981.
Parfin 1958.
Parker & Stange 1965.
Redborg 1982.
Redborg and MacLeod 1985.
Rehn 1939.
Werner & Butler 1965.

BEROTHIDAE

Beaded Lacewings

Figures 33.18, 33.19

Relationships and Diagnosis: The berothids and mantispids form a sister-group whose relationship with other neuropteran families is not clear (*see* Mantispidae). The larvae of both families and dilarid larvae share several characteristics including small heads, relatively straight jaws, and antennae placed far forward. However, berothid larvae can be distinguished by their pigmented heads and bodies (first and third instars), long jaws that exceed half the length of the head (short in Mantispidae), and thorax whose length is at least one-third that of the abdomen (less in Dilaridae). In addition, North American berothid larvae (*Lomamyia*) have 2 pairs of stemmata (one pair in Dilaridae, two in Mantispidae).

The berothids, as do the mantispids, undergo hypermetamorphosis, but only in the Berothidae is the second instar different in form and habits from the first and third. The first instars of both families and the third instars of the berothids are active, pigmented, and campodeiform.

Biology and Ecology: Small clusters of about 8–12 eggs are laid on or near termite-infested logs and are attached to the substrate by 1 to 4 long, flexible, intertwined silken stalks. The first instar is very mobile and apparently locates termite or ant nests; how this is done is not known. Feeding occurs after the larva has incapacitated its termite prey with a neurotoxin (Brushwein 1987) or a termite-specific allomone (Johnson & Hagen 1981). The second instar, unlike the first and third, is a nonfeeding stage, during which the larva hangs downward in a C-shaped position from the roof of the tunnel. This stage lasts only a few days (Tauber & Tauber 1968). All instars appear to have some sort of resistance to attack by termites (Johnson & Hagen 1981). Immature ants and termites apparently constitute the larval diet (Gurney 1947); some species of berothids may be host specific (Johnson & Hagen 1981).

Mature larvae spin sheer, oval cocoons that resemble those of hemerobiids. Adults are largely active at dusk; their feeding habits are unknown. Mating involves the transfer of a large spermatophore which remains protruding from the female bursa for several days (Tjeder 1959; MacLeod & Adams 1967). The season cycle is also unknown.

Description: (largely from Gurney 1947; Toschi 1964; Tauber & Tauber 1968). Mature larvae about 9 mm long, elongate, with large body, small head (fig. 33.18). Head (fig. 33.19) brown, somewhat elongate, two pairs of anterolateral stemmata; antennae longer than jaws, 3-segmented, each with long apical seta. Jaws straight, forward projecting, broad basally, tapering distally to sharp points (first instar) or blunt tips (third instar); maxillae with dorsal sculpturing; labial palpi 3-segmented, longer than jaws.

Thorax with dorsolateral sclerites well developed, pigmented; legs well developed, but small relative to body, approximately equal in size; tarsi and tibiae freely articulated; two pairs equal-sized tarsal claws, no pulvilli; empodium incised apically; hook-like sole below empodium. Abdomen large, with purplish brown transverse bands on dorsum, remainder pale; tip with sucker-like organ used in clinging to surfaces.

Second instar similar to third, but inactive. Body elongate, curved ventrally in C-shape. Antennae and mouthparts grouped together, extending forward; legs short, stubby, directed anteriorly. Head, prothorax, legs pale.

Comments: The Berothidae is a small family with about 60 described species in four subfamilies (MacLeod & Adams 1967). Representatives occur on all continents, but they are largely restricted to warm areas (Tjeder 1959; Aspöck *et al.* 1980). The North American fauna contains one genus, *Lomamyia;* although several species are widespread throughout the region, they are rarely encountered. Larvae of two species in this genus have been described and compared with the first instars of the Australian genus *Spermophorella* (Gurney 1947; Toschi 1964; Tauber & Tauber 1968; Brushwein 1987).

Selected Bibliography

Aspöck *et al.* 1980.
Brushwein 1987.
Gurney 1947.
Johnson & Hagen 1981.
MacLeod & Adams 1967.
Brushwein 1987.
Tauber & Tauber 1968.
Tjeder 1959.
Toschi 1964.

CHRYSOPIDAE

Green Lacewings

Figures 33.20–33.23

Relationships and Diagnosis: The Chrysopidae is the largest and most well-known neuropteran family. It is a monophyletic sister-group of the Hemerobiidae; however, the relationship of the two families to the other Neuroptera is not clear.

Chrysopid larvae are distinguished by their relatively large heads, usually long, toothless, sickle-shaped jaws, relatively long antennae, somewhat reduced first abdominal segment, usually well-developed setigerous tubercles on thorax and abdomen, and trumpet-shaped empodia (all instars).

Biology and Ecology: Chrysopids occur in a wide variety of habitats—forests, fields, deserts, and in association with ant nests. Except in one genus, eggs are laid at the ends of silken stalks—either singly or in clusters. There are three larval instars—all of which are usually similar in structure

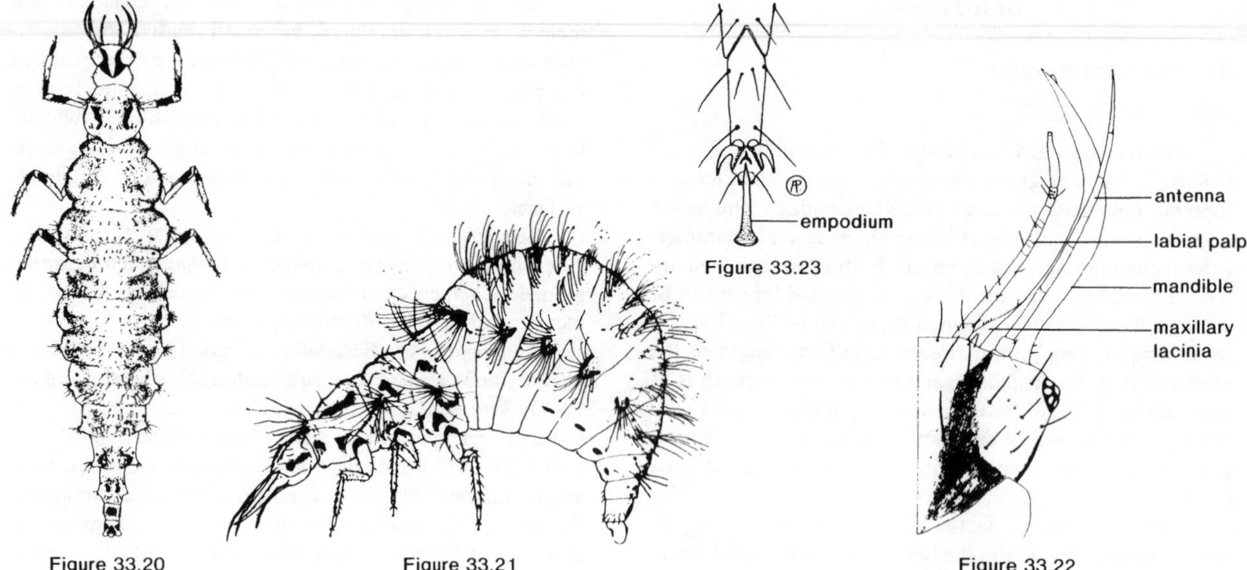

Figure 33.20. Chrysopidae. Mature third instar larva (*Chrysoperla carnea*) (from Tauber 1974). Length = 10 mm.

Figure 33.21. Chrysopidae. Mature third instar larva (*Suarius apache*) (from Tauber 1975).

Figure 33.22. Chrysopidae. Dorsum of third instar head (*Chrysoperla carnea*) (from Tauber 1974).

Figure 33.23. Chrysopidae. Tarsus. (From Peterson, 1951).

and habits. Newly hatched larvae descend their stalks and search for prey. Some species feed on a wide range of diverse soft-bodied arthropod hosts; others have clear preferences for a specific host or range of host species. One larva may consume large numbers of its prey, especially if the hosts are small.

Larvae of many species carry debris such as the skins of their prey, the waxy secretions of aphids, pieces of bark, lichens, or coccid scales in dense or loose packets on their backs. Modified thoracic and abdominal structures, specialized dorsal setae, and specialized behavioral patterns subserve trash carrying, which may serve both as a means of avoiding notice by predators and parasitoids, as well as a defense against aphid-tending ants (Eisner *et al.* 1978). The mature larva spins a 2-layered silken cocoon. Trash-carrying species usually incorporate their trash into the outer layer of the cocoon.

Adults are either predaceous and feed on soft-bodied arthropods (mainly aphids) or they eat honeydew and/or pollen (Hagen 1950; Sheldon & MacLeod 1971; Tauber & Tauber 1974). There can be one to several generations per year (Tauber & Tauber 1982). Overwintering occurs in all stages except the egg. Adult feeding habits and the overwintering stage are characteristic for genera and subgenera.

Description: (largely from Withycombe 1925; Tjeder 1966; Tauber 1974, 1975). Mature larva about 7 to 10 mm long; body broadly to narrowly fusiform, flattened and cucujid-like, or short and stout (figs. 33.20, 33.21). Integument covered with microtrichiae and bearing various types of setae (long, short, straight, hooked, smooth, serrated). Head (fig. 33.22) broad, flattened; 6 pairs of lateral stemmata; antennae longer than jaws, simple, filiform, 3-segmented, with midsegment subdivided, distal segment terminating in a bristle. Jaws slender, curving inward; mandibles pointed; maxillae blunt apically; labial palpi slightly shorter than jaws, 3-segmented, with midsegment subdivided.

Thoracic segments large, each usually with pair of lateral sclerites and a pair of dorsolateral setigerous tubercles that vary from small with short straight setae to elongate or stalked with long, hooked, or modified setae. Legs well developed; tibio-tarsal articulation not very free; tarsi each with 2 claws, long, trumpet-like empodium (fig. 33.23).

Abdomen with first segment smaller than second or third, with rows of setae dorsally, small or no lateral tubercles. Segments 2 to 8 each with setigerous dorso-lateral tubercles of varying lengths, dorsal rows of setae. Anal papilla adhesive.

Comments: The Chrysopidae currently contains three described subfamilies, two of which (Nothochrysinae and Apochrysinae) are relatively small, well defined, and presumably monophyletic (Kimmins 1952, Tjeder 1966; Adams 1967; New 1980). The third, Chrysopinae, which is the largest and which contains the most well-known and common species, is in need of worldwide taxonomic treatment (Hölzel 1970; Adams 1978; Aspöck *et al.* 1980). The New World chrysopine fauna is under investigation (Adams 1978).

No Apochrysinae and only a few species of Nothochrysinae occur in North America (western region) (Adams 1967); about 100 species in several genera of the Chrysopinae are distributed throughout the continent. One genus is restricted to Hawaii (Adams 1962, 1978; Zimmerman 1957).

Diagnostic larval characters have been identified, and larvae of some North American species have been described (Smith 1922; Toschi 1965; Tauber 1969, 1974, 1975). The chrysopids are considered to be among the most valuable natural enemies of agricultural pests, and many species are being studied to increase their efficiency in biological control programs (Hagen *et al.* 1971; Ridgway & Kinzer 1974; New 1975; Tauber & Tauber 1975, 1983; Agnew *et al.* 1981).

Selected Bibliography

Adams 1962, 1967, 1978.
Agnew *et al.* 1981.
Aspöck *et al.* 1980.
Eisner *et al.* 1978.
Hagen 1950.
Hagen *et al.* 1971.
Hölzel 1970.
Kimmins 1952.
New 1975, 1980.
Ridgway & Kinzer 1974.
Sheldon & MacLeod 1971.
Smith 1922.
Tauber 1969, 1974, 1975.
Tauber & Tauber 1974, 1975, 1982, 1983.
Tjeder 1966.
Toschi 1965.
Withycombe 1925.
Zimmerman 1957.

HEMEROBIIDAE

Brown Lacewings

Figures 33.24–33.28

Relationships and Diagnosis: The Hemerobiidae, which now contains the Sympherobiidae, appears to be a monophyletic group. Hemerobiids are most closely related to chrysopids with which they share many larval and adult characters. The relationships of the two families to other Neuroptera are not clear.

Hemerobiid larvae are similar to chrysopid larvae in that they have fairly long, inward-curving jaws without teeth, two terminal claws on each tarsus, and relatively well-developed antennae and labial palpi. They differ from the chrysopids in their consistently fusiform shape, their smooth bodies that lack setigerous tubercles, the similarity in size of abdominal segments 1 through 3, their somewhat shorter and stouter jaws, and the absence of pulvilli between the tarsal claws of the second and third instars.

Biology and Ecology: Eggs are unstalked and usually laid singly or in small groups on leaves, twigs, and bark. There are three larval stages. All genera have active first instars. The second and third instars of most genera are also active, but those of *Boriomyia* and *Sympherobius* are relatively sluggish. Larvae feed on aphids, scales, and other soft-bodied arthropods. Some species show distinct preferences for a particular host species; this becomes apparent when both larvae and adults occur almost exclusively in association with a particular species of plant. Unlike many chrysopid larvae, hemerobiid larvae are not known to carry trash.

The mature larva forms a 2-layered, loosely spun cocoon in a sheltered place in which to pupate. Adults are active at dusk or at night, and in most species they appear to be predaceous. When disturbed, they may feign death.

There may be one to several generations per year. Larvae, pupae, or adults overwinter, depending on the species.

Description: (largely from Withycombe 1925; MacLeod 1960; New & Boros 1983). Mature larva 4 to 7 mm long; fusiform to swollen (figs. 33.24, 33.25). Integument smooth, with very small microtrichiae, few simple setae. Head (figs. 33.26, 33.27) small, somewhat flattened, retracted into prothorax in some genera; eyes with 3 to 6 stemmata; antennae short to long, with three segments; terminal segment short and stout in some genera. Jaws stout, smooth, curved inward; apex of mandible acute, with barbs; apex of maxilla blunt, with minute hairs; labial palpi 3-segmented (although subdivisions may give appearance of more), ranging from shorter to longer than jaws.

Thorax and abdomen soft; thorax with first subsegment short, collar-like to elongate, tapered, with small dorsal sclerites, the largest of which is usually crescent-shaped and on the prothorax; prothoracic sclerites greatly reduced in some genera; legs well developed; tibio-tarsal joint freely articulated; each tarsus with two terminal claws (fig. 33.28); trumpet-shaped empodium present in first instar only. Abdomen varying from elongate and tapered to lobed and physogastric; first segment similar in size to second and third; tenth abdominal segment with pair of eversible appendages.

Comments: The Hemerobiidae is a diverse, worldwide family, containing approximately 40 described genera in two subfamilies: the very small Notiobellinae (Australian Region, Asia, Africa) and the large, but presumably monophyletic Hemerobiinae (worldwide) (Nakahara 1960). There are 60 described species in 8 genera in America north of Mexico (Carpenter 1940; Nakahara 1965); the family is distributed throughout the United States. The Hawaiian fauna contains 28 recorded species of which 24 are endemic (Zimmerman 1957; MacLeod 1964).

Larvae of some hemerobiids have been described (see MacLeod 1960; Nakahara 1954; New & Boros 1983; Gepp 1984), but those of most species are unknown. The Hemerobiidae are very important predators of insect pests in forests and agricultural crops (e.g., Agnew *et al.* 1981). Some species appear to be particularly well adapted to cool conditions and may therefore have a unique role to play in the suppression of pests early in the season or under conditions where other natural enemies are not very effective (Neuenschwander 1976).

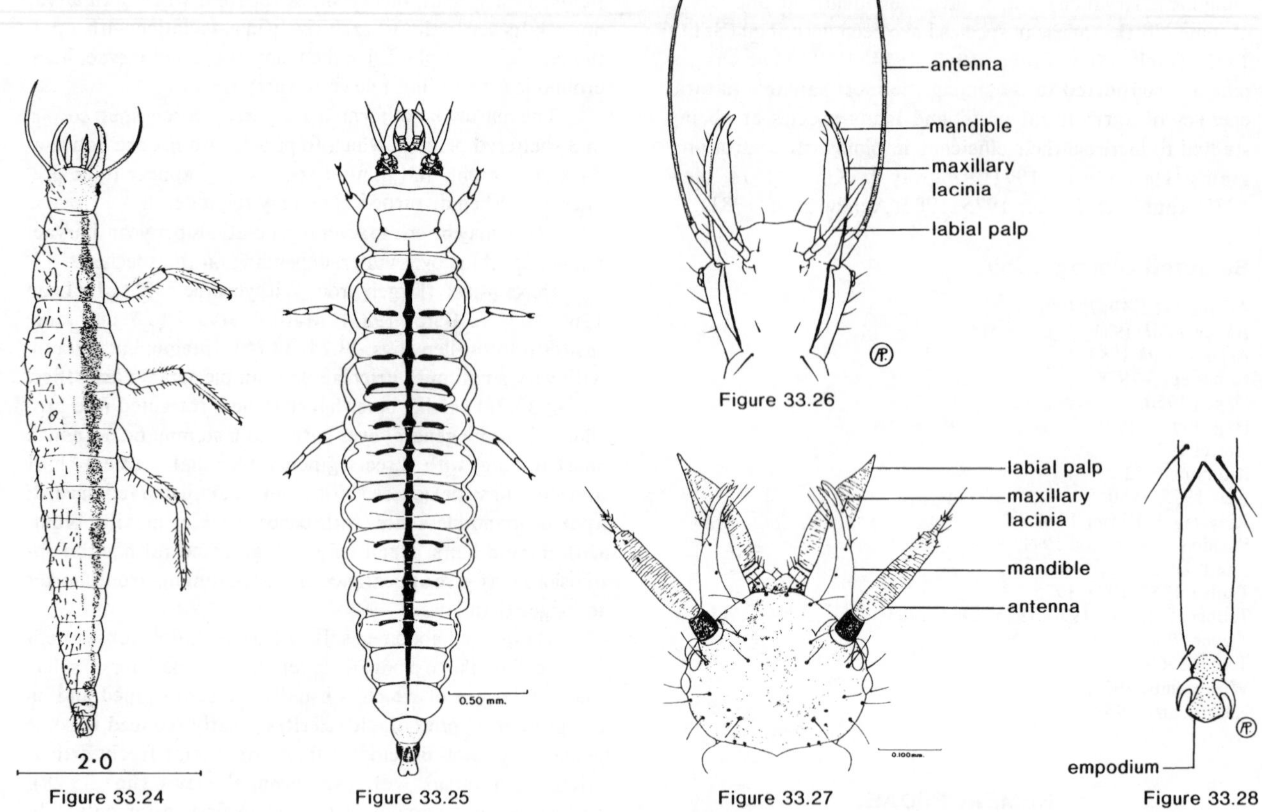

Figure 33.24 Figure 33.25 Figure 33.26 Figure 33.27 Figure 33.28

Figure 33.24. Hemerobiidae. Third instar larva (*Micromus tasmaniae*) (from New & Boros 1983).

Figure 33.25. Hemerobiidae. Third instar larva (*Boriomyia fidelis*) (from MacLeod 1960).

Figure 33.26. Hemerobiidae. Ventral view of head (third instar *Hemerobius*). (From Peterson, 1951).

Figure 33.27. Hemerobiidae. Dorsum of head (first instar *Boriomyia fidelis*) (from MacLeod 1960).

Figure 33.28. Hemerobiidae. Tarsus (*Hemerobius*). (From Peterson, 1951).

Selected Bibliography

Agnew *et al.* 1981.
Carpenter 1940.
Gepp 1984.
MacLeod 1960, 1964.
Nakahara 1954, 1960, 1965.
Neuenschwander 1976.
New & Boros 1983.
Withycombe 1925.
Zimmerman 1957.

MYRMELEONTIDAE

Antlions

Figures 33.29–33.31

Relationships and Diagnosis: The Myrmeleontidae and Ascalaphidae are the only myrmeleontoid families represented in North America; absent are the Psychopsidae (South Africa, Asia, Australia), Nemopteridae (Mediterranean area, Asia, Africa, Australia, South America), and Nymphidae (Australian region). The Stilbopterygidae, previously considered a sixth family within the Myrmeleontoidea, has recently been shown to be polyphyletic and is discounted as a functional taxonomic unit. On the basis of larval and adult characters, most of the species are now included in the Myrmeleontidae; a few are in the Ascalaphidae (New 1982a, b).

Myrmeleontid larve are highly variable, but they share several characters with the other Myrmeleontoidea, specifically: a heavily sclerotized, roughly quadrate head capsule with unique tentorial structure; large, inwardly curving jaws, usually with lateral teeth; relatively small multisegmented antennae with enlarged scapes and small pedicel; labium with pair of large prelabia each bearing a short, usually 2–4 segmented palp; stout, ovoid body; short legs with apical empodia; and dolichasterine (trumpet-shaped) setae over part or most of the body surface (Henry 1976). Among the Myrmeleontoidea, myrmeleontid and ascalaphid larvae are the only ones with metathoracic tibio-tarsal fusion. However, unlike ascalaphid larvae, myrmeleontid larvae have an enlarged metatarsal claw, and the dorsocaudal margin of the

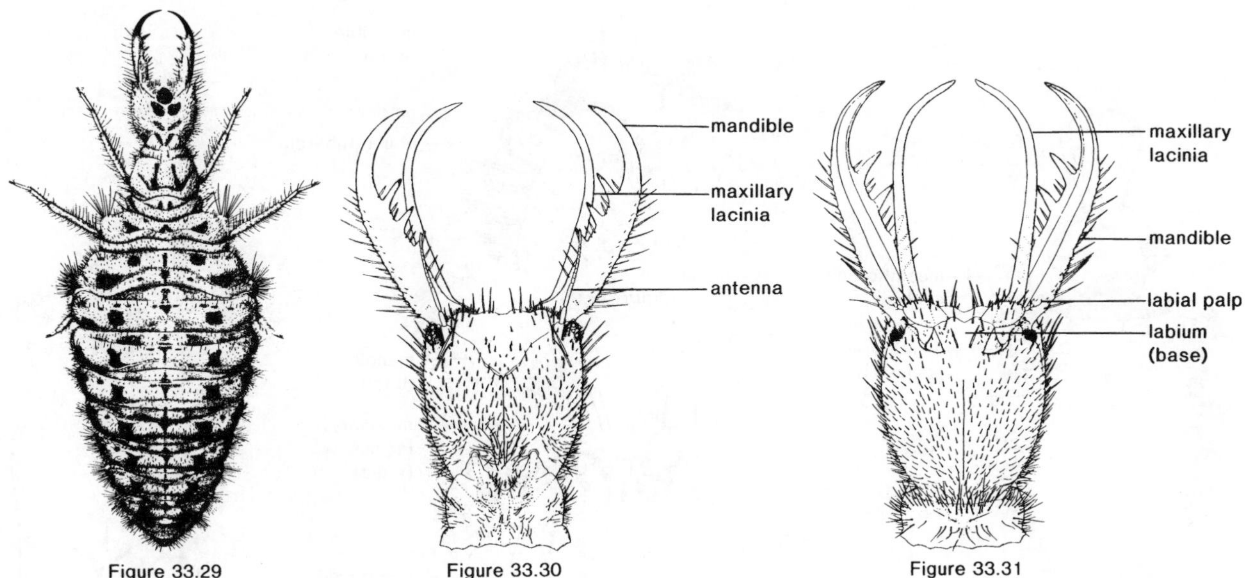

Figure 33.29. Myrmeleontidae. Mature third instar larva (*Myrmeleon inconspicuus*) (from Principi 1943).

Figure 33.30. Myrmeleontidae. Dorsum of head (*Myrmeleon inconspicuus*) (from Principi 1943).

Figure 33.31. Myrmeleontidae. Venter of head (*Myrmeleon inconspicuus*) (from Principi 1943).

head is not bilobed posteriorly. Myrmeleontids are also generally differentiated from ascalaphid larvae by their relatively small thoracic and abdominal lateral projections (scoli) and their variable number of mandibular teeth (1 to 4 in myrmeleontids, three in ascalaphids).

Biology and Ecology: Little is known about myrmeleontid reproductive behavior; some species are known to oviposit single eggs in soil or sand; others lay groups of eggs. The association of myrmeleontid larvae with funnel-shaped pitfall traps applies mainly to one genus (*Myrmeleon*) and a few species in other genera (*Brachynemurus, Hagenomyia*) in the Myrmeleontinae (Withycombe 1925; Stange 1980b; Henry 1982; see also Tuculescu *et al.* 1975). Larvae of most genera move across or under the surface of the sand and apparently search actively for prey. They often leave conspicuous trails on the surface of the sand. Others live in the hollows of trees, in animal burrows, or on logs (Stange 1980a; Miller & Stange 1983, 1985). The larval diet consists of termites, ants, beetle larvae, and other insects; extraintestinal digestion has been demonstrated (Buschinger & Bongers 1969). Running behavior is genus- and species-specific; movement can be fast or slow, backward or forward. Larvae of some species carry trash (Stange 1980b), and those of one species appear to mimic mutillid wasps (Brach 1978).

Larval development is relatively slow, requiring one to two years; at least for some species, any instar can overwinter (Furunishi & Masaki 1982). Mature larvae spin circular, 2-layered silken cocoons on trees or in the soil or sand. In most cases they are covered with sand or debris. Adults of many species are nocturnal and may be encountered at lights. Other species are active during the day or at dusk (Aspöck *et al.* 1980). Adult feeding habits are poorly known; some species are predaceous, and others presumably are pollen feeders.

Description: (largely from Withycombe 1925; Principi 1943, 1947; Stange and Miller 1985). Mature larva medium to large, 10–22 mm long, ovoid with lateral, setae-bearing protuberances (scoli) small, rounded, not finger-shaped (fig. 33.29). Integument with stiff setae, dolichasters, species-specific markings. Head (figs. 33.30, 33.31) without enlarged occipital lobes posterolaterally; jaws large, curving inward at the tip; mandibles with 1 to 4 mesolateral teeth, often with setae; eyes with 7 stemmata: 6 dorsad, and 1 ventrad on ocular lobe that extends forward over base of mandibles; antennae short, arising mesad of eyes; labial palpi short, 3-segmented, arising from large bilobed base.

Thorax and abdomen ovoid; prothorax (in pit-digging species) with median dorsal area raised, often overlapping back of head. Legs relatively long, with numerous stiff setae, with 2, almost straight tarsal claws that often appear as 1; hindlegs with tibiae and tarsi fused; fore- and midlegs without tibio-tarsal fusion; metathoracic claw enlarged; no empodia. Abdomen with 8 visible segments, ninth and tenth retracted.

Interspecific variation includes reduction of ocular lobes, reduction or enlargement of body scoli, antennal and jaw modifications, and specialization of the anal area.

Comments: The Myrmeleontidae is a relatively large family, containing approximately 2000 species on all continents; the group is in need of world-based taxonomic study (Aspöck *et al.* 1980). The family is represented by a large and diverse assemblage of species in America north of Mexico. Most species occur in the South and West; a few are recorded from Hawaii (Zimmerman 1957). In the Western Hemisphere, there are three subfamilies (Palparinae, Acanthclisinae, and Myrmeleontinae) that are further subdivided into somewhat poorly defined tribes (Stange 1970). Although the larvae of most American species are unknown, there is a larval

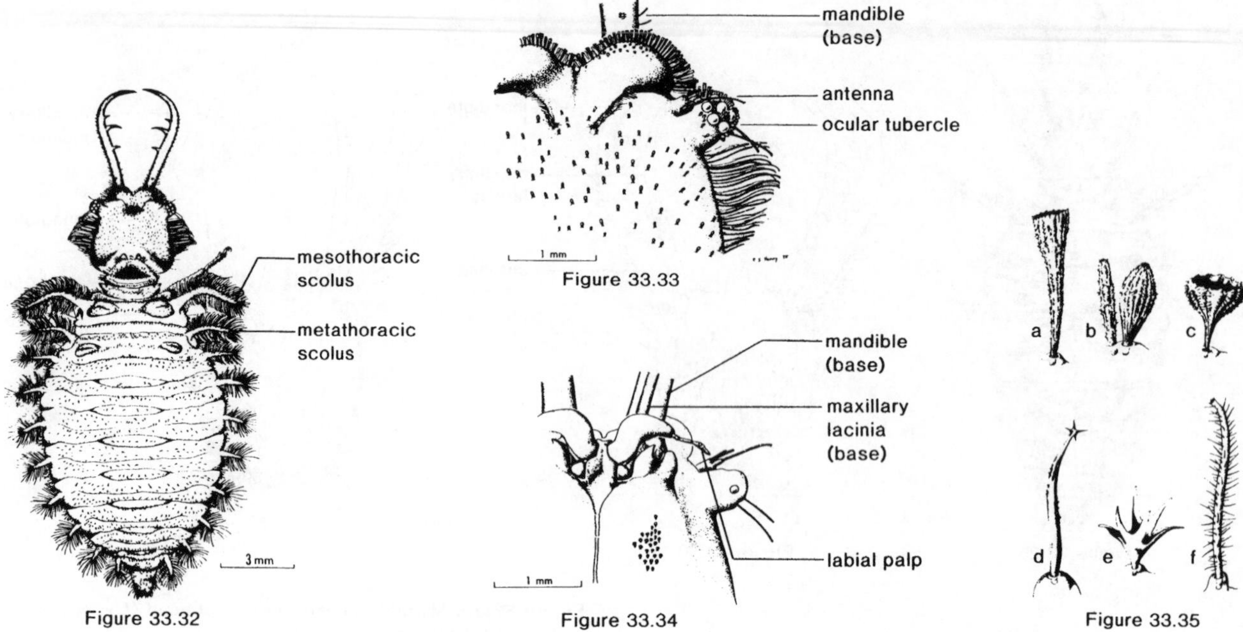

Figure 33.32 Figure 33.33 Figure 33.34 Figure 33.35

Figure 33.32. Ascalaphidae. Mature third instar larva (*Ululodes mexicana*) (from Henry 1976).

Figure 33.33. Ascalaphidae. Dorsal anterior region of head (*Ululodes mexicana*) (from Henry 1976).

Figure 33.34. Ascalaphidae. Ventral anterior region of head (*Ululodes mexicana*) (from Henry 1976).

Figure 33.35. Ascalaphidae. Setal types. a = typical dolichaster, b = clavate dolichasters, c = gobulet-shaped dolichaster, d = stellate-tipped seta, e = scale-like seta, f = plumose seta (from Henry 1976).

key to the myrmeleontid genera of Florida, and the larvae of some species have been described (Stange 1980b; Lucas & Stange 1981; Stange and Miller, 1985). The importance, if any, of the family to biological control is unknown.

Selected Bibliography

Aspöck *et al.* 1980.
Brach 1978.
Buschinger & Bongers 1969.
Furunishi & Masaki 1982.
Henry 1976, 1982.
Lucas & Stange 1981.
Miller & Stange 1983.
New 1982a, 1982b.
Principi 1943, 1947.
Stange 1970, 1980a, 1980b.
Stange and Miller 1985.
Tuculescu *et al.* 1975.
Withycombe 1925.
Zimmerman 1957.

ASCALAPHIDAE

Owlflies

Figures 33.32–33.35

Relationships and Diagnosis: Together, the Myrmeleontidae and Ascalaphidae comprise the myrmeleontoid families with representatives in America north of Mexico. Larvae resemble each other closely in many respects (see Myrmeleontidae). However, most larval ascalaphids can be distinguished by the prominent occipital lobes on the posterolateral margin of the head, consistently 3-toothed mandibles, variously enlarged, flattened or finger-shaped scoli of the thorax and abdomen (fig. 33.32).

Biology and Ecology: Eggs are laid in groups on twigs. Immediately below the egg mass, the female encircles the twig with a fence-like barrier of abortive eggs, called repagula (*see* Henry 1972, 1978b). Upon hatching, the larvae of one species descend the twig, feed on the repagula, return to the empty eggshells, and defend themselves for 7 to 10 days before dispersing. Newly hatched larvae of another species do not feed on the repagula but remain with their siblings on the egg mass for 7 to 10 days before descending to the ground on silken threads.

Larvae tend to conceal themselves under stones or on tree trunks; many species are cryptically colored, others carry trash on their bodies. Feeding apparently involves waiting with the jaws open (in some cases spread up to 280°) until prey come into range.

Mature larvae spin silken cocoons on trees or on the ground; often sand or debris is incorporated into the cocoon. Adults are generally nocturnal, although there are some notable exceptions.

Description: (largely from Rousset 1973; Henry 1976, 1978a; *see* also Pieper & Willman 1980). Mature larva medium to large, 15–20 mm long; ovoid, with long lateral projections (scoli) from thorax and abdomen (fig. 33.32). Integument with numerous types of highly modified dolichastrine and other setae (fig. 33.35). Head roughly quadrate, usually with large occipital lobes posterolaterally; jaws large, curving inward at tips; mandible with three mesolateral teeth; eyes with 7 stemmata: 6 dorsad, 1 ventrad; ocular lobe extending forward over base of mandibles; antennae short, arising mesad of eyes; labial palpi short, 3-segmented, arising from large bilobed prelabium (figs. 33.33, 33.34).

Thorax and abdomen ovoid, with long finger-like or flattened scoli; scoli in single or double rows. Legs short, each with two curved, distinct tarsal claws; no empodia; hindlegs with tibiae and tarsi fused; fore- and mid-legs without tibiotarsal fusion. Abdomen with 9 visible segments, 10th retracted.

Comments: The Ascalaphidae contains two subfamilies (Ascalaphinae and Neuroptynginae), both of which occur in the New World (Aspöck *et al.* 1980, Penny 1981). Each subfamily is represented in the U.S. (most commonly in the southwestern region) by one genus, the larvae of which have been described (Henry 1976; Gepp 1984).

Selected Bibliography

Aspöck *et al.* 1980.
Gepp 1984.
Henry 1972, 1976, 1978a, 1978b.
Penny 1981.
Pieper & Willmann 1980.
Rousset 1973.

BIBLIOGRAPHY

Adams, P. A. 1950. Notes on *Oliarces clara* Banks (Neuroptera, Ithonidae). Pan-Pacific Ent. 26:137–8.

Adams, P. A. 1962. Taxonomy of Hawaiian *Chrysopa* (Neuroptera: Chrysopidae). Proc. Hawaiian Ent. Soc. 18:221–3.

Adams, P. A. 1967. A review of the Mesochrysinae and Nothochrysinae (Neuroptera: Chrysopidae). Bull. Mus. Comp. Zool. 135:215–38.

Adams, P. A. 1970. A review of the New World Dilaridae. Postilla 148:1–30.

Adams, P. A. 1978. Zoogeography of New World Chrysopidae, a progress report. Folia Ent. Mex. 39–40:210–11.

Agnew, C. W., Sterling, W. L., and Dean, D. A. 1981. Notes on the Chrysopidae and Hemerobiidae of eastern Texas with keys for their identification. Southwest. Ent. Suppl. 4:1–20.

Aspöck, H., Aspöck, U. and Hölzel, H. 1980. Die Neuropteren Europas. Goecke and Evers, Krefeld, Vol. I, 495 pp.; Vol. II, 355 pp.

Badgley, M. E., Fleschner, C. A. and Hall, J. C. 1955. The Biology of *Spiloconis picticornis* Banks (Neuroptera: Coniopterygidae). Psyche 62:75–81.

Barnard, P. C. 1981. The Rapismatidae (Neuroptera): montane lacewings of the Oriental region. Syst. Ent. 6:121–36.

Batra, S. W. T. 1972. Notes on the behavior and ecology of the mantispid, *Climaciella brunnea occidentalis*. J. Kansas Ent. Soc. 45:334–40.

Brach, V. 1978. *Brachynemurus nebulosus* (Neuroptera: Myrmeleontidae): A possible Batesian mimic of Florida mutillid wasps (Hymenoptera: Mutillidae). Ent. News 89: 153–6.

Brown, H. P. 1952. The life history of *Climacia areolaris* (Hagen), a neuropterous "parasite" of fresh water sponges. Amer. Midl. Nat. 47:130–160.

Brushwein, J. R. 1987. Bionomics of *Lomamyia hamata* (Neuroptera: Berothidae). Ann. Ent. Soc. Amer. 80:671–79.

Buschinger, A. and Bongers, J. 1969. Zur extraintestinalen Verdauung des Ameisenlöwen (*Euroleon nostras* Four., Myrmeleonidae). Z. vergl. Physiologie 62:205–13.

Carpenter, F. M. 1940. A revision of the Nearctic Hemerobiidae, Berothidae, Sisyridae, Polystoechotidae and Dilaridae (Neuroptera). Proc. Amer. Acad. Arts Sci. 74:193–280.

Carpenter, F. M. 1951. The structure and relationship of *Oliarces* (Neuroptera). Psyche 58:32–40.

Collyer, E. 1951. The separation of *Conwentzia pineticola* End. from *Conwentzia psociformis* (Curt.), and notes on their biology. Bull. Ent. Res. 42:555–64, 2 pl.

Eisner, T., Hicks, K., Eisner, M. and Robson, D. S. 1978. "Wolf-in-sheep's-clothing" strategy of a predaceous insect larva. Science 199:790–94.

Furunishi, S. and Masaki, S. 1982. Seasonal life cycle in two species of ant-lion (Neuroptera: Myrmeleontidae). Japan. J. Ecol. 32:7–13.

Gepp, J. 1984. Erforschungsstand der Neuropteren-Larven der Erde. pp. 183–239. *In* Progress in World's Neuropterology (J. Gepp, H. Aspöck, & H. Hölzel, eds.). Proc. 1st Int. Symp. Neuropterology, Graz. 265 pp.

Gilbert, C. and Rayor, L. S. 1983. First record of mantisfly (Neuroptera: Mantispidae) parasitizing a spitting spider (Scytodidae). J. Kansas Ent. Soc. 56:578–80.

Gilyarov, M. S. 1962. The larva of *Dilar turcicus* Hag. and the position of the family Dilaridae in the suborder Planipennia. Ent. Rev. 41:244–53.

Greve, L. 1974. The larva and pupa of *Heliconis lutea* (Wallengren, 1871) (Neuroptera, Coniopterygidae). Norsk Ent. Tidsskr. 21:19–23.

Grigarick, A. A. 1975. The occurrence of a second genus of spongilla-fly (*Sisyra vicaria* (Walker)) at Clear Lake, Lake County, California. Pan-Pac. Ent. 51:296–7.

Gurney, A. B. 1947. Notes on Dilaridae and Berothidae, with special reference to the immature stages of the Nearctic genera (Neuroptera). Psyche 54:145–69.

Hagen, K. S. 1950. Fecundity of *Chrysopa californica* as affected by synthetic foods. J. Econ. Ent. 43:101–4.

Hagen, K. S., Sawall, E. F. and Tassan, R. L. 1971. The use of food sprays to increase effectiveness of entomophagous insects. pp. 59–81 *In* Proc. Tall Timbers Conf. on Ecol. Anim. Control by Habitat Management, Feb. 1970.

Henry, C. S. 1972. Eggs and repagula of *Ululodes* and *Ascaloptynx* (Neuroptera: Ascalaphidae): a comparative study. Psyche 79:1–22.

Henry, C. S. 1976. Some aspects of the external morphology of larval owlflies (Neuroptera: Ascalaphidae), with particular reference to *Ululodes* and *Ascaloptynx*. Psyche 83:1–31.

Henry, C. S. 1978a. An unusual ascalaphid larva (Neuroptera: Ascalaphidae) from Southern Africa, with comments on larval evolution within the Myrmeleontoidea. Psyche 85:265–74.

Henry, C. S. 1978b. An evolutionary and geographical overview of repagula (abortive eggs) in the Ascalaphidae (Neuroptera). Proc. Ent. Soc. Wash. 80:75–86.

Henry, C. S. 1982. Neuroptera. pp. 470–82 *In* Synopsis and Classification of Living Organisms, Vol. 2 (ed. S. Parker). McGraw-Hill Book Co., Inc., New York, 1232 pp.

Hölzel, H. 1970. Zur generischen Klassifikation der paläarktischen Chrysopinae. Eine neue Gattung und zwei neue Untergattungen der Chrysopidae (Planipennia). Zeit. Arb. Österr. Ent. 22:44–52.

Johnson, J. B. and Hagen, K. S. 1981. A neuropterous larva uses an allomone to attack termites. Nature 289:506–7.

Johnson, V. 1981a. New species of Coniopterygidae (Neuroptera) from North America. Psyche 87:181–92.

Johnson, V. 1981b. Review of the Coniopterygidae (Neuroptera) of North America with a revision of the genus *Aleuropteryx*. Psyche 87:259–98.

Kimmins, D. E. 1952. A revision of the genera of the Apochrysinae (Fam. Chrysopidae). Ann. Mag. Nat. Hist. (ser. 12) 5:929–44.

Kuroko, H. 1961. On the eggs and first-instar larvae of two species of Mantispidae. Esakia 3:25–31.

Lucas, J. R. and Stange, L. A. 1981. Key and descriptions to the *Myrmeleon* larvae of Florida (Neuroptera: Myrmeleontidae). Fla. Ent. 64:207–16.

Lucchese, E. 1955. Ricerche sulla *Mantispa perla* Pallas (Neuroptera Planipennia—Fam. Mantispidae). Ann. Fac. Agraria, Univ. Perugia 11:242–62.

Lucchese, E. 1956. Ricerche sulla *Mantispa perla* Pallas (Neuroptera Planipennia—Fam. Mantispidae). Ann. Fac. Agraria, Univ. Perugia 12:83–213.

MacLeod, E. G. 1960. The immature stages of *Boriomyia fidelis* (Banks) with taxonomic notes on the affinities of the genus *Boriomyia* (Neuroptera: Hemerobiidae). Psyche 67:26–40.

MacLeod, E. G. 1964. The presence of the genus *Hemerobius* in Hawaii with a note on the wing venation of *Nesobiella hospes* (Perkins) (Neuroptera: Hemerobiidae). Pacific Insects 6:1–4.

MacLeod, E. G. 1970. The Neuroptera of the Baltic Amber. I. Ascalaphidae, Nymphidae, and Psychopsidae. Psyche 77:147–80.

MacLeod, E. G. and Adams, P. A. 1967. A review of the taxonomy and morphology of the Berothidae, with a description of a new subfamily from Chile (Neuroptera). Psyche 74:237–65.

MacLeod, E. G. and Redborg, K. E. 1982. Larval platymantispine mantispids (Neuroptera: Planipennia): possibly a subfamily of generalist predators. Neurop. Internat. 2:37–41.

MacLeod, E. G. and Spiegler, P. E. 1961. Notes on the larval habitat and developmental peculiarities of *Nallachius americanus* (McLachlan) (Neuroptera: Dilaridae). Proc. Ent. Soc. Wash. 63:281–6.

McKeown, K. C. and Mincham, V. H. 1948. The biology of an Australian mantispid (*Mantispa vittata* Guérin). Australian Zool. 11:207–24.

Meinander, M. 1972. A revision of the family Coniopterygidae. Acta Zool. Fenn. 136:1–357.

Meinander, M. 1974a. Coniopterygidae from western North America. Ent. Scand. 5:217–32.

Meinander, M. 1974b. The larvae of two North American species of Coniopterygidae (Neuroptera). Notulae Entomologicae 54:12–16.

Meinander, M. 1975. Coniopterygidae from North America (Neuroptera). Notulae Entomologicae 55:28–32.

Meinander, M. 1986. Coniopterygidae of America (Neuroptera). pp. 31–43 *in* Gepp, J., H. Aspöck and H. Hölzel (eds.). Proc. 2nd Int. Symp. neuropterology, Hamburg. 176 pp.

Merritt, R. W. and Cummins, K. W. 1984. An Introduction to the Aquatic Insects of North America, 2nd ed. Kendall/Hunt Publ. Co., Dubuque, Iowa. 722 pp.

Miller, R. B. and Stange, L. A. 1983. The ant-lions of Florida. *Glenurus gratus* (Say) (Neuroptera: Myrmeleontidae). Fla. Dept. of Agric. & Consumer Serv., Div. Plant Industry, Ent. Circ. 251:1–2.

Miller, R. B. and Stange, L. A. 1985. Description of the antlion larva *Navasoleon boliviana* Banks with biological notes (Neuroptera: Myrmeleontidae). Neurop. Internat. 3:119–126.

Monserrat, V. 1977. A systematic and alphabetic list of Neurorthidae and Sisyridae (Neuroptera). Nouv. Rev. Ent. 7:91–6.

Nakahara, W. 1954. Early stages of some Japanese Hemerobiidae including two new species. Kontyû 21:41–6, 6 pl.

Nakahara, W. 1960. Systematic studies on the Hemerobiidae. Mushi 34:1–69, 16 pl.

Nakahara, W. 1965. Contributions to the knowledge of the Hemerobiidae of western North America. Proc. U.S.N.M. 116:205–22.

Neuenschwander, P. 1976. Biology of the adult *Hemerobius pacificus*. Env. Ent. 5:96–100.

New, T. R. 1975. The biology of Chrysopidae and Hemerobiidae (Neuroptera), with reference to their usage as biocontrol agents: a review. Trans. R. Ent. Soc. Lond. 127:115–40.

New, T. R. 1980. A revision of the Australian Chrysopidae (Insecta: Neuroptera). Aust. J. Zool. Suppl. Series. No. 77:1–143.

New, T. R. 1982a. A reappraisal of the status of the Stilbopterygidae (Neuroptera: Myrmeleontoidea). J. Aust. Ent. Soc. 21:71–5.

New, T. R. 1982b. Notes on some early stages of *Stilbopteryx* Newman (Neuroptera, Myrmeleontidae). Neurop. Internat. 2:89–94.

New, T. R. and Boros, C. 1983. The early stages of *Micromus tasmaniae* (Neuroptera: Hemerobiidae). Neurop. Internat. 2:213–17.

Opler, P. A. 1981. Polymorphic mimicry of polistine wasps by a Neotropical neuropteran. Biotropica 13:165–76.

Parfin, S. 1958. Notes on the bionomics of the Mantispidae (Neuroptera: Planipennia). Ent. News 69:203–7.

Parfin, S. I. and Gurney, A. B. 1956. The Spongilla-flies, with special reference to those of the Western Hemisphere (Sisyridae, Neuroptera). Proc. U.S.N.M. 105:421–529, 3 pl.

Parker, F. D. and Stange, L. A. 1965. Systematic and biological notes on the tribe Platymantispini (Neuroptera: Mantispidae) and the description of a new species of *Plega* from Mexico. Can. Ent. 97:604–12.

Penny, N. D. 1981. Review of the generic level classification of the New World Ascalaphidae (Neuroptera). Acta Amaz. 11:391–406.

Peterson, A. 1951. Larvae of Insects. Part II. Coleoptera, Diptera, Neuroptera, Siphonaptera, Mecoptera, Trichoptera. Published by the author, Columbus, Ohio, 416 pp.

Pieper, H. and Willmann, R. 1980. Die Larven griechischer Ascalaphiden-Arten (Ins., Planipennia). Stuttgarten Beitr. Naturle. A (337) 11:1–11.

Poirrier, M. A. 1969. Some fresh-water sponge hosts of Louisiana and Texas spongilla-flies, with new locality records. Amer. Midl. Nat. 81:573–5.

Poirrier, M. A. and Arceneaux, Y. M. 1972. Studies on southern Sisyridae (spongilla-flies) with a key to the third-instar larvae and additional sponge-host records. Amer. Midl. Nat. 88:455–8.

Principi, M. M. 1943. Contributi allo studio dei Neurotteri Italiani. II. *Myrmeleon inconspicuus* Ramb. ed. *Euroleon nostras* Fourcroy. Boll. Ist. Entom. Univ. Bologna XIV:131–92, Figg. I–XXIII.

Principi, M. M. 1947. Contributi allo studio dei Neurotteri Italiani. VI. *Synclisis baetica* Ramb. Boll. Ist Entom. Univ. Bologna XVI:234–53. Figg. I–X.

Pupedis, R. J. 1980. Generic differences among New World Spongilla-fly larvae and a description of the female of *Climacia striata* (Neuroptera: Sisyridae). Psyche 87:305–14.

Redborg, K. E. 1982. Interference by the mantispid *Mantispa uhleri* with the development of the spider *Lycosa rabida*. Ecol. Ent. 7:187–96.

Redborg, K. E. and MacLeod, E. G. 1985. The developmental ecology of *Mantispa uhleri* Banks (Neuroptera: Mantispidae). Illinois Biological Monographs, 53, 131 pp. University of Illinois Press, Urbana and Chicago.

Rehn, J. W. H. 1939. Studies in North American Mantispidae (Neuroptera). Trans. Amer. Ent. Soc. 65:237–263.

Ridgway, R. L. and Kinzer, R. E. 1974. Chrysopids as predators of crop pests. Entomophaga 7:45–51.

Riek, E. F. 1974. The Australian moth-lacewings (Neuroptera: Ithonidae). J. Aust. Ent. Soc. 13:37–54.

Rousset, A. 1966. Morphologie céphalique des larves de Planipennes (Insectes Néuroptéroïdes). Mém. Mus. Nat. Hist. Nat., Ser. A., 42:1–999.

Rousset, A. 1973. Morphologie externe et caractères distinctifs des larves de trois espèces d'Ascalaphes. Bull. Soc. Ent. France 78:164–78.

Sheldon, J. K. and MacLeod, E. G. 1971. Studies on the biology of the Chrysopidae. II. The feeding behavior of the adult of *Chrysopa carnea* (Neuroptera). Psyche 78:107–21.

Smith, R. C. 1922. The biology of the Chrysopidae. Cornell Univ. Agr. Exp. Sta. Mem. 58:1287–1372.

Smithers, C. N. 1979. Gallard 1932: an overlooked paper on the food of *Ithone fusca* Newman (Neuroptera: Ithonidae). Aust. Ent. Mag. 6:74.

Stange, L. A. 1970. A generic revision and catalog of the Western Hemisphere Glenurini with the description of a new genus and species from Brazil (Neuroptera: Myrmeleontidae). Los Angeles Co. Mus., Cont. in Science, 186:1–28.

Stange, L. A. 1980a. The antlions of Florida. I. Genera (Neuroptera: Myrmeleontidae). Fla. Dept. Agric. & Consumer Serv., Div. Plant Industry, Ent. Circ. 215:1–4.

Stange, L. A. 1980b. The ant-lions of Florida. II. Genera based on larvae (Neuroptera: Myrmeleontidae). Fla. Dept. of Agric. & Consumer Serv., Div. Plant Industry, Ent. Circ. 221:1–4.

Stange, L. A. and Miller, R. B. 1985. A generic review of the Acanthaclisine ant-lions based on larvae (Neuroptera: Myrmeleontidae). Insecta Mundi 1:29–42.

Tauber, C. A. 1969. Taxonomy and biology of the lacewing genus *Meleoma* (Neuroptera: Chrysopidae). Univ. Calif. Pub. Ent. 58:1–94.

Tauber, C. A. 1974. Systematics of North American chrysopid larvae: *Chrysopa carnea* group (Neuroptera). Can. Ent. 106:1133–53.

Tauber, C. A. 1975. Larval characteristics and taxonomic position of the lacewing genus *Suarius*. Ann. Ent. Soc. Amer. 68:695–700.

Tauber, C. A. and Tauber, M. J. 1968. *Lomamyia latipennis* (Neuroptera: Berothidae) life history and larval descriptions. Can. Ent. 100:623–9.

Tauber, M. J. and Tauber, C. A. 1974. Dietary influence on reproduction in both sexes of five predacious species (Neuroptera). Can. Ent. 106:921–5.

Tauber, M. J. and Tauber, C. A. 1975. Criteria for selecting *Chrysopa carnea* biotypes for biological control: adult dietary requirements. Can. Ent. 107:589–95.

Tauber, C. A. and Tauber, M. J. 1982. Evolution of seasonal adaptations and life history traits in *Chrysopa:* response to diverse selective pressures. pp. 51–72 *In* Evolution and Genetics of Life Histories (ed. H. Dingle and J. P. Hegmann). Springer-Verlag, N.Y.

Tauber, M. J. and Tauber, C. A. 1983. Life history traits of *Chrysopa carnea* and *Chrysopa rufilabris* (Neuroptera: Chrysopidae): Influence of humidity. Ann. Ent. Soc. Amer. 76:282–5.

Tillyard, R. J. 1922. The life-history of the Australian moth-lacewing, *Ithone fusca* Newman (Order Neuroptera, Planipennia). Bull. Ent. Res. 13:205–23.

Tjeder, B. 1959. Neuroptera—Planipennia. The lacewings of Southern Africa. 2. Family Berothidae. pp. 256–314. *In* South African Animal Life Vol. 6. Statens Naturvetenskapliga Forskningsråd, Stockholm.

Tjeder, B. 1966. Neuroptera—Planipennia. The lacewings of Southern Africa. 5. Chrysopidae. pp. 228–534 *In* South African Animal Life, Vol. 12, Statens Naturvetenskapliga Forskningsråd, Stockholm.

Toschi, C. A. 1964. Observations on *Lomamyia latipennis,* with a description of the first instar larva. Pan-Pacific Ent. 40:21–6.

Toschi, C. A. 1965. The taxonomy, life histories, and mating behavior of the green lacewings of Strawberry Canyon (Neuroptera: Chrysopidae). Hilgardia 36:391–431.

Tuculescu, R., Topoff, H., and Wolfe, S. 1975. Mechanisms of pit construction by antlion larvae. Ann. Ent. Soc. Amer. 68:719–20.

Welch, P. S. 1914. The early stages of the life history of *Polystoechotes punctatus* Fabr. Bull. Brooklyn Ent. Soc. 9:1–6, 1 pl.

Werner, F. G. and Butler, G. D., Jr. 1965. Some notes on the life history of *Plega banksi* (Neuroptera: Mantispidae). Ann. Ent. Soc. Amer. 58:66–8.

Withycombe, C. L. 1923. Notes on the biology of some British Neuroptera (Planipennia). Trans. Ent. Soc. London (1922), pp. 501–94, pl. 38–43.

Withycombe, C. L. 1925. Some aspects of the biology and morphology of the Neuroptera. With special reference to the immature stages and their possible phylogenetic significance. Trans. Ent. Soc. London (1924), pp. 303–411, pl. 39–44.

Zimmerman, E. C. 1957. Insects of Hawaii. Vol. 6. Ephemeroptera, Neuroptera, Trichoptera. Univ. Hawaii Press, Honolulu, pp. 1–175.

Zwick, P. 1967. Beschriebung der aquatischen Larve von *Neurorthus fallax* (Rambur) und Errichtung der neuen Planipennierfamilie Neurorthidae fam. nov. Gewasser und Abwässer 44/45:65–86.

Order Coleoptera

34

John F. Lawrence, coordinator
Division of Entomology, CSIRO

BEETLES*

The Coleoptera is the largest order of insects, with more than 350,000 species worldwide and about 25,000 occurring in America north of Mexico. In the present book, they are placed in 156 families, 31 of which do not occur in the United States or Canada. Larvae of Coleoptera, unlike those of some other orders, have no one common name that can be applied to all forms, and various terms like grub, white grub, wireworm, mealworm, rootworm, glow worm, round-headed borer, flat-headed borer, timber borer, water penny, etc. have been used for individual larval types. Most beetles are terrestrial, but a number of genera and some families are aquatic as larvae, some occur in water throughout most (or rarely all) of the life cycle, and a few are aquatic in the adult stage only. The majority of beetle larve feed on various kinds of living and dead plant tissue (roots, stems, trunks, branches, logs, leaves, flowers, seeds), but many feed on fungi, carrion or dung, some are predaceous, and a few are parasitic. Many Coleoptera cause damage to agricultural crops, forests, or stored products, but very few are of medical importance.

DIAGNOSIS

There is no single feature which will distinguish beetle larvae from those of other insect orders and only a few which may be considered universal within the group. The following list includes most characters considered of diagnostic value.

1. Head capsule well-developed and usually sclerotized.
2. Head usually without paired endocarinae (**adfrontal ridges**) forming a V or Y, and never with adfrontal areas formed between them and the frontal sutures or ecdysial lines; paired endocarinae, when present, very rarely extending to the anterior part of the frontoclypeal region.
3. Antennae almost always with 4 segments or fewer and with a sensorium on the penultimate segment.
4. Number of stemmata on each side always 6 or fewer.
5. Mouthparts almost always of the chewing or orthopteroid type with opposable mandibles moving in a transverse plane, and with normal palp-bearing maxillae and labium; rarely modified to form a sucking tube bearing stylets, or with immovable or non-opposable mandibles.
6. Median labial silk gland or spinneret always absent.
7. Legs usually with 5 or 6 segments (coxa, trochanter, femur, tibia, and either tarsungulus or tarsus and pretarsus, the latter consisting of one or two claws) or absent; occasionally with reduced segmentation.
8. Abdomen usually with 10 segments (occasionally fewer), without articulated appendages (cerci) on segment 10.
9. Paired abdominal prolegs almost always absent; asperity-bearing prolegs occasionally present on sterna 2–4, 3–4 or 2–5; simple prolegs rarely present on sterna 1–8; crochet-bearing prolegs rarely present on sterna 2–7 or 3–7.
10. Respiratory system usually peripneustic, without functional spiracles on metathorax; occasionally amphipneustic, metapneustic, or apneustic.
11. Spiracles often with accessory openings (annular-uniforous, annular-biforous, annular-multiforous), with divided opening (biforous), or with poroid sieve-plate (cribriform); ecdysial scar, if present, never completely enclosed by sieve plate.

Because of the great diversity of form within the Coleoptera, at least some beetle larvae may be confused with immatures belonging to each of the other endopterygote orders and probably those of a few exopterygote groups as well. The triungulinids of **Strepsiptera** are said to resemble the triungulins of Meloidae and Rhipiphoridae, but they differ from them in several respects, including the absence of mandibles, antennae, and labial palps, the lack of trochanters in the legs, and the presence of a pair of long setae on the terminal 10th segment. The endoparasitic forms of Strepsiptera (larvae and adult females) are easily distinguished because of their extremely reduced body which is indistinctly segmented. The free-living last larval instar (puparium) in Mengenillidae has 5-segmented legs with 2 claws but differs from beetle larvae in the presence of compound eyes composed of numerous facets.

Larvae of **Trichoptera, Megaloptera, Raphidioptera,** and **Neuroptera** may be distinguished from most beetle larvae by the presence of 6-segmented legs, which in the Coleoptera occur only in the suborders Archostemata and Adephaga. The distinctive head of Cupedidae and Micromalthidae, with a median endocarina, well-developed mandibular molae, and a sclerotized ligula, separate them from members of the above orders, while the absence of a free labrum will distinguish adephagan larvae from those of Trichoptera, Megaloptera, and Raphidioptera. Neuroptera larvae also have the labrum fused to the head capsule, but they differ from any beetle

*In this section a figure reference such as (Kfig. 121 p190) refers to Key fig. 121, page 190.

larvae in their distinctive feeding apparatus, consisting of a pair of sucking organs formed by the falciform or styliform mandibles and maxillae on each side, and by the complete absence of maxillary palps. The divided, blade-like mandibles of Lycidae or the perforated, sucking mandibles of Lampyridae and Phengodidae might be confused with neuropteran mouthparts, but these groups always have maxillary palps and 5-segmented legs. Six-segmented legs also occur in 2 rare families: Nannochoristidae (**Mecoptera**) and Heterobathmiidae (**Lepidoptera**); larvae of both may be distinguished from those of Coleoptera by the combination of leg segmentation and free labrum, while the former has more than 6 stemmata on each side and panorpoid spiracles (with numerous openings surrounding an ecdysial scar), and the latter has adfrontal ridges, adfrontal areas, and a median labial gland.

Larvae of **Lepidoptera** are easily confused with various beetle larvae, especially those of Chrysomelidae, but all of the former differ in having a median labial gland, which is usually developed into a protruding spinneret, almost all of them differ in having paired adfrontal ridges, with adfrontal areas formed between them and the ecdysial lines, and most of them differ in having paired, crochet-bearing prolegs on abdominal sterna 3 to 6 and 10. The median labial gland may be difficult to see in some primitive forms without distinct spinnerets (Micropterigidae), and the adfrontal ridges are absent in Micropterigidae and Agathiphagidae. A number of groups lack prolegs, while some also lack thoracic legs and may be confused with apodous Coleoptera larvae; all of these groups, however, have the typical lepidopteran head. Several beetle larvae have abdominal prolegs, but these are usually simple or asperate, without crochets, and occur on different abdominal segments than those in most Lepidoptera (1 to 9 in schizopodine Buprestidae; 2 or 3 to 7 in Hydrophilidae; 2 to 4, 3 to 4, or 2 to 5 in Oedemeridae; and 1 to 8 in Curculionidae). Crochets occur on the prolegs of some Hydrophilidae, but these larvae differ from any lepidopteran in having biforous spiracles and the labrum fused to the head capsule. Larval Micropterigidae differ from those of other Lepidoptera in having simple, more or less acute, mecopteran-like prolegs, which are not always distinct, on segments 1 to 8. The lack of lepidopteran head characters makes them particularly difficult to separate from beetle larvae; but the combination of a more or less hexagonal shape in cross-section, a retracted head, the lack of an antennal sensorium, 3-segmented legs, and a vestiture consisting of characteristic thickened setae should distinguish them.

Three-segmented thoracic legs and simple prolegs on segments 1 to 8 will also distinguish larvae of most **Mecoptera** from those of Coleoptera, and, in addition, most Mecoptera larvae have panorpoid spiracles, more than 6 stemmata on each side, and no distinct cardo. Larvae of Boreidae have vestigial spiracles, no prolegs, and only 3 stemmata, but they may be distinguished by their short, clawless, 3-segmented legs, which are very close together on the prothorax and widely separated on the meso- and metathorax.

Various **Hymenoptera** larvae are easily confused with those of beetles, although all possess a median labial gland (not always easy to see) and many have 2 thoracic spiracles. Larvae of Apocrita and some Symphyta (Orussidae) are lightly sclerotized and legless, with a hypognathous head, and are most likely to be confused with some Meloidae, Rhipiphoridae, and Curculionidae; in addition to the labial gland and the presence of a 2nd thoracic spiracle (in some), the former groups have reduced maxillary and labial palps (1-segmented or absent). Larvae of most Symphyta differ from those of Coleoptera in having paired prolegs on abdominal segments 2 to 7 or 8 and usually 10 (Tenthredinidae, Argidae, Cimbicidae, Diprionidae), more than 4 antennal segments (Blasticomidae, Pamphiliidae, Xyelidae), or more than 4 plicae on most abdominal segments (Cimbicidae, Diprionidae). The only beetle larvae with prolegs on the same segments are Hydrophilidae, which have a fused labrum and biforous spiracles, never occurring in Hymenoptera, while those few beetle larvae with more than 4 antennal segments either have long, flagellate antennae, highly modified mouthparts, and a terminal respiratory chamber (Helodidae) or a median endocarina, 6-segmented legs, and mandibular molae (Cupedidae). Larvae of the last group bear a superficial resemblance to larvae of Cephidae, and both have 4 or 5 antennal segments and a median spine (suranal process) on tergum 9. Of those Symphyta larvae with fewer antennal segments, fewer plicae, and no prolegs, the Xiphydriidae and Siricidae both have vestigial legs and a suranal process, much as in larvae of Mordellidae; they differ from mordellids in having elongate-elliptical spiracles. Some surface-feeding Pergidae which lack prolegs differ from any beetle larvae in having a single large stemma on each side and the antennae reduced to dome-like structures; also there is usually a smaller, but obvious, 2nd thoracic spiracle.

The apodous larvae of **Diptera** can often be distinguished from legless beetle larvae on the basis of their incomplete head capsule and/or vertically oriented, non-opposable mandibles; in the Nematocera, however, most larvae have a well-developed head capsule and transverse, opposable mandibles, as in most Coleoptera. Except for some aquatic forms without functional spiracles (Chironomidae, Ceratopogonidae, Simuliidae), most nematoceran larvae may be distinguished from those of beetles by the type and location of their spiracles, as well as by the universal absence of a spiracular closing apparatus (present in all legless Coleoptera). The presence of spiracles on the thorax and 8th abdominal segments only (amphipneustic system, as in Tanyderidae, Psychodidae, Trichoceridae, etc.) and the presence of a single pair of spiracles on segment 8 (metapneustic system of Tipulidae, Culicidae, etc.) are conditions known among Coleoptera, but not in groups with apodous larvae. Bibionidae and a few other groups have a holopneustic system with 2 thoracic and 8 abdominal spiracles on each side, a condition unknown in Coleoptera; and the remaining Nematocera with a peripneustic system usually differ from beetles in having the thoracic and last abdominal spiracles enlarged. The spiracles of Nematocera are either of the panorpoid type, with a ring of small openings surrounding an ecdysial scar, or they have a characteristic cribriform plate with 3 openings; spiracles are never of the biforous type (as in sphaeridiine Hydrophilidae, schizopodine Buprestidae, or Eucnemidae), never have a poroid cribriform plate (as in most Buprestidae), and are never annular, annular-uniforous, or annular-biforous (as in apodous Bothrideridae, Rhipiphoridae, Meloidae, Chrysomeloidea, or Curculionoidea).

Larvae of **Siphonaptera** are also legless, and they differ from apodous beetle larvae in having 2 pairs of thoracic spiracles; these spiracles are very difficult to see, however, and a more practical suite of distinguishing features consists of the fused labrum, 1-segmented but long and prominent antennae, pair of anal struts on segment 10, and transverse rows of stiff setae on the body.

A few Coleoptera larvae (cerylonine Cerylonidae and the leiodid genus *Myrmecholeva*) have the mouthparts modified to form a median tube through which the stylet-like mandibles and maxillae move; both of these larval types have a more or less hypognathous head and might be confused with certain **Homoptera;** members of the last group, however, always lack maxillary and labial palps and often have the head immovably united to the thorax. The general head structure, multisegmented antennae, mouthparts, and aquatic habits of larval Helodidae cause them to resemble the larvae (nymphs) of some **Plecoptera,** but they can be distinguished from that exopterygote group by the lack of wing pads and the structure of the leg, which has a single tarsungulus, rather than a subdivided tarsus and paired claws.

LARVAL MORPHOLOGY

Body Form

Beetle larvae are extremely variable in general form (figs. 34.1–6), and the body may be elongate to short and broad, parallel-sided to almost circular, and highly convex to strongly flattened. Although the body shape is best described in simple geometric terms (cylindrical, elliptical, ovoid), a number of words are commonly used, which have been derived either from Latin and Greek words describing shape or more often from the names of other arthropods having a similar shape. Beetle pupae (figs. 34.7–8) are more uniform in shape and are described later.

The term **campodeiform** (also called **lepismatoid** or **thysanuriform**) refers to a type of larva with an elongate, slightly flattened body, often somewhat tapered (at least posteriorly), and usually moderately heavily sclerotized (at least dorsally), with long, well-developed legs (fig. 34.1). This is an actively moving type of larva, usually inhabiting surfaces or relatively large **interstitial** spaces within leaf litter, and often predaceous. Campodeiform larvae occur throughout the Adephaga (with some modifications in aquatic forms), are common in the Staphylinoidea, and are found in various other groups which are relatively active in the larval stage. The term also applies to the first-instars (triungulins) of Micromalthidae, Rhipiceridae, passandrine Cucujidae, ectoparasitic Bothrideridae, Meloidae, and Rhipiphoridae. The word **fusiform,** referring to a body which tapers at either end, may also be used to describe most campodeiform larvae.

An **eruciform larva** is one which is caterpillar-like, that is, more or less cylindrical with moderately well-developed legs (fig. 34.2). This type of body may be found in surface-active forms which are relatively slow moving and may be protected by cuticular armature or defensive secretions. Among the Coleoptera, many Chrysomelidae are said to have an eruciform type of larva, but the term might apply to a variety of other beetle larvae. Cylindrical larvae which are moderately to heavily and uniformly sclerotized, with short legs, are sometimes called **vermiform,** but are more commonly known as **wireworms** (fig. 34.3). This type of larva, which usually inhabits soil or rotten wood, occurs in Callirhipidae (fig. 34.3), Dryopidae, Elmidae, Elateridae, Cebrionidae, and Tenebrionidae (Alleculinae, Blaptini, Helaeini). Ghilarov (1964b) restricted the word vermiform to the peculiar larvae of cardiophorine Elateridae, which are very long and narrow, lightly sclerotized, and appear to have a large number of abdominal segments, due to the transverse subdivisions of each segment. The term **orthosomatic** may be applied to any larva which is elongate and more or less parallel-sided.

An **onisciform** larva is one which is short and broad, more or less ovate, and moderately to strongly flattened; the word is derived from the name of a terrestrial isopod crustacean. The body itself is often relatively narrow and fusiform, but the onisciform appearance is caused by the presence of lateral tergal outgrowths, which may become closely adapted, so that a continuous oval or circular outline is produced, as in Psephenidae (fig. 34.4). Extremely flattened and ovoid larvae, which have the head and legs completely concealed beneath the body, are sometimes said to be **limpet-like** or **cheloniform** (turtle-like), while the term **disc-like** is used in the larval key. Onisciform larvae are known in Carabidae (Cychrini), Micropeplidae, Scydmaenidae (*Cephennium*), Silphidae (Silphinae), Byrrhidae (*Cytilus*), Elmidae (*Phanocerus*), Psephenidae, Corylophidae (Rypobiini, *Corylophodes*), Cerylonidae (*Murmidius,* Ceryoninae), Discolomidae, Tenebrionidae (Nilioninae), and Chrysomelidae (some Hispinae), but the extreme disc-like larvae occur only among the psephenids, corylophids, and discolomids, and in *Murmidius* and some hispines. This type of body has evolved for different reasons in the various groups; psephenids attach themselves to algae-covered rocks in streams and are subjected to strong currents, while the flattened hispine Chrysomelidae (Cephaloleiini, Arescini, Alurnini) live in leaf axils. The term **cyphosomatic** is used to describe a larva which is strongly convex dorsally but more or less flattened ventrally; it has been applied to larvae of some Chrysomelidae (Chrysomelinae, Criocerinae, Cassidinae).

Scarabaeiform is a term based on the larvae of Scarabaeoidea, which are relatively lightly sclerotized and strongly curved ventrally, forming a C or U (fig 34.5); these larvae have also been called **grubs,** and the term **C-shaped** has been used in the key. Scarabaeiform larvae are characteristic of most scarabaeoids, some Byrrhidae, many Bostrichoidea, Bruchidae, some Chrysomelidae (Sagrinae, Cryptocephalinae, Clytrinae, Chlamisinae, Eumolpinae) and some Curculionoidea. Most of these occur in the soil, but some may be wood-borers or seed-eaters, and larvae of the cryptocephaline group are adapted for living within the confines of a faecal case. A slight ventral curvature may be found in many beetle larvae, and this may be exaggerated with preservation; but these are not considered to be scarabaeiform. In a few internally feeding larvae, the body may be curved dorsally instead

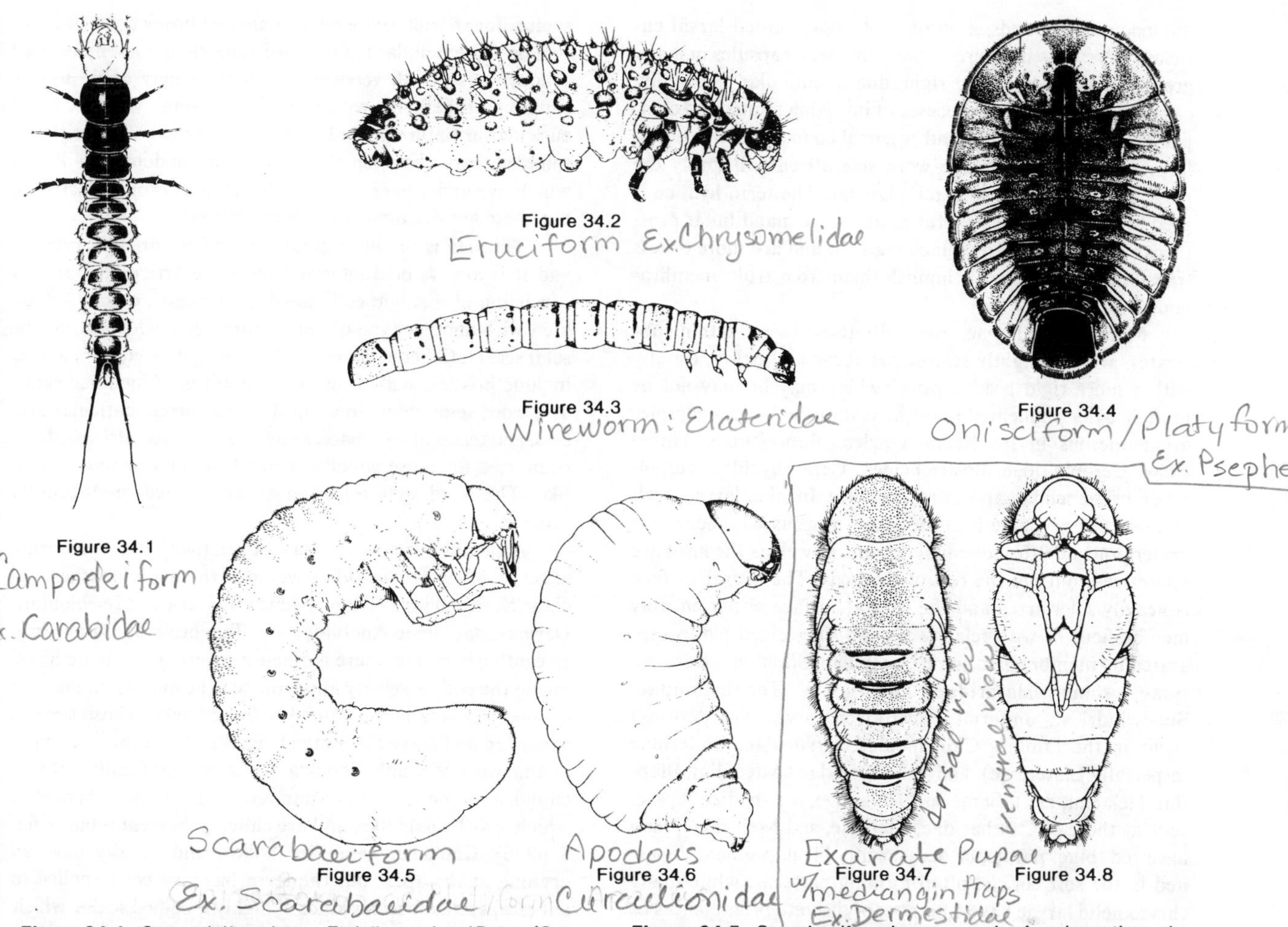

Figure 34.1. Campodeiform larva. *Eudalia macleayi* Bates (Carabidae: Odacanthinae). (From Britton, 1970 *in* The Insects of Australia, Melbourne University Press).

Figure 34.2. Eruciform larva. *Paropsisterna beata* (Newman) (Chrysomelidae: Chrysomelinae). (From Britton, 1970, *in* The Insects of Australia, Melbourne University Press).

Figure 34.3. Wireworm. *Zenoa picea* (Beauvois) (Callirhipidae).

Figure 34.4. Onisciform or platyform larva. *Sclerocyphon* sp. (Psephenidae: Eubriinae). (From Britton, 1970 *in* The Insects of Australia, Melbourne University Press).

Figure 34.5. Scarabaeiform larva or grub. *Anoplognathus pindarus* Carne (Scarabaeidae: Rutelinae). (From Britton, 1970, *in* The Insects of Australia, Melbourne University Press).

Figure 34.6. Apodous larva. *Trigonotarsus rugosus* (Beauvois) (Curculionidae: Rhynchophorinae). (From Britton, 1970, *in* The Insects of Australia, Melbourne University Press).

Figure 34.7. Exarate pupa. *Dermestes maculatus* De Geer (Dermestidae), dorsal view, showing median gin traps. (From Britton, 1970, *in* The Insects of Australia, Melbourne University Press).

Figure 34.8. Same, ventral.

of ventrally; this is known in the puffball-feeding *Pocadius* (Nitidulidae) and in species of *Leiodes* (Leiodidae), which occur in subterranean fungi.

Apodous or legless larvae (fig. 34.6) occur in several large groups living in dense substrates like soil, wood, seeds, etc. Most are lightly sclerotized, and many (Curculionoidea) are scarabaeiform in shape, while the wood-boring Buprestidae, Eucnemidae, and legless Cerambycidae have a straight or orthosomatic body. Most cerambycid and buprestid larvae have the thoracic region enlarged (Kfig. 121 p190) and are sometimes called **clavate** (Ghilarov, 1964b).

Integument

The larval integument or **cuticle** is usually softer, more flexible, and more lightly pigmented than that of an adult beetle, but there is considerable variation. The term **sclerotization** is defined as the stabilization or hardening of cuticle by the addition of aromatic (phenolic and quinonoid) compounds, which cross-link proteins in a polymerization-like process called **tanning.** The formerly used synonym **chitinization** was based on the erroneous assumption that the compound **chitin,** a component of all cuticle, was responsible for the hardening process. Although tanning is usually accompanied by a darkening of the cuticle, due to the presence of quinonoid compounds, a number of unpigmented larval cuticles (those of some "wireworms" and head capsules in many groups) may be hard and rigid, due to cuticular thickening or possibly a hardening process not involving pigmented compounds (Hackman, 1984 and personal communication). Following common usage, the word **sclerotized** will imply the presence of yellowish to black pigment. The term **hyaline** is used to describe rigid structures, like some mandibular prosthecae (*see* below), which lack pigment and are more or less translucent; this is to distinguish them from truly membranous structures.

Many beetle larvae, especially those living within substrates, are very lightly sclerotized above and below, usually with a more rigid head capsule, which may or may not be more heavily pigmented than the rest of the body; examples are numerous in the Scarabaeoidea, Buprestidae, Throscidae, Cerophytidae, Bostrichoidea, Cerambycidae, eumolpine Chrysomelidae, and Curculionoidea. In other larvae, such as those of agrypnine Elateridae and various Cleroidea, the protergum and 9th abdominal tergum, as well as the head are sclerotized, but not the rest of the body. The dorsal surface is heavily sclerotized and the ventral surface either entirely membranous or with relatively small sclerotized plates separated by membrane in some Carabidae, Silphidae, Dermestidae, some Elateridae and some Tenebrionidae. Subcylindrical, uniformly sclerotized larvae (wireworms) occur in the families Callirhipidae, Dryopidae, Elateridae (especially Elaterinae), and Tenebrionidae (Alleculini, Blaptini, Helaeini, etc.). Some surface-active, soft-bodied larvae, such as those of Cantharidae, Cleridae, and Melyridae, may have red, blue, purple, or occasionally black pigment deposited in the soft cuticle, often forming patterns, while many chrysomelid larvae have a number of discrete prominences or plates which are pigmented (fig. 34.2).

In most beetle larvae, the integument is relatively smooth, and the **vestiture** consists of scattered, simple setae, which may be of fixed position within a particular taxon. In some groups, however, the sculpture is more complex and the vestiture may include modified setae of various kinds. Sometimes the entire surface or the dorsal surface is covered with fine **granules,** or these may be more sparsely distributed and form definite patterns, as in Psephenidae, priasilphine Phloeostichidae, Nitidulidae, and *Murmidius* (Cerylonidae). More prominent **setiferous tubercles** (Kfig. 3 p210) occur in a number of families, including Micropeplidae, Eucinetidae, Derodontidae, Hobartiidae, and Languriidae (Xenoscelinae, Cryptophilinae); while complex sculpture is restricted to relatively few groups, such as Erotylidae, Endomychidae, Cerylonidae (Euxestinae), Coccinellidae, and Chrysomelidae. A set of terms for cuticular prominences have been developed for Coccinellidae (Gage, 1920) and includes (from the simple to the complex): **seta,** for an articulated hair arising from a flat surface; **chalaza,** for a minute, pimple-like projection bearing a seta; **verruca,** for a small, mound-like projection bearing several setae; **struma,** for a mound-like projection bearing a few chalazae; **sentus,** for an elongate, unbranched projection bearing a few short setae along its trunk; and **parascolus** and **scolus,** for shorter and longer, branched projections (*see* section on Coccinellidae). The word **tubercle** is sometimes used synonymously with **verruca,** but here it may apply to any simple protuberance, with or without a seta. More localized cuticular armature, including patches or rows of **asperities** or **spinules,** may occur on the thorax and abdomen of larvae which live under bark or bore into relatively solid substrates, but these are discussed in a later section.

The **seta** is the most common type of cuticular process, and it is always characterized by being articulated and by consisting of a single cell called a **trichogen,** closely associated with another type of cell, a **tormogen,** which forms the setal socket. Other articulated structures, derived from a seta, include **bristles, scales,** and various types of modified setae. The word **spine** refers to a multicellular, fixed, cuticular process, and terms like **spinules, asperities,** and **microtrichiae** have been used for those which are much smaller or more toothlike. The word **spur** refers to an articulated, multicellular process.

Although most larvae have a relatively sparse covering of setae, a few are densely covered with long hairs; these include Scarabaeidae (Glaphyrinae), Elateridae (Tetralobini), Dermestidae, some Anobiidae and Tenebrionidae (Lagriini). In cantharid larvae, there is a dense covering of minute hairs, giving the body a velvety appearance. The most common type of modified seta is the so-called **frayed seta,** whose apex is expanded and frayed looking (Kfig. 4 p244); these are found in a number of Staphylinoidea and in several families of Cucujoidea. Some setae have numerous minute protuberances, which may be hair-like, and are called **pubescent setae** (Kfig. 5 p236). **Glandular setae** are hollow and usually have an opening at the apex; but the term has also been applied to pubescent setae. Various kinds of highly modified **scales,** which may be club-like or sometimes flattened, occur in some groups, like *Pseudomorpha* (Carabidae), Brachypsectridae, and a few Cantharidae and Coccinellidae. Setae which are covered with numerous barbs are called **spicisetae** (Kfig. 6 p230), and occur in most Dermestidae and in *Murmidius* (Cerylonidae). A more complex setal type in Dermestidae is the **hastiseta,** which has barbs along the shaft and a pair of 3-pronged structures at the apex (Kfig. 7 p230).

Other cuticular structures include the openings of defensive glands, which may be associated with reservoirs, and luminous or light-producing organs. Cuticular defensive glands are probably widespread among beetle larvae, but they have been reported for only a few groups, partly because the openings and reservoirs may be very small. Paired dorsolateral gland openings on most thoracic and abdominal terga are known in the families Staphylinidae, Cerophytidae, Phengodidae, Cantharidae, Trogossitidae (Trogossitinae), Melyridae, Coccinellidae, Tenebrionidae, and Chrysomelidae, while in Corylophidae, there are large, paired openings on abdominal segments 1 and 8 (Kfig. 138 p258) or 1–7. A single median gland opening occurs on the 8th tergum in most aleocharine Staphylinidae and on the 9th tergum in some Tenebrionidae, while head glands have been reported in Staphylinidae, Pselaphidae and Tenebrionidae (Doyen and

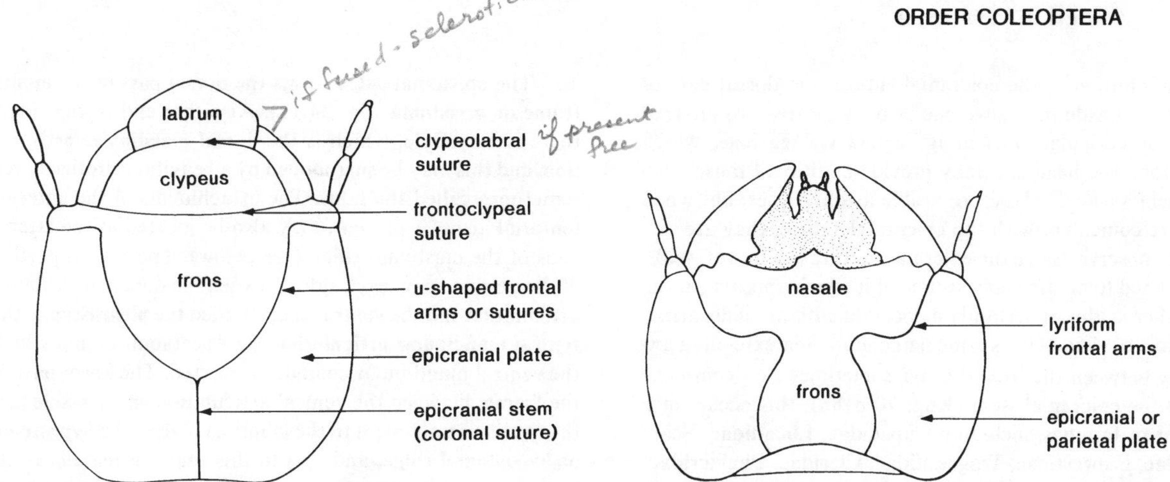

Figure 34.9. Head (Tenebrionidae), dorsal.

Figure 34.10. Head (Elateridae), dorsal.

Lawrence, 1979; Frank and Thomas, 1984; Kasule, 1966). Luminous or light producing organs occur on the head in Phengodidae and on the thorax and/or abdomen in Phengodidae, Lampyridae, Homalisidae, and pyrophorine Elateridae (Sivinski, 1981).

Head Capsule

The head capsule or **cranium** in beetle larvae is always distinct and usually moderately to strongly sclerotized, exceptions occurring in forms with strongly retracted heads and in the protracted heads of some Scarabaeoidea, Anobiidae, Prostomidae, Oedemeridae, Tenebrionidae, Chrysomelidae, and Curculionoidea. There are 2 large openings: the posterior **occipital foramen,** leading into the cervical region or prothorax, and the opening of the **mouth cavity** or **preoral cavity,** which may be anterior or ventral, depending upon the type of head attachment (*see* below).

Larval heads may be of 3 different types, based on the nature of their attachment to the thorax. The most common type is the **protracted** and **prognathous** head (Kfig. 8 p186), which is not or only slightly declined and not deeply retracted into the prothorax. A second type is also protracted but is more strongly declined or **hypognathous** (Kfig. 10 p186), so that the mouthparts are directed ventrally like those of primitive orthopteroid insects. Some strongly hypognathous heads, like those of cerylonine Cerylonidae (Kfig. 28 p200), have the mouthparts directed somewhat posteriorly; these have been called **opisthognathous,** but the term is more suitable for describing the condition in Hemiptera. The third type is prognathous but strongly **retracted** into the prothorax (Kfig. 9 p186). The protracted-prognathous head is widely distributed throughout the order, with the exception of the superfamilies Scarabaeoidea, Byrrhoidea, Bostrichoidea, Chrysomeloidea, and Curculionoidea. Distinctly hypognathous heads occur in Microsporidae, a few Staphylinoidea (Ptiliidae, Silphidae), Dascillidae, all Scarabaeoidea, Byrrhidae, Dermestidae, *Endecatomus* (Bostrichidae), most Anobiidae, Ptinidae, Lymexylidae, Dermestidae, some Endomychidae, Erotylidae, some Coccinellidae, cerylonine Cerylonidae, Mordellidae, later instars of Meloidae and Rhipiphoridae, nilionine Tenebrionidae, and most Chrysomelidae and Curculionoidea. Strongly retracted heads are characteristic of Buprestidae, most Bostrichidae, a few Anobiidae, most Cerambycidae, some Chrysomelidae (Sagrinae, leaf-mining Alticinae and Hispinae), Bruchidae, a few Anthribidae (*Bruchela, Euxenus*), Aglycyderidae, Oxycorynidae, Belidae, and various Curculionidae (Rhynchitini, Allocoryninae, leaf miners). In the following discussions, the head is presumed to be protracted and prognathous unless otherwise noted.

The cranium, as viewed from above, is usually divided by an **ecdysial line** called the **epicranial suture** (fig. 34.9). This "suture" may be V-shaped, U-shaped, Y-shaped, or **lyriform** (shaped like a lyre with sinuous, rather than straight or evenly curved arms, and with or without a stem). The stem, when present, is called the **epicranial stem** or **coronal suture,** while the lateral arms are called **frontal arms** or **frontal sutures.** In some Cerambycidae, the epicranial stem is located within a broad furrow for the attachment of retractor muscles (Kfig. 17 p192). The epicranial suture divides the cranium into three major areas: a **frons (front)** or **frontoclypeal region** in front of and between the two frontal arms, and two **epicranial plates (epicranial halves)** or **parietal plates,** which are behind and lateral to the frontal arms and on either side of the epicranial stem when this is present; when there is no stem, the frontal arms may be contiguous or separated at the base, so that the frons extends posteriorly to the occipital foramen, as in some Carabidae and Nitidulidae. The dorsal part of the combined epicranial plates is sometimes called the **vertex,** as in adult beetles. When complete, the frontal arms extend to the antennal insertions, but in many groups they are incomplete anteriorly, and in some the entire epicranial suture may be indistinct or absent (second instar Histeridae). In a few beetle groups (Pterogeniidae, pisenine Tetratomidae, hallomenine Melandryidae, Chalcodryidae, toxicine Tenebrionidae, anaspidine Scraptiidae), there is an additional transverse ecdysial line joining the two frontal arms near their anterior ends. In beetle larvae there are no **adfrontal areas,** which are characteristic of most Lepidoptera.

In addition to the epicranial suture, the dorsal part of the head capsule may have one or occasionally two internal ridges or cuticular thickenings called **endocarinae,** which strengthen the head and may provide additional muscle attachment surface. These are visible as dark lines, but when they are coincident with the epicranial suture, they are difficult to observe. Since the epicranial suture is a line of weakness related to molting, any section of it which appears darker or thicker is almost certainly associated with an endocarina. The most obvious type is a median endocarina extending anteriorly between the frontal arms, sometimes as a continuation of the epicranial stem (Kfig. 20 p188); this occurs in a number of families, including Cupedidae, Lucanidae, Scarabaeidae, Buprestidae, Trogossitidae, Cleridae, Phalacridae, Anthicidae, Cerambycidae, and Chrysomelidae. The type which is entirely concealed beneath the epicranial stem (Kfig. 19 p216) is more widely distributed in Scarabaeidae, and also occurs in Callirhipidae, Bostrichoidea, Lymexylidae, Languriidae (Languriinae), Melandryidae, Mordellidae, Ciidae, a few Tenebrionidae, Chrysomelidae (Clytrinae and relatives), Bruchidae, and a number of Curculionoidea. A third type of median endocarina is Y-shaped and coincident with both epicranial stem and at least the bases of the frontal arms (Kfig. 23 p220); this type is known in larvae of a few Staphylinidae (Leptochirini), Passalidae, a few Trogossitidae (Peltinae), some Melandryidae, some Zopheridae, Oedemeridae, Cephaloidae (except Stenotrachelinae), and Belidae. Paired endocarinae may occur beneath the bases of the frontal arms, when the epicranial stem is absent (Kfig. 24 p209); these are known in Buprestidae (Trachyini), Eucnemidae, Nitidulidae (*Pocadius*), Cucujidae (Laemophloeinae), Melandryidae (*Zilora*), Colydiidae (*Cicones*), Zopheridae (*Phellopsis*), Salpingidae, Inopeplidae, and a few other groups. Finally there may be paired endocarinae which are not coincident with the frontal arms but lie just mesad of them (Kfig. 25 p243), as in many Buprestidae, a few Eucnemidae, *Thymalus* (Trogossitidae), Monommidae, and Othniidae.

The frontoclypeal region may be divided by a transverse invagination called the **frontoclypeal suture** (fig. 34.9) or **epistomal suture,** represented internally by the **epistomal ridge** (*see* below). The large area behind the suture is called the **frons,** while the smaller area in front of it is the **clypeus;** when the frontoclypeal suture is absent, these areas may be arbitrarily delimited by an imaginary line connecting the dorsal mandibular articulations (*see* below). The clypeus may be further divided into a posterior, more heavily sclerotized **postclypeus** and an anterior, lightly sclerotized **anteclypeus;** the term **epistoma** has been applied by different workers to one or the other subdivision or to the clypeus as a whole. A separate sclerite, the **labrum,** is attached to the anterior edge of the clypeus (*see* below), and the suture between them is called the **clypeolabral suture.**

The epistomal ridge forms the dorsal part of the **mouth frame** or **peristoma** (fig. 34.12). At either end of the ridge, there is a condyle, which is the **dorsal mandibular articulation,** and this may be surrounded by a heavily sclerotized area sometimes called the **talus.** The attachments of the anterior tentorial arms, if present, may also be located at the lateral ends of the epistomal ridge (*see* below). The lateral portion of the peristoma on each side, between the dorsal mandibular articulation and the ventral one, is called the **pleurostoma;** the **ventral mandibular articulation** is an acetabulum into which the ventral mandibular condyle is inserted. The lower part of the frame, between the ventral articulation on each side and the median attachment to the labium, is called the **hypostoma** or **hypostomal ridge,** and it is to this that the maxilla is attached. The hypostomata may be almost transverse in larvae with **protracted ventral mouthparts** (Kfig. 214 p201), so that the maxillary and mandibular articulations are more or less at the same level; in larvae with **retracted ventral mouthparts** (fig. 34.11), however, the maxillary articulations are located well behind those of the mandibles, and the hypostomata form a ventral cavity surrounding the maxillae and labium.

In various families of beetles, the labrum has become partly or completely fused to the clypeus or frontoclypeal region; complete fusion occurs throughout the Adephaga, Hydrophiloidea, Elateroidea (except Artematopidae), and Cantharoidea (except Brachypsectridae), and in the families Scydmaenidae, Pselaphidae, Staphylinidae (Staphylininae, Paederinae, Steninae, Euaesthetinae), Rhipiceridae, Nitidulidae (Meligethinae), Phalacridae (Litochrini), Cucujidae (Passandrinae), Bothrideridae (*Sosylus*), Corylophidae, Cerylonidae (Ceryloninae), Rhipiphoridae, Meloidae (some triungulins), and Chrysomelidae (Cryptocephalinae and relatives), while partial fusion is known in Dascillidae, some Eucinetidae, Nitidulidae (Cybocephalinae), Helotidae, Cucujidae (Uleiotini), Phloeostichidae (Hymaeinae), Phalacridae (Phalacrini), Biphyllidae, Byturidae (*Byturellus*), and Nemonychidae. In most groups with a completely fused labrum, the anterior part of the head forms a median projecting lobe called the **nasale** (fig. 34.10), sometimes with an additional pair of lateral lobes called **adnasalia** (sing. **adnasale**) by carabid workers (Emden, 1942b) and **paranasal lobes** by those working on elaterid larvae (Glen, 1950).

The epicranial plates extend ventrally on each side, and where they continue behind the hypostomata and ventral mouthparts they may be referred to as **postgenae;** each plate bears an **antennal fossa** and may have from 1 to 6 **stemmata** (*see* below). The antennal fossa is usually separated from the pleurostoma by a narrow bar of cuticle (fig. 34.12), but in some larvae (Microsporidae, some Endomychidae, Tenebrionidae: Nilioninae), the antennae are separated from the pleurostoma by a wide strip of cuticle, and in others, there is only a narrow strip of membrane between the two, and the

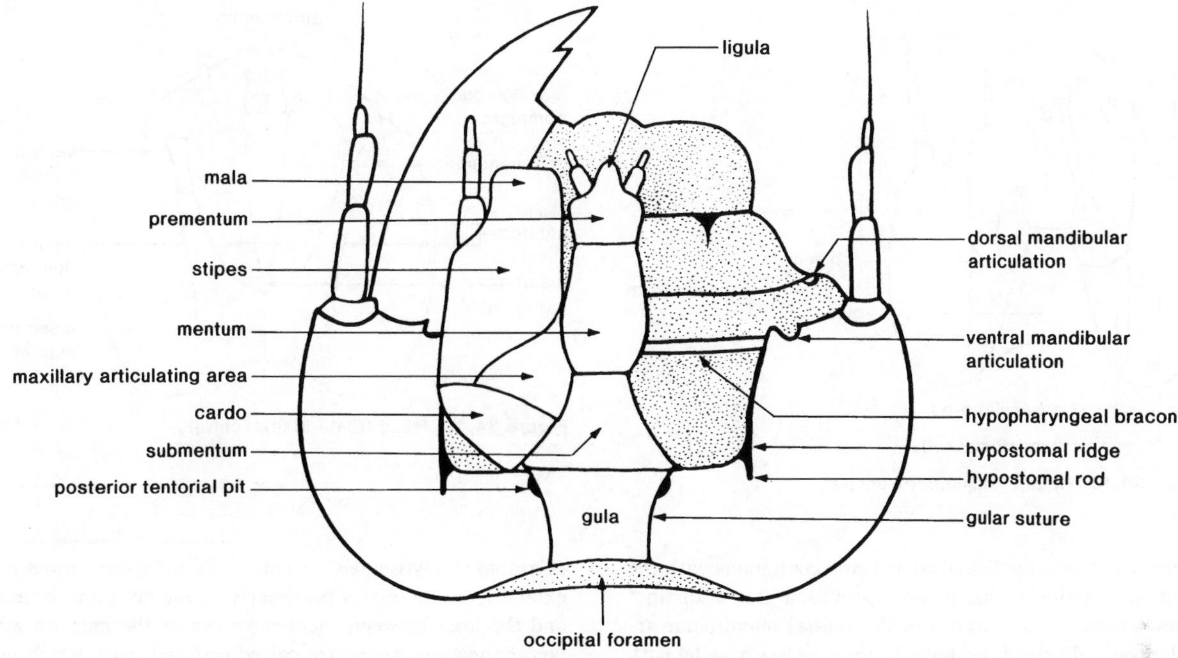

Figure 34.11. Head (composite), ventral, with left mandible and maxilla removed.

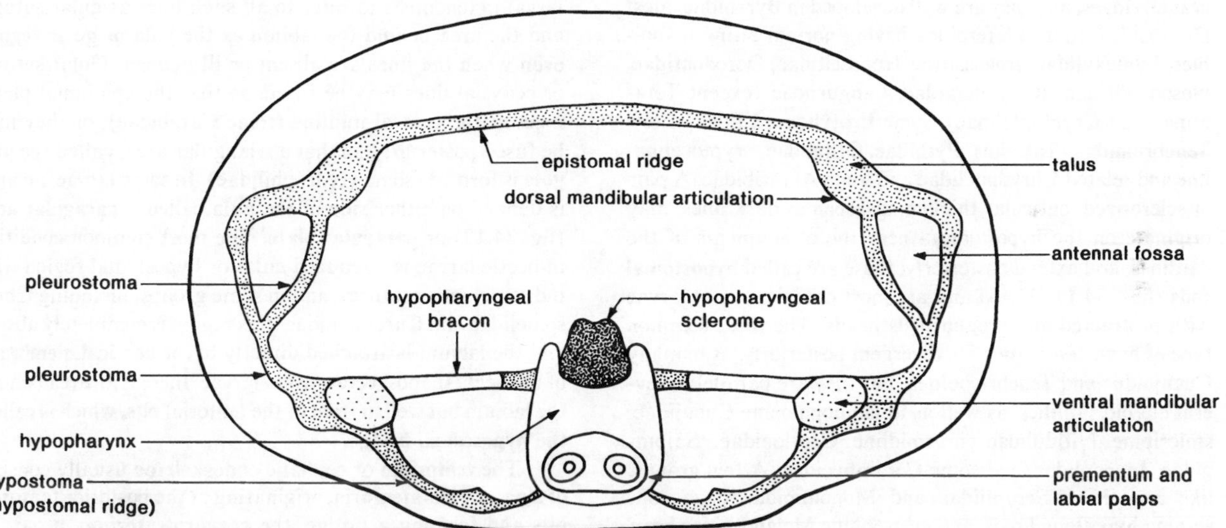

Figure 34.12. Head (Tenebrionidae), anterior, with clypeus, labrum and epipharynx, antennae, mandibles, and maxillae removed.

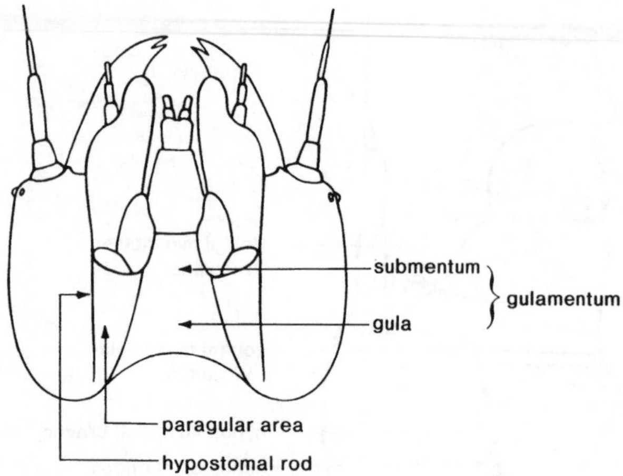

Figure 34.13. Head (Salpingidae), ventral.

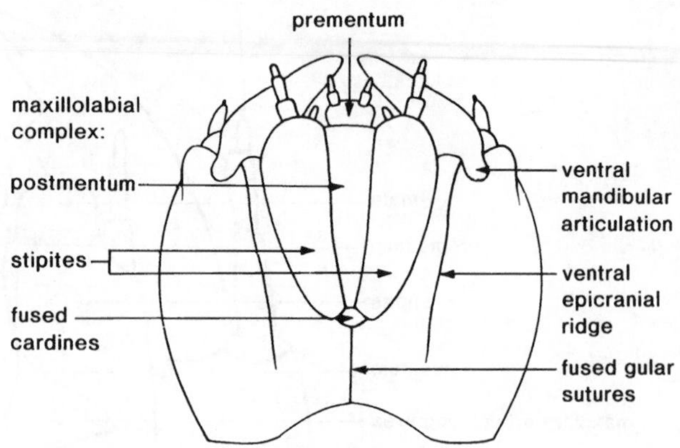

Figure 34.14. Head (Elateridae), ventral.

antennae are usually described as being contiguous with the mandibular articulations. In some groups, a pair of distinct ridges originates just laterad of the ventral mandibular articulations and extends posteriorly, more or less parallel with the hypostomata and forming with them a housing for the ventral mouthparts (fig. 34.14); these are called **ventral epicranial ridges,** and they are well developed in Byrrhidae, most Dryopoidea, those Elateroidea having normal biting mandibles, Lymexylidae, trogossitine Trogossitidae, Derodontidae, Nosodendridae, Rhizophagidae, Languriidae (except Toraminae and Cryptophilinae), some Erotylidae, Ciidae, a few Tenebrionidae (Toxicini), Pythidae, Othniidae, cryptocephaline and related Chrysomelidae and some Anthribidae. A pair of sclerotized, cuticular thickenings, seen as dark lines, may originate on the hypostomata near the attachments of the cardines, and extend posteriorly; these are called **hypostomal rods** (figs. 34.11, 34.13) and are most conspicuous in larvae with protracted and prognathous heads. The most common type of hypostomal rods are divergent posteriorly, as in many Cucujoidea and Tenebrionoidea, but they are parallel in several cleroid families, as well as in laemophloeine Cucujidae, smicripine Nitidulidae, murmidiine Cerylonidae, Salpingidae, Inopeplidae, and some Cerambycidae. A few groups, like anaspidine Scraptiidae and Monommidae have converging hypostomal rods, and eustrophine Melandryidae have two pairs, one parallel and one diverging.

The area behind the ventral mouthparts is sometimes called the **ventral head closure,** but the terms **hypostomal region** and **gular region** have also been used. The term ventral head closure usually implies that the area is at least moderately sclerotized, so that the type of head found in Melyridae, in which this region is very lightly sclerotized, is often said to lack a ventral head closure. The **posterior tentorial pits** (fig. 34.11), representing invaginations of the **tentorium,** usually occur just behind the maxillary articulations, but they may be located well behind these in some prognathous heads (Carabidae, Dytiscidae, Cleridae). Paired **gular sutures** may extend from these pits posteriorly to the occipital foramen, and the area between them is known as the **gula.** In some larvae, however, there are paired ecdysial lines, which may extend anterad of the tentorial pits or may not be associated with them at all (Hinton, 1963). It has been customary among larval taxonomists to refer to all such lines as gular sutures and the area behind the labium as the gula or **gular region,** even when the lines are absent or ill defined. Gular sutures or ecdysial lines may be fused, so that the epicranial plates meet at the ventral midline (some Carabidae), or they may be fused posteriorly, so that a triangular area, called the **pregula** is formed (some Hydrophilidae). In some larvae, an area is defined on either side of the gula called a **paragular area** (fig. 34.13) or **paragular plate.** The most common condition in beetle larvae is a reduced gular or hypostomal region with indistinct gular sutures, and in some groups, including Chrysomelidae and Curculionidae, this region is completely absent and the labium is attached directly to the **cervical membrane** of the neck. In most cerambycid larvae, there is an area behind the labium but well in front of the tentorial pits, which is called the **hypostomal bridge.**

The **tentorium** or cephalic **endoskeleton** usually consists of a pair of **metatentoria,** originating at the posterior tentorial pits and forming a bridge, the **corporotentorium,** a pair of **pretentoria** extending anteriorly to the epistomal ridge, and a pair of **supratentoria** extending dorsally from the pretentoria to a pair of depressions just laterad of the frontal arms on the epicranium. Reductions occur in many groups, especially with respect to the anterior and dorsal branches. The tentorium serves to strengthen the head capsule and acts as an attachment for muscles of the maxillae and labium-hypopharynx (*see* below). Another endoskeletal structure, the **hypopharyngeal bracon** (figs. 34.11–12), extends between the ventral mandibular articulations, passing through the hypopharynx and supporting that structure and its sclerome, when

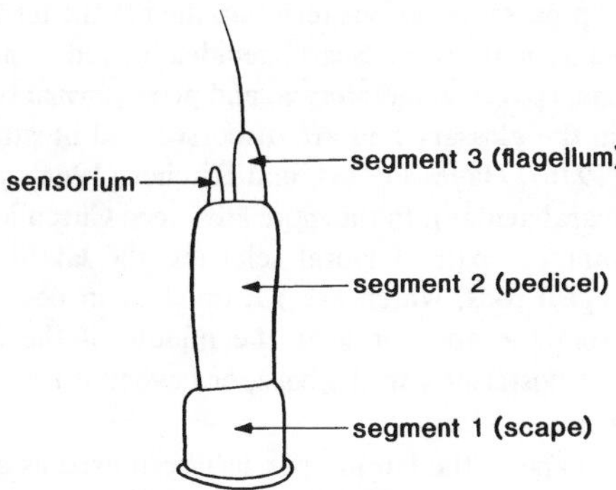

Figure 34.15. Antenna (composite).

present (*see* below). The hypopharyngeal bracon appears to be absent in Adephaga but occurs in Myxophaga and Archostemata. Within the Polyphaga, it is widely distributed, but absent in Hydrophiloidea, Dascillidae, Byrrhidae, Buprestidae, most Dryopoidea, some Cantharoidea and Elateroidea, some Bostrichoidea, and Chrysomeloidea. The presence of the bracon is one of the characters separating curculionoid larvae from those of Bruchidae and Chrysomelidae.

Antennae

The antennae arise from a pair of **antennal fossae** or antennal cavities located on the epicranial plates near the mandibular articulations. The basic number of antennal segments in Coleoptera larvae is apparently three (fig. 34.15): a basal **scape,** which is often short and broad, a **pedicel,** which may be more elongate and which bears at or near its apex one (or occasionally more than one) hyaline vesicle called a **sensorium (sensory appendage, sensory appendix, sensorial appendage, supplementary joint),** and the **flagellum,** which almost always consists of a single segment with a group of sensilla and one or more setae at the apex. Four-segmented antennae occur in the primitive family Cupedidae, as well as in all Adephaga (except a few derived forms of Carabidae); that this represents a derived condition resulting from the division of the scape is suggested by the position of the sensorium on the penultimate segment and the existence of a series of larvae of *Distocupes varians* (Lea), where the early instars have 3 antennal segments, while mature larvae have 4, 5, or 6, the additional segments appearing to be divisions of the scape. Other instances of derived 4-segmented antennae occur within the polyphagan families Leiodidae (*Prionochaeta*), Scydmaenidae (*Mastigus*), Staphylinidae (Staphylininae, Paederinae, Steninae), and Scarabaeidae (Aphodiinae, Scarabaeinae, Ochodaeinae, Glaphyrinae, and all Pleurosticta). The occurrence of more than four segments is known in a few Carabidae, some Dytiscidae, Helodidae, and aphodiine Scarabaeidae, as well as in cupedids. In the case of Aphodiinae, the additional segment appears to be derived from the scape; in Helodidae, the flagellum is divided into numerous segments (occasionally as few as 6); and in Dytiscidae, all the segments may become subdivided, except for the last, which remains reduced and lies side by side with the sensorium. Most beetle larvae have 3 antennal segments, but reductions to 2 or 1 are not uncommon. Two-segmented antennae are known in Microsporidae, Hydroscaphidae, Scydmaenidae, Pselaphidae, Passalidae, Buprestidae, Lycidae, Anobiidae, Cucujidae (Silvanini), Coccinellidae, Discolomidae, Ciidae, Tenebrionidae, Cerambycidae, Chrysomelidae, Belidae, and Aglycyderidae, while antennae consisting of a single segment are known in Rhipiceridae, Callirhipidae, Anobiidae, Ptinidae, Coccinellidae, Mordellidae, Rhipiphoridae, Meloidae, Chrysomelidae, and most families of Curculionoidea. In some groups (Amphizoidae, Buprestidae, Brachypsectridae, some Tenebrionidae), the last segment is highly reduced, so 3-segmented antennae may appear to have only 2 segments. Reduction may take place by the loss of the basal or apical segment or by the fusion of two segments. When the terminal segment is lost, the sensorium comes to lie at the antennal apex, usually accompanied by one or more setae and sensilla representing the true apical segment; **the lack of a seta at the apex of the sensorium distinguishes it from a true segment.** The fusion of apical and penultimate segments may result in the location of the sensorium near the middle of the apical segment, as in Microsporidae. In cases of extreme reduction, as in most Curculionidae, the antenna consists of a membranous dome bearing a few setae and sensilla. A most unusual situation occurs in some Phalacridae (*Phalacrus, Phalacropsis*), where the basal segment has been lost and the apical segment has become subdivided, resulting in a 3-segmented antenna with the sensorium at the apex of the first segment.

The size and shape of the sensorium may be useful in distinguishing certain family groups, but in some families they are extremely variable. In most larvae, the sensorium is elongate-conical or palpiform (sometimes almost setiform), and it may be very short or sometimes (Hydroscaphidae, Heteroceridae, some Cerylonidae and Endomychidae, Ciidae) longer than the terminal segment. Short, dome-like sensoria occur in a number of families, including Rhysodidae, Scydmaenidae, Scarabaeidae, Dascillidae, Lycidae, Lampyridae, Cucujidae, Erotylidae, Tenebrionidae, Pyrochroidae, and Scraptiidae. Bifurcate sensoria occur in some Pselaphidae, and tenebrionid larvae exhibit a variety of complex, dome-like sensoria, which may be C-shaped, sinuate, or broken up into several parts. In some instances, there may be more than one sensorium on the penultimate segment; multiple sensoria are known in a few Hydrophilidae, Scarabaeidae, and Elateridae, and in most Histeridae. In many Scarabaeidae, there are additional vesicles on the terminal segment; these have been called **dorsal sensory spots** by Ritcher (1966). As mentioned above, the sensorium is usually placed at the end of the penultimate segment, except in a few cases where the last two segments have become fused. An exception occurs in most

Staphylinoidea, where the apex of the penultimate segment is strongly oblique and the large sensorium is attached well before the apex and not on a line with the last segment. Some of the large, dome-like sensoria may also be subapical.

Stemmata

The lateral eyes of larval holometabolous insects are often referred to as **ocelli** (sing. **ocellus**), but since that term has also been applied to the median and paired simple eyes of adult insects, the word **stemma** (pl. **stemmata**) is used in this book. The number of stemmata borne on each epicranial plate in Coleoptera larvae varies from 1 to 6 and many larval types lack them altogether. Each stemma may have a well-developed lens and retina, or it may be reduced so that it is little more than a pigment spot; in Carabidae: Cicindelinae, one or two of the stemmata are much larger than the others and contain numerous sensory cells (Paulus, 1979). After clearing in potassium hydroxide, stemmata are particularly difficult to see, and pigment spots may disappear; it is advisable to count the stemmata and make a sketch of their arrangement and position before clearing. A full complement of 6 stemmata occurs in most families of Adephaga, and in Agyrtidae, some Silphidae and Staphylinidae, many Hydrophilidae, Byrrhidae, several groups of Dryopoidea, Derodontidae, Dermestidae (*Dermestes*), a few Cleroidea, many cucujoid families, Cephaloidae, and several subfamilies of Chrysomelidae. The most common number in Polyphaga is 5, but it varies considerably within some families and reduction has occurred several times. Some families, like Nitidulidae and Endomychidae never have more than 4 on each side, while Coccinellidae have 3 or fewer; families of Tenebrionoidea (except some Cephaloidae) do not have more than 5, and Curculionoidea (except some Belidae) have no more than 2. In many substrate-dwelling larvae, stemmata are absent or there are only 1 or 2 pairs of pigment spots; this is the case in Scarabaeoidea, most Bostrichoidea, Bruchidae and most Curculionoidea. Some other groups, like the Elateroidea and Cantharoidea, usually have a single large stemma with a well-developed lens on each side.

Stemmata may be well separated, as in Byrrhidae or Dryopidae, or tightly clustered, as in Elmidae and Ptilodactylidae, and the arrangement may be characteristic for a group; stenine Staphylinidae, for instance, have a circular arrangement, many Carabidae have two curved rows forming an incomplete circle, and many groups have two vertical rows (3 and 3, 4 and 2, 3 and 2).

Labrum and Epipharynx

The area on the underside (**ental** surface or that facing the inside of the mouth cavity) of the labrum is called the **epipharynx,** and the term **labrum-epipharynx** is sometimes used for the entire structure. The base of the labrum bears a pair of lateral sclerites called **tormae,** to which the lateral labral muscles are attached. Tormae vary considerably and may be useful taxonomic characters; they may be longitudinally or transversely oriented, symmetrical or asymmetrical, separate or joined mesally, and simple or complex, with accessory processes. Various terms applied to the tormae and associated structures in Scarabaeoidea include **apotorma, dexiotorma, epitorma, laetotorma,** and **pternotorma;** these are defined in the glossary and are discussed and illustrated by Böving (1936), Hayes (1928), and Ritcher (1966) (*see* section on Scarabaeidae). In the more advanced Curculionoidea, there is another pair of labral sclerites, the **labral rods** or **epipharyngeal rods,** which are paramedian in position and extend from the apex or near the middle of the labrum-epipharynx posteriorly to the base, or beyond it into the clypeal region.

The shape of the labral apex has been used as a taxonomic character in Scarabaeoidea (Kfigs. 35 p268, 36 p273) and various other groups, while the type and distribution of labral setae is important in Tenebrionidae. The surface of the epipharynx (Kfigs. 40–41 p265) may be very complex and is taxonomically useful in some families. The anterior edge usually bears a row of setae or fine hairs, and in addition, there may be patches of setae or spines, groups of sensilla of different sorts, and sclerotized plates or small rods. A terminology for describing the epipharynx has been developed only for the Scarabaeoidea, where there may be six regions—the median **corypha, haptomerum, pedium,** and **haptolachus,** extending from front to back, and the paired lateral **paria;** other terms used for epipharyngeal structures (*see* glossary and references above) include **acanthoparia, acroparia, chaetoparia, clithrum, crepis, epizygum, gymnoparia, helus, nesium, phoba, plegma, plegmatium, proplegmatium, tylus,** and **zygum** (*see* section on Scarabaeidae). The area immediately behind the epipharynx forms the dorsal wall of the **cibarium** and is often provided with a complex armature, including a series of obliquely transverse **cibarial plates,** which are fringed with microtrichia and are located above the microtrichial armature at the base of the mandibles (*see* below).

Mandibles

With few exceptions, Coleoptera larvae have a pair of well-developed, **opposable** mandibles, which move in a horizontal plane and articulate with the dorsal and ventral **mandibular articulations** of the head capsule by means of a **dorsal acetabulum** (figs. 34.55, 34.60) and a **ventral condyle** (figs. 34.45, 34.56–57). Each mandible is moved by a large **adductor muscle,** inserted at the inner angle (just laterad of the mola, if this is present), and a small **abductor muscle,** inserted near the outer edge; both muscles originate on the walls of the cranium, but the former occupies most of the cranial cavity and may be attached as well to an internal ridge or **endocarina.** In exceptional cases, such as mandibles of some Eucnemidae, which are not opposable, the abductor may be larger than the adductor (Ford and Spilman, 1979).

Since mandibles have been used extensively in larval taxonomy at the family level, mandibular structure and function will be dealt with in some detail. In most cases, statements on function are not based on behavioral observations, but on a combination of detailed morphological study, knowledge of food habits, examination of gut contents, and occasionally, manipulation of cleared and bleached specimens. For descriptive purposes, the mandible may be divided into four

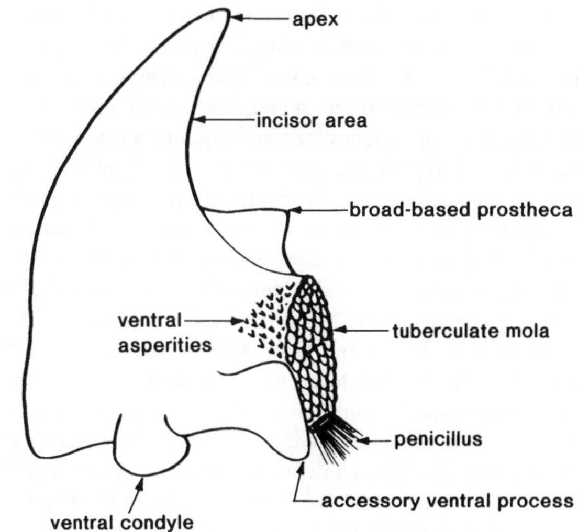

Figure 34.16. Right mandible (Languriidae), ventral.

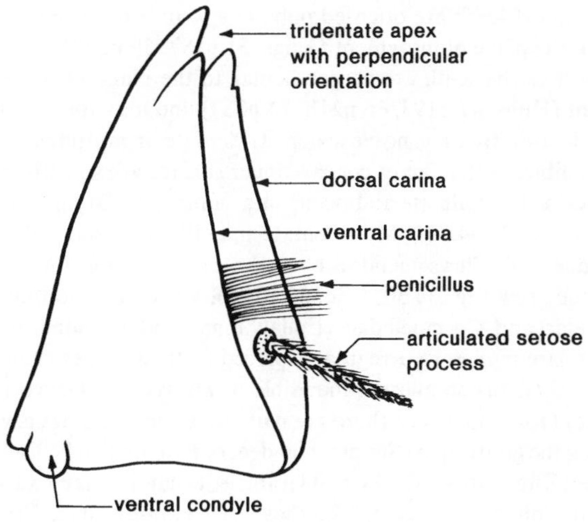

Figure 34.17. Right mandible (Elmidae), ventral.

areas (fig. 34.16): the **apex,** the **incisor area (scissorial area** or **cutting edge)** occupying much of the mesal surface, the **base** including the mesal surface, and the **outer edge.** The mesal portion of the mandibular base is sometimes called the **molar area,** but this term implies the presence of a definite basal structure, the **mola,** which acts in conjunction with the mola of the opposite mandible (and sometimes with a sclerome on the hypopharynx, *see* below) to process and ingest food.

Mandibular structure varies considerably, but the basic type, which is thought to represent that of an ancestral beetle larva, has a relatively wide base with a distinct mola and a narrow, relatively weak apex (figs. 34.16, 34.27, 34.39–40). There are usually two apical teeth, which are oriented obliquely to the plane of movement, sometimes an additional subapical tooth in the incisor area, and a thin, **hyaline** or translucent structure called a **prostheca** located just distal to the mola. The outer edge usually bears a pair of setae of fixed position. The surface of the mola is covered with small **tubercles** or **asperities,** which often continue onto the ventral surface, while the dorsal surface of the mandibular base has a patch of **microtrichia** arranged in oblique rows. This primitive type of mandible usually occurs in larvae which feed on small particles (spores, pollen) or very soft tissue (loose fungal hyphae, highly decayed animal or plant matter); this type of feeding is referred to as **microphagy.** The molae are symmetrical, and their surfaces meet so that the asperities or tubercles interdigitate to move small particles into the mouth cavity, while simultaneously shearing or crushing larger pieces into smaller ones, a process called **comminution.** The ventral asperities may also participate in the process, since they pass over the **hypopharyngeal sclerome** beneath them, while the oblique rows of microtrichia on the dorsal surfaces act together with the fringes on the **cibarial plates** to move small particles caught above the basal edges of the mandibles toward the molae and into the mouth cavity. The ventral part of the mandibular base has, in addition to the ventral condyle, a second, shelf-like projection, the **accessory ventral process** (also **accessory condyle, accessory ventral condyle, second condyle**) (figs. 34.16, 34.51–52), located near the insertion of the adductor muscle and just laterad of the mola; this process it not a condyle, but its true function is not known. This primitive type of mandible occurs in Myxophaga, primitive Staphylinoidea, Eucinetidae, Derodontidae, and many Cucujoidea, while the same basic design with modifications in the mola or elsewhere may be found in Archostemata, Helodidae, Dascillidae, Scarabaeoidea, Lymexylidae, various Cucujoidea, many Tenebrionoidea, and a few primitve Curculionoidea.

Of the many specialized mandibles that can be derived from the primitive type, most fall into two categories: 1) the cutting, biting, gouging or scraping mandible of a **phytophagous** or **xylophagous** larva (figs. 34.29–31), and 2) the grasping, piercing mandible of a larva which is always liquid-feeding (practicing **extraoral digestion**) and usually predaceous (figs. 34.35–36, 34.38, 34.59). In both types the mola has been lost and food processing occurs at the mandibular apex, rather than at the base. Mandibles of the first group usually have a very broad base with a greater cross-sectional area and a gradually narrowed apex, which is sometimes not much narrower than the base; the gap between the mandibular apices is small and much greater force can be exerted at the apex and along the incisor edge. The second group usually includes mandibles which are narrow and **falcate (falciform),** with a narrower base and wide gap; these mandibles can grab and often pierce objects but cannot generate the force needed at the apex to cut through heavily reinforced plant or fungal tissue. Details of these and other specialized mandible types are covered in a later paragraph.

The mandibular apex may be simple, consisting of a single tooth (**unidentate,** Kfigs. 44, 46 p239), but more commonly there are two apical teeth (**bidentate,** Kfigs. 61 p239, 70 p262), and there may be three or more (**tridentate,** figs. 34.17, 34.31, Kfigs. 47 p193, 66 p218; **multidentate,** fig. 34.30, Kfig. 51 p187); each tooth may be acute or rounded, but this often varies with wear in a single individual. In most larvae, the apical teeth are oriented obliquely to or almost parallel with the plane of movement (Kfigs. 51 p187, 59 p239), while in others the teeth are perpendicular to the plane of movement (Kfigs. 47 p193, 66 p218, 74 p257) and may form with the incisor area a concave scoop. Tridentate or multidentate mandibles with a flat or concave inner surface are sometimes described as **palmate** and occur in a number of Dryopoidea (figs. 34.17, 34.31). Multidentate mandibles of some Byrrhidae and Chrysomelidae have the teeth arranged in an oblique row (fig. 34.30, Kfig. 51 p187), while in some Eucinetidae and Coccinellidae (Epilachninae and Psylloborini) there are numerous, irregularly placed teeth or spines (Kfig. 56 p227). In the unusual mandible of *Dasycerus* (Dasyceridae) (Kfig. 55 p241), there are numerous curved spines covering the entire apex. Simple, rounded, or flattened and chisel-like (Kfigs. 46 p187, 48 p234) apices occur in many substrate-inhabiting larvae, like those of Cerambycidae, Bruchidae, and Curculionoidea, while acute apices (Kfigs. 43, 44 p200) are usually found in predators. In some larvae with well-developed molae, the apex may become highly reduced, as in *Lycoperdina* (Endomychidae) (fig. 34.54) or some Lathridiidae.

The incisor area of the mandible often bears one or more subapical teeth, which may be located on a carina extending basally from the dorsalmost apical tooth (fig. 34.17), and sometimes this **dorsal carina** may be serrate. A second **ventral carina** occurs in some mandibles lacking a mola, and a concave, scoop-like area may be formed between the two carinae (fig. 34.17). An extreme form of this occurs in mandibles of certain predaceous larvae, where an open or partly closed groove is formed along the inner surface (Kfig. 44 p200); such a condition was probably prerequisite to the evolution of **perforate mandibles** to be discussed later. Another feature of the incisor area is the presence of a heavy, sclerotized tooth called a **retinaculum** (although that term is sometimes misapplied to the prostheca). A single retinaculum (or occasionally more than one) is usually found on the falcate mandibles of predatory larvae (Carabidae, fig. 34.59, Hydrophilidae, Histeridae, Elateridae, fig. 34.57), but similar structures, called **scissorial teeth,** may also occur in non-predatory larvae of Dascillidae (fig. 34.52) or Scarabaeoidea.

The subapical region of the mandible may also bear a mola-like area, consisting of tubercles, teeth (Kfig. 54 p243) or fine ridges (Kfig. 52 p188); this structure is called a **pseudomola,** but that term has also been applied to a variety of basal or sub-basal molae, which are thought to have evolved secondarily in larvae which have lost the true mola.

The **mola** is the processing area at the base of many mandibles which acts in conjunction with the mola of the opposing mandible to feed small particles into the mouth cavity, to cut and shear larger particles into smaller ones, or sometimes to compact small particles into pellets to be ingested. Primitive molae are symmetrical and relatively narrow, and they extend to the basal edge of the mandible; examples may be found in many Cucujoidea (figs. 34.39, 34.42, Kfigs. 75, 78 p259), as well as in derodontids, eucinetoids, and primitive staphylinoids. From this type have evolved the slightly asymmetrical molae (Kfig. 73 p188) of Mycetophagidae or Archeocrypticidae, and the strongly asymmetrical molae (figs. 34.28, 34.41, Kfig. 74 p257) of the more advanced tenebrionoids (Prostomidae, Oedemeridae, Synchroidae, Pythidae, Pyrochroidae etc.), where each mola consists of numerous transverse or oblique ridges which meet at an angle forming multiple shearing surfaces for handling rotten wood and reinforced fungal hyphae. In most cases, the left mola is more vertical, with a flat or concave surface and often a tooth or lobe at the distal end, while the right one is inclined away from the midline and has a convex surface (figs. 34.28, 34.41, Kfigs. 73, 74 p263). In another derived type of mandible, both molae are concave, usually with smooth surfaces, as in Heteroceridae (fig. 34.32) and some Tenebrionidae (fig. 34.45); these serve to compact loose material, and may also shear larger pieces with the sharp outer edges. The tenebrionid type may act in conjunction with a concave hypopharyngeal sclerome (fig. 34.46). The concave molae in Heteroceridae are used to compact masses of wet sand or mud, which are then swallowed whole for the extraction of diatoms, algae, and other organic matter; the molae are not basal, and are considered to be pseudomolae, since heterocerids have almost certainly evolved from an ancestor lacking a mola. Another molar type, found in the endomychid *Lycoperdina* (fig. 34.54), is broad and finely tuberculate for crushing fungal spores.

The molar surface in more primitive mandibles is covered with numerous, sharp asperities or rounded tubercles, which may form transverse rows and may extend onto the ventral surface of the mandibular base (figs. 34.16, 34.40, 34.43). The molar armature is of taxonomic importance, but care should be exercised when using this character exclusively, since the surface is subject to variation caused by wear. The molar surface may be reinforced by the formation of simple, continuous ridges, which are usually transverse (fig. 34.60) but occasionally longitudinal (Dascillidae, fig. 34.52), or complex ridges, separated by deep channels (figs. 34.63–66). The ridges may become interconnected with the formation of a poroid superstructure (fig. 34.62), which may act as a fluid press, since the pores and channels communicate with the dorsal and ventral surfaces (fig. 34.61).

The dorsal surface of the mandibular base adjacent to the mola often bears a patch of fine microtrichia, which are usually arranged in oblique rows (figs. 34.39, 34.41, 34.60–61, 34.63–64). These are found in many groups having an asperate or transversely ridged mola, and they apparently act in conjunction with the fringed cibarial plates above them to move small particles toward the molar area and away from the articular areas. The ventral surface of the mandibular base may also have patches of microtrichia or asperities, but these are usually not arranged in regular rows. The ventral asperities are assumed to aid in the food processing by acting against the sclerome on the hypopharynx. In some groups of Scarabaeidae (Rutelinae, Dynastinae, Cetoniinae), there is a group of fine transverse ridges at about the middle of the mandibular base or towards the lateral edge; this is a **stridulitrum** (called **stridulatory area**), which acts in conjunction with a **plectrum** consisting of **stridulatory teeth** on the dorsal surface of the maxilla.

There are three types of accessory structures associated with the mola: a **premolar tooth,** a **submolar lobe,** and a **prostheca.** The premolar tooth is a heavily sclerotized process which is actually part of the mola. The submolar lobe is a hyaline area, usually setose, at the base of the mola in Biphyllidae, Anthicidae, Euglenidae, and some Scraptiidae, but in the family Byturidae (Kfig. 76 p279), it is a flexible, setose, projecting lobe. A brush of hairs or **penicillus** may also occur at the molar base, as in some Bostrichidae (Kfig. 72 p187), *Orphilus* (Dermestidae), some cucujoids and some scarabaeoids.

The **prostheca** is a very important taxonomic character in beetle larvae, but there has been some confusion in the use of the term. In the present treatment, the word is restricted to any of the structures located just distad of the mola but not a part of it, which are almost always hyaline, at least basally, or composed of membrane or groups of hairs. The structure is not to be confused with a premolar tooth or with the retinaculum, which is heavily sclerotized, more apical in position, and almost always borne on mandibles lacking a mola. Structures similar to the prostheca may occur in mandibles without a mola, and these are discussed below. Other terms which have been used for the prostheca are **lacinia mobilis** and **lacinia mandibulae.**

The most common type of prostheca is a fixed, hyaline process, which is relatively rigid and more or less acute at the apex; it is often said to be articulated, but rarely fulfills the requirements of an articulated appendage, as defined in the glossary. This type usually has a relatively narrow base and simple, acute apex, as in Agyrtidae, Leiodidae, or Cucujidae (figs. 34.39–40, Kfig. 78 p193), but it may be serrate (Rhizophagidae, Kfig. 82 p270) or bifid (Cryptophagidae, Kfig. 83 p270), or may have a broader base and more or less angulate apex (Languriidae, Kfig. 81 p276). More complex types of prosthecae may be composed of series of fringed membranes (Nitidulidae, fig. 34.53, Kfig. 79 p276), a brush of simple hairs associated with an acute process (Derodontidae), a group of comb-hairs (Biphyllidae, fig. 34.47, Kfig. 80 p276), or a membranous lobe (some Bostrichidae, Kfig. 75 p259; some Endomychidae, fig. 34.54). Other structures which may be called prosthecae are the comb-like process in Helodidae (fig. 34.51, Kfig. 77 p273), or the hyaline area and small, articulated process in Dascillidae (fig. 34.52).

In mandibles lacking a mola, there may be a variety of structures at the base of the mesal edge, and these, like the prosthecae, are useful for identification. Commonly, there is a brush of hairs (Lampyridae; Byrrhidae, Kfig. 63 p233; Histeridae, Kfig. 53 p239); this is usually called a **penicillus,** but that term has also been restricted to a thin "pencil" of hairs. There may an acute, rigid, hyaline process similar to the simple prostheca (and sometimes called a "retinaculum"), as in Paussini (Carabidae), Proteininae (Staphylinidae), Passandrinae (Cucujidae), *Deretaphrus* (Bothrideridae), some Coccinellidae (Kfig. 57 p224), and some Ciidae (Kfig. 58 p239), or a more complex, bifid or serrate process, usually referred to as a **lacinia mobilis** or **lacinia mandibulae,** as in various Cleroidea (Kfig. 61 p239). A group of from 2 to 5 hyaline processes occurs in some Trogossitidae, laemophloeine Cucujidae, some Phalacridae (Kfig. 60 p239), tetratomine Tetratomidae, Perimylopidae, and Salpingidae (fig. 34.55), while a row of hyaline teeth occurs in Monommidae (fig. 34.56); these structures may represent a highly reduced mola or prostheca or may represent independent developments. Articulated processes also occur, and they may be covered with fine hairs, as in *Araeopidius* (Ptilodactylidae) or have a brush of hairs at the apex, as in *Ptilodactyla* (fig. 34.58). Some mandibles have two types of basal structures, such as a brush and articulated, setose process in some Elmidae (fig. 34.17) or a fixed, hyaline lobe and brush of hairs in *Dermestes.* A large, fringed, membranous lobe occurs without a mola in some Erotylidae (Kfig. 64 p239), eustrophine Melandryidae, and *Endomychus* (Endomychidae).

The outer edge of the mandible in most larvae bears a pair of setae of fixed position, but in some groups, like Tenebrionidae, numerous setae may occur and some of these may be stout and spine-like. In some ground-dwelling larvae, the outer edges of the mandible are carinate and serve as wedges used in burrowing; examples are known in Cebrionidae and several groups of Tenebrionidae.

As mentioned above, mandibles lacking a mola usually fall into two groups: the broader, cutting, biting, gouging, or scraping type, and the narrower, grasping or piercing type. These, in turn, are usually associated with two different feeding strategies, namely **macrophagy,** or feeding on large particles (which are often cut or otherwise removed from an even larger food mass and then swallowed whole), and liquid-feeding or **extraoral digestion** (where digestive enzymes are ejected into a large food mass or the surrounding medium, and the breakdown products sucked back into the gut through an **oral filter,** which keeps larger particles out (*see* below)). Heavier mandibles and macrophagous habits almost always occur in larvae which feed on the reinforced tissue of higher plants and advanced basidiomycete fungi, while narrower, falcate mandibles and extraoral digestion usually occur in predators; exceptions are not uncommon, however, and there are a number of liquid-feeding, phytophagous Elateridae and Carabidae, as well as predaceous larvae with relatively broad and heavy mandibles.

Among the macrophagous types of larvae, wood-feeding or **xylophagous** forms are most common in the families Buprestidae, Melandryidae, Mordellidae, Cerambycidae, and Curculionidae, while those specializing on leaves and herbaceous tissue (both fresh and decomposed) occur in Byrrhidae, some dryopoid families, some Languriidae and Coccinellidae, Chrysomelidae, and some Curculionidae. Also falling into this group are those larvae which scrape algae from rock surfaces (Psephenidae, Elmidae). Larvae specializing on the tougher fruiting bodies of higher Basidiomycetes are in the families Trogossitidae (Peltinae, *Calitys*), Erotylidae, Ciidae, and Mordellidae, while seed feeders occur in Bruchidae, Anthribidae, and Curculionidae. Although there is variation in the type of apex, most mandibles in this group are the heavier type with a broad base. Exceptions occur in some Byrrhidae and Chrysomelidae (fig. 34.30), which have a somewhat flattened mandible with a multidentate apex, used to scrape tissue from a plant surface. A similar function is performed by the multispinose apex in phytophagous Coccinellidae (Kfig. 56 p227).

Narrow and falcate mandibles occur in most Adephaga, many Staphylinoidea, most Hydrophiloidea, many Elateridae, all Cantharoidea, and a few cucujoid and tenebrionoid triungulins. One of the main developments in mandibles of this type is the formation of an open groove or closed perforation, which serves to channel digestive fluids into the food mass; the internal perforation is sometimes called a **blood channel.** An open or partly closed groove occurs on the mesal surface of the mandible (Kfig. 44 p212) in larvae of some Noteridae and Dytiscidae, spercheine Hydrophilidae, Homalisidae, Drilidae, Cantharidae and a few Cleridae, while **perforate mandibles** (Kfig. 43 p212) are characteristic of most Dytiscidae, Gyrinidae, Brachypsectridae, Phengodidae (fig. 34.38) and Lampyridae. A different type of perforate mandible, with a broad base and relatively large internal perforation, occurs in larvae of Haliplidae (Kfig. 49 p212), which apparently suck algal cells into the mouth through the perforation (Hickman, 1931). Another unusual type of mandible, apparently associated with liquid feeding, is that found in Lycidae (Kfig. 42 p196); each mandible is split longitudinally into two parts, which fit together to form a channel; the two mandibles are not opposable. **Styliform** mandibles are also non-opposable, but they are extremely narrow and may be capable of anteroposterior movement; they are usually enclosed, together with styliform maxillae, in a median tubular **proboscis** (Cerylonidae: Ceryloninae, Kfig. 27 p224; Leiodidae: Camiarinae), but in *Cerylon* (Kfig. 28 p200) they are enclosed within the mouth cavity **(endognathous).** Non-opposable mandibles also occur in Eucnemidae, Throscidae, Cerophytidae, and cardiophorine Elateridae, but these are not associated with a proboscis. In some eucnemids, they are movable, with laterally projecting teeth; in other Eucnemidae, Throscidae, and Cardiophorinae, they are fused to the head capsule; and in Cerophytidae and *Phyllocerus* (Eucnemidae), they are styliform at the apex. All of these groups are liquid feeders, and all move through relatively compact substrates by using the mandibles, labium, or entire head capsule as a wedge.

Ventral Mouthparts

The **ventral mouthparts** include the paired **maxillae** and unpaired **labium,** whose attachments in a prognathous head are ventrad or mesad of the mandibular articulations; they are also referred to as the **maxillolabial complex,** especially when they are closely coadapted and move as a single unit (*see* below). The ventral mouthparts are said to be strongly **protracted** when the maxillary bases (usually the points of attachment of the cardines) are at about the same level as the ventral mandibular articulations and the hypostomal margins are short and transverse (fig. 34.36, Kfigs. 92, 95 p253). This condition is characteristic of most Adephaga, a few Staphylinidae, most Hydrophiloidea, Cleridae and Chaetosomatidae, smicripine Nitidulidae, laemophloeine and passandrine Cucujidae, Phalacridae, some Corylophidae and Lathridiidae, Prostomidae, most Cerambycidae, Bruchidae, and some Chrysomelidae. When the ventral mouthparts are **retracted,** the attachments of the cardines lie posterior to the mandibular articulations, so that the entire maxillolabial complex is recessed into the head capsule and the hypostomal margin on each side is more or less longitudinal or oblique (figs. 34.33, Kfigs. 85–90, 93 pp240,260); this is the normal condition in beetle larvae. Intermediate situations occur in some groups, like Cantharidae (Kfig. 94 p191), ptilodactyline Ptilodactylidae (fig. 34.34), and Mycteridae, where the maxillae and labium are only slightly retracted. The points of attachment of the maxillae do not necessarily reflect the degree to which the ventral mouthparts extend forward, since the latter can be influenced by the length and orientation of the cardo and the length of the stipes. As mentioned above, a recessed housing for the ventral mouthparts may be formed between the hypostomal margins and ventral epicranial ridges (figs. 34.31, 34.33, Kfigs. 90, 93 p240).

The **maxilla** usually consists of a small basal sclerite, the **cardo** (pl. **cardines**), a larger **stipes** (pl. **stipites**), a segmented **maxillary palp** (or **palpus**), which is borne on a shelf-like projection called the **palpifer,** and a pair of **apical lobes**—the **galea** or **outer lobe** and the **lacinia** or **inner lobe;** in many larvae, there is a membranous region called the **maxillary articulating area (basimaxillary membrane** of Das, 1937), which lies between the junction of stipes and cardo and the base of the labium (*see* below) (fig. 34.33, Kfigs. 85 p191, 89 p197, 105, 106 p200). The cardo may be divided into two distinct sclerites, but most often it is either undivided or bisected by an internal ridge; areas mesad and distad of this ridge have been called the **proxicardo** and **disticardo,** respectively. In most larvae, the cardo is more or less transversely oriented or slightly oblique (Kfigs. 105–106 p200), but in some, including those of most Nitidulidae (Kfig. 91 p200), it is longitudinal. In some larvae with protracted mouthparts (Phalacridae, Kfig. 92 p199; Nitidulidae (Smicripinae); Cucujidae (Laemophloeinae); Histeridae), the cardo may be absent or represented by a membranous area at the maxillary base; in some other groups, like the Coccinellidae, the cardo has become fused to the stipes. The two cardines are usually widely separated, lying on either side of the labium, but in some larvae, they may be approximate, displacing the labium anteriorly (Callirhipidae, Kfig. 99 p216; Heteroceridae), contiguous, meeting behind the labium (Limnichidae, Kfig. 100 p194; some Elateridae), or fused into a single sclerite (some Elateridae, Kfig. 93 p193; Chelonariidae; Brachypsectridae, Kfig. 97 p201). In some Cerambycidae, the entire area at the base of the ventral mouthparts is lightly sclerotized, without distinct separations of the labium, cardines, and stipites.

The stipes is the main body of the maxilla, housing the muscles that operate the palp and apical lobes; it is usually longer than wide, but in some groups (Cleridae; Phalacridae, Kfig. 92 p199) it may be subquadrate or wider than long. The stipes is occasionally divided into a basal and apical section (**proxistipes** and **dististipes,** respectively), and Gilyarov (1964b) has used the term **costipes** for an anterior subdivision in Hydrophiloidea usually called the palpifer and here considered to be the first segment of the maxillary palp (Kfig. 102 p200). Another term, **juxtastipes,** has been used by Böving and Craighead (1931) for an area extending mesally from the base of the stipes. The dorsal surface of the stipes, near the base, in various Scarabaeoidea may have a patch or oblique row of teeth or tubercles called **stridulatory teeth,** but only in the Dynastinae, Cetoniinae, Rutelinae, and some

Melolonthinae is there an opposing stridulatory area on the ventral surface of the mandible (*see* above).

The **maxillary articulating area** is well-developed and broadly triangular in most primitive larval types, loosely joining the base of the maxilla with that of the labium, so that the 2 structures are well separated and the maxilla is capable of considerable independent movement (fig. 34.27). In many larvae, however, the articulating area extends farther forward, and may become narrowly oval, as in Ciidae or Anobiidae, so that maxillary movement is more restricted. The area is usually membranous, but in some groups it may be partly sclerotized, and occasionally it is divided into two parts. In many Ptilodactylidae, and to some degree in the Helodidae and Psephenidae, the area may be well-developed but partly or entirely concealed behind a laterally expanded labium (fig. 34.34, Kfig. 98 p230). In other groups there is no articulating area, because the maxilla and labium have become closely coadapted and more or less connate, with further restriction of independent movement. Further consolidation of the maxillolabial complex through connation and fusion is discussed at the end of this section.

The **palpifer** (or **maxillary palpiger**) is a shelf-like area located externally (laterally) near the apex of the stipes and bears the maxillary palp. There is considerable confusion about the number of palp segments in some groups, because an enlarged palpifer may be counted as an additional segment or the basal palp segment may be considered a palpifer (when the true palpifer is poorly developed or absent). **In constructing the family key, a structure with both inner and outer edges and apparently articulated at the base was always called a palp segment, while a shelf-like extension of the stipes was considered to be a palpifer; in doubtful cases, both character states were included in the matrix.** A 3-segmented maxillary palp is thought to be primitive in Coleoptera larvae, but the number varies from 1 to 5. Four-segmented palps occur in most Adephaga, some advanced Staphylinidae, most Hydrophiloidea (if the so-called palpifer is considered to be a palp segment), Helodidae, almost all Scarabaeoidea, Byrrhidae, almost all Dryopoidea, Elateroidea and Cantharoidea, Dermestidae (except Anthreninae), the bostrichid genus *Endecatomus,* and a few Anobiidae, while 5-segmented palps are restricted to a few Carabidae and some Histeridae. Reduction in the number of palp segments occurs in a number of small larvae and in many which are internal feeders. Two-segmented palps are known in the Hydroscaphidae, a few Leiodidae and Scydmaenidae, *Acalyptomerus* (Clambidae), late instar Rhipiceridae, Buprestidae, *Eubrianax* (Psephenidae), some Eucnemidae, a few Phengodidae, some Bostrichidae, a few Anobiidae, rentoniine Trogossitidae, smicripine and cybocephaline Nitidulidae, passandrine Cucujidae, *Sosylus* (Bothrideridae), some Corylophidae, noviine Coccinellidae, Discolomidae, Mordellidae, first instar Rhipiphoridae, some Chrysomelidae (Sagrinae, Cassidinae, most Hispinae and some Alticinae), some Bruchidae, and most Curculionoidea. Single-segmented palps occur in first instar Rhipiceridae, intermediate and late instar Rhipiphoridae, a few hispine Chrysomelidae, some Bruchidae and some Curculionidae, while maxillary palps may be absent in coarctate Meloidae and some Eucnemidae. The apical segment of the maxillary palp has a group of sensilla at the apex (the "bouquet sensoriel" of Corbière-Tichané, 1973), and at its base there is often one or more **digitiform sensilla,** which are flattened and located in a shallow cavity, where they are appressed to the surface and difficult to see; these structures, which occur on both maxillary and labial palps in larvae and adults, apparently function as tactile mechanoreceptors (Honomichl, 1980; Zacharuk, 1962a; Zacharuk *et al.,* 1977).

It is likely that the maxilla of ancestral beetle larvae had two well-developed **apical lobes,** both relatively narrow and as long as or longer than the stipes: a basally articulated, 2-segmented **galea,** perhaps with a setose apex, and a more or less fixed, 1-segmented **lacinia** appearing as an inner apical extension of the stipes and bearing at least one tooth or spine at the apex. This combination of characters occurs in no known beetle larva, but is present in some modern Plecoptera, where both lobes are basally articulated and 2-segmented. A 2-segmented galea occurs in Carabidae, Gyrinidae, trogine Scarabaeidae, some Byrrhidae, Eulichadidae, some Ptilodactylidae, Artematopidae, various Elateridae, and a few Cantharoidea, but in most of these groups both apical lobes are small and palpiform, while in the Troginae both are falcate with apical teeth. The major evolutionary trends exhibited by the maxillary apex include: 1) the loss of the basal segment of the galea and its basal articulation, resulting in two fixed lobes (Kfigs. 109, 111 p241); 2) the reduction of both lobes, which become more or less palpiform (Kfig. 100 p203); 3) the loss of one of the lobes, usually the lacinia, resulting in a single process, often referred to as the **mala** (Kfig. 103 p205); 4) the basal fusion and then complete fusion of the galea and lacinia, resulting in a combined structure also called the **mala** (Kfigs. 105–106 p200); and 5) the loss of both apical lobes, so that only the palp arises from the apex of the stipes (Kfigs. 101–102 p194). A basally articulated galea and fixed lacinia occur in Archostemata, Micropeplidae, Dascillidae, the more primitive groups of Scarabaeoidea, Byrrhidae, most Dryopoidea and Elateroidea, some Cantharoidea, the bostrichid genus *Endecatomus,* and the Belidae, while an articulated mala or galea, without a lacinia, is characteristic of some Carabidae, Dytiscidae, Noteridae, Amphizoidae, paederine Staphylinidae, Buprestidae, Callirhipidae, some Elateridae, and most Cantharoidea. The complete loss of apical lobes occurs in Dytiscidae, Hygrobiidae, Hydrophiloidea, Eucnemidae, Rhipiphoridae, and Meloidae. Many workers consider the small, palpiform process on the first segment of the maxillary palp in Hydrophiloidea (Kfig. 102 p194) to be a galea, but this interpretation is not followed here.

A maxilla having both galea and lacinia fixed occurs in Haliplidae, the more primitive Staphylinoidea, Eucinetoidea, Rhipiceridae, some Scarabaeidae, Psephenidae, Derodontoidea, Bostrichoidea, and a few Cucujoidea, while a single, fixed mala is known in Rhysodidae, Dytiscidae, Haliplidae, most Staphylinoidea, Clambidae, Scarabaeidae (advanced subfamilies), some Buprestidae, Throscidae, some Cantharidae, Lymexylidae, most Cucujoidea and Tenebrionoidea, Chrysomeloidea, and Curculionoidea. Intermediate conditions occur where the apex of the mala is deeply cleft, and these have usually been coded for both character states in constructing the key.

The galea is usually broader, more rounded or truncate, and more setose than the lacinia, except in those groups where both lobes have become reduced and palpiform, but the structure varies considerably, and it may be falcate like the lacinia in Dascillidae (Kfig. 112 p208) and Scarabaeoidea. In primitive Staphylinoidea, the galea is said to be **fimbriate,** with a dense row of modified setae along the outer edge of the apex (Kfig. 111 p241), while in Helodidae, it is very broad and its apex is covered with specialized comb-hairs (fig. 34.49). The lacinia is often narrower and more falcate that the galea, and may have one or more teeth or spines at the apex; these apical teeth are usually called **unci,** but the term **uncus** has been used for a variety of lobes or teeth at the inner edge of the maxillary apex. A rounded laciniar lobe occurs in some Anobiidae, and a simliar structure in the Ciidae (Kfig. 110 p239) may represent the lacinia or a secondary structure which has developed on the mala. In most Bostrichoidea, the lacinia has become reduced to a simple, spine-like process at the inner edge of the galea (Kfigs. 97a, 215 p218).

The form of the fixed mala also varies considerably, but it is usually rounded or truncate. A falciform mala (Kfig. 105 p241) occurs in some Staphylinidae, spercheine Hydrophilidae, Throscidae, Rhizophagidae, Phloeostichidae, cucujine and silvanine Cucujidae, Cryptophagidae, Languriidae, Biphyllidae, Byturidae, some Bothrideridae, a few Endomychidae and Corylophidae, and donaciine and zeugophorine Chrysomelidae. In *Donacia,* there appear to be two maxillary lobes, but one of them may be a modified seta. A styliform mala (Kfig. 104 p241) occurs in the leiodid genus *Myrmecholeva,* proteinine Staphylinidae, Cerophytidae, the eucnemid genus *Phyllocerus,* and the cerylonine Cerylonidae. In *Myrmecholeva* and most cerylonines, these stylets and the styliform mandibles fit together with the labrum to form a piercing beak or proboscis (Kfig. 27 p224), while in *Cerophytum* they fit into lateral channels in the labial plate. In many groups with a truncate mala, there is a cleft at the apex (Kfig. 106 p249), and within this cleft there may be a **malar sclerome;** it is not certain whether the cleft represents an incomplete fusion of the galea and lacinia or a secondary subdivision of the mala. A cleft mala is most common in the Tenebrionoidea, but it also occurs in Erotylidae, Helotidae, Lymexylidae, and some staphylinoids. The malar apex sometimes has a distinct lobe at its inner angle; this structure, which occurs in Nitidulidae, Trictenotomidae, anaspidine Scraptiidae and a few other groups, is usually called an **uncus**, but the term is also used for laciniar teeth and for the one or more malar teeth on the inner apical angle in many cucujoids and tenebrionoids. Other structures occurring at the malar apex include **spatulate setae,** as in the spore-feeding nitidulids of the genus *Aphenolia* (fig. 34.44) and the **pedunculate seta** (an elongate tubercle with one or more setae at the apex) found in all Cleroidea, except Phloiophilidae and primitive Trogossitidae (Crowson, 1964d). The inner edge of the mala often bears one or more rows of spines which may continue along the edge of the stipes, and its dorsal surface may be armed with spines, setae, or a dense brush of hairs forming part of the oral filter.

The **labium** consists of two major parts: the apical **prementum** and the basal **postmentum** (fig. 34.14), but the latter is often subdivided into a basal **submentum** and a **mentum,** which lies between it and the prementum (fig. 34.11). The submentum is sometimes fused with the gula, so that a combined area called the **gulamentum** (fig. 34.13) is formed, but the limits of the submentum and gula may be arbitrarily defined by an imaginary line extending between the bases of the cardines. These terms have been used in the keys and descriptions, but they do not necessarily reflect true homologies, and various alternative terminologies have been proposed (*see* Anderson, 1936; Crowson, 1981; Das, 1937; Dorsey, 1943). When there are only two labial sclerites, for instance, the basal one may be called the mentum (Crowson, 1981), and when there are three, the first two have be considered subdivisions of the prementum, based on muscle insertions (Anderson, 1936). The presence of only two labial sclerites occurs throughout the order, but it is most commonly met with among the Adephaga, Dryopoidea, Elateroidea, Cantharoidea, Cleroidea, Chrysomeloidea, and Curculionoidea. In various groups (some Adephaga, Hydrophiloidea, Rhipiphoridae, some Curculionidae, etc.), there may be further reduction, so that there is a single labial plate.

The prementum represents the fusion of paired labial sclerites which are serially homologous to the maxillary stipites, and bears a pair of **palpigers** (to which the labial palps are attached), rarely a pair of **glossae,** which are homologous to the maxillary laciniae, and more often a median structure, the **ligula,** which represents fused glossae. **Paraglossae,** found in many other insect orders, may not occur in beetle larvae, although there are paired lobes on either side of the ligula in Staphylinoidea, which have been called paraglossae by Böving and Craighead (1931). In some larvae (Agyrtidae and other primitive staphylinoids, some Eucinetidae, Dascillidae), the ligula is broad and distinctly bilobed, but more often it is undivided, and in various families it is reduced or absent. In Cupedidae, Micromalthidae, and Callirhipidae, the ligula has become heavily sclerotized and wedge-like, forming a **ligular sclerome** (Kfigs. 99 p216, 117 p189).

The usual number of labial palp segments is two, but reduction often occurs, and in a few cases, there has been an increase to 3 (some Carabidae and Dytiscidae, Gyrinidae, some staphylinine Staphylinidae, some Histeridae) or 4 (some Dytiscidae). Single-segmented labial palps occur in the families Cupedidae (*Priacma* first instar), Micromalthidae, Rhysodidae, Hydrophilidae (*Spercheus*), Clambidae (most), Rhipiceridae, Scarabaeidae (Ceratocanthinae), Buprestidae (most, but *see* below), Eucnemidae (many), Bostrichidae (Dinoderinae and Lyctinae), Trogossitidae (Rentoniinae), Nitidulidae, Hobartiidae, Cryptophagidae (Cryptophaginae), Rhizophagidae (Monotominae), Endomychidae (*Lycoperdina* and Merophysiinae), Coccinellidae (*Hyperaspis*), Corylophidae (*Corylophodes*), Lathridiidae (many), Mycetophagidae (*Thrimolus*), Colydiidae (*Nematidium, Pseudendestes*), Meloidae (Tetraonycini triungulins), Mycteridae

(Lacconotinae and Hemipeplinae), Tenebrionidae (Leiochrini), Pedilidae (Cononotinae), Rhipiphoridae (Rhipidiinae), Chrysomelidae (various subfamilies), Anthribidae (some), Brentidae (few), and Curculionidae (some). In Buprestidae, the palps are represented by minute papillae or setose areas, which may not be true palps at all, and labial palps are absent in *Cerophytum*, some Eucnemidae, the coarctate instars of Meloidae, rhipiphorine Rhipiphoridae, Bruchidae, Urodontidae, and the anthribid genus *Euxenus*.

The mentum or postmentum is very broad in some groups, like Helodidae, Ptilodactylidae (fig. 34.34), and Psephenidae, where it may be expanded laterally to conceal the maxillary articulating area. In anchytarsine Ptilodactylidae, it is divided into three parts by longitudinal sutures (Kfig. 98 p230). In some other groups, like trogossitine Trogossitidae and Elateridae, it may be much narrower than the stipes, and in the latter group, it may be narrowed posteriorly and separated from the gular area by contiguous or fused cardines (Kfig. 100 p233).

The ental surface of the labium, posterior to the ligula, is called the **hypopharynx** (fig. 34.12). In some larvae it is a simple, membranous lobe, usually clothed with short hairs, but in others it may have a more complex structure. The hypopharynx is often supported by the **hypopharyngeal bracon** (*see* above) and sometimes by a pair of **hypopharyngeal rods,** extending ventrally on each side to the base of the mentum, and a **hypopharyngeal suspensorium,** consisting of one or more pairs of rods extending dorsally on either side of the cibarium and inserting in its dorsal wall. The detailed structure of the suspensorium and its variation within the order has never been investigated. The hypopharynx is simple in larvae which have the mandibular molae reduced or absent, but in those with a well-developed mola, there is usually a sclerotized bar or a cup-like or molar-like structure called the **hypopharyngeal sclerome** (figs. 34.12, 34.46, Kfig. 119 p189), which acts in conjunction with the mandibles in the processing of food. The development of this sclerome and its form may vary considerably within a family, and the structure has been used as a taxonomic character in Tenebrionidae. The sclerome is particularly well-developed in the Tenebrionoidea but is also widely distributed in the Cucujoidea and occurs in various other groups with mandibular molae. In some families, like Elateridae, there may be a bar-like sclerotization which is not associated with a mola, but its function is not clear. In Dascillidae and Scarabaeoidea, there may be two or more scleromes called **oncyli** (sing. **oncylus**), while in Dascillidae and Helodidae, the hypopharynx may also have two or more **hypopharyngeal combs** (fig. 34.48). According to Beier (1952), these combs in Helodidae are associated with the filter-feeding apparatus; fine particles are gathered by the maxillary comb-hairs (fig. 34.49), which are then brushed over the combs, where food particles are collected, moved posteromesally towards the back of the hypopharynx and compacted into a solid pellet by the hypopharyngeal sclerome and mandibular molae. The compacting function of the hypopharyngeal sclerome is also seen in the mycophagous tenebrionid *Platydema ellipticum* (Fabricius) (figs. 34.45–46), where the larva shears masses of reinforced hyphae from the surface of a bracket fungus and compacts them between the concave molae and the concave sclerome, probably with the aid of the setose maxillae. In addition to the sclerome, the hypopharynx may have well-defined brushes of hairs or lateral lobes, which have been called **maxillulae** by Böving and Craighead (1931). In Oedemeridae, there is a distinct columnar structure, the **prehypopharynx,** between the sclerome and the ligula (Rozen, 1960).

As mentioned above, the more primitive labium is more or less free from the maxillae, except basally, where they are joined by the maxillary articulating area. In several groups of beetles, there has been a consolidation of the maxillae and labium by reduction and loss of the articulating area and connation of the cardines and stipites with the postmentum or mentum. The resulting maxillolabial complex can be moved only as a single unit. This process is usually accompanied by the loss of the mandibular mola and the reduction of the apical maxillary lobes. Connation of maxillae and labium (figs. 34.32, Kfigs. 93, 100 p264) occurs mainly in Dryopoidea, Elateroidea, Cantharoidea, and Cleroidea, but it is also known in a few cucujoids and some Chrysomelidae (especially Hispinae and the clytrine-cryptocephaline group). Partial fusion is known in Dryopidae (fig. 34.31), Brachypsectridae (Kfig. 97 p201), and Lycidae, while complete fusion occurs in Chelonariidae and various Eucnemidae. This consolidation is often associated with liquid feeding, and the ental surface of the maxillolabial complex may be densely clothed with short hairs, forming with a similar brush on the epipharynx an **oral filter** (figs. 34.35–37), which serves to keep solid particles out of the gut. There are no labial silk glands in beetles, as there are in Lepidoptera, but maxillary glands, located between the bases of the maxillae and the labium, have been reported by Srivastava (1959) in Tenebrionidae, Cerambycidae and Chrysomelidae.

Thorax and Legs

The thorax consists of three segments: **prothorax (T1), mesothorax (T2),** and **metathorax (T3),** each of which bears a pair of articulated legs, except in specialized, internal-feeding larvae to be discussed below. In general, the prothorax is largest and most highly modified, while the meso- and metathorax are similar in structure. In larvae of most Buprestidae, some Eucnemidae, and various Chrysomeloidea and Curculionoidea, the prothorax is greatly enlarged and the legs are reduced or absent. The **cervical region** or neck is usually membranous, but sometimes there are dorsal or ventral sclerotizations, the latter of which may be confused with anterior plates of the prothoracic sternum (*see* below). In Cebrionidae, there is a large **cervical membrane,** lined with ridges and asperities, which is capable of being everted, displacing the head dorsally and posteriorly (Kfig. 120 p207).

The three thoracic segments and abdominal segments 1 to 7 or 8 have a similar general structure, except for thoracic modifications associated with the head and leg attachments. The **dorsum** of each segment usually consists of a large plate, called the **tergum,** and one or more pairs of smaller, lateral sclerites, the **laterotergites.** Each tergum is usually divided at the midline by a narrow **ecdysial line,** but in some forms there is a broader division, with the formation of paired, smaller

tergites. Paired tergites occur on all thoracic and most abdominal segments in Sphindidae and on the meso- and metathorax in some Cleridae and Trogossitidae. In some lightly sclerotized larvae (Rhysodidae, Gyrinidae, Histeridae, trogine Scarabaeidae, some Elateridae, many Cleroidea, various Chrysomeloidea and Curculionoidea) only the prothorax bears a tergal plate. Spiracles are usually located between the tergum and laterotergites on each side, but sometimes they are on a laterotergite or on the tergum. The **pleuron** is somewhat reduced, except on the thorax, where it consists of sclerites partly surrounding the leg articulations, while the **venter** usually consists of a single plate, except on the thorax, where complex and poorly understood subdivisions may occur (*see* below).

The structure of the **protergum** is usually simple, consisting of no more than a sclerotized plate, but in some Lymexylidae, Buprestidae, Oedemeridae, and Cerambycidae, there are patches of asperities, in belid larvae there may be a prominent protergal carina, in Platypodidae there is a transverse row of ring-like sclerites, and in some Eucnemidae and Bostrichidae, there are rod-like sclerotizations. Transverse carinae or rows or patches of asperities occur more commonly on the meso- and metatergum, and usually on the anterior abdominal terga as well. Examples of larvae with patches of asperities on the meta- or meso- and metatergum may be found in the families Scarabaeidae, Lucanidae, Anobiidae, Colydiidae (Pycnomerini), Synchroidae, Zopheridae, and Cephaloidae (Stenotrachelinae); while rows of asperities on these segments are known in Rhysodidae, Dascillidae (forming serrulate carinae), Scarabaeidae (Troginae), Colydiidae (*Lasconotus*), Zopheridae, and Cephaloidae (Stenotrachelinae). In some soft-bodied larvae, there may be a folding of the tergal region, so that 2 to 4 **plicae** or folds can be seen from above, while other subdivisions are visible laterally. When the tergum is divided transversely into 2 parts, as on the thorax, the anterior division is sometimes called the **prescutum** and the posterior one the **postscutum;** when 3 folds are present, they may be referred to as the prescutum, **scutum,** and **scutellum.** The thoracic spiracle is usually located in a laterotergite, sometimes called a **spiracular sclerite** (fig. 34.18) or **alar area,** lying between the prothorax and mesothorax; but in some groups the spiracle is on the mesotergum or protergum.

The thoracic pleuron (fig. 34.18) is divided by a short **pleural suture,** extending dorsally or laterally from the coxal articulation and representing an internal apodeme called the **pleural apophysis.** This suture divides the pleuron into an anterior **episternum** (= **prehypopleurum**) and a posterior **epimeron** (= **posthypopleurum**); sometimes there are two additional pleural sclerites, the **precoxale** in front of the episternum, and the **postcoxale** behind the epimeron, but these are often fused with the episternum and epimeron or with adjacent sternal elements. In some Tenebrionidae, the postcoxale on each side is enlarged and extends mesally to fuse with the **sternellum,** forming a crescent-shaped sclerite behind the coxae; this may be called the **postcoxal bridge,** although that term has also been used as a synonym of postcoxale.

The ventral region of the thorax (fig. 34.18) is more lightly sclerotized and its subdivisions usually are neither clearly defined nor clearly separated from the pleural sclerites. In some larvae, there are 2 **sternal pits** lying in the middle of the sternum between the coxal bases. These represent the invaginations of the **sternal apophyses,** and they may be joined by a transverse suture, the **sternacostal suture,** which divides the sternum into an anterior **basisternum** and a posterior **sternellum;** in the absence of a suture or pits, these terms are still used for the sternal areas in front of and behind the coxae. The basisternum may be further subdivided, but the nomenclature concerning these anterior and anterolateral sclerites is confused, and different terms are used by workers dealing with different taxa. There is often a large, sclerotized area on the prothorax, which is set off from the basisternum and extends to the anterior edge; this has been called the **presternum** or **prosternum,** and it may be variously subdivided or joined laterally to the precoxalia. In addition, there may be a smaller, mesal sclerite, which may be folded beneath the head and is thought to be part of the cervical region; this has been called **cervicosternum** by Watt (1970), but the same structure was considered by St. George (1939) to be the presternum. Other terms applied to these sclerites, such as **eusternum, preeusternum,** and **articulating area,** are in the glossary and are defined, discussed, or illustrated by Böving and Craighead (1931), Hyslop and Böving (1935), Glen (1950), St. George (1924, 1939), Wade and St. George (1923), and Watt (1970, 1974b). Sometimes there is a small sclerite lying between the sternellum and the following segment; this is called an **intersternite** by Watt (1970), but was known to earlier workers as a **poststernellum.** Various special structures of the thoracic sterna include patches of asperities on the prothorax in Cupedidae and most Buprestidae, and paired, sclerotized prosternal rods in Throscidae, Cerophytidae, and Eucnemidae.

The basic number of leg segments in larvae of Adephaga and Archostemata (figs. 34.19–20) is 6, including, from base to apex, **coxa, trochanter, femur, tibia, tarsus,** and **pretarsus,** the last consisting of paired claws or a single claw (**ungulus**). In Myxophaga and Polyphaga (fig.34.21), the basic number is 5, including coxa, trochanter, femur, tibia, and **tarsungulus** (also referred to as the **claw**). Peterson (1951) and some other authors exclude the claw in counting leg segments, and thus consider the adephagan leg to be 5-segmented and the polyphagan leg 4-segmented. The homologies of the last 2 or 3 segments in the leg have been the subject of much controversy. Jeannel (1949) considered the apical segment in both leg types to be a tarsus; the penultimate segment was called the tibia in both groups, and the adephagan leg was said to have an extra segment, the **medius,** located between the femur and tibia. Emden (1942b) thought that the apical segment in both groups was a part of the **pretarsus** called an **ungulus** or claw, and that the penultimate segment in the polyphagan leg was a **tibiotarsus,** representing the fusion of the tibia and tarsus. Böving and Craighead (1931) and Crowson (1955, 1964d) considered the apical segment in Polyphaga to represent a fusion of the tarsus and ungulus (thus tarsungulus), and this interpretation is followed here.

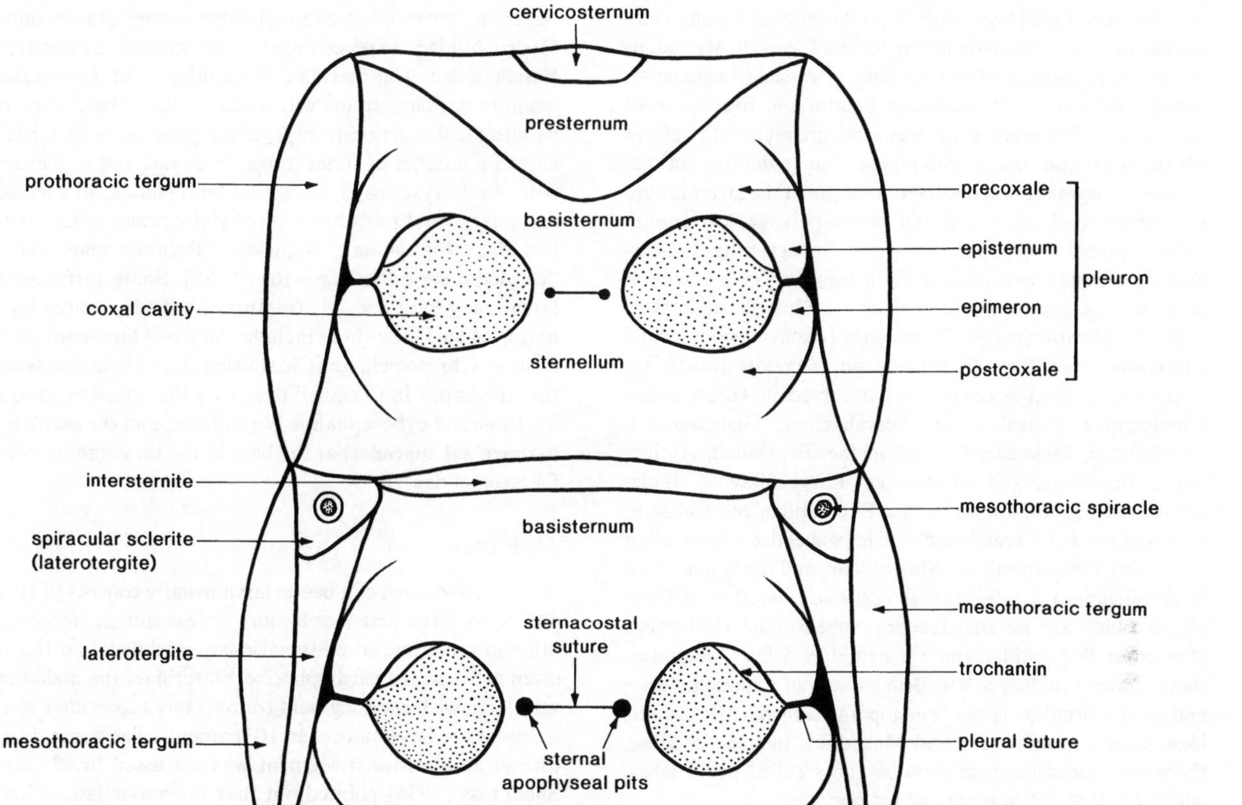

Figure 34.18. Prothorax and mesothorax (composite), ventral, with legs removed.

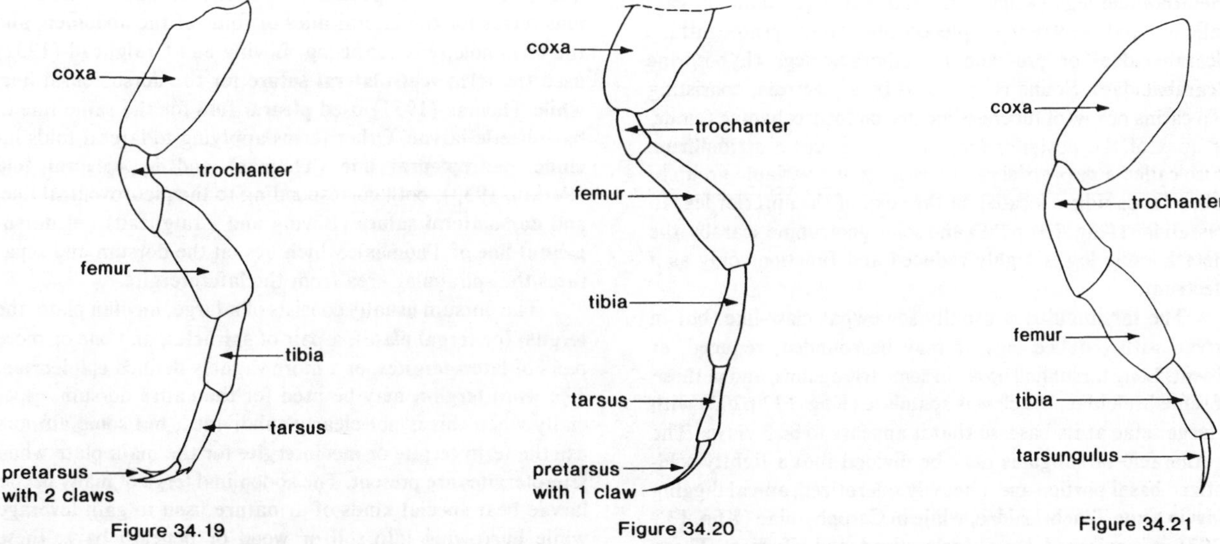

Figure 34.19. Mesothoracic leg (Carabidae).
Figure 34.20. Mesothoracic leg (Dytiscidae).
Figure 34.21. Mesothoracic leg (composite polyphagan).

Six-segmented legs occur in all Archostemata and Adephaga, with the exception of the legless forms of Micromalthidae, the myrmecophilous Carabidae: Paussini (3 segments), and the ectoparasitic Carabidae: Lebiini and Brachinini (4 segments). Five-segmented legs are characteristic of all Myxophaga and many Polyphaga, but reduction in the number of segments has occurred throughout the latter group. Four-segmented legs are found in the polyphagan families Micropeplidae, Hydrophilidae (most Sphaeridiinae), Scarabaeidae (some Geotrupinae), Callirhipidae (few), Eucnemidae (few), Cucujidae (Passandrinae), Rhipiphoridae (late instars), Cerambycidae, Bruchidae (Pachymerinae), and Chrysomelidae (Hispinae, Donaciinae, Megalopodinae). Reduction to 2 or 3 segments has occurred in Georyssidae, Rhipiceridae, Scarabaeidae (Scarabaeinae, Geotrupinae), Mordellidae, Meloidae, Cerambycidae, Bruchidae, Anthribidae, Brentidae, and Ithyceridae. Larvae having single-segmented legs are known in rare Buprestidae and Eucnemidae, and in a few Cerambycidae, Chrysomelidae (leaf-mining Hispinae), Nemonychidae, Anthribidae, and the genus *Cylas* (Curculionidae). Legless larvae occur in a number of Polyphaga which are internal feeders (sphaeridiine Hydrophilidae, most Buprestidae and Eucnemidae, a few Anobiidae, some Cerambycidae, a few Bruchidae and Chrysomelidae, and most Curculionoidea) or ectoparasites (most Bothrideridae, some Rhipiphoridae and Meloidae). In some of these, there are paired thoracic protuberances called **pedal lobes,** which are thought to represent leg remnants.

In many substrate-inhabiting larvae, the prothoracic legs are enlarged and variously modified for digging; examples may be found among the Tenebrionidae (Alleculinae, Tentyriinae, Helaeini, Blaptini) or in the Cebrionidae and Cerophytidae. Prehensile prothoracic legs are formed by the opposition of the femur (Noteridae) or the tibia (Haliplidae, Dytiscidae: Matini) to the apical leg segments. In several groups of Scarabaeoidea, **stridulatory organs** are formed by the meso- and metathoracic legs (Kfigs. 129, 130 p267) (Lucanidae, Passalidae, and geotrupine, pleocomine, and ceratocanthine Scarabaeidae) or pro- and mesothoracic legs (hybosorine Scarabaeidae). Sound is produced by a **plectrum,** consisting of a carina or row of tubercles located on the trochanter, femur, or apex of the posterior leg, rasping against a **stridulitrum** (also called a **pars stridens**), consisting of a variable group of tubercles or ridges located on the coxa of the anterior leg. In Passalidae (Kfig. 131 p258) and some geotrupine scarabs, the metathoracic leg is highly reduced and functions only as a plectrum.

The tarsungulus is usually somewhat claw-like, but in larvae with reduced legs, it may be rounded, reduced, or absent. Long tarsunguli occur in some triungulins, and in those of some Meloidae, the claw is **spatulate** (Kfig. 132 p203) with 2 large setae at its base, so that it appears to be 3 claws. The prothoracic tarsungulus may be divided into a lightly sclerotized basal portion and a heavily sclerotized, apical digging claw in some Tenebrionidae, while in Cerophytidae (Kfig. 133 p192), it is enlarged, heavily sclerotized, and bifurcate. There are usually 2 setae at the base of the tarsungulus, and these are usually placed side by side, but may be in a line, with one distal to the other in certain groups. Plurisetose tarsunguli occur in the families Staphylinidae (some Staphylininae), Hydrophilidae (Spercheinae), Dascillidae, Scarabaeidae, Eulichadidae, Cantharidae, Anobiidae, and Lymexylidae, while in various groups with reduced legs, these setae may be absent. Reduction from 2 tarsungular setae to 1 has occurred a number of times in beetle larvae, and is characteristic of all Myxophaga, almost all Dryopoidea, all Cleroidea, and a large and probably monophyletic group of Cucujoidea including Biphyllidae, Byturidae, Bothrideridae, and the "cerylonid series" of Crowson (1955). Some surface-active larvae may have special structures at the leg apices for adhering to surfaces; these include the basal tarsungular tooth in some Chrysomelidae (Chlamisinae), the **adhesive setae** at the tibial apex in Coccinellidae or on the tarsungulus in meligethine and cybocephaline Nitidulidae, and the **pulvillus** (or **paronychial appendix**) at the base of the tarsungulus in some Chrysomelidae.

Abdomen

The abdomen of a beetle larva usually consists of 10 segments, with the first 7 or 8 more or less similar in form, the 9th and sometimes the 8th variously modified, and the 10th often reduced. In cardiophorine Elateridae, the abdomen is very long, with each segment transversely subdivided, so that it appears to have more than 10 segments. The general structure of an abdominal segment was discussed briefly above. Snodgrass (1935) pointed out that the major lateral line of demarcation on the abdomen is the **dorsopleural line,** which separates the **dorsum** from the **pleuron** and **venter** and continues onto the thorax just above the episternum and epimeron; the abdominal pleuron is that area, just below this line, which corresponds to the plates around the bases of the thoracic legs and which may or may not be separated from the venter by a **pleuroventral line.** In most beetle larvae, there is no abdominal pleuron, so the major lateral line is actually a **tergosternal fold.** Specialists on beetle larvae have used various terms for the lateral lines or folds of the abdomen, and the terminology is confusing. Böving and Craighead (1931) used the term **ventrolateral suture** for the dorsopleural line, while Thomas (1957) used **pleural fold** for the same line in bark beetle larvae. Other terms applying to lateral folds include **ventropleural line** (Thomas) and **hypopleural fold** (Parkin, 1933), both corresponding to the pleuroventral line, and **dorsolateral suture** (Böving and Craighead) (= **dorsopleural line** of Thomas), which lies on the dorsum and separates the spiracular area from the laterotergite.

The dorsum usually consists of a large, median plate, the **tergum** (or **tergal plate**), a pair of **spiracles,** and one or more pairs of **laterotergites,** or a more vaguely defined **epipleurum.** The word tergum may be used for the entire dorsum, especially when this is not clearly subdivided, but some authors use the term **tergite** or **mediotergite** for the main plate when laterotergites are present. The abdominal terga of many beetle larvae bear special kinds of armature used to gain leverage while burrowing into soil or wood or beneath bark; these have proven useful in larval identification. Patches of asperities or spinules (Kfigs. 141–143 pp293,274,295) occur in various groups, including Histeridae, many Scarabaeoidea, lissomine and oestodine Elateridae, most Anobiidae, Lymexylidae, Synchroidae, most Oedemeridae, some Cerambycidae, and rhynchitine Curculionidae. In other larvae, there may be rows of tergal asperities (Kfigs. 144–148 pp251,249,214,296,213), which sometimes form incomplete circles; examples are known in the families Rhysodidae, Cucujidae (Laemophloeinae), Phalacridae (*Litochrus*), Phloeostichidae (Hymaeinae), Monommidae, and Mycteridae. In stenotracheline Cephaloidae, there are both patches and rows of asperities, while in Trictenotomidae there are peculiar patterns formed by longitudinal ridges. Another type of structure serving a similar function is the **ampulla** (Kfig. 149 p262) (also called **ambulacral wart, ambulatory wart, scansorial wart**), a hump-like projection sometimes armed at the apex with asperities; these occur in Cupedidae, Micromalthidae, Histeridae, Trogossitidae (Trogossitinae), Cleridae, Melandryidae (Melandryinae), Mordellidae, Oedemeridae, and some Curculionoidea. In tiger beetle larvae (Carabidae: Cicindelinae) (Kfig. 135 p214) there is a large protuberance bearing 2 or 3 pairs of hooks on the 5th tergum, while in *Sphallomorpha* (Carabidae: Pseudomorphinae) (Kfig. 137 p214), there are rows of stout spines on terga 5–7. In some Eucnemidae (*Fornax*), the abdominal terga (also sterna) have well-defined areas called **matte patches,** which are covered with minute asperities (Gardner, 1935b).

In many soft-bodied larvae, the abdominal terga are transversely divided into 2 to 4 folds or **plicae** (Kfig. 136 p217), which may extend along the sides as well, and the names applied to these vary with the taxon studied (Scarabaeoidea, Ritcher, 1966; Anobiidae, Böving, 1954, Parkin, 1933; Curculionoidea, Anderson, 1952, Thomas, 1957). When there are two plicae, the anterior one may be called **prescutum, prodorsal fold,** or **prenotal fold,** and the posterior one the **scutum, postdorsal fold,** or **postnotal fold.** Three folds in Scarabaeoidea are called **prescutum, scutum,** and **scutellum,** with another fold, the **postscutellum,** behind the scutellum, but visible mainly from the side, and a lateral area, the **subscutum,** wedged between the scutum and scutellum. In Curculionoidea, the folds may be numbered, but the numbering has not been consistent (May, 1967). The area around the spiracle has been called the **alar area, parascutal area, paradorsal area,** or **spiracular sclerite.** Other terms applied to laterotergites or laterotergal areas include: **epipleurite, epipleural area, dorsopleural lobe, pleural lobe, paratergum,** and **paratergite.** Most of the paired lateral processes occurring in beetle larvae are extensions of the laterotergites, and are not pleural in origin. The true abdominal pleuron, which lies immediately below the dorsopleural line, has been called **pleurum, hypopleurum,** and **ventropleural lobe** by various authors. In many groups, it is absent or represented by a narrow membranous region, but distinct pleural sclerites occur in some groups, including Carabidae and various families of Dryopoidea, Elateroidea, and Cantharoidea.

The abdominal venter often consists of a single **sternum** or **sternal plate,** but this may be divided transversely or longitudinally. The mesal sclerite of a longitudinally divided venter is sometimes called the **mediosternite,** and the lateral plates are **laterosternites,** but have also been called **laterosternal folds, pedal areas,** or **coxal lobes.** When the venter is transversely divided, the main plate is the **basisternum** and that posterior to it is the **sternellum,** but as in the thorax, the term **eusternum** has been used inconsistently as synonymous with the basisternum or entire sternum. In Curculionoidea, the terms **mediosternal fold** and **transverse fold** have been used for the two areas, while in Carabidae, there may be a main **ventrite,** with **preventrites** and **postventrites** in front of and behind it. Ambulatory modifications of the sterna are more or less similar to those on the terga, and include patches or rows of asperities, ampullae, plicae, and matte patches. In some larvae, there are distinct **prolegs,** similar to those found in Lepidoptera, but usually without apical hooks or **crochets.** Prolegs with simple or setose apices occur on segments 1 to 7 or 8 in larvae of various Chrysomelidae (Criocerinae, Eumolpinae, Alticinae) and Curculionidae (*Cionus,* Hyperinae), while most Oedemeridae have asperity-bearing prolegs on segments 2 to 3, 4, or 5. Some hydrophilid larvae have prolegs with crochets at the apex on sterna 3–7 (*Enochrus*) or 2–7 (genus near *Coelostoma*). In some larvae the sternal regions are very narrow and may be absent posteriorly; examples of this occur throughout the Dryopoidea.

Abdominal Apex

The last two or three abdominal segments in beetle larvae usually differ from those anterior to them because of special adaptations involving locomotion, defense, or respiration. A type of larval abdomen which may have been ancestral in Coleoptera is characterized by having an unmodified, subterminal 9th segment and a terminal 10th segment, with distinct tergum and sternum bordering a transverse anal opening; this type of apex occurs in some Dascillidae, Scarabaeidae, and Byrrhidae, and with slight modifications in various other families. The major changes which have taken place in the abdominal apex include the following: 1) transformation of segment 10 into a cylindrical or conical **pygopod,** sometimes with hooks, asperities, or eversible holdfast organs at the apex; 2) development of a ventral pygopod (sternum 10) below the anus (Limnichidae and Heteroceridae); 3) posterior production of tergum 10 forming a median process and resulting in the ventral position of the anal region (some Haliplidae); 4) development of **articulated,** terminal or subterminal **urogomphi** on the posterior portion of tergum 9; 5) formation of lobes, pads, setose areas, or sclerites around the anus in soft-bodied, internal feeding larvae; 6) reduction of segment 9 and concealment of its sternum, with retention of a well-developed terminal 10th segment (some Cucujidae); 7) ventral movement of the anal region, so that tergum 9 becomes terminal, extending partly onto the ventral surface, and segment 10 becomes ventrally or posteroventrally oriented; 8) sclerotization of tergum 9 and development of various kinds of armature, including paired, **fixed urogomphi;** 9) transformation of tergum 9 into a hinged sclerotized plate articulating with tergum 8 (often accompanied by the partial enclosure of sternum 9 within an emargination of sternum 8, or sometimes concealment of sternum 9), in flat, subcortical larvae; 10) fusion of tergum and sternum 9, forming a complete sclerotized ring; 11) development of paired **pygopods,** sometimes armed with hooks or asperities, from sternum 10; 12) formation of a dorsally-hinged **operculum** (tergum 9) articulating with tergum 8 and concealing sternum 9 and the anal region; 13) formation of a ventrally-hinged operculum (sternum 10) articulating with the ventral portion of segment 9 and concealing the anal region, in various aquatic and riparian larvae; 14) complete concealment of segment 10 between the edges of tergum and sternum 9; 15) reduction of segments 9 and 10 and movement of tergum 8 into a terminal position, with accompanying modifications; 16) formation of dorsal, posteriorly oriented spiracular process on tergum 8; 17) formation of a terminal **respiratory chamber** from portions of segments 8 and 9; 18) development of gill tufts or osmoregulatory papillae in the anal region. The above transformations represent a series of complex and interrelated evolutionary events. Most of them will be treated further in discussions on variation in each of the segments involved, while the last three will be covered in the section on special respiratory adaptations. Segment 8 in most larvae is not differentiated from the anterior abdominal segments, although it may have sclerotized areas or carinae associated with locomotory adaptations of the segments posterior to it. In certain cases, however, tergum 8 may become terminal, while the last two segments are reduced. Larvae of cassidine Chrysomelidae, for instance, have reduced 9th and 10th segments, and tergum 8 bears a large forked process (Kfig. 155 p221), which accumulates exuviae and other debris, thus concealing the surface-active larvae from predators. Another feature of the 8th tergum is the median defense gland in aleocharine Staphylinidae; these glands are particularly prominent in the Division Bolitocharinea (Seevers, 1978), but they occur in other groups as well (Frank and Thomas, 1984). In the adephagan families Hygrobiidae, Amphizoidae, Noteridae, and Dytiscidae, the 8th segment is more or less terminal and lies above the reduced 9th and 10th segments, the former of which bears a pair of articulated urogomphi (*see* below). In the last three groups, tergum 8 bears a pair of large spiracles at the posterior end (often at the apex of a median process), but in Hygrobiidae, which lacks abdominal spiracles, it forms a long narrow process, similar in structure to the urogomphi. A terminal, spiracle-bearing process on tergum 8 has also evolved in the family Nosodendridae, where segments 9 and 10 are reduced and form a pair of ventrally located anal pads. Other modifications associated with the 8th pair of spiracles are discussed in the next section.

The simplest type of modification of tergum 9 involves its sclerotization, which is often greater than that of preceding abdominal terga; in some Cleridae, Corylophidae, and Chrysomelidae, there may be a distinct tergal plate similar to that on the prothorax; this plate is sometimes referred to as the **pygidium,** especially when it is clearly set off from the rest of tergum 9 (*see* below). The most obvious and consistent feature of the 9th tergum in many larvae is the presence of paired **urogomphi** (sing. **urogomphus**), which may be articulated or fixed at the base and may consist of from 1 to several segments. The term urogomphi has been used for almost any paired prominences on tergum 9, with the exception of setae or articulated spines, and those structures so designated are not necessarily homologous. The term **cerci** (sing. **cercus**) has also been used but is more properly restricted to the appendages of segment 10 in primitive insects. Other terms which have been applied to the urogomphi are **pseudocerci** and **corniculi.** In most active campodeiform larvae, which inhabit surfaces or interstitial spaces (leaf litter, etc.), the urogomphi are relatively long and narrow, project posteriorly or posterodorsally, have relatively little curvature, and bear a number of setae. These are basally articulated and/or segmented (Kfigs. 154–155 pp213,221, 162–163 p195) in some Carabidae (supertribes Nebriitae, Callistitae, Odacanthitae, Lebiitae, etc.), and most Hydradephaga, Staphylinoidea, and Hydrophiloidea, and they appear to be tactile in function. In those larvae which inhabit narrower spaces beneath bark or bore into substrates like rotten wood or fungi, the urogomphi are more solidly built, never articulated or segmented, and often recurved, so that the apex points anteriorly (Kfig. 171 p196); these serve to fix the larva within its burrow and assist in locomotion.

Although most types of urogomphi are simple or at most tuberculate, some may be bifurcate or have a number of accessory branches or spines. Bifurcate urogomphi (Kfigs. 169 p201, 175, 177 p293, 183–184 pp275,249) occur in a number of families, including Elateridae (Athoinae), Melyridae, Nitidulidae, Tetratomidae, Pedilidae, and Salpingidae, while more complex forms (Kfigs. 179, 181 p275) characterize some Trogossitidae (*Calitys, Thymalus*), Nitidulidae (*Epuraea, Cryptarcha*), Languriidae (Xenoscelinae, Cryptophilinae), Cerylonidae (Euxestinae), Pythidae, Othniidae, and Inopeplidae. Sometimes urogomphi may be accompanied by tubercles or teeth which extend onto the surface of the tergum (Kfig. 174 p295) or by smaller, paired processes called **pregomphi** (Kfig. 168 p255). Urogomphi may be approximate or widely separated, and in some cases there may be a median process (Kfig. 172 p215), a pit or cavity (Kfigs. 173–174 pp279,295), or 2 such pits (Kfig. 175 p293) between them.

In some larval types, tergum 9 bears a single median process, which may be spine-like and posteriorly projecting (Kfig. 159 p261), as in Archostemata, Elateridae (Elaterinae), Erotylidae (*Microsternus*), Colydiidae (*Cicones*), and Mordellidae, rounded and posteriorly projecting, as in Melandryidae (Osphyinae) and some Chrysomelidae (Alticinae), or hooked like a typical urogomphus, as in some Dryopidae and Ciidae. A forked median process occurs in some Cucujidae (*Pediacus, Platisus*). In scraptiine Scraptiidae, tergum 9 bears a large, rounded, lightly sclerotized, setose process (Kfig. 158 p261), which is deciduous, so that these larvae are often recognizable, not by the presence of the process, but by the truncate tergal apex from which the process has been removed.

Another distinctive type of 9th tergum is that bearing a concave disc or pygidium, often surrounded by teeth or serrations (Kfig. 160 p261). This feature is known in some Byrrhidae (Syncalyptinae), Lymexylidae, Tenebrionidae (Amarygmini), and Ciidae, and appears to be an adaptation for blocking the larval burrow to prevent the entrance of predators (Lawrence, 1974b). A more complex pygidium, lined with various kinds of teeth or spines, occurs in larvae of strongyliine Tenebrionidae and the pythid genus *Priognathus.* In paussine Carabidae, the 9th tergum forms a complex, multilobed disc of unknown function. In various flattened larvae living under bark (cucujine and laemophloeine Cucujidae; Phalacridae (*Litochrus*), Prostomidae, Mycteridae, Pyrochroidae, techmessine Pythidae), the entire 9th tergum forms a hinged plate, sometimes called a **urogomphal plate** (Young, 1975) (Kfigs. 164–165 pp255,247, 176, 178 p275, 185 p277), articulated to segment 8 and apparently used to wedge the larva between layers of bark or wood. In the Brachypsectridae, the tergum forms an articulated spine, which is capable of extending anteriorly over the back and assists in prey capture (Crowson, 1973b). Another type of tergal plate in Callirhipidae forms a dorsally-hinged operculum, which encloses the anal region (Kfig. 186 p219).

The 9th sternum is usually relatively simple, but in various Tenebrionoidea, there are teeth or asperities variously distributed along the base (anterior edge) or occasionally at the apex. In the Hallomeninae (Melandryidae), Synchroidae, Salpingidae, *Aglenus* (Othniidae), and *Pedilus* (Pedilidae), there is a single tooth at each basal (anterolateral) angle (Kfigs. 183–184 pp275,249), while in the genus *Calopus* (Oedemeridae) and various members of the families Pythidae, Trictenotomidae, Pedilidae, Pyrochroidae, Othniidae, and Inopeplidae, there may be from 2 to several asperities on each side of the anterior edge, sometimes forming a complete basal row (Kfigs. 178–182 p275). Two posteromesal asperities occur in the known larva of Eurygeniinae (Pedilidae) (Kfig. 177 p275), while in *Sphindocis* (Ciidae) and Prostomidae (Kfig. 176 p275), there is an apical (posterior) row of asperities. The 9th sternum is often partly enclosed within an emargination of sternum 8 in those larvae with a hinged tergal plate, but an extreme condition is known in Boridae and Mycteridae (Kfig. 185 p277), where the 9th sternum is deeply set into the 8th, and is U-shaped, almost completely enclosing segment 10 and bearing 2 or more teeth at the posterior edge; a similar condition occurs in some Elateridae (Pleonomini, Semiotini). In some larval types, sternum 9 is reduced, and it may be concealed by sternum 8, as in some Phalacridae (Kfig. 164 p255) or silvanine Cucujidae (Kfig. 167 p271). As mentioned above, the tergum is usually separated from the pleuron or combined pleuron and sternum on each side by a distinct dorsopleural line or fold; in some groups, including many Dryopoidea, however, tergum 9 is completely fused to the pleurosternal region, so that a continuous sclerotized ring is formed.

Modifications of the 10th segment are particularly striking in those larvae using this segment as an ambulatory device (pygopod or paired pygopods) and in those soft-bodied larvae in which the abdominal apex assists locomotion within substrates. In some active larvae with a terminal pygopod, such as those of various Carabidae, Staphylinoidea, and Lampyridae, there is an anal **holdfast organ** composed of a number of eversible, asperated tubes (Kfig. 191 p205) (Brass, 1914; Kemner, 1918). Paired lobes or **pygopods** (fig. 34.22), one on either side of the anal opening, occur in a number of groups. They may be simple (Archostemata, first instar brachinine Carabidae, cardiophorine Elateridae and some Tenebrionidae) or armed with 1 (Byrrhidae) or several (Ptilodactylidae, Kfig. 189 p229; Lymexylidae; Tenebrionidae) asperities, teeth, or hooks. Paired hooks may also occur in larvae without paired pygopods. In Hydroscaphidae (Kfig. 161 p209), Ptiliidae, Hydraenidae (Kfig. 163 p195), some Clambidae, Limnichidae, and some Elateridae, there is a single hook on each side of the single pygopod; gyrinid larvae have 2 pairs of long hooks (Kfig. 157 p213); microsporids have 3 hooks on each side; and larvae of the elaterid genus *Semiotus* have a ring of teeth surrounding segment 10.

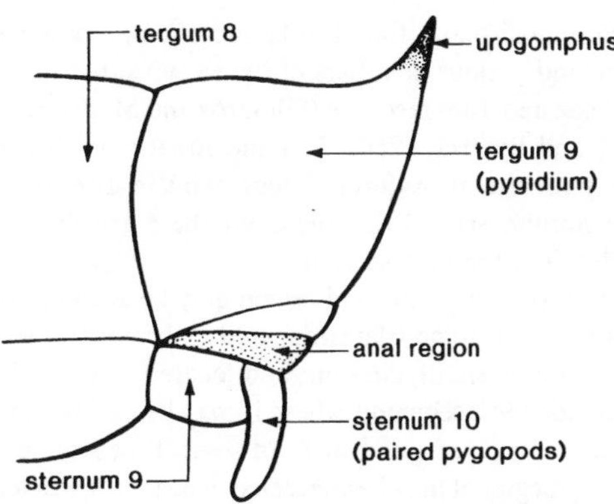

Figure 34.22. Abdominal segments 9 and 10 (Tenebrionidae), lateral.

The anal region may be surrounded by 2 or more membranous lobes, which do not project as pygopods, and in various soft-bodied larvae there are a variety of supporting structures in the vicinity of the anus. In Bostrichidae, Anobiidae, and Ptinidae there is usually a longitudinal groove lying below the anal opening and separating a pair of oval pads or cushions; this structure has been called the **nates** by Böving (1954), who used the term **bow** for the curved sclerite below this. In most Scarabaeoidea, the anal opening may be transverse, Y-shaped, or longitudinal, and is surrounded by a dorsal **anal lobe** and 1 or 2 ventral anal lobes, or a pair of lateral anal lobes; the terms used for these lobes vary even among scarab workers (Ritcher, 1966). In most Lucanidae, the dorsal lobe is usually absent (or highly reduced), and the lateral lobes each may bear a glabrous area surrounded by a fine sclerotized line; this area is called an **anal pad.** In Scarabaeinae and Geotrupinae, the area below the anus may be subdivided forming several ventral lobes. The sternal area immediately anterior to the anal region in Scarabaeoidea often bears a group of hairs, bristles, or hooked setae, which is called a **raster** and is used extensively in the identification of scarab grubs. The raster may consist of a pair of **palidia,** each of which consists of a row of heavy spines or **pali,** surrounding a bare area called the **septula** and sometimes bordered laterally on each side by a **tegillum,** which is a continuous patch of shorter setae; in some cases, there is a single **palidium,** which is curved with the pali facing towards the anus, or the pallidium may be replaced by a **teges,** which is a continuous patch of short, straight or hooked setae. Other associated features include the **campus,** a bare area between the raster and the anterior edge of sternum 10, and a **barbula,** which is a tuft of hairs or short setae at the side of the abdomen near the anus. In the Australian genus *Cephalodesmius* (Scarabaeinae: Scarabaeini), the 10th sternum bears a coarsely striate, longitudinal ridge, which acts as a **plectrum** by rubbing against a finely striate **stridulitrum** on the gular region (Monteith and Storey, 1981).

Other modifications of the anal region include the paired spine-like processes on either side of the vertical anal slit in agriline Buprestidae (Kfig. 190 p191) and the complex lobes and sclerotized plates found in larvae of eumolpine Chrysomelidae and various members of the weevil subfamilies Brachycerinae and Tanymecinae (Ghilarov and Medvedev, 1964; Arnoldi and Byzova, 1964). In some aquatic or semiaquatic larvae, gill tufts or osmoregulatory papillae also arise from the membrane around the anus, but these are discussed in more detail in the next section.

Concealment of the anal region may be accomplished in several ways. In some Elateridae and Cebrionidae, the 10th segment is very small, circular, and located near the base of the undivided 9th segment, where it may be partly concealed beneath the apex of sternum 8. In some Tenebrionidae, the entire 10th segment may be concealed beneath the 9th tergum, and in Callirhipidae it is concealed by the dorsally-hinged operculum formed by the 9th tergum. In a number of dryopoid larvae (araeopidiine Ptilodactylidae, Chelonariidae, Dryopidae, Lutrochidae, Elmidae, eubriine and psephenoidine Psephenidae), a ventrally-hinged **operculum** (Kfigs. 150, 187, 188 pp237,238) is formed from the 10th sternum, which articulates with the ventral part of the 9th segment. The operculum is operated by 1 or 2 sets of muscles, and it may be equipped with hooks (Dryopidae, Lutrochidae, Elmidae) or not. In Elmidae and Psephenidae, the operculum conceals retractable gills, but in the other groups gills are absent. Both the 10th segment and 9th sternum are reduced and at least partly concealed in larvae of some cucujine and laemophloeine Cucujidae, Phalacridae, Brachypsectridae, Nosodendridae, and various Hydradephaga and Hydrophilidae with apical respiratory structures (*see* below).

Spiracles and Special Respiratory Structures

The normal type of respiratory system in beetle larvae is a **peripneustic** one, with functional spiracles on the mesothorax and abdominal segments 1 to 8, but a few larval types (Eulichadidae, Silphidae) have non-functional spiracular remnants on the metathorax. Hinton (1947) has shown that there is considerable variation in the number and position of functional spiracles, and this is complicated by the fact that the final instars of some aquatic larvae differ in this respect from earlier instars. A **hemipneustic** system, with spiracles on the mesothorax and abdominal segments 1–7, is present in the final instars of Hygrobiidae, Haliplidae, and some aquatic Lampyridae, as well as in the coarctate phase of the parasitic nemognathine Meloidae. An **amphipneustic** respiratory system, with metathoracic and 8th abdominal spiracles only, occurs in all instars of the gill-bearing psephenine and eubrianacine Psephenidae, while a **metapneustic** system, with a single pair of functional spiracles on abdominal segment 8, is known in Helodidae, *Araeopidius* (Ptilodactylidae), eubriine Psephenidae, and early and intermediate instars of Noteridae, Amphizoidae, Dytiscidae, and donaciine Chrysomelidae. An **apneustic** system, with no functional spiracles, is characteristic of earlier instars of Gyrinidae, Hygrobiidae, Haliplidae, berosine Hydrophilidae, Elmidae, Lutrochidae, some Ptilodactylidae and some eubriine Psephenidae, and all instars of *Psephenoides* (Psephenidae). In a few beetle groups, the type of larval respiratory system differs from any of those mentioned; examples are final instar Noteridae and all instars of Microsporidae and Torridincolidae, with 8 abdominal spiracles and none on the thorax, Hydroscaphidae with spiracles on the mesothorax and 1st and 8th abdominal segments, last instar Gyrinidae with spiracles on the first 3 abdominal segments, and triungulins of *Tetraonyx* (Meloidae), with an enlarged pair on abdominal segment 1 and minute pairs on 2–5 (Hinton, 1947, 1967a; MacSwain, 1956; Parker and Böving, 1914).

The simplest type of spiracle is an **annular** or **uniforous** type (fig. 34.24), which consists of a circular or oval opening, the edge of which is called the **peritreme,** a spiracular chamber or **atrium,** and a **spiracular closing apparatus,** lying between the atrium and the end of the tracheal trunk and consisting of a sclerotized ring and apodeme to which a muscle attaches. The inside of the atrium may be variously modified with internally projecting hairs or plates, these sometimes forming a filter. The peritreme may be simple or crenulate. During **ecdysis,** the old spiracle is pulled out through the opening of the newly formed one. The term **bilabiate** has been used by Schiödte (1862–1883), Roberts (1930), and others for a type of annular or uniforous spiracle which is elongate-oval or elliptical, as in many Cerambycidae; the term is an unfortunate one, since it can be easily confused with biforous and bicameral (*see* below). An annular type of spiracle with a closing apparatus is thought to be the primitive condition in Coleoptera and occurs in all Archostemata, Adephaga, and Myxophaga; within the Polyphaga, however, it may have been secondarily derived several times from the annular-biforous type mentioned below. Annular spiracles are characteristic of most Staphylinoidea, all Eucinetoidea, Brachypsectridae, most Bostrichoidea, some Cleridae, all Melyridae and many Cucujoidea, Tenebrionoidea, Chrysomeloidea, and Curculionoidea. The distribution of the spiracular closing apparatus in the Coleoptera has been of interest in phylogenetic studies; this structure is widespread throughout the order, but is absent in Dascillidae, Scarabaeidae (with the exception of Troginae and Glaphyrinae), Byrrhidae, Dryopoidea, Artematopidae, Elateridae, Cebrionidae, and Cantharoidea (except Brachypsectridae).

In some types of spiracles, there are one or more **accessory openings (accessory chambers, lateral air tubes, secondary chambers)** connected to the main opening (fig. 34.25). The terms **annular-uniforous, annular-biforous** (also **bicameral**), and **annular-multiforous** are used for those with 1, 2, or more than 2 accessory openings (Crowson, 1981; Roberts, 1930; Steinke, 1919). Annular-uniforous spiracles are known in only a few groups of Polyphaga, including some Bostrichidae and Anobiidae, Phalacridae (*Phalacrus*), a few Melandryidae, and Aglycyderidae. Annular-multiforous spiracles occur in Lymexylidae, a few Anobiidae, several groups of Tenebrionidae, *Calopus* (Oedemeridae), Mycteridae, and some Cerambycidae. The accessory openings are usually distributed around the peritreme, but in Mycteridae they are clustered at one end.

The annular-biforous spiracle may represent the ancestral condition in Polyphaga, since it occurs in a number of primitive groups, like Agyrtidae, Leiodidae, Derodontidae, Nosodendridae, and Phloiophilidae, and is common and

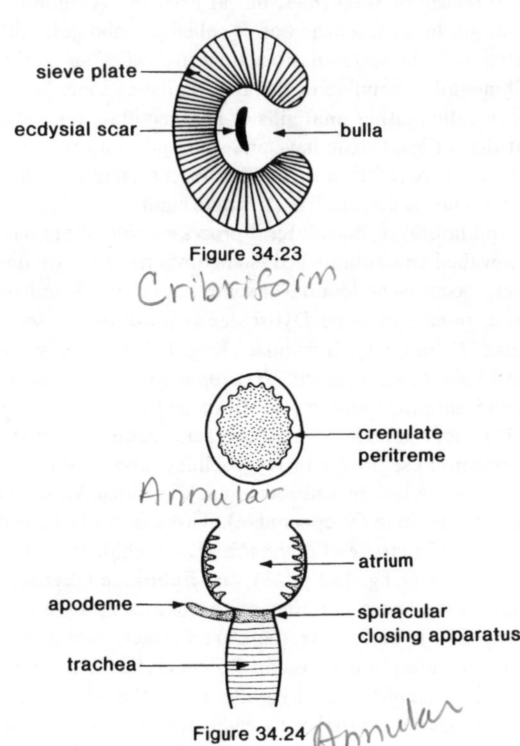

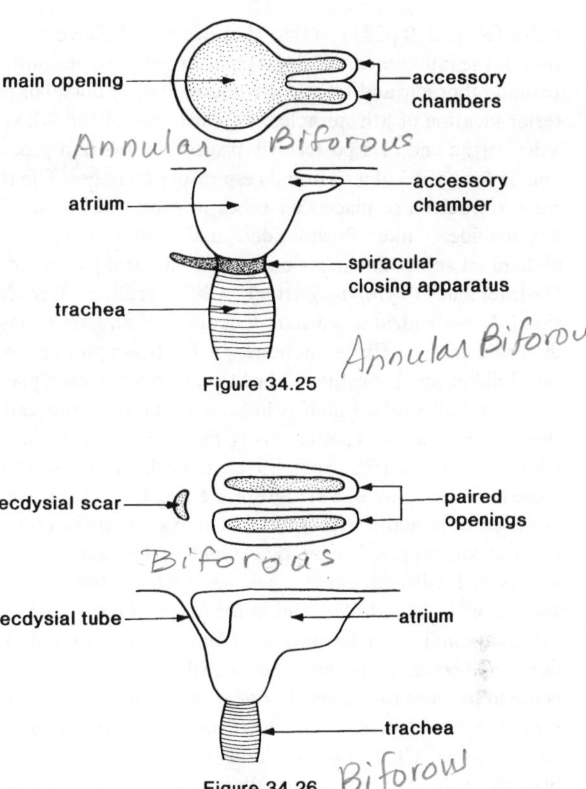

Figure 34.23. Cribriform spiracle (Scarabaeidae).

Figure 34.24. Annular spiracle (composite) and cross-section.

Figure 34.25. Annular-biforous spiracle (composite) and cross-section.

Figure 34.26. Biforous spiracle (composite) and cross-section.

widespread in all cucujiform superfamilies. In some cases, the accessory openings are very small, so that the spiracle looks like the annular type, and in others the main opening is very small and the accessory openings long, so that it resembles the biforous type, but lacks an ecdysial scar (*see* below). In descriptions, these are usually called biforous. In Cerambycidae and Curculionidae, the early instars may have annular-biforous spiracles, while later instars have the annular or annular-multiforous type (Duffy, 1953; Gardner, 1925, 1927; Roberts, 1936).

The **biforous spiracle** has its opening blocked by a median septum, with a narrow, slit-like opening on either side of it (figs. 34.26, 34.50). These two openings are accompanied by an **ecdysial scar** (or **stigmatic scar**), which represents the opening through which the old spiracle is pulled during molting. This specialized type of molting has been called by Hinton (1947) the **elateroid ecdysial process** and is also associated with the cribriform spiracle. Biforous spiracles occur in Hydrophiloidea, late instar Rhipiceridae, *Trox* (Scarabaeidae: Troginae), *Eucanthus* (Scarabaeidae: Geotrupinae), Byrrhidae, *Schizopus* (Buprestidae), all Dryopoidea with functional spiracles, all Elateroidea, and Cantharoidea (except Brachypsectridae).

The **cribriform spiracle** (fig. 34.23) also has the main opening blocked, though not by a single division but by a **sieve plate (respiratory plate, cribriform plate)** bearing numerous perforations. The cuticular area immediately adjoining the sieve plate and usually partly or almost completely surrounded by it is called the **bulla,** and it is here that the **ecdysial scar** is located. Cribriform spiracles with a typical kidney-shaped or **reniform** sieve plate (Kfigs. 205–206, 208 p265) occur in Dascillidae and almost all Scarabaeoidea and Buprestidae, while Heteroceridae have a different type (Kfig. 207 p265). The **undulate spiracle** (Kfig. 209 p281) found in Chelonariidae is of a biforous type with undulated openings which may be partly blocked by cross pieces, forming a transition to the cribriform spiracle. A different sort of cribriform spiracle also occurs on the 8th segment in some eubriine Psephenidae (Hinton, 1955).

The normal location of the thoracic spiracle is on a mesosternal laterotergite that is wedged in between the protergum and mesotergum, but in some soft-bodied larvae (lyctine Bostrichidae, Ptinidae) it may be on the prothorax. The abdominal spiracles vary in size and position, and sometimes the first and last differ in position from those in between. The abdominal spiracles are usually situated on laterotergites or lateral extensions of a single tergal plate.

Many aquatic larvae, as well as those living in wet terrestrial environments subjected to flooding or in wet, sappy areas in wood or under bark, have mechanisms for allowing at least some spiracles access to free air. These include: 1) placement of some or all spiracles at the ends of *spiracular tubes* (Kfig. 210 p281) so that they are raised above the surface of the integument; 2) dorsal placement of some spiracles (usually thoracic and 1st and 8th abdominal; 3) dorsal or posterior location of 8th spiracles; 4) placement of the 8th spiracles at the end of a posteriorly projecting, median process; and 5) formation of a terminal **respiratory chamber.** The thoracic spiracles are placed on tubes in some subcortical Tenebrionoidea, like Prostomidae and Inopeplidae, while abdominal spiracular tubes occur in many groups, including Hydraenidae (*Tympanogaster*), Chelonariidae, Derodontidae, Nosodendridae, various Cucujoidea, and some cassidine and hispine Chrysomelidae. In Hydroscaphidae, some Nitidulidae, and Biphyllidae, the 8th spiracles are each placed at the end of a tube which is located at the posterior end of the segment and faces posteriorly (Kfigs. 161 p261, 168 p255), while in *Donacia* (Chrysomelidae), the 8th spiracles are enclosed within spine-like structures (Kfig. 156 p231), which are used to puncture the stems of aquatic plants in order to tap the air supply. Dorsal placement of thoracic spiracles occurs in Hydroscaphidae, *Tympanogaster* (Hydraenidae), and some Nosodendridae; and in the first and last groups, the 1st abdominal spiracles may also be dorsally situated. The dorsal and posterior movement of the 8th spiracles, which may come to lie close to one another near the midline, is known in some Psephenidae, *Araeopidius* (Ptilodactylidae) (Kfig. 187 p238), some Chelonariidae, and Dryopidae, while their placement at the apex of a median process occurs in Dytiscidae (Kfig. 154 p213), Noteridae, Amphizoidae (Kfig. 153 p213), and Nosodendridae.

The **respiratory chamber** (also called **stigmatic atrium** or **breathing pocket**) of advanced Hydrophilidae is comprised of the 8th and 9th terga and encloses a large, posteriorly projecting pair of spiracles; this chamber may be complexly lobed, and its floor (9th tergum) may have as many as 3 pairs of urogomphus-like processes—the **mesocerci, paracerci,** and **acrocerci.** The mesocerci are usually the most prominent within the chamber, while the acrocerci lie at its posterior edge and are often visible when the chamber is closed. A pair of **procerci** may also be present at the lateral edges of the 8th tergum, above the spiracles. The 10th segment is ventrally oriented, and lies below the floor of the respiratory chamber. A terminal respiratory chamber also occurs in larvae of Helodidae, but it is simpler in structure and also encloses the anal region.

In aquatic, semi-aquatic, and riparian larvae there may be, in addition to or instead of spiracles, a variety of structures which allow oxygen to pass into the tracheal system or directly into the haemocoel when the cuticle is entirely submerged in water. The general term **gill** is used for these structures, and the term **plastron** refers to a specific situation in which a gas layer is held in position by hydrofuge structures which prevent the entry of water under pressure. Most gills are thin-walled processes which project from the venter, the lateral portions of thorax and abdomen, or the anal region. Almost all gills in Coleoptera are **tracheal gills,** that is they contain **tracheae** or **tracheoles,** but at least in Hygrobiidae, the larval gill lacks tracheae and is called a **blood gill.** Gills associated with the spiracular openings are called **spiracular gills.** Thin-walled papillae in the anal region of some larvae have been called either **anal gills** or **anal papillae,** since it is thought that at least some have an osmoregulatory function. Typical gills, here referred to as **gill tufts,** consist of a number of fine branches radiating from a single stem.

Paired lateral or dorsolateral processes, which are usually thin-walled and unbranched (sometimes fringed with fine processes), occur on abdominal or sometimes thoracic and abdominal segments in some Dytiscidae (*Coptotomus*), some Haliplidae (*Peltodytes*), Gyrinidae (Kfig. 157 p213), Hydrophilidae (*Spercheus, Crenitis, Hydrophilus,* various Berosinae), and aquatic Lampyridae. Branched gill tufts (Kfig. 151 p231) occur ventrally on all thoracic segments and the first 3 abdominal segments in Hygrobiidae, and on A1–7 in Euclichadidae, A1–4 in Eubrianacinae (Psephenidae), and A2–6 in Psepheninae (Psephenidae). Three finely-branched **anal gills** occur in larvae of *Hyphalus* (Limnichidae), Lutrochidae, Elmidae (Kfig. 188 p223), and Eubriinae (Psephenidae), and in the last group they are accompanied by a pair of **anal papillae.** In Helodidae, there are 5 larger, unbranched anal gills. In many anchytarsine Ptilodactylidae, the anal region bears a single conical papilla above the anus and 2 smaller papillae associated with each pygopod, but in *Anchytarsus* (Kfig. 189 p229) there is, in addition, a cluster of 9 gills on each side of the anal region.

Spiracular gills are of 3 types: segmented, as in Torridincolidae (Kfig. 152 p257); vesicular, as in Microsporidae and most Hydroscaphidae (Kfig. 211 p208); and tufted, as described by Reichardt (1974) for the hydroscaphid genus *Scaphydra* (Kfig. 212 p208). The first type consists of a 2- or 3-segmented process, with the spiracular opening at the end of segment 1 or 2 and the apical portion of the gill covered with a plastron mesh. A **vesicular gill** consists of a thin-walled extension of the peritreme to form a balloon-like structure with the opening at the apex. The last type looks like a minute gill tuft extending from the peritreme.

The ptilodactylid *Araeopidius monachus* (LeConte) has a very peculiar type of respiratory structure. The thoracic spiracles and abdominal spiracles 1–7 are atrophied and each is partly surrounded by a plate-like structure (Kfig. 213 p251) that contains a fine mesh and whose surface is covered with minute pores (.05–.13 microns); this is called a **plastron plate** and apparently functions as a respiratory organ during periods of submergence in much the same way as the plastron of a housefly egg (Hinton, 1967b). Some aquatic larvae are probably capable of breathing directly through portions of the cuticle with tracheae near the surface. In some Haliplidae, tracheoles actually penetrate minute cuticular processes that have been called **microtracheal gills** (Seeger, 1971).

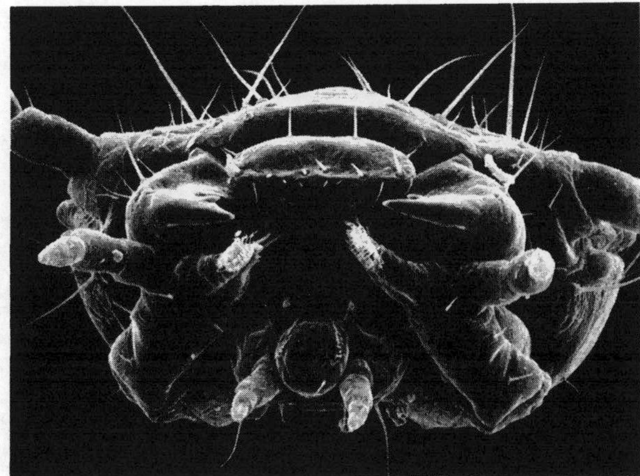

Figure 34.27. *Anisotoma blanchardi* Horn (Leiodidae). Passaconaway, New Hampshire. Head, anterior, showing symmetrical, asperate molae and highly flexible maxillae with fimbriate apices.

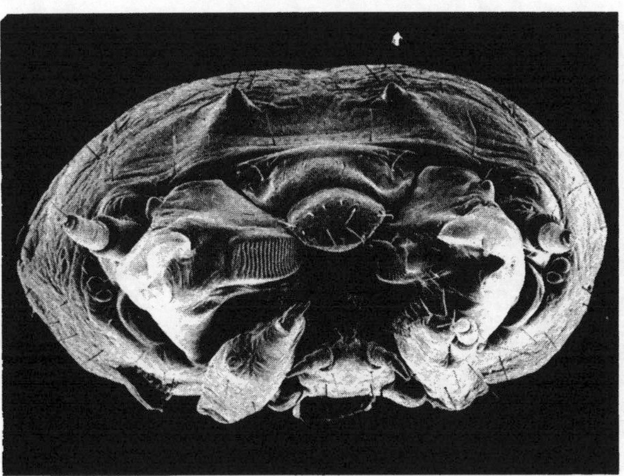

Figure 34.28. *Bolitotherus cornutus* (Panzer) (Tenebrionidae). Frederick, Maryland. Head, anterior, showing asymmetrical, transversely ridged molae.

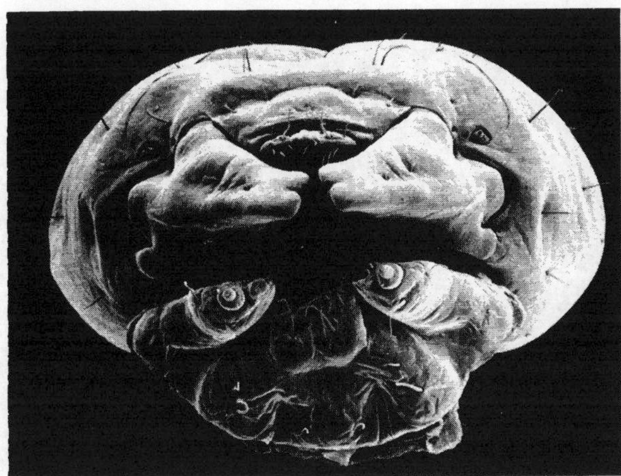

Figure 34.29. *Curculio* sp. (Curculionidae). Ontario. Head, anterior, showing heavy, broad-based, phytophagous mandibles without molae.

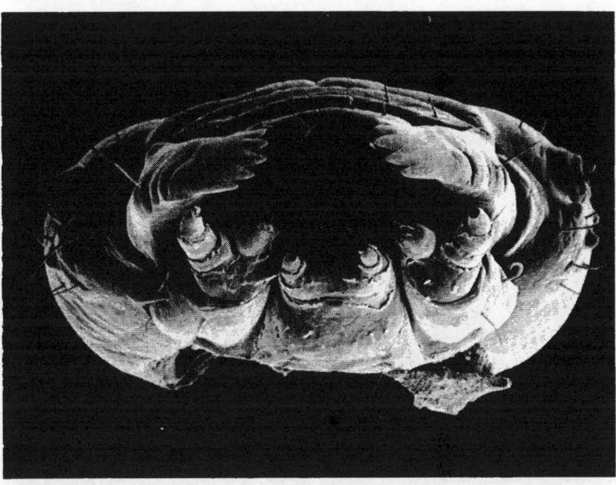

Figure 34.30. *Leptinotarsa lineolata* (Stål) (Chrysomelidae). Gardner Canyon, Arizona. Head, anterior, showing phytophagous mandibles with multidentate apices and no molae.

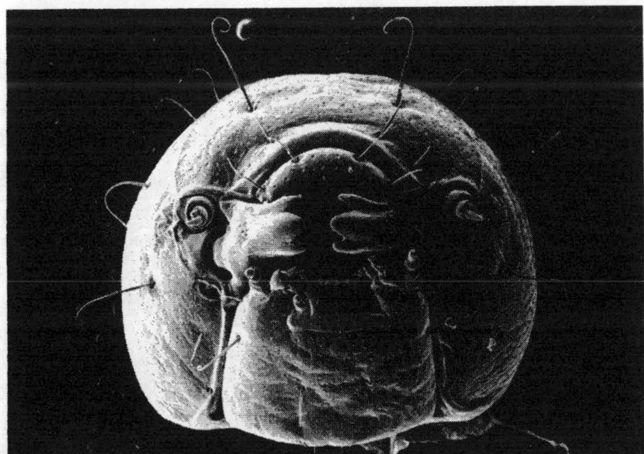

Figure 34.31. *Dryops* sp. (Dryopidae). Barro Colorado Island, Panama. Head, anterior, showing phytophagous type mandibles with tridentate, palmate apices and no molae, and maxillae fused to labium.

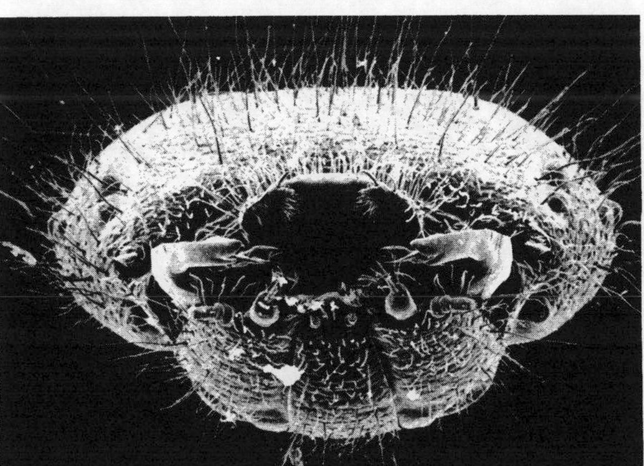

Figure 34.32. *Lanternarius gemmatus* (Horn) (Heteroceridae). Del Norte Co., California. Head, anterior, showing mandibles with concave pseudomolae and consolidated maxillolabial complex.

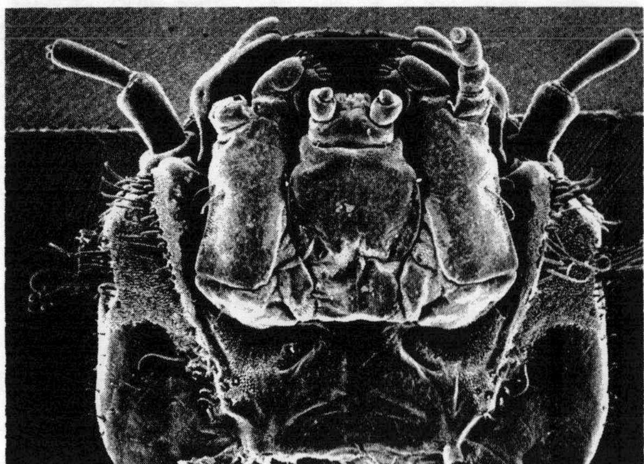

Figure 34.33. *Stenocolus scutellaris* LeConte (Eulichadidae). Richardson Springs, California. Head, ventral, showing ventral epicranial ridges, well-developed maxillary articulating area, and articulated, digitiform galea and lacinia.

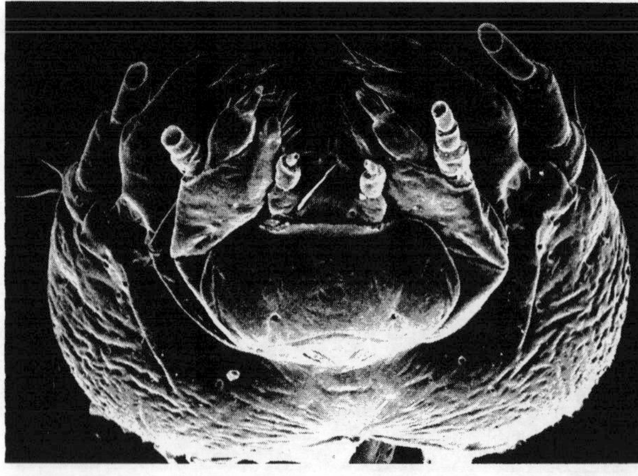

Figure 34.34. *Ptilodactyla* sp. (Ptilodactylidae). 8 mi. S Gatlinburg, Tennessee. Head, anteroventral, showing expanded mentum and concealed maxillary articulating area.

Figure 34.35. *Athous* sp. (Elateridae). Peddler Hill, California. Head, anterior, showing unidentate mandibles without molae and liquid-feeding mouthparts with oral filter blocking mouth cavity.

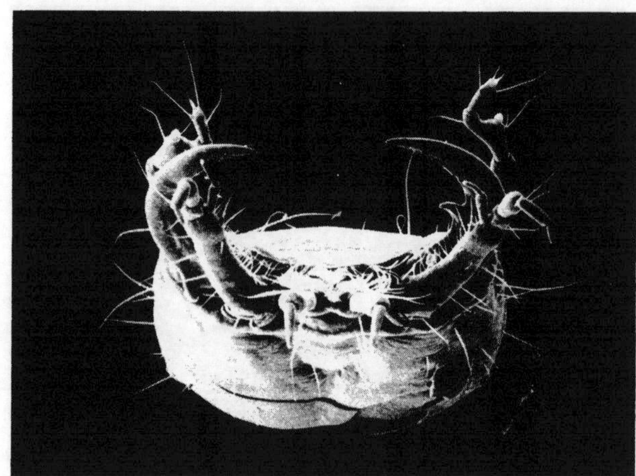

Figure 34.36. *Agonum* sp. (Carabidae). Concord, Massachusetts. Head, anterior, showing falcate mandibles with retinacula, strongly protracted ventral mouthparts, articulated, 2-segmented maxillary mala, and oral filter.

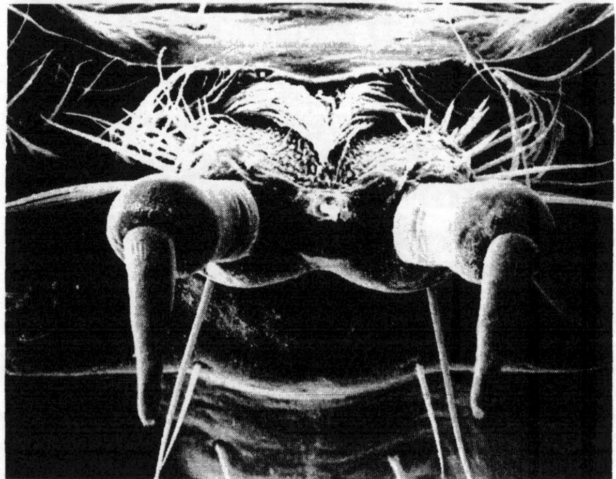

Figure 34.37. *Agonum* sp. Same data. Oral filter at higher magnification, showing barbed hairs.

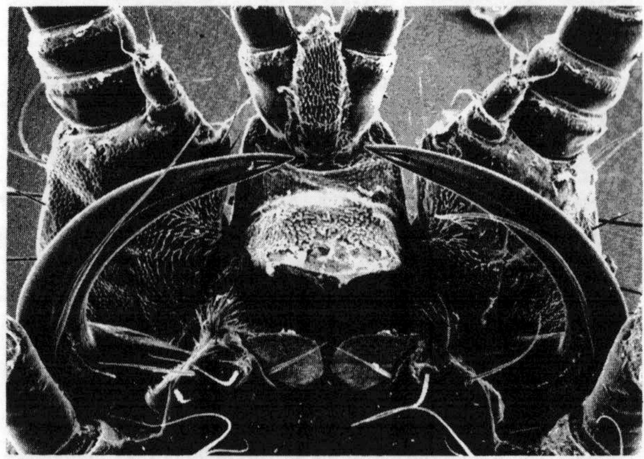

Figure 34.38. *Phengodes* sp. (Phengodidae). Rose Lake, Clinton Co., Michigan. Anterior part of head, anterodorsal view, showing nasale and falcate, perforate mandibles.

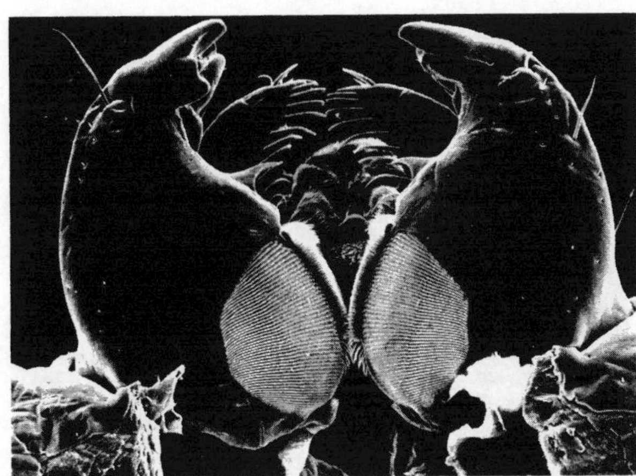

Figure 34.39. *Cucujus clavipes* Fabricius (Cucujidae). 6 mi. E Mineral, California. Anterior part of head, dorsal with frontoclypeal region and labrum-epipharynx removed, showing mandibles in closed position with molar asperities interdigitating, dorsal microtrichial patches, acute prostheca, and falcate, spinose maxillary malae.

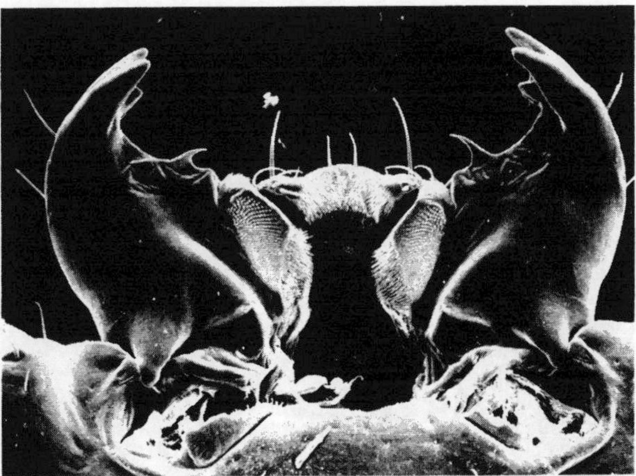

Figure 34.40. *Cucujus clavipes.* Same data. Anterior part of head, ventral with maxillae and labium removed, showing mandibles in open position, ventral condyle, accessory ventral process, asperities on ventral surface of mandibular base, epipharynx, and cibarial plates.

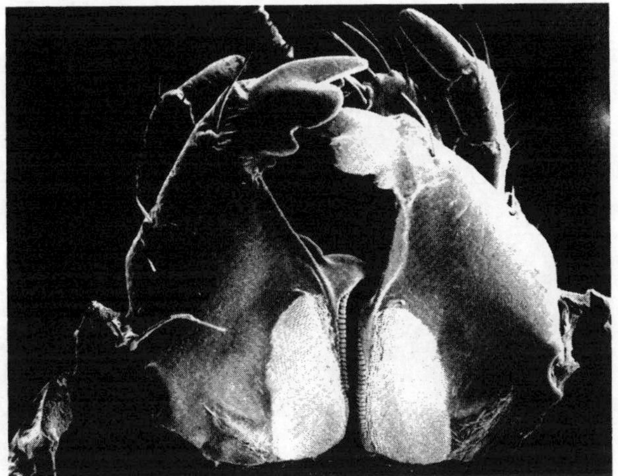

Figure 34.41. *Dendroides* sp. (Pyrochroidae). Phoenix Lake, Marin Co., California. Anterior part of head, dorsal, with frontoclypeal region and labrum-epipharynx removed, showing large, concave/convex, strongly asymmetrical, multiple-shearing molae, premolar tooth on left mola, and dorsal microtrichial patches.

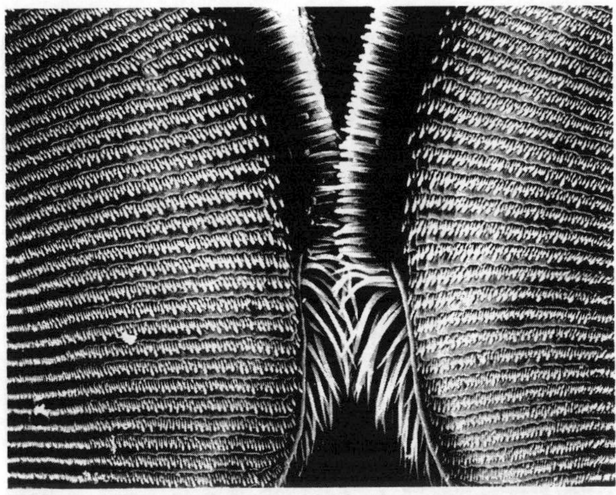

Figure 34.42. *Cucujus clavipes.* Same data as fig. 39. Mandibular bases, dorsal, at higher magnification, showing details of molar surfaces and dorsal rows of microtrichia.

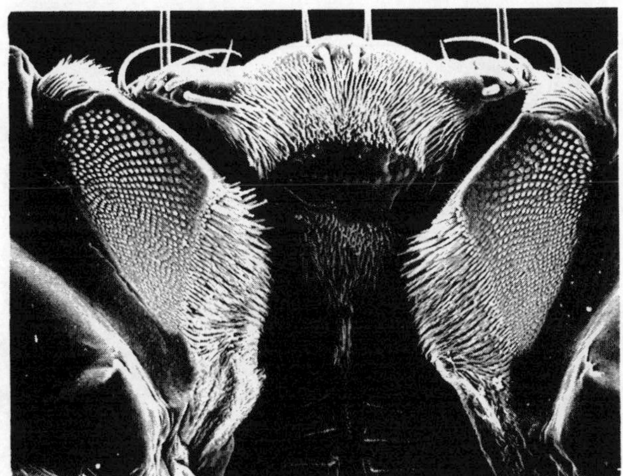

Figure 34.43. *Cucujus clavipes.* Same data as fig. 39. Mandibular bases, ventral, at higher magnification, showing details of ventral asperities, epipharynx, and cibarial plates.

Figure 34.44. *Aphenolia monogama* (Crotch) (Nitidulidae). Wasco Co., Oregon. Apex of right mala, dorsal, showing spatulate setae.

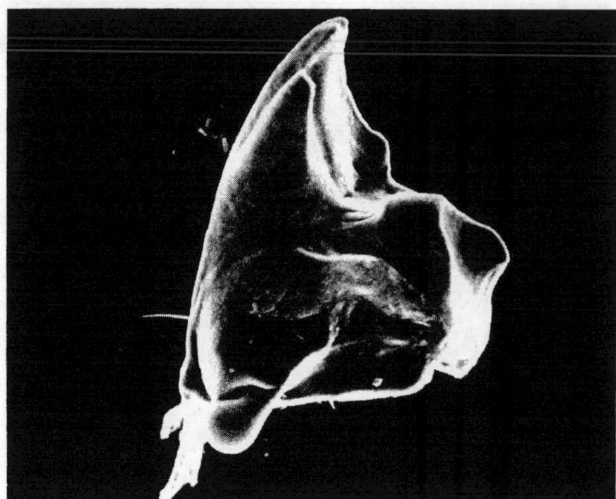

Figure 34.45. *Platydema ellipticum* (Fabricius) (Tenebrionidae). Wakulla Springs, Florida. Right mandible, ventral, showing simple, concave mola, ventral condyle, and dorsal carina with single, weak tooth.

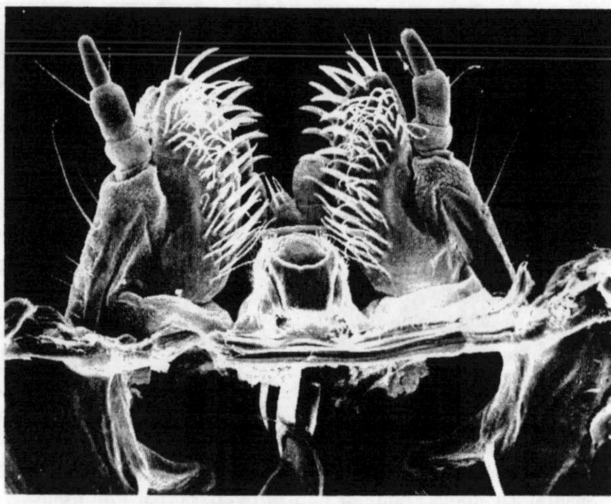

Figure 34.46. *Platydema ellipticum.* Same data. Anterior part of head, dorsal, with frontoclypeal region, labrum-epipharynx, and mandibles removed, showing truncate maxillary mala with setose-spinose inner edge and dorsal surface, hypopharyngeal bracon, and hypopharyngeal sclerome.

Figure 34.47. *Anchorius lineatus* Casey (Biphyllidae). 1 mi. NW Arivaca, Arizona. Mandibular prostheca, ventral, showing comb-hairs.

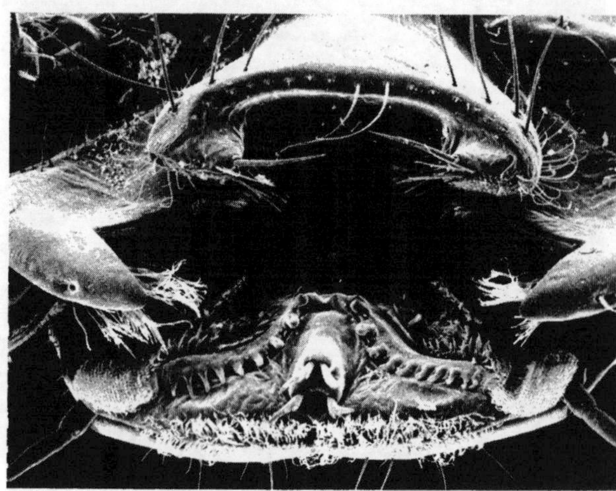

Figure 34.48. *Prionocyphon* sp. (Helodidae). Bedford, Massachusetts. Anterior part of head, with maxillae laterally displaced, showing labrum-epipharynx, mandibular apices, and hypopharynx with series of combs, spines, and setae.

Figure 34.49. *Prionocyphon* sp. Same data. Apex of maxilla, showing comb-hairs.

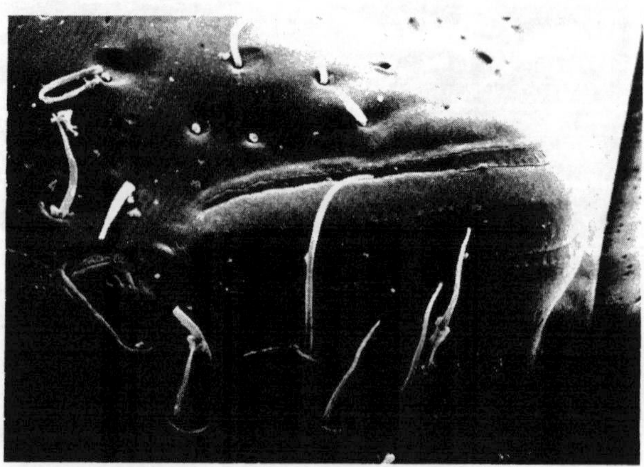

Figure 34.50. *Dryops* sp. (Dryopidae). Barro Colorado Island, Panama. Biforous spiracle on abdominal segment 8, showing 2 openings and ecdysial scar.

Figure 34.51

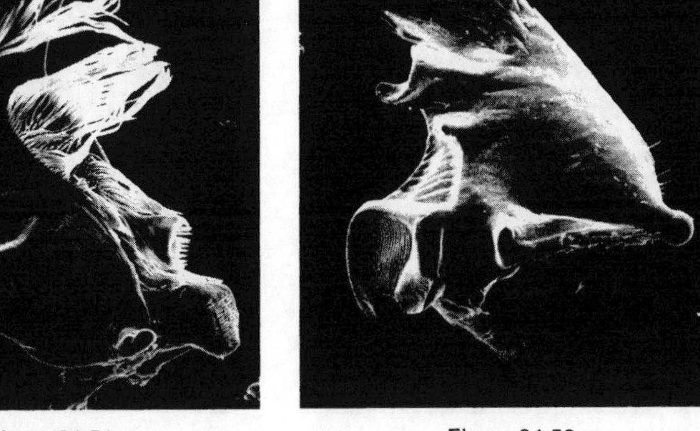

Figure 34.52

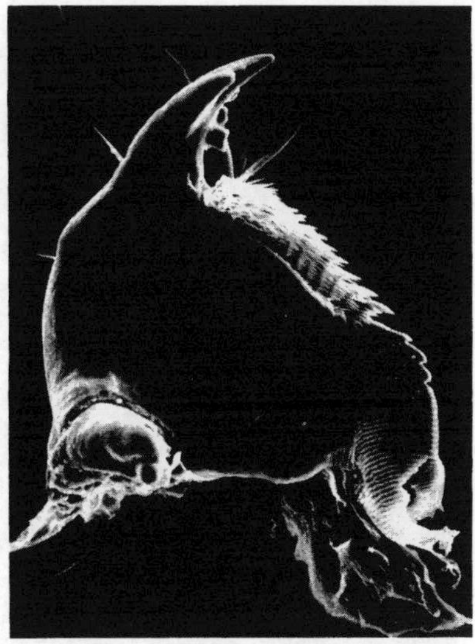

Figure 34.53

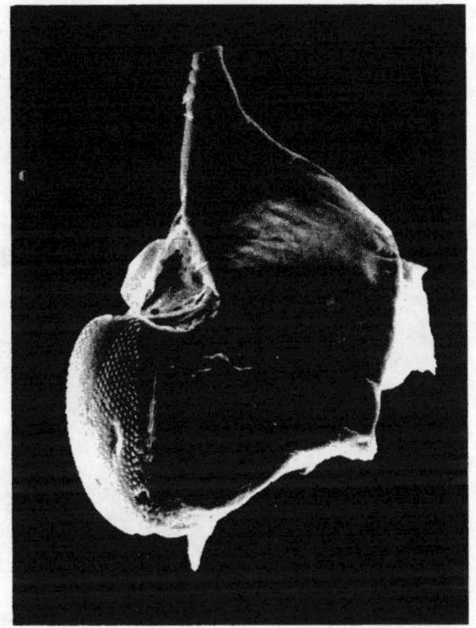

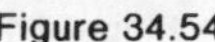

Figure 34.54

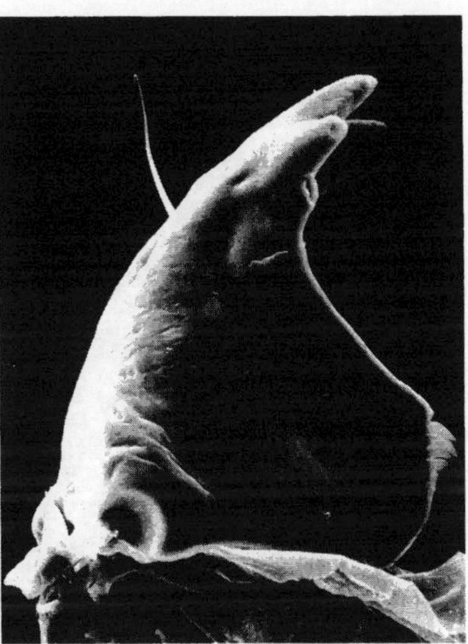

Figure 34.55

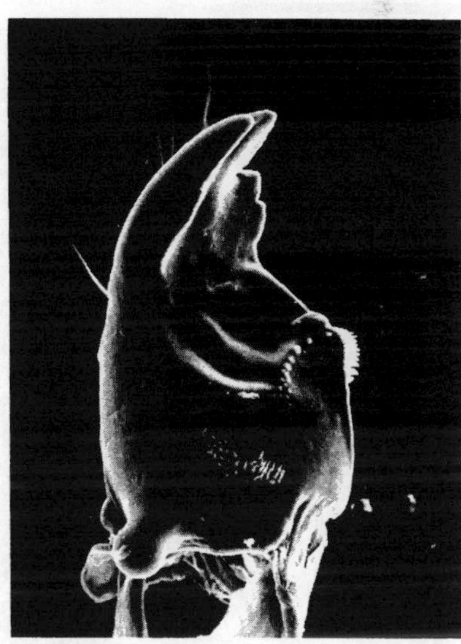

Figure 34.56

Figure 34.51. *Prionocyphon* sp. Same data as fig. 48. Right mandible, ventral, showing transversely ridged mola, comb-like prostheca, small ventral condyle, and accessory ventral process.

Figure 34.52. *Dascillus davidsoni* LeConte (Dascillidae). Novato, California. Left mandible, ventral, showing longitudinally ridged mola, large accessory ventral process, small, articulated prostheca, hyaline area (appearing wrinkled), and large scissorial teeth.

Figure 34.53. *Glischrochilus* sp. (Nitidulidae). Great Smoky Mts. Nat. Park, Tennessee. Right mandible, ventral, showing enlarged ventral condyle, ventral asperities, and prostheca consisting of fringed membranes.

Figure 34.54. *Lycoperdina ferruginea* LeConte (Endomychidae). Great Smoky Mts. Nat. Park, Tennessee. Right mandible, mesodorsal view, showing reduced apex and large, finely tuberculate mola (worn in middle) (prostheca has been damaged).

Figure 34.55. *Rhinosimus* sp. (Salpingidae). Great Smoky Mts. Nat. Park, Tennessee. Left mandible, dorsal, showing dorsal acetabulum and group of hyaline processes at mandibular base.

Figure 34.56. *Hyporhagus gilensis* Horn (Monommidae). Vail, Arizona. Right mandible, ventral, showing dorsal carina with 2 weak teeth and ring of asperities representing reduced mola.

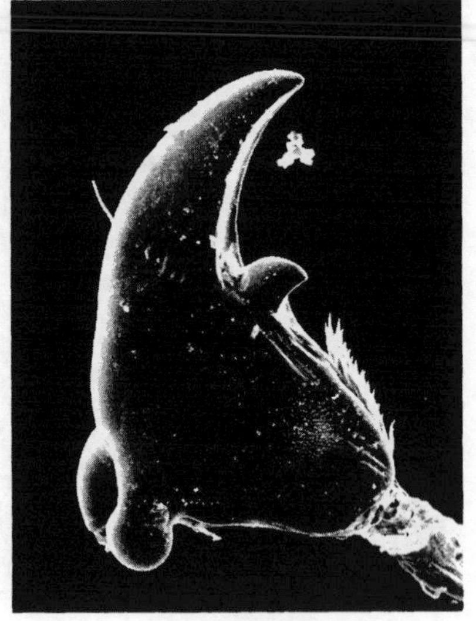
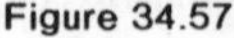

Figure 34.57

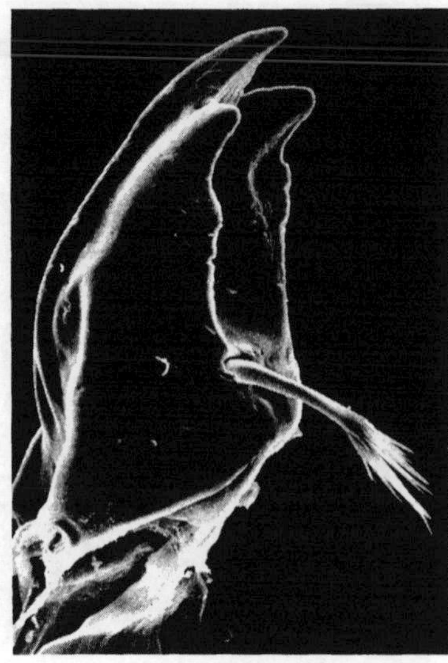

Figure 34.58

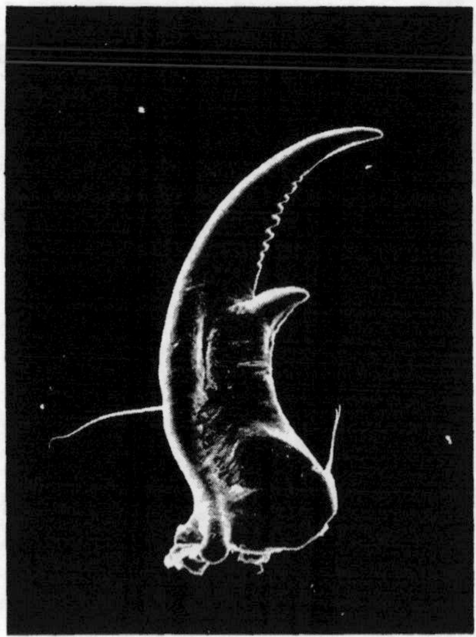

Figure 34.59

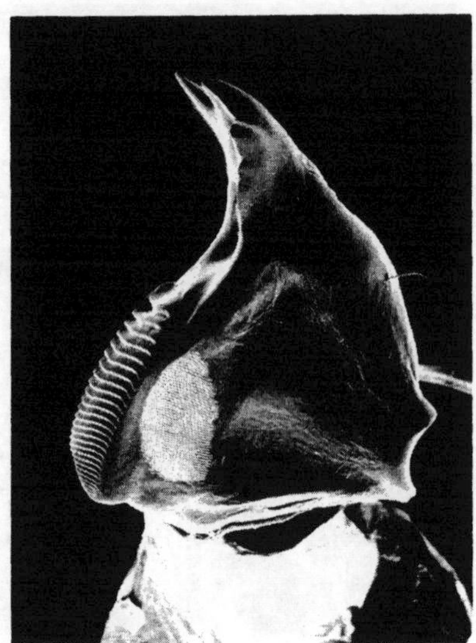

Figure 34.60

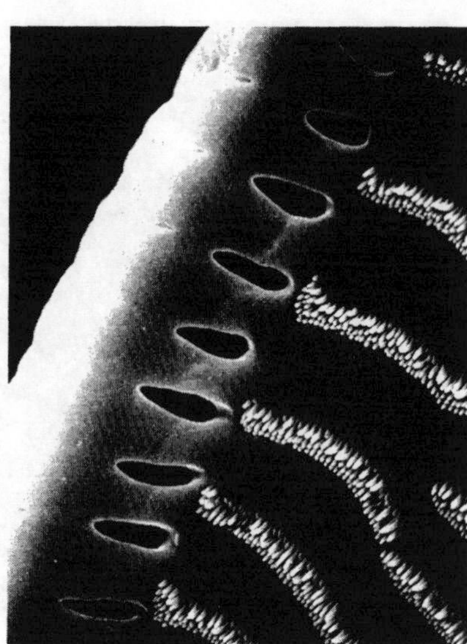

Figure 34.61

Figure 34.62

Figure 34.57. *Athous* sp. (Elateridae). Peddler Hill, California. Right mandible, ventral, showing fairly broad base with penicillus, retinaculum, and unidentate apex.

Figure 34.58. *Ptilodactyla* sp. 8 mi. S Gatlinburg, Tennessee. Right mandible, ventral, showing tridentate apex, concave incisor area, and articulated process with setose apex at mandibular base.

Figure 34.59. *Agonum* sp. (Carabidae). Concord, Massachusetts. Right mandible, ventral, showing narrow base, fine penicillus, retinaculum, unidentate apex, and serrate incisor edge.

Figure 34.60. *Ditylus* sp. (Oedemeridae). Vernon, British Columbia. Right mandible, dorsal, showing dorsal acetabulum (in profile), dorsal microtrichial patch, and mola consisting of simple, transverse ridges.

Figure 34.61. *Pterogenius nietneri* Candeze (Pterogeniidae). Hakgala, Sri Lanka. Part of mesal edge of mandible, dorsal, showing reinforced mola, openings to channels beneath mola, and rows of microtrichia.

Figure 34.62. *Pterogenius nietneri* Candeze. Same data. Surface of mandibular mola, showing ridges and pores.

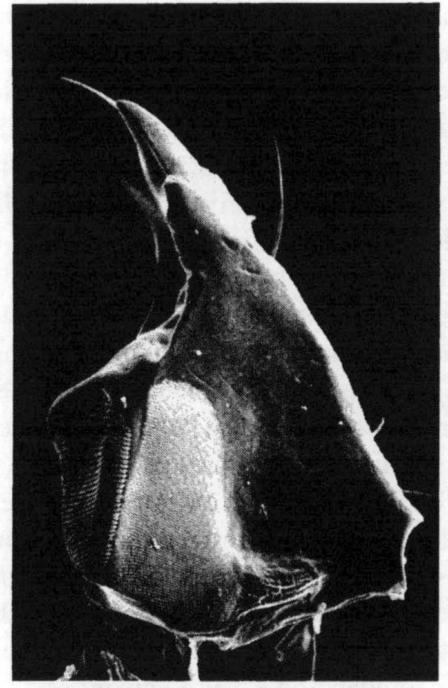

Figure 34.63

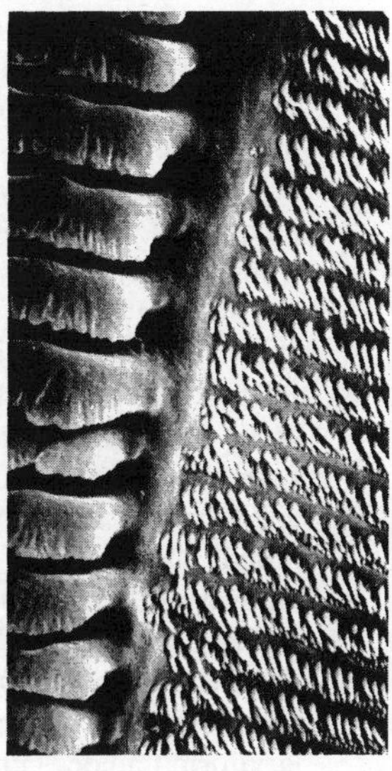

Figure 34.64

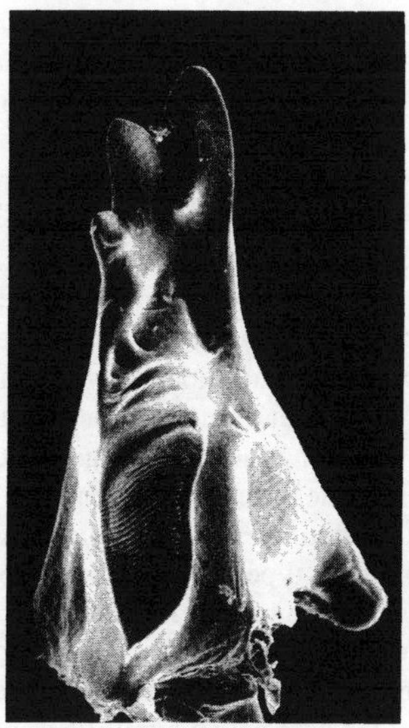

Figure 34.65

Figure 34.66

Figure 34.63. *Synchroa punctata* Newman (Synchroidae). Bear Mountain, New York. Right mandible, dorsal, showing dorsal microtrichial patch and concave mola with complex, transverse ridges separated by deep grooves.

Figure 34.64. *Synchroa punctata*. Same data. Part of mandibular base, dorsal, showing molar ridges and dorsal microtrichia.

Figure 34.65. *Synchroa punctata*. Same data. Left mandible, mesal view, showing complex mola and transverse ridges.

Figure 34.66. *Synchroa punctata*. Same data. Part of molar surface, showing complex ridges and deep grooves.

BIOLOGY AND ECOLOGY

Beetles are holometabolous insects, with a life cycle consisting basically of an egg, larva, pupa, and adult or imago; there are some cases, however, where the egg or pupal stage has been eliminated (*see* below). Truly **viviparous** species, which have eliminated the egg stage, are rare among the Coleoptera, examples occurring in the archostematan genus *Micromalthus* and the termitophilous *Corotoca* (Staphylinidae: Aleocharinae). **Ovoviviparous** forms, in which eggs hatch immediately before or during deposition, are known in some aleocharine Staphylinidae and chrysomeline Chrysomelidae. In some cave-dwelling Leiodidae, such as *Leptodirus*, the eggs are very large and the larva upon hatching is almost fully developed. The size of beetle eggs and the number of eggs produced by a single female vary considerably (*see* Table 2 in Hinton, 1981). Very small beetles, such as Myxophaga, Ptiliidae or Clambidae, produce only 1 or 2 eggs at a time, while very large numbers of small eggs may be found in some Meloidae and Lymexylidae. The morphology of beetle eggs has been covered in detail by Hinton (1981); those with a hard, sculptured **chorion** are rare in the group, but they have been reported in Cupedidae (*Cupes, Priacma*). Eggs are usually deposited singly, but in several groups, such as Coccinellidae and Chrysomelidae, they form characteristic clusters. **Oöthecae** or egg cases occur in several families. Members of the Hydrophilidae and some Hydraenidae enclose their eggs, singly or in batches, within silken cocoons produced by the **colleterial glands.** A number of Chrysomelidae produce oöthecae which may be composed of faeces (**scatoshells**), colleterial gland secretions, or both; these have been reported in Cassidinae and in the so-called Camptosomata (Clytrinae, Cryptocephalinae, Chlamisinae, and Lamprosomatinae), but they also occur in some Australian Sagrinae and Eumolpinae. In the Clytrinae and related groups, the larval and pupal stages are also spent within a faecal case. Parental care, which includes mass provisioning and the guarding of eggs and sometimes larvae, has been reported in a number of beetles, including some Carabidae (*Abax, Molon, Mormolyce*), Silphidae, Staphylinidae (*Platystethus, Bledius*), Scarabaeidae (Geotrupinae, Scarabaeinae), Heteroceridae, Chrysomelidae (Cassidinae), Platypodidae and Scolytidae.

A common feature of newly hatched larvae is the presence of **egg bursters** (also known as **hatching spines**), which are small teeth or spines used to break the chorion upon hatching. Egg bursters appear to be absent in Archostemata and Myxophaga, while those in Adephaga occur on the frontal region. Frontal egg bursters are also known in the polyphagan families Hydraenidae, Hydrophilidae, and Derodontidae (*Laricobius*). More common among Polyphaga are thoraco-abdominal egg bursters, which occur in pairs on some thoracic and abdominal terga; these are known in Staphylinidae, Histeridae, Lucanidae, Passalidae, Scarabaeidae, Rhipiceridae, Byrrhidae, Bostrichidae, Ptinidae, Nitidulidae, Coccinellidae, Tenebrionidae, Pyrochroidae, Melandryidae, Cerambycidae, Chrysomelidae, and Curculionidae, and will probably be discovered in other families as well.

The number of larval instars in Coleoptera varies from 1 to almost 30, but is normally 3 to 5. In Adephaga, Staphylinoidea, and Scarabaeoidea, the number is usually 3, while in Dermestidae it is usually 6 or 7 but may be much higher.

Beetles with only 2 larval instars are relatively rare, occurring in Histeridae and some Leiodidae, and a single instar is known only in some highly evolved cave forms (Leiodidae: Bathysciinae).

First instars differ from later instars in many groups, and sometimes these differences are striking. Differences between first and later instars may be in general form, tergal pigmentation and armature (Sphaerosomatidae, teredine Bothrideridae), form of the 9th tergum (Lymexylidae, Tenebrionidae), or number of antennal segments (Cupedidae, Dytiscidae, Helodidae). Differences may also occur among later instars, especially that immediately preceding the pupal stage, which is often called the **prepupa.** Extreme differences among larval instars, often connected with differences in habits, is called **hypermetamorphosis.** A common type of hypermetamorphosis involves a campodeiform type of first instar, which is involved in active or passive dispersal; this type of larva is usually called a triungulin (based on the apparently 3-clawed first instars of some Meloidae). These active larval types are usually found in beetles which are parasitic, and they may seek out the host themselves (Carabidae, Staphylinidae, Rhipiceridae, Drilidae, Cucujidae, Bothrideridae) or attach themselves to flying adults, in order to be passively transported to the host larvae (Meloidae, Rhipiphoridae). Among non-parasitic groups, active first instars may be found in some groups where eggs are laid on or near the surface of the soil and larvae must seek out subsurface food sources, like roots (schizopodine Buprestidae, vesperine Cerambycidae); but in *Micromalthus,* the triungulin, known as a **caraboid larva,** apparently seeks out wood in the proper state of decay. A unique situation occurs in some Cantharidae, where the first instar is an inactive form called a **prolarva;** this has been termed **foetometamorphosis.** Another type of hypermetamorphosis involves only the last instar or prepupa. In some Eucnemidae, for instance, the prepupa is buprestid-like, with an enlarged prothorax, while earlier instars are elongate and flattened. The most complex types of hypermetamorphosis may involve more than two larval types; these occur in Micromalthidae, Rhipiphoridae, and Meloidae, and are discussed in the sections dealing with those families.

Beetle pupae (figs. 34.7–8) are always **adecticous** (without functional mandibles) and almost always **exarate** (with free appendages). **Obtect** pupae, which are usually more heavily sclerotized, with the appendages more or less fixed to the body, have evolved 5 or more times in the Coleoptera, occurring in Ptiliidae, staphylinine Staphylinidae, Clambidae, Coccinellidae, and Chrysomelidae (especially Hispinae and Cassidinae). Many pupae develop within substrates or in pupal cells of various kinds, but others occur on surfaces, partly enclosed within the last larval exuviae, which may be armed with modified setae, spines, or glandular reservoirs. There is considerable variation in the number of functional pupal spiracles; there are usually 6 in Adephaga, 8 in Cantharoidea and some Elateroidea, 7 in numerous groups, and fewer in some specialized forms. Pupal spiracles have a closing apparatus, even when one is lacking in the larva. Modifications of the pupal prothorax are usually precursors of the same structures of the adult (horns in male scarabaeoids and tenebrionids), but in Pterogeniidae (Lawrence, 1977c), the pupal pronotum bears a spiny, club-like anterior process not found in the adult. Defensive structures on the pupa include long setae or spines, glands (Cantharidae), and pinching devices known as **gin traps** (Hinton, 1946a). Gin traps are formed by sclerotized and often armed portions of two adjacent sclerites, and may involve 2 to 7 abdominal segments. Median, unpaired gin traps are known in Dryopidae, Ptilodactylidae, and Dermestidae; paramedian, paired ones occur in Scarabaeidae (Dynastinae and some Rutelinae), Psephenidae (some Eubriinae), Cucujidae (Passandrinae), and Cerambycidae (some Prioninae); and lateral gin traps are found in a number of Tenebrionidae and a few lamiine Cerambycidae. Most aquatic beetles have terrestrial pupae, but aquatic pupae with plastron-bearing gills are known in Torridincolidae and Psephenidae (Psephenoidinae). Loss of the pupal stage occurs in those beetles which have completely larviform females (Phengodidae and a few Lycidae).

Habitats

Although beetle larvae may be found in almost any type of situation, there are some habitats, like leaf litter, rotten wood, bracket fungi, and the rotten cambial layer beneath the bark of logs or standing dead trees, which are particularly fruitful for collecting. In the following sections, lists are given of the families likely to be found in some major habitat types. Some of these families are not commonly collected and may be dependent upon particular conditions of the microhabitat. Some litter inhabitants, like Ptilodactylidae or Dryopidae, for instance, are restricted to wetter, riparian conditions, while others (Dermestidae, Melyridae) turn up more often in dry litter samples. A subcortical region undergoing bacterial or yeast fermentation will harbor a different fauna than one which has been excavated by bark beetles or one which is relatively dry or covered with a bloom of fungal conidia. The litter and soil habitats are combined, since they grade into one another, and the subcortical habitat is treated together with dead wood for the same reason. Notes have been added on those forms which are more or less restricted to one or another. It is also possible to find inhabitants of bark and wood turning up in "leaf litter" samples, especially when the latter includes sticks and debris under logs, as in most forest areas. Some families, of course, are geographically restricted, and the family composition of samples from the southeastern United States may differ somewhat from that of samples collected in boreal forests. Exotic groups have been marked with an asterisk.

Marine Riparian. Beaches, Mudflats, Intertidal Rocks: Carabidae, Staphylinidae, Limnichidae (Hyphalinae,* Cephalobyrrhinae, Thaumastodinae*), Phycosecidae,* Melyridae, Tenebrionidae (Coniontini, Phaleriini, Opatrini), Salpingidae (Aegialitinae).

Freshwater Riparian: Cyathoceridae,* Microsporidae, Carabidae, Hydraenidae, Staphylinidae (*Bledius,* etc.), Georyssidae, Hydrophilidae, Helodidae, Ptilodactylidae, Chelonariidae, Heteroceridae, Limnichidae, Dryopidae, Curculionidae.

Freshwater Lotic: Amphizoidae, Hydraenidae, Eulichadidae, Ptilodactylidae, Lutrochidae, Elmidae, Psephenidae.

Freshwater Hydropetric: Torridincolidae,* Hydroscaphidae, Hydraenidae (some*), Psephenidae.

Freshwater Lentic: Haliplidae, Hygrobiidae,* Noteridae, Dytiscidae, Gyrinidae, Hydrophilidae, Helodidae, Chrysomelidae (Donaciinae), Curculionidae.

Leaf Litter and Soil: Carabidae, Ptiliidae, Leiodidae, Scydmaenidae, Scaphidiidae, Staphylinidae, Pselaphidae, Hydrophilidae (Sphaeridiinae), Histeridae, Eucinetidae, Clambidae, Dascillidae, Rhipiceridae, Scarabaeidae, Byrrhidae, Buprestidae, Ptilodactylidae, Chelonariidae, Dryopidae, Limnichidae, Elateridae, Cebrionidae, Throscidae, Eucnemidae (*Phyllocerus*),* Lycidae, Phengodidae, Lampyridae, Cantharidae, Dermestidae, Cleridae, Melyridae, Nitidulidae (*Epuraea, Haptoncus, Stelidota, Lasiodactylus,* etc.), Rhizophagidae, Phloeostichidae (Priasilphinae),* Phalacridae, Cucujidae (Silvaninae), Cryptophagidae, Languriidae (Cryptophilinae, Toraminae), Cerylonidae, Discolomidae,* Endomychidae, Corylophidae, Lathridiidae, Archeocrypticidae, Colydiidae, Tenebrionidae, Meloidae, Anthicidae, Euglenidae, Scraptiidae, Cerambycidae, Chrysomelidae (Clytrinae, Cryptocephalinae, Eumolpinae, Galerucinae, Alticinae), Ithyceridae, Curculionidae.

Most of the groups mentioned above live in leaf litter or in the upper layers of the soil, and some represent groups living elsewhere as larvae and dropping into the litter and soil for pupation. Specialized soil-dwelling larvae, grubs and wireworms of various kinds which feed on humus, plant roots, or other soil inhabitants, occur in the families Dascillidae, Rhipiceridae, Scarabaeidae, Buprestidae (Schizopodinae, Julodinae, Stigmoderini), Elateridae, Cebrionidae, Throscidae, Eucnemidae (*Phyllocerus*), Tenebrionidae (Tentyriinae, Phaleriini, Helaeini, Blaptini, Eleodini, Opatrini, Alleculini), Cerambycidae (Vesperinae, Anoplodermininae, Lepturinae, etc.), Chrysomelidae (Eumolpinae, Galerucinae, Alticinae), Ithyceridae, Curculionidae (many). Some leiodids and nitidulids are found in the vicinity of subterranean fungi.

Under Bark and in Dead Wood: Cupedidae, Micromalthidae, Rhysodidae, Carabidae, Ptiliidae, Leiodidae, Scydmaenidae, Scaphidiidae, Staphylinidae, Pselaphidae, Hydrophilidae (Sphaeridiinae), Sphaeritidae, Synteliidae, Histeridae, Eucinetidae, Clambidae, Lucanidae, Passalidae, Scarabaeidae (Ceratocanthinae, Aphodiinae, Rutelinae, Dynastinae, Cetoniinae), Buprestidae, Callirhipidae, Ptilodactylidae, Cerophytidae, Throscidae, Eucnemidae, Elateridae, Brachypsectridae, Lycidae, Derodontidae, Nosodendridae, Dermestidae, Bostrichidae, Anobiidae, Lymexylidae, Trogossitidae, Chaetosomatidae,* Cleridae, Melyridae, Sphindidae, Nitidulidae, Rhizophagidae, Phloeostichidae (Hymaeinae),* Cucujidae, Phalacridae, Helotidae,* Cryptophagidae, Languriidae (some Xenoscelinae), Erotylidae, Biphyllidae, Bothrideridae, Cerylonidae, Discolomidae,* Endomychidae, Coccinellidae, Corylophidae, Lathridiidae, Mycetophagidae, Ciidae, Tetratomidae, Melandryidae, Mordellidae, Rhipiphoridae, Colydiidae, Monommidae, Prostomidae, Synchroidae, Zopheridae, Tenebrionidae, Cephaloidae, Oedemeridae, Mycteridae, Boridae, Trictenotomidae,* Pythidae, Pyrochroidae, Pedilidae, Othniidae, Salpingidae, Inopeplidae, Anthicidae, Euglenidae, Scraptiidae, Cerambycidae, Anthribidae, Belidae,* Aglycyderidae,* Brentidae, Curculionidae, Scolytidae, Platypodidae.

Most of the above groups are collected under the bark of rotten logs and stumps or in rotten wood which is fairly damp and soft; others are found in the vicinity of fungus fruiting bodies on which they feed. Those families with flattened larvae specialized for living in subcortical spaces include Staphylinidae (Piestinae, Osoriinae, Oxytelinae), Buprestidae, Nitidulidae (some), Phloeostichidae (Hymaeinae), Cucujidae (Cucujinae, Laemophloeinae), Phalacridae (some), Tenebrionidae (few), Mycteridae, Boridae, Pythidae (most), Pyrochroidae, Pedilidae (some), Othniidae, Salpingidae, Inopeplidae and Cerambycidae. Larvae of Hydrophilidae (Sphaeridiinae), Sphaeritidae, Synteliidae, Histeridae, Nosodendridae, Nitidulidae (many Carpophilinae and Cryptarchinae, some Nitidulinae) Rhizophagidae (Rhizophaginae), Helotidae, and Biphyllidae, are usually associated with sap flows, slime fluxes, or fermenting areas under bark (some as predators of fly larvae). Others, like Brachypsectridae and some Coccinellidae, represent predators which may occur under loose bark. The flattened larvae of Prostomidae live in pockets of red rotten wood, where they feed in between wood layers, while flattened eucnemid larvae may move between fibers of wood which may be rotten to fairly solid. Those more cylindrical or grub-like larvae which bore through rotten wood are included in the families Cupedidae, Micromalthidae, Rhysodidae, Staphylinidae (Osoriinae), Lucanidae, Passalidae, Scarabaeidae, Callirhipidae, Cerophytidae, Throscidae, Elateridae, Lymexylidae, Melandryidae, Mordellidae, Zopheridae, Tenebrionidae (Phrenapatini, Tenebrionini, Amarygmini, Ulomini, Coelometopini, Cnodalonini, Strongyliini, Alleculini, Helopini, etc.), Cephaloidae, Oedemeridae, Trictenotomidae, Pythidae (*Priognathus*), Cerambycidae and Curculionoidea. Some groups are restricted to relatively sound wood or that which is undergoing dry rot; they often occur in branches and small twigs. These include Buprestidae, Eucnemidae, Dermestidae (Orphilinae), Bostrichidae, Anobiidae, Ciidae, Melandryidae, Mordellidae, Colydiidae, Salpingidae, Cerambycidae, and many Curculionoidea.

Dung: Ptiliidae, Leiodidae, Staphylinidae, Hydrophilidae (Sphaeridiinae), Histeridae, Scarabaeidae (few Geotrupinae, Hybosorinae, Scarabaeinae, Aphodiinae), Ptinidae.

Carrion: Leiodidae, Silphidae, Staphylinidae, Histeridae, Scarabaeidae (Troginae, few Scarabaeinae), Dermestidae (*Dermestes*), Cleridae (some Corynetinae), Phycosecidae,* Melyridae, Nitidulidae (*Omosita, Nitidula*).

Fruiting Bodies of Slime Molds (Myxomycetes): Leiodidae (many Anisotomini), Scaphidiidae (*Baeocera*), Staphylinidae, Eucinetidae, Clambidae, Sphindidae, Lathridiidae (most *Enicmus*).

Fruiting Bodies of Pyrenomycetes (Xylariales, etc.): Rhizophagidae (Monotominae), Cucujidae (Laemophloeinae), Phalacridae (*Litochropus, Acylomus*), Cryptophagidae (some *Cryptophagus*), Biphyllidae, Lathridiidae (some *Enicmus*), Colydiidae (*Bitoma, Cicones,* etc), Mycetophagidae (*Litargus*), Anthribidae.

Fruiting Bodies of Macrofungi (especially Polyporales): Carabidae (especially Lebiini), Ptiliidae (especially Nanosellini), Leiodidae, Eucinetidae, Derodontidae (*Derodontus*), Dermestidae (*Orphilus*), Bostrichidae (*Endecatomus*), Anobiidae (Dorcatominae), Trogossitidae, Cleridae (Thanerocierinae), Nitidulidae, Hobartiidae,* Cryptophagidae, Lamingtoniidae,* Languriidae (*Cryptophilus*), Erotylidae, Sphaerosomatidae,* Discolomidae,* Endomychidae, Corylophidae, Lathridiidae, Mycetophagidae, Pterogeniidae,* Archeocrypticidae, Ciidae, Tetratomidae, Melandryidae (Hallomeninae, Eustrophinae, Orchesiini), Mordellidae (*Mordella*), Colydiidae, Zopheridae, Tenebrionidae (Bolitophagini, Toxicini, Diaperini), Anthribidae (*Euparius*).

Epigean Gasteromycetes (Puffballs etc.): Leiodidae (*Creagrophorus, Nargomorphus*), Nitidulidae (*Pocadius*), Cryptophagidae (some *Cryptophagus*), Endomychidae (*Lycoperdina*).

Hypogean Fungi (Hemenogastrales, Tuberales, etc.): Leiodidae (Hydnobiini, Leiodini, Catopocerinae, Coloninae), Scarabaeidae (some Geotrupinae), Nitidulidae (*Thalycra*).

Mosses and Liverworts: Pselaphidae, Byrrhidae, Artematopidae (Macropogoninae), Elateridae, Cantharidae, Tenebrionidae (Leiochrini*).

Surfaces of Leaves: Phalacridae (*Acylomus, Stilbus*), Coccinellidae, Corylophidae (*Corylophodes*), Lathridiidae (Corticariinae), Chrysomelidae (Criocerinae, Chrysomelinae, Cassidinae, Chlamisinae), Curculionidae (Attelabinae, Gonipterinae, etc.).

Mining in Leaves: Buprestidae (Trachyinae), Nitidulidae (*Xenostrongylus,* Anister**), Chrysomelidae (Zeugophorinae, some Hispinae and Alticinae), Curculionidae (*Orchestes, Prionomerus,* Rhamphus,* Cionus,** etc).

Flowers, Flower Heads, Male Cones: Nitidulidae (Cateretinae, Meligethinae, some Nitidulinae), Phalacridae (*Olibrus*), Languriidae (some Xenoscelinae on cycads), Byturidae, Meloidae (triungulins), Rhipiphoridae (triungulins), Nemonychidae (on gymnosperms), Oxycorynidae (Hydnoraceae),* Curculionidae (Allocoryninae and Antliarrhininae on cycads, many groups on angiosperms).

Seeds, Seed Pods: Nitidulidae, Bruchidae, Urodontidae, Anthribidae, Curculionidae.

TECHNIQUES

Techniques for collecting and preserving larval Coleoptera and preparing them for study are covered in Volume 1 and in a number of general texts or manuals, such as Borror, DeLong and Triplehorn (1981), Cogan and Smith (1974), Emden (1942a), Martin (1977), Oldroyd (1958), Peterson (1951, 1959), Upton and Norris (1980) and Walsh and Dibb (1975), and will not be treated in detail here. Although many larvae are easily collected by hand or with an aspirator, mass collecting techniques are very useful in obtaining specimens which are very small or which live in habitats where they are difficult to observe. When specimens are widely scattered among leaf litter and other debris, it is necessary to first concentrate them and then remove them from the concentrate into preservative or rearing containers. Concentration is accomplished by a **sifter (concentrator** or **extraction sieve)**. This consists of a long cloth cylinder, narrowed at one end, where it is tied off, and with a metal hoop (with a handle) holding the other end open; a second metal hoop (also with a handle) to which an 8 or 10 mm metal sieve is soldered, is sewn into the cloth at some distance from the open hoop. When leaf litter, humus, sticks, bits of wood and bark, etc. are placed between the open hoop and the sieve and vigorously shaken, fine debris passing through the sieve accumulates at the narrow, tied end, and is eventually poured into cloth bags for further processing.

Extraction devices using heat to drive living organisms out of the sample and into preserving fluid are usually called **Berlese funnels** or **Tulgren funnels,** depending upon construction details and heat source; the most common type uses electric light bulbs for a heat source and alcohol as a preserving fluid. Noxious chemicals may be used instead of heat, and alcohol may be replaced by wet towels if living samples are desired. A **Winkler apparatus** is based on the same principle but the concentrate is kept in small, gauze bags, which are allowed to dry over a funnel.

Another collecting device is the **pitfall trap,** which consists of an open container set into the ground, usually covered with a rain roof and filled with a non-volatile liquid, which kills the insects and retards fungal and bacterial decay until the sample can be placed in a more permanent preservative. A mixture of propylene glycol and propylene phenoxetol has a useful combination of fungicidal, bactericidal, humectant, and narcotic properties.

The **elutriator** is a useful device for mass collecting aquatic, littoral or interstitial organisms from sand, gravel, or bottom debris. It consists of a cylindrical container open at one end, with several small openings near the other end, through which air and water are pumped; when the substrate is placed in the container, light organic matter and living organisms float to the top and spill over the edge of the container to be caught in a fine mesh screen (Kingsbury and Beveridge, 1977). Many other special collecting techniques exist for aquatic insects, and these are discussed in some detail in Merritt and Cummins (1984).

Although the above methods will result in large collections of larval Coleoptera, many of which can be identified by comparison with previously determined specimens, it is essential to associate as many larvae as possible with adults to insure correct identifications at the species level. This may be accomplished by 1) collecting larvae alive and attempting to rear them through to the adult stage in the laboratory, 2) collecting pupae associated with larval exuviae, and allowing the pupae to eclose in the laboratory, or 3) collecting living adults and allowing them to produce eggs and larvae in the laboratory. Techniques involved in rearing larvae are numerous, and some are covered in Peterson (1959), or in other chapters of this book.

Fixation of specimens is usually accomplished by boiling in water or by using one of several fixatives containing alcohol and formalin or acetic acid (KAAD, Carnoy's, Kahle's, Pampel's, etc.). It is sometimes desirable to think about just what is required of a fixative. Some fixatives are good for preserving internal structures or for histological studies, but they make some specimens very difficult to macerate for cuticular examination. In most cases, some breakdown of the soft tissue is an advantage in obtaining properly macerated and cleared specimens. It may be necessary to preserve part of a series in an internal fixative and part in a relatively weak acid alcohol.

I have found the following procedure helpful in preparing larval material for study:

1. Sketch the entire larva or make a permanent habitus drawing in case of uniques or short series.
2. Note the number and development of stemmata, which may be difficult to see after the specimen is treated with KOH.
3. Make note of any color patterns, which will also disappear after maceration.
4. Remove the head (with a minutin pin in small specimens).
5. Cut the abdomen at the 4th or 5th segment, so that the larva is now in 3 pieces.
6. Remove a section of the gut and place contents on a slide, if it is desirable to determine food source.
7. Place the 3 pieces of larva in potassium hydroxide (KOH, usually 10% solution), and leave overnight cold, or warm for a few minutes on a hot plate or over a flame. The time varies considerably with the size and condition of the specimen.
8. While the pieces are still in KOH, use forceps or dissecting pins or hooks to remove soft tissue. The head may be squeezed gently with forceps in order to force tissue out of the occipital foramen. At this point note any sclerotizations in the proventriculus or rectum.
9. Remove the mandibles with a hook.
10. It may be necessary to remove the entire maxillolabial complex (ventral mouthparts) in order to get a better view of the epipharynx. The alternative removal of the labrum will destroy the roof of the cibarium in many cases.
11. Transfer pieces to alcohol to which a few drops of acetic acid have been added to neutralize the KOH.
12. Transfer to 80% alcohol.
13. Transfer to glycerine for study.
14. Store in glycerine or make permanent slide(s). Various techniques for making permanent slides may be found in any textbook on histology or microtechnique and in Volume 1. With very small larvae, it may be necessary to carefully clear the entire specimen and make a whole mount permanent slide. The disadvantages of this are that the specimen must be viewed from a single aspect and that some structures are superimposed on others, thus obscuring detail.

CLASSIFICATION

The classification used here represents a compromise between one reflecting this author's present ideas on the phylogeny of the group (based in large part on the works of Crowson and others) and a more conservative treatment, especially when the limits of taxa are in dispute or the opinions of other contributors to this volume are at variance with my own views; in the last case, I have usually adopted the family limits and nomenclature preferred by the author of that particular section. Within Carabidae, the Trachypachinae, Cicindelinae, and Paussinae are often given family status, while several staphylinoid families (Limulodidae, Leptinidae, Scaphidiidae, Micropeplidae, and Dasyceridae) may only be subfamilial groups within the families Ptiliidae, Leiodidae, or Staphylinidae. The recognition of only three families of Scarabaeoidea is in disagreement with the classification used by Crowson and others, where Troginae, Geotrupinae, Ceratocanthinae, and some other groups are given family rank. The Trogossitidae is used in the broad sense, and includes the families Peltidae and Lophocateridae, as defined by Crowson (1964d, 1966a, 1970). The Cucujidae (including Passandrinae, Laemophloeinae, and Silvaninae) certainly would be subdivided in a phylogenetic classification, but the neighboring groups are still in need of study and the limits of various cucujoid families may be fluid for the next few years. Within Curculionoidea, Urodontidae, Scolytidae, Platypodidae, Ithyceridae, and Brentidae are given family rank, while Attelabinae, Rhynchitinae, Allocoryninae, Apioninae, and *Cylas* are all included in a broadly defined Curculionidae; this is not consistent with most recent phylogenetic studies. Rhynchitinae, Allocoryninae and *Cylas* are keyed out separately, while Apioninae, Attelabinae, and Scolytidae are not.

The family names used are, for the most part, those familiar to North American workers. Helodidae has been used, although Scirtidae appears to be the correct name for the group (Pope 1976). The use of Microsporidae instead of Sphaeriidae follows a recent opinion of the International Commission on Zoological Nomenclature (I.C.Z.N., 1985).

The superfamily classification is basically that used by Lawrence and Newton (1982), which, in turn, follows Crowson (1955, 1960b, 1964d, 1971, 1972b, 1973b, 1978, 1981), with the following exceptions: 1) Hydraenidae is placed in Staphylinoidea; 2) Hydrophilidae (*sensu lato*) plus Histeroidea are included in an expanded Hydrophiloidea (as first proposed by Böving and Craighead, 1931); 3) the three families comprising Crowson's Artematopoidea have been moved to Dryopoidea (Callirhipidae), Elateroidea (Artematopidae), and Cantharoidea (Brachypsectridae); 4) Crowson's Dermestoidea is dismembered, Dermestidae being placed in Bostrichoidea and the other 3 families remaining together as Derodontoidea; 5) the sections Clavicornia and Heteromera of Crowson's Cucujoidea have been treated as separate superfamilies, Cucujoidea and Tenebrionoidea; and 6) the family Stylopidae is considered to represent an independent order, the Strepsiptera.

Although the superfamilies have not been grouped into series, as was done by Crowson (1955, 1960b, 1981), some of his terms are useful in general discussions. There may be some confusion, however, between Crowson's series and the informal "lineages" used by Lawrence and Newton (1982). The Staphyliiformia and the staphyliniform lineage are equivalent terms and include the superfamilies Staphylinoidea and Hydrophiloidea (*sensu lato*). The elateriform lineage includes Crowson's Eucinetiformia (Eucinetoidea), Scarabaeiformia (Dascilloidea and Scarabaeoidea), and Elateriformia (Byrrhoidea, Buprestoidea, Dryopoidea, Elateroidea and Cantharoidea in the list below). The cucujiform lineage includes Crowson's Bostrychiformia (*sic*) (Derodontoidea and Bostrichoidea below) and Cucujiformia (Cleroidea, Lymexyloidea, Cucujoidea, Tenebrionoidea, Chrysomeloidea, and Curculionoidea below). In various discussions above and below, the terms cucujiform and elateriform are used in Crowson's original sense to avoid further confusion.

CLASSIFICATION[1]

Order **COLEOPTERA** (25000:350000)
- Suborder Archostemata
  - Ommatidae* (0:5)
    - (= Ommadidae, incl. Tetraphaleridae)
  - Cupedidae (4:26)
    - (= Cupesidae)
  - Micromalthidae (1:1)
- Suborder Myxophaga
  - Cyathoceridae* (0:2)
    - (= Lepiceridae)
  - Torridincolidae (0:27)
  - Microsporidae (3:18)
    - (= Sphaeriidae)
  - Hydroscaphidae (1:14)
- Suborder Adephaga
  - Rhysodidae (8:330)
  - Carabidae (2500:30000)
    - (incl. Brachinidae, Cicindelidae, Omophronidae, Paussidae, Pseudomorphidae, Trachypachidae, etc.)
  - Haliplidae (67:200)
  - Hygrobiidae (0:5)
    - (= Pelobiidae)
  - Amphizoidae (3:4)
  - Noteridae (13:230)
    - (incl. Phreatodytidae)
  - Dytiscidae (446:3000)
  - Gyrinidae (52:1100)
- Suborder Polyphaga
  - Staphylinoidea
    - Hydraenidae (90:900)
      - (= Limnebiidae)
    - Ptiliidae (115:400)
    - Limulodidae (4:28)
      - (= Cephaloplectidae)
    - Agyrtidae (14:63)
    - Leiodidae (200:2300)
      - (= Anisotomidae, Liodidae; incl. Camiaridae, Catopidae, Cholevidae, Colonidae, Leptodiridae)
    - Leptinidae (6:13)
      - (incl. Platypsyllidae)
    - Scydmaenidae (180:3570)
    - Micropeplidae (16:50)
    - Dasyceridae (3:12)
    - Scaphidiidae (60:1120)
    - Silphidae (30:215)
    - Staphylinidae (3500:31200)
      - (incl. Brathinidae, Empelidae)
    - Pselaphidae (500:8300)
      - (incl. Clavigeridae)
  - Hydrophiloidea
    - Hydrophilidae (256:2200)
      - (incl. Helophoridae, Hydrochidae, Sphaeridiidae, Spercheidae)
    - Georyssidae (2:35)
      - (= Georissidae)
    - Sphaeritidae (1:3)
    - Synteliidae (0:5)
    - Histeridae (350:3700)
      - (incl. Niponiidae)
  - Eucinetoidea
    - Eucinetidae (8:30)
    - Clambidae (10:120)
      - (incl. Calyptomeridae)
    - Helodidae (34:600)
      - (= Cyphonidae, Scirtidae)
  - Dascilloidea
    - Dascillidae (5:80)
      - (incl. Karumiidae)
    - Rhipiceridae (5:52)
      - (= Sandalidae)
  - Scarabaeoidea
    - Lucanidae (40:1200)
    - Passalidae (3:500)
    - Scarabaeidae (1400:25000)
      - (incl. Acanthoceridae, Ceratocanthidae, Cetoniidae, Diphyllostomatidae, Geotrupidae, Glaphyridae, Hybosoridae, Pleocomidae, Trogidae, etc.)
  - Byrrhoidea
    - Byrrhidae (40:300)
      - (incl. Syncalyptidae)
  - Buprestoidea
    - Buprestidae (750:15000)
      - (incl. Schizopodidae)
  - Dryopoidea
    - Callirhipidae (1:150)
    - Eulichadidae (1:12)
    - Ptilodactylidae (12:450)
    - Chelonariidae (1:230)
    - Psephenidae (17:120)
      - (incl. Eubriidae, Psephenoididae)
    - Lutrochidae (3:15)
    - Dryopidae (15:230)
    - Limnichidae (28:220)
    - Heteroceridae (31:300)
    - Elmidae (88:1100)
      - (= Elminthidae, Helminthidae)

1. The number of species for America north of Mexico and the world respectively is given in parentheses.

* = larvae not known

Elateroidea
- Artematopidae (9:60)
  - (incl. Eurypogonidae)
- Cerophytidae (2:10)
- Elateridae (844:9000)
  - (incl. Dicronychidae, Lissomidae)
- Cebrionidae (46:250)
  - (incl. Plastoceridae of authors)
- Throscidae (21:190)
  - (= Trixagidae)
- Eucnemidae (71:1300)
  - (= Melasidae; incl. Perothopidae, Phylloceridae)

Cantharoidea
- Brachypsectridae (1:4)
- Cneoglossidae* (0:7)
- Plastoceridae* (0:2)
- Homalisidae (0:10)
- Lycidae (75:3500)
- Drilidae (0:80)
- Phengodidae (23:200)
  - (incl. Rhagophthalmidae)
- Telegeusidae* (3:5)
- Lampyridae (115:1900)
- Omethidae* (7:21)
- Cantharidae (410:5100)
  - (incl. Chauliognathidae)

Derodontoidea
- Derodontidae (9:22)
  - (incl. Laricobiidae, Peltasticidae)
- Nosodendridae (2:48)
- Jacobsoniidae (0:10)
  - (= Sarothriidae)

Bostrichoidea
- Dermestidae (110:880)
  - (incl. Thorictidae, Thylodriidae)
- Bostrichidae (100:700)
  - (= Bostrychidae; incl. Endecatomidae, Lyctidae, Psoidae)
- Anobiidae (332:1600)
- Ptinidae (50:500)
  - (incl. Ectrephidae, Gnostidae)

Lymexyloidea
- Lymexylidae (2:50)
  - (= Lymexylonidae; incl. Atractoceridae)

Cleroidea
- Phloiophilidae (0:2)
- Trogossitidae (69:600)
  - (incl. Lophocateridae, Ostomidae, Peltidae, Temnochilidae)
- Chaetosomatidae (0:9)
- Cleridae (286:4000)
  - (incl. Corynetidae, Korynetidae)
- Phycosecidae (0:4)
- Acanthocnemidae* (0:1)
- Melyridae (456:5000)
  - (incl. Dasytidae, Malachiidae, Prionoceridae, Rhadalidae)

Cucujoidea (= clavicornia)
- Protocucujidae* (0:3)
- Sphindidae (6:35)
  - (incl. Aspidiphoridae)
- Nitidulidae (150:3000)
  - (incl. Brachypteridae, Cateretidae, Cybocephalidae, Smicripidae)
- Rhizophagidae (53:250)
  - (incl. Monotomidae)
- Boganiidae (0:8)
- Phloeostichidae (0:8)
- Helotidae (0:100)
- Cucujidae (140:1200)
  - (incl. Catogenidae, Laemophloeidae, Passandridae, Scalidiidae, Silvanidae)
- Propalticidae* (0:35)
- Phalacridae (122:600)
  - (incl. Phaenocephalidae?)
- Hobartiidae (0:2)
- Cavognathidae (0:5)
- Cryptophagidae (140:600)
  - (incl. Catopochrotidae, Hypocopridae)
- Lamingtoniidae* (0:1)
- Languriidae (38:900)
  - (incl. Cryptophilidae)
- Erotylidae (50:2500)
  - (incl. Dacnidae)
- Biphyllidae (4:200)
  - (= Diphyllidae)
- Byturidae (2:15)
- Bothrideridae (13:300)
- Sphaerosomatidae (0:50)
- Cerylonidae (18:650)
  - (= Cerylidae; incl. Aculagnathidae, Anommatidae, Dolosidae, Euxestidae, Murmidiidae)
- Discolomidae (0:400)
  - (= Notiophygidae)
- Endomychidae (45:1300)
  - (incl. Merophysiidae, Mycetaeidae)
- Coccinellidae (475:4200)
  - (incl. Cerasommatidiidae, Epilachnidae)
- Corylophidae (60:400)
  - (= Orthoperidae)
- Lathridiidae (120:500)

Tenebrionoidea (= heteromera)
- Mycetophagidae (26:200)
- Archeocrypticidae (1:35)
- Pterogeniidae (0:8)
- Ciidae (86:550)
  - (= Cisidae)
- Tetratomidae (9:30)
- Melandryidae (70:450)
  - (= Serropalpidae)
- Mordellidae (200:1200)
- Rhipiphoridae (50:300)
- Colydiidae (75:1000)
  - (incl. Adimeridae, Monoedidae)
- Monommidae (5:225)
- Prostomidae (1:20)
- Synchroidae (2:8)
- Zopheridae (48:125)
  - (incl. Merycidae)
- Perimylopidae (0:8)
- Chalcodryidae (0:6)
- Tenebrionidae (1550:18000)
  - (incl. Alleculidae, Cossyphodidae, Lagriidae, Nilionidae, Rhysopaussidae, Tentyriidae)
- Cephaloidae (10:20)
  - (incl. Nematoplidae, Stenotrachelidae)

Meloidae (350:2800)
(incl. Tetraonycidae)
Oedemeridae (75:1500)
Mycteridae (10:160)
(incl. Hemipeplidae)
Boridae (2:9)
Trictenotomidae (0:12)
Pythidae (9:50)
Pyrochroidae (12:125)
Pedilidae (80:150)
(incl. Cononotidae)
Othniidae (6:50)
(= Elacatidae)
Salpingidae (11:200)
(incl. Aegialitidae, Dacoderidae, Eurystethidae, Tretothoracidae)
Inopeplidae (2:65)
Anthicidae (300:2000)
Euglenidae (40:800)
(= Aderidae, Hylophilidae, Xylophilidae)
Scraptiidae (45:250)
(incl. Anaspididae)

Chrysomeloidea
Cerambycidae (1200:35000)
(incl. Disteniidae, Hypocephalidae, Parandridae, Spondylidae)
Bruchidae (110:1300)
(= Lariidae, Mylabridae)
Chrysomelidae (1500:35000)
(incl. Cassididae, Cryptocephalidae, Megalopodidae, Sagridae, etc.)

Curculionoidea (= Rhynchophora)
Nemonychidae (8:65)
(= Rhinomaceridae)
Anthribidae (86:2500)
(= Platystomidae)
Urodontidae (0:50)
(= Bruchelidae)
Oxycorynidae (0:30)
Belidae (0:150)
Aglycyderidae (0:164)[2]
Ithyceridae (1:1)
Brentidae (6:1200)
(= Brenthidae)
Curculionidae (2500:50000)
(incl. Allocorynidae, Antliarrhinidae, Apionidae, Apoderidae, Attelabidae, Calendridae, Cossonidae, Cyladidae, Pterocolidae, Rhynchitidae, Rhynchophoridae, etc.)
Scolytidae (500:6000)
(= Ipidae)
Platypodidae (7:1500)

2. Many species in Hawaii.

KEY TO THE FAMILIES AND MANY SUBFAMILIES OF COLEOPTERA LARVAE (WORLDWIDE)

John F. Lawrence
Division of Entomology, CSIRO

Introduction

The following key was originally constructed from a 150 × 550 character/taxon matrix, using the DELTA system, devised by M. J. Dallwitz (1974, 1978, 1980, 1984). Its coverage is worldwide, although all taxa not occurring in America north of Mexico are labelled "exotic." The key extends below the family level for many groups, and some families may key out a number of times (24 families more than 5 times; Cerambycidae 18 and Chrysomelidae 22). There are two characteristics of the key which will intimidate beginners: its length and the fact that many sections have 3 or more choices. The length (395 "couplets") is deceiving, since in a computer-generated key, the number of choices required to reach a solution is kept to a minimum by having early "couplets" separate large groups of taxa. This ideal is compromised when particularly useful or obvious characters are given more weight and thus used earlier in the key, or when some sections are revised "by hand", as was done in this case. In spite of this, the user will find that relatively few choices are necessary to key out most taxa.

There are some very useful characters which have been difficult for beginners to interpret correctly during test runs. These include segment counts for antennae, palps, or legs, nature of the endocarina, and various features of the mandible. Although these structures are covered in detail in the morphology section, possible difficulties in determining their character states will be discussed briefly below. The abbreviations T1, T2, T3 (thoracic segments) and A1, A2, etc. (abdominal segments) are frequently used.

The **endocarina** (or paired endocarinae) on the head are cuticular thickenings that extend internally. They are usually thicker and/or darker than the ecdysial lines (epicranial stem, frontal arms) which do not extend internally. If this is not obvious, the head can be cleared in KOH and the internal ridge felt with a micropin.

Errors in determining the correct number of antennal segments usually arise from counting the basal membrane or the sensorium as an additional segment; complete absence of pigmentation, setae, or sensilla should distinguish either of these from a true segment. When the sensorium is apical, there will be no setae at the antennal apex, and those lying at the base of the sensorium usually represent the remnants of a reduced apical segment. When the antenna is highly reduced, it may consist of a membranous area containing a few setae and sensilla; this is considered to be 1-segmented.

Errors in counting palp segments almost always arise from confusion between the basal palp segment and the palpifer or palpiger from which it arises. In the key and in subsequent family coverage, a palpifer (or maxillary palpiger) or palpiger (labial palpiger) is a shelf-like extension of the apicolateral portion of the stipes or prementum, respectively, which is not separated by a distinct line of demarcation and does not have a distinct inner edge. There are some instances where an intermediate condition exists, and these usually have been coded for both character states. **Difficulties in counting leg segments** cannot be avoided in groups with reduced legs, like Cerambycidae or Mordellidae, and these larvae are usually coded for several states. In some larvae, the coxa is not clearly separated from the pleural region, while in others the tarsungulus may appear to be divided. As mentioned above, the pretarsus is considered to be a separate segment consisting of either paired claws or a single claw; thus most Archostemata and Adephaga have 6 leg segments. The polyphagan leg may be 5-segmented or less, and the last segment or tarsungulus may or may not be claw-like. This terminology conflicts with that used by Peterson (1951) where the terminal claws are not counted as leg segments.

Mandibular characters are extremely useful and cannot be avoided in a family key, but they do **require the removal of one or both mandibles**; sometimes it is possible to examine the mandibles in place in well-cleared specimens, but the superimposition of the maxillae may create artifacts. Mandibular characters are discussed in detail in the morphology section above, and only a few comments are needed here. A mandibular **mola** is an enlargement and modification of the mandibular base which meets a similar structure on the opposing mandible for the purpose of ingesting, compacting, or breaking down food material; it is often asperate, tuberculate, or ridged. In some larvae (various Carabidae, Histeridae or Hydrophilidae), the mandibles are slightly enlarged at the base but have no structure which can be called a mola. A **prostheca** is any structure just distad of the mola that is not a sclerotized tooth attached to the mola (premolar lobe), and it may be membranous, hyaline, or partly sclerotized, or may consist of a group of hyaline processes or a brush of hairs. A mandible without a mola cannot have a prostheca, as defined here, and similar structures at the mandibular base are called processes, lobes, or brushes, or are sometimes referred to collectively as a lacinia mobilis in the key or family diagnoses. The term **pseudomola** has been used for a food processing area of the mandible which may be, but is usually not, basal, and which is thought to have evolved secondarily in a mandible type which has lost the true basal mola. Needless to say, this term has little practical value in a key, but it has been used for some subapical, mola-like structures which cannot be confused with a basal mola. A **retinaculum** is a heavily sclerotized tooth on the incisor edge of a mandible lacking a mola. Heavy incisor lobes or teeth in mola-bearing mandibles of dascillids and scarabaeoids are called scissorial teeth.

Urogomphi are paired processes on the 9th tergum; there is usually a single or dominant pair. They are not regarded as homologous structures but as numerous parallel developments, and they vary considerably throughout the order, being simple or complex, fixed or articulated, and of various shapes and orientations. When a second, smaller pair of processes occurs in front of the urogomphi, they are called **pregomphi.**

KEY

—IMPORTANT

A. For ease of use: Read the introduction to this key (p. 184).

B. If the head is partially embedded in the thorax, it *must* be dissected so the rear of the head, cranial sutures and related structures are clearly visible.

C. The presence or absence of a mola can usually be determined without removing the mandibles by cutting on both sides of the head between the mandible and maxilla so the maxillolabial complex can be pried downward, revealing the inner bases of the mandibles.

D. Key Figures: Some figures are duplicated many places in the key. They are usually placed in numerical sequence (e.g. 8, 23, 96, 150, etc.) on the page or facing page where they are used. If the figure is not present, it will be on the previous (←) or following (→) two pages. Check the bottom corners for its location.

1. Thoracic legs absent or represented by non-articulated and non-segmented protuberances (pedal lobes), which are broader than long. All forms lightly sclerotized, except some Eucnemidae, which have the head modified forming a wedge-like plate. Mandibular mola usually absent 4

1′. Thoracic legs present and articulated at base (or if basal articulation not well defined, then legs segmented or narrow and longer than wide) 2

2(1′). Labrum partly (fig. 12) or entirely (fig. 16) fused to head capsule (without or with incomplete clypeolabral suture) 31

2′. Labrum entirely free (fig. 13) (clypeolabral suture complete) 3

3(2′). Mandibular mola absent (figs. 46, 51, 53) 115

3′. Mandibular mola present (figs. 68, 72, 81) 248

4(1). Mouthparts and antennae absent; body very lightly sclerotized and moderately to strongly curved ventrally (C-shaped) (fig. 1). Endoparasitoids of cockroaches. Exotic (Neotropical and Old World) .... **(Rhipidiinae part) *Rhipiphoridae*** p. 509

4′. Mouthparts and antennae present 5

5(4′). Head protracted or slightly retracted into thorax (figs. 8, 10) 6

5′. Head strongly retracted into thorax (fig. 9) 18

6(5). Head highly modified, usually heavily sclerotized, with paired dorsal and ventral endocarinae, or forming wedge-like plate, which is usually apically serrate (fig. 11); antennae minute, 2-segmented; mandibles either fused to head capsule or with non-opposable, divergent apices (fig. 122); body usually strongly flattened or with enlarged prothorax bearing paired, longitudinal or T-shaped rods; spiracles biforous. In dead wood .... **(part) *Eucnemidae*** p. 419

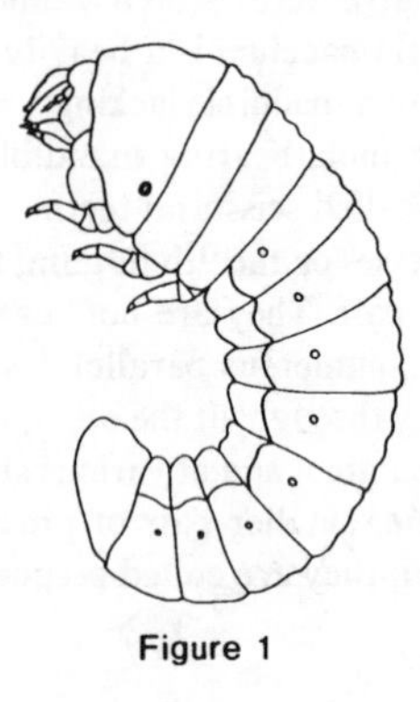
Figure 1

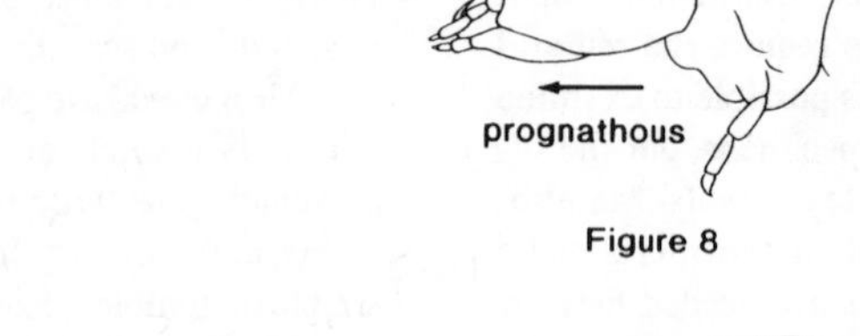

Figure 8

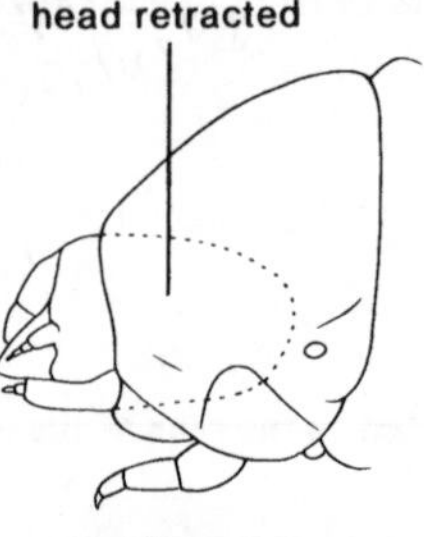

Figure 9

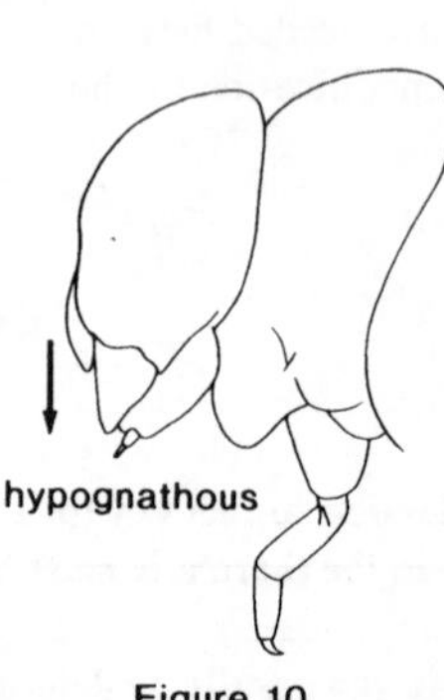

Figure 10

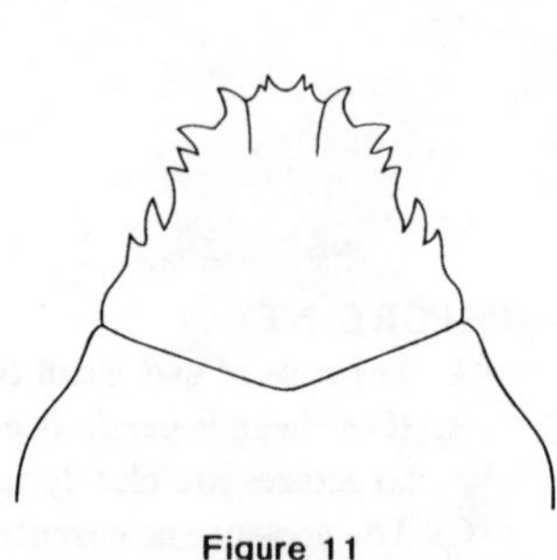
Figure 11

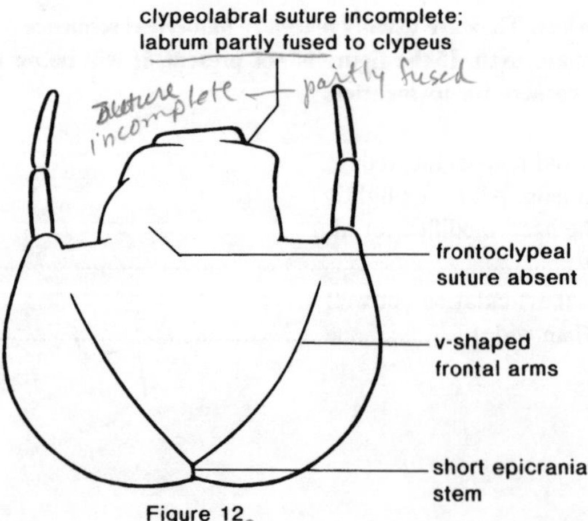

Figure 12

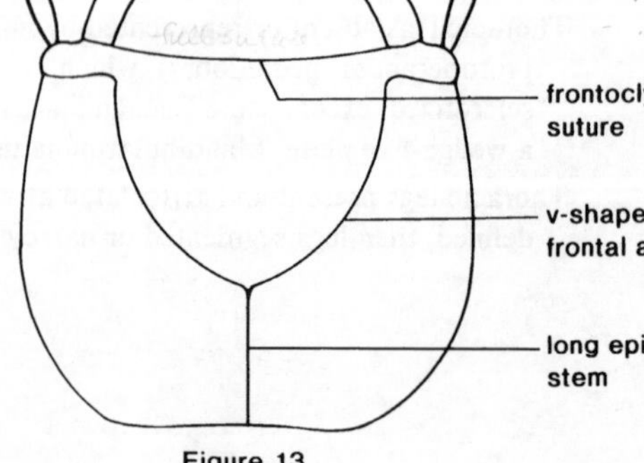

Figure 13

6′.	Head capsule normal, without or with a single, median endocarina; mandibles free, with opposable apices; if spiracles biforous, antennae well-developed and 3-segmented; body form different	7
7(6′).	Antennal segments 1	8
7′.	Antennal segments 2; mandible with broad base and acute apex; body lightly sclerotized, usually with a deep indentation on each side between T2 and T3. Ectoparasites in galleries of ambrosia beetles (Platypodidae) (**see 3rd choice**)	(*Sosylus* part) ***Bothrideridae*** p. 477
7^2.	Antennal segments 3	15
8(7).	With deep indentation on each side between T2 and T3; labrum fused to head capsule; maxilla without apical lobes; maxillary and labial palps absent or represented by minute papillae. In galleries of ambrosia beetles (Platypodidae)	(*Sosylus* part) ***Bothrideridae*** p. 477
8′.	Without thoracic indentations; without other characters in combination	9

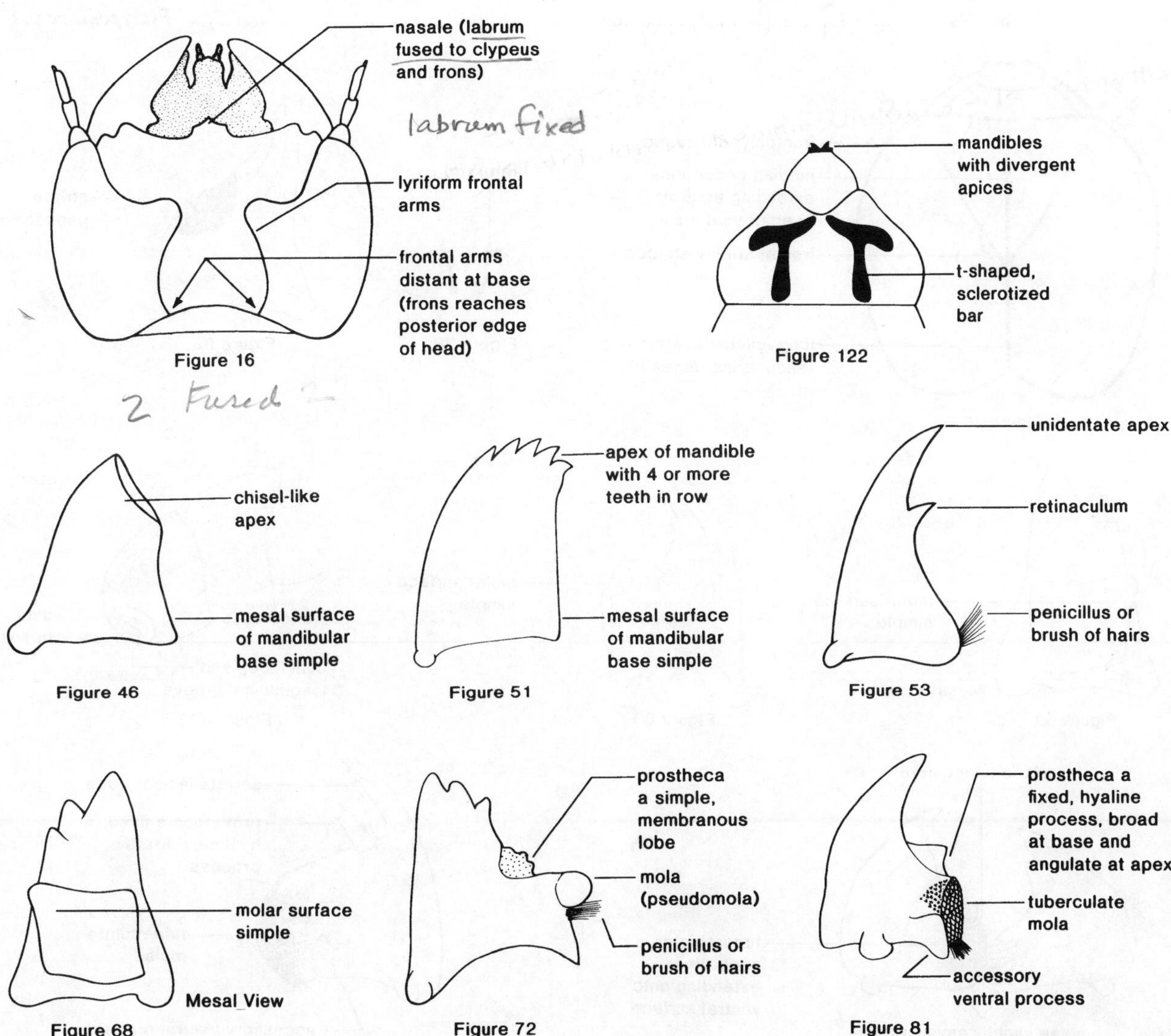

9(8′). Segments in maxillary palp 1 or 2 (fig. 96) 10

9′. Segments in maxillary palp 3 (fig. 105) **(see 3rd choice)** 13

9². Segments in maxillary palp 0. In nests of wild bees .... (part) ***Meloidae*** p. 530

10(9). Body straight, somewhat flattened, distinctly tapering posteriorly; head prognathous (fig. 8), deeply emarginate posteriorly; median endocarina extending anteriorly almost to edge of clypeus. Leafminers .... (part) ***Curculionidae*** p. 594

10′. Body usually moderately to strongly curved ventrally (C-shaped) or if straight, then without other characters in combination 11

11(10′). Labral rods absent; mandibular mola present (fig. 69); hypopharyngeal sclerome present (fig. 119). In seeds, dried fruits, and other stored products .... (part) ***Anthribidae*** p. 586

11′. Labral rods present (fig. 38); mandibular mola absent, but asperate or striate pseudomola (fig. 52) may occur on dorsal surface of mandibular base; hypopharyngeal sclerome absent 12

12(11′). Protergum with pair of sclerotized plates or series of rings joined by fine lines; epipleural and pleural regions of abdomen with 3 or more longitudinal lobes (parallel to long axis of body); clypeus vestigal and labrum enlarged; dorsal surface of mandibular base with series of transverse ridges and a group of asperities forming a pseudomola; maxillary and labial palps usually 1-segmented. Living in fungus-lined (blackened) tunnels in wood (ambrosia beetles) .... ***Platypodidae*** p. 616

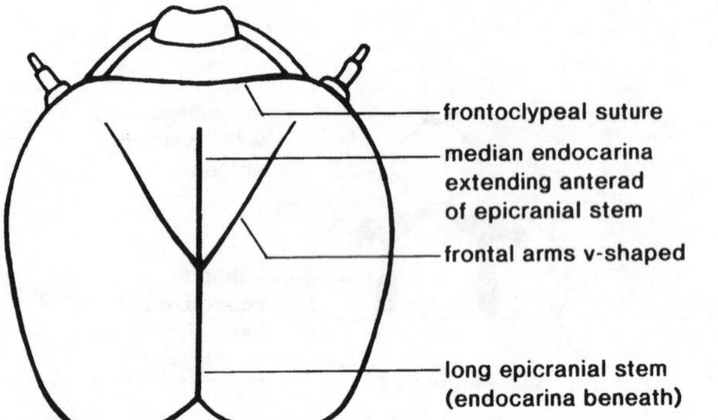

Figure 20

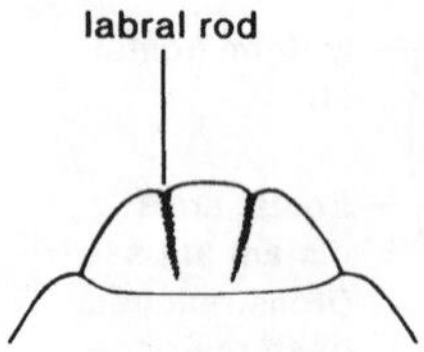

Figure 38

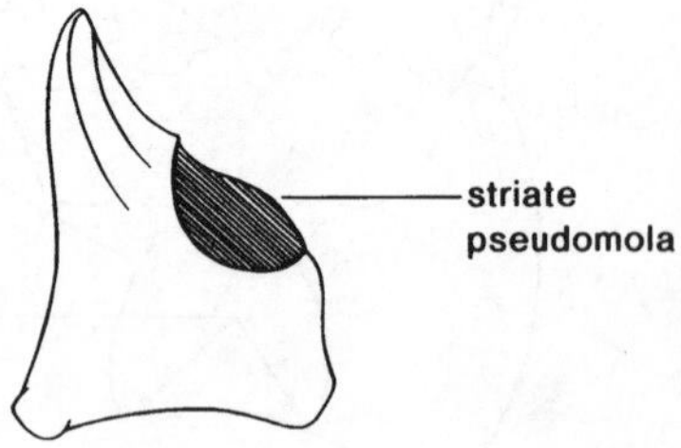

Figure 52

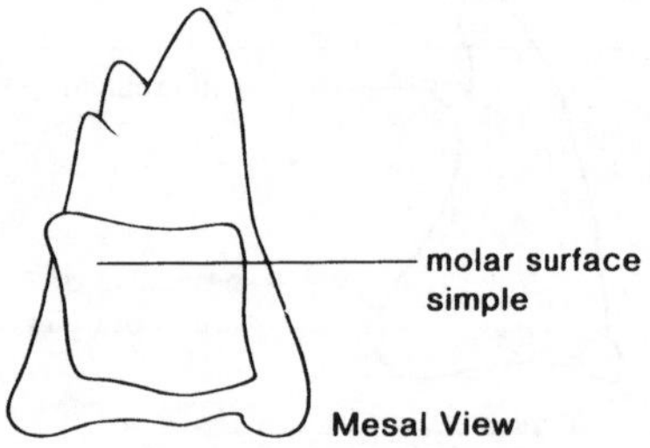

Figure 68

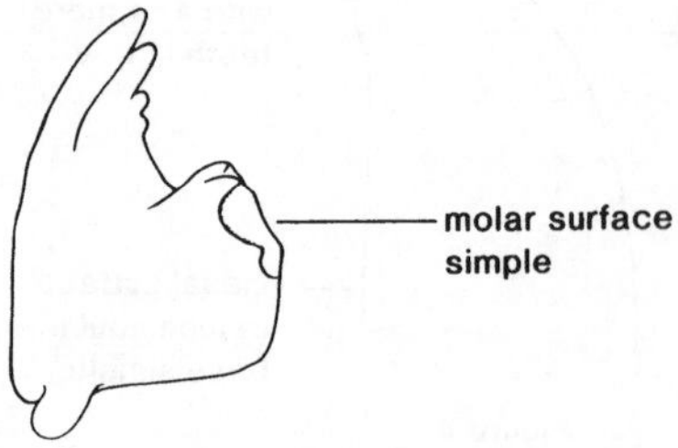

Figure 69

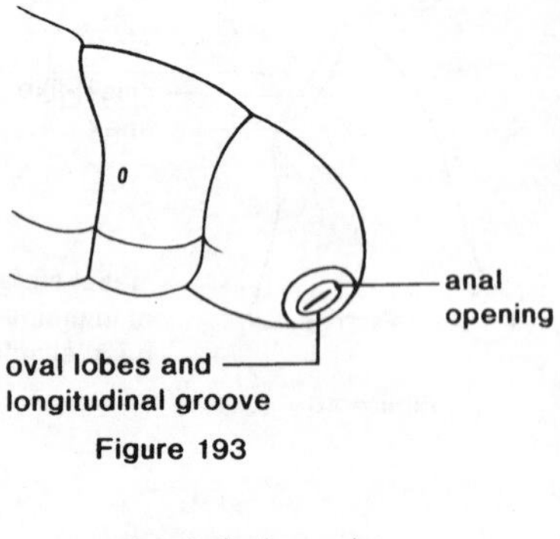

Figure 193

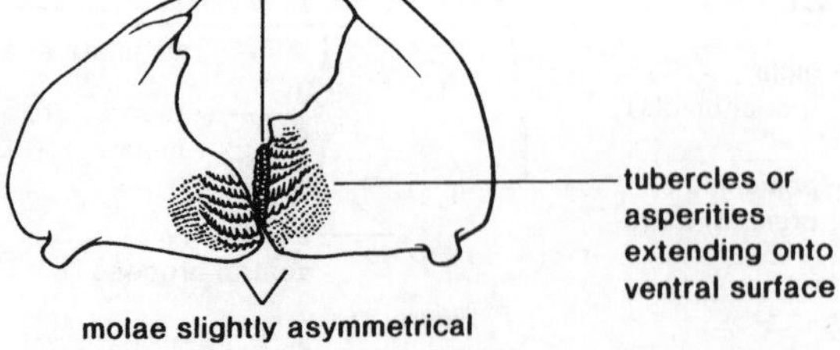

Figure 73

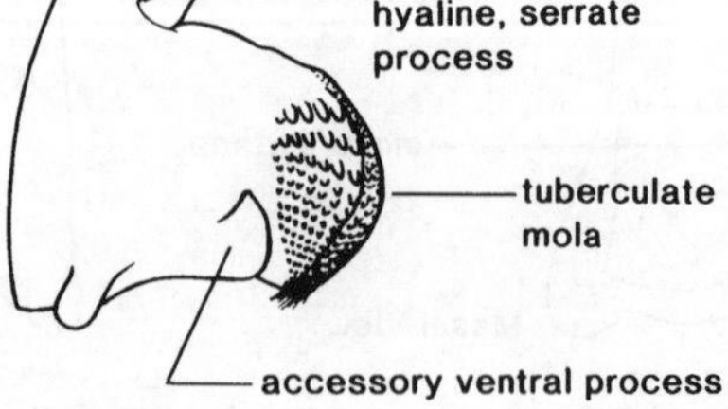

Figure 75

← Figs. 8, 51, 53

12′. Without above characters in combination. Under bark, in various types of living and dead plant tissue .... (part) ***Curculionidae*** p. 594 and ***Scolytidae*** p. 613

13(9′). Maxilla with galea and lacinia (fig. 110); mandibular mola (pseudomola) distinctly tuberculate; maxillary articulating area reduced or absent; segment A10 with pair of oval lobes separated by longitudinal groove (fig. 193); frontal arms absent. In puffballs .... (part) ***Anobiidae*** p. 441

13′. Maxilla with single mala (figs. 88, 96); mandibular mola not tuberculate; maxillary articulating area well-developed; segment A10 without oval lobes separated by longitudinal groove; frontal arms present (fig. 20) 14

14(13′). Mandibles with accessory ventral process (fig. 75); apex of mala straight or slightly emarginate; labium well-developed. In male cones of *Araucaria*. Exotic (Southern Hemisphere) .... (part) ***Nemonychidae*** p. 585

14′. Mandibles without accessory ventral process; apex of mala rounded or acute; labium often reduced in size. In fungus fruiting bodies, dead wood, stems, vines .... (part) ***Anthribidae*** p. 586

15(7^2). Mandibular mola absent (figs. 51, 53); labial palps contiguous or separated by less than width of first palpal segment; ligula absent (fig. 86); stemmata present 16

15′. Mandibular mola present (figs. 68, 73); labial palps separated by more than width of first palpal segment; ligula longer than labial palp, forming a sclerotized, wedge-like structure (fig. 117); stemmata absent 17

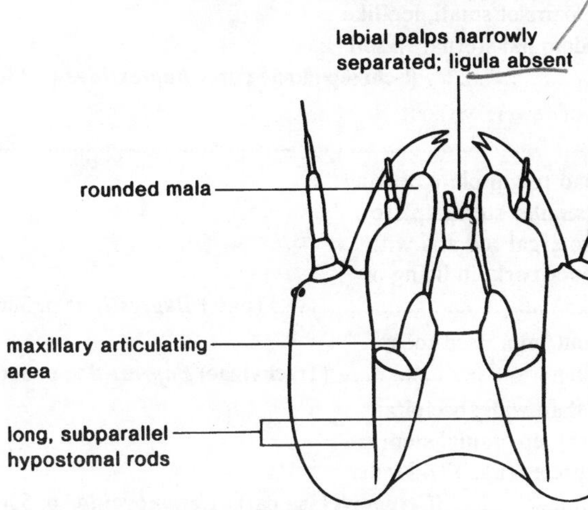

Figure 86

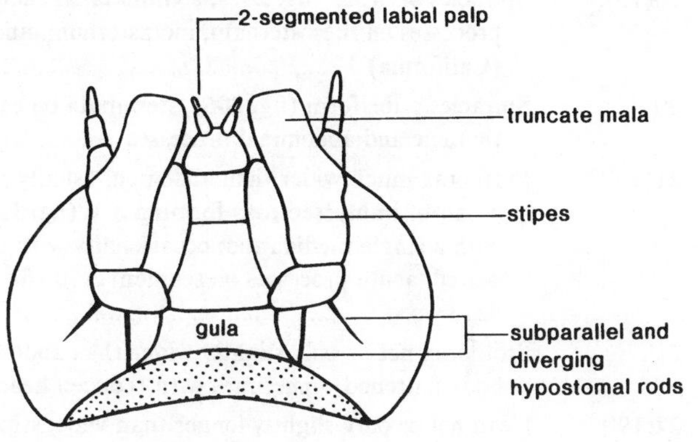

Figure 88

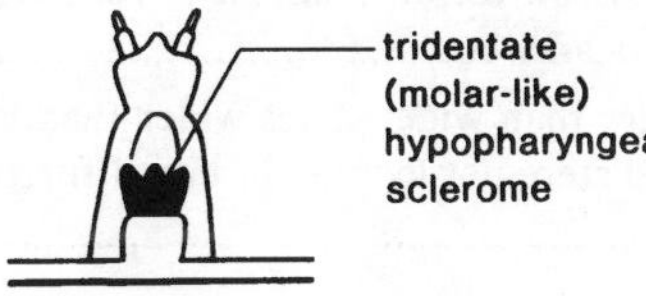

Figure 119

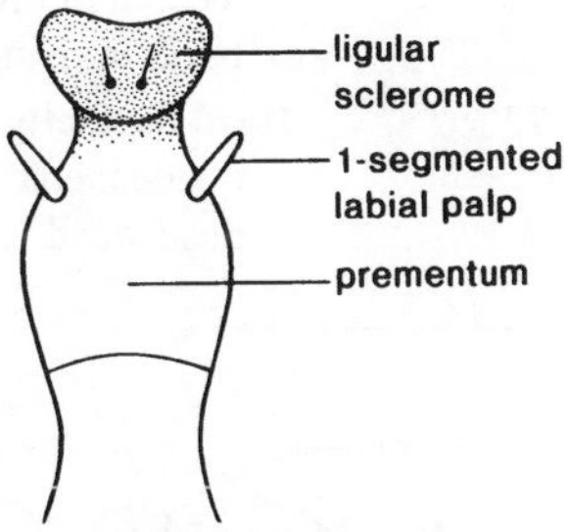

Figure 117

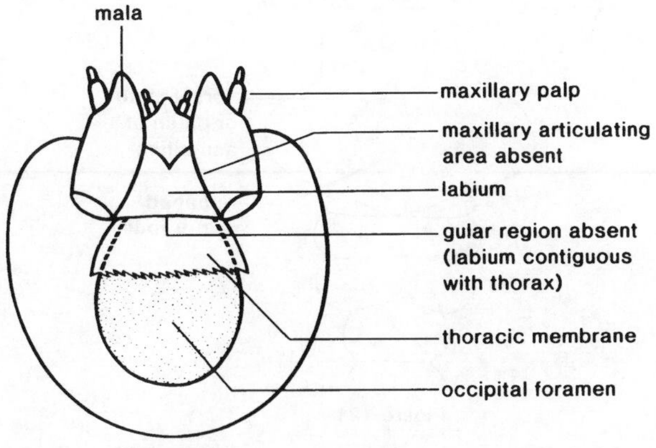

Figure 96

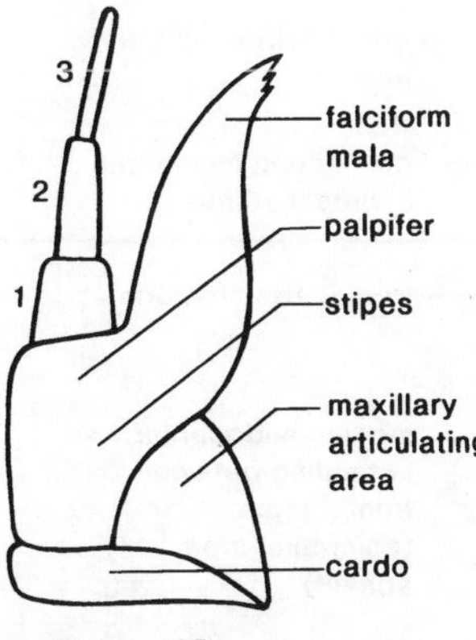

Figure 105

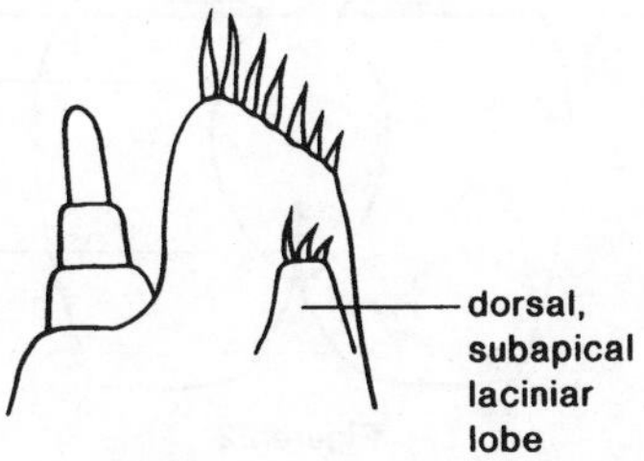

Figure 110

16(15). Median endocarina absent; abdominal apex with respiratory chamber (pocket formed by 8th and 9th terga and enclosing enlarged 8th spiracles); segments in maxillary palp 4; segments in labial palp 2; gular region present (separating labium from thorax) (fig. 94). In rotting vegetation, leaf litter, dung .... (**Sphaeridiinae** part) ***Hydrophilidae*** p. 355

16′. Median endocarina located between frontal arms (fig. 22); abdominal apex without respiratory chamber; segments in maxillary palp 3; segments in labial palp 1; gular region absent (labium contiguous with thoracic membrane) (fig. 96). Leaf miners .... (**Zeugophorinae** and a few **Hispinae**) ***Chrysomelidae*** p. 568

17(15′). Tergum A9 without median process; head narrower than thorax. In rotten wood .... (curculionoid larva and paedogenetic forms) ***Micromalthidae*** p. 300

17′. Tergum A9 with median process (fig. 159); head broader than thorax. In rotten wood .... (cerambycoid larva) ***Micromalthidae*** p. 300

18(5′). Gular region present (separating labium from thorax) (fig. 85); body relatively straight 19

18′. Gular region absent (labium contiguous with thoracic membrane) (fig. 96); body often moderately to strongly curved ventrally (C-shaped) 23

19(18). Labial palps minute and 1-segmented or apparently absent; spiracles cribriform (fig. 206) or rarely biforous (fig. 202) 20

19′. Labial palps well-developed, 2-segmented; spiracles annular, annular-multiforous (fig. 200), or occasionally annular-biforous (fig. 197) 22

20(19). Spiracles biforous (fig. 202); stemmata on each side 3; 2 pairs of small, leg-like processes on mesosternum, metasternum, and most abdominal sterna. In soil (California) .... (**Schizopodinae** part) ***Buprestidae*** p. 386

20′. Spiracles cribriform (fig. 206); stemmata on each side 2 or fewer; without thoracic and abdominal processes 21

21(20′). Prothorax much wider than abdomen, usually with a broad tergal plate bearing a pair of impressed rods forming a V (fig. 121) and a similar sternal plate with a single median rod; occasionally with a single protergal rod and with paired, acute processes on segment A10 (fig. 190). Under bark, in living or dead wood .... (part) ***Buprestidae*** p. 386

21′. Prothorax not or only slightly wider than abdomen, without impressed rods; body flattened, sometimes with reduced head. Leaf miners .... (**Trachyinae**) ***Buprestidae*** p. 386

22(19′). Head not or only slightly longer than wide; stipes longer than wide; occipital foramen divided into 2 parts by tentorial bridge (fig. 95); epicranial stem located in broad furrow for attachment of retractor muscles (fig. 17). Under bark, in living or dead wood .... (**Cerambycinae** part) ***Cerambycidae*** p. 556

22′. Head distinctly longer than wide; stipes wider than long; occipital foramen not divided; epicranial stem not located in broad furrow. Under bark, in living or dead wood .... (**Lamiinae** part) ***Cerambycidae*** p. 556

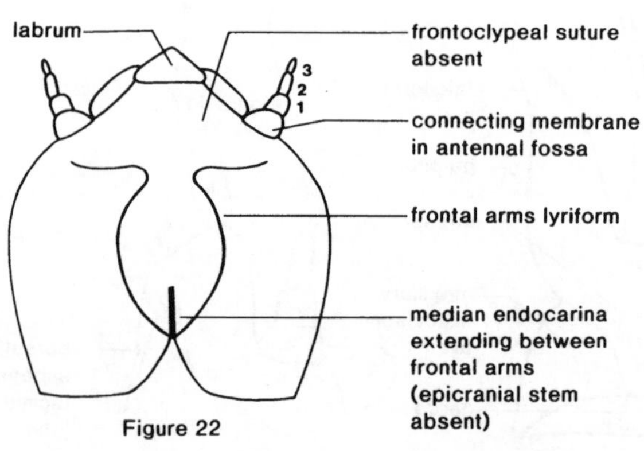

Figure 22

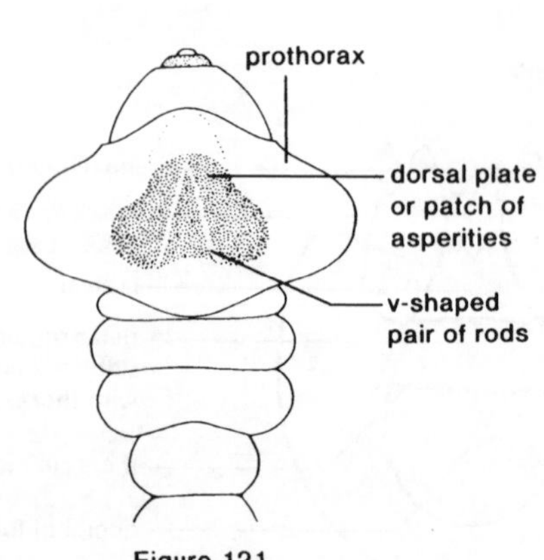

Figure 121

← Figs. 69, 96, 119

23(18′). Labial palps absent 24

23′. Labial palps present 26

24(23). Antennae 2-segmented; hypopharyngeal bracon absent; segments in maxillary palp 1. In seed pods (usually of Fabaceae) .... (part) ***Bruchidae*** p. 561

24′. Antennae 1-segmented; hypopharyngeal bracon present (as in fig. 85); segments in maxillary palp 2 25

25(24′). Mandibular mola (fig. 69) and hypopharyngeal sclerome (fig. 119) present; stemmata absent; body broader and slightly thickened anteriorly without tergal ampullae. In fungus fruiting bodies .... (*Euxenus*) ***Anthribidae*** p. 586

25′. Mandibular mola and hypopharyngeal sclerome absent; head with 1 distinct stemma on each side; body slender, with ampullae on metatergum and abdominal terga 1 to 8. In seed capsules. Exotic (southern Europe and Africa) .... ***Urodontidae*** p. 589

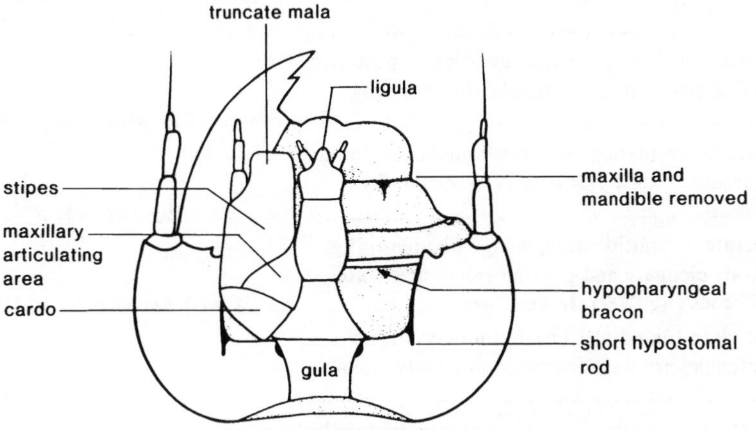

Mouthparts Retracted

Figure 85

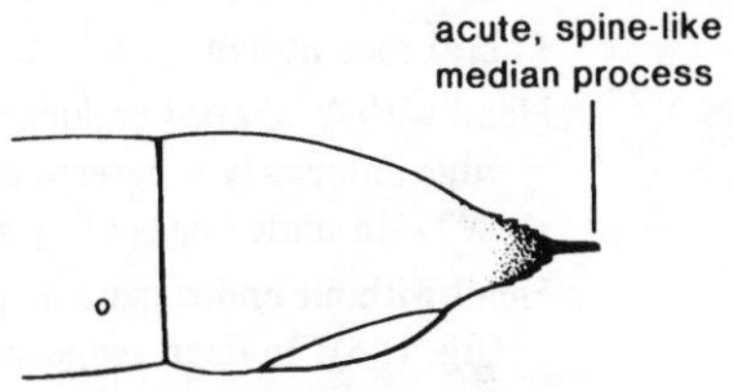

Figure 159

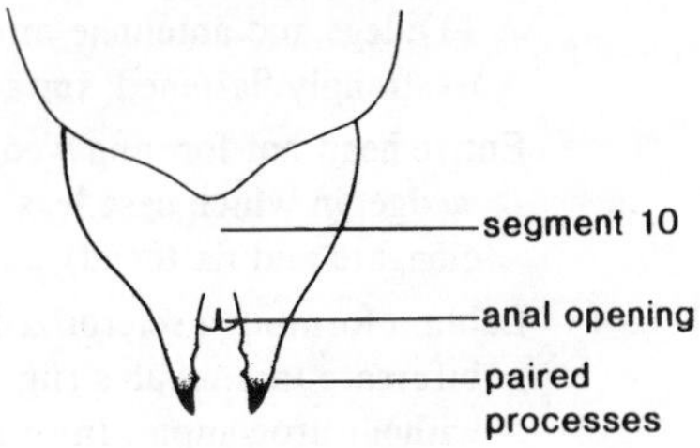

Figure 190

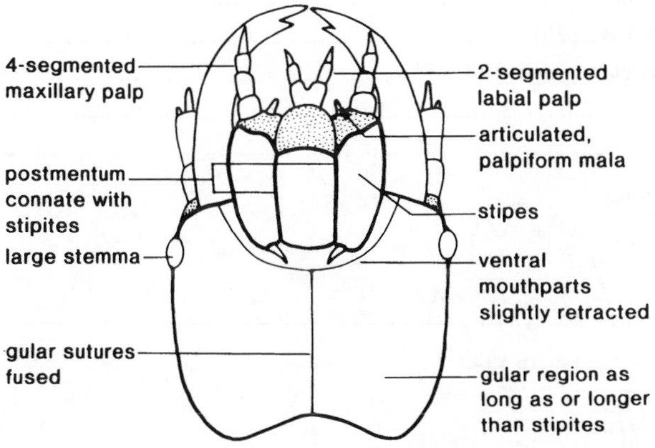

Figure 94

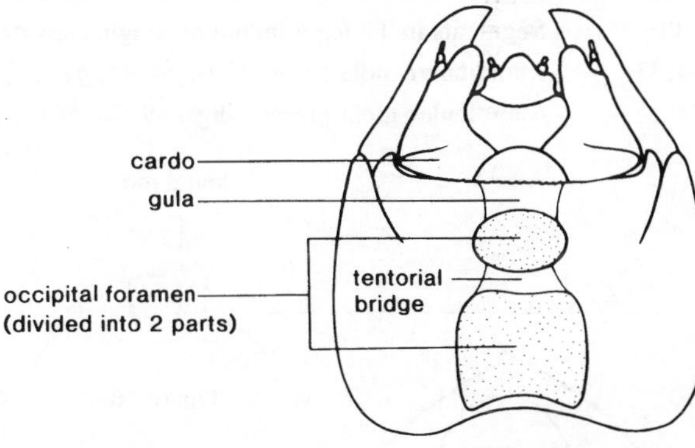

Figure 95

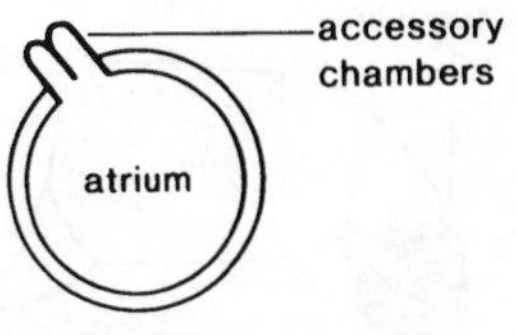

Annular Biforous

Figure 197

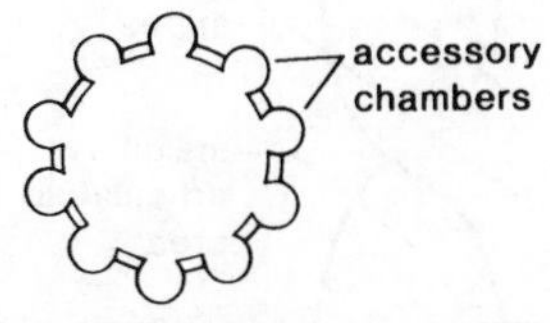

Annular Multiforous

Figure 200

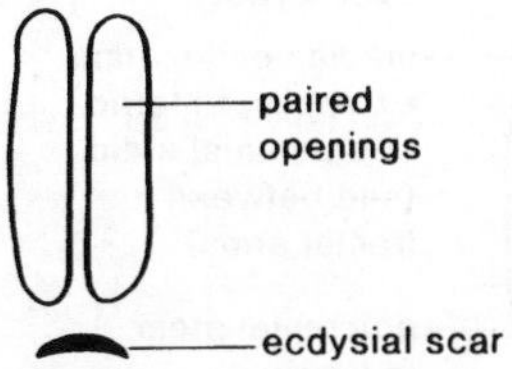

Biforous

Figure 202

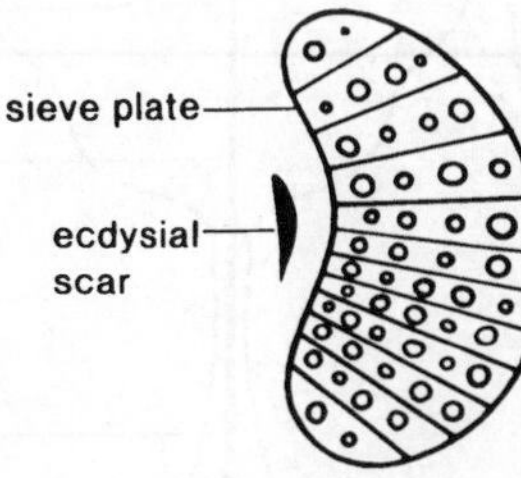

Cribriform

Figure 206

Fig. 17 →

26(23′). Segments in maxillary palp 3 (fig. 105) 27
26′. Segments in maxillary palp 1 or 2 (fig. 96) 29
27(26). Antennae 2-segmented; protergum enlarged and hump-like, with transverse, keeled plate. Boring in living or recently killed twigs and branches. Exotic (Southern Hemisphere) .... ***Belidae*** **p. 589**
27′. Antennae 1-segmented; protergum not enlarged and hump-like, without transverse keel 28
28(27′). Labral rods absent; thoracic and abdominal terga without patches of asperities; protergum with sclerotized plate. In various fruits and cones. Exotic (Southern Hemisphere) .... ***Oxycorynidae*** **p. 589**
28′. Labral rods present (fig. 38); thoracic and abdominal terga with patches of asperities (fig. 142); protergum without sclerotized plate. In flower buds or fruits .... **(Rhynchitinae)** ***Curculionidae*** **p. 594**
29(26′). Labral rods present (fig. 38). In various types of plant tissue .... (part) ***Curculionidae*** **p. 594**
29′. Labral rods absent 30
30(29′). Head with Y-shaped endocarina extending from deep emargination at posterior edge anteriorly to lateral edges of clypeus; spiracles annular-biforous (fig. 197). In male cones of cycads .... **(Allocoryninae)** ***Curculionidae*** **p. 594**
30′. Head without endocarina or posterior emargination; spiracles annular-uniforous (fig. 196). In dead twigs, stems, fronds. Exotic (Hawaii, New Zealand, Pacific) .... ***Aglycyderidae*** **p. 590**
31(2). Entire head forming sclerotized, serrate or multidentate, wedge-like plate (fig. 11); legs and antennae minute; body elongate and parallel-sided, moderately to strongly flattened; spiracles biforous (fig. 202). In dead wood .... (part) ***Eucnemidae*** **p. 419**
31′. Entire head not forming wedge-like plate (mandibles or labium may form a wedge, in which case legs and antennae are well-developed and body is not elongate and flattened) 32
32(31′). Labium forming a sclerotized, 5-dentate plate (fig. 114); T1 legs enlarged with bifurcate tarsungulus (fig. 133); body lightly sclerotized and grub-like, without urogomphi. In rotten wood .... ***Cerophytidae*** **p. 409**
32′. Labium not forming 5-dentate plate; T1 tarsungulus never bifurcate 33
33(32′). Segments in T2 leg 5 or fewer including tarsungulus (fig. 126) 34
33′. Segments in T2 leg 6 including single claw or paired claws (figs. 124, 125) 102
34(33). Mandibular mola absent (figs. 44, 45, 47, 53) 35
34′. Mandibular mola present (figs. 69, 78, 80) 89

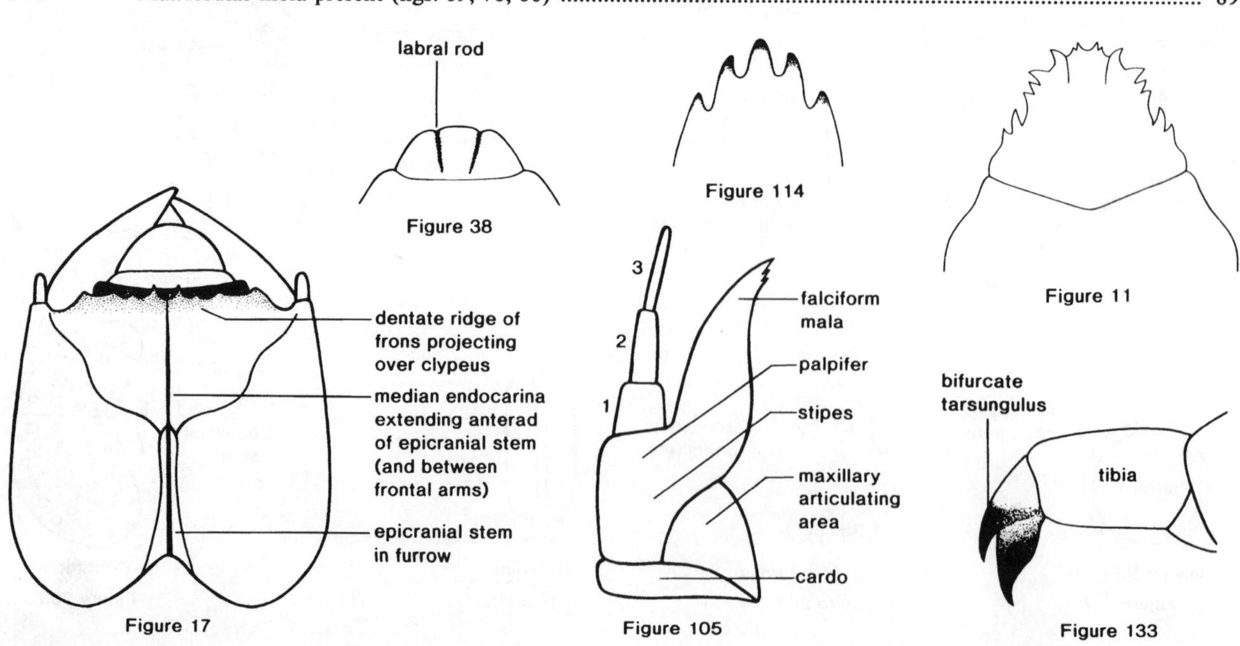

Figure 38

Figure 114

Figure 11

Figure 17

Figure 105

Figure 133

← Figs. 197, 202

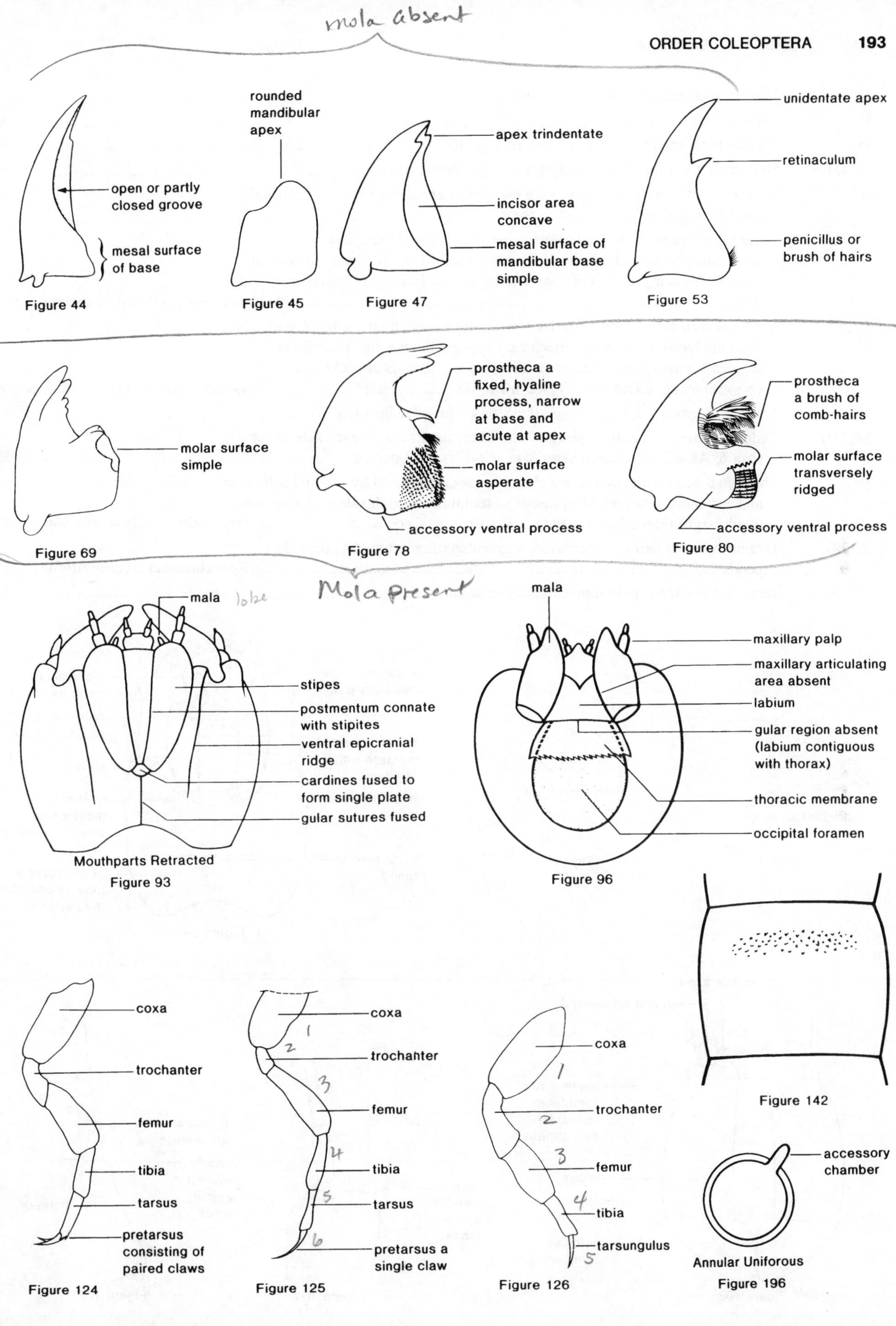

Figure 44

Figure 45

Figure 47

Figure 53

Figure 69

Figure 78

Figure 80

Figure 93

Figure 96

Figure 142

Figure 124

Figure 125

Figure 126

Figure 196

35(34). Maxilla without apical lobes (fig. 101) 36

35′. Maxilla with single mala (figs. 93, 96, 105). **(see 3rd choice)** 48

35^{2}. Maxilla with separate galea and lacinia (fig. 100) 87

36(35). First segment of maxillary palp without digitiform appendage 37

36′. First segment of maxillary palp with digitiform appendage (fig. 102) (usually described as a galea attached to the palpifer) 41

37(36). Antennal segments 1; body strongly curved ventrally (C-shaped); head moderately to strongly declined (hypognathous) (fig. 10); most abdominal segments bearing 2 pairs of cone-like processes. In nests of native bees and wasps .... **(Rhipiphorinae** part) ***Rhipiphoridae*** p. 509

37′. Antennal segments 2; body relatively straight or only slightly curved ventrally; head moderately to strongly declined (hypognathous) (fig. 10); abdominal segments without cone-like processes. Endoparasitoids of cockroaches. Exotic (Neotropical and Old World) **(see 3rd choice)** .... **(Rhipidiinae** part) ***Rhipiphoridae*** p. 509

37^{2}. Antennal segments 3; head prognathous or slightly declined (fig. 8) 38

38(37^{2}). Abdominal apex without respiratory chamber; stemmata on each side fewer than 6; A8 spiracles about same size as others on abdomen 39

38′. Abdominal apex with respiratory chamber (pocket formed by 8th and 9th terga and enclosing enlarged A8 spiracles); stemmata on each side 6; A8 spiracles much larger than others on abdomen. At edges of ponds .... **(Hydrochinae)** ***Hydrophilidae*** p. 355

39(38). Tergum A9 with pair of articulated, segmented urogomphi (fig. 162). In ponds .... **(Epimetopinae)** ***Hydrophilidae*** p. 355

39′. Tergum A9 without paired processes or urogomphi 40

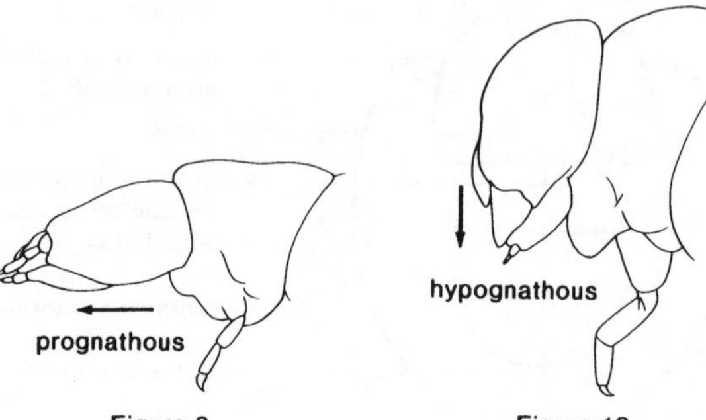

Figure 8

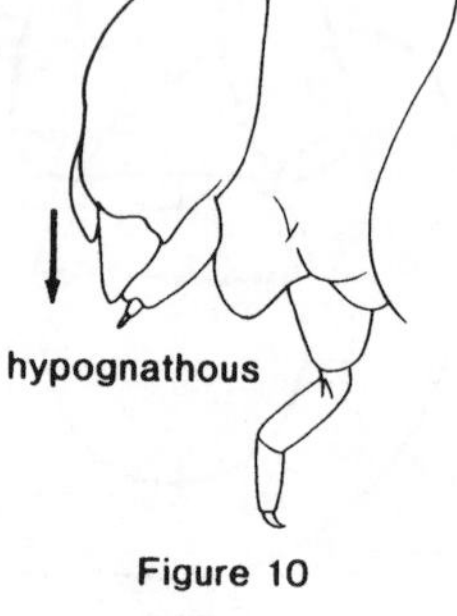

Figure 10

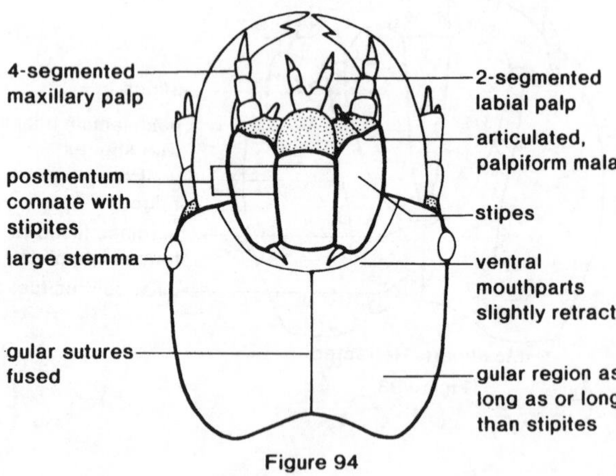

Figure 94

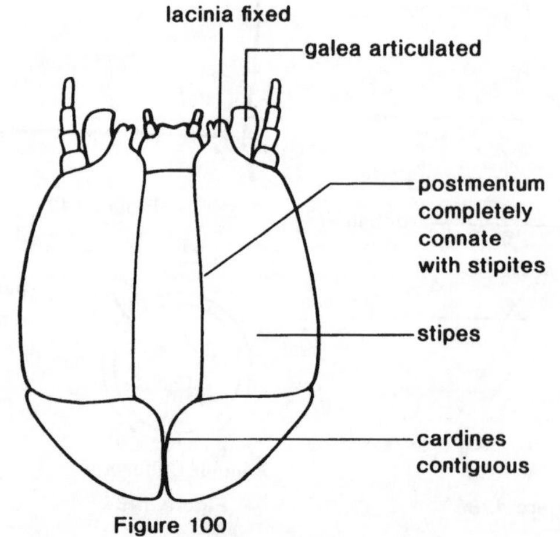

Figure 100

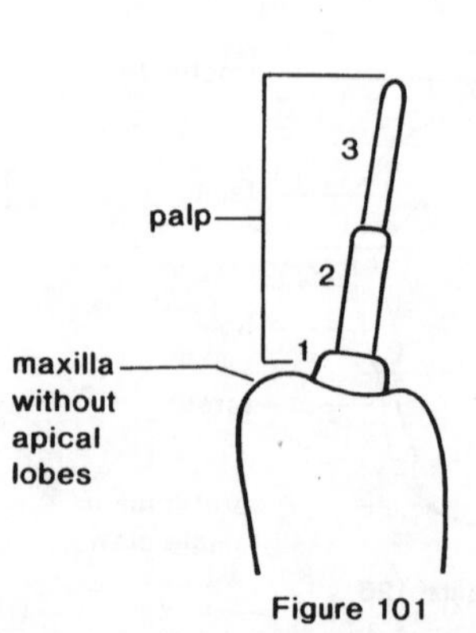

Figure 101

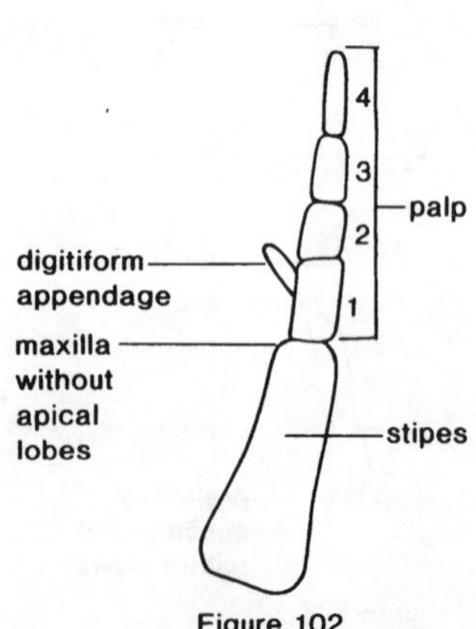

Figure 102

← Figs. 44, 53, 93, 96, 105, 126

40(39′). Larvae minute, length usually about 0.5 mm; head with 4 stemmata on each side; body clothed with moderately long, scattered setae, some of which are heavy and spine-like; without paired dorsal glands .... (triungulins) (**Rhipidiinae** part) ***Rhipiphoridae*** p. 509

40′. Larvae larger, length usually more than 1 mm; head with 1 large stemma on each side (fig. 94); body clothed with very short, fine pubescence, which gives surface a dull appearance; paired dorsal glands (fig. 139) present on thoracic terga and terga A1–8 or A1–9 .... (part) ***Cantharidae*** p. 429

41(36′). Abdominal apex without respiratory chamber 42

41′. Abdominal apex with respiratory chamber (pocket formed by 8th and 9th terga and enclosing enlarged A8 spiracles). In ponds, lakes, slow-moving streams, leaf litter, and dung .... (part) ***Hydrophilidae*** p. 355

42(41). Tergum A9 without paired processes (urogomphi) 43

42′. Tergum A9 with paired processes (urogomphi) (fig. 162) 44

43(42). Mesal surface of mandibular base simple or slightly expanded (fig. 44); segments in labial palp 2; segments in maxillary palp 4; visible abdominal segments 9; stemmata on each side 6; cardo present; segments A1–7 each with pair of long, narrow, lateral processes. In ponds .... (**Berosinae**) ***Hydrophilidae*** p. 355

43′. Mesal surface of mandibular base with brush of hairs (penicillus) (fig. 53); segments in labial palp 3; segments in maxillary palp 5; visible abdominal segments 10; stemmata on each side 1 or 0; cardo apparently absent; abdominal segments without long processes. Under bark, in leaf litter, dung, decaying vegetation, ant nests .... (part) ***Histeridae*** p. 361

44(42′). Mesal surface of mandibular base simple or slightly expanded (fig. 44); T2 leg 3- or 4-segmented; urogomphi 1-segmented. In sand or mud along streams .... ***Georyssidae*** p. 358

44′. Mesal surface of mandibular base with brush of hairs (penicillus) (fig. 53); T2 leg 5-segmented including tarsungulus (fig. 126); urogomphi usually with 2 or more segments (figs. 162, 163) 45

45(44′). Stemmata on each side 1 or 0; maxillary articulating area absent; cardo apparently absent; antennal foramen contiguous with mandibular articulation 46

45′. Stemmata on each side 6; maxillary articulating area present; cardo present; antennal foramen separated from mandibular articulation. In small streams, ponds, wet soil .... (**Helophorinae**) ***Hydrophilidae*** p. 355

46(45). Mentum fused to head capsule; urogomphi with 2 segments or fewer. Under bark, in leaf litter, dung, decaying vegetation, ant nests .... (part) ***Histeridae*** p. 361

46′. Mentum separated from head capsule by suture; urogomphi 4-segmented 47

47(46′). Mandible broadly and abruptly expanded at base. In fermenting sap flows .... ***Sphaeritidae*** p. 359

47′. Mandible gradually expanded at base. Under bark of rotting logs. Exotic (Mexico, Asia) .... ***Synteliidae*** p. 360

48(35′). Antennal segments 2 49

48′. Antennal segments 3 (**see 3rd choice**) 62

48². Antennal segments 4 84

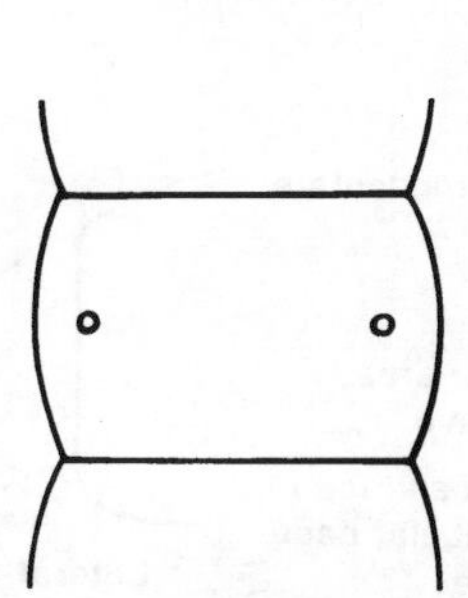
Figure 139

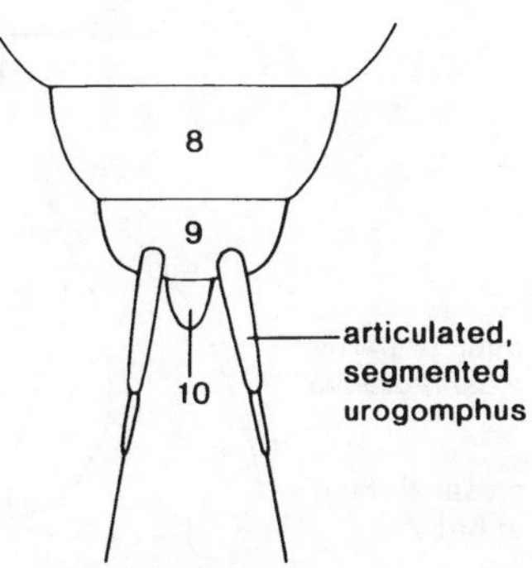

Figure 162

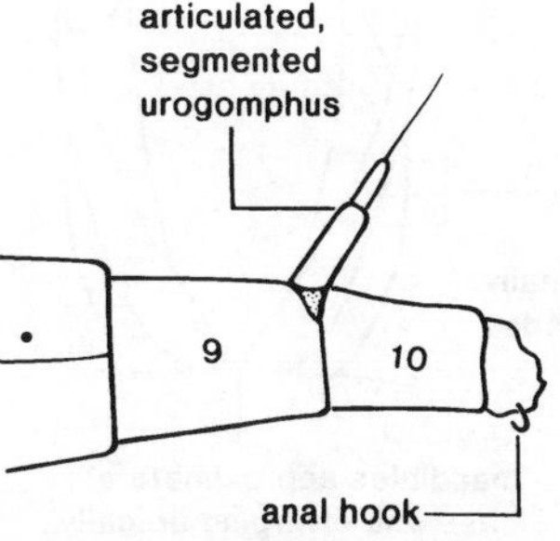

Figure 163

49(48). Tergum A9 without paired processes (urogomphi) 50

49′. Tergum A9 with paired processes (urogomphi) (fig. 171) 58

50(49). Segments in maxillary palp 1; maxillary and labial palps large and tusk-like; body minute, length less than 1 mm. In bark crevices, soil; in association with cicadas .... (triungulins) (part) ***Rhipiceridae*** p. 370

50′. Segments in maxillary palp 2 (fig. 96) **(see 3rd & 4th choices)** 51

50^{2}. Segments in maxillary palp 3 (fig. 105) 53

50^{3}. Segments in maxillary palp 4 (figs. 94, 100, 113) 56

51(50′). Mesal surface of mandibular base simple or slightly expanded (fig. 44); ratio of antennal length to head width more than .5; segments in T2 leg 5 including tarsungulus (fig. 126); hypostomal rods absent 52

51′. Mesal surface of mandibular base with fixed, rigid, hyaline process, sometimes partly sclerotized (fig. 58); ratio of antennal length to head width less than .15; segments in T2 leg 3 or 4; hypostomal rods present (fig. 89). In tunnels of wood-boring beetles, in stored products .... (**Passandrinae** part) ***Cucujidae*** p. 463

52(51). Body not broadly ovate and strongly flattened; head not concealed from above; ratio of length of abdomen to length of thorax more than 2; dorsal surfaces very lightly pigmented or sclerotized. In tunnels of ambrosia beetles (Platypodidae) .... (triungulins) (*Sosylus* part) ***Bothrideridae*** p. 477

52′. Body broadly ovate, strongly flattened, and disc-like (fig. 2); head concealed from above by prothorax (fig. 2); ratio of length of abdomen to length of thorax 1.2 to 2; dorsal surfaces more or less heavily pigmented or sclerotized. In leaf litter .... (part) ***Corylophidae*** p. 495

53(50^{2}). Body relatively straight or slightly curved ventrally; mandibles narrow and falcate (figs. 42, 44); ligula absent; tergum A3 without transverse plicae 54

53′. Body strongly curved ventrally (C-shaped) (fig. 1); mandibles broad and stout or more or less wedge-like (fig. 47); ligula shorter than palp; tergum A3 with 2 or more transverse plicae (as in fig. 136). Case-bearing larvae in ant nests or on leaf surfaces .... (**Clytrinae-Chlamisinae** part) ***Chrysomelidae*** p. 568

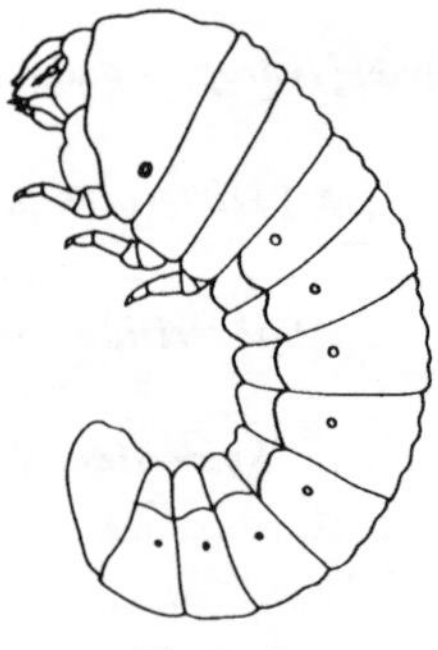

Figure 1

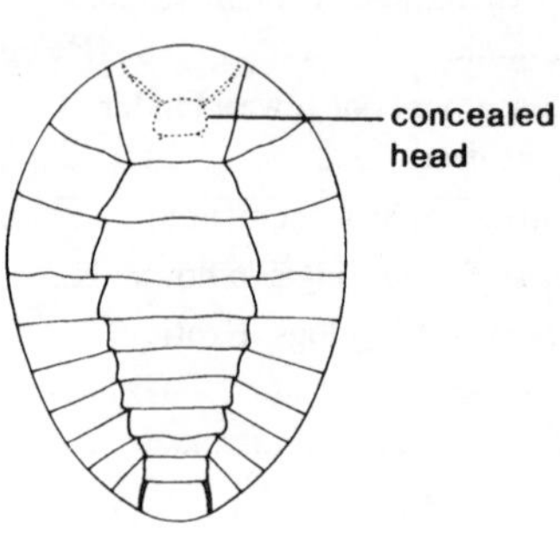

concealed head

Figure 2

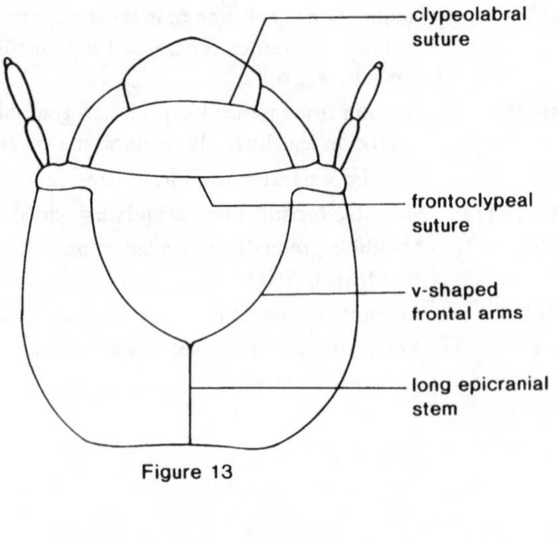

clypeolabral suture
frontoclypeal suture
v-shaped frontal arms
long epicranial stem

Figure 13

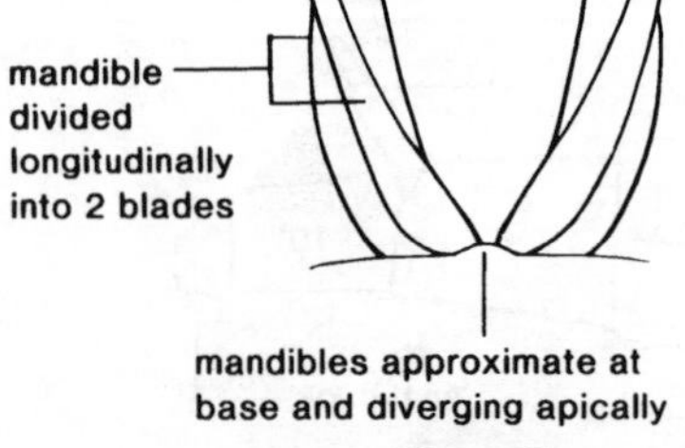

mandible divided longitudinally into 2 blades
mandibles approximate at base and diverging apically

Figure 42

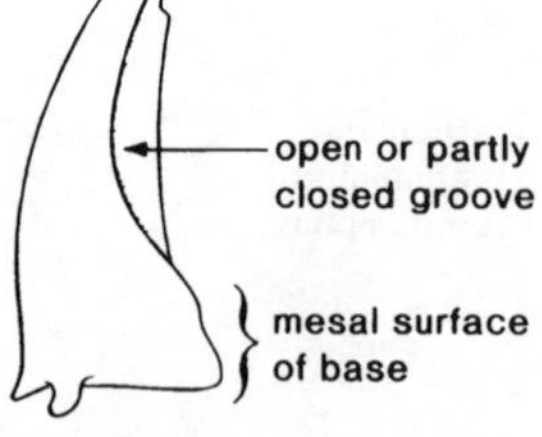

open or partly closed groove
mesal surface of base

Figure 44

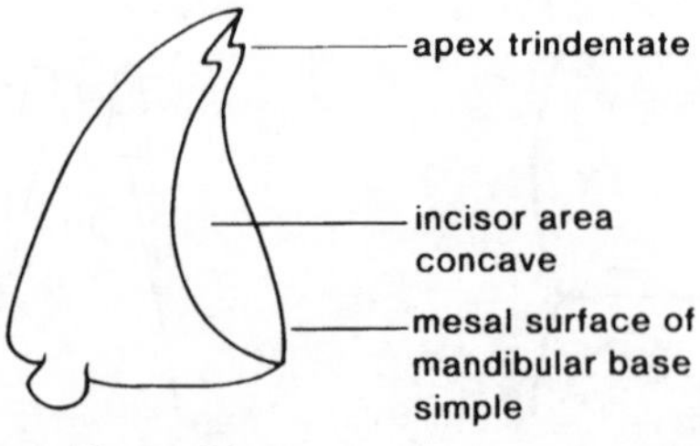

apex trindentate
incisor area concave
mesal surface of mandibular base simple

Figure 47

Lateral

Figure 171

Figs. 8, 10

54(53). Mandibles approximate at base and divided longitudinally so that 2 pairs of diverging blades or stylets are formed (fig. 42); epicranial suture absent .... (part) ***Lycidae*** p. 423

54′. Mandibles distant at base, converging apically, and not divided longitudinally into 2 parts; epicranial suture distinct, with V-shaped frontal arms (fig. 13) 55

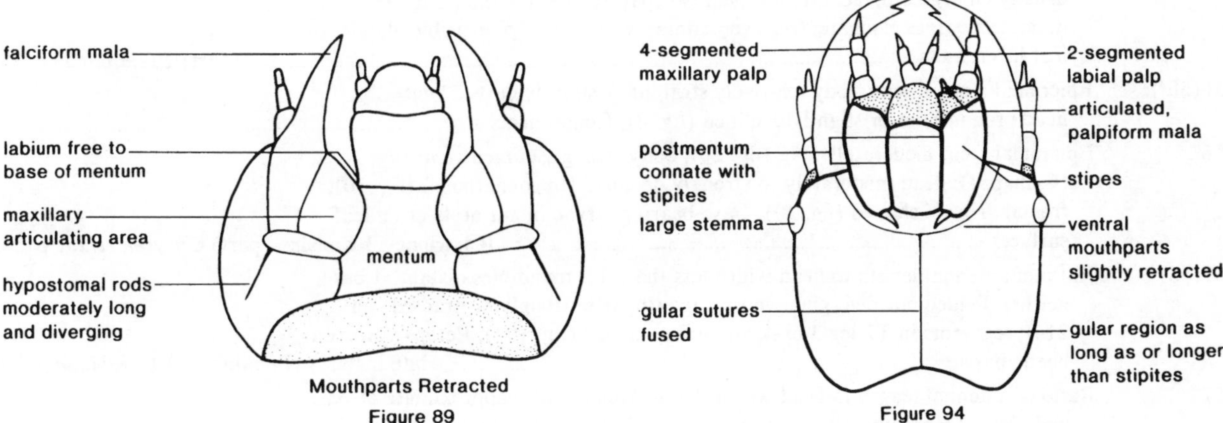

Figure 89

Figure 94

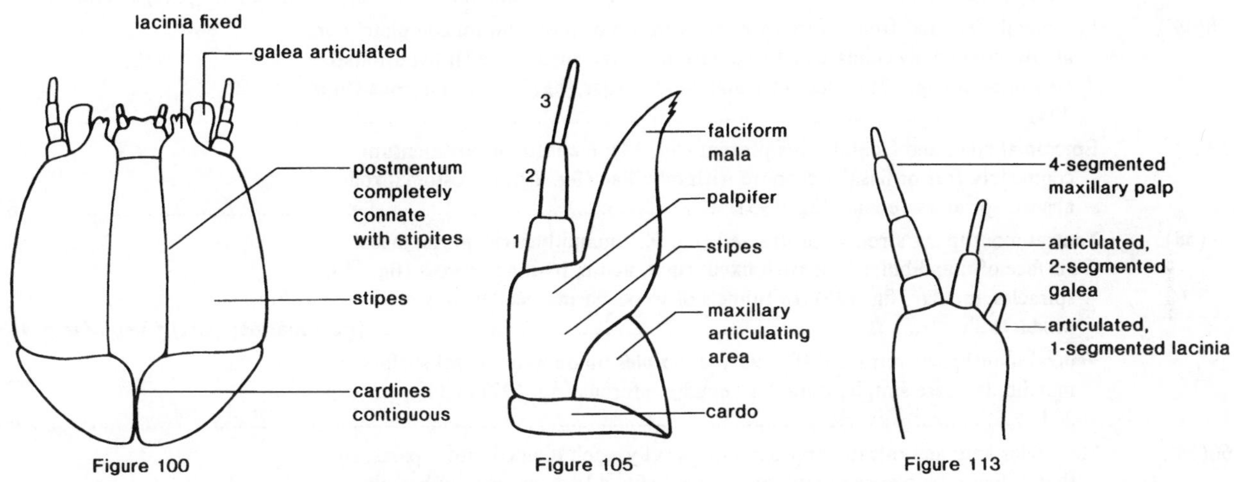

Figure 100

Figure 105

Figure 113

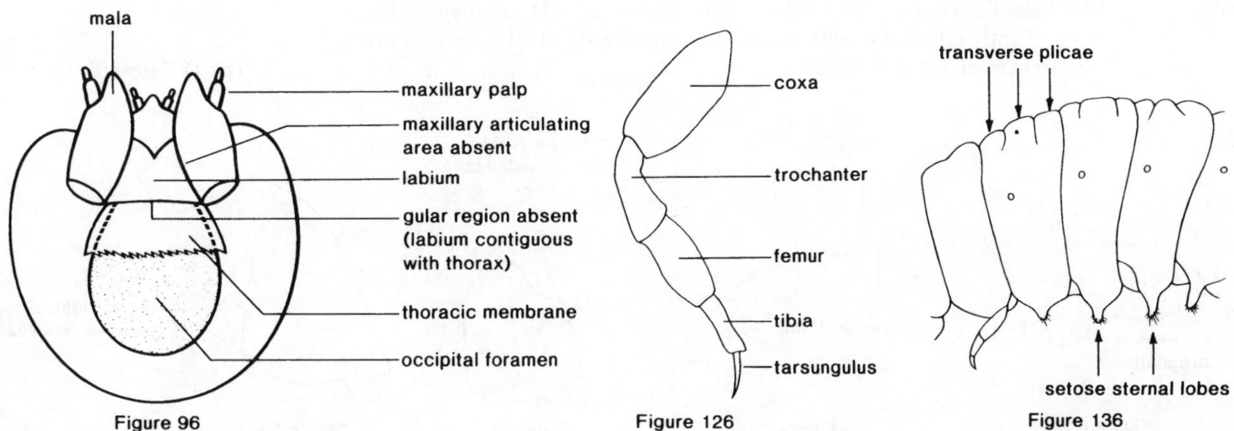

Figure 96

Figure 126

Figure 136

Fig. 58 ⟶

55(54′). Stemmata on each side 1, or 3 forming tight cluster or triangle; antennal sensorium short and dome-like (fig. 31); apex of prementum not emarginate; labial palps separated by more than 2 basal diameters; stipes short and broad; head without protrusible glands. In leaf litter, moss .... (part) ***Scydmaenidae*** p. 330

55′. Stemmata on each side 2 or 3, well-separated and forming a row; antennal sensorium elongate, palpiform (fig. 30) or setiform; apex of prementum usually emarginate; labial palps usually separated by less than 2 basal diameters; stipes elongate; frons sometimes with pair of protrusible glands. In leaf litter, moss .... (part) ***Pselaphidae*** p. 353

56(50[3]). Epicranial stem absent; body relatively straight or slightly curved ventrally; head prognathous or slightly declined (fig. 8); frontal arms absent 57

56′. Epicranial stem moderately long (fig. 20); body strongly curved ventrally (C-shaped): head moderately to strongly declined (hypognathous) (fig. 10); frontal arms V-shaped (fig. 20). Case-bearing larvae in ant nests or on leaf surfaces .... **(Clytrinae-Chlamisinae** part) ***Chrysomelidae*** p. 568

57(56). Ratio of antennal length to head width less than .15; mandibles distant at base, neither divided nor diverging; ventral mouthparts strongly protracted (fig. 214); segments in T2 leg 3 or 4; spiracles annular (fig. 194). Ectoparasitic on beetle pupae .... (late instars) (Lebiini part) ***Carabidae*** p. 305

57′. Ratio of antennal length to head width .15 to .5; mandibles approximate at base and divided longitudinally, forming 2 pairs of diverging blades or stylets (fig. 42); ventral mouthparts retracted (fig. 93); segments in T2 leg 5 including tarsungulus (fig. 126); spiracles biforous (fig. 204). In rotten wood, leaf litter, under bark .... (part) ***Lycidae*** p. 423

58(49′). Epicranial stem and frontal arms absent; mentum or postmentum completely or almost completely connate or fused with maxillae (figs. 92, 93); hypostomal rods present (fig. 92) or spiracles annular-biforous (fig. 197) or biforous (fig. 204) 59

58′. Epicranial stem and frontal arms present (fig. 14); mentum or postmentum completely free or basally connate with maxillae (fig. 89); hypostomal rods absent; spiracles annular (fig. 195) 61

59(58). Ventral mouthparts strongly protracted (fig. 92); mandible tridentate; mesal surface of mandibular base with fixed, rigid, acute, hyaline process (fig. 58); spiracles annular (fig. 195). In tunnels of wood-boring beetles, in stored products .... **(Passandrinae** part) ***Cucujidae*** p. 463

59′. Ventral mouthparts retracted (fig. 93); mandibles unidentate; mesal surface of mandibular base simple; spiracles annular-biforous (fig. 197) or biforous (fig. 204) 60

60(59′). Mandibles long and falcate, approximate basally, each divided into 2 parts, so that 4 diverging blades or styli are formed (fig. 42); body usually heavily sclerotized, with lateral tergal processes on most segments, and without ventral sclerotized rods on T1. In rotten wood, leaf litter, under bark .... (part) ***Lycidae*** p. 423

60′. Mandibles flattened and rounded or subtriangular (fig. 45), not divided; body very lightly sclerotized, with ventral sclerotized rods on T1. In rotten wood, in soil associated with roots .... (part) ***Throscidae*** p. 418

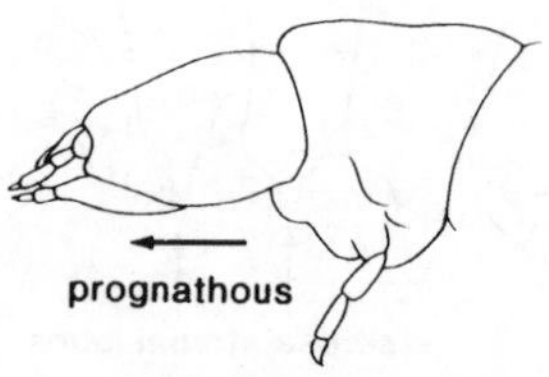

prognathous

Figure 8

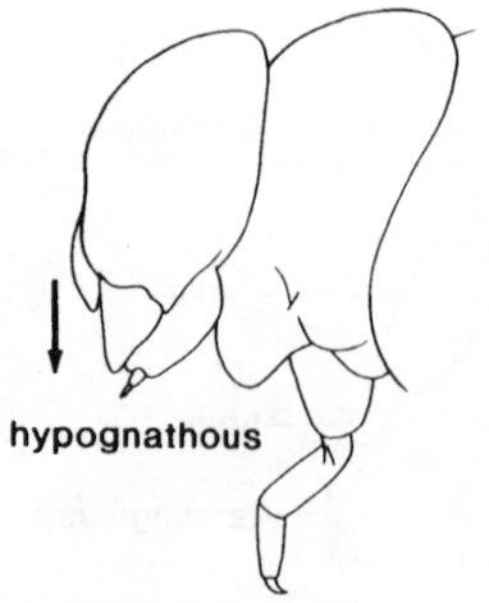

hypognathous

Figure 10

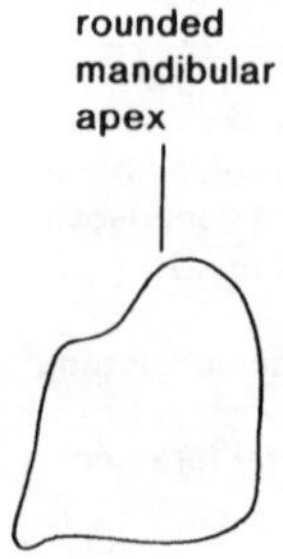

rounded mandibular apex

Figure 45

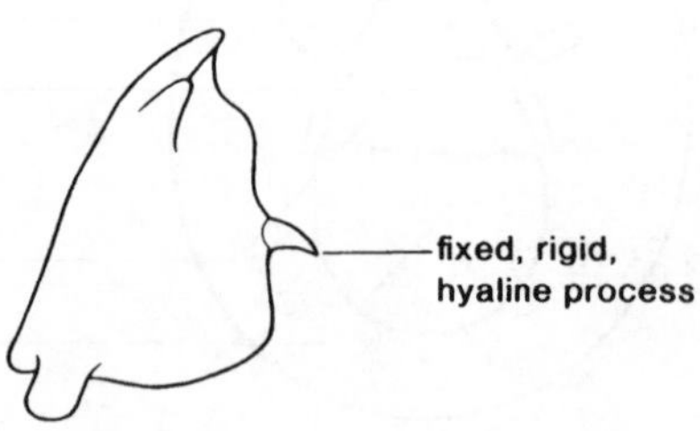

fixed, rigid, hyaline process

Figure 58

← Figs. 42, 89, 126

61(58′). Stemmata absent; antennal sensorium dome-like (fig. 31); apex of prementum not emarginate; labial palps separated by more than 2 basal diameters. In leaf litter .... (Eutheiini part) ***Scydmaenidae*** p. 330

61′. Stemmata on each side 2 or 3; antennal sensorium palpiform (fig. 30), setiform, or sometimes bifid; apex of prementum usually emarginate; labial palps usually separated by less than 2 basal diameters. In leaf litter, moss .... (part) ***Pselaphidae*** p. 353

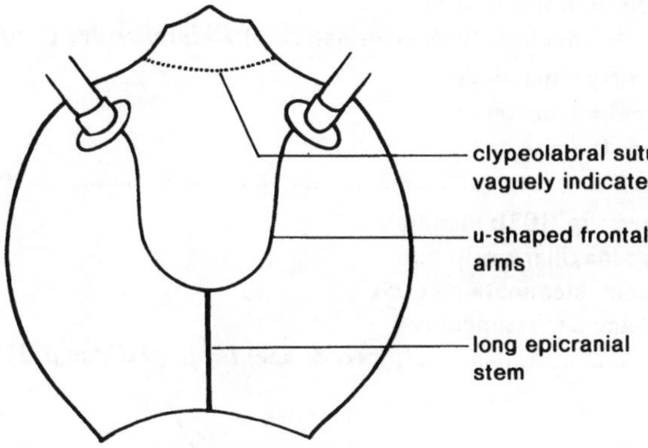

clypeolabral suture vaguely indicated
u-shaped frontal arms
long epicranial stem

Figure 14

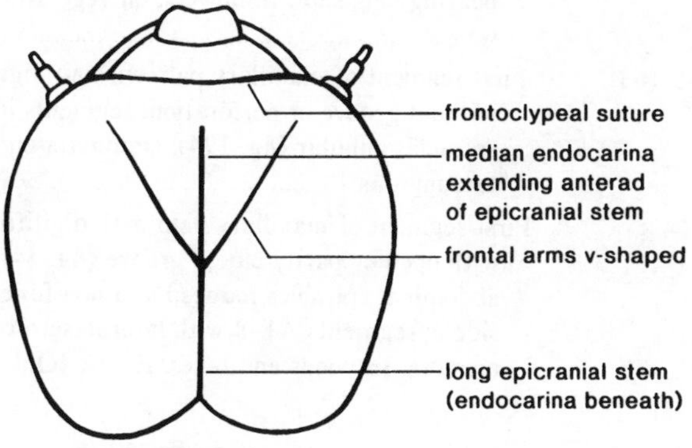

frontoclypeal suture
median endocarina extending anterad of epicranial stem
frontal arms v-shaped
long epicranial stem (endocarina beneath)

Figure 20

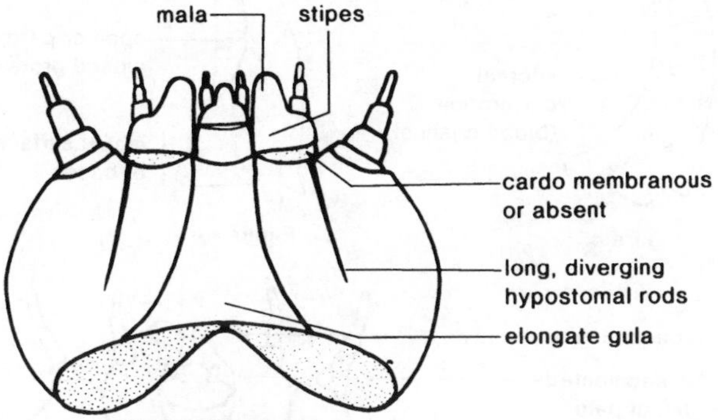

mala
stipes
cardo membranous or absent
long, diverging hypostomal rods
elongate gula

Ventral Mouthparts Strongly Protracted
Figure 92

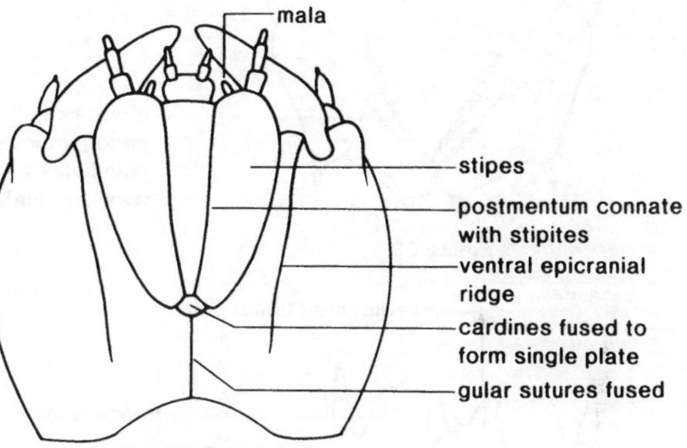

mala
stipes
postmentum connate with stipites
ventral epicranial ridge
cardines fused to form single plate
gular sutures fused

Mouthparts Retracted
Figure 93

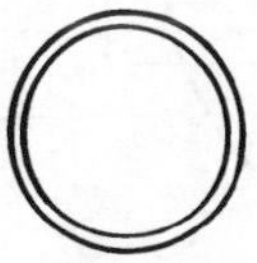

accessory chambers
atrium

Annular
Figure 194

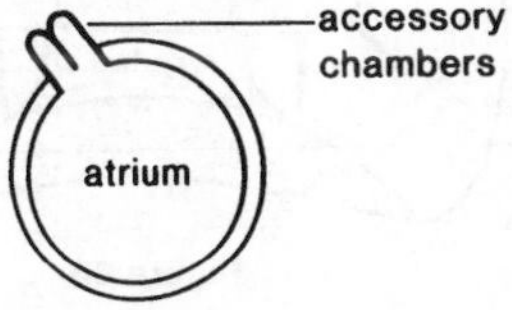

Annular Biforous
Figure 197

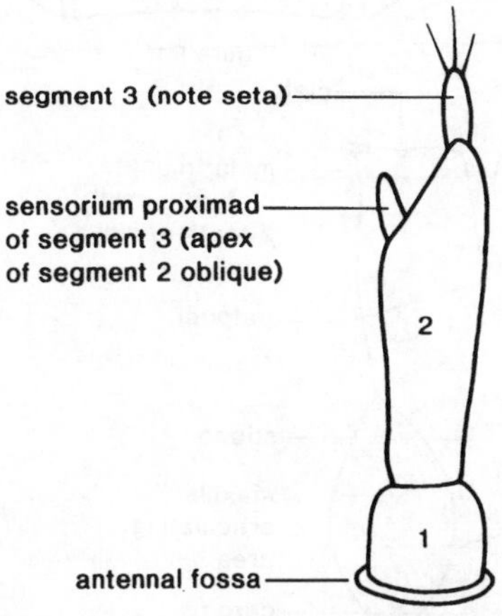

segment 3 (note seta)
sensorium proximad of segment 3 (apex of segment 2 oblique)
2
1
antennal fossa

Figure 30

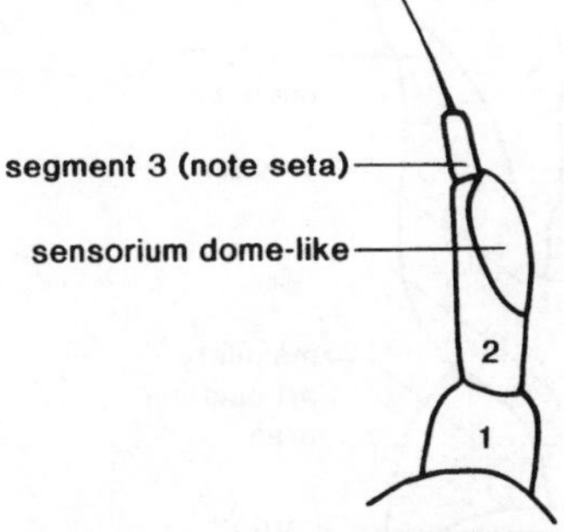

segment 3 (note seta)
sensorium dome-like
2
1

Figure 31

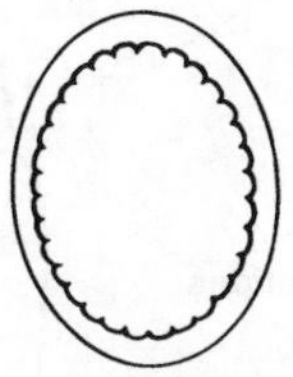

Annular
Figure 195

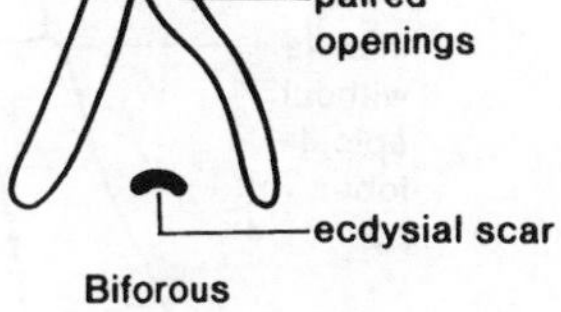

paired openings
ecdysial scar

Biforous
Figure 204

Fig. 214 ⟶

62(48′). Tergum A9 without paired processes (urogomphi) 63

62′. Tergum A9 with paired processes (urogomphi) (figs. 164, 169) 77

63(62). Segments in labial palp 1 (fig. 91) 64

63′. Segments in labial palp 2 (figs. 88, 94, 97) **(see 3rd choice)** 66

63². Segments in labial palp 0; body minute (length less than 1 mm), fusiform; number of stemmata on each side 5; head prognathous or slightly declined (fig. 8); ratio of antennal length to head width more than .5; apex of antenna bearing long seta. In flowers, on vegetation, or attached to native bees or wasps .... (triungulins) (**Rhipiphorinae** part) ***Rhipiphoridae*** p. 509

64(63). First segment of maxillary palp without digitiform appendage; mandible without groove or perforation; segments in maxillary palp 3; abdominal spiracles annular (fig. 194); stemmata on each side 1 or 0; setae on tarsungulus 1 65

64′. First segment of maxillary palp with digitiform appendage (fig. 102); mandible with open or partly closed groove (fig. 44); segments in maxillary palp 4; abdominal spiracles reduced and non-functional or absent; stemmata on each side 6; segments A1–8 with lateral, setose processes; setae on tarsungulus 4 or more. In ponds and lakes. Exotic (Old World) .... **(Spercheinae)** ***Hydrophilidae*** p. 355

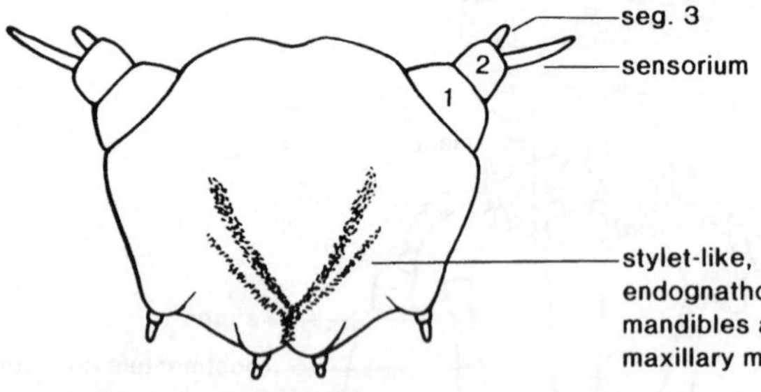

Figure 28

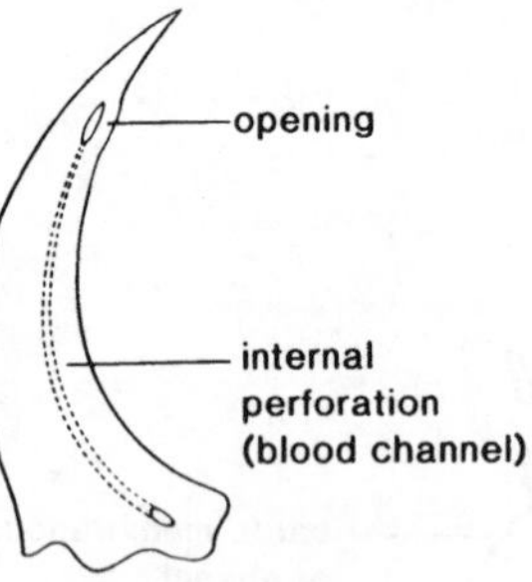

Figure 43

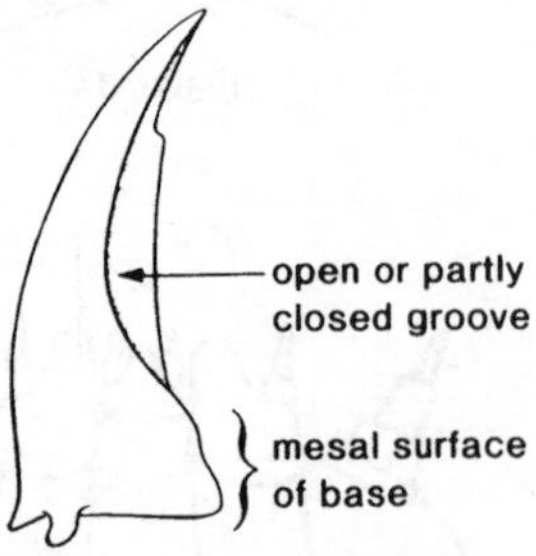

Figure 44

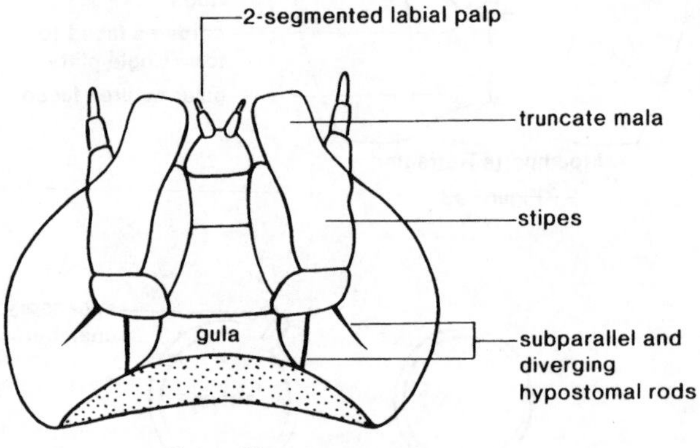

Figure 88

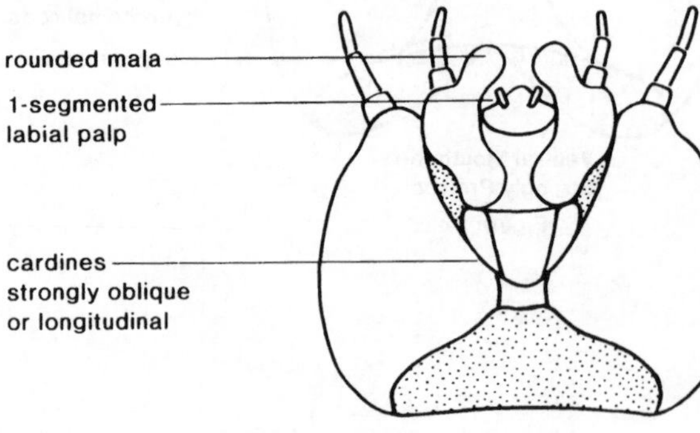

Figure 91

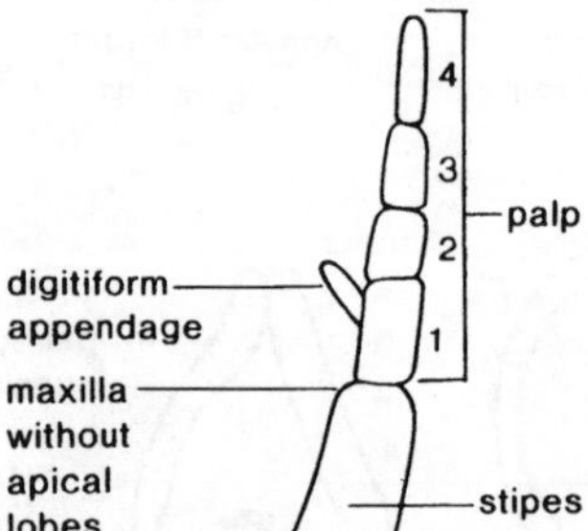

Figure 102

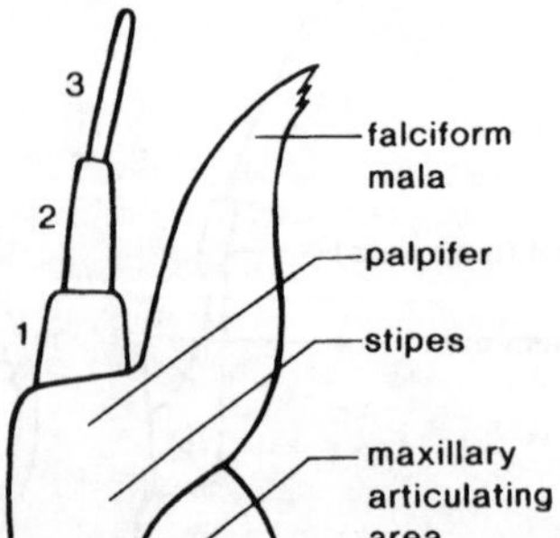

Figure 105

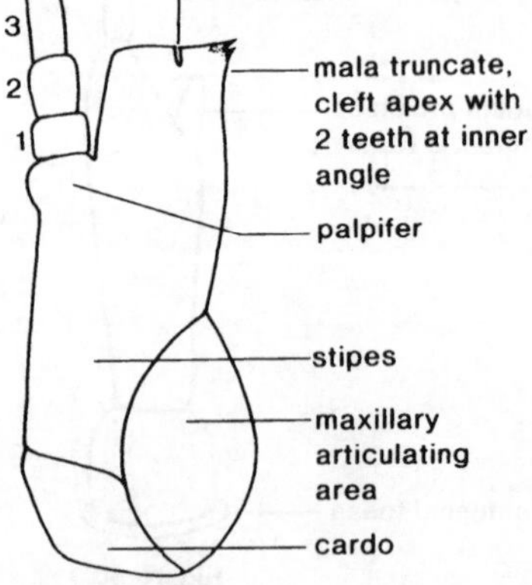

Figure 106

← Figs. 8, 10, 194

65(64). Ratio of antennal length to head width .15 to .5; head moderately to strongly declined (hypognathous) (fig. 10); ratio of length of abdomen to length of thorax more than 2; dorsal surfaces very lightly pigmented or sclerotized; mandibles stylet-like and endognathous (fig. 28); abdominal apex without pair of long setae. Under bark of rotten logs .... (*Cerylon*) (**Cerylonina**e) ***Cerylonidae*** p. 480

65′. Ratio of antennal length to head width more than .5; head prognathous or slightly declined; ratio of length of abdomen to length of thorax less than 1.2; dorsal surfaces more or less heavily pigmented or sclerotized; mandibles falcate (fig. 44), not endognathous; abdominal apex with pair of long setae. On flowers or attached to native bees .... (triungulins) (Tetraonycini part) ***Meloidae*** p. 530

66(63′). Tergum A8 without spiracular processes 67

66′. Tergum A8 with paired processes, each bearing spiracle at apex. On flowers or attached to native bees .... (triungulins) (**Nemognathinae** part) ***Meloidae*** p. 530

67(66). Mandible without groove or perforation 68

67′. Mandible with open or partly closed groove (fig. 44) **(see 3rd choice)** 75

67². Mandible with internal perforation (fig. 43) 76

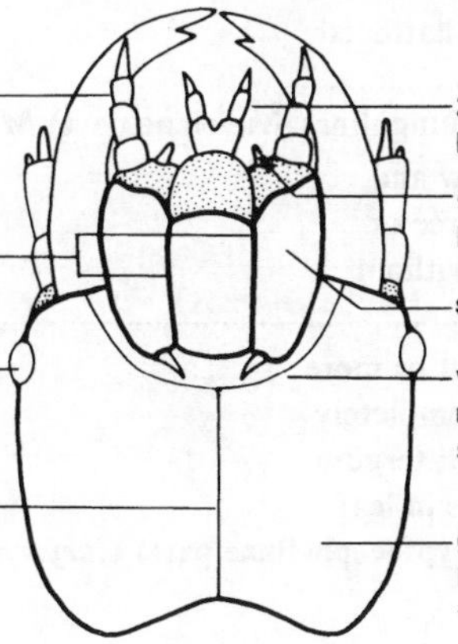

4-segmented maxillary palp
2-segmented labial palp
articulated, palpiform mala
postmentum connate with stipites
stipes
large stemma
ventral mouthparts slightly retracted
gular sutures fused
gular region as long as or longer than stipites

Figure 94

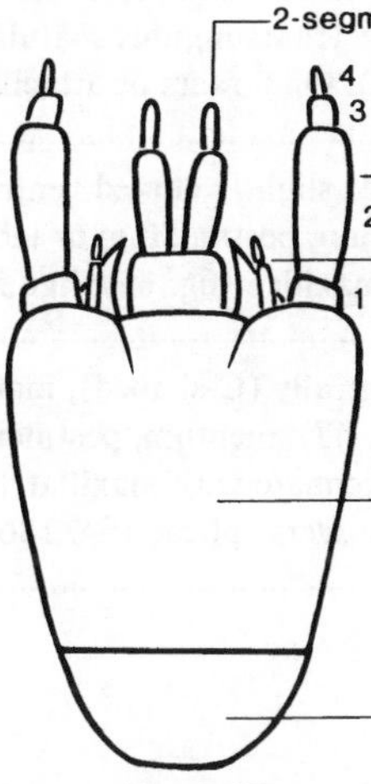

2-segmented labial palp
4
3
4-segmented maxillary palp
2
2-segmented, palpiform galea and blade-like lacinia
1
labium almost completely fused to maxillae
cardines fused, forming single plate

Figure 97

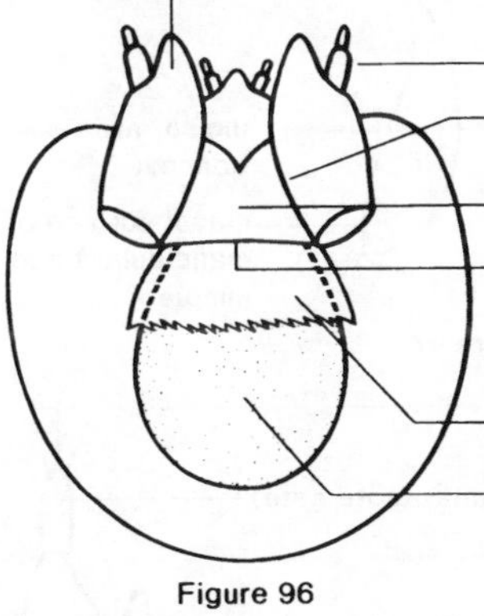

mala
maxillary palp
maxillary articulating area absent
labium
gular region absent (labium contiguous with thorax)
thoracic membrane
occipital foramen

Figure 96

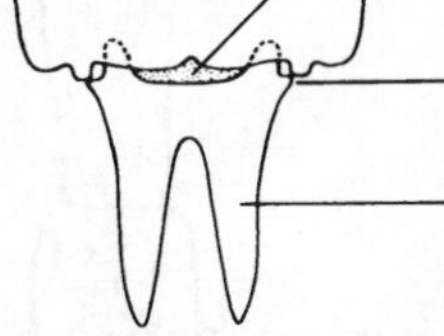

sternum 8
sternum 9 and segment 10 concealed
tergum 9 hinged
tergum 9 (urogomphal plate)
Ventral

Figure 164

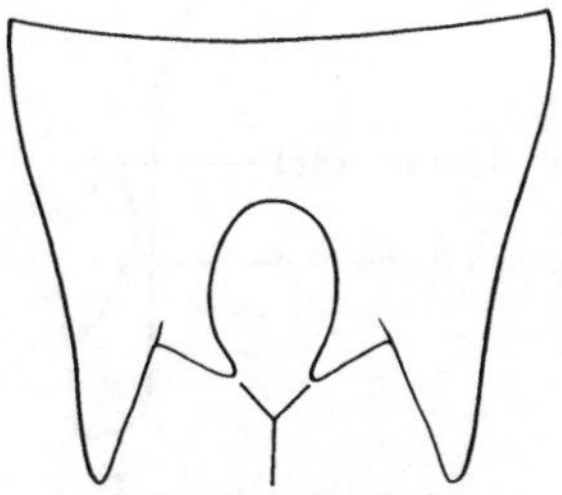

accessory processes
Dorsal

Figure 169

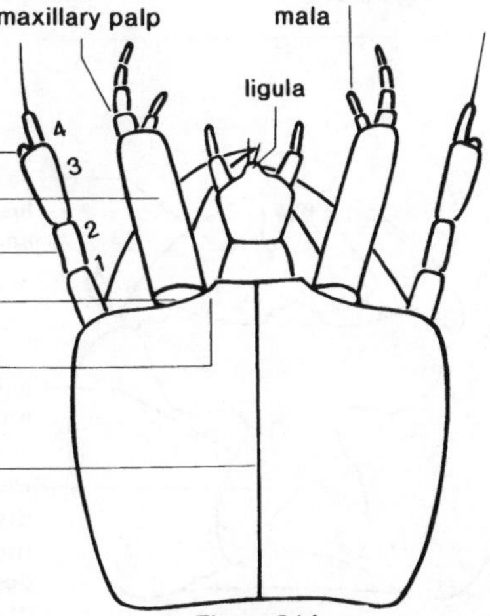

4-segmented maxillary palp
2-segmented mala
ligula
sensorium
4
3
stipes
2
antenna
1
cardo
strongly protracted ventral mouthparts
fused gular sutures

Figure 214

68(67). Segments in maxillary palp 2 (fig. 96) 69

68′. Segments in maxillary palp 3 (figs. 105, 106); mala distinct (**see 3rd choice**) 70

68². Segments in maxillary palp 4 (figs. 94, 100, 113); or if 3, then mala reduced to a brush of hairs 73

69(68). Frontal arms absent; head completely visible from above; body not broadly ovate and strongly flattened; ratio of length of abdomen to length of thorax more than 2; dorsal surfaces very lightly pigmented or sclerotized. In tunnels of ambrosia beetles (Platypodidae) .... (triungulins) (*Sosylus* part) ***Bothrideridae*** p. 477

69′. Frontal arms present (fig. 16); head not visible from above; body broadly ovate, strongly flattened, and disc-like (fig. 2); ratio of length of abdomen to length of thorax 1.2 to 2; dorsal surfaces more or less heavily pigmented or sclerotized. In leaf litter, moss .... (part) ***Corylophidae*** p. 495

70(68′). Ventral mouthparts retracted (fig. 89); ventral surfaces very lightly pigmented; antennae without long apical seta; stemmata on each side more than 1; tarsungulus not spatulate, without long setae at base 71

70′. Ventral mouthparts strongly protracted (fig. 92); ventral surfaces more or less heavily pigmented; antenna with apical seta as long as remainder of antenna; stemmata on each side 1; tarsungulus spatulate, with pair of long, flattened setae at base (fig. 132). On flowers or attached to native bees .... (triungulins) (**Meloinae** part) ***Meloidae*** p. 530

71(70). Body relatively straight or slightly curved ventrally; mandibles narrow and falcate (fig. 44); mentum, postmentum or labial plate completely free or basally connate with maxillae (fig. 89); ligula absent; tergum A3 without transverse plicae 72

71′. Body strongly curved ventrally (C-shaped); mandibles broad and stout or more or less wedge-like (fig. 47); mentum, postmentum or labial plate completely or almost completely connate with maxillae (fig. 96); ligula present; tergum A3 with 2 or more transverse plicae (fig. 136). Case-bearing larvae in leaf litter .... (**Cryptocephalinae** part) ***Chrysomelidae*** p. 568

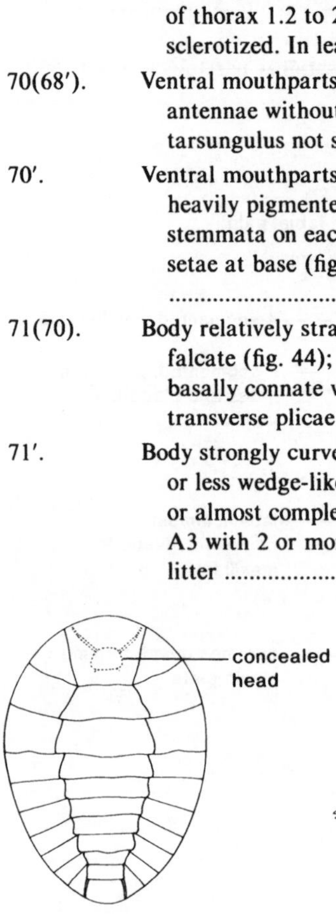

Figure 2

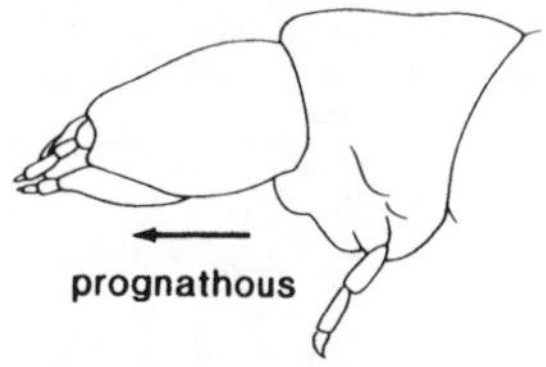

Figure 8

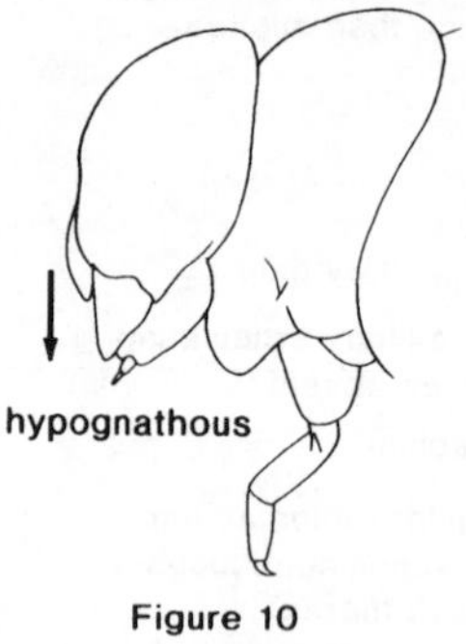

Figure 10

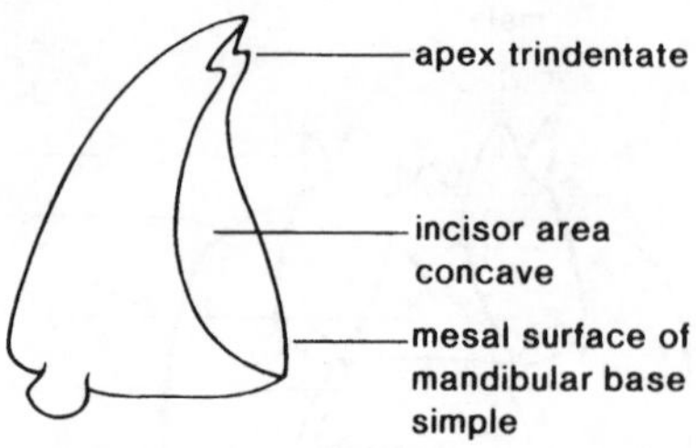

Figure 47

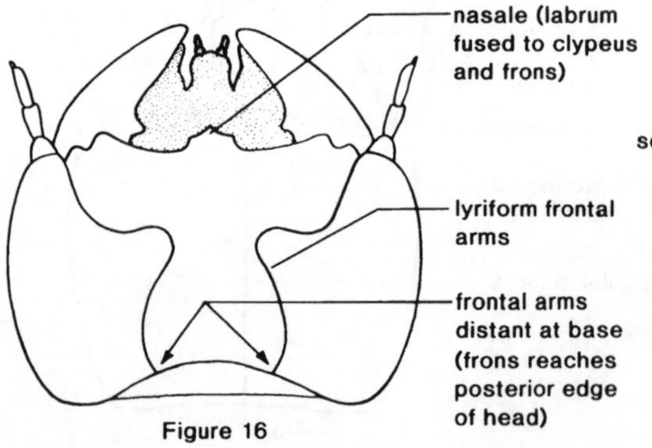

Figure 16

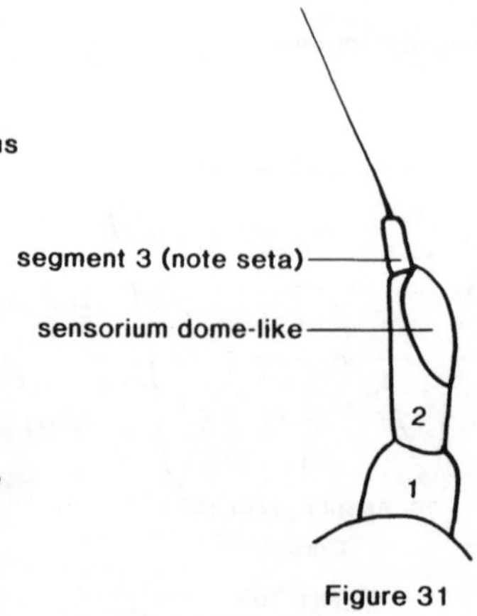

Figure 31

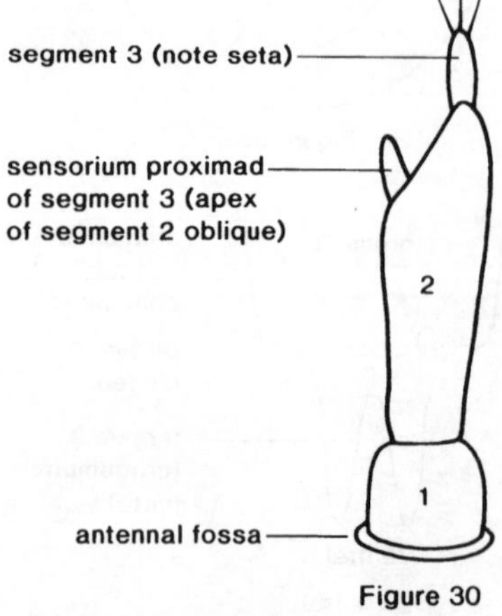

Figure 30

← **Figs. 44, 94, 96, 105, 106**

72(71). Stemmata on each side 3, forming tight cluster or triangle; antennal sensorium short and broad, conical or dome-like (fig. 31); prementum not emarginate; labial palps separated by more than 2 basal diameters; stipes short and broad; thoracic and most abdominal segments with lateral tergal processes. In leaf litter, moss .... (part) ***Scydmaenidae*** p. 330

72′. Stemmata on each side 2 or 3 well-separated and forming a row; antennal sensorium more elongate, palpiform (fig. 30), setiform, or bifid; apex of prementum usually emarginate; labial palps usually separated by less than 2 basal diameters; stipes elongate; thoracic and abdominal segments without lateral tergal processes. In leaf litter, moss .... (part) ***Pselaphidae*** p. 353

73(68²). Epicranial stem moderately long (fig. 20); head moderately to strongly declined (hypognathous) (fig. 10); body strongly curved ventrally (C-shaped); stemmata on each side 5 or 6. Case-bearing larvae in leaf litter .... (**Cryptocephalinae** part) ***Chrysomelidae*** p. 368

73′. Epicranial stem absent (fig. 16) or very short; head prognathous or slightly declined (fig. 8); body relatively straight or slightly curved ventrally; stemmata on each side 1 or 0 74

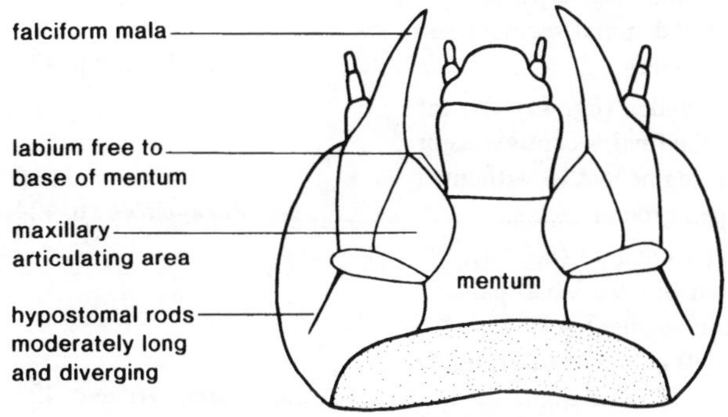

falciform mala
labium free to base of mentum
maxillary articulating area
hypostomal rods moderately long and diverging
mentum

Mouthparts Retracted

Figure 89

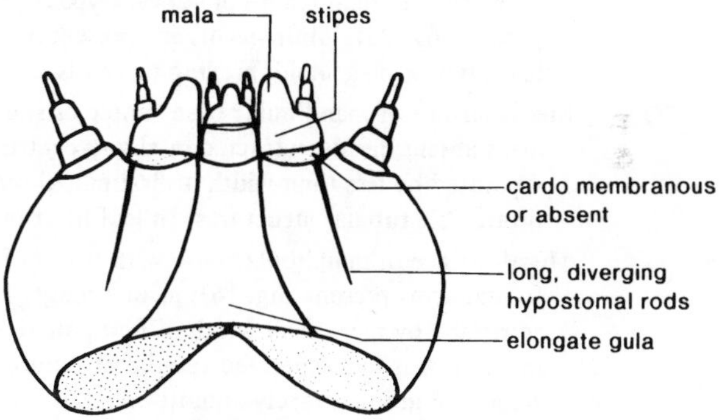

mala
stipes
cardo membranous or absent
long, diverging hypostomal rods
elongate gula

Ventral Mouthparts Strongly Protracted

Figure 92

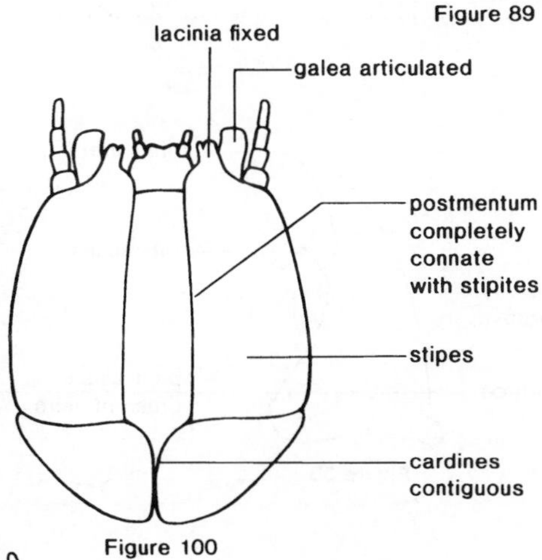

lacinia fixed
galea articulated
postmentum completely connate with stipites
stipes
cardines contiguous

Figure 100

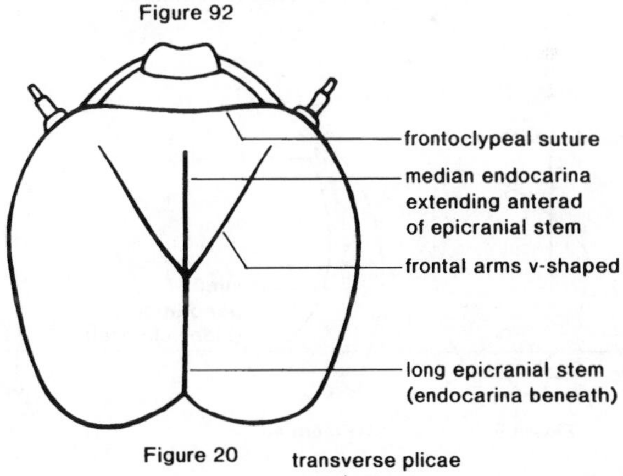

frontoclypeal suture
median endocarina extending anterad of epicranial stem
frontal arms v-shaped
long epicranial stem (endocarina beneath)

Figure 20

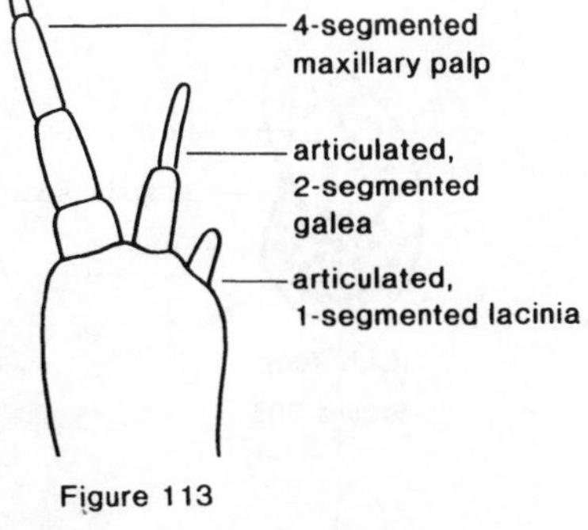

4-segmented maxillary palp
articulated, 2-segmented galea
articulated, 1-segmented lacinia

Figure 113

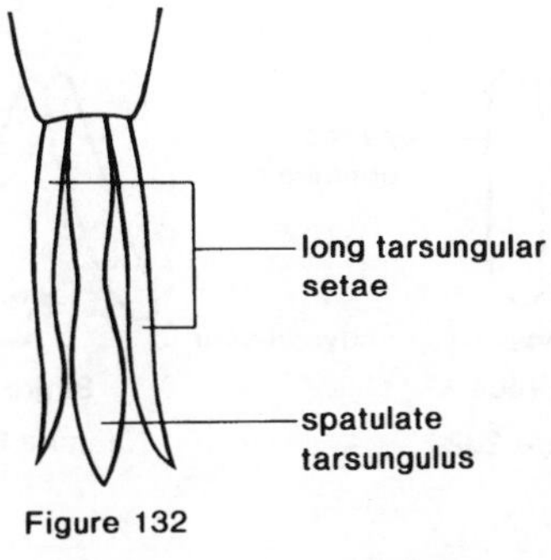

long tarsungular setae
spatulate tarsungulus

Figure 132

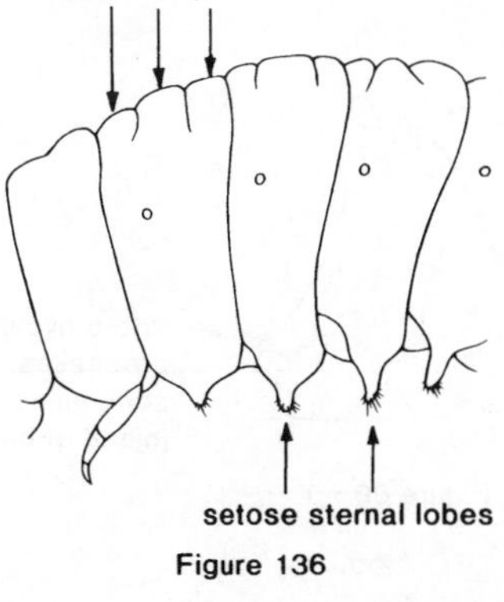

transverse plicae
setose sternal lobes

Figure 136

74(73′). Paired dorsal glands (fig. 139) on thoracic terga and terga A1–8 or A1–9, body lightly sclerotized and flexible (sometimes with dark pigment pattern) and clothed with very short, fine pubescence, which gives the surface a dull appearance; spiracles small and cribriform (fig. 205) or apparently annular; mala consisting of a single segment (fig. 94). In soil and leaf litter, under stones .... (part) ***Cantharidae*** p. 429

74′. Paired dorsal glands absent; body usually more heavily sclerotized and rigid, clothed with setae, or occasionally long, fine hairs; spiracles biforous (figs. 202, 204); mala 2-segmented (fig. 103). In rotten wood, soil, leaf litter .... (part) ***Elateridae*** p. 410

75(67′). Ratio of antennal length to head width less than .5; dorsal surfaces very lightly sclerotized; segment A10 distinct and visible from above; anal region posteriorly or terminally oriented; gular region as long as or almost as long as stipes (fig. 94); clypeus not produced into tapered process; luminous organs absent. In soil and leaf litter, under stones .... (part) ***Cantharidae*** p. 429

75′. Ratio of antennal length to head width more than .5; dorsal surfaces more or less heavily sclerotized; segment A10 concealed from above, distinct or more or less fused to segment A9; anal region ventrally oriented; gular region much shorter than length of stipes; clypeus produced into long, tapered process (fig. 26); luminous organs present at sides of abdominal segments. In leaf litter or on ground, preying on snails. Exotic (Eurasia) .... ***Homalisidae*** p. 422

76(67^{2}). Mesal surface of mandibular base simple or slightly expanded (fig. 43); frontal arms absent; head protracted or slightly retracted; labial palps contiguous or separated by less than width of first palpal segment; segment A10 without protrusible tubular structures. In leaf litter or on open ground .... ***Phengodidae*** p. 424

76′. Mesal surface of mandibular base with brush of hairs (penicillus) (fig. 53); frontal arms present (fig. 16); head strongly retracted (fig. 9); labial palps separated by more than width of first palpal segment; segment A10 with 8 or more protrusible, asperated tubular structures (fig. 191). In leaf litter, under stones, on ground, rarely aquatic .... ***Lampyridae*** p. 427

77(62′). Mesal surface of mandibular base simple or slightly expanded (fig. 44); hypostomal rods absent 78

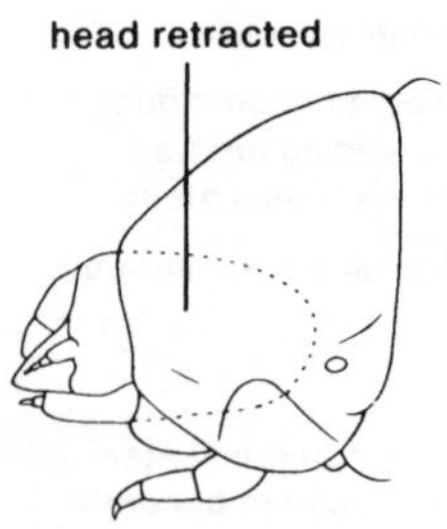

head retracted

Figure 9

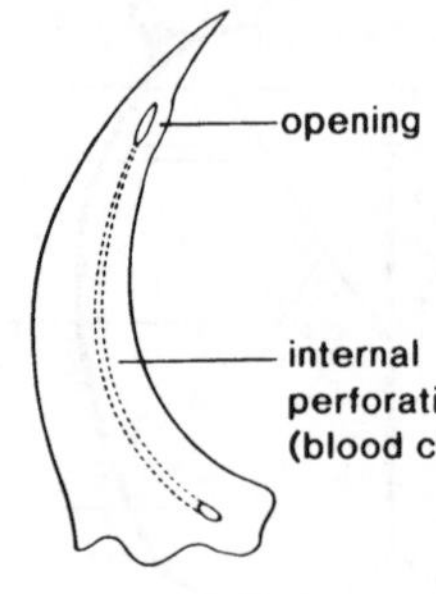

opening
internal perforation (blood channel)

Figure 43

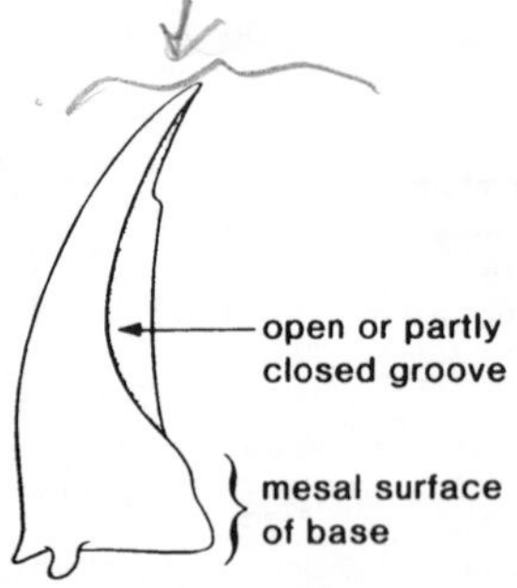

open or partly closed groove
mesal surface of base

Figure 44

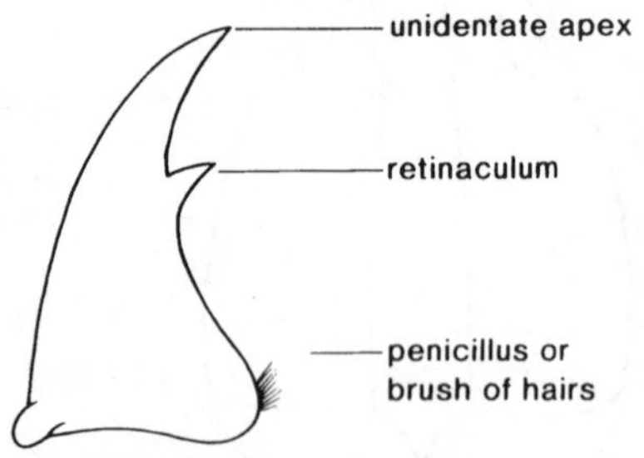

unidentate apex
retinaculum
penicillus or brush of hairs

Figure 53

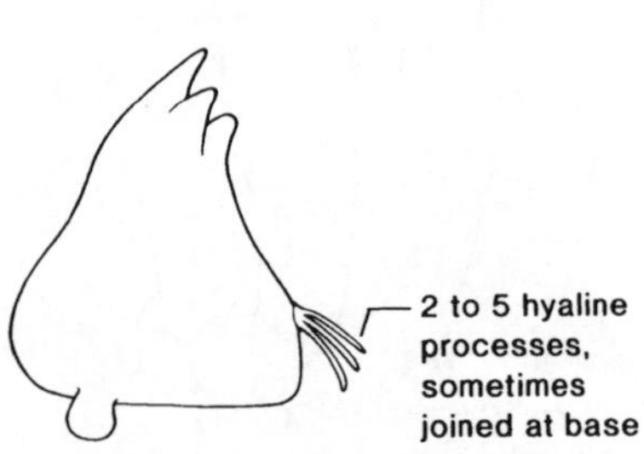

2 to 5 hyaline processes, sometimes joined at base

Figure 60

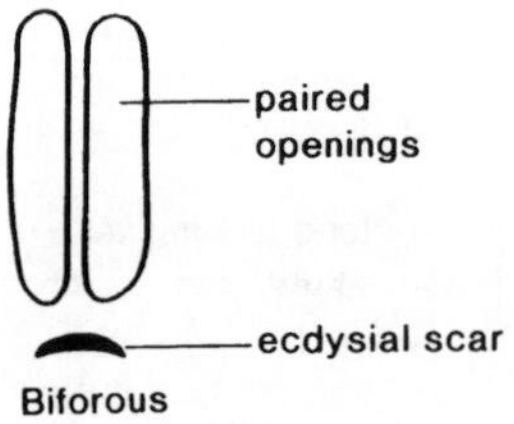

paired openings
ecdysial scar
Biforous

Figure 202

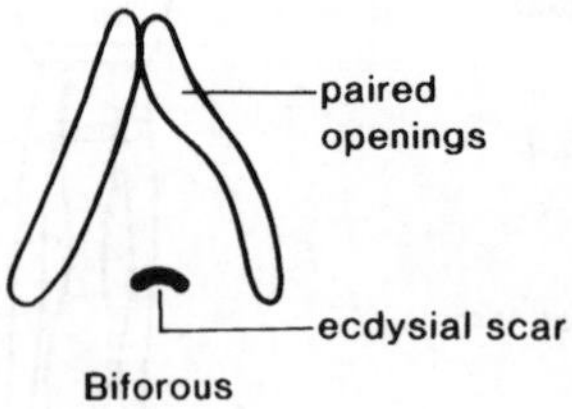

paired openings
ecdysial scar
Biforous

Figure 204

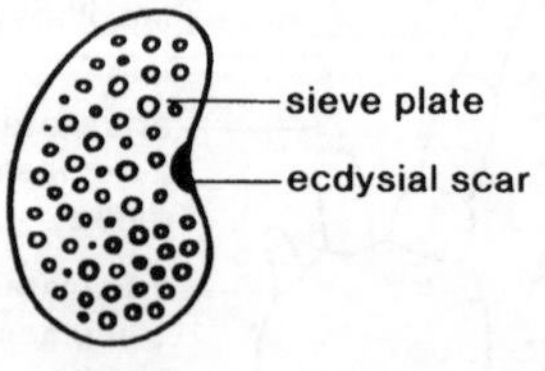

sieve plate
ecdysial scar
Cribriform

Figure 205

77′. Mesal surface of mandibular base with brush of hairs (penicillus) (fig. 53); hypostomal rods absent; ventral epicranial ridges present (fig. 93). In rotten wood, soil, leaf litter (**see 3rd choice**) .... (part) ***Elateridae*** p. 410

77². Mesal surface of mandibular base with one or more hyaline processes (fig. 60); hypostomal rods long and diverging (fig. 92); ventral epicranial ridges absent; ventral mouthparts strongly protracted (fig. 92); cardines absent, indistinct or membranous (fig. 92); stipes wider than long. In flowers, rotten wood, fungus fruiting bodies .... (part) ***Phalacridae*** p. 466

78(77). Epicranial stem absent or very short 79

78′. Epicranial stem moderately long (fig. 14) 82

79(78). Frontal arms absent; cardo absent, indistinct, or membranous 80

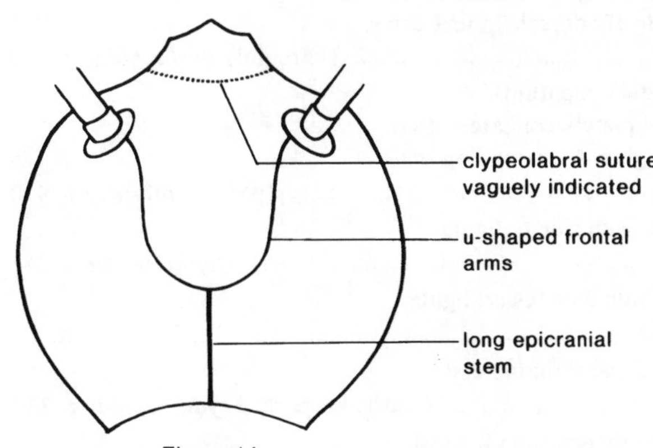

Figure 14

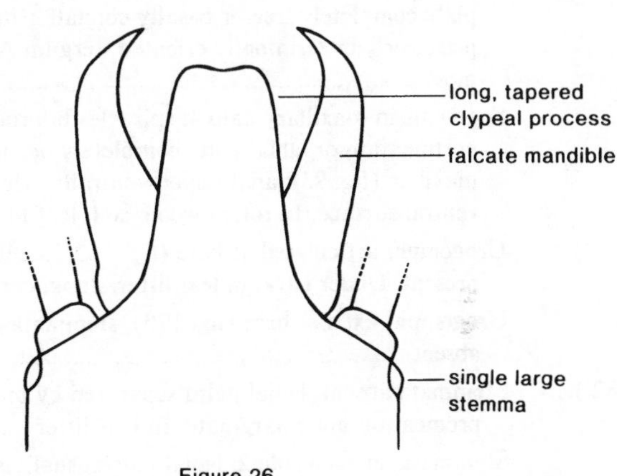

Figure 26

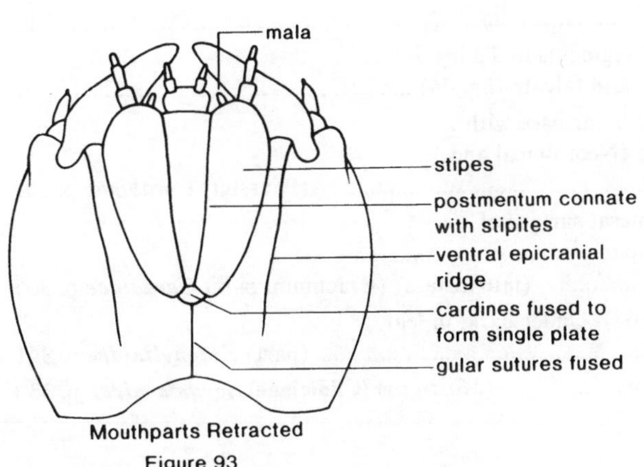

Mouthparts Retracted
Figure 93

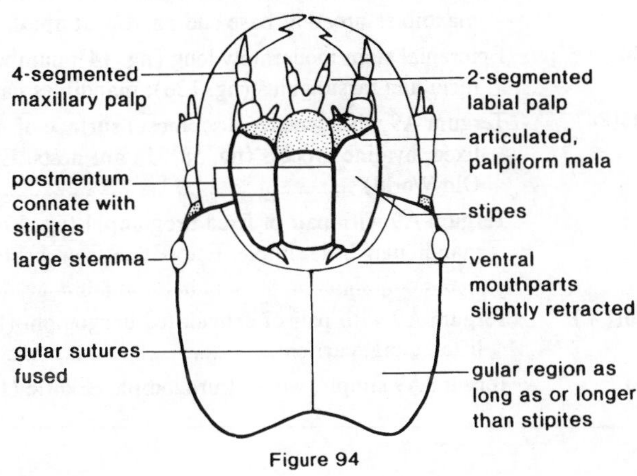

Figure 94

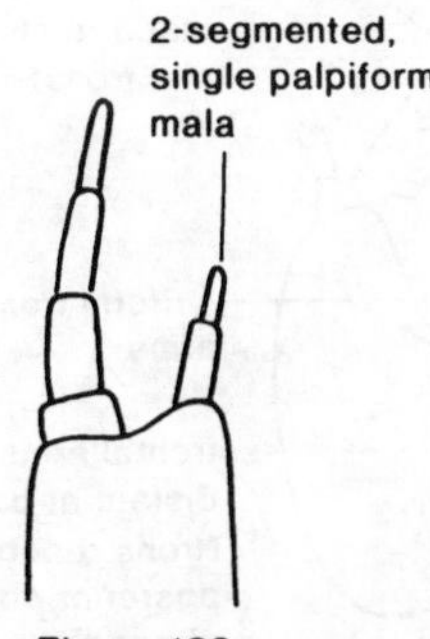

Figure 103

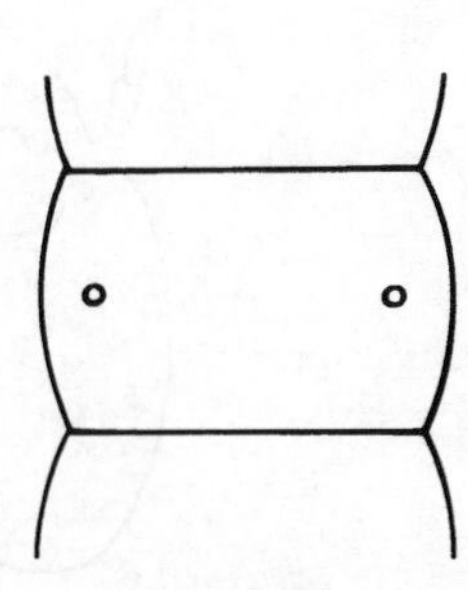

Figure 139

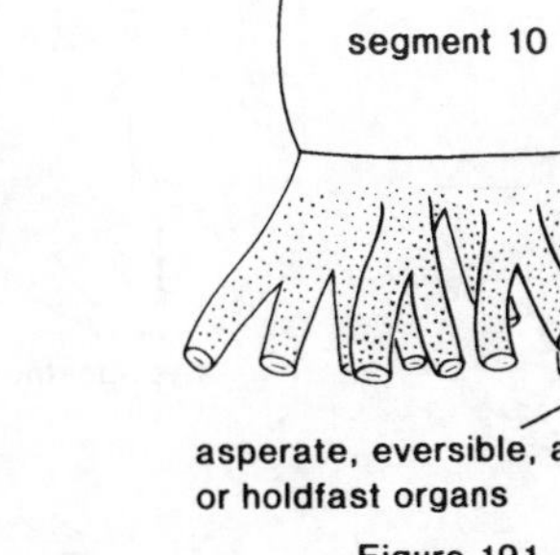

Figure 191

Fig. 92 ⟶

79′. Frontal arms present (fig. 16); cardo distinct and sclerotized (fig. 100) 81

80(79). Ratio of antennal length to head width less than .15; segments in maxillary palp 3; labial palps contiguous or separated by less than width of first palpal segment; mandibles flattened and more or less triangular, without groove (fig. 45); abdomen without lateral processes; stemmata absent. In rotten wood, in soil associated with roots .... (part) ***Throscidae*** p. 418

80′. Ratio of antennal length to head width .15 to .5; segments in maxillary palp 4; labial palps separated by more than width of first palpal segment; mandibles narrow and falcate, with open or partly closed groove (fig. 44); abdomen with well-developed lateral processes on each side; stemmata on each side 1. On ground, in leaf litter (snail predators). Exotic (Old World) .... ***Drilidae*** p. 424

81(79′). Segments in maxillary palp 3; spiracles annular; mentum, postmentum or labial plate completely free or basally connate with maxillae (fig. 89); anal region posteriorly or terminally oriented; tergum A9 completely dorsal. In leaf litter, moss .... (Faronini) ***Pselaphidae*** p. 353

81′. Segments in maxillary palp 4; spiracles biforous (fig. 204); mentum, postmentum or labial plate completely or almost completely connate with maxillae (fig. 93); anal region ventrally oriented; tergum A9 extending onto ventral surface. In rotten wood, soil, leaf litter .... (part) ***Elateridae*** p. 410

82(78′). Urogomphi articulated at base (fig. 162); stemmata on each side 6; ligula present. Under bark, in leaf litter, dung, carrion .... (part) ***Staphylinidae*** p. 341

82′. Urogomphi fixed at base (fig. 170); stemmata on each side 3 or fewer; ligula absent 83

83(82′). Stemmata absent; labial palps separated by more than 2 basal diameters; prementum not emarginate. In leaf litter .... (Eutheiini part) ***Scydmaenidae*** p. 330

83′. Stemmata on each side 2; labial palps usually separated by less than 2 basal diameters; apex of prementum usually emarginate. In leaf litter, moss .... (part) ***Pselaphidae*** p. 353

84(48²). Epicranial stem very short or absent; number of segments in T2 leg 3 or 4; mandibles broad at base and narrow at apex 85

84′. Epicranial stem moderately long (fig. 14); number of segments in T2 leg 5 including tarsungulus (fig. 126); mandibles narrow and falcate (fig. 44) 86

85(84). Tergum A9 with concave disc; mesal surface of mandibular base with fixed, hyaline process (fig. 58). In ant nests. Exotic (Neotropical and Old World) .... (Paussini) ***Carabidae*** p. 305

85′. Tergum A9 with pair of fixed urogomphi (fig. 170); mesal surface of mandibular base simple. Ectoparasitic on beetle pupae .... (late instars) (Brachinini part) ***Carabidae*** p. 305

86(84′). Tergum A9 with pair of articulated urogomphi (fig. 162). Under bark, in leaf litter, dung, carrion .... (part) ***Staphylinidae*** p. 341

86′. Tergum A9 simple, without urogomphi. Exotic (Eurasia) .... (*Mastigus*) (**Clidicinae**) ***Scydmaenidae*** p. 330

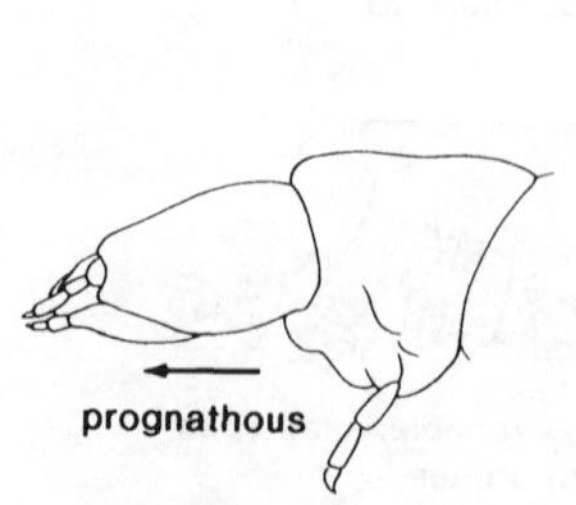

Figure 8

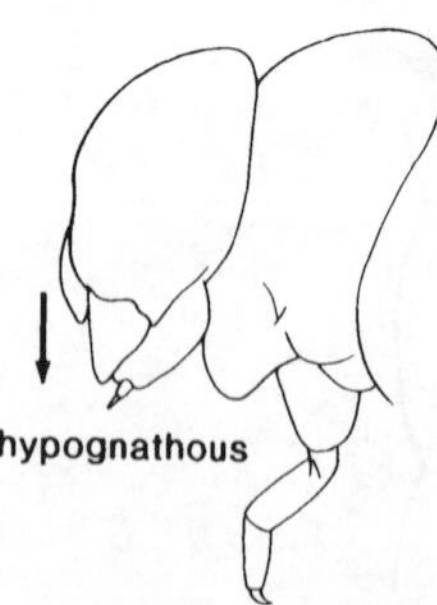

Figure 10

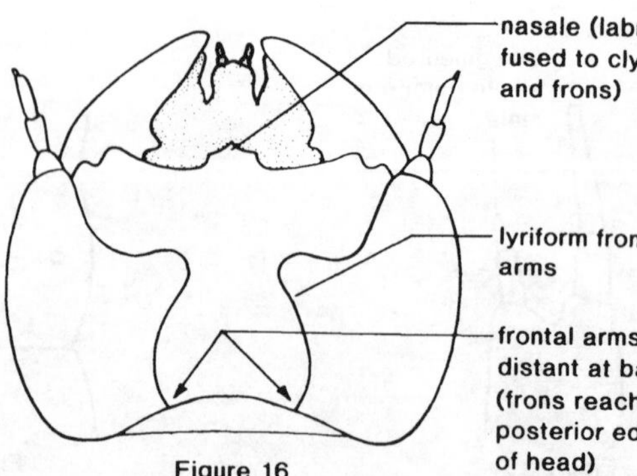

Figure 16

←— **Figs. 14, 44, 93, 204**

87(35²). Antennal segments 1; cardines not distinctly separated from stipites but separated from each other by labium; segments in maxillary palp 2; segments in labial palp 1; ventral epicranial ridges absent. In soil, associated with cicada larvae (ectoparasitic) .... (part) ***Rhipiceridae*** p. 370

87′. Antennal segments 3; cardines distinctly separated from stipites but closely approximate or contiguous with one another (fig. 93), not separated by labium; segments in maxillary palp 4; segments in labial palp 2; ventral epicranial ridges present (fig. 93) 88

88(87′). Head moderately to strongly declined (hypognathous) (fig. 10), globular in shape; outer edges of mandibles forming sharp, slightly raised carina; cervical region of prosternum produced forward and enclosing eversible membrane (fig. 120); body cylindrical, moderately heavily sclerotized, with reduced legs. In soil .... ***Cebrionidae*** p. 418

88′. Head prognathous or slightly declined (fig. 8), usually more or less flattened; outer edges of mandibles not carinate; cervical region of prosternum not strongly produced forward, without eversible membrane. In rotten wood, soil, leaf litter .... (part) ***Elateridae*** p. 410

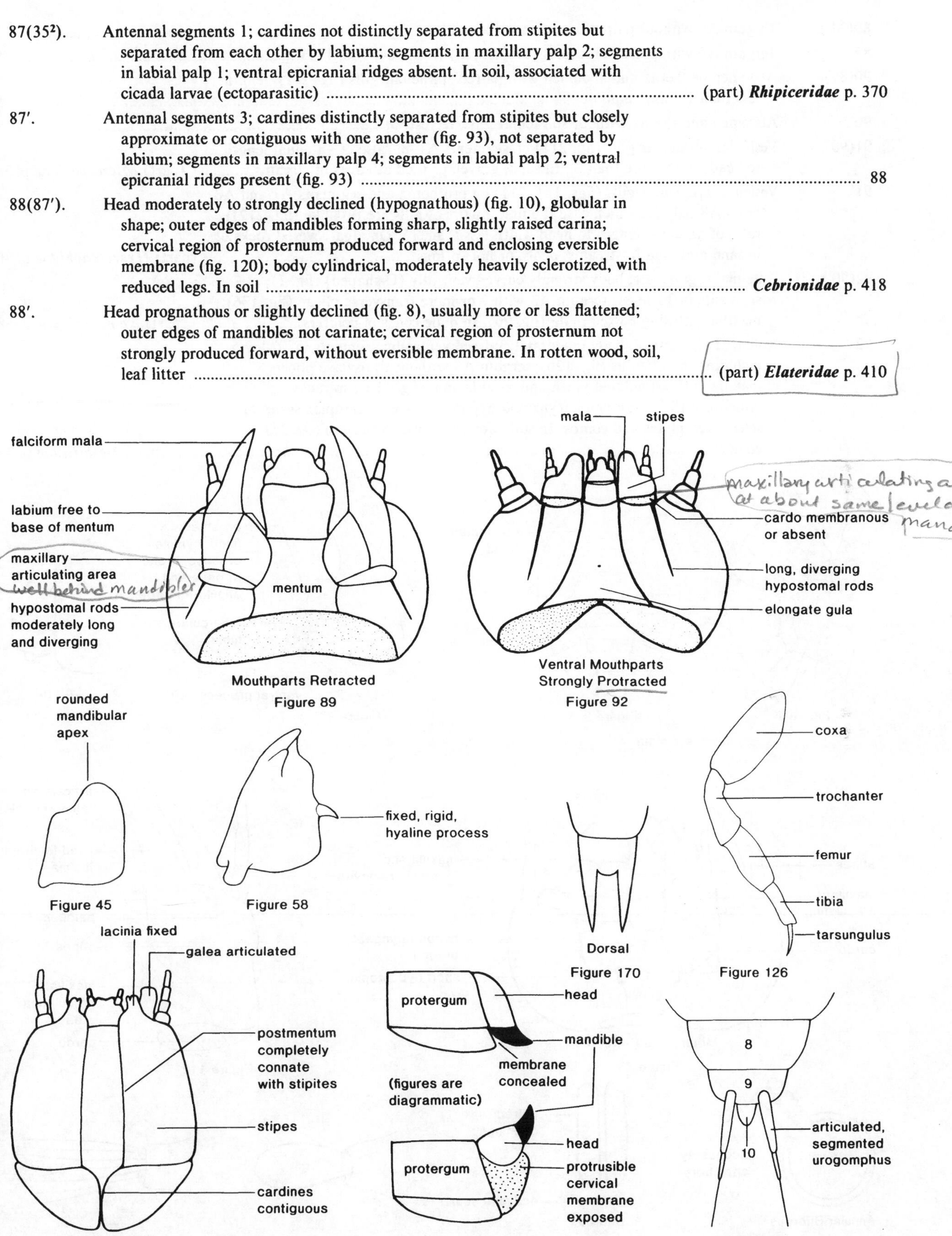

Figure 89

Figure 92

Figure 45

Figure 58

Figure 170

Figure 126

Figure 100

Figure 120

Figure 162

89(34′). Tergum A9 without paired processes (urogomphi) 90

89′. Tergum A9 with paired processes (urogomphi) (figs. 164, 166, 171) 98

90(89). Abdomen or thorax and abdomen with paired spiracular gills (fig. 211); length less than 1.7 mm; aquatic 91

90′. Abdomen and thorax without spiracular gills; terrestrial 92

91(90). Vesicular spiracular gills (fig. 211) on segments A1–8; ratio of antennal length to head width more than .2. In sand, gravel, or mud at edges of streams .... (part) ***Microsporidae*** p. 302

91′. Vesicular spiracular gills (figs. 161, 211) on prothorax and segments A1 and A8 or on A8 only (in which case the first 2 pairs of gills are tuft-like (fig. 212)); ratio of antennal length to head width less than .15. On alga-covered rocks or in sand and mud in streams, pools, or hot springs .... (part) ***Hydroscaphidae*** p. 303

92(90′). Antennal segments 1; body strongly curved ventrally (C-shaped) (fig. 1); segments in T2 leg 1; tergum A3 with 2 or more transverse plicae (fig. 136); maxilla with single fixed mala (fig. 96). In male cones of conifers .... (part) ***Nemonychidae*** p. 585

92′. Antennal segments 2; body moderately curved ventrally; segments in T2 leg 5 including tarsungulus (fig. 126); tergum A3 without transverse plicae; maxilla with articulated galea and fixed lacinia (fig. 112); spiracles cribriform (fig. 206); epipharynx and hypopharynx with complex series of sclerotized plates and combs. In soil (western North America) (**see 3rd choice**) .... (part) ***Dascillidae*** p. 369

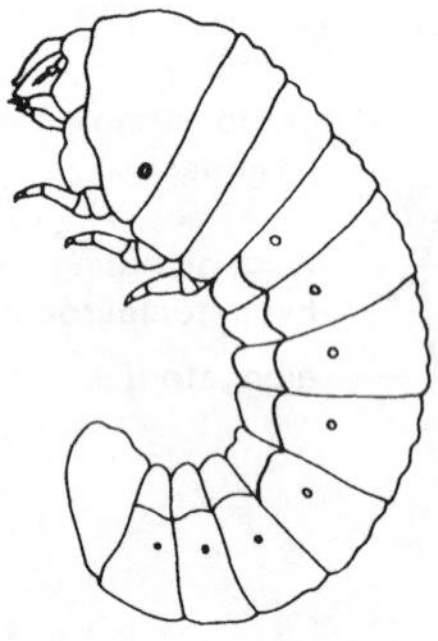

Figure 1

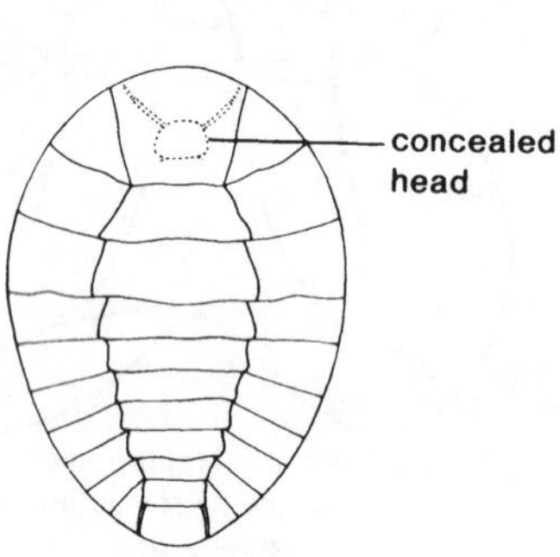

Figure 2

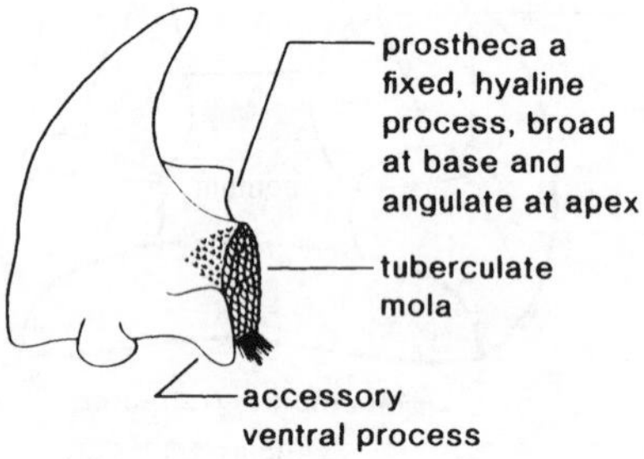

Figure 81

Lateral

Figure 171

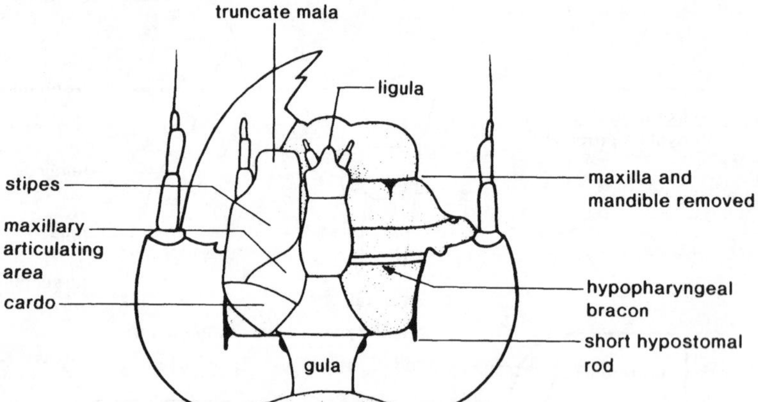

Mouthparts Retracted

Figure 85

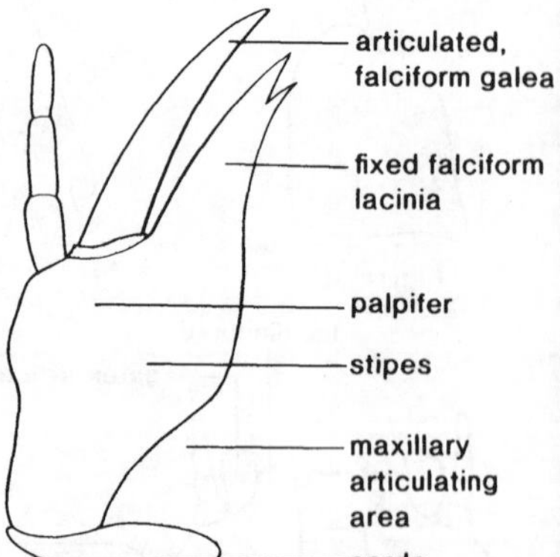

Figure 112

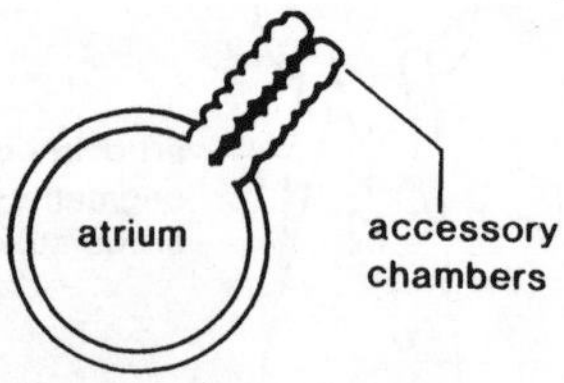

Annular Biforous

Figure 198

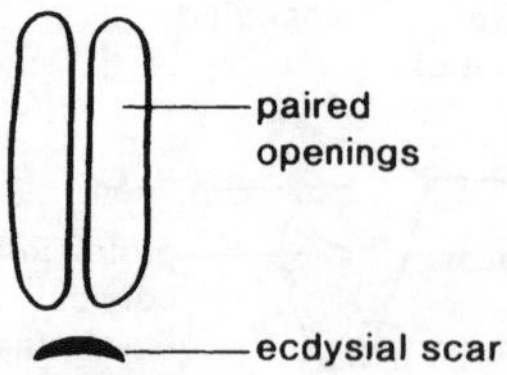

Biforous

Figure 202

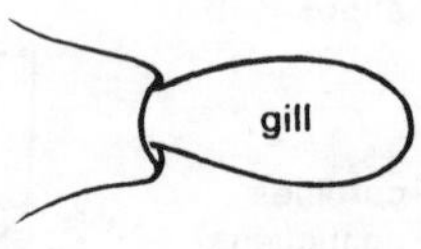

Figure 211

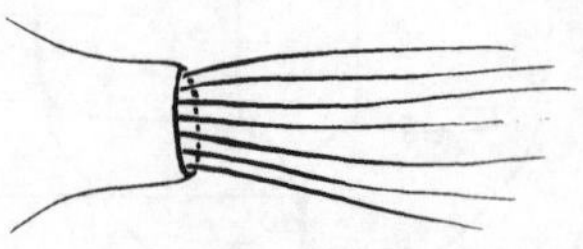

Figure 212

← Figs. 16, 92, 126

92[2]. Antennal segments 3; segments in T2 leg 5 including tarsungulus (fig. 126) 93

93(92[2]). Segments in labial palp 1; maxillary articulating area absent 94

93′. Segments in labial palp 2; maxillary articulating area present (fig. 85) 96

94(93). Body broadly ovate, strongly flattened and disc-like (fig. 2); stemmata on each side 2; ratio of antennal length to head width more than .5; spiracles annular; cardines indistinct or absent. In leaf litter .... (part) ***Corylophidae*** p. 495

94′. Body not broadly ovate and strongly flattened; stemmata on each side 3 or 4; ratio of antennal length to head width less than .5; spiracles annular-biforous (fig. 198) or biforous (fig. 202); cardines distinct, more or less elongate (fig. 91) 95

95(94′.) Bases of frontal arms distinctly separated (fig. 16); prostheca broad with obtusely angulate or rounded apex (fig. 81); hypostomal rods absent. In flowers .... (**Meligethinae** part) ***Nitidulidae*** p. 456

95′. Bases of frontal arms contiguous (fig. 24); prostheca absent; hypostomal rods present (fig. 92). On plant surfaces (feeding on coccids) .... (**Cybocephalinae**) ***Nitidulidae*** p. 456

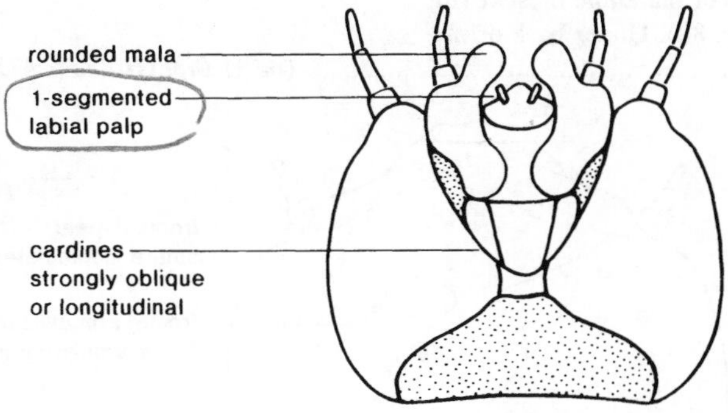

rounded mala
1-segmented labial palp
cardines strongly oblique or longitudinal

Figure 91

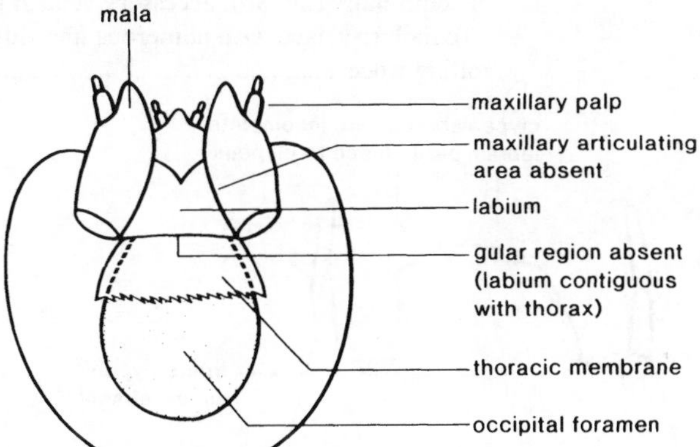

mala
maxillary palp
maxillary articulating area absent
labium
gular region absent (labium contiguous with thorax)
thoracic membrane
occipital foramen

Figure 96

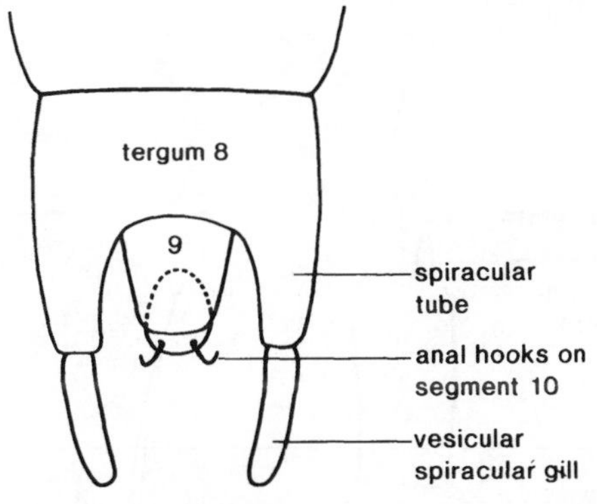

tergum 8
9
spiracular tube
anal hooks on segment 10
vesicular spiracular gill

Figure 161

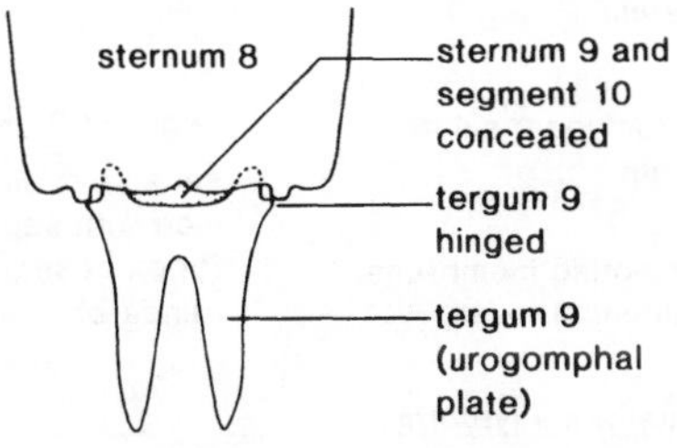

sternum 8
sternum 9 and segment 10 concealed
tergum 9 hinged
tergum 9 (urogomphal plate)

Ventral

Figure 164

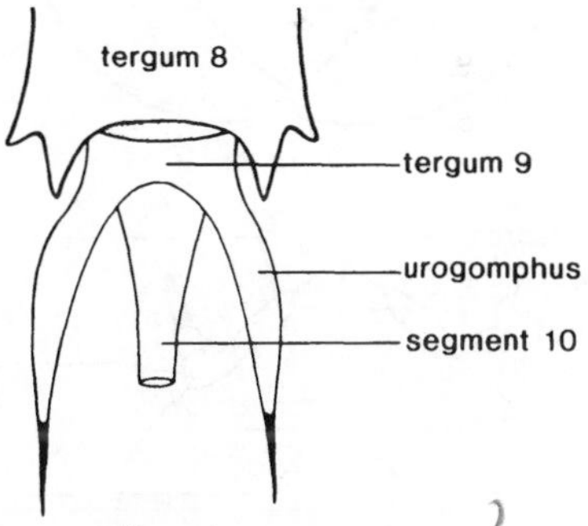

tergum 8
tergum 9
urogomphus
segment 10

Dorsal

Figure 166

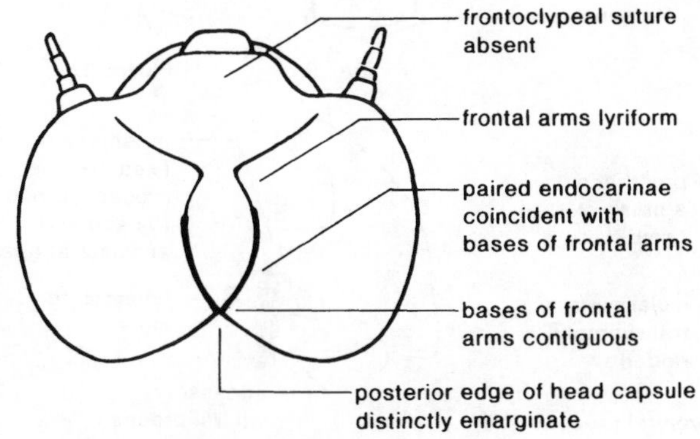

frontoclypeal suture absent
frontal arms lyriform
paired endocarinae coincident with bases of frontal arms
bases of frontal arms contiguous
posterior edge of head capsule distinctly emarginate

Figure 24

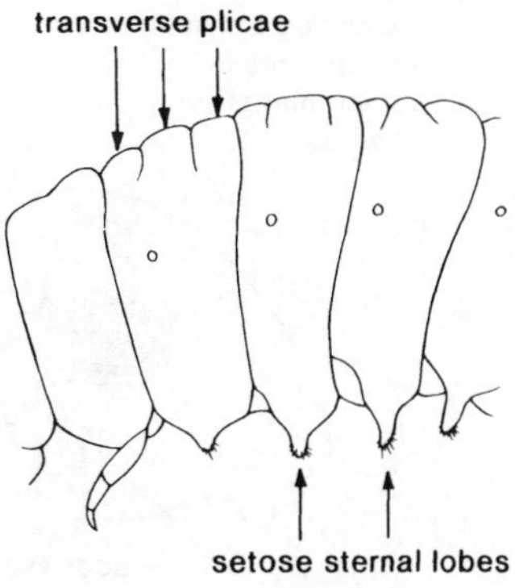

transverse plicae
setose sternal lobes

Figure 136

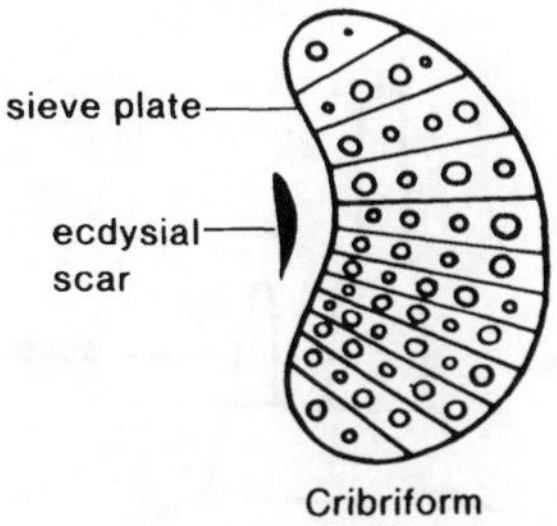

sieve plate
ecdysial scar

Cribriform

Figure 206

96(93′). Spiracles annular; stemmata on each side 5; setae on tarsungulus 2; mandibular apex with numerous teeth and spines; epicranial stem moderately long (fig. 18) and frontal arms lyriform (fig. 22); most dorsal setae borne on elongate tubercles (fig. 3). In rotten wood, sometimes associated with slime molds .... (part) ***Eucinetidae*** p. 364

96′. Spiracles annular-biforous (fig. 198); stemmata on each side 6; setae on tarsungulus 1; mandibular apex bidentate (fig. 80); epicranial stem absent and frontal arms lyriform (fig. 216); dorsal setae not on elongate tubercles **(see 3rd choice)** 97

96². Spiracles cribriform (fig. 206); stemmata on each side 0; setae on tarsungulus 4 or more; mandible unidentate, but with a complex, bifid retinaculum and a small, sclerotized, articulated process; epicranial stem very short and frontal arms V-shaped (fig. 12); dorsal setae not borne on tubercles; epipharynx and hypopharynx with a complex series of sclerotized plates and combs. In soil (western North America) .... (part) ***Dascillidae*** p. 369

97(96′). Body circular in cross-section; ventral epicranial ridges present (fig. 90); A8 spiracles placed at ends of spiracular tubes; prostheca consisting of a brush of comb-hairs (fig. 80); accessory ventral process of mandible present (fig. 80); molar surface with numerous fine ridges (fig. 80). Under bark or in rotting wood .... (part) ***Biphyllidae*** p. 475

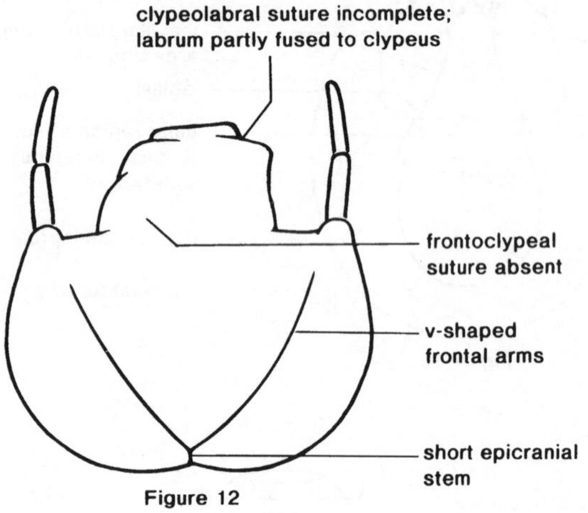

clypeolabral suture incomplete; labrum partly fused to clypeus
frontoclypeal suture absent
v-shaped frontal arms
short epicranial stem

Figure 12

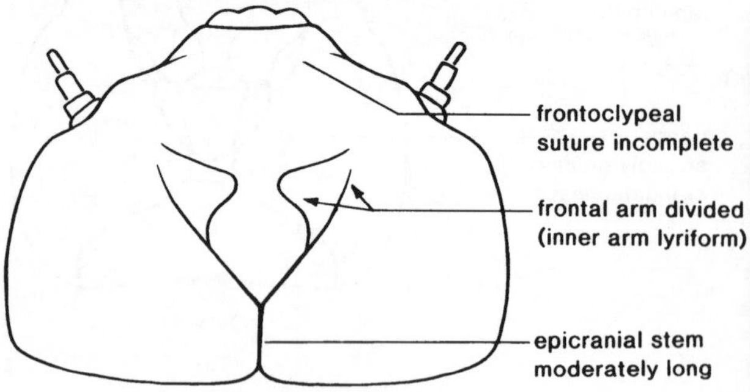

frontoclypeal suture incomplete
frontal arm divided (inner arm lyriform)
epicranial stem moderately long

Figure 18

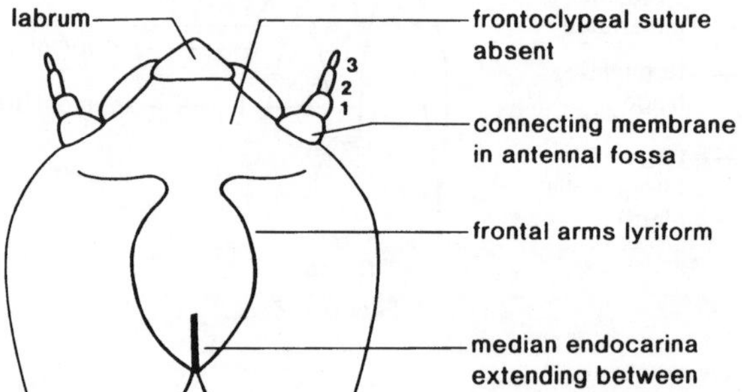

labrum
frontoclypeal suture absent
3
2
1
connecting membrane in antennal fossa
frontal arms lyriform
median endocarina extending between frontal arms (epicranial stem absent)

Figure 22

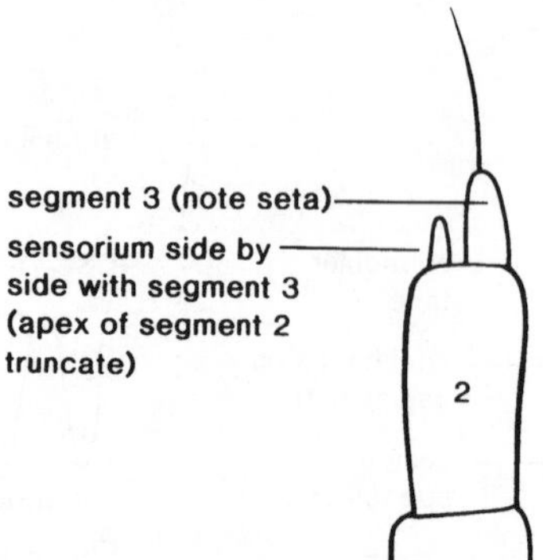

segment 3 (note seta)
sensorium side by side with segment 3 (apex of segment 2 truncate)
2
1
antennal fossa

Figure 29

Figure 3

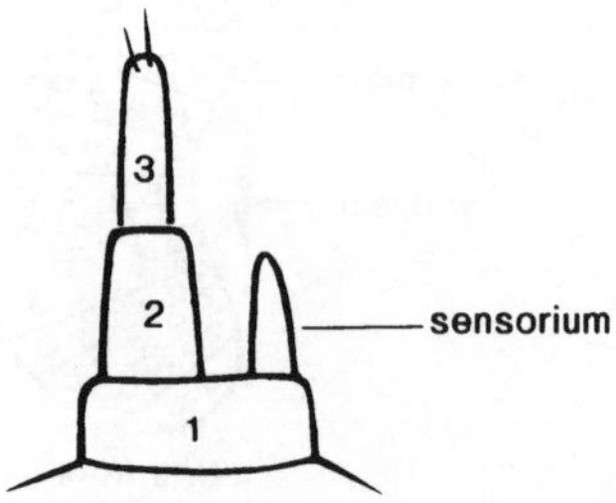

3
2
sensorium
1

Figure 33

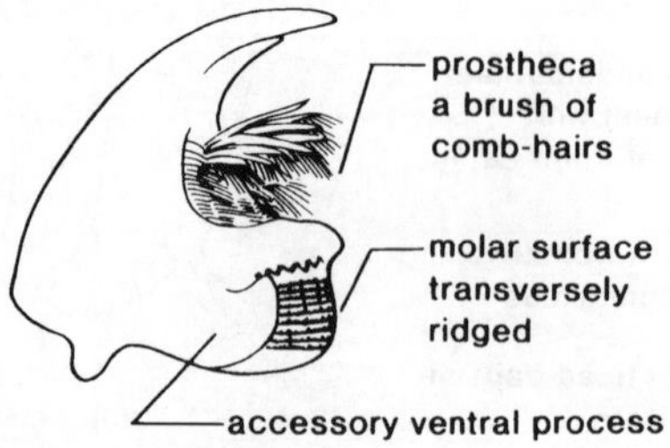

prostheca a brush of comb-hairs
molar surface transversely ridged
accessory ventral process

Figure 80

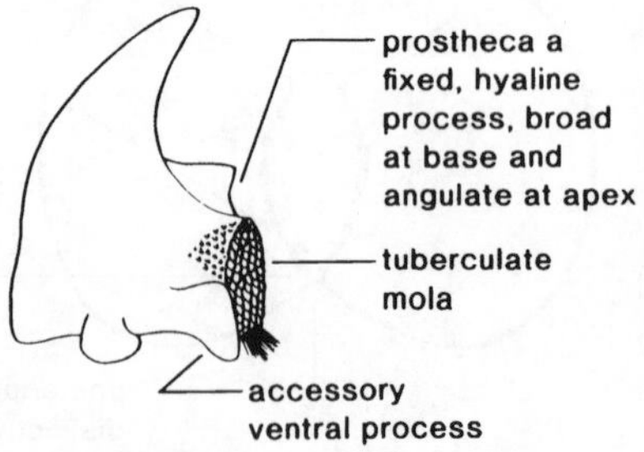

prostheca a fixed, hyaline process, broad at base and angulate at apex
tuberculate mola
accessory ventral process

Figure 81

← Fig. 206

97′. Body slightly flattened; ventral epicranial ridges absent; A8 spiracles not placed at ends of spiracular tubes; prostheca consisting of a brush of simple hairs or several hyaline processes; accessory ventral process of mandible absent; molar surface tuberculate or asperate (fig. 81). In oak catkins (western North America) .... (*Byturellus*) ***Byturidae*** p. 476

98(89′). Abdominal spiracles annular 99

98′. Abdominal spiracles annular-biforous (figs. 197, 198) **(see 3rd choice)** 100

98^{2}. Abdominal spiracles cribriform (fig. 206); antennae 2- or 3-segmented with a very small terminal segment; setae on tarsungulus 4 or more; epipharynx and hypopharynx with a complex series of sclerotized plates and combs. In soil (western North America) .... (part) ***Dascillidae*** p. 369

99(98). Ratio of antennal length to head width less than .15; antennal sensorium on segment 1 and not on segment 2 (fig. 33); ventral mouthparts strongly protracted (fig. 92); stipes wider than long; epicranial stem absent; segment A10 absent or completely concealed from above (fig. 164). On plant surfaces affected by rusts or smuts .... (part) ***Phalacridae*** p. 466

99′. Ratio of antennal length to head width more than .5; antennal sensorium on segment 2 (fig. 29); ventral mouthparts retracted (fig. 90); stipes longer than wide; epicranial stem present; segment A10 distinct and visible from above (fig. 166). Under bark .... (Brontini) ***Cucujidae*** p. 463

100(98′). Segments in labial palp 1; frontal arms V- or U-shaped (fig. 12); stemmata on each side fewer than 6; maxillary articulating area absent. In flowers .... **(Meligethinae** part) ***Nitidulidae*** p. 456

100′. Segments in labial palp 2; frontal arms lyriform (fig. 22); stemmata on each side 6; maxillary articulating area present (fig. 90) 101

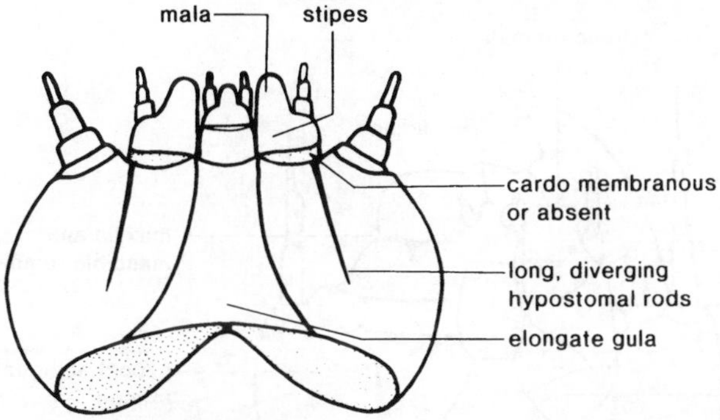

Ventral Mouthparts Strongly Protracted
Figure 92

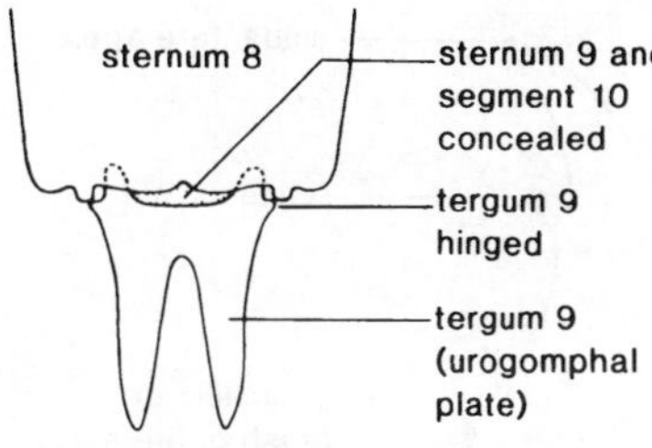

Ventral
Figure 164

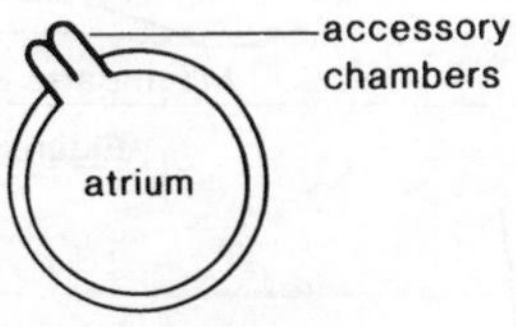

Annular Biforous
Figure 197

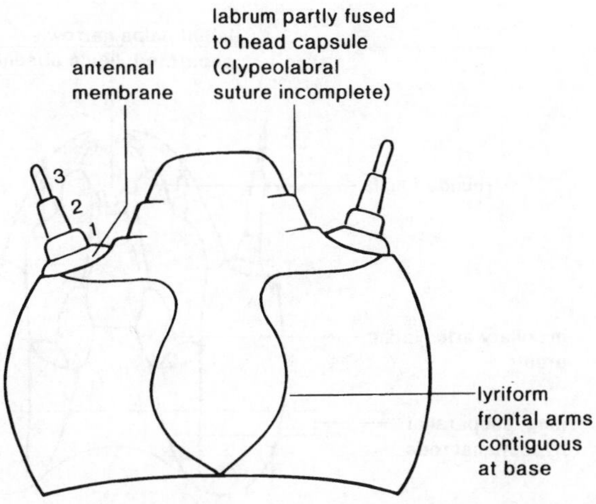

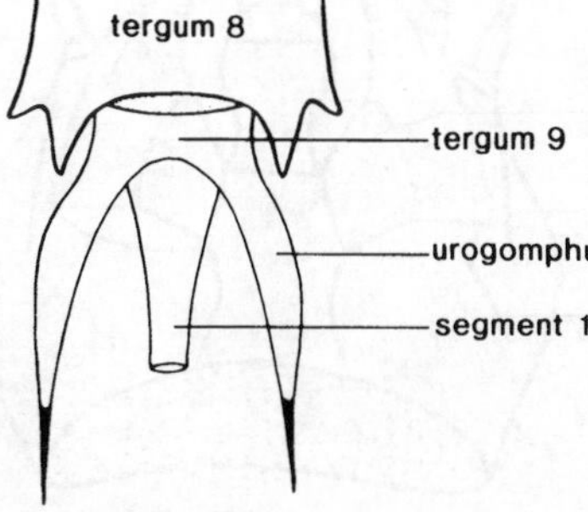

Dorsal
Figure 166

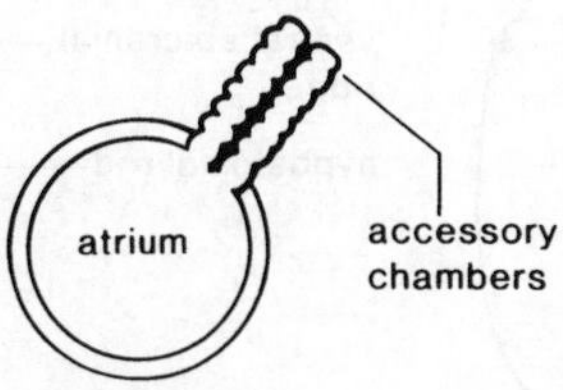

Annular Biforous
Figure 198

Figure 216

Fig. 90 ⟶

101(100′). Ratio of antennal length to head width .15 to .5; ratio of length of abdomen to length of thorax more than 2; body strongly flattened; apex of mala or galea falciform (fig. 90); abdominal terga with paired rows of asperities sometimes forming incomplete rings (fig. 148). Under bark. Exotic (Southern Hemisphere) .... **(Hymaeinae)** ***Phloeostichidae*** p. 462

101′. Ratio of antennal length to head width more than .5; ratio of length of abdomen to length of thorax 1.2 to 2; body slightly flattened; apex of mala or galea rounded or truncate (figs. 85, 86); abdominal terga without rows of asperities. In sap flows or under fermenting bark. Exotic (Africa and Asia) .... ***Helotidae*** p. 463

102(33′). Antennal segments 3; thoracic and abdominal terga produced laterally to form plate-like processes; A8 spiracles dorsally placed (fig. 153); urogomphi 1-segmented, articulated at base (fig. 153). In mountain streams of western North America .... ***Amphizoidae*** p. 313

102′. Antennal segments 4; if thoracic and abdominal terga produced laterally, A8 spiracles laterally placed. **(see 3rd choice)** 103

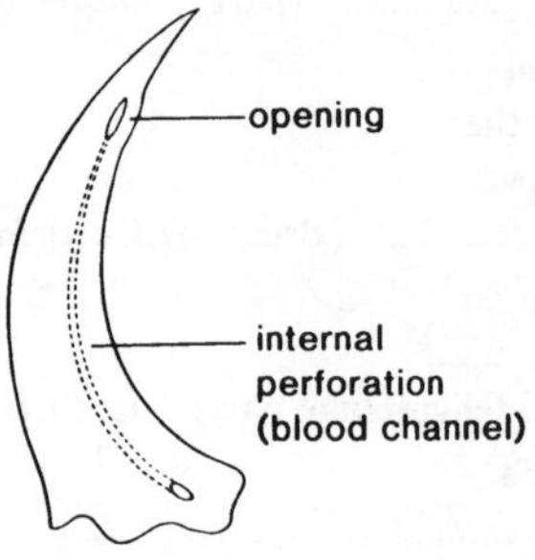

Figure 43

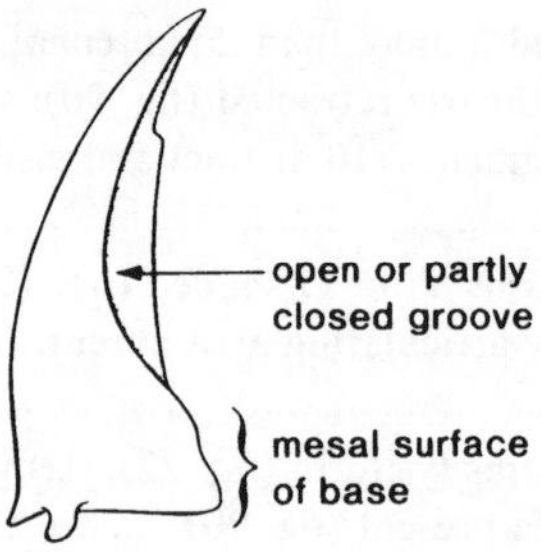

Figure 44

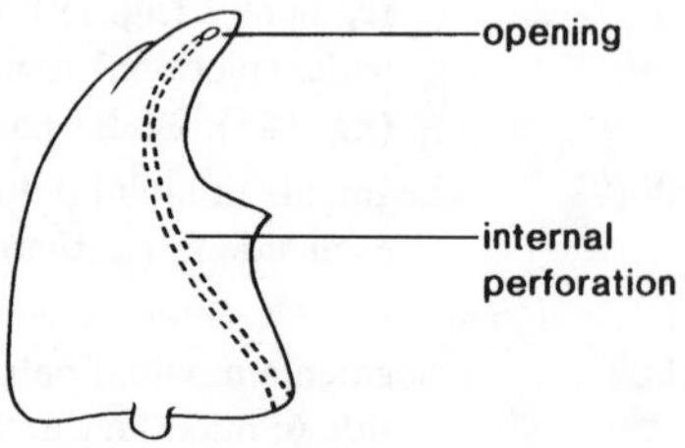

Figure 49

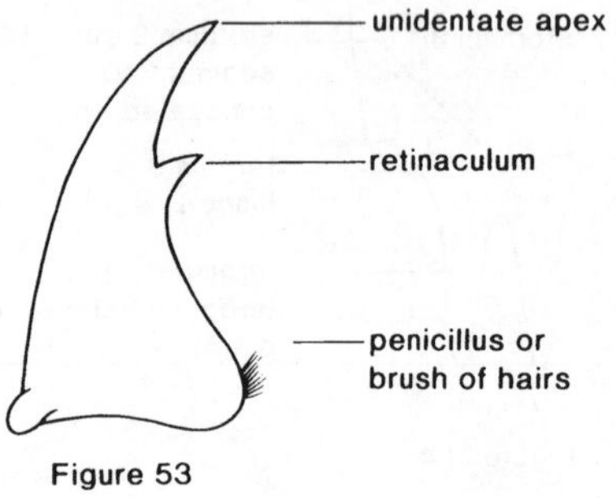

Figure 53

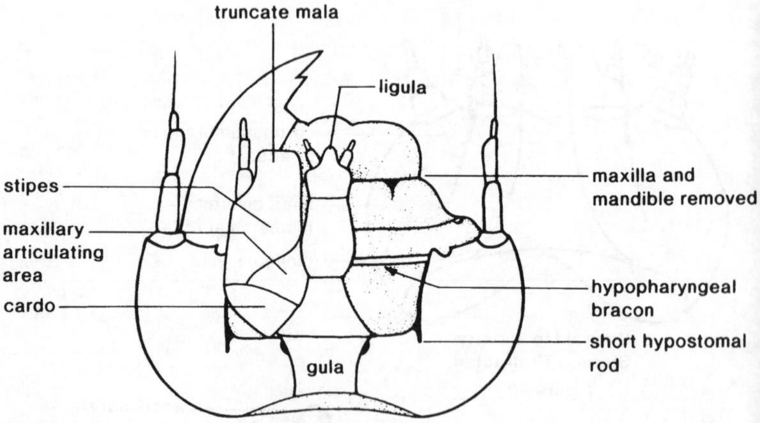

Mouthparts Retracted

Figure 85

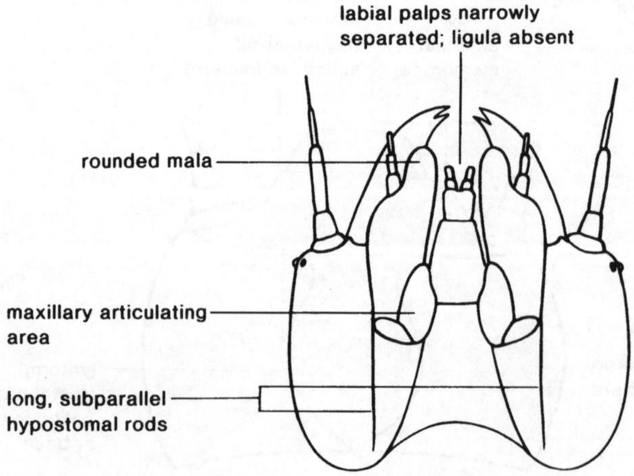

Figure 86

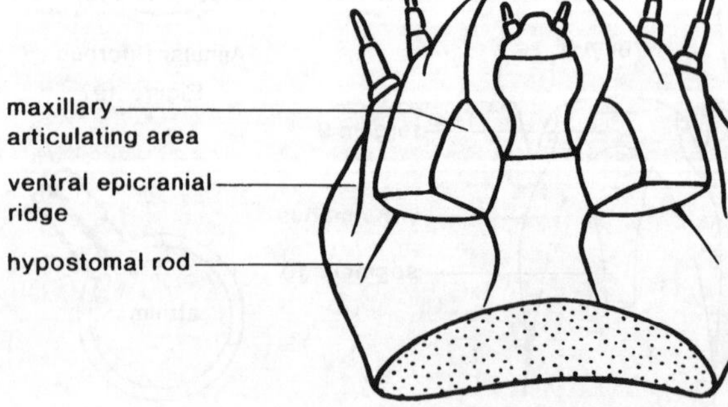

Mouthparts Retracted

Figure 90

102². Antennal segments 5 or more; thoracic and abdominal terga without lateral, plate-like processes. Aquatic 114

103(102′). Tergum A8 without special armature 104

103′. Tergum A8 with single median process, simple at apex; tergum A9 with paired, articulated urogomphi (fig. 153); coxal bases and sterna A1–3 with ventral gill tufts. In ponds or lakes. Exotic (Old World) (**see 3rd choice**) ..... ***Hygrobiidae*** p. 312

103². Tergum A8 with single median process bearing spiracles at apex (fig. 154); articulated urogomphi usually present 113

104(103). Tergum A9 without paired processes (urogomphi) 105

104′. Tergum A9 with paired processes (urogomphi) (figs. 153, 162, 170) 112

105(104). Mandible with internal perforation (figs. 43, 49); dorsal surfaces of body granulate-spinose or abdominal segments with lateral gills. Aquatic 106

105′. Mandible without internal perforation (fig. 53), sometimes with open or partly closed groove (fig. 44); dorsal surfaces of body not granulate or spinose, sometimes with rows or patches of asperities; abdominal segments without lateral gills. Terrestrial 107

106(105). Mandibles narrow and falcate (fig. 43); ratio of antennal length to head width more than .5; segments in maxillary palp 4; segments in labial palp 3; dorsal body surfaces smooth; segment A10 with 2 pairs of hooks (fig. 157); segments A1–8 each with pair of lateral gills, segment A9 with 2 pairs (fig. 157). In ponds or pools ..... ***Gyrinidae*** p. 319

106′. Mandibles broad at base and narrow at apex (fig. 49); ratio of antennal length to head width less than .5; segments in maxillary palp 3; segments in labial palp 2; dorsal body surfaces granulate or spinose; segment A10 without 2 pairs of hooks; abdominal segments without lateral gills. In ponds ..... (part) ***Haliplidae*** p. 311

107(105′). Tergum A9 forming concave, glandular disc, sometimes bearing branched processes. Under bark ..... (Ozaenini) ***Carabidae*** p. 305

107′. Tergum A9 simple, not forming concave disc 108

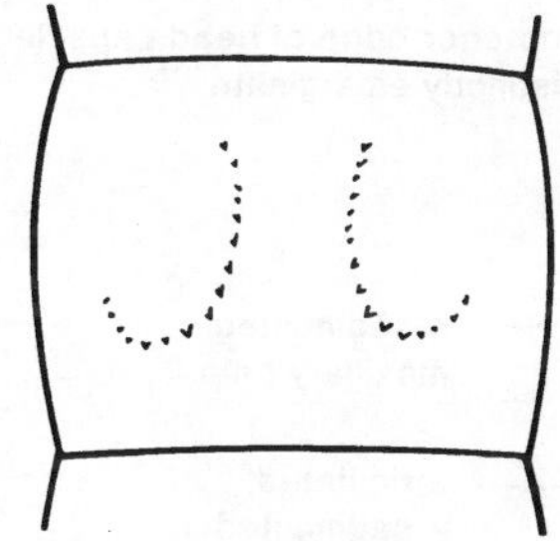
Figure 148

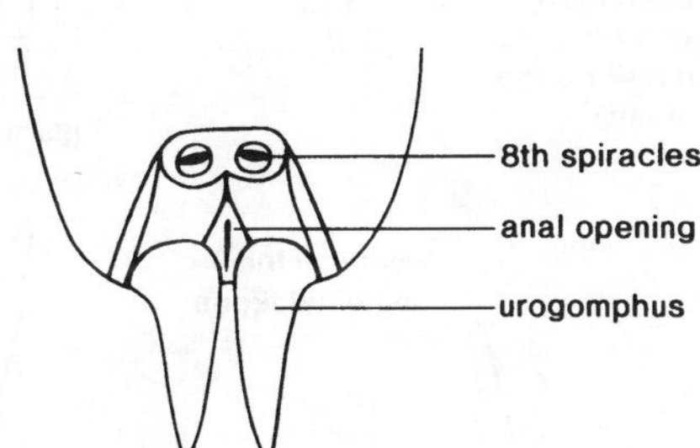

Figure 153

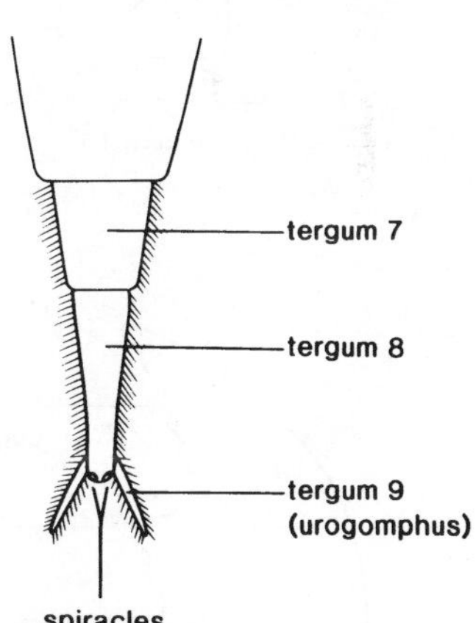

Figure 154

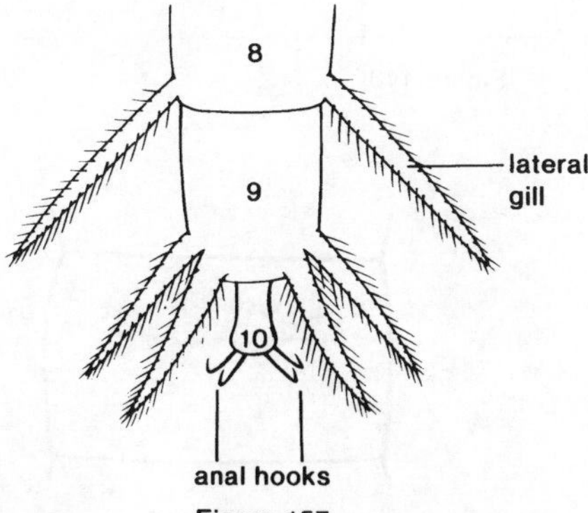

Figure 157

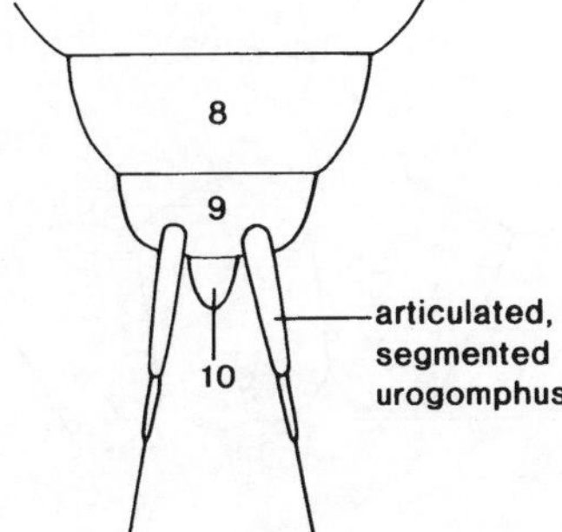

Figure 162

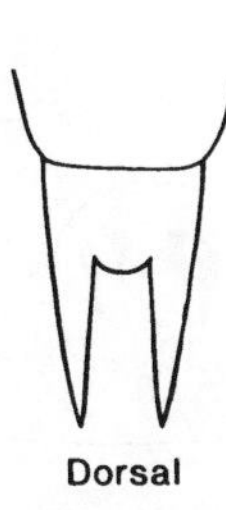

Figure 170

108(107′). Apex of maxilla without distinct lobes, its surface and that of labium sometimes covered with fine fringed membranes; segments in labial palp 1; terga A1 to A7 or A8 each with 2 rows of asperities, which may form incomplete rings (fig. 146). In rotten wood .... ***Rhysodidae*** p. 304

108′. Apex of maxilla with 1 or 2 distinct lobes that may be segmented (figs. 103, 113), without fringed membranes; segments in labial palp 2 109

109(108′). Bases of frontal arms contiguous (fig. 24); tarsal claws on T2 leg 2; tergum A5 or terga A5–7 with hooks or spines (figs. 135, 137). In burrows in ground 110

109′. Bases of frontal arms distinctly separated (fig. 16); tarsal claws on T2 leg 1; terga A5–7 without hooks or spines. In nests of ants or termites 111

110(109). Maxilla with separate galea and lacinia (fig. 113); terga A5–7 each with transverse row of 4 spines (fig. 137). Exotic (Australia) .... (*Sphallomorpha*) (Pseudomorphini) ***Carabidae*** p. 305

110′. Maxilla with single mala (fig. 103); tergum A5 with distinct protuberance bearing 2 or 3 pairs of hooks (fig. 135) .... **(Cicindelinae)** ***Carabidae*** p. 305

111(109′). Head longer than wide, covered with specialized, apically expanded setae; number of tarsal claws on T1 leg 2. In ant nests .... (*Pseudomorpha*) (Pseudomorphini) ***Carabidae*** p. 305

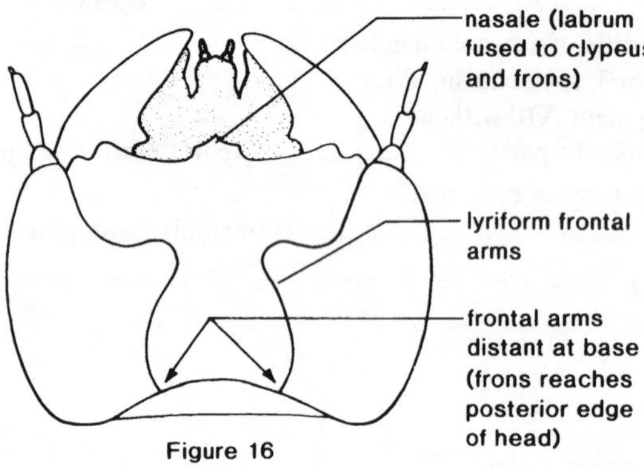

Figure 16

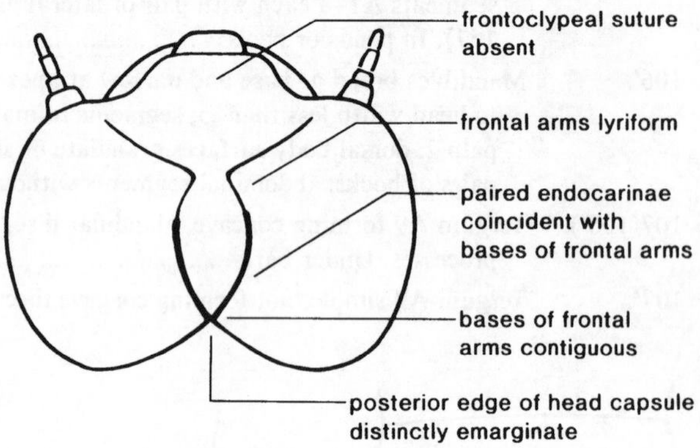

Figure 24

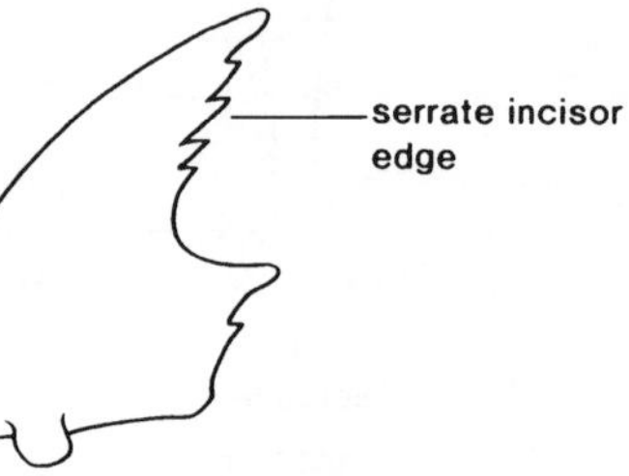

Figure 50

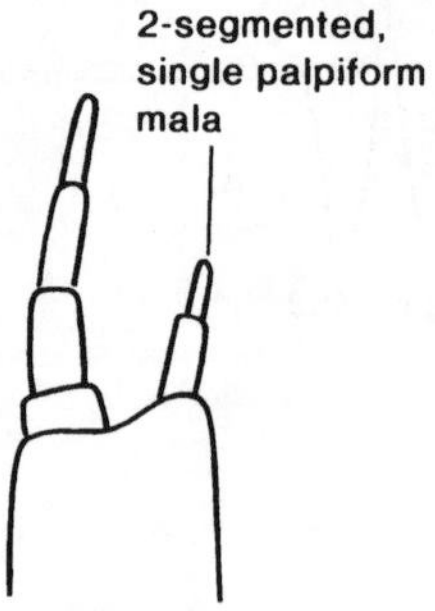

Figure 103

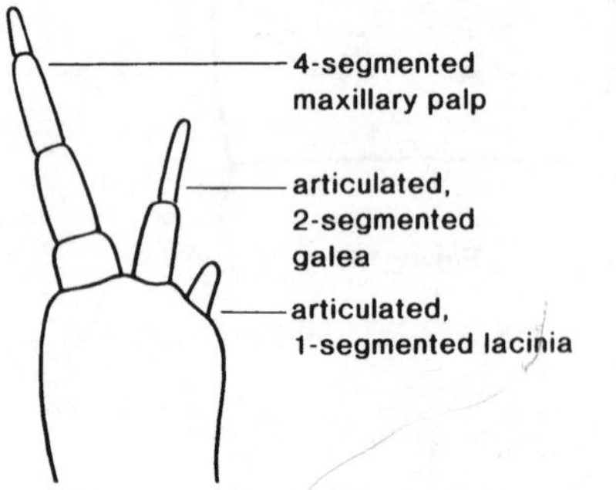

Figure 113

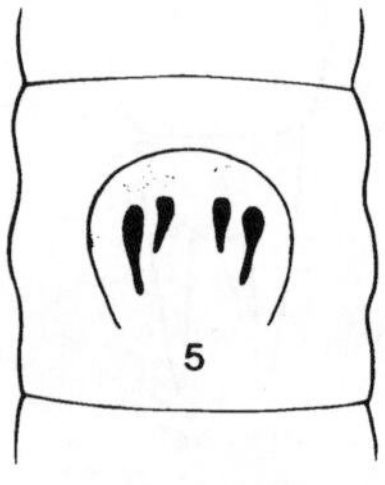

Figure 135

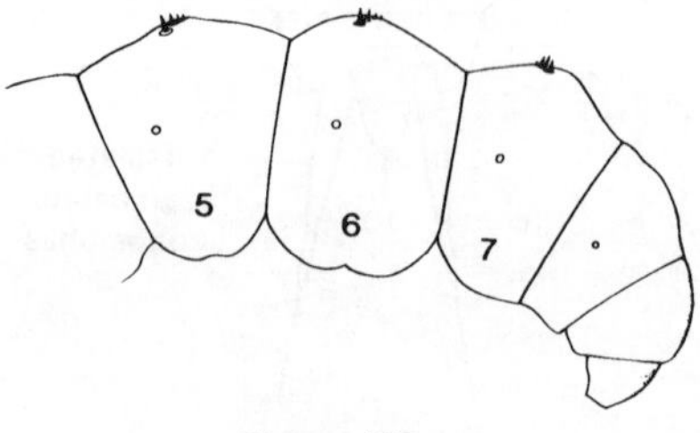

Figure 137

Figure 146

← Figs. 44, 49, 85, 154, 162

111′. Head subquadrate or slightly wider than long, without specialized setae; number of tarsal claws on T1 leg 1; terga A1 to A5 or A6 each with 2 patches of asperities. In termite nests. Exotic (Africa and Asia) .... (Orthogoniini) ***Carabidae*** p. 305

112(104′). Segment A10 distinct and visible from above (fig. 162); sternum A9 exposed; gular sutures fused (fig. 214); ligula almost always present (fig. 214). Usually terrestrial .... (part) ***Carabidae*** p. 305

112′. Segments A9 and A10 absent or completely concealed from above; gular sutures separate (fig. 85); ligula absent. In lakes or ponds .... (part) ***Noteridae*** p. 314

113(103²). Ratio of antennal length to head width less than .5; mandibles broad at base, narrow at apex (fig. 50); incisor edge of mandible serrate (fig. 50). In mud around roots of aquatic plants .... **(Noterinae)** ***Noteridae*** p. 314

113′. Ratio of antennal length to head width more than .5; mandibles narrow and falcate (fig. 44); incisor edge of mandible simple. In streams or ponds .... (part) ***Dytiscidae*** p. 315

114(102²). Tergum A8 without special armature; thoracic and abdominal segments each with 4 or more long, tergal processes; ratio of antennal length to head width less than .5; mandibles broad at base and narrow at apex (fig. 49). In ponds .... (*Peltodytes*) ***Haliplidae*** p. 311

114′. Tergum A8 with single median process bearing spiracles at apex (fig. 154); thorax and abdomen without long tergal processes; ratio of antennal length to head width more than .5; mandibles narrow and falcate (fig. 44). In streams and ponds .... (part) ***Dytiscidae*** p. 315

115(3). Tergum A9 without paired processes or urogomphi 116

115′. Tergum A9 with pair of urogomphi (figs. 162, 164, 172, 184, 217) 193

116(115). Body elongate, parallel-sided, rounded at both ends, strongly flattened, and heavily sclerotized; head forming a sclerotized, rounded plate; mouthparts endognathous; mandibles with narrow, falciform apex. In soil. Exotic (Mediterranean Region, Central Asia, Africa) .... (*Phyllocerus.)* ***Eucnemidae*** p. 419

116′. Without above combination of characters 117

117(116′). Antennae greatly reduced and difficult to see 127

117′. Antennal segments 1 **(see 3rd & 4th choices)** 118

117². Antennal segments 2 127

117³. Antennal segments 3 143

118(117′). Segments in T2 leg 1 or 2 119

118′. Segments in T2 leg 3 or 4 (figs. 127, 128) **(see 3rd choice)** 122

118². Segments in T2 leg 5 including tarsungulus (fig. 126) 124

119(118). Labral rods absent; labium consisting of prementum, mentum, and submentum. In dead wood, stems, vines, fungus fruiting bodies .... (part) ***Anthribidae*** p. 586

119′. Labral rods present (fig. 38); labium consisting of prementum and postmentum 120

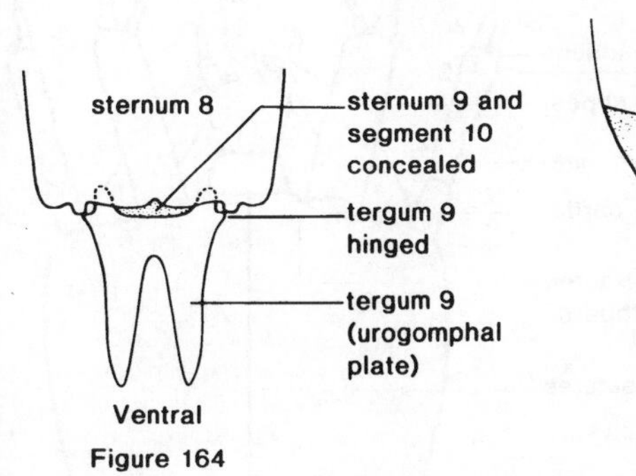

Figure 164

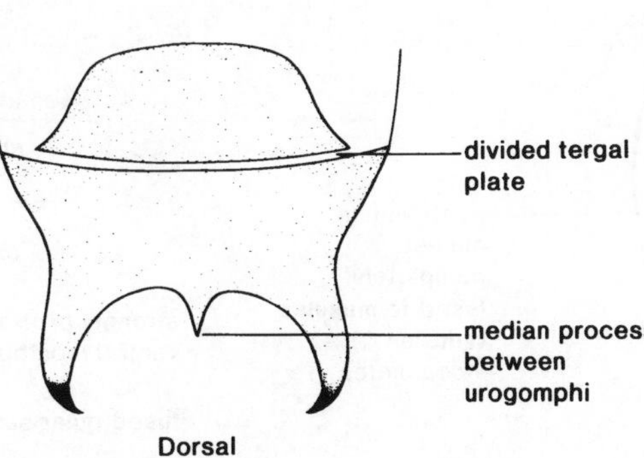

Figure 172

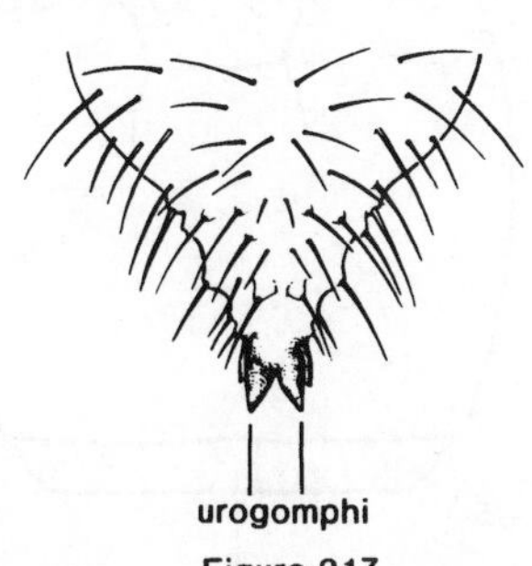

Figure 217

Figs. 38, 126, 127, 128, 184, 214 ⟶

120(119′). Median endocarina coincident with epicranial stem (fig. 19); labral rods not extending beyond posterior edge of clypeus (fig. 38); stemmata absent; patch of asperities (fig. 141) present on each side of mesotergum. In dead wood, under bark .... ***Brentidae*** p. 592

120′. Median endocarina extending anterad of epicranial stem (fig. 20); labral rods extending beyond posterior edge of clypeus; stemmata present; patches of mesotergal asperities absent 121

121(120′). Legs shorter than wide, rounded and lobe-like (but distinctly articulated with at least 1 ring-like segment); transverse plicae on abdominal terga absent or weakly developed; mandibles about as long as wide. In tubers of Convolvulaceae, especially sweet potato .... (*Cylas*) ***Curculionidae*** p. 594

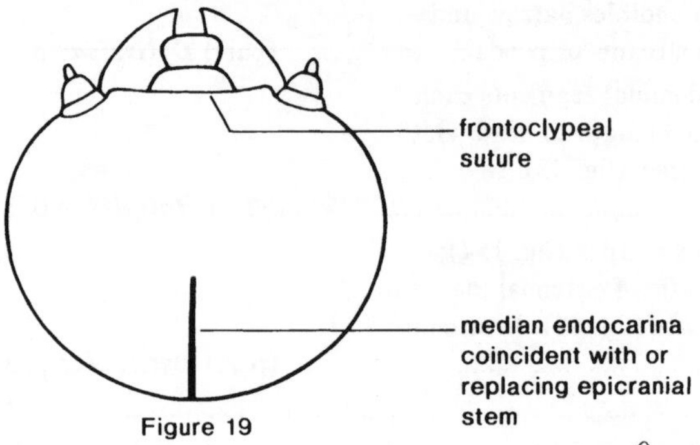

Figure 19

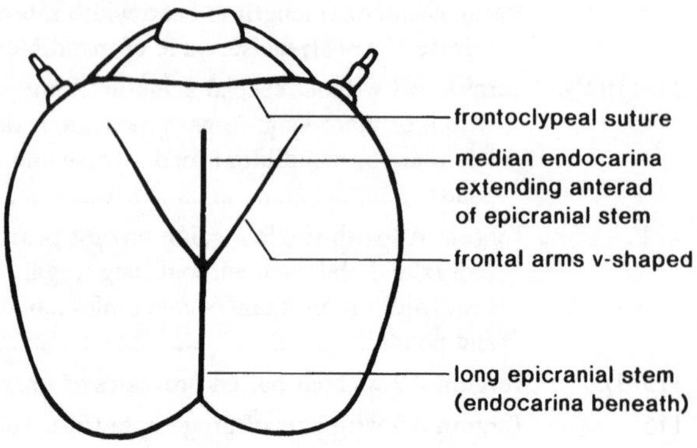

Figure 20

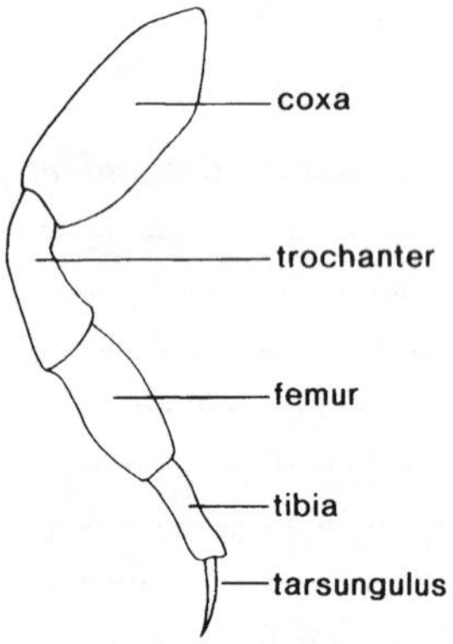

Figure 126

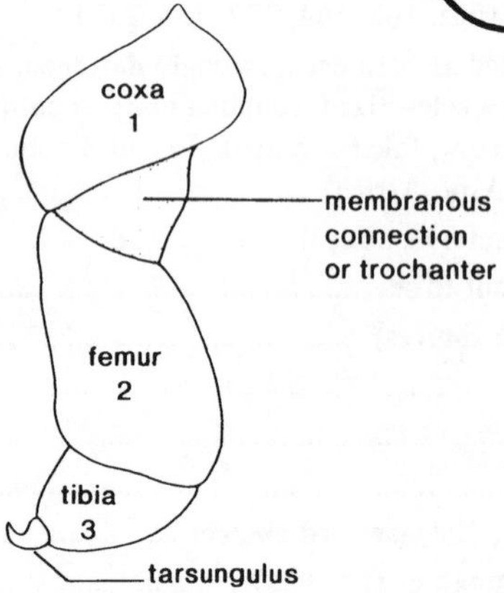

Figure 127

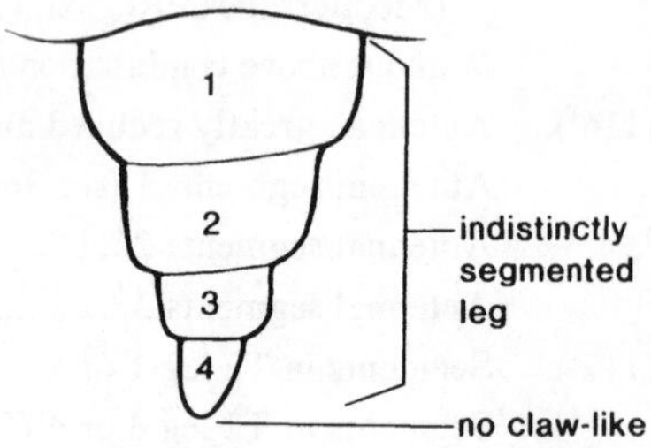

Figure 128

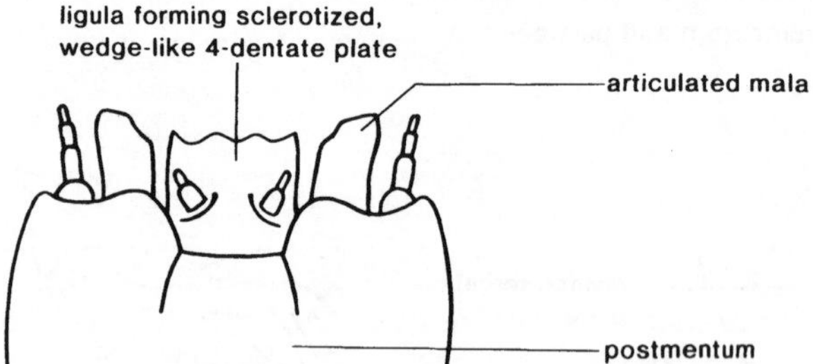

Figure 99

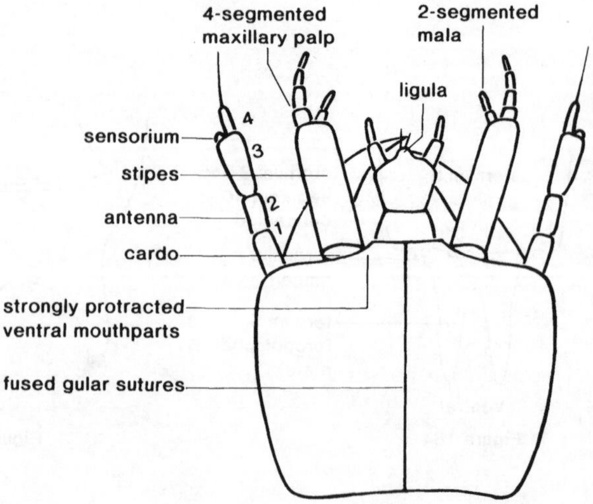

Figure 214

121′. Legs much longer than wide, not lobe-like; transverse plicae on abdominal terga distinct; mandibles distinctly longer than wide. In oak roots .... (part) ***Ithyceridae*** p. 590

122(118′) Body relatively straight or slightly curved ventrally; abdominal spiracles annular (fig. 195); tergum A3 without transverse plicae; labral rods absent. 123

122′. Body strongly curved ventrally (C-shaped); abdominal spiracles annular-biforous (fig. 197); tergum A3 with 2 or more transverse plicae (fig. 136); labral rods present. In oak roots .... (part) ***Ithyceridae*** p. 590

123(122). Segments in labial palp 2; segments in maxillary palp 2 or 3; apex of mandible with single lobe or tooth; labial palps contiguous or separated by less than width of first palpal segment; tergum A9 usually with median process (fig. 159). In rotten wood, stems, fungus fruiting bodies .... (part) ***Mordellidae*** p. 508

123′. Segments in labial palp 1; segments in maxillary palp 4; apex of mandible with 3 or more lobes or teeth (fig. 51); labial palps separated by more than width of first palpal segment; abdominal tergum A9 without median process. In soil .... (Synetinae) ***Chrysomelidae*** p. 568

124(118²). Maxilla with single mala (figs. 96, 99); segment A10 without oval lobes separated by longitudinal groove; frontal arms present (figs. 16, 20); body variably sclerotized and not C-shaped 125

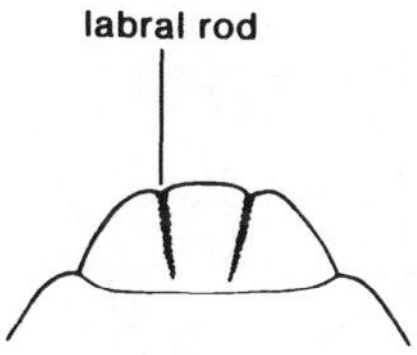

labral rod

Figure 38

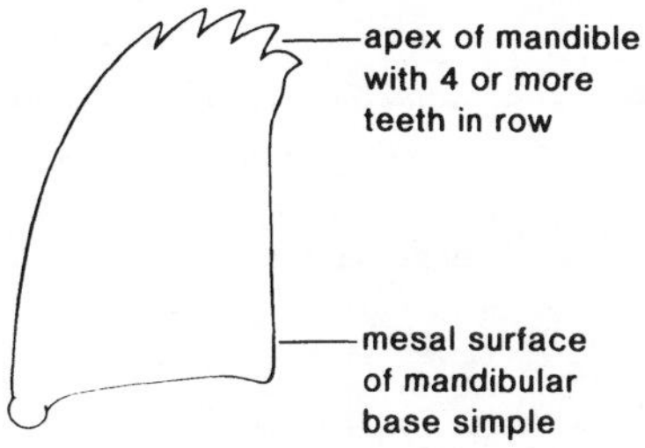

apex of mandible with 4 or more teeth in row
mesal surface of mandibular base simple

Figure 51

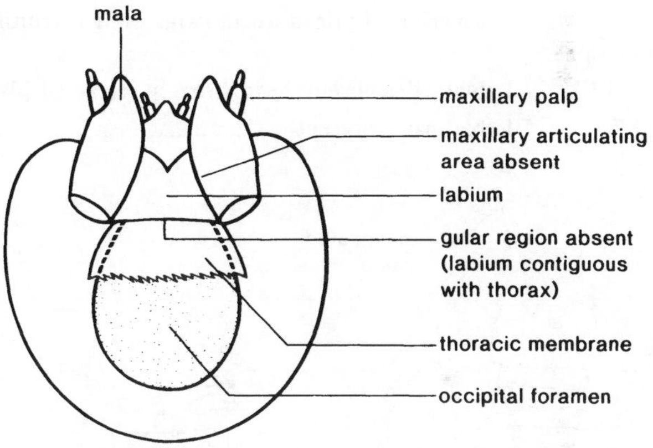

mala
maxillary palp
maxillary articulating area absent
labium
gular region absent (labium contiguous with thorax)
thoracic membrane
occipital foramen

Figure 96

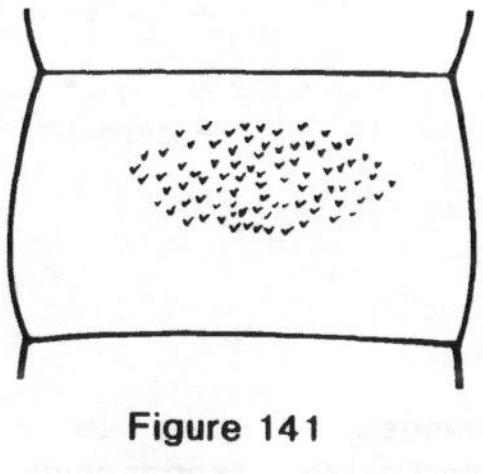

Figure 141

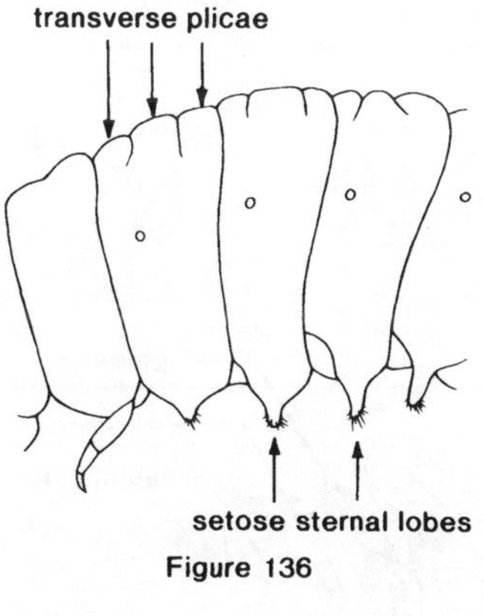

transverse plicae
setose sternal lobes

Figure 136

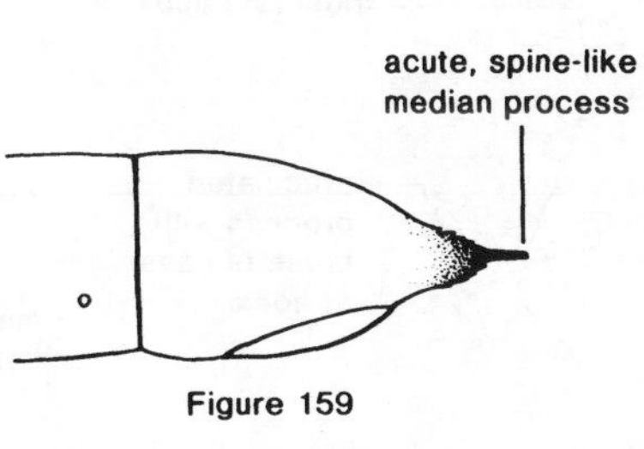

acute, spine-like median process

Figure 159

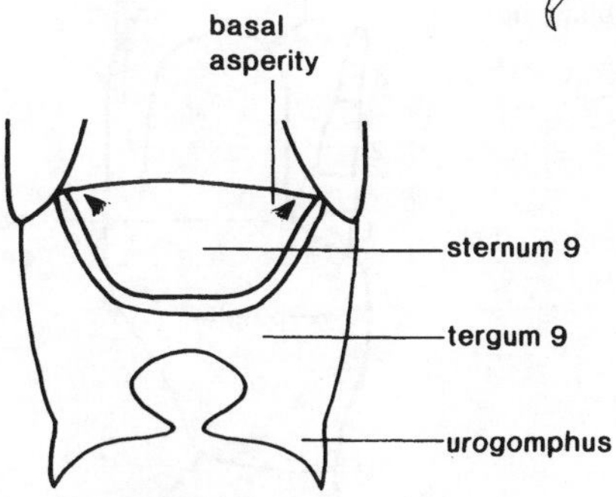

basal asperity
sternum 9
tergum 9
urogomphus

Ventral

Figure 184

Annular

Figure 195

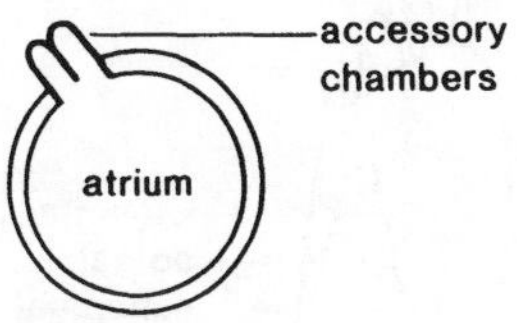

accessory chambers
atrium

Annular Biforous

Figure 197

Fig. 16 on p. 214

124′. Maxilla with separate galea and lacinia (figs. 97a, 110, 215); segment A10 with pair of oval lobes separated by longitudinal groove (fig. 193); frontal arms absent or only vaguely indicated; body lightly sclerotized and strongly curved or C-shaped 126

125(124). Segments in maxillary palp 2; body oblong to ovate; median endocarina absent; apex of mandible bidentate; stemmata on each side 3; spiracles annular; legs not reduced; ligula not sclerotized. On plant surfaces .... (part) ***Coccinellidae*** p. 485

125′. Segments in maxillary palp 3; body elongate and more or less parallel-sided; median endocarina extending anterad of epicranial stem (fig. 20); apex of mandible with 3 or more teeth in a row (fig. 51); stemmata on each side 2 or fewer; spiracles annular; legs not reduced; ligula not sclerotized. In leaf litter, soil, on plant surfaces **(see 3rd choice)** .... **(Galerucinae-Halticinae** part) ***Chrysomelidae*** p. 568

125². Segments in maxillary palp 4; body elongate and cylindrical, heavily sclerotized; median endocarina absent; apex of mandible tridentate (fig. 66); stemmata on each side 0; spiracles biforous (fig. 202); legs reduced; ligula forming sclerotized, 4-dentate, wedge-like plate (fig. 99); abdominal apex with dorsally-hinged operculum (fig. 186). In rotten wood .... ***Callirhipidae*** p. 389

126(124′). Thoracic spiracle at anterior end of T1; abdominal terga without bands of asperities. In dung, animal nests, stored products .... ***Ptinidae*** p. 444

126′. Thoracic spiracle at posterior end of T1; abdominal terga usually with bands of asperities. In dead wood, twigs, fungus fruiting bodies, stored products .... (part) ***Anobiidae*** p. 441

127(117, 117²). Labial palps absent; head strongly retracted (fig. 9) 128

127′. Labial palps present 129

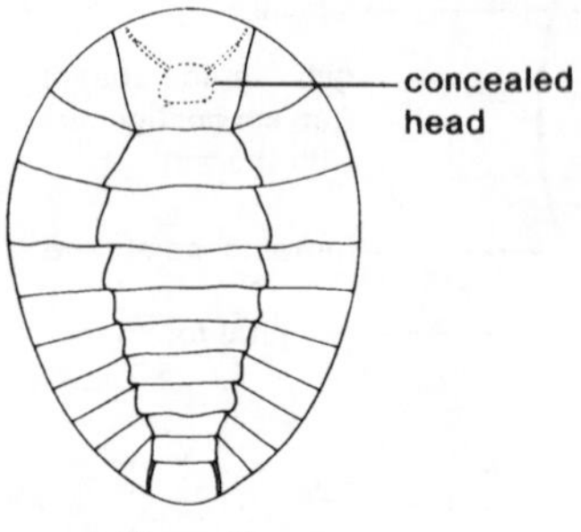

Figure 2

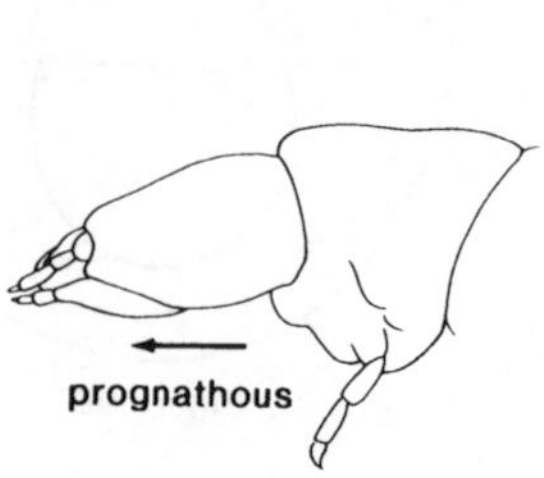

Figure 8

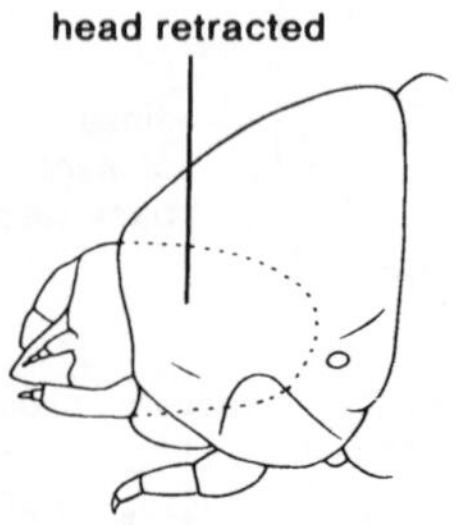

Figure 9

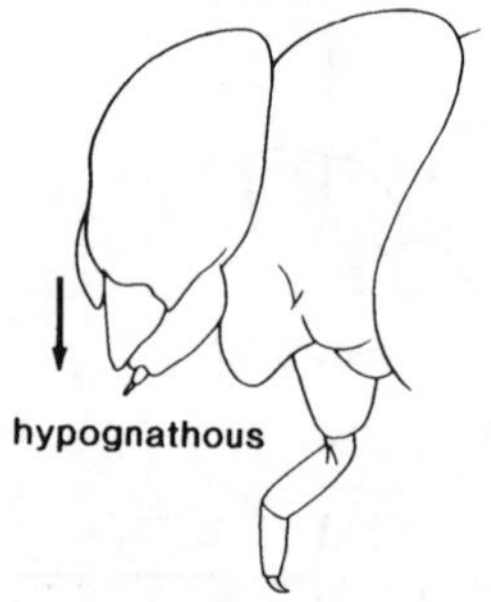

Figure 10

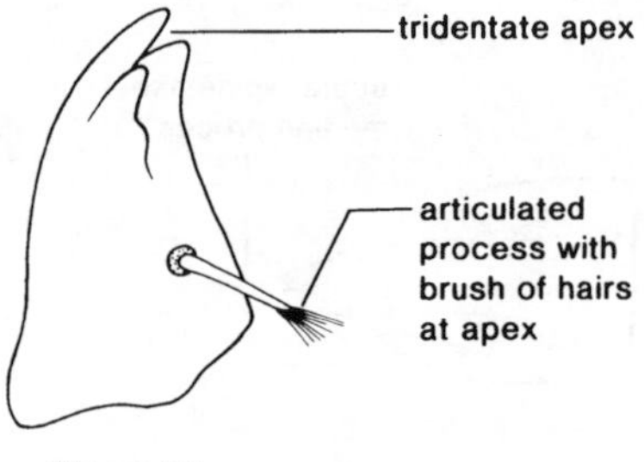

Figure 66

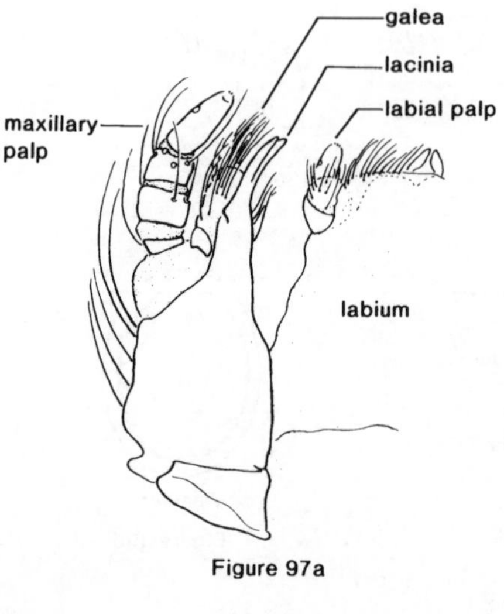

Figure 97a

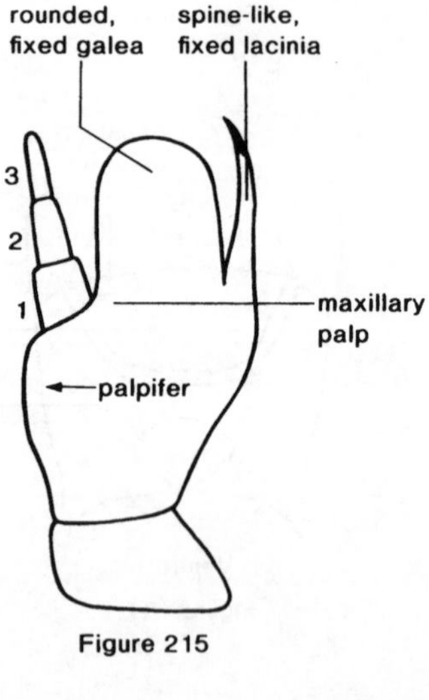

Figure 215

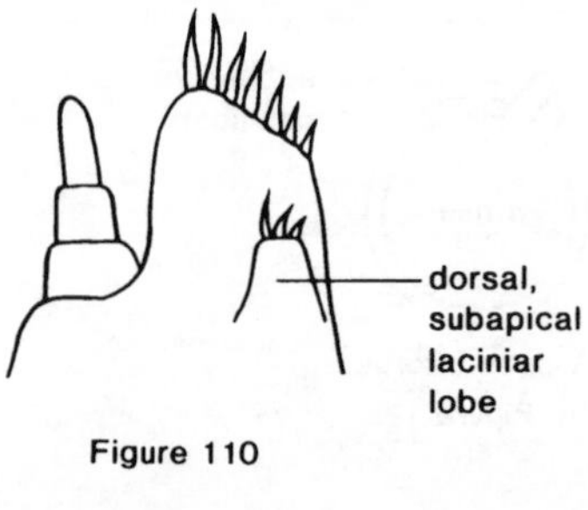

Figure 110

← Figs. 20, 51, 99, 136, 159

128(127). Body strongly curved ventrally (C-shaped); true legs present and T1 never greatly enlarged and flattened. In seeds or pods of various plants .... (part) ***Bruchidae*** p. 561

128′. Body straight or sometimes curved laterally; T1 greatly enlarged and flattened (fig. 121) or true legs absent. Under bark or in wood .... (part) ***Buprestidae*** p. 386

129(127′). Segments in labial palp 1 (fig. 91) (sometimes a minute papilla only) 130

129′. Segments in labial palp 2 (figs. 88, 94, 97) 134

130(129). Head strongly retracted (fig. 9) 131

130′. Head protracted or slightly retracted (figs. 8, 10) 133

131(130). Body broadly ovate, strongly flattened and disc-like (fig. 2); segments in T2 leg 4 or 5. In leaf axils .... (Cephaloleiini) (**Hispinae**) ***Chrysomelidae*** p. 568

131′. Body not broadly ovate and strongly flattened; segments in T2 leg 1 or 2 132

132(131′). True thoracic legs absent (2 pairs of small processes occur on meso- and metasterna, which might be mistaken for legs). In soil .... (**Schizopodinae** part) ***Buprestidae*** p. 386

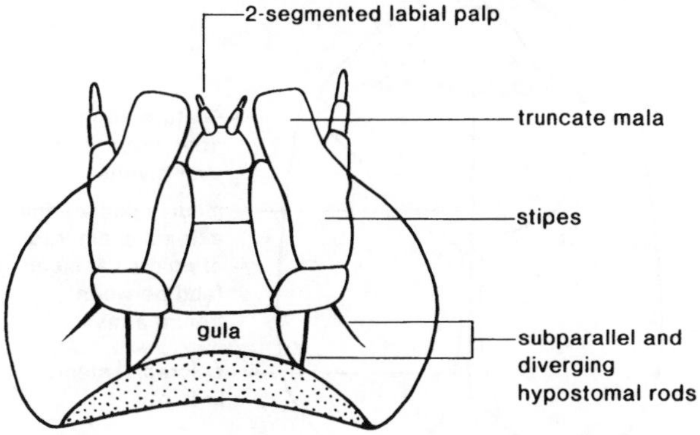

Figure 88

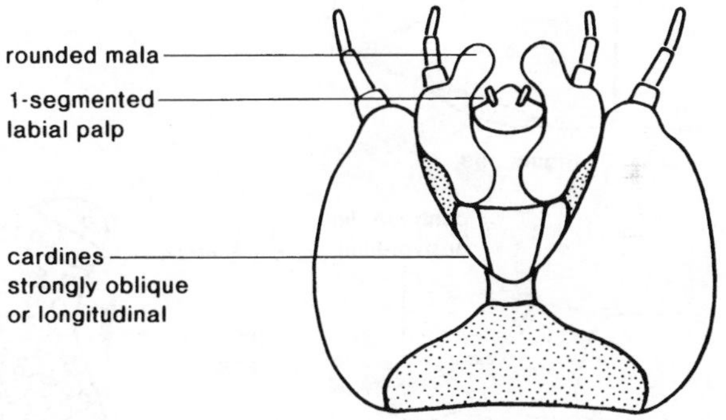

Figure 91

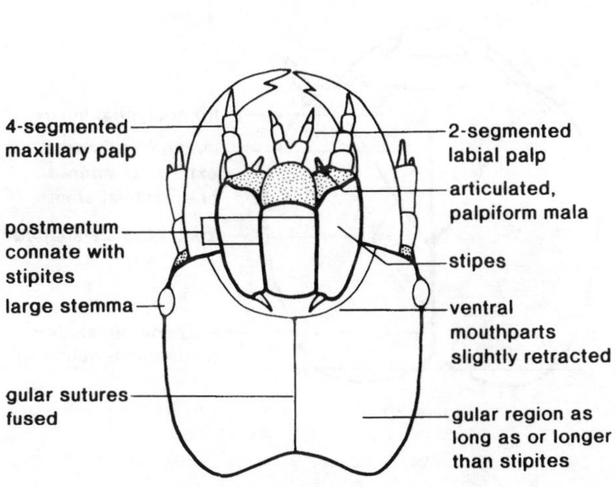

Figure 94

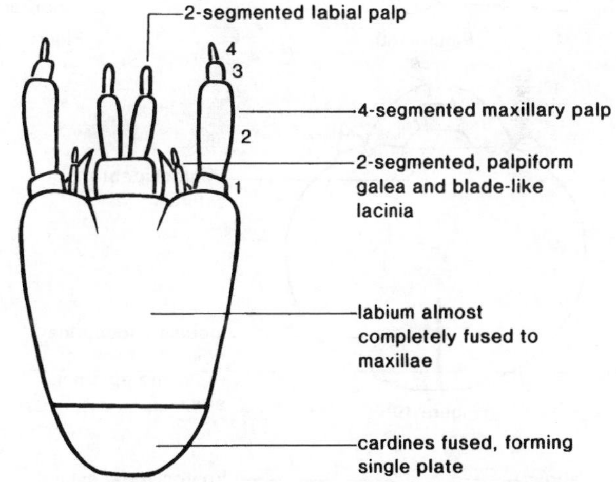

Figure 97

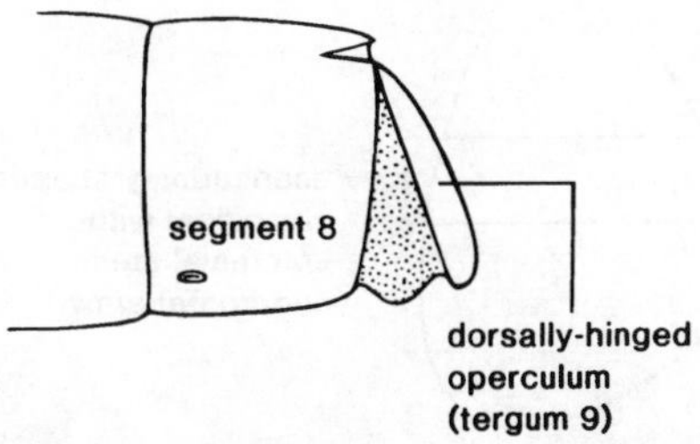

Figure 186

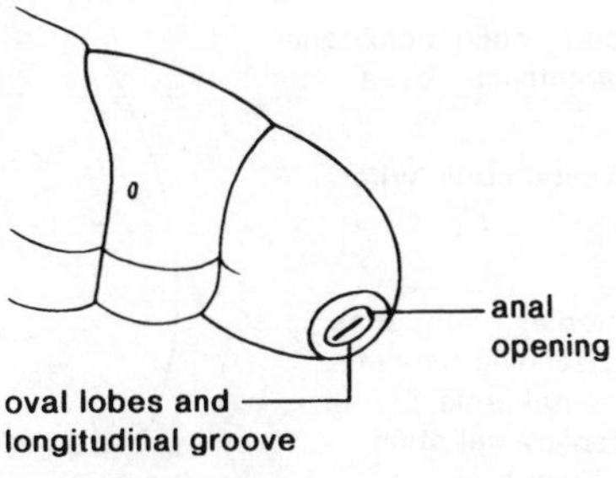

Figure 193

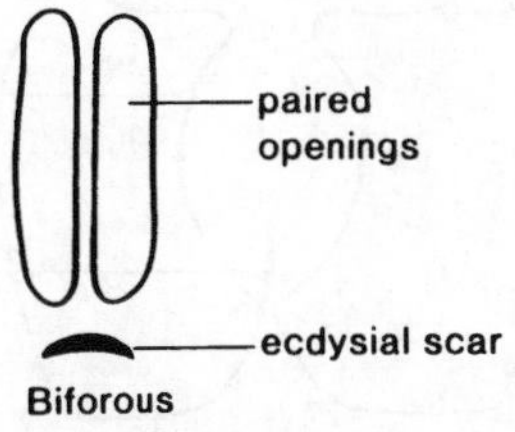

Figure 202

Fig. 121 ⟶

132′. Legs minute with 1 or 2 segments; spiracles cribriform (fig. 206); stemmata on each side 0; T1 much wider than abdomen (fig. 121). Under bark or in wood .... (part) ***Buprestidae*** p. 386

133(130′). Segment A8 terminal, with free posterior edge, usually bearing a long forked process (fig. 155); segments A9 and A10 reduced but more or less visible beneath segment A8; spiracles reduced on segment A8; head moderately to strongly declined (hypognathous) (fig. 10); stemmata on each side 5 or 6. On leaf surfaces .... (**Cassidinae** part) ***Chrysomelidae*** p. 568

133′. Segment A8 not terminal, without forked process and with functional spiracles; body flattened and lightly sclerotized; head prognathous or slightly declined (fig. 8); stemmata on each side 1. Leaf miners .... (**Galerucinae-Halticinae** part) ***Chrysomelidae*** p. 568

134(129′). Median endocarina absent or coincident with epicranial stem (fig. 19) 135

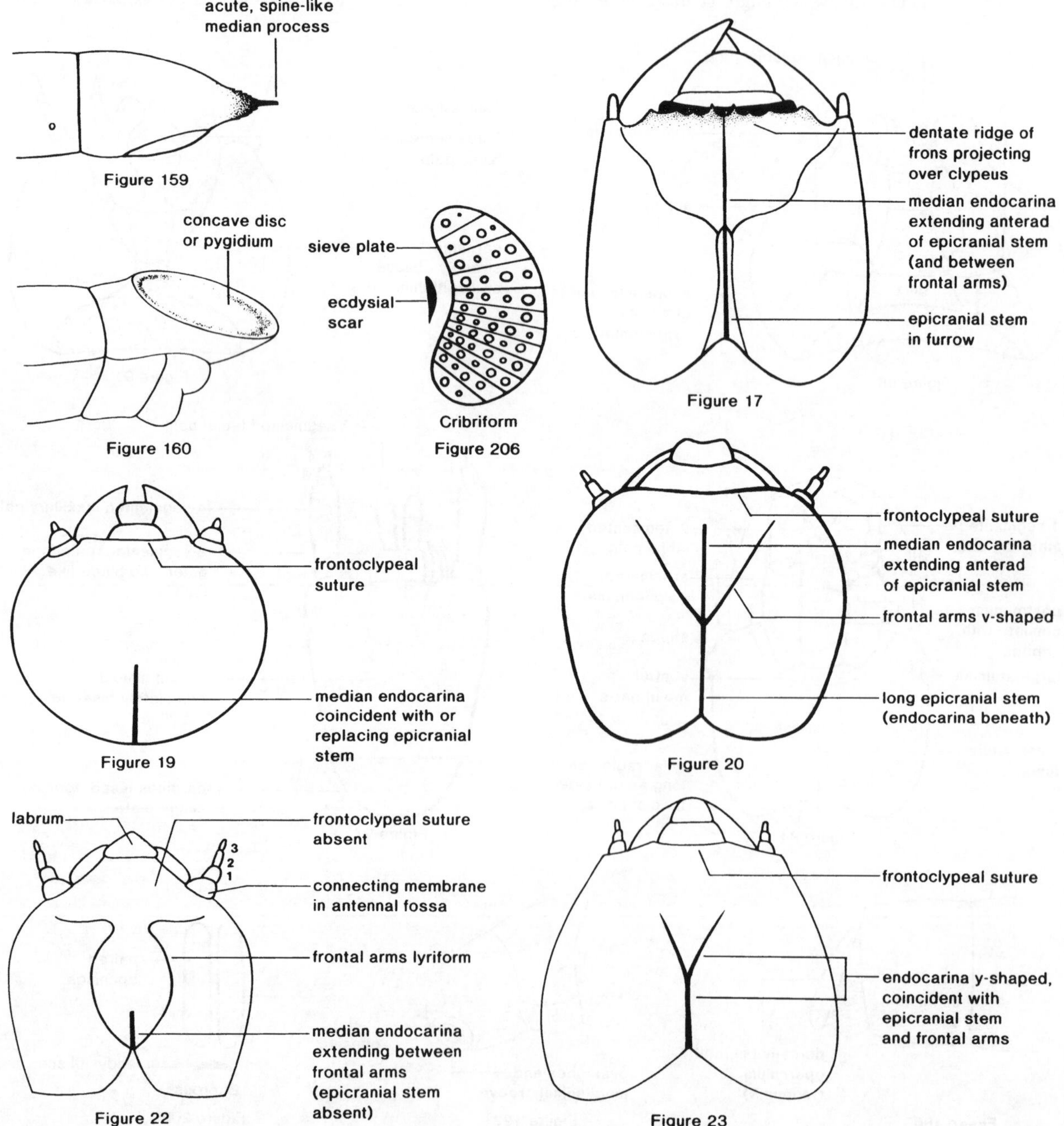

Figure 159

Figure 160

Figure 206

Figure 17

Figure 19

Figure 20

Figure 22

Figure 23

Figs. 2, 8, 10, 88, 97, 193

134′. Median endocarina Y-shaped, coincident with epicranial stem and frontal arms (fig. 23); tergum A9 with median process (fig. 159). In dead wood. **(see 3rd and 4th choices)** .... **(Osphyinae** part) ***Melandryidae*** p. 505

134². Median endocarina extending anterad of epicranial stem (figs. 17, 20) 140

134³. Median endocarina located between frontal arms (fig. 22); epicranial stem absent. Under bark, in living or dead wood .... **(Lepturinae** part) ***Cerambycidae*** p. 556

135(134). Body relatively straight or slightly curved ventrally; tergum A3 without transverse plicae; segment A10 without oval lobes separated by longitudinal groove 136

135′. Body strongly curved ventrally (C-shaped); tergum A3 with 2 or more transverse plicae (fig. 136); segment A10 with pair of oval lobes separated by longitudinal groove (fig. 193). In dead wood, twigs, fungus fruiting bodies, stored products .... (part) ***Anobiidae*** p. 441

136(135). Tergum A9 without median process or terminal disc (entire tergum may form spine) 137

136′. Tergum A9 with median process (fig. 159). **(see 3rd choice)** 139

136². Tergum A9 with concave, terminal disc (fig. 160). In fungus fruiting bodies, under bark, in rotten wood .... **(Ciinae** part) ***Ciidae*** p. 502

137(136). Body not broadly ovate and strongly flattened; cardines if present, separated from each other by labium; ratio of antennal length to head width less than .15; segment A10 exposed; tergum A9 not forming spine 138

137′. Body broadly ovate, strongly flattened, and disc-like (as in fig. 2); cardines completely fused forming single plate (fig. 97); ratio of antennal length to head width more than .5; segment A10 concealed; tergum A9 forming articulated spine. Under bark or stones in dry areas .... (part) ***Brachypsectridae*** p. 421

138(137). Gula absent; body elongate and more or less cylindrical, lightly sclerotized; head moderately to strongly declined (hypognathous) (fig. 10); stemmata absent. In rotten wood, stems, fungus fruiting bodies .... (part) ***Mordellidae*** p. 508

138′. Gula wider than long (fig. 88); body more oblong or ovate and somewhat flattened; dorsal surfaces of abdominal segments usually with transverse row of 6 protuberances which may bear setose or branched processes (fig. 134); head prognathous or slightly declined; stemmata on each side 3. On plant surfaces **(see 3rd choice)** .... (part) ***Coccinellidae*** p. 485

138². Gula longer than wide (fig. 92); body elongate, with enlarged abdomen (physogastric), lightly sclerotized; head prognathous or slightly declined; stemmata on each side 0 or 1. In tunnels of wood-boring beetles .... **(Bothriderinae** part) ***Bothrideridae*** p. 477

139(136′). Segments in T2 leg 3 or 4 (articulations sometimes indistinct) (fig. 128); mesal surface of mandibular base simple or slightly expanded (fig. 46); ventral epicranial ridges absent. In rotten wood, stems, fungus fruiting bodies .... (part) ***Mordellidae*** p. 508

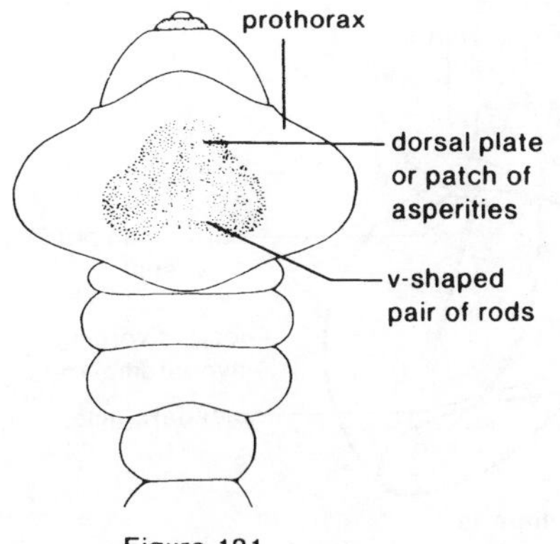

Figure 121

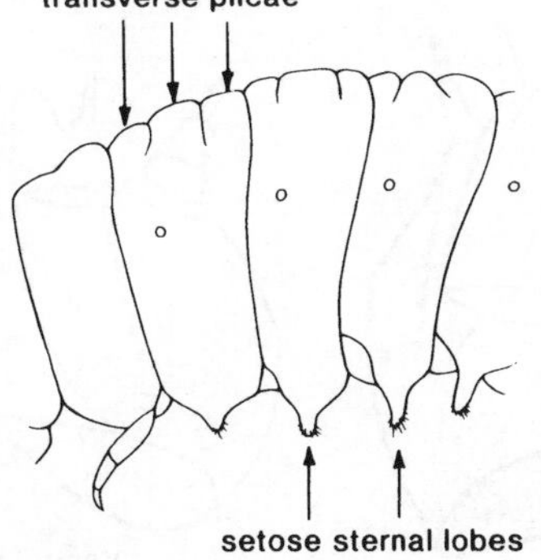

Figure 136

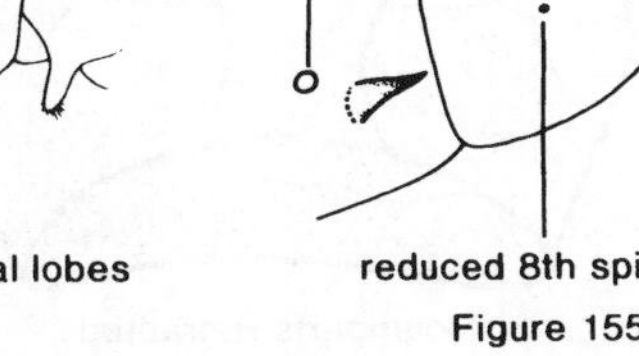

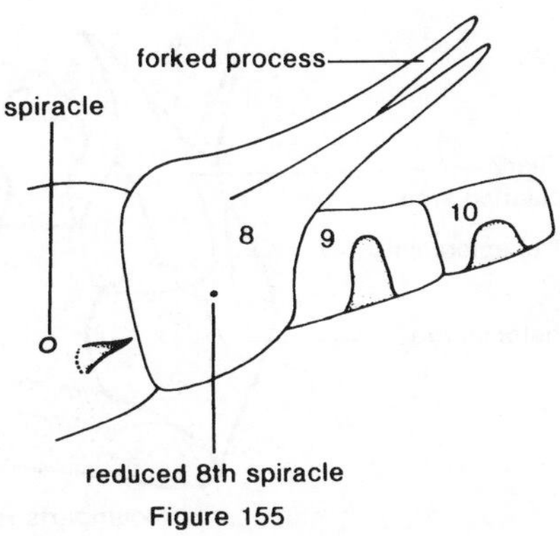

Figure 155

Figs. 46, 92, 128, 134 ⟶

139′. Segments in T2 leg 5 including tarsungulus (fig. 126); mesal surface of mandibular base with fixed, rigid, hyaline process (fig. 58); ventral epicranial ridges present (fig. 90). In fungus fruiting bodies, under bark, in rotten wood .... (Ciinae part) ***Ciidae*** p. 502

140(134^{2}). Segments in T2 leg 1 or 2; head distinctly longer than wide; occipital foramen not divided; epicranial stem not located in broad furrow. Under bark, in living or dead wood .... (**Lamiinae** part) ***Cerambycidae*** p. 556

140′. Segments in T2 leg 3 or 4; head not or only slightly longer than wide; occipital foramen divided into 2 parts by tentorial bridge (fig. 95); epicranial stem located in broad furrow for attachment of retractor muscles (fig. 17). Under bark, in living or dead wood **(see 3rd choice)** .... (**Prioninae** part) ***Cerambycidae*** p. 556

140^{2}. Segments in T2 leg 5 including tarsungulus (fig. 126) 141

141(140^{2}). Segments in maxillary palp 2; maxillary articulating area absent; ratio of length of abdomen to length of thorax 1.2 to 2; tergum A8 with paired processes, or a forked process (fig. 155). On leaf surfaces .... (**Cassidinae** part) ***Chrysomelidae*** p. 568

141′. Segments in maxillary palp 3; maxillary articulating area present (figs. 85); ratio of length of abdomen to length of thorax more than 2; tergum A8 without special armature 142

142(141′). Head protracted or slightly retracted (figs. 8,10); occipital foramen not divided; epicranial stem not located in broad furrow; gular region absent (labium contiguous with thoracic membrane) (fig. 96). In leaf litter, soil, on plant surfaces .... (**Galerucinae-Halticinae** part) ***Chrysomelidae*** p. 568

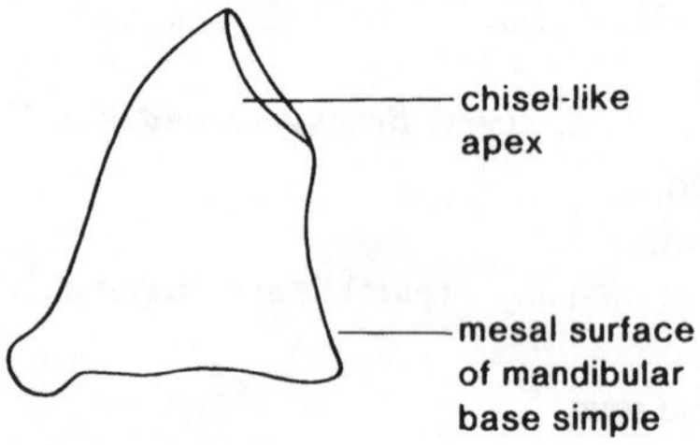

chisel-like apex
mesal surface of mandibular base simple

Figure 46

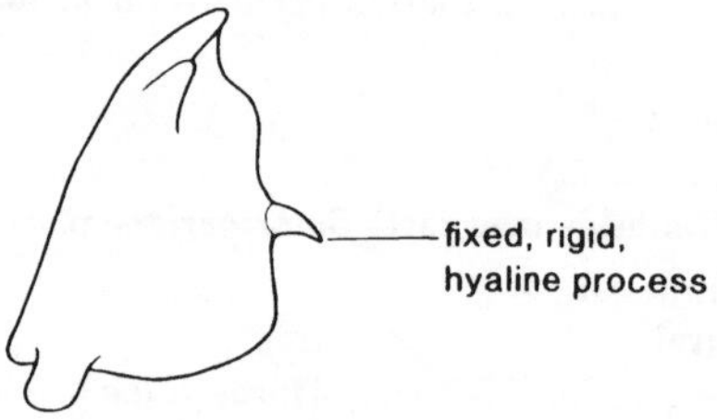

fixed, rigid, hyaline process

Figure 58

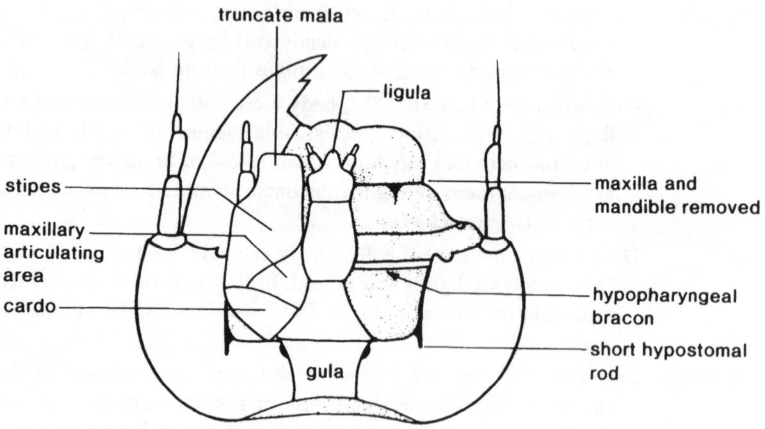

truncate mala
ligula
stipes
maxilla and mandible removed
maxillary articulating area
cardo
hypopharyngeal bracon
short hypostomal rod
gula

Mouthparts Retracted

Figure 85

maxillary articulating area
ventral epicranial ridge
hypostomal rod

Mouthparts Retracted

Figure 90

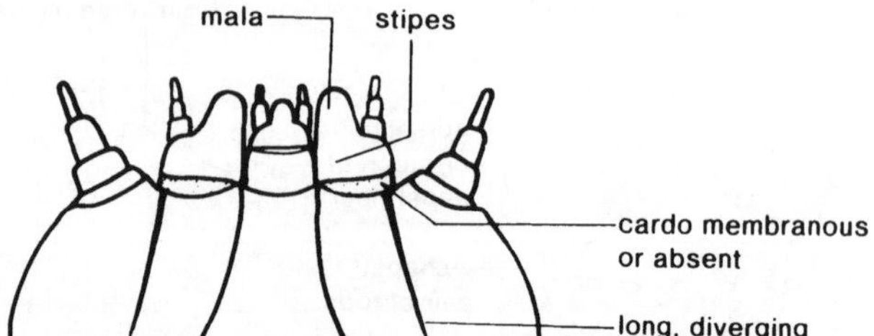

mala
stipes
cardo membranous or absent
long, diverging hypostomal rods
elongate gula

Ventral Mouthparts Strongly Protracted

Figure 92

← Figs. 17. 19. 20. 23. 155

142′. Head strongly retracted (fig. 9); occipital foramen divided into 2 parts by tentorial bridge (fig. 95); epicranial stem located in broad furrow for attachment of retractor muscles (fig. 17); gular region present (separating labium from thorax) (fig. 95). Under bark, in living or dead wood .... (**Prioninae** part) ***Cerambycidae*** p. 556

143(117^3). Abdominal apex without hinged operculum 144

143′. Abdominal apex with ventrally hinged operculum (figs. 150, 188) 189

144(143). Median endocarina absent or coincident with epicranial stem (fig. 19) 145

144′. Median endocarina Y-shaped, coincident with epicranial stem and frontal arms (fig. 23) (**see 3rd and 4th choices**) 174

144^2. Median endocarina extending anterad of epicranial stem (figs. 17, 20); if epicranial stem absent, then endocarina beginning at middle of the head and not extending to the base (fig. 20a) 175

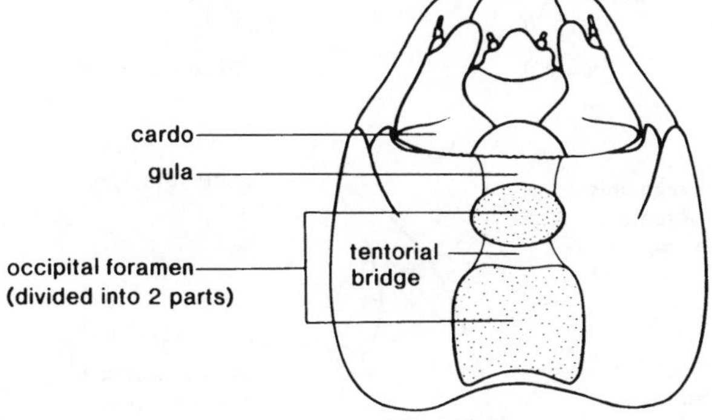

Figure 95

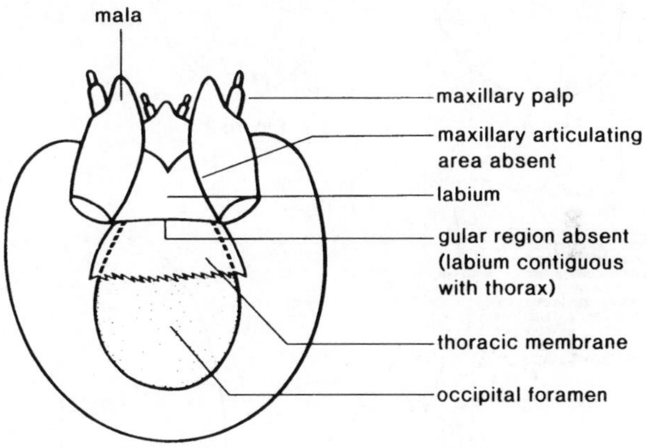

Figure 96

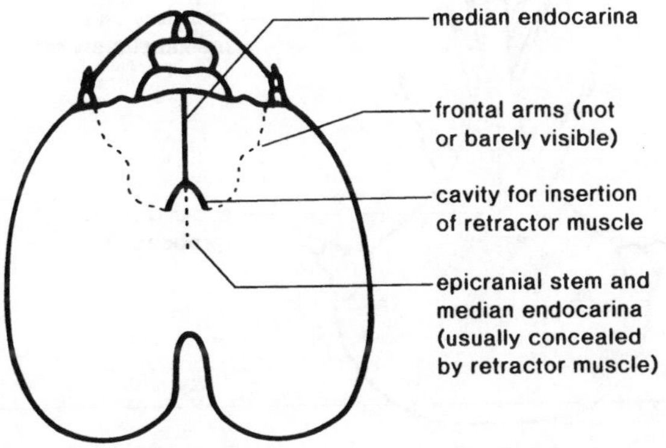

Figure 20a

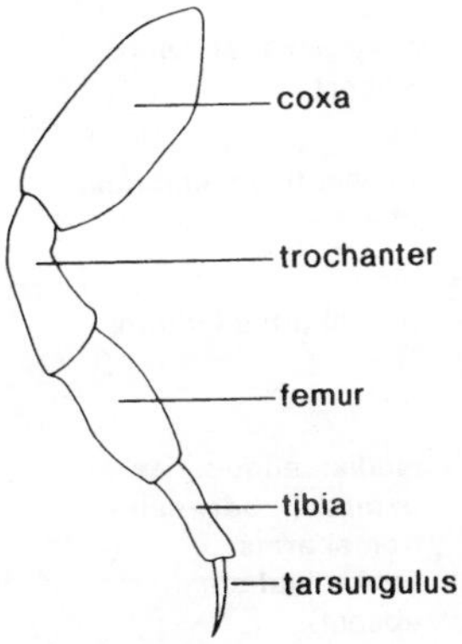

Figure 126

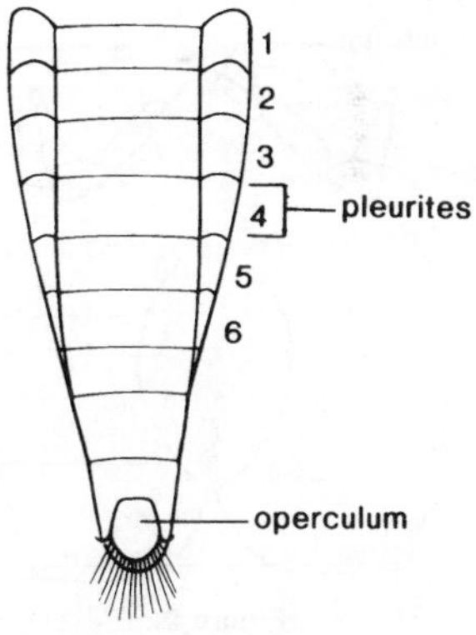

Figure 150

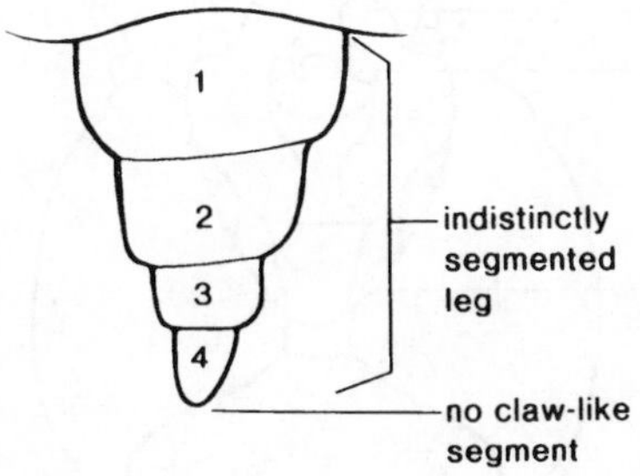

Figure 128

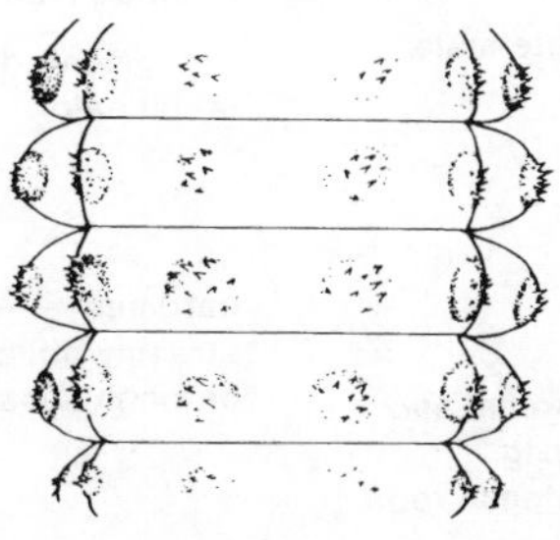

Figure 134

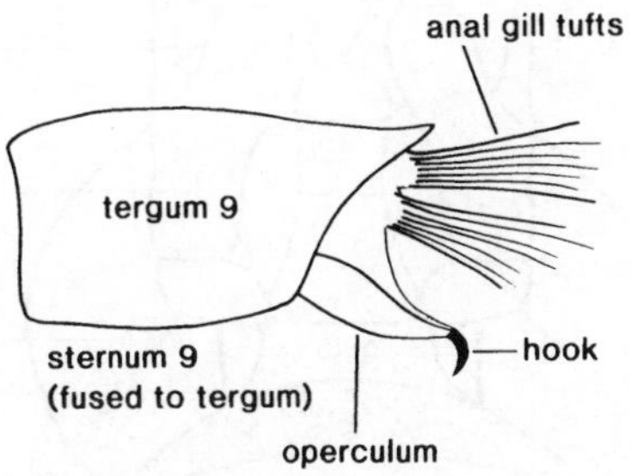

Figure 188

Figs. 8, 9, 10

144³.	Median endocarina located between frontal arms (fig. 22); epicranial stem absent, endocarina beginning at base of the head	186
145(144).	Maxilla without apical lobes (fig. 101); body very lightly sclerotized; number of stemmata on each side 1. In cells of native bees	(part) ***Meloidae*** p. 530
145′.	Maxilla with single mala (figs. 88, 96) (**see 3rd choice**)	146
145².	Maxilla with separate galea and lacinia (figs. 97, 97a, 100, 215)	165

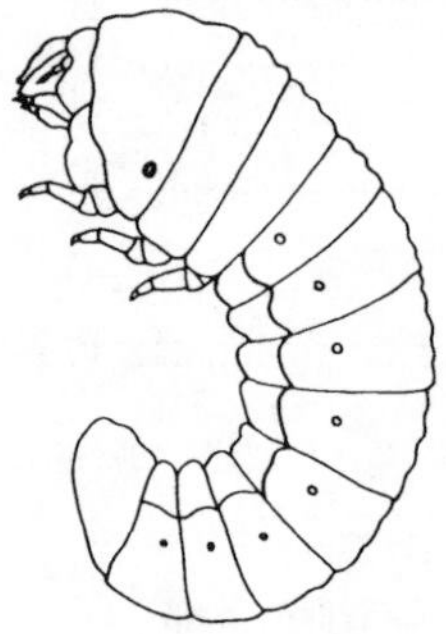

Figure 1 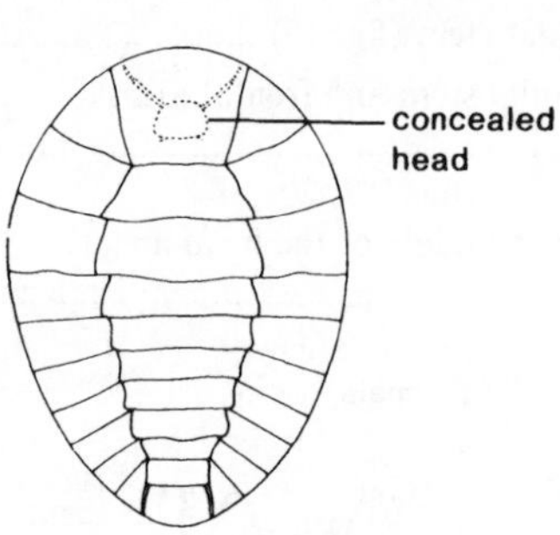 Figure 2 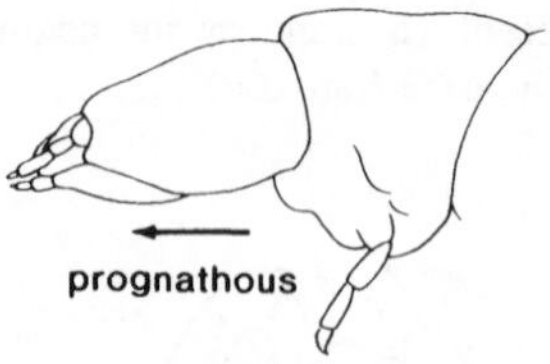 Figure 8 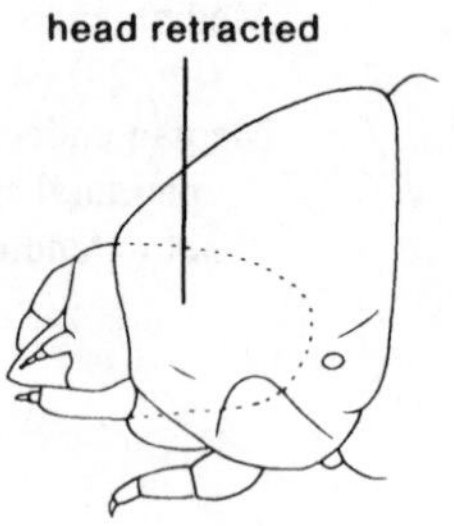 Figure 9

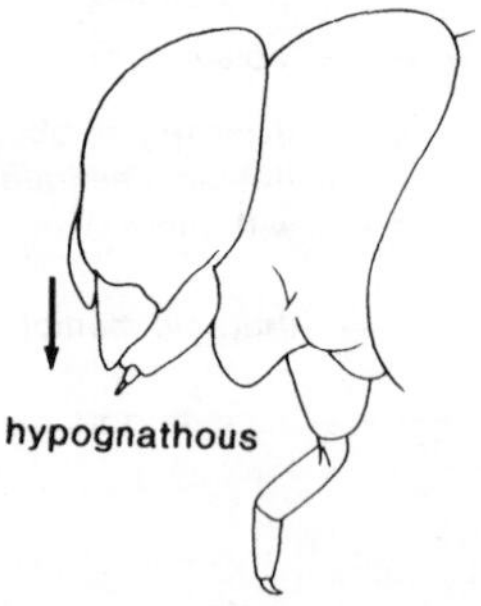

Figure 10 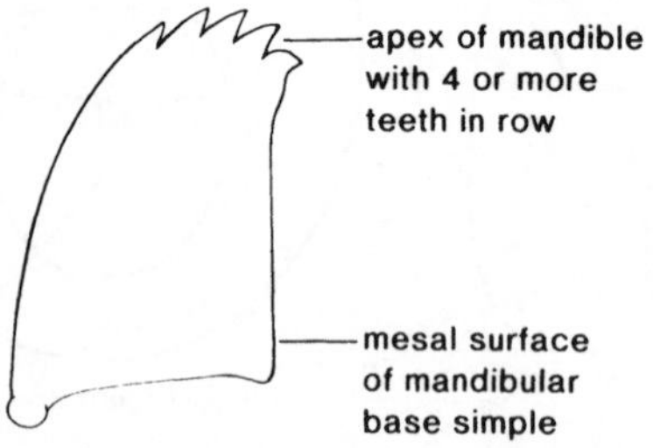 Figure 51 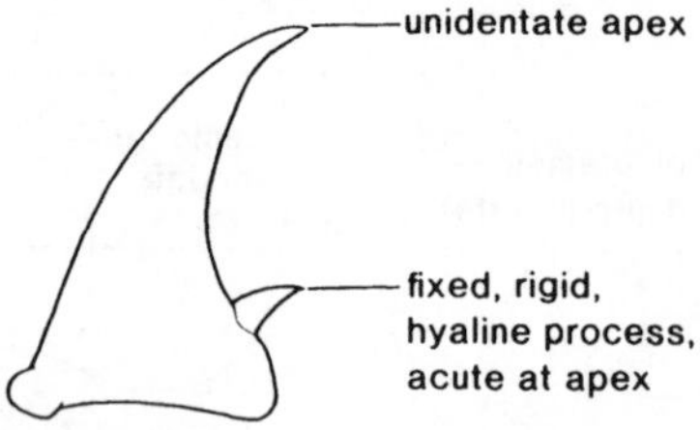 Figure 57

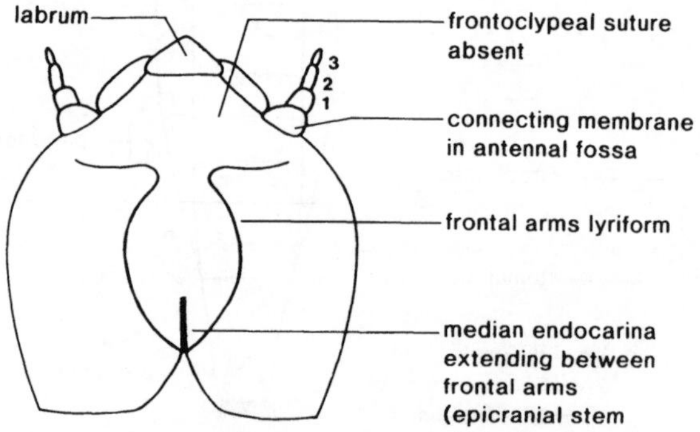

Figure 22 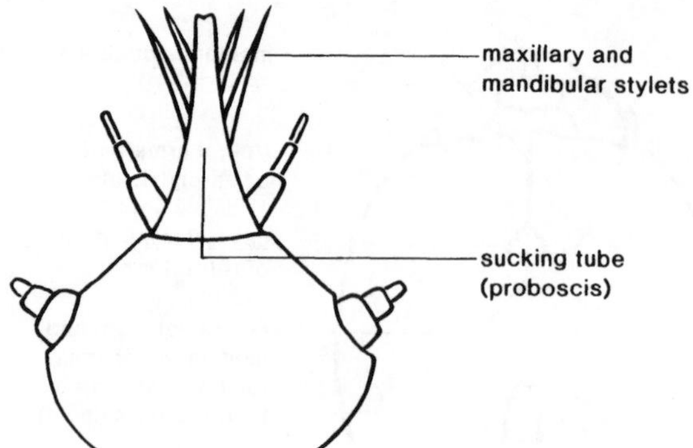 Figure 27

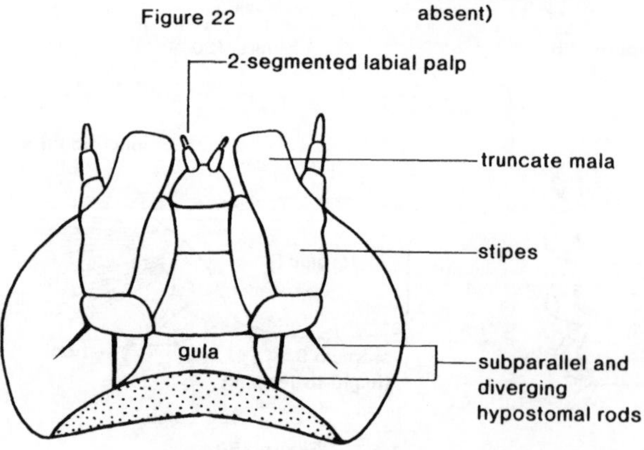

Figure 88 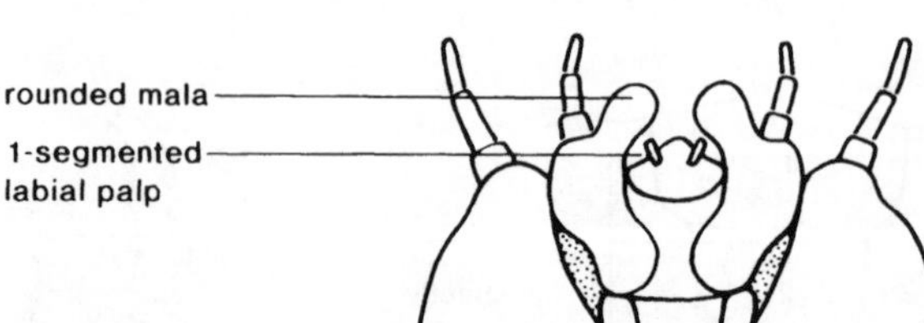 Figure 91

←—— Figs. 85, 96

146(145′). Segments in labial palp 1 (fig. 91) 147

146′. Segments in labial palp 2 (figs. 88, 94, 97) **(see 3rd choice)** 148

146². Segments in labial palp 0; body strongly curved ventrally (C-shaped) (fig. 1) and lightly sclerotized; head strongly retracted (fig. 9). In seeds or pods of various plants .... (part) ***Bruchidae*** p. 561

147(146). Epicranial stem absent or very short; stipes longer than wide; gular region present (separating labium from thorax) (fig. 85); mesal surface of mandibular base with fixed, rigid, hyaline process, sometimes partly sclerotized (fig. 57); apex of mandible with single tooth (fig. 57); stemmata on each side 3. On plant surfaces .... (part) ***Coccinellidae*** p. 485

147′. Epicranial stem moderately long; stipes wider than long; gular region absent (labium contiguous with thoracic membrane) (fig. 96); mesal surface of mandibular base simple or slightly expanded (fig. 51); apex of mandible with 4 or more teeth in row (fig. 51); stemmata on each side 6; body often covered with slimy fecal material. On plant surfaces .... **(Criocerinae)** ***Chrysomelidae*** p. 568

148(146′). Body relatively straight or slightly curved ventrally 149

148′. Body strongly curved ventrally (C-shaped) 163

149(148). Mouthparts not forming sucking tube; head not concealed from above by T1 150

149′. Mouthparts forming sucking tube (fig. 27); head completely concealed from above by T1 (as in fig. 2). In leaf litter, rotten wood, fungi .... **(Ceryloninae** part) ***Cerylonidae*** p. 480

150(149). Tergum A9 without median process or terminal disc (entire tergum may form spine) 151

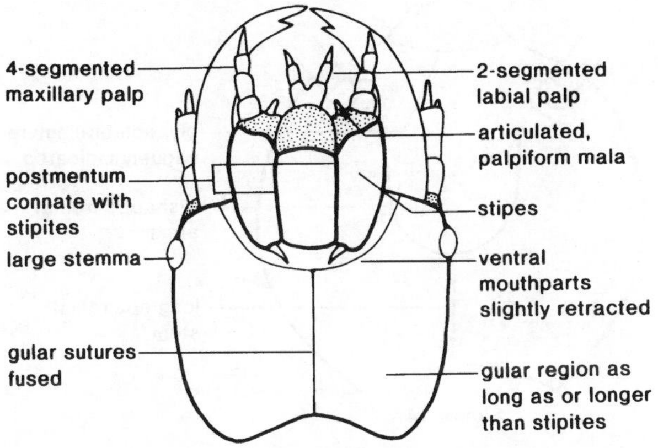

Figure 94

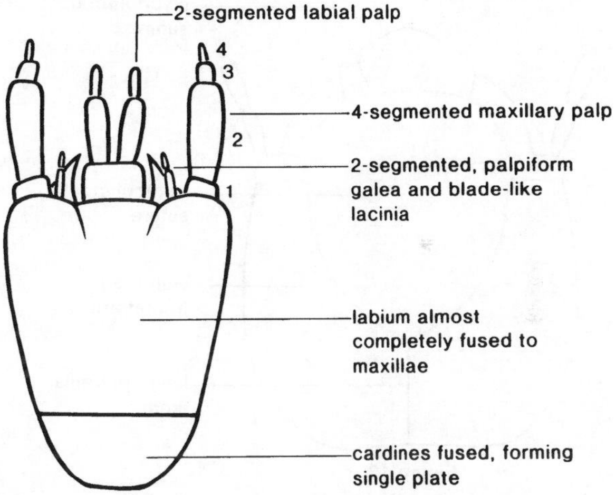

Figure 97

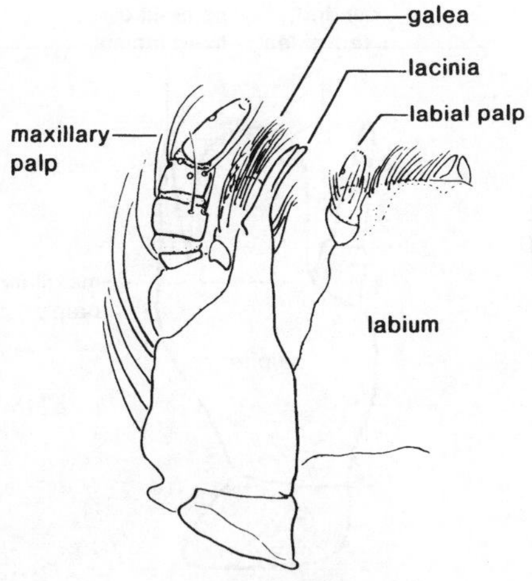

Figure 97a

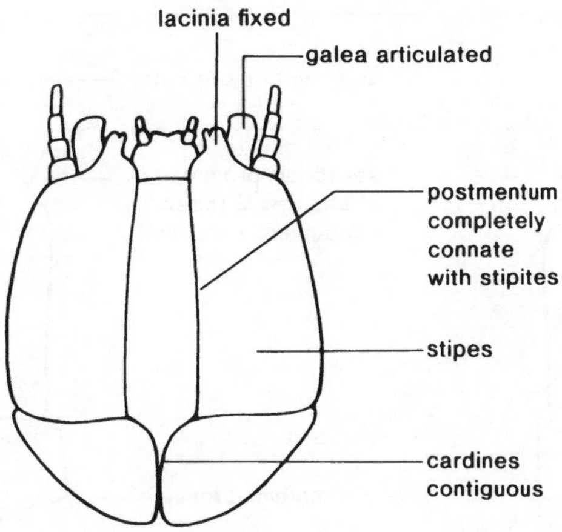

Figure 100

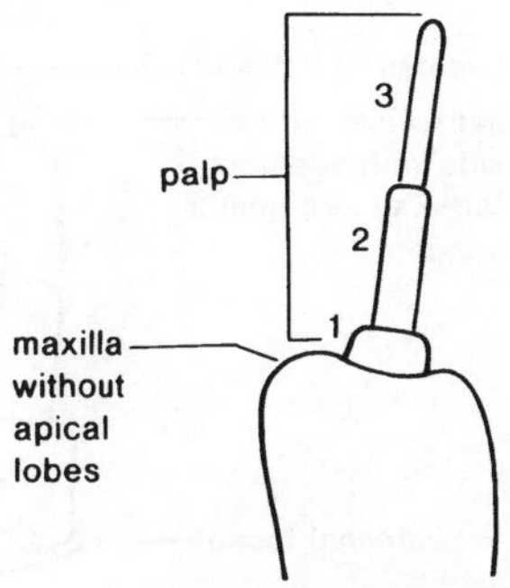

Figure 101

Fig. 215 ⟶

150′. Tergum A9 with median process (fig. 159). In rotten wood, stems, fungus fruiting bodies (**see 3rd choice**) .... (part) ***Mordellidae*** p. 508

150². Tergum A9 with concave, terminal disc (fig. 160). In rotten wood, fungus fruiting bodies (western North America) .... (*Sphindocis* part) (**Sphindociinae**) ***Ciidae*** p. 502

151(150). Head prognathous or slightly declined 152

151′. Head moderately to strongly declined (hypognathous) (fig. 10) 161

152(151). Body not broadly ovate and strongly flattened; cardines if present, separated from each other by labium; visible abdominal segments 10; tergum A9 not forming articulated spine 153

152′. Body broadly ovate, strongly flattened, and disc-like (fig. 2); cardines completely fused forming single plate (fig. 97); visible abdominal segments 9; tergum A9 forming articulated spine; thoracic segments and segments A1–8 each with pair of lateral, branched processes; body covered with scale-like setae; mandibles internally perforated (fig. 43). Under bark or stones in dry areas .... (part) ***Brachypsectridae*** p. 421

153(152). Stipes longer than wide 154

153′. Stipes wider than long 158

154(153). Epicranial stem absent 155

154′. Epicranial stem present (fig. 14) 156

155(154). Hypostomal rods absent; frontal arms absent; setae on tarsungulus 0; stemmata on each side 1; dorsal surfaces of abdominal segments without transverse row of protuberances. In cells of native bees .... (**Nemognathinae** part) ***Meloidae*** p. 530

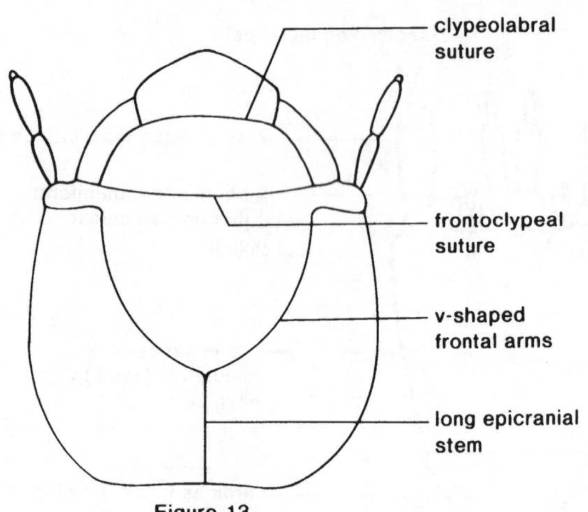

Figure 13

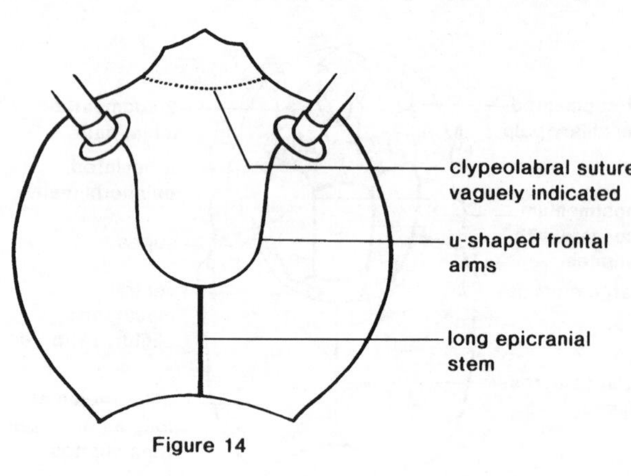

Figure 14

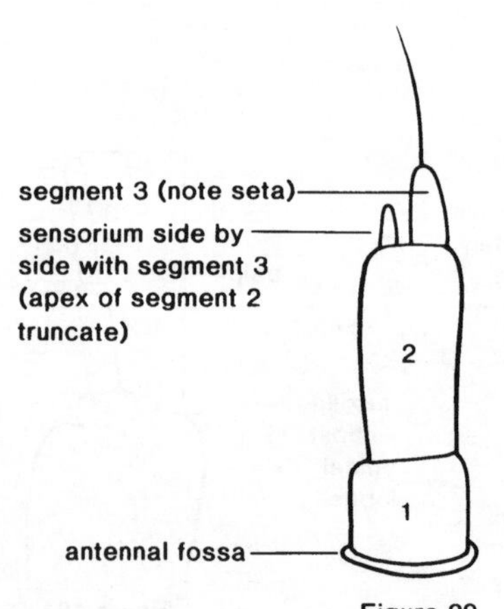

Figure 29

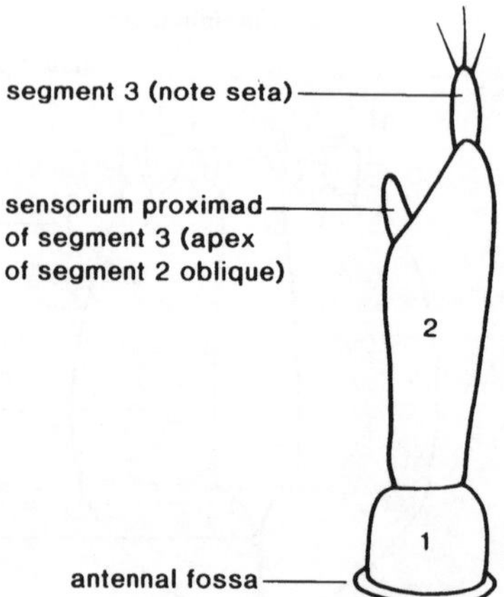

Figure 30

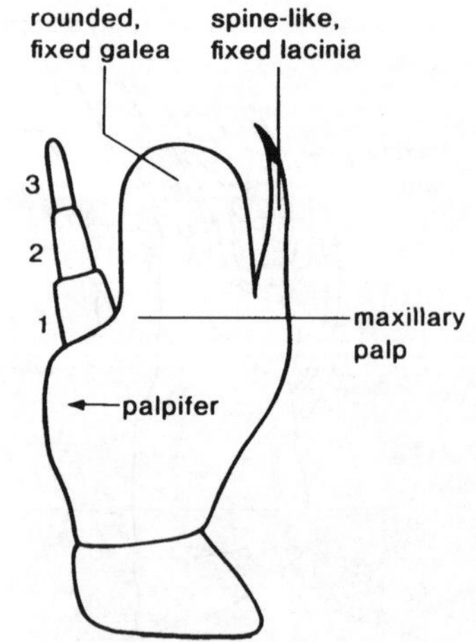

Figure 215

←— Figs. 2, 10, 97

155′. Hypostomal rods present (fig. 89); frontal arms present (fig. 13); setae on tarsungulus 1; stemmata on each side 3; dorsal surfaces of abdominal segments with transverse row of 6 protuberances, which may bear branched or setose processes (fig. 134). On plant surfaces .... (part) ***Coccinellidae*** p. 485

156(154′). Frontoclypeal suture absent or vaguely indicated; hypostomal rods absent; segment A10 distinct and visible from above (fig. 162); apex of mandible not multidentate; dorsal surfaces of abdominal segments without transverse row of protuberances; apex of antennal segment 2 oblique, so that sensorium arises proximad of segment 3 (fig. 30) .... (part) ***Staphylinidae*** p. 341

156′. Frontoclypeal suture distinct (fig. 13); hypostomal rods present (fig. 89) or mandibular apex multidentate (fig. 56); apex of antennal segment 2 truncate, so that sensorium and segment 3 arise together (fig. 29) 157

157(156′). Hypostomal rods present (fig. 89), usually long and diverging; body elongate and more or less parallel-sided, lightly sclerotized, sometimes with paired dorsal and ventral protuberances (ampullae); apex of mandible bidentate; stemmata on each side 5 or 0. In rotten wood, fungus fruiting bodies .... (**Melandryinae** part) ***Melandryidae*** p. 505

157′. Hypostomal rods absent; body oblong or ovate; dorsal surfaces of abdominal segments each with transverse row of 6 protuberances which bear branched processes (fig. 134); apex of mandible multidentate; stemmata on each side 3. On plant surfaces .... (**Epilachninae** part) ***Coccinellidae*** p. 485

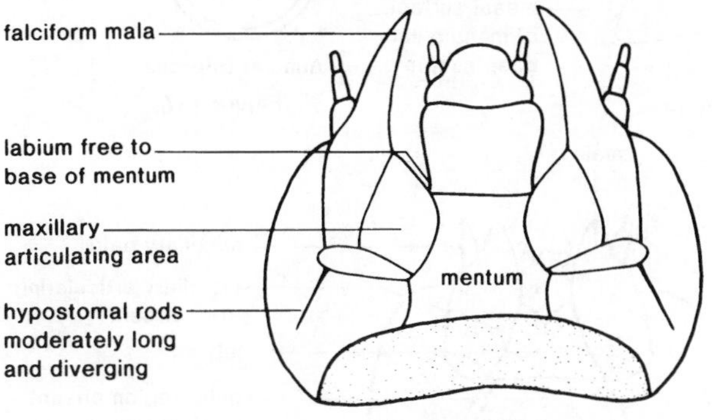

Mouthparts Retracted
Figure 89

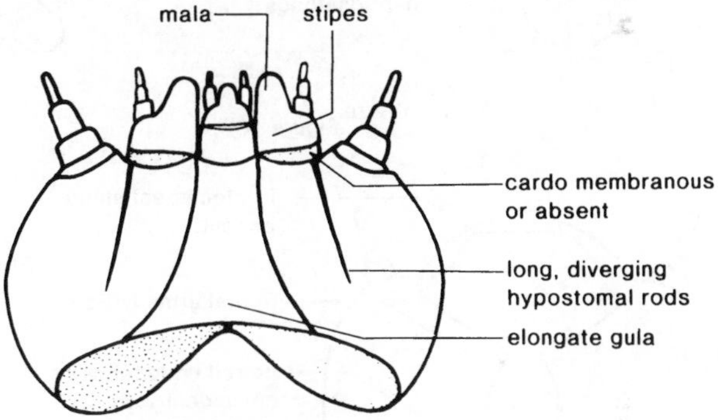

Ventral Mouthparts Strongly Protracted
Figure 92

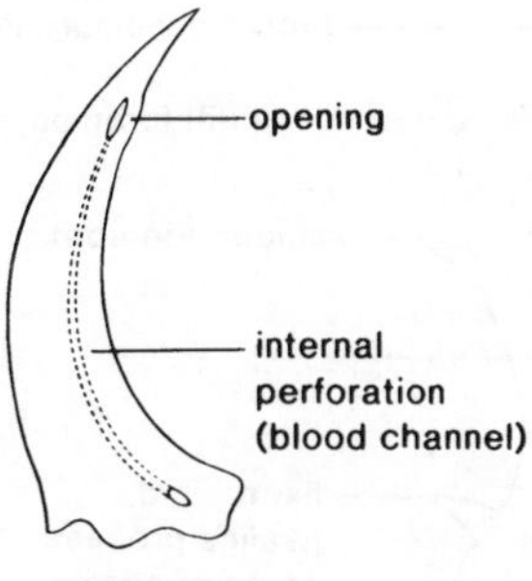

Figure 43

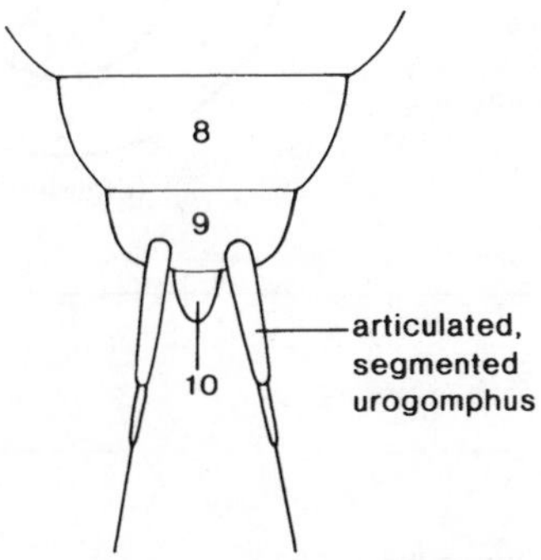

Figure 162

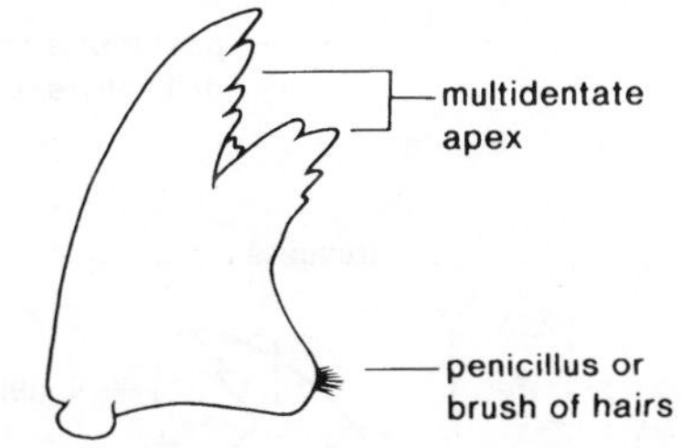

Figure 56

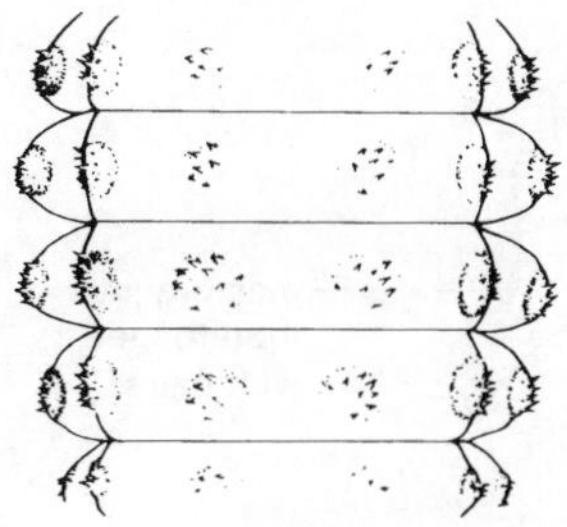

Figure 134

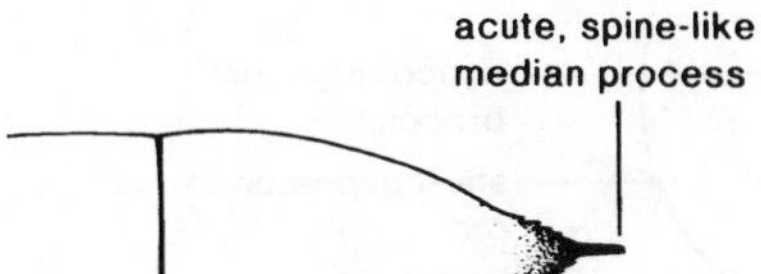

Figure 159

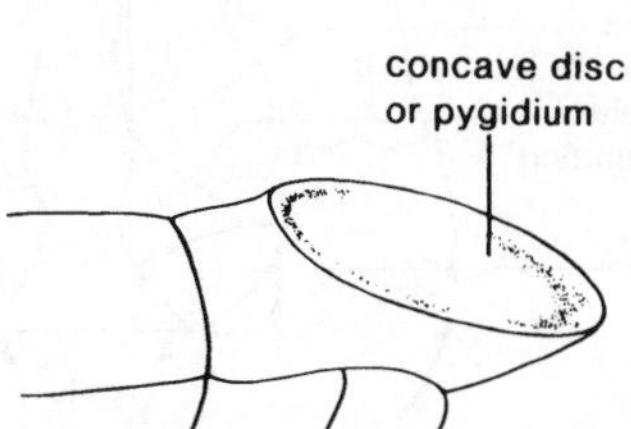

Figure 160

158(153′). Ventral mouthparts strongly protracted (fig. 92); gula much longer than wide (fig. 92); apex of antennal segment 2 truncate, so that sensorium and segment 3 arise close together (fig. 29) 159

158′. Ventral mouthparts retracted (fig. 89); gula not or only slightly longer than wide; apex of antennal segment 2 oblique, so that sensorium arises proximad of segment 3 (fig. 30). In leaf litter, rotten wood, carrion, dung, under bark .... (part) ***Staphylinidae*** p. 341

159(158). Epicranial stem present (fig. 13); paired endocarinae absent; hypostomal rods absent or very short and subparallel 160

159′. Epicranial stem absent and paired endocarinae coincident with frontal arms (fig. 24); hypostomal rods long and diverging (fig. 92); thorax and abdomen without tergal plates. In sooty molds. Exotic (New Zealand) (**see 3rd choice**) .... (*Cyclaxyra*) ***Phalacridae*** p. 466

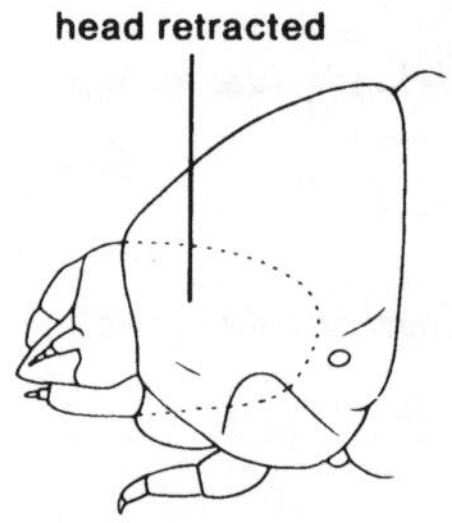

Figure 9

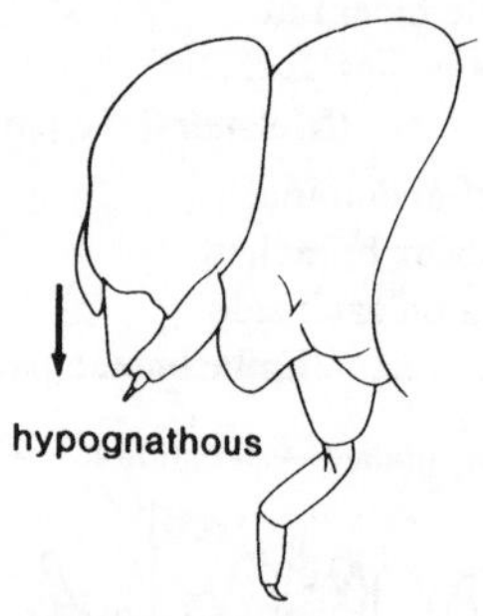

Figure 10

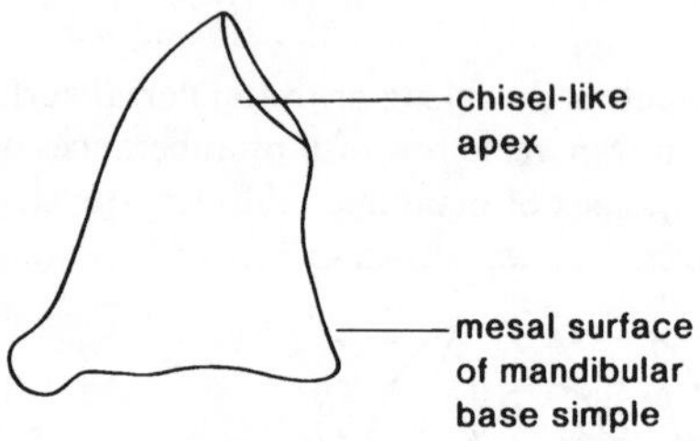

Figure 46

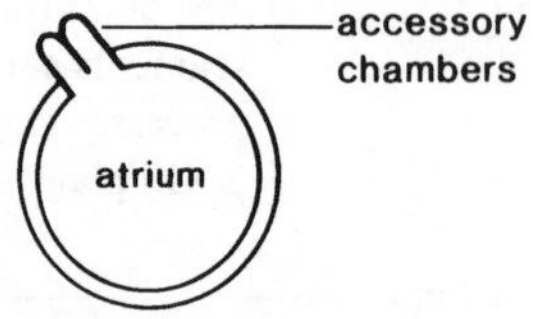

Annular Biforous
Figure 197

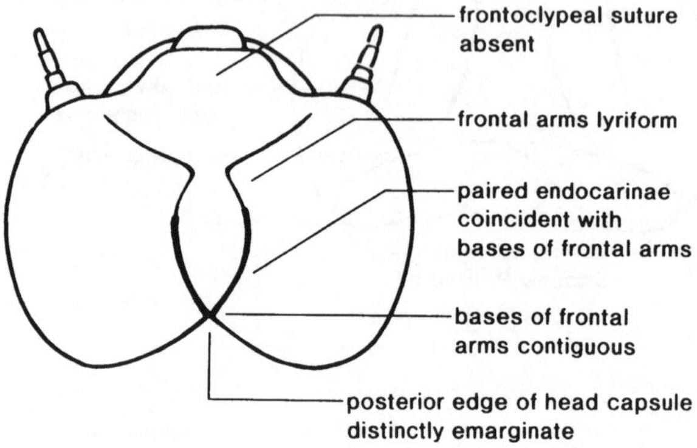

Figure 24

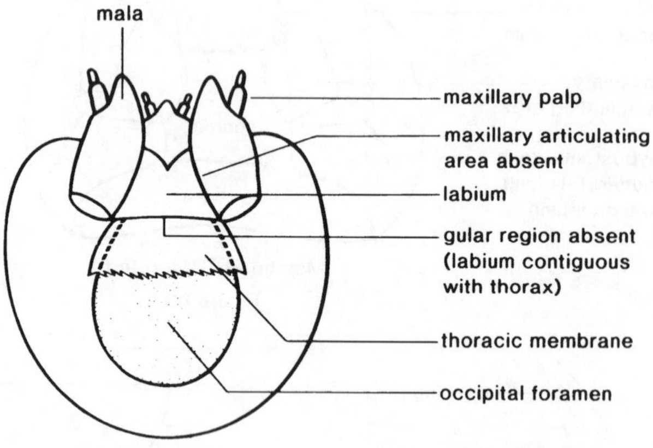

Figure 96

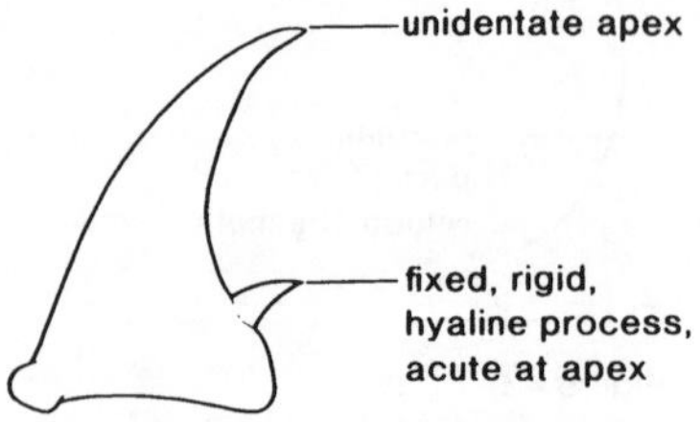

Figure 57

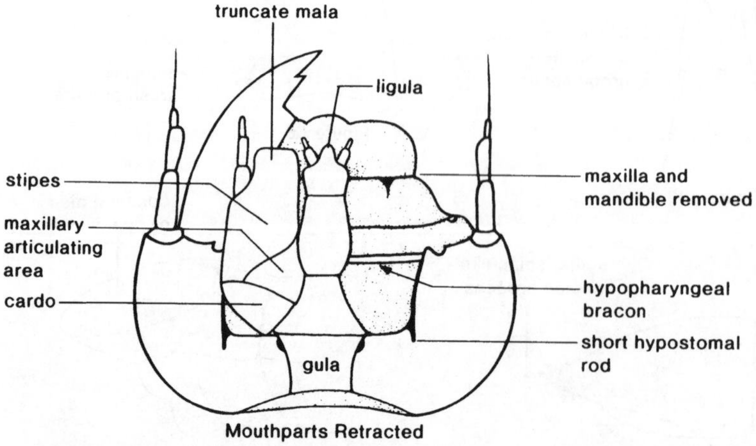

Mouthparts Retracted
Figure 85

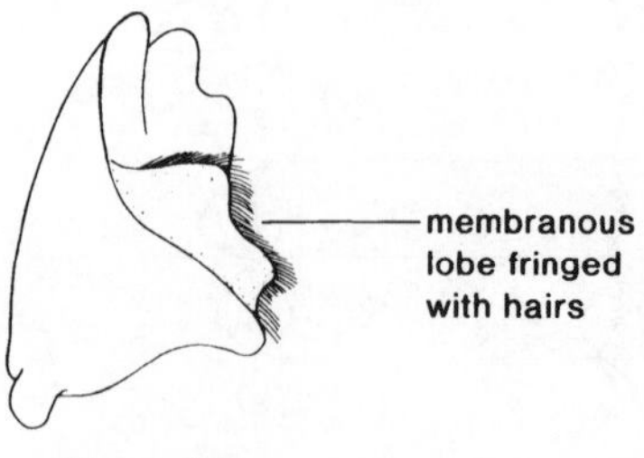

Figure 64

← Figs. 13, 29, 30, 89, 92, 134

159². Epicranial stem and paired endocarinae absent; hypostomal rods absent or short and subparallel; with sclerotized plate on protergum and paired plates on meso- and metaterga. Under bark, in leaf litter or rotten wood .... (**Phyllobaeninae** part) ***Cleridae*** p. 450

160(159). Posterior edge of head capsule distinctly emarginate dorsally; spiracles annular-biforous (fig. 197); anal region posteroventrally oriented; stemmata on each side 5; tergum A9 with sclerotized plate. In fungus fruiting bodies, stored products .... (**Thaneroclerinae**) ***Cleridae*** p. 450

160′. Posterior edge of head capsule not or only slightly emarginate dorsally; spiracles annular; anal region posteriorly or terminally oriented; stemmata on each side 1; tergum A9 with pair of long setae. In soil, on plant surfaces, in Orthoptera egg masses, in bee cells .... (triungulins) (**Lyttinae** part) ***Meloidae*** p. 530

161(151′). Gula absent (fig. 96); mandible broad and stout or more or less wedge-like, with a single lobe or tooth at apex (fig. 46); body elongate and cylindrical, lightly sclerotized. In rotten wood, stems, fungus fruiting bodies .... (part) ***Mordellidae*** p. 508

161′. Gula present (fig. 85); mandible with 2 or more teeth at apex; body more oblong or ovate and slightly flattened 162

162(161′). Mesal surface of mandibular base with membranous lobe (fig. 64); stemmata on each side 4; abdominal segments each with 2 pairs of laterally projecting processes. Under bark, on fungus fruiting bodies (Auriculariaceae) .... (*Endomychus*) (**Endomychinae**) ***Endomychidae*** p. 482

162′. Mesal surface of mandibular base either simple or with fixed, rigid, hyaline process (fig. 57); stemmata on each side 3; abdominal segments each with transverse row of 6 protuberances, which may bear setose or branched processes (fig. 134) .... (part) ***Coccinellidae*** p. 485

163(148′). Epicranial stem absent; segments in T2 leg 3 or 4. In cells of native bees .... (part) ***Meloidae*** p. 530

163′. Epicranial stem moderately long; segments in T2 leg 5 including tarsungulus (fig. 126) 164

164(163′). Head protracted or slightly retracted; head moderately to strongly declined (hypognathous) (fig. 10); gular region present (separating labium from thorax) (fig. 85); ratio of antennal length to head width .15 to .5. In cells of native bees .... (**Lyttinae** part) ***Meloidae*** p. 530

164′. Head strongly retracted (fig. 9); head prognathous or slightly declined; gular region absent (labium contiguous with thoracic membrane) (fig. 96); ratio of antennal length to head width less than .15. Exotic (Asia, Africa, Australia, southern South America) .... (**Sagrinae** part) ***Chrysomelidae*** p. 568

165(145²). Maxillary articulating area well-developed but concealed behind lateral edges of expanded postmentum, which is divided longitudinally into 3 parts (fig. 98); anal hooks 3 or more on each side (fig. 189); gular sutures separate; mesal surface of mandibular base with brush of hairs (fig. 53). In plant debris in or near streams .... (**Anchytarsinae** part) ***Ptilodactylidae*** p. 391

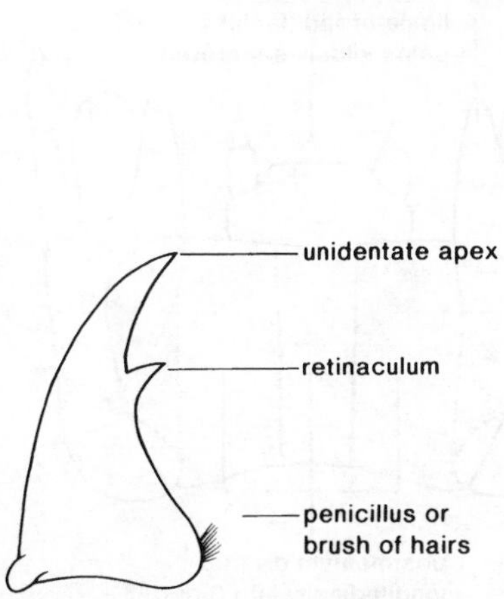

Figure 53

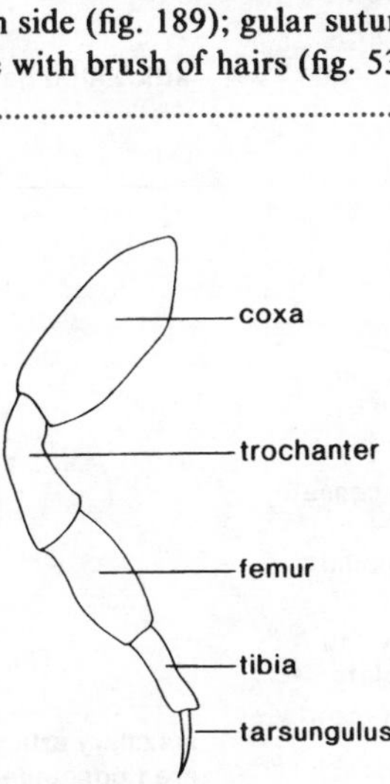

Figure 126

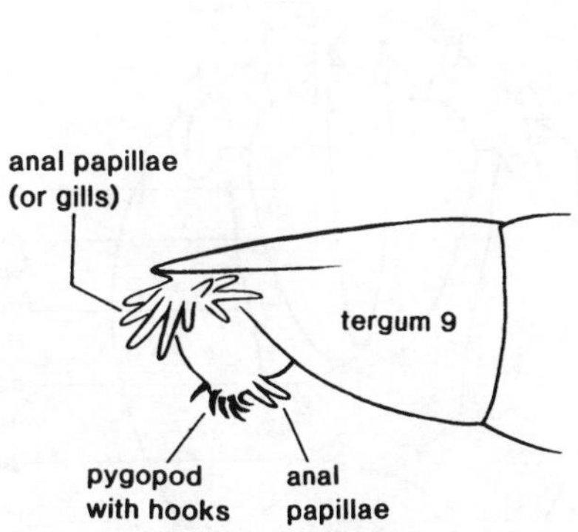

Figure 189

Fig. 98 ⟶

165′. Postmentum not divided longitudinally into 3 parts; if maxillary articulating area concealed behind expanded postmentum, gular sutures are fused (fig. 93) and mesal surface of mandibular base with articulated process (fig. 66) 166

166(165′). Body broadly ovate, strongly flattened and disc-like (fig. 2) 167

166′. Body not broadly ovate and strongly flattened 168

167(166). Ventral abdominal gill tufts absent; tergum A9 forming articulated spine; mandibles narrow and falcate, with internal perforation (fig. 43). Under bark or stones in dry areas .... (part) ***Brachypsectridae*** p. 421

167′. Ventral abdominal gill tufts on segments 2–6 (fig. 151); tergum A9 not forming articulated spine; mandibles neither falcate nor internally perforate. On stones in running water **(see 3rd choice)** .... **(Psepheninae)** ***Psephenidae*** p. 395

167². Ventral abdominal gill tufts on segments 1–4 (fig. 151); tergum A9 not forming articulated spine; mandibles neither falcate not internally perforate. On stones in running water .... ***(Eubrianax)*** **(Eubrianacinae)** ***Psephenidae*** p. 395

168(166′). Anterior abdominal spiracles absent or non-functional; spiracles on segment A8 forming pair of projecting spines (fig. 156); body strongly curved ventrally (C-shaped) and lightly sclerotized; head protracted or slightly retracted. On roots and stems of aquatic plants .... **(Donaciinae)** ***Chrysomelidae*** p. 568

168′. All abdominal spiracles annular or annular-uniforous (fig. 196), those on segment A8 not forming pair of spines **(see 3rd choice)** 169

168². All abdominal spiracles annular-biforous (fig. 197, 198) or biforous (fig. 202), those on segment A8 not forming pair of spines 170

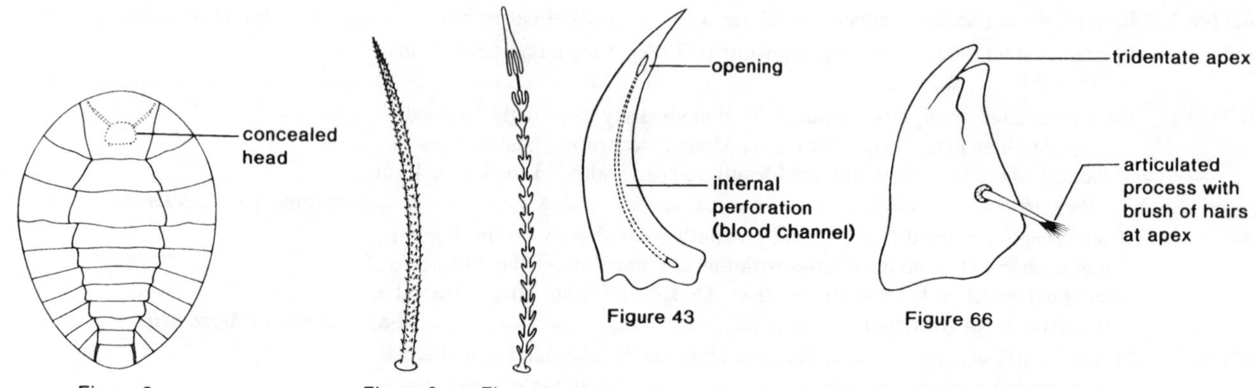

Figure 2 Figure 6 Figure 7 Figure 43 Figure 66

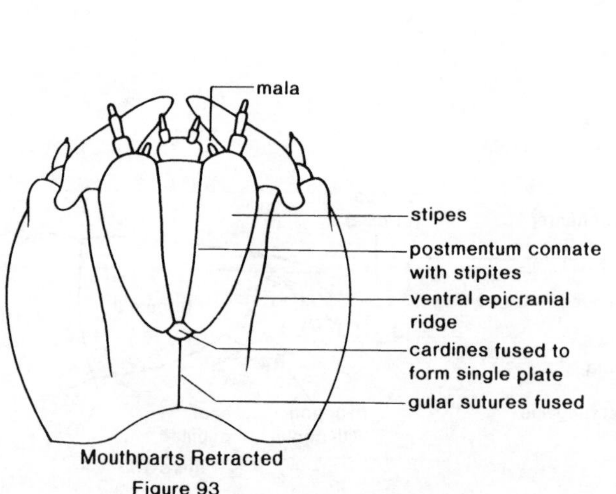

Mouthparts Retracted
Figure 93

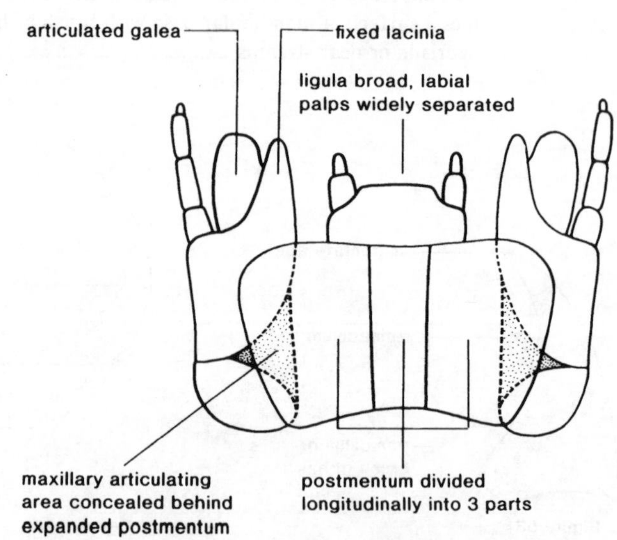

Figure 98

← **Fig. 9**

169(168′). Body strongly curved ventrally (C-shaped); head strongly retracted (fig. 9); tergum A3 with 2 or more transverse plicae (fig. 136); segment A10 with pair of oval lobes separated by longitudinal groove (fig. 193); vestiture consisting of fine hairs or setae only. In dead wood .... (part) ***Bostrichidae*** p. 439

169′. Body relatively straight or only slightly curved ventrally; head protracted or slightly retracted; tergum A3 without transverse plicae; segment A10 without oval lobes; vestiture including specialized setae (barbed spicisetae (fig. 6) or complex hastisetae (fig. 7)). On a variety of animal and plant products .... (part) ***Dermestidae*** p. 434

170(168²). Epicranial stem very short or absent (figs. 12, 16); stemmata on each side 1 or 5–6 forming tight cluster 171

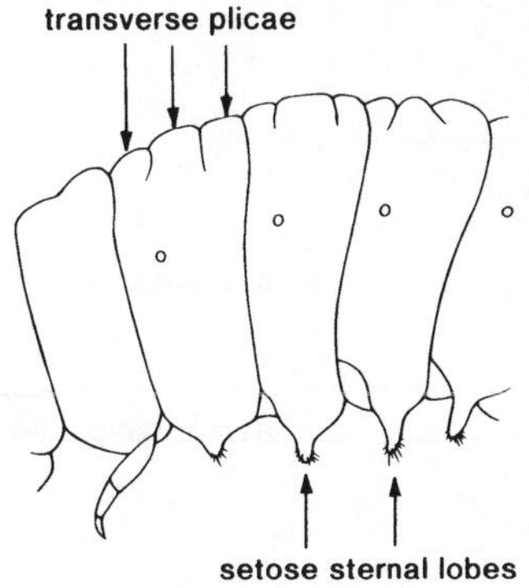

Figure 136

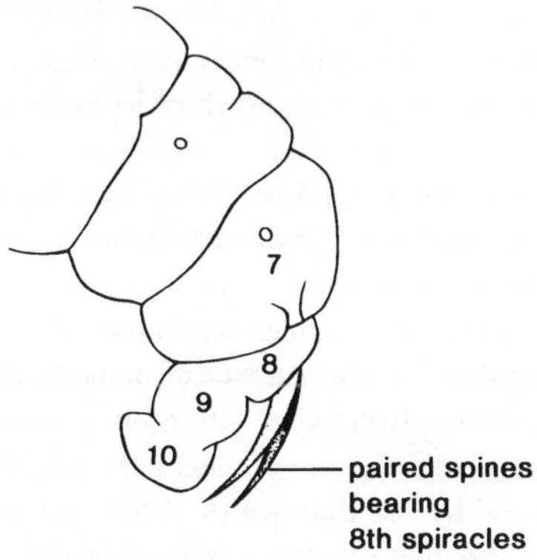

Figure 156

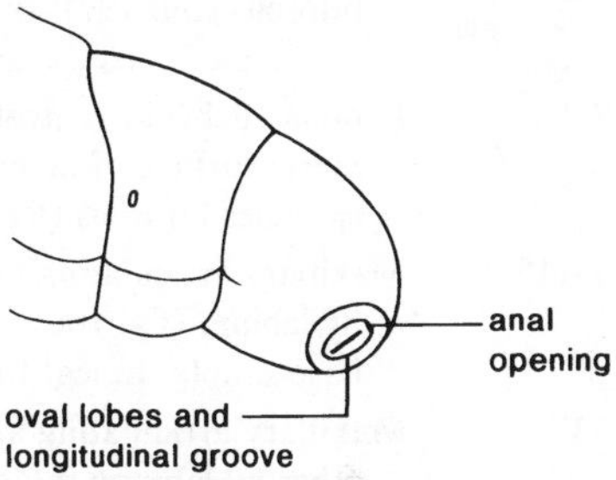

Figure 193

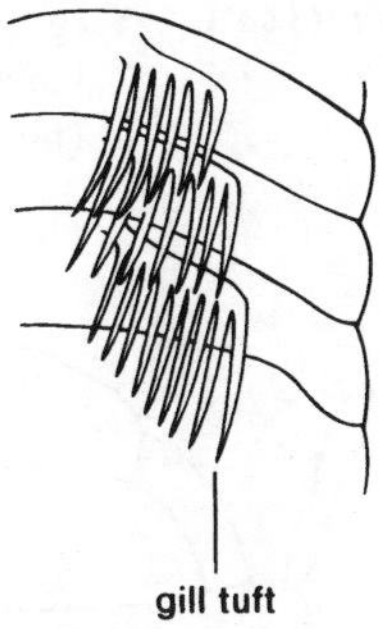

Figure 151

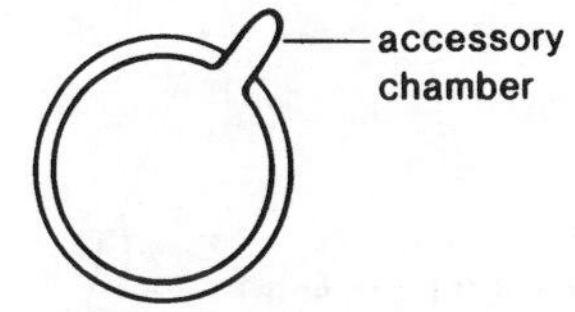

Annular Uniforous
Figure 196

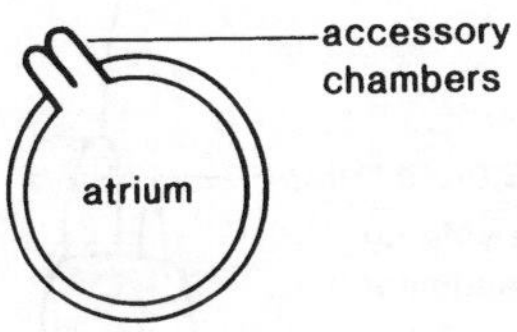

Annular Biforous
Figure 197

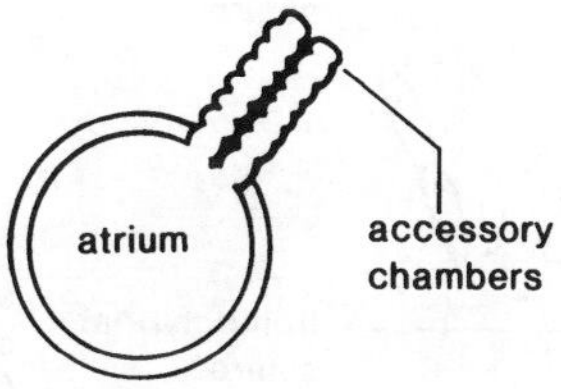

Annular Biforous
Figure 198

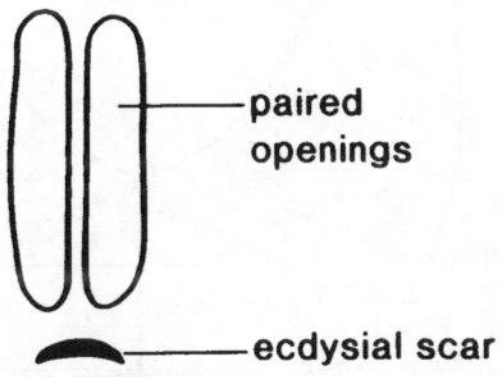

Biforous
Figure 202

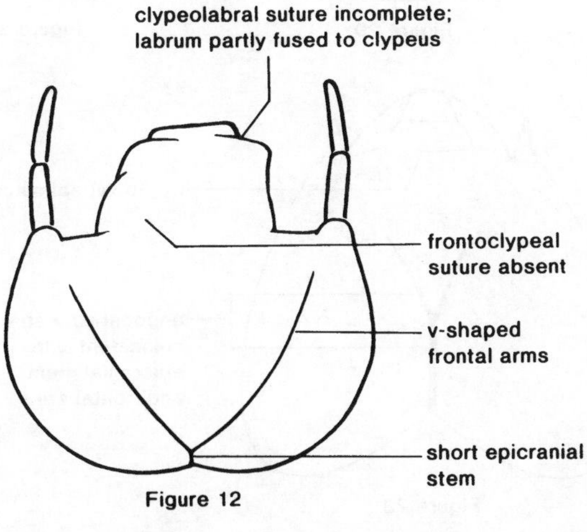

Figure 12

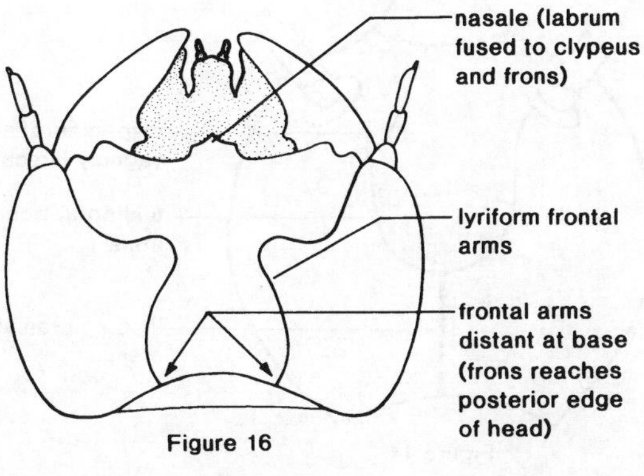

Figure 16

170′. Epicranial stem moderately long (figs. 13, 14); stemmata on each side 5 or 6, well separated and not forming tight cluster 172

171(170). A single large stemma on each side; mesal surface of mandibular base simple (fig. 47); anal hooks absent; gular sutures separate. Under moss or lichens .... ***Artematopidae*** p. 407

171′. A tight cluster of 5 or 6 stemmata on each side; mesal surface of mandibular base with articulated process bearing brush of hairs at apex (fig. 66); anal hooks 3 or more on each side (fig. 189); gular sutures fused (fig. 93). In leaf litter, flood debris, rotten wood .... (**Ptilodactylinae**) ***Ptilodactylidae*** p. 391

172(170′). Lacinia subapical and dorsally situated, so that it is not visible in ventral view (fig. 110); antennal sensorium longer than segment 3; mesal surface of mandibular base with fixed, rigid, hyaline process (fig. 58); spiracles annular-biforous (fig. 197). In rotten wood, fungus fruiting bodies .... (*Sphindocis* part) (**Sphindociinae**) ***Ciidae*** p. 502

172′. Lacinia visible in ventral view; antennal sensorium longer than segment 3; mesal surface of mandibular base simple or with brush of hairs (fig. 63); spiracles biforous (fig. 202) 173

173(172′). Maxillary articulating area absent (fig. 100); cardines contiguous, not separated by labium (fig. 100); setae on tarsungulus 1; mesal surface of mandibular base simple. In leaf litter, soil, flood debris .... ***Limnichidae*** p. 401

173′. Maxillary articulating area present (fig. 85); cardines separated from each other by labium; setae on tarsungulus 2; mesal surface of mandibular base with brush of hairs (fig. 63). In mosses, liverworts, soil, roots .... ***Byrrhidae*** p. 384

174(144′). Tergum A9 without median process; epicranial stem moderately long (fig. 23); labial palps contiguous or separated by less than width of first palpal segment (fig. 86); sensorium on preapical antennal segment conical or palpiform (fig. 29). In rotten wood, fungus fruiting bodies .... (**Melandryinae** part) ***Melandryidae*** p. 505

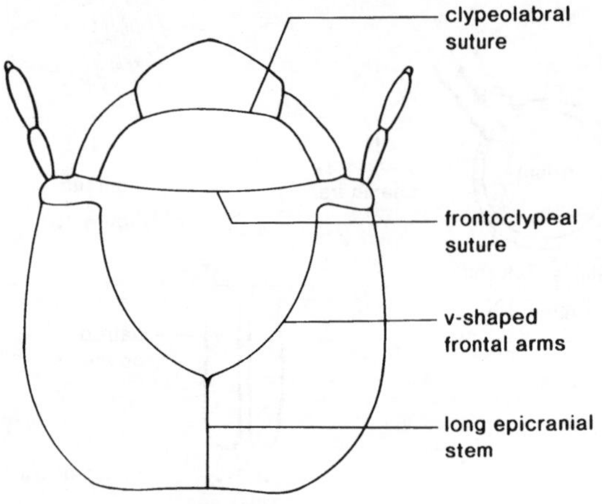

Figure 13

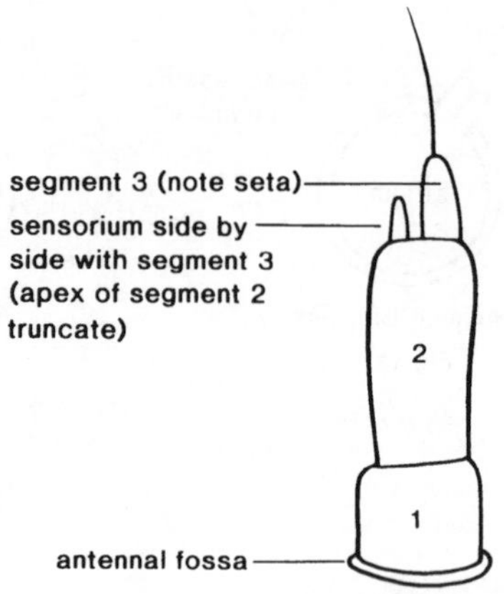

Figure 29

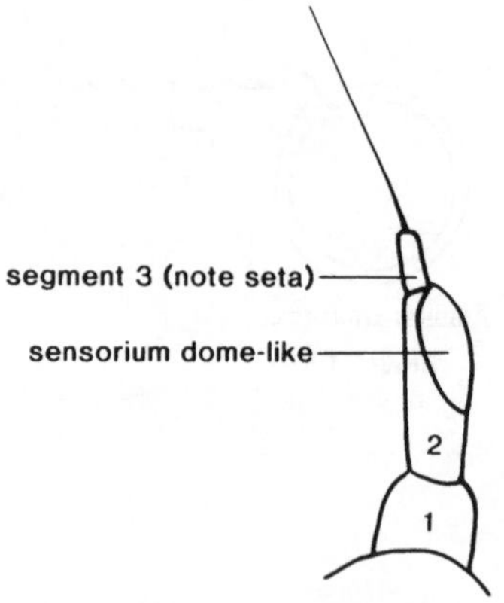

Figure 31

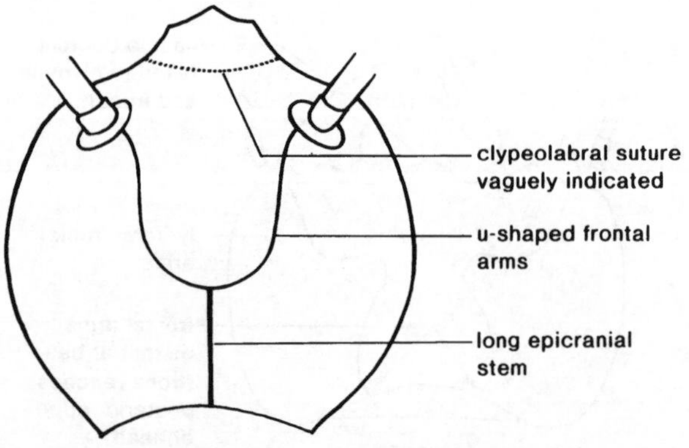

Figure 14

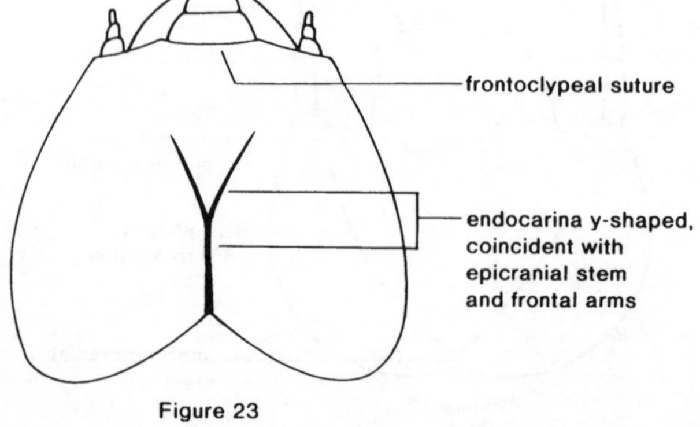

Figure 23

Figs. 66, 93, 197, 202

174′. Tergum A9 with median process (fig. 159); epicranial stem absent or very short; labial palps separated by more than width of first palpal segment; sensorium on preapical antennal segment dome-like (fig. 31). In dead wood .... (**Osphyinae** part) ***Melandryidae*** p. 505

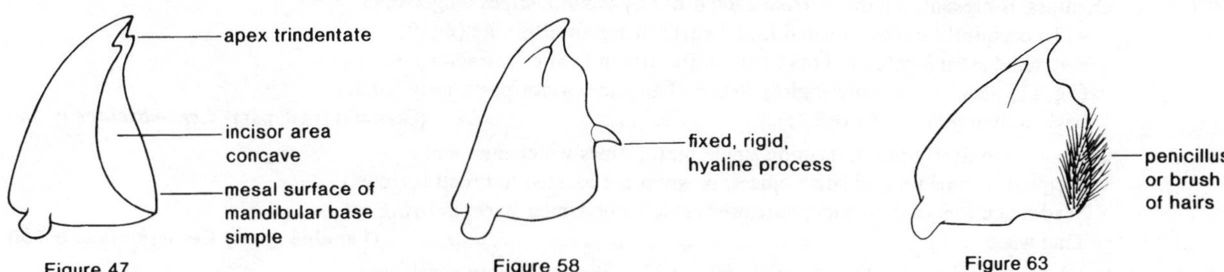

Figure 47 Figure 58 Figure 63

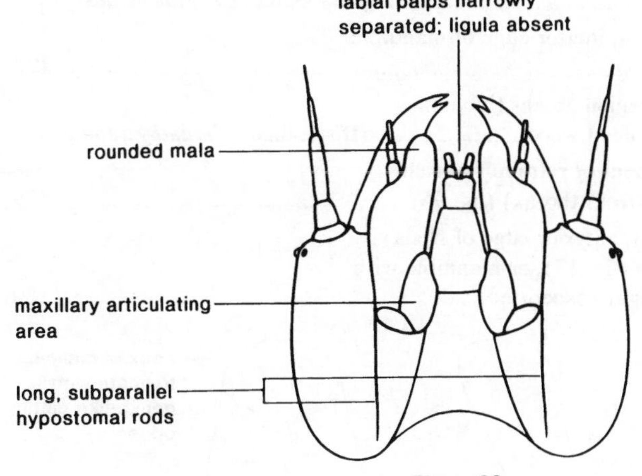

Figure 86

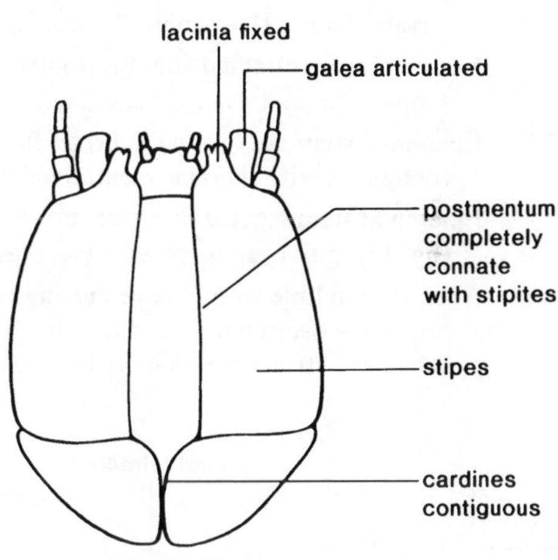

Figure 100

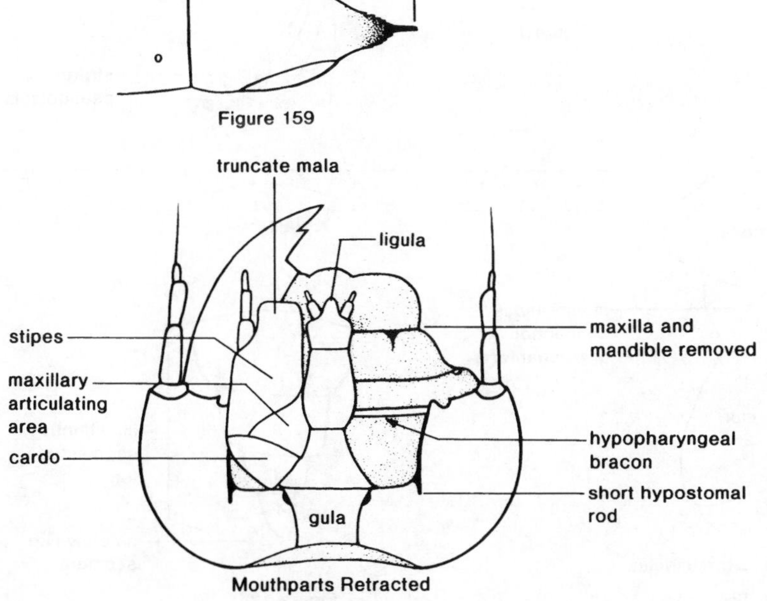

Figure 159

Figure 85

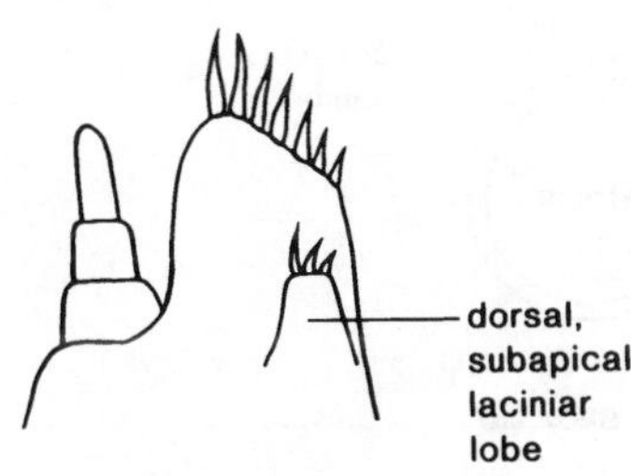

Figure 110

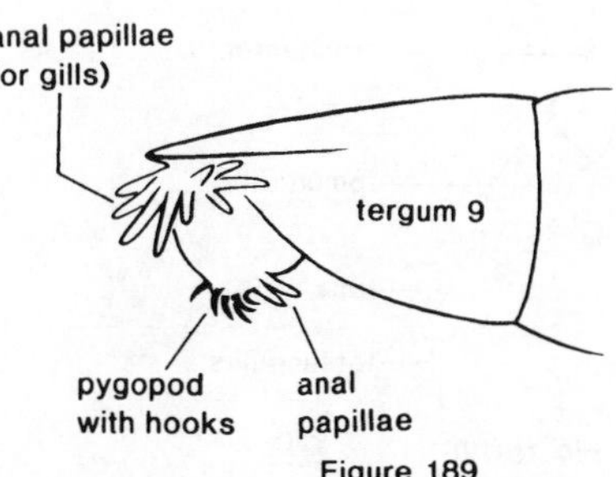

Figure 189

175(144^{2}). Segments in T2 leg 1 or 2; head strongly retracted (fig. 9); body lightly sclerotized 176

175′. Segments in T2 leg 3 or 4 (figs. 127, 128) **(see 3rd choice)** 177

175^{2}. Segments in T2 leg 5 including tarsungulus (fig. 126) 181

176(175). Cardines, if present, separated from each other by labium; stipes longer than wide; occipital foramen divided into 2 parts by tentorial bridge (fig. 95); epicranial stem located in broad furrow for attachment of retractor muscles (fig. 17); head not or only slightly longer than wide, wider posteriorly. Under bark, in living or dead wood .... **(Cerambycinae** part) ***Cerambycidae*** p. 556

176′. Cardines completely fused, forming single plate; stipes wider than long; occipital foramen not divided; epicranial stem not located in broad furrow; head much longer than wide, narrower posteriorly. Under bark, in living or dead wood .... **(Lamiinae** part) ***Cerambycidae*** p. 556

177(175′). Head protracted or slightly retracted; gular region absent (labium contiguous with thoracic membrane) (fig. 96); incisor edge of mandible denticulate or serrate. Exotic (Neotropical) .... **(Megalopodinae)** ***Chrysomelidae*** p. 568

177′. Head strongly retracted (fig. 9); if gular region absent, incisor edge of mandible simple 178

178(177′). Epicranial stem not located in broad furrow; gular region absent (labium contiguous with thoracic membrane) (fig. 96). In dead wood .... **(Disteniinae)** ***Cerambycidae*** p. 556

178′. Epicranial stem located in broad furrow for attachment of retractor muscles (fig. 17); gular region present (separating labium from thorax) (fig. 95) 179

179(178′). Apex of mandible with oblique cutting edge (fig. 52); anterior edge of frons projecting over clypeus and forming dentate ridge (fig. 17), or mandible with subapical, striate pseudomola (fig. 52) and protergum asperate 180

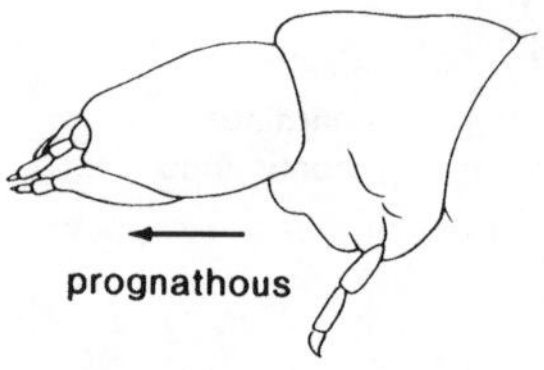

Figure 8

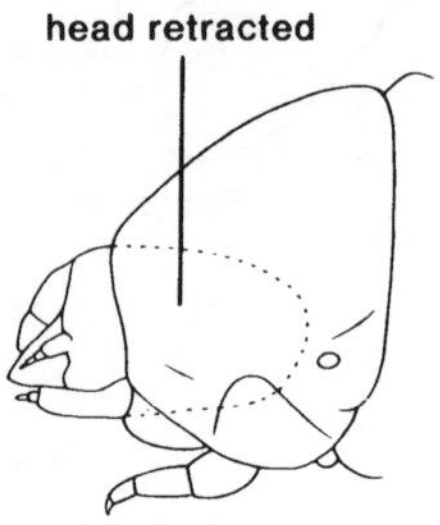

Figure 9

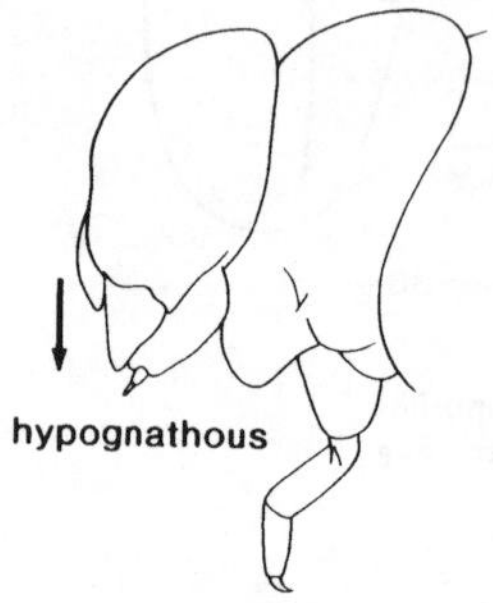

Figure 10

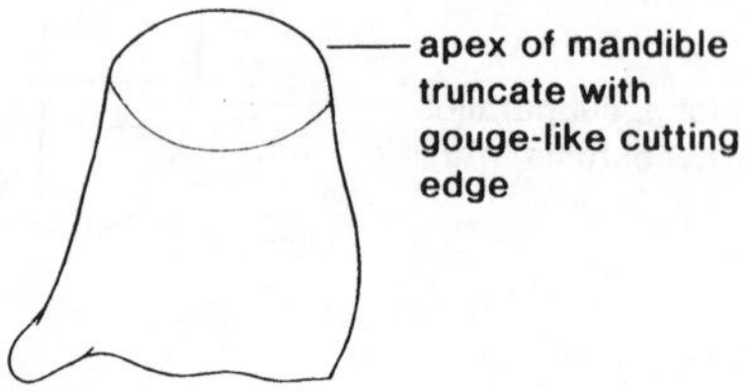

Figure 48

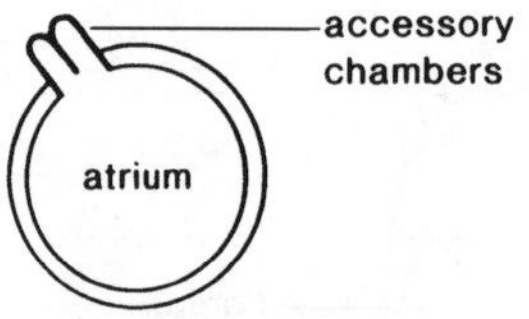

Annular Biforous
Figure 197

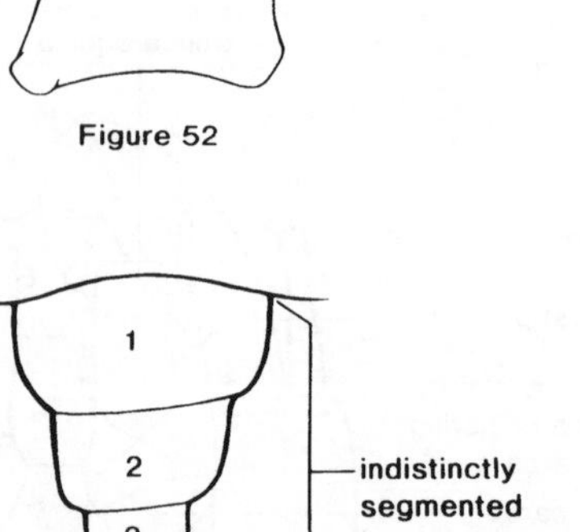

Figure 52

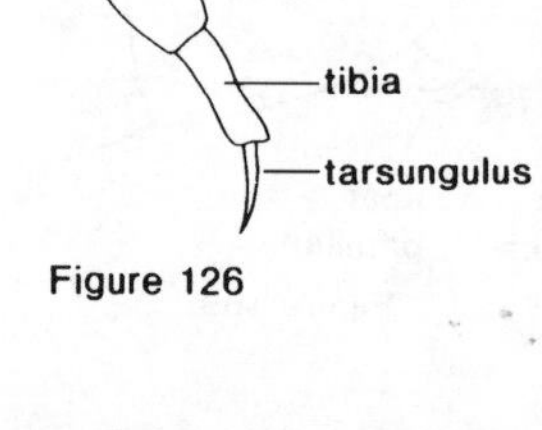

Figure 126

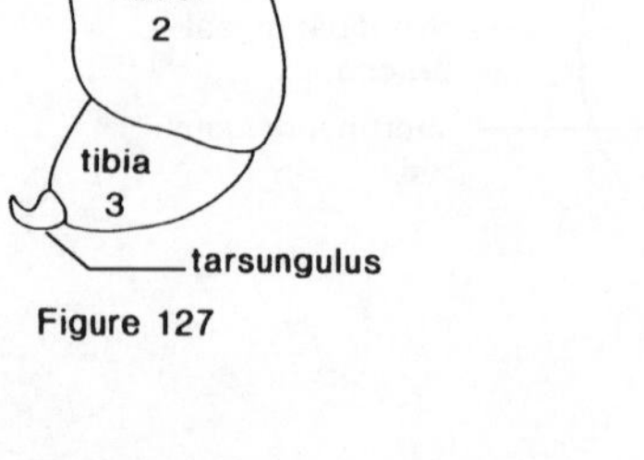

Figure 127

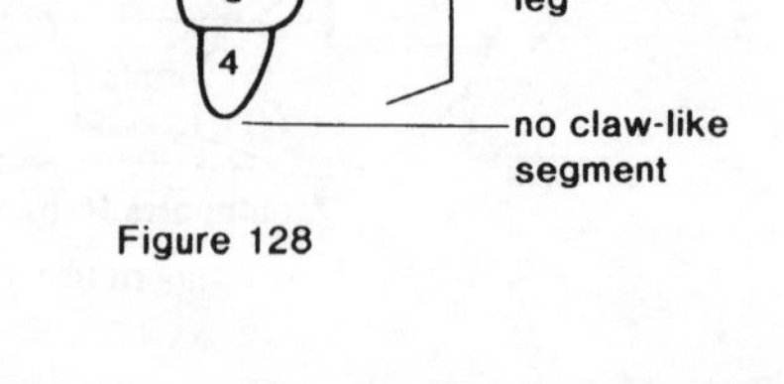

Figure 128

←—— Fig. 159

179′. Apex of mandible with truncate or gouge-like cutting edge (fig. 48); anterior edge of frons never projecting over clypeus and not forming dentate ridge; mandible without striate pseudomola; protergum not asperate. Under bark, in living or dead wood .... (**Cerambycinae** part) ***Cerambycidae*** p. 556

180(179). Mandible without striate pseudomola; protergum without patch of asperities; anterior edge of frons projecting over clypeus and usually forming dentate ridge (fig. 17); ratio of antennal length to head width less than .15. Under bark, in living or dead wood .... (**Prioninae** part) ***Cerambycidae*** p. 556

180′. Mandible with subapical, striate pseudomola (fig. 52); protergum with patch of asperities; anterior edge of frons not projecting over clypeus; ratio of antennal length to head width more than .15. In rotten wood .... (**Parandrinae**) ***Cerambycidae*** p. 556

181(175²). Head protracted or slightly retracted (figs. 8, 10) 182

181′. Head strongly retracted (fig. 9) 185

182(181). Maxillary mala cleft at apex (fig. 106); mesal surface of mandibular base with membranous lobe and brush of hairs; spiracles annular-biforous (fig. 197); body lightly sclerotized, except for apex of tergum A9, which bears a median, spine-like process (fig. 159). In fungus fruiting bodies .... (*Microsternus*) (**Dacninae**) ***Erotylidae*** p. 473

182′. Maxillary mala simple at apex; mesal surface of mandibular base simple or with brush of hairs only; spiracles annular; body usually with pigmented dorsal and lateral plates, and never with sclerotized, spine-like process on tergum A9 183

183(182′). Segments in labial palp 1; dorsal and lateral surfaces very lightly sclerotized, without pigmented plates; abdominal sterna usually with projecting setose lobes (fig. 136). In soil, associated with roots .... (**Eumolpinae**) ***Chrysomelidae*** p. 568

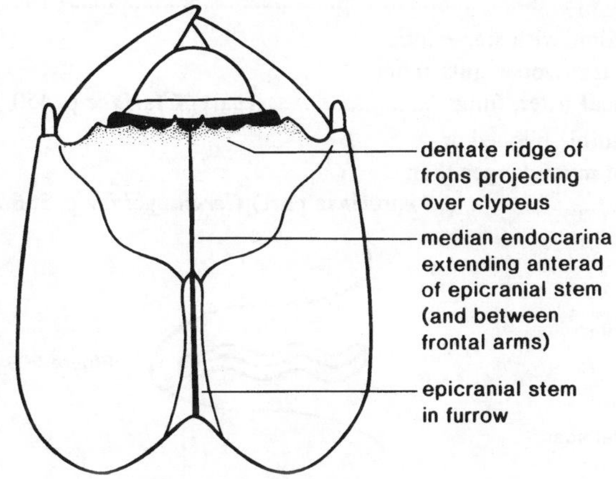

Figure 17

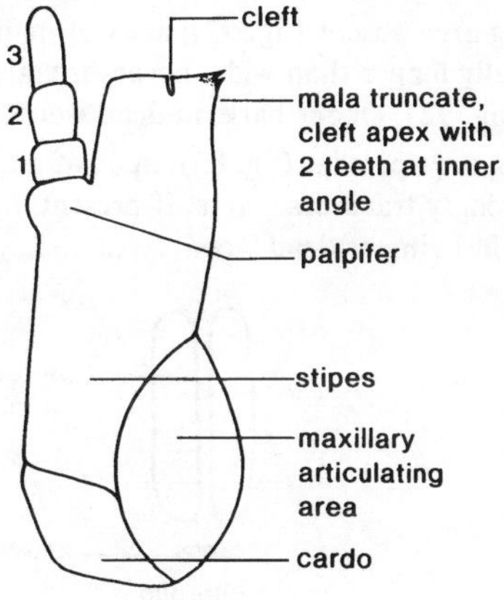

Figure 106

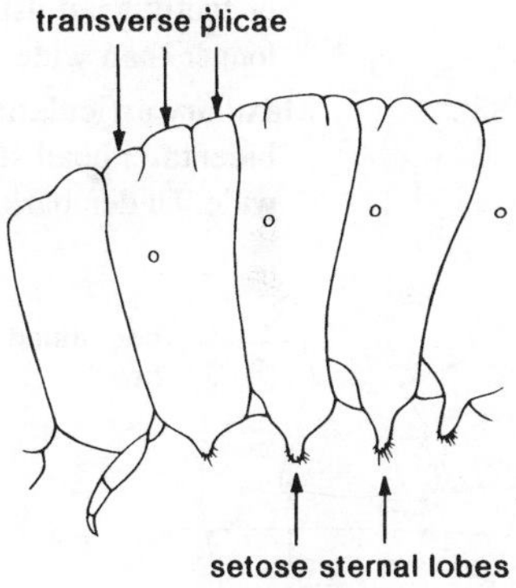

Figure 136

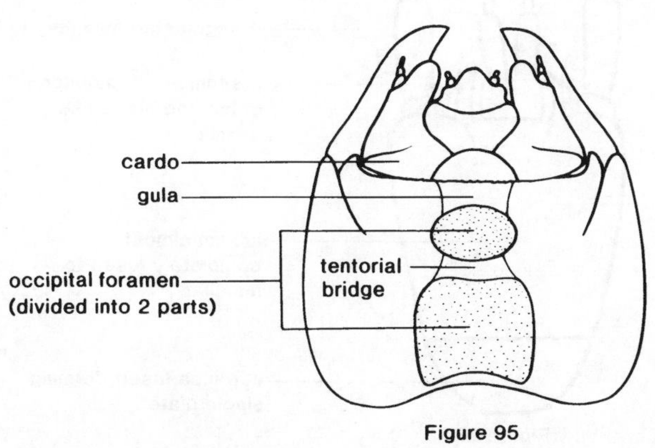

Figure 95

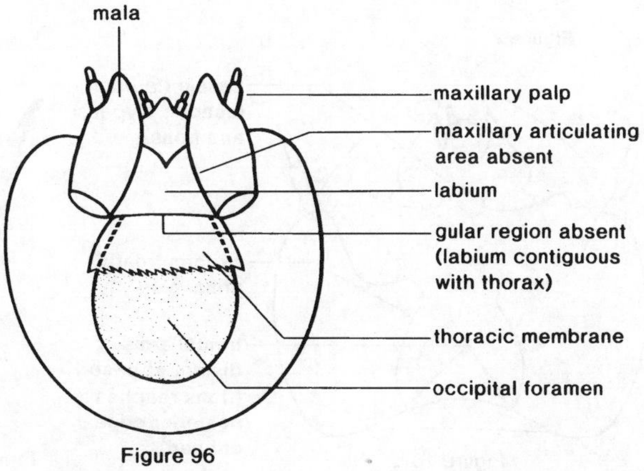

Figure 96

183′. Segments in labial palp 2; dorsal and lateral surfaces often with pigmented plates; abdominal sterna without projecting setose lobes 184

184(183′). Stemmata on each side 1 or 0. In leaf litter, soil, on plant surfaces .... (**Galerucinae-Halticinae** part) ***Chrysomelidae*** p. 568

184′. Stemmata on each side 5 or 6. On leaf surfaces .... (**Chrysomelinae**) ***Chrysomelidae*** p. 568

185(181′). Body relatively straight or slightly curved ventrally; occipital foramen divided into 2 parts by tentorial bridge (fig. 95); epicranial stem located in broad furrow for attachment of retractor muscles (fig. 17); gular region present (separating labium from thorax) (fig. 95). Under bark, in living or dead wood .... (**Prioninae** part) ***Cerambycidae*** p. 556

185′. Body strongly curved ventrally (C-shaped); occipital foramen not divided; epicranial stem not located in broad furrow; gular region absent (labium contiguous with thoracic membrane) (fig. 96). Exotic (Asia, Africa, Australia, southern South America) .... (**Sagrinae** part) ***Chrysomelidae*** p. 568

186(144[3]). Segments in T2 leg 4 or fewer (fig. 127); A8 spiracles dorsally placed; segments in maxillary palp 1 or 2; body flattened and lightly sclerotized. Leaf miners .... (Chalepini and Uroplatini) (**Hispinae**) ***Chrysomelidae*** p. 568

186′. Segments in T2 leg 5 including tarsungulus (fig. 126); A8 spiracles laterally placed; segments in maxillary palp 3 187

187(186′). Stipes longer than wide; ventral mouthparts retracted (fig. 85); labial palps contiguous or separated by less than width of first palpal segment; ratio of length of abdomen to length of thorax 1.2 to 2. On plant surfaces .... (*Microweisea*) (**Sticholotidinae**) ***Coccinellidae*** p. 485

187′. Stipes wider than long; ventral mouthparts strongly protracted (figs. 92, 95); labial palps separated by more than width of first palpal segment; ratio of length of abdomen to length of thorax more than 2 188

188(187′). Maxillary articulating area absent (fig. 92); apex of mandible with single lobe or tooth; head usually longer than wide, never strongly transverse; gula much longer than wide (fig. 92). Under bark, in dead wood, leaf litter, fungi .... (part) ***Cleridae*** p. 450

188′. Maxillary articulating area present (fig. 85); apex of mandible bilobed or bidentate; head strongly transverse; gula, if present, not much longer than wide. Under bark, in living or dead wood .... (**Lepturinae** part) ***Cerambycidae*** p. 556

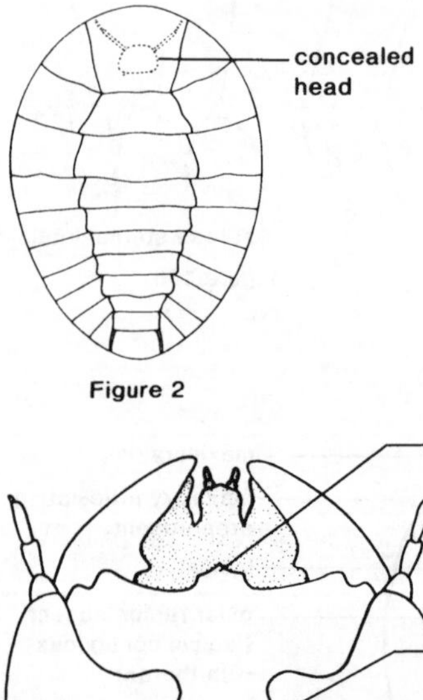

Figure 2

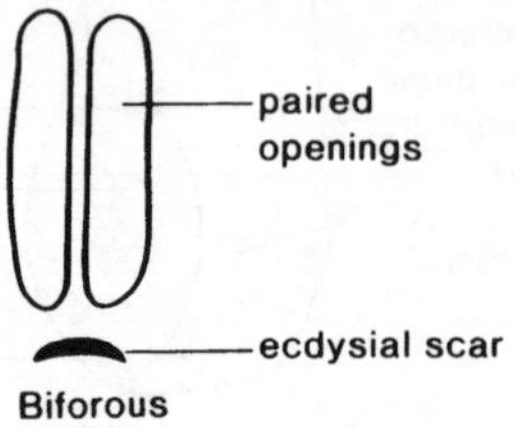

Biforous

Figure 202

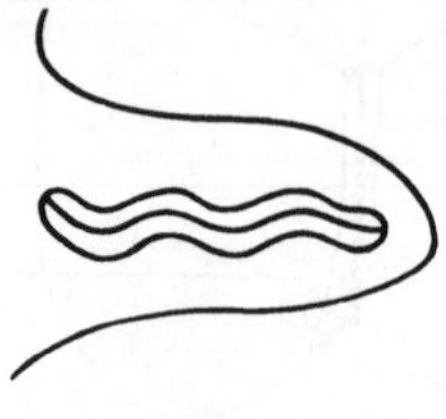

Figure 209

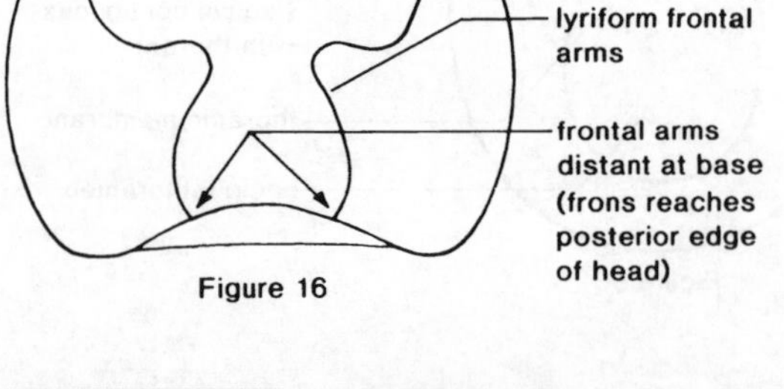

Figure 16

Figure 5

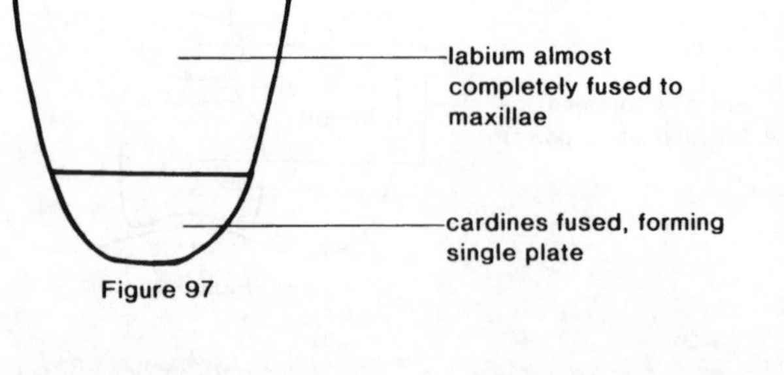

Figure 97

← Figs. 95, 96, 126, 127

189(143′). Body not broadly ovate and strongly flattened; head not concealed from above by T1; maxillary articulating area absent; ventral epicranial ridges present (fig. 93); frontal arms present (fig. 16) 190

189′. Body broadly ovate, strongly flattened, and disc-like (fig. 2); head completely concealed from above by T1; maxillary articulating area present (fig. 85); ventral epicranial ridges absent; frontal arms absent; anal gill tufts present (fig. 188). On stones in running water .... **(Eubriinae)** ***Psephenidae*** p. 395

190(189). Anal gill tufts absent; anal hooks absent; body cylindrical and smooth or spiracles undulate (fig. 209) 191

190′. Anal gill tufts present (fig. 188); anal hooks 1 on each side (fig. 188); body somewhat flattened and more or less granulate dorsally; spiracles biforous (fig. 202) 192

191(190). Cardines, if present, separated from each other by labium; dorsal surfaces generally smooth; vestiture consisting of fine hairs or setae only; spiracles biforous (fig. 202). In leaf litter, flood debris, soil .... ***Dryopidae*** p. 399

191′. Cardines completely fused forming single plate (fig. 97); dorsal surfaces generally granulate or tuberculate; vestiture including pubescent hairs (fig. 5); spiracles undulate (fig. 209). In leaf litter, flood debris, ant refuse heaps .... (part) ***Chelonariidae*** p. 394

192(190′). Abdomen with 4 pairs of pleurites; mesal surface of mandibular base with short brush of hairs, sometimes not visible; stemmata forming loose cluster; cardo absent (fused to stipes). In running water, burrowing in travertine (spring-deposited limestone) .... ***Lutrochidae*** p. 397

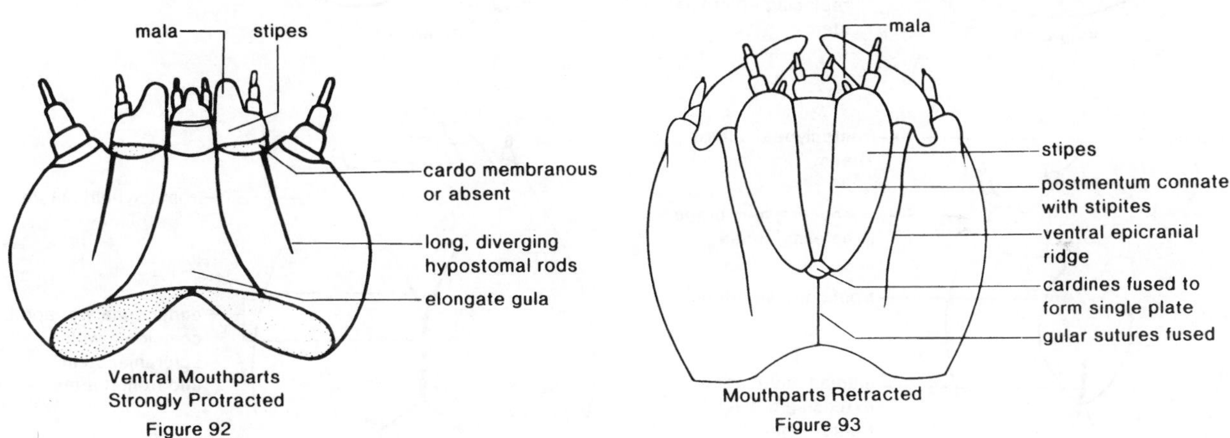

Ventral Mouthparts Strongly Protracted
Figure 92

Mouthparts Retracted
Figure 93

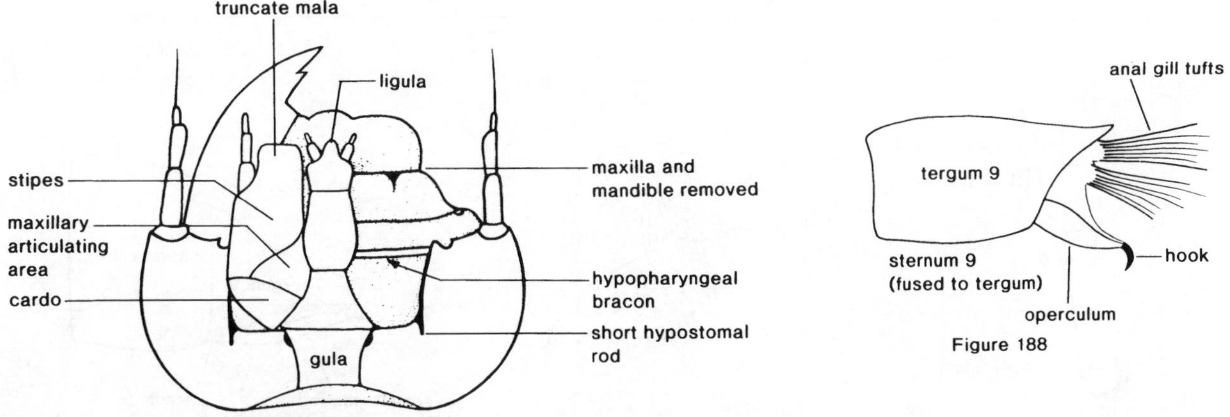

Mouthparts Retracted
Figure 85

Figure 188

192′. Abdomen with 6 or more pairs of pleurites (fig. 150); mesal surface of mandibular base usually with articulated, pubescent process, as well as brush of hairs (fig. 67); stemmata forming tight cluster; cardo distinct and sclerotized. On rocks, submerged wood, debris in streams, sometimes in lakes or pools .... ***Elmidae*** p. 404

193(115′). Median endocarina absent or coincident with epicranial stem (fig. 19) 194

193′. Median endocarina Y-shaped, coincident with epicranial stem and frontal arms (fig. 23) **(see 3rd and 4th choices)** 231

193². Median endocarina extending anterad of epicranial stem (fig. 20) 232

193³. Median endocarina located between frontal arms (fig. 22); epicranial stem absent 238

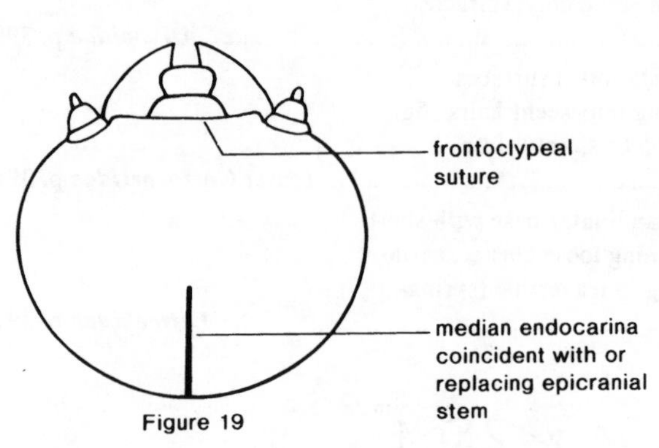

Figure 19

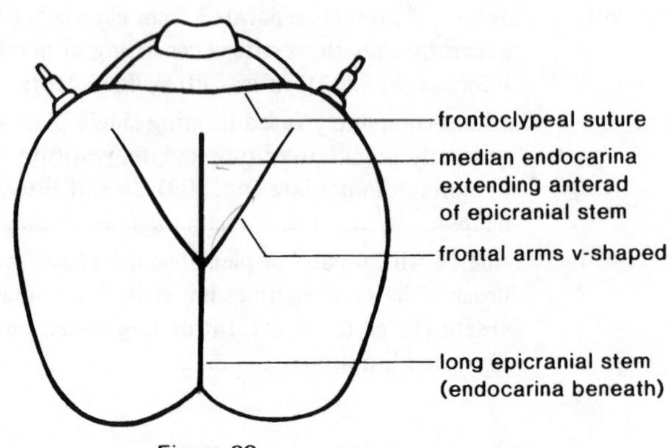

Figure 20

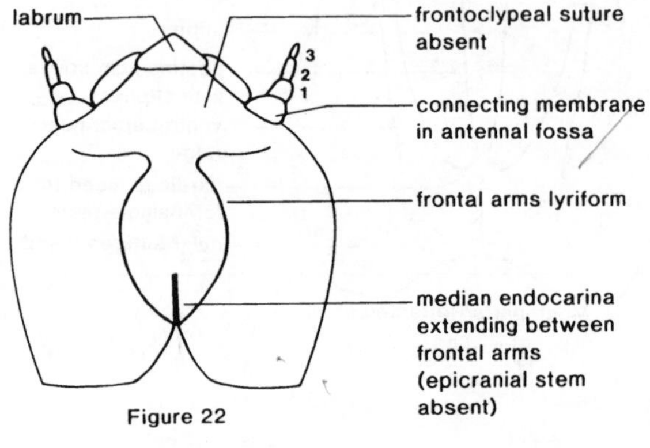

Figure 22

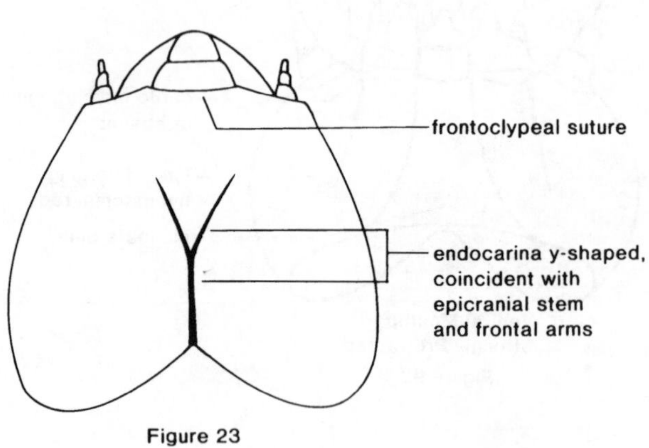

Figure 23

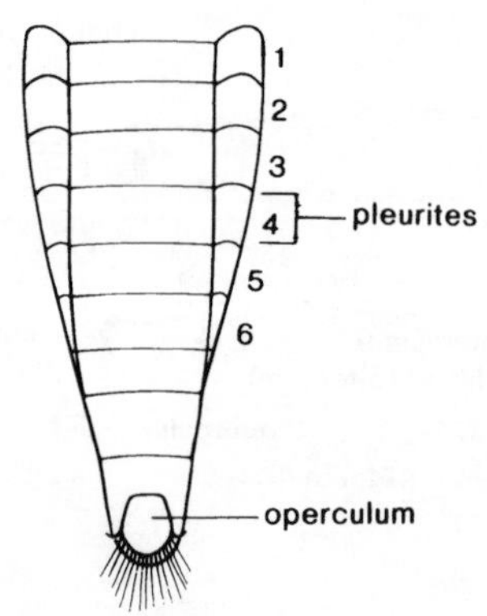

Figure 150

Figure 151

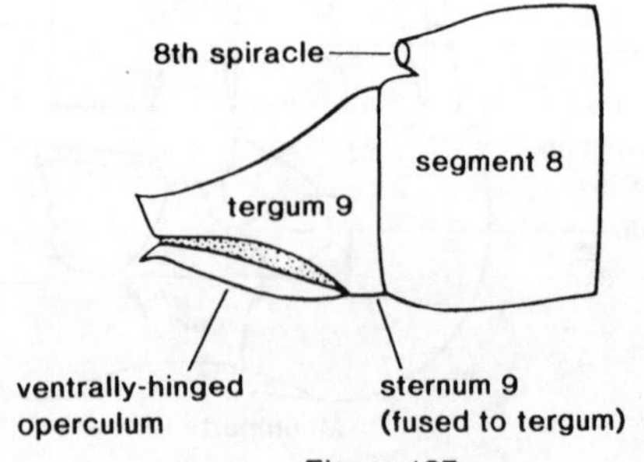

Figure 187

194(193). Abdominal apex without hinged operculum 195
194′. Abdominal apex with ventrally hinged operculum (fig. 187) 230
195(194). Mesal surface of mandibular base simple or slightly expanded (figs. 44, 46, 47) 196
195′. Mesal surface of mandibular base with brush of hairs (penicillus) (fig. 53) **(see next 4 choices)** 208
195². Mesal surface of mandibular base with *articulated,* pubescent process (fig. 67); paired ventral abdominal gill tufts (fig. 151) on segments A1–7. In debris or soil on river or stream bottoms. California .... ***Eulichadidae*** p. 390
195³. Mesal surface of mandibular base with *fixed,* rigid, hyaline process, sometimes partly sclerotized (figs. 58, 61, 82) 209
195⁴. Mesal surface of mandibular base with 2 to 5 hyaline processes, sometimes joined basally (figs. 59, 60) 216
195⁵. Mesal surface of mandibular base with membranous lobe (fig. 64) 226

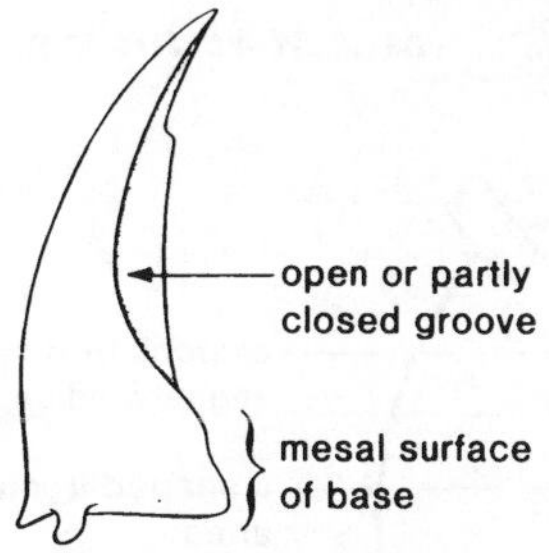

Figure 44

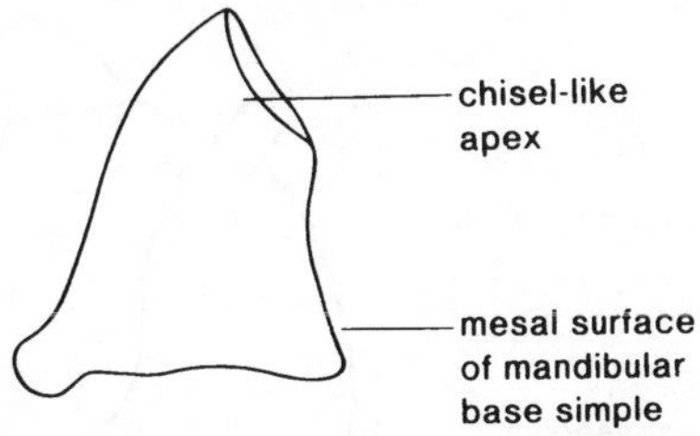

Figure 46

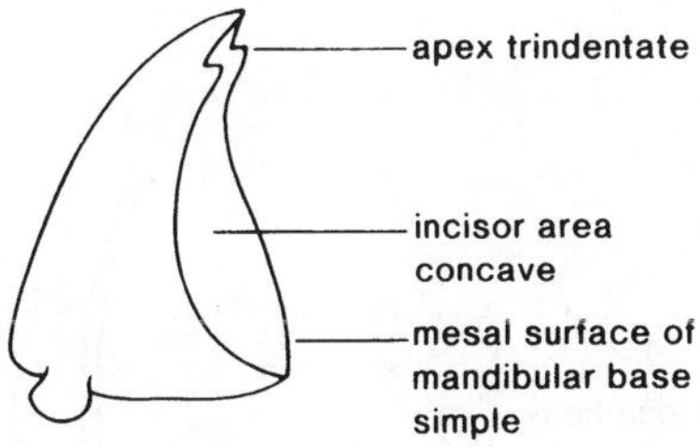

Figure 47

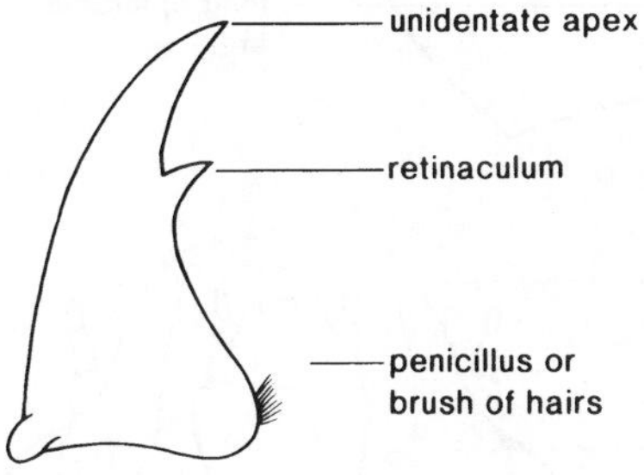

Figure 53

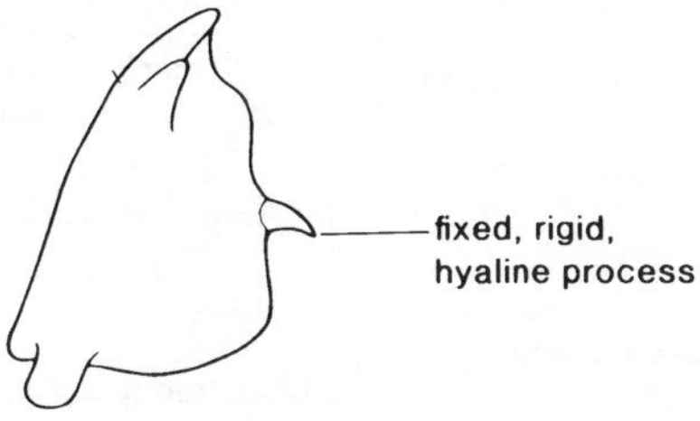

Figure 58

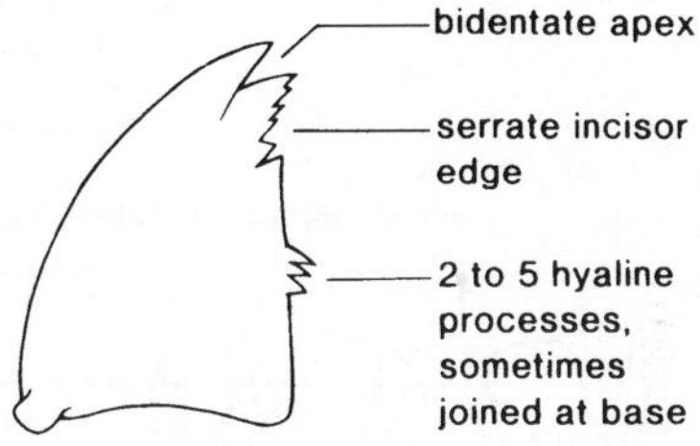

Figure 59

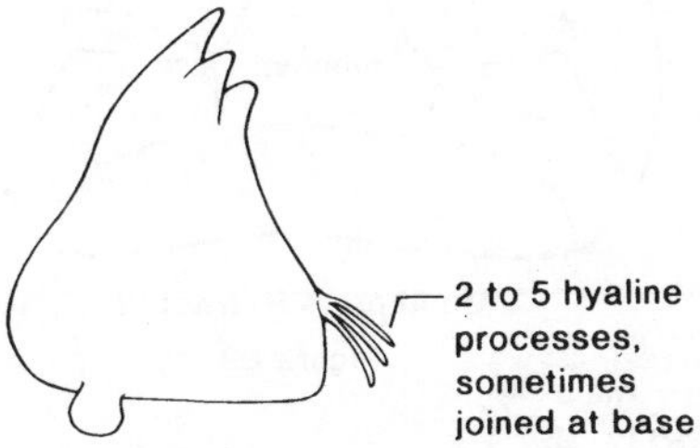

Figure 60

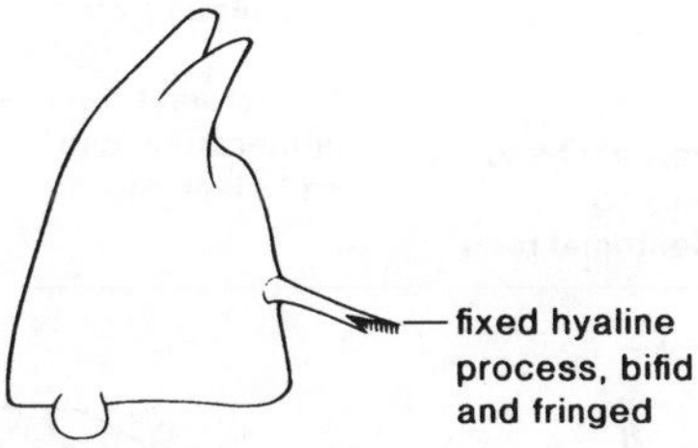

Figure 61

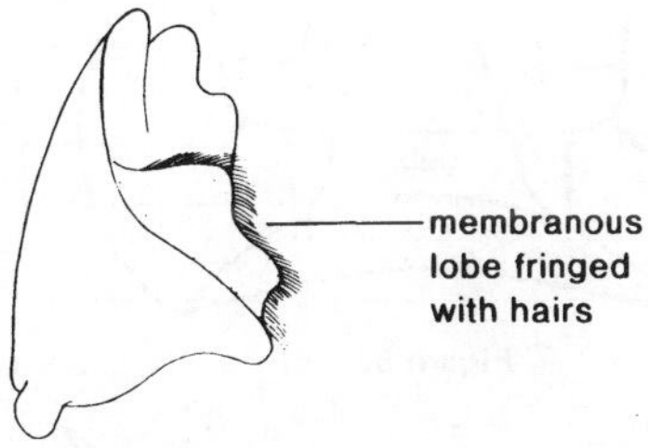

Figure 64

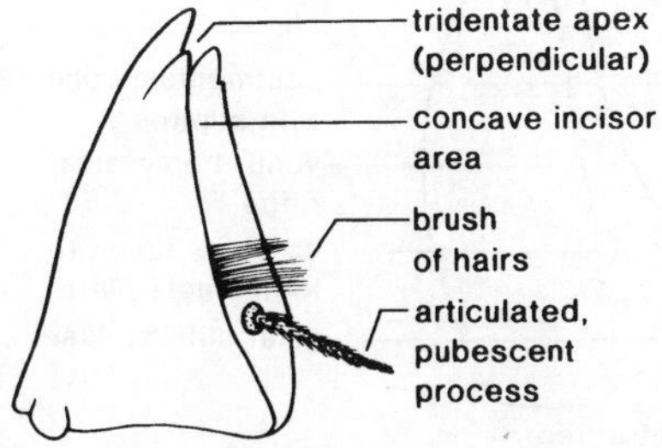

Figure 67

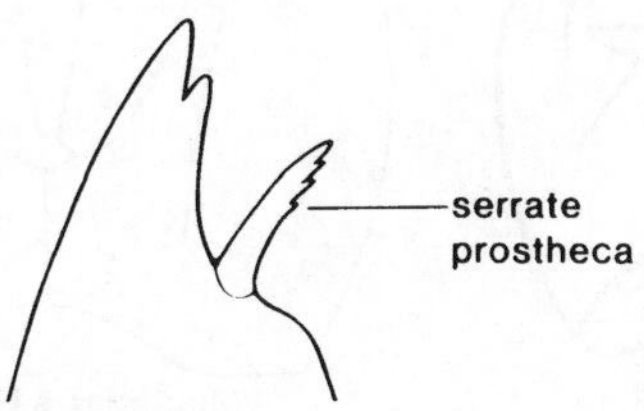

Figure 82

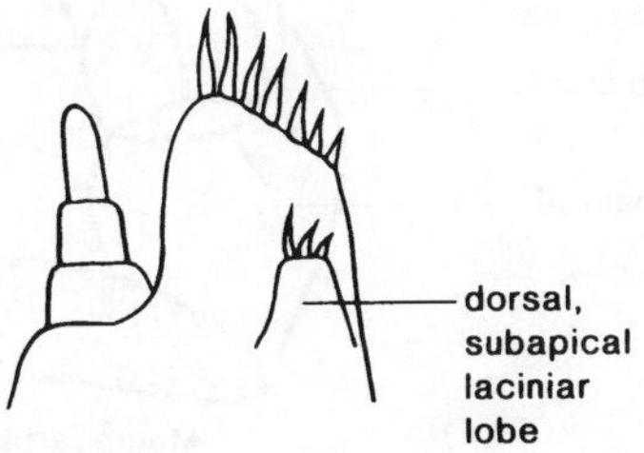

Figure 110

Elmidae

196(195). Apex of mala or galea rounded or truncate (figs. 88, 109, 110) 197

196′. Apex of mala or galea falciform (figs. 89, 105, 111) or stylet-like (fig. 104); urogomphi usually articulated at base (figs. 162, 163) **(see 3rd choice)** 205

196². Apex of mala trilobed (fig. 107); urogomphi articulated at base (fig. 162); labrum more or less fused to clypeus, but with vague indication of clypeolabral suture (fig. 14). In rotten mushrooms .... (*Oxyporus*) **(Oxyporinae)** ***Staphylinidae*** p. 341

197(196). Antennal segments 1 or 2; body elongate and subcylindrical, lightly sclerotized; head moderately to strongly declined (hypognathous) (fig. 10) 198

197′. Antennal segments 3 **(see 3rd choice)** 199

197². Antennal segments 4; urogomphi articulated at base (fig. 162). In leaf litter, rotten wood, carrion, dung, under bark .... (part) ***Staphylinidae*** p. 341

198(197). Segments in T2 leg 3 or 4, the segments indistinctly separated (fig. 128); ventral epicranial ridges absent; antennal sensorium not or only slightly longer than terminal antennal segment. In rotten wood, stems, fungus fruiting bodies .... (part) ***Mordellidae*** p. 508

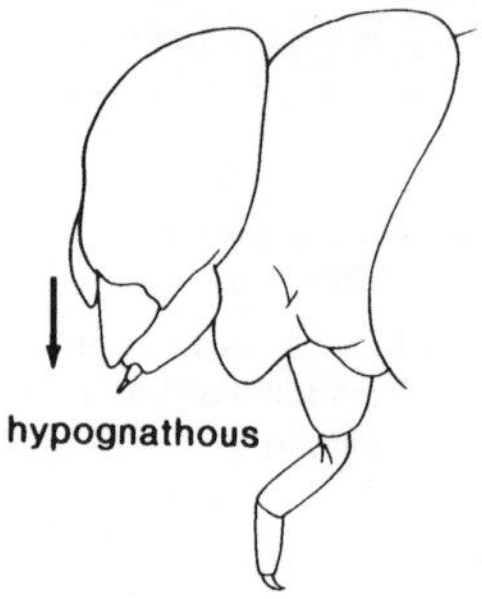

Figure 10

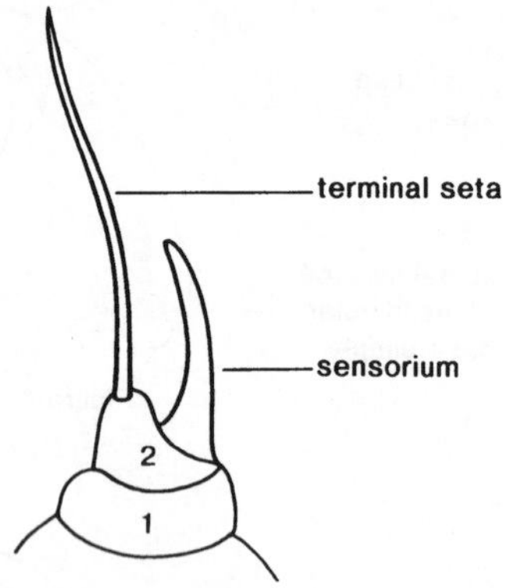

Figure 32

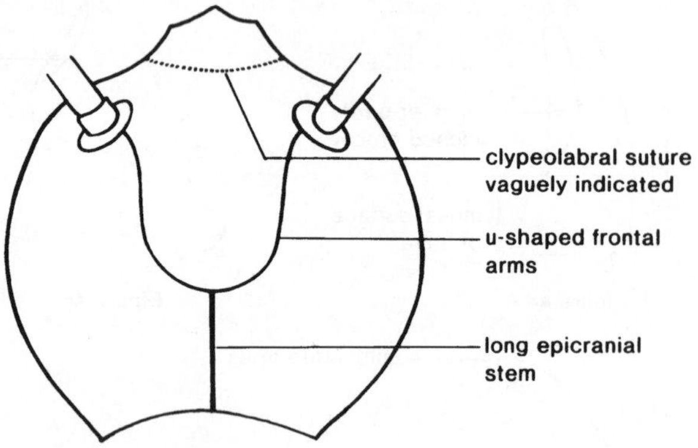

Figure 14

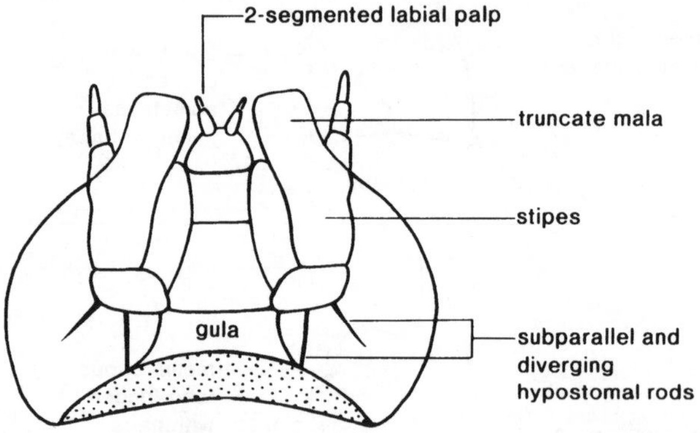

Figure 88

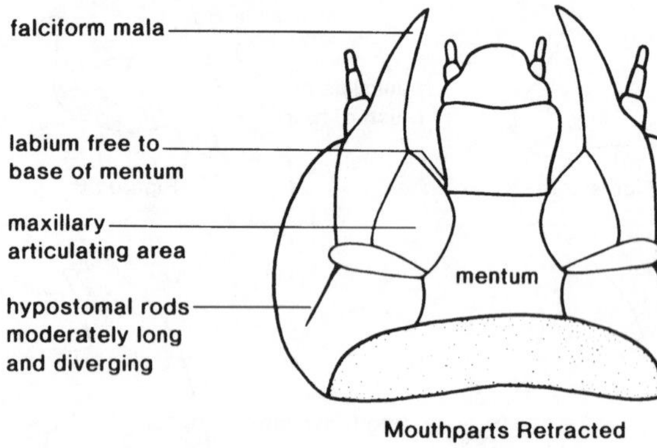

Mouthparts Retracted

Figure 89

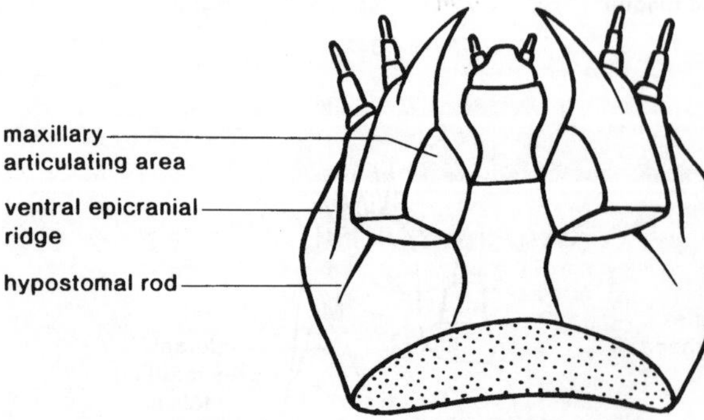

Mouthparts Retracted

Figure 90

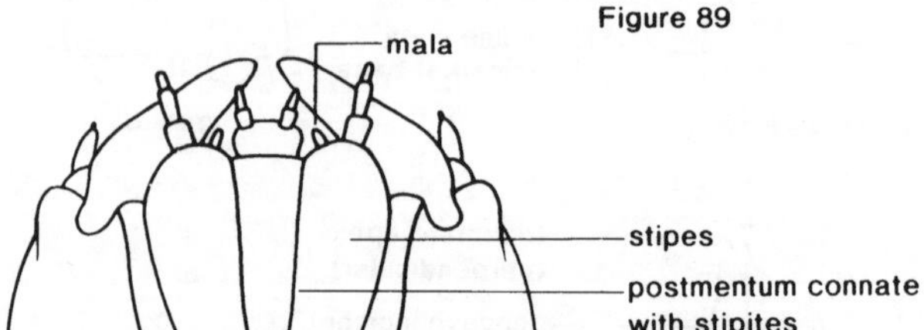

Mouthparts Retracted

Figure 93

← Fig. 110

198′. Segments in T2 leg 5 including tarsungulus (fig. 126); ventral epicranial ridges present (figs. 90, 93); antennal sensorium much longer than terminal (2nd) antennal segment (fig. 32). In fungus fruiting bodies .... (**Ciinae** part) ***Ciidae*** p. 502

199(197′). Maxilla with single mala (figs. 88, 89) 200

199′. Maxilla with separate galea and lacinia (figs. 109, 111) 204

200(199). Apex of mandible truncate and lined with rows of spines (fig. 55); body somewhat flattened and lightly sclerotized; urogomphi fixed at base. On fungus covered logs .... ***Dasyceridae*** p. 335

200′. Apex of mandible not truncate or lined with rows of spines 201

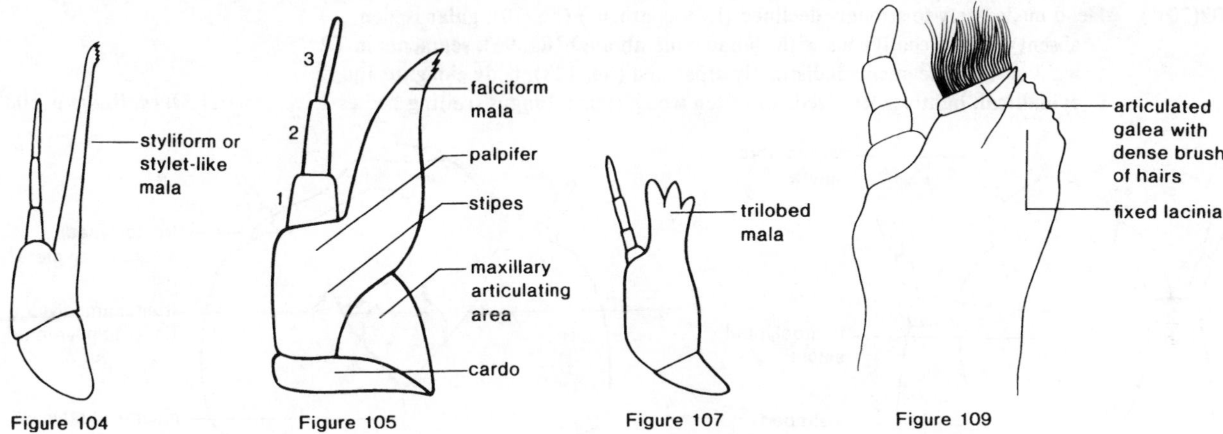

Figure 104 Figure 105 Figure 107 Figure 109

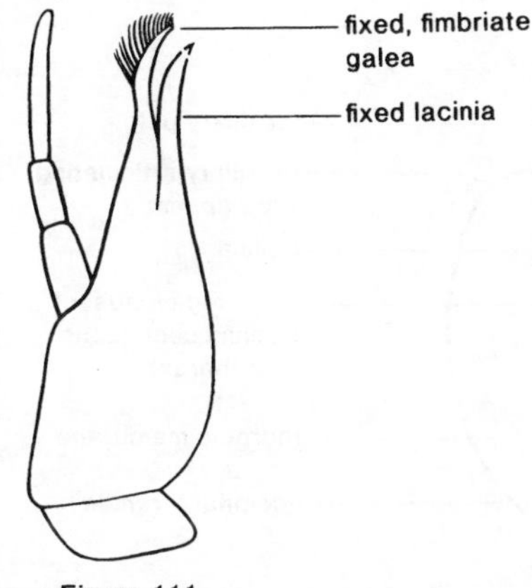

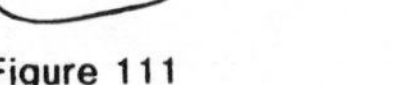
Figure 111

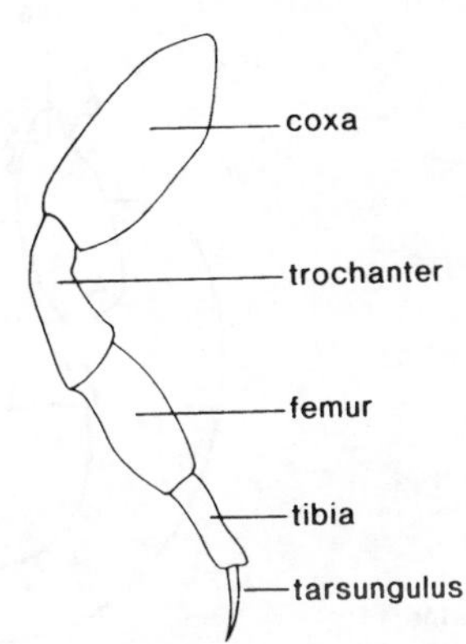

Figure 126

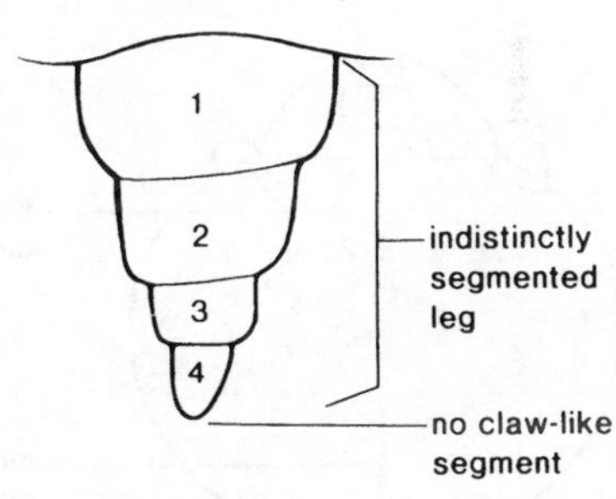

Figure 128

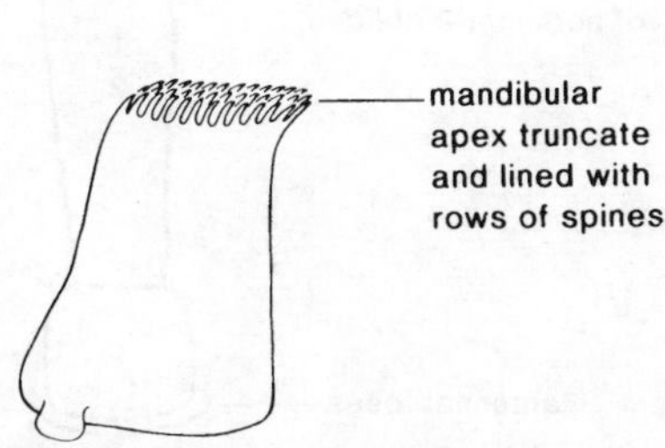

Figure 55

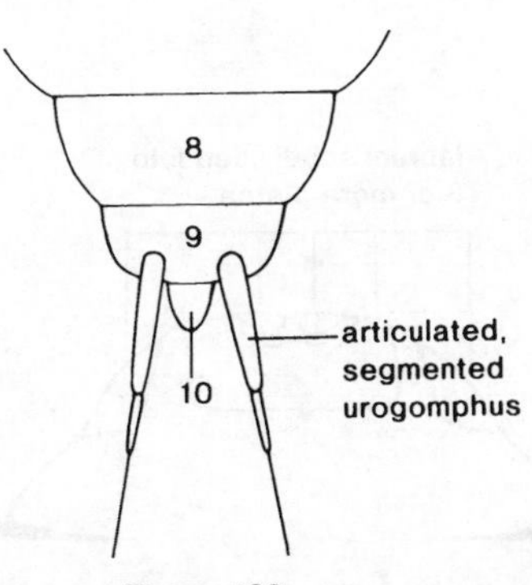

Figure 162

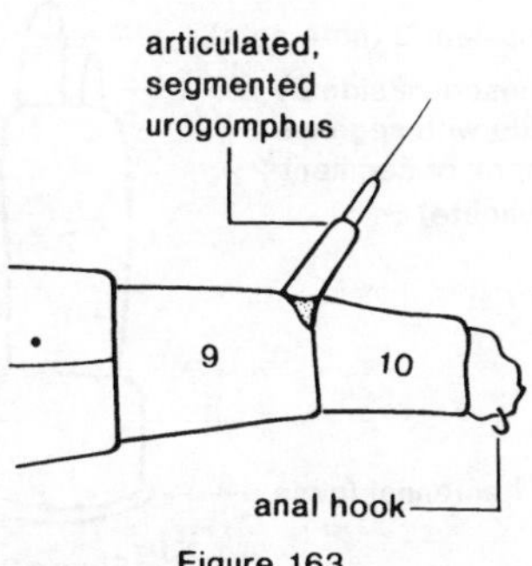

Figure 163

201(200′). Apex of antennal segment 2 oblique, so that sensorium arises proximad of segment 3 (fig. 30); frontoclypeal suture absent; hypostomal rods absent; segment A10 distinct and usually visible from above (fig. 162); urogomphi often articulated at base (fig. 162); abdominal terga usually bearing distinct plates, which are set off from those on sterna and are never asperate. In leaf litter, rotten wood, carrion, dung, under bark .... (part) ***Staphylinidae*** p. 341

201′. Apex of antennal segment 2 truncate, so that sensorium and segment 3 arise together (fig. 29); frontoclypeal suture (figs. 13, 19) and hypostomal rods (fig. 89) present, OR abdominal terga with patches or rows of asperities; segment A10 reduced, not visible from above; urogomphi fixed at base; abdominal terga and sterna not forming distinct plates 202

202(201′). Head moderately to strongly declined (hypognathous) (fig. 10); gular region absent (labium contiguous with thoracic membrane) (fig. 96); segments in T2 leg 3 or 4, the segments indistinctly separated (fig. 128); body elongate and cylindrical, lightly sclerotized. In rotten wood, stems, fungus fruiting bodies .... (part) ***Mordellidae*** p. 508

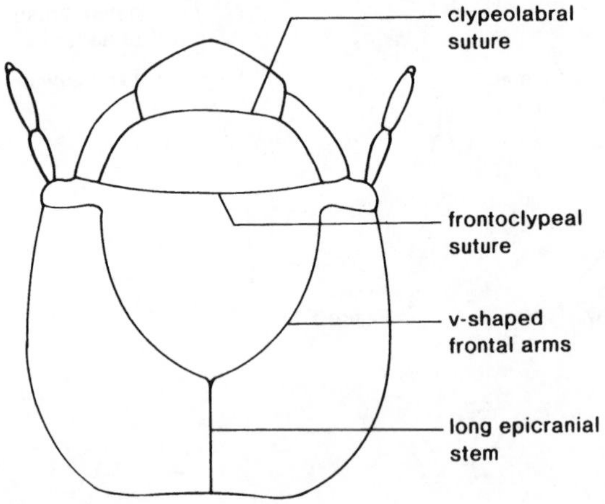

Figure 13

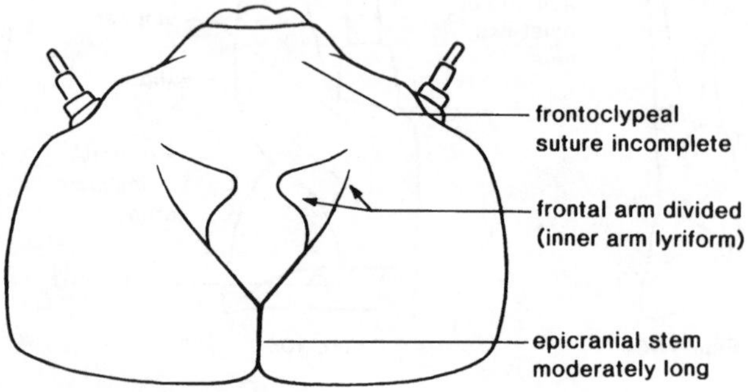

Figure 18

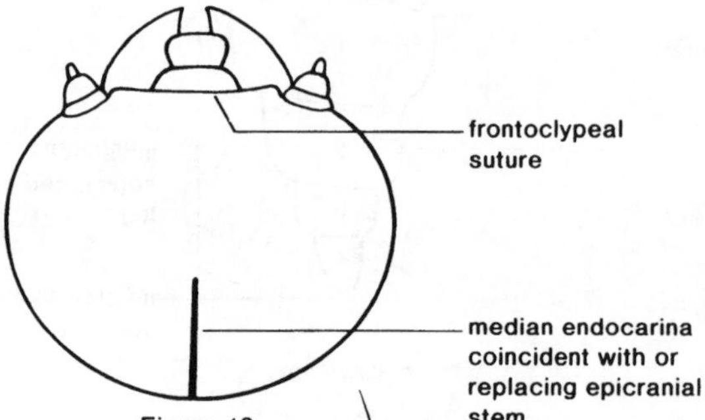

Figure 19

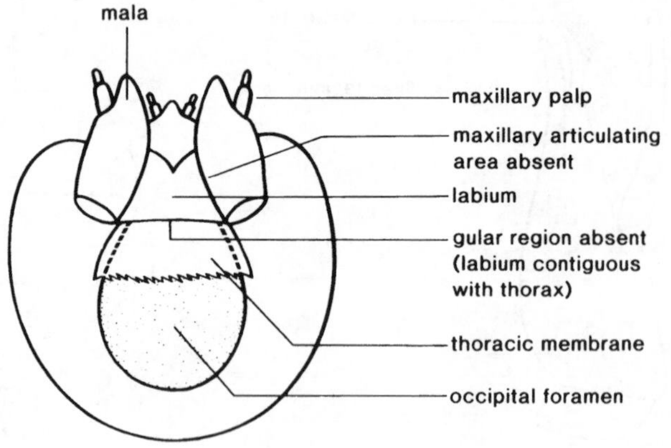

Figure 96

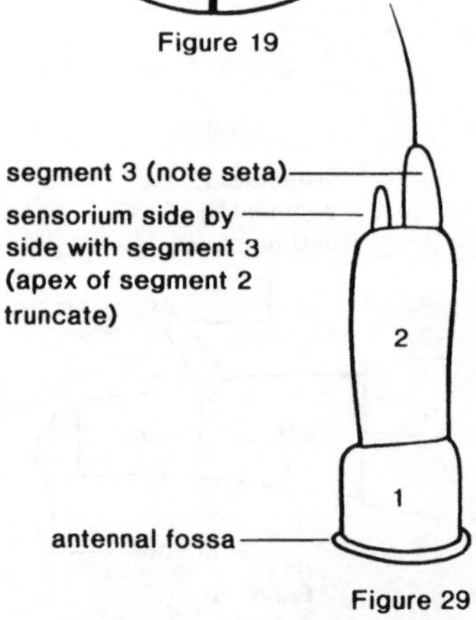

Figure 29

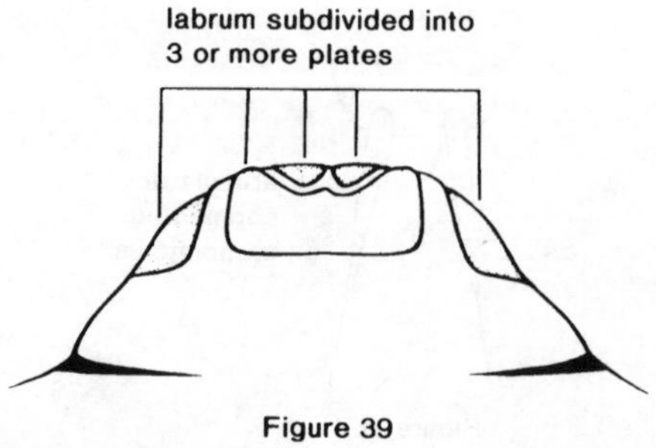

Figure 39

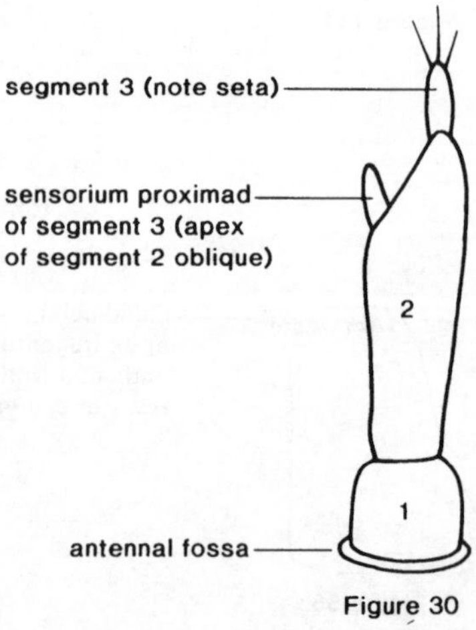

Figure 30

←— Figs. 10, 14, 89, 105, 109, 111, 162

202′. Head prognathous or slightly declined; gular region present (separating labium from thorax) (fig. 85); segments in T2 leg 5, including tarsungulus (fig. 126) 203

203(202′). Frontoclypeal suture present (figs. 13, 19); hypostomal rods almost always present (fig. 89); mesocoxae almost always separated by more than 2 coxal diameters; dorsal surfaces very lightly pigmented and without asperities. In rotten wood, fungus fruiting bodies .... (**Melandryinae** part) ***Melandryidae*** p. 505

203′. Frontoclypeal suture absent (fig. 18); hypostomal rods absent; mesocoxae separated by 1 to 2 coxal diameters; dorsal surfaces with pigmented maculae and rows or patches of asperities. Under bark, in soft fungi .... (*Penthe* part) (**Penthinae**) ***Tetratomidae*** p. 504

204(199′). Epicranial stem absent or very short; mandibles broad at base, not falcate; labrum not subdivided; abdominal segments without lateral plates or spine-like processes; galea without large, dense brush of hairs, but often fimbriate (with fringe of setae) (fig. 111). In leaf litter, carrion, fungi .... (part) ***Leiodidae*** p. 327

204′. Epicranial stem moderately long (fig. 14); mandibles narrow and falcate (fig. 44); labrum subdivided into 3 or more sclerites (fig. 39); abdominal segments produced laterally forming tergal plates or spine-like processes; galea with large, dense brush of hairs (fig. 109). In carrion, decaying vegetation .... ***Silphidae*** p. 339

205(196′). Epicranial stem absent or very short; maxilla with separate galea and lacinia (fig. 111) 206

205′. Epicranial stem moderately long (fig. 14); maxilla with single mala (fig. 105) 207

206(205). Head prognathous or slightly declined; segments in T2 leg 5 including tarsungulus (fig. 126); urogomphi articulated at base (fig. 162); dorsal surfaces generally smooth; thoracic and abdominal segments without lateral tergal processes; mandible without subapical pseudomola. In leaf litter, carrion, fungi .... (part) ***Leiodidae*** p. 327

206′. Head moderately to strongly declined (hypognathous) (fig. 10); segments in T2 leg 3 or 4; urogomphi fixed at base; dorsal surfaces generally spinose or complexly sculptured; thoracic and abdominal segments each with 1 or 2 pairs of lateral tergal processes; mandible with subapical pseudomola consisting of several spines or teeth (fig. 54). In decaying vegetation .... ***Micropeplidae*** p. 334

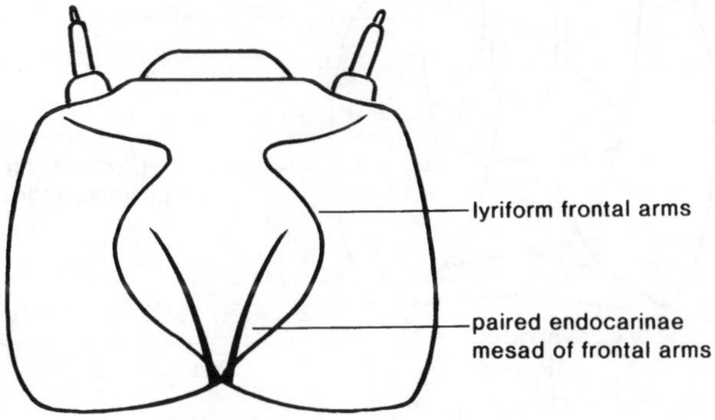

lyriform frontal arms
paired endocarinae mesad of frontal arms

Figure 25

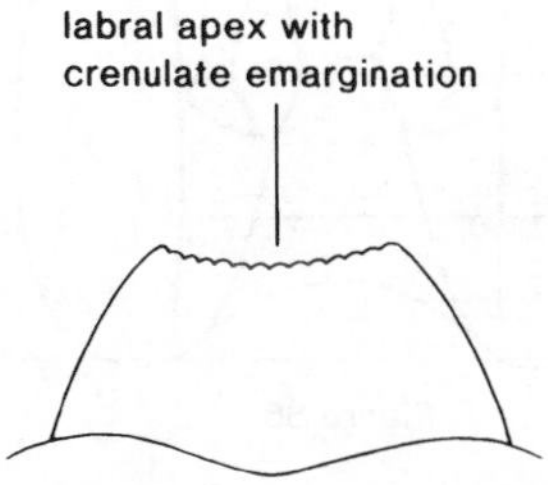

labral apex with crenulate emargination

Figure 37

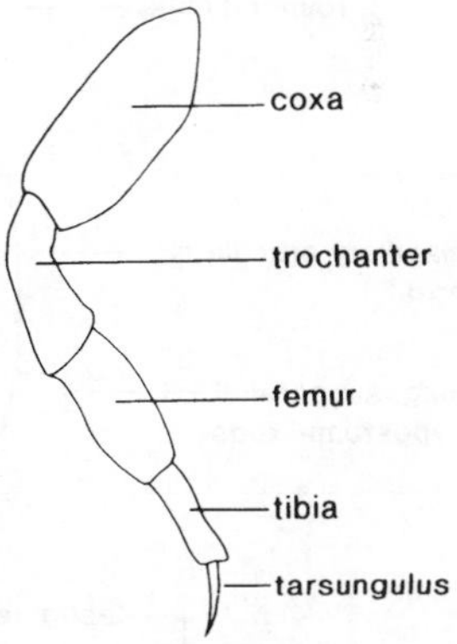

coxa
trochanter
femur
tibia
tarsungulus

Figure 126

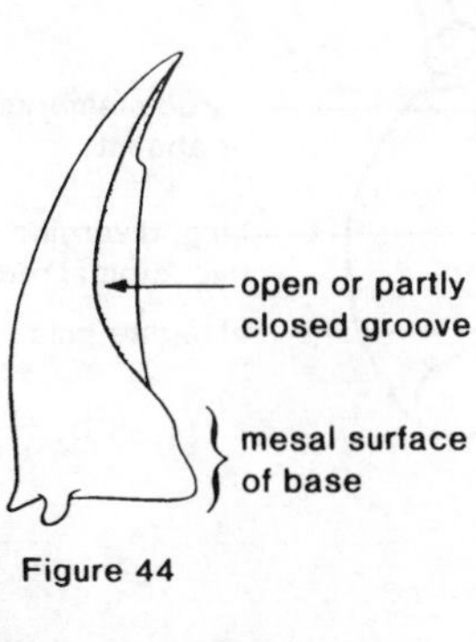

open or partly closed groove
mesal surface of base

Figure 44

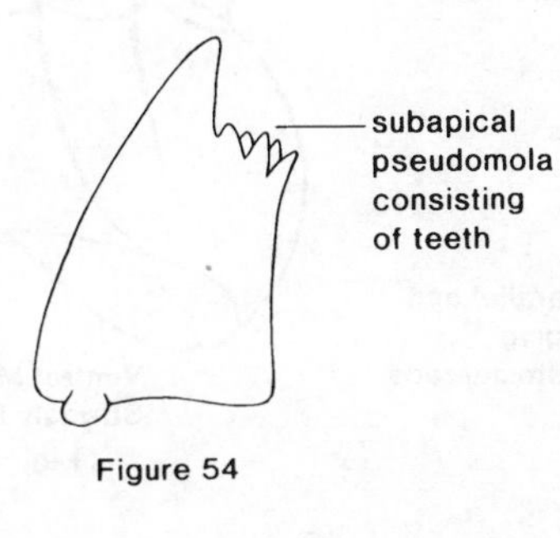

subapical pseudomola consisting of teeth

Figure 54

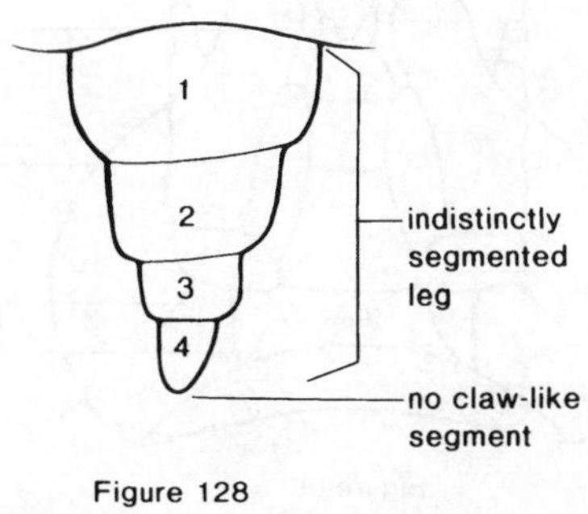

1
2
3
4
indistinctly segmented leg
no claw-like segment

Figure 128

Fig. 85 ⟶

207(205′). Labrum with crenulate emargination (fig. 37). On surfaces of fungi and bark, in leaf litter .... ***Scaphidiidae*** p. 337

207′. Labrum without crenulate emargination. In leaf litter, rotten wood, carrion, dung, under bark .... (part) ***Staphylinidae*** p. 341

208(195′). Mentum or postmentum not divided longitudinally; paired endocarinae located mesad of frontal arms (fig. 25); maxilla with single mala (fig. 88); segments in maxillary palp 3; ventral epicranial ridges absent; maxillary articulating area exposed (fig. 85); anal region without hooks or papillae. Under bark, in rotting stems .... (part) ***Monommidae*** p. 514

208′. Mentum or postmentum divided longitudinally into 3 parts (fig. 98); paired endocarinae absent; maxilla with separate galea and lacinia (fig. 98); segments in maxillary palp 4; ventral epicranial ridges present (fig. 93); maxillary articulating area concealed behind expanded mentum (fig. 98); anal region with several pairs of hooks and 1 or more papillae (fig. 189). In plant debris in or near streams .... (**Anchytarsinae** part) ***Ptilodactylidae*** p. 391

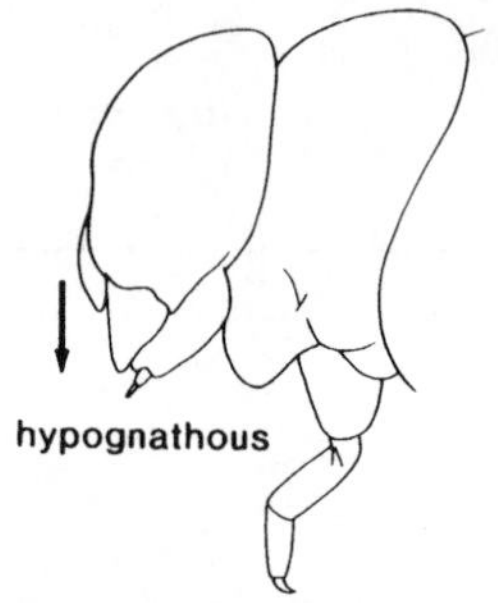

Figure 10

Figure 4

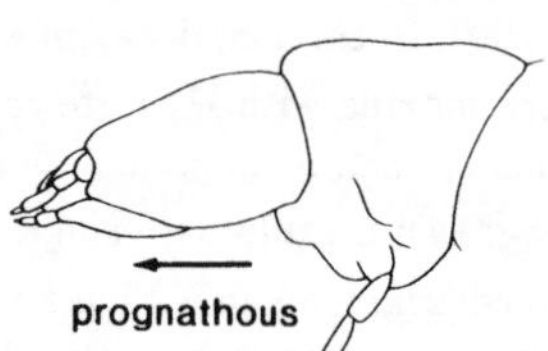

Figure 8

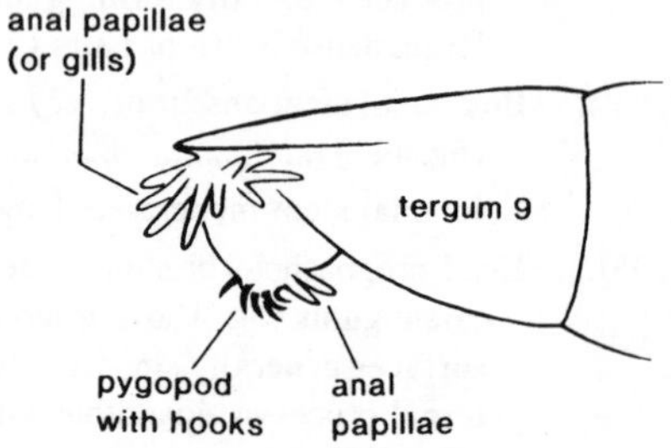

Figure 189

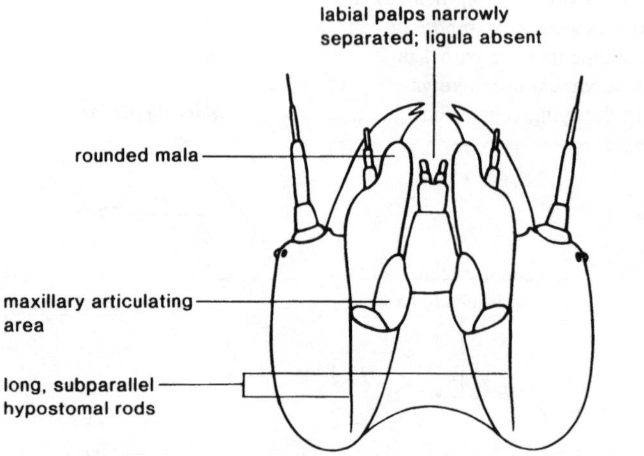

Figure 86

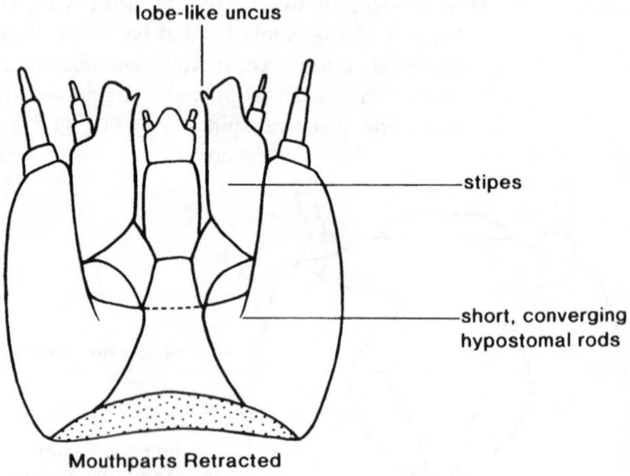

Mouthparts Retracted

Figure 87

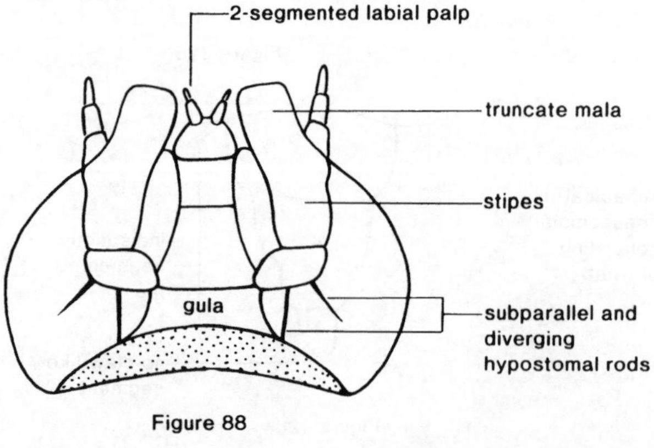

Figure 88

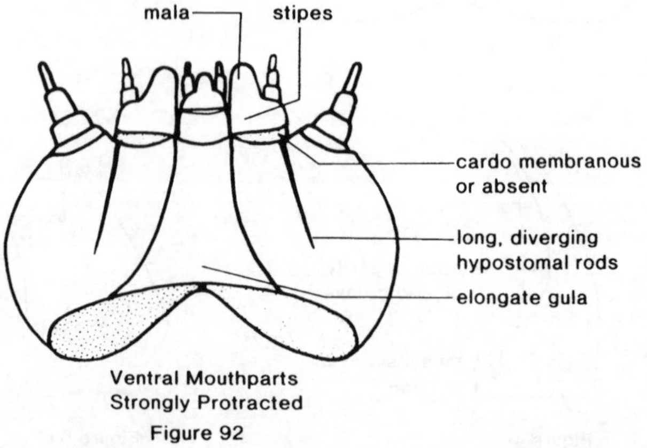

Ventral Mouthparts
Strongly Protracted

Figure 92

← Figs. 25, 37

209(195[3]). Antennal segments 2 210

209′. Antennal segments 3 212

210(209). Head prognathous or slightly declined (fig. 8); epicranial stem absent; labial palps separated by more than width of first palpal segment; ventral epicranial ridges absent; body with enlarged abdomen (physogastric). Usually in tunnels of woodboring insects (ectoparasitic) 211

210′. Head moderately to strongly declined (hypognathous) (fig. 10); epicranial stem moderately long; labial palps contiguous or separated by less than width of first palpal segment; ventral epicranial ridges present (fig. 93); body subcylindrical or slightly flattened, without enlarged abdomen. In fungus fruiting bodies .... (**Ciinae** part) ***Ciidae*** p. 502

211(210). Ventral mouthparts strongly protracted (fig. 92); hypostomal rods distinct, diverging (fig. 92) .... (**Passandrinae** part) ***Cucujidae*** p. 463

211′. Ventral mouthparts retracted (fig. 87); hypostomal rods absent .... (**Bothriderinae** part) ***Bothrideridae*** p. 477

212(209′). Head prognathous or slightly declined (fig. 8); maxilla with single mala (figs. 86, 92, 104); stemmata present 213

212′. Head moderately to strongly declined (hypognathous) (fig. 10); maxilla with separate galea and lacinia (fig. 215); stemmata absent. In leaf litter, stored products, ant nests .... (**Thorictinae**) ***Dermestidae*** p. 434

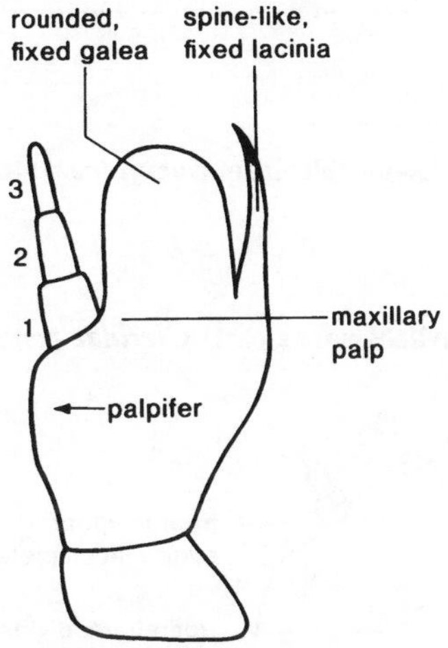

Figure 215

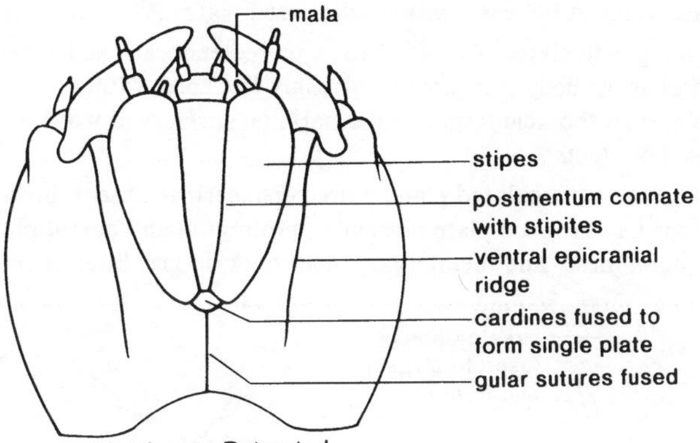

Mouthparts Retracted

Figure 93

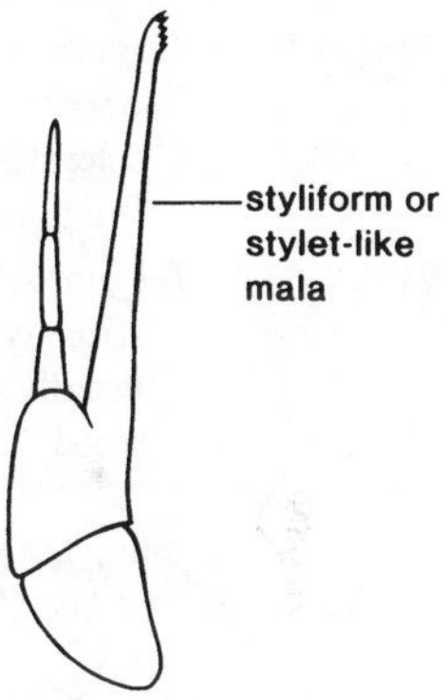

Figure 104

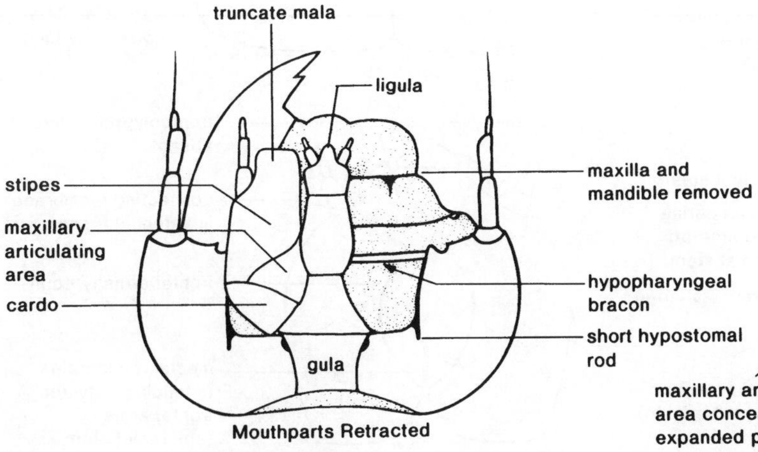

Mouthparts Retracted

Figure 85

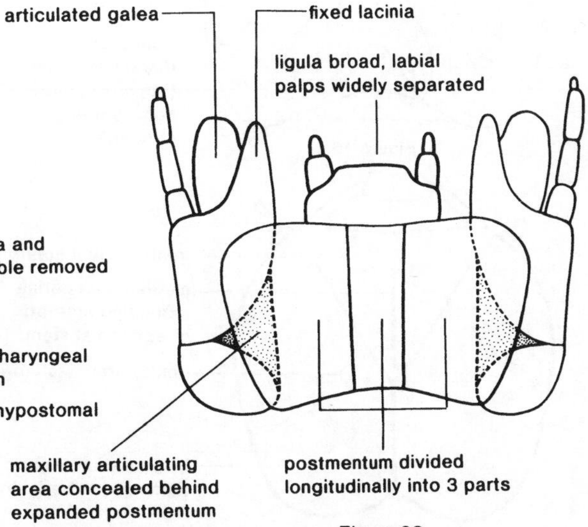

Figure 98

213(212). Urogomphi articulated at base (fig. 162); mala stylet-like (fig. 104); vestiture including frayed setae (fig. 4); stemmata on each side 3 or 6. In leaf litter .... **(Proteininae)** ***Staphylinidae*** p. 341

213′. Urogomphi fixed at base; mala rounded or truncate (figs. 86, 92); vestiture consisting of simple setae 214

214(213′). Epicranial stem absent and frontal arms lyriform (fig. 22); posterior edge of head capsule distinctly emarginate dorsally; body strongly flattened. Under bark, in rotten wood .... (part) ***Inopeplidae*** p. 551

214′. Epicranial stem moderately long and frontal arms V- or U-shaped (fig. 20); posterior edge of head capsule not or slightly emarginate dorsally; body not or only slightly flattened 215

215(214′). Hypostomal rods well-developed and extending almost to posterior edge of head (fig. 86); stemmata on each side 6; paired dorsal abdominal glands absent. In sand dunes. Exotic (Australia, New Zealand, New Caledonia) .... ***Phycosecidae*** p. 452

215′. Hypostomal rods absent; stemmata on each side 5 or fewer; abdominal segments with 1 or more pairs of dorsal glands (fig. 140). In leaf litter, soil, on ground, under bark, in stems .... ***Melyridae*** p. 453

216(195[4]). Epicranial stem absent 217

216′. Epicranial stem moderately long (fig. 18) 224

217(216). Ventral mouthparts strongly protracted; stipes wider than long (fig. 92); frontal arms distant at base (fig. 16); terga A1–8 never with sclerotized plates 218

217′. Ventral mouthparts retracted; stipes longer than wide (fig. 89); if frontal arms distant at base, terga A1–8 each with a sclerotized plate 219

218(217). Tergum A9 forming articulated plate (fig. 165); paired endocarinae located beneath frontal arms; body elongate and strongly flattened, without sclerotized plates on thoracic terga. Under bark, on surfaces of wood or fungi, in stored products .... **(Laemophloeinae)** ***Cucujidae*** p. 463

218′. Tergum A9 not forming articulated plate; paired endocarinae absent; body not strongly flattened; sclerotized plate present on protergum and paired plates usually present on meso- and metaterga. Under bark, in leaf litter or rotten wood .... **(Phyllobaeninae** part) ***Cleridae*** p. 450

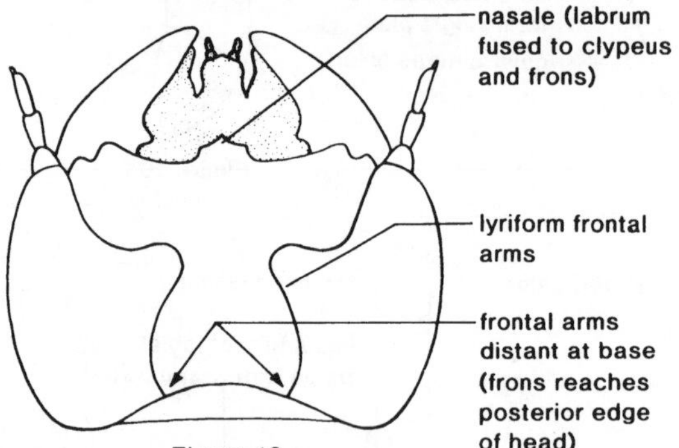

Figure 16

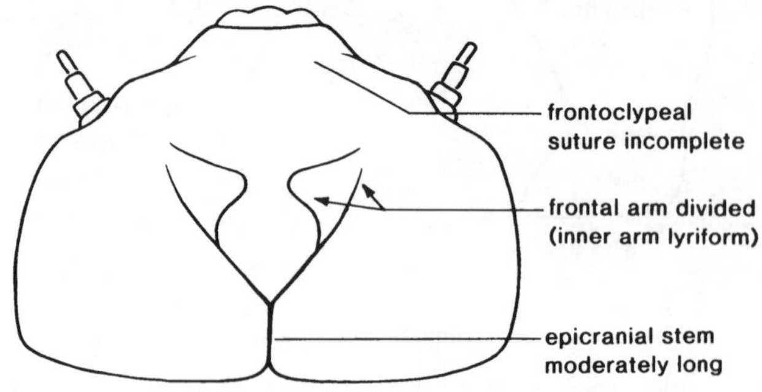

Figure 18

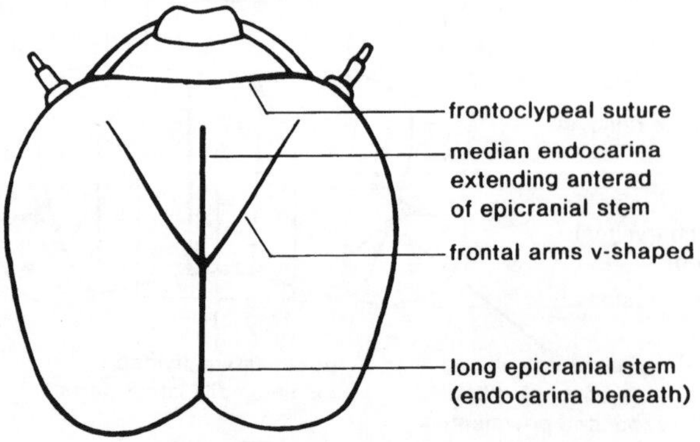

Figure 20

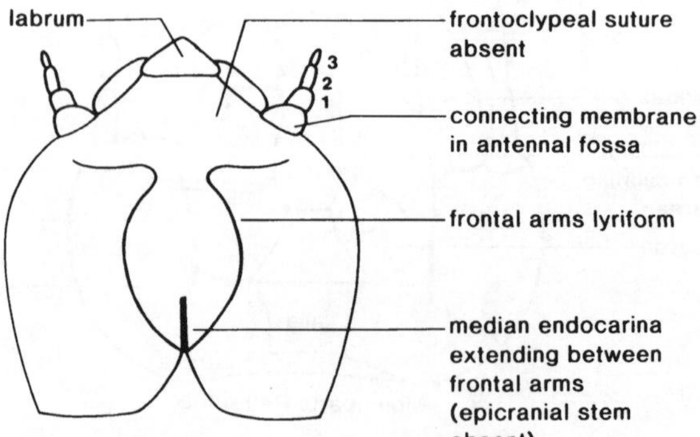

Figure 22

←—— Figs. 4, 86, 87, 92, 104

219(217′). Paired endocarinae absent; frontal arms distant at base; mesal surface of mandibular base with 2 hyaline processes, one acute and the other rounded and pubescent; thoracic terga and abdominal terga A1–9 each with a distinct, pigmented plate. In bird nests. Exotic (South temperate regions) .... (part) ***Cavognathidae*** p. 469

219′. Paired endocarinae present and frontal arms contiguous at base (figs. 24, 25); basal mandibular processes different; without pigmented plates on all thoracic and abdominal terga 220

220(219′). Paired endocarinae coincident with frontal arms (fig. 24); abdominal spiracles annular 221

220′. Paired endocarinae located mesad of frontal arms (fig. 25); abdominal spiracles annular-biforous (fig. 197) 223

221(220). Labial palps contiguous or separated by less than width of first palpal segment; hypostomal rods present (figs. 86, 89); spiracles not on sclerotized plates 222

221′. Labial palps separated by more than width of first palpal segment; hypostomal rods absent; spiracles located on sclerotized plates. In cracks on intertidal rocks (western North America) .... (***Aegialites***) (**Aegialitinae**) ***Salpingidae*** p. 549

222(221). Ratio of antennal length to head width less than .15; hypostomal rods diverging and moderately long (fig. 89); urogomphi simple (fig. 170). In rotten wood, fungus fruiting bodies .... (**Melandryinae** part) ***Melandryidae*** p. 505

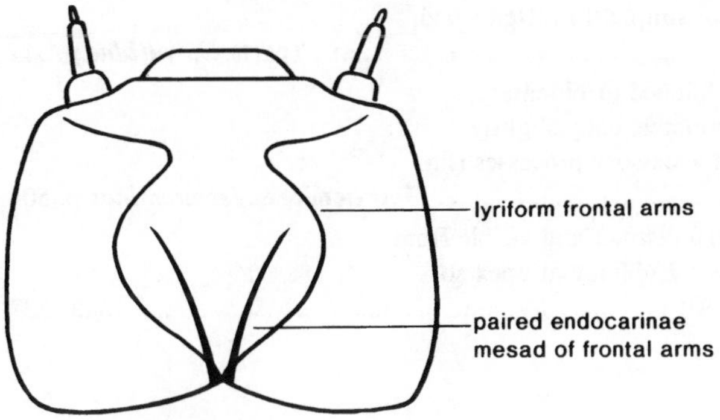

Figure 25

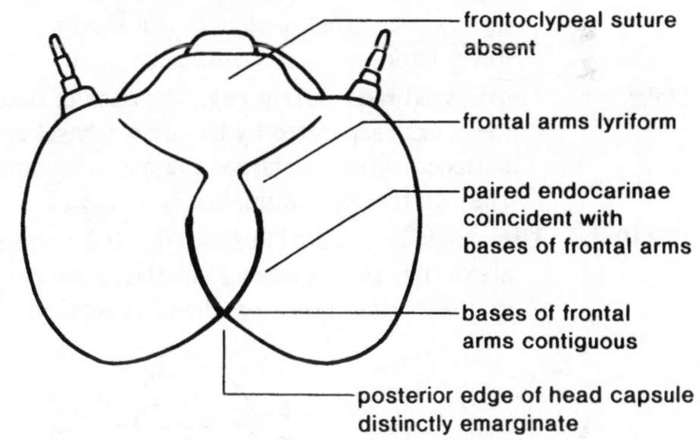

Figure 24

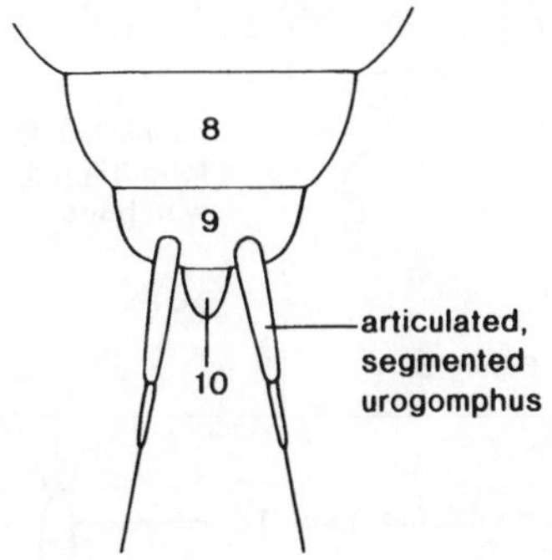

Figure 162

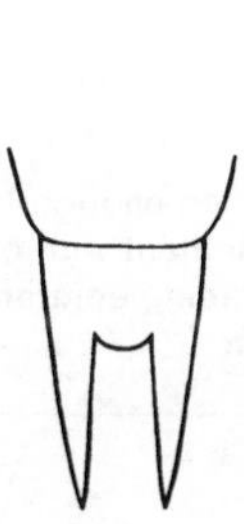
Dorsal
Figure 170

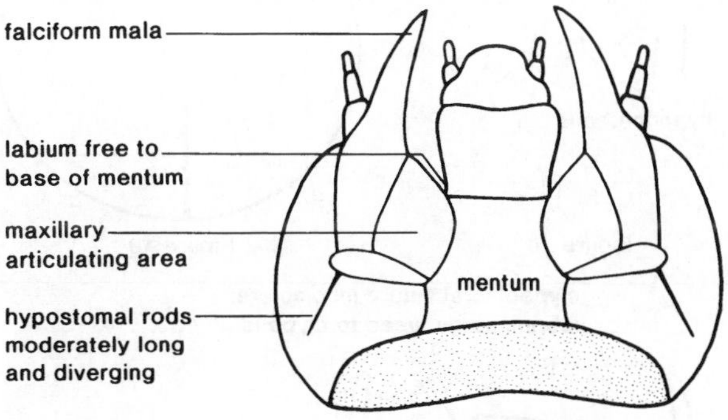

Mouthparts Retracted
Figure 89

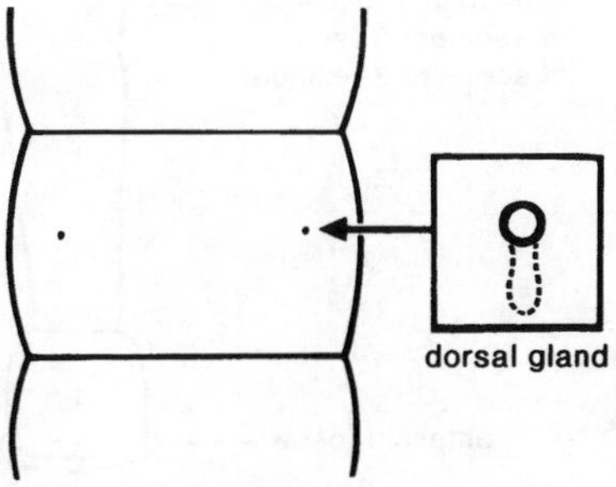

Figure 140

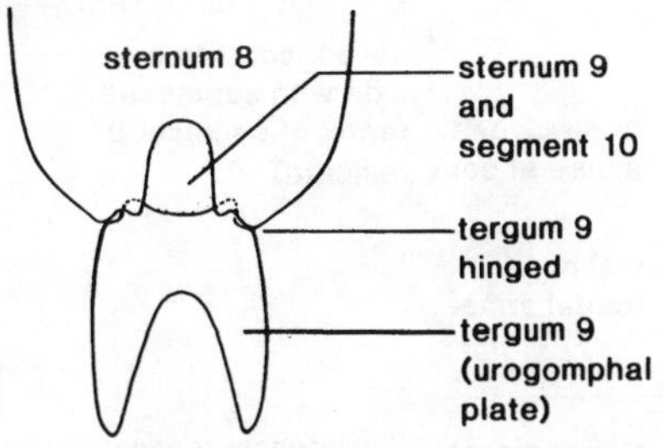

Ventral
Figure 165

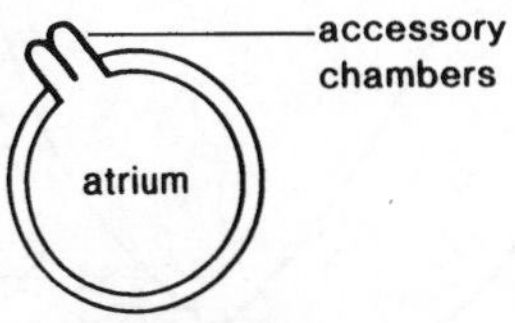

Annular Biforous
Figure 197

Fig. 86 ⟶

222′. Ratio of antennal length to head width .15 to .5; hypostomal rods subparallel and extending almost to posterior edge of head (fig. 86); urogomphi bifurcate (fig. 184). Under bark, in rotten stems, leaf litter .... (**Salpinginae** part) ***Salpingidae*** p. 549

223(220′). Frontal arms lyriform (fig. 25); ratio of antennal length to head width .15 to .5; labial palps separated by more than width of first palpal segment; hypostomal rods converging posteriorly (fig. 87); abdominal terga with paired rows of asperities forming open or closed rings (fig. 145); urogomphi simple with pit between them (fig. 173). Under bark, in rotting stems .... (part) ***Monommidae*** p. 514

223′. Frontal arms V- or U-shaped (fig. 12); ratio of antennal length to head width less than .15; labial palps contiguous or separated by less than width of first palpal segment; hypostomal rods subparallel (fig. 86); dorsal surfaces without asperities; urogomphi complex, with accessory processes (fig. 169) and without pit between them. In fungus fruiting bodies .... (*Thymalus*) (**Peltinae**) ***Trogossitidae*** p. 448

224(216′). Ratio of antennal length to head width .15 to .5; spiracles annular-biforous (fig. 197); setae on tarsungulus 2 225

224′. Ratio of antennal length to head width more than .5; spiracles annular; setae on tarsungulus 1; urogomphi with accessory processes (fig. 181). In tufts of tussock grass. Exotic (South temperate regions) .... ***Perimylopidae*** p. 520

225(224). Hypostomal rods absent; apex of mandible with single lobe or tooth; mesocoxae separated by 1 to 2 basal coxal diameters; body more or less cylindrical and lightly sclerotized, with enlarged thorax; urogomphi simple. In rotten wood, under bark .... (part) ***Colydiidae*** p. 512

225′. Hypostomal rods present (fig. 89); apex of mandible bilobed or bidentate; mesocoxae separated by less than 1 basal coxal diameter; body slightly flattened without enlarged thorax; urogomphi with accessory processes (fig. 169). In fungus fruiting bodies .... (**Tetratominae**) ***Tetratomidae*** p. 504

226(195[5]). Urogomphi articulated at base (fig. 162); segment A10 distinct and visible from above (fig. 162); spiracles annular; antennal segment 2 oblique at apex so that sensorium arises proximad of segment 3 (fig. 30) 227

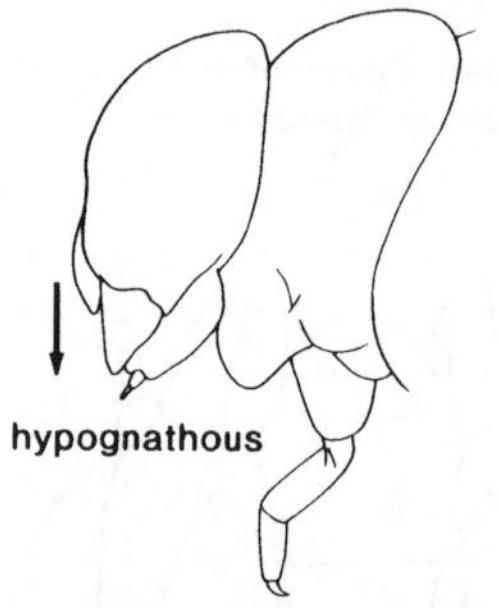

Figure 10

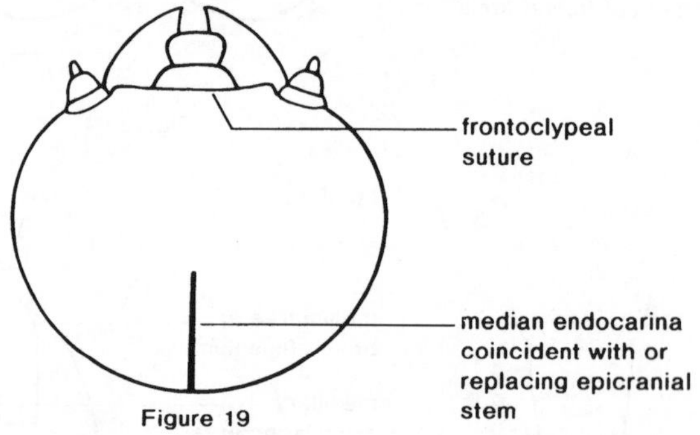

Figure 19

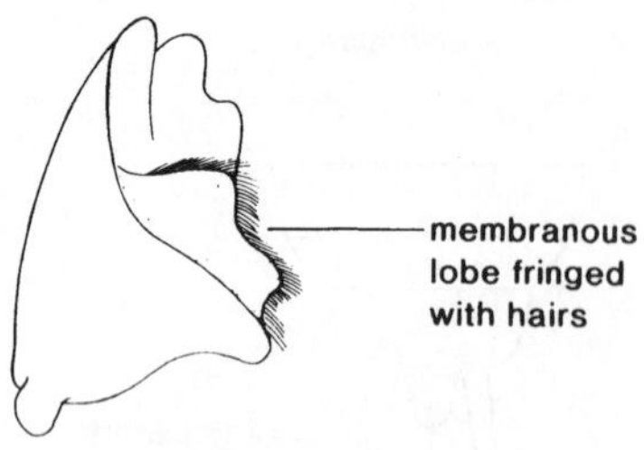

Figure 64

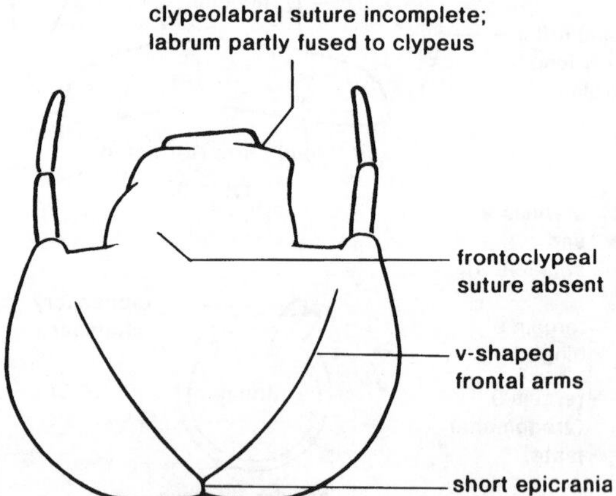

Figure 12

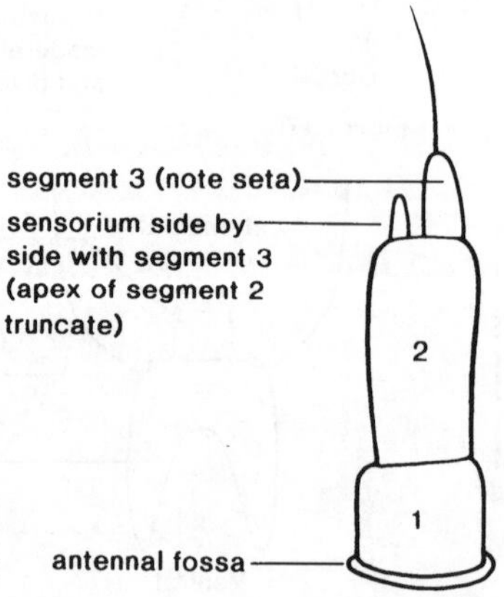

Figure 29

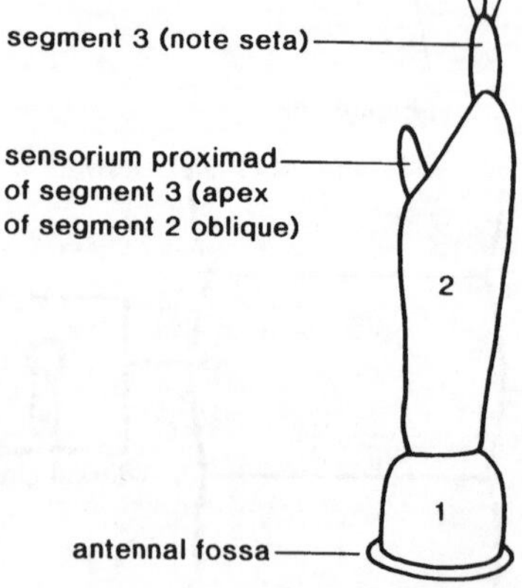

Figure 30

← Figs. 25, 89, 162, 197

226′. Urogomphi fixed at base; segment A10 concealed from above; spiracles annular-biforous (fig. 19) or dorsal surfaces granulate or spinose; antennal segment 2 truncate at apex, so that sensorium and segment 3 arise together (fig. 29) 228

227(226). Maxillary articulating area absent; frontal arms absent; labium consisting of a single plate; head strongly transverse; urogomphi unsegmented. In beaver nests, on beaver pelts .... (*Platypsyllus*) (**Platypsyllinae**) ***Leptinidae*** p. 330

227′. Maxillary articulating area present (fig. 85); frontal arms present; labium consisting of prementum, mentum, and submentum; head only slightly transverse; urogomphi segmented (fig. 162). In leaf litter, rotten wood, fungi .... (part) ***Leiodidae*** p. 327

228(226′). Head moderately to strongly declined (hypognathous) (fig. 10); apex of mala usually cleft (fig. 106); mandible usually tridentate at apex (fig. 64); if spiracles annular-biforous, labial palps separated by more than width of first palpal segment. In or on surfaces of fungus fruiting bodies .... (part) ***Erotylidae*** p. 473

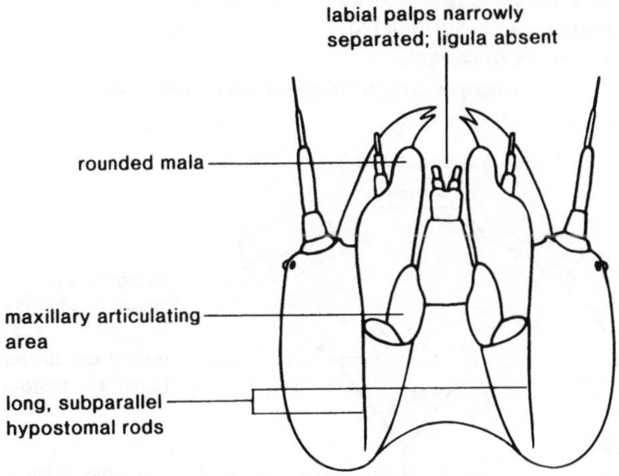

labial palps narrowly separated; ligula absent
rounded mala
maxillary articulating area
long, subparallel hypostomal rods

Figure 86

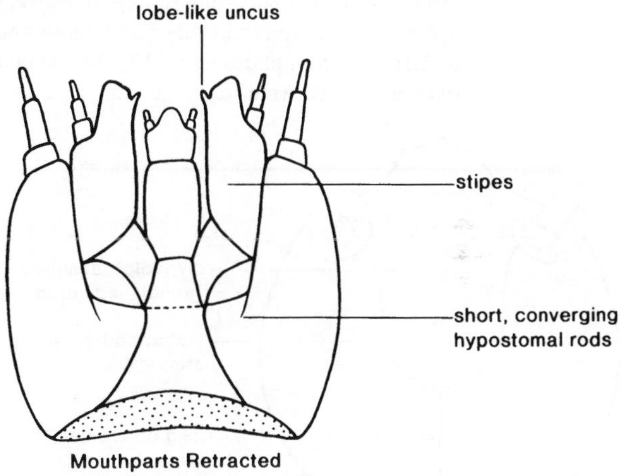

lobe-like uncus
stipes
short, converging hypostomal rods
Mouthparts Retracted

Figure 87

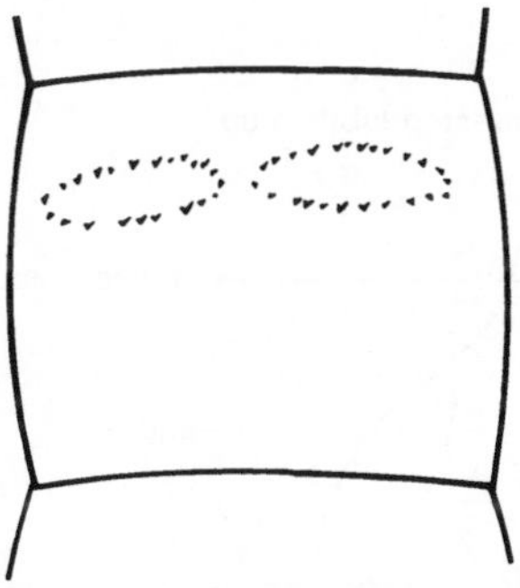

Figure 145

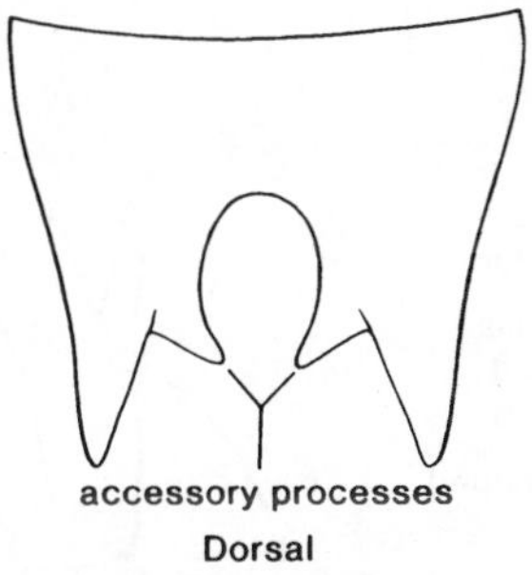

accessory processes
Dorsal

Figure 169

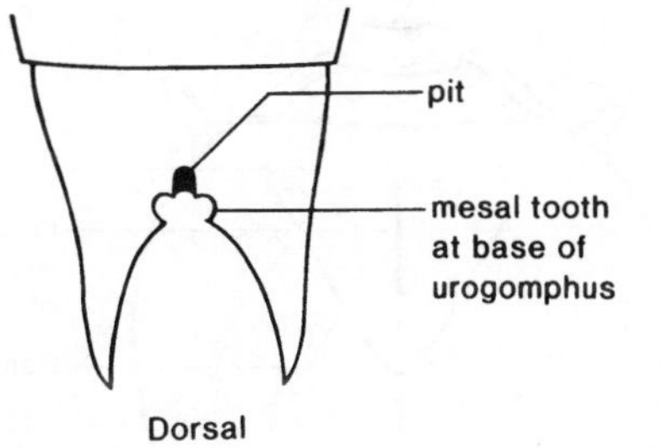

pit
mesal tooth at base of urogomphus
Dorsal

Figure 173

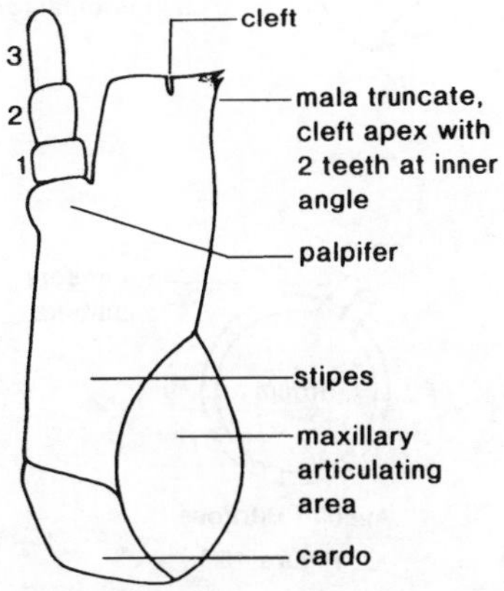

cleft
3
2
1
mala truncate, cleft apex with 2 teeth at inner angle
palpifer
stipes
maxillary articulating area
cardo

Figure 106

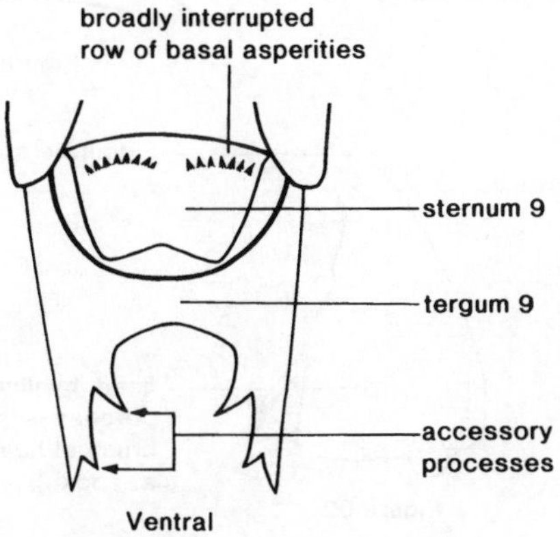

broadly interrupted row of basal asperities
sternum 9
tergum 9
accessory processes
Ventral

Figure 181

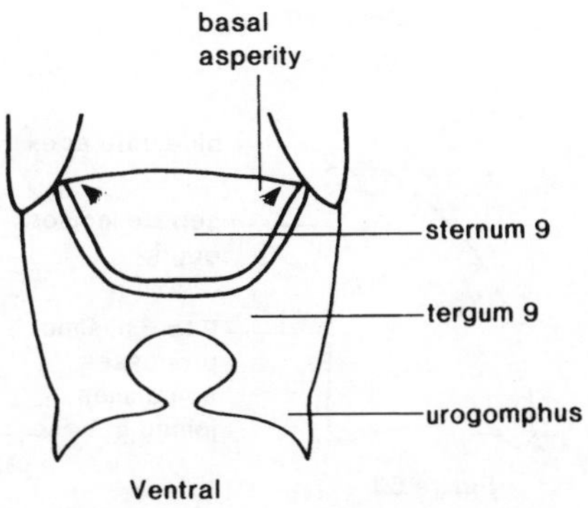

basal asperity
sternum 9
tergum 9
urogomphus
Ventral

Figure 184

Fig. 85: See p. 252

228′. Head prognathous or slightly declined; apex of mala not cleft; mandible bidentate at *apex* (fig. 62 (may have serrate edge below apex as in fig. 59)); spiracles annular-biforous (fig. 197); labial palps separated by less than width of first palpal segment 229

229(228′). Hypostomal rods present (2 pairs) (fig. 88); epicranial stem absent; ratio of antennal length to head width .15 to .5; abdominal terga without asperities. In fungus fruiting bodies, in rotten wood .... **(Eustrophinae** part) ***Melandryidae*** p. 505

229′. Hypostomal rods absent; epicranial stem moderately long (fig. 18); ratio of antennal length to head width less than .15; abdominal terga with patches (fig. 142) or rows (fig. 144) of asperities. Under bark, in soft fungi .... (*Penthe* part) **(Penthinae)** ***Tetratomidae*** p. 504

230(194′). Epicranial stem absent or very short; maxilla with separate galea and lacinia (fig. 100); cardines separated from each other by labium; posterior edge of head capsule distinctly emarginate dorsally; ratio of antennal length to head width .15 to .5; spiracles A1–7 small and non-functional, partly surrounded by large plastron plates (fig. 213). In soil or gravel at edges of streams (western North America) .... (*Araeopidius*) **(Araeopidiinae)** ***Ptilodactylidae*** p. 391

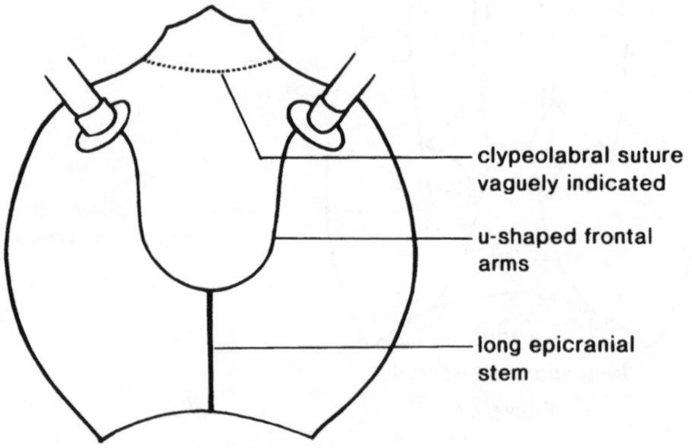

clypeolabral suture vaguely indicated
u-shaped frontal arms
long epicranial stem

Figure 14

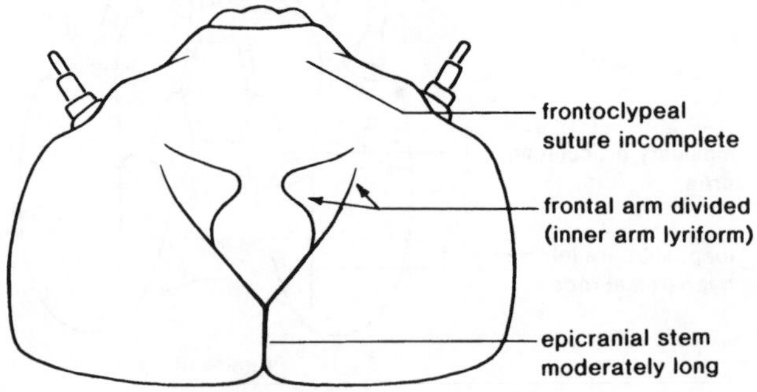

frontoclypeal suture incomplete
frontal arm divided (inner arm lyriform)
epicranial stem moderately long

Figure 18

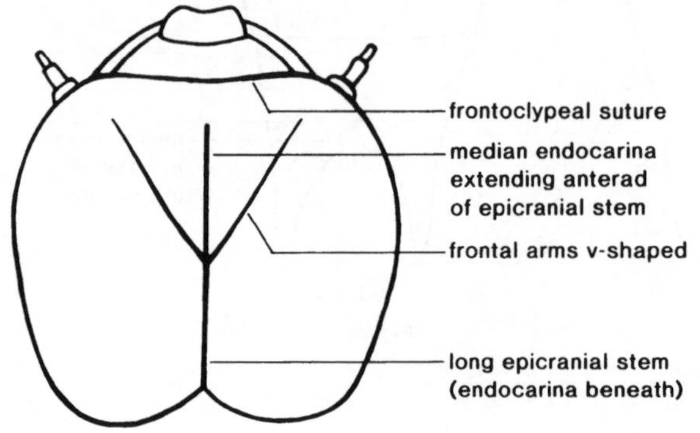

frontoclypeal suture
median endocarina extending anterad of epicranial stem
frontal arms v-shaped
long epicranial stem (endocarina beneath)

Figure 20

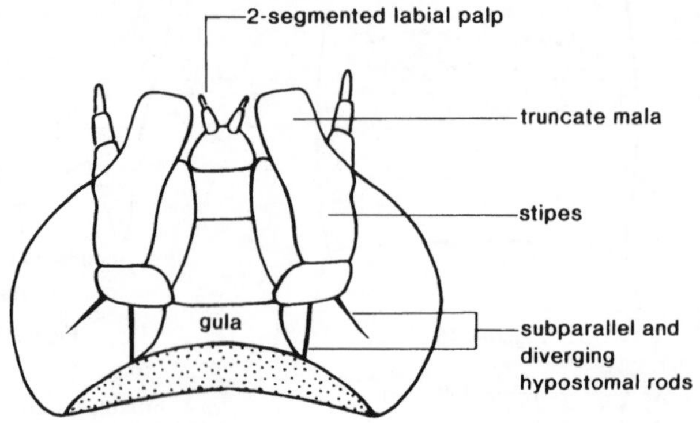

2-segmented labial palp
truncate mala
stipes
gula
subparallel and diverging hypostomal rods

Figure 88

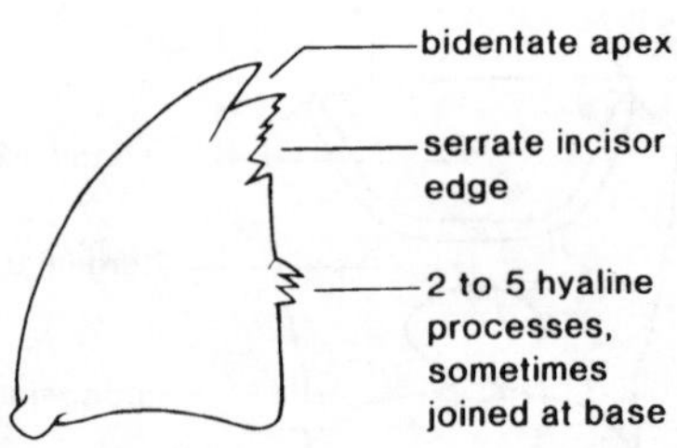

bidentate apex
serrate incisor edge
2 to 5 hyaline processes, sometimes joined at base

Figure 59

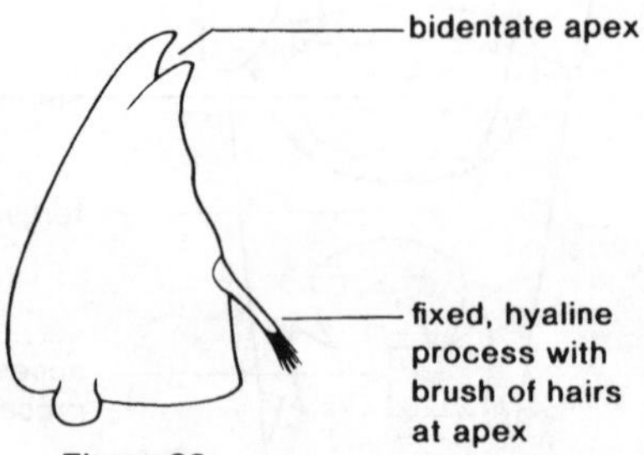

bidentate apex
fixed, hyaline process with brush of hairs at apex

Figure 62

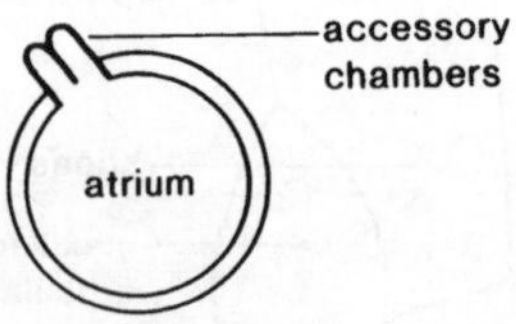

accessory chambers
atrium
Annular Biforous

Figure 197

230′. Epicranial stem moderately long (fig. 20); maxilla with single mala (fig. 99); cardines completely fused forming single plate (fig. 97); posterior edge of head capsule not or only slightly emarginate dorsally; ratio of antennal length to head width less than .15; spiracles A1–7 undulate, placed at the ends of short lateral processes (fig. 209); plastron plates absent. In leaf litter, flood debris, ant refuse heaps .... (part) ***Chelonariidae*** p. 394

231(193′). Epicranial stem moderately long (fig. 14); apex of mala falciform (fig. 105); urogomphi articulated at base (fig. 162); posterior edge of head capsule not or only slightly emarginate dorsally; ratio of antennal length to head width more than .5. In leaf litter, rotten wood, carrion, dung, under bark .... (part) ***Staphylinidae*** p. 341

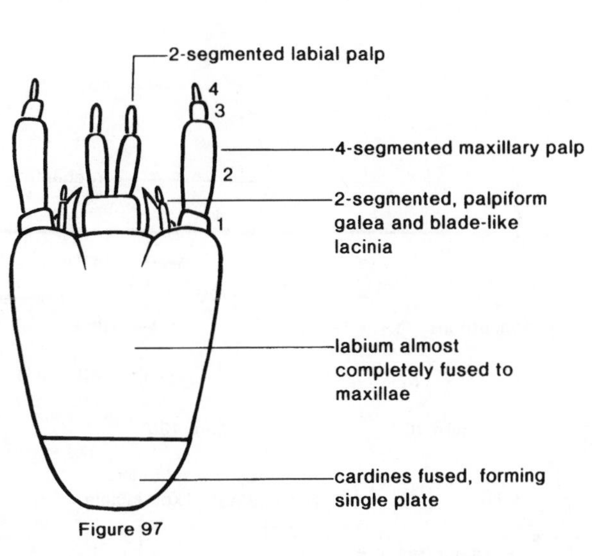

Figure 97

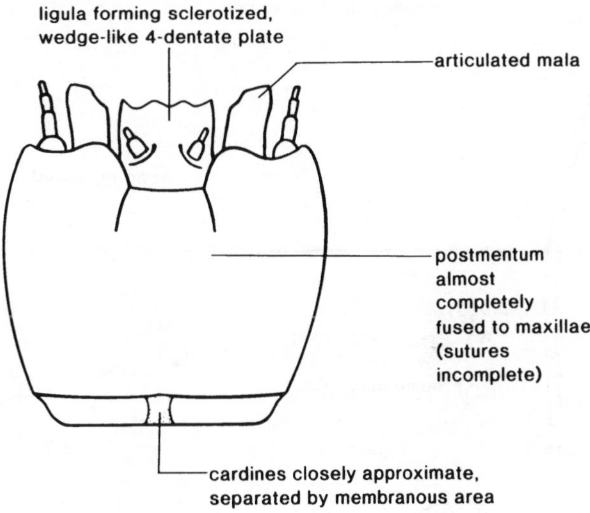

Figure 99

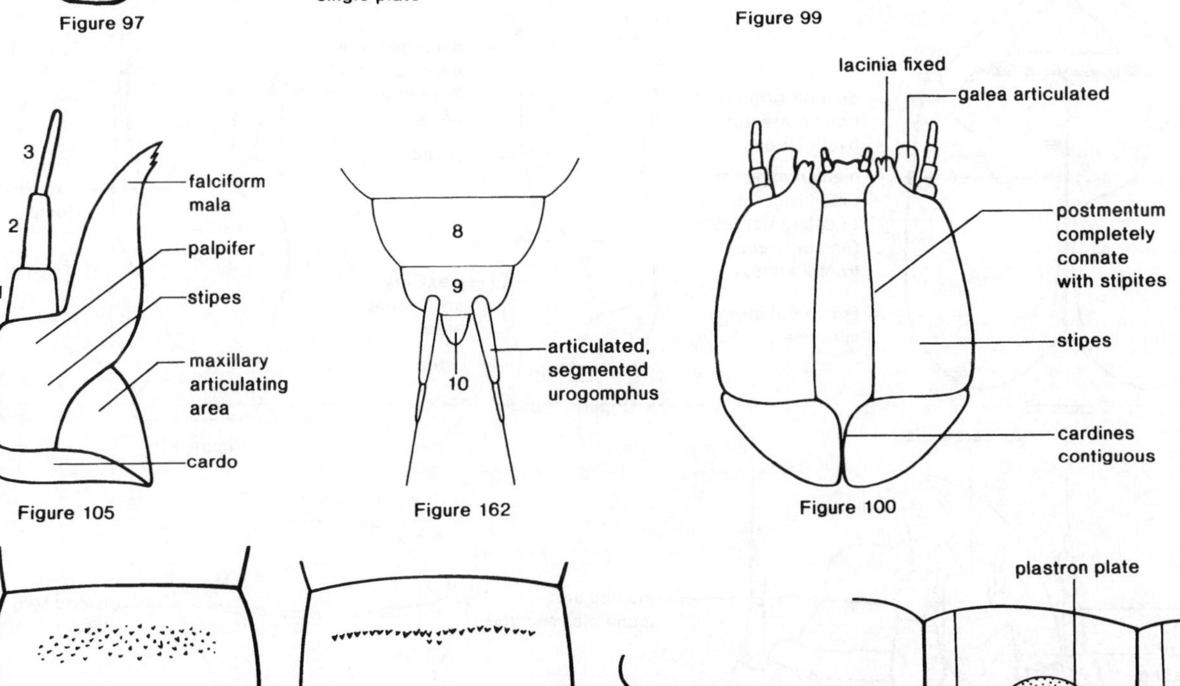

Figure 105

Figure 162

Figure 100

Figure 142

Figure 144

Figure 209

Figure 213

231′. Epicranial stem absent or very short; apex of mala rounded or truncate; urogomphi fixed at base; posterior edge of head capsule distinctly emarginate dorsally; ratio of antennal length to head width less than .15. In fungus fruiting bodies .... (**Peltinae** part) ***Trogossitidae*** p. 448

232(193[2]). Legs more or less reduced with 4 segments or fewer; head strongly retracted (fig. 9) 233

232′. Legs not reduced, with 5 segments including tarsungulus (fig. 126); head protracted or slightly retracted (fig. 8) 235

233(232). Head distinctly longer than wide, narrower posteriorly; epicranial stem not located in broad furrow; cardines completely fused with labium. Under bark, in living or dead wood .... (**Lamiinae** part) ***Cerambycidae*** p. 556

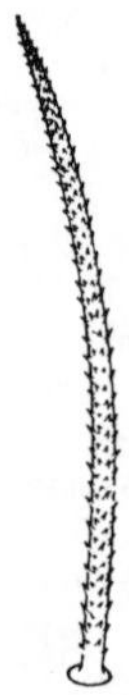

Figure 6

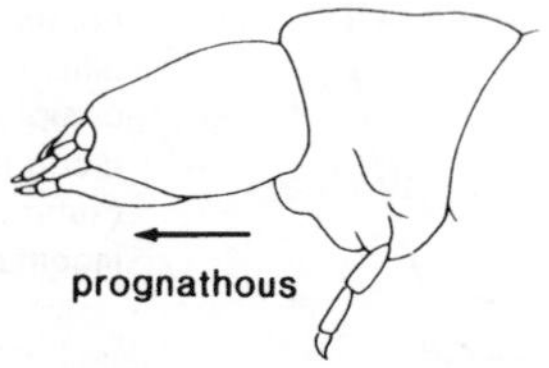

Figure 8

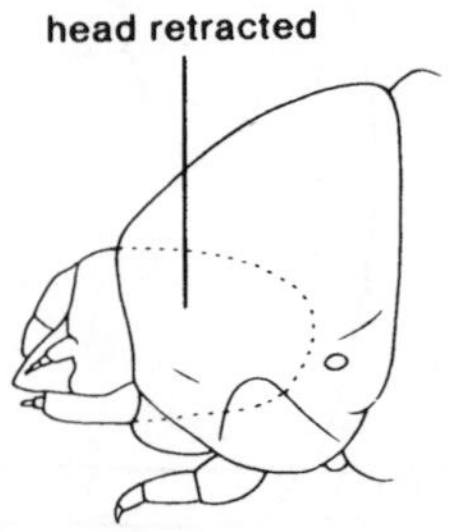

Figure 9

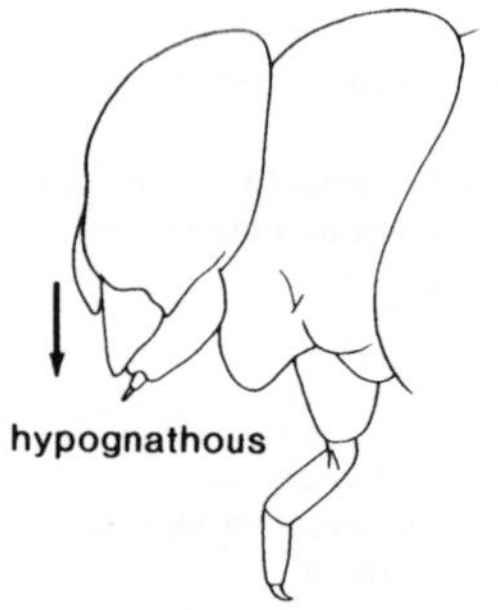

Figure 10

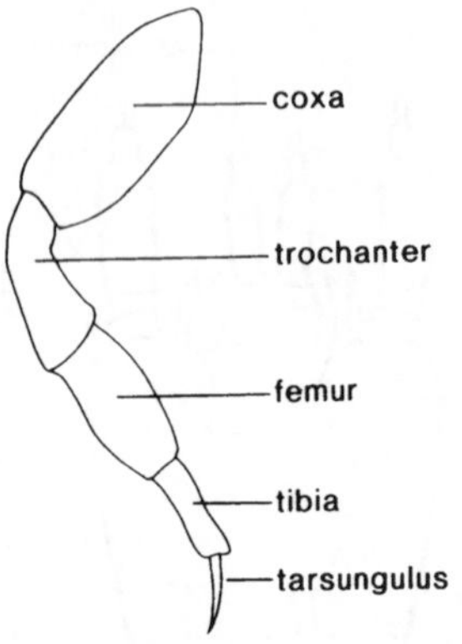

Figure 126

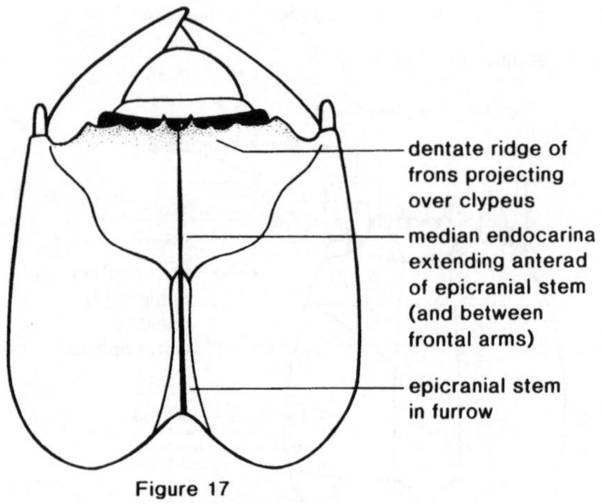

Figure 17

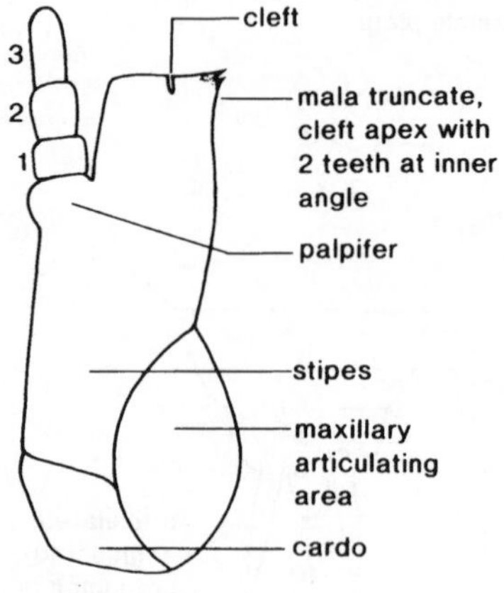

Figure 106

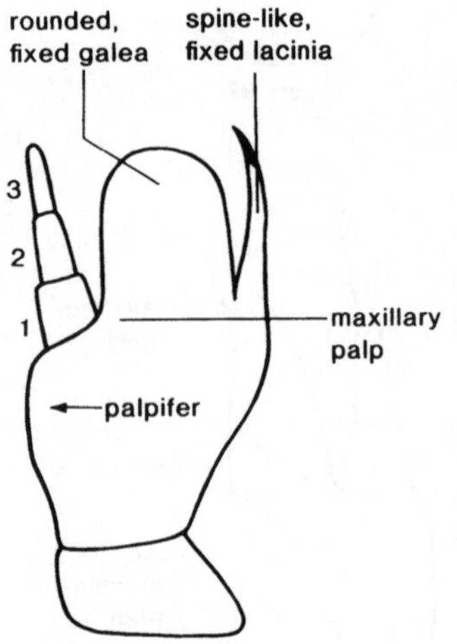

Figure 215

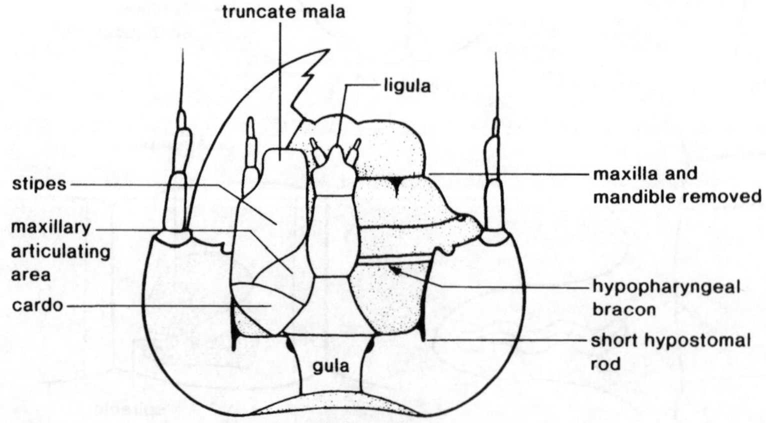

Mouthparts Retracted

Figure 85

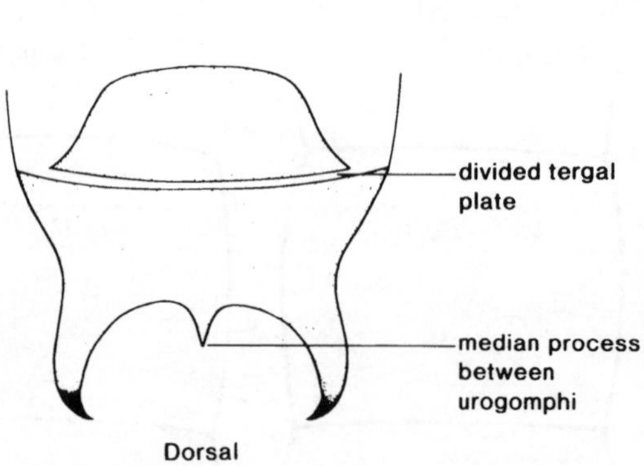

Dorsal

Figure 172

233′. Head not or only slightly longer than wide, not narrowed posteriorly; epicranial stem located in broad furrow for attachment of retractor muscles (fig. 17); cardines separated from each other by labium (fig. 95) 234

234(233′). Occipital foramen not divided; apex of mandible parallel or oblique to plane of movement, its cutting edge straight and not gouge-like. In living or dead conifer wood .... (**Aseminae**) ***Cerambycidae*** p. 556

234′. Occipital foramen divided into 2 parts by tentorial bridge (fig. 95); apex of mandible perpendicular to plane of movement, its cutting edge rounded and gouge-like (fig. 48). Under bark, in living or dead wood .... (**Cerambycinae** part) ***Cerambycidae*** p. 556

235(232′). Head prognathous or slightly declined (fig. 8); ventral mouthparts protracted (fig. 92) or only slightly retracted; stipes wider than long; gula longer than wide (fig. 92); apex of mandible with single lobe or tooth; body surfaces smooth and vestiture consisting of simple setae or hairs 236

235′. Head moderately to strongly declined (hypognathous) (fig. 10); ventral mouthparts retracted (fig. 85); gula wider than long (fig. 85); apex of mandible trilobed or tridentate; dorsal surfaces of body granulate or tuberculate or vestiture including barbed hairs (spicisetae, fig. 6) 237

236(235). Hypostomal rods extending to posterior edge of head. In galleries of wood-boring insects. Exotic (New Zealand, Madagascar) .... ***Chaetosomatidae*** p. 450

236′. Hypostomal rods, if present, not extending to posterior edge of head .... (part) ***Cleridae*** p. 450

237(235′). Maxilla with single mala, sometimes cleft or with 1 or more teeth at inner apical angle (fig. 106); segments in maxillary palp 3; segment A10 concealed from above; vestiture consisting of fine hairs or setae only; anal region posteroventrally oriented; dorsal surfaces usually granulate or tuberculate. In fungus fruiting bodies .... (part) ***Erotylidae*** p. 473

237′. Maxilla with separate galea and spine-like lacinia (as in fig. 215); segments in maxillary palp 4; segment A10 distinct and visible from above; vestiture including barbed hairs (spicisetae, fig. 6); anal region posteriorly or terminally oriented; dorsal surfaces smooth. On a variety of animal and plant products, including carrion .... (**Dermestinae**) ***Dermestidae*** p. 434

238(193³). Stipes longer than wide; ventral mouthparts retracted (fig. 87) 239

238′. Stipes wider than long; ventral mouthparts strongly protracted (fig. 92) 246

239(238). Tergum A9 without divided plate or median process between urogomphi 240

239′. Tergum A9 with transversely divided plate and median process between urogomphi (fig. 172); sclerotized plate on tergum A9 transversely divided (fig. 172). Under bark, in rotten wood, fungus fruiting bodies, stored products .... (**Lophocaterinae**) ***Trogossitidae*** p. 448

240(239). Gular region absent (labium contiguous with thoracic membrane) (fig. 96); antennae very short, with sensorium on segment 2 much longer than reduced 3rd segment; apex of mandible multilobed or multidentate; stemmata on each side 1 .... (**Orsodacninae**) ***Chrysomelidae*** p. 568

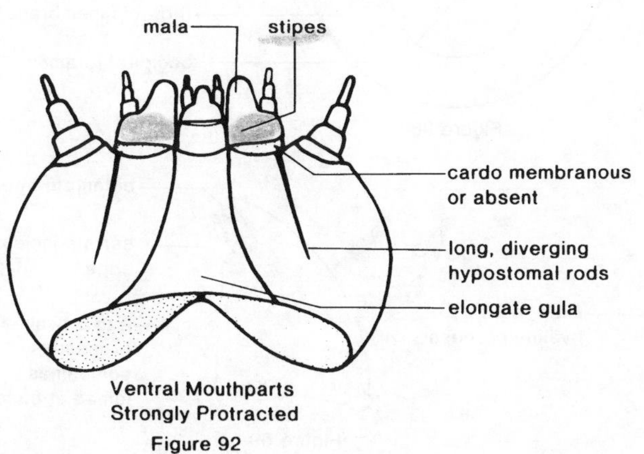

Ventral Mouthparts Strongly Protracted

Figure 92

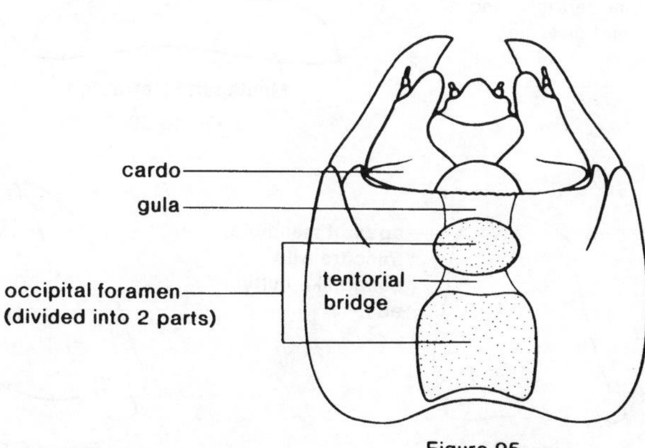

Figure 95

Figs. 48, 87, 96 ⟶

240′. Gular region present (separating labium from thorax) (fig. 85); antennae longer, with segment 3 longer than sensorium; apex of mandible usually bidentate; stemmata on each side 4 or more 241

241(240′). Mesal surface of mandibular base with pubescent, membranous area (fig. 64); 2 pairs of hypostomal rods present, one subparallel and the other diverging (fig. 88); terga T1–3 and terga A1–9 each with distinct pigmented plate. In fungus fruiting bodies .... **(Eustrophinae part)** ***Melandryidae*** p. 505

241′. Mesal surface of mandibular base with 1 or more hyaline processes (figs. 58–61); single pair of hypostomal rods present (fig. 89); thorax and abdomen with tergal plates on some segments only, and these usually paired 242

242(241′). Gula wider than long (fig. 85) 243

242′. Gula longer than wide (fig. 92) 245

243(242). Labial palps 1-segmented; mesal surface of mandibular base with single, acute, hyaline process (fig. 58); urogomphi approximate, arising from elevated common base. In leaf litter. Exotic (South temperate regions) .... **(Rentoniinae)** ***Trogossitidae*** p. 448

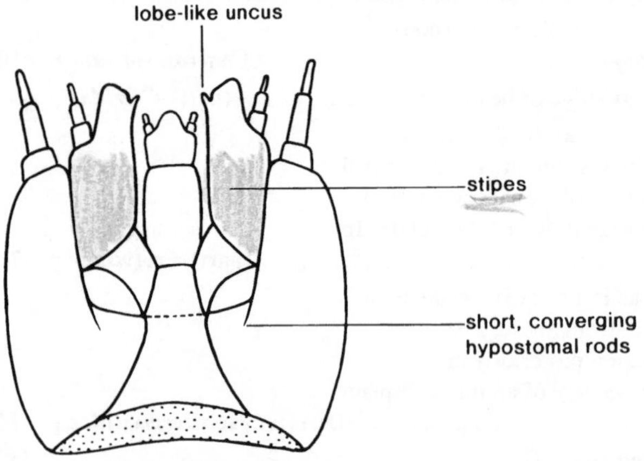

lobe-like uncus
stipes
short, converging hypostomal rods

Mouthparts Retracted
Figure 87

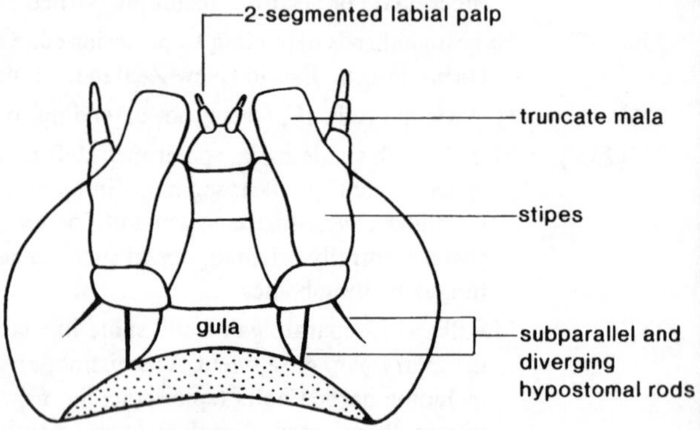

2-segmented labial palp
truncate mala
stipes
gula
subparallel and diverging hypostomal rods

Figure 88

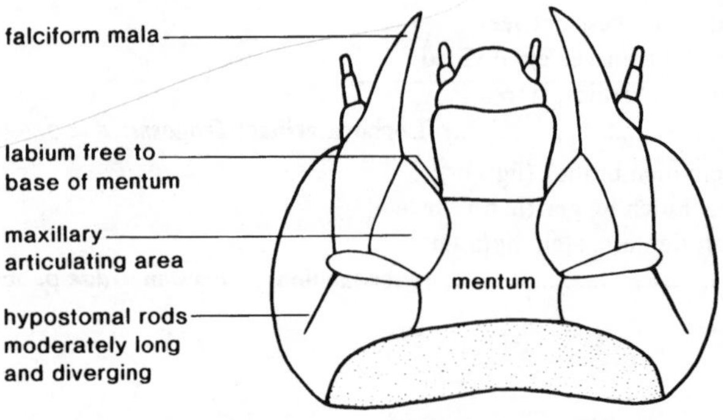

falciform mala
labium free to base of mentum
maxillary articulating area
mentum
hypostomal rods moderately long and diverging

Mouthparts Retracted
Figure 89

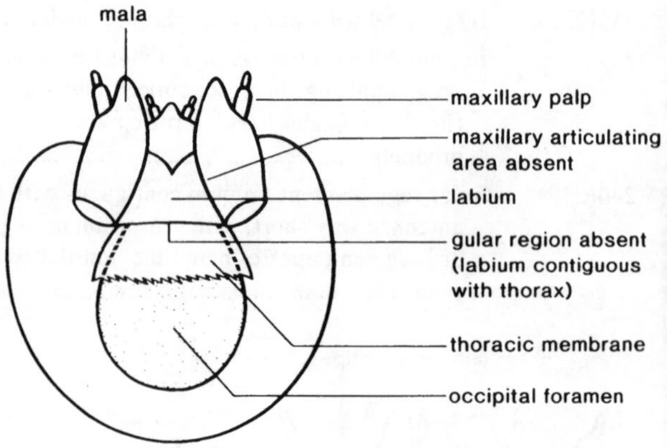

mala
maxillary palp
maxillary articulating area absent
labium
gular region absent (labium contiguous with thorax)
thoracic membrane
occipital foramen

Figure 96

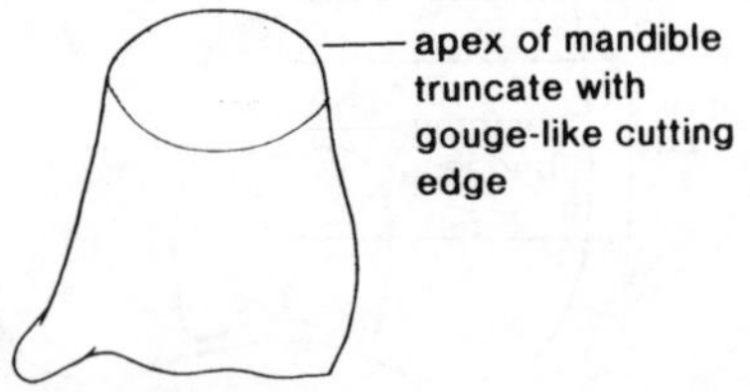

apex of mandible truncate with gouge-like cutting edge

Figure 48

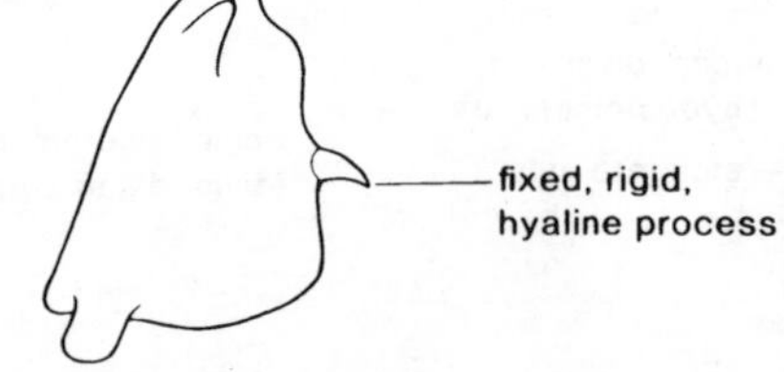

fixed, rigid, hyaline process

Figure 58

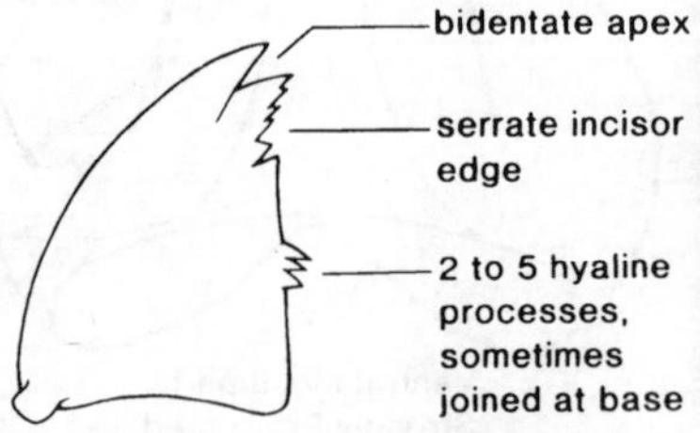

bidentate apex
serrate incisor edge
2 to 5 hyaline processes, sometimes joined at base

Figure 59

243′. Labial palps 2-segmented; mesal surface of mandibular base with 3 acute, hyaline processes (fig. 59); urogomphi distant at base, strongly curved, with apices slightly converging 244

244(243′). Frontal arms distinctly lyriform, each with an acute angle at middle (fig. 25); terga A6–8 without paired sclerotized plates. Under bark or in fungus fruiting bodies. Exotic (Europe) .... ***Phloiophilidae*** p. 447

244′. Frontal arms only slightly, obtusely angulate at middle; terga A6–8 with paired, sclerotized plates. Under bark or in fungus fruiting bodies. Exotic (New Zealand) .... **(Protopeltinae)** ***Trogossitidae*** p. 448

245(242′). Urogomphi with accessory processes (fig. 169); head transverse; tergal sclerites of mesothorax indistinct and separated by more than their width; sclerotized plate on tergum A9 with raised rim. Under bark, in rotten wood, fungus fruiting bodies .... (*Calitys*) **(Calitinae)** ***Trogossitidae*** p. 448

245′. Urogomphi without accessory processes; head subquadrate or longer than wide; tergal sclerites of mesothorax distinct and separated by much less than their width; sclerotized plate on tergum A9 without raised rim. Under bark, in rotten wood, fungus fruiting bodies, stored products .... **(Trogossitinae)** ***Trogossitidae*** p. 448

246(238′). Sternum A9 completely concealed beneath sternum A8 (fig. 164); hypostomal rods long and diverging (fig. 92); apex of mandible tridentate (fig. 60); mesal surface of mandibular base with 1 or more hyaline processes (fig. 60); A8 spiracles located at posterior end of segment and facing posteriorly (fig. 168). In flower heads, rotting wood, fungus fruiting bodies .... (part) ***Phalacridae*** p. 466

246′. Sternum A9 partly or entirely exposed; hypostomal rods, if present, short or subparallel; apex of mandible with fewer than 3 teeth or lobes; A8 spiracles laterally placed and oriented 247

247(246′). Apex of mandible with single lobe or tooth; gula longer than wide (fig. 92); maxillary articulating area absent. Under bark, in dead wood, leaf litter, fungi .... (part) ***Cleridae*** p. 450

247′. Apex of mandible bilobed or bidentate; gula wider than long (fig. 85); maxillary articulating area present (fig. 85). Under bark, in living or dead wood .... **(Lepturinae** part) ***Cerambycidae*** p. 556

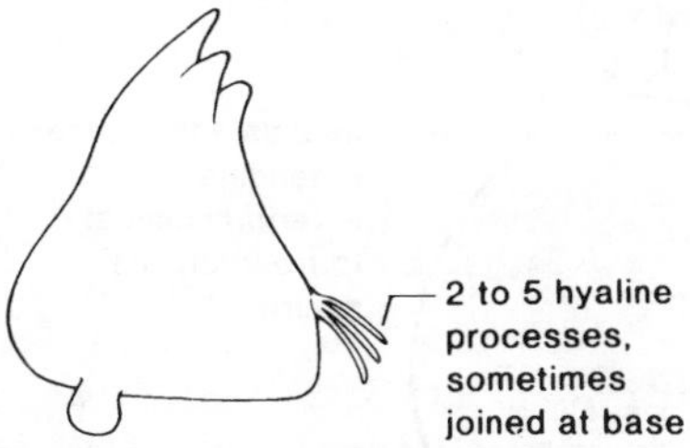

Figure 60

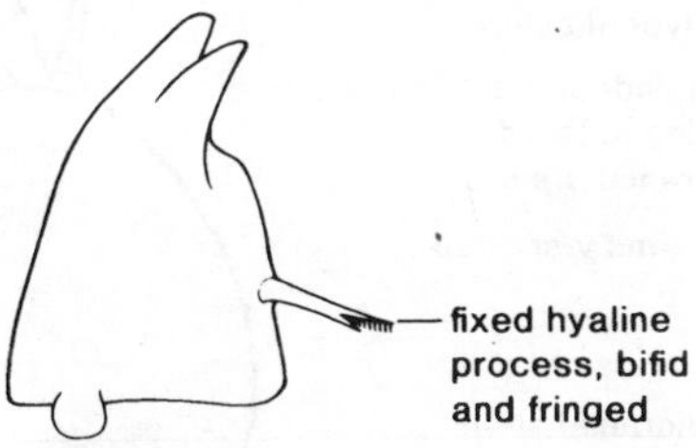

Figure 61

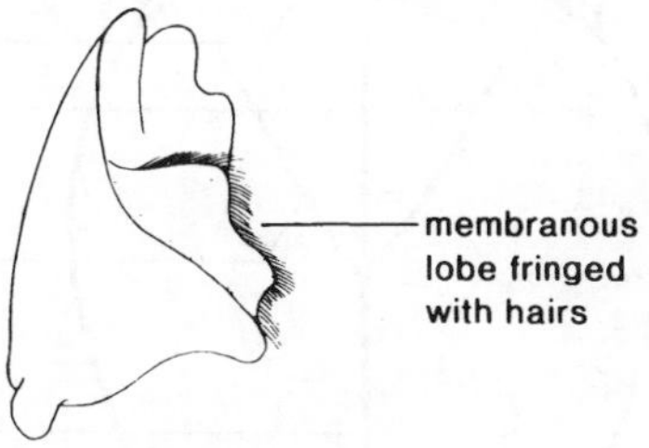

Figure 64

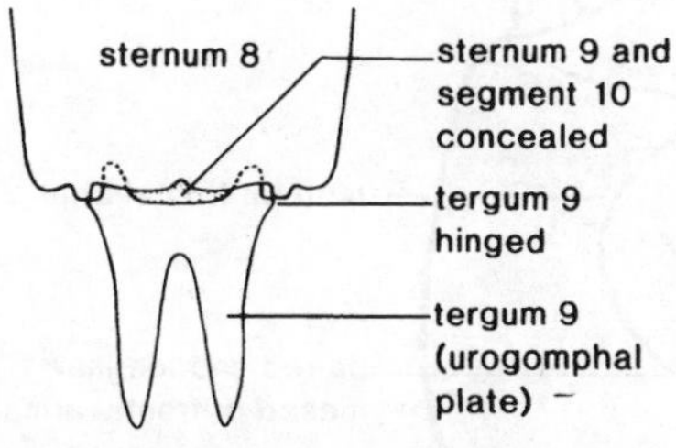

Figure 164

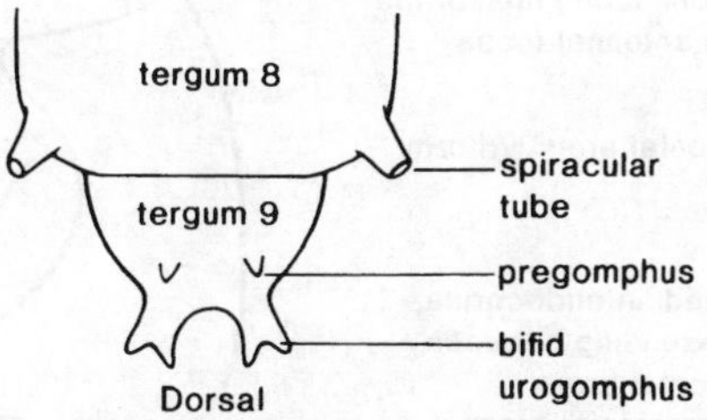

Figure 168

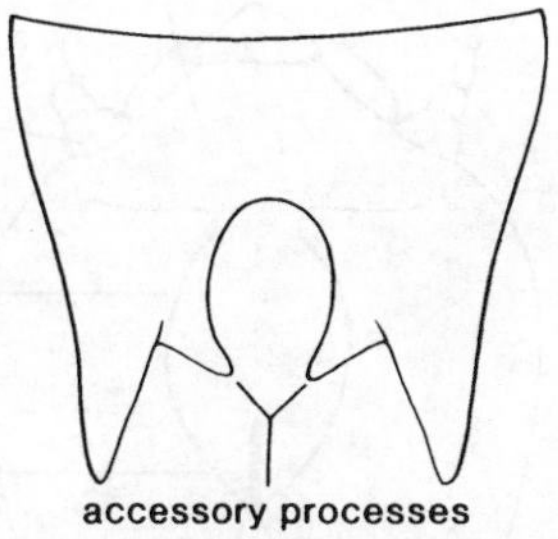

Figure 169

Fig. 25 ⟶

248(3′). Tergum A9 without paired processes (urogomphi) 249

248′. Tergum A9 with paired processes (urogomphi) (figs. 162, 166–185, 217) 311

249(248). Antennal segments 1 250

249′. Antennal segments 2 (**see 3rd and 4th choices**) 253

249^2. Antennal segments 3 (minute antennae may appear only 2-segmented) 261

249^3. Antennal segments 4 or more 307

250(249). Abdominal terga without patches of asperities; frontal arms present (figs. 20, 22); segment A10 without oval lobes separated by longitudinal groove 251

250′. Abdominal terga with patches of asperities on 1 or more segments (fig. 142); frontal arms absent; segment A10 with pair of oval lobes separated by longitudinal groove (fig. 193). In fungus fruiting bodies, rotting wood, stems, vines .... (part) ***Anobiidae*** p. 441

251(250). Legs distinctly 5-segmented (fig. 126); frontal arms lyriform (figs. 22, 25); molar surface with numerous, fine ridges (fig. 74); ratio of antennal length to head width greater than .15; body relatively straight or slightly curved ventrally, flattened, ovate, and heavily pigmented dorsally. On bark or rock surfaces, in moss. Exotic (Old World) .... (Leiochrini) ***Tenebrionidae*** p. 520

251′. Legs with 4 segments or fewer (possibly with an additional coxal lobe); frontal arms, if present, V- or U-shaped (fig. 20); molar surface simple (fig. 69); ratio of antennal length to head width less than .15; body strongly curved ventrally (C-shaped), more or less elongate and lightly pigmented 252

252(251′). Legs 4-segmented; mandible with accessory ventral process (fig. 75). In flowers of *Delphinium* (Ranunculaceae). Exotic (Eurasia and northern Africa) .... (*Nemonyx*) ***Nemonychidae*** p. 585

252′. Legs with 2 segments or fewer; mandible without accessory ventral process. In fungus fruiting bodies, dead wood, stems, vines, etc. .... (part) ***Anthribidae*** p. 586

253(249′). Segments of T2 leg 6 including single claw (fig. 125); head with median endocarina extending anteriorly almost to edge of clypeus (fig. 21); ligula sclerotized and wedge-like (fig. 117). Probably in rotten wood .... (*Priacma*, first instar) ***Cupedidae*** p. 298

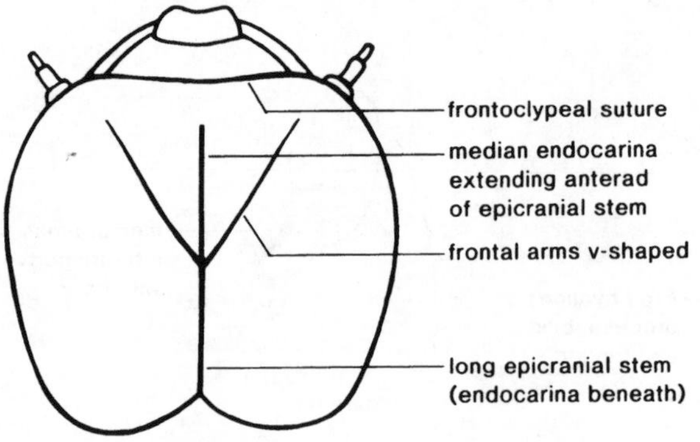

Figure 20

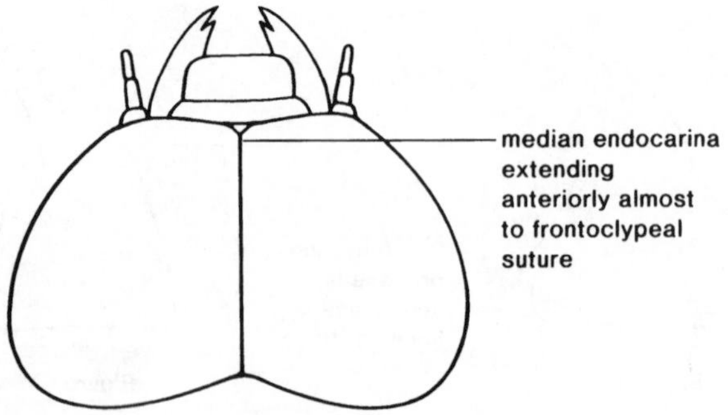

Figure 21

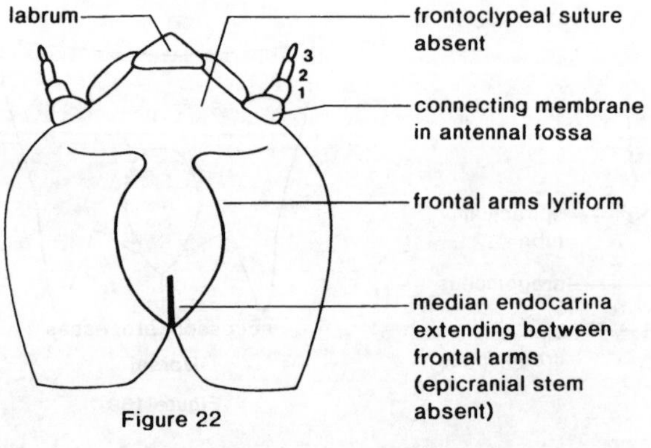

Figure 22

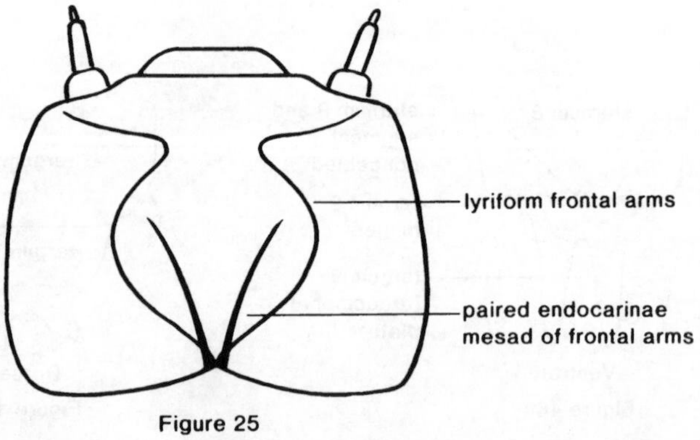

Figure 25

Figs. 166–185: See pages 249, 255, 265, 275

253′. Segments of T2 leg 5 including tarsungulus (fig. 126); median endocarina, if present, not extending far anteriorly; ligula not forming wedge-like sclerome 254

254(253′). Spiracular gills (transparent vesicles, gill tufts, or segmented, spiracle-bearing processes) present on abdomen or thorax and abdomen (figs. 152, 211, 212); length less than 3mm (2mm in North American forms); aquatic 255

254′. Spiracular gills absent; length usually more than 2mm; terrestrial 256

255(254). Segmented spiracular gills (fig. 152) on segments A1–8; ratio of antennal length to head width more than .2. On rocks in streams or near waterfalls. Exotic (Brazil, South Africa, Madagascar) .... (part) ***Torridincolidae*** p. 302

255′. Vesicular spiracular gills (fig. 211) on raised spiracular tubes on A1–8; ratio of antennal length to head width more than .2. In sand, gravel, or mud at edges of streams (**see 3rd choice**) .... (part) ***Microsporidae*** p. 302

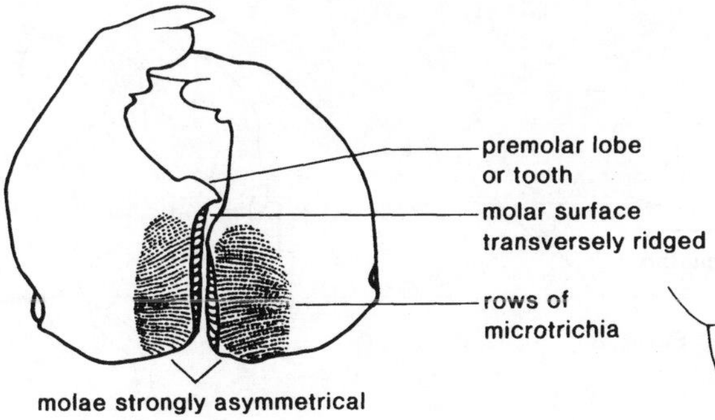

Figure 74

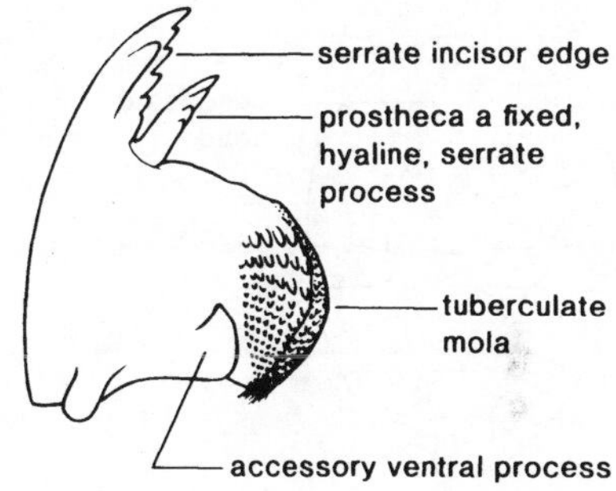

Figure 75

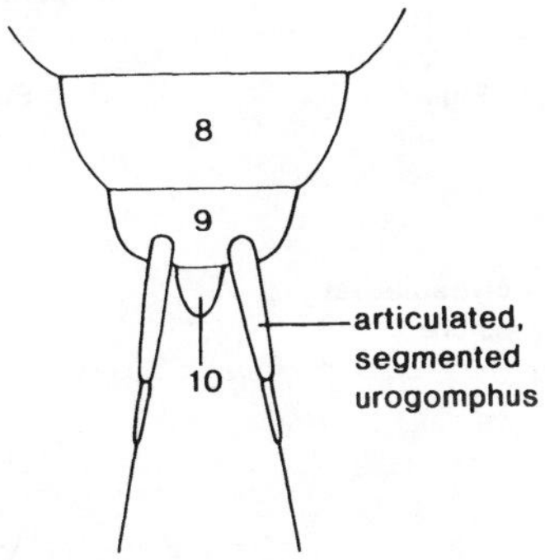

Figure 162

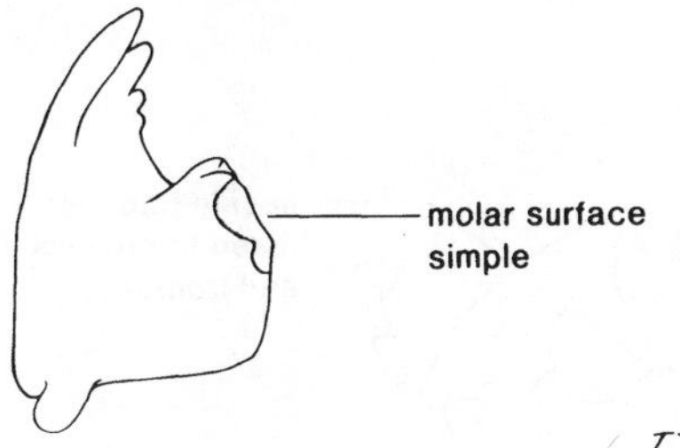

Figure 69

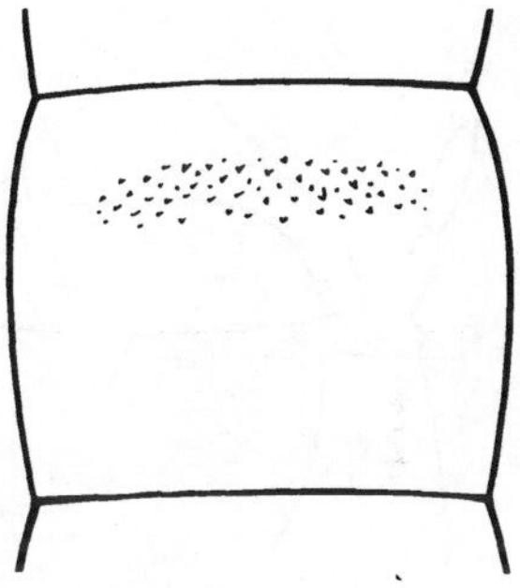

Figure 142

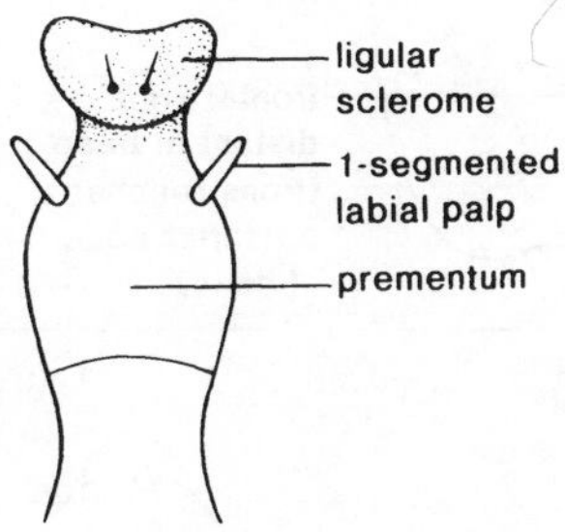

Figure 117

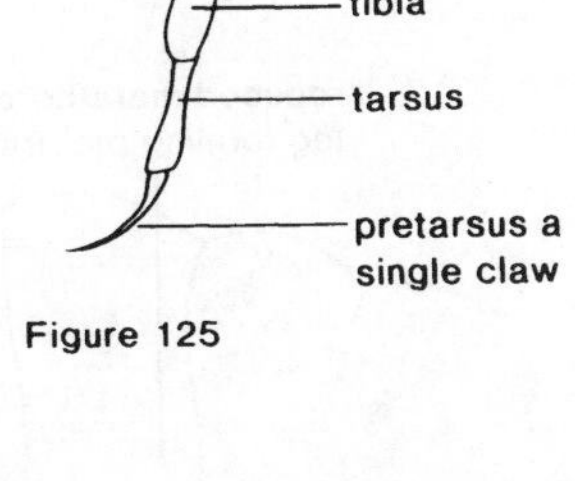

Figure 125

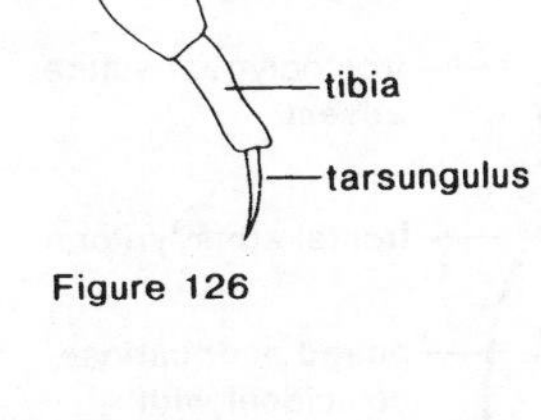

Figure 126

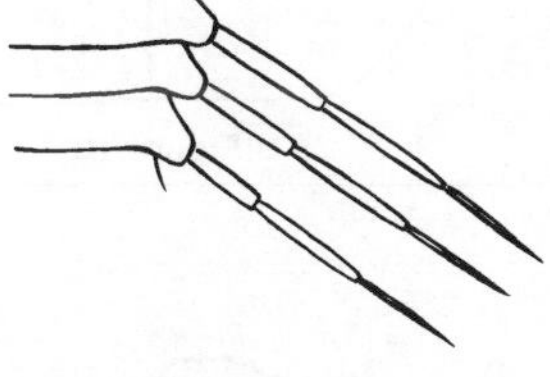

Figure 152

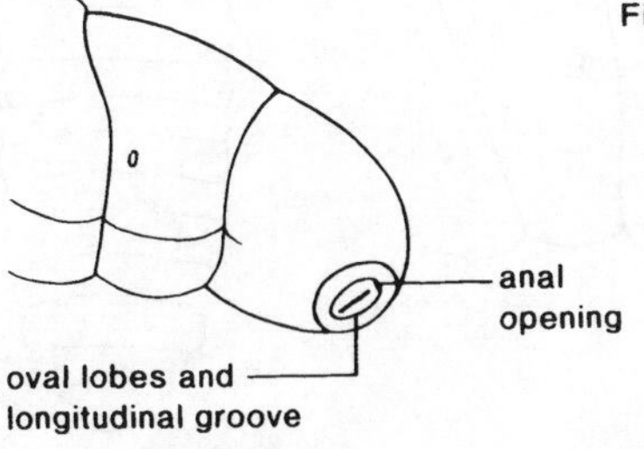

Figure 193

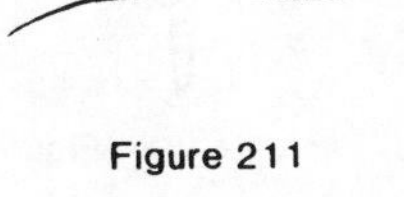

Figure 211

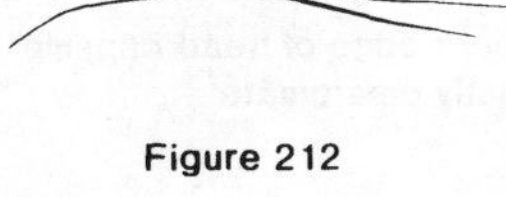

Figure 212

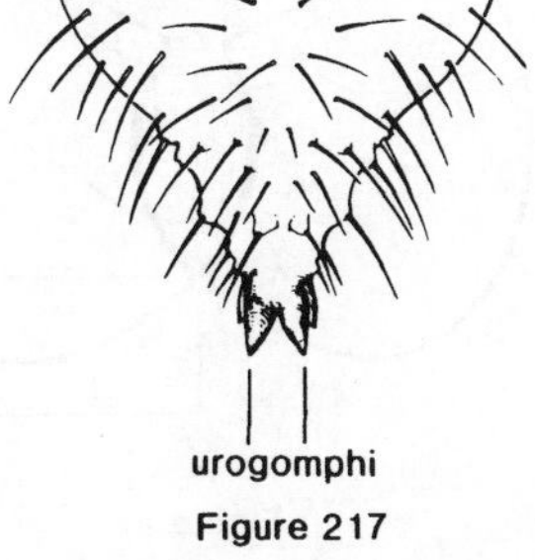

Figure 217

255[2]. Vesicular (fig. 211) or rarely tufted (fig. 212) gills present on raised spiracular tubes on T1, A1 and A8; spiracular tubes on segment A8 long and posteriorly projecting (fig. 161); ratio of antennal length to head width less than .15. On alga-covered rocks or in sand and mud in streams, pools, or hot springs .... (part) ***Hydroscaphidae*** p. 303

256(254′). Body broadly ovate, strongly flattened, and disc-like (as in fig. 2); head concealed from above by T1; segments in maxillary palp 2. Under bark, in fungus fruiting bodies, in leaf litter. Exotic (Neotropical and Old World); possibly in southern United States .... ***Discolomidae*** p. 481

256′. Body more or less elongate; head not concealed from above 257

257(256′). T3 leg reduced and 1-segmented (fig. 131); spiracles cribriform (fig. 208). In rotten wood .... ***Passalidae*** p. 375

257′. T3 leg not reduced; spiracles not cribriform 258

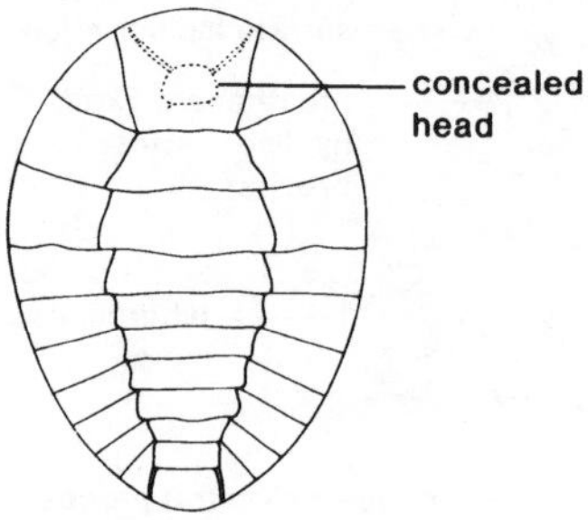

Figure 2

Figure 4

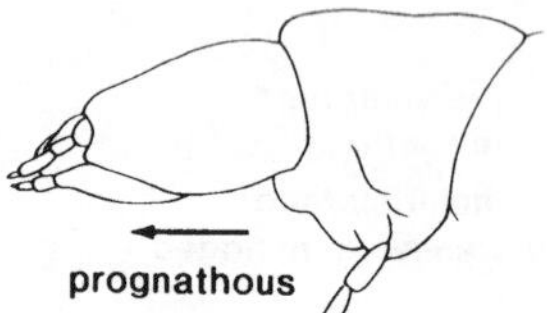

Figure 8

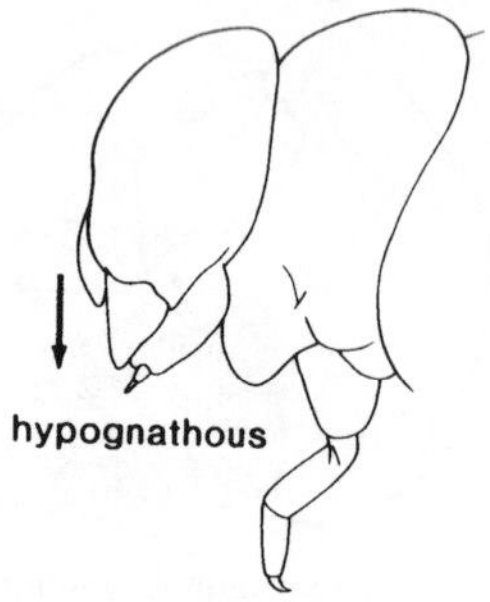

Figure 10

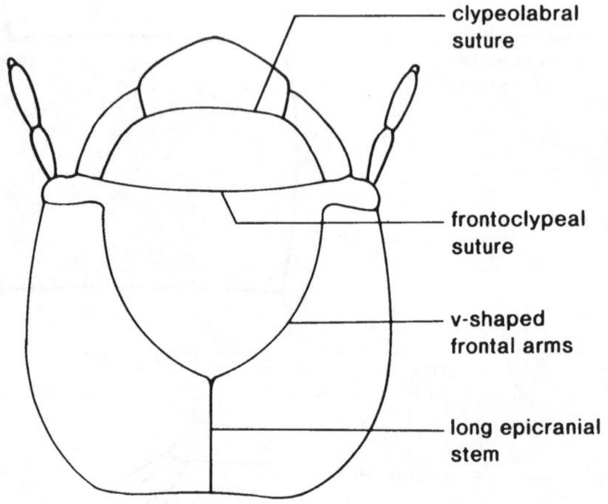

Figure 13

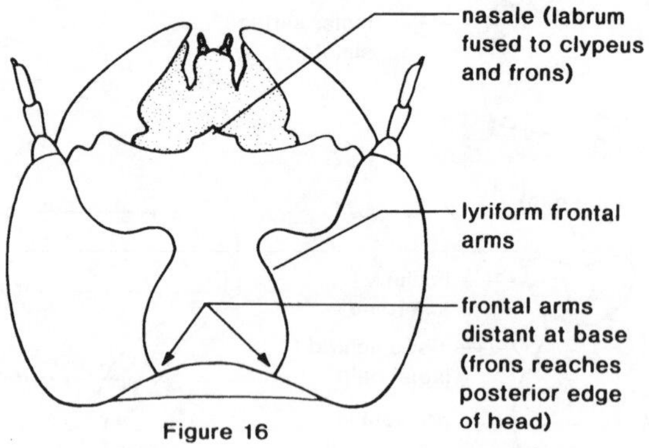

Figure 16

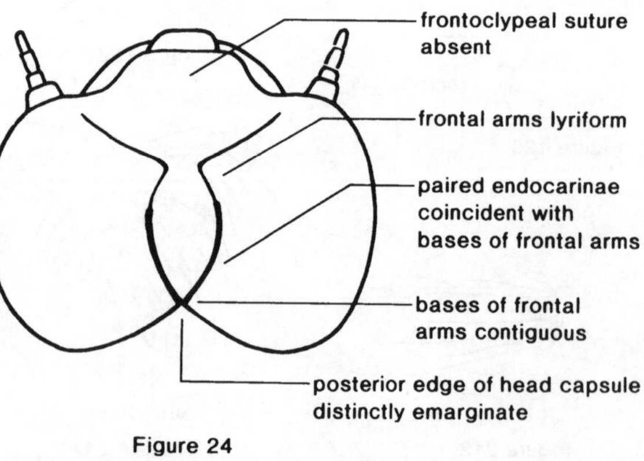

Figure 24

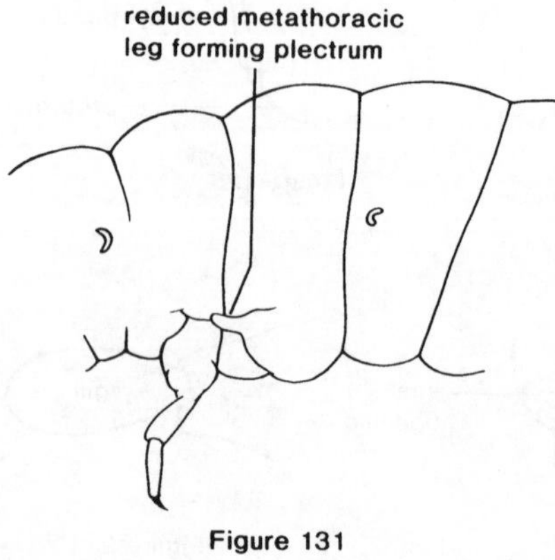

Figure 131

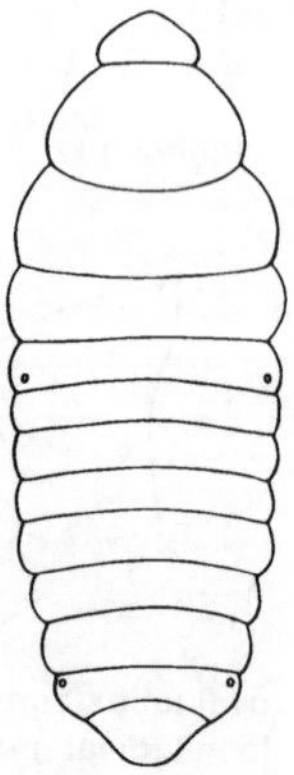

Figure 138

← Figs. 75, 211, 212

258(257′). Head strongly declined (hypognathous) (fig. 10), globular; prothorax enlarged and swollen dorsally, forming a hood-like structure over head; ratio of antennal length to head width less than .15; body elongate and cylindrical. Tunneling in wood .... (part) ***Lymexylidae*** p. 446

258′. Head prognathous or slightly declined (fig. 8); prothorax not enlarged and hood-like; ratio of antennal length to head width more than .15 259

259(258′). Frontoclypeal suture present (fig. 13); hypostomal rods absent. In rotten wood, fungus fruiting bodies, leaf litter .... (part) ***Tenebrionidae*** p. 520

259′. Frontoclypeal suture absent (fig. 24); hypostomal rods long and diverging (figs. 89, 92) 260

260(259′). Frontal arms separated at base (fig. 16); mala rounded or truncate (fig. 92); large paired gland openings on terga A1 and A8 (fig. 138); body more or less oblong, fusiform, vestiture including expanded or clavate setae (fig. 4). In leaf litter, on surfaces of fungi and molds .... (part) ***Corylophidae*** p. 495

260′. Frontal arms contiguous at base (fig. 24); mala falciform (fig. 89); abdomen without paired gland openings; body elongate and more or less parallel-sided; vestiture consisting of simple setae. Under bark, in leaf litter, in stored products .... (Silvanini part) ***Cucujidae*** p. 463

261(249[2]). Prostheca absent 262

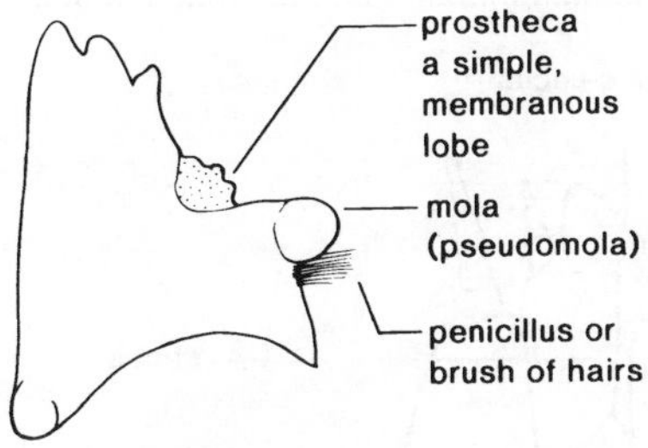

Figure 72

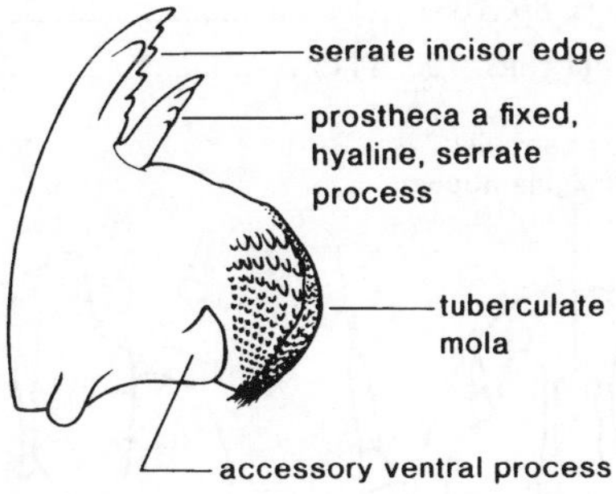

Figure 75

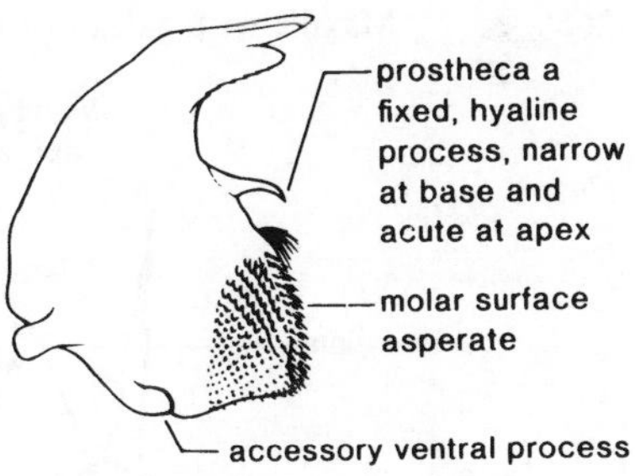

Figure 78

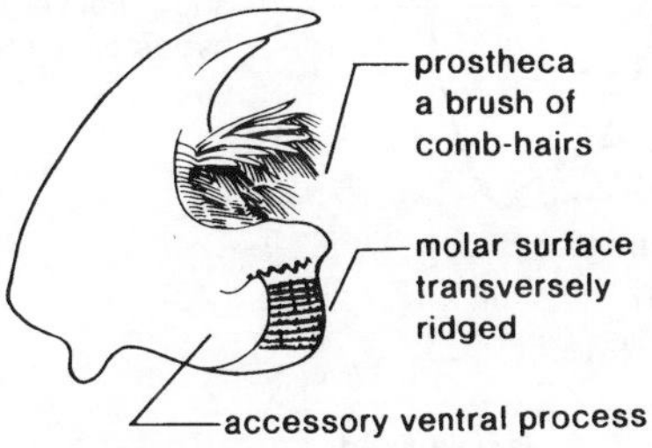

Figure 80

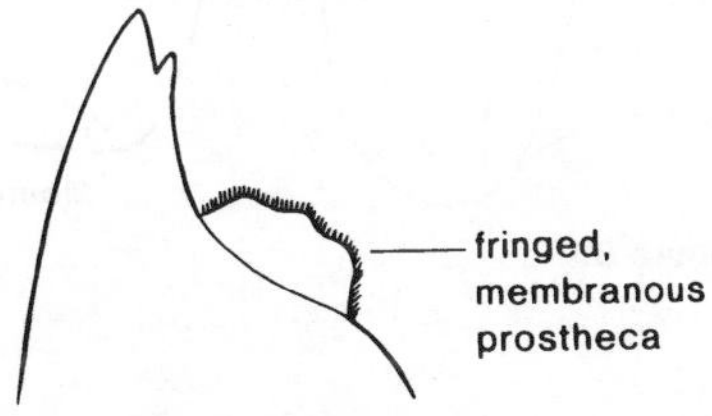

Figure 84

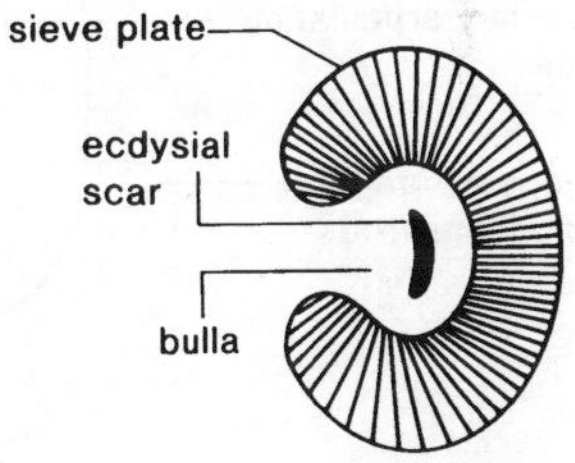

Cribriform
Figure 208

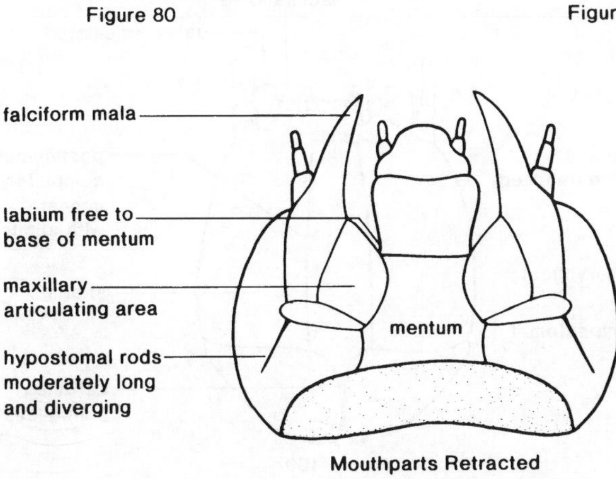

Mouthparts Retracted
Figure 89

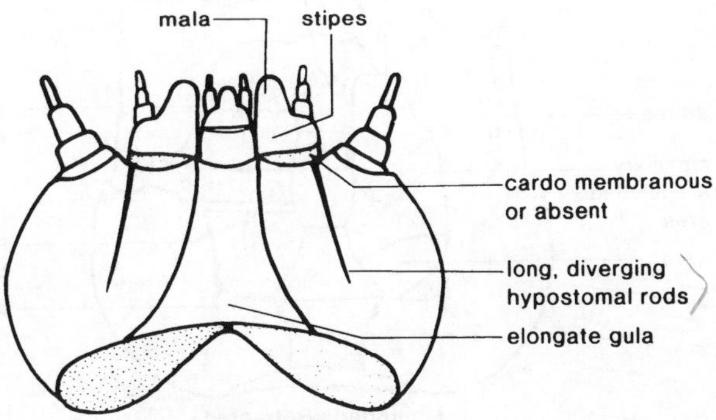

Ventral Mouthparts Strongly Protracted
Figure 92

Fig. 161 ⟶

261′. Prostheca a brush of simple or complex hairs (fig. 80) **(see 3rd thru 6th choices)** 286

261^{2}. Prostheca a fixed, rigid, hyaline process (figs. 75, 78) sometimes partly sclerotized and sometimes appearing articulated 287

261^{3}. Prostheca consisting of 2 hyaline processes; length less than 2.5mm; paired gland openings (fig. 139) on terga A1–7. In leaf litter, on surfaces of fungi .... (part) ***Corylophidae*** p. 495

261^{4}. Prostheca a simple, membranous lobe (fig. 72) 303

261^{5}. Prostheca a membranous lobe fringed with simple or complex hairs (fig. 84) 306

262(261). Tergum A9 simple 265

262′. Tergum A9 with median process (figs. 158, 159) **(see 3rd and 4th choices)** 281

262^{2}. Tergum A9 with concave, terminal disc (fig. 160) 285

262^{3}. Tergum A9 complex, with several sclerotized processes (fig. 180) 263

263(262^{3}). Sternum A9 simple; frontoclypeal suture present (fig. 13). In rotten wood .... (Strongyliini) ***Tenebrionidae*** p. 520

263′. Sternum A9 with asperities at base (figs. 180, 182); frontoclypeal suture absent (fig. 24) 264

264(263′). Hypostomal rods present (fig. 85); sternum A9 with 2 basal asperities on each side. Exotic (Old World) .... **(Trogocryptinae** part) ***Othniidae*** p. 547

264′. Hypostomal rods absent; sternum A9 with doubly curved row of basal asperities (fig. 180). In rotten wood .... **(Pythinae** part) ***Pythidae*** p. 539

265(262). Maxilla with single mala (figs. 86, 106) 266

265′. Maxilla with galea and lacinia (figs. 100, 111) 276

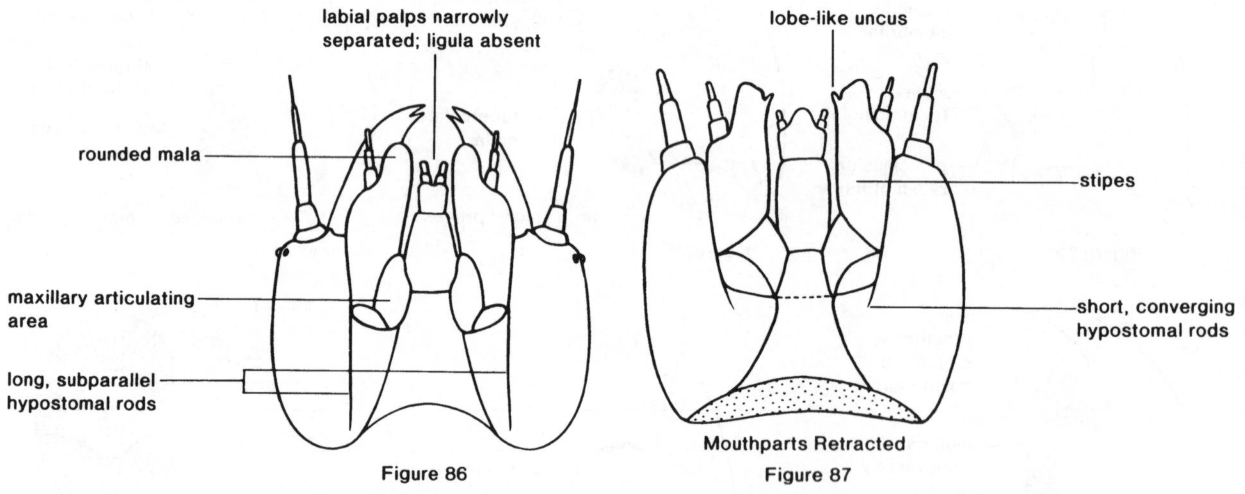

Figure 86

Figure 87

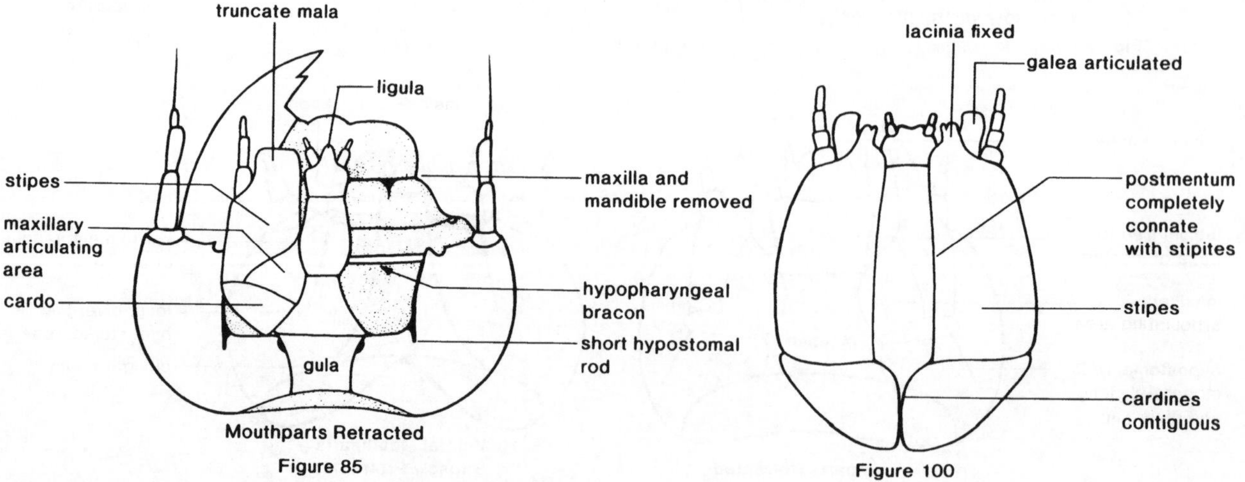

Figure 85

Figure 100

← Figs. 8, 10, 13, 24, 72, 75, 78, 80, 84, 92, 138

266(265). Head moderately to strongly declined (hypognathous) (fig. 10), globular; T1 enlarged and swollen dorsally (hump-like); body very elongate, with abdomen more than 5 times as long as thorax; patches of asperities on lateral portions of all thoracic segments and on apex of tergum A9, which is blunt and rounded. Tunneling in wood .... (part) ***Lymexylidae*** p. 446

266′. Head prognathous or slightly declined (fig. 8); body not as elongate; without other characters in combination 267

267(266′). Ventral mouthparts strongly protracted (fig. 92); gular sutures long, subparallel, with heavily sclerotized internal ridges; large paired gland openings almost always present on terga A1 and A8 (fig. 138), A2 and A8, or A1–7; vestiture always including some modified setae (barbed, frayed, star-shaped). In leaf litter, on surfaces of molds and other fungi .... (part) ***Corylophidae*** p. 495

267′. Ventral mouthparts retracted (fig. 86); gular sutures shorter and without obvious internal ridges; dorsal glands absent or inconspicuous; vestiture almost always consisting of simple setae 268

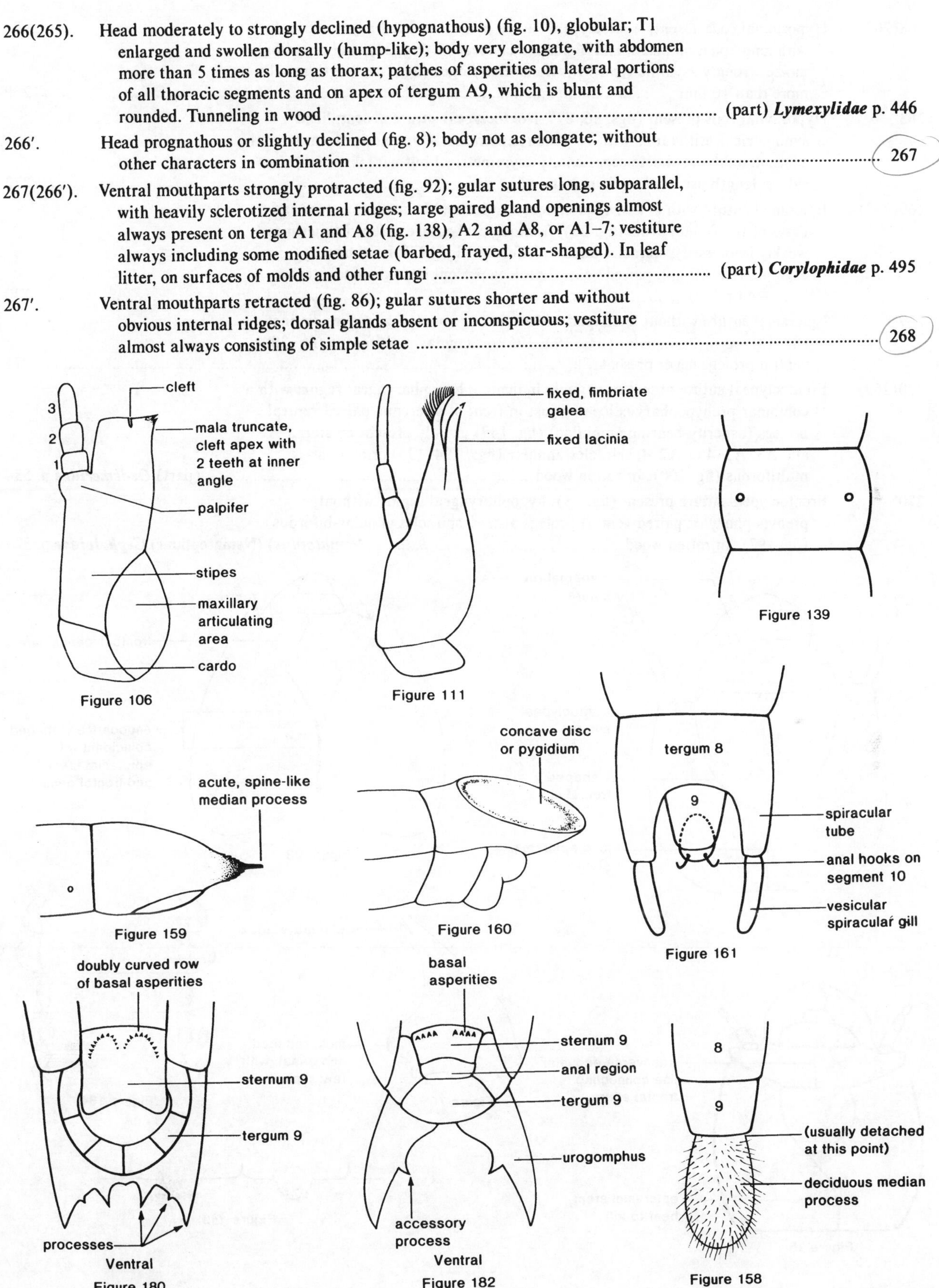

Figure 106
Figure 111
Figure 139
Figure 159
Figure 160
Figure 161
Figure 180
Figure 182
Figure 158

268(267′). Hypostomal rods absent; frontoclypeal suture present or head asymmetrical, with long epicranial stem and Y-shaped endocarina (fig. 23), and mandibular molae strongly asymmetrical with transverse ridges (fig. 74); length usually more than 10 mm 269

268′. Hypostomal rods present (figs. 86, 87); frontoclypeal suture absent; head symmetrical, without or with very short epicranial stem and lyriform frontal arms; mandibular molae occasionally asymmetrical but without transverse ridges; length usually less than 10 mm 272

269(268). Epicranial suture with a Y-shaped endocarina located beneath the stem and bases of the frontal arms (fig. 23); frontoclypeal suture absent (sometimes weakly impressed); spiracles annular-biforous (fig. 197); head usually asymmetrical (if not, paired ventral prolegs present on abdominal sterna 2–3, 3–4 or 2–4) 270

269′. Epicranial suture without Y-shaped endocarina; frontoclypeal suture distinct (fig. 13); spiracles annular (figs. 194, 195) or annular-multiforous (fig. 200); ventral prolegs never present 271

270(269). Frontoclypeal suture absent or vaguely indicated; hypopharyngeal region with a columnar prehypopharynx located just in front of sclerome; paired ventral prolegs (asperity-bearing ampullae) (fig. 149) usually present on sterna A2 and A3, A3–4 or A2–4; spiracles annular (figs. 194–195) or annular-multiforous (fig. 200). In rotten wood .... (part) ***Oedemeridae*** p. 534

270′. Frontoclypeal suture present (fig. 13); hypopharyngeal region without prehypopharynx; paired ventral prolegs absent; spiracles annular-biforous (fig. 197). In rotten wood .... (*Nematoplus*) **(Nematoplinae)** ***Cephaloidae*** p. 529

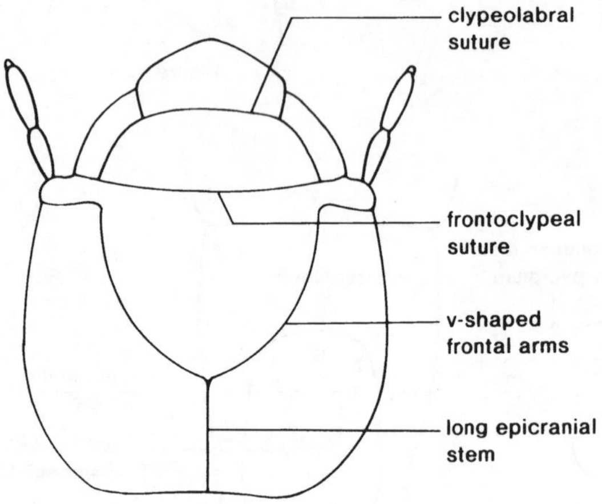

Figure 13

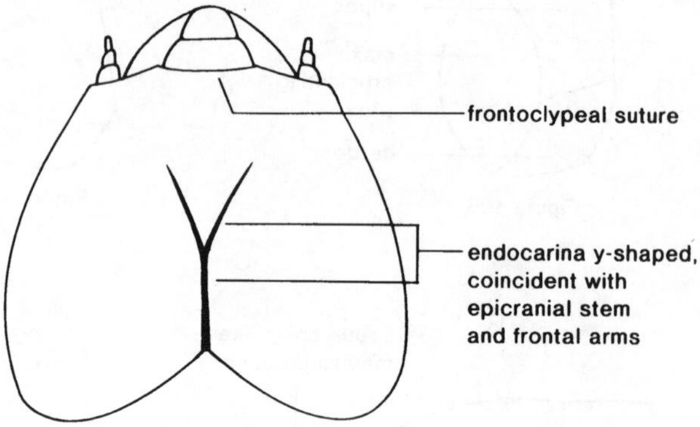

Figure 23

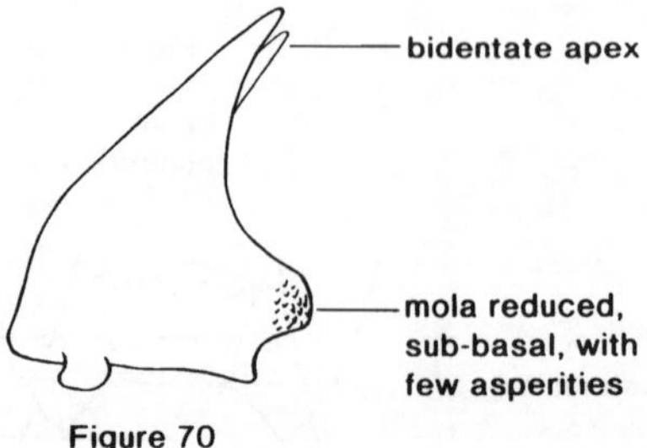

Figure 15

Figure 70

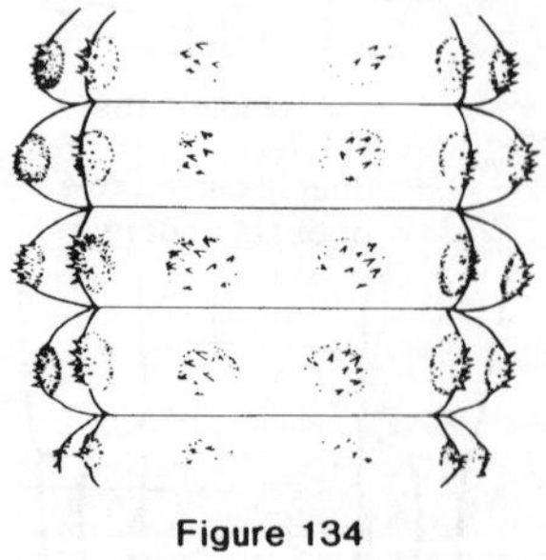

Figure 134

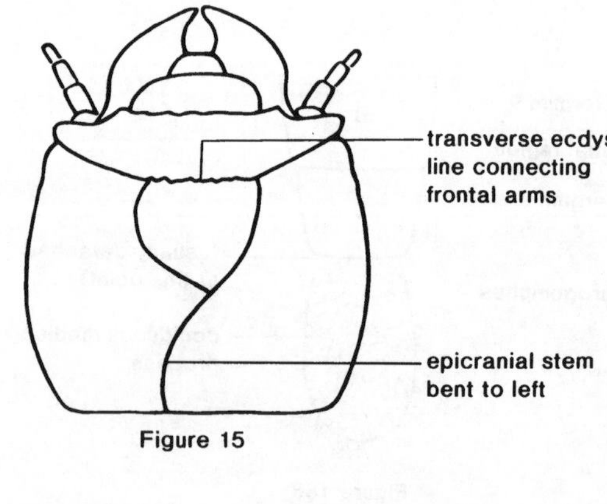

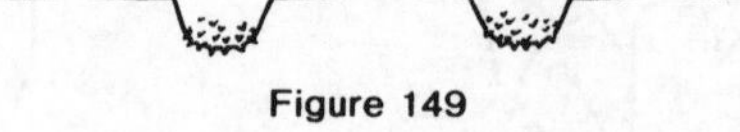

Figure 149

← Figs. 86, 87

271(269′). Abdominal terga with rows of asperities (fig. 144); head heavily sclerotized with 5 well-developed stemmata on each side; frontoclypeal region produced on each side to form a plate; frontal arms joined anteriorly by transverse ecdysial line (fig. 15). On lichens and mosses. Exotic (New Zealand, Australia) .... ***Chalcodryidae*** p. 520

271′. Abdominal terga without rows of asperities; without other characters in combination. Under bark, in rotten wood, soil, leaf litter .... (part) ***Tenebrionidae*** p. 520

272(268′). Segments in labial palp 1 (fig. 91) 273

272′. Segments in labial palp 2 (figs. 88, 94, 97); spiracles annular (fig. 194) 274

273(272). Mandibles asymmetrical; mola well-developed, with tubercles or asperities extending onto ventral surface (fig. 73); setae on tarsungulus 2; spiracles annular. On fungus fruiting bodies .... (*Thrimolus*) ***Mycetophagidae*** p. 498

273′. Mandibles symmetrical; mola reduced, bearing a few tubercles which do not extend onto ventral surface; setae on tarsungulus 1; spiracles annular-biforous (fig. 197). In flowers .... (**Cateretinae**) ***Nitidulidae*** p. 456

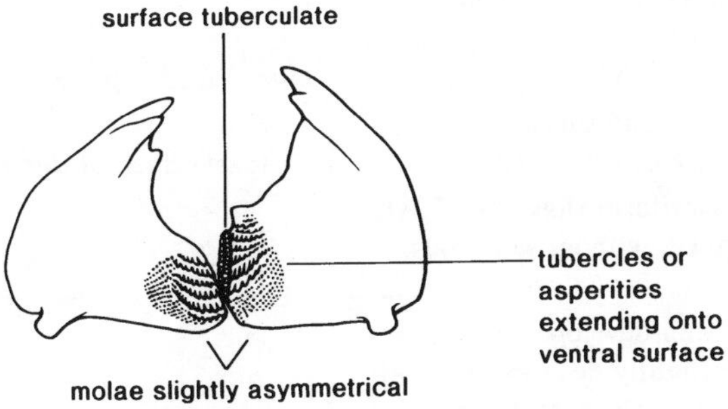

surface tuberculate
tubercles or asperities extending onto ventral surface
molae slightly asymmetrical

Figure 73

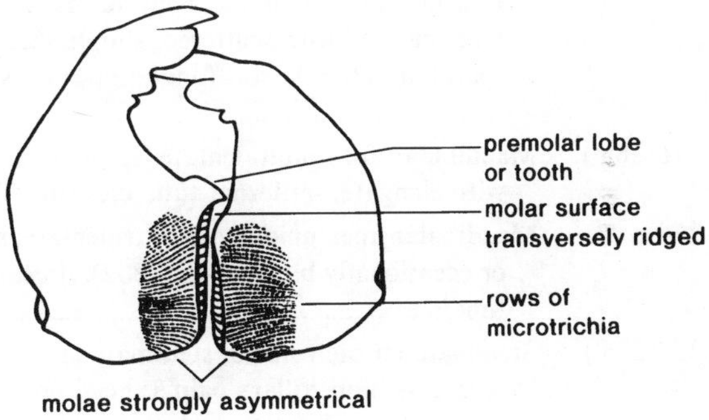

premolar lobe or tooth
molar surface transversely ridged
rows of microtrichia
molae strongly asymmetrical

Figure 74

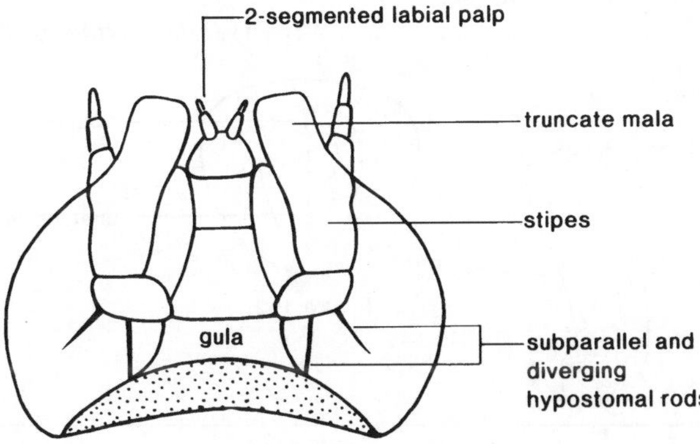

2-segmented labial palp
truncate mala
stipes
gula
subparallel and diverging hypostomal rods

Figure 88

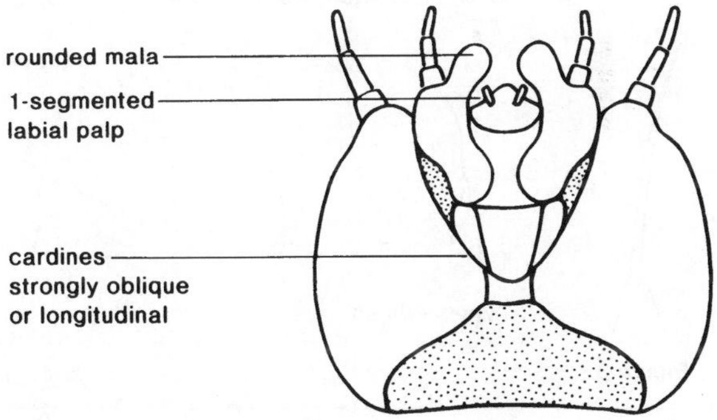

rounded mala
1-segmented labial palp
cardines strongly oblique or longitudinal

Figure 91

Figure 144

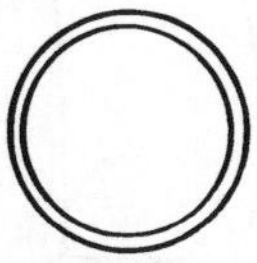

Annular

Figure 194

Annular

Figure 195

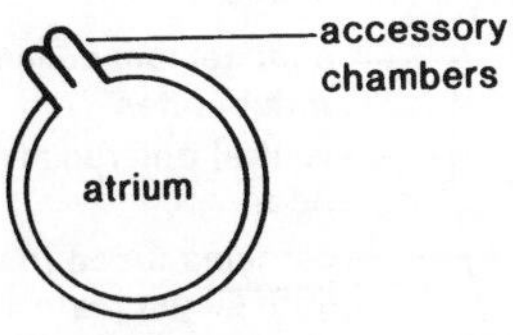

accessory chambers
atrium
Annular Biforous

Figure 197

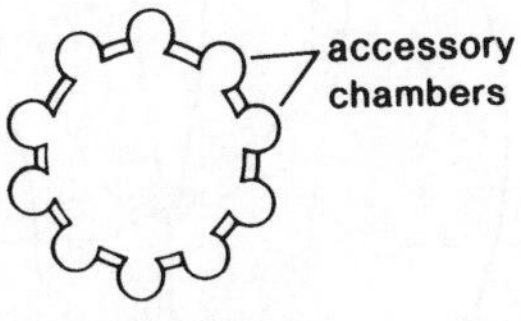

accessory chambers
Annular Multiforous

Figure 200

Figs. 94, 97: See page 225.

274(272′). Stemmata on each side 1, each stemma with a well-developed lens; antennal segment 2 more than 2.5 times as long as segment 1; sensorium dome-like and partly surrounding segment 3; mola sub-basal, reduced, usually with a hyaline lobe or brush of hairs at base; tarsungular setae 2 (1 long and 1 very short); dorsal body surfaces smooth, with vestiture of scattered, long, simple setae .... (part) ***Anthicidae*** p. 552

274′. Stemmata on each side 3 or 0; antennal segment 2 less than 2.5 times as long as segment 1; sensorium conical, often longer than segment 3; mola well-developed and extending to base of mandible; tarsungular setae 1; dorsal body surfaces more complex, usually with transverse rows of protuberances bearing groups of setae or simple or branched processes 275

275(274′). Cardo and stipes fused; mola reduced, sub-basal, and simple or asperate (fig. 70); dorsal surfaces of abdominal segments usually with transverse row of 6 protuberances, which may bear setose or branched processes (fig. 134); tibiae enlarged with group of spatulate setae at apex. On plant surfaces .... (part) ***Coccinellidae*** p. 485

275′. Cardo and stipes distinct; mola well-developed, tuberculate, and extending to base of mandible; dorsal body surfaces smooth, without protuberances of any kind, clothed with scattered, simple setae; tibiae not enlarged, without spatulate setae. In soft fungi, under bark, in leaf litter .... (**Merophysiinae** and **Leiestinae**) ***Endomychidae*** p. 482

276(265′). Mandibular apex multidentate; spiracles annular; dorsal body surfaces covered with elongate, setiferous tubercles (fig. 3). Under bark of rotten logs .... (part) ***Eucinetidae*** p. 364

276′. Mandibular apex unidentate to tridentate; spiracles cribriform (figs. 205–208), or occasionally biforous (fig. 202); dorsal body surfaces without setiferous tubercles 277

277(276′). Stemmata on each side 5; stemmata large, each with well-developed lens; segments in maxillary palp 3; head prognathous or slightly declined; ratio of antennal length to head width less than .15; body relatively straight or slightly curved ventrally, heavily pigmented dorsally; ventral epicranial ridges present (fig. 93); labium almost completely connate with maxillae (fig. 100). In sand or mud along streams .... (part) ***Heteroceridae*** p. 402

Figure 3

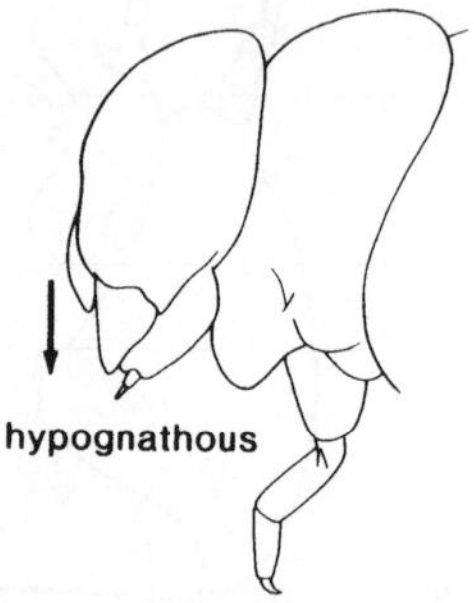

hypognathous

Figure 10

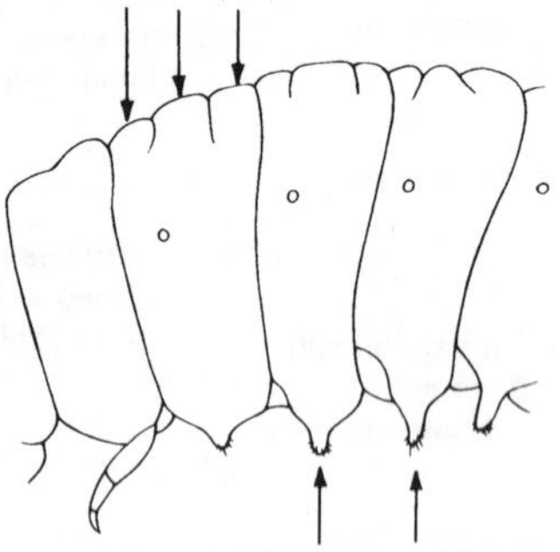

transverse plicae

setose sternal lobes

Figure 136

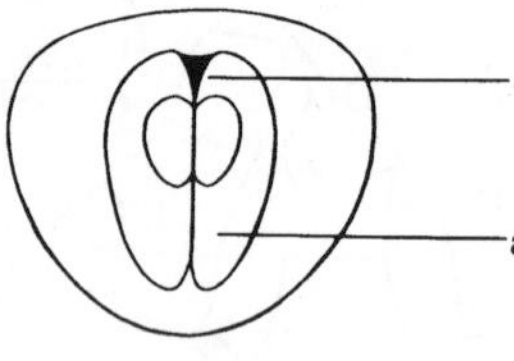

anal opening

anal lobes

Figure 192

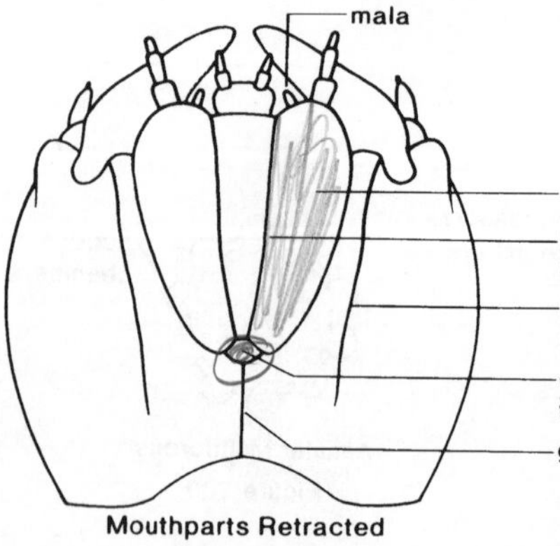

mala

stipes

postmentum connate with stipites

ventral epicranial ridge

cardines fused to form single plate

gular sutures fused

Mouthparts Retracted

Figure 93

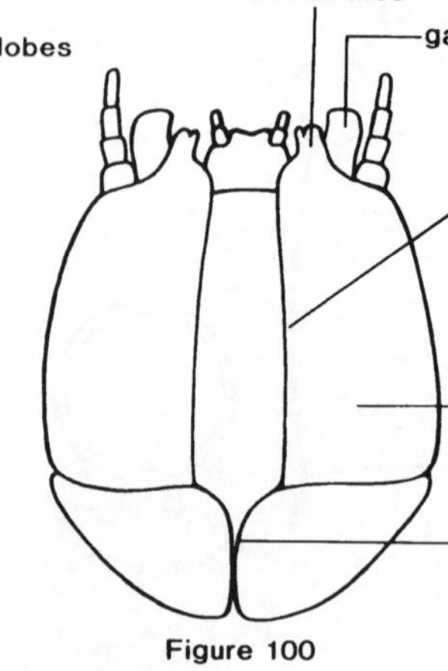

lacinia fixed

galea articulated

postmentum completely connate with stipites

stipes

cardines contiguous

Figure 100

← Figs. 70, 134

277′. Stemmata absent or consisting of single reduced pair; segments in maxillary palp 4; head moderately to strongly declined (hypognathous) (fig. 10); ratio of antennal length to head width greater than .15; body strongly curved ventrally (C-shaped), lightly pigmented; ventral epicranial ridges absent; labium completely free or only basally connate with maxillae (fig. 89) 278

278(277′). Tormae of epipharynx not united mesally (fig. 41); last abdominal segment without fleshy lobes around anus; tergum A3 with 4 transverse plicae bearing patches of short, stout setae. In soil (North America, West Coast) .... **(Pleocominae)** ***Scarabaeidae*** p. 377

278′. Tormae of epipharynx united mesally (fig. 40); last abdominal segment usually with fleshy lobes around anus (fig. 192); tergum A3 with fewer than 4 transverse plicae (fig. 136), with or without patches of stout setae 279

279(278′). Anterior edge of labrum with 3 truncate lobes (fig. 35); 3rd antennal segment almost as long as 2nd, and bearing a sensorium and apical process representing fused 4th segment (fig. 34); stridulatory organs formed from pro- and mesothoracic legs. In soil, dung .... **(Hybosorinae)** ***Scarabaeidae*** p. 377

279′. Anterior edge of labrum without 3 truncate lobes; 3rd antennal segment much shorter than 2nd, which bears 1 or more sensoria (as in fig. 29); stridulatory organs, if present, formed from T2 and T3 legs 280

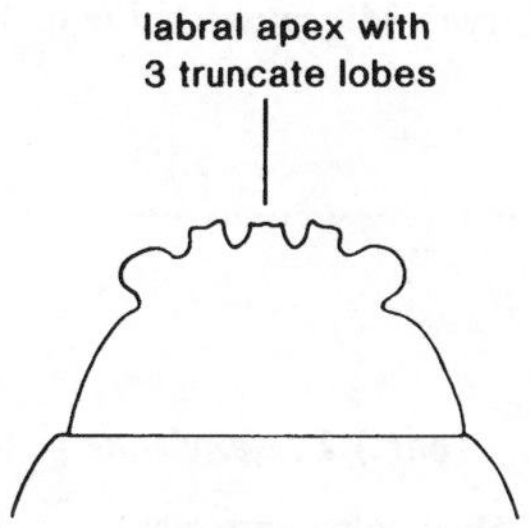

labral apex with 3 truncate lobes

Figure 35

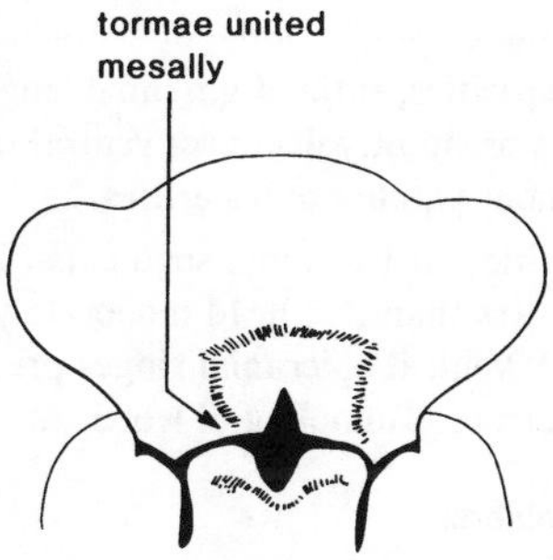

tormae united mesally

Figure 40

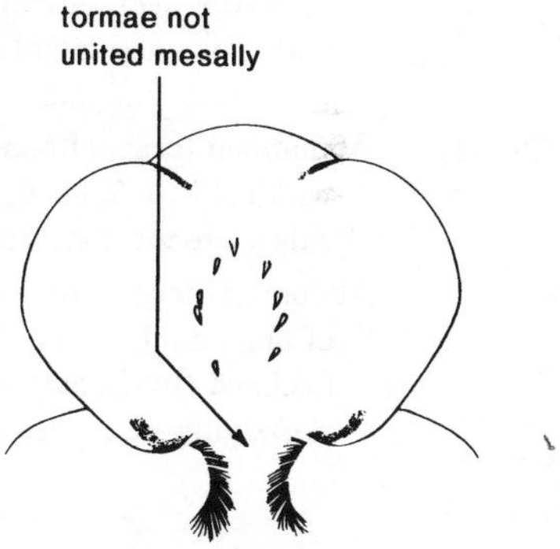

tormae not united mesally

Figure 41

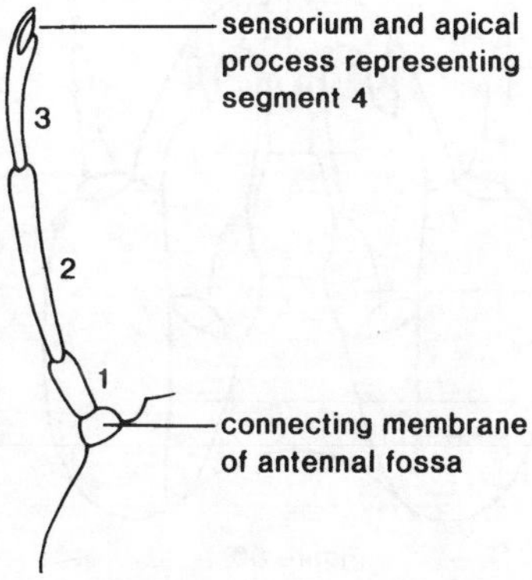

sensorium and apical process representing segment 4
3
2
1
connecting membrane of antennal fossa

Figure 34

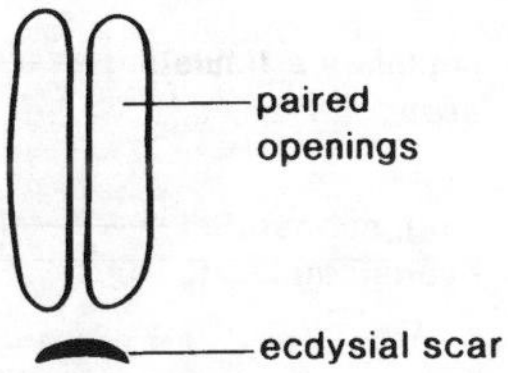

paired openings
ecdysial scar
Biforous

Figure 202

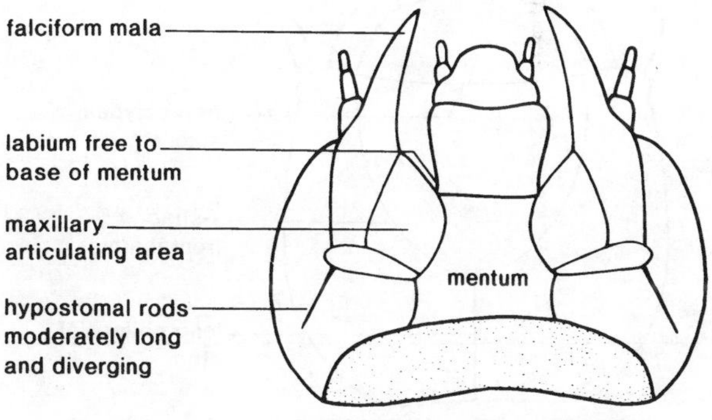

falciform mala
labium free to base of mentum
maxillary articulating area
mentum
hypostomal rods moderately long and diverging
Mouthparts Retracted

Figure 89

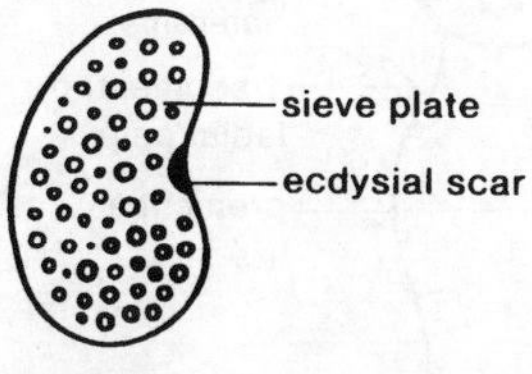

sieve plate
ecdysial scar
Cribriform

Figure 205

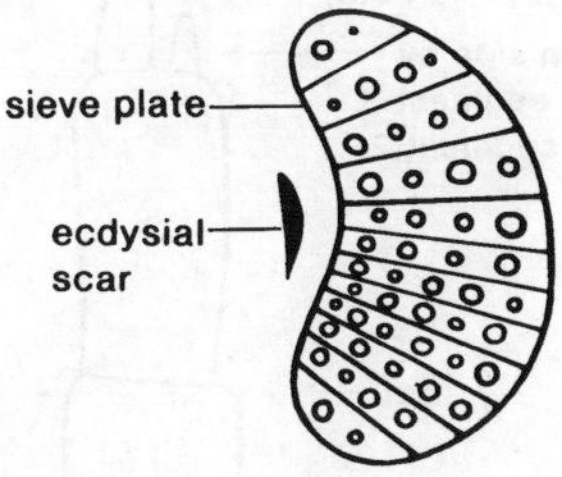

sieve plate
ecdysial scar
Cribriform

Figure 206

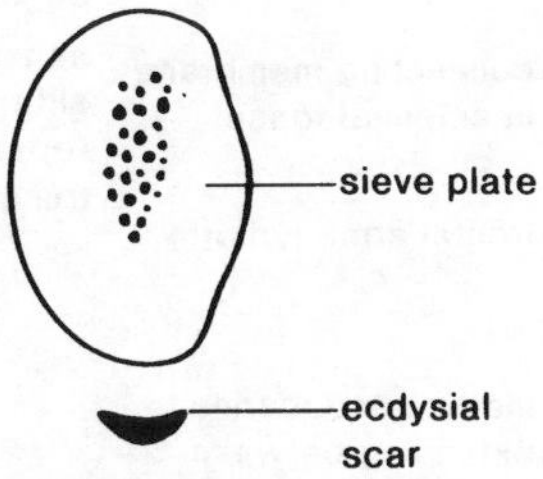

sieve plate
ecdysial scar
Cribriform

Figure 207

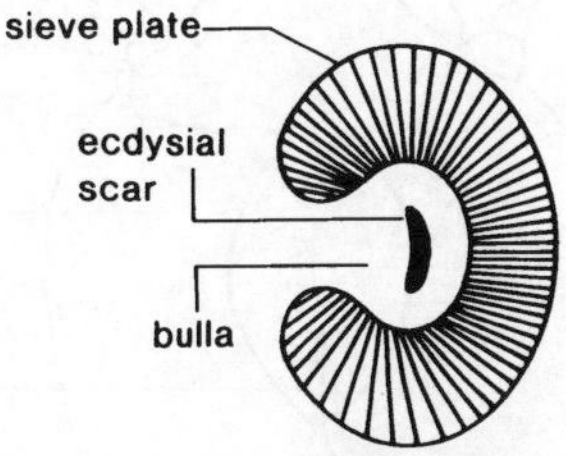

sieve plate
ecdysial scar
bulla
Cribriform

Figure 208

Fig. 29 ⟶

280(279′). Tergum A3 with 3 distinct transverse plicae (fig. 136), each bearing one or more rows of short, stiff setae; T2 and T3 legs not forming stridulatory organs; head much more darkly pigmented than body (except for T1 shield); spiracles cribriform (fig. 205) or biforous (fig. 202). In carcasses .... **(Troginae)** ***Scarabaeidae*** p. 377

280′. Tergum A3 with 2 transverse plicae which do not bear rows or patches of stiff setae or asperities; head and body very lightly pigmented; legs often with fewer than 5 segments, T3 legs sometimes greatly reduced; T2 and T3 legs usually forming stridulatory organs. In burrows in ground **(see 3rd choice)** .... **(Geotrupinae)** ***Scarabaeidae*** p. 377

280². Tergum A3 without distinct transverse plicae; T2 and T3 legs always forming stridulatory organs (asperities on mesocoxa (fig. 129) and plectrum on metatrochanter (fig. 130)); head not much darker than body; spiracles always cribriform (fig. 208); abdominal terga usually with patches of asperities (fig. 141). In rotten wood .... (part) ***Lucanidae*** p. 372

281(262′). Segments in T2 leg 5 or less including tarsungulus (fig. 126); segments in labial palp 2; frontal arms present; ligula without wedge-like sclerome; maxilla with single mala (fig. 86) 282

281′. Segments in T2 leg 6 including paired claws (fig. 124); segments in labial palp 1; frontal arms absent; ligula with wedge-like sclerome (fig. 117); maxilla with galea and lacinia (fig. 100); length less than 2mm. In rotten wood .... (caraboid larva) ***Micromalthidae*** p. 300

282(281). Abdominal terga without rows of asperities; ratio of antennal length to head width .15 to .5; head prognathous or slightly declined; ventral epicranial ridges absent; thoracic terga without patches of asperities 283

282′. Abdominal terga with rows of asperities on 1 or more segments (fig. 144); ratio of antennal length to head width less than .15; head moderately to strongly declined (hypognathous) (fig. 10); ventral epicranial ridges present (fig. 93); protergum with patch(es) of asperities. Tunneling in wood .... (part) ***Lymexylidae*** p. 446

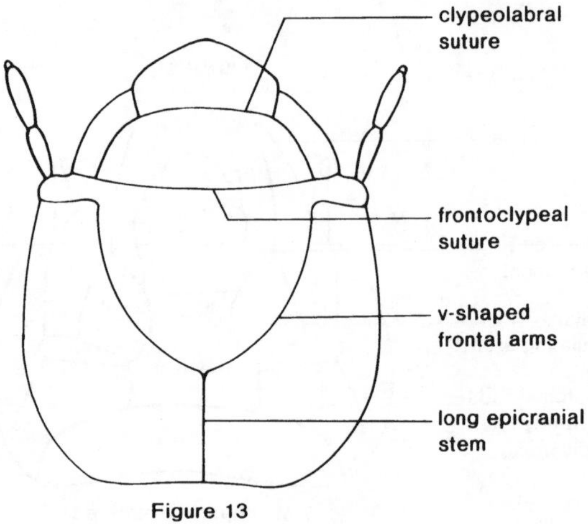

Figure 13

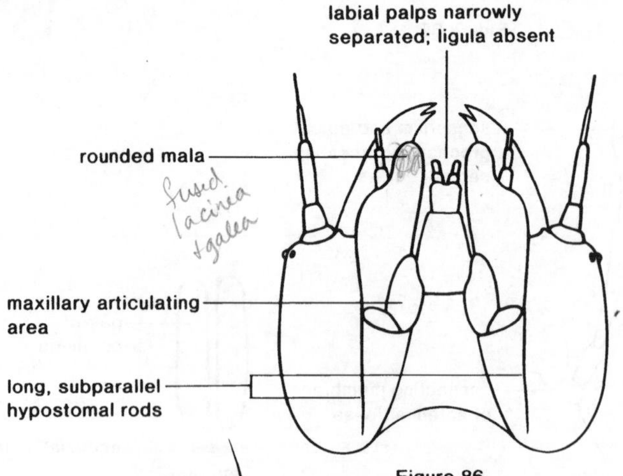

Figure 86

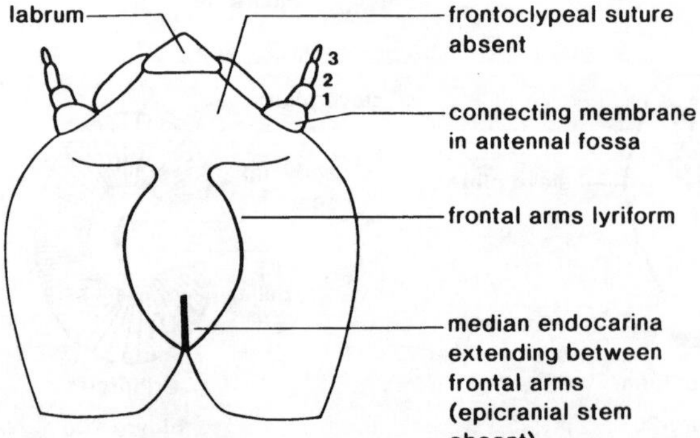

Figure 22

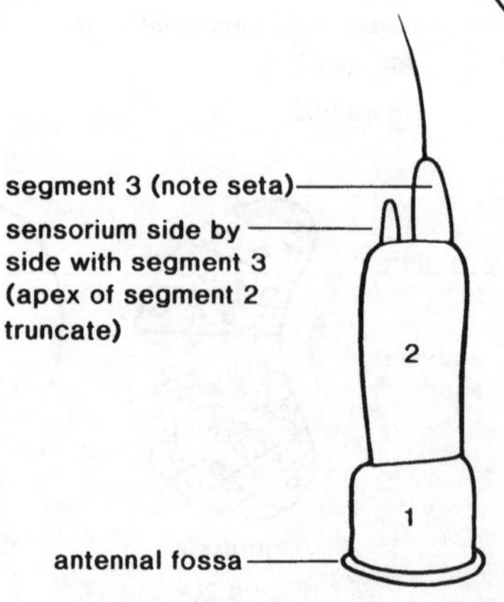

Figure 29

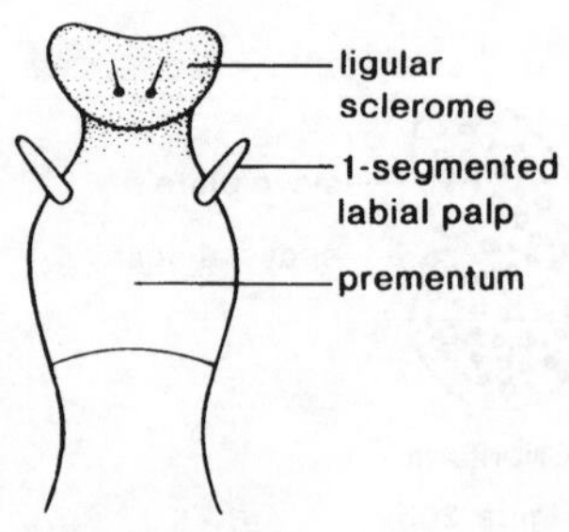

Figure 117

← Figs. 10, 89, 93, 100, 136, 202, 205, 208

283(282). Frontal arms lyriform (fig. 22); posterior edge of head capsule distinctly emarginate dorsally; process on tergum A9 lightly pigmented, rounded, and deciduous (fig. 158) OR hypostomal rods present (fig. 89), frontoclypeal suture absent, and spiracles annular-biforous (fig. 197) 284

283′. Frontal arms V- or U-shaped (fig. 13); posterior edge of head capsule not or only slightly emarginate dorsally; process on tergum A9 not rounded or deciduous; hypostomal rods absent; frontoclypeal suture distinct (fig. 13); spiracles annular or annular-multiforous (fig. 200) .... (part) ***Tenebrionidae*** p. 520

284(283). Paired endocarinae absent; frontoclypeal suture distinct (fig. 13); apex of mala cleft (fig. 106); hypostomal rods absent; abdominal spiracles annular. Median process on tergum A9 lightly sclerotized, rounded apically, pubescent, and deciduous (fig. 158). In leaf litter .... (**Scraptiinae**) ***Scraptiidae*** p. 555

284′. Paired endocarinae present (fig. 24); frontoclypeal suture absent or vaguely indicated; apex of mala simple; hypostomal rods present (fig. 89); abdominal spiracles annular-biforous (fig. 197); median process on tergum A9 sclerotized and more or less acute (fig. 159). Under bark, in rotten wood .... (part) ***Colydiidae*** p. 512

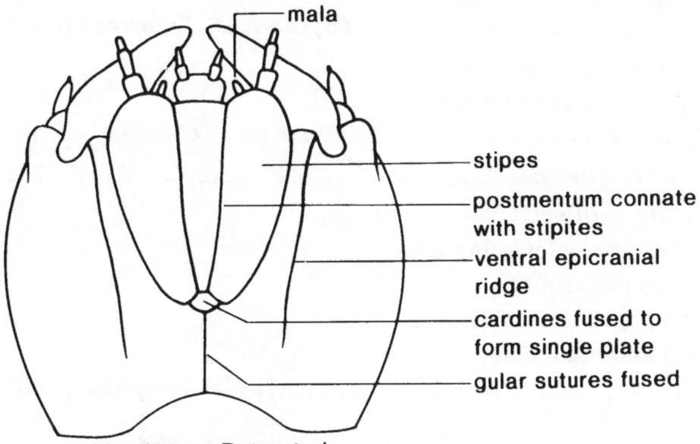

mala
stipes
postmentum connate with stipites
ventral epicranial ridge
cardines fused to form single plate
gular sutures fused
Mouthparts Retracted

Figure 93

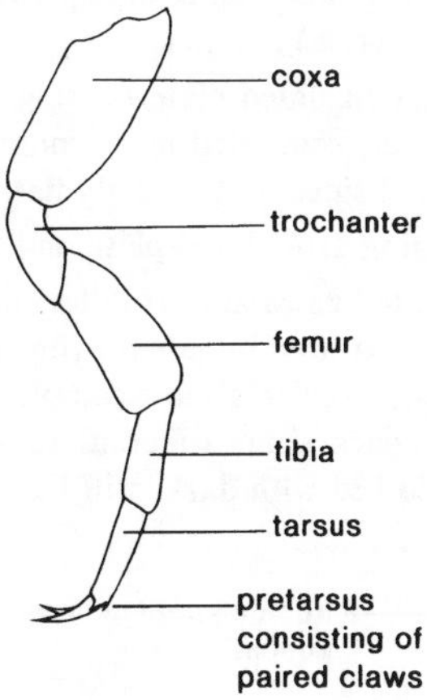

coxa
trochanter
femur
tibia
tarsus
pretarsus consisting of paired claws

Figure 124

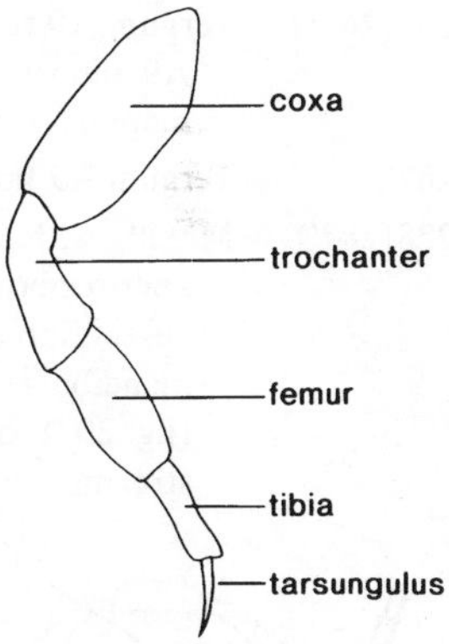

coxa
trochanter
femur
tibia
tarsungulus

Figure 126

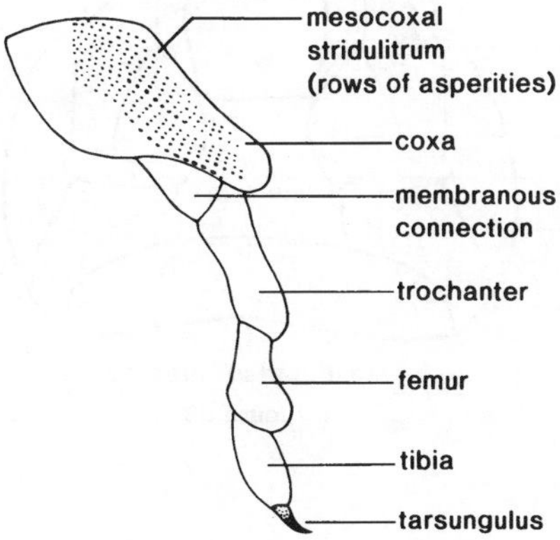

mesocoxal stridulitrum (rows of asperities)
coxa
membranous connection
trochanter
femur
tibia
tarsungulus

Figure 129

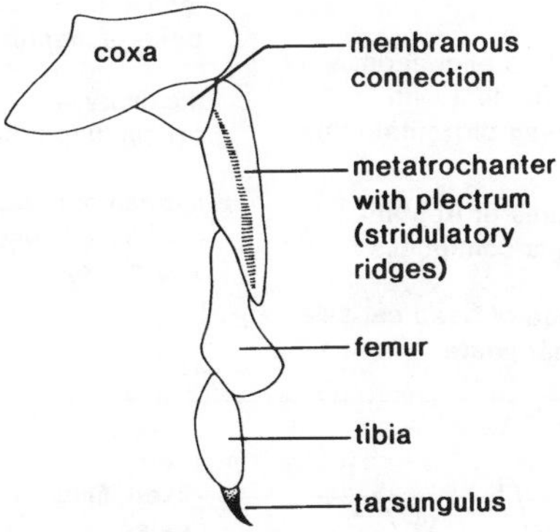

coxa
membranous connection
metatrochanter with plectrum (stridulatory ridges)
femur
tibia
tarsungulus

Figure 130

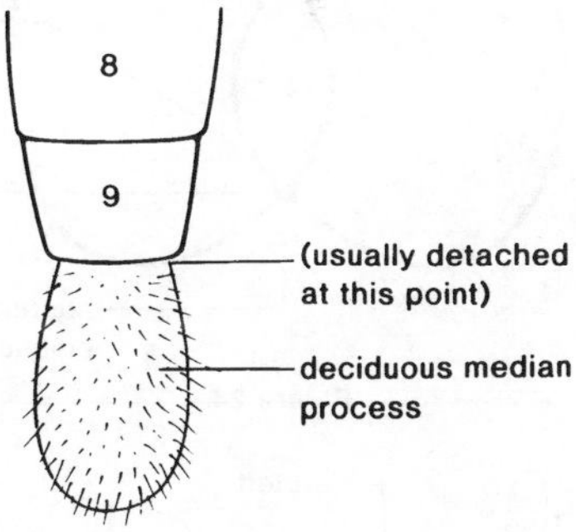

8
9
(usually detached at this point)
deciduous median process

Figure 158

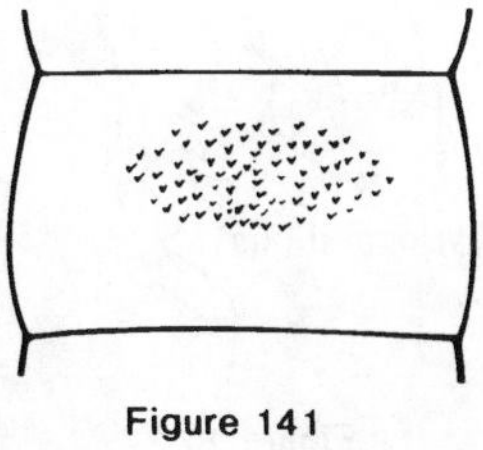

Figure 141

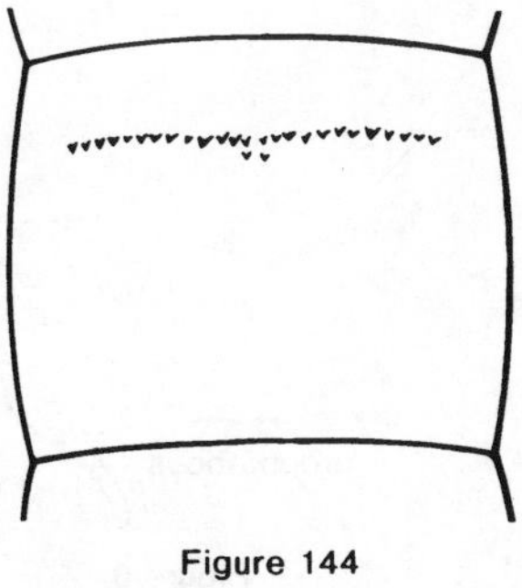

Figure 144

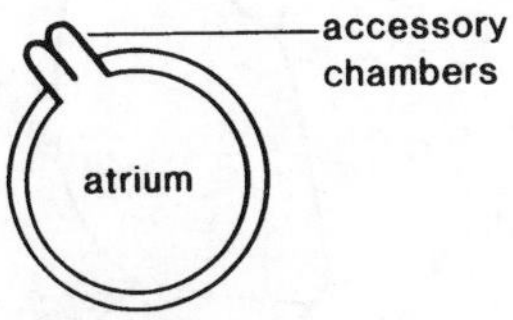

accessory chambers
atrium
Annular Biforous

Figure 197

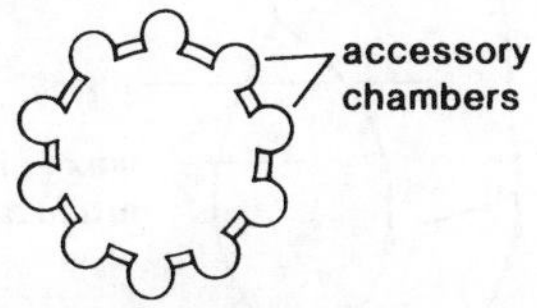

accessory chambers
Annular Multiforous

Figure 200

Figs. 24, 106, 159 ⟶

285(262[2]). Head prognathous or slightly declined (fig. 8); ratio of antennal length to head width .15 to .5; mandibles asymmetrical; thoracic terga without patches of asperities; apex of mala simple. Under bark, in rotten wood and fungus fruiting bodies .... (part) ***Tenebrionidae*** p. 520

285′. Head moderately to strongly declined (hypognathous) (fig. 10); ratio of antennal length to head width less than .15; mandibles symmetrical; protergum with patch(es) of asperities; apex of mala cleft (fig. 106). Tunneling in wood .... (part) ***Lymexylidae*** p. 446

286(261′). Body circular in cross-section; ventral epicranial ridges present (fig. 90); spiracles on A8 located at ends of spiracular tubes; prostheca consisting of a brush of comb-hairs (fig. 80); accessory ventral process of mandible present (fig. 80); molar surface with numerous fine ridges (fig. 80). Under bark or in rotting wood .... (part) ***Biphyllidae*** p. 475

286′. Body slightly flattened; ventral epicranial ridges absent; abdominal spiracles not located at ends of spiracular tubes; prostheca consisting of brush of simple hairs or several hyaline processes; accessory ventral process of mandible absent; molar surface tuberculate or asperate (fig. 81). In oak catkins (western North America) .... (*Byturellus*) ***Byturidae*** p. 476

287(261[2]). Tergum A9 forming articulated plate bearing median, forked process; sternum A9 almost completely concealed by sternum A8 (fig. 164); body elongate, more or less parallel-sided and strongly flattened. Under bark .... **(Cucujinae part) *Cucujidae*** p. 463

287′. Tergum A9 not forming articulated plate and without forked process 288

288(287′). Maxilla with articulated galea and fixed lacinia (fig. 100); cardines approximate, not separated by labium (fig. 100); maxillary articulating area absent; labium and maxillae almost completely connate (fig. 100); mandibular mola concave, not tuberculate or asperate; spiracles cribriform (fig. 207); body clothed with dark, stiff hairs. In sand and mud along streams .... (part) ***Heteroceridae*** p. 402

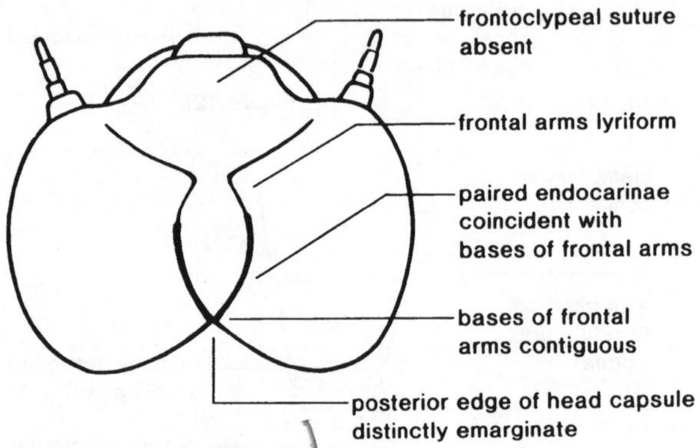

frontoclypeal suture absent
frontal arms lyriform
paired endocarinae coincident with bases of frontal arms
bases of frontal arms contiguous
posterior edge of head capsule distinctly emarginate

Figure 24

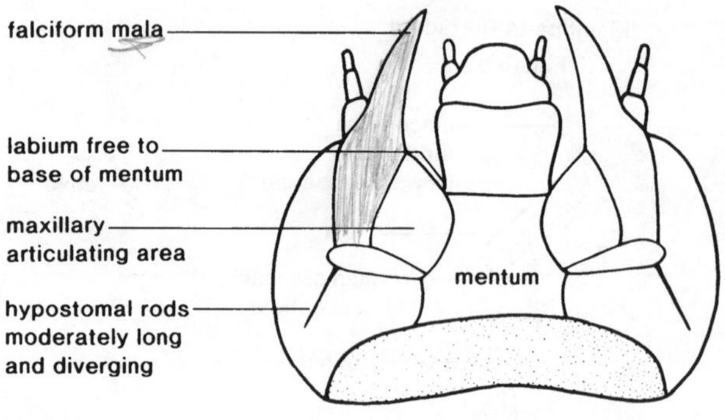

falciform mala
labium free to base of mentum
maxillary articulating area
hypostomal rods moderately long and diverging
mentum
Mouthparts Retracted

Figure 89

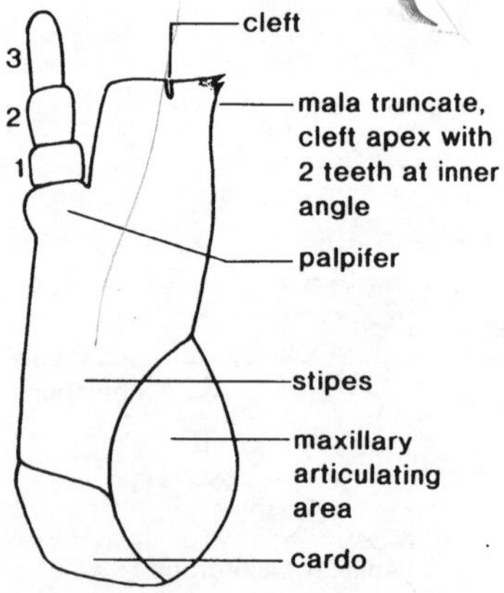

cleft
mala truncate, cleft apex with 2 teeth at inner angle
3
2
1
palpifer
stipes
maxillary articulating area
cardo

Figure 106

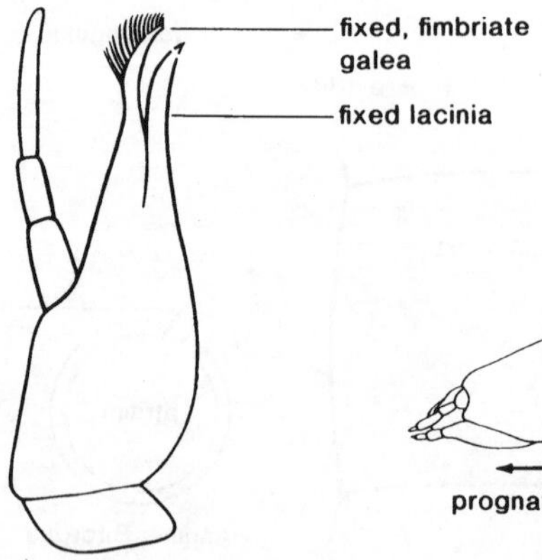

fixed, fimbriate galea
fixed lacinia

Figure 111

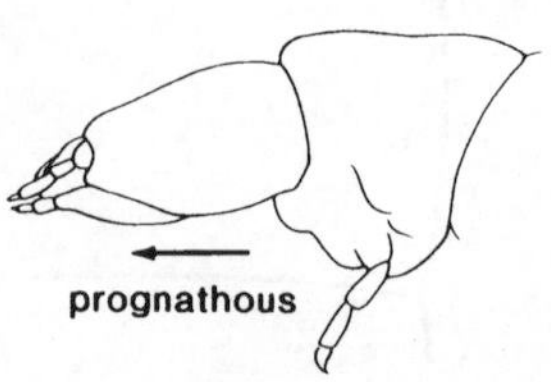

prognathous

Figure 8

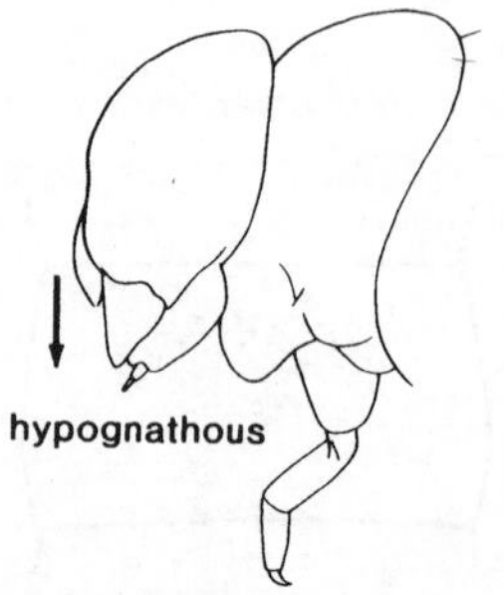

hypognathous

Figure 10

← Figs. 86, 197

288′. Maxilla with fixed galea and lacinia (fig. 111) **(see 3rd choice)** 289

288². Maxilla with single fixed mala (fig. 86) 292

289(288′). Segments in labial palp 2; stemmata on each side 5 or 6 290

289′. Segments in labial palp 1; stemmata on each side 4 or fewer 291

290(289). Antennal segment 2 about 3 times as long as segment 3; head strongly narrowed anteriorly; spiracles annular; mandibular mola well-developed (fig. 78); stemmata on each side 5. In fruiting bodies of slime molds (Myxomycetes), in leaf litter, under bark .... (part) ***Eucinetidae*** p. 364

290′. Antennal segment 2 less than twice as long as segment 3; head not strongly narrowed anteriorly; spiracles annular-biforous (fig. 197); mola somewhat reduced, not extending to base of mandible (fig. 70); stemmata on each side 6. On bark surfaces feeding on woolly aphids (Adelgidae) .... (*Laricobius*) (Laricobiini) ***Derodontidae*** p. 431

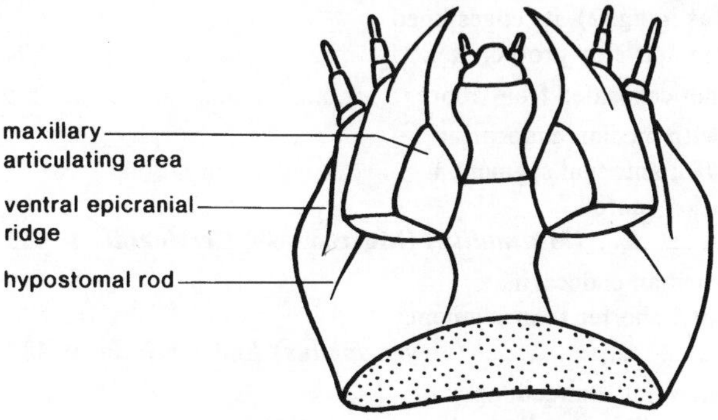

Mouthparts Retracted

Figure 90

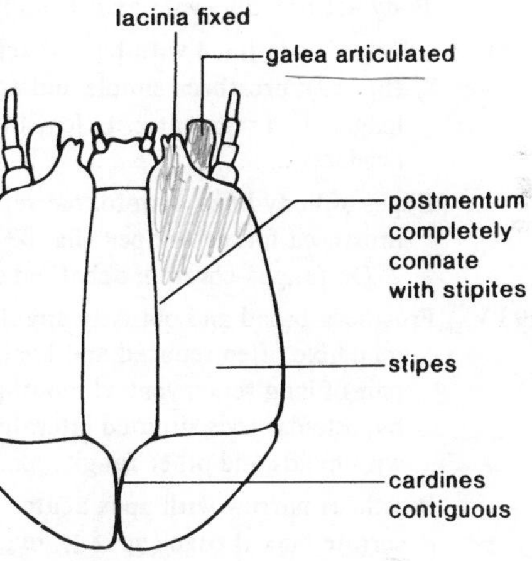

Figure 100

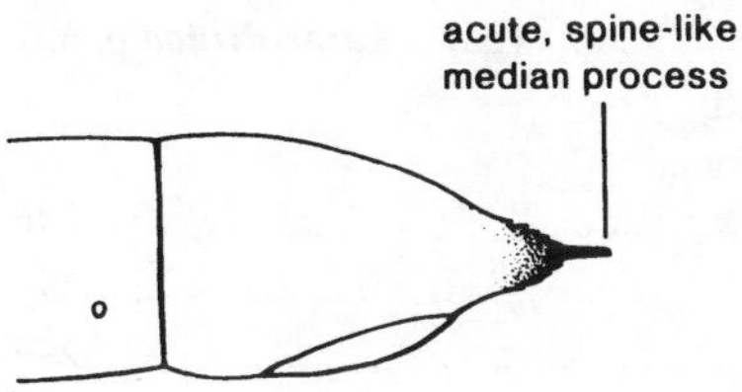

Figure 159

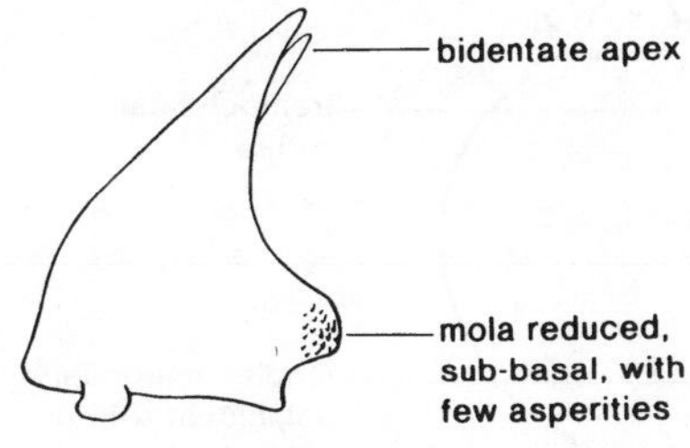

Figure 70

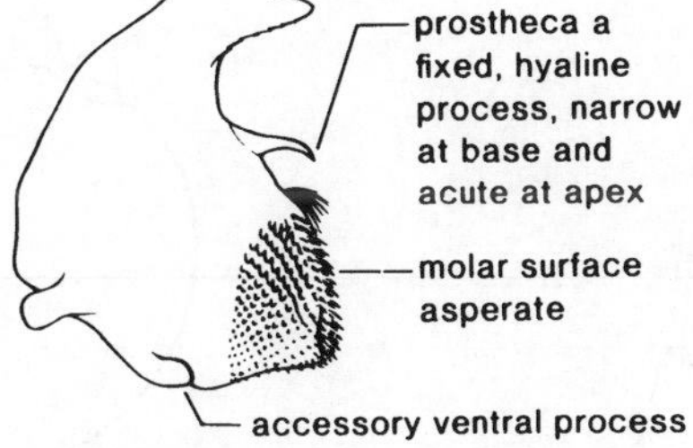

Figure 78

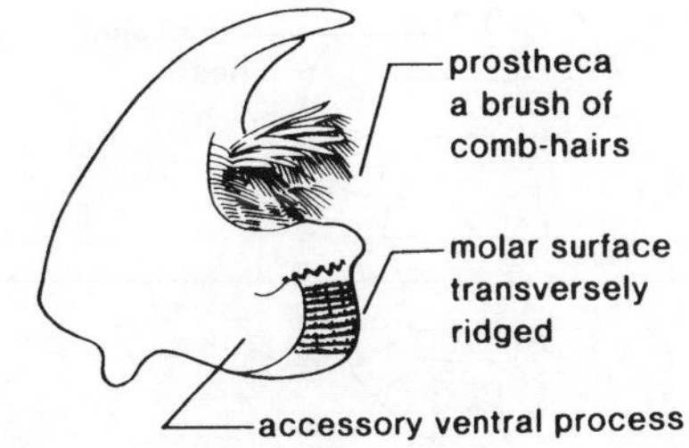

Figure 80

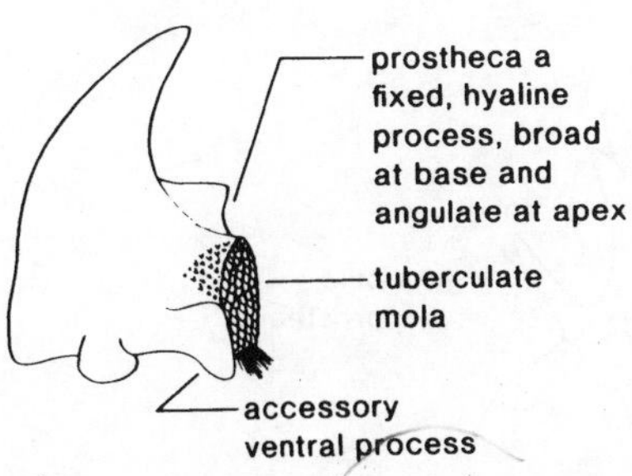

Figure 81

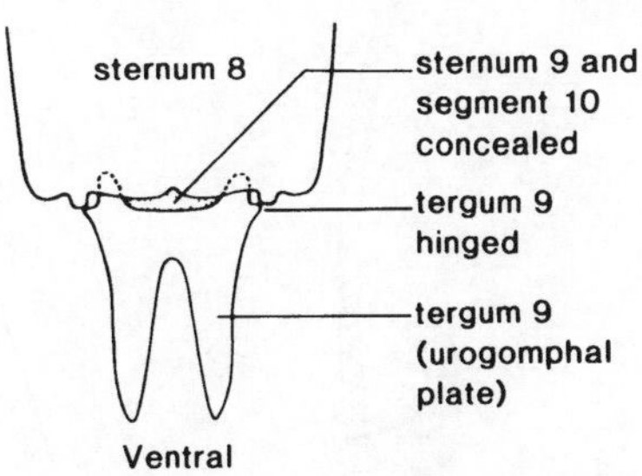

Ventral

Figure 164

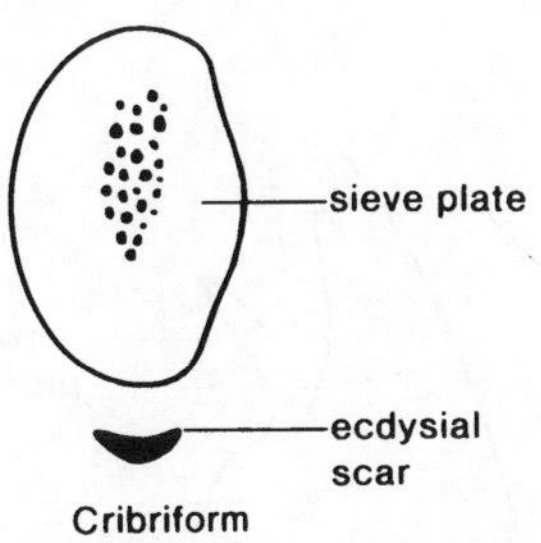

Cribriform

Figure 207

291(289′). Ratio of antennal length to head width less than .2; antennal segment 2 only slightly longer than segment 3; segments in maxillary palp 2; thoracic and abdominal segments with lateral tergal processes; dorsal surfaces covered with setiferous tubercles (fig. 3). In leaf litter, hay stacks, fungi. Exotic (Neotropical and Old World) .... (*Acalyptomerus*) (**Calyptomerinae**) ***Clambidae*** p. 365

291′. Ratio of antennal length to head width more than .5; antennal segment 2 much longer than segment 3; segments in maxillary palp 3; thoracic and abdominal segments without lateral tergal processes; dorsal surfaces clothed with simple setae. On molds .... (*Eufallia*) (**Lathridiinae**) ***Lathridiidae*** p. 497

292(288[2]). Segment A10 with pair of hooks (fig. 163); body minute (length less than 1.2 mm), elongate and cylindrical; head moderately to strongly declined (hypognathous) (fig. 10); stemmata absent. In pore tubes of bracket fungi .... (**Nanosellinae**) ***Ptiliidae*** p. 322

292′. Segment A10 without hooks; without other characters in combination 293

293(292′). Body broadly ovate, strongly flattened and disc-like (as in fig. 2), its edges lined with complex setae; head completely concealed from above by prothorax 294

293′. Body not broadly ovate and strongly flattened; head not concealed from above 295

294(293). Edges of body lined with barbed setae (fig. 6); head with median endocarina (fig. 19); prostheca simple and acute at apex (fig. 78); antennal segment 1 longer than segment 2. In leaf litter, rotting vegetation, stored products .... (*Murmidius*) (**Murmidiinae**) ***Cerylonidae*** p. 480

294′. Edges of body lined with forked setae; head without median endocarina; prostheca forked at apex (fig. 83); antennal segment 1 shorter than segment 2. On fungus-covered, debarked surfaces of trees .... (*Agaricophilus*) ***Endomychidae*** p. 482

295(293′). Prostheca broad and obtusely angulate (fig. 81) or somewhat rounded; apex of mandible often reduced and hyaline; outer edge of mandible usually with pair of long setae; ventral mouthparts often more or less protracted with hypostomal rods situated laterally. In leaf litter, stored products, associated with molds and other fungi .... (part) ***Lathridiidae*** p. 497

295′. Prostheca narrow with apex acute (fig. 78) or bifid (fig. 83), sometimes with serrate buccal edge (fig. 82); mandibles different than above, and hypostomal rods, if present, ventrally situated 296

296(295′). Mala more or less truncate (fig. 88) 297

296′. Mala falciform (fig. 89) 299

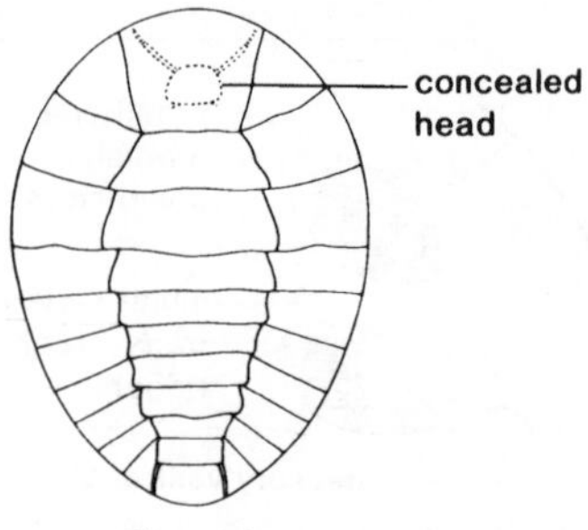

concealed head

Figure 2

Figure 3

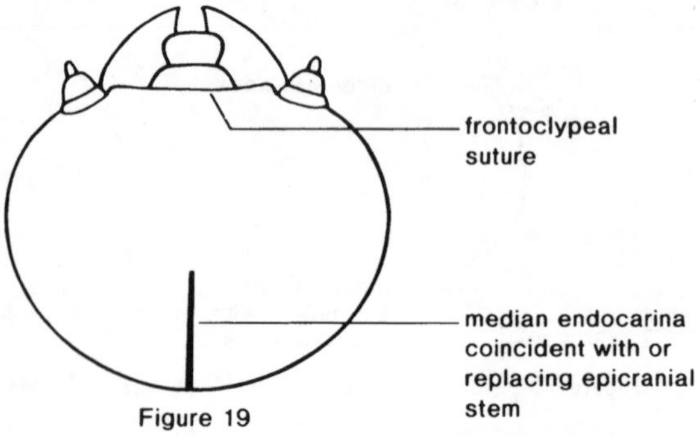

frontoclypeal suture

median endocarina coincident with or replacing epicranial stem

Figure 19

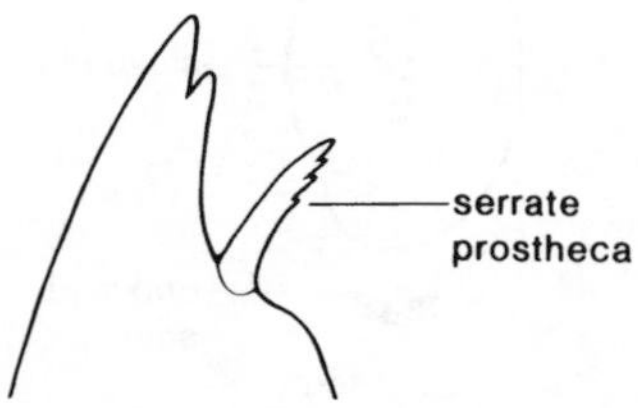

serrate prostheca

Figure 82

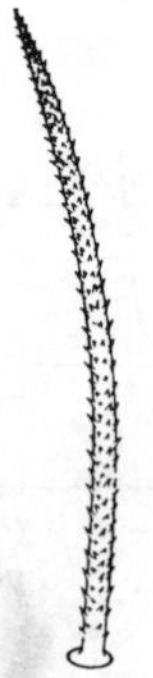

Figure 6

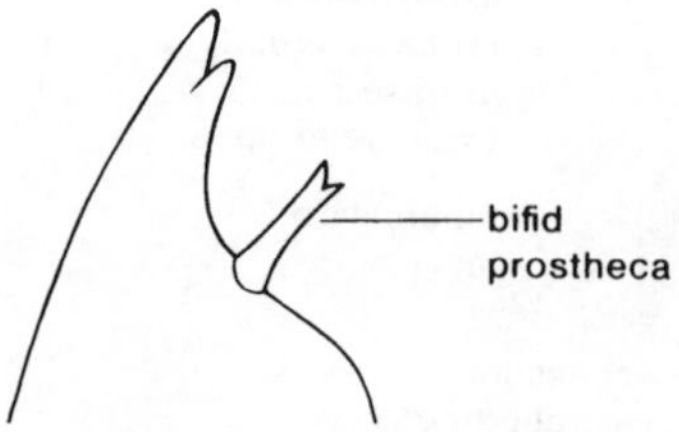

bifid prostheca

Figure 83

← Figs. 8, 10, 78, 81

297(296). Head strongly narrowed anteriorly; stemmata on each side 4; head moderately to strongly declined (hypognathous) (fig. 10); segment A10 ventrally oriented and not visible from above. Under bark, on wood surfaces, associated with fungi .... (**Eumorphinae**) ***Endomychidae*** p. 482

297′. Head not strongly narrowed anteriorly; stemmata on each side 6, occasionally weak or absent; head prognathous or slightly declined (fig. 8); segment A10 posteriorly or terminally oriented and more or less visible from above 298

298(297′). Antennal segment 3 much longer than 2; tarsungulus with single seta. In leaf litter, hay stacks, fruiting bodies of slime molds (Myxomycetes) .... (**Clambinae**) ***Clambidae*** p. 365

298′. Antennal segment 2 much longer than segment 3; tarsungulus with 2 setae. In fruiting bodies of slime molds (Myxomycetes) .... (part) ***Sphindidae*** p. 455

299(296′). Body more or less ovate; thoracic and abdominal terga produced laterally; body densely clothed with long, fine hairs; accessory ventral process of mandible absent. In leaf litter, rotten wood, fungi .... (**Mychotheninae** part) ***Endomychidae*** p. 482

299′. Body elongate and more or less parallel-sided or somewhat narrowed posteriorly; thoracic and abdominal terga not produced laterally; accessory ventral process of mandible present (fig. 78); body not densely clothed with long hairs 300

300(299′). Segment A9 reduced, its sternum usually concealed and its tergum much shorter than that of segment A8 or A10 (fig. 167); antennal segment 3 highly reduced. Under bark, in leaf litter, stored products .... (Silvanini part) ***Cucujidae*** p. 463

300′. Segment A9 not reduced, its sternum well-developed and exposed; antennal segment 3 not reduced 301

301(300′). Antennal segment 3 shorter than sensorium; stemmata on each side 5. In leaf litter, hay stacks .... (*Calyptomerus*) (**Calyptomerinae**) ***Clambidae*** p. 365

301′. Antennal segment 3 much longer than sensorium; stemmata on each side 1 302

302(301′). Prostheca simple at apex; labial palps 1-segmented; spiracles annular-biforous (fig. 197). In flowers .... (*Telmatophilus*) (**Cryptophaginae**) ***Cryptophagidae*** p. 469

302′. Prostheca bifid at apex (fig. 83); labial palps 2-segmented; spiracles annular. In leaf litter, grass piles .... (**Atomariinae** part) ***Cryptophagidae*** p. 469

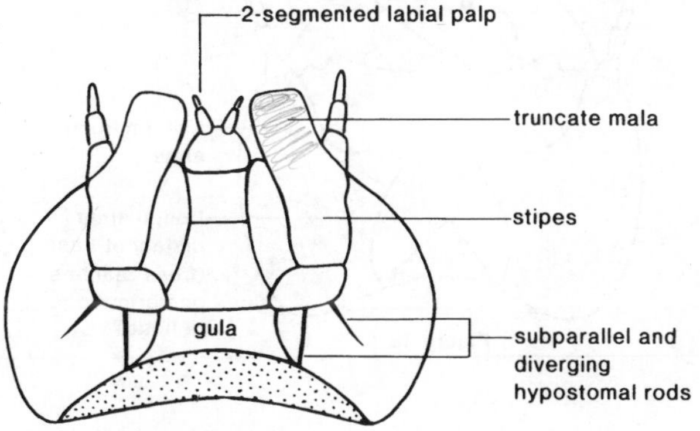

2-segmented labial palp
truncate mala
stipes
gula
subparallel and diverging hypostomal rods

Figure 88

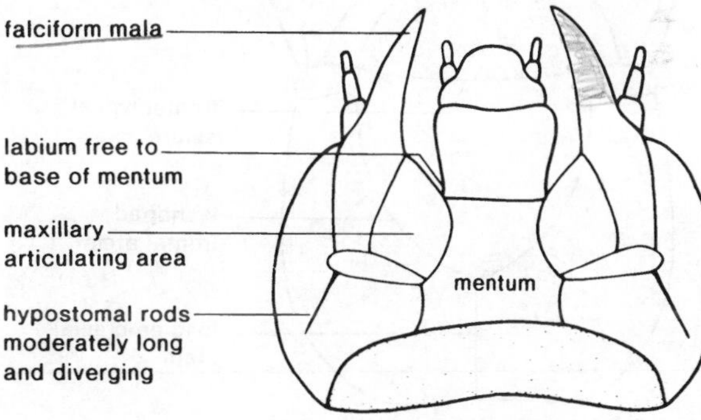

falciform mala
labium free to base of mentum
maxillary articulating area
mentum
hypostomal rods moderately long and diverging
Mouthparts Retracted

Figure 89

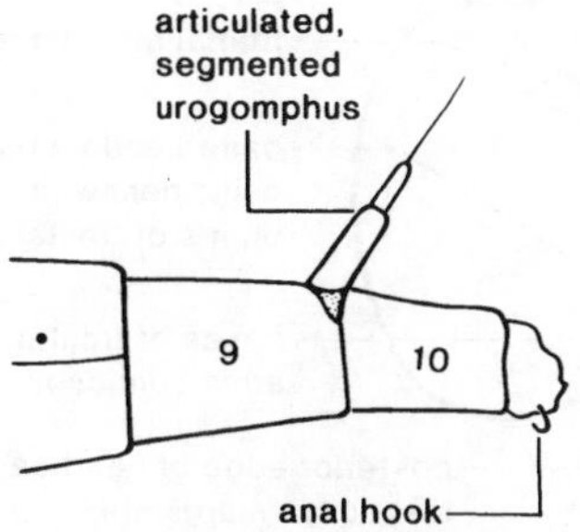

articulated, segmented urogomphus
9
10
anal hook

Figure 163

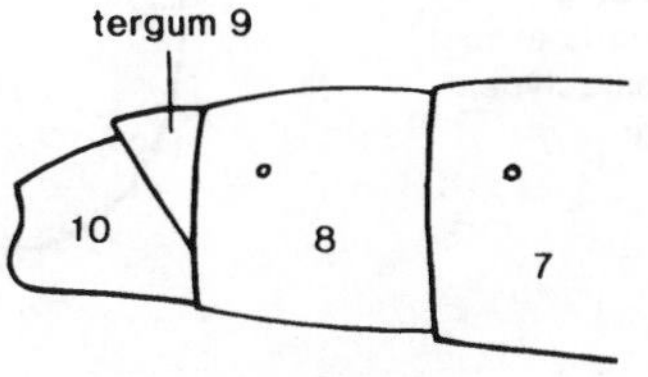

tergum 9
10
8
7

Figure 167

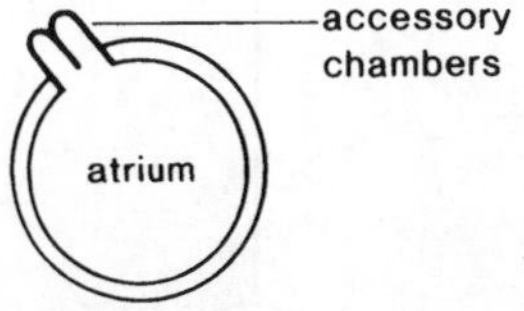

accessory chambers
atrium
Annular Biforous

Figure 197

303(261[4]). Body relatively straight or only slightly curved ventrally; segment A10 without oval lobes separated by longitudinal groove; maxilla with single mala (fig. 86); head prognathous or slightly declined and protracted or slightly retracted (fig. 8). On leaf surfaces, mildew, in leaf litter, stored products .... (part) ***Lathridiidae*** **p. 497**

303′. Body strongly curved ventrally (C-shaped); segment A10 with oval lobes separated by longitudinal groove (fig. 193); maxilla with galea and lacinia (fig. 215); head either hypognathous (fig. 10) or strongly retracted (fig. 9) 304

304(303′). Head moderately to strongly declined (hypognathous) (fig. 10) and protracted or slightly retracted; segments in maxillary palp 4. In fungus fruiting bodies .... (*Endecatomus*) (**Endecatominae**) ***Bostrichidae*** p. 439

304′. Head prognathous or slightly declined and strongly retracted (fig. 9); segments in maxillary palp 3. In dead wood 305

305(304′). Spiracles on segment A8 much larger than those on A1–7 .... (**Lyctinae part**) ***Bostrichidae*** p. 439

305′. Spiracles on segment A8 about same size as those on A1–7 .... (**Dinoderinae, Dysidinae, Lyctinae part**) ***Bostrichidae*** p. 439

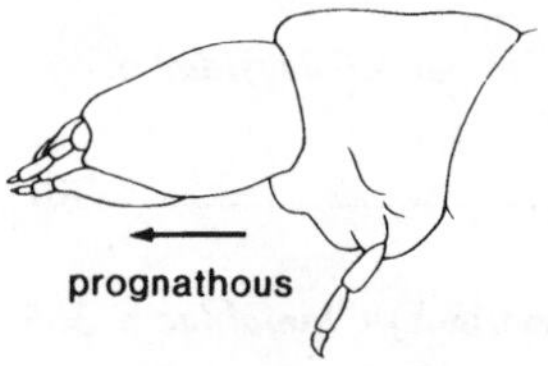

prognathous

Figure 8

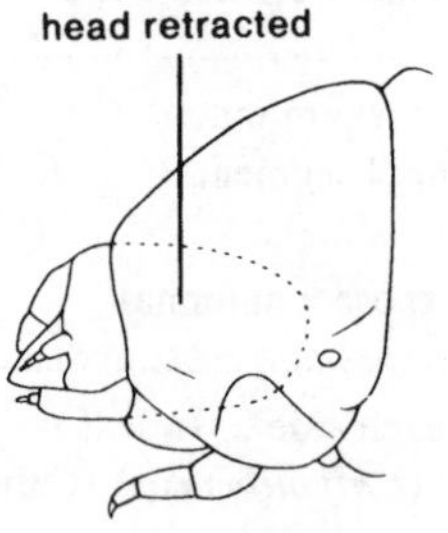

head retracted

Figure 9

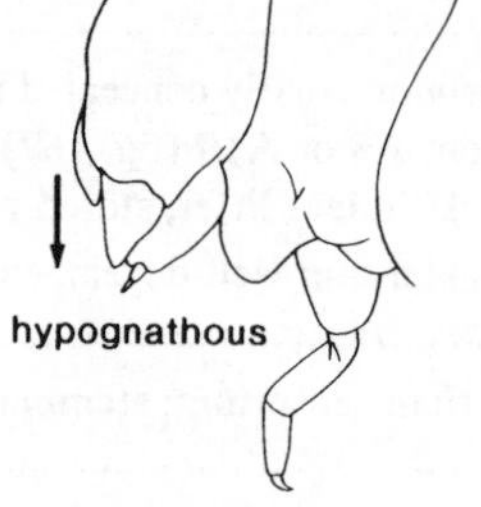

hypognathous

Figure 10

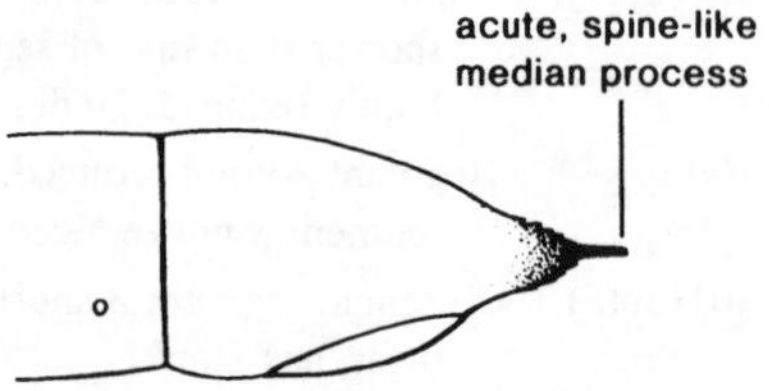

acute, spine-like median process

Figure 159

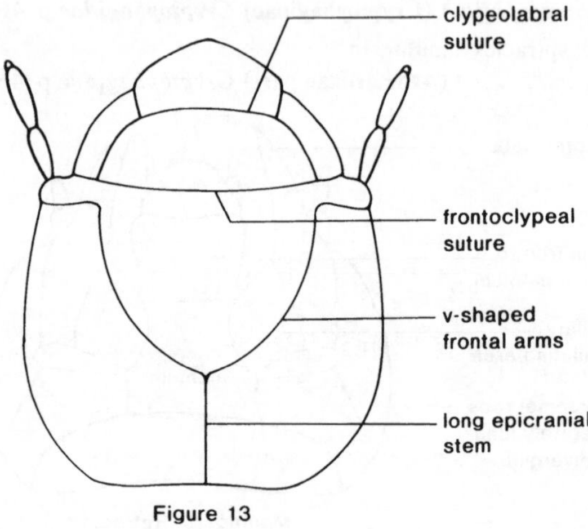

clypeolabral suture

frontoclypeal suture

v-shaped frontal arms

long epicranial stem

Figure 13

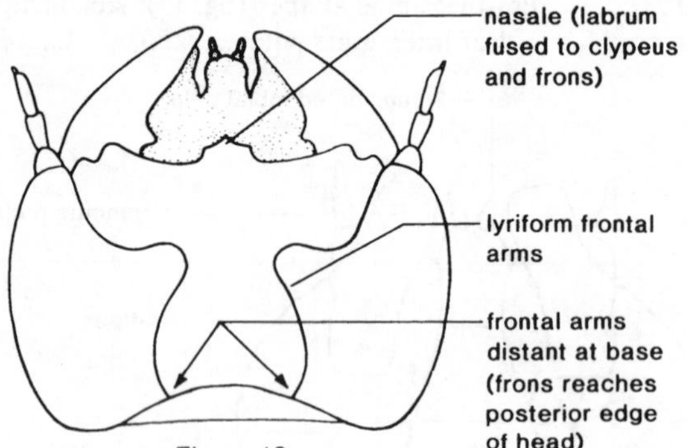

nasale (labrum fused to clypeus and frons)

lyriform frontal arms

frontal arms distant at base (frons reaches posterior edge of head)

Figure 16

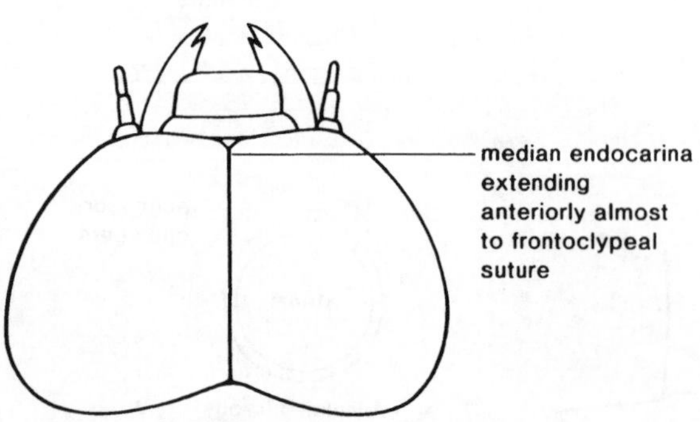

median endocarina extending anteriorly almost to frontoclypeal suture

Figure 21

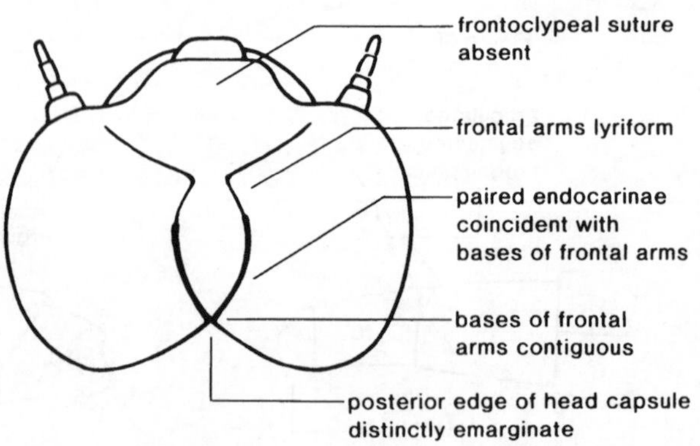

frontoclypeal suture absent

frontal arms lyriform

paired endocarinae coincident with bases of frontal arms

bases of frontal arms contiguous

posterior edge of head capsule distinctly emarginate

Figure 24

← Fig. 89

306(261[5]). Tergum A8 without special armature; bases of frontal arms contiguous (fig. 24); head moderately to strongly declined (hypognathous) (fig. 10); frontoclypeal suture absent or vaguely indicated; hypostomal rods present (fig. 89); stemmata on each side 4 or fewer; tergum A9 dorsal, extending onto ventral surface. Under bark, in fungus fruiting bodies, on bark surfaces .... (part) ***Endomychidae*** p. 482

306′. Tergum A8 forming tapered process bearing spiracles at apex; bases of frontal arms distinctly separated (fig. 16); head prognathous or slightly declined (fig. 8); frontoclypeal suture distinct (fig. 13); hypostomal rods absent; stemmata on each side 5; tergum A9 completely ventral. In slime fluxes of fermenting sap .... ***Nosodendridae*** p. 432

307(249[3]). Tergum A9 without median process; segments in T2 leg 5 or fewer including tarsungulus (fig. 126); segments in maxillary palp 4 (the last occasionally vestigal); prosternum without armature; posterior edge of head capsule not or only slightly emarginate dorsally; median endocarina, if present, not extending to frontoclypeal suture; ligula not sclerotized 308

307′. Tergum A9 with median process (fig. 159); segments in T2 leg 6 including single claw or paired claws (figs. 124, 125); segments in maxillary palp 3; prosternum with 2 patches of asperities; posterior edge of head capsule distinctly emarginate dorsally; median endocarina extending anteriorly almost to edge of clypeus (fig. 21); ligula forming sclerotized wedge (fig. 117); antennal segments 4 or 5. In rotten wood .... (part) ***Cupedidae*** p. 298

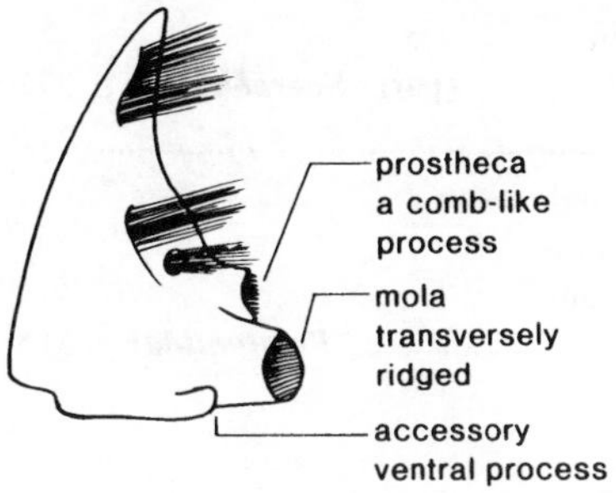

Figure 77

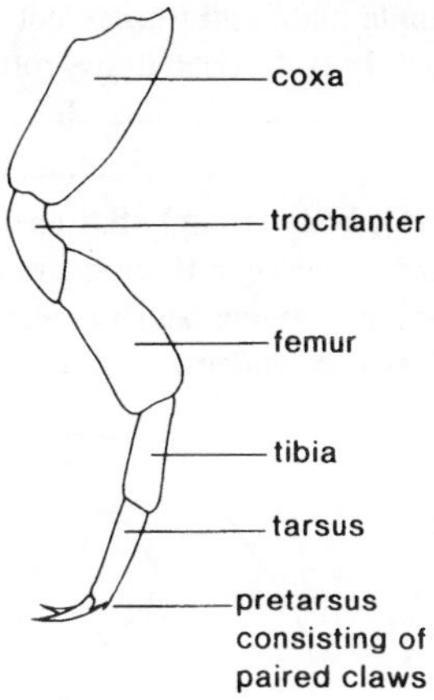

Figure 124

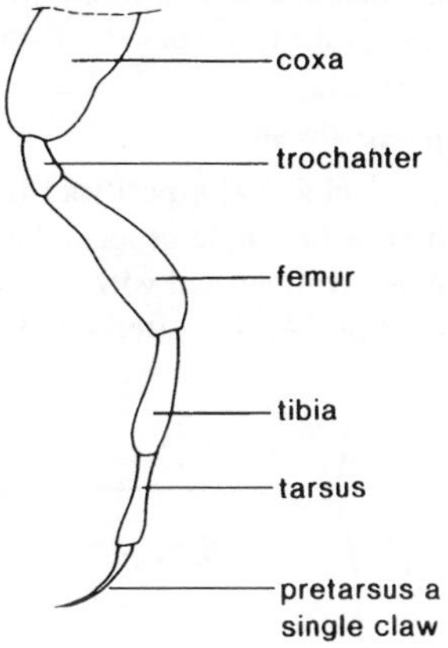

Figure 125

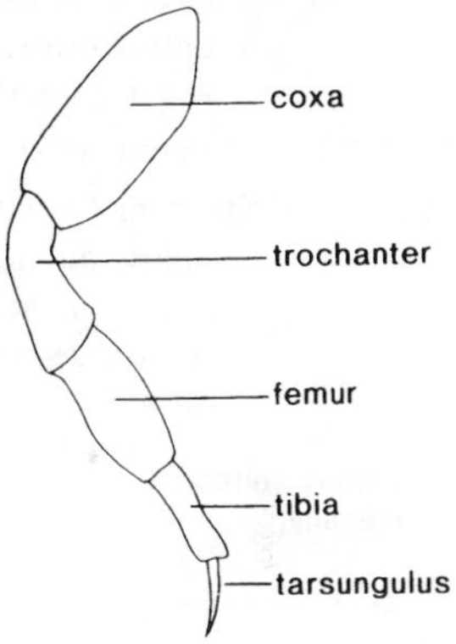

Figure 126

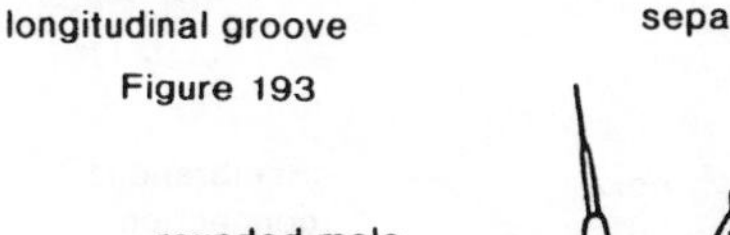

Figure 193

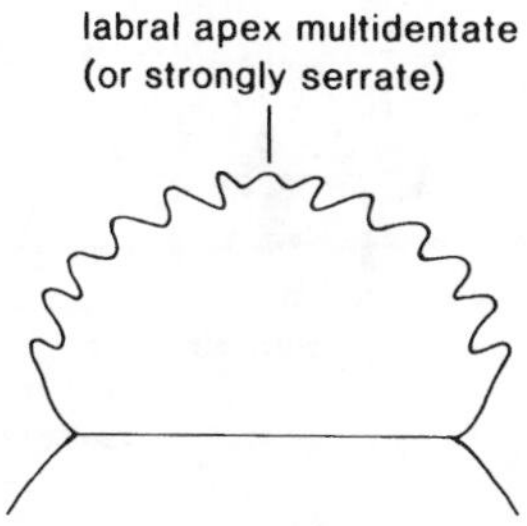

Figure 36

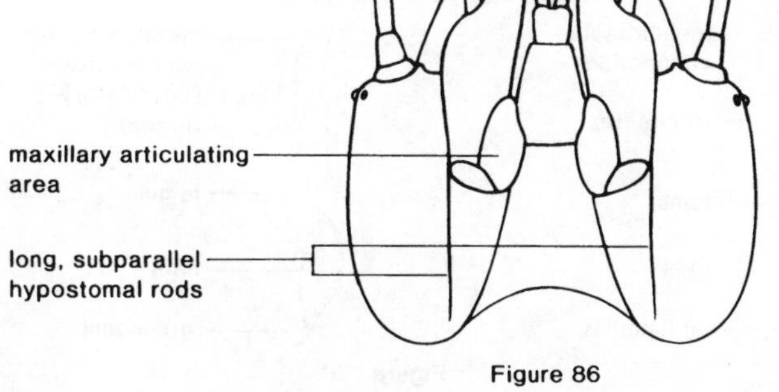

Figure 86

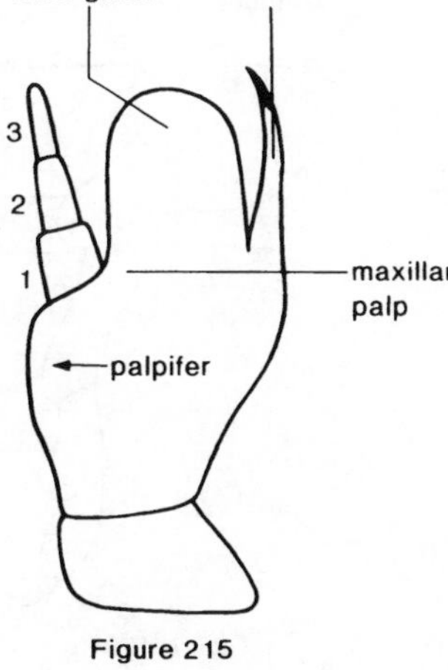

Figure 215

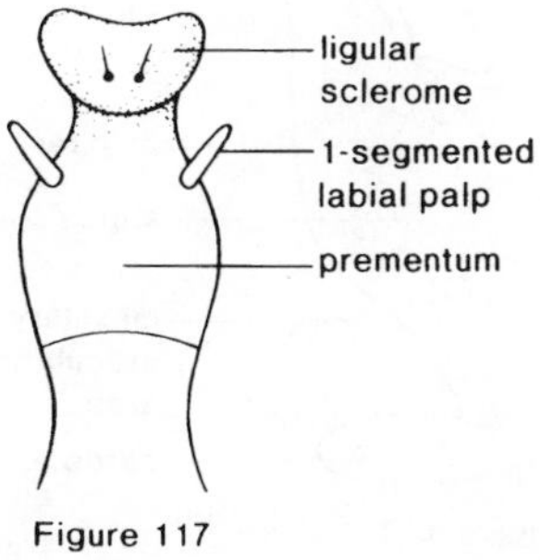

Figure 117

308(307). Antennal segments 6 or more (usually many); body relatively straight or only slightly curved ventrally; visible abdominal segments 8 or 9; abdominal apex with a respiratory chamber (pocket formed by 8th and 9th terga and enclosing enlarged A8 spiracles); prostheca complex, with sclerotized, comb-like process and brush of hairs (fig. 77). In ponds, lakes, tree holes, wet rotten wood .... ***Helodidae*** p. 366

308′. Antennal segments 4 or 5; body strongly curved ventrally (C-shaped); visible abdominal segments 10; abdominal apex without respiratory chamber; prostheca absent 309

309(308′). Labrum with multidentate or strongly serrate apex (fig. 36); segments in labial palp 1; length usually less than 10mm. Under bark, in rotten wood, termite nests .... **(Ceratocanthinae)** ***Scarabaeidae*** p. 377

309′. Labrum rounded, truncate, or weakly trilobed (fig. 40); segments in labial palp 2; length usually more than 10 mm. 310

310(309′). Stridulatory organs (patches or rows of asperities) present on T2 (fig. 129) and T3 (fig. 130) legs; tergum A3 without distinct transverse plicae (sometimes weakly divided into 2 parts); anterior abdominal terga with patches of asperities (fig. 142) or short, stout setae; galea and lacinia distinct (fig. 112); tormae united mesally (fig. 40); anal slit more or less vertical, bordered by 2 or 3 fleshy lobes (fig. 192). In rotten wood .... (part) ***Lucanidae*** p. 372

310′. Stridulatory organs absent on T2 and T3 legs; tergum A3 with 3 distinct, transverse plicae (fig. 136); if anterior abdominal terga have patches of asperities, then galea and lacinia fused into single mala and tormae not united mesally; anal slit transverse or Y-shaped. In soil, roots, dung, rotten wood, animal nests .... (part) ***Scarabaeidae*** p. 377

311(248′). Sternum A9 simple or absent 312

311′. Sternum A9 with row of apical asperities (fig. 176); body strongly flattened and lightly pigmented, with simple urogomphi; head wider than thorax; stemmata absent; mandibles asymmetrical with well-developed molae bearing numerous transverse ridges (fig. 74). In rotten wood **(see next 6 choices)** .... ***Prostomidae*** p. 515

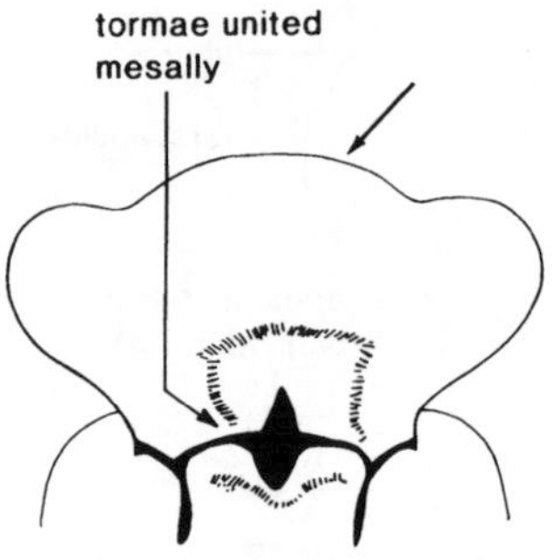

Figure 40

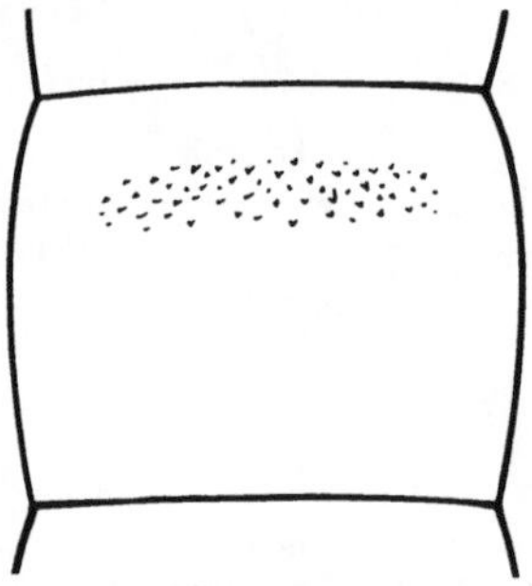

Figure 142

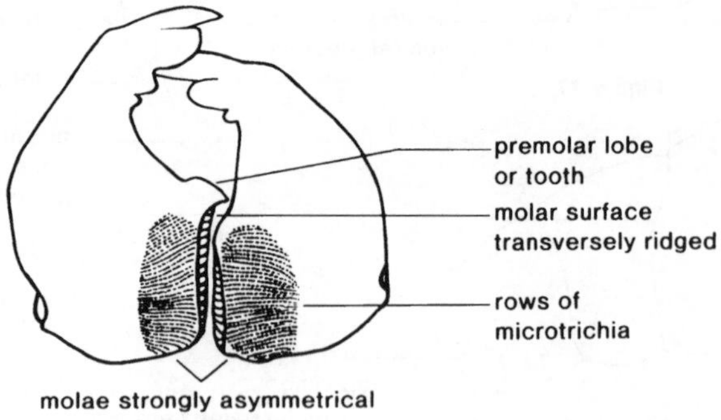

Figure 74

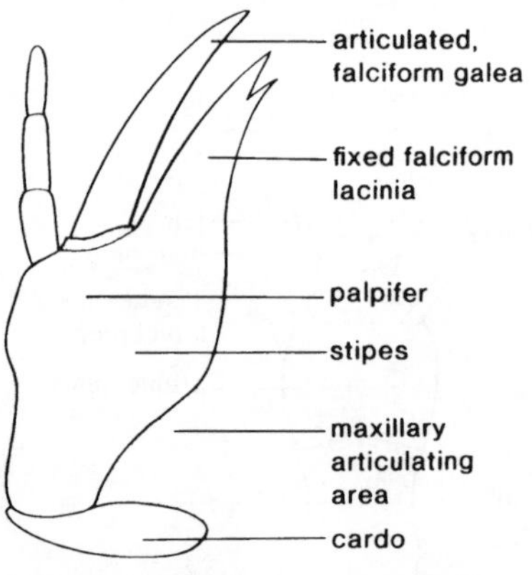

Figure 112

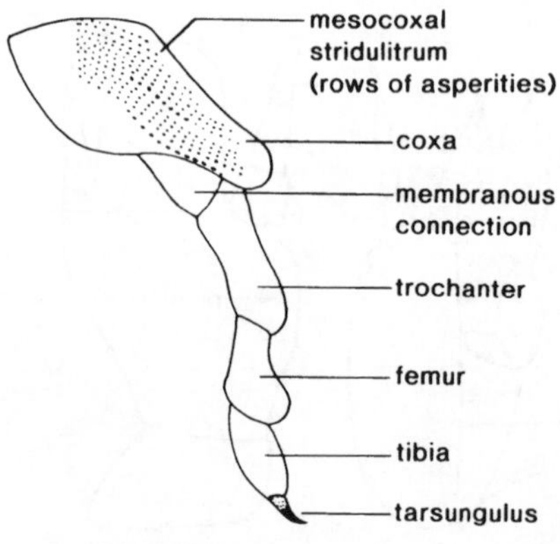

Figure 129

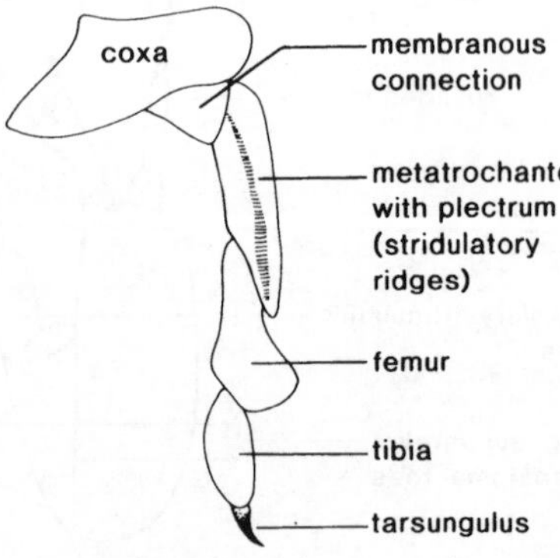

Figure 130

← Figs. 36, 77

311[2]. Sternum A9 with 2 apicomesal asperities (fig. 177); body slightly flattened, lightly pigmented; urogomphi each with an accessory mesal process near base (fig. 177); stemmata absent; mandibular mola with only 2 or 3 ridges. In soil, leaf litter .... **(Eurygeniinae)** ***Pedilidae*** p. 544

311[3]. Sternum A9 with 1 basal asperity on each side (fig. 183) 383

311[4]. Sternum A9 with 2 to 6 basal asperities on each side (fig. 182) 386

311[5]. Sternum A9 with more than 6 basal asperities on each side, forming row (figs. 178–181) 389

311[6]. Sternum A9 with 1 basal asperity on each side, 2 apicomesal asperities, and 2 more mesal asperities near the center; body strongly flattened; hypostomal rods absent; urogomphi diverging, each with several teeth at its base (**see choice 311[7]**) .... (*Catapiestus*) ***Tenebrionidae*** p. 520

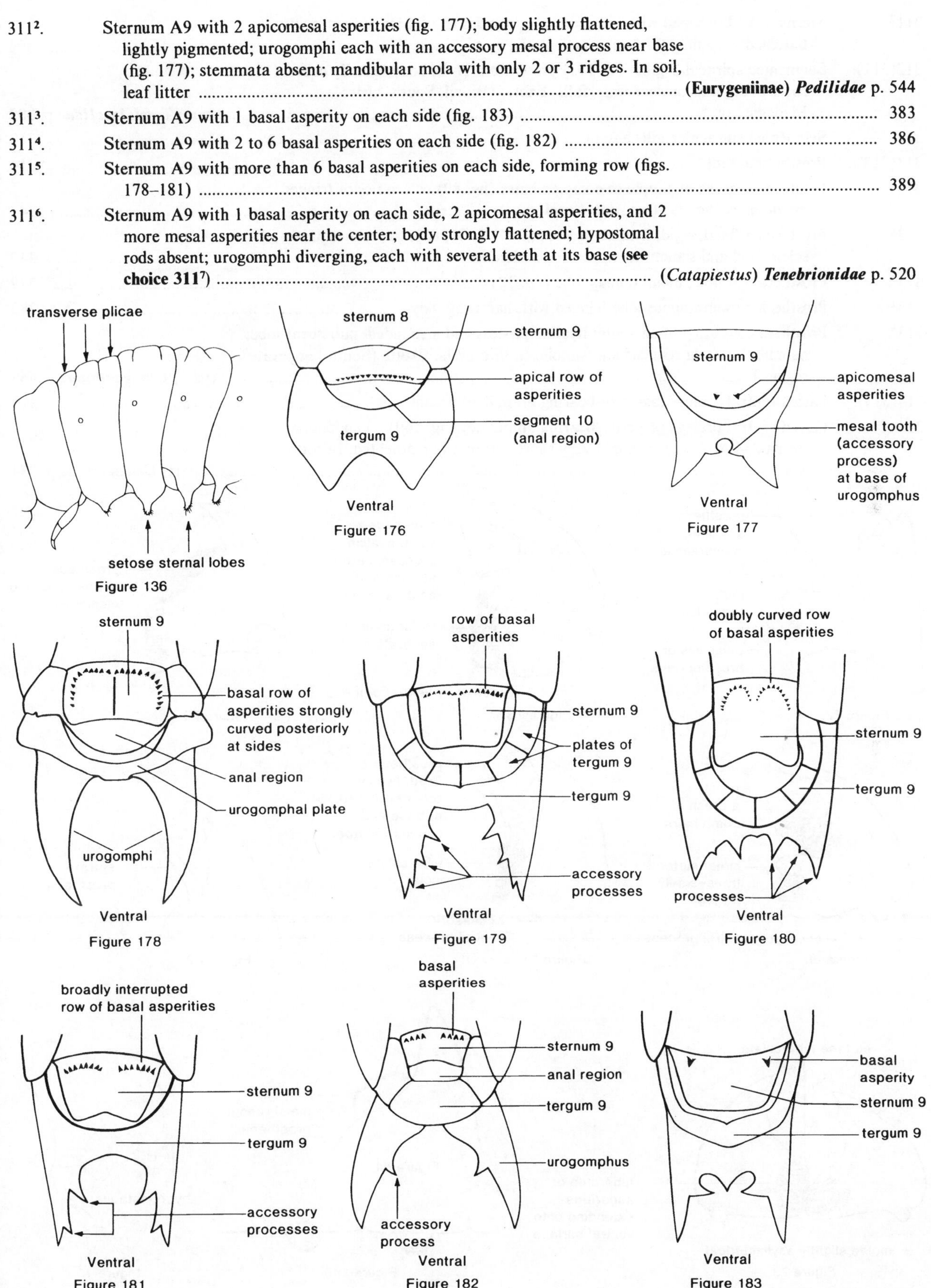

Fig. 192 ⟶

311^{7}. Sternum A9 U-shaped with apicolateral asperities (fig. 185); body strongly flattened; tergum A9 forming articulated plate 394

312(311). Segmented spiracular gills (fig. 152) present on A1–8; body broadly ovate. On rocks in streams or near waterfalls. Exotic (Brazil, South Africa, Madagascar) .... (part) ***Torridincolidae*** p. 302

312′. Segmented spiracular gills absent 313

313(312′). Prostheca absent 314

313′. Prostheca a brush of simple or complex hairs (fig. 80) or a series of fringed membranes (fig. 79) **(see next 4 choices)** 347

313^{2}. Prostheca a fixed, rigid, hyaline process (figs. 78, 81, 82), sometimes partly sclerotized and sometimes appearing articulated 349

313^{3}. Prostheca a simple, membranous lobe (fig. 72) 379

313^{4}. Prostheca a membranous lobe fringed with hairs (fig. 84) 382

313^{5}. Prostheca consisting of an acute, hyaline process and a rounded, pubescent lobe; mandibular mola reduced and simple. In bird nests. Exotic (South temperate regions) .... (part) ***Cavognathidae*** p. 469

314(313). Paired ventral prolegs (asperity-bearing ampullae) absent 315

314′. Paired ventral prolegs (asperity-bearing ampullae) (fig. 149) on abdominal sterna 2–4 or 2–5; urogomphi very short and lightly pigmented. In rotten wood .... (part) ***Oedemeridae*** p. 534

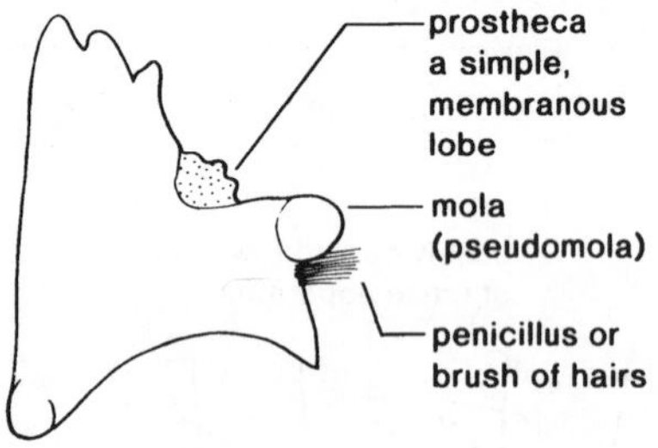

Figure 72

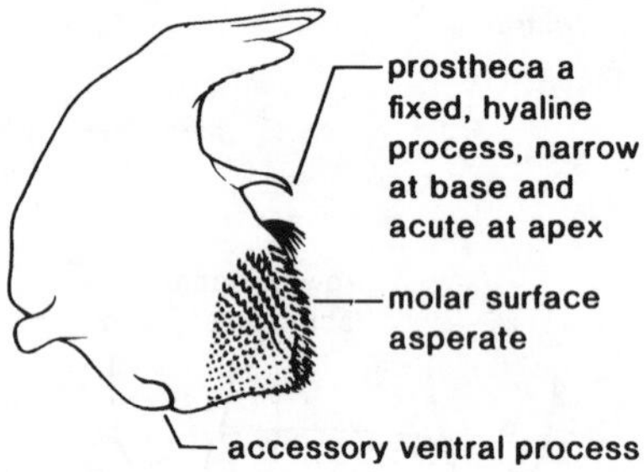

Figure 78

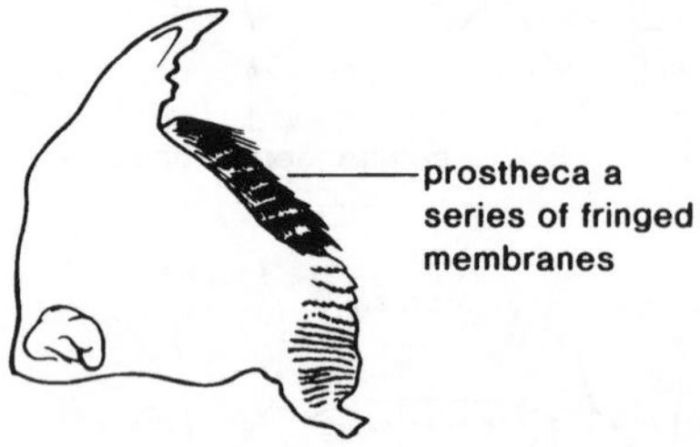

Figure 79

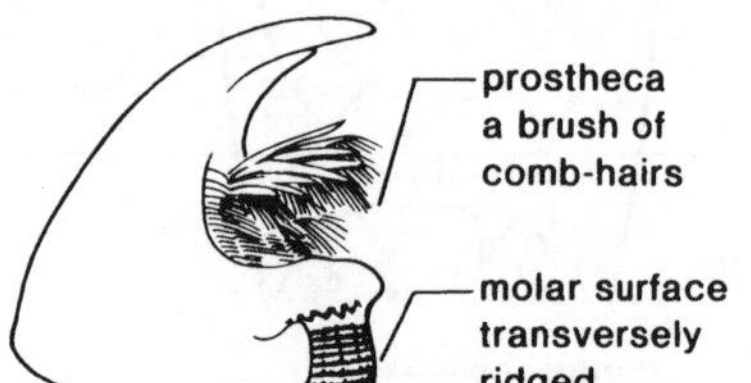

Figure 80

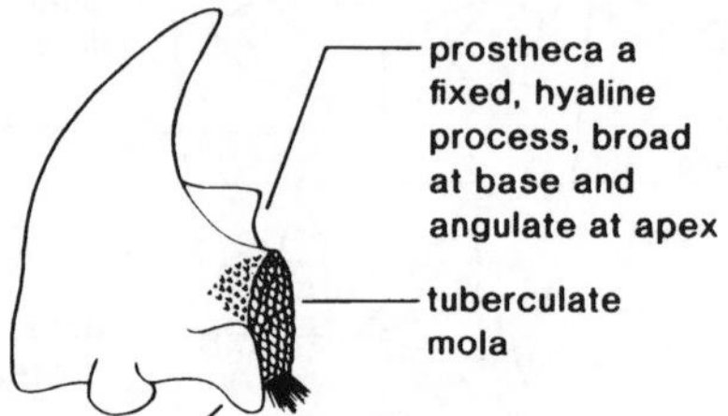

Figure 81

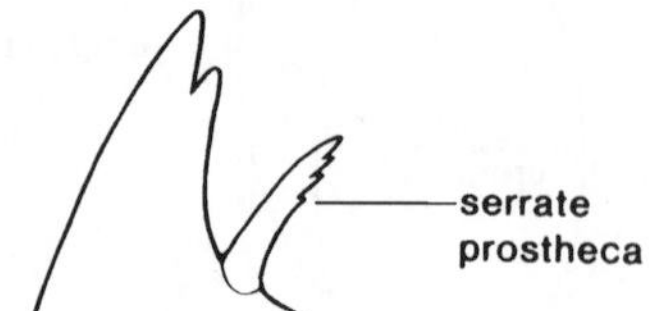

Figure 82

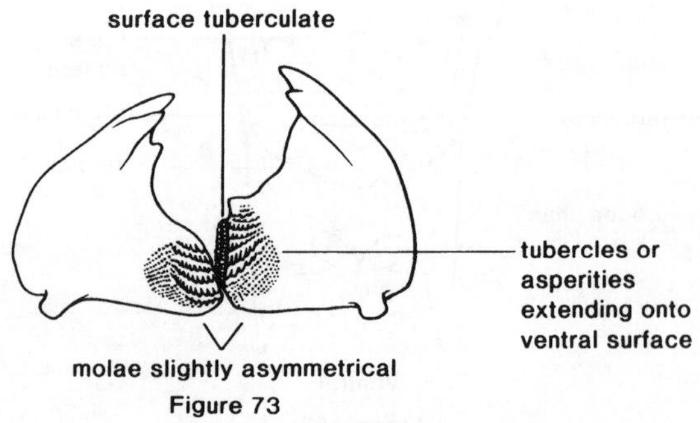

Figure 73

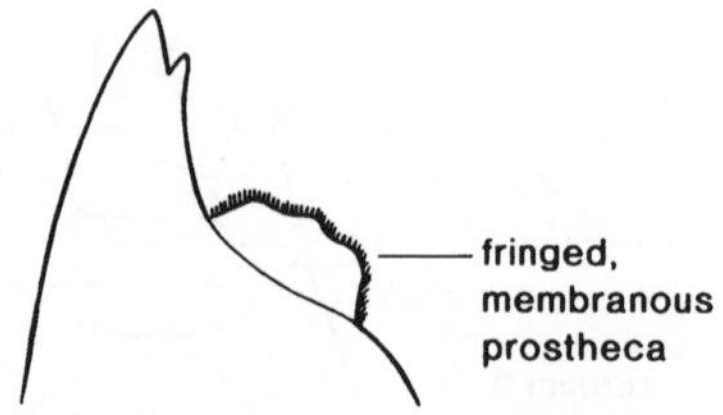

Figure 84

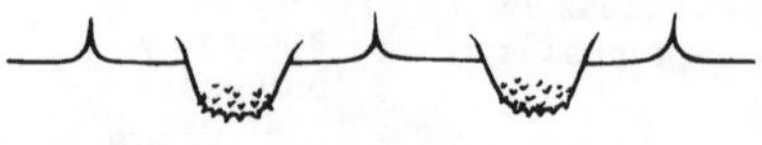

Figure 149

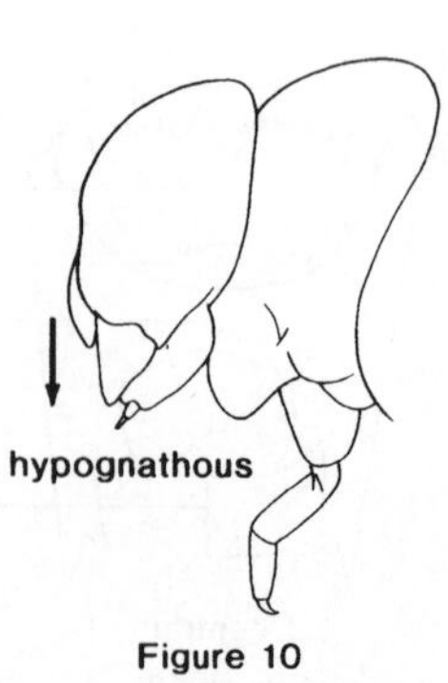

Figure 10

← Fig. 74; Fig. 84 on p. 259

315(314). Antennal segments 2 316

315′. Antennal segments 3 **(see 3rd choice)** 317

315². Antennal segments 4; urogomphi articulated at base (fig. 162). In leaf litter, fungi, carrion .... (part) ***Leiodidae*** p. 327

316(315). Head prognathous or slightly declined (fig. 8); ratio of antennal length to head width more than .15; antennae usually setose, with short dome-like sensorium; frontoclypeal suture distinct (fig. 13); single median endocarina absent; hypopharyngeal sclerome present (fig. 119). Under bark, in rotten wood, soil, leaf litter .... (part) ***Tenebrionidae*** p. 520

316′. Head moderately to strongly declined (hypognathous) (fig. 10); ratio of antennal length to head width less than .15; antennal sensorium palpiform, longer than segment 2 (fig. 32); frontoclypeal suture absent or vaguely indicated; single median endocarina coincident with epicranial stem (fig. 19); hypopharyngeal sclerome absent. In fungus fruiting bodies .... **(Ciinae part)** ***Ciidae*** p. 502

317(315′). Mandibles symmetrical; left and right molae more or less similar in shape 318

317′. Mandibles asymmetrical; left mola differing considerably from right one (figs. 73, 74) 336

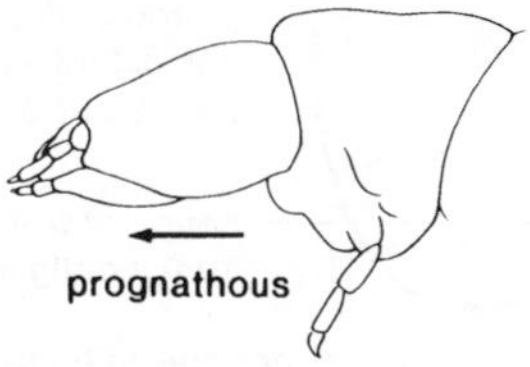

Figure 8

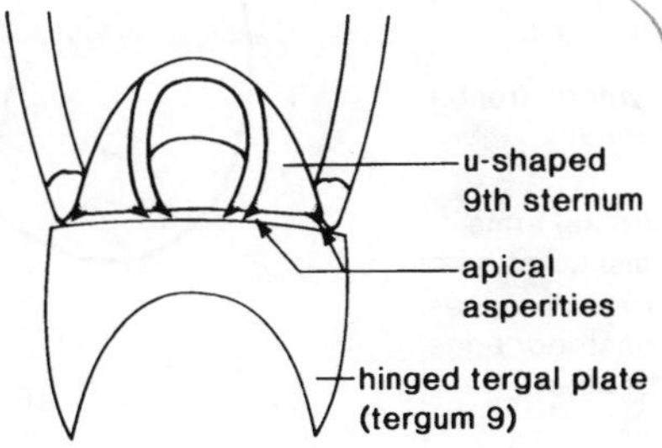

Figure 185

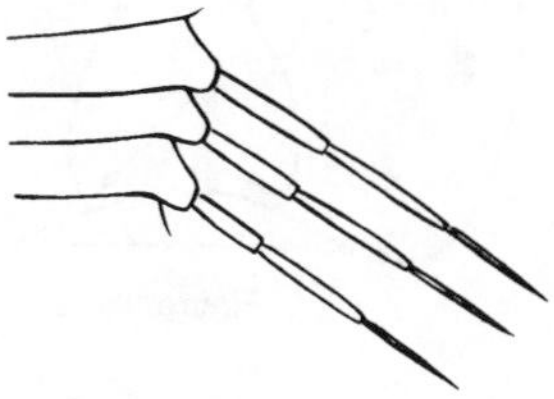

Figure 152

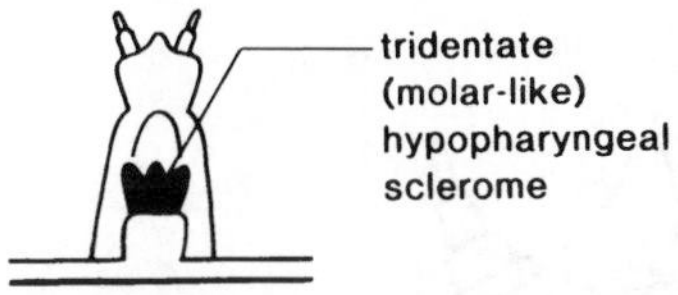

Figure 119

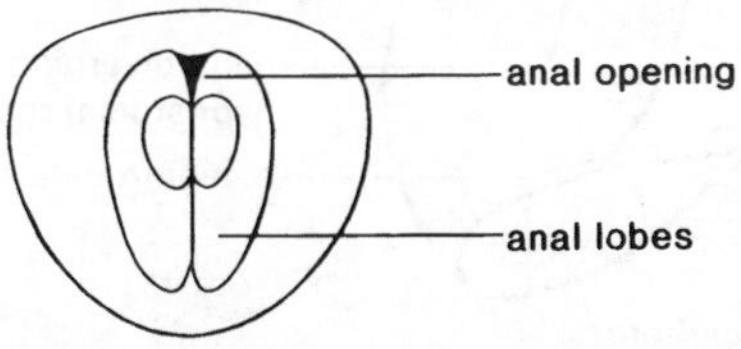

Figure 192

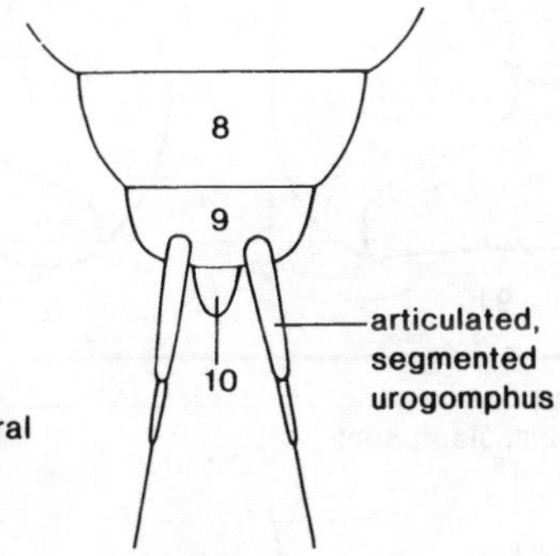

Figure 162

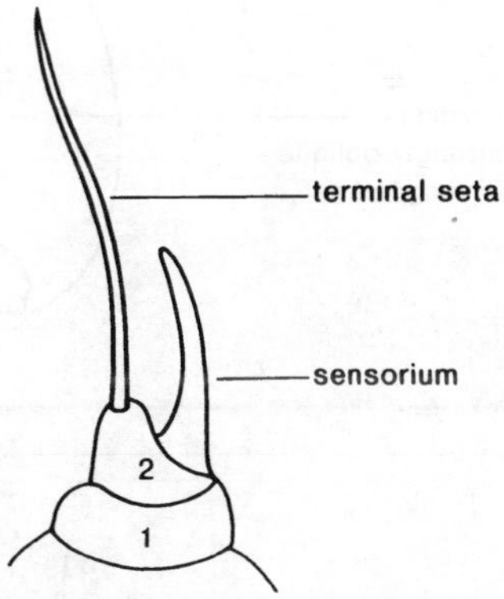

Figure 32

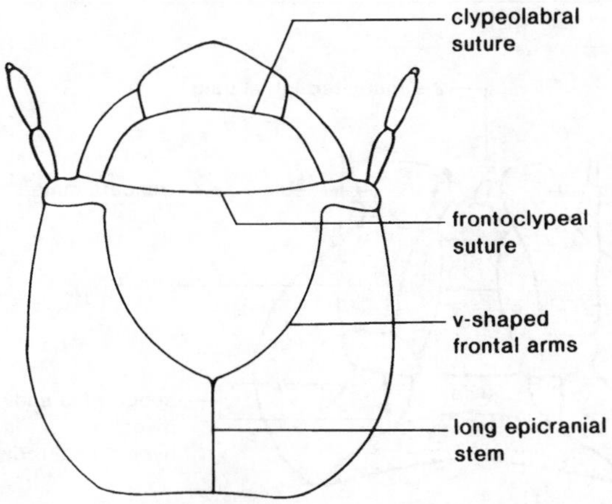

Figure 13

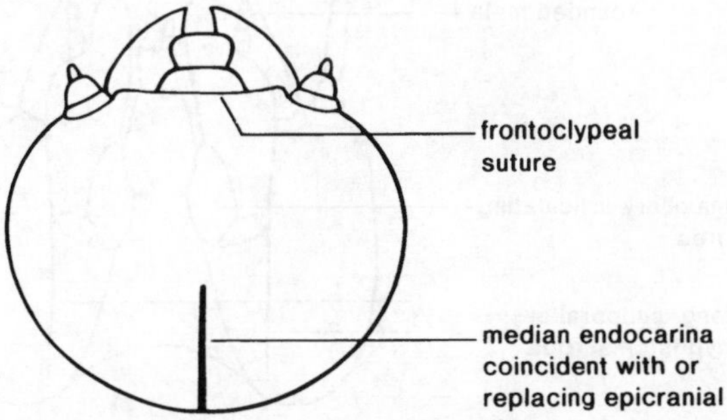

Figure 19

318(317). Segments in labial palp 1 (fig. 91) 319

318′. Segments in labial palp 2 (figs. 88, 94, 97) 320

319(318). Bases of frontal arms contiguous (fig. 24); segments in maxillary palp 3; ratio of antennal length to head width less than .15; labial palps separated by more than width of first palpal segment; hypostomal rods very short (fig. 85); body more or less cylindrical; length more than 3mm. Under bark, in rotten wood .... (part) ***Colydiidae*** p. 512

319′. Bases of frontal arms distinctly separated (fig. 16); segments in maxillary palp 2; ratio of antennal length to head width .15 to .5; labial palps contiguous or separated by less than width of first palpal segment; hypostomal rods extending almost to posterior edge of head (fig. 86); body somewhat flattened; length less than 3mm. In leaf litter, rotting vegetation, under bark .... (*Smicrips*) (**Smicripinae**) ***Nitidulidae*** p. 456

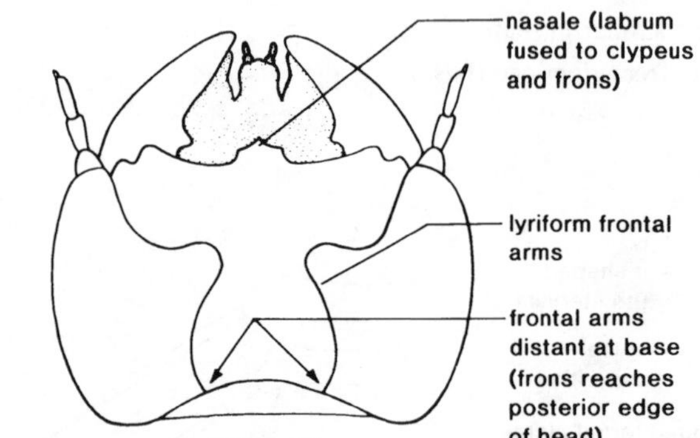

Figure 16

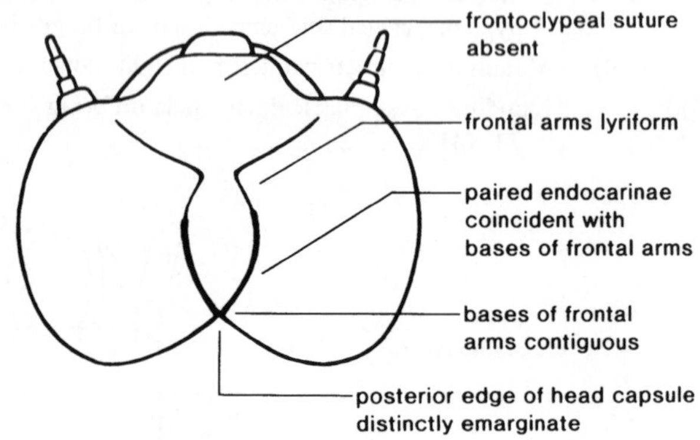

Figure 24

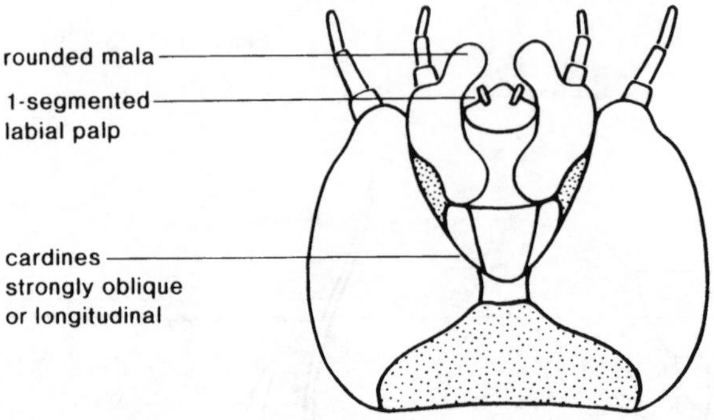

Figure 91

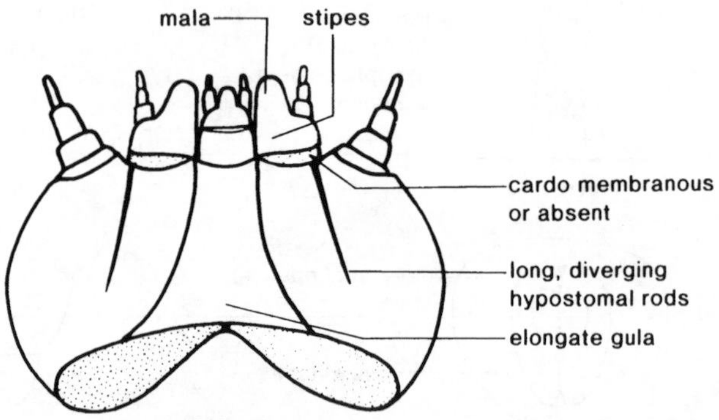

Ventral Mouthparts Strongly Protracted

Figure 92

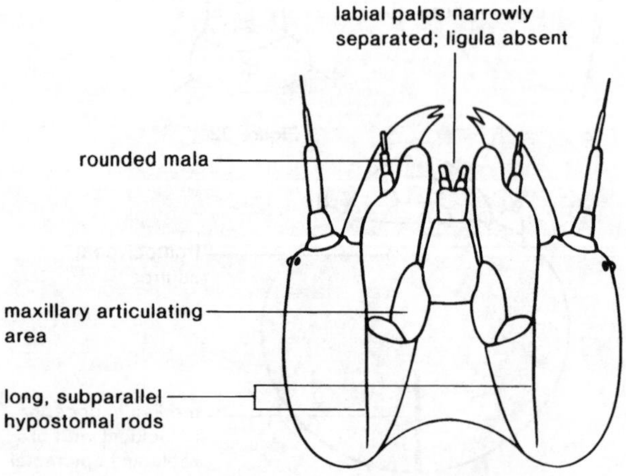

Figure 86

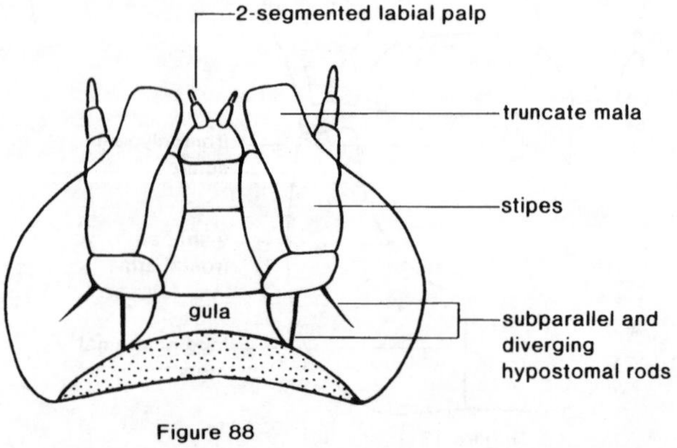

Figure 88

Figs. 94, 97: See page 225.

320(318′). Ventral mouthparts strongly protracted (fig. 92); stipes wider than long; cardo absent or membranous (fig. 92); sternum A9 completely concealed or absent (fig. 164); hypostomal rods long and diverging (fig. 92). In fungus fruiting bodies, smuts, ergots .... (part) ***Phalacridae*** p. 466

320′. Ventral mouthparts retracted (fig. 85); stipes longer than wide; cardo distinct and sclerotized; sternum A9 exposed 321

321(320′). Segment A10 distinct and visible from above (fig. 162); urogomphi narrow, straight, and almost always articulated at base (fig. 162); dorsal body surfaces smooth; mandibular mola without pubescent lobe at base; ratio of antennal length to head width usually more than .5; apex of antennal segment 2 oblique, so that palpiform sensorium arises proximad of segment 3 (fig. 30). In leaf litter, carrion, fungi .... (part) ***Leiodidae*** p. 327

321′. Segment A10 not visible from above, OR, if visible from above, dorsal body surfaces granulate and mandibular mola with pubescent, hyaline lobe at base (fig. 76); urogomphi usually curved and always fixed at base; ratio of antennal length to head width less than .5; apex of antennal segment 2 truncate, so that sensorium and segment 3 arise together (fig. 29), or sensorium dome-like (fig. 31) 322

322(321′). Mandibular mola with hyaline lobe at base (fig. 76) 323

322′. Mandibular mola without hyaline lobe at base 325

323(322). Stemmata on each side 6; dorsal surfaces granulate; segment A10 more or less cylindrical; setae on tarsungulus 1; spiracles annular-biforous (fig. 198). In fruits or berries .... (*Byturus*) ***Byturidae*** p. 476

323′. Stemmata on each side 1 or 0; dorsal surfaces smooth or spiracles annular (fig. 194); segment A10 transverse; setae on tarsungulus 2 324

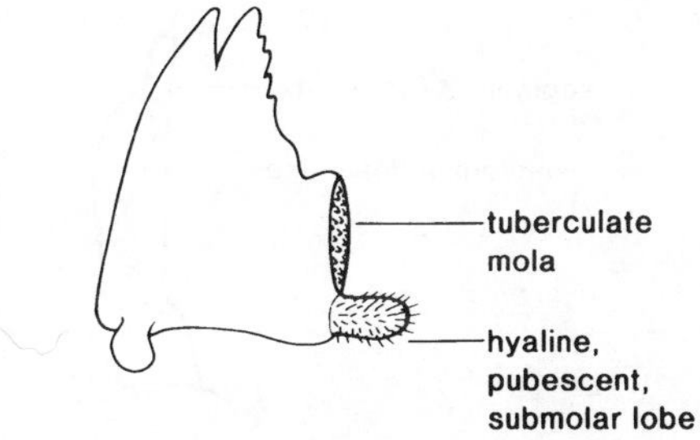

Figure 76

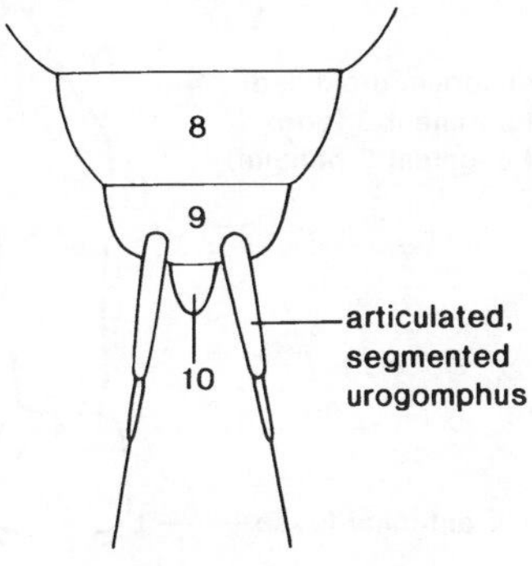

Figure 162

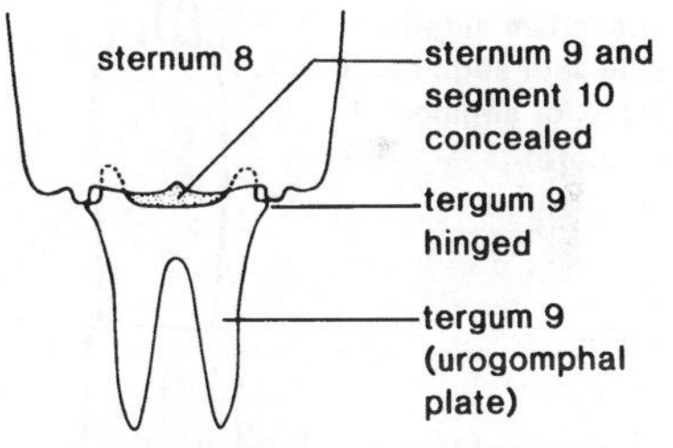

Ventral
Figure 164

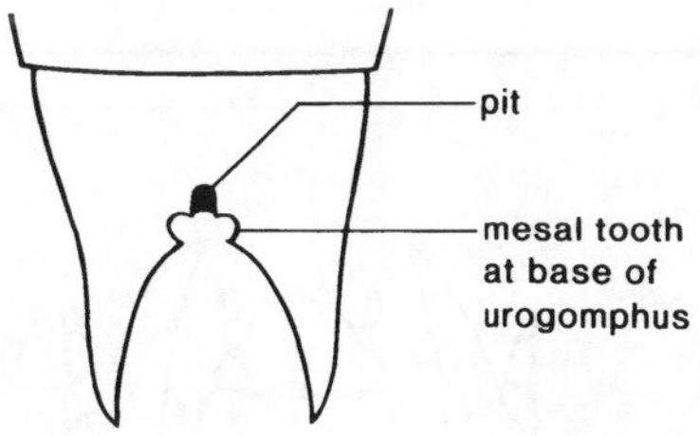

Dorsal
Figure 173

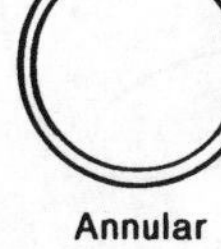

Annular
Figure 194

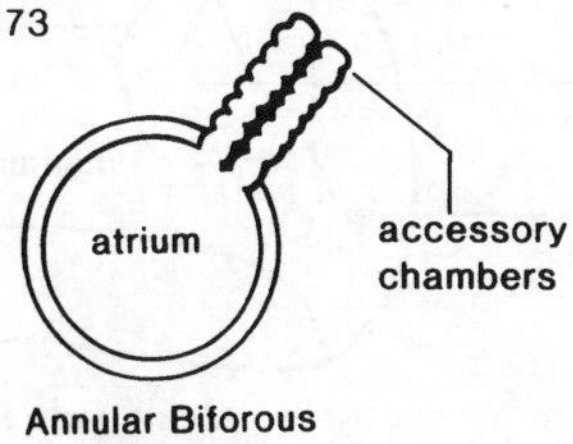

Annular Biforous
Figure 198

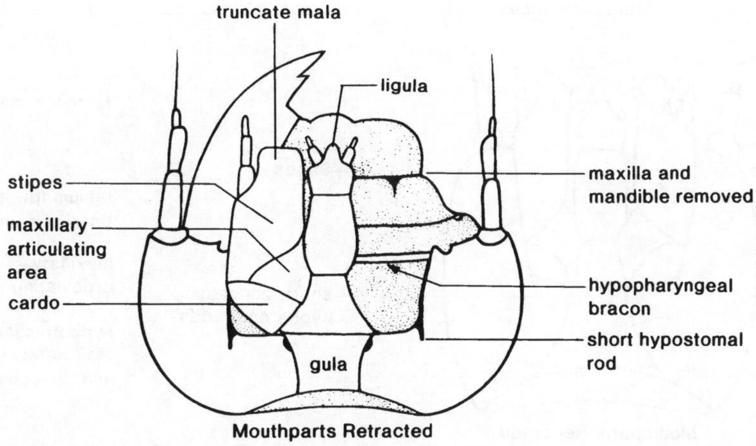

Mouthparts Retracted
Figure 85

Figs. 29, 30, 31 ⟶

324(323′). Median endocarina absent; inner apical angle of mala simple or with 1 or 2 small teeth; urogomphi well separated at base and without pit between them; hypostomal rods absent. In leaf litter, under bark .... (part) ***Euglenidae*** p. 554

324′. Median endocarina absent; inner apical angle of mala with distinct lobe or uncus (fig. 87); urogomphi approximate at base and with pit between them (fig. 173); hypostomal rods short and converging (fig. 87). In leaf litter, under bark **(see 3rd choice)** .... **(Anaspidinae)** ***Scraptiidae*** p. 555

324². Median endocarina extending anterad of epicranial stem (fig. 20); inner apical angle of mala simple or with 1 or 2 small teeth (fig. 106); urogomphi more or less approximate, with or without pit between them; hypostomal rods usually moderately long and converging. Under bark, stones, in leaf litter .... **(Anthicinae)** ***Anthicidae*** p. 552

325(322′). Mandible with accessory ventral process (figs. 78, 80, 81); setae on tarsungulus 1; spiracles usually on long or short tubes (figs. 209, 210) 326

325′. Mandible without accessory ventral process; setae on tarsungulus 2; spiracles not on tubes 329

326(325). Lateral tergal processes present on segments A6–9; urogomphi short, widely separated and strongly upturned, with 2 pairs of truncate lobes between them; stemmata absent; vestiture of simple setae. In rotten wood, leaf litter .... **(Anommatinae)** ***Cerylonidae*** p. 480

326′. Lateral tergal processes, if present, occurring on segments A1–9; urogomphi without 2 pairs of truncate lobes between them; vestiture usually including some expanded or frayed setae 327

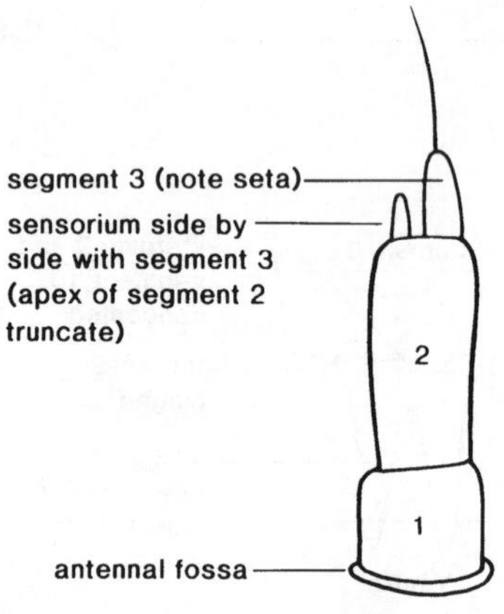

segment 3 (note seta)
sensorium side by side with segment 3 (apex of segment 2 truncate)
2
1
antennal fossa

Figure 29

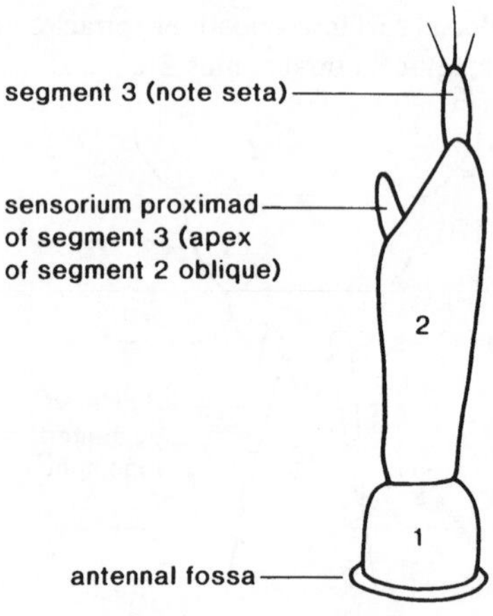

segment 3 (note seta)
sensorium proximad of segment 3 (apex of segment 2 oblique)
2
1
antennal fossa

Figure 30

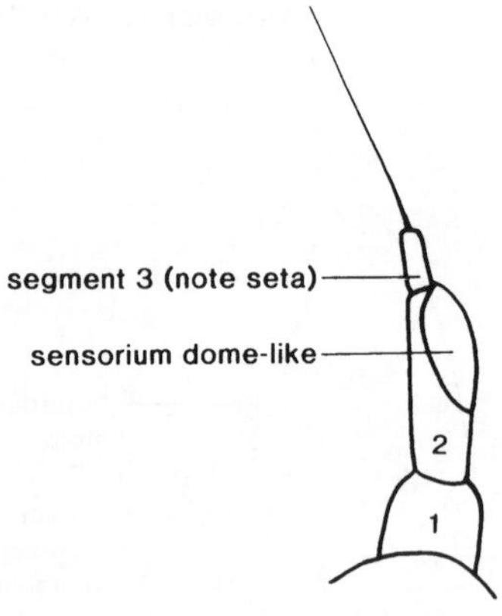

segment 3 (note seta)
sensorium dome-like
2
1

Figure 31

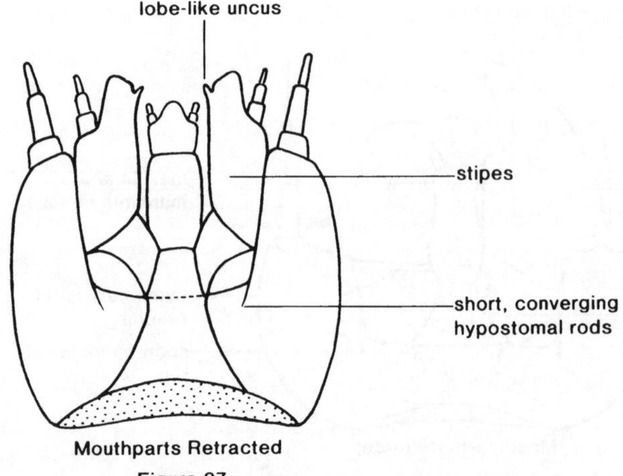

Mouthparts Retracted
Figure 87

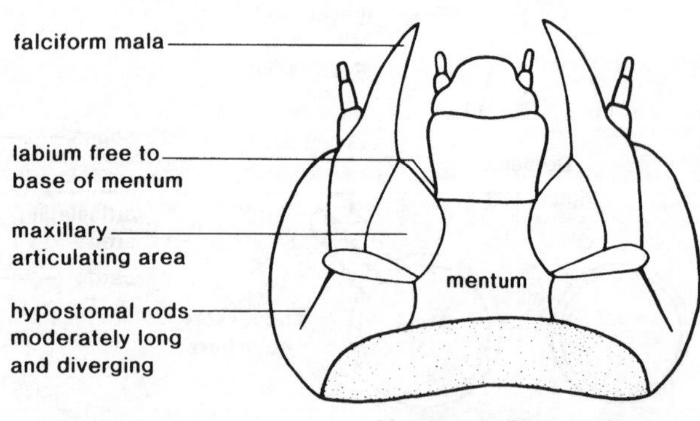

Mouthparts Retracted
Figure 89

← Figs. 24, 85, 86, 173

327(326′). Hypostomal rods absent; stemmata on each side 2 or fewer; urogomphi long and narrow, straight or slightly curved, usually approximate at base and diverging apically, with 2 or more accessory processes. In rotten wood, leaf litter, fungi, ant nests, guano .... **(Euxestinae)** ***Cerylonidae*** p. 480

327′. Hypostomal rods well-developed, diverging (fig. 89); stemmata on each side 5 or 6; urogomphi widely separated at base and strongly curved at apex 328

328(327′). Mala broadly rounded at apex (fig. 86); tergum A8 with pair of conical tubercles near midline; spiracular tubes long and tooth-like, dorsally located. In fungus fruiting bodies. Exotic (Europe) .... ***Sphaerosomatidae*** p. 479

328′. Mala narrowly rounded or subfalciform (fig. 89) at apex; tergum A8 without pair of tubercles (or paired tubercles on several abdominal terga); spiracular tubes shorter, laterally placed. In tunnels of ambrosia beetles (Platypodidae) .... **(Teredinae)** ***Bothrideridae*** p. 477

329(325′). Abdominal terga without rows of asperities 330

329′. Abdominal terga with rows of asperities on one or more segments 334

330(329). Maxillary mala falciform (fig. 105); body elongate, cylindrical, and lightly sclerotized. In stems .... **(Languriinae** part**)** ***Languriidae*** p. 471

330′. Maxillary mala rounded or truncate (fig. 85) 331

331(330′). Frontoclypeal suture present (fig. 13); hypostomal rods absent; frontal arms always V- or U-shaped (fig. 13); spiracles annular (fig. 194, 195), annular-multiforous (fig. 200), or occasionally annular-biforous with very short accessory chambers 332

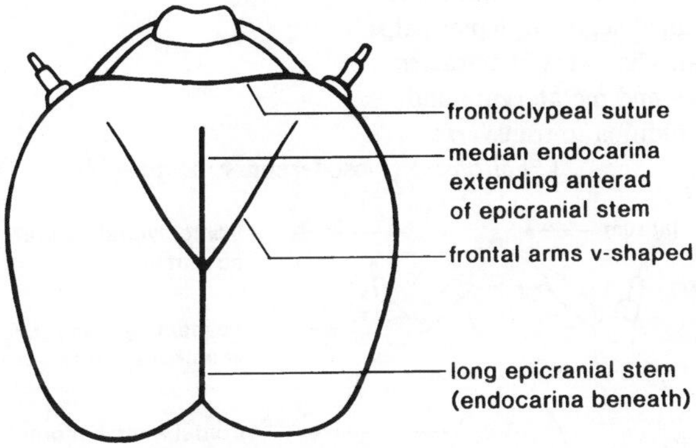

frontoclypeal suture
median endocarina extending anterad of epicranial stem
frontal arms v-shaped
long epicranial stem (endocarina beneath)

Figure 20

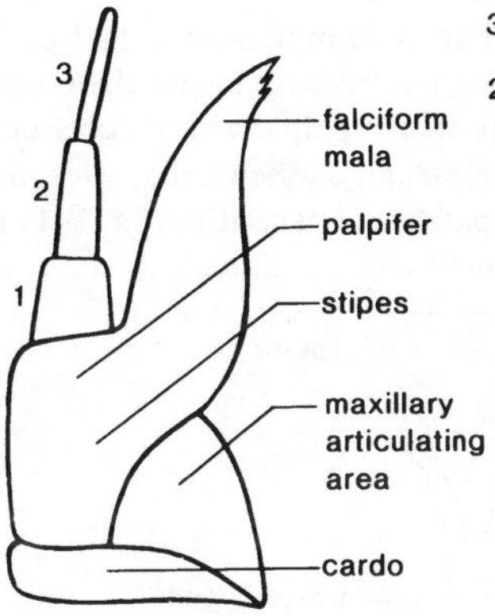

3
2
1
falciform mala
palpifer
stipes
maxillary articulating area
cardo

Figure 105

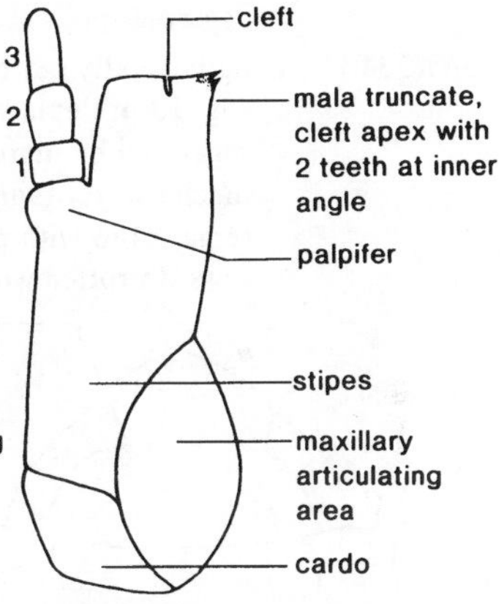

3
2
1
cleft
mala truncate, cleft apex with 2 teeth at inner angle
palpifer
stipes
maxillary articulating area
cardo

Figure 106

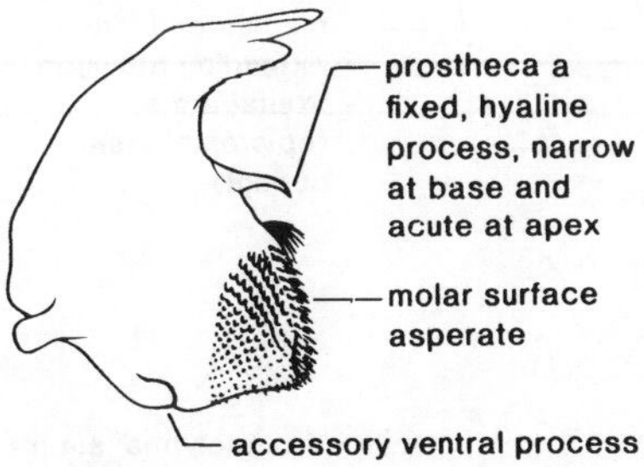

prostheca a fixed, hyaline process, narrow at base and acute at apex
molar surface asperate
accessory ventral process

Figure 78

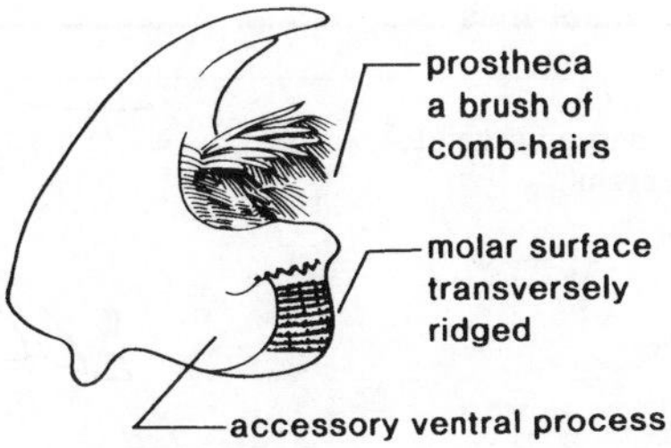

prostheca a brush of comb-hairs
molar surface transversely ridged
accessory ventral process

Figure 80

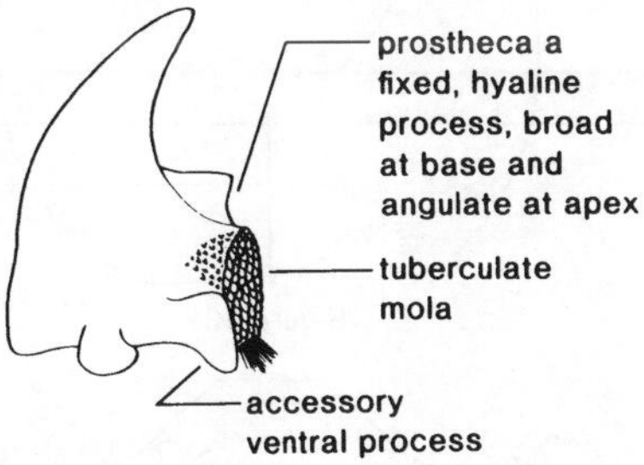

prostheca a fixed, hyaline process, broad at base and angulate at apex
tuberculate mola
accessory ventral process

Figure 81

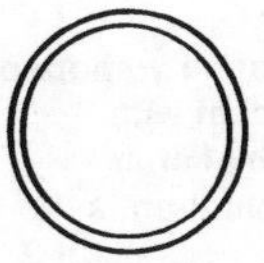

Annular

Figure 194

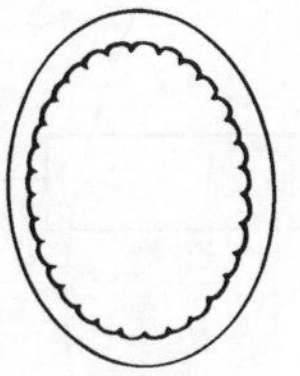

Annular

Figure 195

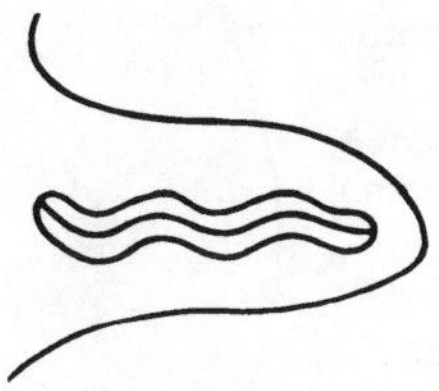

Figure 209

Figure 210

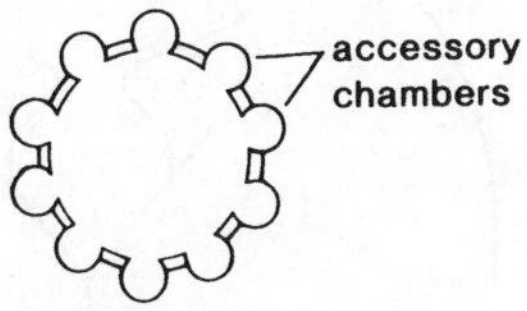

accessory chambers
Annular Multiforous

Figure 200

Fig. 13 ⟶

331′. Frontoclypeal suture absent (fig. 24); hypostomal rods present (figs. 85, 89); frontal arms lyriform (fig. 22); spiracles annular-biforous with well-developed accessory chambers (fig. 198) 333

332(331). Maxillary mala cleft (fig. 106); cardo distinctly divided; median endocarina Y-shaped, coincident with epicranial stem and frontal arms (fig. 23); spiracles annular-biforous with very short accessory chambers; prosternum with patch of asperities .... **(Zopherinae)** ***Zopheridae*** p. 518

332′. Maxillary mala simple (fig. 85); cardo not divided; median endocarina absent; spiracles annular (figs. 194, 195) or annular-multiforous (fig. 200); prosternum without patch of asperities .... (part) ***Tenebrionidae*** p. 520

333(331′). Apex of mala cleft (fig. 106); hypostomal rods (fig. 85) present; tergum A9 distinctly tuberculate; urogomphi without pit between them; ligula as long as or longer than labial palps. Under bark, in fungus fruiting bodies. Exotic (Australia) .... (*Meryx*) **(Ulodinae)** ***Zopheridae*** p. 518

333′. Apex of mala usually simple as in fig. 85, OR, if cleft, then hypostomal rods absent; tergum A9 not tuberculate, or urogomphi with pit between them; ligula much shorter than labial palps. Under bark or in rotten wood .... (part) ***Colydiidae*** p. 512

334(329′). Epicranial stem present (fig. 23); frontal arms V- or U-shaped (fig. 23); hypostomal rods absent; urogomphi small and approximate; transverse rows of asperities, if present, on 5 or fewer abdominal terga. In rotten wood .... **(Nosoderminae** part) ***Zopheridae*** p. 518

334′. Epicranial stem absent and frontal arms lyriform (fig. 216); hypostomal rods present (fig. 85); urogomphi large and well separated; transverse rows of asperities present on terga A1–6 335

335(334′). Length usually more than 10 mm; mala with 1 or 2 teeth at inner apical angle (fig. 106); ligula well-developed, longer than first palpal segment; labial palps separated by more than a palpal width; accessory air chambers of spiracles much shorter than atrium diameter (fig. 197); meso- and metatergum and terga A1–6 with patches of asperities (fig. 141) in addition to transverse rows. In rotten wood .... (*Phellopsis*) **(Nosoderminae)** ***Zopheridae*** p. 518

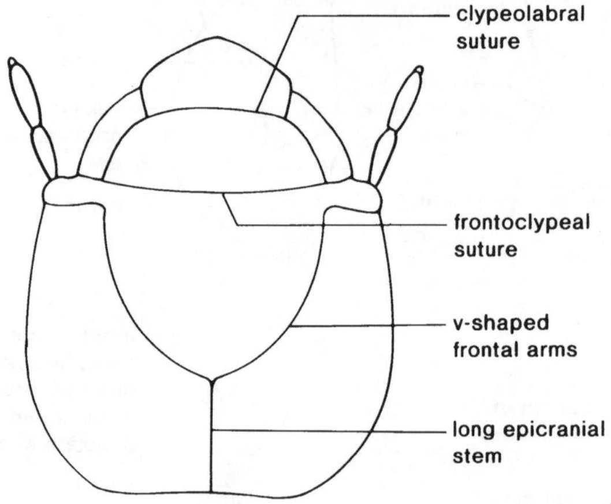

Figure 13

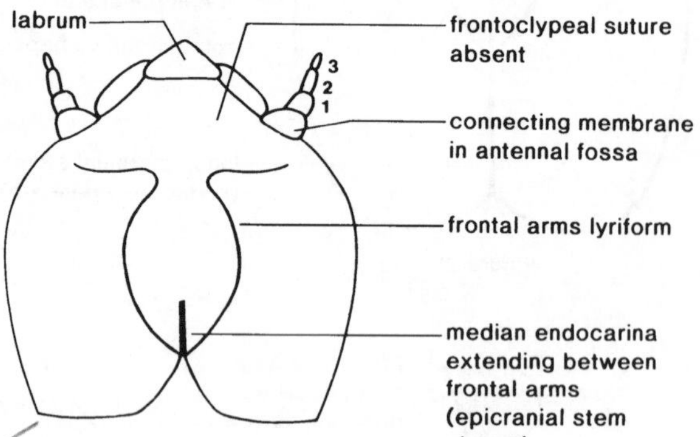

Figure 22

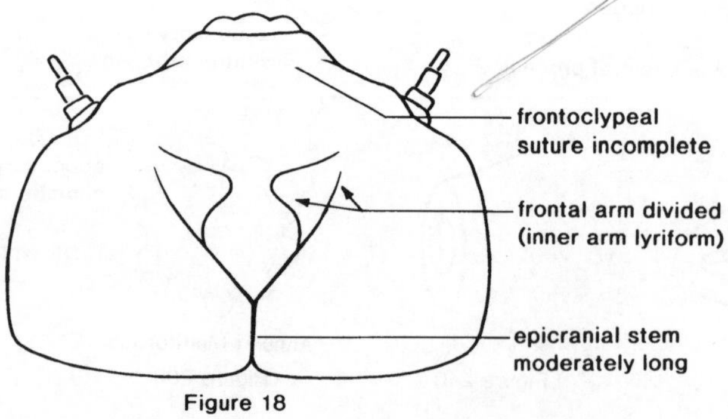

Figure 18

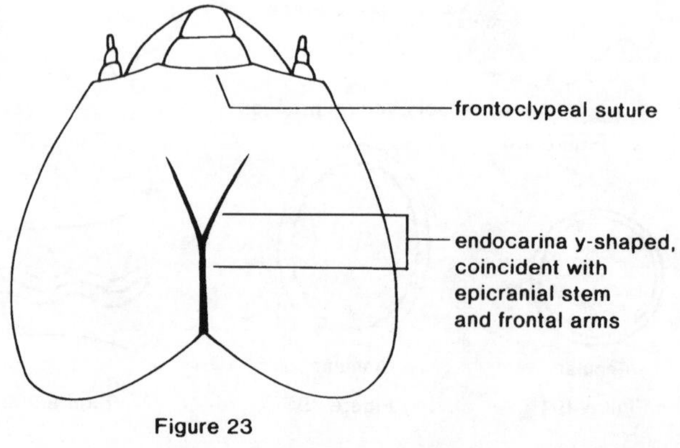

Figure 23

← Figs. 29, 30, 89, 105, 106, 194, ..., 200

335′. Length usually less than 10 mm; mala with inner apical angle simple; ligula reduced; labial palps separated by less than a palpal width; accessory air chambers of spiracles at least as long as atrium diameter (fig. 198); metatergum and terga A1–6 with transverse rows of asperities only (fig. 144). In rotten wood .... (**Usechinae**) ***Zopheridae*** p. 518

336(317′). Frontoclypeal suture absent or vaguely indicated (figs. 18, 22) 337

336′. Frontoclypeal suture distinct (fig. 13) 344

337(336). Hypostomal rods absent 338

337′. Hypostomal rods present (fig. 89) 341

338(337). Urogomphi simple and tergum A9 without additional tubercles or processes; anal region posteriorly or terminally oriented 339

338′. Urogomphi with accessory processes and tergum A9 with several additional spines or tubercles; anal region posteroventrally oriented. In rotten wood. Exotic (Southern Hemisphere) .... (**Ulodinae** part) ***Zopheridae*** p. 518

339(338). Maxillary mala more or less falcate (fig. 105), or if maxilla with 2 apical lobes, then lacinia falcate (fig. 111); antennal segment 2 oblique at apex, so that sensorium arises proximad of segment 3 (fig. 30); urogomphi almost always articulated at base (figs. 112–113). In leaf litter, fungi, carrion .... (part) ***Leiodidae*** p. 327

339′. Maxillary mala truncate (figs. 85, 106) or rounded (fig. 86); antennal segment 2 truncate at apex, so that sensorium and segment 3 arise together (fig. 29); urogomphi fixed at base (figs. 170–171) 340

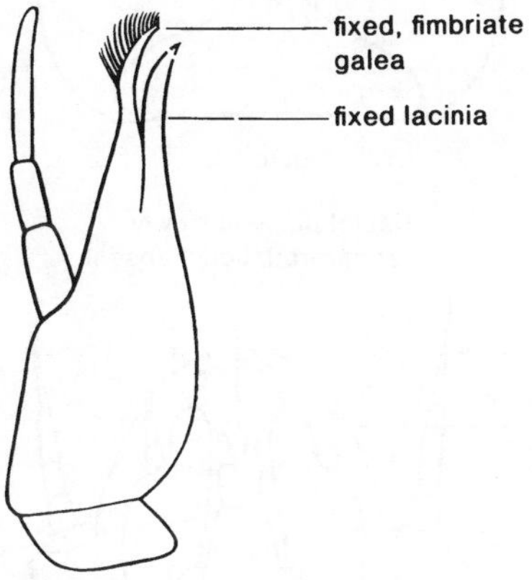

Figure 111

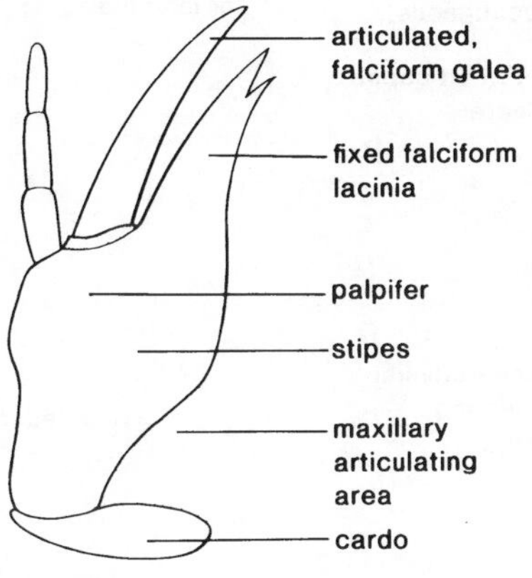

Figure 112

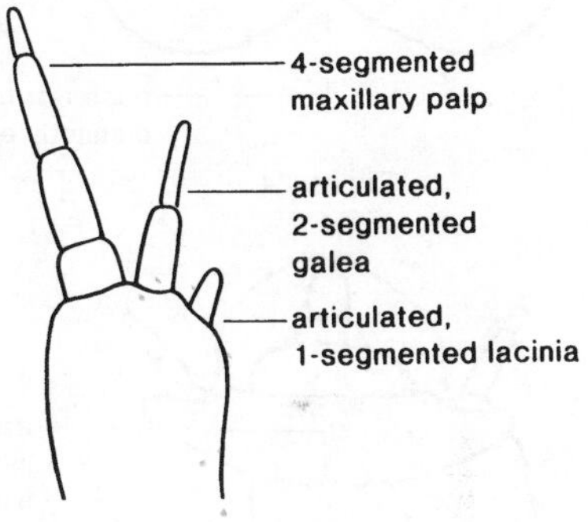

Figure 113

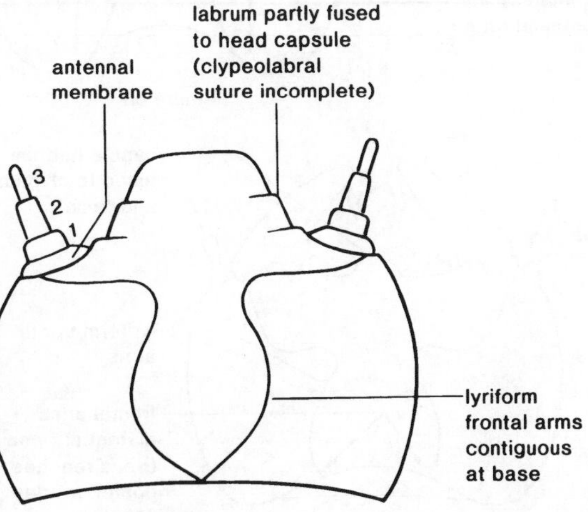

Figure 216

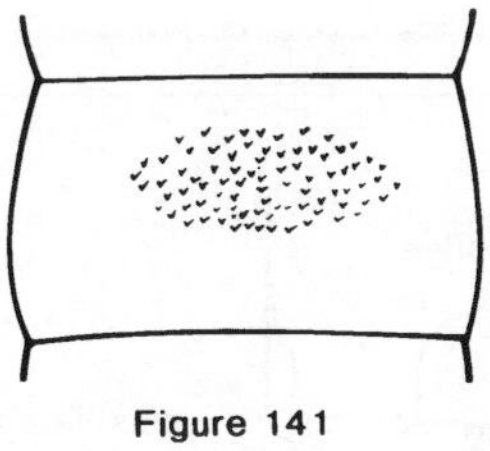

Figure 141

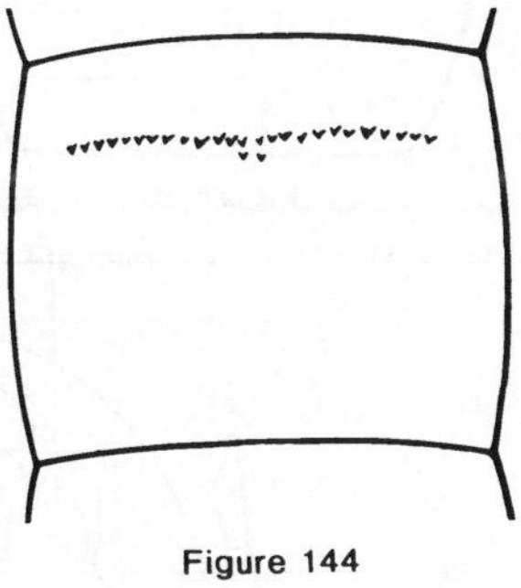

Figure 144

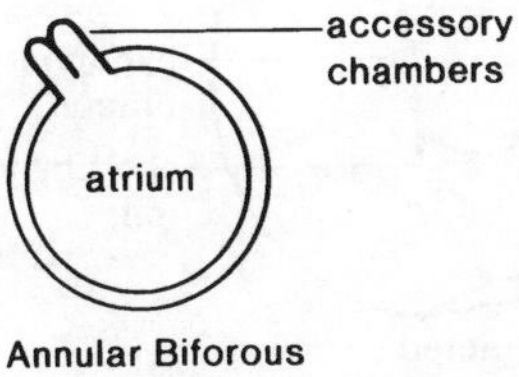

Annular Biforous
Figure 197

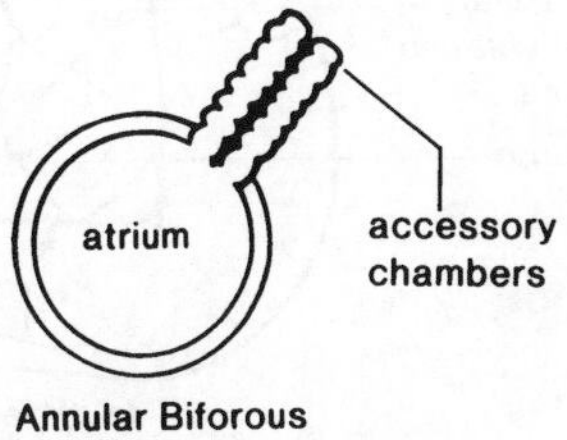

Annular Biforous
Figure 198

Figs. 24, 85, 86, 170, 171 ⟶

340(339′). Stemmata on each side 0; mandibular mola with hyaline lobe at base; body parallel-sided and somewhat flattened; length usually less than 10 mm. In leaf litter, under bark .... (part) ***Euglenidae*** p. 554

340′. Stemmata on each side 5 or 6; mandibular mola without hyaline lobe at base; body somewhat curved ventrally and narrowing posteriorly; length usually more than 10 mm .... (*Cephaloon*) (**Cephaloinae**) ***Cephaloidae*** p. 529

341(337′). Epicranial stem long and bent to the left (fig. 15); frontal arms joined anteriorly by transverse ecdysial line (fig. 15); spiracles on segment A8 much larger than those on A7. In fungus fruiting bodies. Exotic (Asia, East Indies, New Guinea) .... ***Pterogeniidae*** p. 501

341′. Epicranial stem shorter and not bent to the left (fig. 18); frontal arms not joined anteriorly by transverse line; spiracles on segment A8 not larger than those on A7 342

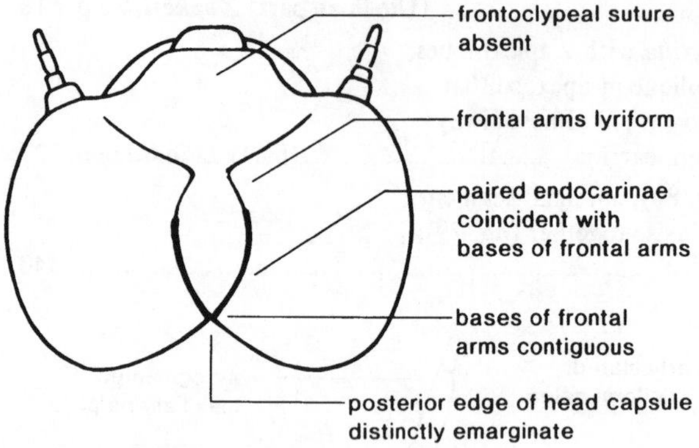

Figure 24

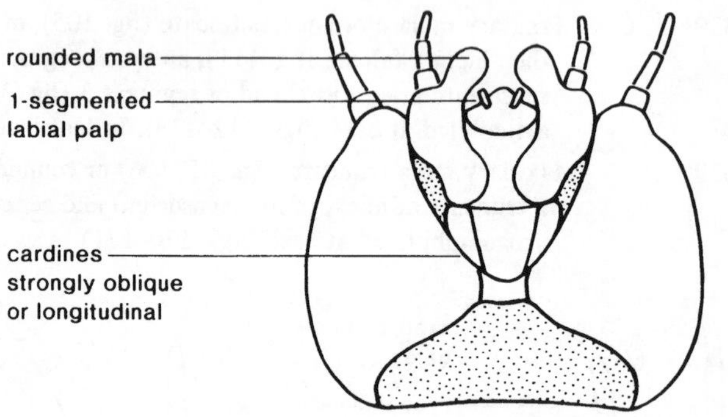

Figure 91

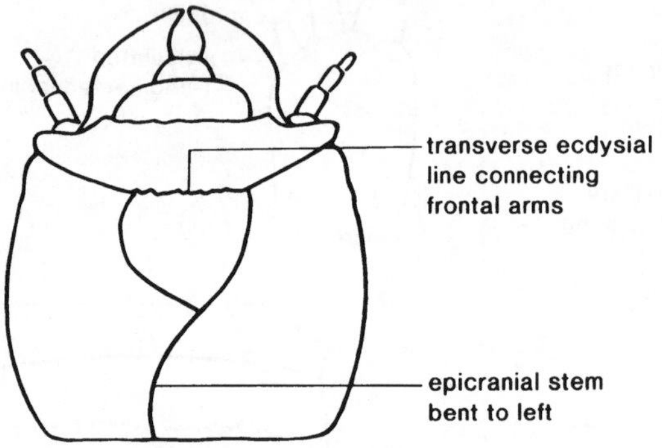

Figure 15

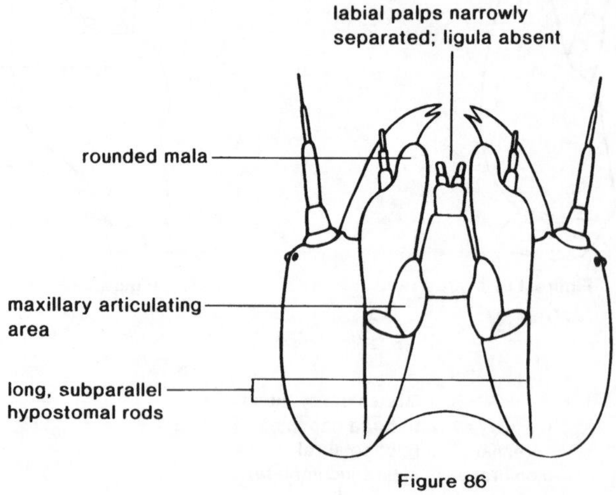

Figure 86

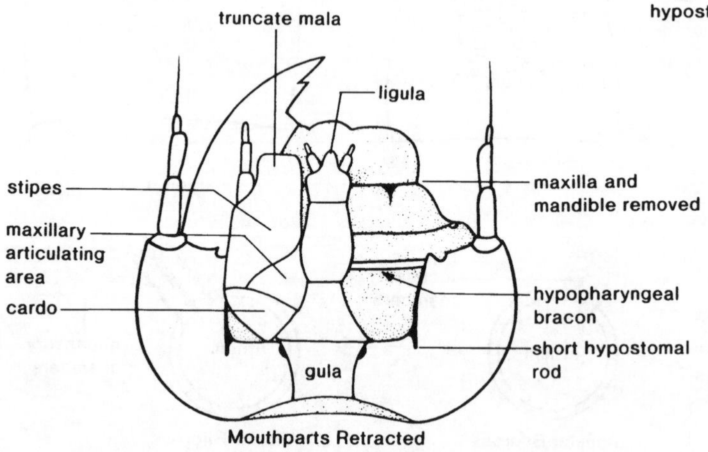

Mouthparts Retracted

Figure 85

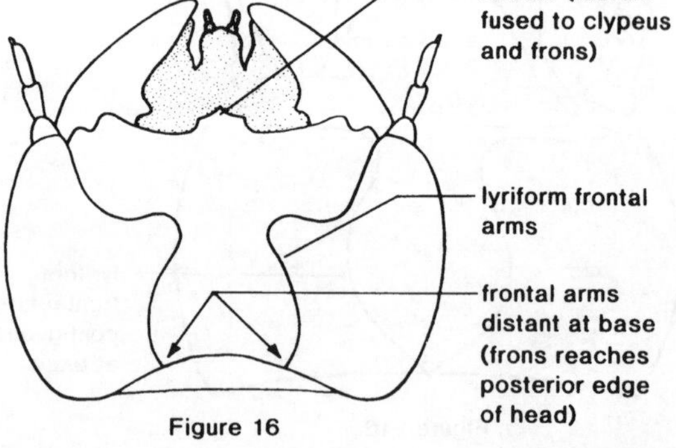

Figure 16

←— **Figs. 18, 197, 198**

342(341′). Anal region posteriorly or terminally oriented; spiracles annular, annular-uniforous (fig. 196), or annular-biforous with short accessory chambers (fig. 197); molae tuberculate, with tubercles extending onto ventral surface (fig. 73). In fungus fruiting bodies, leaf litter, stored products .... (part) ***Mycetophagidae*** p. 498

342′. Anal region posteroventrally or ventrally oriented; spiracles annular-biforous with long accessory chambers (fig. 198); molae transversely ridged, without tubercles on ventral surface 343

343(342′). Urogomphi with accessory processes (fig. 196); epicranial stem moderately long (fig. 18); mala simple and rounded (fig. 86); abdominal terga without rows or patches of asperities; tergum A9 with only a few tubercles and without U-shaped groove. In fungus fruiting bodies .... (**Piseninae** part) ***Tetratomidae*** p. 504

343′. Urogomphi simple; epicranial stem very short or absent (fig. 24); apex of mala truncate and slightly emarginate, with 3 teeth at inner apical angle; abdominal terga each with transverse row of asperities followed by a patch of irregularly distributed asperities; tergum A9 with numerous tubercles and with a U-shaped, transverse groove. In rotten wood .... (**Stenotrachelinae**) ***Cephaloidae*** p. 529

344(336′). Hypostomal rods absent; abdominal spiracles annular (figs. 194, 195) or annular-multiforous (fig. 200). Under bark, in rotten wood, soil, leaf litter .... (part) ***Tenebrionidae*** p. 520

344′. Hypostomal rods present (fig. 89); abdominal spiracles annular-biforous (figs. 197, 198) 345

345(344′). Epicranial stem very short; urogomphi straight, narrow, acute, lightly sclerotized and posteriorly oriented, without accessory processes (fig. 170); hypostomal rods subparallel (fig. 85); ligula absent. In fungus fruiting bodies, leaf litter .... ***Archeocrypticidae*** p. 500

345′. Epicranial stem moderately long; urogomphi with accessory processes (fig. 169); hypostomal rods diverging posteriorly (fig. 89); ligula present 346

346(345′). Apex of mala simple; molar surface with numerous fine ridges (fig. 80); dorsal body surfaces generally smooth. In fungus fruiting bodies, under bark .... (**Piseninae** part) ***Tetratomidae*** p. 504

346′. Apex of mala cleft (fig. 106); molar surface simple (fig. 69); dorsal body surfaces generally granulate or tuberculate. Under bark, in fungus fruiting bodies. Exotic (Southern Hemisphere) .... (**Ulodinae** part) ***Zopheridae*** p. 518

347(313′). Segments in labial palp 1; cardo strongly oblique or longitudinal (fig. 91); apex of mala rounded or truncate (fig. 91); hypostomal rods absent 348

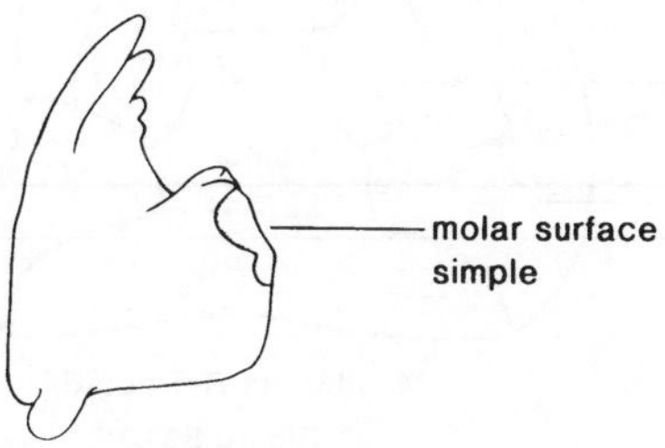

Figure 69

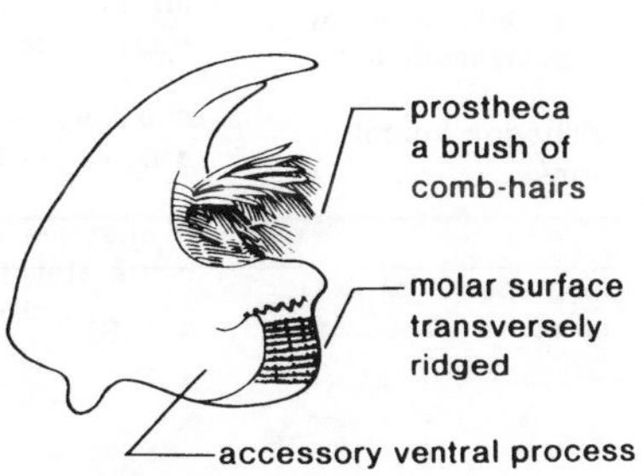

Figure 80

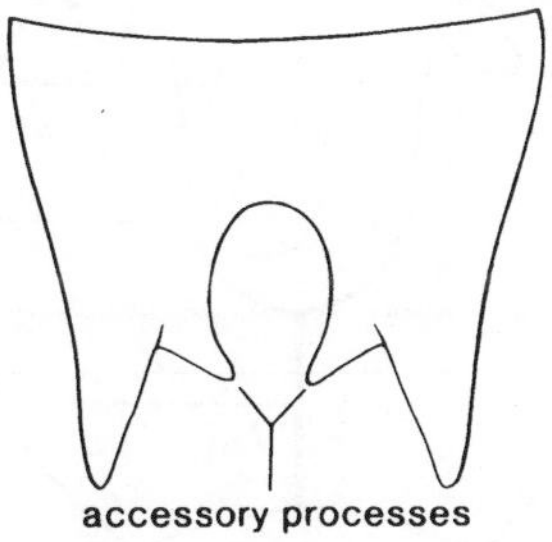

Dorsal

Figure 169

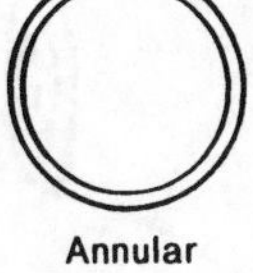

Annular

Figure 194

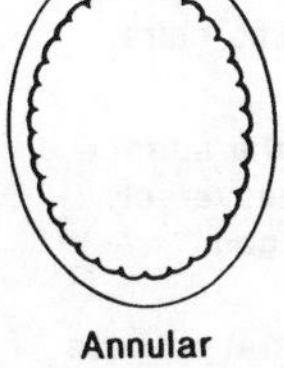

Annular

Figure 195

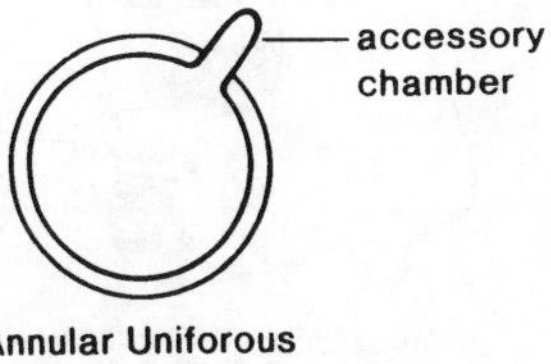

Annular Uniforous

Figure 196

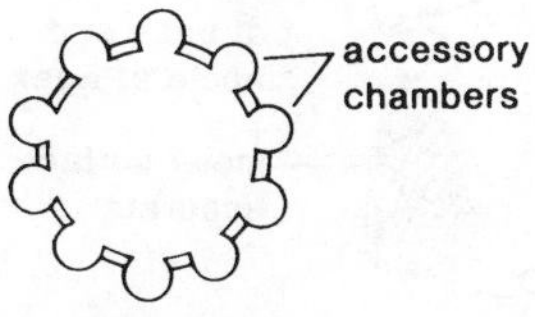

Annular Multiforous

Figure 200

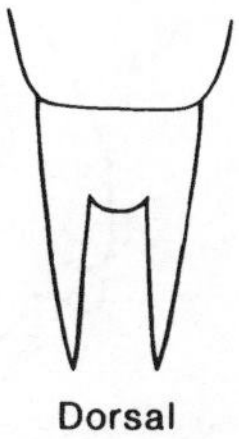

Dorsal

Figure 170

Lateral

Figure 171

Figs. 73, 89, 106 ⟶

347′. Segments in labial palp 2; cardo transverse or slightly oblique; apex of mala falciform (fig. 89); hypostomal rods present (fig. 89). Under fermenting bark, in fungus fruiting bodies .... (part) ***Biphyllidae*** p. 475

348(347). Bases of frontal arms contiguous (fig. 24); paired endocarinae present (fig. 24); ligula absent. In puffballs .... (*Pocadius* part) (**Nitidulinae**) ***Nitidulidae*** p. 456

348′. Bases of frontal arms distinctly separated (fig. 16); paired endocarinae absent; ligula shorter than palp. Under bark, in leaf litter, fungus fruiting bodies, carrion, rotting fruit .... (part) ***Nitidulidae*** p. 456

349(313[2]). Urogomphi articulated at base (fig. 162); maxilla with fixed galea and lacinia (fig. 111), sometimes fused or connate for part of their lengths; galea often with fringe of setae at apex (fimbriate galea) (fig. 111) 350

349′. Urogomphi fixed at base (fig. 170); maxilla usually with single, fixed mala (fig. 105) 355

350(349). Stemmata on each side 6. In decaying vegetation or carrion .... ***Agyrtidae*** p. 324

350′. Stemmata on each side 5 or fewer 351

351(350′). Segment A10 bearing a pair of hooks (fig. 163) 352

351′. Segment A10 without a pair of hooks 353

352(351). Urogomphi 2-segmented; stemmata on each side 3 to 5; epicranial stem present (fig. 14). In sand, on rocks, or in vegetation in or near streams and ponds .... ***Hydraenidae*** p. 320

352′. Urogomphi 1-segmented; stemmata on each side 0 or occasionally 1; epicranial stem absent. In leaf litter, decaying vegetation, rotten wood, fungi .... (part) ***Ptiliidae*** p. 322

353(351′). Stemmata present; epicranial stem present (fig. 14); vestiture often including expanded setae (fig. 4). In leaf litter, fungi, carrion .... (part) ***Leiodidae*** p. 327

353′. Stemmata absent; epicranial stem absent; vestiture of simple setae 354

354(353′). Urogomphi 2-segmented (fig. 162); body usually more than 1.2 mm in length; lacinia scoop-like and bidentate. In mammal nests .... (**Leptininae**) ***Leptinidae*** p. 330

354′. Urogomphi 1-segmented; body minute, length 1.2 mm or less; lacinia not as above. In leaf litter, associated with army ants .... ***Limulodidae*** p. 324

355(349′). Maxilla with galea and lacinia (figs. 111, 215) 356

355′. Maxilla with single mala (fig. 105) 358

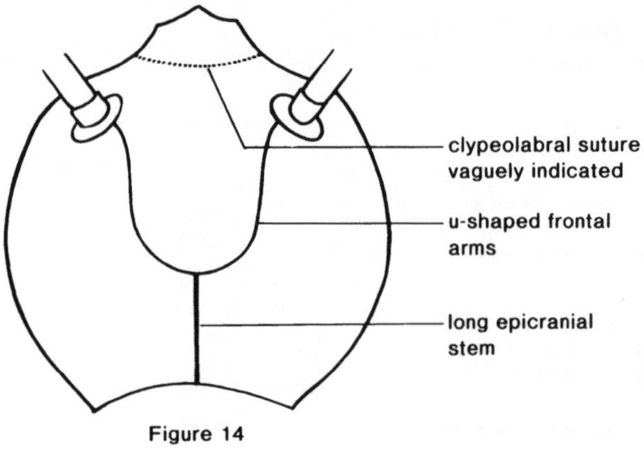

Figure 14

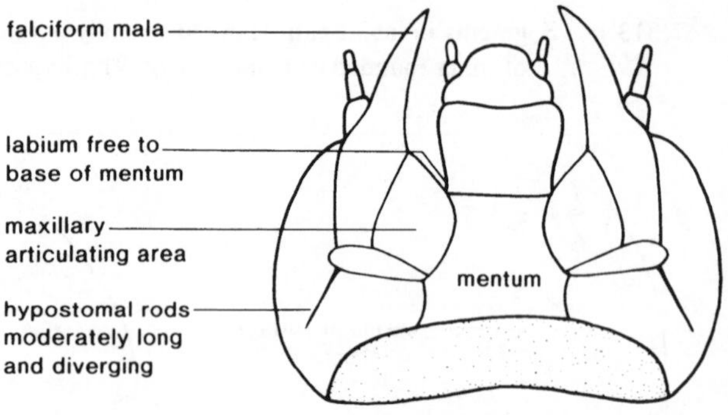

Mouthparts Retracted
Figure 89

Figure 3

Figure 4

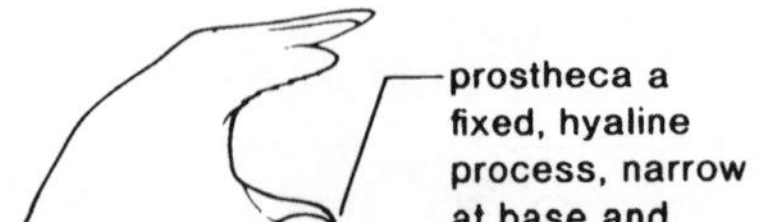

Figure 78

Figure 80

Figure 210

←— Figs. 24, 85, 170

356(355). Dorsal body surfaces coarsely granulate or tuberculate, usually with distinct setiferous processes (fig. 3); stemmata on each side 6; accessory ventral process of mandible present (figs. 78, 80); abdominal spiracles located at ends of spiracular tubes (fig. 210). Under fermenting bark, on fungi .... (part) ***Derodontidae*** p. 431

356′. Dorsal body surfaces smooth or very finely granulate, without setiferous processes; stemmata on each side 5 or fewer; accessory ventral process of mandible absent; abdominal spiracles not located at ends of spiracular tubes 357

357(356′). Dorsal body surfaces more heavily pigmented than ventral ones; ratio of antennal length to head width less than .2; apex of antennal segment 2 truncate, so that sensorium and segment 3 arise together (fig. 29); minute (length usually less then 1.3 mm). Under bark, in rotten wood, leaf litter. Exotic (Neotropical and Old World) .... ***Jacobsoniidae*** p. 433

357′. Dorsal and ventral body surfaces similarly pigmented; ratio of antennal length to head width more than .2; apex of antennal segment 2 oblique, so that sensorium arises proximad of segment 3 (fig. 30). In leaf litter, fungi, carrion .... (part) ***Leiodidae*** p. 327

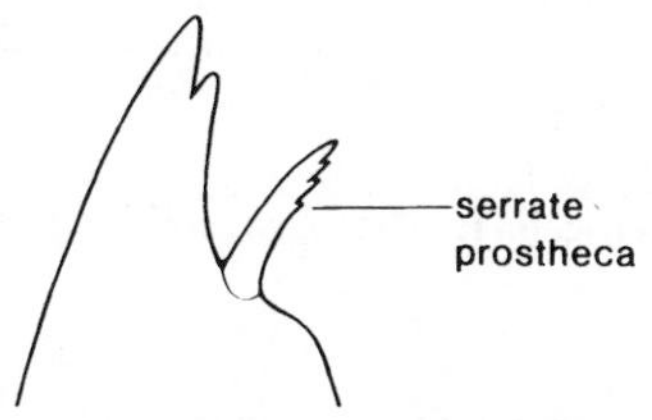

serrate prostheca

Figure 82

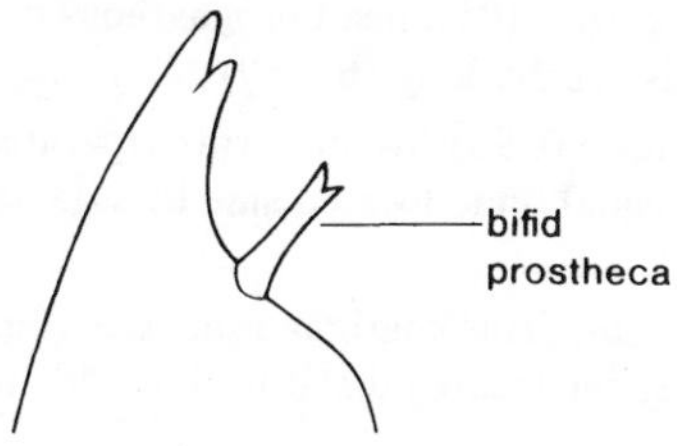

bifid prostheca

Figure 83

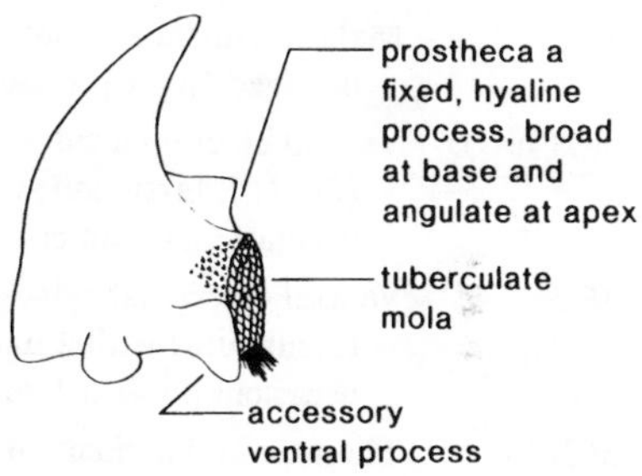

prostheca a fixed, hyaline process, broad at base and angulate at apex

tuberculate mola

accessory ventral process

Figure 81

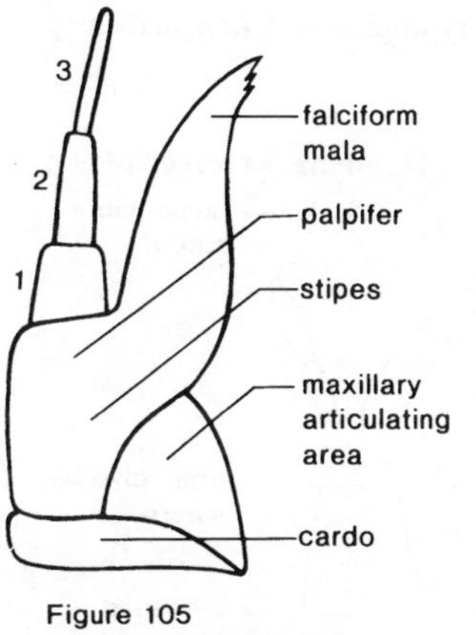

3
2
1
falciform mala
palpifer
stipes
maxillary articulating area
cardo

Figure 105

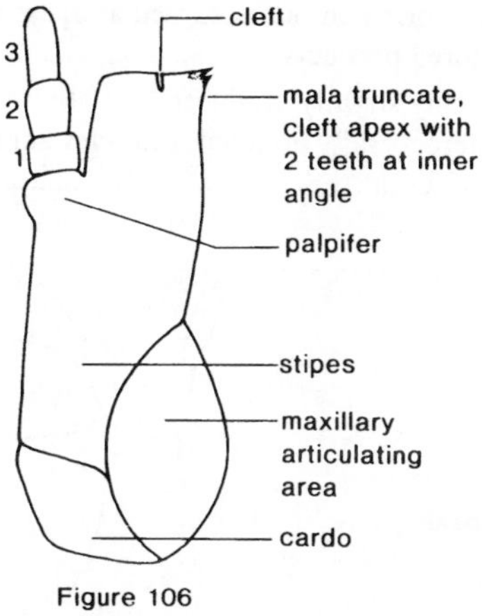

cleft
3
2
1
mala truncate, cleft apex with 2 teeth at inner angle
palpifer
stipes
maxillary articulating area
cardo

Figure 106

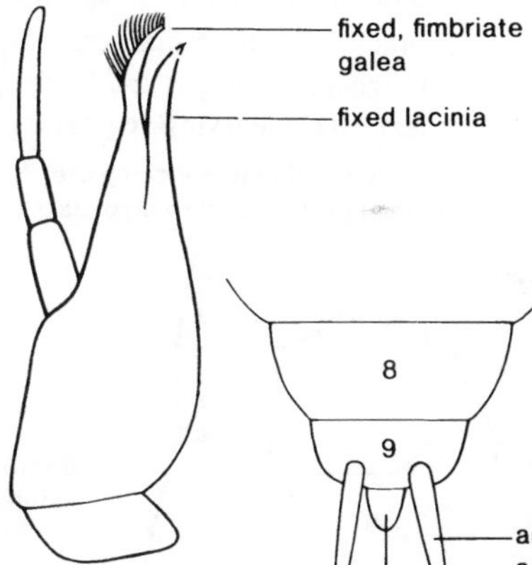

fixed, fimbriate galea
fixed lacinia

Figure 111

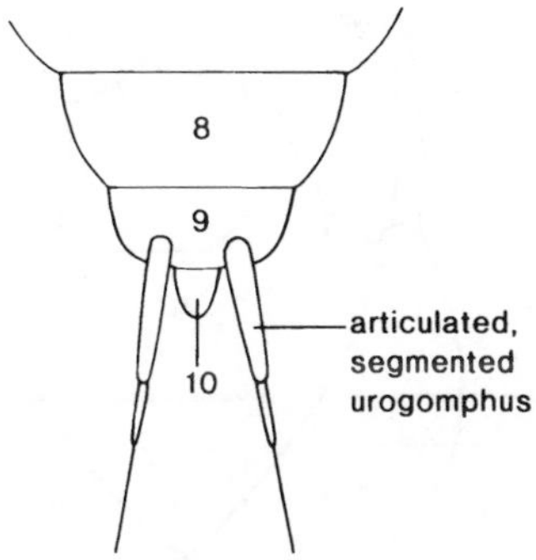

8
9
10
articulated, segmented urogomphus

Figure 162

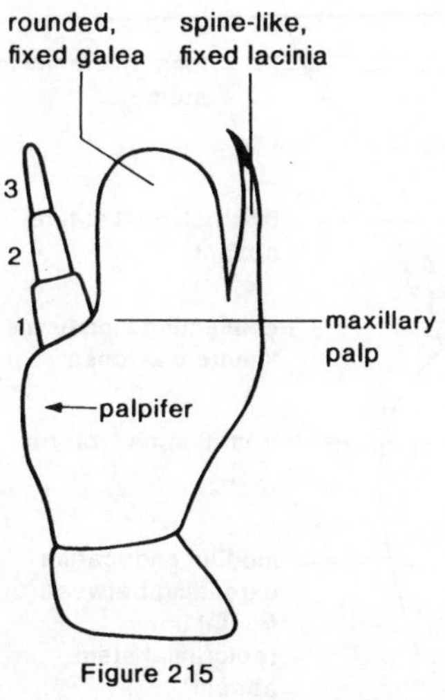

rounded, fixed galea
spine-like, fixed lacinia
3
2
1
maxillary palp
palpifer

Figure 215

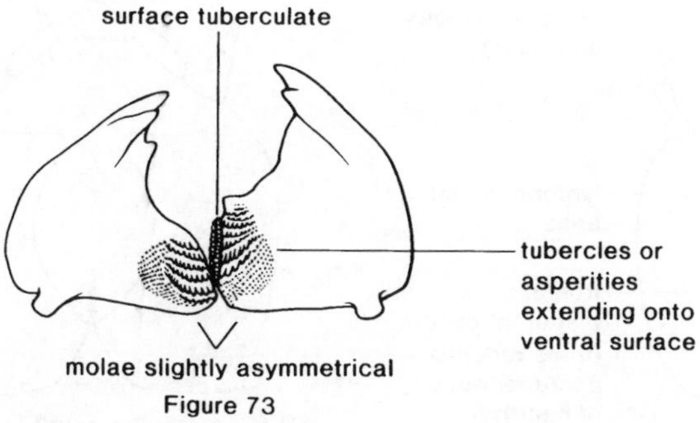

surface tuberculate
tubercles or asperities extending onto ventral surface
molae slightly asymmetrical

Figure 73

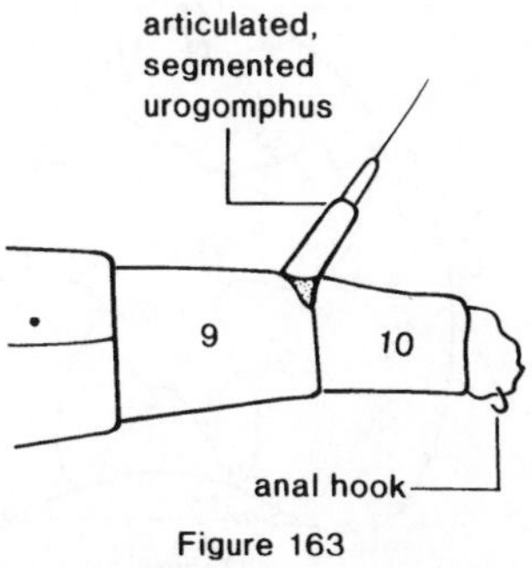

articulated, segmented urogomphus
9
10
anal hook

Figure 163

Figs. 16, 29, 30 ⟶

358(355′). Median endocarina located between frontal arms (fig. 22). In cycad cones, stored products .... (*Pharaxonotha*) (**Xenoscelinae**) ***Languriidae*** p. 471

358′. Median endocarina absent 359

359(358′). Prostheca broad, with apex obtusely angulate (fig. 81), or occasionally rounded 360

359′. Prostheca narrow, with apex acute (fig. 78) or occasionally bifid (fig. 83) or serrate (fig. 82) 364

360(359). Dorsal surfaces smooth; body more or less cylindrical and lightly sclerotized; epicranial stem moderately long (fig. 13); accessory ventral process of mandible absent. In stems .... (**Languriinae** part) ***Languriidae*** p. 471

360′. Dorsal surfaces granulate or tuberculate; body usually slightly flattened; epicranial stem very short (fig. 12) or absent (fig. 24); accessory ventral process of mandible present (figs. 78, 80) 361

361(360′). Mala more or less truncate and cleft at apex (fig. 106); head moderately to strongly declined (hypognathous) (fig. 10); hypostomal rods very short (fig. 85). In fruiting bodies of higher fungi .... (**Dacninae** part) ***Erotylidae*** p. 473

361′. Mala falciform and not cleft at apex (fig. 105); head prognathous or slightly declined (fig. 8); hypostomal rods usually long (fig. 89) 362

362(361′). Ventral epicranial ridges present (figs. 90, 93); frontal arms separated at base (fig. 16); tarsungulus with 2 subequal setae located side by side. In rotting vegetation, cycad cones, leaf axils .... (**Xenoscelinae** part) ***Languriidae*** p. 471

362′. Ventral epicranial ridges absent; frontal arms contiguous at base (fig. 24); tarsungulus with 2 unequal setae, one located distal to the other, or occasionally with 1 seta only 363

363(362′). Thoracic and abdominal terga produced laterally to form flattened processes; urogomphi acute at apex, posteriorly oriented but upturned at apex; spiracles biforous (fig. 202). In leaf litter, stored products .... (**Cryptophilinae**) ***Languriidae*** p. 471

363′. Thoracic and abdominal terga not or only slightly produced laterally; urogomphi straight and blunt, posterodorsally oriented; spiracles annular. In leaf litter, decaying vegetation .... (**Toraminae**) ***Languriidae*** p. 471

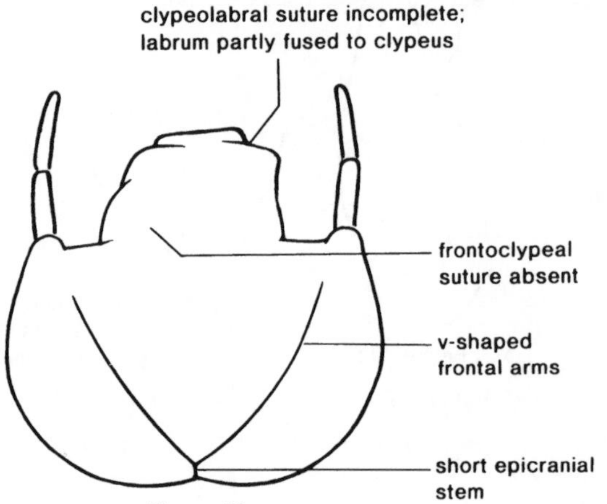

Figure 12

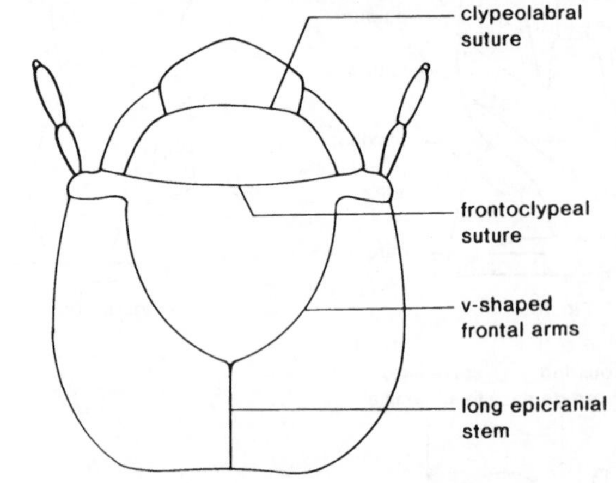

Figure 13

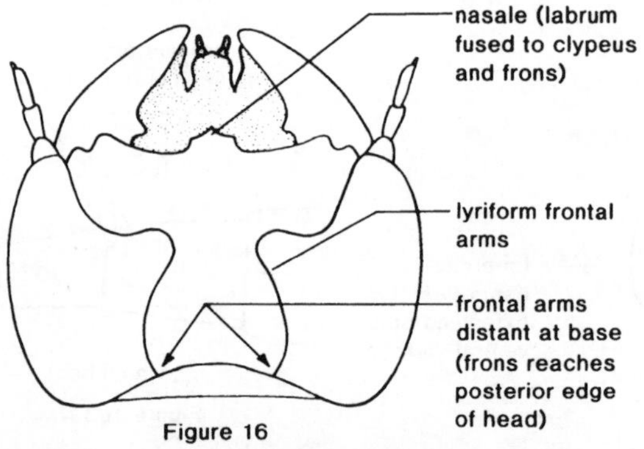

Figure 16

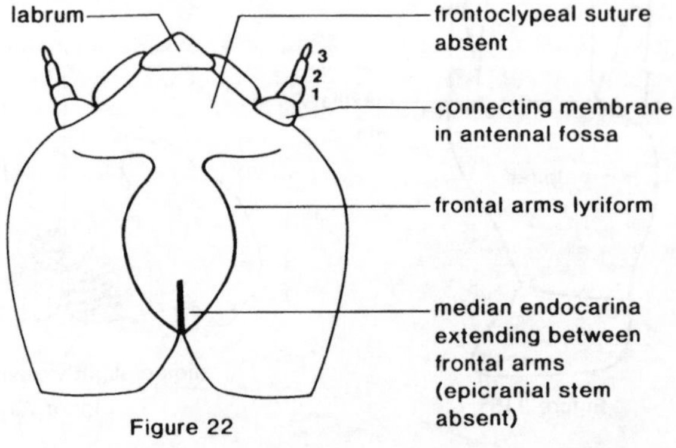

Figure 22

Figs. 4, 78, 80, 81, 82, 83, 89, 105, 106

364(359′). Mala truncate (fig. 88) or rounded (fig. 91) 365

364′. Mala falciform (fig. 105) 369

365(364). Ratio of antennal length to head width less than .15; stemmata on each side 4 or fewer 366

365′. Ratio of antennal length to head width more than .15; stemmata on each side 5 or 6 367

366(365). Abdominal segments with paired tergal processes; vestiture consisting of simple setae only; stemmata on each side 4. In puffballs .... (*Lycoperdina*) (**Eumorphinae**) ***Endomychidae*** p. 482

366′. Abdominal segments without paired processes; vestiture including expanded setae (fig. 4); stemmata on each side 2. On molds, in stored products .... (*Mycetaea*) (**Mycetaeinae**) ***Endomychidae*** p. 482

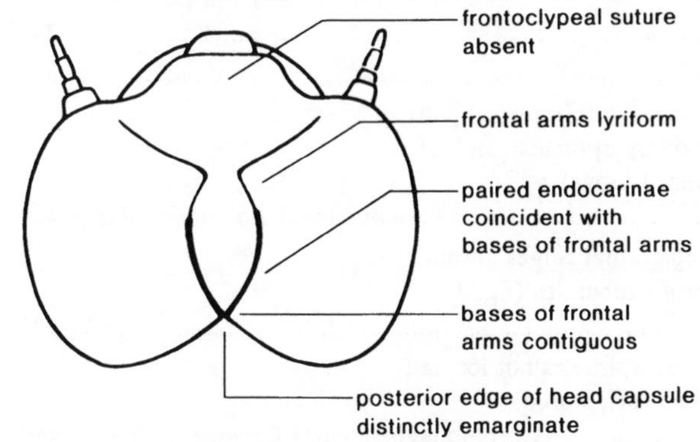

frontoclypeal suture absent
frontal arms lyriform
paired endocarinae coincident with bases of frontal arms
bases of frontal arms contiguous
posterior edge of head capsule distinctly emarginate

Figure 24

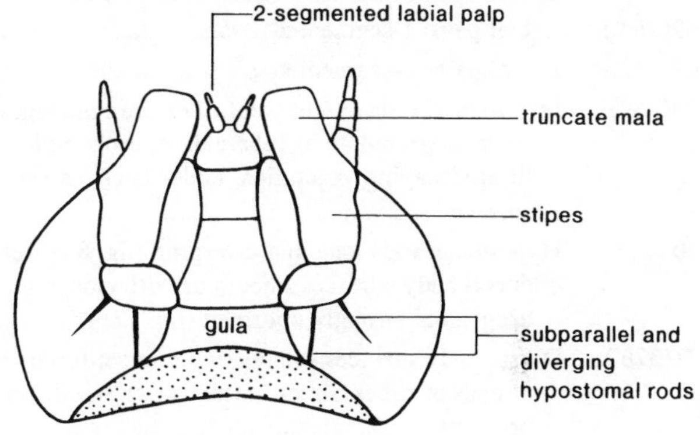

2-segmented labial palp
truncate mala
stipes
gula
subparallel and diverging hypostomal rods

Figure 88

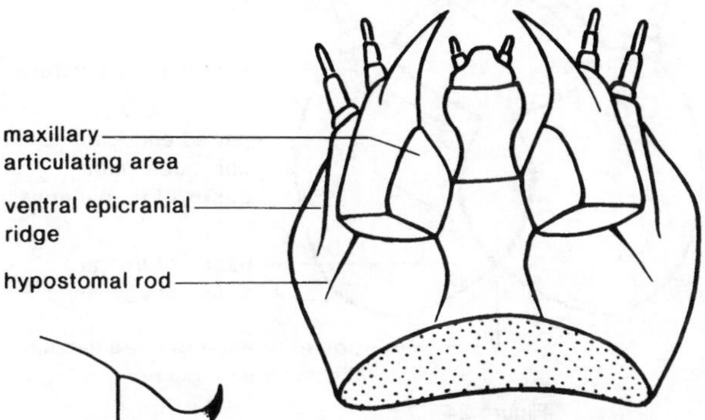

maxillary articulating area
ventral epicranial ridge
hypostomal rod

Mouthparts Retracted

Figure 90

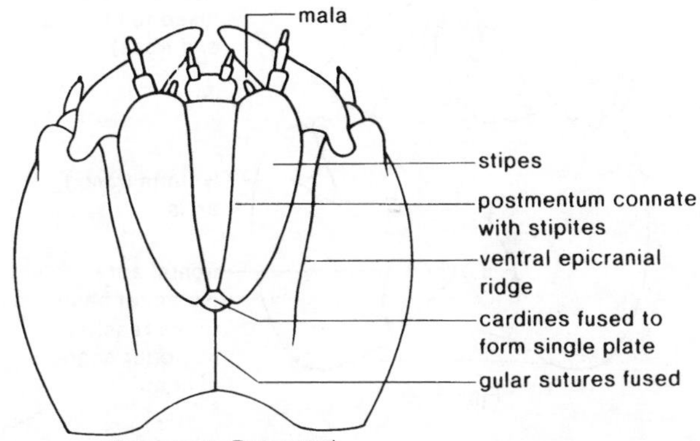

mala
stipes
postmentum connate with stipites
ventral epicranial ridge
cardines fused to form single plate
gular sutures fused

Mouthparts Retracted

Figure 93

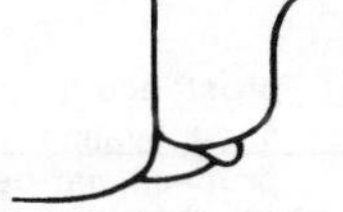

Lateral

Figure 171

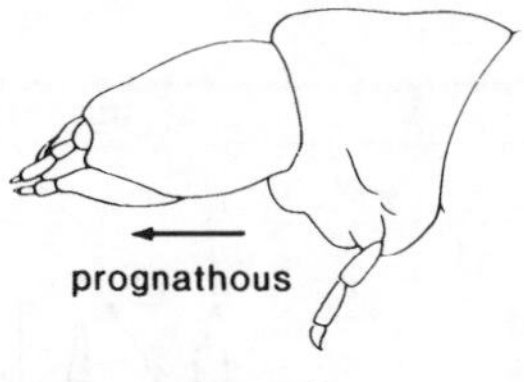

prognathous

Figure 8

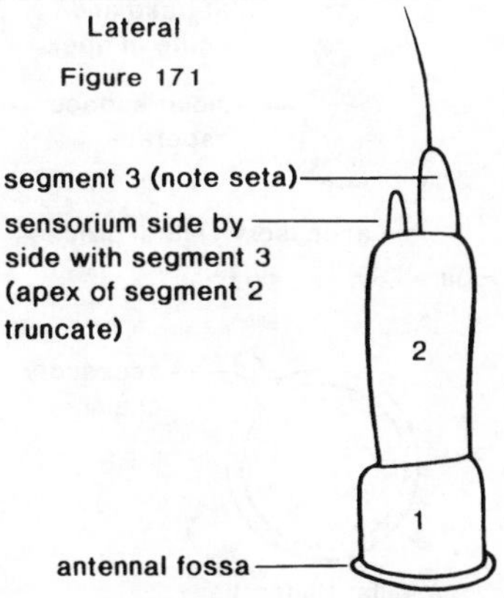

segment 3 (note seta)
sensorium side by side with segment 3 (apex of segment 2 truncate)
2
1
antennal fossa

Figure 29

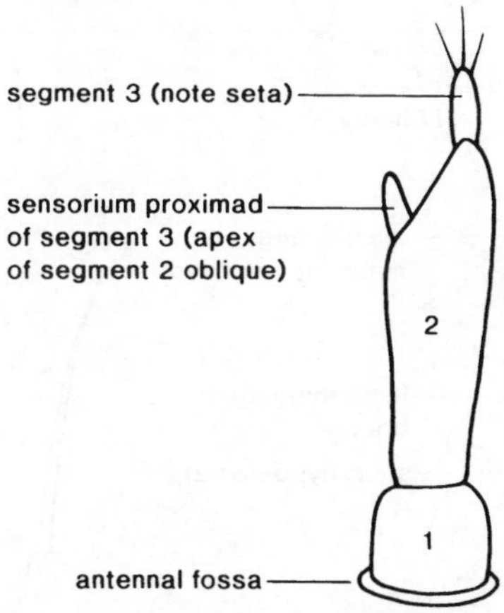

segment 3 (note seta)
sensorium proximad of segment 3 (apex of segment 2 oblique)
2
1
antennal fossa

Figure 30

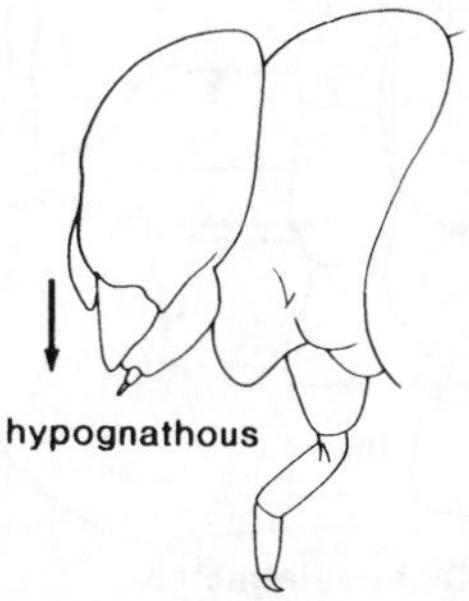

hypognathous

Figure 10

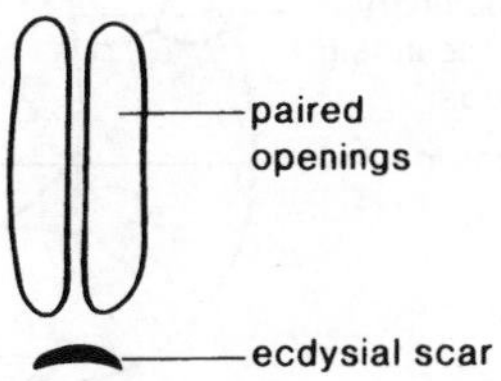

paired openings
ecdysial scar
Biforous

Figure 202

Figs. 85, 91 →

367(365′). Labial palps 1-segmented; hypostomal rods long and diverging (fig. 89); dorsal body surfaces with numerous elongate, setiferous tubercles (fig. 3). In rotting fungus fruiting bodies. Exotic (Southern Hemisphere) .... (part) ***Hobartiidae*** p. 468

367′. Labial palps 2-segmented; hypostomal rods very short (fig. 85); dorsal body surfaces smooth or granulate-tuberculate, but without elongate, setiferous tubercles 368

368(367′). Dorsal body surfaces granulate or tuberculate; head moderately to strongly declined (hypognathous) (fig. 10); spiracles annular-biforous. In fruiting bodies of higher fungi .... (**Dacninae** part) ***Erotylidae*** p. 473

368′. Dorsal body surfaces smooth; head prognathous or slightly declined (fig. 8); spiracles annular or occasionally annular-uniforous (fig. 196). In fruiting bodies of slime molds (Myxomycetes) .... (*Odontosphindus*) (**Sphindinae**) ***Sphindidae*** p. 455

369(364′). Labial palps 1-segmented 370

369′. Labial palps 2-segmented 372

370(369). Hypostomal rods absent; ventral epicranial ridges present (fig. 93); dorsal body surfaces granulate or tuberculate; urogomphi not strongly upturned. In leaf litter, decaying vegetation, under bark, on spore-covered fungal surfaces .... (**Monotominae**) ***Rhizophagidae*** p. 460

370′. Hypostomal rods long and diverging (fig. 89); ventral epicranial ridges absent; dorsal body surfaces smooth or with elongate, setiferous tubercles (fig. 3); urogomphi strongly upturned (fig. 171) 371

371(370′). Dorsal body surfaces smooth, without setiferous tubercles; spiracles not located at ends of tubes. In leaf litter, rotten wood, fungi, bee nests, stored products .... (**Cryptophaginae** part) ***Cryptophagidae*** p. 469

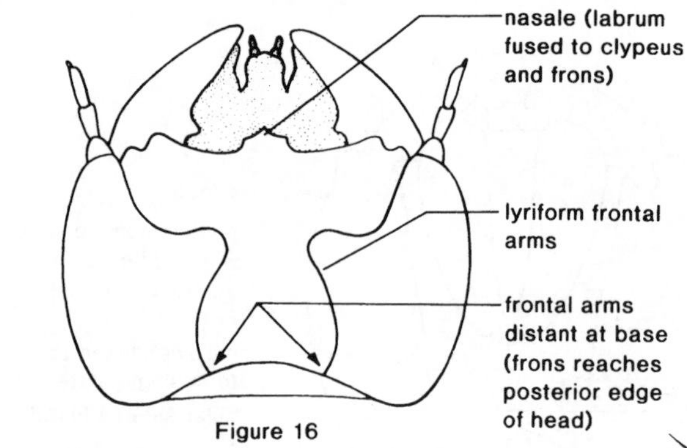

Figure 16

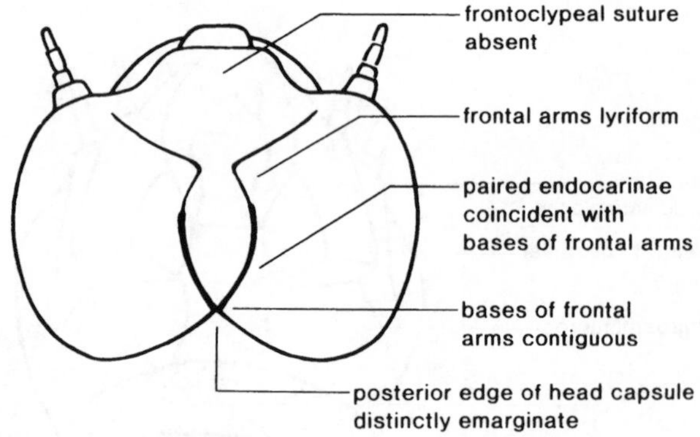

Figure 24

Figure 210

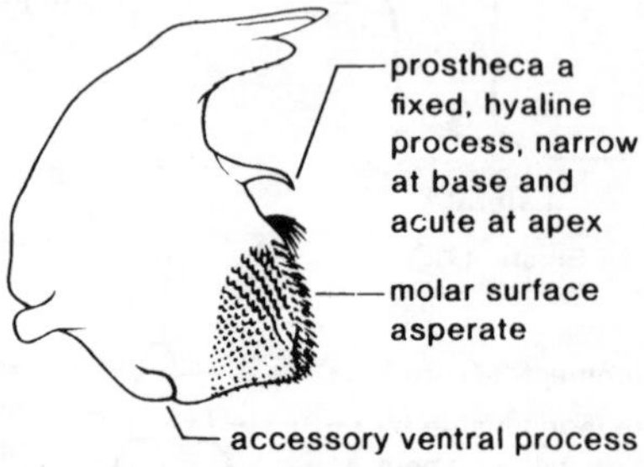

Figure 78

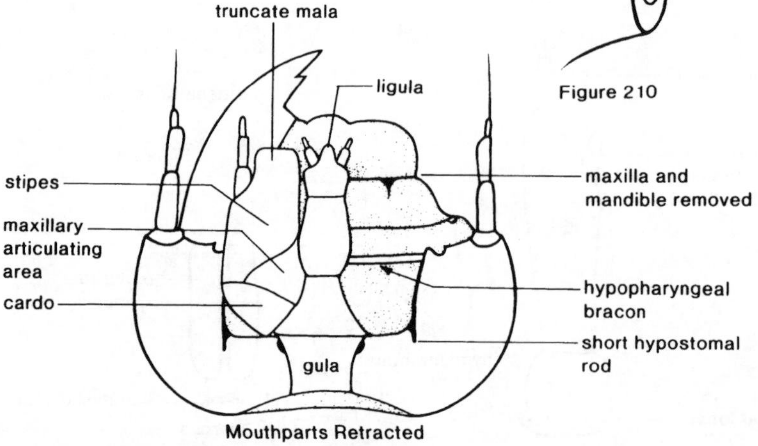

Mouthparts Retracted
Figure 85

Figure 3

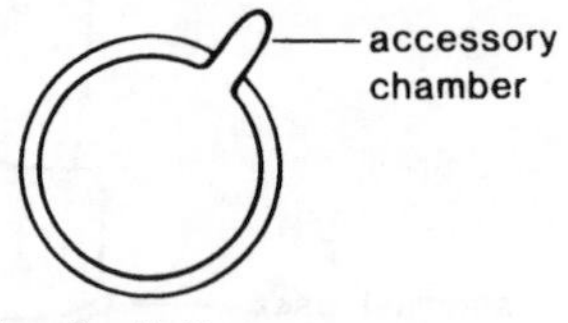

Annular Uniforous
Figure 196

← Figs. 8, 10, 16, 24, 90, 93, 171

371′. Dorsal body surfaces covered with elongate, setiferous tubercles (fig. 3); spiracles located at ends of short tubes (fig. 210). In rotting fungus fruiting bodies. Exotic (Southern Hemisphere) .... (part) ***Hobartiidae*** p. 468

372(369′). Thoracic and abdominal segments bearing lateral tergal processes; body more or less ovate 373

372′. Thoracic and abdominal segments without lateral tergal processes; body elongate and more or less parallel-sided 374

373(372). Abdominal spiracles located at ends of tergal processes; dorsal body surfaces more or less granulate (often covered with dirt); stemmata on each side 5; accessory ventral process of mandible present (fig. 78). In leaf litter. Exotic (Australia and New Zealand) .... **(Priasilphinae)** ***Phloeostichidae*** p. 462

373′. Abdominal spiracles not located at ends of tergal processes; dorsal body surfaces not granulate; stemmata on each side 1 or 0; accessory ventral process of mandible absent. In fungi, rotten wood, leaf litter .... **(Mychotheninae** part) ***Endomychidae*** p. 482

374(372′). Frontal arms distant at base (fig. 16); ventral epicranial ridges present (fig. 90); stemmata on each side 4. Under bark of rotten logs .... (*Rhizophagus*) **(Rhizophaginae)** ***Rhizophagidae*** p. 460

374′. Frontal arms approximate at base (fig. 24); ventral epicranial ridges absent; stemmata on each side 5 or 6 375

375(374′). Tergum A9 (without urogomphi) much shorter than tergum A8 (fig. 166); segment A10 posteriorly oriented; body more or less flattened 376

375′. Tergum A9 well-developed, about as long (without urogomphi) as tergum A8; segment A10 posteroventrally oriented; body usually not flattened 377

376(375). Segment A10 easily visible from above (fig. 166); body usually lightly sclerotized; urogomphi approximate and more or less parallel. In leaf litter, stored products .... (Cryptamorphini) ***Cucujidae*** p. 463

376′. Segment A10 reduced and not visible from above; body heavily sclerotized dorsally and ventrally; urogomphi strongly diverging. Under bark .... (*Cucujus*) **(Cucujinae)** ***Cucujidae*** p. 463

377(375′). Stemmata on each side 6; mala articulated at base; segment A10 with 2 long tubular pygopods. In male cones of cycads. Exotic (Australia) .... (*Paracucujus*) ***Boganiidae*** p. 462

377′. Stemmata on each side 5 or fewer; mala fixed at base; segment A10 without paired pygopods 378

378(377′). Stemmata on each side 1 or 0; urogomphi simple; spiracles annular. In leaf litter, grass piles, stored products, under bark .... **(Atomariinae** part) ***Cryptophagidae*** p. 469

378′. Stemmata on each side 5; urogomphi complex, each with 2 tubercles at base; spiracles annular-biforous (fig. 197). In sooty molds. Exotic (New Zealand) .... (*Agapytho*) **(Agapythinae)** ***Phloeostichidae*** p. 462

379(313³). Head strongly retracted (fig. 9); body strongly curved ventrally (C-shaped); A8 spiracle much larger than others. In dead wood .... (first instar) **(Lyctinae** part) ***Bostrichidae*** p. 439

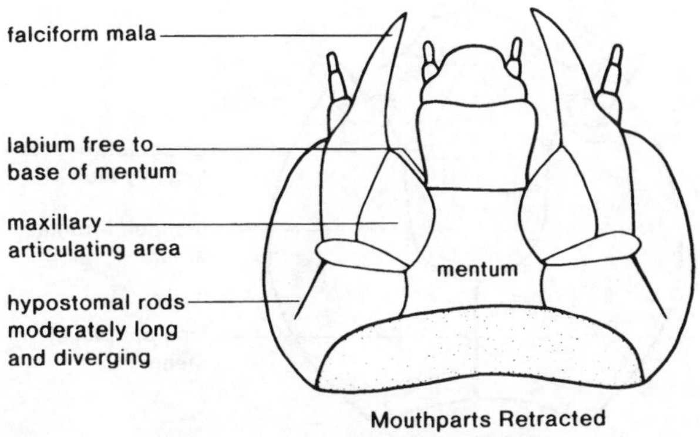

Mouthparts Retracted
Figure 89

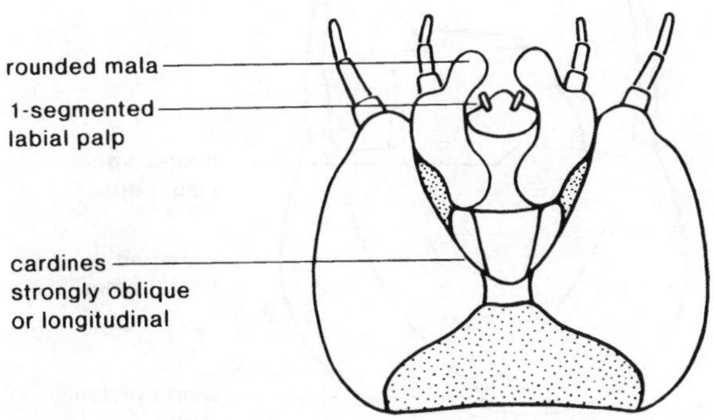

Figure 91

Figs. 9, 166, 197 ⟶

379′. Head protracted or slightly retracted (figs. 8, 10); body not strongly curved ventrally; A8 spiracle about the same size as others 380

380(379′). Epicranial stem moderately long (fig. 14); maxillary palps 4-segmented; hypostomal rods absent; body densely clothed with long, fine hairs. In dead wood .... (*Orphilus*) (**Orphilinae**) ***Dermestidae*** p. 434

380′. Epicranial stem very short (fig. 12) or absent; maxillary palps 3-segmented; hypostomal rods present; body without long, fine hairs; upper surfaces either granulate-tuberculate or with paired processes near midline on all abdominal terga 381

381(380′). Stemmata on each side 4; epicranial stem absent; frontoclypeal suture present; labial palps 1-segmented (or apparently so); spiracles annular; tarsungular setae 1; with paired processes near midline on all abdominal terga. In puffballs .... (*Lycoperdina* part) ***Endomychidae*** p. 482

381′. Stemmata on each side 5 or 6; epicranial stem present; frontoclypeal suture absent; labial palps 2-segmented; spiracles annular-biforous (fig. 197); tarsungular setae 2; without paired tergal processes; upper surface usually granulate or tuberculate. In fungus fruiting bodies (usually Basidiomycetes) .... (part) ***Erotylidae*** p. 473

382(313[4]). Bases of frontal arms contiguous (fig. 24). In puffballs .... (*Pocadius* part) (**Nitidulinae**) ***Nitidulidae*** p. 456

382′. Bases of frontal arms distinctly separated (fig. 16). Under bark, in leaf litter, fungus fruiting bodies, carrion, rotting fruit .... (part) ***Nitidulidae*** p. 456

383(311[3]). Abdominal terga with patches of asperities (fig. 141) on most segments; prothoracic sternum with median tooth (fig. 123); tergum A9 with indistinct pit between urogomphi. Under bark .... ***Synchroidae*** p. 516

383′. Abdominal terga without patches of asperities; prothoracic sternum without median tooth; tergum A9 without pit or with 2 distinct pits between urogomphi 384

384(383′). Tergum A9 with 2 distinct pits between urogomphi (fig. 175). Under bark or in rotting vegetation .... (**Pedilinae**) ***Pedilidae*** p. 544

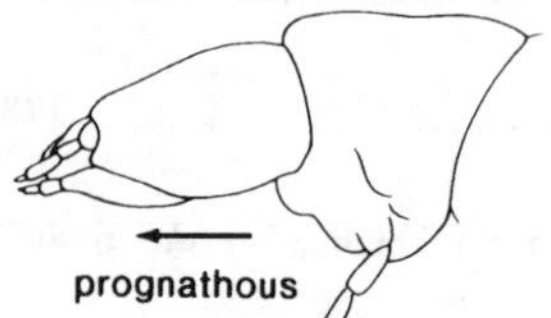

Figure 8

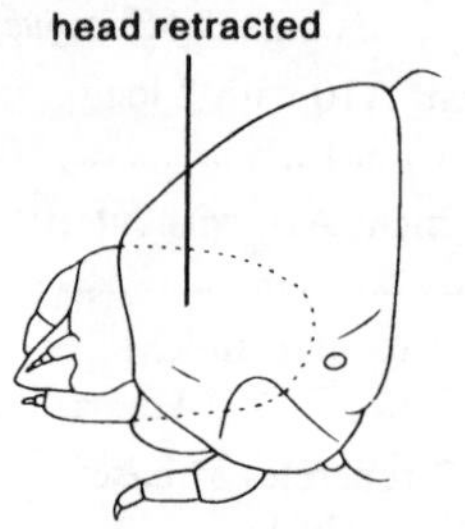

Figure 9

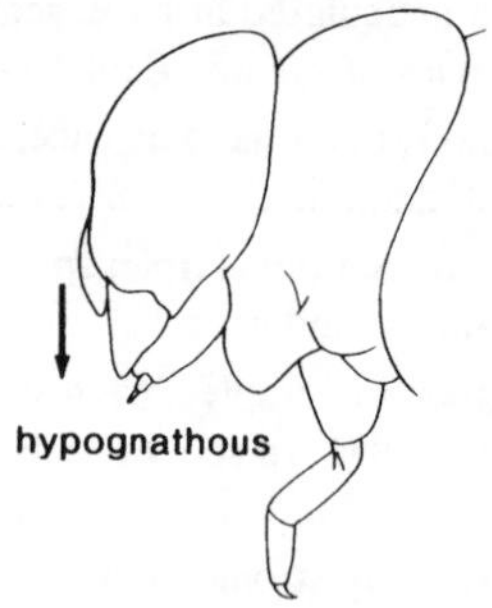

Figure 10

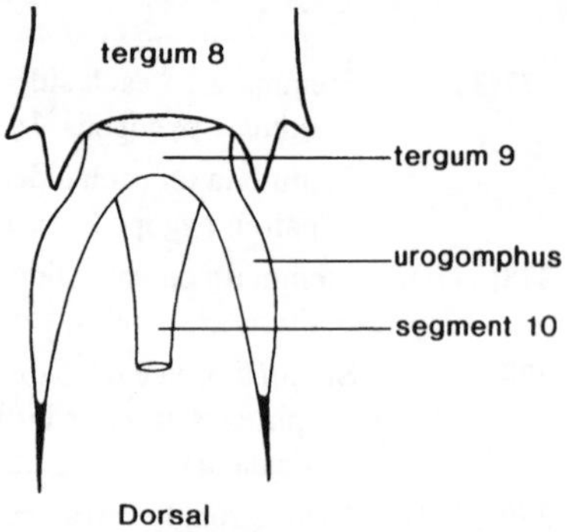

Dorsal

Figure 166

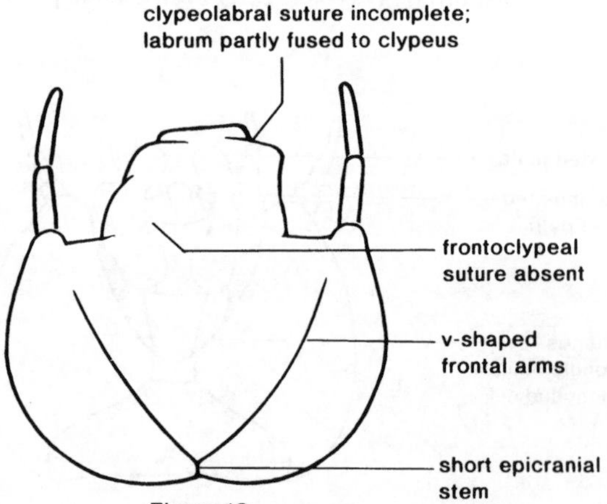

Figure 12

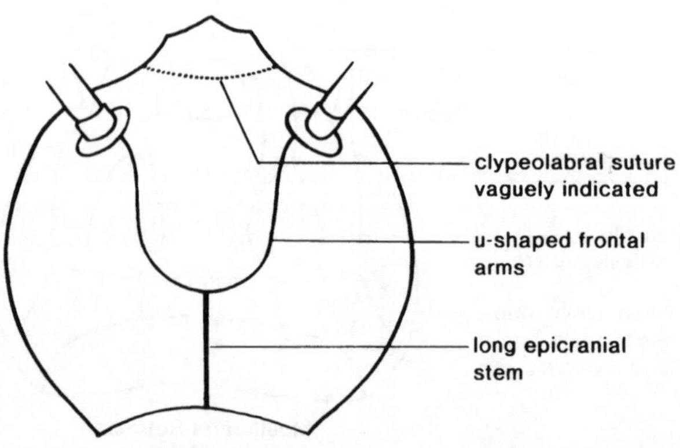

Figure 14

← Figs. 16, 24, 85, 89

384′. Tergum A9 without pit between urogomphi (sometimes the appearance of a pit is caused by the proximity of the 2 small, mesal teeth at the urogomphal bases, as in fig. 177) 385

385(384′). Each urogomphus with small, mesal tooth near base; hypostomal rods very short (fig. 85); epicranial stem absent; paired endocarinae absent; mola well-developed and tuberculate or asperate (fig. 73). In leaf litter, stored products, guano .... (*Aglenus*) (**Agleninae**) ***Othniidae*** p. 547

385′. Urogomphi simple; hypostomal rods moderately long but not extending to posterior edge of head (fig. 89); epicranial stem moderately long (fig. 18); paired endocarinae absent; mola well-developed and tuberculate or asperate (fig. 73). In fungus fruiting bodies (**see 3rd choice**) .... (**Hallomeninae**) ***Melandryidae*** p. 505

385². Urogomphi bifurcate (fig. 184); hypostomal rods extending almost to posterior edge of head (fig. 86); epicranial stem absent; paired endocarinae coincident with frontal arms (fig. 24); mola consisting of a few teeth only (fig. 71). Under bark, in rotten wood, stems, leaf litter .... (**Salpinginae** part) ***Salpingidae*** p. 549

386(311⁴). Paired ventral prolegs (asperity-bearing ampullae) absent; frontoclypeal suture absent or vaguely indicated; abdominal terga without patches of asperities 387

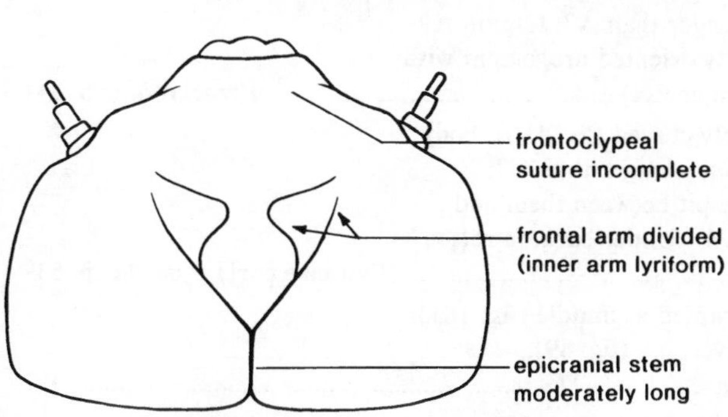

frontoclypeal suture incomplete
frontal arm divided (inner arm lyriform)
epicranial stem moderately long

Figure 18

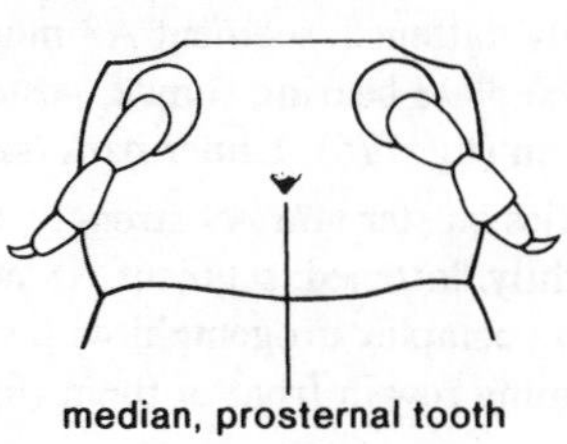

median, prosternal tooth

Figure 123

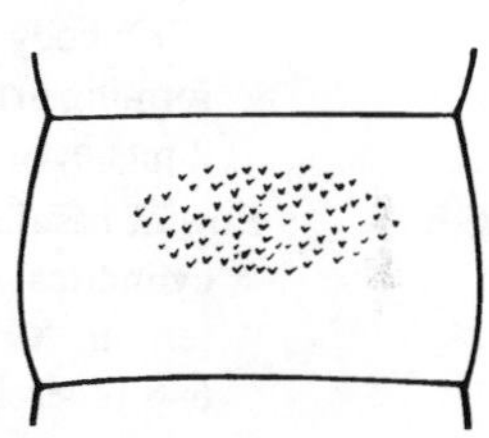

Figure 141

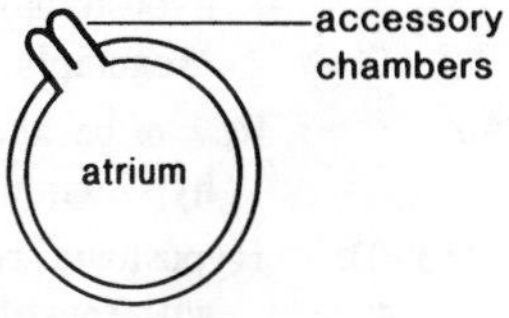

accessory chambers
atrium

Annular Biforous
Figure 197

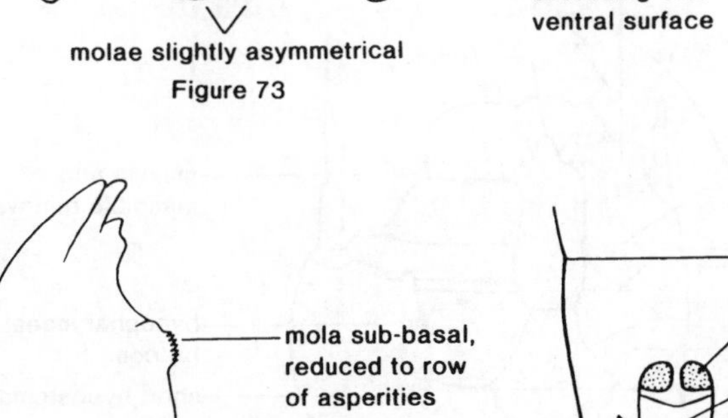

surface tuberculate
tubercles or asperities extending onto ventral surface
molae slightly asymmetrical

Figure 73

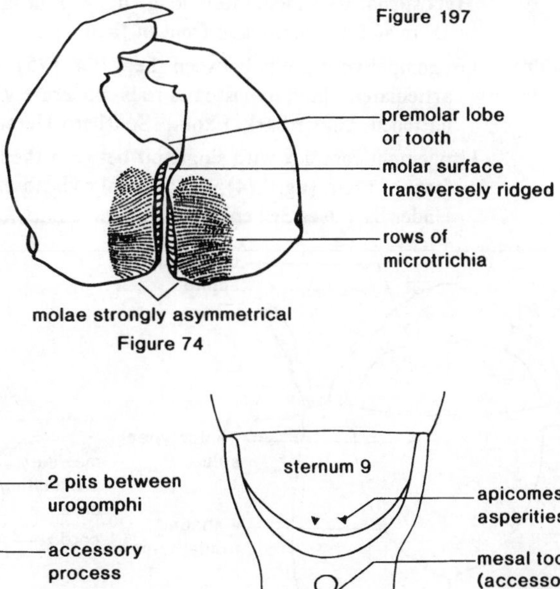

premolar lobe or tooth
molar surface transversely ridged
rows of microtrichia
molae strongly asymmetrical

Figure 74

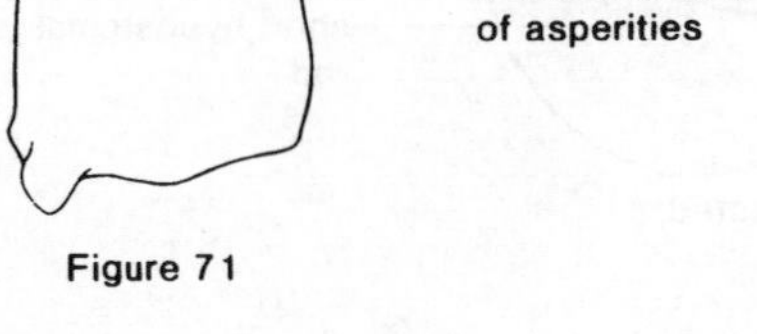

mola sub-basal, reduced to row of asperities

Figure 71

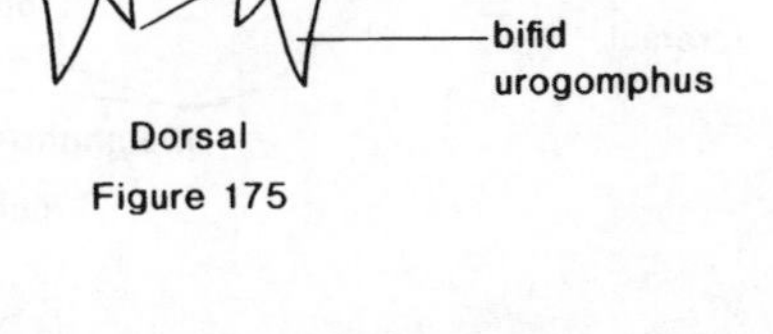

2 pits between urogomphi
accessory process
bifid urogomphus
Dorsal

Figure 175

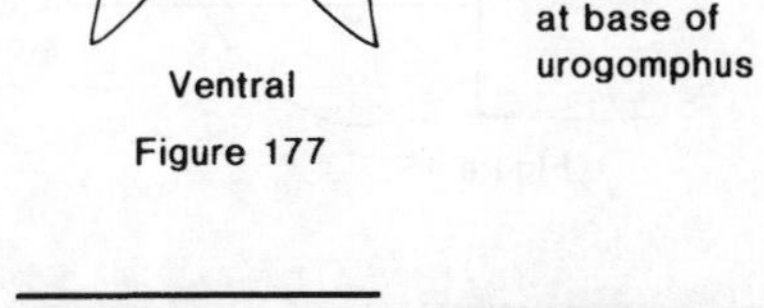

sternum 9
apicomesal asperities
mesal tooth (accessory process) at base of urogomphus
Ventral

Figure 177

Fig. 86: See p. 296. Fig. 184 ⟶

386′. Paired ventral prolegs (asperity-bearing ampullae) (fig. 149) on sterna A2–4 or A2–5; frontoclypeal suture distinct (fig. 13); abdominal terga with patches of asperities (fig. 143) on 1 or more segments. In rotten wood .... (*Calopus*) (**Calopodinae**) ***Oedemeridae*** p. 534

387(386). Urogomphi simple, somewhat converging, with 2 distinct pits between them (fig. 175); stemmata absent; apex of mala cleft (fig. 106); mandibles more or less symmetrical. In decaying vegetation in dry, sandy areas .... (**Cononotinae**) ***Pedilidae*** p. 544

387′. Urogomphi bifid or with accessory processes (fig. 182), without or with a single weakly defined pit between them; stemmata present; apex of mala simple and rounded; mandibles asymmetrical (figs. 73, 74) 388

388(387′). Stemmata on each side 1; ligula well-developed and labial palps distinctly separated (fig. 85) .... (**Trogocryptinae** part) ***Othniidae*** p. 547

388′. Stemmata on each side 2 or 5; ligula absent and labial palps subcontiguous (fig. 86) .... ***Inopeplidae*** p. 551

389(311^5). Row of basal asperities on sternum A9 straight or slightly sinuate and only slightly curved posteriorly at sides (figs. 179, 181) 390

389′. Row of basal asperities on sternum A9 strongly curved posteriorly at sides (fig. 178); body strongly flattened; segment A8 much longer than A7; tergum A9 forming articulated plate bearing simple, posteriorly oriented urogomphi with 2 pits between them (fig. 175). Under bark (**see 3rd choice**) .... ***Pyrochroidae*** p. 541

389^2. Row of basal asperities on sternum A9 strongly, doubly curved (fig. 180); body cylindrical or slightly flattened; segment A8 not much longer than A7; tergum A9 bearing complex urogomphi with single pit between them and 6–8 tubercles forming row in front of them (fig. 174). Under bark, in rotten wood .... (**Pythinae** part) ***Pythidae*** p. 539

390(389). Row of basal asperities on sternum A9 broadly interrupted at middle (fig. 181); hypostomal rods very short (fig. 85) or long and diverging (fig. 89); urogomphi without or with single pit between them 391

390′. Row of basal asperities on sternum A9 more or less continuous (fig. 179); hypostomal rods absent or urogomphi with 2 pits between them (fig. 175) 392

391(390). Hypostomal rods very short (fig. 85); urogomphi with no pit between them and with row of 6 asperities in front of them .... (*Sphalma*) (**Pythinae**) ***Pythidae*** p. 539

391′. Hypostomal rods moderately long (fig. 89); urogomphi with single pit between them and 4 asperities in front of them .... (*Elacatis*) (**Othniinae**) ***Othniidae*** p. 547

392(390′). Urogomphi with 2 pits between them (fig. 175); tergum A9 forming sclerotized, articulated plate; hypostomal rods moderately long (fig. 89); body strongly flattened. Under bark. Exotic (Southern Hemisphere) .... (**Pilipalpinae** = **Techmessinae**) ***Pythidae*** p. 539

392′. Urogomphi complex with single pit between them and 8 or more tubercles in front of them (fig. 174); hypostomal rods absent; body only slightly flattened. Under bark (**see 3rd choice**) .... (*Pytho*) (**Pythinae**) ***Pythidae*** p. 539

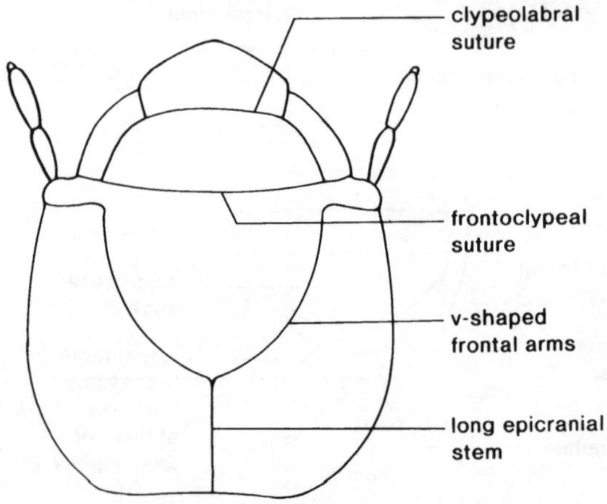

clypeolabral suture
frontoclypeal suture
v-shaped frontal arms
long epicranial stem

Figure 13

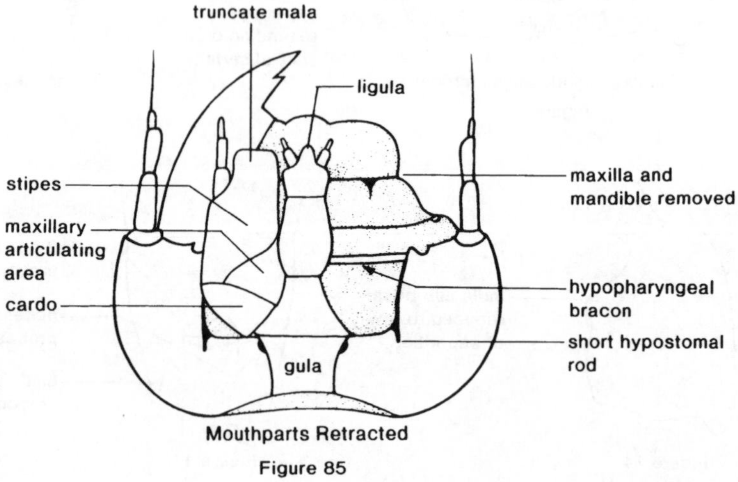

truncate mala
ligula
stipes
maxillary articulating area
cardo
maxilla and mandible removed
hypopharyngeal bracon
short hypostomal rod
gula
Mouthparts Retracted

Figure 85

← Figs. 73, 74

392[2]. Urogomphi without pit between them; hypostomal rods absent. Exotic .... p. 393

393(392[2]). Urogomphi simple, more or less parallel, upturned at apex; meso- and metaterga, terga A1–7 and sterna A2–8 with series of longitudinal ridges. Length of mature larva greater than 100 mm. Exotic (Southeast Asia and East Indies) .... ***Trictenotomidae*** p. 539

393′. Urogomphi diverging, each bearing 5 or more long spines; without longitudinal ridges on thorax and abdomen. Exotic (Japan) .... (*Istrisia*) ***?Salpingidae*** p. 549

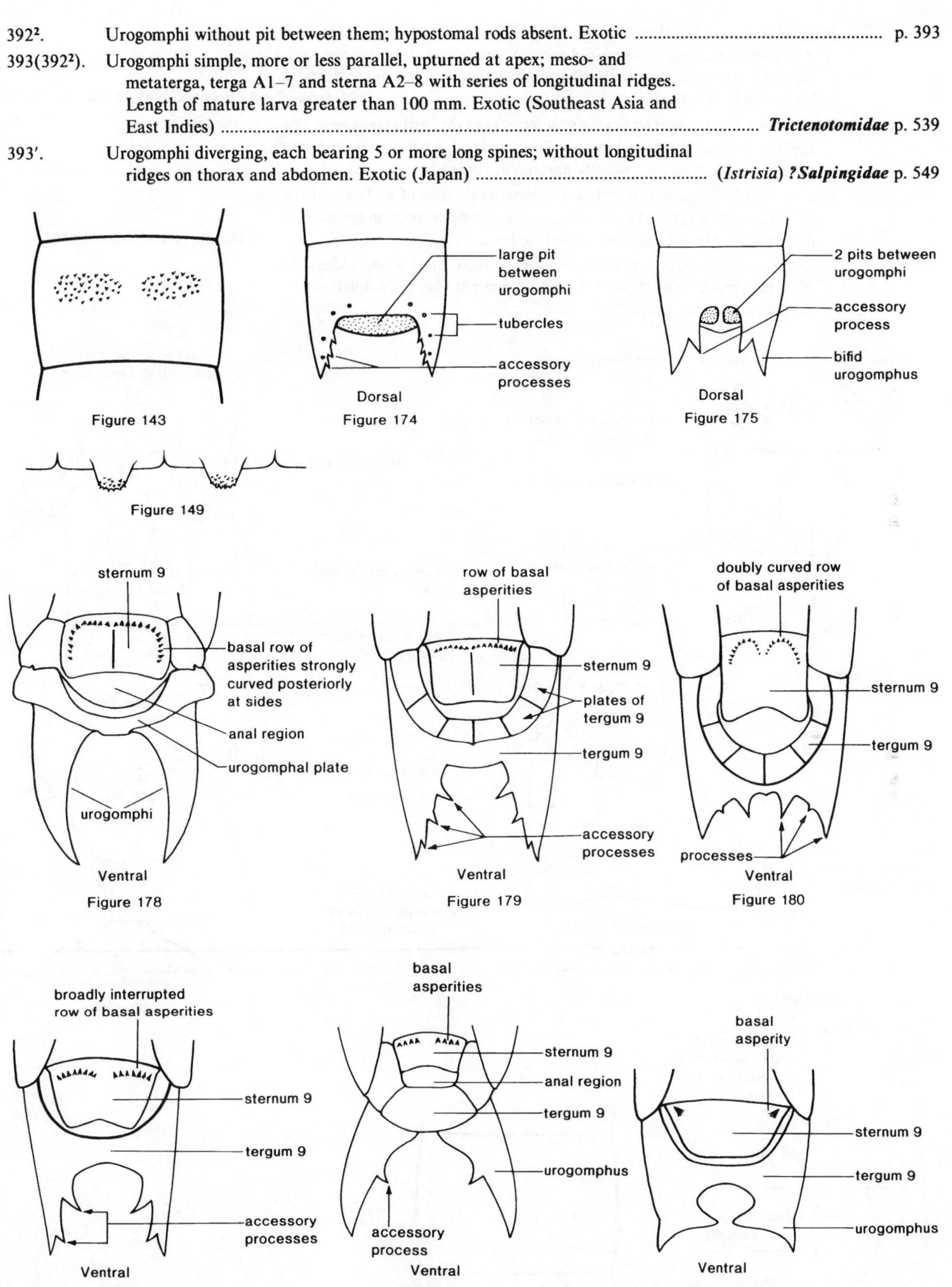

Figure 143

Figure 174

Figure 175

Figure 149

Figure 178

Figure 179

Figure 180

Figure 181

Figure 182

Figure 184

394(311[7]). Ratio of antennal length to head width .15 to .5; cardo strongly oblique (fig. 91); abdominal terga with paired rows of asperities forming incomplete rings (fig. 147); posterior edge of head capsule distinctly emarginate dorsally. Under bark, in leaf axils .... **(Mycterinae** and **Lacconotinae)** ***Mycteridae*** p. 535

394′. Ratio of antennal length to head width more than .5; cardo transverse or slightly oblique; abdominal terga without rows of asperities; posterior edge of head capsule not or only slightly emarginate dorsally 395

395(394′). Segments in labial palp 1; mandibles symmetrical; apex of mala simple (as in fig. 86); tergum A9 without pit between urogomphi; median endocarina present (fig. 22). At bases of cycad fronds .... **(Hemipeplinae)** ***Mycteridae*** p. 535

395′. Segments in labial palp 2; mandibles asymmetrical; apex of mala cleft (fig. 106); tergum A9 with 2 pits between urogomphi (fig. 175); median endocarina absent. Under bark .... ***Boridae*** p. 537

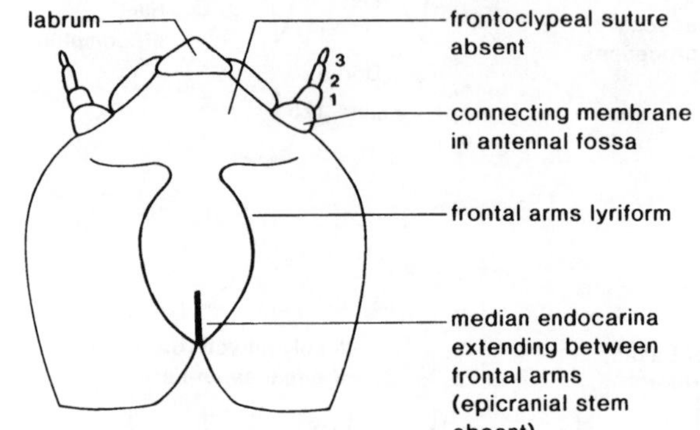

labrum
frontoclypeal suture absent
3
2
1
connecting membrane in antennal fossa
frontal arms lyriform
median endocarina extending between frontal arms (epicranial stem absent)

Figure 22

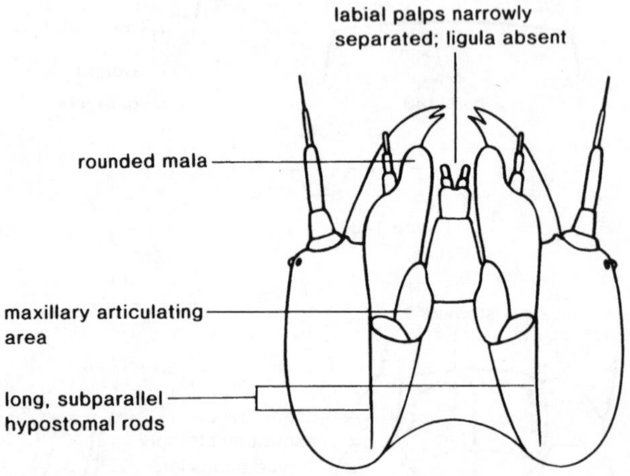

labial palps narrowly separated; ligula absent
rounded mala
maxillary articulating area
long, subparallel hypostomal rods

Figure 86

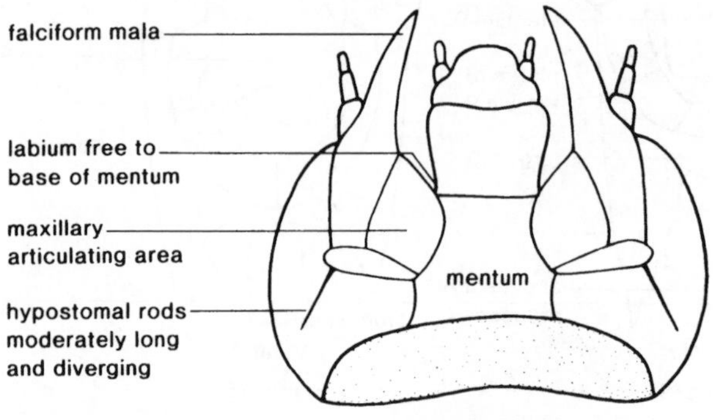

falciform mala
labium free to base of mentum
maxillary articulating area
mentum
hypostomal rods moderately long and diverging

Mouthparts Retracted

Figure 89

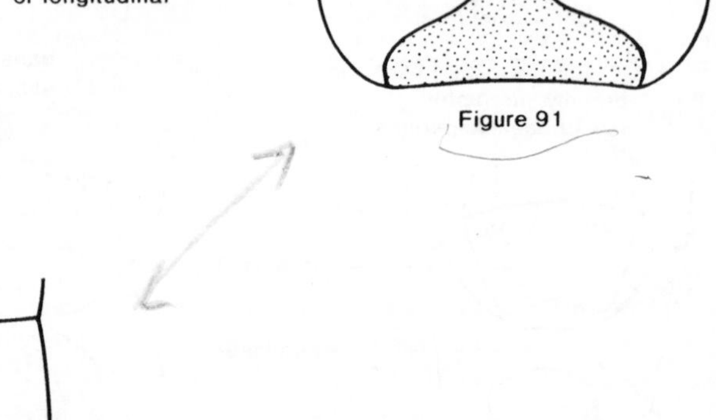

rounded mala
1-segmented labial palp
cardines strongly oblique or longitudinal

Figure 91

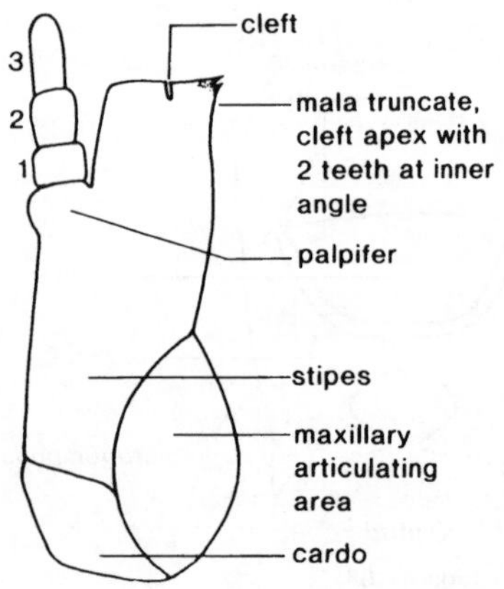

cleft
3
2
1
mala truncate, cleft apex with 2 teeth at inner angle
palpifer
stipes
maxillary articulating area
cardo

Figure 106

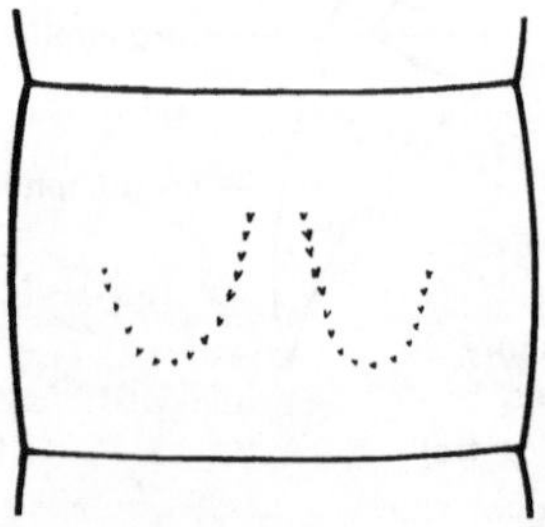

Figure 147

Fig. 175

Selected Bibliography for Introduction

Anderson 1936 (labium).
Anderson 1974b (larval characters in Curculionoidea).
Ashlock and Lattin 1963 (stridulation).
Beier 1949 (morphology of *Helodes* larva).
Beier 1952 (feeding mechanism in larval Helodidae).
Benick 1952 (fungus beetles).
Bitsch 1966 (head capsule).
Böving 1914b (abdominal structure in campodeiform larvae).
Böving 1929b (larval descriptions and illustrations).
Böving 1936 (epipharynx and raster in scarab larvae).
Böving 1954 (morphology of anobiid larvae).
Böving and Craighead 1931 (larval keys and illustrations).
Borror, DeLong and Triplehorn 1981.
Brass 1914 (locomotory function of 10th abdominal segment).
Britton 1974 (larval key).
Burke 1919 (cocoon making in Bothrideridae).
Butcher 1930 (cocoon making in Gyrinidae).
Butt 1951 (feeding mechanism in epilachnine Coccinellidae).
Candeze 1861.
Carpenter and MacDowell 1912 (mouthparts in dascillid and scarabaeid larvae).
Chapuis and Candeze 1853.
Chu 1949 (larval key).
Cogan and Smith 1974 (collection and preservation).
Corbière-Tichané 1973 (sensory structure in larval Leiodidae).
Cornell 1972a (larval bibliography).
Costa et al. 1985.
Crowson 1955 (larval characters in keys).
Crowson 1962 (Cupedidae).
Crowson 1964d (larval characters in Cleroidea; tarsungulus problem).
Crowson 1981 (larval morphology).
Daggy 1947 (pupae of Tenebrionidae).
Dallwitz 1974, 1978, 1980, 1984 (computer keys and descriptions).
Danilevskii 1976 (digestive system in larval Cerambycidae).
Das 1937 (musculature of mouthparts).
Dorsey 1943 (musculature of labrum, labium and pharynx).
Doyen and Lawrence 1979 (larval Tenebrionidae and Zopheridae).
Duffy 1953 (morphology of larval Cerambycidae).
Eidt 1958 (anatomy and histology of an elaterid larva).
Eidt 1959 (feeding mechanism of an elaterid larva).
Emden 1934 (tarsus in Polyphaga).
Emden 1942a (techniques for study and collection).
Emden 1942c (larval key).
Emden 1946 (egg bursters).
Ford and Spilman 1979 (feeding mechanism in Eucnemidae).
Frank and Thomas 1984 (glands in aleocharine Staphylinidae).
Freeman 1980 (stored products pests).
Fuchs 1974 (pollen feeding in beetles).
Gage 1920 (larvae of Coccinellidae; terminology for cuticular armature).
Gardiner 1966 (egg bursters in Cerambycidae).
Gardner 1935b (larval Eucnemidae).
Gardner 1947 (larval bibliography).
Gilyarov 1964b (larval morphology and key).
Glen 1950 (morphology of larval Elateridae).
Goulet 1977 (technique for larval study).
Grandi 1959 (morphological adaptations in various beetle larvae).
Hackman 1984 (cuticle chemistry).
Hayashi 1981, 1986 (larval illustrations).
Hayes 1928 (epipharynx in Scarabaeoidea).
Hickman 1931 (feeding mechanism in Heliplidae).
Hicks 1959, 1962, 1971 (fauna of bird nests).
Hinton 1945b (stored products beetles).
Hinton 1946a (gin traps in beetle pupae), 1946b (pupae).
Hinton 1947 (spiracles and molting in aquatic larvae).
Hinton 1948a (dorsal cranial areas in caterpillars).
Hinton 1949 (pupae).
Hinton 1955 (respiratory adaptations in Psephenidae).
Hinton 1958 (panorpoid orders).
Hinton 1963 (ventral ecdysial lines).
Hinton 1966 (respiratory adaptations in psephenid pupae).
Hinton 1967a (spiracles in Myxophaga).
Hinton 1967c (ecdysial process in Scarabaeoidea).
Hinton 1981 (eggs of Coleoptera).
Hiznay and Krause 1956 (larval head and mouthparts in Passalidae).
Honomichl 1980 (digitiform sensilla in adult beetles).
Hyslop and Böving 1935 (morphology of larval Elateridae).
Kasule 1966 (larval Staphylinoidea).
Kemner 1918 (abdominal apex).
Kingsbury and Beveridge 1977 (elutriator).
Kinzelbach 1971 (Strepsiptera).
Klausnitzer 1978 (larval keys).
Kristensen 1981 (phylogeny of insect orders).
Kristensen 1984 (Lepidoptera: Agathiphagidae).
Kristensen and Nielsen 1983 (Lepidoptera: Heterobathmiidae).
Larsson 1942 (larval key).
Lawrence 1977b (larval mouthparts and feeding mechanisms).
Lawrence 1981 (larval Lucanidae).
Lawrence 1982c (larval descriptions).
Lawrence and Hlavac 1979 (feeding mechanisms in larval Derodontidae).
Leiler 1976 (eucnemid larvae).
Lengerken 1924 (extraoral digestion).
Lepesme 1944 (stored products beetles).
Martin 1977 (collection and preservation).
May 1967 (terminology for weevil larvae).
McAlpine *et al.*, 1981 (Diptera).
Meixner 1935 (larval morphology).
Merritt and Cummins 1984 (aquatic collecting techniques).
Nakamura 1981 (cerambycid pupae).
Nikitsky 1976c, 1980.
Oldroyd 1958 (collection and preservation).
Parkin 1933 (anobiid larvae terminology).
Paulian 1943 (general).
Paulian 1949a (general).
Paulus 1979 (larval eyes).
Perris 1855, 1862, 1876–1877 (larval descriptions).
Peterson 1951 (larval descriptions, illustrations, keys).
Peterson 1959 (collection, preservation, rearing techniques).
Phuoc and Stehr 1974 (pupae of Coccinellidae).
Pilgrim 1972 (Mecoptera: Nannochoristidae).
Riek 1970a (Megaloptera).
Riek 1970b (Neuroptera).
Riek 1970c (Strepsiptera).
Riek 1970d (Mecoptera).
Ritcher 1966 (larvae of Scarabaeoidea).
Roberts 1930 (larval key terminology).
Roberts 1936 (spiracles of first instar weevil larvae).
Ross and Pothecary 1970 (first instar *Priacma*).
Rozen 1963a (pupae of Nitidulidae).
Rozen 1963b (pupae of Archostemata).
St. George 1924 (morphology of tenebrionid larvae).
St. George 1939 (morphology of perimylopid larvae).
Schicha 1967 (functional morphology of mouthparts in adult pollen-feeders).
Schiödte 1864–1883 (larval descriptions).
Schmidt 1971 (mandibles in wood-boring insects).
Scott 1936, 1938 (paedogenesis in Micromalthidae).
Seeger 1971 (morphology and habits of Haliplidae).
Seevers 1957 (termitophilous Staphylinidae).
Sivinski 1981 (bioluminescence).
Smith and Sears 1982 (mandibular morphology and feeding habits).
Snodgrass 1935 (insect morphology).
Snodgrass 1947 (insect cranium).
Srivastava 1959 (maxillary glands in larval beetles).
Steinke 1919 (larval spiracles).
Stickney 1923 (adult head capsule and tentorium).
Striganova 1961a (feeding mechanism and morphology of Helodidae).
Striganova 1964c (mouthparts in phytophagous larvae).

Striganova 1967 (morphology of head and mandibles in burrowing larvae).
Striganova 1971 (morphology of mouthparts of predacious larvae).
Thomas 1957 (morphology of larval Scolytidae).
Thorpe and Crisp 1949 (plastron respiration).
Tillyard 1923 (Lepidoptera: Micropterygidae).
Torre–Bueno 1937 (glossary).
Treherne 1952 (respiration in helodid larvae).
Treherne 1954 (osmotic regulation in helodid larvae).
Upton and Norris 1980 (collection and preservation).
Verhoeff 1923 (larval morphology).
Viedma 1964 (larval key).
Wade 1935 (larval bibliography).
Wade and St. George 1923 (morphology of larval Tenebrionidae).
Walsh and Dibb 1975 (techniques of collecting and study).
Watt 1970 (larvae of Perimylopidae, terminology).
Watt 1974b (larvae of Tenebrionidae).
Wheeler and Blackwell 1984 (mycophagy).
Yasuda 1962 (Lepidoptera: Micropterygidae).
Yuasa 1922 (Hymenoptera: Symphyta).
Zacharuk 1962a (sense organs on head of elaterid larva).
Zacharuk 1962b (larval characters in Elateridae).

OMMATIDAE (ARCHOSTEMATA)

(= OMMADIDAE, INCLUDING TETRAPHALERIDAE)

John F. Lawrence,
Division of Entomology, CSIRO

The family Ommatidae, as delimited by Lawrence and Newton (1982), includes 3 genera: *Omma*, with 2 species in eastern Australia; *Tetraphalerus*, with 2 species in southern Brazil, Argentina, and Bolivia; and *Crowsoniella*, with a single species, *C. relicta* Pace, from central Italy. The group belongs in the Archostemata, along with the Cupedidae and Micromalthidae, and fossils attributable to the family are known from the Lower Jurassic.

Little is known about the biology of ommatids, and no immature forms have been described. Adults of *Omma* are rarely collected from central Queensland to South Australia, not only in the heavily forested areas adjacent to the coast, but also in the more open and arid *Eucalyptus* woodland west of the Great Divide. *Tetraphalerus* species often occur in open, dry areas with little or no tree cover, and adults may be attracted to lights. *Crowsoniella relicta* was found in rather dry and granular, calcareous soil at the bases of old chestnut trees, along with other interstitial beetles, like species of *Leptotyphlus* and *Scotonomus* (Staphylinidae), and *Anommatus* (Cerylonidae). Larval ommatids may be found eventually among the root systems of large shrubs or trees, or possibly in the rotting central cores of old standing trees or stumps.

Selected Bibliography

Crowson 1962 (review of Cupedidae and Ommatidae), 1975 (fossil record), 1976 (*Crowsoniella*).
Lawrence and Newton 1982.
Monros and Monros 1952 (South American Archostemata).
Neboiss 1960 (*Omma* distribution).
Pace 1976 (*Crowsoniella*).
Ponomarenko 1969 (fossil Archostemata).
Vulcano and Pereira 1975 (South American Archostemata).

CUPEDIDAE (ARCHOSTEMATA)

(= CUPESIDAE)

John F. Lawrence,
Division of Entomology, CSIRO

Reticulated Beetles

Figures 34.67a–e

Relationships and Diagnosis: Cupedidae is one of three families comprising the oldest and most primitive suborder of beetles, the Archostemata. The formerly included genera *Omma* and *Tetraphalerus* have been placed in a separate family, Ommatidae, along with the minute, soil-dwelling *Crowsoniella relicta* Pace (Crowson, 1976; Lawrence, 1982c; Lawrence and Newton, 1982).

Cupedid larvae resemble other wood-boring types, like those of some Cephaloidae, Cerambycidae, Lymexylidae, Melandryidae, Oedemeridae, and Tenebrionidae, in that they are lightly sclerotized and more or less cylindrical, with relatively short legs and thoracic and abdominal ampullae. They differ from all of these, however, in having 6-segmented legs, a heavily-sclerotized, wedge-like ligula, and usually 4-segmented antennae. A combination of mandibular mola and separate galea and lacinia separates cupedids from all but Lymexylidae, which lack an anteriorly projecting median endocarina and have a hypognathous head without a gular region. A wedge-like ligula also occurs in Micromalthidae and Callirhipidae, but larvae of the latter are quite unlike cupedids in most respects. *Micromalthus* is the only other archostematan for which the larva is known. It resembles the cupedid larva in many ways, but is much smaller and may be distinguished by a sternal process on segment A9, the lack of legs in the cerambycoid stage, and the long, narrow tarsus with two moveable claws in the caraboid or triungulin stage.

Biology and Ecology: All known larvae bore through firm but fungus-infested wood, with the aid of their thoracic and abdominal ampullae, and apparently feed on wood fibers, fungal hyphae, and fungal biproducts. Larvae of the Japanese *Tenomerga mucida* (Chevrolat) are reported to feed in wood infected with the fungus *Stromatoscypha* (as *Porothelium*) (Basidiomycetes: Schizophyllaceae) (Fukuda, 1941). Some have been found in seasoned timber, but only in those portions attacked by a dry rot. Pupation occurs within the wood. *Priacma* larvae are known only as first instars, which hatched from eggs laid by a female beaten from a hemlock tree (*Tsuga*); it is probable that these larvae also feed on decayed wood (Ross and Pothecary, 1970). Adults are known to feed on pollen and may fly to lights (Crowson, 1962), while males of *Priacma serrata* (LeConte) are known to be attracted to bleach solution (Atkins, 1957); females are rare.

Description: Mature larvae 15 to 35 mm. Body elongate, parallel-sided, straight, subcylindrical, and lightly sclerotized, whitish, except for anterior part of head and tip of terminal process. Surfaces smooth; vestiture of scattered, simple, fine hairs, moderately dense in places.

Head: Protracted and prognathous, broad, transverse, slightly flattened, and deeply emarginate posteriorly. Epicranial stem absent; frontal arms usually indistinct or absent,

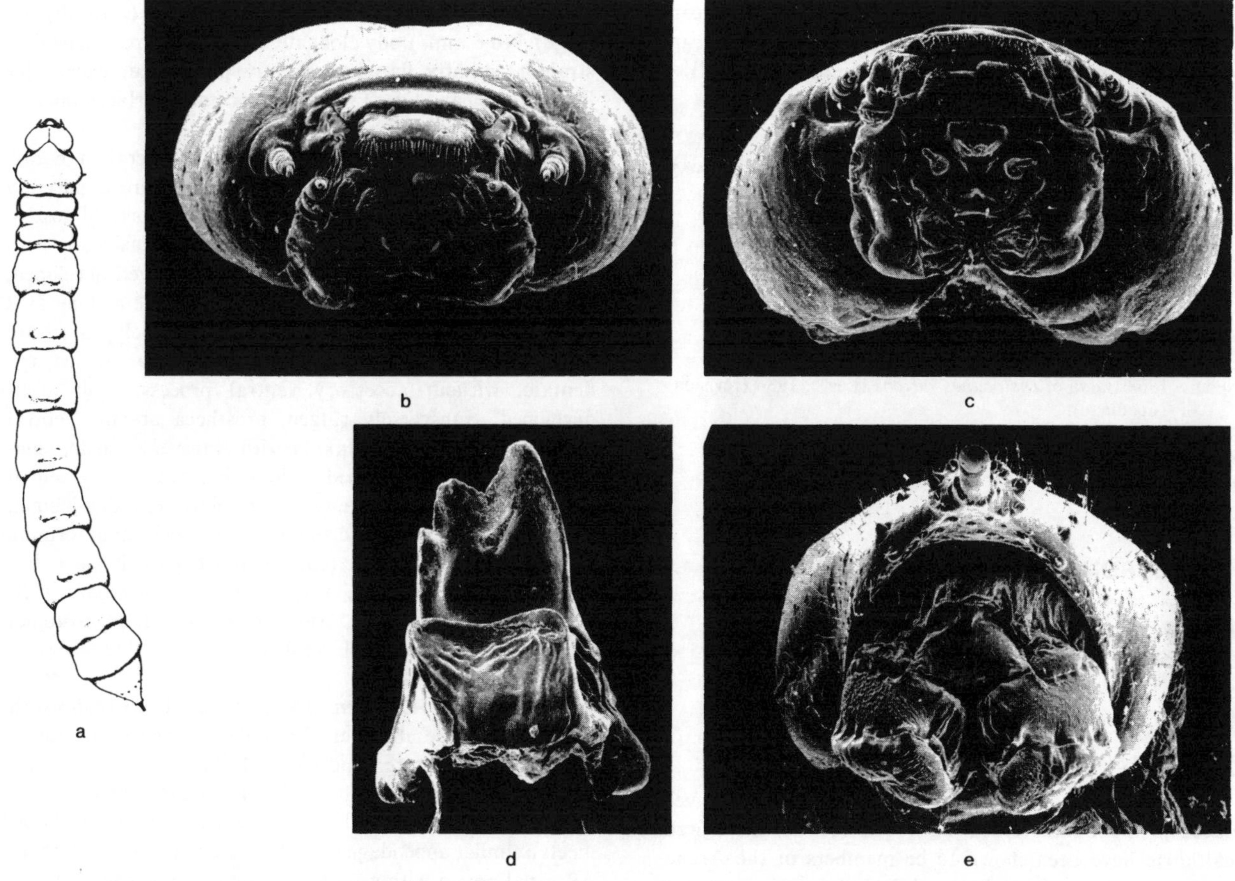

Figures 34.67a–e. Cupedidae. *Tenomerga concolor* (Westwood). East Lansing, Michigan. Length = 34 mm. **a.** Larva, dorsal; **b.** Head, anterior; **c.** Head, anteroventral; **d.** Left mandible, mesal; **e.** Abdominal apex, posteroventral.

sometimes V-shaped (lyriform in first instar of *Priacma*). Median endocarina extending anteriorly almost to edge of clypeus. Stemmata usually absent; occasionally 1 or more small eye spots on each side. Antennae usually well-developed and 4-segmented; rarely with 5 or 6 segments; 3-segmented in early instar *Cupes;* 2-segmented in first instar *Priacma.* Frontoclypeal suture present, sometimes indistinct; labrum free. Mandibles (fig. 34.67d) symmetrical or slightly asymmetrical, robust, tridentate or bidentate with large, subapical tooth, and without accessory ventral process; mola well-developed, simple or with a few tubercles or transverse ridges; prostheca absent. Ventral mouthparts retracted. Maxilla with transverse cardo, elongate stipes, well-developed articulating area, 3-segmented palp with large palpifer, articulated, rounded galea, and fixed, rounded lacinia. Labium with mentum and submentum fused; ligula forming a wedge-like sclerome, extending well beyond labial palps, which are usually 2-segmented (1-segmented in first instar *Priacma*). Hypopharyngeal sclerome well-developed, forming a single unit with ligular sclerome. Hypostomal rods long and slightly diverging at base and converging at apex. Ventral epicranial ridges absent. Gula transverse.

Thorax and Abdomen: Thorax relatively small, less than a fourth the length of the abdomen in later instars. Prothorax slightly larger than meso- or metathorax, with large, paired patches of asperities on sternum. Legs short, widely separated, 6-segmented, including simple or bifurcate claw. Meso- and metaterga, abdominal terga and sterna 1–7 each with transverse ampulla. Tergum A9 with median, sclerotized process, blunt at apex and with asperities around base, or in *Priacma* with median process bearing 2 dorsal and 2 ventral tubercles. Sternum A9 well-developed, simple. Anal region bordered laterally by 2 large, rounded lobes.

Spiracles: Annular, longitudinally oval, with more or less crenulate peritreme.

Comments: The Cupedidae is a small family containing 9 genera and 26 species worldwide, with 4 genera and 4 species in N. America. There are two subfamilies: Priacminae with the monotypic genus *Priacma* from northwestern N. America, and Cupedinae containing the rest of the genera. The Cupedinae have been revised recently (Neboiss, 1984) and now include: *Cupes* (1 species; N. America); *Tenomerga* (10 species; Holarctic, southeast Asia, New Guinea, South Africa); *Distocupes* (1 species; Australia); *Adinolepis* (4 species; Australia); *Ascioplaga* (2 species; New Caledonia);

Rhipsideigma (4 species; East Africa and Madagascar); *Prolixocupes* (2 species; southwestern N. America and southern S. America); and *Paracupes* (1 species; Brazil).

Selected Bibliography

Atkins 1957 (attraction of male *Priacma serrata* to bleach), 1963 (review of family).

Böving 1929b (larva of *Tenomerga concolor* (Westwood)).

Böving and Craighead 1931 (larva of *Tenomerga concolor*).

Crowson 1962 (review of family), 1976 (*Crowsoniella relicta*).

Fukuda 1938 (larva of *Tenomerga mucida*), 1941 (biology of *Tenomerga mucida*).

Lawrence 1982c.

Lawrence and Newton 1982.

Neboiss 1968 (larva of *Distocupes varians* (Lea)), 1984 (revision of Cupedinae).

Ross and Pothecary 1970 (first instar of *Priacma serrata*).

Rozen 1963b (pupa of *Tenomerga concolor*).

Vulcano and Pereira 1975 (S. American Archostemata).

MICROMALTHIDAE (ARCHOSTEMATA)

John F. Lawrence,
Division of Entomology, CSIRO

Telephone-pole Beetles

Figures 34.68a–i

Relationships and Diagnosis: Although tentatively placed in the Cantharoidea by Arnett (1968), the Micromalthidae have been shown to be members of the Archostemata on the basis of both adult and larval characters (Lawrence and Newton, 1982). The larvae are very similar to those of Cupedidae, and the leg-bearing triungulins (caraboid larvae) might be mistaken for first instar cupedids, from which they differ in having a long, narrow tarsus bearing two moveable claws, an apical process on abdominal sternum 9, and a differently formed hypopharyngeal sclerome.

Biology and Ecology: Micromalthid larvae are usually found in localized colonies in red rotten wood of *Quercus* and *Castanea,* but they are capable of infesting rotten building materials of various woods, including *Pinus, Pseudotsuga, Acacia,* and *Eucalyptus,* when used in flooring, furniture, telephone poles, railroad ties, bridge abutments, and mine timbers. The life cycle is more complex than that found in any other insect group. The first instar (caraboid larva) is an active triungulin, which develops into a legless, feeding form (cerambycoid larva), which may undergo three molts. The cerambycoid larva may pupate to become an adult diploid female or may develop into one of three types of larviform reproductives: (1) a thelytokous paedogenetic female, which, in turn, produces viviparously a number of triungulins; (2) an arrhenotokous paedogenetic female, which lays a single egg destined to be a stump-legged curculionoid larva, which, in turn, devours the mother and pupates to form an adult haploid male; and (3) an amphitokous paedogenetic female, which may produce either form. The production of various larval types is apparently affected by environmental conditions. (Barber, 1913a, 1913b; Pringle, 1938a; Scott, 1936, 1938, 1941).

Description: Length of mature larva (cerambycoid stage) 4 to 6 mm. Body elongate, more or less parallel-sided, straight, slightly flattened, lightly pigmented, except for buccal region and tips of terminal processes. Surfaces smooth; vestiture of scattered, simple hairs.

Head: Protracted and prognathous, broader than thorax, transverse, slightly flattened. Epicranial suture apparently absent. Median endocarina extending anteriorly almost to edge of clypeus. Stemmata usually absent; occasionally with a single stemma on each side. Antennae well-developed, 3-segmented, with segment 3 more than twice as long as 2 and narrower than antennal sensorium. Frontoclypeal suture present; labrum free. Mandibles asymmetrical, robust, tridentate, without accessory ventral process; mola well-developed, transversely ridged; prostheca absent. Ventral mouthparts retracted. Maxilla with transverse cardo, elongate stipes, well-developed articulating area, 3-segmented palp, articulated, truncate galea, and fixed, rounded lacinia. Labium with mentum and submentum fused; ligula forming a wedge-like sclerome, extending well beyond labial palps, which are 1-segmented. Hypopharyngeal sclerome well-developed, subtriangular, transversely ridged. Hypostomal rods long and subparallel. Ventral epicranial ridges absent. Gula transverse.

Thorax and Abdomen: Thorax reduced, less than a fifth the length of the abdomen. Legs absent. Meso- and metaterga, abdominal terga and sterna 1–7 each with transverse ampulla. Tergum A9 with median appendage, which curves ventrally and has several teeth at apex, and which almost meets a similar appendage extending from the apex of sternum A9. Anal region with a large, rounded lobe on each side.

Spiracles: Annular, very small.

Triungulin (*caraboid Larva*): Length 1.3 to 2 mm. Similar to cerambycoid larva, but with well-developed, 6-segmented legs, each with a very long and narrow tarsus and with 2 moveable claws with pair of spatulate appendages beneath them.

Curculionoid Larva: More or less grub-like, with narrower head, short, stump-like legs, and no terminal abdominal processes.

Reproductive Forms: More or less similar to legless cerambycoid larva, but with narrower head and no terminal abdominal processes. Developing eggs or viviparous larvae usually can be seen within abdomen.

Comments: This family includes the single species *Micromalthus debilis* LeConte, whose natural range includes the northeastern part of the United States, from Kentucky to Michigan and east to the Atlantic. The total range of the species is far greater, with records from British Columbia; New Mexico; Florida; Oahu, Hawaii; Cuba; the states of Bahia, Minas Gerais, and São Paulo in Brazil; Gibraltar; Victoria and Kowloon, Hong Kong; and Witwatersrand and Johannesburg in South Africa. All peripheral records, however, as well as those from other parts of the world, are based on infestations of human artifacts, usually made from wood imported from North America. The South African specimens, for instance, were found in mines, where the original timbers were made from pitch pine imported from the United States.

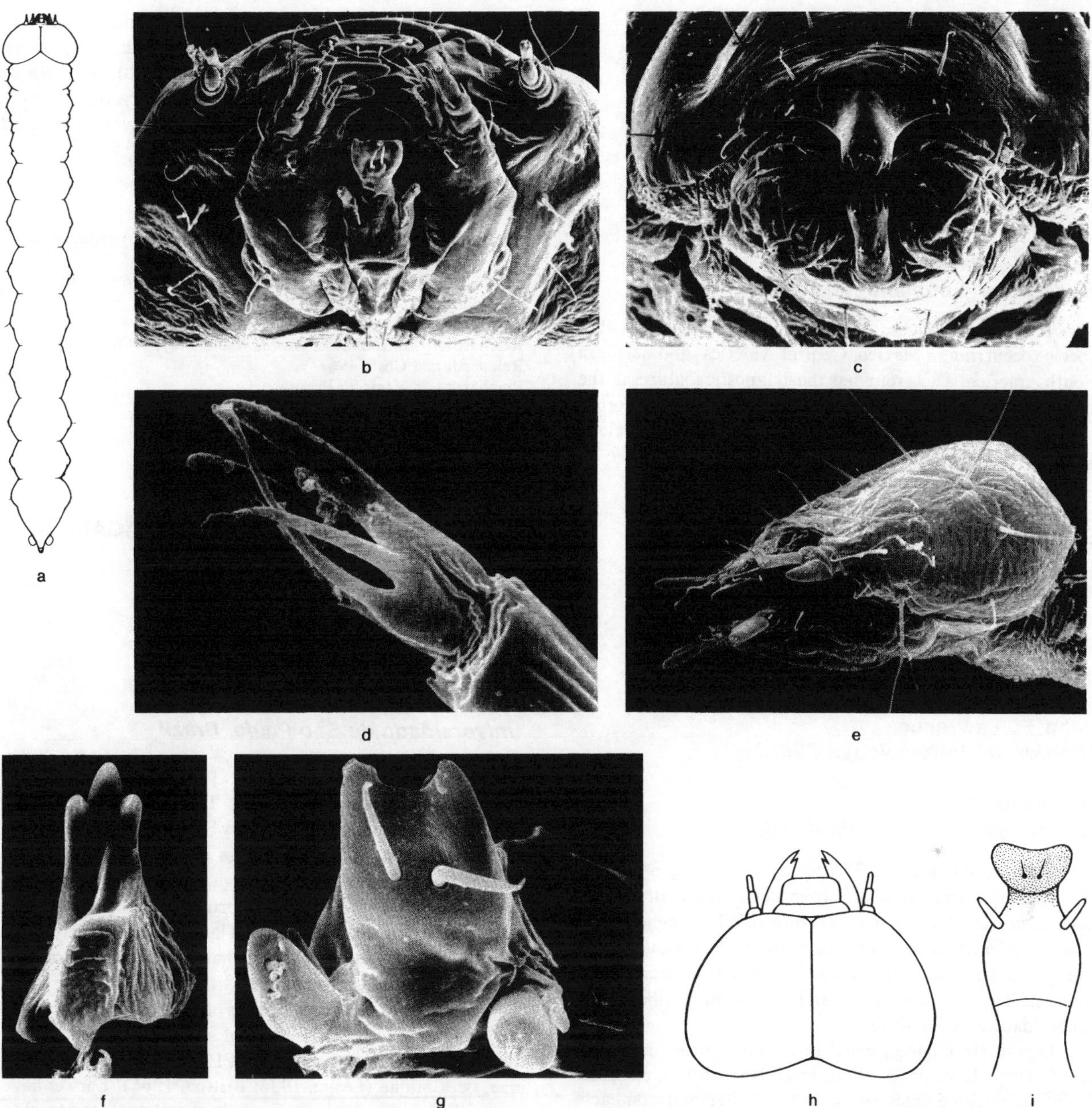

Figures 34.68a-i. Micromalthidae. *Micromalthus debilis* LeConte. **a.** Clinton Co., Ohio. Cerambycoid larva, dorsal, length = 3.7 mm; **b.** Rose Lake, Clinton Co., Michigan. Cerambycoid larva. Head, anteroventral; **c.** Abdominal apex, posterior; **d.** Clinton Co., Ohio. Caraboid larva. Tarsal claws and attached appendages; **e.** Hong Kong. Caraboid larva. Head, anterolateral; **f.** Right mandible, mesal; **g.** Labium, anteroventral; **h.** Head, dorsal; **i** Labium, showing sclerotized ligula.

Although South African specimens are thought to be conspecific with American forms, those from Hong Kong (triungulins without associated adults) could represent a distinct species, since they possess a single pair of stemmata, absent in triungulins of *M. debilis*. (Marshall and Thornton, 1963; Pringle, 1938a; Silvestri, 1941; Swezey, 1940). Fossil triungulins are known from the Miocene of Mexico (Rozen, 1971) and the Oligocene of Europe (Lawrence, unpublished), while a related form is known from the Cretaceous of Lebanon (Crowson, 1981).

Selected Bibliography

Barber 1913a (life cycle), 1913b (life cycle).
Böving 1929b (larva and triungulin).
Böving and Craighead 1931 (larva and triungulin).
Crowson 1981.
Lawrence and Newton 1982.
Marshall and Thornton 1963 (triungulin from Hong Kong).
Peterson 1951 (larva).
Pringle 1938a (occurrence in South Africa; larva; life cycle).
Rozen 1963b (pupa), 1971 (fossil triungulin).

Scott 1936 (haploidy), 1938 (life cycle; paedogenesis), 1941 (production of males and females).
Silvestri 1941 (distribution).
Swezey 1940 (occurrence in Hawaii).

CYATHOCERIDAE (MYXOPHAGA)

(= LEPICERIDAE)

John F. Lawrence,
Division of Entomology, CSIRO

This family includes the single genus *Lepicerus*, with 2 species occurring in Mexico, Central America, and northern South America. This group has the diagnostic features of the suborder Myxophaga, but is somewhat isolated from the other myxophagan families. Adults have been collected along streams in flood debris, but nothing else is known of the biology and larvae are undescribed.

Selected Bibliography

Hinton 1934 (habitat of *Lepicerus*).
Reichardt 1976c (review of family).

TORRIDINCOLIDAE (MYXOPHAGA)

John F. Lawrence,
Division of Entomology, CSIRO

H. Reichardt,*
Universidade de São Paulo, Brazil

This family includes 27 species comprising 6 genera: *Hintonia, Ytu,* and *Claudiella* from southeastern Brazil; *Incoltorrida* from Madagascar; *Torridincola* from central and southern Africa; and *Delevea* from southern Africa and Japan. The group is placed in the suborder Myxophaga, and adults resemble Triassic fossils in the archostematan families Schizophoridae and Catiniidae.

Larvae are minute (less than 2.5 mm), broad and flattened, with a large prognathous head, free labrum, elongate 2- or 3-segmented antennae, and 3 to 5 stemmata on each side. The epicranial stem is short and the frontal arms lyriform. The mandibles have a large, tuberculate mola and a prostheca. The galea and lacinia are more or less fused, and the maxillary palps are 1- or 3-segmented. The labium is broad, with 2-segmented palps. The legs have 5 segments, including the tarsungulus. Abdominal segments 1 to 8 each has a pair of 2- or 3-segmented, lateral, spiracular gills, most of the surfaces of which are covered with plastron mesh. Tergite A9 usually bears a pair of fixed urogomphi. Pupae are obtect and have 2 pairs of long, spiracular gills.

Larvae and adults occur on rock surfaces which are covered with a thin film of moving water (hygropetric habitat) or which are located in the spray zone beneath waterfalls; they feed on algae growing on the rocks. Adults usually possess a plastron on the abdomen, while larvae have a plastron on the gills. According to Spangler (1980b), *Ytu brutus* Spangler and an associated hydrophilid (*Oocyclus* sp.) were literally swimming in the water film running over a wet guard rail across from a waterfall.

Selected Bibliography

Bertrand 1962 (larva of *Delevea*, as an undescribed genus of Hydraenidae).
Hinton 1967a (larval spiracles in Myxophaga), 1969 (adult plastron).
Lawrence and Newton 1982.
Reichardt 1973a (larvae of *Hintonia* and *Ytu*), 1976b.
Reichardt and Costa 1967.
Reichardt and Vanin 1976.
Sato 1982 (Japanese *Delevea*).
Spangler 1980b (larva of *Ytu*).
Steffan 1964 (larva of *Torridincola*).

MICROSPORIDAE (MYXOPHAGA)

(= SPHAERIIDAE)

John F. Lawrence,
Division of Entomology, CSIRO

H. Reichardt,*
Universidade de São Paulo, Brazil

Figures 34.69a, b

Relationships and Diagnosis: The Microsporidae were originally placed in the superfamily Staphylinoidea and some workers (Barlet, 1972) still retain them there, but Forbes (1926) recognized the hydradephagan features of their wings, and Crowson (1955) placed the group in his newly formed suborder Myxophaga. Microsporid larvae are extremely small (0.8 to 1.2 mm) and are unique in having vesicular spiracular gills on abdominal segments 1 to 8 (Hinton, 1967a).

Biology and Ecology: Species of Microsporidae occur in wet sand or gravel at the edges of streams or rivers, where they may occur with Georyssidae, Hydraenidae, Hydrophilidae, or Elmidae (Lesne, 1936; Britton, 1966). Larvae have been collected by allowing a sample of wet sand or gravel to drain through metal gauze into a jar or placing such a sample in an elutriator. Because of their small size, microsporids form part of the interstitial fauna, and their food almost certainly consists of algae occurring among the sand grains. Females were observed to lay large, single eggs (Britton, 1966).

Description: Mature larvae about 1.2 mm. Body elongate, fusiform, narrowed posteriorly, slightly flattened, lightly sclerotized, yellowish. Dorsal surfaces smooth; vestiture of longer and shorter, simple setae, with bands of densely packed, stout, highly refractive setae at posterior ends of trunk segments.

Head: Protracted and hypognathous, broad, narrowed anteriorly. Epicranial stem and frontal arms indistinct. Median

*Deceased

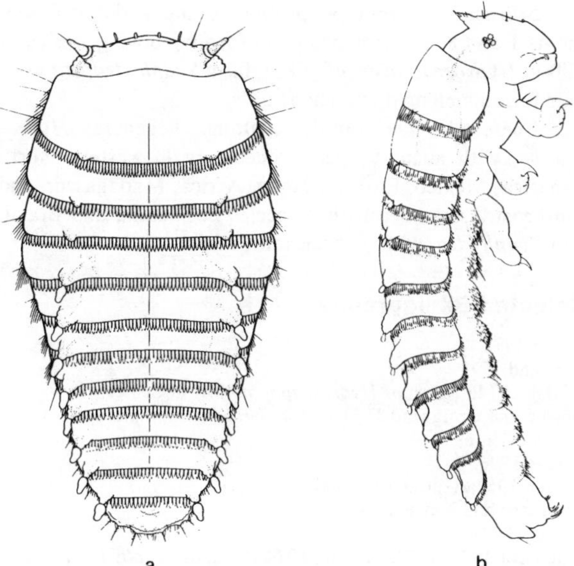

Figures 34.69a,b. Microsporidae. *Microsporus ovensensis* (Oke). **a.** Larva, dorsal; **b.** lateral. Length = 1.2 mm. (from Britton, 1966 and 1974, Supplement to The Insects of Australia, Melbourne University Press).

endocarina absent. Stemmata 4 on each side; lenses well-developed. Antennae short, 3-segmented, with segments 2 and 3 indistinctly separated, giving the appearance of a 2-segmented antenna with a long sensorium arising from middle of apical segment; antennal insertions dorsally placed and well separated from mandibular articulations. Frontoclypeal suture present; clypeal area enlarged; labrum apparently fused to head capsule. Mandibles symmetrical, short and broad, bidentate, without accessory ventral process; mola large and tuberculate; prostheca apparently absent. Ventral mouthparts retracted. Maxilla with oblique cardo, elongate stipes, well-developed articulating area, and falciform mala with 2 or 3 stout setae at inner edge; palp 3-segmented. Labium not distinctly subdivided; ligula broad, bilobed; labial palps 2-segmented, widely separated. Hypostomal rods and ventral epicranial ridges absent. Gula transverse.

Thorax and Abdomen: Thoracic and abdominal segments strongly transverse. Legs moderately large, stout, close together; tarsungulus short, with 1 seta. Abdominal segments 1–8 each with posteriorly directed spiracular tube on each side near posterior edge. Tergum A9 simple, truncate at apex. Segment A10 ventrally oriented, with 3 pairs of hooks.

Spiracles: Present on abdominal segments 1–8; forming vesicular gills, which are more or less ovate.

Comments: The family contains the single genus *Microsporus* with 18 species occurring throughout the Northern Hemisphere and in Australia and Madagascar (Britton, 1966; Lesne, 1936; Paulian, 1949b); two or three species are known from N. America.

Selected Bibliography

Barlet 1972.
Bertrand 1972.
Britton 1966 (larva of *Microsporus ovensensis* (Oke)).
Crowson 1955.
Forbes 1926.
Hinton 1967a (spiracular gills).
I.C.Z.N., 1985 (*Microsporus*).
Lawrence and Newton 1982.
Lesne 1936 (distribution).
Paulian 1949b (occurrence in Madagascar).
Reichardt 1973a.

HYDROSCAPHIDAE (MYXOPHAGA)

John F. Lawrence,
Division of Entomology, CSIRO

H. Reichardt,*
Universidade de São Paulo, Brazil

Figures 34.70 a, b

Relationships and Diagnosis: The Hydroscaphidae are placed by most workers in the primitive suborder Myxophaga (Crowson, 1955, 1960b; Lawrence and Newton, 1982), but they have been included at various times in Hydrophiloidea (Böving, 1914a), Staphylinoidea (Paulian, 1941), or Adephaga (Barlet, 1972; Forbes, 1926). The larvae are easily distinguished by their small size and the presence of spiracular gills on the prothorax and abdominal segments 1 and 8. The gills are almost always vesicular like those of Microsporidae, but they are more elongate and occupy different positions on the body (Hinton, 1967a). The gills of Torridincolidae occur on abdominal segments 1 to 8, but they are segmented. Some hydraenid larvae (*Tympanogaster*) have snorkel-like spiracular tubes on the prothorax, but these do not bear vesicles.

Biology and Ecology: Adults and larvae of Hydroscaphidae are usually found feeding on algae over which a thin film of water is flowing, usually into a stream or river, but some species have been found in hot springs, at the edges of swimming pools, under rocks in fast-flowing streams, or in flood debris. Adults of *Hydroscapha natans* LeConte have been collected in large numbers flying at dusk, along with a species of *Microsporus*. Hydroscaphids pupate among the algae, usually beneath the water, and the pupae have various aquatic adaptations, such as long spiracular processes or plastron-bearing spiracular gills (Böving, 1914a; Reichardt and Hinton, 1976).

Description: Mature larvae about 1.5 mm. Body elongate, fusiform, narrowing posteriorly, slightly flattened, lightly sclerotized or moderately sclerotized dorsally. Dorsal surfaces smooth or somewhat tuberculate; vestiture of longer and shorter, simple setae, sometimes broad and flattened.

Head: Protracted and prognathous, transverse, moderately broad and slightly flattened. Epicranial stem very short; frontal arms V-shaped. Median endocarina absent. Stemmata 5 on each side, each with well-developed lens. Antennae

*Deceased

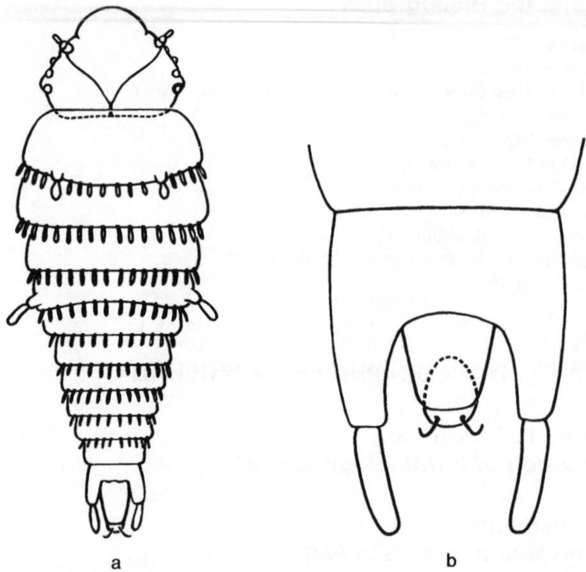

Figures 34.70a,b. Hydroscaphidae. *Hydroscapha natans* LeConte. **a.** larva, dorsal, length = 1.5 mm; **b.** abdominal apex, dorsal. (redrawn from Böving 1914a)

very short, 2-segmented; segment 2 more than 3 times as long as 1; sensorium slightly longer than segment 2. Frontoclypeal suture absent; clypeal area enlarged; labrum indistinct, highly reduced or fused to clypeus. Mandibles more or less symmetrical, with very broad base and reduced apex, which is tridentate; accessory ventral process absent; mola well-developed, tuberculate; prostheca consisting of a fixed, hyaline lobe, bifid or trifid at apex and attached to a membranous lobe. Ventral mouthparts retracted. Maxilla with strongly oblique cardo, slightly elongate stipes, well-developed articulating area, and falciform mala with 2 or 3 teeth at apex; palp 2-segmented with well-developed sensilla on segment 1. Labium with mentum and submentum fused; ligula short and broad; labial palps 2-segmented and widely separated. Hypopharyngeal sclerome present. Hypostomal rods absent; ventral epicranial ridges absent. Gula transverse.

Thorax and Abdomen: Thorax enlarged, more than half as long as abdomen. Posterior edges of thoracic terga and abdominal terga 1–7 in *Hydroscapha* fringed with short, stout setae (becoming finer posteriorly); lateral edges of thoracic and abdominal terga in *Scaphydra* lined with setiferous tubercles. Protergum with short, dorsally-projecting spiracular tube on each side toward the lateral margin near posterior edge (*Hydroscapha*) or laterally-projecting tube near each posterior angle (*Scaphydra*). Legs well-developed, 5-segmented, separated by about 1 coxal diameter; tarsungulus long, with 1 seta. Abdominal segments gradually decreasing in width posteriorly. Tergum A1 with short, laterally-projecting spiracular tube on each side near posterior angle. Tergum A8 with pair of long, posteriorly-projecting, spiracular tubes. Segment A9 with simple tergum and sternum. A10 forming ventrally-hinged operculum, with pair of long hooks at apex (not described for *Scaphydra*).

Spiracles: Present on prothorax and abdominal segments 1 and 8; anterior two pairs forming oblong, vesicular gills in *Hydroscapha* or tuft-like gills in *Scaphydra;* posterior pair forming elongate vesicular gills.

Comments: The family contains 3 genera: *Hydroscapha,* with about 9 species occurring in western North America, Mexico, Eurasia, North Africa, Madagascar, and southeast Asia; *Yara* with 2 species in Panama and Brazil; and *Scaphydra,* with 3 species in Brazil.

Selected Bibliography

Barlet 1972.
Bertrand 1972.
Böving 1914a (larva of *Hydroscapha natans* LeConte).
Böving and Craighead 1931 (larva of *Hydroscapha natans* LeConte).
Forbes 1926.
Hinton 1967a (spiracular gills).
Lawrence and Newton 1982.
Paulian 1941.
Reichardt 1971, 1973a, 1973b, 1974 (larva of *Scaphydra angra* Reichardt).
Reichardt and Hinton 1976.

RHYSODIDAE (ADEPHAGA)

Ross T. Bell, *University of Vermont*

Wrinkled Bark Beetles

Figures 34.71, 72

Relationships and Diagnosis: There are 2 conflicting ideas about the relationships of Rhysodidae. Some regard them as an independent family of possibly primitive Adephaga, while Bell and Bell (1962) consider them to be highly specialized Carabidae. The larvae are recognized as Adephaga by the combination of 6-segmented legs with the absence of a mola on the mandible. They are soft-bodied, slow-moving, and grub-like. They differ from most Carabidae in the complete absence of urogomphi. They can be distinguished from those parasitic or myrmecophilous Carabidae which also lack urogomphi by lacking distinct segmented labial palpi, and by having, at least in the final instar, a transverse row of spinulae on at least some of the thoracic and abdominal tergites.

Biology and Ecology: The larvae are found in short galleries in rotten wood. The gallery behind the larva is tightly packed with wood shavings. They probably feed on slime molds and fungi in the wood, and not on the wood itself. Larvae are usually associated with adults, but the latter can also live in harder wood than the larvae can. In most species at least, the adults do not excavate distinct galleries, but push their way among the rotting wood fibers. Pupation is in the larval gallery.

Description: Small, pale larvae, 9 mm or less in length; body feebly sclerotized, short, depressed, tapered at both ends.

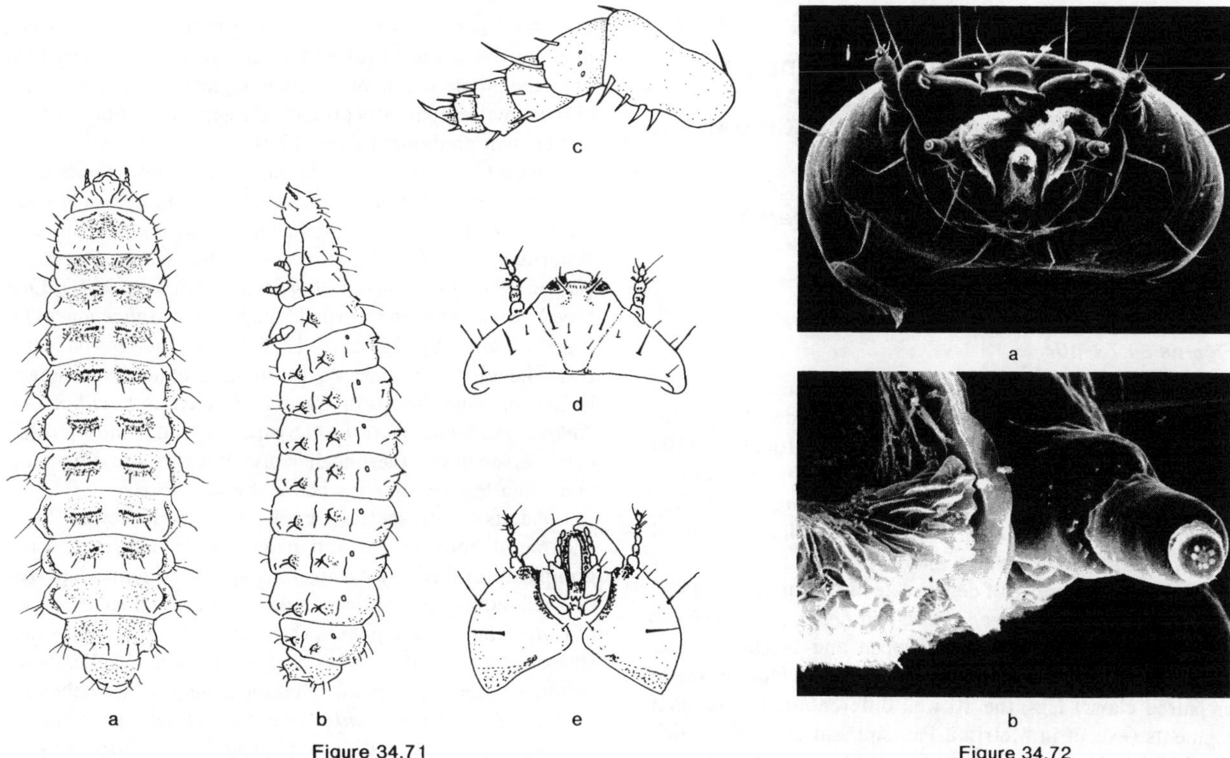

Figures 34.71a-e. Rhysodidae. *Omoglymmius americanus* (Laporte). **a.** larva, dorsal; **b.** lateral; **c.** leg, showing coxa, trochanter, femur, tibia, tarsus, and pretarsus; **d.** head, dorsal; **e.** head, ventral.

Figures 34.72a,b. Rhysodidae. *Omoglymmius hamatus* (LeConte). 1.9 mi. E Crane Flat, Tuolumne Co., Calif. **a.** Head, anteroventral; **b.** detail of maxillary apex.

Head: Prognathous; eyes absent; antennae short, 4-segmented; anterior margin of head with simple median lobe forming nasale; mandibles short, curved; mola absent; maxilla with well developed, 4-segmented palp; maxillary apex reduced, lightly sclerotized, sometimes bearing a series of fringed membranes; labial palps inconspicuous, 1-segmented, located at apex of elongate prementum, which may be clothed with microtrichia or fringed membranes.

Thorax and Abdomen: Prosternum and pronotum more sclerotized than remaining body segments; each leg with six segments, including a simple pointed pretarsus forming a claw; most body segments each with a pair of lightly sclerotized tergal humps, each with a transverse row of spinulae; rows of spinulae always interrupted at midline (in *Clinidium* also interrupted near lateral margin). Spinulae always present on metanotum and on abdominal tergites 1–6 (in *Clinidium,* also present on mesonotum and on tergite 7); ninth abdominal tergite narrow, in dorsal view appearing to form posterior end of body; tenth segment a narrow ring around anus, forming pygopod. Abdominal segments each with a pair of prominent ventrolateral tubercles in *Omoglymmius.* (These very poorly developed in *Clinidium.*)

Spiracles: Annular, on mesothorax and abdominal segments 1–8.

Comments: A poorly known group both taxonomically and ecologically. There are about 130 described species, but Bell and Bell are currently working on a world revision, and have found nearly 200 undescribed species. Eight species are known from North America, two from the western half of the continent and six from the eastern region. They are of no known economic importance, but are of great interest in zoogeography. Rhysodidae are especially prominent in insular faunas. Böving and Craighead (1931) illustrated a larva as *Clinidium sculptile* Newman. It has since been discovered that five species have been confused under this name, so it is uncertain which species they studied. Grandi (1955) and Burakowski (1975a) have illustrated and described the larvae of two European species. No keys are available to North American larvae.

Selected Bibliography

Bell 1970.
Bell and Bell 1962, 1978, 1979, 1982.
Böving and Craighead 1931.
Burakowski 1975a.
Grandi 1955.
Vanin and Costa 1978.

CARABIDAE (ADEPHAGA)

(INCLUDING BRACHINIDAE, CICINDELIDAE, OMOPHRONIDAE, PAUSSIDAE, PSEUDOMORPHIDAE, TRACHYPACHIDAE, ETC.)

Yves Bousquet, *Biosystematics Research Centre, Agriculture Canada*

Ground Beetles, Tiger Beetles

Figures 34.73–104

Relationships and Diagnosis: Carabidae, commonly known as ground beetles, belong to the suborder Adephaga. They form a large and cosmopolitan family with about 1500 genera and 30,000 species (Lawrence 1982c) of which some 2,500 occur in North America. The relationships of the Carabidae with other families of Adephaga, as well as its natural limits, are still challenged by many authors.

With the exception of degenerate instars of ectoparasitic larvae of *Brachinus, Pheropsophus* and *Lebia,* carabid larvae are recognized by the well developed and 2- (usually) or 4-segmented labial palpi, the 6-segmented (including single or paired claws) legs, the 10 well differentiated abdominal segments (except in Metriini, Paussini and Ozaenini which have the last 3 segments markedly modified or reduced), and the sublateral position (between the tergum and the epipleurite) of all spiracles on the abdominal segments.

Carabid larvae are likely to be confused with larvae of Staphylinidae, Silphidae and Histeridae. The 6-segmented legs in carabid larvae instead of 5-segmented legs in the other families will distinguish them.

Biology and Ecology: Only some aspects of the biology of carabids will be considered here. The reader is referred to the publication of Thiele (1977) for more information.

In temperate and boreal regions, the majority of carabid species are terrestrial. Adults and larvae are commonly found under rocks and debris in fields, forests and on shores, in leaf litter and under bark of logs in forest areas, or among gravel along river banks. Some species are hygrophilous, living in bogs, swamps, or marshes where they are collected by treading the vegetation under water. Many species, particularly within the Trechini, are cave inhabiting. Although some species regularly climb trees, bushes or plants, very few, such as species of *Dromius,* are truly arboreal; however, the number of species living on trees increases in the tropics.

At least in the temperate and boreal regions, carabids appear to be univoltine (Thiele 1977). Hibernation occurs in adult and/or larval stages. In the cases of *Amara infima* Duftschmid (Schjøtz-Christensen 1965) and *Sphaeroderus nitidicollis brevoorti* LeConte (Bousquet and Pilon 1980), indirect evidence suggests that winter is spent in the adult and egg stages.

Many species lay their eggs individually in small hollows dug out in the substrate. Others encase their eggs individually in mud cells which are dropped on the substrate or attached to stems or leaves. Some species of Pterostichini have developed parental care; the female lays her eggs in a cell made in the soil or in rotten logs and stays on or near them until the larvae hatch and become pigmented. The presence of the female apparently protects the eggs from fungal attack rather than predators (Löser 1970).

Most Carabidae have 3 larval instars; a few species have 2 (some species of *Amara* (Bílý 1975b), some species of *Harpalus* (Kirk 1974), *Thermophilum sexmaculatum* Fabricius (Paarmann 1979)), 4 (*Eurycoleus macularis* Chevrolat (Erwin 1975)) or 5 instars (*Brachinus* (Erwin 1967)). One larval instar has been recorded for species of *Aphaenops* (Deleurence, cited by Dajoz 1961), but this requires confirmation. Though in many species development of immature stages is fast, in some groups of Carabidae, such as *Scaphinotus, Nebria, Patrobus, Calathus, Synuchus,* and some Pterostichini, larvae have a thermic diapause. Pupation usually occurs in a cell a few centimeters below the substrate.

Adults of most species are polyphagous (Hengeveld 1980) and ingest both animal and plant matter. However, some groups are exclusively carnivorous and often somewhat specialized: adults of *Carabus* feed on worms and snails, those of *Calosoma* on caterpillars, those of Cychrini on snails and slugs, those of *Notiophilus* and *Loricera* on Collembola. Adults of other species are phytophagous and variably specialized: *Zabrus tenebroides* Goeze consumes ripe grains of rye, wheat, barley and corn (Thiele 1977), *Ditomus clypeatus* Rossi feeds on seeds of plantain (Schremmer 1960), and *Carterus calydonius* Rossi those of carrot (Brandmayr and Brandmayr Zetto 1974).

Information on the nutrition of carabid larvae is scarce. However, most species are probably carnivorous in their larval stages and mainly feed on soft bodied insects, snails, and worms. A few species are even very specialized in their feeding habits: larvae of *Orthogonius* live on termites, and those of the South African species, *Arsinoe grandis* Péringuay cling to and suck on the larvae of the tenebrionid *Catamerus revoili* Fairmaire. Ectoparasitic larvae are also known amongst Carabidae: larvae of *Lebia* are parasitic on larvae and pupae of Chrysomelidae, those of *Brachinus* on pupae of aquatic beetles, those of *Pheropsophus* on mole cricket eggs, and those of *Pelecium* on beetle pupae and millipedes. Larvae of some species have been reported to be phytophagous such as those of *Harpalus puncticeps* Stephens and *Carterus calydonius* which feed on carrot seeds (Brandmayr Zetto and Brandmayr 1975, Brandmayr and Brandmayr Zetto 1974). While larvae of most species actively search for their prey, those of others, such as species of Cicindelinae and *Sphallomorpha,* passively wait at the top of their burrow. Under laboratory conditions, cannibalism has been commonly observed.

The economic importance of carabid beetles is moderate. They feed on many injurious insects but are unable to control effectively any pest. Consequently, they should best be considered as valuable natural "auxiliaries" (Thiele 1977). Damage by carabids to crops and stored products is minimal.

Insectivores, bats, rodents, birds, frogs and toads, ants, robber flies and spiders are the most important predators of Carabidae. A number of Nemathelminthes, Acari (especially Podapolipidae), Hymenoptera (Proctotrupoidea, Braconidae, Mutillidae) and Diptera (Tachinidae) are known to be parasitic on adults and/or larvae.

Description: Body length of mature larvae, 2 to 50 mm. General shape campodeiform (figs. 34.73, 74, 76) for most species, rarely onisciform (e.g., Cychrini; fig. 34.75) or physogastric (e.g. some instars of *Brachinus* and *Pheropsophus*, some Pseudomorphini).

Head: Prognathous. Frons, clypeus and labrum fused to form frontale (= frontal piece), its anterior portion (nasale) usually prominent and variable in form. Epicranial suture present, of various shapes (figs. 34.77–84); epicranial stem present (figs. 34.77, 81, 82) or absent (figs. 34.78–80, 83, 84). Parietale in general with 6 stemmata on each side in 2 vertical rows of 3 per row, sometimes with fewer than 6 or without stemmata. Cervical groove present in many groups (e.g., figs. 34.81, 82). Antenna in general nearly as long as or longer than mandible, and 4-segmented, rarely 3-segmented (Anthiini, with second and third segments fused). First antennal segment in some Pterostichini (e.g., *Cyclotrachelus, Molops, Abax*) with circular membranous area near base and therefore seemingly 5-segmented. Third antennal segment usually with 3 small sensilla and 1 large, generally bulbous sensorium anterolaterally (= sensorial appendage of van Emden 1942b).

Mouthparts protracted from anterior part of cephalic capsule. Mandible more or less falcate or subtriangular, symmetrical, without mesal hyaline process (present in Paussini according to van Emden 1942b) or molar region; retinaculum present in nearly all species, consisting of single tooth of various shapes (figs. 34.85–87, 89), occasionally bidentate (*Omophron;* fig. 34.88); penicillus consisting of one to many closely associated setae (figs. 34.85–87, 89), or absent (fig. 34.88). Maxilla with cardo proportionally small; stipes elongate and in general membranous or thinly sclerotized dorsally; lacinia absent (figs. 34.90–92) or developed as a small, rounded or acuminate tubercule (fig. 34.93), or as a protuberance sometimes nearly as long as galea (e.g. *Metrius, Omophron;* fig. 34.94); galea palpiform, 2-segmented (figs. 34.91–95), rarely 1-segmented (e.g., *Metrius, Brachinus;* fig. 34.90); maxillary palp 4-segmented (figs. 34.90–94), rarely 5-segmented (Trechini, with last segment subdivided; fig. 34.95). Labium with prementum in general short, arising from membranous mentum; labial palp 2-segmented (figs. 34.96, 97), or rarely 4-segmented (Trechini, with double subdivision of last segment; fig. 34.98); ligula (usually bearing apical pair of setae) present in many groups as a short (fig. 34.98) or long (e.g., *Metrius, Omophron;* fig. 34.97) protuberance.

Thorax and Abdomen: Thoracic dorsum formed by single sclerite divided medially by narrow membranous area (ecdysial line). Pleural region usually with epimeron and episternum on all 3 segments and also with trochantin and pleurite on meso- and metathorax. Ventral side of thorax in general mostly membranous, except for anterior part of prothorax with large and often more or less triangular, rarely divided (Cicindelinae), prosternite. Legs in general long, 6-segmented (fig. 34.99) including one or two, unequal or subequal, movable claws, rarely 5-segmented (Collyrini, with tarsus and claws fused, according to van Emden 1935b), or 3-segmented (some degenerate instars of Brachinini, according to Erwin 1967).

Abdomen 10-segmented. Segment 9, with rare exceptions (e.g., Cicindelinae, some Cychrini, some Harpalini), with pair of dorso-apical urogomphi; urogomphus articulated (fig. 34.101) or fixed (figs. 34.100, 102–104), segmented or entire, with (fig. 34.104) or without membranous areas (figs. 34.100–103), short (figs. 34.100, 102) or long (figs. 34.101, 103, 104). In Metriini, Ozaenini and Paussini, eighth and ninth segments highly modified (*see* Bousquet 1986). Tenth segment usually tubular, acting as proleg, often with paired eversible vesicles bearing coarse pointed microsculpture. Terga 1–8 usually formed by single sclerite divided by narrow, median membranous area (ecdysial line). In Cicindelinae, fifth tergite modified, bearing 2 or 3 pairs of hooks. Pleural region of segments 1–8 with epipleurite (subdivided in some groups) and hypopleurite (not apparent in first instar of some groups, e.g., *Omophron*). Ventral side of segments 1–7 usually with 7 sclerites: small paired anterior sternites, median sternite (anterior ventrite of van Emden 1942b), and paired inner and outer sternites (postventrites of van Emden 1942b). These sclerites fused or partly fused, often also with hypopleurites on eighth and ninth segments.

Spiracles: Annular, on mesothorax and between tergum and epipleurite of abdominal segments 1–8.

Comments: The first larval instar of most species differs from subsequent instars by having paired egg-bursters located posteriorly on the frontale or rarely basally on the dorsal side of the parietale (some *Bembidion,* D. R. Maddison, pers. comm.). In some Bembidiini, the egg-bursters consist of coarse pointed microsculpture; however, they are usually formed by one or two raised microspines (derived from pointed microsculpture) or a longitudinal series of microspines often fused or partly fused into a carina. Larval instars of most species can also be distinguished by the chaetotaxy. The first instar has primary setae, while the second and third instars often bear, in addition, secondary setae which are in general more numerous in the third instar. The width of the cephalic capsule is often also used to distinguish the larval instars.

The keys of van Emden (1942b) and Thompson (1979a) are useful for the identification of the Nearctic and Palaearctic carabid larvae at tribal level. Larval characteristics of Nearctic species are poorly known, as many taxa are undescribed (larvae of only about 7% of the Nearctic species have been described, according to Thompson 1977) and most descriptions are incomplete or superficial. It is hoped, however, that the rearing and preparation techniques improved by Goulet (1976, 1977) will stimulate research in this field. His study of Elaphrini larvae (Goulet 1983) should stand as a guide.

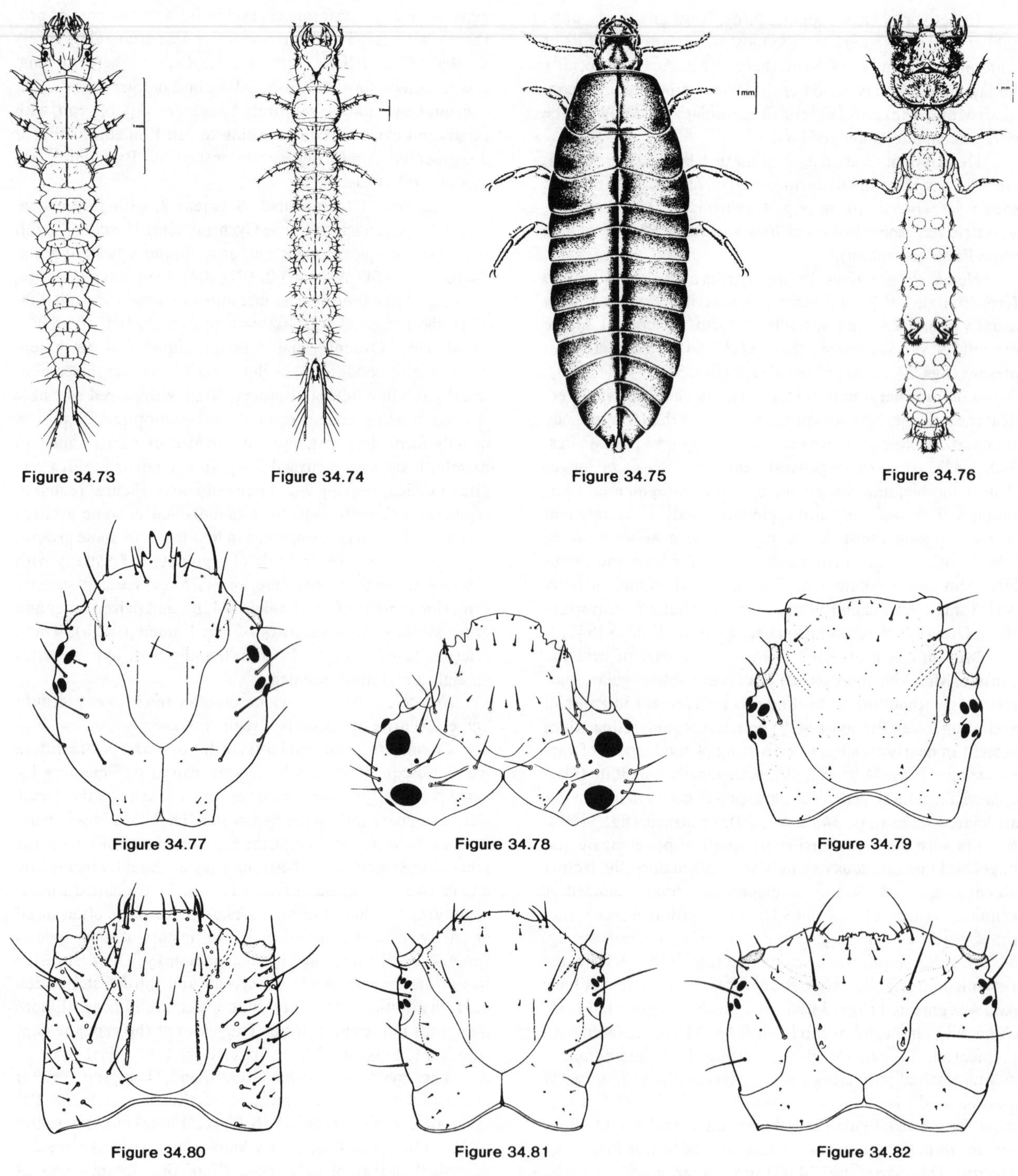

Figure 34.73 Figure 34.74 Figure 34.75 Figure 34.76

Figure 34.77 Figure 34.78 Figure 34.79

Figure 34.80 Figure 34.81 Figure 34.82

Figure 34.73. Carabidae. *Omophron tesselatus* Say, first instar.

Figure 34.74. Carabidae. *Pterostichus diligendus* Chaudoir, third instar.

Figure 34.75. Carabidae. *Sphaeroderus lecontei* Dejean, third instar.

Figure 34.76. Carabidae. *Cicindela* sp., second instar. Cape Breton, Nova Scotia, Canada.

Figures 34.77–34.82. Carabidae. HEAD CAPSULES, dorsal, all first instars.

Figure 34.77. *Nebria* sp.

Figure 34.78. *Cicindela* sp.

Figure 34.79. *Sphaeroderus lecontei* Dejean.

Figure 34.80. *Promecognathus laevissimus* Dejean.

Figure 34.81. *Pterostichus diligendus* Chaudoir.

Figure 34.82. *Anisodactylus nigrita* Dejean.

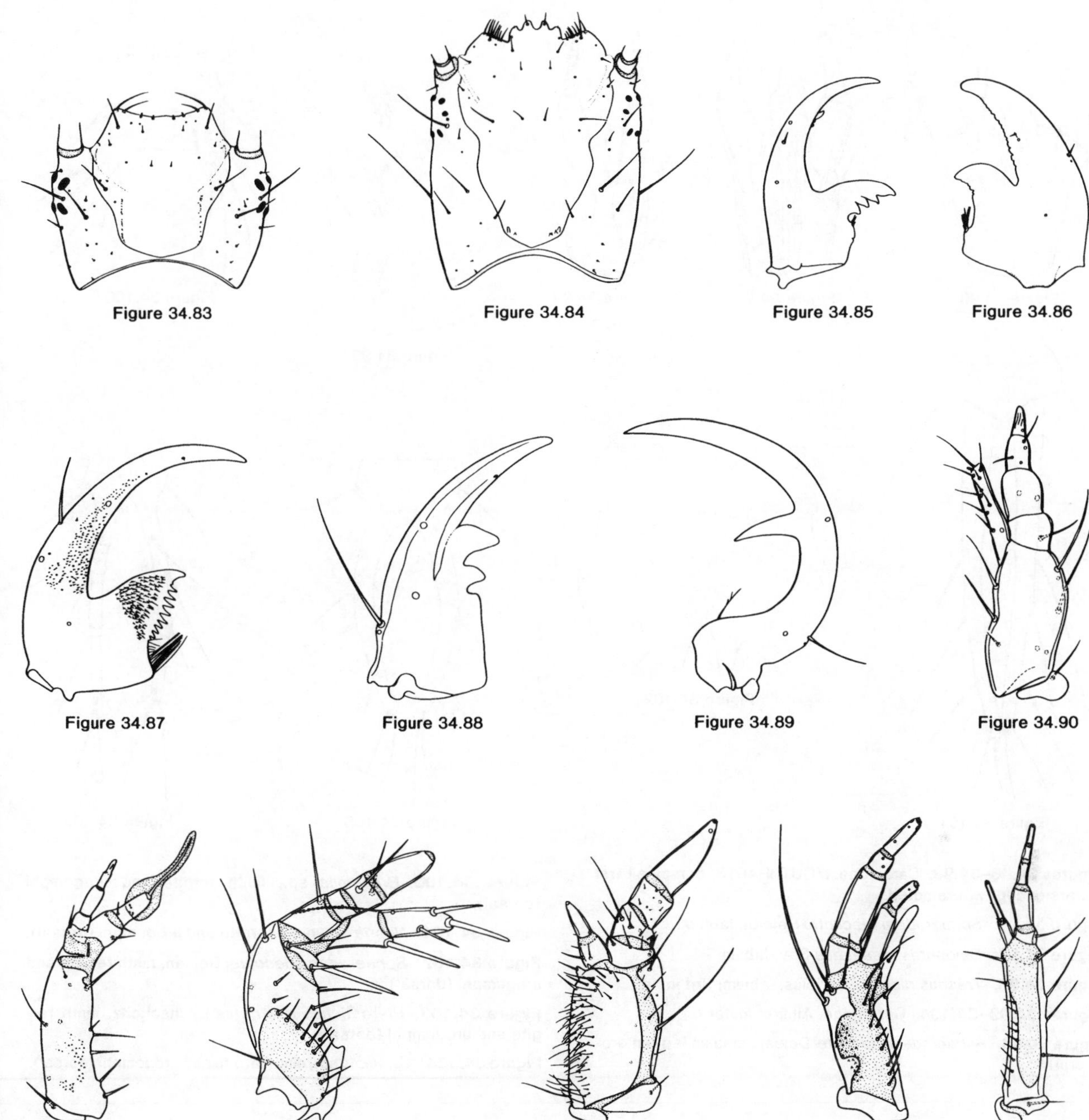

Figures 34.83–34.84. Carabidae. HEAD CAPSULES, dorsal, all first instars.

Figure 34.83. *Diplocheila striatopunctata* LeConte.

Figure 34.84. *Calleida punctata* LeConte.

Figures 34.85–34.95. Carabidae. MOUTHPARTS, dorsal, all first instars except where noted.

Figure 34.85. *Loricera pilicornis* Fabricius, left mandible.

Figure 34.86. *Calosoma frigidum* Kirby, right mandible.

Figure 34.87. *Sphaeroderus lecontei* Dejean, left mandible.

Figure 34.88. *Omophron tesselatus* Say, left mandible.

Figure 34.89. *Abax parallelepipedus* Piller and Mitterpacher, right mandible.

Figure 34.90. *Brachinus* sp., right maxilla.

Figure 34.91. *Loricera pilicornis* Fabricius, left maxilla.

Figure 34.92. *Cicindela* sp., left maxilla.

Figure 34.93. *Sphaeroderus lecontei* Dejean, right maxilla.

Figure 34.94. *Omophron tesselatus* Say, left maxilla.

Figure 34.95. *Trechus rubens* Fabricius, right maxilla, 3rd instar.

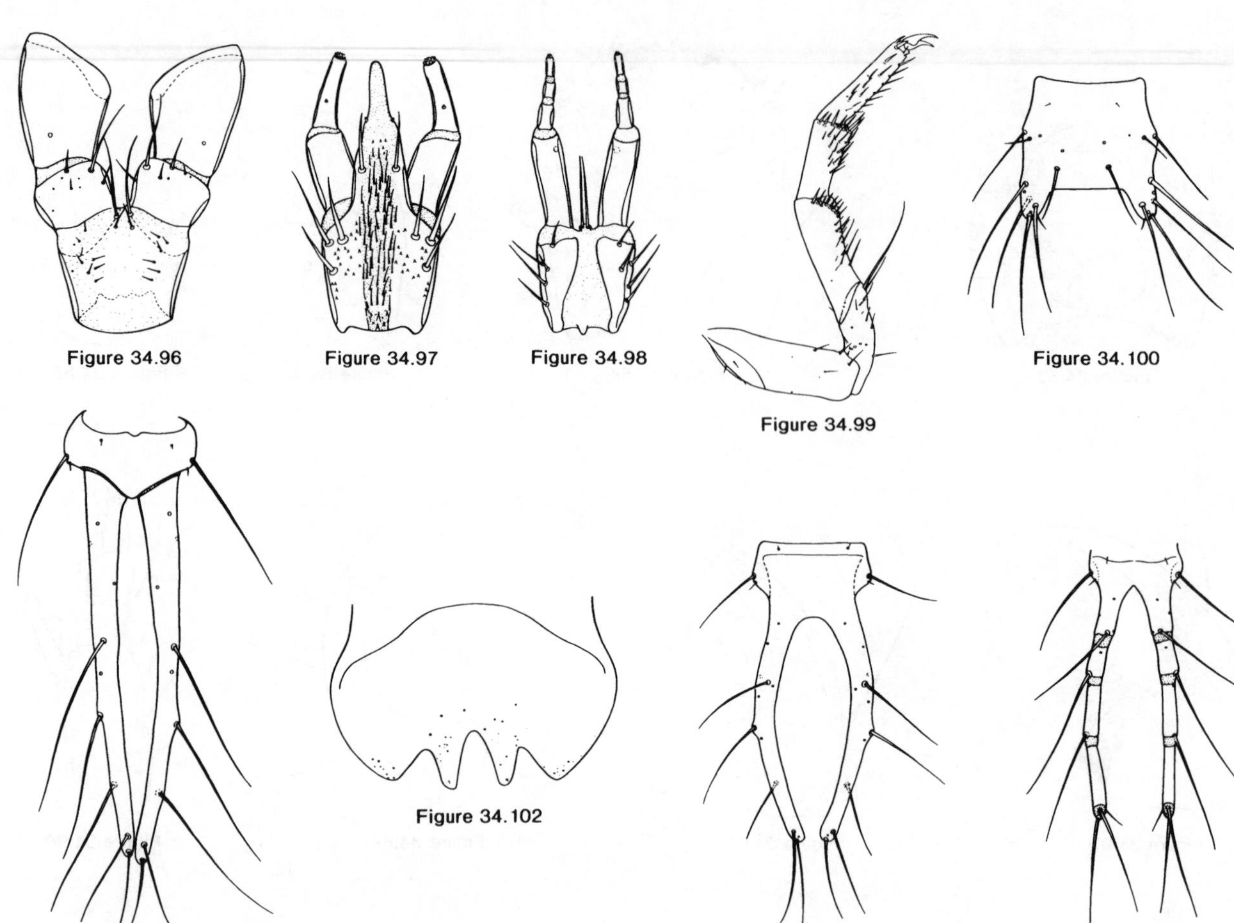

Figures 34.96–34.98. Carabidae. MOUTHPARTS, dorsal, all first instars except where noted.

Figure 34.96. *Sphaeroderus lecontei* Dejean, labium.

Figure 34.97. *Omophron tesselatus* Say, labium.

Figure 34.98. *Trechus rubens* Fabricius, labium, 3rd instar.

Figures 34.99–34.104. Carabidae. All first instars.

Figure 34.99. *Sphaeroderus lecontei* Dejean, median leg (anterolateral).

Figure 34.100. *Brachinus* sp., ninth tergite and urogomphi (dorsal).

Figure 34.101. *Nebria* sp., ninth tergite and urogomphi (dorsal).

Figure 34.102. *Sphaeroderus lecontei* Dejean, ninth tergite and urogomphi (dorsal).

Figure 34.103. *Pterostichus adstrictus* Eschscholtz, ninth tergite and urogomphi (dorsal).

Figure 34.104. *Cymindis* sp., ninth tergite and urogomphi (dorsal).

Selected Bibliography (for Carabidae).

Bily 1975b
Bousquet 1985, 1986, 1987.
Bousquet and Goulet 1984.
Bousquet and Pilon 1980.
Böving and Craighead 1931.
Brandmayr and Brandmayr Zetto 1974, 1982.
Brandmayr Zetto and Brandmayr 1975.
Dajoz 1961.
Emden 1935b, 1936, 1942b.
Erwin 1967, 1975, 1981.
Erwin et al. 1979.
Gardner 1929, 1930b, 1931b, 1933a, 1936b, 1938a.
Goulet 1976, 1977, 1983.
Habu and Sadanaga 1961, 1963, 1965, 1969, 1970.
Hamilton 1925.
Hengeveld 1980.
Houston and Luff 1975.
Hurka and Ducháč 1980a, 1980b.
Hurka and Smrz 1981.
Kirk 1974.
Landry and Bousquet 1984.
Lawrence 1982c.
Lindroth 1960.
Löser 1970.
Luff 1969, 1972, 1976, 1978, 1980, 1985.
Moore 1964, 1965, 1966, 1974.
Paarmann 1979.
Pearson 1988.
Peyrieras 1976.
Schjøtz-Christensen 1965.
Schremmer 1960.
Sharova 1976.
Sharova and Makarov 1984.
Silvestri 1905.
Smrz 1979.
Thiele 1977.
Thompson 1977, 1979a, 1979b.
Vanek 1984.

HALIPLIDAE (ADEPHAGA)

Paul J. Spangler, *Smithsonian Institution*

Crawling Water Beetles

Figures 34.105, 106

Relationships and Diagnosis: The family Haliplidae belongs to the suborder Adephaga; however, unlike all other known adephagan larvae, haliplid larvae have only 1 instead of 2 claws.

Haliplid larvae may be distinguished immediately from other adephagous aquatic beetle larvae by the 9 or 10 abdominal segments and the 6-segmented leg including a single claw. Larvae are of 2 types. Larvae of *Apteraliplus, Brychius,* and *Haliplus* (fig. 34.106) are elongate and taper from head to apex of the last abdominal segment that ends in a subspiniform process that may be bifurcate; they have rough and rigid cuticle, therefore, the body is stiff and can bend very little; gills are absent. Larvae of *Peltodytes* (fig. 34.105) have slender, rather stiff, hairlike gills (prothorax with 3 pairs, meso- and metathorax each with 2 pairs, abdominal segments 1–8 each with 2 pairs, and segments 9 and 10 each with 1 pair); these gills are about as long as or slightly longer than the length of the body which is moderately stiff but can assume a C-shape. Mature larvae range in length from 5.0 to 12.0 mm.

Biology and Ecology: Known haliplid larvae feed on algae. Adults and larvae of various species of *Haliplus* are commonly found on *Chara, Nitella,* and *Ceratophyllum;* species of *Peltodytes* occur more often on *Spirogyra.* Adults and larvae of *Apteraliplus, Haliplus,* and *Peltodytes* normally occur in weedy ditches, ponds, lakes and similar lentic habitats. Species of *Brychius* favor lotic habitats although some have been collected in lakes. Most haliplids occur in shallow water but Hickman (1931) found some species of *Haliplus* 6 feet below the surface. When haliplid larvae are collected they at first feign death and, when they move, they crawl very slowly; therefore, they may be easily overlooked among the vegetation and associated debris in an aquatic net.

The life cycles of some species of *Haliplus* and *Peltodytes* are reasonably well known as a result of Hickman's studies (1930, 1931); however, those of *Apteraliplus* and *Brychius* have not been described. In the United States and Canada, egg laying by *Haliplus* and *Peltodytes* begins in spring and continues through the summer. Adults of *Peltodytes* attach their eggs to the host plants, and females of *Haliplus* insert their eggs inside plant cells. Larvae of both genera molt 3 times before the last instar leaves the water, digs into suitably moist soil, and forms a pupal cell. Adults normally pass the winter underwater. Larvae that attain maturity late in the summer occasionally overwinter in moist soil above the waterline.

Description: Body elongate, cylindrical or subcylindrical, and tapering or not posteriorly. Gills elongate and hairlike (*Peltodytes*) or gills absent (*Apteraliplus, Brychius, Haliplus*). The cuticle of *Apteraliplus, Brychius,* and *Haliplus* is rough because of numerous tuberculate and spinous

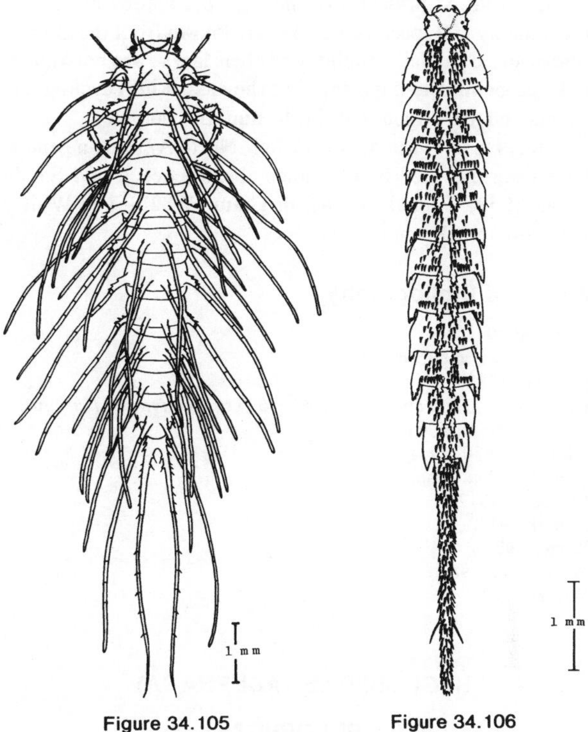

Figure 34.105 Figure 34.106

Figure 34.105. Haliplidae. *Peltodytes* sp., dorsal.

Figure 34.106. Haliplidae. *Haliplus* sp., dorsal.

processes. Integument and sclerites whitish in freshly hatched and freshly molted larvae; sclerotized areas becoming light yellowish brown to reddish brown with aging.

Head: Semiprognathous to hypognathus. Antenna, 4-segmented. Stemmata, 6 pairs. Ecdysial cleavage line and labrum absent. Mandible as wide as long, hook-shaped, hollow; with sharp apical tooth. Maxilla short and broad; cardo small; stipes large and broad; mala with setae; palp, 3-segmented. Labium small; labial palp small, 2-segmented.

Thorax and Abdomen: Prothoracic, meso- and metathoracic terga well sclerotized. Prothorax with 3 pairs of gills; meso- and metathorax each with 2 pairs of gills. Legs, 6 segmented, including single claw; because the fourth segment is produced they are weakly to moderately chelate and serve to grasp algae.

Abdomen of 10 segments. In *Peltodytes,* segments 1–8 each bear 2 pairs of threadlike gills; segments 9 and 10 each bear 1 pair of gills; segment 10 ends bluntly. *Apteraliplus, Brychius,* and *Haliplus* lack gills and segment 10 tapers to a subspiniform process that may be bifurcate or not. Microtracheal "gills" have been reported in *Haliplus* (Seeger, 1971).

Spiracles: Absent, except in last instar, which has annular spiracles on mesothorax and abdominal segments 1–7.

Comments: Haliplids are found throughout the world with most species occurring in the temperate regions. The number of taxa is small with about 200 species in 5 genera in the world. At present 4 of the 5 genera and 67 of the approximately 200 species described are known from the United States and Canada. Haliplids and their larvae are not known to be economically important but they assist in recycling nutrients and are a food item in the aquatic ecosystem.

Keys to larval stages of the four North American genera have been provided by Chandler (1956), Leech and Sanderson (1959), Pennak (1978), Brigham (1982b), and White, Brigham and Doyen (1984).

Selected Bibliography

Bertrand 1972.
Böving and Craighead 1931.
Brigham 1982b.
Chandler *in* Leech and Chandler 1956 (key to larvae of *Haliplus* and *Peltodytes*).
Hickman 1930, 1931.
Leech and Sanderson 1959.
Matheson 1912.
Pennak 1978.
Seeger 1971.
White et al. 1984.
Wilson 1923.

HYGROBIIDAE (ADEPHAGA)

(= PELOBIIDAE)

John F. Lawrence,
Division of Entomology, CSIRO

This family includes the genus *Hygrobia,* with 5 species occurring in western China, western and central Europe, North Africa, and Australia (southeastern part, Northern Territory, and Cape York). The group is generally considered to belong to the complex of aquatic families including Amphizoidae, Noteridae, and Dytiscidae.

Larvae are about 10 mm in length and fusiform, with a large, prognathous head and enlarged thorax. The head has a long epicranial stem, 4-segmented antennae, a labrum fused to the head capsule, and 6 stemmata on each side. Mandibles are falcate, but not perforate, and the protracted ventral mouthparts include maxillae without apical lobes, 4-segmented maxillary palps and 2-segmented labial palps. Gular sutures are completely separated. The legs are relatively long and slender, 6-segmented, including paired, movable claws, and have fringes of swimming hairs on the tibiae and tarsi. Segments A8 and A9 are reduced; segment A8 bears a long, narrow, median process, without spiracles at the apex, and segment A9 bears a pair of long, narrow urogomphi. Segment A10 is reduced and membranous. Paired gill tufts arise from the coxal bases and from abdominal sternites 1–3. Reduced spiracles are present on the mesothorax and abdominal segments 1–7 in the last instar only.

Hygrobiids are bottom-feeding predators in ponds, and adults may be attracted to lights. They occur only in ponds in which the bottom is covered with fine ooze, in which they feed on insect larvae and *Tubifex* worms. The adult maintains an air bubble beneath the elytra and may remain submerged for up to 30 minutes. Fully grown larvae leave the water to pupate within a closed cell in sand or mud.

Selected Bibliography

Balfour-Browne 1922 (biology).
Bertrand 1928 (larva), 1972 (larva).
Böving and Craighead 1931 (larva).
Britton 1981 (Australian *Hygrobia*).
Hammond 1979 (relationships).

AMPHIZOIDAE (ADEPHAGA)

David H. Kavanaugh,
California Academy of Sciences

Trout-Stream Beetles

Figures 34.107a–h

Relationships and Diagnosis: Members of the genus *Amphizoa* LeConte represent a family distinct from but clearly related to the other Caraboidea. They share many characteristics with Carabidae on one hand and Dytiscidae on the other (Edwards, 1951; Horn, 1883; Hubbard, 1892a, 1892b; Leech and Chandler, 1956; Roughley, 1981). Based on a cladistic analysis of both extinct and extant "families" of Adephaga, Kavanaugh (1986) concluded that amphizoids represent the earliest divergent lineage from the common ancestor of all Hydradephaga, excluding Haliplidae, and that this divergence probably occurred in the Triassic.

Larval amphizoids are most similar in size and form to carabid larvae of the tribe Cychrini or to silphid larvae (e.g. genus *Silpha* or *Nicrophorus*); however, their aquatic habits readily distinguish them from members of these other groups. Further, amphizoid larvae are distinguished from larvae of all other North American beetles in having their mandibles sulcate medially, without an internal duct, thoracic legs ambulatory, each 6-segmented, including pretarsus with two movable, terminal claws, the abdomen 8-segmented, without hooks at apex, urogomphi present, short, 1-segmented, and with the spiracles of the eighth abdominal segment located paramedially on the dorsum (basal to urogomphi).

Biology and Ecology: Amphizoids are primarily aquatic, although both adults and larvae are sometimes found out of water; eggs deposited in moist places on land will develop and hatch, and mature larvae leave the water to pupate on land. They inhabit cool or cold fresh water streams (from small rills to large rivers) and (less frequently) lakes, where they are confined to the shallow shorelines. Neither adults nor larvae swim. They crawl over submerged rocks, logs, and vegetation in search of food and shelter, and are often found clinging to floating debris, especially in eddies and backwashes. If dislodged from their substrate, they make feeble walking movements and drift with the current. They are much more agile out of water. Adults have fully formed wings and are no doubt capable of dispersal by flight (Darlington, 1930). Both adults and larvae must come to the surface for air. Larvae assume a posture such that the apex of the abdomen breaks the water surface and the spiracles of the eighth segment are in contact with air.

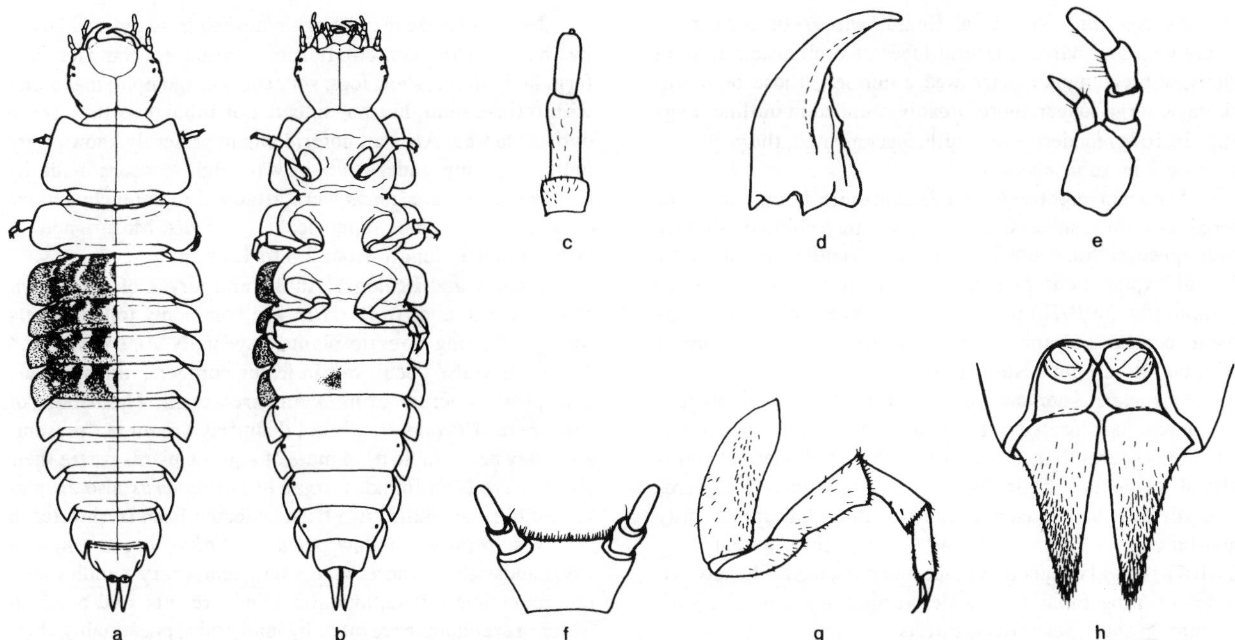

Figures 34.107a-h. Amphizoidae. *Amphizoa insolens* LeConte. **a.** dorsal (pigmentation pattern for left half of metathoracic and abdominal terga 1-3); **b.** ventral (pigmentation pattern for right half of metathorax and abdominal segments 1-3); **c.** left antenna, dorsal; **d.** left mandible, dorsal; **e.** left maxilla, ventral; **f.** labium, ventral; **g.** left metathoracic leg, anterior; **h.** apex of abdomen, posterodorsal oblique aspect, showing spiracles of abdominal segment 8 and urogomphi.

Although previously assumed to feed as scavengers on sluggish or dead and drifting insects. Edwards (1954) found that all stages of larvae and adults studied fed exclusively on living stonefly (Plecoptera) larvae. However, if kept out of water, amphizoids will accept a variety of freshly macerated insects presented as food.

Eggs, which are very large (over 2 mm in length), may be deposited in protected spots under water, such as in cracks on the undersurfaces of floating logs (Edwards, 1954; Leech and Chandler, 1956), or in moist places on land adjacent to water bodies. Oviposition is usually in late August or early September. First instar larvae, also exceptionally large, are found in September through November and again in early spring. This is probably the normal overwintering stage. Mature larvae appear in May through August and leave the water to pupate under stones on adjacent shores in July and August. Adults may be found throughout the year, but teneral (newly-emerged) adults are common only in August.

Description: Mature (i.e. third instar) larvae with total length 12.0 to 17.0 mm, maximum width 3.5 to 4.7 mm. Body form (figs. 34.107a, b), elongate, spindle-shaped, depressed, broadest at mid-length and tapered toward both ends, moderately convex and markedly sclerotized dorsally, flat and only faintly sclerotized ventrally; combined lengths of head and thorax almost equal to length of abdomen; thoracic and abdominal terga broadly explanate laterally as thin, laterally projected lobes; dorsal midline, from vertex of head to apex of abdomen, with a deeply incised longitudinal furrow (actually the ecdysial line of weakness). Dorsum brown (teneral individuals may be testaceous) to piceous, with maculation pattern as in fig. 34.107a but less evident in darkest individuals; venter pale, yellowish white, except undersurface of head and eighth abdominal sternum brown; darkest individuals also with darkened areas along midline on 1 or more abdominal sterna. Surface of dorsal sclerites finely punctulate and/or transversely rugulose, sparsely covered with short, fine, apically hooked, prostrate setae; ventral sclerites smooth and glabrous, except ventrolateral surfaces of head with fine punctures and setae as on dorsum.

Head: Large, protruded, circular in silhouette, narrowed posteriorly, flattened dorsally, moderately convex ventrally, with lateral margins (of genae) carinate. Antennae (fig. 34.107c) short, apparently 3-segmented, but with reduced 4th segment at apex. Three pairs of lateral stemmata present on each side, 1 pair each above, on, and below the lateral margin. Labrum trapezoidal, vertical, glabrous dorsally (anteriorly), except for a fringe of short setae at clypeolabral suture. Mandibles (fig. 34.107d) sickle-shaped, each with a deep, longitudinal groove medially (but without an internal duct), retinaculum minutely dentiform, ventral cutting edge minutely denticulate, without prostheca or penicillus, molar region simple, narrow, unmodified, without denticles. Maxillae (fig. 34.107e) stout, with cardo and stipes fused, lacinia and galea fused as a 2-segmented mala, palpifer present, palp 3-segmented. Labium (fig. 34.107f) broad, transverse, with a fringe of short setae across apical margin, without distinct ligula, palps 2-segmented. Gula absent, gular suture simple, linear in midline.

Thorax and Abdomen: Broad, with prothorax relatively long and narrow; lateral lobes of prothoracic tergum short, subrectangular, narrowed anteriorly, those of pterothoracic terga longer, more broadly rounded in outline. Legs (fig. 34.107g) moderate in length, 6-segmented, the pretarsus bearing 2 movable claws.

Abdomen eight-segmented, explanate lateral lobes of terga 1–7 thin, successively more acutely pointed, slightly overlapped; eighth segment trapezoidal in outline, tergum with lateral explanations present only as lateral carinae. Urogomphi (fig. 34.107h) present, 1-segmented, short but prominent, conical and apically pointed, articulated (not fused) with eighth tergum posterolaterally.

Spiracles: Thoracic spiracles restricted to a single mesothoracic pair located anterolateral to bases of mesocoxae, annular, apparently non-functional. Paired abdominal spiracles of segments 1–7 (fig. 34.107b) located ventrally, anterolaterally near base of each explanate lateral lobe, apparently non-functional; spiracular pair of eighth segment (fig. 34.107h) valvular, functional, located dorsomedially between bases of urogomphi and posteromedial margin of eighth tergum on short, sclerotized turrets.

Comments: The family is very small, with only three species known from North America (Kavanaugh, 1986) and one species from western China. In North America, the family is restricted to the western half of the continent, from the Rocky Mountains west to the Pacific Coast, and from southern Alaska and Yukon Territory to southern California and northern Arizona and New Mexico. The larvae of *Amphizoa insolens* LeConte and *A. lecontei* Matthews have been described and illustrated (Böving and Craighead, 1931; Hubbard, 1892a, 1892b; Peterson, 1951) and their habits discussed (Darlington, 1930; Edwards, 1951, 1953, 1954; Leech and Chandler, 1956). Neither immatures nor adults of this family have apparent economic importance.

Selected Bibliography

Böving and Craighead 1931.
Darlington 1930.
Edwards 1951, 1953, 1954.
Horn 1883.
Hubbard 1892a, 1892b.
Kavanaugh 1986.
Leech and Chandler 1956.
Peterson 1951.
Roughley 1981.

NOTERIDAE (ADEPHAGA)

Paul J. Spangler, *Smithsonian Institution*

The Noterids

Figures 34.108a–e, 34.109

Relationships and Diagnosis: The noterids were formerly considered a subfamily of the Dytiscidae. However, morphological and biological differences readily distinguish them from the dytiscids.

Noterid larvae may be distinguished from dytiscid larvae by the following combination of characters: compact fusiform body shape; short legs; very short urogomphi; mandibles with retinaculum, but not sulcate nor tubular as they are in dytiscid larvae. Also, all noterid pupae presently known were found pupating underwater in watertight cocoons made by the larvae; these cocoons were attached to aerenchymatous cells of aquatic plant stems, leaves, or roots. Mature noterid larvae range in length from 2.0 to 4.5 mm.

Biology and Ecology: Adults and larvae of *Hydrocanthus, Suphis,* and *Suphisellus* are commonly found among roots of floating aquatic plants, especially *Eichhornia* and *Pistia;* they also occur, but in lesser numbers, among emergent plants where floating plants are absent. The habitat of the larvae of *Pronoterus* is not definitely known but presumably they occur in floating mats of aquatic plants where their adults have been found. Larvae of *Notomicrus* also are unknown but the adults have been collected from freshwater in weedy margins of shallow ponds, in *Sphagnum* swamps, in woodlands where there were small temporary puddles containing culicid larvae, in water-filled tire ruts, and brackish water in crabholes excavated by land crabs; presumably, their tiny larvae will be found in the same habitats where the adults live.

The life cycles of most noterids are unknown. However, F. Balfour-Browne and J. Balfour-Browne (1940) described the interesting larval and pupal stage of the European species *Noterus capricornis* Herbst and showed that pupation took place underwater in cocoons attached to aerenchymatous cells of roots of aquatic plants. Spangler (1981, 1982) reported the same pupal habits for the genera *Hydrocanthus, Suphis,* and *Suphisellus.*

Because noterid larvae are present in the United States in the months of June through August, it is presumed that noterids oviposit in late spring or early summer. Because the ovipositor is long and soft, it is assumed that the eggs are laid on aquatic plants or in the mud near the plants. The complete life cycle has not been established for any noterid but the available evidence suggests that it is similar to that of dytiscid beetles except noterid pupation occurs underwater instead of on land.

The food habits of the noterids are poorly known. Wesenberg-Lund (1912) assumed from the shape of the mandibles of a European larva of *Noterus* that it was entirely vegetarian. However, F. Balfour-Browne and J. Balfour-Browne (1940) observed that larvae of *Noterus* feed readily on dead *Chironomus* larvae and dead individuals of their own kind; they also saw the larvae work their mandibles on the surface of roots but seemed not to get anything off. They suggested that the larval mandibles may be a modification of the phytophagous type of mandible and stated, "possibly, therefore, the larva flourishes on a mixed diet." Young (1967) reported that noterid larvae and adults are vegetation-detritus feeders.

Description: Body cylindrical, spindle shaped, or teardrop shaped (*Suphis*); usually tapering strongly posteriorly; subcylindrical in cross section. Cuticle relatively smooth. Integument and sclerotized parts white when freshly hatched but yellow to reddish-brown upon aging. Larvae of some species of *Hydrocanthus* may have dark bands in early instars.

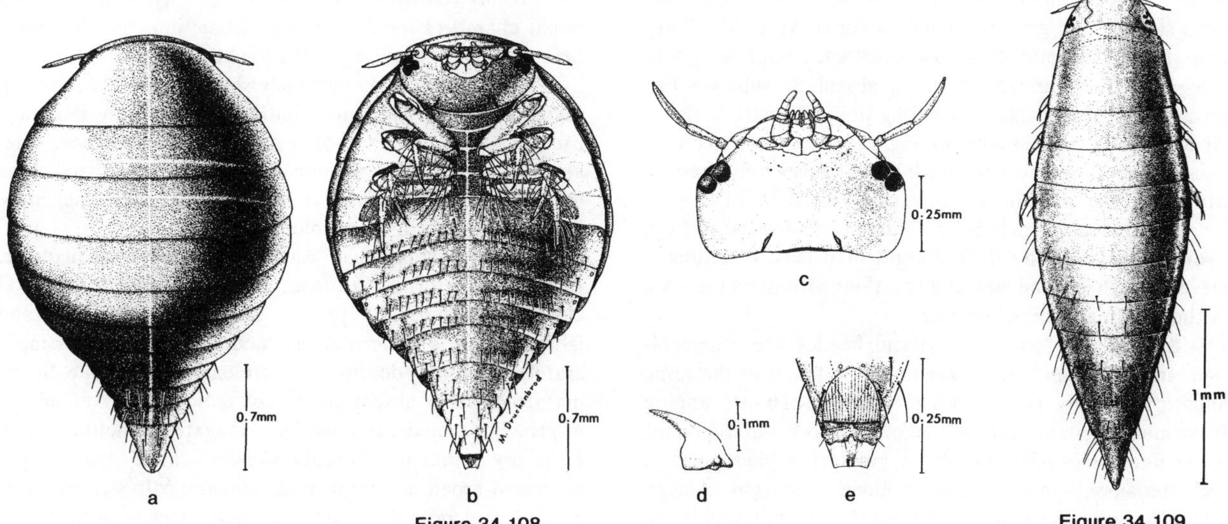

Figures 34.108a-e. Noteridae. *Suphis inflatus* (LeConte). **a.** dorsal; **b.** ventral; **c** head, ventral; **d.** mandible; **e.** last abdominal segment, ventral.

Figure 34.109. Noteridae. *Hydrocanthus* sp., dorsal.

Head: Partially retracted into pronotum, prognathous, globose (fig. 34.108c); without temporal spines; anterior margin arcuate; ecdysial cleavage lines present. Antenna 4-segmented, elongate, slender. Clypeus and labrum fused. Mandible curved, moderately slender; inner margin smooth, serrate, or dentate; neither sulcate nor tubular (fig. 34.108d). Maxillary palp short, 3-segmented; cardo small, stipes long and broad. Labial palp short, stout, 2-segmented. Six pairs of stemmata.

Thorax and Abdomen: Prothoracic, meso-, and metathoracic terga well sclerotized. Prothorax about as long as meso- and metathorax combined. Legs short, 6-segmented, including 2 slender claws. Abdomen of 8 visible segments, urogomphi extremely short (fig. 34.108e).

Spiracles: Annular, present on abdominal segments 1–8, the eighth pair lying together at end of eighth segment and beneath small extension of that tergum. Functional spiracles present on segment 8 only in first and second instars.

Comments: Noterids are cosmopolitan but occur primarily in tropical regions. A few taxa occur widely in temperate regions of the Eastern and Western Hemispheres. The family is small in number of taxa, with 12 described genera and 230 species in the world fauna. Five genera and 13 species have been reported from the United States. Adults and larvae are not known to have any economic importance but Young (1967) pointed out that they are important in recycling nutrients. They also serve as a food item in the diet of various predators in the aquatic ecosystem.

Leech (1956) discussed the biology and immature stages, and Chandler (1956) provided a key to larvae of 2 of the 5 genera known from North America. Spangler and Folkerts (1973b) described the larva of *Suphis inflatus* (LeConte) and included a key to larvae of 3 of the 5 genera found in North America. Bertrand (1972) reviewed and summarized the literature available on noterid biology and immature stages and included a key to larvae of 6 of the 12 genera in the world fauna.

Selected Bibliography

Balfour-Browne and Balfour-Browne 1940.
Bertrand 1972.
Brigham 1982e (biology; key to larvae of *Hydrocanthus, Suphis,* and *Suphisellus*).
Chandler *in* Leech and Chandler 1956.
Crowson 1955.
Leech *in* Leech and Chandler 1956.
Peterson 1951 (description and figure of *Hydrocanthus* larva).
Spangler 1981, 1982, 1986.
Spangler and Folkerts 1973b.
Wesenberg-Lund 1912.
White et al. 1984.
Young 1954, 1967.

DYTISCIDAE (ADEPHAGA)

Paul J. Spangler, *Smithsonian Institution*

Predacious Diving Beetles

Figures 34.110-126

Relationships and Diagnosis: Dytiscid larvae may be distinguished by the following combination of characters. Head prominent, prognathous, exserted; form may be subquadrate, rounded, pyriform, flattened, or subcylindrical; anterior margin moderately to strongly produced; ecdysial cleavage lines present. Antenna, 4-segmented (sometimes with as many as 5 additional small accessory segments), elongate, slender. Clypeus and labrum fused. Mandible curved, usually long, slender, and grooved or hollow for sucking. Maxillary palp slender, elongate, 4- to 10-segmented. Labial palp, 2- to 4-segmented. Stemmata, 6 pairs; absent in subterranean forms. Legs 6-segmented, including pretarsus with 2 claws. Abdomen of 8 visible segments; eighth segment may be very elongate; ninth segment reduced; lateral gills rarely present. Mature larvae range in length from 1.5 to 70.0 mm.

Dytiscid larvae may be immediately distinguished from hydrophilid larvae by their 6-segmented legs, including 2 claws on each leg, and lack of a breathing atrium on the apex of the eighth abdominal segment.

Biology and Ecology: Dytiscid beetles are cosmopolitan, and adults and larvae are normally found in the same aquatic situations. They are found in diverse habitats ranging from microhabitats such as potholes in rock outcroppings, water-filled holes of land crabs, hygropetric niches, springs, and artesian wells to sheltered coves along the margins of large lakes. However, the majority of dytiscid taxa are found in the smaller, shallow, weedy habitats such as the margins of ponds, drainage ditches, gravel pits and stock ponds. In more arid regions, dytiscids may be found in virtually any aquatic habitat available such as pools in intermittent streams, stock tanks, irrigation ditches and overflow areas, and saline and mineral springs or pools. Oftentimes in one of these habitats, large numbers of adults and larvae of a single species will be found and the larvae may be recognized easily by association although it is preferable to rear the larva to the adult to confirm the relationship.

The type of ovipositor generally indicates where the female will lay her eggs. Dytiscid ovipositors are of three major types: (1) Those in which the genital valves are short, blunt, weak, dorsoventrally flattened, obviously not adapted for piercing plant tissues, but setose and apparently tactile—dytiscids with this type of ovipositor, such as *Desmopachria,* generally glue their eggs to surfaces of aquatic plants, drop their eggs at random, or insert them in a muddy substrate (done by many of the Hydroporinae); (2) Those in which the ovipositor is similar to the type described above but is very elongate—undoubtedly the elongate ovipositors allow females to insert their eggs wherever a small, deep, crevice or space narrow enough to conceal the eggs from predators or parasites can be found, such as between or under leaves, sticks, bark, stones, etc.; (3) Those with the genital valves well sclerotized and usually serrate or with toothlike margins—dytiscids with this type of ovipositor, such as *Laccophilus* and *Ilybius,* use it to make incisions in plant tissues in each of which they deposit an egg. Others with piercing valves, such as *Graphoderus* and *Hydaticus,* reportedly place a number of eggs in each hole (Wesenberg-Lund, 1912).

Dytiscids have 3 larval instars. As far as is known, all dytiscid larvae, unlike the Noteridae, leave their aquatic habitat and pupate on land in earthern cells in friable soil where the larvae can burrow beneath the surface a short distance. The pupal chamber is shaped by wriggling movements of the larva which compress the soil so the wall of the chamber will not collapse easily. Some larvae pupate among matted roots of mosses or other plants or under rocks, boards, leaves, logs, etc. Freshly eclosed dytiscid beetles usually remain in their pupal chambers for 3 to 7 days (sometimes much longer) before they leave and enter the aquatic habitat.

Description: Body variously shaped—usually elongate, fusiform, cylindrical, subcylindrical, moderately flattened, usually widest at metathorax or middle of abdomen. Integument white, yellow, greenish, or brown; sclerotized areas becoming yellowish, brown, or reddish brown with aging; may be spotted, striped, or unicolored.

Head: Prognathous; usually exserted; subquadrate, rounded, pyriform, flattened, or subcylindrical; ecdysial cleavage lines usually present and obvious. Clypeus and labrum fused; anterior margin moderately arcuate or moderately to strongly dentiform or prolonged as a nasale. Stemmata 6 pairs or absent (in stygobiontic taxa). Antenna of 4 principal segments, slender, elongate, sometimes with accessory segments. Maxilla slender; cardo small; stipes short and broad or narrow and elongate; palp slender, elongate, 3 or 4 principal segments, sometimes with accessory segments. Labial palp usually 2-segmented; sometimes 3- or 4-segmented. Mandible distinct, curved, usually slender and usually grooved or hollow for sucking blood of hosts.

Thorax and Abdomen: Prothorax usually longer than meso- and metathorax combined. Thoracic terga usually well developed. Legs usually long and slender; some genera with fringes of swimming hairs on femur, tibia, and tarsus; legs 6-segmented including pretarsus consisting of 2 claws.

Abdomen of 8 visible segments; eighth segment may be very elongate; ninth segment reduced; lateral gills rarely present; 8 or 9 pairs of spiracles. Urogomphi usually long and slender but sometimes short and stubby.

Spiracles: Annular, on abdominal segments 1–8 and usually mesothorax in last instar. Early instars with functional spiracles present only on segment 8.

Comments: This family is cosmopolitan, but a greater number of genera and species occur in the tropical regions. There are about 145 genera and 3,000 species of dytiscids known in the world fauna, with 42 genera and about 446 species of Dytiscidae described from America north of Mexico. Larvae of representatives of 28 of these genera have been described.

Most dytiscid larvae are not known to be economically important, but when larvae of the larger dytiscids, such as *Cybister,* become numerous in fish hatcheries, they are destructive to fingerlings (Wilson, 1923). However, most dytiscid larvae probably are beneficial because they feed on larvae of mosquitoes, ceratopogonids, and other noxious insects, as well as other aquatic organisms. Dytiscid larvae are also preyed upon by other insects, birds, and various mammals, and, in this manner, they play an important part in the aquatic food web.

Because of the large number of species, their cosmopolitan distribution, attractive size, color, form, and the accessibility to living specimens, the bionomics of the adult and immature stages of the Dytiscidae are reasonably well known. The most comprehensive studies of dytiscid larvae are those by Bertrand (1928, 1972, 1977). The publications on larval Coleoptera by Böving and Craighead (1931) and Peterson (1951), on dytiscid biologies by Wesenberg-Lund (1912),

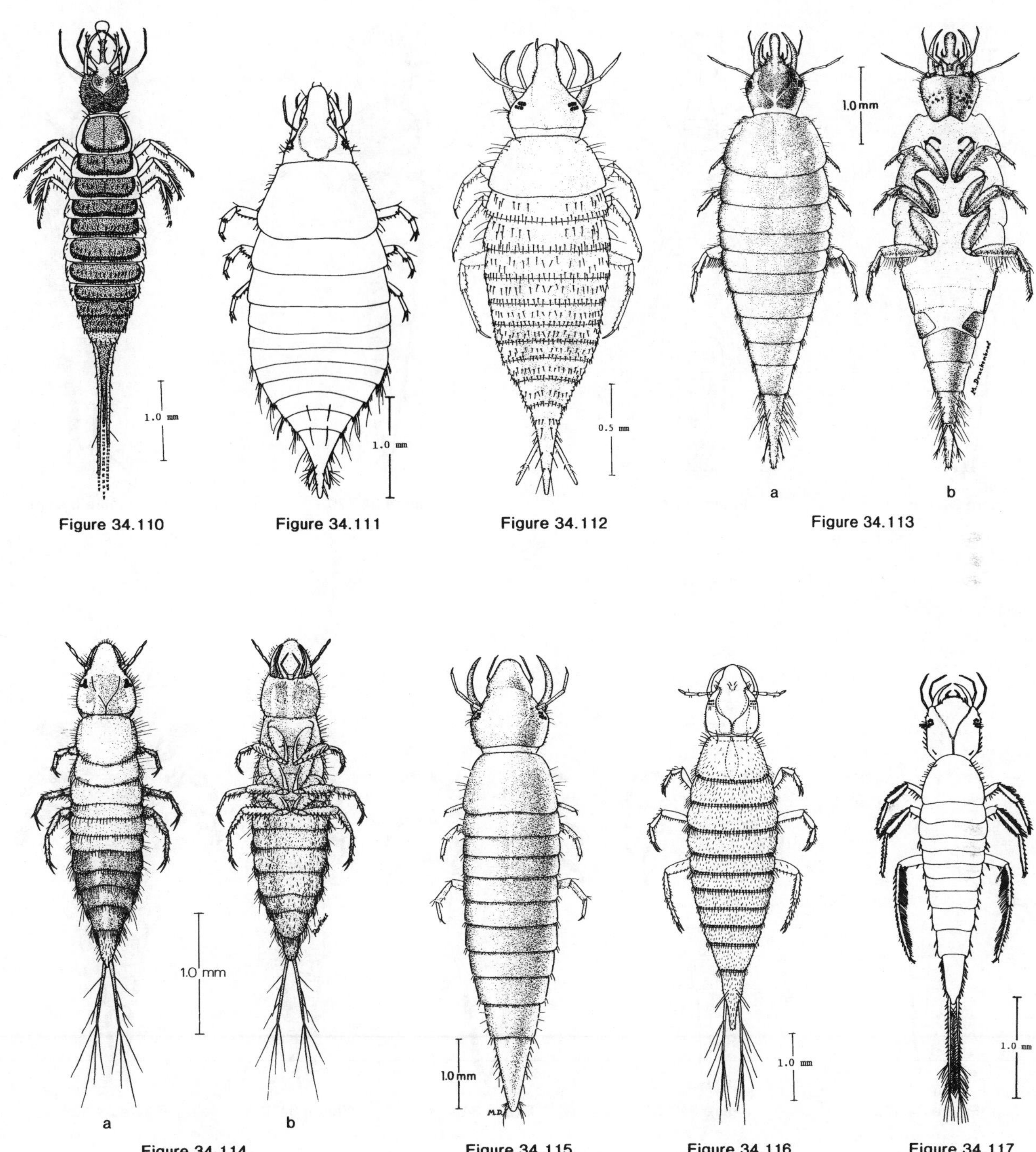

Figure 34.110 Figure 34.111 Figure 34.112 Figure 34.113

Figure 34.114 Figure 34.115 Figure 34.116 Figure 34.117

Figure 34.110. Dytiscidae. *Derovatellus ibarrai* Spangler.

Figure 34.111. Dytiscidae. *Hydrovatus* sp.

Figure 34.112. Dytiscidae. *Desmopachria* sp.

Figures 34.113a,b. Dytiscidae. *Pachydrus princeps* (Blatchley). **a.** dorsal; **b.** ventral.

Figures 34.114a,b. Dytiscidae. *Neoclypeodytes cinctellus* (LeConte). **a.** dorsal; **b.** ventral. [Figures 34.114 a, b. after Perkins, 1981.]

Figure 34.115. Dytiscidae. *Laccornis difformis* (LeConte).

Figure 34.116. Dytiscidae. *Hygrotus sayi* (J. Balfour-Browne).

Figure 34.117. Dytiscidae. *Laccophilus* sp.

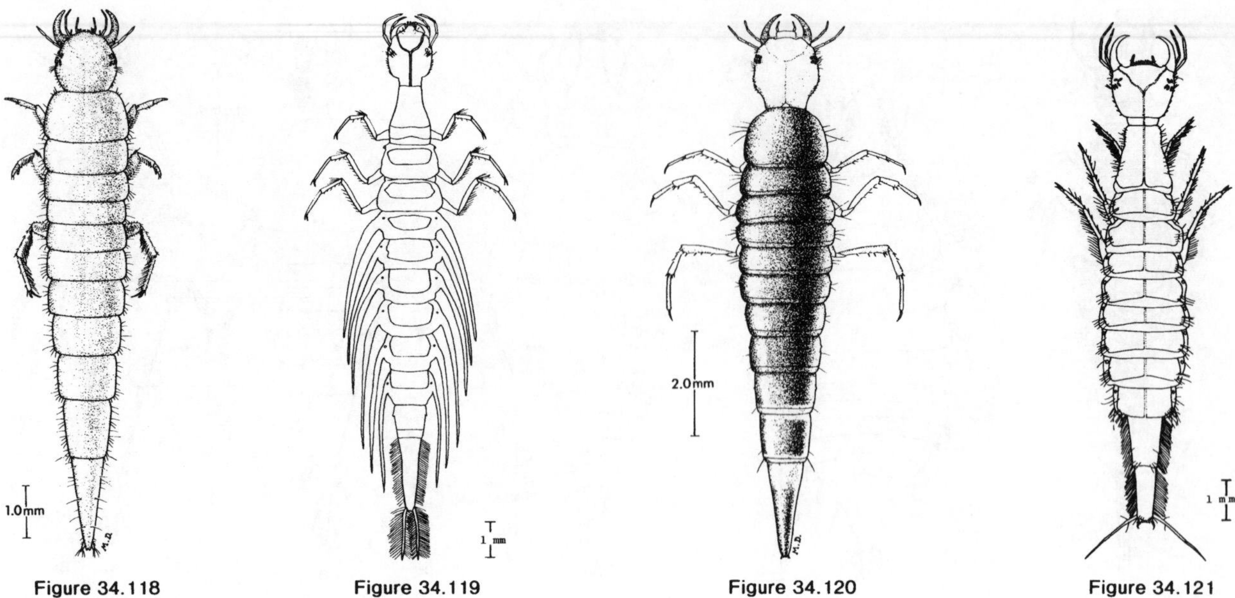

Figure 34.118 Figure 34.119 Figure 34.120 Figure 34.121

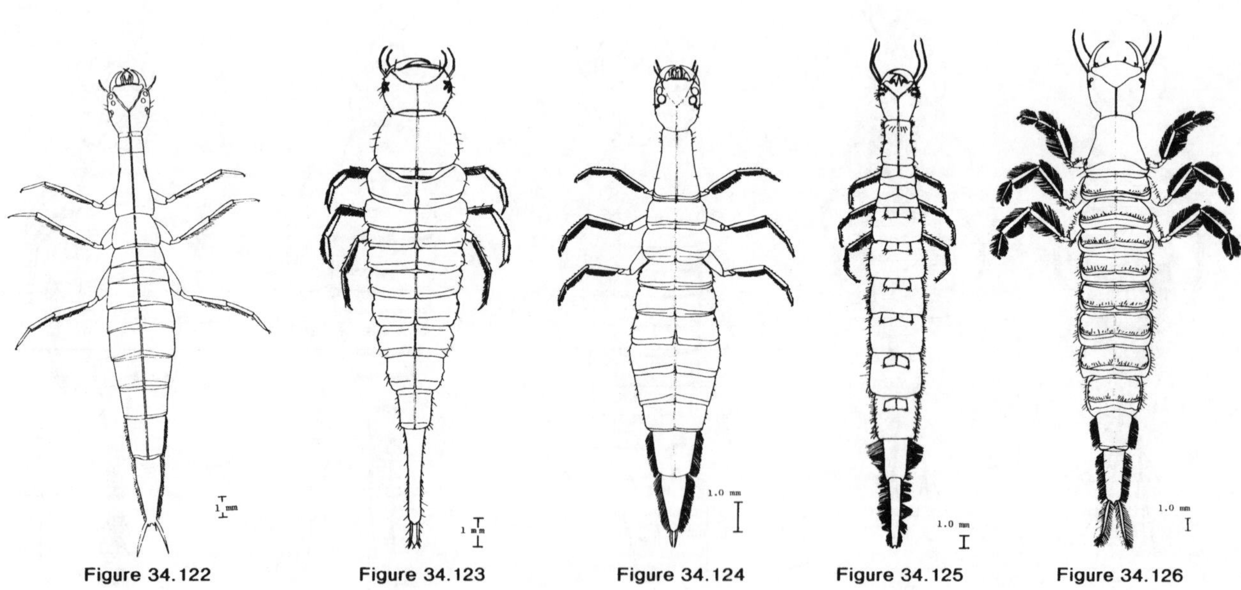

Figure 34.122 Figure 34.123 Figure 34.124 Figure 34.125 Figure 34.126

Figure 34.118. Dytiscidae. *Matus bicarinatus* (Say).

Figure 34.119. Dytiscidae. *Coptotomus* sp.

Figure 34.120. Dytiscidae. *Agabetes acuductus* (Harris).

Figure 34.121. Dytiscidae. *Hydaticus bimarginatus* (Say).

Figure 34.122. Dytiscidae. *Acilius* sp.

Figure 34.123. Dytiscidae. *Rhantus calidus* (Fabricius).

Figure 34.124. Dytiscidae. *Thermonectus basillaris* (Harris).

Figure 34.125. Dytiscidae. *Cybister fimbriolatus* (Say).

Figure 34.126. Dytiscidae. *Dytiscus fasciventris* (Say).

Balduf (1935), and Leech (1956), and the keys to genera by Chandler (1956), Leech and Sanderson (1959), Pennak (1978), Brigham (1982a), and White, Brigham, and Doyen (1984) will be helpful in identifying larvae.

Selected Bibliography

Balduf 1935.
Barman 1973.
Bertrand 1928, 1972, 1977.
Böving and Craighead 1931.
Brigham 1982a.
Chandler *in* Leech and Chandler 1956.
DeMarzo 1979.
Galewski 1971.
Hilsenhoff 1974.
James 1969.
Leech *in* Leech and Chandler 1956.
Leech and Sanderson 1959.
Longley and Spangler 1977.
Matta 1983, 1986.
Matta and Peterson 1985, 1987.
Pennak 1978.
Perkins 1981b.
Peterson 1951.
Spangler 1962a, 1962b, 1966b, 1973, 1974a, 1981, 1982, 1986.
Spangler and Folkerts 1973a.
Spangler and Gillespie 1973.
Spangler and Gordon 1973.
Watts 1970.
Wesenberg-Lund 1912.
White et al. 1984.
Wilson 1923.
Wolfe 1980.
Wolfe and Roughley 1985.
Young 1954.

GYRINIDAE (ADEPHAGA)

Paul J. Spangler, *Smithsonian Institution*

Whirligig Beetles

Figures 34.127–132

Relationships and Diagnosis: Gyrinids are highly adapted to their aquatic habitats. They appear to have been derived from caraboid stock because their larvae have 2 movable claws on each leg, a character unknown outside of the Adephaga.

Gyrinid larvae may be distinguished easily from other aquatic beetle larvae by the following combination of characters: lateral gills on abdominal segments 1–9; 2 claws on each leg; 10 abdominal segments; 4 decurved hooks on tenth abdominal segment (fig. 34.132).

Because of their elongate form, lateral gills, hooked anal feet, and creamy white color, gyrinid larvae resemble freshly molted larvae of some megalopteran genera and the dytiscid genus *Coptotomus*. However, gyrinid larvae may be easily distinguished from the megalopteran and dytiscid genera by their 4 hooked anal feet arising from the tenth abdominal segment. In contrast, megalopteran larvae have 2 hooked anal prolegs or a single long median process arising from the ninth abdominal segment. In addition, gyrinid larvae have lateral gills on abdominal segments 1–9 and larvae of *Coptotomus* have lateral gills only on segments 1–6.

Biology and Ecology: Gyrinid adults are highly adapted for life in the aquatic environment and are the only family of beetles that normally use the surface film for support. They are the fastest and most efficient swimmers of all the aquatic beetles and are equally at home underwater. They are found on the surface of clean lotic and lentic habitats where they may occur singly or in large aggregations composed of 1 to 8 species. Larvae occur in the same habitats as the adults but they live submerged until they leave the water to pupate. Both adults and larvae normally occur in shallow weedy margins of their lentic or lotic habitats.

Gyrinid females in temperate areas begin laying eggs in spring. Eggs are laid in rows or masses on floating or submerged plant stems and leaves. Larvae undergo 3 molts and last-instar larvae leave the water to pupate. Larvae of *Gyrinus* climb emergent vegetation or crawl onto shore to pupate; at this time they gather debris (from the emergent plant stems or the shore), mix the debris with an adhesive substance, crawl to a suitable place, attach the debris to plant stems or other surfaces, dig into the mass, and form a closed chamber by wriggling movements of the body. Larvae that crawl onto shore may make their cocoons from sand or mud and attach them to standing objects or beneath rocks, boards, etc. Pupal cocoons of *Gyrinus* often are common near the top or on top of plants such as *Eleocharis, Typha, Scirpus,* and other emergent plants. Cocoons of *Dineutus* usually are found on shore, near the water, and under objects such as boards or rocks. There are several genera of Hymenoptera that parasitize gyrinid pupae and parasitized pupae may be found quite frequently.

Larval gyrinids, like the adults, are voracious predators; however, adults feed primarily on the surface of their habitats and larvae feed underwater where they seek out soft-bodied larvae of chironomids, tubificids, odonate larvae, etc.; when confined they also feed on their siblings.

Description: Mature larvae, 6.0 to 25.0 mm in length. Body form elongate, narrow, depressed, with lateral gills. Integument white with sclerotized parts yellow to yellow brown and sometimes with dark brown, grey, or black spots on the head capsule.

Head: Prognathous, exserted; subquadrate, depressed; ecdysial cleavage lines present. Anterior margin of nasale truncate or lobed (figs. 34.129–131). Stemmata, 6 pairs. Antenna 4-segmented, slender, elongate. Maxillae slender, elongate; cardo and stipes large; stipes quadrangular, with a galea and lacinia; maxillary palp slender, elongate, 4-segmented. Labial palp elongate, slender, 3-segmented.

Thorax and Abdomen: Prothoracic terga with 2 moderately large sclerites; meso- and metathoracic terga membranous, without sclerites. Legs 6-segmented, including pretarsus with 2 claws.

Abdomen of 10 segments; with a pair of lateral gills on segments 1–8; ninth with 2 pairs of gills; tenth without gills, with 4 stout decurved hooks; some gills may be naked but most have hairlike fringes.

Spiracles: Absent in first and second instars, present on abdominal segments 1–3 in last instar.

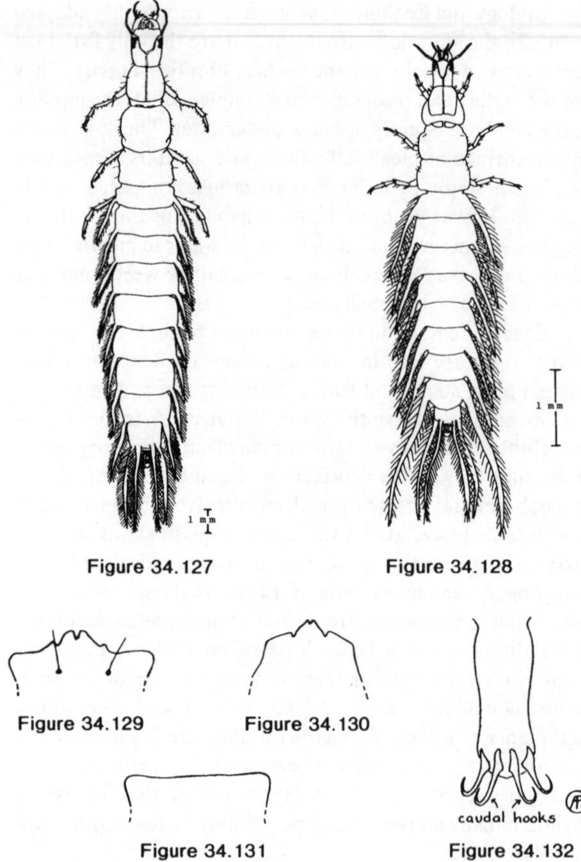

Figure 34.127. Gyrinidae. *Dineutus* sp.

Figure 34.128. Gyrinidae. *Gyrinus* sp.

Figures 34.129–34.131. Gyrinidae. Nasales, dorsal. **(34.129.)** *Dineutus* sp.; **(34.130.)** *Gyrinus* sp.; **(34.131.)** *Gyretes* sp.

Figure 34.132. Gyrinidae. Caudal hooks, *Dineutes* sp. (from Peterson, 1951).

Comments: The family Gyrinidae is cosmopolitan; however, it is richest in taxa pantropically although numerous taxa occur in the temperate regions of the world. The family includes 11 genera and about 1,100 species in the world fauna and is represented in America north of Mexico by 52 species in 4 genera. In North America, the larvae of *Gyretes* and *Spanglerogyrus* remain undescribed; however, *Gyretes* is included in keys to larvae by Chandler (1956), Sanderson (1982), and White, Brigham, and Doyen (1984). Gyrinid beetles are not known to be economically important, but they serve as food for other organisms, and their predatory habits assist in recycling nutrients in aquatic ecosystems.

Very few studies have been published on the biology and immature stages of North American Gyrinidae. However, those that are available are thorough and informative. The most useful of these publications is Wilson's (1923) descriptions of the life cycles and immature stages of *Dineutus* and *Gyrinus*. The publications by Butcher (1930) on the construction of the pupal cocoon and parasitoids of pupae (1933) are also very interesting. In addition, the review of the family by Leech (1956) nicely summarizes the information available for the family. Sanderson (1982) provides a discussion of gyrinid bionomics and includes a key to the larvae of the genera *Dineutus, Gyretes,* and *Gyrinus*. White, Brigham, and Doyen (1984) also include a key to three of the four gyrinid genera of North America. For a summary of information on immature stages of gyrinids of the world *see* Bertrand (1972).

Selected Bibliography

Bertrand 1972.
Böving and Craighead 1931 (key to genera, including *Dineutus, Gyretes,* and *Gyrinus*).
Butcher 1930, 1933.
Carthy and Goodman 1964.
Chandler *in* Leech and Chandler 1956 (key to *Dineutus, Gyretes* and *Gyrinus*).
Folkerts 1979 (new subfamily; adult key to subfamilies, tribes and genera of U.S.).
Hatch 1927.
Leech *in* Leech and Chandler 1956.
Leech and Sanderson 1959.
Sanderson 1982.
Steiner and Anderson 1981.
White et al. 1984.
Wickham 1893, 1894.
Wilson 1923 (larva, pupa and life cycle of *Dineutus* and *Gyrinus*).

SUBORDER POLYPHAGA

HYDRAENIDAE (STAPHYLINOIDEA)

(= LIMNEBIIDAE)

Paul J. Spangler, *Smithsonian Institution*

Hydraenid Beetles

Figures 34.133a,b–135a,b

Relationships and Diagnosis: The hydraenid beetles were originally placed in the family Hydrophilidae and the superfamily Hydrophiloidea. However, as a result of a study of the comparative morphology of Coleoptera larvae, Böving and Craighead (1931) placed the family in the superfamily Staphylinoidea. Coleopterists have different opinions on the proper assignment of the Hydraenidae. Crowson (1955) states that the hydraenid "relationship to the hydrophiloids is indicated by the palpicorn type of antennae and general aquatic adaptations of the adults while the mouthparts are quite primitive and without any indication of the predacious specialisations." Dybas (1976) disagrees with this placement and comments as follows: "Though there has been lack of agreement as to the systematic position of the family Hydraenidae, I regard it as clearly belonging in the Staphylinoidea because of the characters of the larva (particularly the maxilla of *Hydraena*) and because of the close resemblance in numerous features of the dorsum of the abdomen of the adult to that of the generalized ptiliid *Nossidium* (unpublished data)." In the latest discussion of the family assignment, Perkins (1981a) reports the following: "Based on adult antennal form, aquatic habits, and metendosternite, relationships appear to be with

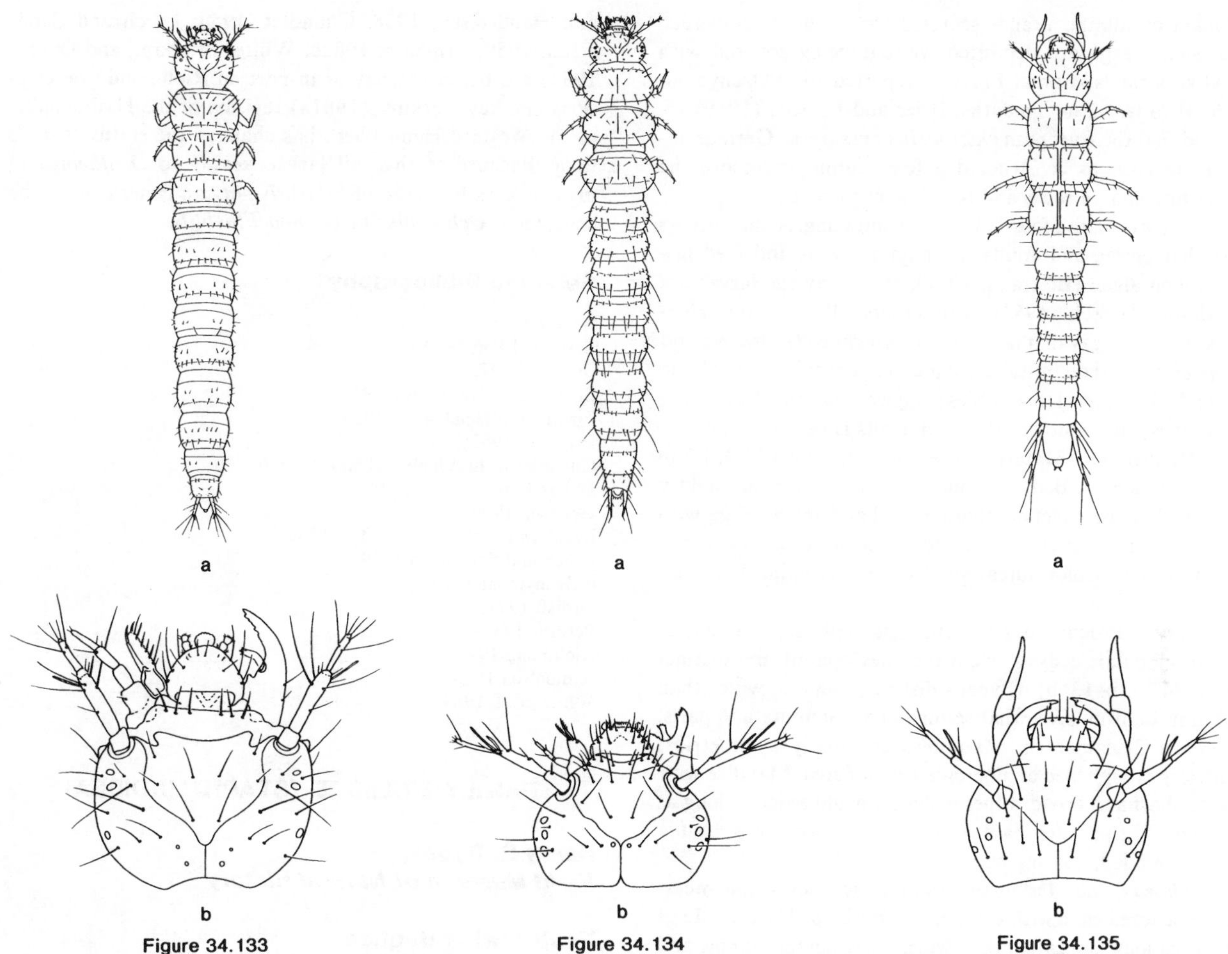

Figure 34.133

Figure 34.134

Figure 34.135

Figures 34.133a,b. Hydraenidae. *Ochthebius impressus* Marsham. **a.** larva; **b.** head.

Figures 34.134a,b. Hydraenidae. *Limnebius* (?)*papposus* Mulsant. **a.** larva; **b.** head.

Figures 34.135a,b. Hydraenidae. *Hydraena pennsylvanica* Kiesenwetter. **a.** larva; **b.** head.

(Figures 34.133a, b and 34.134a, b redrawn from Böving and Craighead, 1931. Figures 135a, b redrawn from Richmond, 1920.)

the Hydrophiloidea. The wings, however, are similar to Staphylinoidea, and larvae markedly resemble those of the Ptiliidae." Perkins concludes that "It is evident that phylogenetic relationships of the Hydraenidae remain equivocal, a fertile area for further research."

The tiny larvae (1.3 to 3.0 mm) superficially resemble those of various smaller Staphylinoidea, such as Ptiliidae, Leiodidae, and Staphylinidae. Hydraenid larvae are distinguished from those of any staphylinid by the presence of a distinct roughened or tuberculate mandibular mola and a pair of recurved hooks on the last abdominal segment. The lack of hooks on segment 10 also distinguishes larvae of Leiodidae, Leptinidae, and Limulodidae from those of hydraenids. Ptiliid larvae have a mandibular mola and recurved hooks at the abdominal apex, but they differ from larvae of Hydraenidae in having urogomphi with only 1 segment (or occasionally no urogomphi) and in lacking stemmata (rarely with 1 pair).

Biology and Ecology: Most hydraenid adults are aquatic and inhabit the margins of cascades, rills, and hygropetric habitats where they occur in leaf packs or matted roots; others occur in splash zones of cascades and waterfalls on moss-covered rocks or other damp or wet plant materials. Adults are also found in sandy margins of streams, potholes, rock outcrops beside streams, in marshy margins of ponds, holes and woodland ponds; others occur on rocky coastlines that are periodically submerged by rising tides. Although most hydraenids are found in fresh water habitats, a few may occur abundantly in brackish to very saline coastal or inland waters. The exotic genus *Meropathus,* lives on moss-covered rocks alongside streams and in grasses and offal around nests of seabirds in their rookeries. All of the larvae known from N. America are terrestrial, but they are usually found along the damp margins of the aquatic habitats of the adults.

Those hydraenids whose egg-laying habits are known deposit their eggs in moist sites out of water, often on leaves, rocks, or algae adjacent to the adult habitat. Eggs are deposited singly and usually secured with a sparse covering or blanket of silklike strands secreted from caudal spinnerets; occasional eggs are deposited without being secured with silklike strands. Larval life was reported by d'Orchymont (1913) to last 2 or 3 months. Beier and Pomeisl (1959) reported that the European *Ochthebius exsculptus* Germar pupates in cocoons constructed a few centimeters above the waterline and on rocks and boulders in torrents.

Reports on the food habits are conflicting. Some authors say that larvae and adults are phytophagous and feed primarily on algae (Bertrand, 1972); others say the larvae are predators (Leech, 1958); still others (Böving and Hendriksen, 1938) say that larvae of "*Ochthebius, Hydraena,* and *Limnebius* feed on infusoria, spores and decaying particles in water." The larval mandibles suggest that the larvae are predators but a study of the food habits is needed.

Description: Mature larvae (figs. 34.133a–135a), 1.3 to 3.0 mm in length. Body form elongate, subcylindrical, slightly flattened in cross section; thoracic and abdominal terga with large sclerites; cuticle with numerous setae; annuliform spiracles present. Color: integument white, becoming light yellowish with age.

Head: Prognathous to semihypognathous; rounded or ovoid, globular; ecdysial cleavage lines present and distinct (figs. 34.133b–135b). Clypeus distinct, arcuate, wider than labrum. Labrum somewhat semicircular. Stemmata, 5 pairs. Antenna 3-segmented. Mandibles essentially symmetrical with large roughened or tuberculate molar area. Maxillae with cardo short and broad; stipes wider than and twice as long as segment-like palpifer; palp closely united with stipes. Ligula short, rounded, bearing papillae.

Thorax and Abdomen: Prothoracic, meso- and metathoracic terga each with a large, well-developed sclerite. Legs about as long as prothorax is wide, 5-segmented; tarsus and pretarsus fused, forming a single claw-like tarsungulus.

Abdomen of 10 segments; 1–8 covered with broad sclerite; 9 bearing a pair of movable, 2-segmented urogomphi; 10 bearing a pair of stout, recurved hooks.

Spiracles: Annular, on mesothorax and abdominal segments 1–8.

Comments: Hydraenids are cosmopolitan with about 900 species known from the world. At present there are 90 species in 5 genera known from America north of Mexico. They are not reported to be economically important, but they often are very abundant in microhabitats where they must be efficient in recycling food items upon which they feed.

Although all hydraenid larvae known from the United States and Canada are terrestrial, some Australian larvae apparently are aquatic. These larvae have a pair of large well-developed prothoracic spiracles borne on dorsal tubercles that appear to function as a snorkel; this suggests that these larvae may live under a thin film of water such as is found in hygropetric habitats. Some of the exotic hydraenid larvae reportedly lack the pair of hooks found on the tenth abdominal segment of the described N. American larvae.

References to the immature stages of hydraenids are few. The most thorough descriptions of North American hydraenid larvae are found in Richmond's (1920) treatise. All subsequent keys to genera of larvae (Bertrand, 1972; Böving and Hendriksen, 1938; Chandler, 1956; Leech and Sanderson, 1959; Brigham, 1982c; White, Brigham, and Doyen, 1984) are based entirely or in part on Richmond's descriptions and key. Perkins' (1981a) revision of the Hydraenidae of the Western Hemisphere has changed the status of some taxa. Because of this, all larvae keying to *Ochthebius* in present keys to larvae of North American genera should be considered *Ochthebius* or *Gymnochthebius*.

Selected Bibliography

Arnett 1968.
Beier and Pomeisl 1959.
Bertrand 1972.
Böving and Craighead 1931.
Böving and Hendriksen 1938.
Brigham 1982c.
Chandler *in* Leech and Chandler 1956.
Dybas 1976.
Hrbácek 1950.
Leech *in* Leech and Chandler 1956.
Leech and Sanderson 1959.
d'Orchymont 1913.
Ordish 1971.
Perkins 1981a.
Richmond 1920.
Samuelson 1964.
White et al. 1984.

FAMILY PTILIIDAE (STAPHYLINOIDEA)

Henry S. Dybas,*
Field Museum of Natural History

Featherwing Beetles

Figures 34.136–139

Relationships and Diagnosis: The family Ptiliidae, which contains the smallest known beetles, is distributed worldwide in both temperate and tropical regions. Larvae are most similar to those of the Limulodidae, but the only known species of the latter lacks anal hooks. The families also show numerous similarities to the Hydraenidae, Leptinidae, and Leiodidae, particularly in the possession (in some members of each family, at least) of a unique and presumably derived structure—the fringed galea of the maxilla (fig. 34.136b). The Hydraenidae, like the Ptiliidae, possess 1 pair of anal hooks. This grouping of the 5 families within the Staphylinoidea essentially corresponds to the "leptinid association" of families recognized by Böving and Craighead (1931) on the basis of other larval characters.

Ptiliidae larvae can be distinguished from larvae of related families of Staphylinoidea by the following characters: unusually small size (ca. 1.0–2.0 mm long), linear form, unpigmented body, lack of eyespots, usually 1-segmented urogomphi, and a pair of anal hooks. Exceptions and additional characters are detailed below.

Biology and Ecology: Adults and larvae occur chiefly in moist, decaying, organic matter in habitats such as the forest floor, tree holes, decaying logs, compost heaps, animal dung, under bark of dead trees, rubbish heaps of ants (e.g., *Atta*

*Deceased

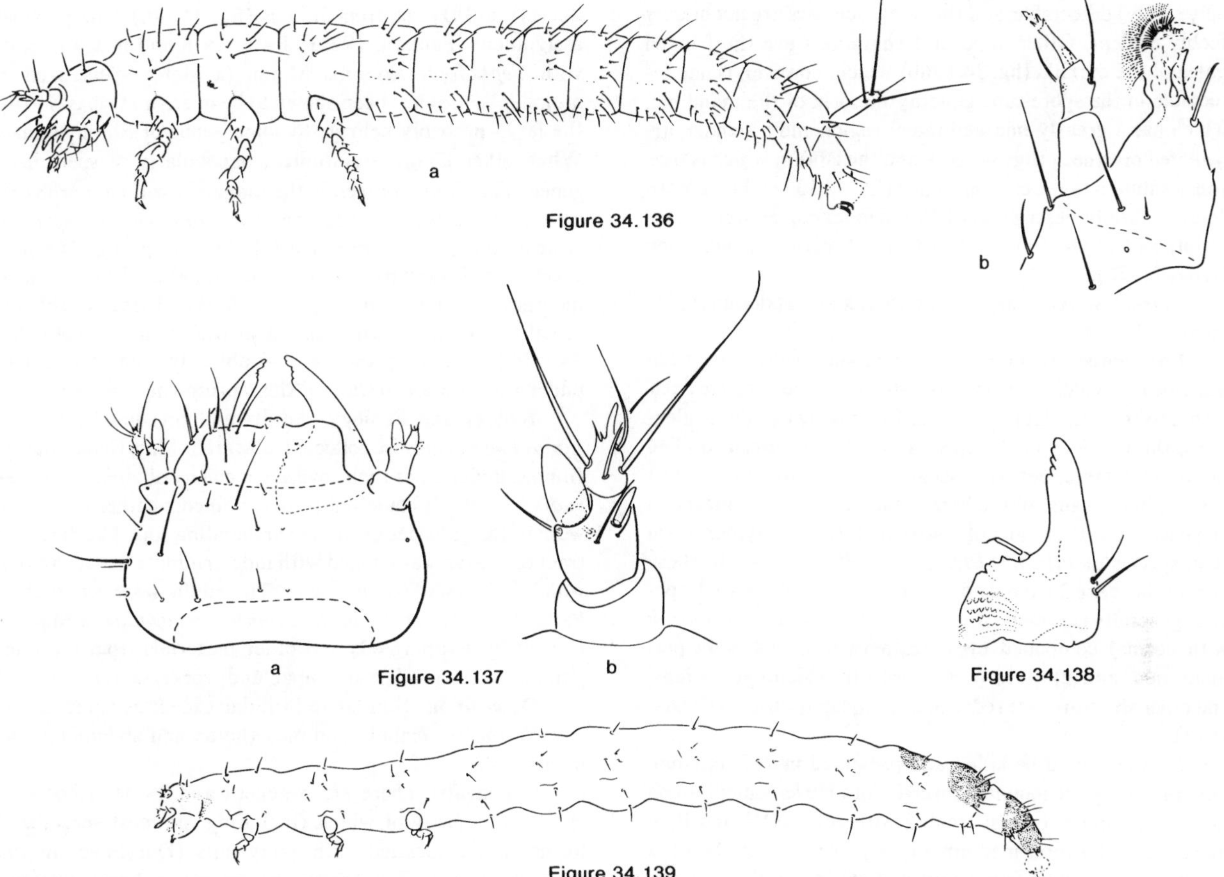

Figures 34.136a,b. Ptiliidae. *Pteryx* sp. **a.** late instar larva, lateral, setal pattern approximate; **b.** maxilla, ventral, showing lacinia and fringed galea united on mala.

Figures 34.137a,b. Ptiliidae. *Nossidium americanum* Motschulsky. **a.** head capsule, dorsal, maxillae not shown; **b.** left antenna, ventral.

Figure 34.138. Ptiliidae. *Nossidium,* left mandible, ventral.

Figure 34.139. Ptiliidae. *Throscoptilium duryi* Barber, lateral, showing yellowish pigmented areas at apex of abdomen (note absence of urogomphi on segment 9).

spp., leaf-cutting ants), decaying piles of seaweed, and similar moist decaying organic materials suitable for the growth of molds and other fungi, where they are frequently the most abundant beetles. These habitats are best sampled by means of the Berlese or Tullgren funnel. Some species of *Actidium,* and a related undescribed genus in the United States, occur on bare sand and gravel bars and flats along water-courses, and can be collected by flotation techniques. One group of ptiliids, *Nanosella* and allies, is found on the underside of shelf fungi (Polyporaceae) where larvae and adults live in the spore tubes and feed on growing spores. This group includes the smallest of all beetles (as small as 0.30 mm long). When the fungus is disturbed, adults and larvae leave the spore tubes and cross the under surface for a short distance before entering other tubes. They can be picked off the under surface with a wetted forceps point.

The main food of both adult and larval Ptiliidae appears to be spores and hyphae of fungi and, probably, soft organic materials containing microorganisms. Under favorable conditions, ptiliids appear to reproduce continuously, as evidenced by finding larvae and teneral and fully hardened adults together at different times of the year. Only a single egg, usually about 1/3–1/2 the length of the female, is matured at a time. Development seems to be fast, 32–45 days from egg to adult at 20° C in 3 species of *Ptinella* (Taylor, 1975). The number of instars in *Ptinella* is 3 (Taylor, 1975). The pupal stage has been adequately described only for *Acrotrichis fascicularis* (Hinton, 1941a).

Description: Ptiliidae larvae are linear, about 1.0–2.0 mm. long (figs. 34.136a, 34.139). Body usually entirely white (unpigmented) or with a yellowish tinge on the more sclerotized regions (mandibles, head capsule), but species of *Actidium* from the exposed riparian habitats are dark. Pigmented eyespots lacking, except in a new species of the generalized genus *Nossidium* (*s.l.*) from Panama. The 10-segmented abdomen possesses a membranous anal vesicle which is furnished with a pair of anal claws or hooks (fig. 34.136a); the ninth abdominal segment has a pair of articulated, 1-segmented urogomphi (fig. 34.136a) that are lacking in species (*Nanosella* and allies) that live in spore tubes of shelf fungi (fig. 34.139). Except in the darkly pigmented species of *Actidium,* the epicranial lines of the head capsule are not (or only rarely) detectable, and the tergal sclerites are not or only feebly evident. Other important characters are the fringed galea of the maxilla (fig. 34.136b) which, however, is not detectable in the spore-tube genera; the form of the mandible, which has a greatly enlarged molar region and a slender, articulated prostheca (fig. 34.138); and the labrum which is free and not united to the clypeal region (fig. 34.137a). For a more detailed family description of Ptiliidae larvae, as well as descriptions and illustrations of 9 North American genera, see Dybas (1976).

Spiracles: Annular, on mesothorax and abdominal segments 1–8.

Comments: About 62 genera and 400 species have been described, of which 23 genera and about 115 species have been recorded from the United States. Judging by existing collections, the majority of the species are still undescribed. The family is notable, not only because it contains the smallest beetles, but because of the high incidence of parthenogenesis (Dybas, 1966, 1978) and of a striking form of polymorphism (e.g. species of *Ptinella, Pteryx,* and *Ptinellodes*). In these genera there are 2 strongly differentiated morphs in each species, generally represented in both sexes: 1) a *normal morph* with normal compound eyes, featherwings, and body pigmentation, and 2) a *vestigial morph* in which eyes, wings, and other structures are reduced or completely absent (Dybas, 1978).

Larvae should be killed and preserved in 70% ethanol. For study they are mounted directly into Hoyer's medium on microscope slides, or first treated with cold KOH and then mounted in Hoyer's medium or in glycerine gel. Hoyer's medium (not a good permanent mount) is soluble in water, and larvae can be soaked off and remounted in different positions. Larvae in glycerine gel can be repositioned by placing the slide on a slide-warming table for a few moments to soften the gel, and then manipulating the specimens with a fine needle. Ptiliidae larvae are best studied with a compound microscope at magnifications of 400× or more, using phase contrast optics (which eliminates the need for staining).

Selected Bibliography

Böving and Craighead 1931.
Dybas 1966, 1976, 1978.
Hinton 1941a.
Taylor 1975.

FAMILY LIMULODIDAE (STAPHYLINOIDEA) (= CEPHALOPLECTIDAE)

Henry S. Dybas,*
Field Museum of Natural History

Limulodids

Figures 34.140a–d

Relationships and Diagnosis: The Limulodidae are closely related to the Ptiliidae. At present the larvae are characterized on the basis of 3 larvae of 1 species, *Limulodes parki* Seevers and Dybas, from Illinois (fig. 34.140). The previous assignment (Paulian, 1941) of a larva found in Costa Rica with *Cephaloplectus mus* Mann, (a highly specialized limulodid beetle) has been shown to be in error (Dybas, 1976); the larva probably belongs to some genus of Staphylinidae. When other larvae are studied, particularly of specialized genera like *Cephaloplectus,* the diagnosis will very probably have to be revised. The larvae of *L. parki* closely resemble typical larvae of the Ptiliidae (cf. *Pteryx* sp., fig. 34.136a) except for the apparent loss of the 2 anal hooks in the anal membrane of abdominal segment 10, the absence of the terminal tuft of the third segment of the maxillary palp (fig. 34.140b), and the presence of 4 obtusely-pointed denticles under the anterior margin of the labrum (fig. 34.140c).

Biology and Ecology: Adults and presumed larvae of *Limulodes parki* were collected in March–May, Cook County, Illinois, under a flat rock covering a colony of *Aphaenogaster rudis s.l.* ants. It could not be determined whether the larvae were in the galleries or in the surrounding soil. The digestive tract of 1 larva was packed with indeterminate matter among which were small soil particles. The adults were reported by Park (1933, under the name *Limulodes paradoxus* Matth.) to feed by scraping oils and other materials from the integument of the ant larvae, pupae and workers.

Description: Similar to Ptiliidae except as noted above.

Spiracles: Annular, on mesothorax and abdominal segments 1–8.

Comments: There are 5 genera and 28 described species, the majority of which (including the most specialized forms) are associated with army ants (Dorylinae) in the American tropics. Four species in 2 genera are known to occur in the United States. All species are found only in association with ants and occur only in the New World and the Australian region. Adults are blind, flightless, compactly built, with a smoothly contoured tear-drop shape reminiscent of *Limulus* (horseshoe crab) from which the family name is derived. A general account of the classification and biology is given in Seevers and Dybas (1943); see also Wilson *et al.,* (1954), Park (1933), and Dybas (1976).

Selected Bibliography

Dybas 1976.
Park 1933.
Paulian 1941.
Seevers and Dybas 1943.
Wilson, Eisner and Valentine 1954.

AGYRTIDAE (STAPHYLINOIDEA)

Alfred F. Newton, Jr.,
Field Museum of Natural History

Figures 34.141–145

Relationships and Diagnosis: The family Agyrtidae has usually been considered a subfamily of Silphidae. Based on studies soon to be published (Newton, in preparation), agyrtids are placed as a distinct family most closely related to Leiodidae rather than to Silphidae and allied families (Lawrence and Newton 1982, Anderson and Peck 1985).

*Deceased

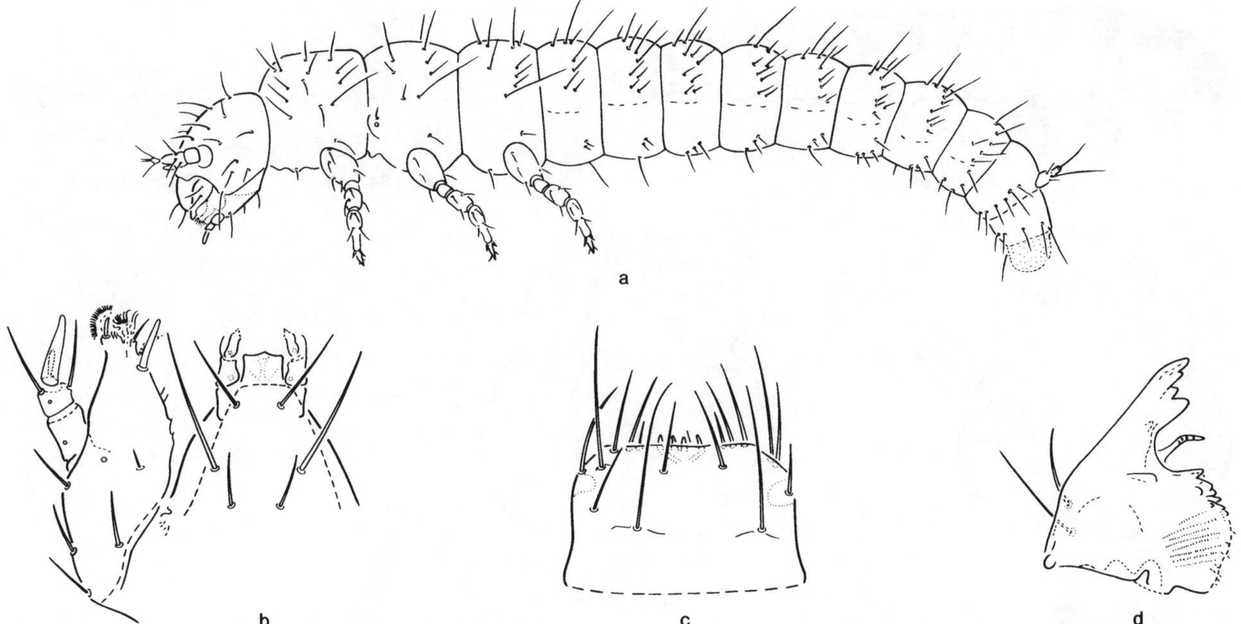

Figures 34.140a-d. Limulodidae. *Limulodes parki* S. & D. **a.** late instar larva, lateral; **b.** right maxilla and labium, ventral; **c.** labrum, dorsal; **d.** right mandible, ventral.

Agyrtid larvae may be distinguished from all other Coleoptera larvae by their possession of a combination of: mandibles with large contiguous molar lobes; 6 stemmata on each side of the head; and articulated, 2-segmented urogomphi. Larvae of Leiodidae, Ptiliidae and Hydraenidae are very similar but have at most 5 stemmata, while larvae of Silphidae and all remaining Staphylinoidea lack molar lobes on the mandibles. In addition, mature larvae of agyrtids are large, about 8 mm or longer, while larvae of related families such as Leiodidae are usually much smaller.

Biology and Ecology: Relatively little is known about the natural history of most agyrtids. *Necrophilus* species are found at small decaying animal carcasses and feces; *N. pettiti* Horn has been reared on mouse feces and decayed squirrel meat (J. A. Payne, unpublished observations) and the European *N. subterraneus* (Dahl) has been reared on decaying snails (Will 1886, Zwick 1981). *Ipelates* species have been found around decaying trees and in forest litter, while *Lyrosoma* species are confined to marine beaches fringing the northern Pacific Ocean. *Pteroloma* and *Apteroloma* species occur under stones and debris in mountainous areas, often above timberline (e.g., Bolívar y Pieltain 1940), and are probably predators as adults (personal observations).

Egg nearly round, yellowish white.

Pupa exarate, with functional abdominal spiracles on segments 1–2.

Description: Mature larvae about 8–20 mm long, elongate and more or less parallel-sided, slightly flattened, straight or slightly curved ventrally. Body surfaces heavily to lightly pigmented and sclerotized, smooth or finely microspinose or granulate, with sparse vestiture of simple setae only or including frayed or bifid setae.

Head: Prognathous or slightly declined, protracted, without differentiated neck. Epicranial stem short, frontal arms V-shaped to lyriform, each arm anteriorly bifurcate; endocarinae absent. Stemmata 6, well separated, on each side of head. Antenna 3-segmented, about 0.6–1.4 times as long as head width, sensorium of preapical segment anterad of apical segment and conical or awl-shaped. Frontoclypeal suture absent. Labrum free, tormae present. Mandibles symmetrical, apex with single tooth, incisor edge with 1 large and several fine subapical teeth; mesal surface of mandibular base with mola bearing numerous fine transverse ridges, distal portion of molar lobe with rounded or tooth-like prostheca and large ventral setose area which extends to mesal edge. Cardines transverse, externally divided, widely separated from each other by submentum. Stipes elongate. Mala long, fixed, divided at apical third to half into fixed galea and lacinia; galea with 2 long setiform sensilla and with apical fringe of 2–5 rows of setae, fringe rarely absent; lacinia falcate, spinose along mesal edge. Maxillary palp 3-segmented. Labium consisting of prementum, mentum and submentum. Ligula shorter or longer than first palpal segment but shorter than palp, apex bilobed. Labial palps 2-segmented, separated by more than width of first palpal segment. Gula transverse. Occipital foramen divided into 2 parts by tentorial bridge.

Thorax and Abdomen: Thoracic terga and abdominal terga and sterna with 1 or more sclerotized plates, without patches or rows of asperities, without lateral tergal processes. Legs long, 5-segmented including bisetose tarsungulus. Abdomen 10-segmented, about 1.5–2 times as long as thorax. Tergum A9 with pair of long 2-segmented urogomphi which may be multiannulate. Segment A10 visible from above, anal region terminally oriented, with membranous anal lobes bearing numerous fine teeth.

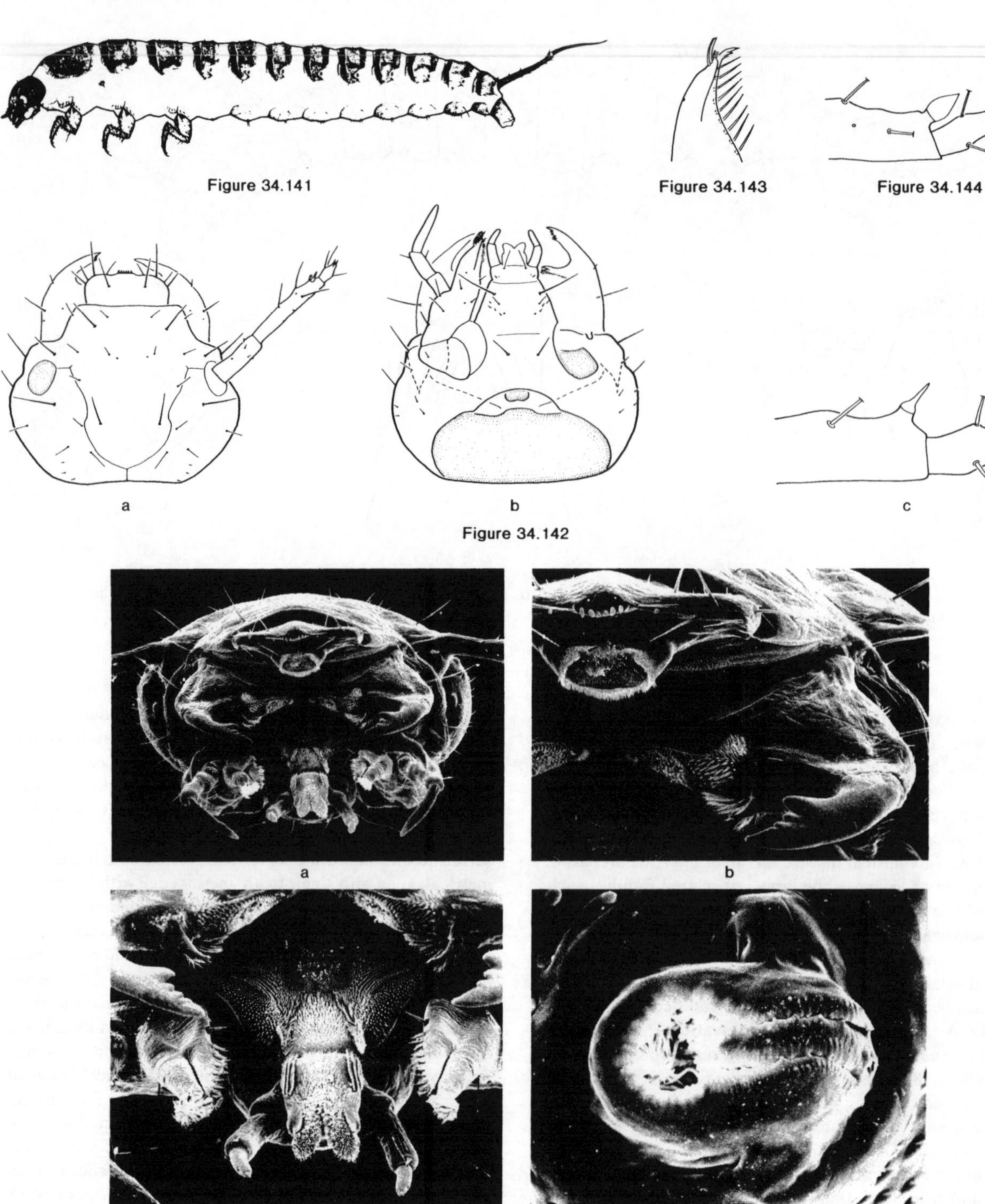

Figure 34.141 Figure 34.143 Figure 34.144

a b c
Figure 34.142

a b c d
Figure 34.145

Figure 34.141. Agyrtidae. *Necrophilus hydrophiloides* Mannerheim. Inverness (1 mile southeast), California; February; on rotten potatoes. Mature larva, lateral. Length = 18 mm.

Figures 34.142a-c. Agyrtidae. *Necrophilus pettiti* Horn. Perry, Georgia; January; reared on feces and carrion. **a.** second instar larva, head, dorsal, left antenna not shown; **b.** head, ventral, left maxilla not shown; **c.** right antenna, dorsal view of apex.

Figure 34.143. Agyrtidae. *?Pteroloma nebrioides* Brown. Cameron Lake, Waterton Park, Alberta; August; moss etc. Right maxilla, ventral view of apex.

Figure 34.144. Agyrtidae. *?Apteroloma tenuicorne* (LeConte). Sagehen Creek, Nevada County, California; October. Right antenna, dorsal view of apex.

Figures 34.145a-d. Agyrtidae. *Necrophilus pettitii.* **a.** head, anterior, mouthparts spread apart; **b.** detail of labrum, left mandible; **c.** detail of labium, maxilla; **d.** abdominal spiracle, dorsal.

Spiracles: Annular, annular-biforous or modified annular with several peripheral chambers to one side; dorsolateral; closing apparatus present.

Comments: The family is a small one of about 8 genera and 60 species. It is nearly confined to the Holarctic region, with the notable exception of *Necrophilus prolongatus* Sharp, found only in New Zealand. Six genera and about 14 species are known from N. America (Anderson and Peck, 1985). The species are seldom encountered without special effort, and none are known to be of any economic importance.

Larvae have not been adequately described. Only 3 species are treated at all in the literature: *Necrophilus subterraneus* (Will 1886, Zwick 1981); *N. prolongatus* (Hudson 1934); and *Apteroloma* sp. (Bolívar y Pieltain 1940). The present treatment is based on larvae of 4 *Necrophilus* species, and on 3 kinds of unreared larvae from western N. America tentatively attributed (based on distribution, habitat and association with adults) to the genera *Ipelates, Pteroloma* and *Apteroloma.*

Selected Bibliography

Anderson and Peck 1985 (classification, adult ecology).
Blaisdell 1901.
Bolívar y Pieltain 1940.
Hudson 1934.
Lawrence and Newton 1982.
Will 1886.
Zwick 1981.

LEIODIDAE (STAPHYLINOIDEA)

(= ANISOTOMIDAE, LIODIDAE, INCLUDING CAMIARIDAE, CATOPIDAE, CHOLEVIDAE, COLONIDAE, LEPTODIRIDAE)

Alfred F. Newton, Jr.,
Field Museum of Natural History

Round Fungus Beetles, Small Carrion Beetles and Allies

Figures 34.146–155

Relationships and Diagnosis: The family is used here in the broad sense to include what are sometimes (e.g., Jeannel 1957) considered to be as many as 4 families: Leiodidae (= Anisotomidae), Camiaridae, Colonidae and Cholevidae (= Catopidae, = Leptodiridae). Leptinidae are very closely related and are included in Leiodidae in some recent concepts of the family (Crowson 1981, Lawrence and Newton 1982). All of these taxa were at one time included in Silphidae and are occasionally still cited under that family, but the only closely related family of Staphylinoidea is Agyrtidae, a recent removal from Silphidae (Lawrence and Newton 1982).

Most leiodid larvae are distinguishable from other beetle larvae by possession of a combination of: 2- or 1-segmented urogomphi; mandibles with molar lobes; 5 or fewer stemmata (in North America 3 or fewer) on each side of the head, rather than 6 stemmata as in otherwise very similar agyrtids; and anal lobes with numerous fine hooks or without hooks rather than with a pair of large hooks as in hydraenids and ptiliids. Those few species that lack well developed molar lobes have the other characters and have the occipital foramen divided into 2 parts by the tentorial bridge, in contrast to Staphylinidae, Silphidae and other families lacking a mola which have an undivided foramen. A few Leiodinae with short fixed urogomphi have the other characters listed above, a dorsally curved abdomen and subterranean feeding habits. Leptinid larvae will fall out with leiodids according to the above diagnosis but differ from leiodids in having either a scoop-like lacinia or mandibles without molar lobes and with ventrally curved apices.

Biology and Ecology: Leiodids are generally, and perhaps primitively, saprophages and scavengers in a variety of habitats including forest litter, dung, carrion, rotting fungi and other decomposing organic matter, and in nests of mammals, birds, ants, termites and stingless bees. Many species, including a majority of the many Palearctic Bathysciini and the isolated North American genus *Glacicavicola,* are obligate cave dwellers. Molds and other fungi may form a part of the diet of larvae and adults of many species, and a number of taxa are obligate mycophages of specific groups of fungi, including Agathidiini and *Neopelatops* on slime molds (Myxomycetes); *Creagrophorus* and *Nargomorphus* on puffballs (Gasteromycetes); Hydnobiini, Leiodini, and possibly Catopocerinae and Coloninae on diverse hypogean fungi; and a few miscellaneous species on Polyporaceae and allied epigean fungi. Larval development is generally rapid, taking as little as 2 days in some *Anisotoma* species feeding on short-lived slime mold fruiting bodies. See Arzone (1970, 1971), Casale (1975), Crowson (1984a), Deleurance-Glaçon (1963), Newton (1984), Wheeler (1979, 1984, 1985) and Zwick (1979).

Egg ovoid, white, smooth.

Pupa exarate, with functional abdominal spiracles on segments 1–2.

Description: Mature larvae about 2–8 mm long, elongate and more or less parallel-sided or (*Neocamiarus*) broadly ovate, slightly to strongly flattened, straight to slightly curved ventrally or (Hydnobiini, Leiodini) curved dorsally. Body surfaces heavily to very lightly pigmented and sclerotized, smooth to microspinose, with vestiture consisting of fine setae only or including expanded or complex setae.

Head: Prognathous, protracted, without differentiated neck. Epicranial stem very short to moderately long, rarely absent; frontal arms V- or U-shaped to lyriform or with bases separate, each arm sometimes anteriorly bifurcate; endocarinae absent. Stemmata 5, 3, 2 or 1 on each side, or absent. Antenna 3- or 4-segmented (*Prionochaeta*), about 0.2–1.2 times as long as head width, sensorium on preapical segment anterad of apical segment and conical or palpiform. Frontoclypeal suture absent. Labrum free, tormae present. Mouthparts forming piercing-sucking tube in *Myrmicholeva* (Australia). Mandibles symmetrical to moderately asymmetrical, broad at base and narrow at apex, or triangular, or (*Myrmicholeva*) stylet-like; apex with single tooth or bifid, incisor edge with 1 or 2 subapical teeth or serrate or simple; mesal surface of mandibular base with mola bearing ridged or asperate surface, or (Camiarini and *Agathidium* (*s. str.*))

with membranous setose lobe, or (*Myrmicholeva*) simple; prostheca consisting of membranous or partly sclerotized lobe or absent. Cardines transverse, divided by internal ridge, widely separated from each other by submentum. Stipes elongate. Mala large, fixed, often divided apically into galea and lacinia; galea falcate or rounded, apex glabrous, setose or bearing up to 3 setal combs; lacinia or mala falcate, spinose. Maxillary palp 3-segmented. Labium consisting of prementum, mentum and submentum. Ligula longer than first palpal segment to as long as palp, apex bilobed, quadrilobed, truncate or complex. Labial palps 2-segmented, separated by more than width of first palpal segment (except *Myrmicholeva*). Gula variable. Occipital foramen divided into 2 parts by tentorial bridge.

Thorax and Abdomen: Thoracic terga and abdominal terga and sterna consisting of 1 or more sclerotized plates, without patches or rows of asperities, without lateral tergal processes. Legs 5-segmented including bisetose tarsungulus. Abdomen 10-segmented, about twice or more as long as thorax. Tergum A9 with pair of urogomphi which may be 2-segmented with very long multiannulate apical segment to short and (some Hydnobiini and Leiodini) 1-segmented or fixed. Segment A10 visible from above, anal region terminally oriented, membranous anal lobes with or without numerous fine teeth.

Spiracles: Annular, annular-biforous or modified annular with several peripheral chambers; lateral or dorsolateral; closing apparatus present.

Comments: The family includes about 300 genera and 2300 species worldwide, especially from northern and southern temperate regions, with about 30 genera and 200 species in America north of Mexico. Most of the genera and species are placed in the 2 large subfamilies Leiodinae and Cholevinae, but the monogeneric subfamilies Coloninae, Catopocerinae and Glacicavicolinae also occur in North America, and the latter 2 subfamilies are confined to this continent. With the exception of some Cholevinae attracted to carrion and dung, the species are not commonly encountered and are of no known economic importance.

Immature stages of Leiodidae are poorly known and no comprehensive treatment exists. Best known are larvae of European genera of Cholevinae, with recent keys provided by Zwick *in* Klausnitzer (1978) for non-Bathysciini and Deleurance-Glaçon (1963) for Bathysciini. The few described leiodine genera are treated by Schiödte (1862, 1864), Saalas (1917), Paulian (1941), Klausnitzer (1978), Wheeler (1979, 1985), Hayashi (1986) and Angelini and DeMarzo (1984). Larvae of the primitive south temperate subfamily Camiarinae have only recently been described (Jeannel 1957, Zwick 1979), and larvae of Coloninae and Glacicavicolinae are unknown. The present treatment is based on previously undescribed larvae of about two dozen genera including *Catopocerus* (Catopocerinae), as well as previously known genera.

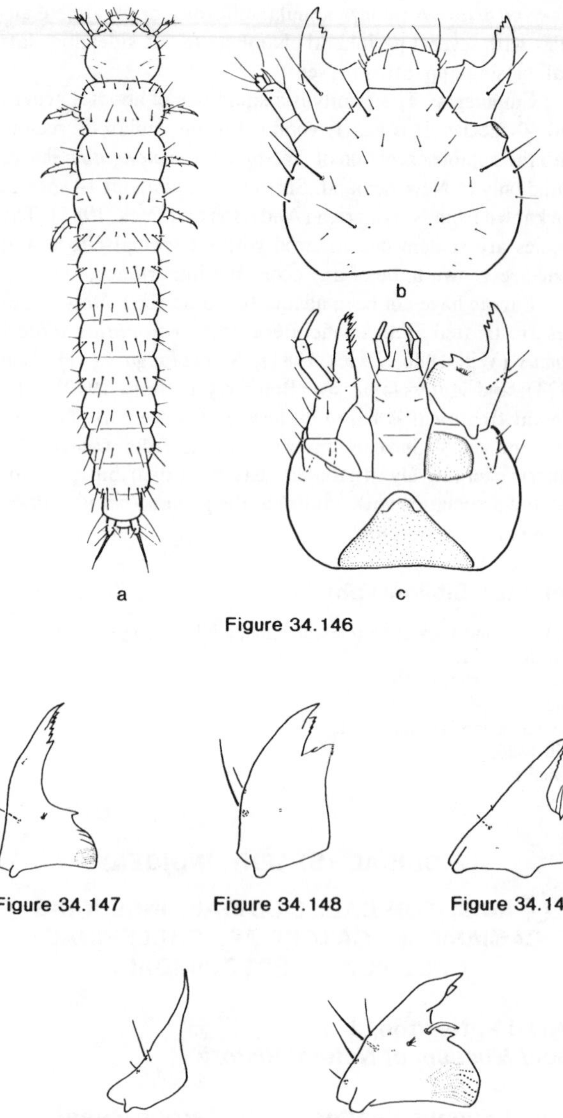

Figure 34.146

Figure 34.147

Figure 34.148

Figure 34.149

Figure 34.150

Figure 34.151

Figures 34.146a-c. Leiodidae. *Catopocerus appalachianus* Peck (Catopocerinae). Ellis Cave, Madison County, Alabama; reared. Length = 3.5 mm. **a.** mature larva, dorsal; **b.** head, dorsal, right antenna not shown; **c.** head, ventral, left maxilla not shown.

Figure 34.147. Leiodidae. *Hydnobius crestonensis* Hatch (Leiodinae). McKenzie Pass vicinity, 5147 feet, Oregon; October; on hypogeous *Gautieria* sp. fungus. Right mandible, ventral.

Figure 34.148. Leiodidae. *Agathidium* (s. str.) *oniscoides* Beauvois (Leiodinae). Bedford, Massachusetts; July; on yellow slime mold plasmodium. Right mandible, ventral.

Figure 34.149. Leiodidae. *Zearagytodes maculifer* (Broun) (Camiarinae). Waipoua State Forest, New Zealand; March; on *Ganoderma* sp. fungus. Right mandible, ventral.

Figure 34.150. Leiodidae. *Myrmicholeva acutifrons* Lea (Camiarinae). Alfred National Park, Victoria, Australia; May; ex rotting logs. Right mandible, ventral.

Figure 34.151. Leiodidae. *Nemadus ?horni* Hatch (Cholevinae). Pine Mountain, New Hampshire; July; in forest floor litter. Right mandible, ventral.

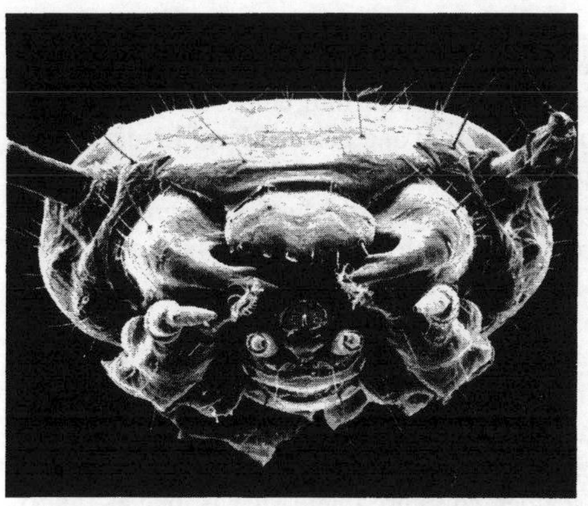

a

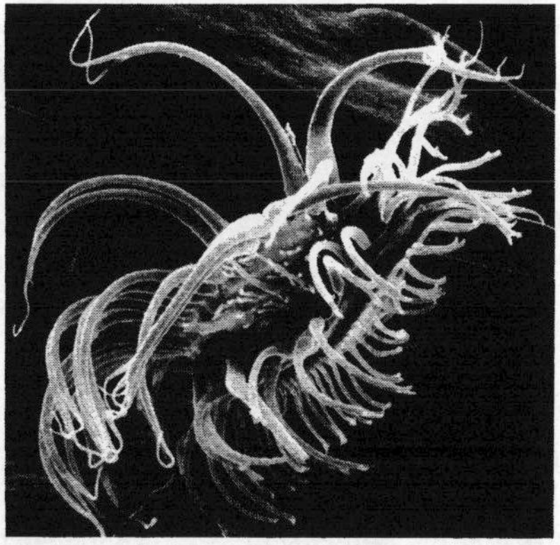

Figure 34.152 b

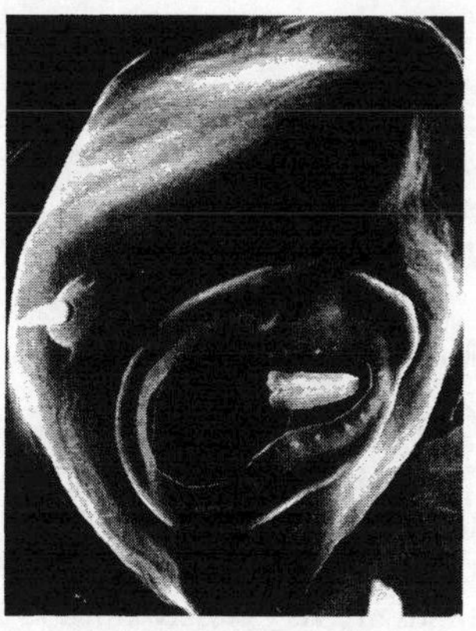

Figure 34.153

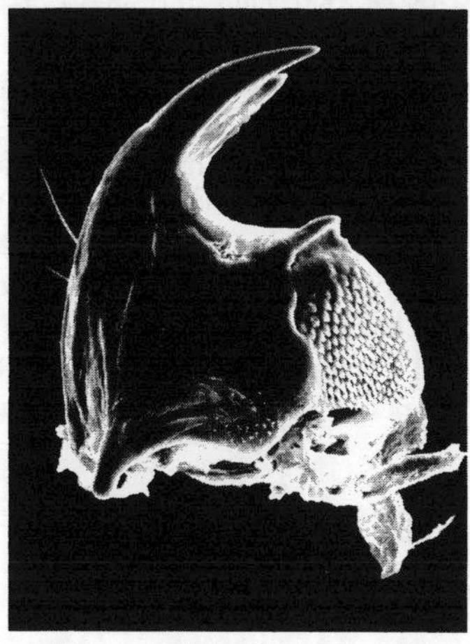

Figure 34.154

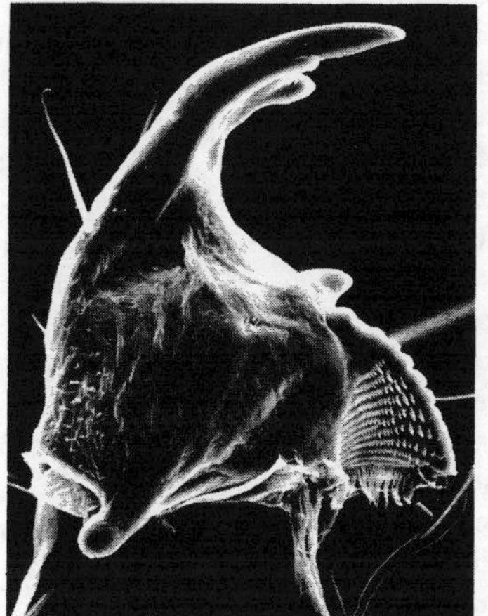

a

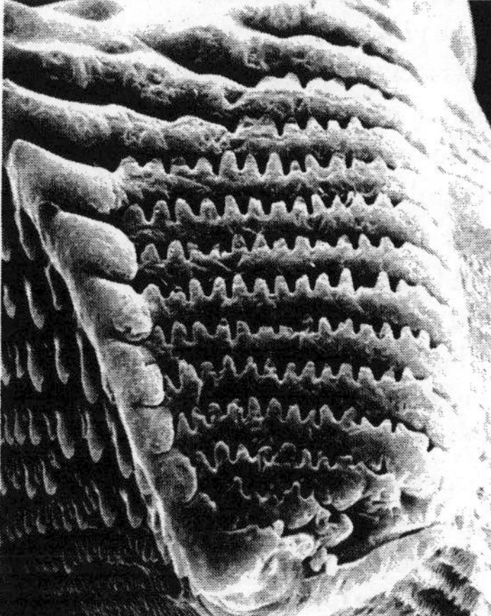

Figure 34.155 b

Figures 34.152a,b. Leiodidae. *Anisotoma errans* Brown (Leiodinae). Longmire (4.7 miles west), Washington; July; on *Stemonitis* sp. fruiting bodies. **a.** head, anterior; **b.** galeal fringe, anterior.

Figure 34.153. Leiodidae. *Zearagytodes maculifer.* Left mesothoracic spiracle, lateral.

Figure 34.154. Leiodidae. *Anisotoma errans.* Right mandible, ventral.

Figures 34.155a,b. Leiodidae. *Colenis impunctata* LeConte (Leiodinae). Rancocas State Park, New Jersey; June; litter under fermenting sap. **a.** right mandible, ventral; **b.** right mandible, mesal view of molar lobe.

Selected Bibliography

Angelini and DeMarzo 1984.
Arzone 1970, 1971.
Benick 1952.
Blas and Vives 1978.
Böving and Craighead 1931.
Casale 1975.
Corbière-Tichané 1973.
Crowson 1981, 1984a.
Deleurance-Glaçon 1963.
Fogel and Peck 1975.
Hayashi 1986 (*Catops, Sphaeroliodes* larvae).
Hayashi et al. 1959 (*Catops* larva).
Jeannel 1957 (*Neocamiarus* larva).
Kistner 1982.
Klausnitzer 1978.
Lawrence and Newton 1980, 1982.
Newton 1984.
Paulian 1941.
Peck 1973 (biology, ecology).
Peyerimhoff 1907.
Rehfous 1955.
Russel 1979.
Saalas 1917 (*Agathidium* larva).
Samuelson 1964 (*Paracatops* larva).
Scheerpeltz and Höfler 1948.
Schiödte 1862, 1864.
Westcott 1968.
Wheeler 1979 (*Creagrophorus* larva), 1984, 1985.
Zwick 1978.
Zwick 1979 (larvae of several Australian genera).

LEPTINIDAE (STAPHYLINOIDEA) (INCLUDING PLATYPSYLLIDAE)

Alfred F. Newton, Jr.,
Field Museum of Natural History

Mammal Nest Beetles

Figures 34.156–158

Relationships and Diagnosis: A close relationship of the louse-like *Platypsyllus* to *Leptinus* and allied genera has long been established (*see* Wood 1965). It is also clear that leptinids are very closely allied to leiodids, and the family is included as a subfamily in some recent, expanded concepts of Leiodidae (Crowson 1981, Lawrence and Newton 1982).

Platypsyllus larvae are easily recognized by their obligate association with beavers (*Castor* spp.) and by their peculiar mandibles with ventrally-directed apices and 1-segmented urogomphi. Larvae of the other 3 leptinid genera closely resemble leiodid larvae but differ in having a scooplike apex of the lacinia. They differ from larvae of other Coleoptera in the same way as Leiodidae (q.v.).

Biology and Ecology: All leptinids are associated with mammal hosts as far as known. In *Leptinus* species the association is a loose one; the life cycle is completed in the nests of a variety of ground-dwelling mammals, but adults are frequently found in other habitats and the species are apparently not host-specific. The other 3 genera are host-specific ectoparasites of semiaquatic mammals, as follows: *Silphopsyllus desmanae* Olsufiev on *Desmana moschata*, a Russian mole; *Leptinillus aplodontiae* Ferris on *Aplodontia rufa*, the mountain beaver; and *L. validus* (Horn) and *Platypsyllus castoris* Ritsema on beavers, *Castor* spp. Adults and larvae are apparently scavengers on the host or in the host nest (Wood 1965, Ising 1969).

Egg ovoid, white, smooth.

Pupa exarate, with functional abdominal spiracles on segments 1–2.

Description: Mature larvae about 2–6 mm long, elongate and more or less parallel-sided, slightly to strongly flattened, relatively straight. Body surfaces very lightly pigmented and sclerotized, smooth, with vestiture consisting of fine setae.

Head: Prognathous, protracted, without differentiated neck. Epicranial stem short, frontal arms lyriform (stem and arms not visible in *Platypsyllus*); endocarinae absent. Stemmata absent. Antenna 3-segmented, about 0.75 or (*Platypsyllus*) 0.25 times as long as head width, sensorium of preapical segment anterad or (*Platypsyllus*) dorsad of apical segment and conical or palpiform. Frontoclypeal suture absent. Labrum free, tormae present. Mandibles symmetrical, apex bilobed, incisor edge simple, mesal surface of base with mola bearing asperate surface, prostheca consisting of partly sclerotized lobe; or (*Platypsyllus*) mandibles with single tooth at ventrally directed apex, incisor edge simple, mesal surface of base with setose lobe, prostheca absent. Cardines transverse, divided by internal ridge, widely separated from each other by submentum. Stipes elongate. Mala large, fixed, divided apically into galea and lacinia, galea bearing pair of setal combs, lacinia with scooplike apex, or (*Platypsyllus*) mala undivided, rounded and setose at apex. Maxillary palp 3-segmented. Labium consisting of prementum, mentum and submentum, or prementum and postmentum. Ligula shorter than first palpal segment and quadrilobed or (*Platypsyllus*) longer than palp and rounded. Labial palps 2-segmented, separated by more than width of first palpal segment. Gula transverse or (*Platypsyllus*) absent. Occipital foramen divided into 2 parts by tentorial bridge.

Thorax and Abdomen: Thoracic terga and abdominal terga and sterna consisting of 1 or more sclerotized plates, without patches or rows of asperities, without lateral tergal processes. Legs 5-segmented including bisetose tarsungulus. Abdomen 10-segmented, more than twice as long as thorax. Tergum A9 with pair of 1- or 2-segmented urogomphi. Segment A10 visible from above, anal region terminally oriented, membranous anal lobes without teeth.

Spiracles: Annular, lateral or dorsolateral, closing apparatus present.

Comments: This small family includes 4 genera: the Holarctic *Platypsyllus* (1 species, *castoris* Ritsema) and *Leptinus* (9 species, 3 in North America); the North American *Leptinillus* (2 species); and *Silphopsyllus desmanae* Olsufiev from the USSR. Larvae of all 4 genera have been described; see Böving and Craighead (1931) and Wood (1965) for North American species. The family is of no known economic importance.

Selected Bibliography

Arnett 1968 (adult classification).
Böving and Craighead 1931.
Casale 1975.
Crowson 1981.
Ising 1969.
Lawrence and Newton 1982.
Neumann and Piechocki 1984, 1985.
Parks and Barnes 1955.
Peck 1982 (*Leptinus* adult classification).
Peterson 1951.
Reid 1942a.
Semenov-Tian-Shansky and Dobzhansky 1927.
Wood 1965.

SCYDMAENIDAE (STAPHYLINOIDEA)

Alfred F. Newton, Jr.,
Field Museum of Natural History

Figures 34.159–167

Relationships and Diagnosis: Scydmaenids have long been recognized as a distinctive family of Staphylinoidea, although their precise relationships remain controversial. They have been considered related to leiodids (e.g., Brown and Crowson 1980), to pselaphids (e.g., Böving and Craighead 1931), and to certain groups of "higher" Staphylinidae (Lawrence and Newton 1982).

The diversity of body forms and absence of any distinctive common characteristics can make scydmaenid larvae difficult to recognize. Most larvae can be recognized by common possession of the following characteristics: size small (under

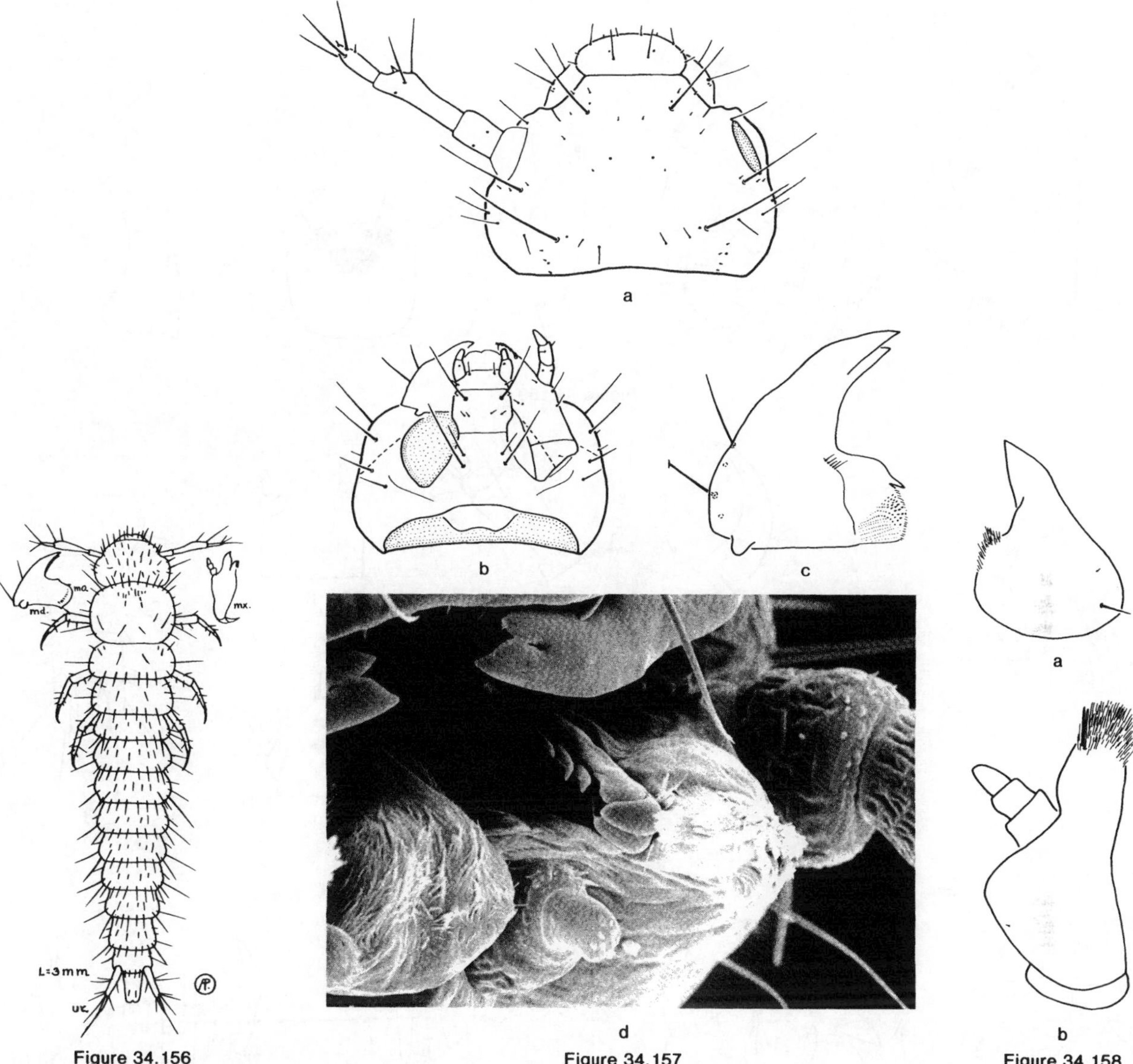

Figure 34.156. Leptinidae. *Leptinus* sp. Larva, dorsal. Length = 3 mm. (From Peterson 1951).

Figures 34.157a-d. Leptinidae. *Leptinillus validus* (Horn). Gatineau Park, Quebec; November; on beaver. **a.** mature larva, head, dorsal, right antenna not shown; **b.** head, ventral, right maxilla not shown; **c.** right mandible, ventral; **d.** left maxilla, anteromesal view of apex of lacinia and galea.

Figures 34.158a,b. Leptinidae. *Platypsyllus castoris* Ritsema. Alameda, California; November; on beaver. **a.** mature (?) larva, left mandible, ventral; **b.** right maxilla, ventral.

5 mm long); head lacking differentiated neck; stemmata 3 or fewer on each side, in close cluster; antenna large, often club-shaped; labrum fused to head capsule; mandible falcate, usually with single mesal edge bearing several, 1, or no teeth; mala large, fixed, rounded or blunt at apex and sometimes bilobed; ligula absent; and urogomphi fixed and very short (except Eutheiini) or absent. They are most easily confused with pselaphid larvae, but lack eversible glandular structures between the antennae, have the principal antennal sensorium conical to domelike and more or less anterior in position rather than palpiform or complex and differently situated, and have tibiae gradually tapered rather than abruptly narrowed apically. Faronine pselaphid larvae, which resemble scydmaenids in these respects, have large fixed urogomphi and a palpiform antennal sensorium while the only scydmaenids with large urogomphi (Eutheiini) have a domelike sensorium.

Biology and Ecology: Scydmaenids are fairly common inhabitants of forest floor litter, mosses, rotting logs, tree holes, sawdust piles and similar habitats, and a few species occur in mammal or ant nests and caves. Adults and larvae are believed to be predatory on mites and other small organisms in

a b c d

Figure 34.159

Figure 34.160

a b c d

Figure 34.161

a b

Figure 34.162

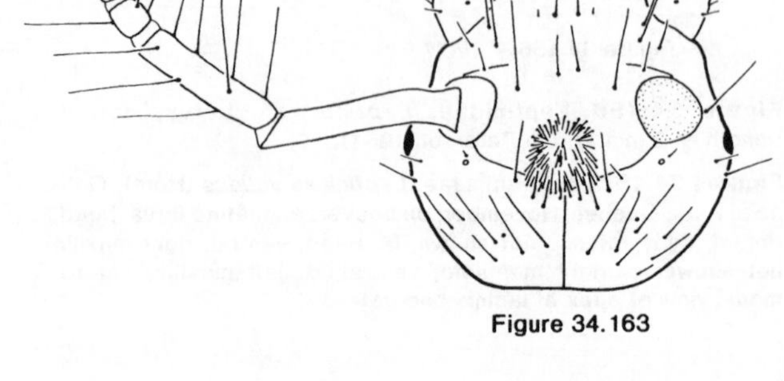

Figure 34.163

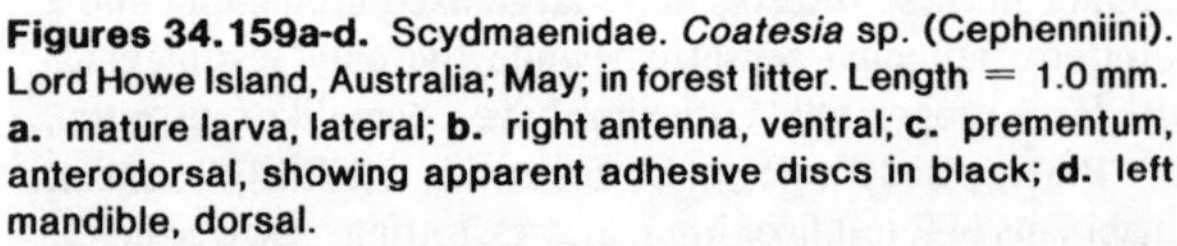

Figures 34.159a-d. Scydmaenidae. *Coatesia* sp. (Cephenniini). Lord Howe Island, Australia; May; in forest litter. Length = 1.0 mm. **a.** mature larva, lateral; **b.** right antenna, ventral; **c.** prementum, anterodorsal, showing apparent adhesive discs in black; **d.** left mandible, dorsal.

Figure 34.160. Scydmaenidae. *Mastigus ruficornis schimitscheki* Mach. (Mastigini). Bolu (25km east), 1000m, Turkey; May. Prementum, ventral.

Figures 34.161a-d. Scydmaenidae. *Veraphis* sp. (Euthiini). Crane Flat (1.9 miles east), 6600 feet, Mariposa County, California; May; in forest litter. **a.** head, dorsal, left antenna not shown; **b.** prementum, ventral; **c.** left maxilla, ventral; **d.** abdominal apex, dorsal.

Figures 34.162a,b. Scydmaenidae. *Scydmaenus* sp. (Scydmaenini). Madden Preserve, Panama; June; in refuse deposit of *Atta* sp. **a.** abdominal apex, dorsal; **b.** head, ventral, right maxilla not shown.

Figure 34.163. Scydmaenidae. *Mastigus ruficornis schimitscheki*. Head, dorsal, right antenna not shown.

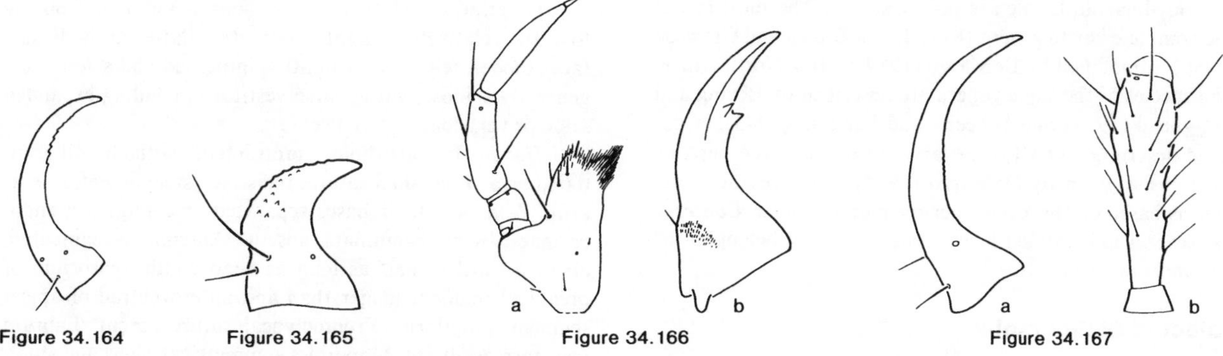

Figure 34.164. Scydmaenidae. *Stenichnus collaris* Müller and Kunze (Stenichnini). Praz-de-Fort, 1200m, Switzerland; June; in mosses. Left mandible, dorsal.

Figure 34.165. Scydmaenidae. *Cephennium thoracicum* Müller (Cephenniini). Veytaux, Switzerland; October; at base of chestnut tree stump. Left mandible, dorsal.

Figures 34.166a,b. Scydmaenidae. *Mastigus ruficornis schimitscheki.* **a.** right maxilla, ventral; **b.** right mandible, ventral.

Figures 34.167a,b. Scydmaenidae. *?Euconnus* sp. (Euconnini). Tallahassee, Florida; March; in forest litter. **a.** left mandible, dorsal; **b.** left antenna, dorsal.

these habitats, but published observations are few. Adults and larvae of at least some species of *Cephennium* are known to specialize on oribatid mites; larvae "stick" the mite to their mouths (possibly with apparently adhesive labial discs described below) and rasp a hole in the mite with the rough outer edges of the mandibles (Schuster 1966a, b). Early instar larvae of *Mastigus pilifer* Kraatz feed only on a secretion left by the female, but final instar larvae leave the oviposition site to feed (DeMarzo, 1983). Egg ovoid, orange, smooth (DeMarzo, 1983). Pupa exarate (DeMarzo, 1984).

Description: Mature larvae about 2–5 mm long. Form very variable, from elongate, parallel-sided, straight and slightly flattened to ovate, ventrally curved and capable of rolling into a ball, to broadly ovate and strongly flattened, disclike. Body surfaces moderately to lightly pigmented and sclerotized, rarely darkly pigmented dorsally; smooth or microspinose; vestiture of fine setae only or including long bristles or complex setae.

Head: Prognathous, protracted and visible from above or (Cepheniini and a few other species) retracted and concealed from above by prothorax; without differentiated neck. Epicranial stem moderately long, frontal arms V- or U-shaped; endocarinae absent or median endocarina present at base of stem. Stemmata on each side 3 (in close triangle or nearly fused) or 1 or absent. Antenna of 2, 3 or (*Mastigus*) 4 segments, about half or more as long as head width, often apically thickened and club-shaped; sensorium of second segment anterad or anterodorsad of apical segment or its unarticulated remnant (in *Mastigus* at apex of fourth segment), conical or domelike. Frontoclypeal suture absent. Labrum fused to head capsule, forming nasale which may bear numerous small teeth. Mandibles symmetrical, narrow and falcate, apex with single tooth; incisor edge simple, serrate, with 1 or 2 small teeth, or with 1 to several teeth on each edge of double mesal edges; mesal surface of base simple, mola and prostheca absent (outer edge of apical half of mandible tuberculate in *Cephennium*). Cardines transverse to strongly oblique, undivided, separated from each other by submentum, or cardines mesally fused to submentum. Stipes elongate to transverse. Mala large, fixed, apically rounded or truncate, setose, sometimes with fixed dorsomesal lobe. Maxillary palp usually 3-segmented but 2- and 4-segmented in *Eutheia* and *Mastigus*, respectively. Labium consisting of prementum, mentum and submentum. Ligula absent. Labial palps 2-segmented, separated by much more than width of first palpal segment. Adoral surface of prementum with 2 or 3 adhesive (?) discs in Cepheniini and some other species. Gular sutures absent or partly or completely fused, gula absent. Occipital foramen not divided by tentorial bridge which, if evident, originates from posterior arms of tentorium.

Thorax and Abdomen: Thoracic and abdominal terga and sterna consisting of 1 or more sclerotized plates, without patches or rows of asperities; with or without lateral tergal processes; sterna rarely membranous. Legs long, 5-segmented including bisetose tarsungulus which in Stenichnini also bears fine subapical spines; tibiae evenly tapered to apex. Abdomen 10-segmented or apparently 9-segmented, slightly longer to more than twice as long as thorax. Tergum A9 variable in shape, sometimes pear-shaped, usually without urogomphi but sometimes with fixed pair of minute and spiniform or (Eutheiini) long and thick urogomphi. Segment A10 distinct and visible from above, anal lobes with or without several hooks (segment 10 invaginated or apparently absent in Cephenniini).

Spiracles: Annular, lateral or dorsolateral, elevated or not, closing apparatus present; spiracles of 1–4 apical abdominal segments atrophied in some species.

Comments: The family is large and widely distributed, with over 3600 known species worldwide, mostly from warm temperate and tropical regions; nearly half of these species are placed in the single broadly defined genus *Euconnus*. About 13 genera and 180 species are found in the United States and Canada. The species are relatively common but of no economic importance.

Scydmaenid larvae are poorly known. The most recent and complete key to genera (8) is that of Brown and Crowson (1980) as modified by DeMarzo (1984); a few North American species of the same genera are described by Böving and Craighead (1931) and Wheeler and Pakaluk (1983). A detailed description of all three larval instars and the pupa of *Mastigus* is given by DeMarzo (1984). The present treatment is based on the known genera plus *Veraphis, Coatesia* and several unidentified larval types probably belonging to *Euconnus* (*sensu lato*).

Selected Bibliography

Arnett 1968 (adult classification).
Böving and Craighead 1931.
Brown and Crowson 1980.
DeMarzo 1983, 1984.
Hayashi 1986 (unidentified larva).
Jeannel and Paulian 1945 (*Scydmaenus* larva).
Kasule 1966 (family characterization).
Kistner 1982.
Klausnitzer 1978.
Lawrence and Newton 1982.
Paulian 1941.
Schuster 1966a, b.
Wheeler and Pakaluk 1983.

MICROPEPLIDAE (STAPHYLINOIDEA)

Alfred F. Newton, Jr.,
Field Museum of Natural History

Figures 34.168–169

Relationships and Diagnosis: Micropeplids have always been closely associated with staphylinids and are often treated as a subfamily of Staphylinidae, placed in the vicinity of Proteininae. Separation of the 2 families is based largely on the unusual (for Staphylinidae) habitus and mouthpart structure of *Micropeplus* larvae.

Known micropeplid larvae have a very distinctive habitus which, in combination with the small size (under 3 mm long) and mouthpart structure (including characteristic mandibles and maxillae) will separate them from other beetle larvae (*see* figures).

Biology and Ecology: Most species of *Micropeplus* and *Pelomicrus* are found in forest floor litter, but a few species of the former genus are apparently restricted to nests of certain mammals such as wood rats and beavers. *Kalissus nitidus* LeConte has been found on the pebbly or muddy margins of lakes, and at least 1 species of *Micropeplus* (*sculptus* LeConte) inhabits swamps and bogs (Campbell 1968). The biology of the family is poorly known, although Hinton and Stephens (1941b) found that larvae and adults of some *Micropeplus* species fed primarily on mold spores and hyphae. These feeding habits may be normal for the family; records from fruiting bodies of higher fungi are probably accidental (Newton 1984).

Egg unknown.

Pupa exarate, with functional abdominal spiracles on segments 1–2.

Description: Mature larvae about 2–3 mm long, oblong to ovate, relatively straight, moderately flattened. Body surfaces moderately to very lightly pigmented and sclerotized, generally spinose, with sparse vestiture including expanded setae or very long tapered setae.

Head: Hypognathous, protracted, without differentiated neck. Epicranial sutures indistinct, stem absent, frontal arms V-shaped, their bases separated or contiguous; endocarinae absent. Stemmata absent. Antenna 3-segmented, about a third to half as long as head width, sensorium of preapical segment longer than and anteroventrad of apical segment, palpiform. Frontoclypeal suture absent. Labrum free, tormae absent. Mandibles symmetrical, short and stout, each with single apical tooth and large subapical pseudomola bearing 2 or more coarse teeth; mesal surface of base simple, mola and prostheca absent. Cardines transverse, divided by internal ridge, widely separated by submentum. Stipes elongate. Mala large, fixed, divided at apex into articulated galea and fixed lacinia; galea with 2 membranous fringed lobes; lacinia falcate, with single apical spur and spinose mesal edge. Maxillary palp 3-segmented. Labium consisting of prementum, mentum and submentum. Ligula longer than first palpal segment, broadly rounded. Labial palps short, 2-segmented, separated by much more than width of first palpal segment. Gular sutures absent. Occipital foramen not divided by tentorial bridge, which arises from posterior margin of head.

Thorax and Abdomen: Thoracic and abdominal terga and sterna consisting of 1 or more sclerotized plates; terga microspinose on disc to coarsely spinose along posterior margins, laterally bearing tergal processes consisting of a single lobe (abdominal segments 1–9) or bifid or trifid lobes (thorax). Legs short, 4-segmented including bisetose tarsungulus, femur and trochanter fused, tibiae with bifid setae. Abdomen 10-segmented, more than twice as long as thorax. Tergum A9 with fixed tergal lobe as in preceding segments, without articlated urogomphi. Segment A10 short, scarcely visible from above, anal region terminally oriented, membranous anal lobes with very fine teeth or spines.

Spiracles: Annular, lateral, at ends of short tubes below tergal lobes, closing apparatus present.

Comments: The family is small, with 4 genera and fewer than 60 species known worldwide: the Holarctic genus *Micropeplus*, including 15 species in North America; *Kalissus*, with a single species *nitidus* LeConte in British Columbia and Washington State; *Peplomicrus* with 7 species in tropical Central and South America and Africa, and *Cerapeplus* with one Oriental species (Campbell 1968, Löbl and Burckhardt 1988). The species are quite rare, and of no economic importance.

The larva of only 1 species, *Micropeplus staphylinoides* (Marsham), has been described (Lubbock 1868; Kasule 1966). Hinton and Stephens (1941b) have described the only known pupa, that of *M. fulvus* Erichson. Larvae of *M. fulvus* and *M. neotomae* Campbell have been examined for this treatment.

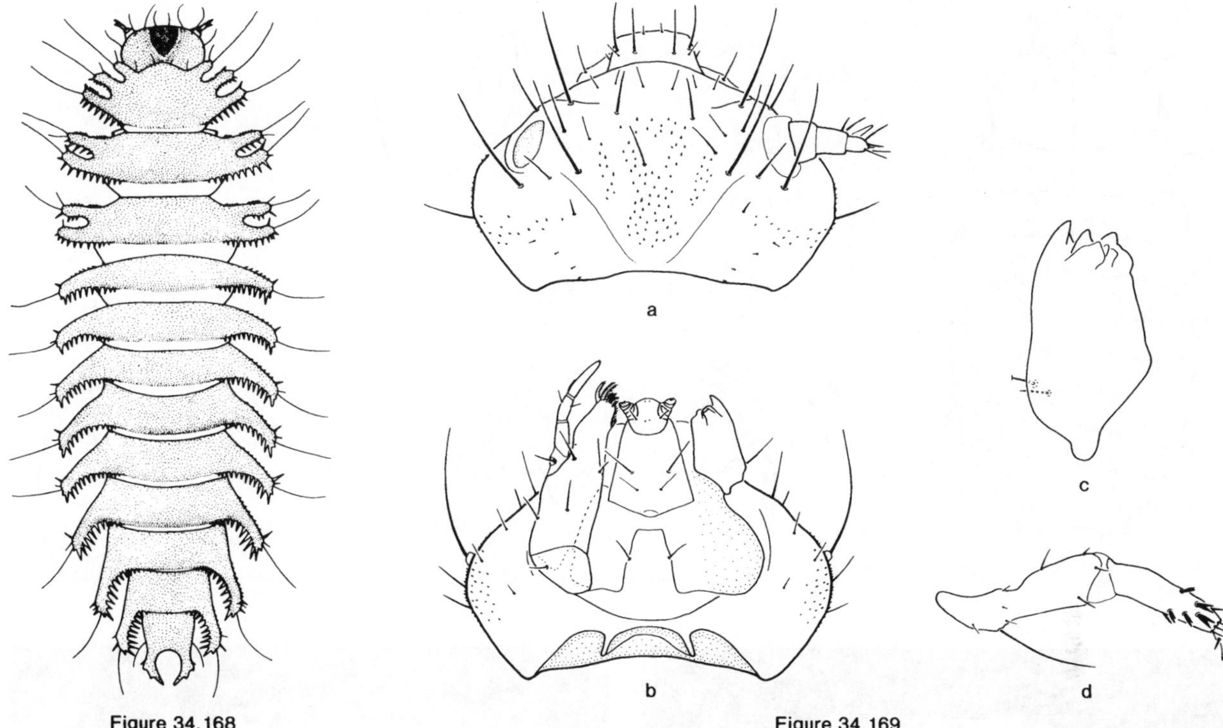

Figure 34.168. Micropeplidae. *Micropeplus neotomae* Campbell. Los Gatos (7.5 miles south), Santa Clara County, California; February; *Neotoma* house. Mature (?) larva, dorsal. Length = 3.0 mm.

Figures 34.169a-d. Micropeplidae. Same species as figure 34.168. Suver Junction (1 mile north), Polk County, Oregon; March; ex fecal material, *Neotoma* nest. **a.** head, dorsal, left antenna not shown; **b.** head, ventral, left maxilla not shown; **c.** right mandible, ventral; **d.** left proleg, anterior.

Selected Bibliography

Campbell 1968 (adult classification, ecology).
Hinton and Stephens 1941b.
Kasule 1966.
Löbl and Burckhardt 1988 (adult classification).
Lubbock 1868.
Newton 1984.
Thayer 1987 (family relationships).
Topp *in* Klausnitzer 1978.

DASYCERIDAE (STAPHYLINOIDEA)

Alfred F. Newton, Jr.,
Field Museum of Natural History

Figures 34.170a-h

Relationships and Diagnosis: The genus *Dasycerus* was included in Lathridiidae until Crowson (1955) placed it tentatively as a family of Staphylinoidea. More recently it has been considered related to certain subfamilies of Staphylinidae (Lawrence and Newton 1982, Thayer 1987).

Larvae are easily recognized by the unique structure of their mandibles.

Biology and Ecology: The species, most of which are flightless, apparently inhabit forest floor litter or decaying trees. The only known large series of adults and larvae, of an undescribed North American species, was from the bark surface of a decaying log (*Aesculus* sp.) covered with *Stereum* sp., an undetermined ascomycete, lichens and moss. Food of adults and larvae remains uncertain but may include molds (Newton 1984, Wheeler 1984b).

Egg and pupa unknown.

Description: Mature larvae about 2–3 mm long, ovate, slightly or not at all flattened, straight. Body surfaces moderately to lightly pigmented and sclerotized, generally smooth or microspinose, with sparse vestiture of simple setae.

Head: Moderately declined, protracted, without differentiated neck. Epicranial stem very short, frontal arms V-shaped; endocarinae absent. Stemmata 6 on each side, widely spaced. Antenna 3-segmented, slightly less than half as long as head width, sensorium of preapical segment anterad of and longer than apical segment, palpiform. Frontoclypeal suture absent. Labrum free, tormae absent. Mandibles symmetrical, short and stout, apex of each truncate and bearing dense array of slender teeth, incisor edge and base simple, mola and prostheca absent. Cardines transverse, divided by internal ridge, widely separated by submentum. Stipes elongate. Mala large, fixed, with rounded setose apex. Maxillary palp 3-segmented. Labium consisting of prementum, mentum and submentum. Ligula nearly as long as palp, with long setae. Labial palps short, 2-segmented, separated by much more than

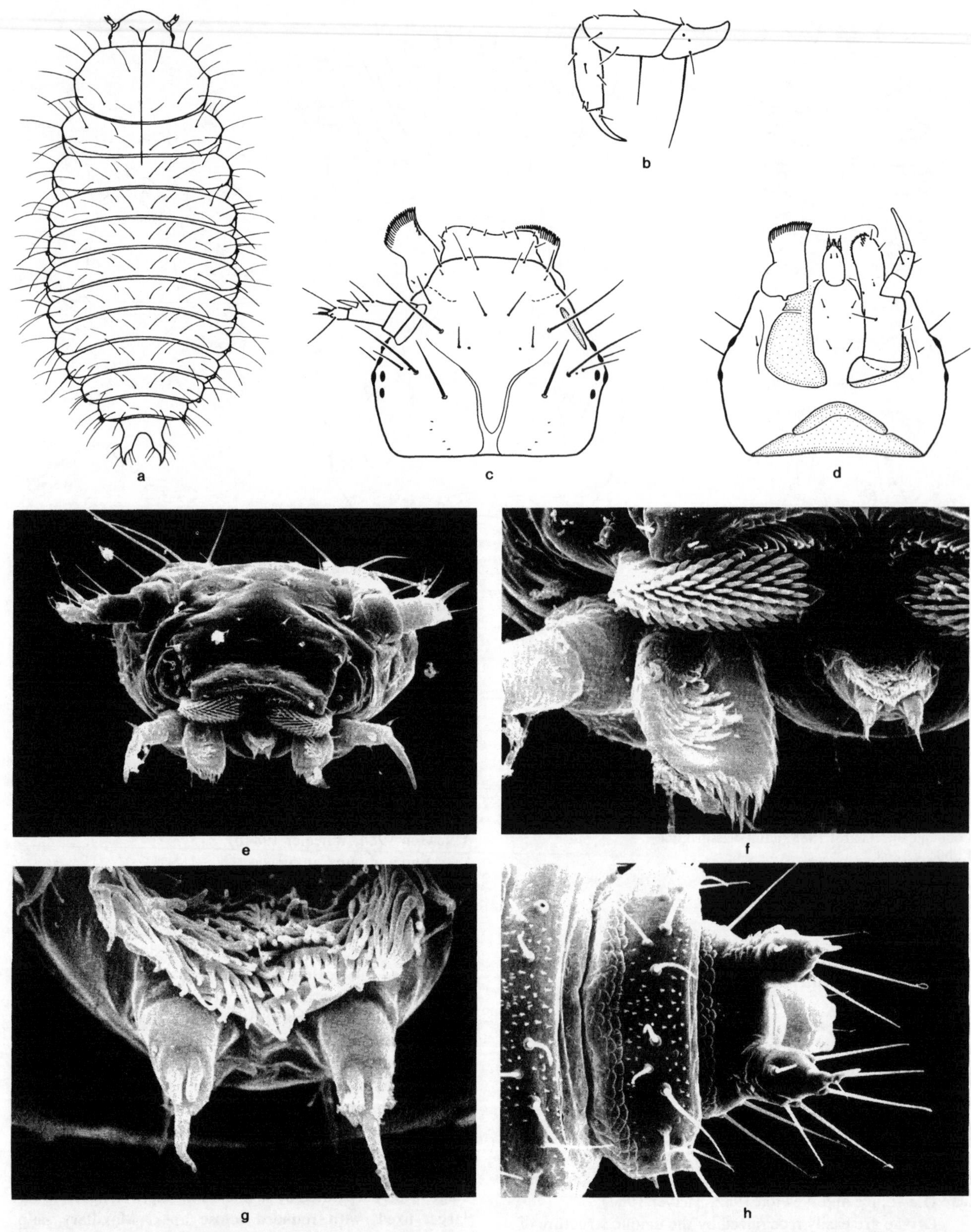

Figures 34.170a-h. Dasyceridae. *Dasycerus* n. sp. Ramsey Cascade Trail, 3900 feet, Great Smoky Mountains National Park, Tennessee; May; on bark of log of *Aesculus* sp. covered with *Stereum* sp., an undetermined ascomycete, lichens and moss. Length = 2.0 mm. **a.** penultimate instar (?) larva, dorsal; **b.** right proleg, anterior; **c.** head, dorsal, right antenna not shown; **d.** head, ventral, right maxilla not shown; **e.** head, anterior; **f.** head, anterior, detail of maxilla, mandible; **g.** head, anterior, detail of labium; **h.** abdominal apex, dorsal.

width of first palpal segment. Gula transverse. Occipital foramen not divided by tentorial bridge, which arises from posterior edge of head.

Thorax and Abdomen: Thoracic and abdominal terga and sterna consisting of 1 or more sclerotized plates; terga microspinose. Legs short, 5-segmented including bisetose tarsungulus. Abdomen 10-segmented, less than twice as long as thorax. Tergum A9 with pair of large fixed urogomphi with spine at apex. Segment A10 scarcely visible from above, anal region terminally oriented, membranous anal lobes without hooks or teeth.

Spiracles: Annular, lateral, slightly elevated, closing apparatus present.

Comments: The family includes the single genus *Dasycerus,* with 10 species in Eurasia and 3 from the United States: 1 from California and 2 (1 undescribed) from the southern Appalachian Mountains. The species are rare and of no economic importance.

Larvae have not been previously described. The present treatment is based on larvae associated with the undescribed Appalachian species and with *D. japonicus* Nakane.

Selected Bibliography

Crowson 1955.
Lawrence and Newton 1982.
Löbl 1977 (revision, adult ecology).
Newton 1984.
Thayer 1987 (family relationships).
Wheeler 1984b.

SCAPHIDIIDAE (STAPHYLINOIDEA)

Alfred F. Newton, Jr.,
Field Museum of Natural History

Shining Fungus Beetles

Figures 34.171–176

Relationships and Diagnosis: Scaphidiids have generally been considered closely related to staphylinids, and have recently been placed in the vicinity of a group of staphylinid subfamilies including Piestinae, Osoriinae and Oxytelinae (Kasule 1966, Lawrence and Newton 1982); Kasule (1966) and Lawrence (1982c), in fact, reduced them to a subfamily of Staphylinidae.

Scaphidiid larvae can usually be recognized by a combination of a toothed or crenulate emargination of the anterior margin of the labrum, mandibles without basal molar lobes, but often with subapical dentate or spinose lobes, and articulated urogomphi which may be minute or (some *Baeocera* species) absent. The crenulate labral emargination is absent from the otherwise similar larvae of certain Staphylinidae. The association with fungi of various kinds and the usual presence of 5 well separated stemmata on each side of the head are also useful recognition traits.

Biology and Ecology: Adults and larvae are associated with and feed on fungi of various kinds, as follows: *Scaphidium, Toxidium,* most *Scaphisoma* and some *Baeocera* species on hyphae of tree fungi (Polyporales); *Cyparium, Scaphium* and some *Scaphisoma* species on hyphae of mushrooms and coral fungi (Agaricales and Clavariaceae); and *Scaphobaeocera* and most *Baeocera* species on fruiting bodies of slime molds (Myxomycetes) (Lawrence and Newton 1980, Newton 1984; Ashe, 1984a). Larval development in *Baeocera* is very rapid, taking only a few days of feeding on the spores of the short-lived host, but life histories have not been worked out in detail for any scaphidiid species.

Egg oval, white, smooth.

Pupa exarate, with functional abdominal spiracles on segments 1–3.

Description: Mature larvae about 2–12 mm long, elongate and more or less parallel-sided to ovate, slightly or not at all flattened, straight or slightly curved ventrally. Body surfaces moderately or lightly pigmented and sclerotized, smooth, with sparse vestiture of simple setae.

Head: Prognathous or slightly declined, protracted or slightly retracted, without differentiated neck. Epicranial stem moderately long, frontal arms V-shaped or lyriform, sometimes joined anteriorly by transverse line; endocarinae absent. Usually 5 well separated stemmata on each side, sometimes 6 or (*Cyparium*) 3. Antenna 3-segmented, about half or more as long as head width, sensorium of preapical segment anterad of apical segment and conical or palpiform. Labrum free, anterior edge with crenulate emargination; tormae absent. Mandibles symmetrical, broad and stout, apex of each bilobed or with a single lobe, incisor edge serrate and sometimes with a subapical pseudomola bearing teeth or spines, mesal surface of base simple, mola and prostheca absent. Cardines transverse, internally divided, separated from each other by submentum or postmentum. Stipes elongate. Mala large, fixed, falcate, apex glabrous and with single large tooth, mesal edge spinose. Maxillary palp 3-segmented. Labium consisting of prementum, mentum and submentum, or prementum and postmentum only. Ligula shorter or slightly longer than first palpal segment, transverse, and rounded, truncate, or obtusely pointed at apex. Labial palps 2-segmented, separated by width of first palpal segment or more. Gular sutures absent. Occipital foramen not divided by tentorial bridge, which arises from posterior arms of tentorium.

Thorax and Abdomen: Thoracic and abdominal terga and sterna consisting of 1 or more sclerotized plates, without patches or rows of asperities. Legs long, 5-segmented including bisetose tarsungulus. Abdomen 10-segmented, about twice or more as long as thorax. Tergum A9 with pair of 1- or 2-segmented urogomphi which may be minute or (some *Baeocera* species) absent. Segment A10 visible from above, anal region terminally oriented, membranous anal lobes bearing numerous fine teeth.

Spiracles: Annular, lateral or dorsolateral, closing apparatus present.

Comments: The family is moderately large, with 50 genera and over a thousand species known from throughout the world, especially from warm temperate and tropical areas. About 60 species, placed in 7 genera, occur in North America. Many species are commonly encountered on fungi; none are of any known economic importance.

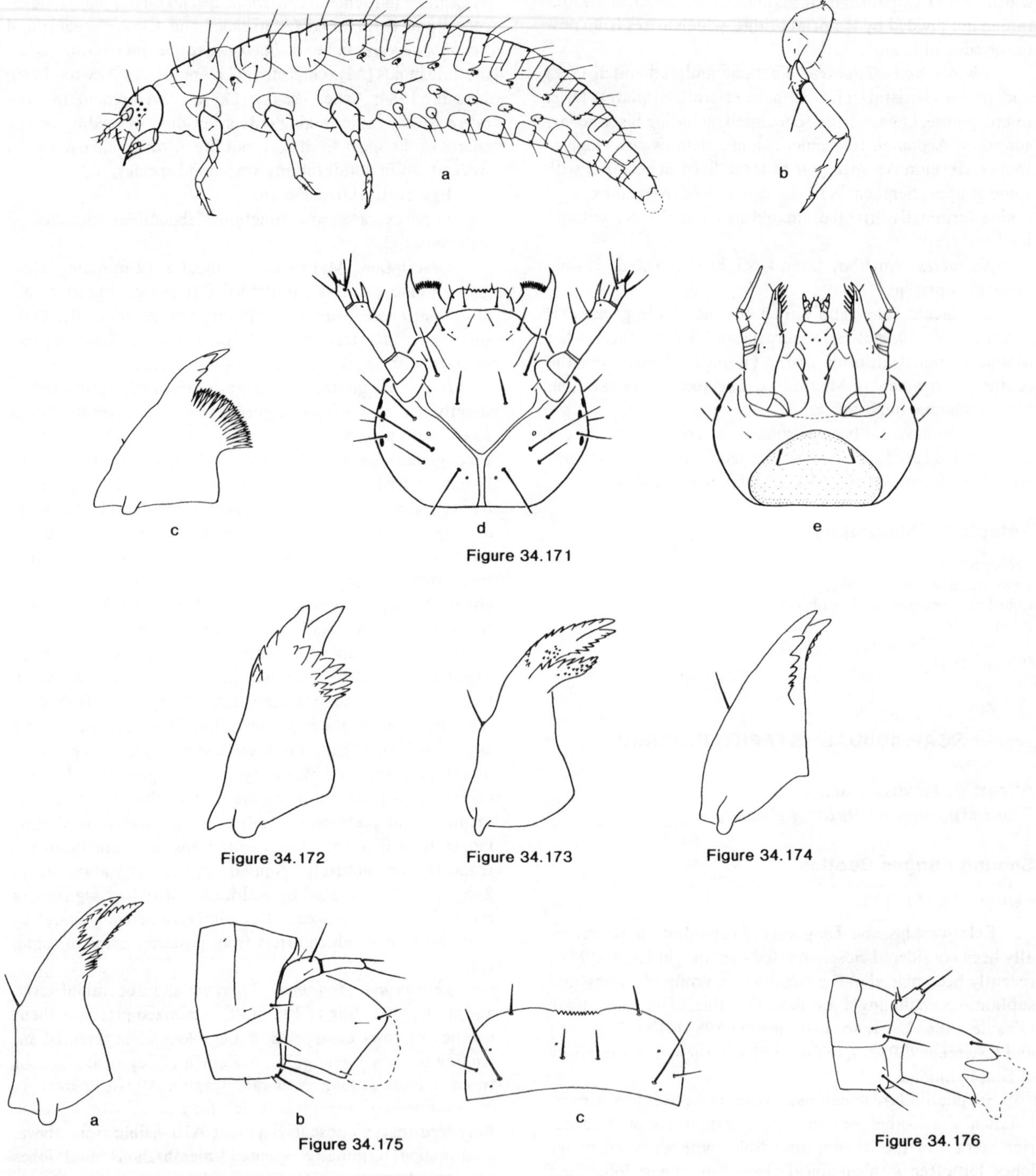

Figures 34.171a–e. Scaphidiidae. *Baeocera picea* Casey. Bedford, Massachusetts; August; on slime mold, *Ceratiomyxa fruticulosa.* Length = 2.6 mm. **a.** mature larva, lateral; **b.** right proleg, posterolateral; **c.** right mandible, ventral; **d.** head, dorsal; **e.** head, ventral.

Figure 34.172. Scaphidiidae. *Scaphisoma* sp. Manchester, Vermont; September; on coral fungus, *Clavaria coronata.* Right mandible, ventral.

Figure 34.173. Scaphidiidae. *Cyparium terminale* Matthews. Tenancingo, Mexico; September; on coral fungus, *?Clavaria* sp. Right mandible, ventral.

Figure 34.174. Scaphidiidae. *Scaphidium* sp. Cerro Azul, Panama; June; on fungusy log. Right mandible, ventral.

Figures 34.175a–c. Scaphidiidae. *Scaphium castanipes* Kirby. George Lake, Alberta; August; on mushroom, *Cortinarius* sp. **a.** right mandible, ventral; **b.** abdominal apex, lateral; **c.** labrum, dorsal.

Figure 34.176. Scaphidiidae. *Cyparium terminale.* Abdominal apex, lateral.

Larval characterization of Scaphidiidae has been unsatisfactory for many years because of the misidentification of larvae of the staphylinid genus *Sepedophilus* as *Scaphisoma* larvae by Böving and Craighead (1931), Paulian (1941), Peterson (1951), Dajoz (1965), and others (*see* Newton 1984). The errors were partly corrected by Kasule (1966), who did not know the identity of the misidentified larvae. Larvae of only 4 genera, *Scaphium, Scaphidium, Scaphisoma,* and *Cyparium* have been adequately described in the literature (Perris 1876, Kasule 1968, Ashe 1984a, Hayashi 1986). The present treatment is also based on larvae of *Baeocera, Scaphobaeocera* and *Toxidium* species.

Selected Bibliography

Arnett 1968 (adult classification).
Ashe 1984a.
Benick 1952.
Böving and Craighead 1931 (misidentified *Scaphisoma* larva).
Dajoz 1965 (misidentified *Scaphisoma* larva).
Hayashi 1986.
Hayashi et al. 1959 (*Scaphidium* larva).
Kasule 1966, 1968.
Lawrence and Newton 1980, 1982.
Newton 1984.
Paulian 1941 (misidentified *Scaphisoma* larva).
Perris 1876 (true *Scaphisoma* larva).
Peterson 1951 (misidentified *Scaphisoma* larva).
Rehfous 1955.
Russell 1979.
Scheerpeltz and Höfler 1948.

SILPHIDAE (STAPHYLINOIDEA)

Alfred F. Newton, Jr.,
Field Museum of Natural History

Carrion Beetles (Silphinae) and Burying Beetles (Nicrophorinae)

Figures 34.177-182

Relationships and Diagnosis: Silphids are closely related to Staphylinidae and allied small families such as Scaphidiidae. The family is here used in a restricted sense, including only the subfamilies (or tribes) Silphinae and Nicrophorinae. Agyrtinae are treated as a separate family, Agyrtidae (q.v.), based on studies soon to be published (Newton, in preparation; also Lawrence and Newton 1982, Anderson and Peck 1985).

Silphid larvae can be recognized by possession of a combination of: mandible without a molar lobe; maxilla with broad, apically cleft mala bearing dense setal brush on outer lobe; and articulated urogomphi present and usually 2-segmented. The large size of mature larvae (12+ mm) and the association of most species with carrion are also useful recognition traits. Each subfamily has a very distinctive larval habitus - grub-like with small terga bearing spines (Nicrophorinae, fig. 34.177), or flattened with lateral tergal lobes (Silphinae, fig. 34.178).

Biology and Ecology: Adults of *Nicrophorus* species, the burying beetles, pair up at a small animal carcass and bury it. Eggs are laid near the burial chamber and the emerging larvae are fed by one or both adults and also feed directly on the carcass. Larval development lasts only a few days, and pupation occurs near the burial chamber. For details see Pukowski (1933), Roussel (1964), Milne and Milne (1976), and Anderson and Peck (1985). Most adults of Silphinae are attracted to large animal carcasses, near which eggs are laid. Larvae, unattended by adults, feed on the carrion until development is complete in 1 to a few weeks. *Silpha* (*sensu stricto*) species (not in North America) are not carrion-feeding and have diverse habits; some species are predatory and a few are specialized snail-feeders. *Aclypea* (formerly *Blitophaga*) species are phytophagous. The most detailed treatment of the biology of Silphinae is given in a series of papers on European species by Heymons and von Lengerken (*see* von Lengerken 1938a and contained references); *see* also Cooley (1917), Ratcliffe (1972), Brewer and Bacon (1975), and Anderson and Peck (1985) for North American species.

Egg oval, whitish, smooth.

Pupa exarate, with functional abdominal spiracles on segments 1–4.

Description: Mature larvae about 12–40 mm long, elongate and more or less parallel-sided to ovate, slightly to strongly flattened, relatively straight or slightly curved ventrally. Body surfaces heavily pigmented and sclerotized (Silphinae) or lightly pigmented and sclerotized (Nicrophorinae), smooth or microspinose, with sparse vestiture of short simple setae only or (rarely) slightly bifid or frayed setae.

Head: Prognathous or slightly declined, protracted, without differentiated neck. Epicranial stem moderately long, frontal arms V- or U-shaped; endocarinae absent. Stemmata 6 (Silphinae) or 1 (Nicrophorinae) on each side. Antenna 3-segmented, a third to twice as long as head width, sensorium of preapical segment extending anterad of apical segment and usually conical or palpiform, rarely platelike. Frontoclypeal suture absent. Labrum free, of 3–5 sclerites, tormae absent. Mandibles symmetrical, narrow and falcate, apex with single tooth or (*Aclypea*) bilobed; incisor edge with 1 large preapical tooth and several smaller ones, ventral edge near apex also serrate in *Aclypea;* mesal surface of base simple, mola and prostheca absent. Cardines transverse, divided by internal ridge, separated from each other by submentum. Stipes elongate. Mala large, fixed, divided at apical fourth into fixed galea and lacinia; galea with dense brush of setae; lacinia falcate, with 2 or 3 apical spurs and spinose mesal edge. Maxillary palp 3-segmented. Labium consisting of prementum, mentum and submentum. Ligula shorter or longer than first palpal segment but shorter than palp, apex bilobed. Labial palps 2-segmented, separated by more than width of first palpal segment. Gular sutures absent. Occipital foramen not divided by tentorial bridge, which arises from posterior arms of tentorium.

Thorax and Abdomen: Thoracic terga and abdominal terga and sterna consisting of 1 or more sclerotized plates, without patches or rows of asperities, each tergum with 1 (Silphinae) or 2 (*Ptomascopus*) lateral tergal processes extending beyond edges of sterna or (*Nicrophorus*) without such

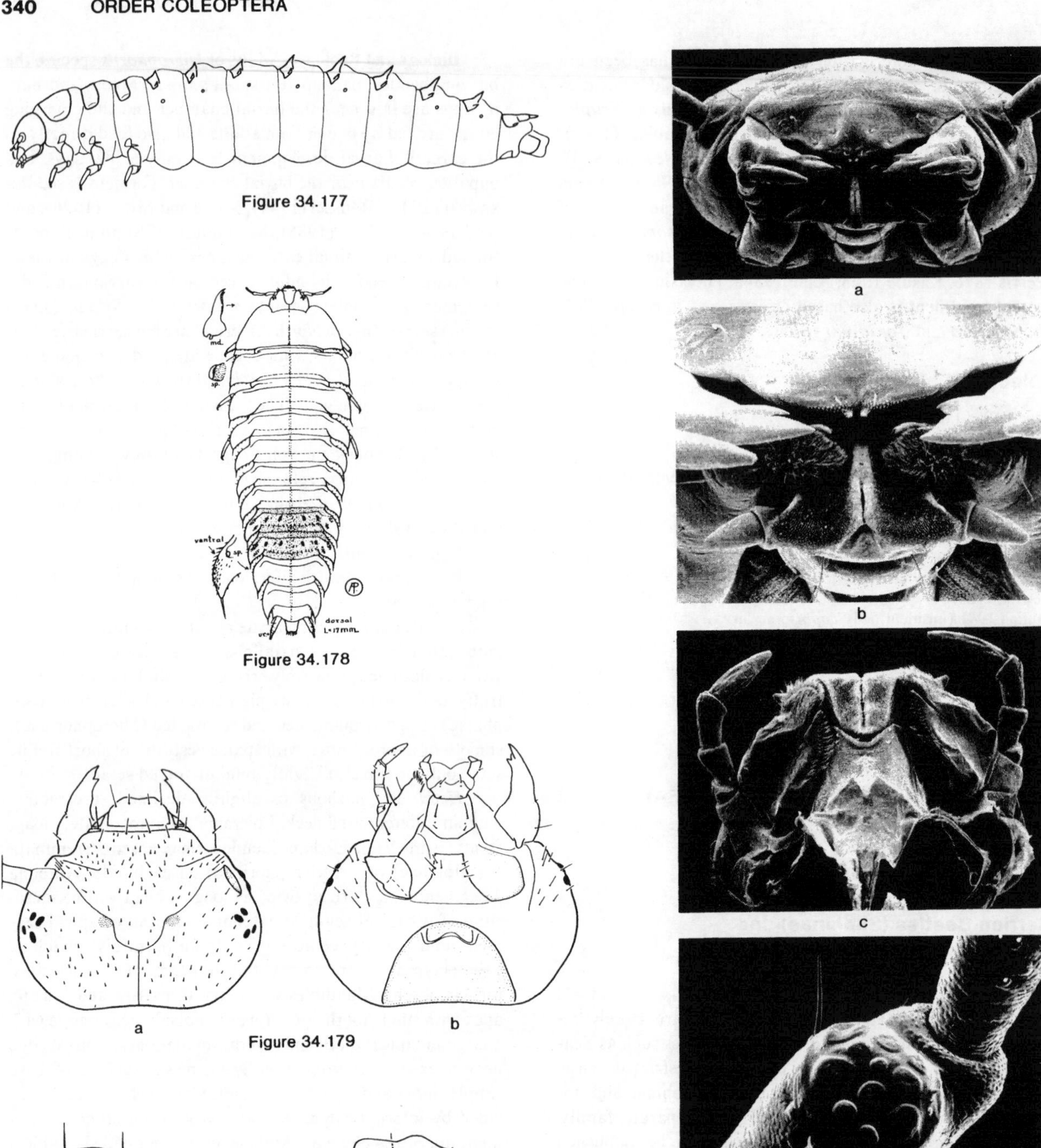

Figure 34.177. Silphidae. *Nicrophorus sayi* Laporte. Basin Pond, New Hampshire; June; reared from adults found at dead chipmunk. Mature larva, lateral. Length = 18 mm.

Figure 34.178. Silphidae. *Oiceoptoma noveboracense* (Forster). Mature larva, dorsal. Length = 17 mm. [From Peterson, 1951.]

Figures 34.179a,b. Silphidae. *Oiceoptoma rugulosum* Portevin. Gainesville, Florida; December; at carrion. **a.** head, dorsal, right antenna not shown; **b.** head, ventral, left maxilla not shown.

Figure 34.180. Silphidae. *Aclypea bituberosa* (LeConte). Whitelaw, Alberta; June; ex rape field. Right mandible, ventral.

Figure 34.181. Silphidae. *Nicrophorus orbicollis* Say. Hurricane Cave, Alabama; July; carrion bait. Right mandible, ventral.

Figures 34.182a-d. Silphidae. *Necrophila americana* (L.). Ozone (2 miles east), Tennessee; July; carrion trap. **a.** head, anterior; **b.** head, anterior, detail; **c.** maxillae and labium (detached), dorsal view; **d.** left antenna, frontal view, detail of apex of second segment.

processes but with 4 spinose projections along posterior margin of abdominal terga. Legs long, 5-segmented including bisetose tarsungulus. Abdomen 10-segmented, more than twice as long as thorax. Tergum A9 with pair of 2- or 1-segmented urogomphi. Segment A10 visible from above, anal region terminally oriented, with membranous anal lobes bearing numerous fine teeth (Silphinae) or without teeth (Nicrophorinae).

Spiracles: Annular, lateral or dorsolateral, closing apparatus present.

Comments: The family Silphidae is small, with fewer than 200 species known worldwide, most of them from the Holarctic region. *Nicrophorus,* with about 85 species worldwide including 15 in North America, forms most of the Nicrophorinae, which also includes the more primitive genus *Ptomascopus* with 2 species in eastern Asia. Silphinae includes from 3 to 22 genera, depending on the authority, with 15 species in the United States and Canada currently placed in 7 genera (Anderson and Peck, 1985). The species are often common and conspicuous members of the carrion microhabitat.

Immature stages of most genera but of fewer than half of all species are known. The most comprehensive larval descriptions are those of Pukowski (1934) and Anderson (1982) for Nicrophorinae, and Karsch (1884), von Lengerken (1938a and preceding parts referred to therein, and 1938b), Dorsey (1940), Paulian (1941), and Anderson and Peck (1985) for Silphinae. One or more species of all genera included in the family have been examined for the present treatment.

Carrion-frequenting species are of no known economic importance, but the phytophagous *Aclypea* species such as the North American Spinach Carrion Beetle, *A. bituberosa* (LeConte), may be pests of spinach, beets and other commercial crops (Cooley 1917, Martin 1945). Larvae of *Aclypea* species are separable from other silphine larvae by the coarsely serrate inner apex of the mandible (fig. 34.180).

Selected Bibliography

Anderson 1982.
Anderson and Peck 1985.
Balduf 1935 (biology).
Böving and Craighead 1931.
Brewer and Bacon 1975.
Cooley 1917.
Dorsey 1940.
Hayashi 1986 (*Nicrophorus, Eusilpha* larvae).
Hayashi et al. 1959.
Karsch 1884.
Klausnitzer 1978.
Lawrence and Newton 1982.
Lengerken, von 1938a, 1938b.
Martin 1945.
Milne and Milne 1976.
Paulian 1941.
Peterson 1951.
Prins 1984.
Pukowski 1933, 1934.
Ratcliffe 1972.
Roussel 1964a, 1964b.
Schiödte 1862.

STAPHYLINIDAE (STAPHYLINOIDEA) (INCLUDING BRATHINIDAE, EMPELIDAE)

J. H. Frank, *University of Florida*

The Rove Beetles

Figures 34.183–287. First instar, figures 3.129h, i (Vol. 1)

Relationships and Diagnosis: Phylogenetic relationships within the superfamily Staphylinoidea are unclear, and there are several interpretations of the confines of the family Staphylinidae. In North America, limits of the family now generally accepted are those put forward by Arnett (1968), with inclusion of *Brathinus* (family Brathinidae, now placed in Omaliinae) and exclusion of Micropeplidae (subfamily Micropeplinae). These are the limits accepted here.

Structures of larvae are diverse and present many similarities to those of related families. No single structure characterizes larvae of the family, and larvae of many genera, some tribes, and some subfamilies (Leptotyphlinae, Megalopsidiinae, Pseudopsinae) are undescribed. Knowledge of them may require modification in the following diagnosis. Larvae are elongate and campodeiform (with an uncertain report of an onisciform larva in *Cilea* (Tachyporinae)). The median epicranial suture is present. Antennae have 3 or 4 segments, with a sensory appendage on the penultimate segment; typically this appendage is on the anterior (= inner) face, but not so in some larvae where the frons, clypeus, and labrum are represented by a single sclerite, the nasale. Mandible without a distinct molar lobe, the base without asperities, the apex not acuminate and rather equally bifid in horizontal plane; only in some Omaliinae, Proteininae and Piestinae is there a process on the inner margin. Maxillary mala typically without a dense tuft of setae, but if a tuft is present, then setae plumose. Leg with claw-like tarsungulus, typically shorter than tibia. Tergal sclerites not extending laterally beyond edges of thoracic and abdominal segments. Spiracles uniforous. Urogomphi typically present and articulated, atypically absent or reduced to bulbous structures.

Biology and Ecology: Larvae are restricted to that part of the adults' habitat which is moist and concealed. Such habitats include banks and shores of rivers, streams, ponds, lakes and the sea, piles of decaying plant material and of animal dung, carrion, nests of vertebrates and social insects, fruiting bodies of fungi, leaf litter, caves, and the bark of fallen trees. Larvae of many species are predatory; although some of these may be exclusively predatory, this has not yet been demonstrated, and many of these larvae will feed in the laboratory on vegetable matter such as pieces of fruit. In contrast, and especially among Oxytelinae, some larvae may be primarily fungivorous, algivorous, or saprophagous. The food ingested by many adults is liquid after pre-oral digestion, and larvae appear to feed in the same way.

Eggs are spherical, spheroidal or, in Steninae, pyriform. Most are white or near-white. Chorionic microsculpture is pronounced in eggs of some Staphylininae. In many cases, eggs are laid singly, dropped on the substrate. In some Aleocharinae, excavations are made in the substrate for placement. In some Tachyporinae, the egg is smeared with dirt by the female. In Oxyporinae and especially Oxytelinae, brood chambers may be made for the eggs, and there may be subsocial behavior by the adults (Hinton, 1944). Hinton's (1981) descriptions of eggs are commendable.

The known pupae of Staphylininae and Xantholininae are obtect, sclerotized and pigmented. Some bear a fringe of fixed pronotal processes, and lateral processes on abdominal segments 7 and 8. Pupae of the remaining subfamilies are exarate, unsclerotized and unpigmented. Pupae of Steninae bear at least 1 pair of abdominal tergal processes. Pupation occurs in cells in the substrate. In Steninae, *Astenus* (Paederinae) and most Aleocharinae, a silken cocoon is spun (Frank and Thomas, 1984).

Most species appear to have 3 larval instars, but some Staphylininae and at least some Paederinae have 2, whereas some Aleocharinae and Oxytelinae are reported to have 4 and 5 instars respectively. A prepupal stage occurs at the end of the final instar. Published descriptions are mainly of the final instar.

Description: Final instar larvae ca. 1 to > 20 mm. Typically, sclerites of head pigmented pale to dark brown, but sclerites of thorax and abdomen feebly or unpigmented. Membrane between sclerites unpigmented.

Head: Typically prognathous, but rather hypognathous in some Osoriinae. With a nuchal constriction in Paederinae, Xantholininae and Staphylininae, and a feeble constriction in Euaesthetinae, but no constriction in remaining subfamilies. Divided dorsally into 3 regions by V-shaped ecdysial lines which separate the frontal area from the epicranial areas. Divided ventrally by more or less distinct median ecdysial line; anterior to this median line in Paederinae, Xantholininae and Staphylininae is an anterior ventral apotome, the "gula"; a small, posterior ventral apotome present in some Staphylininae. Frontal area with a single sclerite, the nasale, in some forms, but in others a labrum may be distinguished from clypeus and frons. Anterior margin of labrum or nasale lobed or serrate in many species. The areas on each side of head, posterior to antennal insertion, bear no stemmata (uncommon), 1 stemma (most Aleocharinae and Xantholininae), a group of 2 (uncommon), or 3, 4, 5 or 6 stemmata. Antennae borne dorso-laterally outside mandibular bases, of 3 or 4, typically elongate segments. Penultimate segment of antenna bears, distally, a sensory appendage, typically located on anterior (= inner) face, but in some species the appendage not evidently so placed, the result either of migration of the appendage or rotation of the antenna; in Euaesthetinae the appendage is dorso-lateral; in Steninae it is morphologically posterior, but directed anteriorly by antennal rotation. Mandibles do not have the distinctly enlarged and in some cases asperate molar base present in some other families of Staphylinoidea; a mesal mandibular process is present only in Proteininae and some Omaliinae and Piestinae; in Paederinae, Xantholininae, Staphylininae and Steninae, mandibles more or less simply falciform; in other subfamilies apex of the mandible may be divided and there is variation in number and position of tooth-like projections of median margin; asymmetry between left and right mandibles is usual. Each maxilla consists of cardo (in some species subdivided), stipes, palpifer (in some species very small), palp, and a single distal median process, the mala; palp of either 3 or 4 segments; mala a fixed structure in all subfamilies except Paederinae, Xantholininae and Staphylininae, where it is finger-like and mobile, resembling "outer lobe" of Adephaga. Labium in less derived forms consists of prementum, mentum and submentum, with a ligula which is broad and rounded, or flattened-acuminate, or cylindrical with rounded apex, or bilobed; in Paederinae, Xantholininae and Staphylininae, mentum and submentum fused indistinguishably to ventral surface of head capsule; labial palp of 2 or 3 segments; paraglossae absent; hypopharyngeal sclerite absent except in Steninae.

Thorax and Abdomen: Prothorax larger or much larger than meso- or metathorax. Each segment dorsally with a pair of tergal sclerites, the narrow, membranous, medial separation serving as an ecdysial line. Each segment ventrally with a number of separate, sclerotized plates in addition to epipleura and hypopleura. The 3 pairs of legs are present even in parasitic forms; coxa, trochanter, femur, tibia and tarsungulus are distinguishable, the last with 2 or more small, articulated spines; the tibia of some Staphylininae with a ctenidium of simple or bifid spines.

Abdomen with 10 segments, of which 1 through 6 or 8 are similar in width and length, the remainder through 9 gradually narrower and in many cases longer. Tergites and sternites represented by single sclerites or, in Staphylininae, Xantholininae and Paederinae, paired similarly to those of thorax. Epipleural and hypopleural sclerites fused to tergites and sternites respectively, or present as separate entities. Segment 9 bears dorsally a pair of articulated urogomphi (cerci), each with 1 to 3 segments and large, conspicuous setae; in a few species each urogomphus is a single bladder-shaped segment, or urogomphi are absent. Segment 10, the anal pseudopod, is an elongate cylinder, a truncated cone, or of intermediate form. The abdomen, as well as head and thorax, bears numerous articulated setae of which several distinct forms exist; thread-like trichobothria occur in a few locations in Paederinae; frayed setae are present in Proteininae, some Piestinae, Omaliinae, Tachyporinae, Staphylininae and Aleocharinae; bifid setae occur in the tibial ctenidium of some Staphylininae; plumose setae are recorded from the maxilla of some Oxytelinae; plain setae occur in numerous positions in all subfamilies.

Spiracles: A single pair of uniforous thoracic spiracles is borne in lobes between the prothorax and mesothorax. On the abdomen eight pairs of uniforous spiracles are present: pair 1 anterolaterally on segment 1, the remaining 7 pairs laterally on segments 2–8, either between the epipleural and tergal sclerites, or incorporated within the edges of the tergites.

Comments: This is one of the largest families of Coleoptera, with more than 3400 species estimated to occur in America north of Mexico (Frank and Curtis, 1979). Large numbers of individuals of some species may be collected by appropriate methods. Collection methods for larvae include pitfall trapping, Berlese funnel extracting (into containers with moist, crumpled paper, so that larvae remain alive), and separation of organic material from the habitat by means of a sieve (or sifter) of appropriate mesh size, allowing larvae to drop onto a tray or sheet. Collection methods for adults described by Smetana (1971) are, in part, appropriate to larvae. There are advantages to collecting adults alive, identifying them, and maintaining them in laboratory colonies to obtain reliably identified eggs, larvae and pupae. To maintain such laboratory colonies, overcrowding should be avoided because of cannibalism, humidity must be kept high, and unconsumed food must be removed scrupulously to prevent infestation by fungi or mites. Vermiculite or peat moss make adequate substrates for cultures of many species.

Knowledge of the natural history and immature stages of species of the region is negligible, so that Moore and Legner (1974) were able to list only about 3% of the species as even partially investigated. Most of our knowledge of the immature stages, and the only broad analysis of their structures, comes from publications by authors in Europe, notably Paulian (1941), Pototskaya (1967), Kasule (1966, 1968, 1970), Steel (1970), Tikhomirova (1973), and Topp (1978). Several hundred parasites, mainly fungi and Nemata, are known from adult staphylinids, few from immature stages (Frank, 1982).

There are few records from any part of the world of species being harmful to agriculture. Economic importance is achieved through the beneficial role of predatory and parasitic species, and has resulted in attempted uses for biological control of agricultural pests. Although laboratory studies have been performed to determine feeding rates of some predators, quantified field evaluations of predation are rare. Special attention has been given to species of *Aleochara*, parasitoids of Diptera puparia, and to species of *Oligota*, predators of mites.

There is yet no documentation of medical importance in North America, but there are many records from other parts of the world. Members of *Paederus* may release a potent toxin which is a DNA inhibitor and causes vesication (Frank and Kanamitsu, 1987). Certain predatory species are believed important in suppressing flea populations in rodent burrows; this is especially important where endemic plague occurs.

KEY TO SUBFAMILIES

NOTE: The "if" statements in couplets 3–8 are an attempt to be all-encompassing when dealing with a poorly known fauna.

1. Maxillary mala fixed (Figs. 34.241–246, 34.249–253) 2

1′. Maxillary mala articulated (Figs. 34.247, 248) 11

2(1). Maxillary mala equal to or exceeding segments 1 + 2 combined of maxillary palp (Figs. 34.241–246, 34.249–251) 3

2′. Maxillary mala shorter than segment 1 of maxillary palp (Figs. 34.252, 253) 10

3(2). Maxillary mala exceptionally long (Fig. 34.241) and urogomphus with 2 segments (Fig. 34.276); with frayed setae (Fig. 34.287) .... *1* **Proteininae**

3′. Maxillary mala shorter as in Figs. 34.242–246, and 34.249–251; or if approaching length of mala in Fig. 34.241, then urogomphus with a single segment (Fig. 34.277); frayed setae present or absent 4

4(3′). Urogomphus with a single segment (Fig. 34.277) and 5 stemmata (Fig. 34.193) or (uncommonly) 0 or (rarely) 6, and each mandible (left and right) with apex undivided (Fig. 34.229); larva not a parasitoid, not an inquiline of nests of social insects, without a prominent dorsal gland on abdominal segment 8 .... *2* **Omaliinae**

4′. Urogomphus with 2 segments (Figs. 34.278, 34.280–281); or if with a single segment (Figs. 34.279, 34.285) then mandibular apex (at least one of the 2 mandibles) divided, or 1,3, or 4 stemmata, or larva parasitic or an inquiline of nests of social insects, or with a prominent dorsal gland 5

5(4′). Urogomphus with 2 segments, 1 longer than anal pseudopod and much longer than 2 (Fig. 34.278) and 4 or 6 stemmata (Fig. 34.194) and ligula of labium broadly rounded (Fig. 34.257) .... *3* **Piestinae**

5′. Urogomphus with a single segment (Figs. 34.279, 34.285); or if with 2 segments (Figs. 34.280, 34.281, 34.284) and 4 or 6 stemmata (Figs. 34.196, 34.200) then ligula of labium cylindrical or conical (Figs. 34.259, 34.261) or segment 1 of urogomphus shorter than anal pseudopod (Fig. 34.284) 6

6(5′). Urogomphus with 2 segments and head with 3 stemmata or fewer; ligula of labium broadly rounded and mandibular apex (at least 1 of the 2 mandibles) trifid .... *4* **Osoriinae** (part)

6′. Urogomphus with a single segment (Fig. 34.285) and stemmata 0, 1, 3, 5, or 6; or if urogomphus with 2 segments and stemmata 3 or fewer, then ligula of labium cylindrical or conical (Figs. 34.259, 34.261) or mandibular apex undivided (not bifid, trifid or quadrifid, at most with a tooth near apex as in Fig. 34.236) 7

7(6′). Urogomphus with a single segment (Fig. 34.279) and mandibles stout (Fig. 34.231) with apex of at least 1 of them bifid, trifid or quadrifid; without a prominent dorsal gland on abdominal segment 8; found in decaying wood, under bark, and among roots of plants .... *4* **Osoriinae** (part)

7′. Urogomphus with 2 segments (Figs. 34.280, 34.281, 34.284); or if urogomphus with a single segment (Fig. 34.285) then mandible with a constriction near midpoint of length (Fig. 34.238) or apex undivided (not bifid, trifid or quadrifid, at most with a tooth near apex), or abdominal segment 8 with a prominent dorsal gland 8

8(7′). Urogomphus with a single segment (Fig. 34.285) and mandibular apex trifid (Fig. 34.238) or (rarely) bifid; 0, 1 (Fig. 34.201) or 3 stemmata; without a prominent dorsal gland on abdominal segment 8 .... *5* **Oxytelinae**

8′. Urogomphus with 2 segments (Figs. 34.280, 34.281, 34.284); or if urogomphus with a single segment, then mandibular apex simple, at most with a tooth near apex (Fig. 34.236) 9

9(8′). Maxillary mala trilobed (Fig. 34.250) and mandible stout and deeply bifid (Fig. 34.237); found in mushrooms .... *6* **Oxyporinae**

9′. Maxillary mala not trilobed (Figs. 34.245, 246, 249) and mandible not stout and deeply bifid (Figs. 34.232, 233, 236) 13

10(2′). Labium bilobed (Fig. 34.265) and head without a nuchal constriction .... *8* **Steninae**

10′. Labium with a conical ligula (Fig. 34.266) and head with a nuchal constriction .... *9* **Euaesthetinae**

11(1′). Labium with tapering ligula with dense setulae (Fig. 34.262); 6 stemmata (Fig. 34.198) or (uncommonly) 5 or (rarely) absent .... *11* **Paederinae**

11′. Labium with conical ligula rarely with more than a pair of setae (Fig. 34.260); 4 or 1 or (rarely) 2 stemmata 12

12(11′). One or (rarely) 2 stemmata .... *12* **Xantholininae**

12′. Four stemmata (Fig. 34.197) .... *13* **Staphylininae**

13(9′). Three stemmata (Fig. 34.195) or 5 or 6 and urogomphus of 2 segments of which apical segment less than 0.5× length of basal (Fig. 34.280) .... *15* **Phloeocharinae**

13′. If 3 or 6 stemmata, then apical segment of urogomphus not less than about 0.7× length of basal (Fig. 34.281) 14

14(13′). Three, 5 or 6 stemmata (Fig. 34.196); abdominal segment 8 without a dorsal gland .... *16* **Tachyporinae**

14′. One stemma (Fig. 34.199), rarely absent; abdominal segment 8 with an evident dorsal gland in more highly derived tribes .... *17* **Aleocharinae**

SUBFAMILY DIAGNOSES (LARVAE)

The following diagnoses may be incomplete or inaccurate because larvae of many genera remain unknown. For this reason, each diagnosis is based on a named genus, usually the type genus of the subfamily, and applies to the final instar larva. Variation provided by known larvae of other genera is noted.

1. **PROTEININAE:** *Proteinus* (examined). Subcylindrical. Head without nuchal constriction; labrum separated by unsclerotized area. Three stemmata (fig. 34.192). Antenna of 3 segments, segment 2 longest, bearing prominent sensory appendage (Fig. 34.215). Mandible with undivided apex, without pre-apical teeth, broad basally and slender apically, with mesal process (fig. 34.228). Maxilla with very long, narrow, slightly curved, fixed mala, and palp of 3 segments, extending anteriorly in advance of head (fig. 34.241). Labium with narrow ligula, and palp with 2 segments (fig. 34.254). Urogomphus with 2 segments, longer than anal pseudopod, segment 2 longest (fig. 34.276). Setation includes frayed setae (fig. 34.287). Variation: mandible with pre-apical teeth, 6 stemmata. The primary habitat is mushrooms, with carrion, caves and mammal burrows as subsidiary habitats.
2. **OMALIINAE:** *Omalium* (from Steel (1970)). Subcylindrical. Head without nuchal constriction; labrum separated by unsclerotized area (fig. 34.206). Five stemmata (fig. 34.193). Antenna of 3 segments, 2 longest bearing prominent sensory appendage (fig. 34.216). Mandible apically undivided, with pre-apical tooth, broad basally, moderately broad apically (fig. 34.229). Maxilla with strap-shaped fixed mala and palp of 3 segments, segment 2 shorter than 1 or 3 (fig. 34.242). Labium with narrow ligula and palp with 2 segments (fig. 34.255). Urogomphus with 1 segment, longer than anal pseudopod (fig. 34.277). Without frayed setae. Variation: 0 (uncommonly) or 6 stemmata (rarely), mandible with 0 or 2 pre-apical teeth and with mesal process, maxillary mala similar to that of Proteininae, ligula of labium broad or

absent, urogomphus shorter than anal pseudopod, frayed setae present. Adults and larvae occur in leaf litter, moss, and under bark of dead trees; adults of several species and larvae of a few also occur in flowers; larvae of most species are predatory, but adults of a few species are known to damage flowers.

3. **PIESTINAE:** *Piestus* (examined). Depressed (fig. 34.183). Head without a nuchal constriction; labrum separated by unsclerotized area (fig. 34.204). Four stemmata (fig. 34.194). Antenna of 3 segments, 2 longest with small sensory appendage (fig. 34.217). Left mandible quadrifid (fig. 34.230), right trifid, slender at middle of length, broadened basally. Maxilla with broad, fixed mala, palp of 3 segments, 2 longer than 1, about as long as 3 (fig. 34.243). Labium with broad ligula and small palp of 2 segments (fig. 34.257). Anal pseudopod short, broad, conical; urogomphus with 2 segments, 1 much longer than 2, much longer than pseudopod (fig. 34.278). Without frayed setae. Variation: 6 stemmata, mandible with a mesal, prostheca-like process, anal pseudopod longer, segment 1 of urogomphus not so much longer than 2, frayed setae present. Adults and larvae occur under bark of dead trees to which habitat they may be adapted by the depressed form; in the tropics they also occur in decaying vegetable matter (banana stems, cacao husks).
4. **OSORIINAE:** *Osorius* (examined). Cylindrical. Head without a nuchal constriction; labrum separated by a suture (fig. 34.205). Stemmata absent. Antenna stout, of 3 segments, 2 longest with small sensory appendage (fig. 34.218). Mandibles stout, left undivided at apex but with large tooth near midpoint, right trifid at apex (fig. 34.231). Maxilla with broad fixed mala, palp of 3 segments, 2 slightly longer than 1, 3 longest (fig. 34.244). Labium with broad ligula with 2 minute lobules, palp of 2 stubby segments (fig. 34.258). Urogomphus with a single segment, short, stout, just longer than the broad, rounded anal pseudopod (fig. 34.279). Anterior tibia stout (fig. 34.269). Without frayed setae. Variation: form depressed as in *Nacaeus* (fig. 34.184), 1, 2, 3, or 4 stemmata, antenna more slender, mandibular apex quadrifid, ligula without minute lobules, urogomphus with 2 segments. Adults and larvae of depressed form occur under bark of dead trees where they seem to feed on fungi; larvae resemble those of Piestinae; adults and larvae of cylindrical form occur in rotten wood and among the roots of plants, and there are reports of damage to lawn grass by *Osorius*.
5. **OXYTELINAE:** *Oxytelus* (examined). Subcylindrical (fig. 34.185). Head without nuchal constriction, labrum separated by suture (fig. 34.210). One stemma (fig. 34.201). Antenna of 3 segments, fairly stout, segment 2 longest and with prominent sensory appendage (fig. 34.225). Mandible broad at base, then constricted, expanded and trifid at apex (fig. 34.238). Maxilla with broad, fixed mala, palp of 3 segments, 1 and 3 about equal in length, 2 shorter (fig. 34.251). Labium with broad, rounded ligula, and palp of 2 stubby segments (fig. 34.264). Urogomphus of a single tapering segment, just longer than the short, stubby anal pseudopod (fig. 34.285). Without frayed setae. Variation: almost onisciform or cylindrical, 0 or 3 stemmata, maxillary mala with an apical tuft of setae, segment 2 of maxillary palp longer than 1, urogomphus falcate or minute. Riparian sites provide a major habitat, while vertebrate dung and decaying plant materials provide others. *Bledius* species live in tunnels in mud in riparian habitats and adults and larvae feed on algae. Some *Carpelimus* species have been accused of damaging vegetables. *Oxytelus* and *Anotylus* adults inhabit dung, and perhaps in response to the need to locate this transient habitat, are highly adept at aerial dispersal.
6. **OXYPORINAE:** *Oxyporus* (examined). Cylindrical or nearly so. Head without a nuchal constriction, labrum fused (fig. 34.211). Six stemmata (fig. 34.200). Antenna of 3 segments, 2nd longest and with a small sensory appendage (fig. 34.224). Mandible stout basally, broad and flat apically, bifid apically, much of the apical margin finely serrulate (fig. 34.237). Maxilla with large, fixed, trilobed mala, palp of 3 segments of which 3 is more slender than 1 and 2 (fig. 34.250). Labium with membranous ligula with 2 spines directed anteriorly, palp of 2 segments directed ventrally (fig. 34.263). Urogomphus of 2 segments, 1 slightly shorter than 2, together shorter than the cylindrical anal pseudopod (fig. 34.284). Without frayed setae. No major variation in this small subfamily. Adults and larvae inhabit mushrooms.
7. **MEGALOPSIDIINAE:** Larvae are undescribed. Adults are found on bracket fungi, though seldom.
8. **STENINAE:** *Stenus* (examined). Subcylindrical. Head narrowed posteriorly but without a nuchal constriction, labrum fused. Six stemmata (fig. 34.202). Antenna of 4 segments, 2 and 3 exceptionally long and slender in some but not all species, with minute sensory appendage at apex of 3 (fig. 34.226). Mandible falciform, without distinct molar lobe, finely denticulate along inner margin, apex undivided (fig. 34.239). Maxilla with small, fixed mala, shorter than segment 1 of palp, segment 3 as long as 1 and 2 combined (fig. 34.252). Labium with broadly bilobed ligula, palp of 2 segments (fig. 34.265). Urogomphus with a single segment about 0.6× length of anal pseudopod, bearing a very long terminal seta (fig. 34.286). Legs long and slender in some species. Frayed setae absent. Variation of length of appendages occurs but there is no major variation in structure. Adults occur in freshwater riparian sites and are predatory on smaller insects; larvae are rarely collected and also predatory; unlike most staphylinids, adults are active in bright daylight, are not lucifugous, and adults of some species climb on vegetation. Adults are remarkable for the protractile mouthparts as well as for the ability to skim on water surfaces (by using a glandular secretion to reduce surface tension).
9. **EUAESTHETINAE:** *Euaesthetus* (examined). Subcylindrical. Head with a feeble nuchal constriction, labrum fused (fig. 34.212). Six stemmata (fig. 34.203). Antenna stout, of 3 segments, 2 the longest and with a sensory appendage with basal annulus (fig. 34.227). Mandible falciform, without a molar lobe, undivided at apex (fig.

34.240). Maxilla with mala fixed and very small, much exceeded in length by palp, segment 3 of palp much longer and more slender than 1 and 2 combined (fig. 34.253). Labium with conical ligula, with palp of 2 segments, 2 longer than 1 (fig. 34.266). Urogomphus of a single segment about equal in length to anal pseudopod, terminated by a seta about 2× length of pseudopod. Larvae of other genera are unknown. Adults and larvae have been collected from wet leaf litter.

10. **LEPTOTYPHLINAE:** Larvae are undescribed. Adults are minute, apterous, eyeless, soil-inhabiting and of very restricted distribution.
11. **PAEDERINAE:** *Paederus* (examined). Subcylindrical. Head with a nuchal constriction, labrum fused, anterior margin with 2 pairs of large teeth and a median pair of small teeth (fig. 34.214). Six stemmata (fig. 34.198). Antenna of 4 segments of which 1 is annular, 2–4 elongate, 3 with a prominent sensory appendage (fig. 34.222). Mandible long, falciform, weakly serrate over much of length of inner margin, apically undivided (fig. 34.235). Maxilla with articulated mala, with palp of 3 segments of which 3 is long and acuminate (fig. 34.248). Labium with tapered ligula, weakly sclerotized, its surface setulose, palp of 2 segments, 1 the longest (fig. 34.262). Urogomphus with segment 1 nodose, about 2× length of segment 2, about 2× length of conical anal pseudopod. Anterior leg with slender, tapering tibia (fig. 34.273). Without frayed setae. With trichobothria (a) behind each group of stemmata, (b) at base of maxillary stipes, (c) at postero-lateral angle of pronotum. Variation: stemmata reportedly 5, maxillary palp with 4 segments, labial palp with 3, the form of the head capsule and arrangement of teeth along its anterior margin provide good diagnostic characters. The subfamily includes many genera (e.g. *Lithocharis* (fig. 34.186)) and species, and these occur in numerous moist habitats. Adults of some *Paederus* species climb on vegetation and are not lucifugous (and in this are similar to Steninae). Adults of the tribe Pinophilini seem adapted for climbing (by greatly expanded, setose anterior tarsi) and perhaps concomitantly for resistance to desiccation (by having abdominal terga fused with sterna to reduce unsclerotized area). Adults and larvae are mainly predatory.
12. **XANTHOLININAE:** *Xantholinus* (examined). Larvae are very similar to those of Staphylininae (below), so only the variation is noted. 1 or (rarely) 2 stemmata. Maxillary and labial palpi with 4 and 3 segments respectively. Tibial brush present. Frayed setae absent. Adults and larvae are predatory. Pupae obtect.
13. **STAPHYLININAE:** *Parabemus* (examined, considered by some authors as a subgenus of *Staphylinus*). Subcylindrical. Head with a nuchal constriction, labrum fused, anterior margin with 9 teeth (fig. 34.213). Antenna of 4 segments, 1 short, 2 and 3 the longest, 3 with a small sensory appendage (fig. 34.221). Four stemmata (fig. 34.197). Mandible falciform, without a molar lobe, undivided at apex (fig. 34.234). Maxilla with an articulated mala, palp with 3 segments (fig. 34.247). Labium with conical ligula with a pair of setae, not densely setulose, palp with 2 segments (fig. 34.260). Urogomphus with 2 segments, 1 longer than 2 and longer than the cylindrical anal pseudopod (fig. 34.282). Anterior tibia with a brush of bifid spines (fig. 34.272). Without frayed setae. Thoracic and abdominal tergites each with 2 pairs of very long setae (similar to trichobothria of Paederinae); head capsule with 2 pairs of setae (behind stemmata and at postero-lateral angle), one present near base of maxillary stipes. Variation: maxillary and labial palpi of 4 and 3 segments respectively, ligula densely setulose, tibial brush absent or of simple setae, urogomphus shorter than anal pseudopod, urogomphus of a single bulbous segment or apparently of 3 segments, frayed setae present. The subfamily includes many genera (e.g. *Creophilus* (fig. 34.187)) and species found in numerous moist habitats. Adults of some species attain sizes of > 20 mm, the largest known in the family. Adults and larvae are mainly predatory. Pupae are obtect.
14. **PSEUDOPSINAE:** Larvae are undescribed. Adults are small, soil-inhabiting and of restricted distribution.
15. **PHLOEOCHARINAE:** *Charhyphus* (examined). Depressed. Head without a nuchal constriction, labrum separated by unsclerotized area (fig. 34.207). Three stemmata (fig. 34.195). Antenna of 3 segments, 2 the longest and bearing a prominent sensory appendage (fig. 34.219). Left mandible falciform (fig. 34.232), right stouter and with preapical cleft and serrulation along apical third. Maxilla with broad, fixed mala, palp of 3 segments of progressive length (fig. 34.245). Labium with conical ligula and palp of 2 segments (fig. 34.259). Urogomphus considered here to have 2 segments although ability of presumed segment 2 to articulate not clear, segment 1 longer than cylindrical anal pseudopod, much longer than 2 (fig. 34.280). Anterior tibia short, about 2× length of tarsungulus (fig. 34.270). Without frayed setae. Variation: stemmata are reported as 5 or 6, segment 1 of urogomphus relatively shorter. Adults and larvae occur under bark of fallen trees and in leaf litter. Larvae are not very distinct from those of Tachyporinae.
16. **TACHYPORINAE:** *Tachyporus* (examined). Subcylindrical. Head without nuchal constriction, labrum separated by a suture (fig. 34.208). Six stemmata (fig. 34.196). Antenna of 3 segments, 2 the longest and with a prominent sensory appendage (fig. 34.220). Mandible narrowed from base to apex, with single pre-apical tooth, apex undivided, inner margin serrulate on either side of tooth (fig. 34.233). Maxilla with broad, fixed mala and palp of 3 segments of progressive length (fig. 34.246). Labium with conical ligula and palp of 2 segments (fig. 34.261). Urogomphus with segment 1 longer than cylindrical anal pseudopod, longer than segment 2 (fig. 34.281). Without frayed setae. Variation: (e.g., *Coproporus* (fig. 34.188)): body reported to be onisciform, 3 or 5 stemmata, antenna with 4 segments, ligula absent or short and obtuse, urogomphus shorter than anal pseudopod. Adults and larvae occur in a variety of moist habitats. Adults of some species climb on plants at night and feed on small, soft-bodied insects such as aphids. Other species are associated with mushrooms.

17. **ALEOCHARINAE:** *Aleochara* (examined). Subcylindrical. Head without a nuchal constriction, labrum separated by a suture (fig. 34.209). One stemma (fig. 34.199). Antenna short, of 3 segments, second with a prominent sensory appendage (fig. 34.223). Mandible broad basally, narrowed apically, undivided apically, with a single pre-apical tooth (fig. 34.236). Maxilla with broad, fixed mala and very small palp of 3 segments, stipes with a long seta (fig. 34.249). Labium with broadly lobed ligula, palp of 2 stubby segments (fig. 34.256). Abdominal segment 10 shaped as a truncated cone, little protruding beyond apex of segment 9, perhaps not functional as an anal pseudopod, urogomphi absent. Anterior tibia broad, less than 2× as long as tarsungulus (fig. 34.274). *Aleochara* is atypical because larvae are parasitoids in Diptera puparia and appendages are reduced (*see* figs. 3.129h,i, Vol. 1). Variation (e.g., *Atheta* (fig. 34.189)): urogomphus present and with 1 or 2 segments, antennae, palpi and legs relatively longer than in *Aleochara,* frayed setae present, stemmata generally 1 but absent in other *Aleochara* and in larvae of some species occurring as inquilines in nests of social insects; abdominal segment 8 in larvae of more highly derived tribes with a prominent dorsal (defensive) gland. Adults and larvae of this huge subfamily occur in almost every conceivable moist habitat. Adults and larvae of some are known to be predatory.

Selected Bibliography

Arnett 1968.
Ashe and Watrous 1984.
Frank 1982.
Frank and Curtis 1979.
Frank and Kanamitsu 1987.
Frank and Thomas 1984.
Hinton 1944, 1981.
Kasule 1966, 1968, 1970.
Kistner 1982.
Lawrence and Newton 1982.
Moore and Legner 1974, 1975.
Paulian 1941.
Pototskaya 1967, 1976a, 1976b.
Seevers 1957, 1965, 1978.
Smetana 1971.
Steel 1970.
Szujecki 1966.
Tikhomirova 1973.
Tikhomirova and Melnikov 1970.
Topp 1978.

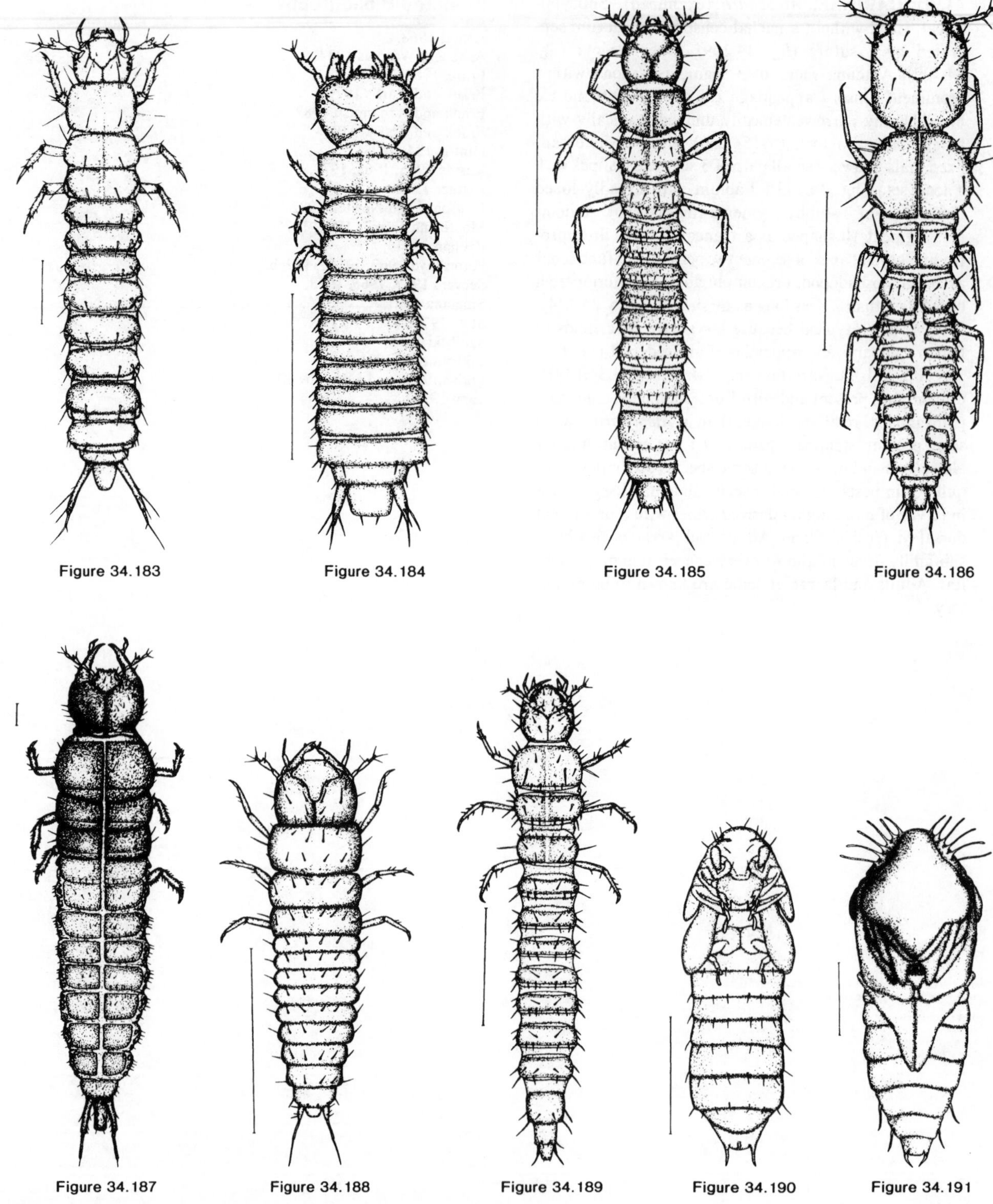

Figures 34.183–34.189. Staphylinidae. Final instar larvae: **(34.183).** *Piestus* (Piestinae), **(34.184).** *Nacaeus* (Osoriinae), **(34.185).** *Oxytelus* (Oxytelinae), **(34.186).** *Lithocharis* (Paederinae), **(34.187).** *Creophilus* (Staphylininae), **(34.188).** *Coproporus* (Tachyporinae), **(34.189).** *Atheta* (Aleocharinae). Scale line = 0.1 mm.

Figures 34.190, 34.191. Staphylinidae. Ventral view of pupa of: **(34.190).** *Oxytelus* (Oxytelinae), **(34.191).** *Belonuchus* (Staphylininae). Scale line = 0.1 mm.

Figures 34.192–34.203. Staphylinidae. Schematic diagrams of stemmata (small circles) and antennal insertion (large circle) of left side of head of larva of: **(34.192).** *Proteinus* (Proteininae), **(34.193).** *Omalium* (Omaliinae), **(34.194).** *Piestus* (Piestinae), **(34.195).** *Charhyphus* (Phloeocharinae), **(34.196).** *Tachyporus* (Tachyporinae), **(34.197).** *Parabemus* (Staphylininae), **(34.198).** *Paederus* (Paederinae), **(34.199).** *Aleochara* (Aleocharinae), **(34.200).** *Oxyporus* (Oxyporinae), **(34.201).** *Oxytelus* (Oxytelinae), **(34.202).** *Stenus* (Steninae), **(34.203).** *Euaesthetus* (Euaesthetinae).

Figures 34.204–34.214. Staphylinidae. Anterior part of upper surface of head of larva of: **(34.204).** *Piestus* (Piestinae), **(34.205).** *Osorius* (Osoriinae), **(34.206).** *Omalium* (Omaliinae), **(34.207).** *Charhyphus* (Phloeocharinae), **(34.208).** *Tachyporus* (Tachyporinae), **(34.209).** *Aleochara* (Aleocharinae), **(34.210).** *Oxytelus* (Oxytelinae), **(34.211).** *Oxyporus* (Oxyporinae), **(34.212).** *Euaesthetus* (Euaesthetinae), **(34.213).** *Parabemus* (Staphylininae), **(34.214).** *Paederus* (Paederinae).

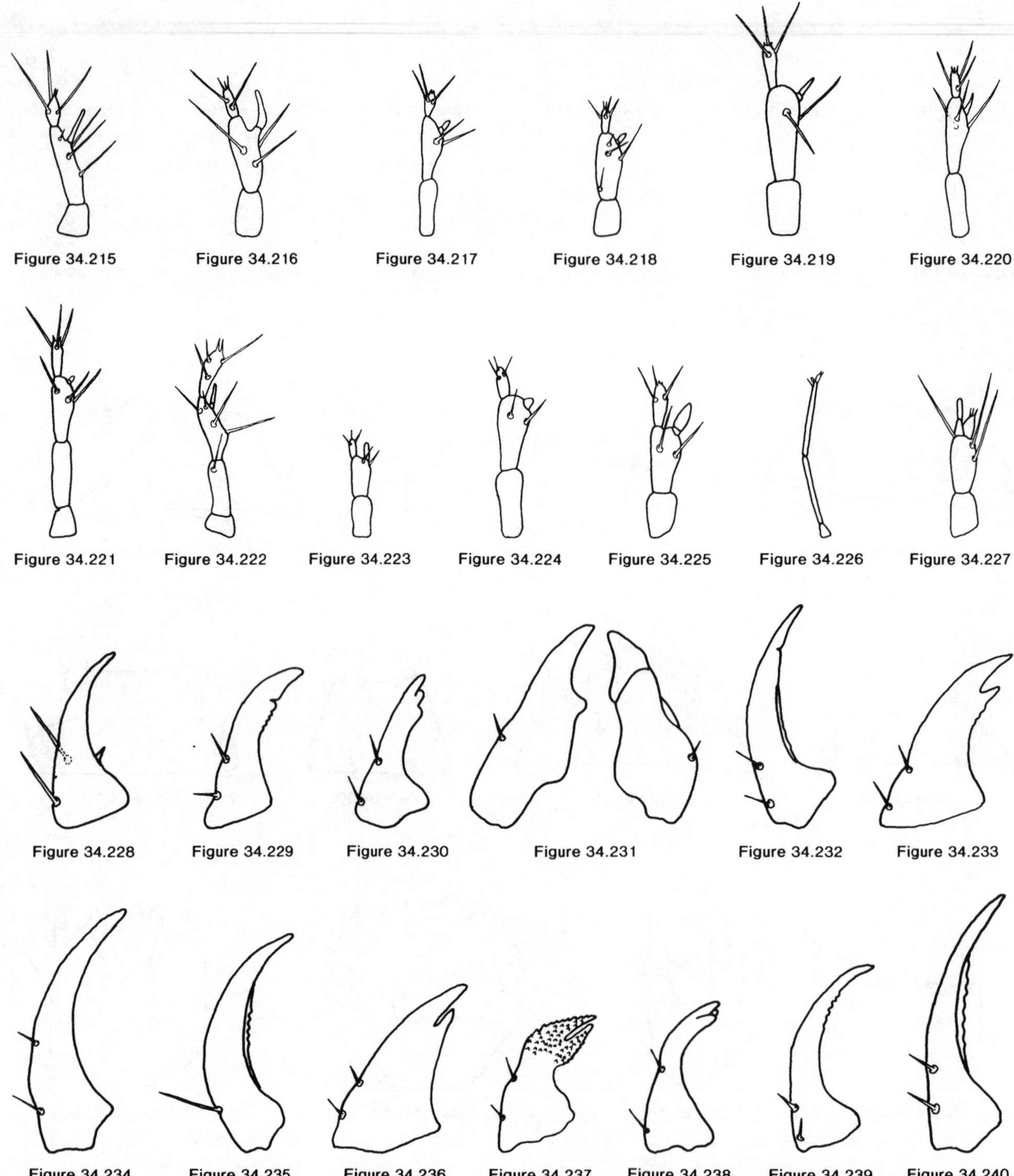

Figures 34.215–34.227. Staphylinidae. Left antenna from above of larva of: **(34.215).** *Proteinus* (Proteininae), **(34.216).** *Omalium* (Omaliinae), **(34.217).** *Piestus* (Piestinae), **(34.218).** *Osorius* (Osoriinae), **(34.219).** *Charhyphus* (Phloeocharinae), **(34.220).** *Tachyporus* (Tachyporinae), **(34.221).** *Parabemus* (Staphylininae), **(34.222).** *Paederus* (Paederinae), **(34.223).** *Aleochara* (Aleocharinae), **(34.224).** *Oxyporus* (Oxyporinae), **(34.225).** *Oxytelus* (Oxytelinae), **(34.226).** *Stenus* (Steninae), **(34.227).** *Euaesthetus* (Euaesthetinae).

Figures 34.228–34.240. Staphylinidae. Left mandible from above of larva of: **(34.228).** *Proteinus* (Proteininae), **(34.229).** *Omalium* (Omaliinae), **(34.230).** *Piestus* (Piestinae), **(34.231).** *Osorius* (Osoriinae) (both mandibles shown), **(34.232).** *Charhyphus* (Phloeocharinae), **(34.233).** *Tachyporus* (Tachyporinae), **(34.234).** *Parabemus* (Staphylininae), **(34.235).** *Paederus* (Paederinae), **(34.236).** *Aleochara* (Aleocharinae), **(34.237).** *Oxyporus* (Oxyporinae), **(34.238).** *Oxytelus* (Oxytelinae), **(34.239).** *Stenus* (Steninae), **(34.240).** *Euaesthetus* (Euaesthetinae).

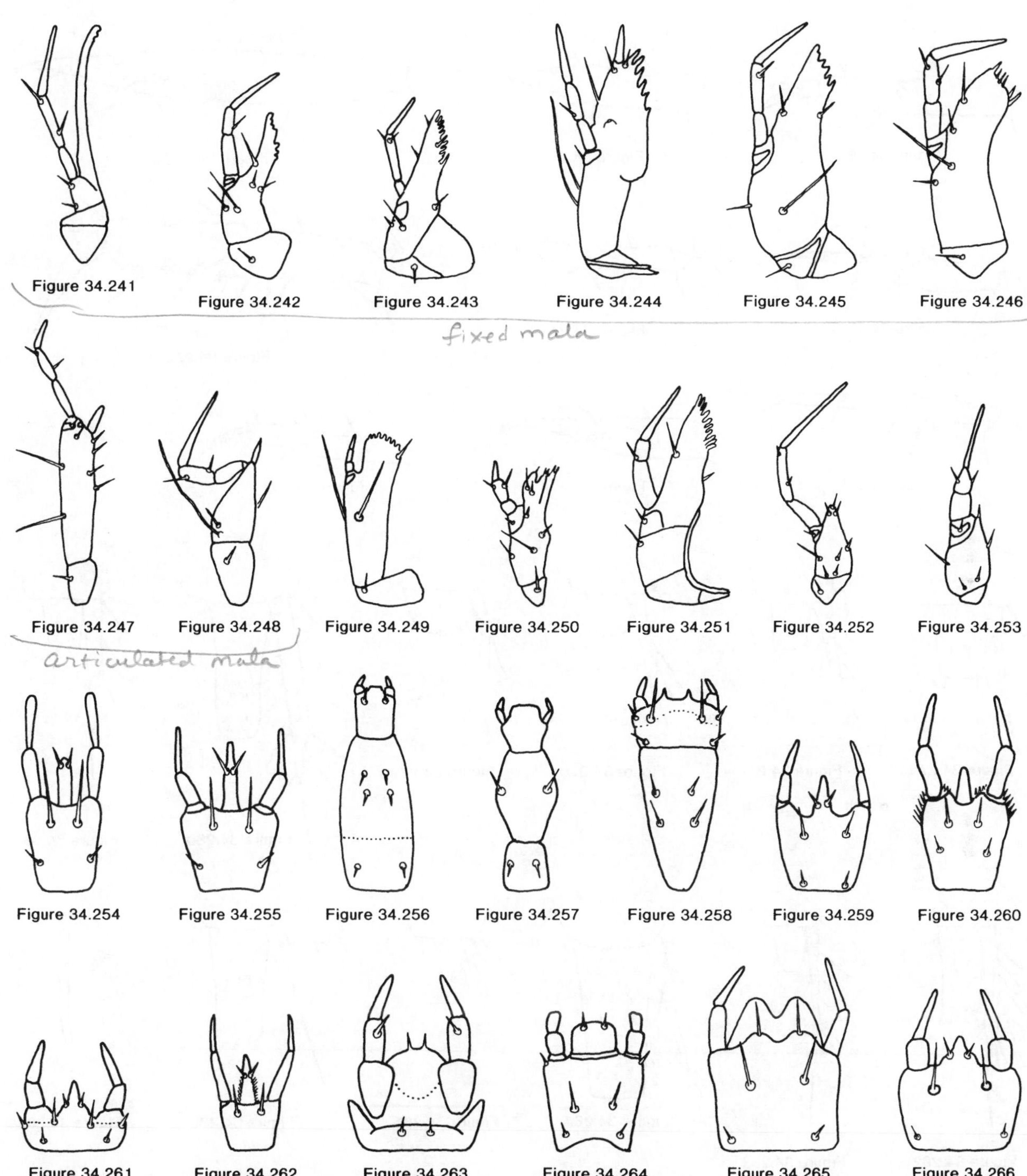

Figures 34.241–34.253. Staphylinidae. Right maxilla from above of larva of: **(34.241).** *Proteinus* (Proteininae), **(34.242).** *Omalium* (Omaliinae), **(34.243).** *Piestus* (Piestinae), **(34.244).** *Osorius* (Osoriinae), **(34.245).** *Charhyphus* (Phloeocharinae), **(34.246).** *Tachyporus* (Tachyporinae), **(34.347).** *Parabemus* (Staphylininae), **(34.248).** *Paederus* (Paederinae), **(34.249).** *Aleochara* (Aleocharinae), **(34.250).** *Oxyporus* (Oxyporinae), **(34.251).** *Oxytelus* (Oxytelinae), **(34.252).** *Stenus* (Steninae), **(34.253).** *Euaesthetus* (Euaesthetinae).

Figures 34.254–34.266. Staphylinidae. Labium and palpi of larva of: **(34.254).** *Proteinus* (Proteininae), **(34.255).** *Omalium* (Omaliinae), **(34.256).** *Aleochara* (Aleocharinae), **(34.257).** *Piestus* (Piestinae), **(34.258).** *Osorius* (Osoriinae), **(34.259).** *Charhyphus* (Phloeocharinae), **(34.260).** *Parabemus* (Staphyininae), **(34.261).** *Tachyporus* (Tachyporinae), **(34.262).** *Paederus* (Paederinae), **(34.263).** *Oxyporus* (Oxyporinae), **(34.264).** *Oxytelus* (Oxytelinae), **(34.265).** *Stenus* (Steninae), **(34.266).** *Euaesthetus* (Euaesthetinae).

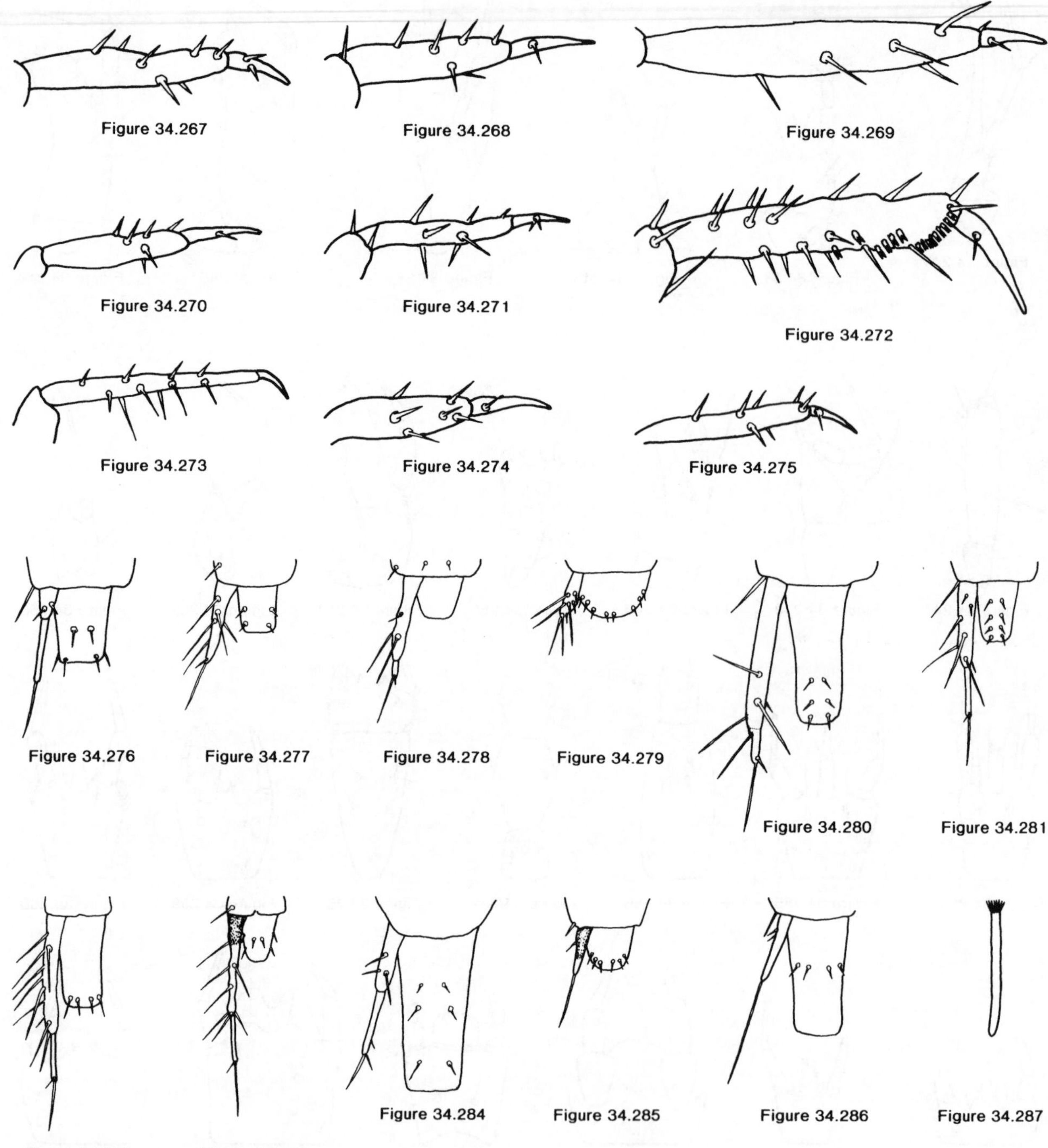

Figures 34.267–34.275. Staphylinidae. Anterior aspect of left protibia and tarsungulus of larva of: **(34.267).** *Omalium* (Omaliinae), **(34.268).** *Piestus* (Piestinae), **(34.269).** *Osorius* (Osoriinae), **(34.270).** *Charhyphus* (Phloeocharinae), **(34.271).** *Tachyporus* (Tachyporinae), **(34.272).** *Parabemus* (Staphylininae), **(34.273).** *Paederus* (Paederinae), **(34.274).** *Aleochara* (Aleocharinae), **(34.275).** *Oxytelus* (Oxytelinae).

Figures 34.276–34.287. Staphylinidae. Dorsal view of apex of abdominal segment 9, of anal pseudopod and left urogomphus (right omitted) of larva of: **(34.276).** *Proteinus* (Proteininae), **(34.277).** *Omalium* (Omaliinae), **(34.278).** *Piestus* (Piestinae), **(34.279).** *Osorius* (Osoriinae), **(34.280).** *Charhyphus* (Phloeocharinae), **(34.281).** *Tachyporus* (Tachyporinae), **(34.282).** *Parabemus* (Staphylininae), **(34.283).** *Paederus* (Paederinae), **(34.284).** *Oxyporus* (Oxyporinae), **(34.285).** *Oxytelus* (Oxytelinae), **(34.286).** *Stenus* (Steninae). Figure **34.287.** Frayed seta as found in *Proteinus* (Proteininae) and larvae of some other subfamilies.

PSELAPHIDAE (STAPHYLINOIDEA) (INCLUDING CLAVIGERIDAE)

Alfred F. Newton, Jr.,
Field Museum of Natural History

Figures 34.288–291

Relationships and Diagnosis: Although usually maintained as a distinct family, pselaphids have universally been considered to be closely related to and probably derived from Staphylinidae. Opinions differ on whether they are most closely related to certain primitive staphylinid subfamilies such as Omaliinae (e.g., Lawrence and Newton 1982, Thayer 1987, and most taxonomic specialists on the family) or to "higher" subfamilies such as Steninae and Euaesthetinae (e.g., Kasule 1966, Naomi 1985). Adults resemble some Staphylinidae but have less flexible abdomens.

Pselaphid larvae may be separated from most other beetle larvae by possession of a combination of: size small (under 5 mm long); head lacking differentiated neck; stemmata 3 or fewer on each side (usually 2, widely spaced in vertical row); labrum fused to head capsule; mandible falcate, with more or less scoop-like mesal surface and double mesal edge, with at least 1 tooth on one of the edges and usually 1 or more on each edge; mala large, fixed, apically blunt and sparsely setose; ligula absent; and urogomphi short and fixed to completely absent. With the exception of Faronini, larvae also have a pair of eversible glandular structures between the antennae; a sensorium of the second antennal segment that is variably and often complexly modified from palpiform, and placed ventrally, posteriorly or apically on the segment; and legs with tibia abruptly differentiated into wide basal and slender apical portions. Faronine larvae, with palpiform anteroventral antennal sensorium, normal legs and no glands, closely resemble certain scydmaenid and staphylinid larvae, but the former have a dome-like sensorium and the latter have a distinct ligula.

Biology and Ecology: Pselaphids are prominent members of the forest floor litter fauna, but also occur in a variety of other microhabitats including rotting logs, mosses, treeholes, caves, mammal nests, and other sources of decaying organic matter. Many species inhabit ant and termite nests, and the large subfamily Clavigerinae is apparently mostly myrmecophilous with many highly modified species that may be fed by the ants. All pselaphids are believed to be predators as adults and larvae, feeding on mites and other minute organisms or (inquilinous species) on host brood. Larvae of *Batrisodes oculatus* (Aubé) capture Collembola by means of a viscous secretion, and spin a silky pupal cocoon with their fore legs (DeMarzo 1986). The biology of most species is very poorly known, however, and no complete life histories have been worked out. See Arnett (1968), Kistner (1982) and Newton and Chandler (1989) and contained references, and Besuchet (1952, 1956b).

Egg unknown.

Pupa exarate, with functional abdominal spiracles on segments 1–3.

Description: Mature larvae about 2–4 mm long, elongate and parallel-sided to ovate, slightly flattened, relatively straight. Body surfaces very lightly pigmented and sclerotized (except some Faronini, with well sclerotized brown terga); smooth or (rarely) microspinose, with vestiture of short to very long simple setae only or including frayed or complex setae.

Head: Prognathous or slightly declined, protracted, without differentiated neck. Epicranial stem short to moderately long, frontal arms V-shaped; endocarinae absent. Stemmata usually 2 on each side, well separated in vertical row; sometimes 3 (in vertical row, or 1 dorsal, 2 ventral), 1, or absent. Antenna 3- or 2-segmented, less than half as long to longer than head width, sensorium of second segment ventrad to posterad of third segment or (some 2-segmented antennae) apical or (Faronini) anteroventral, sensorium palpiform to bifid or complexly modified and usually longer than third segment. Frontoclypeal suture absent. Frons between antennae (except Faronini) with pair of eversible membranous lobes which may bear complex glandular (?) or adhesive (?) structures; each lobe originating from mesad extension of antennal foramen or from base of antenna. Labrum completely fused to head capsule, forming nasale with or without small teeth. Mandibles symmetrical or nearly so, narrow and falcate, apex with single tooth, mesal surface concave between double edges, each edge with one to several teeth (Faronini with single tooth on dorsal edge only); mesal surface of base simple or with one or several fine teeth; mola and prostheca absent. Cardines oblique or longitudinal, undivided, narrowly separated by submentum or membrane. Stipes elongate to transverse. Mala large, fixed, apically rounded or truncate, sparsely setose, sometimes (some Faronini) with dorsal preapical lobe. Maxillary palp 3- or 2-segmented, not much longer than mala. Labium consisting of prementum, mentum and submentum, or prementum only with mentum and submentum indistinct or membranous. Ligula absent. Labial palps 2- or 1-segmented, contiguous to separated by much more than width of first palpal segment. Gular sutures separate, fused or absent; gula, if present, elongate. Occipital foramen not divided by tentorial bridge which, if evident, originates from posterior edge of head and may be very elongate.

Thorax and Abdomen: Thoracic and abdominal terga and most or all sterna consisting of 1 or more sclerotized plates, without patches or rows of asperities. Legs long, 5-segmented including bisetose or glabrous tarsungulus; tibiae differentiated into wide basal and slender apical portions (except Faronini). Abdomen 10-segmented, slightly longer to more than twice as long as thorax. Tergum A9 with pair of fixed urogomphi which may be long and thick, short and spiniform, or minute, or urogomphi absent. Segment A10 usually distinct and visible from above, anal region terminally oriented, membranous anal lobes with or without numerous fine teeth.

Spiracles: Annular, lateral or dorsolateral, slightly elevated, closing apparatus present; spiracles of several apical abdominal segments atrophied in some species.

Comments: This is a large and diverse family of about 1100 genera and well over 8000 species worldwide, especially well represented in tropical and warm temperate areas but

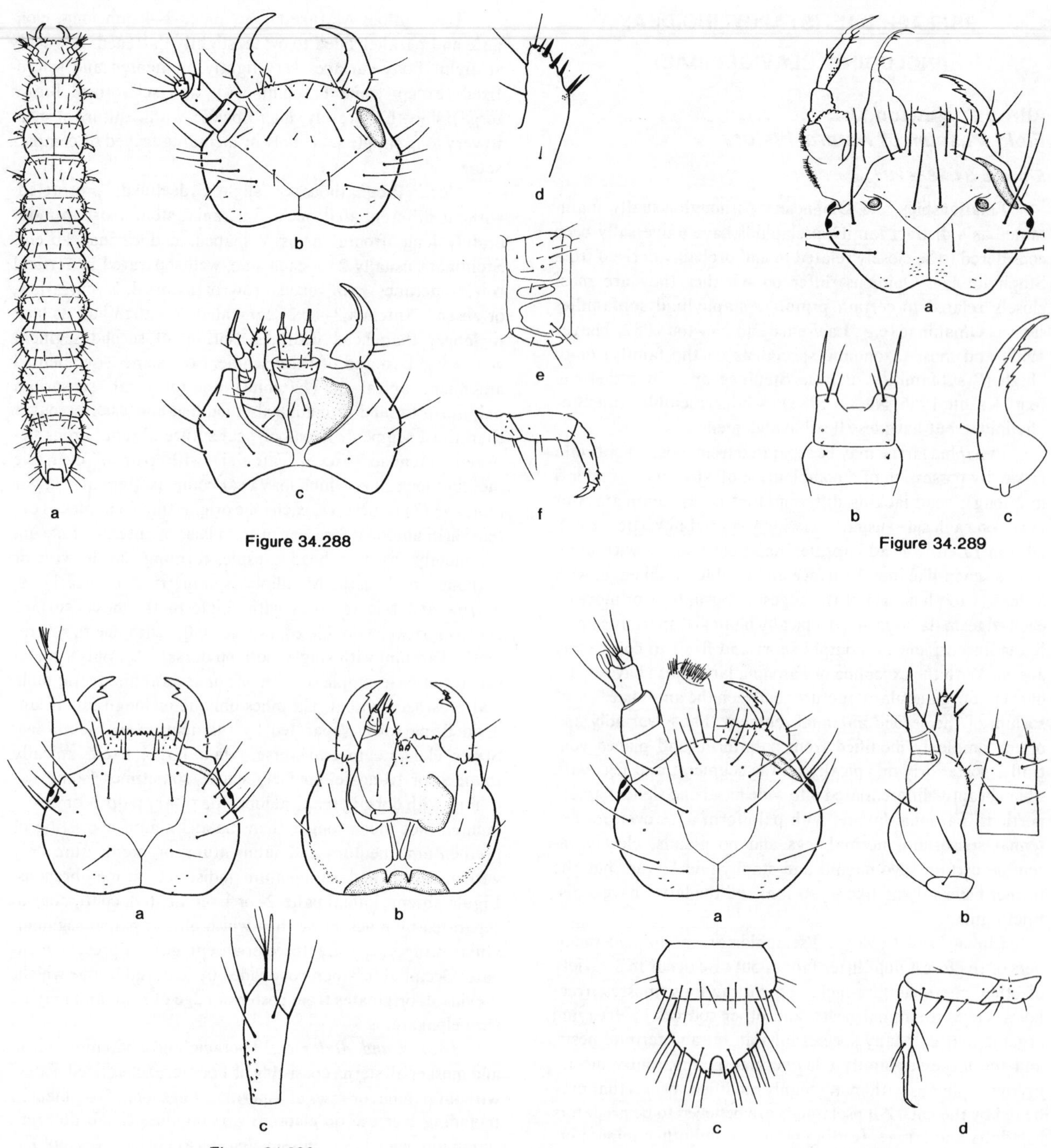

Figures 34.288a-f. Pselaphidae. *Sonoma cavifrons* Casey (Faronini). Inverness (3 miles northwest), California; May; ground litter from *Alnus* grove along stream. Length = 2.3 mm. **a.** mature (?) larva, dorsal; **b.** head, dorsal, right antenna not shown; **c.** head, ventral, left maxilla not shown; **d.** right mala, mesoventral view of apex; **e.** abdominal segment 3, left lateral; **f.** left proleg, anterior.

Figures 34.289a-c. Pselaphidae. Genus "G" (tribe unknown). Cerro Campana, 3200 feet, Panama; June; in leaf litter, cloud forest. **a.** head, dorsal, right antenna and eversible gland not shown; **b.** labium, ventral; **c.** right mandible, ventral.

Figures 34.290a-c. Pselaphidae. Genus "A" (tribe unknown). Tlanchinol (2.5 miles north), 2500 feet, Hidalgo, Mexico; July; under bark of hardwood logs. **a.** head, dorsal, right antenna not shown, eversible glands inverted; **b.** head, ventral, left maxilla not shown; **c.** left antennal segment 2, ventral.

Figures 34.291a-d. Pselaphidae. Genus "B" (?Euplectini). Jefferson Notch, 3000 feet, New Hampshire; September; from ground litter, *Picea* forest. **a.** head, dorsal, right antenna and eversible gland not shown; **b.** labium and right maxilla, ventral; **c.** abdominal apex, dorsal; **d.** right proleg, anterior.

found nearly everywhere. Over 70 genera and 500 species placed in 18 tribes are known from the United States and Canada. Although commonly encountered, the species are of no known economic importance.

Larvae are rarely seen and very poorly known. Klausnitzer (1978) presents the only key, covering 4 genera in 2 tribes (Euplectini, Batrisini). Useful larval descriptions or illustrations are provided by Rosenberg (1925), Böving and Craighead (1931), Besuchet (1952, 1956b), Kasule (1966), and DeMarzo and Vit (1982). The present treatment is based on more than a dozen larval types, none of them reared and many not identifiable to tribe; at least Faronini, Euplectini, Batrisini and Pselaphini are included.

Selected Bibliography

Arnett 1968 (adult classification).
Besuchet 1952, 1956b.
Böving and Craighead 1931.
Brown and Crowson 1980 (possible larva of Faronini).
DeMarzo 1986 (*Batrisodes* larva, biology).
DeMarzo and Vit 1982 (*Batrisodes* larva).
Kasule 1966 (family characterization).
Kemner 1927 (*Euceroncinus*, genus based on larva).
Kistner 1982.
Klausnitzer 1978.
Lawrence and Newton 1982.
Naomi 1985.
Newton and Chandler (1989) (adult classification).
Rosenberg 1925.
Silvestri 1920 (*Ceroncinus*, genus based on pselaphid larva).
Thayer 1987.
Wagner 1975 (*Euplectus* larva).
Wasmann 1918 (possible larva of Clavigerinae).

HYDROPHILIDAE (HYDROPHILOIDEA)

(INCLUDING HELOPHORIDAE, HYDROCHIDAE, SPHAERIDIIDAE, SPERCHEIDAE)

Paul J. Spangler, *Smithsonian Institution*

Water Scavenger Beetles

Figures 34.292–303

Relationships and Diagnosis: The family Hydrophilidae belongs to the suborder Polyphaga. It presently is divided into 9 subfamilies—Berosinae, Chaetarthriinae, Epimetopinae, Hydrochinae, Helophorinae, Hydrobiinae, Hydrophilinae, Spercheinae and Sphaeridiinae. Böving and Craighead (1931) and Bertrand (1972) in their treatment of larvae included several of the subfamilies listed above as valid families. Van Emden (1956), on the basis of the presumed larva of *Georyssus,* also considered the Georyssidae to be a subfamily of the Hydrophilidae, but most subsequent authors have maintained the genus *Georyssus* as a separate family. Members of the subfamily Sphaeridiinae, unlike the other hydrophilid subfamilies, are terrestrial but are associated with moist habitats such as rotting vegetation, rotting fungi, dung, and debris in bird and mammal nests.

Hydrophilid larvae (figs. 34.292–303) are easily distinguished from dytiscid larvae and other adephagan larvae by their 5-segmented legs. In addition, the breathing atrium on the eighth abdominal segment is distinctive for hydrophilid larvae (except larvae of *Berosus* and *Helophorus* which do not have a breathing atrium). Also, the single claw (tarsungulus) of hydrophilid larvae distinguishes them from adephagan larvae (which have 2 claws, except Haliplidae). From larvae of non-adephagan families of aquatic beetles, most hydrophilid larvae may be recognized by the breathing atrium and the conspicuous, usually toothed and/or serrulate inner margins of the mandibles. Larvae of the few genera of terrestrial hydrophilids assigned to the Sphaeridiinae whose larvae resemble the non-adephagan aquatic larvae have their legs reduced or absent. Mature hydrophilid larvae range in length from 1.5 to 60.0 mm.

Biology and Ecology: Hydrophilid adults and larvae inhabit diverse habitats. Most are aquatic and occur in standing or running water. Members of the terrestrial subfamily Sphaeridiinae occur in moist habitats as mentioned above. Females of the plesiomorphic hydrophilids (i.e., the Sphaeridiinae) lay their eggs in damp places singly, in pairs, or in clusters under loosely applied silklike strands laid down by spinnerets arising from anal glands. Females of the more primitive, truly aquatic species also fasten their eggs loosely under a few strands of silklike material in contrast to the derived taxa which construct sturdy, floating, submerged, or partly submerged egg cases. Some of these egg cases are terminated with spikelike masts which stick up above the floating or attached cases. Others end in fobs or longer ribbon-like strips which trail off loosely into the water or are attached to the substrate. The egg cases are usually distinctive for each genus, therefore, many may be identified by their characteristic shapes, sizes, etc. In some genera the species also may be identified by the characteristic shape of their egg cases, masts, ribbons, flanges, and similar adornments. Females of some genera (*Helochares, Helobata, Epimetopus*) carry their eggs in masses attached to the abdomen by a few silklike filaments until the eggs hatch. Females from temperate regions begin laying eggs in early spring and continue throughout late summer. Larvae undergo 3 molts and pupation usually occurs in moist soil. Larvae and adults are abundant throughout the summer, but by late August larvae become scarce. Hydrophilids normally overwinter as adults. Hydrophilid adults are omnivorous and feed on vegetable or animal substances; this is in contrast to the larvae which are mostly voracious predators.

Egg cases and larvae usually may be found in association with adults during the appropriate season. The aquatic larvae usually are found in the shallow weedy margins of ponds, puddles, ditches, stream banks, etc., close to the shoreline.

Description: Mature larvae 1.5 mm to 60.0 mm. Body form variously shaped, elongate, semicylindrical, cylindrical or moderately flattened in cross section; cuticle usually flaccid except for thoracic tergal sclerites and occasionally small sclerites on abdominal terga; tubercles, asperities, and gills may be present on larvae of some genera. Integument white

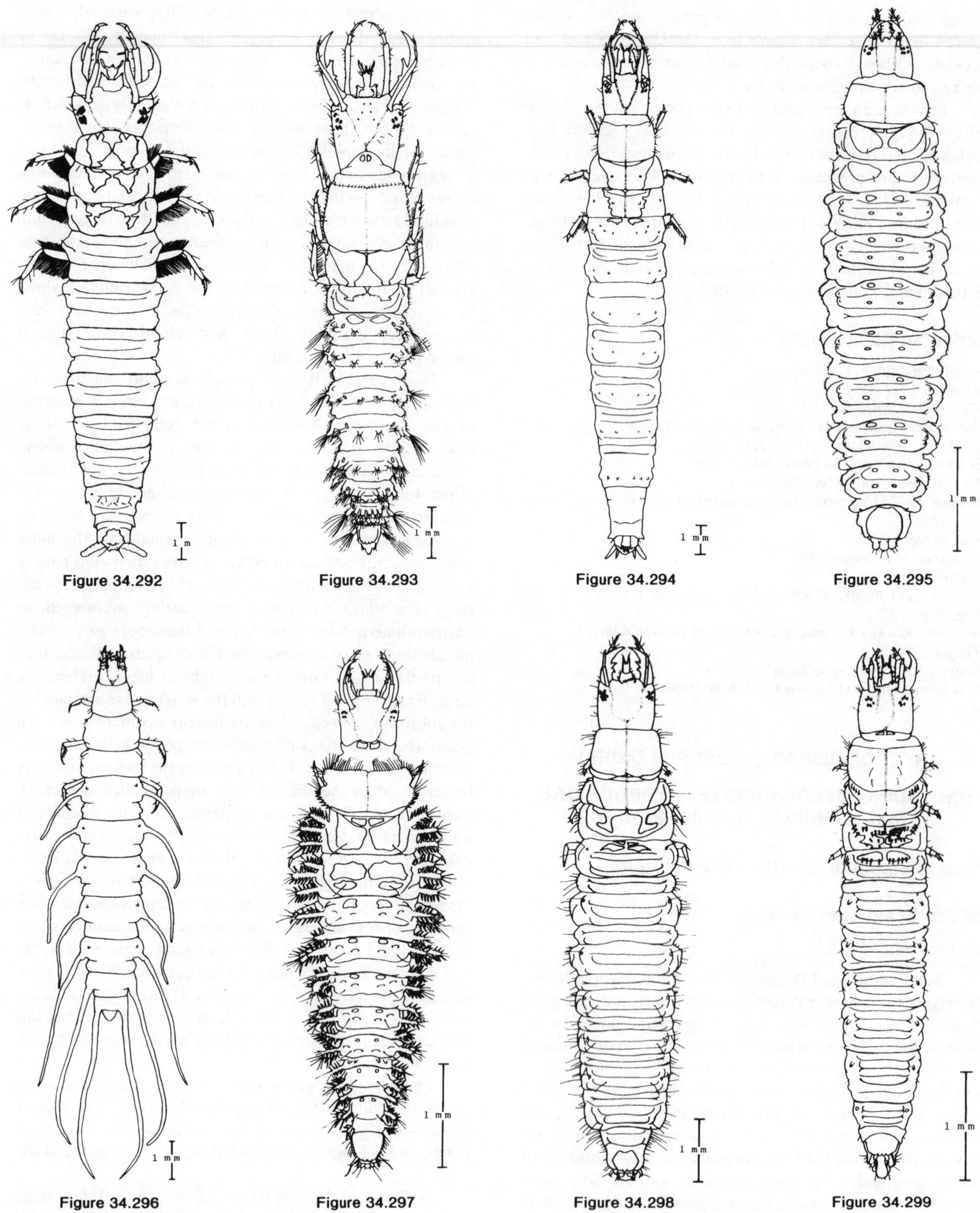

Figure 34.292. Hydrophilidae. *Hydrophilus* sp.

Figure 34.293. Hydrophilidae. *Tropisternus* sp.

Figure 34.294. Hydrophilidae. *Hydrobiomorpha casta* (Say).

Figure 34.295. Hydrophilidae. *Laccobius* sp.

Figure 34.296. Hydrophilidae. *Berosus metalliceps* Sharp.

Figure 34.297. Hydrophilidae. *Derallus* sp.

Figure 34.298. Hydrophilidae. *Helochares* sp.

Figure 34.299. Hydrophilidae. *Enochrus* sp.

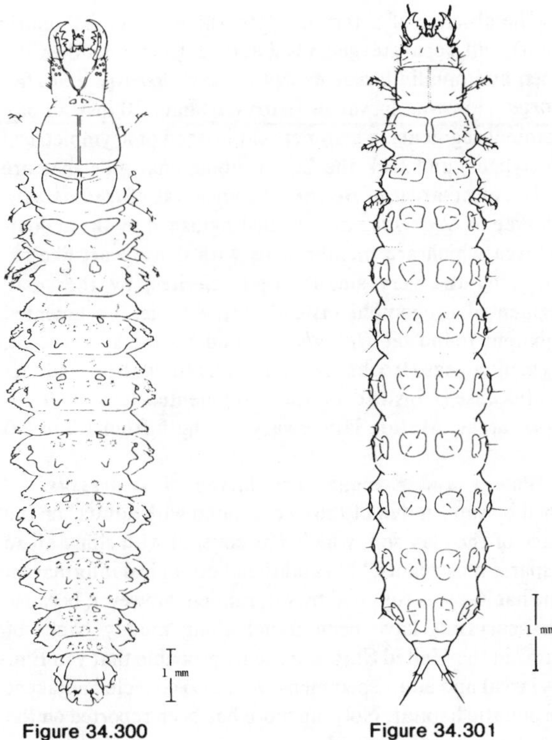

Figure 34.300 Figure 34.301

Figure 34.300. Hydrophilidae. *Sperchopsis tessellata* (Ziegler).
Figure 34.301. Hydrophilidae. *Helophorus* sp.

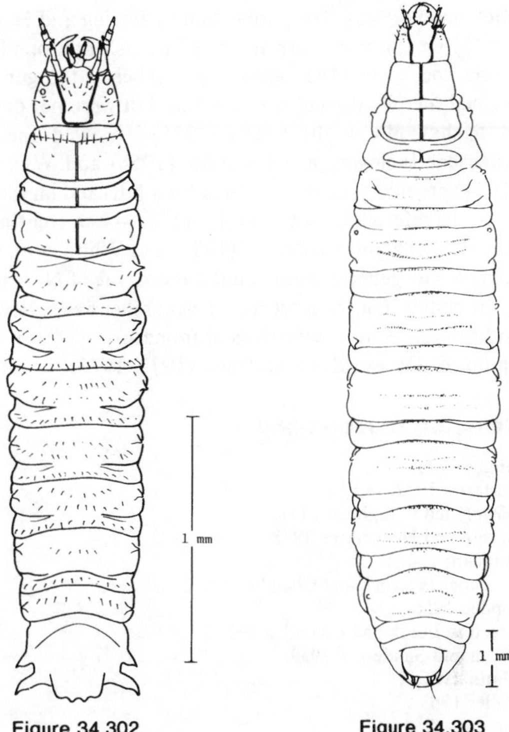

Figure 34.302 Figure 34.303

Figure 34.302. Hydrophilidae. *Sphaeridium scarabaeoides* (L.).
Figure 34.303. Hydrophilidae. *Dactylosternum* sp.

when larva is freshly hatched, then often becoming grey with aging; sclerotized portions become darker yellowish brown with aging.

Head: Prognathous; usually flattened; ecdysial cleavage lines usually present and distinct. Clypeus and labrum fused. Labroclypeus (nasale) toothed or not. Stemmata usually 6 pairs, sometimes fewer. Antennae 3- or 4-segmented, inserted farther from lateral margin of head than mandible. Mandibles large, moderately curved; inner margin usually with 1 to 3 teeth, sometimes serrate. Maxilla with cardo small; stipes usually elongate and slender or robust; palpifer segment-like, bearing small slender process apicomedially. Ligula absent or present.

Thorax and Abdomen: Prothoracic tergal sclerite usually well developed; meso- and metathoracic tergal sclerites diminishing in size. Legs present (most aquatic taxa) or absent; legs, when present, normally 5-segmented, including tarsungulus.

Abdomen of 8–10 segments, usually with 8 visible segments. Gills present or absent. Cuticle usually with numerous folds and moderate lobes, pubescent or leathery. Urogomphi 1- or 2-segmented and short in those genera with 8 abdominal segments; urogomphi 2- or 3-segmented and longer in those genera with 10 abdominal segments. Terminal breathing cavity present except in *Berosus* which has lateral gills, and *Helophorus* which has 9 instead of the usual 8 abdominal segments in other genera.

Spiracles: Biforous, on mesothorax and abdominal segments 1–8, those on segment 8 usually enlarged; spiracles absent in first and second instars of Berosini.

Comments: The family is cosmopolitan but with a greater number of species in tropical areas. At present there are about 1600 species worldwide, with 34 genera and 256 species known from America north of Mexico. Larvae of 25 of the 34 North American genera have been described.

Most species of hydrophilids are not directly economically important, however, adults of some are known to be intermediate hosts of parasites of domestic animals and some are thought to be involved in the transmission of botulism in waterfowl. Predacious larvae are known to feed on noxious snails, and mosquito and ceratopogonid larvae, while phytophagous or omnivorous adults and larvae feed on plant materials. Three species of the genus *Dactylosternum* have been introduced into Hawaii and the Philippines as a biological control agent against the sugar cane borer, and 1 species has been introduced to help control a banana borer in Jamaica. Also, both adults and larvae are food for aquatic birds and various mammals. Thus, hydrophilids play an important role in recycling plant and animal materials in their respective aquatic or terrestrial ecosystems, and to a certain extent as biological control agents against some pests of humans.

The Hydrophilidae is the second largest family of water beetles in North America, and the immature stages of many genera have been described. The most comprehensive study of the immatures of America north of Mexico is that by

Richmond (1920). The publication by Böving and Henriksen (1938) on the immature stages of Danish hydrophilids deserves special mention because a number of the genera described therein also occur in the U.S. but were not described in Richmond's study. Leech (1956) reviewed the North American biologies, and Chandler (1956) and White et al. (1984) provide keys to the genera for adults and larvae. They also cite references to the major publications treating both stages. Leech and Sanderson (1959) and White et al. (1984) also provide generic keys to adults and larvae of N. American hydrophilids. For the most recent comprehensive review, keys, and references to descriptions of immature stages of hydrophilids of the world see Bertrand (1972, 1977).

Selected Bibliography

Angus 1973.
Bertrand 1972, 1977.
Böving and Craighead 1931.
Böving and Hendriksen 1938.
Brigham 1982d.
Chandler *in* Leech and Chandler 1956.
Emden 1956.
Leech *in* Leech and Chandler 1956.
Leech and Sanderson 1959.
Matta 1982.
Miller 1963.
Moulins 1959.
Pennak 1978.
Perkins and Spangler 1981.
Richmond 1920.
Rocha 1967.
Spangler 1961, 1962c, 1966a, 1966c, 1974b.
Spangler and Cross 1972.
White et al. 1984.

GEORYSSIDAE (HYDROPHILOIDEA)

Paul J. Spangler, *Smithsonian Institution*

Minute Mud-Loving Beetles

Figures 34.304a–e

Relationships and Diagnosis: In North America, the georyssids, with the single genus *Georyssus,* are included in the Hydrophiloidea along with the Hydrophilidae (*sensu lato*), Sphaeritidae, Synteliidae and Histeridae. There is some disagreement whether they should be maintained as a separate family or as a subfamily of the Hydrophilidae. Crowson (1955), Leech and Chandler (1956), Arnett (1968), Bertrand (1972), Doyen and Ulrich (1978) and White et al. (1984) treated them as a separate family, Georyssidae. When the presumed larva of the European *Georyssus crenulatus* Rossi was described by van Emden (1956), he considered the georyssids as a subfamily of the Hydrophilidae because of the similar morphological characteristics of the presumed georyssid larva to those of other hydrophilid larvae. In most of its morphological characteristics, the larva (fig. 34.304a) of *Georyssus crenulatus* most closely resembles larvae of members of the Helophorinae (Hydrophilidae).

The absence of a terminal breathing cavity (stigmatic atrium) will separate georyssid larvae from all other described hydrophilid larvae except those of *Berosus* and *Helophorus*. However, larvae of *Georyssus* have 10 instead of 8 abdominal segments, a symmetrical instead of asymmetrical labroclypeus, and lack the lateral abdominal gills that are found on most larvae of *Berosus*. From larvae of *Helophorus,* the larvae of *Georyssus* may be distinguished by the abdominal terga which are membranous with dense wart-like asperities, by the very small tergal sclerites, by the small 1-segmented urogomphi instead of the large, 3-segmented urogomphi found on *Helophorus,* and by the short, stout, 3-segmented legs (trochanter and femur fused, tibia and tarsungulus fused) instead of the 5-segmented legs of *Helophorus* larvae. Mature larvae vary in length from 1.2 to 2.0 mm.

Biology and Ecology: The larvae of *Georyssus* described by van Emden (1956) were found with adults "on the surface of the clay soil, which was covered with slimy algae and sparse mossplants." No additional ecological information on the habitat was supplied by van Emden. However, because adult georyssids have been found along muddy banks of streams in the United States, it seems probable that van Emden's larval and adult specimens were also collected adjacent to an aquatic habitat. Nothing more has been reported on the bionomics of the Georyssidae.

Description: Only 2 larvae of *Georyssus* are reported in the literature (van Emden, 1956); one is 1.21 mm long and the other is 2.05 mm long. Because the georyssid adults are small (1.5 to 3 mm), the larger known larvae may represent the last instar. Head, mouthparts, pronotum, and the small tergal sclerites pale testaceous; the rest of the larva pale brownish with darker brownish asperites. The larvae resemble those of Hydrophilidae.

Head: Prognathous, exserted; subquadrate, moderately flattened; ecdysial cleavage lines present (fig. 34.304b). Clypeus and labrum fused. Labroclypeus (nasale) essentially symmetrical; with a distinct medial dentiform projection; anterolateral angles prolonged anteriorly and bearing numerous dentes on inner margins. Stemmata, 6 pairs. Antenna, 3-segmented; antennal foramen extending farther laterad than the point of articulation of the mandible. Maxilla with small cardo, subquadrate stipes; elongate palpiger bears a small outer lobe and a 3-segmented palp (fig. 34.304c). Labium with 3-segmented palps; ligula absent. Mandibles large, symmetrical; strongly curved; inner margin of each bearing 2 large teeth and finely serrate from those teeth to apex.

Thorax and Abdomen: Pronotal sclerite complete, almost covering entire dorsal surface. Meso- and metathorax each with 2 pairs of sclerites. Legs short and stout, about a third as long as width of mesothorax; prothoracic pair (fig. 34.304d) markedly stouter than mesothoracic pair (fig. 34.304e). Trochanter and femur fused, much wider than long. Tibia and tarsungulus fused (figs. 34.304d,e).

Abdominal segments 1–7 each with 3 pairs of small ovate tergal sclerites; eighth segment with only 2 pairs of tergal sclerites; ninth without paired tergal sclerites, and with a pair of widely separated truncate conical urogomphi; apex of each

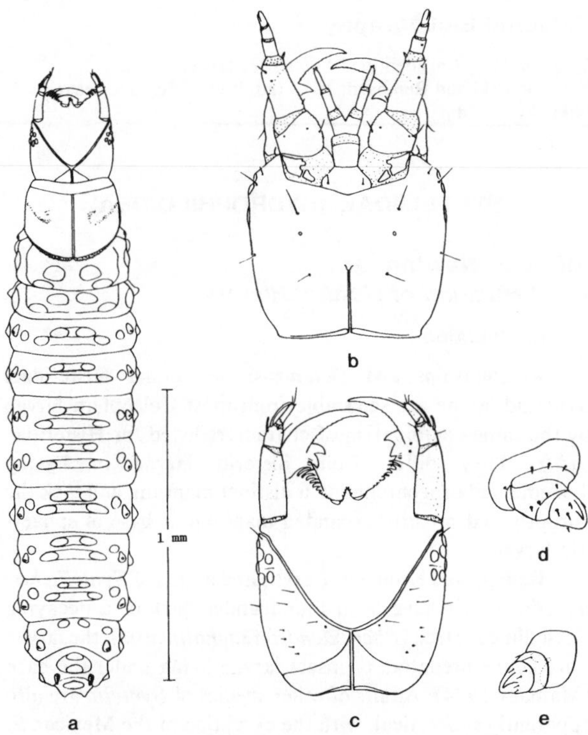

Figures 34.304a-e. Georyssidae. *Georyssus crenulatus* Rossi. **a.** larva, dorsal; **b.** head capsule, ventral; **c.** head capsule, dorsal; **d.** leg, prothoracic; **e.** leg, mesothoracic. (Redrawn from van Emden, 1956.)

urogomphus with a small papilla-like structure, presumably the vestigal second segment; tenth segment broad and short, very distinct.

Spiracles: Biforous, those of mesothorax large and prominent. Abdominal segments 1–8 each with a dorsolateral pair. No terminal breathing cavity.

Comments: This family is cosmopolitan and known from temperate and tropical regions. At present there are about 35 species reported, all in the genus *Georyssus*. Only 2 species (rare in collections) are known from America north of Mexico; the immature stages have not been described. Georyssids are of no known economic importance.

Selected Bibliography

Arnett 1968.
Bertrand 1972.
Crowson 1955.
Doyen and Ulrich 1978.
Emden 1956.
White et al. 1984.

SPHAERITIDAE (HYDROPHILOIDEA)

Alfred F. Newton, Jr.,
Field Museum of Natural History

Figures 34.305a-h

Relationships and Diagnosis: *See* under Histeridae. Sphaeritid larvae are separable from most Coleoptera larvae by the same combination of characters listed for Histeridae (q.v.). They differ from histerid larvae in having 4-segmented urogomphi and a distinct mentum, and from synteliid larvae in having mandibles with an abruptly and broadly expanded base.

Biology and Ecology: Adults of *Sphaerites glabratus* (Fabricius) were found on the sides of birch stumps feeding on exuding sap and mating; females laid eggs in the upper layers of soil impregnated with sap at the base of the stumps (Nikitsky 1976d). First instar larvae were obtained from these eggs, but larval feeding has not been recorded.

Egg large, white.

Pupa unknown.

Description: (Instar 1 of *Sphaerites glabratus* (Fabricius), based on Nikitsky (1976d): Larva elongate, slightly flattened. Body surfaces very lightly to moderately pigmented and sclerotized, generally smooth, with sparse vestiture.

Head: Prognathous, protracted, strongly transverse, without differentiated neck. Epicranial stem absent, frontal arms present, their bases distinctly separated; endocarinae absent. Stemmata apparently absent. Antenna 3-segmented, slightly less than half as long as head width, sensoria of preapical segment posterad of apical segment and conical. Labrum completely fused to head capsule, forming nasale with single median tooth. Mandibles symmetrical, falcate, apex with single tooth, incisor edge with heavily sclerotized tooth, mesal surface of base with penicillus or brush of hairs and with enlarged mola-like lobe. Cardines distinct, transverse, separated from each other by mentum. Stipes elongate. Maxilla without apical lobes. Maxillary palp 4-segmented, first segment with articulated digitiform appendage. Labium consisting of prementum, mentum and submentum. Mentum apparently membranous. Ligula absent. Labial palps 2-segmented, separated by less than width of first palpal segment. Gula elongate.

Thorax and Abdomen: Thoracic terga with 1 or more sclerotized plates, without asperities. Legs 5-segmented including tarsungulus. Abdomen apparently 10-segmented. Terga and sterna A1–8 each with numerous small sclerotized plates and with transverse row of asperities. Tergum A9 much longer than tergum A8, with pair of long 4-segmented urogomphi.

Spiracles: Biforous, lateral.

Comments: The family includes the single rare genus *Sphaerites*, with 1 species (*politus* Mannerheim) widespread in western North America and 2 others known from Europe and China. The only known immature stages are eggs and first instar larvae of the European *S. glabratus* (Fabricius) reared by Nikitsky (1976d), who demonstrated that larvae attributed by Crowson (1974) to this species were incorrectly identified. None of the species are of economic importance.

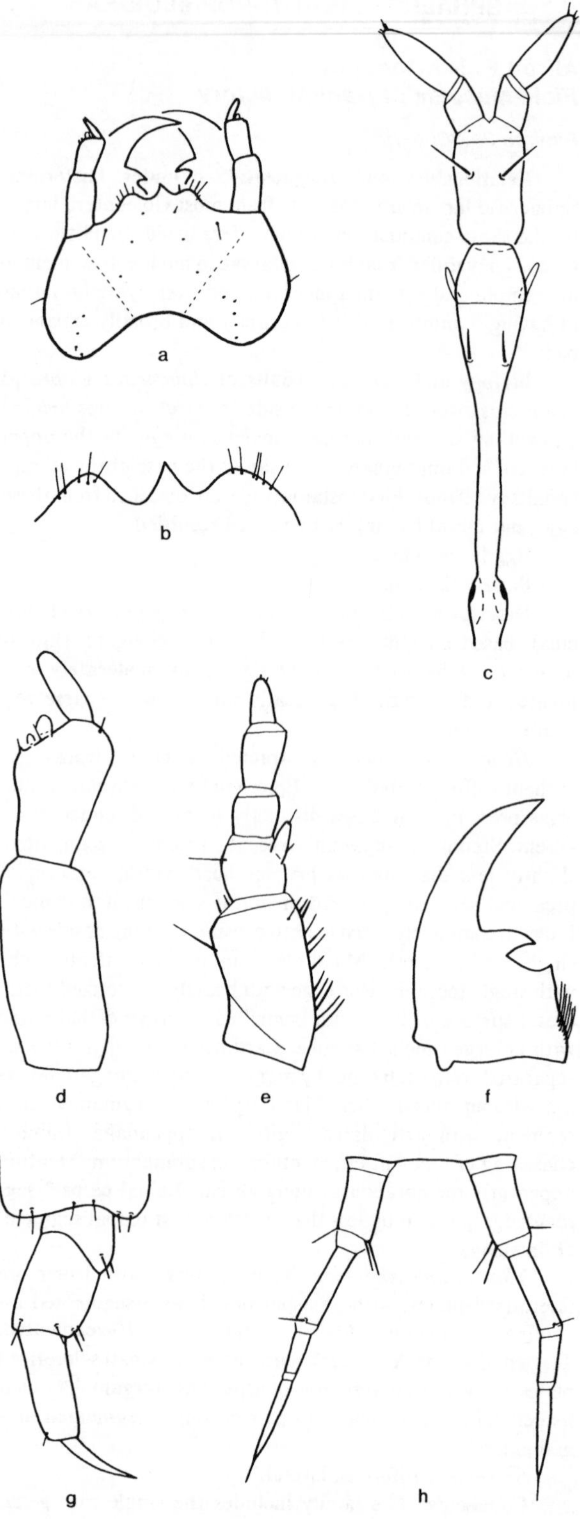

Figures 34.305a-h. Sphaeritidae. *Sphaerites glabratus* (Fabricius). Moscow Province, USSR; June; reared from eggs laid by beetle found in soil impregnated with birch sap near a fresh birch stump. **a.** instar 1, head, dorsal; **b.** nasale; **c.** labium, mentum, submentum; **d.** antenna; **e.** maxilla; **f.** mandible; **g.** mesothoracic leg; **h.** urogomphi. (Redrawn from Nikitsky, 1976d.)

Selected Bibliography

Crowson 1974 (misidentified *Sphaerites* larva).
Kryzhanovski and Reikhardt 1976 (adult classification, ecology).
Nikitsky 1976d.

SYNTELIIDAE (HYDROPHILOIDEA)

Alfred F. Newton, Jr.,
Field Museum of Natural History

Figures 34.306a–c

Relationships and Diagnosis: *See* under Histeridae. Synteliid larvae are separable from most Coleoptera larvae by the same combination of characters listed for Histeridae (q.v.). They differ from histerid larvae in having 4-segmented urogomphi and a distinct mentum, and lack the abruptly and broadly expanded mandibular base of sphaeritid larvae.

Biology and Ecology: Larvae and adults of *Syntelia histeroides* Lewis have been found under bark of a decaying Sakhalin cork tree (*Phellodendron sachalinensis*); the larvae were active predators of insect larvae living under the bark (Mamaev 1974). Adults of other species of *Syntelia* are also apparently subcortical, with the exception of the Mexican *S. westwoodi* Sallé which has only been found on large columnar cacti in a fermentation stage of decay; the latter adults were observed feeding on maggots (personal observations).

Egg unknown.

Pupa exarate, with functional abdominal spiracles on segments (1 or 2?)–6.

Description: Mature larva about 15 mm long, elongate and parallel-sided, straight, moderately flattened. Body surfaces very lightly pigmented and sclerotized, smooth, with sparse vestiture of simple or slightly truncate setae.

Head: Prognathous, protracted, well sclerotized and pigmented, without differentiated neck. Epicranial sutures absent (instar 2) or (instar 1) with short stem and V-shaped frontal arms which end before antennal foramen; endocarinae absent. Stemmata absent. Antenna 3-segmented, about half as long as head width, sensoria of preapical segment posterad of apical segment and conical. Antennal insertion separated from mandibular insertion by narrow strip of membrane only. Labrum completely fused to head capsule, forming nasale with rounded median projection. Mandibles symmetrical, narrow and falcate, apex with single tooth, incisor edge with heavily sclerotized tooth, mesal surface of base with penicillus or brush of hairs, mola absent. Cardo absent (or fused to mentum). Stipes elongate. Maxilla without apical lobes. Maxillary palp 4-segmented, first segment with articulated digitiform appendage. Labium consisting of prementum and mentum, mentum apparently fused laterally to cardines. Ligula absent. Labial palps 2-segmented, separated by about width of first palpal segment. Gular sutures absent, but indistinct medial ecdysial line present. Venter of head about as long and well sclerotized as dorsum. Occipital foramen not divided by tentorial bridge, which is indistinct or absent.

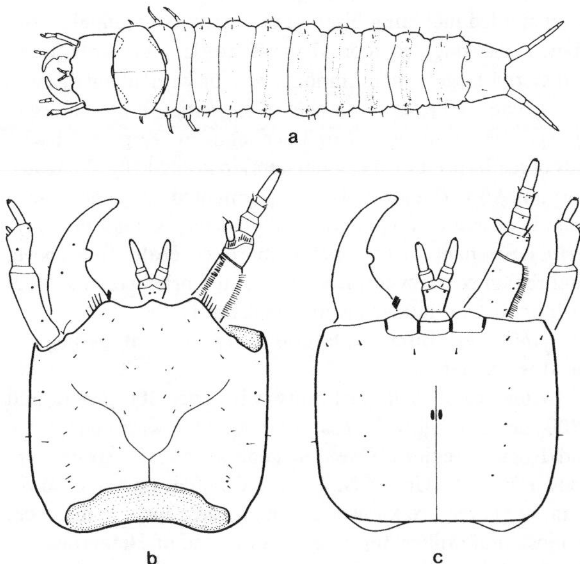

Figures 34.306a-c. Synteliidae. *Syntelia histeroides* Lewis. Kunashir, Kurile Islands, USSR; September; under bark of fallen Sakhalin cork tree (*Phellodendron sachalinensis*). Length (somewhat contracted) = 16 mm. **a.** mature larva, dorsal; **b.** instar 1, head, dorsal, right antenna and mandible not shown; **c.** instar 1, head, ventral, right maxilla and left mandible not shown.

Thorax and Abdomen: Thoracic terga without asperities, pronotum with large sclerotized plate, meso- and metanotum not well sclerotized. Legs short, 5-segmented including bisetose tarsungulus. Abdomen 10-segmented, more than twice as long as thorax. Terga and sterna largely membranous, with 2 transverse rows of asperities on most segments. Tergum A1 with pair of teeth in instar 1. Tergum A9 with pair of long 4-segmented urogomphi. Segment A10 distinct, scarcely visible from above, anal region posteroventrally oriented, anal hooks absent.

Spiracles: Biforous, lateral, closing apparatus present.

Comments: The family contains the single rare genus *Syntelia,* with 2 species in temperate areas of central Mexico and 3 in eastern Asia. Larvae and pupae of the Asian *S. histeroides* Lewis are the only known immature stages (Mamaev 1974, Hayashi 1986). None of the species are of economic importance.

Selected Bibliography

Hayashi 1986.
Kryzhanovski and Reikhardt 1976 (adult classification, ecology).
Mamaev 1974.
Nikitsky 1976d.

HISTERIDAE (HYDROPHILOIDEA)
(INCLUDING NIPONIIDAE)

Alfred F. Newton, Jr.,
Field Museum of Natural History

Hister Beetles (Clown Beetles)

Figures 34.307-315

Relationships and Diagnosis: Histeridae are most closely related to the more primitive families Sphaeritidae and Synteliidae, and these families together are often treated as an isolated superfamily, Histeroidea, of the series Staphyliniformia (Crowson 1974). Increasing evidence, especially larval structure, suggests a close relationship of Histeroidea to Hydrophiloidea and has led to the combining of these superfamilies under the latter name (Böving and Craighead 1931, Nikitsky 1976d, Lawrence and Newton 1982).

Histerid larvae are recognizable by possession of a combination of: head with at most a single stemma on each side, without a free labrum or separate antennal foramen; mandibles without molar lobes but with a penicillus at base; maxilla without cardo and without apical lobes on the stipes, but with an articulated appendage on the first palpal segment; mentum absent; and usually by the presence of 1- or 2-segmented urogomphi, short legs and a soft, largely membranous abdomen. Closely similar larvae of Synteliidae and Sphaeritidae have 4-segmented urogomphi and a distinct mentum, while Hydrophilidae and Georyssidae have a free cardo, antennal foramen externally separated from mandibular foramen, and usually 6 stemmata on each side. Superficially similar larvae of Carabidae, Cantharoidea and some Staphylinidae differ by lacking an articulated appendage on the first palpal segment and (except Staphylinidae) by lacking articulated urogomphi; Carabidae also differ in having 6-segmented legs with (usually) 2 claws.

Biology and Ecology: Histerids are found in a wide variety of microhabitats, especially as follows: under the bark of decaying trees and in the burrows of Scolytidae and other wood-boring insects (most Niponiinae, Abraeinae, Trypeticinae, Trypanaeinae, and many Dendrophilinae, Tribalinae and Histerinae); in dung, carrion, decaying fungi and other decomposing or fermenting organic matter, and in nests of birds and mammals (many Saprininae and Histerinae, and some Dendrophilinae, Tribalinae and Abraeinae); and in nests or colonies of social insects, especially ants (most Hetaeriinae and Chlamydopsinae, and a few others); a few genera are characteristic of forest litter or soil, or sand dunes. Larvae and adults are primarily predators of the immature stages of Diptera and other insects, but the small adults of *Bacanius* are known to feed on fungi (V. Vomero *via* R. Wenzel, personal communication). Larval development is rapid and apparently involves only 2 instars.

Egg white, elongate-oval and slightly curved, smooth.

Pupa exarate, with functional abdominal spiracles on segments 1–4.

Description: Mature larvae about 2–30 mm long, elongate and more or less parallel-sided, slightly to strongly flattened, relatively straight. Body surfaces very lightly or not at all pigmented and sclerotized, smooth or (rarely) granulate or microspinose, with sparse vestiture of simple setae.

Head: Prognathous, protracted, usually well sclerotized and pigmented, without differentiated neck. Epicranial sutures absent (instar 2) or (instar 1) with short to moderately long stem and lyriform or V-shaped frontal arms which end at or near antennal foramen; endocarinae absent. Stemmata absent or 1 on each side. Antenna 3-segmented, about a fourth to three-fourths as long as head width, sensoria of preapical segment posterad of apical segment and conical or palpiform. Antennal insertion separated from mandibular insertion by narrow strip of membrane only. Labrum completely fused to head capsule, forming nasale which is usually asymmetrical and variably toothed. Mandibles symmetrical, narrow and falcate, apex with single tooth, incisor edge with 1 or 2 heavily sclerotized teeth or (rarely) simple or serrate, mesal surface of base with penicillus or brush of hairs and (rarely) small tooth, mola absent. Cardo absent. Stipes elongate. Maxilla without apical lobes. Maxillary palp 4- or 5-segmented, first segment with articulated digitiform appendage. Labium consisting of prementum only. Ligula absent. Labial palps 2- or 3-segmented, usually separated by less than width of first palpal segment. Gular sutures absent, medial ecdysial line sometimes present. Venter of head usually about as long and well sclerotized as dorsum. Occipital foramen not divided by tentorial bridge, which is indistinct or absent.

Thorax and Abdomen: Thoracic terga without asperities and with 1 or more sclerotized plates. Legs short, 5-segmented including bisetose or glabrous tarsungulus. Abdomen 10-segmented, more than twice as long as thorax. Terga and sterna largely membranous, with or without small sclerites, plicae, and rows or patches of asperities; if present, rows of asperities transverse, 1 to 3 per segment. Tergum A1 with pair of teeth (presumed egg bursters) in instar 1 (fig. 34.310b). Tergum A9 with pair of 1- or 2-segmented urogomphi (urogomphi absent in Trypeticinae, Trypanaeinae, and some *Teretrius*). Segment A10 distinct, visible or concealed from above, anal region posteroventrally or ventrally oriented, anal lobes with or without numerous fine hooks or teeth.

Spiracles: Biforous, lateral or dorsolateral, closing apparatus present.

Comments: A large family, with over 300 genera and 3700 species worldwide, best represented in warm temperate and tropical regions. Over 50 genera and 350 species are known from the United States and Canada. Larvae of fewer than 30 genera are known, although these include members of most subfamilies and tribes. No larvae of Hetaeriinae or the Australian Chlamydopsinae are definitely known, but a termitophilous larva described by Böving and Craighead (1931, Plate 21–I) probably belongs to Hetaeriinae. Existing keys to subfamilies and genera are inadequate and largely confined to portions of the European fauna (Schiödte 1864, Perris 1876, Hinton 1945a, Lindner 1967, Nikitsky 1976a, Mamaev et al. 1977, Klausnitzer 1978).

A number of genera that are associated with decaying trees are known or suspected to be of value in controlling bark beetle pests (Nikitsky 1976a, Mateu 1972, Rees 1985, Struble 1930), while *Hister* and other genera inhabiting dung have been used to control flies in Australia and elsewhere (Bornemissza 1968, Summerlin et al. 1981, 1984). The following key will allow subfamily identification of most genera:

1. Maxillary palp 5-segmented, labial palp 3-segmented (fig. 34.311) 4
   Maxillary palp 4-segmented, labial palp 2-segmented (figs. 34.307a,b, 34.312a) 2

2(1). Dorsum of prementum with membranous setose area at base (fig. 34.315a); mesonotum of 5 sclerites above level of spiracles .... **Saprininae**
   Dorsum of prementum (except *Dendrophilus*) without membranous setose area (figs. 34.307c, 34.312b, 34.313); mesonotum of 3 sclerites above level of spiracles (fig. 34.310a) 3

3(2). Prementum without lateral lobes or dorsal teeth (fig. 34.307c) .... **Histerinae**
   Prementum on each side with small lobe which may be acute, obtuse, spiniform or setose; dorsal surface of prementum at base usually with several small teeth (figs. 34.312b, 34.313) .... **Dendrophilinae, Tribalinae**

4(1). First and second antennal segments subequal in length and width (fig. 34.312a); maxilla articulated to head by normal monocondylic joint .... **Abraeinae**
   Second antennal segment about half as long as, and much narrower than first segment (fig. 34.311); maxilla eversible, attached to head by long connecting membrane .... **Trypanaeinae** (New World), **Trypeticinae** (Old World)

The above treatment and key are based on examination of about 3 dozen genera, including previously undescribed larvae of Trypanaeinae (*Trypaneus* sp.), Trypeticinae (*Trypeticus* sp.), and numerous other genera.

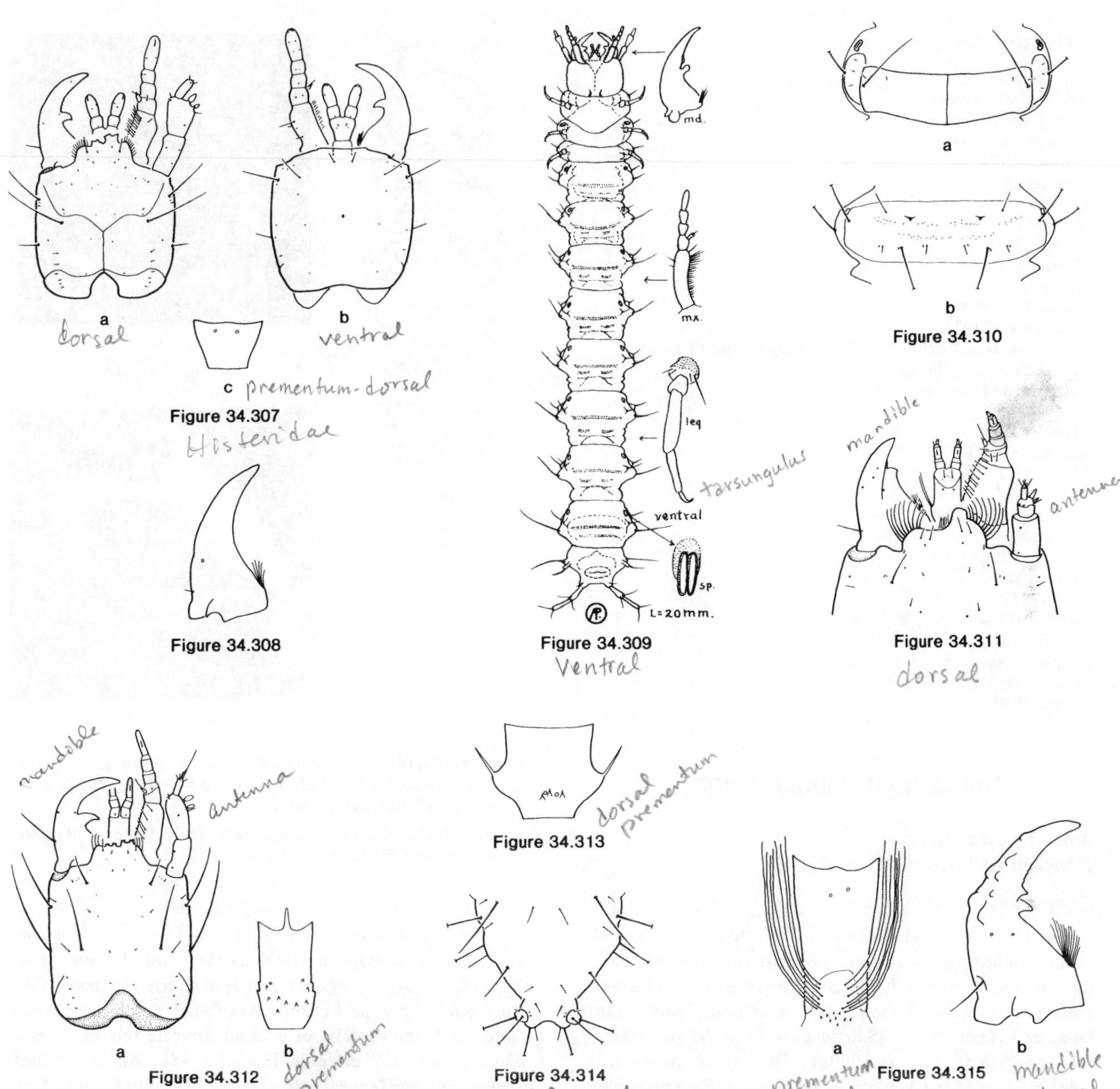

Figure 34.307

Figure 34.308

Figure 34.309

Figure 34.310

Figure 34.311

Figure 34.312

Figure 34.313

Figure 34.314

Figure 34.315

Figures 34.307a-c. Histeridae. *Phelister subrotundus* (Say) (Histerinae). Riley County, Kansas; June; with army ant, *Neivamyrmex opacithorax*. **a.** first instar, head, dorsal, left antenna and right mandible not shown; **b.** head, ventral, right antenna and mandible and left maxilla not shown; **c.** prementum, dorsal.

Figure 34.308. Histeridae. *Teretrius cylindrellus* Casey (Abraeinae). Mill Valley, California; October; from dead branch. Left mandible, dorsal.

Figure 34.309. Histeridae. *Hololepta aequalis* Say (Histerinae). Under bark of dead or dying trees. Final instar larva, ventral, with details of mandible, maxilla, leg and spiracle at right. Length = 20 mm. (From Peterson, 1951.)

Figures 34.310a,b. Histeridae. *Phelister subrotundus* (Say). **a.** mesonotum, dorsal; **b.** abdominal segment 1, dorsal, showing pair of egg bursters.

Figure 34.311. Histeridae. *Trypanaeus* sp. (Trypanaeinae). Barro Colorado Island, Panama; February; under bark of dead tree. Final instar, head, dorsal view of anterior portion, left antenna and right mandible not shown.

Figures 34.312a,b. Histeridae. *Bacanius punctiformis* LeConte (Dendrophilinae). Shades State Park, Indiana; September; from wood fiber pile. **a.** final instar, head, dorsal, left antenna and right mandible not shown; **b.** prementum, dorsal.

Figure 34.313. Histeridae. *Carcinops* sp. (Dendrophilinae). Bedford, Massachusetts; August; from pile of old grass cuttings. Prementum, dorsal.

Figure 34.314. Histeridae. *Aeletes* sp. (Abraeinae). Cambridge, Massachusetts; August; from old wood chip pile. Abdominal segment 9, dorsal.

Figures 34.315a,b. Histeridae. *Hypocaccus estriatus* (LeConte) (Saprininae). Great Salt Lake, Utah; July. **a.** prementum, dorsal; **b.** left mandible, dorsal.

Selected Bibliography

Arnett 1968 (adult classification, ecology).
Balduf 1935 (biology).
Böving and Craighead 1931.
Bornemissza 1968.
Crowson 1974.
Gardner 1930a.
Hayashi 1986.
Hayashi et al. 1959.
Hinton 1945a.
Hudson 1934.
Kistner 1982.
Klausnitzer 1978.
Kryzhanovski 1973.
Kryzhanovski and Reikhardt 1976 (adult classification and ecology, some larvae).
Lawrence and Newton 1982.
Lindner 1967.
Mamaev 1974.
Mamaev et al. 1977.
Mateu 1972.
Mazur 1984.
Nikitsky 1976a, 1976c, 1976d.
Perris 1876.
Peterson 1951.
Prins 1984.
Rees 1985.
Saalas 1917.
Schiödte 1864.
Struble 1930.
Summerlin et al. 1981, 1984.
Vienna 1980.

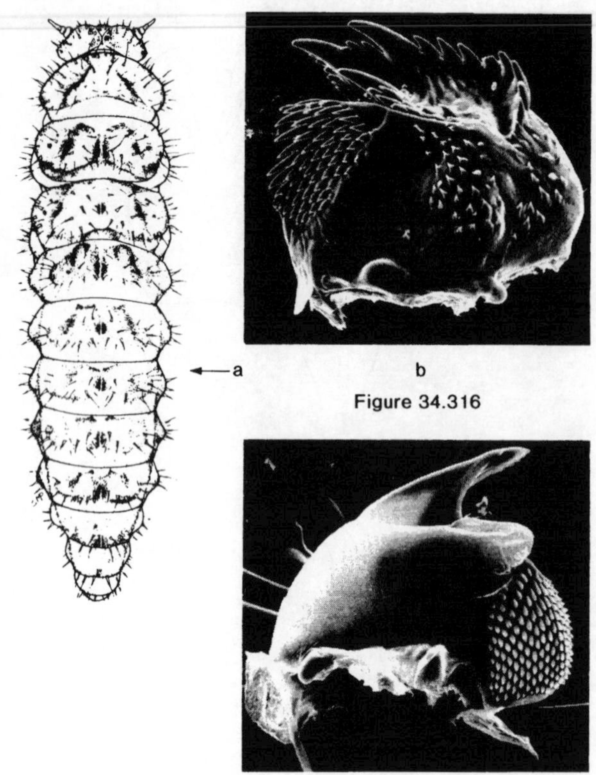

Figure 34.316

Figure 34.317

Figures 34.316a,b. Eucinetidae. *Eucinetus infumatus* LeConte. Caspar, Mendocino Co., California. Length = 7.25 mm. **a.** larva, dorsal; **b.** left mandible, ventral.

Figure 34.317. Eucinetidae. *Eucinetus morio* LeConte. Bedford, Massachusetts. Right mandible, ventral.

EUCINETIDAE (EUCINETOIDEA)

John F. Lawrence,
Division of Entomology, CSIRO

Figures 34.316-317

Relationships and Diagnosis: The eucinetids are small, saltatorial beetles, which appear to be most closely related to the Clambidae. Formerly placed in the Dascilloidea (Arnett, 1968; Crowson, 1955), the family was combined with Clambidae and Helodidae (= Scirtidae) to form the superfamily Eucinetoidea (Crowson, 1960b). The larvae resemble a number of smaller Cucujoidea, such as the Endomychidae, but differ from them in having a separate galea and lacinia. This character also separates them from all Clambidae except *Acalyptomerus,* which has only 2 segments in the maxillary palp.

Biology and Ecology: Eucinetidae are mycophagous and are usually collected in leaf litter, in rotten wood, or under the bark of decaying logs. Several species of *Eucinetus,* including *E. infumatus* LeConte and *E. meridionalis* Castelnau, may be found as aggregations of larvae, pupae, and adults under the loose bark of dead trees in association with fungus growth. *Eucinetus haemorrhous* Duftschmid has been found in drier areas under accumulations of rotting plant remains, in straw piles, or in fungusy roots of *Euphorbia.* The North American *E. morio* LeConte appears to specialize on the spores of slime molds, and has been recorded from the fruiting bodies of *Arcyria, Stemonitis, Fuligo,* and *Tubifera.* *Eucinetus punctulatus* LeConte has been collected on *Paragyrodon sphaerosporus* (Boletaceae) and *Paxillus involutus* (Agaricaceae), where larvae feed on fresh spores or gills. *Eucinetus oviformis* LeConte was found in all stages associated with the mycelia of a basidiomycete fungus, *Coniophora olivacea* (Wheeler and Hoebeke 1984). An undescribed *Jentozkus* from Tennessee has been collected in fungusy bark and in a carrion trap (Bruns, 1984; Hatch, 1962; Horion, 1960; Lawrence and Newton, 1980; Weiss and West, 1921a).

Description: Mature larvae 1 to 6 mm. Body elongate, fusiform, narrowed posteriorly, straight or slightly curved ventrally, slightly flattened, lightly sclerotized, but sometimes with darker maculae on terga and pleura. Dorsal surfaces smooth to tuberculate, sometimes with elongate, setiferous tubercles (*Eucinetus infumatus* LeConte); vestiture of relatively long, simple setae, sometimes arising in dense patches.

Head: Protracted and prognathous or somewhat hypognathous, broad, slightly flattened, strongly narrowed anteriorly. Epicranial stem short to moderately long; frontal arms lyriform. Median endocarina absent. Stemmata 5 on each side. Antennae moderately long, 3-segmented, with segment 2 much longer than 1 or 3; insertions well separated from mandibular articulations. Frontoclypeal suture absent; labrum free or sometimes partly fused to head capsule. Mandibles symmetrical, short and broad, with apex bidentate, sometimes with a number of additional teeth at apex, without accessory ventral process; mola well-developed and asperate; prostheca a blunt, fixed, hyaline lobe (apparently absent in *E. infumatus*). Ventral mouthparts retracted. Maxilla with transverse cardo, elongate stipes, well-developed articulating area, 3-segmented palp, and fixed galea and lacinia; galea falciform and lacinia truncate in *E. infumatus;* galea truncate and lacinia falciform in *E. morio*. Labium more or less free to base of mentum; ligula short and broad, bilobed; labial palps 2-segmented and widely separated. Hypopharyngeal sclerome consisting of a transverse bar. Hypostomal rods and ventral epicranial ridges absent. Gula strongly transverse or absent.

Thorax and Abdomen: Thorax relatively large, more than half as long as abdomen. Legs well-developed, 5-segmented; tarsungulus with 2 setae lying one distal to the other. Abdominal segments gradually decreasing in length and width; segments 1–8 with distinct pleural lobes. Tergum and sternum A9 simple. Segment A10 posteriorly oriented.

Spiracles: Annular, all placed at ends of short tubes; those on abdomen located between pleural lobes and terga.

Comments: The Eucinetidae is a small family with 6 genera and about 30 species worldwide and 3 genera and 8 species occurring in N. America. The genus *Eucinetus* is more or less cosmopolitan in distribution, *Jentozkus* occurs throughout the New World (from the southeastern United States to Brazil), *Euscaphurus* is restricted to northwestern N. America, while the genera *Subulistomella, Tohlezkus,* and *Bisaya* are known from Japan, Turkey, and Central Asia, respectively. Species of *Jentozkus, Tohlezkus,* and *Subulistomella* are all very small (1.2 mm or less) and have piercing-sucking mouthparts (Sakai, 1980; Vit, 1977).

Selected Bibliography

Arnett 1968.
Böving 1929b (larva of *Eucinetus morio*).
Böving and Craighead 1931 (larva of *Eucinetus morio*).
Bruns 1984 (feeding habits of *Eucinetus punctulatus*).
Crowson 1955, 1960b.
Hatch 1962.
Horion 1960 (habitats).
Lawrence and Newton 1980.
Perris 1851 (larva of *Eucinetus meridionalis*).
Sakai 1980.
Vit 1977.
Weiss and West 1921a.
Wheeler and Hoebeke, 1984.

CLAMBIDAE (EUCINETOIDEA) (INCLUDING CALYPTOMERIDAE)

John F. Lawrence,
Division of Entomology, CSIRO

Figures 34.318–319

Relationships and Diagnosis: Clambids are most closely related to the Eucinetidae and form part of the primitive polyphagan superfamily Eucinetoidea (Crowson, 1960b). The larvae closely resemble those of eucinetids but usually differ in having an undivided maxillary mala; in *Acalyptomerus,* the maxilla has a distinct galea and lacinia as in Eucinetidae, but the maxillary palps are only 2-segmented.

Biology and Ecology: Clambidae are spore feeders, which are most commonly collected in samples of leaf litter, rotting straw, or other accumulations of plant debris. *Clambus nigriclavis* Stephens has been found in flood debris or on water-logged sticks near streams; larvae in laboratory cultures were observed to feed on spores and hyphae of molds (Crowson and Crowson, 1955). Hay stacks have been recorded as habitats for *Clambus pubescens* Redtenbacher, *Acalyptomerus asiaticus* Crowson, and *Calyptomerus dubius* Redtenbacher, and the last of these species has been reared on molds in the laboratory. Central American *Acalyptomerus* have been collected on basidiomycete fruiting bodies (*Hirneola mesenterica* and *Hirschioporus sector*), but it is likely that they were feeding on surface molds. Species of *Clambus* from Panama were found breeding in the fruiting bodies of Myxomycetes (*Stemonitis* and *Arcyria*) (Lawrence and Newton, 1980). *Loricaster* species are often found in drier situations; they are usually collected in leaf litter, but adults of *L. testaceus* Müller were found associated with fruiting bodies of *Dalcinia concentrica* (Ascomycetes: Xylariaceae) (Grigarick and Schuster, 1961; Peyerimhoff, 1926). According to Crowson and Crowson (1955), *Clambus nigriclavis* laid fewer, relatively large eggs and had more rapidly developing larvae than *Calyptomerus dubius*. Pupae of Clambidae are obtect and exposed or partly enclosed in the last larval skin.

Description: Mature larvae 1 to 3 mm. Elongate, fusiform, straight, slightly flattened, moderately to very lightly pigmented, sometimes with darker head. Dorsal surfaces smooth, but sometimes with setiferous tubercles as in Eucinetidae; vestiture of scattered long, fine setae.

Head: Protracted and prognathous, moderately broad, slightly flattened. Epicranial stem absent or very short; frontal arms absent or lyriform and contiguous. Median endocarina absent. Stemmata usually 5 or 6 on each side, sometimes 3, 1, or 0. Antennae short to moderately long, 3-segmented, usually with last segment well-developed, but sometimes with segment 3 shorter than antennal sensorium (Calyptomerinae). Frontoclypeal suture absent or indistinct; labrum free. Mandibles symmetrical, bidentate or tridentate, with or without accessory ventral process; mola well-developed, asperate or tuberculate; prostheca a fixed, hyaline lobe, which is simple or bifid at apex. Ventral mouthparts retracted. Maxilla with transverse cardo, elongate stipes, well-developed articulating area and usually a more or less truncate mala, which in *Acalyptomerus* is deeply divided forming fixed galea and lacinia; palp usually 3-segmented (2-segmented in *Acalyptomerus*). Labium with mentum and submentum fused; ligula broad, sometimes longer than labial palps, which are usually 2-segmented (1-segmented in *Acalyptomerus*) and widely separated. Hypopharyngeal sclerome present. Hypostomal rods long and diverging or absent (Calyptomerinae). Ventral epicranial ridges absent. Gula transverse.

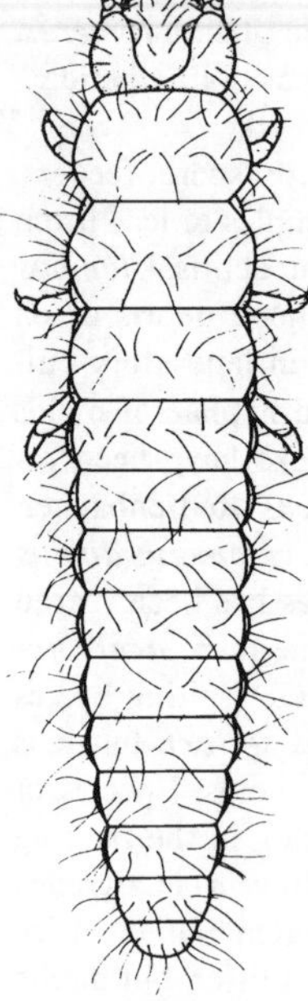

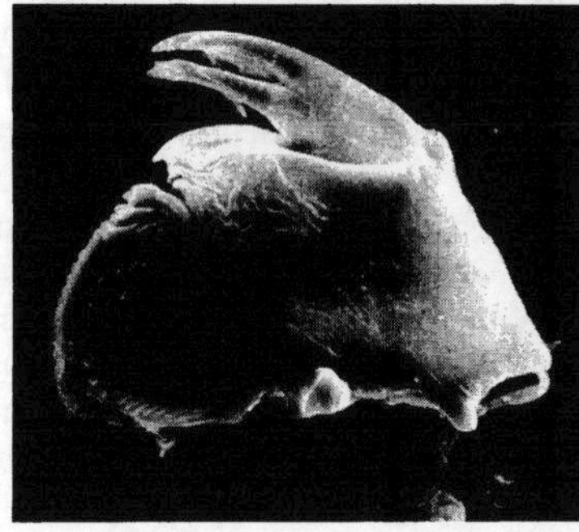

Figure 34.318 Figure 34.319

Figure 34.318. Clambidae. *Clambus* sp. Barro Colorado Is., Canal Zone, Panama. Larva, dorsal. Length = 1.25 mm.

Figure 34.319. Clambidae. *Acalyptomerus* sp. Mona, Jamaica. Left mandible, ventral.

Thorax and Abdomen: Thorax relatively large, more than half as long as abdomen. Thoracic terga and to a lesser extent abdominal terga produced laterally in *Acalyptomerus.* Legs well-developed, 5-segmented; tarsungulus with 2 setae, lying one distal to the other, or with 1 seta only. Tergum A9 simple, usually more or less truncate; segment A10 posteroventrally or posteriorly oriented, occasionally with pair of fine hooks.

Spiracles: Annular, sometimes vestigial on segment A8.

Comments: The Clambidae is a small family with 5 genera and about 70 species, of which 3 genera and 10 species occur in N. America. Crowson (1979) has recognized 3 subfamilies: the monotypic Acalyptomerinae and Calyptomerinae, and the Clambinae with the remaining genera. *Acalyptomerus asiaticus* was described from Ceylon, but specimens have been seen from Japan, Malaysia, India, Australia, Mauritius, Kenya, South Africa, Jamaica, Panama, and Ecuador. These populations may be conspecific, resulting from accidental transport by man, probably in straw (Crowson, 1979). *Calyptomerus* contains 3 species from western N. America and Europe. The European *C. dubius,* which also occurs in hay and straw, has been introduced into Tasmania and South Africa (Endrödy-Younga, 1974). The clambine genera *Loricaster* and *Sphaerothorax* appear to be restricted to the Northern Hemisphere and Southern Hemisphere, respectively, while species of *Clambus* occur throughout the world. Genera formerly included in Clambidae but removed to other families include *Empelus* (to Staphylinidae) (Crowson, 1960b; Lawrence and Newton, 1982) and *Australiodes* (to Trogossitidae) (Crowson, 1966a; Endrödy-Younga, 1960).

Selected Bibliography

Crowson 1955, 1960b, 1966a, 1979 (larvae of *Acalyptomerus asiaticus, Calyptomerus dubius,* and *Clambus nigriclavis*).
Crowson and Crowson 1955 (habits, habitats).
Endrödy-Younga 1959, 1960, 1974, 1981.
Grigarick and Schuster 1961.
Johnson 1966.
Lawrence and Newton 1980 (association with Myxomycetes), 1982.
Perris 1852 (larva of *Calyptomerus dubius*).
Peyerimhoff 1926.
Roberts 1930 (larva of *Calyptomerus dubius* in key).
Wheeler and Hoebeke 1984.

HELODIDAE (EUCINETOIDEA)

(= CYPHONIDAE, SCIRTIDAE)

Laurent LeSage, *Biosystematics Research Centre, Agriculture Canada*

Marsh Beetles

Figures 34.320–322

Relationships and Diagnosis: As defined by Crowson (1955) the Helodidae form a homogenous group both in adult and larval stages. Pope (1976) discussed the validity of the name Helodidae and concluded that the family name Scirtidae should be used instead of Helodidae. However, his suggestion was not followed by many authors. Helodids are members of the superfamily Eucinetoidea (Crowson 1960b) which also includes the Eucinetidae and Clambidae. The superfamily is defined in the larval stage by: a mandibular mola, maxillae with distinct, but not articulated, galea and lacinia, spiracles with normal closing apparatus, urogomphi absent, and abdominal tergites and sternites similarly sclerotized.

Helodid larvae can be recognized by several distinctive features: antennae very long and multisegmented (unique not only among beetle larvae, but also among holometabolous larvae), mandibles with striated mola and complex prostheca, epipharynx and hypopharynx highly specialized and bearing several unusual structures such as the epi- and hypopharyngeal teeth, the filtering apparatus, etc., anal gills present in the ninth abdominal segment.

Biology and Ecology: The larvae of several genera are aquatic and frequent stagnant or flowing water. In general, larvae are found in the marginal vegetation and shallows. *Cyphon* colonizes mainly standing waters; *Microcara* occurs in marshes and bogs; *Scirtes* prefers standing clear waters; *S. tibialis* is associated with ponds and ditches were *Lemna* grows (Kraatz 1918); *Helodes* searches for running clean waters. Many larvae living in standing waters are characterized by the presence of swimming hairs on legs whereas those living in running waters show a peculiar development of spinous setae on legs and body, together with a flattening of the whole body (Bertrand 1972). In the tropics some larvae develop in tree and bamboo stumps or in the leaf bases of Bromeliaceae (Knab 1913). *Prionocyphon* larvae live in still water where leaves are found on the surface (Good 1924) and in tree holes containing water (Osten-Sacken 1862; Striganova 1961a). I have collected *Cyphon* larvae in beaver lodges and in muskrat nests.

Treherne (1952, 1954) showed that the larva of *Helodes* possesses several tracheal air sacs and is able to force a bubble of air out through the last pair of abdominal spiracles, and which is maintained in position by semi-hydrofuge hairs. The air in the bubble is in free communication with that contained in the tracheal system; in extracting, by diffusion, the oxygen dissolved in the water, the air bubble functions as an efficient physical gill. The insect periodically renews its air supply by rising to the surface and taking in air through the 2 abdominal spiracles. The anal papillae or "gills" appear to have little significance as respiratory structures, and one of their main functions is the absorption of salts from the external environment.

According to Bertrand (1972) all helodid larvae are active and phytophagous: *Helodes* larvae are microphagous or eat vegetation detritus, *Scirtes* thrives on *Lemna, Prionocyphon* feeds on epidermal cells of dead leaves and larvae of *Helodes, Microcara,* and *Scirtes* can be reared on dead leaves.

The functioning of the buccal apparatus is complex and has been studied in detail by Beier (1949). The food particles are "sorted" by the comb hairs on the maxillae and then transferred onto the combs of the hypopharynx. The "crushing" and "packing" of the food particles into a ball is accomplished by the hypopharynx moving on the molar area of the mandibles.

Larval growth seems quite rapid and up to 8 stages have been reported (Kraatz 1918). Some species may overwinter as larvae and pupate in spring. Larvae usually pupate in pupal cells in damp soil (Leech and Chandler 1956), but also in moss and dead leaves (Beier 1949). Pupae of *Scirtes* attach to aquatic plants by a secretion for the final metamorphosis (Lombardi 1928).

Description: Size of mature larvae 5–15 mm. General habitus campodeiform, looking superficially like immature cockroaches (figures 34.320a, 321a). All body segments moderately sclerotized and pigmented in brown. The body vestiture consists of numerous scattered small setae and a few large setae regularly arranged.

Head: More or less retractile into the prothorax but always prognathous and visible from above, usually smaller than prothorax, depressed and transverse. Shape variable, narrowed behind, rounded, transverse (figures 34.321b,c), or triangular with anterior angles produced. Epicranial suture present, without or with a short coronal stem and with frontal arms long and strongly divergent.

Stemmata variable in number, 3 or 4 in *Helodes,* 2 in *Prionocyphon* and *Microcara,* reduced and confluent in *Scirtes* and *Cyphon,* but always located on the anterior corners of the head. Antennae filiform and multisegmented; scape 2-segmented and followed by a flagellum which may be made by more than 100 segments, their combined length exceeding the length of the body in some cases (Bertrand 1972).

The buccal cavity shows a very complex organization and the reader is referred to the monograph of Beier (1949) for a detailed discussion. This cavity is broadly opened and has its roof formed by the labrum and the epipharyngeal structures (figure 34.321i); the epipharyngeal teeth vary with genera and are useful for taxonomic purposes. The floor of the mouth is made of a lip bearing dorsally comb-like maxillulae, spinose superlingual lobes and 2 or 4 apical hypopharyngeal setae or teeth (fig. 34.321h); these setae provide good generic characters in most cases. Finally a filtering and grinding apparatus is located at the back of the head and is made of a dorsal cone, a central cavity and 2 ventral toothlike prominences (fig. 34.321h).

Mandible provided with many structures with exact homology not yet known. Mandible very broad at its base, truncated, rounded or pointed at apex (Fig. 34.321f). Shape variable with instars. Mola robust and striated; additional ventral crushing condyle present; mesal margin with a complex prostheca (?) consisting of 2 superposed processes, one fixed and serrate, the other articulated, with a fringe of forked setae; outer edge provided with long robust setae; dorsal aspect usually with a row of short setae located in the basal third. Maxillae located on the far lateral portion of the head, consisting of a small cardo and a distinct stipes, weakly sclerotized on inner part; lacinia movable and bearing an inner fringe of hairs; galea movable as well and subdivided into 2 parts, the subgalea and the galea, the latter being covered with very dense hook- or comb-like hairs (fig. 34.321d). Maxillary palps 4-segmented, but appearing 3-segmented in some cases when the fourth segment is vestigial and reduced to an inconspicuous dome (fig. 34.320b); third segment bearing sensory organs which are usually distinctive for each genus (figs. 34.320b, 34.322a). Labium with a broad transverse submentum (fig. 34.321c); mentum and ligula fused together; labial palps, small, 2-segmented, and located at the outer angles of the labium.

Thorax and Abdomen: Thorax sclerotized, variable in shape, more or less expanded laterally with acute posterior angles in some genera; prothorax larger than either meso- or metathorax. Legs well developed, 5-segmented, bearing natatory or spinous setae (Fig. 34.321g), and including a claw-like tarsungulus.

Usually only 8 abdominal segments visible from a dorsal view. Each segment sclerotized, more or less convex or flattened, sometimes with acute posterior angles. Ninth abdominal segment reduced to a small sinuous plate, hardly or not visible in a dorsal view. Between the dorsal portion of the ninth and the apex of the eighth segment is located a slit which opens in an atrium-like chamber. In the bottom of this chamber are found 2 large annular spiracles at the end of 2 tracheal trunks. The anus is located between the ninth tergite and the ninth sternite which is clearly visible in ventral view. Five retractile anal gills are located on the ninth segment (Fig. 34.321j). They are typically formed of 1 median single tube and 2 pairs of lateral simple tubes. However, in *Prionocyphon* these tubes are extensively branched (fig. 34.322c).

Spiracles: Vestigial or absent in the thoracic and abdominal segments 1–7, but well-developed on abdominal segment 8 (see below).

Comments: The Helodidae is a small family with only 34 species and 7 genera recorded in North America (Arnett 1968). The Nearctic larvae have not been thoroughly studied

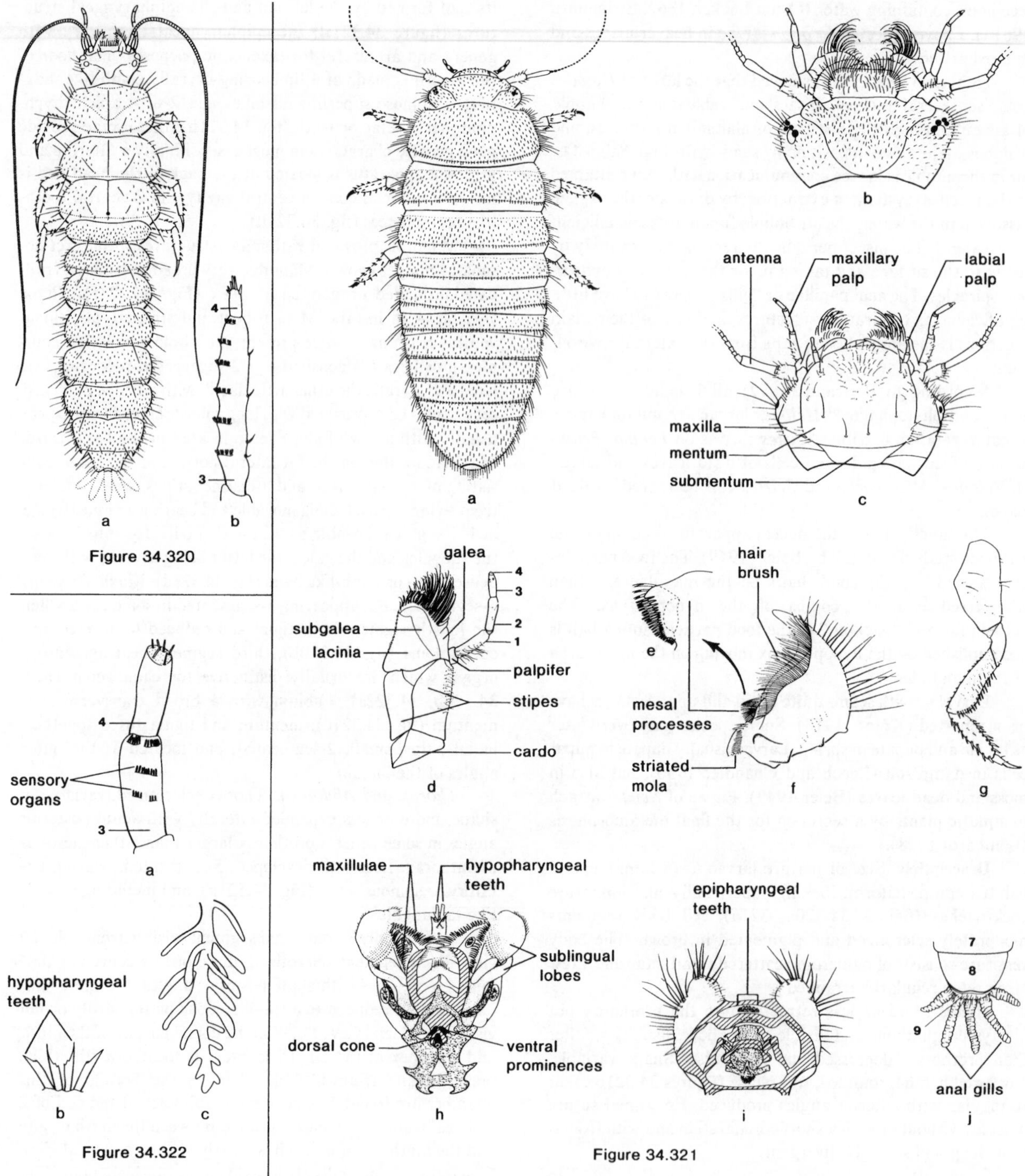

Figures 34.320a,b. Helodidae. *Cyphon* sp. **a.** larva; **b.** maxillary palp, apical segments. 10–12 mm. Body somewhat depressed and more or less uniformly sclerotized; light to reddish brown. Head with 4 large stemmata. Elongate multisegmented, flexible antennae. Body evenly covered by many small setae and a few long setae. Five retractile glands at caudal end. Living in very shallow rust-colored water. (Redrawn from Peterson, 1951.)

Figures 34.322a-c. Helodidae. *Prionocyphon discoideus* Say. **a.** maxillary palp, apical segments; **b.** hypopharyngeal teeth; **c.** anal gills.

Figures 34.321a-j. Helodidae. *Helodes* sp. Body length 8 mm. **a.** larva; **b.** head, dorsal; **c.** head, ventral; **d.** maxilla; **e.** mesal processes of mandible; **f.** mandible; **g.** midleg; **h.** hypopharynx, dorsal; **i.** epipharynx, ventral; **j.** anal gills. Body campodeiform, flattened, sclerotized, almost uniformly brown, except thorax with darker and lighter markings. Stemmata 4, on the far corners of the head. Antennae multisegmented, whip-like. Legs bearing spines. Body evenly covered by numerous small setae and few long lateral setae. Five retractile gills occur in the ninth segment. Mer Bleue, Ontario, in muskrat nest, L. LeSage.

and information concerning them may be found either in general textbooks (Böving and Craighead 1931; Peterson 1951) or in scattered descriptions (Beerbower 1943; Good 1924). However, all common Nearctic genera can be keyed out using recent European literature (Bertrand 1972; Klausnitzer 1975a). *Sarabandus* and *Ora* are still unknown in the larval stage.

Selected Bibliography

Arnett 1968.
Beerbower 1943.
Beier 1949, 1952.
Benick 1924.
Bertrand 1972 (key).
Böving and Craighead 1931.
Brigham et al. 1982 (key).
Crowson 1955, 1960b.
Good 1924.
Klausnitzer 1975a (key).
Knab 1913.
Kraatz 1918.
Leech and Chandler 1956.
Lombardi 1928.
Osten-Sacken 1862.
Pennack 1978 (key).
Peterson 1951.
Pope 1976.
Striganova 1961a.
Treherne 1952, 1954.

DASCILLIDAE (DASCILLOIDEA) (INCLUDING KARUMIIDAE)

John F. Lawrence,
Division of Entomology, CSIRO

Soft-bodied Plant Beetles

Figures 34.323a–g

Relationships and Diagnosis: The Dascillidae, as here understood, includes *Dascillus, Anorus,* and several exotic genera, among which are the highly specialized, apparently termitophilous forms usually treated as a separate family (Karumiidae) (Crowson, 1971; Lawrence, 1982c). The group is currently grouped with the Rhipiceridae in the superfamily Dascilloidea, which, in turn, is often considered to be the sister group of the Scarabaeoidea; the composition of the Dascilloidea and their apparent close relationship to scarabaeoids, however, are still open to question (Crowson, 1971; Lawrence and Newton, 1982).

Dascillid larvae are very similar to those of Lucanidae and Scarabaeidae, in that they are large, grub-like forms, with a large head, long antennae, cribriform spiracles, similarly robust mandibles with well-developed mola and accessory ventral process, separate galea and lacinia, and relatively complex epipharynx and hypopharynx. They differ from all scarabaeoids in having: (1) distinct (though lightly pigmented) tergal plates, most of which bear an anterior carina; (2) a labrum and clypeus at least partly fused together and to the frons; (3) a very short epicranial stem; (4) an articulated process beneath the mandibular retinaculum; (5) heavily sclerotized, comb-like structures on both epipharynx and hypopharynx; (6) a strongly bilobed ligula; (7) a more or less reduced 10th abdominal segment without anal pads; and (8) urogomphi on tergum A9 (except in *Dascillus davidsoni* LeConte). Larvae of Helodidae also resemble those of dascillids in having a very short epicranial stem, complex epipharynx and hypopharynx, and a large labral sclerite, which may represent the fusion of clypeus and labrum; they differ from dascillids in having long, multiannulate antennae, a modified metapneustic respiratory system with atrophied anterior spiracles, enlarged 8th abdominal spiracles, and anal gills or osmoregulatory papillae, and no trace of urogomphi.

Biology and Ecology: Larvae of *Dascillus* are usually found in moist soil and may be collected under rocks in the spring. Like many Scarabaeidae, they probably feed on a variety of plant roots, as well as soil and humus, which are broken down by the sclerotized epipharyngeal and hypopharyngeal plates acting in conjunction with the large mandibles. In California, they have been associated with roots of certain native plants, as well as *Acacia,* apple and cherry (Essig, 1958), while in Europe they are known to feed on roots in boggy meadows and have been reported damaging grasslands and forage crops (Gahan, 1908; Horion, 1955; Zocchi, 1961). The larvae take 2 years to develop, and pupation occurs in the spring (Horion, 1955). Adults may be found clinging to grass stems, and they have been reported to be floricolous (Crowson, 1981). Males of *Anorus* fly to lights in drier areas, while the females are brachypterous and have been taken in the following situations: (1) at the edge of a burrow in hard soil (Blaisdell, 1934); (2) in a house near baseboards recently treated with fog insecticide for termites; and (3) in a dead *Acacia greggii* root in association with the subterranean dry-rot termite (*Paraneotermes simplicicornis* Banks) (W. L. Nutting, personal communication). Adults of *Karumia* were discovered in a shipment of termites from Afghanistan, and it is possible that termitophily is widespread in the Karumiinae (Arnett, 1964).

Description: Mature larvae 10 to 35 mm. Body elongate, more or less parallel-sided, subcylindrical or slightly flattened, and somewhat curved ventrally, lightly sclerotized and yellowish. Surfaces smooth; vestiture of longer and shorter hairs, dense in places, and occurring mainly on lateral and ventral surfaces.

Head: Protracted and moderately declined, large, transverse, strongly rounded laterally, and slightly flattened. Epicranial stem very short or absent and frontal arms V-shaped. Median endocarina absent. Stemmata absent. Antennae relatively long, 3-segmented, but with segment 3 highly reduced, so that they appear 2-segmented. Frontoclypeal suture absent; labrum partly fused to clypeus (suture absent mesally); clypeolabral area asymmetrical; epipharynx complex, with heavily sclerotized plates and combs. Mandibles asymmetrical, robust, bidentate, with well-developed accessory ventral process; incisor area complex, with 2 or 3 scissorial teeth (retinacula) and a small, articulated, sclerotized process; mola well-developed, finely, longitudinally ridged; prostheca, if present, consisting of a flattened, membranous area. Ventral mouthparts only slightly retracted. Maxilla with transverse, distinctly divided cardo, elongate stipes, well-developed articulating area, 3-segmented palp, articulated

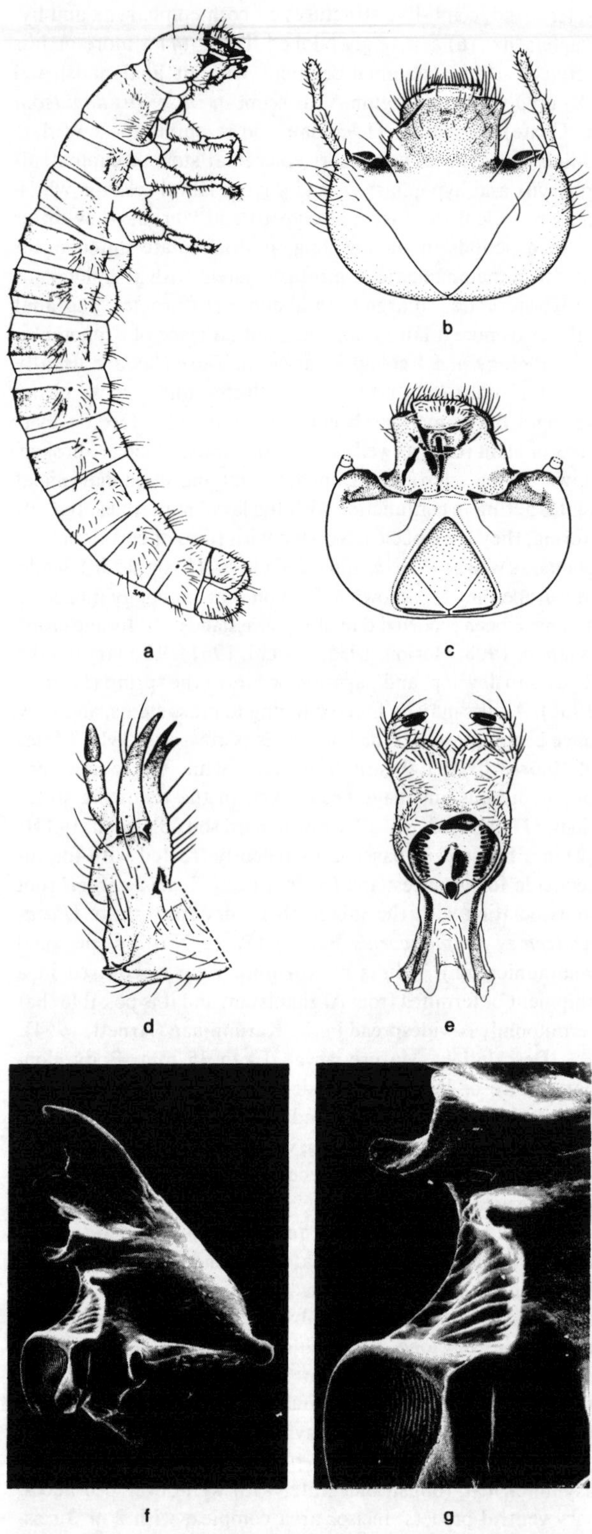

Figures 34.323a–g. Dascillidae. *Dascillus davidsoni* LeConte. Novato, Marin Co., California. Length = 30 mm. **a.** larva, lateral; **b.** head, dorsal; **c.** head, ventral (with ventral mouthparts and mandibles removed); **d.** right maxilla, ventral; **e.** labium-hypopharynx, dorsal; **f.** left mandible, ventral; **g.** details of mandibular mola, prostheca, articulated process, and retinaculum.

falciform galea and fixed, falciform lacinia. Labium free to base of mentum; ligula short, broad, and bilobed; labial palps 2-segmented and widely separated. Hypopharynx complex, with at least 3 scleromes and a pair of heavily sclerotized combs. Hypostomal rods moderately long and strongly diverging. Ventral epicranial ridges absent. Gula transverse.

Thorax and Abdomen: Prothorax slightly larger than meso- or metathorax. Thoracic terga and abdominal terga 1–9 with well-developed, yellowish tergal plates. Meso- and metaterga and abdominal terga 1–8 each with transverse carina near anterior edge; metatergum and abdominal terga 1–4 or 5 sometimes with heavier, asperate carina near posterior edge. Legs well-developed, 5-segmented; tarsungulus usually with 4 setae. Abdomen with 10 visible segments, 1–8 more or less equal; 9 slightly shorter, its tergum dorsal, usually bearing a pair of short, lightly sclerotized, upturned urogomphi (absent in *Dascillus davidsoni* LeConte); 10 reduced, posteriorly oriented; anal opening transverse, with 2 oval lobes beneath it and a transverse lobe above; segments 9 and 10 more or less uniformly covered with short hairs.

Spiracles: Cribriform, large and reniform on thorax and first abdominal segment, becoming smaller and more semicircular posteriorly. Spiracular closing apparatus absent.

Comments: The family Dascillidae contains 15 genera and about 80 species worldwide (2 genera and 5 species in North America). There are 2 subfamilies. *Dascillus,* the African and Asian genera *Coptocera* and *Pseudolichas,* and the Australian *Notodascillus* comprise the Dascillinae, while *Anorus, Genecerus, Drilocephalus,* and a few other genera occurring in the more arid parts of North and South America, Africa, and Central Asia, make up the Karumiinae.

Selected Bibliography

Arnett 1964 (Karumiinae).
Blaisdell 1934 (female of *Anorus piceus* LeConte).
Böving 1929b (larva of *Dascillus davidsoni*).
Böving and Craighead 1931 (larva of *Dascillus davidsoni*).
Crowson 1971, 1981.
Emden 1942b (larva of *Dascillus cervinus* Linnaeus in key).
Essig 1958.
Gahan 1908 (larva of *Dascillus cervinus*).
Horion 1955.
Lawrence 1982c.
Lawrence and Newton 1982.
Zocchi 1961 (larva of *Dascillus cervinus*).

RHIPICERIDAE (DASCILLOIDEA) (= SANDALIDAE)

John F. Lawrence,
Division of Entomology, CSIRO

Figures 34.324, 325

Relationships and Diagnosis: The Rhipiceridae includes only *Rhipicera, Sandalus* and a few related genera. *Callirhipis, Zenoa* and their allies were removed by Crowson (1955) to form a separate group, the Callirhipidae. The placement of Rhipiceridae in the superfamily Dascilloidea is based on substantial evidence from adult structure (Crowson, 1971), but there is little or no larval evidence to support it. There are 2 types of rhipicerid larvae known, a triungulin and an ectoparasitic form, both occurring in the species *Sandalus niger* Koch. The *Sandalus* triungulin differs from those of Meloidae, Rhipiphoridae, and Bothrideridae in having relatively short antennae, with segment 2 shorter than the sensorium, in having long, 1-segmented maxillary and labial palps, and in having no long setae at the apex of the abdomen. The ectoparasitic larva differs from those of meloids and rhipiphorids in having a prognathous head, retracted ventral mouthparts, distinct galea and lacinia, biforous spiracles, and urogomphi. Larvae of Bothrideridae and Cucujidae-Passandrinae have a similar physogastric form and prognathous head, and usually have urogomphi, but they differ from the *Sandalus* larva in having a free labrum, bidentate or tridentate mandible with a fixed, hyaline process at the base, and annular spiracles. Ectoparasitic forms of Carabidae have either the antennae or the maxillary palps 4-segmented and do not have a distinct galea and lacinia.

Biology and Ecology: Rhipicerid larvae are external parasitoids of cicada larvae. According to Elzinga (1977), females deposit large numbers of eggs in holes and cracks in the bark of elms and seem to prefer areas where cicadas have also oviposited. Triungulin larvae drop to the ground or are washed off the bark by rain, and thus enter the soil along with the young cicadas. Adult activity was recorded from late September to late October (Elzinga, 1977), and an unusual abundance of adults was observed in southern Indiana in the year following the emergence of a brood of periodical cicadas (*Magicicada septemdecim* (L.)) (Young, 1956). The ectoparasitic larva was reconstructed from an exuvia associated with a pupa located within a dead cicada which had failed to emerge from its burrow.

Description: Mature larvae 15 to 35 mm. Body elongate, fusiform, with enlarged abdomen (physogastric); lightly sclerotized; vestiture of scattered, short, simple hairs.

Head: Protracted and prognathous, small, slightly flattened. Epicranial stem long and frontal arms V-shaped. Median endocarina absent. Stemmata absent. Antennae very short, 1-segmented, conical. Frontoclypeal suture absent; labrum completely fused to head capsule. Mandibles symmetrical, lightly sclerotized, acute at apex, without accessory ventral process or mola. Ventral mouthparts retracted. Maxilla without distinct cardo, with elongate stipes, no articulating area, 2-segmented palp, fixed, falciform galea, and fixed, rounded lacinia. Labrum with prementum, mentum, and submentum fused into single plate, at least partly fused with stipites; ligula present; labial palps 1-segmented. Hypopharyngeal sclerome absent. Hypostomal rods and ventral epicranial ridges absent. Gula transverse.

Thorax and Abdomen: Legs short, stout, widely separated, 3- or 4-segmented; tarsungulus with 1 seta. Abdominal segments gradually enlarged to 5th or 6th segment, then reduced posteriorly; segment 9 much smaller than 8, its tergum with pair of short, upturned, fixed urogomphi; segment 10 reduced, posteroventrally oriented.

Spiracles: Annular-biforous, with long accessory tubes.

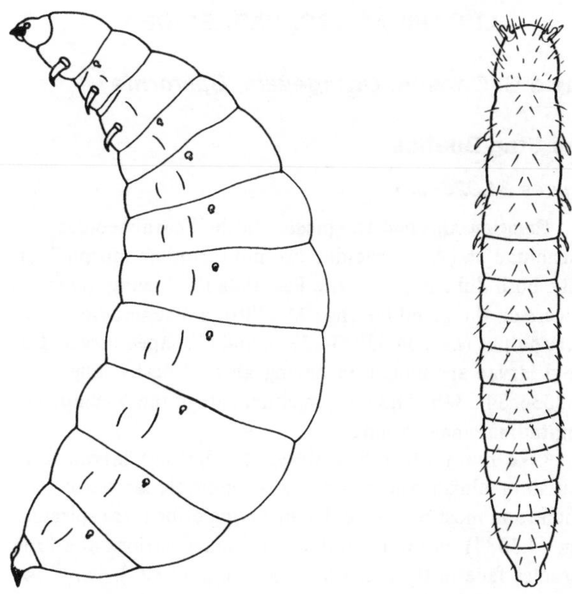

Figure 34.324 Figure 34.325

Figure 34.324. Rhipiceridae. *Sandalus niger* Knoch. Larva, lateral. Length = 25-30 mm. (Redrawn from Craighead, 1921.)

Figure 34.325. Rhipiceridae. *Sandalus niger* Knoch. Columbus, Ohio. Triungulin larva, dorsal. Length = 0.9 mm.

Triungulin: Length less than 1 mm. Body elongate, more or less parallel-sided, somewhat flattened, clothed with scattered setae. Antennae 2-segmented with sensorium on segment 1 longer than segment 2. Mandible flattened, with apex more or less rounded. Maxillary apex with single rounded mala, indistinctly divided into 2 parts; maxillary palp 1-segmented, long and tusk-like. Labial palp 1-segmented, similar to that of maxilla. Legs long and 5-segmented; tarsungulus long and narrow, bearing 1 seta. Paired egg bursters present on abdominal terga 3–7. Tergum A9 without urogomphi. Segment A10 well-developed and posteriorly oriented, with 4 lobes around anus.

Spiracles: Annular, located on mesothorax and abdominal segments 1 and 8.

Comments: The family includes 5 genera and 52 species worldwide: *Rhipicera* (Australia, New Caledonia, southern South America); *Sandalus* (North and South America, Africa, Asia); *Chamoerhipis* (Africa); *Arrhaphipterus* (southern Europe, northern Africa, Asia Minor); and *Polymerius* (Chile). Five species of *Sandalus* occur in N. America.

Selected Bibliography

Böving and Craighead 1931 (ectoparasitic larva of *Sandalus niger* Koch).
Craighead 1921 (ectoparasitic larva of *Sandalus niger*).
Crowson 1955, 1971.
Elzinga 1977 (triungulin of *Sandalus niger*).
Young 1956.

LUCANIDAE (SCARABAEOIDEA)

David C. Carlson, *Orangevale, California*

The Stag Beetles

Figures 34.326–337

Relationships and Diagnosis: In the Scarabaeoidea, the Lucanidae and Scarabaeidae are more similar morphologically, both differing from the Passalidae in having a ventral process on the mandible (fig. 34.328b), a 4-segmented maxillary palpus (figs. 34.330, 34.331), and C-shaped larvae. Lucanid larvae are unique in having an oval fleshy lobe (figs. 34.333e, 34.334b, 34.336b) on either side of the Y-shaped or longitudinal anal opening.

Like many other Scarabaeoidea, lucanid larvae have whitish or bluish bodies. They also resemble larvae of Passalidae and most Scarabaeidae in having cribriform spiracles (fig. 34.333d), but in Lucanidae the emarginations of all the spiracles face in the same direction (fig. 34.327), while the emargination of the prothoracic spiracle in similar-appearing Scarabaeidae and Passalidae faces opposite from the abdominal spiracular emarginations.

Lucanid larvae resemble those of the less specialized Scarabaeidae in having the last antennal segment reduced in size, the epipharynx with tormae united medially (figs. 34.328d, 34.329, 34.332), and in having maxillae with separate galea and lacinia (figs. 34.330, 34.331).

Unlike larvae of Scarabaeidae and Passalidae, lucanid larvae usually lack maxillary stridulatory teeth. All lucanid larvae have stridulatory organs on the mesothoracic (figs. 34.333a, 34.335a) and metathoracic (figs. 34.333b, 34.335b) legs as do the Passalidae and some Scarabaeidae (Geotrupinae). However, all Passalidae and some Geotrupinae have the metathoracic legs reduced in size which lucanids do not.

Biology and Ecology: Lucanidae live in old stumps, in decaying roots, and in or under logs of a wide variety of trees, where all stages may be found (Ritcher, 1966; Furniss and Carolin, 1977). They are most common in woodlands and moist situations. Little is known about the habits of the adults, however, there have been studies of mating behavior in some species (Mathieu, 1969). Adults of several species of *Lucanus* (including *Pseudolucanus*) are frequently attracted to lights while others are found most commonly in or under decaying wood or resting on foliage. Larvae of 1 species of *Lucanus* are known to feed on live roots (Shenefelt and Simkover, 1950).

Description: Full grown larvae are 15 to 40 mm long, subcylindrical, and C-shaped. They have light yellow to reddish-brown heads and bluish or whitish bodies.

Head: (figs. 34.326, 34.327, 34.328c). Hypognathous, exserted, with well-developed mouthparts. Epicranial suture fairly long, frontal arms enclosing a V-shaped or lyriform frons. Stemmata absent. Labrum free, symmetrical, broadly rounded at apex or faintly trilobed. Epipharynx (figs. 34.328d, 34.329, 34.332) with poorly developed haptomerum, often with phobae surrounding the pedium, with tormae fused on midline, with an epitorma, and with 3 nesia. Antennae (fig. 34.326) 3- or 4-segmented, the last segment much reduced in size. Mandibles (figs. 34.328a,b) elongate, with a ventral process and well developed molar region; left mandible with 3 apical teeth. Maxilla (figs. 34.330, 34.331) with separate, falcate galea and lacinia. Maxillary stridulatory teeth absent on stipes of most species (present in *Platycerus,* fig. 34.331). Maxillary palp 4-segmented. Hypopharynx (fig. 34.331) with heavily sclerotized, asymmetrical, oncylus. Labial palpi 2-segmented (fig. 34.331).

Thorax and Abdomen: Thoracic segments (figs. 34.327, 34.328c) distinct, each with a well developed pair of 5-segmented legs (including the tarsungulus which bears 2 or more setae); stridulatory organs present on meso- and metathoracic legs, consisting of a file or patch of granules on the mesocoxae (figs. 34.333a, 34.335a) and a file, patch of granules or rows of granules on the metatrochanters (figs. 34.333b, 34.335b).

Abdomen cylindrical, consisting of 10 segments with the last 2 usually reduced in size. Dorsa of abdominal segments without conspicuous folds (fig. 34.327). Anal opening Y-shaped (figs. 34.334b, 34.336b) or longitudinal (fig. 34.333e), lying between 2 fleshy lobes.

Spiracles: Cribriform (fig. 34.333d), consisting of 1 pair of prothoracic spiracles and 1 pair of spiracles on each of the first 8 abdominal segments. Respiratory plate emarginations of all spiracles facing cephalad or cephaloventrad.

Comments: The Lucanidae is a relatively small family containing approximately 750 species world-wide (Arnett, 1968). Some 40 species representing 9 genera occur in the United States and Canada. In the Eastern United States, the most common genera are *Lucanus* (including *Pseudolucanus*) and *Platycerus.* In the Pacific Northwest the most common genera are *Sinodendron, Platycerus* and *Ceruchus.* In the Southwest *Lucanus* (*Pseudolucanus*) and *Platycerus* are the most common genera. *Diphyllostoma,* known only from California, was transferred to the Scarabaeidae by Holloway (1972). The larva of *Diphyllostoma* is undescribed and may ultimately help resolve the placement of this genus.

The group is of little economic importance since the larvae of most species feed on decaying wood. Larvae and adults of *Ceruchus* occur in dead coniferous wood (Furniss and Carolin, 1977); larvae of most other genera are limited to wood of deciduous trees.

Hayes (1928, 1929) contains keys to larvae of some of the North American genera and Böving and Craighead (1931) published keys to larvae of subfamilies and known genera occurring in the United States. The only larval description of *Nicagus* is in Böving and Craighead (1931). Van Emden (1935a, 1941) also has keys for separating larvae of some of the genera which occur in the United States. Ritcher (1966) summarized the information about lucanid larvae, presented a key to 5 genera and descriptions of species of each of these genera. Since Ritcher (1966), little has been published on North American lucanid larvae. However, works on adult biology, morphology, and systematics have appeared in the literature (Mathieu, 1969; Holloway, 1969, 1972; Hatch, 1971; Howden and Lawrence, 1974). Lawrence (1981) reviewed previous work on larval Lucanidae, discussed relationships among major groups within the family, and presented a key to the primitive genera and more well defined tribes of advanced Lucanidae.

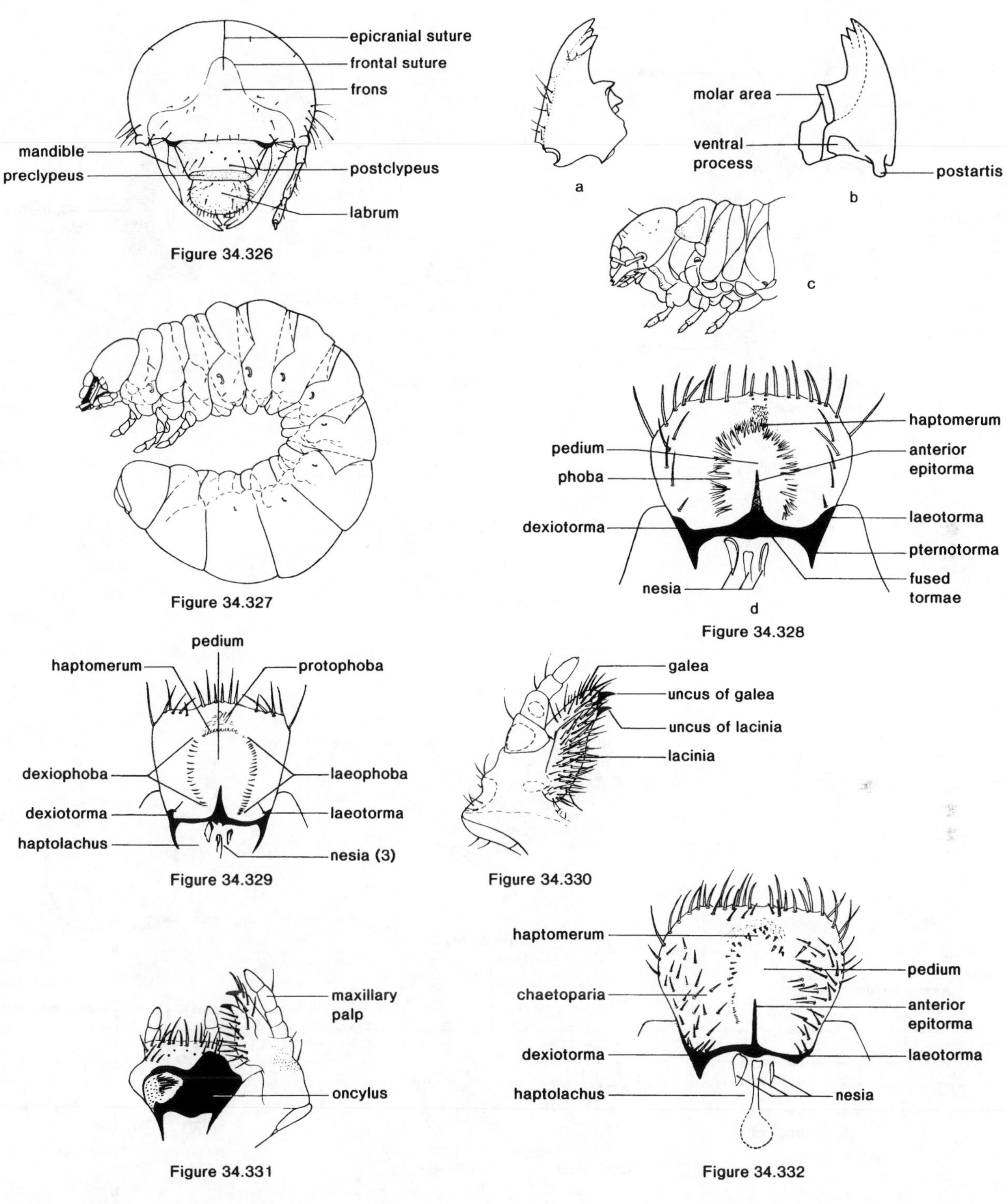

Figure 34.326. Lucanidae. *Lucanus placidus* (Say). Head.

Figure 34.327. Lucanidae. *Dorcus parallelus* (Say). Entire larva (setae omitted).

Figures 34.328a-d. Lucanidae. *Sinodendron rugosum* Mannerheim. **a.** left mandible, dorsal; **b.** left mandible, ventral; **c.** head, thorax, and part of first abdominal segment; left lateral; **d.** epipharynx.

Figure 34.329. Lucanidae. *Ceruchus piceus* Weber. Epipharynx.

Figure 34.330. Lucanidae. *Lucanus capreolus* (L.). Left maxilla, dorsal.

Figure 34.331. Lucanidae. *Platycerus oregonensis* Westwood. Right maxilla and hypopharynx, ental view.

Figure 34.332. Lucanidae. *Dorcus parallelus* (Say). Epipharynx.

(All figures except 34.328b are from Ritcher, 1966, reprinted with permission of the Oregon State University Press.)

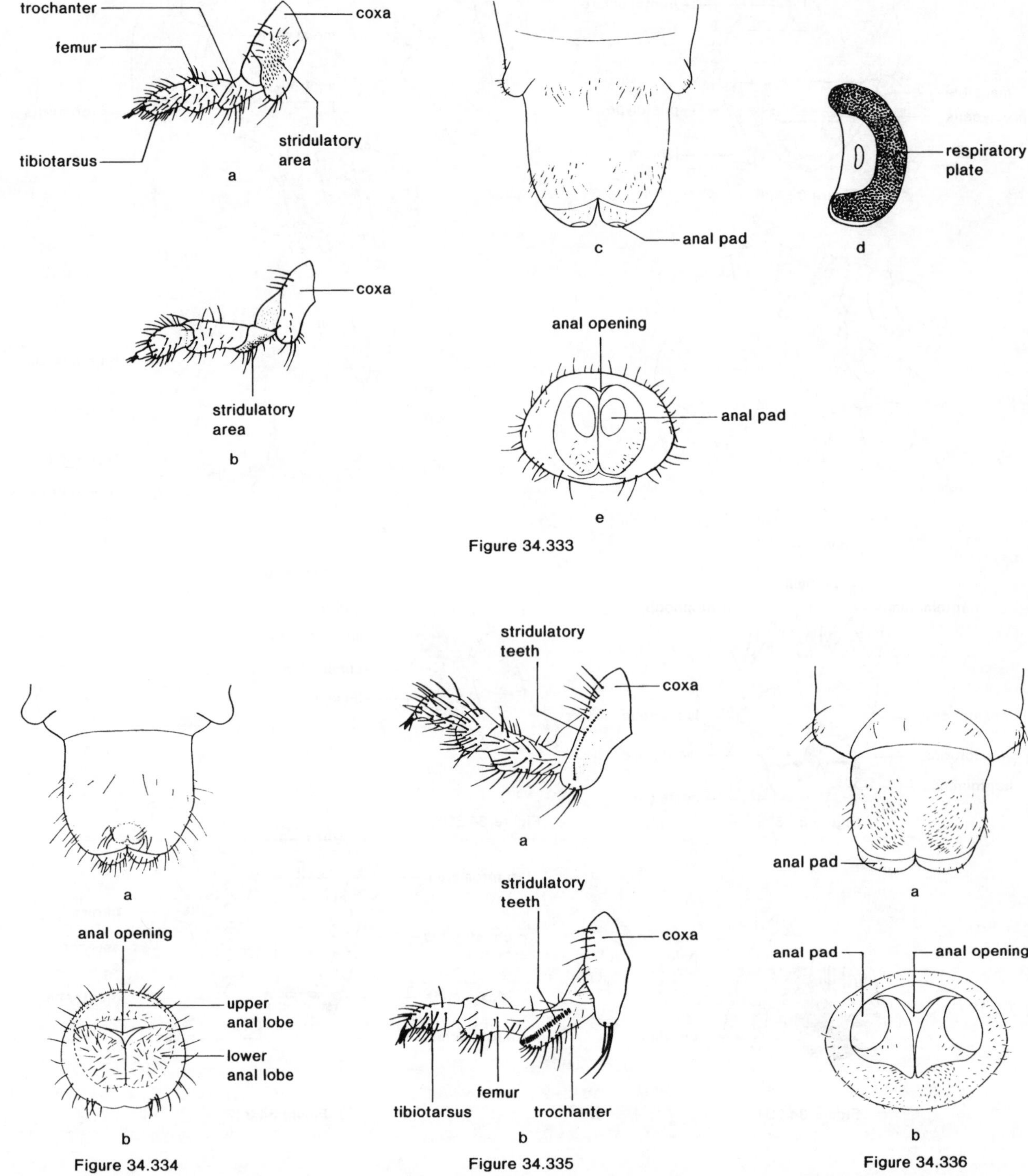

Figures 34.333a-e. Lucanidae. *Sinodendron rugosum* Mannerheim. **a.** right mesothoracic leg; **b.** left metathoracic leg; **c.** venter of last abdominal segment; **d.** abdominal spiracle; **e.** caudal view of last abdominal segment.

Figures 34.334a,b. Lucanidae. *Ceruchus piceus* Weber. **a.** venter of last abdominal segment; **b.** caudal view of last abdominal segment.

Figures 34.335a,b. Lucanidae. *Dorcus parallelus* (Say). **a.** right mesothoracic leg; **b.** left metathoracic leg.

Figures 34.336a,b. Lucanidae. *Playtcerus oregonensis* Westwood. **a.** venter of last abdominal segment; **b.** caudal view of last abdominal segment.

(All figures except 34.333d are from Ritcher, 1966, reprinted with permission of the Oregon State University Press.)

Selected Bibliography

Arnett 1968.
Böving and Craighead 1931.
Emden 1935a, 1941.
Furniss and Carolin 1977.
Hatch 1971.
Hayes 1928, 1929.
Holloway 1969, 1972.
Howden and Lawrence 1974.
Hurka 1978d.
Lawrence 1981.
Mathieu 1969.
Ritcher 1966.
Shenefelt and Simkover 1950.

PASSALIDAE (SCARABAEOIDEA)

David C. Carlson, *Orangevale, California*

The Passalids (Peg or Bess Beetles, Horned Passalids)

Figures 34.338a–h

Relationships and Diagnosis: Among the Scarabaeoidea, Passalidae are more distinct morphologically than Scarabaeidae and Lucanidae. Passalid larvae differ from both lucanids and scarabs in lacking a ventral process on the mandible (fig. 34.338f), having 3-segmented maxillary palpi (fig. 34.338g), and having 2-segmented antennae (fig. 34.338b).

Passalid larvae have whitish or bluish, slightly curved bodies with yellowish-brown heads. The emarginations of the respiratory plates of the prothoracic spiracles face anteriorly and those of the abdominal spiracles face posteriorly (fig. 34.338e). In Lucanidae all spiracle emarginations face in the same direction and in Scarabaeidae the emarginations of prothoracic and abdominal spiracles may face in opposite directions, but the abdominal spiracle emarginations never face posteriorly.

Larvae of Passalidae have a transverse anal slit, maxillary stridulatory teeth (fig. 34.338g), and the metathoracic legs reduced to an unsegmented stub which functions as a stridulatory organ (fig. 34.338c). In Lucanidae, the anal slit is longitudinal or Y-shaped, maxillary stridulatory teeth are usually absent, and the metathoracic legs are not reduced. Larval Scarabaeidae may have a transverse anal slit, the metathoracic legs of some species (Geotrupinae) are reduced, and maxillary stridulatory teeth are usually present.

Passalid larvae lack dorsal folds on the body segments (fig. 34.338e), an oncylus on the hypopharynx (fig. 34.338g) and do not have the tormae united at the midline (fig. 34.338a).

Biology and Ecology: Passalidae live in rotting wood where all stages may be found. The most common habitat is a moist, rotting log in the intermediate stage of decay. They are known to occur in standing, as well as fallen logs (Schuster, 1978) and are most frequently found in moist forest situations. *Odontotaenius disjunctus* (Illiger), the common United States species, has been found in rotting stumps and logs of many species of deciduous trees, but only rarely in coniferous logs (Gray, 1946).

Both adults and larvae possess sound producing organs and occur in the same galleries. They are considered to be sub-social and studies on larval stridulation have been reported by Reyes-Castillo and Jarman (1980).

The adults construct tunnels and line them with finely chewed wood pulp which forms the principal food source for larvae (Gray, 1946). The life cycle of *O. disjunctus* is relatively short (about 3 months) and there is only 1 generation per year. There are 3 larval instars (Gray, 1946).

Description: Full grown larvae are 30 to 40 mm long, elongate, subcylindrical, and slightly curved (fig. 34.338e). They have lightly pigmented, straw colored or yellow-brown heads with bluish bodies which become whitish as larvae mature.

Head: (fig. 34.338b) Exserted, hypognathous, somewhat smaller in diameter than thorax, with well-developed mouthparts. Epicranial suture about 1/3 as long as frontal suture, frons triangular. Stemmata absent. Antennae 2-segmented. Labrum free, symmetrical, slightly trilobed. Epipharynx (fig. 34.338a) symmetrical, haptomerum and haptolachus without conspicuous features; tormae bilaterally symmetrical, not fused at midline. Mandibles (figs. 34.338d,f) nearly symmetrical, strongly sclerotized, each with 3 apical teeth and well-developed molar area; ventral process absent. Maxillae (fig. 34.338g) moveable, with falcate, distinctly separated galea and lacinia; dorsal surface of stipes with patch of conical stridulatory teeth. Maxillary palp 3-segmented. Hypopharynx symmetrical, lacking oncyli. Labial palpi 2-segmented.

Thorax and Abdomen: Thoracic segments distinct, with a few conspicuous, long setae; lacking dorsal folds; first 2 segments with well-developed, 4-segmented legs each terminating in a falcate claw. Metathoracic legs (figs. 34.338e, 34.338c) reduced to 1-segmented stubs, the apical teeth of which rub against a striated, stridulatory area on coxa of mesothoracic legs.

Abdomen subcylindrical, consisting of 10 segments, all segments lacking dorsal folds. Sparse, long setae present on abdominal segments and around anal opening; lower anal lobes without setae (fig. 34.338h). Anal opening transverse.

Spiracles: Cribriform, prothoracic spiracle with respiratory plate emarginations facing anteriorly; abdominal spiracles on first 8 segments with respiratory plate emarginations facing posteriorly; spiracles decreasing in size posteriorly.

Comments: The Passalidae is a small family of approximately 500 species with a primarily pan-tropical distribution. Three species have been recorded from the United States (*Odontotaenius disjunctus* (Illiger), *Passalus puntiger* St. Fargeau et Serville, and *Passalus interruptus* (L.)). However, only 1 of these (*O. disjunctus*) is common and the other 2 records may be in error (Schuster, 1978). Several additional species are known from northern Mexico (Schuster, 1978).

Hayes (1929) and Böving and Craighead (1931) described and figured the larva of *O. disjunctus* and included it in their keys to larval Scarabaeoidea. The life history of this species has been described in detail by Gray (1946).

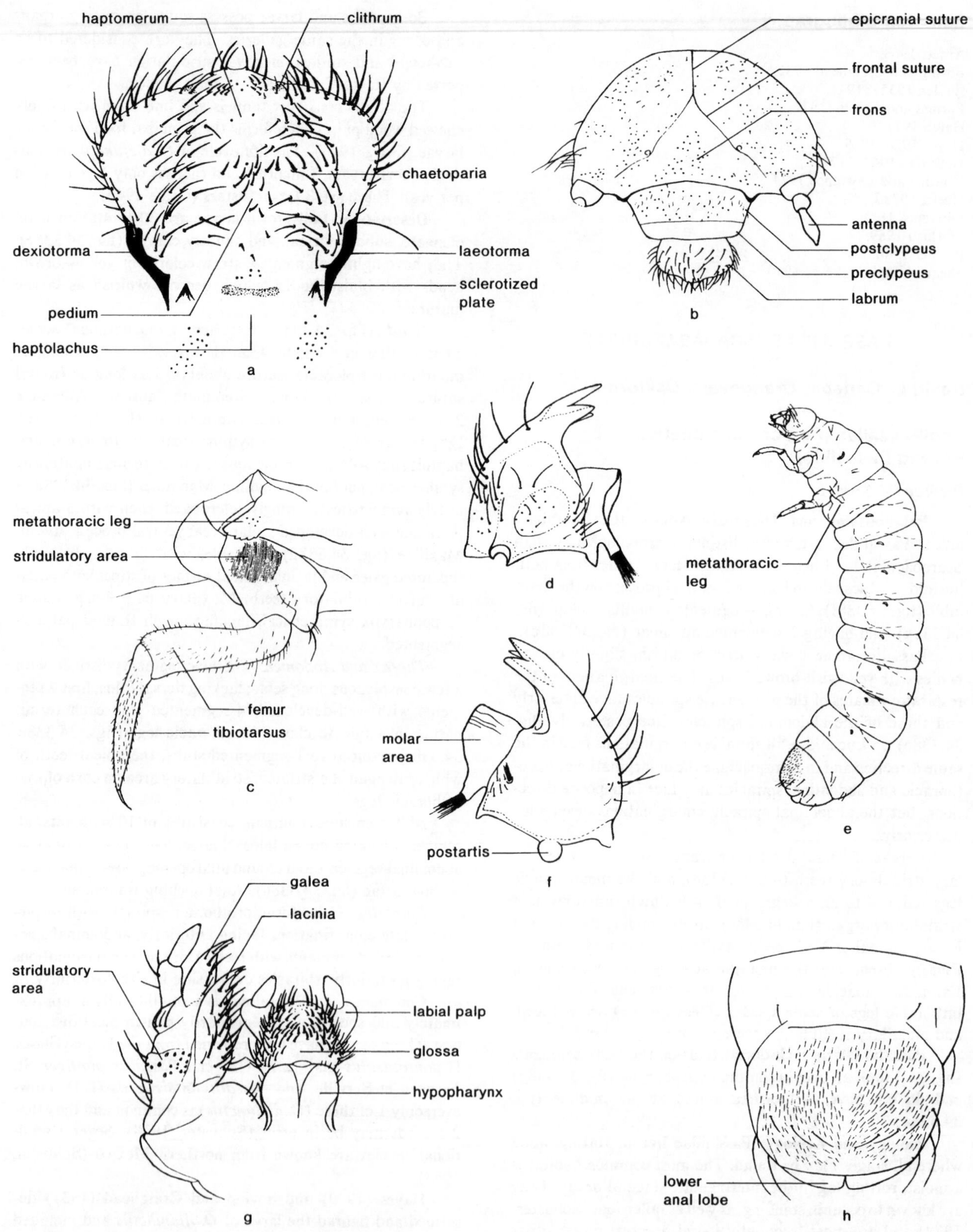

Figures 34.338a-h. Passalidae. *Odontotaenius disjunctus* (Illiger). **a.** epipharynx; **b.** head; **c.** left mesothoracic and metathoracic legs; **d.** left mandible, dorsal; **e.** third-stage larva; **f.** left mandible, ventral; **g.** left maxilla and labium, dorsal; **h.** venter of last 2 abdominal segments. (All figures are from Ritcher, 1966, reprinted with permission of the Oregon State University Press.)

Ritcher (1966) summarized the information on passalid larvae, presented a key to several genera and species, and included descriptions of several species. The most recent comprehensive treatment of passalid larvae is Schuster and Reyes-Castillo (1981). They presented a key to subfamilies, tribes, and 21 New World genera based on 85 species and included information on taxonomic relationships and diagnostic characters of these species. The primary characters used to differentiate taxa included chaetotaxy, antennal form, mandibular dentition, and prothoracic spiracle striation. Reyes-Castillo (1970) published the most recent taxonomic revision of adults of New World species.

The group is of little economic importance since the larvae and adults feed on dead, decaying wood and do not attack living trees. However, they are important factors in the decomposition of logs in forest ecosystems.

Selected Bibliography

Arnett 1968.
Böving and Craighead 1931.
Gray 1946.
Hayes 1929.
Hiznay and Krause 1956.
Reyes-Castillo 1970.
Reyes-Castillo and Jarman 1980.
Ritcher 1966.
Schuster 1978.
Schuster and Reyes-Castillo 1981.

SCARABAEIDAE (SCARABAEOIDEA) (INCLUDING ACANTHOCERIDAE, CERATOCANTHIDAE, CETONIIDAE, DIPHYLLOSTOMATIDAE, GEOTRUPIDAE, GLAPHYRIDAE, HYBOSORIDAE, PLEOCOMIDAE, TROGIDAE, ETC.)

David C. Carlson, *Orangevale, California*

The Scarabs (May Beetles, June Beetles, Rain Beetles, Flower Beetles, Dung Beetles, "Tumble Bugs," and Chafers)

Figures 34.339–385

Relationships and Diagnosis: Scarabaeidae are more similar morphologically to Lucanidae than Passalidae, the other members of the Scarabaeoidea. Larvae of Scarabaeidae are commonly referred to as "white grubs" and differ from Passalidae in having a ventral process on the mandible (figs. 34.379, 34.380b), 3-, 4-, or apparently 5-segmented antennae (figs. 34.349, 34.350, 34.355), a 4-segmented maxillary palp (figs. 34.376a, 34.378, 34.381b, 34.382), and C-shaped (sometimes "hump-backed") larvae (figs. 34.339, 34.340, 34.342–445, 34.347). Scarabaeidae differ from Lucanidae in having a transverse (figs. 34.371, 34.373), angulate (fig. 34.369), or Y-shaped (fig. 34.374) anal opening and lacking oval fleshy lobes on either side of the anal opening. Lucanidae may have a Y-shaped anal opening, but have fleshy lobes on both sides.

Scarab larvae have white, yellow, orange, or bluish bodies, with yellowish or reddish-brown heads. The emarginations of the thoracic and abdominal spiracles may face in similar (figs. 34.339, 34.347) or opposite (figs. 34.340, 34.342) directions, but never with the abdominal emarginations facing posteriorly. In Lucanidae, all emarginations face in the same direction, while in Passalidae thoracic and abdominal emarginations face in opposite directions, but the abdominal emarginations face posteriorly.

In Scarabaeidae, maxillary stridulatory teeth are usually present (figs. 34.376a, 34.381b), stridulatory organs may be present on the legs (figs. 34.383, 34.384a,b, 34.385), and the legs may be reduced in size (figs. 34.347, 34.385). Most Lucanidae lack maxillary stridulatory teeth.

Biology and Ecology: Scarabaeidae exhibit a wide variety of habits. The biology of this family was reviewed by Ritcher (1958) and there are excellent accounts of the habits (adult and larval) of various groups. References to some of these are included in the selected bibliography.

In general, the larvae are saprophagous or phytophagous. Saprophagous species are known that feed on carrion, dung, humus, decaying vegetable matter, duff, litter, and wood. Phytophagous species usually feed on the roots of seed plants. Larvae of Scarabaeinae (Coprinae) are coprophagous or necrophagous. Larvae of Aphodiinae are mostly coprophagous on mammal dung, but a few species attack living roots in the soil. Some species of Aphodiinae and Cetoniinae are found in ant or termite nests. Larvae of Glaphyrinae and some Geotrupinae, Dynastinae, Melolonthinae, Aphodiinae, and Cetoniinae feed on various kinds of decaying plant material. Larvae of some Rutelinae, Cetoniinae, and Dynastinae feed in decaying wood. Larvae of Geotrupinae, Scarabaeinae, and some Dynastinae feed on saprophagous material provisioned for them by adults. Nidification behavior (nest building) in Scarabaeinae has been studied extensively (Halffter and Matthews, 1966; Halffter and Edmonds, 1981, 1982). Larvae of Troginae feed on animal remains and are often found beneath mammal or bird carcasses or in the nests of birds and burrows of mammals where they feed on feathers and fur (Baker, 1968). The larvae of many species of Melolonthinae, Pleocominae, Rutelinae, and Dynastinae feed on living roots and are often destructive.

Relatively little is known of the biology of members of the Acanthocerinae, Hybosorinae, and Ochodaeinae in North America. Observations of adult and larval habits of these groups are often conflicting.

The life cycles of Scarabaeidae vary considerably and are generally longer in temperate regions. They range in duration from less than 1 year (bivoltine) in some Aphodiinae to 8 or more years in the Pleocominae. Most species have life cycles ranging from 1 to 3 years. In most species there are 3 larval instars. The pupa is usually found in a cell constructed by the last larval instar. Most often the pupal cell is found in the soil, however, it may be constructed of old dung, duff, wood fragments, or humus-like material.

Since habits and life cycles vary considerably, it is not possible to present a very representative generalized life cycle. However, for univoltine species in northern latitudes, eggs are laid in the soil in July and August and hatch in a few weeks.

The first stage larvae commence feeding and continue until colder weather forces them deeper (below the frost line) where they overwinter. In April the larvae resume feeding activity near the soil surface, pass through 2 more instars and pupate in June. Adults emerge a few weeks later. In species with longer life cycles, the larvae may only molt once each year or the number of instars may be greater (Pleocominae). Some species may overwinter as pupae or adults.

Description: Full grown larvae are 10 to 125 mm long along the curved dorsal surface, C-shaped (figs. 34.340, 34.342, 34.343, 34.347), and subcylindrical with fleshy bodies. Some species are "hump-backed" (Scarabaeinae (figs. 34.339, 34.344, 34.345)). They have yellow to reddish–brown or black heads and lightly sclerotized whitish, yellowish, bluish or orange bodies. The caudal end of abdomen may appear blackish due to accumulated feces.

Head: (figs. 34.349–359). Hypognathous, exserted, with well developed mouthparts. Epicranial suture shorter than frontal arms; V-shaped or lyriform frons enclosed by frontal arms. Stemmata sometimes present (fig. 34.351). Antennae 3- or 4-segmented (figs. 34.349, 34.350), last segment reduced in size in more generalized subfamilies (figs. 34.358, 34.359). Labrum free, sometimes asymmetrical, distal margin entire, bilobed or trilobed. Epipharynx (figs. 34.360–367) often with well developed haptomeral teeth (fig. 34.361), chaetoparia (fig. 34.363), and haptolachus (figs. 34.360, 34.361, 34.363); tormae fused on midline in more generalized subfamilies (figs. 34.362, 34.365, 34.367), 1–3 nesia (SC or SP) present (figs. 34.361, 34.366). Mandibles (figs. 34.376b, 34.379, 34.380a,b) asymmetrical, with ventral articulating process (fig. 34.379), well developed scissorial and molar areas, the scissorial area toothed (fig. 34.376b) or blade-like (fig. 34.380a); ventral surface in Dynastinae, Rutelinae and Cetoniinae with oval patch of transverse striae (fig. 34.379) opposite the stridulatory teeth on maxillary stipes (figs. 34.376a, 34.381b). Maxillae moveable, with separate galea and lacinia in generalized forms (fig. 34.376a); with mala in some specialized forms (Melolonthinae, Dynastinae, Rutelinae, and Cetoniinae) (figs. 34.381b, 34.382); galea and lacinia with apical teeth (unci). Maxillary palp 4-segmented.

Thorax and Abdomen: Thoracic segments distinct, each with 2–3 dorsal lobes and pair of well-developed 2- to 5-segmented legs. Legs with claws; stridulatory organs on pro- and mesothoracic legs in Hybosorinae (figs. 34.384a,b), meso- and metathoracic legs in Pleocominae (fig. 34.383), Acanthocerinae, and Geotrupinae (fig. 34.385), lacking in other subfamilies.

Abdomen subcylindrical, of 10 segments (figs. 34.339, 34.342); segments 9 and 10 fused in some Cetoniinae (fig. 34.348); surface weakly sclerotized; usually with 3 dorsal folds on first 6 segments, folds often with numerous setae. Entire body hairy in Glaphyrinae. Venter of last abdominal segment often with distinctive setal arrangement (raster) (figs. 34.370–375). Anal opening Y-shaped (fig. 34.374), V-shaped (fig. 34.369), or transversely curved (fig. 34.373); conspicuous fleshy lobes on last abdominal segment, near anus, in Aphodiinae, Scarabaeinae, and Geotrupinae (figs. 34.339, 34.343).

Spiracles: Usually cribriform (fig. 34.341), biforous (fig. 34.346) in some Troginae and Geotrupinae, consisting of 1 pair of prothoracic spiracles and 1 pair on abdominal segments 1–8. Respiratory plate emarginations of thoracic spiracles facing posteriorly, ventrally, or caudoventrally; emarginations of abdominal spiracles facing cephalad, ventrad, or cephaloventrad.

Comments: The Scarabaeidae is the largest of the 3 families comprising the superfamily Scarabaeoidea. It has a world-wide distribution and contains some 13,000 species ranging in size from a few to over a hundred millimeters in length. Approximately 1400 species representing some 120 genera and 13 subfamilies occur in the United States and Canada (Arnett, 1968; Blackwelder, 1944).

The immature stages of some groups are fairly well known, often because they are of economic importance (e.g. *Phyllophaga*) or have unusual habits (e.g. *Trox*). There are several early and significant works dealing with scarabaeid larvae (Hayes, 1928; Böving and Craighead, 1931; Böving, 1936, 1942), however, the most recent comprehensive treatment of the larvae is Ritcher (1966). Ritcher's monograph contains descriptions and keys to subfamilies, genera, and species, including many species of economic importance. Subsequently, the immature stages of numerous species and a few genera have been described. References to some of these papers are included in the selected bibliography. Tashiro (1987) provides excellent coverage of the turfgrass pests.

The immature stages of many N. American genera and species are yet to be described. Approximately 65% of the genera which occur in the United States and Canada have had the larvae of 1 or more species described (Ritcher, 1966). In many cases, however, relatively large genera are represented by the description of only 1 or a few species (e.g. *Coenonycha,* 30+ species, larvae of 1 species described; *Onthophagus,* about 40 N. American species, larvae of 6 species described; *Ataenius,* 63 U.S. and Canadian species, larvae of 8 species described).

The larvae of many species are economically important as agricultural pests of various crops. Some of the most destructive species belong to the subfamilies Melolonthinae, Rutelinae, and Dynastinae. Among the Melolonthinae, larvae of *Phyllophaga, Serica, Maladera, Diplotaxis,* and *Rhizotrogus* are serious pests of grasses, corn, wheat, strawberries, and tree seedlings. The Ruteline genera *Anomala, Cotalpa,* and *Popillia* contain species whose larvae cause serious damage to turf, vegetable crops, corn, small grains, strawberries, and nursery stock. The larvae of *Cyclocephala* and other Dynastine genera are often pests of sod, corn, small grains, sugar cane, rice, and vegetable crops. The adults of many scarab species often cause serious economic damage by attacking the leaves, flowers, and fruit of various plants.

Not all economically significant Scarabaeidae are detrimental. Many species of Scarabaeinae are beneficial in that they incorporate organic material (dung) into the soil and in so doing increase soil fertility and destroy breeding places for important livestock pests.

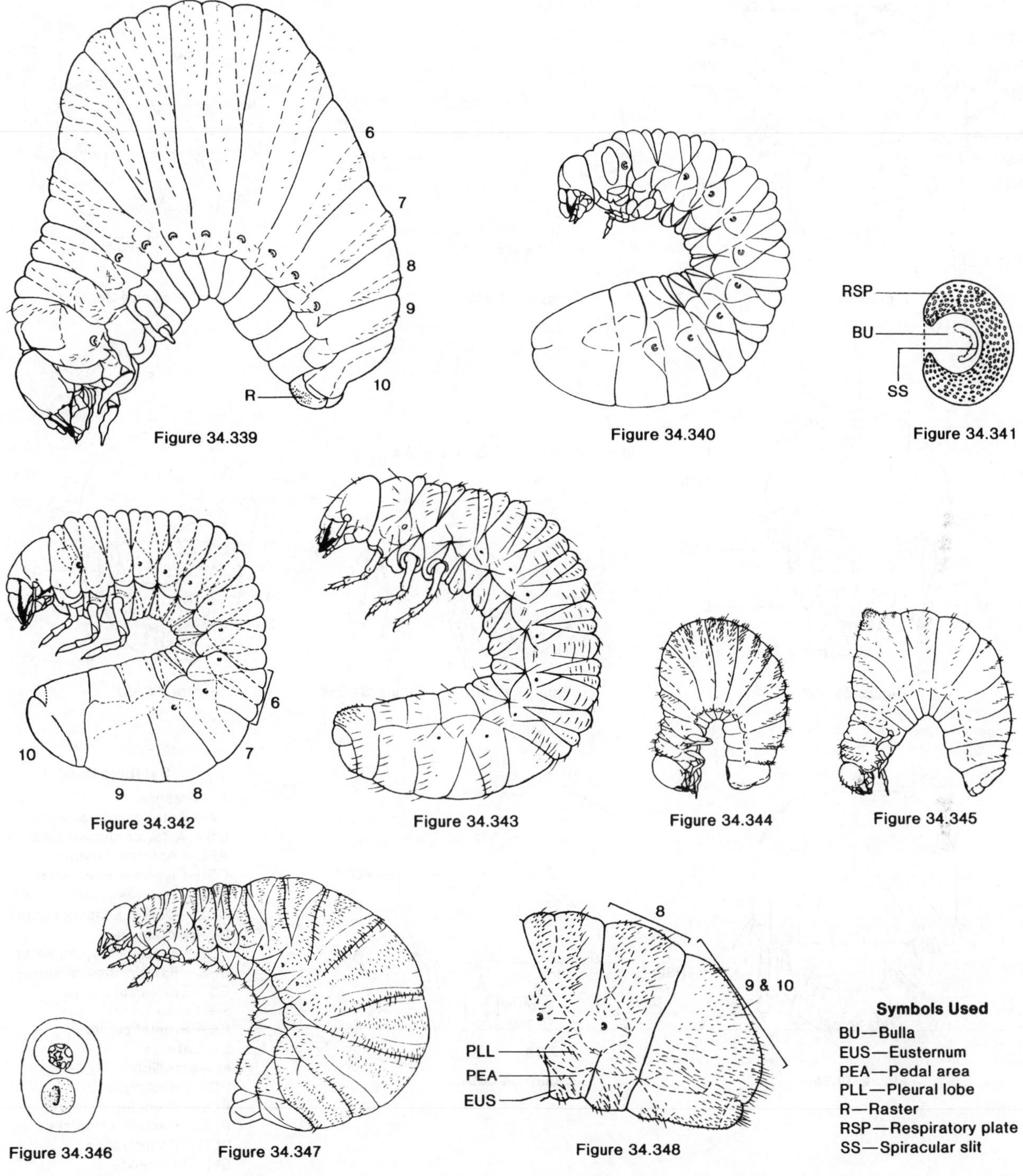

Figure 34.339 Figure 34.340 Figure 34.341

Figure 34.342 Figure 34.343 Figure 34.344 Figure 34.345

Figure 34.346 Figure 34.347 Figure 34.348

Figures 34.339–34.348. Scarabaeidae.

Figure 34.339. *Dichotomius carolinus* (L.). Third-stage larva.

Figure 34.340. *Euphoria inda* (L.). Entire larva (setae omitted).

Figure 34.341. *Anomala innuba* (Fabricius). Abdominal spiracle.

Figure 34.342. *Cyclocephala immaculata* Olivier. Third-stage larva.

Figure 34.343. *Aphodius sparsus* LeConte. Third-stage larva.

Figure 34.344. *Ateuchus histeroides* Weber. Third-stage larva.

Figure 34.345. *Onthophagus hecate hecate* (Panzer). Third-stage larva.

Figure 34.346. *Bolboceras simi* (Wallis). Thoracic spiracle.

Figure 34.347. *Peltotrupes youngi* Howden. Third-stage larva.

Figure 34.348. *Osmoderma eremicola* Knoch. Abdominal segments 8-10, lateral.

(All figures from Ritcher, 1966, reprinted with permission of the Oregon State University Press.)

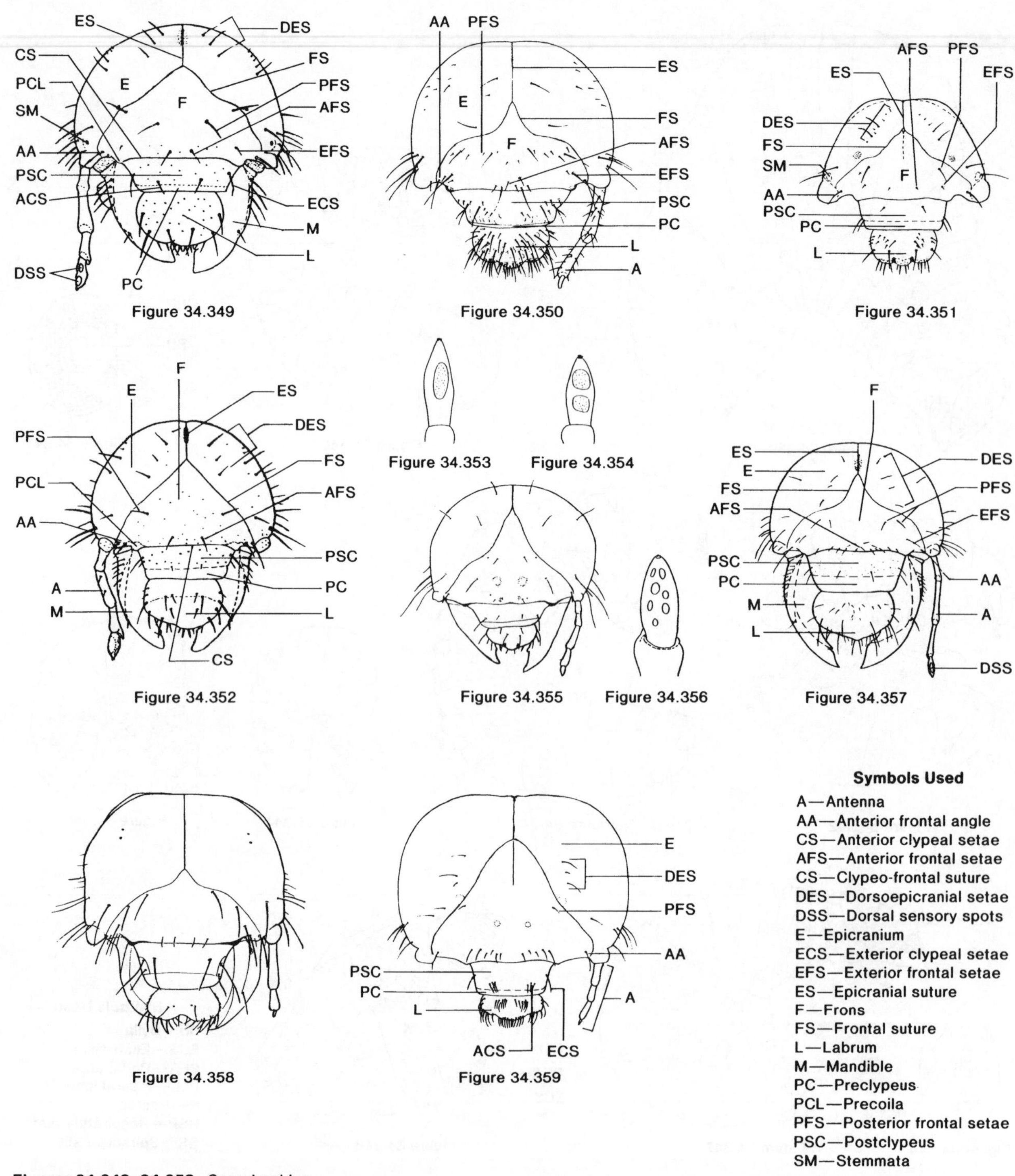

Figures 34.349–34.359. Scarabaeidae.

Figure 34.349. *Cyclocephala immaculata* Olivier. Head.

Figure 34.350. *Pleocoma hirticollis vandykei* Linsley. Head.

Figure 34.351. *Cotinis nitida* (L.). Head.

Figure 34.352. *Anomala innuba* (Fabricius). Head.

Figure 34.353. *Aphonus densicauda* Casey. Distal segment of antenna, dorsal.

Figure 34.354. *Bothynus gibbosus* (De Geer). Distal segment of antenna, dorsal.

Figure 34.355. *Aphodius hamatus* Say. Head.

Figure 34.356. *Dynastes tityus* (L.). Distal segment of antenna, dorsal.

Figure 34.357. *Phyllophaga horni* (Smith). Head.

Figure 34.358. *Trox suberosus* Fabricius. Head.

Figure 34.359. *Dichotomius carolinus* (L.). Head.

(All figures from Ritcher, 1966, reprinted with permission of the Oregon State University Press.)

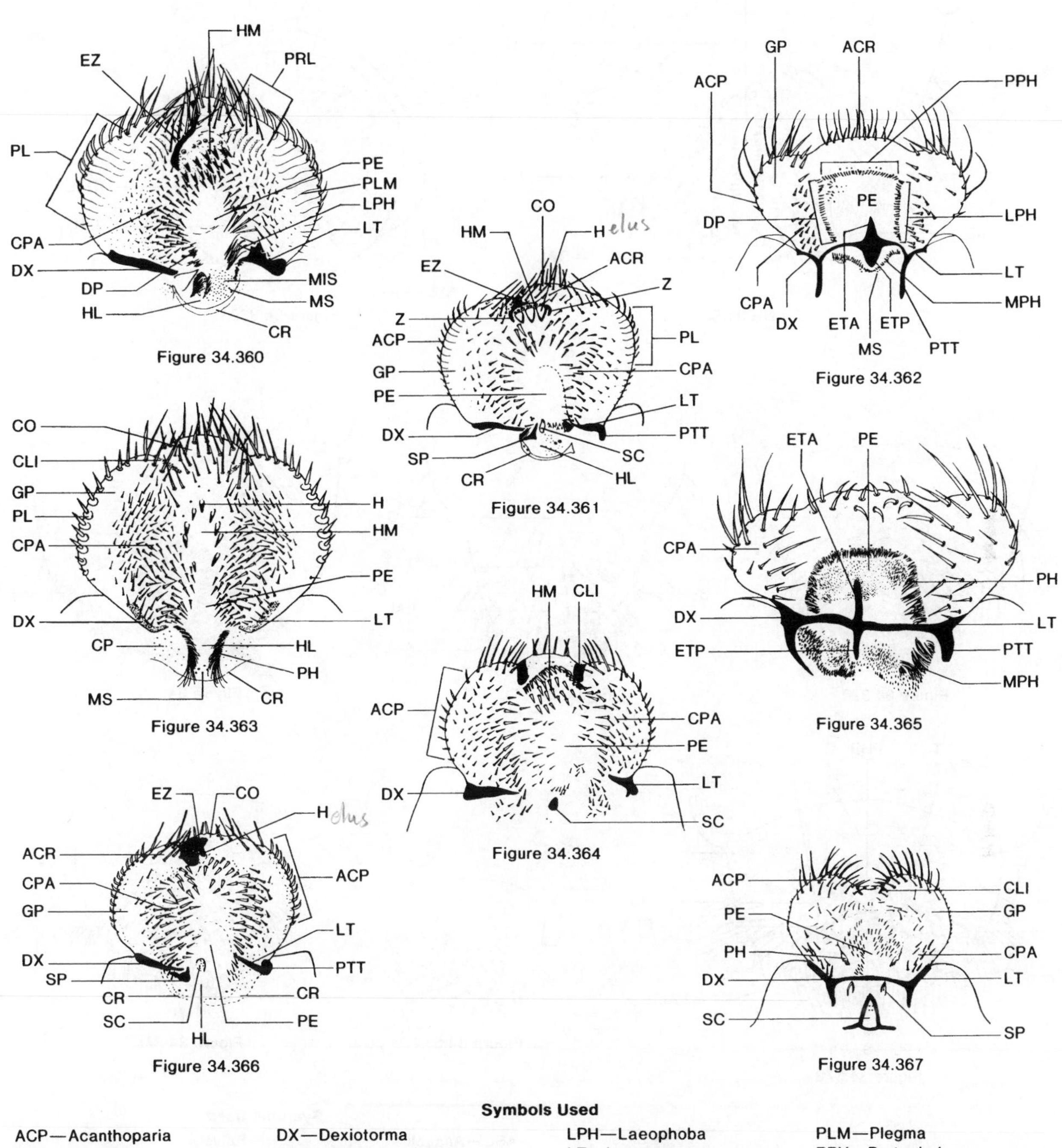

Symbols Used

ACP—Acanthoparia
ACR—Acroparia
CLI—Clithrum
CO—Corypha
CP—Crepidal punctures
CPA—Chaetoparia
CR—Crepis
DP—Dexiophoba
DX—Dexiotorma
ETA—Anterior epitorma
ETP—Posterior epitorma
EZ—Epizygum
GP—Gymnoparia
H—Helus
HL—Haptolachus
HM—Haptomerum
LPH—Laeophoba
LT—Laeotorma
MIS—Microsensilla
MPH—Mesophoba
MS—Macrosensilla
PE—Pedium
PH—Phoba
PL—Plegmatium
PLM—Plegma
PPH—Protophoba
PRL—Proplematium
PTT—Pternotorma
SC—Sense cone
SP—Sclerotized plate
Z—Zygum

Figures 34.360–34.367. Scarabaeidae. Epipharynges.

Figure 34.360. *Phyllophaga hirticula* (Knoch).

Figure 34.361. *Anomala innuba* (Fabricius).

Figure 34.362. *Dichotomius carolinus* (L.).

Figure 34.363. *Pleocoma hirticollis vandykei* Linsley.

Figure 34.364. *Cotinis nitida* (L.).

Figure 34.365. *Geotrupes blackburnii excrementi* Say.

Figure 34.366. *Cyclocephala immaculata* Olivier.

Figure 34.367. *Trox suberosus* Fabricius.

(All figures from Ritcher, 1966, reprinted with permission of the Oregon State University Press.)

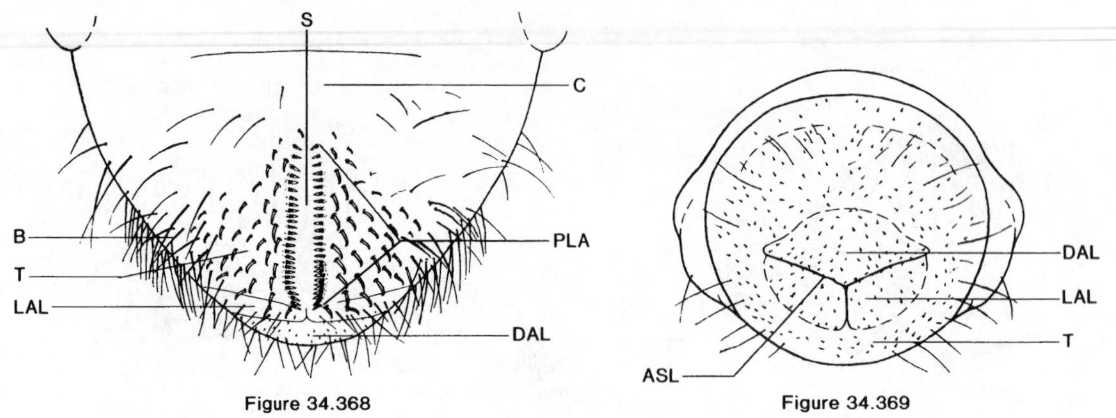

Figure 34.368

Figure 34.369

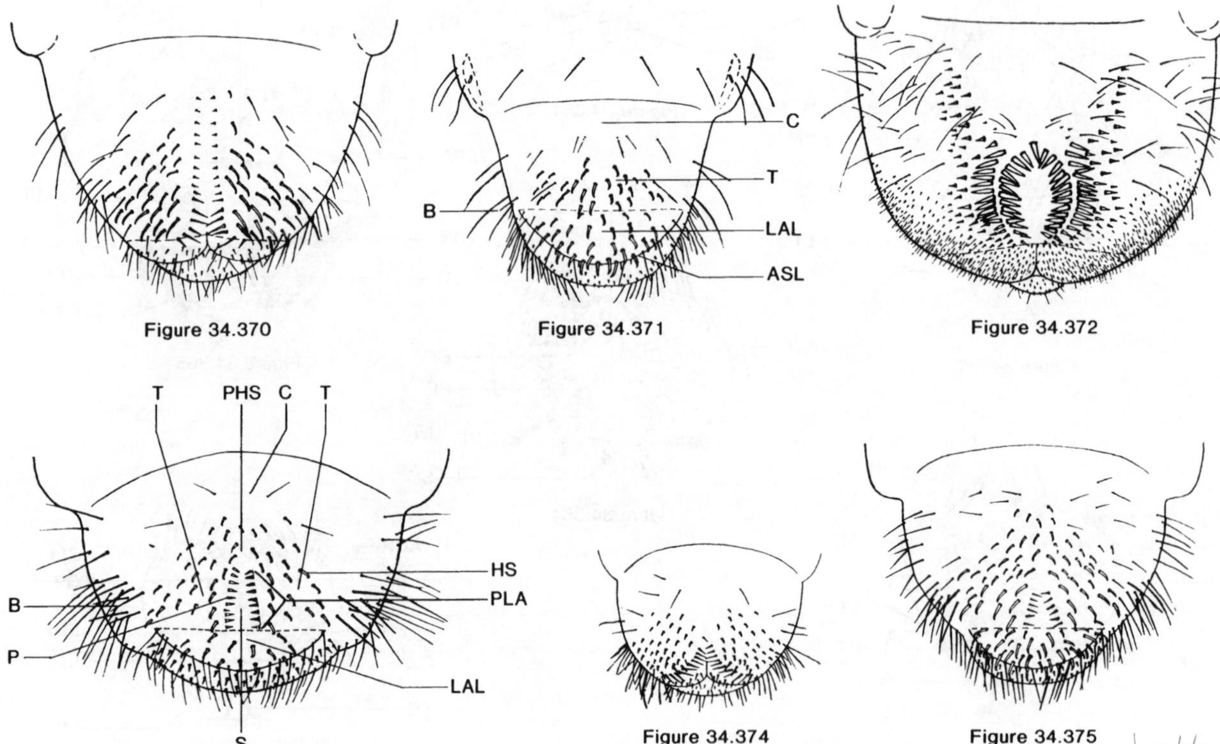

Figure 34.370

Figure 34.371

Figure 34.372

Figure 34.373

Figure 34.374

Figure 34.375

Symbols Used

ASL—Anal slit
B—Barbula
C—Campus
DAL—Dorsal anal lobe
HS—Hamate setae
LAL—Lower anal lobe
P—Palus
PHS—Preseptular hamate setae
PLA—Palidia
S—Septula
T—Tegilla

Figures 34.368–34.375. Scarabaeidae. Venters of last abdominal segment (except fig. 34.369).

Figure 34.368. *Phyllophaga hirticula* (Knoch).

Figure 34.369. *Pleocoma hirticollis vandykei* Linsley. Caudal view.

Figure 34.370. *Rhizotrogus (Amphimallon) majalis* (Razoumowsky).

Figure 34.371. *Cyclocephala immaculata* Olivier.

Figure 34.372. *Phobetus comatus sloopi* Barrett.

Figure 34.373. *Anomala innuba* (Fabricius).

Figure 34.374. *Diplotaxis* sp.

Figure 34.375. *Popillia japonica* Newman.

(All figures from Ritcher, 1966, reprinted with permission of the Oregon State University Press.)

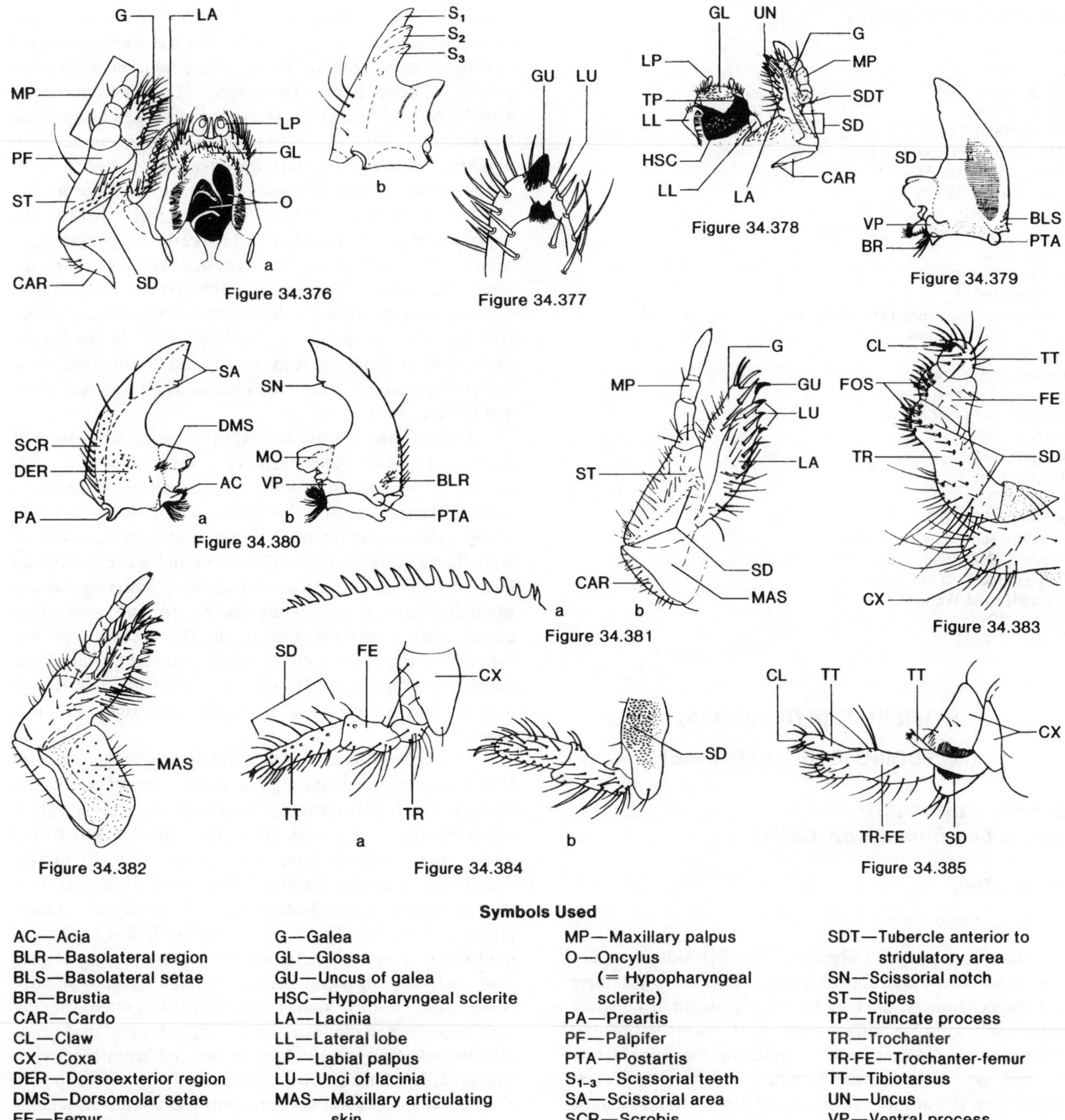

Symbols Used

AC—Acia
BLR—Basolateral region
BLS—Basolateral setae
BR—Brustia
CAR—Cardo
CL—Claw
CX—Coxa
DER—Dorsoexterior region
DMS—Dorsomolar setae
FE—Femur
FOS—Fossorial setae
G—Galea
GL—Glossa
GU—Uncus of galea
HSC—Hypopharyngeal sclerite
LA—Lacinia
LL—Lateral lobe
LP—Labial palpus
LU—Unci of lacinia
MAS—Maxillary articulating skin
MO—Mola bearing point
MP—Maxillary palpus
O—Oncylus (= Hypopharyngeal sclerite)
PA—Preartis
PF—Palpifer
PTA—Postartis
S_{1-3}—Scissorial teeth
SA—Scissorial area
SCR—Scrobis
SD—Stridulatory teeth or area of striae
SDT—Tubercle anterior to stridulatory area
SN—Scissorial notch
ST—Stipes
TP—Truncate process
TR—Trochanter
TR-FE—Trochanter-femur
TT—Tibiotarsus
UN—Uncus
VP—Ventral process

Figures 34.376–34.385. Scarabaeidae

Figures 34.376a,b. *Dichotomius carolinus* (L.). **a.** left maxilla, labium, and hypopharynx, ental view; **b.** left mandible, dorsal.

Figure 34.377. *Plusiotis woodi* Horn. Distal part of left maxilla.

Figure 34.378. *Cyclocephala immaculata* Olivier. Right maxilla, labium, and hypopharyngeal sclerome, ental view.

Figure 34.379. *Anomala innuba* (Fabricius). Left mandible, ventral.

Figures 34.380a,b. *Phyllophaga horni* (Smith). **a.** left mandible, dorsal; **b.** left mandible, ventral.

Figures 34.381a,b. *Polyphylla decemlineata* (Say). **a.** maxillary stridulatory teeth of right maxilla; **b.** left maxilla, dorsal.

Figure 34.382. *Phyllophaga crenulata* (Froelich). Right maxilla, ventral.

Figure 34.383. *Pleocoma hirticollis vandykei* Linsley. Right metathoracic leg.

Figures 34.384a,b. *Hybosorus orientalis* West. **a.** right mesothoracic leg; **b.** left prothoracic leg.

Figure 34.385. *Geotrupes blackburnii excrementi* Say. Left metathoracic and mesothoracic legs, lateral.

(All figures from Ritcher, 1966, reprinted with permission of the Oregon State University Press.)

Selected Bibliography

Arnett 1968.
Baker 1968.
Blackwelder 1944.
Blume 1981.
Böving 1936, 1942.
Böving and Craighead 1931.
Carlson and Ritcher 1974.
Cornell 1967, 1972b.
Edmonds 1967.
Edmonds and Halffter 1972, 1978.
Erwin 1970.
Fellin 1975, 1981.
Gardner 1935a.
Halffter and Edmonds 1981, 1982.
Halffter and Matthews 1966.
Hatch 1971.
Hayes 1928, 1929.
Helgesen and Post 1967.
Howden 1955, 1964, 1982.
Hurka 1978c.
Hurpin 1962.
Jerath 1960.
Moron 1977.
Ratcliffe 1976, 1977.
Ratcliffe and Chalumeau 1980.
Ritcher 1958, 1966, 1967, 1973.
Ritcher and Duff 1971.
Rosander and Werner 1970.
Tashiro 1987.
Woodruff 1973.

BYRRHIDAE (BYRRHOIDEA) (INCLUDING SYNCALYPTIDAE)

John F. Lawrence,
Division of Entomology, CSIRO

Pill Beetles

Figures 34.386–391

Relationships and Diagnosis: The Byrrhidae are isolated from other Elateriformia by a combination of primitive and unique features, and they are usually placed in a distinct superfamily. Larvae differ from most members of the dryopoid-elateroid-cantharoid complex by the combination of large, hypognathous head with distinct ventral epicranial ridges; free labrum; 6 well-separated stemmata; Y-shaped epicranial suture with the frontal arms forming an acute angle; well-developed lacinia, articulated galea, and maxillary articulating area; and bisetose tarsungulus. The C-shaped body and hypognathous head of some byrrhid larvae resemble those of Dascillidae and Scarabaeoidea, but larvae of the latter lack stemmata and have a well-developed mandibular mola and usually cribriform spiracles. Larvae of Chrysomelidae and Byrrhidae are often similar in body form, and both lack a mandibular mola and gular region; chrysomelid larvae, however, always lack a distinct galea and lacinia, and they usually have a median endocarina extending anteriorly between the frontal arms.

Biology and Ecology: Most Byrrhidae feed on the leaves and rhizoids of mosses and liverworts (Bryophyta), but some have been recorded from lichens and others are known to attack the roots of higher plants. *Cytilus* larvae were found in moss-covered sand on the banks of a river, and fragments of rhizoids were found in the gut, while Johnson and Russell (1978) collected *Exomella pleuralis* (Casey) from mosses growing on boulders. Other records of moss-feeding are cited by Horion (1955) and Watt (1971). The Japanese *Lamprobyrrhulus nitidus* (Schaller) was taken under liverworts of the genus *Marchantia* (Hayashi, 1972), and *Epichorius* species have been found on both liverworts and lichens in the subantarctic region (Watt, 1971). Species of *Byrrhus* have been found in soil around the roots of seedling trees in nurseries (Peterson, 1951), while *Amphicyrta* in California is destructive to the roots of wild grasses, clover, oats, and weeds (Essig, 1958). A species of "*Pedilophorus*" in the Snowy Mountains of Australia feeds on the rosettes of *Cardamine* and is responsible for the transmission of a tymovirus (Guy and Gibbs, 1981).

Description: Mature larvae 2 to 15 mm. Body form of 2 general types: (1) elongate, more or less parallel-sided, subcylindrical, straight to strongly curved ventrally (C-shaped), without laterally expanded tergites, so that pleura are more or less exposed, usually more lightly sclerotized in middle of body than at either end; and (2) oblong and fusiform, slightly flattened, and moderately curved ventrally, with tergal plates expanded laterally concealing the pleura, and with entire dorsal surface heavily sclerotized. The second type less common, occurring in *Cytilus, Lioon,* and many South Temperate "*Pedilophorus.*" Dorsal surfaces smooth; vestiture of long or short, simple hairs, variously distributed and sometimes dense.

Head: Protracted and moderately to strongly declined (hypognathous), globular. Epicranial stem long; frontal arms V-shaped or slightly lyriform (*Amphicyrta*), forming a characteristic acute angle with each other. Median endocarina often present beneath base of epicranial stem. Stemmata usually 6 on each side, forming 2 groups of 3. Antennae short to moderately long, 3-segmented. Frontoclypeal suture present; labrum free. Mandibles symmetrical, short and broad, unidentate or multidentate, without accessory ventral process; mola absent; mesal surface of mandibular base with dense brush of hairs. Ventral mouthparts retracted. Maxilla with transverse cardo, elongate stipes, well-developed articulating area, 4-segmented palp, articulated, palpiform galea, and fixed, falciform or truncate lacinia. Labium free to base of mentum; ligula absent or very short and broad; labial palps 2-segmented and well separated. Hypopharyngeal sclerome absent. Hypostomal rods absent. Ventral epicranial ridges well-developed and long, extending well past the beginning of occipital foramen. Gular region absent (labium contiguous with thoracic membrane).

Thorax and Abdomen: Legs well-developed, 5-segmented, moderately widely separated; tarsungulus with 2 setae lying side by side. Abdominal segments with distinct pleural and sternal lobes, sometimes bearing plates, usually with groups of setae, and often divided into 2 or more parts. Tergum A8 sometimes with distinct, transverse carina or raised, microasperate area. Tergum A9 without urogomphi, usually more or less strongly rounded, sometimes flattened,

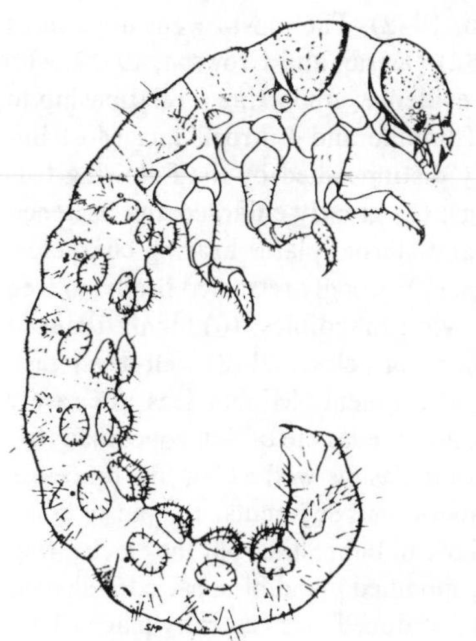

Figure 34.386

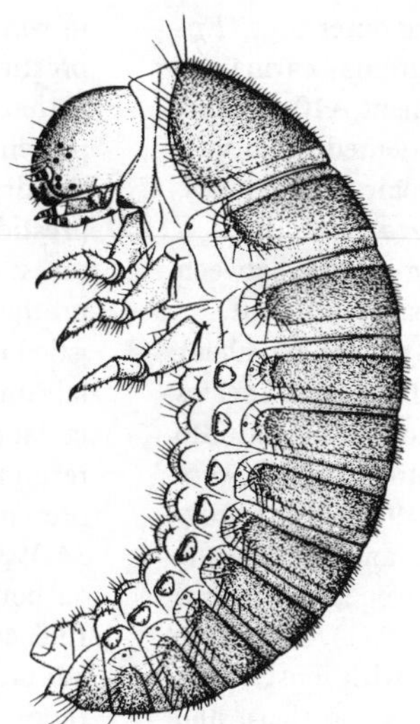

Figure 34.387

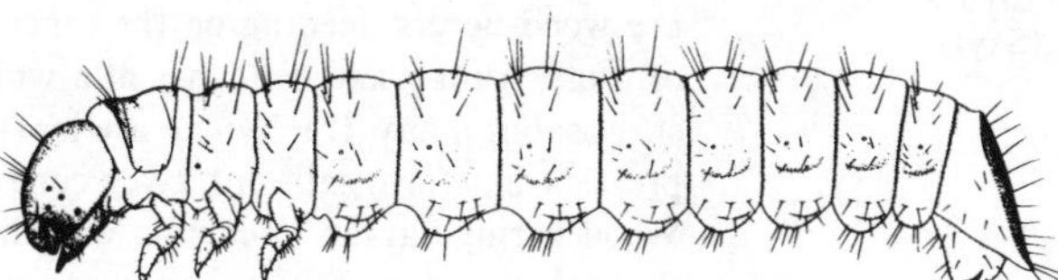

Figure 34.388

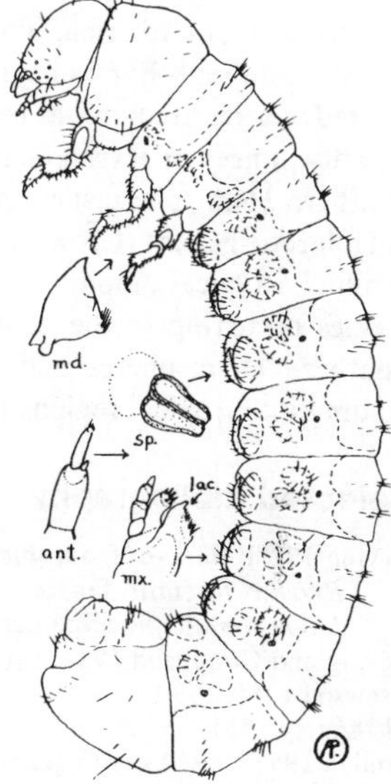

Figure 34.389

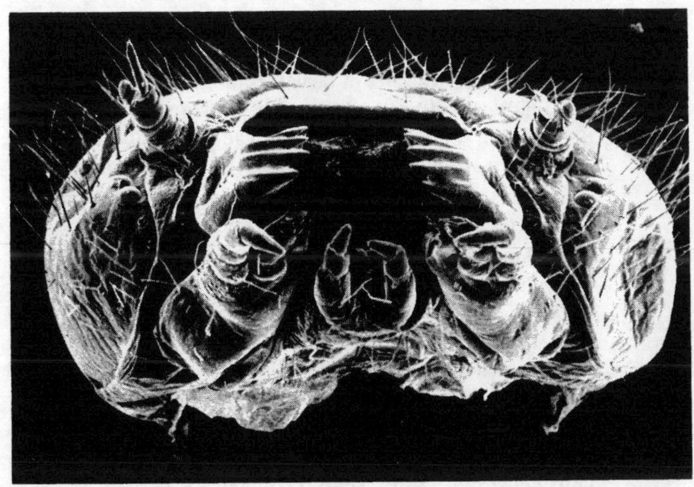

Figure 34.390

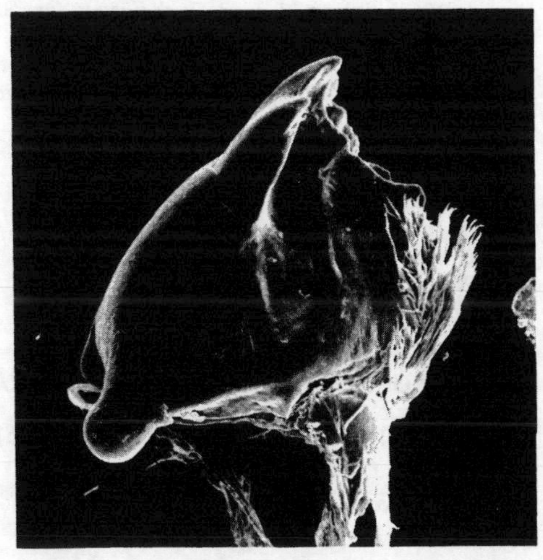

Figure 34.391

Figure 34.386. Byrrhidae. *Amphicyrta* sp. Sebastopol, Sonoma Co., California. Larva. Length = 23 mm.

Figure 34.387. Byrrhidae. *Cytilus alternatus* (Say). Bedford, Middlesex Co., Massachusetts. Larva. Length = 5 mm.

Figure 34.388. Byrrhidae. *Microchaetes* sp. Lower Gordon River, Tasmania. Larva. Length = 3 mm.

Figure 34.389. Byrrhidae. *Byrrhus* sp. Larva. Length = 17 mm. (From Peterson, 1951.)

Figure 34.390. Byrrhidae. *Lioon* sp. 7 mi. S Port Angeles, Washington. Head, anteroventral.

Figure 34.391. Byrrhidae. *Cytilus alternans* (Say). Thornton, New Hampshire. Right mandible, ventral.

with incomplete, semi-circular carina around outer edge (*Epichorius*), and sometimes with complete, circular carina enclosing concave plate (Syncalyptinae). Segment A10 reduced but distinct, posteriorly or posteroventrally oriented, often with distinct tergal and sternal plates, and with pair of anal lobes, each of which may bear a hook (*Amphicyrta*).

Spiracles: Biforous, those on abdomen located between pleural lobes and tergal plates; closing apparatus absent.

Comments: The Byrrhidae includes 28 genera and about 300 species worldwide, with 14 genera and about 40 species occurring in N. America. The group is usually divided into two subfamilies—Byrrhinae and Syncalyptinae, which were treated as separate families by El Moursy (1961). The family is badly in need of revision on a world basis, and many of the Southern Hemisphere species have been erroneously placed in Holarctic genera (Crowson, 1980; Watt, 1971). The group exhibits a typical amphipolar distribution, with most of the species occurring in the cooler parts of the Northern and Southern Hemispheres and only a few Syncalyptinae extending into tropical regions (Crowson, 1980; Paulus, 1972).

Selected Bibliography

Böving 1929b (larvae of *Amphicyrta chrysomelina* Erichson, *Byrrhus fasciatus* Forster, *Cytilus alternatus* (Say), and *Lioon simplicipes* (Mannerheim)).
Böving and Craighead 1931 (same as above).
Crowson 1955, 1980.
El Moursy 1961.
Emden 1958 (larval key to genera).
Essig 1958.
Gressitt and Samuelson 1964b (larva of *Epichorius sorenseni* (Brookes)).
Guy and Gibbs 1981 (virus transmission by *Pedilophorus* on *Cardamine*).
Hayashi 1962b (larva of *Lamprobyrrhulus nitidus* (Schaller)).
Horion 1955.
Hudson 1934 (larva of *Pedilophorus coruscans* Pascoe).
Johnson and Russell 1978 (habitat of *Exomella pleuralis* (Casey)).
Lawrence and Newton 1982.
LeSage 1983 (larva of *Cytilus alternatus* (Say)).
Paulus 1972.
Peterson 1951 (larva of *Byrrhus*).
Striganova 1964b (larval key to genera).
Watt 1971 (*Epichorius* larvae).

BUPRESTIDAE (BUPRESTOIDEA) (INCLUDING SCHIZOPODIDAE)

John F. Lawrence,
Division of Entomology, CSIRO

Metallic Wood-Boring Beetles; Flat-Headed Wood-Borers

Figures 34.392–399

Relationships and Diagnosis: The Buprestidae are normally placed in a separate superfamily, and their phylogenetic relationships have been the subject of much controversy. The family is sometimes considered to have evolved directly from a dascilloid ancestor, partly because of the similarities in wing venation between Dascillidae and schizopodine Buprestidae (Forbes, 1926, 1942). The most recent hypothesis is that of Crowson (1982) (Kasap and Crowson, 1975), who presented considerable evidence supporting a relationship to the dryopoid families Elmidae and Lutrochidae. Most buprestid larvae are easily distinguished by the following features: (1) absence of legs; (2) greatly enlarged and flattened prothorax, with tergal and sternal plates bearing characteristic impressed markings; (3) strongly retracted head; (4) free labrum; (5) normal chewing mandibles; (6) highly reduced labial palps; (7) cribriform spiracles; and (8) well-developed, terminal 10th abdominal segment. Minute legs are rarely present in a few species, and the larvae of Schizopodinae (fig. 34.393) have biforous spiracles, as well as leg-like processes on both thoracic and abdominal segments. Larvae of some Eucnemidae resemble those of buprestids, but they lack a free labrum, and have highly modified mandibles, paired T-shaped thoracic markings, and a reduced and ventrally placed 10th segment. Those legless cerambycid larvae with a retracted head and enlarged prothorax differ in having annular, annular-multiforus, or annular-biforous spiracles and well-developed labial palps and lacking the prothoracic markings.

Biology and Ecology: The majority of buprestid larvae are wood-borers, feeding on the inner bark and outer wood of roots, trunks, and branches of a wide variety of trees and shrubs; but many Trachyinae are leaf-miners, and other buprestids may form galls or feed on cones or herbaceous plants. Wood-boring larvae excavate winding, flattened galleries through sound or decayed sapwood or heartwood, and pupation occurs within the gallery. Although most species attack weakened or dead plants, bark-borers are capable of girdling and killing healthy trees, while wood-borers may be destructive to lumber. Adult Buprestidae are usually collected at flowers, where they feed on pollen and nectar, but some species feed on twigs or foliage. Eggs are usually laid in cracks or wounds near the site of larval feeding, but in some species (especially Julodinae and probably Schizopodinae), they are deposited in soil and first instar larvae are capable of moving through the soil to the root system. Buprestidae may be serious pests of forest, shade, ornamental, and fruit trees, and some attack building materials. Some common pest species in N. America include the bronze birch borer (*Agrilus anxius* Gory, fig. 34.397), the flatheaded apple tree borer (*Chrysobothris femorata* (Olivier), fig. 34.396), species of *Buprestis* and *Melanophila,* which attack conifers, and species of *Brachys,* which mine the leaves of oaks and elms, but whose damage is more aesthetic. In Europe, a number of Sphenopterinae are destructive to fruit and shade trees. (Drooz, 1985; Balachowsky *et al.,* 1962; Furniss and Carolin, 1977; Gardner, 1944; Lesne, 1898; Peterson, 1951; Rees, 1941).

Description: Mature larvae 2 to 100 mm, usually 50 mm or less. Body elongate, usually with abdomen more or less parallel-sided, straight, and slightly flattened, and with thorax (especially prothorax) greatly enlarged and moderately to strongly flattened; body strongly flattened and widest at middle, without enlarged thorax in specialized leaf miners, subcylindrical without enlarged thorax in Schizopodinae. Surfaces lightly sclerotized, whitish or cream in color, except for buccal region and sometimes portions of prothorax and abdominal apex. Vestiture of simple hairs, usually scattered, sometimes dense, long and dense in early instars of Julodinae.

Head: Retracted and prognathous, relatively small, slightly to strongly flattened, usually deeply emarginate posteriorly (except in Schizopodinae and Julodinae). Epicranial stem absent; frontal arms usually V-shaped, sometimes indistinct, apparently absent in Schizopodinae and Julodinae. Median endocarina usually extending anteriorly to frontoclypeal suture; in Julodinae and Schizopodinae endocarina not reaching suture and continuous anteriorly with cavity for attachment of retractor muscle; in Trachyini absent. Paired endocarinae usually present mesad of frontal arms, and other internal support structures may by present as well. Stemmata usually absent or represented by 1 or 2 eye spots on each side; 3 well-developed stemmata present on each side in Schizopodinae. Antennae very short, 2- or 3-segmented, but often with only 1 apparent segment; segment 2 often reduced, forming cup-like structure, fringed with hairs and bearing within its cavity a conical sensorium and sometimes a minute 3rd segment (more often represented by a setiferous tubercle only); in some Trachyini, antennae with as many as 3 sensoria and a distinct 3rd segment. Frontoclypeal suture present; clypeal region (epistoma of authors) usually heavily sclerotized (membranous in Schizopodinae and Julodinae, which have anterior portion of frons sclerotized); labrum free. Mandibles symmetrical, robust, usually bidentate or tridentate, but sometimes with truncate apex, without mola or accessory ventral process; mesal surface of mandibular base usually simple, occasionally with brush of hairs. Ventral mouthparts protracted. Maxilla usually without distinct cardo, with elongate stipes, no articulating area, 2-segmented palp, and single, articulated or fixed mala. Labium with mentum and submentum fused or with all sclerites fused into single plate; ligula broad and truncate or slightly bilobed; labial palps either absent or represented by minute, 1-segmented papillae or setose areas. Hypopharyngeal sclerome absent. Hypostomal rods usually long and subparallel or diverging, absent in Schizopodinae. Ventral epicranial ridges absent. Gula subquadrate or elongate.

Thorax and Abdomen: Prothorax usually greatly enlarged and flattened, meso- and metathorax moderately so; protergum sometimes with large patch of asperities and often with 2 anteriorly converging endocarinae (represented on surface as grooves or sclerotized rods), which form a V; sometimes with single median endocarina. Prosternum sometimes with similar patch of asperities and usually with single endocarina, which may be Y-shaped; Schizopodinae and Trachyini without asperities or endocarinae. Legs almost always absent or represented by minute papillae or setose spots only; occasionally minute, with 1 or 2 segments; Schizopodinae (fig. 34.393) with 2 pairs of leg-like processes on mesothorax, metathorax, and abdominal segments 1–7, as well as a single pair on segments 8 and 9. Abdominal segments 2–9 sometimes with paired, longitudinally oval or elliptical impressions on terga and sterna. In Schizopodinae and Julodinae abdominal terga 1–8 each with up to 4 transverse plicae, and meso- and metaterga with 2 or 3. Tergum A9 simple, without urogomphi. Segment A10 well-developed or somewhat reduced, usually posteriorly or terminally oriented, with pair of vertically oval, fleshy lobes bordering anal opening; sometimes with pair of acute, posteriorly projecting processes (Agrilini and some first instars); A10 posteroventrally oriented and without lobes or acute processes in Schizopodinae.

Spiracles: Almost always cribriform (biforous in Schizopodinae), those on thorax usually larger and reniform, those on abdomen usually more oval in shape.

Comments: The family is a large one, with as many as 400 genera and 15000 species worldwide, with 44 genera and about 750 species occurring in America north of Mexico. The group is presently divided into 13 subfamilies: Julodinae, Schizopodinae, Thrincopyginae, Mastogeniinae, Acmaeoderinae, Polycestinae, Chalcophorinae, Chrysobothrinae, Buprestinae, Sphenopterinae, Agrilinae, Trachyinae, and Cylindromorphinae (Cobos, 1980; Nelson, 1981). Of these, most are widely distributed, but Julodinae, Sphenopterinae, and Cylindromorphinae are native to the Old World (Eurasia and Africa), while Schizopodinae and Thrincopyginae occur in the southwestern part of North America.

Selected Bibliography

Alexseev 1960 (larva of European *Agrilus*).

Drooz 1985 (forest pests; key to some genera in eastern North America).

Balachowsky *et al.* 1962 (agricultural pests).

Benoit 1964 (*Chrysobothris* larvae), 1965 (*Agrilus* larvae), 1966a (*Melanophila* larvae), 1966b (*Chrysobothris* larvae).

Bilý 1972a (larva of *Dicera berolinensis* (Herbst)), 1972b (larva of *Ptosima flavoguttata* (Illiger)), 1975a (*Chrysobothris* larvae), 1978, 1983 (larvae of *Julodis variolaris freygessneri* Obenberger and *Paracylindromorphus transversicollis* (Reitter)).

Böving and Craighead 1931 (various larvae).

Burke 1917 (forests pests), 1918 (*Buprestis* larvae), 1928 (larva of *Trachykele blondeli* Marseul).

Burke and Böving 1929 (larva of *Chrysobothris mali* Horn).

Chapman 1915 (larva of *Agrilus bilineatus* (Weber)), 1923 (larva of *Taphrocerus gracilis* (Say)).

Chittenden 1922 (larva of *Agrilus ruficollis* (Fabricius)).

Cobos 1980, 1981.

Costa and Vanin 1984a (larva of *Euchroma gigantea* (Linnaeus)).

Crowson 1955, 1960b, 1982.

Dumbleton 1932 (some New Zealand buprestid larvae).

Evans 1964, 1966.

Falcoz 1923 (larvae of *Coroebus sinuatus* Creutzer).

Forbes 1926, 1942.

Furniss and Carolin 1977 (forest pests).

Gardner 1929 (key to larvae of Indian genera), 1930b (larva of *Buprestis* sp.), 1944 (Indian *Sternocera* larvae).

Gravely 1916.

Hawkeswood 1985 (larva of *Diadoxus erythrurus* (White)).

Hudson 1934 (larva of *Nascioides enysii* (Sharp)).

Kasap and Crowson 1975.

Lawrence and Newton 1982.

Lesne 1898 (larva of *Julodis*).

Levey 1978 (putative larva of *Prospheres aurantiopicta* (Laporte and Gory)).

Nelson 1981.

Peterson 1951 (various North American larvae).

Rees 1941 (first instar larvae of *Buprestis rusticorum* (Kirby) and *Schizopus sallei* Horn).

Schaefer 1953 (larva of *Paracylindromorphus subuliformis* (Mannerheim)).

Synder 1919 (damage of *Chrysobothris tranquebarica* (Gmelin)).

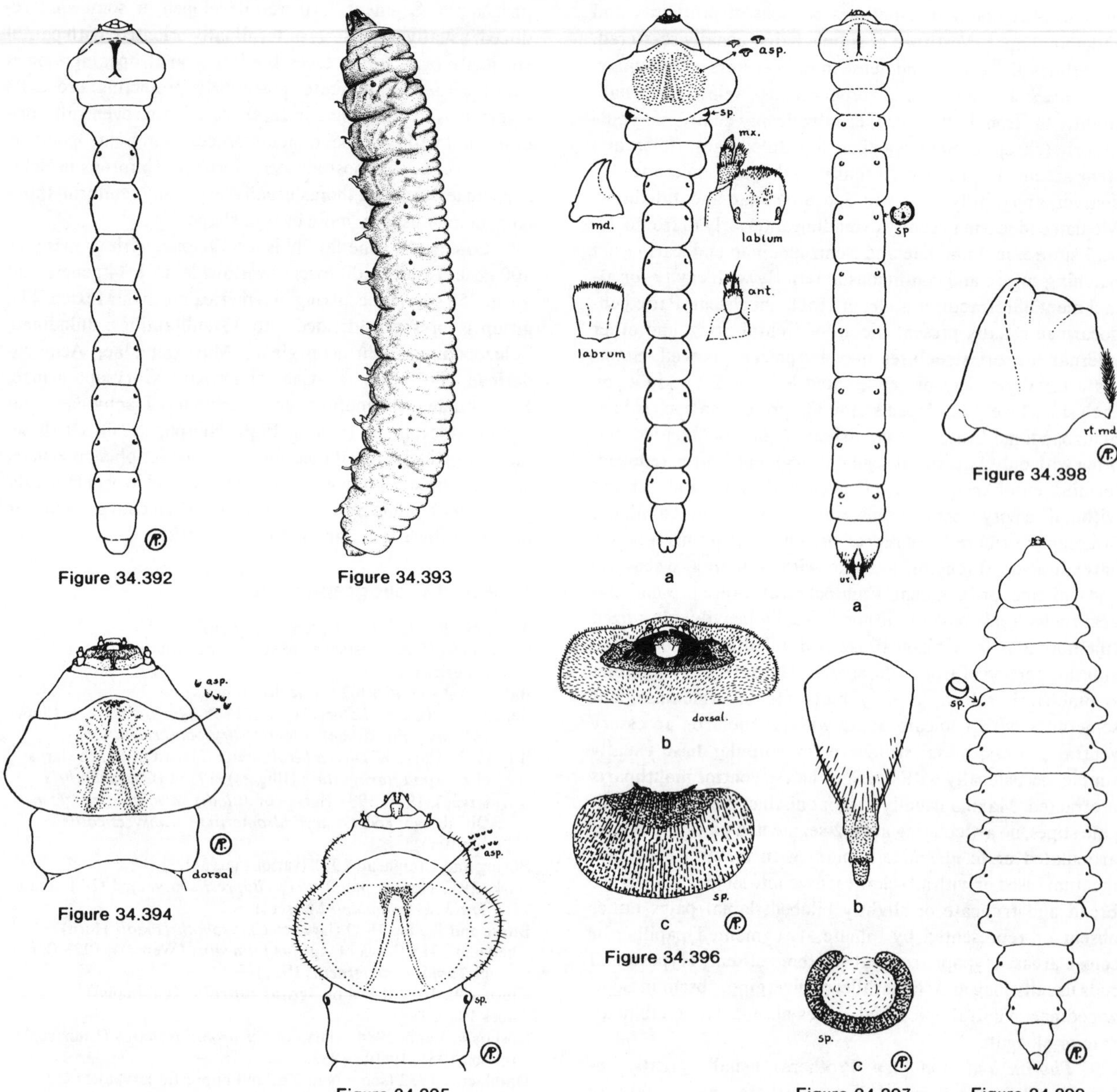

Figure 34.392. Buprestidae. *Chrysophana* sp. Larva, dorsal. Length = 25 mm. Infests twigs, branches and wood of conifers, especially pines and firs. (From Peterson, 1951.)

Figure 34.393. Buprestidae. *Dystaxia?* sp. Warner's Hot Springs, San Diego Co., California. Larva, lateral. Length = 27 mm.

Figure 34.394. Buprestidae. *Chalcophora* sp. Head and thorax, dorsal. (From Peterson, 1951.)

Figure 34.395. Buprestidae. *Dicera lurida* (Fabricius). Head and thorax, dorsal. (From Peterson, 1951.)

Figure 34.396a-c. Buprestidae. *Chrysobothris femorata* (Olivier). Flatheaded apple tree borer. Length = 26 mm. **a.** larva, dorsal (plus details); **b.** head, dorsal; **c.** prothoracic spiracle. Near white, head retracted with no stemmata; abdominal spiracles each with a broad, oval, respiratory plate possessing 20 plus openings. Attacks many kinds of fruit, woodland, and shade trees and bushes. Most destructive to nursery trees, newly set, or old weakened trees, especially on the sunny side. (From Peterson, 1951.)

Figures 34.397a-c. Buprestidae. *Agrilus anxius* Gory. Bronze birch borer. **a.** larva, dorsal, length = 20 mm; **b.** abdominal apex, lateral; **c.** abdominal spiracle. Attacks most birch species, especially European white birch and other species under stress in yards. Excavates galleries which wind back and forth in the cambium area, eventually girdling the tree, which dies from the top down. Adults emerge through D-shaped exit holes. (From Peterson, 1951.)

Figure 34.398. Buprestidae. *Agrilus bilineatus* (Weber). Two-lined chestnut borer. Right mandible, ventral. Length = 15–20 mm. Infests trunks (cambium) of various hardwoods, especially chestnut, but most commonly found in oaks since the demise of chestnut. (From Peterson, 1951.)

Figure 34.399. Buprestidae. *Pachyschelus laevigatus* (Say). Larva, dorsal. Length = 3.5 mm. A leafminer in *Lespedeza, Trifolium* and others. (From Peterson, 1951.)

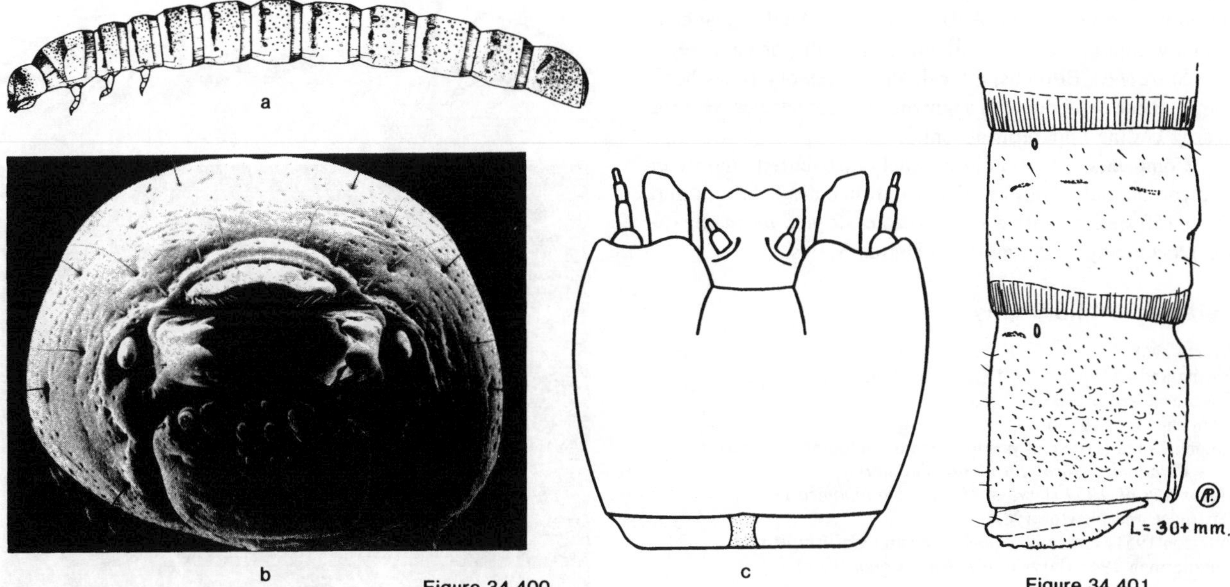

Figure 34.400

Figure 34.401

Figures 34.400a-c. Callirhipidae. *Zenoa picea* (Beauvois). Florida Caverns State Park, Florida. Length = 20 mm. **a.** larva, lateral; **b.** head, anterior; **c.** maxillo-labial complex, ventral.

Figure 34.401. *Callirhipis* sp. Abdominal apex, lateral. (From Peterson, 1951.)

CALLIRHIPIDAE (DRYOPOIDEA)

John F. Lawrence,
Division of Entomology, CSIRO

Figures 34.400–401

Relationships and Diagnosis: In most classifications (Arnett, 1968), Callirhipidae have been included in the same family with Rhipiceridae (= Sandalidae), but Crowson (1955) separated them and later moved Callirhipidae to the superfamily Artematopoidea, along with Artematopidae and Brachypsectridae. Lawrence and Newton (1982) included the family in an expanded Dryopoidea. Larvae of Callirhipidae resemble those of some Elateridae, Cebrionidae, and Dryopidae in having a well sclerotized, cylindrical body and globular head with highly consolidated ventral mouthparts, but they are easily distinguished from any other larval type by the dorsally-hinged operculum formed from the 9th abdominal tergum.

Biology and Ecology: Callirhipid larvae occur in old wood which is punky and heavily penetrated by white rot fungi. The larvae live for 2 years or more, while the adults are short-lived and may be attracted to lights.

Description: Mature larvae 10 to 50 mm. Body elongate and cylindrical, straight, heavily sclerotized and uniformly dark reddish-brown, or yellowish-brown with darker head, thoracic terga, and abdominal apex. Vestiture consisting of a few scattered, short setae.

Head: Protracted and slightly to moderately declined, globular. Epicranial stem long; frontal arms V-shaped. Median endocarina present beneath epicranial stem. Stemmata absent. Antennae very short, 1-segmented. Frontoclypeal suture present; labrum free. Mandibles symmetrical, robust, broad and wedge-like, tridentate, or appearing 5-dentate because of subapical tooth on either side of broadly concave, scoop-like mesal surface; mola and accessory ventral process absent. Ventral mouthparts retracted. Maxilla with strongly transverse cardo, elongate stipes, no articulating area, 4-segmented palp, and articulated, truncate mala, notched at inner apical angle. Cardines closely approximate, with membranous area between them, not separated by labium, which may also be closed off posteriorly by approximate stipital bases. Labium with mentum and submentum fused; postmentum completely or almost completely connate with and sometimes partly fused to stipites; ligula forming a sclerotized, wedge-like, 4-dentate plate; labial palps 2-segmented and narrowly separated. Hypopharyngeal sclerome absent. Hypostomal rods absent. Ventral epicranial ridges present. Gula longer than wide.

Thorax and Abdomen: Prothorax much longer than meso- or metathorax. Legs very short, usually 5-segmented, including tarsungulus, occasionally with only 4 segments; tarsungulus without a seta; coxae subcontiguous. Protergum, mesotergum, and anterior part of metatergum sometimes with transverse asperities or rugosities. Abdominal segments 1–8 forming complete rings, without sutures separating terga, pleura, and sterna. Mesotergum, metatergum, and abdominal terga 1–8 often with paired striate impressions, or these occurring only on tergum 8, where they open posteriorly, forming 2 slits above operculum. Abdominal terga 2–7 usually with paired cavities having sclerotized openings; cavities sometimes absent or present on tergum 7 only. Tergum A9 forming dorsally-hinged operculum; urogomphi absent; sternum A9 and segment A10 not distinct. Anal region concealed within operculum, without hooks, gills, or papillae.

Spiracles: Biforous, lateral, with accessory tubes horizontal; those on abdominal segments 3–8 sometimes reduced in size; closing apparatus absent.

Comments: The family is widely distributed, mainly in tropical and subtropical regions, and includes 8 genera and about 150 species, with a single species, *Zenoa picea* (Beauvois), occurring in the southern United States.

Selected Bibliography

Arnett 1968.
Böving and Craighead 1931 (larva of *Zenoa picea*).
Costa and Vanin 1985.
Crowson 1955, 1973b.
Emden 1932a (larvae of various genera, including *Callirhipis, Celadonia, Horatocera,* and *Simianus*).
Hayashi *et al.* 1959 (larva of *Horatocera niponica* Lewis).
Lawrence and Newton 1982.
Peterson 1951 (larvae of *Zenoa picea* and *Callirhipis* sp.).
Zimmerman 1942 (larva of *Callirhipis onoi* Blair).

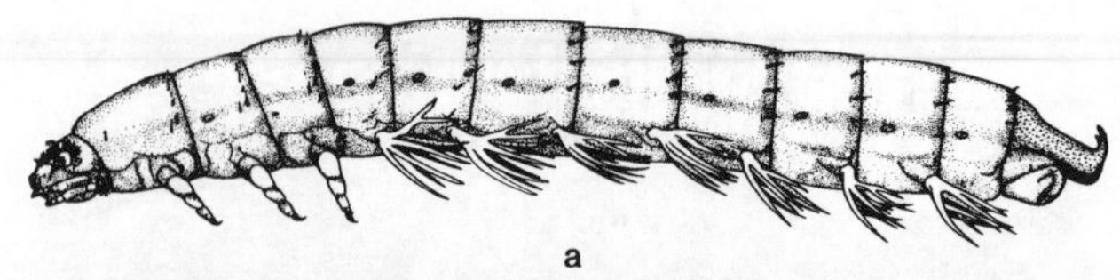

a

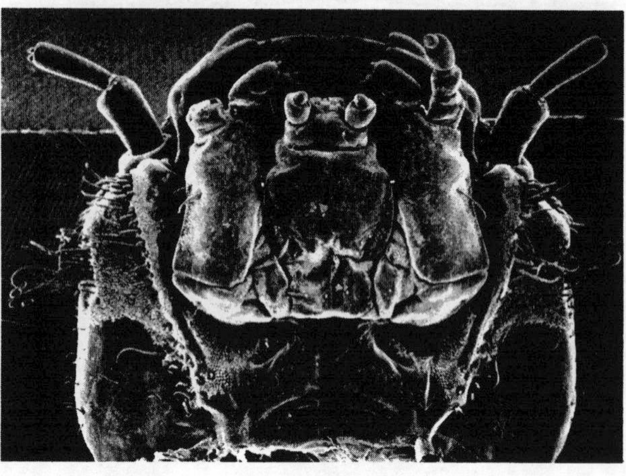

b

Figures 34.402a,b. Eulichadidae. *Stenocolus scutellaris* LeConte. Richardson Springs, north of Chico, California. Length = 43 mm. **a.** larva, lateral; **b.** head, ventral.

EULICHADIDAE (DRYOPOIDEA)

(= LICHADIDAE)

John F. Lawrence,
Division of Entomology, CSIRO

Figures 34.402a,b

Relationships and Diagnosis: The Eulichadidae have usually been treated as part of the Ptilodactylidae (Crowson, 1955) or Dascillidae (Arnett, 1968); their relationship to Callirhipidae was recognized by Forbes (1926), who first placed them in a distinct family (Lichadidae), and their systematic position has been further discussed by Crowson (1978) and Lawrence and Newton (1982). They are considered to be one of the most primitive groups of Dryopoidea. Eulichadid larvae are easily distinguished from any other beetle larvae by their relatively large size and elongate body with paired abdominal gills on segments 1–7.

Biology and Ecology: Larvae of Eulichadidae occur in streams or small rivers, where they burrow among detritus or in the substratum and apparently feed on decaying vegetation or roots (Leech and Chandler, 1956).

Description: Mature larvae 15 to 50 mm. Body elongate, parallel-sided, straight, slightly flattened. Brown or gray dorsally, slightly lighter ventrally; dorsal surfaces microgranulate and dull, with numerous shallow pits or punctures; vestiture of longer and shorter hairs, often arising in tufts.

Head: Protracted and prognathous, slightly flattened, deeply emarginate posteriorly. Epicranial stem absent; frontal arms usually not visible, V-shaped, not contiguous at base. Median endocarina absent. Stemmata consisting of a single large lens on each side, beneath which are 2 to 5 pigment spots. Antennae well-developed, 3-segmented, with last segment reduced. Frontoclypeal suture present; anterior part of frons densely clothed with short, stout bristles. Labrum free, strongly upturned, and bearing 4 lobes at apex. Mandibles symmetrical, tridentate, without mola or accessory ventral process; mesal surface of mandibular base with an articulated, setose process. Ventral mouthparts retracted. Maxilla with transverse cardo, elongate stipes, narrow articulating area, slightly concealed behind expanded mentum, 4-segmented palp, rounded, articulated galea, and rounded, articulated lacinia. Labium more or less free to base of mentum; ligula very short and broad; labial palps 2-segmented and separated by slightly more than the width of a basal segment. Hypopharyngeal sclerome absent. Hypostomal rods absent. Ventral epicranial ridges present. Gula transverse.

Thorax and Abdomen: Legs well-developed, clothed with patches of bristles; tarsungulus with 2 setae lying side by side or with several setae. Abdominal sterna 1–7 each with a pair of gills, each having 2 main branches and 10 to 25 filaments. Tergum A9 with pair of fixed urogomphi, approximate at base, slightly diverging and strongly upturned at apex; dorsal surface of tergum A9 usually concave, lateral edges sometimes lined with dense fringe of long hairs. Sternum A9 simple. Segment A10 more or less circular, ventrally or posteroventrally oriented, with 2 lobes bordering anus.

Spiracles: Biforous, lateral, highly reduced and nonfunctional; remnant of second thoracic spiracle on metathorax. Closing apparatus absent.

Comments: The family includes only 2 genera: *Eulichas,* with about 11 species in southeast Asia, and *Stenocolus,* with a single species, *S. scutellaris* LeConte, occurring in streams entering the San Joaquin and Sacramento Valleys of California up to an altitude of 4000 feet.

Selected Bibliography

Arnett 1968.
Böving and Craighead 1931 (larva of *Eulichas* sp., as Asiatic Ptilodactylidae).
Crowson 1955, 1978.
Forbes 1926.
Leech and Chandler 1956 (larva of *Eulichas* sp., as *Stenocolus* sp.).

PTILODACTYLIDAE (DRYOPOIDEA)

John F. Lawrence,
Division of Entomology, CSIRO

Figures 34.403–409

Relationships and Diagnosis: The Ptilodactylidae, as defined here, includes 2 groups of genera (Anchytarsini and Haploglossini of Champion, 1897, and "Cneoglossini" of Arnett, 1968) formerly placed in Dascillidae but transferred to the present family by Crowson (1955), and excludes the genus *Cneoglossa* (Crowson 1972b; *see* Cneoglossidae) and the genera *Eulichas* and *Stenocolus* (Crowson, 1978; *see* Eulichadidae). The genus *Araeopidius* has a distinctive type of wing-folding (Forbes, 1926) and a unique larval type; it probably should be placed in a separate family (Lawrence and Newton, 1982). In both larval and adult features, *Araeopidius* appears to be intermediate between Ptilodactylidae and Chelonariidae. Ptilodactylid larvae are of 3 major types: those of *Araeopidius* and *Ptilodactyla,* and a third type found in all remaining groups (Anchytarsinae). The main character distinguishing Anchytarsinae and Ptilodactylinae from other Elateriformia is the presence of a well-developed maxillary articulating area which is partly or wholly concealed behind an expanded mentum. In *Araeopidius,* the maxillolabial complex is more or less consolidated, with the reduction of the articulating area and the connation of maxillae with stipites, as in many other dryopoids, elateroids, and cantharoids; it may be distinguished by the combination of free labrum, vertically-clustered stemmata, dorsally-placed terminal spiracles, and ventral operculum, and by the unique plastron plates surrounding the non-functional thoracic and anterior abdominal spiracles. Eulichadid larvae are distinguished from those of ptilodactylids by the well-developed, exposed maxillary articulating area, the paired gill tufts on abdominal sterna 1 to 7, and the bisetose or multisetose tarsungulus.

Biology and Ecology: Larvae of Ptilodactylidae usually occur in wetter habitats and may be riparian or aquatic; they feed on rotting vegetation, including leaves, roots, or dead wood. Anchytarsine larvae appear to be truly aquatic. According to LeSage and Harper (1976b), the larvae of *Anchytarsus* feed on submerged decaying wood, require 3 years to complete their growth, and pupate out of the water in chambers built within decaying vegetation in cracks in rocks (as in many Elmidae); similar habits have been observed for *Tetraglossa* and various other anchytarsines. *Araeopidius monachus* (LeConte) lives in gravel or sand at the edges of streams, and the larval adaptations (dorsally-placed terminal spiracles and plastron plates) probably have evolved in response to periodic flooding. Ptilodactylinae larvae occur in leaf litter or in rotting wood and appear to be entirely terrestrial. Several species have been recorded as pests of plants in greenhouses and nurseries. *Ptilodactyla exotica* Chapin has been found in rosehouses in the United States (Chapin, 1927), and Süss and Puppin (1976) reported the same species feeding on roots of *Aechmea, Dracaena,* and *Anthurium* in European greenhouses. Spilman (1961) recorded *Ptilodactyla serricollis* (Say) in soil of potted *Ficus elastica,* and Horion (1955) mentioned an Indonesian species *P. luteipes* Pic associated with banana plants in Europe. Pupae of both *Ptilodactyla* and *Anchytarsus* are known to have gin-traps on the abdomen (LeSage and Harper, 1976b; Spilman, 1961). Adult ptilodactylids are nocturnal and may be attracted to lights.

Description: Mature larvae 3 to 25 mm, usually 15 mm or less. Body elongate, parallel-sided, subcylindrical or slightly flattened, straight or slightly curved ventrally, sometimes (Ptilodactylinae) with abdominal apex strongly curved ventrally. Dorsal surfaces usually yellowish or yellowish-brown, with ventral surfaces lighter; occasionally (*Araeopidius*) uniformly dark reddish-brown. Surfaces usually smooth (microgranulate and densely punctate in *Araeopidius*); vestiture of scattered long, simple hairs.

Head: Protracted and prognathous, slightly flattened. Epicranial stem very short or absent; frontal arms V-shaped and contiguous. Median endocarina absent. Stemmata 3, 5 or 6 on each side, usually closely clustered, so that they appear as a single stemma. Antennae well-developed, 3-segmented, relatively long, with segments 1 and 2 much longer than 3. Frontoclypeal suture present, sometimes vaguely indicated; labrum free. Mandibles symmetrical, tridentate, without mola or accessory ventral process; incisor edge usually broadly excavate; mesal surface of mandibular base with articulated process, setose at apex, or with 1 or 2 brushes of hairs. Ventral mouthparts retracted (less so in Ptilodactylinae). Maxilla with transverse or oblique cardo, subquadrate or slightly elongate stipes, 4-segmented palp, articulated galea and fixed or articulated lacinia; galea occasionally 2-segmented. Maxillary articulating area usually well-developed but concealed behind laterally expanded mentum (reduced or absent in *Araeopidius*). Labium usually free to base of mentum, partly connate with stipites in *Araeopidius;* mentum divided longitudinally into 3 parts in Anchytarsinae; ligula short and broad; labial palps 2-segmented, moderately to widely separated. Hypopharyngeal sclerome absent. Hypostomal rods absent. Ventral epicranial ridges present. Gula usually subquadrate or slightly elongate; gular sutures fused in Ptilodactylinae.

Thorax and Abdomen: Legs well-developed, 5-segmented (shorter with dense patches of bristles in *Araeopidius*); tarsungulus with 1 seta. Coxae narrowly to broadly separated (contiguous in *Araeopidius*). Paired eversible glands present on mesothorax and abdominal segment 8 in Ptilodactylinae. Thoracic segments and abdominal segments 1–8 each with 2 pairs of elongate, striate impressions on each side, and mesothorax and abdominal segments 1–7 each with paired plastron plates surrounding spiracles in *Araeopidius.* Distinct pleurites and sternites present on abdominal segments 1–7 in Ptilodactylinae; sternites only present on segments 8 and 9; pleurites absent in Anchytarsinae and *Araeopidius;*

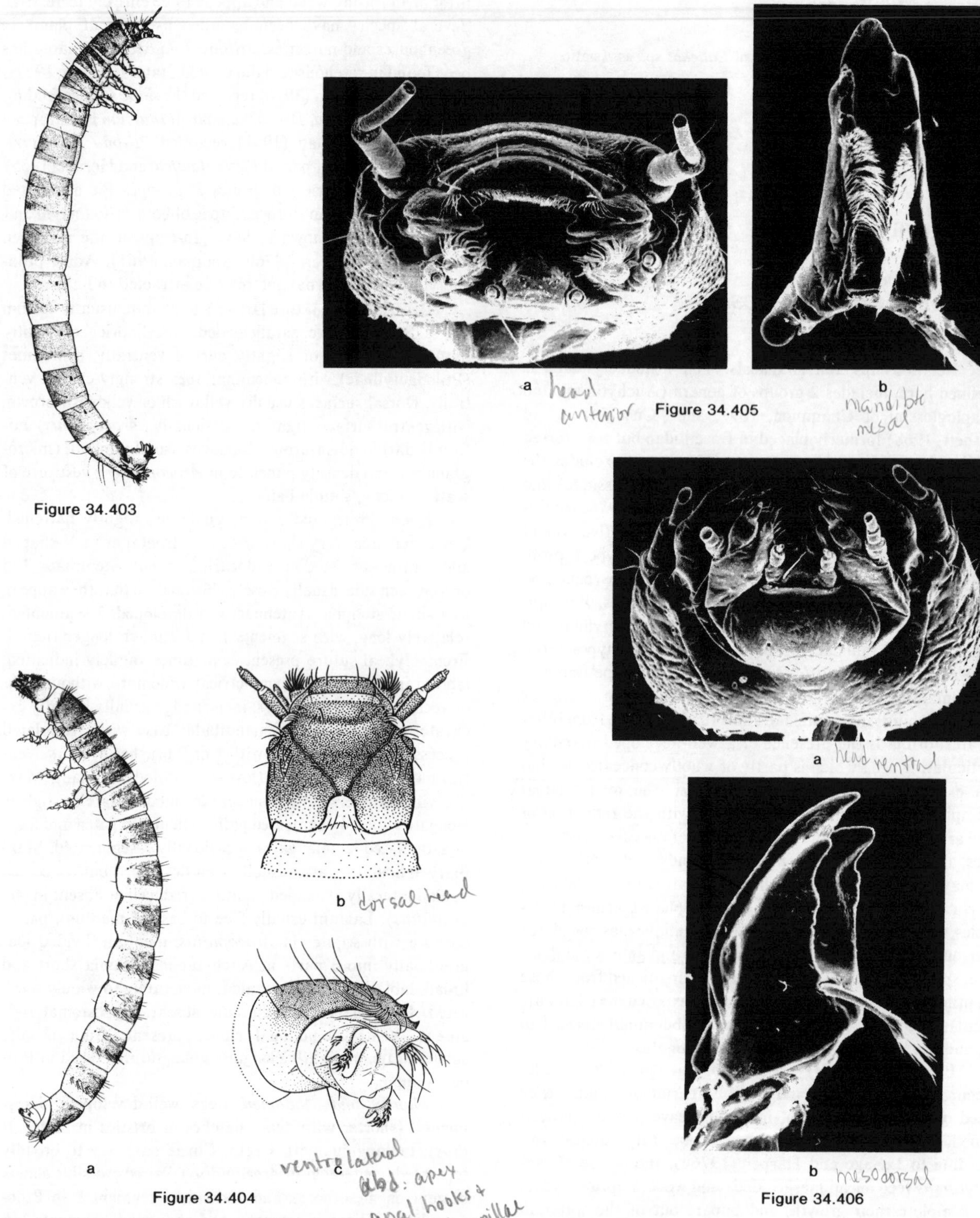

Figure 34.403. Ptilodactylidae. *Anchytarsus bicolor* (Melsheimer). Elk Garden, West Virginia. Larva, lateral. Length = 20.3 mm.

Figures 34.404a-c. Ptilodactylidae. *Anchycteis velutina* Horn. Alder Campground, Tehama Co., California. Length = 24.5 mm. **a.** larva, lateral; **b.** head, dorsal; **c.** abdominal apex, postero-ventrolateral view, showing anal hooks and anal papillae.

Figures 34.405a,b. Ptilodactylidae. *Tetraglossa* sp. Barro Colorado Is., Canal Zone, Panama. **a.** head, anterior; **b.** right mandible, mesal.

Figures 34.406a,b. Ptilodactylidae. *Ptilodactyla* sp. Great Smoky Mts. National Park, Tennessee. **a.** head, anteroventral; **b.** left mandible, mesodorsal.

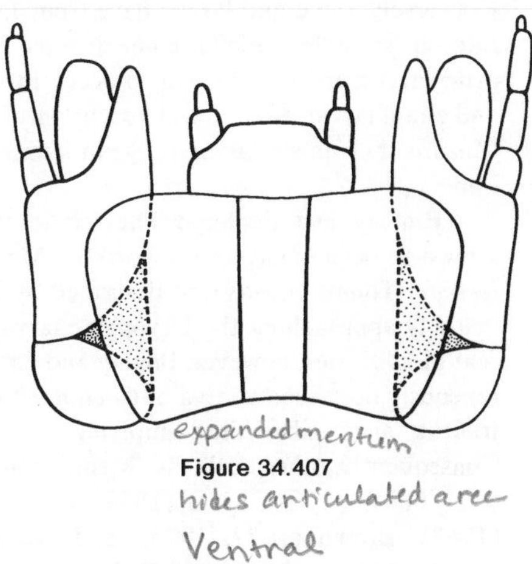

Figure 34.407

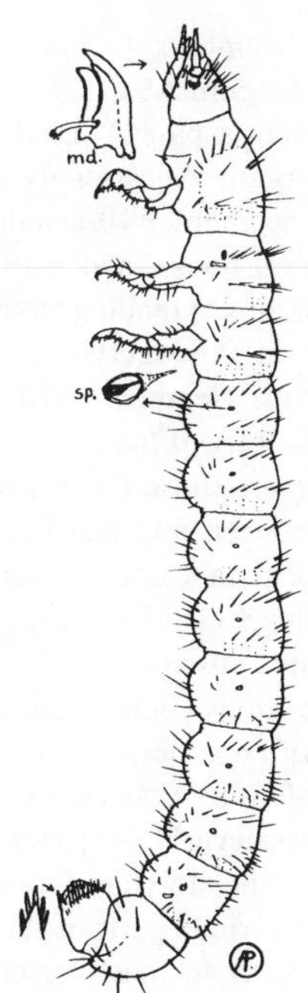

Figure 34.408

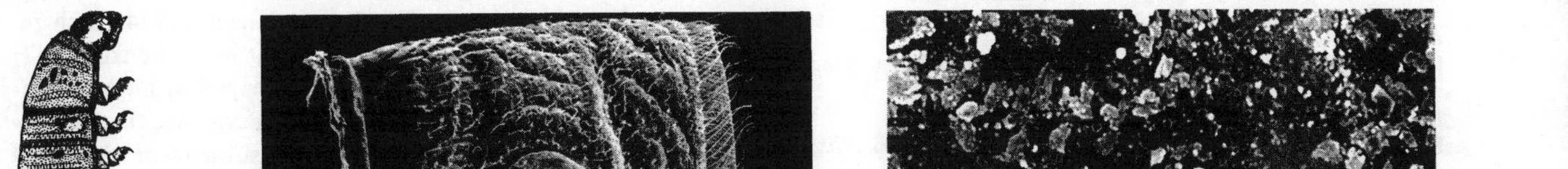

a

b

c

d

e

Figure 34.409

Figure 34.407. Ptilodactylidae. Anchytarsinae. Maxillo-labial complex, ventral (showing partly concealed maxillary articulating area).

Figure 34.408. Ptilodactylidae. *Ptilodactyla serricollis* Say. Larva, lateral. Length = 14 mm. (From Peterson, 1951.)

Figures 34.409a-e. Ptilodactylidae. *Araeopidius monachus* (LeConte). 4 mi. W Forest Glen, Trinity Co., California. Length = 15.3 mm. **a.** larva, lateral; **b.** abdominal segment 4, lateral (left side, showing plastron plate); **c.** same (right side); **d.** same. Details of plastron plate; **e.** abdominal apex, lateral (showing dorsally-placed 8th spiracles and ventrally-hinged operculum).

sternites present on segments 1–7 or 8 in Anchytarsinae, 1–5 in *Araeopidius*. Tergum A9 usually more or less concave dorsally, with or without a pair of small urogomphi at apex; in Ptilodactylinae tergum A9 strongly convex and simple; sternum A9 simple. Segment A10 usually with pair of anal lobes, each with several hooks, and with 3 to several osmoregulatory papillae or gills extending from membrane around anus; papillae absent in Ptilodactylinae; hooks and papillae absent in *Araeopidius*. Segment A10 forming ventrally-hinged operculum in *Araeopidius*.

Spiracles: Biforous, those on mesothorax and abdominal segments 1–7 reduced and non-functional in *Araeopidius*, all non-functional in early instars of Anchytarsinae. Spiracles on segment A8 usually lateral, dorsal in *Araeopidius*. Closing apparatus absent.

Comments: The family contains about 40 genera and 450 species worldwide, with 6 genera and 12 species occurring in N. America. The subfamily Araeopidiinae, with a single species, occurs in northwestern N. America, from central California to Alaska. The Ptilodactylinae are abundant in tropical America but occur in various parts of the Old World as well. The Anchytarsinae occur in most regions of the world, including New Zealand (*Byrrocryptus*). In N. America, the subfamily is represented by the genera *Anchytarsus, Anchycteis,* and *Odontonyx*.

Selected Bibliography

Arnett 1968.
Bertrand 1966 (larva of *Anchycteis velutina* Horn; key to genera), 1972 (larvae of various genera).
Böving 1929b (larva of *Ptilodactyla serricollis*).
Böving and Craighead 1931 (larvae of *Ptilodactyla serricollis* and *Anchytarsus bicolor* (Melsheimer)).
Britton 1970 (larva of *Epilichas* sp.).
Chapin 1927.
Champion 1897.
Crowson 1955, 1972b, 1978.
Horion 1955.
Hudson 1934 (larvae of *Byrrocryptus urquharti* Broun).
Leech and Chandler 1956 (larva of *Araeopidius monachus*, as *Helichus* sp.).
LeSage and Harper 1976b (biology of *Anchytarsus bicolor*).
Peterson 1951 (larva of *Ptilodactyla serricollis*).
Spangler 1983 (larva of *Tetraglossa palpalis* Champion).
Spilman 1961 (pupa of *Ptilodactyla serricollis*).
Stribling 1986.
Süss and Puppin 1976 (larva of *Ptilodactyla exotica* Chapin).

CHELONARIIDAE (DRYOPOIDEA)

Paul J. Spangler, *Smithsonian Institution*

The Chelonariids

Figures 34.410 a–h

Relationships and Diagnosis: The chelonariids are presently assigned to the superfamily Dryopoidea, where they were placed by Böving (1929b) based on larval characters. They appear to be most closely related to the elmids, and the larvae of these 2 families closely resemble each other in general facies although the adults do not.

Larvae may be distinguished easily from elmid larvae by the absence of a pair of anal hooks and the absence of anal gills under the operculum, and by differences in the spiracles; chelonariids have 9 pairs of large projecting and cribriform or biforous spiracles that are of a deviating sinuous type. Conversely, the elmid larvae have from 1 to 9 pairs of small biforous spiracles which are not sinuous. In addition, chelonariid larvae are now known to occur in terrestrial habitats and elmid larvae are known to occur only in aquatic habitats. Mature chelonariid larvae range in length from 12.0 to 15.0 mm.

Biology and Ecology: The larvae of the only species known to occur in America north of Mexico, *Chelonarium lecontei* Thomson, was first described by Böving (1929b) as "closely approaching the Dryopidae larvae" but having "no real gills." Later, however, Böving and Craighead (1931) erroneously described a larva of *Chelonarium* sp. as having retractile anal gills, thus implying that it was aquatic. Consequently, other authors—Mequignon (1934), Fleutiaux et al. (1947), Costa Lima (1953), Crowson (1955), Arnett (1968), Brown (1972, 1975), and Bertrand (1977)—followed Böving and Craighead's lead and repeated that chelonariid larvae were aquatic. However, Spangler (1980a) reported that chelonariid larvae lack anal gills, are not aquatic, frequently occur in packing materials around the roots of orchids (where ants often nest), in termite galleries in branches, in the trunk of "sweet orange," and under the bark of a dead tree. Previously, Lenko (1967) and Janzen (1974) each reported finding chelonariid larvae in ant nests. Lenko (1967) also found live chelonariid adults and pupae with the chelonariid larvae in ant nests; thus providing evidence that chelonariids are truly myrmecophilous. The subsequent reports by Janzen (1974) and Spangler (1980a) verify the myrmecophilous relationships of chelonariid larvae. Chelonariid larvae are now known from nests of ants in epiphytic plants growing in trees, from mound-building ant nests, and circumstantially from ant or termite galleries beneath bark of dead trees.

Nothing about the egg-laying habits or the number of larval instars has been reported. However, Lenko (1967) reported that pupation occurred within the last larval exuvia and the pupal period lasted 45 days. Spangler (1980a) suggests that the fleshy filaments arising on the carinae on the abdominal segments of the larvae provide a secretion used by the ants and may provide the chelonariid larvae immunity from attack. Coats and Selander (1979) described the unusual behavior of the adult male of *Chelonarium lecontei* which uses the aedeagus to right itself.

Description: Mature larvae, 12.0 to 15.0 mm (fig. 34.410a); body subcylindrical, fusiform; integument hard and rigid. Integument and sclerotized areas moderately dark to dark reddish brown; fleshy filaments creamy white.

Head: Hypognathous. Ecdysial cleavage lines present (fig. 34.410b). 1 stemma on each side (fig. 34.410b). Antenna short, 3-segmented (fig. 34.410e). Mandible (figs. 34.410f–h) without mola, apices broadly dentiform, never perforate nor deeply cleft. Labium subquadrangular (fig. 34.410d); gular area present (fig. 34.410c).

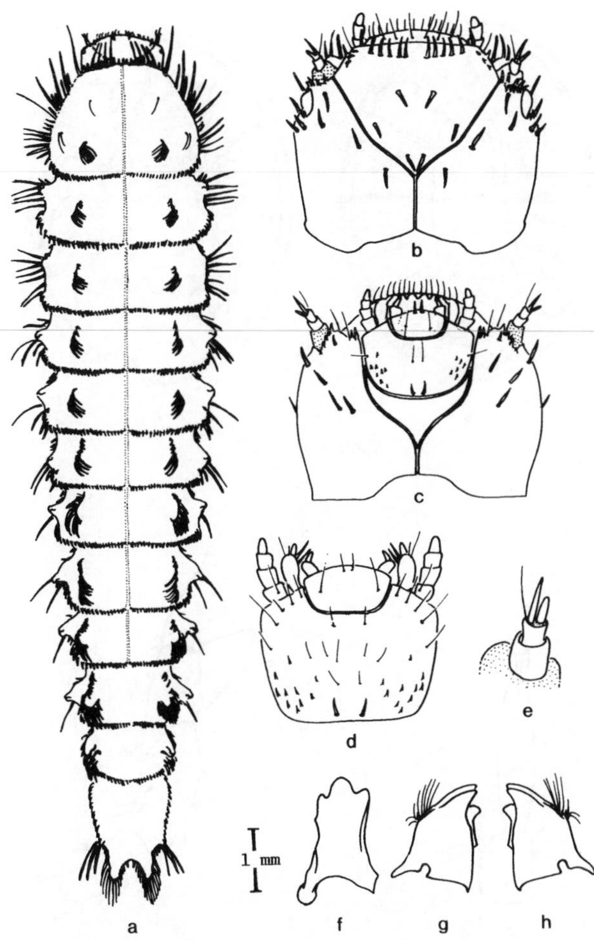

Figures 34.410a-h. Chelonariidae. *Chelonarium* sp. **a.** dorsal; **b.** head capsule, dorsal; **c.** head capsule, ventral; **d.** labium; **e.** antenna; **f.** mandible, medial surface; **g.** left mandible; **h.** right mandible.

Thorax and Abdomen: Pronotum, mesonotum, and metanotum each with a large sublateral carina on each side, with each carina bearing a row of long, almost fleshy filaments. Legs 5-segmented, ambulatory; tarsungulus claw-like.

Abdomen of 9 apparent segments, each with a large sublateral carina on each side and each carina bearing a row of long, almost fleshy filaments. Ninth abdominal segment ventrally with movable operculum covering cloacal chamber. Without gills or internal hooks under operculum. Urogomphi absent.

Spiracles: Of the undulate or sinuous type, biforous, usually with the openings partially blocked and so appearing cribriform. Location lateral on mesothorax and abdominal segments 1–8, except for those on segment 8, which may be dorsal.

Comments: The Chelonariidae is a small family with 2 genera and about 230 species known primarily from tropical regions. It is especially rich in the Neotropics from where 215 species and subspecies have been described. The single species known from the United States, *Chelonarium lecontei* Thomson, is known from Virginia to Florida and westward to Oklahoma and Texas. Chelonariid beetles are of no known economic importance.

Selected Bibliography

Arnett 1968.
Bertrand 1977.
Böving 1929b.
Böving and Craighead 1931.
Brown 1972, 1975.
Coats and Selander 1979.
Costa Lima 1953.
Crowson 1955.
Fleutiaux *et al.* 1947.
Janzen 1974.
Lenko 1967.
Mequignon 1934.
Spangler 1980a.

PSEPHENIDAE (DRYOPOIDEA)

(INCLUDING EUBRIIDAE, PSEPHENOIDIDAE)

Harley P. Brown,
University of Oklahoma, Norman

Water Pennies

Figures 34.411–414

Relationships and Diagnosis: As employed here, the family Psephenidae includes not only the Psepheninae, but also the Eubrianacinae and Eubriinae, which are either given family status or placed in the family Dascillidae by some authors (e.g., Arnett 1968, Bertrand 1972). The flattened, onisciform water pennies bear a superficial resemblance to larvae of the corylophid and cerylonid fungus beetles, young whiteflies (Homoptera: Aleyrodidae), and pupae of the psychodid fly, *Maruina*. From all of these they may be readily distinguished by the presence of abdominal gills, either as paired external structures arising from the sterna or in the form of a retractile caudal brush protected by an anal operculum. Mature water penny larvae are also much larger, commonly measuring over 5 mm in length.

The discoidal shape and ventral head, with paired ventral gills arising from several abdominal segments, serve to separate the larvae of Psepheninae and Eubrianacinae from those of all other beetles. In the Eubrianacinae, the eighth abdominal segment has lateral expansions like those of the preceding segments, whereas in the Psepheninae this segment has none. The onisciform shape and ventral head, which normally cannot be seen from above, will distinguish larvae of the Eubriinae from other larvae that have retractile cloacal gills.

Biology and Ecology: All are aquatic, typically occurring on the undersides of stones in moving water, either in streams or along rocky wave-washed lake shores. As daylight fades, they move out to graze upon the day's growth of encrusting algae. Their flattened bodies enable them to hug the rocks almost as tenaciously as limpets, rendering them relatively safe from predators and from being swept away by the current. When a stone is lifted from the water and turned over, the water pennies soon begin to move toward the under side, often with surprising speed. Water pennies vary greatly in coloration, from pale translucent amber to almost black,

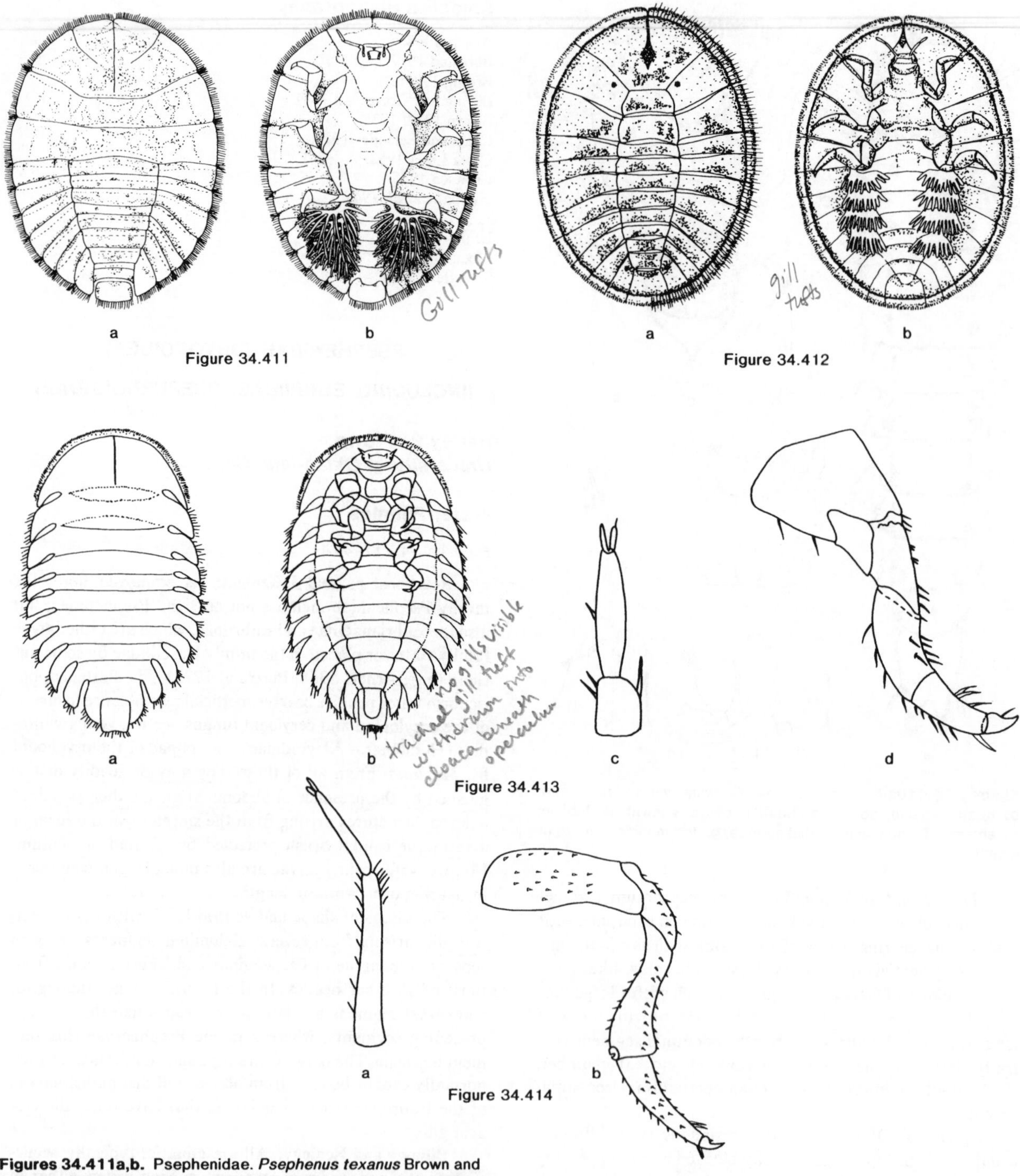

Figures 34.411a,b. Psephenidae. *Psephenus texanus* Brown and Arrington (Psepheninae). Length = 6–10 mm. **a.** dorsal. Note that abdominal segment 8 has no lateral or pleural extension and does not reach the margin; **b.** ventral. Gill tufts arise from the posterior edges of abdominal segments 2–6. Segment 9 bears no operculum or tracheal gill.

Figures 34.412a,b. Psephenidae. *Eubrianax edwardsii* (LeConte) (Eubrianacinae). Length 6–10 mm. **a.** dorsal. As in the Psepheninae, the lateral expansions of adjacent segments are fused together. *Eubrianax* differs from *Psephenus* in that abdominal segment 8 has lateral expansions extending to the body margin; **b.** ventral. Gill tufts arise from abdominal segments 1–4. Segment 9 bears no operculum or tracheal gill.

Figures 34.413a-d. Psephenidae. *Ectopria* sp. (Eubriinae). Length 4–6 mm. **a.** dorsal. Lateral expansions of adjacent segments are separated, unlike those of *Psephenus* and *Eubrianax;* **b.** ventral. The tracheal gill tuft is withdrawn into the cloacal chamber beneath the operculum, so no gills are visible; **c.** antenna, 2nd segment longest; **d.** hind leg.

Figures 34.414a,b. Psephenidae. *Psephenus texanus.* **a.** antenna, basal segment longest; **b.** hind leg.

often with attractive patterns, but tending to match their background substrate and making them very difficult to see while they are immobile. They overwinter as larvae, the adults being short-lived.

Eggs of *Psephenus* are usually bright yellow or orange, and laid in a flat mass on the underside of a stone in a shallow riffle. Oviposition habits of *Eubrianax* are similar to those of *Psephenus*, but are undescribed for the Eubriinae.

Pupation normally occurs in a protected site near water, beneath the carapace of the last larval skin. For other details, including keys, see Brown (1972, 1976, 1980, 1987), Brown and Murvosh (1974), Leech and Chandler (1956), Murvosh (1971), and West (1929).

Description: Mature larvae 3–10 mm. Body oval to subcircular in outline, conspicuously flattened, most segments with flat lateral extensions bordered by a marginal fringe of partially fused hairs. Cuticle of dorsum rather smooth, often with a labyrinthine pattern of micropunctures. The lateral extensions of adjacent segments are fused in the Psepheninae and Eubrianacinae, but free and slightly separated in the Eubriinae.

Head: Ventral, about as wide as long. Antennae 3-segmented, the apical segment being very small and setiform; long and slender in Eubrianacinae and Psepheninae, less so in Eubriinae. Labrum broader than long. Mandibles well-sclerotized, either short and triangular or somewhat elongate (both types present in the same species), with a brushlike basal process but no mola. Stemmata closely clustered to form a single eye on each side. Maxillary palps 2-segmented in Eubrianacinae, 4-segmented in Psepheninae, and 3- or 4-segmented in Eubriinae. Labium with submentum, mentum, and ligula fused; palps 3-segmented and well-developed in Eubriinae, but small and 2-segmented in Psepheninae and Eubrianacinae.

Thorax and Abdomen: Thorax broadly arcuate anteriorly; completely covering head; attaining a width about 3 to 8 times broader than head, and forming from one-third to one-half of entire body. Legs not visible from above, 5-segmented, the apical segment being a rather short tarsungulus bearing a mesal spine; coxae widely separated; hind legs largest. Spiracles, if present (*Eubrianax*), on dorsal surface of mesothorax.

Abdomen with 9 segments visible from above; in the Eubriinae the sternum of the 9th segment bears an operculum for the cloacal chamber into which the filamentous tracheal gills may be withdrawn. In Psepheninae and Eubrianacinae, there is no such operculum or anal gill; instead, pairs of feathery tracheal gills arise from the ventral surfaces of segments 1–4 (*Eubrianax*) or 2–6 (*Psephenus*). In *Eubrianax* and *Psephenus* the spiracles of the 8th segment are on the dorsal surface; in Eubriinae, the spiracles are situated at the apices of the lateral expansions of this segment.

Spiracles: Functional biforous spiracles on mesothorax and 8th abdominal segment in Psepheninae and Eubrianacinae; all other spiracles non-functional. Functional spiracles cribriform and present only on 8th abdominal segment in Eubriinae, and with a special spiracular brush on the 9th abdominal segment to keep the spiracle clean. Functional spiracles absent in all instars in the exotic Psephenoidinae.

Comments: The family is small. Only 1 of the world's 27 known species of Eubrianacinae occurs in N. America—in the Pacific coastal states. Seven species of *Psephenus* represent our share of the world's 31 species of Psepheninae. Of the Eubriinae, we have 9 of the 55 described species; 1 in the genus *Dicranopselaphus* (exceedingly rare), 4 in *Acneus* in the Pacific coastal states, 3 in *Ectopria* in the central and eastern states, and 1 in *Alabameubria*, presently known only from a few larvae taken in northern Alabama and Tennessee. Perhaps only the common eastern water penny is sufficiently abundant and widespread (Georgia to Oklahoma and north into Canada) to be considered familiar to the average aquatic biologist. None are of appreciable economic significance.

Until the 1950's the larvae of *Dicranopselaphus* and *Ectopria* were widely considered those of the dryopid genus *Helichus*, rendering family characterizations invalid.

Selected Bibliography

Arnett 1968 (key to adult beetles).
Artigas 1963 (larva of *Tychepsephus*).
Bertrand 1972 (most extensive coverage of larvae and pupae of world fauna).
Böving and Craighead 1931.
Brown 1972, 1976 (figures of each genus; keys to both adults and larvae), 1980, 1981a, 1987 (biology).
Brown and Murvosh 1974 (includes reproduction of West's figures).
Hinton 1955 (respiratory systems of larvae and pupae; subfamily distinctions), 1966.
Leech and Chandler 1956 (keys to all genera but *Dicranopselaphus*).
Murvosh 1971.
West 1929.

LUTROCHIDAE (DRYOPOIDEA)

Harley P. Brown,
University of Oklahoma, Norman

Travertine Beetles

Figures 34.415–416

Relationships and Diagnosis: The Lutrochidae comprise a small New World family of aquatic and semi-aquatic beetles recently removed from the subfamily Limnichinae of the Limnichidae. Prior to the work of Hinton (1939), they had long been considered members of the Dryopidae, in which family Bertrand (1972) still places them. Ecologically, as well as in many anatomical features, they are closest to the Elmidae (Brown 1987). Larval lutrochids differ from limnichids and dryopids in possessing well-developed tracheal gills like those of elmids, which are completely retractile within a cloacal chamber covered by a ventral operculum on the 9th abdominal segment. The body also tapers apically, like most elmids, rather than being parallel-sided as are the elateroid limnichid and dryopid larvae. They differ from elmids in having no more than 4 pairs of abdominal pleura, whereas elmids have 5 or more, and in having the apex of the 9th abdominal segment broadly rounded while it is typically emarginate in elmids.

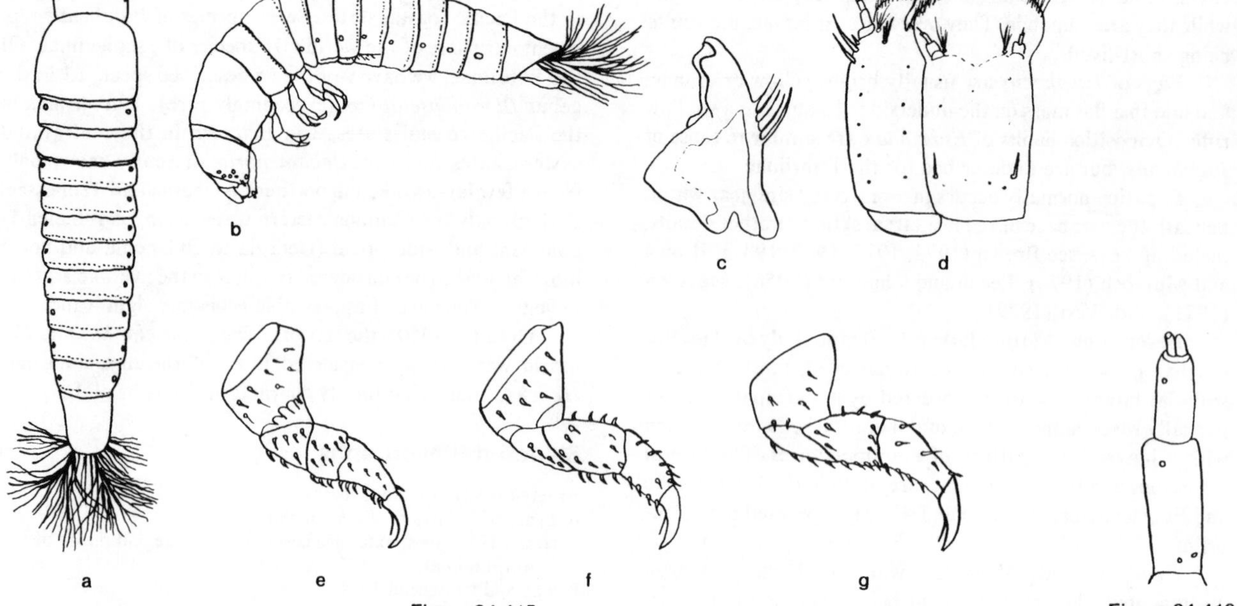

Figures 34.415a-g. Lutrochidae. *Lutrochus luteus* LeConte. **a.** dorsal, with anal gills extended; **b.** lateral; **c.** mandible; **d.** right maxilla and labium, ventral; **e.** front leg; **f.** middle leg; **g.** hind leg.

Figure 34.416. Lutrochidae. *Lutrochus* sp. Antenna.

Lutrochid larvae are separable from all others by the following combination of characters: aquatic, with retractile anal tracheal gills; head large and visible from above, though partially retractile into prothorax; body robust, elongate, tapering posteriorly to broadly rounded apex; abdomen with 9 segments, segments 5–8 without pleura.

Biology and Ecology: Lutrochid larvae are aquatic, but unlikely to be collected in routine dip-netting. They typically burrow in travertine (the hard, calcareous encrustation that is produced by certain algae in shallow rapids or cascades where the calcium content of the water is high). Their food consists largely of the filamentous algae, diatoms, and other organic matter incorporated in the travertine, but they may also burrow in and feed upon decaying wood. Their slow movements and mottled camouflage coloration, as well as their burrowing habit, make them unlikely to be noticed in their natural habitat.

Our lutrochids overwinter as larvae, which pupate above the water line in the spring, in small chambers in rotten wood, travertine, or soil. Recently-emerged adults fly quite ably, and are attracted to lights. Thereafter, adults of some species (e.g., *Lutrochus luteus* LeConte) may enter the water and remain submerged permanently; as with many elmids, a heavy mineral encrustation often develops on body and legs, precluding any possibility of further flight. In other species (e.g., *L. laticeps* Casey), adults are essentially riparian, looking and behaving very much like overgrown limnichids. Oviposition occurs where the adults are feeding, eggs being inserted in the travertine substrate by means of bladelike ovipositors. In temperate regions, there is probably but a single generation per year.

Description: Mature larvae 4–7 mm. Body robust, elongate, with 3 pairs of 5-segmented legs (the terminal segment a tarsungulus). Head large, visible from above, and partially retractile into prothorax; prognathous. Mesothorax and abdominal segments 1–8 each with a pair of biforous lateral spiracles. Thoracic sterna largely membranous; thoracic pleurites slender, parallel to tergal margins. Abdomen 9-segmented; first 3 or 4 segments with pleura as well as distinct sterna. Dorsal cuticle typically mottled in coloration, and bearing numerous tubercles, setae, and marginal spines or setae.

Head: Large, exposed, slightly broader than long, partially retractile. Antennae 3-segmented, small, slender; partially retractile; basal segment longest; apical segment small, subequal in size to sensory peg beside it. Eyes on each side composed of a loose cluster of 5 stemmata, with a 6th stemma directed ventrally at base of mandible. Clypeus much broader than long, and bearing arborescent setae. Labrum transverse, with broadly rounded angles. Mandibles symmetrical, strong, with 3 apical (and sometimes 2 subapical) teeth; without mola; mesal surface of mandibular base with a patch of parallel bristles (obscured in mesal concavity and not visible in aspect figured). Maxilla without differentiated palpifer; palp 4-segmented; galea separate from and much larger than lacinia; stipes elongate; cardo largely membranous or absent. Labium with postmentum undivided; ligula broader than long; palps 2-segmented but with well-developed palpiger. In S. American lutrochids, the labium is fused with the stipes of each maxilla, but there is no such fusion in N. American larvae examined.

Thorax and Abdomen: Prothorax subequal in length to mesothorax and metathorax combined. Median thoracic sternites lacking in N. American lutrochids. Legs well developed but relatively short and robust; not differing greatly in size, though increasing slightly in length posteriorly; each tarsungulus interiorly with both a long spine and, alongside it, a very short peg.

Abdomen with 9 segments, tapering gradually. Segments 1–4 with pleura diminishing in size posteriorly, those on segment 4 being very small and triangular. (Neotropical lutrochids have no trace of pleura on segment 4, and only sternopleural sutures on segment 3.) Segments 5–8 (4–8 in neotropical forms) have no sutures dividing sterna from terga, the cuticle of each segment forming a complete sclerotized ring. Segment 9 with a flattened, oval, ventral operculum bearing on its dorsal surface 2 prominent curved claws, each lined interiorly by a series of spines or setae. Retractile anal tracheal gills prominent when extended, composed of both simple and branched filaments, with a total of about 80 terminal filaments in mature larvae. Without urogomphi.

Spiracles: Functional spiracles absent except in last instar, which has lateral biforous spiracles on mesothorax and abdominal segments 1–8.

Comments: This small family is primarily Neotropical. Only 1 genus, represented by several species, occurs in the United States. Of these, 1 is in Arizona, 1 in Texas and bordering areas, and 1 is rather widely distributed in the East and Midwest. None are known to be of economic importance.

Selected Bibliography

Bertrand 1972 (included in Dryopidae).
Brown 1976 (included in Limnichidae; larval figures; distribution and habitat), 1987 (biology).
Brown and Murvosh 1970 (new species; ecology and behavior).
Hinton 1939 (larva of Neotropical species; included in Limnichidae).

DRYOPIDAE (DRYOPOIDEA)

(= PARNIDAE)

Harley P. Brown,
University of Oklahoma, Norman

Figures 34.417-421

Relationships and Diagnosis: For many years the dryopids, elmids, and often the psephenids were placed in the family Parnidae. Even now, some authors (e.g., Bertrand, 1972) include the elmids and lutrochids within the family Dryopidae. Unfortunately, the larvae of various eubriine psephenids (Eubriidae of Bertrand 1972) were mistakenly identified as larvae of dryopids, and from at least the 1800's until the 1950's the larva of *Ectopria* was figured and described as that of *Helichus*. The larva of *Dicranopselaphus* was also figured as that of *Helichus*. Another eubriine larva from Panama was thought to be *Pelonomus* by Böving and Craighead (1931). The larva considered to be that of *Helichus* by Leech and Chandler (1956), Leech and Sanderson (1959) and Brown (1976) is apparently that of the ptilodactylid, *Araeopidius*. The actual larvae of *Helichus* and *Dryops*, first reared and described by Beling (1882), bear little resemblance to eubriine or other psephenid larvae, which are conspicuously flattened. All known dryopid larvae are elateriform, rather than onisciform. Furthermore, instead of being aquatic, they typically occur in humid soil or decaying vegetation.

The following combination of characters will distinguish dryopid larvae from all others: head rounded, partly retractile, and almost as wide as thorax; mandibles without molar process or prostheca; body cylindrical; abdominal segments with narrow sterna and no pleura; caudal segment bluntly rounded as seen from above, without urogomphi or anal tracheal gills, but with a ventral operculum; all spiracles functional.

Biology and Ecology: Unlike their elmid relatives, dryopid larvae are not aquatic, but occur typically in damp soil or decomposing vegetation, feeding primarily upon decaying plant tissue but also upon tender roots. This is quite unusual, for the adults are closely associated with water, *Dryops* and *Pelonomus* being mostly riparian, but *Helichus* being so thoroughly aquatic that it need not come to the surface for air and is a characteristic riffle inhabitant. Although many aerial insects have aquatic larvae, very few insects are aquatic as adults but terrestrial as larvae.

In contrast with elmids and psephenids, most female dryopids have well developed blade-like ovipositors which are apparently used to insert the eggs into soil or plant tissue. The complete life cycle requires only a few months in warm climates, but 2 years in cooler regions. Adults are long-lived in the genus *Helichus*. Despite the fact that adults are locally abundant, larvae have rarely been found in this hemisphere, and Ulrich (1986) has made the only detailed observations on them. Adults of *Dryops* and *Pelonomus* are most likely to be taken at lights, sometimes in great numbers. Some species of *Helichus* are also commonly taken at lights.

Description: Mature larvae ca. 7–12 mm, ca. 6 times longer than wide. Body cylindrical, rounded at both ends. Cuticle horny, glabrous, yellow to brown. The darker, striate posterior margins of the thoracic and first 8 abdominal segments often give the larva a banded appearance.

Head: About as wide as long, rounded, retractile to about half its length into prothorax. Epicranial and frontal sutures usually complete but obscure except in cleared specimens. Antennae short, 3-segmented, with basal segment usually longest. Eyes, when present, represented on each side by a dorsolateral cluster of 2 stemmata, a lateral cluster of 3 stemmata behind base of antenna, and a single stemma directed ventrally near base of mandible. Labrum and clypeus transverse. Mandible strong, with 3 rather blunt apical teeth; without mola or plumose basal process. Maxilla with 4-segmented palp; galea and lacinia with apical tufts of spines; suture between cardo and stipes not conspicuous or cardo small and separated from stipes by membrane. Labium with palpiger and short, 2-segmented palp.

Thorax and Abdomen: Thorax not much wider than head. Prothorax as long as meso- and metathorax combined. Legs short, ventral, 5-segmented, the apical segment being a tarsungulus with a small interior spine; coxae not widely separated.

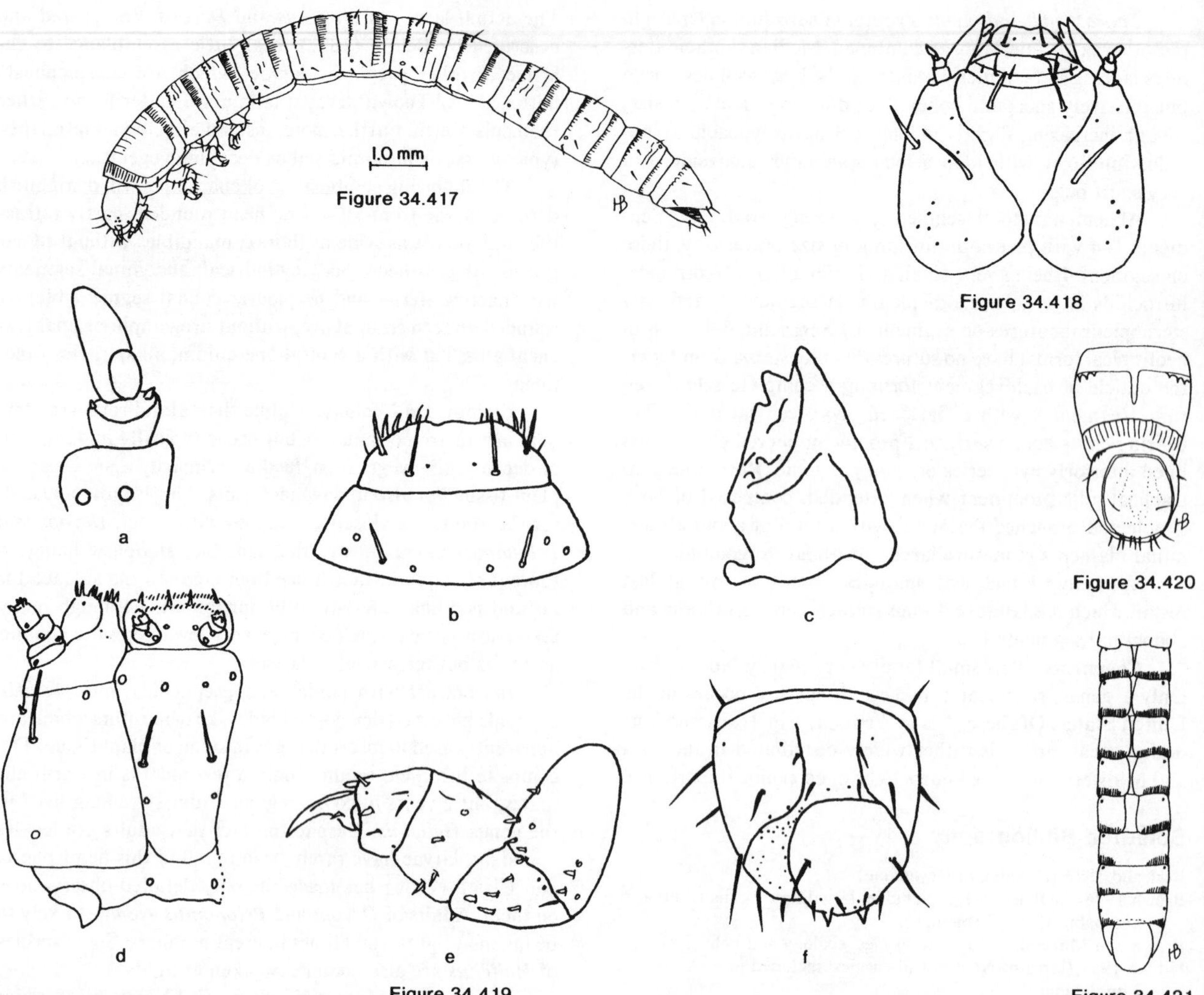

Figure 34.417. Dryopidae. *Pelonomus obscurus* LeConte. Lateral.

Figure 34.418. Dryopidae. *Dryops rudolfi* Rusek. Head, dorsal.

Figure 34.419a-f. Dryopidae. **a.** antenna; **b.** labrum and clypeus, dorsal; **c.** mandible; **d.** right maxilla and labium, ventral; **e.** middle leg; **f.** ventral aspect of 9th abdominal segment with operculum partially open.

(Figures 34.418, 34.419a-f are redrawn from Rusek.)

Figure 34.420. Dryopidae. *Pelonomus.* Ventral aspect of 8th and 9th abdominal segments with operculum slightly open.

Figure 34.421. Dryopidae. *Helichus suturalis* LeConte. Ventral aspect of abdomen, showing narrow sternum on first 5 segments but no sternum on segments 6–8. In at least this species of *Helichus* the spiracles of segment 8 are lateral and visible even in ventral aspect, but in most dryopids they are dorsal on this segment.

Abdomen cylindrical, sides parallel and subequal in width to thorax; caudal segment tapering roundly to a blunt apex. Segments 1–5 with a narrow sternum but no pleura. Segments 6–8 forming complete rings without distinct sutures or pleura or with small and indistinct sterna. Segment 9 bearing a ventral, rather flattened operculum (sternite 10?) but no tracheal gills.

Spiracles: All instars with functional, small, biforous spiracles with delicate transverse partitions, lateral on mesothorax and abdominal segments 1–7 but more or less dorsolateral on segment 8.

Comments: The family is worldwide in distribution, with over 230 species in 17 genera. Only 3 genera are known from the United States. One species of *Pelonomus* occurs from Texas and Florida as far north as Canada, but appears to be common only in marshy subtropical regions. *Dryops* is represented by 1 species described from Arizona, 1 Mexican species extending into southern Texas, and possibly an introduced European species in Maine (*Dryops viennensis* Heer has become established along the St. Lawrence River in the Canadian province of Quebec). We have 11 species of the cosmopolitan genus *Helichus,* several of which are widespread and locally abundant. The genus occurs throughout the continental United States and adjoining Canadian provinces, chiefly in shallow streams. None of our species appear to be of economic importance.

Selected Bibliography

Beling 1882 (larvae and pupae of *Dryops auriculatus* (Geoffroy) and *Helichus substriatus* (Müller)).
Berjon 1964 (*Dryops* larvae as potential pests on roots of grain crops).
Bertrand 1972 (keys to genera of known larvae and pupae; figures of *Dryops* larva).
Böving and Craighead 1931.
Brown 1976 (adults keyed to species, larvae to genera) 1981a, 1987 (biology).
Hinton 1936 (detailed figures of larva).
Leech and Chandler 1956 (key to genera).
Leech and Sanderson 1959 (key to genera).
Olmi 1972 (notes on ecology).
Rusek 1973 (detailed figures of adult and larva; larvae as dominant decomposers in soil of wet Czechoslovakian meadows).
Ulrich 1986 (larvae of 2 species of *Helichus*).

LIMNICHIDAE (DRYOPOIDEA)

Harley P. Brown,
University of Oklahoma, Norman

Minute Marsh-Loving Beetles

Figures 34.422–423

Relationships and Diagnosis: The limnichids are closest to such fellow dryopoids as the Dryopidae, Lutrochidae, Elmidae, and Psephenidae. They are also more like the Chelonariidae than like Heteroceridae and Ptilodactylidae. For many years they were associated with the Byrrhidae. They also resemble some of the small, convex dermestids. Hinton (1939) removed *Lutrochus* from the Dryopidae and placed it in this family, where it remained until recently. In all recent American literature dealing with limnichids, unfortunately, the treatment of immature stages is based upon *Lutrochus*, and is consequently quite misleading. It was not until 1970 that bona fide limnichid larvae and pupae were described in Europe by Paulus. Although the genera he described (*Limnichus* and *Pelochares*) are not known to occur in North America, his work enables us to recognize the larvae of such American genera as *Limnichites*, upon which our present treatment is largely based. Limnichid larvae differ from lutrochids in being terrestrial and lacking anal gills; nor do they have pleura on abdominal segments 1–4; abdominal segments 1–7 have a clearly-defined sternum whereas in lutrochids segments 5–8 are encased in complete sclerotized rings with no lateral sutures or identifiable sterna. Limnichid larvae are more like those of dryopids, but differ in exhibiting broad, membranous sterna on abdominal segments 1–7, in contrast to the armored abdomens of the dryopids which show small sterna on several anterior segments but often none at all on the remaining segments.

Limnichid larvae are characterized by the following combination of features: body elongate and subcylindrical; head large but partially retractile within prothorax; antennae and mouthparts shorter than head; mandibles without molar process or prostheca; with 9 pairs of small biforous spiracles; terrestrial and without tracheal gills; abdominal segments 1–7 with broad, membranous sterna; without urogomphi; caudal segments deflexed; legs 5-segmented, including tarsungulus.

Biology and Ecology: So far as known, limnichid larvae live in damp soil or humus near streams or other bodies of water, apparently feeding chiefly upon decaying plant matter. Adults live near by, along or near the water margin, often in trash left above the water level as floods subside. When disturbed, they may fall onto the water surface, from which they fly readily, as do many psephenids, dryopids, and lutrochids. Adults are sometimes taken in great numbers at lights. No complete life histories have been described, but pupation occurs in the larval habitat.

Description: Mature larvae small, most being less than 5.0 mm long and little more than 0.5 mm in width. Body elongate and subcylindrical, terminating rather bluntly or broadly at both ends, pale testaceous to brownish in color. Posterior margins of thoracic and first 8 abdominal segments with longitudinally striated borders.

Head: Exposed but partially retractile into prothorax, prognathous. Antennae short, 3-segmented; segment 3 with an apical spine; apical sensory peg of segment 2 hyaline, broad and conical, with rounded tip. Eyes represented on each side by a dorsolateral quadrangle of stemmata behind antennal base, a ventrally-directed stemma just behind base of mandible, and in *Limnichites* another small ventral stemma a short distance behind the first and below the posterior lateral stemma. (The posterior ventral stemma is apparently absent in *Limnichoderus* as well as *Limnichus* and *Pelochares*.) Epicranial sutures complete. Labrum and clypeus transverse. Mandible rather strong, with two apical teeth; excavated or hollowed medially; without mola or basal process. Maxilla with 4-segmented palp that extends barely, if at all, beyond tip of galea; suture between cardo and stipes distinct; galea and lacinia stubby, with apical clusters of short spines; without palpifer. Labium several times longer than wide; with short, small, 2-segmented palps.

Thorax and Abdomen: Thorax little wider than head. Prothorax about as long as mesothorax and metathorax combined. Legs rather short, 5-segmented, the apical segment a tarsungulus with a small interior spine.

Abdomen cylindrical, sides parallel and subequal in width to metathorax; rounded at apex; decurved posteriorly to the extent that the dorsal surface of segment 9 is almost at a 90° angle from body axis. Segments 1–7 with broad, membranous sterna but without pleura. Sternum of segment 8 narrower. Segment 9 without tracheal gills; sternum either forming a feeble operculum or appearing to constitute a 10th segment.

Comments: Although the family is cosmopolitan, the N. American fauna is indigenous. None of our genera are known from the Old World, though some of the Neotropical species belong to Old World genera. Limnichids occur throughout the United States and neighboring provinces, but

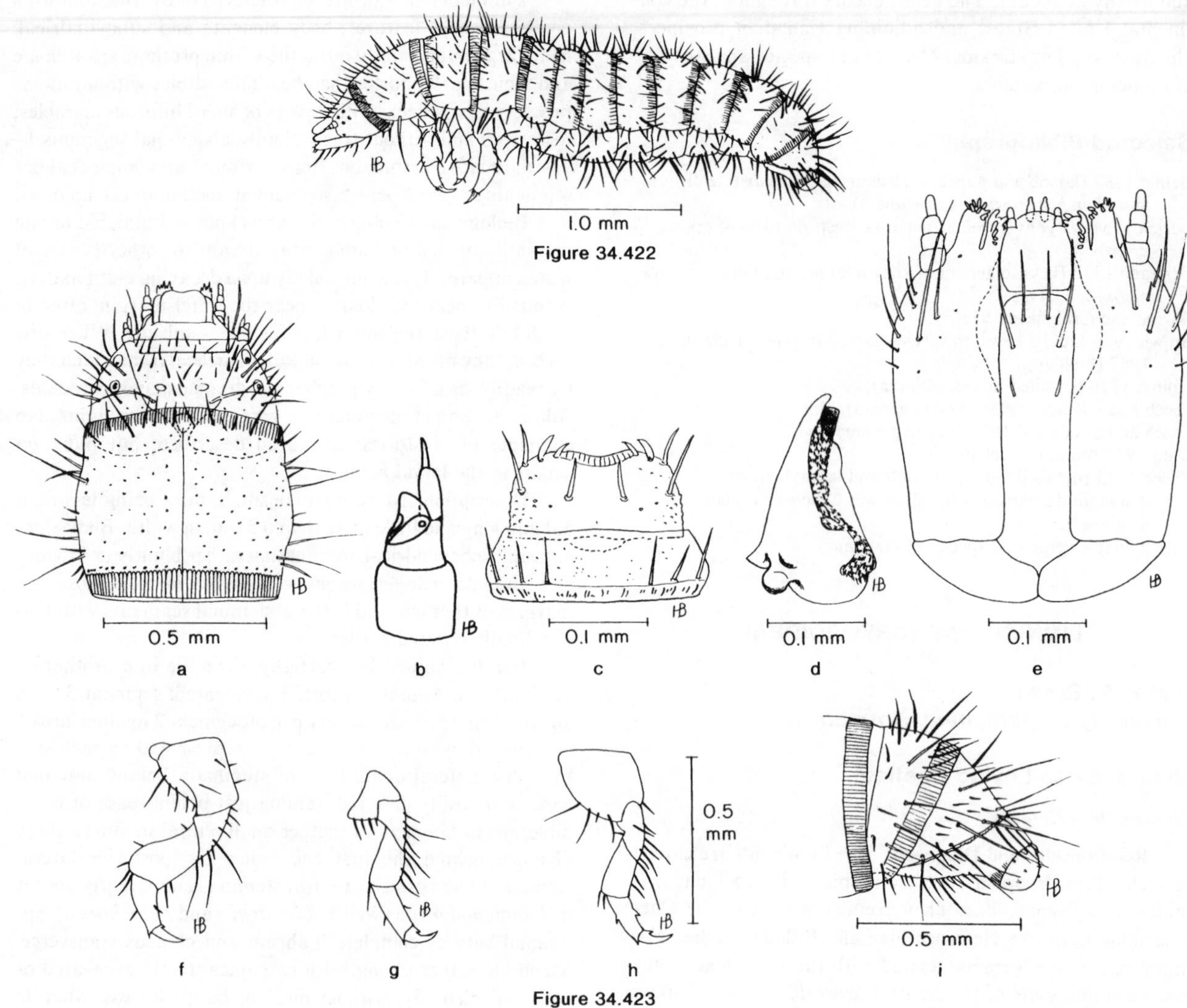

Figure 34.422. Limnichidae. *Limnichites foraminosus* Casey. Lateral.

Figures 34.423a-i. Limnichidae. *Limnichites.* **a.** head and pronotum, dorsal; **b.** antenna; **c.** labrum and clypeus, dorsal; **d.** mandible; **e.** labium and maxillae, ventral; **f.** front leg; **g.** middle leg; **h.** hind leg; **i** posterior end, left side.

all are small and inconspicuous, and none are of known economic importance. Of our 6 genera, 1 (*Throscinus*) is confined to warm seashores, 1 (*Physemus*) to the arid Southwest, and 1 (*Lichminus*) to the Pacific states. The other 3 are widely distributed and sometimes locally abundant. Although *Limnichoderus* is quite drab, specimens of *Eulimnichus* and *Limnichites* are often veritable gems because of their iridescent pubescence. Very few larvae have been collected and identified.

Selected Bibliography

Britton 1971 (intertidal species).
Casey 1912 (descriptions of adults; nothing on larvae).
Hinton 1939 (treatment of limnichid larvae based on *Lutrochus*, now placed in separate family).
Paulus 1970 (larvae and pupae of *Pelochares* and *Limnichus*).
Spilman 1959, 1972.
Wooldridge 1975 (key to New World genera based on adults).

HETEROCERIDAE (DRYOPOIDEA)

John F. Lawrence,
Division of Entomology, CSIRO

Variegated Mud-Loving Beetles

Figures 34.424-426

Relationships and Diagnosis: The Heteroceridae were formerly associated with the Dascillidae because of the presence of a mandibular mola and cribriform spiracles in the larva (Böving, 1929b), but in all more recent classifications they are placed in the Dryopoidea and are thought to be most closely related to Limnichidae and Dryopidae (Crowson, 1978). Heterocerid larvae are easily distinguished from those of any other family by the combination of very short antennae, large, scattered stemmata, strongly consolidated

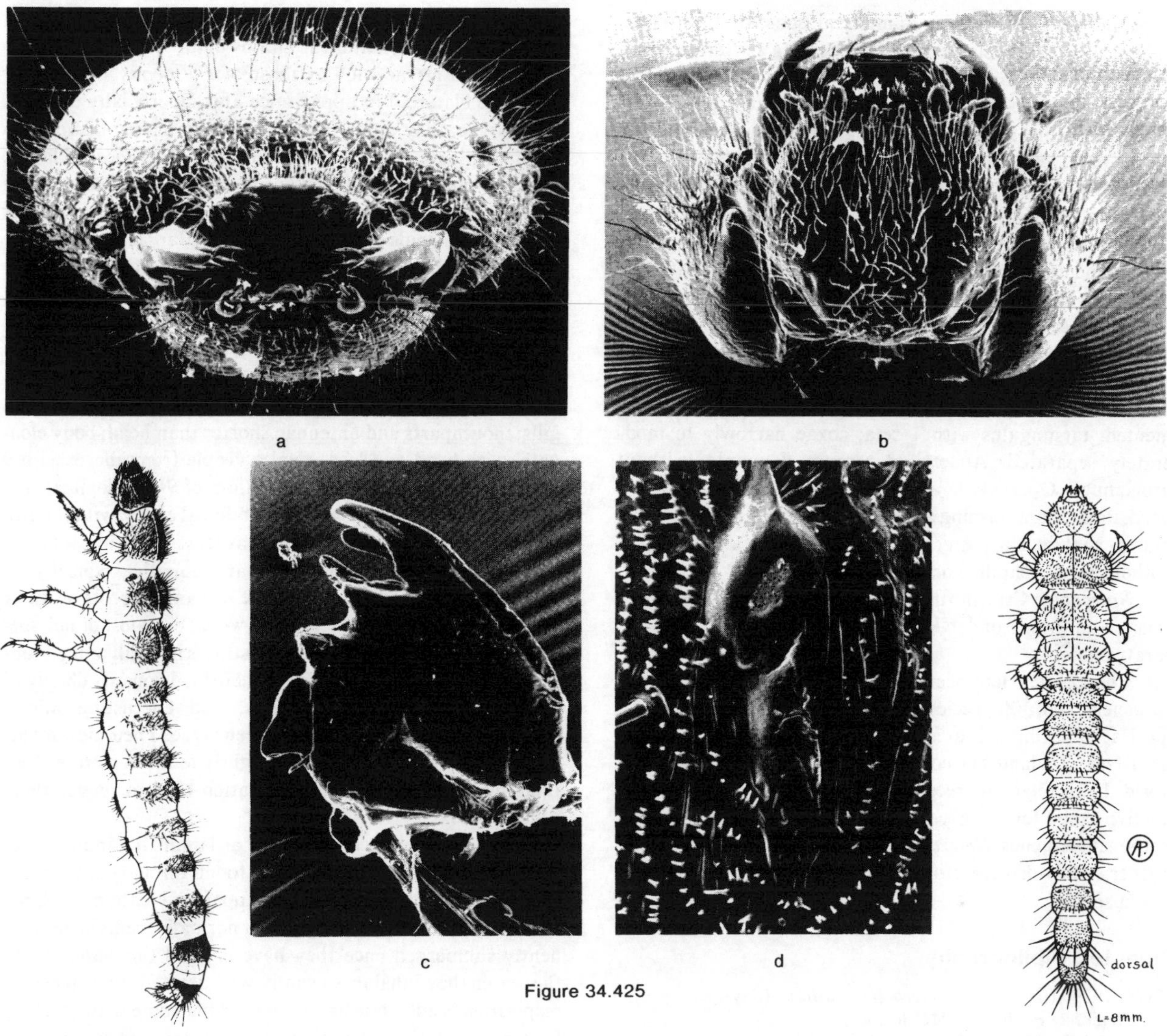

Figure 34.424

Figure 34.425

Figure 34.426

Figure 34.424. Heteroceridae. *Lanternarius gemmatus* (Horn). Rancheria Creek at mouth Yale Creek, Mendocino Co., California. Larva, lateral. Length = 11.8 mm.

Figures 34.425a-d. Heteroceridae. Same as figure 34.424. Del Norte Co., California. **a.** head, anterior; **b** head, ventral; **c.** right mandible, dorsal; **d.** abdominal spiracle.

Figure 34.426. Heteroceridae. *Heterocerus?* sp. Larva, dorsal. Length = 8 mm. (From Peterson, 1951.)

maxillo-labial complex, a distinctive mandible with complex incisor area and concave pseudomola, cribriform spiracles of a unique type, and a 10-segmented abdomen without hooks, gills, or operculum.

Biology and Ecology: Heterocerids are found in wet sand, mud, or clay along the banks of ponds, streams, or other bodies of fresh, brackish, or salt water, but adults are often attracted to lights. Adults burrow into the sand to lay eggs, and larvae may be found in superficial tunnels, where they ingest quantities of substrate from which they extract algae, diatoms, and other organic material. Pupation takes place in a mud cell.

Description: Mature larvae 2 to 10 mm. Body elongate, widest in thoracic region, tapering posteriorly, convex to slightly flattened. Dorsal surfaces darkly pigmented, smooth; vestiture of short and long hairs, the latter denser and more lightly pigmented on head and thorax, sparser and darkly pigmented on abdomen.

Head: Protracted and prognathous, broad, somewhat flattened. Epicranial stem absent; frontal arms V- or U-shaped, contiguous at base. Median endocarina absent. Stemmata 5 on each side, well separated and with well-developed lenses. Antennae very short, 3-segmented, with large, bulbous sensorium on segment 2 which is longer than reduced 3rd segment. Frontoclypeal suture present; labrum large and free. Mandibles symmetrical, somewhat flattened, bidentate, with complex incisor area bearing a sclerotized retinaculum and hyaline process, and without accessory ventral process; mola sub-basal, elongate, divided into 2 parts, concave with simple surface, bearing an articulated process at its base. Ventral mouthparts retracted. Maxilla with transverse cardo, elongate stipes, no articulating area, 3-segmented palp, articulated, rounded galea, and fixed, rounded lacinia; cardines approximate. Labium with mentum and submentum fused; postmentum connate with stipites for more than half its length; ligula short and broad; labial palps 2-segmented, widely separated. Hypopharyngeal sclerome absent. Hypostomal rods absent. Ventral epicranial ridges present. Gula transverse.

Thorax and Abdomen: Legs well-developed, 5-segmented; tarsungulus with 1 seta; coxae narrowly to moderately separated. Abdominal tergum 9 simple, without urogomphi. Operculum absent. Segment A10 well-developed, its sternum forming a conical pygopod; tergum A10 reduced; anal region posteriorly or posteroventrally oriented, without hooks, papillae or gills.

Spiracles: Cribriform; of a unique type appearing annular-uniforous under lower magnifications. Closing apparatus absent.

Comments: Heteroceridae is a small family with about 15 genera and 300 species worldwide and 9 genera and 31 species in the United States and Canada. The family is absent from New Zealand but common in most other areas of the world. Until relatively recently, all species except a few distinctive exotic forms (*Elythomerus, Micilus*) were included in the single genus *Heterocerus;* Pacheco (1964, 1978) has further subdivided the group, primarily on the basis of New World species.

Selected Bibliography

Böving 1929b (larva of *Neoheterocerus pallidus* (Say), as *Heterocerus ventralis* Melsheimer).
Böving and Craighead 1931 (larva of *N. pallidus*).
Clarke 1973.
Claycomb 1919 (habits).
Pacheco 1964, 1978.
Peterson 1951 (larva of *Heterocerus* sp.).
Pierre 1945 (larva of *Heterocerus aragonicus* Kiesenwetter).

ELMIDAE (DRYOPOIDEA)

(= ELMINTHIDAE, HELMINTHIDAE)

Harley P. Brown,
University of Oklahoma, Norman

Riffle Beetles

Figures 34.427–431

Relationships and Diagnosis: The Elmidae constitute the largest family of the Dryopoidea. Members of the 2 subfamilies Elminae and Larinae differ appreciably in ecology, appearance, and adult life style. It is only in recent years that they have been separated from the dryopids and placed within a single family, most convincingly by Hinton (1939). Bertrand (1972) still retains the elmids in the family Dryopidae. Some elmid larvae such as those of *Cylloepus* and *Narpus* are quite elateroid in appearance and superficially like dryopid larvae; they differ from dryopids in having distinct abdominal pleura and tracheal anal gills. *Phanocerus* larvae resemble eubriine psephenid larvae in being flattened and onisciform, but the head is readily visible from above whereas that of psephenids is ventral. Lutrochid larvae not only look like elmids, but live in the same habitat; they differ from elmids in having only 4 pairs of abdominal pleura while elmids have 5 or more; the broadly rounded apex of the last segment of lutrochids also separates them from most elmids.

Elmid larvae are distinguished by the following combination of characters: aquatic, with retractile anal tracheal gills; mouthparts and antennae shorter than head; body elongate, with head and legs usually visible from above; with 9 pairs of biforous spiracles; operculum of 9th abdominal segment with a well-developed pair of dorsal claws; without urogomphi; abdomen with pleura on at least segments 1–5.

Biology and Ecology: All are aquatic, typically in shallow, moving fresh water. Some live among roots or water moss, others in or on waterlogged wood, but most of our species occur on rocky or gravelly substrates in small or medium-sized, relatively clean streams. Their food consists chiefly of decaying plant matter, detritus, and encrusting micro-organisms such as diatoms and green algae. Their movements are slow, and they usually cling tightly to the substrate. Neither their actions nor their coloration tend to render them conspicuous.

Some species overwinter only as larvae, but in most species both adults and larvae may be found throughout the year, usually in the same habitat and apparently eating the same food. Adults are long-lived, and normally remain permanently submerged once they have entered the water. Since the water they inhabit is usually well-aerated, their plastron respiration is adequate and they need not come to the surface for air. Copulation takes place under water, and eggs are deposited singly or in small groups upon the substrate, often in crevices. Mature larvae either crawl out of the water or wait until the water level recedes and leaves them on the shore. After pupation in a protected site, the adults may fly at night before entering the water. Many species are collected at lights. Adults of the subfamily Larinae are usually not aquatic, but riparian (Brown 1987).

Description: Mature larvae 3–16 mm, most less than 8 mm. Body elongate, with 3 pairs of well-developed, 5-segmented legs (the terminal segment a tarsungulus). Head partially retractile into thorax, but usually visible from above. Antennae and mouthparts shorter than head. Thoracic segments with or without median sterna. Mesothorax and abdominal segments 1–8 each with a pair of spiracles. At least the first 5 abdominal segments with pleura as well as terga and sterna. Ninth abdominal segment with ventral operculum and retractile brush of tracheal anal gills.

Head: Usually exposed but partially retractile into prothorax, and about as wide as long. Antennae 3-segmented, shorter than head; second segment usually longest; third segment small and setose, sometimes smaller than apical seta or

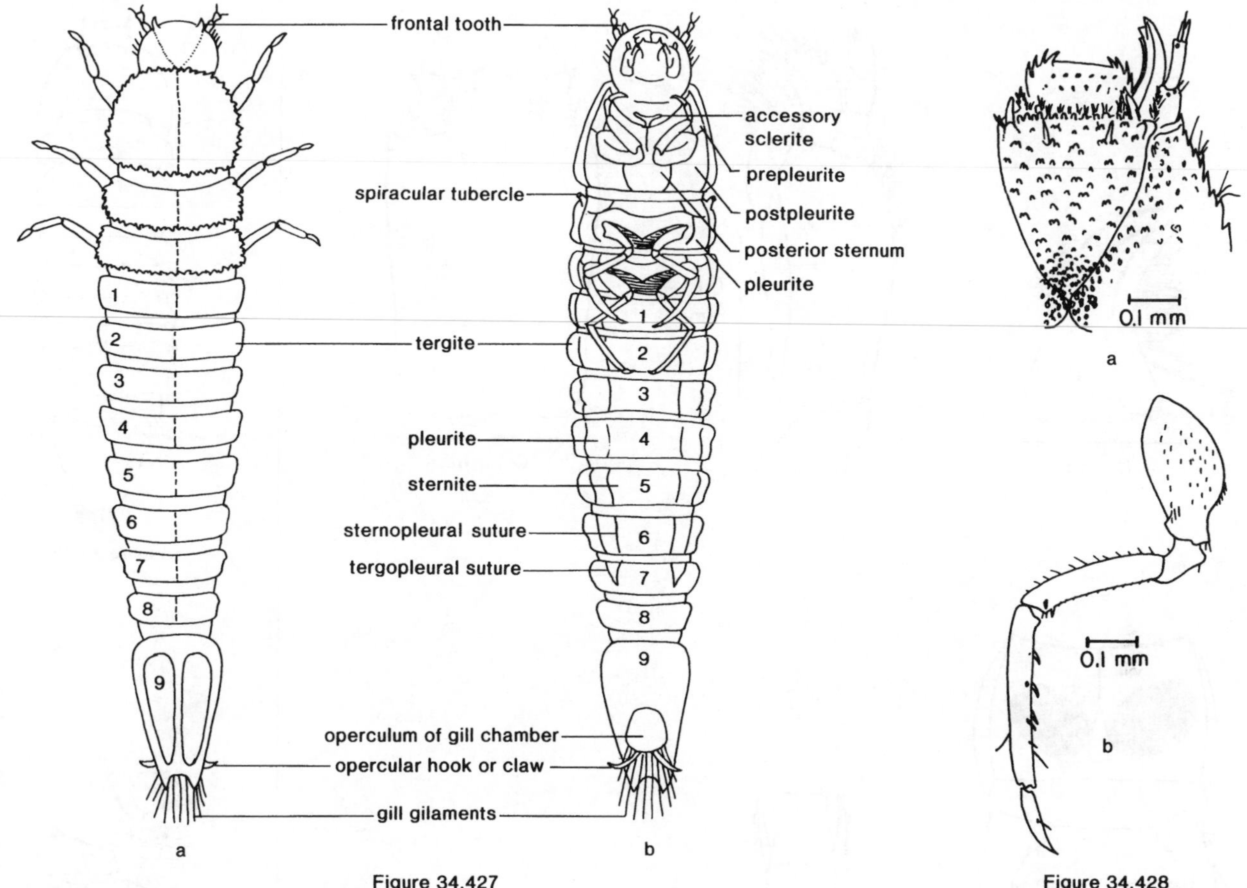

Figure 34.427

Figure 34.428

Figures 34.427a,b. Elmidae. *Neocylloepus boeseli* Brown. **a.** diagrammatic representation of larva in dorsal aspect to show characteristic features of elmid larvae; **b.** ventral aspect with operculum slightly open and filaments of tracheal gill partially extended.

Figures 34.428a,b. Elmidae. *Neocylloepus boeseli* Brown. **a.** enlarged portion of head in dorsal aspect, showing details of ornamentation; **b.** right middle leg, ventral.

spine of second segment. Stemmata normally clustered to form a single "eye" on each side. Labrum transverse. Mandibles usually with 3 apical teeth; without mola; with basal pubescent or plumose process. Maxilla without distinct articulating area; stipes without palpifer; palp 4-segmented; galea and lacinia separate; cardo distinct. Labial palp usually 2-segmented and sometimes with palpiger; postmentum undivided. Gula well developed.

Thorax and Abdomen: Prothorax often as long as meso- and metathorax combined. Mesothorax with a pair of lateral spiracles. Median thoracic sterna either present or absent. Legs well developed, 5-segmented, the tarsungulus with an inner spine or seta; hind legs often longest; all legs commonly visible from above when in use.

Abdomen with 9 segments, the last few often tapering in width and the ninth commonly emarginate (with a shallow to deep notch) at apex. Segment 9 with a ventral operculum bearing a pair of well-developed claws on its inner surface; retractile tuft of filamentous tracheal anal gills. Sterna usually distinct; pleura present on at least segments 1–5.

Spiracles: Functional spiracles absent in first and second instars, biforous spiracles on mesothorax and abdominal segments 1–8 in last instar.

Comments: The family is cosmopolitan. Of about 100 known species of Larinae, only 3 occur in the United States, but of around 1,000 species of Elminae, we have over 80 representing 23 genera. All are apparently indigenous, and a few are quite localized. *Stenelmis,* which occupies about the eastern three-quarters of N. America, does not extend far south of us, but is well represented in most of the Old World. Two other genera are shared with Europe and several with eastern Asia, but a larger proportion appear to be primarily Neotropical. Several are known only from N. America. None are of direct economic importance, but some are of increasing significance as indicators of water quality, and for this purpose generic identification is essential. Since adults of many insects are necessary for specific identification, a major advantage of elmids over other indicator groups is the fact that adults are generally present (along with the larvae) throughout the year. Of course, other factors contribute to

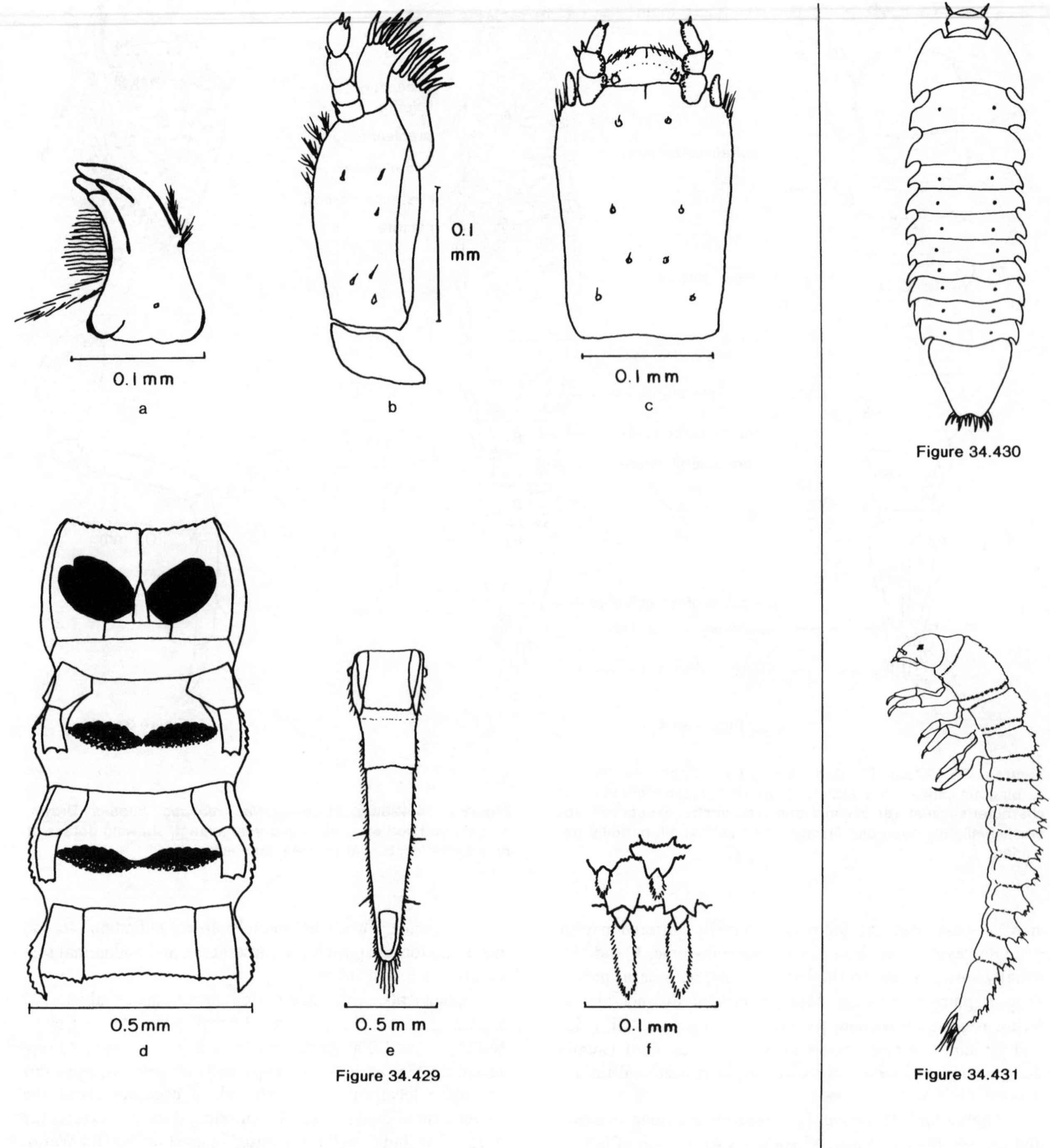

Figure 34.429

Figure 34.430

Figure 34.431

Figures 34.429a-f. Elmidae. *Hexacylloepus ferrugineus* (Horn). **a.** mandible; **b.** right maxilla, ventral; **c.** labium, ventral; **d.** sclerites of thorax and first abdominal segment, ventral; **e.** abdominal segments 7-9, ventral aspect. Some species of *Dubiraphia* have larvae with even more elongate and slender apical segments than this; **f.** tubercles and tufted setae of second abdominal sternite near middle of posterior margin. The details of such cuticular features are of taxonomic value at both specific and generic levels.

Figure 34.430. Elmidae. *Phanocerus clavicornis* Sharp. Dorsal. This flattened, onisciform larva resembles those of eubriine psephenids, but is relatively devoid of lateral setae and has the head readily visible from above.

Figure 34.431. Elmidae. *Gonielmis dietrichi* (Musgrave). Lateral, with tracheal gills partially extruded from beneath cloacal operculum. Other elmid larvae such as *Ampumixis* and *Promoresia* also exhibit dorsal and / or lateral bulges, especially on abdominal segments, but most of our elmid larvae have a much smoother profile.

their usefulness in such work: they are abundant, almost ubiquitous, convenient to sample, relatively indestructible, and can be identified with reasonable certainty.

Selected Bibliography

Bertrand 1972 (keys to known larvae and pupae of world; elmids included in family Dryopidae).
Brown 1976 (keys to species of adults; notes on distribution and habitat; key to genera of larvae with figures), 1981a, 1981b, 1987 (biology).
Crowson 1978 (relationships).
Hilsenhoff 1977 (water quality indicators).
Hinton 1939 (important taxonomic paper), 1940 (most exhaustive work on elmids of this hemisphere).
Leech and Chandler 1956 (keys to genera for adults and larvae).
Leech and Sanderson 1959 (keys to genera for adults and larvae).
LeSage and Harper 1976a (figures of pupae of 5 genera), 1977 (life histories of representatives of 5 genera).
Sanderson 1953 (cornerstone of modern elmid taxonomy; larvae play an important role), 1954 (continuation of above).
White et al. 1984.

ARTEMATOPIDAE (ELATEROIDEA)
(= EURYPOGONIDAE)

Kenneth W. Cooper,
University of California, Riverside

Figures 34.432–434

Relationships and Diagnosis: The artematopids are a family of elateriforms with strong affinities to the eulichadids and to the elaterids. The morphologies of the few species of larval artematopids that have been reared lend support to the inferred relationships, as do those of a small number of related larvae from the United States and one from Mexico for which generic identities are unknown.

At first glance, known larval artematopids may be confused with some larvae of distantly related tenebrionids, and with many larvae of closely related elaterids, but not with the eulichadids even though the similarities to them are numerous and striking. The lack of filaments ("gills") on the ventral segments of the abdomen sets known artematopid larvae apart from those of eulichadids. Artematopid larvae lack both a well-defined frontoclypeal suture and mandibular mola, two features characteristic of tenebrionid larvae, and they have a differentiated galea and lacinia in contrast to the tenebrionid mala. Finally, the free, movable labrum of artematopid larvae contrasts sharply with the nasale of elaterid larvae, providing an easy, clean-cut separation of them.

Biology and Ecology: Larvae of Macropogoninae have been found repeatedly in mats of mosses growing on granitic boulders, often of huge size and along streams, or in open woods that receive winter snow cover or rains and are moist in early spring. Larvae tunnel the sods of the moss, and the transforming larva presses out its pupal cell amid the rhizoids or between sod and rock, often along joints or cracks in the boulder. The pupa is exarate, and wholly free of the larval exuvia.

Macropogon sequoiae Hopping has been taken as larvae and pupae associated with some 6 genera of mosses occurring frequently in mixed mats, and it is still uncertain which, or how many, of these mosses may serve as food. Larvae of *M. rufipes* Horn and *Eurypogon harrisi* (Westwood) occur in beds of the moss *Paraleucobryum longifolium* on granitic boulders, and they have been reared from early larval instars to adult upon that moss alone. The larvae exit from their tunnels to feed upon the leaflets of moss, filtering their intake through the long filamentous hairs and brushes of parts associated with the mouth so that the gut contains a soup almost entirely comprised of fine particles. Whether all macropogonine larvae feed exclusively or primarily on moss cannot be decided at this time. Two specimens preserved in the U.S. National Museum collections are compatible with other regimes [as is Crowson's association of a larva and pupa of *Eurypogon niger* (Melsheimer) with the lichen *Umbilicara*], but do not require them. One, probably the larva of *M. testaceipennis* Motschulsky, is recorded as "in soil with Klamath weed" (Siskiyou Co., Cal.). The other artematopid larva, almost certainly *not* a macropogonine, is from Mexico, bearing on its label the notation "Orchid Plants."

The Brazilian *Artematopus discoidalis* Pic (Artematopinae) has been reared from larvae found in soil just beneath forest litter (Costa, Casari Chen & Vanin 1985). Because larvae "readily ate" cut-up tenebrionid larvae and termites, although they did not attack the insects when presented with forceps, it was concluded (perhaps wrongly) that *Artematopus* larvae are carnivorous. The pupa is exarate.

Adult artematopids have been collected by sweeping grasses, shrubs and lower limbs of trees; little is known of their habits or food.

Description: (Subfamily Macropogoninae) Known mature larvae less than 15 mm long, body elongate, nearly cylindrical, black-brown, often with a bronzy glint, or mottled above; primary setation of head and body sparse, moderately long.

Head: Prognathous, cordate from above, wedge-shaped in profile, proximal fifth or so enclosed by the anterior margin of the pronotum. Labrum free, deeply, broadly incised, with long filamentous hairs projecting from below. Clypeus transverse, slightly wider than frons; frontoclypeal suture obsolete. Frontal arms of ecdysial lines convergent and transversely connected behind, epicranial stem very short or absent; no endocarina. Antennae anterolateral, short, 3-segmented; sensorium ventrolateral, acorn-shaped, half or more as long as cylindrical apical joint. A single pair of stemmata immediately posterior to antennal sockets; lens convex, subcircular in outline.

Mandibles strong, nearly symmetrical, apices pointed, with a dorsal and 1 or 2 ventral subapical teeth; no mola or prostheca, a scant penicillus, a shallow scrobe bearing a long basal seta, a short lateral seta near mandibular midpoint.

Ventral mouthparts reaching nearly to posterior two-thirds of head capsule, protrusible; gula distinct, elongated. Maxillae with 4-segmented palps, 2-segmented galea; lacinia with a complex, dense brush of spines, setae and filamentous hairs dorsally; stipes elongate, articulating area nearly absent;

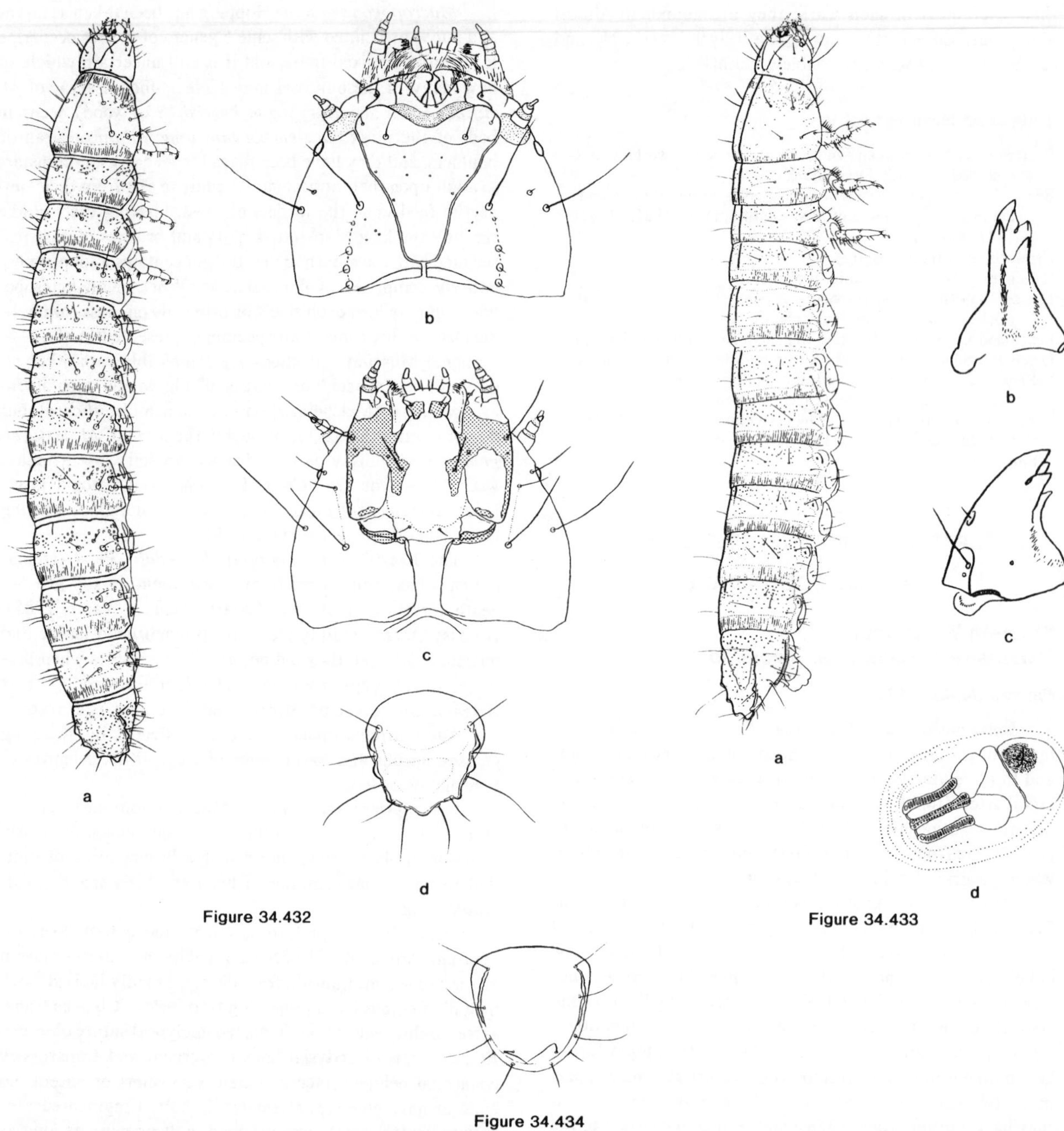

Figure 34.434

Figures 34.432a-d. Artematopidae. *Macropogon sequoiae* Hopping **a.** larva, right dorsolateral, length about 13 mm (head exserted); **b.** head capsule, dorsal; **c.** ventral; **d.** ninth abdominal tergite, dorsal. Brown to deep brown above, with bronzy sheen, creamy yellow to white below. Main setae strikingly rimmed by large, pale, unsclerotized insertions (to greater than 2 stemmatal widths in diameter). Dorsal punctation also coarse, sparse. Alder Creek (near Glennville), Greenhorn Mts., Kern County, California, from clump of mixed mosses, dominated by a species of *Grimmia*, growing on woodland dioritic boulder.

Figures 34.433a-d. Artematopidae. *Macropogon rufipes* Horn. **a.** larva, right dorsolateral. Length about 12 mm, and similar in a general way to *M. sequoiae* in color, many proportions, and setation. However, *M. rufipes* is easily set apart by its finer dorsal punctation, by the unsclerotized bases of its main body setae not exceeding a stemma in diameter, and by the ninth tergite which is longer than broad and more conspicuously concave; **b.** right mandible, oral surface; **c.** left mandible, dorsal; **d.** abdominal spiracle, right side. Mt. Cardigan, Grafton County, New Hampshire, from the moss *Paraleucobryum longifolium* on a granitic boulder in open woods.

Figure 34.434. Artematopidae. Ninth abdominal tergite of *Eurypogon harrisi* (Westwood), dorsal. Larval length 8 mm, brownish black to black, similar in overall structure to *Macropogon* species (see also figures in Böving 1929, Böving & Craighead 1931 "*E. niger*"); the known *Eurypogon* are at once separated from *Macropogon* by abdominal tergite 9 which is parabolic in outline in *Eurypogon*, laterally margined but not scalloped, and dorsally convex. Mt. Cardigan, Grafton County, New Hampshire, from the moss *Paraleucobryum longifolium*, on granitic boulder in open woods.

cardo undivided. Posterior tentorial pit small, subadjacent and medial to insertion of cardo, near junction of hypostomal ridges and gular sutures. Labium with prementum, mentum, and short submentum (not always sclerotized); ligula small, with 2 pairs of short subapical setae; labial palps 2-segmented. Hypopharynx terminated by a dense, somewhat triangular sclerome supported laterally by a strong hypopharyngeal bracon originating at the pleurostomal ridge adjacent and posterior to the pleurostomal mandibular acetabulum.

Thorax and Abdomen: Thorax with pronotum anteriorly and posteriorly margined by a broad, longitudinally striated transverse band; meso- and metanotum, and abdominal tergites 1–8 each with a posterior striated marginal band only. Legs 5-segmented, fossorial, strong, stout, bearing hair-like, spinous, and peg-like setae; tarsungulus bisetose.

Abdominal tergites 1–8 similar in general appearance and setation to each other and to meso- and metathorax. Without urogomphi as usually understood. In *Macropogon* tergite A9 is elongate, conspicuously but shallowly crenated laterally, with a median, terminal, blunt tooth. In *Eurypogon* tergite A9 is elongate, weakly margined but regular in outline, without an apical tooth. The sternite and tergite of A10 enclose a protrusible anal papilla. (Two quite different additional larval forms, unassigned to genera, have the ninth abdominal tergite either slopingly rounded off, unmargined and not crenated, or inflated into a pronounced dome. Otherwise, the morphologies of these larvae are concordant with the general description given.)

Spiracles: Biforous, located laterally and anteriorly on mesothorax and abdominal segments 1–8; a *tenth*, atrophied spiracle (or spiracular scar) usually present on metathorax (as in eulichadids). Spiracles without closing apparatus.

(Subfamily Artematopinae) From the figures and accompanying descriptions by Costa *et al.* (1985), artematopine and macropogonine larvae are closely similar in size, elongate body, and other features, but artematopines are different as follows. Body dorsoventrally flattened, white or transparent, head and prothorax somewhat darkened; primary setation relatively shorter. Head less deeply set into pronotum. Labrum with median tooth and acute lateral angles. Ecdysial lines broadly U-shaped. Sensory appendix of antenna nearly equal in length to third segment. Mandibles with a strong retinaculum and penicillus. Basal segment of maxillary palp nearly equal to distal three together. Ventral mouthparts reaching to posterior fifth or sixth of head capsule. Transverse bands of thoracic and abdominal segments not longitudinally striated. Tarsungulus *not* setose.

Comments: The Artematopidae, predominantly inhabiting the New World tropics and subtropics, is a small family of 60 or so species of which none is known to be of economic significance. It is represented in America north of Mexico by 3 or 4 species of *Macropogon*, 4 of *Eurypogon*, and 1 of *Allopogonia*. Of these, larvae have been reared to adults for *M. rufipes*, *M. sequoiae*, and *E. harrisi*. The probable larva of *E. niger* has been described, figured, and commented upon (Böving 1929b: pl. VIII A–H; Böving and Craighead 1931: pl. 69 I–S; Crowson 1973b). Two additional forms of artematopid larvae are known (*see* above), but have not been associated with their adults. The larva of *Artematopus discoidalis* is most notable (and least primitive) by its nonsetose claws and striking mandibular retinaculum. Unlike reared macropogonines, which are moss-feeders, *Artematopus* larvae may be scavengers (or possibly carnivores).

Selected Bibliography

Böving 1929b.
Böving and Craighead 1931.
Costa et al. 1985.
Crowson 1973b.
Glen 1950.
Pic 1914.

CEROPHYTIDAE (ELATEROIDEA)

John F. Lawrence,
Division of Entomology, CSIRO

Figures 34.435a,b

Relationships and Diagnosis: The Cerophytidae are usually considered to be members of the superfamily Elateroidea, although Hlavac (1975) placed them in a redefined Artematopoidea because of the absence of the specialized elateroid propleurocoxal mechanism (Lawrence and Newton, 1982). The larva is similar in several respects to those of Throscidae and Eucnemidae, but has certain unique features, such as the sclerotized 5-toothed labial plate, enlarged front legs with bifurcate tarsunguli, and paired thoracic and abdominal glands.

Biology and Ecology: *Cerophytum* larvae in Europe are known to inhabit old rotten elm trees, where they feed in layers of decomposing wood (Buysson, 1910), but they have also been recorded from beech, oak, poplar, willow, maple, linden, birch, walnut, and horse-chestnut (Horion, 1955). According to Mamaev (1978), larvae of *Cerophytum elateroides* Latreille were found, along with those of Oedemeridae, in the darkened fungus-softened xylem of an elm withering on the root. Adults may be attracted to lights; they have a unique method of jumping, which is still not well understood.

Description: Mature larvae 6 to 9 mm. Body moderately elongate, broadly subcylindrical, slightly curved ventrally, lightly sclerotized, with dense covering of hairs.

Head: Protracted and prognathous, small, flattened. Epicranial suture absent. Median endocarina absent. Stemmata 1 on each side; lens well-developed. Antennae moderately well-developed, 3-segmented, with broadly conical sensorium. Frontoclypeal suture absent; labrum fused to head capsule, forming nasale which is lightly sclerotized and deeply emarginate anteriorly. Mandibles symmetrical, flattened, broad at base but narrowing to form stylet-like apex; mola absent. Ventral mouthparts retracted, without distinct cardo or articulating area; mala stylet-like and divided longitudinally (partially fused galea and lacinia); palp 3-segmented with well-developed palpifer. Labium consisting of postmentum, which is completely connate with stipites, and prementum, which forms complex sclerotized plate with 5 teeth at apex and a groove at each side, into which mala and mandibular apex are inserted; labial palps absent.

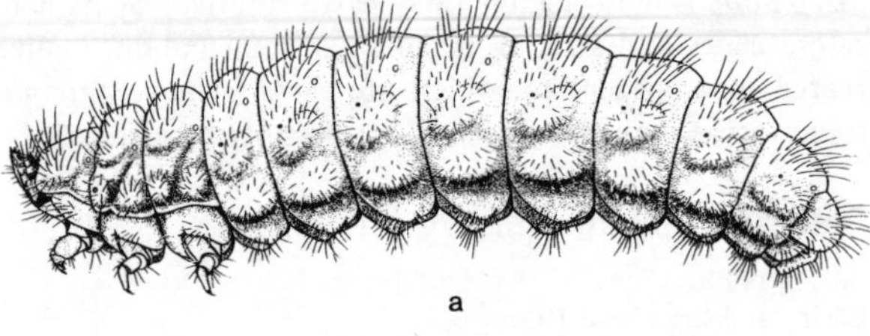

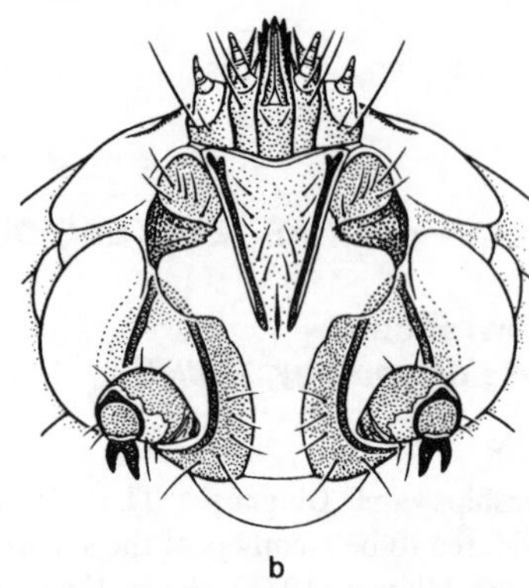

Figures 34.435a,b. Cerophytidae. *Cerophytum elateroides* Latreille. Voronezh Region, U.S.S.R. Length = 10 mm. **a.** larva, lateral; **b.** anteroventral view of head and prothorax.

Thorax and Abdomen: Prothorax with 2 sclerotized sternal rods converging posteriorly. Legs well-developed, 5-segmented; front legs enlarged, with heavily sclerotized, unevenly bifurcate tarsungulus, which is bisetose; front coxae approximate; middle and hind coxae widely separated. All 3 thoracic segments and abdominal segments 1–8 each with conspicuous pair of lateral gland openings near posterior edge of tergum. Abdominal segments 1–8 each with 2 well-developed pleural lobes on each side. Tergum A9 reduced, rounded, without urogomphi; sternum A9 simple. Segment A10 small, posteroventrally oriented, with vertical anal opening.

Spiracles: Biforous, those on abdomen located just above slight indentation in upper pleural lobe; closing apparatus present.

Comments: The family Cerophytidae includes the single genus *Cerophytum* with 10 species in Europe, N. America, and the Neotropical Region; 2 species occur in the United States (Golbach, 1983; Soares and Perracchi, 1964). The genus *Anischia,* normally included in this family, belongs in the Elateridae, in spite of the lack of metacoxal excavations in the adult.

Selected Bibliography

Buysson 1910 (habitat).
Golbach 1983.
Hlavac 1975.
Horion 1955.
Lawrence and Newton 1982.
Mamaev 1978 (larva of *Cerophytum elateroides*).
Rey 1887 (larva of *C. elateroides*).
Soares and Perracchi 1964.

ELATERIDAE (ELATEROIDEA) (INCLUDING DICRONYCHIDAE, LISSOMIDAE)

Edward C. Becker, *Biosystematics Research Centre, Agriculture Canada,* With a Key to 32 Genera of Larvae by **James R. Dogger, *Systematic Entomology Laboratory, ARS, USDA***

Click Beetles, Wireworms

Figures 34.436–451

Relationships and Diagnosis: The Elateridae, as here understood, includes *Drapetes* and *Lissomus* (formerly placed in Throscidae) and excludes the so-called plastocerids *Aphricus, Euthysanius,* and *Octinodes* (now placed in Cebrionidae). It is the largest group in the superfamily Elateroidea, which also includes Artematopidae, Cerophytidae, Cebrionidae, Throscidae, and Eucnemidae. Larval Elateroidea are usually characterized by having distinct frontal sutures, relatively stout and non-perforate mandibles which lack a basal mola, no stemmata or 1 large one on each side of the head, a labrum which is solidly fused to the head capsule forming a nasale (except in Artematopidae), a consolidated maxillo-labial complex, with the stipites either connate with or fused to the labium and with an articulated galea or mala, 2 tarsungular setae, and biforous spiracles with a distinct ecdysial scar; many are subcylindrical and heavily sclerotized, giving them the name "wireworm." Cantharoid larvae share some of the above features with elateroids, but they usually have mandibles which are longitudinally split, grooved, or perforate, they sometimes lack frontal sutures, and they are often soft-bodied. The groups are closely related.

Wireworms, as the name implies, are readily recognized by the usually heavily sclerotized, cylindrical body, although some are lightly sclerotized and have somewhat depressed bodies; by the absence of a free labrum; by the biforous spiracles; by the well-developed legs, each with a claw-like tarsungulus; and by the small, ventrally placed tenth abdominal segment. Frequently elaterid larvae are confused with those of tenebrionids, but the lack of the labrum and similarly sized legs distinguish elaterid larvae, whereas in the tenebrionids the labrum is present and the front legs are frequently larger than the others.

Elaterid larvae are distinguished from those of Cebrionidae by lacking the elongate prothorax and eversible cervical membrane, and from those of Artematopidae by having a fused labrum. Larvae of Cerophytidae differ from elaterid larvae in having a 5-dentate labial plate and enlarged prothoracic legs with bifurcate tarsunguli. Larvae of Throscidae and Eucnemidae have highly reduced or no legs and modified mandibles, which are either fused to the head capsule or outwardly curving and non-opposable.

Biology and Ecology: Adult Elateridae (click beetles) usually are active in the late afternoon in the spring and early summer, although species of a few genera (*Athous, Elathous, Hemicrepidius,* etc.) are active in mid to late summer. Adults can be found crawling on vegetation or can be taken by beating or sweeping. Some species, especially in southern areas, are readily attracted to light or fall into pitfall traps. Very little is known about the food habits of the adults. Apparently they do some minor damage by feeding on tender plant tissues, such as the leaf and flower buds, but this is minimal (*see* Thomas 1940).

Elaterid larvae (wireworms) are found in a variety of habitats, such as leaf litter, soil, moss, rotten wood, and under rocks. In general, some found in leaf litter and in rotten wood are predaceous, while most in soil feed on roots. A list of the genera found in one or more of these habitat types follows, courtesy of J. R. Dogger.

Known Larval Habitats of U.S. and Canadian Elaterid Genera

In Wood, Under Bark	In Soil	In Litter, Under Moss, Under Rocks
Agriotella	*Aeolus*	*Agriotella*
Alaus	*Agriotella*	*Athous*
Ampedus	*Agriotes*	*Ctenicera*
Athous	*Agrypnus*	*Limonius*
Ctenicera	*Athous*	*Sericus*
Denticollis	*Cardiophorus*	
Dicrepidius	*Conoderus*	
Dipropus	*Ctenicera*	
Elater	*Dalopius*	
Hemicrepidius	*Dipropus*	
Hemirhipus	*Glyphonyx*	
Lacon	*Hemicrepidius*	
Limonius	*Horistonotus*	
Megapenthes	*Hypolithus*	
Melanotus	*Limonius*	
Neotrichophorus	*Melanotus*	
Orthostethus	*Negastrius*	
Parallelostethus	*Sericus*	
Pityobius		
Sericus		

Most elaterid larvae take three years to mature, although some of the smaller species may complete their life cycle in one or two years. If conditions are not suitable, the life cycle may be extended more than three years. Normally there are three instars. In most genera, the adults either emerge from the pupal cell in the early fall and then hibernate under bark, boards, rocks, etc., or they remain in the pupal cell. In either case, mating and egg laying take place in the spring and pupation is in late summer of the second year; these adults will overwinter before mating, thus completing the 3-year life cycle. Species in some genera (*Athous, Elathous, Hemicrepidius,* etc.) overwinter as mature larvae and pupate the following spring; adults then emerge in midsummer. For these species, mating and egg laying occur in late summer, but the larvae still usually require three years to complete the life cycle.

Description: Full grown larvae of most species are 15 to 35 mm in length, but smaller species are about 10 mm, and larger ones up to 60 mm. The general shape is *cylindrical,* although some are somewhat depressed and slightly wider in the middle. Many species are light to dark brown, but some are whitish and others may have a darker head and darker ninth abdominal segment. The vestiture consists of a few scattered setae; these frequently are of significance at the generic or specific level.

Head: Distinct, deeply pigmented, somewhat depressed and prognathous. Epicranial or frontal suture lyre-shaped and surrounding frontoclypeal region; the shape of the frontoclypeal region is important. Labrum absent (fused with frons and forming the nasale). Nasale unidentate or tridentate. Antenna 3-segmented, second segment with one to several conical sensory papillae at apical end. Mandible robust, usually conical, usually with retinaculum and sometimes with additional teeth; mandible of Cardiophorinae deeply cleft, with dorsal lobe toothed. Maxillae and labium well developed. Submentum triangular (Pyrophorini and Pityobiini) or narrowly rectangular.

Thorax and Abdomen: Thorax with three subequal segments. Legs well developed, each 5-segmented, terminating in a claw-like tarsungulus. Prosternum as one sclerite or subdivided.

Abdomen with nine distinct segments dorsally (Cardiophorinae has secondary segmentation on segments 1–8, thus giving the appearance of having three segments for each true one).

Segment 9 quite variable, yet very important taxonomically. Rounded or pointed (most Elaterinae), flattened dorsoventrally (Melanotini in Elaterinae), bifurcate (Pyrophorinae). Caudal notch large or small. Each urogomphus usually divided into inner and outer prong, sometimes undivided (*Eanus, Negastrius*). Dorsum of segment 9 with tubercles, pits, or setae; in Pyrophorinae, outer margin with several rounded or pointed protuberances. Pair of "muscular impressions" sometimes present (most *Agriotes*).

Segment 10 on ventral side of segment 9, large or small (Elaterini), sometimes with anal armature consisting of sclerotized tooth or teeth on each side (Pyrophorini); anal papillae sometimes present (Cardiophorinae).

Segments 1–8 sometimes with striate impressions on each side near posterior margin (Ampedini, Melanotini). Segments frequently with impressed "L"-shaped line on each side near anterior margin. Sides and dorsum sometimes with large rounded or semicircular pits. Various setae and shape and size of spiracles sometimes useful.

Pleural areas concealed or reduced (Elaterinae) or visible with one or more sclerites.

Spiracles: Biforous, without closing apparatus.

Comments: Elaterid larvae are among the most commonly collected Coleoptera larvae. In light of this, a key to the larvae of 32 genera is provided below.

KEY TO THIRTY-TWO GENERA OF LARVAL ELATERIDAE[1]

By James R. Dogger

1. A9 with a median caudal notch (Pyrophorinae) 3
1'. A9 without a median caudal notch 2
2(1'). Abdomen entirely soft-skinned; abdominal segments subdivided into 2 or 3 ringlike divisions (fig. 34.438a) (Cardiophorinae) 33
2'. Abdomen partially or completely sclerotized; abdominal segments not subdivided as above (Elaterinae) 19
3(1). Submentum triangular; bases of stipites contiguous (fig. 34.441d) 4
3'. Submentum not triangular; bases of stipites well-separated (Athoini=Lepturoidini) 11
4(3). Anal hooks present on A10 (Pyrophorini) 5
4'. Anal hooks absent on A10; armature consisting of a narrow ridged band on each side of A10 (Pityobiini) .... *Pityobius*
5(4). Nasale consisting of one triple-pointed tooth (as in fig. 34.437b) .... *Hemirhipus*
5'. Nasale tridentate (fig. 34.441b) 6
6(5'). Dorsum of head with a prominent longitudinal groove extending from near frontal suture to hind margin of head on each side (fig. 34.441b); with short hooks in addition to anal hooks on A10 .... *Alaus*
6'. Dorsum of head without extensive longitudinal grooves, sometimes with short grooves or rows of setae posteriorly; without short hooks in addition to anal hooks on A10 7
7(6'). Anal hooks long and prominent, extending to or beyond tip of A10 (subtribe Agrypnina) 8
7'. Anal hooks short and inconspicuous, not extending to tip of A10 (subtribe Conoderina) 10
8(7). Dorsum of A9 without tubercles .... *Agrypnus* (=*Colaulon*)
8'. Dorsum of A9 with tubercles 9
9(8'). Dorsum of A9 covered with small tubercles; margin with 4 or fewer lateral protuberances .... *Lacon*
9'. Dorsum of A9 with few tubercles; margin with more than 4 lateral protuberances (Neotropical) .... *Dilobitarsus*
10(7'). Frons tapering to a blunt point at base .... *Aeolus*
10'. Frons truncate or rounded at base .... *Conoderus*
11(3'). A1–8 conspicuously sculptured or pitted 16
11'. A1–8 smooth; sparsely or finely punctured 12
12(11'). Nasale tridentate (similar to fig. 34.441b), or if consisting of one triple-pointed tooth, with a dorsal seta on the head on each side of and adjacent to the posterior portion of the fronto-clypeal area 17
12'. Nasale consisting of one single- or triple-pointed tooth (fig. 34.437b), without dorsal setae adjacent to the posterior portion of the frontoclypeal area 13
13(12'). Presternum of prothorax consisting of one large triangular sclerite (fig. 34.441d) 14
13'. Presternum of prothorax divided into 2 or more sclerites .... (part) *Ctenicera*
14(13). Lateral margins of A9 with 2 or more prominent projections on each side; caudal notch variable 15
14'. Lateral margins of A9 sinuate with not more than one prominent projection on each side; caudal notch small .... *Limonius*
15(14). Dorsum red-brown to black; urogomphi with outer prongs longer than inner .... *Denticollis*
15'. Dorsum yellow-brown or amber; urogomphi variable .... (part) *Ctenicera*

1. Larvae of *Crepidomenus, Eanus, Elathous, Oestodes* and *Melanactes* have been described, but not studied by me, and are not included in the key.

16(11). Stemmata present; intermediate abdominal segments with crescent-shaped sculpturing or conspicuous pits .... *Athous*

16′. Stemmata absent; intermediate abdominal segments with crescent-shaped sculpturing .... *Hemicrepidius*

17(12). Lateral margins of A9 with prominent projections; anterolateral impressions on tergites of A2–8 reaching or approaching mid-dorsal line .... (part) *Ctenicera*

17′. Lateral margins of A9 smooth or hardly protuberant; anterolateral impressions absent or more abbreviated (subtribe Negastrina = Hypnoidina) 18

18(17′). Nasale consisting of one triple-pointed tooth; urogomphi two-pronged .... *Hypolithus*

18′. Nasale tridentate; urogomphi simple, undivided .... *Negastrius*

19(2′). A9 sternum narrower anteriorly than the adjacent portion of the A8 sternum; tip of A9 broadly rounded 20

19′. A9 sternum as broad or broader anteriorly than the adjacent portion of the A8 sternum; tip of A9 bluntly or sharply pointed or scalloped 24

20(19). Mandible with several teeth; second antennal segment with a single papilla; nasale of one single- or triple-pointed tooth; usually dark colored .... *Sericus*

20′. Mandible with a single tooth; second antennal segment with 5 or more papillae; nasale tridentate or quadrate with 3 equally projecting teeth on a quadrate base; color usually shades of yellow or red-brown 21

21(20′). Base of nasale with an arc of 10 or more closely set setae extending to either side 22

21′. Base of nasale without such closely set setae 23

22(21). Second antennal segment with 6 or 7 papillae .... *Orthostethus*

22′. Second antennal segment with 8 to 13 papillae .... *Parallelostethus*

23(21′). Second antennal segment with 6 or 7 papillae; tergites smooth and finely punctate .... *Elater*

23′. Second antennal segment with 8 or more papillae; tergites sometimes with transversely oriented sculpturing .... *Neotrichophorus*

24(19′). Dorsum of A9 flattened and with margin scalloped (Melanotini) .... *Melanotus*

24′. Dorsum of A9 neither flattened nor scalloped. 25

25(24′). Intermediate abdominal segments with striate impressions or coarsely punctured; nasale variable 29

25′. Intermediate abdominal segments without striate impressions and not coarsely punctured; nasale consisting of one triple-pointed tooth (Agriotini) 26

26(25′). A9 with a pair of central dorso-tergal setae .... *Dalopius*

26′. A9 without central dorso-tergal setae 27

27(26′). A1–8 tergites with 3 or more prominent setae in a transverse line anteriorly (additional short setae may be present) 28

27′. A1–8 tergites with only 2 prominent setae in a transverse line anteriorly (additional short setae may be present) .... *Agriotella*

28(27). Submentum approximately 4 times as long as average width; the most dorsad of the large anterior setae on A1–8 tergites with small setae on either side transversely .... *Glyphonyx*

28′. Submentum not more than 3 times as long as average width; the most dorsad of the large anterior setae on A1–8 tergites without the small setae on either side transversely .... *Agriotes*

29(25). Nasale consisting of one triple-pointed tooth; A9 rounded to a short, sharp tip (subtribe Dicrepidiina) 30

29′. Nasale consisting of one single-pointed tooth or tridentate; A9 with tip attenuated or tri-pointed (subtribe Ampedina) 31

30(29). A9 tergite with tubercles larger than punctures; striae of abdominal striate impressions frequently appearing bent or broken .... *Dicrepidius*

30′. A9 tergite with tubercles smaller than punctures; striae of abdominal striate impressions straight and parallel .... *Dipropus* (=*Ischiodontus*)

31(29′). Nasale consisting of one single-pointed tooth; abdominal segments conspicuously sculptured or with pits or coarse punctures .... *Ampedus*
31′. Nasale tridentate; abdominal segments variable 32
32(31′). Intermediate abdominal segments with striate impressions; abdominal segments sparsely or finely punctate .... *Megapenthes*
32′. Intermediate abdominal segments without striate impressions; abdominal segments with coarse punctures .... *Ectamenogonus (insignis)*
33(2). With a single stemma at base of each antenna .... *Cardiophorus*
33′. Without stemmata .... *Horistonotus*

Comments: In the following discussion the supergeneric classification of Hyslop (1917) has been used, mainly because of its simplicity and the fact that it is based primarily on larvae. The following subfamilies and tribes are recognized (some renamed based on current literature):

Pyrophorinae
- Pyrophorini
- Pityobiini
- Athoini (= Lepturoidini of Hyslop)
- Oestodini (including *Drapetes*)

Elaterinae
- Elaterini (= Steatoderini of Hyslop)
- Agriotini
- Ampedini (= Elaterini of Hyslop)
- Physorhinini
- Melanotini

Cardiophorinae

There are 844 species of Elateridae in N. America (Arnett 1968). The larvae of fewer than 10% of these are known and most of these are in the larger genera, such as *Conoderus, Limonius, Ctenicera, Athous, Ampedus,* and *Melanotus*. On the other hand, there are some genera in N. America in which the larvae remain unknown, including:

Pyrophorinae—*Lanelater, Meristhus, Oistus, Bladus, Perissarthron, Oedostethus, Anthracopteryx, Melanactes.*[2]

Elaterinae—*Diplostethus, Oxygonus, Anchastus, Drasterius? debilis, Blauta, Physorhinus.*

Cardiophorinae—*Coptostethus, Aptopus, Eniconyx,*[3] *Aphricus,*[3] *Esthesopus.*

Mature larvae are readily identifiable to subfamily. Larvae of Pyrophorinae are somewhat depressed and not too heavily sclerotized, have the ninth abdominal segment bifurcate (sometimes the first instar lacks the bifurcation), and have the pleural areas membranous and visible.

Larvae of Elaterinae are cylindrical and heavily sclerotized; have the ninth abdominal segment rounded or pointed, never bifurcate, and have the pleural areas decidedly reduced and concealed.

Larvae of Cardiophorinae are cylindrical and membranous, have each abdominal segment subdivided so it appears to be triple-segmented (fig. 34.438a), have deeply cleft mandibles (fig. 34.438b), and have anal papillae (these last three characters are unique among the Elateridae).

2. The phosphorescent larva generally attributed to *Melanactes piceus* (*see* Blatchley 1910) is certainly not in this genus and probably is not even an elaterid.

3. *Eniconyx* and *Aphricus* have been placed in the plastocerids, but they belong in Cardiophorinae and are allied to *Aptopus*.

Within the **Pyrophorinae** there are four tribes. The larvae of Pyrophorini (*Lacon, Alaus, Chalcolepidius, Hemirhipus, Pyrophorus, Conoderus, Aeolus,* etc.) have the submentum triangular (fig. 34.441d), the mandible without teeth on the inner surface, and the tenth abdominal segment armed with teeth (fig. 34.441c). Larvae of Pityobiini (*Pityobius*) also have the submentum triangular and armature on the tenth abdominal segment, but the mandible has 3 teeth on the inner surface. Larvae of Athoini (*Athous, Denticollis, Limonius, Hypolithus, Negastrius, Ctenicera, Hemicrepidius,* etc.) have the submentum rectangular, the mandible with a tooth on the inner surface, and lack armature on the tenth abdominal segment. Larvae of Oestodini (*Oestodes, Drapetes*) are very distinctive with a pair of heavy downwardly or inwardly curved spines on the ninth abdominal segment (fig. 34.436).

Within the **Elaterinae** there are five tribes. Larvae of Elaterini (*Sericus, Neotrichophorus, Orthostethus,* etc.) have the ninth abdominal segment smoothly ellipsoidal and a small tenth segment (fig. 34.444). Larvae of Agriotini (*Agriotes, Dalopius,* etc.) have the nasale tridentate and the ninth abdominal segment more or less bluntly pointed (fig. 34.443). Larvae of Ampedini (*Ampedus, Megapenthes, Dicrepidius, Dipropus,* etc.) usually have the nasale unidentate, the ninth abdominal segment terminating in a short but conspicuous spine (fig. 34.448), and have conspicuous transverse muscular impressions on the abdominal tergites. Larvae of Physorhinini (*Anchastus, Physorhinus,* etc.) are not known, but should fall within this subfamily. Larvae of Melanotini (*Melanotus*) are very distinctive because the ninth abdominal segment is flattened dorsoventrally and has a trilobed apex (fig. 34.445); they may also have conspicuous muscular impressions on the abdominal tergites.

The larvae of **Cardiophorinae** (*Cardiophorus, Horistonotus,* etc.) are the most distinctive subfamily with the apparent triple segmentation of the abdominal segments, cleft mandibles, anal papillae, etc.

Some wireworms are predacious, although no studies have been made to determine how beneficial they may be. At least one species has been used as a biological control agent in the tropics. On the other hand, several plant feeding species are quite destructive, particularly in irrigated land in western areas. Most pest species are in the genera *Limonius, Ctenicera, Conoderus, Hypolithus, Melanotus,* and *Agriotes*. Most pest wireworms feed on roots, especially those of newly germinating seeds, and they can destroy areas of newly planted fields; others tunnel into potatoes, carrots, etc. Riley and Keaster (1981) provide a pictorial field key to wireworms attacking corn in the area east of the 104th meridian (= the North Dakota/Montana border).

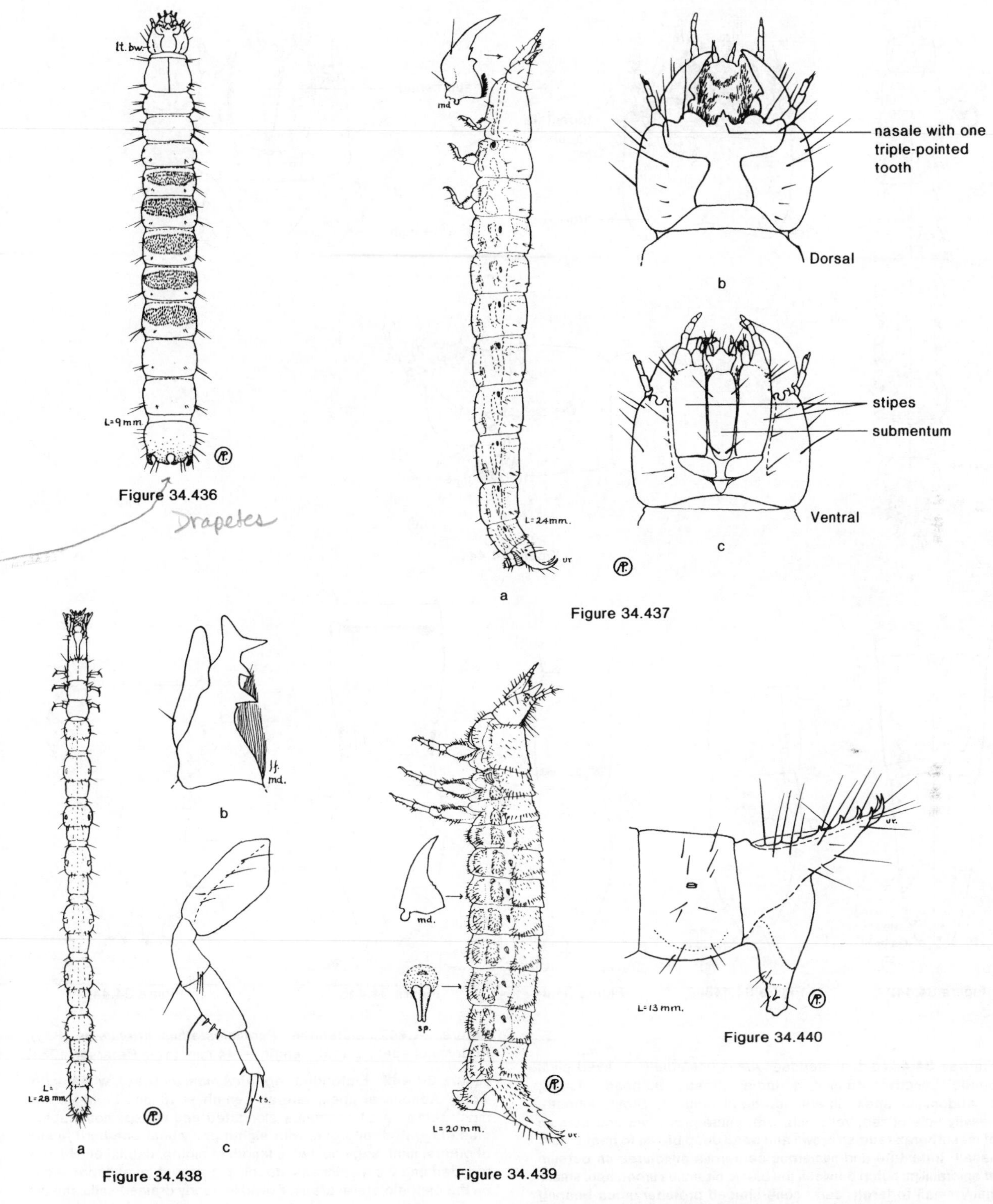

Figure 34.436

Figure 34.437

Figure 34.438

Figure 34.439

Figure 34.440

Figure 34.436. Elateridae. *Drapetes* sp. Maryland. Larva, dorsal. Length = 9 mm. (from Peterson 1951).

Figures 34.437a-c. Elateridae. *Ctenicera* sp. Length = 24 mm. **a.** larva, lateral; **b.** head, dorsal; **c.** head, ventral. (from Peterson 1951)

Figures 34.438a-c. Elateridae. *Cardiophorus* sp. Length = 28 mm. **a.** larva, dorsal; **b.** left mandible, dorsal; **c.** mesothoracic leg. (from Peterson 1951)

Figure 34.439. Elateridae. *Hemirhipis fascicularis* (Fab.). Alabama. Larva, lateral. Length = 20 mm. (from Peterson 1951).

Figure 34.440. Elateridae. *Aeolus mellillus* (Say). Abdominal apex, lateral. (from Peterson 1951).

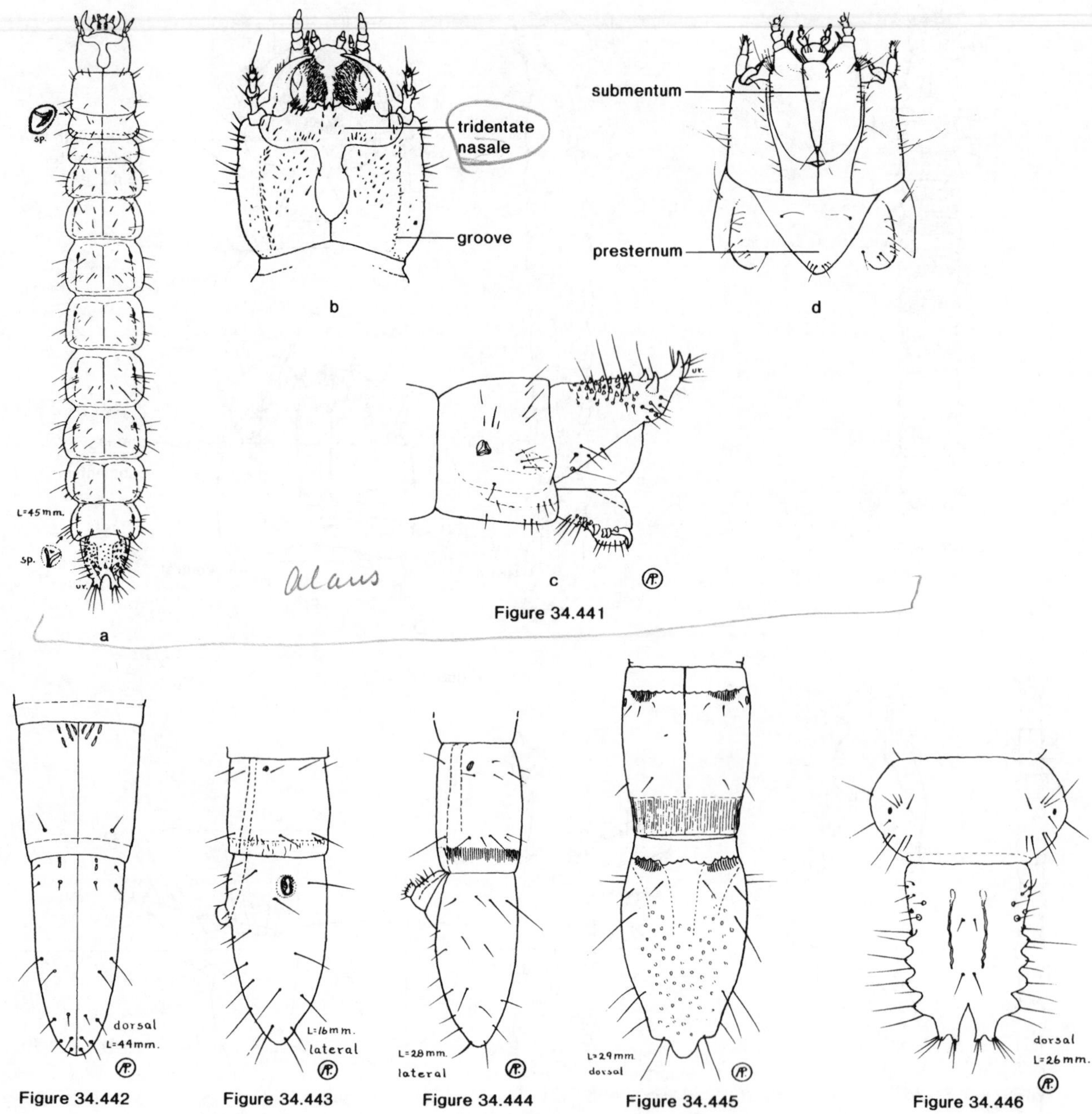

Figures 34.441a-d. Elateridae. *Alaus oculatus* (L.), eyed click beetle Length = 45 mm. **a.** larva, dorsal; **b.** head, dorsal; **c.** abdominal apex, lateral; **d.** head, ventral. Stout, smooth, heavily sclerotized, yellowish, with entire prothorax and dorsum of mesothorax reddish brown and head deep brown to near black; nasale tridentate and numerous setiferous punctures on dorsum of epicranium; notch between the black, biramous urogomphi small; many small to large, dark, cone-shaped protuberances bearing setae on dorsum and lateral aspects of ninth segment; tenth segment with two anal hooks caudolaterad of the anus and 10 to 12 asperities and setae cephalolaterad of the anus; spiracles biforous, V-shaped, located on the lateroventral aspect of the mesothorax and the dorsolateral aspects of abdominal segments 1 to 8. Lives in decayed wood, especially moist stumps, predacious on other insects. (from Peterson 1951, except d, drawn by Peter Carrington).

Figure 34.442. Elateridae. *Parallelostethus attenuatus* (Say). Abdominal apex, dorsal. Length = 44 mm. (from Peterson 1951)

Figure 34.443. Elateridae. *Agriotes mancus* (Say), wheat wireworm. Abdominal apex, lateral. Length = 16 mm. Light yellow; most setae on all segments elongated and conspicuous; spiracles elongated, biforous, with eighth pair about one-third length of others; ninth segment twice length of eighth, caudal end bluntly rounded and 2 conspicuous, deeply pigmented, oval depressions on the cephalolateral areas. Found in poorly drained soils and old meadows. Glen, et al., 1943. (from Peterson 1951)

Figure 34.444. Elateridae. *Sericus* sp. Michigan. Abdominal apex, lateral. Length = 28 mm. (from Peterson 1951)

Figure 34.445. Elateridae. *Melanotus communis* (Gyllenhal). Abdominal apex, dorsal. Length = 29 mm. (from Peterson 1951)

Figure 34.446. Elateridae. *Conoderus lividus* (De Geer). Abdominal apex, dorsal. Length = 26 mm. (from Peterson 1951)

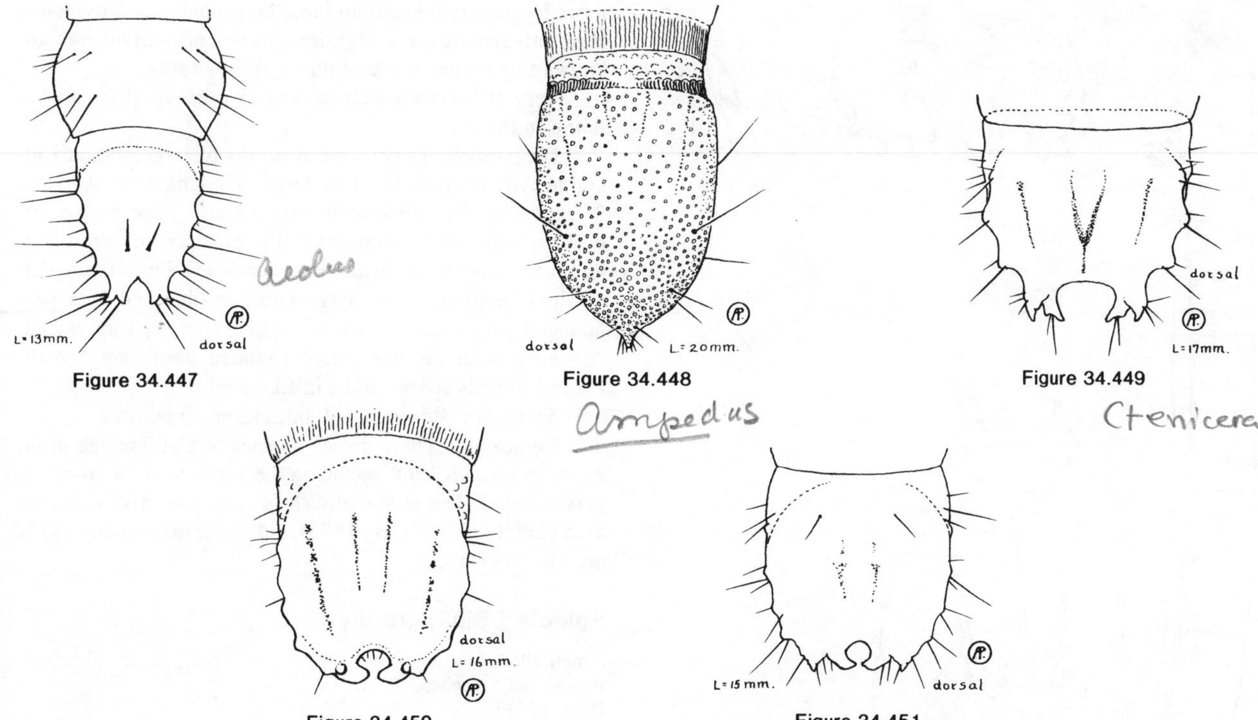

Figure 34.447 Figure 34.448 Figure 34.449

Figure 34.450 Figure 34.451

Figure 34.447. Elateridae. *Aeolus mellillus*. Abdominal apex, dorsal. Length = 13 mm. (from Peterson 1951)

Figure 34.448. Elateridae. *Ampedus* sp. Abdominal apex, dorsal. Length = 20 mm. (from Peterson 1951)

Figure 34.449. Elateridae. *Ctenicera aeripennis destructor* (Brown), prairie grain wireworm. Abdominal apex, dorsal. Length = 17 mm. Larva brownish-yellow, shiny and nearly smooth; stemmata (eye spots) present; nasale unidentate with a row of 8 or 9 short, sharp, subnasal (ventral) processes (may be dulled by wear); ninth abdominal segment longer than eighth, flattened on dorsum, tapering caudad, caudal notch wide, V-shaped urogomphi bifurcate with inner prong larger, more ventrad and distally turning inward beyond the knob on its lateral margin, outer prong shorter and projecting caudodorsolaterad; tenth abdominal segment tubular, moderate size with 2 whorls of 10 short setae and no anal armature. A soil dweller and pest of wheat. (from Peterson 1951)

Figure 34.450. Elateridae. *Limonius agonus* (Say), eastern field wireworm. Abdominal apex, dorsal. Length = 16 mm. Larva reddish-brown; head depressed, oval and no stemmata (eye spots); nasale tridentate with median prong largest; ninth segment half again as long as eighth and nearly as broad, flattened on dorsum and terminating in rounded knob-like urogomphi which nearly meet on the meson and possess a distinct, rounded dorsal projection on the basal portion of each urogomphus. Found in soil, injurious to roots and tubers of vegetables. (from Peterson 1951)

Figure 34.451. Elateridae. *Limonius californicus* (Mannerheim), sugar-beet wireworm. Abdominal apex, dorsal. Length = 15 mm. Larva reddish-brown with lighter, wide, caudal bands on each segment and areas between sclerites on lateroventral aspects; head somewhat depressed; no stemmata (eye spots); nasale prominent, terminating in 3 subequal prongs; ninth segment subequal in length and width to eighth; urogomphi projecting mesad, almost completing a circle, each with a prominent, horny protuberance at its base projecting caudodorsad. A pest of sugar beets in western states. (from Peterson 1951)

Selected References

Arnett 1968.
Blatchley 1910.
Böving and Craighead 1931.
Burakowski 1973, 1976.
Costa 1971, 1977.
Crowson 1961b.
Dietrich 1945.
Dogger 1959.
Dolin 1978.
Emden, F. I. 1945.
Emden, H. F. 1956.
Gaedike 1969, 1975, 1979.
Gardner 1936c.
Glen 1950.
Glen et al. 1943.
Gur'yeva 1969.
Hawkins 1936.
Hyslop 1917.
Jewett 1946.
Lanchester 1939, 1946.
McDougall 1934.
Ôhira 1962.
Peterson 1951.
Riley and Keaster 1979, 1981? (no date on publication).
Rudolph 1974, 1982.
Stibick 1979.
Thomas 1940.
Zacharuk 1962a, 1962b.

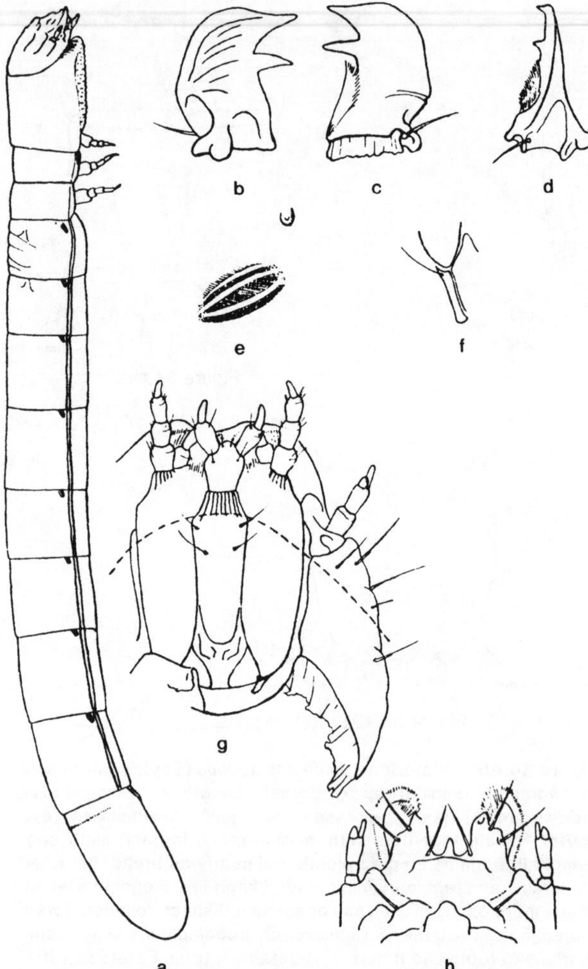

Figures 34.452a-h. Cebrionidae. *Aplastus speratus* LeConte. **a.** larva, lateral; **b.** right mandible, ventral; **c.** right mandible, dorsal; **d.** right mandible, lateral; **e.** abdominal spiracle; **f.** cardo; **g.** head, ventral; **h.** anteromesal portion of head, dorsal view, showing nasale. (From Hyslop, 1923.)

CEBRIONIDAE (ELATEROIDEA)

(INCLUDING PLASTOCERIDAE OF AUTHORS, NOT CROWSON)

Edward C. Becker, *Biosystematics Research Centre, Agriculture Canada*

Figures 34.452a-h

Relationships and Diagnosis: Currently *Selonodon* (= *Cebrio* of American authors), *Scaptolenus*, and the so-called plastocerids (*Aplastus, Octinodes,* and *Euthysanius*) belong to the Cebrionidae. The larvae are readily recognized by the elongate prothorax, which is longer than the other 2 thoracic segments combined, and the enlarged and eversible cervical membrane on the prosternum.

Biology and Ecology: All species occur in the southern areas, particularly in the southwestern states. Males are much more frequently taken than females, probably because most, if not all females, are flightless. Males of *Scaptolenus* are known to emerge *en masse* during heavy rains.

Very little is known about the larvae except that they are found in the soil.

Description: Larvae are medium to large, cylindrical, and heavily sclerotized. Head large, hypognathous. Antenna 3-segmented. Mandible with carinate outer edge. Stipes and postmentum mostly concealed by prosternum. Prothorax longer than meso- and metathorax combined. Prosternum with cervical membrane very large and eversible and with pronounced projection anteriorly. Legs rather well developed. Abdomen with ninth segment rounded, tenth very small. Pleural regions concealed as in Elaterini.

Spiracles: Biforous, without closing apparatus.

Comments: There are 46 species of Cebrionidae in N. America (Arnett 1968, including the 3 genera of the so-called plastocerids). The larvae of *Cebrio* from the Mediterranean area (Lefébure de Cerisy 1853) and *Aplastus* (Hyslop 1923) are also known.

Selected Bibliography

Arnett 1968.
Böving and Craighead 1931.
Hyslop 1923.
Lefébure de Cerisy 1853.

THROSCIDAE (ELATEROIDEA)

(= TRIXAGIDAE)

Edward C. Becker, *Biosystematics Research Centre, Agriculture Canada*

Figures 34.452x-z

Relationships and Diagnosis: The compact body of adult throscids is reminiscent of some buprestids, however, at least some throscids have the ability to click (Burakowski 1975b, Muona 1981). In N. America the family contains only 2 genera, *Trixagus* (= *Throscus*) and *Aulonothroscus*. Two other genera were formerly in the throscids; Crowson (1961b) transferred *Drapetes* to Elateridae, but left *Lissomus* in Throscidae, and Burakowski (1973) placed both genera in a separate family Lissomidae (*see* Elateridae). Larvae of Throscidae are small; lightly sclerotized; with reduced antennae, mandibles, and legs; and with sclerotized rods on the ventral part of the prothorax (fig. 34.452y).

Biology and Ecology: Little is known about the habits of adults in N. America. They have been taken at light, in Berlese samples, and occasionally during grain elevator inspections. In Europe adults are found on old oaks, alders, herbs of low growth, in sand and gravel pits, near entrances to rabbit burrows, etc. (Burakowski 1975b).

In Europe, according to Burakowski (1975b), larvae are found in various types of soil, but always near stumps of trees and usually just under the litter; they feed on the "external

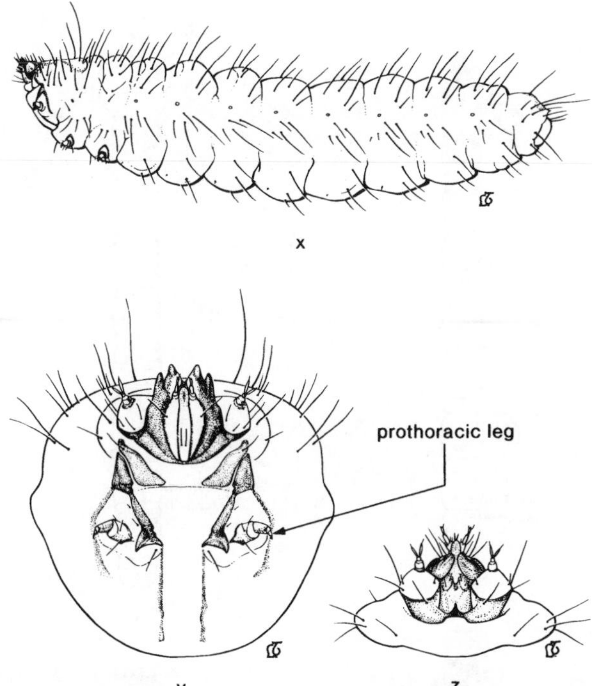

Figures 34.452x-z. Throscidae. *Trixagus* (possibly *Aulonothroscus*). x. larva, lateral; y. head and thorax, ventral; z. head, dorsal.

stratum of ectotrophic mycorrhizal roots of birch, beech, elm, and even pine." The specimen illustrated was collected in decaying pine boards in Fairfax Co., Virginia (U.S. Nat. Mus.).

Description: Larvae of *Trixagus* are very small, elongate, broadest near thorax, and slightly sclerotized.

Head: Small, depressed, prognathous. Antenna 2- or 3-segmented. Mandibles reduced, upturned and more or less flattened.

Thorax and Abdomen: Prothorax with paired sclerotized rods ventrally. Legs small, very short. Abdomen with 9 visible segments dorsally, segment 9 bearing a minute pair of fixed urogomphi, segment 10 reduced.

Spiracles: Biforous, with closing apparatus.

Comments: There are only 21 species of Throscidae in N. America (Arnett 1968, excluding *Drapetes*). Böving and Craighead's (1931) *Trixagus* (= *Throscus*) larva could easily be that of an *Aulonothroscus*, since there is very little difference in the adults of these genera.

Selected Bibliography

Arnett 1968.
Böving and Craighead 1931.
Burakowski 1973, 1975b.
Cobos 1961, 1966.
Crowson 1961b.
Muona 1981.

EUCNEMIDAE (ELATEROIDEA)

(= MELASIDAE, INCLUDING PEROTHOPIDAE, PHYLLOCERIDAE)

Edward C. Becker, *Biosystematics Research Centre, Agriculture Canada*

Figures 34.453-456

Relationships and Diagnosis: Eucnemid larvae are easily recognized by the lack of legs and by the mandibles curving outwardly (curved inwardly in all other Elateroidea).

Biology and Ecology: Adult eucnemids are rare in collections and little is known about their habits. Like the elaterids, they are active in late afternoon in spring and early summer and may be taken by sweeping or beating. All species have the clicking mechanism (prosternal spine and mesosternal cavity), but many can not be provoked into clicking. Muona (1981) gave a list of the species that have been reported to click. Most adults have cylindrical bodies and long antennae, which they use to help right themselves; therefore some may have "lost" the ability to click.

Larvae of eucnemids are found in decaying stumps or logs, usually deciduous ones, although Peterson (1951) noted that they were found in newly dead trees. The larvae bore across the grain of the wood. Only a few species are known in the larval stage.

Description: Larvae small to medium, lightly to deeply pigmented, and distinctly sclerotized. Two general types occur: (1) Body subcylindrical, lightly sclerotized, prothorax wider than other segments (similar to larvae of buprestids) with T-shaped, sclerotized rods dorsally or dorsally and ventrally, and mouthparts exposed dorsally (*Melasis, Isorhipis*); (2) Body somewhat depressed, distinctly sclerotized, prothorax about same width as rest of body, and mouthparts not exposed dorsally (*Fornax, Palaeoxenus*).

Head: Prognathous; depressed; highly specialized, lacking epicranial suture, frons, clypeus, labrum, and stemmata. Antenna vestigial or reduced to 2-segmented rudiment. Mandible very distinctive; curved and moving outwardly, sometimes fused with head or vestigial; frequently with 2 outwardly-curved teeth. Maxilla and labium nearly vestigial.

Thorax and Abdomen: Legless or with very small unsegmented protrusions. Thorax and abdomen frequently with sclerotized, rugose, median, oval areas dorsally and/or ventrally (presumably these assist in movement and those on prothorax may be used as an anchor for movement of mouthparts). Ninth abdominal segment bluntly rounded at apex, without cerci (except in *Palaeoxenus*).

Spiracles: Biforous, with closing apparatus.

Comments: There are 71 species of Eucnemidae in N. America (Arnett 1968). According to Gardner (1935b), larvae of the Indian species can not be arranged to conform with the adult classification. As with the elaterids, discovery of more larvae should help with the classification.

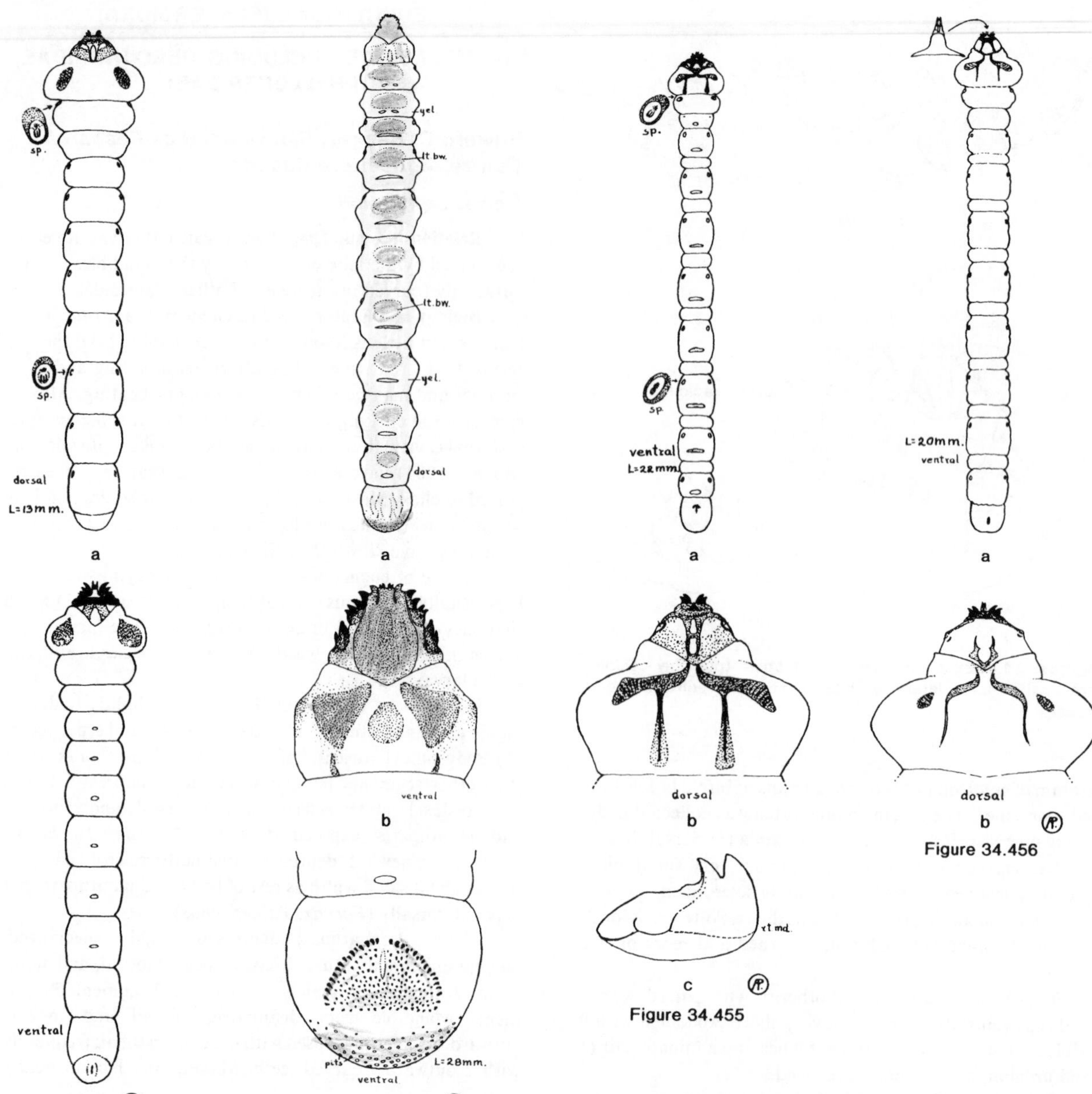

Figures 34.453a,b. Eucnemidae. *Hylochares nigricornis* (Say). Hocking Co., Ohio. Length = 13 mm. **a.** larva, dorsal; **b.** ventral. (from Peterson 1951)

Figures 34.454a-c. Eucnemidae. *Fornax* sp. Length = 30 mm. **a.** larva, dorsal; **b.** head and prothorax, ventral; **c.** abdominal apex, ventral. Larva yellowish, depressed, orthosomatic and legless; prothorax no wider than other segments but more depressed and with paired, deeply pigmented areas that are irregularly quadrangular on the tergum and triangular on the sternum; deep yellow to light brown circular areas and laterally elongate ovals located on the meson of the sternum and tergum of abdominal segments 1-8; caudal segment with numerous pigmented pits on distal margin and semi-circular group of asperities near center of the sternum. Under bark of fallen logs, also deep in decayed beech. (from Peterson 1951)

Figures 34.455a-c. Eucnemidae. *Melasis pectinicornis* Melsheimer. Ohio. Length = 22 mm. **a.** larva, ventral; **b.** head and prothorax, dorsal; **c.** right mandible, ventral. Larva orthosomatic, slender, legless, smooth, near white and hairless; head depressed, diamond-shaped, distinctly pigmented with a few setae on the non-sclerotized areas; vestigial antennae, no stemmata, labrum or maxillae; mandible with 2 prominent teeth that project cephalolaterad; labium reduced to 2 inconspicuous, lightly sclerotized palpi and a median pointed ligula. Each abdominal segment 2-8 with a narrow annulet and a broad annulet with the spiracles on the latter; sclerotized rings, circular or oval, located ventrally on the metathorax through the eighth abdominal segment, restricted to segments 6-8 dorsally, all spiracles inconspicuously annular-biforous and surrounded by sclerotized rings. Found under bark of fallen hardwood trees, including maple and black gum. (from Peterson 1951)

Figures 34.456a,b. Eucnemidae. *Isorhipis obliqua* (Say). Ohio. Length = 20 mm. **a.** larva, ventral; **b.** head and prothorax, dorsal. (from Peterson 1951)

Selected Bibliography

Arnett 1968.
Böving and Craighead 1931.
Cobos 1964.
Ford and Spilman 1979.
Gardner 1930b, 1935b, 1936c.
Ghilarov 1979.
Leiler 1976.
Lucht 1981.
Mamaev 1976b.
Muona 1981.
Palm 1959.

BRACHYPSECTRIDAE (CANTHAROIDEA)

John F. Lawrence,
Division of Entomology, CSIRO

Figures 34.457a-d

Relationships and Diagnosis: The position of the Brachypsectridae is still open to question; the group has been included in Dascilloidea, Elateroidea, and Cantharoidea by various authors, and it was combined with Artematopidae and Callirhipidae in a new superfamily, Artematopoidea, by Crowson (1973b). The family is here considered to be a primitive member of the Cantharoidea, following a more recent treatment (Kasap and Crowson, 1975), but it would make more sense phylogenetically to combine Elateroidea, Cantharoidea, Brachypsectridae and Artematopidae into a single superfamily (Lawrence and Newton, 1972). The unique larva of *Brachypsectra* was first discovered by Barber in 1905, but it was not associated with the adult until 25 years later (Blair, 1930). It may be distinguished by the broad, flattened form (like that of the Psephenidae and some Chrysomelidae-Hispinae), with branched processes along the sides and a covering of scale-like setae; also the 9th abdominal tergum is formed into a long spine, the antennae are club-like and apparently 2-segmented, and the mandibles are falciform and perforate.

Biology and Ecology: *Brachypsectra* larvae are slow-moving, ambush predators, which occur under bark, on the ground under pieces of wood, or in rock crevices in arid regions. Larvae remain motionless until a spider or other small arthropod (possibly attracted by the row of glands on the dorsal surface) walks onto the flattened body, when the flexible head and tail spine are abruptly raised, pinning the prey between the point of the spine and the piercing-sucking mandibles. Although spiders appear to be the normal prey, a larva has been observed with a hymenopteran, and others were successfully reared on small roaches in the laboratory. Prior to pupation, the larvae spin a coarsely woven, silk cocoon. Adults are very rare and are probably short-lived (Crowson, 1973b; Ferris, 1927).

Description: Mature larvae 6 to 10 mm. Body broadly ovate and moderately to strongly flattened. Dorsal surfaces yellowish with brown plates and maculae, covered with granules, tubercles, and modified scales.

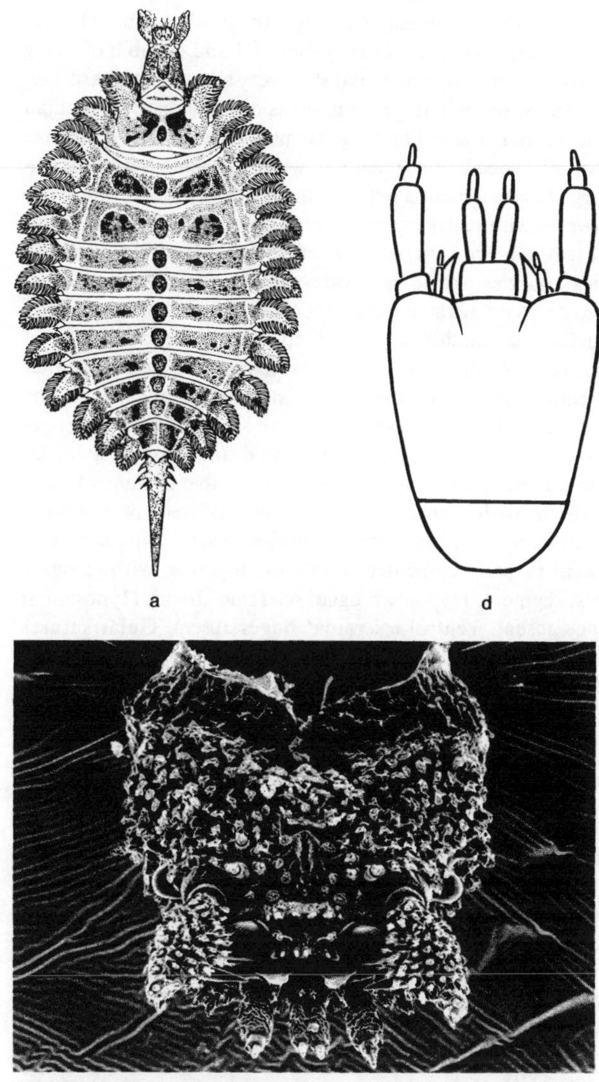

Figures 34.457a-d. Brachypsectridae. *Brachypsectra fulva* LeConte. Mexicali, Imperial Co., California. Length = 12.8 mm. **a.** larva, dorsal; **b.** head, anterodorsal; **c.** head, ventral; **d.** maxillo-labial complex, ventral.

Head: Protracted and prognathous, elongate and more or less parallel-sided, usually elevated and capable of being strongly elevated. Epicranial stem very short; frontal arms U-shaped or somewhat lyriform, occasionally indistinct. Median endocarina absent. One large stemma on each side. Antennae well-developed, 3-segmented, with segment 2 greatly enlarged and segment 3 reduced and not easily visible; antennae covered with scales like most of the head. Frontoclypeal suture absent; labrum free but reduced in size. Mandibles symmetrical, narrow and falcate, internally perforate, widely separated at base, without mola or accessory ventral process; mesal surface of mandibular base simple. Ventral mouthparts retracted. Maxillae with cardines completely fused together forming single, trapezoidal plate; stipes elongate; articulating area absent; palp 4-segmented, with segment 2 much longer than others; galea small, articulated and 2-segmented; lacinia fixed, falciform and not easily visible in ventral view. Labium with mentum and submentum fused; postmentum completely fused to stipites (sutures absent); ligula absent; labial palps 2-segmented, very close together, with elongate first segment. Hypopharyngeal sclerome absent. Hypostomal rods absent. Ventral epicranial ridges absent. Gular sutures fused; gular region (ventral head closure) longer than maxillolabial complex.

Thorax and Abdomen: Thoracic segments each with 2 pairs of lateral, branched processes, which are similar in size; abdominal segments 1–8 each with 2 pairs of lateral, branched processes, the anterior of which are highly reduced. Thoracic terga and abdominal terga 1–7 each with a median, elevated, gland-like structure. Legs well-developed, 5-segmented; tarsungulus with 2 setae lying side by side. Tergum A9 modified to form a long, tapered, acute spine, which articulates with tergum A8 and is capable of being flexed dorsally; base of spine with few pairs of lateral, branched processes; sternum A9 simple, located ventrally beneath spine; segment A10 not visible.

Spiracles: Thoracic spiracles biforous, located ventrally on 2 sclerites placed near anterior edge of mesothorax. Abdominal spiracles reduced and apparently annular, located dorsally on terga. Closing apparatus present.

Comments: The genus *Brachypsectra* contains 3 described species from southwestern N. America, southern India, and Malaysia, and an apparently undescribed species (known from larvae only) from northwestern Australia.

Selected Bibliography

Barber 1905 (larva of *Brachypsectra fulva* LeConte).
Blair 1930 (larvae of *B. fulva* and *B. lampyroides* Blair).
Böving and Craighead 1931 (larva of *B. fulva*).
Crowson 1973b (habits).
Ferris 1927 (habitat).
Kasap and Crowson 1975.
Lawrence and Newton 1982.

CNEOGLOSSIDAE (CANTHAROIDEA)

John F. Lawrence,
Division of Entomology, CSIRO

This family contains the genus *Cneoglossa* with 7 species from Central and South America. These are small, somewhat flattened beetles, with the head concealed beneath the prothorax. They were placed in Cantharoidea by Crowson (1972b). Nothing is known about the biology or immature stages.

Selected Bibliography

Crowson 1972b.

PLASTOCERIDAE (CANTHAROIDEA)

John F. Lawrence,
Division of Entomology, CSIRO

This family includes the genus *Plastocerus* (= *Ceroplastus*), with 2 species: *P. angulosus* (Germar) from Turkey and Asia Minor, and *P. thoracicus* Fleutiaux from Southeast Asia. Adults resemble various Elateridae, but they differ in having exposed prothoracic trochantins and free abdominal sternites; Crowson (1972b) has placed the group in Cantharoidea. Nothing is known of the biology of plastocerids and immature stages have not been described.

Selected Bibliography

Crowson 1972b.
Lawrence and Newton 1982.

HOMALISIDAE (CANTHAROIDEA)

John F. Lawrence,
Division of Entomology, CSIRO

This family contains the genus *Homalisus* with about 10 species known from southern and eastern Europe. The adult female has shortened elytra and no hindwings, but is in most other respects similar to the male; luminous organs are lacking in both sexes. The group is apparently related to Lycidae.

Larvae are elongate, narrow, and flattened, with well-sclerotized terga. The head is prognathous, with long, 3-segmented antennae, a long, narrow median lobe formed by the clypeus and fused labrum, a single, large stemma on each side, and a narrow, but distinct ventral closure. The mandibles are long, narrow and falcate, approximate at the base, and each has a deep longitudinal groove and transparent inner lamella. The mentum and stipites are closely joined but not fused, the cardines are indistinct, and the galea is 1-segmented. The legs are 5-segmented with a bisetose tarsungulus. Tergum A9 lacks urogomphi, and segment 10 is reduced and ventral. Luminous organs occur on the sides of the abdomen.

Larvae of *Homalisus* occur in leaf litter; they have been reported to feed on snails, but there is little evidence to support this (Magis, 1977).

Selected Bibliography

Bertkau 1891 (larva and female of *Homalisus fontisbellaquei* (Fourcroy), as *H. suturalis* (Olivier)).
Crowson 1972b (larval head of *H. fontisbellaquei*).
Magis 1977.
Ochs 1949 (female).

LYCIDAE (CANTHAROIDEA)

John F. Lawrence,
Division of Entomology, CSIRO

Net-Winged Beetles

Figures 34.458a,b

Relationships and Diagnosis: The Lycidae are a distinctive group of cantharoid beetles, whose larvae are easily distinguished by the unique mandibles, which are divided longitudinally forming 2 pairs of basally approximate, apically diverging, non-opposable, blade-like structures. Other distinguishing features include the 2-segmented antennae, absence of an epicranial suture, and complete or almost complete fusion of the maxillae with the labium. Closely related families include the Drilidae, Homalisidae, and Lampyridae.

Biology and Ecology: Although lycid larvae were once thought to be predaceous (Britton, 1970; Crowson, 1955), it has been fairly well established that they feed on soft or fluid material associated with rotting wood. Withycombe (1926) observed larvae of *Calopteron tropicum* (L.) (as *C. fasciatum* (Fabricius)) feeding on fermenting pulp under the bark of a legume (*Erythrina*) in Trinidad, while Mjöberg (1925) noted that Asian "trilobite" larvae (*Duliticola*) suck juices from wet pieces of wood and that the guts were filled with dark masses of "decayed woody products." Larvae of *Platycis sculptilis* (Say) burrow into relatively soft and wet, decaying wood, pushing through the spaces between fibers (rather than constructing a permanent tunnel by comminution of the wood) and "bleeding" the wood for the juices on which they feed (McCabe and Johnson, 1979a). McCabe and Johnson (1979b) offered various prey items to larvae of *Calopteron terminale* (Say), but the larvae were induced to feed only on aphid honeydew (although they avoided live aphids). An unidentified Mexican lycid larva was found feeding on a slime mold fruiting body at night, and other larvae in Arizona were found in old conifer wood penetrated by a bright yellow plasmodium (*Fuligo?* sp.). Since many slime molds are colorless in the plasmodial stage, it is possible that plasmodia represent the food source of other lycid species. The so-called "trilobite" larvae of southeast Asia represent a unique group of lycids in which the females are larviform and may be up to 80 mm in length. Adult lycids are short-lived and many are nectar feeders. They are usually active during the day

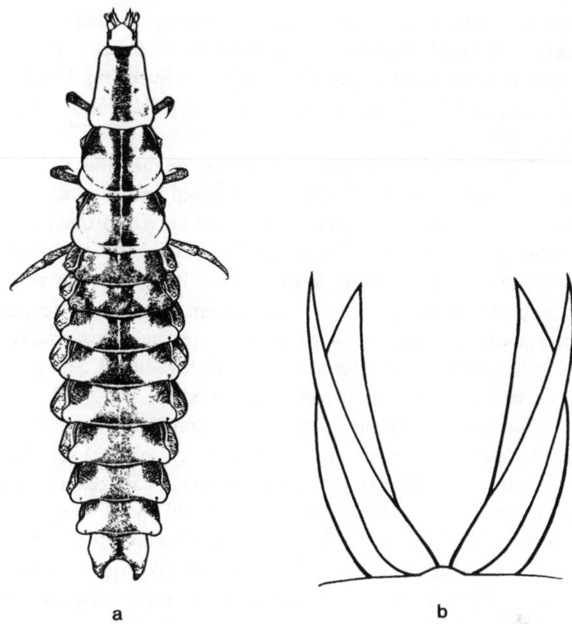

Figures 34.458a,b. Lycidae. *Calopteron* sp. 8 mi. NW Sontecopaman, Veracruz, Mexico. Length = 18.6 mm. **a.** larva, dorsal; **b.** mandibles, dorsal.

and may be abundant on flowers, but some species are also attracted to lights. Most species are aposematically colored and may be involved in mimicry complexes.

Description: Mature larvae 6 to 25 mm (with some exotic forms reaching 80 mm), usually 15 mm or less. Body elongate and parallel-sided or wider in middle and tapering anteriorly and posteriorly, slightly to strongly flattened. Dorsal or both dorsal and ventral surfaces often heavily sclerotized and variously colored, usually smooth (occasionally tuberculate) and glabrous or with scattered, short, simple setae.

Head: Prognathous, usually protracted, short and broad, and wider anteriorly; occasionally elongate and retracted; usually somewhat flattened. Epicranial stem and frontal arms absent. Median endocarina absent. Stemmata absent, or with single stemma on each side. Antennae short to moderately long, 2-segmented, with dome-like, variously shaped, sensorium at apex of segment 2. Frontoclypeal suture absent; labrum completely fused to head capsule. Mandibles symmetrical, narrow and falcate, without mola or accessory ventral process, approximate at base and diverging apically, divided longitudinally into 2 parts, one of which may be stylet-like; mesal surface of mandibular base simple, without brush of hairs. Ventral mouthparts retracted. Maxilla with cardo distinct or fused to stipes, no articulating area, subquadrate or elongate stipes, and mala closely associated with palpifer, both of which are separated from stipes by membranous area; palp 3-segmented or apparently 4-segmented, due to enlarged palpifer; mala fixed or articulated, and acute, rounded, or truncate at apex. Labium with mentum and submentum fused; postmentum completely or almost completely fused to stipites; ligula absent; labial palps 2-segmented, usually narrowly separated. Hypopharyngeal sclerome absent. Hypostomal rods long and subparallel to slightly diverging. Ventral epicranial ridges absent. Gular region very short and indistinct.

Thorax and Abdomen: Thoracic segments and abdominal segments 1–8 often with paired tergal and pleural processes, in the form of flat plates, knobs, or narrow projections, extending laterally and dorsolaterally. Legs well-developed, 5-segmented; tarsungulus with 2 setae lying side by side. Tergum A9 usually without urogomphi, sometimes deeply emarginate at apex or with pair of distinct, posteriorly-projecting urogomphi; sternum A9 simple, reduced. Segment A10 short and circular, ventrally oriented.

Spiracles: Biforous. Abdominal spiracles located in upper pleural sclerite or upper part of single pleural sclerite.

Comments: The family contains about 150 genera and 3500 species worldwide, with 17 genera and about 75 species occurring in N. America. The group has been divided into a number of tribes based on adult features (Bourgeois, 1891; Kleine, 1933), but larval characters have rarely been used in higher classification.

Selected Bibliography

Böving and Craighead 1931 (larvae of *Caeniella dimidiata* (Fabricius), *Calopteron reticulatum* (Fabricius), and *Eros humeralis* (Fabricius)).
Bourgeois 1891.
Britton 1970.
Bugnion 1907 (larva of *Ditonectes pubicornis* (Walker)).
Crowson 1955, 1972b.
Gardner 1946 (larvae of *Calochromus* spp., *Lycostomus similis* (Hope), *Plateros* sp.).
Gravely 1915 (larva of *Lypropaeus biguttatus* Gorham).
Kleine 1933.
Korschefsky 1951 (larvae of *Dictyopterus* spp. and *Lygistopterus sanguineus* (Linnaeus)).
McCabe and Johnson 1979a (larva of *Platycis sculptilis* (Say)), 1979b (larva of *Calopteron terminale* (Say)).
Mjöberg 1925 (larva and female of *Duliticola paradoxa* Mjöberg).
Peterson 1951 (larva of *Platycis?* sp., as *Calopteron* sp.; *see* McCabe and Johnson, 1979a).
Rosenberg 1943 (larvae of *Lycostomus sanguineus* (Fabricius), *Metriorrhynchus?* sp.; also several "trilobite" larvae).
Withycombe 1926 (feeding habits of *Calopteron tropicum* (Linnaeus)).

DRILIDAE (CANTHAROIDEA)

John F. Lawrence,
Division of Entomology, CSIRO

This family, as redefined by Crowson (1972b) includes only *Drilus, Selasia, Malacogaster,* and a few related genera, with about 80 species occurring in southern Eurasia, Asia Minor, Africa, and India. The group is thought to be related to Phengodidae and Lampyridae.

Larvae are elongate and somewhat flattened, strongly setose, except for the last instar, and usually with well-developed lateral tergal and pleural processes on most abdominal segments and a pair of fixed urogomphi on tergum 9. The head lacks an epicranial suture and has a narrow but distinct ventral closure (gular region); antennae are 3-segmented, the labrum is fused to the head capsule, and there is a single large stemma on each side. The mandibles are falcate, incompletely perforate and widely separated at the base. The maxillae and labium are closely joined but not fused, the cardo is indistinct, and the galea is 1-segmented. The legs are 5-segmented and the tarsungulus bears 2 long and spatulate setae. Spiracles are biforous and located in the membrane between terga and pleura. Luminous organs are absent.

Drilid adults are short-lived and apparently diurnal in habits. The females have the head and legs of an adult, but the thorax and abdomen are larviform, without elytra or wings. Larvae are predators of gastropods; each of the three larval instars feeds on a different individual snail, and pupation takes place within the snail shell.

Selected Bibliography

Balduf 1935.
Barker 1969 (larva and life cyle of *Selasia unicolor* (Guérin)).
Böving and Craighead 1931 (larvae of *Selasia* sp. and *Drilus concolor* Ahrens).
Cros 1925 (habits of *Drilus mauritanicus* Lucas and *Malacogaster passerinii* Bassi), 1926 (habits of *D. mauritanicus*), 1930 (habits of *M. passerinii*).
Crowson 1972b.
Korschefsky 1951 (larvae of *Drilus* spp.).
Rosenberg 1909 (larva and feeding habits of *Drilus concolor*).

PHENGODIDAE (CANTHAROIDEA) (INCLUDING RHAGOPHTHALMIDAE)

Laurent LeSage,
Biosystematics Research Centre, Ottawa

Glowworms
Figures 34.459–460

Relationships and Diagnosis: The Phengodidae or "glowworms", allied to the Lampyridae and Drilidae (Crowson 1972b), is a group of luminescent beetles with larviform females. These insects are found in both North and South America and are usually not abundant in any locality (Crowson, 1972b, also includes the Asian Rhagophthalminae). Their general appearance, the luminescence of the body, and the channelled or grooved mandibles will separate the Lampyridae-Phengodidae larvae from all other beetle larvae (some elaterid larvae have luminescence but no channelled mandibles). The families are distinguished readily by the epicranial suture which is absent in the Phengodidae but well-developed in Lampyridae, and by the retinaculum of the mandible which is present only in Lampyridae.

Biology and Ecology: The early observations on the ecology of glowworms were reported in short notes (Murray 1870; Riley 1886a, 1886b, 1887; Barber 1905; Dahlgren 1917). The excellent photographs and original observations of Tiemann (1967) on *Zarhipis integripennis* (LeConte) constitute the first complete account of the natural history of phengodid beetles, and are summarized here.

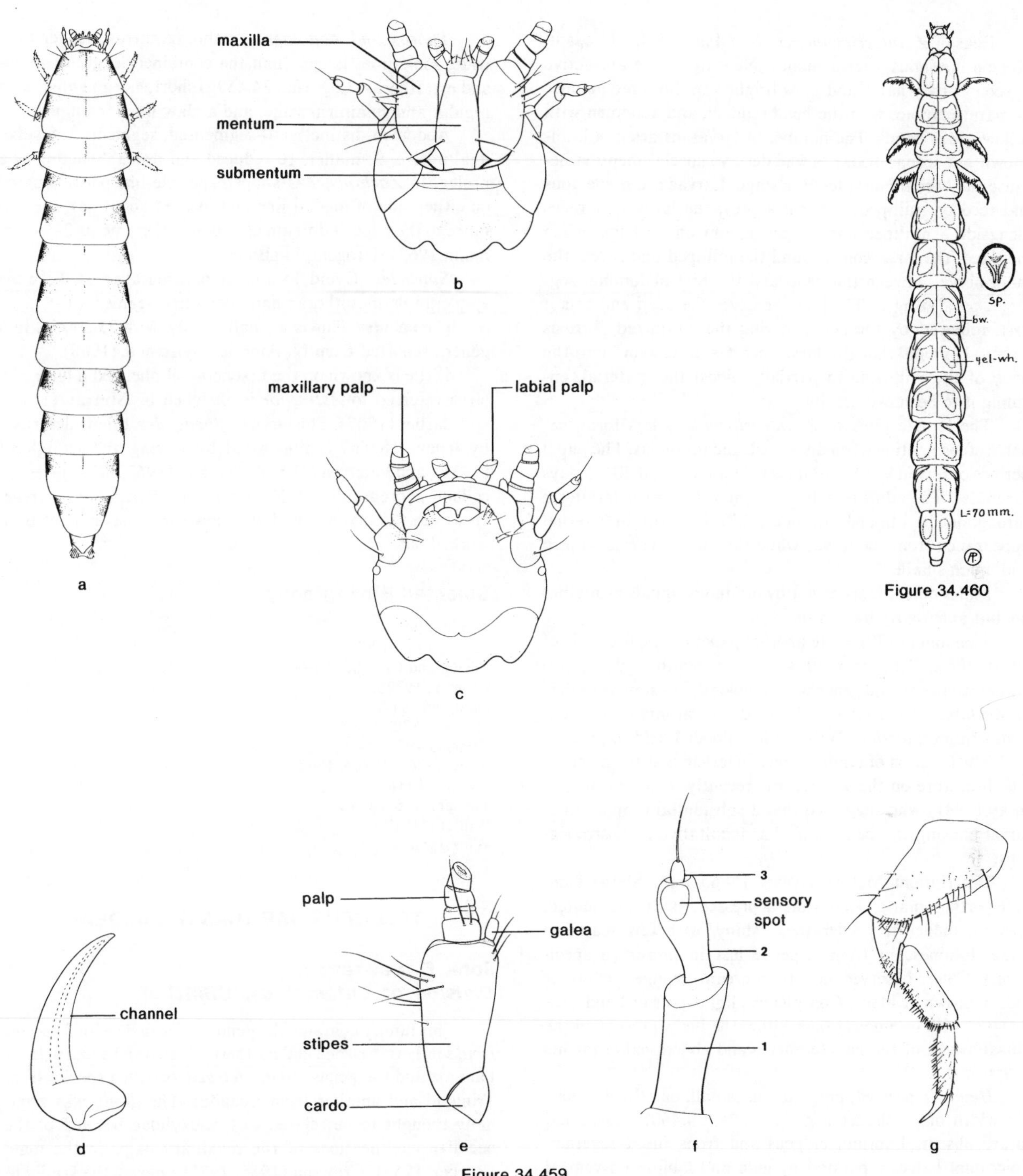

Figures 34.459a-g. Phengodidae. *Phengodes* sp. Length = 62 mm. **a.** larva, dorsal; **b.** head, ventral; **c.** head, dorsal; **d.** mandible; **e.** maxilla; **f.** antenna; **g.** midleg.

Figure 34.460. Phengodidae. *Phengodes laticollis* LeConte. Length = 65-70 mm. Elongate, subcylindrical, venter flattened; cream to lemon yellow with dorsum of all segments dark brown, each possessing on each lateral half a large yellow spot; head small, prognathous without epicranial suture, clypeus or labrum; one pair of stemmata; setae absent except a few on head, many on legs and a few on ventral aspect and caudal end; spiracles bilabiate or a modified biforous type with heavy annular ring about each. Larvae are luminescent. Found on or in ground under debris in woodland areas or in garden soil. (From Peterson, 1951.)

Eggs of *Z. integripennis* are round and oval, white, and become luminous after a month. Neonate larvae are active as soon as they hatch and glow brightly in the same manner as mature larvae, with the head reddish, and abdomen with yellow-green bands. The number of larval instars is still unknown but second instar larvae deserve special mention because they are the first feeding stage. Larvae are predacious and feed on millipeds. To kill a prey, the larva first races alongside a milliped until it can mount on its back; when mounted, the larva coils around the milliped and severs the main nerve between the head and the first abdominal segment, paralyzing it. The larva removes the head and eats it first, followed by the body, leaving the chitinized portions empty. It seems that the larva secretes an enzyme into the body of the millipede to partially digest the material, enabling it to be sucked easily.

The female pupa of *Z. integripennis* is larviform, capable of locomotion when disturbed, and luminous. The pupal periods observed last 12–13 days for females, and 20–35 days for males. The adult female of *Z. integripennis* is larviform throughout its life and luminous, differing little in external appearance from the larva, while the adult male is winged and much smaller.

Males detect females not by luminous signals as fireflies do, but apparently by a faint scent.

Tiemann (1970) wrote another paper on the life cycle of *Phrixothrix*, "the railroad worm", a South and Central American phengodid genus with a biology very similar to that of *Zarhipis*. Wing (1984) observed several larvae of *Phengodes ?nigromaculata* Wittmer in a flooded field of grass.

The function of luminescence in larvae is still a mystery. The literature on the subject was recently reviewed by Sivinsky (1981) who suggested that a substantial proportion of larval glowing can be explained as facultative defensive signals.

Description: Mature larvae, 15–65 mm. Shape elongate, orthosomatic, somewhat depressed. Body pigmented, smooth, moderately sclerotized, shiny, with few scattered setae, luminescent (luminescence lost in preserved specimens). Colors observed vary from cream, orange, yellow, or red, to brown or black. Conspicuous light spots or bands, associated with luminous tissue within the body, present on the dorsal aspect of thoracic segments and abdominal segments 1–9.

Head: Depressed, prognathous, small, one third to half the width of prothorax (figs. 34.459a, 34.460). Epicranial suture absent. Labrum, clypeus and frons fused together. Epicranial halves separated by gula and labium on ventral aspect of the head. Antenna (fig. 34.459f) conspicuous, 3-segmented, third segment short in *Phengodes*, located close to the base of the mandibles. One pair of stemmata on each side of the head. Mandible (fig. 34.459d) robust, falciform, and channelled from the tip to the base; mola and retinaculum absent. Maxilla (fig. 34.459e) with distinct stipes and cardo; maxillary palp 3- or 4-segmented; galea appendage-like. Labium with small submentum, distinct prementum and ligula; labial palp short, 2-segmented.

Thorax and Abdomen: Prothorax narrow, longer than wide, sometimes longer than the combined length of meso- and metathorax. Legs (fig. 34.459g) short, 4-segmented, with regular and spiniform setae, and a claw-like tarsungulus.

Abdomen distinctly 10-segmented, segments 1–8 subequal in size, 9 smaller, 10 reduced and probably acting as a proleg. In *Zarhipis*, 2 U-shaped spiracle-like pores situated on either side of medial line and located about half the distance to the edge of dorsum of abdominal segments 2–9 (Tiemann 1967). Urogomphi absent.

Spiracles: Ovoid, located on parascutal areas above the epipleura on mesothorax and abdominal segments 1–8.

Comments: This is a small family, with 23 species in 6 genera reported from N. America by Arnett (1968).

Little is known on the taxonomy of phengodid larvae. A larva referred to *Astraptor* is sketched by Murray (1870), and Barber (1907). The larva of *Phengodes*, briefly described by Riley (1886a) is illustrated by Böving and Craighead (1931), and Peterson (1951). Tiemann (1967, 1970) gives excellent photographs of *Zarhipis* and *Phrixothrix* larvae. However, the taxonomy of the glowworms has not yet been worked out.

Selected Bibliography

Arnett 1968.
Barber 1905, 1907.
Böving and Craighead 1931.
Crowson 1972b.
Dahlgren 1917.
Murray 1870.
Peterson 1951.
Riley, 1886a, 1886b, 1887.
Sivinsky 1981.
Tiemann 1967, 1970.
Williams 1917.
Wing 1984.

TELEGEUSIDAE (CANTHAROIDEA)

John F. Lawrence,
Division of Entomology, CSIRO

This family contains the genus *Telegeusis* with 3 species from southern Arizona and northern Mexico and a fourth from Panama and the genus *Pseudotelegeusis* with a species from Trinidad and another from Ecuador. The group was originally thought to be related to Lymexylidae because of the peculiar modifications of the maxillary palps in the male (Barber, 1952). Crowson (1955, 1972b) moved the family to Cantharoidea and placed it next to the Phengodidae. The female and larva are unknown.

Selected Bibliography

Allen and Hutton 1969 (key to species).
Barber 1952.
Crowson 1955, 1972b.
Wittmer 1976 (*Pseudotelegeusis*).

LAMPYRIDAE (CANTHAROIDEA)

Donna M. LaBella and J. E. Lloyd,
University of Florida

Fireflies, Lightningbugs

Figures 34.461–464

Relationships and Diagnosis: The Lampyridae are members of the Cantharoidea as defined by Crowson (1972b). Larval characteristics of this superfamily include: falcate mandibles which may be cleft longitudinally or channelled, and which lack a molar region; a reduction in the articulated area of the maxillae; an ill-defined labrum which may be contained in a nasale; a pygopod which aids in locomotion; and legs that are 5-jointed with tarsus and claw fused into a tarsungulus (Böving and Craighead 1931, Gardner 1946).

The lampyrids can be distinguished from other cantharoids by the presence of an epicranial suture which is absent in similar species, and a photogenic organ which is usually situated on the venter of abdominal segment 8. *Pterotus obscuripennis,* a California species, has its photogenic organ situated on abdominal segments 7 and 8 and can easily be mistaken for a phengodid because of its subparallel-shaped body. However, N. American phengodids have paired luminous organs on abdominal segments 1–8, and lack a retinaculum which characterizes the channelled mandible of the Lampyridae. Lycids can be distinguished by having longitudinally split mandibles and no epicranial suture. The cantharids differ from all of these by having the head fused ventrally behind the mouthparts, and by a velvety pubescence.

Biology and Ecology: Lampyrid larvae are generally found in damp places, but especially in the western United States a number of dry-habitat species occur. Larvae are found under stones, in ground depressions, under leaf litter, in vegetation along streams and ponds, in rotten logs, and on floating and emergent vegetation. Though larvae are able to withstand temporary dryness by burrowing into the soil, high microenvironmental humidity is essential for life (McClean et al. 1972). Some, such as *Photinus,* spend most of their time underground. Others are terrestrial (*Photuris*) or semiaquatic (*Pyractomena*). All are predaceous, feeding on snails and other soft-bodied forms—*Photinus* may specialize on earthworms. Larvae probably find prey via chemical clues. Schwalb (1960) reported that the European glowworm *Lampyris noctiluca* follows old slime trails to attack the snails that made them. Upon contacting a prey, larvae use their falcate, hollowed mandibles to pierce the prey and inject a poison that stuns it (Balduf 1935, Schwalb 1960). A strong case has been made for a facultative warning function of larval luminescence (Sivinski 1981). Larvae undergo 4 to 5 molts, and usually take from a few months to 2 years to reach pupation (Williams 1917, Hess 1920, Buschman 1977, 1984, 1988).

Pupae are exarate. Some species pupate underground in an excavated cell, others in a covered excavation at the surface, and others in natural cavities in dead logs. Pupae are generally milky white with a little yellow or pink pigmentation, and darken before eclosion. However, *Pyractomena* pupae hang like Lepidoptera chrysalids from vegetation, probably as a result of their semi- or transiently-aquatic habitats. Most pupae luminesce from larval light organs when mechanically stimulated. Pupal duration varies from about a week (*Pyropyga, Pyractomena*) to 2, 3, or more (*Photuris*) (Williams 1917, Hess 1920, Buschman 1977, 1984, 1988).

Eggs are usually smooth, spherical, and may be dull creamy white, pale yellow, or colorless. They are deposited singly or in groups, in or on the ground or at the base of grasses (*Pyractomena*). The incubation period lasts from 13 to 27 days (Williams 1917, Buschman 1977, 1984, 1988).

Though most adults are typical soft-winged cantharoid beetles, females of some species are short-winged (brachypterous), wingless (apterous), or larviform (*Pleotomus*). Many species are active at night and use bioluminescent sexual signals; others are diurnal and use pheromones (Lloyd 1972). Bioluminescent communications and mimicry interactions are complex (Lloyd 1971, 1978, 1981, 1983).

Description: Mature larvae 17–50 mm in length. Body either onisciform (*Photuris,* fig. 34.462) and somewhat dorsoventrally flattened, or fusiform and elongate (*Photinus, Pyractomena,* figs. 34.461, 34.464), or subparallel (*Pterotus*). Cuticle of the dorsum shiny or dull, usually granulose and punctate with fine pubescence. Color varies and may be black, testaceous, castaneous, or variegated. There are 12 terga, with an ochraeous median dorsal vitta or sulcus generally present through abdominal tergum 7. The posterolateral tergal margins may be ochraeous or pinkish, and may be concave. The venter may be smokey gray, pink, or pale. The photogenic organ appears as 2 distinct spots on the epipleura of abdominal segment 8. The pygopod bears retractible anal filaments which may have hooks.

Head: Small, usually longer than wide, subcylindrical; smooth and colored testaceous or castaneous. Head not fused ventrally and retractile within the prothorax. Occiput sometimes concealed by the muscular sheath which envelops the head. Epicranial sutures present, varying in length of stem and its forward and rearward extensions. Labrum and clypeus indistinct; a nasale suggested (not always formed in the early instar larvae). Mandibles falcate and channelled, opening at the distal margin just before the apex. Retinaculum present, forming a distinct subapical or median tooth. Maxillae and labium with reduced articulation; cardo, small and subquadrate; stipes broad and elongate. Maxillary palpi 3- or 4-segmented with the last segment either blunt or tapering. Galea 2-segmented, and tapering with a long seta at its apex; lacinia reduced, appearing at the base of the galea as a brush border opening to the buccal cavity. Labium with small mentum and elongate submentum; 2-segmented palpi. The mouthparts are characterized by many fine setae, and a few spines on the submentum and stipes. Antennae, retractible and 3-segmented with globular-shaped accessory structures on the third joint. Stemmata situated just caudad of the antennal base.

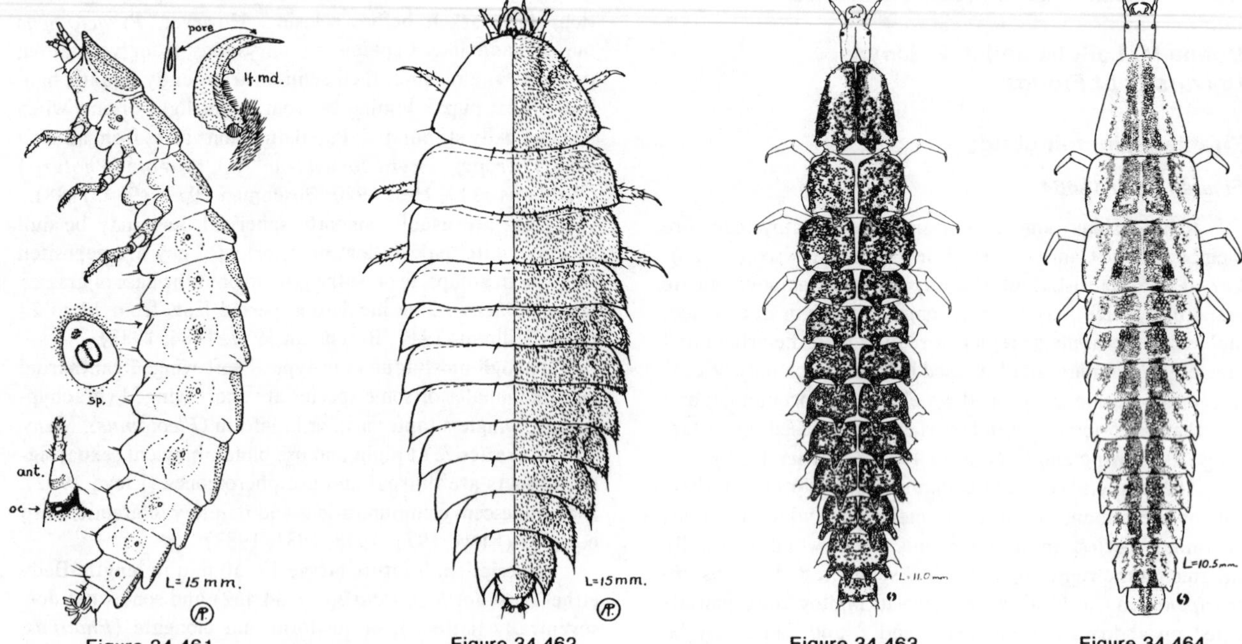

Figure 34.461 Figure 34.462 Figure 34.463 Figure 34.464

Figure 34.461. Lampyridae. *Photinus* sp. Larva, lateral (with insets of mandible, spiracle, and antenna). Length = 15 mm. (From Peterson, 1951)

Figure 34.462. Lampyridae. *Photuris* sp. Larva, dorsal. Length = 15 mm. (From Peterson, 1951)

Figure 34.463. Lampyridae. *Micronaspis floridana* Green. Larva, dorsal.

Figure 34.464. Lampyridae. *Pyractomena* sp. Larva, dorsal.

Thorax and Abdomen: Pronotum longer than wide or subelliptical, narrowed anteriorly with antero-lateral margins rounded; base straight except for *Micronaspis*. Meso- and metathorax subquadrate, slightly wider than long. Thoracic terga divided mid-dorsally with longitudinal carinae usually present. Posterolateral margins may be rounded, angulate, or may protrude (*Micronaspis*, fig. 34.463). Three pairs of 5-segmented legs present, may be stout or slender, and have a tarsungulus.

Abdomen with 9 visible terga; 1 through 7 broader than long and divided mid-dorsally, margins the same as thoracic margins. Body widest at about the second or third abdominal segment, and may narrow more bluntly anteriorly than posteriorly. The 10th abdominal segment (pygopod) is situated ventrally and bears retractile filaments which aid in locomotion or serve as a cleaning apparatus (Paiva 1919, McClean et al. 1972).

Spiracles: Functional annular biforous spiracles occur in the mesepipleura and in the epipleura of abdominal segments 1–8.

Comments: The family is large, consisting of approximately 1900 species in 92 genera and subgenera worldwide (McDermott 1966). About 20 genera, and 150 species occur in America north of Mexico. Representatives of the tribe Photini—*Photinus, Ellychnia, Pyropyga,* and *Lucidota*—are difficult to distinguish in the larval stage, though a better understanding of reared material may reduce this problem. Also, the semi-aquatic species of *Pyractomena* do not possess lateral abdominal gills as do some Asian *Luciola*. Finally, two European species of *Lamprohiza* (photine) have the photogenic organ situated in the venter of abdominal segments 2 and 6 or 2 through 6.

Selected Bibliography

Balduf 1935.
Böving and Craighead 1931.
Buschman 1977, 1984, 1988.
Crowson 1972b.
Gardner 1946.
Hess 1920.
Lloyd 1971, 1972, 1978, 1981, 1983.
McDermott 1964, 1966.
McClean *et al.* 1972.
Paiva 1919.
Schwalb 1960.
Sivinski 1981.
Williams 1917.

OMETHIDAE (CANTHAROIDEA)

John F. Lawrence,
Division of Entomology, CSIRO

This family includes the genus *Drilonius* (formerly placed in Drilidae), with about 10 species in East Asia, the genera *Matheteus* and *Ginglymocladus* (formerly placed in Lampyridae), with 1 and 2 species, respectively, from California, and the subfamily Omethinae (formerly placed in

Cantharidae), with 5 genera and 8 species from N. America and Japan. The group is thought to be most closely related to Cantharidae. Both sexes are winged, but the wings are occasionally reduced in females. Nothing is known of the biology of immature stages.

Selected Bibliography

Crowson 1972b.
Fender 1975.
Magis and Wittmer 1974.
Wittmer 1970.

CANTHARIDAE (CANTHAROIDEA) (INCLUDING CHAULIOGNATHIDAE)

Laurent LeSage,
Biosystematics Research Centre, Ottawa

Soldier Beetles

Figures 34.465-467

Relationships and Diagnosis: The Cantharidae belong to the superfamily Cantharoidea whose larvae are characterized by the grooved or channelled mandibles without a molar part, maxillae with articulated galea, biforous spiracles without a closing apparatus, and the lack of articulated urogomphi. In the larvae, starting from forms with the maxilla-labium complex deeply retractile in the subfacial sinus and with all the typical parts present, one can trace parallel progressions involving the loss of the separate parts and the shallowing of the subfacial sinus; in these respects, lampyrid larvae seem among the most primitive and lycids and cantharids the most advanced (Crowson 1955).

Cantharid larvae are readily distinguished from all others by their unique velvety appearance due to a covering of fine hairs, and the fully exserted head with a single large stemma on each side. Another distinctive feature is the presence of at least 1 pair of dorsal repugnatorial gland pores on all thoracic segments and abdominal segments 1–8 or 9 (fig. 34.466a).

Biology and Ecology: Adult Cantharidae, or "soldier beetles," are diurnal, frequent the foliage and flowers of trees, shrubs and herbs, and as a whole are in part predatory and also feed to some extent on plants, nectar, and pollen.

The larvae are primarily nocturnal, preying chiefly on small arthropods living in the ground debris. They have also been recorded to feed on caterpillars and syrphid larvae (Payne 1916). The larvae of some *Cantharis* are considered carnivorous (Balduf 1935) while those of *Silis* with stout mandibles may be predominantly herbivorous (Striganova 1962). Larvae may be found under stones, in leaf litter, under bark, in moss, in damp soil, etc. Hibernation occurs in the ground as mature larvae. Pupation takes place in spring in earthen cells in moist soil below the surface, and lasts 10–15 days, most species making their appearance in May and June. So far as known eggs are deposited in the summer in masses on the ground, under stones, and in cavities. In the subfamily Cantharinae, the newly hatched larvae of some species are feebly developed and are called "prolarvae"; they are soft-bodied and their legs are imperfectly developed. These larvae stay congregated and, about 2 days after hatching, they shed the prolarval skin and become active in the usual form of the species (Balduf 1935). The exact number of molts is still unknown but seems relatively large, 8 to 10 in *Cantharis* and *Rhagonycha* (Janssen 1963). Depending on the species, 1 or 2 generations develop in a year. Some larvae have a defensive mechanism in addition to that of the repugnatorial glands, and when annoyed they produce a blackish distasteful liquid from the mouth (Balduf 1935).

Description: Size in mature larvae 5 to 20 mm, sometimes reaching 30 mm. Shape campodeiform, orthosomatic, about the same diameter throughout, except for smaller caudal segments (Fig. 34.466g). Vestiture unique, velvet-like, consisting of numerous fine setae covering the entire cuticle of the body. Most species are deeply colored with brown, reddish, purple, or black.

Head: Well developed, pigmented, sclerotized, exserted, prognathous, depressed, and densely covered with fine setae (figs. 34.466b,c). Epicranial suture absent; frons, clypeus and labrum fused into a large area covering entire dorsal surface of the head. The form of the nasale (fig. 34.467a) is of taxonomic value and is usually characteristic of each genus although there is wide intraspecific variation in large genera. Epicranial halves fused on the ventral aspect, forming a continuous sclerotized area with gular sutures obsolete or more usually absent (fig. 34.466b). Antennae 3-segmented and well developed (figs. 34.466h, 34.467c); apex appearing 2-jointed, consisting of a short dome-shaped or cupola-like sensory appendage located in the membranous area of the second segment, and of a longer spine-like, usually setose, third segment. The structure of the second segment is of generic value, its general shape being characteristic of each subfamily: apex simple in Cantharinae (fig. 34.466h), deeply indented in Malthininae (fig. 34.467c). One pair of large stemmata present on each side of the head. The mouthparts are protracted and arise from the anterior portion of the head. Mandibles falciform, grooved, with a series of fine hairs on their mesal margin (fig. 34.466e); longitudinal channel usually well developed and open, almost closed in *Malthinus;* molar area absent, retinaculum well developed, tooth-like, generally present, but virtually absent in *Malthinus*. Maxilla with small cardo, strongly developed stipes, and 3- or 4-segmented palp, missing in *Chauliognathus*; lacinia densely hairy. Labium with prementum and mentum well-developed; labial palps 2-segmented, ligula absent (fig. 34.466b).

Thorax and Abdomen: Thoracic segments subequal or gradually decreasing in size from prothorax to metathorax, usually bearing a pair of pigmented darker spots. A pair of repugnatorial gland pores present dorsally on each segment. Legs 5-segmented (fig. 34.466k), well developed, with a claw-like tarsungulus which is armed with setae, a feature useful for the separation of several genera (figs. 34.465c, 34.466f, 34.467d).

Abdominal segments subequal in size and similar in structure, densely velvety, usually deeply pigmented. Tenth segment small and sometimes hardly visible from dorsal view. One or 2 pairs of round glandular openings (when 2, one pair

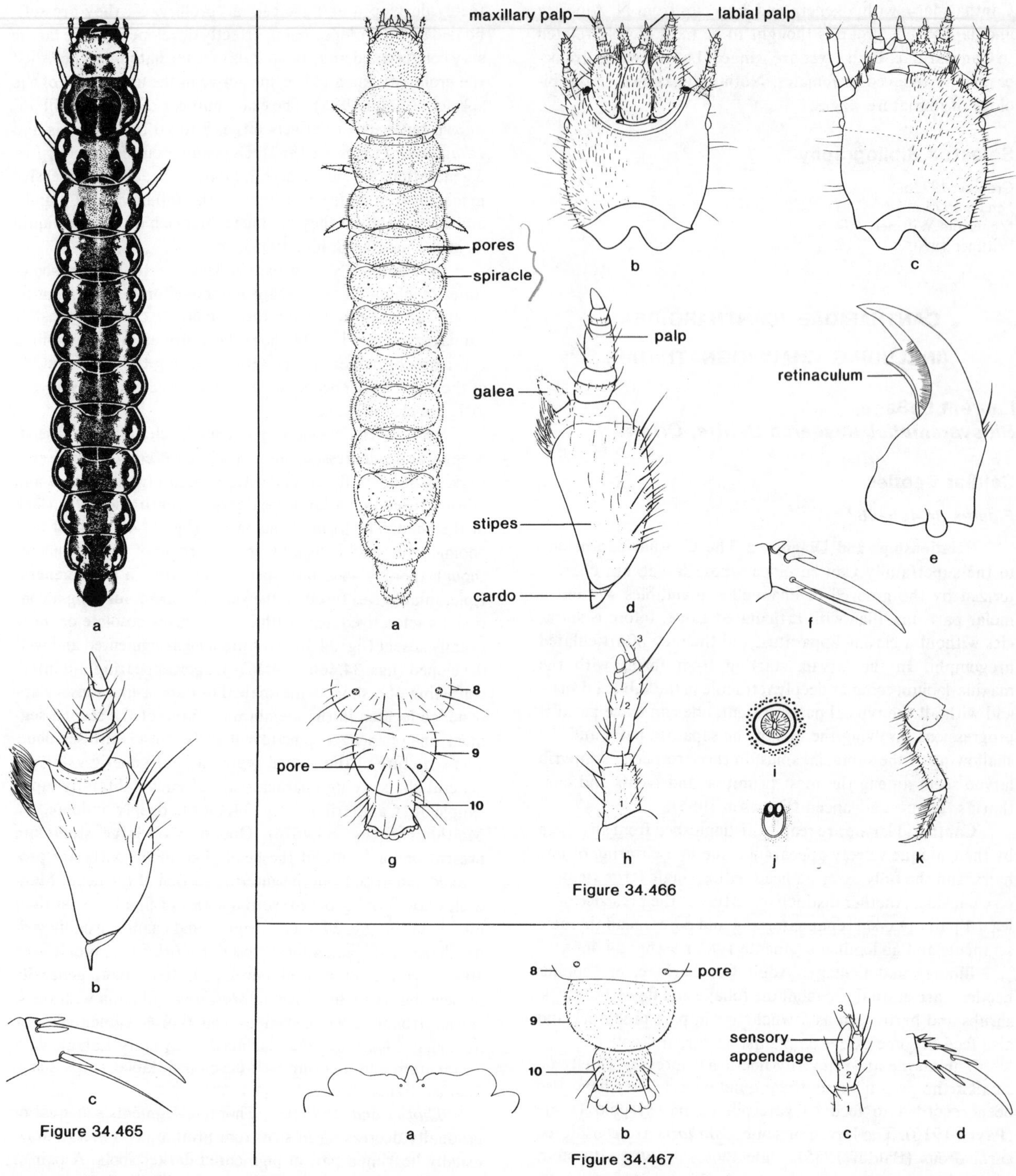

Figures 34.465a-c. Cantharidae. (Chauliognathinae). Length = 35 mm. **a.** larva; **b.** maxilla; **c.** claw-like tarsungulus. Pores of repugnatorial glands large, located on meso- and metathorax and on abdominal segments 1-8. Spiracles small, oval and cephaloventrad of gland pores. Predacious, living in moist habitats, under debris. Red Oak, Latimer Co., Oklahoma, Oct.-Nov. 1976, K. Stephan. (Drawn by S. Poulakis).

Figures 34.466a-k. Cantharidae (Cantharinae). *Cantharis* sp. Length = 20 mm. **a.** larva; **b.** head, ventral; **c.** head, dorsal; **d.** maxilla; **e.** mandible; **f.** claw-like tarsungulus; **g.** caudal segments, dorsal; **h.** antenna; **i.** pore of repugnatorial glands; **j.** spiracle; **k.** midleg. One pair of large dorsal pores of repugnatorial glands on thoracic segments and on abdominal segments 1-9; spiracles ovoid, biforous, located laterally and ventrad of repugnatorial glands. Pike River, Quebec, 7-IX-1970, trapped in an alfalfa field, G. Mailloux.

Figures 34.467a-d. Cantharidae (Malthininae). *Malthodes* sp. **a.** nasale, young instar; **b.** caudal segments, dorsal; **c.** antenna; **d.** claw-like tarsungulus.

is very small) of the repugnatorial glands (fig. 34.466i) on dorsum of segments 1–8 in Malthininae (fig. 34.467b) and segments 1–9 in Cantharinae (fig. 34.466g). They may be mistaken for the spiracles which are located ventrad of the glandular pores.

Spiracles: Usually elliptical (fig. 34.466j), biforous and located on the ventrolateral portion of the mesothorax and abdominal segments 1–8.

Comments: The Cantharidae is a relatively small family with 215 species recorded so far from N. America (Arnett 1968). The larval taxonomy of Nearctic cantharids is poorly known, and keys do not exist. However, several cantharid genera are common to both Europe and N. America, and the works of European authors are useful for the study of the Nearctic fauna (Fitton 1975; Gardner 1946; Janssen 1963; Striganova 1962; Verhoeff 1917, 1923). Additional information is scattered through general works (Böving & Craighead 1931; Peterson 1951) or in short papers dealing with biology (Riley 1868; Wilson 1913).

Selected Bibliography

Arnett 1968.
Balduf 1935 (biology).
Böving and Craighead 1931.
Brancucci 1980.
Crowson 1955.
Fitton 1975 (key).
Gardner 1946.
Jackson and Crowson 1969.
Janssen 1963 (biology).
Klausnitzer 1978 (key).
Payne 1916 (biology).
Peterson 1951.
Riley 1868.
Striganova 1962, 1964c.
Verhoeff 1917, 1923 (key).
Wilson 1913.

DERODONTIDAE (DERODONTOIDEA)

(INCLUDING LARICOBIIDAE, PELTASTICIDAE)

John F. Lawrence,
Division of Entomology, CSIRO

Figures 34.468–469

Relationships and Diagnosis: The Derodontidae is a primitive family of cucujiform beetles, which has apparent affinities with both Cucujoidea and Bostrichoidea (Dermestidae). The family has been placed in its own superfamily or variously combined with Dermestidae, Nosodendridae, and Jacobsoniidae (Crowson 1959, 1960b; Lawrence and Hlavac, 1979; Lawrence and Newton, 1982). In general form and in the granulate or tuberculate dorsal surface, most derodontid larvae resemble those of various cucujoid families, such as Nitidulidae, Rhizophagidae, and Languriidae, from which they differ in having a distinct galea and lacinia. Eucinetid larvae may resemble those of *Laricobius,* but they differ in having annular, rather than annular-biforous spiracles, and a well-developed, asperate mola. The body form, mandibular type, and habitat of *Laricobius* are similar to those of some Coccinellidae, which lack a distinct galea and lacinia and have three stemmata or fewer. The closely-related family Nosodendridae have larvae which are very different from derodontids in having the eighth spiracles placed at the end of a long, tapered, terminal process. Jacobsoniid larvae are minute and have a fringed maxillary mala, while those of Dermestidae either lack a mandibular mola or have a simple one similar to that in Bostrichidae.

Biology and Ecology: Most Derodontidae are mycophagous, but members of the genus *Laricobius* have become predators of woolly aphids (Homoptera: Adelgidae). Species of *Peltastica* occur in fermenting sap flows or in areas under bark where decay and fermentation are taking place; gut contents include a variety of cell types, including spores and hyphae. *Derodontus* species attack a wide variety of higher Basidiomycetes (including Tricholomataceae, Hericiaceae, and Polyporaceae), where they feed on the softer tissue of fresh or rotting fruiting bodies. Species of *Nothoderodontus* appear to feed exclusively on the black hyphal masses of sooty molds (Ascomycetes: Capnodiaceae) which grow on tree trunks. It is thought that the predaceous *Laricobius* may have evolved from ancestors with similar habits. The European *Laricobius erichsonii* Rosenhauer has been introduced into the northwestern United States and Maritime Provinces of Canada to control the balsam woolly aphid (*Adelges picea* (Ratzeburg)). (Crowson, 1959; Clark and Brown, 1958; Franz, 1958; Fukuda, 1963; Lawrence and Hlavac, 1979; Shepard, 1976).

Description: Mature larvae 2 to 6.5 mm. Body elongate, more or less parallel-sided, or slightly fusiform, slightly flattened. Dorsal surfaces sclerotized and usually granulate or tuberculate, sometimes darkly pigmented; vestiture of scattered, short, simple setae.

Head: Protracted and prognathous, moderately broad and slightly flattened. Epicranial stem very short or absent; frontal arms lyriform and contiguous at base. Median endocarina absent. Stemmata 6 on each side (occasionally 5 in *Nothoderodontus*). Antennae well-developed, 3-segmented. Frontoclypeal suture absent or vaguely indicated; labrum free. Mandibles symmetrical, bidentate, with accessory ventral process and usually with serrate incisor edge (simple in *Laricobius*). Mola usually well-developed and asperate; prostheca usually consisting of fixed, acute, hyaline process accompanied by a brush of hairs; mola and prostheca reduced in *Laricobius*. Ventral mouthparts retracted. Maxilla with transverse cardo, elongate stipes, and 3-segmented palp, and usually with well-developed articulating area, a fixed, narrowly rounded or truncate galea, and fixed, falciform lacinia, which may be bifid or trifid at apex; in *Laricobius,* articulating area absent and galea and lacinia reduced and broadly rounded, having the appearance of a single, cleft mala. Labium usually free to base of mentum, more or less connate with stipites in *Laricobius;* ligula broad, not bilobed; labial palps 2-segmented and widely separated. Hypopharyngeal sclerome represented by transverse bar. Hypostomal rods absent. Ventral epicranial ridges usually present (absent in *Laricobius*). Gula much wider than long.

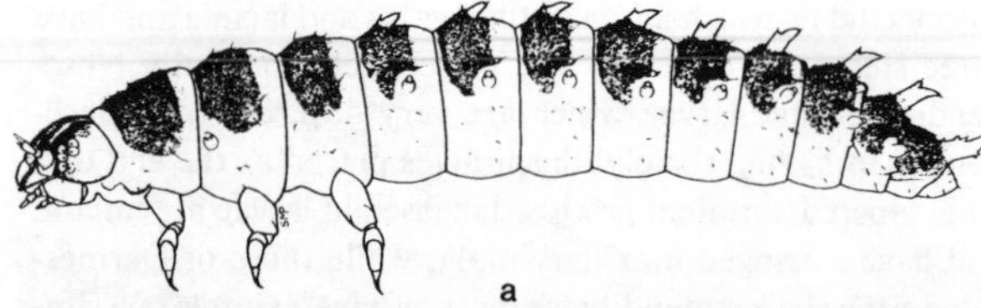

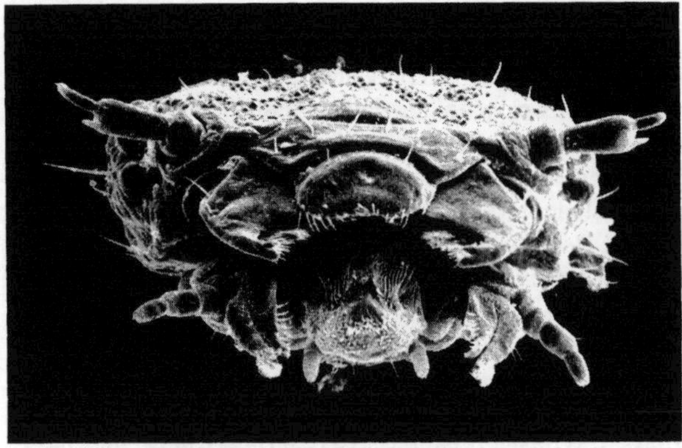

Figure 34.468

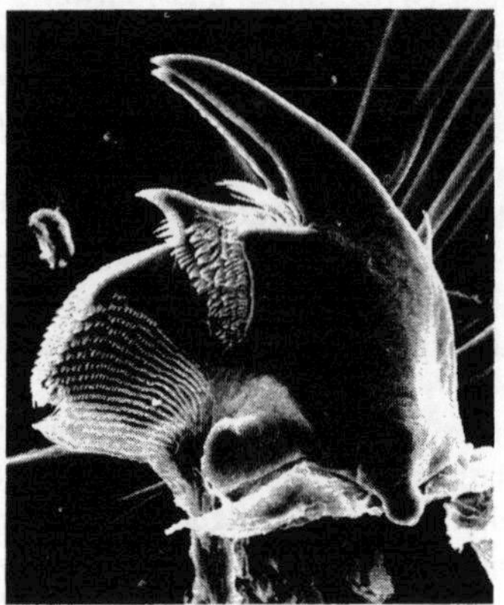

Figure 34.469

Figures 34.468a,b. Derodontidae. *Peltastica tuberculata* Mannerheim. Blodgett Forest, near Georgetown, El Dorado Co., California. Length = 6.5 mm. **a.** larva, lateral; **b.** head, anterior.

Figure 34.469. *Derodontus* sp. Rock Creek Park, Maryland. Left mandible, ventral.

Thorax and Abdomen: Legs well-developed, 5-segmented; tarsungulus with 2 setae lying side by side. Thoracic terga and abdominal terga 1–9 sometimes with paired, pigmented plates. Pleural lobes usually well-developed. Abdominal terga 1–8 with 1 or more pairs of setiferous processes or a median, forked process (absent in *Laricobius*). Tergum A9 usually with pair of fixed urogomphi (absent in *Laricobius*). Sternum A9 simple. Segment A10 posteriorly oriented, cylindrical, sometimes sclerotized dorsally.

Spiracles: Annular-biforous, lateral; both thoracic and abdominal spiracles borne on tubular, cone-like processes.

Comments: The family includes 4 genera and 20 species, with 3 genera and 9 species occurring in N. America. The group has an amphipolar distribution with *Peltastica, Derodontus,* and *Laricobius* occurring in the Holarctic Region and *Nothoderodontus* restricted to the cooler parts of the Southern Hemisphere (Chile, Australia, New Zealand) (Crowson, 1980; Lawrence, 1985a).

Selected Bibliography

Böving and Craighead 1931 (larva of *Derodontus* sp.).

Clark and Brown 1958 (*Laricobius erichsonii* as *Adelges* predator in North America), 1960 (*L. rubidus* as *Pineus* predator in N. America).

Crowson 1959 (larva of *Nothoderodontus gourlayi* Crowson), 1980 (amphipolar distribution).

Franz 1958 (larva and life history of *L. erichsonii*).

Fukuda 1963 (larva of *Peltastica reitteri* Lewis).

Lawrence 1985a (larval characters in *Nothoderodontus*).

Lawrence and Hlavac 1979 (larval key; feeding habits; larva of *Peltastica tuberculata* Mannerheim).

Shepard 1976.

NOSODENDRIDAE (DERODONTOIDEA)

John F. Lawrence,
Division of Entomology, CSIRO

Figures 34.470a-c

Relationships and Diagnosis: Nosodendridae is one of the more primitive cucujiform families, sharing features with both Derodontidae and Dermestidae. The larvae are very distinctive, with the eighth abdominal spiracles located at the apex of a long, tapered, terminal process, a condition found elsewhere only in distantly-related, aquatic or subaquatic groups. The larval head resembles that of a nitidulid, with the frontal arms distant at base and the mandibular prostheca complex and fringed with hairs; the maxilla, however, is more like that of Derodontidae, with a distinct galea and spur-like lacinia.

Biology and Ecology: Nosodendrids appear to be restricted to slime flux, which form when tree wounds are attacked by various yeasts (Ascoideaceae, Endomycetaceae); the beetles have been reported to be predators of fly larvae (Arnett, 1968; Sokoloff, 1959), but it is more likely that they feed on fungi and fungal biproducts.

Description: Mature larvae 5 to 12 mm. Body elongate, fusiform, somewhat flattened. Dorsal surfaces sclerotized and darkly pigmented, reddish-brown, distinctly granulate and tuberculate, more or less glabrous, except for tufts of moderately long, yellowish hairs along lateral edges of thorax and abdomen. Body often covered with debris.

Head: Protracted and prognathous, short and broad, somewhat flattened, Epicranial stem absent and frontal arms weakly lyriform and distant at base (frontal arms contiguous and joined to short epicranial stem only in New Zealand species *N. ovatum* Broun). Median endocarina absent. Stemmata 5 on each side. Antennae well-developed, 3-segmented, with very small 3rd segment. Frontoclypeal suture present; clypeus very short; labrum free, moderately large. Mandibles

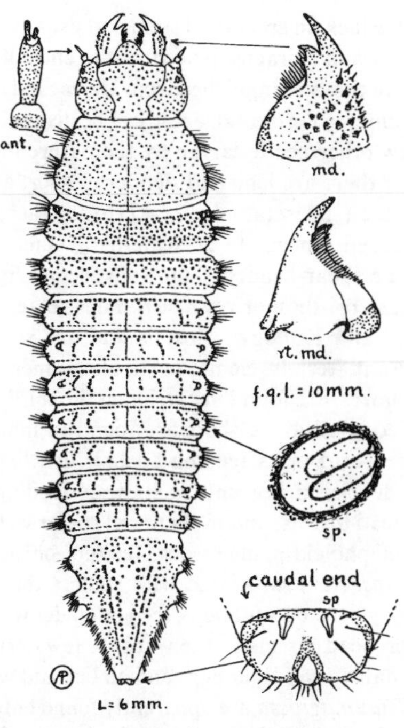

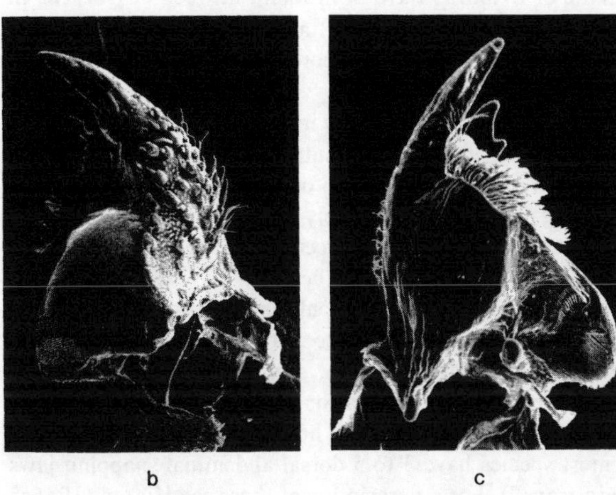

Figures 34.470a–c. Nosodendridae. *Nosodendron unicolor* Say. Length = 6 mm. **a.** larva, dorsal (plus details); **b.** right mandible, dorsal; **c.** right mandible, ventral. Great Smoky Mountains National Park, Tennessee. (figure "a" from Peterson 1951)

symmetrical, unidentate, with bifid, sclerotized retinaculum on incisor edge, and with accessory ventral process; outer edge of mandible tuberculate; mola well-developed, transversely ridged, prostheca consisting of hyaline lobe fringed with hairs and a series of fringed membranes; ventral surface of mola with extensive microtrichial patch. Ventral mouthparts retracted. Maxilla with transverse cardo, elongate stipes, well-developed articulating area, 3-segmented palp, truncate, articulated galea, and fixed, falciform lacinia with bifid apex. Labium free to base of mentum; ligula short and broad, bilobed; labial palps 2-segmented and widely separated. Hypopharyngeal sclerome transversely ridged. Hypopharyngeal, epipharyngeal, and cibarial regions with complex series of microtrichial patches and fringed membranes. Hypostomal rods absent. Ventral epicranial ridges present. Gula transverse.

Thorax and Abdomen: Legs relatively short, 5-segmented, armed with short spines; tarsungulus with 2 very short setae lying side by side. Abdominal terga 1–7 sometimes with blunt, lateral tergal processes lined with long hairs. Segment A8 produced and tapered posteriorly forming a terminal process, with a pair of spiracles at the apex and sometimes a pair of sclerotized knobs at base; tergum usually lined with long hairs. Sternum A8 divided longitudinally. Segments A9 and A10 reduced to a pair of longitudinal anal pads, located ventrally at base of terminal process.

Spiracles: Annular-biforous, located on short, tubular processes. Thoracic spiracles usually lateral (dorsal in *N. unicolor* Say). Abdominal spiracles on segment 1 usually dorsal and those on 2–7 lateral or dorsal (those on 1–7 rarely all lateral). Spiracles on segment 8 contiguous, at end of terminal process.

Comments: The family includes the single genus *Nosodendron* which occurs in most regions of the world, including Australia and New Zealand (Reichardt, 1976a).

Selected Bibliography

Böving 1929b (larvae of *Nosodendron californicum* Horn and *N. unicolor* Say).
Böving and Craighead 1931 (larvae of *N. californicum* and *N. unicolor*).
Crowson 1959 (larva of *N. ovatum* Broun).
Hayashi *et al.* 1959 (larva of *N. asiaticum* Lewis).
Hayes and Chu 1946 (larvae of *N. californicum* and *N. unicolor*).
Klausnitzer 1971 (larva of *N. fasciculare* (Olivier)).
Osborne and Kulhavy 1975.
Peterson 1951 (larva of *N. unicolor*).
Reichardt 1976a.
Sokoloff 1959.

JACOBSONIIDAE (DERODONTOIDEA)

John F. Lawrence,
Division of Entomology, CSIRO

This family includes about 10 described species, which have been placed in the following genera: *Derolathrus* (Hawaii, Guadeloupe), *Gomya* (Fiji, Mauritius, Reunion, Sri Lanka), *Lathridiomorphus* (Canary Islands), *Saphophagus* (New Zealand), and *Sarothrias* (= *Jacobsonium*) (Sumatra, India, and the Seychelles); however undescribed *Sarothrias* have been seen from Micronesia, Fiji, and Australia, while species of the *Derolathrus-Gomya* complex have been seen from Australia, New Zealand, Solomon Islands, southern and central Africa, Panama, and Brazil. The systematic position of the group is still uncertain, and it is usually placed near Derodontidae and Nosodendridae.

Larvae are minute (less than 2.5 mm, and usually less than 1 mm), elongate, more or less parallel-sided and slightly flattened, with evenly sclerotized terga. The head is somewhat flattened, without an epicranial stem, median endocarina, or stemmata, and with lyriform frontal arms, distant at base, and short, broad antennae. The mandibles are unidentate, with a serrate incisor lobe, slender prostheca, and tuberculate mola. The galea and lacinia are distinct, the former bearing a tuft of serrate hairs, and the maxillary palps are 3-segmented. The ligula is bilobed, and the labial palps are 2-segmented. Legs are short and the tarsungulus is unisetose. Tergum A9 has a pair of short, fixed urogomphi, and segment A10 is more or less circular, forming a complete ring around the anus, which is ventrally oriented. Spiracles are annular.

Saphophagus adults and larvae have been found under moist, tight-fitting bark, on a darkened wood surface, where they were associated with species of *Picrotus* (Cryptophagidae) and *Lenax* (Rhizophagidae). *Sarothrias* has been collected in leaf litter. *Lathridiomorphus anophthalmus* Franz was taken from deep in the soil around rotten roots of a laurel stump. *Derolathrus atomus* Sharp was found with its larva beneath the bark of *Acacia koa*. Species of *Gomya* have been collected in bat guano and by sifting *Pandanus* trash while these and other related forms have turned up in leaf litter samples from various parts of the world.

Selected Bibliography

Crowson 1955, 1959 (larva of *Saphophagus*).
Dajoz 1972 (*Gomya*), 1978 (*Sarothrias* from India).
Franz 1968 (*Lathridiomorphus*).
Lawrence 1982c.
Lawrence and Hlavac 1979 (relationships).
Lawrence and Newton 1982 (relationships).
Sen Gupta 1979 (Derolathrinae in Merophysiidae).

DERMESTIDAE (BOSTRICHOIDEA)

(INCLUDING THORICTIDAE, THYLODRIIDAE)

Richard S. Beal, *Colorado Christian College*

Carpet Beetles, Hide Beetles, Larder Beetles

Figures 34.471–482.

Relationships and Diagnosis: Crowson (1955) placed the Dermestidae with Nosodendridae and Derodontidae in the superfamily Dermestoidea, partly on the basis of such larval characters as the presence of setose tergal plates, a free labrum, and a galea distinctly separated from the lacinia by a deep or shallow cleft; however the affinities of Dermestidae with the bostrichoid families (Bostrichidae, Anobiidae, Ptinidae) have also been pointed out by Crowson (1959, 1961a), Lawrence and Hlavac (1979) and Lawrence and Newton (1982). The lack of a tuberculate mola and the presence of a long epicranial stem and spur-like lacinia are larval characters shared with Bostrichoidea, but typical bostrichoid larvae are highly modified wood-borers which differ from all dermestids in having a lightly sclerotized, more or less C-shaped body. Dermestid larvae are readily distinguished from those of Nosodendridae, since the latter have a spine-like, terminal 8th abdominal segment; in Dermestidae abdominal segment 9 is always distinct although sometimes quite short. Derodontidae lack an epicranial stem and usually have a tuberculate mola and spiracles placed at the ends of projecting tubes, and are thus distinguished from Dermestidae. All commonly encountered dermestid larvae can be distinguished from all but a few other beetle larvae by the nature of the dorsal setae. Either there are long and slender or short and clubbed spinulate setae (spicisetae, fig. 34.471b), or there are longitudinally ribbed lanceolate or scale-like setae. In addition there may be spear-headed setae (hastisetae, fig. 34.471c) forming dense brushes, or rows of ramous setae.

Biology and Ecology: Most larvae are scavengers on dried protein materials. Some, particularly members of the genus *Dermestes,* are found in carrion in the third, or butyric, stage of decomposition. Others are found in mammal nests and sheltered bird nests feeding on hair, feathers and proteinaceous debris, in bee and wasp nests feeding on pollen stores and dead insects, and in sheltered spider webbing, such as masses of pholcid spider webs, feeding on dead insects. Larvae of *Apsectus,* associated with spiders that live under stones in dry, protected situations, feed on spider webbing itself as well as on dead insects in the webs. A few dermestid species are predatory on spider eggs and on bee and wasp larvae. Larvae of *Thaumaglossa* are apparently found only in mantid oöthecae, a frequent habitat as well for larvae of *Orphinus.* Larvae of *Orphilus* have been found in wood of *Quercus* in North Africa (Paulian, 1942) and in dead branches of *Arbutus* infested with the polypore *Trametes sepium* in California (Beal, 1985).

Eggs are deposited singly in or on the material on which the larvae will feed. First instar larvae of some species are noteworthy for their ability to penetrate packaging material enclosing food stores, probably an adaptation in nature to movement from one food source to another, such as from one mud dauber wasp cell to another.

The pupal stage is commonly spent within the last larval skin. Most dermestids pupate within the larval food site. Larvae of *Dermestes,* however, tunnel into wood, plaster, or even stone to construct pupation chambers. In such cases the larval skin is shed and used to help close the chamber. Pupae of most species have 3 to 5 dorsal abdominal snapping jaws ("gin-traps"). For a description of these see Hinton (1946a) and Beal (1964).

Description: Mature larvae minute to medium in length, 2–30 mm. Body elongate, fusiform, oblong or oval, straight or somewhat cyphosomatic, somewhat depressed to subcylindrical. Cuticle smooth, punctate, minutely papillate, or with row of small tubercles behind each tergal carina (often called antecostal suture). Dorsum with spicisetae or longitudinally ribbed setae except in *Thorictodes* and *Orphilus;* hastisetae or ramous setae often present. Spiracles annular, oval, or slit-like, not evidently biforous, not on projecting tubes; 9 pairs present; abdominal spiracles on tergum or on membrane lateral to tergum.

Head: Free, hypognathous and subglobular; epicranial and frontal sutures present. Frons triangular. Stemmata none, 3 or 6. Antenna 3-segmented; second segment with apical or subapical sensorium. Labrum free; labral-epipharyngeal margin (figs. 34.481, 34.482) with lateral series of sharply or slightly curved lanceolate setae; middle of margin with series

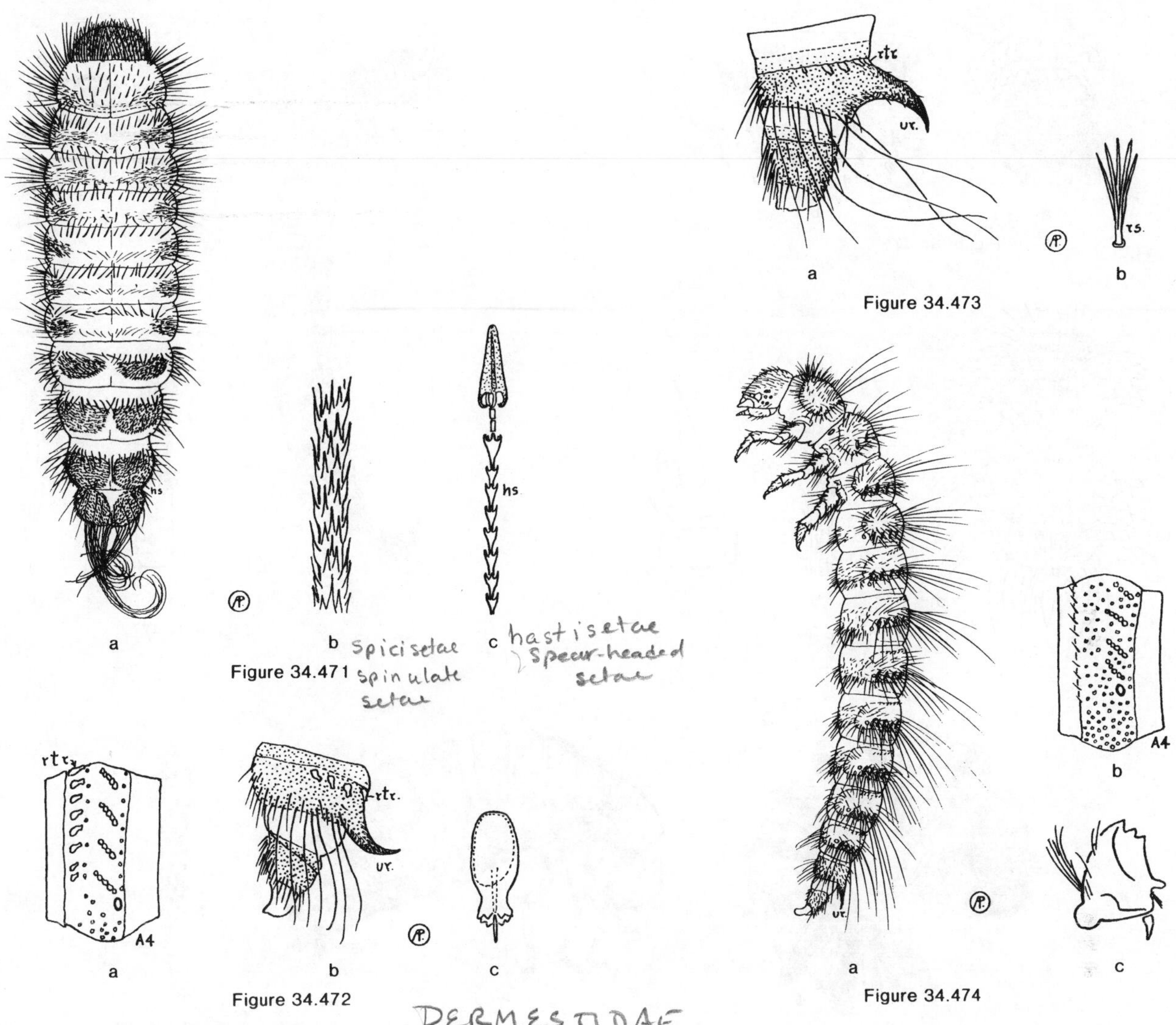

Figure 34.471

Figure 34.472

Figure 34.473

Figure 34.474

Figures 34.471a-c. Dermestidae. *Trogoderma inclusum* LeConte. Length = 7-8 mm. **a.** larva, dorsal; **b.** short section of midsection of long spiciseta; **c.** tip of hastiseta. Setiferous, orthosomatic, dorsum covered with long and short spicisetae and hastisetae; hastisetae forming dense clumps on posterior terga, particularly on abdominal segments 6, 7, and 8; abdominal segment 9 short and bearing brush of long, straight spicisetae. Often found as scavenger in wasp and bee nests. Feeds on dried proteinaceous materials, particularly dead insects; often infests dried milk and dry animal feeds. (figures 34.471a, c from Peterson 1951)

Figures 34.472a-c. Dermestidae. *Dermestes maculatus* De Geer, hide beetle. Length = 14+ mm. **a.** denuded tergum of abdominal segment 6, lateral; **b.** abdominal segments 9 and 10, lateral; **c.** retrorse tubercle. Members of genus recognized by presence of urogomphi, numerous long spicisetae, and sclerotized ring completely encircling abdominal segment 10. Mature larvae of this species with broad, median, yellowish stripe extending posteriorly from anterior margin of pronotum; retrorse tubercles on abdominal terga 4-9, tubercles with acute wartlike processes and terminal seta; frons with pair of prominent tubercles, one on each side. Commonly found beneath carrion. Often infesting stored hides, hams, dried and salted fish, etc. (Figures 34.472a, b, c from Peterson, 1951.)

Figures 34.473a,b. Dermestidae. *Dermestes lardarius* L., larder beetle. Length = 14 to 15 mm. **a.** segments 9 and 10 of abdomen, lateral; **b.** ramous seta. Similar to species above but with dorsum uniformly pigmented except for fine cleavage line. Head lacking tubercle on frons; abdominal terga 3-10 with retrorse tubercles; terga 3-8 each with transverse series of ramous setae immediately posterior to retrorse tubercles. Often found in carrion and in birds' nests, particularly those of martins, swallows, and house sparrows. Infests a wide variety of stored products, including furs, leather, bacon, hams, cheese, etc. May be serious pests of poultry houses and pigeon lofts, attacking and killing young birds. (from Peterson 1951)

Figures 34.474a-c. Dermestidae. *Dermestes ater* De Geer, black larder beetle. Length = 15+ mm. **a.** larva, lateral; **b.** denuded tergum of abdominal segment 6, lateral view; **c.** mandible. Similar to preceding species but very setiferous; dorsally with broad, discontinuous, median, yellowish stripe extending posteriorly from anterior margin of mesonotum; pronotum incompletely divided by stripe; head lacking tubercles on frons; abdominal terga each with series of slender, upright setae immediately posterior to antecostal suture; retrorse tubercles lacking. Commonly found in carcases of dead birds and mammals. Often a pest of dried fish, hides, hog bristles, and the like. Occasionally a predator attacking cocoons of silkworms and is known to prey on larvae of the gypsy moth. (Figures 34.474a, b from Peterson, 1951.)

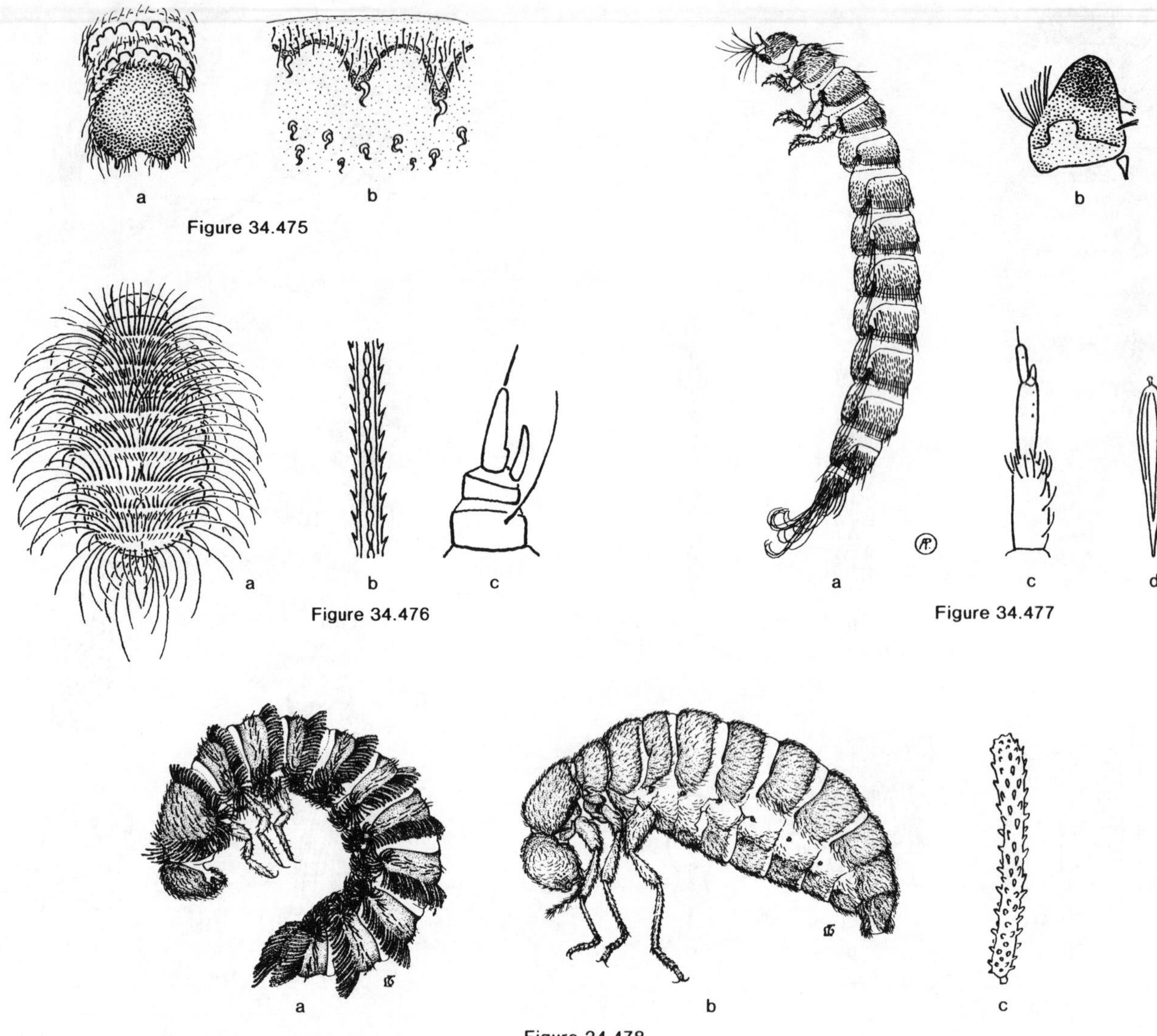

a b c

Figure 34.478

Figures 34.475a,b. Dermestidae. *Orphilus subnitidus* LeConte. Length = 3 mm. **a.** posterior segments of abdomen, dorsal; **b.** anterior half of short section of abdominal segment. Elongate, orthosomatic, creamy, with abdominal segment 8 heavily pigmented and dark reddish brown; mandible with well-developed mola; dorsal setae consisting of fine, simple hairs becoming longer at lateral margins and short, tapering, twisted hairs; antecostal sutures of nota and terga forming sclerotized, crenulate carinae; each tooth of carina with twisted hair at apex; segment 8 with pair of short, peg-like fixed urogomphi. Probably feeds on fungus.

Figures 34.476a-c. Dermestidae. *Apsectus araneorum* Beal. Length = 2 mm. **a.** larva, dorsal; **b.** short section of shaft of long spiciseta; **c.** antenna, dorsal. Body elongate oval, cyphosomatic, dusky brown with long and short quadrate spicisetae; antenna short with accessory papilla half as long as terminal segment; urogomphi absent; caudal brush absent. Found in spider webbing in dry, sheltered situations. Feeds on webbing and on dead insects in webs.

Figures 34.477a-d. Dermestidae. *Attagenus unicolor* (Brahm), black carpet beetle. Length = 7-8 mm. **a.** mature larva, lateral; **b.** mandible; **c.** antenna; **d.** lanceolate seta from tergite of abdominal segment 1. Orthosomatic (although usually appearing somewhat cyphosomatic in preserved specimens), dark reddish brown; dorsal surface covered with long and short spicisetae and longitudinally ribbed lanceolate setae; caudal brush of long, straight spicisetae arising on very short abdominal segment 9 (these curled in alcoholic specimens); urogomphi absent. Commonly found in sheltered birds' nests such as swallow, English sparrow, and *Sayornis* spp. and in rodent nests in hollow trees. Common pest of granaries and stored woolens, feathers, silk, fur, and other animal products. (figure "a" from Peterson 1951)

Figures 34.478a-c. Dermestidae. *Thylodrias contractus* Motschulsky, the odd beetle. Length = 2.3 mm. **a.** mature larva; **b.** larviform adult female; **c.** clavate spiciseta from abdominal tergum 1. Oval, strongly cyphosomatic, light golden brown; posterior margin of each notum and abdominal terga 1-4 with row of closely spaced, clavate spicisetae of form shown; anterior margin of pronotum, lateral margins of each notum and tergum and posterior margin of posterior abdominal terga with longer, tapering spicisetae; antenna short; 3 large subequal stemmata on each side of head; urogomphi absent; caudal brush of setae absent. Feeds on feathers, fur, carrion in last stages of decomposition, dead insects, and other proteinacious materials in dry situations. Not uncommon in houses; occasional pest in insect, bird and other dried animal collections.

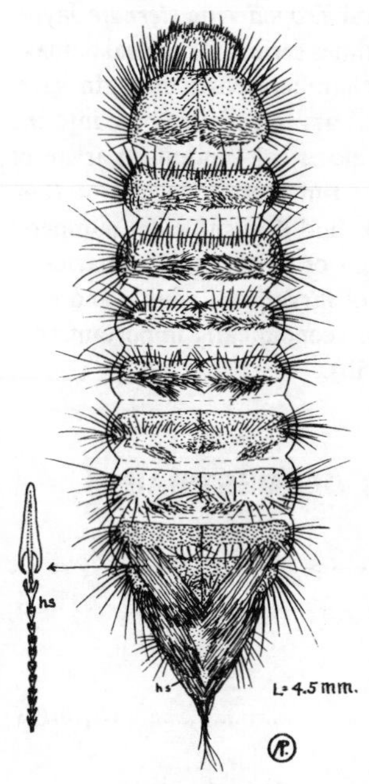

Figure 34.479

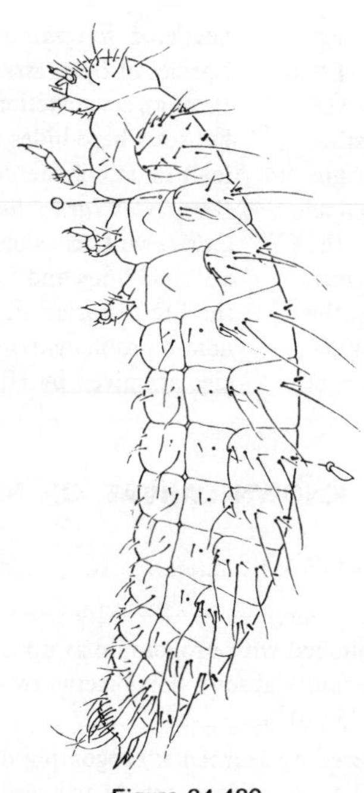

Figure 34.480

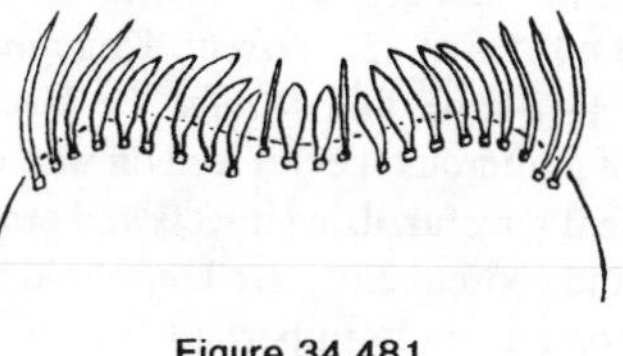

Figure 34.481

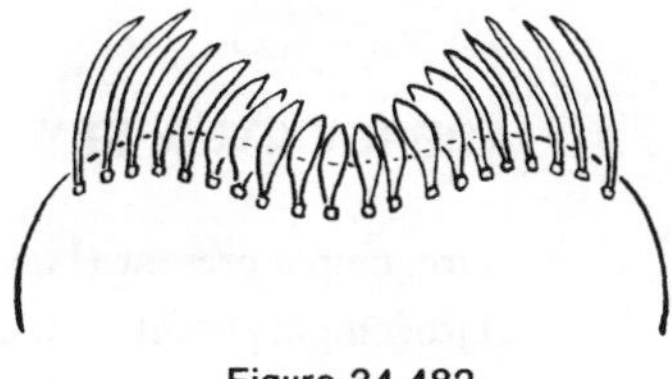

Figure 34.482

Figure 34.479. Dermestidae. *Anthrenus verbasci* (L.), varied carpet beetle, dorsal, with detail of tip of hastiseta. f.g.l. 4–5 mm. Brown to black, with converging brushes of hastisetae on abdominal segments 5, 6 and 7 inserted on each side of membrane behind the tergum. Body compact and broadest at segments 4–6. Feeds on many animal products, including fur, feathers, hides, skins and leather, wool, silk and insect collections. Also found in bird nests. (from Peterson 1951)

Figure 34.480. Dermestidae. *Thorictodes heydeni* Reitter. Length = 2 mm. Setae simple, with the pair near the middorsal line on each thoracic and abdominal segments 1–8 short and clublike (*see* detail in figure 34.480); urogomphi present, slightly decurved. Originally from the Mediterranean region, now widespread in Europe and Asia; transported in rice, peanuts, wheat and other foodstuffs (Van Emden, 1951). Recorded from a rice mill in Texas, and Nebraska and Mexico (from Anderson, 1949).

Figure 34.481. Dermestidae. *Trogoderma* sp., or *Reesa* sp. Labral-epipharyngeal margin.

Figure 34.482. Dermestidae. *Megatoma* sp. Labral-epipharyngeal margin.

of lanceolate and/or linear setae; epipharynx with 1 or 2 series of transverse sensory pores with or without distal cluster or row of sensory papillae; row of setae or spines lacking except in rare Thaumaphrastini; labral rods moderately long, straight, sinuate, curved or (in *Orphilus*) somewhat Y-shaped. Mandible distally rounded with 2 or 3 apical teeth or with pointed or rounded, medially concave cusp; membranous retinaculum usually present; mola absent except in *Orphilus;* basal setal brush present or absent. Maxilla with 3- or 4-segmented palp; galea simple, setiferous; lacinia with an elongate, strong, apically bifid spur; maxillary articulating area reduced. Labium with 2-segmented palp; ligula bilobed (except in *Thorictodes?*). Gula present.

Thorax and Abdomen: Thorax relatively short, 1/4 to 2/5 times as long as length of abdomen. Legs 5-segmented with tarsus and claw fused into single, claw-shaped, tarsungulus.

Abdomen with 9 or 10 segments evident; segment 10 encircled by sclerotized and pigmented ring (*Dermestes*) and used as pygopod, or sometimes unsclerotized. Most terga with distinct transverse carina near anterior edge; this may be straight or (in *Orphilus*) deeply scalloped. Segment 9 with or without urogomphi; when present (Dermestinae, Orphilinae) urogomphi fixed; in many species segment 9 quite short and bearing caudal brush of long, slender setae.

Comments: The distribution of the family is worldwide with a total of about 880 described species. About 110 species are found in the continental United States representing 7 of the 9 tribes. Knowledge of the larvae is fairly complete for economically important genera with the exception of *Anthrenus*. Nearctic representatives of this genus need taxonomic revision. Larvae and larval habitats of the abundant flower-visiting species of *Cryptorhopalum* remain mostly unknown.

Many species are of considerable economic importance. The khapra beetle, *Trogoderma granarium* Everts, once eradicated from the United States but reintroduced to the environs of Baltimore, is considered the world's second most important depredator of stored grains. *T. glabrum* (Herbst) and *T. variabile* Ballion are less important pests of stored grains but along with *T. inclusum* LeConte (figs. 34.471a–c) are significant pests in dried milk plants and malt stores. Stored, salted, smoked and dried fish and meats, hides, furs, bristles, feathers, and cheese are attacked by many species, most notably by *Dermestes maculatus* De Geer (fig. 34.472a–c), *D. frischi* Kugelmann, *D. lardarius* L. (fig. 34.474a,b), and *D. ater* De Geer (figs. 34.473a–c). Larvae of numerous species feed on woolens, silks, horsehair, bristles, feathers, fur, dried insects and other materials of animal origin and consequently are household and museum pests as well as commercially important insects. Chief among these are the black carpet beetle, *Attagenus unicolor* (Brahm) (figs. 34.497a–d) (= *A. megatoma* F. and *A. piceus* Olivier), the varied carpet beetle, *Anthrenus verbasci* (L.) (fig. 34.479), the furniture carpet beetle, *A. flavipes* LeConte, the carpet beetle, *A. scrophulariae* (L.) and *Trogoderma sternale* Jayne. Species of *Dermestes* cause at times serious structural damage through construction of pupal chambers, particularly in warehouses where hides are stored. Larvae will also bore into the lead of telephone cables and electrical conduits. Larvae of *Dermestes lardarius* L. and at times other species of *Dermestes* will occasionally attack, and if in sufficient numbers, kill ducklings and young pigeons or chickens. Summaries of the life histories, descriptions of larvae and adults, and economic problems posed by most economically important species are given by Hinton (1945b).

ABBREVIATED KEY TO GENERA OF KNOWN LARVAE OF NEARCTIC DERMESTIDAE

1. Urogomphi present (Figs. 34.472a, 34.473a, 34.475a) 2

   Urogomphi absent 4

2. Segment A9 heavily sclerotized, posteriorly bilobed with urogomphi at apparent posterior apex of insect; segment A10 apparently absent; dorsal setae twisted; tergal carinae deeply scalloped (Figs. 34.475a,b) .... *Orphilus*

   Segment A9 sclerotized to same extent as preceding segments; urogomphi dorsal; segment A10 present but may or may not be sclerotized; tergal carinae more or less straight (Figs. 34.472a, b, 34.474) 3

3. Abdominal segment 10 completely encircled by sclerotized ring; spicisetae present and forming 8 distinct, oblique series across each abdominal tergum (Figs. 34.472a, 34.474b); ramous setae (Fig. 34.473b) and retrorse tubercles (Fig. 34.472c) sometimes present; mature larvae up to 1 or 2 cm long (Figs. 34.472b, 34.473a, 34.474a) .... *Dermestes*

   Abdominal segment 10 membranous, consisting mostly of lobes around anus; all setae simple; terga without tubercles; mature larvae minute, about 2.0 mm long (Fig. 34.480) (very rare) .... *Thorictodes*

4. Maxillary palp 4-segmented; body without hastisetae 5

   Maxillary palp 3-segmented; body with hastisetae, some of which form tufts on or behind 1 or more posterior abdominal segments (Fig. 34.471a) 7

5. 4 or 5 stemmata on each side of the head; caudal brush of long spicisetae present; some lanceolate setae (Fig. 34.477d) or scale-like setae present on terga; body elongate and somewhat tapering posteriorly (Fig. 34.477a) .... *Attagenus, Novelsis*

   3 or 6 stemmata on each side of head; no caudal brush of long spicisetae present; spicisetae (Figs. 34.471b, 34.476b, 34.478c) present on terga but no lanceolate or scale-like setae present; body somewhat compact, C-shaped, broadest near middle 6

6. Each tergum with a posterior, comb-like row of coarse, short, clavate spicisetae (Fig. 34.478c); 3 large subequal stemmata on each side of head .... *Thylodrias*

   Terga without comb-like series of clavate setae but with long spicisetae that are quadrate in cross section, some half as long as body or longer (Figs. 34.471b, 34.476b); 6 stemmata on each side of head in 3 groups of 2 each .... *Apsectus*

7. Abdominal segment 7 and often 5 and 6 with brush of hastisetae inserted on each side of membrane behind tergum 8

   All hastisetae and hastisetal brushes inserted on sclerotized areas of terga, never on membranes behind terga (in *Trogoderma primum* brush of hastisetae on abdominal segment 7 inserted on semidetached sclerite behind tergum) 9

8. Brushes of hastisetae present behind abdominal segments 5, 6 and 7; body compact, broadest at abdominal segments 4–6 .... *Anthrenus*

   Brushes of hastisetae present behind abdominal segment 7 only; body more elongate, broadest at or before abdominal segment 2 .... *Cryptorhopalum, Orphinus, Thaumaglossa*

9. Setae of tarsungulus unequal in length, one about half as long as other; middle setae of labral-epipharyngeal margin consisting of 2 broad inner and 2 very narrow outer setae (Fig. 34.481) .... *Trogoderma*
   Setae of tarsungulus subequal in length; middle setae of labral-epipharyngeal margin various 10
10. Middle 4 setae of labral-epipharyngeal margin consisting of 2 broad inner and 2 very narrow outer setae .... *Reesa*
   Middle 4 setae of labral-epipharyngeal margin consisting of 4 broad (spatulate) setae (Fig. 34.482) .... *Megatoma*

Selected Bibliography

Anderson 1949 (larva of *Thorictodes* as *Thaumaphrastus*).
Beal 1959a, 1959b, 1960 (*Trogoderma* larvae), 1964, 1970, 1975, 1985 (*Orphilus* larva).
Crowson 1955, 1959, 1961a.
Emden 1924 (larva of *Thorictodes*), 1951 (*Thorictodes* = *Thaumaphrastus*).
Hinton 1945b (stored products pests), 1946a (gin traps).
Lawrence and Hlavac 1979.
Lawrence and Newton 1982.
Mroczkowski 1968, 1975.
Paulian 1942.
Rees 1943.
Zhantiev 1976.

BOSTRICHIDAE (BOSTRICHOIDEA)

(= BOSTRYCHIDAE, INCLUDING ENDECATOMIDAE, LYCTIDAE)

John F. Lawrence,
Division of Entomology, CSIRO

Powderpost Beetles, Twig and Wood-Borers

Figures 34.483–487

Relationships and Diagnosis: The Bostrichidae, as defined by Crowson (1961a), includes *Endecatomus,* as well as those genera formerly comprising the family Lyctidae. Bostrichid larvae differ from those of other Bostrichoidea in having well-developed, 3-segmented antennae, and, except for those of *Endecatomus,* they all differ from ptinid or anobiid larvae in having a strongly retracted head. A similar body form and retracted head are present in Bruchidae and sagrine Chrysomelidae, whose larvae differ from bostrichids in lacking the longitudinally-oval anal pads and in having the ventral mouthparts more retracted. *Endecatomus* larvae resemble those of Anthribidae, but the latter also lack the anal pads and have highly reduced antennae.

Biology and Ecology: Most Bostrichidae are wood-borers, and some will attack hard, well-seasoned timber. Adults are capable of boring vertically into hard wood and then excavating horizontal galleries in which they lay their eggs; after egg-laying the adults often die within the entrance tunnel, thus blocking it and preventing the entrance of predators or parasitoids. Larvae excavate galleries parallel to the wood surface, with each generation boring somewhat deeper into the wood, and then proceeding toward the surface for pupation. Many species prefer smaller twigs, vines, or branches, but others attack larger stumps, trunks, roots, or bark. Many Lyctinae and Dinoderinae prefer to attack monocotyledons, such as bamboos or canes. The lesser grain borer, *Rhyzopertha dominica* (Fabricius) (fig. 34.486), has become a cosmopolitan pest of stored grains and a wide variety of cereal products. Members of the Lyctinae may be serious pests of wood products composed of sapwood and are particularly destructive to articles made of cane and bamboo. Species of *Endecatomus* are atypical for the family, since they feed as both adults and larvae in the fruiting bodies of bracket fungi; they have been associated with several polypore species, including *Bjerkandera adusta, Phellinus gilvus, Inonotus radiatus,* and *Fomes fomentarius.*

Description: Mature larvae 3 to 60 mm, usually less than 20 mm. Body elongate, subcylindrical or slightly flattened, almost always moderately to strongly curved ventrally (C-shaped) (first instars sometimes straight), lightly sclerotized, except for buccal region, whitish to cream or yellow; surfaces smooth; vestiture usually consisting of a few, scattered, simple setae dorsally, sometimes with dense patches of long, fine hairs laterally and ventrally; entire body covered with long, fine hairs in *Endecatomus.*

Head: Usually retracted and prognathous, elongate and somewhat flattened (protracted, hypognathous, and subglobular in *Endecatomus*). Epicranial stem long; frontal arms usually indistinct or absent, sometimes V-shaped. Median endocarina coincident with epicranial stem. Stemmata absent or 1 on each side. Antennae usually moderately long, occasionally very short, 3-segmented or 2-segmented due to fusion of segments 2 and 3 (in which case sensorium located in middle of terminal segment). Frontoclypeal suture present; clypeus transverse; labrum free, sometimes enlarged with pair of epipharyngeal rods; pharyngeal sclerome (fulcrum) sometimes present. Mandibles symmetrical, robust, unidentate to tridentate, sometimes wedge-like, without accessory ventral process. Mola absent in Bostrichinae and Euderiinae, which have mandibular base simple or sometimes with an articulated process associated with a membranous lobe; remainder of family with simple mola (pseudomola), membranous prostheca, and brush of hairs at base of mola. Ventral mouthparts retracted. Maxilla with transverse cardo, elongate stipes, narrow or indistinct articulating area, and 2- or 3-segmented palp borne on distinct palpifer, so that it may appear to have an additional segment. Galea broadly rounded or truncate, fixed or articulated, and lacinia falciform, with sclerotized spur at apex. Labium sometimes with mentum and submentum fused; ligula present or absent; labial palps 1- or

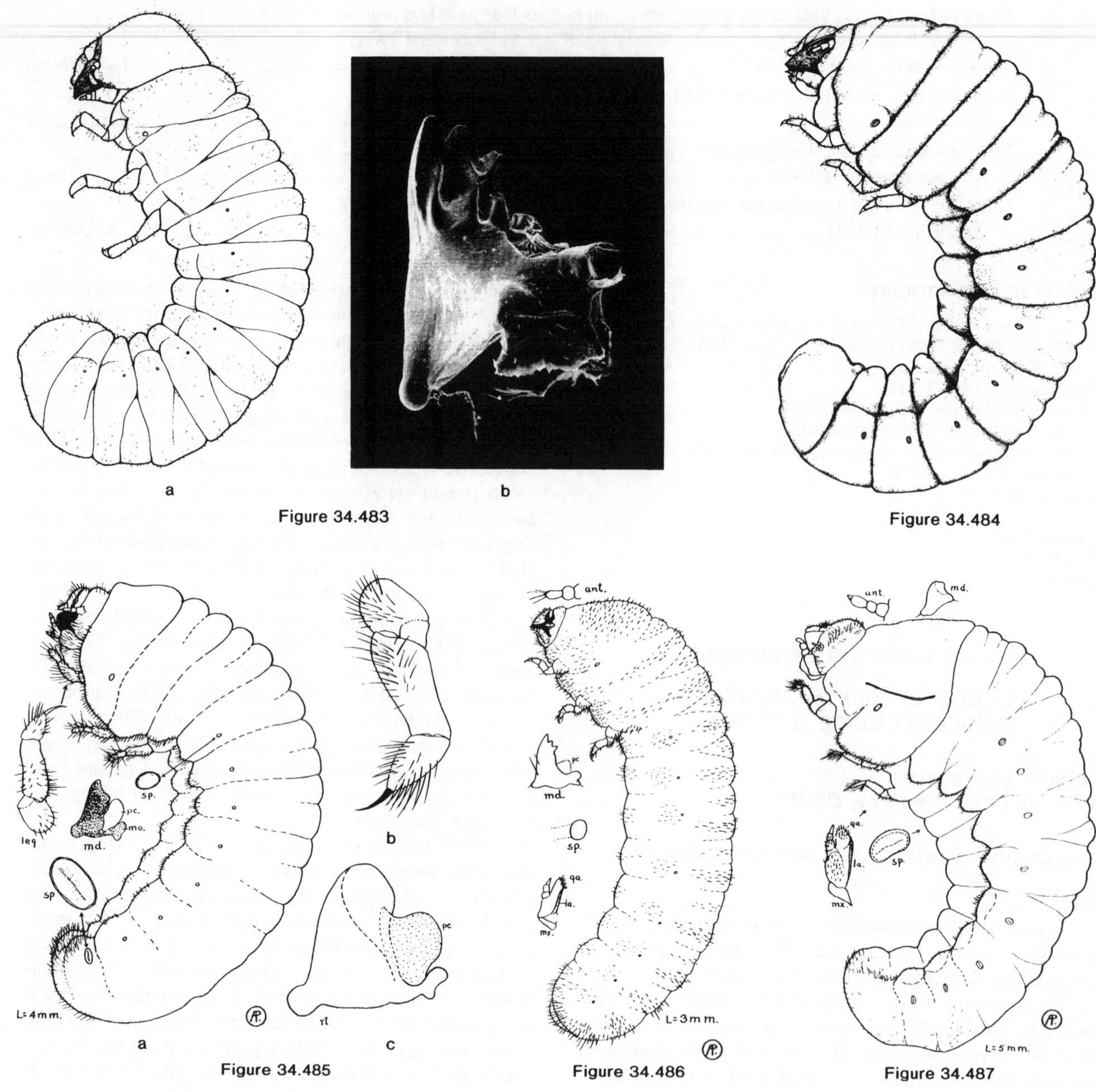

Figures 34.483a,b. Bostrichidae. *Endecatomus rugosus* Randall. Cedar Rapids, Linn Co., Iowa. Length = 5.8 mm. **a.** larva, lateral; **b.** right mandible, ventral.

Figure 34.484. *Psoa maculata* LeConte. San Luis Obispo Co., California. Larva, lateral. Length = 13.4 mm.

Figures 34.485a-c. Bostrichidae. *Lyctus planicollis* LeConte, Southern lyctus beetle. Length = 4 mm. **a.** larva, lateral (plus details); **b.** prothoracic leg; **c.** right mandible, ventral. Near white, including head except the area adjacent to the mouthparts; short, colorless, setae confined largely to the head, legs and ventrolateral aspects of each segment; labrum densely setiferous; prothoracic legs similar to those on mesothorax and metathorax but larger, more setiferous and with distinct, elongated, curved spine-like claws; spiracles on eighth abdominal segment much larger and more nearly oval than those on segments 1-7. Infests many kinds of dry or seasoned hardwood, especially ash, hickory and oak. (Figures 34.485a-c from Peterson, 1951)

Figure 34.486. Bostrichidae. *Rhyzopertha dominica* (Fabricius), lesser grain borer. Larva, lateral (plus details). Length = 3 mm. Near white except for the pigmented mouthparts, portions of the partially retracted head, and claws; a few, short light-colored setae on most parts of the body and head, longest and most abundant at caudal end and on ventral aspect; spiracles annuliform and very inconspicuous. Infests nearly all stored grains, especially wheat, and other products, namely cork, dried roots, drugs, leather goods and seeds. Also many penetrate paper boxes containing food. (From Peterson, 1951.)

Figure 34.487. Bostrichidae. *Xylobiops texanus* (Horn). Larva, lateral (plus details). Length = 5 mm. Infests dry mesquite wood. (from Peterson 1951)

2-segmented. Hypopharyngeal sclerome present or absent. Hypostomoal rods absent or very short. Ventral epicranial ridges absent. Gular region absent.

Thorax and Abdomen: Thorax usually enlarged with prothorax larger than meso- or metathorax (not so in *Endecatomus*). Prothorax in Bostrichinae with an oblique, sclerotized rod on each side. Legs relatively small, 5-segmented; tarsungulus slender, without setae or with 2 setae lying side by side. Meso- and metatergum and abdominal terga 1 to 6 or 7 each with 2 to 4 transverse plicae, but without patches of asperities. Abdominal segments 1–8 usually with well-developed pleural lobes. Segment A9 enlarged, almost always simple (small urogomphi or median process present in some first instars). Segment A10 reduced, usually with pair of longitudinal oval anal pads, but in Lyctinae with transverse pad as well.

Spiracles: Annular or annular-uniforous; thoracic spiracles located on posterior part of prothorax or on mesothorax; spiracles on segment A8 enlarged in Lyctinae.

Comments: The family includes about 90 genera and 700 species distributed worldwide, with 35 genera and about 100 species in N. America. There are 7 recognized subfamilies, which appear to form distinct groups: Endecatominae, Dysidinae-Psoinae, Dinoderine-Lyctinae, and Euderiinae-Bostrichinae. The genus *Endecatomus* is Holarctic; Dysidinae includes the S. American *Dysides* and the Oriental *Apoleon;* Psoinae are widespread, with individual groups in the Holarctic Region, tropical America, southeast Asia, and Chile; and *Euderia* is restricted to New Zealand. The Dinoderinae, Lyctinae, and Bostrichinae are all widely distributed.

Selected Bibliography

Anderson 1939 (larval key).
Baker, N. 1971 (habits of *Dinapate wrighti* Horn).
Beeson and Bhatia 1936.
Böving and Craighead 1931 (larvae of *Lyctus cavicollis* LeConte, *Polycaon stouti* (LeConte), *Stephanopachys substriatus* (Paykull), and *Scobicia declivis*.
Burke, H. E. *et al.* 1922 (biology of *Scobicia declivis*) (LeConte).
Chararas and Balachowsky 1962 (agricultural pests).
Crowson 1961a (larval key; larvae of *Endecatomus rugosus* (Randall) and *Euderia squamosa* Broun).
Drooz 1985.
Emden 1943 (larval key).
Fisher 1950 (North American revision, except Lyctinae).
Furniss and Carolin 1977 (forest pests).
Gardner 1933b (keys and descriptions of various larvae).
Gerberg 1957 (revision of North American Lyctinae).
Iablokov 1940 (mycophagy in *Endecatomus*).
Iwata and Nishimoto 1981 (larva and pupa of *Lyctus brunneus* (Stephens)).
Kojima 1932 (biology of *Lyctus linearis* (Goeze)).
Kompantsev 1978 (larva of *Endecatomus lanatus* Lesne).
Lesne 1934 (*Euderia squamosa* Broun).
Mateu 1967 (larva and biology of *Xylomedes rufocoronata* (Fairmaire)).
Mathur 1956 (keys and descriptions of various larvae).
Matthewman and Pielou 1971 (mycophagy in *Endecatomus*).
Peterson 1951 (larvae of *Rhizopertha dominica* and *Xylobiops texanus* (Horn)).
Pringle 1938b (larvae of *Lyctus brunneus, Xylopsocus sellatus* (Fabricius), *Heterobostrychus brunneus* Murray, *Xylion adustus* (Fåhraeus), and *Enneadesmus forficula capensis* Lesne).
Snyder 1916 (egg and oviposition in *Lyctus planicollis* LeConte).
Weiss and West 1920 (mycophagy in *Endecatomus*).

ANOBIIDAE (BOSTRICHOIDEA)

John F. Lawrence,
Division of Entomology, CSIRO

Manuel G. de Viedma,*
Universidad Politecnica de Madrid, Spain

Deathwatch Beetles, Furniture Beetles, Cigarette Beetle, Drugstore Beetle

Figures 34.488–495

Relationships and Diagnosis: Anobiidae, like Ptinidae and Bostrichidae, are characterized in the larval stage by the lightly sclerotized and more or less C-shaped or scarabaeiform body, usually with well-developed legs, and with a pair of longitudinally oval pads beneath the anus. Anobiid larvae may be distinguished from those of Bostrichidae by the combination of protracted and hypognathous head and very short antennae, which usually have only 1 evident segment. They differ from larvae of the closely related Ptinidae by the location of the thoracic spiracle posteriorly on the prothorax or between the prothorax and mesothorax, and usually by the presence of transverse bands of asperities on the abdominal terga. Scarabaeoid larvae differ from those of anobiids in having longer, 3- or 4-segmented antennae, a well-developed mandibular mola and accessory ventral process, and cribriform (or occasionally biforous) spiracles. Larvae of some Anthribidae resemble anobiid larvae in several respects but lack the longitudinal anal pads and tergal asperities; the same 2 features will separate legless *Caenocara* larvae from those of other curculionoids.

Biology and Ecology: Larvae of most Anobiidae feed on dead or dying wood, but some species attack conifer cones, twigs, vines, seeds, galls, fungus, fruiting bodies, or stored products of both animal and plant origin. Xylophagous species occur in all subfamilies, and include serious pests of furniture and structural timber, such as *Ernobius mollis* (L.), *Anobium punctatum* (De Geer), and *Xestobium rufovillosum* (De Geer). Larvae of the last species, as they tunnel through wood, produce a tapping sound once considered to be an omen of death; this is the origin of the common name deathwatch beetle. Within the Ernobiinae, *Ozognathus* species occur in stems, twigs, and galls, while some *Ernobius* attack the cones of conifers. The Dorcatominae include some wood-damaging forms, especially in the genus *Calymmaderus,* but the majority of species are mycophagous. In North America, species of *Dorcatoma* and *Byrrhodes* occur in the larger, more durable fruiting bodies of *Fomes, Ganoderma, Inonotus,* and *Phellinus,* while species of *Caenocara* are apparently restricted to puffballs (*Calvatia, Lycoperdon, Scleroderma,* etc.) (Benick, 1952; Lawrence, 1973; Matthewman and Pielou, 1971; Weiss, 1922; Weiss and West, 1920, 1921b). The subfamily Tricoryninae includes most of the seed-feeding anobiids, but *Tricorynus herbarius* (Gorham) damages books, leather, and stored foods, while *T. tabaci* (Guérin-Méneville)

*Deceased

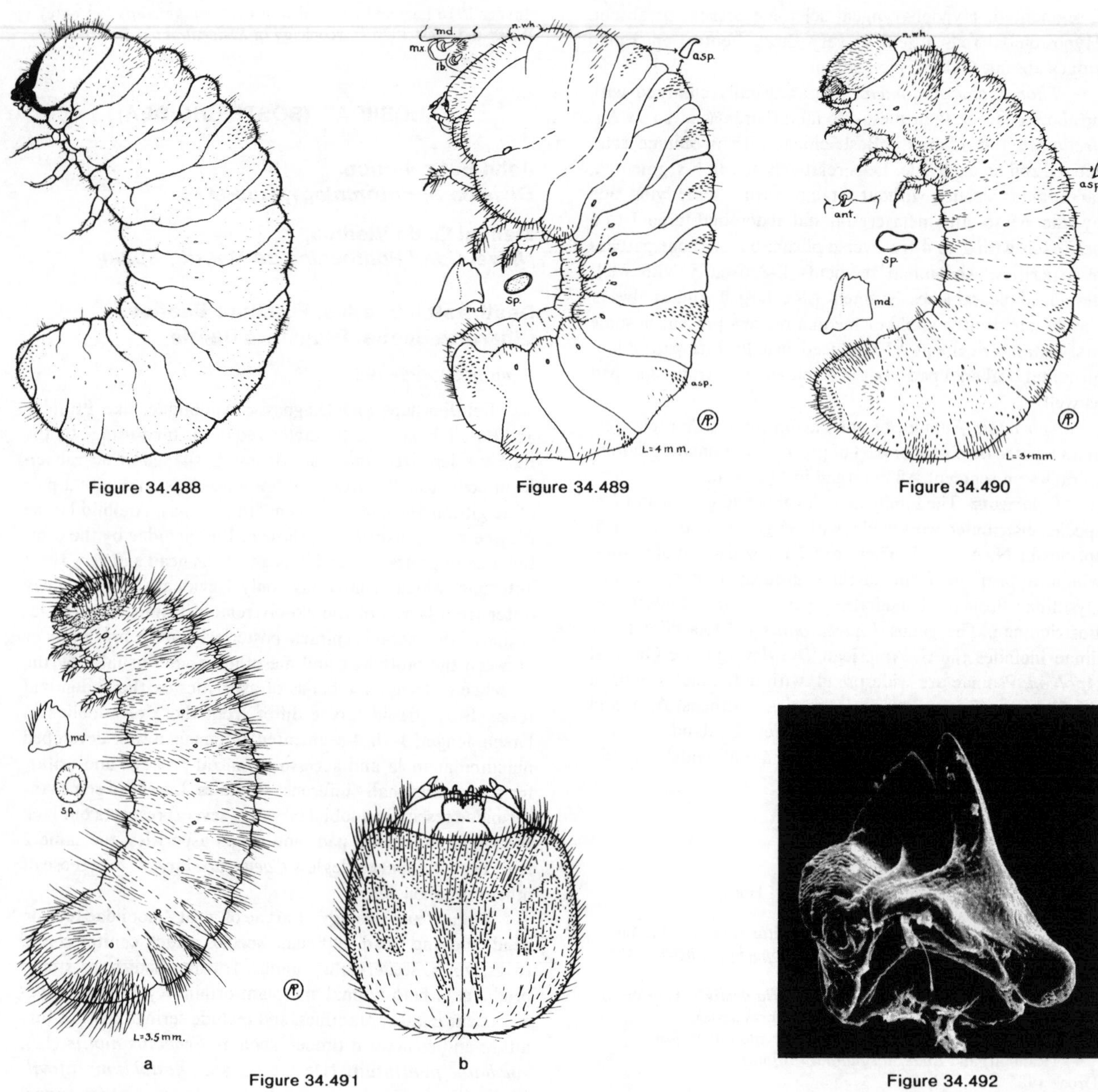

Figure 34.488. Anobiidae. *Dorcatoma* sp. Rensselaerville, Albany Co., New York. Larva, lateral. Length = 8.3 mm.

Figure 34.489. Anobiidae. *Hemicoelus carinatus* (Say). Larva, lateral (plus details). Length = 4 mm. Near white except cream head and moderately to deeply pigmented epistoma, mandibles, asperities and tarsi; short or long, light setae on all parts of body, most numerous on head, caudal segments and ventrad of spiracles on all segments; thorax enlarged; groups of short, curved asperities on dorsa of metathorax to abdominal segment 7, and on caudolateral apsects of caudal segment; spiracles oval and inconspicuous. Infests dry dead wood in buildings and elsewhere. Recorded chiefly from deciduous trees. (From Peterson, 1951).

Figure 34.490. Anobiidae. *Stegobium paniceum* (L.), drugstore beetle. Larva, lateral (plus details). Length = 3 mm. All segments subequal in diameter except the smaller prothorax. Near white except mouthparts and ventral margin of head capsule; numerous short to long, light colored setae on all areas of head, body and legs; inconspicuous, colored asperities present as single or narrow bands on dorsa of all abdominal segments; inconspicuous peanut-shaped spiracles present. Attacks a wide variety of dry plant products including book covers containing glue, paste or casein, leather, and bamboo. (From Peterson, 1951)

Figures 34.491a,b. Anobiidae. *Lasioderma serricorne* (Fabricius) cigarette beetle. **a.** larva, lateral (plus details), length = 3.5 mm; **b.** head, anterior. Wrinkled, segments subequal in diameter; near white including head capsule except for yellowish-brown pigmented areas; very numerous, long, light setae on all parts of body, head and legs; spiracles annuliform to oval with thoracic pair on prothorax. Infests mainly dried vegetable products. Destructive to cured leaf tobacco or tobacco products, especially in storage. Also known to infest dried yeast cakes, seeds, dried botanical specimens, dried fish, leather goods, rugs, tapestry, and upholstered furniture. (From Peterson, 1951).

Figure 34.492. Anobiidae. *Byrrhodes* sp. Concord, Massachusetts. Left mandible, mesoventral.

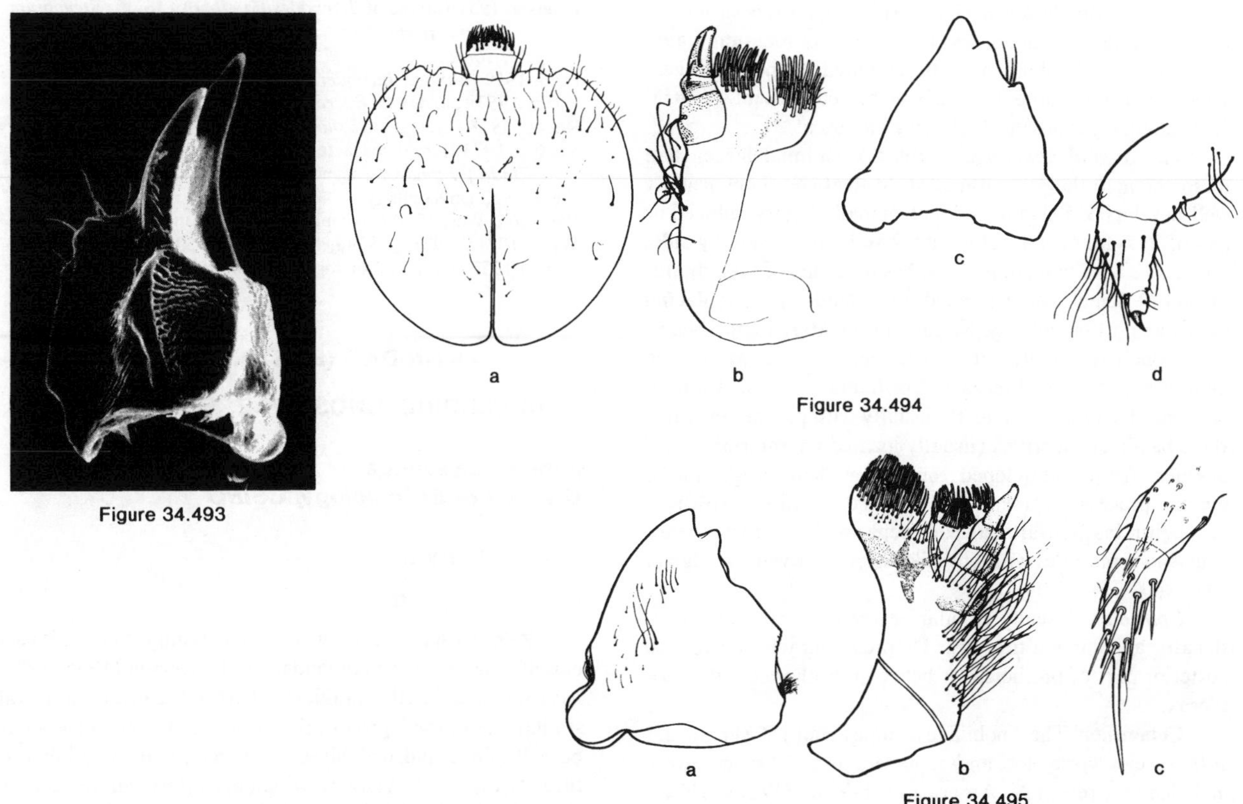

Figure 34.493

Figure 34.493. Anobiidae. *Caenocara* sp. Gloucester, Massachusetts. Left mandible, mesal.

Figures 34.494a-d. Anobiidae. *Grynobius planus* (Fabricius). **a.** head, dorsal; **b.** right maxilla, ventral; **c.** left mandible, dorsal; **d.** apex of prothoracic leg.

Figures 34.495a-c. Anobiidae. *Nicobium villosum* (Brullé). **a.** left mandible, dorsal; **b.** right maxilla, dorsal; **c.** apex of leg.

is found in various stored herbs, bulbs, and seeds (White, 1982). The 2 most common and widespread pests of stored products are the drugstore beetle, *Stegobium paniceum* (L.) (fig. 34.490) (Anobiinae) and the cigarette beetle, *Lasioderma serricorne* (Fabricius) (fig. 34.491a,b) (Xyletininae). These species damage a wide variety of plant and animal products, including tobacco, drugs, spices, seeds, grains, and leather (White, 1982). Many anobiids exhibit an advanced type of digestive physiology, in which intracellular symbionts play a large part in breaking down cellulose and providing B vitamins (Blewett and Fraenkel, 1944; Buchner, 1965; Parkin, 1940); however little is known about the physiology of the primitive Hedobiinae and Dryophilinae. Hedobiinae are apparently unique in forming a silk-like cocoon for pupation (actually formed of chitinous threads produced by the peritrophic membrane); the fact that *Ptinus* species construct a similar cocoon has been used to support the union of the Anobiidae and Ptinidae (Crowson, 1981; Hickman, 1974; Viedma, 1973).

Description: Mature larvae 2 to 12 mm. Body elongate, more or less parallel-sided, subcylindrical to slightly flattened, moderately to strongly curved ventrally (C-shaped), lightly sclerotized, except for buccal region, whitish to yellowish; vestiture consisting of short, simple setae or longer, fine hairs, often dense in places.

Head: Protracted and hypognathous (rarely retracted and prognathous), globular to somewhat flattened. Epicranial stem long; frontal arms usually indistinct or absent, sometimes V-shaped. Median endocarina coincident with epicranial stem. Stemmata absent or 1 on each side. Antennae very short, usually 2-segmented, with sensorium on segment 1 longer than segment 2, sometimes 1-segmented with sensorium at apex. Frontoclypeal suture present; labrum free; clypeolabral region narrow. Mandibles usually symmetrical, robust, unidentate to tridentate (occasionally 4-dentate), without accessory ventral process; mesal surface of mandibular base usually simple or with brush of hairs (penicillus); occasionally a mola (pseudomola) present which may be ridged or tuberculate. Ventral mouthparts retracted. Maxilla with transverse cardo, elongate stipes, and articulating area narrow, indistinct, or absent; maxillary palp usually 3-segmented, occasionally 2-segmented, or 4-segmented where palpifer is completely articulated; galea and lacinia fixed; galea rounded or truncate; lacinia rounded to falciform, sometimes with sclerotized spur or spine at apex, occasionally vestigial. Labium more or less free to base of mentum; ligula short and broad; labial palps 2-segmented. Hypopharyngeal sclerome usually absent. Hypostomal rods absent. Ventral epicranial ridges absent. Gular region absent.

Thorax and Abdomen: Prothorax not much larger than meso- or metathorax, without sclerotizations. Legs usually well-developed, 5-segmented, occasionally highly reduced or absent; tarsungulus usually with 2 setae lying side by side, sometimes with numerous setae; basal portion of tarsungulus (arolium) sometimes expanded and bladder-like, replacing claw-like portion entirely. Meso- and metaterga and abdominal terga 1–7 usually with 2 transverse plicae. Meso- and metaterga, abdominal terga 1–7 or 8, and sometimes lateral portions of segments 9 and 10, usually with patches or transverse bands of asperities (usually located on anterior plica). Segment A9 well-developed, rounded, without urogomphi or median process. Segment A10 reduced, terminal, with pair of longitudinally oval anal pads separated by groove below anus and sometimes with bow-like sclerite at ventral edge of anal pads.

Spiracles: Usually annular or annular-uniforous; occasionally annular-multiforous. Thoracic spiracle located on posterior part of prothorax or between prothorax and mesothorax.

Comments: The Anobiidae include about 150 genera and 1600 species worldwide, and 52 genera and 332 species were included in a recent N. American catalogue (White, 1982). The family is currently divided into 9 subfamilies on the basis of adult characters (White, 1974b): Hedobiinae, Dryophilinae, Anobiinae, Ptilininae, Alvarenganiellinae, Xyletininae, Dorcatominae, and Tricoryninae; except for the Hedobiinae, however, none of these divisions have been characterized using larval features (Böving, 1954; Viedma, 1973, 1977). The major work on anobiid larvae is that of Böving (1954), which includes representatives of more than 40 genera.

Selected Bibliography

Blewett and Fraenkel 1944 (intracellular symbiosis and vitamin requirements).
Böving 1954 (larvae of many species; larval key), 1955 (corrections to 1954).
Böving and Craighead 1931 (larvae of *Hedobia imperialis* (Linnaeus), *Trichodesma klagesi* Fall, *Lasioderma serricorne*, and *Nevermannia dorcatomoides* Fisher).
Buchner 1965 (intracellular symbiosis).
Crowson 1955, 1961a, 1981.
Drooz 1985.
Eidmann and Heqvist 1958 (larva of *Dryophilus pusillus* (Gyllenhal)).
Espanõl 1970.
Frankie 1973 (biology of *Ernobius conicola* Fisher).
Furniss and Carolin 1977 (forest pests).
Gardner 1937b (larvae of several species; larval key).
Hayashi 1951 (*Dorcatoma* larvae and pupae).
Hickman 1974.
Lawrence 1973 (mycophagy).
Matthewman and Pielou 1971 (mycophagy).
Palm 1959.
Parkin 1933 (larvae of wood-boring species; larval key), 1940 (digestive enzymes).
Peterson 1951 (larvae of *Trichodesma gibbosa* (Say), *Stegobium paniceum*, *Hemicoelus carinatus* (Say), and *Lasioderma serricorne*).
Prota 1959 (larva of *Lasioderma baudii* Schilsky).
Rozen 1957 (biology of *Eucrada humeralis* (Melsheimer)).
Ruckes 1958 (biology of *Ernobius* spp.).
Viedma 1973 (larval characters of Hedobiinae), 1977 (larva of *Clada fernandezi* Espanõl).
Weiss 1922 (mycophagy).
Weiss and West 1920 (mycophagy), 1921b (mycophagy).
White 1971a, 1971b, 1974a, 1974b, 1982 (catalogue).
Zocchi 1957 (larva and biology of *Ernobius abietis* (Fabricius)).

PTINIDAE (BOSTRICHOIDEA) (INCLUDING GNOSTIDAE, ECTREPHIDAE)

John F. Lawrence,
Division of Entomology, CSIRO

Spider Beetles

Figures 34.496–497

Relationships and Diagnosis: The family Ptinidae is very closely related to the Anobiidae, and Crowson (1955, 1981) has suggested that the families be merged because of the larval similarities in the 2 groups, the presence of a pupal cocoon in both Ptinidae and hedobiine Anobiidae, and the adult features of the genus *Xylodes*, which are intermediate between the 2 groups. Ptinid larvae may be distinguished from those of anobiids mainly by the position of the thoracic spiracle near the anterior edge of the prothorax, but also by the absence of tergal asperities present in most anobiid larvae.

Biology and Ecology: Ptinidae are primarily scavengers and feed on a wide variety of animal and plant products. It is thought that wood-boring may be the ancestral habit of the group (Crowson, 1955), and at least a few species of *Ptinus*, such as *P. palliatus* Perris and *P. lichenum* Marsham, do occur in wood (Belles, 1980; Hinton, 1941e). It is much more common, however, for Ptinidae to feed on feces, hair, feathers, arthropod cuticle, or other organic debris, especially that accumulating in caves or in the nests of mammals, birds, spiders or insects. In arid regions, caves and animal nests provide them with access to drinking water, without which they cannot produce eggs (Howe, 1959). Some species, such as *Ptinus sexpunctatus* Panzer and *P. californicus* Pic, which occur in the nests of bees, have become depredators, feeding on the pollen stores (Linsley, 1944; Linsley and MacSwain, 1942). According to Hickman (1974), the Australian *Ptinus exulans* Erichson lives in the nests of a salticid spider (*Plexippus validus* Urquhart), where both larvae and adults feed on dead arthropods and spider eggs. Myrmecophily has evolved at least 5 times in the family, and the more highly specialized forms, such as the Australian Ectrephinae, are probably fed by the ants, at least as adults (Lawrence and Reichardt, 1969). The best known ptinids are those which have become pests of stored products and have been spread worldwide by human transport. Examples are *Mezium americanum* Laporte, *Gibbium psylloides* (Czenpinski), *Niptus hololeucus* (Falderman),

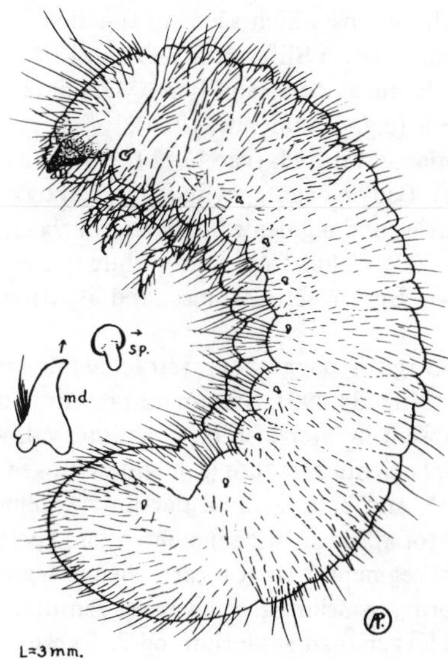

Figure 34.496. Ptinidae. *Ptinus brunneus* Duftschmid. Illinois. Larva, lateral (plus details). Length = 3 mm. Collected from tankage in Illinois. (from Peterson 1951)

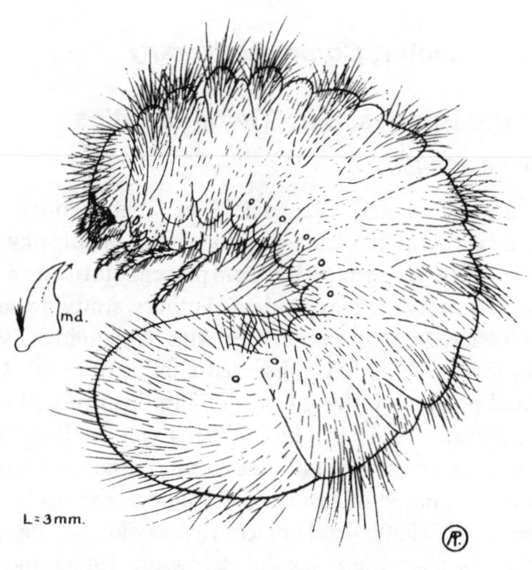

Figure 34.497. Ptinidae. *Mezium americanum* Laporte. Cleveland, Ohio. Larva, lateral (plus details). Length = 3 mm. Reared from rat pellets. (from Peterson 1951)

Ptinus fur (Linnaeus), *P. villiger* Reitter, and *P. tectus* Boieldieu. According to Howe (1959), ptinids cause only minor damage to most goods, either by the accumulation of feces or silk, or by the larvae boring into materials to pupate. The "silken" cocoon formed for pupation has been shown to be composed of chitin and produced by the peritrophic membrane (Crowson, 1981).

Description: Mature larvae 2 to 6 mm. Body elongate, more or less parallel-sided, subcylindrical to slightly flattened, moderately to strongly curved ventrally (C-shaped), lightly sclerotized, whitish to yellowish; vestiture of moderately long, fine hairs, densely and more or less uniformly distributed.

Head: Protracted and hypognathous. Epicranial stem long; frontal arms absent. Median endocarina coincident with epicranial stem. Stemmata absent. Antennae very short, 1-segmented, with well-developed sensorium at apex. Frontoclypeal suture present; labrum free; clypeolabral region narrow. Mandibles symmetrical, robust, wedge-like, with simple, unidentate apex, oblique incisor edge, and no accessory ventral process; mesal surface of mandibular base simple; mola absent. Ventral mouthparts retracted. Maxilla with transverse cardo, elongate stipes, narrow or indistinct articulating area, and 3-segmented palp; galea truncate and lacinia falciform with sclerotized, spine-like apex. Labium free to base of mentum; ligula absent; labial palps 2-segmented. Hypopharyngeal sclerome absent. Hypostomal rods absent. Ventral epicranial ridges absent. Gular region absent.

Thorax and Abdomen: Prothorax not enlarged and without sclerotizations. Legs well-developed, 5-segmented; tarsungulus usually with 1 seta, sometimes with 2. Meso- and metaterga and abdominal terga 1–7 with 2 transverse plicae; tergal asperities absent. Segment A9 well-developed, rounded, without urogomphi or median process. Segment A10 reduced, terminal, with pair of longitudinally oval anal pads separated by groove below anus and with U-shaped or V-shaped sclerite (preanal sclerite) at ventral edge of pads.

Spiracles: Annular or annular-uniforous. Thoracic spiracles located near anterior end of prothorax.

Comments: The family contains about 50 genera and 500 species, with 11 genera and about 50 species occurring in America north of Mexico. The subfamily and tribal classification is still in doubt. Belles (1981) has divided the family into 9 informal groups based on the genera *Xylodes, Gynopterus, Maheoptinus, Ptinus, Niptus, Casopus, Trigonogenius, Sphaericus,* and *Gibbium*. This breakdown does not include the myrmecophilous groups, such as *Gnostus* and *Fabrasia* in the New World, *Diplocotidus* in South Africa, or the Australian Ectrephinae.

Selected Bibliography

Belles 1980 (biology of *Ptinus lichenum*), 1981.
Böving 1956 (larva of *Ptinus californicus*).
Böving and Craighead 1931 (larvae of *Niptus* sp. and *Ptinus fur*).
Braune 1930 (biology of *Niptus hololeucus*).
Crowson 1955, 1981.
Hall and Howe 1953 (key to larvae in stored products).
Hickman 1974 (biology of *Ptinus exulans*).
Hinton 1941e (species of economic importance).
Howe 1959 (biology of the Ptinidae).
Lawrence and Reichardt, 1969 (myrmecophily).
Linsley 1944 (natural reservoirs for stored products species).
Linsley and MacSwain 1942 (biology of *Ptinus californicus*).
Manton 1945 (key to larvae in stored products).
Mathur 1957a (larva of *Gibbius psylloides*).
Peterson 1951 (larvae of *Ptinus clavipes* Panzer and *Mezium americanum*).
Viedma 1973.

LYMEXYLIDAE (LYMEXYLOIDEA)

Quentin Wheeler, *Cornell University*

Lymexylid Beetles, Ship-Timber Beetles

Figures 34.498-501

Relationships and Diagnosis: The Lymexylidae is a primitive family of the Cucujiformia of uncertain phylogenetic relationship, reflected by their present status as a separate superfamily, Lymexyloidea. Authors studying larval characters have noted similarities with Cucujoidea, while those studying adults have observed similarities with Cleroidea (and previously "Malacoderm" families) (e.g., Böving and Craighead, 1931; Bright, 1979). Lymexyloidea is regarded here as a monofamilial superfamily, although Crowson (1981) has proposed inclusion of Stylopidae (Strepsiptera). Crowson (1955, 1960b, 1981) places Lymexylidae in the series Cucujiformia near Cleroidea (cf. Crowson, 1964) and Cucujoidea (Clavicornia + Heteromera). This situation is presently an unresolved trichotomy. Wheeler (1986) discusses phylogenetic relationships of world genera and recognizes 3 subfamilies: Hylecoetinae (*Hylecoetus*), Lymexylinae (*Lymexylon, Atractocerus*), and Melittomminae (a new subfamily for *Melittomma* and related genera *Protomelittomma, Melottommopsis,* and *Australymexylon*). North American Lymexylidae include *Hylecoetus lugubris* and *Melittomma sericeum* in the U.S. and Canada, and *Atractocerus brasiliensis* in Mexico.

The large hoodlike pronotum, elongate body form, lateral folds, and highly modified ninth abdominal segment (figs. 34.498–500) distinguish the Lymexylidae from most other Coleoptera. Three major structural configurations of the ninth abdominal segment exist: (1) a long, heavily sclerotized, serrated swordlike form in Hylecoetinae (fig. 34.499); (2) a large, bulbous, membranous to lightly sclerotized form in Lymexylinae (fig. 34.500); and a truncate, cylindrical, heavily sclerotized (at least apically) and often toothed form in Melittomminae (sensu Wheeler 1986; cf. fig. 34.498). Some lymexylids (*Hylecoetus* and *Atractocerus*) have plurisetose tarsunguli (fig. 34.501), and late-instar larvae lack stemmata.

Biology and Ecology: All Lymexylidae are wood-boring as larvae, and believed to be symbiotic associates of fungi and microbes. *Hylecoetus dermestoides* L., a Palearctic species, is best understood, and the North American *H. lugubris* is apparently identical in habits. *Hylecoetus* is fungus-growing (Wilson, 1971), cultivating *Ascoidea hylecoeti* in tunnels in many hardwoods and most softwoods (cf. Batra & Francke-Grosmann, 1961; Batra, 1967; Francke-Grosmann, 1967; Lyngnes, 1958). Spores are transmitted to eggs in a sticky substance, and host spores are stored by the female in fungus-pouches of the genitalia (cf. Francke-Grosmann, 1967; Wheeler, 1986). First instar larvae differ from later instars in body proportion, having stemmata, and sometimes bearing trichobothria. Pupation occurs in the wood. Some larvae may live for as many as 6 years or more, and have an indeterminate number of instars. The short-lived adults have bizarre maxillary palp organs which seem to function in chemoreception (Germer, 1912; Slifer et al., 1975). Efforts to determine the symbiotic associates of lymexylids have sometimes proven difficult (e.g., Simmonds, 1956).

Description: Mature larvae 12–20 mm (tropical species up to 50 mm). Body elongate, cylindrical, orthosomatic, definitely sclerotized and pigmented with lateral abdominal folds (figs. 34.498–500). Colors from nearly white to creamy yellow to dark brown. Sparse setae, spines, and asperities on integument.

Head: Large, hypognathous, retracted into large, hoodlike pronotum (fig. 34.499). Stemmata present in first instar larvae, but absent in later instars. Cranium oval, with short postero-medial epicranial suture (sometimes broken or hidden by pronotum), and pale lines on dorsum. Antennae small, 3-segmented (or apparently 2-segmented), and posterodorsal to mandibles; segment 2 (first clearly visible segment) with large digitiform sensorium and few other sensilla; segment 3 only slightly larger than sensorium on 2. Labrum elongate, simple. Epipharynx membranous, with dense setae. Mandible broad, short, and heavily sclerotized; with distinct molar part covered by transverse grooves or minute spines; prostheca absent. Maxilla with distinct cardo; stipes and mala fused; mala partially divided at apex in some taxa, and mesal surface with double row of long setae and membranous area between; maxillary palp 3-segmented. Labium small, lightly sclerotized; labial palp 2-segmented; ligula broad, membranous.

Thorax and Abdomen: Prothorax large, hoodlike, partially covering head; with sparse spines and asperities, especially anteriorly. Meso- and metathorax smaller than prothorax. Three pairs of small but well-developed legs; legs with many spines and setae, terminating in heavily sclerotized tarsungulus which is bisetose or plurisetose (*Hylecoetus* and *Atractocerus*).

Abdomen elongate, subcylindrical, with 9 large segments, each with lateral fold (fig. 34.500), and a minute tenth segment inferior to ninth and restricted to area surrounding anus. Ninth segment with 3 structural types: long, heavily sclerotized, swordlike and spinose (e.g., *Hylecoetus,* fig. 34.499); large, bulbous, lightly sclerotized, usually with dense asperities (e.g., *Atractocerus,* fig. 34.500); and cylindrical, truncate, and often with large teeth around apical rim (e.g., *Melittomma,* fig. 34.498).

Spiracles: Present on abdominal segments 1–8; lateral, annular or annular-multiforous, elliptical.

Comments: The Lymexylidae includes about 7 genera and 50 species worldwide. The family is primarily tropical, with only two species established in the United States and Canada (*Hylecoetus lugubris* Say and *Melittomma sericeum* Harris), easily identified by the form of the ninth abdominal segment (cf. figs. 34.498, 34.499). *Atractocerus brasiliensis* is a widespread species in Mexico and Central and South America. In Europe, *Hylecoetus dermestoides* is sometimes a serious pest of stored lumber, but *H. lugubris* has only rarely been a problem in N. America. *Melittomma sericeum* used to infest up to 90% of the trees in local stands of chestnut in N. America, but is today an uncommon species. Although many lymexylid larvae have been described

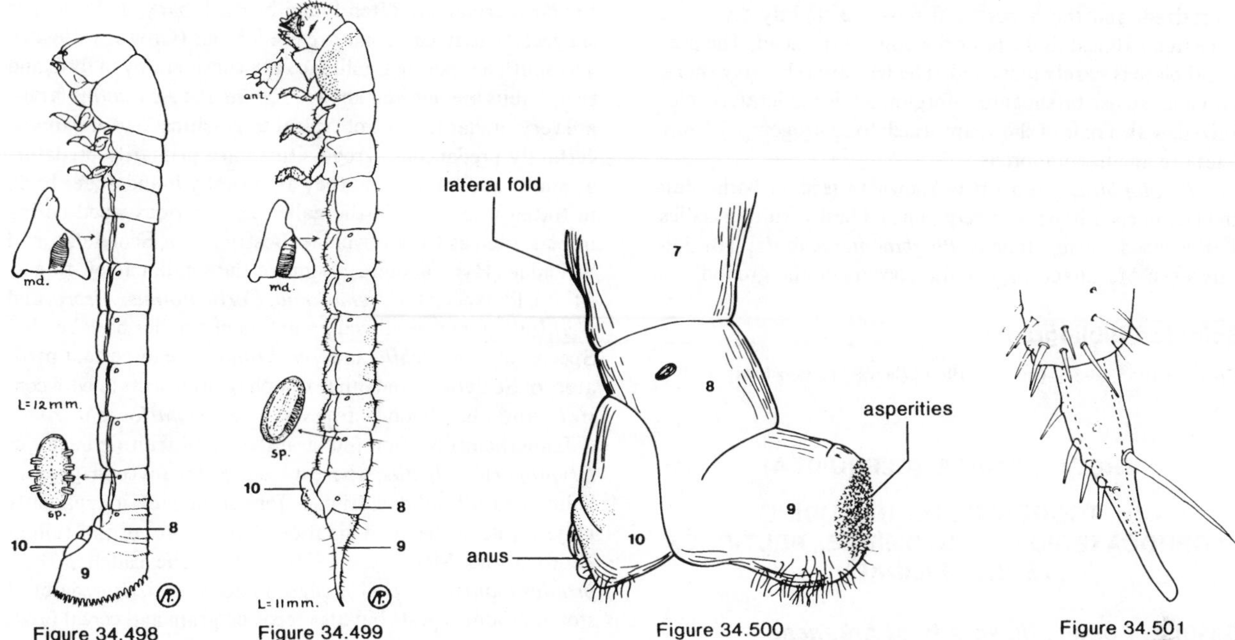

Figure 34.498 Figure 34.499 Figure 34.500 Figure 34.501

Figure 34.498. Lymexylidae. *Melittomma sericeum* Harris, chestnut timberworm. Larva, lateral. Formerly a serious pest of chestnut, now uncommon, chiefly in white oak. (From Peterson, 1951)

Figure 34.499. Lymexylidae. *Hylecoetus lugubris* Say. Larva, lateral. Infests assorted softer hardwoods, including poplar, birch, tulip-poplar, basswood, walnut and buckeye. (From Peterson, 1951)

Figure 34.500. Lymexylidae. *Atractocerus brevicornis* Palisot de Beauvois. Apex of abdomen (lateral) showing lymexyline type of ninth segment. (From Wheeler, 1986)

Figure 34.501. *Hylecoetus dermestoides* L. Tarsungulus, showing plurisetose condition. (From Wheeler, 1986)

(*see* review in Wheeler, 1986.), there remains much work to be done before the full range of structural and biological variability is known. The genera have recently been revised (Wheeler 1986).

Selected Bibliography

Batra 1967 (ambrosia fungi).
Batra and Francke-Grosmann 1961 (ambrosia fungi).
Böving and Craighead 1931 (larvae of *Hylecoetus lugubris* Say and *Melittomma sericeum* (Harris)).
Bright 1979.
Brown 1954 (biology and control of *P. insulare* Fairmaire).
Clark 1925 (larva of *Atractocerus kreusleri* Pascoe), 1931 (larva of *A. crassicornis* Clark).
Crowson 1955, 1960b, 1964d, 1981.
Emden 1943 (larvae of *Lymexylon navale* (Linnaeus) and *H. dermestoides* (Linnaeus)).
Francke-Grosmann 1967 (ectosymbiosis).
Fulmek 1930 (larva, 1st instar, pupa of *A. emarginatus* Castlenau).
Gahan 1908 (larva of *P. insulare*).
Gardner 1926 (larva of *A. emarginatus*), 1929 (larva of *A. reversus* Walker), 1935d (larvae of *Atractocerus* spp.), 1937b (larva of *M. albitarsis* Blair).
Germer 1912.
Grandi 1960 (larva of *H. dermestoides*).
Hayashi *et al.* 1959 (larva of *H. cossis* Lewis).
Lyngnes 1958.
Peterson 1951 (larvae of *H. lugubris* and *M. sericeum*).
Simmonds 1956 (biological control of *M. insulare*).
Slifer *et al.* 1975.
Vesey-Fitzgerald 1941 (larva and economic damage of *P. insulare*).
Wheeler 1986.
Wilson 1971.

PHLOIOPHILIDAE (CLEROIDEA)

John F. Lawrence,
Division of Entomology, CSIRO

This family includes the single genus *Phloiophilus*, containing the European species *P. edwardsi* Stephens and a second species from North Africa. The group is considered to be the most primitive member of the Cleroidea, and it shares a number of features with protopeltine Trogossitidae.

Larvae are 4 to 4.5 mm in length, and are elongate, slightly flattened, and lightly sclerotized, except for the head and tergum A9. The head is transverse, strongly rounded laterally, without an epicranial stem, and with lyriform frontal arms and a long median endocarina. There are 6 stemmata on each side, and the antennae are relatively short. The mandibles lack a mola, but each has 3 short, acute processes at the base. The ventral mouthparts are retracted, with the stipes longer than the cardo, and the maxillary mala bears a pedunculate seta. The gular and paragular regions are lightly sclerotized, and the hypostomal rods are slightly diverging and extend almost to the posterior edge of the head. The protergal plate is barely indicated. The legs are relatively short, with a unisetose tarsungulus. Tergum A9 is moderately sclerotized, with a pair of short, upturned, fixed urogomphi. Spiracles are annular-uniforous.

Phloiophilus edwardsi is known to feed as both adult and larva beneath the relatively thin and fleshy fruiting bodies of the wood-rotting fungus *Phlebia merismoides* (Basidiomycetes: Meruliaceae); pupation occurs on the ground.

Selected Bibliography

Crowson 1955, 1964a (habits), 1964d (larva; relationships).

TROGOSSITIDAE (CLEROIDEA)

(= TROGOSITIDAE; INCLUDING LOPHOCATERIDAE, OSTOMIDAE, PELTIDAE, TEMNOCHILIDAE)

David E. Foster, ***University of Arkansas***

John F. Lawrence,
Division of Entomology, CSIRO

Bark-Gnawing Beetles, Cadelles

Figures 34.502–504

Relationships and Diagnosis: The family Trogossitidae is used here in its broadest sense, encompassing the families Peltidae and Lophocateridae, as defined by Crowson (1964d, 1966a, 1970). This family and the related Phloiophilidae include the more primitive types of larval Cleroidea, which are similar to those of some cucujoids, as well as the predaceous, clerid-like larvae of Trogossitini. Trogossitid larvae may be distinguished from those of Cleridae and Chaetosomatidae by having the ventral mouthparts retracted, and from those of Melyridae and Phycosecidae by lacking a long epicranial stem and having the gular and paragular regions as heavily sclerotized as the epicranium.

Biology and Ecology: Although most of the better known Trogossitidae are predatory, several are mycophagous and 2 have become stored products pests. The New Zealand *Protopeltis* is apparently a fungus feeder, like the related Phloiophilidae. Little is known about the south temperate Rentoniinae; the single presumed larva was collected in leaf litter but assumed to have dropped from an arboreal habitat (Crowson, 1966a). Adults of the S. American Decamerinae are apparently pollen feeders, as are those of the N. American *Eronyxa*. Members of the Peltinae and Calitinae feed on the fruiting bodies of various Polyporaceae. *Thymalus marginicollis* Chevrolat breeds mainly in the birch bracket fungus (*Piptoporus betulinus*), but has been found in other polypores as well. *Ostoma pippingskoeldi* (Mannerheim) and *Calitys scabra* (Thunberg) are usually associated with *Fomitopsis pinicola*, a common brown rot of conifers, while *O. ferruginea* (Linnaeus) and *C. minor* Hatch were collected together in *Tyromyces leucospongia* brackets and *O. columbiana* Casey was found in *Hapalopilus alboluteus*. The Lophocaterinae are often found beneath bark, and evidence suggests that at least some are predaceous (Crowson, 1964d). The south temperate Egoliinae vary considerably in form and their habits are not well known; *Egolia* and *Acalanthis* larvae are very similar to those of typical trogossitines and are almost certainly predaceous. Trogossitinae are primarily predators as adults and larvae, and they are usually found under bark, in rotten wood, and in the galleries of various wood-boring insects, such as Cerambycidae, Bostrichidae, Scolytidae, and Siricidae (Hymenoptera). Some of the small, narrow trogossitines, like species of *Nemosoma, Corticotomus, Airora,* and *Cylidrella,* are specialized predators of smaller bark beetles. Species of *Temnochila* and *Tenebroides* are important predators of Scolytidae and other xylophagous insects in both conifer and hardwood forests. *Temnochila chlorodia* (Mannerheim) is often found in the small fruiting bodies of *Cryptoporus volvatus,* where it feeds on the mycophagous inhabitants (Nitidulidae, Ciidae, Tenebrionidae); it apparently plays a role in the dissemination of the spores of this fungus (Borden and McClaren, 1970, 1972). The cadelle, *Tenebroides mauritanicus* (L.) (figs. 34.502a–c), is a predator of stored products pests but also feeds on grain and cereal products.

Description: Mature larvae 2 to 50 mm (usually 10 to 30 mm). Body elongate, subcylindrical to slightly flattened, occasionally with enlarged thoracic region. Usually lightly sclerotized, except for head, protergum, and part of tergum A9 (occasionally without sclerotizd areas); paired sclerotizations often present on meso- and metaterga (and rarely terga A6–8); color usually cream or white, but varying to purple in some surface-active forms. Surfaces smooth; vestiture consisting of shorter and longer, scattered setae.

Head: Protracted and prognathous, slightly elongate and parallel-sided in Trogossitinae, transverse with strongly rounded sides in Peltinae, with intermediate conditions occurring in *Calitys* and Lophocaterinae. Epicranial stem usually absent (very short in some Peltinae); frontal arms V-shaped. Median endocarina extending between frontal arms in most species; a Y-shaped endocarina coincident with the short epicranial stem and the frontal arms in most Peltinae; paired endocarinae located between frontal arms in *Thymalus.* Stemmata usually 2 or 5 on each side; sometimes absent and occasionally 3, 4, or 6. Antennae well-developed, 3-segmented, very short in Peltinae. Frontoclypeal suture absent except in some Peltinae; labrum free. Mandibles heavy and subtriangular, bidentate, without a mola or accessory ventral process; inner edge of mandibular base bearing 1 to several hyaline processes forming a lacinia mobilis (also called lacinia mandibulae). Ventral mouthparts more or less retracted, the maxillolabial complex at least as long as gula; cardo transverse, stipes elongate, maxillary articulating area usually absent, and mala simple and rounded or truncate, with or without a pedunculate seta; maxillary palp almost always 3-segmented (2-segmented in Rentoniinae). Labium with mentum and submentum fused; postmentum elongate in Trogossitinae; labial palps almost always 2-segmented (1-segmented in Rentoniinae); ligula usually absent. Gula longer than wide, except in Peltinae. Hypostomal rods usually subparallel and extending to posterior edge of head, delimiting sclerotized paragular plates; hypostomal rods absent

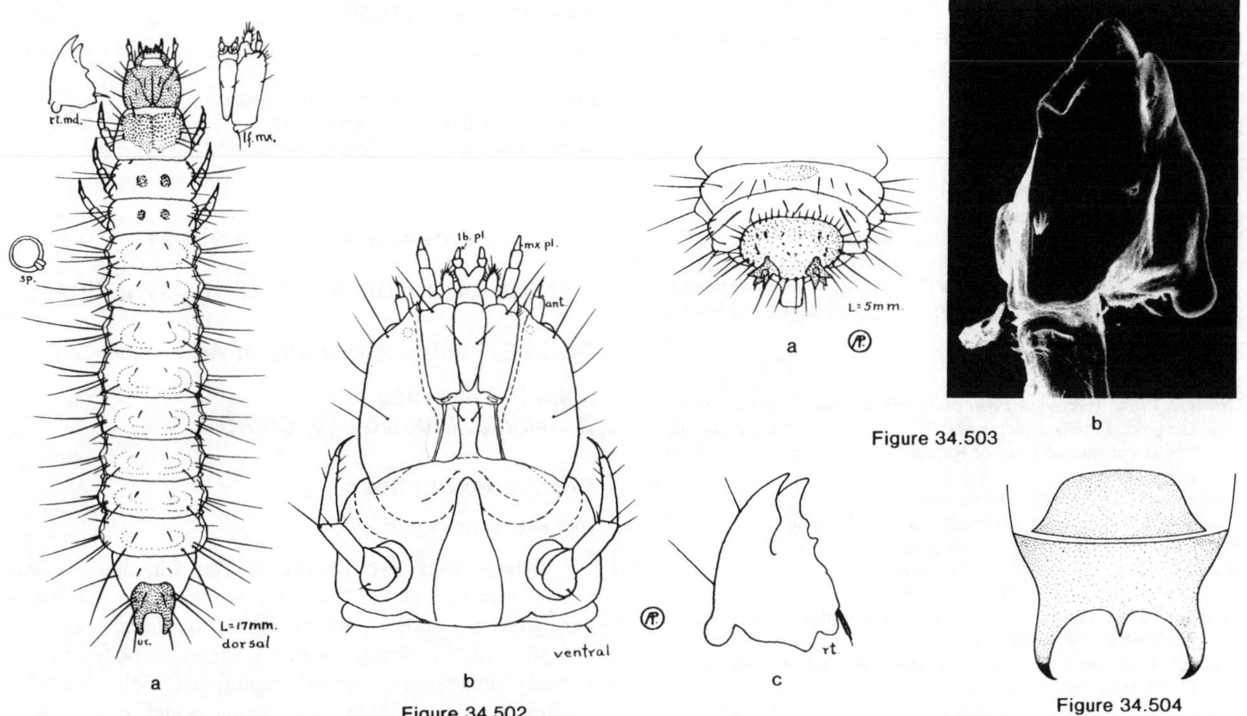

Figures 34.502a-c. Trogossitidae. *Tenebroides mauritanicus* (Linnaeus), the cadelle. **a.** larva, dorsal, length = 17 mm; **b.** head and prothorax, ventral; **c.** right mandible, ventral. Dirty white, with head, pronotum, anal plate and urogomphi deep brown; long, colorless setae on all segments, especially the lateral aspects; anal plate bearing 2, thick, rounded urogomphi with small pointed tips. Infests stored grains, rice, flour, meal, and may injure paper and wood containers, silk bolting cloth and similar materials associated with food products. (From Peterson, 1951)

Figures 34.503a,b. Trogossitidae. *Thymalus marginicollis* Chevrolat. **a.** abdominal apex, dorsal; **b.** left mandible, mesal. Concord, Massachusetts. (Figure 34.503a from Peterson, 1951)

Figure 34.504. Trogossitidae. (Lophocaterinae). Abdominal apex, dorsal.

in some Peltinae. Ventral epicranial ridges present in Trogossitinae and *Calitys*.

Thorax and Abdomen: Protergum with sclerotized plate in *Thymalus, Calitys,* and Trogossitinae; mesotergum with single plate in Trogossitini but with paired plates in *Calitys* and Egoliini; metatergum usually with paired plates in Trogossitini. Prothorax enlarged in *Ostoma* and related genera. Ventral regions of thorax with various sclerotizations in Trogossitinae. Legs moderately well-developed, somewhat reduced in some Peltinae, widely separated; tarsungulus with 1 seta. Abdominal terga 1–6 or 7 sometimes with paired ampullae. Paired, coiled, tubular glands (not easily visible) often present on abdominal terga 1–8. Tergum A9 usually with sclerotized basal plate (absent in most Peltinae) and always with paired urogomphi, which bear accessory processes in *Calitys* and *Thymalus*. In Lophocaterinae, basal plate transversely divided and median, posteriorly projecting tooth present between urogomphi. Sternum A9 well-developed; segment A10 ventrally located, small, circular.

Spiracles: Annular-biforous or apparently annular.

Comments: The family includes about 60 genera and 600 species worldwide, with 14 genera and about 60 species known from America north of Mexico. The classification has been subject to dispute in recent years. Barron (1971, 1975) divided the family along traditional lines, with 2 groups based on adult structure: Peltinae, with a more or less broadly oval body and explanate elytra, and Trogossitinae, with a narrower, more elongate body. The first group, with the exception of *Calitys* also has open procoxal cavities. Crowson (1964d, 1966a, 1970) divided the family into 8 groups, which he recognized as 3 separate families: Peltidae, with the subfamilies Decamerinae (including *Eronyxa*), Protopeltinae, Rentoniinae, and Peltinae (including *Ostoma* and *Thymalus*); Lophocateridae (including *Lophocateres, Lycoptis,* and *Grynocharis*); and Trogossitidae, with the subfamilies Calitinae (*Calitys*), Egoliinae, and Trogossitinae (equivalent to that of Barron). Crowson's 8 groups are all considered subfamilies here. Decamerinae are primarily S. American, with the western N. American *Eronyxa* somewhat doubtfully included; Protopeltinae, Rentoniinae, and Egoliinae are restricted to the Southern Hemisphere; Lophocaterinae and Trogossitinae are worldwide; and Peltinae and Calitinae are Holarctic.

There are 2 economically important species: the cadelle (*Tenebroides mauritanicus,* figs. 34.502a–c), which is a primary pest of stored grains and cereal products (in spite of its predatory habits), and the Siamese grain beetle, (*Lophocateres pusillus* (Klug), which is a secondary grain pest, especially common in rice.

Selected Bibliography

Aitken 1975 (stored products pests).
Balduf 1935 (biology of predators).
Barron 1971 (North American revision), 1975.
Böving and Craighead 1931 (larvae of *Ostoma ferruginea, Thymalus marginicollis, T. limbatus* (Fabricius), *Calitys scabra, Temnochila virescens* (Fabricius), *Tenebroides nanus* (Melsheimer), *Airora cylindrica* (Serville), and *Corticotomus cylindricus* (LeConte)).
Borden and McClaren 1970, 1972.
Craighead 1950 (forest species).
Crowson 1964d (larvae of *Protopeltis viridescens* (Broun); key to larvae of Lophocaterinae and Trogossitinae in New Zealand), 1966a (presumed larva of Rentoniinae; subfamily key), 1970 (larval keys).
Furniss and Carolin 1977 (forest species).
Gardner 1931b (larva of *Melambia cardoni* Leveille).
Halstead 1968b (biology of *Lophocateres pusillus*).
Hatch 1962 (Pacific Northwest species).
Kline and Rudinsky 1964 (forest species).
Mamaev 1976c (larvae of *Tenebroides, Temnochila, Nemosoma, Melambia,* and *Lepidopteryx*).
Palm 1951 (larvae of *Ostoma ferruginea, Grynocharis oblonga* (Linnaeus), and *Thymalus limbatus*).
Peterson 1951 (larvae of *Thymalus marginicollis* and *Tenebroides mauritanicus*).

CHAETOSOMATIDAE (CLEROIDEA)

John F. Lawrence,
Division of Entomology, CSIRO

The family Chaetosomatidae includes the monotypic New Zealand genera *Chaetosoma* and *Chaetosomodes*, and the Madagascar genera *Malgassochaetus* and *Somatochaetus*, with 5 and 2 species, respectively. The family is thought to be most closely related to Cleridae.

Larvae are elongate, parallel-sided, slightly flattened, white or pink, and lightly sclerotized, except for head, protergal plate, A9 tergum, and lightly sclerotized plates on meso- and metatergum. The head is elongate, with a short epicranial stem, V-shaped frontal arms, and a median endocarina. The base of the mandible has no mola but bears 2 slender spines and a seta. Ventral mouthparts are strongly protracted, with the cardo longer than the stipes, and the mala has a pedunculate seta. The hypostomal rods are subparallel and extend to the posterior edge of the head. Gular and paragular plates are well-sclerotized. Legs are 5-segmented, with a unisetose tarsungulus. Paired gland openings are absent from the abdominal terga, and tergum A9 is concave, with a pair of upturned urogomphi. Spiracles are annular-biforous.

Chaetosomatid larvae live in the galleries of wood-inhabiting beetles and are apparently predaceous. A larva of *Chaetosoma scaritides* Westwood has been found in the galleries of a cossonine weevil.

Selected Bibliography

Crowson 1964d (relationships; larva of *Chaetosoma*), 1970 (key to cleroid larvae).
Ekis and Menier 1980 (Madagascar species).
Hudson 1934 (larva of *Chaetosoma*).
Menier and Ekis 1982 (Madagascar species).

CLERIDAE (CLEROIDEA)

(INCLUDING CORYNETIDAE, KORYNETIDAE)

David E. Foster, *University of Arkansas*

John F. Lawrence,
Division of Entomology, CSIRO

Checkered Beetles

Figures 34.505-508

Relationships and Diagnosis: The Cleridae are advanced members of the Cleroidea, whose larvae are distinguished from all other members of the superfamily (with the exception of the closely related Chaetosomatidae) by the strongly prognathous ventral mouthparts with transverse stipites and the elongate gular region, which is as heavily sclerotized as the epicranial region. Surface-active clerid larvae are usually reddish or mottled with shades ranging from blue to red, so that they superficially resemble melyrid larvae occurring in the same habitats. Melyrid larvae differ from those of most clerids, however, by having a long epicranial stem and no median endocarina, and from those of all clerids by having the ventral mouthparts more retracted with elongate stipites and the gular region more lightly sclerotized than the epicranium (head lacking a ventral closure). Gallery-inhabiting clerid larvae are subcylindrical and lightly pigmented, and they may have dorsal ampullae similar to those of some wood-boring, xylophagous beetles on which they prey; but the features of the head capsule and mouthparts are still diagnostic in these forms. The larva of *Chaetosoma* differs from those of Cleridae in having hypostomal rods which extend to the posterior edge of the head capsule.

Biology and Ecology: With the exception of some species of *Necrobia*, all Cleridae are predaceous as adults and larvae. The majority are xylophilous and prey on a variety of wood-boring and subcortical insects, but some are restricted to other habitats and others are wide-ranging predators. Members of the Thaneroclerinae usually prey on small mycophagous insects and are found under bark or in fungus fruiting bodies; one of these, *Thaneroclerus buqueti* (Lefevre), is associated with stored products, where it feeds on various pest species, especially *Lasioderma serricorne* (Fabricius) (Anobiidae). Xylophilous clerid larvae fall into 2 groups: those which are relatively unspecialized and capable of moving rapidly on surfaces or in tunnels or galleries, and those which have become specially adapted for living within the tunnels of wood-boring insects. The second type has evolved several times in the subfamilies Tillinae (*Cylidrus,*

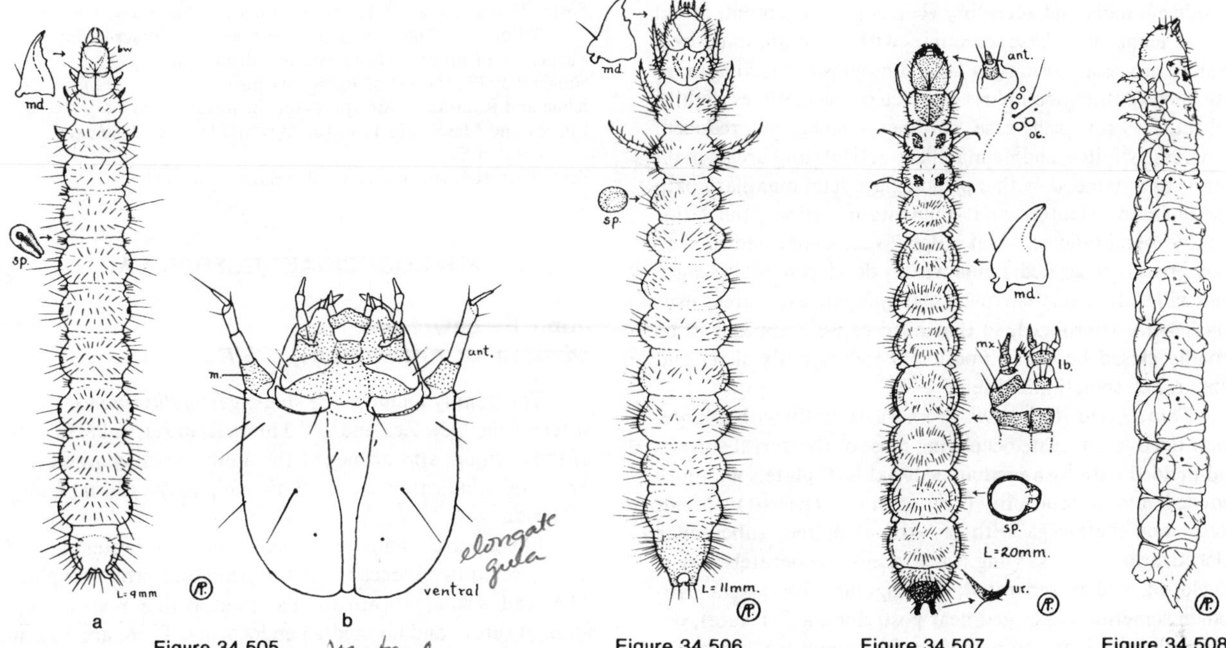

Figure 34.505 Figure 34.506 Figure 34.507 Figure 34.508

Figures 34.505a,b. Cleridae. *Necrobia rufipes* (De Geer), the redlegged ham beetle. Length = 9 mm. **a.** larva, dorsal; **b.** head, ventral. *See* "Comments." (from Peterson 1951)

Figure 34.506. Cleridae. *Cymatodera* sp. Arizona. Larva, dorsal, length = 11 mm. (from Peterson 1951)

Figure 34.507. Cleridae. *Enoclerus* sp. Massachusetts. Larva, dorsal, length = 20 mm. (from Peterson 1951)

Figure 34.508. Cleridae. *Priocera castanea* (Newman). Larva, lateral, length = 13 mm. (from Peterson 1951)

Monophylla), Epiphloeinae (*Neichnea, Phlogistosternus*), Enopliinae (*Pelonium, Chariessa, Cregya, Orthopleura, Tenerus*), Tarsosteninae (*Tarsostenus, Paratillus*), and Clerinae (*Priocera*).

There may be strong preferences for habitat or prey type. Some genera like *Chariessa* are associated with deciduous trees, while *Thanasimus* are restricted to conifers. Species of *Monophylla* and *Cylidrus* are predators of Bostrichidae; *Paratillus carus* (Newman) and *Tarsostenus univittatus* (Rossi) are associated with lyctine Bostrichidae; and *Thanasimus* species prefer bark beetles (Scolytidae). Within the genus *Enoclerus, E. sphegeus* (Fabricius) and related species occur on conifers, where several are important bark beetle predators, *E. eximus* (Mannerheim) occurs in riparian situations and preys on anobiids in alder and willow, *E. coccineus* (Schenklin) and *E. acerbus* Wolcott occur in the roots of rabbit brush, where they prey on cerambycid larvae, and *E. abdominalis* (Chevrolat) preys on yucca moth larvae in the pods and weevil larvae in the stems of *Yucca.* Many *Cymatodera* species are apparently associated with wood borers, but they have also been reported from cynipid galls and the cells of megachilid bees. Species of *Phyllobaenus* are also wide-ranging predators, and may occur not only in woody habitats, but in leaf litter, galls, or the webbing of Lepidoptera. *Lecontella cancellata* has been recorded from the cells of 7 species of aculeate Hymenoptera. Species of *Trichodes* occur primarily in the cells of bees, but *T. oregonensis* Barr and *T. nutalli* (Kirby) prey on the eggs of slantfaced grasshoppers. *Aulicus terrestris* Linsley feeds on the eggs of lubber grasshoppers.

Within the subfamily Corynetinae, *Corynetes coeruleus* occurs in wood and preys on Anobiidae, but some species of *Necrobia* frequent carrion and are scavengers as well as predators (*see* "Comments").

Description: Mature larvae 5 to 25 mm. Body elongate and parallel-sided to somewhat fusiform or clavate and subcylindrical to slightly flattened; lightly sclerotized, except for head (only rarely as lightly sclerotized as rest of body), prothorax, which bears a protergal plate and usually paired meso- and metatergal plates, and tergum A9, which usually bears a sclerotized basal plate and/or urogomphi. Color white or cream in gallery-inhabiting forms, but often with varying amounts of red or blue pigments in free-living species. Surfaces smooth, and vestiture of numerous short, fine hairs or longer, scattered hairs.

Head: Protracted and prognathous, often elongate and flattened, with subparallel sides, but sometimes subconical or globular, with modified upper surface. Posterior edge of head sometimes emarginate or notched. Epicranial stem usually absent (well-developed in Thaneroclerinae and some exotic Enopliinae); frontal arms V-shaped, usually contiguous at base, sometimes narrowly separated, widely separated in Phyllobaeninae. Median endocarina extending between frontal arms in all but Thaneroclerinae. Stemmata varying from 1 to 5 on each side or occasionally absent, usually arranged in 2 subvertical rows. Antennae well-developed, 3-segmented. Frontoclypeal suture usually distinct; labrum free. Mandibles heavy, subtriangular, unidentate, with apex acute or somewhat rounded; retinaculum, if present, weakly developed; mola and accessory ventral process absent; mesal edge of mandibular base sometimes with 2 or more spine-like, hyaline processes forming a lacinia mobilis (or lacinia mandibulae). Ventral mouthparts protracted or sometimes slightly retracted, with transverse cardines, sometimes reduced, transverse stipites, and no maxillary articulating areas; mala simple and rounded, with a pedunculate seta; maxillary palp 3-segmented. Mentum and submentum distinct, the latter usually membranous; labial palps 2-segmented and moderately broadly separated; ligula weakly developed. Gula longer than wide, often very narrow; gular and paragular areas usually heavily sclerotized, so that head capsule appears to be broadly closed behind. Hypostomal rods usually short and subparallel, sometimes absent.

Thorax and Abdomen: Prothorax usually with a large, sclerotized tergal plate occupying most of the dorsal surface and divided only by a median ecdysial line; plate sometimes reduced and occasionally (*Orthopleura, Tenerus*) absent. Meso- and metaterga with or without paired, subcircular sclerotizations of varying size. Legs moderately well-developed, widely separated; tarsungulus with 1 seta. Abdomen sometimes enlarged near posterior end (clavate), occasionally very long and narrow; paired ampullae sometimes present on abdominal terga 2–8. Tergum A9 almost always with a sclerotized basal plate and usually with a pair of urogomphi (reduced or absent in *Phyllobaenus* and some Thaneroclerinae), which vary from long, hooked structures to short, blunt projections, and may be widely separated or occasionally joined at base. Segment A10 reduced, ventrally or posteroventrally situated, and oval or subcircular.

Spiracles: Usually annular or annular-biforous with very small accessory openings; occasionally (Korynetinae) with well-developed accessory openings.

Comments: The family includes about 150 genera and 4000 species worldwide, with 37 genera and about 300 species occurring in America north of Mexico. Crowson (1964d) gave a key to the 8 subfamilies based on adults, and later (1972a) discussed their phylogenetic relationships based on studies of the alimentary canal made by Ekis and Gupta (1971).

Two species of Cleridae have been recorded as stored products pests: *Necrobia ruficollis* (Fabricius), the red shouldered ham beetle, and *N. rufipes* (De Geer) the red legged ham beetle (figs. 34.505a,b). The former occurs mainly on animal matter, and is relatively scarce in plant products, while the latter is a common insect in copra, oilseeds, oilcake, and cocoa beans, where it feeds on other pests, as well as on the products themselves.

Selected Bibliography

Aitken 1975 (species in stored products).
Balduf 1935 (biology of various predators).
Barr 1961 (review of Pacific Northwest species).
Beeson 1926 (biology of Indian species).
Böving and Champlain 1921 (North American larvae; 32 species in 15 genera).
Böving and Craighead 1931.
Craighead 1950 (predators of forest pests).
Crowson 1964d, 1972a.
Ekis 1977 (*Perilyphus* revision; larval description).
Ekis and Gupta 1971.
Foster 1976a (larvae of Thaneroclerinae), 1976b (larvae of Tillinae), 1976c (*Trichodes* revision; larval description).
Furniss and Carolin 1977 (predators of forest pests).
Gardner 1937a (larvae of Indian species).
Kline and Rudinsky 1964 (predators of forest pests).
Linsley and MacSwain 1943 (biology and larva of *Trichodes ornatus* Say).
Scott 1919 (biology and larva of *Necrobia ruficollis*).

PHYCOSECIDAE (CLEROIDEA)

John F. Lawrence,
Division of Entomology, CSIRO

This family includes the single genus *Phycosecis,* with 1 species from New Zealand and 3 more from Australia. In spite of the cucujoid appearance of the adult, larval characters of *Phycosecis* indicate a close relationship to the cleroid family Melyridae.

Larvae are elongate, narrow, slightly flattened, and lightly sclerotized, except for the head and protergal plate. The head is subquadrate and flattened, with a Y-shaped epicranial suture and no median endocarina. There are 6 stemmata on each side and the mandible lacks a mola but has a plumose process at the base. The ventral mouthparts are retracted with the stipes much longer than the cardo, and the mala bears a pedunculate seta. The gular and paragular regions are lightly sclerotized, so that the head capsule appears open behind, and the hypostomal rods extend to the posterior edge of the head. The protergal plate is weakly developed, and thoracic and abdominal terga lack paired gland openings. Tergum A9 is slightly more heavily sclerotized than the preceding terga and has a pair of subparallel, posteriorly projecting urogomphi, slightly curved at apex. Spiracles are annular.

Phycosecids live along coastlines among sand dunes, and adults may be active during the day but are difficult to see because of their covering of grayish scales or scale-like setae. Larvae and adults are scavengers and have been found on dead birds and fishes along the beach. Crowson (1964d) found insect parts in the larval gut of *P. limbata* (Fabricius), but I have reared *P. littoralis* Pascoe through an entire generation in a closed container with fresh fish, lettuce, and bran (larvae were usually clustered beneath the rotting pieces of fish).

Selected Bibliography

Crowson 1964d (relationships, larva), 1970 (key to larval Cleroidea).
Hudson 1934 (habits of *P. limbata*).

ACANTHOCNEMIDAE (CLEROIDEA)

John F. Lawrence,
Division of Entomology, CSIRO

This family is based on a single species, *Acanthocnemus nigricans* Hope, which is native to Australia, but has been introduced into southern Europe, North Africa, southern

Africa, Madagascar, India, Burma, Thailand, and New Caledonia. The group is considered to have affinities with both Trogossitidae and Melyridae. An unassociated first instar larva has been described by Crowson (1970), but this is not described here because of the questionable identification. *Acanthocnemus* adults often fly to lights and one series was taken from the hot ashes of a campfire; in addition, they have been found under bark on several occasions. Crowson's larva was found in litter around the base of a *Eucalyptus*. An unusual adult feature which may provide a clue to the niche occupied by this widespread species is the presence on each pronotal hypomeron of a deep cavity, which is almost totally blocked by a thin cuticular flap, attached at one point to the rim of the cavity.

Selected Bibliography

Champion 1922 (distribution).
Crowson 1970 (relationships, putative larva).

MELYRIDAE (CLEROIDEA)

(INCLUDING DASYTIDAE, MALACHIIDAE, PRIONOCERIDAE, RHADALIDAE)

David E. Foster, *University of Arkansas*

John F. Lawrence, *Division of Entomology, CSIRO*

Soft-winged Flower Beetles

Figures 34.509–510

Relationships and Diagnosis: The family Melyridae is one of the more derived groups of Cleroidea and appears to be most closely related to Acanthocnemidae and Phycosecidae, both of which are endemic to the Australian Region. Melyrid larvae (and also those of Phycosecidae) may be distinguished from those of other cleroids by the distinctive head capsule, which is somewhat parallel-sided and flattened, with a long-stemmed, Y-shaped epicranial suture and no endocarina, and is well-sclerotized dorsally with membranous gular and paragular regions ventrally; this type of head capsule is sometimes described as lacking a ventral closure. Additional features distinguishing melyrid larvae from those of Cleridae and Chaetosomatidae are the retracted ventral mouthparts with elongate stipites, and shorter, broader gula. Phycosecid larvae differ from those of melyrids in having well-developed hypostomal rods and 6 stemmata on each side.

Biology and Ecology: Adult Melyridae are active and usually diurnal inhabitants of surfaces, often occurring in flowers or on foliage. The large subfamily Malachiinae contains many brightly colored, aposematic species, which have eversible defense glands; this same group is characterized by having a variety of glandular structures on the head, antennae, or elytra of the male, which are used in courtship (Matthes, 1970a, 1970b). Adult melyrids may feed on pollen or other insects; Crowson (1964d) found pollen grains in the guts of representatives of all 5 subfamilies and insect fragments in at least some Haplocneminae, Dasytinae, and Malachiinae.

Larval melyrids occur in a variety of habitats where they are either scavengers or predators. Larvae of species of *Dasytes, Anthocomus, Malachius,* and *Hypebaeus* have all been collected under bark or in the galleries of wood-boring beetles, but melyrids in general are less common than Cleridae and Trogossitidae in this habitat. A number of larvae occur on the ground or in soil and leaf litter, where they feed on living or dead animals (insects, molluscs, worms, carrion). They are particularly common in sandy soils and on beaches; *Amecocerus* larvae were found in soil beneath dune vegetation. *Endeodes* species occur in the debris accumulated at the high tide mark on beaches along the Pacific Coast (Moore, 1956; Moore and Legner, 1977). A larva belonging to a species of *Rhadalus* was collected from rotting *Yucca* stems in Arizona. Adults of the European *Psilothrix viridicaeruleus* (Geoffroy) deposit eggs in dead flower heads or inflorescences, and the first instar is immobile and non-feeding; later instars become scavengers on the ground, and the last instar bores into the stems of herbaceous plants where it pupates (Fiori, 1971). *Anthocomus horni* (Fall) has been collected under pine bark, but also in tephritid galls on sagebrush and in the nests of black-billed magpies (Foster and Antonelli, 1973). *Malachius bipustulatus* (L.) has been collected in *Artemisia* stems inhabited by larvae of *Mordellistena* (Mordellidae), and *M. aeneus* (L.) is known to be a predator on insect larvae in grasses. Larvae of *Collops* species are predaceous in a variety of situations, including grasses, beaches, termite nests, and refuse chambers of harvester ants. At least 3 species prey on important economic pests: *C. bipunctatus* (Say) on the alfalfa weevil (*Hypera postica* (Gyllenhal)), *C. hirtellus* LeConte on the pea leaf weevil (*Sitona lineata* (Linnaeus)), and *C. vittatus* (Say) on the alfalfa caterpillar (*Colias erytheme* Boisduval). *Dicranolaius villosus* (Lea) occurs in soil in Australia, where it is an egg pod predator of the Australian plague locust (*Choroicetes terminifera* (Walker)) (Farrow, 1974).

Description: Mature larvae 2.5 to 25 mm (usually 5 to 10 mm). Body elongate and usually narrow, sometimes wider at middle, subcylindrical to slightly flattened. Lightly sclerotized except for head, usually tergum A9, and paired plates (or pigmented areas) on thoracic terga; color cream to pink, red, or purple; vestiture of fine hairs, usually short and scattered but sometimes long and dense.

Head: Protracted and prognathous, subquadrate or slightly elongate, parallel-sided and more or less flattened. Epicranial stem long and frontal arms V-shaped; median endocarina absent. Stemmata usually 4 or 5 on each side, arranged in 2 subparallel, vertical rows, sometimes fewer (2 in Haplocneminae). Antennae relatively short, 3-segmented, with first 2 segments usually subequal and the last much narrower and usually longer; sensorium broadly conical. Mandibles broadly triangular, bidentate, without a mola and with a serrate and/or bifurcate lacinia mobilis at base of mesal edge. Ventral mouthparts retracted. Maxilla with transverse, subtriangular cardo, elongate stipes, no articulating area, and

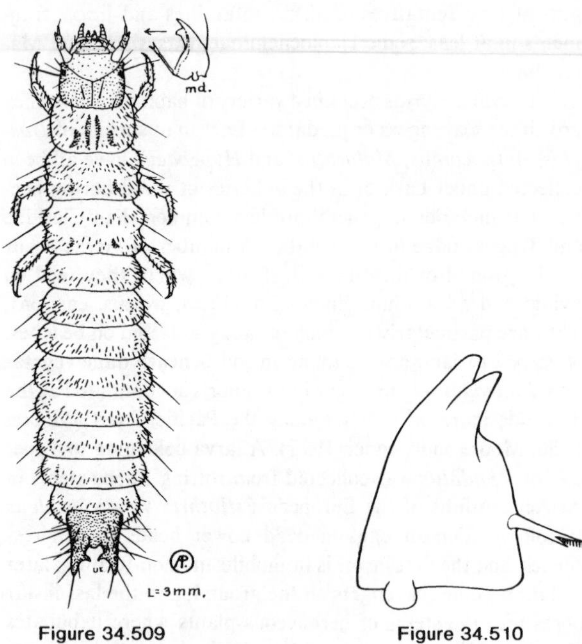

Figure 34.509 Figure 34.510

Figure 34.509. Melyridae. *Collops* sp. Larva, dorsal, length = 3 mm. (From Peterson, 1951)

Figure 34.510. Melyridae. (Malachiinae). Mandible, ventral.

simple, rounded mala; maxillary palp 3-segmented. Labium with submentum more or less distinct and mentum connate with stipites for most of its length; labial palps 2-segmented, broadly to narrowly separated. Gula subquadrate to slightly elongate; gular and paragular areas lightly sclerotized, in contrast with dorsal surface, so that head capsule appears to be "open" ventrally (lacking a sclerotized ventral closure). Hypostomal rods absent; ventral epicranial ridges occasionally present. Hypopharyngeal bracon present; hypopharyngeal sclerome absent.

Thorax and Abdomen: Protergum often with 1 to 3 pairs of sclerites or pigmented areas, a larger, elongate median pair, and 1 or 2 oval or crescent-shaped lateral ones; rarely with single prothoracic plate. Meso- and metaterga often with 1 pair of small plates or pigmented areas (rarely with 2). Legs well-developed, widely separated; tarsungulus with 1 seta. Thoracic terga and abdominal terga 1–8 or A1 and A8 often with 1 or 2 pairs of gland openings. Abdominal terga 1–8 rarely (Prionocerinae) with paired plates or pigmented areas. Tergum A9 usually with sclerotized basal plate and almost always with paired urogomphi, which may be arcuate or hooked and sometimes have mesally projecting teeth near the base. Basal plate or urogomphal bases occasionally with 1 (Haplocneminae) or 2 (Prionocerinae) pairs of membranous, tubular appendages. Sternum A9 simple. Segments A10 ventrally or posteroventrally situated, transversely oval.

Spiracles: Annular.

Comments: The family includes about 200 genera and 5000 species worldwide, with 47 genera and about 450 species known from America north of Mexico. Crowson (1964d, 1970) divided the group into 5 subfamilies: Haplocneminae (including *Rhadalus*), Melyrinae (including *Melyrodes* and *Eurelymis*), Prionocerinae (exotic), Dasytinae, and Malachiinae. Melyrinae are most diverse in S. America and Africa, Prionocerinae occur in southern Eurasia and tropical Africa, Dasytinae are particularly abundant in the Mediterranean region and southwestern N. America, and the Malachiinae is the largest and most widespread group and the only one to have radiated extensively in Australia. There are no economically important species, but *Zygia oblonga* (Fabricius) occurs in stored carob, barley, dried fruit, and lentils in the Mediterranean region.

Selected Bibliography

Aitken 1975 (stored products records).
Balduf 1935.
Böving and Craighead 1931 (larvae of *Malachius bipustulatus* (Linnaeus), *Anthocomus auritus* (LeConte), *Collops nigriceps* (Say), and *Dasytes caeruleus* (De Geer)).
Crowson 1964d, 1970.
Farrow 1974 (larva of *Laius villosus* Lea).
Fiori 1963, 1971 (larva of *Psilothrix viridicoeruleus* (Fourcroy)).
Foster and Antonelli 1973 (larva of *Anthocomus horni* (Fall)).
Gardner 1929 (larvae of *Idgia* spp.), 1931b (larva of *Carphurus almorensis* (Champion).
Matthes 1970a (courtship in *Malachius bipustulatus* (L.), 1970b (courtship in *Troglops albicans* (L.)).
Medvedev and Galata 1969 (larva of *Simoderus reflexicollis* Gebl)
Moore 1956 (habitat and larvae of *Endeodes*).
Moore and Legner 1977 (larvae and pupae of *Endeodes*).
Vinson 1957 (larva of *Pelecophora pikei* Vinson).

PROTOCUCUJIDAE (CUCUJOIDEA)

John F. Lawrence,
Division of Entomology, CSIRO

This family is based on the genus *Ericmodes* (= *Protocucujus*), which includes 2 species from Chile and a third from Australia, but several species and probably an additional genus remain to be described from these regions. It is one of the more primitive cucujoid families and is thought to be closely related to the Sphindidae.

Adult protocucujids have been collected in leaf litter, in flight intercept traps, and at lights, but little is known of the biology and immature states have not been described. Tillyard (1926) reported *Ericmodes australis* Grouvelle from *Uromycladium* rust galls on *Acacia.*

Selected Bibliography

Crowson 1955.
Sen Gupta and Crowson 1969a, 1979.
Tillyard 1926.

SPHINDIDAE (CUCUJOIDEA) (INCLUDING ASPIDIPHORIDAE)

John F. Lawrence,
Division of Entomology, CSIRO

Figure 34.511

Relationships and Diagnosis: The family Sphindidae is one of the more primitive groups of Cucujoidea, and may be related to Protocucujidae, Boganiidae, or Phloeostichidae. Larvae may be distinguished by the 6 stemmata, blunt mala, relatively narrow, hyaline prostheca, and annular spiracles, and most of them are further distinguished by lacking urogomphi and having paired tergal plates on the thoracic and abdominal segments.

Biology and Ecology: Both larval and adult Sphindidae appear to be restricted to the fruiting bodies of slime molds, where they feed on the spore mass. They have been recorded from a number of Myxomycetes, including species of *Arcyria, Fuligo, Lycogala, Reticularia,* and *Stemonitis* (Lawrence and Newton, 1980; Russell, 1979; Sen Gupta and Crowson, 1979).

Description: Mature larvae 2 to 5 mm. Body elongate, parallel-sided or somewhat fusiform, straight, slightly flattened; dorsal surfaces smooth, usually with head plus paired thoracic and abdominal tergal plates darkly pigmented; body occasionally very lightly pigmented, with darker head and a few small, tergal maculae. Vestiture usually consisting of scattered, simple setae; body sometimes densely clothed with long, fine hairs.

Head: Protracted and prognathous, moderately broad, slightly flattened. Epicranial stem short or absent; frontal arms lyriform and contiguous at base. Median endocarina absent. Stemmata 6 on each side. Antennae well-developed, 3-segmented, with segment 2 slightly to much longer than 1 or 3. Frontoclypeal suture absent; labrum free. Mandibles symmetrical, unidentate or bidentate, sometimes with serrate incisor edge, and with accessory ventral process; mola well-developed, tuberculate or asperate, prostheca a narrow, fixed, hyaline process, acute at apex. Ventral mouthparts retracted. Maxilla with transverse cardo, elongate stipes, well-developed articulating area, 3-segmented palp, and broad and obtuse mala. Labium free to base of mentum; ligula short and broad; labial palps 2-segmented, widely separated. Hypopharyngeal sclerome a transverse bar. Hypostomal rods usually long and slightly to strongly diverging, occasionally very short or absent. Ventral epicranial ridges absent. Gula very short and indistinct.

Thorax and Abdomen: Thoracic terga and abdominal terga 1–8 with large, paired, pigmented plates in *Sphindus* and *Protosphindus*. Thoracic terga expanded laterally in *Aspidiphorus*. Legs well-developed, 5-segmented; tarsungulus with 2 setae lying one distal to the other; coxae moderately to widely separated. Abdominal segments 1–8 each with 2 pairs of lateral processes in *Aspidiphorus*. Tergum A9 unpigmented or with a few small maculae, usually simple, occasionally (*Odontosphindus* and *Protosphindus*) with pair of slightly upturned, fixed urogomphi; sternum A9 well-developed, simple. Segment A10 short, circular, posteriorly oriented.

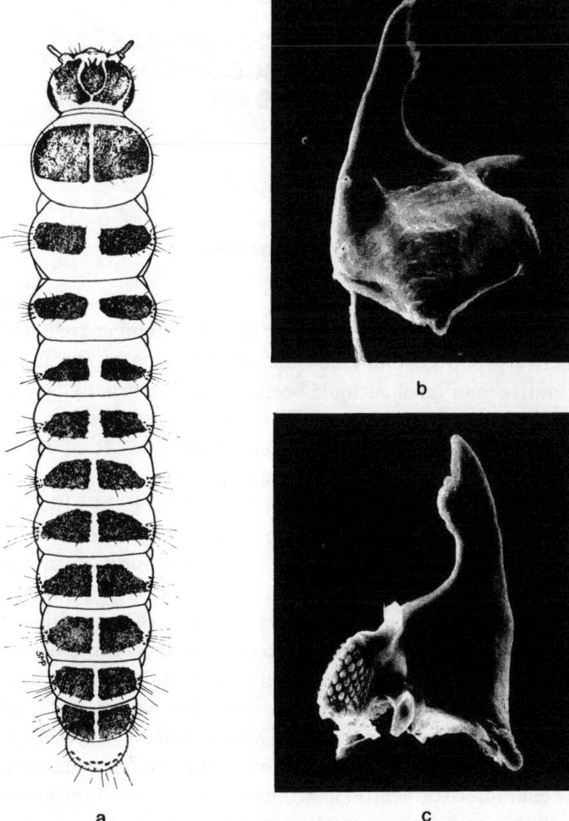

Figures 34.511a-c. Sphindidae. *Sphindus americanus* LeConte. Bedford, Middlesex Co., Massachusetts. Length = 3.2 mm. **a.** larva, dorsal; **b.** left mandible, dorsal; **c.** left mandible, ventral.

Spiracles: Usually annular and located at ends of short, conical tubes; thoracic spiracles in *Odontosphindus* annular-uniforous.

Comments: The family includes 6 genera and about 35 species worldwide, with 3 genera and 6 species occurring in America north of Mexico. *Odontosphindus* and *Sphindus* are both found in the Holarctic Region, but the latter extends into S. America, India, tropical Africa, and Madagascar; *Eurysphindus* and *Aspidiphorus* are restricted to the New World and Old World, respectively, while *Protosphindus* is endemic to Chile and *Sphindophorus* to South Africa. Undescribed genera have been seen from Australia and tropical America.

Selected Bibliography

Böving and Craighead 1931 (larva of *Sphindus americanus* LeConte).
Emden 1943 (larvae of *Aspidiphorus orbiculatus* (Gyllenhal) and *Sphindus dubius* (Gyllenhal)).
Lawrence and Newton 1980 (slime mold associations).
Perris 1877 (larvae of *Aspidiphorus lareyniei* Jacquelin DuVal and *Sphindus dubius*).
Peyerimhoff 1921 (larvae of *A. orbiculatus* and *S. dubius*).
Russell 1979 (slime mold associations).
Sen Gupta and Crowson 1979 (larva of *Odontosphindus clavicornis* Casey; larval key).

NITIDULIDAE (CUCUJOIDEA)

(INCLUDING BRACHYPTERIDAE, CATERETIDAE, CYBOCEPHALIDAE, SMICRIPIDAE)

John F. Lawrence,
Division of Entomology, CSIRO

Sap Beetles, Dried Fruit Beetles

Figures 34.512–522

Relationships and Diagnosis: The Nitidulidae are a distinctive group generally considered to be one of the more primitive families of cucujoid beetles. They are usually placed near the Rhizophagidae on the basis of adult features, but there are considerable differences between larvae of the 2 groups. Almost all nitidulid larvae may be distinguished by the unique ventral head structure (figs. 34.516b, 34.519c), in which the hypostomal ridges are strongly convergent posteriorly, the retracted mouthparts are pushed forward by the longitudinal cardines, the maxillary mala is blunt, and the labial palps are 1-segmented. In addition, the mandibular prostheca is usually complex, consisting of comb-hairs or fringed membranes (fig. 34.514c), the frontal arms are usually distant at base, the spiracles are often placed on raised tubes, and the urogomphi often have a pair of tubercles or pregomphi in front of them. The only nitidulid larvae lacking the characteristic ventral mouthparts are species of the genus *Smicrips,* which is usually excluded from the family (Crowson, 1955; Kirejtshuk, 1982). In the *Smicrips* larva, the mouthparts are truly protracted, without distinct cardines, as in Phalacridae and Cucujidae-Laemophloeinae; it may be distinguished from phalacrid larvae by the long, parallel hypostomal rods and from those of Laemophloeinae by the structure of the abdominal apex, in which segment 8 is not elongate and tergum 9 does not form a hinged plate. Larvae of the Cybocephalinae, which have also been removed from the family by various authors (Böving and Craighead 1931; Kirejtshuk, 1982), do not differ significantly from those of other Nitidulidae. Rhizophagid larvae differ from those of nitidulids in having retracted ventral mouthparts, transverse cardines, a falciform mala, a narrow mandibular prostheca, which may be acute or serrate, and the usual presence of ventral epicranial ridges.

Biology and Ecology: Nitidulids have one of the largest feeding repertoires among the Coleoptera. In general, they are saprophagous, feeding on decaying or fermenting plant tissue or occasionally carrion, but many species feed on fungal fruiting bodies or spores, and some may be predaceous or may attack healthy leaves, flowers, or fruits. The most characteristic habit of nitidulids is an association with yeasts and other fungi causing fermentation in tree wounds, under bark, or among decaying leaves, fruits, and flowers. Species of *Brachypeplus, Cillaeus, Colopterus, Epuraea, Prometopia, Soronia, Cryptarcha,* and *Glischrochilus* occur most commonly in rancid sap or under fermenting bark, while *Amphicrossus* may be restricted to slime fluxes, where they occur with *Nosodendron. Haptoncus, Lasiodactyla,* and *Stelidota* are particularly abundant in leaf litter and other accumulations of rotting vegetation, *Camptodes* has been found breeding in rotting cactus (*Opuntia*), and decaying flowers form the food source for some *Conotelus, Macrostola, Mystrops,* and *Smicrips.* Fermenting fruits are attractive to many nitidulids, especially species of *Carpophilus, Epuraea, Haptoncus, Aethina,* and *Lobiopa.* The dried-fruit beetle, *Carpophilus hemipterus* (L.) (figs. 34.514a–c), the corn sap beetle, *Carpophilus dimidiatus* (Fabricius), the dusky sap beetle, *Carpophilus lugubris* Murray, and several related forms are serious pests of stored fruits, grains, and other plant products (Böving and Rozen, 1962; Bondar, 1940a; Connell, 1956; Hayashi, 1978; Hinton, 1945b; Horion, 1960; Lepesme, 1944; Schedl, 1962).

Several species occurring in the habitats mentioned have been implicated in the transmission of pathogenic fungi. Some *Colopterus, Carpophilus, Haptoncus,* and *Glischrochilus* are involved in the transmission of *Ceratocystis* (Ascomycetes: Ophiosomatales) in oak, aspen, and sugarcane, while *Glischrochilus* are known to transmit *Fusarium* (Hyphomycetes) in corn. Some species of *Epuraea* are attracted to the nectar exuding from the spermagonia of rusts, such as *Cronartium* and *Peridermium* (Basidiomycetes: Uredinales), and they may aid in the transmission of these diseases in pines. (Chang and Jensen, 1974; Crowson, 1984a; Dorsey and Leach 1956; Dorsey *et al.,* 1953; Hinds, 1972; Jewell, 1956; Parsons, 1969b; Powell, 1971; Powell *et al.,* 1972; Windels *et al.,* 1976).

Mycophagy in the Nitidulidae involves most of the major fungal groups, including those mentioned above. Other ascomycete specialists include some species of *Epuraea,* which feed on conidia of molds (Eurotiales), *Prometopia,* which has been associated with an *Hypoxylon* (Xylariales) (Lawrence, 1977a), and several New Zealand species, including *Soronia hystrix* Broun, which feed on the black hyphal masses of sooty molds (Dothideales: Capnodiaceae) growing on tree trunks.

Some nitidulids occur on the softer fruiting bodies of higher Basidiomycetes (Agaricales and Aphyllophorales), especially those which are decaying, and certain genera appear to be restricted to this habitat. *Phenolia grossa* (Fabricius) has been collected in *Pleurotus, Bondarzewia, Laetiporus,* and *Hydnum,* while *Pallodes, Neopallodes, Cychramus,* and *Cyllodes* are found in gilled fungi, such as *Armillaria, Lampteromyces, Pleurotus, Pholiota,* and *Russula.* Most of these species feed on hyphal tissue, but species of *Aphenolia* feed on the spores of the polypore *Cryptoporus volvatus* (Böving and Rozen, 1962; Gillogly and Gillogly, 1954; Hayashi, 1978; Palm, 1959; Weiss and West, 1921a).

Gasteromycetes serve as hosts for certain Nitidulidae. Species of *Psilopyga,* for instance, are found on the fruiting structures of Phallaceae, while *Pocadius* species feed on the spores of puffballs (*Calvatia, Lycoperdon,* and *Scleroderma*). The genus *Thalycra* is restricted to the hypogean fruiting bodies of Hymenogastraceae (*Rhizopogon* and *Gautieria*), and it is likely that the Australian *Thalycrodes* has similar habits. (Donisthorpe, 1935; Fogel and Peck, 1975; Hayashi, 1978; Howden, 1961; Palm, 1959; Rehfous, 1955).

Although some nitidulids feed on rotting flowers, pollen- and nectar-feeding occur primarily in the subfamilies Cateretinae and Meligethinae. Cateretinae occur on a variety of angiosperms, such as Papaveraceae (*Amartus, Brachyleptus*), Agavaceae (*Anthonaeus*), Scrophulariaceae (*Brachypterolus*), Compositae, Urticaceae, and Umbelliferae (*Brachypterus*), or Rosaceae and Caprifoliaceae (*Cateretes* and *Heterhelus*). Species of *Meligethes* attack many types of flowers, including those of Rosaceae, Cruciferae, Ranunculaceae, Caprifoliaceae, Campanulaceae, and Labiatae, while *Pria* have been recorded from solanaceous flowers. Adults of *Conotelus* (Carpophilinae) feed and oviposit in live solanaceous flowers, but larvae develop after the flowers begin to rot. Floricolous habits are much less common in the Nitidulinae, but the Australian genera *Circopes* and *Macroura* breed in the male cones of cycads and the flowers of Malvaceae (*Hibiscus* and *Alyogyne*), respectively, while the S. American *Neopocadius nitiduloides* Grouvelle occurs in the flowers of *Prosopanche* (Hydnoraceae), which give off the odor of carrion (Böving and Rozen, 1962; Bruch, 1923; Hayashi, 1978; Horion, 1960; Jourdheuil, 1962; Peyerimhoff, 1926). Other associates of living plants include members of the genera *Xenostrongylus* and *Anister*, which are leafminers of various Brassicaceae (Hinton, 1942; Peyerimhoff, 1910).

Predatory habits occur in members of the Cybocephalinae, which attack various mealybugs and scale insects (Homoptera: Coccoidea) (Horion, 1960; Silvestri, 1910) and the nitiduline *Cychramptodes murrayi* Reitter, which is a predator of the wattle tick scale (*Cryptes baccatus* (Maskell)) in Australia. In the United States *Cybocephalus nipponicus* has been established, and both larvae and adults are effective predators against euonymus scale, *Unaspis euonymi* (Comstock), feeding beneath the scale covers (R. M. Hendrickson, personal communication to FWS). Species of *Epuraea, Glischrochilus,* and *Pityophagus* have been recorded as predators of scolytid eggs and larvae (Beaver, 1967; Kleine, 1908, 1909, 1944; Horion, 1960), but it is doubtful that any of these are obligate predators. Carrion feeding occurs in the genera *Omosita* and *Nitidula* (Böving and Rozen, 1962; Hayashi, 1978; Horion, 1960), while *Brachypeplus auritus* Murray and certain species of *Epuraea* are known to inhabit the nests of bees (Böving and Rozen, 1962; Murray, 1864). The European *Amphotis marginata* (Fabricius) is an associate of the ant *Lasius fuliginosus* (Latreille); adults induce food-laden ants to regurgitate food by tapping the ant's labium and are protected from the ant's attack by the flattened body form (Hölldobler, 1971; Horion, 1960).

Description: Mature larvae 1 to 20 mm, usually less than 15 mm. Body elongate, parallel-sided or fusiform, straight or slightly curved ventrally, occasionally curved dorsally (*Pallodes, Pocadius*), subcylindrical to strongly flattened. Dorsal surfaces smooth to granulate or tuberculate, usually lightly pigmented, except for head, protergum, and tergum A9, occasionally with single or paired plates on all exposed terga. Vestiture of scattered, simple setae.

Head: Protracted and prognathous, usually transverse and slightly flattened. Epicranial stem almost always absent (present in *Pocadius*); frontal arms lyriform or V-shaped, distant at base except in *Pocadius,* some Cateretinae and Cybocephalinae. Median endocarina absent, except in some Cateretinae, where it is simple or in *Pocadius,* where it is forked (Y-shaped) and coincident with epicranial stem and bases of frontal arms. Stemmata 2 to 4 on each side. Antennae usually well developed and always 3-segmented. Frontoclypeal suture absent, except in Cybocephalinae; labrum almost always free, partly fused to head capsule in Cybocephalinae, completely so in Meligethinae. Mandibles symmetrical, unidentate or bidentate, sometimes with serrate incisor edge, without accessory ventral process; mola well developed, usually with tubercles or asperities, which often form transverse rows; basal armature usually extending onto both dorsal and ventral surfaces; prostheca usually consisting of a brush of simple or complex hairs or several fringed membranes; in Meligethinae it is a simple membranous lobe, and in Cateretinae, Cybocephalinae and Smicripinae it is absent. Ventral mouthparts almost always retracted, in that their bases are located well behind the mandibular articulations, but appear protracted because of the longitudinally oblique cardines, which displace the stipes and mala farther forward; ventral mouthparts truly protracted in Smicripinae. Cardines absent in Smicripinae, longitudinally oblique in all other nitidulids. Stipes usually elongate, wider than long in Smicripinae, Cybocephalinae, and some Cateretinae; articulating area more or less reduced or absent; maxillary palp usually 3-segmented, 2-segmented in Smicripinae and Cybocephalinae. Mala obtuse, usually with mentum and submentum fused; postmentum partly connate with maxillae; ligula present or absent; labial palps 1-segmented. Hypopharyngeal sclerome usually a transverse bar, sometimes tooth-like. Hypostomal rods usually absent, long and parallel in Smicripinae, long and diverging in Cybocephalinae and Cateretinae. Ventral epicranial ridges absent. Gula indistinctly separated from labium in most nitidulids, longer than wide in Smicripinae.

Thorax and Abdomen: Protergum usually with pigmented plate; meso- and metaterga with or without plates or maculae. Meso- and metathorax and abdominal segments 1–8 sometimes with paired dorsal or lateral tergal processes, or paired pleural processes. Legs well developed, 5-segmented, usually not very long, tarsungulus usually with 2 setae, which may be unequal, with one or both forming adhesive lobe, sometimes with single seta; coxae usually moderately widely separated. Tergum A9 usually with paired, fixed urogomphi, which may be simple, bifid, or complex, and which often have a pair of pregomphi in front of them; tergum A9 simple in Cateretinae, Meligethinae, and Cybocephalinae. Segment A10 short, circular, ventrally or posteroventrally oriented.

Spiracles: Biforous or annular biforous, placed at the ends of short to moderately long tubes in Carpophilinae, Cryptarchinae and most Nitidulinae.

Comments: The family includes about 160 genera and 3000 species worldwide, with 36 genera and about 150 species occurring in N. America. In addition to the Smicripinae and Cybocephalinae, the family is usually divided into 5 more subfamilies—Cateretinae, Carpophilinae, Meligethinae, Nitidulinae, and Cryptarchinae, and Kirejtshuk (1982) has proposed an additional subfamily for the genus *Calonecrus.* Most of the subfamilies are widely distributed, but Calonecrinae are restricted to the Philippines and East Indies, Smicripinae occur in the warmer parts of the New World, and Cateretinae are found in the temperate parts of the Northern and Southern Hemispheres.

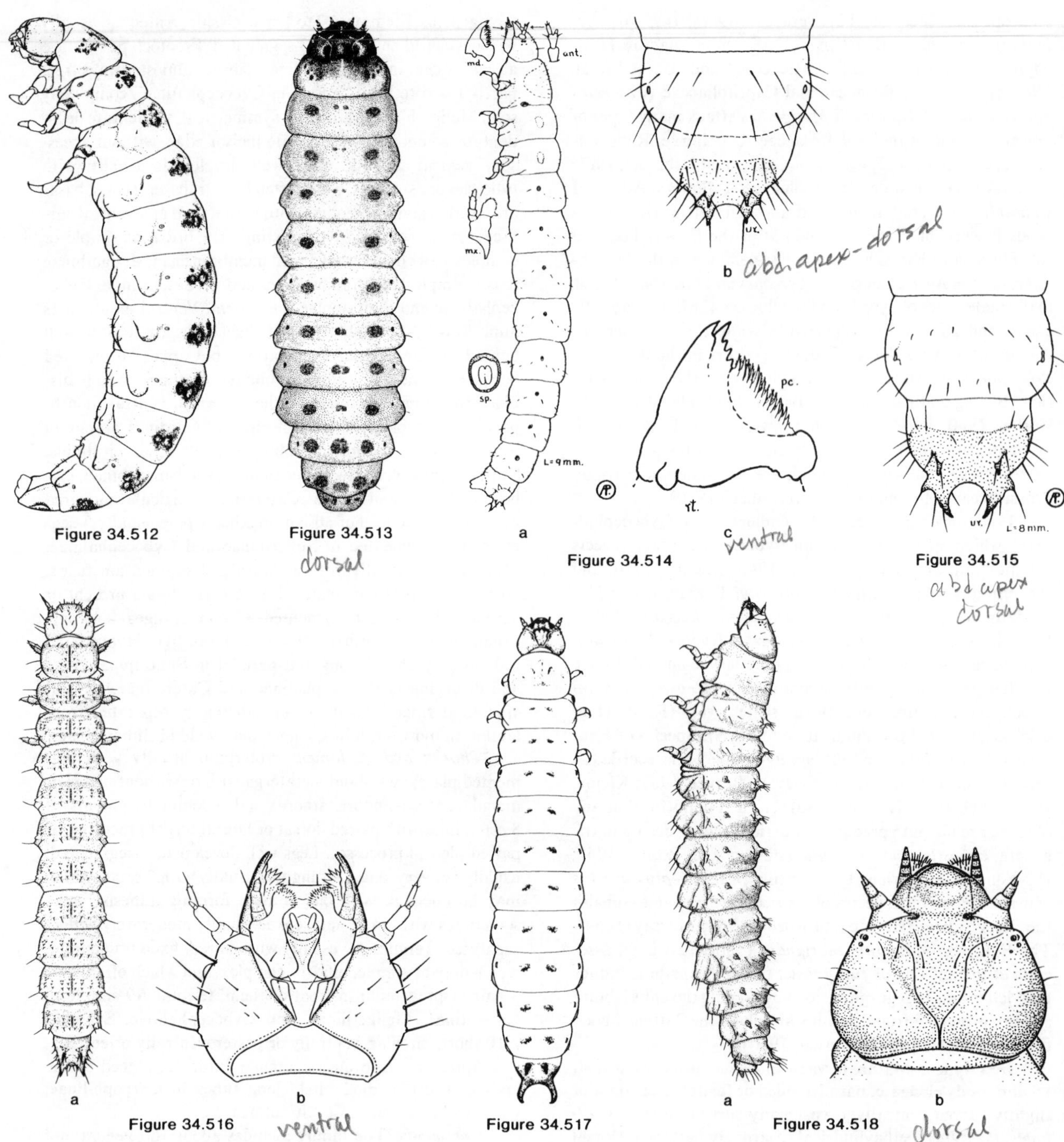

Figure 34.512. Nitidulidae. *Amartus rufipes* LeConte. Point Reyes National Seashore, Marin Co., California. Larva, lateral. Length = 6.3 mm.

Figure 34.513. Nitidulidae. *Meligethes* sp. El Cerrito, California. Larva, dorsal. Length = 3.7 mm.

Figures 34.514a-c. Nitidulidae. *Carpophilus hemipterus* (L.). Length = 9 mm. **a.** larva, lateral view (plus details); **b.** abdominal apex, dorsal; **c.** right mandible, ventral. Infests various kinds of ripe and dried stored fruits, especially raisins, prunes, figs, dates, apples and other stored food products; also found in decayed fruits. (From Peterson, 1951)

Figure 34.515. Nitidulidae. *Carpophilus niger* (Say). Abdominal apex, dorsal. It and *C. lugubris* (Murray) commonly infest insect-damaged sweet corn and other decaying vegetation. (From Peterson, 1951)

Figures 34.516a,b. Nitidulidae. *Epuraea* sp. Mt. Rainier, Pierce Co., Washington. **a.** larva, dorsal; **b.** head, ventral. Length = 4.6 mm.

Figure 34.517. Nitidulidae. *Phenolia grossa* (Fabricius). Dorset, Bennington Co., Vermont. Larva, dorsal. Length = 12.8 mm.

Figures 34.518a,b. Nitidulidae. *Pocadius* sp. Tamborine Mountain, Queensland. **a.** larva, lateral; **b.** head, dorsal. Length = 3.3 mm.

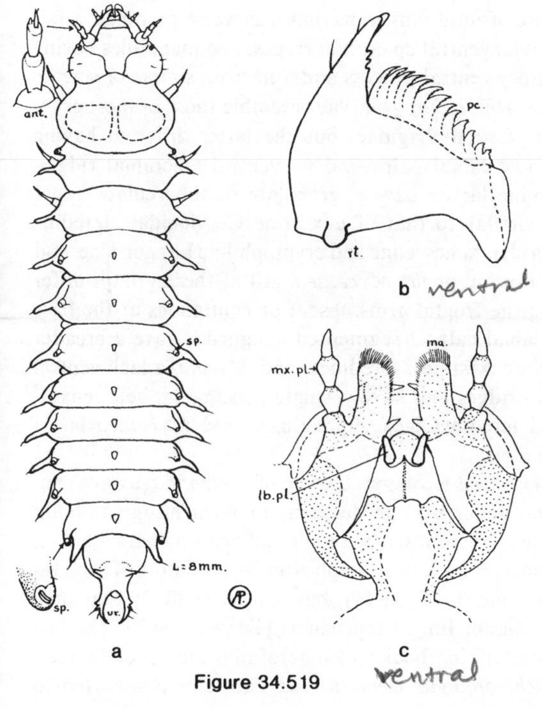

Figure 34.519

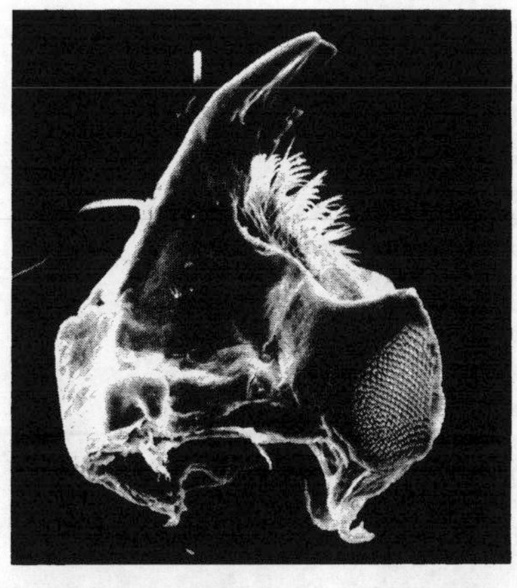

a

b

Figure 34.521

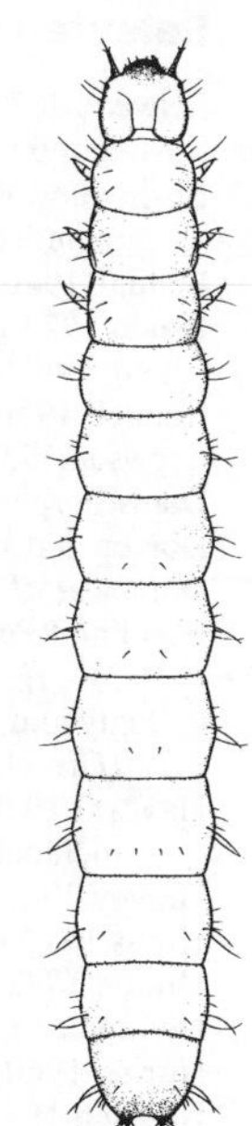

Figure 34.522

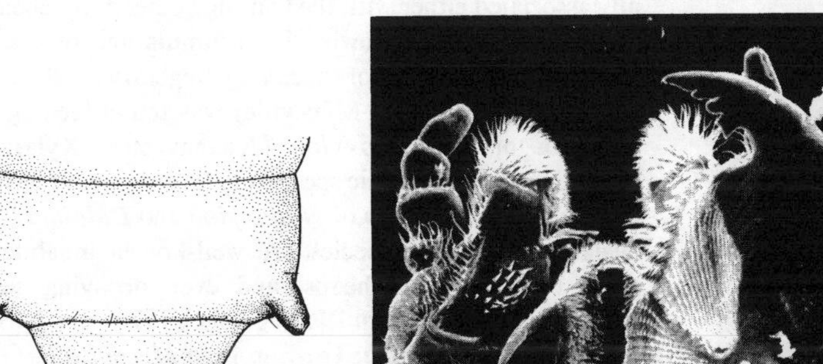

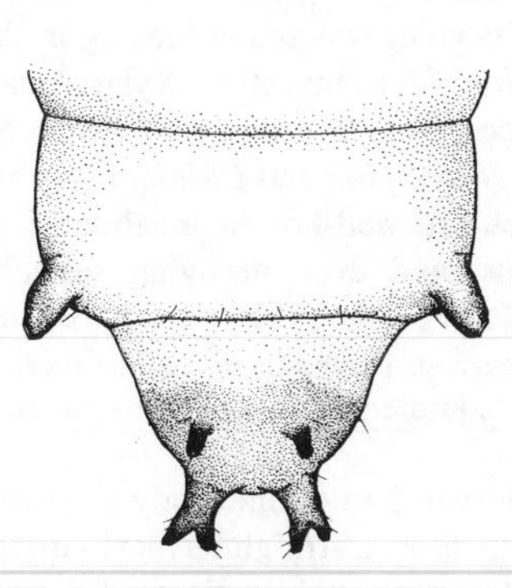

a

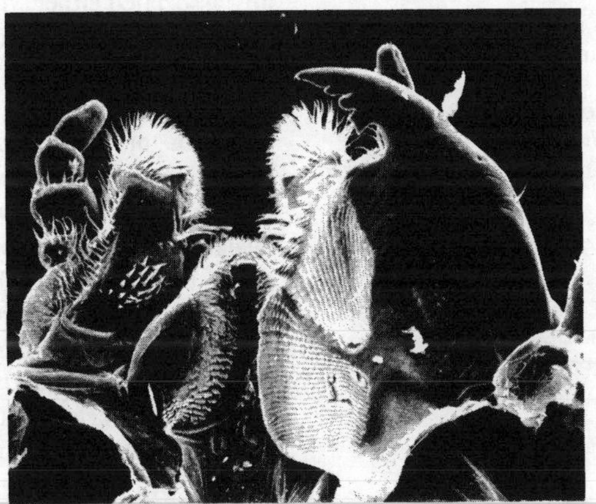

b

Figure 34.520

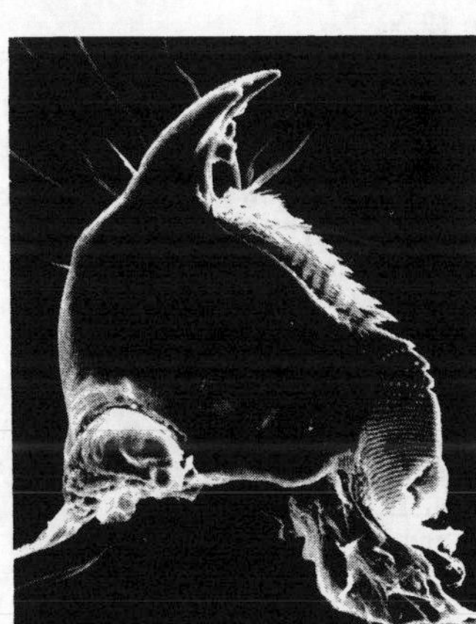

c

Figures 34.519a-c. Nitidulidae. *Cryptarcha ampla* Erichson. **a.** larva, dorsal (plus details); **b.** right mandible, ventral; **c.** maxillo-labial complex, ventral. Length = 8 mm. (from Peterson 1951)

Figures 34.520a-c. Nitidulidae. *Glischrochilus* sp. Great Smoky Mountains National Park, Tennessee. **a.** abdominal apex, dorsal; **b.** right mandible and maxillo-labial complex, dorsal; **c.** right mandible, ventral. Adult *G. quadrisignatus* (Say) and *G. fasciatus* (Olivier) are the common, black with yellow spots, "picnic" beetles.

Figures 34.521a,b. Nitidulidae. *Aphenolia monogama* (Crotch). Wasco Co., Oregon. **a.** right mandible, ventral; **b.** apex of maxillary mala.

Figure 34.522. Nitidulidae. *Smicrips* sp. Madera Canyon, Santa Rita Mts., Arizona. Larva, dorsal. Length = 3.1 mm.

Selected Bibliography

Beaver 1967 (predaceous species).
Böving and Craighead 1931 (larvae of various species).
Böving and Rozen 1962 (larvae of many species in key; 27 genera included).
Bondar 1940b (larva of *Mystrops palmarum* Bondar).
Bruch 1923 (larva of *Neopocadius nitiduloides* Grouvelle).
Chang and Jensen 1974 (transmission of pineapple disease).
Connell 1956 (larvae of *Carpophilus* spp.; key).
Crowson 1955, 1984a.
Donisthorpe 1935 (mycophagy).
Dorsey and Leach 1956 (association with oak wilt).
Dorsey *et al.* 1953 (transmission of oak wilt).
Fogel and Peck 1975 (association of *Thylacra* spp. with hypogeous fungi).
Gillogly and Gillogly 1954 (mycophagy in *Aphenolia monogama* (Crotch)).
Hayashi 1978 (larvae of many species; larval key; 24 genera included).
Hervey 1927 (larva *Brachypterolus pulicarius* (Linnaeus)).
Hinds 1972 (transmission of *Ceratocystis*).
Hinton 1942 (leaf mining species), 1945b (stored products pests).
Hölldobler 1971 (myrmecophily).
Horion 1960 (habits and habitats).
Howden 1961 (larva of *Thalycra sinuata* Howden).
Jelinek 1975.
Jewell 1956 (transmission of oak wilt).
Jourdheuil 1962 (agricultural pests).
Kirejtshuk 1982 (subfamily classification).
Kleine 1908 (predaceous species), 1909 (predaceous species), 1944 (predaceous species).
Lawrence 1977a (association with Ascomycetes).
Lepesme 1944 (stored products pests).
Lesne 1938.
Murray 1864 (monograph).
Okumura and Savage 1974 (species in dried fruits).
Osborne 1965 (larvae of *Meligethes* spp.).
Palm 1959.
Parsons 1969b (association of *Epuraea* with rusts).
Peterson 1951 (larvae of *Carpophilus* spp. and *Cryptarcha ampla* Erichson).
Peyerimhoff 1910 (leaf-mining larva of *Xenostrongylus lateralis* Chevrolat), 1916 (larva of *Lasiodactylus chevrolati* Reitter), 1926 (biology of phytophagous species).
Powell 1971 (association with blister rust).
Powell *et al.* 1972 (association with stem rust).
Rehfous 1955 (mycophagy).
Rozen 1963a (pupae of several species).
Saalas 1917 (biology of forest species).
Schedl 1962 (associations with bark beetles).
Silvestri 1910 (larva of *Cybocephalus rufifrons* Reitter).
Verhoeff 1923.
Weiss and West 1920 (mycophagy).
Windels *et al.* 1976 (transmission of *Fusarium* to corn).

RHIZOPHAGIDAE (CUCUJOIDEA) (INCLUDING MONOTOMIDAE)

John F. Lawrence,
Division of Entomology, CSIRO

Figures 34.523–525

Relationships and Diagnosis: The family Rhizophagidae is usually considered to be closely related to the Nitidulidae on the basis of adult similarities, but larval characters indicate a closer relationship to other primitive cucujoids, such as Cryptophagidae or Biphyllidae. Larval rhizophagids differ from those of nitidulids in having transverse cardines, a falciform mala, ventral epicranial ridges, and mandibles having an accessory ventral process and a narrow, simple or serrate prostheca. *Rhizophagus* larvae resemble those of most cryptophagine Cryptophagidae, but the latter differ in having 1-segmented labial palps and no ventral epicranial ridges. Monotomine larvae have a granulate or tuberculate upper surface, similar to that of euxestine Cerylonidae, teredine Bothrideridae, xenosceline and cryptophiline Languriidae, and the endomychid genus *Mycetaea.* All of these groups differ in having the frontal arms absent or contiguous at the base and the labial palps 2-segmented; languriids have a broader prostheca; euxestines, teredines, and *Mycetaea* lack ventral epicranial ridges and have a single tarsungular seta; euxestines and teredines lack a prostheca: and *Mycetaea* has a truncate mala.

Biology and Ecology: Larvae of *Rhizophagus* are usually found under bark in decaying or fermenting cambium and often in association with the galleries of bark beetles. Several species have been implicated as predators of scolytid larvae (Kleine, 1909, 1944), but it is more likely that they feed on fungi or fungal biproducts (Beaver, 1967). The few records on fruiting bodies of macrofungi are probably incidental. *Rhizophagus bipustulatus* Fabricius is reported to damage wine corks (Lepesme, 1944). Larvae of *Shoguna* sp. have been taken under bark of logs in the fermenting stage in Australia, and *S. conradti* Grouvelle has been collected in the galleries of ambrosia beetles (Platypodidae and *Xyleborus*) in Africa (Schedl, 1962). *Lenax* adults have been taken among fine debris under the bark of *Nothofagus* logs. Monotomine larvae may also occur under bark, but they are usually associated either with the fruiting bodies of pyrenomycete fungi or with fungi growing in accumulations of grass cuttings or other types of decaying vegetation. *Bactridium ephippigerum* (Guérin-Méneville) was found feeding in the stromata of an *Hypoxylon* (Ascomycetes: Xylariaceae) (Lawrence, 1977a), while species of *Hesperobaenus* have been taken in fruiting bodies of *Hypoxylon* and *Daldinia* (Xylariaceae). *Monotoma* species are well-known inhabitants of grass piles, compost heaps, and even decaying seaweed (Chandler, 1983; Horion, 1960; Kuschel, 1979), and have been widely dispersed by man. The European *M. conicollis* Guérin-Méneville is apparently a commensal in the nests of ants (*Formica*) (Horion, 1960).

Description: Mature larvae 2 to 6 mm. Body elongate, parallel-sided or somewhat fusiform, straight, subcylindrical to slightly flattened. Dorsal surfaces lightly pigmented, usually whitish, sometimes with weakly developed yellowish tergal plates in *Rhizophagus,* usually granulate or tuberculate (smooth in *Rhizophagus*). Vestiture consisting of scattered, long, simple setae, sometimes mixed with stout, expanded setae.

Head: Protracted and prognathous, moderately broad and transverse, sometimes sharply constricted posteriorly (*Monotoma*). Epicranial stem absent; frontal arms lyriform, separated at base. Median endocarina absent. Stemmata usually 2 or 4 on each side, sometimes absent. Antennae well-developed, 3-segmented. Frontocylpeal suture absent; labrum

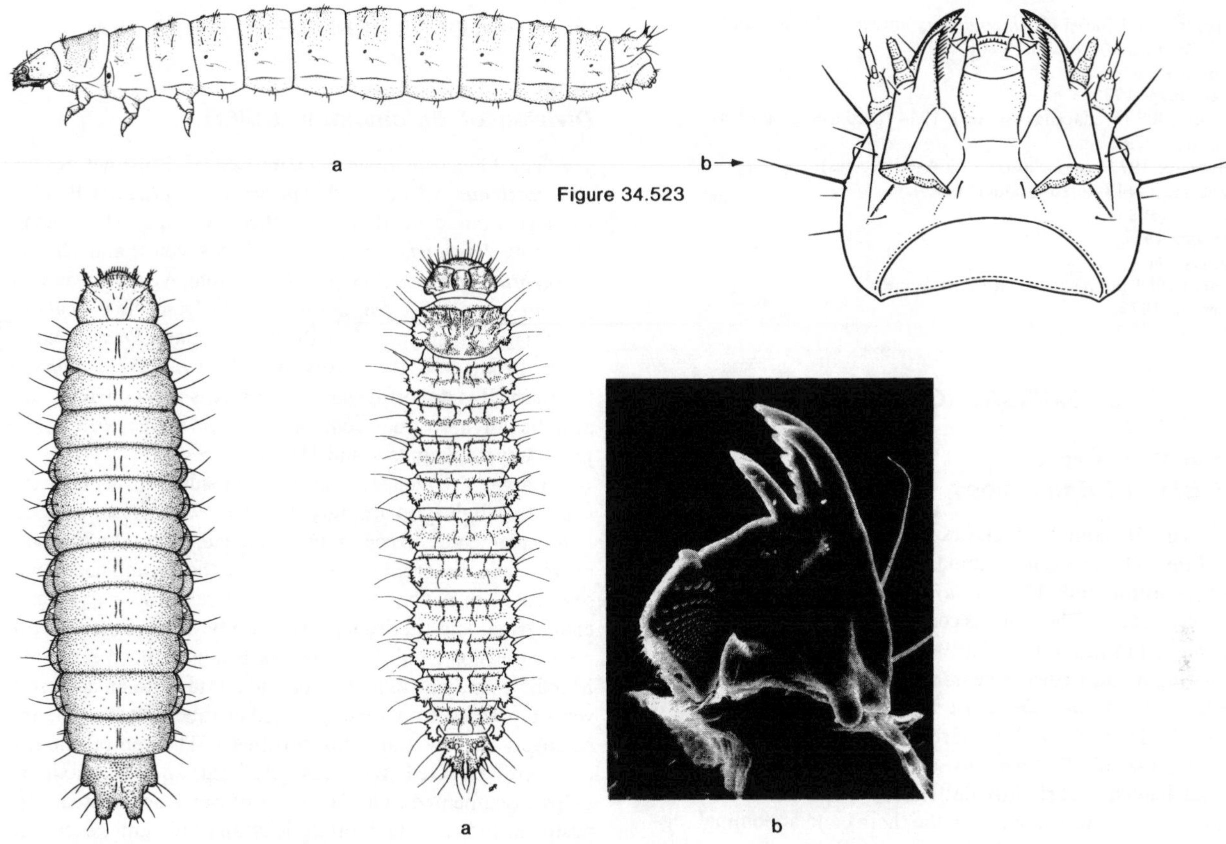

Figures 34.523a,b. Rhizophagidae. *Rhizophagus* sp. West side Mt. Rainier, Pierce Co., Washington. **a.** larva, lateral, length = 5.6 mm; **b.** head, ventral.

Figure 34.524. Rhizophagidae. *Bactridium epphipigerum* (Guérin-Méneville). Savannah, Georgia. Larva, dorsal. Length = 2.4 mm.

Figures 34.525a,b. Rhizophagidae. *Monotoma americana* Aube. Bedford, Middlesex Co., Massachusetts. **a.** larva, dorsal, length = 3.1 mm; **b.** left mandible, ventral.

free. Mandibles symmetrical, bidentate, sometimes with serrate incisor edge, with accessory ventral process; mola well-developed, asperate or tuberculate; prostheca a fixed, narrow, hyaline process, which is acute, bifid, or serrate. Ventral mouthparts retracted. Maxilla with transverse cardo, elongate stipes, well-developed articulating area, 3-segmented palp, and falciform mala. Labium free to base of mentum; ligula short and broad; labial palps usually 1-segmented (2-segmented in *Rhizophagus*), widely separated. Hypopharyngeal sclerome a transverse bar. Hypostomal rods usually short to moderately long and diverging (absent in *Monotoma*). Ventral epicranial ridges present. Gula strongly transverse.

Thorax and Abdomen: Thoracic terga and abdominal terga 1–9 each with a pair of lateral processes in *Monotoma*. Legs well-developed, 5-segmented; tarsungulus with 2 setae lying side by side; coxae moderately to widely separated. Tergum A9 with pair of urogomphi, which are usually complex, with accessory tubercles; sternum A9 well-developed, simple. Segment A10 short, circular, posteroventrally oriented.

Spiracles: Biforous or annular-biforous, usually raised on short, conical tubes.

Comments: The Rhizophagidae includes about 20 genera and 250 species, of which 9 genera and 53 species occur in America north of Mexico. The bulk of the genera are placed in the subfamily Monotominae, which occurs in most regions of the world. The subfamily Rhizophaginae includes the Holarctic genus *Rhizophagus*, and the Thioninae (*Thione* and *Shoguna*) occur in the warmer parts of the New and Old Worlds. The fourth subfamily, Lenacinae, contains the New Zealand genus *Lenax*, which is unknown in the larval stage and somewhat doubtfully included in the family. A larva described as *Myrmechixenus* by Klausnitzer (1975b) is almost certainly a misidentified *Monotoma* (*see* Doyen and Lawrence, 1979, and Nikitsky, 1983a).

Selected Bibliography

Beaver 1967 (predaceous species).
Böving and Craighead 1931 (larvae of *Hesperobaenus* sp. and *Rhizophagus grandis* Gyllenhal).
Chandler 1983 (larva of *Monotoma producta* LeConte).
Doyen and Lawrence 1979.

Hayashi 1980 (larva of *Rhizophagus japonicus* Reitter and *Mimemodes monstrosus* Reitter).
Horion 1960.
Klausnitzer 1975b.
Kleine 1909 (predaceous species), 1944 (predaceous species).
Kuschel 1979.
Lawrence 1977a (association with Ascomycetes).
Lepesme 1944 (stored products pests).
Nikitsky 1983a.
Peacock 1977.
Saalas 1917.
Schedl 1962 (association with bark beetles).
Verhoeff 1923.

BOGANIIDAE (CUCUJOIDEA)

John F. Lawrence,
Division of Entomology, CSIRO

The Boganiidae includes the Australian genera *Boganium* and *Paracucujus,* and the South African genera *Afroboganium* and *Metacucujus* (Endrödy-Younga and Crowson 1986). The group is considered to be one of the more primitive famiies of Cucujoidea, and *Paracucujus rostratus* Sen Gupta and Crowson bears a marked resemblance to the Jurassic fossil *Parandrexis parvula* Martynov from Central Asia (Rodendorf and Ponomarenko, 1962).

Larvae of *P. rostratus* are elongate, more or less parallel-sided, and slightly flattened, with a sclerotized head and pigmented tergal plates on the thorax and abdominal segments 1 to 9. The head lacks an epicranial stem or median endocarina and has lyriform frontal arms; antennae are moderately long and 3-segmented, with a small sensorium, and there are 6 stemmata on each side. Mandibles are unidentate, without an accessory ventral process and with a tuberculate mola and acute prostheca. Ventral mouthparts are retracted; the maxilla has a relatively short stipes, an articulated, falciform mala, and a 3-segmented palp with well-developed palpifer; and the labium has a broadly rounded ligula and long, 2-segmented palps. Hypostomal rods are short and slightly diverging. Legs are well-developed but relatively thin and widely separated, with a bisetose tarsungulus. Tergal plates are smooth and even, and those on the meso- and metathorax and abdominal segments 1 to 8 bear a transverse carina near the anterior edge. Tergum A9 bears a pair of short, posteriorly projecting urogomphi. Segment A10 is posteriorly oriented and bears a pair of long, tubular pygopods. Spiracles are small, annular-biforous, and not placed at the ends of tubes.

Paracucujus rostratus occurs in the male cones of the cycad *Macrozamia riedlei,* where both larvae and adults feed on pollen, while the South African *Metacucujus encephalarti* feeds on the pollen of another cycad, *Encephalartos lanatus.* Species of *Afroboganium* occur on flowering shrubs, but little is known of the habits of *Boganium.*

Selected Bibliography

Crowson 1981.
Endrödy-Younga and Crowson 1986.
Rodendorf and Ponomarenko 1962.
Sen Gupta and Crowson 1966, 1969a.

PHLOEOSTICHIDAE (CUCUJOIDEA)

John F. Lawrence,
Division of Entomology, CSIRO

The Phloeostichidae is comprised of 4 distinct groups: *Phloeostichus* with a single species (*P. denticollis* Redtenbacher) from central and southern Europe; Hymaeinae, including *Hymaea,* with 3 Australian species, and *Rhopalobrachium clavipes* Boheman from Chile; Agapythinae with a single species (*A. foveicollis* Broun) from New Zealand; and Priasilphinae, including the New Zealand *Priasilpha obscura* Broun and the Tasmanian *Priastichus tasmanicus* Crowson. It is doubtful that this family is monophyletic, and affinities of individual components seem to be with Cucujidae, Cryptophagidae, and Helotidae.

Larvae are elongate and parallel-sided to fusiform and subovate, slightly to strongly flattened, with dorsal surfaces lightly pigmented, except in Priasilphinae, which also has long, curved, spinose tergal processes on the thorax and abdominal segments 1 to 8. The head lacks an epicranial stem or median endocarina, and the frontal arms are lyriform; antennae are moderately long and there are 5 or 6 stemmata on each side. Mandibles may be bidentate or tridentate, with an accessory ventral process, transversely ridged or tuberculate mola, and relatively narrow and acute prostheca. The ventral mouthparts are retracted, the mala falciform, and the maxillary palps 3-segmented. The labial palps are 2-segmented. Hypostomal rods are moderately long and diverging, except in Priasilphinae, where they are very short or absent. Weak ventral epicranial ridges occur in Priasilphinae. The legs are widely separated, with a bisetose tarsungulus. Dorsal surfaces are covered with elongate, setiferous tubercles forming patterns in Priasilphinae; the anterior 5 or 6 abdominal terga and sterna in Hymaeinae each has a pair of semicircular bands of asperities; surfaces are smooth in *Agapytho,* except for a pair of elongate, setiferous tubercles in front of each urogomphus. Tergum A9 has a pair of urogomphi, widely separated and upcurved in Hymaeinae, approximate, diverging, and upcurved in Agapythinae, and straight and posteriorly oriented in Priasilphinae. Spiracles are biforous and not or only slightly raised, except in Priasilphinae, where they are annular and placed at the ends of the long lateral processes. Segment A10 is ventrally or posteroventrally oriented and somewhat transverse.

Phloeostichus and the Hymaeinae occur under the bark of dead trees and probably feed on fungal hyphae and conidia or dead cambium. *Agapytho* occurs on sooty molds (Ascomycetes: Capnodiaceae, etc.). Larvae and adults of *Priasilpha* and *Priastichus* occur in leaf litter and moss; their feeding habits are unknown.

Selected Bibliography

Cekalovic 1976 (larva of *Rhopalobrachium*).
Crowson 1973a (Priasilphinae).
Sen Gupta and Crowson 1966 (larva of *Hymaea*), 1969a (larvae of *Agapytho* and *Rhopalobrachium*).
Weise 1897 (larva of *Phloeostichus*).

HELOTIDAE (CUCUJOIDEA)

John F. Lawrence,
Division of Entomology, CSIRO

This family includes the single genus *Helota,* with about 100 species occurring in the warmer parts of the Old World from Japan to India and tropical Africa. It is usually placed with the more primitive cucujoid families and may be related to Phloeostichidae.

Larvae of helotids are of moderate size (10 to 20 mm), elongate and slightly flattened, with lateral processes on the meso- and metathorax and abdominal segments 1 to 8. The head is broad, without an epicranial stem or median endocarina, the frontal arms are lyriform and distant at base, the antennae are moderately long, and there are 6 stemmata on each side; the labrum is fused to the clypeus, except laterally. The mandibles are bidentate, with a large, tuberculate mola, an accessory ventral process, and a narrow, acute prostheca. The ventral mouthparts are retracted and the mala is obliquely truncate at apex and partly divided, with a dense fringe of hairs along the inner edge and several curved spines at the apex; maxillary palps are 3-segmented and labial palps are 2-segmented. Hypostomal rods are absent and ventral epicranial ridges are present. Tergum A9 has a pair of long, diverging, spinose urogomphi. Spiracles are annular-biforous.

Helotids feed as adults and larvae on the sap exuding from trees.

Selected Bibliography

Crowson 1973a.
Hayashi *et al.* 1959 (larva).
Olliff 1882 (larva).
Roberts 1958 (larva in key).

CUCUJIDAE (CUCUJOIDEA)

(INCLUDING CATOGENIDAE, LAEMOPHLOEIDAE, PASSANDRIDAE, SILVANIDAE)

John F. Lawrence,
Division of Entomology, CSIRO

Flat Bark Beetles

Figures 34.526–533

Relationships and Diagnosis: The Cucujidae includes a diverse assemblage of beetles which have been placed in several families by Crowson (1955, 1981) and by Böving and Craighead (1931). Larval characters support the recognition of 4 distinct groups: Passandrinae, Cucujinae, Laemophloeinae, and Silvaninae. Closely related families include the Phalacridae and Phloeostichidae. Diagnostic features of cucujid larvae include the reduction of the 9th abdominal segment, usually with the loss or concealment of the sternum, the somewhat flattened, elongate body, more or less protracted ventral mouthparts, relatively broad head, and well-developed legs and antennae; none of these characters, however, occur in all members of the family.

Flattened, subcortical larvae in the tenebrionoid (heteromeran) families Prostomidae, Pythidae, Pedilidae, Pyrochroidae, Othniidae, Inopeplidae, Salpingidae, Mycteridae, and Tenebrionidae may be confused with some cucujids, but most of these larvae have a well-developed 9th sternum bearing at least 1 pair of basal asperities, while those tenebrionids with a reduced sternum have a distinct frontoclypeal suture and lack both the falciform mala and symmetrical mandibles of the Cucujinae and Uleiotini and the prognathous mouthparts of the Laemophloeinae. The exotic Hymaeinae Phloeostichidae have the abdominal rings of asperities like those in Laemophloeinae and mouthparts similar to those of cucujines and brontines, but the frontal arms are distant at the base, and the 9th abdominal segment has a well-developed sternum. Some flattened Nitidulidae differ from cucujids in having unique, longitudinal cardines, distant frontal arms and a complex prostheca. Smicripine nitidulids have prognathous mouthparts and long, parallel hypostomal rods as in Laemophloeinae, but they lack the paired endocarinae and articulated 9th tergum of that group. Some Phalacridae (*Litochrus* and their allies) resemble laemophloeine cucujids in general shape, presence of a sclerotized, articulated 9th tergum, long hypostomal rods, protracted ventral mouthparts, and mandibular structure; these phalacrids differ in having a median endocarina, diverging hypostomal rods, and a completely concealed 9th sternum. Some lophocaterine Trogossitidae are also cucujid-like, but have V-shaped frontal arms, a median endocarina, and a transversely divided plate on tergum A9.

Biology and Ecology: Cucujidae are commonly found under bark, in leaf litter, or in stored products; they are primarily mycophagous, but some can be reared on grains or cereals, and others have become predaceous or ectoparasitic.

Members of the Laemophloeinae usually feed on fungal spores or hyphae growing under bark or in the galleries of bark beetles. Several laemophloeines have been associated with the fruiting bodies of Ascomycetes, apparently feeding on the sporulating surfaces; larvae of *Laemophloeus biguttatus* (Say) were found feeding on conidia produced by a species of *Hypoxylon, Placonotus illustris* Casey was collected on *Daldinia* stromata, and similar habits have been reported for several European species (Dajoz, 1966; Lawrence, 1977a; Palm, 1959). A number of laemophloeines are known to be stored products pests; *Laemophloeus janeti* (Grouvelle) attacks cocoa beans and a *Planolestes* has been reported from legume pods, but the most serious economic pests are species of *Cryptolestes* (fig. 34.529), including the rusty grain beetle, *C. ferrugineus* (Stephens), which cause major damage to stored grains. Some species of *Cryptolestes* may also be facultative predators (Bräuer, 1970; Howe and Lefkovitch, 1957; Lefkovitch, 1957a, 1957b, 1958, 1959, 1962a, 1962b, 1962c, 1964). Laemophloeine larvae are unique among the Coleoptera in producing silk for pupation from paired prothoracic glands (Bishop, 1960; Roberts and Rilett, 1953).

The Cucujinae are found exclusively under bark, and at least some of them appear to be predaceous. Smith and Sears (1982) studied the larval gut contents of *Cucujus clavipes* (Fabricius) (figs. 34.528a–d) and found only animal matter, including insect parts; in my own dissections, insect remains were also common, but these were mixed with some plant

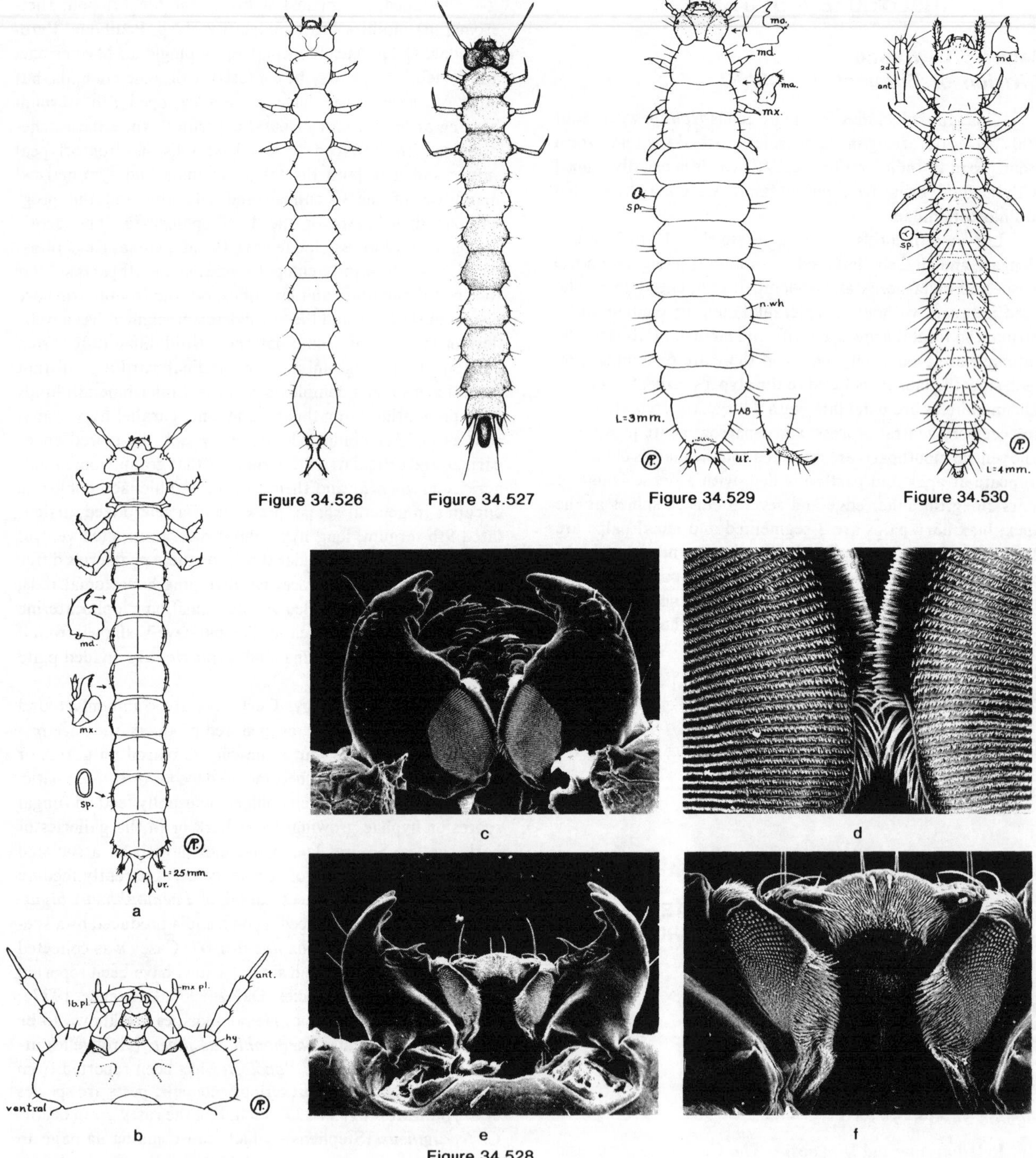

Figure 34.526 Figure 34.527 Figure 34.529 Figure 34.530

a b c d e f

Figure 34.528

Figure 34.526. Cucujidae. *Pediacus* sp. Peddler Hill, Amador Co., California. Larva, dorsal. Length = 6 mm.

Figure 34.527. Cucujidae. *Dendrophagus americanus* Mannerheim. Blodgett Forest, near Georgetown, El Dorado Co., California. Larva, dorsal. Length = 10.3 mm.

Figures 34.528a-f. Cucujidae. *Cucujus clavipes* Fabricius. **a.** larva, dorsal (plus details), length = 25 mm; **b.** head, ventral; **c.** anterior portion of head capsule, dorsal, with clypeolabral region removed, showing mandibular molae engaged; **d.** details of mandibular bases; **e.** anterior portion of head capsule, ventral, with maxillo-labial complex removed and mandibles spread, showing epipharynx and pharynx; **f.** details of mandibular bases, epipharynx and pharynx. Lives under bark of logs and dead trees. (Figures 34.528a, b from Peterson, 1951)

Figure 34.529. Cucujidae. *Cryptolestes* sp. Larva, dorsal (plus details). Length = 3 mm. (From Peterson, 1951)

Figure 34.530. Cucujidae. *Oryzaephilus surinamensis* (L.), saw-toothed grain beetle. Larva, dorsal (plus details). Length = 4 mm. Infests grains and cereal products, dried fruits, nuts, yeast, candy, tobacco, dried meats and many other food products. (From Peterson, 1951)

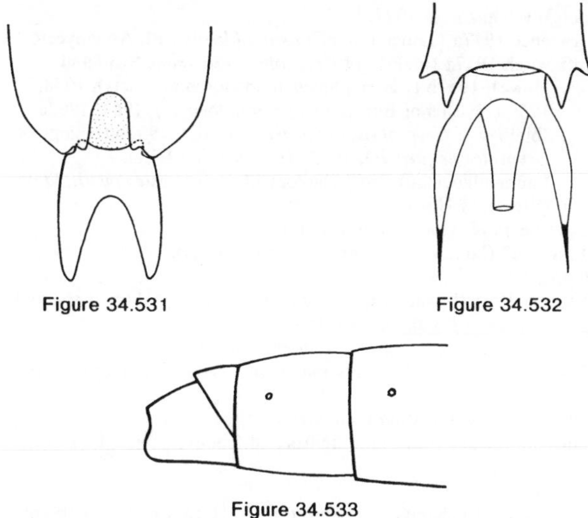

Figure 34.531 Figure 34.532

Figure 34.533

Figure 34.531. Cucujidae. Laemophloeinae. Abdominal apex, ventral.

Figure 34.532. Cucujidae. Uleiotini. Abdominal apex, dorsal.

Figure 34.533. Cucujidae. Silvanini. Abdominal apex, lateral.

fragments, fungal hyphae, and hyphomycete conidia. *Cucujus* larvae are probably facultative predators, whose feeding apparatus is still of the microphagous type, unlike those of most predaceous forms, which practice liquid feeding. The Passandrinae are true obligate predators with ectoparasitic larvae, and are usually found in the tunnels of wood-boring insects. *Catogenus rufus* (Fabricius) is known to attack larvae of a braconid wasp and cerambycid beetles (Baker, 1977; Wheeler, 1921a), while *Hectarthrum trigeminum* (Newman) has been associated with a cryptorrhynchine weevil (*Mechistocerus*) (Gravely, 1916). *Laemotmetus rhizophagoides* (Walker) is known to inhabit rattan canes and bamboo, where it feeds on larvae of Lyctinae and *Dinoderus* (Bostrichidae) (Aitken, 1975; Beeson, 1941).

Among the Silvaninae, members of the tribe Uleiotini are subcortical fungus feeders like the laemophloeines, but the larvae are larger and more active. Crowson and Ellis (1969) showed that *Dendrophagus crenatus* (Paykull), which lives under the bark of old conifers, carries spores of the fungi on which it feeds in the mandibular cavities of the adult; larvae were reared through to the adult stage on *Ceratocystis* grown from spores obtained from these mandibular cavities. Although some other Silvaninae also occur under bark, the Cryptamorphini, Psammoecini, and many Silvanini are found in leaf litter and decaying vegetation, and a number of species have become pests of stored products. Species of economic importance include the foreign grain beetle, *Ahasverus advena* (Waltl), the squarenecked grain beetle, *Cathartus quadricollis* (Guérin-Méneville), the sawtoothed grain beetle, *Oryzaephilus surinamensis* (L.) (fig. 34.530), and the merchant grain beetle, *O. mercator* (Fauvel). A unique habit occurs in the so-called social beetles of the Neotropical Region; *Coccidotrophus socialis* Schwarz and Barber and *Eunausibius wheeleri* Schwartz and Barber live in the enlarged petioles of the ant-plant (*Tachigalia*), where they feed not only on the nutritive parenchyma, but also on the honeydew produced by a species of *Pseudococcus* (Homoptera) (Wheeler, 1921a).

Description: Mature larvae 2 to 30 mm, usually less than 20 mm. Body elongate, usually more or less parallel-sided (physogastric in Passandrinae), straight, slightly to strongly flattened, usually uniformly lightly pigmented (with heavily pigmented abdominal apex in Laemophloeinae and some Cucujinae) or uniformly heavily pigmented in *Cucujus;* surfaces smooth; vestiture of scattered, simple setae.

Head: Protracted and prognathous, slightly to strongly transverse, sometimes strongly flattened, and usually as wide as or wider than thorax. Epicranial stem very short or absent; frontal arms usually lyriform and contiguous at base, subparallel and distant at base in Laemophloeinae, absent in Passandrinae. Paired, subparallel endocarinae present in Laemophloeinae. Stemmata usually 5 or 6 on each side, sometimes absent. Antennae usually long and 3-segmented, sometimes longer than head width, very short in Passandrinae; terminal segment reduced or antenna 2-segmented in some Silvanini. Frontocylpeal suture absent; labrum usually free, sometimes partly (Uleiotini) or completely (Passandrinae) fused to head capsule. Mandibles symmetrical, bidentate or tridentate, with accessory ventral process; mola usually well-developed, tuberculate or asperate, with slender, acute, fixed, hyaline prostheca; mola replaced by single hyaline process in Passandrinae or by several processes in Laemophloeinae. Ventral mouthparts usually retracted, strongly protracted in Laemophloeinae and Passandrinae. Maxilla usually with transverse cardo, elongate stipes, well-developed articulating area, 3-segmented palp, and falciform mala; cardo and articulating area absent, stipes short and broad, and mala obtuse in Laemophloeinae and Passandrinae, palp reduced in latter. Labium more or less free to base of mentum; ligula short and broad; labial palps 2-segmented and widely separated. Hypopharyngeal sclerome usually a tooth-like structure. Hypostomal rods short to moderately long and usually diverging, long and subparallel in Laemophloeinae. Ventral epicranial ridges absent. Gula usually transverse, sometimes elongate.

Thorax and Abdomen: Legs usually well-developed, 5-segmented, and widely separated (reduced and indistinctly segmented in Passandrinae); tarsungulus usually with 2 setae lying side by side. Abdomen with 9 or 10 segments, with segment 8 sometimes enlarged and segment 9 usually reduced. Abdominal terga and sterna 1–7 in Laemophloeinae with paired rows of asperities forming incomplete rings which are longitudinally oriented. Tergum A9 with pair of urogomphi in some Passandrinae, *Cucujus,* Laemophloeinae, Uleiotini (fig. 34.532) and Cryptamorphini; urogomphi absent in Silvanini (fig. 34.533), Psammoecini, and some Passandrinae, and replaced by median forked process in some Cucujinae. Tergum A9 in Cucujinae and Laemophloeinae forming sclerotized, articulated plate. Sternum A9 reduced, sometimes completely concealed. Segment A10 well-developed, more or less tubular, posteriorly oriented, and visible from above in Passandrinae and Silvaninae, reduced and concealed from above in Cucujinae and Laemophloeinae; sternum A9 and segment A10 in Laemophloeinae (fig. 34.531) membranous and located in deep emargination of sternum A8.

Spiracles: Usually annular and raised on short tubes, annular-biforous in Laemophloeinae; spiracles on segment A8 more or less posteriorly placed, except in Silvanini and Passandrinae.

Triungulin (Passandrinae): Passandrine triungulins have not been examined but apparently resemble those of Bothrideridae, except for the shorter 3rd antennal segment, biforous spiracles, and presence of 5 or 6 stemmata on each side (Crowson, *in litt.*).

Comments: The family includes about 75 genera and 1200 species, of which 17 genera and 140 species occur in North America. The subfamily Passandrinae contains the N. American genera *Catogenus* and *Scalidia,* plus various exotic groups, like *Passandra, Hectarthrum,* and *Laemotmetus. Cucujus, Pediacus,* and the exotic *Platisus* belong in the Cucujinae. Laemophloeinae includes *Laemophloeus* and a number of related genera, such as *Cryptolestes, Placonotus, Narthecius, Lathropus, Sysmerus,* and *Rhinomalus.* The Silvaninae is divided into 4 tribes: Uleiotini (*Uleiota, Dendrophagus*), Psammoecini (*Psammoecus, Telephanus*), Cryptamorphini (*Cryptamorpha*), and Silvanini (the bulk of the genera, including *Silvanus, Oryzaephilus, Cathartus, Monanus, Ahasverus,* and *Nausibius*). Several genera have been removed to other families: these include *Hemipeplus* (to Mycteridae), *Prostomis* (to Prostomidae), *Inopeplus* (to Inopeplidae), and *Hypocoprus* (to Cryptophagidae) (Böving and Craighead, 1931; Crowson, 1955, 1973a, 1981; Crowson and Viedma, 1964; Lefkovitch, 1961, 1962c).

Selected Bibliography

Aitken 1975.
Beeson 1941.
Bishop 1960 (larvae of *Cryptolestes ferrugineus, C. minutus* (Olivier), and *C. turicus* (Grouvelle)).
Böving 1921 (larvae of *Coccidotrophus socialis, Eunausibius wheeleri, Nausibius clavicornis* (Kugelann), *Ahasverus advena, Telephanus* sp., and *Scalidia linearis* LeConte).
Böving and Craighead 1931 (larvae of *Cucujus clavipes* Fabricius, *Brontes* sp., *Oryzaephilus surinamensis,* and those listed for Böving, 1921).
Bräuer 1970 (biology of *Cryptolestes*).
Cekalovic and Quezada 1972 (larva of *Uleiota chilensis* (Blanchard)).
Crowson 1955, 1973a, 1981.
Crowson and Ellis 1969 (mycangia in *Dendrophagus crenatus* (Paykull)).
Crowson and Viedma 1964.
Cutler 1971 (larvae of *Ahasverus advena, Cathartus quadricollis, Oryzaephilus surinamensis,* and *O. mercator*).
Dajoz 1966 (association with Ascomycetes).
Drooz 1985 (habits of *Catogenus rufus*).
Emden 1931 (larva of *Telephanus costaricensis* Nevermann).
Gravely 1915 (larva of *Uleiota indica* Arrow; key to *Uleiota* larvae), 1916 (larva of *Hectarthrum trigeminum* (Newman)).
Halstead 1973 (revision of *Silvanus* and relatives), 1980 (revision of *Oryzaephilus* and relatives).
Hayashi 1980 (larvae of *Cucujus coccinatus* Lewis, *Uleiota arborea* Reitter, *Dendrophagus longicornis* Reitter, *Pediacus japonicus* Reitter, *Silvanus lateritius* Reitter, and *Laemophloeus* sp.).
Howe and Lefkovitch 1957.
Hudson 1934 (larvae of *Cryptamorpha brevicornis* (White) and *Dendrophagus capito* Pascoe (as *Parabrontes setiger* Broun; *see* Lefkovitch, 1961)).
Iablokov-Khnzorian 1977.
Lawrence 1977a (association of *Laemophloeus* with Ascomycetes).
Lefkovitch 1957a (biology of *Cryptolestes ugandae* Steel and Howe), 1957b (Laemophloeinae in stored products), 1958, 1959 (revision of European Laemophloeinae), 1961, 1962a (biology of *Cryptolestes capensis* (Waltl)), 1962b (biology of *Cryptolestes turicus*), 1962c (revision of African Laemophloeinae), 1964 (biology of *Cryptolestes pusilloides* (Steel and Howe)).
Lepesme 1944 (stored products pests).
Olliff 1882 (larva of *Cucujus coccinatus* Lewis).
Palm 1959.
Peterson 1951 (larvae of *Cucujus clavipes, Laemophloeus* sp., and *Oryzaephilus surinamensis*).
Reid 1942b (stored products laemophloeines).
Roberts and Rillett 1953 (silk glands in *Cryptolestes ferrugineus*).
Saalas 1917.
Schedl 1962 (association with bark beetles).
Smith and Sears 1982 (mouthparts and feeding habits of *Cucujus clavipes*).
Thomas 1984.
Wheeler 1921a (habits and habitats), 1921b (association with ant-plants).
Williams 1931 (larva of *Cryptamorpha desjardinsi* (Guérin-Meneville)).

PROPALTICIDAE (CUCUJOIDEA)

John F. Lawrence,
Division of Entomology, CSIRO

This family includes the genus *Propalticus* with about 25 described species from Asia, the East Indies, the Pacific Region, Australia, Madagascar, and Africa, and *Discogenia* with about 10 species occurring only in Africa. Adults are minute, broad and strongly flattened, and are capable of jumping using the front pair of legs. On the basis of adult structure, the group appears to be related to Cucujidae-Laemophloeinae, but a relationship to Languriidae-Cryptophilinae was suggested by Crowson and Sen Gupta (1969), based on larvae collected in fungusy bark of a dead tree with adults of an Australian *Propalticus.*

Selected Bibliography

Crowson 1955.
Crowson and Sen Gupta 1969 (putative larva; relationships).
John 1960.
Sen Gupta and Crowson 1971.

PHALACRIDAE (CUCUJOIDEA)

John F. Lawrence,
Division of Entomology, CSIRO

Shining Flower Beetles

Figures 34.534-538

Relationships and Diagnosis: The systematic position of the Phalacridae is still in doubt, but the family is generally considered to be one of the more primitive groups of Cucujoidea, with possible affinities with the Nitidulidae and Cucujidae. In general, the group is well defined, but the inclusions

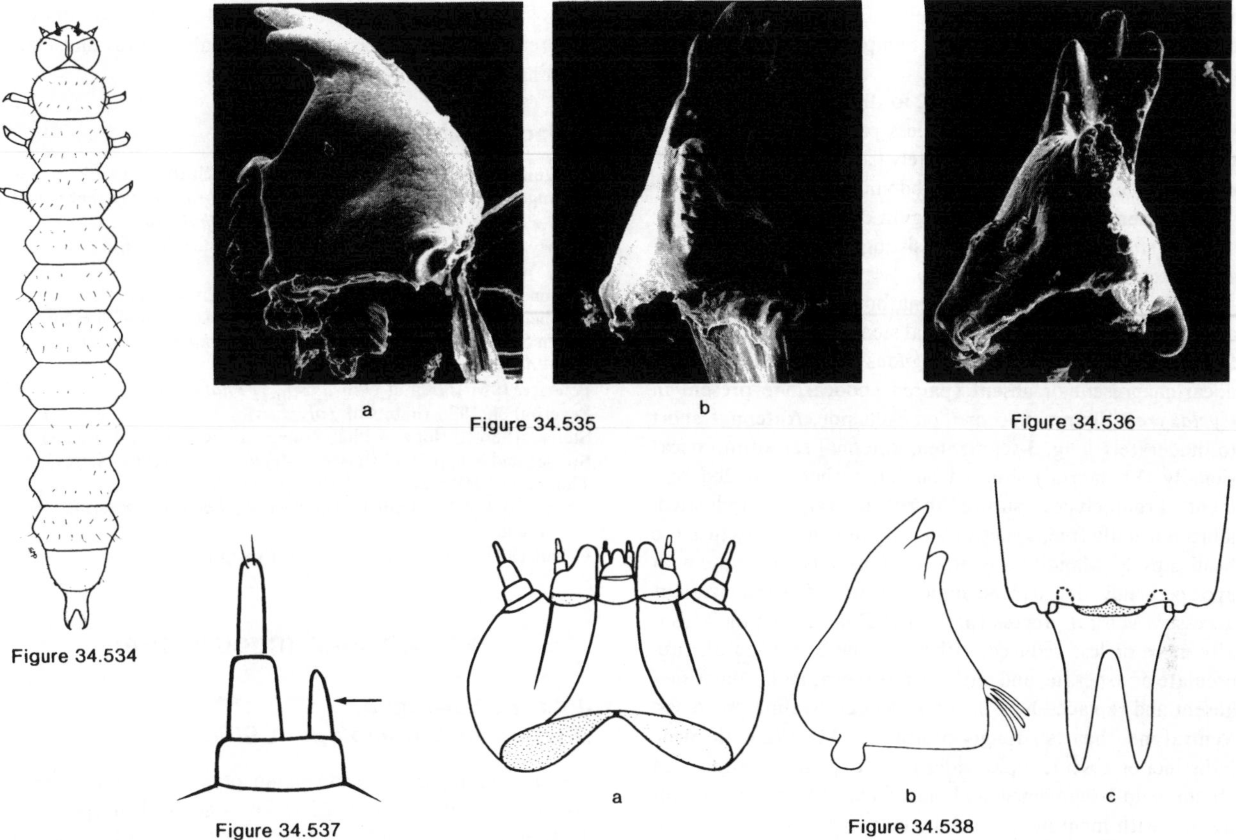

Figure 34.534

Figure 34.535

Figure 34.536

Figure 34.537

Figure 34.538

Figure 34.534. Phalacridae. *Litochropus clavicornis* Casey. Welder Wildlife Refuge, near Sinton, Texas. Larva, dorsal. Length = 4.8 mm.

Figures 34.535a,b. Phalacridae. *Phalacropsis dispar* (LeConte). Lee Vining, California. **a.** right mandible, dorsal; **b.** left mandible, mesal.

Figure 34.536. Phalacridae. *Acylomus pugetanus* Casey? Bedford, Massachusetts. Left mandible, mesal.

Figure 34.537. Phalacridae. *Phalacrus* sp. Antenna (showing sensorium on segment 1).

Figures 34.538a-c. Phalacridae. *Litochrus* sp. **a.** head, ventral; **b.** right mandible, ventral; **c.** abdominal apex, ventral.

of the Australian *Phalacrinus* and the New Zealand *Cyclaxyra* are still open to question (Crowson, 1955; Sen Gupta and Crowson, 1966). Phalacrid larvae are usually distinguished from those of other cucujoids by the combination of protracted ventral mouthparts, reduced cardines, long, diverging hypostomal rods, enlarged and posteriorly oriented 8th spiracles, and concealed 9th sternite and anal region. Protracted ventral mouthparts also occur in laemophloeine Cucujidae, the nitidulid genus *Smicrips*, and the family Cleridae, but these groups do not have lyriform frontal arms which are contiguous at the base or a concealed anal region (in Laemophloeinae it is exposed by a notch in sternum 8). Antennae of the Phalacrinae (*Phalacrus* and *Phalacropsis*) are unique in the Coleoptera, in that the sensorium is located on the first and not the second segment.

Biology and Ecology: Most Phalacridae are associated with ascomycete and primitive basidiomycete fungi, including molds, mildews, sooty molds, ergots, smuts, and rusts, but some feed in rotten wood or bark and others occur in the flowers heads of Compositae. Species of *Phalacrus* are often found in the spore masses of smuts (*Ustilago, Tilletia, Cintractia*) infesting various monocots, such as *Carex*, rye, wheat, corn, and sugarcane, but at least 2 Australian species feed on spores formed on the galls of *Uromycladium* rusts. The related genus *Phalacropsis* is associated with the aecia of pine stem rusts (*Peridermium* spp.). Most species of *Acylomus* and *Stilbus* feed on the spores of molds and mildews occurring on dead or dying leaves, seed clusters, etc. An exception is *A. pugetanus* Casey, whose larvae feed within the sclerotia (ergots) of *Claviceps purpurea*. Species of *Litochropus* feed on spores and stromatal tissue of *Daldinia* spp. (Xylariaceae), while Australian species of *Litochrus* occur in rotten wood or bark. Species of *Olibrus* breed in the flower heads of various Compositae (*Aster, Cirsium, Eupatorium, Solidago*), and the European *Tolyphus* apparently have similar habits. *Olibrus* larvae feed on fresh flowers, and no spores or other fungal material have been found in the gut. Pupation may occur on surfaces (*Acylomus*), within the food mass (*Phalacrus, Acylomus pugetanus, Litochropus, Litochrus*), or in the ground (*Olibrus*) (d'Aguilar, 1944; Crowson, 1981,

1984a; Lambert and McIlveen, 1976; Löben Sels, 1934; Peyerimhoff, 1926; Steiner, 1984; Thompson, 1958; Thompson and Marshall, 1980).

Description: Mature larvae 2 to 10 mm, usually less than 5 mm. Body elongate, more or less parallel-sided or somewhat fusiform, slightly to moderately flattened; usually lightly pigmented, except for head and abdominal apex, occasionally with pigmented plates on protergum or on all visible terga; surfaces smooth and vestiture consisting of scattered or dense, simple setae or longer hairs.

Head: Protracted and prognathous, moderately broad and somewhat flattened. Epicranial stem very short or absent; frontal arms lyriform and contiguous at base. Median endocarina present or absent (paired endocarinae present in *Cyclaxyra*). Stemmata 5 or 6 on each side. Antennae short to moderately long, 3-segmented; antennal sensorium occasionally (Phalacrini) situated on 1st, rather than 2nd segment. Frontoclypeal suture absent or vaguely indicated; labrum usually free, sometimes partly or completely fused to head capsule. Mandibles symmetrical, usually tridentate with apex perpendicular to plane of movement, and usually without accessory ventral process (present in Phalacrini); mola usually more or less reduced, either narrow or sub-basal, tuberculate or asperate, and without prostheca; mola sometimes absent and replaced by 2 or more slender, hyaline processes. Ventral mouthparts strongly protracted. Maxilla with cardo indistinct or absent, stipes wider than long, articulating area absent, palp 3-segmented, and mala fixed and obtuse. Labium usually with mentum and submentum fused; ligula present; labial palps 2-segmented and widely separated. Hypopharyngeal sclerome absent or consisting of transverse bar. Hypostomal rods long and diverging. Ventral epicranial ridges absent. Gula elongate.

Thorax and Abdomen: Legs short to moderately long, 5-segmented, widely separated; tarsungulus with single seta or 2 setae of unequal lengths. Abdominal segments 1–7 occasionally with paired rows of asperities forming incomplete rings on terga and sterna. Segment A9 much narrower and usually shorter than segment A8; tergum A9 almost always with pair of urogomphi (absent in *Cyclaxyra*), sometimes forming sclerotized plate which articulates with A8. Sternum A9 and segment A10 more or less concealed beneath sternum A8 (except in *Cyclaxyra*).

Spiracles: Annular-biforous, those on segment A8 usually enlarged and posteriorly situated.

Comments: The family is comprised of about 55 genera and 600 species worldwide, with 13 genera and 122 species occurring in N. America. Although there is no generally accepted subfamilial classification, the Phalacrinae (*Phalacrus* and *Phalacropsis*), *Tolyphus, Phalacrinus,* and *Cyclaxyra* form distinctive groups on the basis of adult characters, and in the case of Phalacrinae and *Cyclaxyra* this is supported by features of the larvae. The remaining phalacrids, whose larvae are known, fall into 2 groups: those having a median endocarina and lacking a mandibular mola (with 2 or more hyaline processes at the mandibular base) (*Litochrus, Litochropus, Olibrus*); and those lacking a median endocarina but having a mandibular mola (*Acylomus, Stilbus*). Steiner (1984) illustrated mandibles of several genera and discussed the relationship between morphological features and larval feeding habits.

Selected Bibliography

d'Aguilar 1944 (larva of *Phalacris caricis* Sturm; larval key).
Böving and Craighead 1931 (larvae of *Olibrus aeneus* (Fabricius), *Phalacrus politus* Melsheimer, and *Phalacrus* sp.).
Crowson 1955, 1981, Crowson, 1984a (association with Ascomycetes).
Emden 1928 (larva of *Phalacrus grossus* Erichson).
Lambert and McIlveen 1976 (*Acylomus* associated with ergot).
Löben Sels 1934 (larva of *Olibrus* sp., as *Phalacrus politus* Melsheimer).
Peterson 1951 (larva of *Olibrus* sp., as *Phalacrus*).
Peyerimhoff 1926 (habits of *Tolyphus*).
Steiner 1984 (biology of Phalacridae; larvae of several species).
Steiner and Singh 1987 (larva of *Acylomus pugetanus* Casey).
Thompson 1958 (review of British species).
Thompson and Marshall 1980 (larva of *Phalacrus uniformis* (Blackburn)).
Urban 1926 (biology, larva of *Olibrus* spp.).

HOBARTIIDAE (CUCUJOIDEA)

John F. Lawrence,
Division of Entomology, CSIRO

This family is based on two monotypic Australian genera, *Hobartius* and *Hydnobioides,* but undescribed species of *Hobartius* are known from Australia, Chile, and Argentina.

Larvae are elongate, somewhat fusiform, and slightly flattened, with a relatively lightly pigmented upper surface which is granulate and/or tuberculate. The head lacks an epicranial stem and median endocarina and has lyriform frontal arms, distant at base, 5 stemmata on each side, and well-developed antennae. The mandibles have a simple or serrate incisor edge, a well-developed tuberculate mola, accessory ventral process, and narrow, acute, hyaline prostheca (reduced in *Hobartius*). The ventral mouthparts are retracted, the mala falciform or narrowly rounded, and the maxillary palps 3-segmented. The labial palps are only 1-segmented. The hypostomal rods are short to moderately long and diverging, and there are no ventral epicranial ridges. The legs have a bisetose tarsungulus. Abdominal terga are armed with transverse rows of short to very long setiferous processes, and tergum 9 has a pair of upturned, fixed urogomphi; segment 10 is circular and more or less ventrally oriented. Spiracles are biforous and borne on short spiracular tubes.

Hobartiids are known to breed in soft, rotting fruiting bodies of some basidiomycete fungi, including Tricholomataceae (*Pleurotus*), Polyporaceae (*Grifola*), and Clavariaceae, but adults may be attracted in large numbers to carrion traps. The larva described as *Hydnobioides pubescens* by Sen Gupta and Crowson (1969a) was misidentified.

Selected Bibliography

Crowson 1981.
Sen Gupta and Crowson 1966, 1969a.

CAVOGNATHIDAE (CUCUJOIDEA)

John F. Lawrence,
Division of Entomology, CSIRO

This family includes 5 described species: *Taphropiestes fusca* Reitter from Chile, *Cavognatha pullivora* Crowson from Australia, *Neocercus electus* Broun from New Zealand, and the New Zealand genus *Zeonidicola* with 2 species; other forms have been seen from Brazil, Argentina, and Chile. The group is usually placed near the Cryptophagidae.

Cavognathid larvae are elongate, more or less parallel-sided, and slightly flattened, with a darkly pigmented head and tergal plates on the thorax and abdominal segments 1 to 9. The head is short and broad, with moderately long antennae, 6 stemmata on each side, lyriform frontal arms, which are distant at base, and no epicranial stem or median endocarina. The mandibles are bidentate, with an additional subapical tooth, and the mola is absent, being replaced by 2 hyaline processes, one rounded and setose and the other narrow and acute. The ventral mouthparts are retracted, the mala is falciform, maxillary palps are 3-segmented and the labial palps 2-segmented. Hypostomal rods are moderately long and diverging. The legs are moderately widely separated and the tarsungulus bears 2 setae, one distal to the other. Tergum A9 has a pair of posteriorly projecting urogomphi, and segment A10 is circular and more or less posteriorly oriented. Spiracles are biforous and placed at the ends of very short tubes.

Cavognathids occur in the nests of birds and may be attached to nestlings. They have been associated with a number of bird species. *Z. chathamensis* has been found in the nests of the pipit (Motacillidae), sooty shearwater, fairy prion, giant petrel, and royal albatross (Procellaridae) on the Chatham Islands; *Z. dumbletoni* has been associated with the spotted shag (Phalacrocoracidae); a species identified as "*Taphropiestes* sp." was found with the weebill (Acanthizidae); and *C. pullivora* has been found in the nests of the magpie (Cractidae), white-winged triller (Campephagidae), regent honeyeater (Meliphagidae), and a finch. Cavognathids are widely distributed in Australia but they appear to be more common in the arid inland regions.

Selected Bibliography

Chisholm 1952 (bird nest record).
Crowson 1964b (larva of *Cavognatha pullivora*), 1973a.
Sen Gupta and Crowson 1966, 1969a.
Watt 1980 (larva of *Zeonidicola dumbletoni* Crowson).

CRYPTOPHAGIDAE (CUCUJOIDEA)

(INCLUDING CATOPOCHROTIDAE, HYPOCOPRIDAE)

John F. Lawrence,
Division of Entomology, CSIRO

Silken Fungus Beetles

Figures 34.539-542

Relationships and Diagnosis: Because of recent studies on cucujoid phylogeny (Crowson, 1955, 1980; Sen Gupta, 1967, 1968a; Sen Gupta and Crowson, 1971), the constitution of the Cryptophagidae has undergone considerable change, and of those North American genera listed by Arnett (1968), *Cryptophilus, Toramus, Loberus, Hapalips,* and *Pharaxonotha,* have been transferred to Languriidae, while the genus *Hypocoprus* (Rhizophagidae in Arnett) has been added to the family. The group is a relatively primitive one, sharing plesiomorphic features with Cavognathidae and Phloeostichidae, but differing from all basal cucujoids in the adult stage by the outward closure of the mesocoxal cavities by the sterna. Cryptophagid larvae may be distinguished from those of other cucujoid families by the combination of falciform mala, narrow-based, hyaline prostheca (which may be simple, bifid, or serrate), 2 tarsungular setae, no ventral epicranial ridges, and either 1-segmented labial palps or annular spiracles.

Biology and Ecology: Cryptophagidae are usually found under bark or in leaf litter, compost heaps, animal nests, or stored products, where they feed on fungal spores and hyphae, but a few are known to feed on pollen or fern spores, and others have been associated with the fruiting bodies of higher fungi. Hinton and Stephens (1941a) found the adults of *Cryptophagus acutangulus* Gyllenhal fed on spores of many fungi, but larvae were reared on molds (*Penicillium* and *Botrytis*). Many of the species recorded from the larger fruiting bodies of Basidiomycetes were probably feeding on hyphae and conidia of surface molds. Exceptions include the Californian *Cryptophagus maximus* Blake and the closely related Japanese species, *C. enormis* Hisamatsu, which feed on the spore mass of the pouch fungus, *Cryptoporus volvatus* (Hisamatsu, 1962); *Pteryngium* species, which are consistently associated with *Fomitopsis pinicola* (Scheerpeltz and Höfler, 1948); and *Crosimus* species, which have been taken on a variety of trametoid Polyporaceae (Weiss and West, 1920). *Cryptophagus lycoperdi* (Scopoli) is a frequent inhabitant of the gasteromycete *Scleroderma vulgare,* but it has also been recorded from the hypogean fruiting body of *Choiromyces meandriniformis* (Tuberales) (Donisthorpe, 1935). *Cryptophagus ruficornis* Stephens was collected on the early stage, sporulating stromata of *Daldinia* sp. (Ascomycetes: Xylariaceae), while *C. dentatus* (Herbst) was associated with older fruiting bodies (Hingley, 1971). A number of *Cryptophagus* species have been taken from the nests of birds (Hicks, 1959, 1962, 1971), mammals, and social insects; species of *Antherophagus* are apparently restricted to *Bombus* nests (Horion, 1960; Meer Mohr and Lieftinck, 1947); and *Emphylus glaber* (Gyllenhal) occurs in nests of various *Formica* species (Eichelbaum, 1907; Horion, 1960). Some atomariines, such as *Anchicera ruficornis* (Marsham) and *A. lewisi* (Reitter), occur in compost heaps and grass cuttings, and have been widely distributed by man (Evans, 1961). *Atomaria linearis* Stephens has been recorded as a pest of mangels and some other food plants (Newton, 1932). Species apparently belonging to an undescribed genus were collected on sporangia of a fern (*Cystopteris fragilis*) on Juan

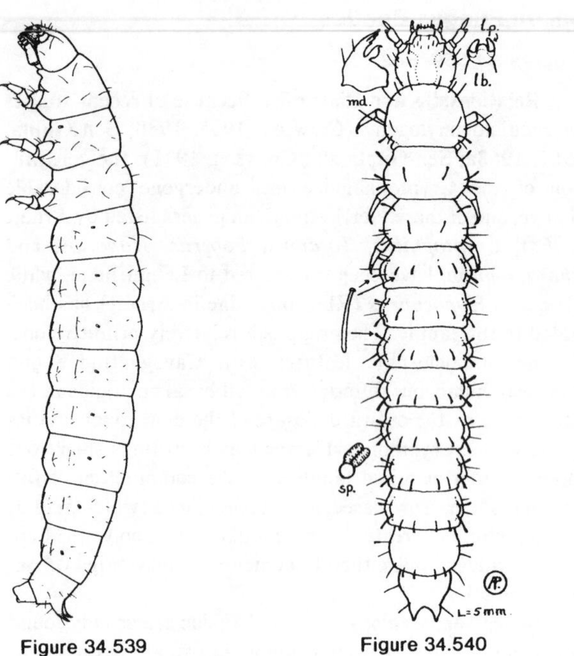

Figure 34.539

Figure 34.540

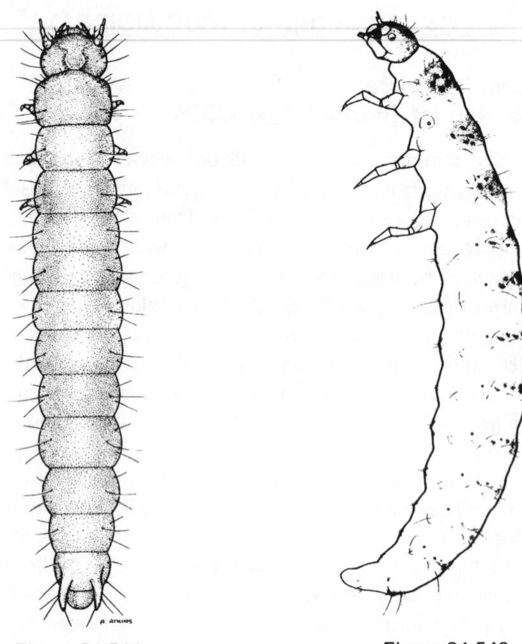

Figure 34.541

Figure 34.542

Figure 34.539. Cryptophagidae. *Antherophagus* sp. Manotick, Ontario. Larva, lateral. Length = 9.5 mm.

Figure 34.540. Cryptophagidae. *Cryptophagus* sp. Larva, dorsal (plus details). Length = 5 mm. (From Peterson, 1951)

Figure 34.541. Cryptophagidae. *Atomaria longipennis* (Casey). 7.5 mi. S Los Gatos, Santa Clara Co., California. Larva, dorsal. Length = 3.1 mm.

Figure 34.542. Cryptophagidae. *Anchicera lewisi* Reitter. Bedford, Middlesex Co., Massachusetts. Larva, lateral. Length = 3.3 mm.

Fernandez; sporangia were found on the mouthparts and fern spores in the gut. The N. American and European species of *Telmatophilus* feed in the flower heads of aquatic monocots, such as *Carex, Typha,* and *Sparganium* (Hatch, 1962; Horion, 1960). *Henoticus californicus* (Mannerheim), and several species of *Cryptophagus* and *Anchicera* are known to be pests in stored products (Hinton, 1945b).

Description: Mature larvae 2 to 6 mm. Body elongate, more or less parallel-sided or slightly wider in middle, straight, subcylindrical to slightly flattened; surfaces lightly pigmented, occasionally with yellowish-brown tergal plates which may be maculate, smooth, and clothed with scattered, simple, setae.

Head: Protracted and prognathous, moderately broad and slightly flattened. Epicranial stem absent; frontal arms lyriform, usually contiguous at base. Median endocarina absent. Stemmata usually indistinct or absent, sometimes 1, 2, or 5 or each side. Antennae well-developed, 3-segmented. Frontocylpeal suture absent; labrum free. Mandibles symmetrical, bidentate, with accessory ventral process and often with serrate incisor edge; mola well-developed, tuberculate or asperate, sometimes with brush of hairs at base; prostheca consisting of slender, fixed, hyaline process, which may be acute, bifid, or serrate. Ventral mouthparts retracted. Maxilla with transverse cardo, elongate stipes, well-developed articulating area, 3-segmented palp, and falciform mala. Labium free to base of mentum; ligula present; labial palps 1-segmented (Cryptophaginae) or 2-segmented (Atomariinae), widely separated. Hypopharyngeal sclerome usually tooth-like. Hypostomal rods moderately long and diverging. Ventral epicranial ridges absent. Gula transverse.

Thorax and Abdomen: Legs well-developed, 5-segmented; tarsungulus with 2 setae, either side by side or with one distal to the other; coxae moderately close together. Tergum A9 usually with pair of strongly upturned, fixed urogomphi, simple and sclerotized at apex only, occasionally reduced or absent. Sternum A9 well-developed, simple. Segment A10 well-developed, circular, posteriorly or posteroventrally oriented.

Spiracles: Annular or annular-biforous, not raised on tubes.

Comments: The family includes about 30 genera and 600 species for the world, and 16 genera and 140 species occur in the United States and Canada. The subfamily and tribal classification is currently being revised by Sen Gupta and Crowson, and several genera remain to be described, especially from the Southern Hemisphere. Crowson (1980) has divided the group into 4 subfamilies: Hypocoprinae (the Holarctic *Hypocoprus*), Alfieriellinae (the Palearctic *Alfieriella*), Cryptophaginae and Atomariinae; the last 2 groups have been reconstituted by Crowson, so that *Caenoscelis* and related genera fall into the former, along with the New Zealand *Pirotus* and *Thortus,* while *Ephistemus* and its allies, as well as various Southern Hemisphere forms related to

Salltius are included in the latter (Crowson, 1980). The family exhibits an amphipolar distribution, with tribes restricted to either North Temperate (Cryptophaginae: Cryptophagini and Caenoscelini; Atomariinae: Atomariini) or South Temperate (Cryptophaginae: Cryptosomatulini and Picrotini; Atomariinae: Salltiini) Regions. Exceptions occur in the cryptophagine genera (*Mnioticus* (East Africa) and *Micrambe* (extending into tropical Africa and Asia).

Selected Bibliography

Arnett 1968.
Böving and Craighead 1931 (larvae of *Antherophagus* sp., *Cryptophagus saginatus* Sturm, *Telmatophilus typhae* Fall, and *Henoticus californicus* (Mannerheim), as *H. germanicus*).
Coombs and Woodroffe 1955.
Crowson 1955, 1967, 1980 (amphipolar distribution).
Donisthorpe 1935 (mycophagy).
Eichelbaum 1907 (larva of *Emphylus glaber* (Gyllenhal)).
Evans 1961 (larva of *Anchicera ruficornis*).
Falcoz 1922 (larva of *Henoticus californicus* (Mannerheim)), 1924 (larvae of *Cryptophagus* spp.).
Hatch 1962.
Hingley 1971 (association with Ascomycetes).
Hinton 1945b (stored products pests).
Hinton and Stephens 1941a (larva of *Cryptophagus acutangulus* Gyllenhal).
Hisamatsu 1962 (association with *Cryptoporus*, a polypore).
Horion 1960.
Klippel 1952 (association with molds; larvae and pupae of *Cryptophagus* spp.).
Meer Mohr and Lieftinck 1947 (larva of *Antherophagus ludekingi* Grouvelle).
Newton 1932 (larva of *Atomaria linearis* Stephens).
Peterson 1951 (larva of *Cryptophagus* sp.).
Scheerpeltz and Höfler 1948 (mycophagy).
Sen Gupta 1967, 1968a.
Sen Gupta and Crowson 1971.
Verhoeff 1923.
Weiss and West 1920 (mycophagy).

LAMINGTONIIDAE (CUCUJOIDEA)

John F. Lawrence,
Division of Entomology, CSIRO

This family is based on *Lamingtonium binnaburrense* Sen Gupta and Crowson, which was collected under the bark of a dead standing tree in southern Queensland, Australia. On the basis of adult features, it was considered to be a link between primitive Cucujoidea and the family Languriidae (Sen Gupta and Crowson, 1969b). Larvae have not been described.

Selected Bibliography

Lawrence and Newton 1982.
Sen Gupta and Crowson 1969b.

LANGURIIDAE (CUCUJOIDEA) (INCLUDING CRYPTOPHILIDAE)

John F. Lawrence,
Division of Entomology, CSIRO

Figures 34.543–547

Relationships and Diagnosis: The Languriidae, as defined by Sen Gupta and Crowson (1971), includes not only the typical languriines, but also a number of genera formerly placed in Cryptophagidae (Arnett, 1968), such as *Pharaxonotha, Hapalips, Loberus, Cryptophilus,* and *Toramus*. The family appears to be closely related to Erotylidae, and some members of the Xenoscelinae are difficult to distinguish in the larval stage from dacnine erotylids. Larvae of both Erotylidae-Dacninae and Languriidae may be distinguished by the presence of a well-developed, tuberculate or asperate, mandibular mola, accompanied by a characteristic, fixed, hyaline prostheca, which is broad at base and more or less angulate at the apex; languriid larvae differ from those of dacnines in having a falciform, rather than a truncate maxillary mala. Xenosceline and cryptophiline larvae usually have a more or less granulate or tuberculate upper surface, resembling the condition found in many other cucujoids, such as *Monotoma* (Rhizophagidae), *Epuraea* (Nitidulidae), *Teredolaemus* (Bothrideridae), or *Mycetaea* (Endomychidae), but in all of these groups the prostheca is differently formed (narrow and acute, serrate, complex) or absent.

Biology and Ecology: Members of the subfamily Xenoscelinae have a variety of feeding habits but are usually associated with decaying plant tissue. The European *Eicolyctus brunneus* (Gyllenhal) feeds in red-rotten wood of maple, alder and birch (Horion, 1960; Lundberg, 1973; Sen Gupta and Crowson, 1967). The widespread *Leucohimatium arundinaceum* (Forskål) has been recorded in oats and barley, and was found breeding in a smut (*Ustilago*) on *Spinifex* in Australia, while *Leucohimatops javanus* Heller was commonly found in tea shipments in Java (Aitken, 1975; Heller, 1923). Several xenoscelines feed on the pollen of cycadaceous gymnosperms; *Pharaxonotha floridana* (Casey) occurs on *Zamia* in N. America, *Pharaxonotha* sp. on *Encephalartos* in South Africa, and *Xenocryptus tenebroides* Arrow and *Hapalips* spp. on *Macrozamia* in Australia (Roberts, 1939; Sen Gupta and Crowson, 1971). Various other *Hapalips* species have been recorded from corn stalks, banana leaves, cactus flowers, and palm leaf bases, while the stored products pest *Pharaxonotha kirschi* Reitter is known to cause damage to cotton bolls, corn meal, edible tubers, stored maize, wheat, and beans (Grouvelle, 1914; Hinton, 1945b; Sen Gupta, 1968a; Sen Gupta and Crowson, 1971). Perhaps the most bizarre habits are those of the Mexican *Loberopsyllus traubi* Martinez and Barrera, a blind and apterous commensal of the volcano mouse (*Neotomodon*), which attaches itself to the fur and feeds on dead skin and other organic matter (Barrera, 1969).

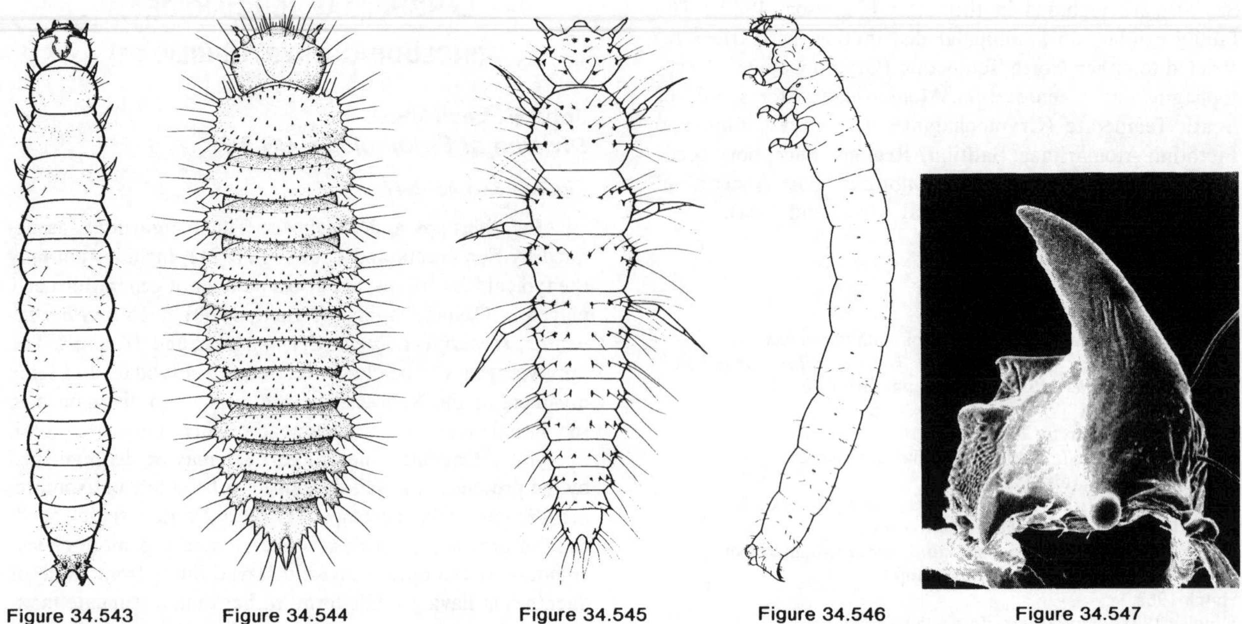

Figure 34.543. Languriidae. *Pharaxonotha floridana* (Casey). Crescent City, Florida. Larva, dorsal. Length = 8.3 mm.

Figure 34.544. Languriidae. *Cryptophilus* sp. Shipton's Flat, North Queensland. Larva, dorsal. Length = 2.7 mm.

Figure 34.545. Languriidae. *Loberoschema* sp. Barro Colorado Is., Canal Zone, Panama. Larva, dorsal. Length = 1.2 mm.

Figure 34.546. Languriidae.. *Acropteroxys gracilis* (Newman). Bedford, Middlesex Co., Massachusetts. Larva, lateral. Length = 17.5 mm.

Figure 34.547. Languriidae. *Hapalips* sp. Galapagos Is. Left mandible, ventral.

The only member of the Setariolinae, *Setariola sericea* (Mulsant and Rey), occurs under bark and in dead leaves and vegetable detritus in southern Europe (Falcoz, 1921). The closely related groups Cryptophilinae and Toraminae are also found in decaying vegetation and may be associated with molds. *Cryptophilus integer* (Heer) has been implicated as a stored products pest because of its occurrence in cellars and warehouses, sometimes in association with dried fruit or cereals (Aitken, 1975; Hinton, 1945b), and Peyerimhoff (1919) reared the species from larvae found on the underside of a bracket fungus (*Fomes fomentarius*). *Cryptophilus seriatus* Casey has been found on rotten fruiting bodies of *Polyporus squamosus* and *Steccherinum septentrionale,* where it was apparently feeding on surface molds. *Cryptophilus* species are also commonly collected in leaf litter. Adults and larvae belonging to the toramine genera *Loberoschema* and *Empocryptus* were taken in numbers in the rotting fallen flowers of a tropical tree (*Pseudobombax*) in Panama. The subfamily Languriinae includes those species which feed in the living stems of various herbaceous angiosperms, and a number of them may be pests of garden vegetables or forage crops (Hinton, 1945b; Lawrence and Vaurie, 1983; Vaurie, 1948).

Description: Mature larvae 2 to 25 mm. Body moderately to very elongate, more or less parallel-sided or sometimes narrowed posteriorly, straight, and cylindrical to slightly flattened. Dorsal surfaces usually granulate or tuberculate (smooth in Languriinae and *Leucohimatium*), lightly pigmented, sometimes with darker head and tergal plates; vestiture of long, fine, scattered hairs, occasionally mixed with expanded or frayed setae.

Head: Protracted and prognathous, moderately broad, slightly flattened. Epicranial stem usually absent (long in Languriinae); frontal arms usually lyriform (V-shaped in Languriinae) and contiguous at base (separated at base in some Loberini); epicranial suture sometimes indistinct. Median endocarina extending between frontal arms in some Xenoscelinae (*Eicolyctus, Leucohimatium, Pharaxonotha*). Stemmata usually 5 or 6 on each side, occasionally 2 or 0. Antennae moderately long, 3-segmented. Frontoclypeal suture absent; labrum free. Mandibles symmetrical, bidentate or tridentate, usually with accessory ventral process (absent in Languriinae); mola well-developed, tuberculate or asperate, with tubercles or asperities sometimes forming transverse rows, sometimes with hyaline lobe at base; prostheca usually a fixed, hyaline process, broad at base and obtusely angulate at apex (broadly triangular), sometimes reduced. Ventral mouthparts retracted. Maxilla with transverse cardo, elongate stipes, well-developed articulating area, 3-segmented palp, and falciform mala. Labium more or less free to base of mentum; ligula short and broad; labial palps 2-segmented and widely separated. Hypopharyngeal sclerome tooth-like. Hypostomal rods moderately to very long and diverging. Ventral epicranial ridges absent, except in Cryptophilinae and Toraminae.

Thorax and Abdomen: Thoracic and abdominal terga sometimes with weak (Toraminae) or strong (Cryptophilinae) lateral processes. Legs well-developed, 5-segmented, moderately widely separated; tarsungulus usually with 2 setae lying more or less side by side, sometimes (Toraminae and Cryptophilinae) with 2 unequal setae lying one distal to the other. Tergum A9 well-developed, with pair of fixed urogomphi, which are strongly upturned (not so in Toraminae and Cryptophilinae) and unpigmented or pigmented at apex only; accessory setiferous tubercles or pregomphi often present in front of urogomphi (absent in Languriinae). Segment A10 usually more or less circular and posteroventrally oriented (somewhat transverse in Languriinae).

Spiracles: Biforous or annular-biforous, not raised on tubes.

Comments: The family contains about 85 genera and 900 species, with 9 genera and 38 species occurring in America north of Mexico. Five subfamilies are currently recognized: Xenoscelinae, with the tribes Xenoscelini (= Pharaxonothini), Loberini, and Loberonothini; Setariolinae, with the European genus *Setariola;* Languriinae, with the tribes Languriini, Cladoxenini, and Thallisellini; Cryptophilinae, with the tribes Cryptophilini and Xenoscelinini; and Toraminae (Lawrence and Vaurie, 1983; Sen Gupta, 1967, 1968a, 1968b; Sen Gupta and Crowson, 1967, 1969b, 1971). Unlike the Cryptophagidae, with which the group was formerly confused, Languriidae are widely distributed throughout the warmer parts of the world, and are not as common in cool temperate regions. The larvae described by Sen Gupta and Crowson (1971) as *Cryptophilus integer* and *Xenoscelinus australiensis* Sen Gupta and Crowson, were apparently misidentified.

Selected Bibliography

Aitken 1975.
Arnett 1968.
Barrera 1969 (habits of *Loberopsyllus traubi*).
Böving and Craighead 1931 (larvae of *Languria angustata* (Palisot de Beauvois) and *Pharaxonotha kirschi*).
Falcoz 1921 (relationships of *Setariola sericea* (Mulsant)).
Gardner 1931b.
Grouvelle 1914 (habitat of *Hapalips*).
Heller 1923 (*Leucohimatops javanus* in tea).
Hinton 1945b (stored products pests).
Horion 1960.
Kingsolver 1973.
Lawrence and Vaurie 1983 (North American catalogue).
Lundberg 1973 (larva of *Eicolyctus brunneus*).
Martins and Pereira 1966.
Peterson 1951 (larva of *Languria mozardi* Latreille).
Peyerimhoff 1919 (habits of *Cryptophilus integer*).
Roberts 1939 (larvae of *Pharaxonotha floridana, Bolerus angulosus* Arrow, *Hapalips* sp., and several Languriinae), 1958 (various genera in key).
Sen Gupta 1967 (Toraminae), 1968a (Loberini; larva of *Hapalips prolixus* Sharp), 1968b (Cladoxenini).
Sen Gupta and Crowson 1967 (larva of *Eicolyctus brunneus*), 1969b, 1971 (family revision).
Vaurie 1948 (North American Languriinae).
Villiers 1943 (Languriinae).

EROTYLIDAE (CUCUJOIDEA) (INCLUDING DACNIDAE)

John F. Lawrence,
Division of Entomology, CSIRO

Pleasing Fungus Beetles

Figures 34.548-553

Relationships and Diagnosis: Larvae of the Erotylidae may be distinguished from those of the closely related Languriidae by the presence of a truncate, rather than a falciform maxillary mala, and by the absence of a mandibular mola in all groups but the Dacninae. Dacnine larvae may be separated from those of most other groups (Languriidae excluded) by the presence of a characteristic prostheca, which consists of a fixed, hyaline lobe with a broad base and more or less angulate apex. Some erotylid larvae without a mola may be confused with those of certain Tenebrionoidea, such as Tetratomidae (*Penthe*) and Melandryidae (Eustrophinae), but they differ from those of either group in having either a median endocarina or a hypognathous head, cleft mala, and tridentate mandibles.

Biology and Ecology: The Erotylidae appear to feed exclusively on the larger fruiting bodies of basidiomycete fungi in the orders Aphyllophorales and Agaricales. Most larvae are internal feeders on the supportive (context) tissue, but those of the New World Erotylinae feed externally on the soft hymenial or spore-bearing surfaces. Those occurring under bark are usually associated with basidiomycete hyphal masses. Many species of *Dacne, Thallis, Tritoma,* and *Triplax* feed on the fruiting bodies of wood-inhabiting mushrooms (*Pleurotus, Panus,* etc.) or on the softer bracket fungi, like *Polyporus, Tyromyces, Hapalopilus, Laetiporus,* and *Piptoporus;* others occur in woodier sporophores, like those of *Phellinus* (*Microsternus* spp.) or *Ganoderma* (*Megalodacne* spp.). Some species of *Dacne* are known to be pests in commercial dried mushrooms. Pupation usually occurs within the fungal matrix or in the soil, but at least some erotylines, such as *Cypherotylus californicus* (Lacordaire), pupate on the surface, partly enclosed within the larval skin. Adults of most species and larvae of Erotylinae are aposematically colored with red and yellow pigments and probably contain defensive compounds. The surface-grazing erotyline larvae also have a covering of spines.

Description: Mature larvae 3 to 25 mm. Body elongate, more or less parallel-sided or fusiform, subcylindrical to slightly flattened, straight or slightly curved ventrally. Dorsal surfaces usually lightly pigmented but covered with granules or setiferous tubercles, sometimes with darkly pigmented tergal plates and/or complex, branched spines.

Head: Protracted and moderately to strongly declined (hypognathous), globular to slightly flattened. Epicranial stem usually long, absent in Triplacinae and some Dacninae; frontal arms lyriform or V-shaped, contiguous at base; epicranial suture sometimes indistinct or absent. Median endocarina

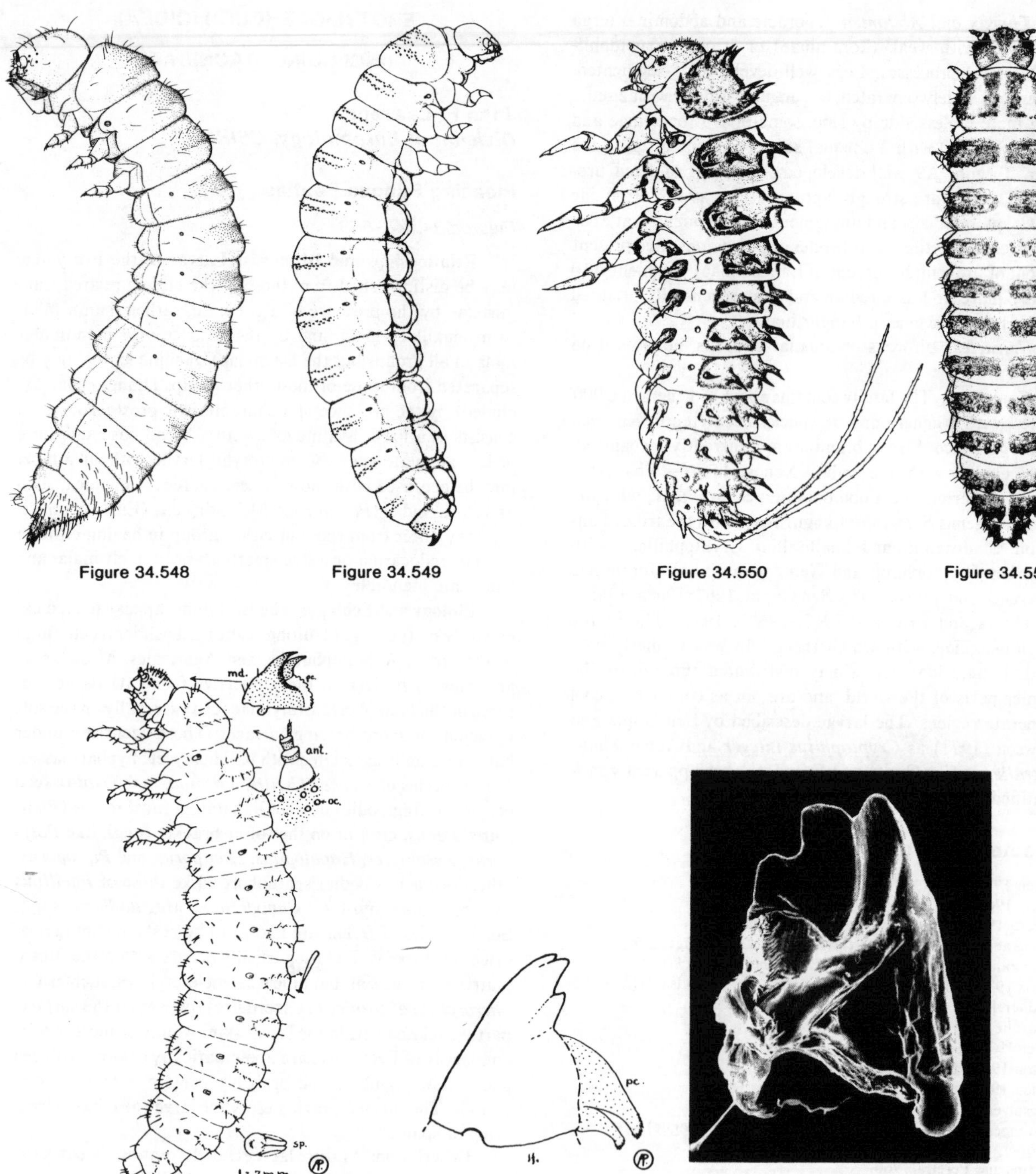

Figure 34.548. Erotylidae. *Microsternus ulkei* (Crotch). Cincinnati, Ohio. Larva, lateral. Length = 7.5 mm.

Figure 34.549. Erotylidae. *Megalodacne* sp. Barro Colorado Is., Canal Zone, Panama. Larva, lateral. Length = 25 mm.

Figure 34.550. Erotylidae. *Cypherotylus californicus* (Lacordaire). Sunnyside Canyon, Huachuca Mts., Arizona. Larva, lateral. Length = 17 mm.

Figure 34.551. Erotylidae. *Ischyrus quadripunctatus* (Olivier). Dead Lake, Chipola Park, Calhoun Co., Florida. Larva, dorsal. Length = 11 mm.

Figure 34.552. Erotylidae. *Tritoma* sp. Larva, lateral (plus details). Length = 7 mm. (From Peterson 1951)

Figures 34.553a,b. Erotylidae. *Cypherotylus californicus* (Lacordaire). **a.** left mandible, dorsal; **b.** left mandible, mesoventral. Sunnyside Canyon, Huachuca Mts., Arizona. (Figure 34.553a from Peterson, 1951)

sometimes present between frontal arms and extending anterad of epicranial stem, when the latter is present; endocarina absent in some Dacninae and many Erotylinae. Stemmata 5 or 6 on each side. Antennae short to moderately long, 3-segmented. Frontoclypeal suture absent or vaguely indicated; labrum free. Mandibles symmetrical, usually tridentate, and usually without accessory ventral process (present in *Cryptodacne, Dacne,* and *Thallis*); mola usually absent, replaced by membranous setose lobe and sometimes an additional brush of hairs; mola in *Cryptodacne, Dacne,* and *Thallis* well-developed, armed with asperities or tubercles, which may form transverse rows, and accompanied by fixed, hyaline prostheca, which is broadly triangular. Ventral mouthparts retracted. Maxilla with transverse cardo (which may be divided), elongate stipes, and 3-segmented palp; articulating area present or absent; mala blunt, often cleft, and variously armed with teeth, spines, and setae. Labium sometimes with mentum and submentum fused, usually more or less free to base of mentum; ligula usually short and broad, sometimes absent; labial palps 2-segmented and narrowly to widely separated. Hypopharyngeal sclerome usually absent, sometimes consisting of a transverse bar or tooth-like structure. Hypostomal rods usually absent, occasionally short and diverging. Ventral epicranial ridges present or absent. Gula transverse.

Thorax and Abdomen: Legs short to moderately long, 5-segmented, and narrowly separated; tarsungulus with 2 setae lying side by side. Thoracic and abdominal terga sometimes with granulate plates, rows or patches of asperities, or 2 to several pairs of simple or branched spines. Tergum A9 usually with pair of fixed urogomphi, which are usually short and strongly upturned, but may be long and narrow (some Erotylinae), sometimes with single median, sclerotized process (some Dacninae); sternum A9 well-developed, simple. Segment A10 circular, posteroventrally oriented.

Spiracles: Usually annular-biforous, sometimes (Erotylinae) annular.

Comments: The family contains about 30 genera and 2500 species, of which 10 genera and 50 species occur in America north of Mexico. There are at least 4 distinct groups: Dacninae (including *Cryptodacne*), Megalodacninae, Triplacinae, and Erotylinae. The Dacninae occur in the North and South Temperate Regions, while the Megalodacninae are pantropical, but most diverse in the Indo-Australian Region. The Triplacinae are usually found in the warmer parts of the New and Old Worlds, while the Erotylinae are primarily Neotropical, with a single species of *Cypherotylus* extending into the southwestern United States.

Selected Bibliography

Böving and Craighead 1931 (larvae of *Cypherotylus aspersus* Gorham, *Homoeotelus confusus, Megalodacne* sp., and *Tritoma unicolor* Say).

Boyle 1956 (North American revision).

Chujo 1969 (Japanese fauna).

Graves 1965 (habits of *Cypherotylus californicus*).

Hayashi and Takanaka 1965 (larva of *Encaustes praenobilis* Lewis).

Hayashi *et al.* 1959 (larvae of several Japanese species).

Nobuchi 1954 (biology and larvae of various Japanese species), 1955 (biology and larvae of various Japanese species).

Park and Sejba 1935 (nocturnal behavior of *Megalodacne heros* (Say)).

Peterson 1951 (larvae of *Tritoma* sp. and *Cypherotylus californicus*).

Rehfous 1955 (fungus hosts).

Roberts 1939 (family relationships; subfamily key), 1958 (larvae of many species, representing 16 genera).

Scheerpeltz and Höfler 1948 (fungus hosts).

Sen Gupta 1969 (larvae of *Thallis crichsoni* Crotch and *Cryptodacne synthetica* Sharp).

Weiss 1920b (fungus hosts).

Weiss and West 1920 (fungus hosts).

BIPHYLLIDAE (CUCUJOIDEA)

John F. Lawrence,
Division of Entomology, CSIRO

Figures 34.554a–c

Relationships and Diagnosis: The family Biphyllidae is a sharply defined group, which has been associated with the cucujoid families Erotylidae (Roberts, 1958), Languriidae (Cryptophilinae) (Crowson, 1955; Sen Gupta and Crowson, 1971), and Byturidae (Falcoz, 1926), but has also been placed in the Tenebrionoidea by Crowson (1960b) and Abdullah (1966). Several features of both larva and adult also indicate a relationship to the Bothrideridae and possibly the euxestine Cerylonidae. Falcoz (1926) presented a reasonable argument for the close relationship of Biphyllidae and Byturidae, but biphyllid larvae differ from those of byturids in having an accessory ventral process on the mandible, a basal hyaline mandibular lobe, which is neither projecting nor setose, well-developed ventral epicranial ridges, and enlarged, posteriorly oriented 8th spiracles, which may be raised on short tubes. The complex prostheca, consisting of comb-hairs, distinguishes byphyllid larvae from those of most other cucujoid families; a similar type of prostheca may be found in Nitidulidae, which have longitudinally oblique cardines and a truncate mala. Although distinct urogomphi are present in the 2 European larvae which have been previously described, these structures are absent in *Anchorius,* several Neotropical *Goniocoelus,* and Australian species presently included in the genera *Biphyllus* and *Diplocoelus.*

Biology and Ecology: Biphyllidae are mycophagous and appear to be associated with various types of Ascomycetes. Larvae of *Anchorius lineatus* Casey were found under the fermenting bark of mesquite (*Prosopis*) in Arizona, while several Neotropical biphyllids (*Goniocoelus* spp.) have also been found under bark in the fermenting stage. The European species *Biphyllus lunatus* (Fabricius) and *Diplocoelus fagi* Guérin-Méneville feed on the spores and stromata of the xylariaceous Ascomycetes *Daldinia concentrica* and *Tubercularia confluens,* respectively (Donisthorpe, 1935; Hingley, 1971; Horion, 1960; Palm, 1959), and several Australian species of *Biphyllus* and *Diplocoelus* appear to have similar habits. Some Australian species have been found on the rotting flower stalks of a monocot *Xanthorrhoea* and in rotting cycad cones, where they may be feeding on molds or other fungi.

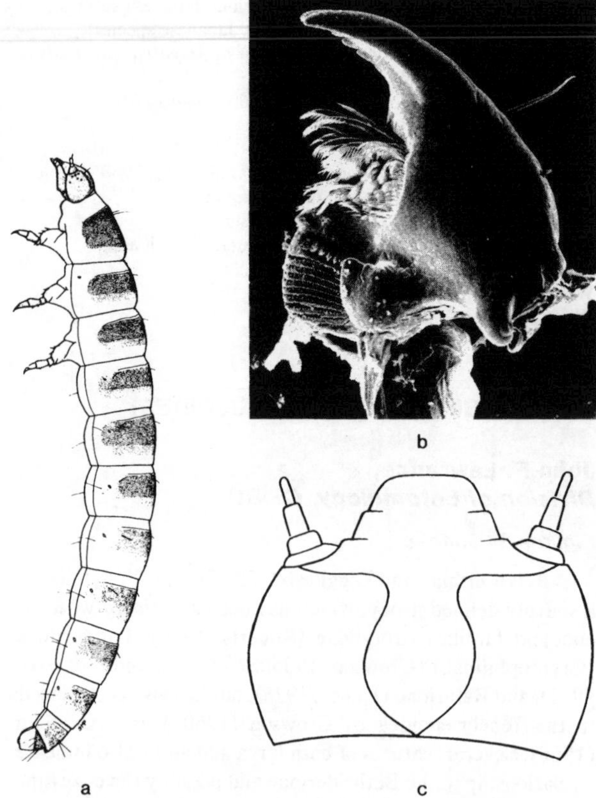

Figures 34.554a-c. Biphyllidae. *Anchorius lineatus* Casey. 1 mi. NW Arivaca, Pima Co., Arizona. **a.** larva, lateral view. Length = 7 mm; **b.** left mandible, ventral; **c.** head, dorsal.

Description: Mature larvae 2.5 to 10 mm. Body elongate, parallel-sided, cylindrical to very slightly flattened, straight or slightly curved ventrally; dorsal surfaces lightly to moderately heavily pigmented, smooth, with scattered, simple setae.

Head: Protracted and prognathous, moderately broad, slightly flattened. Epicranial stem absent or very short; frontal arms lyriform and contiguous at base. Median endocarina absent. Stemmata 6 on each side. Antennae well-developed, 3-segmented. Frontoclypeal suture absent; labrum partly fused to head capsule. Mandibles symmetrical, bidentate, with accessory ventral process and sometimes with serrate incisor edge; mola well-developed, transversely ridged, with hyaline lobe at base; prostheca consisting of a brush of complex comb-hairs. Ventral mouthparts retracted. Maxilla with transverse cardo, elongate stipes, well-developed articulating area, 3-segmented palp, and falciform mala. Labium free almost to base of mentum; ligula short; labial palps 2-segmented, widely separated. Hypopharyngeal sclerome consisting of a transverse bar. Hypostomal rods short and diverging. Ventral epicranial ridges present. Gula transverse.

Thorax and Abdomen: Legs well-developed, 5-segmented, moderately close together, clothed with short spines; tarsungulus with 1 seta. Mesotergum, metatergum, and abdominal terga 1–8 each with a transverse carina near anterior edge. Tergum A9 usually without urogomphi; sometimes with a pair of short, fixed urogomphi; sternum A9 well-developed, simple. Segment A10 more or less cylindrical, posteriorly or posteroventrally oriented.

Spiracles: Annular-biforous, those on segment A8 raised on short tubes and posteriorly or posterodorsally placed.

Comments: The family includes 6 genera and about 200 species worldwide, with *Anchorius lineatus* and 3 species of *Diplocoelus* occurring in America north of Mexico.

Selected Bibliography

Abdullah 1966.
Crowson 1955, 1960b, 1981.
Donisthorpe 1935 (mycophagy).
Falcoz 1926 (relationships).
Hingley 1971 (association with Ascomycetes).
Horion 1960.
Nikitsky 1983b (larva of *Biphyllus flexuosus* Reitter).
Palm 1959.
Perris 1851 (larva of *Biphyllus lunatus*).
Roberts 1958 (larvae of *Biphyllus lunatus* and *Diplocoelus fagi* in key).
Sen Gupta and Crowson 1971 (relationships).

BYTURIDAE (CUCUJOIDEA)

John F. Lawrence,
Division of Entomology, CSIRO

Fruit Worms

Figures 34.555–556

Relationships and Diagnosis: The family Byturidae is generally considered to be most closely related to the Biphyllidae (Falcoz, 1926), and both groups are now placed in the superfamily Cucujoidea (Crowson, 1981; Lawrence and Newton, 1982). Byturid larvae differ from those of Biphyllidae in lacking an accessory ventral process on the mandible, ventral epicranial ridges, and modified 8th abdominal spiracles, and in having a projecting and setose, hyaline process at the base of the mandibular mola.

Biology and Ecology: Species of *Byturus* are known to breed in the fruits of *Rubus* species, and may be pests of blackberries and raspberries, while larvae of *Byturellus grisescens* (Jayne) feed in the catkins of oak (*Quercus* species) in California (d'Aguilar, 1962a; Barber, 1942; Schöning, 1953).

Description: Mature larvae 4 to 10 mm. Body elongate, parallel-sided or slightly wider in middle, subcylindrical to slightly flattened, straight or slightly curved ventrally; dorsal surfaces moderately heavily pigmented, somewhat granulate and tuberculate, with vestiture of scattered, simple setae.

Head: Protracted and prognathous. Epicranial stem very short or absent; frontal arms lyriform and contiguous at base. Median endocarina absent. Stemmata almost always 6 on each side (rarely 5). Antennae well-developed, 3-segmented. Frontoclypeal suture absent; labrum free or partly fused to head capsule. Mandibles symmetrical, with bidentate apex, serrate incisor edge and no accessory ventral process; mola sub-basal, tuberculate, with hyaline, setose lobe at base; prostheca absent or consisting of a few hyaline processes or brush

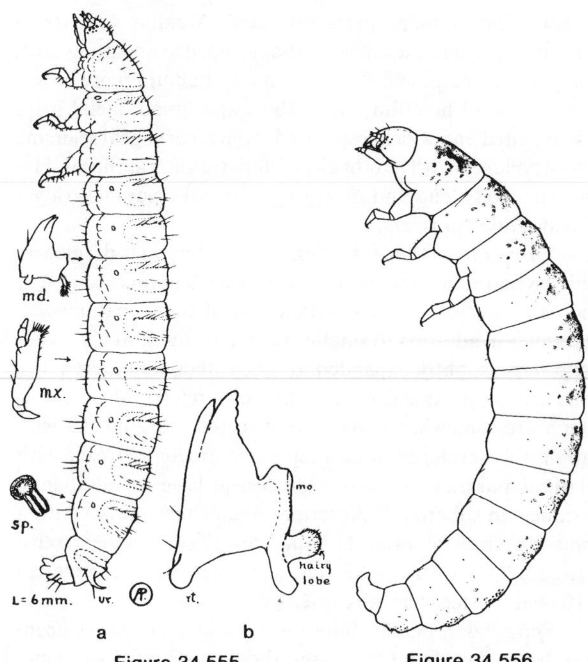

Figure 34.555 Figure 34.556

Figures 34.555a,b. Byturidae. *Byturus* sp. **a.** larva, lateral (plus details), length = 6 mm; **b.** right mandible, ventral. (From Peterson, 1951)

Figure 34.556. Byturidae. *Byturellus grisescens* (Jayne). Strawberry Canyon, Berkeley, California. Larva, lateral. Length = 6 mm.

of hairs. Ventral mouthparts retracted. Maxilla with transverse cardo, elongate stipes, well-developed articulating area, 3-segmented palp, and falciform mala, which may be somewhat obtuse with 2 spines at inner angle. Labium free almost to base of mentum; ligula present; labial palps 2-segmented, widely separated. Hypopharyngeal sclerome a transverse bar or tooth-like structure. Hypostomal rods moderately long and diverging. Ventral epicranial ridges absent. Gula transverse.

Thorax and Abdomen: Legs well-developed, 5-segmented, moderately close together; tarsungulus with 1 seta. Tergum A9 usually with pair of upturned, fixed urogomphi (absent in *Byturellus*); sternum A9 well-developed, simple. Segment A10 circular, posteriorly or posteroventrally oriented.

Spiracles: Annular, biforous, not raised on tubes, those on segment A8 laterally placed.

Comments: The family includes 7 genera: the Holarctic *Byturus*, *Byturellus* from California, *Byturodes* from western China, *Xerasia* and *Terobyturus* from Japan, and *Platydascillus* and *Dascillocyphon* from Sumatra.

Selected Bibliography

Abdullah 1966.
d'Aguilar 1962b (agricultural pests; larva of *Byturus tomentosus* (De Geer)).
Barber 1942 (revision).
Böving and Craighead 1931 (larvae of *Byturus tomentosus* and *B. unicolor* Say).
Crowson 1981.
Falcoz 1926 (relationships).
Lawrence and Newton 1982.
Peterson 1951 (larva of *Byturus* sp.).
Schöning 1953 (larvae of *B. fumatus* (Fabricius) and *B. tomentosus*).
Springer and Goodrich 1983 (North American revision).
Verhoeff 1923 (larva of *B. tomentosus*).

BOTHRIDERIDAE (CUCUJOIDEA)

John F. Lawrence,
Division of Entomology, CSIRO

Figures 34.557-558

Relationships and Diagnosis: Although Craighead (1920) treated this group as a distinct family, subsequent authors (Arnett, 1968; Crowson, 1955, 1981) considered it to be a subfamily of Colydiidae. Lawrence (1980) presented evidence for placing Bothrideridae in the Cucujoidea, but their relationships within that superfamily are still unclear, and it is not at all certain that the Teredinae and Bothriderinae, as delimited below, form a monophyletic unit. Teredine larvae have a number of features in common with Sphaerosomatidae and some Cerylonidae (Euxestinae), while adult teredines are reminiscent of Metacerylinae (Cerylonidae). The Biphyllidae must also be considered as a possible sister group of the Bothrideridae. The ectoparasitic larvae of Bothriderinae are similar to those of Cucujidae: Passandrinae, but differ in having retracted ventral mouthparts and lacking hypostomal rods.

Biology and Ecology: All species of Bothrideridae appear to be associated with the galleries or tunnels of wood-boring insects; especially Coleoptera, and most are ectoparasitoids of larvae or pupae. Larvae of Teredinae are apparently free-living, and do not show any of the modifications associated with ectoparasitism; those of *Oxylaemus, Teredolaemus,* and *Teredomorphus* are associated with the tunnels of ambrosia beetles (Platypodidae and some Scolytidae), while *Teredus* larvae live in the galleries of a number of wood-borers, including Scolytidae (*Dryocoetes*), Anobiidae (*Ptilinus, Xestobium, Anobium*), and Cerambycidae (*Callidium*) (Dajoz, 1977; Horion, 1961; Schedl, 1962). *Teredolaemus leae* (Grouvelle) was reared from tunnels of *Platypus subgranosus* Schedl in a *Nothofagus* log from Tasmania; the pupa was not enclosed in a cocoon (Lawrence, 1985b). Species of *Sosylus* also inhabit the tunnels of Platypodidae and other ambrosia beetles, but they are hypermetamorphic, with active triungulins and later instars modified as ectoparasitoids, which construct a waxen chamber for pupation. The related *Asosylus rhysodoides* (Grouvelle) has similar habits, but its host is a brentid (*Carcinopisthius kolbei* Senna), which is, in turn, a nest parasite of Platypodidae (Browne, 1962; Roberts, 1968, 1969, 1980; Schedl, 1962). Larvae of *Dastarcus, Deretaphrus,* and the genera of Bothriderini are all ectoparasitoids of wood-borers (usually excluding ambrosia beetles), and they spin a silken cocoon for pupation. *Dastarcus* larvae have been associated with a cerambycid host (Lieu, 1944), but they have also been found parasitizing larvae of carpenter bees (*Xylocopa*) in Asia (Piel, 1938). *Deretaphrus* larvae are known

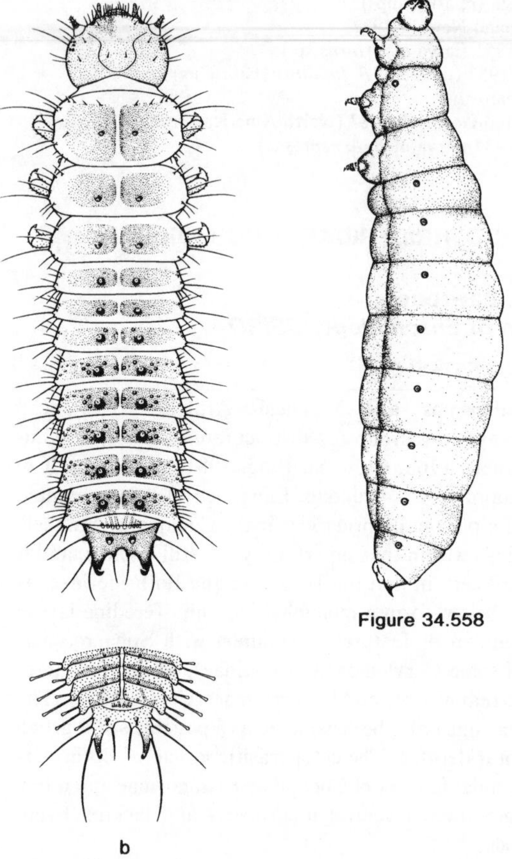

Figures 34.557a,b. Bothrideridae. *Teredolaemus leae* (Grouvelle). Arve Valley, Tasmania. **a.** larva, late instar, dorsal, length = 2.6 mm. **b.** first instar, abdominal apex, dorsal.

Figure 34.558. *Deretaphrus oregonensis* Horn. Giant Forest, California. Larva, lateral. Length = 15 mm.

to attack Cerambycidae, Buprestidae, and some of the larger Scolytidae (*Dendroctonus*), while bothriderine larvae are often found in the tunnels of Bostrichidae, but may also feed on larvae or pupae of Cerambycidae, Buprestidae, Anobiidae, Curculionidae, or Scolytidae (Burke, 1919; Craighead, 1920; Dajoz, 1977; Horion, 1961; Rasmussen, 1967; Schedl, 1962).

Description: *Free-living Larvae (Teredinae).* Mature larvae 3–4 mm. Body elongate, more or less parallel-sided and moderately to strongly flattened. Dorsal surfaces moderately pigmented and granulate-tuberculate; vestiture of longer and shorter, simple setae.

Head: Protracted and prognathous, moderately broad and somewhat flattened. Epicranial stem very short; frontal arms lyriform. Median endocarina absent. Stemmata usually 5 on each side (occasionally with an indistinct 6th stemma). Antennae short, 3-segmented. Frontoclypeal suture absent; labrum free. Mandibles symmetrical, bidentate, with 2 to several teeth along incisor edge and with large accessory ventral process. Mola well-developed and asperate; prostheca absent. Ventral mouthparts retracted. Maxilla with transverse cardo, elongate stipes, well-developed articulating area, 3-segmented palp, and falciform mala. Labium more or less free to base of mentum; ligula short and broad; labial palps 2-segmented and widely separated. Hypopharyngeal sclerome well-developed with pair of characteristic anterior horns. Hypostomal rods long and diverging. Ventral epicranial ridges absent. Gula transverse.

Thorax and Abdomen: Legs moderately well-developed, widely separated; tarsungulus with 1 seta. Thoracic terga and abdominal terga 1–9 each with a pair of sharp, paramedian tubercles, in addition to smaller tubercles and granules. Thoracic terga slightly expanded at sides; abdominal terga 1–8 more strongly expanded laterally forming tergal processes which are somewhat posteriorly directed. Tergum A9 with pair of well-developed, widely separated urogomphi, each with a lateral and mesal accessory process at base (in addition to paramedian tubercles). Anterior edge of tergum A9 with a transverse row of internal processes (visible under transmitted light). Sternum A9 well-developed, simple. Segment A10 oval, posteroventrally oriented.

Spiracles: Annular biforous, with long accessory openings, borne on tubular processes; those on segment A8 somewhat larger than others.

First Instar: Body broader and flatter, with smoother and more heavily pigmented dorsal surfaces, and with several pleural and sternal pigmented plates. Vestiture including modified setae which are expanded at apex and probably glandular. Lateral tergal processes on thorax and first 8 abdominal segments more well-developed, forming flattened plates. Paramedian tubercles on thorax and abdomen reduced, except for those on tergum A9. Accessory basal processes of urogomphi reduced. Spiracles located at the ends of longer tubes.

Ectoparasitic Larvae (Bothriderinae): Mature larvae 3 to 15 mm. Body elongate, subcylindrical, but with abdomen enlarged (physogastric), very slightly sclerotized, smooth, with vestiture of short, scattered setae.

Head: Protracted and prognathous, reduced in size, somewhat flattened. Epicranial stem absent; frontal arms usually indistinct or absent, sometimes V-shaped and distant at base. Median endocarina absent. Stemmata absent or 1 on each side. Antennae very short, usually 2-segmented, with long sensorium on segment 1, occasionally 1-segmented. Frontoclypeal suture absent; labrum usually free. Mandibles symmetrical, unidentate or bidentate, without accessory ventral process or mola; mesal surface of mandibular base simple or with fixed, hyaline or partly sclerotized, acute process. Ventral mouthparts retracted. Maxilla usually with transverse or oblique cardo, slightly elongate stipes, well developed articulating area, 2-segmented palp (sometimes with distinct palpifer and appearing 3-segmented), and a blunt but narrowly rounded mala; in *Sosylus*, cardo, stipes and articulating area indistinct, mala sometimes absent, and palp sometimes absent or represented by minute papilla. Labium usually free to base of mentum, with ligula longer than labial palps which are 2-segmented; in *Sosylus*, labium may not be subdivided, ligula is absent, and labial palps may be absent or represented by minute papillae. Hypopharyngeal sclerome absent. Hypostomal rods absent. Ventral epicranial ridges absent. Gula longer than wide.

Thorax and Abdomen: Thorax relatively short and narrower than abdomen, sometimes (*Sosylus*) strongly narrowed in between mesothorax and metathorax. Legs usually reduced, but 5-segmented, widely separated; tarsungulus usually with 1 seta; legs absent in *Sosylus*. Tergum A9 usually much shorter than A8, without urogomphi or with short, strongly upturned pair; in *Sosylus*, tergum A9 not reduced, without urogomphi; sternum A9 well-developed, simple. Segment A10 circular, posteriorly or posteroventrally oriented.

Spiracles: Annular, usually with heavily sclerotized peritreme, not raised on tubes; absent on thorax and segment A8 in *Sosylus*.

Triungulin (Sosylus): About 1 to 1.5 mm long. Body elongate, fusiform, strongly flattened, more heavily sclerotized than mature larvae, with vestiture of stout spines. Head strongly transverse. Labrum apparently fused to head capsule. Stemmata well-developed, 1 on each side. Antennae 2-segmented, with segment 2 much longer than 1, longer than sensorium, and bearing long seta at apex. Mandibles narrow and falcate, strongly curved and sickle-shaped, without mola. Ventral mouthparts retracted. Maxilla with longitudinally oblique cardo, elongate stipes, indistinct articulating area, 2-segmented palp, and broadly rounded mala. Labium with mentum and submentum fused; labial palps 2-segmented. Legs well-developed, 5-segmented, with long, narrow tibiae. Segment A9 not much shorter than A8, without urogomphi, with a pair of long setae at apex. Segment A10 circular, posteriorly oriented. Spiracles annular, present on mesothorax and abdominal segments 1–8.

Comments: The family includes about 30 genera and 300 species worldwide, and 6 genera and 13 species are known from America north of Mexico. The generic and suprageneric concepts are in need of revision, and current studies are in progress by Slipinski, Crowson, Sen Gupta, Pal, and Lawrence. The family probably should be divided into 2 groups: Teredinae, with obvious trochanters in the adult and free-living larvae; and Bothriderinae (including Dastarcini and Deretaphrini), with reduced and partly concealed trochanters and ectoparasitic larvae. Klausnitzer (1975b) described a single larva purported to be that of *Oxylaemus* sp.; since the specimen was not reared and nothing in the short description suggests a relationship to known bothriderid larvae, it is possible that this is a misidentification.

Selected Bibliography

Arnett 1968.
Böving and Craighead 1931 (larvae of *Bothrideres geminatus* (Say) and *Deretaphrus oregonensis* Horn).
Browne 1962 (biology and larval forms of *Sosylus spectabilis* Grouvelle).
Burke, 1919 (cocoon making in *Deretaphrus oregonensis*).
Craighead, 1920 (larvae of *D. oregonensis*, *B. geminatus*, and *Lithophorus succineus* (Pascoe); cocoons).
Crowson, 1955, 1981.
Dajoz 1977.
Horion 1961.
Klausnitzer 1975b (putative larva of *Oxylaemus* sp.).
Lawrence 1980 (relationships), 1985b (biology of *Teredolaemus leae*).
Lieu 1944 (biology of *Dastarcus helophoroides* Fairmaire).
Pal and Lawrence 1986.
Piel 1938 (biology of *Dastarcus helophoroides*).
Rasmussen, 1967 (association of *Bothrideres contractus* (Fabricius) with *Hylotrupes*).
Roberts, 1968 (association of *Sosylus* with Platypodidae), 1969 (biology of African *Sosylus* spp.), 1980 (larval forms of *Sosylus* spp. and *Asosylus rhysodoides*).
Schedl, 1962 (association with bark beetles).
Sen Gupta and Crowson, 1971 (larva of *Teredolaemus* sp., as *Cryptophilus* and *Xenoscelinus*), 1973.

SPHAEROSOMATIDAE (CUCUJOIDEA)

John F. Lawrence,
Division of Entomology, CSIRO

This family contains only *Sphaerosoma*, with about 50 species distributed from southern Europe to North Africa and Asia Minor. Although the genus was formerly included in the Endomychidae, its true affinities appear to be with the families Bothrideridae (Teredinae) or perhaps Cerylonidae (Euxestinae).

Larvae are minute (less than 2.5 mm), elongate and slightly flattened, with the dorsal surfaces heavily pigmented in earlier instars but lightly so in later ones, with a vestiture of scattered, simple setae. The head lacks an epicranial stem or median endocarina, and has lyriform frontal arms, approximate at base, 5 stemmata on each side, and short, broad antennae. Mandibles are tridentate, with a tuberculate mola and no prostheca. The ventral mouthparts are retracted, with an obtuse mala, 3-segmented maxillary palps and 2-segmented labial palps. The hypostomal rods are short and slightly diverging. The tarsungulus is unisetose. Tergum A8 is armed with a pair of dorsal tubercles and tergum A9 has a pair of upturned, slightly convergent urogomphi; segment A10 is circular and ventrally oriented. Spiracles are biforous, elongate and elliptical, and borne at the ends of well-developed spiracular tubes.

Sphaerosomatids feed on the fruiting bodies of a variety of mushrooms (Agaricales); they have been recorded from species of *Armillaria, Collybia, Hypholoma, Russula, Mycena, Panus, Pholiota, Schizophyllum,* and *Tricholoma.*

Selected Bibliography

Crowson 1981.
Lawrence 1982c.
Peyerimhoff 1913 (larva of *Sphaerosoma algiricum* (Reitter)), 1926 (habits).
Rehfous 1955 (fungus hosts).
Scheerpeltz and Höfler 1948 (fungus hosts).
Sen Gupta and Crowson 1971 (relationships).

CERYLONIDAE (CUCUJOIDEA)

(= CERYLIDAE; INCLUDING ACULAGNATHIDAE, ANOMMATIDAE, DOLOSIDAE, EUXESTIDAE, MURMIDIIDAE)

John F. Lawrence,
Division of Entomology, CSIRO

Figures 34.559-562

Relationships and Diagnosis: The Cerylonidae are a diverse group of cucujoid beetles, which probably should be placed in two or more families. Larvae of the subfamily Ceryloninae are easily distinguished by the highly modified mandibles and maxillae, which are stylet-like and either endognathous or enclosed in a tubular proboscis. Adults of Anommatinae and Metaceryloninae exhibit characters indicating a relationship to Bothrideridae, while larvae of Anommatinae and Metaceryloninae are not easily distinguished from free-living forms of the bothriderid subfamily Teredinae (Pal and Lawrence 1986). The positions of the Anommatinae and Murmidiinae are not clear, and larvae of the latter are disc-like, resembling those of the family Discolomidae. Current studies in progress by Crowson, Sen Gupta, Lawrence, Pal, and Slipinski may shed light on the phylogenetic relationship of these groups.

Biology and Ecology: Cerylonidae occur commonly in leaf litter, in rotten wood, or under bark, and probably feed on fungal hyphae and spores. *Hypodacne punctata* LeConte has been found in the galleries of carpenter ants (*Camponotus*) in rotten oak and elm logs (Stephan, 1968), while Australian species of this genus are collected in samples of leaf litter. *Euxestus erithacus* Chevrolat was found breeding in bat guano in a cave in Jamaica, and the African *Elytrotetrantus chappuisi* (Jeannel and Paulian) has been collected in the nests of mole rats. Species of *Cerylon, Philothermus,* and *Mychocerus* are most common under bark, while species of *Lapethus* are characteristic inhabitants of leaf litter and have been taken in the refuse deposits of leaf-cutter ants (*Atta*). *Murmidius ovalis* (Beck) occurs in granaries and warehouses, where it has been associated with stored products, such as corn, wheat, rice, flour, hay, pepper, ginger, and various fruits and seeds; it also occurs in dead leaves and cut grass and is probably a mold feeder (Halstead, 1968a; Hinton, 1945b). *Anommatus duodecimstriatus* (Müller) is known from damp soil beneath railroad ties, in grass roots, grass cuttings, tree holes, and leaf litter; it has apparently been distributed by man in soil samples (Kuschel, 1979; Lawrence and Stephan, 1975).

Description: Mature larvae 1 to 5 mm. Body oblong and fusiform to broadly ovate, slightly to strongly flattened. Dorsal surfaces usually lightly sclerotized, sometimes with pigmented tergal plates, often granulate or tuberculate; vestiture often including frayed or clavate setae, or occasionally (*Murmidius*) barbed setae.

Head: Protracted and prognathous to strongly hypognathous or opisthognathous, sometimes completely concealed from above by prothorax, slightly flattened. Epicranial stem usually absent, occasionally short; frontal arms usually absent, sometimes lyriform and contiguous at base. Median endocarina present in *Murmidius*. Stemmata usually absent, sometimes 1 to 3 on each side. Antennae short to moderately long, 3-segmented, usually with sensorium longer than segment 3. Frontoclypeal suture usually absent; labrum usually free, forming long, tapered, tubular proboscis in most Ceryloninae, apparently fused to head capsule in *Cerylon*. Mouthparts in Ceryloninae highly modified, either enclosed within proboscis or endognathous (*Cerylon*); mandibles andmaxillae both forming narrow stylets; maxillary palps 3-segmented and labial palps usually 2-segmented (1-segmented in *Cerylon*). Mouthparts in unspecialized forms (Euxestinae, *Murmidius, Anommatus*) as follows: mandibles symmetrical, bidentate or tridentate, with well-developed mola, which is tuberculate, asperate, or transversely ridged, usually with an accessory ventral process, and occasionally (*Murmidius*) with a fixed, hyaline prostheca; ventral mouthparts retracted; maxilla with transverse or oblique cardo, elongate stipes, well-developed articulating area, 3-segmented palp, and blunt mala; labium free to base of mentum, with ligula and 2-segmented palps; hypopharyngeal sclerome a transverse bar or tooth-like structure. Hypostomal rods usually absent (short and subparallel in *Murmidius*). Gula transverse.

Thorax and Abdomen: Thoracic and abdominal terga usually with lateral or dorsal and lateral processes. Legs well-developed, 5-segmented; tarsungulus with 1 seta. Tergum A9 usually simple, with short, complex urogomphi in *Anommatus,* and with long urogomphi bearing accessory tubercles or processes in Euxestinae; sternum A9 relatively short, simple. Segment A10 circular, ventrally oriented.

Spiracles: Annular or annular-biforous, occasionally raised on tubes, located laterally beneath lateral tergal processes, when present.

Comments: The family, as here defined, includes about 55 genera and 650 species worldwide, with 10 genera and 18 species occurring in N. America. The subfamily Euxestinae is worldwide and includes such genera as *Euxestus, Hypodacne,* and *Elytrotetrantus.* The Anommatinae are native to Europe, but the species *Anommatus duodecimstriatus* (Müller) has been introduced into N. America, South Africa, Chile, and Australia. Species of Ostomopsinae, Metaceryloninae and Murmidiinae occur in the warmer parts of the Old and New Worlds, but *Murmidius ovalis* has been widely introduced with stored products. The Ceryloninae is the largest and most widespread subfamily, and includes such genera as *Cerylon, Lapethus,* and *Philothermus.* The genus *Eidoreus* (= *Eupsilobius*) was transferred to the family Endomychidae by Sen Gupta and Crowson (1973). The larva described and illustrated by Peterson (1951) as *Murmidius ovalis* is actually a corylophid.

Selected Bibliography

d'Aguilar 1937 (larva of *Anommatus duodecimstriatus*).
Arnett 1968.
Böving and Craighead 1931 (larva of *Murmidius ovalis*).
Dajoz 1963, 1968 (larva of *Anommatus duodecimstriatus*).
Halstead 1968a (biology of *M. ovalis*).
Hinton 1945b (stored products pests).

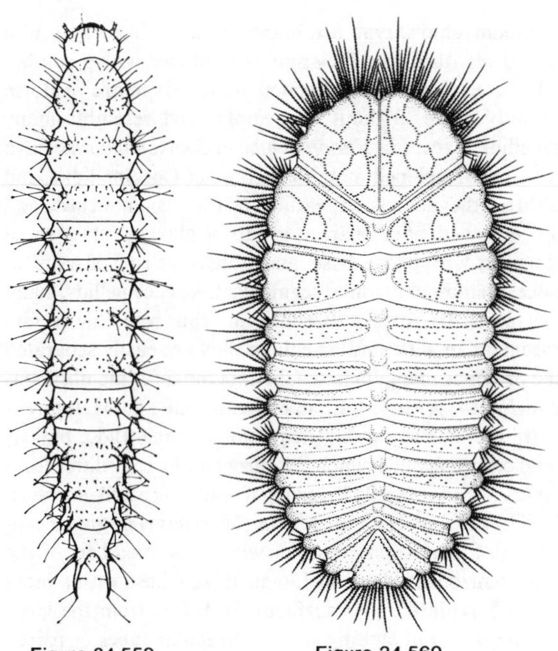

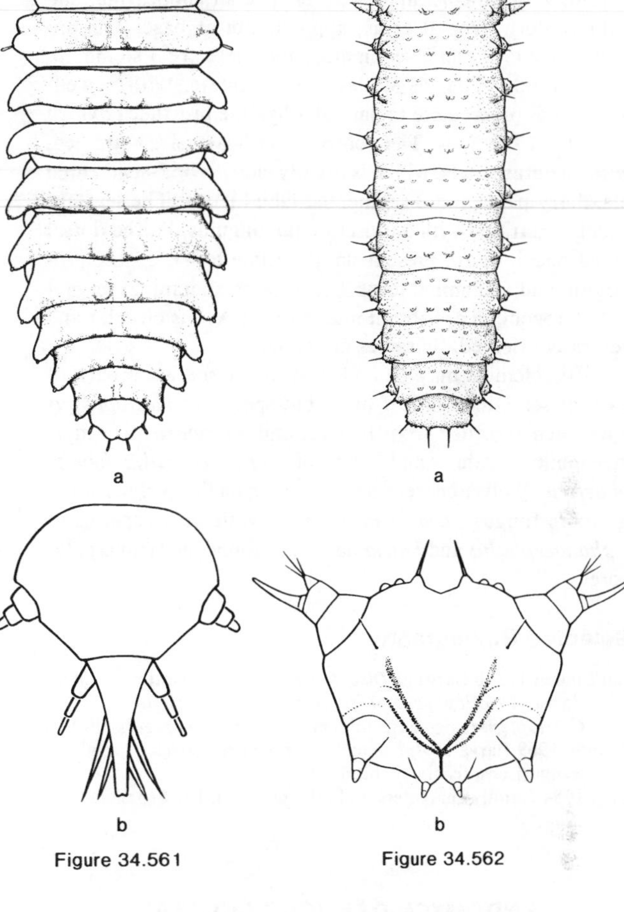

Figure 34.559. Cerylonidae. *Euxestus erithacus* Chevrolat. Bandon Hill Cave, St. James Parish, Jamaica. Larva, dorsal. Length = 4.9 mm.

Figure 34.560. Cerylonidae. *Murmidius ovalis* (Beck). Slough, England (in culture). Larva, dorsal. Length = 1.9 mm.

Figures 34.561a,b. Cerylonidae. *Lapethus* sp. 21 km. W El Hato del Volcan, Panama. **a.** larva, dorsal, length = 2.3 mm; **b.** head, dorsal (anteroventral view).

Figures 34.562a,b. Cerylonidae. *Cerylon unicolor* (Ziegler). 5.4 mi. N Shady Lake Recreation Area, Polk Co., Arkansas. **a.** larva, dorsal. Length = 2.4 mm; **b.** head, ventral.

Jeannel and Paulian 1945 (larva of *Elytrotetrantus chapuisi*).
Kuschel 1979 (*Anommatus* in New Zealand).
Lawrence 1982a (North American catalogue).
Lawrence and Stephan 1975 (North American review).
Nikitsky and Belov 1979 (larva of *Hypodacne edithae* (Reitter)).
Oke 1932.
Pal and Lawrence 1986.
Peterson 1951.
Sen Gupta and Crowson 1973 (larval characters and key; larvae of *Hypodacne bivulneratus* (Lea), *Euxestus* sp., *Cerylon histeroides* (Fabricius), and *Philothermus bicavus* Sharp).
Slipinski 1984 (larva of *Lapethus ferrugineus* (Hinton and Ancona)).
Stephan 1968 (habits of *Hypodacne punctata*).

DISCOLOMIDAE (CUCUJOIDEA)

(= NOTIOPHYGIDAE)

John F. Lawrence,
Division of Entomology, CSIRO

The family Discolomidae is comprised of 18 genera and about 400 species distributed throughout the warmer parts of the world, but is much more diverse in the Eastern Hemisphere. Of the 5 subfamilies, Notiophyginae and Aphanocephalinae have representatives in the Neotropical, Ethiopian, and Indo-Australian Regions, Discolominae occur in tropical Africa and America, Cephalophaninae are restricted to Southeast Asia and the East Indies, and Pondonatinae occur only in Africa. Both larval and adult features indicate a possible relationship with *Murmidius* and its relatives (Cerylonidae).

Larvae are small to minute, broadly ovate, strongly flattened, and disc-like, with the dorsal surfaces pigmented, granulate, and clothed with various kinds of modified setae or scales, and the lateral edges lined with scale-like setae. The head is prognathous but concealed from above by the pronotum, with no epicranial stem or median endocarina, and with lyriform frontal arms, approximate at base. The antennae are long and 2-segmented, and there are 3 stemmata on each side. The mandibles are tridentate, with a well-developed, tuberculate mola and a hyaline prostheca, which is acute at the apex. The ventral mouthparts are retracted, with an obtuse mala, which is usually cleft, 2- or 3-segmented maxillary palps and 2-segmented labial palps. The legs are widely separated with a unisetose tarsungulus. Paired dorsal gland openings are present on all visible terga, on the protergum and abdominal terga 1 to 4, or on abdominal terga 4 to 8. Urogomphi are absent and segment A10 is circular and ventrally oriented. Spiracles are annular.

Discolomids may be collected under the loose bark of trees or sometimes in leaf litter, but species of *Notiophygus* have been recorded from lichens, and *Katoporus* are myrmecophilous. Adults and larvae of *Aphanocephalus hemisphaericus* Wollaston were found feeding on the surface of the polypore fungus *Favolus arcularius,* while other species of *Aphanocephalus* and *Fallia* have been found on various polypores.

Selected Bibliography

van Emden 1932b (larva of *Discoloma cassideum* Reitter), 1938a (larva of *Notiophygus hessei* John), 1957 (larvae of *Cassidoloma angolense* John and *Notiophygus piger* John).

Fukuda 1969 (larva of *Aphanocephalus hemisphaericus;* larval comparisons; feeding habits).

John 1954 (family characters and relationships), 1959 (genera), 1964.

ENDOMYCHIDAE (CUCUJOIDEA)

(INCLUDING MEROPHYSIIDAE, MYCETAEIDAE)

John F. Lawrence,
Division of Entomology, CSIRO

Handsome Fungus Beetles

Figures 34.563-569

Relationships and Diagnosis: Endomychidae, as defined here, excludes the European genus *Sphaerosoma* (Sphaerosomatidae), but includes *Agaricophilus reflexus* Motschulsky (removed by Mamaev, 1977) and *Eidoreus* (= *Eupsilobius*) (formerly placed in Cerylonidae), as well as those genera (*Merophysia, Coluocera, Holoparamecus,* etc.) currently comprising the family Merophysiidae (Crowson, 1955, 1981; Lawrence, 1982c; Sasaji, 1971; Sen Gupta and Crowson, 1973). The family forms part of Crowson's cerylonid series and is thought to be related to Sphaerosomatidae and Coccinellidae.

Endomychid larvae are highly variable in form, sculpture, and vestiture, and they may be confused with a number of other cucujoids. Typical endomychid larvae are fusiform to broadly ovate, without urogomphi, and resemble many coccinellids. Except in *Endomychus,* endomychid larvae have a more well-developed mola than those of Coccinellidae, and in addition they lack such features as the fusion of cardo and stipes, enlarged tibia bearing group of clavate setae, small tarsungulus with broad base, and transverse row of 6 protuberances on most abdominal segments. Like coccinellids, many endomychids are surface dwellers, and thus resemble certain Chrysomelidae with similar habits; they are easily separated by the presence of a gula and a distinct mola (or membranous lobe in *Endomychus*). Larvae of Leiestinae and Merophysiinae are more or less cylindrical and resemble those biphyllids and cryptophagids which lack urogomphi; they differ from members of either family in having no prostheca and an obtuse mala. The larva of *Mycetaea hirta* (Marsham) might be confused with Sphaerosomatidae or with some members of the families Rhizophagidae and Languriidae (those with a more or less tuberculate upper surface). It differs from the larva of *Sphaerosoma* in lacking raised spiracular tubes or paired tubercles on tergum A8, and from both rhizophagids and languriids in having an obtuse mala and small, annular spiracles. Larvae of Cerylonidae: Ceryloninae resemble endomychids in general body form, but they have highly modified mouthparts, which are either endognathous or enclosed within a tubular beak. *Murmidius ovalis* (Beck) has a disc-like larva resembling that of *Agaricophilus,* but the latter has a much shorter basal antennal segment and lacks a median endocarina and barbed setae. Those corylophid larvae which have a disc-like body differ from *Agaricophilus* in having paired gland openings on abdominal segments 1 and 8.

Biology and Ecology: Endomychidae are basically mycophagous, feeding on a wide variety of fungal types, but some occur with ants or termites, a few are predators, and others are pests on stored products. Most endomychids are associated with rotten wood, and they may occur under bark, in rotting wood, or more often on wood or bark surfaces (especially at night). Their specific food source usually consists of the less conspicuous cryptograms and is thus difficult to determine; *Aphorista morosa* LeConte was found in association with a yellow plasmodium of a slime mold (Myxomycetes), while a Neotropical species of *Amphix* was found feeding on spore capsules of an ascomycete fungus. The more obvious Basidiomycetes are also fed upon by some endomychids. *Endomychus* species are known to breed in the fruiting bodies of *Hirneola mesenterica* (Auriculariaceae) and *Schizophyllum commune* (Schizophyllaceae); species of *Symbiotes* have been collected in *Pleurotus* mushrooms (Tricholomataceae); and species of *Phymaphora, Rhanidea, Mycetina, Bystus,* and *Eumorphus* have been associated with various bracket fungi (Polyporaceae) (Donisthorpe, 1935; Weiss and West, 1920). The last group represents several different larval types, some (*Phymaphora, Rhanidea*) feeding internally in a relatively soft matrix, and others (*Eumorphus*) feeding on surface tissue. Species of *Lycoperdina* have distinctive, internally feeding larvae, which occur in puffballs

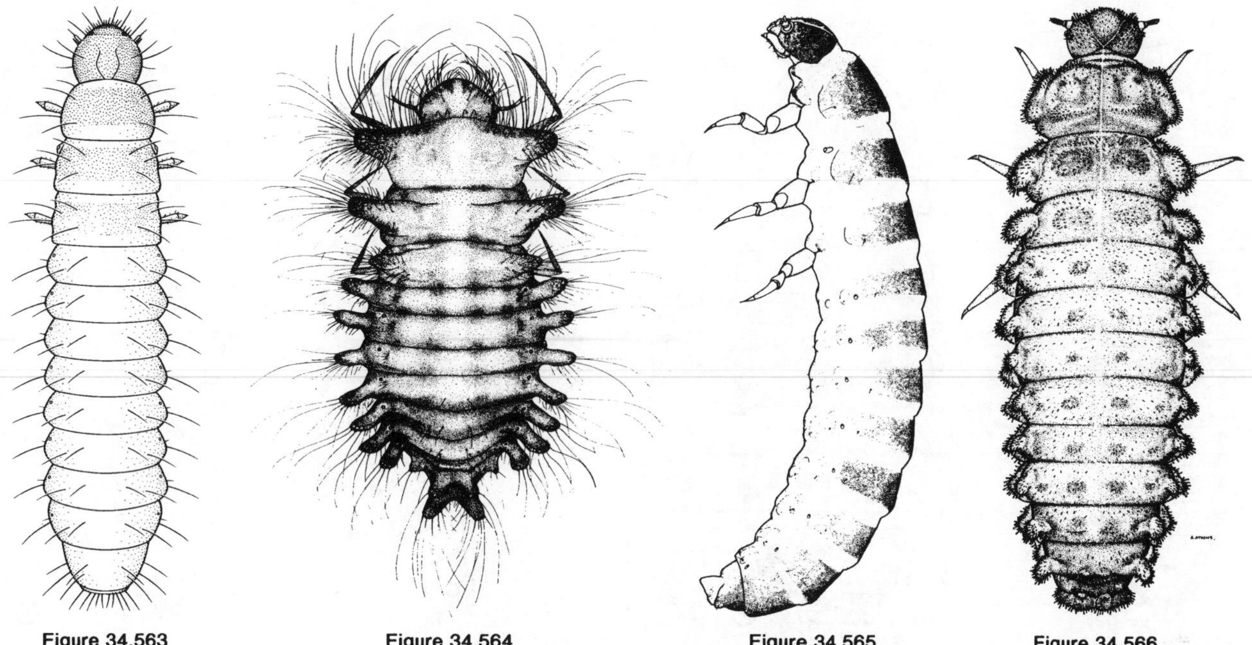

Figure 34.563. Endomychidae. *Holoparamecus* sp. Island MUD2, Monroe Co., Florida. Larva, dorsal. Length = 1.9 mm.

Figure 34.564. Endomychidae. *Bystus* sp. Barro Colorado Is., Canal Zone, Panama. Larva, dorsal. Length = 4.7 mm.

Figure 34.565. Endomychidae. *Rhanidea unicolor* (Ziegler). Rancocas, Burlington Co., Vermont. Larva, lateral. Length = 3.3 mm.

Figure 34.566. Endomychidae. *Aphorista morosa* LeConte. Mt. Bigelow, Santa Catalina Mts., Arizona. Larva, dorsal. Length = 8.9 mm.

of the genus *Lycoperdon* (Donisthorpe, 1935; Weiss and West, 1920). Various endomychids turn up regularly in leaf litter samples and feed on molds and other fungi on decaying plant products. Among the mold feeders, *Mycetaea hirta* (Marsham) and various species of *Holoparamecus* have become stored products pests, occurring in granaries and warehouses and causing minor damage to a variety of stored goods (Aitken, 1975; Hinton, 1945b; Klippel, 1952). Species of *Trochoideus* occur in termite nests, while *Coluocera* and *Merophysia* are myrmecophiles, but little is known of the food habits of either group (Kemner, 1924; Silvestri, 1912). *Saula japonica* Gorham has developed predaceous habits and feeds on scale insects and phytophagous mites (Sasaji, 1978b).

Description: Mature larvae 2 to 20 mm, usually 10 mm or less. Body elongate and subcylindrical, oblong, or fusiform and slightly flattened, or broadly ovate and strongly flattened. Dorsal surfaces lightly to darkly pigmented, sometimes with distinct tergal plates, smooth or sometimes granulate, with vestiture variable, consisting of short, fine setae, long, fine hairs, stout spines, or specialized (barbed, clavate, frayed) setae or scales.

Head: Protracted and prognathous or slightly declined to strongly hypognathous, usually broad and sometimes narrowed anteriorly. Epicranial stem very short or absent; frontal arms usually lyriform (V-shaped in *Mycetaea* and contiguous at base (separated at base in *Endomychus*), sometimes indistinct or absent. Median endocarina absent. Stemmata usually 2 to 4 on each side, 1 or 0 in Mychotheninae and Merophysiinae. Antennae usually well-developed, 3-segmented, occasionally short and broad, often with segment 2 very long and 3 reduced; antennal insertions often distant from mandibular articulations. Frontoclypeal suture present or absent; labrum free. Mandibles symmetrical, unidentate to tridentate or occasionally 4-dentate, with incisor edge sometimes serrate and accessory ventral process usually absent; mola almost always well-developed, bearing tubercles or asperities, which may form transverse rows or ridges, rarely (*Endomychus*) reduced and replaced by membranous lobe; prostheca usually a narrow, fixed, hyaline process, which may be acute or bifid, sometimes a fringed membrane, occasionally absent. Ventral mouthparts retracted. Maxilla usually with transverse cardo, elongate stipes, well-developed articulating area, 3-segmented palp, and obtuse mala, which is sometimes densely fringed with setae at apex; mala falciform in Mychotheninae. Labium with mentum and submentum fused; ligula present, often longer than labial palps, which are usually 2-segmented (1-segmented in *Lycoperdina* and Merophysiinae). Hypopharyngeal sclerome usually consisting of a distinct tooth-like structure. Hypostomal rods usually long and diverging, sometimes almost parallel. Ventral epicranial ridges absent. Gula transverse.

Thorax and Abdomen: Abdominal or thoracic and abdominal terga and sometimes abdominal pleura with lateral processes, which may be dehiscent; *Lycoperdina* with paired,

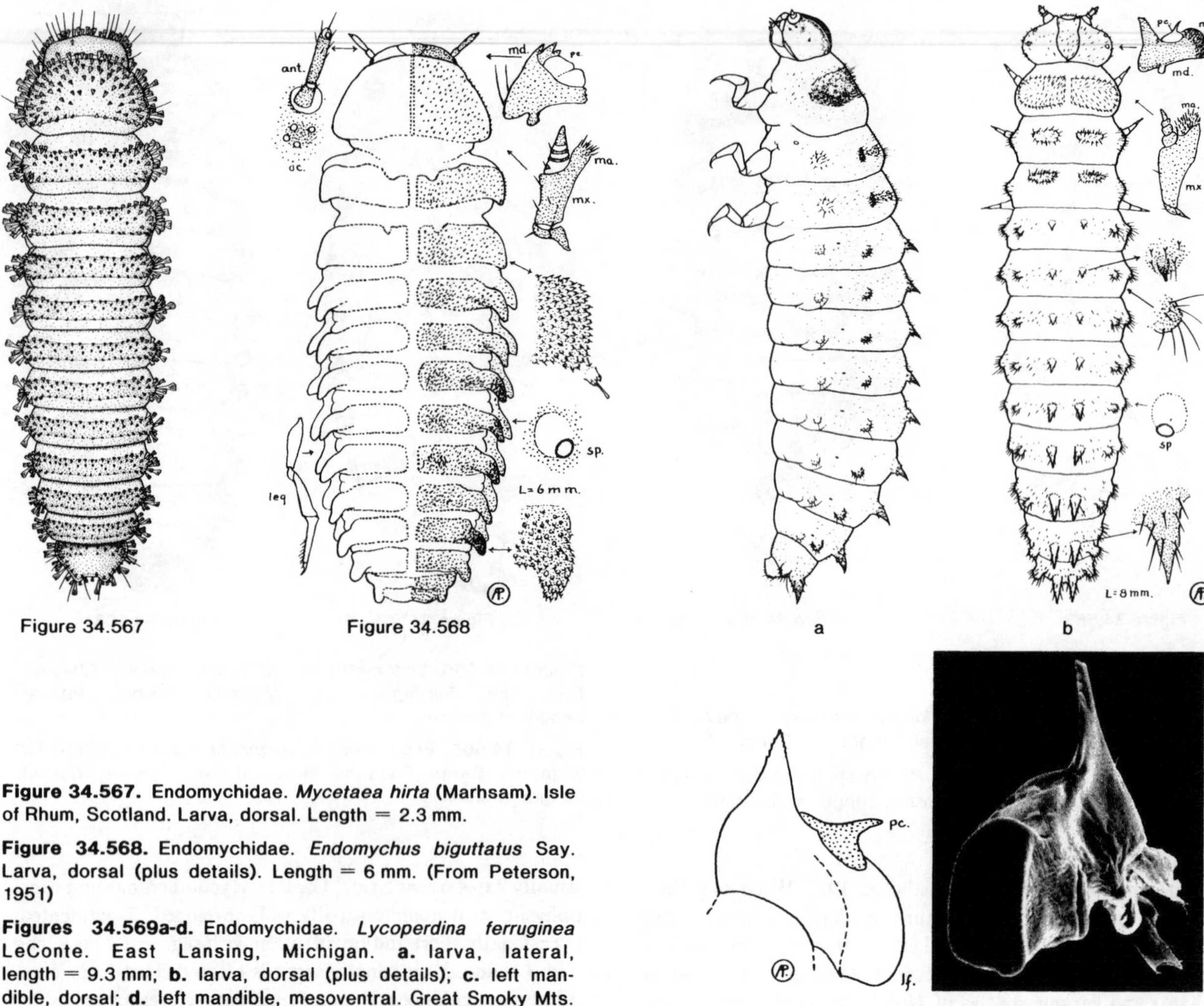

Figure 34.567. Endomychidae. *Mycetaea hirta* (Marhsam). Isle of Rhum, Scotland. Larva, dorsal. Length = 2.3 mm.

Figure 34.568. Endomychidae. *Endomychus biguttatus* Say. Larva, dorsal (plus details). Length = 6 mm. (From Peterson, 1951)

Figures 34.569a-d. Endomychidae. *Lycoperdina ferruginea* LeConte. East Lansing, Michigan. **a.** larva, lateral, length = 9.3 mm; **b.** larva, dorsal (plus details); **c.** left mandible, dorsal; **d.** left mandible, mesoventral. Great Smoky Mts. National Park, Tennessee. (figures 34.569b, c from Peterson, 1951)

acute, dorsal processes on terga A1–9, the last being the urogomphi. Legs well-developed, 5-segmented; tarsungulus with 1 seta. Tergum A9 usually without urogomphi (except in *Lycoperdina, Mycetaea,* and some Mychotheninae); sternum A9 short, simple. Segment A10 circular, ventrally or posteroventrally oriented.

Spiracles: Annular, not raised on tubes (except in some Mychotheninae).

Comments: The family includes about 120 genera and 1300 species worldwide, with 19 genera and about 45 species occurring in America north of Mexico. The subfamily classification currently in use is that of Strohecker (1953, and in Arnett, 1968), but more recent changes have been made by Crowson and Sen Gupta (1973), Lawrence (1982c, and this work), and Sasaji (1978a). The following groups are presently included in the family, but their constitution, rank, and arrangement are only tentative: Leiestinae (*Leiestes, Phymaphora, Rhanidea,* etc.); Holoparamecinae (*Holoparamecus, Lycoperdinella,* etc.); Merophysiinae (*Merophysia, Coluocera,* etc.); Eupsilobiinae (*Eidoreus*); Mycetaeinae (*Mycetaea*); Mychotheninae (*Mychothenus, Bystus, Symbiotes, Clemmus,* etc.); Agaricophilinae (*Agaricophilus*); Xenomycetinae (*Xenomycetes*); Trochoideinae (*Trochoideus*); Eumorphinae (*Eumorphus, Lycoperdina, Mycetina, Aphorista, Amphix,* etc.); Stenotarsinae (*Stenotarsus, Danae, Epipocus,* etc.); and Endomychinae (*Endomychus,* etc.).

Selected Bibliography

Aitken 1975.
Arnett 1968.
Arrow 1920.
Bates 1861.
Böving and Craighead 1931 (larvae of *Bystus ulkei* (Crotch), *Amphix laevigata* Gerstaecker, *Aphorista vittata* (Fabricius), *Endomychus coccineus* (Linnaeus), *E. biguttatus* Say, *Lycoperdina ferruginea* LeConte, *L. succincta* (Linnaeus), *Mycetaea hirta* and *Stenotarsus hispidus* (Herbst)).
Bugnion 1909 (larva of *Eumorphus pulchripes* Gerstaecker).
Coquerel 1848 (larva of *Holoparamecus kunzei* (Aubé)).

Crowson 1955, 1981.
Donisthorpe 1935 (fungus hosts).
Gardner 1931b.
Gorham 1887.
Hayashi *et al.* 1959 (larvae of *Ancylopus melanocephalus* (Olivier), *Endomychus gorhami* Lewis, *Lycoperdina mandarinea* Gerstaecker, and *L. dux* Gorham).
Hinton 1945b (stored products species).
Kemner 1924 (larva of *Trochoideus termitophilus* Roepke).
Klippel 1952 (larva of *M. hirta*).
Lawrence 1982c.
Mamaev 1977 (larva of *Agaricophilus reflexus* Motschulsky).
Pakaluk 1984 (larva and biology of *Lycoperdina ferruginea*).
Peterson 1951 (larvae of *Endomychus biguttatus* and *Lycoperdina ferruginea*).
Sasaji 1971, 1978a (larva of *Mychothenus asiaticus* Sasaji), 1978b (larva of *Saula japonica*).
Sen Gupta and Crowson 1973 (position of *Eidoreus*).
Silvestri 1912 (larva of *Coluocera formicaria* Motschulsky).
Strohecker 1953.
Verhoeff 1923 (larva of *M. hirta*).
Weiss and West 1920 (fungus hosts).

COCCINELLIDAE (CUCUJOIDEA)

Laurent LeSage, *Biosystematics Research Centre, Agriculture Canada*

The Lady Beetles, Lady Birds

Figures 34.570-624

Relationships and Diagnosis: The Coccinellidae, commonly known as ladybird beetles, belong to the superfamily Cucujoidea, and are closely related to the Endomychidae and Corylophidae (van Emden 1949; Crowson 1955). There are about 490 genera and 4200 world species (Sasaji 1971) of which about 425 are known from the United States and Canada. Coccinellid larvae usually have a reduced mola on the mandible (absent in Epilachninae) (figs. 34.593–34.599), a gular area between the labium and the thorax (fig. 34.586), and the median epicranial stem (coronal suture) is absent in most genera (figs. 34.583–34.585). The most distinctive characteristics of coccinellid larvae are a great development of the body armature into setose processes in most tribes, a campodeiform and usually brightly colored body, and mandibles of a predaceous type in most tribes (acute at apex). Except in Epilachnini and Psylloborini, they are usually very active predators.

Many galerucine and some alticine larvae in the Chrysomelidae superficially resemble larvae of coccinellids. However, chrysomelid larvae have short legs, none or only 1 pair of stemmata, no setose processes on the body (rarely with processes covered by fine setae) and the mandibles lack a mola, whereas coccinellid larvae which may be mistaken for chrysomelid larvae have long legs, 3 pairs of stemmata, large setose body processes covered by robust setae, and a mandibular mola.

Biology and Ecology: It is not possible to treat here in detail the bionomics and ecology of Coccinellidae. See the excellent reviews by Balduf (1935), Hagen (1962), and Hodek (1967, 1973).

The eggs, usually oval or spindle-shaped, vary in colour from yellowish to reddish orange, and are mostly laid in clusters on the underside of leaves or in bark crevices in the vicinity of prey.

In general, there are 4 larval instars which last about 10 days each, but with great variation according to species and ecological factors. A unique feature among coccinellid larvae is the presence of secretory structures that produce a visible coating of waxy threads in the larvae of several tribes; these probably have primarily a defensive role against predators (Pope 1979). Cannibalism is frequent in coccinellid larvae and increases the chances of survival when there is a very low density of prey (Hodek 1973; Dimetry 1976). Larvae perceive their prey only by contact (Fleschner 1950; Putman 1955a, 1955b; Dixon 1959; Kaddou 1960; Frazer et al. 1981). With the development of biological control programs, attempts have been made to rear larvae on artificial diets or dried food. Several recent experiments gave excellent results (Smith 1960; Fisher 1963; Shands et al. 1966; Hodek 1973 (review); Kariluoto 1980).

Different types of pupae occur in Coccinellidae. Coccinellinae and Sticholotini have naked pupae attached by the cauda to the substrate. Pupae of the Chilochorini and Noviini are partly covered by the skin of the last larval instar, and the Hyperaspini and Scymnini have pupae completely covered by larval skins. The pupa is not entirely immobile; if irritated, the head region is raised several times by upward jerks of the body.

The number of generations varies greatly according to species and latitudes; types of voltinism were summarized by Hagen (1962). Perhaps the most fascinating phenomenon coccinellid adults display is the formation of aggregations. Species involved usually feed mostly on aphids, exhibit long dormancy or diapause periods, and mate at the aggregation site before the beetles disperse or migrate (Hagen 1962; Hodek 1967, 1973; Benton & Crump 1979; Lee 1980).

The role of coccinellids in natural control has been demonstrated many times. Various interrelated factors affect the ability of coccinellids to check pest infestations. However, it seems that the most important factor involved, temperature, was probably also the most neglected in theories of insect predation (Baumgaertner et al. 1981).

Since the bionomics, general habitus and food preference are generally distinctive for each tribe, a key to tribes is presented below, followed by a short synopsis of each tribe.

Body Armature: Gage's terminology (Gage 1920) is generally followed for different structures on the body of larvae. A *seta* (fig. 34.610) is situated directly on the body surface; a *chalaza* is a seta mounted on a small base (fig. 34.611). A *verruca* or *tubercle* is a small protuberance covered by setae instead of chalazae (fig. 34.612). A *struma* appears to be a mound-like projection of the body-wall upon which are situated a few chalazae (fig. 34.613). A *parascolus* (fig. 34.614) is an elongate process covered by chalazae, but less than 3 times as long as wide. A *scolus* (fig. 34.616) is a branched projection, usually more than 5 times as long as wide; each branch bears at its distal end a single stout seta. A *sentus* (fig. 34.615) is a projection of the body-wall which is not branched like a scolus but bears stout setae on its trunk. Larvae may have structures intermediate between scoli, senti, parascoli or strumae.

KEY TO TRIBES OF NORTH AMERICAN LARVAE OF COCCINELLIDAE

1. Body with scoli (figs. 34.570, 34.616); mandible without mola (fig. 34.597) .... Epilachnini

   Body without scoli (figs. 34.572–34.575); mandible with mola (figs. 34.593–34.596) 2

2(1). Epicranial suture present (figs. 34.582, 34.583, 34.585) 3

   Epicranial suture absent (fig. 34.584) 11

3(2). Epicranial suture V-shaped (fig. 34.585) 4

   Epicranial suture U-, Y-, or lyre-shaped (figs. 34.582, 34.583) 6

4(3). Antennae large, conspicuous, second and third segments elongate (fig. 34.600) .... Scymnillini

   Antennae small, inconspicuous, second and third segments short (fig. 34.585) 5

5(4). Mature larvae small, less than 3 mm (fig. 34.575); tibiotarsi with a pair of apical flattened setae (fig. 34.618); maxillary palps 3-segmented (fig. 34.608) .... Sticholotini

   Mature larvae larger, more than 6 mm (fig. 34.574); tibiotarsi with several apical clavate setae (fig. 34.617); maxillary palps 2-segmented (fig. 34.609) .... Noviini

6(3). Pores of repugnatorial glands present in the coria between abdominal segments on the antero-lateral margin (fig. 34.587); body always with long senti (figs. 34.587, 34.615) .... Chilochorini

   Pores of repugnatorial glands absent, body usually with strumae (fig. 34.613) or parascoli (fig. 34.614), rarely with senti (fig. 34.615) 7

7(6). Apex of mandible simple (fig. 34.594) 8

   Apex of mandible bidentate (figs. 34.596, 34.599) 10

8(7). Body densely covered with fine hairs and long setae (fig. 34.624); tibiotarsi slender and narrowing apically .... Serangiini

   Body covered with few large setae located on tubercles or strumae (figs. 34.623, 34.573); tibiotarsi short, stout, and truncated apically (fig. 34.588) 9

9(8). Three pairs of conspicuous, pigmented, sclerotized plates (strumae) on abdominal segments 1–8 (fig. 34.623); body not covered by wax-like secretions .... Stethorini

   Three pairs of inconspicuous, not pigmented, tubercles on abdominal segments 1–8; body covered by wax-like secretions .... (in part) Scymnini

10(7). Few large chalazae on disk or posterior margin of abdominal segment 9 (fig. 34.622); body dull yellowish; third antennal segment always well-developed and cupola-like (fig. 34.602) .... Coccidulini

   Numerous setae and/or small chalazae on abdominal segment 9 (fig. 34.621); body brightly colored with black, brown, red, yellow or orange; third antennal segment usually much reduced, antenna appearing 2-segmented (fig. 34.606) .... Coccinellini

11(2). Apex of mandible multidentate (fig. 34.595); body without wax-like secretions .... Psylloborini

   Apex of mandible simple (fig. 34.594); body covered with wax-like secretions 12

12(11). Labial palp very small, dome-shaped, 1-segmented (fig. 34.592) .... Hyperaspini

   Labial palp normal, 2-segmented (fig. 34.591) .... (in part) Scymnini

Tribal Information

Unlike other coccinellids, the **Epilachnini** (Epilachninae) have an unusual porcupine-like appearance (fig. 34.570), and are phytophagous. Furthermore, in this tribe the mandible (fig. 34.597) lacks a mola and has a multidentate apex. *Epilachna borealis* (Fabricius), the squash beetle, attacks squash and pumpkins, and *E. varivestis* (Mulsant), the Mexican bean beetle, is a serious pest of beans including soybeans (Guyon & Knull 1925). The European alfalfa beetle, *Subcoccinella vigintiquatuorpunctata* (L.) was discovered in 1972 in Pennsylvania (Annonymous 1974). While an important pest of alfalfa and clover in Europe, it has been found feeding only on bouncing bet, *Saponaria officinalis*, campion, *Lychnus alba*, and oatgrass, *Arrhenatherum elatius* in the United States (Annonymous 1974).

Chilochorini (Chilochorinae) superficially resemble Epilachnini when senti are well-developed as in *Chilochorus* (fig. 34.571), or some Coccinellini when senti are more reduced. However, the presence of large pores of repugnatorial glands (fig. 34.587) on the abdomen will separate them easily from both. Chilochorini feed primarily on aphids and scales, therefore are used for biological control (Huffaker & Doutt 1965). For example, *Exochomus flavipes* Thungerg, indigenous to South Africa (Geyer 1947a, 1947b), was successfully used in the United States against mealybugs infesting commercial greenhouses (Doutt 1951). *E. quadripustulatus* (L.) was released against the wooly aphid *Adelges piceae* (Ratz). The twice-stabbed lady beetle, *Chilochorus stigma* (Say), is an important predator of the Florida red scale, *Chrysomphalus aonidum* (L.) which infests citrus groves (Muma 1955a, 1955b).

Coccinellini (Coccinellinae) are the best known coccinellid larvae because they live exposed, are very active, relatively large, and brightly coloured. The body armature is very diverse in this tribe (figs. 34.572, 34.579–34.581) and all structures are represented except scoli. These larvae can be distinguished from those of other tribes by the lyre-shaped epicranial suture of the head (fig. 34.583), the bidentate apex of the mandible (fig. 34.599), the reduced, inconspicuous third antennal segment (fig. 34.606) and the well-developed body armature. All native Coccinellini are beneficial and several foreign species have been introduced to aid in control of pests (DeBach 1964; Hodek 1967, 1973). Some species are widely distributed and well known. *Anatis mali* (Say), the eye-spotted lady beetle, bears senti on the body similar to fig. 34.580, and occurs on conifers where it is able to survive at low prey densities (Smith 1965; Watson 1976). The spotted ladybird, *Coleomegilla maculata* (De Geer) (fig. 34.621), eats pollen as well as aphids and is usually found on herbaceous plants, wild and cultivated, where its food is abundant (Smith 1965). *Hippodamia* species are important aphid predators (Cuthright 1924; Hodek 1973) and a common species, *Hippodamia convergens* Guérin, the convergent lady beetle (fig. 34.579), can keep aphids in check in alfalfa fields (Cooke 1963). *Adalia bipunctata* (L.) (fig. 34.581), the 2-spotted ladybird, is a widespread polymorphic species (Hodek 1973) which prefers trees above 2 m. Consequently, it is especially beneficial in orchards and groves where it is the most important coccinellid aphid predator (Smith 1958; Putman 1964; Hodek 1973). *Coccinella* species (fig. 34.572) are known as aphid predators (Palmer 1914; Clausen 1916; McMullen 1967), and some have become established after repeated releases over large areas; others like *C. undecimpunctata* L., are becoming well established on their own, along with the aid of man's commerce (Watson 1979; Wheeler & Hoebeke 1981).

Coccidulini (Coccidulinae) much resemble Coccinellini but differ by the features of the last abdominal segment (fig. 34.622) and antenna (fig. 34.602), their dull coloration, and the presence of a thin powdery coating of wax (Pope 1979). Their biology is not well known. *Coccidula* live in wet habitats. *Rhyzobius ventralis* (Erichson) has been introduced from Australia to California and Hawaii for control of scale insects (Pope 1981; Richards 1981).

Scymnini and **Hyperaspini** larvae are strikingly different from others because of their thick coating of wax (fig. 34.577) which is absent or inconspicuous in other tribes.

Scymnini (Scymninae) larvae have very sharp unidentate mandibles (fig. 34.594) and very small tubercles on the abdomen (fig. 34.573). *Scymnus* species feed mainly on aphids; some are useful predators in red pine plantations (Gagné & Martin 1968), cotton fields (Davidson 1921b) or sugarbeet fields (Buntin & Tamaki 1980); others attack psyllids and are beneficial in pear orchards (Westigard et al. 1968) while a few feed on mealybugs in citrus groves (Muma 1955a), or on phylloxera on wild grape (Wheeler & Jubb 1979).

Hyperaspini (Scymninae) larvae are separated from all others by the unique dome-shaped, 1-segmented labial palps (fig. 34.592). *Hyperaspis* (figs. 34.576, 34.577) species are known as efficient predators of scale insects (Simanton 1916; Böving 1917; Phillips 1963).

Noviini (Coccidulinae) is the only tribe where the larvae have only 2 pairs of sclerotized tubercles and 1 pair of soft lateral projections on the abdominal segments (fig. 34.574). *Rodolia cardinalis* Mulsant, the vedalia lady beetle, is a famous classic example of successful use of coccinellids in biological control of coccids (DeBach 1964).

Psylloborini (Coccinellinae) larvae are immediately recognized by the multidentate apex of their mandibles (fig. 34.595). They differ from other tribes in that they are mycophagous and feed on mildew. They are beneficial because they eat destructive fungi (Davidson 1921a).

Larvae of the 4 remaining tribes are usually overlooked because of their small size. *Microweisea* larvae (fig. 34.575) in the tribe **Sticholotini** (Sticholotinae) are easily identified by the 2 large flattened setae at the apex of the tibiae (fig. 34.618). They are beneficial scale feeders (Burgess & Collins 1912; Muma 1955a; Sharma & Martel 1972).

Serangiini (Sticholotinae) larvae have the body densely covered with fine setae (fig. 34.624) and their tibiae are unusually slender (fig. 34.590), characters which distinguish them from all others. *Delphastus* species in this tribe are predators of Aleyrodidae (Muma 1955a, 1955b).

Stethorini (Scymninae) larvae resemble superficially the Serangiini but are separated from them and all other tribes by the few large setae fixed on small tubercles covering the body (figs. 34.578, 34.623) and their short tibiae, apically truncated (fig. 34.588). They feed chiefly on mites, and many *Stethorus* species are active predators of these pests (Fleschner 1950; Robinson 1953; Putman 1955a, 1955b; Putman & Herne 1966; Tanigoshi & McMurtry 1977).

Scymnillini (Coccidulinae) larvae are distinguished by their unusual, large, antennae (fig. 34.600). Their biology is poorly known. *Scymnillus aterrimus* Horn has been reported as an incidental predator of scale insects in citrus groves (Muma 1955b).

Description: Coccinellid larvae are extremely diverse as described in the previous section, and illustrated in the family key, where they key out at several couplets. Therefore, they cannot be distinguished altogether by only 1 or 2 characters as in many beetle families. On the other hand, larvae of each tribe show a distinctive general habitus and have morphological features which are shared by all members of the tribe. The striking flattened larvae of the Palaearctic Platynaspini do not occur in North America.

Head: Hypognathous, usually rounded (fig. 34.582), sometimes elongate (fig. 34.585) as in *Microweisea*, or transverse as in Hyperaspini, Platynaspini, and some Scymnini (fig. 34.584). In most species, the head is completely sclerotized, but sometimes it may be partly or only very slightly sclerotized. The epicranial suture is usually distinct, V-shaped (fig. 34.585), Y-shaped (fig. 34.582), lyre-shaped (fig. 34.583), or absent (fig. 34.584) but the epicranial stem is usually absent.

The antenna of the typical form of coccinellid larvae consists of 3 sclerotized segments (figs. 34.600, 34.602, 34.603), and bears a large spine-like seta on the membranous apical area of the second segment. However, they may appear to be 2-segmented (figs. 34.601, 34.605, 34.606) or even 1-segmented (fig. 34.604) when the third, and sometimes the second and third segments are reduced and not sclerotized; in those cases the homology of the segments apparently missing is often difficult to establish (Sasaji 1968b). The labrum is distinct and transverse (fig. 34.582).

The mandible is either apically simple and acute (figs. 34.593, 34.594), bidentate (figs. 34.596, 34.599), or multidentate in plant feeders (figs. 34.595, 34.597); the mola is usually present but reduced (figs. 34.595, 34.596 34.598, 34.599), highly reduced in *Microweisea* (fig. 34.593) and absent in *Epilachna* (fig. 34.597); a retinaculum may be either absent (figs. 34.593, 34.594), developed with 1 tooth (fig. 34.596), or multidenticulate (fig. 34.595). The maxillary palps are generally 3-segmented, but are 2-segmented in Noviini (fig. 34.609). The labial palps are either 1- or 2-segmented (figs. 34.591, 34.592). The labium has the submentum fused with the ligula (fig. 34.586).

Thorax and Abdomen: Pronotum with 2 or 4 plates. Meso- and metanotum each with 2 plates and distinct armature. Legs usually long and slender (figs. 34.589, 34.590), short in Hyperaspini and Stethorini (fig. 34.588), consisting of 5 segments (coxa, trochanter, femur, tibia and claw-like tarsungulus). Tarsungulus curved (fig. 34.620), with a robust quadrangular tooth in some species (fig. 34.619). The apex of the tibiae usually bears clavate or flattened setae which are important in taxonomy (figs. 34.588, 34.617, 34.618).

Abdomen 10-segmented, widest basally, tapering to caudal end, dorsally with distinct armature, usually characteristic for each tribe. The tenth segment may be modified as a proleg or a sucking disk. Pores of repugnatorial glands may occur on each antero-lateral margin of terga in the coria between segments (fig. 34.587).

Spiracles: Small, annular, and located on abdominal segments 1–8 (fig. 34.587).

Comments: The larval stage of most coccinellid species consists of 4 instars. The first instar can be recognized by the paired egg-bursters on pronotum, and differences in proportions in size of head, abdomen, legs, setae, etc. Changes occur in coloration, proportions and armature of the body between successive instars. Larvae of the first and second instars are monochrome, with sclerotization and armature less developed than in older instars. Larvae of the third and fourth instars are usually brightly coloured and well sclerotized.

Palearctic coccinellid larvae are now fairly well known with the recent contributions of several authors (van Emden 1949; Savoiskaya 1957, 1960, 1962, 1964a, 1964b; Kamiya 1965; Sasaji 1968a, 1968b; Klausnitzer 1970; Savoiskaya & Klausnitzer 1973). The taxonomy of the Nearctic coccinellid larvae has not been comprehensively studied; only the early works of Böving (1917) and Gage (1920) provide a general treatment of the family. With the field key of Storch (1970) one can identify the larvae of 5 common native species. Phuoc and Stehr (1974) studied the morphology and the phylogenetic relationships of coccinellid pupae based on their morphology and suggested the need for a similar study of the larvae.

Selected Bibliography

Annonymous 1974.
Balduf 1935.
Baumgaertner *et al.* 1981.
Benton and Crump 1979.
Böving 1917. (key)
Buntin and Tamaki 1980.
Burgess and Collins, 1912.
Clausen 1916.
Cooke 1963.
Crowson 1955.
Cuthright 1924.
Davidson 1921a, 1921b.
DeBach 1964.
Dimetry 1976.
Dixon 1959.
Doutt 1951.
Emden 1949. (key)
Fisher 1963.
Fleschner 1950.
Frazer *et al.* 1981.
Gage 1920. (key)
Gagné and Martin 1968.
Geyer 1947a, 1947b.
Guyon and Knull 1925.
Hagen 1962.
Hodek 1967, 1973.
Huffaker and Doutt 1965.
Kaddou 1960.
Kamiya 1965.
Kariluoto 1980.
Klausnitzer 1970. (key)
Lee 1980.
McMullen 1967.
Muma 1955a, 1955b.
Palmer 1914.
Phillips 1963.
Phuoc and Stehr 1974. (key, pupae)
Pope 1979, 1981.
Putman 1955a, 1955b, 1964.
Putman and Herne 1966.
Richards 1981.
Robinson 1953.
Sasaji 1968a, 1968b, 1971.
Savoiskaya 1957 (key), 1960, 1962, 1964a, 1964b.
Savoiskaya and Klausnitzer 1973. (key)
Shands, Holmes and Simpson 1966.
Shands et al. 1972.
Sharma and Martel 1972.
Simanton 1916.
Smith 1958, 1960, 1965.
Storch 1970. (key)
Tanigoshi and McMurtry 1977.
Watson 1976, 1979.
Westigard *et al.* 1968.
Wheeler and Hoebeke 1981.
Wheeler and Jubb 1979.

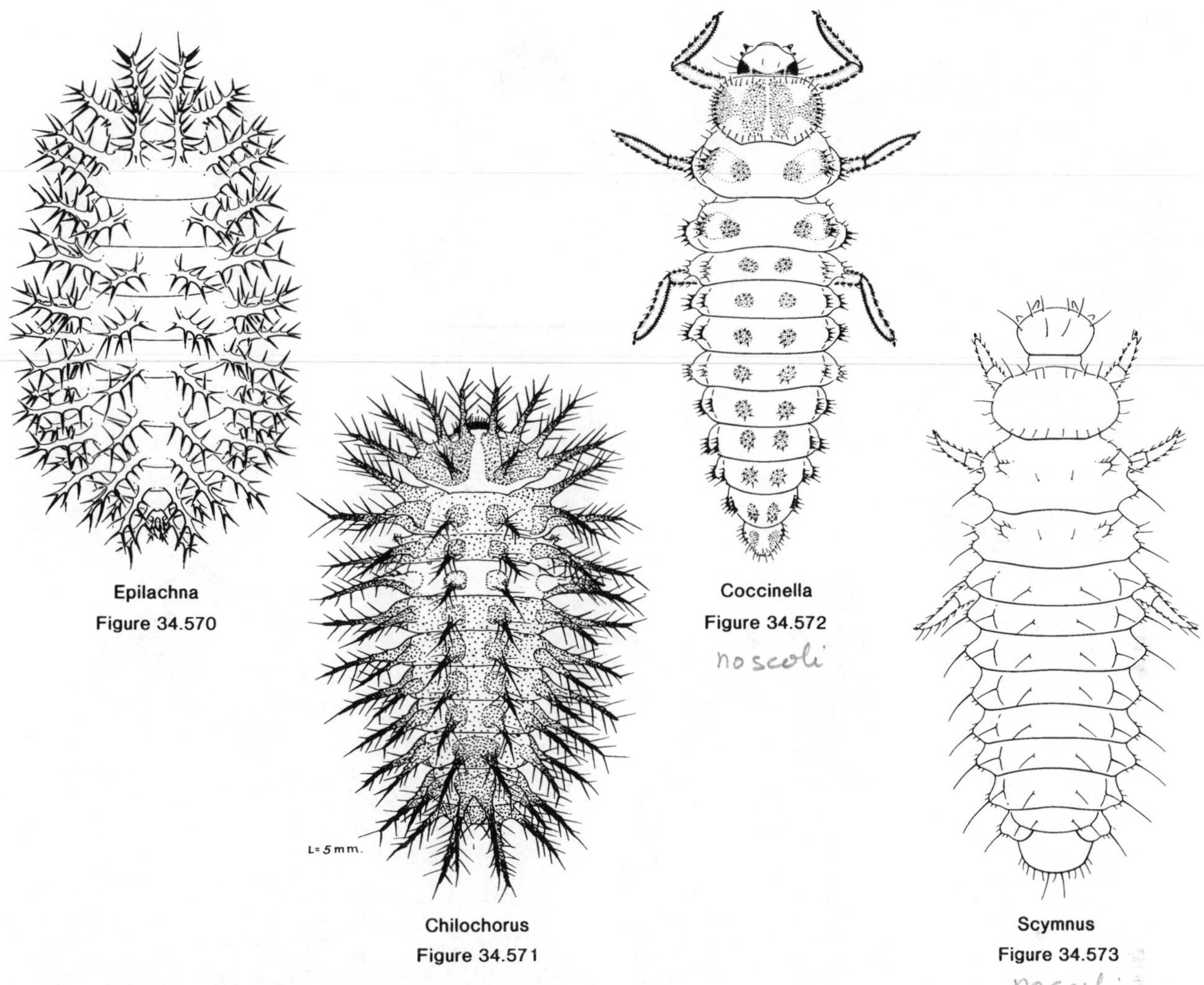

Epilachna
Figure 34.570

Chilochorus
Figure 34.571

Coccinella
Figure 34.572
no scoli

Scymnus
Figure 34.573
no scoli

Figure 34.570. *See* also figures 34.597, 34.601, 34.616. Coccinellidae. *Epilachna varivestis* Mulsant. Mexican bean beetle. (Epilachnini) Length 11 mm; cyphosomatic, yellow, brownish around stemmata and tips of scoli; abdominal segments 1-8 each bearing 6 scoli; 4 scoli on prothorax, 6 on both meso- and metathorax; light slender setae on head, legs, and verrucae of all segments; 1 pair of verrucae on sterna of each thoracic and abdominal segment 1; 3 pairs on abdominal segments 2-7, 2 pairs on 8; caudal segment in the form of a sucker-like protuberance; spiracles inconspicuous, annular, and ventrad of supraspiracular scoli. It can be a serious defoliator of many kinds of beans, including soybeans. Jerome, Idaho, 25-VII-1960, Gibson & Evans.

Figure 34.571. Coccinellidae. *Chilocorus stigma* (Say), twice-stabbed lady beetle. (Chilochorini) Length = 5 ± mm. Cyphosomatic, oval, and covered with prominent senti; color brownish with most senti and pinacula near black, mid-dorsal line and dorsum of first abdominal segment yellow to near white; prothorax with 5 pairs of prominent senti, mesothorax and metathorax with 4 pairs of senti; pinacula of dorsal senti on abdominal segments 1-5 separated, contiguous on segments 6-8; all senti deep brown to near black except the yellow to near white dorsal and supraspiracular senti on the first abdominal segment; all abdominal segments with 3 pairs of senti except the eighth where lateral senti are wanting; 7 pairs of conspicuous, circular openings to glands are found in the coriae between abdominal segments 1-8; less conspicuous circular spiracles occur on the mesothorax and abdominal segments 1-8 cephaloventrad of the supraspiracular senti. Feeds on scale insects, especially soft bodied scales and immatures. (From Peterson, 1951)

Figure 34.572. Coccinellidae. *Coccinella transversoguttata* Faldermann. (Coccinellini) Transverse lady beetle, an aphid predator. Length = 11 mm. Fusiform with ground color bluish-gray, all processes on abdominal segments black except for lateral and dorsolateral ones on abdominal segments 1 and 4; basal portion of head black, labrum and frons cream to white; epicranial suture lyre-shaped; medial plates on prothorax separated by narrow yellowish stripe, sclerotized plates on both meso- and metathorax well-separated; dorsal and dorsolateral aspects of abdominal segments 1-8 each provided with parascoli or strumae, ventral aspect with verrucae or chalazae; legs well-developed, robust, and black; basal portion of claw with a distinct rectangular tooth. Ottawa, Ontario, 20-VIII-1980, on potatoes, L. LeSage.

Figure 34.573. Coccinellidae. *Scymnus hemorrhous* LeConte, a predator. (Scymnini) Length = 3 mm. Shape fusiform; head, body and legs yellowish; epicranial suture absent; prothorax transverse, ovoid, with a row of marginal setae; meso- and metathorax each with a pair of dorsal sclerotized tubercles and 2 pairs of lateral and moderately developed tubercles; abdominal segments 1-8 each with 3 pairs of weakly developed tubercles, each bearing a large seta and a few small setae; legs relatively short and rather stout. Baton Rouge, Louisiana, 15-VIII-1952, (reared), O.L.C.

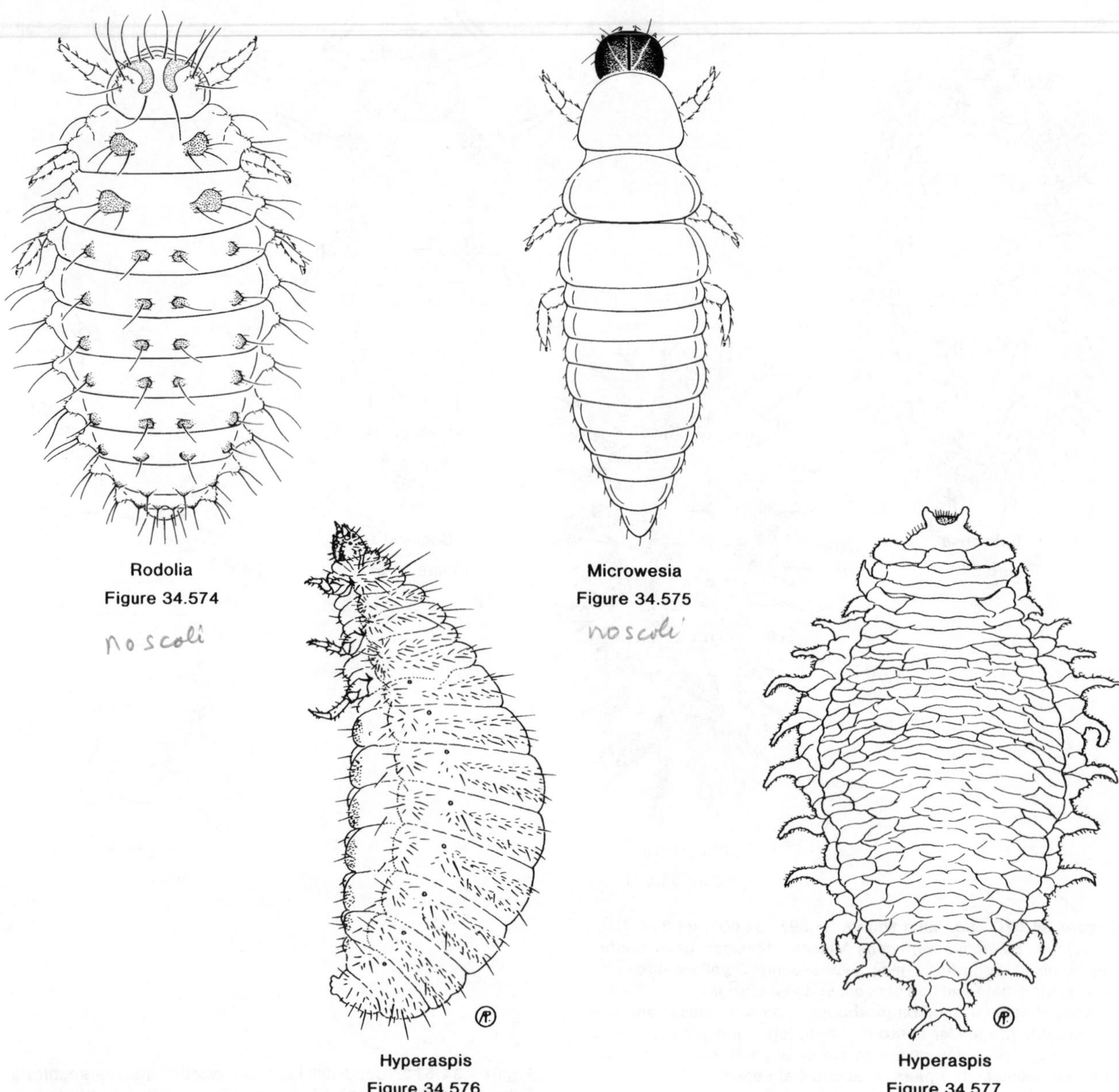

Figure 34.574. Coccinellidae. *Rodolia cardinalis* (Mulsant). (Noviini) Vedalia lady beetle, a predator on the cottony-cushion scale, *Icerya purchasi* Raaskell. Length = 7 mm. Shape elliptical and weakly convex above; dorsum dark red with brown sclerotized plates and tubercles; head black, subquadrate; epicranial suture V-shaped; maxillary palpi 2-segmented; each thoracic segment with a pair of sclerotized plates, meso- and metathorax each bearing 2 additional pairs of soft lateral projections; abdominal segments 1-8 each with 2 pairs of dorsal plates and 1 pair of soft lateral projections bearing 2 long setae. Los Angeles, California, July 1892, D. V. Coquillet.

Figure 34.575. Coccinellidae. *Microweisea* sp. (Stilochotini) Length = 3 mm. Shape fusiform, yellowish throughout except the dark brown head; head elongate with a V-shaped epicranial suture; prothorax trapezoidal, meso- and metathorax transverse; lateral margin of each thoracic segment with a fringe of inconspicuous fine setae; legs well-developed; tibiotarsi with a pair of large flattened setae at the apex; abdominal segments 1-8 similar, but becoming successively smaller, bearing a few inconspicuous small setae. New Orleans, Louisiana, 13-VII-1923, Quaintance.

Figures 34.576, 34.577. Coccinellidae. *Hyperaspis signata* Olivier. (Hyperaspini) Length = 5 mm. Cyphosomatic, greatest width near mid-abdominal region and entire dorsal aspect covered with a near white, cottony, wax covering; body cream to greenish color (may be pinkish in preserved specimens), head mottled brown and legs brown especially the 2 distal segments; numerous short, brown setae scattered over lateral and dorsal aspects of all segments and also on the head, only a few setae on venter of abdomen; legs small, well developed but not projecting beyond sides of body; caudal segment with an eversible sucking disk; inconspicuous circular spiracles on lateral aspects dorsad of lateral ridge. Larvae feed on mealybugs, and soft scales, especially immatures. (From Peterson, 1951).

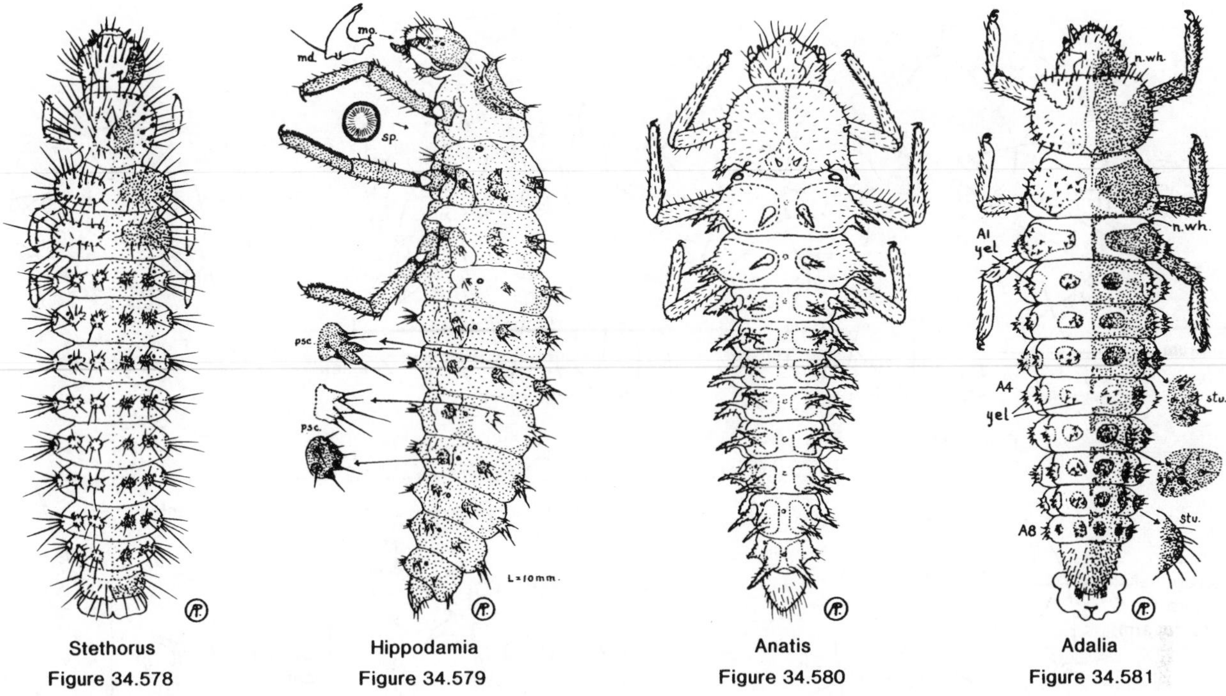

Stethorus
Figure 34.578

Hippodamia
Figure 34.579

Anatis
Figure 34.580

Adalia
Figure 34.581

Figures 34.578, and 34.623. Coccinellidae. *Stethorus punctum* LeConte. (Stethorini) Length = 2.5-3 mm. Somewhat fusiform with greatest diameter in metathorax and first abdominal segments; light gray to light brown with brown to gray verrucae and brown to gray areas on thorax and head; head light brown laterally and mottled on frons; solid brown to mottled brown pigment areas laterad of the meson on all thoracic segments; 6 verrucae on the dorsal and lateral aspects of abdominal segments 1-8; numerous, light colored, elongate setae on the head, dorsum of thorax, on all verrucae and venter of the abdomen; well developed legs possess a few setae and are partially pigmented; spiracles circular and located ventrad of supraspiracular verrucae. Feeds on plant-infesting mites, common on apple. (From Peterson, 1951)

Figure 34.579. Coccinellidae. *Hippodamia convergens* Guérin-Méneville, convergent lady beetle. (Coccinellini) Length = 10-11 mm. Somewhat fusiform, subcylindrical with greatest diameter in region of metathorax; dark brown to black with a bluish cast, light areas ranging from orange, to yellow to near-white; prothorax oval, wider than long and with 4 longitudinal dark areas with light yellow areas between, cephalad, and caudad of dark areas. Chalazae on the cephalic and lateral portions and on the pigmented areas; parascoli on caudolateral margin of metathorax cream colored; each abdominal segment 1-8 with 3 pairs of parascoli, subdorsal, supraspiracular and subspiracular, all deeply pigmented except the supraspiracular and subspiracular on the first and fourth abdominal segments and the subdorsal on the fourth segment; the light colored parascoli and the areas about them plus the areas between the subdorsal and supraspiracular parascoli on the 6th and 7th segments are yellow to deep orange; sterna on segments 2-8 with transverse rows of 6 verrucae; legs well developed, elongate and tarsal claws without appendiculate teeth. Feeds on aphids and soft bodied insects. (From Peterson, 1951)

Figure 34.580. Coccinellidae. *Anatis quindecimpunctata* Olivier. (Coccinellini) Length = 17-18 mm. Elongate, widest at metathorax and tapering toward both ends; color on dorsal half a deep brown except for light spots and median line, ventral half near white to yellow; head one-half diameter of prothorax, flattened, light on frontal area and dark on caudolateral portions; prothorax with parascoli on caudolateral margin and a light spot on the caudomeson bearing 2 chalazae; mesothorax and metathorax bearing 2 pairs of senti and 1 extreme, lateral pair of parascoli; abdominal segments 1-8 with 2 pairs of senti dorsad of the spiracles and 1 pair of senti or parascoli immediately ventrad of spiracles; thoracic sterna similar, each bearing a pair of verrucae adjacent to the meson; thoracic legs long, slender, nearly 1.5 times as long as the metathorax is wide. Feeds chiefly on aphids. (From Peterson, 1951)

Figure 34.581. Coccinellidae. *Adalia bipunctata* (L.), two-spotted lady beetle. (Coccinellini) Length = 9 ± mm. Fusiform with greatest width in region of 2nd to 4th abdominal segments; dark brown to bluish-gray, mottled with yellow to cream spots, dorsal half of head deeply pigmented and ventral portion of frons and clypeus cream to white; prothorax with a medium yellow stripe and 2 cephalolateral yellow areas; each abdominal segment 1-8 with 2 pairs of strumae dorsad of the spiracles and 1 pair ventrad; each struma may have 3-8 chalazae and additional setae; 2 of the lateral strumae on first abdominal segment cream to yellow, and on the fourth abdominal segment the most lateral strumae and the pair adjacent to the meson are light colored or yellow; ninth segment deeply pigmented, setiferous and giving rise to a large, eversible, fleshy protuberance. Feeds chiefly on aphids. (From Peterson, 1951)

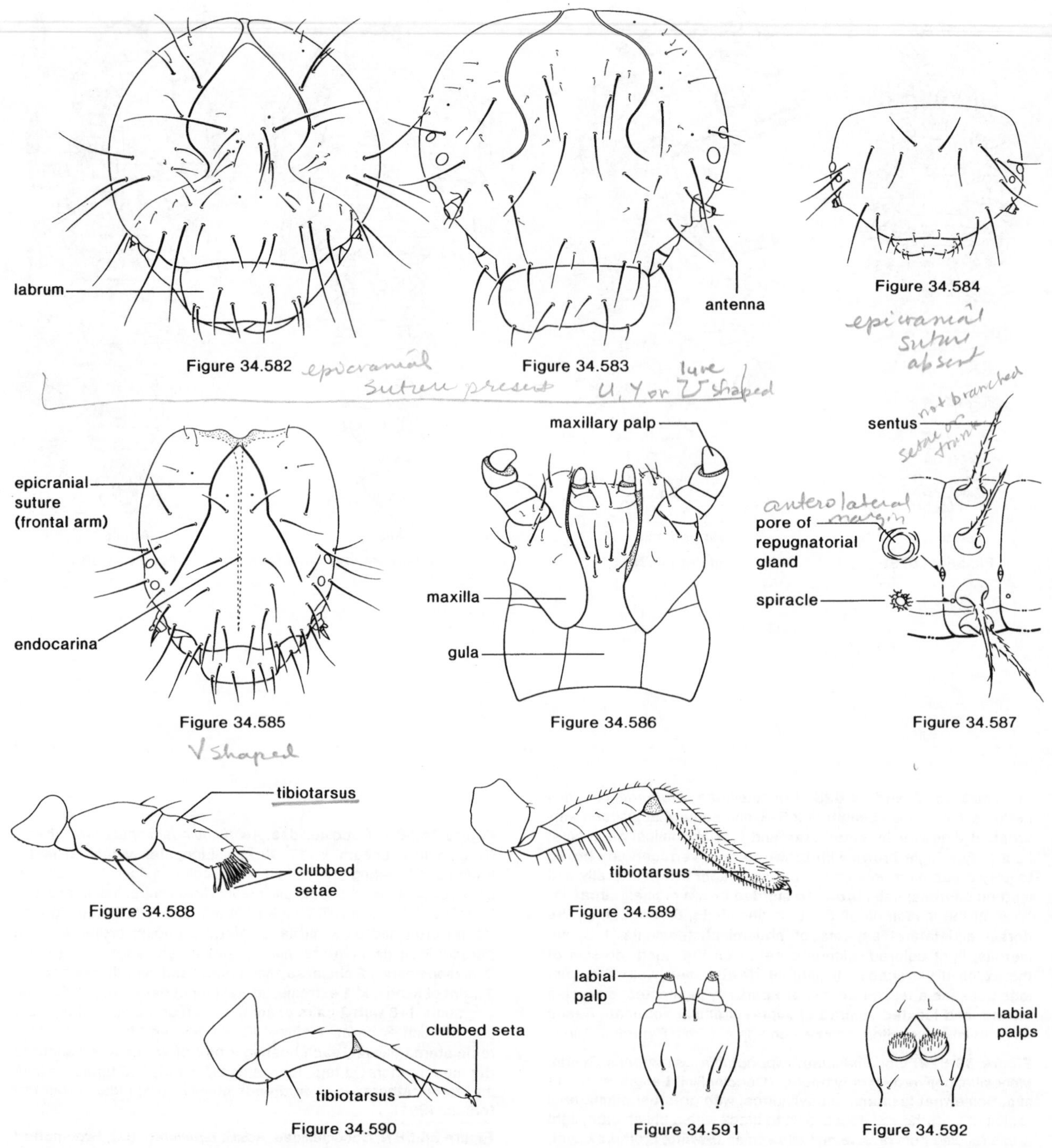

Figures 34.582–34.592. Coccinellidae. Structures of larvae.

Figure 34.582. *Chilochorus cacti* (L.), head capsule, dorsal. (Chilocorini)

Figure 34.583. *Coccinella transversoguttata* Faldermann, head capsule, dorsal. (Coccinellini)

Figure 34.584. *Scymnus creperus* Mulsant, head capsule, dorsal. (Scymnini)

Figure 34.585. *Microweisea* sp., head capsule, dorsal. (Stichotini)

Figure 34.586. *Hippodamia tredecimpunctata* (L.), head capsule, ventral. (Coccinellini)

Figure 34.587. *Chilochorus cacti* (L.), third abdominal segment, lateral. (Chilochorini)

Figure 34.588. *Stethorus histrio* Chazeau, foreleg. (Stethorini)

Figure 34.589. *Coccinella transversoguttata* Faldermann, foreleg. (Coccinellini)

Figure 34.590. *Delphastus sonoricus* Casey, foreleg. (Serangiini)

Figure 34.591. *Scymnus collaris* Melsheimer, labium. (Scymnini)

Figure 34.592. *Hyperaspis binotata* Say, labium. (Hyperaspini)

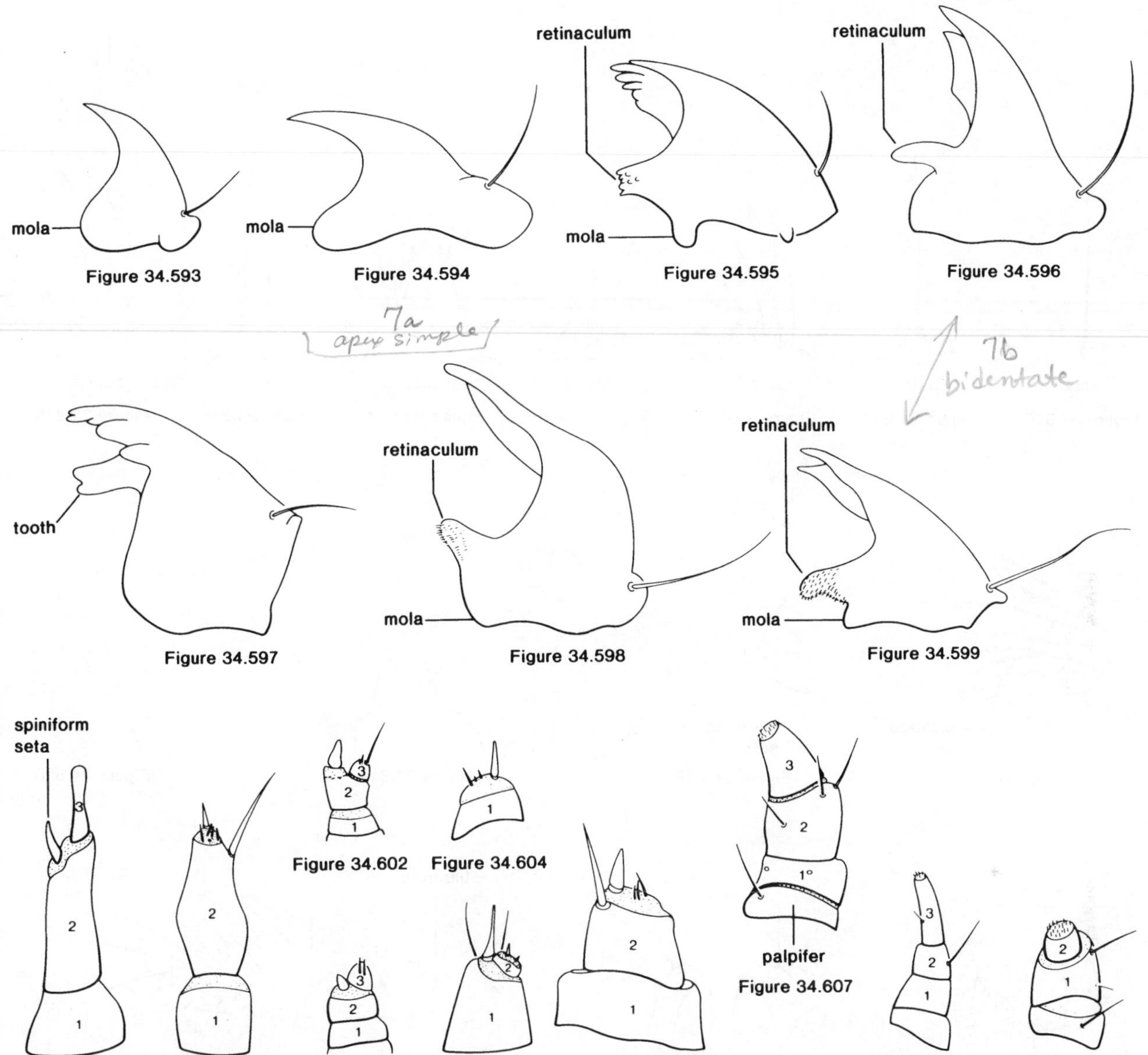

Figures 34.593–34.609. Coccinellidae. Structures of larvae.

Figure 34.593. *Microweisea*, mandible. (Stilochotini)

Figure 34.594. *Scymnus creperus* Mulsant, mandible. (Scymnini)

Figure 34.595. *Psyllobora* sp., mandible. (Psylloborini)

Figure 34.596. *Lindorus lophantae* Blaisdell, mandible. (Coccidulini)

Figure 34.597. *Epilachna varivestis* Mulsant, mandible. (Epilachnini)

Figure 34.598. *Hyperaspis binotata* Say, mandible. (Hyperaspini)

Figure 34.599. *Coccinella* sp., mandible (Coccinellini)

Figure 34.600. *Zagobla ornata* (Horn), antenna (Scymnillini)

Figure 34.601. *Epilachna varivestis* Mulsant, antenna. (Epilachnini)

Figure 34.602. *Lindorus lophantae* Mulsant, antenna. (Coccidulini)

Figure 34.603. *Scymnus creperus* Mulsant, antenna. (Scymnini)

Figure 34.604. *Chilochorus cacti* (L.), antenna. (Chilochorini)

Figure 34.605. *Rodolia cardinalis* (Mulsant), antenna. (Noviini)

Figure 34.606. *Coccinella transversoguttata* Faldermann, antenna. (Coccinellini)

Figure 34.607. *Coccinella transversoguttata* Faldermann, maxillary palp. (Coccinellini)

Figure 34.608. *Microweisea* sp., maxillary palp. (Stilochotini)

Figure 34.609. *Rodolia cardinalis* (Mulsant), maxillary palp. (Noviini)

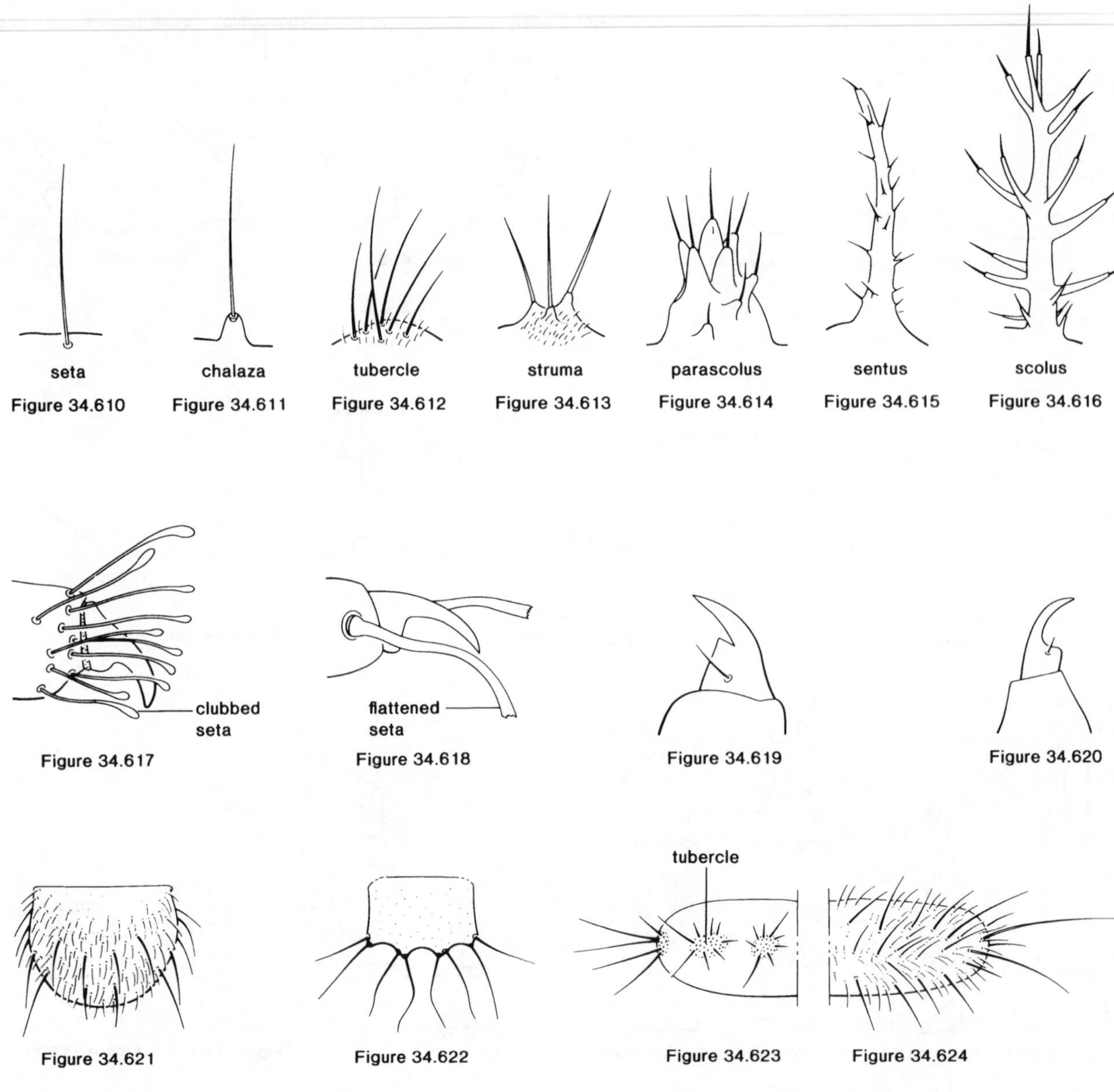

Figures 34.610–34.624. Coccinellidae. Structures of larvae.

Figure 34.610. *Hippodamia tredecimpunctata* (L.), seta. (Coccinellini)

Figure 34.611. *Hippodamia tredecimpunctata* (L.), chalaza. (Coccinellini)

Figure 34.612. *Microweisea* sp., verruca. (Stilochotini)

Figure 34.613. *Coleomegilla maculata* (De Geer), struma. (Coccinellini)

Figure 34.614. *Coccinella transversoguttata* Faldermann, parascolus. (Coccinellini)

Figure 34.615. *Chilochorus cacti* (L.), sentus. (Chilorchorini)

Figure 34.616. *Epilachna varivestis* Mulsant, scolus. (Epilachnini)

Figure 34.617. *Rodolia cardinalis* (Mulsant), apex of tibiotarsus. (Noviini)

Figure 34.618. *Microweisea* sp., apex of tibiotarsus. (Stilochotini)

Figure 34.619. *Coccinella transversoguttata* Faldermann, apex of tibiotarsus. (Coccinellini)

Figure 34.620. *Coleomegilla maculata* (De Geer), apex of tibiotarsus. (Coccinellini)

Figure 34.621. *Coleomegilla maculata* (De Geer), ninth abdominal tergite. (Coccinellini)

Figure 34.622. *Lindorus lophantae* Mulsant, ninth abdominal tergite. (Coccidulini)

Figure 34.623. *Stethorus punctum* (LeConte), third abdominal tergite, left half. (Stethorini)

Figure 34.624. *Delphastus sonoricus* (LeConte), third abdominal tergite, right half. (Serangiini)

CORYLOPHIDAE (CUCUJOIDEA)

(= ORTHOPERIDAE)

John F. Lawrence,
Division of Entomology, CSIRO

Minute Fungus Beetles

Figures 34.625–629

Relationships and Diagnosis: Corylophidae is a family whose relationships are somewhat obscured by the structural modifications accompanying small size. The group has been included in the "cerylonid" series (Crowson, 1955) and is usually placed near the Coccinellidae, but there are some larval features resembling Lathridiidae, and the lathridiid genus *Hyplathrinus* is actually a corylophid (Pakaluk, 1985). Larvae are highly variable, some being broadly ovate and strongly flattened like those of *Murmidius* (Cerylonidae), *Agaricophilus* (Endomychidae), and Discolomidae, and others more elongate and resembling some Lathridiidae, Endomychidae, and Coccinellidae. All corylophid larvae may be distinguished by the presence of paired dorsal or dorsolateral gland openings on abdominal segments 1–7, 1 and 8, or 2 and 8.

Biology and Ecology: Corylophids are primarily spore feeders in both larval and adult stages, and they are usually collected in leaf litter, refuse heaps, or on surfaces covered with sporulating fungi of various kinds. Species of *Sericoderus* and *Orthoperus* are often associated with molds. Hinton (1945b) reared *S. lateralis* (Gyllenhal) in the laboratory on spores of *Mucor mucedo* (Zygomycetes: Mucorales) and conidia of *Penicillium glaucum* (Ascomycetes: Eurotiales), and the same species has been associated with *Penicillium* occurring in vegetable detritus (Peyerimhoff, 1919, 1921). *Orthoperus scutellaris* LeConte has been collected on *Pencillium*, and Chandler (1983) found the species in decaying seaweed feeding on *Alternaria* and *Helminthosporium* conidial types; other *Orthoperus* have been found in lawn clippings and in sorghum heads. *Corylophodes marginicollis* LeConte was found breeding beneath the leaves of European horse chestnut on a powdery mildew, *Uncinula flexuosa* (Ascomycetes: Erysiphales) (Morrill, 1903). *Arthrolips obscurus* (Sahlberg) was found by Peyerimhoff (1919) feeding on spores falling out of the pore tubes of *Fomes fomentarius* (Basidiomycetes: Polyporaceae) and being caught on the viscous filaments produced by the mycetophilid fly, *Ceroplatus tipuloides* Bosc. *Arthrolips aequalis* Wollaston was reported feeding on the stromata of *Nummulariola bulliardii* (Ascomycetes: Xylariaceae), and various other corylophids have been observed on xylariaceous fruiting bodies. Paulian (1950) noted the occurrence of certain African species in the infloresences of *Lobelia* and *Senecio* and others associated with scale insects (*Chionaspis* and *Lecanium*). *Foadia maculata* Pakaluk has been collected under the bark of mangroves in the Florida Keys (Pakaluk, 1985). Other corylophids have been collected in rotten reeds (*Rypobius* and *Corylophus*), tussock grass (*Holopsis*), and tree ferns (*Holopsis*) (Endrödy-Younga, 1964; Horion, 1949). Pupae of Corylophidae are obtect and are attached to the substrate, enclosed basally by the larval exuvia, as in most Coccinellidae (Hinton, 1941b; Morrill, 1903).

Description: Mature larvae 1 to 3 mm. Body elongate, somewhat fusiform and slightly flattened, or broadly ovate and strongly flattened. Dorsal surfaces usually lightly pigmented, except for head and maculae on protergum and abdominal tergum 9, sometimes entirely pigmented or maculate, smooth to granulate, with vestiture of fine simple setae, usually mixed with various types of expanded setae or scales.

Head: Protracted and prognathous, but sometimes concealed from above by prothorax, elongate to strongly transverse, slightly to strongly flattened. Epicranial stem absent; frontal arms, if present, V-shaped and distant at base. Median endocarina absent. Paired, subparallel endocarinae sometimes present. Stemmata absent or 2 on each side. Antennae moderately to very long, 2- or 3-segmented, sometimes with very long 2nd segment. Frontoclypeal suture absent; labrum free or occasionally completely fused to head capsule. Mandibles symmetrical, usually broad at base and narrow at apex, bidentate or tridentate, without accessory ventral process and with tuberculate or asperate mola and prostheca absent or consisting of 2 hyaline processes; mandibles occasionally narrow and falcate, endognathous. Ventral mouthparts usually protracted. Maxilla with cardo, stipes, and articulating area usually indistinctly separated; mala obtuse or falciform; maxillary palp 2- or 3-segmented, with apical segment much longer than penultimate. Labium usually consisting of single plate; labial palps 2-segmented with apical segment much longer than basal, or occasionally 1-segmented. Hypopharyngeal area often with complex internal skeleton. Hypostomal rods usually absent, sometimes long and diverging. Ventral epicranial ridges absent. Gula sometimes longer than wide; gular sutures usually well-marked by parallel internal ridges.

Thorax and Abdomen: Legs well-developed, 5-segmented, moderately widely separated; tarsungulus with 1 seta, usually clavate. Thoracic and abdominal terga sometimes broadly expanded at edges forming contiguous flat plates, so that body has a disc-like form. Paired dorsal gland openings present on abdominal terga 1 and 8, or sometimes 2 and 8 or 1–7. Tergum A9 simple, without urogomphi, sometimes with pigmented plate or macula; sternum A9 short, simple. Segment A10 circular, ventrally oriented.

Spiracles: Annular, those on abdomen dorsally placed.

Comments: The family includes about 35 genera and 400 species worldwide, with 12 genera and about 60 species occurring in America north of Mexico. The subfamilial, tribal, and generic concepts are badly in need of revision. On the basis of larval characters, Böving and Craighead (1931) divided the group into Arthrolipinae (*Arthrolips* and *Orthoperus*) with gland openings on segments A1–7, and Corylophinae (*Corylophodes, Molamba, Sacium,* and *Sericoderus*) with openings on A1 and A8 only. This is in conflict with the adult classification proposed by Matthews (1899) and Casey (1900) and modified by Paulian (1950), where *Arthrolips* falls into the same group as *Sacium* and *Molamba*.

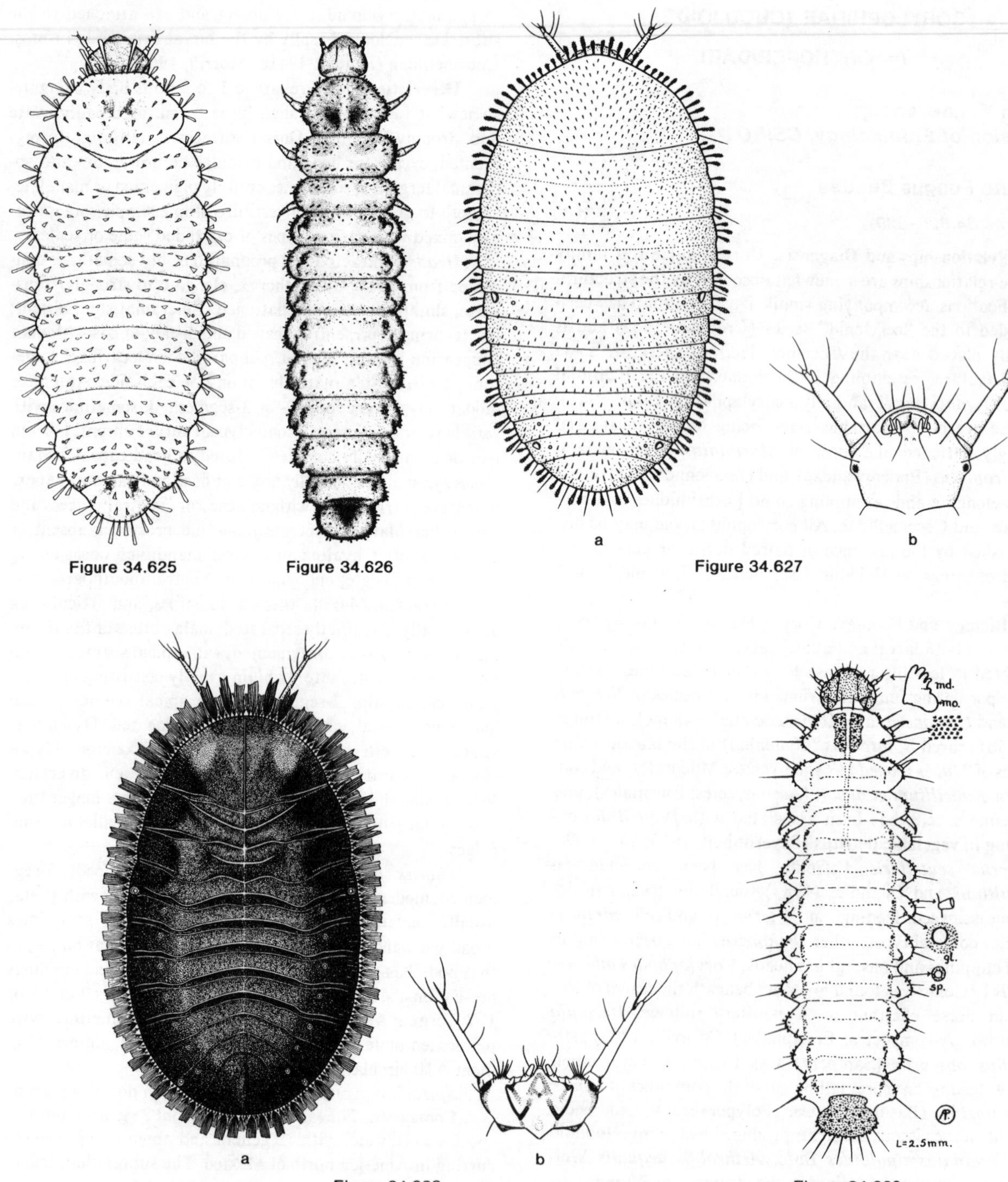

Figure 34.625. Corylophidae. *Sericoderus lateralis* (Gyllenhal). Bedford, Middlesex Co., Massachusetts. Larva, dorsal. Length = 1.7 mm.

Figure 34.626. Corylophidae. *Arthrolips* sp. Ysleta, Texas. Larva, dorsal. Length = 2.3 mm.

Figures 34.627a,b. Corylophidae. *Corylophodes* sp. Mt. Gingera, Australian Capital Territory. **a.** larva, dorsal, length = 1.2 mm; **b.** head, ventral.

Figures 34.628a,b. Corylophidae. Genus? Lamington National Park, Queensland. **a.** larva, dorsal, length = 1.9 mm; **b.** head, ventral.

Figure 34.629. Corylophidae. *Arthrolips* sp. Larva, dorsal (plus details). Length = 2.5 mm. (From Peterson, 1951)

It is possible, however, that Böving's illustrations of *Molamba lunata* LeConte and *Sacium* sp. were based on misidentifications, since there is no information available on how the determinations were made. Paulian recognized 4 subfamilies: Peltinoidinae (*Peltinoides* from North Africa), Corylophinae (including Peltinini, Orthoperini, Corylophodini, Corylophini, and Rypobiini), Sericoderinae and Saciinae (including *Arthrolips*); Casey's Aenigmaticini (Phanerocephalina of Matthews, 1899) was included in Orthoperini. A more thorough study of corylophid larvae and the discovery of the larval forms of critical genera, such as *Rypobius* and *Corylophus* would greatly help in understanding this difficult group.

Selected Bibliography

Böving and Craighead 1931 (larvae of *Arthrolips* sp., *Molamba lunata* (LeConte), *Sacium* sp., and *Corylophodes marginicollis*).
Casey 1900.
Chandler 1983 (larva of *Orthoperus scutellaris*).
Crowson 1955.
Endrödy-Younga 1964.
Gressitt and Samuelson 1964a (larvae of *Holopsis oblongus* Endrödy-Younga).
Hinton 1941b (larva of *Sericoderus lateralis*).
Horion 1949.
Matthews 1899.
Morrill 1903 (larva of *Corylophodes marginicollis*).
Pakaluk 1985.
Paulian 1950.
Peterson 1951 (larva misidentified as *Murmidius ovalis* (Beck)).
Peyerimhoff 1919 (habits of *S. lateralis*), 1921 (larva of *S. lateralis*).

LATHRIDIIDAE (CUCUJOIDEA)

John F. Lawrence,
Division of Entomology, CSIRO

Minute Brown Scavenger Beetles

Figures 34.630-631

Relationships and Diagnosis: The Lathridiidae, as defined by Belon (1909) and Arnett (1968), contains several genera which belong to other families. These include *Dasycerus* (considered to be a staphylinoid family Dasyceridae by Crowson, 1955), *Merophysia, Holoparamecus,* and their allies (placed in Merophysiidae by Crowson, 1955, but here considered as part of the family Endomychidae), and *Hyplathrinus,* which is closely related to the corylophid genus *Aenigmaticum* on the basis of both larval and adult characters. Lathridiid larvae are usually distinguished by their small size, 4 stemmata or fewer on each side of the head, truncate or rounded maxillary mala, tuberculate mandibular mola with a broad, hyaline prostheca, reduced mandibular apex which is partly or entirely unsclerotized, moderately to strongly protracted mouthparts, with hypostomal rods more laterally placed than is usual, single tarsungular seta, annular spiracles, and lack of urogomphi or sclerotized abdominal tergites.

Biology and Ecology: Lathridiids are often found in leaf litter, decaying vegetation, animal nests, or human habitations, or on the surfaces of leaves, bark, or wood, where they feed on the spores of a variety of fungi. Spores of Myxomycetes are fed upon by *Revelieria californica* Fall and by several species of *Enicmus* in N. America, Europe, Australia, and New Zealand (Andrews, 1976a; Crowson and Hunter, 1964; Dajoz, 1960; Lawrence and Newton, 1980; Russell, 1979). Most lathridiids feed on spores or conidia of Zygomycetes (Mucorales) and Ascomycetes (Eurotiales, Erysiphales, Pyrenomycetes, and their imperfect stages). Species of *Metophthalmus, Lithostygnus, Adistemia, Cartodere, Dienerella,* and *Corticaria* have been reared on various "molds", including *Mucor, Aspergillus, Penicillium,* and *Botrytis* (Andrews, 1976b; Hammad, 1953; Hinton, 1941c, 1945b; Kerr and McLean, 1956; Klippel, 1952). *Enicmus maculatus* LeConte was reared on *Nodulisporium* type conidia produced by a pyrenomycete fungus (*Hypoxylon*) (Lawrence, 1977a), while 2 European *Enicmus* were recorded from *Nummulariola bulliardii* (Dajoz, 1966). Chandler (1983) collected large numbers of *Corticaria valida* Fall in rotting seaweed and found conidia of the *Helminthosporium* and *Alternaria* types in their guts. Several *Corticaria* and *Melanophthalma* may be collected by sweeping leaves and other vegetation, where they are probably feeding on mildews. Gordon (1938) recorded *Dienerella filum* (Aube) from herbarium specimens of smuts (*Ustilago* and *Tilletia*) and puffballs (*Lycoperdon*), while a Californian species of *Corticaria* is known to breed in the spore masses produced by a polypore (*Cryptoporus volvatus*). Many of the mold-feeding lathridiids are commonly found in cellars, warehouses, and granaries, either on moldy walls or in stored grains or other foods; these stored products species include *Aridius nodifer* (Westwood), *Lathridius minutus* (Linnaeus), *Cartodere constricta* (Gyllenhal), *Dienerella argus* (Reitter), *Corticaria pubescens* (Gyllenhal) and *Migneauxia orientalis* Reitter (Aitken, 1975; Hinton, 1941c, 1945b; Klippel, 1952). One species of lathridiid, *Eufallia seminivea* Motschulsky, has been reported biting humans in Florida; these bites apparently produced red, itching lesions (Parsons, 1969a).

Description: Mature larvae 1 to 3 mm. Body elongate and fusiform or elongate-oval and slightly flattened. Dorsal surfaces usually very lightly pigmented, sometimes with darker head and protergal macula, smooth, with vestiture of scattered long setae, occasionally mixed with expanded or frayed hairs.

Head: Protracted and prognathous, usually more heavily sclerotized dorsally than ventrally, slightly flattened. Epicranial stem usually short to moderately long; frontal arms usually V-shaped; epicranial suture sometimes absent. Median endocarina absent. Stemmata 1 to 5 on each side (usually 3 or 4), sometimes absent. Antenna well-developed, often long, 3-segmented, usually with 2nd segment much longer than 1st or 3rd. Frontoclypeal suture absent; labrum free. Mandibles symmetrical, usually with 1 to 4 teeth at apex, sometimes with apical portion rounded and hyaline, without accessory ventral process, and sometimes with 2 long setae arising from outer edge; mola well-developed and tuberculate or asperate;

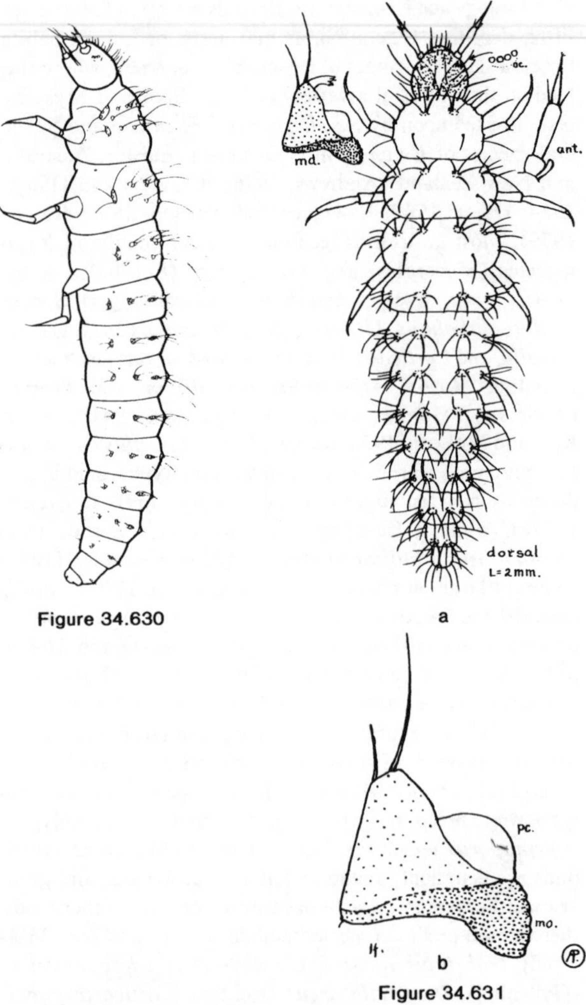

Figure 34.630

a

b

Figure 34.631

Figure 34.630. Lathridiidae. *Lathridius minutus* (L.). Lake of the Clouds, Mt. Washington, New Hampshire. Larva, lateral. Length = 3.3 mm.

Figures 34.631a,b. Lathridiidae. *Cartodere* sp. Poland. **a.** larva, lateral, length = 2 mm; **b.** left mandible, dorsal. (From Peterson, 1951)

prostheca consisting of a broad, rounded or acute, fixed hyaline process. Ventral mouthparts retracted or strongly protracted. Maxilla with cardo present or absent, stipes elongate or broader than long, articulating area present or absent, palp 3-segmented, and mala usually obtuse, sometimes falciform, and occasionally cleft at apex. Labium usually with mentum and submentum fused, sometimes consisting of a single plate only; ligula present or absent; labial palps 1- or 2-segmented and widely separated. Hypopharyngeal sclerome a transverse bar. Hypostomal rods usually long and diverging, sometimes absent. Ventral epicranial ridges absent. Gula sometimes elongate.

Thorax and Abdomen: Legs well-developed, 5-segmented; tarsungulus with 1 seta. Tergum A9 without urogomphi; sternum A9 well-developed. Segment A10 circular and posteriorly or terminally oriented.

Spiracles: Annular, not raised on tubes.

Comments: The family includes about 25 genera and 500 species worldwide, and appears to be most diverse in the northern and southern temperate regions; in America north of Mexico, there are 15 genera and about 120 species. There are 2 well-defined subfamilies—Lathridiinae and Corticariinae, but the generic concepts have been relatively unstable, and several well-known genera and species have undergone recent name changes (Walkley, 1952; Pope, 1977). Widespread human introduction into new territory has added to the taxonomic confusion in the group.

Selected Bibliography

Aitken 1975.
Andrews 1976a (association with Myxomycetes), 1976b.
Arnett 1968.
Belon 1909 (genera).
Böving and Craighead 1931 (larvae of *Corticaria dentigera* LeConte, *Dienerella costulata* (Reitter), *Eufallia seminiveus* Motschulsky, and *Melanophthalma chamaeropis* Fall).
Chandler 1983 (larva of *Corticaria valida* LeConte).
Crowson 1955.
Crowson and Hunter 1964.
Dajoz 1960 (association with Myxomycetes), 1966 (association with Ascomycetes).
van Emden 1942b (larval mandible of *Aridius nodifer*).
Gordon 1938 (feeding habits of *Dienerella filum*).
Hammad 1953 (larva of *Lithostygnus serripennis* Broun).
Hinton 1941c (stored products species; larval key; larvae and pupae of *Aridius nodifer, Lathridius minutus, Adistemia watsoni* (Wollaston), *Dienerella filum, D. filiformis* (Gyllenhal), and *Corticaria fulva* (Comolli)), 1945b (stored products species; larvae as in 1941c).
Kerr and McLean 1956 (biology, control).
Klippel 1952 (association with molds; larvae of *Aridius nodifer* and *Corticaria fulva*).
Lawrence 1977a (association with Ascomycetes).
Lawrence and Newton 1980 (association with Myxomycetes).
Lefkovitch 1960 (*Adistemia watsoni* in bird nest).
Parsons 1969a (*Eufallia seminivea* biting humans).
Peterson 1951 (larvae of *Dienerella?* sp., as *Cartodere* sp.).
Pope 1977.
Russell 1979 (association with Myxomycetes).
Saalas 1923 (biology of various species; larvae).
Verhoeff 1923 (larva of *Lathridius minutus*).
Walkley 1952 (generic concepts).

MYCETOPHAGIDAE (TENEBRIONOIDEA)

John F. Lawrence,
Division of Entomology, CSIRO

Hairy Fungus Beetles

Figures 34.632-633

Relationships and Diagnosis: The Mycetophagidae is one of the more primitive tenebrionoid families whose affinities appear to be with the Tetratomidae and Archeocrypticidae (Crowson, 1964c, 1966b; Lawrence and Newton, 1982; Miyatake, 1960; Watt, 1974b). Larvae of *Penthe* and the Tetratominae differ from those of mycetophagids in lacking a mandibular mola (replaced by membranous lobe or group of hyaline processes), while larvae of the pisenine tetratomids

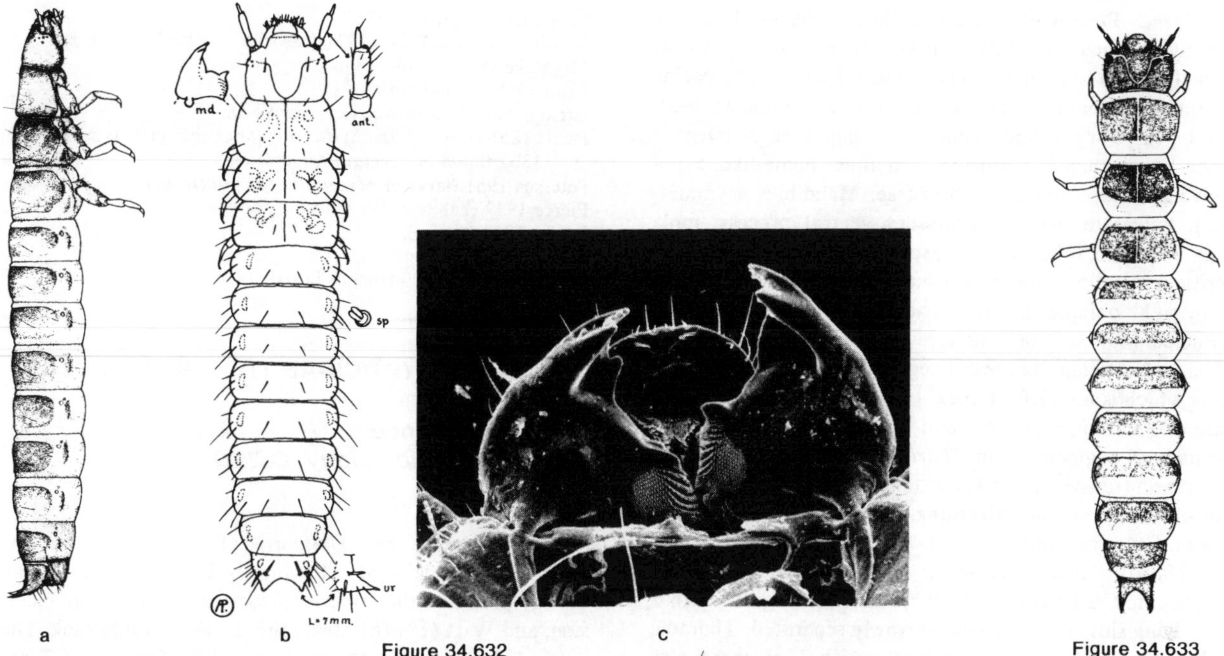

Figure 34.632

Figure 34.633

Figures 34.632a–c. Mycetophagidae. *Mycetophagus punctatus* Say. Tyringham, Massachusetts. **a.** larva, lateral, length = 8.6 mm; **b.** larva, dorsal (plus details); **c.** anterior part of head capsule, with ventral mouthparts removed, showing asymmetrical mandibles. (Figure 34.632b from Peterson, 1951)

Figure 34.633. Mycetophagidae. *Litargus sexpunctatus* (Say). Savannah, Georgia. Larva, dorsal. Length = 4.7 mm.

have accessory processes on the urogomphi, a more ventrally-oriented, pygopod-like anal region, and longer accessory spiracular openings. The transverse and posteriorly oriented anal region of mycetophagids is also found in Archeocrypticidae and most members of the anthicid-scraptiid complex; the former group differs in having a distinct frontoclypeal suture and characteristic urogomphi, which are straight, posteriorly oriented, and lightly pigmented, while the latter usually have reduced mandibular molae, a dome-like antennal sensorium, or a median endocarina.

Biology and Ecology: Almost all mycetophagids are mycophagous, feeding on the softer and often rotting tissue of various hymenomycete fruiting bodies or the spores and sterile tissue of molds and xylariaceous Ascomycetes. Species of *Triphyllus, Mycetophagus, Pseudotriphyllus,* and *Litargus* has been recorded from a variety of Hymenomycetes, including *Pleurotus ostreatus* (Tricholomataceae), *Fistulina hepatica* (Fistulinaceae), *Climacodon septentrionale* (Hydnaceae), *Polyporus squamosus, Piptoporus betulinus, Laetiporus sulphureus, Bjerkandera adusta, Daedaleopsis confragosa,* and *Fomes fomentarius* (Polyporaceae) (Crowson, 1960a; Donisthorpe, 1935; Matthewman and Pielou, 1971; Palm, 1959; Weiss and West, 1920). The European *Mycetophagus atomarius* Fabricius, at least 1 *Esarcus,* and a number of species of *Litargus* appear to prefer the fruiting bodies of Xylariaceae, such as *Hypoxylon, Daldinia,* or *Trichoderma* (conidial stage) (Crowson, 1960a; Dajoz, 1964; Hingley, 1971; Lawrence, 1977a; Palm, 1959). A number of mycetophagids are associated with molds, and may occur in hay stacks, decaying vegetation, or moldy fruits, grains, and cereal products in granaries, warehouses, or domestic premises. The most commonly encountered stored products species are *Litargus balteatus* LeConte, *Mycetophagus quadriguttatus* Müller, and *Typhaea stercorea* (Linnaeus) (Aitken, 1975; Crowson, 1960a; Hinton, 1945b).

Members of the genus *Berginus* have different habits from most of the family. *Berginus tamarisci* Wollaston has been collected in flowers of tamarisk (Tamaricaceae), is known to breed in the male cones of *Pinus maritimus,* and has been found feeding on pollen and stamens of pharmaceutical cactus flowers (*Cereus grandifloreus*) imported into Germany (Hinton, 1945b; Horion, 1961; Perris, 1862). *Berginus californicus* Pierce has been recorded from flowers of a dodder (*Cuscuta*) in California (Pierce, 1939), while *B. maindroni* Grouvelle is known to be a predator of the lac scale (*Tachardia*) in India (Imms and Chatterjee, 1915). Another unusual habit is found in a Chilean species presently placed in the genus *Mycetophagus* (*M. chilensis* Philippi); specimens from Juan Fernandez Island were collected on ferns, and fern spores were found in the gut.

Description: Mature larvae 1.5 to 8 mm. Body elongate, more or less parallel-sided, straight, slightly flattened. Head and all visible terga usually pigmented, brown or yellow in color; pleural areas sometimes with distinct sclerotized plates as well; surfaces smooth; vestiture consisting of long and short, fine, simple hairs, sometimes densely distributed.

Head: Protracted and prognathous, moderately broad, slightly flattened. Epicranial stem very short or absent; frontal arms lyriform and contiguous at base. Median endocarina absent. Stemmata 4 or 5 on each side. Antennae well-developed, 3-segmented, sometimes long, with segment 2 much longer than 1; sensorium sometimes dome-like. Frontoclypeal suture absent; labrum free. Mandibles asymmetrical, bidentate, without accessory ventral process; mola well-developed, tuberculate or asperate, with tubercles or asperities extending onto ventral surface, left mola almost vertical, right oblique; prostheca absent. Ventral mouthparts retracted. Maxilla with transverse cardo, elongate stipes, well-developed articulating area, 3-segmented palp, and rounded mala which is not cleft at apex. Labium more or less free to base of mentum; ligula present; labial palps usually 2-segmented (1-segmented in *Thrimolus*). Hypopharyngeal sclerome a transverse bar. Hypostomal rods moderately long, subparallel or slightly diverging. Ventral epicranial ridges absent. Gula transverse.

Thorax and Abdomen: Legs moderately long, 5-segmented, usually somewhat spinose; tarsungulus with 2 setae lying side by side; coxae narrowly separated. Thoracic and abdominal segments sometimes with 1 or more pigmented pleurites. Tergum A9 usually with pair of well-developed, posteriorly oriented, slightly upturned, simple urogomphi (absent in *Thrimolus*); sternum A9 simple. Segment A10 transverse, posteriorly oriented.

Spiracles: Annular, annular-uniforous, or annular-biforous, with short accessory tubes.

Comments: The family includes 8 genera and about 200 species worldwide, with 5 genera and 26 species occurring in America north of Mexico (Parsons, 1975). The group may be divided into three subfamilies: Bergininae (*Berginus* with species in Europe, North Africa, Madagascar, India, Ceylon, North and Central America and the West Indies); Esarcinae (*Esarcus* with species from southern Europe and North Africa); and Mycetophaginae, including the remaining genera. The family occurs in most parts of the world, but the greatest diversity appears to be in the temperate parts of the Northern and Southern Hemispheres.

Selected Bibliography

Aitken 1975.
Böving and Craighead 1931 (larvae of *Litargus balteatus, L. connexus* (Fourcroy), *L. sexpunctatus* (Say), *Mycetophagus obsoletus* (Melsheimer), *M. punctatus* Say, *Thrimolus duryi* Casey, and *Typhaea stercorea*).
Byzova 1958 (larva of *Mycetophagus?* sp., as *Diaperis boleti* (Linnaeus)).
Crowson 1960a (habits, fungus hosts), 1964c (relationships), 1966b (relationships).
Dajoz 1964 (*Esarcus*).
Donisthorpe 1935 (fungus hosts).
Hayashi 1971 (larvae of *Mycetophagus* spp., *Parabaptistes* sp., *Triphyllioides seriatus* (Reitter), and *Typhaea stercorea*).
Hingley 1971 (association with Ascomycetes).
Hinton 1941d (larva and pupa of *Mycetophagus quadripustulatus* (Linnaeus)), 1945 (stored products species; larva of *T. stercorea*).
Horion 1961.
Imms and Chatterjee 1915 (biology of *Berginus maindroni*, predator of lac-scale).
Lawrence 1977a (association with Ascomycetes).
Lawrence and Newton 1982.
Matthewman and Pielou 1971 (association with Polyporaceae).
Miyatake 1960 (relationships).
Palm 1959 (fungus hosts).
Parsons 1975 (North American revision).
Perris 1851 (larva of *Triphyllus bicolor* (Fabricius), as *punctatus*), 1862 (larva of *Berginus tamarisci*).
Peterson 1951 (larva of *Mycetophagus punctatus*).
Pierce 1939 (habits of *Berginus californicus*).
Saalas 1923 (larvae).
Watt 1974b (relationships).
Weiss and West 1920 (fungus hosts).

ARCHEOCRYPTICIDAE (TENEBRIONOIDEA)

John F. Lawrence,
Division of Entomology, CSIRO

Figures 34.634a–d

Relationships and Diagnosis: The Archeocrypticidae were orignally included in the family Tenebrionidae. Kaszab (1964) proposed the tribe Archeocrypticini for their inclusion, and Watt (1974b) raised the taxon to family rank. The group belongs among the more primitive families of Tenebrionoidea and appears to be most closely related to Mycetophagidae and Pterogeniidae (Lawrence, 1977c; Lawrence and Newton, 1982). Archeocrypticid larvae resemble those of Mycetophagidae in general form, structure of the head and mouthparts, and form of the anal region, which is transverse and more or less posteriorly oriented, but the urogomphi are unique in being long, subparallel, lightly pigmented, and posteriorly oriented, and the frontoclypeal suture is distinct.

Biology and Ecology: Archeocrypticids are most commonly collected in leaf litter and other decaying plant material, but some of them feed in the softer fruiting bodies of certain Basidiomycetes. *Enneboeus caseyi* Kaszab has been collected in leaf litter, old pine cones, and rotting flowers. Several Australian species in the genera *Enneboeus* and *Enneboeopsis* have been found breeding in the fruiting bodies of *Piptoporus portentosus* and *Grifola berkeleyi;* the larvae are active like those of Mycetophagidae and are not adapted for boring into harder types of fungi.

Description: Mature larvae 2 to 6 mm. Body elongate, more or less parallel-sided, straight, slightly flattened. Dorsal surfaces lightly pigmented or with weakly developed, yellow tergal plates, smooth, with vestiture of longer and shorter, simple hairs, the latter sometimes densely distributed.

Head: Protracted and prognathous, moderately broad, slightly flattened. Epicranial stem short; frontal arms lyriform. Median endocarina absent. Stemmata 5 on each side. Antennae moderately long, 3-segmented, with short, conical sensorium. Frontoclypeal suture present; labrum free. Mandibles asymmetrical, bidentate or tridentate, without accessory ventral process; mola well-developed, tuberculate, with tubercles forming transverse ridges; left mola more vertical, produced into a tooth apically; prostheca absent. Ventral mouthparts retracted. Maxilla with transverse cardo, elongate stipes, well-developed articulating area, 3-segmented palp, and truncate mala which is not cleft. Labium free to base of mentum; ligula absent; labial palps 2-segmented and

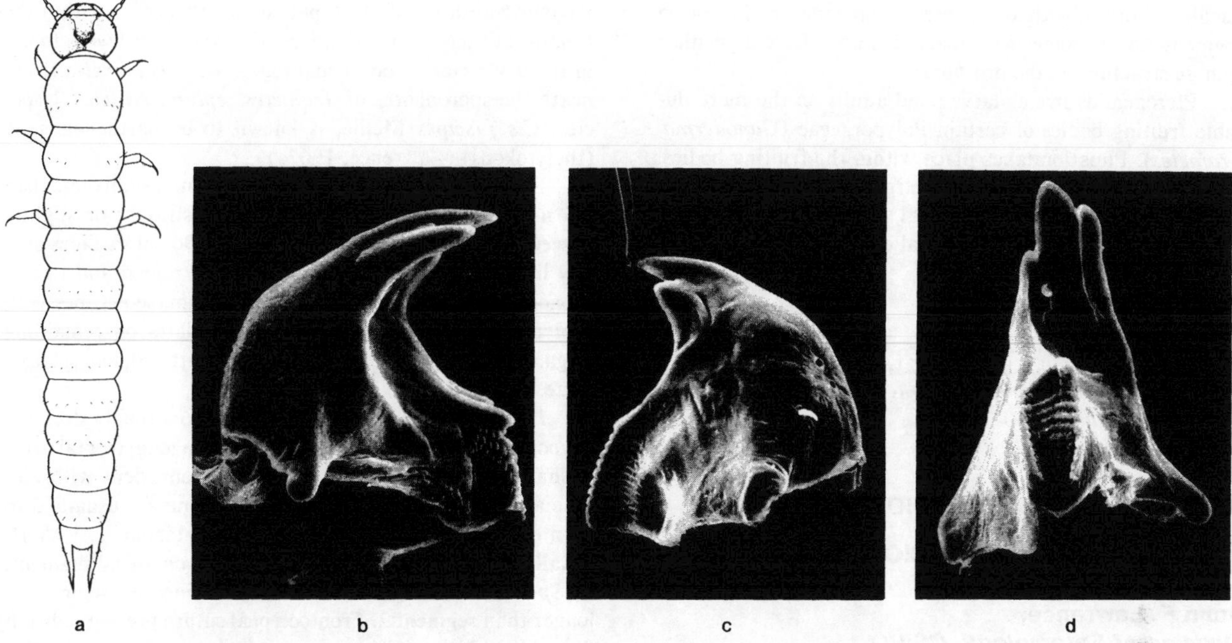

Figures 34.634a-d. Archeocrypticidae. *Enneboeus caseyi* Kaszab. Barro Colorado Is., Canal Zone, Panama. **a.** larva, dorsal, length = 3.6 mm; **b.** right mandible, ventral; **c.** right mandible, dorsal; **d.** left mandible, mesal.

narrowly separated. Hypopharyngeal sclerome consisting of a tooth-like structure. Hypostomal rods very short and subparallel or slightly diverging. Ventral epicranial ridges absent. Gula transverse.

Thorax and Abdomen: Legs well-developed, 5-segmented; tarsungulus with 2 setae lying one distal to the other; coxae moderately narrowly separated. Tergum A9 with pair of posteriorly projecting, narrow, straight, lightly pigmented urogomphi which are widely separated, subparallel and acute at apex; sternum A9 simple. Segment A10 transverse, posteriorly oriented.

Spiracles: Annular-biforous.

Comments: The family contains 8 genera and 35 species worldwide with the single species *Enneboeus caseyi* Kaszab (= *Uloporus ovalis* Casey, not *Enneboeus ovalis* Waterhouse) occurring in America north of Mexico. The genus *Enneboeus,* as defined by Kaszab (1981), extends from the southern United States to Chile, and is also found in Australia. Other genera include the Neotropical *Pseudenneboeus, Archeocrypticus* from southern S. America and New Zealand, *Sivacrypticus* with a number of species in Africa and the Indomalayan Region, and the Australian genera *Enneboeopsis, Parenneboeus, Neboissianus,* and *Wattianus.*

Selected Bibliography

Kaszab 1964, 1969a, 1969b, 1979, 1981.
Lawrence 1977c (relationships), 1982c (larval characters).
Lawrence and Newton 1982 (relationships).
Triplehorn and Wheeler 1979 (relationships).
Watt 1974b (larval characters).

PTEROGENIIDAE (TENEBRIONOIDEA)

John F. Lawrence,
Division of Entomology, CSIRO

This family includes 2 genera: *Pterogenius,* with 2 described species from Sri Lanka, and *Histanocerus,* with 6 described species and several undescribed, extending from Southeast Asia and the East Indies to New Guinea and the Solomon Islands. The group is considered to be among the more primitive heteromerous families, with relationships to Mycetophagidae, Tetratomidae, Ciidae, and Archeocrypticidae.

Larvae are elongate and subcylindrical, lightly sclerotized, except for the head and sometimes the last 2 terga, with smooth surfaces and a vestiture of long, simple setae. The head is subquadrate, as seen from above, with a long epicranial stem, bent to the left, and lyriform frontal arms, joined anteriorly by a transverse ecdysial line. There are 4 or 5 stemmata on each side, and the antennae are short and 3-segmented, with the sensorium as long as or longer than segment 3. The mandibles are highly asymmetrical, with large, transversely ridged molae and no prostheca. The ventral mouthparts are retracted, with a truncate mala, 3-segmented maxillary palps, 2-segmented labial palps, and a well-developed ligula. Hypostomal rods are short and divergent and ventral epicranial ridges are absent. The legs are close together and the tarsungulus is bisetose. Tergum A9 bears a pair of strongly upturned urogomphi, which may be simple or bifurcate; sternum A9 is simple. Segment A10 is transversely oval and posteroventrally oriented. The spiracles are annular-biforous, with long accessory openings, and those on segment A8 are somewhat enlarged. Pupae have a peculiar spinose structure on the prothorax.

Pterogeniids live as larvae and adults on the more durable fruiting bodies of certain Polyporaceae (*Ganoderma, Trametes*). Pupation takes place within the fruiting bodies. Adults are dimorphic, and the enlarged and broadened head (*Pterogenius*) or enlarged antennal scape (*Histanocerus*) in the male is probably used in sexual combat.

Selected Bibliography

Crowson 1955.

Lawrence 1977c (relationships; habits; larvae of *Pterogenius nietneri* Candeze and *Histanocerus pubescens* Motschulsky; pupa of *H. pubescens*).

CIIDAE (TENEBRIONOIDEA)

(= CISIDAE, CIOIDAE)

John F. Lawrence,
Division of Entomology, CSIRO

Minute Tree Fungus Beetles

Figures 34.635-639

Relationships and Diagnosis: The family Ciidae was traditionally placed in the Bostrichoidea or Cleroidea, based primarily on larval characters (presence of lacinia, reduction of maxillary articulating area, lack of true mandibular mola), but later authors have treated the group as either a member of the Cucujoidea (Crowson, 1955) or a primitive family of Tenebrionoidea (Crowson, 1960b; Lawrence, 1971, 1974b). The acute hyaline process at the base of the larval mandible and the inner (dorsal) lobe of the maxilla are difficult to interpret; the former may represent the cucujoid prostheca (unknown in other tenebrionoids) or a remnant of the mola (as in some Tetratomidae and Melandryidae), while the latter may be the true lacinia or a secondary development. When both adult and larval characters are taken into consideration, the most logical placement of the family is among the primitive Tenebrionoidea, near Tetratomidae and Mycetophagidae. Ciid larvae may be distinguished by the subcylindrical form and more or less hypognathous head, the lack of a true mola (several have anobiid-like pseudomolae), the long epicranial stem and V-shaped frontal arms, the dorsally situated laciniar lobe, and the reduced antennae which lack a third segment (except in *Sphindocis*, where segment 3 is shorter than the sensorium).

Biology and Ecology: Ciidae appear to feed almost exclusively on the fruiting bodies or vegetative hyphae of Basidiomycetes. Some ciids, such as species of *Orthocis*, may be associated with the primitive Auriculariales or found in rotten branches or vines, but the majority of species occur in the more durable fruiting bodies of the Polyporaceae and Hymenochaetaceae. Most species exhibit a certain degree of host preference, as has been shown by Lawrence (1973) and Paviour-Smith (1960a). Pupation usually occurs within the fungus, but pupae of *Sphindocis denticollis* Fall were found in the dry rotten wood of madrone (*Arbutus menziesii*) beneath the sporophores of *Trametes sepium*. At least 1 species, *Cis fuscipes* Mellié, is known to be parthenogenetic (thelytokous) (Lawrence, 1967a).

Description: Mature larvae 1 to 7 mm, usually less than 3.5 mm. Body elongate, parallel-sided, straight or slightly curved ventrally, usually subcylindrical. Dorsal surfaces usually lightly pigmented except for buccal region and tips of urogomphi or entire 9th abdominal tergum, sometimes with protergal plate and rarely with weak plates on remaining terga. Surfaces smooth and vestiture of scattered, long, simple setae.

Head: Protracted and moderately to strongly declined (hypognathous), globular. Epicranial stem long; frontal arms V-shaped. Median endocarina usually coincident with epicranial stem (absent in *Sphindocis*). Stemmata usually 3 to 5 on each side, sometimes 2, 1, or 0. Antennae very short, usually 2-segmented with long sensorium on apical segment; in *Sphindocis* 3-segmented with sensorium on segment 2 longer than segment 3. Frontoclypeal suture present; labrum free. Mandibles usually asymmetrical, broad and stout, bidentate, without accessory ventral process; mola usually absent, often replaced by narrow, acute, fixed, hyaline process on one or both mandibles; occasionally with well-developed pseudomola bearing transverse ridges. Ventral mouthparts retracted. Maxilla with transverse cardo, elongate stipes, 3-segmented palp, and reduced articulating area (well-developed in *Sphindocis*); galea rounded or truncate; lacinia truncate, reduced, and subapical, not visible in ventral view, so that maxilla appears to have single mala. Labium free to base of mentum or partly connate with maxillae; ligula short, sometimes absent; labial palps 2-segmented, narrowly separated. Hypopharyngeal sclerome absent. Hypostomal rods absent. Ventral epicranial ridges present. Gula transverse.

Thorax and Abdomen: Legs relatively short and broad; tarsungulus with 2 setae lying side by side; coxae moderately close together. Tergum A9 usually with pair of small, upturned urogomphi, occasionally with single median process, more than 2 processes, or concave disc (*Sphindocis* and some *Cis*); sternum A9 usually simple, rarely (*Sphindocis*) with apical row of fine asperities. Segment A10 transversely oval, posteroventrally oriented.

Spiracles: Usually annular, rarely annular-uniforous or (*Sphindocis*) annular-biforous with long accessory tubes.

Comments: The family is presently divided into 2 subfamilies: Sphindociinae, with the single species *Sphindocis denticollis* from the northern coastal region of California, and the Ciinae, including about 40 genera and 550 species worldwide and 13 genera and 85 species in America north of Mexico (Lawrence, 1971, 1974b, 1982b). The Holarctic fauna is fairly well understood, but the group requires a world revision and many genera and species remain to be described from the Equatorial Region and Southern Hemisphere.

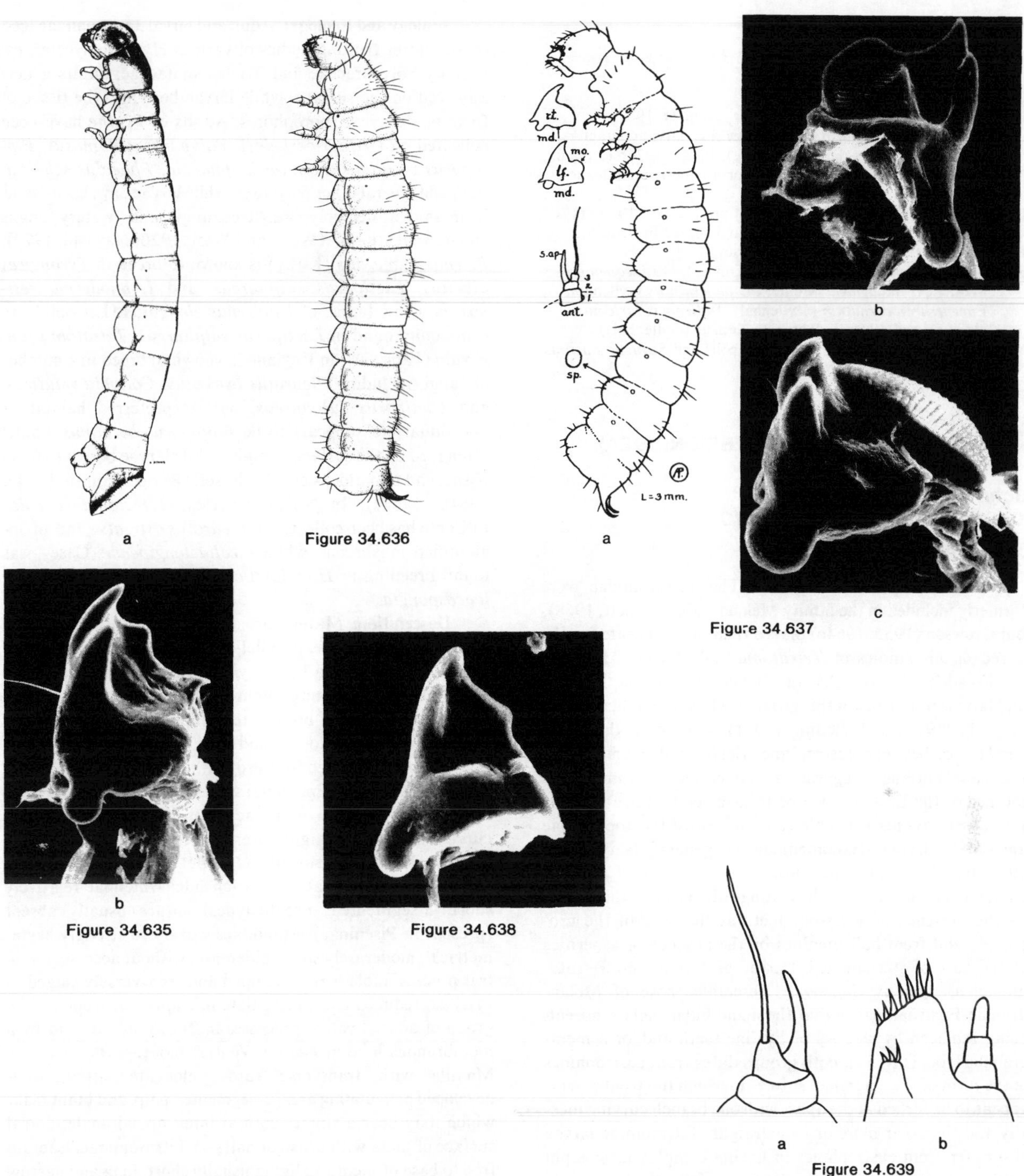

Figures 34.635a,b. Ciidae. *Sphindocis denticollis* Fall. 4 mi. W. Forest Glen, Trinity Co., California. **a.** larva, lateral view, length = 7.6 mm; **b.** right mandible, ventral. Alpine Lake, Marin Co., California.

Figure 34.636. Ciidae. *Cis vitulus* Mannerheim. Big Sur, Monterey Co., California. Larva, lateral. Length = 3.9 mm.

Figures 34.637a-c. Ciidae. *Cis levettei* (Casey). **a.** larva, lateral (plus details). Length = 3 mm; **b.** left mandible, ventral; **c.** right mandible, ventral. Great Smoky Mts. National Park, Tennessee. (Figure 34.637a from Peterson, 1951)

Figure 34.638. Ciidae. *Malacocis brevicollis* (Casey). 3 mi. E Bloomingdale, New York. Right mandible, mesoventral.

Figures 34.639a,b. Ciidae. Ciinae. **a.** antenna; **b.** maxillary apex, dorsal.

Selected Bibliography

Böving and Craighead 1931 (larvae of *Cis fuscipes* and *Ceracis* sp., as *Ennearthron*).
Crowson 1955, 1960b.
Lawrence 1967a (parthenogenesis in *Cis fuscipes*), 1967b (*Ceracis*), 1971 (North America revision; larval characters), 1973 (fungus hosts; host preference), 1974a, 1974b (larva of *Sphindocis denticollis*), 1982b (North Amerian catalogue), 1982c (larval characters).
Paviour-Smith 1960a (fungus hosts: host preference), 1960b (*Cis bilamellatus* Fowler, introduction and spread in Britain).
Peterson 1951 (larva of *Cis levettei* (Casey)).
Saalas 1923 (larvae of *Cis bidentatus* (Olivier), *Cis jacquemarti* Mellié, *Cis quadridens* Mellié, *Cis punctulatus* Gyllenhal, *Ennearthron cornutum* (Gyllenhal). *Dolichocis laricinus* (Mellié), and *Rhopalodontus perforatus* (Gyllenhal)).
Verhoeff 1923 (larvae of *Cis boleti* (Scopoli) and *Sulucacis affinis* (Gyllenhal)).

TETRATOMIDAE (TENEBRIONOIDEA)

John F. Lawrence,
Division of Entomology, CSIRO

Figures 34.640–642

Relationships and Diagnosis: The Tetratomidae were formerly included in the family Melandryidae (Arnett, 1968), but Crowson (1955, 1964c) placed them in a separate family, based on his studies of *Tetratoma* and *Penthe*. Miyatake (1960) added the tribe Pisenini and commented on the adult and larval similarities of this group to Mycetophagidae, while Hayashi (1975) and Viedma (1971) both noted the larval similarities between eustrophine Melandryidae and Tetratominae/Penthinae, suggesting that either the families be merged or the Eustrophinae be transferred to Tetratomidae. Larvae of *Pisenus* resemble those of both Mycetophagidae and Melandryidae-Hallomeninae in general body form, structure of the epicranial suture, and presence of a tuberculate mola, but they may be distinguished from either group by the presence of accessory teeth at the base of the urogomphi and from hallomenines by the absence of asperities at the base of sternum A9. Larvae of *Penthe* and Tetratominae, as well as *Eupisenus*, resemble those of Melandryidae-Eustrophinae in that the mandibular mola is absent, being replaced by a series of hyaline teeth and/or a membranous lobe. However, both groups differ from eustrophines in the form of the epicranial suture, in which the frontal arms appear to be forked at the apex, with one branch curving mesally and the other more or less straight. Tetratomine larvae also differ from eustrophines in having complex urogomphi (with accessory teeth), while the larva of *Penthe* has simple urogomphi but differs in lacking hypostomal rods and having a long, narrow ligula. The family Ciidae is also closely related to the Tetratomidae, but may be distinguished by the combination of more or less cylindrical body, globular head, long epicranial stem with V-shaped frontal arms, absence of a mola (sometimes replaced by a single hyaline process), and presence of a distinct laciniar lobe on the maxilla.

Biology and Ecology: Adult and larval Tetratomidae feed on the softer fruiting bodies of various Hymenomycetes, especially Polyporaceae and Tricholomataceae; adults generally feed on the surface, while larvae bore into the tissue of fresh or decaying sporophores. Adults of *Penthe* have been collected on *Grifola berkeleyi, Polyporus squamosus, Piptoporus betulinus, Fomitopsis pinicola, Phaeolus schweinitzii,* and several other polypores, while larvae have been found in an unidentified polypore, "tree fungi," and "watery fungus on chestnut stump" (Weiss and West, 1920; Hayashi, 1972). *Pisenus humeralis* (Kirby) is known to breed in *Tyromyces albellus, Laetiporus sulphureus,* and *Ischnoderma resinosum,* while larvae of *Eupisenus elongatus* (LeConte) are commonly found in *Laetiporus sulphureus. Tetratoma fungorum* (Fabricius) in England is known to breed in a number of fungi, including *Pleurotus ostreatus, Collybia velutipes,* and *Laetiporus sulphureus,* but its preferred habitat or "headquarters" appears to be *Piptoporus betulinus;* adults often feed on the surface at night, while larvae bore into fresh tissue, and pupation occurs in the soil (Paviour-Smith, 1964a, 1964b, 1964c). In North America, *Tetratoma concolor* LeConte has been collected in *Pleurotus ostreatus* and an unidentified mushroom, while *Incolia longipennis* Casey was found breeding in *Hapalopilus alboluteus* and *Tyromyces leucospongia.*

Description: Mature larvae 3 to 17 mm, usually 12 mm or less. Body elongate, parallel-sided or fusiform, straight or slightly curved ventrally, subcylindrical or slightly flattened. Head and all visible terga lightly sclerotized or with weakly developed, yellowish-brown, tergal plates; dorsal surfaces smooth; vestiture of longer and shorter, simple setae.

Head: Protracted and prognathous, moderately broad, somewhat flattened. Epicranial stem short to moderately long; frontal arms lyriform or characteristically forked, each with curved inner and straight outer branch. Median endocarina usually absent, occasionally (*Penthe*) coincident with epicranial stem. Stemmata 5 on each side. Antennae relatively short, 3-segmented. Frontoclypeal suture usually absent (present in Piseninae). Mandibles weakly to strongly asymmetrical, moderately stout, bidentate, without accessory ventral process; mola well-developed and transversely ridged in *Pisenus,* reduced and tuberculate in *Eupisenus,* replaced by group of acute hyaline processes in Tetratominae, and by a membranous lobe in *Penthe.* Ventral mouthparts retracted. Maxilla with transverse cardo, elongate stipes, well-developed articulating area, 3-segmented palp, and blunt mala which may bear a single tooth at inner apical angle; dorsal surface of mala with brush of hairs on Tetratominae. Labium free to base of mentum; ligula usually short, long and narrow in *Penthe;* labial palps 2-segmented, usually narrowly separated. Hypopharyngeal sclerome usually absent or represented by transverse bar, tooth-like in *Pisenus.* Hypostomal rods usually short to moderately long and diverging, absent in *Penthe.* Ventral epicranial ridges absent. Gula transverse to almost as long as wide.

Thorax and Abdomen: Legs well developed, 5-segmented; tarsungulus with 2 setae, one distal to the other. Meso- and metatergum and abdominal terga 1–8 in *Penthe* with

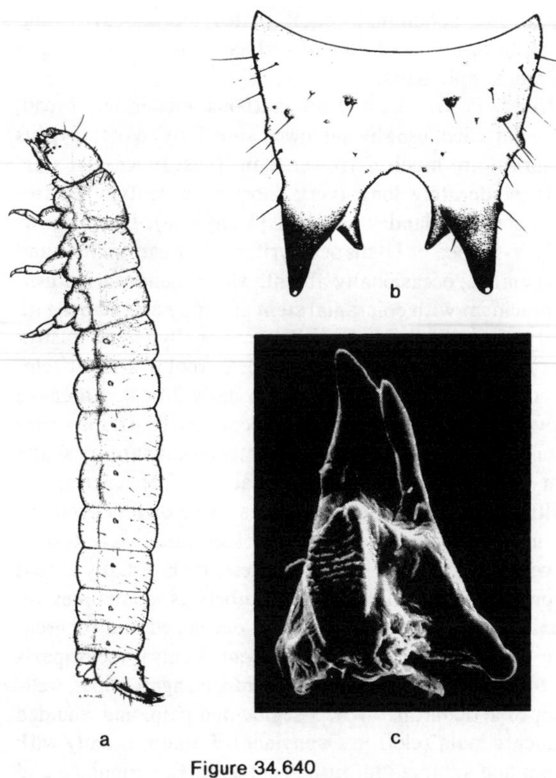

Figure 34.640

Figures 34.640a-c. Tetratomidae. *Pisenus humeralis* (Kirby). Florida Caverns State Park, Jackson Co., Florida. **a.** larva, lateral, length = 6 mm; **b.** abdominal apex, anterodorsal; **c.** left mandible, mesal.

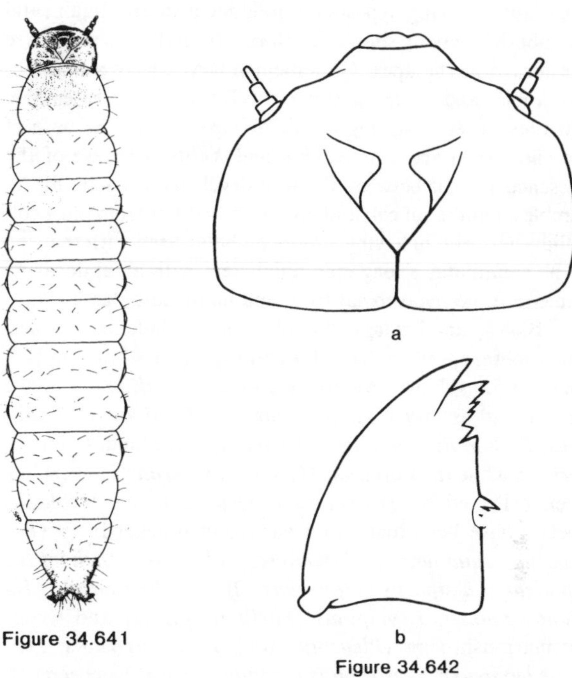

Figure 34.641

Figure 34.642

Figure 34.641. Tetratomidae. *Incolia longipennis* Casey. Paintbrush Canyon, Grand Teton Mts., Wyoming. Larva, dorsal. Length = 5 mm.

Figures 34.642a,b. Tetratomidae. *Tetratoma* sp. **a.** head, dorsal; **b.** right mandible, ventral.

patches or rows of asperities, in addition to maculations. Tergum A9 with pair of urogomphi, which are usually bifid, or with an accessory process at the base of each (simple in *Penthe*); sternum A9 simple. Segment A10 transversely oval, posteroventrally oriented.

Spiracles: Annular-biforous, with moderately long accessory tubes.

Comments: The family occurs throughout the Holarctic Region, with species of *Penthe* extending into India and the East Indies, and includes 6 genera and about 30 species. There are three recognized subfamilies: Piseninae (*Pisenus* and *Eupisenus*), Tetratominae (*Tetratoma, Abstrulia,* and *Incolia*), and Penthinae (*Penthe*). *Eupisenus* is only tentatively included in the first group, since both larvae and adults exhibit features more or less intermediate between Piseninae and Tetratominae.

Selected Bibliography

Arnett 1968.

Böving and Craighead 1931 (larva of *Penthe pimelia* (Fabricius)).

Crowson 1955, 1964c (larvae of *Tetratoma fungorum* and *T. desmaresti* Latreille).

Hayashi 1972 (larvae of *Pisenus rufitarsis* (Reitter), *P. insignis* (Reitter), and *Penthe japana* Marseul), 1975 (relationships).

Miyatake 1960 (relationships).

Paviour-Smith 1964a (biology of *Tetratoma fungorum;* habitat), 1964b (larva of *Tetratoma fungorum*), 1964c (biology of *T. fungorum*).

Viedma 1971 (relationships).

MELANDRYIDAE (TENEBRIONOIDEA) (= SERROPALPIDAE)

John F. Lawrence,
Division of Entomology, CSIRO

Figures 34.643-649

Relationships and Diagnosis: The family Melandryidae appears to be closely related to the Tetratomidae and the mordellid-rhipiphorid lineage, but the adult similarities between melandryids and anaspidine Scraptiidae (Crowson, 1966b) are probably a result of convergence, since anaspidine larvae are of an entirely different type. Larvae of the melandryid subfamilies Melandryinae and Osphyinae differ from those of Tetratomidae in lacking a mola or any other mesal structure at the base of the mandible, and usually by having a Y-shaped epicranial suture and an endocarina which extends beneath each frontal arm. Eustrophine larvae have a membranous lobe and/or a series of hyaline teeth at the mandibular base as in larvae of Tetratominae and Penthinae, but they differ from the former in having simple urogomphi, from the latter in having hypostomal rods and a shorter ligula, and from both in having simple (lyriform) frontal arms which are not forked at the apex. Hallomenine larvae have a mandibular mola, and resemble larvae of Tetratomidae-Piseninae, but they differ in having a pair of asperities at the base of sternum A9. Larvae of *Rushia* and *Xylita*, because of the presence of a pit between the well-developed urogomphi, resemble a number of colydiid larvae; they differ from most colydiids, however, in having a more globular head with reduced or no stemmata, a long epicranial stem with an endocarina beneath it, no hypostomal rods, and no mandibular mola.

Biology and Ecology: Members of the Hallomeninae and Eustrophinae feed on the softer fruiting bodies of various Hymenomycetes. North American species of *Hallomenus* have been found as larvae in sporophores of *Laetiporus sulphureus, Dichomitus squalens, Osteina abducta*, and *Trametes serialis*, while the European *Mycetoma suturale* (Panzer) has been collected on *Ischnoderma resinosum*. *Eustrophinus* species have been found in a variety of bracket fungi (*Polyporus squamosus, Tyromyces albellus, Dichomitus squalens, Laetiporus sulphureus, Bjerkandera adusta, Inonotus munzii, I. vulpinus, Phellinus gilvus*) and wood-rotting mushrooms (*Pleurotus sapidus, Panus rudis, Lentinus lepideus*); *Synstrophus repandus* (Horn) has been collected on *Pleurotus ostreatus, Phaeolus schweinitzii*, and *Coriolus versicolor*, and *Eustrophus tomentosus* Say has been found in *Polyporus squamosus*. Larvae of Osphyinae and most Melandryinae occur in dead wood of varying consistency (relatively sound to soft and spongy), but often that which has been infested with a white rot fungus, such as *Coriolus versicolor, Hirschioporus abietinus*, or *Phellinus gilvus*. When they are associated with sporophores, the larvae feed in the rotting wood beneath; some species appear to prefer conifers or hardwoods, but this may be an artifact of collecting. Members of the tribe Orchesiini exhibit a wide variety of feeding habits, and some have been found in rotten wood, moss, and sooty molds, but a number *Orchesia* feed as larvae and pupate within relatively hard and durable polypore fruiting bodies. In eastern North America 3 species of *Orchesia* exhibit a definite preference for the type of sporophore in which they will breed: *O. castanea* (Melsheimer) prefers the thin, leathery, and light colored brackets of species of *Coriolus* and *Lenzites* (*sensu stricto*) (Polyporaceae); *O. gracilis* (Melsheimer) breeds in the hard, yellowish brown conks of the hymenochaetaceous fungus *Phellinus gilvus;* and *O. cultriformis* Laliberte occurs in the softer, dark brown fruiting bodies of *Inonotus* species (Hymenochaetaceae) (Dorn, 1936; Hayashi, 1975; Hoebeke and McCabe, 1977; Horion, 1956; Iablokoff, 1944; Viedma, 1966; Weiss, 1919; Weiss and West, 1920).

Description: Mature larvae 2.5 to 30 mm. Body elongate, more or less parallel-sided or slightly wider in middle, straight or slightly curved ventrally, subcylindrical to slightly flattened; usually very lightly sclerotized, except for buccal region and urogomphi, if present; occasionally with head and thoracic and abdominal tergal plates more heavily pigmented, brown in color; dorsal surfaces smooth; vestiture of scattered, simple hairs.

Head: Protracted and prognathous, moderately broad, slightly flattened, usually narrowed anteriorly so that clypeus and labrum are much narrower than frons. Epicranial stem usually moderately long (very short or absent in Eustrophinae, some Melandryinae, and Osphyinae); frontal arms usually V-shaped or U-shaped, lyriform in Eustrophinae and Hallomeninae, occasionally absent. Median endocarina usually coincident with epicranial stem and may extend beneath frontal arms as well (Y-shaped); occasionally (some Eustrophinae) extending anteriorly between frontal arms. Stemmata usually 5 on each side, occasionally 2 or 0. Antennae often very short, 3-segmented, or occasionally (Osphyinae) with only 2 apparent segments. Frontoclypeal suture usually absent (present in Eustrophinae); labrum free, emarginate apically in Eustrophinae. Mandibles more or less symmetrical, usually stout, unidentate or bidentate, rarely tridentate, without accessory ventral process; mola usually absent or represented by a few teeth or tubercles, sometimes replaced by a membranous lobe; well-developed and tuberculate in Hallomeninae; prostheca absent. Ventral mouthparts retracted. Maxilla with transverse cardo, elongate stipes, well-developed articulating area, 3-segmented palp, and rounded or truncate mala (cleft in Osphyinae). Labium usually with mentum and submentum fused (not so in Eustrophinae and Hallomeninae), usually free to base of mentum or postmentum; ligula usually present, often longer than labial palps, sclerotized in Osphyinae, absent or very short in Eustrophinae; labial palps 2-segmented and usually narrowly separated. Hypopharyngeal sclerome usually absent, represented by transverse bar in Eustrophinae, tooth-like in Hallomeninae. Hypostomal rods usually short to moderately long and diverging, 2 pairs present in Eustrophinae, absent in some Melandryinae (*Xylita, Rushia*). Ventral epicranial ridges absent. Gula usually transverse, longer than wide in Osphyinae.

Thorax and Abdomen: Meso- and metatergum and anterior abdominal terga and sterna sometimes with distinct ampullae. Legs well developed or short and broad, 5-segmented; tarsungulus with 1 or 2 setae. Tergum A9 usually without or with reduced urogomphi, sometimes with well-developed urogomphi (Eustrophinae, Hallomeninae, *Xylita, Rushia*), or with median process (Osphyinae); a pit present between urogomphi in *Xylita* and *Rushia*. Sternum A9 usually simple, with a single asperity on each side at base in Hallomeninae. Segment A10 transverse to almost circular, posteroventrally oriented.

Spiracles: Usually annular-biforous with very short accessory tubes, occasionally annular, annular-uniforous, or annular-biforous with long accessory tubes.

Comments: The family contains about 80 genera and 450 species worldwide, with 28 genera and about 70 species occurring in America north of Mexico. The subfamily Hallomeninae includes only the genera *Hallomenus* from N. America, Eurasia, and Africa, and *Mycetoma* from Europe. Eustrophinae includes several genera from N. America, tropical America, Eurasia, Africa, and Madagascar. Osphyinae

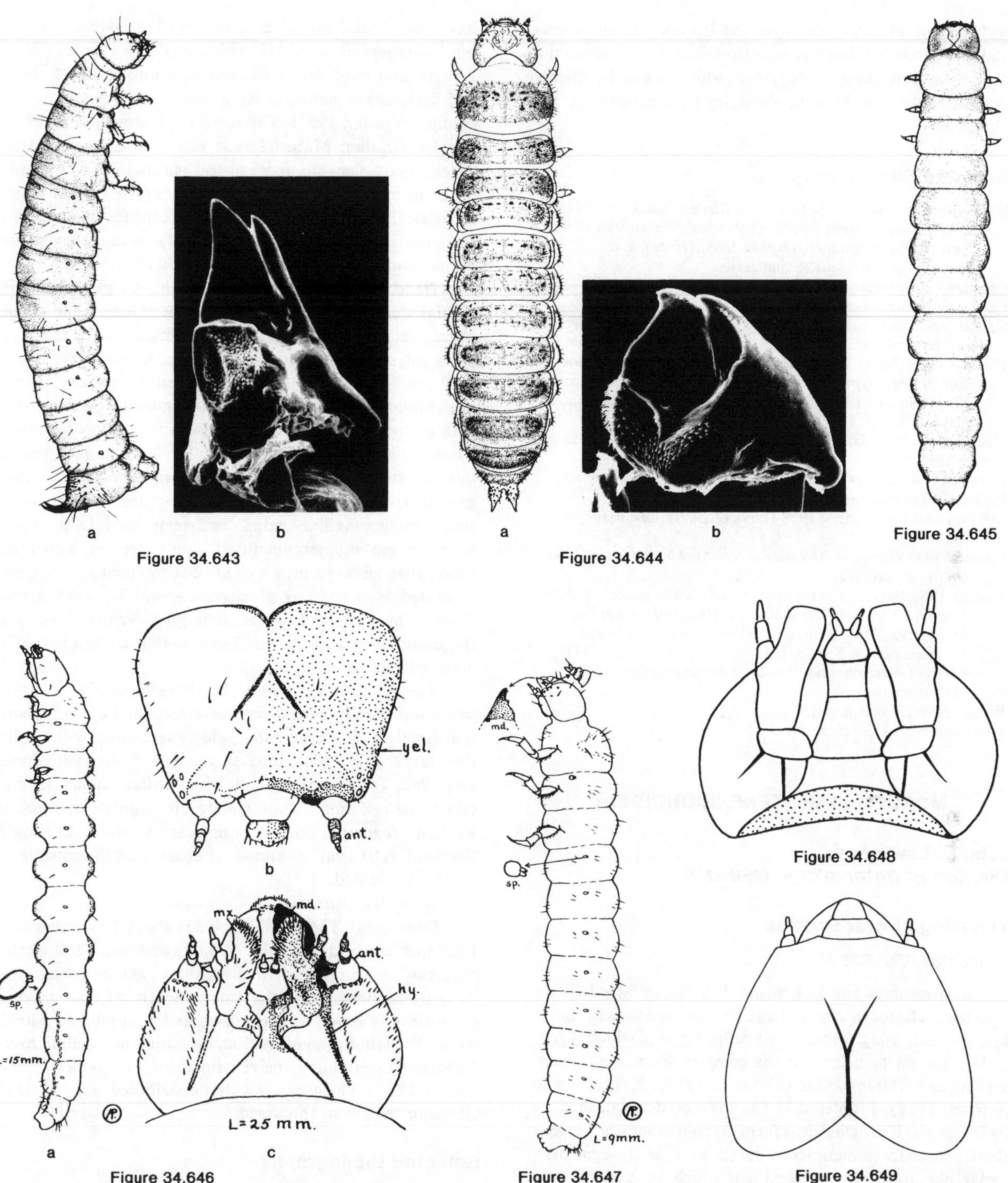

Figures 34.643a,b. Melandryidae. *Hallomenus* sp. Blue Hills, Norfolk Co., Massachusetts. **a.** larva, lateral, length = 6.9 mm; **b.** right mandible, mesal.

Figures 34.644a,b. Melandryidae. *Eustrophinus bicolor* (Fabricius). 2 mi. S Newbury, Essex Co., Massachusetts. **a.** larva, dorsal, length = 9.6 mm; **b.** left mandible, ventral.

Figure 34.645. Melandryidae. *Zilora occidentalis* Mank. Mt. Lemmon, Santa Catalina Mts., Arizona. Larva, dorsal. Length = 15 mm.

Figures 34.646a-c. Melandryidae. *Melandrya striata* Say. **a.** larva, lateral view, length = 15 mm; **b.** head, dorsal; **c.** head, ventral. (From Peterson, 1951)

Figure 34.647. Melandryidae. *Orchesia castanea* Melsheimer. Larva, lateral (plus details). Length = 9 mm. (From Peterson, 1951)

Figure 34.648. Melandryidae. (Eustrophinae). Head, ventral.

Figure 34.649. Melandryidae. (Melandryinae). Head, dorsal.

includes the Holarctic *Osphya*, the European *Conopalpus*, and 2 apparently related genera from Brazil. The remaining genera fall into the Melandryinae, which is usually divided into the tribes Orchesiini, Melandryini, Serropalpini, and Hypulini.

Selected Bibliography

Böving and Craighead 1931 (larvae of *Dircaea quadrimaculata* (Say), *Eustrophinus bicolor* (Fabricius), *Melandrya striata* Say, *Orchesia castanea*, *Osphya lutea* (Horn), and *Serropalpus barbatus* (Schaller)).

Crowson 1966b (relationships).

Dorn 1936 (fungus hosts).

Gressitt and Samuelson 1964b (larva of *Orchesia rennelli* Gressitt and Samuelson).

Hayashi 1975 (larvae of 14 species included in genera *Abdera*, *Euryzilora*, *Hypulus*, *Melandrya*, *Orchesia*, *Phloeotrya*, *Phryganophilus*, and *Serropalpus*), 1980 (same comments as Hayashi 1975).

Hoebeke and McCabe 1977 (larva of *Serropalpus coxalis* Mank).

Horion 1956.

Hudson 1934 (larva of *Mecorchesia brevicornis* Broun).

Iablokoff 1944 (fungus hosts).

Mamaev and Kompantzev 1978 (larva of *Euryzilora ussurica* Mamaev).

Peterson 1951 (larvae of *Melandrya striata*, *Melandrya* sp., and *Orchesia castanea*).

Viedma 1966 (larvae of many species included in genera *Abdera*, *Carida*, *Clinocara*, *Conopalpus*, *Hallomenus*, *Hypulus*, *Melandrya*, *Mycetoma*, *Osphya*, *Orchesia*, *Phloeotrya*, *Phyrganophilus*, *Serropalpus*, *Xylita*, and *Zilora*), 1971 (larva of *Eustrophinus bicolor* and relationships to Tetratomidae).

Weiss 1919b (fungus hosts).

Weiss and West 1920 (fungus hosts).

MORDELLIDAE (TENEBRIONOIDEA)

John F. Lawrence,
Division of Entomology, CSIRO

Tumbling Flower Beetles

Figures 34.650–653

Relationships and Diagnosis: The family Mordellidae, with the exclusion of Anaspidinae (transferred to Scraptiidae by Crowson, 1955), forms a well-defined group, which is generally thought to belong to the complex including Melandryidae and Rhipiphoridae (Crowson, 1966b; Lawrence and Newton, 1982). Mordellid larvae may be distinguished by their more or less cylindrical form, hypognathous head, very short antennae (sometimes reduced to 1 or 2 segments), wedge-like mandibles, reduced legs which have only 3 or 4 indistinct segments, and usually the presence of a median, spine-like process on tergum A9.

Biology and Ecology: Adult mordellids are commonly found on flowers, where they apparently feed on nectar and pollen; the common name, tumbling flower beetles, is based on the habit of rapidly moving the hindlegs back and forth, when disturbed, causing the wedge-shaped body to tumble off its resting surface. Mordellid larvae are highly adapted for boring into relatively compact and often woody substrates. Many species feed in rotten wood, but some *Mordella* have been found breeding in the fruiting bodies of *Gloeophyllum saepiarium* and *G. trabeum* (Polyporaceae) (Weiss, 1920a), and most *Mordellistena* bore into stems of shrubs and herbaceous plants, causing damage to some cultivated species (Aguilar, 1962b; Crowson, 1955; Franciscolo, 1974).

Description: Mature larvae 3 to 18 mm, usualy 10 mm or less. Body elongate, more or less parallel-sided or slightly wider in middle, straight or slightly curved ventrally, subcylindrical. Very lightly sclerotized except for buccal region and sometimes apex of tergum A9; surfaces smooth; vestiture of fine, simple hairs, sometimes densely distributed.

Head: Protracted and prognathous, globular. Epicranial stem long; frontal arms absent. Median endocarina coincident with epicranial stem. Stemmata usually indistinct or absent, sometimes 1 to 3 on each side. Antennae very short, 1-, 2-, or 3-segmented. Frontoclypeal suture present; labrum free. Mandibles symmetrical, robust, more or less wedge-like with undivided apex and no accessory ventral process; mola absent; mesal surface of mandibular base simple. Ventral mouthparts retracted. Maxilla with transverse cardo, elongate stipes, well-developed articulating area, and rounded, simple mala; maxillary palp 2- or 3-segmented. Labium with mentum and submentum fused; ligula present, sometimes longer than labial palps, which are 2-segmented and narrowly separated. Hypopharyngeal sclerome absent. Hypostomal rods short to moderately long and diverging. Ventral epicranial ridges absent. Gular region absent so that labium is contiguous with thorax.

Thorax and Abdomen: Prothorax usually enlarged, sometimes with ridge or asperities on tergum. Legs very short, indistinctly 3- or 4-segmented, widely separated, without distinct tarsungulus. Abdominal segments 1–6 often with dorsal ampullae. Tergum A9 usually with median, spine-like process or pair of small, approximate urogomphi (occasionally without urogomphi or median process); sternum A9 simple. Segment A10 oval or almost circular, posteroventrally or ventrally oriented.

Spiracles: Annular.

Comments: The family includes about 100 genera and 1200 species worldwide, with 6 genera and about 200 species occurring in America north of Mexico. There are 2 subfamilies: Ctenidiinae, with the single South African species, *Ctenidia mordelloides* Castelnau, and Mordellinae with 5 tribes (Stenaliini, Reynoldiellini, Conaliini, Mordellini, Mordelistenini) containing the remainder of the species (Franciscolo, 1957). The group is widely distributed, and occurs in all major regions of the world.

Selected Bibliography

d'Aguilar 1962c (agricultural pests).

Böving and Craighead 1931 (larvae of *Tomoxia bidentata* (Say) and *Mordellistena* sp.).

Crowson 1955, 1966b.

Ermisch 1956.

Franciscolo 1952, 1957, 1974 (larva of *Mordellistena ghanii* Franciscolo).

Hayashi 1980 (larvae of *Falsomordellistena katoi* Nomura, *Glipa fasciata* Kono, *Glipostena pelecotomidea* (Pic), *Hoshihananomia perlata* (Sulzer), and *Mordella truncatoptera* (Nomura)).

Lawrence and Newton 1982.

Peterson 1951 (larvae of *Mordellistena* sp. and *Tomoxia* sp.).

Weiss 1920a (biology of *Mordella marginata* Melsheimer; fungus host).

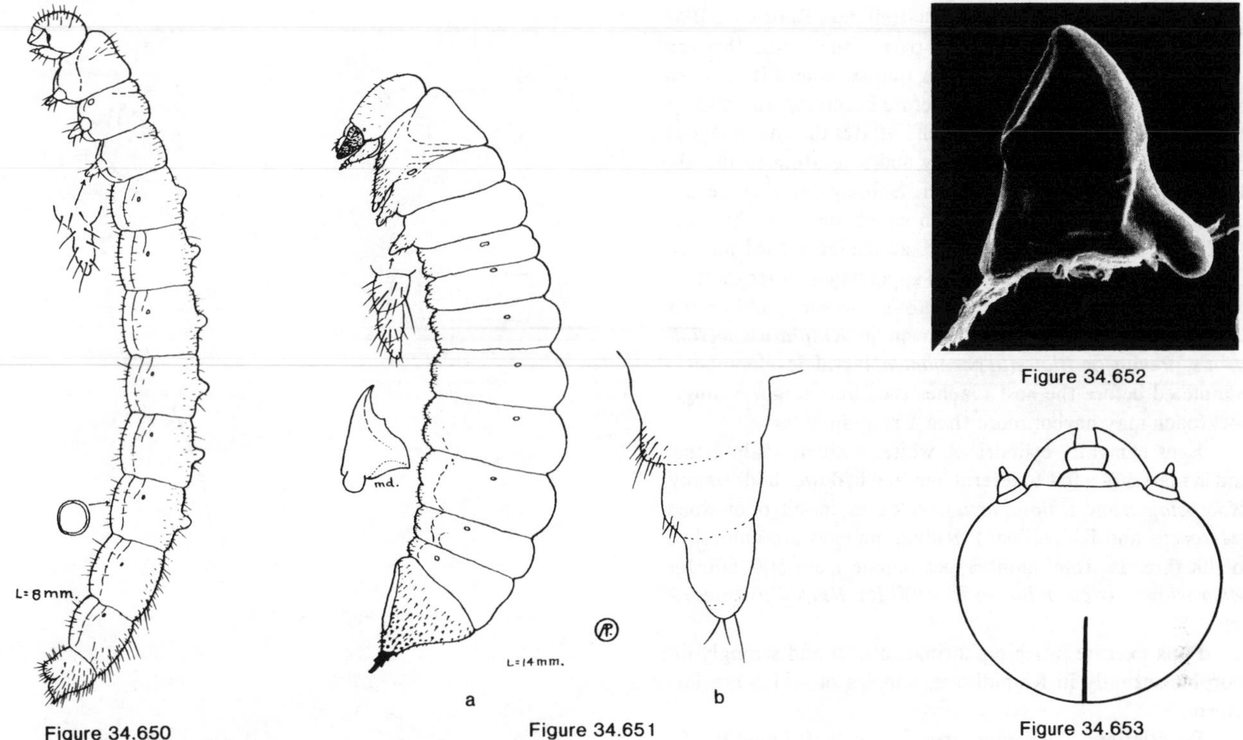

Figure 34.650. Mordellidae. *Mordellistena* sp. Larva, lateral (plus details). Length = 8 mm. (From Peterson, 1951)

Figures 34.651a,b. Mordellidae. *Tomoxia* sp. **a.** larva, lateral (plus details), length = 14 mm; **b.** metathoracic leg. (From Peterson, 1951)

Figure 34.652. Mordellidae. *Mordella* sp. Red Mountain, Colorado. Left mandible, ventral.

Figure 34.653. Mordellidae. Head, dorsal.

RHIPIPHORIDAE (TENEBRIONOIDEA)

Richard B. Selander, ***University of Illinois***

Rhipiphorid Beetles

Figures 34.654–657

Relationships and Diagnosis: Rhipiphorids are hypermetamorphic parasitoids of immature Hymenoptera and Blattodea. [Reports of attacks on scale insects (Peyerimhoff, 1942; Tucker, 1952) are of questionable validity.] Development involves both external and internal feeding stages. Because it is entomophagous and has a triungulin-type first instar larva, the family has been associated with the Meloidae and Stylopoidea (Strepsiptera) (Böving and Craighead, 1931; Arnett, 1968). However, evidence from adult anatomy indicates that the three groups are convergent (Crowson, 1955; Selander, 1957).

Rhipiphorid and meloid triungulins differ from those of the order Strepsiptera by having distinct antennae, mandibles, maxillary palpi, and trochanters and by the fact that the caudal setae arise from abdominal segment 9 (not 10). Triungulins of Rhipiphoridae and Meloidae are distinguished in the section on Meloidae. In later instars Rhipiphorinae become grotesque grubs feeding externally on wasp and bee larvae, whereas Rhipidiinae, feeding internally in cockroach larvae (nymphs), are degenerate, apodous, sac-like larvae initially, but later develop distinct appendages. There is no larval phase resembling the coarctate phase of Meloidae. Unlike Strepsiptera, both sexes of Rhipiphoridae abandon the host to pupate and are free-living as adults.

Biology and Ecology: No significant bionomic information is available for Nephritinae, Ptilophorinae, or Micholaeminae. In Pelecotominae the larva of *Rhipistena* has been described as an eruciform, apparently free-living predator of cerambycid larvae (*Prionoplus*), quite unlike other rhipiphorids (Hudson, 1934). In Rhipiphorinae the triungulin of *Macrosiagon* and *Metoecus* attaches to an adult wasp (Eumenidae, Scoliidae, Tiphiidae, Vespidae) and that of *Rhipiphorus* to an adult bee (Apidae, Halictidae) by grasping a body hair with the mandibles and is thereby carried to a provisioned nesting cell. After waiting, if necessary, for the host larva to hatch, the triungulin burrows into its thorax and eventually (sometimes after overwintering) becomes enormously distended (endophagous stage). When the host larva approaches maturity, the rhipiphorid makes an opening in the mesothorax and, after ecdysis, emerges to feed through 5 short instars by wrapping itself around the venter of the host's thorax (ectophagous stage). After devouring the host, the larva pupates in the cell. In Rhipidiinae (*Rhipidius, Rhipidioides,* and relatives) (fig. 34.656) the triungulin attaches directly to a young cockroach (Blattellidae, Blattidae, Blaberidae) and inserts the head and part of the thorax through a membrane on the venter of the thorax, where it remains 2–3 weeks (ectophagous stage) before becoming an apodous 2nd instar larva (2nd phase) which initiates the endophagous stage by entering the host's body and migrating to the abdomen, where it may overwinter. Subsequently, there are several instars of rapid growth in which the larva has recognizable, although unsegmented, appendages (3rd phase). Ultimately, it develops segmented appendages, emerges from between the last tergites of the host's abdomen, and crawls away to pupate (4th phase). Except in *Rhipidius pectinicornis,* preying on *Blattella germanica,* larval development is completed before the host reaches the adult stage. A single cockroach may harbor more than 1 rhipidiine larva.

Eggs elongate, cylindrical, white, without sculpturing, laid in masses of a few to several hundred in flower buds (many *Macrosiagon* and *Rhipiphorus*), on leaves, in soil, or on wood (*Metoecus* and Rhipidiinae). Rhipidiine eggs are enmeshed in silk threads. Total number per female from 500–600 for *Macrosiagon tricuspidatum* to 2200 for *Rhipidius quadriceps.*

Pupa exarate in Rhipiphorinae, obtect and strongly dimorphic sexually in Rhipidiinae, females of which are larviform.

Description: Triungulin larva (1st instar): Length: .45–.95 mm. Body heavily sclerotized; form navicular before feeding, crescentic, with intersegmental membranes enormously enlarged when replete; spiracles on mesothorax and abdominal segments 1–8 (Macrosiagonini), 7–8 (Rhipidiinae), or 8 (Rhipiphorini); definite pattern of setae or setae and spinules (Rhipidiinae) on body (chaetotaxy sparser than in Meloidae); lacking line of dehiscence on body.

Head: Prognathous, rounded in front, with or without median dorsal cleft or anterior notch, with basal transverse ridge. Epicranial suture absent. Four (Rhipidiinae) or 5 stemmata on each side of head, behind antennal foramina, which are well behind middle of head. Antennae 2- (Rhipidiinae) or 3-segmented; segment 2 bifid in Rhipidiinae, with spiniform apical sensory appendix in Rhipiphorinae; last segment with long terminal seta. Labrum fused with frontoclypeus. Mandibles working more or less vertically, crossed in repose, falciform distally, smooth, with large basal lobe. Maxillae with stipes large and elongate or (Rhipidiinae) vestigial; cardo fused with stipes; palps 2- or 3-segmented, conspicuous; last segment with long terminal seta in Rhipidiinae. Labium much reduced; palps a pair of papillae.

Thorax and Abdomen: Thoracic nota well developed. Legs slender, elongate, 5-segmented; tibiae very long in Rhipidiinae; tarsungulus with 0–2 small claws and conspicuous pulvillus. Abdomen with 10 distinct segments; 10th tubular (may be subdivided in Rhipidiinae). Abdominal tergites and sternites well developed, with single transverse row of setae (or setae/spinules); pleurites distinct or not. Urogomphi absent; segment 9 generally with pair of enlarged caudal setae (shorter pair sometimes present on 8).

Later Instars: Rhipiphorinae (2nd–6th instars): Ectophagous. Body crescentic, lightly sclerotized, sparsely, minutely setate; spiracles on mesothorax and abdominal segments 1–7.

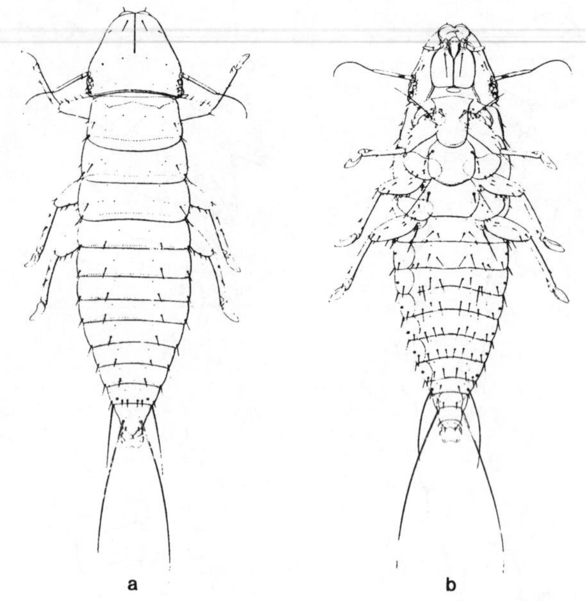

Figure 34.654

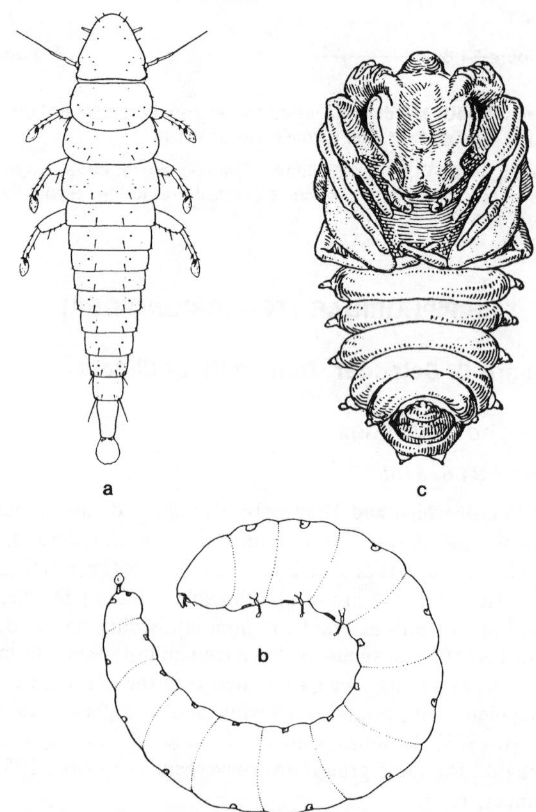

Figure 34.655

Figures 34.654a,b. Rhipiphoridae. *Macrosiagon ferrugineum flabellatum* Fabricius. **a.** triungulin, dorsal; **b.** ventral (after Grandi, 1936).

Figures 34.655a-c. Rhipiphoridae. *Rhipiphorus smithi* Linsley & MacSwain. **a.** triungulin, dorsal; **b.** replete triungulin, lateral; **c.** pupa, ventral (after Linsley *et al.*, 1952).

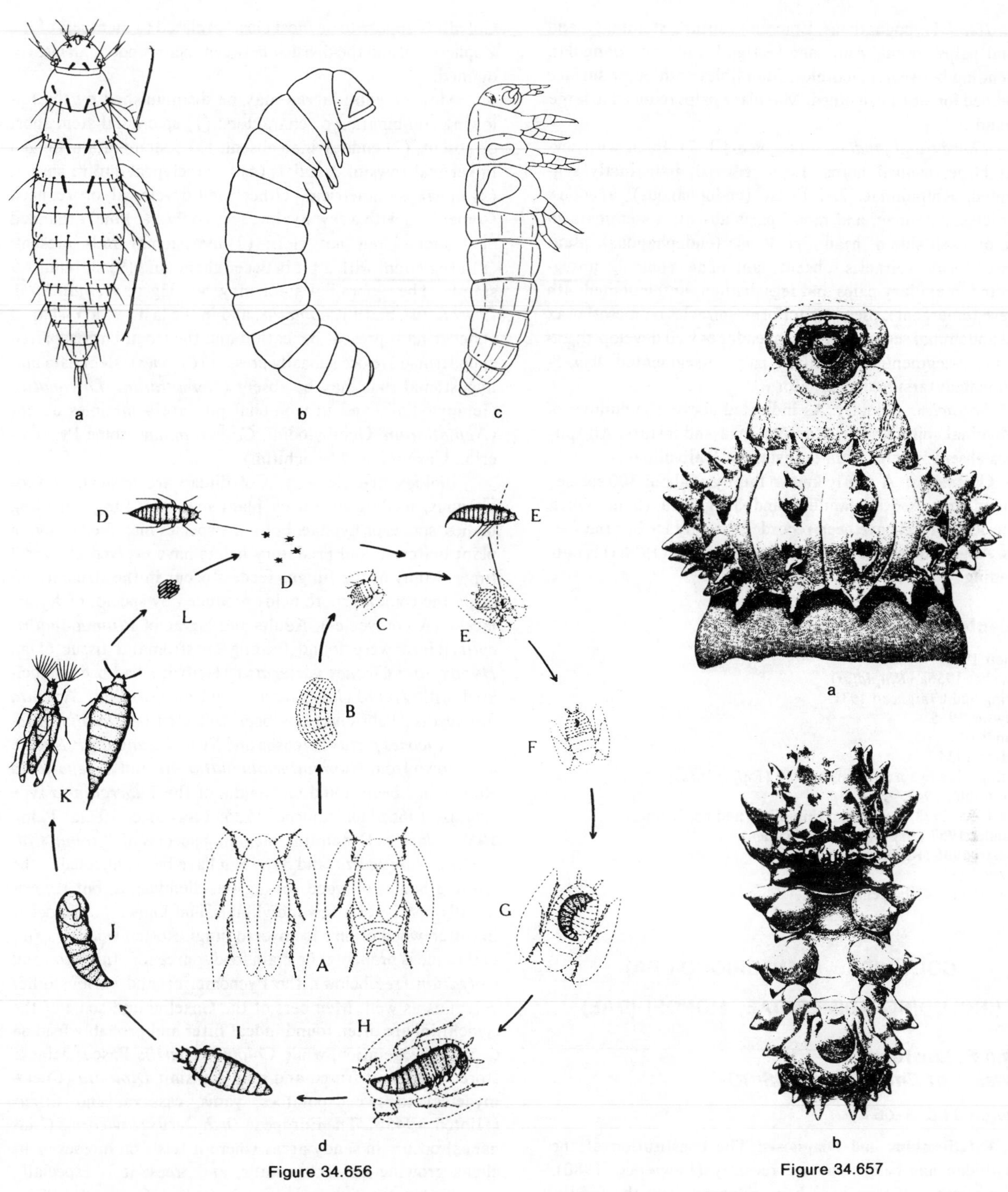

Figures 34.656a-d. Rhipiphoridae. *Rhipidius quadriceps* Abeille de Perrin. **a.** triungulin, dorsal; **b.** 3rd phase larva, lateral; **c.** 4th phase larva; **d.** life cycle. A, host cockroaches (*Ectobius*) adults; B, host oötheca; C, host larva; D, triungulin; E, triungulin attached to host; F, 2nd phase; G, 3rd phase; H, 4th phase, before emergence; I, 4th phase, emerged; J, pupa; K, adults; L, egg (after Besuchet, 1956).

Figures 34.657a,b. Rhipiphoridae. *Macrosiagon ferrugineum flabellatum*. **a.** ectophagous phase larva, on thorax of larval wasp; **b.** ventral view (after Grandi, 1936).

Head: Hypognathous. Epicranial suture, stemmata, and labial palps absent. Antennae vestigial. Labrum triangular, extending between mandibles. Mandibles with outer surface modified for cutting, toothed. Maxillary palps reduced to large mounds.

Thorax and Abdomen (segments 1–7). Each with several large, conical horns. Legs reduced, indistinctly segmented. Rhipidiinae: 2nd Phase (endophagous): apodous; spiracles, antennae, and mouthparts absent; 5 stemmata in line on each side of head. 3rd Phase (endophagous): pseudoeruciform; spiracles absent; antennae conical, unsegmented; maxillary palps and legs distinct, unsegmented. 4th Phase (emergent): pseudoeruciform; spiracles on mesothorax and abdominal segments 1–6; appendages well developed; antennae 2-segmented, maxillary palps 3-segmented, legs 5-segmented; tarsungulus falciform.

Spiracles: Annular. As indicated above, the number of abdominal spiracles varies among taxa and instars. All spiracles absent in intermediate instars of Rhipidiinae.

Comments: A poorly known family of about 300 species divided among 6 subfamilies and 38 genera (5 in North America). Hosts have been recorded for only 15% of the species. Among bionomic works that of Besuchet (1956a) is outstanding.

Selected Bibliography

Arnett 1968.
Besuchet 1956a (*Rhipidius*).
Böving and Craighead 1931.
Crowson 1955.
Grandi 1936.
Hudson 1934.
Linsley, MacSwain and Smith 1952 (*Rhipiphorus*).
Peyerimhoff 1942.
Riek 1955, 1973 (Australian rhipidiines and nephritines).
Selander 1957 (*Nephrites*, classification).
Silvestri 1906 (Rhipidiinae).
Tucker 1952.
Viana 1971 (Micholaeminae).

COLYDIIDAE (TENEBRIONOIDEA)
(INCLUDING ADIMERIDAE, MONOEDIDAE)

John F. Lawrence,
Division of Entomology, CSIRO

Figures 34.658–661

Relationships and Diagnosis: The constitution of the Colydiidae has been reviewed recently (Lawrence, 1980), and a number of taxa have been removed from the family, including the following groups listed in Arnett (1968): *Anchomma* (Tenebrionidae); *Aglenus* (Othniidae, Trogocryptinae); Deretaphrini and Bothriderini (Bothrideridae); Cerylonini, Euxestinae, and Murmidiinae (except *Eupsilobius*) (Cerylonidae); and *Eupsilobius* (= *Eidoreus*) (Endomychidae) (Sen Gupta and Crowson, 1973). The family is usually placed near the base of the Tenebrionoidea because of various "clavicorn" features of the adult, including abruptly clubbed antennae and reduced tarsal segmentation. Colydiids appear to be most closely related to members of the Zopheridae, and the division between them is not at all clearly defined.

Most colydiid larvae may be distinguished by the following combination of characters: (1) epicranial stem short or absent; (2) endocarinae absent; (3) 5 stemmata arranged in vertical rows of 3 and 2; (4) frontoclypeal suture absent; (5) molae symmetrical, either well-developed or reduced (sometimes with a few hyaline teeth only); (6) mala rounded or truncate and not cleft; (7) hypostomal rods present; (8) urogomphi with a pit between them; and (9) sternum A9 simple. The epicranial stem may be long in *Nematidium, Lasconotus,* and *Pycnomerus,* and in the last two a Y-shaped endocarina is present beneath it and the frontal arms; paired endocarinae are occasionally present (*Cicones*); stemmata and hypostomal rods may be absent (*Nematidium, Lasconotus,* Gempylodini); and urogomphal pits are sometimes absent (*Nematidium,* Gempylodini, *Colydium,* and some Pycnomerini, Coxelini, and Synchitini).

Biology and Ecology: Colydiidae are basically mycophagous, feeding on rotten plant material, fungal fruiting bodies, spores or hyphae, but some species may feed on living plant material, and predatory habits have evolved at least 3 times. Many of the fungus feeders occur in the stromata of, or on the conidial spore fields produced by species of Xylariaceae (Ascomycetes). Adults and larvae of *Bitoma quadricollis* Horn were found feeding on stromatal tissue of an *Hypoxylon; Cicones variegatus* (Hellwig) has been associated with *Hypoxylon deustum* and an *Ustulina; Synchita humeralis* (Fabricius) has been collected on *Daldinia tuberosa; Cicones pictus* Erichson and *Synchita angularis* Abeille are known from *Nummulariola bulliardii;* and *S. separanda* Reitter has been found on conidia of the *Tubercularia* type (Dajoz, 1966; Donisthorpe, 1935; Lawrence, 1977a; Palm, 1959). Various Synchitini, including species of *Bitoma, Colydodes, Namunaria,* and *Synchita* have been collected in the fruiting bodies of wood-rotting Basidiomycetes, but records usually consist of a few individuals, and known larvae occur in rotten wood or on ascomycete fungi. Rotten wood or cambial tissue is probably the main food source for *Aulonium* and *Colydium* (*see* below), the Pycnomerini, and various other colydiids as well. Members of the Coxelini and some of the Synchitini are often found in leaf litter and probably feed on decaying vegetation, while *Colobicus parilis* Pascoe attacks stored roots and fruits, and may transmit *Diplodia* (Coelomycetes) to sweet potatoes, yams, cassava, and *Citrus* (Hinton, 1945b). The European *Orthocerus clavicornis* (Linnaeus) occurs in sandy areas where it feeds on mosses or lichens growing on rocks, walls, and trees; it is especially common on the lichen *Peltigera canina* (Crowson, 1984a; Horion, 1961). Species of *Dryptops* occur on tree surfaces and may also be lichen feeders; they are peculiar in having an epicuticular growth of cryptogams, which gives the surface a green color (Samuelson, 1966). Other plant-associated colydiids include species of *Monoedus* whose larvae have been collected in the pith of stems (Craighead, 1920), and the Australian *Todima* which feed on the flower spikes of grass trees (*Xanthorrhoea*).

Almost all cases of predatory habits in the Colydiidae concern members of the Colydiini and Gempylodini and the synchitine genus *Lasconotus*, but it is possible that *Nematidium* is also predaceous. Members of the genus *Aulonium* are often abundant in the subcortical galleries of bark beetles, and are generally considered to be predators (Dajoz, 1977; Horion, 1961). Although larvae of *A. longum* LeConte have been observed feeding on larval *Dendroctonus* (Dahlsten, 1970), both *A. trisulcum* (Fourcroy) and *A. tuberculatum* Kraus have been reared entirely on plant material (Craighead, 1920; Marshall, 1978). Species of *Colydium* have also been recorded as bark beetle predators (Blackman and Stage, 1924; Dahlsten, 1970; Horion, 1961), but there is a lack of convincing evidence. A number of African gempylodines (*Aprostoma* and *Mecedanum*) have been reported to be predators in the galleries of ambrosia beetles (Platypodidae and *Xyleborus*) (Hinton, 1948c; Roberts, 1968; Schedl, 1962), as have species of *Pseudendestes* and *Munaria* in the Australian Region (Lawrence, 1980), and at least some of these records are based on detailed feeding observations. All known larvae of Colydiini and Gempylodini lack any obvious modifications usually associated with predaceous habits (molar reduction, loss of microtrichial patches on mandibles). Predatory behavior has been well established in the genus *Lasconotus,* and the larvae have highly modified mandibles. Hackwell (1973) noted that, although early instars fed primarily on fungi, third stage larvae of *L. subcostulatus* Kraus were aggressive predators on larvae and pupae of *Ips*.

Description: Mature larvae 2 to 20 mm, usually less than 10 mm. Body elongate, more or less parallel-sided, straight, and subcylindrical to slightly flattened. Usually lightly pigmented except for head and A9 tergum, sometimes with all visible terga pigmented; dorsal surfaces smooth; vestiture of scattered, simple setae.

Head: Protracted and prognathous, slightly flattened. Epicranial stem short to moderately long, sometimes absent; frontal arms lyriform or V-shaped, contiguous at base. Median endocarina usually absent, occasionally coincident with epicranial stem or with both stem and frontal arms. Paired endocarinae occasionally present beneath frontal arms. Stemmata usually 5 on each side, sometimes absent. Antennae usually well developed, occasionally very short, 3-segmented, with elongate, palpiform or conical sensorium. Frontoclypeal suture absent; labrum free. Mandibles symmetrical, usually bidentate, occasionally unidentate (*Lasconotus*) or tridentate, sometimes with serrate incisor edge, without accessory ventral process; mola either well-developed and tuberculate, asperate, or transversely ridged, or reduced, sometimes consisting of a few asperities or hyaline teeth only; prostheca absent. Ventral mouthparts retracted. Maxilla with transverse cardo, elongate stipes, well-developed articulating area, 3-segmented palp, and rounded or truncate mala, sometimes with tooth at inner apical angle. Labium usually free to base of mentum; ligula present (except in *Nematidium*); labial palps usually 2-segmented (1-segmented in *Nematidium* and Gempylodini). Hypopharyngeal sclerome usually present, variable. Hypostomal rods short to moderately long and diverging, sometimes absent. Ventral epicranial ridges absent. Gula usually transverse, occasionally longer than wide.

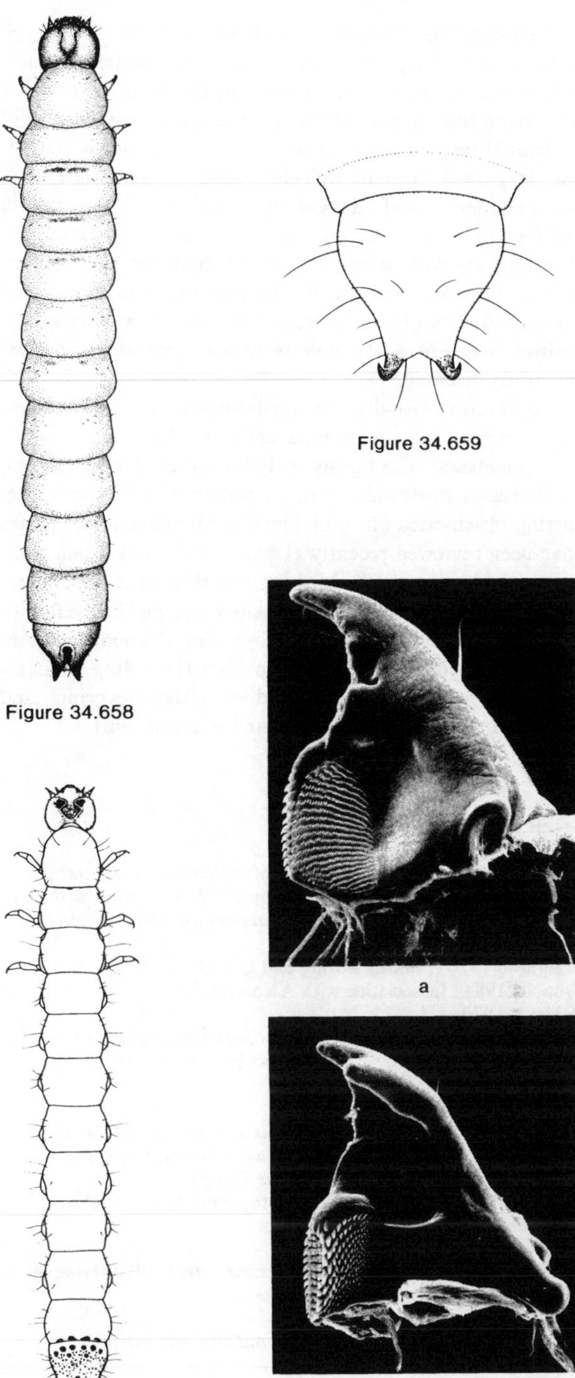

Figure 34.658. Colydiidae. *Pycnomerus* sp. Lake St. Clair, Tasmania. Larva, dorsal. Length = 8.7 mm.

Figure 34.659. Colydiidae. *Pycnomerus* sp. Monga State Forest, New South Wales. Abdominal apex, dorsal.

Figure 34.660. Colydiidae. *Namunaria pacifica* (Horn). Phoenix Lake, Marin Co., California. Larva, dorsal. Length = 7.5 mm.

Figures 34.661a,b. Colydiidae. *Aulonium longum* LeConte. Chuska Mts., Arizona. **a.** right mandible, dorsal; **b.** left mandible, ventral.

Thorax and Abdomen: Prothorax sometimes enlarged (*Lasconotus*). Legs well-developed, 5-segmented, moderately to widely separated; tarsungulus almost always with 2 setae lying side by side. Meso- and metatergum and anterior abdominal terga sometimes with rows or patches of asperities. Tergum A9 usually well-developed and sometimes variously pigmented and tuberculate, occasionally (*Nematidium* and Gempylodini) reduced, much shorter than tergum A8, almost always with paired, upturned, simple urogomphi, often with pit between them, rarely (*Cicones*) with median spine. Sternum A9 usually simple, sometimes with row of apical asperities. Segment A10 transversely oval, ventrally or posteroventrally oriented.

Spiracles: Usually annular-biforous with short accessory openings, sometimes apparently annular.

Comments: The family includes about 150 genera and 1000 species worldwide, with 18 genera and 75 species occurring in America north of Mexico. The Palearctic species have been reviewed recently (Dajoz, 1977), but much work remains to be done in establishing a worldwide tribal and generic classification. The tribes presently included in the family are Nematidiini, Gempylodini, Colydiini, Pycnomerini, Acropini, Monoedini, Synchitini, Coxelini (including Megataphrini), Diodesmini, Langelandiini, Rhopalocerini, and Orthocerini (including Corticini and Rhagoderini).

Selected Bibliography

Arnett 1968.
Blackman and Stage 1924.
Böving and Craighead 1931 (larvae of *Aulonium tuberculatum* LeConte, *Bitoma crenata* (Fabricius), *Nematidium filiforme* LeConte, *Phloeonemus catenulatus* Horn, and *Synchita fuliginosa* Melsheimer).
Craighead 1920 (same as Böving and Craighead 1931).
Crowson 1984a (association with Ascomycetes).
Dahlsten 1970.
Dajoz 1968 (larva of *Langelandia anophthalma* Aubé), 1971 (larva of *Pycnomerus fuliginosus* Erichson), 1977 (biology; Palaearctic revision).
Donisthorpe 1935 (fungus hosts).
Hackwell 1973 (biology of *Lasconotus subcostulatus* Kraus).
Hayashi 1972 (larvae of *Cicones hayashii* Sasaji, *Penthelispa vilis* (Sharp), and *Sympanotus pictus* Sharp).
Hinton 1945b (*Colobicus parilis* in stored products), 1948c (predators of ambrosia beetles).
Horion 1961.
Lawrence 1977a (association with Ascomycetes), 1980 (larva of *Pseudendestes robertsi* Lawrence).
Mamaev 1975.
Marshall 1978 (biology and larva of *Aulonium trisulcum* (Fourcroy)).
Nikitsky and Belov 1980a, 1980b (key to larvae of 19 species (16 genera); larvae of *Cicones pictus* Erichson, *Lastrema verrucollis* Reitter, *Niphopelta imperialis* Reitter, *Pycnomerus terebrans* Olivier, and *Synchita mediolanensis* Villa).
Palm 1959 (association with Ascomycetes).
Peterson 1951.
Roberts 1968 (predators of ambrosia beetles).
Samuelson 1966 (epizoic symbiosis).
Saalas 1923 (larvae of *Lasconotus jelskii* (Wankow), *Bitoma crenata* (Fabricius)).
Schedl 1962 (association with bark and ambrosia beetles).
Sen Gupta and Crowson 1973.
Verhoeff 1923 (larva of *Bitoma crenata*).

MONOMMIDAE (TENEBRIONOIDEA)

John F. Lawrence,
Division of Entomology, CSIRO

Figures 34.662–664

Relationships and Diagnosis: The Monommidae are closely related to Colydiidae and Zopheridae, but their larvae differ from those of either group in having a reduced mandibular mola, paired endocarinae mesad of the frontal arms, and double rows of asperities forming incomplete transverse rings on abdominal terga 2 to 6 (Crowson, 1955, 1966b).

Biology and Ecology: Monommids feed on decaying plant material and are usually found in soft and highly decayed stems, but may also occur under bark of rotten logs, and have been found in yams and *Yucca* pods. *Hyporhagus gilensis* Horn was found in *Yucca* stems in Arizona, *Monomma brunneum* Thomson in papaya stems in India (Fletcher, 1916), and *Inscutomonomma hessei* Freude in the stems of *Euphorbia* in Africa (Freude, 1958). Pupation takes place in the soil (Freude, 1958).

Description: Mature larvae 5 to 15 mm, usually 10 mm or less. Body elongate, more or less parallel-sided, straight, slightly flattened. Dorsal surfaces lightly pigmented, smooth except for head and A9 tergum; vestiture of scattered, fine, simple setae.

Head: Protracted and prognathous, moderately broad, somewhat flattened. Epicranial stem absent; frontal arms lyriform and contiguous at base. Median endocarina absent; paired endocarinae located mesad of frontal arms. Stemmata 5 on each side. Antennae well developed, 3-segmented. Frontoclypeal suture absent; labrum free. Mandibles symmetrical, somewhat longer than wide, bidentate, without accessory ventral process; mola reduced and sub-basal, represented by a row of 10 or more hyaline teeth. Ventral mouthparts retracted. Maxilla with transverse cardo, elongate stipes, well-developed articulating area, 3-segmented palp, and truncate mala which is cleft at apex and has 2 teeth at inner apical angle. Labium free to base of mentum; ligula present; labial palps 2-segmented and moderately widely separated. Hypopharyngeal sclerome absent. Hypostomal rods moderately long and subparallel or slightly diverging (converging again at apex). Ventral epicranial ridges absent. Gula transverse.

Thorax and Abdomen: Metatergum and abdominal tergum 1 sometimes with transverse row of asperities (broken at midline). Legs short and spinose, 5-segmented, widely separated; tarsungulus with 2 setae lying side by side. Abdominal terga 2 to 6 each with double row of asperities forming transverse, complete or incomplete ring or with paired rings. Tergum A9 with pair of urogomphi, between which is a heavily sclerotized pit; sternum A9 simple. Segment A10 transversely oval, ventrally oriented.

Spiracles: Annular-biforous.

Comments: The family includes 12 genera and about 225 species worldwide, with *Apasthines* (1 species) and *Hyporhagus* (4 species) occurring in America north of Mexico. The group occurs in tropical and subtropical regions of both hemispheres, has its greatest diversity in Africa and Madagascar.

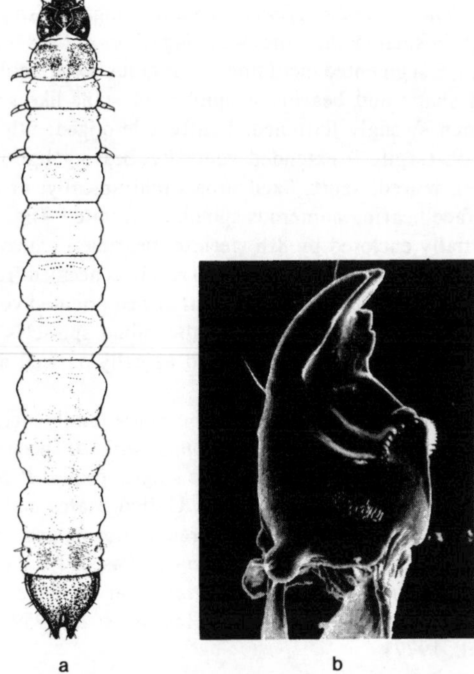

Figure 34.662

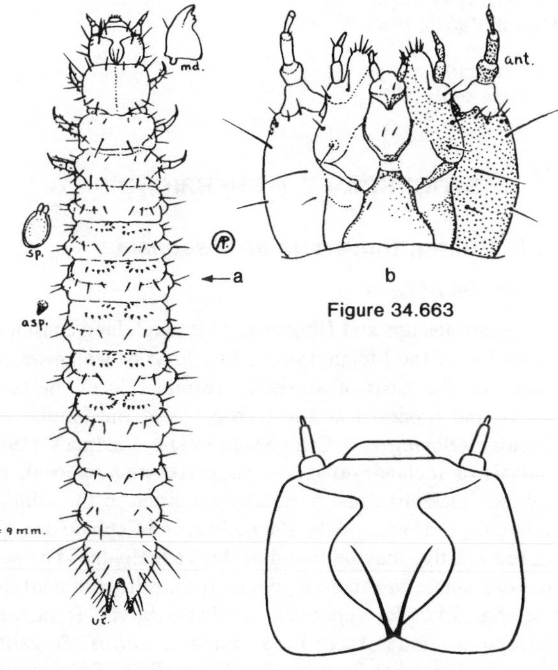

Figure 34.663

Figure 34.664

Figures 34.662a,b. Monommidae. *Hyporhagus gilensis* Horn. Vail, Pima Co., Arizona. **a.** larva, dorsal, length = 18.9 mm; **b.** right mandible, ventral.

Figures 34.663a,b. Monommidae. *Hyporhagus texanus* Linnell. El Paso, Texas. **a.** larva, dorsal (plus details), length = 9 mm; **b.** head, ventral. (From Peterson, 1951)

Figure 34.664. Monommidae. *Hyporhagus* sp. Head, dorsal.

Selected Bibliography

Crowson 1955, 1966b.
Fletcher 1916 (larva of *Monomma brunnea* Thomson).
Freude 1958 (larva of *Inscutomonomma hessei* Freude).
Peterson 1951 (larva of *Hyporhagus texanus* Linnell).

PROSTOMIDAE (TENEBRIONOIDEA)

D. K. Young, *University of Wisconsin*

Figures 34.665-666

Relationships and Diagnosis: The composition of the Prostomidae is essentially that proposed by Böving and Craighead (1931) on the basis of critical examination of the larval stages. Crowson (1955) enumerated several characters of both larvae and adults which would support placement in the Heteromera, but treated it as a subfamily of Cucujidae until some years later (Crowson, 1967). The closest relative may be among the Inopeplidae, Salpingidae or Othniidae, all of which appear to be closely related. However, Lawrence (1977c) stated that the prostomids might also be derived from Crowson's (1966b) synchroid-zopherid-tenebrionid group. The asymmetrical cranium, with the right side larger than the left (fig. 34.665b), and transverse row of asperities on the distal margin of the ninth abdominal sternite (fig. 34.665c) characterize prostomid larvae.

Biology and Ecology: Larvae and adults are associated with dead, decaying trees. Typically, they are found within a characteristic mud- or clay-like material between layers of the decaying wood. Adult *Prostomis mandibularis* (Fabricius) have been collected beneath loose bark and within decaying wood of various conifers in western North America. Larvae of *P. mandibularis* and the Japanese *P. latoris* Reitter have been found beneath bark and within decaying wood of pines; in the case of *P. mandibularis*, the wood was in the red-rotten stage of decay (Hayashi, 1969b; Lawrence, Young, unpublished notes).

Description: Mature larvae attain lengths of 8–9 mm and widths of 1.3–2 mm. Body strongly flattened, smooth, lightly sclerotized, creamy white, widest slightly beyond the middle and narrowing posteriorly and anteriorly; vestiture sparse, consisting of short setae.

Head: (fig. 35.665b) Prognathous, exserted from prothorax, conspicuously asymmetrical with right side larger than left. Epicranial suture poorly defined, frontal arms lyriform, stem short or absent; endocarinae absent. Symmetrical labrum anterad of fused frons and clypeus. Stemmata absent. Antennal insertions fully exposed, antennae elongate, 3-segmented, sensorium of segment 2 conical, 0.5× length of apical segment. Mouthparts protracted, supported ventrally by well developed, posteriorly divergent hypostomal rods. Mandibles (figs. 34.666a,b) heavily sclerotized, movable, asymmetrical; left mandible with prominent molar tooth; apices of mandibles tridentate. Maxilla with 1-segmented cardo; undivided, pad-like maxillary articulating area; maxillary mala shallowly cleft subapically; 3-segmented palp. Labium free to base of mentum and possessing apically rounded ligula and 2-segmented palps. Hypopharyngeal sclerome transversely rectangular; gular sutures separate.

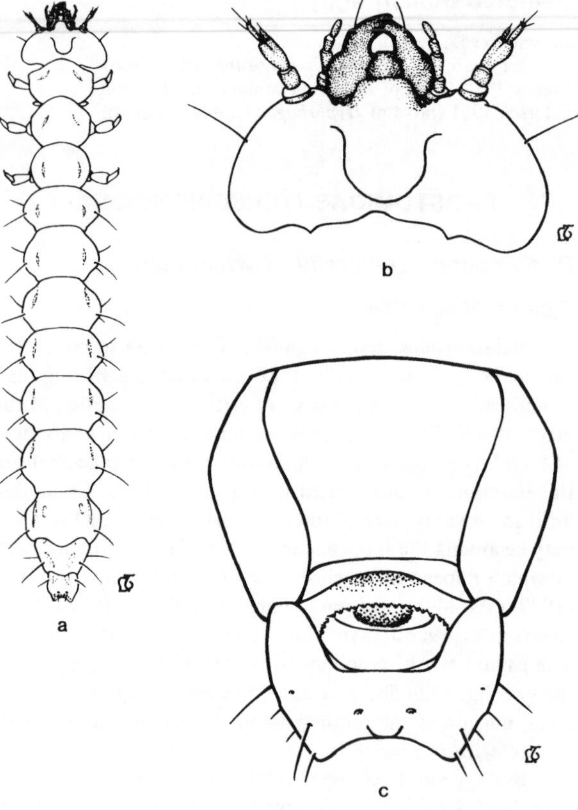

Figure 34.665

a b

Figure 34.666

Figures 34.665a-c. Prostomidae. *Prostomis mandibularis* (Fabricius). Pacific House, El Dorado Co., California, elev. 3400 feet, under bark of red rotten pine. **a.** larva, dorsal; **b.** head capsule, dorsal; **c.** apex of abdomen, ventral.

Figures 34.666a,b. Prostomidae. *Prostomis* sp. East slope of Mt. McKinley, Mindanao, Philippine Islands. **a.** left mandible, oroventral view; **b.** right mandible, dorsal.

Thorax and Abdomen: Thorax elongate with prothorax slightly smaller than meso- or metathorax. Legs well developed, 5-segmented including tarsungulus. Legs similar in size and shape and bearing a number of spine-like setae. Abdomen strongly flattened, lightly sclerotized; 9th segment small; tergite 9 extended ventrally, bearing lightly sclerotized, paired, short, fixed urogomphi apically; urogomphal surface bearing numerous spinulae; sternite 9 (fig. 34.665c) partially enclosed by 8th sternite, bearing a transverse row of apical asperities. Segment 10 small, concealed from above by projection of 9th tergite; anal orifice oriented ventrally.

Spiracles: Thoracic and abdominal spiracles annular-biforous, those of thorax placed apically on well developed spiracular tubes.

Comments: This small and obscure family includes only 2 genera, *Dryocora* and *Prostomis,* with 20 species distributed throughout the world. The single species, *P. mandibularis* (Fabricius), occurs in the United States and Canada. Larvae of this species were figured by Böving and Craighead (1931) along with those of *D. howitti* Pascoe from New Zealand. Several additional *Prostomis* larvae have been described (Hayashi, 1969b, 1980; Hayashi, et al., 1959; Mamaev et al., 1977).

Selected Bibliography

Böving and Craighead 1931.
Crowson 1955, 1966b, 1967.
Hayashi 1969b, 1980.
Hayashi *et al.* 1959.
Lawrence 1977c.
Mamaev *et al.* 1977.

SYNCHROIDAE (TENEBRIONOIDEA)

D. K. Young, *University of Wisconsin*

Figures 34.667a–e

Relationships and Diagnosis: Although long considered a member of the Melandryidae, *Synchroa* Newman was excluded on the basis of a critical examination of the larval stages, and proposed as the type genus of the family Synchroidae (Böving and Craighead, 1931). Viedma's (1966) analysis of melandryid larvae supported the removal, and Crowson (1966b) drew particular attention to the affinities between synchroids and the Zopheridae and Cephaloidae. The nearest relative may be found in the Cephaloidae. The presence of a single pair of asperities on the ninth abdominal sternite (fig. 34.667c) separates synchroid larvae from other Coleoptera larvae except for Pedilidae (*Pedilus, Pergetus*), Boridae (*Lecontia*), Salpingidae (*Aegialites, Rhinosimus*), Othniidae (*Aglenus*) and certain Melandryidae (e.g. *Hallomenus*). The urogomphi of synchroids, *Lecontia* and *Hallomenus* are simple; in those Pedilidae, Salpingidae, and Othniidae which have but a single pair of caudoventral asperities the urogomphi are branched. Other diagnostic features of synchroid larvae include the moderately long epicranial stem and small mesal dentiform process associated with the prothoracic sternum (fig. 34.667b).

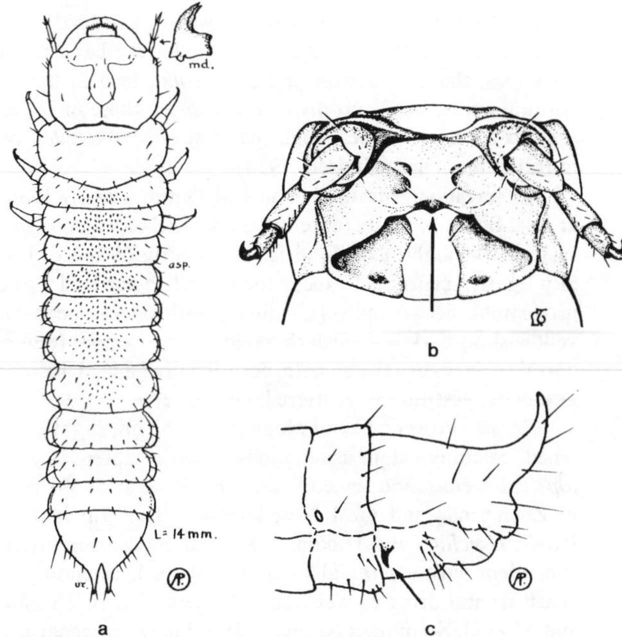

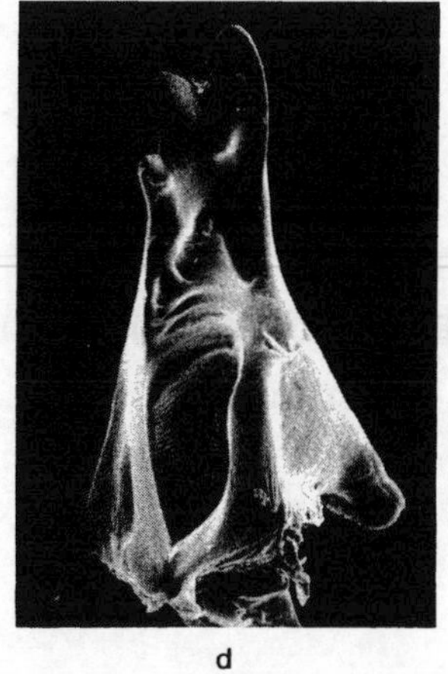

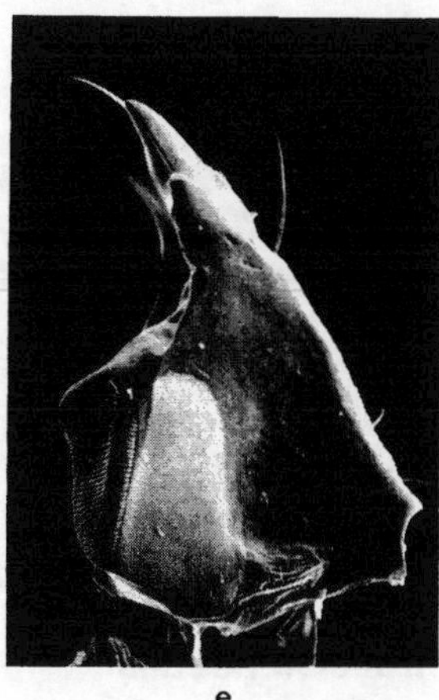

Figures 34.667a-e. Synchroidae. *Synchroa punctata* Newman. Commonly collected in the Midwest and northeastern United States beneath somewhat loose bark of decaying hardwoods. **a.** larva, dorsal; **b.** prothorax, ventral; **c.** apex of abdomen, lateral; **d.** left mandible, oral; **e.** right mandible, dorsal. (Figures 34.667a,c from Peterson, 1951)

Biology and Ecology: Adults of *Synchroa punctata* Newman have been collected at lights and while running over the surfaces of dead deciduous trees at night. Larvae are associated with somewhat cool, moist conditions beneath slightly loose bark of decaying deciduous trees, especially wild cherry, *Prunus serotina*. Both woody and fungal materials have been extracted from the gut, but fungi appear to play a significant role in nutrition of larvae. From 1 to several years may be required for larval development depending upon the diet, and several instars commonly occur together beneath the bark of a given log. The pupal stage is spent beneath bark in a frass-walled ovate chamber prepared by the larva (Payne, 1931; Peterson, 1951; Young, unpublished notes).

Description: Mature larvae attain lengths of 15–18 mm and widths of 3–4 mm. Body orthosomatic and subcylindrical, subparallel throughout length, lightly sclerotized except for head and apices of urogomphi; vestiture of fine setae. Dorsum of meso- and metathorax and abdominal segments 1–5 beset with patches of small asperities, body otherwise generally smooth. Head and urogomphal apices yellowish-brown to piceous, remainder of body yellowish-white.

Head: Prognathous, exserted from prothorax. Epicranial suture with moderately elongate stem and lyriform frontal arms that are complete to near the antennal insertions; endocarinae absent. Labrum symmetrical, separated from fused frons and clypeus by suture. Stemmata on each side 5, 3 in anterior and 2 in posterior group. Antennal insertions fully exposed; antennae elongate, 3-segmented, small palpiform sensorium associated with segment 2. Mouthparts retracted. Mandibles (figs. 34.667d,e) heavily sclerotized, movable, asymmetrical, molar region of right mandible more prominent than that of left; apices of mandibles bidentate with single subapical tooth. Maxilla with cardo divided by internal fold and thus appearing 2-segmented, small pad-like maxillary articulating area; maxillary mala distinctly cleft subapically on adoral margin; 3-segmented palp. Labium free to base of mentum and possessing elongate, apically rounded ligula and 2-segmented palps. Hypopharyngeal sclerome molar-like with anterior surface concave, posterior rim provided with dense pile of stout, flattened setae. Mentum slightly wider than long with anterior margin shallowly emarginate; submentum rectangular, longer than wide; gular region transverse.

Thorax and Abdomen: Thorax elongate with sides subparallel; prosternum bearing a distinct mesal dentiform process (fig. 34.667b). Legs well developed, 5-segmented including tarsungulus. Legs similar in size and shape and bearing numerous stout, spine-like setae. Abdomen subcylindrical; tergite 9 extended ventrally, possessing single, shallow pit between bases of paired, heavily sclerotized, fixed urogomphi; urogomphal apices strongly curved upward; 9th sternite with single, stout asperity near each anterolateral margin (fig. 34.667c). Segment 10 more or less fused to 9th, visible ventrally surrounding anal orifice.

Spiracles: Thoracic and abdominal spiracles annular-biforous.

Comments: The 2 genera and 8 species have been found in Indonesia, Japan, and eastern North America. Worldwide, 7 species of *Synchroa* are currently recognized along with *Mallodrya subaenea* Horn. Two species, *M. subaenea* and *S. punctata* Newman, occur in North America. Larvae of *S.*

punctata and *S. melanotoides* Lewis have been described or illustrated (Böving and Craighead, 1931; Hayashi, 1980; Hayashi, *et al*, 1959; Peterson, 1951).

Selected Bibliography

Böving & Craighead 1931.
Crowson 1966b.
Hayashi 1980.
Hayashi *et al.* 1959.
Payne 1931.
Peterson 1951.
Viedma 1966.

ZOPHERIDAE (TENEBRIONOIDEA) (INCLUDING MERYCIDAE)

John F. Lawrence,
Division of Entomology, CSIRO

Ironclad Beetles

Figures 34.668-671

Relationships and Diagnosis: The Zopheridae was originally based on a few genera comprising the tenebrionid tribes Zopherini and Nosodermini (Böving and Craighead, 1931), but more recently the group has been expanded and redefined (Crowson, 1955; Doyen and Lawrence, 1979; Kamiya, 1963; Watt, 1967b, 1974a, 1974b). Possible sister groups include Synchroidae, Cephaloidae, Colydiidae-Monommidae, Perimylopidae, Chalcodryidae and Tenebrionidae (Crowson, 1966b; Doyen and Lawrence, 1979; Lawrence and Newton, 1982; Watt, 1967a, 1974a, 1974b). Although most zopherids were first described as tenebrionids, the group is easily distinguished from Tenebrionidae in both adult and larval stages. Larvae of Zopheridae differ from those of Tenebrionidae in having an internally divided cardo, cleft maxillary mala, annular-biforous spiracles (sometimes with highly reduced accessory openings), usually lyriform frontal arms, no frontoclypeal suture (occasionally vaguely indicated), and well-developed tenth abdominal segment without pygopods or spines. Larvae of *Phellopsis* are very similar to those of Synchroidae and stenotracheline Cephaloidae, from which they differ in having more or less symmetrical mandibles without transversely ridged molae and in lacking basal asperities on sternum A9 and a transverse groove on tergum A9. A cleft mala and divided cardines usually distinguish zopherid larvae from those of Colydiidae and Monommidae, and in addition zopherid mandibles resemble those of tenebrionids more than those of either group. The southern temperate family Perimylopidae differ from Zopheridae in having annular spiracles and a reduced mandibular mola.

Biology and Ecology: Adult Zopherinae and Usechinae have been recorded from the fruiting bodies of various fungi, but all known larvae appear to feed under bark or in dead, rotten wood, especially that which has been attacked by white rot (delignifying) fungi. Species of *Cotulades* in Australia are often found on lichen-covered surfaces at night, but their larvae also feed within white rotten wood. Among the Ulodinae, larvae of *Meryx* are common under the bark of fungusy logs, those of *Ulodes* and *Dipsaconia* feed in the soft fruiting bodies of *Pleurotus* or *Piptoporus*, those of *Brouniphylax* occur in woodier fungi. and those of *Syrphetodes* feed in rotten branches (Hudson, 1934).

Description: Mature larvae 4 to 45 mm, usually 30 mm or less. Body elongate, more or less parallel-sided or slightly fusiform, straight, subcylindrical or slightly flattened. Usually lightly sclerotized except for buccal region and tips of urogomphi; occasionally (Ulodinae) with darker head and yellowish to dark brownish thoracic and abdominal tergites; dorsal surfaces usually smooth, occasionally (some Ulodinae) granulate; vestiture of scattered, simple setae.

Head: Protracted and prognathous, broad, slightly flattened, epicranial stem usually long (short or absent in *Phellopsis, Usechus,* and some Ulodinae); frontal arms V-shaped in Zopherinae and most Nosoderminae, lyriform in *Phellopsis, Usechus,* and Ulodinae. Median endocarina usually coincident with epicranial stem or Y-shaped, extending beneath frontal arms as well (absent in *Phellopsis, Usechus,* and *Meryx*). Stemmata on each side 5, 3 or 0. Antennae usually well-developed, 3-segmented, sometimes very short. Frontoclypeal suture usually absent (present in *Dipsaconia* and *Zopherus*); labrum free. Mandibles symmetrical or slightly asymmetrical, robust, bidentate or tridentate, without accessory ventral process; mola present, sometimes reduced in size, usually tuberculate or transversely ridged; prostheca absent. Ventral mouthparts retracted. Maxilla with transverse cardo, often distinctly divided, elongate stipes, well-developed articulating area, 3-segmented palp, and truncate mala, which is always cleft at apex and often has 1 or more teeth at inner apical angle. Labium more or less free to base of mentum; ligula present, sometimes longer than labial palps, which are 2-segmented and narrowly or broadly separated. Hypopharyngeal sclerome almost always tooth-like. Hypostomal rods absent in Zopherinae and most Nosoderminae, otherwise short to moderately long and diverging. Gula transverse.

Thorax and Abdomen: Legs well-developed, 5-segmented, sometimes short and spinose; tarsungulus usually with 2 setae lying side by side: coxae narrowly to widely separated. Meso- and metatergum and first 5 to 8 abdominal terga with patches, rows, or ridges of asperities. Tergum A9 with smooth surface and small, approximate urogomphi in Zopherinae and most Nosoderminae; *Phellopsis, Usechus* and Ulodinae with granulate and/or tuberculate 9th tergum and with larger urogomphi, which may have accessory processes (rare); sternum A9 simple. Segment A10 transverse, ventrally or posteroventrally oriented.

Spiracles: Annular-biforous, with long to very short accessory tubes (in the latter case appearing annular).

Comments: The family includes about 26 genera and 125 species worldwide, with 7 genera and 48 species in N. America. Three subfamilies are currently recognized (Doyen and Lawrence, 1979): Usechinae (*Usechus* and *Usechimorpha* from North America and Japan); Zopherinae (*Zopherus, Nosoderma, Phellopsis,* and 6 other genera from North, Central and South America, Africa, and Australia);

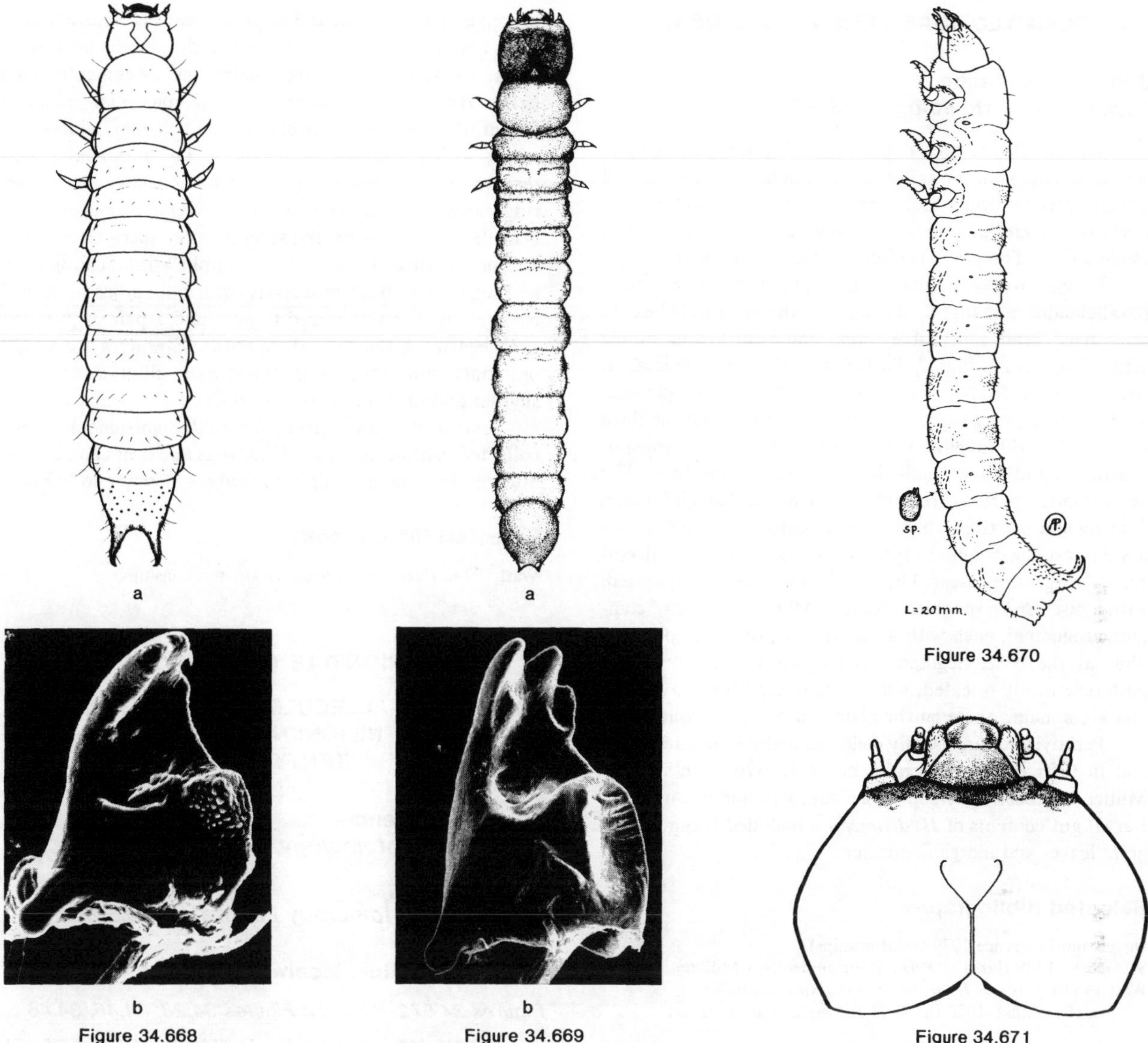

Figure 34.668

Figure 34.669

Figure 34.670

Figure 34.671

Figures 34.668a,b. Zopheridae. *Usechus lacerta* Motschulsky. Santa Clara Co., California. **a.** larva, dorsal, length = 8.5 mm; **b.** right mandible, mesoventral.

Figures 34.669a,b. Zopheridae. *Nosoderma* sp. 10 mi. S Las Vigas, Veracruz, Mexico. **a.** larva, dorsal, length = 15.5 mm; **b.** right mandible, mesoventral.

Figure 34.670. Zopheridae. *Phellopsis obcordata* (Kirby). Larva, lateral (plus details). Length = 20 mm. (From Peterson, 1951)

Figure 34.671. Zopheridae. *Zopherus nodulosus* Solier. Head, dorsal.

and Ulodinae (bulk of the genera from Chile, New Zealand, and Australia). The Australian genus *Meryx* is usually placed in a distinct family on the basis of adult features (Crowson, 1955); larvae of *Phellopsis* are quite distinct from other Zopherinae (Doyen and Lawrence, 1979); and the systematic position of several southern temperate groups is still uncertain.

Selected Bibliography

Böving and Craighead 1931 (larvae of *Phellopsis obcordata* (Kirby)).
Crowson 1955, 1966b.
Doyen 1976 (larva of *Phloeodes diabolicus* LeConte).
Doyen and Lawrence 1979 (larvae of *Brouniphylax varius* (Broun), *Dipsaconia pyritosa* Pascoe, *Meryx rugosa* Latreille, *Nosoderma* sp., *Syrphetodes punctatus* Broun, *Usechus lacerta* Motschulsky, *Zopherus granicollis* Horn, and *Z. nodulosus* Solier; key to New World genera).
Hudson 1934 (larvae of *Brouniphylax varius* and *Syrphetodes punctatus*).
Kamiya 1963.
Keleinikova and Mamaev 1971 (larva of *Phellopsis amurensis* Heyden).
Lawrence and Newton 1982.
Peterson 1951 (larva of *Phellopsis obcordata*).
Watt 1967b, 1974a, 1974b.

PERIMYLOPIDAE (TENEBRIONOIDEA)

John F. Lawrence,
Division of Entomology, CSIRO

This family includes 4 genera and 8 species occurring in southern Patagonia and Tierra del Fuego (Argentina and Chile) (*Hydromedion* and *Chanopterus*), the Falkland Islands (*Darwiniella*), and South Georgia (*Hydromedion* and *Perimylops*). The group is closely related to Zopheridae.

Larvae are of moderate size (7–15 mm), elongate, parallel-sided, and slightly flattened, with moderately heavily sclerotized head and visible terga, and vestiture of simple setae. The head is slightly flattened, with a short epicranial stem, lyriform frontal arms, 5 stemmata on each side, and long antennae, with segments 1 and 2 subequal and the third reduced. Mandibles have 3 or 4 teeth at the apex, no mola or prostheca, and a few small, hyaline teeth at the base. The ventral mouthparts are retracted, with a rounded, cleft mala, 3-segmented maxillary palps, 2-segmented labial palps, and a well-developed ligula. Hypostomal rods and ventral epicranial ridges are absent. The legs are long, widely separated, with a bisetose tarsungulus. Tergum A9 bears a pair of complex urogomphi, each with an accessory process or pregomphus at the base. Segment A10 is transversely oval and posteroventrally oriented, with a sclerotized tergum. Spiracles are annular, those on the abdomen being subequal.

Perimylopids are usually collected under stones and moss and in tufts of tussock grass, but *Perimylops antarcticus* Müller has been taken in large numbers under carcasses. Larval gut contents of *Hydromedion* included fragments of grass leaves and inorganic matter.

Selected Bibliography

Doyen and Lawrence 1979 (relationships).
St. George 1939 (larva of *Perimylops antarcticus* Müller).
Watt 1967b (larva of *Hydromedion sparsutum* Müller; relationships), 1970 (larva of *Perimylops antarcticus*).

CHALCODRYIDAE (TENEBRIONOIDEA)

John F. Lawrence,
Division of Entomology, CSIRO

This family includes the Tasmanian species *Sirrhas limbatus* Champion and 5 New Zealand species in the genera *Chalcodrya, Onysius,* and *Philpottia.* Adult chalcodryids have a number of features in common with Zopheridae and Perimylopidae, but the larvae resemble those of certain tenebrionids.

Larvae are moderately large (up to 30 mm), elongate, parallel-sided, and subcylindrical or slightly flattened, with a darkly pigmented head and lightly pigmented, brownish-yellow terga. The head is slightly flattened and evenly tuberculate, with a long epicranial stem, no median endocarina, V-shaped frontal arms, joined anteriorly by a transverse ecdysial line, 5 stemmata on each side, relatively short antennae, and a pair of sharp ridges extending anteriorly from the frontoclypeal region. The mandibles are bidentate, with a simple, concave mola and no prostheca. The ventral mouthparts are retracted, with a divided cardo, simple and rounded mala, 3-segmented maxillary palps and 2-segmented labial palps. Hypostomal rods and ventral epicranial ridges are absent, the legs are relatively long and slender, widely separated, with a bisetose tarsungulus. The meso- and metatergum and abdominal terga 1 to 8 each have a transverse ridge near the anterior edge, a transverse row of asperities near the posterior edge, and several more asperities of varying size on the disc. Tergum A9 is simple, without urogomphi, and segment A10 is transversely oval, more or less terminally oriented, with a sclerotized tergum and a pair of pygopods, each bearing a few lateral asperities. Spiracles are annular or annular-multiforous, larger and more elliptical anteriorly, smaller and more circular posteriorly.

Larvae of *Chalcodrya variegata* Redtenbacher have been collected in or beaten from dead twigs and branches, but they apparently emerge at night to feed on mosses and lichens.

Selected Bibliography

Watt 1974a (larva of *Chalcodrya* sp.; relationships).

TENEBRIONIDAE (TENEBRIONOIDEA)

(INCLUDING ALLECULIDAE, COSSYPHODIDAE, LAGRIIDAE, NILIONIDAE, RHYSOPAUSSIDAE, TENTYRIIDAE)

John F. Lawrence,
Division of Entomology, CSIRO

T. J. Spilman,
Sytematic Entomology Laboratory, USDA

Darkling Beetles, Mealworms, False Wireworms

Figures 34.672–693, and Figures 34,28, 34.45, 34.46

Relationships and Diagnosis: The limits of the family Tenebrionidae as used here conform to those of Watt (1967a) and Doyen and Lawrence (1979), with the inclusion of the Alleculidae, Lagriidae, Nilionidae, Cossyphodidae, and several genera, like *Myrmechixenus,* formerly placed in other families. The group is the largest in the superfamily, and is relatively easily distinguished in both larval and adult stages from other Tenebrionoidea. Diagnostic larval characters found in all members of the family include: the presence of a frontoclypeal suture, simple malar apex which is not cleft, simple 9th sternum, and annular or annular-multiforous spiracles; and the absence of an endocarina, mandibular prostheca, hypostomal rods, ventral prolegs, and patches or rows of tergal asperities. In addition, most species have a characteristic antennal sensorium, which is almost always flat and dome-like and usually C-shaped, so that it forms an incomplete ring around the terminal segment (or its fused remnant in species with 2-segmented antennae). Soil-dwelling tenebrionid larvae superficially resemble wireworms (larvae of many Elateridae) and have thus been called false wireworms; elaterid larvae are easily distinguished from those of Tenebrionidae, however, by their more flattened head with a fused labrum, the presence of 1 large stemma on each side (or the absence of stemmata), the lack of a mandibular mola, and the presence of biforous spiracles with a distinct ecdysial scar. The phylogenetic relationships of Tenebrionidae are not clear, and possible related groups include Chalcodryidae, Perimylopidae, Zopheridae (and the related colydiid-monommid complex), Synchroidae, Cephaloidae, and Oedemeridae. Among these groups, a frontoclypeal suture is found only in Chalcodryidae, zopherine and a few ulodine Zopheridae, *Nematoplus* (Cephaloidae), and *Calopus* (Oedemeridae); most of these have a cleft or otherwise modified malar apex, *Nematoplus* has a highly asymmetrical head, *Calopus* has distinct ventral prolegs, and *Chalcodrya* has rows of tergal asperities and a transverse ecdysial line joining the frontal arms anteriorly. Annular-biforous spiracles will also distinguish all but Oedemeridae, Perimylopidae and Chalcodryidae.

Biology and Ecology: The majority of adult Tenebrionidae are relatively large, heavily sclerotized, dark in color, long-lived, and active at night, either on the ground or on the surfaces of logs or tree trunks. In some groups (Phrenapatini, some Pycnocerini, some Opatrini, Trachyscelini, Ulomini) the adults are adapted for burrowing into substrates, occupying the same habitats as their larvae; while in others (Lagriini, Cyphaleini, Alleculini, Cnodalonini, and Strongyliini) the adults are relatively soft-bodied, short-lived, and often diurnal and brightly colored. Brachyptery and aptery are common, and characterize almost all members of certain large taxa, like Tentyriinae and Adeliini. Many ground-inhabiting tenebrionids inhabit deserts and exhibit a number of morphological, physiological, and behavioral adaptations to arid, sandy environments (subelytral cavity, sand-walking and sand-swimming modifications, water catchment devices) (Cloudsley-Thompson, 1965; Dizer, 1955; Koch, 1962; Medvedev, 1965; Seely, 1976; Seely and Hamilton, 1976). A major feature of many adult tenebrionids is the presence of defense glands located near the abdominal apex and sometimes on the prothorax; the morphology of these glands, the chemistry of the secretions, and the types of defensive behavior (posturing, stridulation, distribution of secretions) associated with the gland system vary greatly and have been the subject of several recent studies (Eisner *et al.*, 1964, 1974; Kendall, 1968, 1974; Tschinkel, 1975a, 1975b, 1975c; Tschinkel and Doyen, 1980).

Tenebrionids are primarily saprophagous, feeding on a variety of dead plant and animal matter, including humus, leaf litter, rotten wood and cambium, wind-blown detritus, carrion, and dung; some soil-dwelling larvae, however, will feed on living plant roots or seedlings, while some forest-dwellers have become specialized fungus-feeders or facultative predators. Larvae may be divided into 2 groups according to major habitat: 1) xylophilous larvae, which occur in rotten wood and associated cambium and subcortical spaces, and 2) geophilous larvae, which occur in the soil and leaf litter. These will be discussed separately below.

Those larvae which bore directly into wood which has been softened and chemically altered by the action of various fungi include all or most Phrenapatini, Zolodinini, Tenebrionini, Cyphaleini, Amarygmini, Ulomini, Coelometopini, Cnodalonini, and Strongyliini, and some Adeliini, Diaperini, Gnathidiini, Alleculini, and Helopini. The tribes Alphitobiini, Triboliini, Hypophloeini, and Diaperini, include a number of smaller tenebrionids which occur under bark or in the galleries of bark beetles, ambrosia beetles, or wood borers, where they are primarily mycophagous. Larvae of the diaperine *Adelina plana* (Fabricius) have been found in a yeast-rich, fermenting tree wound along with larvae of *Anchorius* (Biphyllidae) and *Cryptarcha* (Nitidulidae). Species of *Corticeus* (Hypophloeini) are common under bark in the galleries of Scolytidae; although the majority are mycophagous, most will feed on larval, pupal or teneral bark beetles, and a few are normally predaceous, if not obligately so (Nikitsky, 1976b; Parker and Davis, 1971; Schedl, 1962; Struble, 1930). Species of *Lyphia* (Triboliini) occur in bamboo and palm fronds, where they prey on larvae of lyctine and dinoderine Bostrichidae (Hinton, 1948b); some *Tribolium* and *Palorus* have also been recorded as semipredators.

Those mycophagous Tenebrionidae which have become specialized for feeding on the reinforced hyphae comprising the relatively large and durable fruiting bodies of Basidiomycetes (especially Polyporaceae, Hymenochaetaceae, and Tricholomataceae) belong to the tribes Bolitophagini (including Rhipidandrini), Diaperini, Dysantini, and Toxicini. Most bolitophagines occur in the large conks of *Ganoderma* and *Fomes* species, but *Eleates* and *Megeleates* are commonly associated with *Fomitopsis pinicola*, a brown rot of conifers, while some species of *Rhipidandrus* and *Eledona* frequent dried mushrooms or smaller polypores. Within the diaperine genus *Platydema*, the degree of host preference varies, and species like *P. oregonense* LeConte and *P. americanum* Laporte and Brullé breed in a variety of fungal types, while *P. neglectum* Triplehorn is restricted to *Cryptoporus volvatus*, a peculiar polypore on conifers, and *P. ellipticum* (Fabricius) breeds only in *Phellinus gilvus* and related Hymenochaetaceae. The 2 species of *Neomida* in N. America appear to have nonoverlapping niches, in that *N. ferruginea* (LeConte) occurs in the larger, brownish fruiting bodies of *Ganoderma* and *Fomes* species, while *N. bicornis* (Fabricius) prefers the smaller, light-colored brackets of *Coriolus versicolor*, *Lenzites betulina*, and related forms. Other fungus-feeding Diaperini include *Alpitophagus bifasciatus* (Say) and species of *Diaperis*, *Pentaphyllus*, and *Ceropria*.

Surface-grazing on algae, lichens, and mosses growing on wood, bark, and occasionally rock surfaces, has been described for *Cylindronotus laevioctostriatus* (Goeze) (Helopini) (Brendell, 1975) but is probably of much wider occurrence within the family. The habit is not common in larvae, but it has been observed in Nilionini, Leiochrini, *Titaena* (Cyphaleini), and a species of *Amarygmus* (Amarygmini) (Jorge, 1974; Watt, 1974b).

Various xylophilous tenebrionids have come to inhabit the nests of birds, ants, or termites, and often the same groups have become stored products pests. Termitophily is known in the Rhysopaussini (probably a specialized group of Amarygmini) and in *Pseudeba novica* Blackburn; myrmecophily occurs in *Tribolium myrmecophilum* Lea; various *Tribolium* are known to inhabit the nests of wild bees; and species of *Tenebrio*, *Alphitobius*, and *Palembus* have been recorded from bird nests. Other invasions of these habitats have involved geophilous species (*see* below). All of the important stored products pests in this family have been derived from xylophilous members of the tribes Tenebrionini, Triboliini, Alphitobiini, and Diaperini; these are listed under *Comments* below.

The geophilous Tenebrionidae include the assemblage of tribes (Tentyriini, Asidini, Pimeliini, etc.) often forming the subfamily Tentyriinae, the Goniaderini, most Adeliini and Lagriini, Phaleriini, most Helaeini and Nyctozoilini, Scaurini, Blaptini and Eleodini, Opatrini (in the broad sense), Trachyscelini, Crypticini, Apocryphini, some Alleculini and some Helopini. The group may be further divided into the surface and litter inhabitants, like Adeliini, Lagriini, Goniaderini, some Opatrini and Apocryphini, and those whose larvae burrow through the soil. Many of the latter group occur in the steppes and deserts, and some, like Phaleriini, Trachyscelini, and some Opatrini, are restricted to littoral regions, where they occur in the soil beneath dune plants or in seaweed, carrion, or other organic debris. Some members of this group, like the Stenosini, Cossyphodini, and some Opatrini (*Scleron, Notibius*) have become associated with ants, while others (some Opatrini) may inhabit nests of ground-nesting birds. Some members of the Eleodini (*Embaphion*) and the Australian helaeine group (*Pterohelaeus* and *Brises*) have invaded caves and probably feed on bat guano. Crop pests in Europe and America occur mainly in the tribes Opatrini, Eleodini, Blaptini, and Alleculini (*see Comments*).

Description: Mature larvae 5 to 70 mm. Body usually elongate and cylindrical to slightly flattened, occasionally short and broad, fusiform, or strongly flattened. Head and all trunk segments usually evenly, lightly to heavily sclerotized, occasionally with head and 9th tergum more heavily sclerotized than rest of body. Surfaces usually smooth, without patches or rows of asperities, sometimes with large, flat-bottomed punctures, occasionally with transverse carinae; abdominal apex sometimes more heavily sculptured; vestiture usually consisting of scattered short and/or long setae (body densely setose in Lagriini, Nilionini, and some groups of Tentyriinae).

Head: Protracted and prognathous to slightly declined, globular or slightly flattened. Epicranial stem usually long, occasionally short, and rarely absent; frontal arms almost always V- or U-shaped. Median endocarina absent. Stemmata 5 or fewer on each side, often inconspicuous or absent. Antennae usually 3-segmented, with elongate 2nd segment and with segment 3 small and much narrower than 2; sometimes with segment 3 highly reduced or absent, so that antennae are only 2-segmented (Lagriinae); rarely with basal segment also reduced, so that only a single segment is apparent (some Leiochrini). Sensorium usually flattened and dome-like, often forming an incomplete ring around the base of segment 3, occasionally elongate and conical or complex (sinuous or subdivided). Antennal insertions lateral, usually adjacent to the mandibular articulations, separated from them by a narrow strip of membrane; sometimes separated by a sclerotized bar which may be broad (Leiochrini, Nilionini). Frontoclypeal suture distinct. Labrum usually transverse, with few to many setae; tormae usually with transverse mesal arms which may be united at midline. Epipharynx usually with an anterior, subanterior, and posterior group of sensilla, a centrally located pair of short setae, an asymmetrical pair of posterior plates, and lateral setose patches. Mandibles usually more or less asymmetrical, short, stout, and subtriangular, with 1 to 3 apical teeth and a sharp incisor edge, sometimes bearing an additional tooth; molae well-developed and usually concave, irregularly tuberculate, or coarsely ridged, sometimes finely transversely ridged, the left one often with a projecting premolar lobe or tooth; prostheca and accessory ventral process absent; basolateral portion of dorsal surface sometimes with an elevated, setose, membranous area; outer edge occasionally carinate. Ventral mouthparts retracted; maxilla with transverse, subtriangular, undivided cardo, slightly elongate stipes, well-developed articulating area, 3-segmented palp, and rounded or truncate mala, which is not cleft, but sometimes slightly notched and only rarely with an uncus at the inner apical angle. Labium almost always with distinct mentum and submentum and usually with the latter separated by a suture from the gula; labial palps 2-segmented, broadly separated to subcontiguous; ligula usually present. Hypopharyngeal bracon present; hypopharyngeal sclerome well-developed, often bicuspidate or tricuspidate, often with setose, membranous elevation in front of it. Hypostomal rods always absent, ventral epicranial ridges rarely present (Toxicini). Gula well-developed, usually subquadrate to elongate (rarely transverse or undifferentiated).

Thorax and Abdomen: Prothorax usually slightly larger than meso- or metathorax. Legs well-developed, 5-segmented; prothoracic legs often much stouter and/or longer, with larger and denser setae or spines; legs usually more or less contiguous, but sometimes moderately widely separated; tarsungulus, especially that of prothoracic leg, sometimes very large and heavily sclerotized, and occasionally divided into 2 parts; 2 tarsungular setae usually unequal in length. Thoracic and abdominal terga usually simple (except for median ecdysial line) or occasionally with transverse carinae, tergum A8 sometimes with well-developed carina or 2 posteriorly-projecting tubercles. Thoracic and abdominal segments sometimes with paired or occasionally median gland openings located on terga or sterna. Tergum A9 usually terminal, extending onto ventral surface, its armature diagnostic at various levels; apex of tergum A9 simple and rounded, triangular, or bearing an acute median process, pair of urogomphi, concave plate or more complex armature; tergum A9 more rarely restricted to dorsal surface so that segment A10 is terminal. Sternum A9 long to very short, and sometimes highly reduced and not clearly separated from segment A10, almost always simple, without asperities. Segment A10 reduced, posteriorly to ventrally located, often with a pair of projecting pygopods, which may be spinose; sternum A9 and segment A10 occasionally concealed beneath the abutting edges of sternum A8 and tergum A9.

Spiracles: Annular, with circular or oval peritreme, which may be crenulate or lined with very small accessory chambers.

Comments: The family includes about 1700 genera and 18,000 species worldwide, with about 200 genera and 1550 species occurring in America north of Mexico. The subfamily and tribal classification has been the subject of recent studies (Doyen, 1972, 1984, 1985; Doyen and Lawrence, 1979; Doyen and Tschinkel, 1982; Watt, 1974b) and is undergoing further revision. Since the subfamilial and tribal concepts used in the discussions above are based on these recent studies, they do not correspond to those given in Arnett (1968). The Lagriinae refers to the family Lagriidae plus the genera *Anaedus, Paratenetus* and *Prateus* in N. America. Tentyriinae includes the tribes usually placed in Tentyriinae and Asidinae plus the Coniontini (*sensu* Doyen, 1972), *Alaephus, Eupsophulus, Alaudes,* and *Cnemeplatia.* Opatrini is used in the broad sense to include most of the Opatrini and Pedinini. Ulomini is restricted to *Uloma* and related genera, such as *Eutochia* and *Ulosonia,* while the rest of the ulomine genera listed in Arnett are placed in Alphitobiini (*Alphitobius*), Triboliini (*Tribolium, Lyphia, Latheticus, Palorus, Mycotrogus,* and *Tharsus*), Diaperini (*Gnatocerus, Cyaneus, Sitophagus, Adelina* (= *Doliema*), and *Doliopines*), or Hypophloeini (*Corticeus*). Goniaderini includes only *Anaedus* in the N. American fauna. The Tenebrionini, as understood here, includes *Tenebrio, Neatus, Zophobas, Scotobaenus, Centronopus,* and *Bius,* while the rest of the genera usually included in that tribe are placed in Coelometopini.

The Tenebrionidae include a number of economically important species which damage crops or more commonly stored products. Those larvae living in the soil and damaging cultivated plants are usually called false wireworms; they include *Eleodes* species in North America, species of *Blaps* (Blaptini), *Opatrum* and *Gonocephalum* (Opatrini), and *Omophlus* and *Podonta* (Alleculini) in Eurasia and North Africa, and species of *Pterohelaeus* (Helaeini) and *Gonocephalum* in Australia. Those species occurring in stored products are secondary pests of varying importance primarily in grains and cereal products. They include: the broad-horned flour beetle, *Gnatocerus cornutus* (Fabricius); slender-horned flour beetle, *G. maxillosus* (Fabricius); *Sitophagus hololeptoides* (Laporte); large black flour beetle, *Cyaneus angustus* (LeConte); 2-banded fungus beetle, *Alphitophagus bifasciatus* (Say); *Palembus ocularis* Casey; yellow mealworm, *Tenebrio molitor* L.; dark mealworm, *T. obscurus* Fabricius; confused flour beetle, *Tribolium confusum* Jacquelin duVal; red flour beetle, *T. castaneum* (Herbst); black flour beetle, *T. audax* Halstead; long-headed flour beetle, *Latheticus oryzae* Waterhouse; depressed flour beetle, *Palorus subdepressus* (Wollaston); small-eyed flour beetle, *P. ratzeburgi* (Wissmann); lesser mealworm, *Alphitobius diaperinus* (Panzer); and black fungus beetle, *A. laevigatus* (Fabricius). Some of these, such as *Tenebrio molitor* and *Alphitobius diaperinus* occur on a wide variety of plant and animal materials, and the latter is commonly found breeding in poultry manure; others like the species of *Gnatocerus, Palorus, Tribolium,* and *Latheticus* are most important as grain pests. Species of *Tribolium* are probably second only to those of *Drosophila* in their importance as study animals in genetics and population ecology.

Selected Bibliography

Aguilar 1962a (agricultural pests).
Aitken 1975 (stored products pests).
Allsopp 1979 (false wireworms in Australia).
Arnett 1968.
Artigas and Brañas-Rivas 1973 (larva and pupa of *Praocis curta* Solier).
Böving and Craighead 1931 (larvae of 17 species (17 genera)).
Brendell 1975 (British species).
Brown 1973 (larvae of *Philolithus densicollis* (Horn) and *Stenomorpha puncticollis* (LeConte)).
Byzova 1958 (larvae of *Belopus procerus* Mulsant, *Cryphaeus cornutus* (Fischer), and *Laena starki* Reitter).
Byzova and Gilyarov 1956 (larvae of 7 species (4 genera) of Helopini).
Byzova and Keleinikova 1964 (larvae of 75 species (53 genera) in key).
Cekalovic and Morales 1974 (larva of *Oligocara nitida* (Solier)).
Cekalovic and Quezada 1973 (larva and pupa of *Emmallodera multipunctata curvidens* Kulzer), 1982 (larva of *Nycterinus abdominalis* Eschscholtz).
Cloudsley-Thompson 1965 (subelytral cavity).
Costa and Vanin 1981 (larvae of *Goniadera* spp.).
Cotton and St. George 1929 (larvae of *Tenebrio molitor* and *T. obscurus*).
Daggy 1947 (pupae of 14 species (12 genera) in key).
Dizer 1953 (larvae of 7 species (2 genera) of Platyscelini), 1955 (subelytral cavity).
Doyen 1972 (classification; larval characters), 1974 (larvae of *Bothrotes plumbeus* (LeConte) and *Lobometopon fusiforme* Casey), 1976a (larvae of *Coelus* spp.), 1984, 1985.
Doyen and Kitayama 1980 (larva and pupa of *Apocrypha anthicoides* Eschscholtz).
Doyen and Lawrence 1979 (larvae of *Alaephus pallidus* Horn, *Phrenapates bennetti* Kirby, *Corticeus praetermissus* (Fall), *Neanopidium simile* Dajoz; notes on larvae of Penetini and Gnathidiini).
Doyen and Tschinkel 1973, 1982 (classification).
Eisner *et al.* 1964.
Eisner *et al.* 1974 (defense glands and stridulation in *Adelium*).
Emden 1947 (first-instars in 7 genera keyed; larvae of 30 species (22 genera) in key; notes on non-British larvae).
Fontes 1979 (larva and pupa of *Uloma misella* Gebien).
Gardner 1929 (larvae of *Derosphaerus crenipennis* Motschulsky and *Setenis* spp.), 1931b (larvae of *Encyalesthus exularis* Gebien, *Strongylium* spp., and *Uloma rubripes* Gebien), 1932c (larva of *Catapiestus indicus* Fairmaire).
Gebien 1922 (larva of *Pseudhadrus seriatus* Kiobe).
Ghilarov 1964a (larva of *Largia hirta* (Linnaeus)).
Ghilarov and Svetova 1963 (larva of *Hedyphanes seidlitzi* Reitter).
Hayashi 1964 (larvae of 4 species (4 genera) of Lagriinae), 1966 (larvae of 58 species (40 genera); larval key), 1968 (larvae of 10 species (10 genera), including *Heterotarsus carinula* Maresul).
Hinton 1948b (biology of *Lyphia* spp.).
Hyslop 1915 (larva and pupa of *Meracantha contracta* (Beauvois)).
Jorge 1974 (larva and pupa of *Nilio varius* Ihering).
Keleinikova 1959 (larvae of 5 species (3 genera) of Tentyriini), 1961a (larvae of 1 species of Platyopini and 4 species (3 genera) of Pimeliini in key), 1961b (larvae of 2 species (1 genus) of Akidini), 1961c (larvae of 5 species (5 genera) of Opatrini in key), 1961d (larva of *Cyphogenia aurita* (Pallas)), 1962 (larvae of 2 species (2 genera) of Erodiini), 1963 (larval types in Tenebrionidae), 1966 (larvae of 10 species (7 genera) of Pedinini in key; 6 described), 1968 (larvae of 10 species (6 genera) of Platyopini and 13 species (10 genera) of Platyscelini), 1970 (larvae of *Colposphaena karelini* Menetries and 4 species (4 genera) of Tentyriini),

1971 (first instars of tentyrioid and tenebrioid Tenebrionidae), 1976 (larvae of *Stenosis fausti* Reitter and *Dichillus reitteri* Semenov).

Kendall 1968, 1974 (defense glands in Lagriini and Alleculinae).

Koch 1962 (adult modifications in desert species).

Korschefsky 1943 (larvae of 39 species (32 genera)).

Lawrence and Medvedev 1982 (larva of *Hyocis occidentalis* Blackburn).

Liles 1956 (biology, larva, pupa of *Bolitotherus cornutus* (Panzer)).

Marcuzzi and Cravera 1981 (larvae of 22 species (7 genera) from West Indies).

Marcuzzi, Cravera, and Faccini 1980 (larvae of 10 species (8 genera)).

Marcuzzi and Floreani 1962.

Marcuzzi and Rampazzo 1960 (larvae of 19 species (13 genera)).

Matthews 1986 (revision of *Brises;* classification).

Mattoli 1974 (larva and pupa of *Neomida haemorrhoidalis* (Fabricius)).

Medvedev 1965 (adult leg modifications in desert and steppe species), 1968 (larvae of 12 genera of Opatrinae in key).

Moore 1974 (larva of *Phaleria rotundata* LeConte).

Nikitsky 1976b (larvae of *Corticeus* spp.), 1983a (larva of *Myrmechixenus subterraneus* Chevrolat).

Ogloblin and Znoiko 1950 (larvae of 11 species (6 genera) of Alleculini in key).

Pace 1967 (habits and life history of *Bolitotherus cornutus*).

Parker and Davis 1971 (feeding habits of *Corticeus substriatus*).

Peterson 1951 (larvae of 10 species (8 genera)).

Pierre and Balachowsky 1962 (agricultural pests).

St. George 1924 (larva and pupa of *Merinus laevis* (Olivier); larvae of 14 genera in key), 1926 (larvae of 4 species (2 genera)), 1930 (larva of *Leichenum variegatum* Küster), 1950 (larvae of 21 genera in key).

Schedl 1962 (tenebrionids associated with bark beetles).

Schulze 1962 (larvae of *Lepidochora discoidalis* Gebien and *Onymacris rugatipennis* Haag; larvae of 15 species (5 genera) of Adesmiini and 14 species (10 genera) of Eurychorini in keys), 1963 (larvae of 8 species (7 genera) of Opatrini), 1964a (larvae of *Bantodemus zulu* Koch, *Quadrideres femineus* (Lesne), and *Zophodes fitzsimonsi* Koch), 1964b (larvae of 11 species of *Onymacris* in key), 1968 (larvae of 10 species (4 genera) of Helopinini), 1969 (larvae of *Carchares macer* Pascoe and *Herpiscius sommeri* Solier; comparisons with Opatrini), 1978 (larvae of 5 species of *Gonopus*).

Seely 1976 (fog basking in *Onymacris*).

Seely and Hamilton 1976 (construction of fog catchment trenches by *Lepidochora* spp.).

Skopin 1959 (larva of *Adesmia* sp.), 1960a (larva of 31 species (6 genera) of Blaptini), 1960b (larva of 7 species (4 genera) of Akidini), 1961 (larvae of *Diaphanidius semenowi* Reitter and *Arthrodosis lobicollis* Reitter), 1962 (larvae of 83 species (33 genera) of Pimeliini), 1964 (larvae of 20 species (12 genera) of Pycnocerini), 1978 (larvae of 60 genera in keys).

Spilman 1966 (larva and pupa of *Amarygmus morio* (Fabricius)), 1979 (larvae and pupae of *Centronopus* spp.), in press (larvae of 23 species (13 genera) in stored products in key).

Striganova 1961b, 1964a (larvae of 14 species (9 genera) in keys).

Struble 1930 (biology, larva, and pupa of *Corticeus substriatus* (LeConte)).

Tschinkel 1975a, 1975b, 1975c.

Tschinkel and Doyen 1980.

Wade and St. George 1923 (larva and biology of *Eleodes suturalis* Say).

Watt 1967a (classification), 1971 (larvae of *Pseudhelops* spp.), 1974b (larval characters; keys to subfamilies and some tribes; larvae of *Zolodinus zeelandicus* Blanchard, *Menimus* spp., *Archaeoglenes costipennis* Broun, *Lepispilus* spp., and *Nyctoporis* spp.; pupae of *Z. zeelandicus* and *Nyctoporis* spp.).

Young 1976a (larva of *Dioedus punctatus* LeConte).

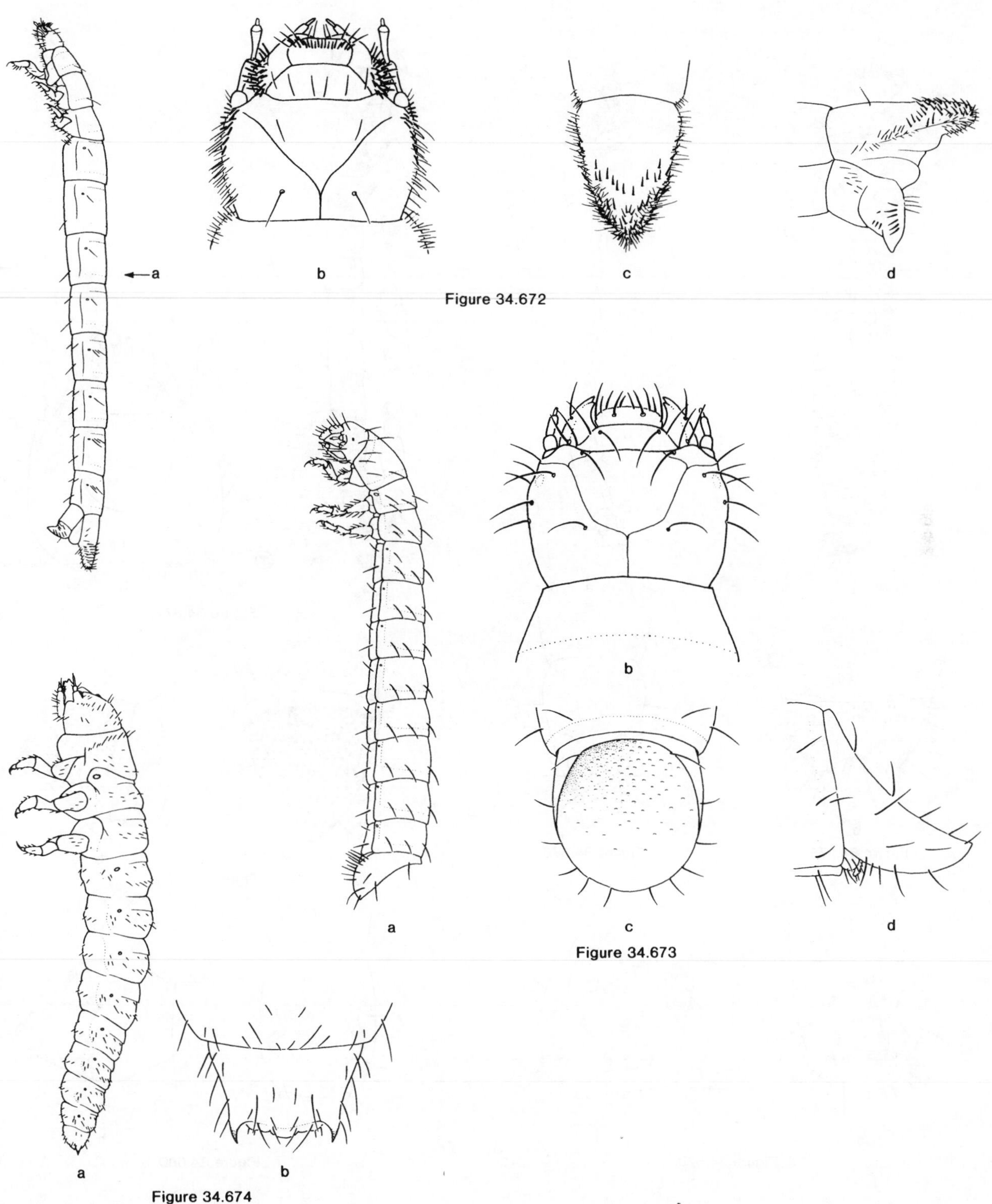

Figures 34.672a-d. Tenebrionidae. *Epitragus* sp. (Tentyriini). **a.** larva, lateral; **b.** head, dorsal; **c,d.** abdominal apex, dorsal and lateral.

Figures 34.673a-d. Tenebrionidae. *Meracantha contracta* (Beauvois) (Amarygmini). **a.** larva, lateral; **b.** head, dorsal; **c,d.** abdominal apex, dorsal and lateral.

Figures 34.674a,b. Tenebrionidae. *Bolitotherus cornutus* (Panzer) (Bolitophagini). **a.** larva, lateral; **b.** abdominal apex, dorsal. Also *see* figure 34.28 for anterior view of head. Feeds in bracket fungi.

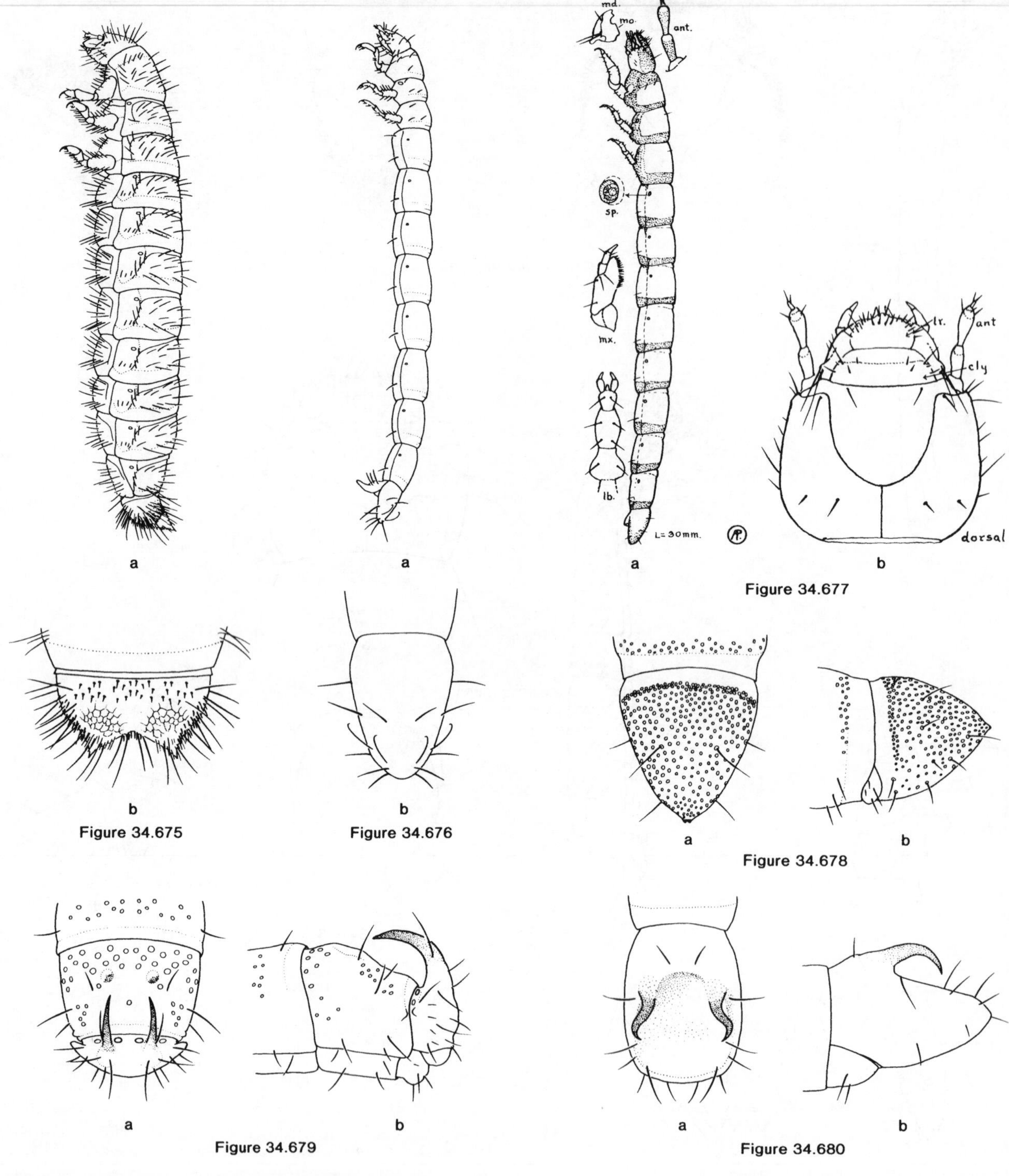

Figures 34.675a,b. Tenebrionidae. *Arthromacra aenea* (Say) (Lagriini). **a.** larva, lateral; **b.** abdominal apex, dorsal.

Figures 34.676a,b. Tenebrionidae. *Hymenorus niger* (Melsheimer) (Alleculini). **a.** larva, lateral; **b.** abdominal apex, dorsal.

Figures 34.677a,b. Tenebrionidae. *Capnochroa fuliginosa* Melsheimer (Alleculini). **a.** larva, lateral. Length = 30 mm; **b.** head, dorsal. Under bark of dead or dying trees and in decayed vegetation. (From Peterson, 1951)

Figures 34.678a,b. Tenebrionidae. *Uloma imberbis* LeConte (Ulomini). Abdominal apex, dorsal and lateral.

Figures 34.679a,b. Tenebrionidae. *Helops pernitens* LeConte (Helopini). Abdominal apex, dorsal and lateral.

Figures 34.680a,b. Tenebrionidae. *Dioedus punctatus* LeConte (Phrenapatini). Abdominal apex, dorsal and lateral.

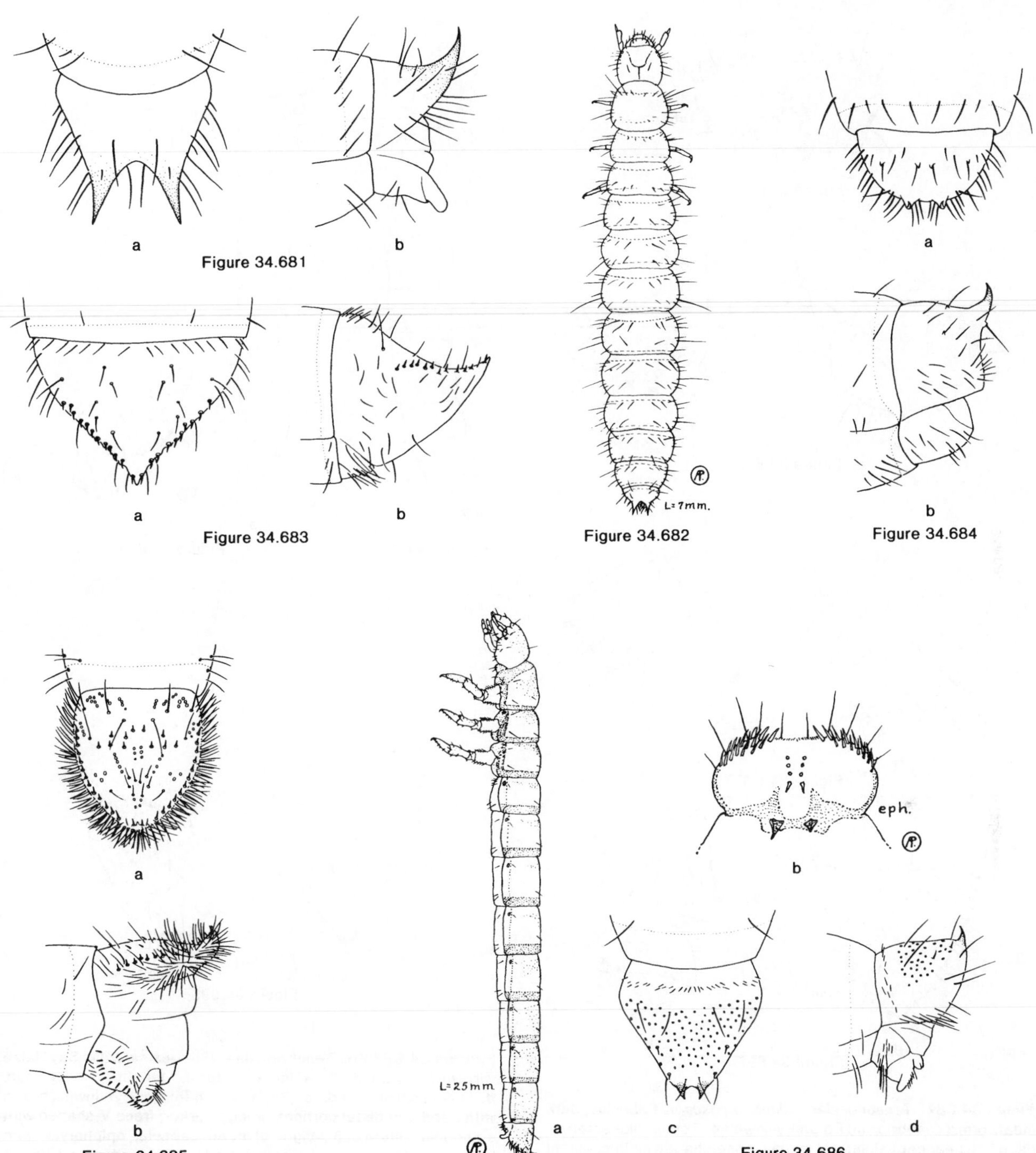

Figures 34.681a,b. Tenebrionidae. *Tribolium castaneum* (Herbst), red flour beetle (Triboliini). Abdominal apex, dorsal and lateral. Habits similar to *T. confusum* (see below).

Figure 34.682. Tenebrionidae. *Tribolium confusum* (Olivier), confused flour beetle. Larva, dorsal. Length = 7 mm. A common pest worldwide in grains, grain products and a variety of other dried foods. (from Peterson 1951)

Figures 34.683a,b. Tenebrionidae. *Asidopsis polita* (Say) (Asidini). Abdominal apex, dorsal and lateral.

Figures 34.684a,b. Tenebrionidae. *Alobates pennsylvanica* (De Geer) (Coelometopini). Abdominal apex, dorsal and lateral.

Figures 34.685a,b. Tenebrionidae. *Coniontis nemoralis* Eschscholtz (Coniontini). Abdominal apex, dorsal and lateral.

Figures 34.686a-d. Tenebrionidae. *Tenebrio molitor* L., yellow mealworm. (Tenebrionini) **a.** larva, lateral, length = 25 mm; **b.** epipharynx, **c,d.** abdominal apex, dorsal and lateral. Yellowish-brown, with dorsum, especially head and caudal segments, more reddish-brown; epipharynx with 12 ± setae near laterocephalic margin of each lateral half; ninth segment cone-shaped and terminating in 2 prominent, curved, sharp-pointed urogomphi which project dorsad. Infests refuse grain, coarse cereal bran, and mill products, especially if somewhat moist. Commonly reared for fish bait and pet food. (Figures a,b from Peterson, 1951)

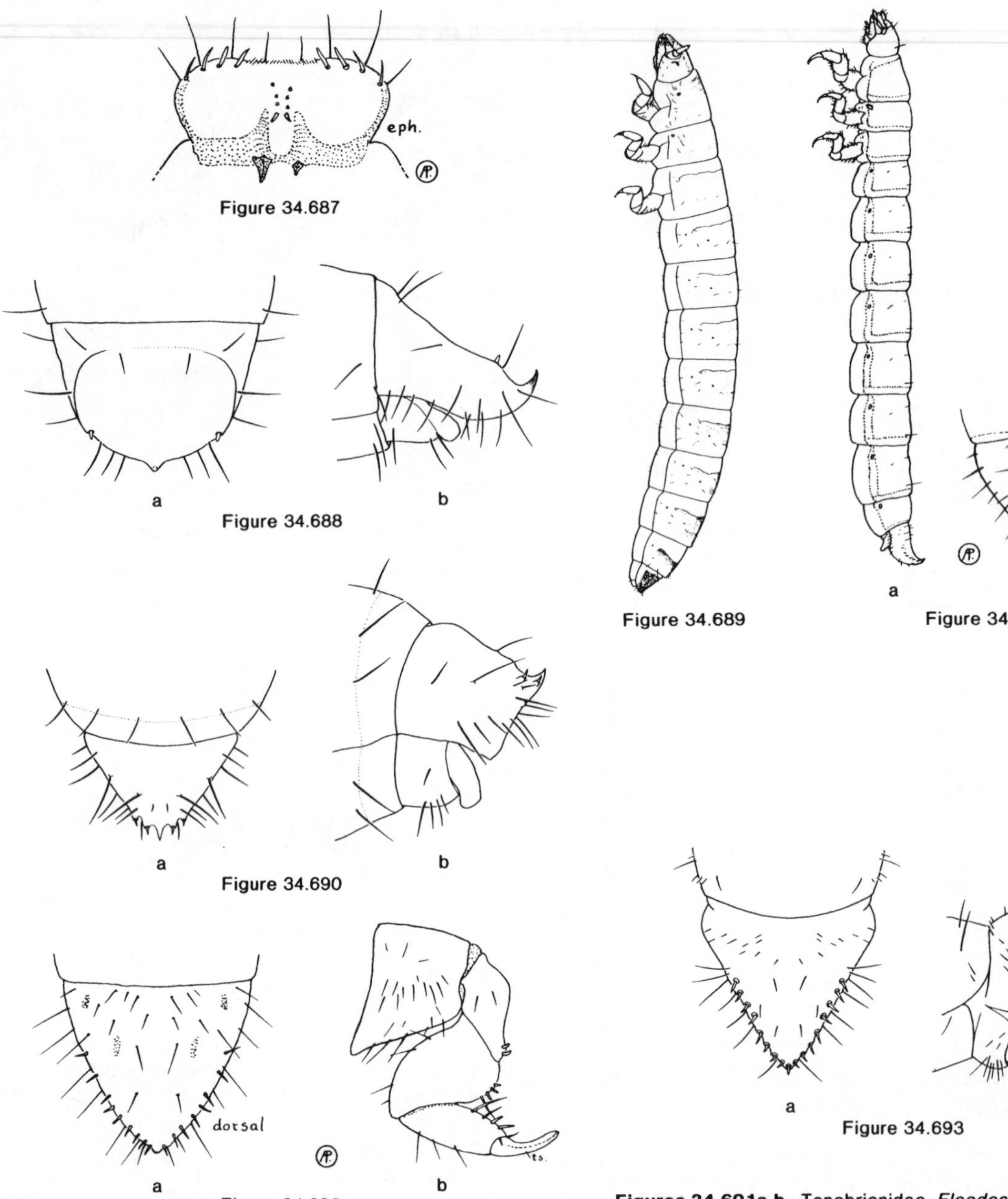

Figure 34.687

Figure 34.688

Figure 34.689

Figure 34.691

Figure 34.690

Figure 34.692

Figure 34.693

Figure 34.687. Tenebrionidae. *Tenebrio obscurus* Fabricius, dark mealworm. (Tenebrionini) Epipharynx with 4–5 spine-like setae on each half near cephalolateral margin; cone-shaped ninth segment projects caudad and terminates in 2 short, pointed urogomphi which project caudodorsad; also possesses two short spines on each lateral aspect cephalad of the urogomphi. Food habits similar to *T. molitor* L. (From Peterson, 1951)

Figures 34.688a,b. Tenebrionidae. *Phaleria rotundata* LeConte (Phaleriini). Abdominal apex, dorsal and lateral.

Figure 34.689. Tenebrionidae. *Corticeus praetermissus* (Fall) (Hypophleoini). Larva, lateral. (From Doyen and Lawrence, 1979).

Figures 34.690a,b. Tenebrionidae. *Platydema ellipticum* (Fabricius) (Diaperini). Abdominal apex, dorsal and lateral. *See* also figure 34.46, anterior part of head, dorsal, with frontoclypeal region, labrum-epipharynx, and mandibles removed, showing hypopharyngeal sclerome, and figure 34.45, right mandible, ventral.

Figures 34.691a,b. Tenebrionidae. *Eleodes suturalis* Say, false wireworm. (Eleodini) **a.** larva, lateral, length = 24 mm; **b.** abdominal apex, dorsal. Surface corneous and yellowish brown with head and distal portions of legs darker; frons V-shaped with epicranial suture .25 length of head capsule; epipharynx with broad, caudal, transverse, sclerotized band possessing 1 pair of stub-like, short, teeth near meson; prothoracic legs distinctly larger than others; lateral margins of upturned abdominal apex with 7 to 9 short spine-like setae. Infests soil planted to wheat or other small grains west of Mississippi River. (From Peterson, 1951)

Figures 34.692a,b. Tenebrionidae. *Eleodes opacus* (Say), plains false wireworm. (Eleodini) **a.** abdominal apex, dorsal. **b.** prothoracic leg. Yellowish with surface glabrous except ventral surface of head and thorax, legs and pygidium; prothoracic legs almost spineless but others spinose; abdominal apex bearing 2 small spines and lateral margins with 6 or 7 blunt spines. Found in soil of wheat west of Mississippi River. (From Peterson, 1951)

Figures 34.693a,b. Tenebrionidae. *Eleodes obsoletus* (Say) (Eleodini). Abdominal apex, dorsal and lateral.

CEPHALOIDAE (TENEBRIONOIDEA)

(INCLUDING NEMATOPLIDAE, STENOTRACHELIDAE)

John F. Lawrence,
Division of Entomology, CSIRO

Figures 34.695–697

Relationships and Diagnosis: The limits of the family are still in dispute, and the inclusion of Stenotrachelinae by Crowson (1955, 1966b) and Arnett (1968) has been rejected by Mamaev (1973), who considered this group and also *Nematoplus* to be in separate families, and by Hayashi (1963) who placed *Stenocephaloon* in the Melandryidae. The larvae of *Stenotrachelus* and *Stenocephaloon* closely resemble those of *Synchroa* (Synchroidae) and *Phellopsis* (Zopheridae); the former differ in having a pair of basal asperities on sternum A9 and a median tooth on the prosternum, while the latter differ in having more symmetrical mandibles and no transverse groove on tergum A9. Larvae of *Nematoplus* and *Cephaloon* bear a strong resemblance to those of Oedemeridae and bolitophagine Tenebrionidae in the lightly sclerotized body, asymmetrical head capsule, and highly asymmetrical, transversely ridged, mandibular molae, and both groups have been considered likely relatives (Lawrence and Newton, 1982; Mamaev, 1973). Oedemerid larvae differ from those of cephaloids in having either abdominal "prolegs" (asperity-bearing ampullae) or patches of tergal asperities, while bolitophagine tenebrionids have annular spiracles. The family Meloidae has also been considered close to Cephaloidae (Abdullah, 1965; Lawrence and Newton, 1982), but all meloid larvae lack a mandibular mola.

Biology and Ecology: Little is known about the habits of adult cephaloids, but they are relatively rare and probably short-lived. Some have been collected from flowers (Arnett, 1968) and the mouthparts are consistent with pollen- or nectar-feeding, but Mamaev (1973) suggested that adult feeding in *Nematoplus* might not be necessary for egg maturation. Larvae of all species feed on decaying wood. Those of *Stenotrachelus aeneus* Paykull are found under bark or in the wood of dead angiosperms, including *Salix, Populus, Alnus,* and *Betula,* and are only rarely associated with conifiers (Palm, 1959; Saalas, 1913). *Nematoplus* larvae feed in highly decomposed wood of stumps or logs which are infested with brown rots (fungi which selectively destroy cellulose); the wood is usually somewhat reddish and may be referred to as being in the red rot stage. At least some species of *Cephaloon,* like *C. variabilis* Motschulsky and *C. ungulare* (LeConte), have similar habits, but according to Mamaev (1973), *C. pallens* Motschulsky is usually found in large branches of deciduous trees.

Description: Mature larvae 10 to 25 mm. Body elongate, more or less parallel-sided, straight, subcylindrical to slightly flattened, lightly sclerotized except for buccal region and occasionally (Stenotrachelinae) tips of urogomphi; dorsal surfaces usually smooth; vestiture of scattered, simple, setae.

Head: Protracted and prognathous, broad, usually asymmetrical, sometimes broader than thorax, slightly flattened. Epicranial stem long and frontal arms V-shaped in *Cephaloon* and *Nematoplus;* epicranial stem absent and frontal arms lyriform and contiguous at base in Stenotrachelinae. Median endocarina usually Y-shaped, coincident with epicranial stem and frontal arms (absent in Stenotrachelinae). Stemmata 5 or 6 on each side, sometimes absent. Antennae well-developed, 3-segmented. Frontoclypeal suture present in *Nematoplus* only. Mandibles strongly asymmetrical, tridentate, without accessory ventral process; molae large and transversely ridged, the left one more or less parallel to long axis and produced at apical end, the right one strongly oblique; prostheca absent. Ventral mouthparts retracted. Maxilla with transverse cardo, elongate stipes, well-developed articulating area, 3-segmented palp, and rounded or truncate mala, which may be weakly emarginate at apex and may bear 1 or more teeth at inner apical angle. Labium free to base of mentum; ligula well-developed; labial palps 2-segmented. Hypopharyngeal sclerome tooth-like. Hypostomal rods absent in *Cephaloon* and *Nematoplus,* short and diverging in Stenotrachelinae. Gula transverse.

Thorax and Abdomen: Legs well-developed, 5-segmented; tarsungulus with 2 setae lying side by side. Meso- and metatergum and abdominal terga 1–6 with rows and patches of asperities in Stenotrachelinae. Tergum A9 granulate or tuberculate in Stenotrachelinae and bearing a pair of upturned urogomphi, sclerotized at apex; tergum A9 smooth in *Cephaloon* and *Nematoplus,* the former with a pair of posteriorly projecting, straight, lightly sclerotized urogomphi, and the latter without urogomphi; sternum A9 simple. Segment A10 transverse, more or less posteriorly oriented in *Cephaloon* and *Nematoplus,* ventrally or posteroventrally oriented in Stenotrachelinae.

Spiracles: Annular-biforous, with long or short accessory tubes.

Comments: The family is a small one, containing about 20 species and 7 genera restricted to the cooler parts of the Holarctic Region; the N. American fauna consists of 4 genera and 10 species. Three distinct subfamilies are recognized: Cephaloinae (*Cephaloon*), Nematoplinae (*Nematoplus, Pedilocephaloon*), and Stenotrachelinae (*Stenotrachelus, Stenocephaloon, Anelpistus,* and *Scotodes*).

Selected Bibliography

Abdullah 1965.
Arnett 1953, 1968.
Böving and Craighead 1931 (larva of *Cephaloon lepturides* Newman).
Crowson 1955, 1966b.
Hayashi 1963 (larva of *Cephaloon pallens* (Motschulsky) and *Stenocephaloon metallicum* Pic).
Lawrence and Newton 1982.
Mamaev 1973 (larvae of *Cephaloon pallens, C. variabilis* Motschulsky, and *Nematoplus semenovi* Nikitskii).
Palm 1959.
Peterson 1951 (larva of *Cephaloon lepturides*).
Saalas 1913 (larva of *Stenotrachelus aeneus* (Fabricius)).

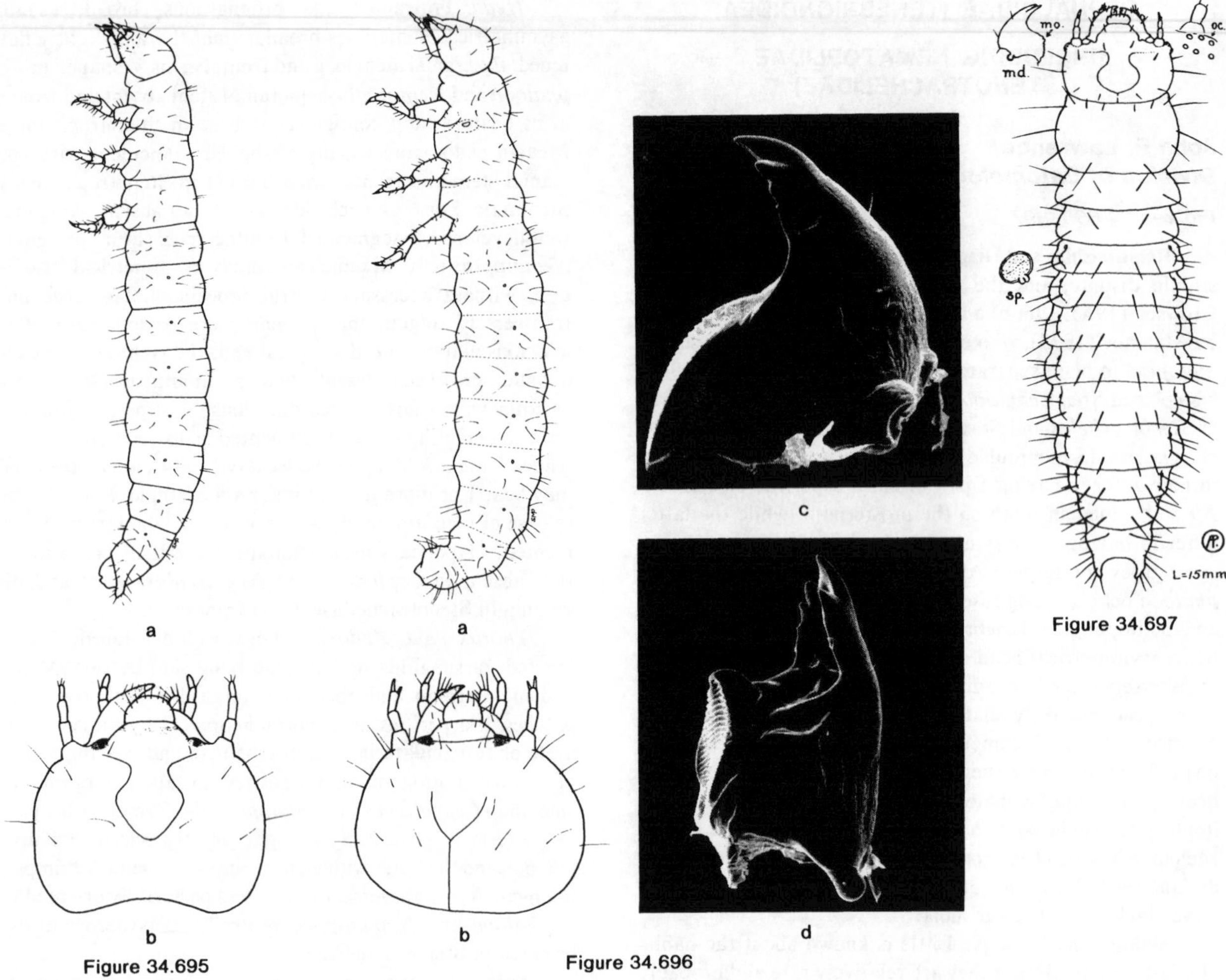

Figure 34.695

Figure 34.696

Figure 34.697

Figures 34.695a,b. Cephaloidae. *Cephaloon ungulare* (LeConte). Jefferson Notch, Coos Co., New Hampshire. **a.** larva, lateral, length = 22.6 mm; **b.** head, dorsal.

Figures 34.696a-d. *Cephaloidae. Nematoplus collaris* LeConte. Jefferson Notch, Coos Co., New Hampshire. **a.** larva, lateral, length = 17.4 mm; **b.** head, dorsal; **c.** right mandible, dorsal; **d.** left mandible, ventral.

Figure 34.697. *Cephaloidae. Cephaloon* sp. Larva, dorsal (plus details). Length = 15 mm. (From Peterson, 1951)

MELOIDAE (TENEBRIONOIDEA)

Richard B. Selander, *University of Illinois*

Blister Beetles, Oil Beetles

Figures 34.698–709

Relationships and Diagnosis: The Meloidae and Rhipiphoridae were isolated from other Heteromera (= Tenebrionoidea) as Meloidea by Böving and Craighead (1931) because of anatomical and behavioral similarities in the first larval instar; some authors would add the Stylopoidea (= Strepsiptera). Meloid larvae are specialized predators of insect eggs, larvae and provisions, whereas rhipiphorid and stylopoid larvae are parasitoids, each attacking a single insect host and spending at least part of the feeding period within its body. Moreover, on the basis of adult anatomy, the meloids and rhipiphorids are more similar to "non-parasitic" Anthicidae and Mordellidae-Scraptiidae, respectively, than to each other, and the Stylopoidea are distinctive enough to warrant assignment to a separate superfamily (Crowson, 1955; Selander, 1964) or an independent order Strepsiptera as is done here (Lawrence and Newton, 1982).

Meloidae exhibit hypermetamorphosis, the larva typically passing through 4 phases (fig. 34.698): triungulin (T), first grub (FG), coarctate (C), and second grub (SG). T larvae are campodeiform or navicular, heavily sclerotized, and active. Phoretic triungulins, in particular, may be confused with first instar rhipiphorids but have 1 or 2 (rather than several) stemmata on each side of the head, well developed labial palpi (except in *Tetraonyx*), a line of dehiscence on the thorax, and lack pulvilli. FG larvae superficially resemble larval Scarabaeidae, from which they differ in having the antennae

3-segmented, the mandibles similar to each other and lacking a definite molar area, the tarsungulus with 2 (1 in *Tetraonyx*) setae, and the abdomen 9-segmented and lacking a raster. C larvae, which are heavily sclerotized and immobile and have the mouthparts and legs reduced to stubs, are more likely to be mistaken for Diptera puparia than for beetle larvae. SG larvae are generally similar to FG larvae (*see* below).

Biology and Ecology: So far as known (Eleticinae have not been studied), larval meloids are predatory. Most Meloinae and all Nemognathinae attack the nesting cells of bees, where they consume both the immature host and the provisions in 1 or more cells. Most species of the subtribes Epicautina and Mylabrina (Meloinae) prey on grasshopper eggs. Some Epicautina, at least, are evidently predators of meloid eggs (Selander, 1981). Phoresy of T larvae on adult bees is characteristic of the Meloini (in the Meloinae) and all Nemognathinae.

The ontogenetic cycle of Meloidae is diagrammed in fig. 34.698. Typically, the larva, after beginning to feed in the first instar, grows rapidly through 4 instars of the FG phase (FG_{2-5}), passes a more or less extended period of inactivity in the C phase (C_6), and finally reverts to a grublike form (SG_7) before becoming a pupa. In most Meloinae the C larva and pupa occupy chambers in the soil excavated during the preceding instar. In Nemognathinae the FG_5 and C_6 exuviae are not cast off by the larva but rather encapsulate it.

In response to high temperature, many *Epicauta* develop directly from the FG phase to the pupal stage. In addition, larvae of this genus may have an extra instar (FG_6) in the FG phase. Presumably in response to adverse conditions, larvae of some Lyttini may revert to the C phase after reaching the SG phase. Rarely, some of the subtribe Lyttina skip the SG phase. Most meloids pass the winter or dry season as C larvae but a few do so by diapausing in the egg, T larval phase, or adult stage.

Eggs cylindrical, white to orange, without sculpturing, laid in masses of a few score to a few thousand in soil (Meloinae), on plants (especially bracts) (most Nemognathinae), or near or in nests of host bees (degenerate nemognathines such as *Tricrania* and *Hornia*).

Pupa typical coleopteran, with (Meloinae, fig. 34.709) or without bracing spines on dorsum.

Description: The larval phases are so distinctive that it is convenient to treat them separately.

T Phase: (figs. 34.699–34.705) Length 0.6–4.5 mm. Body heavily sclerotized, campodeiform or (Nemognathinae) navicular, with conspicuous spiracles on mesothorax and abdominal segments 1–5 (Tetraonycini) or 1–8, a definite pattern of setae on body and legs, and generally a line of dehiscence on thorax, often extending onto abdomen.

Head: Prognathous, often with a basal transverse ridge in phoretic larvae. Epicranial suture well developed, with frontal sutures shortened, reduced to coronal suture, or entirely absent. Gula well developed, with 2 setae (in Meloinae). One or (Nemognathini) 2 stemmata on each side of head. Antennae 3-segmented, inserted before middle of head, with sensory organ (often cone-shaped) at apex of segment 2; segment 3 with long terminal seta. Labrum distinct, closely applied ventrally to head capsule (subtribe Pyrotina), or fused with frontoclypeus to form a nasale (Nemognathinae). Mandibles working horizontally in non-phoretic larvae, more or less vertically in phoretic ones, smooth, dentate, or (many Nemognathinae) with transverse ridges. Maxillae with stipes large, cardo distinct or not; palps 3-segmented, conspicuous. Labium with mentum (1st and 2nd prementa) weakly sclerotized; palps 2-segmented (absent in Tetraonycini).

Thorax: Nota well sclerotized, sterna weakly so in most Meloinae, strongly so in Nemognathinae. Legs slender, elongate, 5-segmented; tarsungulus conical, set with 1 (Tetraonycini), 2, or several (some of the subtribe Mylabrina) slender setae, or spatulate and with 2 spatulate setae, forming a trident (*e.g.*, many *Meloe*, fig. 34.701). Pulvilli absent.

Abdomen: With 9 segments of more or less uniform length and a much reduced tenth segment; terga heavily sclerotized, sterna variably so; tergites with an anterior median group of minute setae and median transverse and posterior marginal rows of setae; pleurites distinct or not. Spiracles of segment 8 on hooklike elevations in Nemognathini, those of segment 1 on large lateral extensions in Tetraonycini. Urogomphi absent; end of abdomen with a pair of large caudal setae except in Nemognathini.

FG Phase: (figs. 34.706, 34.707a) Maximum length 5–25 mm. Differs from T phase as follows: Thorax and abdomen membranous, pale, the body progressively more scarabaeiform with growth. Body setae often numerous in later instars. Spiracles present on mesothorax, metathorax, and abdominal segments 1–8, often dorsal in position in early instars.

Head: Initially prognathous, becoming hypognathous with growth. Epicranial suture well developed, fine. Stemmata replaced by subcuticular black eye spots. Antennae with sensory organ small; segment 3 becoming shorter in successive instars, lacking a long terminal seta. Labrum fleshy, exposed. Mandibles massive, dentate or not. Labial palps always present, 1 or 2-segmented.

Thorax: Pronotal and prosternal plates present in later instars. Tarsungulus conical, bearing 2 short setae.

Abdomen: Segments lobed, with a distinct ridge lateroventrally. Spiracles small, round.

C Phase: (figs. 34.707b, 34.708) Body less C-shaped than in late instar FG larva; cuticle thick, very heavily sclerotized, leathery, brown, glabrous. Body segments fused. Appendages reduced to unsegmented stubs, fused to body. Spiracles normally oval, elevated, those on abdominal segment 8 vestigial in Nemognathinae. Oral cavity and anus closed. Musculature vestigial.

SG Phase: (fig. 34.707c) Much like late-instar FG larva, but with body setae shorter and somewhat sparser, head less strongly sclerotized, labial palps often vestigial, legs shorter and thicker, tarsungulus vestigial or absent, and spiracles oval.

Spiracles: *See* each phase above.

Comments: A large family (about 2500 species) of considerable ecological importance because of its larval biology. In addition, adults of some species (especially of *Epicauta* and *Mylabris*) are crop pests. T (triungulin) anatomy is extensively employed taxonomically, due largely to the pioneering work of Cros (1940) and MacSwain (1956); other larval phases have been largely neglected in this regard.

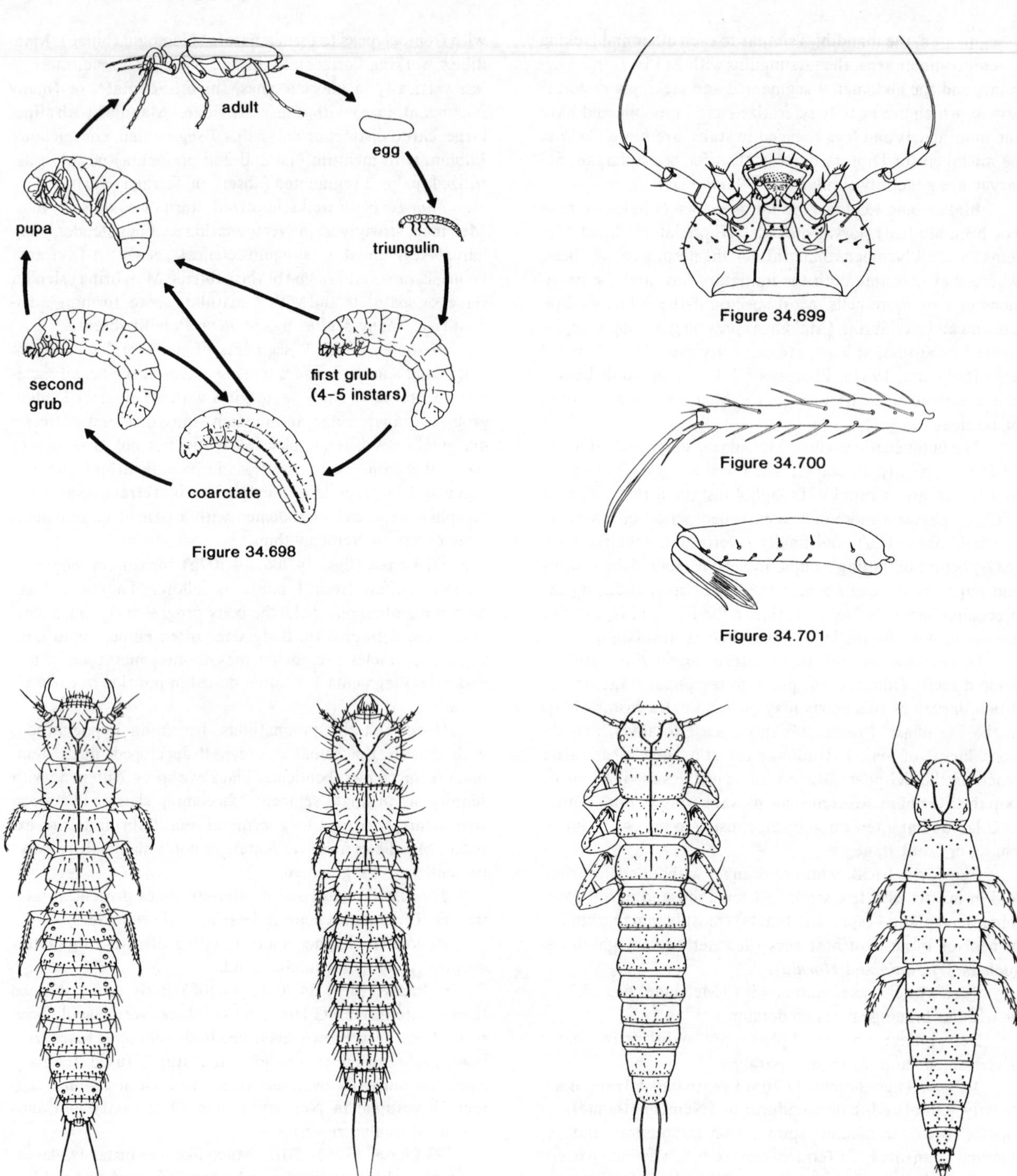

Figures 34.698–34.705. Meloidae.

Figure 34.698. Ontogenetic cycle. (Redrawn from Selander & Weddle, 1969, reprinted with permission from the Entomological Society of America.)

Figure 34.699. *Meloe americanus* Leach, T_1, head, ventral (after Pinto & Selander, 1970). © 1970, reprinted with permission of the Board of Trustees, University of Illinois.

Figure 34.700. *Megetra cancellata* (Brandt & Erichson), T_1, hind tibia and tarsungulus.

Figure 34.701. *Meloe angusticollis* Say, T_1, hind tibia and tarsungulus. Note the "3-clawed" (triungulin) appearance.

Figure 34.702. *Lytta mutilata* Horn, T_1.

Figure 34.703. *Epicauta segmenta* (Say), T_1.

Figure 34.704. *Meloe dianella* Pinto & Selander, T_1.

Figure 34.705. *Nemognatha piezata* (Fabricius), T_1.

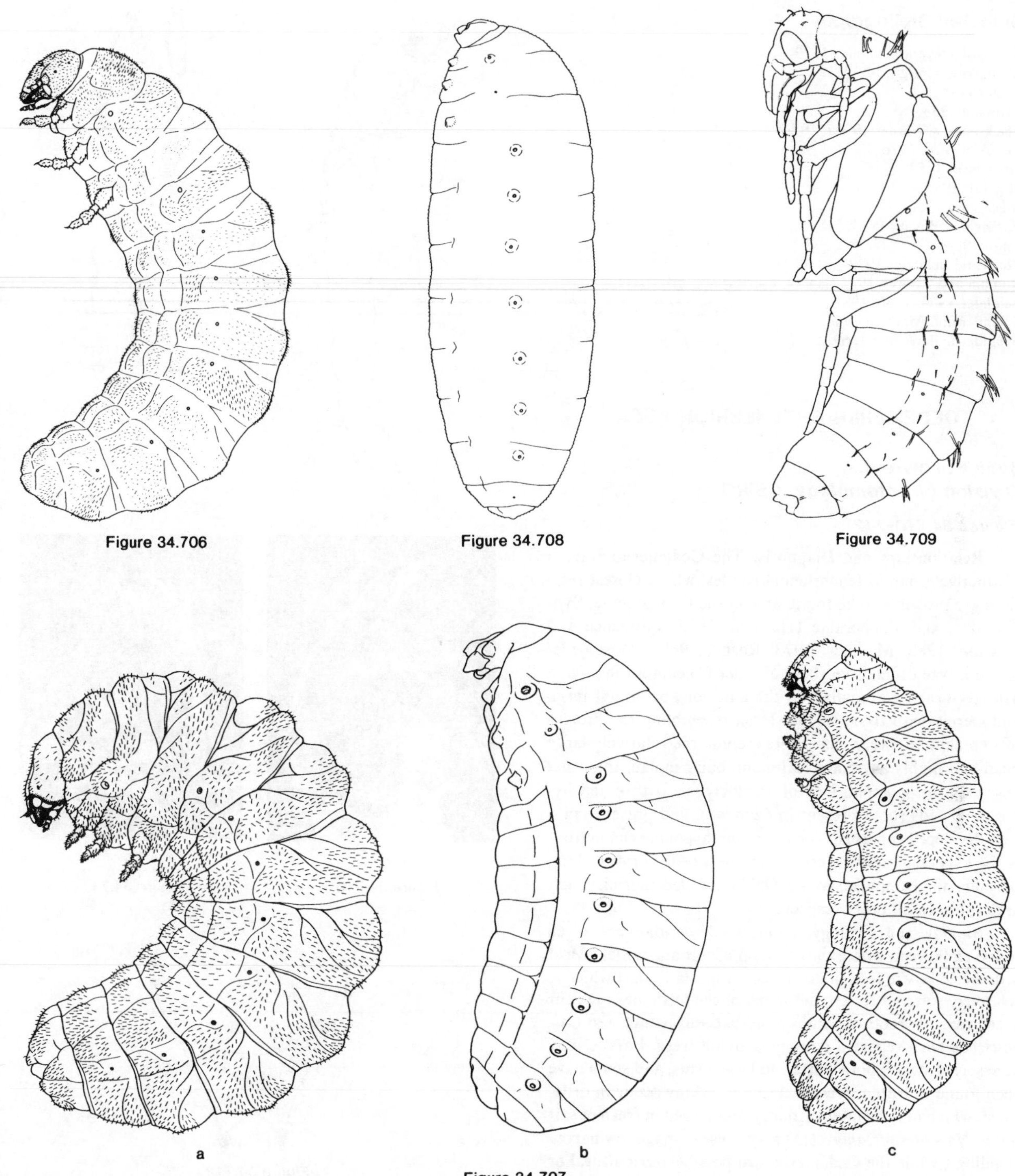

Figure 34.706. Meloidae. *Meloe dianella*, FG_5 (after Pinto & Selander, 1970). © 1970, reprinted with permission of the Board of Trustees, University of Illinois.

Figures 34.707a-c. Meloidae. *Epicauta segmenta*, **a.** FG_5; **b.** C_6; **c.** SG_7.

Figure 34.708. Meloidae. *Nemognatha lutea* LeConte, C_6.

Figure 34.709. Meloidae. *Pyrota palpalis* Champion, pupa (after Selander & Mathieu, 1964).

Selected Bibliography

Berrios-Ortiz and Selander 1979.
Böving and Craighead 1931.
Cros 1940.
Crowson 1955.
Erickson and Werner 1974a, 1974b.
Erickson *et al.* 1976.
Greathead 1963.
Horsfall 1943.
MacSwain 1956.
Parker and Böving 1914.
Pinto 1984.
Pinto and Selander 1970.
Rempel and Church 1972 (last of a series).
Selander 1964, 1966, 1981.
Selander and Mathieu 1964.
Selander and Weddle 1969, 1972.

OEDEMERIDAE (TENEBRIONOIDEA)

John F. Lawrence,
Division of Entomology, CSIRO

Figures 34.710-712

Relationships and Diagnosis: The Oedemeridae are a distinctive group of tenebrionoid beetles, whose closest relatives are probably to be found among the Cephaloidae, Synchroidae, and Zopheridae (Hayashi, 1975; Lawrence and Newton, 1982; Mamaev, 1973; Rozen, 1960). Most oedemerid larvae differ from those of other Coleoptera in having asperity-bearing ampullae on the anterior abdominal terga and sterna, with those on the sterna resembling the prolegs of Lepidoptera; other characters include the relatively large head and highly asymmetrical mandibular molae, the usual absence of urogomphi, lack of frontoclypeal suture and hypostomal rods (both present in *Calopus*), and presence of a well-developed and usually columnar prehypopharynx in front of the hypopharyngeal sclerome. The only oedemerids lacking the abdominal ampullae are the Old World Oedemerini, which are typical in all other respects.

Biology and Ecology: Adult Oedemeridae are often found on flowers, where they feed on nectar and pollen. Most oedemerid larvae feed in dead wood, especially that which is relatively soft and rotten, but many of the Oedemerini occur in the stems or roots of shrubs or herbaceous plants, and *Calopus* has been reported damaging living trees. Larvae may be associated with both white and brown rots, and some have been found in rotten structural timber. Many occur in driftwood, which may be intermittently submerged in fresh or salt water. *Nacerdes melanura* (L.) may cause damage to wharves or pilings, while the Oedemerini are possible agricultural or horticultural pests because of their stem and root boring habits (Arnett, 1968; Burke, 1906; Horion, 1956; Rozen, 1958, 1960).

Description: Mature larvae 10 to 40 mm, usually 25 mm or less. Body elongate, parallel-sided, straight or slightly curved ventrally, and subcylindrical, always very lightly pigmented, except for buccal region and occasionally tips of urogomphi, color nearly white: surface smooth and vestiture of scattered, simple setae.

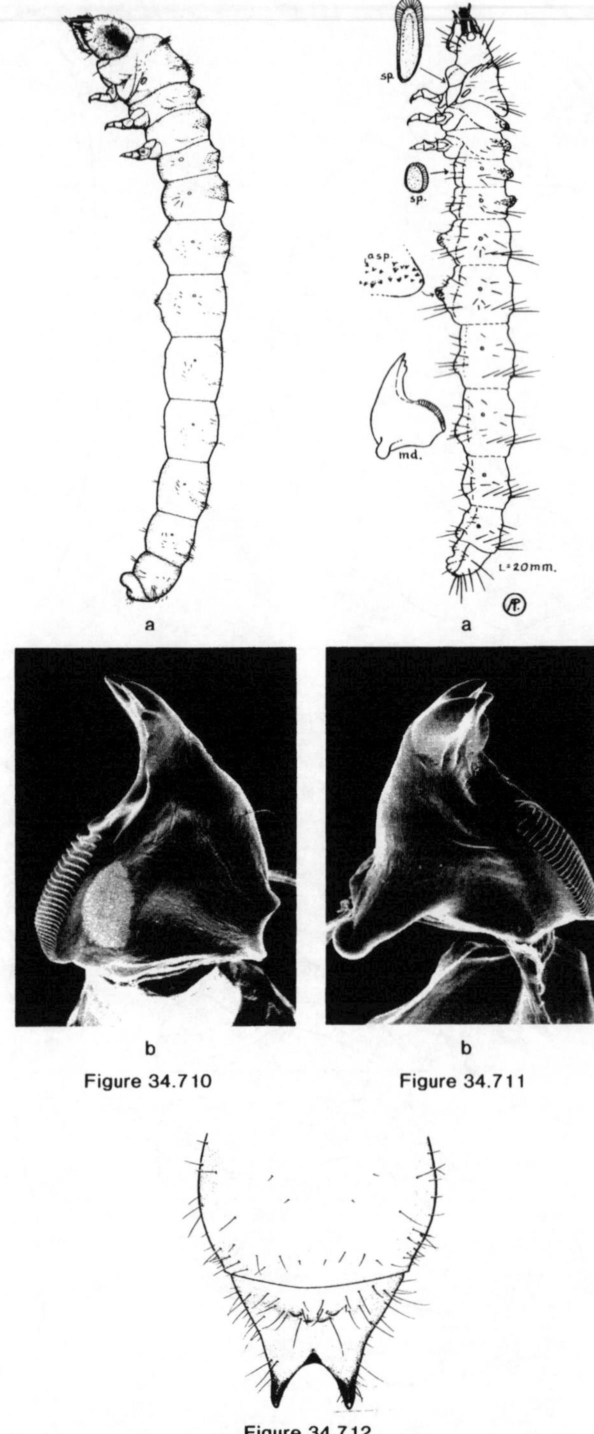

Figure 34.710

Figure 34.711

Figure 34.712

Figures 34.710a,b. Oedemeridae. *Ditylus* sp. Edge of Deep Lake, Vernon, British Columbia. **a.** larva, lateral, length = 38.6 mm; **b.** right mandible, dorsal.

Figures 34.711a,b. Oedemeridae. *Nacerdes melanura* (L.). **a.** larva, lateral view (plus details), length = 20 mm; **b.** right mandible, ventral. St. John, New Brunswick. (Figure 34.711a from Peterson, 1951)

Figure 34.712. Oedemeridae. *Calopus angustus* LeConte. Vancouver, British Columbia. Abdominal apex, dorsal.

Head: Protracted and prognathous, broad, sometimes asymmetrical, slightly flattened. Epicranial stem moderately to very long; frontal arms usually V-shaped (lyriform in *Calopus*). Median endocarina Y-shaped, coincident with epicranial stem and frontal arms. Stemmata usually absent, sometimes 2 or 5 on each side. Antennae well developed, 3-segmented, with reduced 3rd segment. Frontoclypeal suture usually absent (present in *Calopus*); labrum free. Mandibles strongly asymmetrical, bidentate or tridentate, without accessory ventral process; molae large and transversely ridged, the left one more or less parallel to long axis and produced apically, the right one strongly oblique; prostheca absent. Ventral mouthparts retracted. Maxilla with transverse cardo, sometimes distinctly divided, elongate stipes, well-developed articulating area, 3-segmented palp, and rounded mala which is sometimes broadly emarginate at apex and usually has a single tooth at inner apical angle (2 or 3 teeth in *Calopus*). Labium free to base of mentum; ligula well developed or sometimes reduced; labial palps 2-segmented. Hypopharyngeal sclerome a distinct tooth-like structure, in front of which is a distinct and usually columnar prehypopharynx. Hypostomal rods usually absent (moderately long and slightly diverging, but not well sclerotized in *Calopus*). Ventral epicranial ridges absent. Gula transverse.

Thorax and Abdomen: Thorax relatively short, but usually somewhat wider than abdomen. Legs short, moderately to widely separated; tarsungulus with 2 setae lying side by side. Thorax usually with paired patches of asperities on all terga, occasionally absent or present on meso- and metatergum only. Meso- and metatergum and abdominal terga 1–2 or 1–3 (1–5 in *Calopus*) with asperity-bearing ampullae (absent in Oedemerini). Abdominal sterna 2–3, 3–4 or 2–4 usually with paired, asperity-bearing ampullae (on 2–5 in *Calopus*, absent in Oedemerini). Tergum A9 usually without urogomphi or with a very small, lightly sclerotized pair; in *Calopus* with well-developed, upturned urogomphi with pit between them; sternite A9 usually simple, in *Calopus* with 3–4 asperities on each side at base. Segment A10 transverse, posteriorly or posteroventrally oriented.

Spiracles: Annular or annular-multiforous.

Comments: The family contains about 100 genera and 1500 species worldwide, with 17 genera and about 75 species occurring in America north of Mexico. Three subfamilies are usually recognized: Calopodinae (*Calopus, Sparedrus, Sparedropsis,* and *Ocularium*), distributed throughout the Holarctic Region, and extending into southeast Asia and Central America; Nacerdinae (*Nacerdes, Xanthochroa,* and a few other genera); and Oedemerinae, which is further divided into the tribes Ditylini, Asclerini, and Oedemerini (Arnett, 1951, 1961, 1968; Rozen, 1960). This classification is badly in need of world revision.

Selected Bibliography

Arnett 1951, 1961, 1968.
Burke 1906 (larva of *Calopus angustus* LeConte).
Böving and Craighead 1931 (larvae of *Calopus angustus, Copidita thoracica* (Fabricius), and *Alloxacis dorsalis* (Melsheimer)).
Duffy 1952b (larvae of *Sessinia livida* (Fabricius)).
Gardner 1929.
Hayashi 1980 (larvae of *Asclera nigrocyanea* Lewis, *Eobia cinereipennis* Motschulsky, and *Xanthochroa* sp.).
Horion 1956.
Hudson 1975 (larval characters).
Lawrence and Newton 1982.
Liebenow 1978.
Mamaev 1973 (larvae of *Calopus serraticornis* (Linnaeus), *Oedemera virescens* (L.), and *Chrysanthia viridissima* (Linnaeus)).
Peterson 1951 (larva of *Nacerdes melanura*).
Rozen 1958 (larva of *Nacerdes melanura*); 1959 (pupae representing 8 genera); 1960 (larval key; larvae of many species representing 19 genera).

MYCTERIDAE (TENEBRIONOIDEA) (INCLUDING HEMIPEPLIDAE)

John F. Lawrence,
Division of Entomology, CSIRO

Figures 34.713–714

Relationships and Diagnosis: The Mycterinae and Laconotinae, traditionally placed in a broadly defined Salpingidae or Pythidae (Arnett, 1968; Seidlitz, 1920; Spilman, 1952), and the Hemipeplinae, formerly treated as a separate family (Arrow, 1930) or a subfamily of Cucujidae (Arnett, 1968), were combined into a single heteromerous family by Crowson and Viedma (1964), following an earlier suggestion by van Emden (1942b). The family appears to be most closely related to the Boridae and to the complex of tenebrionoid families including Pythidae, Pyrochroidae, Inopeplidae, Othniidae, and Salpingidae. Mycterid larvae may be distinguished from those of most other Coleoptera by the unique structure of the abdominal apex, in which the 9th tergum forms a hinged, terminal plate, the 10th segment is partly enclosed by the U-shaped 9th sternum, and the latter is partly contained within an emargination of sternum 8. Larvae of Boridae have a similar anal region, but they differ from mycterid larvae in having more strongly retracted ventral mouthparts, a well-developed articulating area, well-developed, asymmetrical mandibular molae, and no median endocarina; among the Mycteridae, the mandibular molae are reduced in all forms except Hemipeplinae which have a median endocarina.

Biology and Ecology: Adults of *Mycterus* are commonly collected on flowers; *M. curculionoides* has been taken on *Cistus* in Europe, *M. canescens* Horn and *M. concolor* LeConte both occur on *Ceanothus* blossoms, and *M. quadricollis* Horn is known to frequent flowers of *Yucca* in the American Southwest (Banks, 1912; Crowson and Viedma, 1964; Hopping, 1935). Some mycterids have been recorded as adults from various palms; *Mycteromimus insularis* Champion is known only from *Phoenicophorium,* while *Hemipeplus marginipennis* (LeConte) occurs on *Sabal palmetto* in the southeastern United States (Schwarz, 1878; Scott, 1933). Mycterid larvae occur under bark or in the leaf axils or dead fronds of monocotyledonous plants; they are phytophagous, and plant material has been found in the gut of several species. Larvae of *Mycterus curculionoides* were

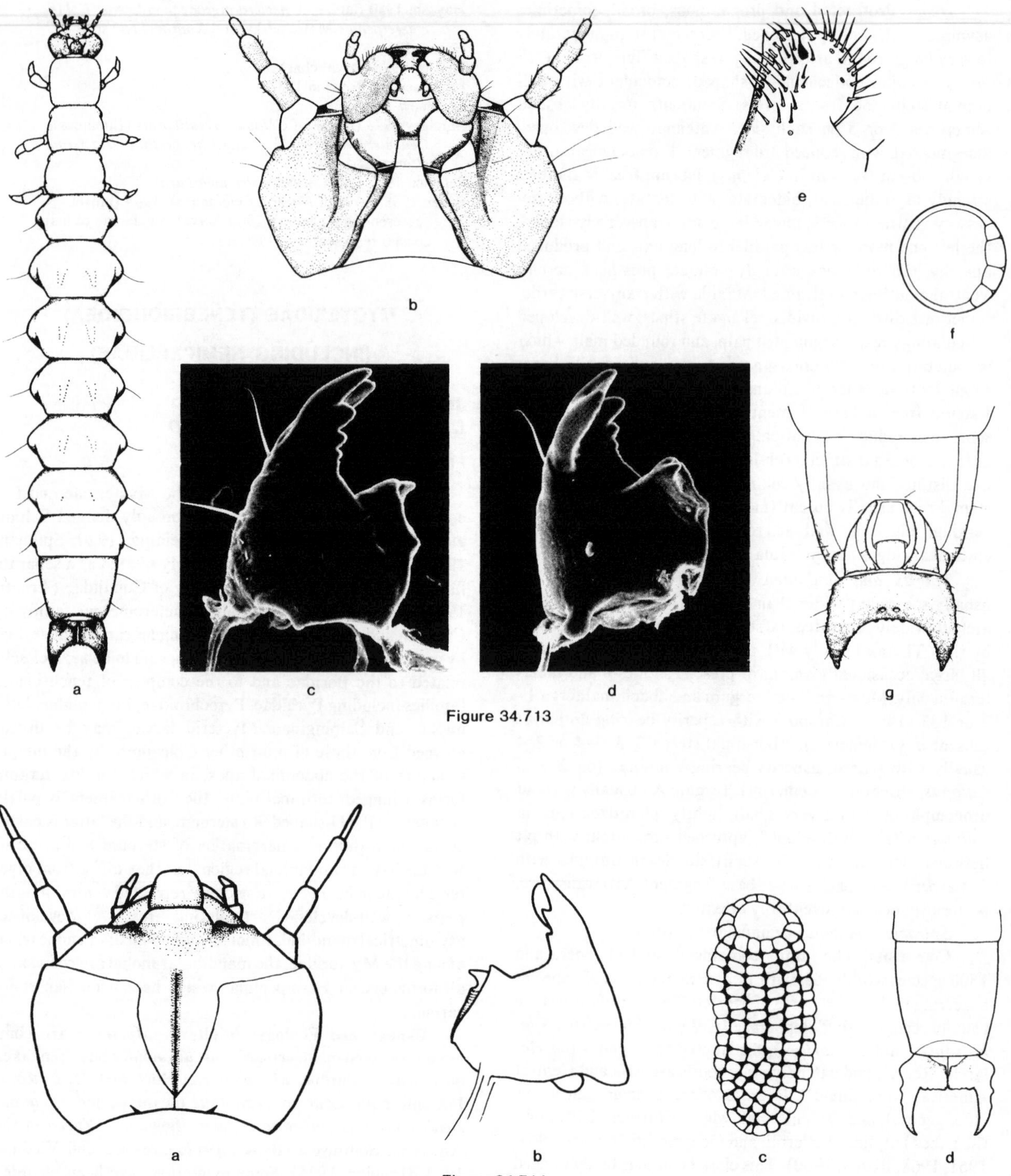

Figure 34.713

Figure 34.714

Figures 34.713a-g. Mycteridae. *Lacconotus pinicola* Horn. Chuska Mts., Apache Co., Arizona. **a.** larva, dorsal, length = 18 mm; **b.** head, ventral; **c.** left mandible, dorsal; **d.** right mandible, ventral; **e.** apex of maxillary mala, ventral; **f.** abdominal spiracle; **g.** abdominal apex, ventral.

Figures 34.714a-d. Mycteridae. *Hemipeplus marginipennis* (LeConte). Indian River, Florida. **a.** head, dorsal; **b.** right mandible, dorsal; **c.** abdominal spiracle; **d.** abdominal apex, dorsal.

taken under the bark of *Pinus*, those of *Lacconotus pinicola* Horn were found under bark of *Abies* and *Populus*, and larvae thought to be *L. pallidus* Van Dyke were found under *Quercus* bark (Crowson and Viedma, 1964). Larvae of *Eurypus muelleri* Seidlitz were collected in dead palm leaves, while those of *Eurypus rubens* Kirby and *Hemipeplus nuciferae* Arrow represent minor pests of the coconut palm; *E. rubens* was found in leaf axils and formed a pupal case from the fibrous material of the leaf, while *H. nuciferae* was observed feeding on young, unexpanded coconut leaves (Arrow, 1930; Bondar, 1940a; Costa and Vanin, 1977).

Description: Mature larvae 5 to 30 mm, usually 20 mm or less. Body elongate, more or less parallel-sided, straight, and strongly flattened. Lightly sclerotized except for head and tergum A9; surfaces smooth; vestiture of scattered, simple setae.

Head: Protracted and prognathous, broad, transversely oval, flattened. Epicranial stem short or absent; frontal arms lyriform and contiguous at base. Median endocarina usually absent, extending anterad of epicranial stem in Hemipeplinae. Stemmata 5 or 2 on each side. Antennae well developed, 3-segmented, usually with a short, dome-like sensorium. Frontoclypeal suture absent; labrum free. Mandibles symmetrical or weakly asymmetrical, bidentate or tridentate, without accessory ventral process; mola usually reduced and tuberculate or simple (well developed and transversely ridged in Hemipeplinae); prostheca absent. Ventral mouthparts slightly retracted. Maxilla with more or less longitudinally oblique cardo, short and broad stipes, well-developed articulating area, 3-segmented palp, and rounded mala, which is cleft at apex, sometimes with a tooth at inner apical angle. Labium free to base of mentum; ligula longer than labial palps, which are usually 1-segmented (2-segmented in *Mycterus*). Hypopharyngeal sclerome a transverse bar. Hypostomal rods long and diverging. Gula transverse.

Thorax and Abdomen: Thorax usually somewhat narrower than abdomen. Legs relatively short, 5-segmented, widely separated; tarsungulus usually with 2 setae lying side by side (2 setae lying one distal to the other in Hemipeplinae). Abdominal terga and sterna 2–6 each with paired rows of asperities forming incomplete rings which are longitudinally oriented. Segment A8 much longer than A7; segment A9 short. Tergum A9 heavily sclerotized and forming an articulated plate, bearing a pair of urogomphi, sometimes with median process or 2 posteriorly oriented pits between them; tergum A9 often longitudinally divided. Sternum A8 posteriorly excavated and partly enclosing sternum A9. Sternum A9 deeply excavate, forming U-shaped sclerite (or combination of sclerites), which encloses segment A10 and bears 1 to 3 asperities or teeth at each end. Segment A10 transversely oval or circular, ventrally oriented.

Spiracles: Annular-multiforous, with a series of accessory openings around peritreme or clustered at one end.

Comments: The family includes 30 genera and about 160 species worldwide, with 5 *Mycterus*, 3 *Lacconotus*, and 2 *Hemipeplus* species occurring in America north of Mexico. Mycterinae occur primarily in the drier parts of N. America, Eurasia, and North Africa, while Lacconotinae and Hemipeplinae are widespread but most diverse in the tropics.

Selected Bibliography

Arnett 1968.
Arrow 1930.
Banks 1912.
Böving and Craighead 1931 (larva of *Hemipeplus* sp.).
Bondar 1940a (larva of *Eurypus rubens* Kirby).
Costa and Vanin 1977 (larva of *Eurypus muelleri* Seidlitz), 1984b.
Crowson and Viedma 1964 (relationships; larva of *Mycterus curculioides* (Fabricius)).
Emden 1942c (larval characters of *Hemipeplus*).
Hopping 1935.
Schwarz 1878.
Scott 1933.
Seidlitz 1920.
Spilman 1952.
Thomas 1985.

BORIDAE (TENEBRIONOIDEA)

D. K. Young, *University of Wisconsin*

Figures 34.715–716

Relationships and Diagnosis: Except for the addition of *Lecontia* (Young, 1985b), the scope of the Boridae is essentially the same as that outlined by Crowson (1955). There is little doubt that the closest relatives are the Mycteridae and Pythidae, even though early classifications based upon adults associated *Boros* Herbst with the Tenebrionidae as well as the Pythidae. Larvae and adults of lacconotine mycterids bear numerous similarities to the borids. The structure of abdominal segment 9 distinguishes borid larvae from those of most other Coleoptera. The ninth tergite forms a hinged plate similar to that of Inopeplidae, Mycteridae, Prostomidae and Pyrochroidae. Additional diagnostic features include the well developed, pad-like maxillary articulating areas, a pair of posteriorly directed, dentiform structures along the posteromesal margin of the pronotum (*Lecontia*) or anteromesal aspect of the mesonotum (*Boros*), and the presence of 2 urogomphal pits. Larvae of Pyrochroidae and Pedilidae (*Cononotus, Pedilus*) also have 2 urogomphal pits, but lack dentiform processes on the thorax. Mycteridae usually lack urogomphal pits and have a reduced maxillary articulating area. Structures similar to the dentiform processes on the dorsum of the pro- and mesothorax of Boridae also occur in the Pythidae, but pythids generally have 0–1 urogomphal pits; they also possess a series of 12 or more preanal asperities along the anterior margin of sternite 9.

Biology and Ecology: Adults of *Boros unicolor* Say and *Lecontia discicollis* (LeConte) have been collected from beneath bark of conifers and by sifting through forest litter. Larvae of *B. unicolor* and the European *B. schneideri* (Panzer) have been collected from beneath the somewhat loose bark of dead pines (St. George, 1931, 1940).

Description: Mature larvae attain lengths of 17–23 mm (*Boros*) or 38–45 mm (*Lecontia*) and widths of 2–2.5 mm (*Boros*) or 3.7–5.5 mm (*Lecontia*). Body strongly flattened, subparallel throughout, lightly sclerotized except for head and abdominal segment 9. Sinuate parabasal ridges associated with thoracic and abdominal tergites; vestiture consisting of

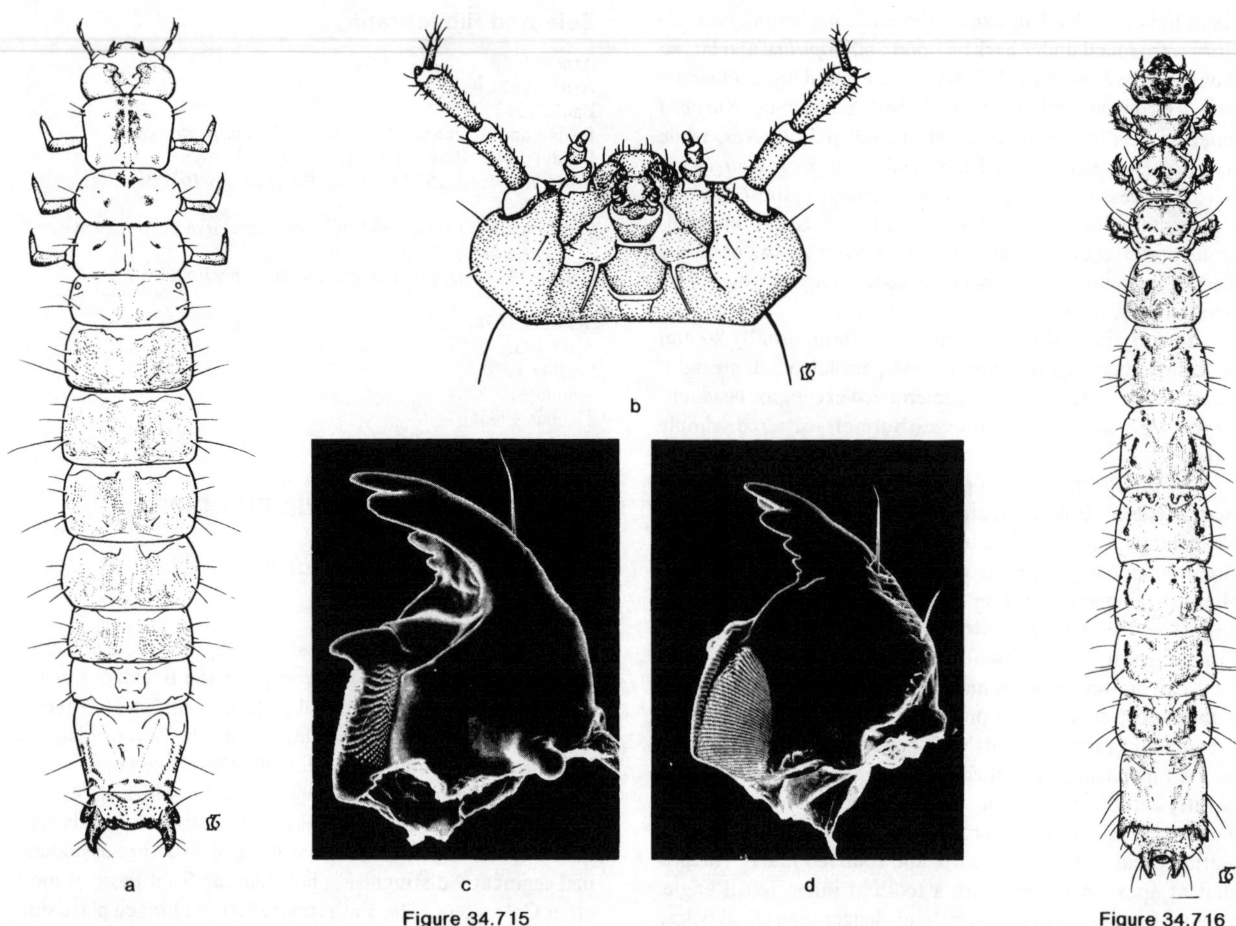

Figures 34.715a-d. Boridae. *Boros unicolor* Say. Figures **a,b:** Grayling, Crawford Co., Michigan, beneath bark of *Pinus;* Figures **c,d:** Miles Standish Forest, Massachusetts. **a.** larva, dorsal; **b.** head capsule, ventral; **c.** left mandible, ventral; **d.** right mandible, dorsal.

Figure 34.716. Boridae. *Lecontia discicollis* (LeConte). Chippewa Co., Michigan. Larva, dorsal.

scattered setae; surface of body otherwise generally smooth. Dorsum of body yellowish-brown, venter yellowish; head and abdominal segment 9 reddish-brown to nearly black.

Head: Prognathous, exserted from prothorax. Epicranial suture with stem short, anterior arms lyriform and complete to posterior margins of antennal insertions; endocarinae absent. Symmetrical labrum anterad of fused frons and clypeus. Stemmata on each side 5, 3 in anterior and 2 in posterior group (*Boros*) or entirely lacking (*Lecontia*). Antennal insertions fully exposed, antennae elongate, 3-segmented, a small conical sensorium associated with segment 2. Mouthparts retracted, supported ventrally by well developed, posteriorly divergent hypostomal rods (fig. 34.715b). Mandibles (figs. 34.715c,d) heavily sclerotized, movable, asymmetrical; left mandible bearing prominent molar tooth; apices of mandibles tridentate. Maxilla with cardo divided by internal fold and thus appearing 2-segmented, also with well developed, pad-like maxillary articulating area, undivided maxillary mala, and 3-segmented palp. Labium free to base of mentum, possessing elongate, distally rounded ligula and 2-segmented palps. Hypopharyngeal sclerome transversely rectangular. Mentum slightly wider than long with anterior margin shallowly emarginate; submentum trapezoidal (*Boros*) or quadrate (*Lecontia*), gular region transverse.

Thorax and Abdomen: Thorax elongate with sides subparallel, cervicosternum divided into 3 plates, posterior margin of pronotum (*Lecontia*) or anterior margin of mesonotum (*Boros*) bearing 2 posteriorly directed, flat, dentiform processes along meson. Legs well developed, 5-segmented including tarsungulus, legs similar in size and shape, bearing numerous fine setae. Abdomen strongly flattened; tergite 9 heavily sclerotized, hinged, extending ventrally to form the entire terminal region or urogomphal plate; ventral aspect divided along meson by longitudinal suture (*Boros*) or deep, wide sulcus (*Lecontia*); caudoventral margin bearing 2 shallow urogomphal pits between the paired, fixed urogomphi; 9th sternite broadly U-shaped (*Boros*) or transversely rectangular (*Lecontia*) and recessed into emargination of 8th sternite, produced distolaterally into a series of small, apically dentiform plates (*Boros*) or simple and bearing a single asperity on each side (*Lecontia*). Segment A10 reduced, visible ventrally surrounding anal orifice.

Spiracles: Thoracic spiracle annular-biforous (*Boros*) or ovate with peritreme partially crenulated (*Lecontia*); abdominal spiracles annular or annular-biforous.

Comments: The Boridae contains 2–3 Holarctic genera and 5–6 species inhabiting coniferous forests. In N. America, borids are represented by *Boros unicolor* Say and *Lecontia discicollis* (LeConte). Larvae have been described or illustrated for *Boros* (Böving and Craighead, 1931; Hayashi, 1980; Mamaev *et al,* 1977; St. George, 1931, 1940) and *Lecontia* (Young, 1985b).

Selected Bibliography

Böving and Craighead 1931.
Crowson 1955.
Hayashi 1980.
Mamaev *et al.* 1977.
St. George 1931, 1940.
Young 1985b.

TRICTENOTOMIDAE (TENEBRIONOIDEA)

John F. Lawrence,
Division of Entomology, CSIRO

This family includes 2 genera: *Autocrates* with 2 species in western China and the Himalayas, and *Trictenotoma,* with 10 species distributed from central China and India to Sri Lanka and the East Indies. The group is thought to be closely related to the family Pythidae.

Larvae are very large (over 100 mm), elongate and parallel-sided and slightly flattened, yellowish-white, with a darker head. The head is large and slightly flattened, with a short epicranial stem, lyriform frontal arms, no stemmata or median endocarina, and relatively long antennae, with a reduced third segment. The mandibles are large, asymmetrical, and tridentate, with a coarsely ridged mola and no prostheca. Ventral mouthparts are retracted, with a truncate mala, broadly excavate at apex and with a distinct uncus at the inner edge; maxillary palps are 3-segmented and labial palps 2-segmented. Hypostomal rods are absent and ventral epicranial ridges are present. The thorax is very short and broad, the legs are short, stout, and widely separated. The mesotergum, metatergum, and abdominal terga 1 to 7 bear a series of short, irregular, longitudinal ridges, which form obliquely transverse rows on thorax and longitudinal rows on abdomen. Sterna A2–8 with short, longitudinal ridges arranged in transverse rows. Tergum A9 with a pair of approximate, posteriorly projecting urogomphi, turned up at the apex; ventral portion of tergum A9 divided into plates; sternum A9 with a continuous row of asperities at the base. Segment A10 transverse and ventrally oriented. Spiracles are annular, with an elliptical peritreme.

Trictenotomid larvae live in rotten wood.

Selected Bibliography

Crowson 1955.
Gahan 1908 (larva of *Trictenotoma childreni* Gray).

PYTHIDAE (TENEBRIONOIDEA)

D. K. Young, *University of Wisconsin*

Figures 34.717–719

Relationships and Diagnosis: As defined here, the North American Pythidae conform well with Arnett's (1968) salpingine tribe Pythini, with the addition of *Sphalma* Horn (as suggested by Young, 1976b) and the exclusion of *Boros* Herbst and *Lecontia* which are referred to the closely related family Boridae. Possible affinities of the monotypic *Trimitomerus* Horn with the family Trictenotomidae were alluded to by Crowson (1955), and discovery of the larva of *T. riversi* Horn could provide meaningful insight into its systematic position. A number of genera related to the Chilean *Pilipalpus* Fairmaire and Australian *Techmessa* Bates have been variously associated with Anthicidae or Pyrochroidae (Abdullah, 1964 and Paulus, 1971). While the pilipalpines appear to be better placed in the Pythidae, there is good evidence to support the contention that pythids and pyrochroids are closely related. The Pythidae also appear to be fairly closely related to the Mycteridae by way of the Boridae. The series of 12 or more preanal asperities along the anterior margin of the ninth abdominal sternite (figs. 34.717d, 34.719b) distinguishes pythid larvae from those of most other families. Asperities in pyrochroid larvae form a single, continuous arch, while pythids and othniids have a double arch. The double arch is nearly continuous mesally in Pythidae, whereas in othniids it is broadly discontinuous along the meson, with the distance between arches greater than twice that between adjacent asperities of a single arch. Larvae of Prostomidae also possess a row of asperities on the ninth sternite, but it is located near the posterior margin.

Biology and Ecology: Adults are found on and beneath the bark of dead logs or by sweeping foliage. Larvae are associated with dead logs; those of *Pytho* are typically found beneath somewhat loose bark of conifers where they feed on the decaying cambial-phloem layer; *Priognathus* larvae are usually found within decaying wood of conifers, and larvae of *Sphalma quadricollis* Horn have been collected from beneath bark of decaying black cottonwood, *Populus trichocarpa*. Larvae of the aberrant *Trimitomerus riversi* Horn are unknown; adults have been taken at lights in Arizona (Anderson and Nilssen, 1978; Hatch, 1965; Hayashi, 1969b; Peterson, 1951; Smith and Sears, 1982; Young, 1976b, unpublished notes).

Description: Mature larvae attain lengths of 8.5–30 mm and widths of 1.5–4 mm. Body orthosomatic and subcylindrical to slightly flattened (*Priognathus, Sphalma*) or strongly flattened (*Pytho*), creamy-yellow to yellowish-brown, lightly sclerotized except for head and abdominal segment 9; sides

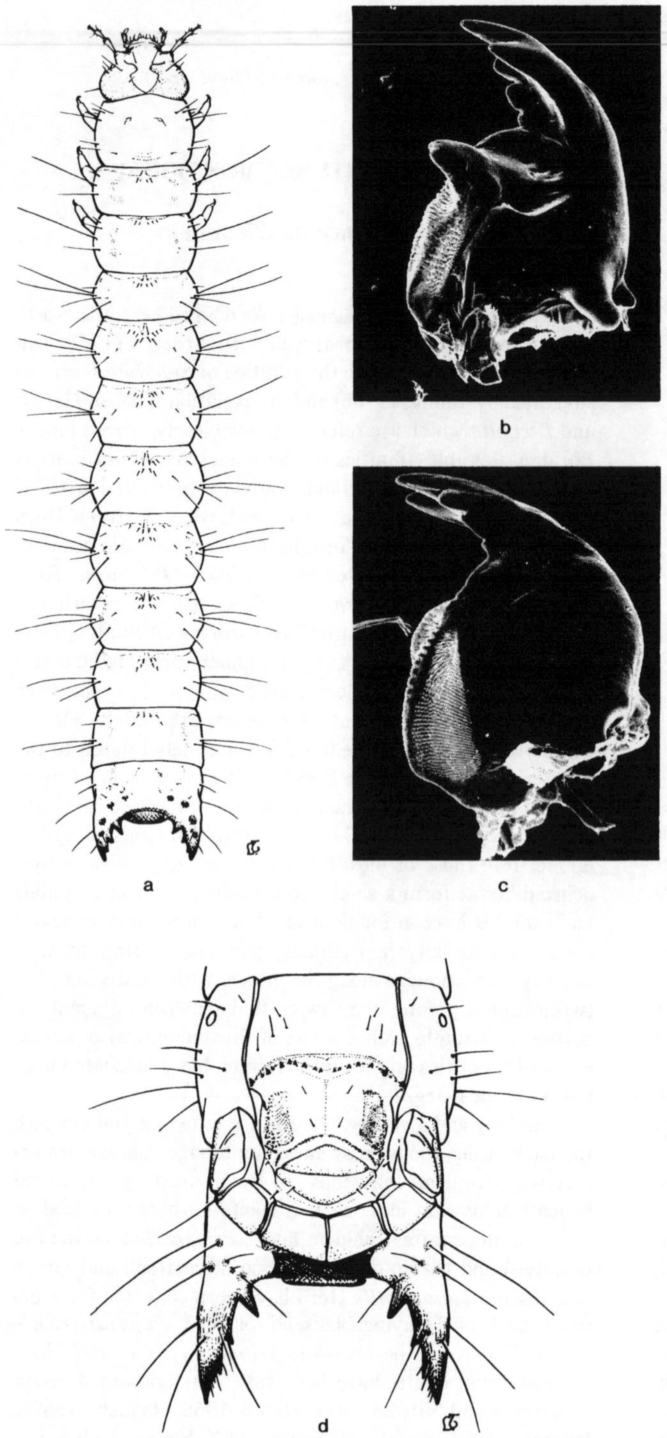

a b c

d

Figure 34.717

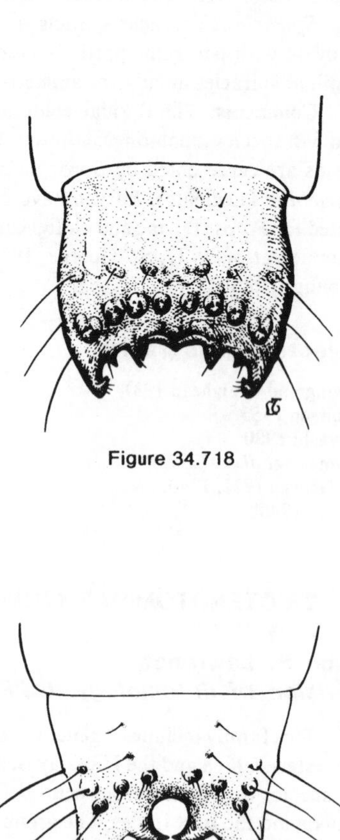

Figure 34.718

a

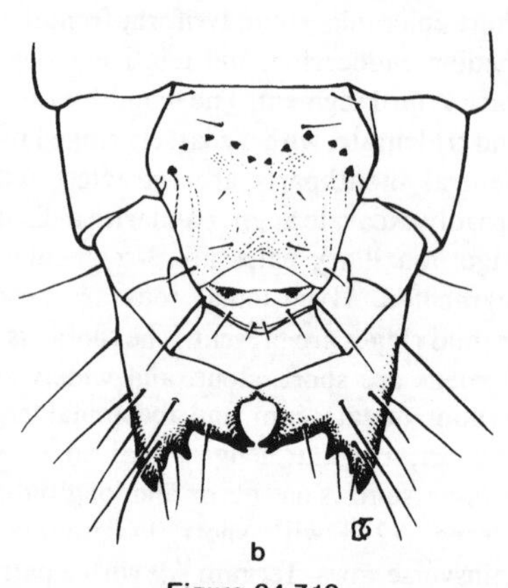

b

Figure 34.719

Figures 34.717a-d. Pythidae. *Pytho* sp. Figures **a,d:** Van Riper State Park, Marquette Co., Michigan, beneath moist bark of *Pinus;* Figures **b,c:** Jefferson Notch, New Hampshire. **a.** larva, dorsal; **b.** left mandible, ventral; **c.** right mandible, dorsal; **d.** apex of abdomen, ventral.

Figure 34.718. Pythidae. *Priognathus monilicornis* (Randall). Grayling, Crawford Co., Michigan, from dead log. Urogomphal plate, dorsal.

Figures 34.719a,b. Pythidae. *Sphalma quadricollis* Horn. Summerdale Campground, Sierra National Forest, Mariposa Co., California, from beneath bark of *Populus trichocarpa,* primarily on underside of log adjacent to soil. **a.** urogomphal plate, dorsal; **b.** apex of abdomen, ventral.

of body subparallel. Parabasal ridges frequently associated with thoracic and abdominal tergites; surface of body otherwise generally smooth with scattered fine setae.

Head: Prognathous, exserted from or slightly retracted within prothorax. Epicranial suture with stem short or absent, frontal arms lyriform and complete to antennal insertions or nearly so; endocarinae absent. Symmetrical labrum anterad of fused frons and clypeus. Stemmata on each side 5, 3 in anterior and 2 in posterior line. Antennal insertions fully exposed, antennae elongate, 3-segmented, sensorium of segment 2 conical. Mouthparts retracted, supported ventrally by pair of epicranial ridges which arise laterad of hypostomal margin; hypostomal rods lacking. Mandibles (figs. 34.717b,c) heavily sclerotized, movable, slightly (*Sphalma*) to conspicuously (*Priognathus, Pytho*) asymmetrical; left mandible bearing prominent molar tooth; molar region of right mandible slightly more prominent than that of left; mandibles with 2–3 apical and usually 2 subapical teeth. Maxilla with cardo divided by internal fold and thus appearing 2-segmented, and a well developed, pad-like maxillary articulating area; maxillary mala shallowly cleft subapically along adoral margin; 3-segmented palp. Labium free to base of mentum and possessing short (*Sphalma*) or elongate (*Priognathus, Pytho*), apically rounded ligula and 2-segmented palps. Hypopharyngeal sclerome transverse, molar-like. Mentum widest distally, apical margin shallowly emarginate, submentum trapezoidal, widest proximally; gular region transverse, lightly sclerotized.

Thorax and Abdomen: Thorax elongate with sides subparallel, cervicosternum divided into 3 plates, anterior margin of metanotum and occasionally mesonotum sometimes bearing 2 posteriorly directed, flat, dentiform processes along meson (many *Pytho* spp.). Legs well developed, 5-segmented including tarsungulus, legs subequal in size and shape, bearing numerous stout, spine-like setae. Abdomen subcylindrical to slightly flattened (*Priognathus, Sphalma*) or strongly flattened (*Pytho*); anterior margins of tergites 1–8 sometimes bearing 2 posteriorly directed, flat, dentiform processes along meson (many *Pytho* spp.). Tergite 9 heavily sclerotized, extending ventrally to form the entire terminal segment or urogomphal plate (figs. 34.717a, 34.718, 34.719a), usually possessing a single shallow urogomphal pit along caudoventral margin between the paired, fixed urogomphi; urogomphi with secondary branching (*Sphalma*), or stout spine-like structures (*Priognathus*), or unbranched and possessing smaller spine-like projections (*Pytho*). Sternite 9 (figs. 34.717d, 34.719b) beset with a double arch of asperities near anterior margin, each arch composed of 6 or more asperities, arches separated mesally by a distance subequal to that between adjacent asperities of a single arch. Segment A10 small, concealed from above by projection of tergite 9, visible ventrally surrounding anal orifice.

Spiracles: Thoracic and abdominal spiracles annular (*Priognathus, Pytho*) or annular-biforous (*Sphalma*).

Comments: As currently defined to include the Pilipalpinae, the Pythidae include 15 genera and about 50 species which are largely restricted to Holarctic coniferous forests and temperate parts of the Southern Hemisphere. The North American fauna is comprised of 3–4 genera and 8–9 species. Larvae have been described or illustrated for *Priognathus* (Peterson, 1951, under the name *Lecontia*), *Pytho* (Böving and Craighead, 1931; Hayashi, 1969b, 1980; Mamaev *et al.* 1977), and *Sphalma* (Young, 1976b).

Selected Bibliography

Abdullah 1964.
Anderson and Nilssen 1978.
Arnett 1968.
Böving and Craighead 1931.
Crowson 1955.
Hatch 1965.
Hayashi 1969b, 1980.
Mamaev *et al.* 1977.
Paulus 1971.
Peterson 1951.
Smith and Sears 1982.
Young 1976b.

PYROCHROIDAE (TENEBRIONOIDEA)

D. K. Young, *University of Wisconsin*

Fire-Colored Beetles

Figures 34.720–725

Relationships and Diagnosis: The family has recently been revised for North America (Young, 1975); some of the problems encountered while attempting to arrive at a sound definition for the group at the world level were discussed. Young (1985a) described the larva of *Ischalia vancouverensis* Harrington and discussed the systematic position of the genus, recommending that *Ischalia* be transferred to the Anthicidae. The closest relatives of pyrochroids are among the Pythidae, the Pedilidae (especially Pedilinae) and perhaps to some extent the Trictenotomidae and Othniidae (especially through similarities of the larvae although some of these may be symplesiomorphic). The well developed mandibular mola and series of several preanal asperities associated with the ventroanterior margin of abdominal sternite 9 separate pyrochroid larvae from all but the Othniidae, some Salpingidae (e.g., *Istrisia*), Pythidae (*Pytho, Priognathus* and *Sphalma*), and Oedemeridae (*Calopus*). Additional diagnostic characters include the presence of divergent hypostomal rods, 2 well developed urogomphal pits, and asperities in a continuous single arch which extends well around the lateral margin of the ninth sternite. The Othniidae also possess divergent hypostomal rods, but have only a single poorly developed urogomphal pit or none at all, and possess a mesally discontinuous double arch of asperities; the Salpingidae and Pythidae that are somewhat similar lack hypostomal rods (though short ventral epicranial ridges are often present), possess 0–1 urogomphal pits, and possess a continuous double arch of asperities. Larvae of *Calopus* (Oedemeridae) lack hypostomal rods and urogomphal pits, and possess a transverse row of asperities at the base of the ninth sternite which does not extend around the lateral margins.

Biology and Ecology: Adults of *Dendroides* and *Neopyrochroa* have been collected at lights and appear to be largely nocturnal. Adult *Neopyrochroa* have also been taken in the evening at fermenting baits such as beer and molasses. Beating foliage is a good collecting technique, suggesting that adults hide in trees and shrubs during the day. Males of *Schizotus cervicalis* Newman, *N. flabellata* (Fabricius) and *N. femoralis* (LeConte) have been taken at cantharidin bait traps being used to attract *Pedilus* and *Notoxus* adults. Adults of the Palearctic genus *Pyrochroa* have been reported to occur on flowers as well as on stumps and logs where oviposition takes place.

Larvae are generally associated with somewhat cool, moist conditions beneath slightly loosened bark and to some extent within decaying wood of deciduous and coniferous trees. While both woody and fungal materials have been extracted from the gut, fungi probably play the most significant dietary role. Under adverse conditions such as overpopulation larvae can become cannibalistic, but they are not normally predaceous.

From 1 to several years are spent as a larva; several instars commonly occur together at any given time. Pupation is commonly beneath bark in an ovate frass-walled chamber prepared by the larva; the duration is 1–2 weeks. In N. America, adults are active primarily from late April to early August, with only the larval stage overwintering. (Buck, 1954; Champlain and Kirk, 1926; Chararas *et al.* 1979; Duffy, 1946; Frankenberg, 1942; Palm, 1959; Payne, 1931; Procter, 1938; Smith and Sears, 1982; Young, 1975, 1984a, 1984b, 1984c and unpublished notes.)

Description: Mature larvae attain lengths of 14–35 mm and widths of 2–5 mm. Body orthosomatic with sides subparallel throughout or slightly wider posteriorly, lightly sclerotized except for head and urogomphal plate, with scattered setae. Parabasal ridges associated with abdominal tergites. Head and body yellowish-brown to amber, or [*Schizotus cervicalis* Newman and sometimes *Neopyrochroa femoralis* (LeConte)] dirty-yellowish with olive-green tint, pigmentation much darker in areas of heavy sclerotization such as mandibles and urogomphal plate.

Head: (figs. 34.720a,b) Prognathous, flattened, exserted from prothorax. Epicranial suture lyre-shaped with stem short and indistinct or absent, frontal arms complete to antennal insertions; endocarinae absent. Symmetrical labrum anterad of fused frons and clypeus. Stemmata on each side 4, in 2 groups of 2. Antennal insertions fully exposed, antennae elongate, 3-segmented, sensorium of segment 2 short. Mouthparts retracted. Mandibles (fig. 34.721a) strong, movable, asymmetrical, molar area of right mandible well developed, left mandible bearing prominent molar tooth; apex of right mandible tridentate, left bidentate, both frequently bearing smaller subapical teeth. Maxilla (fig. 34.721b) with 1-segmented cardo which is diagonally folded upward upon itself toward the stipes and thus appearing 2-segmented, a well developed, undivided, pad-like maxillary articulating area; mala which bears uncus (heavily sclerotized and dentiform in *Neopyrochroa,* slightly less so in *Schizotus* and *Dendroides*); 3-segmented palp. Labium with mentum subquadrate, broadest distally with apical margin shallowly emarginate, submentum elongate with sides shallowly sinuate basally, apical margin convexly rounded; ligula well developed; palp 2-segmented. Hypopharyngeal sclerome molar-like; gular sutures separate.

Thorax and Abdomen: Thorax elongate with sides parallel, cervicosternum divided into 3 plates. Legs (fig. 34.720e) well developed, 5-segmented including tarsungulus, devoid of stout spine-like setae; coxae large, separated by 2–3 coxal diameters. Abdomen flattened, moderately sclerotized, tergites 1–7 and 9 (excluding urogomphi) subequal in length; tergite 8 usually more than twice as long as seventh (slightly less than twice in *Schizotus*). Sternite 8 deeply emarginate apically. Tergite 9 (figs. 34.720a, 34.721c–34.725) hinged, capable of considerable dorsolongitudinal movement, extending ventrally to form the entire terminal segment or urogomphal plate, possessing ledge-like lip (well developed in *Neopyrochroa,* slightly reduced in *Dendroides,* more so in *Schizotus*) and 2 heavily sclerotized pits on apical margin between heavily sclerotized, fixed urogomphi. Sternite 9 (fig. 34.720d) broadly U-shaped, divided longitudinally, partially recessed into emargination of eighth sternite, possessing continuous semicircular arch of asperities anteriorly. Segment A10 much reduced, visible ventrally surrounding anal opening.

Spiracles: Thoracic spiracle ovate, situated on laterotergite. Circular spiracles subequal in size and located on ventrolateral margins of laterotergites 1–7; spiracles of laterotergite 8 slightly larger, circular or ovate, located about midway along its length (*Schizotus* and *Neopyrochroa*) or near the posterior margin (*Dendroides*).

Comments: The family Pyrochroidae is widely distributed, containing 12 genera and 125 species. Diversity is greatest in temperate regions and in mountainous, temperate-like areas within the tropics. Three genera and 12 species occur in the United States and Canada.

The significance of fungi in the larval diet was investigated by Payne (1931) who carried out experiments with *D. canadensis* and the synchroid *Synchroa punctata* Newman. Both Böving and Craighead (1931) and Peterson (1951) figured larvae which were thought to be *N. femoralis* (LeConte). However, proportions of the urogomphal plate indicate that the specimens were probably *N. flabellata.* Spilman and Anderson (1961) presented keys to the species of N. American larvae and to the genera of pupae known to them. In his revision, Young (1975) redescribed the larvae at the family level and provided a description of the pupae. Also included are keys to the known larvae and pupae of N. America and illustrations of salient features. Emden (1943) and Rossem (1945) figured and provided keys to the 3 European larvae; Emden's work also included head capsule measurements which indicated 4 larval instars. A number of Japanese pyrochroid larvae have been described or illustrated (Hayashi, 1963, 1969a, 1980; Hayashi, *et al;* 1959; Kôno and Nishio, 1943; Osawa, 1947). The only other known treatment of pyrochroid larvae is that of Viswanathan (1945). He described and figured the larvae of 3 Indian species. The larvae of *Pseudopyrochroa testaceitarsis* Pic were taken from beneath moist bark of *Cinnamoum zeylanicum* and were observed to make long sinuous tunnels, feeding both on the inner bark and outer soft wood. An analysis of gut contents revealed only coarse woody materials. Larvae of *Pyrochroa subcostulata* Fairmaire were taken from beneath bark of girdled *Betula utilis.*

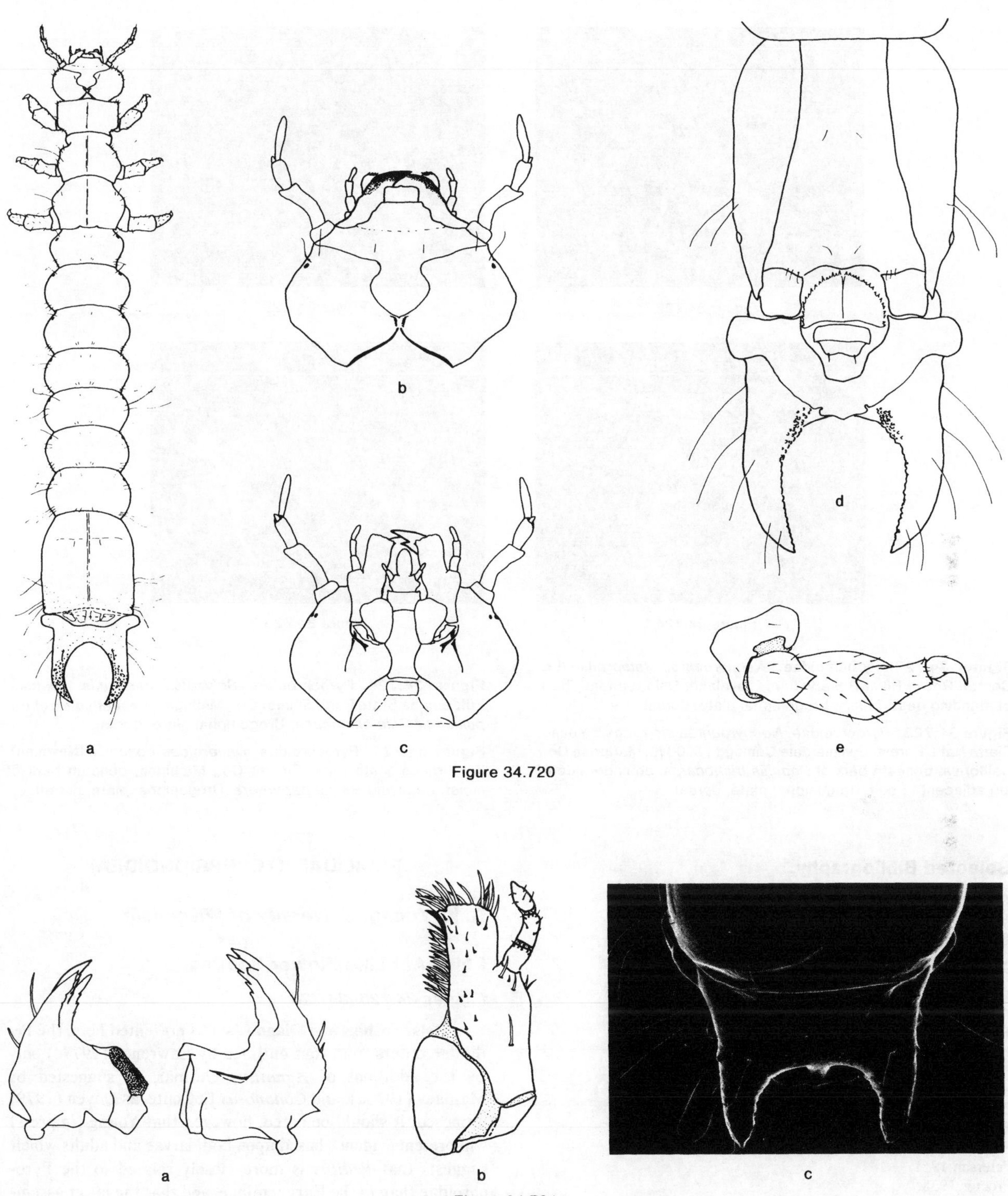

Figures 34.720a–e. Pyrochroidae. *Dendroides canadensis* Latreille. East Lansing, Michigan, beneath bark of *Fagus grandifolia* log on upper aspect. **a.** larva, dorsal; **b.** head capsule, dorsal; **c.** head capsule, ventral; **d.** abdominal apex, ventral; **e.** leg, inner aspect.

Figures 34.721a–c. Pyrochroidae. *Neopyrochroa flabellata* (Fabricius). Pettibone Lake, Newaygo Co., Michigan, beneath bark of *Quercus* sp., on underside of log adjacent to soil. **a.** mandibles, ventral; **b.** left maxilla, ventral; **c.** urogomphal plate, dorsal.

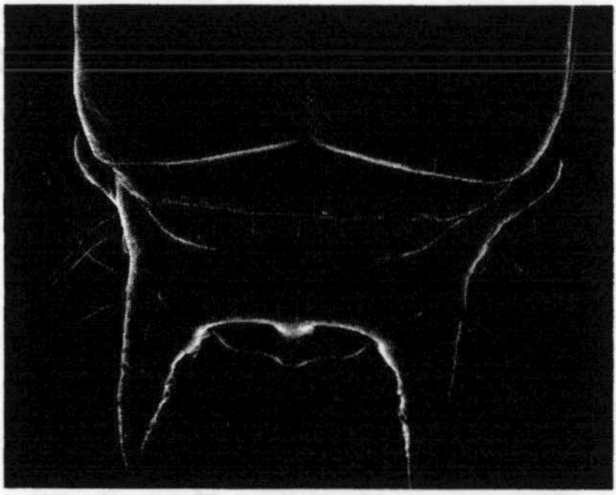

Figure 34.722

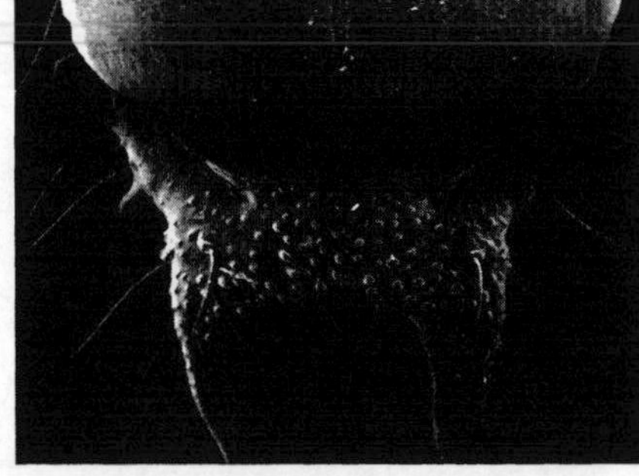

Figure 34.723

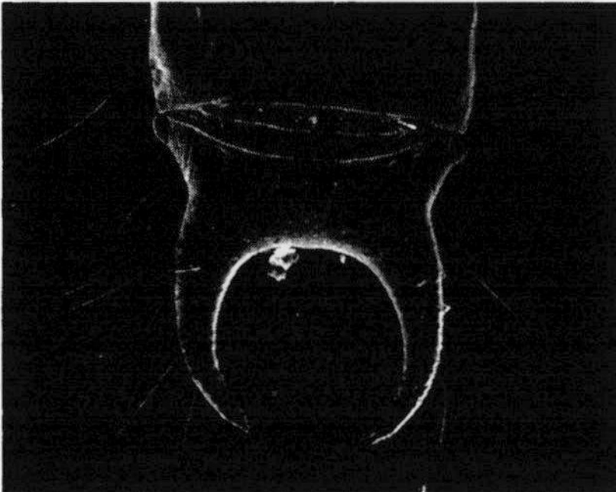

Figure 34.724

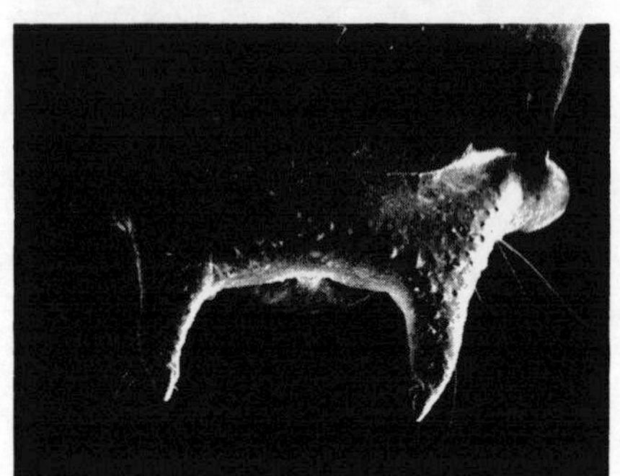

Figure 34.725

Figure 34.722. Pyrochroidae. *Neopyrochroa femoralis* (LeConte). Rocky River Reservation, Cleveland, Ohio, beneath bark of standing dead *Ulmus.* Urogomphal plate, dorsal.

Figure 34.723. Pyrochroidae. *Neopyrochroa sierraensis* Young. Sierra Nat'l. Forest, Summerdale Campgd., 5000 ft., Mariposa Co., California, beneath bark of *Populus trichocarpa,* on underside of log adjacent to soil. Urogomphal plate, dorsal.

Figure 34.724. Pyrochroidae. *Schizotus cervicalis* Newman. Wilderness State Park, Emmet Co., Michigan, beneath bark of *Populus* and *Tilia americana.* Urogomphal plate, dorsal.

Figure 34.725. Pyrochroidae. *Dendroides concolor* (Newman). Wilderness State Park, Emmet Co., Michigan, beneath bark of moist, decaying *Betula papyrifera.* Urogomphal plate, dorsal.

Selected Bibliography

Böving and Craighead 1931.
Blair 1914.
Buck 1954.
Champlain and Kirk 1926.
Chararas *et al.* 1979.
Duffy 1946.
Emden 1943.
Frankenberg 1942.
Hayashi 1963, 1969a, 1980.
Hayashi *et al.* 1959.
Kôno and Nishio 1943.
Osawa 1947.
Palm 1959.
Payne 1931.
Peterson 1951.
Procter 1938.
Rossem 1945.
Smith and Sears 1982.
Spilman and Anderson 1961.
Viswanathan 1945.
Young 1975, 1984a, 1984b, 1984c, 1985a.

PEDILIDAE (TENEBRIONOIDEA)

D. K. Young, *University of Wisconsin*

False Ant-Like Flower Beetles

Figures 34.726–34.728

Relationships and Diagnosis: As presented here, the Pedilidae differs from that outlined by Lawrence (1977c) only by the additions of *Agnathus* Germar, as suggested by Mamaev (1976a), and *Cononotus* LeConte, as Doyen (1979) proposed. It should be noted, however, that Young (in press) will present evidence based upon both larvae and adults which suggests that *Pedilus* is more closely related to the Pyrochroidae than to the Eurygeniinae, and that the latter assemblage of genera is, on the basis of adults, more closely related to the Anthicidae. While the highly specialized adults of *Agnathus* and *Cononotus* are difficult to compare to other heteromerous taxa, the larvae suggest a relationship to those of *Pedilus,* Pyrochroidae, and thus to the pythid-borid complex. Pedilid larvae may be distinguished from those of most other

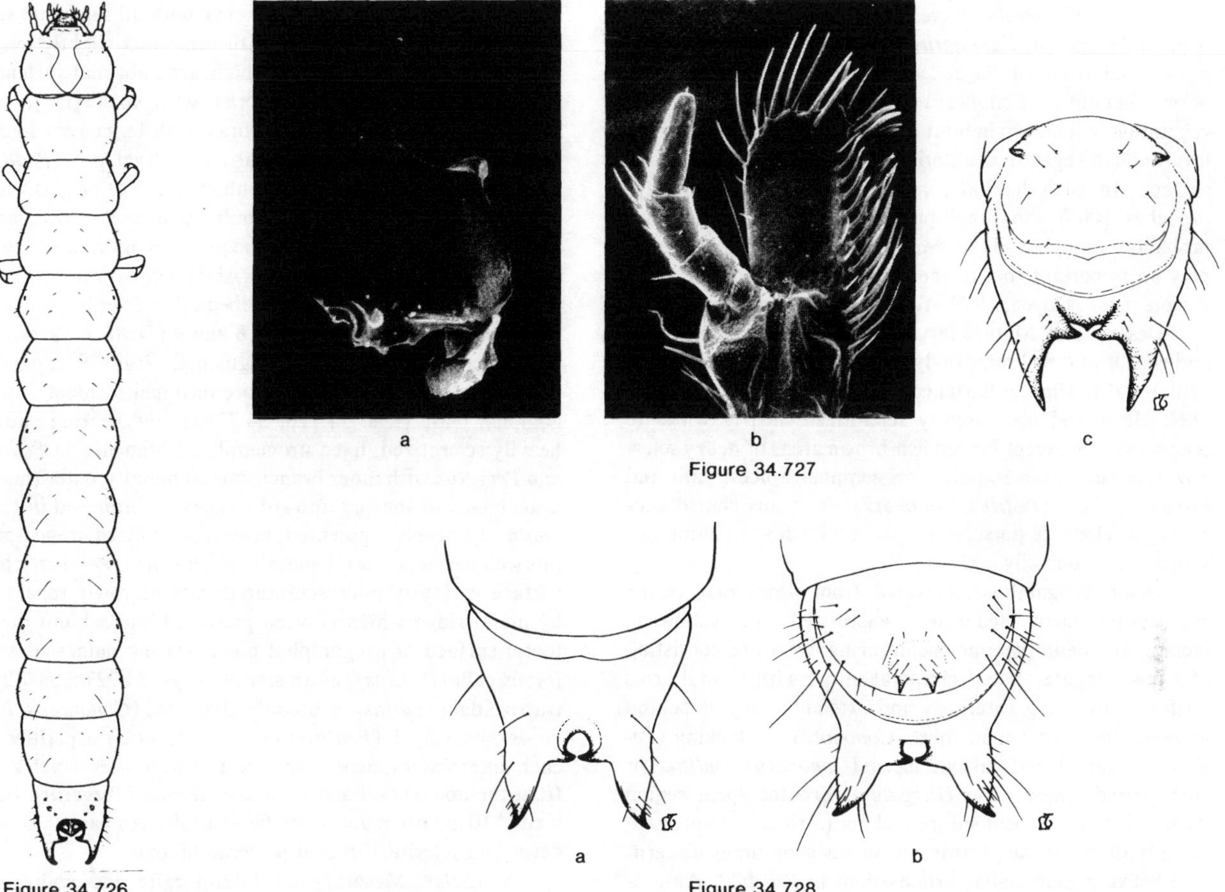

Figure 34.726

Figure 34.727

Figure 34.728

Figure 34.726. Pedilidae. *Pedilus* sp. Turkey Run Recreation Area, Fairfax Co., Virginia, from litter associated with *Liriodendron* and other hardwoods. Larva, dorsal.

Figures 34.727a-c. Pedilidae. *Pedilus lugubris* (Say). NE. corner Sanford woodlot, Michigan State University, Ingham Co., Michigan, collected 1-2 cm beneath underside surface of decaying log. **a.** left mandible, dorsal; **b.** right maxilla, ventral; **c.** apex of abdomen, ventral.

Figures 34.728a,b. Pedilidae. *Pergetus campanulatus* LeConte. Bandon, Oregon, from old cranberry bog. **a.** urogomphal plate, dorsal; **b.** apex of abdomen, ventral.

Coleoptera by structures associated with abdominal segment 9. Larvae of *Pergetus* (Eurygeniinae) possess a pair of stout asperities on the apical margin of the ninth sternite near the meson (fig. 34.728b). While the presence of a single pair of asperities on the ninth sternite is also characteristic of *Pedilus* (fig. 34.727c) as well as the Synchroidae, many Salpingidae (*Aegialites, Rhinosimus*), Othniidae (*Aglenus*) and certain Melandryidae (e.g., *Hallomenus*), the asperities in these taxa are associated with the anterolateral margins of the sternite. The presence of urogomphal pits separates larvae of *Pedilus* (fig. 34.726) and *Cononotus* from all but the Boridae, Othniidae, Pyrochroidae and Pythidae. In *Pedilus, Cononotus,* Boridae and Pyrochroidae, 2 pits are present whereas Othniidae and Pythidae possess but a single, often poorly defined, pit. *Pedilus* and *Cononotus* have 1–2 pairs of stout asperities associated with the anterolateral aspect of the ninth abdominal sternite; the asperities of pyrochroids are numerous and form a continuous transverse arch. Borid larvae possess a pair of posteriorly directed, dentiform structures on the pronotum (*Lecontia*) or mesonotum (*Boros*); such processes are not found in *Pedilus* and *Cononotus*.

Biology and Ecology: Adult Eurygeniinae are most commonly encountered by sweeping vegetation or by use of lights at night. *Cononotus* adults have been found beneath rocks, stones and decaying vegetation; those of *C. bryanti* Van Dyke were taken in numbers from beneath dried cow dung along coastal sand dunes. Most of the associations for adult *Pedilus* refer to diurnal activity at flowers and foliage although several specimens of *P. flabellatus* (Horn) have been taken at lights. Adults of many N. American species of *Pedilus* have been collected in association with meloid beetles or at cantharidin, the compound produced by meloid and some oedemerid beetles as a chemical defense mechanism. With few exceptions, all cantharidin-orienting *Pedilus* have been males.

Larvae of *Pergetus* were taken from an old cranberry bog, while those of *Cononotus* were reared in containers of moist sand from adults collected beneath cow dung. The normal habitat for *Cononotus* larvae is presumably decaying vegetation. All known habitat associations for *Pedilus* larvae involve dead vegetative materials, on or just beneath the soil surface. In such habitats, relative humidity usually approaches 100% and fungi proliferate. *Pedilus* larvae are probably phytophagous or omnivorous, and fungi may also play an important role in their nutrition. (Abdullah, 1969; Arnett, 1968; Doyen, 1979; Young, 1984b, and in press).

Description: Mature larvae attain lengths of 5–15 mm and widths of 0.8–3 mm. Body subcylindrical anteriorly and cylindrical to slightly flattened posteriorly, its sides subparallel. Head and body lightly sclerotized, nearly white to creamy yellow except for reddish-brown areas of heavy sclerotization such as mandibles, urogomphal apices, calli and urogomphal pits (*Pedilus, Cononotus*). Vestiture sparse, consisting of elongate dorsolateral setae with few to numerous shorter setae dorsally and ventrally.

Head: Prognathous, exserted from prothorax, vertex slightly flattened; moderately sclerotized and yellowish, lacking any definitive microsculpturing; vestiture consisting of a few elongate setae. Epicranial suture with frontal arms lyriform, diverging anteriorly and extending to just behind antennal insertions; stem short (*Cononotus*) or lacking (*Pedilus, Pergetus*); endocarinae absent (*Cononotus, Pedilus*) or with paired endocarinae (*Pergetus*). Frontoclypeal region flattened; frontoclypeal suture lacking (a weakly impressed, mesally discontinuous transverse sulcus sometimes discernable between mandibular articulations in *Pedilus*). Anterolateral aspects of head with 4 (rarely 3) pigmented stemmata on each side directed posterad of antennal insertions (*Pedilus*) or stemmata lacking (*Cononotus, Pergetus*). Antennae elongate, 3-segmented, segment 2 bearing a broad sensorium. Mouthparts supported posteroventrally by hypostomal ridges (*Pedilus*); hypostomal rods present (*Cononotus*) or lacking (*Pedilus, Pergetus*); gula nearly as broad posteriorly as its mesal width, strongly narrowed anteriorly. Mandibles (fig. 34.727a) slightly asymmetrical, large, heavily sclerotized and wedge-shaped, with triangular base and 3 apical teeth; molar region of left mandible bearing a prominent distal tooth, that of right mandible lacking tooth, bearing much coarser serrulations. Maxilla (fig. 34.727b) movable, free to base of mentum, with 1-segmented cardo which is diagonally creased upward toward stipes, thus appearing 2-segmented; lightly sclerotized, undivided pad-like maxillary articulating area present; with maxillary mala, and 3-segmented palp. Labium with submentum trapezoidal and longer than wide; mentum slightly longer than wide, broadest distally; prementum subrectangular, well developed ligula projecting anteromesally beyond palps (*Cononotus, Pedilus*) or somewhat shorter than length of palps (*Pergetus*); palps 2-segmented. Hypopharyngeal sclerome transversely subrectangular, with associated setae (*Pedilus*) or not (*Cononotus, Pergetus*); anterior aspect of hypopharynx bearing numerous blunt, fleshy denticles laterally and apically, and a transverse row of setae just anterad of the hypopharyngeal sclerome (*Pedilus*) or lacking denticles and setae (*Cononotus, Pergetus*).

Thorax and Abdomen: Thorax with all segments subequal in length (*Pedilus*) or with prothorax slightly longer than meso- and metathorax which are subequal in length (*Cononotus, Pergetus*); prothorax with coxae large and subglobular; meso- and metathorax with terga very lightly pigmented, each with parabasal ridge developed (*Pedilus, Pergetus*). Legs well developed, about 0.5× as long as width of mesothorax, each bearing a number of fine setae; coxa large and conical; femur and tibia subequal in length; tarsungulus well sclerotized and pigmented. Abdomen with tergites 1–8 and 9 (excluding urogomphi) subequal in length; parabasal ridge well developed on tergites 8 and 9 (*Pedilus, Pergetus*), inconspicuous on remaining segments; tergite 9 expanded posteroventrally to form the entire terminal segment or urogomphal plate (figs. 34.726, 34.728a), and bearing paired, heavily sclerotized, fixed urogomphi (2-branched in *Pedilus* and *Pergetus* with inner branch curved mesally, outer (main) branch curved sharply upward at apex; unbranched in *Cononotus*); 2 deeply pigmented, heavily sclerotized urogomphal pits well developed caudomesally (*Cononotus, Pedilus*), their surface variously microsculptured; pits narrowly separated by mesal ridge which is often produced into a blunt mesal tooth; surface of urogomphal plate bearing numerous setiferous calli (*Pedilus*); ninth sternite (figs. 34.727c, 34.728b) trapezoidal (*Pedilus*) or broadly U-shaped (*Cononotus, Pergetus*), bearing 1 (*Pedilus*) or 2 (*Cononotus*) asperities on each anterolateral aspect, or a pair of asperities which arise from the apicomesal aspect of the sternite (*Pergetus*). Segment A10 greatly reduced, visible ventrally as a narrow, poorly developed sclerite, directed posterad of anus.

Spiracles: Mesothoracic laterotergite and abdominal segments bearing annular spiracles with (*Cononotus, Pedilus*) or without (*Pergetus*) accessory chambers.

Comments: The family contains about 20 genera and 150 species distributed worldwide. Fourteen genera and approximately 80 species occur in the United States and Canada.

The larva of *Pergetus* (= *Eurygenius*) *campanulatus* LeConte was figured by Böving and Craighead (1931) and reevaluated by Doyen (1979) along with his description of the larva of *Cononotus bryanti* Van Dyke. The first published account of the larval stages of *Pedilus* was provided by Mamaev's (1976a) brief description of an unidentified *Pedilus* from Kirgiz USSR. Lawrence (1977c) further characterized the larvae at the generic level. The first species description of a *Pedilus* larva belongs to Wharton (1979), who reared larvae of *P. inconspicuus* (Horn) from El Dorado County, California. Young (in press) will offer a composite generic description of the larval stages of *Pedilus* based upon all known records. The larval stages of *Agnathus decoratus* Germar have also been described (Mulsant and Rey, 1856; Mamaev, 1976). The present treatment is based largely upon larvae of *Pedilus inconspicuus, P. lugubris* (Say), and Böving's specimens of *Pergetus campanulatus*.

Selected Bibliography

Abdullah 1969.
Arnett 1968.
Böving and Craighead 1931.
Doyen 1979.
Lawrence 1977c.

Mamaev 1976a.
Mulsant and Rey 1856.
Wharton 1979.
Young 1984b, in press.

OTHNIIDAE (TENEBRIONOIDEA)

D. K. Young, *University of Wisconsin*

False Tiger Beetles

Figures 34.729-730

Relationships and Diagnosis: (Trogocryptinae not included in diagnosis.) To a great extent, the present treatment follows the outline presented by Lawrence (1977c) with the inclusion of Trogocryptinae as well as Othniidae. To this assemblage, *Aglenus* Erichson is added. As Lawrence (1980) pointed out in his discussion of *Aglenus,* "There is little doubt that members of the Salpingidae, Inopeplidae, and Othniidae are closely related, and perhaps all should be treated as a single family with *Aglenus* forming an independent subgroup." Lawrence (1982c) appears to have followed up on this notion in his broad view of the Salpingidae. The presence of a well developed mandibular mola and a series of preanal asperities associated with the anterior margin of abdominal sternite 9 separates most othniid larvae from all but the Pyrochroidae, some Salpingidae (*Istrisia*) and Pythidae (*Pytho, Priognathus,* and *Sphalma*). Additional diagnostic characters for most othniids include the presence of divergent hypostomal rods, urogomphal pit poorly defined or entirely lacking, and asperities in a mesally discontinuous double arch. The Pyrochroidae also possess divergent hypostomal rods, but have 2 well developed urogomphal pits and possess a continuous single arch of asperities which curves well around the lateral margins of the sternite; the pertinent salpingid taxa lack hypostomal rods (although short ventral epicranial ridges are often present), and possess a continuous double arch of asperities.

Larvae of the aberrant othniid *Aglenus brunneus* (Gyllenhal) possess a single stout asperity on each anterolateral margin of abdominal sternite 9, as do larvae of Pedilidae (*Pedilus*) and most Salpingidae. However, *Pedilus* larvae have a pair of well developed urogomphal pits and salpingids typically lack a true mola (except *Istrisia,* as noted above).

Biology and Ecology: Adults have been taken from recently fallen logs in Queensland, Australia, and in other areas from foliage and various decomposing vegetative materials. Feeding habits of the larvae are poorly known, but label data indicate an association with a variety of decaying vegetative materials. A series of *Elacatis umbrosus* from the Colorado Rockies was taken from beneath loose bark of a *Pinus ponderosa* log. Other host records include *Pinus palustris,* decaying oak logs, banana leaves, beneath bark of fallen twigs, and rotting leaves and cacti. Larvae and adults of *A. brunneus* have been collected from manure and decaying vegetation. (Brooks, 1965; Champion, 1888; Fukuda, 1962; Hayashi, 1969b; Lawrence, 1980; Young, unpublished notes).

Description: (Exclusive of Trogocryptinae.) Mature larvae attain lengths of 4–13 mm and widths of 0.4 mm to about 1 mm. Body moderately flattened to subcylindrical and subparallel throughout, lightly sclerotized except for head and urogomphal plate, with scattered elongate setae. Transverse parabasal ridges associated with abdominal tergites (*Elacatis*), or not (*Aglenus*). Head and body white or yellowish-white, darker in areas of heavy sclerotization such as tips of mandibles and urogomphi.

Head: (figs. 34.729b,c) Prognathous, somewhat depressed, exserted from prothorax. Epicranial suture with stem absent, frontal arms lyriform and complete to antennal insertions; endocarina absent. Symmetrical labrum present anterad of fused frons and clypeus. Stemmata 5 on each side, in groups of 3 anteriorly and 2 posteriorly (*Elacatis*) or lacking (*Aglenus*). Antennal insertions fully exposed, antennae elongate, 3-segmented, sensorium on segment 2 small. Mouthparts retracted, supported ventrally by well developed, posteriorly divergent hypostomal rods. Mandibles (figs. 34.729d, 34.730a) short, movable, asymmetrical, molar area of right mandible well developed, left mandible bearing prominent molar tooth; apices of mandibles with 3 large and 2 or more smaller interlocking teeth. Maxillae (fig. 34.729e) movable, consisting of 1-segmented cardo, lightly sclerotized, undivided, pad-like maxillary articulating area, mala which bears small dentiform uncus, and 3-segmented palps. Labium with mentum subquadrate, sides convexly rounded, broadest distally with apical margins shallowly emarginate; submentum with sides shallowly sinuate, proximal and distal margins convexly rounded; ligula short and apically rounded; palps 2-segmented. Hypopharyngeal sclerome molar-like.

Thorax and Abdomen: Thorax elongate with sides parallel, cervicosternum divided into 3 plates. Legs (fig. 34.729f) well developed, possessing scattered fine setae (*Aglenus*) or stout, spine-like setae (*Elacatis*), terminating with well developed tarsungulus; coxae large, separated by 1.5-2 coxal diameters. Abdomen somewhat flattened to subcylindrical, lightly sclerotized, tergites 1–9 subequal in length, 10th much reduced, visible ventrally anterad of anus; ninth tergite (figs. 34.729a, 34.730b) heavily sclerotized, extending ventrally, possessing (*Elacatis*) or lacking (*Aglenus*) a single poorly defined inflection on apical margin between fixed urogomphi; urogomphi 2-branched with outer (= main) branch upcurved and simple (*Aglenus*) or apically bifurcate (*Elacatis*); 9th sternite (fig. 34.729g) bearing a single asperity near each anterolateral margin (*Aglenus*) or possessing a double arch of mesally discontinuous asperities (*Elacatis*) (distance between arches greater than twice that between adjacent asperities of a single arch).

Spiracles: Thoracic spiracle annular (*Aglenus*) to ovate (*Elacatis*), situated on laterotergite; abdominal spiracles annular, subequal in size.

Comments: The othniids are a small but widely distributed family comprised of roughly 13 genera and 50 species. Diversity is greatest in the tropics and in the warm parts of temperate regions. Two genera and about 6 species occur in the United States.

The larva of *E. umbrosus* (LeConte) was figured by Böving and Craighead (1931); Peterson (1951) also made reference to this species in his discussion and illustrations of the Othniidae. Several additional species of *Elacatis* have been described (Gardner, 1931b; Fukuda, 1962; Hayashi, 1969b,

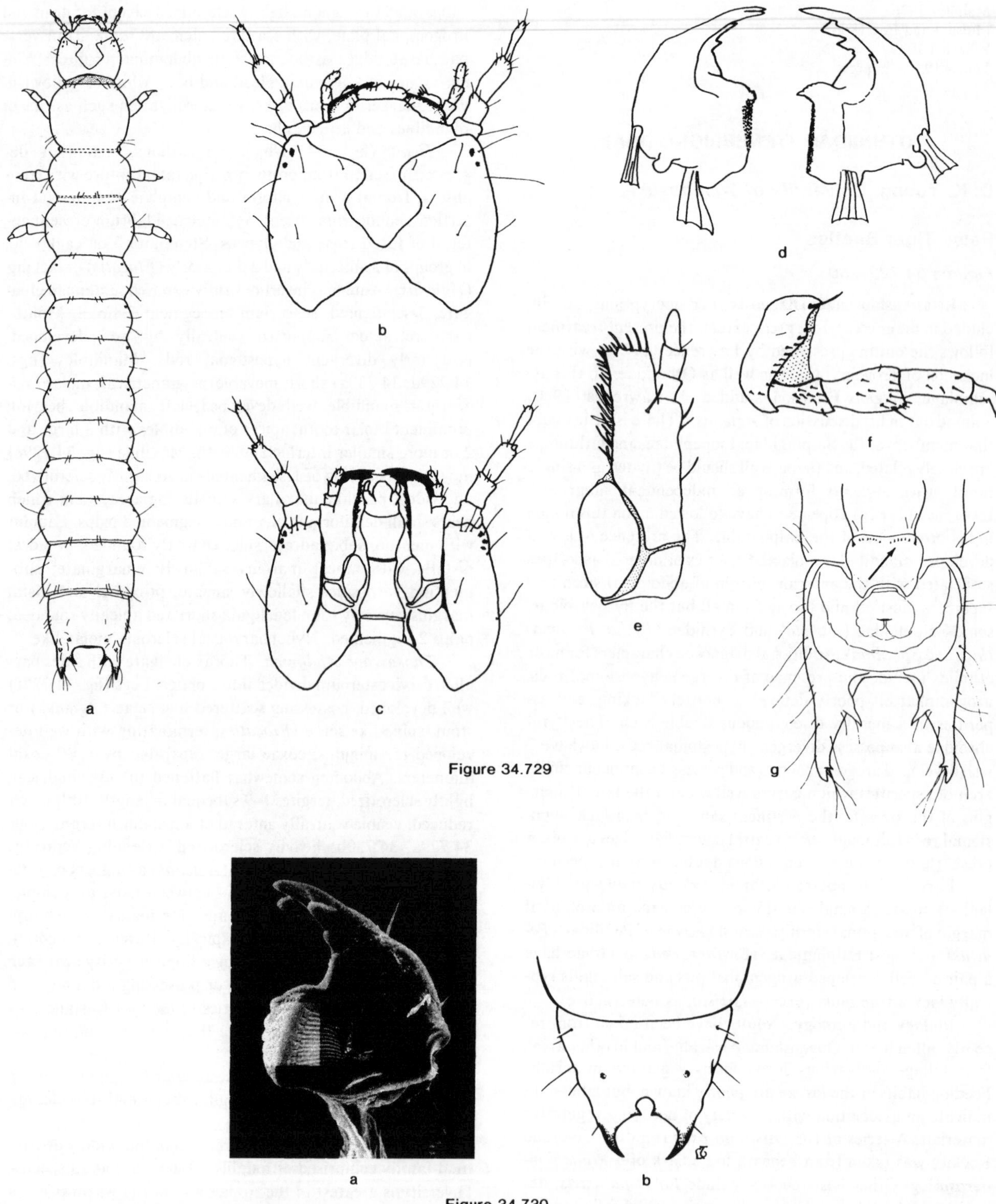

Figures 34.729a-g. Othniidae. *Elacatis umbrosus* (LeConte). Rocky Mtn. Nat'l Park, Deer Ridge Junction, 8930 ft, Larimer Co, Colorado, beneath bark of *Pinus ponderosa*. **a.** larva, dorsal; **b.** head capsule, dorsal; **c.** head capsule, ventral; **d.** mandibles, ventral; **e.** left maxilla, ventral; **f.** leg, inner aspect; **g.** apex of abdomen, ventral.

Figures 34.730a,b. Othniidae. *Aglenus brunneus* (Gyllenhal). Carter Cave State Park, Carter Co., Kentucky, from bat guano in cave. **a.** right mandible, dorsal; **b.** urogomphal plate, dorsal.

1980). Within the Trogocryptinae, the larvae of 2 *Prostominia* have been described (Peyerimhoff *in* Grouvelle, 1914; Hayashi, 1969b, 1980). Lawrence (1980) discussed the systematic position of *Aglenus,* noting several characteristics of the larva. The present discussion is based on larvae of *E. umbrosus* and *A. brunneus.*

Selected Bibliography

Böving and Craighead 1931.
Brooks 1965.
Champion 1888.
Fukuda 1962.
Gardner 1931b.
Hayashi 1969b, 1980.
Lawrence 1977c, 1980, 1982c.
Peterson 1951.
Peyerimhoff *in* Grouvelle 1914.

SALPINGIDAE (TENEBRIONOIDEA)

D. K. Young, *University of Wisconsin*

Narrow-Waisted Bark Beetles

Figures 34.731-733

Relationships and Diagnosis: The present treatment partially follows the outline presented by Lawrence (1977c) with the tentative inclusion of the Dacoderidae (of Watt, 1967b). However, *Aglenus* Erichson is included (along with the Trogocryptinae) in the Othniidae. As Lawrence (1977c: 43) noted, larvae of the Japanese salpingid genus *Istrisia* Lewis (*see* Hayashi, 1969b) provide a link between Pythidae and Othniidae; the salpingids are also closely linked to othniids through the Trogocryptinae, a group of genera referred to the Othniidae (*sensu* Crowson, 1955) or perhaps Salpingidae (*sensu* Lawrence, 1977c). In addition to the othniids, the family Salpingidae is also closely related to the Inopeplidae. These relationships were graphically portrayed in Lawrence's (1982c) discussion of Coleoptera, in which all 3 taxa were treated under the name Salpingidae. The presence of branched urogomphi (figs. 34.731, 34.733) and a single stout asperity associated with each anterolateral margin of the ninth abdominal sternite (fig. 34.732c) distinguishes most salpingid larvae from those of most other families. The Japanese salpingid *Istrisia rufobrunnea* Lewis is exceptional in having a transverse row of asperities across the anterior region of the ninth abdominal sternite; it also has asymmetrical mandibles and a well developed mola unlike other salpingids. Some Pedilidae (*Pedilus*) and Othniidae (*Aglenus*) also have branched urogomphi and a single asperity associated with each anterolateral margin of the ninth sternite. However, larvae of *Pedilus* also possess 2 well developed urogomphal pits caudomesally between the bases of the urogomphi that are entirely lacking in salpingid larvae. Asymmetrical mandibles and a distinct molar region on the right mandible will separate *Aglenus* larvae from those of the Salpingidae (symmetrical mandibles lacking a true mola).

Biology and Ecology: Adult salpingids have been found on flowers, foliage, or in association with decaying twigs and logs. The few known larvae have been collected beneath bark and within decaying twigs and logs. Larvae and adults of the European salpingids, *Lissodema quadripustulatum* Marsham and *Rhinosimus planirostris* Fabricius, have been reported to prey on scolytids. *Rabocerus mutilatus* Beck, another European salpingid, feeds on the chermid, *Adelges piceae,* and also on the fungus *Cucurbitaria pithyophila.* Larvae of *R. mutilatus* were noted to feed either on the chermid or the fungus, whereas the adults fed only on the fungus. The inner bark of red alder, *Alnus rubra,* was the observed food of *Rhinosimus viridiaeneus* Randall in British Columbia.

Both the larvae and adults of the Aegialitinae are found in rock crevices of the intertidal zone of several land masses bordering the Pacific Ocean. Examination of gut contents has demonstrated that *Aegialites* adults feed, at least in part, upon mites and perhaps other small invertebrates. Gut contents of *A. stejnegeri sugiharai* Kôno larvae indicate that it feeds phytophagously; only amorphous materials have been isolated from the digestive systems of other *Aegialites* larvae (Buck, 1954; Franz, 1955; Hatch, 1965; Howden & Howden, 1982; Kleine, 1909; Spilman, 1967).

Description: (Excluding *Istrisia—see* Hayashi, 1969b.) Mature larvae attain lengths of 3–6 mm and widths of 0.4–0.8 mm. Body moderately flattened, subparallel throughout, lightly sclerotized except for head and abdominal segment 9. Parabasal ridge associated with 9th tergite (*Aegialites, Rhinosimus*), 9th tergite with numerous wart-like callosities (*Aegialites,* fig. 34.733) or not (*Rhinosimus,* fig. 34.731), surface of body otherwise generally smooth. Vestiture consisting of scattered, moderately elongate setae. Dorsum of body yellowish-white (*Rhinosimus*) or yellowish-brown (*Aegialites*); venter creamy white to yellowish-white.

Head: Prognathous, exserted from or slightly retracted within prothorax. Epicranial suture with stem absent, frontal arms lyriform and complete to posterior margins of antennal insertions; paired endocarinae present. Symmetrical labrum present anterad of fused frons and clypeus. Stemmata 5 on each side, 3 forming an anterior band and 2 in a posterior line. Antennal insertions fully exposed; antennae elongate, 3-segmented, palpiform sensorium associated with segment 2. Mouthparts retracted, supported ventrally by well developed, subparallel hypostomal rods. Mandibles (figs. 34.732a,b) heavily sclerotized, movable, symmetrical, with basal pectinate hyaline lobe; mandibles with 2–3 apical and 1–2 subapical teeth. Maxilla with undivided cardo, pad-like maxillary articulating area and undivided maxillary mala, 3-segmented palp. Labium free to base of mentum and possessing short, distally rounded ligula (*Aegialites*) or lacking ligula (*Rhinosimus*); labial palps 2-segmented. Hypopharyngeal sclerome transverse, lacking definitive sclerotization; gular sutures separate.

Thorax and Abdomen: Thorax elongate with sides subparallel, prothorax much longer than meso- or metathorax which are subequal in length. Legs well developed, 5-segmented including tarsungulus; legs similar in size and

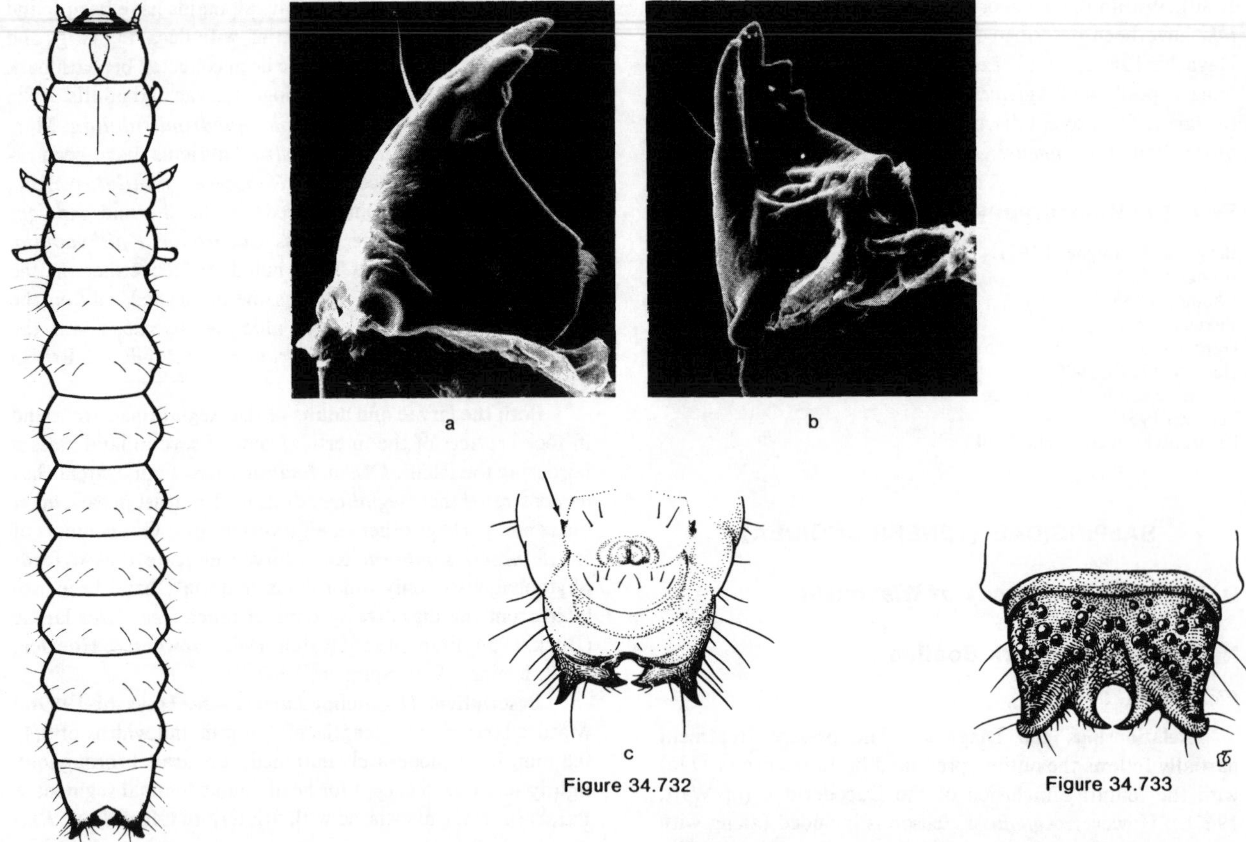

Figure 34.731

Figure 34.731. Salpingidae. *Rhinosimus viridiaeneus* Randall. Point Reyes National Seashore, Marin Co., California, from dead stems of *Lupinus*. Larva, dorsal.

Figures 34.732a-c. Salpingidae. *Rhinosimus* sp. Ramsey Cascade Trail, Great Smoky Mts. National Park, Tennessee, elev. 3200-3400 feet, beneath bark of *Acer saccharum*. **a.** left mandible, dorsal; **b.** right mandible, oroventral view; **c.** apex of abdomen, ventral.

Figure 34.733. Salpingidae. *Aegialites subopacus* (Van Dyke). Tomales Point, Marin Co., California. Urogomphal plate, dorsal.

shape, bearing scattered fine setae. Abdomen moderately flattened; tergite 9 (figs. 34.731, 34.733) heavily sclerotized, extending ventrally, possessing paired, fixed urogomphi caudally; urogomphi 2-branched with outer (= main) branch strongly upcurved; sternite 9 (fig. 34.732c) bearing a single asperity near each anterolateral margin. Segment A10 concealed from above by projection of 9th tergite, visible ventrally surrounding the anal orifice.

Spiracles: Thoracic spiracle annular, inserted on apex of short spiracular tube (*Aegialites*) or not (*Rhinosimus*); abdominal spiracles annular or annular-biforous.

Comments: The Salpingidae includes about 25 genera and 200 species distributed throughout most of the major faunal regions. Although relatively uncommon, the greatest species richness appears to be associated with the temperate region. Four genera and 11 species occur in the United States and Canada.

Wickham (1904) described and figured the larva and pupa of *Aegialites californicus* (Motschulsky); the species was also figured by Böving and Craighead (1931) and Chu (1949) under the generic name *Eurystethus*. Additional treatments of aegialitine larvae include those of Meixner (1935), Kôno (1938), Paulian (1949a), Leech and Chandler (1956) and Spilman (1967). The only salpingine larva described from N. America is that of *Rhinosimus viridiaeneus* Randall (Howden & Howden, 1982). Several additional species of Salpinginae have been described (Böving and Craighead, 1931; Franz, 1955; Hayashi et al., 1959; Hayashi, 1969b, 1980; Klausnitzer, 1978). For the present treatment, larvae of *Rhinosimus* sp. and *Aegialites subopacus* were examined.

Selected Bibliography

Böving and Craighead 1931.
Buck 1954.
Chu 1949.
Crowson 1955.
Franz 1955.
Hatch 1965.
Hayashi 1969b, 1980.
Hayashi *et al.* 1959.
Howden and Howden 1982.
Klausnitzer 1978.
Kleine 1909.

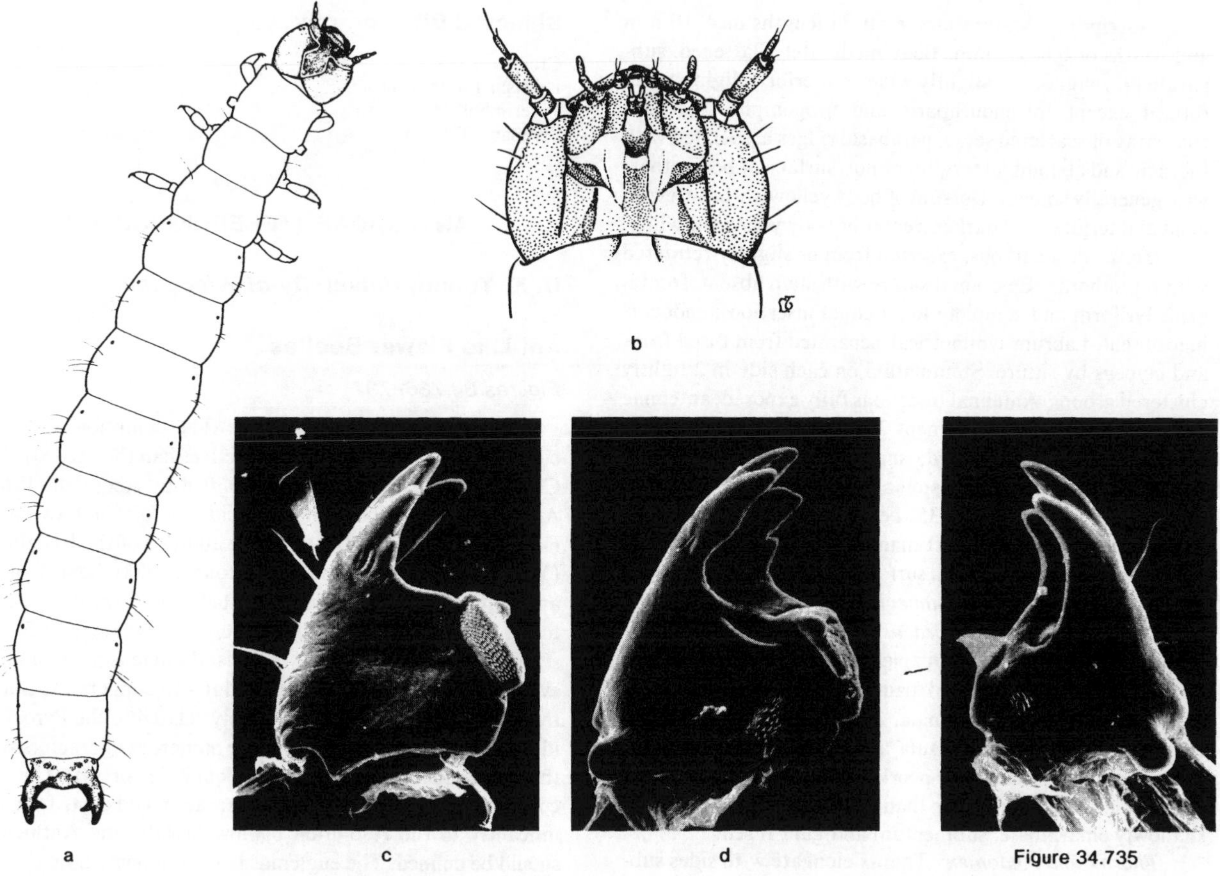

Figure 34.734

Figure 34.735

Figures 34.734a-d. Inopeplidae. *Inopeplus* sp. (*quadrinotatus* form). 15.1 mi. S. Valle Nacional, Oaxaca, Mexico, elev. 4300 feet, under bark of hardwoods. **a.** larval, dorsal; **b.** head capsule, ventral; **c.** left mandible, dorsal; **d.** right mandible, ventral.

Figure 34.735. Inopeplidae. *Inopeplus* sp. (*praeustus* form). 6.6 mi. W. El Bosque, Mexico. Left mandible, ventral.

Kôno 1938.
Lawrence 1977c, 1982c.
Leech and Chandler 1956.
Meixner 1935.
Paulian 1949a.
Spilman 1967.
Watt 1967b.
Wickham 1904.

INOPEPLIDAE (TENEBRIONOIDEA)

D. K. Young, *University of Wisconsin*

Figures 34.734-735

Relationships and Diagnosis: As interpreted here, the family Inopeplidae is the same as that defined primarily on the basis of an examination of adults by Crowson (1955). Although long associated with the Cucujidae, evidence from both larvae and adults strongly suggests affinities with the Heteromera. Nearest relatives of the inopeplids are the closely related Othniidae and Salpingidae. Inopeplid larvae may be distinguished from those of most other Coleoptera by the structure of the ninth abdominal tergite, which forms a hinged, terminal plate (fig. 34.734a). Other heteromerous families with a hinged, plate-like ninth tergite include the Boridae, Mycteridae, Prostomidae and Pyrochroidae. Borids and pyrochroids possess 2 well developed urogomphal pits caudomesally between the urogomphi; inopeplids lack such modifications of the urogomphal plate. The asymmetrical cranium of prostomid larvae differentiates them from inopeplids. In mycterids, the head possesses a short, curved incision along the posterior edge on each side of the midline, the maxillary articulating area is reduced, and the hypostomal rods are posteriorly divergent. Inopeplid larvae lack incisions along the posterior margin of the cranium, possess a well developed, pad-like maxillary articulating area (fig. 34.734b), and have the mouthparts supported ventrally by hypostomal rods which are subparallel throughout their length (fig. 34.734b).

Biology and Ecology: Very little is known about the bionomics of inopeplids. Larvae and adults have been collected from beneath bark of decaying deciduous trees. Larvae of *Inopeplus quadrinotatus* (Gorham) were collected from beneath bark of decaying oak in Japan (Hayashi, 1969b).

Description: Mature larvae attain lengths of 4–10 mm and widths of 0.8–1.5 mm. Body moderately flattened, subparallel throughout or slightly wider posteriorly, lightly sclerotized except for mouthparts and urogomphi. Vestiture consisting of scattered setae, parabasal ridges associated with thoracic and abdominal tergites or not, surface of body otherwise generally smooth. Dorsum of body yellowish-brown with head and tergites 8–9 darker, venter of body yellowish-white.

Head: Prognathous, exserted from or slightly retracted within prothorax. Epicranial suture with stem absent, frontal arms lyriform and complete to antennal insertions; endocarinae absent. Labrum symmetrical, separated from fused frons and clypeus by suture. Stemmata 5 on each side in 2 tightly clustered groups. Antennal insertions fully exposed; antennae elongate, 3-segmented, segment 2 bearing palpiform sensorium. Mouthparts retracted, supported ventrally by well developed, subparallel hypostomal rods (fig. 34.734b). Mandibles (figs. 34.734c,d, 34.735) heavily sclerotized, movable, somewhat asymmetrical, left mandible bearing more or less prominent molar tooth; mesal surface of mandibular base with well developed mola (e.g. *Inopeplus quadrinotatus*) or with simple hyaline lobe (e.g. *I. praeustus*); mandibles bearing 2–3 apical teeth and single subapical tooth. Maxilla with undivided cardo, well developed, pad-like maxillary articulating area, undivided maxillary mala, 3-segmented palp. Labium free to base of mentum, ligula absent, palps 2-segmented. Hypopharyngeal sclerome poorly defined or transversely rectangular. Mentum longer than wide with anterior margin shallowly emarginate; submentum and gula fused.

Thorax and Abdomen: Thorax elongate with sides subparallel, prothorax slightly longer than meso- or metathorax which are subequal in length. Legs well developed, 5-segmented including tarsungulus; legs similar in size and shape, bearing numerous fine setae. Abdomen moderately flattened; tergite 9 heavily sclerotized, hinged, extended ventrally to form the entire terminal segment or urogomphal plate that bears paired, fixed, distally branched urogomphi. Base of urogomphus with (*I. quadrinotatus*) or without (*I. praeustus*) 2 stout setiferous calli; 9th sternite partially recessed within emargination of 8th sternite, devoid of asperities (*I. praeustus*) or with 2 asperities near each anterolateral margin (*I. quadrinotatus*). Segment A10 reduced, visible ventrally surrounding anal orifice.

Spiracles: Thoracic spiracle annular-biforous, placed apically on well developed spiracular tube; abdominal spiracles annular.

Comments: The Inopeplidae is comprised of 2–3 genera and about 65 species widely distributed throughout both temperate and tropical regions. In N. America, they are represented by 12 species of *Inopeplus*.

The larva of *Inopeplus quadrinotatus* (Gorham) was described and illustrated by Hayashi (1969b, 1980) and that of *I. praeustus* Chevrolat by Peyerimhoff (1902) and Spilman (1971). For the present account, one of Hayashi's *I. quadrinotatus* was examined and several unidentified *Inopeplus* larvae from Mexico.

Selected Bibliography

Crowson 1955.
Hayashi 1969b, 1980.
Peyerimhoff 1902.
Spilman 1971.

ANTHICIDAE (TENEBRIONOIDEA)

D. K. Young, *University of Wisconsin*

Ant-Like Flower Beetles

Figures 34.736–737

Relationships and Diagnosis: The definition of Anthicidae followed here is more restrictive than that proposed by Crowson (1955) and further amplified by Abdullah (1969). As presently conceived, the anthicids conform to Lawrence's (1977c) outline, with the tentative addition of the Ischaliinae (Young, 1985a) and Lemodinae (Young, 1978), both of which appear to be far more closely related to the Anthicidae than to the Pyrochroidae. Although the pedilids are given family status in the present work, there is an increasing amount of evidence from both larvae and adults to support the notion that *Pedilus* Fischer is more closely related to the Pyrochroidae while the Eurygeniinae share numerous characters with the Anthicidae (Doyen, 1979; Young, in press). The discovery of additional eurygeniine larvae should contribute significantly to the resolution of how broadly the Anthicidae should be defined. The euglenids have commonly been treated as close relatives of the anthicids although, as Crowson (1955) pointed out, a number of structures in euglenid adults are more suggestive of Clavicornia than Heteromera. An analysis of known larval forms suggests a close relationship between the Anthicidae, Euglenidae and Scraptiidae, particularly with the subfamily Anaspidinae.

Anthicid larvae may be characterized by the following diagnostic character set: head with 1 pair of stemmata and usually a single median endocarina (absent in *Ischalia*); mandibles (fig. 34.737) possessing a well developed penicillus or brush of spine-line setae at base of mola; tergite 9 lacking urogomphal pits; sternite 9 without asperities. Within the Heteromera, larvae of Othniidae (*Prostominia*) and Scraptiidae (Anaspidinae) also possess a single pair of stemmata. Othniids possess a series of preanal asperities along the anterior margin of the ninth abdominal sternite and lack both a median endocarina and a mandibular penicillus. Scraptiids also lack both a median endocarina and a mandibular penicillus.

Biology and Ecology: Adults of many species are found in association with decaying vegetation on the surface of the ground; some appear to be limited to very specific habitats such as sand dunes and margins of fresh water, and salt and alkali lakes. They can also be found at times on flowers and foliage, and many species can be collected in large numbers at light. Numerous *Notoxus* adults as well as several species of *Acanthinus, Anthicus, Formicilla, Formicomus, Mecynotarsus, Sapintus, Tomoderus* and *Vacusus* have been taken

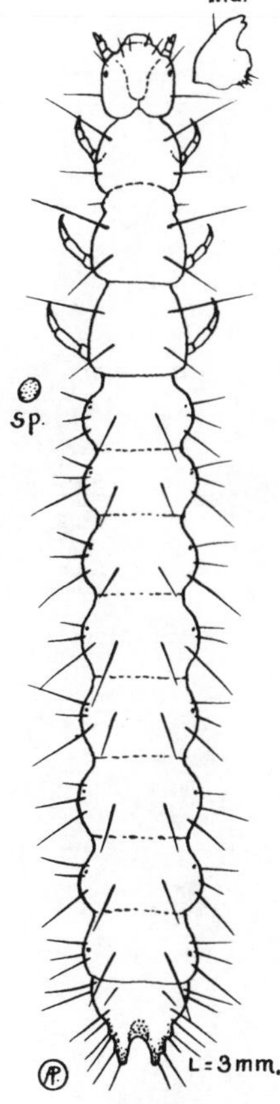

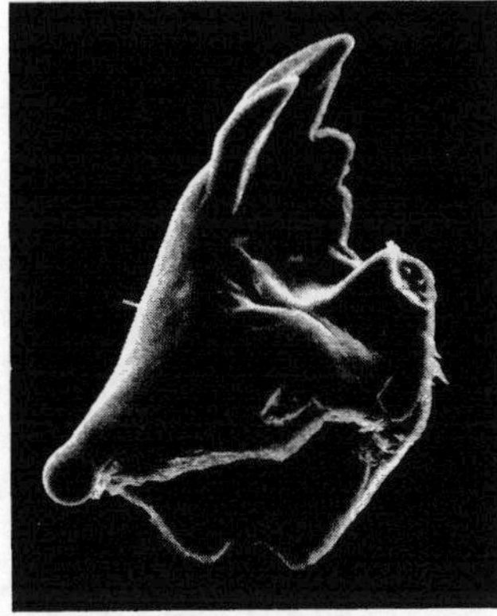

Figure 34.736 Figure 34.737

Figure 34.736. Anthicidae. *Notoxus* sp. Collected from soil in peach orchard. Larva, dorsal. (From Peterson, 1951)

Figure 34.737. Anthicidae. *Anthicus heroicus* Casey. 9 mi. E. Rogers, Benton Co., Arkansas, taken from dobsonfly egg masses (Corydalidae). Right mandible, oroventral view.

in association with meloid beetles or cantharidin, the chemical defense compound produced by meloid and some oedemerid beetles. The adults of most species are probably omnivorous although some are predaceous and perhaps a few subsist exclusively on nectar and pollen.

Larvae are generally associated with decaying vegetative materials where most are omnivorous or mycetophagous. However, larvae of *Anthicus heroicus* Casey have been observed feeding on egg masses of *Corydalus cornutus*, and *Malporus cinctus* (Say) larvae were found feeding on Diptera puparia (Abdullah, 1969; Arnett, 1968; Buck, 1954; Chandler, 1976; Davidson and Wood, 1969; Hatch, 1965; Hinton, 1945b; Howard, 1896; Kitayama, 1982; Landwehr, 1977; Young, 1984a, 1984b).

Description: (Exclusive of the aberrant genus *Ischalia* which is represented in N. America by 3 uncommon species.) Mature larvae attain lengths of 4–7 mm and widths of 0.3–0.8 mm. Body orthosomatic and subcylindrical or slightly flattened, subparallel throughout, lightly sclerotized except for head and urogomphi. Vestiture consisting of fine setae, surfaces of body generally smooth, color pale yellowish-white.

Head: Prognathous, exserted from prothorax. Epicranial suture with stem short, frontal arms lyriform and complete to antennal insertions; single median endocarina extending anterad of epicranial stem. Symmetrical labrum anterad of fused frons and clypeus. Single stemma near base of antenna. Antennal insertions fully exposed; antennae elongate, 3-segmented, segment 2 bearing conical sensorium, segment 3 with an elongate terminal seta. Mouthparts retracted. Mandibles (fig. 34.737) heavily sclerotized, movable, asymmetrical, molar area of right mandible usually more prominent than that of left, molar tooth of left mandible usually well developed; apices of mandibles bidentate; penicillus or brush of spine-like setae well developed at base of mola. Maxilla with 1-segmented cardo, a well developed, undivided, pad-like maxillary articulating area, maxillary mala, and 3-segmented palp. Labium free to base of mentum and possessing short, apically rounded ligula and 2-segmented palps. Hypopharyngeal sclerome transversely rectangular or cup-shaped. Gular sutures separate. Hypostomal rods lacking.

Thorax and Abdomen: Thorax elongate with sides subparallel, prothorax slightly longer than meso- or metathorax which are subequal. Legs well developed, 5-segmented including tarsungulus, similar in size and shape, and bearing numerous fine setae. Abdomen subcylindrical or slightly flattened, lightly sclerotized; tergite 9 extending ventrally, bearing paired, heavily sclerotized, fixed urogomphi apically. Urogomphi upcurved distally, with or without short inner secondary branch; sternite 9 small, not enclosed by 8th sternite. Segment A10 reduced, concealed from above by projection of 9th tergite and more or less fused to 9th segment. Anal orifice oriented posteroventrally.

Spiracles: Thoracic and abdominal spiracles annular.

Comments: The anthicids are a widely distributed family of about 35 genera and 2000 species. About 25 genera and 300 species have been recorded in the United States and Canada.

The larva of *Omonadus* (=*Anthicus*) *floralis* (L.) was briefly described by Rey (1887) and further discussed by Hinton (1945b). An account of the habits and general description of *Anthicus heroicus* Casey was provided by Howard (1896); the species was also figured by Böving and Craighead (1931) and Davidson and Wood (1969). A number of additional anthicid larvae have been described (Böving and Craighead, 1931; Peterson, 1951; Bonadona, 1958; Hayashi et al; 1959; Hayashi, 1980). Kitayama (1982) described and illustrated the larvae of 7 N. American anthicids. He also briefly commented on the larva of *Notoxus robustus* Casey, and summarized general information available relative to the systematics and bionomics of anthicid larvae.

Selected Bibliography

Abdullah 1969.
Arnett 1968.
Bonadona 1958.
Böving and Craighead 1931.
Buck 1954.
Chandler 1976.
Chu 1949.
Crowson 1955.
Davidson and Wood 1969.
Doyen 1979.

Hatch 1965.
Hayashi 1980.
Hayashi *et al.* 1959.
Hinton 1945b.
Howard 1896.
Kitayama 1982.
Landwehr 1977.
Lawrence 1977c.
Peterson 1951.
Rey 1887.
Young 1978, 1984b, 1984c, 1985a.

EUGLENIDAE (TENEBRIONOIDEA)

D. K. Young, *University of Wisconsin*

Ant-Like Leaf Beetles

Figure 34.738

Relationships and Diagnosis: Relative to most heteromerous families, the Euglenidae appears to be rather sharply defined and the intrafamilial classifications of such recent authors as Crowson (1955) and Arnett (1968) are essentially the same. However, a number of characters displayed by adults are suggestive of the Clavicornia rather than the Heteromera, thus tending to obscure interfamilial relationships. Most authors have placed the euglenids next to the Anthicidae; characters of *Escalerosia rubrivestis* (Marseul), the only euglenid larva described in detail (Hayashi, 1972), tend to support such a relationship. Euglenids may be distinguished from most other larvae by the following character combination: mandibles with well developed mola and fleshy hyaline lobe basad of molar region; stemmata, median endocarina, hypostomal rods and frontoclypeal suture lacking; antennal segment 2 bearing large, conical sensory appendix; segment 3 of antennae bearing elongate apical seta; abdominal tergite 9 bearing several elongate setae and a pair of short, fixed, strongly upcurved urogomphi; sternite 9 lacking asperities.

Biology and Ecology: Very little is known regarding the bionomics of euglenids. Adults have been collected on foliage, occasionally on flowers and at malaise and windowpane traps. Large numbers of adult *Elonus basalis* (LeConte) and *E. nebulosus* (LeConte) have been observed mating on the undersides of leaves in a hardwood forest in Michigan. Euglenid larvae are probably associated with decaying wood; one brief account made particular reference to wood in the red-rotten stage of decay. Hayashi's (1972) larvae of *E. rubrivestis* were collected from rotting wood in Japan (Buck, 1954; Hayashi, 1972; Young, unpublished notes).

Description: Mature larvae attain lengths of about 6 mm and widths of about 0.8 mm. Body subcylindrical to flattened, somewhat wider posteriorly, lightly sclerotized except for mouthparts and apices of urogomphi; surface of body generally smooth, creamy white to yellowish-white. Vestiture consisting of scattered, elongate setae.

Head: Prognathous to slightly deflexed, exserted from prothorax. Epicranial suture with stem short to lacking, frontal arms lyriform, ending as distinct sutures prior to attaining antennal insertions; endocarinae absent. Symmetrical labrum anterad of fused frons and clypeus. Stemmata absent. Antennal insertions fully exposed; antennae elongate, 3-segmented; sensorium of segment 2 large, conical; segment 3 bearing elongate, terminal seta. Mouthparts retracted. Mandibles heavily sclerotized, movable, asymmetrical; left mandible with prominent molar tooth; mandibles with simple hyaline lobe basad of mola; apices of mandibles tridentate. Maxilla with 1-segmented cardo, well developed, divided, padlike maxillary articulating area, maxillary mala, and 3-segmented palp. Labium free to base of mentum and possessing short to moderately elongate ligula and 2-segmented palps. Hypopharyngeal sclerome transverse. Gular sutures separate, slightly divergent posteriorly. Hypostomal rods lacking.

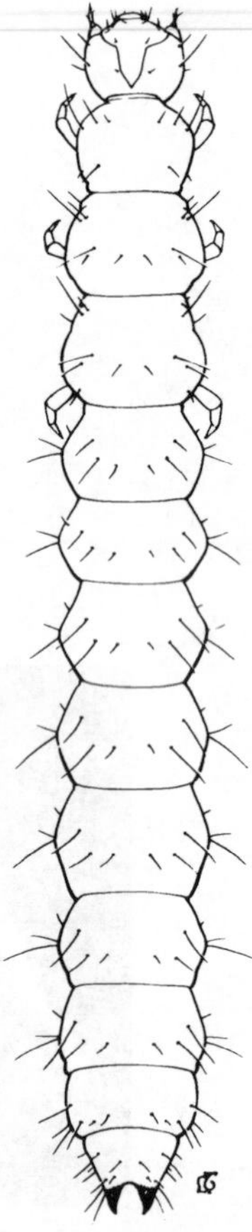

Figure 34.738. Euglenidae. Unassociated euglenid. Chattahoochie State Park, Houston Co., Alabama, from litter associated with log. Larva, dorsal.

Thorax and Abdomen: Thorax elongate, widest posteriorly, prothorax slightly longer than meso- or metathorax which are subequal in length. Legs well developed, 5-segmented including tarsungulus, similar in size and shape and bearing scattered, fine setae. Abdomen subcylindrical to flattened, lightly sclerotized; tergite 9 extending posteroventrally, bearing several elongate, fine setae and possessing paired, fixed urogomphi apically; urogomphi strongly upcurved distally. Sternite 9 lacking asperities, not enclosed by 8th sternite. Segment A10 small, concealed from above by projection of 9th tergite and more or less fused to 9th segment. Anal orifice oriented caudally.

Spiracles: Thoracic and abdominal spiracles annular.

Comments: The Euglenidae includes about 25 genera and 800 species worldwide, but is perhaps best represented in the tropics. In the United States and Canada, the euglenids are represented by 13 genera and about 40 species, most of which are uncommonly encountered.

The only positively associated euglenid larva is that of *Escalerosia rubrivestis* (Marseul), from Japan (Hayashi, 1972, 1980). Larvae similar to that of *E. rubrivestis* have been collected from dead logs and forest litter. These presumed euglenid larvae were relied upon, in part, for the present treatment.

Selected Bibliography

Arnett 1968.
Buck 1954.
Crowson 1955.
Hayashi 1972, 1980.

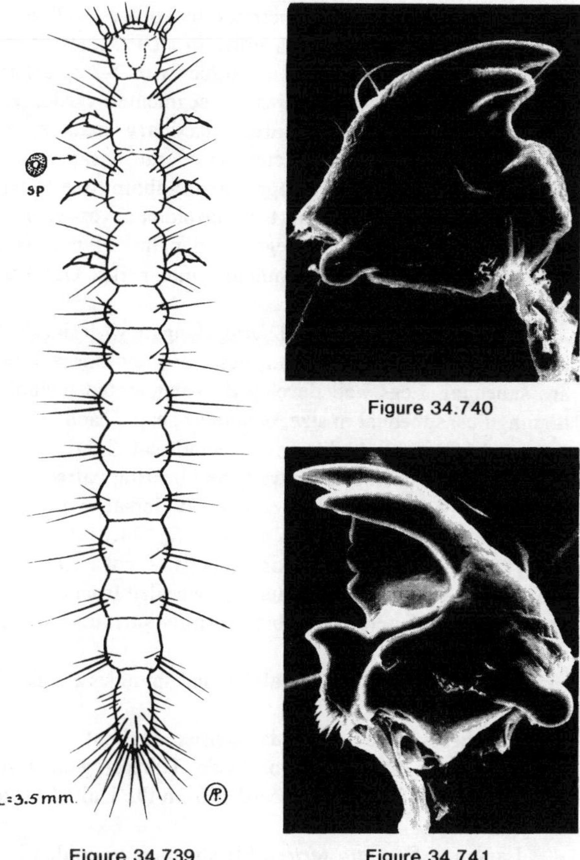

Figure 34.739. Scraptiidae. *Scraptia* sp. Collected from woodland ground cover (duff). Larva, dorsal. (From Peterson, 1951)

Figure 34.740. Scraptiidae. *Canifa* sp. Pickman Area, Bedford, Massachusetts, taken from black mold. Right mandible, ventral.

Figure 34.741. Scraptiidae. *Nassipa rufilabris* (Gyllenhal). Ermelunden, Denmark. Left mandible, ventral.

SCRAPTIIDAE (TENEBRIONOIDEA)

D. K. Young, *University of Wisconsin*

Figures 34.739–741

Relationships and Diagnosis: The present treatment follows the definitions of Scraptiidae proposed by Franciscolo (1972) and Lawrence (1977c) on the basis of both larvae and adults. Relying primarily upon characters of the adults, a number of coleopterists have considered the Scraptiinae as melandryids and the Anaspidinae as mordellids. Crowson (1955) included the anaspidines with the scraptiids, but considered them to be derived from a melandryid-like ancestor. Both Böving and Craighead (1931) and Lawrence (1977c) presented evidence suggesting that the closest relatives of the scraptiids are in the Anthicidae; the relationship between the Anaspidinae and anthicids appears to be especially close based on characters of the larvae. Scraptiine larvae are readily distinguished by the presence of a unique oblong, dehiscent process attached to the caudomesal aspect of the ninth abdominal tergite (fig. 34.739). A distinct frontoclypeal suture and lack of stemmata further characterize larvae of this subfamily. A single pair of stemmata, short hypostomal rods, and a mandibular mola with a brush of stout spines at the base (fig. 34.741) adequately characterizes anaspidine larvae. They also possess a pair of fixed urogomphi on the caudal margin of the ninth abdominal tergite and have sternite 9 simple, lacking asperities.

Biology and Ecology: Adult Scraptiinae and Anaspidinae appear to be largely diurnal and have been collected on foliage and flowers; many anaspidine species can be taken in large numbers from flowers of Apiaceae and Rosaceae growing near marshes and stream margins. The few known scraptiid larvae have been collected from beneath bark and among decaying woody fibers of dead logs, and from lichens. The Japanese anaspidine, *Anaspis funagata* Kôno, was found in association with the lichen, *Parmelia subaurulenta,* growing on a rock (Buck, 1954; Crowson, 1955; Hayashi, 1962a).

Description: Mature larvae attain lengths of 3–10 mm and widths of 0.4–1.3 mm. Body orthosomatic and subcylindrical, subparallel throughout, lightly sclerotized except for mouthparts and urogomphi. Surface of body smooth, pale yellowish-white. Vestiture consisting of fine setae.

Head: Prognathous, exserted from or slightly retracted within the prothorax. Epicranial suture with stem short (Anaspidinae) or absent (Scraptiinae); frontal arms lyriform, complete to antennal insertions; endocarinae absent. Labrum symmetrical, frons and clypeus more or less fused (Anaspidinae) or separated by distinct frontoclypeal suture (Scraptiinae). Stemmata 1 on each side (Anaspidinae) or absent (Scraptiinae). Antennal insertions fully exposed; antennae 3-segmented, segment 3 small, bearing an elongate terminal seta; segment 2 the longest, bearing a small, dome-like sensorium. Mouthparts retracted, supported ventrally by short hypostomal rods (Anaspidinae) or not (Scraptiinae). Mandibles (figs. 34.740, 34.741) heavily sclerotized, movable, nearly symmetrical to asymmetrical; molar area well developed, base of mola beset with brush of stout spines (Anaspidinae) or not (Scraptiinae); mandibles with 2–3 apical and 1–2 subapical teeth. Maxilla with 1-segmented cardo, padlike maxillary articulating area; maxillary mala simple (Scraptiinae) or subapically cleft on adoral margin (Anaspidinae); maxillary palps 3-segmented. Labium free to base of mentum, possessing short (Anaspidinae) or elongate (Scraptiinae) ligula and 2-segmented palps. Hypopharyngeal sclerome transverse, commonly molar-like. Gular sutures separate.

Thorax and Abdomen: Thorax elongate with sides subparallel, prothorax longer than meso- or metathorax which are subequal. Legs well developed, 5-segmented including tarsungulus, subequal in size, of similar shape, each bearing scattered fine setae. Abdomen subcylindrical, lightly sclerotized. Tergite 9 extended ventrally and bearing paired, fixed urogomphi (Anaspidinae) or completely dorsal, with large, oblong dehiscent caudomesal process (fig. 34.739) (Scraptiinae); sternite 9 simple, lacking asperities, not enclosed by 8th sternite. Segment A10 reduced, concealed from above by projection of 9th tergite, visible ventrally surrounding anal orifice.

Spiracles: Thoracic and abdominal spiracles annular or annular-biforous.

Comments: The Scraptiidae consists of about 25 genera and 250 species distributed worldwide. Thirteen genera and about 45 species have been recorded from the United States and Canada.

Larvae of *Scraptia sericea* Melsheimer, an unidentified *Anaspis,* and the European *A. frontalis* (L.) were illustrated by Böving and Craighead (1931). Emden (1942c), Peterson (1951) and Hayashi (1980) figured larvae of *Scraptia* spp., and *S. sericea* was also referred to in Chu (1949). Crowson (1955) made reference to larvae of *Scraptia* and *Anaspis* in his discussion of the definition of Scraptiidae. The only other known descriptions of scraptiid larvae come from Japan (Hayashi *et al.* 1959; Hayashi, 1962a, 1980).

Selected Bibliography

Böving and Craighead 1931.
Buck 1954.
Chu 1949.
Crowson 1955.
Emden 1942c.
Franciscolo 1972.
Hayashi 1962a, 1980.
Hayashi *et al.* 1959
Lawrence 1977c.
Peterson 1951.

CERAMBYCIDAE (CHRYSOMELOIDEA)

(INCLUDING DISTENIIDAE, HYPOCEPHALIDAE, OXYPELTIDAE, PARANDRIDAE, SPONDYLIDAE, VESPERIIDAE)

John F. Lawrence,
Division of Entomology, CSIRO

Long-Horned Beetles, Round-headed Wood-Borers

Figures 34.742–758

Relationships and Diagnosis: The Cerambycidae, as understood here, retains the subfamilies Philinae, Vesperinae, Oxypeltinae, and Disteniinae, which have been removed from the family in some recent works (Crowson, 1981; Linsley, 1961; Mann and Crowson, 1981; Švacha and Danilevsky, 1987). The great majority of cerambycid larvae differ from those of other beetle groups in lacking a mandibular mola and in having more or less protracted ventral mouthparts, with a well-developed maxillary articulating area and simple, fixed mala, and a broad gular or hypostomal region between the labium and thorax; in addition, the body is usually relatively straight, the legs are either reduced or absent, the body is lightly sclerotized, usually with lateral or dorsal swellings or ampullae, the head is often retracted, the stemmata are poorly developed or absent, and the spiracles are usually annular and never cribriform. In Disteniinae, the head is strongly retracted and the gular or hypostomal region is absent, the thoracic membrane being attached directly to the labium. Buprestid larvae may resemble those of some cerambycids, but they lack a maxillary articulating area and have an articulated, palpiform mala and cribriform spiracles (except for *Schizopus,* which has biforous spiracles and fingerlike processes on the thoracic and most abdominal segments).

Biology and Ecology: Cerambycids are phytophagous, and their larvae almost always feed internally on a variety of plant materials, with most species attacking the cambium, sapwood, and heartwood of trees and shrubs. Among those which feed under the bark or in the wood of trunks or larger branches, some occur in living trees (*Tetropium, Megacyllene, Glycobius, Goes, Plectrodera, Saperda*), some prefer recently killed conifers (*Arhopalus, Asemum, Rhagium, Callidium, Monochamus*) or hardwoods (*Knulliana, Xylotrechus, Neoclytus*), and others occur only in older dead wood, which is moist or in contact with the ground (*Parandra, Orthosoma, Ergates, Leptura*). Dry, seasoned timber is attacked by species of *Hylotrupes* and *Callidium* (conifer wood), *Chlorophorus* (bamboo), and *Smodicum* (hardwood). A number of cerambycid larvae feed internally on root systems (*Prionus* on a variety of living and dead trees, *Spondylis, Judolia, Ulochaetes* on conifers, *Desmocerus* on *Sambucus* (elder), *Dorcadion* on grasses), while others live in the soil, feeding externally on roots (*Tetraopes* on *Asclepias* (milkweed), *Homaesthesis* on sod-forming grasses). Some groups are restricted to twigs and smaller branches, which they girdle and cause to drop off; this girdling may be caused by adults (*Oberea, Oncideres*) or larvae (*Elaphidionoides, Aneflomorpha, Xylotrechus*). Other species may form galls (*Saperda, Oberea*) or bore into bark (*Acanthocinus, Elaphidion, Anoplodera*). Unusual larval feeding sites include green conifer cones (*Chlorophorus*), dead conifer cones on the ground (*Paratimia, Stenidea*), and mangrove seeds (*Ataxia, Leptostylus*). A number of cerambycids specialize in living herbaceous plants; these include many Lamiinae, such as species of *Moneilema* (cacti), *Phytoecia* (Apiaceae), and *Agapanthia*. Live stem borers appear to be much more common in tropical regions, where there is severe competition from termites in dead wood habitats. Larvae vary considerably in their degree of host specificity, the outstanding example of polyphagy being the Indian *Stromatium barbatum* (Fabricius), which has been recorded from over 300 host plants. Pupation usually takes place in cells in the wood, but some root feeders may pupate in earthen cells, and calcareous cells are constructed by a few species of Cerambycinae.

Adult feeding habits are also quite variable. Floricolous habits are common in Lepturinae, but a number of cerambycines and lamiines also feed on blossoms. Bark, stem, and leaf feeding are usually restricted to the Lamiinae, with conifer needle and cone feeding occurring mainly in the genus *Monochamus*. European *Dorcadion* feed as adults on the same grass roots which form the larval food. A number of Prioninae and Aseminae, and *Hylotrupes* and *Stromatium* among the Cerambycinae do not seem to feed as adults, although some may be attracted to sugar baits.

The main economic damage caused by Cerambycidae is that to forest products, tropical and subtropical fruit and nut trees, and vegetable and field crops; damage to temperate forest trees is relatively minor, and the major ecological role of this group in temperate forests is the decomposition of slash. *Hylotrupes bajulus* (L.) is a major pest of seasoned, coniferous timber in both N. America and Europe, while *Stromatium barbatum* (Fabricius) causes much damage to furniture and structural timber in Asia. Species of *Vesperus* injure alfalfa, potatoes, and various root vegetables in the Mediterranean Region, while *Philus* is known to destroy roots of sugarcane, and *Dorcadion* damages roots of maize. Other pests of herbaceous crops include species of *Agapanthia* and *Phytoecia*. The most important cerambycid pests are those attacking cacao, coffee, citrus, fig and other tree crops in tropical and subtropical regions; included among these pests are species in the lamiine genera *Anoplophora* and *Apriona*.

Description: Mature larvae 5 to 220 mm, usually 80 mm or less. Body elongate, parallel-sided or enlarged anteriorly, straight, cylindrical to strongly flattened, lightly sclerotized, except for buccal region and occasionally tergum A9. Surfaces smooth; vestiture of scattered, simple setae or occasionally a dense covering of hairs.

Head: Prognathous and usually retracted into prothorax (protracted in Lepturinae). Epicranial stem usually long, sometimes located in broad furrow for attachment of retractor muscle, absent in Lepturinae; frontal arms usually V-shaped and contiguous at base, sometimes indistinct. Median endocarina present, extending anterad of epicranial stem or between frontal arms (when stem is absent). Stemmata usually indistinct or absent, sometimes with 1 to 5 on each side. Antennae usually very short, 2- or 3-segmented. Frontoclypeal suture present or absent; labrum free; anterior edge of frons sometimes projecting over clypeus and forming sclerotized, dentate ridge. Mandibles usually symmetrical, robust, unidentate or bidentate, often with truncate apex, which may be oblique or rounded and gouge-like, and without accessory ventral process; mola absent, a subapical pseudomola sometimes present (Parandrinae, some Lepturinae); prostheca absent; mesal surface of mandibular base simple. Ventral mouthparts protracted or slightly retracted. Maxilla with cardo transverse or indistinct and fused with labium to form single plate or membranous area, stipes elongate to transverse, articulating area present or absent, palp 3-segmented, and mala simple and rounded. Labium sometimes with mentum and submentum fused; ligula well-developed; labial palps 2-segmented. Hypopharyngeal sclerome absent. Hypostomal rods usually present, short or long, converging or diverging. Ventral epicranial ridges absent. Gular region almost always present behind labium (absent in Disteniinae, Oxypeltinae, and Vesperinae, where labium is contiguous with thorax). Occipital foramen sometimes (Parandrinae, Prioninae, Cerambycinae) divided into 2 parts by tentorial bridge.

Thorax and Abdomen: Prothorax usually enlarged; protergum and prosternum sometimes with patches of asperities. Legs variable, often with fewer than 5 segments, sometimes highly reduced or absent, always widely separated; tarsungulus usually without setae, sometimes with 1. Abdominal segments 1 to 6 or 7 often bearing dorsal and ventral ampullae; abdominal terga sometimes with patches of asperities. Tergum A9 usually simple, occasionally with small urogomphi, median process, or spinose, sclerotized plate; sternum A9 simple. Segment A10 more or less oval, usually transverse, almost always posteriorly oriented; anal opening usually Y-shaped, occasionally transverse.

Spiracles: Usually annular or annular-multiforous, occasionally annular-biforous or annular-uniforous.

Comments: The family includes about 4000 genera and 35,000 species worldwide, with about 300 genera and 1200 species occurring in America north of Mexico. The group may be divided into as many as 11 subfamilies, with the 3 largest, Prioninae, Cerambycinae, and Lamiinae, occurring in all major regions of the world. The subfamily Disteniinae includes the genus *Distenia*, with species from North to South America, Asia and Africa, and several other genera from Asia, Madagascar, and South America. Philinae includes *Philus* and *Doesus* from Asia and Africa, while Oxypeltinae includes the 2 monotypic Chilean genera *Oxypeltis* and *Cheloderus*. Vesperinae includes the genus *Vesperus* from southern Europe, North Africa, and Asia Minor; adults resemble lepturines, but the females are wingless and there are 2 distinct larval types. The South American Anoplodermінae includes not only *Anoploderma* and its relatives, but also the peculiar *Hypocephalus armatus* Desmarest, which is a wingless, burrowing form, with short antennae and digging legs. The subfamily Parandrinae includes *Erichsonia* from Mexico and Central America and the widespread genus *Parandra*. The Aseminae includes about a dozen genera from N. America, Eurasia, and Madagascar, while the Lepturinae comprises over 100 genera, also dominant in the Holarctic Region.

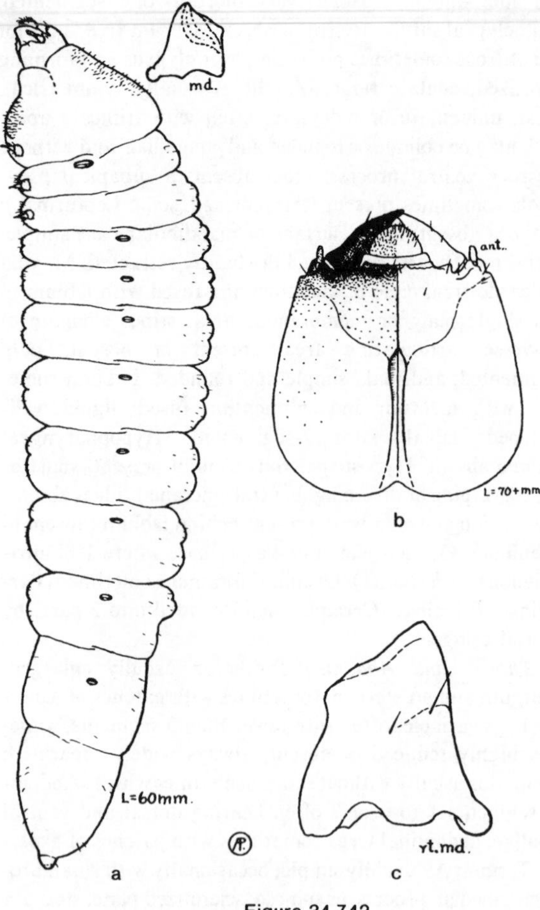

Figure 34.742

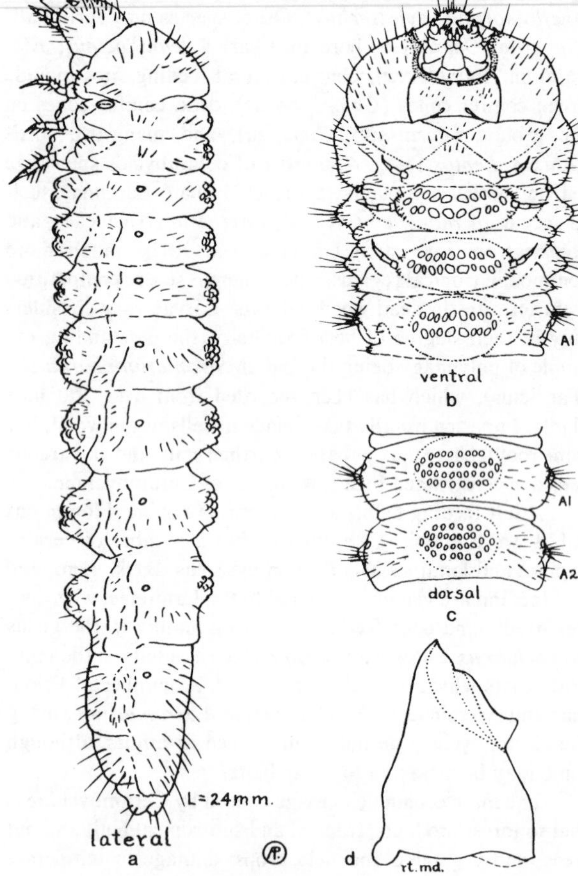

Figure 34.745

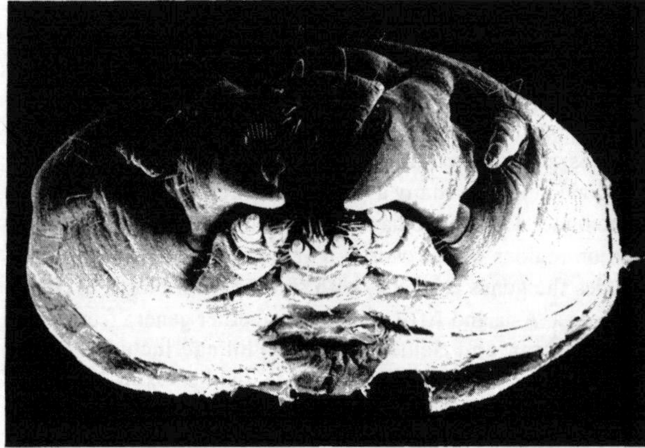

Figure 34.743

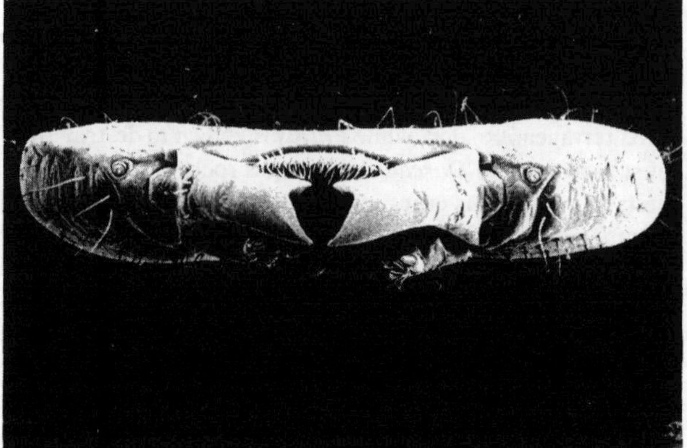

Figure 34.744

Figures 34.742a-c. Cerambycidae. *Orthosoma brunneum* (Forster) (Prioninae). **a.** larva, lateral (plus detail), length = 60-75 mm; **b.** head, dorsal; **c.** right mandible, dorsal. Cylindrical, smooth and shining, yellowish. Antennae prominent, distal segment elongate and cylindrical. Infests dead and decaying logs of hardwoods and conifers, also destructive to cross-ties and timbers in contact with ground. (From Peterson, 1951)

Figure 34.743. Cerambycidae. *Parandra brunnea* (Fabricius) (Parandrinae). Head, anteroventral. Bores in dry, dead pine.

Figure 34.744. Cerambycidae. *Rhagium inquisitor* (L.) (Lepturinae). Head, anterior.

Figures 34.745a-d. Cerambycidae. *Leptura* sp. (Lepturinae). **a.** larva, lateral, length = 24 mm; **b.** head and thorax, ventral; **c.** abdominal segments 1 and 2, dorsum; **d.** right mandible, ventral. (From Peterson, 1951)

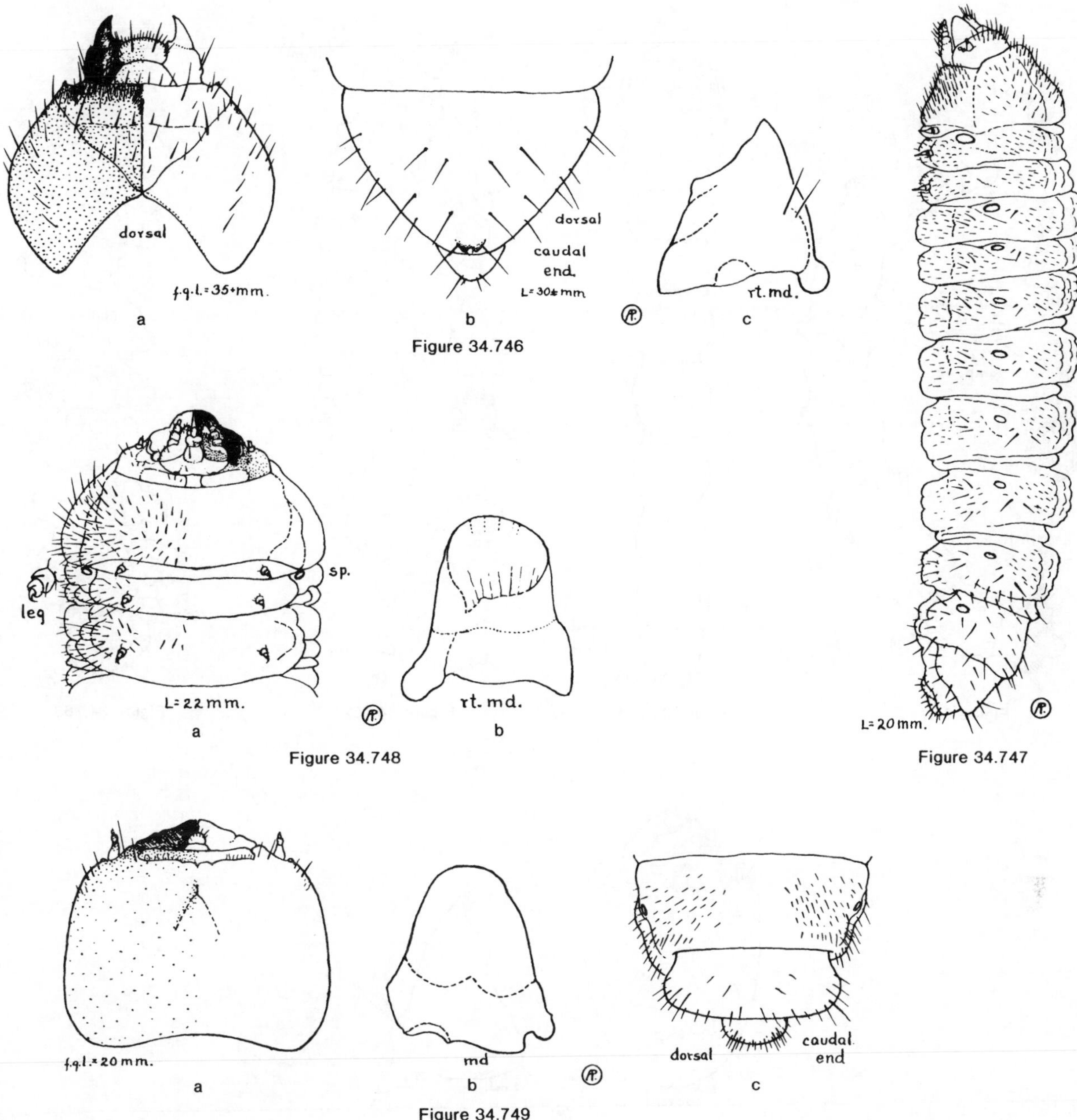

Figures 34.746a-c. Cerambycidae. *Anthophylax hoffmani* Beutenmueller (Lepturinae). **a.** head, dorsal; **b.** abdominal apex, dorsal; **c.** right mandible. Bores in logs of spruce and fir. (From Peterson, 1951)

Figure 34.747. Cerambycidae. *Elaphidionoides villosus* (Fabricius) (Cerambycinae). Twig pruner. Larva, lateral. Length = 20 mm. Long yellowish setae scattered over body. Attacks smaller branches of a wide variety of hardwoods; burrows down center of stem until grown when it cuts off the branch, leaving a thin shell of bark broken by the wind. (From Peterson, 1951)

Figures 34.748a,b. Cerambycidae. *Hylotrupes bajulus* (L.) (Cerambycinae). Old house borer. **a.** head and thorax, ventral; **b.** right mandible, mesal. Robust, slightly depressed with cuticula thin, shining and sparsely covered with long yellowish setae; mandibles broad, basally light brown and distally black with a deep longitudinal impression; prothorax rectangular, depressed, with oblique pronotum a little wider than long, with cephalic portion bearing many setae and caudal portion shining with a few, irregular, indistinct striae and an impressed median suture; metanotum with an inverted v-shaped impression; legs short; dorsal ampullae prominent, shining, nearly tuberculate, with 2 lateral and 2 transverse impressions; spiracles small and broadly oval with thin peritremes. Attacks dry seasoned wood of conifers, especially *Pinus* and *Picea*. Can reinfest wood and cause serious structural damage. (From Peterson, 1951)

Figures 34.749a-c. Cerambycidae. *Megacyllene robiniae* (Forster) (Cerambycinae). Locust borer. **a.** head, dorsal; **b.** left mandible, mesal; **c.** abdominal apex, dorsal. A serious pest of living black locust trees, especially in plantations and under stressed conditions. (From Peterson, 1951)

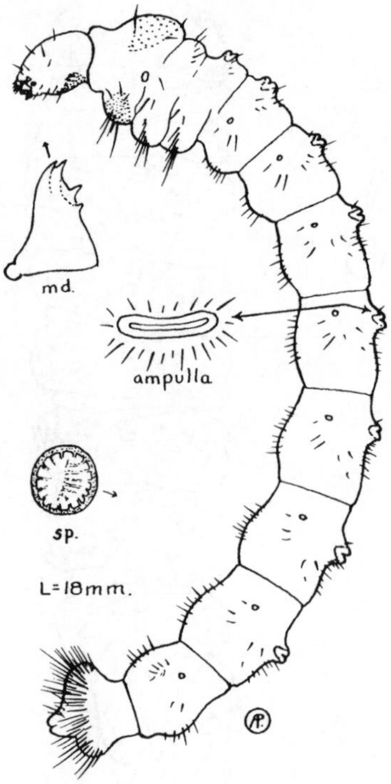

Figure 34.750

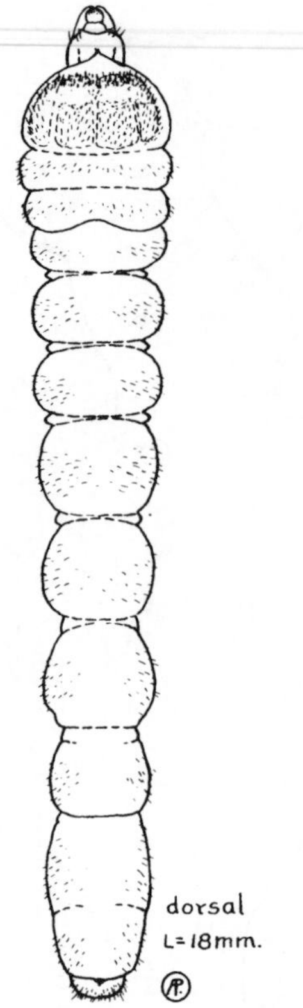

Figure 34.751

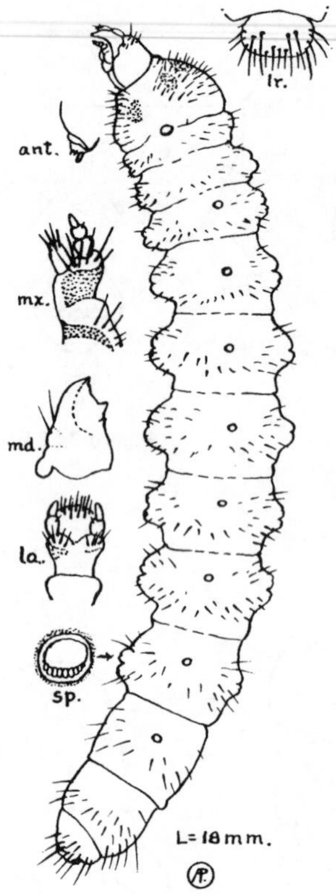

Figure 34.752

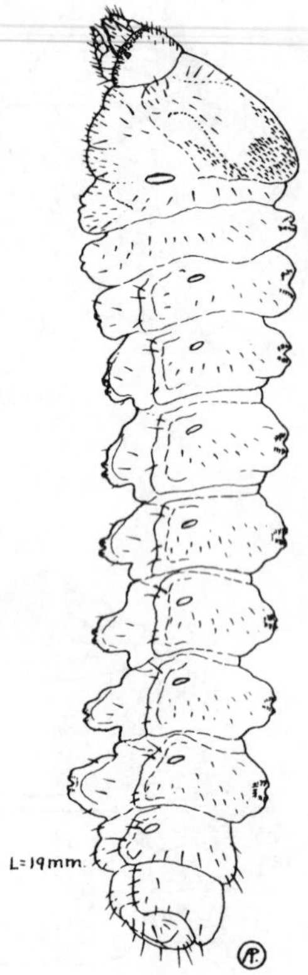

Figure 34.753

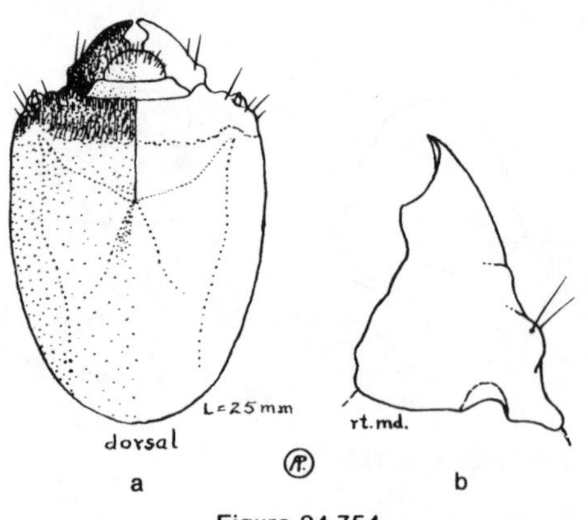

Figure 34.754

Figure 34.750. Cerambycidae. *Hippopsis lemniscata* (Fabricius) (Lamiinae). Larva, lateral (plus details). Length = 18 mm. Infests the stems of ragweed (*Ambrosia*). (From Peterson, 1951)

Figure 34.751. Cerambycidae. *Graphisurus* sp. (Lamiinae). Larva, dorsal. Length = 18 mm. Feeds beneath the bark of dead hardwoods. (From Peterson, 1951)

Figure 34.752. Cerambycidae. *Dectes spinosus* (Say) (Lamiinae). Larva, lateral (plus details). Length = 18 mm. Attacks the stems of large weeds, especially ragweed (*Ambrosia*), Joe-Pye-weeds and other *Eupatorium* spp. and *Xanthium* spp. (From Peterson, 1951)

Figure 34.753. Cerambycidae. *Oberea tripunctata* (Swederus) (Lamiinae). Dogwood twig borer. Larva, lateral. Length = 14 mm. Bores down the center of twigs, making a series of frass-expulsion holes and cutting off parts of the twig behind it. Attacks dogwood, elm, laurel, azalea, sourwood, *Viburnum* and various fruit trees. (From Peterson, 1951)

Figures 34.754a,b. Cerambycidae. *Monochamus scutellatus* (Say) (Lamiinae). Whitespotted sawyer. **a.** head, dorsal; **b.** right mandible, dorsal. Infests recently dead conifers, especially sawlogs and pulpwood. White pine a favorite; but commonly also attacks red and jack pine, balsam fir, larch, and black, white and red spruce. (From Peterson, 1951)

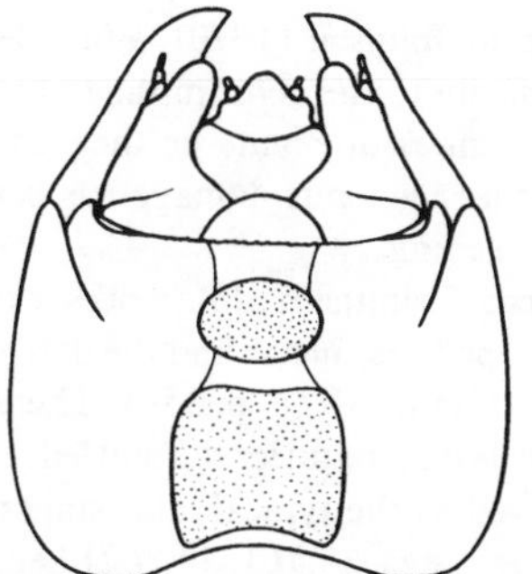

Figure 34.755

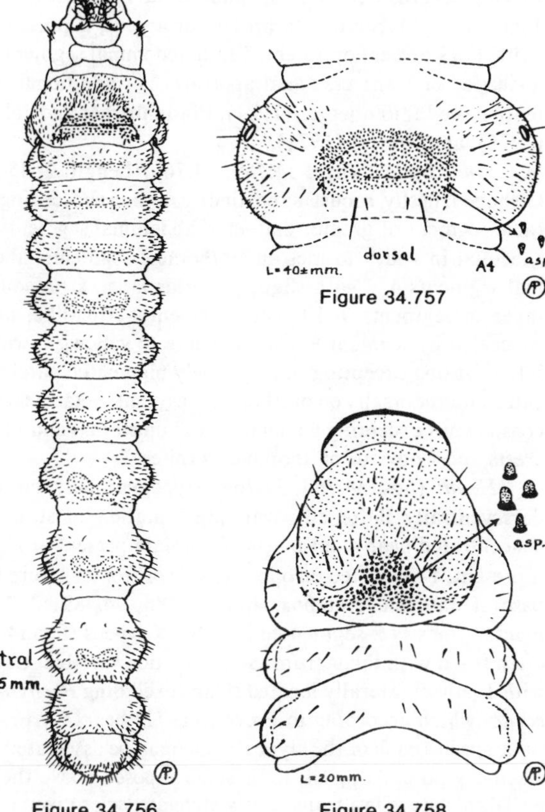

Figure 34.757

Figure 34.756

Figure 34.758

Figure 34.755. Cerambycidae. Prioninae. Head, ventral (showing occipital foramen divided by tentorial bridge).

Figure 34.756. Cerambycidae. *Saperda vestita* Say (Lamiinae). Linden borer. Larva, ventral. Length = 25 mm. Usually attacks weakened or stressed linden (basswood) (*Tilia*), but may attack poplar, boring beneath the bark or deep into the wood. (From Peterson, 1951)

Figure 34.757. Cerambycidae. *Saperda calcarata* Say (Lamiinae). Poplar borer. Abdominal segment 4, dorsum. Attacks poplar and willow, boring in the inner bark, sapwood and heartwood, sometimes causing severe injury or death in plantations, shade trees or forest stands. (From Peterson, 1951)

Figure 34.758. Cerambycidae. *Oberea bimaculata* (Olivier) (Lamiinae). Raspberry cane borer. Head and thorax, dorsal. Also bores in blackberry canes, and reported from plum, quince and apple. (From Peterson, 1951)

Selected Bibliography

Aitken 1975.

Balachowsky 1962 (agricultural pests).

Beeson and Bhatia 1939.

Böving and Craighead 1931 (larvae of a few species, showing subfamily characters).

Cameron and Real 1974 (larva of *Chelodérus childreni* Gray).

Chemsak and Linsley 1983.

Craighead 1923 (larvae of many North American species, representing about 125 genera).

Crowson 1955, 1960b, 1981.

Demelt 1978.

Drooz 1985 (forest pests).

Duffy 1952a (first-instar larva of *Agapanthia villosoviridescens* (De Geer)), 1953 (larvae of 110 species (75 genera) and pupae of 76 species (58 genera) occurring in Britain (native and introduced)), 1957 (larvae of 203 species (137 genera) and pupae of 83 species (61 genera) occurring in Africa), 1960 (larvae of 160 species (123 genera) and pupae of 78 species (67 genera) occurring in the Neotropical Region), 1963 (larvae of 100 species (77 genera) and pupae of 44 species (39 genera) occurring in the Australasian Region), 1968 (larvae of 186 species (124 genera) and pupae of 46 species (40 genera) occurring in the Oriental Region), 1980 (supplement to Duffy, 1957).

Durr 1957 (morphology and biology of *Hylotrupes bajulus*).

Emden 1939 (larvae of British species), 1940 (larvae of British species).

Forchhammer 1981.

Furniss and Carolin 1977 (forest pests).

Gardiner 1966 (egg bursters).

Gardner 1925 (larvae of Indian species), 1927 (larvae of Indian species), 1930b, 1931a (larvae of Indian species), 1931b (larvae of Indian species).

Linsley 1959 (cerambycid biology).

Linsley 1961, 1962a, 1962b, 1963, 1964 (North American monographs).

Linsley and Chemsak 1972, 1976 (North American monographs).

Mamaev and Danilevskii 1973 (larvae of Aseminae).

Mann and Crowson 1981 (family relationships).

Mateu 1963 (larva of *Macrotoma palmata* (Fabricius)), 1966 (larva of *Derolus mauritanicus* (Buquet).

Mathur 1958 (larvae of Indian species).

Nakamura 1981 (pupae of Japanese species, representing more than 100 genera).

Nakamura *et al.* 1976 (larva of *Parandra shibatai* Hayashi; larval subfamily key).

Paulian and Villiers 1940 (larval characters; key to genera).

Peterson 1951 (larvae of various North American species).

Synder 1910 (damage to telephone poles by *Parandra brunnea* (Fabricius)).

Villiers 1978 (monograph of French species; larval key to genera).

BRUCHIDAE (CHRYSOMELOIDEA)

(= Lariidae, Mylabridae)

Gary S. Pfaffenberger,
Eastern New Mexico University

The Bruchids

Figures 34.759–789

Relationships and Diagnosis: The bruchids have been placed in the superfamily Chrysomeloidea along with the chrysomelids and cerambycids. However, adult-based taxonomic schemes do not always depict morphological relationships among included larval forms. For example, the larvae of bruchids and chrysomelids share numerous morphological similarities, particularly in the case of the questionable chrysomelid genus *Rhaebus* (Kingsolver and Pfaffenberger 1980). However, the bruchids and cerambycids appear quite distinct. Furthermore, bruchid larvae share structural as well as morphological similarities with the curculionids, members of the closely related superfamily Curculionoidea.

According to Crowson (1955) representatives of two genera of the family Curculionidae exhibit seed boring habits, a trait exhibited by all Bruchidae and unlike known forms of the Chrysomelidae. Despite their behavioral similarities curculionids and bruchids are readily distinguished by characteristics of the labium. Snout weevil larvae possess 2-segmented labial palps which are absent among larval bruchids (Peterson 1951). Moreover, Böving and Craighead (1931) use the presence of a hypopharyngeal bracon among the Curculionoidea to separate them from the Chrysomeloidea.

Due to the variability of larval characteristics which appear within the superfamily Chrysomeloidea, there are few salient features which may be used to separate the bruchids from the chrysomelids. Possible exceptions may be the presence of 1–3 pair(s) (Pfaffenberger 1982; Pfaffenberger et al. 1984; Pfaffenberger 1984), or perhaps an odd number (Pfaffenberger and Janzen 1985) of short, decurved, median epipharyngeal setae (fig. 34.766) and 1 pair of median submental setae (fig. 34.777). The most noticeable feature of bruchids is, however, the conspicuous absence of labial palps. In addition, all first instar larvae characteristically bear a dorsal X- or H-shaped prothoracic plate (fig. 34.765) (Pfaffenberger and Johnson 1976).

Biology and Ecology: Bruchids are notorious seed weevils which are known to ravage the seeds of economically important legumes under cultivation or in storage. Larvae successfully parasitize the seeds of 30 other families of plants as well (Johnson 1970). Members of the largest subfamily, Bruchinae, are especially common in the southwestern U.S. extending into central Mexico and some even into Central and South America. Many species of this subfamily are commonly collected from a few prevalent plant genera which characterize this geographical region. These genera include *Astragalus, Prosopis, Acacia* and *Mimosa.* For a comprehensive treatment of plant hosts *see* Center and Johnson (1974; 1976) and Johnson (1979; 1981a; 1983).

The overwintering adults copulate and the female oviposits on the pod or seed surface. The larva penetrates the seed, with (Pfaffenberger 1981) or without (Pfaffenberger 1979) the assistance of the egg chorion, and subsequently devours all or a portion of the seed contents. One to several larvae may mature within a single seed or one larva may use several seeds that it glues together (Center and Johnson 1973). In either case the larva will eventually excavate a chamber within which it will pupate. The quarried pupation chambers are near the surface and may be identified by a small, circular, opaque area in the seed integument which was weakened peripherally by the last larval instar prior to pupation. Upon emergence the adult escapes through this area and may begin mating within 15 minutes to an hour (Singh et al. 1979).

According to Johnson (1970), adult feeding habits are restricted primarily to the consumption of pollen and nectar from diverse families of plants or they do not feed at all. Hence, little or no economic damage has been observed as a result of adult feeding.

Description: Prominent differences exist between the dimorphic, oligopod first instar and the derived, apodous subsequent instars (Imms 1977, p. 359). Therefore, identification of bruchid larvae requires a knowledge of the aberrant first instar as well as the typical later stages.

First Instar: (figs. 34.761, 34.762) Length 0.5–1.5 mm, caraboid with slight ventral curvature. Cuticle white to yellowish, unsclerotized except for conspicuous, dorsal prothoracic plate (fig. 34.765). Chaetotaxy rather uniform with varying lengths (cf. figs. 33 and 34 *in* Pfaffenberger and Johnson 1976). Stemmata present or absent, if present 1–3 pairs. Legs present or absent. Tenth abdominal segment with Y-shaped or transverse anal aperture. For additional information see Pfaffenberger (1981), Pfaffenberger and Johnson (1976), and Prevett (1971).

Final Instar: (figs. 34.759, 34.760) Length 2–45 mm, C-shaped, fleshy, robust, subcylindrical, each succeeding thoracic segment of greater diameter, abdominal segments 1–3 subequal in width to meso-metathoracic segments, abdominal segments 4–6 with slight posterior taper, a pronounced taper in segments 7–10, 10th segment button-like, nearly concealed by segment 9. Cuticle white to yellowish, without sclerotization excepting those vaguely pigmented patches located anterodorsally on prothoracic segment. Setae nearly inconspicuous excepting spinulate areas on sternal and pleural crests, most obvious on thoracic sternites.

Head: (figs. 34.763, 34.764) Hypognathous, retracted, dorsoventrally flattened, oval, pigmentation most heavily concentrated near mouthparts. Stemmata present or absent, if present, 1–3 located on opposite side of frontal suture from base of antenna. Antenna (figs. 34.788, 34.789) 2–3 segmented (mostly 2-segmented in U.S.). Clypeus (fig. 34.783) often fused with the labrum. Sclerotization variable. Usually with 1 pair of laterally located setae (excluding *Pachymerus cardo,* which according to Prevett (1971) bears 2 pairs). In some species each of the clypeal setae may be associated with a sensory pore. Adjacent to or superimposed above the clypeolabral suture is a transversely elongate, sclerotized region usually bearing a pair of laterally located setae and associated sensory pore. Labrum (fig. 34.783) with anterior margin generally semi-elliptical, supporting several elongate setae and sensory pores that may be concealed by dense mat of microtrichia. Epipharynx (fig. 34.766) typically bearing 1–3 pair(s) of median epipharyngeal setae. Other features variable such as tormae and medial longitudinal groove. Mandible (fig. 34.770) prognathous, monocondylic, cutting surface concave, smooth molar surface. Denticulations appear in *Conicobruchus strangulatus* and *Tuberculobruchus natalensis* (Prevett 1971). Maxilla (fig. 34.780) with degree of sclerotization and number and arrangement of setae varying greatly throughout family but of value in subfamilial organization. Cardo usually moderately developed, stipes largely membranous, semiquadrate with sclerite extended postero-laterally, sclerite of palpifer well developed. Spatulate setae on lacinia appear consistently throughout family but vary in number from 3–8. According to available information, maxillary palp is 1-segmented in subfamilies Bruchinae and Eubaptinae (Teran 1967); known members of subfamily Amblycerinae possess 2-segmented palps with exception of *Zabrotes* sp. with 1-segmented palps; Pachymerinae appear to be a heterogeneous assemblage since *Caryoborus* has 1-segmented palps, *Pachymerus* 2-segmented palps, *Caryedon* and *Caryobruchus* possess 2–3 segmented palps; maxillary palps appear to be 1-segmented in the subfamily Kytorhininae (*Kytorhinus quadriplagiatus*, unpublished information). Labium (fig. 34.777) with submentum transversely elongate with contrasting degrees of sclerotization. According to Prevett (1971) this sclerite is lacking in *Tuberculobruchus* sp., as it is in *Callosobruchus maculatus* (fig. 34.771) and *C. chinensis* (fig. 34.772). Submentum bearing 1 pair of median setae and separated from mentum by membranous area which supports 2 pairs of lateral setae. Mentum with longitudinal, ovoid appearance, bearing prong-like, anterolateral projections. One pair of medio-lateral setae present, may be isolated in membranous islets. Variable number of setae and sensory pores between bases of glossae and paraglossae. Paraglossae sometimes elongate, fleshy, and membranous.

Thorax and Abdomen: Distance between paired legs increasing with each succeeding segment. Development varies from a papilla (*Zabrotes subfasciatus*) to 5-segmented, fleshy lobes in *Pachymerus* sp. (Pfaffenberger 1974). Chaetotaxy variable. Anal sulcus transverse or Y-shaped.

Spiracles: Slightly oval to round, unicameral or bicameral, atrial fine structure variable. Mesothoracic and 8 abdominal pairs present.

Comments: According to Bottimer (1968) this small family includes 5 subfamilies and 58 genera (the latter according to J. M. Kingsolver, personal communication) containing approximately 1300 species world-wide (Johnson 1970). They are more abundantly represented in tropical and subtropical areas than in temperate areas of both the Old and New Worlds. Roughly 600 species have been described from the Nearctic and Neotropical regions. Members of the subfamily Kytorhininae are unique in being restricted to the northern Palearctic and Nearctic regions.

Taxonomic interest in bruchid larvae had been restricted primarily to the economic forms until Prevett (1971) published his monograph which included larval descriptions of 28 species. Subsequently major contributions include those of Arora (1978) and Pfaffenberger and Johnson (1976). Currently larvae of about 121 species have been described (Pfaffenberger 1985b).

KEY TO FINAL INSTARS OF ECONOMIC SPECIES OF NORTH AMERICAN BRUCHIDAE

1. Submentum of labium absent (fig. 34.771) 2
   Submentum of labium at least partially sclerotized (figs. 34.773–34.778) 3

2(1). Three pairs of setae surrounding base of mentum (fig. 34.771). Sclerite of mentum trilobed anteriorly, with small median lobe. Maxillary stipes with 7 setae (fig. 34.780). Palpifer not continuous with lacinia .... cowpea weevil, *Callosobruchus maculatus* (Fabricius)
   Two pairs of setae surrounding base of mentum (fig. 34.772). Sclerite of mentum bilobed anteriorly. Maxillary stipes with 11 setae. Palpifer continuous with lacinia (fig. 34.779) .... southern cowpea weevil, *Callosobruchus chinensis* (L.)

3(1). Submentum entire (fig. 34.777) 5
   Submentum vestigial, represented by isolated sclerites (fig. 34.774) 4

4(3). Median vestige of submentum present (fig. 34.773). Clypeus lacking postero-lateral sclerotized plates (fig. 34.782). Labrum with 3 equidistantly spaced setae along posterior border of dense setiferous mat (fig. 34.782). Thoracic appendage 4-segmented .... vetch bruchid, *Bruchus brachialis* Fahraeus
   Submentum represented only by lateral vestiges (fig. 34.774). Clypeus with postero-lateral sclerotized plates each bearing a seta and sensory pore (fig. 34.783). Labrum with 3 rows of well developed setae (fig. 34.783): 1st row with 2 tall, blunt setae, 2nd row with 5 short, blunt setae and 3rd row with 2 lateral short, sharp setae. Thoracic appendage papilla-like .... Mexican bean weevil, *Zabrotes subfasciatus* (Boheman)

5(3). Clypeus lacking posterior, transverse, sclerotized plate (fig. 34.784). Epipharynx with semicircular, sclerotized plate along each posterolateral margin of median groove (fig. 34.766) .... broadbean weevil, *Bruchus rufimanus* Boheman
   Clypeus with posterior, transverse, sclerotized plate (fig. 34.785). Sclerotized plates absent on epipharynx 6

6(5). Maxillary palp 2-segmented (fig. 34.781). Transverse, sclerotized clypeal plate with 2 pairs of lateral setae (fig. 34.785). Rectangular, sclerotized plate of clypeolabral suture with 2 pairs of lateral setae (fig. 34.785). Y-shaped anal sulcus. Epipharynx with 2 median and 1 lateral pairs of setae (fig. 34.768). Dense setiferous mat located anteriorly on epipharynx. Ligula with 3 pairs of setae (fig. 34.776) .... palm seed weevil, *Caryobruchus gleditsiae* (L.)

Maxillary palp 1-segmented. Transverse anal sulcus 7

7(6). Epipharynx with ()-shaped tormae (fig. 34.767). Second row of 3 labral setae pointed and equidistantly spaced (fig. 34.786). Submental sclerite reaching opposite sides of labium (fig. 34.777). Distal antennal segment (fig. 34.788) twice as long as basal segment .... pea weevil, *Bruchus pisorum* (L.)

Epipharynx lacking tormae. Second row of labral setae (fig. 34.787) with 1 median and 1 lateral pair. Submentum reduced, not reaching lateral extremities of labium (fig. 34.778). Antenna apparently consisting of 2 equal segments, but actually 1-segmented .... bean weevil, *Acanthoscelides obtectus* (Say)

Acanthoscelides obtectus (Say). The common bean weevil (figs. 34.769, 34.778, 34.787, 34.789). *A. obtectus* is the most notorious bean weevil in the U.S. as well as in many other countries. Its undisputed reputation has led to the publication of volumes of material and foremost with respect to its biology and larval morphology are Johnson (1970; 1981b) and Pfaffenberger (1985a), respectively. It attacks a variety of host plants in the field (Johnson 1983) or in storage. Some hosts of economic importance are: California black-eyed beans (black-eyed peas), navy beans, pinto beans, kidney beans, fava beans, lima beans, cowpeas, peas, lentils and others. Larvae bore out of the egg and into the underlying seed and feed upon the cotyledons. One host seed is sufficient to rear numerous adults.

Bruchus brachialis Fahraeus. The vetch bruchid (figs. 34.773, 34.782). Larvae successfully attack seeds in the field and in storage providing the seeds are not dry. Most species of vetch are subject to attack with the possible exception of Hungarian and common vetch. This species seems to prefer the purple, wooly podded, and smooth or hairy vetches (Pfaffenberger 1977). In storage, multiple generations are produced annually, whereas in the field a single annual generation is produced. Two or more larvae may develop in each seed.

Bruchus pisorum (L.). The pea weevil (figs. 34.760, 34.767, 34.777, 34.786, 34.788). The female of this species will oviposit only on the pods of pea plants (*Pisum* sp.). This pest inflicts most of its damage upon peas in the field. Rarely, if ever, is there damage to stored peas which may be due to the hardness of the dried seed. Cannibalism appears to exist since no more than one adult develops per seed (Pfaffenberger 1977). One generation is produced each year.

Bruchus rufimanus Boheman. The broadbean weevil (figs. 34.766, 34.775, 34.784). This species closely resembles the pea weevil but is readily distinguished on the basis of host preference and numbers of larvae maturing per seed. This generalist will infest the seeds of peas and vetch as well as the pod and seed of the European broadbean (horse bean) in the field. Moreover, at least two or three larvae develop per seed (Pfaffenberger 1977).

The genus *Callosobruchus* Pic. Much confusion exists in the literature as to the preferred common name of *Callosobruchus maculatus* and its close relative *C. chinensis*. In 1898, Chittenden referred to *C. chinensis* as "the cowpea weevil" and *C. maculatus* as "the four-spotted bean weevil." Then in 1919, F. B. Paddock decided that *C. maculatus* should be the cowpea weevil since it, rather than *C. chinensis*, was collected more frequently from cowpeas throughout the state of Texas. Later, Larson and Fisher (1938) referred to *C. maculatus* as "the southern cowpea weevil" alluding to its southern distribution. Added perplexity exists in more recent works such as Metcalf, Flint and Metcalf (1962) where *C. chinensis* is identified as the southern cowpea weevil. To avoid additional confusion I have decided to utilize the name associations defined by Metcalf, Flint and Metcalf (1962).

Callosobruchus maculatus (Fabricius). The cowpea weevil (figs. 34.771, 34.780). This weevil is an important pest of many host plants (Booker 1967; Vats 1974a) but of particular interest, in the U.S., are its effects upon cowpeas. Although its destruction is more pronounced upon stored cowpeas, the adult female will oviposit upon pods in the field. In storage one or more larvae may develop within a single pea. It will also infest peas and beans (mung beans) of other varieties, but with less vigor.

Callosobruchus chinensis (L.). The southern cowpea weevil (figs. 34.772, 34.779). Collection records indicate intense competition and geographical overlap between this species and *C. maculatus*. Also, the disparity in numbers of individuals collected indicates that *C. maculatus* may be much more competitive. Therefore, *C. chinensis* is only of moderate economic importance. Like *C. maculatus*, infestation of cowpeas begins in the field and will continue in storage with one to several larvae infesting the same seed. This species may also infest the seeds of other host plants (Vats 1974a). It is not uncommon to find this species co-infesting the same seed with *C. maculatus* and *Acanthoscelides obtectus*.

Caryobruchus gleditsiae (L.). The palm seed weevil (figs. 34.759, 34.768, 34.776, 34.781, 34.785). This species is easily recognized by its size, being the largest of the family. Within the U.S. its range extends primarily from the southern tip of Texas eastward to Florida and northward along the Atlantic coast into North Carolina (Woodruff 1968). The state most heavily affected is Florida. Outside the U.S. the extent of its range and its preferred host plants are not as well known but larvae have been obtained from the seeds of the stonenut palm, *Sabal uresana*, which were collected near San Carlos Bay,

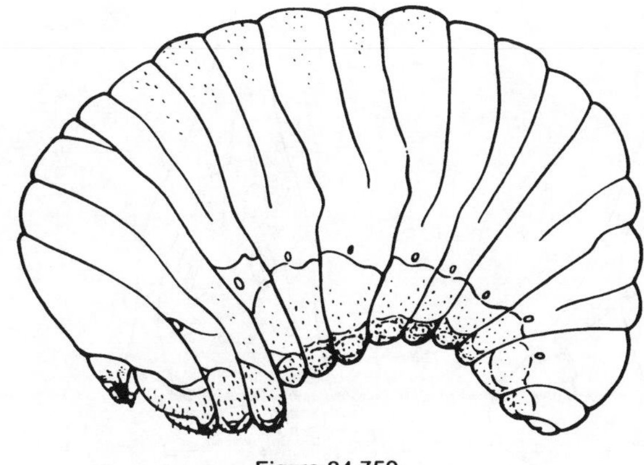
Figure 34.759

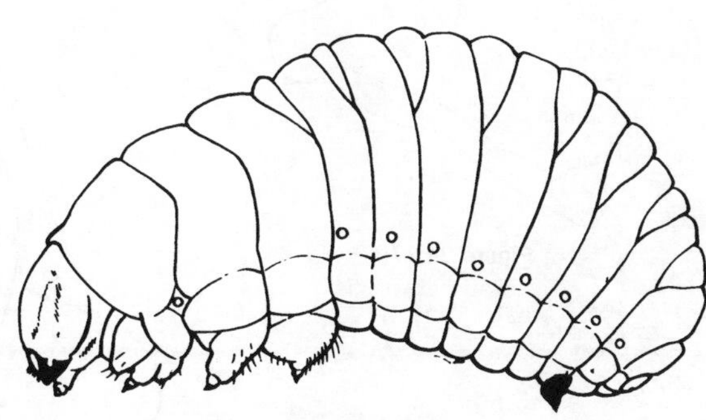
Figure 34.760

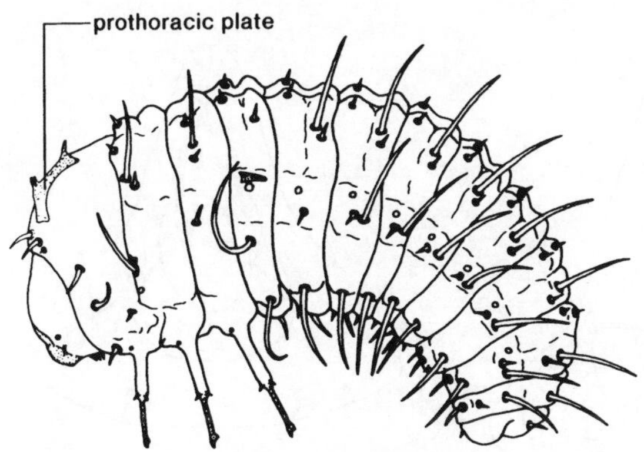

Figure 34.761

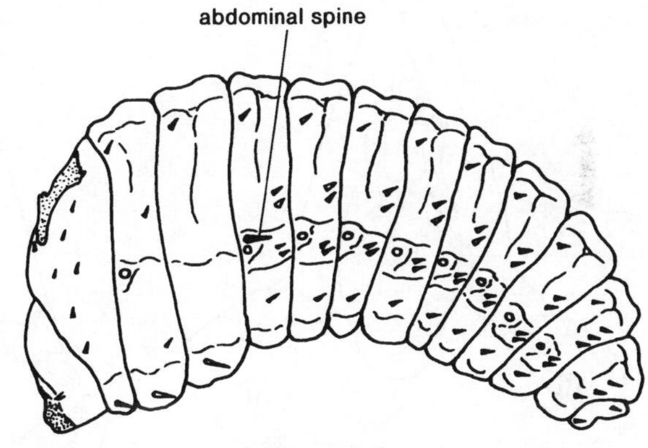

Figure 34.762

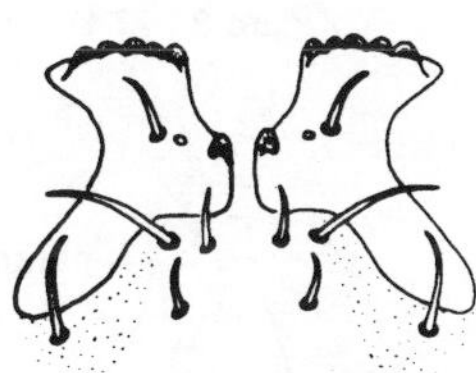
Figure 34.765

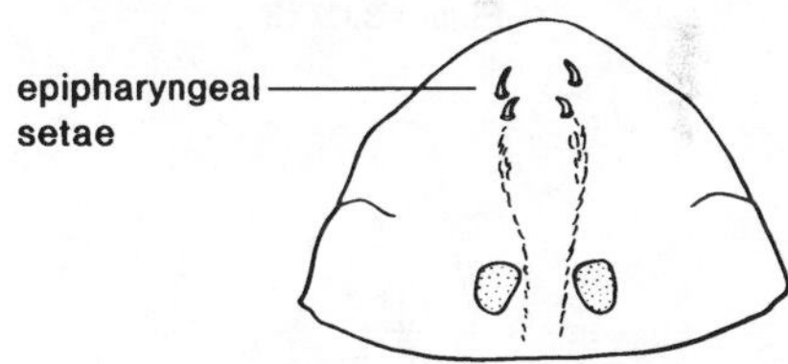

Figure 34.766

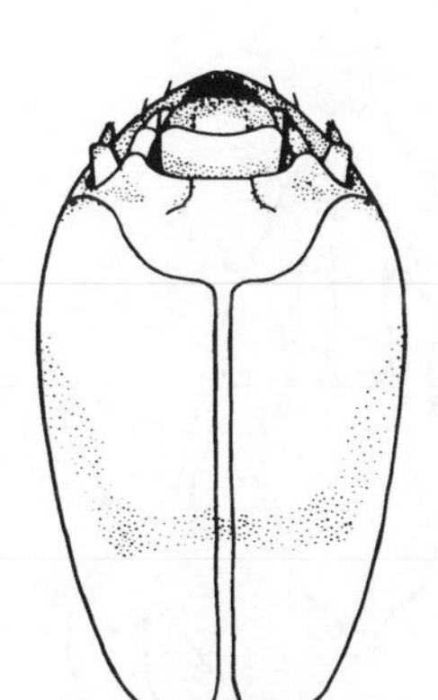
Figure 34.763

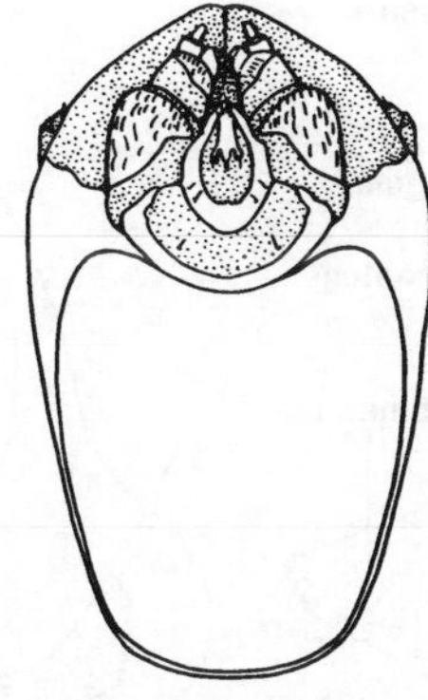
Figure 34.764

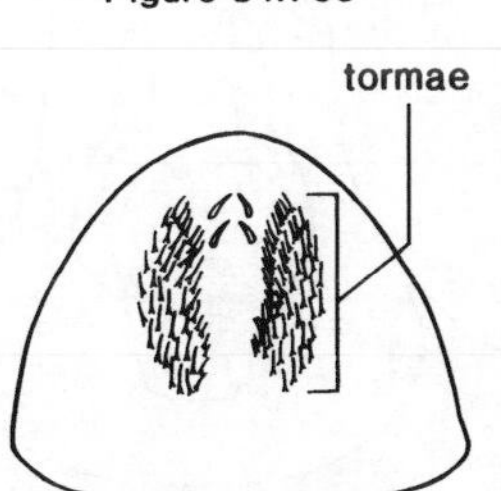
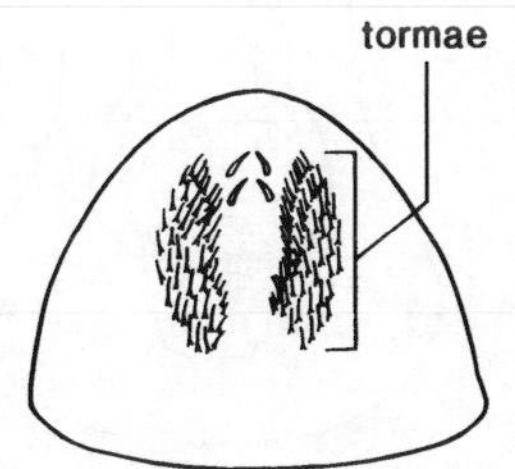

Figure 34.767

Figure 34.768

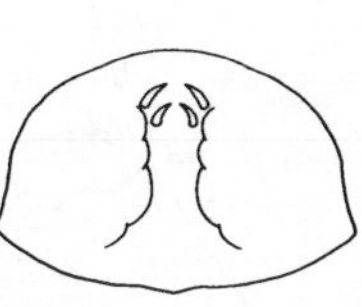
Figure 34.769

Figures 34.759–34.769. Bruchidae.

Figure 34.759. *Caryobruchus gleditsiae* (L.).

Figure 34.760. *Bruchus pisorum* (L.).

Figure 34.761. First stage, *Algarobius prosopis* (LeConte).

Figure 34.762. First stage, *Stator* sp.

Figure 34.763. Head, dorsal.

Figure 34.764. Head, ventral.

Figure 34.765. Prothoracic plate of first stage larva.

Figure 34.766. Epipharynx, *Bruchus rufimanus* Boheman.

Figure 34.767. Epipharynx, *Bruchus pisorum* (L.).

Figure 34.768. Epipharynx, *Caryobruchus gleditsiae* (L.).

Figure 34.769. Epipharynx, *Acanthoscelides obtectus* (Say).

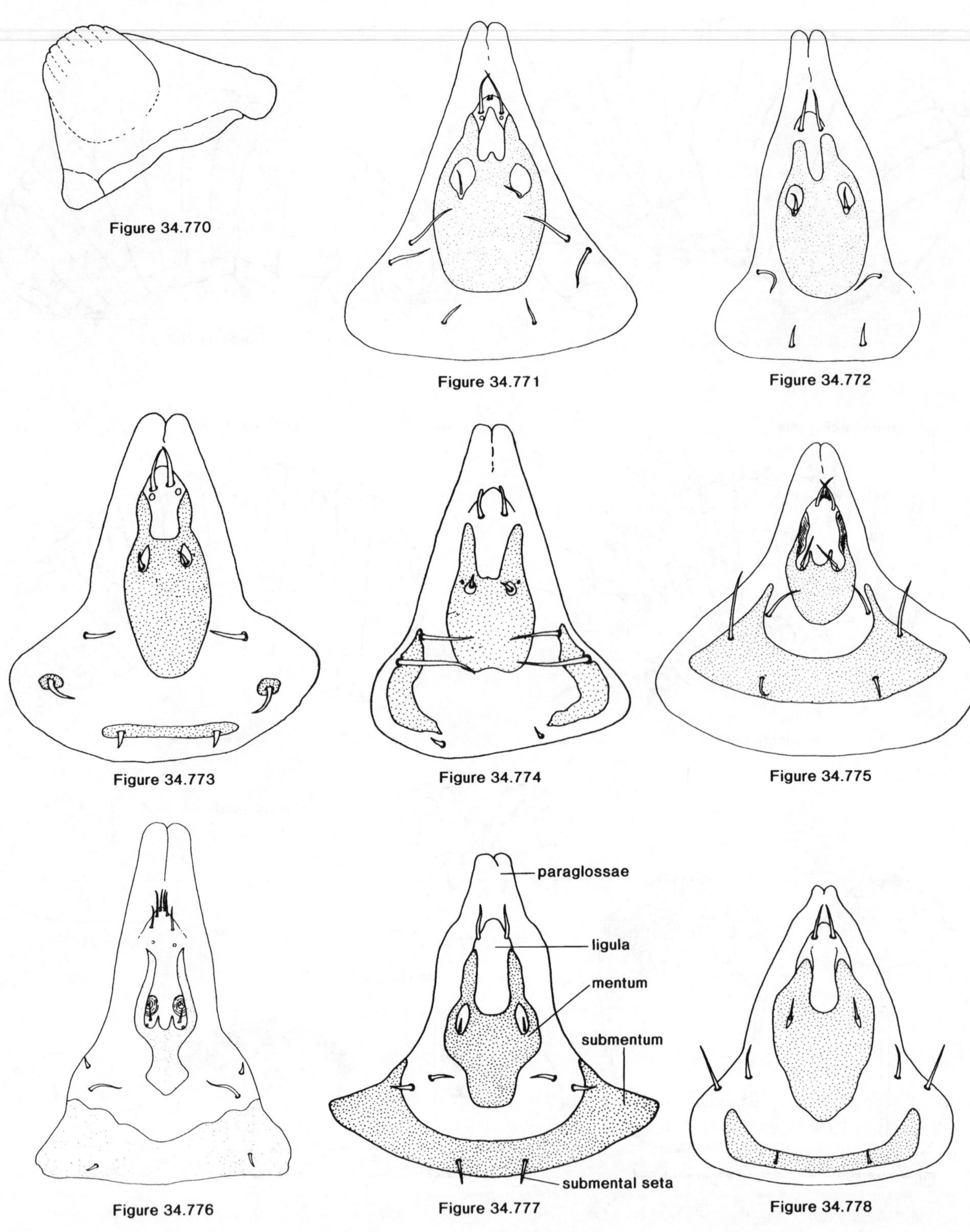

Figure 34.770

Figure 34.771

Figure 34.772

Figure 34.773

Figure 34.774

Figure 34.775

Figure 34.776

Figure 34.777

Figure 34.778

Figures 34.770–34.778. Bruchidae.

Figure 34.770. Mandible.

Figure 34.771. Labium, *Callosobruchus maculatus* (Fabricius).

Figure 34.772. Labium, *Callosobruchus chinensis* (L.).

Figure 34.773. Labium, *Bruchus brachialis* Fahraeus.

Figure 34.774. Labium, *Zabrotes subfasciatus* (Boheman).

Figure 34.775. Labium, *Bruchus rufimanus* Boheman.

Figure 34.776. Labium, *Caryobruchus gleditsiae* (L.).

Figure 34.777. Labium, *Bruchus pisorum* (L.).

Figure 34.778. Labium, *Acanthoscelides obtectus* (Say).

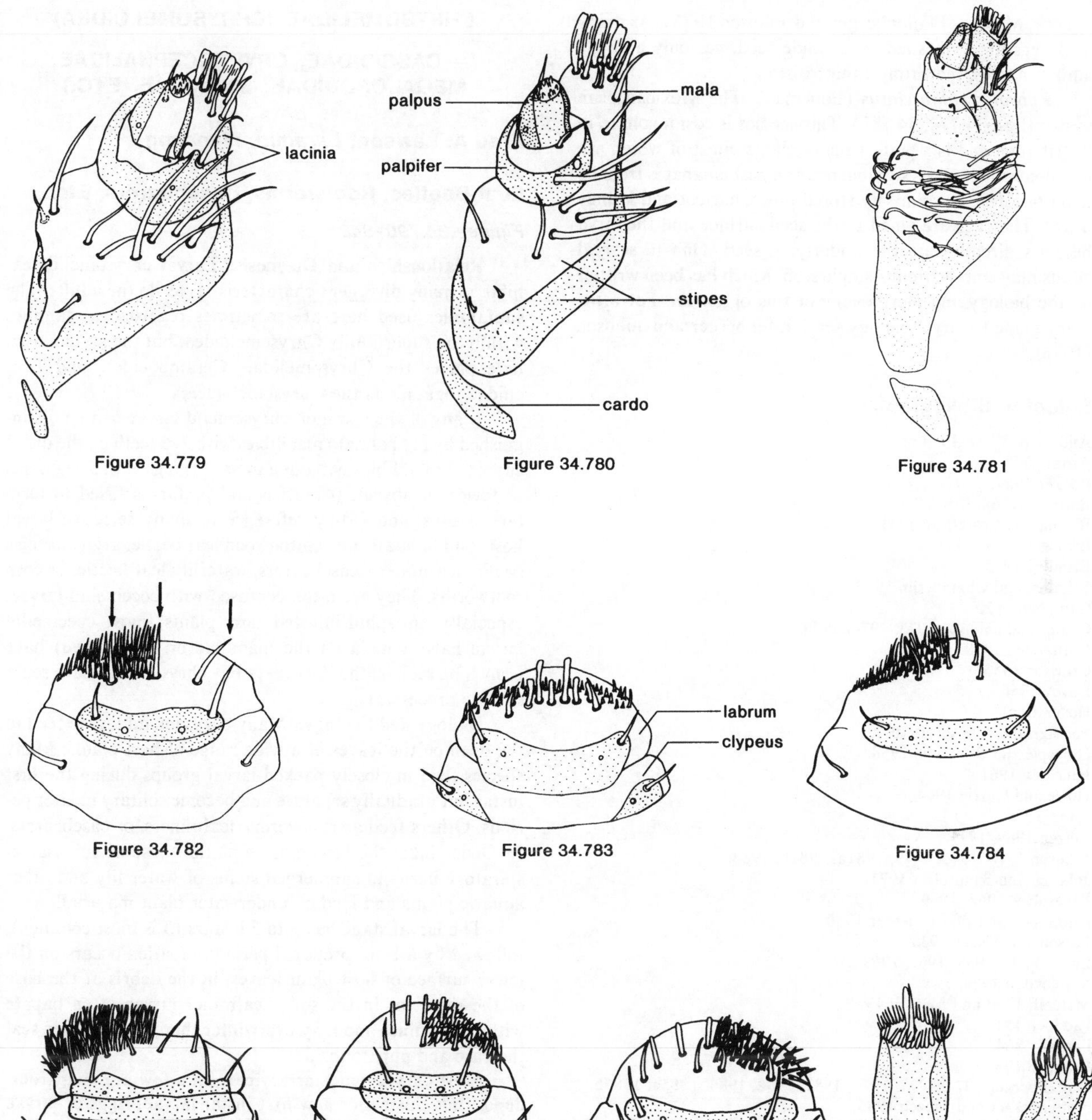

Figure 34.779 Figure 34.780 Figure 34.781

Figure 34.782 Figure 34.783 Figure 34.784

Figure 34.785 Figure 34.786 Figure 34.787 Figure 34.788 Figure 34.789

Figures 34.779–34.789. Bruchidae.

Figure 34.779. Right maxilla, *Callosobruchus chinensis* (L.).

Figure 34.780. Right maxilla, *Callosobruchus maculatus* (Fabricius).

Figure 34.781. Right maxilla, *Caryobruchus gleditsiae* (L.).

Figure 34.782. Clypeolabrum, *Bruchus brachialis* Fahraeus.

Figure 34.783. Clypeolabrum, *Zabrotes subfasciatus* (Boheman).

Figure 34.784. Clypeolabrum, *Bruchus rufimanus* Boheman.

Figure 34.785. Clypeolabrum, *Caryobruchus gleditsiae* (L.).

Figure 34.786. Clypeolabrum, *Bruchus pisorum* (L.).

Figure 34.787. Clypeolabrum, *Acanthoscelides obtectus* (Say).

Figure 34.788. Antenna, *Bruchus pisorum* (L.).

Figure 34.789. Antenna, *Acanthoscelides obtectus* (Say).

Sonora, Mexico (Pfaffenberger and Johnson 1976). As many as 4 eggs are oviposited on a single seed, yet only a single adult emerges, indicating cannibalism.

Zabrotes subfasciatus (Boheman). The Mexican bean weevil (figs. 34.774, 34.783). This species is cosmopolitan in distribution and is a pest of many plants, most of which are non-domesticated. Its significance as a pest emanates from its habit of infesting primarily stored pinto, haricot and kidney beans. The eggs are glued to the seed surface and the larva burrows directly into the underlying seed. One to several adults may emerge from a single seed. Much has been written on the biology and morphology of this organism. For a literary guide to larval biology see Pfaffenberger and Johnson (1976).

Selected Bibliography

Anderson, W. H. 1943.
Arora 1978.
Booker 1967.
Bottimer 1968.
Böving and Craighead 1931.
Bridwell 1929, 1933.
Brindley 1933.
Brindley and Chamberlin 1952.
Campbell 1920.
Center and Johnson 1973, 1974, 1976.
Chittenden 1898.
Crowson 1955.
Davey 1958.
Doyer 1929.
Forister and Johnson 1970.
Gillespie and Randolph 1958.
Hinckley 1961.
Howe and Currie 1964.
Imms 1977.
Janzen 1969, 1971.
Johnson 1970, 1975, 1979, 1981a, 1981b, 1983.
Johnson and Kingsolver 1971.
Kingsolver 1968, 1969.
Kingsolver and Pfaffenberger 1980.
Larson and Fisher 1938.
Luca, Y. de. 1956, 1967, 1968.
Marcovitch 1934.
Metcalf, Flint and Metcalf 1962.
Paddock 1919.
Parnell 1964.
Peterson 1951.
Pfaffenberger 1974, 1977, 1979, 1981, 1982, 1984, 1985a, 1985b.
Pfaffenberger and Janzen 1985.
Pfaffenberger and Johnson 1976.
Pfaffenberger, et al. 1984.
Prevett 1971.
Randolph and Gillespie 1958.
Rybalko 1966.
Singh, Kaur and Saini 1979.
Smith and Michelbacher 1944.
Steffan 1945, 1946.
Steinhausen 1966.
Teran 1962, 1967.
Vats 1972, 1974a, 1974b.
Whitehead 1930.
Woodruff 1968.

CHRYSOMELIDAE (CHRYSOMELOIDEA)

(= CASSIDIDAE, CRYPTOCEPHALIDAE, MEGALOPODIDAE, SAGRIDAE, ETC.)

Fred A. Lawson, *Laramie, Wyoming*

Leaf Beetles, Rootworms, Casebearers, Etc.

Figures 34.790-843

Relationships and Diagnosis: Larval chrysomelids exhibit as many divergent characteristics as do the adults. The subfamilies used here are sometimes regarded as families within the superfamily Chrysomeloidea, but the system used here places the Chrysomelidae, Cerambycidae, and Bruchidae together as the Chrysomeloidea.

Many of the common chrysomelid larvae can be distinguished by (1) palmate mandibles with 3–5 teeth on the distal edge; (2) mandibles without a mola; (3) legs with 5 segments or fewer, or absent; (4) tarsus and pretarsus fused to form tarsungulus; and (5) by reference to many selective larval hosts and habitats: i.e., cottonwood leaf beetle, argus tortoise beetle, leafminers, casebearers, waterlily leaf beetle, or corn rootworms. They are often confused with coccinellid larvae, especially on aphid-infested host plants; most coccinellid larvae have a mola on the mandible or (*Epilachna*) have branching scoli on the *dorsum* (a few chrysomelids have some spinose processes).

Biology and Ecology: Many chrysomelid larvae feed in the open on the leaves of a wide range of host plants. Many species feed in closely packed larval groups during the first instar but gradually separate and become solitary in later periods. Others feed as rootworms, leafminers, or casebearers. In Donaciinae, the larvae are aquatic; these attach via respiratory horns to submerged stems of water lily and other aquatic plants and feed on underwater plant material.

The larval stage has 3 to 5 instars (3 is most common), followed by a brief prepupal period; pupation occurs on the lower surface of host plant leaves, in the debris at the base of the plant, or in the soil. Leafminer larvae often pupate within the mines and may overwinter there; casebearers seal the case and pupate inside.

Most chrysomelids are oviparous; a few are viviparous, depositing clusters of new first instar larvae. Eggs or larvae are deposited on the lower surface of leaves of the host plant, on small twigs or stems early in the season, in small excavations of the lower leaf surface (*Mantura*), in the soil at the base of the host plant, on the lower edge of floating leaves, or (casebearers) within a tightly fitting coat (case) of partially digested plant material, which is tied to a host leaf or twig (*Coscinoptera*) or dropped to the substrate (*Pachybrachis, Cryptocephalus*).

Some species have 2 or more generations each year; many others consistently have only 1. A few can produce repeated generations when fed fresh young host plant material; if old host plant material is provided, adults feed sparingly and go into hibernation; the larvae die very soon after eating the same old host material. Most species spend the winter in the adult stage.

Description: Larvae of the chrysomelid subfamilies differ so much that a complete discussion in concise form is not possible; definitive subfamily characteristics for the larvae are therefore summarized in Table 1. Some general features of leaf beetle larvae are as follows.

Head: Mostly hypognathous, rounded, moderately to heavily sclerotized; epicranial suture short to absent, frontal sutures elongate, strongly divergent; in some leafminers, the head is flattened, prognathous, and lightly sclerotized. Antennae very small to elongate, usually peg-shaped, with 1–3 ring-like segments; they arise in a membranous socket at or near the ends of the frontal sutures. The socket membrane is scarcely wider than the first antennal segment in galerucine larvae; it is 2–5 times wider than the first antennal segment in the alticines. Minute accessory structures (sensoria) may occur apically on the second antennal segment, near the base of the third. Stemmata absent, or 1–6 on each side of the head. The head is sparsely, usually inconspicuously setose in most; some casebearers may have prominent elongate or capitate setae on the frons. Longer head setae occur in some flea beetle larvae.

Mandibles with mesal surface widened and concave (= palmate) in most, with 3–5 teeth on distal edge; in others, the mesal surface is narrowed and bears only a cutting edge or 1–2 small teeth apically. A few are just narrow, tapered gouges. Mola absent. Maxilla usually with a lightly sclerotized stipes, a very short, blunt mala bearing stout apical bristles, and palps with 1–4 ring segments. Labium usually with a small fused basal plate, a very small ligula, and inconspicuous 1–2 segmented palps. The labio-maxillary complex is retracted and relatively inconspicuous in most, but is protracted and prominent in a few.

Thorax and Abdomen: Thorax rounded in most, flattened in some leafminers; cuticle glabrous, asperate or microsetose; setae mostly minute to inconspicuous, but elongate and prominent in casebearers and most leafminers; stout setae or elongate seta-bearing scoli occur in some flea beetle larvae.

Prothoracic shield lightly to heavily sclerotized; size variable, from narrow, transverse mid-dorsal plates to an enlarged saddle extending down each side, sometimes to base of legs. Legs conspicuous in most, often stout, but may be very small and widely separated; usually 5-segmented (including tarsungulus), occasionally 4-segmented (many Hispinae, Donaciinae), 1- or 2-segmented (some leafminers), or absent (Zeugophorinae). In many, the middle and hind legs are spaced successively farther apart on the venter. The legs of casebearers are elongate, slender, and prominent; they project out of the case beside the head when larvae are moving or feeding.

White, membranous, eversible scent glands (eversible pockets) are housed in paired, sclerotized, usually black dorsolateral cones on the meso- and metathorax.

Abdomen with segments rounded in cross section in most, but distinctly flattened in many leafminers. Cuticle smooth, rugose, asperate, minutely spinose (microsetose), or bearing scoli in some. A few larvae are stout-bodied and grublike in the middle segments; casebearer larvae have the middle to apical segments enlarged and strongly curved ventrad and then cephalad, apparently useful in supporting and retaining the case during locomotion and feeding. Many of the remainder are orthosomatic, while some chrysomelines are greatly enlarged dorsally and laterally in the middle abdominal segments; when disturbed, these often curl the venter, becoming cyphosomatic (c-shaped).

Eight abdominal segments usually visible; in some, 9 or 10. The 10th may serve as an anal pseudopod and often bears an eversible anal disc. In Cassidinae, the 9th and 10th segments are invaginated but can be everted and manipulated for placing feces on the dorsum. Segments A1–7 (rarely A8) are subdivided transversely on the dorsum, forming plicae in some, but they are entire in others. Segment A8 of most Cassidinae has a dorsal fork which lies forward over the abdomen and is used to hold exuviae and fecal matter on the dorsum.

Eversible scent glands may occur in paired, dorsolateral sclerotized dark cones on abdominal segments 1–7; in others, an eversible gland occurs middorsally in the A7–8 intersegmental membrane.

Ambulatory lobes (and/or hairs) occur ventrally on some or all abdominal segments of legless forms, which may also bear stout dorsal and lateral hairs; ambulatory lobes may also occur in a few species having thoracic legs. Casebearers usually have stout setae on abdominal segments 1–4.

The anal opening is caudal or caudoventral in most species; it is dorsal in criocerines, allowing wet fecal material to be placed on the dorsum. The anal opening of Cassidinae and of one fleabeetle larva can be manipulated in the same fashion with the same result.

Spiracles: These form a lateral line of mostly annular openings which are set in small sclerites in many species; they vary from conspicuous to only faintly discernible and may occur without sclerites. The opening may also have paired dorsal or dorsolateral accessory openings (annular-biforous type). The spiracles on segment 8 are vestigial in cassidine larvae. In Donaciinae, the spiracles are vestigial on segments 1–7; on 8 they have migrated into a membranous area at the bases of the sclerotized respiratory horns.

Comments: Wilcox (1975) lists 17 subfamilies, 182 genera, and about 1500 species for America north of Mexico. Three of these subfamilies are so poorly known that they are not included in the table (Aulacoscelinae, Lamprosomatinae, Megascelinae); these, and two others (Orsodacninae, Synetinae) are not illustrated because of lack of specimens. Much of the early work on chrysomelid larvae was devoted to intensive studies of pest species; expanded coverage has been a more recent trend. Many rare forms and larger groups with wide geographic range and varied host plants remain to be studied.

Certain chrysomelid larvae can be pests of field crops and shade trees/ornamentals as defoliators, leafminers, or rootworms. Some of these are the corn rootworms, Colorado potato beetle, cereal leaf beetle, elm leaf beetles, cottonwood leaf beetles, strawberry rootworm, flea beetles, and grape rootworm (see illustrations). Several general feeders live on a variety of crop/plant hosts and weeds; many others are limited to 1 host. Many native chrysomelid larvae feed entirely on the leaves or roots of various weeds, but apparently do not become a limiting factor to a given host except rarely on a local basis. The Klamath weed beetle, *Chrysolina quadrigemina* (Suffrian), is a beneficial chrysomelid which was introduced for weed control in the Western U.S.

Table 34.1. Summary of Chrysomelidae–Larval Characteristics

SUBFAMILY	Mandibles: Shape	Mandibles: No. Teeth	Antennae: Segments	Antennae: Relative Size	Legs: Absent – Present +	Legs: Shape	Labrum: State	Labrum: Emargination +=deep –=slight	Max. Palp: No. Segments	Max. Palp: Size	Labial Palp: No. Segments	Labial Palp: Size Pos.	Stemmata: Pairs	Habitats: C-Casebearer, R-Rootfeeder, LM-Leaf Miner, S-Surface,leaf	Abdomen: No. Segments	Abdomen: Shape*
ALTICINAE	Palmate	4	1–2	small	+	stout; or reduced	Free	+ to slightly rounded emarginate	3–4	+	2	+	0, 1, or 2	S, R, LM	10	O, Mostly straight
CASSIDINAE	Palmate	4–5	1–2	small	+	stout	Free	+	2	+	1	+	5 or 6	S	8–Top (9–10 retracted)	8th with dorsal forks Spinose laterally
CHLAMISINAE	Palmate	3	2	small	+	long, thin	Fused	–	3–4	+	2	–	6	C, S	10 (9–10 small)	Hook
CHRYSOMELINAE	Palmate	4–5	3	medium	+	stout	Free	+	3–4	–	2	–	5 or 6	S	10 (10th = disk)	C, O, Cu
CLYTRINAE	+ – Palmate	3	2	small	+	long, thin	Fused	–	3–4	–	2	–	5	C, S	10 (9–10 small)	Hook
CRIOCERINAE	Palmate	4–5	3	small	+	stout	Free	+	3–4	–	1	–	5 or 6	S	9 or 10	C
CRYPTOCEPHALINAE	Gouge	0	3	large	+	long, thin	Fused	–	4	+	2	–	6	C, S	10	Hook
DONACIINAE	Pointed	3–5, on apex	3	stout	+	very stout	Free	none	3	–	1	– –	5	Aquatic	10	C, Cu
EUMOLPINAE	Narrow, Not Palmate	3	3	small	+	stout	Free	+	3–4	+	1	–	none	S	10	Cu
GALERUCINAE	Palmate	4–5	1–2	small	+	stout	Free	+ to –	3–4	+	2	+	1 or none	S, R,	9 or 10 (10th = disk)	O
HISPINAE	Conical, cutting edge	1–3 on apex	3	small	+ or –	stout	Free	–	2–3	–	1	–	4 or 6	S, LM	8	O, ± flat
ORSODACNINAE	Palmate	0	3	small	+	stout	Free	none	3	–	2	?	1	R	10	O, flat
SYNETINAE	Single, Excavated	1 point at apex	3	medium	+	stout	Free	–	3	–	2	Apart	none	S	10	Cu
ZEUGOPHORINAE	Transverse, cutting	0	3	very small	–	–	Free	none	3	+	1	Close at base minute	0 –	LM, S ?	9	O, flat

*C=cyphosomatic O=orthosomatic Cu=curved

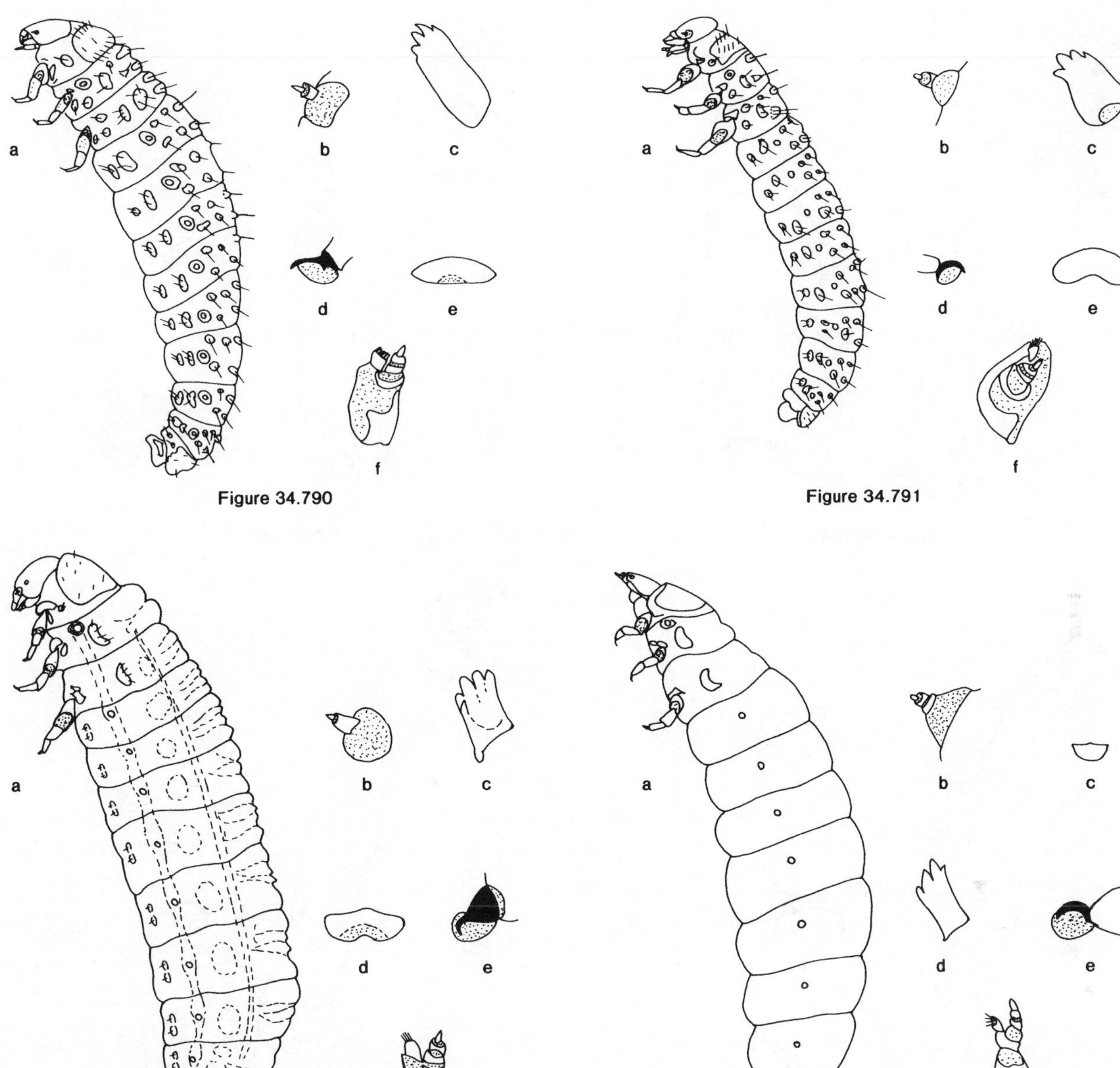

Figures 34.790a-f. Chrysomelidae. *Altica chalybea* Illiger. Grape flea beetle. (Alticinae) Length 6 mm; head, legs dark brown; body tan, tubercles slightly darker, prominent; on buds and leaves of grapes. **a.** larva, lateral; **b.** antenna; **c.** mandible; **d.** tarsungulus; **e.** labrum; **f.** maxilla.

Figures 34.791a-f. Chrysomelidae. *Altica corni* Woods. A dogwood flea beetle. (Alticinae) Length 6 mm; head dark brown, legs medium brown; body reddish-tan, tubercles much darker, prominent; on dogwood. **a.** larva, lateral; **b.** antenna; **c.** mandible; **d.** tarsungulus; **e.** labrum; **f.** maxilla.

Figures 34.792a-f. Chrysomelidae. *Blepharida rhois* Forster. A sumac flea beetle. (Alticinae) Length 8 mm; head, legs, abdomen dark brown to black; pronotum yellow; body yellowish-tan, speckled by setal rings; faint stripes along spiracular and dorsolateral lines of new specimens; body stout, slightly eruciform, swollen caudally; surface faintly, densely rugose; A9 with multilobate ventral pseudopod; A10 with eversible tube which directs feces onto dorsum; living larvae usually with slimy coating; on terminal growth of sumac; may defoliate. **a.** larva, lateral; **b.** antenna; **c.** mandible; **d.** labrum; **e.** tarsungulus; **f.** maxilla.

Figures 34.793a-f. Chrysomelidae. *Dibolia borealis* Chevrolat. A plantain flea beetle. (Alticinae) Length 5 mm; head flattened, prognathous, light brown; pronotal plate reduced; legs gray-white, small, set far apart; body slightly flattened, near white; tubercles indistinct; leaf miner in broad leaf plantain. **a.** larva, lateral; **b.** antenna; **c.** labrum; **d.** mandible; **e.** tarsungulus; **f.** maxilla.

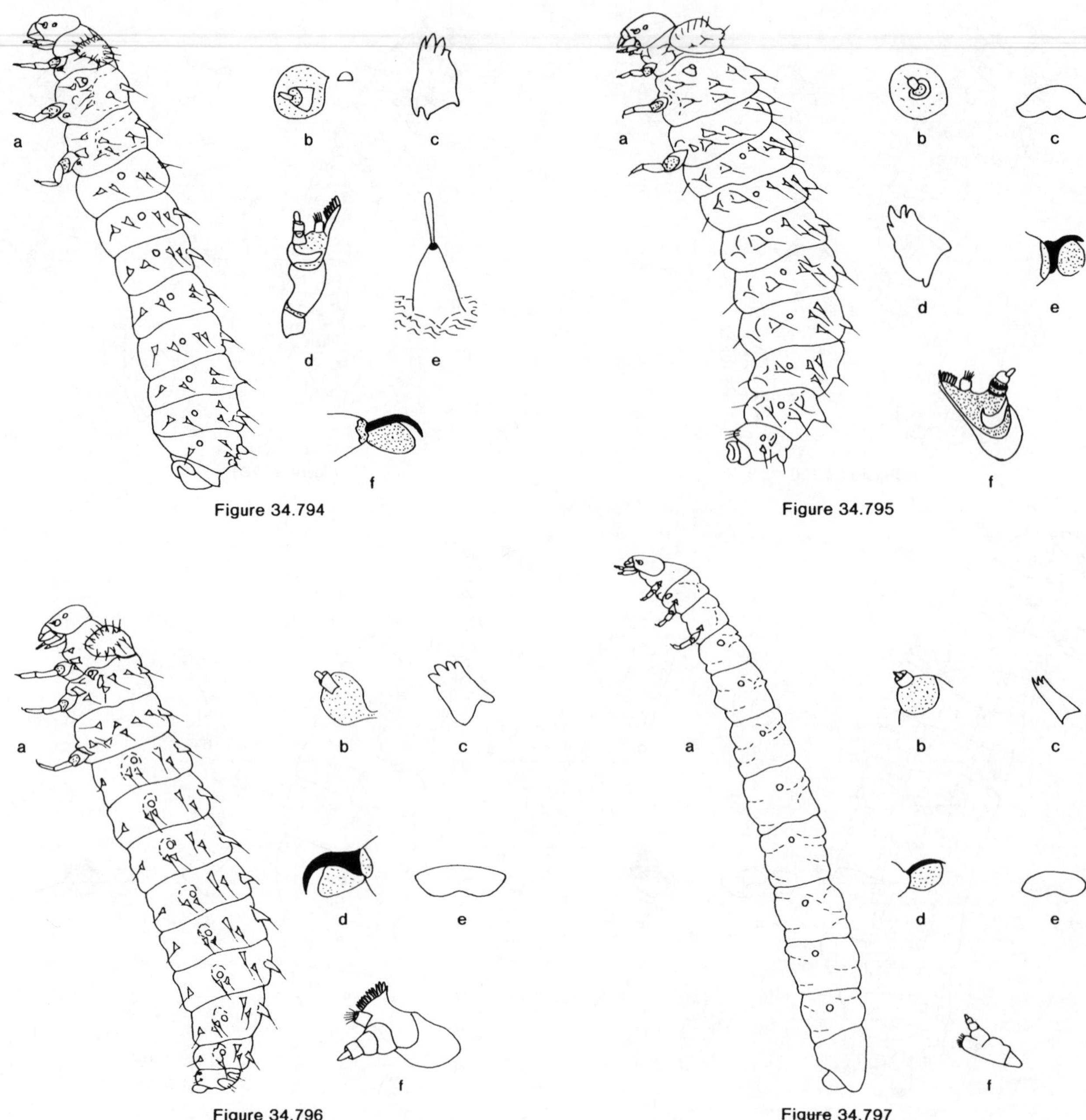

Figures 34.794a-f. Chrysomelidae. *Disonycha xanthomelas* (Dalman). Spinach flea beetle. (Alticinae) Length 6 mm; head, legs light to medium brown; body white, with prominent scoli, each with a brown seta in a dark ring at base; A10 with dorsal plate and caudodorsal row of stout setae; spiracles distinct; on spinach, others. **a.** larva, lateral; **b.** antenna; **c.** mandible; **d.** maxilla; **e.** scolus with seta; **f.** tarsungulus.

Figures 34.795a-f. Chrysomelidae. *Disonycha triangularis* (Say). Threespotted flea beetle. (Alticinae) Length 12 mm; head, legs, dark brown; body light brown, wrinkled above and down to spiracles; scoli creamy white, elongate, setae inconspicuous; A10 short, sclerotized; venter tan, seta bases distinct; on sugar beet, spinach, *Amaranthus*. **a.** larva, lateral; **b.** antenna; **c.** labrum; **d.** mandible; **e.** tarsungulus; **f.** maxilla.

Figures 34.796a-f. Chrysomelidae. *Disonycha alternata* Illiger. A willow flea beetle. (Alticinae) Length 10 mm; head, legs dark brown; body dark brown to black above, down to spiracles; dorsum may appear banded; scoli light, with prominent apical ring and seta; venter tan, seta bases prominent; spiracles distinct; on *Salix*. **a.** larva, lateral; **b.** antenna; **c.** mandible; **d.** tarsungulus; **e.** labrum; **f.** maxilla.

Figures 34.797a-f. Chrysomelidae. *Epitrix cucumeris* (Harris). Potato flea beetle. (Alticinae) Length 5 mm; head tan, epicranial suture short, darker; body pinkish to near white, segments subdivided dorsally; spiracles indistinct; on roots, underground parts of *Solanum*. **a.** larva, lateral; **b.** antenna; **c.** mandible; **d.** tarsungulus; **e.** labrum; **f.** maxilla.

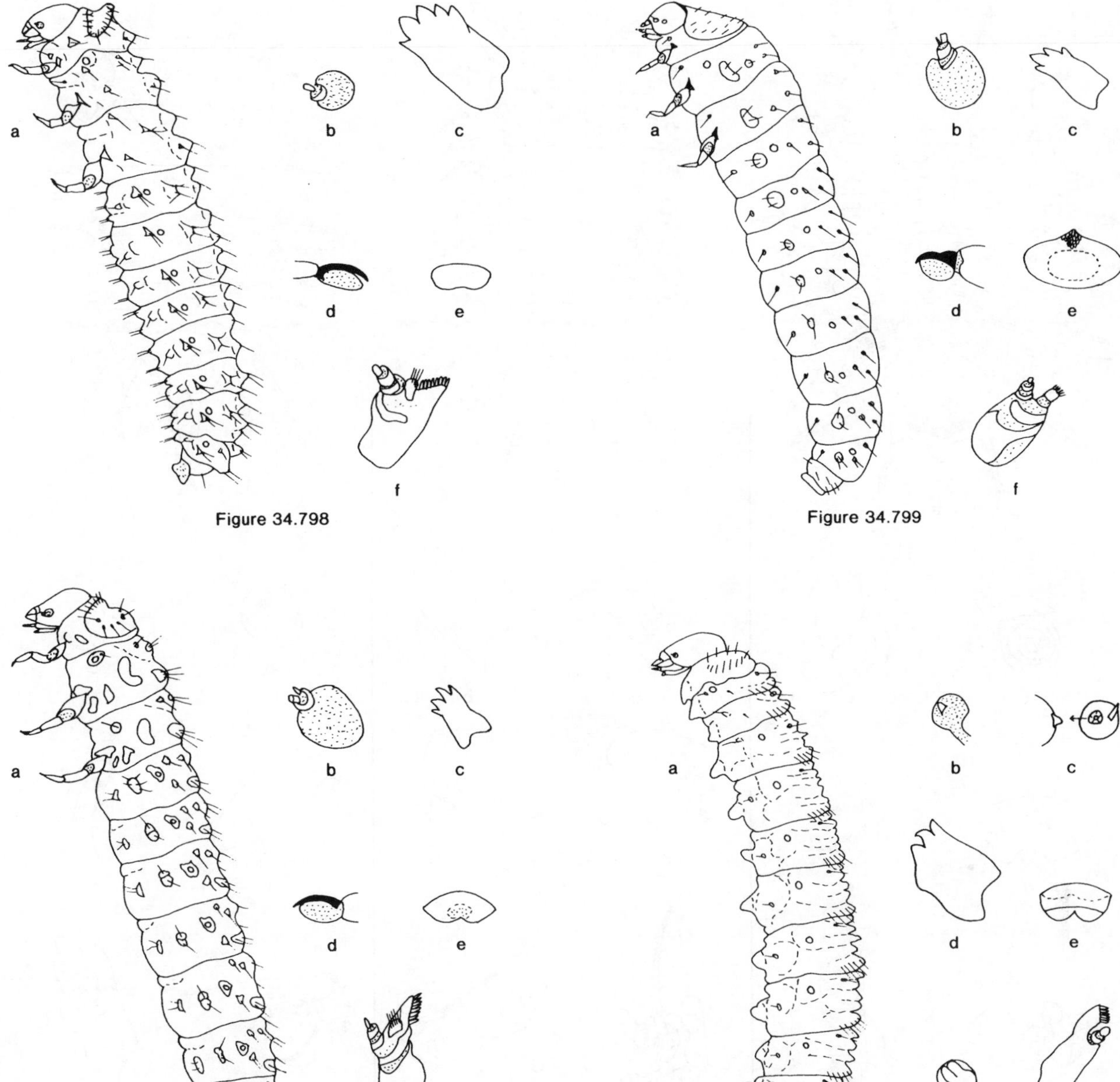

Figure 34.798

Figure 34.799

Figure 34.800

Figure 34.801

Figures 34.798a-f. Chrysomelidae. *Kuschelina gibbitarsa* Say. A mint flea beetle. (Alticinae) Length 6 mm; head brown, legs light brown to tan; body pale pink to near white, with many spinose gibbosities; stemmata absent; on foliage of mint. **a.** larva, lateral; **b.** antenna; **c.** mandible; **d.** tarsungulus; **e.** labrum; **f.** maxilla.

Figures 34.799a-f. Chrysomelidae. *Mantura chrysanthemi floridana* Crotch. A dock flea beetle. (Alticinae) Length 4.5 mm; head, legs dark brown; pronotal plate medium brown, prominent; body white, robust, densely, finely asperate; setae short, in small dark sclerites; spiracles distinct; leaf miner in *Rumex*. **a.** larva, lateral; **b.** antenna; **c.** mandible; **d.** tarsungulus; **e.** labrum; **f.** maxilla.

Figures 34.800a-f. Chrysomelidae. *Macrohaltica ambiens* LeConte. An alder flea beetle. (Alticinae) Length 11 mm; head dark brown to black; legs dark brown; body dark above, lighter below, densely asperate; tubercles darker, with single prominent seta; each spiracle in a dark sclerite; on foliage of alder. **a.** larva, lateral; **b.** antenna; **c.** mandible; **d.** tarsungulus; **e.** labrum; **f.** maxilla.

Figures 34.801a-g. Chrysomelidae. *Systena blanda* Melsheimer. Palestriped flea beetle. (Alticinae) Length 6 mm; head dark brown; minute stumps in place of legs; paired ambulatory lobes on abdominal segments; body moderately C-shaped, white, segments subdivided on top; setal rings and spiracles distinct; general feeder on garden plants, weeds. **a.** larva, lateral; **b.** antenna; **c.** thoracic "leg"; **d.** mandible; **e.** labrum; **f.** spiracle; **g.** maxilla.

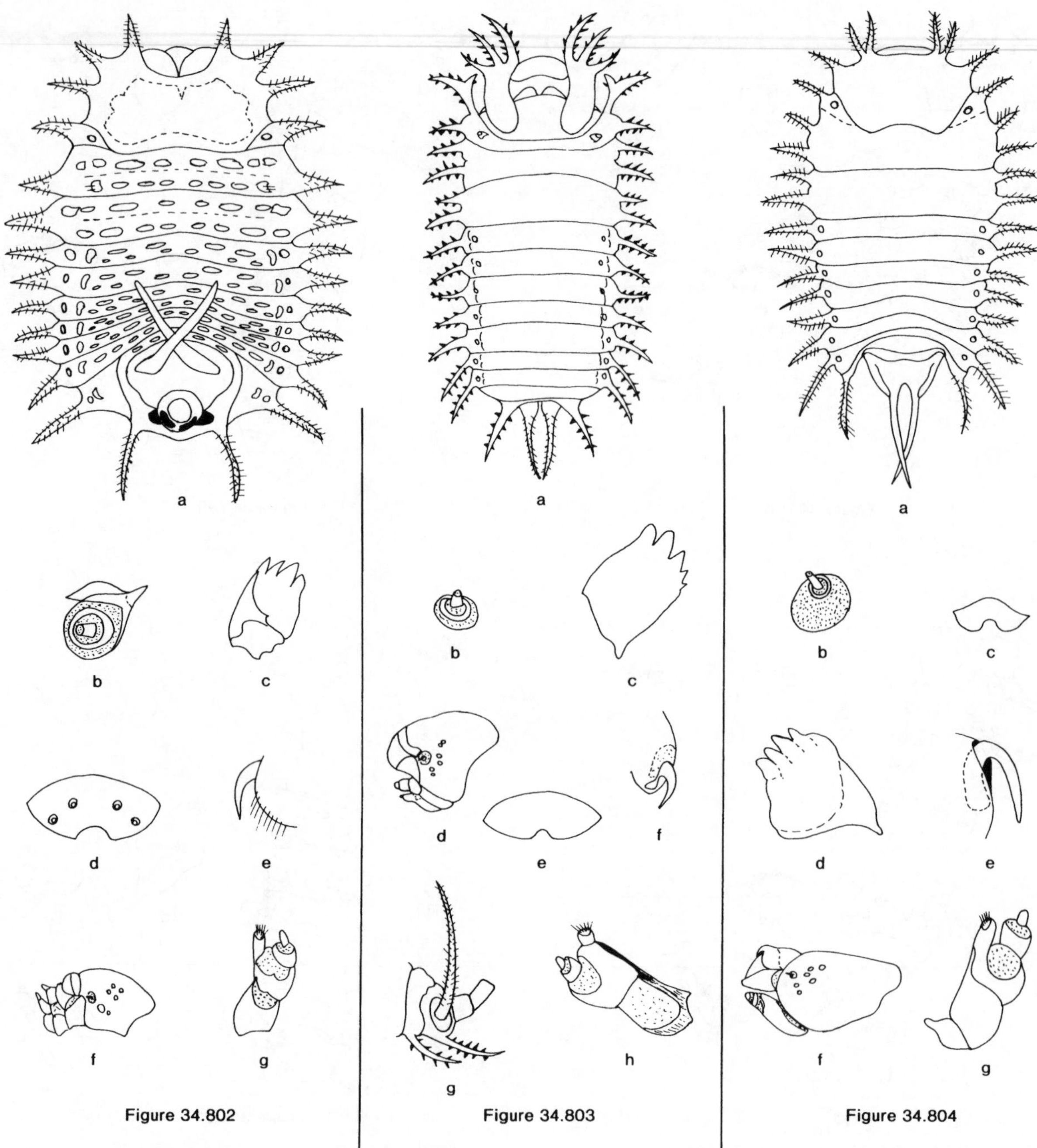

Figures 34.802a-g. Chrysomelidae. *Chelymorpha cassidea* (Fabricius). Argus tortoise beetle. (Cassidinae) Length 8 mm; head brown, hypognathous, mostly hidden; legs stout, reddish brown; pronotal shield expanded laterally and back into T2; body tan, tubercles darker, prominent; venter densely setose, dorsum less so; lateral scoli stout, dark brown to black, spinose; fork (A8) brown, darker apically, tips may be light; spiracles conspicuous; on vine plants (sweetpotato, bindweed, morning glory). **a.** larva, dorsal; **b.** antenna; **c.** mandible; **d.** labrum; **e.** tarsungulus; **f.** head, lateral; **g.** maxilla.

Figures 34.803a-h. Chrysomelidae. *Johnthonota nigripes* (Olivier). Blacklegged tortoise beetle. (Cassidinae) Length 8 mm; head brown, prognathous, mostly hidden; legs stout, light; pronotal shield light, with darkened oval on each half, extends back into T2; body grayish to near white, with irregular darker dorsolateral pattern, and with faint lateral longitudinal stripe; lateral scoli with stout spines, anterior pronotal scoli with 3 branches; fork slender, smooth; body surface densely, finely asperate; spiracles on abdomen inconspicuous; on vine plants. **a.** larva, dorsal; **b.** antenna; **c.** mandible; **d.** head, lateral; **e.** labrum; **f.** tarsungulus; **g.** terminal abdominal segments; **h.** maxilla.

Figures 34.804a-g. Chrysomelidae. *Plagiometriona clavata* Fabricius. A tortoise beetle. (Cassidinae) Length 6 mm; head yellow-tan, mostly hidden; legs stout, light; pronotum pale, with large, oval, shallow depressions; body tan, densely microsetose; lateral scoli short, densely setose; spiracles in elevated tubules; on vine plants. **a.** larva, dorsal; **b.** antenna; **c.** labrum; **d.** mandible; **e.** tarsungulus; **f.** head, lateral; **g.** maxilla.

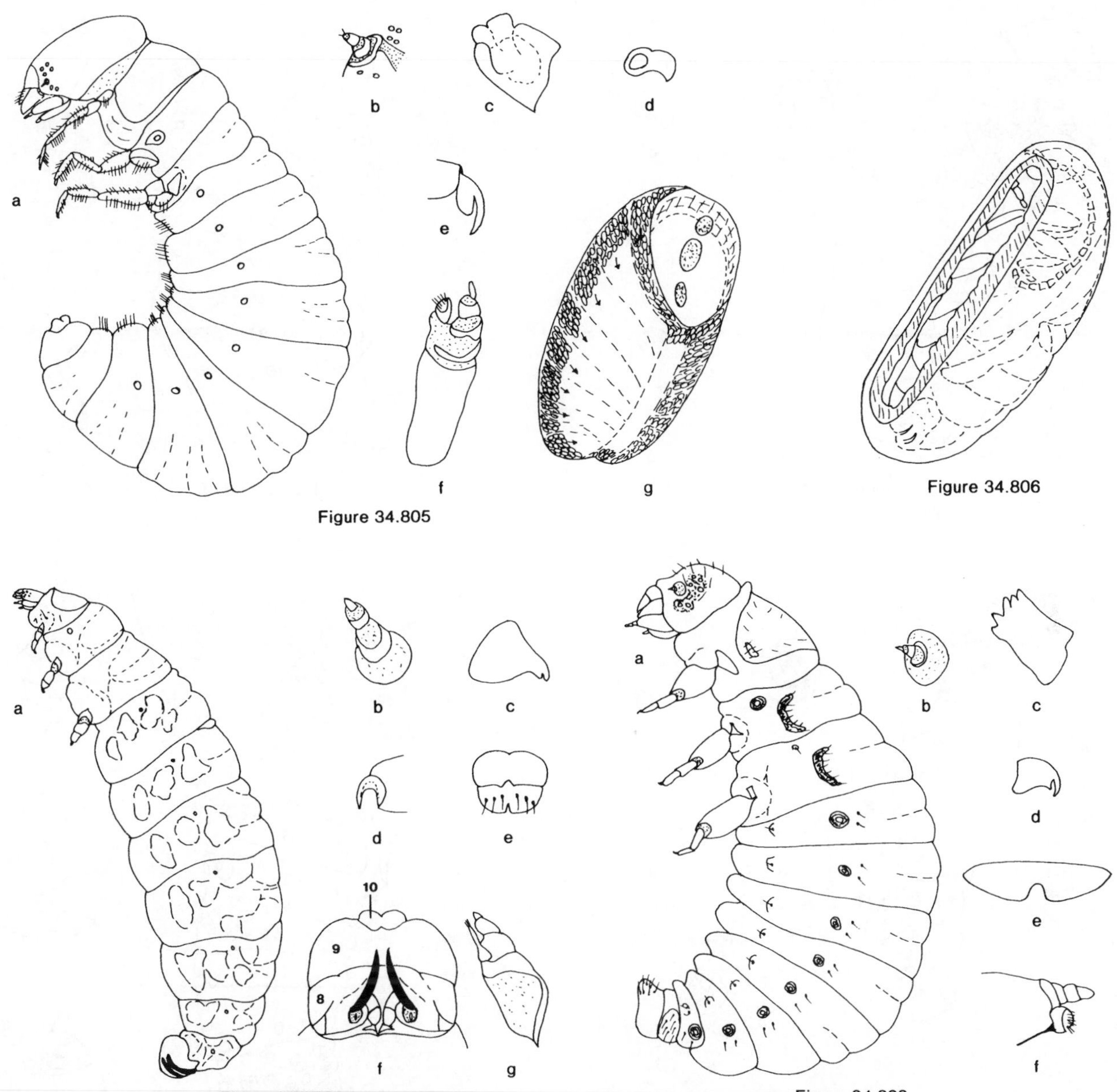

Figures 34.805a-g. Chrysomelidae. *Neochlamysis platani* (Brown). A case bearer. (Chlamasinae) Length 6+ mm; head dark brown to black, with faint lateral stripe to antennal base; legs elongate, dark brown, first and second often held beside head; pronotal shield transverse, extends to base of forelegs; body cream to near white, finely setose, segments subdivided dorsally; mid-abdominal segments expanded dorsally, reduced ventrally, forming hook shape; A9, A10 enlarged dorsally; on wild rose. **a.** larva, lateral; **b.** antenna; **c.** mandible; **d.** spiracle + sclerite; **e.** tarsungulus; **f.** maxilla; **g.** case, ventrolateral. Length 8 mm; collar, midline on venter light; remainder dark or banded light and dark; composed of small fecal pellets; carried by larva, sealed for pupation.

Figure 34.806. Chrysomelidae. Pupal case (pupa inside) of *Donacia quadricollis* Say. (Donaciinae) Length 9-10 mm; tan, semitransparent silk coccoon, with faint impressions of body segments; flattened surface attached to underwater plant parts; overwintering site of some.

Figures 34.807a-g. Chrysomelidae. *Donacia quadricollis* Say. A long horn leaf beetle. (Donaciinae) Length 10 mm; head light brown; legs very short, far apart; body stout, eruciform, faintly yellowish white; pronotal shield lightly sclerotized; lateral spiracles (A1-7) minute; spiracles of A8 open in oval membranous area in base of curved, dark, stout respiratory horns; on underwater parts of water lily, others. **a.** larva, lateral; **b.** antenna; **c.** mandible; **d.** tarsungulus; **e.** labrum; **f.** terminal segments of abdomen; **g.** maxilla.

Figures 34.808a-f. Chrysomelidae. *Calligrapha verrucosa* Suffrian. A willow leaf beetle. (Chrysomelinae) Length 9 mm; head black; legs, lateral thoracic sclerites dark brown to black; body cyphosomatic, reddish, yellow, or faintly yellowish to near white alive; surface densely, finely asperate, with venter more deeply pigmented and strongly setose; pronotum elevated, yellowish, with submarginal lateral spots and narrow median band dark brown; T2 moderately constricted; on short, floodplain *Salix* sp. **a.** larva, lateral; **b.** antenna; **c.** mandible; **d.** tarsungulus; **e.** labrum; **f.** maxilla.

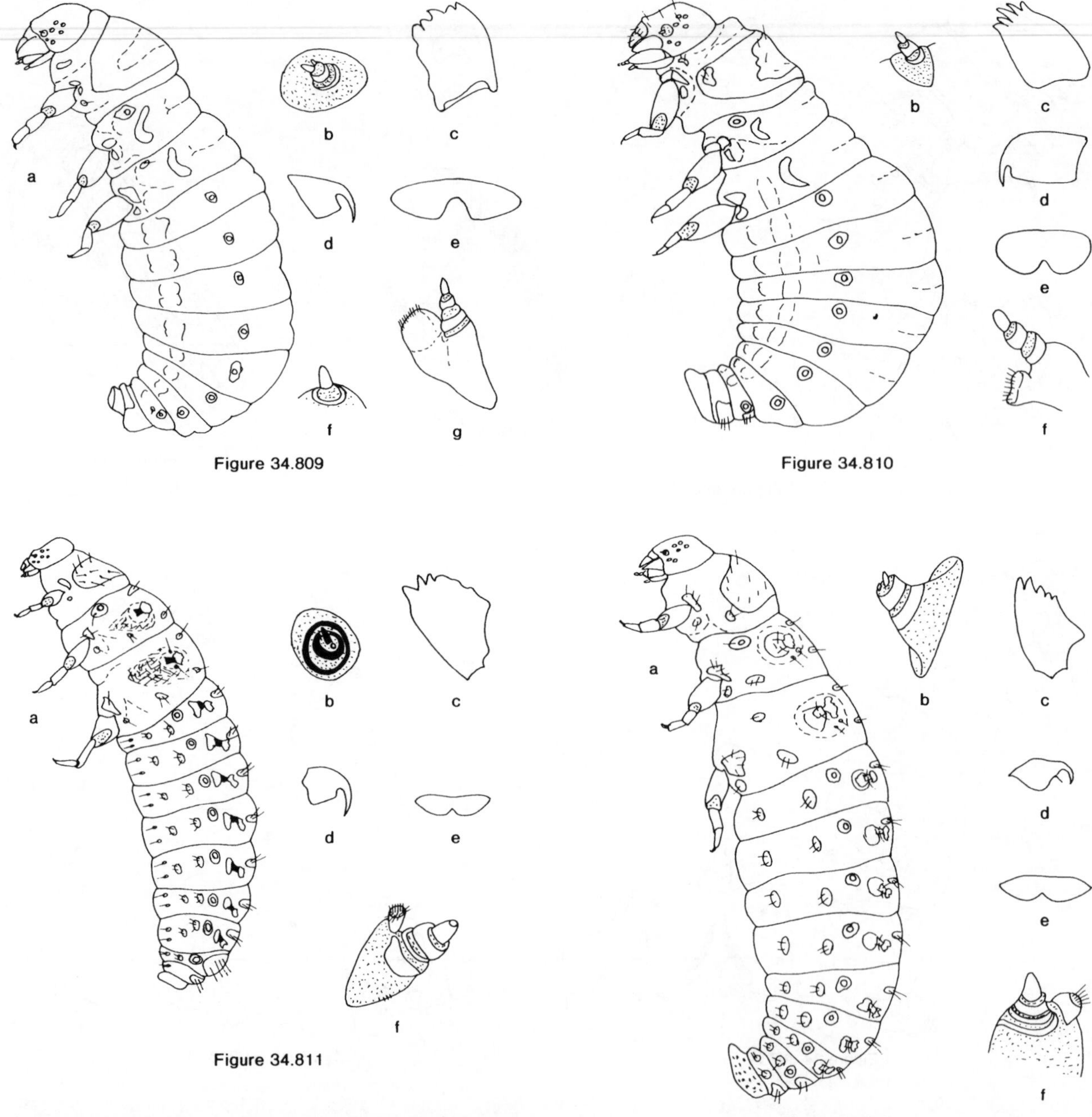

Figure 34.809

Figure 34.810

Figure 34.811

Figure 34.812

Figures 34.809a-g. Chrysomelidae. *Calligrapha multipunctata* Say. A willow leaf beetle. (Chrysomelinae) Length 8 mm; head yellow, black amid stemmata, faintly reticulated vertex; palpi and antennae dark; legs light, black at tarsi and joints, tibiae light tan; body cyphosomatic, white, small black lateral (spiracular) and thoracic sclerites; pronotum with submarginal black spot and small scattered median spots; surface densely microspinose; on taller *Salix* sp., near water. **a.** larva, lateral; **b.** antenna; **c.** mandible; **d.** tarsungulus; **e.** labrum; **f.** labial palpus; **g.** maxilla.

Figures 34.810a-f. Chrysomelidae. *Calligrapha spiraea* Say. A ninebark leaf beetle. (Chrysomelinae) Length 8 mm; head tan, vertex mottled; legs tan, darker at joints; lateral sclerite line reddish brown, sclerites minute to small; pronotum with light brown submarginal lateral spots and single brown median spot; body light tan to pale yellowish, surface finely rugose; pronotum elevated, T2 constricted; abdomen more setose below than above; on ninebark, *Physocarpus opulifolius.* **a.** larva, lateral; **b.** antenna; **c.** mandible; **d.** tarsungulus; **e.** labrum; **f.** maxilla.

Figures 34.811a-f. Chrysomelidae. *Chrysomela scripta* Fabricius. Cottonwood leaf beetle. (Chrysomelinae) Length 11 mm; head, legs near black; pronotal shield, all spiracular, setal, and glandiferous tubercles brown; body gray to yellowish, slightly eruciform, surface densely asperate; eversible scent glands on T2, 3, and A1-7; often has multiple generations on cottonwood or willow, particularly young growth. **a.** larva, lateral; **b.** antenna; **c.** mandible; **d.** tarsungulus; **e.** labrum; **f.** maxilla.

Figures 34.812a-f. Chrysomelidae. *Chrysomela lineatopunctata* Forster. A cottonwood leaf beetle. (Chrysomelinae) Length 8 mm; head, legs dark brown to black; tubercles, sclerites dark brown; eversible scent glands prominent on T2-3, A1-7; body eruciform, yellowish, surface finely, densely asperate; setae longer than in *C. scripta;* on cottonwood (1 generation only). **a.** larva, lateral; **b.** antenna; **c.** mandible; **d.** tarsungulus; **e.** labrum; **f.** maxilla.

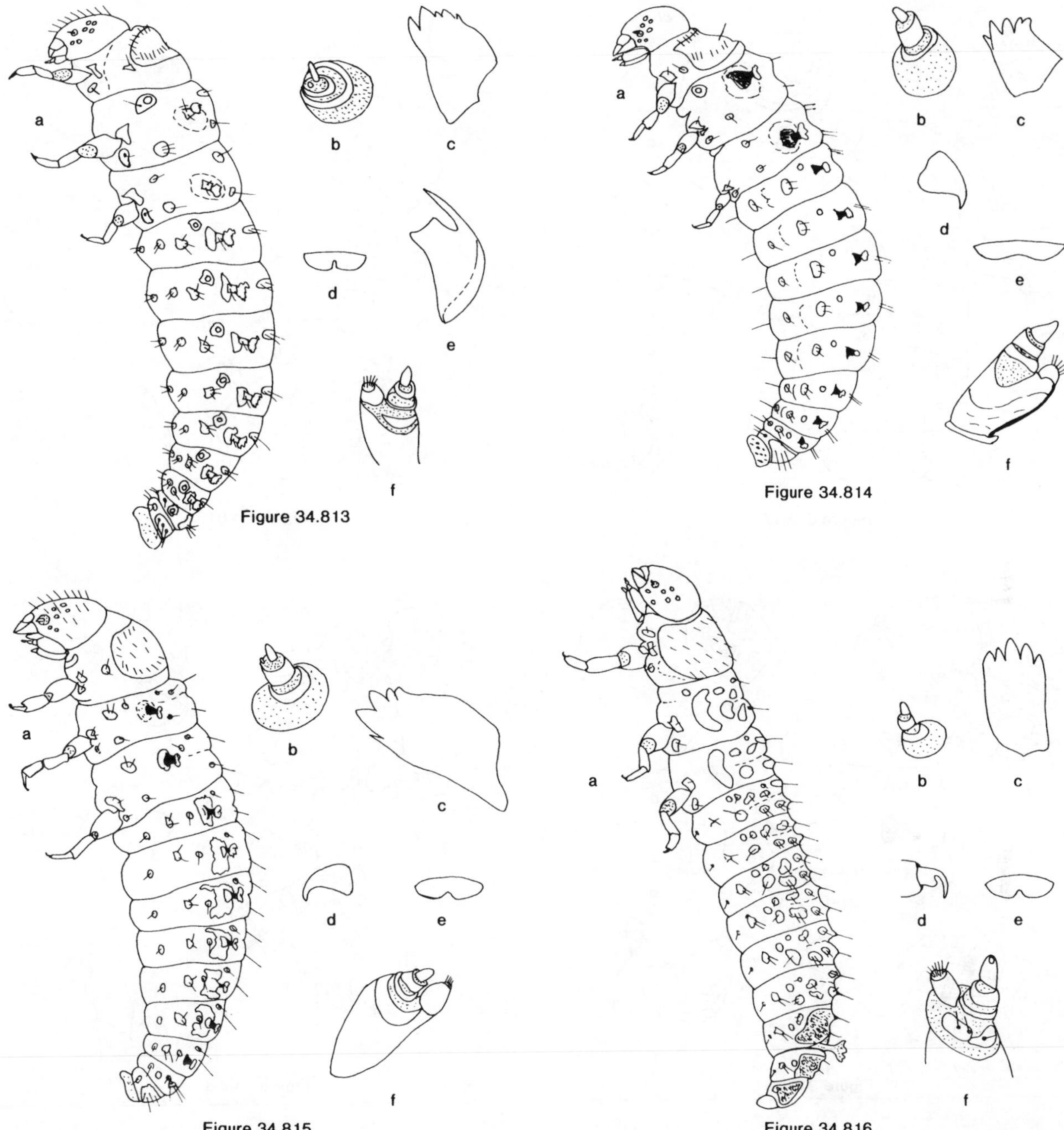

Figures 34.813a-f. Chrysomelidae. *Chrysomela knabi* Brown. A willow-cottonwood leaf beetle. (Chrysomelinae) Length 10 mm; head, legs black; tubercles, sclerites dark brown to black; body eruciform, yellowish, surface densely, darkly asperate; tubercles larger and darker, setae longer than in either *C. scripta* or *C. lineatopunctata;* eversible scent glands prominent; on willow, cottonwood. **a.** larva, lateral; **b.** antenna; **c.** mandible; **d.** labrum; **e.** tarsungulus; **f.** maxilla.

Figures 34.814a-f. Chrysomelidae. *Gastrophysa cyanea* Melsheimer. A dock leaf beetle. (Chrysomelinae) Length 6 mm; head black, legs dark brown to black; body stout, slightly eruciform, dark brown to black, lighter below; surface densely, very darkly asperate; setae, pronotal plate, thoracic scent glands prominent; eversible scent glands on T2-3 and A1-7; on *Rumex.* **a.** larva, lateral; **b.** antenna; **c.** mandible; **d.** tarsungulus; **e.** labrum; **f.** maxilla.

Figures 34.815a-f. Chrysomelidae. *Gastrophysa polygoni* L. A smartweed leaf beetle. (Chrysomelinae) Length 5 mm; head near black; body robust, light, distinct rows of darker tubercles; eversible glands prominent; sclerites, tubercles form distinct transverse segmental rows; on *Polygonum.* **a.** larva, lateral; **b.** antenna; **c.** mandible; **d.** tarsungulus; **e.** labrum; **f.** maxilla.

Figures 34.816a-f. Chrysomelidae. *Gonioctena americana* (Schaeffer). American aspen beetle. (Chrysomelinae) Length 8 mm; head dark brown; pronotal shield covering entire top of segment, medium brown; body segments dark, heavily sclerotized above, lighter below; mid-dorsal scent glands between A7 and A8; on aspen foliage. **a.** larva, lateral; **b.** antenna; **c.** mandible; **d.** tarsungulus; **e.** labrum; **f.** maxilla.

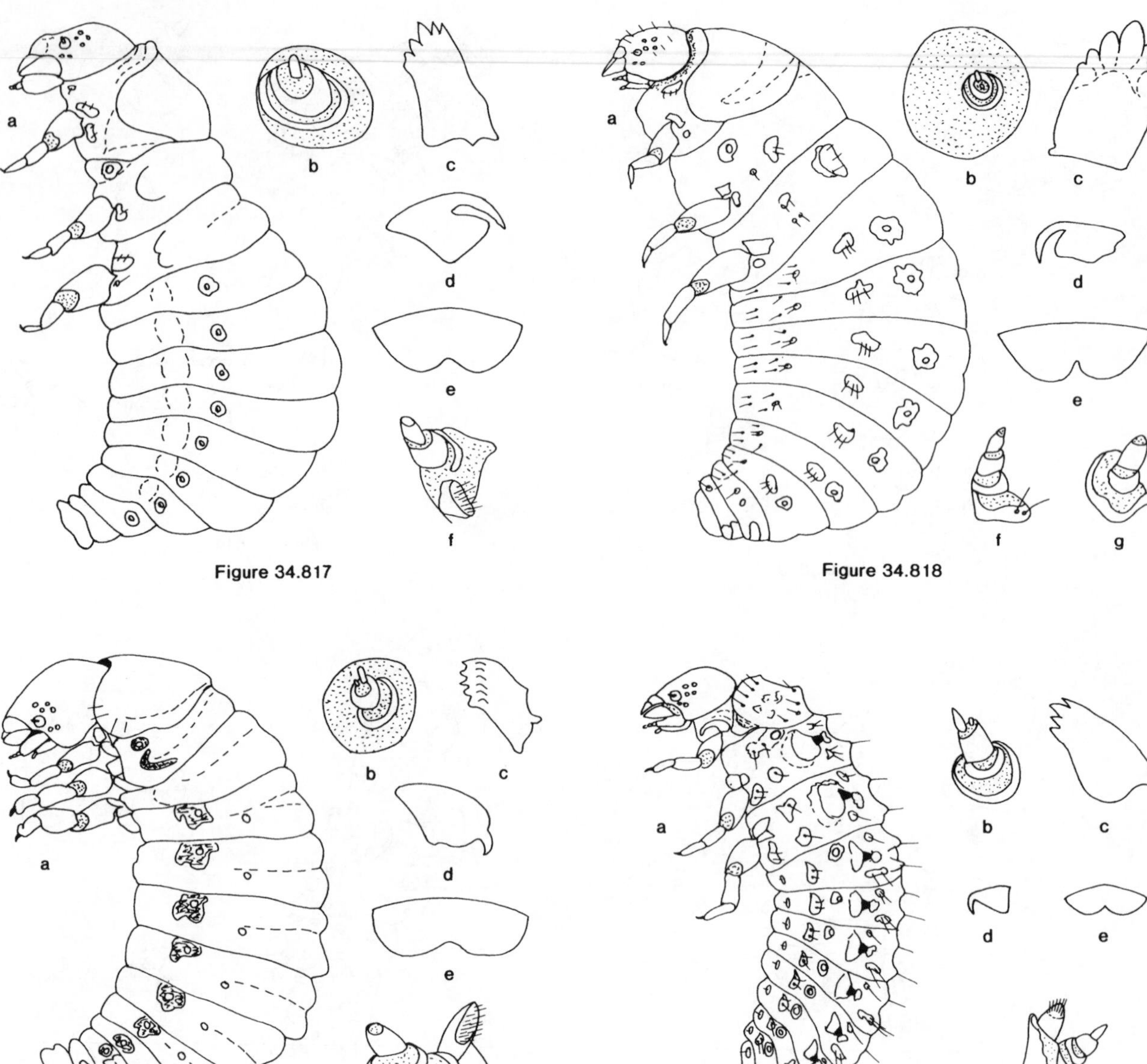

Figure 34.817

Figure 34.818

Figure 34.819

Figure 34.820

Figures 34.817a-f. Chrysomelidae. *Labidomera clivicollis* Kirby. A swamp milkweed leaf beetle. (Chrysomelinae) Length 11 mm; head, legs, pronotal plate tan; head reticulated; body strongly cyphosomatic, abdominal terga greatly expanded; spiracles black circles in irregular small brown sclerites; spiracles on A8 reduced; A7,8 with paired ambulatory lobes; A8,9 with tan dorsal plates; abdomen glabrous above, lightly setose below; on milkweeds. **a.** larva, lateral; **b.** antenna; **c.** mandible; **d.** tarsungulus; **e.** labrum; **f.** maxilla.

Figures 34.818a-g. Chrysomelidae. *Leptinotarsa decemlineata* (Say). Colorado potato beetle. (Chrysomelinae) Length 10 mm; head dark brown; pronotal shield part dark, part yellow (variable); body cream to yellowish to orange, with prominent dark spiracular sclerites and a row of dark subspiracular tubercles; mesothoracic spiracles enlarged; on *Solanum* foliage. A serious pest of potatoes. **a.** larva, lateral; **b.** antenna; **c.** mandible; **d.** tarsungulus; **e.** labrum; **f.** maxilla; **g.** labial palpus.

Figures 34.819a-f. Chrysomelidae. *Leptinotarsa peninsularis* Horn. (Chrysomelinae) Length 8 mm; head tan, with a pair of dark frontal spots; body strongly eruciform, light, with near black spiracular spots; intersegmental lines and a median mid-dorsal line near black; host unknown. **a.** larva, lateral; **b.** antenna; **c.** mandible; **d.** tarsungulus; **e.** labrum; **f.** maxilla.

Figures 34.820a-f. Chrysomelidae. *Phaedon americanus* Horn. (Chrysomelinae) Length 4 mm; head dark brown to black; legs brown; body eruciform, surface densely, finely asperate; body lightly rugose, tan to dark brown; pronotal plate covers entire dorsum of T1; eversible glands more prominent on thorax; all tubercles, sclerites, and setae prominent; feeds on foliage of *Veronica* (speedwell). **a.** larva, lateral; **b.** antenna; **c.** mandible; **d.** tarsungulus; **e.** labrum; **f.** maxilla.

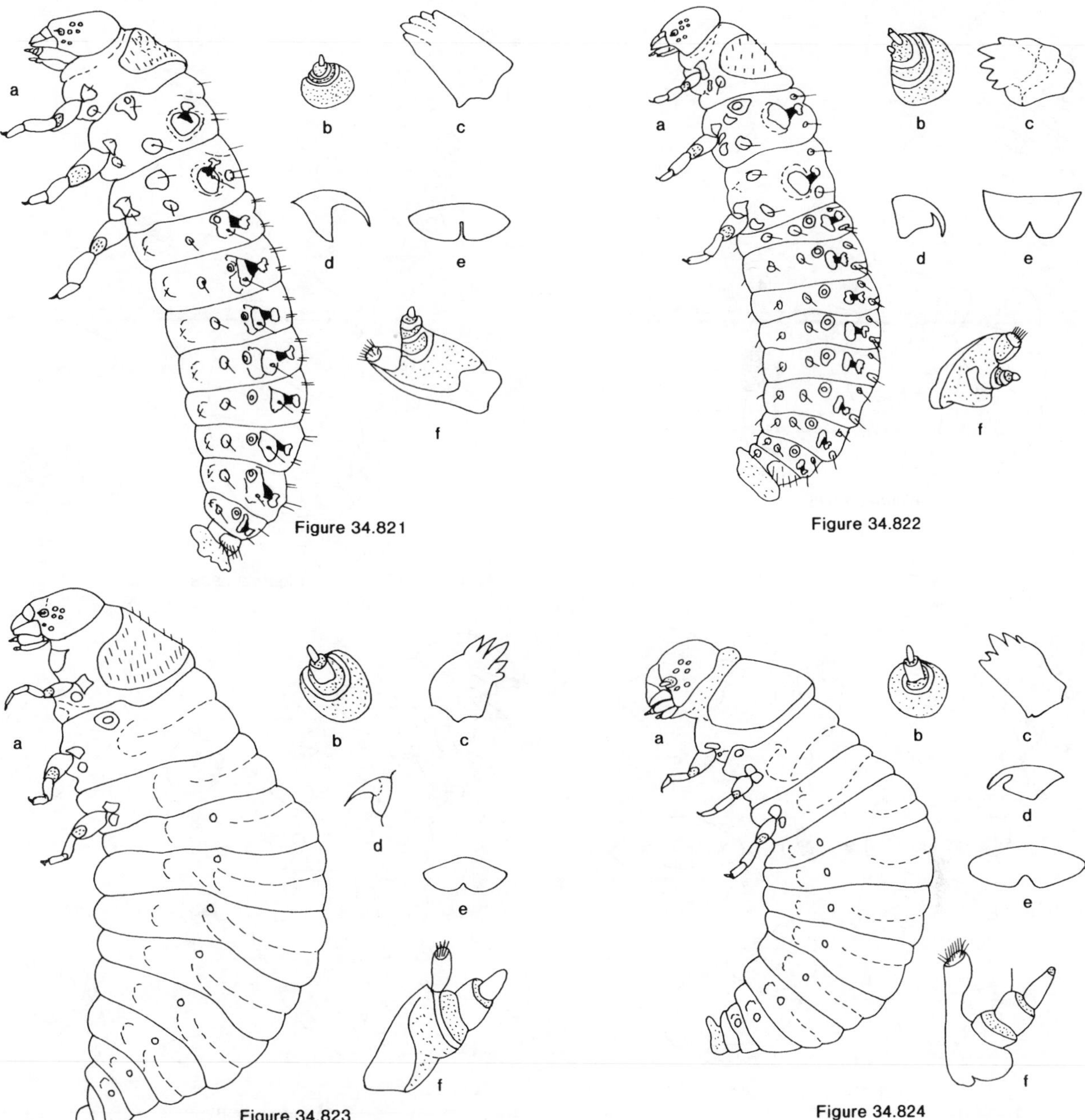

Figures 34.821a-f. Chrysomelidae. *Phratora americana* Schaeffer. A willow, cottonwood leaf beetle. (Chrysomelinae) Length 6 mm; head, legs dark brown; pronotal shield, tubercles, and sclerites dark dorsally and laterally, inconspicuous on venter; eversible gland bases enlarged on thorax, small on abdomen; body near white to gray, surface densely asperate around tubercles, less so elsewhere; setae elongate, spiracles brown circles in small tan tubercles; on willow and cottonwood. **a.** larva, lateral; **b.** antenna; **c.** mandible; **d.** tarsungulus; **e.** labrum; **f.** maxilla.

Figures 34.822a-f. Chrysomelidae. *Plagiodera versicolora* (Laicharting). Imported willow leaf beetle. (Chrysomelinae) Length 5 mm; head red to brown, faintly reticulated; legs tan; T1 plate brown in center, edges tan; sclerites, tubercles light brown; eversible glands bigger on thorax; body light brown, densely asperate on dorsum down to spiracles, sparsely asperate on venter except for a near glabrous midline; on *Salix* (willow). **a.** larva, lateral; **b.** antenna; **c.** mandible; **d.** tarsungulus; **e.** labrum; **f.** maxilla.

Figures 34.823a-f. Chrysomelidae. *Zygogramma exclamationis* (Fabricius). Sunflower beetle. (Chrysomelinae) Length 9 mm; head, pronotal shield light brown, head reticulated; body, legs yellowish; body strongly cyphosomatic, abdominal segments wider dorsally; setae conspicuous; spiracles very small brown circles; on sunflower foliage. **a.** larva, lateral; **b.** antenna; **c.** mandible; **d.** tarsungulus; **e.** labrum; **f.** maxilla.

Figures 34.824a-f. Chrysomelidae. *Zygogramma suturalis* Fabricius. A ragweed leaf beetle. (Chrysomelinae) Length 7 mm; head light tan, faintly reticulated; body light tan, strongly cyphosomatic; setae prominent; spiracles very small brown circles; on ragweed, goldenrod. **a.** larva, lateral; **b.** antenna; **c.** mandible; **d.** tarsungulus; **e.** labrum; **f.** maxilla.

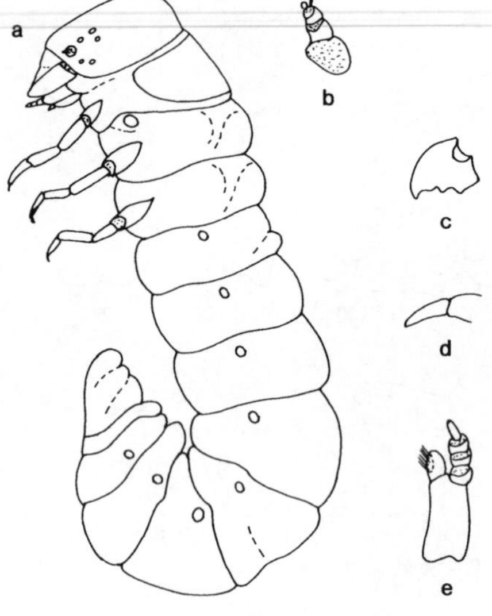

Figure 34.825

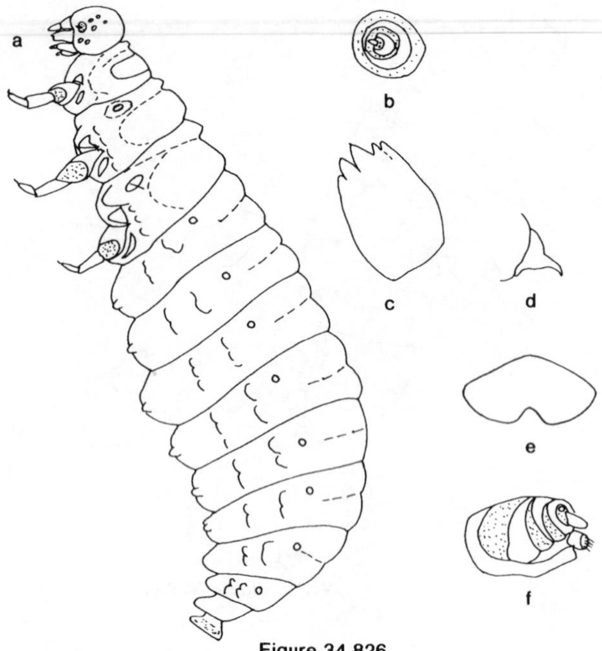

Figure 34.826

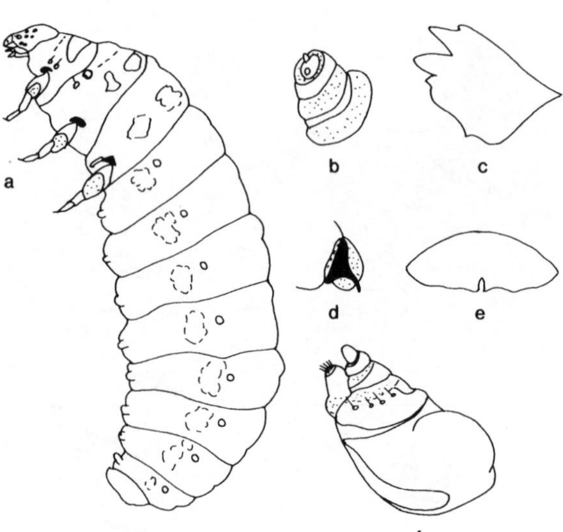

Figure 34.827

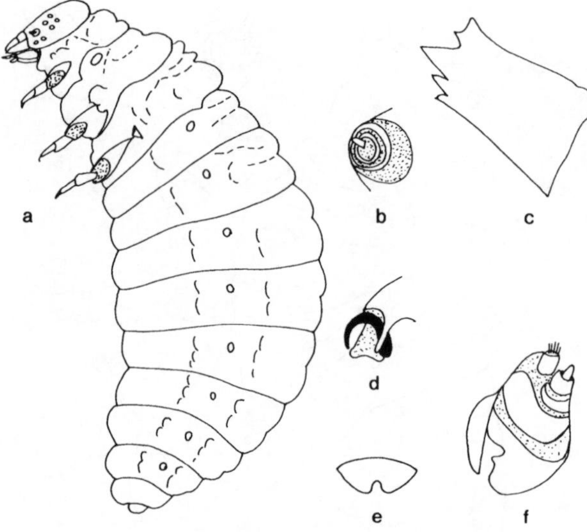

Figure 34.828

Figures 34.825a-e. Chrysomelidae. *Coscinoptera* sp. (?) A casebearer. (Clytrinae) Length 6 mm; head dark brown to black; pronotum, legs gray to black; body white; abdomen hooked, segments wider on dorsum, intersegmental lines indistinct in middle; setae on head and legs elongate, tapered; setae on body shorter; spiracles indistinct; on willow, wild rose, others. **a.** larva, lateral; **b.** antenna; **c.** mandible; **d.** tarsungulus; **e.** maxilla.

Figures 34.826a-f. Chrysomelidae. *Criocerus asparagi* (L.). Asparagus beetle. (Criocerinae) Length 9 mm; head dark brown; legs, pronotal shield light brown; body tan to light tan; slightly eruciform; most segments subdivided once across dorsum; surface finely, densely asperate; spiracles biforous brown circles; paired ambulatory lobes on A1-8 and A10; on asparagus ferns. **a.** larva, lateral; **b.** antenna; **c.** mandible; **d.** tarsungulus; **e.** labrum; **f.** maxilla.

Figures 34.827a-f. Chrysomelidae. *Criocerus duodecempunctata* (L.). Spotted asparagus beetle. (Criocerinae) Length 8 mm; head light brown; pronotal shield dark, divided longitudinally on meson; body robust, off white, moderately eruciform; paired ambulatory lobes on abdominal segments; surface finely, densely asperate; spiracles cribrate, light brown, ovate; inside asparagus fruits. **a.** larva, lateral; **b.** antenna; **c.** mandible; **d.** tarsungulus; **e.** labrum; **f.** maxilla.

Figures 34.828a-f. Chrysomelidae. *Lema trilineata* (Olivier). Threelined potato beetle. (Criocerinae) Length 6 mm; head dark brown; legs, pronotal shield brown; body moderately to strongly eruciform, light tan, with abdominal ambulatory lobes; cuticle densely, finely asperate; spiracles conspicuous, annular-biforous, tan; anal opening dorsal; dorsum of living larvae covered with fecal exudate; on foliage of solanaceous plants; pupates in a silken coccoon in debris at base of host. **a.** larva, lateral; **b.** antenna; **c.** mandible; **d.** tarsungulus; **e.** labrum; **f.** maxilla.

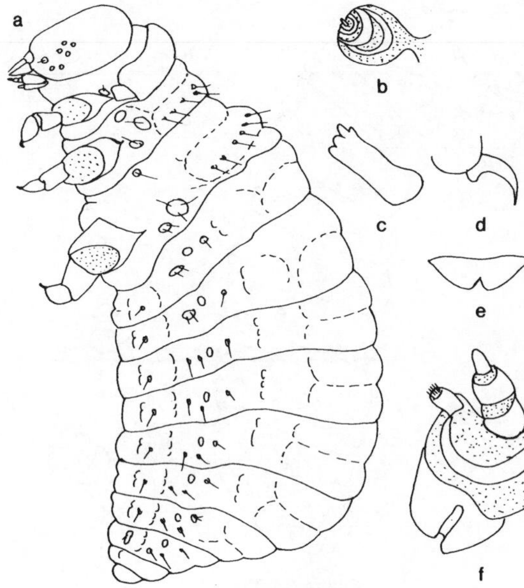

Figure 34.829

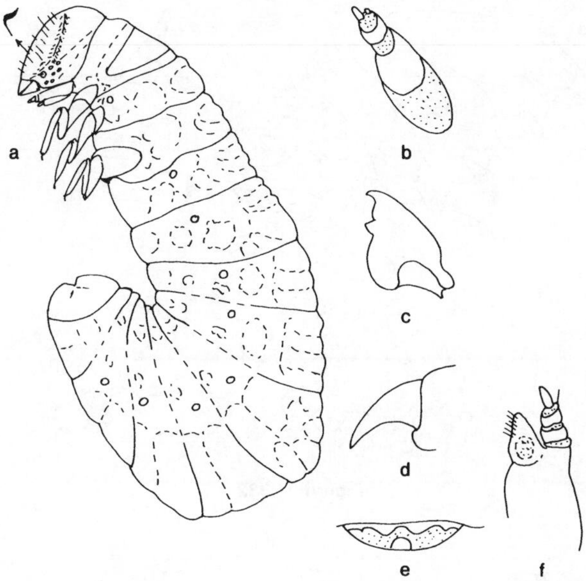

Figure 34.830

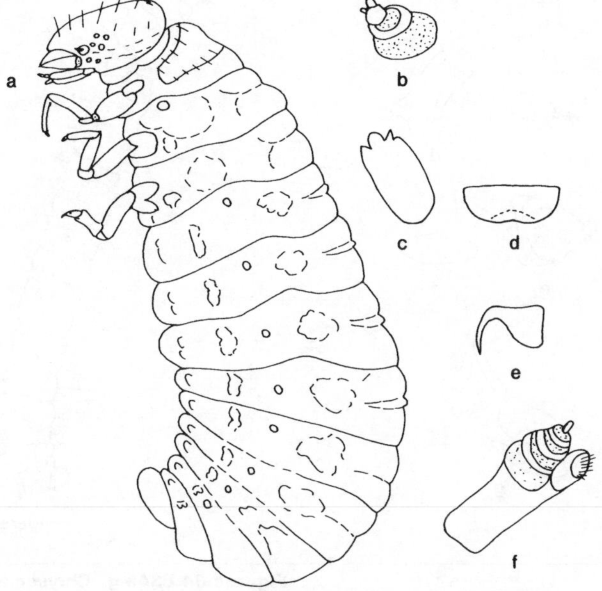

Figure 34.831

Figures 34.829a-f. Chrysomelidae. *Oulema melanopus* (L.). Cereal leaf beetle. (Criocerinae) Length 5 mm; head dark to medium brown, with pale stripe circling cranium to antennae; legs light brown, short; body near white to yellow, robust, dorsal surface strongly arched, venter slightly curved; spiracles minute but distinct; surface lightly microsetose; setae inconspicuous, but stronger on A8–9; feeding instars covered with liquid fecal material; on foliage of oats, other small grains. European origin. **a.** larva, lateral; **b.** antenna; **c.** mandible; **d.** tarsungulus; **e.** labrum; **f.** maxilla.

Figures 34.830a-f. Chrysomelidae. *Pachybrachis bivittatus* (Say). A willow casebearer. (Cryptocephalinae) Length 8 mm; head hypognathous, brown to near black; pronotal shield narrow, transverse; body near white, strongly hooked, apex of abdomen directed forward; spiracles inconspicuous; legs elongate, slender; hairs on frons clubbed and reflexed; live and feed from within cases, on foliage of willow. **a.** larva, lateral; **b.** antenna; **c.** mandible; **d.** tarsungulus; **e.** labrum-clypeus (fused); **f.** maxilla.

Figures 34.831a-f. Chrysomelidae. *Cryptocephalus confluentus* Say. A casebearer. (Cryptocephalinae) Length 8 mm; head hypognathous, dark brown to black, with lateral light stripe to bases of antennae; body white, robust, cylindrical in front, mid-abdominal segments enlarged dorsally, shortened ventrally, forming hook shape; T1 with expanded dorsal and small ventral plates; other segments membranous; legs long, dark brown, placed forward on segments; forelegs often held forward on either side of head; head hairs elongate, slender, pointed; body lightly setose; live and feed in cases on rabbitbrush (*Chrysothamnus*), sagebrush. **a.** larva, lateral; **b.** antenna; **c.** mandible; **d.** labrum-clypeus (fused); **e.** tarsungulus; **f.** maxilla.

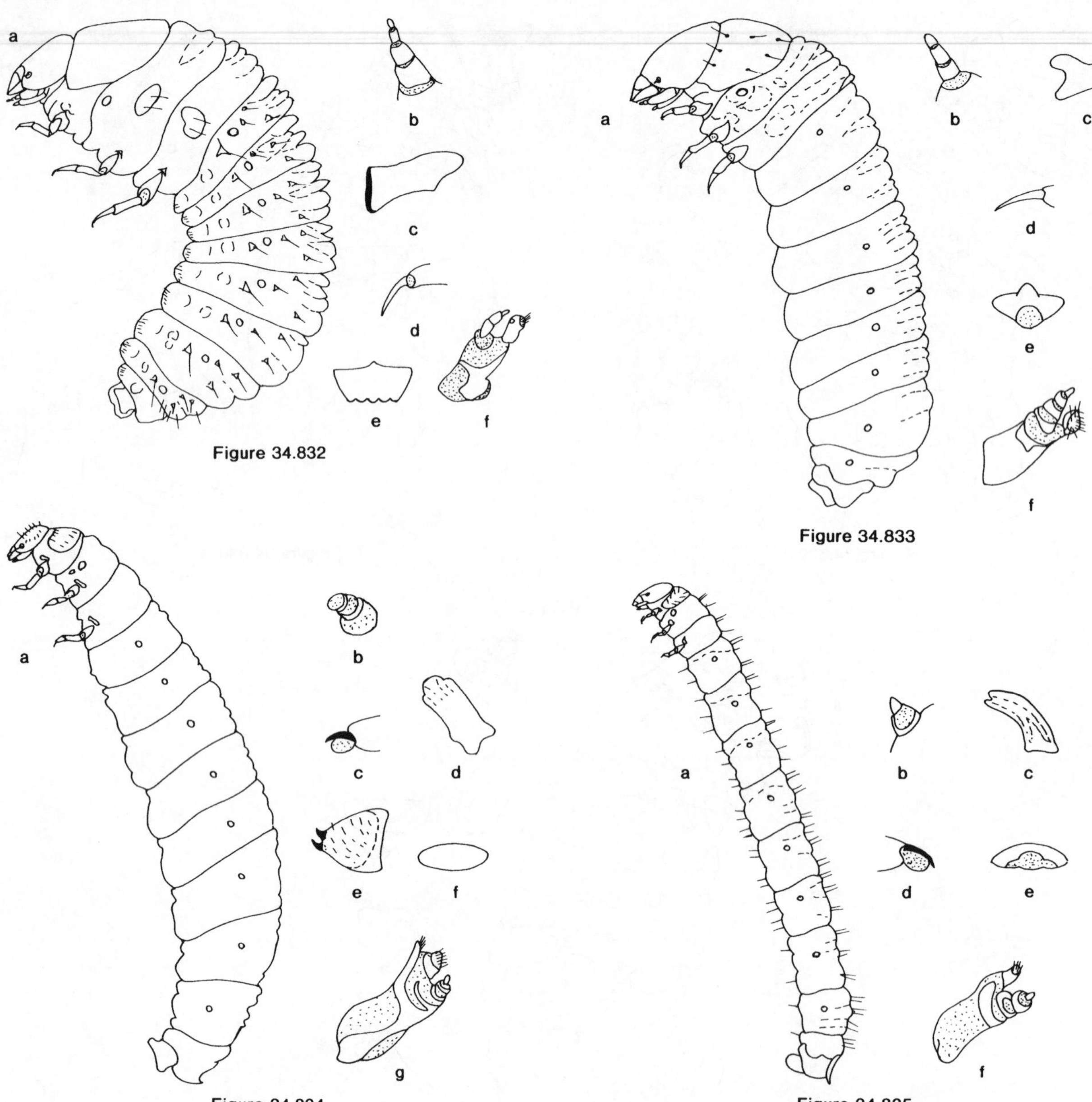

Figures 34.832a-f. Chrysomelidae. *Fidia viticida* Walsh. Grape rootworm. (Eumolpinae) Length 5 mm; head, pronotal shield light yellow; mandibles prominent, darker; no stemmata; body near white, segments subdivided on dorsum, moderately C-shaped, robust, surface densely microrugose; spiracles in tan rings; setae prominent; on roots of grape. **a.** larva, lateral; **b.** antenna; **c.** mandible; **d.** tarsungulus; **e.** labrum; **f.** maxilla.

Figures 34.833a-f. Chrysomelidae. *Paria fragariae* Wilcox. Strawberry rootworm. (Eumolpinae) Length 3 mm; head yellow, mandibles and frontoclypeal ridge darker; body near white; legs long, pale except for dark tarsungulus; body slightly curved, segments subdivided, setae elongate; abdomen with continuous ventrolateral lobate margin; ventral surface of abdomen may be transversely ridged; on roots of strawberry. **a.** larva, lateral; **b.** antenna; **c.** mandible; **d.** tarsungulus; **e.** labrum; **f.** maxilla.

Figures 34.834a-g. Chrysomelidae. *Acalymma vittatum* (Fabricius). Striped cucumber beetle. (Galerucinae) Length 10 mm; head brown; legs, body white; body elongate, cylindrical, slightly curved; pronotal shield small, tan; setae and spiracles inconspicuous; surface sparsely microsetose; A1-8 faintly subdivided dorsally; A9 with sclerotized dorsal plate (anal plate) bearing 2 minute urogomphi near midline; A10 resembles an anal proleg; on underground parts of cucurbits. **a.** larva, lateral; **b.** antenna; **c.** tarsungulus; **d.** mandible; **e.** "anal plate"; **f.** labrum; **g.** maxilla.

Figures 34.835a-f. Chrysomelidae. *Diabrotica undecempunctata* Mannerheim. Spotted cucumber beetle, Southern corn rootworm. (Galerucinae) Length 11 mm; head, anal plate brown; pronotum light brown; remainder of body near white; body elongate, slender, cylindrical, slightly curved, segments subdivided dorsally, slightly thicker in segments A4-7; setae minute, spiracles indistinct; surface densely microasperate; A10 forms an anal proleg; on roots, underground stems of crop plants, weeds. **a.** larva, lateral; **b.** antenna; **c.** mandible; **d.** tarsungulus; **e.** labrum; **f.** maxilla.

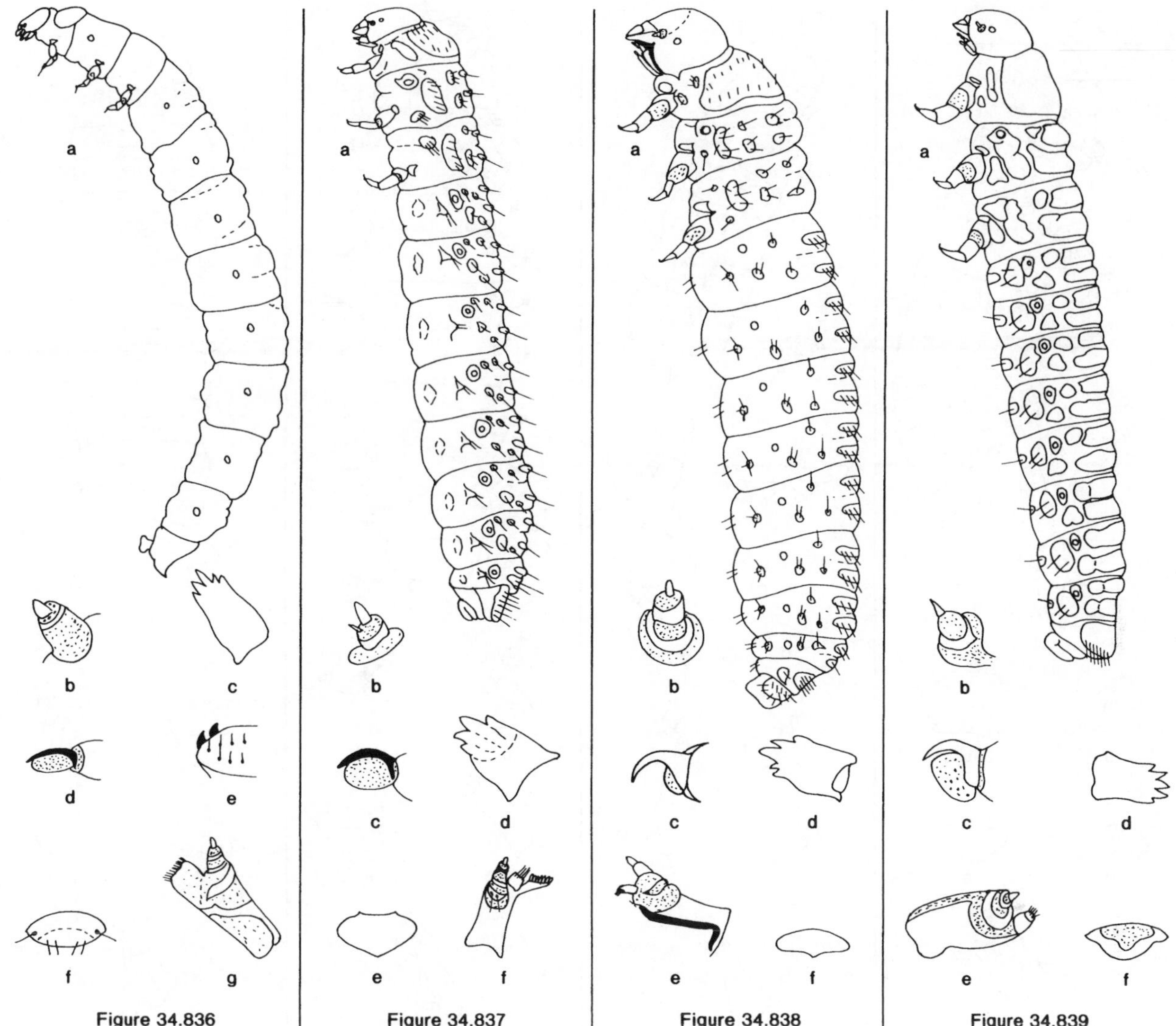

Figures 34.836a-g. Chrysomelidae. *Diabrotica virgifera* LeConte. Western corn rootworm. (Galerucinae) Length 11 mm; head brown, partially reticulated; body near white, elongate, cylindrical, segments subdivided dorsally; setae conspicuous; spiracles indistinct; A1-9 have one small lateral, one larger ventrolateral lobe on each side; A9 has dorsal brown, rugose (anal) plate with minute scattered darker spots; surface minutely, densely rugose; on roots of corn. **a.** larva, lateral; **b.** antenna; **c.** mandible; **d.** tarsungulus; **e.** "anal plate"; **f.** labrum; **g.** maxilla. Larvae of the northern corn rootworm, *Diabrotica longicornis barberi* Smith and Lawrence, are similar in overall appearance, but have 4 minute setae arranged in 2 laterally placed pairs on the narrow, posterior, middorsal A8 sclerite; in the western and southern corn rootworms these 4 setae form an evenly spaced transverse row.

Figures 34.837a-f. Chrysomelidae. *Pyrrhalta luteola* (Müller). Elm leaf beetle. (Galerucinae) Length 10 mm; head, legs, dorsolateral, ventrolateral rows of tubercles black; median dorsal, lateral, ventral tubercles light tan to gray; pronotum brown in center, black edge or dark spots on sides; body yellowish, appears striped in dorsal aspect; surface darkly, densely asperate about tubercles of dorsolateral row, enhancing striped effect; remainder of surface glabrous to sparsely microsetose; on foliage of elm. **a.** larva, lateral; **b.** antenna; **c.** tarsungulus; **d.** mandible; **e.** labrum; **f.** maxilla.

Figures 34.838a-f. Chrysomelidae. *Monoxia pallida* Blake. Goosefoot leaf miner. (Galerucinae) Length 4 mm; head, legs black to dark brown; body pink with prominent gray tubercles and sclerites; spiracles brown, in minute gray sclerites; body slightly curved, surface densely asperate; setae prominent; leaf miner in *Chaenopodium* (goosefoot). **a.** larva, lateral; **b.** antenna; **c.** tarsungulus; **d.** mandible; **e.** maxilla; **f.** labrum.

Figures 34.839a-f. Chrysomelidae. *Galerucella nymphaeae* L. A waterlily leaf beetle. (Galerucinae) Length 6 mm; head, legs, pronotum black; lateral rows of tubercles black; median dorsal sclerites dark brown to near black; ventral tubercles, sclerites small, gray; dorsal, lateral sclerites and tubercles greatly expanded, leaving little exposed yellow to near white intermediate cuticle, imparting an extremely dark tone; dark parts with metallic sheen; setae dark, inconspicuous; ventral surface mostly membranous, yellow in fresh material; on leaves of water smartweed (*Polygonum amphibium*), and water or pond lilies. **a.** larva, lateral; **b.** antenna; **c.** tarsungulus; **d.** mandible; **e.** maxilla; **f.** labrum.

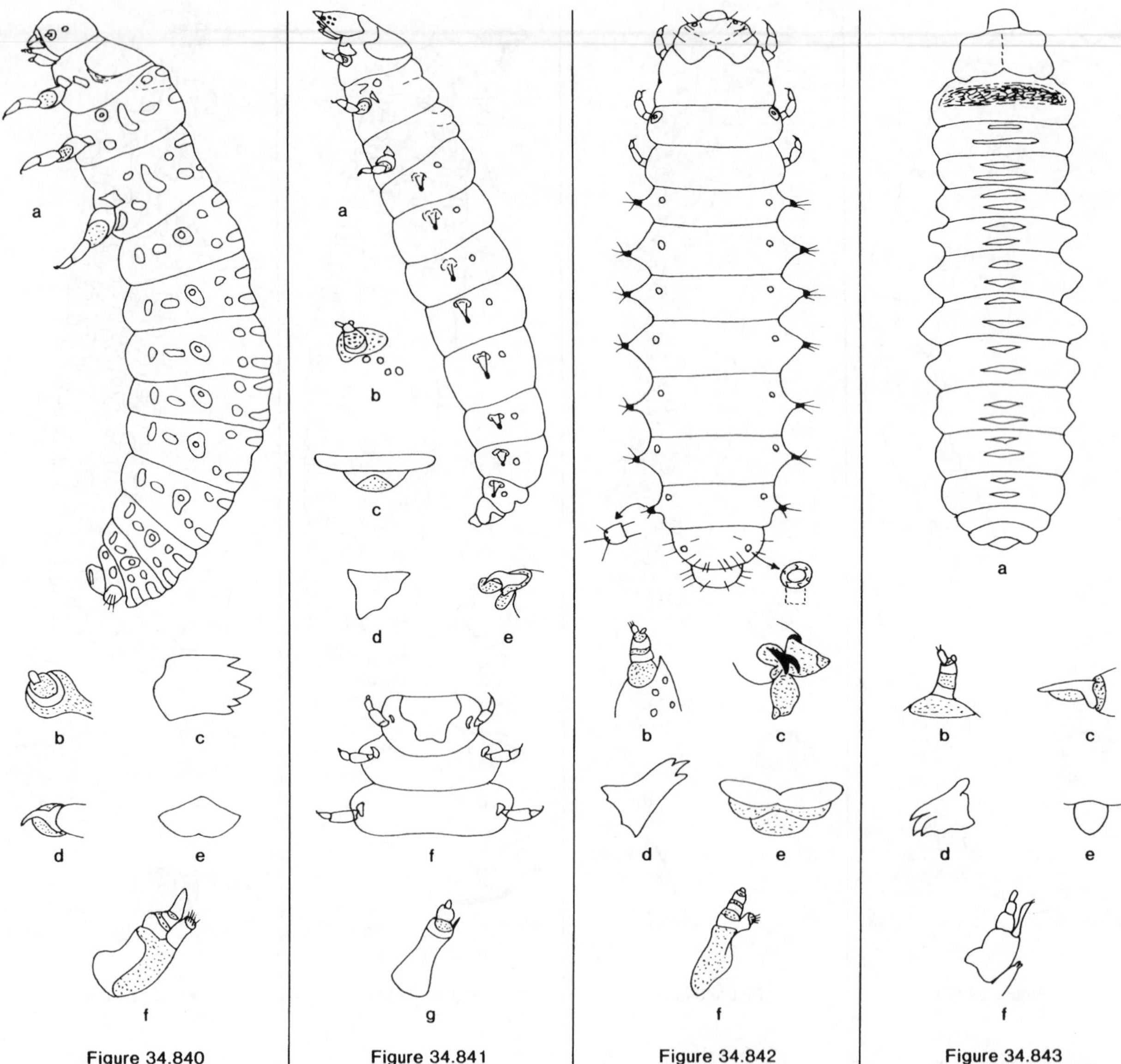

Figures 34.840a-f. Chrysomelidae. *Trirhabda attenuata* Say. A rabbitbrush leaf beetle. (Galerucinae) Length 11 mm; head dark brown to black, one stemma; mandibles elongate; body brownish, mildly cyphosomatic, sclerites and tubercles darker, with metallic sheen; legs dark brown; setae prominent; surface densely, darkly asperate; midline and ventral margin of frons, labrum, and tips of mandibles black; on foliage of rabbitbrush (*Chrysothamnus* spp.) **a.** larva, lateral; **b.** antenna; **c.** mandible; **d.** tarsungulus; **e.** labrum; **f.** maxilla.

Figures 34.841a-g. *Chrysomelidae. Odontota dorsalis* (Thunberg). Locust leafminer. (Hispinae) Length 8 mm; head, pronotal, prosternal shields light to dark brown; legs dark brown above, pale beneath; body orthosomatic, flattened, head semi-prognathous, color pinkish to white; segments A1-7 with triangular lateral extension on each side, these with black capped, trisetose apex; spiracles brown circles, slightly elevated; surface finely, densely punctate; miners in leaves of black locust. **a.** larva, lateral; **b.** antenna and stemmata; **c.** clypeus and labrum; **d.** mandible; **e.** tarsungulus; **f.** thorax, ventral; **g.** maxilla.

Figures 34.842a-f. Chrysomelidae. *Microrhopala vittata* Fabricius. A goldenrod leafminer. (Hispinae) Length 8 mm; head flattened, prognathous, tan to yellowish, vertex deeply notched, epicranial suture absent; frontal sutures prominent, in a wide V; deep groove down middle of clypeus; body white, robust, moderately elongate, depressed; T1 with expanded dorsal and shield-shaped ventral plates; T2-A7 membranous; A8 with small dorsal plate; A1-7 with lateral triangular projection bearing black cap on each side; cuticle densely microsetose; spiracles with prominent black rings; legs dark, long, widely spaced, each with 2 large pulvilli beneath tarsungulus; leafminer in goldenrod. **a.** larva, dorsal; **b.** antenna and stemmata; **c.** tarsungulus; **d.** mandible; **e.** clypeus and labrum; **f.** maxilla.

Figures 34.843a-f. Chrysomelidae. *Zeugophora scutellaris* Suffrian. A cottonwood leafminer. (Zeugophorinae) Length 3 mm; head, T1 strongly declivent to front; head prognathous, wide, flattened; body slender, off white, depressed, tapered caudally; T2-A8 with small dorsal and ventral midline sclerites, also with small, rounded, setose lateral lobes; T1 with expanded dorsal and ventral plates; labial palpi minute, bases adjacent; leafminers in cottonwood, poplar. **a.** larva, dorsal; **b.** antenna; **c.** tarsungulus; **d.** mandible; **e.** labrum; **f.** maxilla and labial palpi.

Selected Bibliography

Anderson, W. H. 1938 (larvae of *Chaetocnema*).
Böving 1927 (Galerucinae-Alticinae), 1929a (larvae of Galerucinae).
Cox 1981 (larva of *Orsodacne*).
Crowson 1955, 1981 (classification; biology).
Erber 1969 (biology of Cryptocephalinae and Clytrinae).
Ford and Cavey 1985 (larval Hispinae).
Grison *et al.* 1962 (agricultural species in Europe).
Hoffman 1939 (larvae of Donaciinae).
Kasap and Crowson 1976 (larva of *Oomorphus*, relationships of Cryptocephalinae, etc.).
Kaufman 1967 (larvae of *Lema*).
Kimoto 1962 (immature stages of Chrysomelinae).
Kurcheva 1967 (larvae of Eumolpinae and *Syneta*).
LeSage 1982 (immature stages of *Exema*), 1984a (immature stages of *Lexiphanes*), 1984b (immature stages of *Neochlamisus*).
Lindquist and Davis 1971 (biology of *Phratoria hudsonia*).
Mann and Crowson 1981 (relationships of *Orsodacne* and *Syneta*).
Marshall 1979 (larvae of *Chrysolina*), 1980 (larvae of British Galerucinae and Alticinae).
Maulik 1931, 1932, 1933 (larvae of Hispinae).
McCauley 1938 (*Microrhopala*).
Medvedev, L. N. 1962.
Medvedev and Zaitzev 1978 (chrysomelid larvae of Siberia and Far East).
Monros 1949 (larva of *Lamprosoma*), 1954 (larva of *Megalopus*), 1955 (larva of *Atalasis*), 1959 (genera of Chrysomelidae).
Patterson 1931 (larvae of various Chrysomelidae).
Peterson 1951.
Sailsbury 1943 (larvae of Criocerinae).
Sanderson 1948 (larva of *Physonota*).
Seeno and Wilcox 1982 (genera of Chrysomelidae).
Steinhausen 1966, 1978.
Takizawa 1972 (larvae of Galerucinae), 1976 (larvae of *Gonioctena*), 1980 (larvae of Cassidinae).
White and Day 1979 (biology of *Lema*).
Wilcox 1975 (North and Central American checklist).
Woods 1918, 1924 (*Altica;* blueberry leafbeetle).
Zaitzev and Dang 1982.

NEMONYCHIDAE (CURCULIONOIDEA)

D. M. Anderson,
Systematic Entomology Laboratory, USDA

The Nemonychid Weevils

Figures 34.844a–e

Relationships and Diagnosis: Larvae of this small family resemble those of other curculionoid families in having a soft body with legs reduced, no terminal armature on the abdomen, hypopharyngeal bracon present, gular sutures absent, and antennae reduced to a single segment. They closely resemble larvae of Anthribidae in having rudimentary legs, body setae numerous, two dorsal folds on most abdominal segments, frontal sutures reaching the anterior margin of the head capsule, and mandible with molar area produced for grinding, but differ from anthribid larvae in having no labral tormae, clypeus not separated from the frons, and labrum not entirely separated from the clypeus. On the basis of larval characters, these weevils were placed in the Anthribidae by van Emden (1938b) and by W. H. Anderson (1947a), but have been excluded from that family and recognized as a separate family by Crowson (1955, 1981, 1985), Kuschel (1983), and Ter-Minasyan (1984), all of whom considered them to be the most primitive family of the Curculionoidea.

Biology and Ecology: All Nemonychidae are believed to be phytophagous, infesting cones or flowers, where both adults and larvae feed primarily on pollen of their hosts, which are mostly gymnosperms (Pinaceae, Araucariaceae, Podocarpaceae) and a few primitive angiosperms (Fagaceae and Ranunculaceae) according to Kuschel (1983) and Ter-Minasyan (1984). Larvae hatching from eggs laid in cones or flowers feed on pollen there, and some cone-infesting larvae have been observed crawling from one cone to another. Mature larvae drop from the cones or flowers and pupate in the soil, where they may remain for 1 to 7 months or may enter a diapause of from 1 to 2 years before emerging as adults (Kuschel, 1983). *See* also Thomas and Herdy (1961).

Description: Based on the descriptions of the larvae of *Neocimberis pilosus* (LeConte) by W. H. Anderson (1947a) and of *Nemonyx lepturoides* (Fabricius) by Ter-Minasyan (1984). Unless noted otherwise, the terminology used here follows that of W. H. Anderson (1947b) (*see* Curculionidae in this volume). Mature larva (fig. 34.844a) about 4.0 to 4.8 mm long, of moderate thickness throughout its length, strongly C-shaped. Body white, covered with a mixture of long and short setae. Minute legs present on thorax.

Head: (fig. 34.844b) Hypognathous, rounded at sides, pigmented, with few to many setae on frons and epicranium. Frontal sutures complete, reaching articulating membrane of mandible. Clypeus not distinguishable from frons and incompletely separated from labrum. One pair of anterior stemmata. Labrum short, its anterior margin rounded, bearing 4 pairs of setae. Antenna of a single membranous segment bearing an accessory appendage. Epipharynx (fig. 34.844c) with 2 pairs of anterolateral setae, 1 pair of anteromedian setae, 2 pairs of median setae (median spines of Anderson, 1947b), and 1 pair of median sensilla, but without labral tormae. Mandible (fig. 34.844d) with 2 apical teeth, an obtuse protuberance on cutting edge, a distinctly produced molar area with a flattened grinding surface, and 1 pair of setae. Hypopharyngeal bracon present. Maxilla (fig. 34.844e) with palp 3-jointed, palpiger present or absent, and setae as figured on stipes and mala. Labial palps of 2 articles. Premental sclerite present, may be divided medially.

Thorax and Abdomen: Pronotal sclerite transverse, lightly pigmented or unpigmented, sparsely covered with setae. Legs very small, subconical, 2 or 3-jointed, with or without a terminal claw (fig. 34.844a).

First 8 segments of abdomen with 2 dorsal folds and bearing annular or bicameral spiracles. Anal opening terminal.

Spiracles: Annular or bicameral (*see* above).

Comments: In an extensive analysis of the systematics, evolution, biology, and distribution of the Nemonychidae, Kuschel (1983) stated that the family consists of 65 species in 22 genera, worldwide. Only 8 species in 2 genera occur in N. America, according to O'Brien and Wibmer (1982). Larvae of these species are rarely collected, but if found (probably in male pine cones), they should be recognizable as Nemonychidae by comparison with the descriptions and figures presented here.

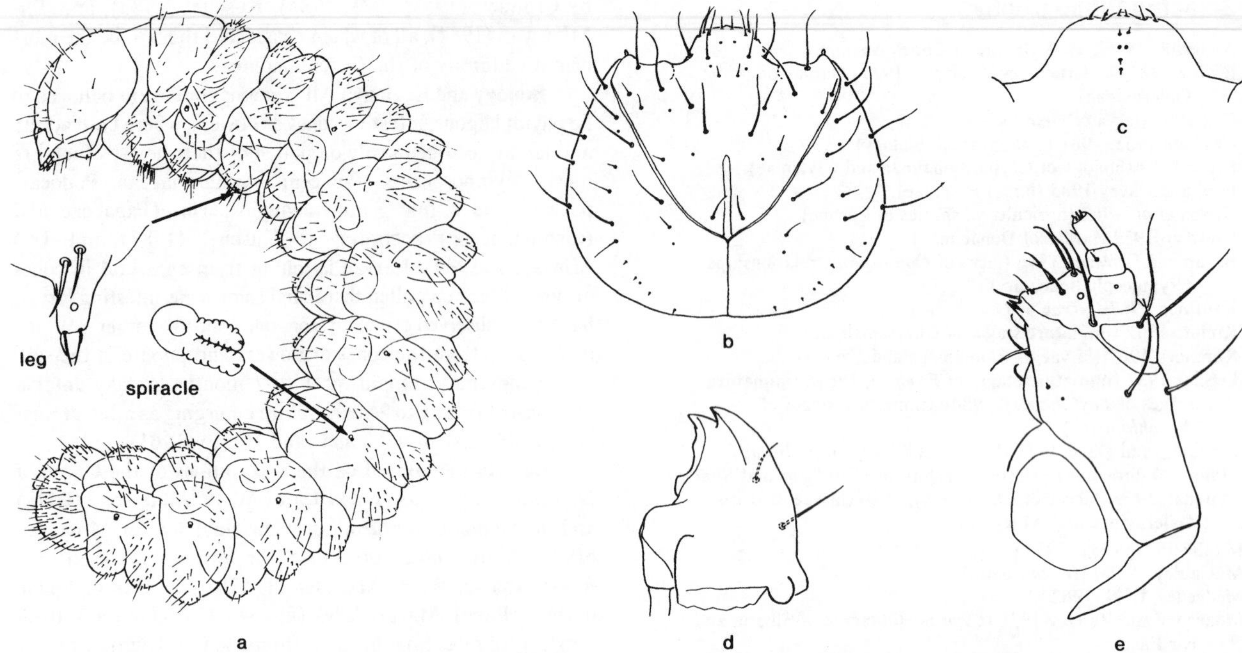

Figures 34.844a-e. Nemonychidae. *Neocimberis pilosus* (LeConte). **a.** larva, lateral, with enlarged metathoracic leg and abdominal spiracle; **b.** head, without mouthparts, dorsal; **c.** epipharynx; **d.** left mandible, ventral; **e.** left maxilla, ventral (from Anderson, 1947a). Larvae live in male cones of pines in eastern N. America and pupate in trash on the ground. For a more detailed description and notes on the 1st instar (as *Cimberis pilosus*), *see* Anderson (1947a).

Selected Bibliography

Anderson, W. H. 1947a, 1947b.
Crowson 1955, 1981, 1985.
Emden 1938b.
Kuschel 1959, 1983.
O'Brien and Wibmer 1982.
Ter-Minasyan 1984.
Thomas and Herdy 1961.

ANTHRIBIDAE (CURCULIONOIDEA)

D. M. Anderson,
Systematic Entomology Laboratory, USDA

The Fungus Weevils

Figures 34.845–847

Relationships and Diagnosis: The Anthribidae belong to the superfamily Curculionoidea (= Rhynchophora), which also includes the Nemonychidae, Ithyceridae, Brentidae, Curculionidae (broad sense), Scolytidae, Platypodidae, and several smaller families. Curculionoid larvae are soft bodied, with legs absent or greatly reduced and without terminal claws, antennae reduced to 2 or fewer segments, hypopharyngeal bracon usually present, gular sutures absent, and 9th abdominal segment without paired urogomphi or a terminal spine. The following combination of characters will separate larvae of Anthribidae from those of the other curculionoid families according to Anderson (1947a): ultimate and penultimate segments of maxillary palp each bearing a short seta (figs. 34.846f, 34.847e); labral rods (= epipharyngeal rods) absent; labral tormae present (fig. 34.845c); and legs usually present (fig. 34.847f), but sometimes reduced or absent. The preceding diagnostic characters and the detailed family description of larvae of the Anthribidae presented later are based upon the same genera (some now under different names) upon which W. H. Anderson (1947a) based his family description, with the exception of *Bruchela* and *Cimberis,* which were excluded from the Anthribidae by Crowson (1955, 1984b), Valentine (1960) and others. The larval terminology used here follows that used by W. H. Anderson (1947b).

Biology and Ecology: Larvae of N. American Anthribidae whose habits are known feed on plant materials. They can be found in the dead twigs and branches of trees, in woody fungi, and under bark of dead and dying trees, particularly those infested with fungi. According to Crowson (1984a), a number of anthribids feed on the stromata of certain ascomycete fungi (Xylariaceae and related families), but species of the genus *Euparius* apparently specialize on the fruiting bodies of higher Basidiomycetes (*see* below). The coffee bean weevil (*Araecerus fasciculatus* (De Geer)) lives in seeds and in other dried plant products. According to Blatchley and Leng (1916) and Arnett (1968), larvae of some American species feed in the living stems or receptacles of certain weeds. Arnett (1968) mentioned that the larvae of the European genus *Anthribus* (formerly *Brachytarsus*) feed on eggs of scale insects of the subfamily Lecaniinae, but this is a strong deviation from the usual phytophagous habits of the Anthribidae.

Description: Mature larvae small to moderately large, about 2.5 mm to 15 mm in length. Body subcylindrical in cross section and C-shaped. Legs usually present, sometimes barely discernible. Body setae sparse (fig. 34.846a) to abundant (fig. 34.845a), at least moderately abundant ventrally on prothorax.

Head: Hypognathous; exserted (fig. 34.845a) or partly retracted into prothorax (fig. 34.846a). Epicranium and frons bearing sparse to abundant setae (figs. 34.845b, 34.846d, 34.847a). Anterior stemma present or absent; posterior stemma absent. Antenna consisting of 1 membranous segment, which bears a conical to subconical sensorium and a few minute setae (fig. 34.846e). Frontal sutures complete (not terminating at antennae) anteriorly (fig. 34.846e), indistinct in some species. Labrum with 1 pair of basal sensilla, without anterior or median sensilla, and with 4 or more pairs of setae (fig. 34.846d). Labral tormae present (fig. 34.845c). Epipharynx with 2 pairs of anteromedian sensilla and 1 pair of peg-shaped sensilla. Mandible usually with a mola (fig. 34.845e), and with 2 or more setae on outer surface (fig. 34.846c). Hypopharyngeal bracon present. Hypopharyngeal sclerome usually present. Maxillary palp with 2 or 3 segments; ultimate and penultimate segments each with a short seta; palpifer absent. Stipes with several to numerous setae (fig. 34.847e). Mala bearing setae, the dorsal setae when present not arranged in a row and usually bearing a thornlike lacinia near middle of inner margin (fig. 34.847e). Labial palps, when present, with 1 or 2 segments, (fig. 34.847d) or absent (fig. 34.846f). Posterior margin of prementum straight or nearly straight; premental sclerite usually present. One to several pairs of premental setae present (fig. 34.847d). Submentum broad, usually with more than 3 pairs of setae (fig. 34.847d).

Thorax and Abdomen: Meso- and metathorax each with 2 dorsal folds (figs. 34.845a, 34.846a). Legs present, of 1 or 2 segments (fig. 34.847f), or indicated by clumps of setae (fig. 34.845a), or absent (fig. 34.846a).

Abdominal segments 1–7 with 2 dorsal folds (figs. 34.845a, 34.846a). Sternellum absent from all segments.

Spiracles: Situated in mesothorax or between prothorax and mesothorax, and on abdominal segments 1–8; annular-biforous, annular-uniforous, or without air tubes.

Comments: Larvae are known for few of the 86 described species in America north of Mexico. The most comprehensive account is by W. H. Anderson (1947a), who described the characters of larvae representing 13 of the 23 genera known to occur north of Mexico. However, the descriptions are based on larvae of 1 or a few species of each genus, and no attempt was made to describe or separate species. Larvae of 3 N. American species, representing 3 genera, are illustrated here as examples of the 2 subfamilies found in N. America. Head width measurements are from W. H. Anderson (1947a).

With the exception of the coffee bean weevil, the Anthribidae are of little economic importance, but their larvae are sometimes collected along with similar larvae of other insects infesting woody vegetation and thus must be identified.

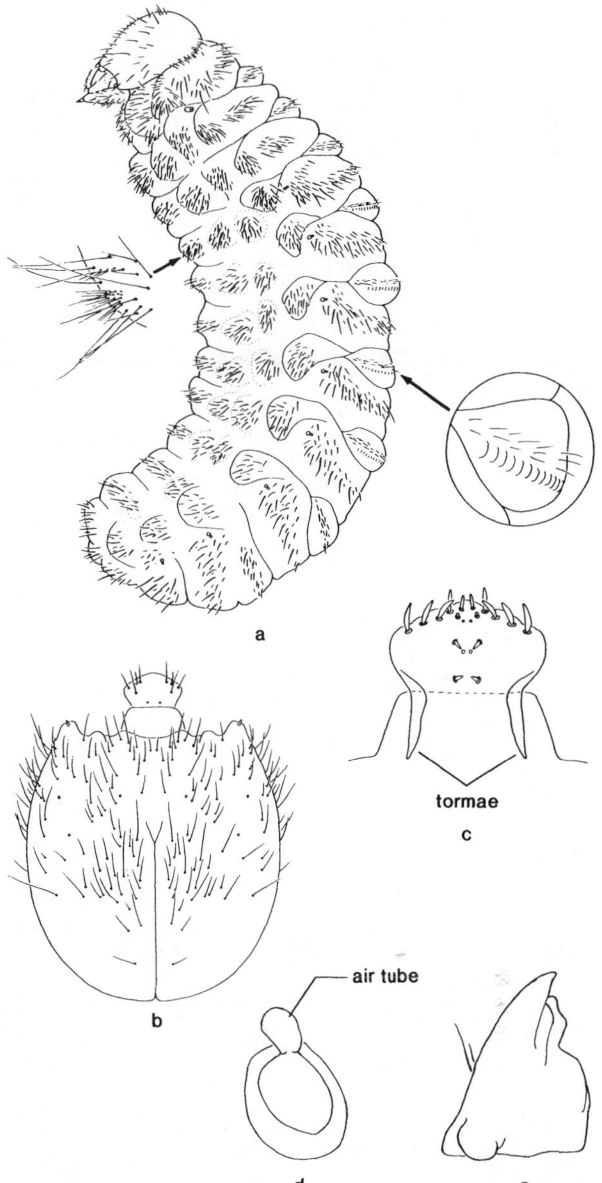

Figures 34.845a-e. Anthribidae. *Araecerus fasciculatus* (De Geer) (Choraginae), the coffee bean weevil. **a.** larva, lateral, with pedal area of metathorax, showing a cluster of setae in place of a leg, and 1st dorsal fold of abdominal segment 4, enlarged, showing row of longitudinal ridges; **b.** head, front, without mandibles or ventral mouthparts; **c.** epipharynx; **d.** abdominal spiracle; **e.** right mandible, ventral. Several authors, including Böving and Craighead (1931), El Sayed (1940), Anderson (1947), and White (1967) have described and/or figured the larva of this rather well-known, nearly cosmopolitan species. It feeds in many types of dried plant products in addition to coffee beans, including corn, dried fruits, and dry stalks or twigs. Diagnosis: Body and head cream colored except for dark mouthparts, and bearing numerous setae. Head width up to 0.95 mm. One stemma present on each side of head near base of mandible. Legs reduced, indicated only by clusters of setae. Thoracic spiracle bicameral; abdominal spiracles unicameral. First dorsal fold on abdominal segments 2-4 with a transverse row of short longitudinal ridges (figure 34.845a). Labial palpi 1-segmented. Features of epipharynx and shape of mandible also helpful in recognizing this species. (Figures 34.845a, b, c original, by C. Feller; d, e, from Anderson, 1947).

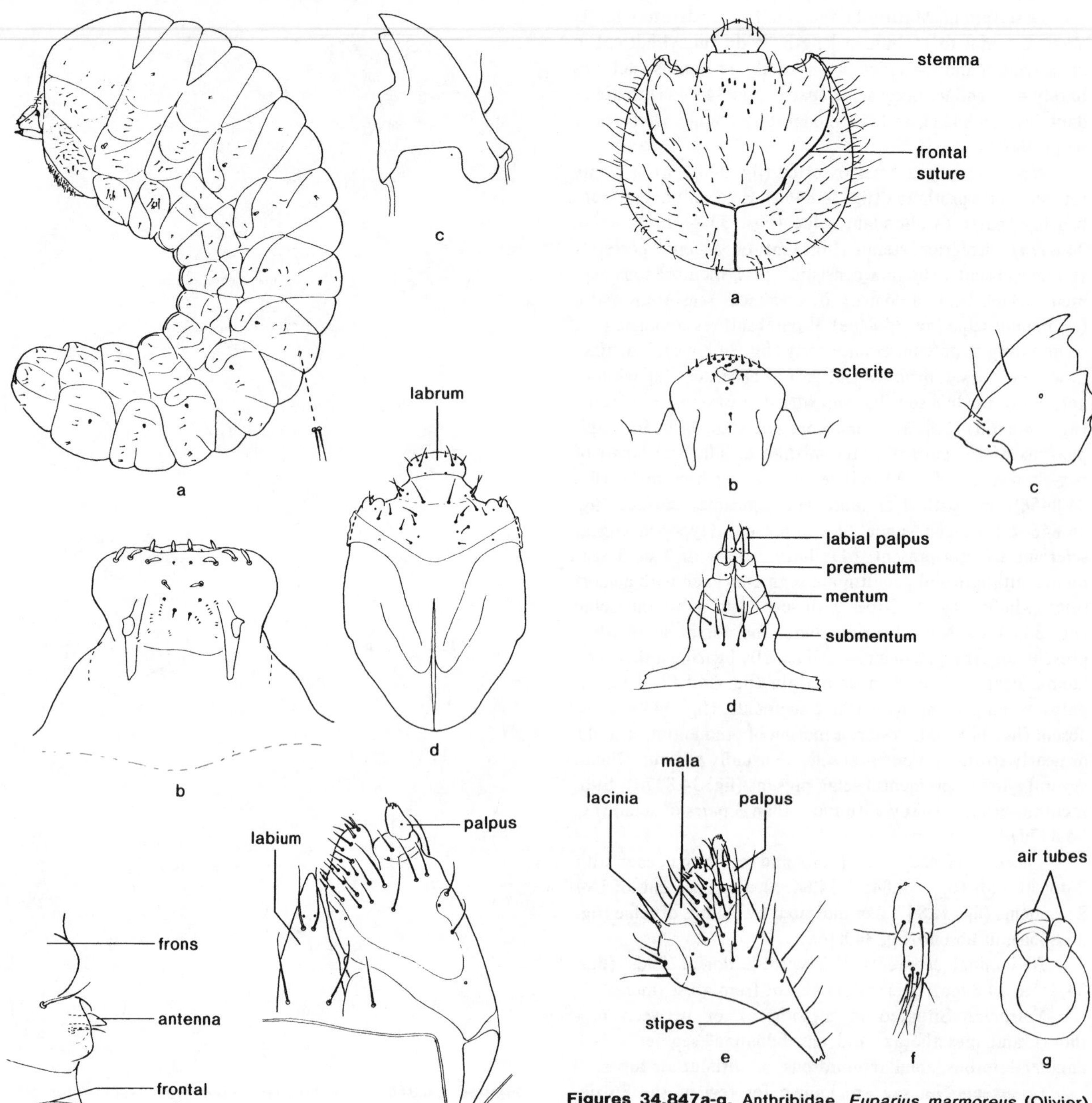

Figures 34.846a-f. Anthribidae. *Euxenus* sp. **a.** larva; **b.** epipharynx, lateral; **c.** left mandible, ventral; **d.** head, dorsal, without mandibles or ventral mouthparts; **e.** antennal area at anterior margin of frons, lateral; **f.** labium and left maxilla, ventral. The larva of this species was described under the name *Holostilpna nitens* (LeConte) by Anderson (1943, 1947). However, as explained by Valentine (1960) the larvae used in that description actually belonged to an undescribed species of *Euxenus*. The only biological information was provided by Anderson (1943, 1947), who collected the larvae from [an ascomycete] fungus, *Hypoxylon atropunctatum*, growing in the bark of dead black oaks in Maryland. Diagnosis: Head up to 0.47 mm wide, deeply retracted into prothorax. Legs absent. Body setae rather sparse, except on underside of prothorax. All spiracles unicameral. Features of epipharynx, shape of mandible, and antenna as illustrated. Labial palpi absent. Mala without a lacinia. (All figures from Anderson 1943).

Figures 34.847a-g. Anthribidae. *Euparius marmoreus* (Olivier) (Anthribinae). **a.** head, front; **b.** epipharynx; **c.** left mandible, dorsal; **d.** labium, ventral; **e.** left maxilla, ventral; **f.** prothoracic leg; **g.** abdominal spiracle. Previous descriptions of the larva of this species include the illustrations (including a complete lateral view) by Böving and Craighead (1931) and the detailed description, with some illustrations, by Anderson (1947). The biology of *E. marmoreus* is not well known. Anderson (1947) stated that the numerous specimens he examined were "from a variety of woody plants but probably always associated with fungus." According to J. F. Lawrence (personal communication), *E. marmoreus* has been collected in the fruiting bodies of at least 12 species of polypore fungi and 1 gilled fungus (*Panus* rudis) (30 records), and larvae were found in 5 species belonging to the genera *Coriolus, Funalia, Bjerkandera, Trametes,* and *Phellinus.* Diagnosis: Head up to 1.6 mm in width, reddish brown, unretracted, with numerous setae, and with 1 pair of stemmata. A small sclerite present in front of median setae on epipharynx. Thornlike lacinia on mala prominent. Mandible shaped as illustrated and bearing 5 setae. Labial palpi 1-segmented. Legs small, 2-segmented. All spiracles bicameral. Setae numerous on most body folds.

Selected Bibliography

Anderson, W. H. 1943, 1947a, 1947b.
Arnett 1968.
Blatchley and Leng 1916.
Böving and Craighead 1931.
Crowson 1955, 1984a (association with Ascomycetes), 1984b (position of *Bruchela*).
El Sayed 1940.
Emden 1938b.
Gardner 1932a, 1936a, 1937b.
Mathur 1957b.
Valentine 1960.
White 1967.

URODONTIDAE (CURCULIONOIDEA) (= BRUCHELIDAE)

John F. Lawrence,
Division of Entomology, CSIRO

This family includes the two genera *Bruchela* Dejean (= *Urodon* Schönherr) and *Cercomorphus* Perris, which together contain about 50 species occurring in Europe, Asia Minor, North Africa, and South Africa. *Bruchela* was included in the family Anthribidae by W. H. Anderson (1947a) and Crowson (1955), but excluded from that family by Valentine (1960) and placed in Bruchidae by van Emden (1938b), Wolfram (1953), and others. Hoffman (1945) proposed a separate family for the group, and Crowson (1984b) has recently supported this.

Larvae are oval to subcircular in cross section, slightly curved ventrally, lightly sclerotized with a darker head, and sparsely to densely setose. The head is strongly retracted, with a long epicranial stem, complete, V-shaped frontal arms, a distinct endocarina extending anterad of epicranial stem (between frontal arms), 1 stemma on each side, and 1-segmented antennae. The frontoclypeal suture is only vaguely defined and the labrum is free, with 2 basal sensilla and no labral rods. The mandibles are bidentate with 1 or 2 subapical teeth and no mola or accessory ventral process. The ventral mouthparts are retracted, with well-developed maxillary articulating areas, fused mentum and submentum, 2-segmented maxillary palps with palpifer, and no labial palps. The maxillary mala may be simple and rounded or may have a spine-like uncus extending from the inner edge. Hypostomal rods are long and subparallel. The hypopharyngeal bracon is present and the hypopharyngeal sclerome is absent. Thoracic legs are absent. All thoracic and most abdominal terga have 2 transverse plicae, while the metathorax and most abdominal segments have bifid, tergal ampullae. Spiracles are annular-uniforous or annular-biforous on the thorax and annular-uniforous on the abdomen.

Bruchela larvae are known to feed in the seed capsules of dicotyledonous plants in the families Resedaceae (*Reseda*) and Brassicaceae (*Iberis, Hirschfeldia*) in Europe and North Africa, but have been reported from *Gladiolus* (Iridaceae) seeds in South Africa. Adult *Cercomorphus* have been associated with *Genista* (Fabaceae) and *Frankenia* (Frankeniaceae). The New Zealand genus *Xenanthribus* Broun, included in this group by Wolfram (1929), is a true anthribid (Holloway, 1982).

Selected Bibliography

Anderson, W. H. 1947a (larvae of *Bruchela* spp.).
Crowson 1984b.
Emden 1938b.
Hoffmann 1945.
Holloway 1982 (*Xenanthribus*).
Peyerimhoff 1911 (*Bruchela* host plants), 1919 (*Cercomorphus* host plants).
Urban 1913 (larva of *Bruchela rufipes* (Olivier)).
Valentine 1960.
Wolfram 1929, 1953.

OXYCORYNIDAE (CURCULIONOIDEA)

John F. Lawrence,
Division of Entomology, CSIRO

This family includes about 30 species comprising 7 genera: *Metrioxena* from the East Indies, *Afrocorynus* and *Hispodes* from South Africa, and *Oxycorynus, Alloxycorynus, Hydnorobius,* and *Oxycraspedus* from southern South America. This is considered to be one of the more primitive curculionoid families, with features closely resembling the Jurassic Eobelidae.

Larvae are short, broad, ventrally curved, and lightly sclerotized. The head is strongly retracted, subrectangular, and slightly convex, with very short, 1-segmented antennae, and no labral rods. Mandibles are bidentate, with a well-developed subapical tooth and no mola or prostheca. Ventral mouthparts are retracted, with a simple, rounded mala, 3-segmented maxillary palps with a well-developed palpifer, and 2-segmented labial palps. Legs are absent. The first 3 abdominal terga each have 2 transverse plicae. Segment A9 is simple and segment A10 reduced. Spiracles are annular-biforous, with the thoracic spiracle located on the mesothorax.

Species of *Oxycraspedus* have been associated with the cones of *Araucaria*. Larvae of *Hydnorobius* species breed in the fleshy flowers of root parasites in the genus *Prosopanche* (Dicotyledonae: Hydnoraceae), and *Alloxycorynus bruchi* (Heller) has also been associated with influorescences of a parasitic plant (Balanophoraceae: *Ombrophytum*). Species of *Metrioxena* have been collected on palm fruits.

Selected Bibliography

Arnoldi and Zherikin 1977 (fossil Eobelidae).
Bruch 1923 (larva of *Hydnorobius hydnorae* (Pascoe)).
Crowson 1955.
Kuschel 1959 (review of family; habits).
Voss 1957.

BELIDAE (CURCULIONOIDEA)

John F. Lawrence,
Division of Entomology, CSIRO

This family includes 13 genera and about 150 species occurring in New Zealand (*Agathinus* and *Pachyurinus*), Australia (*Belus, Rhinotia, Pachyura,* etc.), and southern

South America (*Homalocerus, Dicordylus, Trichophthalmus,* etc.). The family is considered to be one of the most primitive groups of Curculionoidea.

Larvae are broad, slightly flattened, ventrally curved, lightly sclerotized, and hairy, with an enlarged, strongly declivous prothorax. The head is elongate and strongly retracted, with an indistinct median endocarina, 2 to 3 stemmata on each side, and 2-segmented antennae; the frons sometimes bears a median spine, and labral rods are absent. The mandibles are unidentate, without a mola or prostheca. Ventral mouthparts are retracted, with a simple, rounded mala, 3-segmented maxillary palps with a well-developed palpifer, and 2-segmented labial palps. The protergum has a sclerotized, keeled plate on the posterior half. Legs are absent. The abdominal terga usually have 2 indistinct, transverse plicae. Tergum A9 is simple and the anus is terminal and T-shaped. Spiracles are annular and the thoracic spiracles are located between the prothorax and mesothorax.

Belid larvae feed internally on twigs and branches. Larvae of Australian *Belus* have been collected in *Acacia* twigs, adult *Pachyura* have been associated with *Hakea* and other Proteaceae, *Agathinus* and *Pachyurinus* are found on Podocarpaceae, and *Homalocerus* and *Trichophthalmus* have been associated with ferns. Adults of Australian *Rhinotia* and South American *Homalocerus* both mimic lycid beetles.

Selected Bibliography

Crowson 1955.
Emden 1938b (larval characters).
Hudson 1934 (habits).
Kuschel 1959 (review of family; relationships; habits).
Vanin 1976 (South American species).

AGLYCYDERIDAE (CURCULIONOIDEA)

(= PROTERHINIDAE)

John F. Lawrence,
Division of Entomology, CSIRO

This family includes *Aglycyderes setifer* Westwood from the Canary Islands, *Platycephala* with 3 species from New Zealand and New Caledonia, and *Proterhinus* with about 160 species found mainly in Hawaii, but also occurring in the Marquesas, Fiji, Samoa, the Society and Austral Islands and Enderbury Island. This represents another primitive, relictual group.

Larvae are elongate straight or slightly curved ventrally, lightly sclerotized, and setose. The head is elongate, subquadrate, and strongly retracted, without an epicranial suture, and with 1-segmented antennae, 1 stemma on each side, and no labral rods. Ventral mouthparts are retracted, with a simple, blunt mala, 2-segmented maxillary palps and 2-segmented labial palps. Legs are absent. Abdominal terga 1 to 6 each have 4 transverse plicae, and tergum 7 has 3. Tergum A9 is shortened, simple, and the anus is terminal and T-shaped. Spiracles are annular-uniforous, and the thoracic spiracle is located on the mesothorax.

Aglycyderes has been found in *Euphorbia* stems, *Platycephala* have been associated with tree ferns (*Cyathea*), and species of *Proterhinus* occupy a number of different niches, breeding in fronds and stems of ferns, under bark, or in dead leaves, and in one case mining leaves.

Selected Bibliography

Anderson, W. H. 1941b (larvae of *Proterhinus* spp.).
Böving and Craighead 1931 (larva of *Proterhinus anthracias* Perkins).
Crowson 1955.
Hudson 1934 (habits of *Platycephala*).
Paulian 1944 (review of family).
Zimmerman 1948 (habits and distribution of *Proterhinus*).

ITHYCERIDAE (CURCULIONOIDEA)

John F. Lawrence,
Division of Entomology, CSIRO

New York Weevil

Figures 34.848a–d

Relationships and Diagnosis: This family includes the single species *Ithycerus noveboracensis* (Forster), the New York weevil, which appears to be most closely related to members of the Brentidae and the Apioninae (here included in Curculionidae). The presence of thoracic legs distinguishes *Ithycerus* larvae from those of most Curculionoidea, while the presence of labral rods and the absence of a mandibular mola and hypopharyngeal sclerome separates them from the leg-bearing larvae of most Anthribidae and Nemonychidae. They differ from larvae of Brentidae by having an endocarina which extends in front of the epicranial stem (between the frontal arms), 3 stemmata on each side of the head, and labral rods extending beyond the posterior edge of the clypeus, and by lacking patches of asperities on the thorax.

Biology and Ecology: Adult *Ithycerus* feed on the shoots, leaf petioles, leaf buds, and acorn buds of various dicotyledonous trees in the families Fagaceae (*Fagus, Castanea, Quercus*), Juglandaceae (*Juglans, Carya*), and Betulaceae (*Carpinus, Betula*). Oviposition occurs in the soil, and the larvae feed on the roots of the host plant. At least 7 instars have been observed in laboratory cultures. More information may be found in Sanborne (1981).

Description: Mature larvae 20 to 25 mm. Body relatively short and broad, subcylindrical, strongly curved ventrally (C-shaped), and lightly sclerotized, except for head, which is yellowish with heavily sclerotized mouth frame and mandibles. Surfaces smooth; vestiture consisting of moderately long, scattered setae.

Head: Protracted and hypognathous, about as long as wide. Epicranial stem moderately long; frontal arms slightly lyriform, forming narrow angle at base, complete to mandibular articulations. Median endocarina extending well in front of epicranial stem, between frontal arms. Stemmata 3 on each side, weakly developed. Antennae 1-segmented, consisting of a lightly sclerotized, dome-like segment bearing a sensorium,

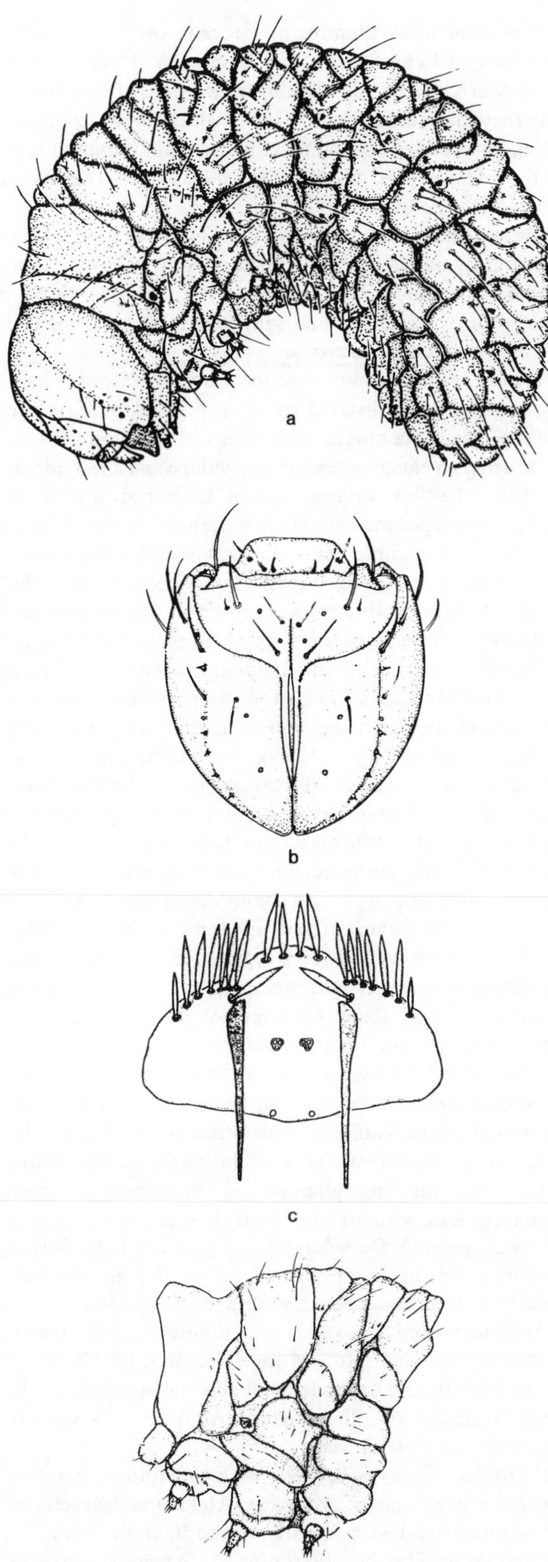

Figures 34.848a-d. Ithyceridae. *Ithycerus noveboracensis* (Forster). **a.** larva, 1st instar, lateral, length = 3 mm; **b.** head capsule, dorsal; **c.** epipharynx; **d.** larva, 6th instar, thorax, lateral. (From Sanborne, 1981).

a bifurcate appendage, and several setae. Frontocylpeal suture distinct; clypeus with 2 sensilla and 3 pairs of setae. Labrum free, with 1 sensillum and 4 pairs of setae; labral rods long, extending beyond posterior edge of clypeus; epipharynx with 7 pairs of anterolateral setae, 1 pair of submarginal, lateral setae, 4 anteromedian setae, 2 groups of centrally located sensory pores, and 1 pair of pores near posterior edge. Mandibles slightly elongate, subtriangular, with 2 weakly developed apical teeth, and 3 blunt teeth on incisor edge; accessory ventral process and mola absent. Ventral mouthparts retracted. Maxilla with transverse, undivided cardo, elongate stipes, well-developed articulating area, 2-segmented palp with palpifer, and rounded, setose mala, without an uncus. Labium with mentum and submentum fused; postmentum partly free from stipites; prementum forming a subtriangular plate. Labial palps 2-segmented, widely separated; ligula weakly developed and slightly emarginate. Hypostomal rods moderately long and subparallel. Hypopharyngeal bracon present; hypopharyngeal sclerome absent. Gula absent.

Thorax and Abdomen: Protergum with moderately sclerotized, yellowish plate; meso- and metaterga each with 2 transverse plicae. Thoracic legs small, widely separated, 2-segmented or apparently 3-segmented; segment 1 short and broad, segment 2 slightly elongate, with a subapical ring of setae, beyond which is a rounded lobe appearing like a 3rd segment. Abdominal terga 1 to 4 each with 3 transverse plicae, 5 to 8 with 2 plicae. Segments A9 and A10 shortened, their terga not subdivided; anal opening pi-shaped, so that 1 dorsal and 3 ventral lobes are formed. Epipleural and pleural lobes well-developed.

Spiracles: Annular-biforous. Thoracic spiracle relatively large, apparently located between prothorax and mesothorax, its opening subtriangular and partly blocked by a sclerotized, 3-lobed valve; accessory chambers very small in later instars. Abdominal spiracles circular with relatively long accessory chambers in early instars, elongate-oval with small accessory chambers in later instars.

Comments: *I. noveboracensis* occurs in Canada and the United States, from southern Ontario and Quebec to Mississippi and Georgia and from the eastern coast to about the 95th meridian. There is some disagreement among specialists on the relationships of *Ithycerus*, the Apioninae (including *Cylas* and Eurhinini), and the Antliarrhininae to Brentidae and the true weevils. The topic is discussed in detail by Sanborne (1981) and in the other references cited below.

Selected Bibliography

Crowson 1955, 1960b.
Kissinger 1968.
Lawrence and Newton 1982.
Morimoto 1976.
Sanborne 1981.

BRENTIDAE (CURCULIONOIDEA)

D. M. Anderson,
Systematic Entomology Laboratory, USDA

The Timberworms (Primitive Weevils, Straight-Snouted Weevils)

Figures 34.849a-i

Relationships and Diagnosis: Larvae of Brentidae (sometimes spelled Brenthidae) resemble those of other families in the superfamily Curculionoidea (Anthribidae, Nemonychidae, Curculionidae (broad sense), Scolytidae, and Platypodidae) in having a soft body without a terminal spine or urogomphi on the abdomen, legs rudimentary or absent, a hypopharyngeal bracon present, gular sutures absent, and antennae reduced to a single segment. However, the few described brentid larvae can be distinguished from those of the other curculionoid families by the following combination of characters: body elongate, subcylindrical, slightly expanded at posterior end (fig. 34.849a); legs usually present; stemmata absent; labral rods rather than tormae (extensions of posterolateral corners of labrum) present (fig. 34.849f); mandible without a mola; labium not fused to stipes of maxilla; 3 or 4 dorsal folds present on abdominal segments 1–8 (fig. 34.849a). This combination of larval characters and the adult characters (as in Arnett, 1968) are distinctive, but do little to indicate the closest relatives of the Brentidae. For example, brentid larvae resemble those of the Anthribidae and Nemonychidae in usually having legs present, the labium not fused with the maxillary stipes, and the cardo divided, but they also resemble larvae of most Curculionidae, Scolytidae, and Platypodidae in having labral rods present, no mola on the mandible, and 3 or more dorsal folds on typical abdominal segments.

The preceding discussion and the description presented later are based on publications by Heller (1904), Böving and Craighead (1931), Gardner (1935c), van Emden (1938b), W. H. Anderson (1947 a&b), unpublished notes by W. H. Anderson (with his permission), and from original observations. Except as noted, the larval terminology follows Anderson (1947b).

Biology and Ecology: The life histories and habits of Brentidae in general are poorly known. The larvae of the few N. American species for which habits have been recorded are wood borers that make their tunnels deep into the heartwood of dying or recently felled hardwood trees, according to LeConte and Horn (1876), Blatchley and Leng (1916), Peterson (1951), Buchanan (1960), Arnett (1968), and labels with larval specimens in the National Museum of Natural History. Arnett (1968) stated that larvae feed on wood and on fungus mycelia. Eggs are laid singly, in deep holes that females chew into the wood, according to Blatchley and Leng (1916), Buchanan (1960), and Arnett (1968). Gardner (1935c) stated that females of some Indian Brentidae oviposit in the galleries of Scolytidae and Platypodidae. Brentid larvae pupate in their tunnels, but there are apparently no published descriptions of the pupal chambers. Some species are myrmecophilous, according to Arnett (1968) and Gardner (1935c), but their larvae are unknown. Descriptions of the habits of some adult brentids by LeConte and Horn (1876), Blatchley and Leng (1916), and Buchanan (1960) indicate that they are usually found on or under bark of tree trunks, where they maintain a relationship in which males attend ovipositing females, fighting off other males that attempt to mate with those females and assisting the females to free their rostra (beaks) from the oviposition holes.

Description: Mature larvae (fig. 34.848a) moderate to large, slightly under 10 mm to about 30 mm in length. Body elongate, subcylindrical, slightly curved; thickest through thorax, slightly expanded and subtruncate at posterior end; small legs present; setae sparse.

Head: Hypognathous, exserted; setae of epicranium and frons usually long, distributed in a regular pattern (fig. 34.849b); stemmata absent. Antennae of 1 membranous segment bearing an elongate sensory appendage and a few minute setae (fig. 34.849e). Frontal sutures U-shaped, faintly defined in some species, extending anteriorly to articulating membrane of mandible (fig. 34.849b). Endocarina absent. Clypeus with 3 pairs of basal setae but without sensilla (fig. 34.849h). Labrum with 3 pairs of setae and 1 basal sensillum (fig. 34.849h). Labral rods (= epipharyngeal rods) present (fig. 34.849f); tormae absent. Epipharynx with 1 or 2 pairs of anterolateral setae, a row of 4–6 anteromedian setae, 1 or more pairs of median setae, and various sensory pores (fig. 34.849f). Mandible (fig. 34.849g) without a mola. Hypopharyngeal bracon present. Hypopharyngeal sclerome absent or reduced to a thin plate. Maxillary palp (fig. 34.849c) 2-segmented, with 1 ventral seta on basal segment; palpifer absent; stipes with few setae, not fused with lateral margins of labium; mala bearing a long comb-like row of dorsal setae and a few ventral setae; cardo divided into 2 pieces. Labium (fig. 34.849i) with palps 1- or 2-segmented; prementum distinct, defined posteriorly by a premental sclerite, and bearing 1 or more pairs of setae; mentum not distinct from postmentum, which bears 3 pairs of setae.

Thorax and Abdomen: (fig. 34.849a) Prothorax transverse; pronotal plate distinct. Mesothorax and metathorax divided into pro- and postdorsal folds. Prodorsal fold lenticular on mesothorax; subtriangular on metathorax, and extending forward into an emargination of mesothorax. Short, 2-segmented legs, without a terminal claw, present (except in some Asian genera). Dorsolateral angles of postnotal fold on mesothorax bearing a conspicuous patch of pigmented asperities (except in some Asian genera) (fig. 34.849a).

Abdomen elongate, smaller in diameter than thorax; segments 1–8 divided into 3–4 folds dorsally; terminal segment usually slightly expanded and subtruncate posteriorly. Sternellum absent on all segments. Anal opening ventral, subterminal, variable in shape.

Spiracles: Thoracic spiracle with or without air tubes; situated at rear margin of prothorax. Abdominal spiracles on first 8 segments and with or without marginal air tubes.

Comments: The brentid fauna of America north of Mexico consists of 6 species representing 5 genera, although there are over 1,200 species worldwide according to Arnett (1968). Of the few species north of Mexico, the larva of only 1, *Arrhenodes minutus* (Drury) has been described (Böving and Craighead, 1931; Peterson, 1951). A description of the

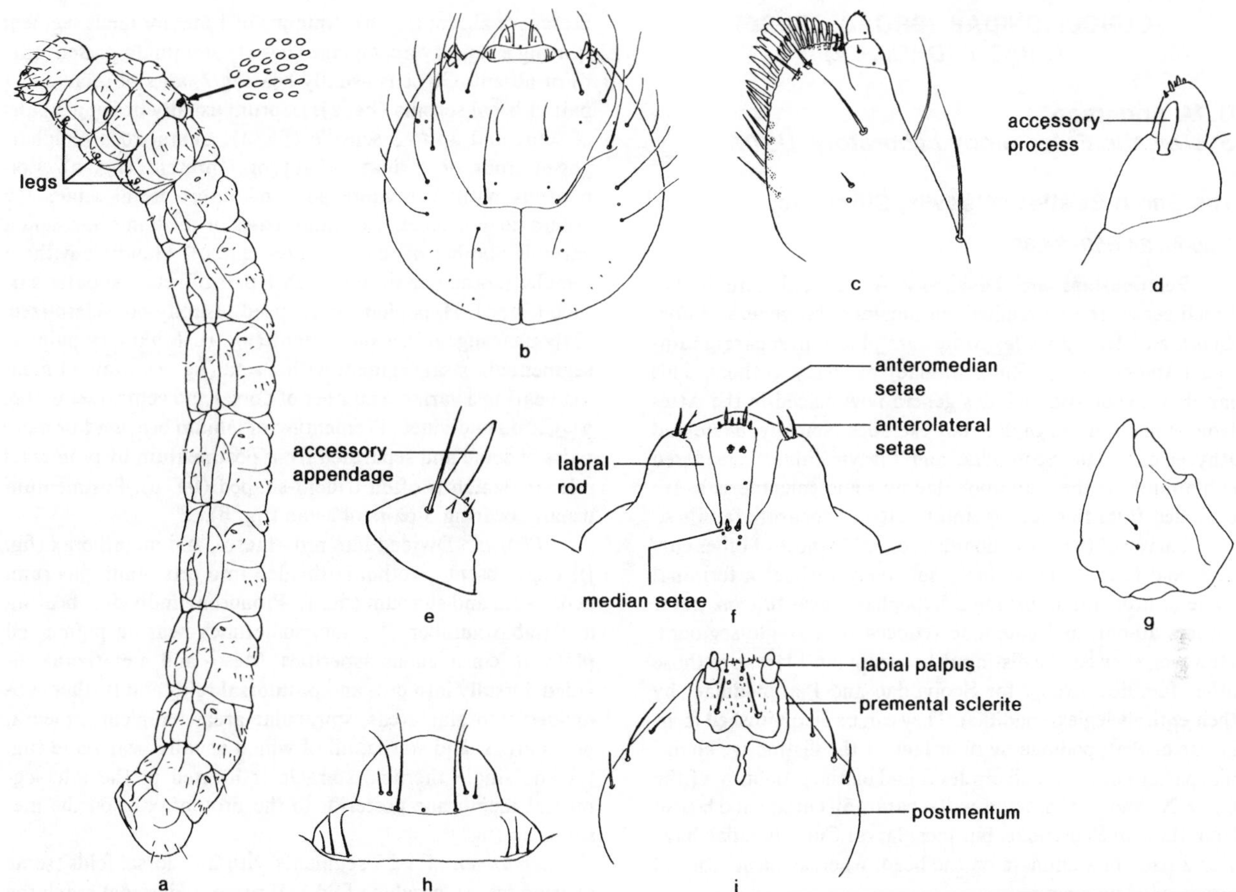

Figures 34.849a-i. Brentidae. *Arrhenodes minutus* (Drury), the oak timberworm. **a.** larva, with asperities of mesothorax; **b.** head, front; **c.** malar area of left maxilla, ventral; **d.** maxillary palpus, dorsal; **e.** antenna, dorsal; **f.** epipharynx; **g.** left mandible, dorsal; **h.** clypeus and labrum, dorsal; **i.** labium, ventral. The larva was illustrated by Böving and Craighead (1931) and briefly described and illustrated by Peterson (1951). A detailed manuscript description by W. H. Anderson exists in the files of the Systematic Entomology Laboratory, U.S.D.A. Most of the diagnostic description and illustrations presented here were taken from that work, with the permission of its author. *A. minutus* is the only brentid found in forest trees in eastern N. America north of Florida. Its hosts are dying or recently felled hardwoods, primarily oak, beech, and poplar, according to Craighead (1950). Buchanan (1960) reported that females he observed preferred to oviposit in recently wounded areas on trunks of oaks. Larvae hatching from the eggs made their tunnels directly into the heartwood, where they remained for 2-4 years before emerging as adults. The damage caused by the larval boring results in pinholes in lumber cut from the infested trees. This makes the wood unfit for special uses such as cooperage, and thus causes lower value. Figures a, b, c, d, e, f, h are original, by W. H. Anderson; g, i and asperities are original, by the author.

larvae of a species of ***Brentus,*** a genus also present in the United States, was published by Heller (1904). Larvae of Brentidae are scarce in collections. At this writing, the only N. American brentid larvae available were those of *Arrhenodes minutus,* the oak timberworm, and of an unidentified species of *Brentus* from southern California.

Selected Bibliography

Anderson, W. H. 1947a, 1947b.
Arnett 1968.
Blatchley and Leng 1916.
Böving and Craighead 1931.
Buchanan 1960.
Craighead 1950.
Emden 1938b.
Gardner 1935c.
Heller 1904.
LeConte and Horn 1876.
Peterson 1951.
Sanborne 1981.

CURCULIONIDAE (BROAD SENSE) (CURCULIONOIDEA)

D. M. Anderson,
Systematic Entomology Laboratory, USDA

The Snout Beetles (Weevils, Curculios)

Figures 34.850–34.884

Relationships and Diagnosis: As treated here in the broad sense, the Curculionidae includes the genera *Apion, Cylas,* and *Merhynchites,* which are placed in separate families (Apionidae and Rhynchitidae) by many authors. This family concept also includes genera now placed in the Attelabidae and Allocorynidae, but excludes Nemonychidae and Ithyceridae. The Scolytidae and Platypodidae, considered subfamilies of the Curculionidae by some coleopterists, are excluded from this section and treated as separate families.

Larvae of the Curculionidae resemble those of other curculionoid families in having a soft body without a terminal spine or urogomphi, having a hypopharyngeal bracon, gular sutures absent, and antennae reduced to a single segment. However, they can be distinguished from most larvae in those other families, except for Scolytidae and Platypodidae, by their entirely legless condition. They can be distinguished from larvae of Platypodidae by their lack of the distinctive chain-like pattern of loops and circles found on the pronotum of the latter. No known characters will separate all curculionid larvae from those of Scolytidae, but most larval Curculionidae have 1 or 2 pairs of stemmata on the head, whereas most scolytid larvae have no stemmata.

The above information and the family description have been drawn from Emden (1938b, 1952), W. H. Anderson (1947b), unpublished notes of W. H. Anderson, and research experience of the author. Except as noted, the terminology follows that of Anderson (1947b) and is illustrated in figs. 34.850a–k.

Biology and Ecology: The life history patterns and ecological adaptations within this family show great variation, although almost all species are phytophagous. Larvae of most species live within their host plants, but those of some genera live externally, others live in the soil, and still others are aquatic. Almost every conceivable type of plant material is utilized as food and/or living space by some curculionid species.

Larvae that live internally usually develop in one part of the plant, i.e., in the roots, stems, fruits, seeds, leaves, galls, etc. Many species are of economic importance as pests of crops or as biocontrol agents against weeds. Examples of these various biological categories can be found among the species treated below.

Description: (figures 34.850a–k) Mature larvae vary in size, from less than 2.0 mm to over 35.0 mm in length. Body often C-shaped but varies from elongate, nearly straight to stout and convex. Legs absent. Setae usually sparse, variable among taxa in length, shape, and arrangement.

Head: Hypognathous or prognathous; completely exserted to deeply retracted within prothorax. Epicranial suture and frontal sutures usually present (fig. h), sometimes obscure or absent, variable in shape. Median endocarina present or absent (fig. h). Antenna of 1 membranous segment bearing a sensory appendage (fig. f). Stemmata present (fig. h) or absent. Clypeus usually bearing 2 pairs of setae and 1 pair of basal sensilla (fig. a). Labrum usually bearing 3 pairs of setae and up to 3 sensilla (fig. a). Labral rods (epipharyngeal rods or tormae of authors) present (fig. c). Epipharynx with 2 or more pairs of anterolateral setae, 2–6 anteromedian setae, 1 or more pairs of median setae, and a variable number of sensory pores (fig. c). Mandible without a mola, toothed or simple, with 1 pair of setae on outer surface (fig. d). Hypopharynx reduced, usually not sclerotized. Hypopharyngeal bracon present (fig. i). Maxillary palpi 2-segmented, basal segment with 1 seta (fig. g). Mala of maxilla bearing a variable number of dorsal and ventral setae (fig. g). Cardo undivided. Prementum of labium bearing 1 or more pairs of setae and separated from postmentum by premental sclerite, which is often trident-shaped (fig. b). Postmentum usually bearing 3 pairs of setae (fig. b).

Thorax: Divided into pro-, meso-, and metathorax (fig. j). Legs absent. Prothorax divided into pronotum, pleurum, pedal area and sternum (fig. j). Pronotum undivided, bearing a variable number of setae, sometimes bearing pigmented plates or conspicuous asperities. Meso- and metathorax divided dorsally into pro- and postdorsal folds and further subdivided into alar areas, spiracular areas, epipleura, pleura, pedal areas, and sterna, all of which usually bear setae (fig. j). The single thoracic spiracle is located in the intersegmental membrane posterior to the prothorax or on the mesothorax (fig. j).

Abdomen: First 7 segments with 2–5 dorsal folds (some bearing setae), of which fold 1, if present, does not reach the midline (figs. j, k). Segment 8 usually with 2 or 3 dorsal folds, and segment 9 with only 1 dorsal fold (fig. k). Segment 10 reduced to anal lobes (fig. l). First 8 segments bearing 1 pair of spiracles, and subdivided laterally into epipleura, pleura, and pedal areas, and ventrally into eusterna and sternella, all of which bear setae (figs. j, k). Ninth segment without spiracles, pleura, pedal areas, or sternella (fig. k). Anal opening variable in shape, surrounded by a variable number of anal lobes.

Spiracles: Annular-biforous (fig. e), annular-uniforous, or without marginal air tubes.

Comments: The immature stages of only a relatively small percentage of the genera and species of this huge family have been studied. According to Burke and Anderson (1976), larvae and/or pupae of only 256 species (representing 123 genera) out of the approximately 2,500 species (representing 390 genera) of the Curculionidae (broad sense) of America north of Mexico have been described well enough to be identifiable. Although this leaves an enormous amount of descriptive work to be done before it will be possible to identify the larvae of all of the Curculionidae in our area to genus, some representatives of all of the major subfamilies (e.g. Anthonominae, Cryptorhynchinae, Rhynchophorinae) and of most smaller subfamilies (e.g. Rhynchaeninae, Magdalinae) have now been described, and among the larvae that have been described are those of many economically important species.

The diagnostic descriptions and illustrations which follow are presented as examples of larvae of the major curculionid subfamilies in N. America and as aids in the identification of some of the economically important species in that area. The successful use of these descriptions will require careful reference to figures 34.850a–k, in which the anatomical terms are illustrated.

Most of these larvae must be identified on the basis of a combination of characters, plus host plant information, rather than by a few distinctive characters. Therefore, it is important to examine all of the features mentioned in the descriptions and to take note of any information available on the habits.

Selected Bibliography

Ahmad and Burke 1972.
Ainslie 1920.
Anderson and Anderson 1973.
Anderson, W. H. 1941a, 1941b, 1947b, 1948a, 1948b, 1952.
Arant 1938.
Balachowsky and Hoffmann 1962.
Balduf 1959.
Barrett 1930.
Boyce 1927.
Böving and Craighead 1931.
Burke 1968.
Burke and Anderson 1976.
Cotton 1921.
Craighead 1950.
Davies 1928.
Emden 1938b, 1950, 1952.
Gardner 1934a, 1938b.
Gibson 1969, 1985.
Hamilton 1980.
Hamilton and Kuritsky 1981.
Herron 1953.
High 1939.
Hoffmann *et al.* 1962.
Houser 1923.
Isely and Schwardt 1934.
Johnson 1944.
Keen 1952.
Keifer 1933.
Kirk 1965.
Kissinger 1964.
Kuschel 1959.
Lalone and Clarke 1981.
Lekander 1967.
LeConte and Horn 1876.
Maier 1980.
Mamaev and Krivosheina 1976.
Manglitz *et al.* 1963.
Martel *et al.* 1976.
Mathur 1954.
May 1966, 1967, 1977.
Metcalf *et al.* 1962.
Morimoto 1962a, 1962b, 1976.
Muniz and Barrera 1969.
O'Brien and Wibmer 1982.
Pepper 1942.
Peterson 1951.
Roberts 1936.
Sanborne 1981.
Scherf 1964, 1978.
Thomas 1968.
Tuttle 1954.
Warner 1966.

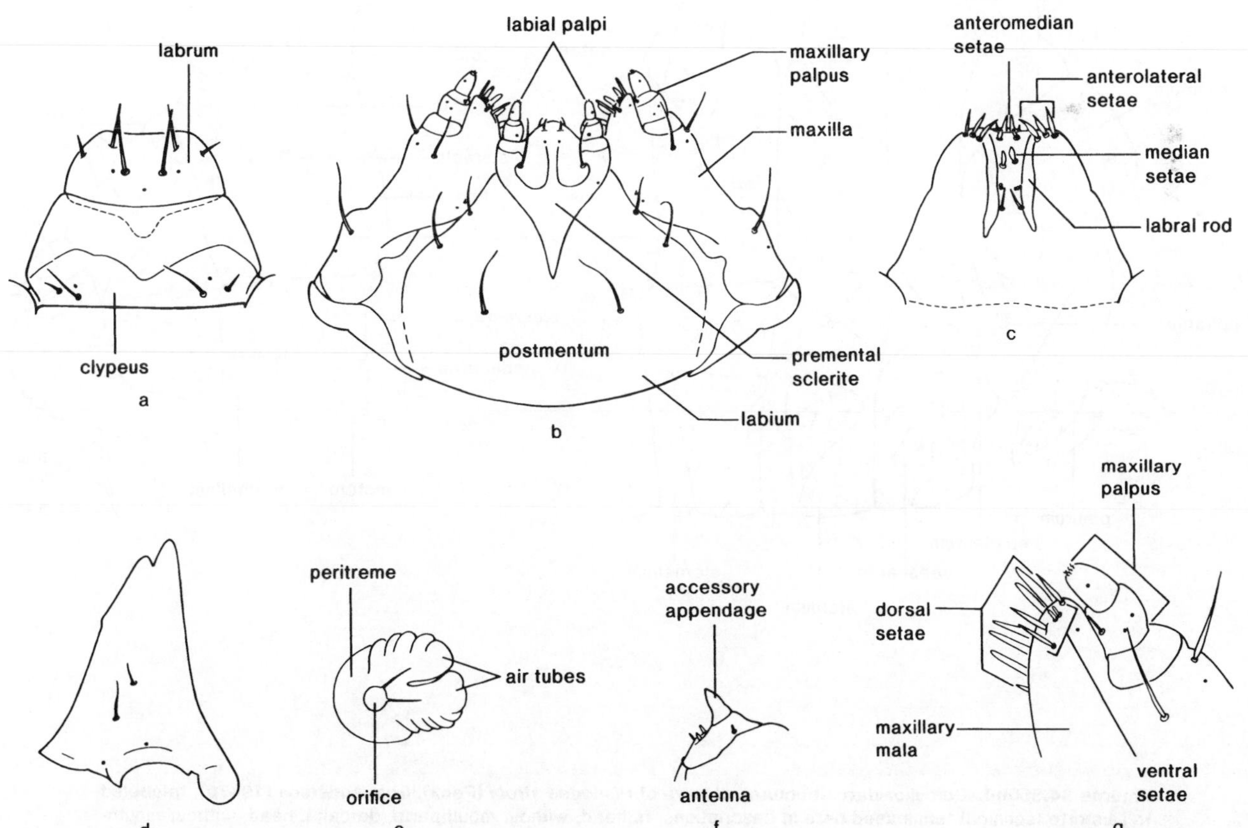

Figures 34.850a-g. Curculionidae. Structures of larva of *Pissodes strobi* (Peck), from Anderson (1947b), relabeled to illustrate technical terms used here in descriptions. **a.** clypeus and labrum; **b.** labium and maxillae, ventral; **c.** epipharynx; **d.** right mandible, outer surface, lateral; **e.** abdominal spiracle; **f.** antenna; **g.** malar area of left maxilla;

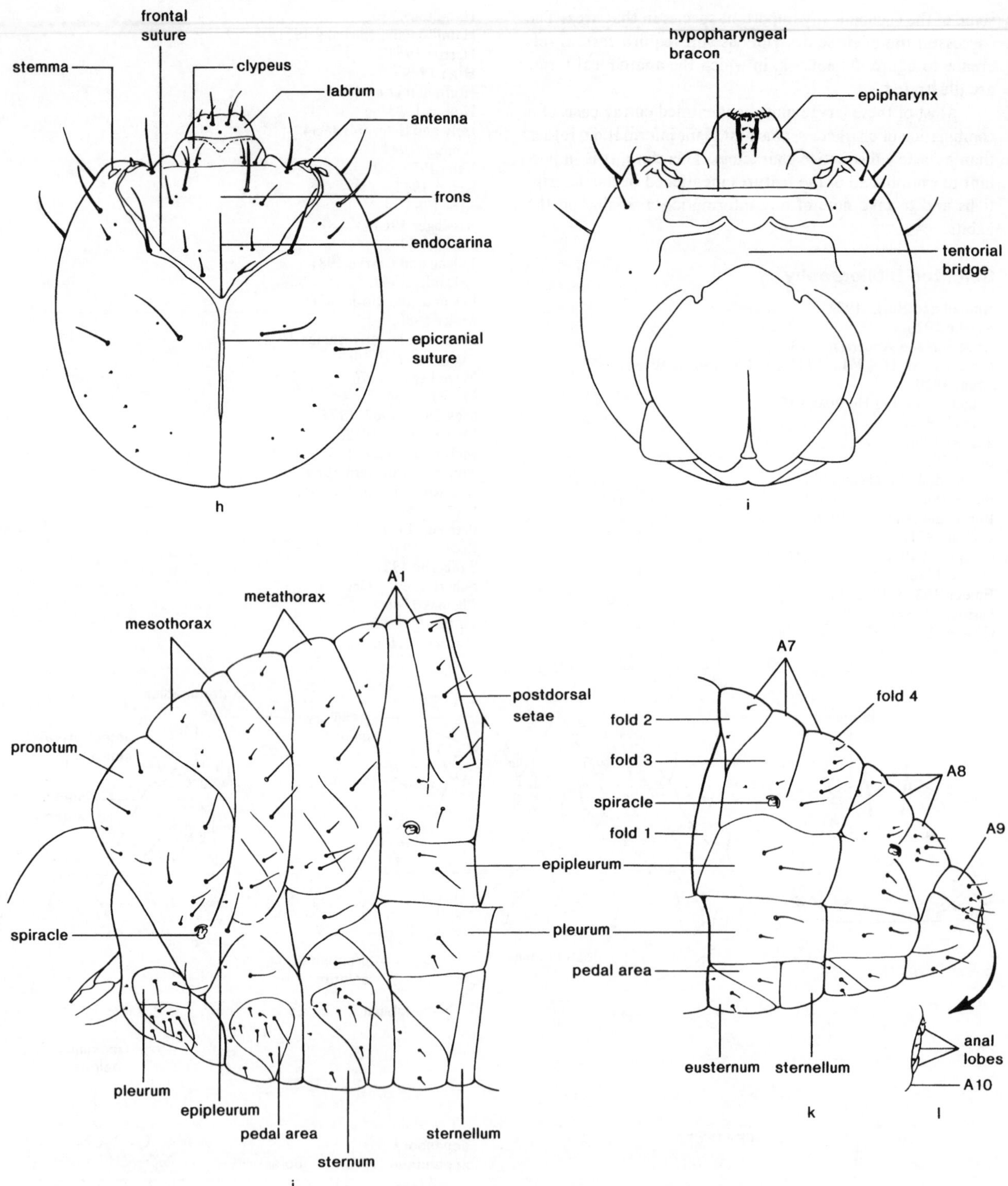

Figures 34.850h-l. Curculionidae. Structures of larva of *Pissodes strobi* (Peck), from Anderson (1947b), relabeled to illustrate technical terms used here in descriptions. **h.** head, without mouthparts, dorsal; **i.** head, without mouthparts, ventral. **j.** thorax and 1st abdominal segment, lateral; **k.** abdomen, segments 7-10, lateral, and **l.** segment 10, enlarged, lateral.

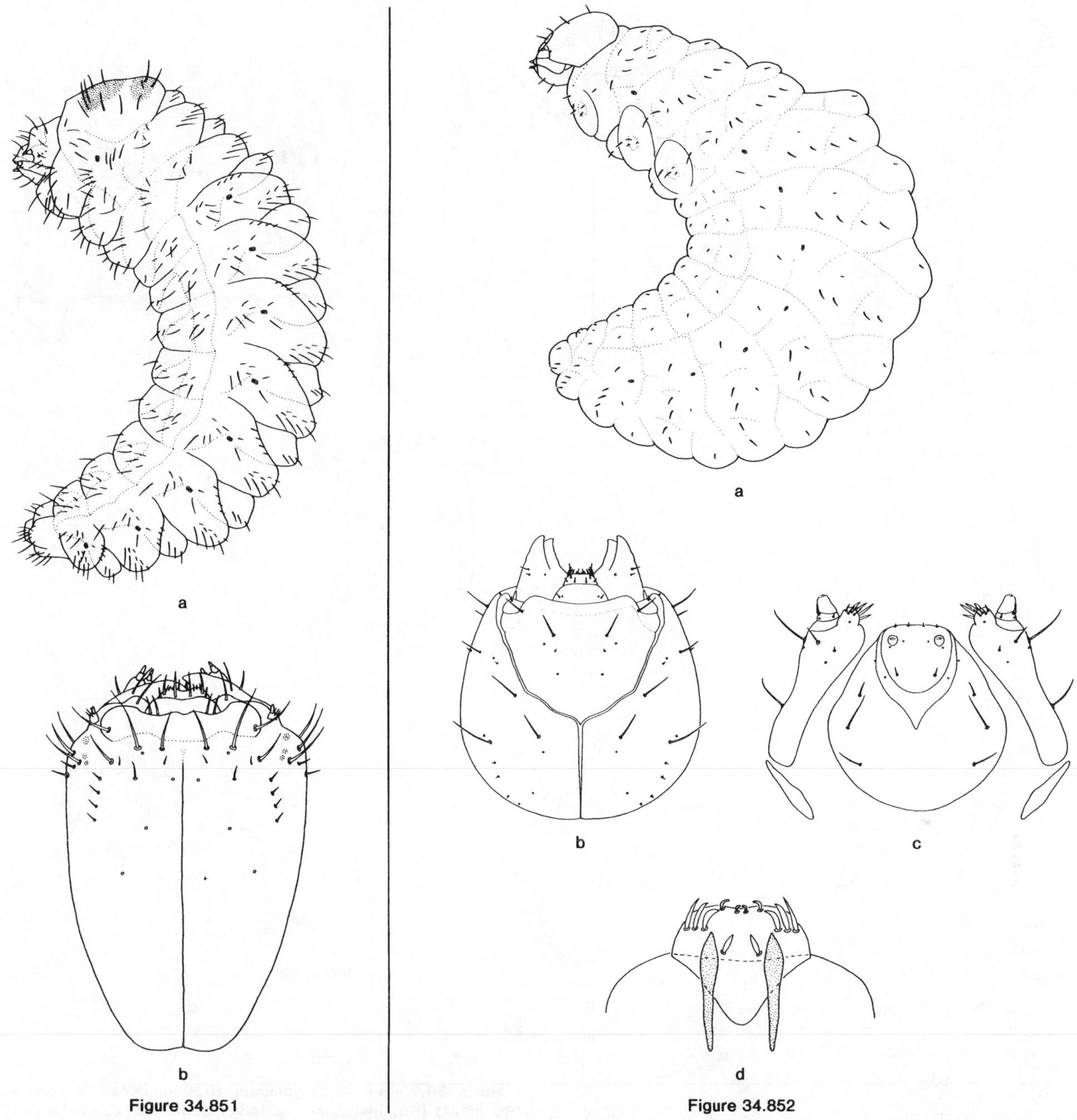

Figure 34.851

Figure 34.852

Figures 34.851a-c. Curculionidae. ***Merhynchites bicolor*** **(Fabricius) (Rhynchitinae), rose curculio. a. larva; b. head. Larvae develop only in the fruit of *Rosa* species. Detailed descriptions of larva and pupa, with a summary of the literature, published by Hamilton and Kuritsky (1981). Balduf (1959) described life cycle, habits, and immature stages of the eastern U.S. population (as *Rhynchites bicolor*). Mature larvae approximately 6 mm long. Body moderately stout, distinctly C-shaped, bearing numerous prominent setae, and tapered at tip of abdomen. Spiracles all bicameral. Head longer than wide, deeply retracted into prothorax and unpigmented except for brownish frontal area and dark brown mouthframe and mandibles. Frontal sutures indistinct. One pair of dark stemmata and 2 pairs of subcutaneous pigment spots. Mandibles bifid at tips. Setae of head as illustrated. Prothorax bearing lightly pigmented pronotal shield. Abdominal segments 1-8 with 2 prominent dorsal folds.**

Figures 34.852a-d. Curculionidae. ***Apion griseum*** **Smith (Apioninae). a. larva; b. head; c. labium and maxilla, ventral; d. epipharynx. The larval hosts are green pods of certain legumes, including stringbeans and lima beans, which are attacked in the United States and Mexico. Tuttle (1954) published notes on the habits, and the larvae are distinguished from those of some other species of *Apion* in a key by van Emden (1938). Mature larvae approximately 2.5 mm long. Body pale white, stout, strongly curved, tapered at both ends. Spiracles small, all unicameral. Setae as illustrated. Head unretracted; unpigmented except for dark mouthframe, mandibles, and single pair of stemmata. Frontal sutures distinct, reaching basal membrane of mandibles. Labial palpi of 1 short, stout segment. Labium and maxillae as illustrated (note crescentic shape of premental sclerite). Epipharynx with labral rods straight, slightly club-shaped, and pigmented; setae as figured. Mandibles bifid at tips. Abdominal segments 1-8 with 2 dorsal folds.**

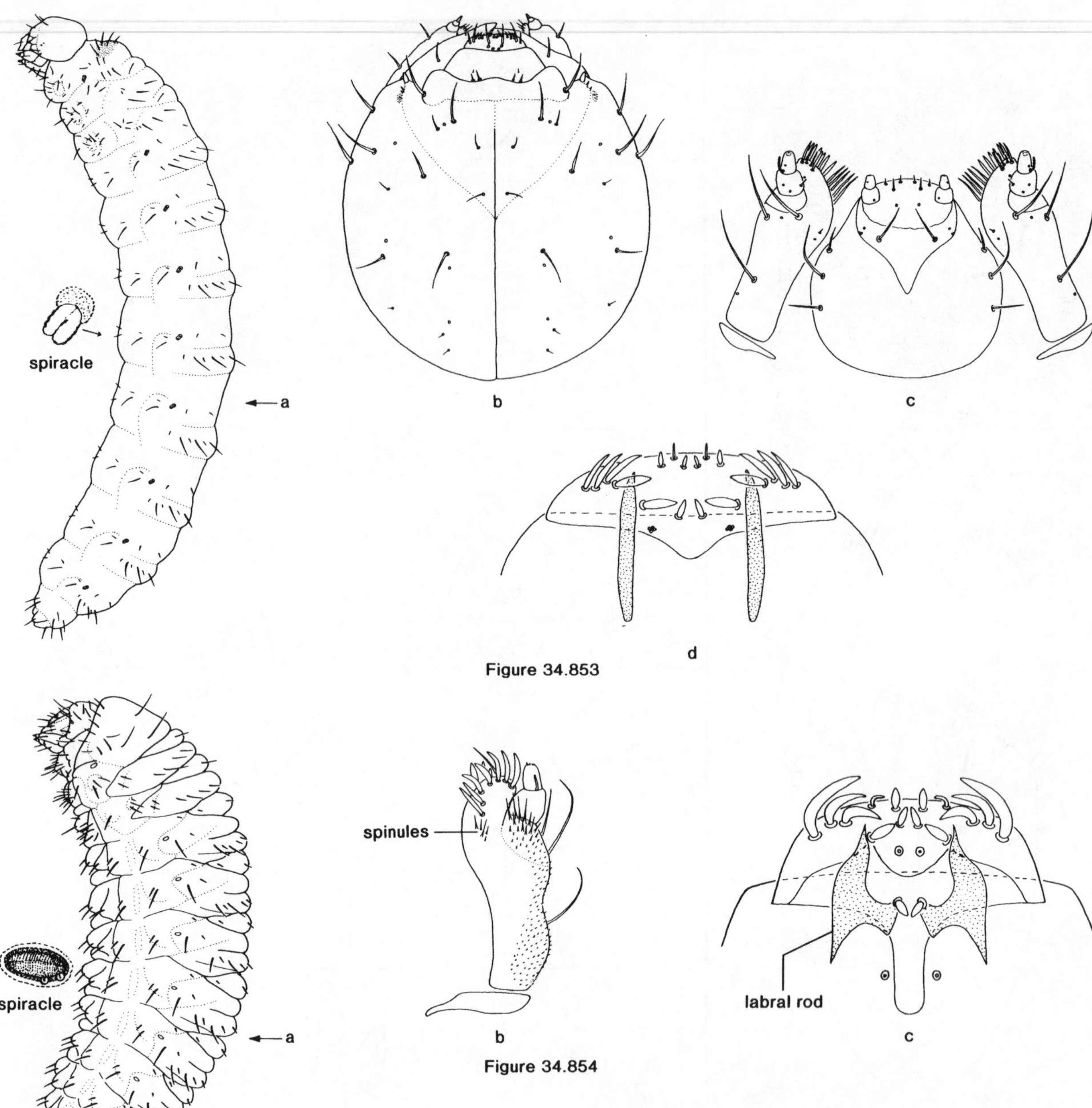

Figures 34.853a-d. Curculionidae. *Cylas formicarius elegantulus* (Summers) (Apioninae), sweetpotato weevil. **a.** larva; **b.** head; **c.** labium and maxilla, ventral; **d.** epipharynx. Larvae and adults feed only on tubers of sweetpotato and some other species of *Ipomoea*. They will attack sweetpotatoes in the field and in storage. The life history and habits were summarized by Metcalf, et al. (1962). Larva briefly described and figured by Peterson (1951) and by Pierce (1918). Mature larva approximately 8 mm long. Body white, elongate, slightly curved, tapered at posterior. Head unretracted, pale brown with darker mandibles; setae as figured. Frontal sutures distinct, reaching basal membrane of mandible. One pair of stemmata, each containing 2 contiguous pigment spots. Epipharynx, labium, and maxillae as figured. Mandibles bifid at tips. Abdominal segments 1-7 with 2 dorsal folds. Body setae as figured.

Figures 34.854a-c. Curculionidae. *Graphognathus leucoloma* (Boheman) (Brachyderinae). **a.** larva; **b.** maxilla; **c.** epipharynx. The 4 N. American species of this genus are known as whitefringed beetles. Their larvae live in the soil, feeding on roots of a wide variety of plants, including such crops as peanuts, corn, and soybeans. All are introduced and their range in N. America is restricted to the southeastern United States. A key to separate larvae of whitefringed beetles from similar soil-inhabiting weevil larvae in the southeastern states was published by Anderson and Anderson (1973). Mature larvae approximately 12-14 mm long. Body stout, slightly curved, yellowish white, with some prominent setae on all segments as figured. Head white, except dark brown mandibles and mouthframe, elongate, partly retracted. Antenna with accessory appendage broader than long. Mandibles undivided at tips. Frontal sutures indistinct. Stemmata absent. Epipharynx and maxilla as figured (note shape of labral rods and presence of a group of spinules at posterior end of dorsal row of setae on mala of maxilla). Segments 1-7 of abdomen with 3 dorsal folds. Spiracles large, elliptical, with 2 small unannulated air tubes on posterior margin.

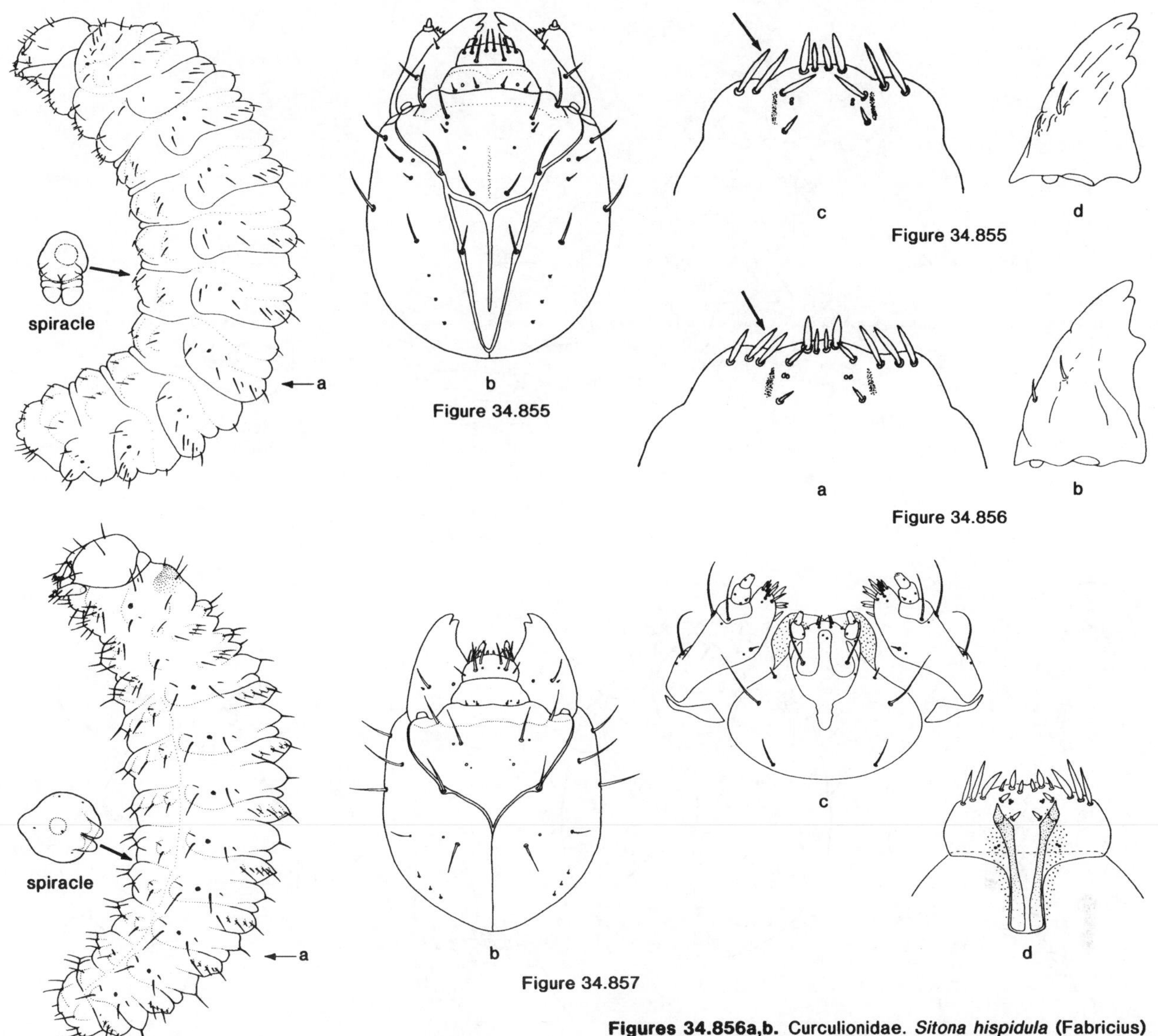

Figures 34.855a-d. Curculionidae. *Sitona cylindricollis* Fahraeus (Brachyderinae), sweetclover weevil. **a.** larva; **b.** head; **c.** epipharynx; **d.** mandible. Primarily associated with sweetclover, but also known to attack alfalfa and various clover species. Larvae feed externally on the roots, including nitrogen-fixing nodules, of their hosts. Herron (1953) described life history in the U.S. Larval feeding habits and taxonomic characters described by Manglitz, et al. (1963). Mature larva approximately 6 mm long, distinctly C-shaped, with a few prominent setae. Head unretracted, pale yellowish except for brown mandibles. Stemmata absent. Accessory appendage of antenna broader than long. Epipharynx and mandible as figured. Abdominal segments 1-7 with 3 dorsal folds. Differs from larva of *Sitona hispidula,* the other species likely to be found feeding on sweetclover roots, in having 3 evenly spaced teeth on the mandible (figure 34.855d) and 2 pairs of anterolateral setae on the epipharynx (figure 34.855c) vs 3rd tooth on mandible widely separated from the 2nd (figure 34.856b) and 3 pairs of anterolateral setae on the epipharynx (figure 34.856a) for *hispidula.* (Figures c-d from Manglitz, et al., 1963.)

Figures 34.856a,b. Curculionidae. *Sitona hispidula* (Fabricius) (Brachyderinae) clover root curculio. **a.** epipharynx; **b.** mandible. Known for its attacks on clover roots, but it also feeds on roots of some other forage crop legumes, including alfalfa, sweetclover, and trefoil. Larvae feed externally on the roots of their hosts. Habits in the U.S. reported by Herron (1953) and by Manglitz, et al. (1963). Larvae resemble those of *Sitona cylindricollis,* but can be distinguished from them by the characters indicated above for that species. (From Manglitz et al., 1963)

Figures 34.857a-d. Curculionidae. *Otiorhynchus sulcatus* (Fabricius) (Otiorhynchinae), black vine weevil. **a.** larva; **b.** head; **c.** labium and maxillae, ventral; **d.** epipharynx. Larvae live in soil and attack the roots or crowns of a wide variety of plants, including strawberry and yew. Well known as a pest in greenhouses, where it feeds on cyclamen, geranium, gloxina, primula, and other ornamentals. Keys separating the larvae from those of some other *Otiorhynchus* species published by van Emden (1952) and by Keifer (1933). Larval development reported by LaLone and Clarke (1981). Mature larva approximately 10 mm long. Body white, crescent-shaped, with some prominent setae on each segment as figured. Head yellowish brown, mandibles darker brown; setae and other features as figured. Endocarina absent. Mandibles bifid at tip. Accessory appendage of antenna broader than long, flattened. Stemmata absent. Epipharynx, labium, and maxillae as figured. Abdominal segments 1-7 with 3 dorsal folds. Spiracles bicameral, marginal air tubes not annulated.

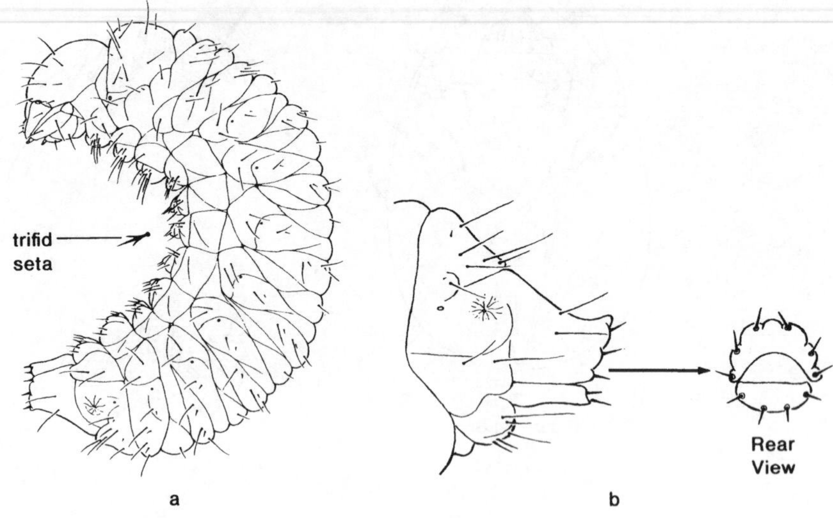

Figure 34.858

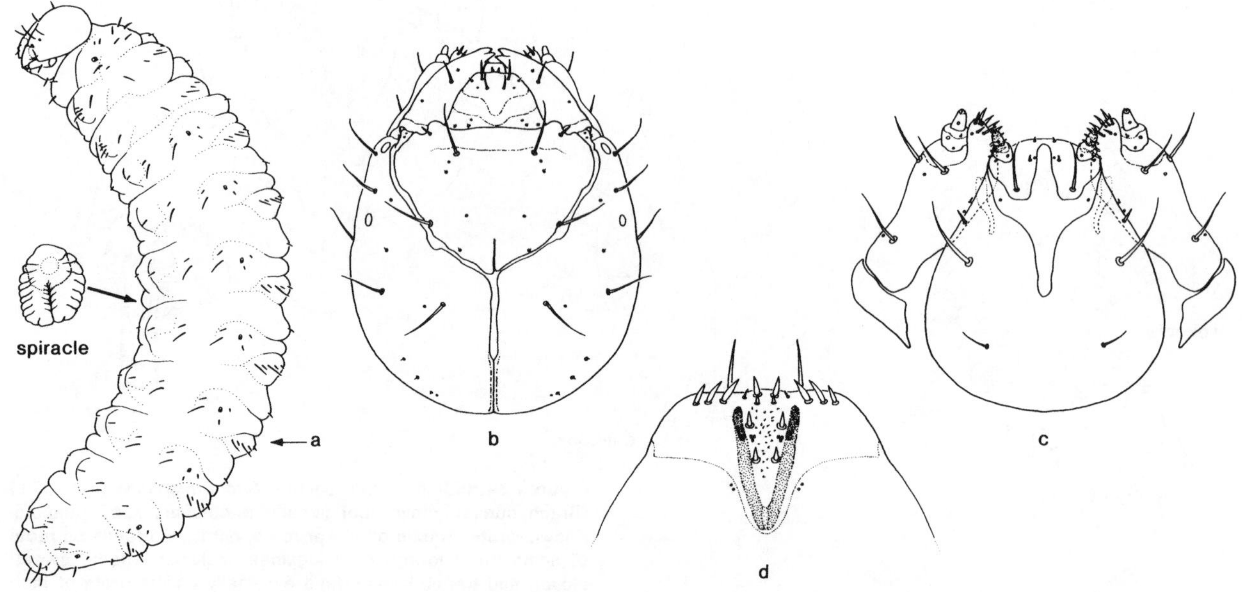

Figure 34.859

Figures 34.858a,b. Curculionidae. *Calomycterus setarius* Roelofs (Eremninae), imported longhorned weevil. **a.** larva; **b.** tip of abdomen. Larvae feed on small roots of many plants, including asters, clovers, alfalfa, and goldenrod. Adults do the principal damage, feeding on foliage of many species, including forage and vegetable crops, ornamentals, shrubs, trees, and weeds. This introduced species is widely established in the eastern and midwestern U.S. Johnson (1944) described life history in detail, and larva described by van Emden (1952). Full grown larva approximately 5 mm long, white or grayish white, moderately curved, bearing some prominent setae on all segments. Head evenly light yellowish except for reddish-brown mandibles, which are bifid at tips. Endocarina absent. Stemmata absent. Accessory process of antenna flattened, broader than long. Body setae as figured. Abdominal segments 1-7 with 3 dorsal folds. Trifid setae (figured) on ventral folds of most abdominal segments. Epipleurum of 8th abdominal segment marked with a flattened, wrinkled disc-shaped area, and terminal segment distinctively subconical and truncate, as figured. (Figures by W. H. Anderson)

Figures 34.859a-d. Curculionidae. *Listronotus oregonensis* (LeConte) (Rhytirhininae), carrot weevil. **a.** larva; **b.** head; **c.** labium and maxillae; **d.** epipharynx. Larvae bore in the stems and roots of carrot, but also attack celery, parsnip, parsley, dill, wild carrot, plaintain, and dock. Pupation in the soil. Detailed accounts of life history of *L. oregonensis* (as *L. latiusculus* (Boheman)) published by Boyce (1927), Pepper (1942), and (as *L. oregonensis*) by Martel, *et al.* (1976). Larva described by Peterson (1951). Mature larva approximately 10 mm long, nearly white, thickest in mid-section. Head light yellowish-brown, with 2 pairs of stemmata and with setae as figured. Accessory appendage of antenna short, subconical. Mandibles bifid at tip. Epipharynx, maxillae, and labium as figured. Abdominal segments 1-7 with 3 dorsal folds; epipleural folds prominent on segments 8 and 9. Spiracles bicameral; air tubes directed posteriorly on abdomen. Body setae not prominent, distributed as figured. May be confused with larvae of *Listronotus texanus* (Stockton), which also attack carrots in southern Texas.

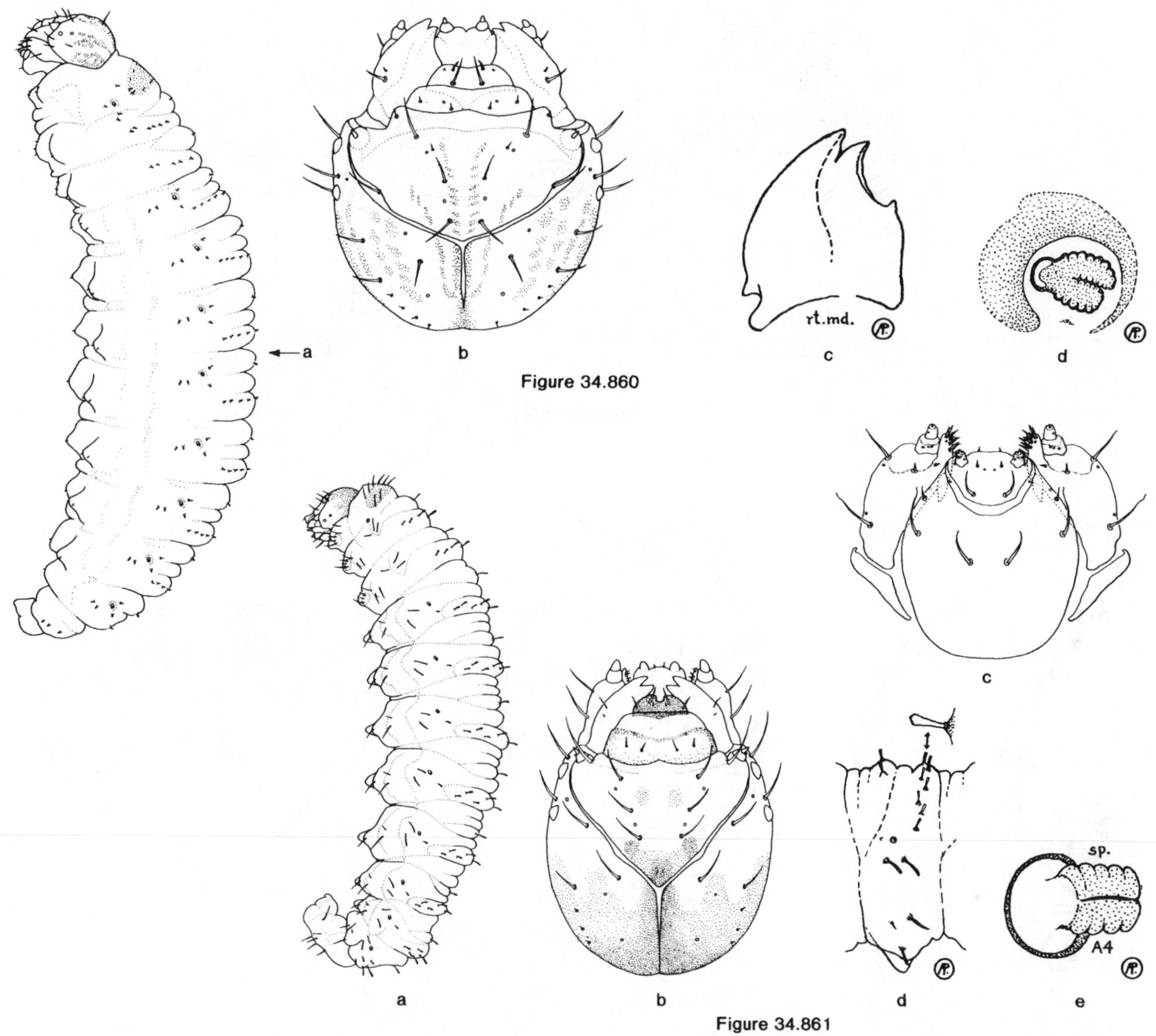

Figures 34.860a-d. Curculionidae. *Listroderes costirostris obliquus* (Klug) (Rhytirhininae), vegetable weevil. **a.** larva; **b.** head; **c.** mandible; **d.** A4 spiracle. This weevil has a very wide host range, but it frequently attacks carrots, turnips, spinach, tomatoes, and other garden crops. Larvae feed externally on all parts of plants, above or below soil surface. Adults feed mostly on foliage of the same plants, sometimes on roots and stems. Pupates in earthen cells near the host. Occurs in the southeastern U.S. and California. Detailed description of life history published by High (1939) and larva described by Barrett (1930), Peterson (1951), and May (1966 & 1977). Mature larva about 15 mm long, robust, slightly curved, and pale to dark green, without stripes, when alive. Green is lost in preserved specimens. Head pale to medium brown, mottled with darker brown in pattern figured, bearing 2 pairs of stemmata and setae as figured. Accessory appendage of antenna elongate, subconical. Endocarina absent. Mandibles bifid at tip. Premental sclerite divided. Pronotal shield darkly pigmented. Abdominal segments 1-7 with 3 dorsal folds; segments 2-6 with paired proleg-like ampullae; anal pygopod present. All spiracles partly enclosed by a dark, crescent-shaped sclerite, figured. Body setae short, flattened at apex, distributed as figured. (Figures c, d from Peterson, 1951)

Figures 34.861a-e. Curculionidae. *Hypera postica* (Gyllenhal) (Hyperinae), alfalfa weevil. **a.** larva; **b.** head; **c.** labium and maxillae, ventral; **d.** abdominal segment 4; **e.** spiracle A4. Best known for its destructiveness to alfalfa although it also infests some other forage crop legumes, including yellow steetclover and the true clovers. Larvae feed externally on foliage and upper stems of their hosts and spin cocoons there before pupating. The life history and economic importance of *H. postica* have been summarized by Metcalf, et al. (1962). The larva was separated from those of most other North American and European species in a key by Anderson (1948b) and described by Peterson (1951). Mature larva approximately 7 mm long, slightly curved, green with a white median dorsal stripe when alive. Green lost in preserved specimens. Body caterpillar-like, with most abdominal segments bearing paired ventral proleg-like ampullae, and with a terminal pygopod. Head unretracted, mottled dark brown to black posteriorly; light orange, except for black mandibles, anteriorly. Two pairs of large stemmata. Front margin of labrum with deep median excavation. Endocarina absent. Setae of head as figured. Mandibles bifid at tip. Premental sclerite of labium crescent-shaped. Body setae mostly short, flattened and expanded at apex, distributed as figured. Asperities prominent, rounded, giving body a pebbled appearance. Abdominal segments 1-7 with 3 dorsal folds; 1-6 with 2 auxilliary dorsal folds between segments. (Figures d, e from Peterson, 1951)

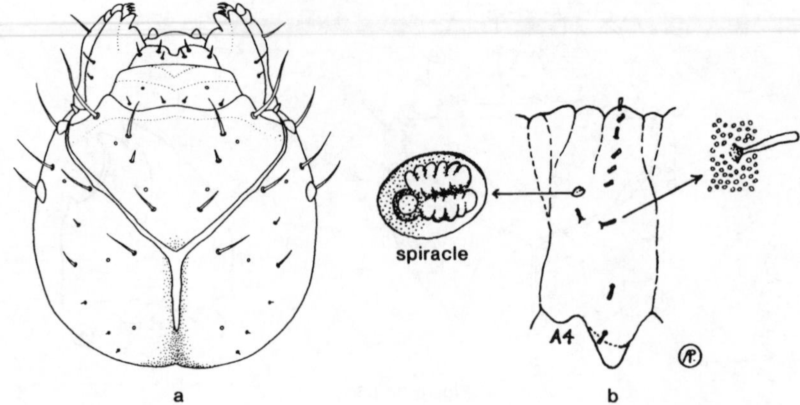

Figure 34.862

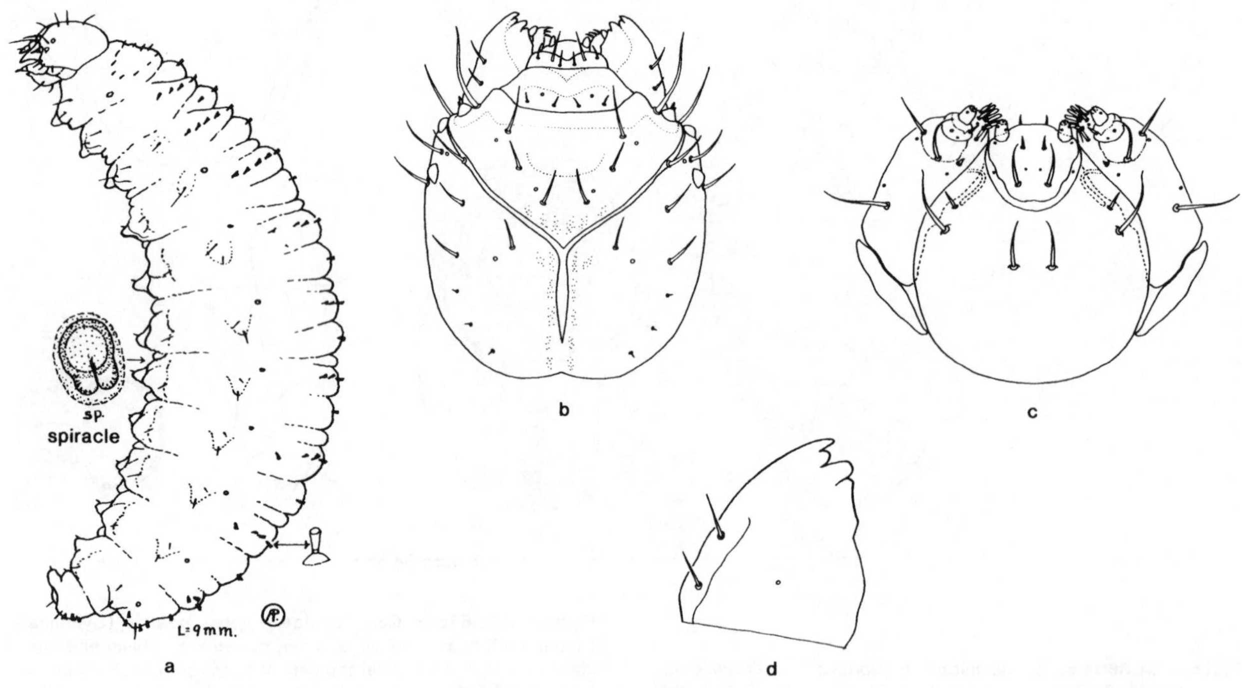

Figure 34.863

Figures 34.862a,b. Curculionidae. *Hypera nigrirostris* (Fabricius) (Hyperinae), lesser clover leaf weevil. **a.** head; **b.** abdominal segment 4. Larvae feed externally, primarily on the foliage of red clover, but will also feed on all common clover species, alfalfa, and sweetclover. Details of life history similar to those outlined above for *Hypera postica,* and summarized by Metcalf, et al. (1962). Characters separating it from other N. American *Hypera* species are in the key by Anderson (1948b). Mature larvae approximately 6.5 mm long, slightly curved, nearly white (rather than green, as in *H. postica*) when alive or preserved. Body covered with convex pale brown asperities. Head light orange-yellow, except for a narrow brownish band along posterior margin and epicranial suture. Other features, including prominent stemmata, length and shape of body setae, dorsal folds of abdomen (figured), and proleg-like ventral abdominal ampullae very similar to *H. postica.* (Figure b from Peterson, 1951)

Figures 34.863a-d. Curculionidae. *Hypera punctata* (Fabricius) (Hyperinae), clover leaf weevil. **a.** larva; **b.** head; **c.** labium and maxillae; **d.** mandible. Larvae feed on the foliage of various legumes, including true clovers, sweetclover, alfalfa, and some bean species. Life history similar to that described above for *Hypera postica* and summarized by Metcalf, et al. (1962). Larva is separated from those of other N. American and European *Hypera* species in the key by Anderson (1948b) and described by Peterson (1951). Mature larvae 9-14 mm long. Form, color, and other characters similar to those of *H. postica,* above, but mandibles have 4, rather than 2 teeth and mature larva is larger. Head with prominent median convexity on the frons; dark mottling on head confined to a broad dorsomedian stripe. No auxilliary dorsal folds between abdominal segments. Also resembles larva of the vegetable weevil described above, but the latter has a much larger mottled area on the head, has 2 rather than 4 teeth on the mandible, and has a dark crescent-shaped area partly enclosing each spiracle. (Figure a from Peterson 1951)

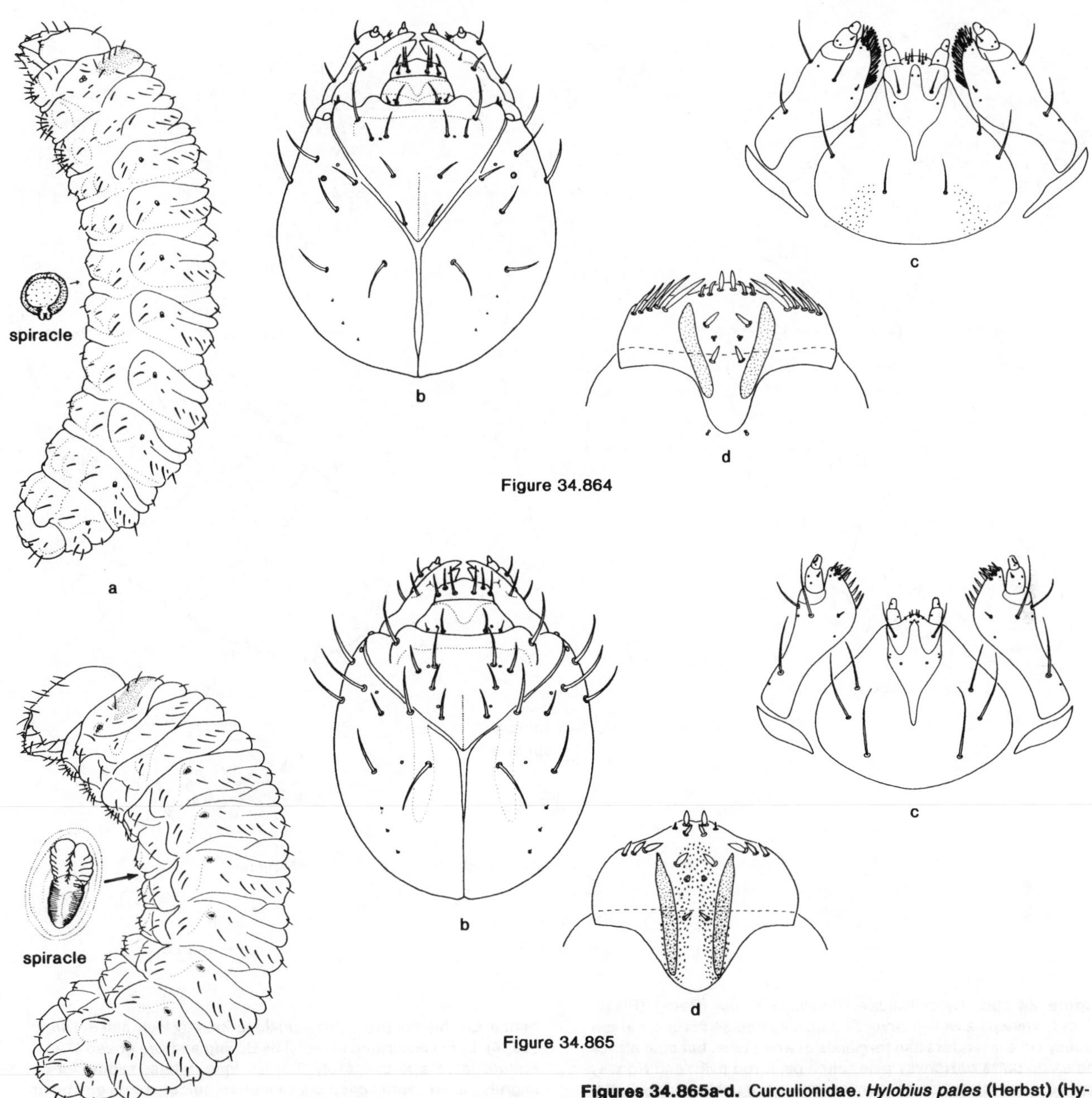

Figure 34.864

Figure 34.865

Figures 34.864a-d. Curculionidae. *Lixus concavus* Say (Cleoninae), rhubarb curculio. **a.** larva; **b.** head; **c.** labium and maxillae, ventral; **d.** epipharynx. Actually lives in the stems of wild dock, *Rumex crispus,* rather than in rhubarb. Adults will lay eggs in flower stalks and other parts of rhubarb plants, but larvae fail to develop there. Pupation within stems of dock. Larva briefly described by Peterson (1951). Mature larva approximately 17 mm long, slightly curved, subcylindrical. Body white, except for brownish pronotal shield, setae as figured. Head light brown, with some irregular pale areas and dark mandibles; setae as figured. One pair of anterior stemmata and a smaller, darker pair of posterior stemmata. Epipharynx with 5 pairs of anterolateral setae and stout, dark labral rods, as figured. Maxilla with unusually dense row of dorsal malar setae. Mandibles bifid at tip. Premental sclerite of labium distinctly trident-shaped. Abdominal segments 1-7 with 3 dorsal folds.

Figures 34.865a-d. Curculionidae. *Hylobius pales* (Herbst) (Hylobiinae), pales weevil. **a.** larva; **b.** head; **c.** labium and maxillae, ventral; **d.** epipharynx. Larvae live in cambium of bark on freshly cut or dying trucks and stumps of several species of pine. Other recorded hosts include tamarack, balsam fir, red spruce, Norway spruce, eastern hemlock, red cedar, white cedar, Douglas-*fir, and* common juniper. Pupation takes place in chip cocoons under bark. Ranges widely in the eastern U.S. and Canada. Taxonomy, life history, and distribution reviewed by Warner (1966). No detailed description of the larva exists, but the pupa was described by Thomas (1968). Mature larva approximately 12 mm long, distinctly C-shaped, cream white, with a few prominent setae (figured) on each segment. Head orange-brown with 2 pale paramedian dorsal stripes, 1 pair of stemmata, several prominent setae, as figured. Epipharynx, maxillae and labium as figured. Mandibles dark brown, bifid at tip. Pronotal shield divided at midline, lightly pigmented. Abdominal segments 1-7 with 3 dorsal folds. All spiracles encircled by a raised, pigmented sclerotized ridge. Last (8th) abdominal spiracle larger than other 7. These characters will serve to identify larvae of the genus *Hylobius* but will not separate *H. pales* from other *Hylobius* species.

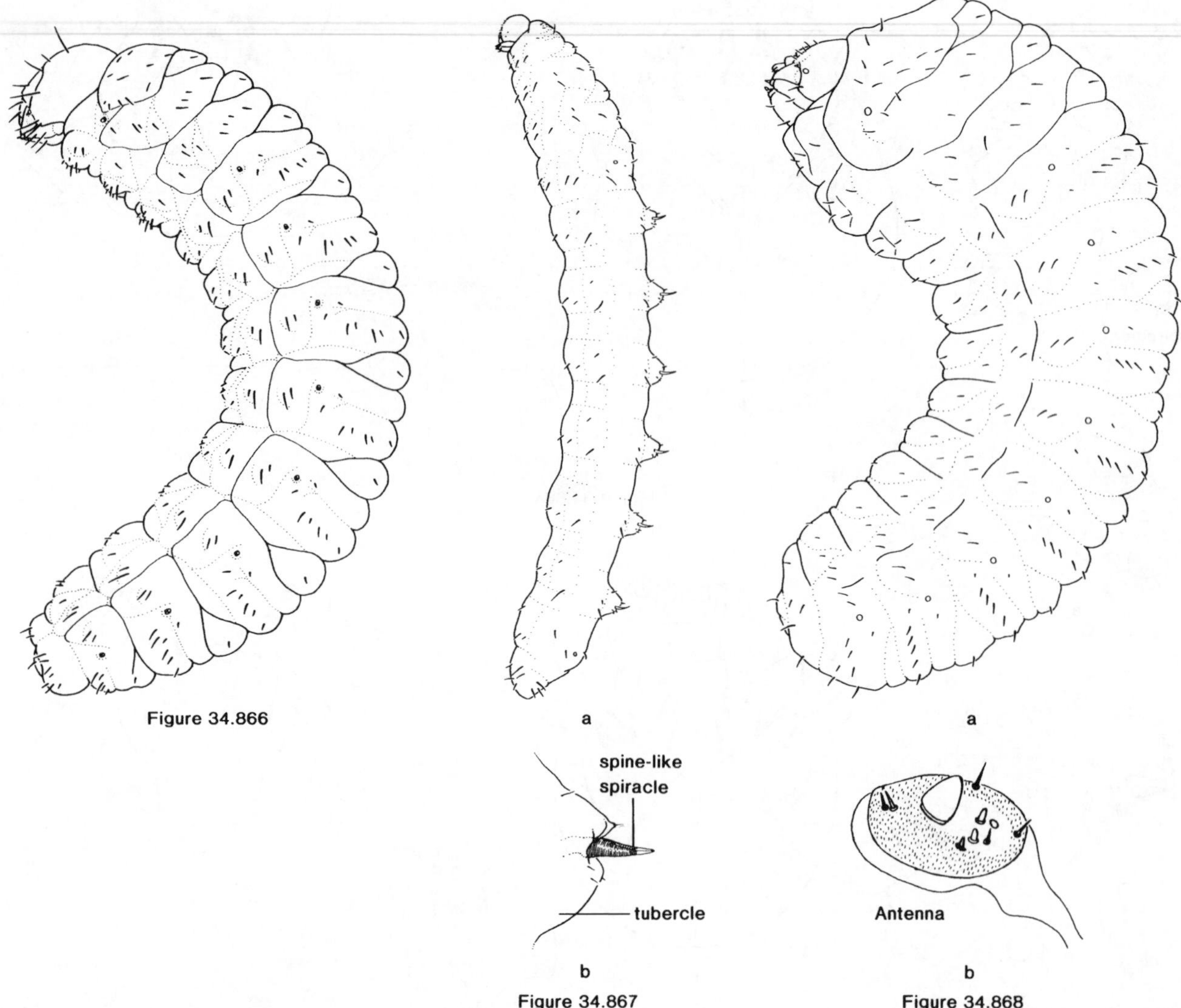

Figure 34.866

a

b

Figure 34.867

a

b

Figure 34.868

Figure 34.866. Curculionidae. *Pissodes strobi* (Peck) (Pissodinae), white pine weevil larva. *See* also figures 34.850a–l. Larvae usually bore in leaders and terminals of white pine, but also attack the same parts of Norway pine, pitch pine, red pine, and Norway spruce. Life history and economic importance published by Craighead (1950). Larval characteristics of *Pissodes* defined by Anderson (1947) and *P. strobi* briefly described by Peterson (1951). Full grown larva approximately 7 mm long, cream white, C-shaped, subtruncate at posterior end. Head yellowish-brown, with 1 pair of stemmata and several pairs of prominent setae, as illustrated. Mandibles bifid at tip. Epipharynx, labium, and maxillae as illustrated in figures 34.850b,c. Pronotal shield lightly pigmented. Abdominal segments 1–7 with 3 dorsal folds; terminal segment subcylindrical, without pleural folds. Body setae as illustrated. The blunt angulation on outer margin of maxilla and rounded subtruncate terminal abdominal segment are important generic characters.

Figures 34.867a,b. Curculionidae. *Lissorhoptrus oryzophilus* Kuschel (Erirhininae), rice water weevil. **a.** larva; **b.** spiracular tubercle. Larvae are aquatic and feed externally on submerged portions of rice plants, using sharp, spine-like abdominal spiracles to tap internal air spaces of plants for oxygen. Pupation takes place in mud-covered cocoons attached to bases of rice plants and supplied with air through a hole chewed into the hollow center of the stem by the larva. Adults also swim, and feed on upper parts of rice. Life history (as *L. simplex*) described by Isely and Schwardt (1934). Larva illustrated in detail by Böving and Craighead (1931). Mature larva approximately 3.5 mm long, white, subcylindrical, slightly curved, with conspicuous mid-dorsal tubercles on abdominal segments 2–7. Head pale, unpigmented except for brownish mouthparts, convex, unretracted, with 2 pairs of small stemmata. Abdominal segments 2–7 each bearing a pair of spine-like spiracles arising from a mid-dorsal tubercle, as illustrated. Body setae mostly short, inconspicuous.

Figures 34.868a,b. Curculionidae. *Magdalis aenescens* LeConte (Magdalinae), bronze appletree weevil. **a.** larva; **b.** antenna. Larvae mine under bark of apple and alder trees in Pacific northwestern N. America. Pupation occurs in a cell under the bark. No detailed descriptions of the larva or pupa are available, but for other species see Scherf (1964) and Leckander (1967). Mature larva approximately 6 mm long, stout, strongly curved, thickest through the thorax, cream white. Head deeply retracted, unpigmented except for brownish mouthframe and dark brown mandibles. Basal membrane of antenna covered with microspicules, as figured. Mandible with oblique cutting edge without teeth. One pair of stemmata, barely evident. Pronotum without sclerotized shield. Abdominal segments 1–7 with 3 dorsal folds. Body setae distributed as figured. All spiracles nearly round, without marginal air tubes.

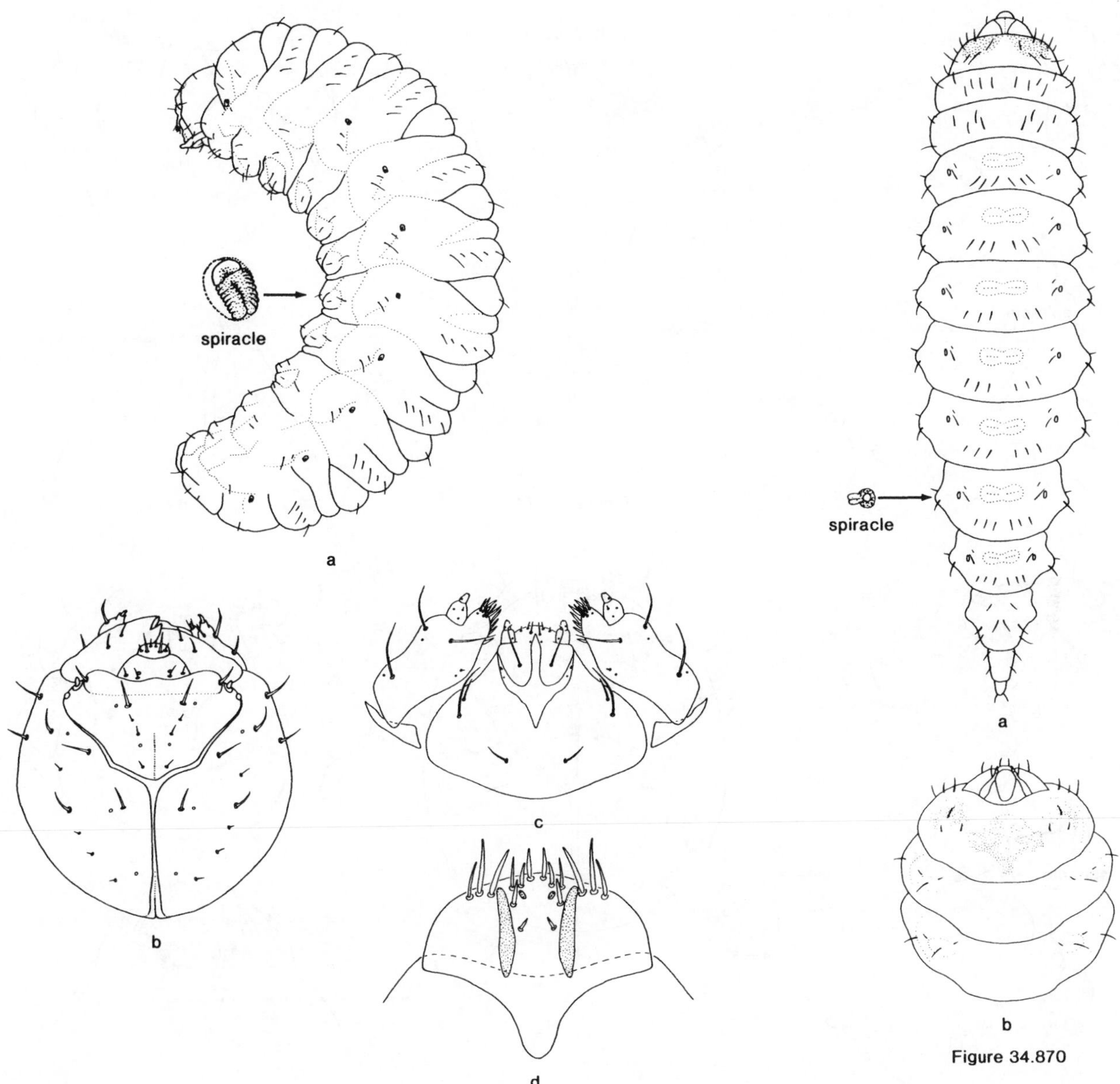

Figure 34.869

Figure 34.870

Figures 34.869a-d. Curculionidae. *Curculio caryae* (Horn) (Curculioninae), the pecan weevil. **a.** larva; **b.** head; **c.** labium and maxilla, ventral; **d.** epipharynx. Larvae live only in nuts of pecan and hickory. When full grown they emerge from the nuts and enter the soil, where they pupate in earthen cells. They are occasionally found in homes, having emerged from nuts. Taxonomy, biology, and distribution reviewed by Gibson (1969), and larva briefly described by Peterson (1951). Mature larva approximately 15 mm long, robust, distinctly C-shaped, yellowish-white, with some prominent pigmented setae. Head medium yellow-brown, unretracted, with 1 pair of stemmata and several pairs of setae, as figured. Epipharynx with dark, stout labral rods. Premental sclerite of labium dark, trident-shaped. Mandibles dark, bifid at tip. Abdominal segments 1-7 with 3 dorsal folds. Anal opening distinctly X-shaped. Body setae as figured. Distinguished from larvae of *Conotrachelus* species also found in pecan and hickory nuts by having a more robust, more strongly C-shaped body form, and in lacking a pigmented pronotal shield (compare figures 34.869a and 34.880a). See also the *Curculio* key by Gibson (1985).

Figures 34.870a,b. Curculionidae. *Rhynchaenus pallicornis* (Say) (Rhynchaeninae), apple flea weevil. **a.** larva; **b.** head and thorax. Larvae make winding mines in leaves of cultivated apple, hawthorn, wild crabapple, chokecherry, quince, hazelnut, winged elm and American elm. Pupation takes place in an inflated portion of the mine at the edge of the leaf. Adults feed on leaves of the same hosts. Life history described by Houser (1923) and summarized by Metcalf, et al. (1962). Peterson (1951) briefly described the larva. Mature larva approximately 4.5 mm long, flattened, broadest in middle third, white except for pair of dark pronotal plates and 3 dark ventral prothoracic plates, as figured. Head dark brown, deeply retracted into prothorax, with 1 pair of small stemmata. Abdomen strongly tapered posteriorly; segments 1-7 with 2 vaguely defined dorsal folds. Body setae as figured. Larvae of other *Rhynchaenus* species are similar but mine leaves of other hosts.

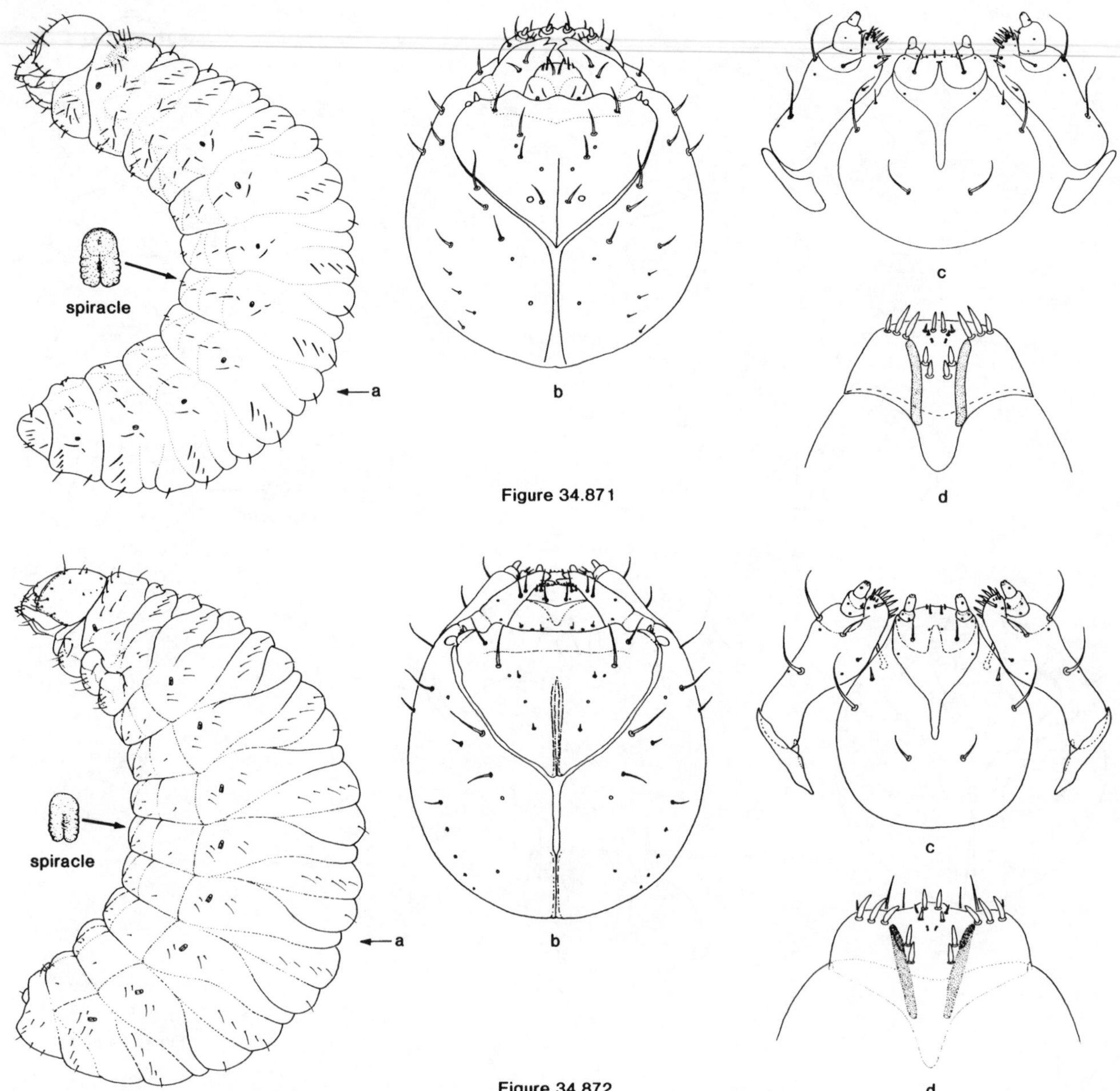

Figures 34.871a-d. Curculionidae. *Anthonomus grandis* Boheman (Anthonominae), boll weevil. **a.** larva; **b.** head; **c.** labium and maxillae, ventral; **d.** epipharynx. Well known for its destructiveness to flower buds (squares) and fruit (bolls) of cultivated cotton. Larvae develop in those parts of cotton and *Cienfuegosia* spp., *Thespesia populnea*, *Hibiscus* sp. (in Mexico), and *Hampea* spp. A detailed technical description of the larva and a key separating it from other larval Anthonomini was published by Ahmad and Burke (1972). Description of the pupa and key separating it from pupae of other Anthonomini published by Burke (1968). Mature larva 5.6-8.1 mm long, robust, thickest through middle abdominal segments, white, distinctly curved, tapered toward posterior end. Head light yellowish-brown, unretracted, with 1 pair of stemmata and several pairs of setae, as figured. Labium with a trident-shaped premental sclerite, 2-segmented palpi, and setae as figured. Maxillae with 1st segment of palpi much stouter than 2nd, as figured. Pronotal shield lightly pigmented. Abdominal segments 1-7 with 3 dorsal folds. Terminal abdominal segment subconical, anal opening ventral. Body setae as figured.

Figures 34.872a-d. Curculionidae. *Anthonomus quadrigibbus* Say (Anthonominae), apple curculio. **a.** larva; **b.** head; **c.** labium and maxillae, ventral; **d.** epipharynx. Larvae develop in apples and in fruit of wild crab, hawthorn, quince, pear, and shadbush (*Amelanchier*). Pupation within the fruit. Other details of life history summarized by Metcalf, et al. (1962). Detailed description of larva published by Ahmad and Burke (1972) and pupa described by Burke (1968). Peterson (1951) briefly described the larva. Mature larva approximately 4.6-6.3 mm long, white, robust, distinctly curved. Head light yellowish-brown, unretracted, with 1 pair of stemmata and several setae, as figured. Epipharynx, labium, and maxillae as figured. Pronotal shield unpigmented. Abdominal segments 1-7 with 3 distinct dorsal folds. Air tubes of abdominal spiracles directed posteroventrally. Anal opening subterminal and ventral. Body setae as figured.

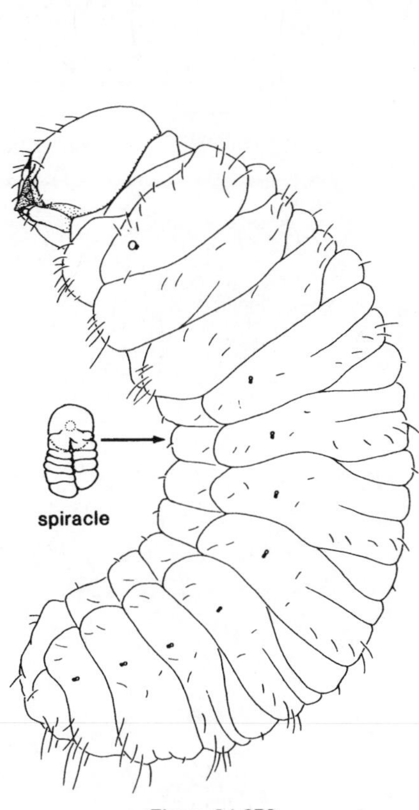

Figure 34.873

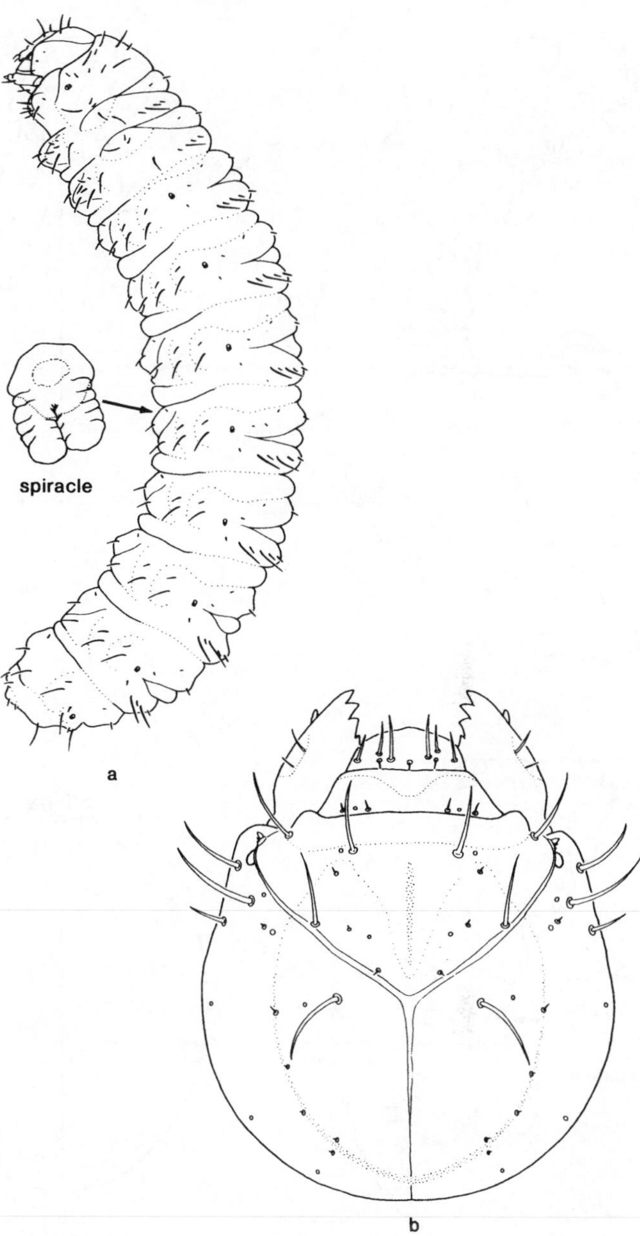

Figure 34.874

Figure 34.873. Curculionidae. *Caulophilus oryzae* (Gyllenhal) (Cossoninae), broadnosed grain weevil. Infests stored grain, particularly corn, and other dried seeds, such as avocado pits, in the southeastern U.S. Larvae hatch from eggs laid in soft or damaged grain and feed on the softer inner part until full grown, then pupate within the grain. Egg, larva, and pupa described in detail by Cotton (1921) and larva by Anderson (1952). Mature larva about 2.5 mm long, white, stout, curved, with a few prominent setae, as figured. Head pale orange-brown, with 1 pair of stemmata and several pairs of setae. Mandibles with 2 apical teeth and a rounded projection on inner margin. Labral rods of epipharynx nearly straight, convergent posteriorly. Premental sclerite of labium trident-shaped. Pronotal shield unpigmented. Abdominal segments 1-7 with 3 dorsal folds. Spiracles bicameral, air tubes of abdominal spiracles directed posteriorly.

Figures 34.874a,b. Curculionidae. *Centrinaspis penicilla* (Herbst) (Baridinae), corn stalk weevil. **a.** larva; **b.** head. Little is known of the life history in association with its native hosts (various grasses), but it is frequently found attacking cornstalks in the southeastern U.S. From eggs laid in the tassel, larvae bore down into first 2-4 nodes of the stalk. Mature larvae emerge from the stalk, drop to the ground and pupate in the soil. Kirk (1965) described life history and briefly described larva and pupa. Böving (*in* Ainslie, 1920) described larva under *Centrinus penicellus.* Mature larva approximately 10 mm long, slightly curved, tapered at ends, ivory white. Head unretracted, light brown; epicranium marked with a darkly pigmented dorsal crescent, as figured; setae stout, distributed as figured. One pair of stemmata. Mandibles with 4 teeth at apex. Pronotal shield not pigmented. Abdominal segments 1-7 with 3 distinct dorsal folds. Spiracles bicameral, with marginal air tubes of abdominal spiracles oriented posteriorly. Body setae as figured.

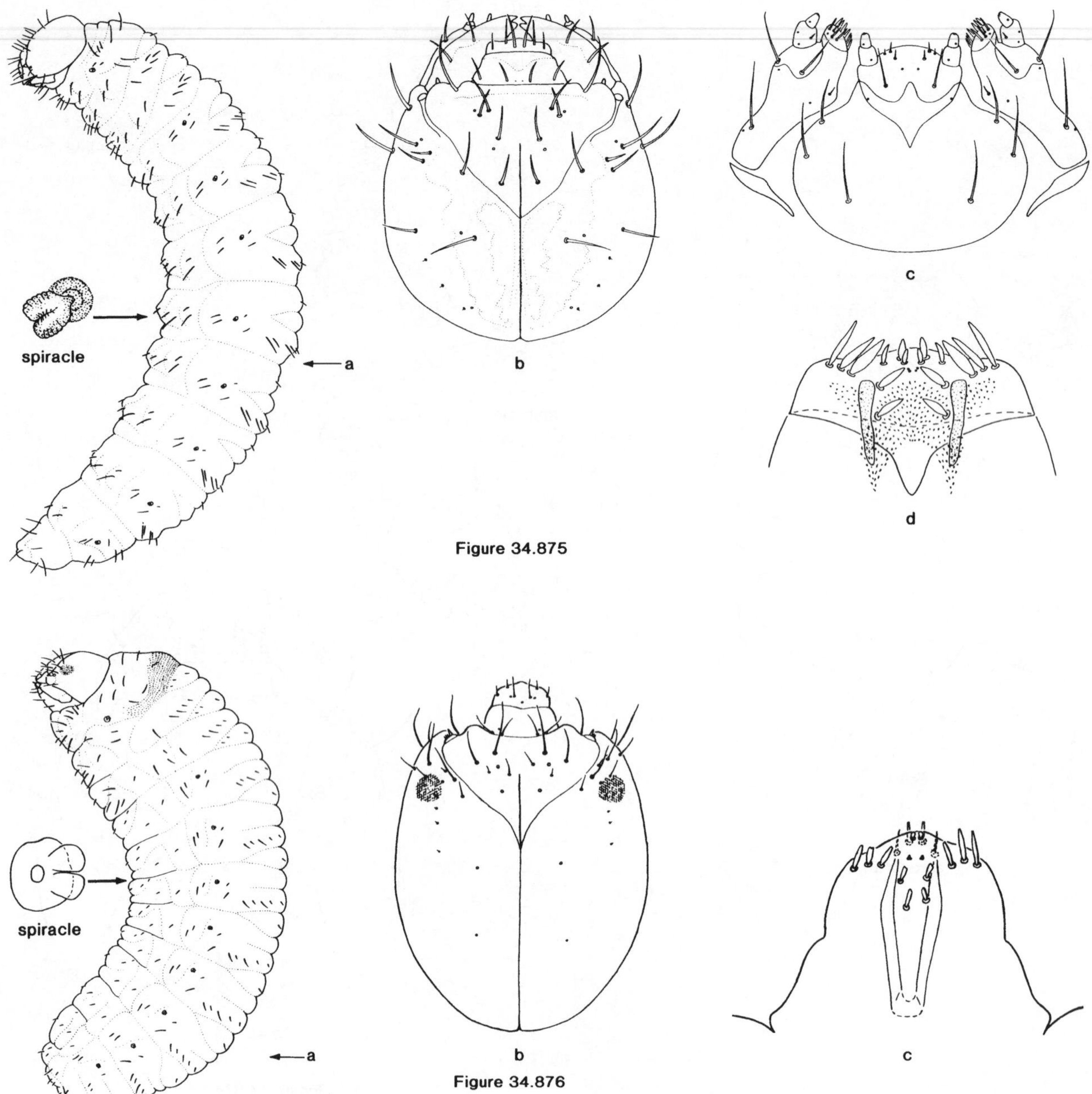

Figures 34.875a-d. Curculionidae. *Trichobaris trinotata* (Say) (Baridinae), potato stalk borer. **a.** larva; **b.** head; **c.** labium and maxillae, ventral; **d.** epipharynx. Hosts include potato, eggplant, tomato, Jimson weed, horse nettle, and ground cherry. Larvae hatch from eggs laid in stems or leaf petioles and bore inside stems, causing plants to wilt and die. Weevils sometimes become abundant enough to do major damage to potato crops. Details of life history summarized by Metcalf, *et al.* (1962) and larva described by Peterson (1951). Mature larva about 10 mm long, white to yellowish, distinctly curved, tapered toward both ends. Head unretracted, light brown, marked with 2 pale irregular paramedian areas on epicranium and bearing several pairs of prominent setae, as figured. One pair of stemmata. Mandibles bifid at tip. Epipharynx, labium, and maxillae as figured. Abdominal segments 1-7 with 3 dorsal folds. Spiracles bicameral, those of abdomen with marginal air tubes oriented posteroventrally. Body setae as figured.

Figures 34.876a-c. Curculionidae. *Cylindrocopturus furnissi* Buchanan (Zygopinae), Douglas-fir twig weevil. **a.** larva; **b.** head; **c.** epipharynx. Attacks and kills small branches on young Douglas-fir in the western U.S. Larvae hatch from eggs laid in the bark and bore into the wood, where they feed until mature and pupate there after overwintering. Details of life history summarized by Keen (1952). Larva and pupa described and larva separated from larva of 3 other *Cylindrocopturus* species by Anderson (1941). Mature larva about 3.3 mm long, stout, curved, subcylindrical. Head (figured) deeply retracted into prothorax, unpigmented except for orange-brown mouthframe, dark mandibles, and large, dark, rounded spot on each side. One pair of stemmata. Epipharynx with labral rods fused posteriorly, as figured. Mandibles bifid at tip. Pronotum with dense band of asperities on posterior half, as figured. Abdominal segments 1-7 with 3 dorsal folds. Spiracles bicameral, marginal air tubes not annulated. Body setae inconspicuous, distributed as figured. (Figures b, c from Anderson, 1941)

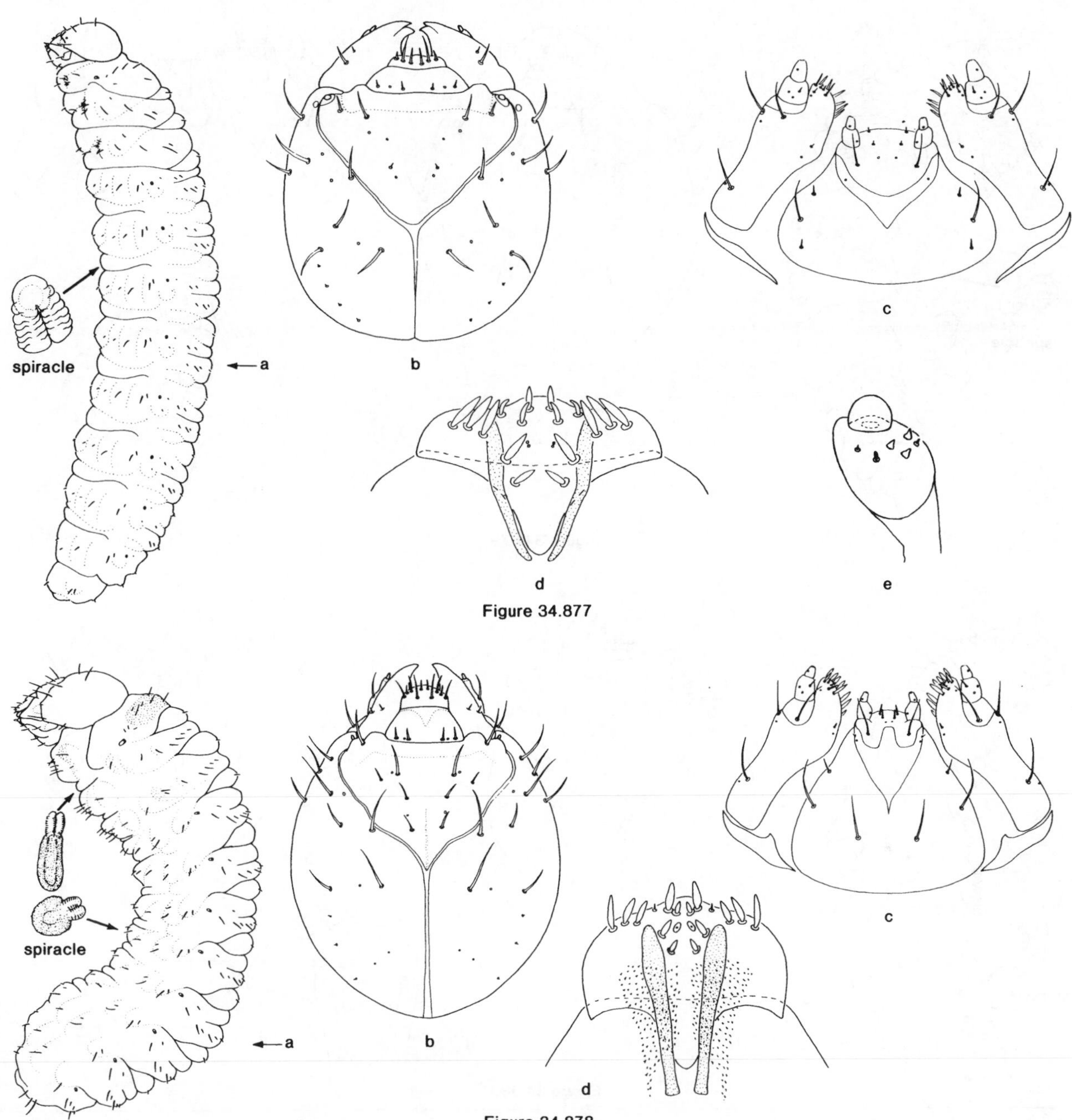

Figures 34.877a-e. Curculionidae. *Ceutorhynchus rapae* Gyllenhal (Ceutorhynchinae), cabbage curculio. **a.** larva; **b.** head; **c.** labium and maxillae, ventral; **d.** epipharynx; **e.** right antenna. A pest of cabbage, turnip, horseradish, cauliflower, and other cruciferous plants throughout most of the U.S. Larvae hatch from eggs laid in the stems, in which they mine until full grown, then pupate in earthen cocoons. Adults feed on the same plants. Details of life history summarized and larva and pupa described by Scherf (1964). Mature larva 6-7 mm long, white, slender, slightly curved. Head unretracted, light orange-brown, no endocarina, 1 pair of stemmata and several pairs of setae. Accessory appendage of antenna subglobular, as figured. Labral rods of epipharynx (figured) elongate, convergent posteriorly. Premental sclerite of labium (figured) angulate posteriorly, not trident-shaped. Abdominal segments 1-7 with 3 dorsal folds. Spiracles bicameral, marginal air tubes directed posterodorsally. Body setae as figured.

Figures 34.878a-d. Curculionidae. *Cryptorhynchus lapathi* (L.) (Cryptorhynchinae), poplar-and-willow borer. **a.** larva; **b.** head; **c.** labium and maxillae, ventral; **d.** epipharynx. Attacks living poplar and willow, and some species of alder and birch. Larvae hatch from eggs laid in the bark and tunnel into the cambium, then into the hard wood, sometimes honeycombing part of a branch or trunk, before making pupal cells in the wood. Details of life history summarized by Craighead (1950) and larva described by Peterson (1951) and by Scherf (1964). Full grown larva approximately 12 mm long, curved, white, except for lightly pigmented pronotal shield, with some prominent dorsal setae near end of abdomen, as figured. Head unretracted, yellowish-brown, with 2 pale dorsal stripes, 1 pair of stemmata and several pairs of setae. Labral rods of epipharynx elongate, convergent posteriorly, as figured. Labium and maxillae as figured. Abdominal segments 1-7 with 3 dorsal folds. Spiracles bicameral, air tubes directed dorsally.

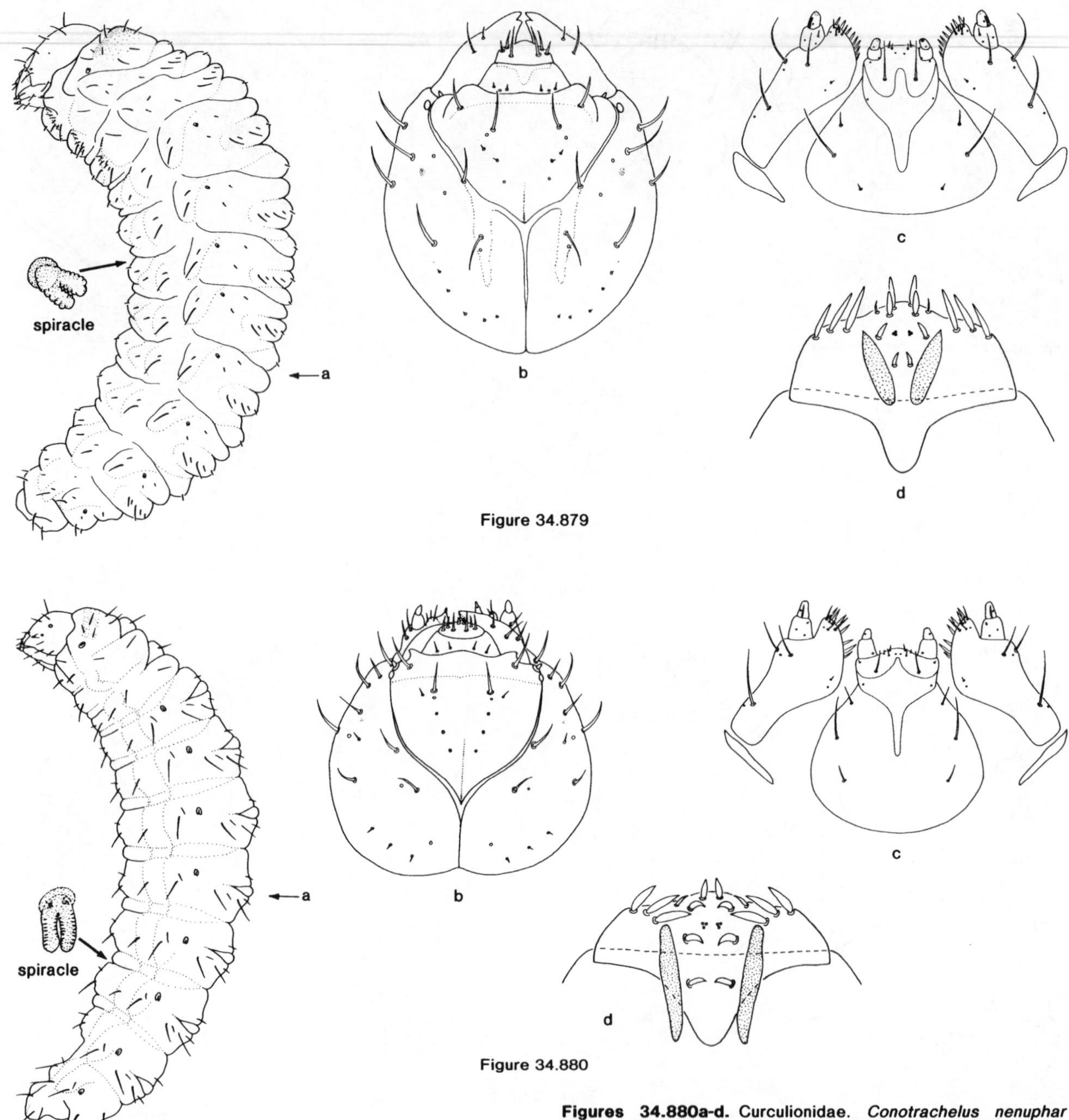

Figures 34.879a-d. Curculionidae. *Chalcodermus aeneus* Boheman (Cryptorhynchinae), cowpea curculio. **a.** larva; **b.** head; **c.** labium and maxillae, ventral; **d.** epipharynx. Larvae develop in seeds of cowpeas, stringbeans, and some wild legumes. Pupation takes place in the soil. Life history studied in detail by Arant (1938) and larva described by Peterson (1951). Mature larva about 7 mm long, yellowish-white except for light brown pronotal shield, thickest through 1st abdominal segment, tapered toward posterior. Head unretracted, light brown, with 2 short pale dorsal stripes on epicranium, 1 pair of stemmata, 1 pair of subcutaneus pigment spots, and setae as figured. Labium and maxillae as figured. Mandibles bifid at tip. Abdominal segments 1-7 with 3 dorsal folds. Spiracles bicameral, marginal air tubes directed posterodorsally. Body setae as figured.

Figures 34.880a-d. Curculionidae. *Conotrachelus nenuphar* (Herbst) (Cryptorhynchinae), plum curculio. **a.** larva; **b.** head; **c.** labium and maxillae, ventral; **d.** epipharynx. Well known as a pest of many cultivated fruits, including plum, peach, cherry, apricot, nectarine, apple, pear, and quince in the eastern U.S. and Canada. Larvae hatch from eggs laid under skin of the fruit and feed within the fruit until full grown, then emerge and pupate in the soil. Details of life history summarized by Metcalf, et al. (1962) and larva described by Peterson (1951). Full grown larva about 9 mm long, dirty white to yellow except for pigmented pronotal shield; slender, thickest through middle of abdomen, with a few prominent setae on each segment as figured. Head unretracted, yellowish brown, with 1 pair of stemmata and several pairs of setae. Labral rods of epipharynx (figured) stout, dark, straight. Mandibles bifid at tip. Labium and maxillae as figured. Abdominal segments 1-7 with 3 dorsal folds. Spiracles bicameral, marginal air tubes longer than width of peritreme, directed posteriorly on abdomen.

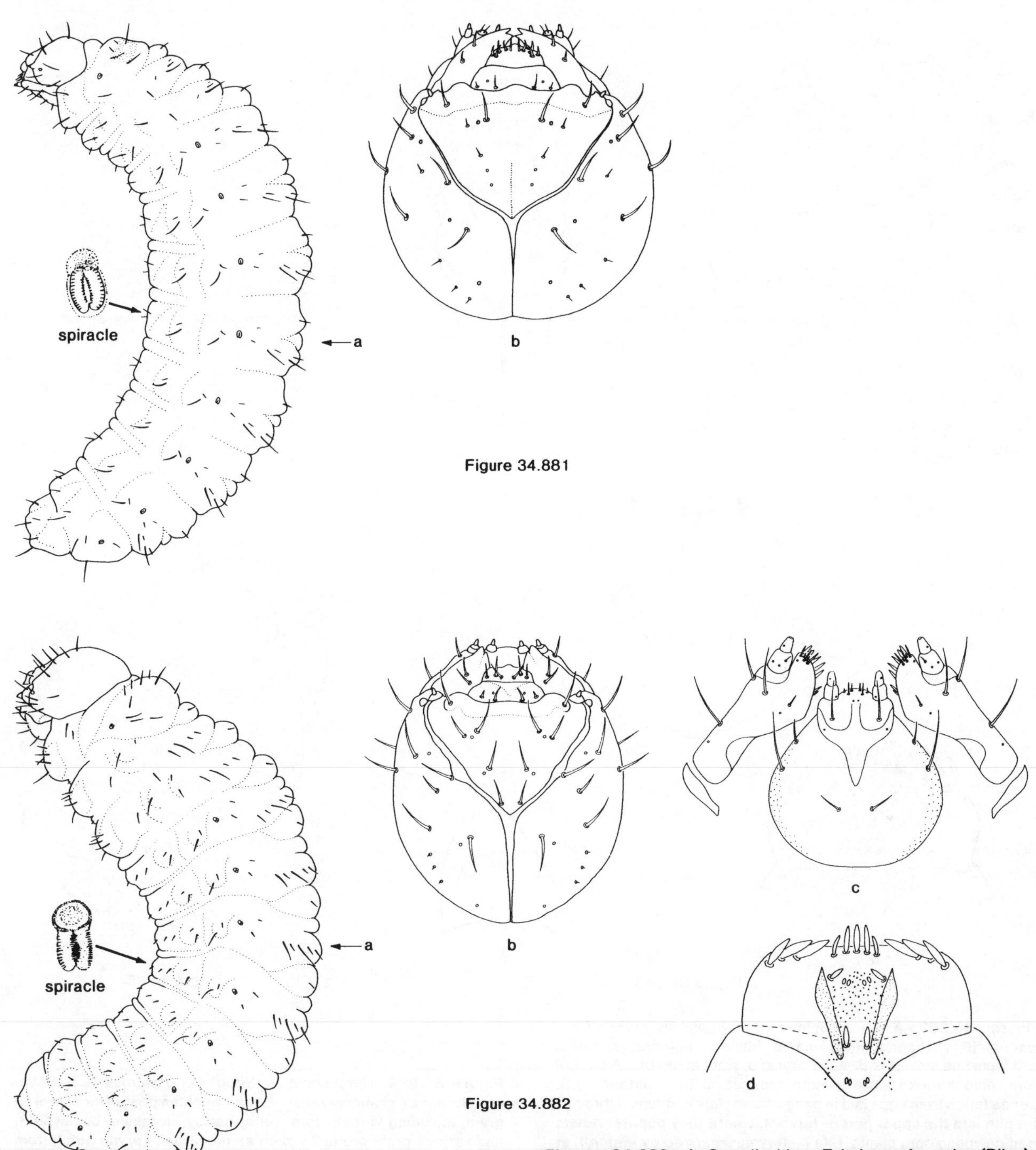

Figures 34.881a,b. Curculionidae. *Conotrachelus crataegi* Walsh (Cryptorhynchinae), quince curculio. **a.** larva; **b.** head. Larvae feed in fruit of quince, pear, apple, and hawthorn; emerge when mature and enter soil, where they overwinter before pupating. Adults feed on same fruit, causing deformation. Life history summarized by Metcalf, et al. (1962) and Maier (1980). Larva described by Peterson (1951). Mature larva about 10 mm long, very similar to that described above for *C. nenuphar* but more robust, with vertex of head (figured) not as strongly indented, with 3 pairs vs. 2 pairs of setae on frons, and with narrower, more widely separated labral rods.

Figures 34.882a-d. Curculionidae. *Tyloderma fragariae* (Riley) (Cryptorhynchinae), strawberry crown borer. **a.** larva; **b.** head; **c.** labium and maxillae; **d.** epipharynx. A pest of cultivated strawberries in the eastern U.S.; also infests wild strawberries and cinquefoil. Larvae hatch from eggs laid in leaf bases and bore down into crown, killing or stunting the plants. Life history briefly summarized by Metcalf, et al. (1962) and larva of this species and *T. foveolata* Say described by Peterson (1951). Mature larva about 7 mm long, white, C-shaped, with inconspicuous setae, as figured. Head unretracted, light amber, with 1 pair of stemmata and several pairs of setae, as figured. Mandibles bifid at tip. Labral rods of epipharynx (figured) stout, dark, convergent posteriorly. Labium and maxillae as figured. Pedal areas of prothorax strongly developed. Abdominal segments 1-7 with 3 dorsal folds. Spiracles bicameral, marginal air tubes directed posterodorsally.

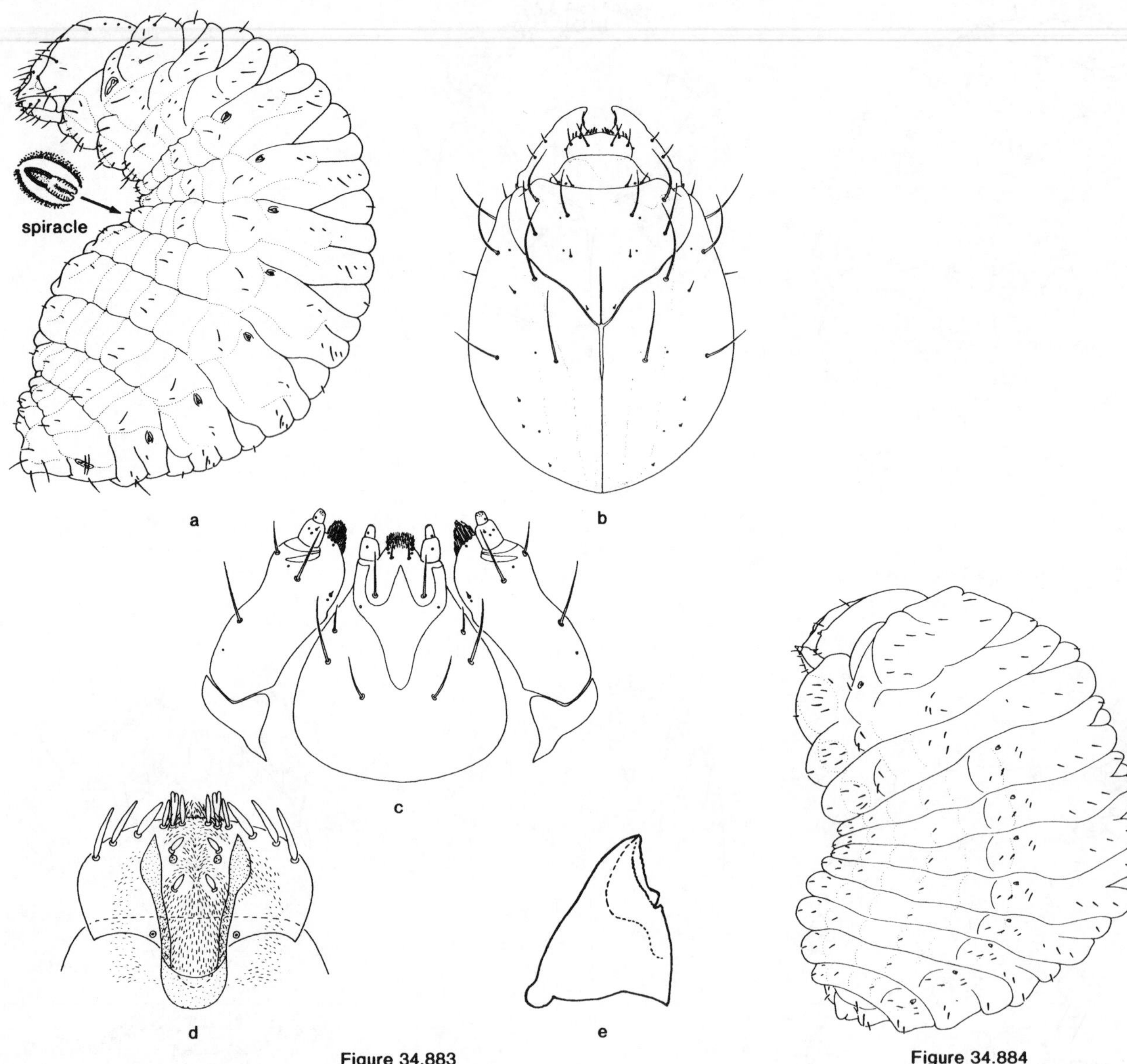

Figure 34.883

Figure 34.884

Figures 34.883a-e. Curculionidae. *Sphenophorus maidis* Chittenden (Rhynchophorinae), maize billbug. **a.** larva; **b.** head; **c.** labium and maxillae; **d.** epipharynx; **e.** right mandible. A pest of corn, also attacks large swamp grasses in the southern U.S. Larvae hatch from eggs laid in base of cornstalk and tunnel through the pith into the upper part of taproot, where they pupate. Adults feed on young corn plants. Life history summarized by Metcalf, et al. (1962) and larva described by Peterson (1951). A generic description of *Sphenophorus* (as *Calendra*) was published by Anderson (1948a). Full grown larva about 17 mm long, white, stout, with a conspicuously swollen area in abdominal segments 4-6, setae inconspicuous except on caudal segments of abdomen, as figured. Head dark reddish brown, with 2 vaguely defined pale dorsal stripes, and several pairs of prominent setae. Mandibles undivided at tip, with cutting edge concave. Epipharynx (figured) with labral rods fused posteriorly and with 1 pair of anteromedian setae branched. Labium and maxillae as figured. Abdominal segments 1-7 with 3 dorsal folds; 1-8 with pleura subdivided into 4 lobes, as figured. Spiracles large, with long air tubes directed dorsally, except on 8th abdominal segment, where air tubes point posteriorly. Terminal abdominal segment flattened, bearing 4 pairs of long setae.

Figure 34.884. Curculionidae. *Sitophilus granarius* (L.) (Rhynchophorinae), granary weevil. A cosmopolitan pest of most stored grain, including wheat, corn, oats, barley, sorghum, buckwheat, and various grain products such as macaroni. Larvae hatch from eggs laid in seeds or other foods, in which they feed until full grown and pupate. Adults feed on same materials. Details of life cycle and habits summarized by Metcalf, et al (1962). Egg, larva, and pupa described by Cotton (1921) and by Scherf (1964); larva distinguished from larvae of other common species of *Sitophilus* by Mathur (1954). Mature larva about 2.75 mm long, white, very robust, strongly convex dorsally. Head partially retracted, light brown with dark mouthframe and mandibles, with 2 vaguely defined pale dorsal stripes, and 1 pair of stemmata. Premental sclerite of labium trident-shaped. Labral rods dark, elongate, convergent posteriorly. Abdominal segments 1-4 with 3 dorsal folds, 5-7 with 2 dorsal folds. Pleura subdivided into 3 folds on abdominal segments 1-7. Spiracles small, bicameral, air tubes directed dorsally, except on terminal abdominal segment, where they point posteriorly. Body setae small, distributed as figured.

SCOLYTIDAE (CURCULIONOIDEA)

D. E. Bright, *Biosystematics Research Centre, Agriculture Canada*

Bark Beetles

Figures 34.885-896

Relationships and Diagnosis: The Scolytidae, Platypodidae; Curculionidae, Anthribidae, Brentidae and several related small families comprise the enormous superfamily Curculionoidea. The Scolytidae were reduced to subfamily rank in the Curculionidae by Crowson (1955) but Morimoto (1962a), Wood (1973) and other specialists have retained family status for the group. The larvae of Scolytidae and Curculionidae are presently impossible to distinguish at the family level, but *see* the diagnosis for Curculionidae. Both are characterized by the C-shaped, legless body with a prominent sclerotized head capsule. Differences at the generic level can sometimes be found.

Two subfamilies of Scolytidae are currently recognized, Scolytinae and Hylesininae. The Hylesininae are more primitive and their larvae are unique in having 3 or 4 lobes or incisoral teeth on each mandible (figs. 34.890, 34.891) while in the Scolytinae the mandibles are bilobed (fig. 34.889) or bear only 1 broad chisel-like tooth. In addition, Lekander (1968) found that the submental setae may be arranged either in a straight line or obliquely to one another. He stated that these 2 types are very constant and divided the larvae of bark beetles into 2 large groups, Ipinae (Scolytinae) and Hylesininae. Also, the Hylesininae have 3 or 4 setae on each pedal lobe, while the Scolytinae apparently always have 2 (Lekander, 1968).

Biology and Ecology: Larvae and adults are strictly phytophagous. Most species in N. America feed and reproduce under the bark or deep in the wood of a variety of coniferous and deciduous trees and in cones, shrubs, etc. Most attack trees that are recently dead or dying or trees weakened by some environmental stress such as disease, drought, air pollution, fire, etc. Some species, however, are capable of attacking and killing healthy, vigorous trees. Extensive damage is frequently recorded, especially in western N. America.

Based on mode of attack and biology, the family is commonly divided into 2 groups. The *true bark beetles* construct galleries under the bark and feed on the phloem and surrounding tissue. The *ambrosia beetles* bore deep into the sapwood where they construct galleries and feed on a fungus that grows on the gallery walls.

A vast amount of literature is available on the biology of bark beetles. For further details, see Bright and Stark (1973), Bright (1976) and Wood (1982).

Description: Mature larvae 2–10 mm. Body C-shaped, subcylindrical, usually white, legless.

Head: Amber or light brown, usually free but slightly retracted into prothorax, concealing cervical membrane; slightly longer than wide to subequal, sides subparallel, continuing into a broadly rounded caudal region; cranium divided by Y-shaped ecdysial suture into a frontal and 2 parietal areas (fig. 34.885b). Frons with posterior margin acutely angled to broad or rounded, usually as broad or broader than long, surface usually smooth but may bear a few small tubercles (figs. 34.893a, 34.895); setae as in figures. Antennae 1-segmented, with up to 6 minute setae and several sensilla located on membranous area surrounding central, cone-like segment. Clypeus (fig. 34.893b) with angular sides and broadly emarginate anterior margin, usually with a pair of setae on each side anterior to frontoclypeal suture. Labrum (fig. 34.893b) attached to clypeus by clypeo-labral suture, posterior edge acuminate or acute, extending below clypeus, visible as a darkly pigmented area; anterior margin rounded, weakly acuminate or emarginate, surface bearing 2 groups of 3 setae; labral rods attached to anterior, inner surface. Maxilla amber to light brown; palp 2-segmented, apical segment with 1 or more sensilla and papillae; stipes longer than broad, margins broadly emarginate, lacinial lobe with about 6–10 dorsal setae at apex (fig. 34.886c). Labial palp 2-segmented except in *Pityophthorus, Conophthorus* and *Gnathotrichus* in which proximal segment is indistinct. Mandible roughly triangular in outline, 2-lobed in Scolytinae (fig. 34.889), 3-to 4-lobed in Hylesininae (figs. 34.890, 34.891).

Thorax and Abdomen: Three thoracic and 10 abdominal segments (fig. 34.887a), delimited by intersegmental lines or grooves, each segment composed of a number of integumental folds. Integument covered with sclerotized dorsal plates, covered with fine, minute, backward-projecting spines or asperities. Tenth abdominal segment represented by indistinct anal folds, segments 9 and 10 sometimes with sclerotized dorsal plates (figs. 34.892a,b). Pedal lobes distinct or indistinct on thorax, with 2–4 setae (fig. 34.887a). Body setae variable in placement.

Spiracles: Variable, located on prothorax and first 8 abdominal segments, of 2 types, 1) circular, or 2) a circular orifice with a pair of annulated air tubes lateral to orifice; prothoracic spiracle always larger.

Comments: About 500 species of Scolytidae occur in N. America. All are phytophagous and a number are of considerable economic importance to forestry. Economic species are covered in Drooz (1985) and Furniss and Carolin (1977).

Selected Bibliography

Bright 1976.
Bright and Stark 1973.
Browne 1961.
Crowson 1955.
Drooz 1985.
Furniss and Carolin 1977.
Gardner 1934b.
Hopkins 1909.
Kaston 1936.
Lekander 1968.
Morimoto 1962b.
Peterson 1951.
Schedl 1978.
Thomas 1957, 1965.
Wood 1961, 1973, 1982.

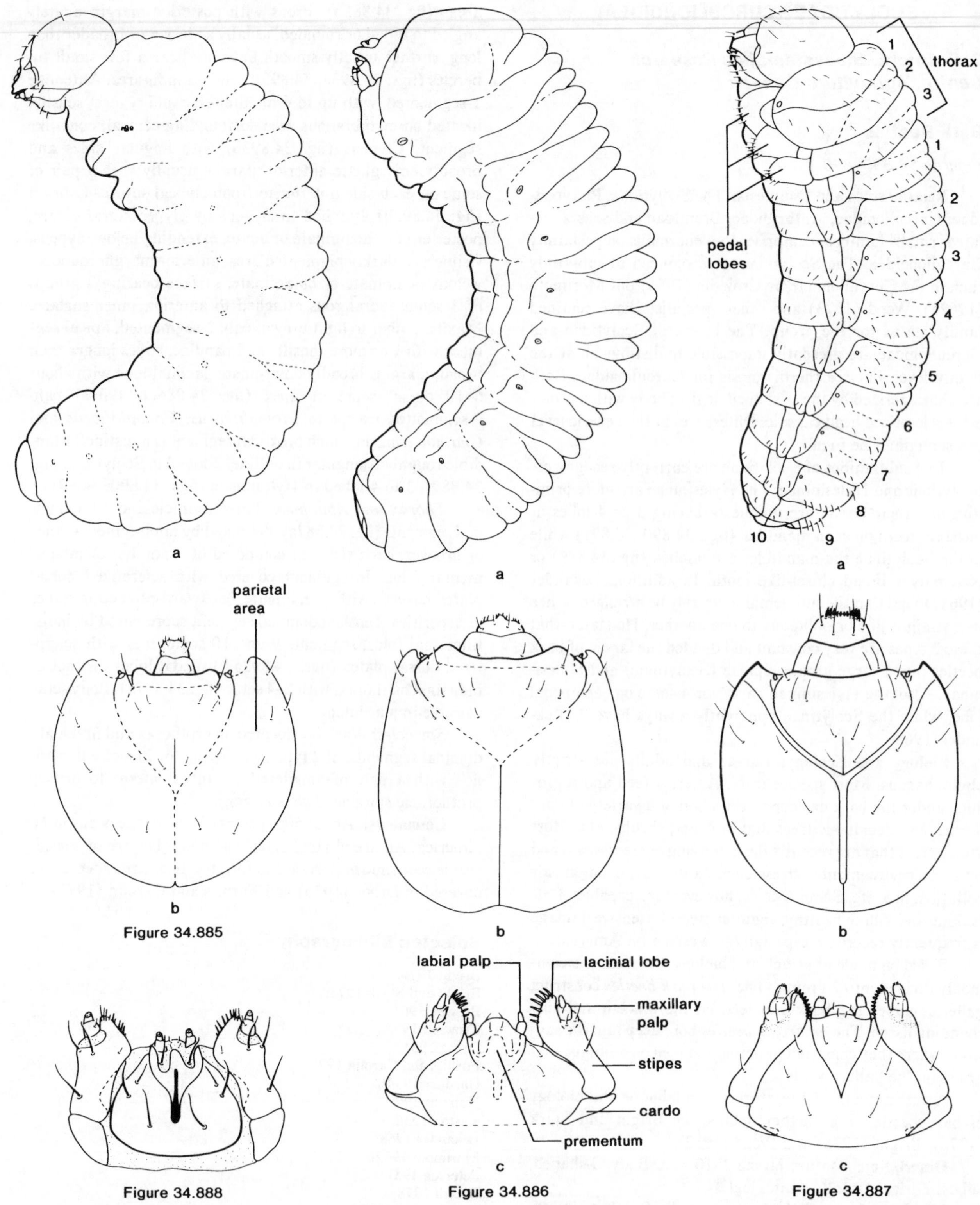

Figures 34.885a,b. Scolytidae. *Hylastinus obscurus* (Marsham), clover root borer. **a.** larva; **b.** head capsule.

Figures 34.886a-c. Scolytidae. *Scolytus mali* (Bechstein), larger shothole borer. **a.** larva; **b.** head capsule; **c.** maxillae and labium.

Figures 34.887a-c. Scolytidae. *Hylurgopinus rufipes* (Eichhoff), native elm bark beetle. **a.** larva; **b.** head capsule; **c.** maxillae and labium.

Figure 34.888. Scolytidae. *Hylurgops pinifex* (Fitch), maxillae and labium. (Figures 34.885a,b, 34.886a redrawn from Peterson, 1951)

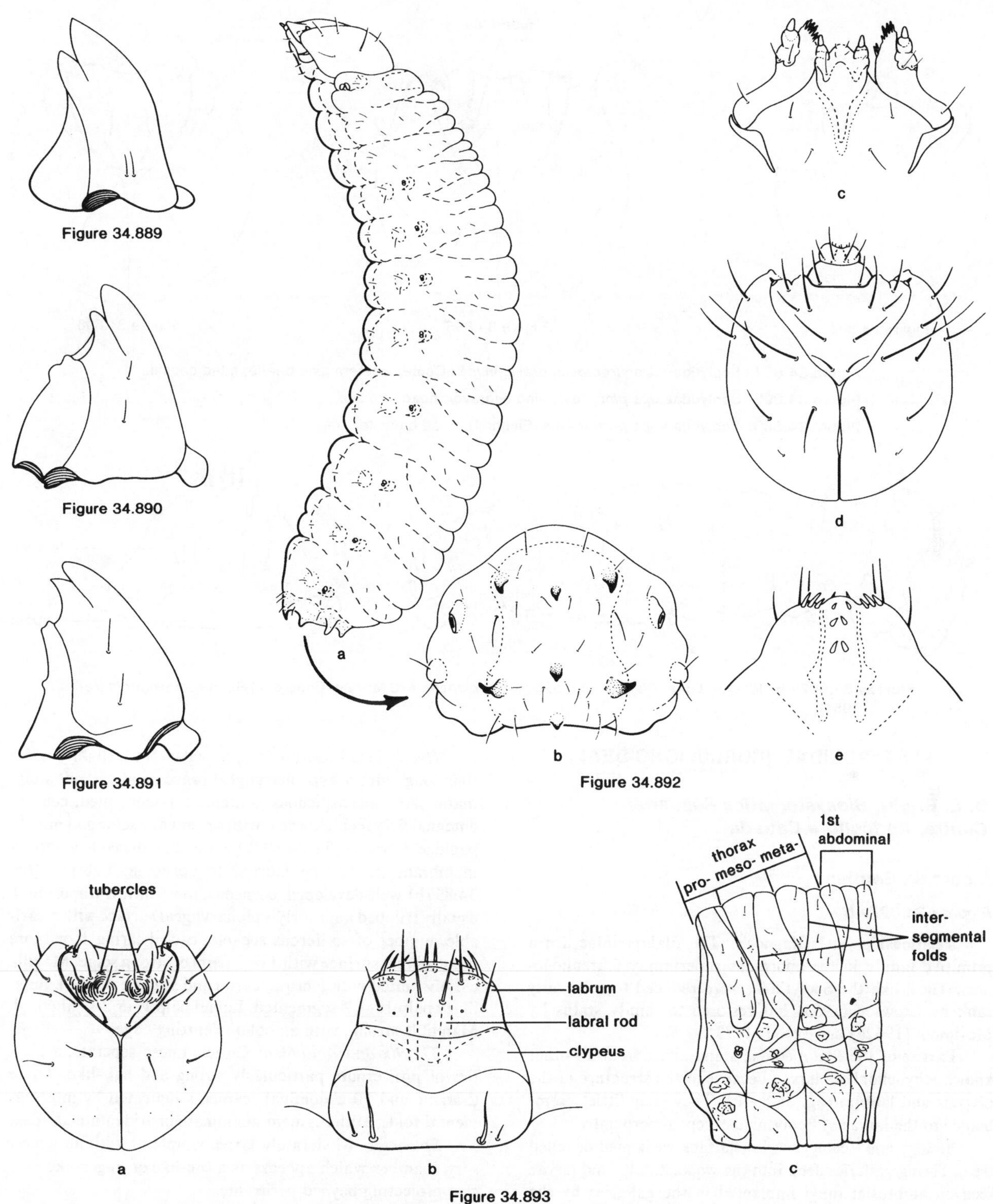

Figures 34.889–34.891. Scolytidae. Mandibles. (34.889) *Scolytus piceae* (Swaine); (34.890) *Dendroctonus ponderosae* Hopkins, mountain pine beetle; (34.891) *Hylurgopinus rufipes* (Eichhoff).

Figures 34.892a-e. Scolytidae. *Dendroctonus terebrans* (Olivier), black turpentine beetle. **a.** larva; **b.** eighth and ninth tergal plates; **c.** maxillae and labium; **d.** head capsule; **e.** epipharynx. (Redrawn from Peterson 1951)

Figures 34.893a-c. Scolytidae. *Dendroctonus ponderosae* Hopkins. **a.** head capsule; **b.** clypeus and labrum; **c.** plural region, thorax and first abdominal segment.

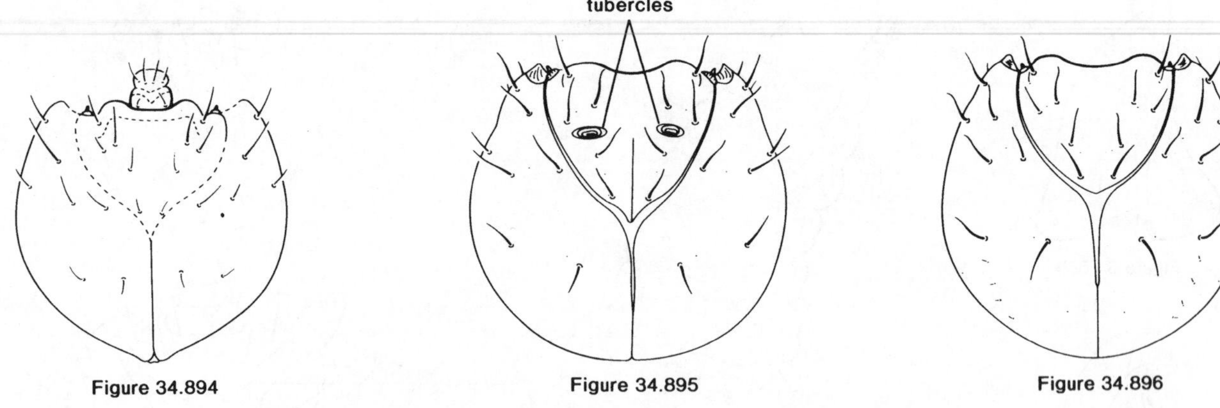

Figure 34.894. Scolytidae. *Dendroctonus brevicomis* LeConte, western pine beetle, head capsule.

Figure 34.895. Scolytidae. *Ips pini* (Say), pine engraver, head capsule.

Figure 34.896. Scolytidae. *Ips perturbatus* (Eichhoff), head capsule.

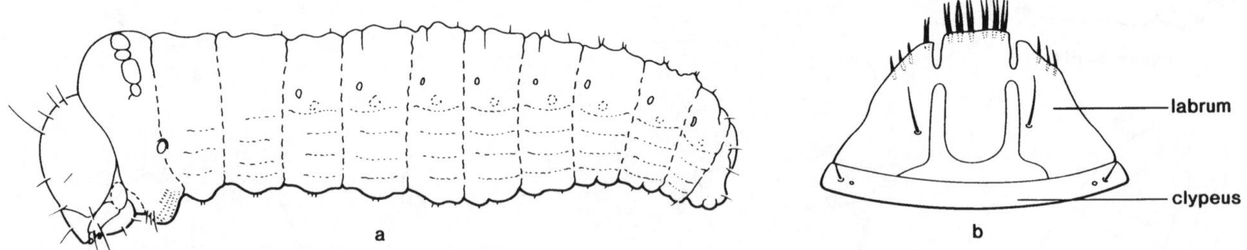

Figures 34.897a,b. Platypodidae. *Platypus* sp. **a.** larva; **b.** clypeus and labrum. (Figure 34.897a redrawn from Peterson, 1951)

PLATYPODIDAE (CURCULIONOIDEA)

D. E. Bright, *Biosystematics Research Centre, Agriculture Canada*

Ambrosia Beetles

Figures 34.897a,b

Relationships and Diagnosis: The Platypodidae are a primitive family in the enormous superfamily Curculionoidea. They, like the Scolytidae, were reduced to subfamily rank by Crowson (1955) but restored to family status by Morimoto (1962a) and Wood (1973).

Larvae of the Platypodidae stand apart from all other known Rhynchophora larvae because of the structure of the clypeus and labrum (fig. 34.897b). They bear little resemblance to the larvae of Scolytidae except superficially.

Biology and Ecology: Adults attack weakened or felled trees, boring galleries deep into the wood. Adults and larvae feed on ambrosial fungi implanted in the galleries by the female which has special organs for storing and transporting the fungal spores. Pupation occurs in special cells constructed along the gallery walls. Attacks by these beetles may cause the downgrading of lumber made from infested logs. Extensive damage is frequently recorded in tropical countries. Damage in N. America is much less severe.

Description: Mature larvae 4–12 mm. Body cylindrical, whitish, sometimes armed with corneous tubercles, particularly on apical segments of abdomen.

Head: Head capsule free, hypognathous, slightly wider than long, with a hypopharyngeal bracon and without stemmata, setae inconspicuous. Antennae 1-segmented, conical, antennal field feebly convex, with several fine setae and minute papillae. Clypeus (fig. 34.897b) vestigial, reduced to a narrow membrane connecting labrum to epistoma. Labrum (fig. 34.897b) well-developed, corneous, more or less trapezoidal, usually trilobed anteriorly, epipharyngeal surface with a variable number of setiferous sockets, each bearing 1 or more setae; dorsal surface with 1 or 2 pairs of strong setae. Maxilla closely united with labrum, cardo and stipes not fused, maxillary palp 1- or 2-segmented. Labial palp 1- or 2-segmented. Mandible strong, with an oblique cutting edge.

Thorax and Abdomen: Cuticle micro-asperate, asperities of prosternum particularly strong and hair-like. Three thoracic and 10 abdominal segments indicated by intersegmental folds, 10th segment acuminate or tridentate at apex.

Spiracles: Moderately large, simple or with an accessory chamber which appears as a sac-like or finger-like process projecting beyond peritreme.

Comments: This family is mainly represented in the tropical regions of the world, with about 1500 species known. Seven species occur in N. America.

Selected Bibliography

Browne 1961, 1972.
Crowson 1955.
Gardner 1932b.
Morimoto 1962b.
Roberts 1968.
Schedl 1972, 1978.
Wood 1973.

COLEOPTERA BIBLIOGRAPHY

Abdullah, M. 1964. A revision of the genus *Pilipalpus* (Coleoptera, Anthicidae: Pedilinae). *Beiträge zur Entomologie* 14(1/2):3–9.

Abdullah, M. 1965. The genus *Nematoplus* LeConte (Coleoptera, Cephaloidae) and its affinities with Meloidae. *Entomologist* 1965:54–59.

Abdullah, M. 1966. Byturidae and Biphyllidae (Coleoptera), two primitive families of the Heteromera not the Clavicornia—a new interpretation of some old observations. *Entomological News* 77:63–69.

Abdullah, M. 1969. The natural classification of the family Anthicidae with some ecological and ethological observations (Coleoptera). *Deutsche Entomologische Zeitschrift* (N.F.)16:323–366, 3 pls.

d'Aguilar, J. 1937. Description de la larve de *Anommatus basalis*. *Bulletin Assoc. Col. Seine* 2:7–8.

d'Aguilar, J. 1944. Contribution à l'étude des Phalacridae. *Annales des epiphyties* (2)10:85–91, pl. 1.

d'Aguilar, J. 1962a. Famille des Alleculidae, pp. 368–373 *in* A. S. Balachowsky, ed., *Entomologie appliquée à l'agriculture. Tome I. Coléoptères. Premier volume.* Paris: Masson, 564 pp.

d'Aguilar, J. 1962b. Famille des Byturidae, pp. 335–340 *in* A. S. Balachowsky, ed., *Entomologie appliquée à l'agriculture. Tome I. Coléoptères. Premier volume.* Paris: Masson, 564 pp.

d'Aguilar, J. 1962c. Famille des Mordellidae, pp. 353–357 *in* A. S. Balachowsky, ed., *Entomologie appliquée à l'agriculture. Tome I. Coléoptères. Premier volume.* Paris: Masson, 564 pp.

Ahmad, M. and Burke, H. R. 1972. Larvae of the weevil tribe Anthonomini (Coleoptera: Curculionidae). *Miscellaneous Publications of the Entomological Society of America* 8(2):31–81, 121 figs.

Ainslie, G. G. 1920. The cornpith weevil (*Centrinus penicellus* Herbst). *Journal Economic Entomology* 13(3):271–280.

Aitken, A. D. 1975. Insect Travellers. Volume I. Coleoptera. *Ministry of Agriculture, Fisheries and Food. Pest Infestation Control Laboratory. Technical Bulletin,* 31, xvi + 191 pp.

Alexseev, A. V. 1960. On the morphology and systematics of larvae of some species of the genus *Agrilus* Curt. in the European part of the U.S.S.R. (In Russian). *Zoologicheskii Zhurnal* 39:1497–1510.

Allen, R. T. and Hutton, R. S. 1969. A new species of Telegeusidae (Leng) from Panama (Coleoptera: Cantharoidea). *Coleopterists Bulletin* 23:109–112.

Allsopp, P. G. 1979. Identification of false wireworms (Coleoptera: Tenebrionidae) from southern Queensland and northern New South Wales. *Journal of the Australian Entomological Society* 18:277–286.

Anderson, D. M. and Anderson, W. H. 1973. A key to separate larvae of white-fringed beetles, *Graphognathus* species, from larvae of closely related species. *United States Department of Agriculture Cooperative Economic Insect Report* 23(49–52):797–800, 10 figs.

Anderson, J. and Nilssen, A. C. 1978. The food selection of *Pytho depressus* L. (Col., Pythidae). *Norwegian Journal of Entomology* 25:225–226.

Anderson, R. S. 1982. Burying beetle larvae: Nearctic *Nicrophorus* and Oriental *Ptomascopus morio* (Silphidae). *Systematic Entomology* 7:249–264.

Anderson, R. S. and Peck, S. B. 1985. *The carrion beetles of Canada and Alaska. Coleoptera: Silphidae and Agyrtidae. The insects and arachnids of Canada.* Part 13, 121 pp. (Publication No. 1778) Research Branch, Canada Department of Agriculture, Ottawa, Canada.

Anderson, W. H. 1936. A comparative study of the labium of coleopterous larvae. *Smithsonian Miscellaneous Collections* 95(13):1–29, 8 pls.

Anderson, W. H. 1938. Description of the larvae of *Chaetocnema denticulata* (Illiger) and *Chaetocnema pulicaria* Melsheimer (Coleoptera: Chrysomelidae). *Proceedings of the Entomological Society of Washington* 40:161–169.

Anderson, W. H. 1939. A key to the larval Bostrichidae in the U.S. National Museum (Coleoptera). *Journal of the Washington Academy of Sciences* 29:382–391.

Anderson, W. H. 1941a. The larva and pupa of *Cylindrocopturus furnissi* Buchanan (Coleoptera: Curculionidae). *Proceedings of the Entomological Society of Washington* 43(7):152–155, 12 figs.

Anderson, W. H. 1941b. On some larvae of the genus *Proterhinus* (Coleoptera: Aglycyderidae). *Proceedings of the Hawaiian Entomological Society* 11(1):25–35.

Anderson, W. H. 1943. The larva of *Holostilpna nitens* (LeC.) and its relationships (Coleoptera: Anthribidae). *Proceedings of the Entomological Society of Washington* 45:171–175.

Anderson, W. H. 1947a. Larvae of some genera of Anthribidae (Coleoptera). *Annals of the Entomological Society of America* 40(3):489–517.

Anderson, W. H. 1947b. A terminology for the anatomical characters useful in the taxonomy of weevil larvae. *Proceedings of the Entomological Society of Washington* 49(5):123–132, 13 figs.

Anderson, W. H. 1948a. Larvae of some genera of Calendrinae (= Rhynchophorinae) and Stromboscerinae (Coleoptera: Curculionidae). *Annals of the Entomological Society of America* 41:413–437, 19 figs.

Anderson, W. H. 1948b. A key to the larvae of some species of *Hypera* Germar, 1817 (= *Phytonomus* Schoenherr, 1823). *Proceedings of the Entomological Society of Washington* 50(2):25–34, 30 figs.

Anderson, W. H. 1949. Larval description and transfer of *Thaumaphrastus karanisensis* from Colydiidae to a new subfamily of Dermestidae (Coleoptera). *Bulletin of the Brooklyn Entomological Society* 44:121–127.

Anderson, W. H. 1952. Larvae of some genera of Cossoninae (Coleoptera: Curculionidae). *Annals of the Entomological Society of America* 45:281–309, 22 figs.

Andrews, F. G. 1976a. A new species of *Fuchsina* Fall with notes on some California Lathridiidae (Coleoptera). *Pan-Pacific Entomologist* 52:339–347.

Andrews, F. G. 1976b. A revision of the North American species of *Metophthalmus* (Coleoptera: Lathridiidae). *Coleopterists Bulletin* 30:37–56.

Angelini, F. and L. DeMarzo. 1984. Morfologia della larva matura e della pupa in *Agathidium varians* Beck (Coleoptera, Leiodidae, Anisotomini). *Entomologica* 19:51–60.

Angus, R. B. 1973. The habits, life histories and immature stages of *Helophorus* F. (Coleoptera: Hydrophilidae). *Transactions of the Royal Entomological Society of London* 125:1–26.

Anonymous. 1974. European alfalfa beetle in the U.S. 1974. *United States Department of Agriculture Cooperative Economic Insect Report* 24(36):731–734.

Arant, F. S. 1938. Life history and control of the cowpea curculio. *Alabama Agricultural Experiment Station Bulletin* 246, 34 pp., 14 figs.

Arnett, R. H. 1951. A revision of the Nearctic Oedemeridae (Coleoptera). *American Midland Naturalist* 45(2):257–391.

Arnett, R. H. 1953. A review of the beetle family Cephaloidae. *Proceedings of the United States National Museum* 103:155–161.

Arnett, R. H. 1961. Contribution towards a monograph of the Oedemeridae 14. A key to and notes on the New World genera. *Coleopterists Bulletin* 15:49–64.

Arnett, R. H. 1964. Notes on Karumiidae (Coleoptera). *Coleopterists Bulletin* 18:65–68.

Arnett, R. H. 1968. *The Beetles of the United States. (A Manual for Identification).* Ann Arbor: American Entomological Institute, xii + 1112 pp.

Arnoldi, L. V. and Byzova, Yu. B. 1964. Family Curculionidae—Weevils, pp. 535–573. *In* Gilyarov, M. S. (edit.), *Determination Key to Soil-dwelling Insect Larvae.* Moscow: "Nauka", 918 pp. (in Russian).

Arnoldi, L. V. and Zherikin, V. V. 1977. Rhynchophora, pp. 142–182. *In* Arnoldi, L. V., Zherikin, V. V., Nikritkin, L. M. and Ponomarenko, A. G. (eds.) *Mesozoic Coleoptera.* Moscow: "Nauka," 204 pp. (in Russian).

Arora, G. L. 1978. Taxonomy of the Bruchidae (Coleoptera) of Northwest India Part II. Larvae. *Oriental Insects.* Supp. 8:48 pp.

Arrow, G. J. 1920. A contribution to the classification of the coleopterous family Endomychidae. *Transactions of the Entomological Society of London* 1920:1–83, 1 pl.

Arrow, G. J. 1930. A new family of heteromerous Coleoptera (Hemipeplidae), with descriptions of a new genus and a few new species. *Annals and Magazine of Natural History* (10)5:225–231.

Artigas, J. N. 1963. Descripcion de la larva de un psefenido Chileno (Coleoptera-Psephenidae). *Gayana (Zoologia)* 8:3–8, 1 pl.

Artigas, J. N. and Brañas-Rivas, P. 1973. Descripcion de la larva y pupa de *Praocis curta* Solier (Coleoptera-Tenebrionidae). *Boletin de la Sociedad de Biologia de Concepcion* 46:65–74.

Arzone, A. 1970. Reperti ecologici ed etologici di *Liodes cinnamomea* Panzer vivente su *Tuber melanosporum* Vittadini. *Annali della Facoltà di Scienze Agrarie della Università degli Studi di Torino* 5:317–357.

Arzone, A. 1971. Nuovi reperti sulla biologia di *Liodes cinnamomea* Panzer in *Tuber magnatum* Pico (Coleoptera: Staphylinoidea). *Allionia* 17:121–129.

Ashe, J. S. 1984a. Description of the larva and pupa of *Scaphisoma terminata* Melsh. and the larva of *Scaphium castanipes* Kirby with notes on their natural history (Coleoptera: Scaphidiidae). *Coleop. Bull.* 38:361–73.

Ashe, J. S. 1984b. Major features of the evolution of relationships between gyrophaenine staphylinid beetles (Coleoptera: Staphylinidae: Aleocharinae) and fresh mushrooms. Pp. 227–255 *in* Wheeler, Q. D. and Blackwell, M. (eds.) *Fungus/Insect Relationships: Perspectives in Ecology and Evolution.* New York; Columbia University Press, 512 pp.

Ashe, J. S. 1986. Structural features and phylogenetic relationships among larvae of genera of gyrophaenine staphylinids (Coleoptera: Staphylinidae: Aleocharinae). *Fieldiana Zoology n.s.* 30:i–iv, 1–60.

Ashe, J. S. and Watrous, L. E. 1984. Larval chaetotaxy of Aleocharinae (Staphylinidae) based on a description of *Atheta coriaria* Kraatz. *Coleopterists Bulletin* 38:165–169.

Ashlock, P. D. and Lattin, J. D. 1963. Stridulatory mechanisms in the Lygaeidae, with a new American genus of Orsillinae (Hemiptera: Heteroptera). *Annals of the Entomological Society of America* 56:693–703.

Atkins, M. D. 1957. An interesting attractant for *Priacma serrata* (LeC.) (Cupesidae: Coleoptera). *Canadian Entomologist* 89:214–219.

Atkins, M. D. 1963. The Cupedidae of the world. *Canadian Entomologist* 95:140–162.

Baker, C. W. 1968. Larval taxonomy of the Troginae in North America with notes on biologies and life histories (Coleoptera: Scarabaeidae). *Bulletin of the United States National Museum* 279:1–80.

Baker, N. W. 1971. Observations on the biology of the giant palm-boring beetle, *Dinapate wrighti* Horn (Coleoptera: Bostrichidae). *Journal of the New York Entomological Society* 79:31–42.

Baker, W. L. 1972. Eastern Forest Insects. *United States Department of Agriculture Forest Service Miscellaneous Publications* 1175, 642 pp.

Balachowsky, A. S. 1962. Famille des Cerambycidae, pp. 394–434 *in* A. S. Balachowsky, ed., *Entomologie appliquée à l'agriculture. Tome I. Coléoptères. Premier volume.* Paris: Masson, 564 pp.

Balachowsky, A. S. Davatchi, A., and Descarpentries, A. 1962. Famille des Buprestidae, pp. 235–300 *in* A. S. Balachowsky, ed., *Entomologie appliquée à l'agriculture. Tome I. Coléoptères. Premier volume.* Paris: Masson, 564 pp.

Balachowsky, A. S. and Hoffmann, A. 1962. Famille des Attelabidae, pp. 1202–1237 *in* A. S. Balachowsky, ed., *Entomologie appliquée à l'agriculture. Tome I. Coléoptères. Second volume.* Paris: Masson, pp. 567–1391.

Balduf, W. V. 1935. *The Bionomics of Entomophagous Coleoptera.* St. Louis: John S. Swift, 220 pp.

Balduf, W. V. 1959. *Obligatory and Facultative Insects in Rose Hips: their Recognition and Bionomics.* Urbana, Illinois: University of Illinois Press, 194 pp.

Balfour-Browne, F. 1922. The life-history of the water beetle *Pelobius tardus* Herbst. *Proceedings of the Zoological Society of London* 1922:79–97.

Balfour-Browne, F. and Balfour-Browne, J. 1940. An outline of the habits of the water-beetle, *Noterus capricornis* Herbst (Coleopt.). *Proceedings of the Royal Entomological Society of London* (A)15:105–112.

Banks, N. 1912. At the *Ceanothus* in Virginia. *Entomological News* 23:102–110.

Barber, H. S. 1905. Notes on *Phengodes* in the vicinity of Washington, D.C. *Proceedings of the Entomological Society of Washington* 7:196–197.

Barber, H. S. 1907. The glow-worm *Astraptor. Proceedings of the Entomological Society of Washington* 9:41–43.

Barber, H. S. 1913a. Observations on the life history of *Micromalthus debilis* Lec. (Coleoptera). *Proceedings of the Entomological Society of Washington* 15:31–38, pl. 3.

Barber, H. S. 1913b. The remarkable life-history of a new family (Micromalthidae) of beetles. *Proceedings of the Biological Society of Washington* 26:185–190.

Barber, H. S. 1942. Raspberry fruitworms and related species. *United States Department of Agriculture Miscellaneous Publications,* No. 468, 32 pp.

Barber, H. S. 1952. Notes on *Telegeusis* and some relatives. *Pan-Pacific Entomologist* 28(3):163–170.

Barker, J. F. 1969. Notes on the life cycle and behavior of the drilid beetle *Selasia unicolor* (Guerin). *Proceedings of the Royal Entomological Society of London* (A)44(10–12):169–172, 1 pl.

Barlet, J. 1972. Sur le thorax de certain Myxophaga Crowson (Coléoptères). *Bulletin de l'Institut royale de sciences naturelles de Belgique* 48(14):1–6.

Barman, E. H., Jr. 1973. Biology and immature stages of *Desmopachria convexa* (Aubé). *Proceedings of the Entomological Society of Washington* 75:233–239.

Barr, W. F. 1961. Family Cleridae, pp. 105–112 *in* Hatch, M. H., *The Beetles of the Pacific Northwest. Part III: Pselaphidae and Diversicornia I.* Seattle: Washington University Press, 503 pp.

Barrera, A. 1969. Notes on the behavior of *Loberopsyllus traubi,* a cucujoid beetle associated with the volcano mouse, *Neotomodon alstoni* in Mexico. *Proceedings of the Entomological Society of Washington* 71:481–486.

Barrett, R. E. 1930. A study of the immature forms of some Curculionidae (Coleoptera). *University of California Publications in Entomology* 5(5):89–104, 28 figs.

Barron, J. R. 1971. A revision of the Trogositidae of America north of Mexico (Coleoptera: Cleroidea). *Memoirs of the Entomological Society of Canada* 75:1–143.

Barron, J. R. 1975. A review of the genus *Lycoptis* Casey (Coleoptera: Trogositidae). *Canadian Entomologist* 107:1117–1122.

Bates, H. W. 1861. On the Endomychidae of the Amazon Valley. *Journal of Entomology* 1:158–172, pl. 11.

Batra, L. R. 1967. Ambrosia fungi: a taxonomic revision, and nutritional studies of some species. *Mycologia* 59:976–1017.

Batra, L. R. and Francke-Grosmann, H. 1961. Contributions to our knowledge of ambrosia fungi I. *Ascoidea hylecoeti* sp. nov. (Ascomycetes). *American Journal of Botany* 48:453–456.

Bauer, T. 1982. Prey-capture in a ground beetle larva. *Animal Behavior* 30:203–208.

Baumgaertner, J. V., Frazer, B. D., Gilbert, N., Gill, B., Gutierrez, A. P., Ives, P. M., Nealis, V., Raworth, D. A., and Summers, C. G. 1981. Coccinellids (Coleoptera) and aphids (Homoptera). The overall relationships. *Canadian Entomologist* 113:975–980.

Beal, R. S., Jr. 1959a. Notes on the biology and systematics of the dermestid beetle genus *Apsectus* with descriptions of two new species. *Annals of the Entomological Society of America* 52(2):132–137.

Beal, R. S., Jr. 1959b. A key to the Nearctic genera of Dermestidae. *Coleopterists Bulletin* 13:99–101.

Beal, R. S., Jr. 1960. Descriptions, biology, and notes on the identification of some *Trogoderma* larvae. *United States Department of Agriculture Technical Bulletin,* No. 1228, 26 pp.

Beal, R. S., Jr. 1964. A new superficially cryptic species of *Trogoderma* from the southwestern United States. *Proceedings of the Entomological Society of Washington* 66(2):79–84.

Beal, R. S., Jr. 1970. A taxonomic and biological study of species of Attagenini (Coleoptera: Dermestidae) in the United States and Canada. *Entomologica Americana* 45(3):141–235.

Beal, R. S., Jr. 1975. Description of adult and larval stages of a new species of *Cryptorhopalum* from Arizona and Mexico (Coleoptera: Dermestidae). *Proceedings of the Entomological Society of Washington* 77(2):228–233.

Beal, R. S., Jr. 1985. Review of the Nearctic species of *Orphilus* (Coleoptera: Dermestidae) with description of the larva of *O. subnitidus* LeConte. *Coleopterists Bulletin* 39(3):265–271.

Beaver, R. A. 1967. Notes on the fauna associated with elm bark beetles in Wytham Wood, Berks.—I. Coleoptera. *Entomologists Monthly Magazine* 102:163–170.

Beerbower, F. V. 1943. Life history of *Scirtes orbiculatus* Fabricius (Coleoptera: Helodidae). *Annals of the Entomological Society of America* 36:672–680.

Beeson, C. F. C. 1926. Notes on the biology of the Cleridae. *Indian Forest Records* 12:217–231.

Beeson, C. F. C. 1941. *The Ecology and Control of the Forest Insects of India and the Neighboring Countries.* Dehra Dun: C. F. C. Beeson, 1007 pp.

Beeson, C. F. C. and Bhatia, B. M. 1936. On the biology of the Bostrychidae (Coleopt.). *Indian Forest Records. New Series. Entomology* 2(12):223–323. 3 pls.

Beeson, C. F. C. and Bhatia, B. M. 1939. On the biology of the Cerambycidae (Coleopt.). *Indian Forest Records. New Series. Entomology* 5(1):i–iv + 1–235, 8 pls.

Beier, M. 1949. Körperbau und Lebensweise der Larve von *Helodes haussmanii* Gredler (Col. Helodidae). *Eos* 25:49–100.

Beier, M. 1952. Bau and Funktion der Mundwerkzeuge bei den Helodiden-Larven (Col.). *Transactions of the Ninth International Congress of Entomology, Amsterdam* 1:135–138.

Beier, M. and Pomeisl, E. 1959. Einiges über Körperbau und Lebenweise von *Ochthebius exsculptus* Germ. und seiner Larve (Col. Hydroph. Hydraen.) *Zeitschrift fur Morphologie und Ökologie der Tiere* 48:72–88.

Beling, T. 1882. Beitrag zur Biologie einiger Käfer aus den Familien Dascyllidae und Parnidae. *Verhandlungen der Zoologisch-Botanischen Gesellschaft in Wien* 32:435–442.

Bell, R. T. 1966. *Trachypachus* and the origin of the Hydradephaga (Coleoptera). *Coleopterists Bulletin* 20:107–112.

Bell, R. T. 1970. The Rhysodini of North America, Central America and the West Indies. *Miscellaneous Publications of the Entomological Society of America* 6(6):289:324.

Bell, R. T. and Bell, J. R. 1962. The taxonomic position of the Rhysodidae. *Coleopterists Bulletin* 16:99–106.

Bell, R. T. and Bell, J. R. 1978. Rhysodini of the world. Part I. A new classification of the tribe, and a synopsis of *Omoglymmius* subgenus *Nitiglymmius,* new subgenus (Coleoptera: Carabidae or Rhysodidae). *Quaestiones Entomologicae* 14:43–88.

Bell, R. T. and Bell, J. R. 1979. Rhysodini of the world. Part II. Revisions of the smaller genera (Col: Carabidae or Rhysodidae). *Quaestiones Entomologicae* 15:377–446.

Bell, R. T. and Bell, J. R. 1982. Rhysodini of the world. Part III. Revision of *Omoglymmius* Ganglbauer (Col: Carabidae or Rhysodidae) and substitutions for preoccupied generic names. *Quaestiones Entomologicae* 18:127–259.

Belles, X. 1980. *Ptinus* (*Pseudoptinus*) *lichenum* Marsham, ptinido perforador de Madera (Col. Ptinidae). *Boletin de la Estacion Central de Ecologia* 9(18):89–91.

Belles, X. 1981. Idees sobre la classificacio supragenerica de la familia Ptinidae (Coleoptera). *II Sessio Conjunta d'Entomologia, Barcelona,* pp. 61–65.

Belon, R. P. 1909. *Genera Insectorum de P. Wytsman. Fascicule 3. Coleoptera Claviornia. Fam. Lathridiidae.* Crainhem, Belgium: Genera Insectorum, 40 pp., 1 pl.

Benick, L. 1924. Zur Biologie der Käferfamilie Helodidae. *Mitteilungen der Geologischen Gesellschaft des Naturhistorischen Museums in Lubeck* 29:47–78.

Benick, L. 1952. Pilzkäfer und Käferpilze. Ökologische und statistische Untersuchungen. *Acta Zoologica Fennica* 70:1–250.

Benoit, P. 1964. Comparative morphology of some *Chrysobothris* larvae (Coleoptera: Buprestidae) of eastern Canada. *Canadian Entomologist* 96:1107–1117.

Benoit, P. 1965. Morphologie larvaire des *Agrilus liragus* Barter et Brown et *Agrilus anxius* Gory (Coleoptera: Buprestidae). *Canadian Entomologist* 97:768–773.

Benoit, P. 1966a. Description de la larve du *Melanophila acuminata* De Geer et de quelques caractères distinctifs du *Melanophila fulvoguttata* (Harris) (Coleoptera: Buprestidae). *Canadian Entomologist* 98:1208–1211.

Benoit, P. 1966b. Descriptions of some *Chrysobothris* larvae (Coleoptera: Buprestidae) occurring in the United States and Mexico. *Canadian Entomologist* 98:324–330.

Benton, A. H. and Crump, A. J. 1979. Observations on aggregation and overwintering in the coccinellid beetle *Coleomegilla maculata* (DeGeer). *Journal of the New York Entomological Society* 87:154–159.

Berjon, J. 1964. Presence dans des sols du pays basque de populations importantes de *Dryops ernesti* Goz. (Col. Dryopidae). *Bulletin de la Société entomologique de France* 69:62–63.

Berrios-Ortiz, A. and Selander, R. B. 1979. Skeletal musculature in larval phases of the beetle *Epicauta segmenta* (Coleoptera, Meloidae). The Hague: W. Junk, 112 pp.

Bertkau, P. 1891. Beschreibung der Larve und des Weibchens von *Homalisus suturalis. Deutsche Entomologische Zeitschrift* 1891(1):37–44.

Bertrand, H. 1928. Les larves et nymphes des dytiscides, hygrobiides, et haliplides. *Encyclopédie entomologique,* Ser. A, 10:1–366, 33 pls.

Bertrand, H. 1962. Contribution à l'étude des premiers états des Coléoptères aquatiques de la région éthiopienne (4me note). *Bulletin de l'Institut francais d'Afrique noire* 24(A):1065–1114.

Bertrand, H. 1966. Les premiers états des Ptilodactylidae (Col.) aquatiques. *Bulletin du Museum National d'Histoire Naturelle* (2)38:143–150.

Bertrand, H. 1972. *Larves et nymphes des coléoptères aquatiques du globe.* Paris: F. Paillart, 804 pp., 561 figs.

Bertrand, H. 1977. *Larves es nymphes des coléoptères aquatiques du globe. Errata et Addenda.* Abbéville: F. Paillart, 19 pp.

Besuchet, C. 1952. Larves et nymphes de *Plectoploeus* (Col. Pselaphidae). *Mitteilungen der Schweizerischen Entomologischen Gesellshaft* 25:251–256.

Besuchet, C. 1956a. Biologie, morphologie et systématique des *Rhipidius* (Col. Rhipiphoridae). *Mitteilungen der Schweizerischen Entomologischen Gesellschaft* 29(2):73–144.

Besuchet, C. 1956b. Larves et nymphes de Pselaphidae (Coléoptères). *Revue suisse de zoologie* 63:697–705.

Bily, S. 1972a. The larva of *Dicera* (*Dicera*) *berolinensis* (Herbst) (Coleoptera, Buprestidae), and a case of prothetely in this species. *Acta Entomologica Bohemoslovaka* 69:266–269.

Bily, S. 1972b. The larva of *Ptosima flavoguttata* (Illiger), (Coleoptera, Buprestidae). *Acta Entomologica Bohemoslovaka* 69:18–22.

Bily, S. 1975a. Larvae of European species of the genus *Chrysobothris* Eschsch. (Coleoptera, Buprestidae). *Acta Entomologica Bohemoslovaka* 72:418–424.

Bily, S. 1975b. Larvae of the genus *Amara* (subgenus *Celia* Zimm.) from Central Europe (Coleoptera, Carabidae). *Studie a Prameny Ceskoslovenske Akademie Vedecky Prague* 13:1–74.

Bily, S. 1978. Buprestidae (partim), pp. 154–165 *in* Klausnitzer, B., (ed.) *Ordnung Coleoptera* (*Larven*). The Hague: W. Junk, vi + 378 pp.

Bily, S. 1983. Larvae of *Julodis variolaris freygessneri* Obenberger and *Paracylindromorphus transversicollis* (Reitter) (Coleoptera, Buprestidae). *Acta Entomologica Bohemoslovaka* 80:65–70.

Bishop, G. W. 1960. Taxonomic observations on the larvae of the three American *Cryptolestes* (Coleoptera: Cucujidae) that infest stored grain. *Annals of the Entomological Society of America* 53:8–11.

Bitsch, J. 1966. L'Evolution des structures céphaliques chez les larves de Coléoptères. *Annales de la Société entomologique de France* (N.S.)II(2):255–324.

Blackman, M. W. and Stage, H. H. 1924. On the succession of insects living in the bark and wood of dying, dead, and decaying hickory. *New York State College of Forestry at Syracuse University. Technical Publication*, No. 17(Vol. 24, No. 22), 218 pp., 14 pls.

Blackwelder, R. E. 1944. Checklist of the coleopterous insects of Mexico, Central America, The West Indies, and South America. Part II. *Bulletin of the United States National Museum* 185:189–341.

Blair, K. G. 1914. A revision of the family Pyrochroidae (Coleoptera). *Annals and Magazine of Natural History* 8(13):310–326.

Blair, K. G. 1930. *Brachypsectra* LeC.—The solution of an entomological enigma. *Transactions of the Royal Entomological Society of London* 78:45–50, 1 pl.

Blaisdell, F. E. 1901. Pupa of *Necrophilus hydrophiloides*. *Entomological News* 12:263–265.

Blaisdell, F. E. 1934. Rare North American Coleoptera. *Transactions of the American Entomological Society* 60:317–326.

Blas, M. and Vives, E. 1978. A proposito de la biologia y desarrollo de *Cholevinus pallidus* (Ménétries, 1832) (Col. Catopidae). *Miscellanea Zoologica* 4:147–159.

Blatchley, W. S. 1910. *An Illustrated Descriptive Catalogue of the Coleoptera or beetles (exclusive of the Rhynchophora) known to occur in Indiana*. Indianapolis, Indiana, Nature. 1386 pp.

Blatchley, W. S. and Leng, C. W. 1916. *Rhynchophora or Weevils of North Eastern America*. Indianapolis: Nature Publishing Co., 682 pp.

Blewett, M. and Fraenkel, G. 1944. Intracellular symbiosis and vitamin requirements of two insects, *Lasioderma serricorne* and *Sitodrepa panicea*. *Proceedings of the Royal Society* (B)132:212–221.

Blume, R. R. 1981. *Glaphyrocanthon viridis viridis* (Beauvois): description of larva and notes on biology (Coleoptera: Scarabaeidae). *Coleopterists Bulletin* 35(2):235–238.

Böving, A. G. 1914a. Notes on the larva of *Hydroscapha* and some other aquatic larvae from Arizona. *Proceedings of the Entomological Society of Washington* 16:169–174, pls. 17–18.

Böving, A. G. 1914b. On the abdominal structure of certain beetle larvae of the campodeiform type. A study of the relationships between the structure of the integument and the muscles. *Proceedings of the Entomological Society of Washington* 16:55–61.

Böving, A. G. 1917. A generic synopsis of the coccinellid larvae in the United States National Museum with a description of the larva of *Hyperaspis binotata* Say. *Proceedings of the United States National Museum* 51:621–650.

Böving, A. G. 1921. The larvae and pupae of the social beetles *Coccidotrophus socialis* (Schwarz and Barber) and *Eunausibius wheeleri* (Schwarz and Barber) with remarks on the taxonomy of the family Cucujidae. *Zoologica* 3:197–213, pls. 7–10.

Böving, A. G. 1927. Descriptions of larvae of the genera *Diabrotica* and *Phyllobrotica*, with a discussion of the taxonomic validity of the subfamilies Galerucinae and Halticinae (Coleoptera: Chrysomelidae). *Proceedings of the Entomological Society of Washington* 29:193–205, 1 pl.

Böving, A. G. 1929a. Beetle larvae of the subfamily Galerucinae. *Proceedings of the United States National Museum* 72:1–48.

Böving, A. G. 1929b. On the classification of beetles according to larval characters. *Bulletin of the Brooklyn Entomological Society* 24:55–80.

Böving, A. G. 1936. Description of the larva of *Plectris aliena* Chapin and explanation of new terms applied to the epipharynx and raster. *Proceedings of the Entomological Society of Washington* 38(8):169–185, pls. 9–10.

Böving, A. G. 1942. A classification of larvae and adults of the genus *Phyllophaga* (Coleoptera: Scarabaeidae). *Memoirs of the Entomological Society of Washington* 2:1–95.

Böving, A. G. 1954. Mature larvae of the beetle family Anobiidae. *Danske Biologiske Meddelelser* 22(2):1–298.

Böving, A. G. 1955. A correction. *Proceedings of the Entomological Society of Washington* 57:202.

Böving, A. G. 1956. A description of the mature larva of *Ptinus californicus* Pic and a discussion about the justification of considering the Ptinidae as a valid family of beetles. *Entomologiske Meddelelser* 27:229–241, 2 pls.

Böving, A. G. and Champlain, A. B. 1921. Larvae of North American beetles of the family Cleridae. *Proceedings of the United States National Museum* 57:575–649.

Böving, A. G. and Craighead, F. C. 1931. An illustrated synopsis of the principal larval forms of the Order Coleoptera. *Entomologica Americana* (*New Series*) 11:1–351.

Böving, A. G. and Hendriksen, K. 1938. The developmental stages of the Danish Hydrophilidae. *Videnskabelige Meddelelser fra Dansk Naturhistorisk Forening i Khobenhavn* 102:27–162.

Böving, A. G. and Rozen, J. G. 1962. Anatomical and systematic study of the mature larvae of the Nitidulidae (Coleoptera). *Entomologiske Meddelelser* 31:265–299.

Bohác, J. 1982. The larval characters of Czechoslovak species of the genera *Abemus*, *Staphylinus* and *Ocypus*. *Studie Ceskoslovenske Akademie Ved* 4:1–96, pl. 1–27.

Bolívar y Pieltain, C. 1940. Sobre algunos insectos alpinos de los volcanes de Mexico. *Revista de la Sociedad Mexicana de Historia Natural* 1:175–177.

Bonadona, P. 1958. *Faune de Madagascar. VI. Insectes Coléoptères Anthicidae*. Tananarive-Tsimbazaza: Institut de Recherche Scientifique, 153 pp.

Bondar, G. 1940a. Insectos nocivos e moles tias do coqueiro (*Cocos nucifera*) no Brasil. *Boletin do Instituto Central de Fomento Economica da Bahia* 8:1–160.

Bondar, G. 1940b. Notas entomologicas da Bahia. V. *Revista de Entomologia* 11:211–214.

Booker, R. H. 1967. Observations on three bruchids associated with cowpea in northern Nigeria. *Journal of Stored Products Research* 3(1):1–15.

Borden, J. H. and McClaren, M. 1970. Biology of *Cryptoporus volvatus* (Peck) Shear (Agaricales, Polyporaceae) in southwestern British Columbia: distribution, host species, and relationship with subcortical insects. *Syesis* 3:145–154.

Borden, J. H. and McClaren, M. 1972. Biology of *Cryptoporus volvatus* (Peck) Shear (Agaricales, Polyporaceae) in southwestern British Columbia: life history development and arthropod infestation. *Syesis* 5:67–72.

Bornemissza, G. F. 1968. Studies on the histerid beetle *Pachylister chinensis* in Fiji, and its possible value in the control of Buffalo-fly in Australia. *Australian Journal of Zoology* 16:673–688.

Borror, D. J., D.M. DeLong and C. A. Triplehorn. 1981. *An Introduction to the Study of Insects.* Fifth edition. Saunders College Publishing. New York, 827 pp.

Bottimer, J. J. 1968. Notes on Bruchidae of America North of Mexico with a list of world genera. *Canadian Entomologist* 100(10):1009–1048.

Bourgeois, J. 1891. Etudes sur la distribution géographique des Malacodermes. I. Lycides. *Annales de la Société entomologique de France* 11:337–364, 1 pl.

Bousquet, Y. 1983. Brood care in *Pterostichus* (*Monoferonia*) *diligendus* Chaudoir (Coleoptera: Carabidae). *Coleopterists Bulletin* 37(4):307–308.

Bousquet, Y. 1985. Morphologie comparée des larves de Pterostichini (Coleoptera: Carabidae): descriptions et tables de détermination des espèces du nord-est de l'Amérique du Nord. *Naturaliste canadien* 112:191–251.

Bousquet, Y. 1986. Description of first-instar larva of *Metrius contractus* Eschscholtz (Coleoptera: Carabidae) with remarks about phylogenetic relationships and ranking of the genus *Metrius* Eschscholtz. *Canadian Entomologist* 118:373–388.

Bousquet, Y. 1987. Description of the larva of *Helluomorphoides praeustus bicolor* Harris with comments on the relationships of the Helluonini (Coleoptera: Carabidae). *Canadian Entomologist* 119:921–930.

Bousquet, Y. and Goulet, H. 1984. Notation of primary setae and pores on larvae of Carabidae (Coleoptera: Adephaga). *Canadian Journal of Zoology* 62:573–588.

Bousquet, Y. and Pilon, J.-G. 1980. Habitat et cycle biologique des *Sphaeroderus* du Québec (Coleoptera: Carabidae: Cychrini). *Naturaliste Canadien* 107:175–184.

Böving—*See* after Blume.

Boyce, A. M. 1927. A study of the biology of the parsley stalk-weevil *Listronotus latiusculus* Boheman, Coleoptera: Curculionidae. *Journal of Economic Entomology* 20(6):814–821, 9 figs.

Boyle, W. W. 1956. A revision of the Erotylidae of America north of Mexico (Coleoptera). *Bulletin of the American Museum of Natural History* 110:61–172, pl. 8.

Brancucci, M. 1980. Morphologie comparée, évolution et systématique des Cantharidae. (Insecta: Coleoptera). *Entomologica Basiliensia* 5:215–388.

Brandmayr, P. 1974. Le cure parentali di *Carterus* (*Sabienus*) *calydonius* Rossi (Coleoptera: Carabidae). *Atti della Academia delle Scienze di Torino I.—Classe di Scienze Fisiche, Matematiche e Naturali* 108(1973–4):811–818.

Brandmayr, P. and Brandmayr Zetto, T. 1974. Sulle cure parentali e su altri aspetti della biologia di *Carterus* (*Sabenius*) *calydonius* Rossi, con alcune considerazioni sui fenomeni di cura della prole sino ad oggi riscontrati in Carabidi (Coleoptera: Carabidae). *Redia* 55:143–175.

Brandmayr, P. & Brandmayr Zetto T. 1982. Identificazione di larve del genere *Ophonus* Dejean, 1821 (*sensu novo*) e note bionomiche (Coleoptera Carabidae). *Memorie Societa entomologica italiana* 60:67–103.

Brandmayr Zetto, T. and Brandmayr, P. 1975. Biologia di *Ophonus puncticeps* Steph. Cenni sulla fitofagia delle larve e loro etologia (Coleoptera, Carabidae). *Annali della Facolta di Scienze Agraria della Universita degli Studi di Torino* 9:421–430.

Brass, P. 1914. Das 10. Abdominalsegment der Käferlarven als Bewegungsorgan. *Zoologische Jahrbücher Abteilung für Systematik* 37:65–122, pls. 4–7.

Bräuer, G. 1970. Bedeutung der Leistenkopfplattkäfer (*Cryptolestes* Gangl.; Coleopt.; Cucujidae) für die Vorratshaltung von Getreide und Getreideerzeugnissen. *Nachrichtenblatt des deutschen Pflanzenschutzdienstes* (*Braunschweig*) 24:216–222.

Braune, R. 1930. Untersuchungen an *Niptus hololeucus* Fald. Teil I: Morphologie und Biologie. *Zeitschrift für Morphologie und Ökologie der Tiere* 16:234–370.

Brendell, M. J. D. 1975. *Handbooks for the Identification of British Insects. Vol. 5. Part 10. Coleoptera. Family Tenebrionidae.* London: Royal Entomological Society of London, 22 pp.

Brewer, J. W. and Bacon, T. R. 1975. Biology of the carrion beetle *Silpha ramosa* Say. *Annals of the Entomological Society of America* 68:786–790.

Bridwell, J. C. 1929. The cowpea bruchid (Coleoptera) under another name—a plea for one kind of entomological specialist. *Proceedings of the Entomological Society of Washington* 31(2):39–44.

Bridwell, J. C. 1933. The hairy-vetch bruchid, *Bruchus brachialis* Fahraeus, in the United States. *Journal of Agricultural Research* 46(8):739–751.

Brigham, W. U. 1982a. Dytiscidae, pp. 10.47–10.72 *in* Brigham, A. R., Brigham, W. U. and Gnilka, A. (eds.) *Aquatic Insects and Oligochaetes of North and South Carolina.* Mahomet, Illinois: Midwest Aquatic Enterprises, 837 pp.

Brigham, W. U. 1982b. Haliplidae, pp. 10.38–10.44 *in* Brigham, A. R., Brigham, W. U. and Gnilka, A. (eds.) *Aquatic Insects and Oligochaetes of North and South Carolina.* Mahomet, Illinois: Midwest Aquatic Enterprises, 837 pp.

Brigham, W. U. 1982c. Aquatic Coleoptera, Hydraenidae, pages 10.72–10.75 *in* Brigham, A. R., W. U. Brigham, and A. Gnilka (eds.). *Aquatic Insects and Oligochaetes of North and South Carolina.* Midwest Aquatic Enterprises, Mahomet, Illinois. 837 pp.

Brigham, W. U. 1982d. Hydrophilidae, p. 10.75–10.96. *In* Brigham, A. R., W. U. Brigham, and A. Gnilka (eds.), *Aquatic Insects and Oligochaetes of North and South Carolina.* Midwest Aquatic Enterprises, Mahomet, Illinois. 837 pp.

Brigham, W. U. 1982e. Noteridae, pp. 10.44–10.47 *in* Brigham, A. R., Brigham, W. U. and Gnilka, A. (eds.) *Aquatic Insects and Oligochaetes of North and South Carolina.* Mahomet, Illinois: Midwest Aquatic Enterprises, 837 pp.

Bright, D. E. 1976. *The Insects and Arachnids of Canada. Part 2. The Bark Beetles of Canada and Alaska* (*Coleoptera: Scolytidae*). Ottawa, Canada: Canada Department of Agriculture, Research Branch (Publication No. 1576), 241 pp.

Bright, D. E. 1979. Cleroidea, p. 377 *in* Danks, H. V. (edit.), Canada and its insect fauna. *Memoirs of the Canadian Entomological Society* 108:1–573.

Bright, D. E. and Stark, R. W. 1973. The bark and ambrosia beetles of California (Coleoptera: Scolytidae and Platypodidae). *Bulletin of the California Insect Survey* 16, 169 pp.

Brindley, T. A. 1933. Some notes on the biology of the pea weevil *Bruchus pisorum* L. (Coleoptera: Bruchidae) at Moscow, Idaho. Journal Economic Entomology 26:1058–1062.

Brindley, T. A. and Chamberlin, J. C. 1952. The pea weevil, pp. 530–537 *in Insects. The Yearbook of Agriculture. 1952.* Washington, D.C.: United States Department of Agriculture, xviii + 780 pp., 72 pls.

Britton, E. B. 1966. On the larva of *Sphaerius* and the systematic position of the Sphaeriidae. *Australian Journal of Zoology* 14:1193–1198.

Britton, E. B. 1970. Coleoptera (Beetles), pp. 495–621 *in* C.S.I.R.O., *The Insects of Australia* Melbourne: Melbourne University Press, 1029 pp.

Britton, E. B. 1971. A new intertidal beetle (Coleoptera: Limnichidae) from the Great Barrier Reef. *Journal of Entomology* (B)40:83–91.

Britton, E. B. 1974. Coleoptera, pp. 62–89 *in* C.S.I.R.O., *The Insects of Australia. Supplement 1974.* Melbourne: Melbourne University Press, viii + 146 pp.

Britton, E. B. 1981. The Australian Hygrobiidae (Coleoptera). *Journal of the Australian Entomological Society* 20:83–86.

Brooks, J. G. 1965. New records of Coleoptera in Australia. *Journal of the Entomological Society of Queensland* 4:85.

Brown, C. and Crowson, R. A. 1980. Observations on scydmaenid (Col.) larvae with a tentative key to the main British genera. *Entomologist's Monthly Magazine* 115:49–59.

Brown, E. S. 1954. The biology of the coconut pest *Melittomma insulare* (Col., Lymexylonidae), and its control in the Seychelles. *Bulletin of Entomological Research* 45:1–66, 6 pls.

Brown, H. P. 1972. *Aquatic Dryopoid Beetles (Coleoptera) of the United States. Biota of Freshwater Ecosystems Identification Manual No. 6* Washington DC: Environmental Protection Agency, 82 pp.

Brown, H. P. 1975. A distributional checklist of North American genera of aquatic dryopoid and dascilloid beetles. *Coleopterists Bulletin* 29:149–160.

Brown, H. P. 1976. Same as Brown 1972, but issued by Cincinnati, Ohio office of EPA.

Brown, H. P. 1980. A new genus and species of water beetle from Alabama (Psephenidae: Eubriinae). *Transactions of the American Microscopical Society* 99:187–192.

Brown, H. P. 1981a. A distributional survey of the world genera of aquatic dryopoid beetles (Coleoptera: Dryopidae, Elmidae, and Psephenidae *sens latu*). *Pan-Pacific Entomologist* 57:133–148.

Brown, H. P. 1981b. Key to the world genera of Larinae (Coleoptera, Dryopoidea, Elmidae), with descriptions of new genera from Hispanola, Colombia, Australia, and New Guinea. *Pan-Pacific Entomologist* 57:76–104.

Brown, H. P. 1987. Biology of riffle beetles. *Annual Review Entomology* 32:253–73.

Brown, H. P. and Murvosh, C. M. 1970. *Lutrochus arizonicus* new species, with notes on ecology and behavior (Coleoptera, Dryopoidea, Limnichidae). *Annals of the Entomological Society of America* 63(4):1030–1035.

Brown, H. P. and Murvosh, C. M. 1974. A revision of the genus *Psephenus* (water penny beetles) of the United States and Canada (Coleoptera, Dryopoidea, Psephenidae). *Transactions of the American Entomological Society* 100(3):289–340.

Brown, K. W. 1973. Description of immature stages of *Philolithus densicollis* and *Stenomorpha puncticollis* with notes on their biology (Coleoptera, Tenebrionidae, Tentyriinae). *Postilla* 162:1–28.

Browne, F. G. 1961. The biology of Malayan Scolytidae and Platypodidae. *Malayan Forest Records,* No. 22, xi + 255 pp.

Browne, F. G. 1962. *Sosylus spectabilis* Grouvelle (Coleoptera, Colydiidae), a predator and parasite of African ambrosia beetles. *Report of the West African Timber Borer Research Unit 5* (1961–62):91–96.

Browne, F. G. 1972. Larvae of the principal Old World genera of Platypodinae (Coleoptera: Platypodidae). *Transactions of the Royal Entomological Society of London* 124:167–190.

Bruch, C. 1923. Coleopteros fertilizadores de *"Prosopanche Burmeisteri"* De Bary. *Revista de la Sociedad Argentina de Ciencias Naturales* 7:82–88.

Bruns, T. D. 1984. Insect mycophagy in the Boletales: fungivore diversity and the mushroom habitat. pp. 91–129 *in* Wheeler, Q. D. and M. Blackwell (eds.). *Fungus/insect relationships: perspectives in ecology and evolution.* New York: Columbia University Press. 514 pp.

Buchanan, W. D. 1960. Biology of the oak timberworm *Arrhenodes minutus. Journal of Economic Entomology* 53(4):510–513.

Buchner, P. 1965. *Endosymbiosis of Animals with Plant Microorganisms. Revised English Version.* New York: Interscience, xvii + 909 pp.

Buck, F. D. 1954. *Handbooks for the Identification of British Insects. Vol. V. Part 9. Coleoptera (Lagriidae, Alleculidae, Tetratomidae, Melandryidae, Salpingidae, Pythidae, Mycteridae, Oedemeridae, Mordellidae, Scraptiidae, Pyrochroidae, Rhipiphoridae, Anthicidae, Aderidae and Meloidae).* London: Royal Entomological Society of London, 30 pp.

Bugnion, E. 1907. Les métamorphoses du *Ditonectes pubicornis* Walk. *Annales de la Société entomologique de France* 76:118–122.

Bugnion, E. 1909. Les métamorphoses de l'*Eumorphus pulchripes* Gerst., de Ceylon. *Annales de la Société entomologique de France* 78:282–286, pl. 11.

Buntin, L. A. and Tamaki, G. 1980. Bionomics of *Scymnus marginicollis* (Coleoptera: Coccinellidae). *Canadian Entomologist* 112:675–680.

Burakowski, B. 1973. Immature stages and biology of *Drapetes biguttatus* (Piller) (Coleoptera, Lissomidae). *Annales Zoologici, Warsaw* 30(10):335–347.

Burakowski, B. 1975a. Descriptions of larva and pupa of *Rhysodes sulcatus* F. and notes on the bionomics of this species. *Annales Zoologici, Warsaw* 32(12):271–287.

Burakowski, B. 1975b. Development, distribution, and habits of *Trixagus dermestoides* (L.), with notes on the Throscidae and Lissomidae (Coleoptera, Elateroidea). *Annales Zoologici, Warsaw* 32(17):375–405.

Burakowski, B. 1976. Post-embryonic development and bionomics of *Quasimus minutissimus* (Germar) (Coleoptera, Elateridae). *Annales Zoologici, Warsaw* 33(15):235–259.

Burgess, A. F. and Collins, C. W. 1912. The value of predaceous beetles in destroying insect pests, pp. 453–466 *in Yearbook of the United States Department of Agriculture. 1911.* Washington, D.C.: United States Department of Agriculture, 732 pp.

Burke, H. E. 1906. Notes on the larva of *Calopus angustus* Lec. *Proceedings of the Entomological Society of Washington* 8:64–66.

Burke, H. E. 1917. Flat-headed borers affecting forest trees in the United States (Buprestidae). *United States Department of Agriculture Bulletin* 437:1–8, 9 pls.

Burke, H. E. 1918. Biological notes on some flat-headed wood borers of the genus *Buprestis. Journal of Economic Entomology* 11:334–338.

Burke, H. E. 1919. Notes on a cocoon-making colydiid. *Proceedings of the Entomological Society of Washington* 21:123–124.

Burke, H. E. 1928. The western cedar pole borer or powder worm, *Trachykele blondeli* Mars. (Buprestidae). *United States Department of Agriculture Technical Bulletin* 48:1–15.

Burke, H. E. and Böving, A. G. 1929. The Pacific flathead borer, *Chrysobothris mali* Horn (Buprestidae). *United States Department of Agriculture Technical Bulletin* 83:1–36.

Burke, H. E., Hartman, R. D., and Snyder, T. E. 1922. The lead-cable borer or "short-circuit beetle" in California. *United States Department of Agriculture Bulletin* 1107, 56 pp., 9 pls.

Burke, H. R. 1968. Pupae of the weevil tribe Anthonomini (Coleoptera: Curculionidae). *Texas Agricultural Experiment Station Technical Monograph* 5, 92 pp., 68 figs.

Burke, H. R. and Anderson, D. M. 1976. Systematics of larvae and pupae of American Curculionoidea: status report, historical review and bibliography. *Southwestern Entomologist* 1(2):56–73.

Buschman, L. L. 1977. Biology and bioluminescence of selected fireflies in three genera: *Pyractomena, Photinus,* and *Photuris. Ph.D. Thesis, University of Florida, Gainesville,* 278 pp.

Buschman, L. L. 1984. Larval biology and ecology of *Photinus* fireflies (Coleoptera: Lampyridae) in northcentral Florida. *Journal Kansas Entomologist Society* 57:7–16.

Buschman, L. L. 1988. Larval development and its photoperiodic control in the firefly, *Pyractomena lucifera* (Coleoptera: Lampyridae). *Annals Entomological Society America* 81:82–90.

Butcher, F. G. 1930. Notes on the cocooning habits of *Gyrinus. Journal of the Kansas Entomological Society* 3:64–66.

Butcher, F. G. 1933. Hymenopterous parasites of Gyrinidae with descriptions of new species of *Hemiteles. Annals of the Entomological Society of America* 26:76–85.

Butt, F. H. 1951. Feeding habits and mechanism of the Mexican bean beetle. *Cornell University Agricultural Experiment Station Memoir* 306, 32 pp.

Buysson, H. du 1910. Matériaux pour servir à l'histoire des insectes de l'aulne. *Annales de la Société entomologique de France* 79:105–128.

Byzova, Yu. B. 1958. Tenebrionid larvae of some tribes of the subfamily Tenebrioninae (Coleoptera). *Zoologicheskii Zhurnal* 37(12):1823–1830.

Byzova, Yu. B. and Ghilarov, M. S. 1956. Soil-dwelling tenebrionid larvae of the tribe Helopini (Coleoptera, Tenebrionidae). *Zoologicheskii Zhurnal* 35:1493–1508. (In Russian).

Byzova, Yu. B. and Keleinikova, S. I. 1964. Family Tenebrionidae—Darkling beetles, pp. 463–496 *in* Ghilarov, M. S. (edit.), *Determination Key to Soil-dwelling Insect Larvae.* Moscow: "Nauka," 918 pp.

Campbell, J. M. 1968. A revision of the New World Micropeplinae (Coleoptera: Staphylinidae) with a rearrangement of the world species. *Canadian Entomologist* 100:225–267.

Campbell, R. E. 1920. The broad-bean weevil. *U.S.D.A. Bull.* No. 807. 23 pp.

Cameron, S. W. and Real, P. 1974. Contribucion a la biologia del coleoptero de la luma, *Cheloderus childreni* Gray (Coleoptera: Cerambycidae). *Revista Chilena de Entomologia* 8:123–132.

Candeze, E. 1861. Histoire des métamorphoses de quelques Coléoptères exotiques. *Mémoires de la Société royale des sciences de Liège* 16:325–410, 6 pls.

Carlson, D. C. and Ritcher, P. O. 1974. A new genus of Ochodaeinae and a description of the larva of *Pseudochodaeus estriatus* (Schaeffer). *Pan-Pacific Entomologist* 50(2):99–110.

Carpenter, G. H. and MacDowell, M. C. 1912. The mouth-parts of some beetle larvae (Dascillidae and Scarabaeidae), with especial reference to the maxillulae and hypopharynx. *Quarterly Journal of Microscopical Science* 57:373–396, pls. 35–37.

Carthy, J. D. and Goodman, L. G. 1964. An electrophysical investigation of the divided eye of *Gyrinus bicolor* F. *Journal of Insect Physiology* 10:431–436.

Casale, A. 1975. Ciclo biologico e morfologia preimmaginale di Coleoptera Staphylinoidea delle famiglie Leptinidae e Catopidae. *Redia* 56:199–230.

Casey, T. L. 1900. Review of the American Corylophidae, Cryptophagidae, Tritomidae and Dermestidae, with other studies. *Journal of the New York Entomological Society* 8:51–172.

Casey, T. L. 1912. Descriptive catalogue of the American Byrrhidae. *Memoirs on the Coleoptera* 3:1–69.

Cekalovic K. T. 1976. Nuevo hallazgo de un insecto poco conocido para Chile (Coleoptera, Phloeostichidae). *Boletin de la Sociedad de Biologia de Concepcion* 50:209–211.

Cekalovic K. T. and Morales A. E. 1974. Descripcion de la larva de *Oligacara nitida* Solier, 1848 (Coleoptera, Tenebrionidae). *Boletin de la Sociedad de Biologia de Concepcion* 48:173–177.

Cekalovic K. T. and Quezada Q. A. E. 1972. Distribucion geografica de *Uleiota chilensis* (Blanchard), 1851 y descripcion de la larva (Coleoptera-Cucujidae). *Boletin de la Sociedad de Biologia de Concepcion* 44:17–22.

Cekalovic K. T. and Quezada Q. A. E. 1973. Descripcion de la larva, pupa y distribucion geografica de *Emmallodera multipunctata curvidens* Kulzer, 1955 (Coleoptera-Tenebrionidae). *Anales del Instituto de la Patagonia* 4(1–3):417–422.

Cekalovic K. T. and Quezada Q. A. E. 1982. Descripcion de la larva de *Nycterinus* (*Eunycterinus*) *abdominalis* Eschscholtz, 1829 (Coleoptera, Tenebrionidae). *Boletin de la Sociedad de Biologia de Concepcion* 53:47–51.

Center, T. D. and C. D. Johnson, 1973. Comparative life histories of *Sennius* (Coleoptera: Bruchidae). *Environmental Entomology* 2(4):669–672.

Center, T. D. and Johnson, C. D. 1974. Coevolution of some seed beetles (Coleoptera: Bruchidae) and their hosts. *Ecology* 55(5):1096–1103.

Center, T. D. and Johnson, C. D. 1976. Host plants and parasites of some Arizona seed feeding insects. *Annals of the Entomological Society of America* 69(2):195–201.

Champion, G. C. 1888. Fam. Othniidae, pp. 465–469 *in* Godman, F. D. and Salvin, O., (eds.), *Biologia Centrali-Americana. Insecta, Coleoptera. Vol. 4, Part 1.* London: Porter, 572 pp., 23 pls.

Champion, G. C. 1897. Family Dascillidae, pp. 586–661, pls. 26–27 *in* Godman, F. D. and Salvin, O., (eds.), *Biologia Centrali-Americana. Insecta. Coleoptera. Vol. 3, Part 1.* London: Porter, 690 pp., 27 pls.

Champion, G. C. 1922. The geographical distribution and synonymy of the dasytid-beetle *Acanthocnemus nigricans* Hope (= *ciliatus* Perris). *Entomologist's Monthly Magazine* 58:77–79.

Champlain, A. B. and Kirk, H. B. 1926. Bait pan insects. *Entomological News* 37:288–291.

Chandler, D. S. 1976. Use of cantharidin and meloid beetles to attract Anthicidae (Coleoptera). *Pan-Pacific Entomologist* 52:179–180.

Chandler, D. S. 1983. Larvae of wrack Coleoptera in the families Corylophidae, Rhizophagidae, and Lathridiidae. *Psyche* 90(3):287–296.

Chang, V. C. S. and Jensen, L. 1974. Transmission of the pineapple disease organism of sugarcane by nitidulid beetles in Hawaii. *Journal of Economic Entomology* 67:190–192.

Chapin, E. A. 1927. The North American species of *Ptilodactyla* (Coleoptera: Helodidae). *Transactions of the American Entomological Society* 8:90–91, pl.1.

Chapman, R. N. 1915. Observations on the life history of *Agrilus bilineatus* Weber (Buprestidae). *Journal of Agricultural Research* 3:283–294.

Chapman, R. N. 1923. Observations on the life history of *Taphrocerus gracilis* Say (Buprestidae). *Memoirs of the Cornell Agricultural Experiment Station* 67:1–13, 4 pls.

Chapuis, M. F. and Candèze, M. E. 1853. Catalogue des larves des Coléoptères, connues jusqu'à ce jour avec la description de plusieurs espèces nouvelles. *Mémoires de la Société royale des sciences de Liège* 8:341–653, pls. 1–9.

Chararas, C. and Balachowsky, A. S. 1962. Famille des Bostrychidae, pp. 304–315 *in* A. S. Balachowsky, (ed.), *Entomologie appliquee à l'agriculture. Tome I. Coléoptères. Premier volume.* Paris: Masson, 564 pp.

Chararas, C., Chipoulet, J.-M., and Courtois, J.-E. 1979. Etude du preferendum alimentaire et des oxidases de *Pyrochroa coccinea* (Coleoptera Pyrochroidae). *Comptes rendus des séances de la Société de biologie et de ses filiales* 173(1):42–46.

Chemsak, J. A. and Linsley, E. G. 1983. *Checklist of the Disteniidae and Cerambycidae of North America, Central America, and the West Indies (Coleoptera). Checklist of the Beetles of North and Central America and the West Indies. Volume 7*. Gainesville, Florida: Flora and Fauna Publications, 138 pp.

Chisholm, A. H. 1952. Bird-insect associations in Australia. *Ibis* 94:395–405.

Chittenden, F. H. 1898. Insects injurious to beans and peas, pp. 233–260 *in Yearbook of Agriculture, 1898*. Washington, D.C.: United States Department of Agriculture, 768 pp.

Chittenden, F. H. 1922. The red-necked raspberry cane-borer, *Agrilus ruficollis* Fab. (Buprestidae). *United States Department of Agriculture Farmers' Bulletin* 1286.

Chu, H. F. 1949. *How to Know the Immature Insects*. Dubuque, Iowa: W. C. Brown, 234 pp.

Chujo, M. 1969. *Fauna Japonica. Erotylidae (Insecta: Coleoptera)* Tokyo: Academic Press of Japan, 316 pp., 23 pls.

Clark, J. 1925. Forest pests. The pin-hole borer (*Atractocerus kreusleri* Pasc.). *Journal of the Department of Agriculture of Western Australia* 2:138–142.

Clark J. 1931. A new species of Lymexylonidae (Coleoptera). *Proceedings of the Royal Society of Victoria* (N.S.)43:120–122.

Clark, R. C. and Brown, N. R. 1958. Studies of predators of the balsam woolly aphid, *Adelges piceae* (Ratz.) (Homoptera: Adelgidae) V. *Laricobius erichsonii* Rosen (Coleoptera: Derodontidae), an introduced predator in eastern Canada. *Canadian Entomologist* 90:657–672.

Clark, R. C. and Brown, N. R. 1960. Studies of predators of the balsam woolly aphid, *Adelges piceae* Ratz. (Homoptera: Adelgidae) VII. *Laricobius rubidus* Lec. (Coleoptera: Derodontidae), a predator of *Pineus strobi* (Htg.) (Homoptera: Adelgidae). *Canadian Entomologist* 92:237–240.

Clarke, R. O. S. 1973. *Handbooks for the Identification of British Insects. Vol. V. Part 2(c). Coleoptera. Heteroceridae*. London: Royal Entomological Society of London, 15 pp.

Clausen, C. P. 1916. Life-history and feeding records of a series of California Coccinellidae. *University of California Publications in Entomology* 1:251–299.

Claycomb, G. B. 1919. Notes on the habits of *Heterocerus* beetles. *Canadian Entomologist* 51:24–25.

Cloudsley-Thompson, J. L. 1965. On the function of the sub-elytral cavity in desert Tenebrionidae (Col.). *Entomologist's Monthly Magazine* 100:148–151.

Coats, J. R. and Selander, R. B. 1979. Notes on *Chelonarium lecontei* Thomson (Coleoptera: Chelonariidae), including description of an unusual righting behavior. *Coleopterists Bulletin* 33:57–59.

Cobos, A. 1961. Sobre la posición sistemática del género *Potergus* Bonvouloir y revisión de las categorias supragenérica de la familia Throscidae (Coleoptera). *Bulletin de l'Institut Royale de Sciences Naturalles de Belgique* 37(35):1–6.

Cobos, A. 1964. Materiales para el estudio de la familia Eucnemidae. Primera parte. (Coleoptera). *Eos* 40:289–435.

Cobos, A. 1966. Estudios sobre Throscidae, II (Col. Sternoxia). *Eos* 42:311–351.

Cobos, A. 1980. Ensayo sobre los generos de la subfamilia Polycestinae (Coleoptera, Buprestidae). Parte I. *Revista Espanola de Entomologia* 54:15–94.

Cobos, A. 1981. Ensayo sobre los generos de la subfamilia Policestinae (Coleoptera, Buprestidae). *Eos* 55–56 (1979–1980):23–94.

Cogan, B. and Smith, K. G. V. 1974. *Insects: Instructions for Collectors. No. 4a. 5th Edition*. London: British Museum (Natural History) (Publication No. 705), vi + 169 pp.

Connell, W. A. 1956. Nitidulidae of Delaware. *University of Delaware Agricultural Experiment Station Technical Bulletin* 318:1–67.

Cooke, W. C. 1963. Ecology of the pea aphid in the Blue Mountain area of eastern Washington and Oregon. *United States Department of Agriculture Technical Bulletin* 1287, 48 pp.

Cooley, R. A. 1917. The spinach carrion beetle *Silpha bituberosa* LeC. *Journal of Economic Entomology* 10:94–102.

Coombs, C. W. and Woodroffe, G. E. 1955. A revision of the British species of *Cryptophagus* (Herbst) (Coleoptera: Cryptophagidae). *Transactions of the Royal Entomological Society of London* 106:237–282.

Coquerel, C. 1848. Observations entomologiques sur divers insectes recueillis à Madagascar (Ire partie). *Annales de la Société entomologique de France* (2)6:177–190.

Corbière-Tichané, G. 1973. Sur les structures sensorielles et leurs fonctions chez la larve de *Speophyes lucidulus*. *Annales de spéléologie* 28(2):247–265.

Cornell, J. F. 1967. Description of the larva of *Aegialia browni* Saylor (Coleoptera: Scarabaeidae: Aphodiinae). *Pan-Pacific Entomologist* 43:189–192.

Cornell, J. F. 1972a. Larvae of the families of Coleoptera: a bibliographic survey of recent papers and tabular summary of 7 selected English language contributions. *Coleopterists Bulletin* 26:81–96.

Cornell, J. F. 1972b. A taxonomic review of the beetle genus *Dichelonyx* Harris in North America (Scarabaeidae: Melolonthinae). *Ph.D. Dissertation, Oregon State University, Corvallis*, 253 pp.

Costa, C. 1971. Descrição de fases imaturas de Pyrophorinae Neotropicais (Coleoptera, Elateridae). *Revista Brasiliera de Entomologia* 15(4):21–30.

Costa, C. 1977. Studies on Elateridae (Coleoptera). Biological notes on Neotropical larvae. *Papeis Avulsos de Zoologia* 31(2):7–18.

Costa, C., Casari Chen, S. A., and Vanin, S. A. 1985. Larvae of Neotropical Coleoptera. XII. Artematopoidea. Artematopidae. *Revista Brasileia de Entomologia* 29:309–314.

Costa, C. and Vanin, S. A. 1977. Larvae of Neotropical Coleoptera. I: Mycteridae, Lacconotinae. *Papeis Avulsos de Zoologia* 31:163–168.

Costa, C. and Vanin, S. A. 1981. Larvae of Neotropical Coleoptera. IV: Tenebrionidae, Lagriinae, Adeliini. *Papeis Avulsos de Zoologia* 34:165–178.

Costa, C. and Vanin, S. A. 1984a. Larvae of Neotropical Coleoptera. VII: Buprestidae, Buprestinae, Chalcophorini. *Papeis Avulsos de Zoologia* 35(10):117–124.

Costa, C. and Vanin, S. A. 1984b. Larvae of Neotropical Coleoptera. X: Mycteridae, Lacconotinae. *Revista Brasileira de Zoologia* 2(2):71–76.

Costa, C. and S. A. Vanin. 1985. Larvae of Neotropical Coleoptera. XI: Callirhipidae, Artematopoidea. *Revista Brasileira de Zoologia*. 2(6):351–355.

Costa, C., Vanin, S. A. and Casari-Chen, S. A. (1988). *Larvas de Coleoptera do Brasil*. São Paulo: Museu de Zoologia da Universidade de São Paulo 282 pp., 165 pls.

Costa Lima, A. da 1953. Insectos do Brasil. 8º Tomo. Coleopteros. 2º Parte. *Escola Nacional de Agronomia, Serie Didactica*, No. 10, 323 pp.

Cotton, R. T. 1921. Four Rhynchophora attacking corn in storage. *Journal of Agricultural Research* 20(8):605–614, 4 pls.

Cotton, R. T. and St. George, R. A. 1929. The meal worms. *United States Department of Agriculture Technical Bulletin* 95:23–35.

Cox, M. L. 1981. Notes on the biology of *Orsodacne* Latreille with a subfamily key to the larvae of the British Chrysomelidae (Coleoptera). *Entomologists's Gazette* 32:123–135.

Craighead, F. C. 1920. Biology of some Coleoptera of the families Colydiidae and Bothrideridae. *Proceedings of the Entomological Society of Washington* 22:1–13.

Craighead, F. C. 1921. Larva of the North American beetle *Sandalus niger* Knoch. *Proceedings of the Entomological Society of Washington* 23:44–48.

Craighead, F. C. 1923. North American cerambycid larvae. A classification and the biology of North American cerambycid larvae. *Technical Bulletin. Department of Agriculture, Canada* 27:1–239, 44 pls.

Craighead, F. C. 1950. Insect enemies of eastern forests. *United States Department of Agriculture Miscellaneous Publication* No. 657, 679 pp.

Cros, A. 1925. Sur les moeurs du *Drilus mauritanicus* Lucas et du *Malacogaster passerinii* Bassi. *Bulletin de la Société d'histoire naturelle de l'Afrique du Nord* 16:300–302.

Cros, A. 1926. Moeurs et évolution du *Drilus mauritanicus* Lucas. *Bulletin de la Société d'Histoire Naturelle de l'Afrique du Nord* 17:181–206, pl. 10.

Cros, A. 1930. *Malacogaster passerinii:* moeurs, évolution. *Bulletin de la Société d'histoire naturelle de l'Afrique du Nord* 21:133–160.

Cros, A. 1940. Essai de classification des *Meloidae* algériens. *Sixth International Congress of Entomology, Madrid, 1935* 311–338.

Crowson, R. A. 1955. *The Natural Classification of the Families of Coleoptera.* London: N. Lloyd, 187 pp.

Crowson, R. A. 1959. Studies on the Dermestoidea (Coleoptera), with special reference to the New Zealand fauna. *Transactions of the Royal Entomological Society of London* 111:81–94.

Crowson, R. A. 1960a. Observations on Scottish Mycetophagidae (Col.). *Entomologist's Monthly Magazine* 96:244.

Crowson, R. A. 1960b. The phylogeny of Coleoptera. *Annual Review of Entomology* 5:111–134.

Crowson, R. A. 1961a. Considerations on the genera *Endecatomus* Mellie and *Euderia* Broun (Coleoptera: Bostrychidae), with descriptions of their larvae. *Proceedings of the Royal Entomological Society of London* (B)30:113–120.

Crowson, R. A. 1961b. On some new characters of classificatory importance in adults of Elateridae (Coleoptera). *Entomologist's Monthly Magazine* 96(1960):158–161.

Crowson, R. A. 1962. Observations on the beetle family Cupedidae, with descriptions of two new fossil forms and a key to the recent genera. *Annals and Magazine of Natural History*(13)5:147–157, pls. 3–4.

Crowson, R. A. 1964a. The habits and life cycle of *Phloiophilus edwardsi* Steph. (Coleoptera: Phloiophilidae). *Proceedings of the Royal Entomological Society of London* (A)39:151–152.

Crowson, R. A. 1964b. A new genus of Australian clavicorn Coleoptera, probably a new family. *Proceedings of the Linnean Society of New South Wales* 89:241–245.

Crowson, R. A. 1964c. Observations on British Tetratomidae (Col.), with a key to the larvae. *Entomologist's Monthly Magazine* 94:82–86.

Crowson, R. A. 1964d. A review of the classification of Cleroidea (Coleoptera), with descriptions of two new genera of Peltidae and of several new larval types. *Transactions of the Royal Entomological Society of London* 116:275–327.

Crowson, R. A. 1966a. Further observatons on Peltidae (Coleoptera: Cleroidea), with definitions of a new subfamily and of four new genera. *Proceedings of the Royal Entomological Society of London* (B)35(9–10):119–127.

Crowson, R. A. 1966b. Observations on the constitution and subfamilies of the family Melandryidae. *Eos* 41:507–513.

Crowson, R. A. 1967. The natural classification of the families of Coleoptera. Addenda and corrigenda. *Entomologist's Monthly Magazine* 103:209–214.

Crowson, R. A. 1970. Further observations on Cleroidea (Coleoptera). *Proceedings of the Royal Entomological Society of London* (B)39(1–2):1–20.

Crowson, R. A. 1971. Observations on the superfamily Dascilloidea (Coleoptera: Polyphaga), with the inclusion of Karumiidae and Rhipiceridae. *Zoological Journal of the Linnean Society of London* 50:11–19.

Crowson, R. A. 1972a. On the systematic value of the alimentary canal in Cleridae. *Systematic Zoology* 21(3):339–340.

Crowson, R. A. 1972b. A review of the classification of Cantharoidea (Coleoptera), with the definition of two new families, Cneoglossidae and Omethidae. *Revista de la Universidad de Madrid* 21(82):35–77.

Crowson, R. A. 1973a. Further observations on Phloeostichidae and Cavognathidae, with definitions of new genera from Australia and New Zealand. *Coleopterists Bulletin* 27:54–62.

Crowson, R. A. 1973b. On a new superfamily Artematopoidea of polyphagan beetles, with the definition of two new fossil genera from the Baltic amber. *Journal of Natural History* 7:225–238.

Crowson, R. A. 1974. Observations on Histeroidea, with descriptions of an apterous larviform male and of the internal anatomy of a male *Sphaerites. Journal of Entomology* (B)42:133–140.

Crowson, R. A. 1975. The evolutionary history of Coleoptera as documented by fossil and comparative evidence. *Atti del X Congresso Nazionale Italiano di Entomologia. Sassari, 20–25 Maggio 1974,* pp. 47–90.

Crowson, R. A. 1976. The systematic position and implications of *Crowsoniella. Bolletino del Museo Civico di Storia Naturale, Verona* 2(1975):459–463.

Crowson, R. A. 1978. Problems of phylogenetic relationships in Dryopoidea (Coleoptera). *Entomologia Germanica* 4:250–257.

Crowson, R. A. 1979. Observations on Clambidae (Coleoptera), with descriptions of a new genus and species and of several larvae. *Revue Suisse de Zoologie* 86:611–623.

Crowson, R. A. 1980. On amphipolar distribution patterns in some cool climate groups of Coleoptera. *Entomologia Generalis* 6:281–292.

Crowson, R. A. 1981. *The Biology of Coleoptera.* Academic Press: London xii + 802 pp.

Crowson, R. A. 1982. On the dryopoid affinities of Buprestidae. *Coleopterists Bulletin* 36(1):22–25.

Crowson, R. A. 1984a. The associations of Coleoptera with Ascomycetes. pp. 256–285 *in* Wheeler, Q. D. and Blackwell, M., (eds.), *Insect/Fungus Relationships: Perspectives in Ecology and Evolution.* New York: Columbia University Press, 512 pp.

Crowson, R. A. 1984b. On the systematic position of *Bruchela* Dejean (*Urodon* auctt.) (Coleoptera). *Coleopterists Bulletin* 38(1):91–93.

Crowson, R. A. 1985. The systematic position of *Nemonyx* Redtenbacher. *Entomologists Monthly Magazine* 121:144.

Crowson, R. A. and Crowson, E. A. 1955. Some observations on beetles of the family Clambidae. *Glasgow Naturalist* 17:205–206.

Crowson, R. A. and Ellis, I. 1969. Observations on *Dendrophagus crenatus* (Paykull) (Cucujidae) and some comparisons with piestine Staphylinidae (Coleoptera). *Entomologist's Monthly Magazine* 104:161–169.

Crowson, R. A. and Hunter, F. A. 1964. Some Coleoptera associated with old trees in Grimsthorpe Park, Lincs. *Entomologist's Monthly Magazine* 100:198–200.

Crowson, R. A. and Sen Gupta, T. 1969. The systematic position of Propalticidae and of *Carinophloeus* Lefkovitch (Coleoptera, Clavicornia) with description of a new species of *Propalticus* and of its supposed larva. *Proceedings of the Royal Entomological Society of London* 38:132–140.

Crowson, R. A. and Viedma, M. 1964. Observations on the relationships of the genera *Circaeus* Yablok. and *Mycterus* Clairv. with a description of the presumed larva of *Mycterus* (Col. Heteromera). *Eos* 40:99–107.

Cuthright, C. R. 1924. Bionomics of *Hippodamia tredecimpunctata* L. *Annals of the Entomological Society of America* 17:188–192.

Cutler, J. R. 1971. A key for distinguishing the larvae of *Ahasverus advena* (Waltl), *Cathartus quadricollis* (Guer.), *Oryzaephilus surinamensis* (L.) and *Oryzaephilus mercator* (Fauv.) (Coleoptera: Silvanidae). *Journal of Stored Products Research* 7:125–127.

Daggy, T. 1947. Notes on the ecology and taxonomy of certain pupae of the family Tenebrionidae (Coleoptera). *Proceedings of the Indiana Academy of Sciences* 56:253–260.

Dahlgren, U. 1917. The production of light by animals. *Journal of the Franklin Institute* 183:593–624.

Dahlsten, D. L. 1970. Section 8. Parasites, predators, and associated organisms reared from western pine beetle infested bark samples. pp. 75–79 *in* Stark, R. W. and Dahlsten, D. L. (edits.) *Studies on the Population Dynamics of the Western Pine Beetle,* Dendroctonus brevicornis *LeConte (Coleoptera: Scolytidae)*. Berkeley: University of California Division of Agricultural Sciences, 174 pp.

Dajoz, R. 1960. Les Coléoptères mycétophiles de la Forêt de la Massane (Pyrénées-Orientales). Note préliminaire. *Vie et milieu* 11:195–208.

Dajoz, R. 1961. Etude analytique et critique des travaux récents sur les Carabiques de la faune de France (Coléoptères). *Cahiers des Naturalistes* 17:1–48.

Dajoz, R. 1963. *Dolosus leleupi* n. g., n. sp. et *Dolosus basilewskyi* n. sp. types d'une famille nouvelle de Cucujoidea (Coléoptères). *Revue de zoologie et botanique africaines* 67:91–96.

Dajoz, R. 1964. Contribution à l'étude des Coléoptères Mycetophagidae. I. Revision du genre *Esarcus* et description de trois espèces nouvelles d'Espagne. *Entomologiste* 20:7–15.

Dajoz, R. 1965. Morphologie et biologie de la larve de *Scaphosoma assimile* Er. *Bulletin mensuel de la Société linnéenne de Lyon* 34:105–110.

Dajoz, R. 1966. Ecologie et biologie des Coléoptères xylophages de la hêtraie. *Vie et milieu* (C)17:525–763.

Dajoz, R. 1968. Révision des Colydiidae anophthalmes de la faune paléarctique (Col.) IV. Etudes sur les genres *Anommatus* et *Langelandia. Annales de la Société entomologique de France* (N.S.)4:975–988.

Dajoz, R. 1971. Coléoptères Colydiidae nouveaux ou peu connus de la région paléarctique. *Entomologiste* 27:83–101.

Dajoz, R. 1972. Coléoptères Lathridiidae de Madagascar et des Mascareignes. *Bulletin du Museum National d'Histoire Naturelle* (3)85 (Zoologie 64):1049–1055.

Dajoz, R. 1977. *Faune de l'Europe et du Bassin Méditerranéen. 8. Coléoptères Colydiidae et Anommatidae paléarctiques.* Paris: Masson, 275 pp.

Dajoz, R. 1978. Une espèce nouvelle de l'Inde du genre *Sarothrias* Grouvelle (Coleoptera, Sarothriidae). *Bulletin de la Société Linnéenne de Lyon* 47:322–324.

Dallwitz, M. J. 1974. A flexible program for generating identification keys. *Systematic Zoology* 23:50–57.

Dallwitz, M. J. 1978. User's guide to KEY—a computer program for generating identification keys. *Commonwealth Scientific and Industrial Research Organization, Australia, Division of Entomology Report* 4, 16 pp.

Dallwitz, M. J. 1980. A general system for coding taxonomic descriptions. *Taxon* 29:41–46.

Dallwitz, M. J. 1984. User's guide to the DELTA system—a general system for coding taxonomic descriptions. Second edition. *Commonwealth Scientific and Industrial Research Organization, Australia, Division of Entomology Report* 13, 93 pp.

Danilevskii, M. L. 1976. Comparative anatomy of the digestive system of long-horned beetle larvae (Coleoptera, Cerambycidae), pp. 123–135 *in* Mamaev, B. M. (edit.), *Evolutionary Morphology of Insect Larvae.* Moscow: "Nauka," 204 pp. (in Russian).

Darlington, P. J., Jr. 1930. Notes on the habits of *Amphizoa. Psyche* 36(1929):383–385.

Das, G. M. 1937. The musculature of the mouth-parts of insect larvae. *Quarterly Journal of Microscopical Science* (N.S.)80:39–80, 12 pls.

Davey, P. M. 1958. The groundnut bruchid, *Caryedon gonagra* (F.). *Bulletin of Entomological Research* 49(2):385–404.

Davidson, J. A. and Wood, F. E. 1969. Description and biological notes on the larva of *Anthicus heroicus* Casey (Coleoptera: Anthicidae). *Coleopterists Bulletin* 23:5–8.

Davidson, W. M. 1921a. Observations on *Psyllobora taedata* LeConte, coccinellid attacking mildews (Col.). *Entomological News* 32:83–89.

Davidson, W. M. 1921b. Observations of *Psyllobora tredecimpunctata* L. *Annals of the Entomological Society of America* 17:188–192.

Davies, W. M. 1928. The bionomics of *Apion ulicis* Först. (gorse weevil), with special reference to its role in the control of *Ulex europaeus* in New Zealand. *Annals of Applied Biology* 15(2):263–286, pls. 15–17.

DeBach, P. 1964. *Biological Control of Insect Pests and Weeds.* London: Chapman and Hall, 844 pp.

Deleurance-Glaçon, S. 1963. Recherches sur les Coléoptères troglobies de la sous-famille des Bathysciinae. *Annales des sciences naturelles. Zoologie* (12)5:1–172.

DeMarzo, L. 1979. Studi sulle larve dei coleotteri ditiscidi. X. Anatomia e funzionamento dell'apparato succhiante cibario-faringeo in alcune forme larvali delle subf. Dytiscinae, Colymbetinae, Laccophilinae e Hydroporinae. *Entomologica* 15:5–72.

DeMarzo, L. 1983. Observazioni sulla ovideposizione e sul ciclo larvale in *Mastigus pilifer* Kraatz (Coleoptera, Scydmaenidae). *Entomologica* 18:125–136.

DeMarzo, L. 1984. Morfologia della larve e della pupa in *Mastigus pilifer* Kraatz (Coleoptera, Scydmaenidae). *Entomologica* 19:61–74.

DeMarzo, L. 1986. Osservazione etologiche sulle larve di *Batrisodes oculatus* Aubé (Coleoptera: Pselaphidae). *Frustula Entomologica* (N.S.) 7–8:501–506.

DeMarzo, L. and Vit, S. 1982. Nota sulle presenza di *Bastrisodes oculatus* Aubé (Coleoptera, Pselaphidae) in una grotta di Puglia. *Entomologica* 17:149–162.

Demelt, C. V. 1978. Bestimmungstabelle für die Larven der Cerambycidae (partim), pp. 335–336 *in* Klausnitzer, B. (ed.) *Ordnung Coleoptera (Larven).* The Hague: W. Junk, vi + 378 pp.

Dietrich, H. 1945. The Elateridae of New York State. *Cornell University Agricultural Experiment Station Memoir* 269, 79 pp.

Dimetry, N. Z. 1976. Studies on the cannibalistic behavior of the predatory larvae of *Adalia bipunctata* L. (Col. Coccinellidae). *Zeitschrift für Angewandte Entomologie* 81:156–163.

Dixon, A. F. G. 1959. An experimental study of the searching behavior of the predatory coccinellid beetle (*Adalia decempunctata* (L.)). *Journal of Animal Ecology* 28:259–281.

Dizer, Yu. B. 1953. Morphological distinctions among larvae of some darkling beetles of the subfamily Platyscelinae and their significance for the systematics of the group. *Zoologicheskii Zhurnal* 32(3):457–466 (in Russian).

Dizer, Yu. B. 1955. On the physiological role of the elytra and sub-elytral cavity of steppe and desert Tenebrionidae. *Zoologicheskii Zhurnal* 34:319–322 (in Russian).

Dogger, J. R. 1959. The Elateridae of Wisconsin. I. A list of the species found in Wisconsin and keys to the identification of genera of adults and larvae. *Wis. Acad. Sci., Arts and Letters.* 48:103–120.

Dolin, V. G. 1978. *Classification Key to the Larvae of Click Beetles in the Fauna of the U.S.S.R., Kiev:* "Urozhai," 124 pp. (in Russian).

Donisthorpe, H. 1935. The British fungicolous Coleoptera. *Entomologist's Monthly Magazine* 71:21–31.

Dorn, K. 1936. Über *Mycetoma suturale* Pz. und *Derodontus macularis* Fuss. (Coleopt.). *Mitteilungen der Entomologischen Gesellschaft in Halle* 14:29–35.

Dorsey, C. K. 1940. A comparative study of the larvae of six species of *Silpha. Annals of the Entomological Society of America* 33:120–139.

Dorsey, C. K. 1943. The musculature of the labrum, labium, and pharyngeal region of adult and immature Coleoptera. *Smithsonian Miscellaneous Collections* 103(7):1–42.

Dorsey, C. K. and Leach, J. G. 1956. The bionomics of certain insects associated with oak wilt with particular reference to the Nitidulidae. *Journal of Economic Entomology* 49:219–230.

Dorsey, C. K., F. F. Jewell, J. G. Leach and R. P. True 1953. Experimental transmission of oak wilt by four species of Nitidulidae. *Plant Disease Reporter* 37:419–420.

Doutt, R. L. 1951. Biological control of mealybugs infesting commercial greenhouse gardenias. *Journal of Economic Entomology* 44:37–40.

Doyen, J. T. 1972. Familial and subfamilial classification of the Tenebrionoidea (Coleoptera) and a revised generic classification of the Coniontini (Tentyriidae). *Quaestiones Entomologica* 8:357–376.

Doyen, J. T. 1974. Larvae of *Bothrotes plumbeus* and *Lobometopon fusiforme* (Coleoptera: Tentyriidae: Epitragini). *Coleopterists Bulletin* 28:159–165.

Doyen, J. T. 1976a. Biology and systematics of the genus *Coelus* (Coleoptera: Tentyriidae). *Journal of the Kansas Entomological Society* 49(4):595–624

Doyen, J. T. 1976b. Description of the larva of *Phloeodes diabolicus* LeConte (Coleoptera: Zopheridae). *Coleopterists Bulletin* 30:267–272.

Doyen, J. T. 1979. The larva and relationships of *Cononotus* LeConte (Coleoptera: Heteromera). *Coleopterists Bulletin* 33:33–39.

Doyen, J. T. 1984. Reconstitution of the Diaperini of North America, with new species of *Adelina* and *Sitophagus* (Coleoptera: Tenebrionidae). *Proceedings of the Entomological Society of Washington* 86(4):777–789.

Doyen, J. T. 1985. Reconstitution of the tenebrionid tribes Ulomini and Triboliini for North and Central America (Coleoptera). *Proceedings of the Entomological Society of Washington* 86(4):777–789.

Doyen, J. T. and Kitayama, C. 1980. Review of the North American species of *Apocrypha* Eschscholtz, with a description of the immature stages of *Apocrypha anthicoides* (Coleoptera: Tenebrionidae). *Pan-Pacific Entomologist* 56(2):121–136.

Doyen, J. T. and Lawrence, J. F. 1979. Relationships and higher classification of some Tenebrionidae and Zopheridae (Coleoptera). *Systematic Entomology* 4:333–377.

Doyen, J. T. and Tschinkel, W. F. 1973. Population size, microgeographic distribution and habitat separation in some tenebrionid beetles (Coleoptera). *Annals of the Etnomological Society of America* 67(4):617–626.

Doyen, J. T. and Tschinkel, W. R. 1982. Phenetic and cladistic relationships among tenebrionid beetles (Coleoptera). *Systematic Entomology* 7:127–183.

Doyen, J. T. and Ulrich, G. 1978. Aquatic Coleoptera, pp. 203–231 *in* Merritt, R. W. and Cummins, K. W. (eds.), *An Introduction to Aquatic Insects*. Dubuque, Iowa: Kendall/Hunt, xiv + 441 pp.

Doyer, L. C. 1929. Aantasting von boonen dar *Bruchus obtectus* Say. *Tijdschrift over Plantenziekten* 35(10):257–263.

Drooz, A. T. (ed.) 1985. Insects of eastern forests. United States Department of Agriculture, Forest Service. Misc. Publ. 1428, 608 pp.

Duffy, E. A. J. 1946. Notes on the British species of *Pyrochroa* (Col. Pyrochroidae) with a key to their first-stage larvae. *Entomologist's Monthly Magazine* 82:92–93.

Duffy, E. A. J. 1952a. Dorsal prolegs and extreme cephalic modification in the first-instar larva of *Agapanthia villosoviridescens* Deg. (Col. Cerambycidae). *Entomologist's Monthly Magazine* 87:313–318.

Duffy, E. A. J. 1952b. The immature stages of *Sessinia livida* (Fabricius). *Proceedings of the Hawaiian Entomological Society* 14:379–383.

Duffy, E. A. J. 1953. *A Monograph of the Immature Stages of British and Imported Timber Beetles* (*Cerambycidae*). London: British Museum (Natural History), 350 pp., 8 pls.

Duffy, E. A. J. 1957. *A Monograph of the Immature Stages of African Timber Beetles (Cerambycidae)*. British Museum (Natural History): London. 338 pp., 10 pls.

Duffy, E. A. J. 1960. *A Monograph of the Immature Stages of Neotropical Timber Beetles (Cerambycidae)*. British Museum (Natural History): London. 327 pp., 13 pls.

Duffy, E. A. J. 1963. *A Monograph of the Immature Stages of Australasian Timber Beetles*. London: British Museum (Natural History), 235 pp., 11 pls.

Duffy, E. A. J. 1968. *A Monograph of the Immature Stages of Oriental Timber Beetles (Cerambycidae)*. London: British Museum (Natural History), 434 pp., 18 pls.

Duffy, E. A. J. 1980. *A Monograph of the Immature Stages of African Timber Beetles (Cerambycidae) Supplement*. London: Commonwealth Institute of Entomology, 186 pp., 4 pls.

Dumbleton, L. J. 1932. Early stages of New Zealand Buprestidae (Coleoptera). *Stylops* 1:41–48.

Durr, H. J. R. 1957. Morphology and bionomics of the European houseborer, *Hylotrupes bajulus* (Coleoptera: Cerambycidae). *Union of South Africa Department of Agriculture. Entomology Memoirs* 4(1):1–136.

Dybas, H. S. 1966. Evidence for parthenogenesis in the featherwing beetles, with a taxonomic review of a new genus and eight new species. *Fieldiana Zoology* 51:11–52.

Dybas, H. S. 1976. The larval characters of featherwing and limulodid beetles and their family relationship in the Staphylinoidea (Coleoptera: Ptiliidae and Limulodidae). *Fieldiana Zoology* 70:29–78.

Dybas, H. S. 1978. Polymorphism in featherwing beetles, with a revision of the genus *Ptinellodes* (Coleoptera: Ptiliidae). *Annals of the Entomological Society of America* 71:695–714.

Edmonds, W. D. 1967. The immature stages of *Phanaeus* (*Coprophaneus*) *jasius* Olivier and *Phanaeus* (*Metallophanaeus*) *saphirinus* Sturm (Coleoptera: Scarabaeidae). *Coleopterists Bulletin* 21(4):97–105.

Edmonds, W. D. and Halffter, G. 1972. A taxonomic and biological study of the immature stages of some New World Scarabaeinae (Coleoptera: Scarabaeidae). *Anales de la Escuela Nacional de Ciencias Biologicas, Mexico City* 19:85–122.

Edmonds, W. D. and Halffter, G. 1978. Taxonomic review of immature dung beetles of the subfamily Scarabaeinae. *Systematic Entomology* 3:307–331.

Edwards, J. G. 1951. Amphizoidae (Coleoptera) of the world. *Wasmann Journal of Biology* 8(1950):303–332.

Edwards, J. G. 1953. The real source of *Amphizoa* secretions. *Coleopterists Bulletin* 7:4.

Edwards, J. G. 1954. Observations on the biology of Amphizoidae. *Coleopterists Bulletin* 8:19–24.

Eichelbaum, F. 1907. Die Larven von *Cis festivus* Panz. und von *Emphylus glaber* Gyll. *Zeitschrift für wissenschaftliche Insekten-Biologie* 3:25–30.

Eidmann, H. and Heqvist, K. J. 1958. Die Larve von *Dryophilus pusillus* Gyll. (Col., Anobiidae). *Entomologisk Tidskrift* 79:38–40.

Eidt, D. C. 1958. Anatomy and histology of the full-grown larva of *Ctenicera aeripennis destructor* (Brown) (Coleoptera: Elateridae). *Canadian Journal of Zoology* 36:317–361.

Eidt, D. C. 1959. Mode of feeding of the larva of *Ctenicera aeripennis destructor* (Brown) (Coleoptera: Elateridae). *Canadian Entomologist* 91:97–101.

Eisner, T., Aneshansley, D., Eisner, M., Rutowski, R., Chong, B. and Meinwald, J. 1974. Chemical defense and sound production in Australian tenebrionid beetles (*Adelium* spp.). *Psyche* 81(1):189–208.

Eisner, T., McHenry, F. and Salpeter, M. M. 1964. Defense mechanisms of arthropods. XV. Morphology of the quinone-producing glands of a tenebrionid beetle (*Eleodes longicollis* Lec.). *Journal of Morphology* 115(3):355–399.

Ekis, G. 1977. Classification, phylogeny, and zoogeography of the genus *Perilypus* (Coleoptera: Cleridae). *Smithsonian Contributions to Zoology* 227, 138 pp.

Ekis, G. and Gupta, A.P. 1971. Digestive system of Cleridae (Coleoptera). *International Journal of Insect Morphology and Embryology* 1(1):51–86.

Ekis, G. and Menier, J. J. 1980. Discovery of Chaetosomatidae in Madagascar. Systematics of the new genus *Malagassochaetus* (Col. Cleroidea). *Annales de la Société Entomologique de France* (N.S.)16(2):197–208.

El Moursy, A. A. 1961. A tentative classification of and a key to the North American genera of the family Byrrhidae (new sense) and family Syncalyptidae (new status). *Coleopterists Bulletin* 15:9–15.

El Sayed, M. T. 1940. The morphology, anatomy, and biology of *Araecerus fasciculatus* DeGeer. *Bulletin de la Societe Fouad Ier d'Entomologie* 24:82–151.

Elzinga, R. J. 1977. Observations on *Sandalus niger* Knoch (Coleoptera: Sandalidae) with a description of the triungulin larva. *Journal of the Kansas Entomological Society* 50:324–328.

Emden, F. I. van 1924. Zur Biologie von *Thorictodes heydeni* Reitt. (Col. Thorictidae). *Treubia* 6(1):1–7.

Emden, F. I. van 1928. Die Larve von *Phalacrus grossus* Er. und Bemerkungen zum Larvensystem der Clavicornia. *Entomologische Blätter* 24:8–20, pl. 1.

Emden, F. I. van 1931. Beschreibung der Larve von *Telephanus costaricensis* Neverm. *Stettiner Entomologische Zeitung* 92:113–117.

Emden, F. I. van 1932a. Die Larven der Callirhipini, eine mutmassliche *Cerophytum*-Larve und Familien-Bestimmungstabelle der Larven der Malacodermata-Sternoxia-Reihe (Coleoptera). *Bulletin et Annales de la Societe Royale Entomologique de Belgique* 72:199–259.

Emden, F. I. van 1932b. Die Larven von *Discoloma cassideum* Reitt. (Col. Colyd.) und *Skewarraia paradoxa* Lac. (Col. Chrysom.). *Zoologischer Anzeiger* 101:1–17.

Emden, F. I. van 1934. Sind Polyphaga-Larven mit selbständigem Tarsus bekannt? (Col.). *Stettiner Entomologische Zeitung* 95:61–64.

Emden, F. I. van 1935a. Die Gattungsunterschiede der Hirschkäferlarven, ein Beitrag zum natürlichen System der Familie (Col. Lucan.). *Stettiner Entomologische Zeitung* 96:178–200.

Emden, F. I. van 1935b. Die Larven der Cicindelinae I. Einleitendes und alocosternale Phyle. *Tijdschrift voor Entomologie* 78:134–183.

Emden, F. I. van 1936. Eine interessante, zwischen Carabidae und Paussidae vermittelinde Käferlarve. *Arbeiten uber physiologische und angewandte Entomologie aus Berlin-Dahlem* 3:250–256.

Emden, F. I. van 1938a. Beschreibung der Larve von *Notiophygus hessei* John (Coleoptera). *Arbeiten über morphologische und taxonomische Entomologie aus Berlin-Dahlem* 5:132–134, pl.1.

Emden, F. I. van 1938b. On the taxonomy of Rhynchophora larvae (Coleoptera). *Transactions of the Royal Entomological Society of London* 87(1):1–37, 108 figs.

Emden, F. I. van 1939. Larvae of British beetles—I. Cerambycidae. (part) *Entomologist's Monthly Magazine* 75:257–273.

Emden, F. I. van 1940. Larvae of British beetles—I. Cerambycidae. (part) *Entomologist's Monthly Magazine* 76:7–13.

Emden, F. I. van 1941. Larvae of British beetles—II. A key to the British lamellicornia larvae. *Entomologist's Monthly Magazine* 77:117–192.

Emden, F. I. van 1942a. The collection and study of beetle larvae. *Entomologist's Monthly Magazine* 78:73–79.

Emden, F. I. van 1942b. A key to the genera of larval Carabidae (Col.). *Transactions of the Royal Entomological Society of London* 92:1–99.

Emden, F. I. van 1942c. Larvae of British beetles—III. Keys to families. *Entomologist's Monthly Magazine* 78:206–226, 253–272.

Emden, F. I. van 1943. Larvae of British beetles. IV. Various small families. *Entomologist's Monthly Magazine* 79:209–270.

Emden, F. I. van 1945. Larvae of British beetles. V. Elateridae. *Entomologist's Monthly Magazine* 81:13–37.

Emden, F. I. van 1946. Egg-bursters in some more families of polyphagous beetles and some general remarks on egg-bursters. *Proceedings of the Royal Entomological Society of London* (A)21:89–97.

Emden, F. I. van 1947. Larvae of British beetles. VI. Tenebrionidae. *Entomologist's Monthly Magazine* 83:154–171.

Emden, F. I. van 1949. Larvae of British beetles. VII. Coccinellidae. *Entomologist's Monthly Magazine* 85:265–283.

Emden, F. I. van 1950. Eggs, egg-laying habits and larvae of short-nosed weevils. *Proceedings of the 8th International Congress of Entomology, Amsterdam*, pp. 365–372.

Emden, F. I. van 1951. On the genus *Thaumaphrastus* Blaisdell (Coleoptera: Thorictidae). *Bulletin of the Brooklyn Entomological Society* 46(2):39–41.

Emden, F. I. van 1952. On the taxonomy of Rhynchophora larvae: Adelognatha and Alophinae (Insecta: Coleoptera). *Proceedings of the Zoological Society of London* 122:651–795, 153 figs.

Emden, F. I. van 1956. The *Georyssus* larva—a hydrophilid. *Proceedings of the Royal Entomological Society of London* (A)31:20–24.

Emden, F. I. van 1957. The larvae of *Cassidoloma angolense* John and *Notiophygus piger* John (Col. Colyd.). *Publicacoes Culturais da Companhia de Diamontes de Angola* 34:27–31.

Emden, F. I. van 1958. Über die Larvenmerkmale einiger deutscher Byrrhidengattungen. *Mitteilungen der Deutschen Entomologischen Gesellschaft* 17:39–40.

Emden, H. F. van 1956. Morphology and identification of the British larvae of the genus *Elater* (Col., Elateridae). *Entomologist's Monthly Magazine* 92:167–188.

Endrödy-Younga, S. 1959. Systematischer Überblick über die Familie Clambidae (Col.). *Opuscula Entomologica* 24:81–116.

Endrödy-Younga, S. 1960. Neue Angaben zur Klärung des Systems der Familie Clambidae und Beschreibung einer neuen Liodiden-Gattung (Coleoptera). *Annales Historico-Naturales Musei Nationalis Hungarici. Pars Zoologica* 52:239–245.

Endrödy-Younga, S. 1964. Insects of Campbell Island. Coleoptera: Orthoperidae. *Pacific Insects Monographs* 7:408–409.

Endrödy-Younga, S. 1974. A revision of the described Australian and New Zealand species of the family Clambidae (Coleoptera) with description of a new species. *Records of the South Australian Museum* 17(1):1–10.

Endrödy-Younga, S. 1981. The American species of the family Clambidae (Coleoptera: Eucinetoidea). *Entomologia Generalis* 7:33–67.

Endrödy-Younga, S. and R. A. Crowson, 1986. Boganiidae, a new beetle family for the African fauna (Coleoptera: Cucujoidea). *Annals of the Transvaal Museum* 34(12):253–273.

Erber, D. 1969. Beitrag zur Entwicklungsbiologie mitteleuropäischer Clytrinen und Cryptocephalinen (Coleoptera, Chrysomelidae). *Zoologische Jahrbücher, Abteilung für Systematik, Ökologie and Geographie der Tiere* 96:453–477.

Erickson, E. H. and Werner, F. G. 1974a. The bionomics of Nearctic bee-associated Meloidae (Coleoptera); life histories and nutrition of certain Meloinae. *Annals of the Entomological Society of America* 67:394–400.

Erickson, E. H. and Werner, F. G. 1974b. Bionomics of Nearctic bee-associated Meloidae (Coleoptera); life histories and nutrition of certain Nemognathinae. *Annals of the Entomological Society of America* 67:401–406.

Erickson, E. H., W. R. Enns and F. G. Werner. 1976. Bionomics of bee-associated Meloidae (Coleoptera); bee and plant hosts of some Nearctic meloid beetles—a synopsis. *Annals of the Entomological Society of America* 69:959–970.

Ermisch, K. von. 1956. Mordellidae, pp. 269–328 *in* Horion, A. (ed.), *Faunistik der mitteleuropäischen Käfer. Band V: Heteromera.* Tutzing bei München: Museum G. Frey, xv + 336 pp.

Erwin, T. L. 1967. Bombardier beetles (Coleoptera, Carabidae) of North America: Part II. Biology and behavior of *Brachinus pallidus* Erwin in California. *Coleopterists Bulletin* 21:41–55.

Erwin, T. L. 1970. A description of the larva of *Thyce harfordi* Casey (Scarabaeidae: Melolonthinae). *Psyche* 77(1):50–53.

Erwin, T. L. 1975. Relationships of predaceous beetles to tropical forest wood decay. Part I. Descriptions of the immature stages of *Eurycoleus macularis* Chevrolat (Carabidae: Lebiini). *Coleopterists Bulletin* 29:297–300.

Erwin, T. L. 1981. A synopsis of the immature stages of Pseudomorphini (Col. Car.) with notes on tribal affinities and behavior in relation to life with ants. *Coleopterists Bulletin* 35:53–68.

Erwin, T. L., Ball, G. E., Whitehead, D. R., and Halpern, A. L., (edits.). 1979. *Carabid Beetles: Their Evolution, Natural History, and Classification.* The Hague: W. Junk, 635 pp.

Español, F. 1970. Notas sobre Anobidos (Col.).XLVI. Contribucion al conocimiento de la subfamilia Hedobiinae. *Memorias de la R. Academia de Ciencias y Artes de Barcelona* 40:457–482.

Essig, E. O. 1958. *Insects and Mites of Western North America.* New York: Macmillan, xiii + 1050 pp.

Evans, M. E. G. 1961. The life history of *Atomaria ruficornis* (Marsh.) (Col., Cryptophagidae). *Transactions of the Society for British Entomology* 14:207–222.

Evans, W. G. 1964. Infra-red receptors in *Melanophila acuminata* DeGeer. *Nature* 4928:211.

Evans, W. G. 1966. Morphology of the infrared sense organs of *Melanophila acuminata* (Buprestidae: Coleoptera). *Annals of the Entomological Society of America* 59:873–877.

Falcoz, L. 1921. Etudes sur les Cryptophaginae (Coléoptères Erotylides). I. Morphologie et affinités systématiques de *Setaria sericea* Muls. *Annales de la Société linnéenne de Lyon* 68:25–40.

Falcoz, L. 1922. Etudes sur les Cryptophaginae (Coléoptères Erotylides). II. *Henoticus californicus* Mannerheim, espèce americaine en voie d'acclimatation européenne. *Annales de la Société linnéenne de Lyon* 69:167–183.

Falcoz, L. 1923. Description de la larve de *Coroebus sinuatus* Creutzer (Col. Buprestidae). *Annales de la Societe Entomologique de France* 92:247–252, pl. 3.

Falcoz, L. 1924. Etudes sur les Cryptophaginae (Coléoptères Erotylides). IV. Essai sur les larves de *Cryptophagus. Annales de la Société linnéenne de Lyon* 71:120–127.

Falcoz, L. 1926. Position systématique des genres *Diphyllus* Stephens et *Diplocoelus* Guerin. *Encyclopédie entomologique.* Série B. I. Coleoptera. 1:69–74.

Farrow, R. A. 1974. The larva of *Laius villosus* (Coleoptera: Melyridae) feeding on the egg pods of the Australian plague locust, *Chortoicetes terminifera* (Orthoptera: Acrididae). *Journal of the Australian Entomological Society* 13:185–188.

Fellin, D. G. 1975. Feeding habits of *Pleocoma* larvae in coniferous forests of Western Oregon. *Northwest Science* 49(2):71–86.

Fellin, D. G. 1981. *Pleocoma* spp. in Western Oregon coniferous forests: observations on adult flight habits and on egg and larval biology (Coleoptera: Scarabaeidae). *Pan-Pacific Entomologist* 57(4):461–484.

Fender, K. M. 1975. Notes and descriptions of some North American Omethinae (Coleoptera: Omethidae). *Pan-Pacific Entomologist* 51:298–302.

Ferris, G. F. 1927. Notes on an entomological enigma. *Canadian Entomologist* 59:279–281.

Fiori, G. 1963. Alcuni appunti sulla sistematica dei Coleotteri Malachiidi e Dasitidi a livello delle famiglie e sulla loro etologia. *Atti della Academia delle Scienze di Torino* 97:265–288.

Fiori, G. 1971. Contributi alla conoscenza morfologica ed etologica dei Coleotteri. IX. *Psilothrix viridicaeruleus* (Geoffr.) (Melyridae: Dasytinae). *Studi Sassaresi. Sezione III* 19:1–70.

Fisher, T. W. 1963. Mass culture of *Cryptolaemus* and *Leptomastix,* natural enemies of citrus mealybug. *Bulletin of the California Agricultural Experiment Station* 797:1–39.

Fisher, W. S. 1950. A revision of the North American beetles belonging to the family Bostrichidae. *United States Department of Agriculture Miscellaneous Publications,* No. 698, 157 pp.

Fitton, M. G. 1975. The larvae of the British genera of Cantharidae (Coleoptera). *Journal of Entomology* 44:243–254.

Fleschner, C. A. 1950. Studies on searching capacity of three predators of citrus red mite. *Hilgardia* 20:233–265.

Fletcher, T. B. 1916. One hundred notes on Indian insects. *Bulletin. Agricultural Research Institute, Pusa* 59:1–39.

Fleutiaux, E., Legros, C., Lepesme, P. and Paulian, R. 1947. *Faune de l'Empire Francais. VII. Coléoptères des Antilles (Volume I).* Paris: Museum National d'Histoire Naturelle, 239 pp., 1 pl.

Fogel, R. and Peck, S. B. 1975. Ecological studies of hypogeous fungi. I. Coleoptera associated with sporocarps. *Mycologia* 67:741–747.

Folkerts, G. W. 1979. *Spanglerogyrus albiventris,* a primitive new genus and species of Gyrinidae (Coleoptera) from Alabama. *Coleopterists Bulletin* 33:1–8.

Fontes, L. R. 1979. On the ontogeny and taxonomy of Brazilian *Uloma* (Coleoptera, Tenebrionidae). *Papeis Avulsos de Zoologia* 32(20):233–241.

Forbes, W. T. M. 1926. The wing folding patterns of the Coleoptera. *Journal of the New York Entomological Society* 34:42–68, 91–139.

Forbes, W. T. M. 1942. The wing of the Schizopini (Coleoptera: Dascillidae). *Entomological News* 53:101–102.

Forchhammer, P. 1981. Evolution of diurnal activity in some of the higher taxa of long-horned beetles (Coleoptera, Cerambycidae) in relation to continental drift. Contributions to the knowledge of Ethiopian Cerambycidae (Coleoptera). Part I. *Natura Jutlandica* 19:91–106.

Ford, E. J. and Cavey, J. F. 1985. Biology and larval descriptions of some Maryland Hispinae (Coleoptera: Chrysomelidae). *Coleopterists Bulletin* 39(1):36–59.

Ford, E. J. and Spilman, T. J. 1979. Biology and immature stages of *Dirrhagofarsus lewisi,* a species new to the United States (Coleoptera, Eucnemidae). *Coleopterists Bulletin* 33(1):75–83.

Forister, G. W. and Johnson, C. D. 1970. Bionomics of *Merobruchus julianus* (Coleoptera: Bruchidae). *Coleopterists Bulletin* 24:84–87.

Foster, D. E. 1976a. North American Thaneroclerine larvae. *Coleopterists Bulletin* 30:75–80.

Foster, D. E. 1976b. A review of North American Tillini larvae, pp. 133–138 *in* Barr, W. E. (edit.), *50 Year Anniversary Publication. Department of Entomology. University of Idaho,* Moscow, Idaho: University of Idaho Press and Idaho Agricultural Experiment Station, 138 pp.

Foster, D. E. 1976c. Revision of North American *Trichodes. Special Publications of the Museum of Texas Technical University* 11:1–86.

Foster, D. E. and Antonelli, A. L. 1973. Larval description and notes on the biology of *Anthocomus horni. Pan-Pacific Entomologist* 49:56–59.

Franciscolo, M. F. 1952. On the systematic position of the genus *Ctenidia* Castelnau, 1840 (Coleoptera). (Contribution xxxiv to the knowledge of the Mordellidae). *Proceedings of the Royal Entomological Society of London* (B)21:155–163.

Franciscolo, M. F. 1957. Coleoptera: Mordellidae. A monograph of the South African genera and species. 1. Morphology, subfamily Ctenidiinae and tribe Stenaliini. *South African Animal Life* 4:207–291.

Franciscolo, M. 1972. Su alcuni generi poco noti de Anaspidinae (47o contributo allaconoscenza degli Scraptiidae). *Memorie della Societa Entomologica Italiana* 51:123–155.

Franciscolo, M. 1974. New and little-known *Mordellistena* Costa from Pakistan and India (Coleoptera: Mordellidae). *Oriental Insects* 8:71–84.

Francke-Grosmann, H. 1967. Ectosymbiosis in wood-inhabiting insects, pp. 141–205 *in* Henry, M. S. (edit.), *Symbiosis, Vol. 2.* New York: Academic Press, 443 pp.

Frania, H. E. 1986. Larvae of *Eustilicus* Sharp, *Rugilus* Leach, *Deroderus* Sharp, *Stilocharis* Sharp, and *Medon* Stephens (Coleoptera: Staphylinidae: Paederinae: Paederini), and their phylogenetic significance. *Canadian Journal of Zoology* 64:2543–2557.

Frank, J. H. 1982. The parasites of the Staphylinidae (Coleoptera). *University of Florida Agricultural Experiment Stations Bulletin,* No. 824, vii + 118 pp.

Frank, J. H. and Curtis, G. A. 1979. Trend lines and the number of species of Staphylinidae. *Coleopterists Bulletin* 33(2):133–149.

Frank, J. H. and K. Kanamitsu. 1987. *Paederus* sensu lato: natural history and medical importance (Coleoptera: Staphylinidae). *Journal Medical Entomology* 24:155–191.

Frank, J. H. and Thomas, M. C. 1984. Cocoon-spinning and the defensive function of the median gland in larvae of Aleocharinae (Coleoptera, Staphylinidae): a review. *Quaestiones Entomologicae* 20:7–23.

Frankenberg, G. von 1942. Fleischfresser im Holz. *Natur und Volk* 72:91–100.

Frankie, G. W. 1973. Feeding habits and seasonal history of *Ernobius conicola* in cones of Monterey Cypress with notes on cohabiting insects (Coleoptera: Anobiidae). *Pan-Pacific Entomologist* 49:102–109.

Franz, H. 1968. Eine neue Gattung und Art aus der Familie Colydiidae von den Kanarischen Inseln (Coleoptera). *Eos* 44:135–139.

Franz, J. 1955. Tannenstammläuse (*Adelges piceae* Ratz.) unter einer Pilzdecke von *Cucurbitaria pithyophila* (Kze. et Schm.) De Not., nebst Beobachtungen an *Aphidoletes thompsoni* Möhn (Dipt., Itonididae) und *Rabocerus mutilatus* Beck (Col., Pythidae) als Tannenlausfeinde. *Zeitschrift für Pflanzenkrankheiten (Pflanzenpathologie) und Pflanzenschutz* 62:49–61.

Franz, J. M. 1958. Studies on *Laricobius erichsonii* Rosenh. (Coleoptera: Derodontidae), a predator on chermesids. *Entomophaga* 3:109–196.

Frazer, B. D., Gilbert, N., Ives, P. M., and Raworth, D. A. 1981. Predation of aphids by coccinellid larvae. *Canadian Entomologist* 113:1043–1046.

Freeman, P. (edit.). 1980. *Common Insect Pests of Stored Products. A Guide in their Identification. 6th Edition.* London: British Museum (Natural History) (Economic Series, No. 15), 69 pp.

Freude, H. 1958. Die Monommidae der Afrikanischen Region (Coleoptera). (IV. Teil der Monommiden der Welt mit Zusammenfassung der Ergebnisse). *Annales du Musee Royal du Congo Belge Tervuren (Belgique). Série in 8o. Sciences Zoologiques* 61:1–115.

Fuchs, G.-V. 1974. Die Gewinnung von Pollen und Nektar bei Käfern. *Natur und Museum* 104:45–54.

Fukuda, A. 1938. Description of the larva and pupa of *Cupes clathratus* (in Japanese). *Transactions of the Natural History Society of Formosa* 28:390–393.

Fukuda, A. 1941. Some ecological studies on *Cupes clathratus. Transactions of the Natural History Society of Formosa* 31:394–399.

Fukuda, A. 1962. Description of the larva of *Elacatis kraatzi* Reitter (Elacatidae, Coleoptera). *Kontyu* 30:17–20, 1 pl.

Fukuda, A. 1963. Studies on the larva of *Peltastica reitteri* Lewis with comments on the classification of Derodontidae based on larval characters (Coleoptera, Derodontidae). *Kontyu* 31:189–193, pl. 10.

Fukuda, A. 1969. Description of larva of *Aphanocephalus hemisphericus* Wollaston, with the relationship of the related genera. *Kontyu* 37:20–26.

Fulmek, L. 1930. Zur Kenntnis der Entwicklung von *Atractocerus emarginatus* Cast. (Coleopt.-Lymexylonidae). *Treubia* 12:389–394.

Furniss, R. L. and Carolin, V. M. 1977. Western Forest Insects. *United States Department of Agriculture Forest Service Miscellaneous Publications* 1339, 654 pp.

Gaedike, R. 1969. Bibliographie der Elateridenlarven-Literatur der Welt. *Beiträge zur Entomologie* 19:159–266.

Gaedike, R. 1975. Bibliographie der Elateridenlarven-Literatur der Welt (1968–1972). *Beiträge zur Entomologie* 25:85–98.

Gaedike, R. 1979. Bibliographie der Elateridenlarven-Literatur der Welt (1973–1977). *Beiträge zur Entomologie* 29:307–317.

Gage, J. H. 1920. The larvae of the Coccinellidae. *Illinois Biological Monographs* 6:232–294.

Gagné, W. C. and Martin, J. L. 1968. The insect ecology of red pine plantations in central Ontario. V– The Coccinellidae (Coleoptera). *Canadian Entomologist* 100:835–846.

Gahan, C. J. 1908. On the larvae of *Trictenotoma childreni* Gray, *Melittomma insulare* Fairm., and *Dascillus cervinus* L. *Transactions of the Entomological Society of London* 1908:275–282.

Galewski, K. 1971. A study on morphobiotic adaptations of European species of the Dytiscidae. *Polskie Pismo Entomologiczne* 41:479–702.

Gardiner, L. M. 1966. Egg bursters and hatching in the Cerambycidae (Coleoptera). *Canadian Journal of Zoology* 44:199–212.

Gardner, J. C. M. 1925. Identification of immature stages of Indian Cerambycidae. I. Cerambycini. *Indian Forest Records (Entomology Series)* 12:89–105, 3 pls.

Gardner, J. C. M. 1926. Description of the early stages of *Fornax gardneri* Fleut. (Melasidae, Col.). Description of the larva of *Atractocerus emarginatus* Cast. (Lymexylonidae, Col.). *Indian Forest Records* 12:273–282, 1 pl.

Gardner, J. C. M. 1927. Identification of immature stages of Indian Cerambycidae. II. *Indian Forest Records. (Entomology Series)* 13:31–61, 4 pls.

Gardner, J. C. M. 1929. Immature stages of Indian Coleoptera (6). *Indian Forest Records. (Entomology Series)* 14(4):103–132, 6 pls.

Gardner, J. C. M. 1930a. The early stages of *Niponius andrewesi,* Lew. (Co. Hist.). *Bulletin of Entomological Research* 21:15–17, 1 pl.

Gardner, J. C. M. 1930b. Immature stages of Indian Coleoptera (7). *Indian Forest Records. (Entomology Series)* 14(13):279–286, 3 pls.

Gardner, J. C. M. 1931a. Immature stages of Indian Coleoptera (8) (Cerambycidae—contd.). *Indian Forest Records. (Entomology Series)* 16:49–89.

Gardner, J. C. M. 1931b. Immature stages of Indian Coleoptera (9). *Indian Forest Records. (Entomology Series)* 16(4):91–111, 5 pls.

Gardner, J. C. M. 1932a. Immature stages of Indian Coleoptera (10) (Anthribidae). *Indian Forest Records. (Entomology Series)* 16(11):327–333, 1 pl.

Gardner, J. C. M. 1932b. Immature stages of Indian Coleoptera (11) (Platypodidae). *Indian Forest Records. (Entomology Series)* 17(3):1–10, 2 pls.

Gardner, J. C. M. 1932c. The larva of *Catapiestus indicus* Fairm. (Coleoptera: Tenebrionidae). *Proceedings of the Entomological Society of Washington* 34:142–145.

Gardner, J. C. M. 1933a. Immature stages of Indian Coleoptera (12) (Carabidae contd.). *Indian Forest Records. (Entomology Series)* 17(8):1–12, 2 pls.

Gardner, J. C. M. 1933b. Immature stages of Indian Coleoptera (13) (Bostrychidae). *Indian Forest Records. (Entomology Series)* 18(9):1–19, 4 pls.

Gardner, J. C. M. 1934a. Immature stages of Indian Coleoptera (14) (Curculionidae). *Indian Forest Records. (Entomology Series)* 20(2):1–42, 6 pls.

Gardner, J. C. M. 1934b. Immature stages of Indian Coleoptera (15) (Scolytidae). *Indian Forest Records. (Entomology Series)* 20(8):1–17, 2 pls.

Gardner, J. C. M. 1935a. Immature stages of Indian Coleoptera (16) (Scarabaeoidea). *Indian Forest Records. New Series. Entomology* 1(1):1–33, 4 pls.

Gardner, J. C. M. 1935b. Immature stages of Indian Coleoptera (17) (Eucnemidae). *Indian Forest Records. New Series. Entomology* 1(4):79–93, 2 pls.

Gardner, J. C. M. 1935c. Immature stages of Indian Coleoptera (18) (Brenthidae). *Indian Forest Records. New Series. Entomology* 1(7):139–148, 2 pls.

Gardner J. C. M. 1935d. A new Indian species of *Atractocerus* (Col. Lymexylonidae). *Stylops* 4:69–70.

Gardner, J. C. M. 1936a. Immature stages of Indian Coleoptera (19) Anthribidae. *Indian Forest Records. New Series. Entomology* 2(2):99–113, 2 pls.

Gardner, J. C. M. 1936b. Immature stages of Indian Coleoptera (20), Carabidae). *Indian Forest Records. New Series. Entomology* 2(9):181–201, 4 pls.

Gardner, J. C. M. 1936c. A larva of the subfamily Balginae (Col., Elateridae). *Proceedings of the Royal Entomological Society of London* (B)5:3–5.

Gardner, J. C. M. 1937a. Immature stages of Indian Coleoptera (21) Cleridae. *Indian Forest Records. New Series. Entomology* 3(2):31–47, 2 pls.

Gardner, J. C. M. 1937b. Immature stages of some Indian Coleoptera (22). *Indian Forest Records. New Series. Entomology* 3(6):127–140, 2 pls.

Gardner, J. C. M. 1938a. Immature stages of Indian Coleoptera (23) (Carabidae—contd.). *Indian Forest Records. New Series. Entomology* 3(8):149–157, 2 pls.

Gardner, J. C. M. 1938b. Immature stages of Indian Coleoptera (24), Curculionidae contd.) *Indian Forest Records. New Series. Entomology* 3(12):227–261, 6 pls.

Gardner, J. C. M. 1944. On some coleopterous larvae from India. *Indian Journal of Entomology* 6:111–116.

Gardner, J. C. M. 1946. Larvae of Cantharoidea (Coleoptera). *Indian Journal of Entomology* 8:121–129.

Gardner, J. C. M. 1947. A list of described immature stages of Indian Coleoptera. *Indian Forest Records. New Series. Entomology* 7(5):163–191.

Gebien, H. 1922. The Percy Sladen trust expedition to the Indian Ocean in 1905. Vol. VII. No. 5. Coleoptera, Heteromera: Tenebrionidae. *Transactions of the Linnean Society of London. Series 2. Zoology* 18:261–324, pl. 23.

Gerberg, E. J. 1957. A revision of the New World species of powder-post beetles belonging to the family Lyctidae. *United States Department of Agriculture Technical Bulletin* 1157:1–55.

Germer, F. 1912. Untersuchungen über den Bau und die Lebensweise der Lymexyloniden, speziell des *Hylecoetus dermestoides* L. *Zeitschrift für Wissenschaftliche Zoologie, Abteilung A* 101:683–735.

Geyer, J. W. C. 1947a. A study of the biology and ecology of *Exochomus flavipes* Thunb. (Coccinellidae, Coleoptera) Part I. *Journal of the Entomological Society of South Africa* 9:219–234.

Geyer, J. W. C. 1947b. A study of the biology and ecology of *Exochomus flavipes* Thunb. (Coccinellidae, Coleoptera) Part II. *Journal of the Entomological Society of South Africa* 10:64–109.

Gibson, L. P. 1969. Monograph of the genus *Curculio* in the New World (Coleoptera: Curculionidae) Part I. United States and Canada. *Miscellaneous Publications of the Entomological Society of America* 6(5):241–285, 140 figs.

Gibson, L. P. 1985. Description and Key to larvae of *Curculio* spp. of Eastern United States and Canada (Coleoptera: Curculionidae). *Proceedings of Entomological Society of Washington.* 87:554–563.

Gillespie, B. B. and Randolph, N. M. 1958. Notes on the biology of *Bruchus brachialis* Fahr. *Journal of Economic Entomology* 51(3):401–402.

Gillogly, L. R. and Gillogly, G. M. 1954. Notes on the biology of *Epuraea monogama* Crotch (Coleoptera: Nitidulidae). *Coleopterists Bulletin* 8:63–67.

Ghilarov, M. S. 1964a. Family Lagriidae—Hairy beetles, pp. 455–456 *in* Ghilarov, M. S. (edit.), *Determination Key to soil-dwelling Insect Larvae.* Moscow: Izdatel'stvo "Nauk", 918 pp. (in Russian).

Ghilarov, M. S. 1964b. Order Coleoptera—Beetles, pp. 49–71 *in* Ghilarov, M. S. (edit.), *Determination Key to soil-dwelling Insect Larvae.* Moscow: Izdatel'stvo "Nauk", 918 pp. (in Russian).

Ghilarov, M. S. 1979. The larva of *Phyllocerus* and the position of this genus in the system of Elateroidea (Coleoptera). *Zoologicheskii Zhurnal* 58(5):655–663 (in Russian).

Ghilarov, M. S. and Medvedev, L. N. 1964. Family Chrysomelidae-Leaf Beetles, pp. 507–530 *in* Ghilarov, M. S. (edit.), *Determination Key to soil-dwelling Insect Larvae.* Moscow: "Nauka", 918 pp. (in Russian).

Ghilarov, M. S. and Svetova, J. A. 1963. Die Larve von *Hedyphanes seidlitzi* Reitter und die Unterschiede der Larven einiger Gattungen der paläarktischen Helopini. *Beitrage zur Entomologie* 13(3/4):327–334.

Glen, R. 1950. Larvae of the elaterid beetles of the tribe Lepturoidini (Coleoptera: Elateridae). *Smithsonian Miscellaneous Collections* 111(11):1–246 (#3987).

Glen R., King, K. M., and Arnason, A. P. 1943. The identification of wireworms of economic importance in Canada. *Canadian Journal of Research* 21(D):358–387.

Golbach, R. 1983. Primera cita de la familia Cerophytidae (Coleoptera) para Paraguay, Bolivia y la Argentina. *Acta Zoologica Lilloana* 37(1):131–137.

Good, H. G. 1924. Notes on the life history of *Prionocyphon limbatus* LeC. (Helodidae, Coleoptera). *Journal of the New York Entomological Society* 32:79–85.

Gordon, H. D. 1938. Note on a rare beetle, *Cartodere filum* Aubé, eating fungus spores. *Transactions of the Btitish Mycological Society* 21:193–197.

Gorham, H. S. 1887. Revision of the Japanese species of the coleopterous family Endomychidae. *Proceedings of the Zoological Society of London* 1887:642–653, pl. 53.

Gorham, J. R. 19 . Insect and mite pests in food: an illustrated key. *USDA. Agr. Handbook,* in press.

Goulet, H. 1976. A method for rearing ground beetles (Coleoptera: Carabidae). *Coleopterists Bulletin* 30:33–36.

Goulet, H. 1977. Technique for the study of immature Coleoptera in glycerine. *Coleopterists Bulletin* 31:381–382.

Goulet, H. 1983. The genera of Holarctic Elaphrini and species of *Elaphrus* Fabricius (Coleoptera: Carabidae): classification, phylogeny and zoogeography. *Quaestiones Entomologicae* 19:219–482.

Grandi, G. 1936. Morfologia ed etologia comparate di insetti a regime specializzato. XII. *Macrosiagon ferrugineum flabellatum* F. *Bolletino dell'Instituto di Entomologia della Universita degli Studi di Bologna* 9:33–64.

Grandi, G. 1955. *Rhysodes germari* Ganglb. Documenti morfologici ed eto-ecologici. *Bolletino dell'Istituto di Entomologia della Universita degli Studi di Bologna* 21:179–195.

Grandi, G. 1959. The problems of "morphological adaptation" in insects. *Smithsonian Miscellaneous Publications* 137:203–230.

Grandi, G. 1960. Campagna di richerche dell'Istituto di Entomologia dell'Universita di Bologna nelle Alpi trentine. I. *Hylecoetus dermestoidea* L. (Coleoptera Lymexylonidae). Morfologia larvale—la larva neonata. *Bolletino dell'Istituto di Entomologia della Universita degli Studi di Bologna* 24:39–51.

Gravely, F. H. 1915. The larvae and pupae of some beetles from Cochin. *Records of the Indian Museum* 11:353–366, pls. 20–21.

Gravely, F. H. 1916. Some lignicolous beetle-larvae from India and Borneo. *Records of the Indian Museum* 12:137–175, pls. 20–22.

Graves, R. C. 1965. Observations on the ecology, behavior and life cycle of the fungus-feeding beetle, *Cypherotylus californicus*, with a description of the pupa (Coleoptera: Erotylidae). *Coleopterists Bulletin* 19:117–122.

Gray, I. E. 1946. Observations on the life history of the horned *Passalus*. *American Midland Naturalist* 35:728–746.

Greathead, D. J. 1963. A review of the insect enemies of Acridoidea (Orthoptera). *Transactions of the Royal Entomological Society of London* 114:437–517.

Gressitt, J. L. and Samuelson, G. A. 1964a. Insects of Campbell Island. Coleoptera: Orthoperidae (larva). *Pacific Insects Monographs* 7:410–411.

Gressitt, J. L. and Samuelson, G. A. 1964b. Insects of Campbell Island. Coleoptera: Hydraenidae, Ptiliidae, Leptodiridae, Byrrhidae, Lathridiidae, Melandryidae. *Pacific Insects Monographs* 7:376–390.

Grigarick, A. A. and Schuster, R. O. 1961. A new species of *Loricaster* from California (Coleoptera: Clambidae). *Pan-Pacific Entomologist* 37:161–164.

Grison, P., Labeyrie, V., Jourdheuil, P., Remaudiere, G. and Balachowsky, A. S. 1962. Famille des Chrysomelidae, pp. 567–873 *in* A. S. Balachowsky, (ed.), *Entomologie à l'appliquee à l'agriculture. Tome I. Coléoptères. Second Volume.* Paris: Masson, pp. 567–1391.

Grouvelle, A. 1914. The Percy Sladen Trust expedition to the Indian Ocean in 1905. Coleoptera: Cucujidae, Cryptophagidae. (Avec une description de la larve et de la nymphe de *Prostominia convexiuscula* Grouvelle (Cucujidae) par P. de Peyerimhoff). *Transactions of the Linnean Society of London* 17:140–159.

Gur'yeva, Ye. L. 1969. Some trends in the evolution of click beetles (Coleoptera, Elateridae). *Entomologicheskoye Obozreniye* 48:263–272 (in Russian; translation in *Entomological Review* 48:154–159).

Guy, P. and Gibbs, A. 1981. A tymovirus of *Cardamine* sp. from alpine Australia. *Australasian Plant Pathology* 10:12–13.

Guyon, T. L. and Knull, J. N. 1925. Mexican bean beetle in Pennsylvania. *Bulletin of the Pennsylvania Department of Agriculture* 8:1–6.

Habu, A. and Sadanaga, K. 1961. Illustrations for identification of larvae of the Carabidae found in cultivated fields and paddy-fields (I) (in Japanese). *Bulletin of the National Institute of Agricultural Sciences, Ser. C* 13:207–248.

Habu, A. and Sadanaga, K. 1963. Illustrations for identification of larvae of the Carabidae found in cultivated fields and paddy-fields (II) (in Japanese). *Bulletin of the National Institute of Agricultural Sciences, Ser. C* 16:151–179.

Habu, A. and Sadanaga, K. 1965. Illustrations for identification of larvae of the Carabidae found in cultivated fields and paddy-fields (III) (in Japanese). *Bulletin of the National Institute of Agricultural Sciences, Ser. C* 19:81–216.

Habu, A. and Sadanaga, K. 1969. Illustrations for identification of larvae of the Carabidae found in cultivated fields and paddy-fields (Suppl. I) (in Japanese). *Bulletin of the National Institute of Agricultural Sciences, Ser. C* 23:113–143.

Habu, A. and Sadanaga, K. 1970. Descriptions of some larvae of the Carabidae found in cultivated fields and paddy-fields (I) (II) (in Japanese). *Kontyu* 38:9–23;24–41.

Hackman, R. H. 1984. Chapter 30. Arthropoda: Cuticle: Biochemistry, pp. 583–610 *in* Bereiter, J., Matolsky, A. G., and Richards, K. S. (edits.), *Biology of the Integument. Volume 1: Invertebrates.* Berlin: Springer-Verlag, xvi + 841 pp.

Hackwell, G. A. 1973. Biology of *Lasconotus subcostulatus* (Coleoptera: Colydiidae) with special reference to feeding behavior. *Annals of the Entomological Society of America* 66:62–65.

Hagen, K. S. 1962. Biology and ecology of predaceous Coccinellidae. *Annual Review of Entomology* 7:289–326.

Halffter, G. and Edmonds, W. D. 1981. Evolucion de la nidification y de la cooperacion bisexual en Scarabaeinae (Ins.: Col.). *Anales de la Escuela Nacional de Ciencias Biologicas, Mexico City* 25:117–144.

Halffter, G. and Edmonds, W. D. 1982. *The Nesting Behavior of Dung Beetles (Scarabaeinae): an Ecological and Evolutive Approach.* Mexico, D. F.: Instituto de Ecologia, 176 pp.

Halffter, G. and Matthews, E. G. 1966. The natural history of dung beetles of the subfamily Scarabaeinae (Coleoptera, Scarabaeidae). *Folia Entomologica Mexicana* 12–14:1–312.

Hall, D. W. and Howe, R. W. 1953. A revised key to the larvae of the Ptinidae associated with stored products. *Bulletin of Entomological Research* 44:1–216.

Halstead, D. G. H. 1968a. Observations on the biology of *Murmidius ovalis* (Beck) (Coleoptera: Cerylonidae). *Journal of Stored Products Research* 4:13–21.

Halstead, D. G. H. 1968b. Some observations on the biology of *Lophocateres pusillus* (Klug) (Coleoptera: Trogositidae). *Journal of Stored Products Research* 4:197–202.

Halstead, D. G. H. 1973. A revision of the genus *Silvanus* Latreille (S. L.) (Coleoptera: Silvanidae). *Bulletin of the British Museum (Natural History). Entomology* 29(2):37–112.

Halstead, D. G. H. 1980. A revision of the genus *Oryzaephilus* Ganglbauer, including descriptions of related genera (Coleoptera: Silvanidae). *Zoological Journal of the Linnean Society* 69:271–374.

Hamilton, C. C. 1925. Studies on the morphology, taxonomy and ecology of the larvae of Holarctic tiger-beetles (family Cicindelidae). *Proceedings of the United States National Museum* 65(17):1–87, pls. 1–12.

Hamilton, R. W. 1980. Notes on the biology of *Eugnamptus collaris* (Fabr.) (Coleoptera: Rhynchitidae), with descriptions of the larva and pupa. *Coleopterists Bulletin* 34(2):227–236.

Hamilton, R. W. and Kuritsky, S. S. 1981. Description of the larva and pupa of *Merhynchites bicolor* (Fabricius) (Coleoptera: Rhynchitidae). *Coleopterists Bulletin* 35(2):189–195, 11 figs.

Hammad, S. M. 1953. The immature stages of *Metophthalmus serripennis* Broun (Coleoptera: Lathridiidae). *Proceedings of the Royal Entomological Society of London* (A)28:133–138.

Hammond, P. M. 1979. Wing-folding mechanisms of beetles, with special reference to investigations of adephagan phylogeny (Coleoptera), pp. 113–180 *in* Erwin, T. L., Ball, G. E., Whitehead, D. R. and Halpern, A. L. (edit.), *Carabid Beetles: their Evolution, Natural History, and Classification.* The Hague: W. Junk, 635 pp.

Hatch, M. H. 1927. Notes on the biology of *Dineutus* (Gyrinidae). *Bulletin of the Brooklyn Entomological Society* 22:27–28.

Hatch, M. H. 1961. *The Beetles of the Pacific Northwest. Part III: Pselaphidae and Diversicornia I.* Seattle: University of Washington Press, ix + 503 pp.

Hatch, M. H. 1965. *The Beetles of the Pacific Northwest. Part IV: Macrodactyles, Palpicornes and Heteromera.* Seattle, Washington: University of Washington Press, 268 pp., 28 pls.

Hatch, M. H. 1971. *The Beetles of the Pacific Northwest. Part V: Rhipiceroidea, Sternoxi, Phytophaga, Rhynchophora, and Lamellicornia.* Seattle: University of Washington Press, 602 pp.

Hawkeswood, T. J. 1985. The larva of *Diadoxus erythrurus* (White) (Coleoptera: Buprestidae). *Australian Entomological Magazine* 12(2):23–28.

Hawkins, J. H. 1936. The bionomics and control of wireworms in Maine. *Maine Agricultural Experiment Station Bulletin* 381, 146 pp.

Hayashi, N. 1951. Studies on the Japanese species of the genus *Dorcatoma* II. The classification of the larvae and pupae. (Studies of Dorcatominae, Anobiidae, I). *The Chuho*, No. 4, 12 pp.

Hayashi, N. 1962a. The larval form of *Anaspis (Anaspis) funagata* Kono (Scraptiidae). (Studies on coleopterous larvae XIV). *Entomological Review of Japan* 15:19–21.

Hayashi, N. 1962b. Notes on the immature stages of *Lamprobyrrhus nitidus* Schaller (Byrrhidae) (Studies on coleopterous larvae XI). *Entomological Review of Japan* 14:48–50, pls. 8–9.

Hayashi, N. 1963. On the larvae of three species of Cephaloidae, Melandryidae and Pyrochroidae occurring in Japan (Coleoptera: Cucujoidea). *Insecta Matsumurana* 26:108–114.

Hayashi, N. 1964. On the larvae of Lagriidae occurring in Japan (Coleoptera: Cucujoidea). *Insecta Matsumurana* 27:24–30.

Hayashi, N. 1966. A contribution to the knowledge of the larvae of Tenebrionidae occurring in Japan (Coleoptera: Cucujoidea). *Insecta Matsumurana, Supplement* 1:1–41.

Hayashi, N. 1968. Additional notes on the larvae of Lagriidae and Tenebrionidae occurring in Japan (Coleoptera: Cucujoidea). *Insecta Matsumurana, Supplement* 3:1–12.

Hayashi, N. 1969a. On the larvae of Pyrochroidae occurring in Japan (Coleoptera: Cucujoidea). *Kontyu* 37:444–452.

Hayashi, N. 1969b. On the larvae of some species of small families of Cucujoidea in Japan (Coleoptera). *Insecta Matsumurana, Supplement* 7, 9 pp., 6 pls.

Hayashi, N. 1971. On the larvae of Mycetophagidae occurring in Japan (Coleoptera: Cucujoidea). *Kontyu* 39:361–367.

Hayashi, N. 1972. On the larvae of some species of Colydiidae, Tetratomidae and Aderidae occurring in Japan (Coleoptera: Cucujoidea). *Kontyu* 40(2):100–111.

Hayashi, N. 1975. On the larvae of Melandryidae (Coleoptera, Cucujoidea) and some related families occurring in Japan. *Kontyu* 43:147–169.

Hayashi, N. 1978. A contribution to the knowledge of the larvae of Nitidulidae occurring in Japan (Coleoptera: Cucujoidea). *Insecta Matsumurana* (N.S.)14:1–97.

Hayashi, N. 1980. Illustrations for identification of larvae of the Cucujoidea (Coleoptera) found living in dead trees in Japan. (In Japanese). *Memoirs of the Education Institute for Private Schools in Japan* 72:95–147, 53 pls.

Hayashi, N. 1981. Illustrations for identification of the coleopterous larvae living in dead trees. (In Japanese). *Memoirs of the Education Institute for Private Schools in Japan* 81:83–96, 12 pls.

Hayashi, N. 1986. Larvae, pp. 202–218 and plates 1–113 *in* Morimoto, K. and Hayashi, N. (edits.) *The Coleoptera of Japan in Color*, Vol. I. Osaka: Hoikusha Pub. Co., vi + 323 pp.

Hayashi, N., Fukuda, A., and Kurosa, K. 1959. Coleoptera, pp. 392–545 *in* Esaki, T., Yuasa, Ishii, T., and Motoki, T. (edits.) *Illustrated Insect Larvae of Japan*. Tokyo: Hokuryukan, 712 + 50 pp.

Hayashi, N. and Takenaka, H. 1965. Notes on the immature stages of *Encaustes praenobilis* Lewis (Coleoptera: Erotylidae). *Mikado* 1:35–39.

Hayes, W. P. 1928. The epipharynx of lamellicorn larvae (Coleop.) with a key to common genera. *Annals of the Entomological Society of America* 21(2):282–306.

Hayes, W. P. 1929. Morphology, taxonomy and biology of larval Scarabaeoidea. *Illinois Biological Monographs* 12(2):1–119.

Hayes, W. P. and Chu, H. F. 1946. The larvae of the genus *Nosodendron* Latr. (Coleoptera, Nosodendridae). *Annals of the Entomological Society of America* 39:69–79.

Helgesen, R. G. and Post, R. L. 1967. Saprophagous Scarabaeidae (Coleoptera) of North Dakota. *North Dakota Insects* 7:1–60.

Heller, K. M. 1904. Brazilianische Käferlarven gesammelt von Dr. Fr. Ohaus. *Stettiner Entomologische Zeitung* 65(1):381–401, 2 pls.

Heller, K. M. 1923. Ein neuer Cryptophagine (Coleopt.) aus Java. *Treubia* 3:275–276.

Hengeveld, R. 1980. Polyphagy, oligophagy and food specialization in ground beetles (Coleoptera, Carabidae). *Netherlands Journal of Zoology* 30:564–584.

Herman, L. H., Jr. 1986. Revision of *Bledius*. Part IV. Classification of species groups, phylogeny, natural history, and catalogue (Coleoptera, Staphylinidae, Oxytelinae). *Bulletin of the American Museum of Natural History* 184:1–367.

Herron, J. C. 1953. Biology of the sweetclover weevil and notes on the biology of the clover root curculio. *Ohio Journal of Science* 53(3):105–112.

Hervey, G. E. R. 1927. A European nitidulid, *Brachypterolus pulicarius* L. (Coleoptera, family Nitidulidae), *Journal of Economic Entomology* 20:809–814.

Hess, W. N. 1920. Notes on the biology of some common Lampyridae. *Biological Bulletin. Marine Biological Laboratory, Woods Hole, Mass.* 38:39–76.

Hickin, N. E. 1963. *The Insect Factor in Wood Decay*. London: Hutchinson, 336 pp.

Hickman, J. R. 1930. Life-histories of Michigan Haliplidae (Coleoptera). *Papers of the Michigan Academy of Science, Arts, and Letters* 11(1929):399–424, 59 figs.

Hickman, J. R. 1931. Contribution to the biology of the Haliplidae (Coleoptera). *Annals of the Entomological Society of America* 24:129–142.

Hickman, V. V. 1974. Notes on the biology of *Ptinus exulans* Erichson (Coleoptera: Ptinidae). *Journal of the Entomological Society of Australia (New South Wales)* 8:7–14.

Hicks, E. A. 1959. *Check-list and Bibliography on the Occurrence of Insects in Birds' Nests*. Ames, Iowa: Iowa State College Press, 681 pp.

Hicks, E. A. 1962. Check-list on the occurrence of insects in birds' nests. Supplement I. *Iowa State Journal of Science* 36(3):233–348.

Hicks, E. A. 1971. Check-list on the occurrence of insects in birds' nests. Supplement II. *Iowa State Journal of Science* 46(2):123–338.

High, M. M. 1939. The vegetable weevil. *United States Department of Agriculture Circular* 530, 25 pp, 12 figs.

Hilsenhoff, W. L. 1974. The unusual larva and habitat of *Agabus confusus* (Dytiscidae). *Annals Entomological Society of America* 67:703–705.

Hilsenhoff, W. L. 1977. Use of arthropods to evaluate water quality of streams. *Department of Natural Resources, Madison, Wisconsin, Technical Bulletin*, No. 10, 15 pp.

Hinckley. A. D. 1961. Comparative ecology of two beetles established in Hawaii: an anthribid, *Araecerus levipennis*, and a bruchid, *Mimosestes sallaei*. *Ecology*. 42(3):526–532.

Hinds, T. E. 1972. Insect transmission of *Ceratocystis* species associated with aspen cankers. *Phytopathology* 62:221–225.

Hingley, M. R. 1971. The ascomycete fungus, *Daldinia concentrica*, as a habitat for animals. *Journal of Animal Ecology* 40:17–32.

Hinton, H. E. 1934. Two coleopterous families new to Mexico. *Pan-Pacific Entomologist* 9(4):160–162.

Hinton, H. E. 1936. Notes on the biology of *Dryops luridus* Erichson (Coleoptera, Dryopidae). *Transactions of the Society for British Entomology* 3:67–78.

Hinton, H. E. 1939. An inquiry into the natural classification of the Dryopoidea, based partly on a study of their internal anatomy (Col.). *Transactions of the Royal Entomological Society of London* 89(7):133–184.

Hinton, H. E. 1940. A monographic revision of the Mexican water beetles of the family Elmidae. *Novitates zoologicae* 42:217–396.

Hinton, H. E. 1941a. The immature stages of *Acrotrichis fascicularis* (Herbst) (Col. Ptiliidae). *Entomologist's Monthly Magazine* 77:245–250.

Hinton, H. E. 1941b. The immature stages of *Sericoderus lateralis* (Gyllenhal) 1827 (Coleoptera, Corylophidae). *Entomologist* 74:198–202.

Hinton, H. E. 1941c. The Lathridiidae of economic importance. *Bulletin of Entomological Research* 33:191–247.

Hinton, H. E. 1941d. Notes on the internal anatomy and immature stages of *Mycetophagus quadripustulatus* (Linnaeus) (Coleoptera, Mycetophagidae). *Proceedings of the Royal Entomological Society of London* (A)16:39–48.

Hinton, H. E. 1941e. The Ptinidae of economic importance. *Bulletin of Entomological Research* 31:331–381.

Hinton, H. E. 1942. A new leaf-mining nitidulid (Coleoptera). *The Entomologist* 75:126–129.

Hinton, H. E. 1944. Some general remarks on sub-social beetles, with notes on the biology of the staphylinid *Platystethus arenarius* (Fourcroy). *Proceedings of the Royal Entomological Society of London* (A)19:115–128.

Hinton, H. E. 1945a. The Histeridae associated with stored products. *Bulletin of Entomological Research* 35:309–340.

Hinton, H. E. 1945b. *A Monograph of the Beetles Associated with Stored Products. Volume 1* London: British Museum (Natural History). 443 pp.

Hinton, H. E. 1946a. The "gin traps" of some beetle pupae: a protective device which appears to be unknown. *Transactions of the Royal Entomological Society of London* 97:473–496.

Hinton, 1946b. A new classification of insect pupae. *Proceedings of the Zoological Society of London* 116:282–328.

Hinton, H. E. 1947. On the reduction of functional spiracles in the aquatic larvae of the Holometabola, with notes on the moulting process of spiracles. *Transactions of the Royal Entomological Society of London* 98:449–473.

Hinton, H. E. 1948a. The dorsal cranial areas of caterpillars. *Annals and Magazine of Natural History* (11)14:843–852.

Hinton, H. E. 1948b. On two species of *Lyphia* introduced with stored products into Britain (Coleoptera, Tenebrionidae). *The Entomologist* 81:15–19.

Hinton, H. E. 1948c. A synopsis of the genus *Mecedanum* Erichson (Coleoptera, Colydiidae). *Novitates Zoologicae* 42:475–484.

Hinton, H. E. 1949. On the function, origin, and classification of pupae. *Proceedings of the South London Entomological and Natural History Society* 1947–1948:111–154.

Hinton, H. E. 1955. On the respiratory adaptations, biology, and taxonomy of the Psephenidae, with notes on some related families (Coleoptera). *Proceedings of the Zoological Society of London* 125:543–568.

Hinton, H. E. 1958. The phylogeny of the panorpoid orders. *Annual Review of Entomology* 3:181–206.

Hinton, H. E. 1963. The ventral ecdysial lines of the head of endopterygote larvae. *Transactions of the Royal Entomological Society of London* 115:39–61.

Hinton, H. E. 1966. Respiratory adaptations of the pupae of beetles of the family Psephenidae. *Philosophical Transactions of the Royal Society of London Series B. Biological Sciences* 251:211–245.

Hinton, H. E. 1967a. On the spiracles of the larvae of the suborder Myxophaga (Coleoptera). *Australian Journal of Zoology* 15:955–959.

Hinton, H. E. 1967b. The respiratory system of the egg-shell of the common housefly. *Journal of Insect Physiology* 13:647–651.

Hinton, H. E. 1967c. Structure and ecdysial process of the larval spiracles of the Scarabaeoidea, with special reference to those of *Lepidoderma*. *Australian Journal of Zoology* 15:947–953.

Hinton, H. E. 1969. Plastron respiration in adult beetles of the suborder Myxophaga. *Journal of Zoology* 159:131–137.

Hinton, H. E. 1981. *Biology of Insect Eggs*. Oxford: Pergamon Press, xxiv + 1125 pp. (3 volumes).

Hinton, H. E. and Stephens, F. L. 1941a. Notes on the biology and immature stages of *Cryptophagus acutangulus*, Gyll. (Col., Cryptophagidae). *Bulletin of Entomological Research* 32:135–143.

Hinton, H. E. and Stephens, F. L. 1941b. Notes on the food of *Micropeplus*, with a description of the pupa of *M. fulvus* Erichson (Coleoptera, Micropeplidae). *Proceedings of the Royal Entomological Society of London* (A)16:29–32.

Hisamatsu, S. 1962. A new species of *Cryptophagus* (Col.: Cryptophagidae). *Niponius. Acta Coleopterologica* 1(20):1–3.

Hiznay, P. A. and Krause, J. B. 1956. The structure and musculature of the larval head and mouthparts of the horned passalus beetle, *Popilius disjunctus* Illiger. *Journal of Morphology* 97:55–70.

Hlavac, T. F. 1975. The prothorax of Coleoptera: (except Bostrichiformia-Cucujiformia). *Bulletin of the Museum of Comparative Zoology* 147:137–183.

Hodek, I. 1967. Bionomics and ecology of predaceous Coccinellidae. *Annual Review of Entomology* 12:79–104.

Hodek, I. 1973. *Biology of Coccinellidae*. The Hague: Academia, W. Junk, 260 pp.

Hoebeke, E. R. and McCabe, T. L. 1977. The life history of *Serropalpus coxalis*, with a description of the larva and pupa (Coleoptera: Melandryidae). *Coleopterists Bulletin* 31(1):57–63.

Hölldobler, B. 1971. Communication between ants and their guests. *Scientific American* 1971:86–93.

Hoffman, C. E. 1939. Morphology of the immature stages of some northern Michigan Donaciini (Chrysomelidae: Coleoptera). *Papers of the Michigan Academy of Science, Arts, and Letters* 25:243–292.

Hoffmann, A. 1945. *Faune de France. 44. Coléoptères Bruchides et Anthribides*. Paris: Office Central de Faunistique, 184 pp.

Hoffmann, A., Jourdheuil, P., Grison, P., Chevalier, M., Steffan, J., Cuillé, Vilardebo, A. and Balachowsky, A. S. 1962. Famille des Curculionidae, pp. 874–1202 *in* A. S. Balachowsky, (ed.), *Entomologie appliquée a l'agriculture*. Tome I. Coléoptères. Second volume. Paris: Masson, pp. 567–1391.

Holloway, B. A. 1969. Further studies on generic relationships in Lucanidae (Insecta: Coleoptera) with special reference to the ocular canthus. *New Zealand Journal of Science* 12:958–977.

Holloway, B. A. 1972. The systematic position of the genus *Diphyllostoma* Fall (Coleoptera: Scarabaeoidea). *New Zealand Journal of Science* 15:31–38.

Holloway, B. A. 1982. Anthribidae (Insecta: Coleoptera). *Fauna of New Zealand* 3, 264 pp.

Honomichl, K. 1980. Die digitiformen Sensillen auf dem Maxillarpalpus von Coleoptera. I. Vergleichend-topographische Untersuchung des kutikulären Apparates. Zoologischer Anzeiger 204(1/2):1–12.

Hopkins, A. D. 1909. Contributions toward a monograph of the scolytid beetles. *United States Department of Agriculture Bureau of Entomology, Technical Series* 17(1), 164 pp.

Hopping, R. 1935. Revision of the genus *Mycterus* Clairv. (Coleoptera, Pythidae). *Pan-Pacific Entomologist* 11:75–78.

Horion, A. 1949. *Faunistik der mitteleuropäischen Käfer. Band III. Palpicornia-Staphylinoidea (ausser Staphylinidae)*. Frankfurt am Main: Klosterman, xxiii + 388 pp.

Horion, A. 1955. *Faunistik der mitteleuropäischen Käfer. Band IV: Sternoxia (Buprestidae), Fossipedes, Macrodactylia, Brachymera*. Tutzing bei München: Museum G. Frey, xxii + 280 pp.

Horion, A. 1956. *Faunistik der mitteleuropäishcen Käfer. Band V: Heteromera*. Tutzing bei München: Museum G. Frey, xv + 336 pp.

Horion, A. 1960. *Faunistik der mitteleuropäischen Käfer. Band VII: Clavicornia. 1. Teil (Sphaeritidae bis Phalacridae)*. Überlingen: A. Feyel, viii + 346 pp.

Horion, A. 1961. *Faunistik der mitteleuropäischen Käfer. Band VIII: Clavicornia. 2. Teil (Thorictidae bis Ciidae), Teredilia, Coccinellidae.* Überlingen: A. Feyel, xvi + 375 pp.

Horn, G. H. 1883. Miscellaneous notes and short studies of North America Coleoptera. *Transactions of the American Entomological Society* 10:269–312.

Horsfall, W. R. 1943. Biology and control of common blister beetles in Arkansas. *University of Arkansas Agricultural Experiment Station Bulletin* 436, 55 pp.

Houser, J. S. 1923. The apple flea weevil, *Orchestes pallicornis* Say (Curculionidae). *Ohio Agricultural Experiment Station Bulletin* 372:395–434, 15 figs.

Houston, W. W. K. and Luff, M. L. 1975. The larvae of the British Carabidae (Coleoptera). III. Patrobini. *Entomologist's Gazette* 26:59–64.

Howard, L. O. 1896. A coleopterous enemy of *Corydalis cornutus. Proceedings of the Entomological Society of Washington* 3:310–313.

Howden, A. T. and Howden, H. F. 1982. The larva and adult biology of *Rhinosimus viridiaeneus* (Coleoptera: Salpingidae). *Canadian Entomologist* 113:1055–1060.

Howden, H. F. 1955. Biology and taxonomy of North American beetles of the subfamily Geotrupinae with revisions of the genera *Bolbocerosoma, Eucanthus, Geotrupes,* and *Peltotrupes* (Scarabaeidae). *Proceedings of the United States National Museum* 104:151–319.

Howden, H. F. 1961. A revision of the New World species of *Thalycra* Erichson, with a description of a new genus and notes on generic synonymy (Coleoptera: Nitidulidae). *Canadian Entomologist, Supplement* 25:1–61.

Howden, H. F. 1964. The Geotrupinae of North and Central America. *Memoirs of the Entomological Society of Canada* 39:1–91.

Howden, H. F. 1982. Larval and adult characters of *Frikius* Germain, its relationship to the Geotrupini, and a phylogeny of some major taxa in the Scarabaeoidea (Insecta: Coleoptera). *Canadian Journal of Zoology* 60(11):2713–2724.

Howden, H. F. and Lawrence, J. F. 1974. The New World Aesalinae, with notes on the North American lucanid subfamilies (Coleoptera, Lucanidae). *Canadian Journal of Zoology* 52:1505–1510.

Howe, R. W. 1959. Studies on beetles of the family Ptinidae. XVII.—Conclusions and additional remarks. *Bulletin of Entomological Research* 50:287–326.

Howe, R. W. and J. E. Currie. 1964. Some laboratory observations on the rates of development, mortality and oviposition of several species of Bruchidae breeding in stored pulses. *Bulletin Entomological Research* 55(3):437–477.

Howe, R. W. and Lefkovitch, L. P. 1957. The distribution of the storage species of *Cryptolestes* (Col., Cucujidae). *Bulletin of Entomological Research* 48:795–809.

Hrbacek, J. 1950. On the morphology and function of the antennae of the central European Hydrophilidae. *Transactions of the Royal Entomological Society of London* 101:239–256.

Hubbard. H. G. 1892a. Notes on the larva of *Amphizoa. Insect Life* 5:19–22.

Hubbard, H. G. 1892b. Description of the larva of *Amphizoa lecontei. Proceedings of the Entomological Society of Washington* 2:341–346, pl. 3.

Hudson, G. V. 1934. *New Zealand Beetles and their Larvae.* Wellington: Ferguson & Osborn, 236 pp., 17 pls.

Hudson, L. 1975. A systematic revision of the New Zealand Oedemeridae (Coleoptera, Insecta). *Journal of the Royal Society of New Zealand* 5:227–274.

Huffaker, C. B. and R. L. Doutt. 1965. Establishment of the coccinellid *Chilocorus bipustulatus* Linnaeus in California in olive groves. *Pan-Pacific Entomologist* 44:61–63.

Hurka, K. 1978a. Cicindelidae, p. 51 *in* Klausnitzer, B. (ed.), *Ordnung Coleoptera (Larven).* The Hague: W. Junk, vi + 378 pp.

Hurka, K. 1978b. Carabidae, pp. 51–69 *in* Klausnitzer, B. (ed.), *Ordnung Coleoptera (Larven).* The Hague: W. Junk, vi + 378 pp.

Hurka, K. 1978c. Scarabaeidae, pp. 103–113 *in* Klausnitzer, B. (ed.), *Ordnung Coleoptera (Larven).* The Hague: W. Junk, vi + 378 pp.

Hurka, K. 1978d. Lucanidae, pp. 114–115 *in* Klausnitzer, B. (ed.), *Ordnung Coleoptera (Larven).* The Hague: W. Junk, vi + 378 pp.

Hurka, K. and Duchàč, V. 1980a. Larvae and the breeding type of the Central European species of the subgenera *Bradytus* and *Pseudobradytus* (Coleoptera, Carabidae, *Amara*). *Vestnik ceskoslovenske spolechnosti zoologicke* 44:166–182.

Hurka, K. and Duchàč, V. 1980b. Larval descriptions and the breeding type of the central European species of *Amara* (*Curtonotus*) (Coleoptera, Carabidae). *Acta entomologica bohemoslovaca* 77:258–270.

Hurka, K. and Smrz, J. 1981. Diagnosis and bionomy of unknown *Agonum, Batenus, Europhilus* and *Idiochroma* larvae (Col., Carabidae, *Platynus*). *Vestnik ceskoslovenske spolecnosti zoologicke* 45:255–276.

Hurpin, B. 1962. Super-famille des Scarabaeoidea, pp. 24–204 *in* A. S. Balachowsky, (ed.), *Entomologie appliquée à l'agriculture. Tome I. Coléoptères. Premier Volume.* Paris: Masson, 564 pp.

Hyslop, J. A. 1915. Observations on the life history of *Meracantha contracta* (Beauv.). *Psyche* 22(2):44–48, pl. 4.

Hyslop, J. A. 1917. The phylogeny of the Elateridae based on larval characters. *Annals of the Entomological Society of America* 10:241–263.

Hyslop, J. A. 1923. The present status of the coleopterus family Plastoceridae. *Proceedings of the Entomological Society of Washington* 25:156–160, pl. 12.

Hyslop, J. A. and Böving, A. G. 1935. Larva of *Tetrigus fleutiauxi* Van Zwaluwenburg. *Proceedings of the Hawaiian Entomological Society* 9(1):49–61.

Iablokoff, E. 1944. Observation sur quelques Melandryidae. *Revue Francaise d'Entomologie* 10:119–120.

Iablokov, A. 1940. Notes sur l'*Hendecatomus reticulatus* Herbst. *Revue Francaise d'Entomologie* 7:34–35.

Iablokov-Khnzorian, S. M. 1977. Beetles of the tribe Laemophloeini (Coleoptera, Cucujidae) in the Soviet fauna. *Communication* 1. *Entomological Review* 56(3):88–98 (translation).

Imms, A. D. 1977. *A General Textbook of Entomology. Tenth Edition* (revised by O. W. Richards and R. G. Davies). London: Methuen, 2 vols., 1354 pp.

Imms, A. D. and Chatterjee, N. C. 1915. On the structure and biology of *Tachardia lacca* Kerr, with observations on certain insects predaceous or parasitic upon it. *Indian Forest Memoirs. Forest Zoology Series* 3(1):41 pp., 8 pls.

International Commission on Zoological Nomenclature. 1985. Opinion 1331. Sphaeriidae Jeffreys, 1862 (1820) (Mollusca, Bivalvia) and Microsporidae Reichardt, 1976 (Insecta, Coleoptera): placed on the Official List. *Bulletin of Zoological Nomenclature* 42(3):230–232.

Isely, D. and Schwardt, H. H. 1934. The rice water weevil, *Lissoroptrus* simplex (Say) (Curculionidae). *Arkansas Agricultural Experiment Station Bulletin* 299, 44 pp.

Ising, E. 1969. Zur Biologie des *Leptinus testaceus* Müller, 1817. *Zoologische Beitrage* 15:393–456.

Iwata, R. and Nishimoto, K. 1981. Observations on the external morphology and the surface structure of *Lyctus brunneus* (Stephens) (Coleoptera, Lyctidae) by scanning electron microscopy. I. Larvae and pupae. *Kontyu* 49:542–557.

Jackson, G. J. and Crowson, R. A. 1969. A comparative anatomical study of the digestive, excretory and central nervous systems of *Malachius viridis* F. (Col., Melyridae) and *Rhagonycha usta* Gemm. (Col., Cantharidae), with observations on their diet and taxonomy. *Entomologist's Monthly Magazine* 105:93–98.

James, H. G. 1969. Immature stages of five diving beetles (Coleoptera, Dytiscidae), notes on their habits and life history and a key to aquatic beetles of vernal woodland pools in southern Ontario. *Proceedings of the Entomological Society of Ontario* 100:52–97.

Janssen, W. 1963. Untersuchungen zur Morphologie, Biologie und Ökologie von *Cantharis* L. und *Rhagonycha* Eschsch. (Cantharidae, Col.). *Zeitschrift fur wissenschaftliche Zoologie* 169:115–202.

Janzen, D. H. 1969. Seed-eaters versus seed size, number, toxicity and dispersal. *Evolution* 23:1–27.

Janzen, D. H. 1971. The fate of *Scheelea rostrata* fruits beneath the parent tree: predispersal attack by bruchids. *Principles* 15(3):89–101.

Janzen, D. H. 1974. Epiphytic myrmecophytes in Sarawak: mutualism through the feeding of plants by ants. *Biotropica* 6:237–259.

Jeannel, R. 1949. Ordre des Coléoptères (Coleoptera Linné, 1758). Partie Générale, pp. 771–891 *in* Grassé, P., (ed.), *Traité de Zoologie: Anatomie, Systématique, Biologie. Tome IX. Insectes: Paléontologie, Géonémie, Aptérygotes, Ephéméroptères, Odonatoptères, Blattoptéroïdes, Orthoptéroïdes, Dermaptéroides, Coléoptères.* Paris: Masson, 1117 pp.

Jeannel, R. 1957. Sur quelques Catopides, Leiodides et Camiarides du Chili. *Revista Chilena de Entomologia* 5:41–65.

Jeannel, R. and Paulian, R. 1945. Mission scientifique de l'Omo. Fauna des terriers des rats-taupes IV. Coléoptères. *Mémoires du muséum national d'histoire naturelle* (N.S.)19:51–147.

Jelinek, J. 1975. New genus of Oriental Meligethinae with notes on the supergeneric classification of Nitidulidae (Coleoptera, Nitidulidae). *Annotationes Zoologicae et Botanicae* 102:1–9.

Jerath, M. L. 1960. Notes on the larvae of nine genera of Aphodiinae in the United States (Coleoptera: Scarabaeidae). *Proceedings of the United States National Museum* 111:43–94.

Jewell, F. F. 1956. Insect transmission of oak wilt. *Phytopathology* 46:244–257.

Jewett, H. H. 1946. Identification of some larval Elateridae found in Kentucky. *Kentucky Agricultural Experiment Station Bulletin,* No. 489, 40 pp.

John, H. 1954. Familiendiagnose der Notiophygidae (= Discolomidae). Ihr Verwandtschaftsverhältnis zu den Colydiidae und Bemerkungen zu einigen systematisch noch nicht eingeordneten Gattungen. *Entomologische Blätter* 50:9–75.

John, H. 1959. *Genera Insectorum de P. Wytsman. Fascicule 213. Coleoptera Clavicornia. Fam. Discolomidae (= Notiophygidae).* Crainhem, Belgium: Genera Insectorum, 56 pp., 8 pls.

John H. 1960. Eine Übersicht über die Familie Propalticidae. *Pacific Insects* 2(2):149–170.

John, H. 1964. Discolomidae Col. aus Angola (*Aphanocephalus* Woll. und *Cassidoloma* Kolbe). *Publicaoes Culturais da Companhia de Diamontes de Angola.* 68:89–93.

Johnson, C. 1966. The Stephensian species of the family Clambidae (Col.), together with a revised British list of the family. *Entomologists Monthly Magazine* 101:185–188.

Johnson, C. D. 1970. Biosystematics of the Arizona, California, and Oregon species of the seed beetle genus *Acanthoscelides* Schilsky (Coleoptera: Bruchidae). *University of California Publications in Entomology* 59:1–116.

Johnson, C. D. 1975. Ecology and redescription of the Arizona grape bruchid, *Amblycerus vitis* (Coleoptera). *Coleopterists Bulletin* 29(4):321–331.

Johnson, C. D. 1979. New host records in the Bruchidae (Coleoptera). *Coleopterists Bulletin* 33(1):121–124.

Johnson, C. D. 198la. Seed beetle host specificity and the systematics of the Leguminosae, p. 995–1027. *In* R. M. Polhill and P. H. Raven (ed.) *Advances in legume systematics.* Part 2. Royal Botanic Gardens, Kew, England.

Johnson, C. D. 1981b. Relations of *Acanthoscelides* with their plants hosts. pp. 73–81. *In* V. Labeyrie (ed.) *The ecology of bruchids attacking legumes (pulses).* Dr. W. Junk Publ. Boston.

Johnson, C. D. 1983. Ecosystematics of *Acanthoscelides* (Coleoptera: Bruchidae) of southern Mexico and Central America. *Misc. Publ. Ent. Soc. Amer.* 56:1–370.

Johnson, C. D. and Kingsolver, J. M. 1971. Descriptions, life histories, and ecology of two new species of Bruchidae infesting Guacima in Mexico. *Journal of the Kansas Entomological Society* 44(2):141–152.

Johnson, J. P. 1944. The imported long-horned weevil, *Calomycterus setarius* Roelofs. *Connecticut Agricultural Experiment Station Bulletin* 479:121–142, 16 figs.

Johnson, P. and Russell, L. K. 1978. Notes on the rediscovery, habitat, and classification of *Exomella pleuralis* (Casey) (Coleoptera: Byrrhidae). *Coleopterists Bulletin* 32:159–160.

Jorge, M. E. 1974. Immature stages of Nilioninae: a contribution toward the taxonomic position of the family (Coleoptera). *Revista Brasiliera de Entomologia* 18(4):123–128.

Jourdheuil, P. 1962. Famille des Nitidulidae, pp. 318–330 *in* Balachowsky, A. S. (edit.) *Entomologie appliquée à l'agriculture. Tome I. Coléoptères. Premier volume.* Paris: Masson, xxvii + 564 pp.

Jurzitza, G. 1979. The fungi symbiotic with anobiid beetles, pp. 65–76 *in* Batra, L. R. (edit.), *Insect-Fungus Symbiosis. Nutrition, Mutualism, and Commensalism.* Montclair, New Jersey: Allanheld, Osmun, ix + 276 pp.

Kaddou, I. K. 1960. The feeding behavior of *Hippodamia quinquesignata* (Kirby) larvae. *University of California Publications in Entomology* 16:181–232.

Kamiya, H. 1963. On the systematic position of the genus *Usechus* Motschulsky, with a description of a new species from Japan (Coleoptera). *Mushi* 37:20–26.

Kamiya, H. 1965. Comparative morphology of larvae of the Japanese Coccinellidae, with special reference to the tribal phylogeny of the family (Coleoptera). *Memoirs of the Faculty of Liberal Arts, Fukui University* 14:83–100.

Kariluoto, K. T. 1980. Survival and fecundity of *Adalia bipunctata* (Coleoptera: Coccinellidae) and some other predatory insect species on an artificial diet and a natural prey. *Annales Entomologici Fennici* 46:101–106.

Karsch, F. 1884. Bestimmungstabellen von Insecten-Larven. *Entomologische Nachrichten Berlin* 10:221–229.

Kasap, H. and Crowson, R. A. 1975. A comparative anatomical study of Elateriformia and Dascilloidea (Coleoptera). *Transactions of the Royal Entomological Society of London* 126:441–495.

Kasap, H. and Crowson, R. A. 1976. On systematic relations of *Oomorphus concolor* (Sturm) (Col., Chrysomelidae), with descriptions of its larva and of an aberrant Cryptocephaline larva from Australia. *Journal of Natural History* 10:99–112.

Kaston, J. 1936. The morphology of the elm bark beetles, *Hyluropinus rufipes* (Eichhoff). *Connecticut Agricultural Experiment Station Bulletin* 387:613–650.

Kasule, F. K. 1966. The subfamilies of the larvae of Staphylinidae (Coleoptera) with keys to the larvae of the British genera of Steninae and Proteininae. *Transactions of the Royal Entomological Society of London* 118:261–283.

Kasule, F. K. 1968. The larval characters of some subfamilies of British Staphylinidae (Coleoptera) with keys to the known genera. *Transactions of the Royal Entomological Society of London* 120:115–138.

Kasule, F. K. 1970. The larvae of Paederinae and Staphylininae (Coleoptera: Staphylinidae) with keys to the known British fauna. *Transactions of the Royal Entomological Society of London* 122:49–80.

Kaszab, Z. 1964. The zoological results of Gy. Topal's collectings in South Argentina. 13. Coleoptera—Tenebrionidae. *Annales Historico-naturalis Musei Nationalis Hungarici* 56:353–387.

Kaszab, Z. 1969a. The scientific results of the Hungarian soil zoological expeditions to South America. 26. Tenebrionidae aus Arentinen und Brasilien. *Opuscula Zoologica, Budapest* 9(1):127–132.

Kaszab, Z. 1969b. The scientific results of the Hungarian soil zoological expeditions to South America. 17. Tenebrionidae aus Chile (Coleoptera). *Opuscula Zoologica, Budapest* 9(2):291–337.

Kaszab, Z. 1979. Die Arten der Gattung *Sivacrypticus* Kaszab, 1964 (Coleoptera, Tenebrionidae). *Annales Historico-naturalis Musei Nationalis Hungarici* 71:185–204.

Kaszab, Z. 1981. Die Gattungen und Arten der Tribus Archeocrypticini (Col. Tenebrionidae). *Folia Entomologica Hungarica* 42:95–115.

Kaufman, D. L. 1967. Notes on the biology of three species of *Lema* (Coleoptera: Chrysomelidae) with larval descriptions and key to described United States species. *Journal of the Kansas Entomological Society* 40:361–372.

Kavanaugh, D. H. 1986. A systematic review of amphizoid beetles (Amphizoidae: Coleoptera) and their phylogenetic relationships to other Adephaga. *Proc. Calif. Acad. Sci.* 44:67–109.

Keen, F. P. 1952. Insect enemies of western forests. *United States Department of Agriculture Miscellaneous Publication* 273, revised edition, 280 pp., 111 figs.

Keifer, H. H. 1933. Some Pacific Coast otiorhynchid weevil larvae. *Entomologica Americana* 13(2):45–85, 52 figs.

Keleinikova, S. I. 1959. Larvae of tenebrionids of the tribe Tentyriini. *Zoologicheskii Zhurnal* 38:1835–1843. (in Russian).

Keleinikova, S. I. 1961a. Larvae of darkling beetles of the subfamily Pimeliinae (Coleoptera, Tenebrionidae) from western Kazakhstan. *Entomologicheskoye Obozreniye* 40:371–384. (In Russian; translation in *Entomological Review* 40:195–200.)

Keleinikova, S. I. 1961b. Toward knowledge of larvae of the tribe Akidini (Col., Tenebrionidae). *Sbornik Trudov Zoologicheskovo Muzeia Moskovskovo Universiteta* 8:151–157. (In Russian).

Keleinikova, S. I. 1961c. Toward knowledge of larvae of the subfamily Opatrinae (Coleoptera, Tenebrionidae) of western Kazakhstan. *Sbornik Trudov Zoologicheskovo Muzeia Moskovskovo Universiteta* 8:159–164. (In Russian).

Keleinikova, S. I. 1961d. *Cyphogenia aurata* Pall. larva (Tenebrionidae, Coleoptera). *Zoologicheskii Zhurnal* 40:776–777. (In Russian).

Keleinikova, S. I. 1962. On larvae of the tribe Erodiini (Coleoptera, Tenebrionidae). *Zoologicheskii Zhurnal* 41:459–463. (In Russian).

Keleinikova, S. I. 1963. On the larval types of darkling beetles (Coleoptera, Tenebrionidae) of the Palearctic. *Entomologicheskoye Obozreniye* 42:539–549. (In Russian; translation in *Entomological Review* 42:292–296.)

Keleinikova, S. I. 1966. Descriptions of larvae of some Palearctic genera of darkling beetles of the tribe Pedinini (Coleoptera, Tenebrionidae). *Entomologicheskoye Obozreniye* 45:589–598. (In Russian; translation in *Entomological Review* 45:335–339).

Keleinikova, S. I. 1968. Soil-inhabiting darkling beetle larvae (Coleoptera, Tenebrionidae) in the Soviet fauna. I. *Sbornik Trudov Zoologicheskovo Muzeia Moskovskovo Universiteta* 11:205–239. (In Russian).

Keleinikova, S. I. 1970. Larvae of tenebrionids of Palearctic tribes of the subfamily Tentyriinae (Coleoptera, Tenebrionidae). *Entomologicheskoye Obozreniye* 49:409–422. (In Russian; translation in *Entomological Review* 49:245–253.)

Keleinikova, S. I. 1971. Morphological peculiarities of the first-stage larvae of Tenebrionidae (Coleoptera). *Proceedings of the 13th International Congress of Entomology, Moscow, 1968* 1:154–155.

Keleinikova, S. I. 1976. Darkling beetle (Coleoptera, Tenebrionidae) larvae of the tribe Stenosini. *Entomogicheskoye Obozreniye* 55:101–104. (In Russian; translation in *Entomological Review* 55:71–74).

Keleinikova, S. I. and Mamaev, B. M. 1971. *Phellopsis amurensis* Heyd. (Coleoptera, Zopheridae)—a specific wood destroyer in the Southern Primoriye. *Byulletin Moskovskovo Obshchestva Ispytatelei Prirody Novaya Seriya. Otdel Biologicheskii* 76(4):124–128. (In Russian).

Kemner, N. A. 1918. *Vergleichende Studien über das Pygopodium einiger Koleopteren-Larven.* Uppsala: Wretsman.

Kemner, N. A. 1924. Über die Lebensweise und Entwicklung des angeblich myrmecophilen oder termitophilen Genus *Trochoideus* (Col. Endomych.), nach Beobachtungen über *Trochoideus termitophilus* Roepke aus Java. *Tijdschrift voor Entomologie* 67:180–194.

Kemner, N. A. 1927. *Termitosuga* and *Euceroncinus,* zwei seltsame termitophile Käferlarven aus Java. *Arkiv voor Zoologi* 18A(29), 33 pp.

Kendall, D. A. 1968. The structure of the defense glands in Alleculidae and Lagriidae (Coleoptera). *Transactions of the Royal Entomological Society of London* 120(5):139–156.

Kendall, D. A. 1974. The structure of defense glands in some Tenebrionidae and Nilionidae (Coleoptera). *Transactions of the Royal Entomological Society of London* 124(4):437–487.

Kerr, T. W. and McLean, D. L. 1956. Biology and control of certain Lathridiidae. *Journal of Economic Entomology* 49(2):269–270.

Kimoto, S. 1962. A phylogenetic consideration of Chrysomelinae based on immature stages of Japanese species. *Journal of the Faculty of Agriculture, Kyushu University* 12:67–88.

Kingsbury, P. D. and Beveridge, J. G. 1977. A simple bubbler for sorting bottom fauna samples by elutriation. *Canadian Entomologist* 109:1265–1268.

Kingsolver, J. M. 1968. One previously described and one new species of South American Bruchidae injurious to commercial legume seed crops. *Proceedings of the Entomological Society of Washington* 70(4):318–322.

Kingsolver, J. M. 1969. A new species of neotropical seed weevil affecting pigeon peas, with notes on two closely related species. *Proceedings of the Entomological Society of Washington* 71(1):50–55.

Kingsolver, J. M. 1973. New synonymy in Languriidae (Coleoptera). *Proceedings of the Entomological Society of Washington* 75:247.

Kingsolver, J. M. and Pfaffenberger, G. S. 1980. Systematic relationship of the genus *Rhaebus* (Coleoptera: Bruchidae). *Proceedings of the Entomological Society of Washington* 82(2):293–311.

Kinzelbach, R. K. 1971. Morphologische Befunde an Fächerflüglern und ihre phylogenetische Bedeutung (Insecta: Strepsiptera). *Zoologica* 119, 256 pp.

Kirejtshuk, A. G. 1982. Systematic position of the genus *Calonecrus* J. Thomson and notes on the phylogeny of the family Nitidulidae (Coleoptera). *Entomologicheskoye Obozreniye* 61:117–129. (In Russian, translation in *Entomological Review* 61:109–122).

Kirk, V. M. 1965. Observations on the corn stalk weevil *Centrinaspis penicellus. Journal of Economic Entomology* 58(4):796–798, 1 fig.

Kirk, V. M. 1974. Biology of a ground beetle, *Harpalus erraticus. Annals of the Entomological Society of America* 67:24–28.

Kissinger, D. G. 1964. *Curculionidae of America North of Mexico.* South Lancaster, Mass.: Taxonomic Publications, 143 pp.

Kissinger, D. G. 1968. *Curculionidae Subfamily Apioninae of North and Central America with Reviews of the World Genera of Apioninae and World Subgenera of* Apion *Herbst.* South Lancaster, Mass.: Taxonomic Publications, vii + 559 pp.

Kistner, D. H. 1982. The social insects' bestiary, pp. 1–244 *in* Hermann, H. R. (edit.) *Social Insects. Vol. 3.* New York: Academic Press, xiii + 459 pp.

Kitayama, C. Y. 1982. Biosystematics of anthicid larvae (Coleoptera: Anthicidae). *Coleopterists Bulletin* 36(1):76–95.

Klausnitzer, B. 1970. Zur Larvalsystematik der mitteleuropäischen Coccinellidae (Col.). *Entomologische Abhandlungen* 38:55–110.

Klausnitzer, B. 1971. Zur Biologie einheimischer Käferfamilien. 6. Nosodendridae. *Entomologische Berichte, Berlin* 1971:71–73.

Klausnitzer, B. 1975a. Zur Kenntnis der Larven der mitteleuropäischen Helodidae. *Deutsche Entomologische Zeitschrift,* N.F., 22:61–65.

Klausnitzer, B. 1975b. Zur Kenntnis der Larven von *Myrmecoxenus* Chevrolat und *Oxylaemus* Erichson. XIX. Beitrag zur Kenntnis der mitteleuropäischen Coleopteren-Larven. (Coleoptera: Colydiidae). *Beiträge zur Entomologie* 25:209–211.

Klausnitzer, B. 1978. *Ordnung Coleoptera (Larven).* The Hague: W. Junk, 378 pp.

Kleine, R. 1908. Die europäischen Borkenkäfer und ihre Feinde aus den Ordnungen der Coleopteren und Hymenopteren. *Entomologische Blätter* 4:205–208, 225–227.

Kleine, R. 1909. Die europäischen Borkenkäfer und ihre Feinde aus den Ordnungen der Coleopteren und Hymenopteren. *Entomologische Blätter* 5:41–50, 76–79, 120–122, 140–141.

Kleine, R. 1933. Pars. 128. Fam. Lycidae. *In* Junk, W. and Schenkling, S., (edit.), *Coleopterorum Catalogus.* Berlin: W. Junk, 145 pp.

Kleine, R. 1944. Die europäischen Borkenkäfer und ihre Feinde aus den Ordnungen der Coleopteren und Hymenopteren. *Entomologische Blätter* 44:68–133.

Kline, L. N. and Rudinsky, J. A. 1964. Predators and parasites of the Douglas-fir beetle. *Oregon Experiment Station Technical Bulletin* 79, 52 pp.

Klippel, R. 1952. Beiträge zur Kenntnis der Schimmel- und Moderkäfer. *Zeitschrift fur hygienische Zoologie und Schädlingsbekämpfung* 40:65–85.

Knab, F. 1913. Larvae of Cyphonidae in Bromeliaceae. *Entomologist's Monthly Magazine* 49:54–55.

Koch, C. 1962. Zur Ökologie der Dünen-Tenebrioniden der Namibwüste Südwest-Afrikas und Angolas. *Lunds Universitets Arsskrift N.F.* Avd. 2, 58(5), 25 pp.

Kojima, T. 1932. Beiträge zur Kenntnis von *Lyctus linearis* Goeze. *Zeitschrift für angewandte Entomologie* 19:325–495.

Kompantzev, A. V. 1978. Description of larva of the powder-post beetle *Endecatomus lanatus* Lesne and the systematic position of the genus *Endecatomus* Mellie (Col. Bostrychidae). *Entomologicheskoye Obozreniye* 57:121–123. (In Russian; translation in *Entomological Review* 57:82–84).

Kôno, H. 1938. An inter-tidal rock-dwelling beetle, *Aegialites stejnegeri,* and its one new subspecies. *Entomological World* 6(46):1–5.

Kôno, H. and Nishio, Y. 1943. Die Beschreibung der Larven von *Pseudopyrochroa rufula* (Motschulsky), und *P. brevitarsis* (Lewis). *Transactions of the Natural History Society of Formosa* 33:566–569.

Korschefsky, R. 1943. Bestimmungstabelle der bekanntesten deutschen Tenebrioniden- u. Alleculiden-Larven (Coleoptera). *Arbeiten über physiologische und angewandte Entomologie aus Berlin-Dahlem* 10:58–68.

Korschefsky, R. 1951. Bestimmungstabelle der bekanntesten deutschen Lyciden-, Lampyriden- und Drilidenlarven (Coleoptera). *Beiträge zur Entomologie* 1:60–64, pl. 1.

Kraatz, W. C. 1918. *Scirtes tibialis,* Guér. (Coleoptera: Dascillidae), with observations on the life history. *Annals of the Entomological Society of America* 11:393–401.

Kristensen, N. P. 1981. Phylogeny of insect orders. *Annual Review of Entomology* 26:135–157.

Kristensen, N. P. 1984. The larval head of *Agathiphaga* (Lepidoptera, Agathiphagidae) and the lepidopteran ground plan. *Systematic Entomology* 9:63–81.

Kristensen, N. P. and Nielsen, E. S. 1983. The *Heterobathmia* life history elucidated: immature stages contradict assignment to suborder Zeugloptera (Insecta, Lepidoptera). *Zeitschrift für zoologische Systematik und Evolutionsforschung* 21(2):101–124.

Kryzhanovski, O. L. 1973. Directions of evolution in Histeridae. *Zoologicheskii Zhurnal* 52:870–875 (in Russian).

Kryzhanovski, O. L. 1976. An attempt at a revised classification of the family Carabidae (Coleoptera). *Entomologicheskoye Obozreniye* 55:80–91 (in Russian; translated in *Entomological Review* 55:56–64).

Kryzhanovski, O. L. and Reikhardt, A. N. 1976. *Fauna of the U.S.S.R. (New Series). Coleoptera. Vol. V. Part 4. Beetles of the Superfamily Histeroidea (Families Sphaeritidae, Histeridae, Synteliidae).* Leningrad: "Nauka", 434 pp. (in Russian).

Kurcheva, G. F. 1967. Leaf beetle larvae of the subfamily Eumolpinae and of *Syneta betulae* F. (Coleoptera-Chrysomelidae). *Entomologicheskoye Obozreniye* 46:222–233. (In Russian; translation in *Entomological Review* 46:361–372.)

Kuschel, G. 1959. Nemonychidae, Belidae y Oxycorynidae de la Fauna Chilena, con algunas consideraciones biogeograficas. *Investigaciones Zoologicas Chilenas* 5:229–271.

Kuschel, G. 1979. The genera *Monotoma* Herbst (Rhizophagidae) and *Anommatus* Wesmael (Cerylidae) in New Zealand (Coleoptera). *New Zealand Entomologist* 7:44–48.

Kuschel, G. 1983. Past and present of the relict family Nemonychidae (Coleoptera: Curculionoidea). *GeoJournal* 7.6:499–504.

LaLone, R. S. and Clarke, R. G. 1981. Larval development of *Otiorhynchus sulcatus* and effects of larval density on larval mortality and injury to Rhododendron. *Environmental Entomology* 10(2):190–191.

Lambert, D. H. and McIlveen, W. D. 1976. *Acylomus* sp. infesting ergot sclerotia. *Annals of the Entomological Society of America* 69:34.

Lanchester, H. P. 1939. The external anatomy of the larva of the Pacific Coast wireworm. *United States Department of Agriculture Technical Bulletin* 693, 40 pp.

Lanchester, H. P. 1946. Larval determination of six economic species of *Limonius* (Coleoptera: Elateridae). *Annals of the Entomological Society of America* 39:619–626.

Landry, J.-F. and Bousquet, Y. 1984. The genus *Omophron* Latreille (Coleoptera: Carabidae): redescription of the larval stage and phylogenetic considerations. *Canadian Entomologist* 116:1557–1569.

Landwehr, V. R. 1977. *Ischyropalpus nitidulus* (Coleoptera: Anthicidae), a predator of mites associated with Monterey pine. *Annals of the Entomological Society of America* 70:81–83.

Larson, A. O. and Fisher, C. K. 1938. The bean weevil and the southern cowpea weevil in California. *United States Department of Agriculture Technical Bulletin* 593, 70 pp.

Larsson, S. G. 1942. Danske Billelarver. Bestemmelsesnogle til Familie. *Entomologiske Meddelelser* 22:239–259.

Lawrence, J. F. 1967a. Biology of the parthenogenetic fungus beetle *Cis fuscipes* Mellie (Coleoptera: Ciidae). *Breviora* 258:1–14.

Lawrence, J. F. 1967b. Delimitation of the genus *Ceracis* (Coleoptera: Ciidae) with a revision of North America species. *Bulletin of the Museum of Comparative Zoology* 136(6):91–144.

Lawrence, J. F. 1971. Revision of the North American Ciidae (Coleoptera). *Bulletin of the Museum of Comparative Zoology* 142:419–522.

Lawrence, J. F. 1973. Host preference in ciid beetles (Coleoptera: Ciidae) inhabiting the fruiting bodies of Basidiomycetes in North America. *Bulletin of the Museum of Comparative Zoology* 145:163–212.

Lawrence, J. F. 1974a. The ciid beetles of California (Coleoptera: Ciidae). *Bulletin of the California Insect Survey* 17:1–41.

Lawrence, J. F. 1974b. The larva of *Sphindocis denticollis* Fall and a new subfamily of Ciidae (Coleoptera: Heteromera). *Breviora* 424:1–14.

Lawrence, J. F. 1977a. Coleoptera associated with an *Hypoxylon* species (Ascomycetes: Xylariaceae) on oak. *Coleopterists Bulletin* 31:309–312.

Lawrence, J. F. 1977b. Extraordinary images show how beetles have adapted to live off plants, and each other. *Horticulture* 55:8–13.

Lawrence, J. F. 1977c. The family Pterogeniidae, with notes on the phylogeny of the Heteromera. *Coleopterists Bulletin* 31:25–56.

Lawrence, J. F. 1980. A new genus of Indo-Australian Gempylodini with notes on the constitution of the Colydiidae (Coleoptera). *Journal of the Australian Entomological Society* 19:293–310.

Lawrence, J. F. 1981. Notes on larval Lucanidae (Coleoptera). *Journal of the Australian Entomological Society* 20:213–219.

Lawrence, J. F. 1982a. A catalog of the Coleoptera of America north of Mexico. Family: Cerylonidae. *United States Department of Agriculture. Agriculture Handbook* 529–95, 10 pp.

Lawrence, J. F. 1982b. A catalog of the Coleoptera of America north of Mexico. Family: Ciidae. *United States Department of Agriculture.* Agriculture Handbook 529–105, 18 pp.

Lawrence, J. F. 1982c. Coleoptera, pp. 482–553 *in* Parker, S. P. (edit.) *Synopsis and Classification of Living Organisms. Vol. 2.* New York: McGraw-Hill, 1232 pp.

Lawrence, J. F. 1985a. The genus *Nothoderodontus* (Coleoptera: Derodontidae), with new species from Australia, New Zealand, and Chile, pp. 68–83 *in* Ball, G. E. (edit.). *Taxonomy, Phylogeny, and Zoogeography of Beetles and Ants. A Volume Dedicated to the Memory of Philip Jackson Darlington, Jr. 1904–1983.* The Hague: W. Junk.

Lawrence, J. F. 1985b. The genus *Teredolaemus* Sharp (Coleoptera: Bothrideridae) in Australia. *Journal of the Australian Entomological Society* 24:205–206.

Lawrence, J. F. and Hlavac, T. F. 1979. Review of the Derodontidae (Coleoptera: Polyphaga) with new species from North America and Chile. *Coleopterists Bulletin* 33:369–414.

Lawrence, J. F. and Medvedev, G. S. 1982. A new tribe of darkling beetles from Australia and its systematic position (Coleoptera, Tenebrionidae). *Entomological Review* 61(3):85–107.

Lawrence, J. F. and Newton, A. F., Jr. 1980. Coleoptera associated with the fruiting bodies of slime molds (Myxomycetes). *Coleopterists Bulletin* 34:129–143.

Lawrence, J. F. and Newton, A. F., Jr. 1982. Evolution and classification of beetles. *Annual Review of Ecology and Systematics* 13:261–290.

Lawrence, J. F. and Reichardt, H. 1969. The myrmecophilous Ptinidae (Coleoptera), with a key to Australian species. *Bulletin of the Museum of Comparative Zoology* 138:1–28.

Lawrence, J. F. and Stephan, K. 1975. The North American Cerylonidae (Coleoptera; Clavicornia). *Psyche* 82:131–166.

Lawrence, J. F. and Vaurie, R. 1983. A catalog of the Coleoptera of America north of Mexico. Family: Languriidae. *United States Department of Agriculture. Agriculture Handbook* No. 529–92, 13 pp.

LeConte, J. L. and Horn, G. H. 1876. The Rhynchophora of America north of Mexico. *Proceedings of the American Philosphical Society* 15:1–455.

Lee, R. E. 1980. Aggregation of lady beetles on the shores of lakes (Coleoptera: Coccinellidae). *American Midland Naturalist* 104:295–304.

Leech, H. B. and Chandler, H. P. 1956. Chapter 13. *Aquatic Coleoptera, pp. 293–371 in* Usinger, R. L., ed., *Aquatic Insects of California,* Berkeley: University of California Press, 508 pp.

Leech, H. B. and Sanderson, M. W. 1959. Chapter 38. Coleoptera, pp. 981–1023 *in* Edmondson, W. T. (edit.), *Fresh Water Biology,* 2nd Edition. New York: Wiley, 1248 pp.

Lefébure de Cerisy, L. C. 1853. Observations sur les metamorphoses des Coléoptères du genre *Cebrio. Revue et Magasin de Zoologie* (2)5:214–225.

Lefkovitch, L. P. 1957a. The biology of *Cryptolestes ugandae* Steel & Howe (Coleoptera, Cucujidae), a pest of stored products in Africa. *Proceedings of the Zoological Society of London* 128(3):419–429.

Lefkovitch, L. P. 1957b. Further records of Laemophloeinae (Col., Cucujidae) in stored products. *Entomologist's Monthly Magazine* 93:239.

Lefkovitch, L. P. 1958. A new genus and species of Laemophloeinae (Col., Cucujidae) from Africa. *Entomologist's Monthly Magazine* 93:271–273.

Lefkovitch, L. P. 1959. A revision of the European Laemophloeinae (Coleoptera: Cucujidae). *Transactions of the Royal Entomological Society of London* 111:95–118.

Lefkovitch, L. P. 1960. *Adistemia watsoni* (Wollaston) (Coleoptera: Lathridiidae) in a bird's nest in Hamburg. *Entomologische Mitteilungen aus dem zoologischen Staatsinstitut und zoologischen Museum Hamburg* 27:12–13.

Lefkovitch, L. P. 1961. Notes on New Zealand Cucujidae and Silvanidae (Coleoptera). *Entomologist's Monthly Magazine* 97:143–144.

Lefkovitch, L. P. 1962a. The biology of *Cryptolestes capensis* (Waltl) (Coleoptera: Cucujidae). *Bulletin of Entomological Research* 53:529–535.

Lefkovitch, L. P. 1962b. The biology of *Cryptolestes turicus* (Grouvelle) (Coleoptera, Cucujidae), a pest of stored and processed cereals. *Proceedings of the Zoological Society of London* 138(1):23–35.

Lefkovitch, L. P. 1962c. Revision of African Laemophloeinae (Coleoptera: Cucujidae). *Bulletin of the British Museum (Natural History). Entomology* 12(4):167–245.

Lefkovitch, L. P. 1964. The biology of *Cryptolestes pusilloides* (Steel & Howe) (Coleoptera, Cucujidae), a pest of stored cereals in the southern hemisphere. *Bulletin of Entomological Research* 54(4):649–656.

Leiler, T. -E. 1976. Zur Kenntnis der Entwicklungsstadien und der Lebensweise nord- und mitteleuropäischen Eucnemiden. *Entomologische Blätter* 72:10–50.

Lekander, B. 1967. A description of two *Magdalis* larvae (Col. Curc.) and a comparison between these and the *Scolytus* larvae (Col. Scol.). *Entomolgisk Tidskrift* 88(3–4):123–129, 2 figs.

Lekander, B. 1968. Scandinavian bark beetle larvae. Descriptions and classification. *Department of Forest Zoology, Royal College of Forestry, Research Notes* 4:1–186.

Lengerken, H. von 1924. Extraintestinale Verdauung. *Biologisches Zentralblatt* 44:273–295.

Lengerken, H. von 1938a. Studien über die Lebenserscheinungen der Silphini (Coleopt.) XI–XIII. *Thanatophilus sinuatus* F., *rugosus* L. und *dispar* Hrbst. *Zeitschrift für Morphologie ünd Okologie der Tiere* 33:654–666.

Lengerken, H. von 1938a. Studien über die Lebenserscheinungen der Silphini (Coleopt.) XI–XIII. *Thanatophilus sinuatus* F., *rugosus* L. und *dispar* Hrbst. *Zeitschrift fur Morphologie und Ökologie der Tiere* 33:654–666.

Lenko, K. 1967. *Chelonarium semivestitum* inquilino da formiga *Camponotus rufipes* (Coleoptera: Chelonariidae). *Studia entomologica* 10:433–438.

Lepesme, P. 1944. Les Coleopteres des denrees alimentaires et des produits industriels entreposes. *Encyclopedie Entomologique* (A)22:1–335, 12 pls.

LeSage, L. 1982. The immature stages of *Exema canadensis* Pierce (Coleoptera: Chrysomelidae). *Coleopterists Bulletin* 36(2):318–327.

LeSage, L. 1983. The larva and pupa of *Cytilus alternatus* Say, with a key to the known genera of Nearctic byrrhid larvae. *Coleopterists Bulletin* 37(2):99–105.

LeSage, L. 1984a. Egg, larva, and pupa of *Lexiphanes saponatus* (Coleoptera: Chrysomelidae: Cryptocephalinae). *Canadian Entomologist* 116:537–548.

LeSage, L. 1984b. Immature stages of Canadian *Neochlamisus* Karren (Coleoptera: Chrysomelidae). *Canadian Entomologist* 116:383–409.

LeSage, L. and Harper, P. 1976a. Description de nymphes d'Elmidae néarctiques. *Canadian Journal of Zoology* 54:65–73.

LeSage, L. and Harper, P. 1976b. Notes on the life history of the toed-winged beetle *Anchytarsus bicolor* (Melsheimer) (Coleoptera: Ptilodactylidae). *Coleopterists Bulletin* 30(3):233–238.

LeSage, L. and Harper, P. 1977. Life cycles of Elmidae (Coleoptera) inhabiting streams of the Laurentian Highlands, Quebec. *Annales de limnologie* 12(2):139–174.

Lesne, P. 1898. Description de la larve et adulte du *Julodis albopilosa* Chevr. et remarques sur divers caracteres des larves de Buprestides. *Bulletin de la Société entomologique de France* 1898:69–75.

Lesne, P. 1934. Note sur un Bostrychide Néo-Zélandais, l'*Euderia squamosa* Broun. *Annales de la Société entomologique de France* 103:389–393.

Lesne, P. 1936. Nouvelles données sur les Coléoptères de la famille des Sphaeriidae, pp. 241–248 *in Livre jubilaire de M. E.-L. Bouvier.* Paris.

Lesne, P. 1938. Sur un nitidulide mycétophage nouveau et sur quelques caractères de la famille (Coleoptera). *Revue Française d'Entomologie* 5:158–168, pls. 1–2.

Levey, B. 1978. A taxonomic revision of the genus *Prospheres* (Coleoptera: Buprestidae). *Australian Journal of Zoology* 26:713–726.

Liebenow, K. 1978. Oedemeridae, pp. 106–209 *in* Klausnitzer, B. (ed.), *Ordnung Coleoptera (Larven).* The Hague: W. Junk, vi + 378 pp.

Lieftinck, M. A. and Wiebes, J. T. 1968. Notes on the genus *Mormolyce* Hagenbach (Coleoptera, Carabidae). *Bijdragen tot de Dierkunde* 38:59–68, 2 pls.

Lieu, K. O. V. 1944. A preliminary note on the colydiid parasite of a willow-branch cerambycid. *Indian Journal of Entomology* 6:125–128.

Liles, M. P. 1956. A study of the life history of the forked fungus beetle, *Bolitotherus cornutus* (Panzer) (Coleoptera: Tenebrionidae). *Ohio Journal of Science* 56(6):329–337.

Lindner, W. 1967. Ökologie und Larvalbiologie cinheimischer Histeriden. *Zeitschrift für Morphologie und Ökologie der Tiere* 59:341–380.

Lindquist, O. H. and Davis, C. N. 1971. The biology of a birch leaf beetle, *Phratora hudsonia* (Coleoptera: Chrysomelidae), with a larval key to forest Chrysomelidae in Ontario. *Canadian Entomologist* 103:622–626.

Lindroth, C. H. 1960. The larvae of *Trachypachus* Mtsch., *Gehringia* Darl., and *Opisthius* Kby. (Col. Carabidae). *Opuscula Entomologica* 25:30–42.

Linsley, E. G. 1944. Natural sources, habitats, and reservoirs of insects associated with stored food products. *Hilgardia* 16(4):187–224.

Linsley, E. G. 1959. Ecology of Cerambycidae. *Annual Review of Entomology* 4:99–138.

Linsley, E. G. 1961. The Cerambycidae of North America. Part I. Introduction. *University of California Publications in Entomology* 18:1–97.

Linsley, E. G. 1962a. The Cerambycidae of North America. Part II. Taxonomy and classification of the Parandrinae, Prioninae, Spondylinae, and Aseminae. *University of California Publications in Entomology* 19:v + 102 pp., 1 pl.

Linsley, E. G. 1962b. The Cerambycidae of North America. Part III. Taxonomy and classification of the subfamily Cerambycinae, tribes Opsimini through Megaderini. *University of California Publications in Entomology* 20:xi + 188 pp.

Linsley, E. G. 1963. The Cerambycidae of North America. Part IV. Taxonomy and classification of the subfamily Cerambycinae, tribes Elaphidionini through Rhinotragini. *University of California Publications in Entomology* 21:ix + 165 pp.

Linsley, E. G. 1964. The Cerambycidae of North America. Part V. Taxonomy and classification of the subfamily Cerambycinae, tribes Callichromini through Ancylocerini. *University of California Publications in Entomology* 22:viii + 197 pp., 1 col. pl.

Linsley, E. G. and Chemsak, J. A. 1972. Cerambycidae of North America. Part VI, No. 1. Taxonomy and classification of the subfamily Lepturinae. *University of California Publications in Entomology* 69:viii + 138 pp., 2 pls.

Linsley, E. G. and Chemsak, J. A. 1976. Cerambycidae of North America. Part VI, No. 2. Taxonomy and classification of the subfamily Lepturinae. *University of California Publications in Entomology* 80:ix + 186 pp.

Linsley, E. G. and MacSwain, J. W. 1942. The bionomics of *Ptinus californicus,* a depredator in the nests of bees. *Bulletin of the Southern California Academy of Sciences* 40:126–137, pls. 11–14.

Linsley, E. G. and MacSwain, J. W. 1943. Observations on the life history of *Trichodes ornatus* (Coleoptera, Cleridae), a larval predator in the nests of bees and wasps. *Annals of the Entomological Society of America* 36:589–601.

Linsley, E. G., MacSwain, J. W. and Smith, R. F. 1952. The life history and development of *Rhipiphorus smithi* with notes on their phylogenetic signficance (Coleoptera, Rhipiphoridae). *University of California Publications in Entomology* 9(4):291–314, pls. 7–12.

Lloyd, J. E. 1971. Bioluminescent communication in insects. *Annual Review of Entomology* 16;97–122.

Lloyd, J. E. 1972. Chemical communication in fireflies. *Environmental Entomology* 1:265–266.

Lloyd, J. E. 1978. Insect bioluminescence. *In* Herring, P. (edit.), *Bioluminescence in Action.* New York: Academic Press, 520 pp.

Lloyd, J. E. 1981. Mimicry in the sexual signals of fireflies. *Scientific American* 245(1):139–145.

Lloyd, J. E. 1983. Bioluminescence and communication in insects. *Annual Review of Entomology* 28:131–160.

Löben Sels, E. von 1934. Some observations on *Phalacrus politus* and other inhabitants of the heads of the New England aster. *Journal of the New York Entomological Society* 42:319–327.

Löbl, I. 1977. Beitrag zur Kenntnis der Gattung *Dasycerus* Brongniart (Coleoptera, Dasyceridae). *Mitteilungen der Schweizerischen Entomologischen Gesellschaft* 50:95–106.

Löbl, I. and Burckhardt, D. 1988. *Cerapeplus* gen. n. and the classification of micropeplids (Coleoptera: Micropeplidae). *Systematic Entomology* 13:57–66.

Löser, S. 1970. Brutfürsorge und Brutpflege bei Laufkäfern der Gattung *Abax. Zoologischer Anzeiger. Supplementband. (Verhandlungen der deutschen zoologischen Gesellschaft* 1969) 33:322–326.

Lombardi, D. 1928. Contributo alla conoscenza dello *Scirtes hemisphaericus* L. (Coleoptera: Helodidae). *Bolletino del Laboratorio di Entomologia del R. Istituto Superiore Agrario di Bologna* 1:236–268.

Longley, G., and Spangler, P. J. 1977. The larva of a new subterranean water beetle, *Haideoporus texanus* (Coleoptera: Dytiscidae: Hydroporinae). *Proceedings of the Biological Society of Washington* 90:532–535.

Lubbock, J. 1868. On the larva of *Micropeplus staphylinoides. Transactions of the Entomological Society of London* 1868:275–277.

Luca, Y. de. 1956. Contributions a l'étude morphologique et biologique de *Bruchus lentis* Froh. Essais de lutte. *Annales de L'Institut Agricole* 10(1):1–94.

Luca, Y. de. 1967. Ethological notes on oviposition and the newly hatched larva of *Bruchidius ater. Bulletin de la Societe Entomologique de France* 72:16–20.

Luca, Y. de. 1968. La larve neonate de *Bruchidius atrolineatus* (Pic) (Coleoptera:Bruchidae). *Bull. de l'Inst. F.A.N.T.* 30:589–592.

Lucht, W. 1981. Die Präimaginalstadien von *Hypocoelus olexai* Palm (Col., Eucnemidae) nebst Bestimmungstabelle der Larven nord- und mitteleuropäischer *Hypocoelus*-Arten. *Entomologische Blätter für Biologie und Systematik der Käfer* 77:61–74.

Lundberg, S. 1973. Bidrag till kännedom om Svenska skalbaggar, 14. *Entomologisk Tidskrift* 94:28–33.

Luff, M. L. 1969. The larvae of the British Carabidae (Coleoptera). I. Carabini and Cychrini. *Entomologist* 102:245–263.

Luff, M. L. 1972. The larvae of the British Carabidae (Coleoptera). II. Nebriini. *Entomologist* 105:161–179.

Luff, M. L. 1976. The larvae of the British Carabidae (Coleoptera). IV. Notiophilini and Elaphrini. *Entomologist's Gazette* 27:51–67.

Luff, M. L. 1978. The larvae of the British Carabidae (Coleoptera). V. Omophronini, Loricerini, Scaritini and Broscini. *Entomologist's Gazette* 29:265–287.

Luff, M. L. 1980. The larvae of the British Carabidae (Coleoptera). VI. Licinini, Panagaeini, Chlaeniini and Oodini. *Entomologist's Gazette* 31:177–194.

Luff, M. L. 1985. The larvae of the British Carabidae (Coleoptera). VII. Trechini and Pogonini. *Entomologist's Gazette* 36:301–316.

Lyngnes, R. 1958. Studier over *Hylecoetus dermestoides* L. under et angrep pa bjorkestokker pa Sunnmore 1954–1955. *Norsk Entomologisk Tidskrift* 10:221–235.

MacSwain, J. W. 1956. A classification of the first instar larvae of Meloidae (Coleoptera). *University of California Publications in Entomology* 12, 182 pp.

Magis, N. 1977. *Catalogue des Coléoptères de Belgique. Fascicule VI. Catalogue raisonné des Cantharoidea. Premiere partie. Homalisidae, Drilidae, Lampyridae et Lycidae.* Bruxelles, Société royale belge d'entomologie, 60 pp.

Magis, N. and Wittmer, W. 1974. Nouvelle repartition des genres de la sous-famille des Chauliognathinae (Coleoptera, Cantharoidea: Cantharidae). *Bulletin de la Société royale des sciences* 43(1–2):78–95.

Maier, C. T. 1980. Quince curculio, *Conotrachelus crategi* Walsh (Coleoptera: Curculionidae), developing in apple, a new host, in southern New England. *Proceedings of the Entomological Society of Washington* 82(1):59–62.

Mamaev, B. M. 1973. The morphology of the larva of the beetle genus *Nematoplus* Lec. and the phylogenetic connections of some families of Heteromera (Coleoptera, Cucujoidea). *Entomologicheskoye Obozreniye* 52:586–598. (In Russian; translation in *Entomological Review* 52:388–395.)

Mamaev, B. M. 1974. The immature stages of the beetle *Syntelia histeroides* Lewis (Synteliidae) in comparison with certain Histeridae (Coleoptera). *Entomologicheskoye Obozreniye* 53:866–871 (In Russian; translated in *Entomological Review* 53:98–101.

Mamaev, B. M. 1975. Morphological and ecological characteristics of xylophilous cylindrical bark beetles in the fauna of the European part of the U.S.S.R. based on larval indices (Coleopt.: Colydiidae). *Nauchnye Doklady Vysshei Shkoli. Biologicheskiye Nauki* 18(12):16–22. (In Russian.)

Mamaev, B. M. 1976a. Larval morphology of the genus *Agnathus* Germ. (Coleoptera, Pedilidae) and the position of the genus in the system of the Coleoptera. *Entomologicheskoye Obozreniye* 55:642–645. (In Russian; translation in *Entomological Review* 55:97–99.)

Mamaev, B. M. 1976b. Morphological types of eucnemid larvae (Coleoptera, Eucnemidae) and their evolutionary significance, pp. 136–155 *in* Mamaev, B. M. (edit.), *Evolutionary Morphology of Insect Larvae.* Moscow: "Nauka," 204 pp. (In Russian).

Mamaev, B. M. 1976c. Review of larvae of the family Trogossitidae (Coleoptera) in the fauna of the U.S.S.R. *Zoologicheskii Zhurnal* 55(11):1648–1658 (in Russian).

Mamaev, B. M. 1977. Larval morphology of *Agaricophilus reflexus* Motsch. as evidence for assignment of the genus *Agaricophilus* Motsch. to the family Cerylonidae (Coleoptera). *Doklady (Proceedings) of the Academy of Sciences of the U.S.S.R. Zoology* 236:456–458 (translation).

Mamaev, B. M. 1978. Morphology of the larvae of *Cerophytum elateroides* Latr. and the phylogenetic ties of the Cerophytidae (Coleoptera) family. *Doklady (Proceedings) of the Academy of Sciences of the U.S.S.R. Zoology* 238:1007–1008 (translation).

Mamaev, B. M. and Danilevskii, M. L. 1973. New data on systematic status of the subfamily Aseminae (Col. Cerambycidae) with reference to the morphology of larvae. *Zoologicheskii Zhurnal* 52:1257–1261 (In Russian).

Mamaev, B. M., and Kompantzev, A. V. 1978. New data on the mycetophilous melandryids comprising the genus *Euryzilora* Lew. (Coleoptera, Melandryidae). *Nauchnye Doklady Vysshei Shkoli. Biologicheskiye Nauki* 21(3):46–49. (In Russian)

Mamaev, B. M. and Krivosheina, N. P. 1976. Morphology of weevil larvae (Coleoptera, Curculionidae) and the ecological pathways of their adaptations for living in wood, pp. 81–122 *in* Mamaev, B. M. (edit.), *Evolutionary Morphology of Insect Larvae.* Moscow: "Nauka," 204 pp. (In Russian).

Mamaev, B. M., Krivosheina, N. P., and Potoskaya, V. A. 1977. *Classification Key of the Larvae of Predatory Insects-Entomophaga of Trunk Pests.* Moscow: "Nauka", 391 pp.

Manglitz, G. R., Anderson, D. M., and Gorz, H. J. 1963. Observations on the larval feeding habits of two species of *Sitona* (Coleoptera: Curculionidae) in sweetclover fields. *Annals of the Entomological Society of America* 56:831–835, 3 figs.

Mann, J. S. and Crowson, R. A. 1981. The systematic position of *Orsodacne* Latr. and *Syneta* Lac. (Coleoptera: Chrysomelidae), in relation to characters of the larvae, internal anatomy and tarsal vestiture. *Journal of Natural History* 15:727–749.

Manton, S. M. 1945. The larvae of the Ptinidae associated with stored products. *Bulletin of Entomological Research* 35:341–365.

Marcovitch, S. 1934. Control of weevils in stored beans and cowpeas. *Tenn. Agric. Exp. Sta. Bull.* 150:3–8.

Marcuzzi, G. and Cravera, C. 1981. Illustrazione di larve di coleotteri tenebrionidi dell'area Caraibica. *Quaderni di Ecologia Animale* 17:1–11, 33 pls.

Marcuzzi, G., Cravera, C., and Faccini, E. 1980. III Contributo alla conoscenza delle forme larvali dei Tenebrionidi (Col. Heteromera). *Eos* 54:167–206.

Marcuzzi, G. and Floreani, L. F. 1962. Contributo allo studio dei sensilli delle larve dei coleotteri tenebrionidi. *Monitore Zoologico Italiano* 69:22–38.

Marcuzzi, G. and Rampazzo, L. 1960. Contributo alla conoscenza delle forme larvali dei tenebrionidi. *Eos* 36:63–117.

Marshall, A. T. and Thornton, I. W. B. 1963. *Micromalthus* (Coleoptera: Micromalthidae) in Hong Kong. *Pacific Insects* 5:715–720.

Marshall, J. E. 1978. The larva of *Aulonium trisulcum* (Fourcroy) (Coleoptera: Colydiidae) and its association with elm bark beetles (*Scolytus* spp.) *Entomologist's Gazette* 29:59–69.

Marshall, J. E. 1979. The larvae of the British species of *Chrysolina* (Chrysomelidae). *Systematic Entomology* 4: 409–417.

Marshall, J. E. 1980. A key to some larvae of the British Galerucinae and Halticinae (Coleoptera: Chrysomelidae). *Entomologist's Gazette* 31:275–283.

Martel, P., Svec, H. J., and Harris, C. R. 1976. The life history of the carrot weevil, *Listronotus oregonensis* (Coleoptera: Curculionidae) under controlled conditions. *Canadian Entomologist* 108(9):931–934, 1 fig.

Martin, H. 1945. Contribution à l'étude des Silphes de la betterave en Suisse. *Landwirtschaftliches Jahrbuch der Schweiz* 59: 757–819.

Martin, J. E. H. 1977. *The Insects and Arachnids of Canada. Part 1. Collecting, Preparing, and Preserving Insects, Mites, and Spiders.* Ottawa, Canada Department of Agriculture, Research Branch (Publication No. 1943), 182 pp.

Martins, U. and Pereira, F. S. 1966. Revisão dos Languriinae neotropicais. *Arquivos de Zoologia* 13:139–300.

Mateu, J. 1963. Notes sur la biologie de *Macrotoma palmata* F. (Coleoptere Prionitae). *Annales des Sciences Naturelles. Zoologie et Biologie Animale* (12)5:793–806.

Mateu, J. 1966. Notes sur la biologie de *Derolus mauritanicus* Buq. (Col. Cerambycidae). *Eos* 41:597–606, pls. 11–12.

Mateu, J. 1967. Notes sur la biologie de *Xylomedes rufocoronata* (Coleoptera: Bostrychidae). *Annales de la Société entomologique de France* (N.S.)3:885–891.

Mateu, J. 1972. *Les insectes xylophages des Acacia dans les régions sahariennes.* PubliCacões do Instituto de Zoologia "Dr. Augusto Nobre" No. 116, 714 pp.

Matheson, R. 1912. The Haliplidae of America north of Mexico. *Annals of the New York Entomological Society* 20:156–193, pls. 10–15.

Mathieu, J. M. 1969. Mating behavior of five species of Lucanidae (Coleoptera: Insecta). *Canadian Entomologist* 101:1054–1062.

Mathur, R. N. 1954. Immature stages of Indian Coleoptera (25) (Curculionidae). *Indian Forest Records. New Series. Entomology* 8(9):227–231, 2 pls.

Mathur, R. N. 1956. Immature stages of Indian Coleoptera (26) (Bostrychidae). *Indian Forest Records. New Series. Entomology* 8(10):233–245, 2 pls.

Mathur, R. N. 1957a. Immature stages of Indian Coleoptera (27) *Gibbium psyllioides* Czempinski (Family: Ptinidae). *Indian Forest Records. New Series. Entomology* 9(4):123–125, 1 pl.

Mathur, R. N. 1957b. Immature stages of Indian Coleoptera (28). Anthribidae. *Indian Forest Records. New Series. Entomology* 9(5):127–129, 1 pl.

Mathur, R. N. 1958. Immature stages of Indian Coleoptera (29) Cerambycidae. *Indian Forest Records. New Series. Entomology* 9(9):175–181, 3 pls.

Matta, J. F. 1982. The bionomics of two species of *Hydrochara* (Coleoptera: Hydrophilidae) with descriptions of their larvae. *Proceedings of the Entomological Society of Washington* 84(3):461–467.

Matta, J. F. 1983. Description of the larva of *Uvarus granarius* (Aubé) with a key to nearctic Hydroporinae larvae. *Coleopterists Bulletin* 37(3):203–207.

Matta, J. F. 1986. *Agabus* (Coleoptera: Dytiscidae) larvae of Southeastern United States. *Proceedings of the Entomological Society of Washington* 88:515–520.

Matta, J. F., and D. E. Peterson. 1985. The larvae of six nearctic *Hydroporus* of the subgenus *Neoporus* (Coleoptera: Dytiscidae). *Proc. Acad. Nat. Sci. Philadelphia* 137(1):53–60.

Matta, J. F. and D. E. Peterson. 1987. The larvae of two North American diving beetles of the genus *Acilius* (Coleoptera: Dytiscidae). *Proceedings of the Entomological Society at Washington* 89:440–443.

Matthes, D. 1970a. *Malachius bipustulatus* (Malachiidae): Balz und Kopulation. *Encyclopaedia Cinematographica* E1567/1969, 10 pp., 4 figs.

Matthes, D. 1970b. *Troglops albicans* (Malachiidae): Balz and Kopulation. *Encyclopaedia Cinematographica* E1639/1970, 13 pp., 8 figs.

Matthewman, W. G. and Pielou, D. P. 1971. Arthropods inhabiting the sporophores of *Fomes fomentarius* (Polyporaceae) in Gatineau Park, Quebec. *Canadian Entomologist* 103:775–847.

Matthews, A. 1899. *A Monograph of the Coleopterous Families Corylophidae and Sphaeriidae.* London: Janson, 220 pp., pls. A + 1–8.

Matthews, E. G. 1986. A revision of the troglobitic genus *Brises* Pascoe, with a discussion of the Cyphaleini (Coleoptera, Tenebrionidae). *Records of the South Australian Museum* 19(6):77–90.

Mattoli, D. 1974. Note di morfologia e di biologia su *Hoplocephala haemorrhoidalis* Fabr. (Coleoptera, Tenebrionidae). *Entomologica Bari* 10:9–30.

Maulik, S. 1931. On the structure of larvae of hispine beetles. *Proceedings of the Zoological Society of London* 1931:1137–1162.

Maulik, S. 1932. On the structure of larvae of hispine beetles.—II. *Proceedings of the Zoological Society of London* 1932:293–322.

Maulik, S. 1933. On the structure of larvae of hispine beetles.—III. *Proceedings of the Zoological Society of London* 1933:669–680.

May, B. M. 1966. Identification of the immature forms of some common soil-inhabiting weevils, with notes on their biology. *New Zealand Journal of Agricultural Research* 9(2):286–316, 2 pls, 18 figs.

May, B. M. 1967. Immature stages of Curculionidae. 1. Some genera in the tribe Araucariini (Cossoninae). *New Zealand Journal of Science* 10:644–660.

May, B. M. 1977. Immature stages of Curculionidae: larvae of the soil-dwelling weevils of New Zealand. *Journal of the Royal Society of New Zealand* 7(2):189–228, 162 figs.

Mazur, S. 1984. A world catalog of Histeridae. *Polskie Pismo Entomologiczne* 54:1–379.

McAlpine, J. F., B. F. Peterson, G. E. Shewell, H. J. Teskey, J. R. Vockeroth and D. M. Wood (Coordinators) (edits.) 1981. *Manual of Nearctic Diptera. Volume 1.* Hull, Quebec: Canadian Government Publishing Centre, vi + 674 pp. Vol. 2. 1987, pp. 675–1332.

McCabe, T. L. and Johnson, L. M. 1979a. The biology of *Platycis sculptilis* (Say) (Coleoptera: Lycidae). *Coleopterists Bulletin* 33(3):297–302.

McCabe, T. L. and Johnson, L. M. 1979b. Larva of *Calopteron terminale* (Say) with additional notes on adult behavior (Coleoptera: Lycidae). *Journal of the New York Entomological Society* 87:283–288.

McCauley, R. H, Jr. 1938. A revision of the genus *Microrhopala* in North America, north of Mexico. *Bulletin of the Brooklyn Entomological Society* 33(4):145–168.

McClean, M., Buck, J., and Hanson, F. E. 1972. Culture and larval behavior of photurid fireflies. *American Midland Naturalist* 87(1):133–145.

McDermott, F.A. 1964. The taxonomy of the Lampyridae (Coleoptera). *Transactions of the American Entomological Society* 90:1–72.

McDermott, F. A. 1966. Pars 9 (Editio secunda). Lampyridae. *In* Steel, W. O. (edit.), *Coleopterorum Catalogus Supplementa.* The Hague: W. Junk, 149 pp.

McDougall, W. A. 1934. The determination of larval instars and stadia of some wireworms (Elateridae). *Queensland Agricultural Journal* 42:43–70.

McMullen, R. D. 1967. A field study of diapause in *Coccinella novemnotata* (Coleoptera: Coccinellidae). *Canadian Entomologist* 99:42–49.

Medvedev, G. S. 1965. Types of adaptation of the structure of legs of the desert darkling beetles (Coleoptera, Tenebrionidae). *Entomologicheskoye Obozreniye* 44:803–826 (in Russian).

Medvedev, G. S. 1968. *Fauna of USSR. Coleoptera. Vol. 19, No. 2. Darkling Beetles (Tenebrionidae), Subfamily Opatrinae. Tribes Platynotini, Dendarini, Pedinini, Dissonomini, Pachypterini, Opatrini (part), and Heterotarsini.* Leningrad:

"Nauka", 285 pp. (In Russian; translation published for U.S. Department of Agriculture by Indian National Scientific Documentation Centre, New Delhi, 386 pp.).

Medvedev, L. N. 1962. Systematics and biology of the larvae of the subfamily Clytrinae (Col. Chrysomelidae). *Zoologicheskii Zhurnal* 41:1334–1344 (in Russian).

Medvedev, L. N. and Galata, L. P. 1969. On the larva of *Simoderus reflexicollis* Gebl (Coleoptera, Melyridae), *Polskie Pismo Entomolgiczne* 39(2):331–338.

Medvedev, L. N. and Zaitzev, Yu. M. 1978. *The larvae of chrysomelid beetles of Siberia and the Far East*. Moscow: "Nauka," 182 pp. (in Russian).

Meer Mohr, J. C. van der and Lieftinck, M. A. 1947. Over de biologie van *Antherophagus ludekingi* Grouv. (Col.) in hommelnesten (*Bombus* Latr.) op Sumatra. *Tijdschrift voor Entomologie* 88 (1945):207–214.

Meixner, J. 1935. Achte Überordnung der Pterygogenea: Coleopteroidea [part], pp. 1245–1340 *in* Kükenthal W. and Krumbach T., edit., *Handbuch der Zoologie. Vierter Band, zweite Halfte. Insecta 2, zweite Lieferung*. Berlin: W. de Gruyter.

Menier, J. J. and Ekis, G. 1982. Les Chaetosomatidae Malgaches. II. Description d'un genre nouveau et de quatre especes nouvelles (Coleoptera, Cleroidea). *Annales de la Société entomologique de France* (N.S.)18(3):343–348.

Mequignon, A. 1934. Les *Chelonarium* de l'Amérique continentale. *Annales de la Société entomologique de France* 103:199–256.

Merritt, R. W. and Cummins, K. W. (edits.). 1984. *An Introduction to the Aquatic Insects of North America. 2nd Edition.* Dubuque, Iowa: Kendall-Hunt, 722 pp.

Metcalf, C. L., Flint, W. P., and Metcalf, R. L. 1962. *Destructive and Useful Insects. 4th Edition.* New York: McGraw-Hill, xii + 1087 pp.

Miller, D. C. 1963. The biology of the Hydrophilidae. *Biologist* 45:33–38.

Milne, L. J. and Milne, M. 1976. The social behavior of burying beetles. *Scientific American* 235:84–89.

Miyatake, M. 1960. The genus *Pisenus* Casey and some notes on the family Tetratomidae (Coleoptera). *Transactions of the Shikoku Entomological Society* 6:121–135.

Mjöberg, E. 1925. The mystery of the so called "trilobite larvae" or "Perty's larvae" definitely solved. *Psyche* 32:119–154, pls. 3–4.

Monros, F. 1949. Descripcion de las metamorfosis de *Lamprosoma chorisiae* Monros y consideraciones taxonomicas sobre Lamprosominae (Col. Chrysomelidae). *Acta Zoologica Lilloana* 7:449–466.

Monros, F. 1954. *Megalopus jacobyi,* nueva plaga de solanaceas en el noroeste Argentino, con notas sobre biologia y taxonomia de Megalopinae (Col. Chrysomelidae). *Revista Agronomica del Noroeste Argentino* 1:167–179.

Monros, F. 1955. Biologia y descripcion de la larva de *Atalasis sagriodes. Revista Agronomica del Noroeste Argentina* 1:275–281.

Monros, F. 1959. Los generos de Chrysomelidae (Coleoptera). *Opera Lilloana* 3: 1–337.

Monros, F. and Monros, M. 1952. Les especies Argentinas de Cupedidae (Coleoptera). *Anales de la Sociedad Cientifica Argentina* 153:19–41.

Monteith, G. G. and Storey, R. I. 1981. The biology of *Cephalodesmius,* a genus of dung beetles which synthesizes "dung" from plant material (Coleoptera: Scarabaeidae: Scarabaeinae). *Memoirs of the Queensland Museum* 20(2):253–277.

Moore, B. P. 1964. Australian larval Carabidae of the subfamilies Broscinae, Psydrinae and Pseudomorphinae (Coleoptera). *Pacific Insects* 6(2):242–246.

Moore, B. P. 1965. Australian larval Carabidae of the subfamilies Harpalinae, Licininae, Odacanthinae and Pentagonicinae (Coleoptera). *Proceedings of the Linnean Society of New South Wales* 90:157–163.

Moore, B. P. 1966. The larva of *Pamborus* and its systematic position. *Proceedings of the Royal Entomological Society of London* (B)35:1–4.

Moore, B. P. 1974. The larval habits of two species of *Sphallomorpha* Westwood (Coleoptera, Carabidae, Pseudomorphinae). *Journal of the Australian Entomological Society* 13:179–183.

Moore, B. P. 1979. Chemical defense in carabids and its bearing on phylogeny. pp. 193–203 *in* Erwin, T. L., Ball, G. E., Whitehead, D. R., and Halpern, A. L., (edits.), *Carabid Beetles: Their Evolution, Natural History, and Classification.* The Hague: W. Junk, 635 pp.

Moore, I. 1956. Notes on some intertidal Coleoptera with descriptions of the early stages (Carabidae, Staphylinidae, Malachiidae). *Memoirs of the San Diego Society of Natural History* 12:207–230.

Moore, I. 1974. Notes on *Phaleria rotundata* LeConte with description of the larva (Coleoptera: Tenebrionidae). *Great Lakes Entomologist* 7(4):99–102.

Moore, I. and Legner, E. F. 1974. A catalogue of the taxonomy, biology and ecology of the developmental stages of the Staphylinidae (Coleoptera) of America north of Mexico. *Journal of the Kansas Entomological Society* 47:469–478.

Moore, I. and Legner, E. F. 1975. A catalogue of the Staphylinidae of America north of Mexico (Coleoptera). *Special Publications of the Division of Agricultural Sciences, University of California,* No. 3015, 514 pp.

Moore, I. and Legner, E. F. 1977. The developmental stages of *Endeodes* LeConte (Coleoptera: Melyridae). *Proceedings of the Entomological Society of Washington* 79(2):172–175.

Morimoto, K. 1962a. Comparative morphology and phylogeny of the superfamily Curculionoidea of Japan (Comparative morphology, phylogeny and systematics of the superfamily Curculionoidea of Japan. I.). *Journal of the Faculty of Agriculture, Kyushu University* 11:331–373.

Morimoto, K. 1962b. Key to the families, subfamilies, tribes and genera of the superfamily Curculionoidea of Japan excluding Scolytidae, Platypodidae and Cossoninae (Comparative morphology, phylogeny and systematics of the superfamily Curculionoidea of Japan. III.). *Journal of the Faculty of Agriculture, Kyushu University* 12:21–66.

Morimoto, K. 1976. Notes on the family characters of Apionidae and Brentidae (Coleoptera), with key to the related families. *Kontyu* 44:469–476.

Moron, M. A. 1977. Description of the third-stage larva of *Megasoma elephas occidentalis* Bolivar y Pieltain et al. (Scarabaeidae: Dynastinae). *Coleopterists Bulletin* 31(4):339–345.

Morrill, A. W. 1903. Notes on the early stages of *Corylophodes marginicollis* LeC. *Entomological News* 14(4):135–138, pl. 6.

Moulins, M. 1959. Contribution a la connaissance de quelques types larvaire d'Hydrophilidae. *Travaux du Laboratoire de Zoologie et de la Station Aquicole Grimaldi de la Faculte des Sciences de Dijon,* No. 30, 46 pp.

Mroczkowski, M. 1968. Distribution of the Dermestidae (Coleoptera) of the world with a catalogue of all known species. *Annales Zoologici* (Warsaw) 4:1–163.

Mroczkowski, M. 1975. *Fauna Polski. 4. Dermestidae. Skornikowate.* Zoological Institute, Polish Academy of Science, Panstwowe Wydawnictwo Naukowe. Warsaw, 168 pp.

Mulsant, E. and Rey, C. 1856. Notes pour servir a l'histoire de l'*Agnathus decoratus.* Description de la larve et de la nymphe de l'*Agnathus decoratus. Annales de la Societe Linneenne de Lyon* 3:114–118.

Muma, M. H. 1955a. Lady beetles (Coccinellidae: Coleoptera) found on citrus in Florida. *Florida Entomologist* 38:117–124.

Muma, M. H. 1955b. Some ecological studies on the twice-stabbed lady beetle *Chilocorus stigma* (Say). *Annals of the Entomological Society of America* 48:493–498.

Muniz, R. and Barrera, A. 1969. *Rhopalotria dimidiata* Chevrolat, 1878: estudio morfologico del adulto y descripcion de la larva (Ins. Col. Curcul.: Oxycorynidae). *Revista de la Sociedad Mexicana de Historia Natural* 30:205–222.

Muona, J. E. 1981. Classification and nomenclature of the beetle family Eucnemidae. *Ph.D. Thesis, University of California at Davis,* 118 pp.

Murray, A. 1864. A monograph of the family of Nitidulariae. *Transactions of the Linnean Society of London* 24:211–414, pls. 32–36.

Murray, F. L. S. 1870. On an undescribed light-giving coleopterous larva (provisionally named *Astraptor illuminator*). *Journal of the Linnean Society. Zoology* 10:74–82.

Murvosh, C. M. 1971. Ecology of the water penny beetle *Psephenus herricki* (DeKay). *Ecological Monographs* 41(1):79–96.

Nakamura, S. 1981. Morphological and taxonomic studies of the cerambycid pupae of Japan (Coleoptera: Cerambycidae). *Miscellaneous Reports of the Hiwa Museum for Natural History* 20:1–159, 75 pls.

Nakamura, S., Kojima, K., and Okajima, S. 1976. Notes on the larva of *Parandra shibatai* Hayashi (Coleoptera, Cerambycidae), with key to the subfamilies of Japanese cerambycid larvae. (In Japanese). *Kontyu* 44:228–233.

Naomi, S. I. 1985. The phylogeny and higher classification of the Staphylinidae and their allied groups (Coleoptera, Staphylinoidea). *Esakia* 23:1–27.

Neboiss, A. 1960. On the family Cupedidae, Coleoptera. *Proceedings of the Royal Society of Victoria* 72(1):12–20, pls. 4–5.

Neboiss, A. 1968. Larva and pupa of *Cupes varians* Lea, and some observations on its biology (Coleoptera; Cupedidae). *Memoirs of the National Museum of Victoria* 28:17–19, pl. 7.

Neboiss, A. 1984. Reclassification of *Cupes* Fabricius (*s. lat.*), with descriptions of new genera and species (Cupedidae: Coleoptera). *Systematic Entomology* 9:443–477.

Nelson, G. H. 1981. A new tribe, genus, and species of North American Buprestidae with consideration of subfamilial and tribal categories. *Coleopterists Bulletin* 35:431–450.

Neumann, V. and Piechocki, R. 1984. Die Entwicklungsstadien der Familie Leptinidae (Coleoptera). *Entomologische Nachrichten und Berichte* 28:237–244.

Neumann, V. and Piechocki, R. 1985. Morphologische und histologische Untersuchungen an den Larvenstadien von *Platypsyllus castoris* Ritsema (Coleoptera, Leptinidae). *Entomologische Abhandlungen, Staatliches Museum fur Tierkunde Dresden* 49:27–34.

Newton, A. F., Jr. 1984. Mycophagy in Staphylinoidea (Coleoptera). pp. 302–353 *in* Wheeler, Q. D. and Blackwell, M. (edit.) *Fungus/Insect Relationships: Perspectives in Ecology and Evolution.* New York: Columbia University Press, 512 pp.

Newton, A. F., Jr. and Chandler, D. S. (1989). World catalog of the genera of Pselaphidae (Coleoptera). *Fieldiana: Zoology* n.s. 53:1–93.

(Coleoptera, Cryptophagidae) and its larval stages. *Annals of Applied Biology* 19:87–97.

Nikitsky, N. B. 1976a. Larval morphology and mode of life of Histeridae (Coleoptera) in bark-beetle passages. *Entomologicheskoye Obozreniye* 55:875–888 (in Russian; translated in *Entomological Review* 55:102–111).

Nikitsky, N. B. 1976b. Morphology of larvae and ecology of beetles of the genus *Hypophloeus* F. (Coleoptera, Tenebrionidae). *Zoologicheskii Zhurnal* 55:41–51 (in Russian).

Nikitsky, N. B. 1976c. Morphology of predatory larvae associated with bark beetles in the southwestern Caucasus, pp. 175–201 *in* Mamaev, B. M (edit.), *Evolutionary Morphology of Insect Larvae.* Moscow: "Nauka," 204 pp. (in Russian).

Nikitsky, N. B. 1976d. The morphology of *Sphaerites glabratus* larva and the phylogeny of Histeroidea. *Zoologicheskii Zhurnal* 55:531–537 (in Russian).

Nikitsky, N B. 1980. *Insect Predators of Bark Beetles and their Ecology.* Moscow: "Nauka", 237 pp. (in Russian).

Nikitsky, N. B. 1983a. Morphology of the *Myrmechixenus subterraneus* Chevr. larva and some remarks on systematics of the genus *Myrmechixenus* Chevr. *Byulletin Moskovskovo Obshchestva Ispytatelei Prirody Novaya Seriya. Otdel Biologicheskii* 88(2):59–63.

Nikitsky, N. B. 1983b. Species of the genus *Biphyllus* (Coleoptera, Biphyllidae) in the East Palearctic. *Zoologicheskii Zhurnal* 62(5):695–706. (in Russian).

Nikitsky, N. B. and Belov, V. V. 1979. New and poorly known species of Clavicornia (Coleoptera) from Talysh. *Zoologicheskii Zhurnal* 58:849–854 (in Russian).

Nikitsky, N. B. and Belov, V. V. 1980a. Larvae of cylindrical bark-beetles (Coleoptera, Colydiidae) of the European part of the USSR and Caucasus, with comments on taxonomy of the family. 1. *Zoologicheskii Zhurnal* 59:1040–1053 (in Russian).

Nikitsky, N. B. and Belov, V. V. 1980b. Larvae of cylindrical bark-beetles (Coleoptera, Colydiidae) of the European part of the USSR and Caucasus, with comments on taxonomy of the family. 2. *Zoologicheskii Zhurnal* 59:1328–1333 (in Russian).

Nobuchi, A. 1954. Morphological and ecological notes of fungivorous insects. (II). On the larvae of erotylid-beetles from Japan (Erotylidae, Coleoptera) (Part 1). *Kontyu* 22:1–6.

Nobuchi, A. 1955. Morphological and ecological notes of fungivorous insects. (II). On the larvae of erotylid-beetles from Japan (Erotylidae, Coleoptera) (Part 2). *Kontyu* 23:53–60, pls. 8–9.

O'Brien, C. W. and Wibmer, G. J. 1982. Annotated checklist of the weevils (Curculionidae *sensu lato*) of North America, Central America, and the West Indies (Coleoptera: Curculionoidea). *Memoirs of the American Entomological Institute* 34, ix + 382 pp.

Ochs, J. 1949. Un staphylinide nouveau pour la faune française et notes diverses. *Bulletin mensuel de la Sociêtê linnêenne de Lyon* 18:203–205.

Ogloblin, D. A. and Znoiko, D. V. 1950. *Fauna of USSR. Coleoptera. Vol. 18, No. 5. Pollen Eaters (Fam. Alleculidae, Part 2. Subfamily Omophlinae)* Moscow: Academia Nauk SSSR, 134 pp.

Ôhira, H. 1962. *Morphological and Taxonomic Study on the Larvae of Elateridae in Japan (Coleoptera).* Okazaki City, Japan: Ôhira, 179 pp.

Oke, C. 1932. Aculagnathidae. A new family of Coleoptera. *Proceedings of the Royal Society of Victoria* (N.S.)44:22–24, pl.2.

Okumura, G. T. and Savage, I. E. 1974. Nitidulid beetles most commonly found attacking dried fruits in California. *National Pest Control Operators News* 34:3–7.

Oldroyd. H. 1958. *Collecting, Preserving and Studying Insects.* London: Hutchinson, 327 pp.

Olliff, A. S. 1882. Descriptions of two larvae and new genera and species of clavicorn Coleoptera, and a synopsis of the genus *Helota,* MacLeay. *Cistula Entomolgica* 3(26):49–65.

Olmi, M. 1972. The palearctic species of the genus *Dryops* (Olivier) (Coleoptera, Dryopidae). *Bolletino del Museo di Zoologia dell'Universita di Torino* 5:69–132.

d'Orchymont, A. 1913. Contribution à l'étude des larves hydrophilides. *Annales de biologie lacustre* 6:173–214.

Ordish, R. G. 1971. Entomology of the Auckland Islands and other islands south of New Zealand: Coleoptera: Hydraenidae. *Pacific Insect Monographs* 27:185–192.

Osawa, S. 1947. Larva and pupa of *Pseudopyrochroa japonica* Heyden. *Insect Ecology* 2:7–10.

Osborne, H. L. and Kulhavy, D. L. 1975. Notes on *Nosodendron californicum* Horn on slime fluxes of grand fir, *Abies grandis* (Douglas) Lindley, in northern Idaho (Coleoptera: Nosodendridae). *Coleopterists Bulletin* 29:71–73.

Osborne, P. 1965. Morphology of the immature stages of *Meligethes aeneus* (F.) and *M. viridescens* (F.) (Coleoptera, Nitidulidae). *Bulletin of Entomological Research* 55:747–759, pl. 8.

Osten-Sacken, C. R. 1862. Description of some larvae of North American Coleoptera. *Proceedings of the Entomological Society of Philadelphia* 1:105–130.

Paarmann, W. 1979. A reduced number of larval instars, as an adaptation of the desert carabid beetle *Thermophilum (Anthia) sexmaculatum* F. (Coleoptera, Carabidae) to its arid environment. pp. 113–117 *in* Boer, P. J. den, Thiele, H. U., and Weber, F. (edits.), *On the Evolution of Behavior in Carabid Beetles.* Wageningen, The Netherlands: Landbouwhogeschool Wageningen (Miscellaneous Papers, No. 18), 222 pp.

Pace, A. E. 1967. Life history and behavior of a fungus beetle, *Bolitotherus cornutus* (Tenebrionidae). *Occasional Papers of the Museum of Zoology, University of Michigan* 653:1–15.

Pace, R. 1976. An exceptional endogenous beetle: *Crowsoniella relicta* n. gen. n. sp. of Archostemata Tetraphaleridae from central Italy. *Bolletino del Museo Civico di Storia Naturale, Verona* 2(1975):445–458.

Pacheco, F. 1964. Sistematica, filogenia y distribucion de los heteroceridos de America (Coleoptera: Heteroceridae). *Monografias del Colegio de Post-graduados, Escuela Nacional de Agricultura, Chapingo, Mexico* 1:1–155.

Pacheco, F. 1978. A catalog of the Coleoptera of America north of Mexico. Family: Heteroceridae. *United States Department of Agriculture, Agriculture Handbook,* No. 529–47, 8 pp.

Paddock, F. B. 1919. The cowpea weevil. *Texas Agricultural Experiment Station Bulletin* 256:1–51, 90–92.

Paiva, C. A. 1919. Notes on the Indian glowworm (*Lamprophorus tenebrosus* Walk.). *Records of the Indian Museum* 16(1):19–28.

Pakaluk, J. 1984. Natural history and evolution of *Lycoperdina ferruginea* (Coleoptera: Endomychidae) with descriptions of immature stages. *Proceedings of the Entomological Society of Washington* 86(2):312–325.

Pakaluk, J. 1985. New genus and species of Corylophidae (Coleoptera) from Florida, with a description of its larva. *Annals of the Entomological Society of America* 78:406–409.

Pal, T. K. and Lawrence, J. F. 1986. A new genus and subfamily of mycophagous Bothrideridae (Coleoptera: Cucujoidea) from the Indo-Australian Region, with notes on related families. *Journal of the Australian Entomological Society* 25:185–210.

Palm, T. 1951. Anteckningar om svenska skalbaggar. VI. *Entomologisk Tidskrift* 72:39–53.

Palm, T. 1959. Die Holz- und Rinden-Käfer der süd- und mittelschwedischen Laubbäume. *Opuscula Entomologica Supplementa* 16:1–374, 93 figs.

Palm, T. 1960. Zur Kenntnis der früheren Entwicklungsstadien schwedischer Käfer. 1. Bisher bekannte Eucnemiden-Larven. *Opuscula Entomologica* 25:157–169.

Palmer, M. A. 1914. Some notes on life history of lady beetles. *Annals of the Entomological Society of America* 7:213–237.

Park, O. 1933. Ecological study of the ptiliid myrmecocole, *Limulodes paradoxus* Matthews. *Annals of the Entomological Society of America* 26:357–360.

Park, O. and Sejba, O. 1935. Studies in nocturnal ecology, IV. *Megalodacne heros. Ecology* 16:164–172.

Parker, D. L. and Davis, D. W. 1971. Feeding habits of *Corticeus substriatus* (Coleoptera: Tenebrionidae) associated with the mountain pine beetle in lodgepole pine. *Annals of the Entomological Society of America* 64(1):293–294.

Parker, J. B. and Böving, A. G. 1914. The blister beetle *Tricrania sanguinipennis*—biology, description of different stages, and systematic relationships. *Proceedings of the United States National Museum* 64(23):1–49, 5 pls.

Parkin, E. A. 1933. The larvae of some wood-boring Anobiidae (Coleoptera). *Bulletin of Entomological Research* 24:33–68.

Parkin, E. A. 1940. The digestive enzymes of some wood-boring beetle larvae. *Journal of Experimental Biology* 17:364–377.

Parks, J. J. and Barnes, J. W. 1955. Notes on the family Leptinidae including a new record of *Leptinillus validus* (Horn) in North America (Coleoptera). *Annals of the Entomological Society of America* 48:417–421.

Parnell, J. R. 1964. The external morphology of the larvae and notes on the pupae of *Bruchidius ater* (Marsh.) (Col., Bruchidae) and *Apion fuscirostre* F. (Col., Curculionidae). *Ent. Mon. Mag.* 100:83–87.

Parsons, C. T. 1969a. A lathridiid beetle reported to bite man. *Coleopterists Bulletin* 23:15.

Parsons, C. T. 1969b. North American Nitidulidae (Coleoptera). V. Species of *Epuraea* related to *corticina* Erichson. *Coleopterists Bulletin* 23:62–72.

Parsons, C. T. 1975. Revision of the Nearctic Mycetophagidae (Coleoptera). *Coleopterists Bulletin* 29(2):93–108.

Patterson, N. F. 1931. Studies on the Chrysomelidae. Part II. The bionomics and comparative morphology of the early stages of certain Chrysomelidae (Coleoptera: Phytophaga). *Proceedings of the Zoological Society of London* 2:879–949.

Paulian, R. 1941. Les premiers états des Staphylinoidea (Coleoptera). Etude de morphologie comparée. *Mémoires du Muséum national d'histoire naturelle* (N.S.)15:1–361.

Paulian, R. 1942. The larvae of the sub-family Orphilinae and their bearing on the systematic status of the family Dermestidae (Col.). *Annals of the Entomological Society of America* 35:393–396.

Paulian, R. 1943. *Les Coléoptères: Formes—Moeurs—Rôle.* Paris: Payot, 396 pp.

Paulian, R. 1944. Les Aglycyderidae, une famille relicte (Col.). *Revue francaise d'entomologie* 10:113–119.

Paulian, R. 1949a. Ordre des Coléoptères (Coleoptera Linné, 1758). Partie Systématique. Premier sous-ordre.-Heterogastra Jeannel et Paulian, 1944, pp. 890–989, *in* Grassé, P., (edit.), *Traité de Zoologie: Anatomie, Systématique, Biologie. Tome IX. Insectes: Paléontologie, Géonémie, Aptérygotes, Ephémeroptères, Odonatoptères. Blattoptéroïdes, Orthoptéroïdes, Dermaptéroïdes, Coléoptères.* Paris: Masson, 1117 pp.

Paulian, R. 1949b. Recherches sur les insectes d'importance biologique de Madagascar. VI. Deux familles de coléoptères nouvelles pour la fauna malgache. *Mémoires de l'Institut scientifique de Madagascar* (a)3:371–374, figs. 17–18.

Paulian, R. 1950. Les Corylophidae d'Afrique (Coleoptera). *Mémoires de l'Institut Français d'Afrique Noire* 12:1–126.

Paulian, R. and Villiers, A. 1940. Les larves des Cerambycidae français. *Revue française d'entomologie* 7:202–217.

Paulus, H. F. 1970. Zur Morphologie und Biologie der Larven von *Pelochares* Mulsant & Rey (1869) and *Limnichus* Latreille (1829) (Coleoptera: Dryopoidea: Limnichidae). *Senckenbergiana Biologica* 51(1/2):77–87.

Paulus, H. F. 1971. Neue Pyrochroidae aus Nepal (Coleoptera, Heteromera), mit einer Diskussion der verwandtschaftlichen Verhältnisse der Familie. *Zeitschrift der Arbeitsgemeinschaft Oesterreichischer Entomologen* 23:75–85.

Paulus, H. F. 1972. Der Stand unserer Kenntnis über die Familie Byrrhidae (Col.). *Folia Entomologica Hungarica* (S.N.)25:335–348.

Paulus, H. F. 1979. Eye structure and the monophyly of Arthropoda, pp. 299–383 *in* Gupta, A. P. (edit.), *Arthropod Phylogeny.* New York: Van Nostrand Reinhold, xx + 762 pp.

Paviour-Smith, K. 1960a. The fruiting-bodies of macrofungi as habitats for beetles of the family Ciidae (Coleoptera). *Oikos* 11:1–71.

Paviour-Smith, K. 1960b. The invasion of Britain by *Cis bilamellatus* Fowler (Coleoptera: Ciidae). *Proceedings of the Royal Entomological Society of London* (A)35(10–12):145–155.

Paviour-Smith, K. 1964a. Habitats, headquarters and distribution of *Tetratoma fungorum* F. (Col., Tetratomidae). *Entomologist's Monthly Magazine* 100:71–80.

Paviour-Smith, K. 1964b. The life history of *Tetratoma fungorum* F. (Col., Tetratomidae) in relation to habitat requirements, with an account of eggs and larval stages. *Entomologist's Monthly Magazine* 100:118–134.

Paviour-Smith, K. 1964c. The night-day activity rhythm of *Tetratoma fungorum* F. (Col., Tetratomidae). *Entomologist's Monthly Magazine* 100:234–240.

Payne, N. M. 1931. Food requirements for the pupation of two coleopterous larvae, *Synchroa punctata* Newn. and *Dendroides canadensis* LeC. (Melandryidae, Pyrochroidae). *Entomological News* 42:13–15.

Payne, O. G. M. 1916. On the life-history and structure of *Telephorus lituratus* Fallén (Coleoptera). *Journal of Zoological Research* 1:4–35.

Peacock, E. R. 1977. *Handbooks for the Identification of British Insects. Vol. V. Part 5(a). Coleoptera. Rhizophagidae.* London: Royal Entomological Society of London, 23 pp.

Pearson, D. L. 1988. Biology of tiger beetles. *Annual Review Entomology* 33:123–47.

Peck, S. B. 1973. A systematic revision and the evolutionary biology of the *Ptomaphagus (Adelops)* beetles of North America (Coleoptera; Leiodidae; Catopinae), with emphasis on cave-inhabiting species. *Bulletin of the Museum of Comparative Zoology* 145(2):29–162.

Peck, S. B. 1982. A review of the ectoparasitic *Leptinus* beetles of North America (Coleoptera: Leptinidae). *Canadian Journal of Zoology* 60:1517–1527.

Pennak, R. W. 1978. *Fresh Water Invertebrates of the United States.* 2nd Edition. New York: Wiley, 803 pp.

Pepper, B. B. 1942. The carrot weevil, *Listronotus latiusculus* (Bohe.), in New Jersey and its control. *New Jersey Agricultural Experiment Station Bulletin* 693, 20 pp.

Perkins, P. D. 1981a. Aquatic beetles of the family Hydraenidae in the Western Hemisphere: classification, biogeography and inferred phylogeny (Insecta: Coleoptera). *Quaestiones Entomologicae* 16(1–2):1–554.

Perkins, P. D. 1981b. Larval and pupal stages of a predaceous diving beetle, *Neoclypeodytes cinctellus* (LeConte) (Dytiscidae: Hydroporinae: Bidessini). *Proceedings of the Entomological Society of Washington* 82:474–481.

Perkins, P. D. and Spangler, P. J. 1981. A description of the larva of *Helocombus bifidus* (Coleoptera: Hydrophilidae). *Pan-Pacific Entomologist* 57:52–56.

Perris, E. 1851. Quelques mots sur les métamorphoses de Coléoptères mycétophages, le *Triphyllus punctatus* Fab.; le *Diphyllus lunatus* Fab.; l'*Agathidium seminulum* Linn., et l'*Eucinetus(Nycteus)* Latr.) *meridionalis* de Castelnaù. *Annales de la Société entomolgique de France* (2)9:39–53, pl. 2(II–V).

Perris, E. 1852. Histoire des métamorphoses du *Clambus enshamensis. Annales de la Société entomologique de France* (2)10:574–578, pl. 14.

Perris, E. 1855. Histoire des metamorphoses de divers insectes. *Mémoires de la Société royal des sciences de Liège* 10:233–280, pl. 5.

Perris, E. 1862. Histoire des insectes du pin maritime. Supplément aux Coléoptères et rectifications. *Annales de la Société entomologique de France* (4)2:173–243, pls. 5–6.

Perris, E. 1876. Larves de Coléoptères. (Part). *Annales de la Société linnéenne de Lyon* 22(1875):259–418.

Perris, E. 1877. Larves de Coléoptères. (Part). *Annales de la Société linnéenne de Lyon* 23(1876):1–430, 14 pls.

Peterson, A. 1951. *Larvae of Insects. An Introduction to Nearctic Species. Part II. Coleoptera, Diptera. Neuroptera, Siphonaptera, Mecoptera, Trichoptera.* Columbus, Ohio: A. Peterson, 416 pp.

Peterson, A. 1959. *Entomological Techniques: How to Work with Insects. 9th Edition.* Ann Arbor, Michigan: A. Peterson, v + 435 pp.

Peyerimhoff, P. de 1902. Descriptions des larves de trois Coléoptères exotiques. *Annales de la Société entomologique de France* 71:710–718.

Peyerimhoff, P. de 1907. Deux types nouveaux de larves Silphidae (Col.). *Annales de la Société entomologique de France* 76:83–88.

Peyerimhoff, P. de 1910. Un nouveau type de larves mineuses appartenant au genre *Xenostrongylus* (Col. Nitidulidae). *Bulletin de la Société entomologique de France* 1910: 266–268.

Peyerimhoff, P. de 1911. Notes sur la biologie de quelques Coléoptères phytophages du Nord-Africain (premiere série). *Annales de la Société entomologique de France* 80:283–314.

Peyerimhoff, P. de 1913. Lalarve de *Sphaerosoma algiricum* Reitt. (Col. Endomychidae) et ses deux formes successives. *Bulletin de la Société entomologique de France* 1913:199–204.

Peyerimhoff, P. de 1916. Description de la larve de *Lasiodactylus chevrolati* Reitt. (Coleoptera, Nitidulidae). *Records of the Indian Museum* 12:109–113.

Peyerimhoff, P. de 1919. Notes sur la biologie de quelques Coléoptères phytophages du Nord-Africain (troisième série). *Annales de la Société entomologique de France* 88:169–258.

Peyerimhoff, P. de 1921. Etudes sur les larves des Coléoptères. I. Introduction.—II. Corylophidae.—III. Sphindidae. *Annales de la Société entomologique de France* 90:97–110, pls. 3–4.

Peyerimhoff, P. de 1926. Notes sur la biologie de quelques Coléoptères phytophages du Nord. Africain (quatrième série). *Annales de la Société entomologique de France* 95:319–390.

Peyerimhoff, P. de 1942. Les *Rhipidius* peuvent-ils parasiter les chenilles? *Bulletin de la Société entomologique de France* 1942:172–177.

Peyrieras, A. 1976. Insectes Coléoptères Carabidae Scaritinae. II. Biologie. *Faune de Madagascar* 41:1–161.

Pfaffenberger, G. S. 1974. Comparative morphology of the final larval instar of *Caryobruchus buscki* and *Pachymerus* sp. (Coleoptera: Bruchidae: Pachymerinae). *Annals of the Entomological Society of America* 67(4):691–694.

Pfaffenberger, G. S. 1977. Comparative descriptions of the final larval instar of *Bruchus brachialis, B. rufimanus* and *B. pisorum* (Coleoptera: Bruchidae). *Coleopterists Bulletin* 31(2):133–142.

Pfaffenberger, G. S. 1979. Comparative description and bionomics of the first and final larval stages of *Amblycerus acapulcensis* Kingsolver and *A. robiniae* (Fabricius) (Coleoptera: Bruchidae). *Coleopterists Bulletin* 33(2):229–238.

Pfaffenberger, G. S. 1981. A comparative description and phenetic analysis of the first instar larvae of seven *Stator* species (Coleoptera: Bruchidae). *Coleopterists Bulletin* 35(1):255–268.

Pfaffenberger, G. S. 1982. Description and phylogenetic comments on the final larval instar of *Caryobruchus veseyi* (Horn) (Coleoptera: Bruchidae). *Pan-Pac. Ent.* 58(3):240–244:

Pfaffenberger, G. S. 1984. Description of first instar larva of *Caryedon palaestinicus* Southgate, new status (Coleoptera: Bruchidae). *Coleopterists Bulletin* 38(3):220–226.

Pfaffenberger, G. S. 1985a. Description, differentiation, and biology of the four larval instars of *Acanthoscelides obtectus* (Coleoptera: Bruchidae). *Coleop. Bull.* 39:239–256.

Pfaffenberger, G. S. 1985b. Checklist of selected species of described first and/or final larval instars (Bruchidae: Coleoptera). *Coleopterists Bulletin.* 39:(1)1–6.

Pfaffenberger, G. S. and D. H. Janzen. 1985. Life history and morphology of first and last larval instars of Costa Rican *Caryedes brasiliensis* Thunberg (Coleoptera: Bruchidae). *Coleopterists Bulletin.* 38(3):267–281.

Pfaffenberger, G. S. and Johnson, C. D. 1976. Biosystematics of the first stage larvae of some North American Bruchidae (Coleoptera). *United States Department of Agriculture Technical Bulletin* 1525, 75 pp.

Pfaffenberger, G. S., Muruaga de L'Aregentier, S. M., and Teran, A. L. 1984. Morphological descriptions and biological and phylogenetic discussions of the first and final instars of four species of *Magacerus* larvae (Bruchidae). *Coleopterists Bulletin* 38(1):1–26.

Phillips, J. H. H. 1963. Life history and ecology of *Pulvinaria vitis* (L.) (Hemiptera: Coccoidea), the cottony scale attacking peach in Ontario. *Canadian Entomologist* 95:372–407.

Phuoc, D. T. and Stehr, F. W. 1974. Morphology and taxonomy of the known pupae of Coccinellidae (Coleoptera) of North America, with a discussion of phylogenetic relationships. *Contributions of the American Entomological Institute (Ann Arbor)* 10:1–125.

Pic, M. 1914. Pars 58. Dascillidae, Helodidae, Eucinetidae. *In* Junk, W. and S. Schenkling (eds.) *Coleopterorum Catalogus.* Berlin: W. Junk, 65 pp.

Piel, O. 1938. Note sur le parasitisme de *Dastracus helophoroides* Fairmaire (Coléoptère: Colydiide). *Notes d'entomologie chinoise* 4:1–15, pls. 1–2.

Pierce, W. D. 1939. The dodder and its insects. *Bulletin of the Southern California Academy of Sciences. Los Angeles* 38:43–53.

Pierre, F. 1945. La larve d'*Heterocerus aragonicus* Kiesw. et son milieu biologique (Col. Heteroceridae). *Revue fransçaise d'entomologie* 12:166–174.

Pierre, F. and Balachowsky, A. S. 1962. Famille des Tenebrionidae, pp. 374–393 *in* A. S. Balachowsky, (ed.), *Entomologie appliquée a l'agriculture. Tome I. Coléoptères. Premier volume.* Paris: Masson, 564 pp.

Pilgrim, R. L. C. 1972. The aquatic larva and the pupa of *Choristella philpotti* Tillyard, 1917 (Mecoptera: Nannochoristidae). *Pacific Insects* 14(1):151–168.

Pinto, J. D. 1984. Cladistic and phenetic estimates of relationship among genera of eupomphine blister beetles (Coleoptera: Meloidae). *Systematic Entomology* 9:165–182.

Pinto, J. D. and Selander, R. B. 1970. The bionomics of blister beetles of the genus *Meloe* and a classification of the new world species. *Illinois Biological Monographs* 42, 222 pp.

Ponomarenko, A. G. 1969. Historical development of the Coleoptera—Archostemata. *Trudi Paleontologicheskovo Instituta* 125:1–240, pls. 1–15.

Pope, R. D. 1976. Nomenclatural notes on the British Scirtidae (= Helodidae) (Col.). *Entomologist's Monthly Magazine* 111:186–187.

Pope, R. D. 1977. *In Kloet and Hincks. A Check List of British Insects Second Edition. Part 3: Coleoptera and Strepsiptera.* London: Royal Entomological Society, 105 pp.

Pope, R. D. 1979. Wax production by coccinellid larvae (Coleoptera). *Systematic Entomology* 4:171–196.

Pope, R. D. 1981. *Rhyzobius ventralis* (Coleoptera: Coccinellidae), its constituent species, and their taxonomy and historical roles in biological control. *Bulletin of Entomological Research* 71:19–31.

Pototskaya, V. A. 1967. *An Identifier for the Larvae of Brachelytrous Beetles (Staphylinidae) of the European part of the U.S.S.R.* (In Russian). Moscow: Academiya Nauk, 120 pp.

Pototskaya, V. A. 1976a. Morphology of the larva of *Trigonurus asiaticus* Reiche and relationships of the tribe Trigonurini to the other tribes in the subfamily Piestinae (Coleoptera, Staphylinidae), pp. 13–21 *in* Mamaev, B. M. (edit.), *Evolutionary Morphology of Insect Larvae.* Moscow: "Nauka," 204 pp. (in Russian).

Pototskaya, V. A. 1976b. Staphylinid larvae (Coleoptera, Staphylinidae) developing in wood, pp. 156–174 *in* Mamaev, B. M. (edit.), *Evolutionary Morphology of Insect Larvae.* Moscow: "Nauka," 204 pp. (in Russian).

Powell, J. M. 1971. The arthropod fauna collected from the commandra blister rust, *Cronartium comandrae,* on lodgepole pine in Alberta. *Canadian Entomologist* 103:908–918.

Powell, J. M., Wong, H. R. and Melvin, J. C. E. 1972. Arthropods collected from stem rust cankers of hard pine in western Canada. *Information Report, Northern Forest Research Centre, Edmonton, Alberta* NOR—X-42, 19 pp.

Prevett, P. R. 1971. The larvae of some Nigerian Bruchidae (Coleoptera). *Transactions of the Royal Entomological Society of London* 123(3):247–312.

Pringle, J. A. 1938a. A contribution to the knowledge of *Micromalthus debilis* LeC. (Coleoptera). *Transactions of the Royal Entomological Society of London* 87:271–286.

Pringle, J. A. 1938b. Observations on certain wood-boring Coleoptera occurring in South Africa. *Transactions of the Royal Entomological Society of London* 87:247–270.

Prins, A. J. 1984. Morphological and biological notes on some South African arthropods associated with decaying organic matter. Part 2. The predatory families Carabidae, Hydrophilidae, Histeridae, Staphylinidae and Silphidae (Coleoptera). *Ann. South African Mus.* 92:295–356.

Procter, W. 1938. Biological survey of the Mount Desert Region, Part IV: the insect fauna with references to methods of capture, food plants, the flora and other biological features. *Memoirs of the Wistar Institute of Anatomy and Biology. Philadelphia* 1938, 496 pp.

Prota, R. 1959. Ricerche sull'entomofauna del Carciofo (*Cynara cardunculus v. scolymus* L.). II. *Lasioderma baudii* Schils. (Coleoptera: Anobiidae). *Studi Sassaressi. Sezione III* 6:210–255, 5 pls.

Pukowski, E. 1933. Ökologische Untersuchungen an *Necrophorus* F. *Zeitschrift für Morphologie und Ökologie der Tiere* 27:518–586.

Pukowski, E. 1934. Zur Systematik der *Necrophorus*-Larven (Col.). *Stettiner Entomologische Zeitung* 95:53–60.

Putman, W. L. 1955a. Bionomics of *Stethorus punctillum* Weise (Coleoptera: Coccinellidae) in Ontario. *Canadian Entomologist* 87:9–33.

Putman, W. L. 1955b. The immature stages of *Stethorus punctillum* Weise (Coleoptera: Coccinellidae). *Canadian Entomologist* 87:506–508.

Putman, W. L. 1964. Occurrence and food of some coccinellids (Coleoptera) in Ontario peach orchards. *Canadian Entomologist* 96:1149–1155.

Putman, W. L. and Herne, D. H. C. 1966. The role of predators and other biotic agents in regulating the population density of phytophagous mites in Ontario peach orchards. *Canadian Entomologist* 98:808–820.

Randolph, N. M. and B. B., Gillespie. 1958. Notes on the biology of *Bruchus brachialis* Fahr. *J. Econ. Ent.* 51(3):401–402.

Rasmussen, S. 1967. *Hylotrupes* (Col., Cerambycidae) in dead trees on Farön, a Swedish island. *Entomologiske Meddelelser* 35:223–226.

Ratcliffe, B. C. 1972. The natural history of *Necrodes surinamensis* (Fabr.) (Coleoptera: Silphidae). *Transactions of the American Entomological Society* 98:359–410.

Ratcliffe, B. C. 1976. Notes on the biology of *Euphoriaspis hirtipes* (Horn) and descriptions of the larva and pupa (Coleoptera: Scarabaeidae). *Coleopterists Bulletin* 30(3):217–225.

Ratcliffe, B. C. 1977. Description of the larva and pupa of *Osmoderma subplanata* (Casey) and *Cremastocheilus wheeleri* LeConte (Coleoptera: Scarabaeidae). *Journal of the Kansas Entomological Society* 50(3):363–370.

Ratcliffe, B. C. and Chalumeau, F. 1980. *Strategus syphax* (Fabr.): a description of the third instar larva and pupa (Coleoptera: Dynastinae). *Coleopterists Bulletin* 43(1):85–93.

Rees, B. E. 1941. First-instar larvae of *Buprestis rusticorum* (Kby.) and *Schizopus sallei* Horn, with notes on the classification of *Schizopus. Proceedings of the Entomological Society of Washington* 43:210–222, plts. 22–23.

Rees, B. E. 1943. Classification of the Dermestidae (larder, hide, and carpet beetles) based on larval characters, with a key to the North American genera. *United States Department of Agriculture Miscellaneous Publications,* No. 511, 18 pp.

Rees, D. P. 1985. Life history of *Teretriosoma nigrescens* Lewis (Coleoptera: Histeridae) and its ability to suppress populations of *Prostephanus truncatus* (Horn) (Coleoptera: Bostrichidae). *Journal of Stored Products Research* 21:115–118.

Rehfous, M. 1955. Contribution à l'étude des insectes des champignons. *Mitteilungen der Schweizerischen entomologischen Gesellschaft* 28:1–106.

Reichardt, H. 1971. Three new Hydroscaphidae from Brazil (Coleoptera, Myxophaga). *Entomologist* 104:290–292, pl. 4.

Reichardt, H. 1973a. A critical study of the suborder Myxophaga, with a taxonomic revision of the Brazilian Torridincolidae and Hydroscaphidae (Coleoptera). *Arquivos de Zoologia* 24:73–162.

Reichardt, H. 1973b. More on Myxophaga: on the morphology of *Scaphydra angra* (Reichardt, 1971) (Coleoptera). *Revista Brasiliera de Entomologia* 17:109–110.

Reichardt, H. 1974. Relationships between Hydroscaphidae and Torridincolidae, based on larvae and pupae, with the description of the immature stages of *Scaphydra angra* (Coleoptera, Myxophaga). *Revista Brasiliera de Entomologia* 18:117–122.

Reichardt, H. 1976a. Monograph of the New World Nosodendridae, and notes on the Old World forms (Coleoptera). *Papeis Avulsos de Zoologia* 29:185–220.

Reichardt, H. 1976b. A new African torridincolid (Coleoptera, Myxophaga). *Revue Zoologique Africaine* 90(1):209–214.

Reichardt, H. 1976c. Revision of the Lepiceridae (Coleoptera, Myxophaga). *Papeis Avulsos de Zoologia* 30:35–42.

Reichardt, H. and Costa, C. 1967. *Ptyopteryx britskii*, a new Neotropical genus and species of the hitherto Ethiopian Torridincolidae (Coleoptera, Myxophaga). *Papeis Avulsos de Zoologia* 21:13–19.

Reichardt, H. and Hinton, H. 1976. On the New World beetles of the family Hydroscaphidae. *Papeis Avulsos de Zoologia* 30:1–24.

Reichardt, H. and Vanin, S. A. 1976. Two new Torridincolidae from Serra do Cipo, Minas Gerais, Brazil (Coleoptera, Myxophaga). *Studia Entomologica* 19(1–4):211–218.

Reid, J. A. 1942a. A note on *Leptinus testaceus* Müller (Coleoptera: Leptinidae). *Proceedings of the Royal Entomological Society of London* (A)17:35–37.

Reid, J. A. 1942b. The species of *Laemophloeus* (Coleoptera: Cucujidae) occuring in stored foods in the British Isles. *Proceedings of the Royal Entomological Society of London* (A)17:27–32.

Rempel, J. G. and Church, N. S. 1972. The embryology of *Lytta viridana* LeConte (Coleoptera: Meloidae). VIII. The respiratory system. *Canadian Journal of Zoology* 50:1547–1554.

Renner, K. 1976. Seltene Käfer aus westfälischen Schillerporlingen. *Natur und Heimat* 36:84–86.

Rey, C. 1887. Essai d'études sur certaines larves de Coléoptères et descriptions de quelques espèces inédites ou peu connues. *Annales de la Société linnéenne de Lyon* (2)33:131–260, pls. 1–2.

Reyes-Castillo, P. 1970. Coleoptera, Passalidae: morfologia y division en grandes grupos; generos americanos. *Folia Entomologica Mexicana* 20–22:3–240.

Reyes-Castillo, P. and Jarman, M. 1980. Some notes on larval stridulation in Neotropical Passalidae (Coleoptera: Lamellicornia). *Coleopterists Bulletin* 34:263–270.

Richards, A. M. 1981. *Rhyzobius ventralis* (Erichson) and *R. forestieri* (Mulsant) (Coleoptera: Coccinellidae), their biology and value for scale insect control. *Bulletin of Entomological Research* 71:33–46.

Richmond, E. A. 1920. Studies on the biology of the aquatic Hydrophilidae. *Bulletin of the American Museum of Natural History* 42(1):1–94, 16 pls.

Riek, E. F. 1955. The Australian rhipidiine parasites of cockroaches (Coleoptera: Rhipiphoridae). *Australian Journal of Zoology* 3(1):71–94.

Riek, E. F. 1970a. Megaloptera (Alderflies), pp. 465–471 *in* C.S.I.R.O., *The Insects of Australia.* Melbourne: Melbourne University Press, 1029 pp.

Riek, E. F. 1970b. Neuroptera (Lacewings), pp. 472–494 *in* C.S.I.R.O., *The Insects of Australia.* Melbourne: Melbourne University Press, 1029 pp.

Riek, E. F. 1970c. Strepsiptera, pp. 622–635 *in* C.S.I.R.O., *The Insects of Australia.* Melbourne: Melbourne University Press, 1029 pp.

Riek, E. F. 1970d. Mecoptera (scorpion-flies), pp. 636–646 *in* C.S.I.R.O., *The Insects of Australia.* Melbourne: Melbourne University Press, 1029 pp.

Riek, E. F. 1973. Rhipiphorid beetles of the subfamily Nephritinae (Coleoptera: Rhipiphoridae). *Journal of the Australian Entomological Society* 12:261–276.

Riley, C. V. 1868. A friend unmasked. *American Entomologist* 1:51–52.

Riley, C. V. 1886a. Notes on *Phengodes* and *Zarhipis. Proceedings of the Entomological Society of Washington* 1:62–63.

Riley, C. V. 1886b. Further notes on *Phengodes* and *Zarhipis. Proceedings of the Entomological Society of Washington* 1:86–89.

Riley, C. V. 1887. On the luminous larviform females of the Phengodini. *Entomologist's Monthly Magazine* 24:148–149.

Riley, T. J. and A. J. Keaster. 1979. Wireworms associated with corn identification of larvae of nine species of *Melanotus* from the North Central States. *Ann. Ent. Soc. Amer.* 72:408–14. (updated by Riley. 1983. *ibid.* 76:999–1001).

Riley, T. J. and Keaster, A. J. 1981. A pictorial field key to wireworms attacking corn in the Midwest. *University of Missouri Miscellaneous Publications* 517, 18 pp.

Ritcher, P. O. 1958. Biology of Scarabaeidae. *Annual Review of Entomology* 3:311–334.

Ritcher, P. O. 1966. *White Grubs and their Allies, a Study of North American Scarabaeoid Larvae.* Corvallis: Oregon State University Press, 219 pp.

Ritcher, P. O. 1967. Keys for identifying larvae of Scarabaeoidea to the family and subfamily. *California Department of Agriculture Bureau of Entomology. Occasional Papers* 10:1–8.

Ritcher, P. O. 1973. A description of the larva of *Hypothyce mixta* Howden (Coleoptera: Scarabaeidae: Melolonthinae). *Coleopterists Bulletin* 27(3):113–116.

Ritcher, P. O. and Duff, R. 1971. A description of the larva of *Ceratophyus gopherinus* Cartwright with a revised key to the larvae of North American Geotrupini and notes on the biology (Coleoptera: Scarabaeidae). *Pan-Pacific Entomologist* 47(2):158–163.

Roberts, A. W. R. 1930. A key to the principal families of Coleoptera in the larval stage. *Bulletin of Entomological Research* 21:57–72.

Roberts. A. W. R. 1936. Observations on the spiracles of curculionid larvae in the first instar, pp. 283–286 *in Livre jubilaire de M. E.-L. Bourvier.* Paris.

Roberts, A. W. R. 1939. On the taxonomy of Erotylidae (Coleoptera), with special reference to the morpological characters of the larvae. *Transactions of the Royal Entomological Society of London* 88:89–117.

Roberts, A. W. R. 1958. On the taxonomy of Erotylidae (Coleoptera), with special reference to the morphological characters of the larvae. II. *Transactions of the Royal Entomological Society of London* 110:245–285.

Roberts, H. 1968. Notes on the biology of ambrosia beetles of the genus *Trachyostus* Schedl (Coleoptera: Platypodidae) in West Africa. *Bulletin of Entomological Research* 58:325–352.

Roberts, H. 1969. A note on the Nigerian species of the genus *Sosylus* Erichson (Col. Colydiidae), parasites and predators of ambrosia beetles. *Journal of Natural History* 3:85–91.

Roberts, H. 1980. Description of the developmental stages of *Sosylus* spp. (Coleoptera: Colydiidae) from New Guinea, parasites and predators of ambrosia beetles (Coleoptera: Platypodidae). *Bulletin of Entomological Research* 70:245–252.

Roberts, R. H. and Rillett, R. O. 1953. Silk glands of the rusty grain beetle, *Laemophloeus ferrugineus* (Steph.). *Transactions of the American Microscopical Society* 72:264–270.

Robinson, A. G. 1953. Notes on *Stethorus punctum* (LeC.) (Coleoptera: Coccinellidae), a predator of tetranychid mites in Manitoba. *Annual Report of the Entomological Society of Ontario* 1952:24–26.

Rocha, A. A. 1967. Biology and first instar larva of *Epimetopus trogoides. Papeis Avulsos de Zoologia* 20:223–228.

Rodendorf, B. B. and Ponomarenko, A. G. 1962. Order of Coleoptera or Beetles, pp. 241–267 *in* Rodendorf, B. B. (edit.) *Fundamentals of Paleontology. Reference Book for the Paleontology and Geology of the U.S.S.R. Tracheate and Chelicerate Arthropoda.* Moscow: Academia Nauk S.S.S.R., 560 pp., 22 pls. (in Russian).

Rosander, R. W. and Werner, F. G. 1970. Larvae of some Arizona species of *Phyllophaga* (Coleoptera: Scarabaeidae). *Annals of the Entomological Society of America* 63(4):1136–1142.

Rosenberg, E. C. 1909. Bidrag til Kundskaben om Billernes Udvikling, Levevis og Systematik. II. *Drilus concolor* Ahr.: Hunnens Forvandling i Skallen af *Helix hortensis. Entomologiske Meddelelser* (2)3:227–240, 2 pls.

Rosenberg, E. C. 1925. Contributions to the knowledge of the life-habits, development and systematics of the Coleoptera. IV. On the larva of *Batrisodes venustus* Reichenb., with remarks on the life-habits of other so-called myrmecophile Coleoptera. *Entomologiske Meddelelser* 14:374–388.

Rosenberg, E. C. 1943. Neue Lyciden-Larven. (Beitrag zur Kenntnis der Lebensweise, Entwicklung und Systematik der Käfer. V.). *Entomologiske Meddelelser* 24:1–42, 8 pls.

Ross, D. A. and Pothecary, D. D. 1970. Notes on adults, eggs, and first-instar larvae of *Priacma serrata* (Coleoptera: Cupedidae). *Canadian Entomologist* 102:346–348.

Rossem, G. 1945. De larven van *Pyrochroa* (Col. Pyrochroidae). *Tijdschrift voor Entomologie* 88:524–530.

Roughley, R. E. 1981. Trachypachidae and Hydradephaga (Coleoptera): a monophyletic unit? *Pan-Pacific Entomologist* 57:273–285.

Roussel, J.-P. 1964a. Le développement larvaire de *Necrophorus vespillo* L. *Bulletin de la Société Zoologique de France* 89:102–110.

Roussel, J. P. 1964b. Le dévelopment larvaire de *Necrophorus fossor* Er. *Bulletin de la Société zoologique de France* 89:111–117.

Rozen, J. G. 1957. Biological notes on *Eucrada humeralis* (Melsheimer) (Anobiidae). *Coleopterists Bulletin* 11:53–54.

Rozen, J. G. 1958. The external anatomy of the larva of *Nacerdes melanura* (Linnaeus) (Coleoptera: Oedemeridae). *Annals of the Entomological Society of America* 51:222–229.

Rozen, J. G. 1959. Systematic study of the pupae of the Oedemeridae (Coleoptera). *Annals of the Entomological Society of America* 52:299–303.

Rozen, J. G. 1960. Phylogenetic-systematic study of larval Oedemeridae. *Miscellaneous Publications of the Entomological Society of America* 1:35–68.

Rozen, J. G. 1963a. Preliminary systematic study of the pupae of the Nitidulidae (Coleoptera). *American Museum Novitates* 2124:1–13.

Rozen, J. G. 1963b. Two pupae of the primitive suborder Archostemata (Coleoptera). *Proceedings of the Entomological Society of Washington* 65:307–310.

Rozen, J. G. 1971. *Micromalthus debilis* LeConte from amber of Chiapas, Mexico (Coleoptera, Micromalthidae). *University of California Publications in Entomology* 63:75–76.

Ruckes, H., Jr. 1958. Observations on two species of pine cone feeding deathwatch beetles in California (Coleoptera: Anobiidae). *Annals of the Entomological Society of America* 51(2):186–188.

Rudolph, K. 1974. Beitrag zur Kenntnis der Elateridenlarven der Fauna der DDR und der BRD (eine morphologisch-taxonomische Studie). *Zoologische Jahrbücher, Abteilung für Systematik, Ökologie und Geographie der Tiere* 101:1–151.

Rudolph, K. 1982. Beiträge zur Insektenfauna der DDR: Coleoptera—Elateridae. *Faunistische Abhandlungen* 10:1–109.

Rusek, J. 1973. *Dryops rudolfi* sp. n. und seine Larve (Coleoptera, Dryopidae). *Acta Entomologica Bohemoslovaka* 70:86–97.

Russell, L. K. 1979. Beetles associated with slime molds (Mycetozoa) in Oregon and California (Coleoptera: Leiodidae, Sphindidae, Lathridiidae). *Pan-Pacific Entomologist* 55:1–9.

Rybalko, A. D. 1966. Ecological-faunistic survey of Bruchidae in the steppe zone of the Ukraine. *Zoologicheskii Zhurnal* 45(10):1493–1503. (in Russian)

Saalas, U. 1913. Die Larven der *Stenotrachelus aeneus* Payk. und *Upis ceramboides* L. Sowie die Puppe der letzteren. *Acta Societatis pro Fauna et Flora Fennica* 37(8):1–12, 2 pls.

Saalas, U. 1917. Die Fichtenkäfer Finnlands. I. Allgemeiner Teil und Spezieller Teil 1. *Annales Academiae Scientiarum Fennicae* (A)8:547 pp.

Saalas, U. 1923. Die Fichtenkäfer Finnlands. II. Spezieller Teil 2 und Larvenbestimmungstab. *Annales Academiae Scientiarum Fennicae* (A)22:746 pp.

Sailsbury, M. B. 1943. The comparative morphology and taxonomy of some larval Criocerinae (Coleoptera, Chrysomelidae). *Bulletin of the Brooklyn Entomological Society* 38(4):129–139.

St. George, R. A. 1924. Studies on the larvae of North American beetles of the subfamily Tenebrioninae with a description of the larva and pupa of *Merinus laevis* (Olivier). *Proceedings of the United States National Museum* 65:1–22, pls. 1–3.

St. George, R. A. 1926. Taxonomic studies of the larvae of the genera *Tenebrio* and *Neatus* LeConte (Coleoptera: Tenebrionidae). *Proceedings of the Entomological Society of Washington* 28:102–111.

St. George, R. A. 1930. The discovery of what is possibly the larva of an introduced tenebrionid, *Leichenum variegatum* Küst. *Proceedings of the Entomological Society of Washington* 32(7):122–123, pl. 8.

St. George, R. A. 1931. The larva of *Boros unicolor* Say and the systematic position of the family Boridae Herbst. *Proceedings of the Entomological Society of Washington* 33:103–115.

St. George, R. A. 1939. The larva of *Perimylops antarcticus* (Müller) and the systematic position of the family Perimylopidae (Coleoptera). *Proceedings of the Entomological Society of Washington* 41(7):207–214, pl. 27.

St. George, R. A. 1940. A note concerning the larva of a beetle, *Boros schneideri* (Panzer), a European species. *Proceedings of the Entomological Society of Washington* 42:68–73.

St. George, R. A. 1950. Family Alleculidae (= Cistelidae). Family Tenebrionidae. Family Lagriidae. pp. 215–220 *in* Craighead, F. C., Insect enemies of eastern forests. *United States Department of Agriculture Miscellaneous Publications,* No. 657, 679 pp, 197 figs.

Sakai, M. 1980. A new genus of Eucinetidae from Japan (Coleoptera). *Transactions of the Shikoku Entomological Society* 15:83–85.

Samuelson, G. A. 1964. Insects of Campbell Island. Appendix. Coleoptera: Hydraenidae, Leptodiridae (larvae). *Pacific Insects Monograph* 7:624–627.

Samuelson, G. A. 1966. Epizoic symbiosis: a new papuan colydiid beetle with epicuticular growth of cryptogamic plants (Coleoptera: Colydiidae). *Pacific Insects* 8(1):290–293.

Sanborne, M. 1981. Biology of *Ithycerus noveboracensis* (Forster) (Coleoptera) and weevil phylogeny. *Evolutionary Monographs* 4:1–80.

Sanderson, M. W. 1948. Larval and adult stages of North American *Physonota* (Chrysomelidae). *Annals of the Entomological Society of America* 41(4):468–477.

Sanderson, M. W. 1953. A revision of the Nearctic genera of Elmidae (Coleoptera). [Part]. *Journal of the Kansas Entomological Society* 26(4):148–163.

Sanderson, M. W. 1954. A revision of the Nearctic genera of Elmidae (Coleoptera). [Part]. *Journal of the Kansas Entomological Society* 27(1):1–13.

Sanderson, M. W. 1982. Gyrinidae, pp. 10.29–10.38 *in* Brigham, A. R., Brigham, W. U. and Gnilka, A. (edits.) *Aquatic Insects and Oligochaetes of North Carolina.* Mahomet, Illinois: Midwest Aquatic Enterprises, 837 pp.

Sasaji, H. 1968a. Description of the coccinellid larvae of Japan and the Ryukyus (Coleoptera). *Memoirs of the Faculty of Education, Fukui University Series II (Natural Sciences)* 18:93–136.

Sasaji, H. 1968b. Phylogeny of the family Coccinellidae (Col.). *Etizenia* 35:1–37.

Sasaji, H. 1971. *Fauna Japonica. Coccinellidae (Insecta: Coleoptera).* Tokyo: Academic Press of Japan, ix + 340 pp., 16 pls.

Sasaji, H. 1978a. Notes on the Japanese Endomychidae, with an establishment of a new subfamily (Coleoptera). *Memoirs of the Faculty of Education, Fukui University. II (Natural Sciences)* 28:1–31.

Sasaji, H. 1978b. On the larva of a predaceous endomychid, *Saula japonica* Gorham (Coleoptera). *Kontyu* 46:24–28.

Sato, M. 1982. Discovery of Torridincolidae (Coleoptera) in Japan. *Annotationes Zoologicae Japonenses* 55(4):276–283.

Savoiskaya, G. I. 1957. The experience of composition of key of coccinellid larvae. *Doklady VII. Nauchnoy Konferencii Tomskovo Universiteta* 3:62–63 (in Russian).

Savoiskaya, G. I. 1960. Morphology and taxonomy of coccinellid larvae from South-east Kasakhstan. *Entomologicheskoye Obozreniye* 29:122–133 (in Russian).

Savoiskaya, G. I. 1962. Coccinellidae of the tribe Chilochorini (Col. Coccinellidae). *Akademiya Nauk Kazakhskoi SSR. Trudy Instituta Zoologii* 8:189–200 (in Russian).

Savoiskaya, G. I. 1964a. Materials on morphology and taxonomy of larvae of the tribe Coccinellini (Col. Coccinellidae). *Trudy Nauchno Issledovatelskovo Instituta Zashiti Rastenii* 8:310–357 (in Russian).

Savoiskaya, G. I. 1964b. On some larvae of the tribe Coelopterini and Hyperaspini, with a description of a new species (Col. Coccinellidae). *Trudy Nauchno Issledovatelskovo Instituta Zashiti Rastenii* 8:358–370.

Savoiskaya, G. I. and Klausnitzer, B. 1973. 2. Morphology and taxonomy of the larvae with keys for their identification, pp. 36–55 *in* Hodek, I., *Biology of Coccinellidae.* The Hague: Academia, W. Junk, 260 pp.

Schaefer, L. 1953. Les premiers états du *Paracylindromorphus subuliformis* (Col. Bupr.). *Miscellanea entomologica* 47:66–69, pl. 4, figs. 1–5.

Schedl, K. E. 1962. Forstentomologie Beiträge aus dem Kongo. Räuber und Kommensalen. *Entomologische Abhandlungen und Berichte aus dem staatlichen Museum für Tierkunde in Dresden* 28:37–84.

Schedl, K. E. 1972. *Monographie der Familie Platypodidae, Coleoptera.* The Hague: W. Junk, 322 pp.

Schedl, K. E. 1978. Evolutionszentren bei den Scolytoidea. *Entomologische Abhandlungen und Berichte aus dem Staatlichen Museum für Tierkunde in Dresden* 41:311–323.

Scheerpeltz, O. and Höfler, K. 1948. *Käfer und Pilze.* Vienna: Verlag für Jugend und Volk, 351 pp., 9 pls.

Scherf, H. 1964. Die Entwicklungsstadien der Mitteleuropaischen Curculioniden (Morphologie, Bionomie, Ökologie). *Abhandlungen herausgegeben von der Senckenbergischen naturforschenden Gesellschaft* 506, 335 pp., 497 figs.

Scherf, H. 1978. Bestimmungtabelle für die Larven der Curculionidae (partim), pp. 344–367 *in* Klausnitzer, B., *Ordnung Coleoptera (Larven).* The Hague: W. Junk, vi + 378 pp.

Schicha, E. 1967. Morphologie und Funktion der Malachiiden-mundwerkzeuge unter besonderer Berücksichtigung von *Malachius bipustulatus* L. (Coleopt., Malacodermata). *Zeitschrift für Morphologie und Okologie der Tiere* 60:376–433.

Schiödte, J. C.1862. De metamorphosi Eleutheratorum observations: bidrag til insekternes udviklingshistorie. *Naturhistorisk Tidsskrift* (3)1:193–232, pls. 1–10.

Schiödte, J. C. 1864. De metamorphosi Eleutheratorum observations: bidrag til insekternes udviklingshistorie. *Naturhistorisk Tidsskrift* (3)3:131–224, pls. 1–12.

Schiödte, J. C. 1866. De metamorphosi Eleutheratorum observations: bidrag til insekternes udviklingshistorie. *Naturhistorisk Tidsskrift* (3)4: pls. 12–14.

Schiödte, J. C. 1867. De metamorphosi Eleutheratorum observations: bidrag til insekternes udviklingshistorie. *Naturhistorisk Tidsskrift* (3)4:415–552, pls. 15–22.

Schiödte, J. C. 1869. De metamorphosi Eleutheratorum observations: bidrag til insekternes udviklingshistorie. *Naturhistorisk Tidsskrift* (3)6:353–378, pls. 1–2.

Schiödte, J. C. 1870. De metamorphosi Eleutheratorum observations: bidrag til insekternes udviklingshistorie. *Naturhistorisk Tidsskrift* (3)6:467–536, pls. 3–10.

Schiödte, J. C. 1872. De metamorphosi Eleutheratorum observations: bidrag til insekternes udviklingshistorie. *Naturhistorisk Tidsskrift* (3)8:165–226, pls. 1–9.

Schiödte, J. C. 1873. De metamorphosi Eleutheratorum observations: bidrag til insekternes udviklingshistorie. *Naturhistorisk Tidsskrift* (3)8:545–564, pls. 18–20.

Schiödte, J. C. 1874. De metamorphosi Eleutheratorum observations: bidrag til insekternes udviklingshistorie. *Naturhistorisk Tidsskrift* (3)9:227–376, pls. 8–19.

Schiödte, J. C. 1876. De metamorphosi Eleutheratorum observations: bidrag til insekternes udviklingshistorie. *Naturhistorisk Tidsskrfit* (3)10:369–458, pls. 12–18.

Schiödte, J. C. 1878. De metamorphosi Eleutheratorum observations: bidrag til insekternes udviklingshistorie. *Naturhistorisk Tidsskrift* (3)11:479–598, pls. 5–12.

Schiödte, J. C. 1880. De metamorphosi Eleutheratorum observations: bidrag til insekternes udviklingshistorie. *Naturhistorisk Tidsskrift* (3)12:513–598, pls. 14–18.

Schiödte, J. C. 1883. De metamorphosi Eleutheratorum observations: bidrag til insekternes udviklingshistorie. *Naturhistorisk Tidsskrift* (3)13:415–426, pl. 18.

Schjøtz-Christensen, B. 1965. Biology and population studies of Carabidae of the Corynephoretum. *Natura Jutlandica* 11:1–173.

Schmidt, H. 1971. Struktur und Funktion der Mandibeln holzminierender Insekten. *Proceedings of the 13th International Congress of Entomology* 1:296–297.

Schöning, R. V. 1953. Biologisch-ökologische Untersuchungen an *Byturus tomentosus* Fabr. und *fumatus* Fabr. *Beiträge zur Entomologie* 3:627–652.

Schremmer, F. 1960. Beitrag zur Biologie von *Ditomus clypeatus* Rossi, eines körnersammelnden Carabiden. *Zeitschrift der Arbeitsgemeinschaft Oesterreichischer Entomologen* 12:140–145.

Schulze, L. 1962. The Tenebrionidae of southern Africa. XXXIII. Descriptive notes on the early stages of *Onymacris rugatipennis* Haag and *Lepidochora discoidalis* Gebien and keys to genera and species of Adesmiini and Eurychorini. *Annals of the Transvaal Museum* 24(2–3):161–180, pls. 37–41.

Schulze, L. 1963. The Tenebrionidae of Southern Africa. XXXVIII. On the morphology of the larvae of Stizopina (Coleoptera: Opatrini). *Scientific Papers of the Namib Desert Research Station* 19:1–23.

Schulze, L. 1964a. The Tenebrionidae of Southern Africa. XXXIV.—Descriptive notes and key to the larvae of *Zophodes fitzsimonsi* Koch, *Bantodemus zulu* Koch and *Quadrideres femineus* (Lesne) (Coleoptera: Platynotina s.str.). *Journal of the Entomological Society of South Africa* 26:441–451.

Schulze, L. 1964b. The Tenebrionidae of Southern Africa. XXXIX.—A revised key to the larvae of *Onymacris* Allard (Coleoptera: Adesmiini). *Scientific Papers of the Namib Desert Research Station* 23:1–7.

Schulze, L. 1968. The Tenebrionidae of Southern Africa. XLI. Descriptive notes on the early stages of four genera of the Drosochrini (Coleoptera). *Annals of the Transvaal Museum* 26:31–51.

Schulze, L. 1969. The Tenebrionidae of Southern Africa. Part XLII: Description of the early stages of *Carchares macer* Pascoe and *Herpiscius sommeri* Solier with a discussion of some phylogenetic aspects arising from the incongruities of adult and larval systematics. *Scientific Papers of the Namib Desert Research Station* 53:139–149.

Schulze, L. 1978. The Tenebrionidae of Southern Africa. XLV. Description of some larvae of the subgenera *Gonopus* and *Agonopus* of the genus *Gonopus* (Coleoptera). *Annals of the Transvaal Museum* 31:1–16.

Schuster, J. C. 1978. Biogeographical and ecological limits of New World Passalidae (Coleoptera). *Coleopterists Bulletin* 32:21–28.

Schuster, J. C. and Reyes-Castillo, P. 1981. New World genera of Passalidae (Coleoptera): a revision of larvae. *Anales de la Escuela Nacional de Ciencias Biologicas, Mexico City* 25:79–116.

Schuster, R. 1966a. Über den Beutefang des Ameisenkäfers *Cephennium austriacum* Reitter. *Naturwissenschaften* 53:113.

Schuster, R. 1966b. Scydmaeniden-Larven als Milben-Räuber. *Naturwissenschaften* 53:439–440.

Schwalb, H. H. 1960. Beiträge zur Biologie der einheimischen Lampyriden *Lampyris noctiluca* Geoffr. und *Phausis splendidula* LeC. und experimentelle Analyse ihres Beutefang- und Sexualverhaltens. *Zoologische Jahrbücher, Abteilung für Systematik, Ökologie und Geographie der Tiere* 88:399–550.

Schwarz, E. A. 1878. The Coleoptera of Florida. *Proceedings of the American Philosophical Society* 17:353–472.

Scott, A. C. 1936. Haploidy and aberrant spermatogenesis in a coleopteran, *Micromalthus debilis* LeConte. *Journal of Morphology* 59:485–515.

Scott, A. C. 1938. Paedogenesis in the Coleoptera. *Zeitschrift für Morphologie und Ökologie der Tiere* 33:633–653.

Scott, A. C. 1941. Reversal of sex production in *Micromalthus*. *Biological Bulletin* 81:420–431.

Scott, H. 1919. Notes on the biology of *Necrobia ruficollis* Fabr. (Coleoptera, Cleridae). *Annals of Applied Biology* 6(23):101–115.

Scott, H. 1933. General conclusions regarding the insect fauna of the Seychelles and adjacent islands. The Percy Sladen trust expedition to the Indian Ocean in 1905 . . . Vol. 8, No. 12. *Transactions of the Linnean Society of London. Series 2. Zoology* 19(3):307–309, pls. 17–23.

Seeger, W. 1971. Morphologie, Bionomie und Ethologie von Halipliden, unter besonderer Berücksichtigung funktions-morphologischer Gesichtspunkte (Haliplidae; Coleoptera). *Archiv für Hydrobiologie* 68:400–435.

Seely, M. K. 1976. Fog basking by the Namib Desert beetle, *Onymacris unguicularis*. *Nature* 262:284–285.

Seely, M. K. and Hamilton, W. J. 1976. Fog catchment sand trenches constructed by tenebrionid beetles, *Lepidochora*, from the Namib Desert. *Science* 193:484–486.

Seeno, T. N. and Wilcox, J. A. 1982. Leaf beetle genera (Coleoptera: Chrysomelidae). *Entomography* 1:1–221.

Seevers, C. H. 1957. A monograph of the termitophilous Staphylinidae (Coleoptera). *Fieldiana Zoology* 40:1–334.

Seevers, C. H. 1965. The systematics, evolution and zoogeography of staphylinid beetles associated with army ants (Coleoptera, Staphylinidae). *Fieldiana Zoology* 47:139–351.

Seevers, C. H. 1978. A generic and tribal revision of the North American Aleocharinae (Coleoptera, Staphylinidae). *Fieldiana Zoology* 71:i-vi, 1–289.

Seevers, C. H. and Dybas, H. S. 1943. A synopsis of the Limulodidae (Coleoptera): a new family proposed for myrmecophiles of the subfamilies Limulodinae (Ptiliidae) and Cephaloplectinae (Staphylinidae). *Annals of the Entomological Society of America* 36:546–586.

Seidlitz, G. von 1920. *Naturgeschichte der Insecten Deutschlands. Abteilung I., Coleoptera. Band 5 (part 2). Lief 3.* Berlin: Nicolai, pp. 969–1206.

Selander, R. B. 1957. The systematic position of the genus *Nephrites* and the phylogenetic relationships of the higher groups of Rhipiphoridae (Coleoptera). *Annals of the Entomological Society of America* 50:88–103.

Selander, R. B. 1964. Sexual behavior in blister beetles (Coleoptera: Meloidae). I. The genus *Pyrota*. *Canadian Entomologist* 96:1037–1082.

Selander, R. B. 1966. A classification of the genera and higher taxa of the meloid subfamily Eleticinae (Coleoptera). *Canadian Entomologist* 98:449–481.

Selander, R. B. 1981. Evidence for a third type of larval prey in blister beetles (Coleoptera, Meloidae). *Journal of the Kansas Entomological Society* 54(4):757–783.

Selander, R. B. and Mathieu, J. M. 1964. The ontogeny of blister beetles (Coleoptera, Meloidae). I. A study of three species of the genus *Pyrota*. *Annals of the Entomological Society of America* 57:711–732.

Selander, R. B. and Weddle, R. C. 1969. The ontogeny of blister beetles (Coleoptera, Meloidae). II. The effects of age of triungulin larvae at feeding and temperature on development in *Epicauta segmenta*. *Annals of the Entomological Society of America* 62:27–39.

Selander, R. B. and Weddle, R. C. 1972. The ontogeny of blister beetles (Coleoptera, Meloidae). III. Diapause termination in coarctate larvae of *Epicauta segmenta*. *Annals of the Entomological Society of America* 65:1–17.

Semenov-Tian-Shansky, A. and Dobzhansky, T. 1927. Die Larve von *Silphopsyllus desmanae* Ols., Parasit der Moschuratte, als Kriterium seiner genetischen Beziehungen und seiner systematischen Stellung. *Revue russe d'entomologie* 21:8–16.

Sen Gupta, T. 1967. A new subfamily of Languriidae (Coleoptera) based on four genera, with a key to the species of *Toramus*. *Proceedings of the Royal Entomological Society of London* (B)36:167–176.

Sen Gupta, T. 1968a. Review of the genera of the tribe Loberini (Coleoptera: Languriidae). *Breviora* 303:1–27.

Sen Gupta, T. 1968b. Revision of the genera of Cladoxenini (= Cladoxeninae Arrow) and Thallisellini trib. nov. of the family Languriidae (Coleoptera: Clavicornia). *Journal of Natural History* 2:463–475.

Sen Gupta, T. 1969. On the taxonomy of Erotylidae (Insecta: Coleoptera: Clavicornia), with descriptions of two new larvae. *Proceedings of the Zoological Society of Calcutta* 22:97–107.

Sen Gupta, T. 1979. A new subfamily of Merophysiidae (Clavicornia: Coleoptera) and descriptions of two new species of *Gomya* Dajoz and its larva. *Revue suisse de zoologie* 86:691–698.

Sen Gupta, T. and Crowson, R. A. 1966. A new family of cucujoid beetles, based on six Australian and one New Zealand genera. *Annals and Magazine of Natural History* (13)9:61–85.

Sen Gupta, T. and Crowson, R. A. 1967. The systematic position of *Eicolyctus* Sahlberg (Coleoptera: Languriidae). *Proceedings of the Royal Entomological Society of London* (B)36:87–93.

Sen Gupta, T. and Crowson, R. A. 1969a. Further observations on the family Boganiidae, with definition of two new families Cavognathidae and Phloeostichidae. *Journal of Natural History* 3:371–590.

Sen Gupta, T. and Crowson, R. A. 1969b. On a new family of Clavicornia (Coleoptera) and a new genus of Languriidae. *Proceedings of the Royal Entomological Society of London* (B)38:125–131.

Sen Gupta, T. and Crowson, R. A. 1971. A review of classification of the family Languriidae (Coleoptera: Clavicornia) and the place of Languriidae in the natural system of Clavicornia. *Memoirs of the Zoological Survey of India* 15(2):1–42.

Sen Gupta, T. and Crowson, R. A. 1973. A review of the classification of Cerylonidae (Coleoptera, Clavicornia). *Transactions of the Royal Entomological Society of London* 124:365–446.

Sen Gupta, T. and Crowson, R. A. 1979. The coleopteran family Sphindidae. *Entomologist's Monthly Magazine* 113:177–191.

Shands, W. A., R. L. Holmes and G. W. Simpson. 1966. Techniques for massproducing *Coccinella septempunctata. Journal of Economic Entomology* 59:102–103

Shands, W. A., G. W. Simpson, H. E. Wave, and C. C. Gordon. 1972. Importance of arthropod predators in controlling aphids on potatoes in northeastern Maine. *Technical Bulletin of University of Maine at Orono* 54:1–49.

Sharma, M. L. and Martel, P. 1972. Some notes on *Microweisea marginata* (LeConte) (Coleoptera: Coccinellidae), predator of the pine needle scale *Phenacaspis pinifoliae* (Fitch.). *Folia entomologica mexicana* 23–24:122–123.

Sharova, I. K. 1976. Fundamental forms of ground beetle larvae (Coleoptera, Carabidae) and their evolutionary interrelationships, pp. 56–80 *in* Mamaev, B. M. (edit.), *Evolutionary Morphology of Insect Larvae.* Moscow: "Nauka," 204 pp. (in Russian).

Sharova, I. K. and Makarov, K. V. 1984. On the larvae of the carabid genus *Broscus* Panz. (Coleoptera, Carabidae) (in Russian). *Entomologicheskoe Obozrenie* 63:742–750.

Shenefelt, R. D. and Simkover, H. G. 1950. White grubs in Wisconsin forest tree nurseries. *Journal of Forestry* 48:429–434.

Shepard, W. D. 1976. Records and notes concerning *Derodontus maculatus* (Mels.) (Coleoptera: Derodontidae). *Southwestern Entomologist* 1:168–170.

Silvestri, F. 1905. Contribuzione alla conoscenza della metamorfosi e dei costumi della *Lebia scapularis* Fourc. con descrizione dell'apparato sericiparo della larva. *Redia* 2(1904):68–84, pls. 3–7.

Silvestri, F. 1906. Descrizione di un nuovo genere di Rhipiphoridae. *Redia* 3(2):315–324, pl. 20.

Silvestri. F. 1910. Metamorfosi del *Cybocephalus rufifrons* Reitter e notize sui suoi costumi. *Bolletino del Laboratorio di Zoologia Generale e Agraria della R. Scuola Superiore d'Agricoltura* 4:221–227.

Silvestri, F. 1912. Contribuzione alla conoscenza dei mirmecofili. II. Di alcuni mirmecofili dell'Italia meridionale e della Sicilia. *Bolletino del Laboratorio di Zoologia Generale e Agraria della R. Scuola Superiore d'Agricoltura* 6:222–245.

Silvestri, F. 1920. Contribuzione alla conoscenza dei termitidi e termitofili dell'Africa occidentale. II. Termitofili. Parte seconda. *Bolletino del Laboratorio di Zoologia Generale e Agraria della Facolta Agraria in Portici* 14:265–319.

Silvestri, F. 1941. Distribuzione geografica del *Micromalthus debilis* LeConte (Coleoptera, Micromalthidae). *Bolletino della Societa Entomologica Italiana* 73:1–2.

Simanton, F. L. 1916. *Hyperaspis binotata,* a predatory enemy of terrapin scale. *Journal of Agricultural Research* 6:197–203.

Simmonds, F. J. 1956. An investigation of the possibilities of biological control of *Melittomma insulare* Fairm. (Coleoptera, Lymexylonidae), a serious pest of coconut in the Seychelles. *Bulletin of Entomological Research* 47:685–702.

Singh, T., Kaur, I., and Saini, M. S. 1979. Biology of *Zabrotes subfasciatus* (Boh.) (Bruchidae: Coleoptera). *Entomon* 4(2):201–203.

Sivinsky, J. 1981. The nature and possible functions of luminescence in coleopterous larvae. *Coleopterists Bulletin* 35:167–179.

Skopin, N. G. 1959. Über die Larven der Gattung *Adesmia* Fisch. (Coleoptera, Tenebrionidae). *Acta Zoologica Academiae Scientiarum Hungaricae* 5:393–400. (In Russian, German summary).

Skopin, N. G. 1960a. Material on the morphology and ecology of larvae of the tribe Blaptini. *Trudy Instituta Zoologii, Akademia Nauk Kazakhskoi SSR* 11:36–71, 13 pls. (In Russian).

Skopin, N. G. 1960b. Über die Larven der Tribus Akidini (Coleoptera, Tenebrionidae). *Acta Zoologica Academiae Scientiarum Hungaricae* 6:149–165.

Skopin, N. G. 1961. Über die Larven der Unterfamilie Erodiinae (Coleoptera, Tenebrionidae). *Annales Historico-Naturales Musei Nationalis Hungarici* 53:407–413. (In Russian, German summary).

Skopin, N. G. 1962. Larvae of the subfamily Pimeliinae (Coleoptera, Tenebrionidae). *Trudy Nauchno-Issledovatelskovo Instituta Zachity Rastenii (Kazakhstan)* 7:191–298. (In Russian).

Skopin. N. G. 1964. Die Larven der Tenebrioniden des Tribus Pycnocerini (Coleoptera: Heteromera). *Annales du Musee Royal de l'Afrique Centrale, Tervuren, Serie in 8, Sciences Zoologiques* 127:1–35, 16 pls.

Skopin, N. G. 1978. Tenebrionidae, pp. 223–248 *in* Klausnitzer, B., *Ordnung Coleoptera (Larven).* The Hague: W. Junk, vi + 378 pp.

Slifer, E. H., Gruenwals, T. F. J., and Sekhon, S. S. 1975. The maxillary palp organs of a wood-boring beetle, *Melittomma sericeum* (Coleoptera, Lymexylonidae). *Journal of Morphology* 147:123–136.

Slipinski, S. A. 1984. Notes on the Lapethini with a revision of the World *Lapethus* Casey (Coleoptera, Cerylonidae), including descriptions of related genera. *Polskie Pismo Entomologiczne* 54:3–104.

Smetana, A. 1971. Revision of the tribe Quediini of America north of Mexico (Coleoptera: Staphylinidae). *Memoirs of the Entomological Society of Canada* 79:1–303.

Smith. B. C. 1958. Notes on relative abundance and variation in elytral patterns of some common coccinellids in the Belleville district (Col.: Coccinellidae). *Report of the Entomological Society of Ontario* 888:59–60.

Smith, B. C. 1960. A technique for rearing coccinellid beetles on dry foods, and influence of various pollens on development of *Coleomegilla maculata lengi* Timb. (Coleoptera: Coccinellidae). *Canadian Journal of Zoology* 38:1047–1049.

Smith, B. C. 1965. Differences in *Anatis mali* Auct. and *Coleomegilla maculata lengi* Timberlake to changes in the quality of larval food (Coleoptera: Coccinellidae). *Canadian Entomologist* 97:1159–1166.

Smith, D. B. and Sears, M. K. 1982. Mandibular structure and feeding habits of three morphologically similar coleopterous larvae: *Cucujus clavipes* (Cucujidae), *Dendroides canadensis* (Pyrochroidae), and *Pytho depressus* (Salpingidae). *Canadian Entomologist* 114:173–175.

Smith, G. L. and E. A. Michelbacher. 1944. The pea weevil. *Calif. Agr. Expt. Sta. Leaflet.* 140:1–2.

Smrz, J. 1979. Über die Larven der *Bembidion (Peryphus)*-Arten (Coleoptera, Carabidae, Bembidiinae). *Acta entomologica bohemoslovaca* 76:244–254.

Snodgrass, R. E. 1935. *Principles of Insect Morphology.* New York: McGraw-Hill, ix + 667 pp.

Snodgrass, R. E. 1947. The insect cranium and the "epicranial suture." *Smithsonian Miscellaneous Collections* 107(7), 52 pp.

Snyder, T. E. 1910. Insects injurious to forests and forest products. Damage to chestnut telephone and telegraph poles by wood-boring insects. *United States Department of Agriculture, Bureau of Entomology Bulletin* 94(1):1–12, pls. 1–2.

Snyder, T. E. 1916. Egg and manner of oviposition of *Lyctus planicollis*. *Journal of Agricultural Research* 6:273–276, pls. 28–31.

Snyder, T. E. 1919. Injury to Casuarina trees in southern Florida by the mangrove borer *Chrysobothris tranquebarica* Gmelin, (Buprestidae). *Journal of Agricultural Research* 16:155.

Soares, B. A. M. and Perracchi, A. L. 1964. Sobre a presença de cerofitidas no Brasil, com a descricão de duas novas especies (Insecta: Coleoptera, Cerophytidae). *Anais do II Congresso Latino-Americano de Zoologia, S. Paulo,* 1962 1:127–134.

Sokoloff, A. 1959. The habitat-niche of American Nosodendridae. *Coleopterists Bulletin* 13:97–98.

Southgate, B. J. 1979. Biology of the Bruchidae. *Annual Review of Entomology* 24:449–474.

Spangler, P. J. 1961. Notes on the biology and distribution of *Sperchopsis tessellatus* (Ziegler) (Coleoptera: Hydrophilidae). *Coleopterists Bulletin* 15:105–112.

Spangler, P. J. 1962a. Biological notes and descriptions of the larva and pupa of *Copelatus glyphicus* (Say) (Coleoptera: Dytiscidae). *Proceedings of the Biological Society of Washington* 75:19–24.

Spangler, P. J. 1962b. Description of the larva of *Hydrovatus cuspidatus pustulatus* Melsheimer (Coleoptera: Dytiscidae). *Coleopterists Bulletin* 17:97–100.

Spangler, P. J. 1962c. Description of the larva and pupa of *Ametor scabrosus* (Horn) (Coleoptera: Hydrophilidae). *Coleopterists Bulletin* 16:16–20.

Spangler, P. J. 1966a. A description of the larva of *Derallus rudis* Sharp (Coleoptera: Hydrophilidae). *Coleopterists Bulletin* 20:97–103.

Spangler, P. J. 1966b. A new species of *Derovatellus* from Guatemala and a description of its larva (Coleoptera: Dytiscidae). *Coleopterists Bulletin* 20:11–18.

Spangler, P. J. 1966c. Results of the Catherwood Foundation Peruvian–Amazon Expedition. Insects, Part XIII, The aquatic Coleoptera (Dytiscidae; Noteridae; Gyrinidae; Hydrophilidae; Dascillidae; Helodidae; Psephenidae; Elmidae). *Monographs of the Academy of Natural Sciences of Philadelphia* 14:377–443.

Spangler, P. J. 1973. The bionomics, distribution, and immature stages of the rare predaceous water beetle, *Hoperius planatus* (Coleoptera: Dytiscidae). *Proceedings of the Biological Society of Washington* 86:423–434.

Spangler, P. J. 1974a. A description of the larva of *Celina angustata* Aubé (Coleoptera: Dytiscidae). *Journal of the Washington Academy of Sciences* 63(1973):165–168.

Spangler, P. J. 1974b. A description of the larva of *Hydrobiomorpha casta* (Coleoptera: Hydrophilidae). *Journal of the Washington Academy of Sciences* 63(1973):160–164.

Spangler, P. J. 1980a. Chelonariid larvae, aquatic or not? *Coleopterists Bulletin* 34:105–114.

Spangler, P. J. 1980b. A new species of *Ytu* from Brazil (Coleoptera: Torridincolidae). *Coleopterists Bulletin* 34(2):145–158.

Spangler, P. J. 1981. Coleoptera, pp. 129–220 *in* Hurlbert, S. H., Rodriquez, G., and Santos, N. D. (eds).) *Aquatic Biota of Tropical South America, Part 1. Arthropoda*. San Diego, California: San Diego State University, xii + 323 pp.

Spangler, P. J. 1982. Coleoptera, pp. 328–397 *in* Hurlbert, S. H. and Villalobos-Figueroa, A. (edits.) *Aquatic Biota of Mexico, Central America, and the West Indies*. San Diego, California: San Diego State University, xv + 529 pp.

Spangler, P. J. 1983. Immature stages and biology of *Tetraglossa palpalis* Champion (Coleoptera: Ptilodactylidae). *Entomological News* 94(5):161–175.

Spangler, P. J. 1986. Insecta: Coleoptera, pp. 622–631. *In* Botosaneanu, Lazare (ed.). *Stygofauna Mundi. A faunistic, distributional, and ecological synthesis of the world fauna inhabiting subterranean waters (including the marine interstitial)*. E. J. Brill/Dr. W. Backhuys: Leiden.

Spangler, P. J. and Cross, J. L. 1972. A description of the egg case and larva of the water scavenger beetle *Helobata striata* (Coleoptera: Hydrophilidae). *Proceedings of the Biological Society of Washington* 85:413–418.

Spangler, P. J. and Folkerts, G. W. 1973a. The larva of *Pachydrus princeps* (Coleoptera: Dytiscidae). *Proceedings of the Biological Society of Washington* 86:351–356.

Spangler, P. J. and Folkerts, G. W. 1973b. Reassignment of *Colpius inflatus* and a description of its larva (Coleoptera: Noteridae). *Proceedings of the Biological Society of Washington* 86:301–310.

Spangler, P. J. and Gillespie, J. M. 1973. The larva and pupa of the predaceous water beetle *Hygrotus sayi* (Coleoptera: Dytiscidae). *Proceedings of the Biological Society of Washington* 86:143–152.

Spangler, P. J. and Gordon, R. D. 1973. Descriptions of larvae of some predaceous water beetles (Coleoptera: Dytiscidae). *Proceedings of the Biological Society of Washington* 86:261–278.

Spilman, T. J. 1952. The male genitalia of the Nearctic Salpingidae. *Coleopterists Bulletin* 6:9–13.

Spilman, T. J. 1959. A study of the Thaumastodinae, with one new genus and two new species (Limnichidae). *Coleopterists Bulletin* 13:111–122.

Spilman, T. J. 1961. On the immature stages of the Ptilodactylidae (Coleoptera). *Entomological News* 72:105–107.

Spilman, T. J. 1966. Larva and pupa of *Amarygmus morio* from Hawaii (Coleoptera: Tenebrionidae). *Proceedings of the Hawaiian Entomological Society* 19(2):297–301.

Spilman, T. J. 1967. The heteromerous intertidal beetles (Coleoptera: Salpingidae: Aegialitinae). *Pacific Insects* 9(1):1–21.

Spilman, T. J. 1971. Bredin-Archbold-Smithsonian biological survey of Dominica: Bostrichidae, Inopeplidae, Lagriidae, Lyctidae, Lymexylonidae, Melandryidae, Monommidae, Rhipiceridae, and Rhipiphoridae (Coleoptera). *Smithsonian Contributions to Zoology,* No. 70, 10 pp.

Spilman, T. J. 1972. A new genus and species of jumping shore beetle from Mexico (Coleoptera: Limnichidae). *Pan-Pacific Entomologist* 48:108–115.

Spilman, T. J. 1979. Larvae and pupae of *Centronopus calcaratus* and *Centronopus suppressus* (Coleoptera: Tenebrionidae), with an essay on wing development in pupae. *Proceedings of the Entomological Society of Washington* 81(4):513–521.

Spilman, T. J. (In press). Darkling beetles (Tenebrionidae, Coleoptera), *in* Gorham, J. R., Insect and mite pests in food: an illustrated key. *United States Department of Agriculture, Agriculture Handbook*.

Spilman, T. J. and Anderson, W. H. 1961. On the immature stages of North American Pyrochroidae. *Coleopterists Bulletin* 15:38–40.

Springer, C. A. and Goodrich, M. A. 1983. A revision of the family Byturidae for North America. *Coleopterists Bulletin* 37(2):183–192.

Srivastava, U. S. 1959. The maxillary glands of some Coleoptera. *Proceedings of the Royal Entomological Society of London* (A)34:57–62.

Steel, W. O. 1970. The larvae of the genera of Omaliinae (Coleoptera: Staphylinidae) with particular reference to the British fauna. *Transactions of the Royal Entomological Society of London* 122:1–47.

Steffan, J. R. 1945. Contribution à l'étude de *Zabrotes subfasciatus* Boheman. *Mem. de Mus. Nat. D'Histoire Naturelle.* 21(2):55–84.

Steffan, J. R. 1946. The primary larva of *Bruchidius fasciatus* and its relationship with several neonatal larvae of Bruchidae. *Soc. Ent. de France, Bull.* 51:12–16.

Steffan, W. 1964. Torridincolidae, Coleopterorum nova familia e regione aethiopica. *Entomologische Zeitschrift* 74:193–200.

Steiner, W. E. 1984. A review of the biology of phalacrid beetles (Coleoptera). pp. 424–445 *in* Wheeler, Q. D. and Blackwell, M. (eds.) *Fungus/insect relationships: perspectives in ecology and evolution.* New York: Columbia University Press, 514 pp.

Steiner, W. E. and Anderson, J. J. 1981. Notes on the natural history of *Spanglerogyrus albiventris* Folkerts, with a new distribution record (Coleoptera: Gyrinidae). *Pan-Pacific Entomologist* 57:124–132.

Steiner,W. E. and Singh, B. P. 1987. Redescription of an ergot beetle, *Acylomus pugetanus* Casey, with immature stages and biology (Coleoptera: Phalacridae). *Proceedings of the Entomological Society of Washington.* 89:744–758.

Steinhausen, W. R. 1966. Vergleichende Morphologie des Labrum von Blattkäferlarven. *Deutsche Entomologische Zeitschrift,* (N.F.) 13:313–322.

Steinhausen, W. 1978. Bestimmungstabelle für die Larven der Chrysomelidae (partim), pp. 336–343 *in* Klausnitzer, B. (ed.), *Ordnung Coleoptera (Larven).* The Hague: W. Junk, vi + 378 pp.

Steinke, G. 1919. Die Stigmen der Käferlarven. *Archiv für Naturgeschichte* 85(A)7:1–58, 2 pls.

Stephan, K. 1968. Notes on additional distribution and ecology of *Euxestus punctatus* LeC. (Coleoptera: Colydiidae). *Coleopterists Bulletin* 22:19.

Stibick, J. N. L. 1979. Classification of the Elateridae (Coleoptera). Relationships and classification of the subfamilies and tribes. *Pacific Insects* 20:145–186.

Stickney, F. S. 1923. The head-capsule of Coleoptera. *Illinois Biological Monographs* 8(1), 104 pp., 8 pls.

Storch, R. H. 1970. Field recognition of the larvae of native Coccinellidae common to the potato fields of Aroostook county. *Maine Agricultural Experiment Station Technical Bulletin* 43:1–16.

Stribling, J. B. 1986. Revision of *Anchytarsus* (Coleoptera: Dryopoidea) and a key to the New World genera of Ptilodactylidae. *Annals of the Entomological Society of America* 79(1):219–234.

Striganova, B. R. 1961a. Morpho-functional characters of *Prionocyphon serricornis* Müll. (Coleoptera, Helodidae) larvae in relation to their inhabitation of water. *Entomologicheskoye Obozreniye* 40:577–583. (in Russian; translation in *Entomological Review* 40:314–317).

Striganova, B. R. 1961b. Morphological peculiarities of and identification key to alleculid-larvae of the subfamily Alleculinae (Coleoptera). *Zoologicheskii Zhurnal* 40:193–200 (in Russian).

Striganova, B. R. 1962. The larva of *Podabrus alpinus* L. and several morphological peculiarities of cantharid larvae (Col.). *Zoologicheskii Zhurnal* 41:546–551 (in Russian).

Striganova, B. R. 1964a. Family Alleculidae—Pollen Beetles, pp. 457–462 *in* Ghilarov, M. S. (edit.), *Determination Key to soil-dwelling Insect Larvae.* Moscow: "Nauka", 918 pp. (in Russian).

Striganova, B. R. 1964b. Family Byrrhidae—Pill Beetles, pp. 336–339 Ghilarov, M. S. (edit.), *Determination Key to soil-dwelling Insect Larvae.* Moscow: "Nauka", 918 pp. (in Russian).

Striganova, B. R. 1964c. Peculiarities of the structure of the mouth apparatus of the plant-eating coleopterous larvae. *Zoologicheskii Zhurnal* 43:560–571 (in Russian).

Striganova, B. R. 1967. Morphological adaptations of the head and mandibles of some coleopterous larvae burrowing solid substrates (Coleoptera). *Beiträge zur Entomologie* 17:639–649.

Striganova, B. R. 1971. Ways of the morphological specialization of the mouth apparatus in predatory coleopterous larvae. *Proceedings of the 13th International Congress of Entomology, Moscow, 1968* 1:308–309.

Strohecker, H. F. 1953. *Genera Insectorum de P. Wytsman. Fascicule 210 Coleoptera. Fam. Endomychidae.* Crainhem, Belgium: Genera Insectorum, 140 pp., 5 pls.

Struble, G. H. 1930. The biology of certain Coleoptera associated with bark beetles in western yellow pine. *University of California Publications in Entomology* 5:105–134.

Summerlin, J. W., Bay, D. E., Harris, R. L. and Russell, D. J. 1981. Laboratory observations on the life cycle and habits of two species of Histeridae (Coleoptera): *Hister coenosus* and *H. incertus. Annals of the Entomological Society of America* 74:316–319.

Summerlin, J. W., Bay, D. E., Stafford, K. C., III and Hunter, J. S., III. 1984. Laboratory observations on the life cycle and habits of *Hister abbreviatus* (Coleoptera: Histeridae). *Annals of the Entomological Society of America* 77:543–547.

Süss, L. and Puppin, O. 1976. Osservationi sulla morfologia e sulla biologia di *Ptilodactyla exotica* Chapin (Coleoptera, Ptilodactylidae) nelle serre della Lombardia e contributo bibliografico allo studio del gruppo. *Bolletino di Zoologia Agraria e di Bachicoltura* (2)13:143–165.

Svacha, P. and Danilevsky, M. L. 1987. Cerambycoid larvae of Europe and Soviet Union (Coleoptera, Cerambycoidea). Part I. *Acta Universitatis Carolinae Biologica* 30(1986):1–176.

Swezey, O. H. 1940. *Micromalthus debilis* LeConte in Hawaii (Coleoptera: Micromalthidae). *Proceedings of the Hawaiian Entomological Society* 10:459.

Szujecki, A. 1966. Notes on the appearance and biology of eggs of several Staphylinidae (Coleoptera) species. *Bulletin de l'Academie Polonaise des Sciences, Classe 2. Serie des Sciences Biologiques* 14:169–175.

Takizawa, H. 1972. Descriptions of larvae of glanduliferous group of Galerucinae in Japan, with notes on subdivisions of the subfamily (Coleoptera: Chrysomelidae). *Insecta Matsumurana. Supplement* 10, 14 pp., 8 pls.

Takizawa, H. 1976. Larvae of the genus *Gonioctena* (Coleoptera, Chrysomelidae): descriptions of Japanese species and the implications of larval characters for the phylogeny. *Kontyu* 44(4):444–468.

Takizawa, H. 1980. Immature stages of some Indian Cassidinae (Coleoptera, Chrysomelidae). *Insecta Matsumurana,* n.s., 21:19–48.

Tanigoshi, L. K. and McMurtry, J. A. 1977. The dynamics of predators of *Stethorus picipes. Hilgardia* 45:237–288.

Tashiro, H. 1987. *Turfgrass insects of the United States and Canada.* Cornell Univ. Press, Ithaca, N.Y. 391 pp.

Taylor, V. A. 1975. The biology of feather-winged beetles of the genus *Ptinella* with particular reference to co-existence and parthenogenesis. *Ph.D. Thesis. University of London,* 296 pp.

Teran, A. L. 1962. Observaciones sobre Bruchidae (Coleoptera) Del Noroeste Argentino. *Acta Zool. Lilloana.* 28:211–242.

Teran, A. L. 1967. Consideraciones sobre *Eubaptus palliatus* Lac., *Bruchus scapularis* Pic, y descripcion de los setados preimaginales de *Eubaptus rufithorax* (Pic). *Acta Zoologica Lilloana* 21:71–89.

Ter-Minasyan, M. E. 1984. A review of the weevil family Nemonychidae (Coleoptera, Rhynchophora) of the fauna of the USSR. *Entomologicheskoye Obozreniye* 63(1):105–110 (in Russian; translation in *Entomological Review,* 63(1):89–95).

Thayer, M. K. 1987. Biology and phylogenetic relationships of *Neophonus bruchi,* an anomalous south Andean staphylinid (Coleoptera: Staphylinidae). *Systematic Entomology* 12:389–404.

Thiele, H. U. 1977. Carabid beetles in their environments: a study on habitat selection by adaptations in physiology and behavior. *Zoophysiology and Ecology* 10, 369 pp.

Thomas, C. A. 1940. The biology and control of wireworms. Review of literature. *Pennsylvania State College of Agriculture and Experiment Station Bulletin* 392, 90 pp.

Thomas, H. A. 1968. A description of the pupa of the pales weevil, *Hylobius pales,* and a method for identifying its sex. *Canadian Entomologist* 100(4):434–437, 2 figs.

Thomas, J. B. 1957. The use of larval anatomy in the study of bark beetles (Coleoptera: Scolytidae). *Canadian Entomologist. Supplement* No. 5, 45 pp.

Thomas, J. B. 1965. The immature stages of Scolytidae: The genus *Dendroctonus* Erichson. *Canadian Entomologist* 97(4):374–400.

Thomas, J. B. and Herdy, H. 1961. A note on the life history of *Cimberis elongatus* (LeC.) (Coleoptera: Anthribidae). *Canadian Entomologist* 93(5):406–408.

Thomas, M. C. 1984. A new species of apterous *Telephanus* (Coleoptera: Silvanidae) with a discussion of phylogenetic relationships of Silvanidae. *Coleopterists Bulletin* 38(1):43–55.

Thomas, M. C. 1985. The species of *Hemipeplus* Latreille (Coleoptera: Mycteridae) in Florida, with a taxonomic history of the genus. *Coleopterists Bulletin* 39(4):365–375.

Thompson, R. G. 1977. A synoptic list of the described ground beetle larvae of North America (Coleoptera: Carabidae). *Proceedings of the Biological Society of Washington* 90:99–107.

Thompson, R. G. 1979a. Larvae of North American Carabidae with a key to the tribes. pp. 209–291 *in* Erwin, T. L., Ball, G. E., Whitehead, D. R. and Halpern, A. L., (edits.), *Carabid Beetles: Their Evolution, Natural History, and Classification.* The Hague: W. Junk, 635 pp.

Thompson, R. G. 1979b. A systematic study of larvae in the tribes Pterostichini, Morionini, and Amarini (Coleoptera: Carabidae). *Arkansas Agricultural Experiment Station Bulletin* 837:1–105.

Thompson, R. T. 1958. *Handbooks for the Identification of British Insects. Vol. V. Part 5(b). Coleoptera. Phalacridae.* London: Royal Entomological Society of London, 17 pp.

Thompson, R. T. and Marshall, J. E. 1980. A taxonomic study of *Phalacrus uniformis* (Coleoptera: Phalacridae), an Australian beetle now established in New Zealand. *New Zealand Journal of Zoology* 7:407–416.

Thorpe, W. H. and Crisp, D. J. 1949. Studies on plastron respiration. IV. Plastron respiration in the Coleoptera. *Journal of Experimental Biology* 26(3):219–260, pl. 7.

Tiemann, D. L. 1967. Observations on the natural history of the western banded glowworm *Zarhipis integripennis* (LeConte) (Coleoptera: Phengodidae). *Proceedings of the California Academy of Sciences* 35:235–264.

Tiemann, D. L. 1970. Nature's toy train, the railroad worm. *National Geographic* 138(1):58–67.

Tikhomirova, A. L. 1973. *Morpho-ecological characterstics and phylogeny of Staphylinidae (with a catalogue of the fauna of the U.S.S.R.).* (In Russian). Moscow: Academiya Nauk, 191 pp.

Tikhomirova, A. L. and Melnikov. O. A. 1970. The late embryogenesis of Staphylinidae and nature of aleocharo- and staphylino-morphous larvae. *Zoologischer Anzeiger* 184:76–87.

Tillyard, R. J. 1923. On the larva and pupa of the genus *Sabatinca* (Order Lepidoptera, Family Micropterygidae). *Transactions of the Entomological Society of London* 1922:437–453.

Tillyard, R. J. 1926. *The Insects of Australia and New Zealand.* Sydney: Angus and Robertson, 560 pp.

Topp, W. 1978. Bestimmungstabelle für die Larven der Staphylinidae, pp. 304–334 *in* Klausnitzer, B., *Ordnung Coleoptera (Larven).* The Hague: W. Junk, vi + 378 pp.

Torre-Bueno, J. R. de la 1937. *A Glossary of Entomology.* Brooklyn: Brooklyn Entomological Society, 336 pp., 9 pls.

Treherne, J. R. 1952. The respiration of the larvae of *Helodes minuta* (Col.). *Transactions of the IXth International Congress of Entomology, Amsterdam* 1:311–314.

Treherne, J. R. 1954. Osmotic regulation in the larvae of *Helodes* (Coleoptera: Helodidae). *Transactions of the Royal Entomological Society of London* 105:117–130.

Triplehorn, C. A. and Wheeler, Q. D. 1979. Systematic placement and distribution of *Uloporus ovalis* Casey (Coleoptera: Heteromera: Archeocrypticidae). *Coleopterists Bulletin* 33:245–250.

Tschinkel, W. R. 1975a. A comparative study of the chemical defense system of tenebrionid beetles: Chemistry of the secretions. *Journal of Insect Physiology* 21:753–783.

Tschinkel, W. R. 1975b. A comparative study of the chemical defense system of tenebrionid beetles. Defensive behavior and ancillary features. *Annals of the Entomological Society of America* 63(3):439–453.

Tschinkel, W. R. 1975c. A comparative study of the chemical defense system of tenebrionid beetles. III. Morphology of the glands. *Journal of Morphology* 145:355–370.

Tschinkel, W. R. and Doyen, J. T. 1980. Comparative anatomy of the defensive glands, ovipositors and female genital tubes of tenebrionid beetles (Coleoptera). *International Journal of Insect Morphology and Embryology* 9:321–368.

Tucker, R. W. E. 1952. The insects of Barbados. *Journal of Agriculture of the University of Puerto Rico* 36:330–363.

Tuttle, D. M. 1954. Notes on the bionomics of six species of *Apion* (Curculionidae, Coleoptera). *Annals of the Entomological Society of America* 47(2):301–307.

Ulrich, G. W. 1986. The larvae and pupae of *Helichus suturalis* LeConte and *Helichus productus* LeConte (Coleoptera: Drypoidae), *Coleopterists Bulletin* 40:325–334.

Upton, M. S. and Norris, K. R. 1980. *The Collection and Preservation of Insects and other Terrestrial Arthropods.* Brisbane, Australia: Australian Entomological Society (Miscellaneous Publication No. 3), 34 pp.

Urban, C. 1913. Beiträge zur Lebensgeschichte der Käfer. *Entomologische Blätter* 9(3/4):57–63.

Urban, C. 1926. Über die *Olibrus*-Larve (Col. Phalacr.). *Deutsche Entomologische Zeitschrift* 5:401–412.

Valentine, B. D. 1960. The genera of the weevil family Anthribidae north of Mexico (Coleoptera). *Transactions of the American Entomological Society* 86:41–85.

van Emden—*See* Emden

Vanek, S. 1984. Larvae of the Palaearctic species *Clivina collaris* and *Clivina fossor* (Coleoptera, Carabidae, Scaritini). *Acta entomologica bohemoslovaca* 81:99–112.

Vanin, S. A. 1976. Taxonomic revision of the South American Belidae (Coleoptera). *Arquivos de Zoologia* 28:1–75.

Vanin, S. A. and Costa, C. 1978. Larvae of Neotropical Coleoptera. II: Rhysodidae. *Papeis Avulsos de Zoologia* 31:195–201.

Vats, L. K. 1972. Tracheal system in the larvae of the Bruchidae (Coleoptera: Bruchidae). *Journal of the New York Entomological Society* 80:12–17.

Vats, L. K. 1974a. Distinctive characters of the larvae of three species of *Callosobruchus* Pic (Bruchidae: Coleoptera), together with a key for their identification. *Indian Journal of Entomology* 36(1):17–22.

Vats, L. K. 1974b. Distinctive characters of the larvae of two species of *Bruchidius* Schilsky (Bruchidae: Coleoptera). *Indian Journal of Entomology* 36(2):113–117.

Vaurie, P. 1948. A review of the North American Languriidae. *Bulletin of the American Museum of Natural History* 92:119–156.

Verhoeff, K. W. 1917. Zur Entwicklung, Morphologie und Biologie der Vorlarven und Larven der Canthariden. *Archiv für Naturgeschichte* (A)83:102–140.

Verhoeff, K. W. 1923. Beiträge zur Kenntnis der Coleopteren-Larven mit besonderer Berücksichtigung der Clavicornia. *Archiv für Naturgeschichte* (A)89:1–109, 7 pls.

Vesey-Fitzgerald, D. 1941. *Melittomma insulare* Fairm., a serious pest of coconut in the Seychelles. *Bulletin of Entomological Research* 31:383–398.

Viana, M. J. 1971. Micholaeminae nueva subfamilia de Rhipiphoridae y *Micholaemus gerstaeckeri* nuevo genero y especie de la Republica Argentina (Insecta, Coleoptera). *Revista de la Sociedad Entomologica Argentina* 33(1–4):69–76.

Viedma, M. G. de 1964. Larvas de Coleopteros. *Graellsia* 20:245–275.

Viedma, M. G. de 1966. Contribucion al conocimiento de las larvas de Melandryidae de Europa (Coleoptera). *Eos* 41:483–506.

Viedma, M. G. de 1971. Redescripcion de la larva de *Eustrophinus bicolor* y consideraciones acerca de la posicion sistematica del genero *Eustrophinus* (Col. Melandryidae). *Annales de la Societe Entomologique de France* (N.S.)7:729–733.

Viedma, M. G. de 1973. Definition of the subfamily Hedobiinae based on larval characteristics (Coleoptera: Anobiidae). *Great Lakes Entomologist* 6:57–58.

Viedma, M. G. de 1977. Descripcion de la larva de *Clada fernandezi* Espanol y consideraciones acerca de la subfamilia Hedobiinae (Col. Anobiidae). *Miscellanea Zoologica Barcelona* 4(1):143–146.

Vienna, P. 1980. *Fauna d'Italia, Vol. 16. Coleoptera: Histeridae.* Bologna Edizioni Calderini, ix + 386 pp.

Villiers, A. 1943. Etude morphologique et biologique des Languriidae. *Publications du Muséum national d'histoire naturelle* 6:1–98.

Villiers, A. 1978. *Faune des Coléoptères de France. I. Cerambycidae.* Paris: Lechevalier (Encyclopédie entomologique 42), xxvii + 611 pp.

Vinson, J. 1957. On two species of *Pelecophora* from Round Island with the description of a larva of this genus (Col. Dasytidae). *The Mauritius Institute Bulletin* 5(1):1–6.

Viswanathan, T. P. 1945. On the larvae of three species of Pyrochroidae (Coleoptera). *Indian Journal of Entomology* 6:121–124.

Vit, S. 1977. Contribution a la connaissance des Eucinetidae (Coleoptera). *Revue suisse de Zoologie* 84:917–935.

Voss, E. 1957. *Archimetrioxena electrica* Voss und ihre Beziehungen zu rezenten Formenkreisen. *Deutsche Entomologische Zeitschrift* (N.F.) 4(I/II):95–102.

Vulcano, M. A. and Pereira, F. S. 1975. Cupesidae (Coleoptera). *Arquivos do Instituto Biologico* 42:31–68.

Wade, J. S. 1935. *A Contribution to a Bibliography of the Described Immature Stages of North American Coleoptera.* Washington, D.C.: United States Department of Agriculture, 114 pp.

Wade, J. S. and St. George, R. A. 1923. Biology of the false wireworm *Eleodes suturalis* Say. *Journal of Agricultural Research* 36(11):547–566, 2 pls.

Wagner, J. A. 1975. Review of the genera *Euplectus, Pycnoplectus, Leptoplectus,* and *Acolonia* (Col.: Pselaphidae) including Nearctic species north of Mexico. *Entomologica Americana* 49:125–207.

Walkley, L. M. 1952. Revision of the Lathridiini of the state of Washington. *Proceedings of the Entomological Society of Washington* 54:217–235.

Walsh, G. B. and Dibb, J. R. (eds.) 1975. *A Coleopterist's Handbook. Second Edition* (revised by J. Cooter and P. W. Cribb). Feltham: Amateur Entomologist's Society, 142 pp.

Warner, R. E. 1966. A review of the *Hylobius* of North America, with a new species injurious to slash pine (Coleoptera: Curculionidae). *Coleopterists Bulletin* 20(3):65–81, 48 figs.

Wasmann, E. 1918. Über die von v. Rothkirch 1912 in Kamerun gesammelten Myrmekophilen. *Entomologische Mitteilungen* 7:135–149.

Watson, W. Y. 1976. A review of the genus *Anatis* Mulsant (Coleoptera: Coccinellidae). *Canadian Entomologist* 108(9):935–944.

Watson, W. Y. 1979. North American distribution of *Coccinella u. undecimpunctata* L. (Coleoptera: Coccinellidae). *Coleopterists Bulletin* 33:85–86.

Watt, J. C. 1967a. A review of classifications of Tenebrionidae (Coleoptera). *Entomologist's Monthly Magazine* 102:80–86.

Watt, J. C. 1967b. The families Perimylopidae and Dacoderidae (Coleoptera, Heteromera). *Proceedings of the Royal Entomological Society of London* (B)36:109–118.

Watt, J. C. 1970. Subantarctic entomology, particularly of South Georgia and Heard island. Coleoptera: Perimylopidae of South Georgia. *Pacific Insects Monographs* 23:243–253.

Watt, J. C. 1971. Entomology of the Aucklands and other islands south of New Zealand: Coleoptera: Scarabaeidae, Byrrhidae, Ptinidae, Tenebrionidae. *Pacific Insects Monographs* 27:193–224.

Watt, J. C. 1974a. Chalcodryidae: a new family of heteromerous beetles (Coleoptera: Tenebrionoidea). *Journal of the Royal Society of New Zealand* 4:19–38.

Watt, J. C. 1974b. A revised subfamily classification of Tenebrionidae (Coleoptera). *New Zealand Journal of Zoology* 1:381–452.

Watt, J. C. 1980. *Zeonidicola* (Coleoptera: Cavognathidae)—Beetles inhabiting birds' nests. *Journal of the Royal Society of New Zealand* 10:331–339.

Watts, C. H. S. 1970. The larvae of some Dytiscidae (Coleoptera) from Delta, Manitoba. *Canadian Entomologist* 102:716–728.

Weise, J. 1897. Biologische Mitteilungen. *Deutsche Entomologische Zeitschrift* 1897:389–395.

Weiss, H. B. 1919a. *Catorama nigritulum* Lec., and its fungus host. *Canadian Entomologist* 51:255–256.

Weiss, H. B. 1919b. Notes on *Eustrophinus bicolor* Fabr., bred from fungi (Coleoptera). *Psyche* 26:132–133.

Weiss, H. B. 1920a. *Mordella marginata* Melsh., bred from fungus (Coleopt.). *Entomological News* XXX:67–68.

Weiss, H. B. 1920b. Notes on *Ischyrus quadripunctatus* Oliv., bred from fungus. *Canadian Entomologist* 52:14–15.

Weiss, H. B. 1922. Notes on the puffball beetle, *Caenocara oculata* (Say). *Psyche* 29:92–94.

Weiss, H. B. and West, E. 1920. Fungous insects and their hosts. *Proceedings of the Biological Society of Washington* 33:1–20.

Weiss, H. B. and West, E. 1921a. Additional fungous insects and their hosts. *Proceedings of the Biological Society of Washington* 34:59–62.

Weiss, H. B. and West, E. 1921b. Additional notes on fungous insects. *Proceedings of the Biological Society of Washington* 34:167–172.

Wesenberg-Lund, C. 1912. Biologische Studien über Dytisciden. *Internationale Revue der Gesamten Hydrobiologie und Hydrographie. Biologisches Supplement* 5:1–129, 9 pls.

West, L. S. 1929. Life history notes on *Psephenus lecontei* Lec. (Coleoptera; Dryopoidea; Psephenidae). *Battle Creek College Bulletin* 3(1):2–20, 3 pls.

Westcott, R. L. 1968. A new subfamily of blind beetles from Idaho ice caves with notes on its bionomics and evolution (Col. Leiodidae). *Contributions to Sciences. Los Angeles County Museum of Natural History* 141:1–14.

Westigard, P. M., Gentser, L. G. and Berry, D. W. 1968. Present status of biological control of the pear psylla in southern Oregon. *Journal of Economic Entomology* 61:740–743.

Wharton, R. A. 1979. Description and habits of larval *Pedilus inconspicuous* (Horn) (Coleoptera: Pedilidae). *Coleopterists Bulletin* 33:27–31.

Wheeler, A. G. and Hoebeke, E. R. 1981. A revised distribution of *Coccinella undecimpunctata* L. in eastern and western North America (Coleoptera: Coccinellidae). *Coleopterists Bulletin* 35:213–216.

Wheeler, A. G. and Jubb, G. L. 1979. *Scymnus cervicalis* Mulsant, a predator of grape phylloxera, with notes on *S. brullei* Mulsant as a predator of woolly aphids on elm (Coleoptera: Coccinellidae). *Coleoperists Bulletin* 33:199–204.

Wheeler, Q. D. 1979. Revision and cladistics of the Middle American genus *Creagrophorus* Matthews (Coleoptera: Leodidae). *Quaestiones Entomologicae* 15:447–479.

Wheeler, Q. D. 1984. Evolution of slime mold feeding in leiodid beetles. pp. 446–477 *in* Wheeler, Q. D. and Blackwell, M. (edit.). *Fungus/insect relationships: perspectives in ecology and evolution.* New York: Columbia University Press, 514 pp.

Wheeler, Q. D. 1984b. Notes on host associations and habitats of Dasyceridae (Coleoptera) in the southern Appalachian Mountains. *Coleopterists Bulletin* 38:227–231.

Wheeler, Q. D. 1985. Larval characters of a Neotropical *Scotocryptus* (Coleoptera: Leodidae), a nest associate of stingless bees (Hymenoptera: Apidae). *J.N.Y. Ent. Soc.* 93:1082–1088.

Wheeler, Q. D. 1986. Revision of the genera of Lymexylidae (Coleoptera: Cucujiformia). *Bulletin American Museum Natural History* 183:113–210.

Wheeler, Q. D. and M. Blackwell (eds.) 1984. *Fungus/insect relationships: perspectives in ecology and evolution.* N.Y.: Columbia Univ. Press. 514 pp.

Wheeler, Q. D. and Hoebeke, E. R. 1984. A review of mycophagy in the Eucinetoidea (Coleoptera), with notes on an association of the eucinetid beetle, *Eucinetus oviformis,* with a Coniophoraceae fungus (Basidiomycetes: Aphyllophorales). *Proceedings of the Entomological Society of Washington* 86(2):274–277.

Wheeler, Q. D. and Pakaluk, J. 1983. Descriptions of larval *Stenichnus (Cystoscydmus): S. turbatus* and *S. conjux,* with notes on their natural history (Coleoptera: Scydmaenidae). *Proceedings of the Entomological Society of Washington* 85:86–97.

Wheeler, W. M. 1921a. Notes on the habits of European and North American Cucujidae *(sens. auct.). Zoologica* 3(5):173–183.

Wheeler, W. M. 1921b. A study of some social beetles in British Guiana and of their relations to the ant-plant *Tachigalia. Zoologica* 3:35–126.

White, D. S., W. U. Brigham, and J. T. Doyen. 1984. Aquatic Coleoptera, pages 361–437. *In* Merritt, R. W. and K. W. Cummins (eds.). *An introduction to aquatic insects of North America, second edition.* Kendall/Hunt Publ. Co., Dubuque, Iowa, 722 pp.

White, R. E. 1967. Identification of the coffee bean weevil, *Araecerus fasciculatus* (DeGeer). *United States Department of Agriculture Cooperative Economic Insect Report* 17(23):496.

White, R. E. 1971a. Key to North American genera of Anobiidae, with phylogenetic and synonymic notes (Coleoptera). *Annals of the Entomological Society of America* 64(1):179–191.

White, R. E. 1971b. A new subfamily in Anobiidae (Coleoptera). *Annals of the Entomological Society of America* 64:1301–1304.

White, R. E. 1974a. The Dorcatominae and Tricoryninae of Chile (Coleoptera: Anobiidae). *Transactions of the American Entomological Society* 100:191–253.

White, R. E. 1974b. Type-species for world genera of Anobiidae (Coleoptera). *Transactions of the American Entomological Society* 99:415–475.

White, R. E. 1982. A catalog of the Coleoptera of America north of Mexico. Family: Anobiidae. *United States Department of Agriculture. Agriculture Handbook* No. 529–70, 58 pp.

White, R. E. and Day, W. H. 1979. Taxonomy and biology of *Lema trivittata* Say, a valid species with notes on *L. trilineata* (Oliv.) (Coleoptera: Chrysomelidae). *Entomological News* 90(5):209–217.

Whitehead, F. E. 1930. The pea weevil problem. *Journal of Economic Entomology* 23(2):399–401.

Wickham, H. F. 1893. Description of the early stages of several North American Coleoptera. *Bulletin of Natural History, University of Iowa* 2:330–344, pl. 9.

Wickham, H. F. 1894. On some aquatic larvae, with notice of their parasites. *Canadian Entomologist* 26:39–41.

Wickham, H. F. 1904. The metamorphosis of *Aegialites. Canadian Entomologist* 36:57–60.

Wilcox, J. A. 1975. *Checklist of the Beetles of North and Central America and the West Indies. Volume 8. Family 129. Chrysomelidae. The Leaf Beetles.* Gainesville, Florida: Flora and Fauna Publications, 166 pp.

Will, F. 1886. Zur Entwicklung des *Necrophilus subterraneus* Dahl, spec. ord. col. *Entomologische Nachrichten* 12:209–213.

Williams, F. X. 1917. Notes on the life history of some North American Lampyridae. *Journal of the New York Entomological Society* 25:11–33.

Williams. F. X. 1931. *Handbook of the Insects and other Invertebrates of Hawaiian Sugar Cane Fields.* Honolulu: Hawaiian Sugar Planters' Association Experiment Station, 400 pp.

Wilson, C. B. 1923. Water beetles in relation to pondfish culture, with life histories of those found in fishponds at Fairport, Iowa. *Bulletin of the United States Bureau of Fisheries* 39:231–345.

Wilson, E. O. 1971. *The Insect Societies.* Cambridge, Massachusetts: Harvard University Press, 548 pp.

Wilson, E. O., Eisner, T., and Valentine, B. D. 1954. The beetle genus *Paralimulodes* Bruch in North America, with notes on morphology and behavior. *Psyche* 61:154–161.

Wilson, H. F. 1913. Notes on *Podabrus pruinosus. Journal of Economic Entomology* 6:457–459.

Wing, S. R. 1984. A spate of glowworms (Coleoptera: Phengodidae). *Ent. News* 95:55–57.

Windels, C. E., Windels, M. B., and Kommedahl, T. 1976. Association of *Fusarium* species with picnic beetles on corn ears. *Phytopathology* 66:328–331.

Withycombe, C. I. 1926. The biology of lycid beetles in Trinidad. *Proceedings of the Royal Entomological Society of London* 1:32.

Wittmer, W. 1970. On some Cantharidae (Coleoptera) of the United States. *Coleopterists Bulletin* 24(2):42–46.

Wittmer, W. 1976. Eine neue Gattung der Familie Telegeusidae (Col.). *Mitteilungen der schweizerischen entomologischen Gesellschaft* 49:293–296.

Wolfe, G. W. 1980. The larva and pupa of *Acilius fraternus fraternus* (Coleoptera: Dytiscidae) from the Great Smoky Mountains, Tennessee. *Coleopterists Bulletin* 34:121–126.

Wolfe, G. W., and R. E. Roughley. 1985. Description of the pupa and mature larva of *Matus ovatus ovatus* Leech (Coleoptera: Dytiscidae) with a chaetotaxal analysis emphasizing mouthparts, legs, and urogomphus. *Proc. Acad. Nat. Sci. Philadelphia* 137(1):61–79.

Wolfram, P. 1929. Pars 102. Anthribidae. *Coleopterorum Catalogus.* The Hague: W. Junk, 145 pp.

Wolfram, P. 1953. Part 102. Anthribidae. *In* Hincks. W. D. (edit.), *Coleopterorum Catalogus Supplementa.* The Hague: W. Junk, 63 pp.

Wood, D. M. 1965. Studies on the beetles *Leptinillus validus* (Horn) and *Platypsyllus castoris* Ritsema from beaver. *Proceedings of the Entomological Society of Ontario* 95:33–63.

Wood, S. L. 1961. A key to the North American genera of Scolytidae. *Coleopterists Bulletin* 15:41–48.

Wood, S. L. 1973. On the taxonomic status of Platypodidae and Scolytidae (Coleoptera). *Great Basin Naturalist* 33:77–90.

Wood, S. L. 1982. The bark and ambrosia beetles of North and Central America (Coleoptera: Scolytidae), a taxonomic monograph. *Great Basin Naturalist Memoirs* 6, 1359 pp.

Woodruff, R. E. 1968. The palm seed "weevil", *Caryobruchus gleditsiae* (L.), in Florida (Coleoptera: Bruchidae). *Florida Department of Agriculture Entomology Circular* 73, 2 pp.

Woodruff, R. E. 1973. *The Scarab Beetles of Florida (Coleoptera: Scarabaeidae). Part I. The Laparosticti (Subfamilies: Scarabaeinae, Aphodiinae, Hybosorinae, Ochodaeinae, Geotrupinae, Acanthocerinae). Arthropods of Florida and Neighboring Land Areas, Volume 8* Gainesville, Florida: Florida Department of Agriculture and Consumer Services, 220 pp.

Woods, W. C. 1918. The biology of Maine species of *Altica. Maine Agricultural Experiment Station Bulletin* 273:149–199.

Woods, W. C. 1924. The blueberry leaf beetle and some of its relatives. *Maine Agricultural Experiment Station Bulletin* 319:93–141.

Wooldridge, D. P. 1975. A key to the New World genera of the beetle family Limnichidae. *Entomological News* 86(1/2):1–4.

Yasuda, T. 1962. On the larva and pupa of *Neomicropteryx nipponensis* Issiki, with its biological notes (Lepidoptera, Micropterygidae). *Kontyu* 30:130–136, pls. 6–8.

Young, D. K. 1975. A revision of the family Pyrochroidae (Coleoptera: Heteromera) for North America based on the larvae, pupae, and adults. *Contributions of the American Entomological Institute* 11(3):1–39.

Young, D. K. 1976a. Description of the larva of *Dioedus punctatus* LeConte (Coleoptera: Tenebrionidae). *Great Lakes Entomologist* 9:79–81.

Young, D. K. 1976b. The systematic position of *Sphalma quadricollis* Horn (Coleoptera: Salpingidae: Pythini) as clarified by discovery of its larva. *Coleopterists Bulletin* 30:227–232.

Young, D. K. 1978. A new species of *Lagriomorpha*, with observations on the systematic position of the genus (Coleoptera: Anthicidae). *Pacific Insects* 18:105–109.

Young, D. K. 1984a. Field studies of cantharidin orientation by *Neopyrochroa flabellata* (Coleoptera: Pyrochroidae). *Great Lakes Ent.* 17:133–135.

Young, D. K. 1984b. Cantharidin and insects: An historical review. *Great Lakes Ent.* 17:187–194.

Young, D. K. 1984c. Field records and observations of insects associated with cantharidin. *Great Lakes Ent.* 17:195–199.

Young, D. K. 1985a. Description of the larva of *Ischalia vancouverensis* Harrington (Coleoptera: Anthicidae: Ischalinae), with observations on the systematic position of the genus. *Coleopterists Bulletin.* 39:201–206.

Young, D. K. 1985b. The true larva of *Lecontia discicollis* (LeConte), and change in the systematic position of the genus (Coleoptera: Boridae). *Great Lakes Ent.* 18:97–101.

Young, D. K. (In press). Bionomics and systematics of the North American species of *Pedilus* Fischer (Coleoptera: Pyrochroidae). *Knull Series, Ohio State University.*

Young, F. N. 1954. The water beetles of Florida. *University of Florida Studies. Biological Science Series* 5(1), ix + 238 pp.

Young, F. N. 1956. Unusual abundance of *Sandalus* in southern Indiana. *Coleopterists Bulletin* 9:74.

Young, F. N. 1967. A possible recycling mechanism in tropical forests. *Ecology* 48(3):506.

Yuasa, H. 1922. A classification of the larvae of the Tenthredinoidea, *Illinois Biological Monographs* 7(4):1–172.

Zacharuk, R. Y. 1962a. Sense organs of the head of larvae of some Elateridae (Coleoptera): their distribution, structure and innervation. *Journal of Morphology* 111:1–33.

Zacharuk, R. Y. 1962b. Some new larval characters for the classification of Elateridae (Coleoptera) into major groups. *Proceedings of the Royal Entomological Society of London* (B)31:29–32.

Zacharuk, R. Y., Albert, P. J. and Bellamy, F. W. 1977. Ultrastructure and function of digitiform sensilla on the labial palp of a larval elaterid (Coleoptera). *Canadian Journal of Zoology* 55:569–578.

Zaitzev, Yu. M. and Dang Thi Dap 1982. The larva of *Sagra femorata* (Coleoptera, Chrysomelidae) from Vietnam. *Zoologicheskii Zhurnal* 61:458–460.

Zhantiev, R. D. 1976. Hide beetles (Family Dermestidae) of the fauna of the U.S.S.R. Moscow: Moscow University, 182 pp (in Russian).

Zimmerman, E. C. 1942. Rhipiceridae of Guam. *Bulletin of the Bernice P. Bishop Museum,* Honolulu 172:45–46, pl. 1.

Zimmerman, E. C. 1948. *Insects of Hawaii. Volume 1. Introduction.* Honolulu: University of Hawaii Press, xx + 206 pp.

Zocchi, R. 1957. Contributi alla conoscenza degli insetti delle piante forestali. III. Note morfobiologiche sull'*Ernobius abietis* F. (Coleoptera Anobiidae). *Redia* 42:291–348.

Zocchi, R. 1961. Contributo alla conoscenza della morfologia larvale del Coleottero Dascillidae *Dascillus cervinus* L. *Studi Sassaresi* (3)9:430–444.

Zwick, P. 1978. Bestimmungstabelle für die Larven der Catopidae, pp. 302–304 *in* Klausnitzer, B. (ed.), *Ordnung Coleoptera (Larven).* The Hague: W. Junk, vi + 378 pp.

Zwick, P. 1979. Contributions to the knowledge of Australian Cholevidae (Catopidae auct.: Coleoptera). *Australian Journal of Zoology, Supplementary Series,* No. 70, 56 pp.

Zwick, P. 1981. Die Jugendstadien des Käfer *Necrophilus subterraneus* (Col. Silph. Agyrtinae). *Beiträge zur Naturkunde Osthessens* 17:133–140.

Order Strepsiptera

35

Ed. Luna de Carvalho
Instituto de Investigação Cientifica Tropical, Lisbon, Portugal

Marcos Kogan
Illinois Natural History Survey and University of Illinois

The Strepsiptera are a small group of peculiar insects, remarkably adapted to endoparasitic life during most of their developmental stages. Their development is hypermetabolic and they present a striking sexual dimorphism (fig. 35.1). There are about 400 species described from all continents.

Most Strepsiptera specialists tend to keep the group as an independent order with affinities to the Coleoptera. There have been, however, many attempts to identify phylogenetic relationships to other taxonomic groups (Pierce 1936, Bohart 1941, Jeannel 1945, Brown 1965, Blazejewski 1969, Riek 1970). According to Lawrence and Newton (1982) the position of Strepsiptera is still unclear but its close relationship to the beetles is based on the following evidence: a) presence of metathoracic flight wings, b) free prothorax, with closely associated meso- and metathoraces, c) abdomen with sternites more heavily sclerotized than tergites, d) structure of the metendosternite, and e) triungulin type larvae similar to those of the Meloidae and Rhipiphoridae. It seems that there are compelling reasons to maintain it as an independent order [for instance Kinzelbach (1971a, c)], despite the vehement arguments to the contrary by Abdullah (1978).

DIAGNOSIS

The best taxonomic characters to separate the families are found in the free-living males. The females have regressed to a very rudimentary external morphology by virtue of remaining partially endoparasitic, except in some genera of the family Mengenillidae in which the adult female is larviform but free-living.

BIOLOGY AND ECOLOGY

Host Relationships: The Strepsiptera, among the insects, approach the definition of a true parasite in that they seldom, if ever, kill the host. The parasitized host, said to be "stylopized", may present changes in developmental rates, morphological malformations, such as changes in wing venation, and acquisition of secondary sexual characters of the opposite sex. These stylopized insects are detectable by the protruding cephalothorax of the female or the cephalic end of the male puparium of the strepsipteran (figs. 35.2a–e). When many parasites occur in one host the host's abdomen may become greatly distorted. Hosts, which are characteristic for the families, have been observed in the Thysanura, Orthoptera, Hemiptera, Homoptera, Diptera and Hymenoptera (more detailed host associations are given below).

Many of the earlier specialists assumed as a rule that the Strepsiptera were highly host specific. As a consequence there was a tendency to describe a new species for each new host record. This led to an excessive fractionation of the order. For example, Pierce (1918) subdivided the then described 166 species into 5 superfamilies, 11 families, and 49 genera. Later revisions, mainly by Bohart (1941) and Kinzelbach (1971c, 1978), have reduced the number of families and genera. Host associations are quite consistent within families but individual species may have many hosts within a given family. Thus, *Elenchus tenuicornis* (Kirby) has been recorded from 55 leafhopper species (Baumert 1959).

Life Cycle: The typical life cycle of Strepsiptera involves a viviparous, free-living first instar known as a triungulinid, by analogy to the triungulins of the Meloidae. The strepsipteran first instar larvae do not have true tarsal claws. They are dispersed by phoresy by the primary parasitized host, and remain on a suitable habitat until a new host is encountered. If the species is parasitic on a hemimetabolous insect (Homoptera, Hemiptera or Orthoptera), the larva may directly penetrate the body of the host and continue its development. In *Corioxenos antestiae* Blair, parasitic on the pentatomids *Antestia* spp. in East Africa, the triungulinid seems to wait for a host to molt, at which time it penetrates mainly through the thoracic integument of a soft, newly molted host (Kirkpatrick 1937). If, however, they are parasites of a hymenopteran, the triungulinid attaches itself to the body of a foraging wasp or bee and is transported to the nest where it moves toward the immature stage of the host, penetrating its soft body to continue development. Penetration is aided by the sharp edge of the head; apparently the mandibles play no role in this process. Soon after penetration the triungulinid undergoes a first molt and becomes a secondary larva. The endoparasitic development of the secondary larva proceeds with several additional molts. In one instance up to 7 molts have been recorded and authors seem to be able to identify a male series and a female series of larvae in some species.

The ensuing phases of development of males and females are strikingly different. The female remains as a neotenic individual and matures within the skin of the last instar (tertiary) larva. In one family (Myrmecolacidae) it seems that males are exclusively parasitic on ants, whereas the females are parasitic on orthopterans (Ogloblin 1939, Luna de Carvalho 1959). The life cycle of these dual-host species is illustrated after Luna de Carvalho (1959) (fig. 35.3).

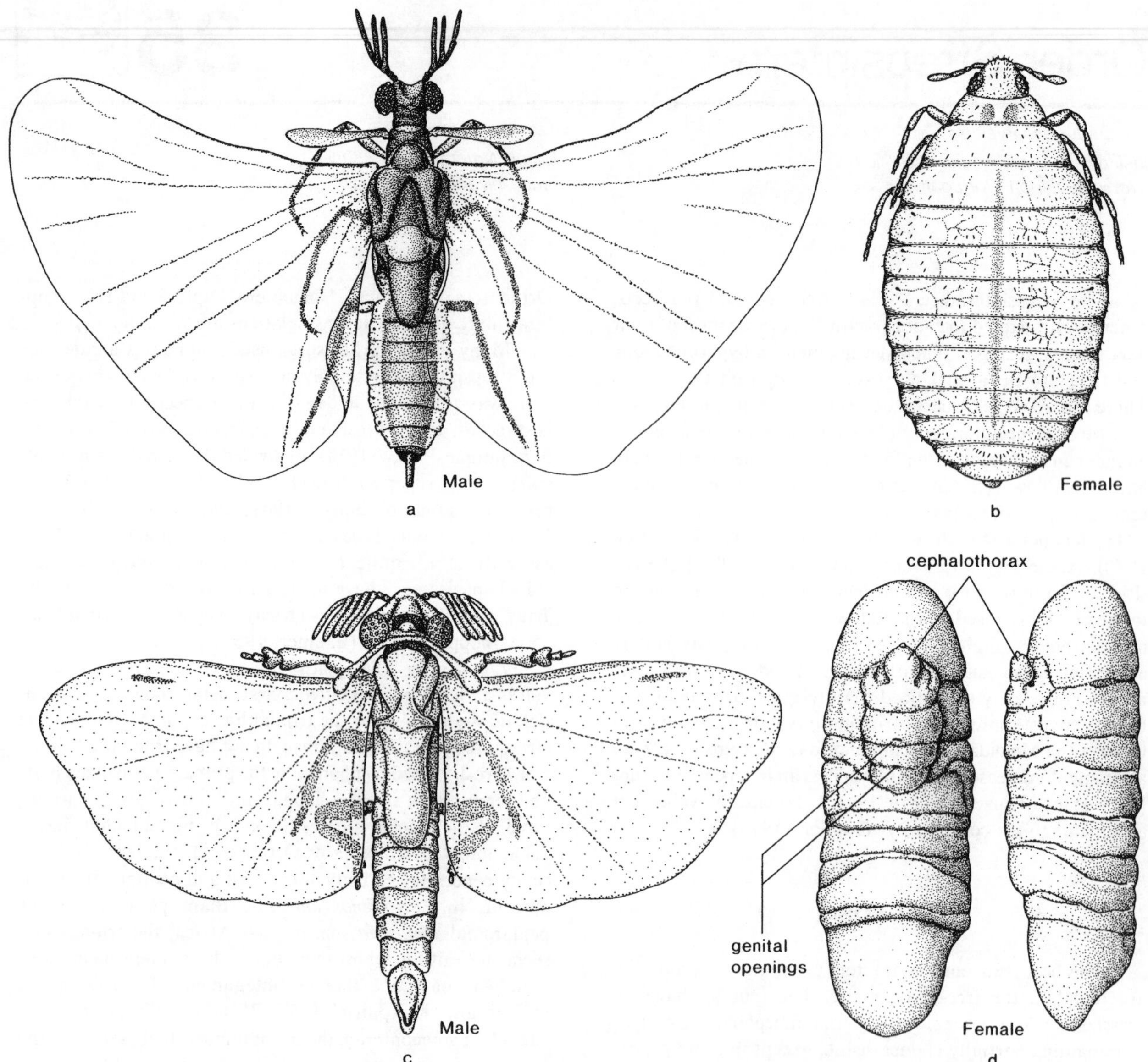

Figure 35.1a-d. Typical adult Strepsiptera. **a, b.** *Eoxenos laboulbenei* Peyerimhoff (Mengenillidae), redrawn from Parker and Smith (1933): **a,** adult male, dorsal; **b,** free living neotenic female, dorsal, **c,d.** *Halictophagus oncometopiae* (Pierce) (Halictophagidae), redrawn from Pierce (1918): **c,** adult male, dorsal; **d,** endoparasitic female, lateral and ventral (or outer view in relation to the host's body).

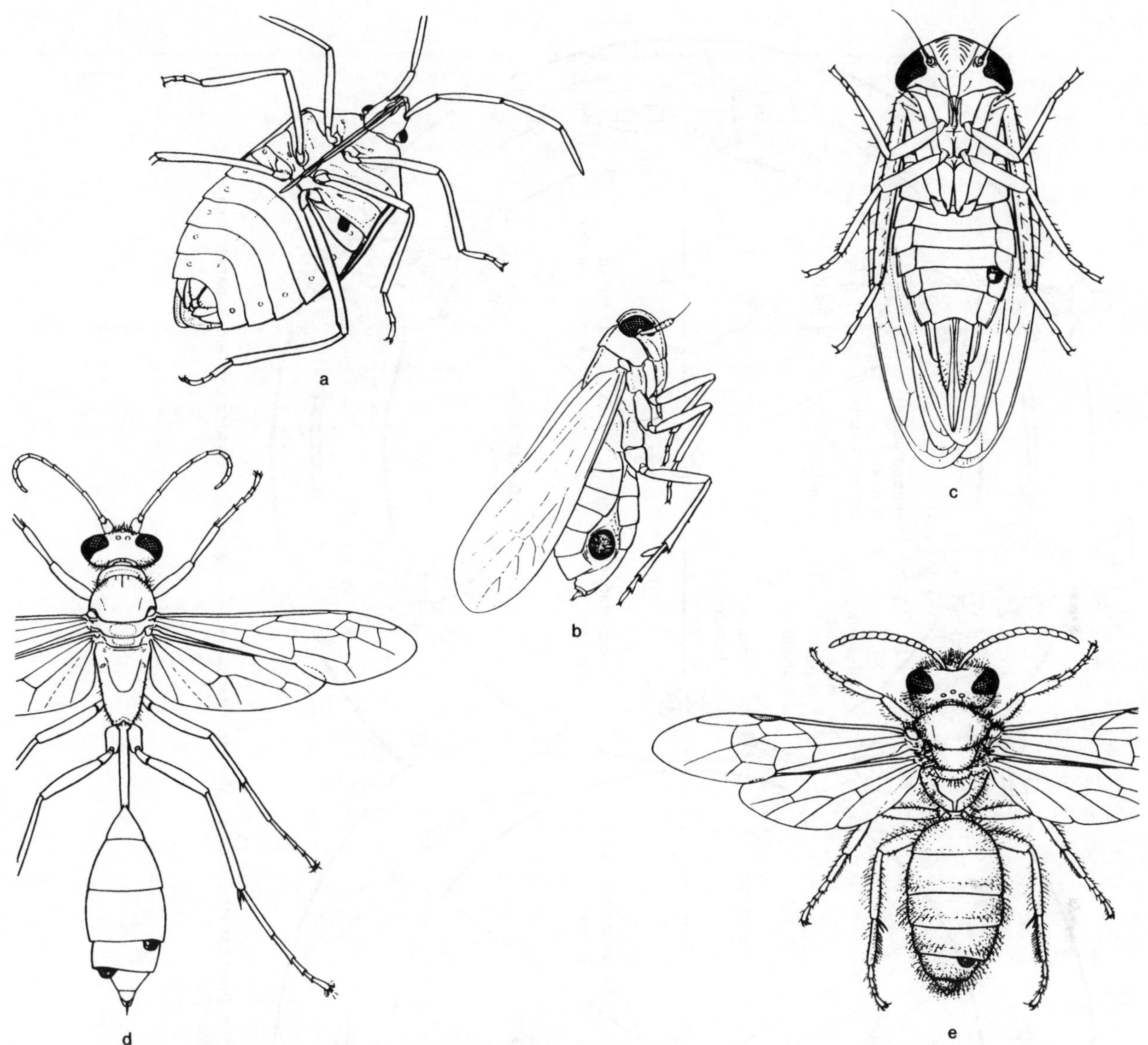

Figure 35.2a-e. Stylopized insects: **a,** Halictophagidae, *Halictophagus* sp., female in pentatomid bug; **b,** Elenchidae, *Elenchus* sp., male puparium in delphacid; **c,** Halictophagidae, *Halictophagus* sp., in female jassine; **d,** Stylopidae, *Pseudoxenos* sp., male puparium and female in *Sceliphron laetum* (Sphecidae); **e,** Stylopidae, *Hylecthrus* sp., female in *Euryglossa* sp. (Colletidae). **a-e** from The Insects of Australia (Melbourne University Press). [M. Quick, artist]

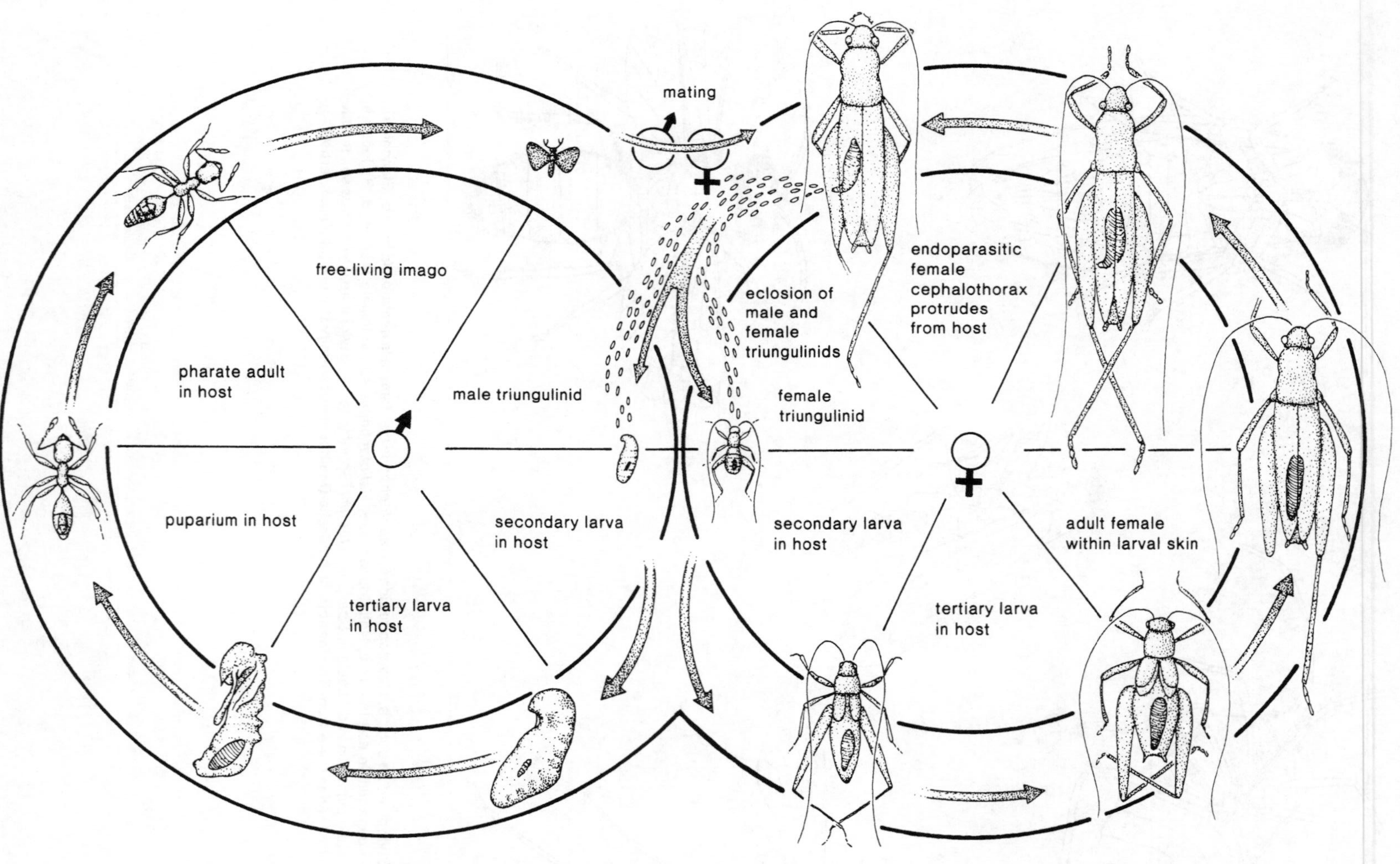

Figure 35.3. Probable life cycle of a dual-host strepsipteran: Myrmecolacidae—with male parasitic on ants and female parasitic on an orthopteran (After Luna de Carvalho 1959).

Description of Developmental Stages

The complex metamorphoses of the Strepsiptera can be defined by the identification of 4 or 5 post-embryonic stages. In the male these stages are: (a) the free-living primary larva (triungulinid), (b) the endoparasitic secondary larva which undergoes several molts, turning into a recognizable tertiary larva, (c) the prepupa, and (d) the pupa, the latter two remaining within the puparium, consisting of the skin of the last secondary or tertiary larval instar. The female matures within the skin of the tertiary larval instar without undergoing pupation (Kinzelbach 1971a). These developmental stages are illustrated in fig. 35.4.

Primary Larvae: Triungulinids (fig. 35.4) are 80–350μ long, not including the caudal bristles, varying in color from pale brown to black. They lack well-formed antennae, mandibles or maxillae, and they bear on each side of the head 3 to 5 stemmata. The head contains a complex tentorium that can be observed under high magnification in cleared preparations; this tentorium seems to provide species specific characteristics (fig. 35.5). The legs are slender, without trochanters and with expanded pulvilli. Thoracic segments are subequal. The abdomen has 10 visible segments, the last one has a caudal plate bearing 2 long caudal bristles. These bristles enable the larvae to jump. The other abdominal segments usually have rows of short hairs ventrally and pairs of longer hairs laterally.

Secondary Larvae: (fig 35.4) The parasitic larva is legless and grub- or maggot-like. It is uncertain whether they undergo additional molts (Riek 1970), although Kirkpatrick (1937) has recorded 6 parasitic instars in *Corioxenos*, and 4–8 molts have been described for *Halictophagus, Elenchus* and *Xenos*. Many authors recognize a tertiary larva, the skin of which forms the male puparium and the outer integument of the female body (fig. 35.4).

Puparium: The skin of the last secondary larval instar (or tertiary larva) forms the puparium. When the pupa is formed within it the puparium is heavily sclerotized anteriorly, and small cephalic protuberances mark the position of the antennae, mandibles and ocellar areas of the pupa.

Pupa: (fig. 35.4) The pupa has a thin, transparent cuticle, through which the pharate adult body can be clearly perceived. The antennae are folded ventrally and directed caudad; the femora and tibiae are bent close together with the tarsi of the legs of the same pair almost touching each other ventrally, and oriented posteriorly. The wings wrap the thorax and reach sternally the coxae of the hind legs. When the imago is entirely formed it can be easily removed from the puparium and the pupal skin. Many type specimens are such unhatched adults thus obtained from stylopized hosts. The mode of adult emergence from the puparium has been recently described for *Elenchus tenuicornis* (Kirby) by Kathirithamby (1983).

Adults: Females are typically neotenic. In the Mengenillidae they have legs and are free living (fig. 35.1b). In all other families they are partially endoparasitic and deprived of legs and of most sensorial organs (fig. 35.1d). They are reduced to a large sac-like abdomen that remains embedded in the host's body and a much smaller, heavily sclerotized cephalothorax that protrudes from the intersegmental membrane of the host's abdomen. The cephalic region is separated from the thorax by a lateral suture and ventrally by the intercephalothoracic membrane that serves both as the sexual opening and the exit window for the triungulinids. The ventral side of the female has up to 5 genital openings. These lead to the brood canal, a structure formed by the juxtaposition of the puparial skin (or the skin of the tertiary larva enveloping the female body) and the adult skin (fig. 35.6). Mature triungulinids migrate through the genital openings into the brood canal and from there to the intercephalothoracic membrane area.

Adult males are free living, with functional metathoracic wings, and twisted, coriaceous forewings. The head bears typical large, faceted eyes and flabellate antennae (*see* figs. 35.1a and 35.1b).

Taxonomic Subdivisions: The most recent taxonomic systems recognize one fossil family and 8 contemporary families (Luna de Carvalho 1978, Kinzelbach 1978). Diagnosis of families is more easily done with male characters, although keys based on female characters or a combination of male and female characters are also available. Since the only immature stage with enough diagnostic characteristics is the triungulinid, a key to the families based on triungulinid characters is also provided. However, most encounters with Strepsiptera result from collections of stylopized hosts which often contain females still full of triungulinids and an occasional male puparium, that, with luck, may contain a pharate adult. Therefore, the family definition is presented with the complementary adult information, since the system of classification currently available is based on those forms, and since there is always the possibility of collecting adult forms associated with the larva one is interested in identifying.

CLASSIFICATION[1]

Order **STREPSIPTERA**[2]
- Suborder Mengenillidia
  - Mengeidae (1 fossil)
  - Mengenillidae (9)
- Suborder Stylopidia
  - Corioxenidae (12)
  - Halictophagidae (63)
  - Callipharixenidae (2)
  - Bohartillidae (1)
  - Elenchidae (19)
  - Myrmecolacidae (53)
  - Stylopidae (214)

1. The classification is based on Kinzelbach (1971c, 1978). It excludes the superfamilies and infraorders proposed by him.

2. The numbers of species given are for the world (Kinzelbach 1971c). Perhaps 35 have been described since then.

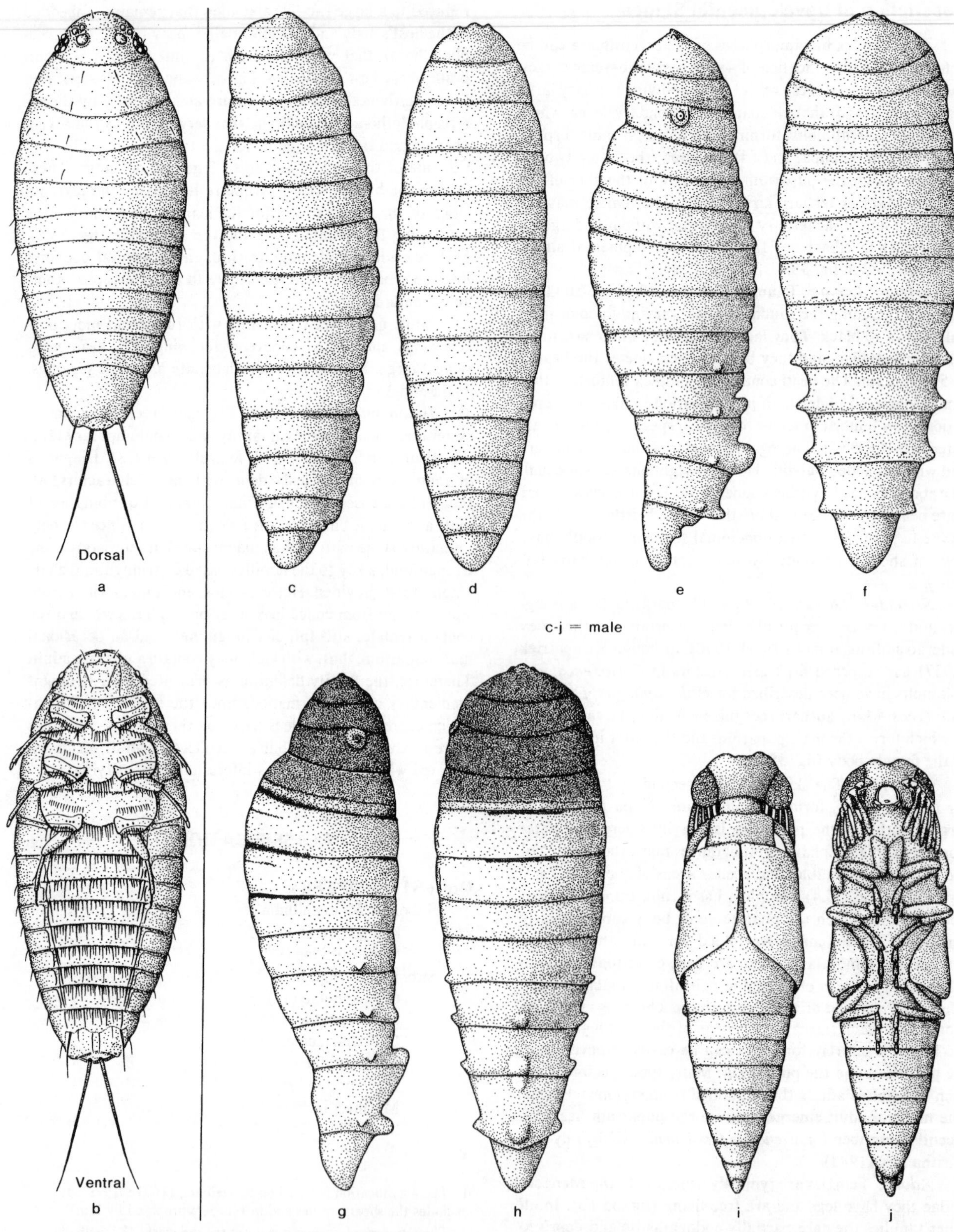

Figures 35.4a-j. Developmental stages of *Halictophagus tettigometrae* Silvestri (Halictophagidae): **a,b:** triungulinid (primary larvae), dorsal and ventral. **c-j:** Stages of male development: **c,d:** secondary larva, lateral and dorsal; **e,f:** tertiary larva, lateral and ventral; **g,h:** puparium, lateral and dorsal; **i,j:** pupa, dorsal and ventral.

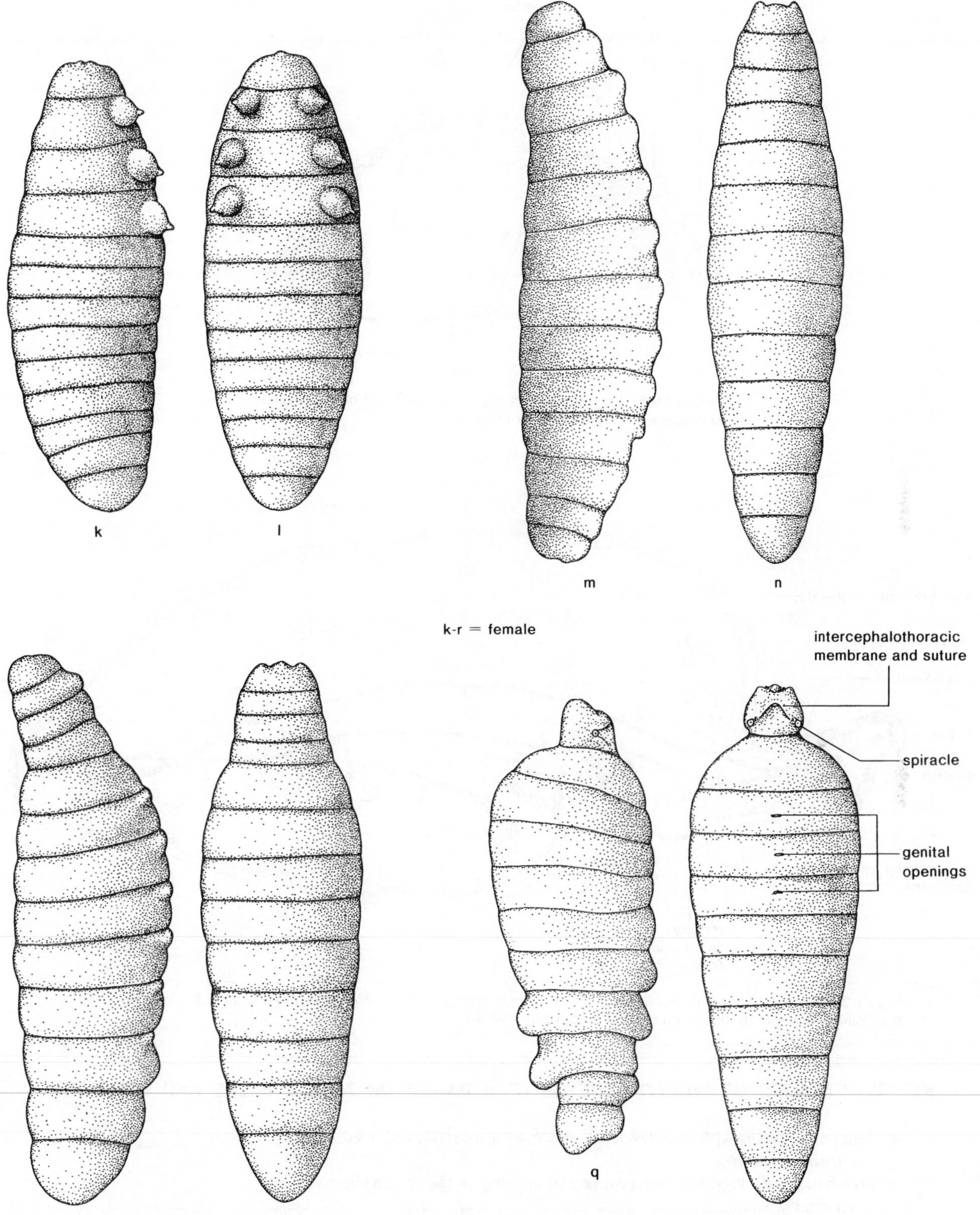

Figures 35.4k-r. Developmental stages of *Halictophagus tettigometrae* Silvestri (Halictophagidae): **k-r:** Stages of female development. **k,l:** primary larva, lateral and ventral; **m,n:** secondary larva (1st instar), lateral and dorsal; **o,p:** secondary larva (2nd instar), lateral and dorsal; **q,r:** tertiary larva and puparium, lateral and ventral. (redrawn after Silvestri 1941b, by omitting details of internal anatomy).

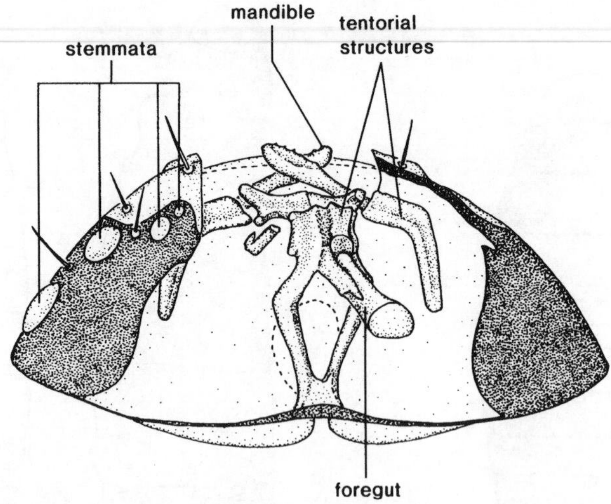

Figure 35.5. Stylopidae. Partially dissected head of triungulinid of *Stylops*. (Redrawn after Borchert 1963).

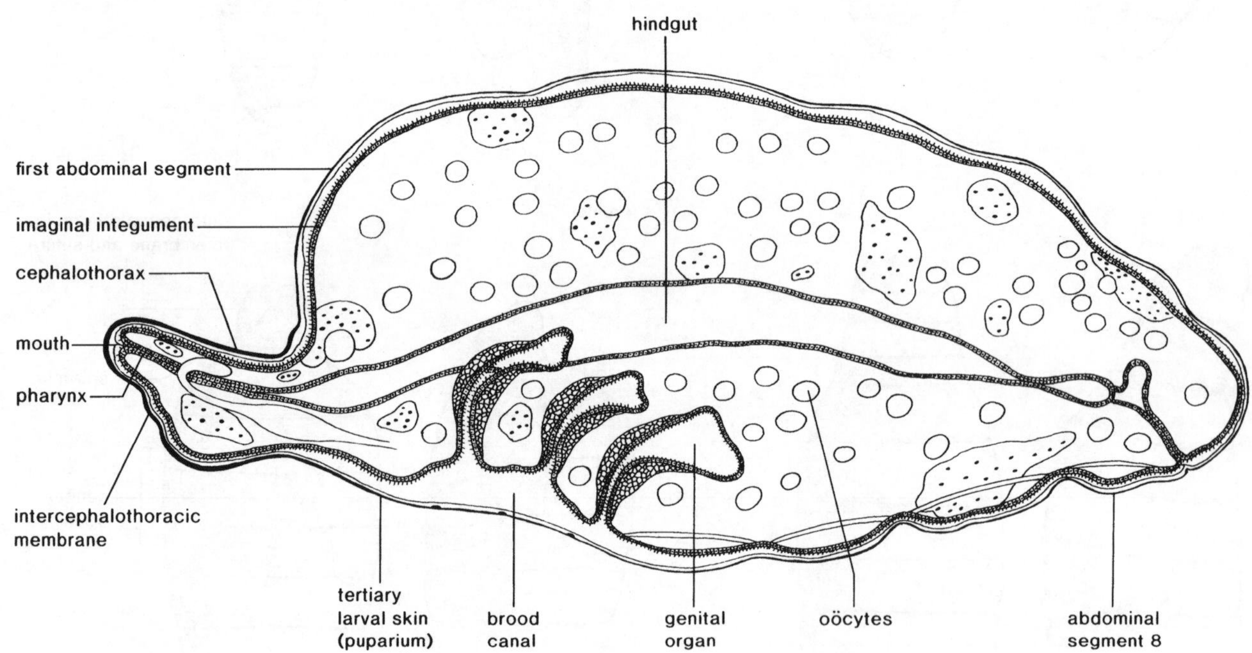

Figure 35.6. Halictophagidae. Adult female *Halictophagus tettigometrae:* morphology in a sagittal section (Redrawn from Kinzelbach 1971b, based on original by Silvestri 1941b).

KEY TO FAMILIES OF WORLD STREPSIPTERA BASED ON ADULT MALES AND FEMALES[3]

1. ♂: Pentameric tarsus with well developed claws; with spear-shaped, medially crossing mandibles.
   ♀: Free-living; with legs, but apterous; genital opening on the 7th abdominal segment 2

1′. ♂: Pretarsus without claws; when (rudimentary) claws present, then mandibles shorter than wide or absent.
   ♀: Partially endoparasitic imago remains within puparial skin; anterior tagmata only, produced into a sclerotized cephalothorax 3

3. Based on Kinzelbach 1978, Luna de Carvalho 1978.

2(1). ♂: Antennae with 7 segments; labial palpi present (fossil).
♀: Unknown .... ***Mengeidae*** (p. 668)

2′. ♂: Antennae with 6 segments; labial palpi absent
♀: Free-living females, with functional legs .... ***Mengenillidae*** (p. 668)

3(1′). ♂: Pretarsus with rudimentary claws, mandibles absent; first tarsomere with dorsal sensorial pit.
♀: With vestigial eye-structure; ventral side of cephalothorax lying adjacent to host's surface .... ***Corioxenidae*** (p. 668)

3′. ♂: Pretarsus never with claws; strong mandibles present; first tarsomere often with discernible ventral sensorial pit.
♀: No eye-structures; ventral side of cephalothorax away from host's surface 4

4(3′). ♂: Antennae with 7 segments; mandibles cone-shaped, not crossing in the median line.
♀: Head more than half the length of cephalothorax; dorsal side not membranous .... ***Halictophagidae*** (p. 668)

4′. ♂: Antennae with fewer than 7 segments; mandibles blade-like, crossing in the median line.
♀: Head less than half the length of cephalothorax or, cephalothorax with membranous dorsal side 5

5(4′). ♂: Maxillary palpus about 1/6 the length of maxillary base.
♀: Unknown .... ***Bohartillidae*** (p. 669)

5′. ♂: Maxillary palpus longer than 1/6 the length of maxillary base.
♀: Head always less than half the length of cephalothorax or cephalothorax with dorsal side membranous and inflated 6

6(5′). ♂: Maxillary palpi always longer than 1/6 the maxillary base.
♀: Metathoracic stigmatal structures rarely present, and never functional; round or bell-shaped cephalothorax 7

6′. ♂: Unknown.
♀: With functional metathoracic stigmata; cephalothorax elongate .... ***Callipharixenidae*** (p. 668)

7(6). ♂: Spoon-shaped subalar sclerite present in metathorax.
♀: Dorsal side of flattened cephalothorax sclerotized; intercephalothoracic membrane slitlike .... ***Stylopidae*** (p. 669)

7′. ♂: Metathorax without distinguishable subalar sclerite.
♀: Dorsal side of cephalothorax membranous and swollen; intercephalothoracic membrane widely expanded 8

8(7′). ♂: With 2 tarsal segments.
♀: With 1–5 genital openings in abdominal segments 2–4 .... ***Elenchidae*** (p. 669)

8′. ♂: With 4 tarsal segments.
♀: With more than 5 genital openings in abdominal segments 2–4; very large .... ***Myrmecolacidae*** (p. 669)

KEY TO FAMILIES OF WORLD STREPSIPTERA BASED ON CHARACTERS OF PRIMARY LARVAE (TRIUNGULINIDS)

Primary larvae of representative species belonging to 7 of the 8 contemporary families of Strepsiptera have been reasonably well described and illustrated. Existing keys use a combination of morphological characters and host records to establish the dichotomies. Obviously, the rare, but possible encounter of an isolated triungulinid without host record will create a problem in the use of such keys. Given, however, the paucity of detailed comparative morphological studies on triungulinids, the present key is the best that is currently available. The following key is adapted mainly from Bohart 1941, and Luna de Carvalho (unpublished).

1. All tarsi long, setiform 2

1′. At least the front tarsi pulvilliform 4

2(1). Posterior margin of abdominal sternites smooth, without bristles (parasites of Pentatomidae) .... ***Callipharixenidae*** (p. 668)

2′. Posterior margin of abdominal sclerites with bristles directed caudad 3

3(2′). Posterior margin of abdominal sclerites with a fringe of long and short bristles; ventral side of head with short bristles on posterior margin (parasites of Homoptera, Gryllidae, Blattidae and Diptera) .... ***Halictophagidae*** (p. 668)

3′. Posterior margin of abdominal sclerites with a pair of bristles directed caudad along the sagittal line; ventral side of head with long bristles reaching the front coxae (parasites of Delphacidae) .... ***Elenchidae*** (p. 669)

4(1′). Front and mid tarsi pulvilliform 5

4′. Only front tarsi pulvilliform; mid and hind tarsi setiform (♂ parasites of Formicidae, ♀ parasites of Orthoptera) .... ***Myrmecolacidae*** (p. 669)

5(4). Posterior margin of abdominal sclerites dentate but without bristles; ventral side of head with two long bristles directed caudad on posterior margin, reaching front coxae (parasites of Hemiptera) .... ***Corioxenidae*** (p. 668)

5′. Posterior margin of abdominal sclerites fringed with hairs of various lengths; ventral-side of head with bristles (if present) very short, not reaching front coxae 6

6(5′). Ventral side of head without hairs (parasites of Hymenoptera Aculeata) .... ***Stylopidae*** (p. 669)

6′. Ventral side of head with short hairs directed caudad (parasites of Lepismatidae) .... ***Mengenillidae*** (p. 668)

MENGEIDAE

A family described from a single adult male preserved in Baltic amber. Seven specimens have been found (Kinzelbach 1978). Fossil.

Selected Bibliography

Kinzelbach 1978.
Menge 1866.
Ulrich 1927.

MENGENILLIDAE

Figures 35.1a,b, 35.7a,b

Parasites of Thysanura (Lepismatidae) with free-living females (fig. 35.1b). Secondary larvae leave the host, undergo an additional molt and shed the puparial exuvia, living in ant nests under rocks. Recorded from tropical and subtropical regions of Europe, Africa, Asia and Australia.

Selected Bibliography

Bolivar y Pieltain 1926.
Kinzelbach 1971c, 1972, 1978.
Luna de Carvalho 1953.
Parker and Smith 1933, 1934.
Silvestri 1941a, 1943b, 1946.

CORIOXENIDAE

Figures 35.7c,d

Parasites of Hemiptera: (Pentatomidae, Cydnidae, Miridae). Females with opening of brood-canal terminal, and a distinctly segmented thorax; genital organs in abdominal segment 3 or 4–7. Representative species have been recorded from North, Central and South America, West and East Africa, Cape Verde Islands, Australia, New Guinea, Philippines, China, Japan, Arabia and Malaysia.

Selected Bibliography

Blair 1936.
Kinzelbach 1971c, 1978, 1980.
Luna de Carvalho 1956, 1985.
Pierce 1909, 1918.
Kogan 1958.

HALICTOPHAGIDAE

Figures 35.1c,d; 35.2a,c; 35.4a–r; 35.6, 35.7f,g

A large and rather polymorphic family with representative species recorded as parasitic on: Homoptera: Auchenorrhyncha (Cicadellidae, Cercopidae, Fulgoridae, Dyctiopharidae, Delphacidae, Membracidae); Hemiptera (Pentatomidae, Coreidae); Orthoptera (Tridactylidae); Blattodea and Diptera. Females have a relatively long head in comparison to the thorax, with 1–4 abdominal genital organs (but generally with three). Cosmopolitan distribution.

Selected Bibliography

Bohart 1943.
Kinzelbach 1971c, 1978.
Perkins 1905, 1907.
Pierce 1918.
Silvestri 1934, 1941b, 1943a.
Yang 1964a,b.

CALLIPHARIXENIDAE

Figure 35.7e

Monotypic family known only from female specimens parasitic on Hemiptera (Scutelleridae, *Calliphara* and *Chrysocoris*), in Southeast Asia (Thailand and Moluccas). Cephalothorax very elongate and flattened, and with a pair of stigmata on the 1st abdominal segment as large as the metathoracic stigmata. With four genital organs.

Selected Bibliography

Blair 1936.
Kinzelbach 1971c, 1978.
Pierce 1918.

BOHARTILLIDAE

Monotypic family known only from male specimens of *Bohartilla* collected in Honduras. Host unknown.

Selected Bibliography

Kinzelbach 1969.

ELENCHIDAE

Figures 35.2b, 35.7l

Parasitic on Homoptera (Fulgoroidea, Cercopoidea). Females with membranous ventral thorax; genital organs in abdominal segments 2–9. Some species are parasitic on important rice pests and have received considerable attention in the last decade as biological control agents (Otake et al. 1976). Reported from Australia, New Guinea, Borneo, Philippines, Japan, Angola, Liberia, Mexico, Central America, central Europe.

Selected Bibliography

Esaki and Hashimoto 1931.
Kinzelbach 1971c, 1978, 1979.
Luna de Carvalho 1956.
Perkins 1905.
Pierce 1908, 1918, 1961.

MYRMECOLACIDAE

Figures 35.3, 35.7h-k, 35.8, 35.9

Originally described, and for many years known only from male specimens collected in light traps or puparia on stylopized ants of the genera *Camponotus, Eciton, Pheidole, Pseudomyrma, Solenopsis* and *Crematogaster* (Oglobin 1939, Luna de Carvalho 1959). Females are currently believed to be parasites of Orthoptera (Gryllidae, Gryllotalpidae, Tettigoniidae) and Mantodea. Until recently these female "species" constituted a different family (Stichotrematidae) now considered a synonym of Myrmecolacidae (Luna de Carvalho 1956). The triungulinids sometimes display a marked polymorphism with males 1/2–3/4 the size of the females (fig. 35.8). The adult female is large (up to 3 cm) with the thoracic end of the cephalothorax very reduced. Genital organs in segments 2–4. Reported from North, Central and South America, Africa and Oceania.

Selected Bibliography

Bohart 1951.
Kifune and Hirashima 1980.
Kinzelbach 1971c, 1978.
Kogan and Oliveira 1964.
Luna de Carvalho 1956, 1959, 1967.
Ogloblin 1939.
Pierce 1909.
Riek 1970.

STYLOPIDAE

Figures 35.2d,e, 35.5, 35.7m-p

This is by far the largest family, including well over half the total number of described species. They are all parasites of Hymenoptera Aculeata (Apoidea, Vespoidea, Sphecoidea, and Scolioidea). Representative species occur in all zoogeographic regions. Female cephalothorax sturdy, flattened and rounded forward; cephalic region small. Genital organs in abdominal segments 2–6 or fewer. The family is usually subdivided into three subfamilies: Xeninae with three genera, Paraxeninae with one, and Stylopinae with 7.

Selected Bibliography

Bohart 1941, 1943.
Kifune and Maeta 1965.
Kinzelbach 1971c, 1978.
Kogan and Oliveira 1966.
Luna de Carvalho 1978, 1979.
Pierce 1908, 1909, 1911a, b, 1918.

BIBLIOGRAPHY

Abdullah, M. 1978. Coleoptera including Strepsiptera syn. nov.: Entomophaga suborder nov. with Stylopiformia series nov. for Stylopoidea comb. nov. Deutsch. Entomol. Zeit., N.F. 25(1–3):129–130.

Baumert, D. 1959. Mehrjahrige Zuchten einheimischer Strepsipteren. 2. Imagines, Lebenszyklus and Artbestimmung von *Elenchus tenuicornis* Kirby. Zool. Beitr. N.F. 4(3):343–409.

Blair, K. G. 1936. A new genus of Strepsiptera. Proc. R. Entomol. Soc. London (B)5:113–117.

Blazejewski, F. 1969. The position of the Strepsiptera in the classification of insects (in Polish with English summary). Przeglad Zoologiezny 13(1):38–42.

Bohart, R. M. 1941. A revision of the Strepsiptera with special reference to the species of North America. Univ. Calif. Publs. Ent. 7(6):91–160.

Bohart, R. M. 1943. New species of *Halictophagus* with a key to the genus in North America (Strepsiptera, Halictophagidae). Ann. Entomol. Soc. Amer. 36(3):341–359.

Bohart, R. M. 1951. The Myrmecolacidae of the Philippines (Strepsiptera). Wasmann J. Biol. 9(1):83–103.

Bolivar y Pieltain, C. 1926. Estudio de un nuevo Mengenillidae de España (Streps. Meng.). Eos 2:5–13.

Borchert, H. M. 1963. Comparative morphological investigations of the Berlin Stylops-L_1 (Strepsipt.) for the purpose of deciding the two specificity questions: 1. Are there more Stylops species in our spring Andrenidae (Hymenoptera, Apidae) and 2. Are there host specificities? Zool. Beitr. (n.s.)8(3):331–445. [in German]

Brown, W. L., Jr. 1965. Numerical taxonomy, convergence, and evolutionary reduction. Systematic Zool. 14(2):101–109.

Esaki, T., and S. Hashimoto. 1931. Report on the leafhoppers injurious to the rice plant and their natural enemies. (In Japanese with English Summary). Publ. Ent. Lab. Kyushu University 2:39–52.

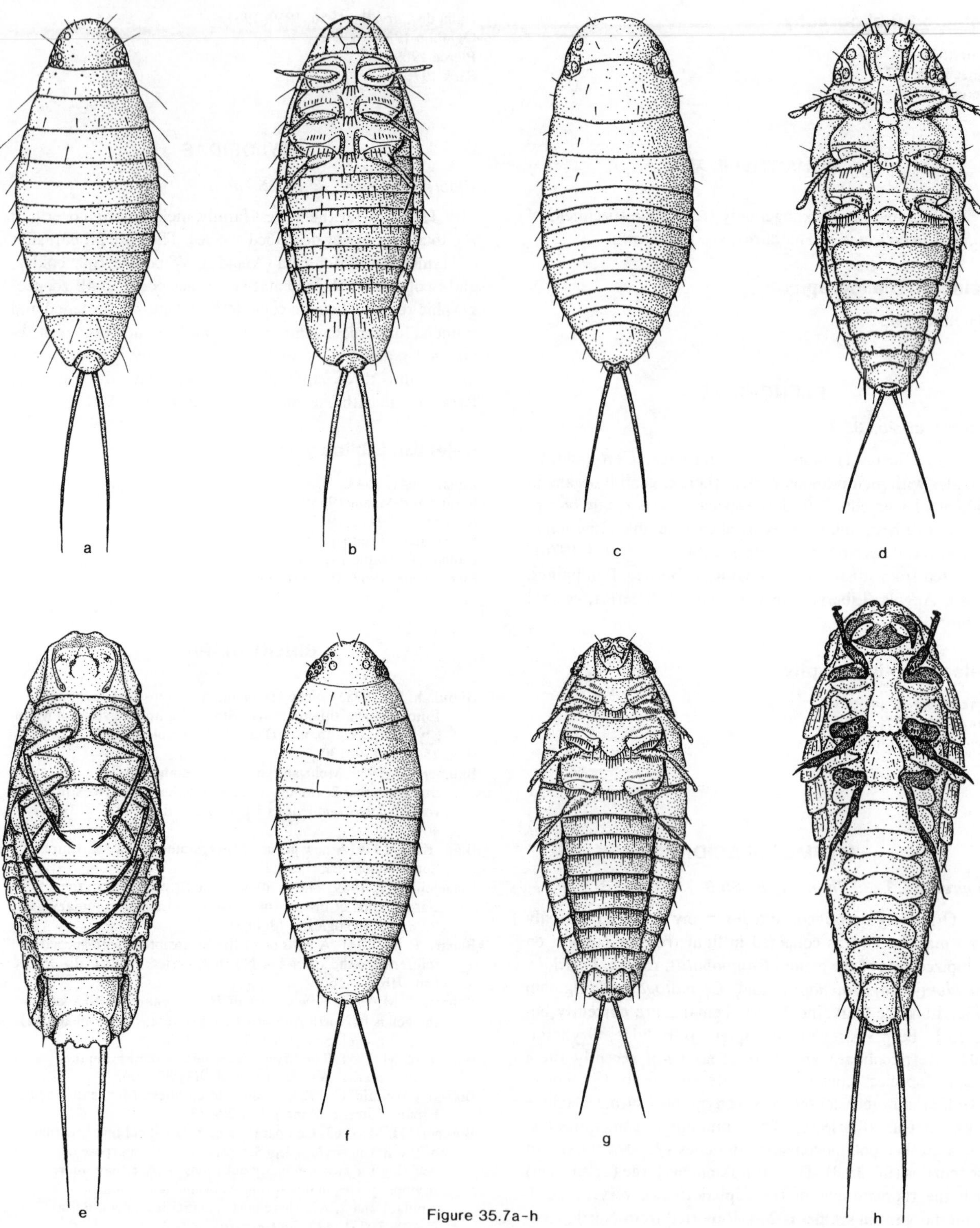

Figures 35.7a-h. Representative triungulinids of the various families of Strepsiptera. **a,b:** Mengenillidae. *Mengenilla chabouti,* dorsal and ventral (redrawn from Silvestri 1943b). **c,d:** Corioxenidae. *Triozocera macrocysti,* dorsal and ventral (redrawn from Esaki and Miyamoto 1958). **e:** Callipharixenidae. *Callipharixenos muiri,* ventral. **f,g:** Halictophagidae. *Halictophagus tettigometrae,* dorsal and ventral (redrawn from Kinzelbach 1971b). **h-k:** Myrmecolacidae. *Stichotrema dallatorreanum* Hofeneder. **h:** female larva.

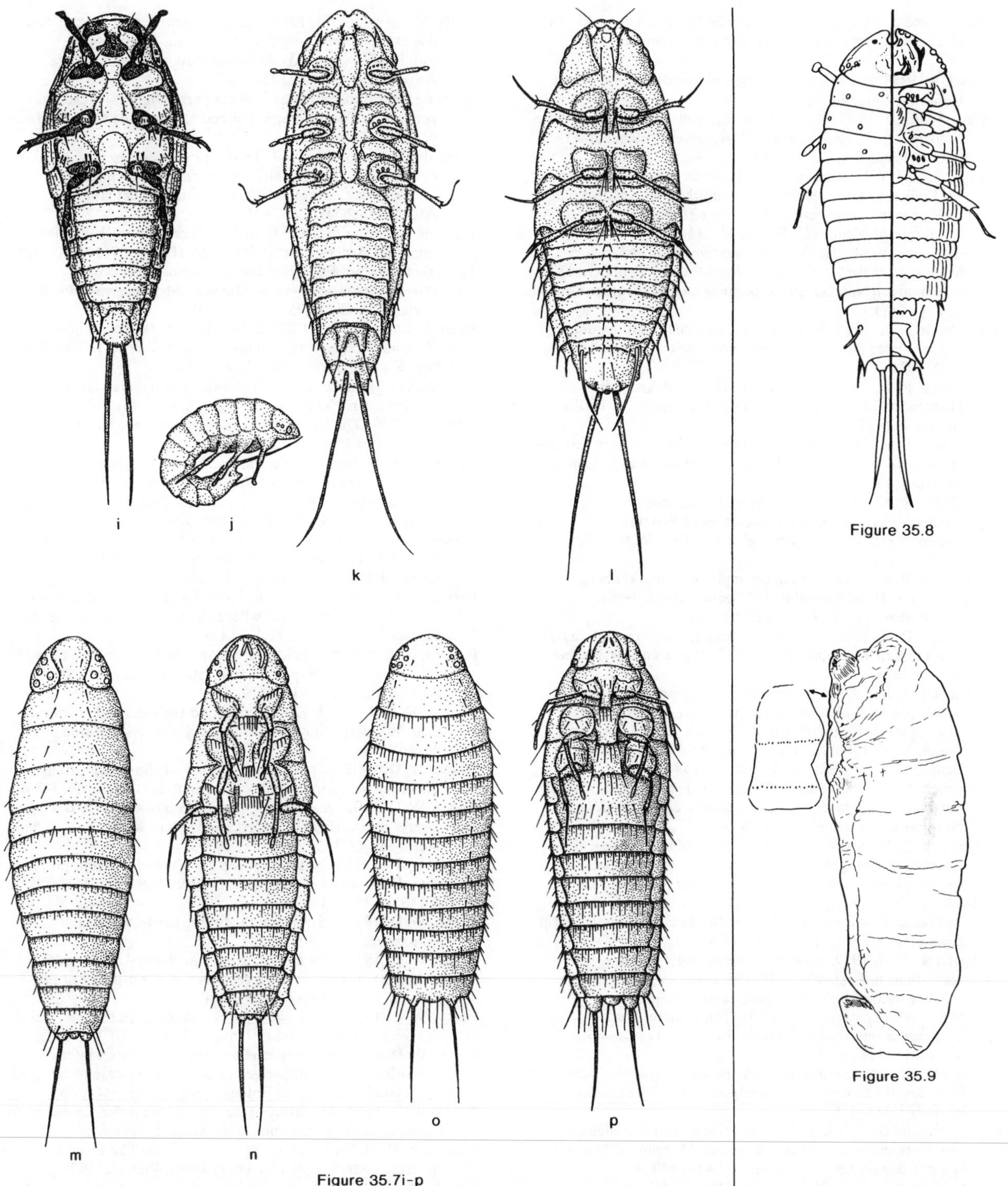

Figure 35.7i-p. Representative triungulinids of the various families of Strepsiptera. **i:** male larva; **j:** lateral view of triungulinid in the jumping position (redrawn after Luna de Carvalho 1967); **k:** Myrmecolacidae. *Caenocholax* sp., ventral (redrawn after Kinzelbach 1971b). **l:** Elenchidae. *Elenchus tenuicornis*, ventral (redrawn after Kinzelbach 1971b). **m,n:** Stylopidae. *Hylecthrus californicus*, dorsal and ventral (redrawn after Bohart 1941). **o,p:** Stylopidae. *Eurystylops desertorum*, dorsal and ventral (redrawn after Bohart 1941).

Figure 35.8. Myrmecolacidae. *Myrmecolax ogloblini*, dorsal and ventral triungulinid. From Argentina.

Figure 35.9. Myrmecolacidae. *Stichotrema barrosmachadoi*, mature 3rd instar and genital openings (left).

Esaki, T., and S. Miyamoto. 1958. The Strepsiptera parasitic on Heteroptera. Proc. Int. Congr. Entomol., Montreal, 1:375–381.

Jeannel, R. 1945. Sur la position systematique des Strepsiptères. Revue Fr. Entomol. 11(3):111–118.

Kathirithambi, J. 1983. The mode of emergence of the adult male *Elenchus tenuicornis* (Kirby) (Strepsiptera: Elenchidae) from its puparium. Zool. J. Linnean Soc. 77:97–102.

Kifune, T., and Y. Hirashima. 1980. Records of the Strepsiptera of Sri Lanka in the collection of the Smithsonian Institution, with descriptions of seven new species (Notulae Strepsipterologicae–VI). Esakia 15:143–159.

Kifune, T., and Y. Maeta. 1965. A tentative list of the Strepsiptera. Part I. The genus *Pseudoxenos* Saunders 1872 (Stylopidae)(Notulae Strepsipterologicae–I). Tohoku Konchu Kenkyu 2(1):1–10.

Kinzelbach, R. K. 1969. Bohartillidae, eine neue Familie der Facherflugler (Insecta, Strepsiptera). Beitr. Neotrop. Fauna 6:92–102.

Kinzelbach, R. K. 1971a. Strepsiptera (Facherflugler). In: Handbuch der Zoologie. IV. Band: Arthropoda, 2. Halfte: Insecta 2(24):1–61.

Kinzelbach, R. K. 1971b. Redeskription und Revision der Gattung *Hylecthrus* Saunders 1850 (Insecta, Strepsiptera). Angew. Parasitol. 12(4):204–220.

Kinzelbach, R. K. 1971c. Morphologische Befunde an Facherfluglern and ihre phylogenetische Bedeutung (Strepsiptera: Insecta). Zoologica 41(119; 1,2):I–XIII, 1–256.

Kinzelbach, R. K. 1972. Die Facherflugler des Senckenberg-Museums. II. Mengenillidae (Insecta: Strepsiptera). Senckenbergiana Biol. 53(5/6):403–409.

Kinzelbach, R. K. 1978. Insecta: Flacherflugler (Strepsiptera). *In:* Die Tierwelt Deutschlands, part 65, 166 p. Gustav Fischer Verlag, Jena.

Kinzelbach, R. K. 1979. The first neotropical fossil of the order Strepsiptera (Amber collection Stuttgart: Insecta, Strepsiptera). Stuttgarter Beitr. Naturk. B 52:1–14.

Kinzelbach, R. K. 1980. Insects of Saudi Arabia. Strepsiptera (2 Beitrag). Fauna of Saudi Arabia, pp. 159–163.

Kinzelbach, R. K. 1983. Flächerflügler aus dem dominikanischen Bernstein (Insecta: Strepsiptera: Myrmecolacidae). Verh. Naturwiss. Ver. Hamburg 26:29–36.

Kinzelbach, R. K., and H. Lutz. 1985. Stylopid larva from the Eocene—a spotlight on the phylogeny of the stylopids (Strepsiptera). Ann. Entomol. Soc. Amer. 78:600–602.

Kirby, W. 1813. VI. Strepsiptera, a new order of insects proposed; and the characters of the order, with those of its genera, laid down. Trans. Linn. Soc. Lond. 11:86–123.

Kirkpatrick, T. W. 1937. Studies on the ecology of coffee plantations in East Africa. II. The autecology of *Antestia* spp. (Pentatomidae) with a particular account of a Strepsipterous parasite. (Part II: The bionomics of *Corioxenos antestiae*). Trans. R. Ent. Soc. London 86:247–343 (281–341).

Kogan, M. 1958. A new species of the genus *Triozocera* Pierce from Brazil (Mengeidae, Strepsiptera). Studia Entomol. (Brazil) 1:421–426.

Kogan, M., and S. J. de Oliveira. 1964. New Guinean Mengeidae and Myrmecolacidae of the American Museum of Natural History. Studia Entomol. (Brazil) 7:459–470.

Kogan, M., and S. J. de Oliveira. 1966. Brazilian Xeninae parasitizing *Polybia* wasps with the description of a new genus and six new species. Rev. Brasil. Biol. 26:345–360.

Lawrence, J. F., and A. F. Newton, Jr. 1982. Evolution and classification of beetles. Ann. Rev. Ecol. Syst. 13:261–290.

Luna de Carvalho, Ed. 1953. Contribuição para o conhecimento dos estrepsípteros de Portugal (III nota). Mems. Estud. Mus. Zool. Univ. Coimbra 215:5–13.

Luna de Carvalho, Ed. 1956. Primeira contribuição para o estudo dos estrepsípteros angolenses (Insecta, Strepsiptera). Publicaçōes Culturais da Companhia de Diamantes da Angola (Lisbon). 29:11–54.

Luna de Carvalho, Ed. 1959. Segunda contribuição para o estudo dos estrepsípteros angolenses (Insecta, Strepsiptera). Publicaçōes Culturais da Companhia de Diamantes da Angola (Lisbon). 41:125–154.

Luna de Carvalho, Ed. 1961. Tabela para a determinação dos generos de estrepsipteros (Insecta). Garcia de Orta (Lisbon). 9(4):691–698.

Luna de Carvalho, Ed. 1967. Terceira contribuição para o estudo dos estrepsípteros angolenses (Insecta, Strepsiptera). Publicaçōes Culturais da Companhia de Diamantes da Angola (Lisbon). 77:17–56.

Luna de Carvalho, Ed. 1978. African Strepsiptera (Ethiopian region). Garcia de Orta, Ser. Zool. (Lisbon) 7(1–2):41–106.

Luna de Carvalho, Ed. 1979. Esboço monográfico dos estrepsípteros de Portugal (Insecta, Strepsiptera). Bol. Soc. Portuguesa Cien. Nat. 19:165–195.

Luna de Carvalho, Ed. 1985. Estrepsípteros da Macaronésia. (Insecta Strepsiptera). Actas II Congr. Ibérico Entomol., Bol. Soc. Port. Entomol. Supl. 1, 2:63–73.

Menge, A. 1866. Über ein Rhipidopteron und einige Helminthen im Bernstein. Schr. Naturf. Ges. Danzig 2(1):1–8.

Ogloblin. 1939. The Strepsiptera parasites of ants. Congr. Int. Ent., Berlin 1938, 2:1277–1284.

Otake, A., P. H. Somasundaram, and M. B. Abeyiloon. 1976. Studies on populations of *Sogatella furcifera* Horvath and *Nilaparvata lugens* Stal (Hemiptera: Delphacidae) and their parasites in Sri Lanka. Appl. Ent. Zool. 11(3):284–294.

Parker, H. L., and H. D. Smith. 1933. Additional notes on the Strepsipteron *Eoxenos laboulbenei* Peyerimhoff. Ann. Entomol. Soc. Amer. 26(2):217–233.

Parker, H. L., and H. D. Smith. 1934. Further notes on *Eoxenos laboulbenei* Peyerimhoff with a description of the male. Ann. Entomol. Soc. Amer. 27:468–479.

Perkins, R. C. L. 1905. Leafhoppers and their natural enemies (Pt. III. Stylopidae). Rep. Wk. Exp. Sta. Hawaii Sugar Plant. Assoc., Bull. 1(3):90–111.

Perkins, R. C. L. 1907. Leafhoppers and their natural enemies. Rep. Wk. Exp. Sta. Hawaii Sugar Plant. Assoc., Bull. 4(7):58–59.

Pierce, W. D. 1908. A preliminary review of the classification of the order Strepsiptera. Proc. Entomol. Soc. Wash. 9:75–83.

Pierce, W. D. 1909. A monographic revision of the twisted winged insects comprising the order Strepsiptera Kirby. Bull. U.S. Nat. Mus. 66:1–232.

Pierce, W. D. 1911a. Notes on insects of the order Strepsiptera, with descriptions of new species. Proc. U.S. Nat. Mus. 40(1834):487–511.

Pierce, W. D. 1911b. Strepsiptera. *In:* Wytsman: Genera Insectorum, fasc. 121:1–54.

Pierce, W. D. 1918. The comparative morphology of the order Strepsiptera together with records and descriptions of insects. Proc. U.S. Nat. Mus. 54:391–501.

Pierce, W. D. 1936. The position of Strepsiptera in the classification of insects. Entomol. News 47(10):257–263.

Pierce, W. D. 1961. A new genus and species of Strepsiptera parasitic on a leafhopper vector of a virus disease of rice and other Gramineae. Ann. Entomol. Soc. Amer. 54(4):467–474.

Pierce, W. D. 1964. The Strepsiptera are a true order, unrelated to Coleoptera. Ann. Entomol. Soc. Amer. 57:603–605.

Riek, E. F. 1970. Strepsiptera. pp. 622–635. *In:* The Insects of Australia. Melbourne University Press, Victoria. 1079 p.

Rossius, P. 1793. Observation de M. Rossi sur un nouveau genre d'insecte, voisin des Ichneumons. Bull. Soc. Philomatique 1:49.

Silvestri, F. 1934. *Halictophagus tettigometrae* Silv. di *Tettigometra impressifrons* Muls. et Rey. Comp. Ent. Applic. 1:366–367.

Silvestri, F. 1941a. Studi sugli Strepsiptera (Insecta). I. Ridescrizione e ciclo dell'*Eoxenos laboulbenei* Peyerimoff (sic!). Boll. Lab. Zool. Gen. Agr.: Fac. Agr. Portici 31:311–341.

Silvestri, F. 1941b. Studi sugli Strepsiptera (Insecta). II. Descrizione, biologia e sviluppo postembrionale dell'*Halictophagus tettigometrae* Silv. Boll. Lab. Zool. Gen. Agr.: Fac. Agr. Portici 32:11–48.

Silvestri, F. 1943a. Morfologia dell'ovo, maturazione e primi stadi di sviluppo dell'*Halictophagus* (Insetti Strepsipteri). Commentat. Pontif. Acad. Sci. 7(17):523–530.

Silvestri, F. 1943b. Studi sugli Strepsiptera (Insecta). III. Descrizione e biologia di 6 specie italiane di *Mengenilla*. Boll. Lab. Zool. Gen. Agr.: Fac. Agr. Portici. 32:197–282.

Silvestri, F. 1946. Identita dei generi di Mengenillidae (Strepsiptera) *Austrostylops* Lea et *Mengenilla* Hofeneder. Boll. Lab. Ent. Agr. Portici 6:15–16.

Ulrich, W. 1927. Über das bisher einzige Strepsipteron aus dem baltischen Bernstein und über eine Theorie der Mengeidenbiologie. Z. Wiss. Biol. A. Morph. Okol. Tiere) 8:45–62.

Yang, C. K. 1964a. Notes on the new subfamily Tridactylophaginae (Strepsiptera: Halictophagidae). Acta Ent. Sin. 13(4):606–613. (in Chinese with English trans.)

Yang, C. K. 1964b. Notes on the genus *Halictophagus* of China (Strepsiptera: Halictophagidae). Acta Zootaxon. Sinica (Beijing) 1:76–83. (in Chinese with English Abstract)

Order Siphonaptera

36 

Robert E. Elbel
University of Utah

FLEAS

Adult fleas are blood-sucking parasites of birds and mammals, including humans, but most larvae are free living. Based on larval and adult characters, fleas are considered to be descended from winged mecopteran-like ancestors (Rothschild, 1975). However, George W. Byers (personal communication) believes that fleas arose from nematocerous Diptera since legless larvae with well-developed head capsules and adults without mandibles are shared by both taxa. Riek (1970) reported a flea-like male from Lower Cretaceous siltstone of Australia with unusually long, non-jumping, hind legs and with nematocerous-type antennae which he suggested supported the conclusion that fleas evolved from a nematocerous-type ancestor. However, Smit (1972) and Traub (1980) refute the claim that the specimen is a flea.

DIAGNOSIS

Flea larvae (fig. 36.10) are distinguished from all other larvae by being eyeless and legless, with chewing mouthparts consisting of labrum, mandibles, maxillae and labium, with maxillary and labial palps projecting down from the ventral surface of the head (fig. 36.1). The one-segmented antenna arises from an elevated mound ringed with three large processes which alternate with smaller ones; the tip of the antenna has a single seta flanked by four minute setae.

BIOLOGY

Fleas mate frequently and eggs either fall from the host or are laid in the nest. Eggs are dry if laid on the host or viscous if laid in the nest. They are laid singly, the number laid at one time varying per species, averaging three for cat fleas, six for rat fleas and 18 for squirrel fleas. The oval, white eggs usually have smooth, polished surfaces and hatch in 3–10 days; newly hatched larvae have an egg breaker which is lost at the first molt. There are three larval instars except for two in some Tungini (Tunginae, Pulicidae). The active larvae feed on flea feces and other organic matter except for the non-feeding *Tunga monositus* Barnes and Radovsky. Third instars stop feeding and go through a defecation period in which all contents of the alimentary canal are expelled; the larva enters a prepupal stage, becomes waxy white and spins a silken cocoon to which grains of sand or dust adhere and in which the prepupa rests, bent double, before molting to a pupa. The pupal period lasts for only a few days and if the flea rests in the cocoon, it does so as a prepupa or as an adult. The entire life cycle usually can be completed in 3–4 weeks above 24 C and 84% RH but lower temperatures and humidities prolong development. When the adult emerges, it tends to find the same host species as that upon which the parents were parasitic.

MORPHOLOGY

The 13-segmented cylindrical body has three thoracic (T) and 10 abdominal (A) segments (fig. 36.18). The A10 or terminal anal segment (fig. 36.14) bears downward, and backward-directed anal struts that aid in locomotion. In moving, the anal struts are turned down onto the surface, the body is elongated, the raised head slides forward to a point where the labrum hooks onto the surface, and the body is drawn forward. Anterior to the anal struts are a few large ventrolateral setae, usually similar in length, with the anal comb above as a single or double row of fine setae (figs. 36.11, 36.14). Anterior to the ventrolateral setae is the smaller anteroventral seta. At the base of the anal struts, there are a number of fine strut setae. On the first 12 body segments, setae are arranged in transverse rows with one row of short setae in the Tungini (Tunginae, Pulicidae) (fig. 36.6) and two rows in other larvae, an anterior row of short setae, and a posterior row of long setae (fig. 36.18). In the anterior row on A1–8, the third seta is usually in front of the others (fig. 36.21, D). The anterior row as used here is equivalent to the middle row of Kirjakova (1964, 1965) since she considers the anterior row in Ceratophyllidae and some Hystrichopsyllidae as a row of five minute spinules on the anterior part of the first 12 segments. In the posterior row, setae arise from dorsal, ventral and lateral plates; on A1–6, the two ventral setae consist of a long and a short seta arising close together from the same plate.

Since the chaetotaxy of flea larvae is bilaterally symmetrical and there are no median dorsal or median ventral setae, descriptions are written for one lateral surface. The major rows of setae on the head are designated as anterior and posterior (Elbel, 1951) (fig. 36.1). Most European workers follow Kirjakova (1961) in illustrating the dorsal and ventral surfaces. They designate Elbel's ventral seta of the anterior head row as belonging to the posterior row. I follow Elbel's arrangement as do Stark et al. (1976). When dorsal and ventral surfaces are illustrated, marginal setae are frequently duplicated on both surfaces, which causes confusion in counting the number of setae per row, and often makes it difficult to distinguish dorsal from ventral setae. I have experienced little difficulty in assigning setae to appropriate rows in lateral view except for strut setae (fig. 36.14) which are found on lateral, dorsal and ventral surfaces of each strut. For the present study, the total number of strut setae on all surfaces was counted in lateral view and divided by two, the number being verified by counts on either side of a dorsal-ventral view.

TECHNIQUES

Descriptions are based most often on larvae obtained by rearing eggs deposited by known gravid females or by placing identified adults on flea-free hosts in a container with clean bedding (Stark et al., 1976; Elbel, in prep.). However, larvae are described from nest material when the host association is such that only a single undescribed larval species would be expected.

The following method of mounting flea larvae is modified from suggestions of W. T. Atyeo, University of Georgia, and A. W. Grundmann, University of Utah. Place larvae in 10% KOH overnight, followed by distilled water for 1/2 hour and chlorolactophenol with a few drops of a 1% aqueous solution of acid fuchsin for four hours. The chlorolactophenol consists of 2 parts chloral hydrate, 1 part melted phenol and 1 part lactic acid. After 15 minutes each in 40%, 70% and 95% ethanol, then at least two hours in Cellosolve, mount larvae inverted in a drop of Cellosolve in center of slide with head facing right (when viewed through compound microscope, larva faces left); add Canada balsam to edge of cover slip and place slide overnight in drying oven at about 60 C. Place host label on right and flea identification label on left.

LARVAL CLASSIFICATION FOR AMERICA NORTH OF MEXICO[1]

Order **SIPHONAPTERA (264)**
- Pulicidae (14)
  - Tunginae
    - Tungini: *Tunga, Echidnophaga* (adult = Pulicinae, Pulicini)
  - Pulicinae
    - Pulicini: *Pulex*
    - Archaeopsyllini—Spilopsyllini—Xenopsyllini: *Ctenocephalides, Actenopsylla,*[2] *Cediopsylla, Euhoplopsyllus, Hoplopsyllus,*[2] *Xenopsylla*
- Rhopalopsyllidae (3)
  - Rhopalopsyllinae: *Polygenis, Rhopalopsyllus*
- Vermipsyllidae (5): *Chaetopsylla*[2]
- Hystrichopsyllidae (103)
  - Ctenophthalminae
    - Carterettini: *Carteretta*[2]
    - Ctenophthalmini: *Ctenophthalmus*
  - Rhadinopsyllinae
    - Corypsyllini: *Corypsylla,*[2] *Nearctopsylla*[2]
    - Rhadinopsyllini: *Paratyphloceras, Rhadinopsylla, Trichopsylloides*
  - Hystrichopsyllinae—Stenoponiinae: *Atyphloceras,*[2] *Hystrichopsylla, Stenoponia, Mioctenopsylla* (adult = Ceratophyllidae, Ceratophyllinae)
  - Anomiopsyllinae
    - Jordanopsyllini: *Jordanopsylla*[2]
    - Anomiopsyllini: *Anomiopsyllus, Callistopsyllus,*[2] *Megarthroglossus,*[2] *Stenistomera*[2]
  - Neopsyllinae
    - Neopsyllini: *Neopsylla,*[2] *Tamiophila*[2]
    - Phalacropsyllini: *Catallagia, Delotelis,*[2] *Epitedia, Meringis,*[2] *Phalacropsylla*[2]
  - Doratopsyllinae
    - Doratopsyllini: *Corrodopsylla,*[2] *Doratopsylla*[2]
- Ischnopsyllidae (8)
  - Ischnopsyllinae: *Myodopsylla, Nycteridopsylla, Sternopsylla*[2]
- Leptopsyllidae (19)
  - Amphipsyllinae
    - Amphipsyllini: *Amphipsylla,*[2] *Ctenophyllus,*[2] *Odontopsyllus*[2]
    - Dolichopsyllini: *Dolichopsyllus*[2]
    - Ornithophagini: *Ornithophaga*[2]
  - Leptopsyllinae
    - Leptopsyllini: *Leptopsylla, Peromyscopsylla, Conorhinopsylla* (adult = Hystrichopsyllidae, Anomiopsyllinae, Anomiopsyllini)
- Ceratophyllidae (112)
  - Ceratophyllinae: *Amphalius,*[2] *Ceratophyllus, Dasypsyllus,*[2] *Diamanus, Jellisonia,*[2] *Malaraeus, Megabothris, Monopsyllus, Nosopsyllus, Opisocrostis, Opisodasys, Orchopeas, Oropsylla, Pleochaetis,*[2] *Tarsopsylla,*[2] *Thrassis*
  - Dactylopsyllinae: *Dactylopsylla,*[2] *Foxella*[2]

1. Modified from that of the adults (Lewis, 1972–75). The arrangement of families, subfamilies and tribes is based on larval similarity. The classification of larvae used here differs from that of the adults; hence, families, subfamilies, tribes and included genera are given. The number of species north of Mexico is given in parentheses after each family.

2. Larvae have not been examined.

KEY TO FAMILIES, SUBFAMILIES AND TRIBES OF KNOWN NORTH AMERICAN FLEA LARVAE

Note: For one lateral surface, setae are numbered from mid-dorsal to mid-ventral line:

1.	First 12 body segments with short setae (fig. 36.6)	Tungini, **Tunginae, *Pulicidae*** (p. 676)
1′.	First 12 body segments with long setae (fig. 36.10)	2
2(1′).	Anal comb with single row of setae (fig. 36.11)	3
2′.	Anal comb with double row of setae (figs. 36.13, 36.14)	5

3(2). Posterior row on A9 with six setae (fig. 36.21) .... Ctenophthalmini, **Ctenophthalminae,** ***Hystrichopsyllidae*** (p. 682)

3′. Posterior row on A9 with 7–8 setae (figs. 36.8, 36.10) 4

4(3′). Anal segment with 4–6 (usually five) ventrolateral setae (fig. 36.9) .... Pulicini, **Pulicinae,** ***Pulicidae*** (p. 680)

4′. Anal segment with three ventrolateral setae (fig. 36.11) .... Archaeopsyllini-Spilopsyllini-Xenopsyllini, **Pulicinae,** ***Pulicidae*** (p. 680)

5(2′). Posterior row on A9 with 7–8 setae (fig. 36.19) 6

5′. Posterior row on A9 with six setae (fig. 36.21) 7

6(5). Anal segment with four ventrolateral setae, first extremely long (fig. 36.12) .... *Polygenis gwyni,* **Rhopalopsyllinae,** ***Rhopalopsyllidae*** (p. 682)

6′. Anal segment with three ventrolateral setae (fig. 36.13) .... **Hystrichopsyllinae-Stenoponiinae,** ***Hystrichopsyllidae*** (p. 683)

7(5′). Mandible broad, curved, with large distal tooth and 5–7 minute blunt teeth (fig. 36.2) .... Rhadinopsyllini, **Rhadinopsyllinae,** ***Hystrichopsyllidae*** (p. 683)

7′. Mandible narrow with five or more teeth (figs. 36.4, 36.5) 8

8(7′). Anal segment with two ventrolateral setae (fig. 36.15) .... Anomiopsyllini, **Anomiopsyllinae,** ***Hystrichopsyllidae*** (p. 684)

8′. Anal segment with three ventrolateral setae (fig. 36.14) 9

9(8′). Anterior row of short setae on A1–5 with setae 2–5 in straight line (fig. 36.20, C) .... Phalacropsyllini, **Neopsyllinae,** ***Hystrichopsyllidae*** (p. 684)

9′. Anterior row of short setae on A1–5 with seta three in front of others (fig. 36.21, D) 10

10(9′). Anal segment with just one seta in anterior row of anal comb (fig. 36.13) .... Leptopsyllini, **Leptopsyllinae,** ***Leptopsyllidae*** (p. 686)

10′. Anal segment with two or more setae in anterior row of anal comb (fig. 36.14) 11

11(10′). Posterior row on A8 with six setae (fig. 36.17) .... **Ischnopsyllinae,** ***Ischnopsyllidae*** (p. 685)

11′. Posterior row on A8 with five setae (fig. 36.21) .... **Ceratophyllinae,** ***Ceratophyllidae*** (p. 686)

Selected Bibliography

Bacot 1914.
Bacot and Ridewood 1914.
Cotton 1970.
Elbel 1951, 1952.
Henderson 1928.
Karandikar and Munshi 1952.
Kirjakova 1961, 1964, 1965.
Mitzmain 1910.
Riek 1970.
Rothschild 1975.
Sharif 1937.
Smit 1972.
Stark et al. 1976.
Traub 1980.

PULICIDAE

Figures 36.3, 36.4, 36.6–36.11

Tunginae: Tungini

Figures 36.6, 36.7

Relationships and Diagnosis: The short setae on the first 12 body segments and the anal struts that are reduced or absent (fig. 36.6) distinguish the Tungini from all other larvae. The anal comb has a double row of short setae (fig. 36.7). Included here is the sticktight flea, *Echidnophaga gallinacea* (Westwood).

Biology and Ecology: The adult chigoe, *Tunga penetrans* (L.), infests many large wild and domestic mammals in southern United States and much of S. America; the chigoe attacks humans, often burrowing into their feet. *T. monositus* Barnes and Radovsky of Baja, California is found as an adult on the deer mouse, *Peromyscus maniculatus,* other *Peromyscus,* and wood rats, *Neotoma.* The adult sticktight flea is a pest of poultry in temperate and semitropical climates throughout the world, attacking often on the head; it parasitizes many mammals, including humans.

The behavior of Tungini adults differs considerably from that of most fleas since the females remain attached to the host. *T. penetrans* bores into the horny epidermis and becomes completely buried in the skin, *E. gallinacea* which does not penetrate under the horny epidermis is anchored only by the proboscis, and *T. monositus* attaches to the dorsal surface of the deer mouse ear with only the caudal disc exposed through a circular opening in the host skin. Copulation takes place on the host and females continuously drop eggs which hatch in the soil where larvae develop. Both *Tunga* species have only two larval instars. *T. monositus* larvae have non-feeding mouthparts and are quiescent except for construction of the cocoon; the relatively large egg contains nutrients sufficient to provide for larval development and survival of the non-feeding adult male. *T. penetrans* larvae develop better on a mixed diet but *E. gallinacea* larvae will die within four days without flea feces or dried chicken blood.

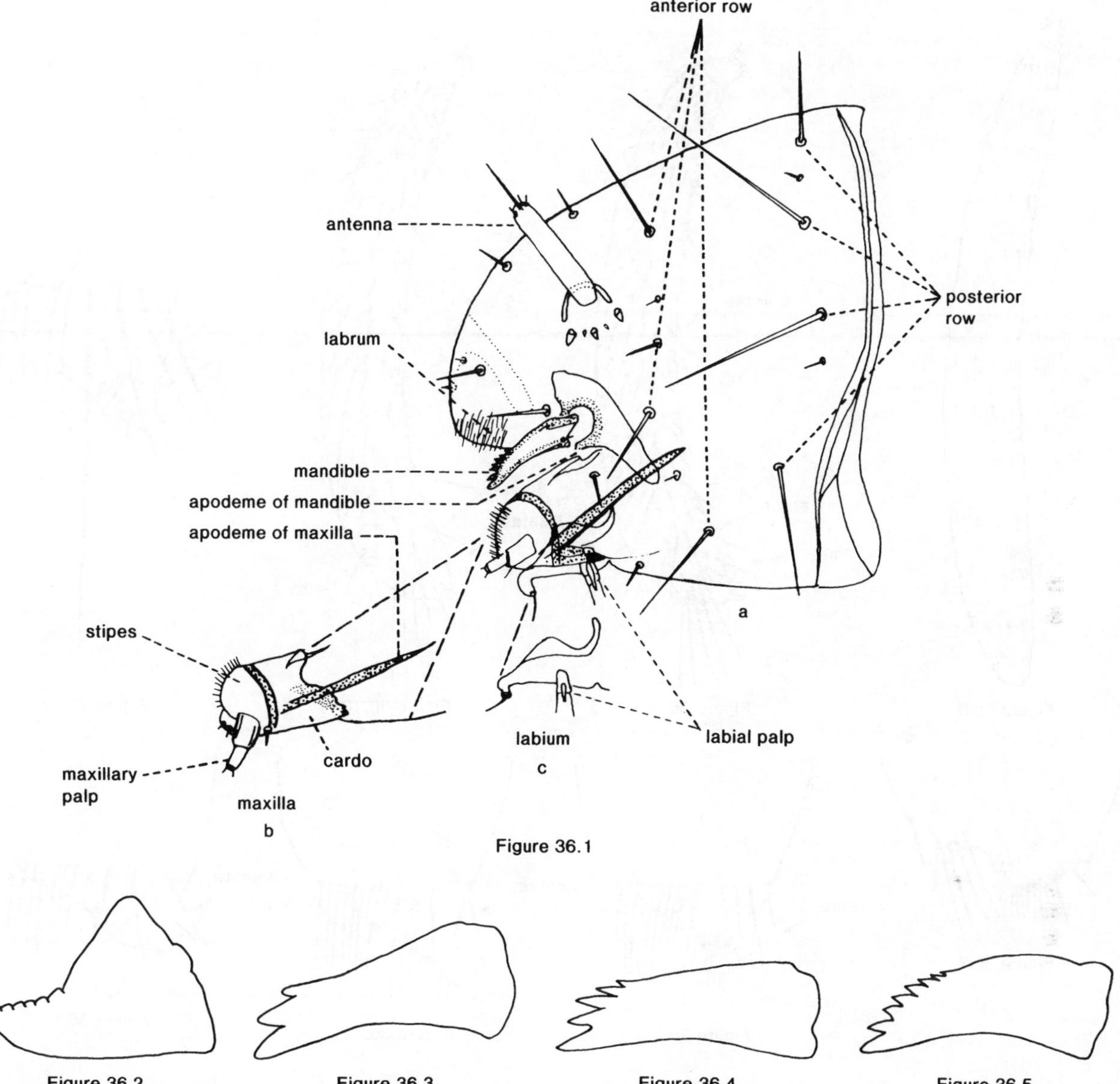

Figure 36.1a-c. Siphonaptera. Ceratophyllidae. *Monopsyllus wagneri* (Baker). **a.** head, lateral; **b.** maxilla; **c.** labium. (Figure 36.1 reprinted with permission of the editor from Stark et al. (1976), J. Med. Ent. 13:107–111.)

Figures 36.2–36.5. Mandibles. **(36.2.)** Hystrichopsyllidae. *Rhadinopsylla pentacantha* (Rothschild). **(36.3.)** Hystrichopsyllidae. *Ctenophthalmus orientalis* (Wagner). **(36.4.)** Pulicidae. *Ctenocephalides felis* (Bouché). **(36.5.)** Ceratophyllidae. *Ceratophyllus gallinae* (Schrank).

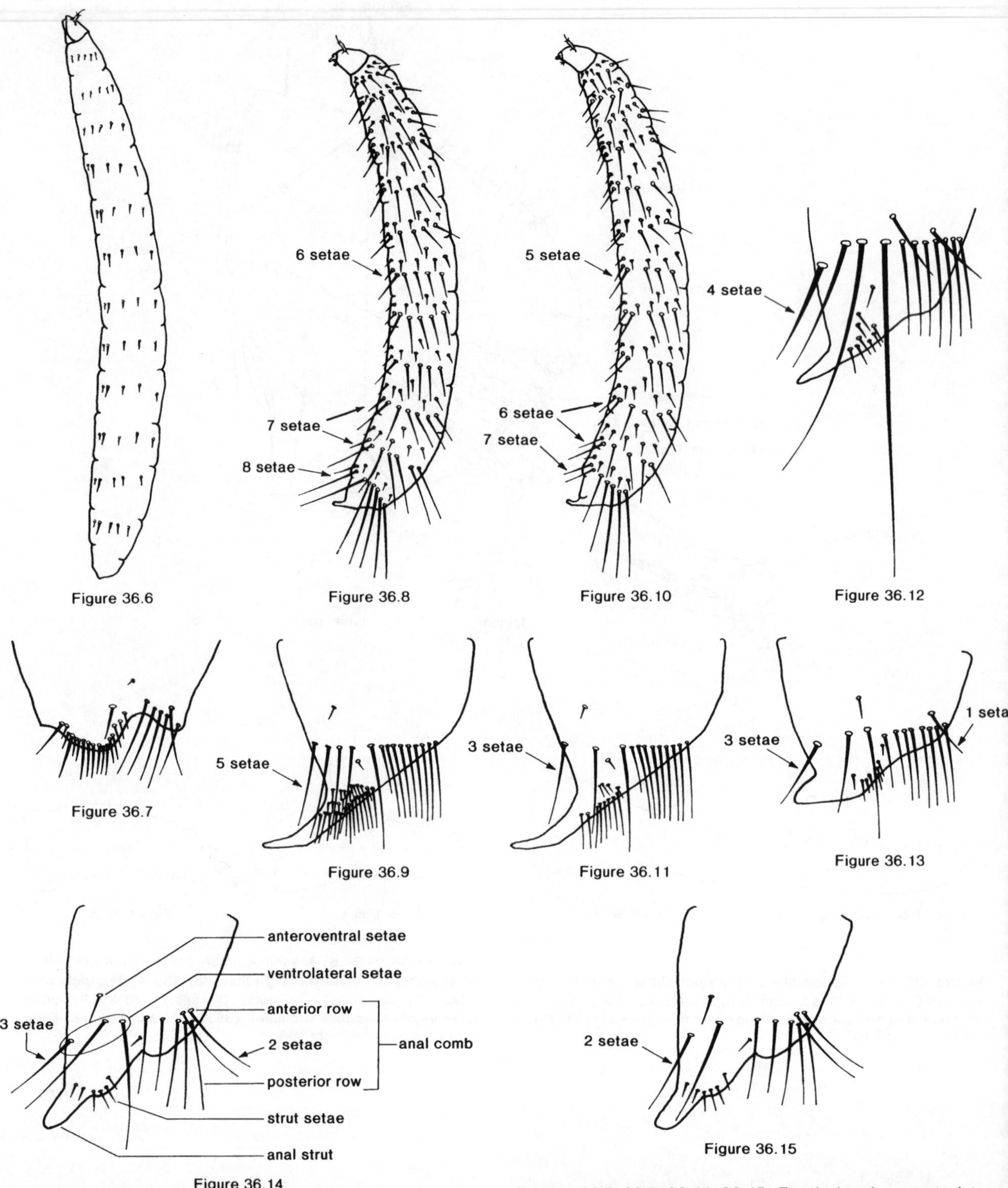

Figure 36.6

Figure 36.8

Figure 36.10

Figure 36.12

Figure 36.7

Figure 36.9

Figure 36.11

Figure 36.13

Figure 36.14

Figure 36.15

Figures 36.6, 36.8, 36.10. Siphonaptera. Entire larvae, lateral. **(36.6.)** Pulicidae. *Echidnophaga gallinacea* (Westwood). **(36.8.)** Pulicidae. *Pulex irritans* L. **(36.10.)** Pulicidae. *Xenopsylla cheopis* (Rothschild).

Figures 36.7, 36.9, 36.11–36.15. Terminal anal segments, lateral. **(36.7.)** Pulicidae. *Echidnophaga gallinacea* (Westwood). **(36.9.)** Pulicidae. *Pulex irritans* L. **(36.11.)** Pulicidae. *Xenopsylla cheopis* (Rothschild). **(36.12.)** Rhopalopsyllidae. *Polygenis gwyni* (C. Fox). **(36.13.)** Leptopsyllidae. *Leptopsylla segnis* (Schönherr). **(36.14.)** Ceratophyllidae. *Orchopeas leucopus* (Baker). **(36.15.)** Hystrichopsyllidae. *Anomiopsyllus hiemalis* Eads and Menzies. (Figures 36.6–36.15 modified from Elbel (1951, 1952) with permission of the editor, J. Parasitology.)

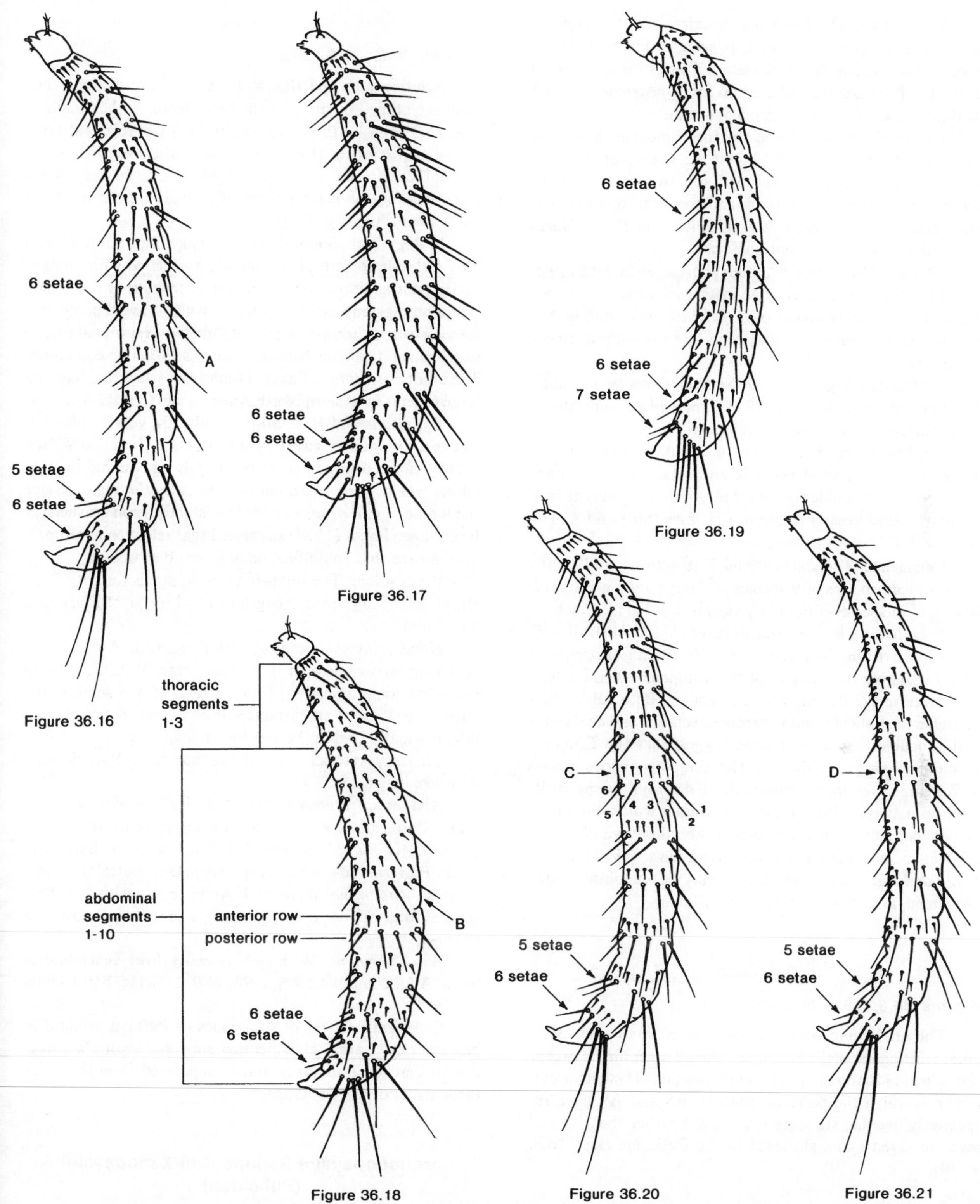

Figures 36.16–36.21. Siphonaptera. Entire larvae, lateral. **(36.16.)** Hystrichopsyllidae. *Anomiopsyllus hiemalis* Eads and Menzies; A, short seta 2 of posterior body row. **(36.17.)** Ischnopsyllidae. *Myodopsylla insignis* (Rothschild). **(36.18.)** Leptopsyllidae. *Leptopsylla segnis* (Schönherr); B, short seta 2 of posterior body row. **(36.19.)** Rhopalopsyllidae. *Polygenis gwyni* (C. Fox). **(36.20.)** Hystrichopsyllidae. *Epitedia wenmanni* (Rothschild); C, short setae 2–5 in straight line in anterior body row. **(36.21.)** Ceratophyllidae. *Orchopeas sexdentatus* (Baker); D, short seta 3 in front of others in anterior body row. (Figures 36.16–36.21 modified from Elbel (1951, 1952) with permission of the editor, J. Parasitology.)

Description: The following description is based on *E. gallinacea*, *T. monositus* and *T. penetrans*. For both *Tunga* species, average length of first instars 2 mm, of second instars 2 mm for *T. monositus* and 2.5 mm for *T. penetrans*. Length of third instar of *E. gallinacea* 3.5–4 mm.

Head: Mandible reduced and inconspicuous in *T. monositus* but *T. penetrans* with 12 teeth, six in each of two rows. *E. gallinacea* with large distal tooth followed by 4–6 smaller teeth in single row. Both anterior and posterior rows with two short setae except *T. monositus* with five and *E. gallinacea* with three short setae in anterior row.

Thorax: With one row of four short setae in both *Tunga* species but with five short setae in *E. gallinacea*. In the anterior part of each segment of *E. gallinacea* (including A1–9) a few very minute setae may represent the anterior row of other larvae.

Abdominal Segments 1–9: With one row of five short setae except for six setae on A9 of *T. penetrans*, ventralmost seta shortest for each species in each row.

Anal Segment: Each species with number of short setae more or less in form of ventrolateral setae, strut setae and anal comb with double row of setae. Anal struts absent in *T. monositus* and greatly reduced in *T. penetrans* and *E. gallinacea*.

Comments: *T. monositus* and *T. penetrans* are the only N. American species of 9 species of *Tunga* and only *E. gallinacea* of 21 species of *Echidnophaga* is found in N. America. The *E. gallinacea* larva as well as larvae of the Australian *E. myrmecobii* Rothschild and *E. perilis* Jordan (Pilgrim, in prep.) resemble the Tungini of the Tunginae but the adults are placed in the Pulicini of the Pulicinae. Both subfamilies belong to the Pulicidae but the reduced setation and anal comb with a double row of short setae suggest that the Tunginae larvae should have family rank. However, the African species of Tunginae, *Neotunga euloidea* Smit, destroys the sharp differences between adult pulicids and tungids which, therefore, can no longer be recognized as a family; the larva of *N. euloidea* has not been examined. Kirjakova (1971) places *Echidnophaga* larvae in the family Ctenophthalmidae (= Ctenophthalminae, Hystrichopsyllidae).

Pulicinae

Figures 36.3, 36.4, 36.8–36.11

The anal comb with a single row of setae (figs. 36.9, 36.11) distinguishes the Pulicinae from all other larvae except the Ctenophthalmini (Ctenophthalminae, Hystrichopsyllidae) in which the posterior rows on A8 and A9 have respectively five and six setae (fig. 36.21) rather than six and seven or seven and eight setae in the Pulicinae (figs. 36.8, 36.10).

Pulicini

Figures 36.3, 36.8, 36.9

Relationships and Diagnosis: Larvae are distinguished from the other Pulicinae by the three teeth on the mandible (fig. 36.3) and by the 4–6 (usually five) ventrolateral setae on the anal segment (fig. 36.9). Also, in the posterior row on A1–6, there are six setae; A7 and A8 have seven and A9 has eight setae in the posterior row (fig. 36.8). There are 25 or more strut setae (fig. 36.9).

Biology and Ecology: The cosmopolitan adult human flea, *Pulex irritans* L., is frequently associated with humans under poor sanitary conditions but *P. irritans* is commonly found on pigs and large carnivores which are probably preferred hosts. *P. irritans* transmits plague experimentally, so may transmit plague between humans in major epidemics. The adult *P. simulans* Baker is found on carnivores, deer and large rodents in western North America and Central America.

P. irritans and *P. simulans* adults are very similar but can be differentiated easily by the males, so Haas and Wilson (1967) isolated gravid females in vials and reared eggs to adults which were separated to species by the males. Since the larvae of *P. irritans* can feed satisfactorily on crushed rat feces alone, Bacot (1914) suggested that selection might produce a race that could feed on any possible rubbish.

Description: The following description is for both *P. irritans* and *P. simulans*. Length of third instar of *P. irritans* 4.6–6 mm.

Head: Narrow mandible with three teeth. Anterior and posterior rows each with three long setae. R. L. C. Pilgrim (personal communication) found a total of 4–5 setae in the anterior row, since an extremely minute seta, not easily visible, is usually present between each long seta.

Thorax: Anterior row with five short setae. Posterior row with five long setae.

Abdominal segments 1–6: Anterior row with six short setae. Posterior row with six long setae, ventralmost seta shortest. *Abdominal segments* 7–8: Anterior row with six short setae. Posterior row with seven long setae, ventralmost seta shortest. *Abdominal segment* 9: Anterior row with 4–5 short setae. Posterior row with eight long setae, ventralmost seta shortest.

Anal Segment: With 4–6 (usually five) ventrolateral setae. Anal comb with single row of 9–11 setae. Strut setae, 25–34.

Comments: Three of six species of *Pulex* are found in N. America. Larval classification supports adult classification in that the Pulicini should be separated from the other three tribes of the Pulicinae.

Archaeopsyllini-Spilopsyllini-Xenopsyllini (Pulicidae)

Figures 36.4, 36.10, 36.11

Relationships and Diagnosis: Larvae are distinguished from the Pulicini by more than three teeth on the mandible (fig. 36.4) and by the three ventrolateral setae on the anal segment (fig. 36.11). Also, in the posterior row on A1–6, there are five setae; A7 and A8 have six and A9 has seven setae in the posterior row (fig. 36.10). There are fewer than 25 strut setae (fig. 36.11).

Biology and Ecology: In the Archaeopsyllini, adult cat fleas, *Ctenocephalides felis* (Bouché), and dog fleas, *C. canis* (Curtis), are almost cosmopolitan but absent on pets in areas of low humidity. They infest many wild and domestic carnivores and human annoyance is common. Adults rarely leave their carnivore hosts so their dry eggs constantly fall to the bedding, ground or floor where the larvae develop.

In the Spilopsyllini, the adult rabbit flea, *Cediopsylla simplex* (Baker) of eastern United States, infests rabbits and hares but will attack humans. *Euhoplopsyllus glacialis glacialis* (Taschenberg) is known from Greenland and northern Canada as an adult parasite of hares and their predators. *E. g. lynx* (Baker) extends from Alaska to northern United States and the adult is a parasite of the lynx, *Lynx canadensis.* Rothschild and Ford (1972) showed that *C. simplex* is dependent on the reproductive cycle of its host since the unmatured female flea develops mature ovaries and chorionated eggs after transfer to a rabbit in the last week of pregnancy but the ovaries regress rapidly and developing oocytes are resorbed if the fleas are returned to the estrous doe. Freeman and Madsen (1949) found the larvae of *E. g. glacialis* living in the fur of the arctic hare, *Lepus arcticus,* which has no permanent nest or resting place; all three instars were found from May through July, the breeding season of the hare. In July 1964, I observed the same situation for *E. g. lynx* on *L. americanus* in Alaska.

In the Xenopsyllini, the adult oriental rat flea, *Xenopsylla cheopis* (Rothschild), a parasite of house rodents, has been transported throughout the world by humans but it is scarce or absent in northern areas. It attacks humans freely and is the principal vector of bubonic plague and murine typhus. Adults of *X. vexabilis* Jordan are parasites of Hawaiian field rodents, *Rattus exulans* and *Mus musculus. X. cheopis* apparently develops successfully on food of low nutritional quality because it has a large egg, weighing 1/12 as much as the adult, suggesting that larvae receive important factors via the egg (Pausch and Fraenkel, 1966). Since the species has adapted to grain stores frequented by *R. rattus,* larvae can develop solely on grain in the store or in grain sacks in transit (Hirst, 1923).

Description: The following description is based on *Cediopsylla simplex, Ctenocephalides canis, C. felis, E. glacialis, X. cheopis* and *X. vexabilis.* Length of first instars 1.5–3 mm and of third instars 3–5.2 mm. *C. canis* and *C. felis* larvae cannot be differentiated but according to Bacot (1911), they differ in egg size, *C. canis* measuring at least 0.54 × 0.36 mm and *C. felis* at most 0.51 × 0.32 mm.

Head: Narrow mandible with 4–6 teeth. Anterior row with five setae and posterior row with three setae.

Thorax: Anterior row with five short setae. Posterior row with five long setae, ventralmost seta in Spilopsyllini shortest.

Abdominal segments 1–6: Anterior row with six short setae. Posterior row with four long setae and with ventralmost short seta which is nearly as long as other setae in Xenopsyllini. *Abdominal segments* 7–8: Anterior row with six short setae. Posterior row with five long setae and with ventralmost short seta which is nearly as long as other setae in Xenopsyllini. *Abdominal segment* 9: Anterior row with five short setae in Xenopsyllini but with four short setae in Archaeopsyllini and Spilopsyllini. Posterior row of all six species with seven long setae, ventralmost seta shortest except in Xenopsyllini which has two ventral setae slightly shorter than other setae.

Anal Segment: With three ventrolateral setae. Anal comb with single row of 5–13 setae. Strut setae, 8–24.

Comments: In the Archaeopsyllini, most of the eight species of *Ctenocephalides* are confined to the Ethiopian Region and only *C. felis* and *C. canis* are found in N. America. In the Spilopsyllini, the N. American species are two of four species of *Cediopsylla* and one of three species of *Euhoplopsyllus.* In the Xenopsyllini, most of 70 species of *Xenopsylla* are restricted to the Ethiopian and Palaearctic Regions but *X. cheopis* and *X. vexabilis* are found in the United States. Larval classification supports adult classification in that these four genera should be separated from the Pulicini but there are not enough differences in larval characters to suggest that these four genera should belong to more than one tribe.

Selected Bibliography

Pulicidae
Tunginae: Tungini
Barnes and Radovsky 1969.
Elbel 1952.
Hicks 1930.
Kirjakova 1968a, 1971.
Lavoipierre et al. 1979.
Lewis 1972.
Suter 1964.
Pulicinae: Pulicini
Bacot 1914.
Bacot and Ridewood 1914.
Haas and Wilson 1967.
Kirjakova 1968a.
Lewis 1972.
Wang 1956.
Pulicinae: Archaeopsyllini-Spilopsyllini-Xenopsyllini
Bacot 1911.
Bacot and Ridewood 1914.
Elbel 1951, 1952.
Freeman and Madsen 1949.
Harwood and James 1979.
Hirst 1923.
Kirjakova 1961, 1968a.
Klein 1964.
Lewis 1972.
Pausch and Fraenkel 1966.
Rothschild and Ford 1972.
Sikes 1930.
Srivastava and Perti 1971.
Wang 1956.

RHOPALOPSYLLIDAE, HYSTRICHOPSYLLIDAE, ISCHNOPSYLLIDAE, LEPTOPSYLLIDAE AND CERATOPHYLLIDAE

Figures 36.13, 36.14, 36.21

The anal comb with a double row of setae (figs. 36.13, 36.14) distinguishes these larvae from the Pulicinae (Pulicidae) except for the Ctenophthalmini (Ctenophthalminae,

Hystrichopsyllidae) which have an anal comb with a single row of setae. However, in the posterior row on A8 and A9 there are respectively five and six setae in Ctenophthalmini (fig. 36.21) instead of six and seven or seven and eight setae in the Pulicinae (figs. 36.8, 36.10).

RHOPALOPSYLLIDAE: Rhopalopsyllinae

Figures 36.4, 36.12, 36.19

Relationships and Diagnosis: On the anal segment, the four ventrolateral setae, the first being extremely long (fig. 36.12), distinguish *Polygenis gwyni* (C. Fox) (Rhopalopsyllinae) from all other larval groups. In the posterior row on A1–8, there are six setae and A9 has seven setae (fig. 36.19), characters shared with some hystrichopsyllids (Hystrichopsyllinae and Stenoponiinae). *Paratyphloceras oregonensis* (Rhadinopsyllini, Rhadinopsyllinae, Hystrichopsyllidae), the ischnopsyllids (Ischnopsyllinae) and some leptopsyllids (Leptopsyllinae, Leptopsyllini) agree in setal number on A1–8 but have only six setae in the posterior row on A9 (fig. 36.17). The anal comb has a double row of setae with just one seta in the anterior row (fig. 36.12), a character shared with the Leptopsyllini.

Biology and Ecology: The adult *P. gwyni* of southeastern United States is mainly a parasite of the cotton rat, *Sigmodon hispidus,* but also infests the rice rat, *Oryzomys palustris,* wood rats, *Neotoma,* and opossums, *Didelphis.*

Description: The following description is based on *P. gwyni*. Length of third instar about 3.6 mm.

Head: Narrow mandible with five teeth. Anterior and posterior rows each with three long and three short setae.

Thorax: Anterior row with five short setae. Posterior row with five long setae, ventralmost seta shortest.

Abdominal segments 1–8: Anterior row with six short setae. Posterior row with six long setae, ventralmost seta slightly shorter than other setae. *Abdominal segment* 9: Anterior row with five short setae. Posterior row with seven long setae, two ventral setae slightly shorter than other setae.

Anal Segment: With four ventrolateral setae, first extremely long. Anal comb with double row, one seta in anterior row and 6–8 setae in posterior row. Elbel (1952) thought that the anteroventral seta was absent and the anterior row of the anal comb consisted of two setae. However, R. L. C. Pilgrim (personal communication) has pointed out that the second seta of Elbel's anterior row should be considered the anteroventral seta since it is situated nearly in front of the first ventrolateral seta (fig. 36.12). This would agree with other flea larvae which all have an anteroventral seta, including larval specimens of *Rhopalopsyllus cacicus* (Rhopalopsyllinae) (Pilgrim, in prep.). Strut setae, 8–13.

Comments: Most of the species in the Rhopalopsyllinae are confined to Central and South America but two of 41 species of *Polygenis* invade the United States and one of six species of *Rhopalopsyllus* is found here. Larval classification suggests that the rhopalopsyllids belong between the pulicids and hystrichopsyllids which agrees with the arbitrary arrangement of Lewis (1973).

Selected Bibliography

Elbel 1952.
Lewis 1973.

HYSTRICHOPSYLLIDAE

Figures 36.2, 36.3, 36.11, 36.13, 36.15-36.21

Ctenophthalminae: Ctenophthalmini

Figures 36.3, 36.11, 36.21

Relationships and Diagnosis: The three large rounded teeth on the mandible (fig. 36.3) distinguish the Ctenophthalmini from all other larvae except the Pulicini (Pulicinae, Pulicidae). In the posterior row on A8 and A9, there are respectively seven and eight setae in the Pulicini (fig. 36.8) but five and six setae in the Ctenophthalmini (fig. 36.21), characters shared with the ceratophyllids (Ceratophyllinae). The anal comb has a single row of setae (fig. 36.11), a character shared with the Pulicinae.

Biology and Ecology: Adult specimens of *Ctenophthalmus levanticus levanticus* Lewis are found in Lebanon on the mole rat, *Spalax ehrenbergi*. The British *C. nobilis nobilis* (Rothschild) adult infests the wood mouse, *Apodemus sylvaticus,* the red-backed mouse, *Clethrionomys glareolus,* long-tailed shrews, *Sorex,* and moles, *Talpa,* but the Russian *C. orientalis* (Wagner) adult is a specific parasite of the ground squirrel, *Citellus citellus*.

Description: The following description is based on *C. levanticus, C. nobilis* and *C. orientalis*. Length of first instars 1.8–2.3 mm and of third instars 3–5.5 mm. The similarity in characteristics of these 3 species suggests that the following description will apply equally well to the one N. American species, *C. pseudagyrtes* Baker, when the larva is known.

Head: Narrow mandible with three large rounded teeth, rarely a small fourth basal tooth. Anterior and posterior rows each with six setae.

Thorax: Anterior row with five short setae. Posterior row with four long setae and with ventralmost short seta.

Abdominal segments 1–7: Anterior row with six short setae. Posterior row with five long setae and with ventralmost short seta. *Abdominal segment* 8: Anterior row with six short setae. Posterior row with five long setae, two ventral setae shortest. *Abdominal segment* 9: Anterior row with five short setae. Posterior row with six long setae, two ventral setae shortest.

Anal Segment: With three ventrolateral setae. Anal comb with single row of 7–10 setae. For *C. orientalis* Kirjakova (1964) showed the anal comb as a double row with one seta in the anterior row, but R. L. C. Pilgrim (personal communication) found only a single row. In addition, the larvae of five other species of *Ctenophthalmus* in his collection agree in having the single row. Strut setae, 6–8.

Comments: Of 116 species of *Ctenophthalmus,* only *C. pseudagyrtes* is found in N. America. Adult *Ctenophthalmus* are hystrichopsyllids but the larvae represent a distinct group between the pulicids and ceratophyllids.

Rhadinopsyllinae: Rhadinopsyllini (Hystrichopsyllidae)

Figures 36.2, 36.13, 36.17, 36.18, 36.21

Relationships and Diagnosis: The broad, curved mandible with large distal tooth and 5–7 minute blunt teeth (fig. 36.2) distinguishes the Rhadinopsyllini from all other larvae. In the posterior row on A1–5 of *Trichopsylloides oregonensis* Ewing, the second seta is at most 1/4 the length of the first and third setae (fig. 36.18, B), a character shared with the leptopsyllids (Leptopsyllinae, Leptosyllini). In the posterior row on A8 and A9, there are respectively five and six setae (fig. 36.21), characters shared with ceratophyllids (Ceratophyllinae), except in *Paratyphloceras oregonensis* Ewing which has six setae on both segments (fig. 36.17), characters shared with the ischnopsyllids (Ischnopsyllinae) and some Leptopsyllini. The anal comb has a double row of setae with just one seta in the anterior row (fig. 36.13), a character shared with the Leptopsyllini.

Biology and Ecology: Adults of *P. oregonensis* and *T. oregonensis* are specific parasites of the mountain beaver, *Aplodontia rufa,* from British Columbia to western California. Adult *Rhadinopsylla integella* Jordan and Rothschild in Europe and western USSR are normal parasites of red-backed mice, *Clethrionomys.* The British *R. pentacantha* (Rothschild) adults infest microtine rodents.

Description: The following description is based on *P. oregonensis, R. integella, R. pentacantha* and *T. oregonensis.* For *P. oregonensis, R. integella* and *T. oregonensis,* length of first instars 2.4–5.5 mm and of third instars 4.2–7.5 mm. *P. oregonensis* and *T. oregonensis* were characterized by courtesy of R. L. C. Pilgrim.

Head: Broad, curved mandible with large distal tooth and 5–7 minute blunt teeth. Anterior and posterior rows each with six setae.

Thorax: Anterior row with five short setae. Posterior row with four long setae and with ventralmost short seta.

Abdominal segments 1–7: Anterior row with six short setae. Posterior row with ventral most short seta and with five long setae except for *T. oregonensis* with second seta in A1–5 at most 1/4 length of first and third setae. *Abdominal segment* 8: Anterior row with six short setae. Posterior row with five long setae, two ventral setae shortest except *P. oregonensis* with six long setae, two ventral setae shortest. *Abdominal segment* 9: Anterior row with five short setae. Posterior row with six long setae, two ventral setae shortest.

Anal Segment: With three ventrolateral setae, first longest. Anal comb with double row, one seta in anterior row and 5–9 setae in posterior row. For *R. pentacantha,* Cotton (1970) showed the anal comb as a single row but R. L. C. Pilgrim (personal communication) found a double row with one seta in the anterior row. Strut setae, 5–7.

Comments: *Rhadinopsylla* has 54 species of which nine are found in N. America. *Parathphloceras* and *Trichopsylloides* are N. American monotypic genera. Although adult Rhadinopsyllini are hystrichopsyllids, the larvae represent a distinct group. Kirjakova (1971) places *Rhadinopsylla* larvae in the family Ctenophthalmidae (=Ctenophthalminae, Hystrichopsyllidae).

Hystrichopsyllinae and Stenoponiinae (Hystrichopsyllidae)

Figures 36.13, 36.16, 36.19

Relationships and Diagnosis: In the posterior row on A1–8, there are 6–8 setae and A9 has 7–8 setae (fig. 36.19). Included here is *Mioctenopsylla.* This chaetotaxy distinguishes these larvae from all others except the rhopalopsyllid, *Polygenis gwyni* (Rhopalopsyllinae), which is unique in having four ventrolateral setae on the anal segment (fig. 36.12). *Paratyphloceras oregonensis* (Rhadinopsyllini, Rhadinopsyllinae, Hystrichopsyllidae), the ischnopsyllids (Ischnopsyllinae) and some leptopsyllids (Leptopsyllinae, Leptopsyllini) agree in setal number on A1–8 but have only six setae in the posterior row on A9 (fig. 36.17). For the posterior row on A1–4 of *Hystrichopsylla talpae talpae* (Curtis), Cotton (1970) shows the second seta less than 1/4 the length of the first and third setae in the third instar, a character of the Leptopsyllini (fig. 36.18, B). Oudemans (1913) correctly shows this seta as this short only in the first instar but in the third instar the second seta on A1–5 is 1/2–3/4 the length of the first and third setae (R. L. C. Pilgrim, personal communication). This condition occurs in all instars of the Anomiopsyllini (Anomiopsyllinae, Hystrichopsyllidae) (fig. 36.16, A) as well as first instars of the Phalacropsyllini (Neopsyllinae, Hystrichopsyllidae) and *Opisocrostis tuberculatus cynomuris* (Ceratophyllinae, Ceratophyllidae). Pilgrim further notes that in *H. schefferi* Chapin, the length of the setae in the first and third instars is as for *H. talpae* except that the third seta may be the short one or both the third and fourth setae may be short. Except for *Mioctenopsylla* with two setae in the anterior row of the anal comb (fig. 36.14), the Hystrichopsyllinae and Stenoponiinae have just one seta in this row (fig. 36.13), a character shared with the Leptopsyllini.

Biology and Ecology: The adult of *H. schefferi* is a specific parasite of the mountain beaver, *Aplodontia rufa,* from British Columbia to western California. *H. talpae* extends from Britain to the USSR, the adult occurring on insectivores and microtine rodents. *Stenoponia americana* (Baker) adults are parasites of *Peromyscus* and *Microtus* east of the Great Plains of North America. Most of the 13 subspecies of *Stenoponia tripectinata* were described by Jordan and most adults infest gerbils of the genera *Meriones* and *Gerbillus* in the Middle East. Both adult species, *Mioctenopsylla arctica arctica* Rothschild and *M. traubi* Holland and Jellison, were collected with the larvae in soggy, wet and cold nests of the Kittiwake gull (*Rissa*) in the Pribilof Islands, Alaska (W. L. Jellison, personal communication).

Description: The following description is based on *H. schefferi, H. talpae, S. americana, S. tripectinata* and *Mioctenopsylla.* Length of first instars 3.8–8 mm and of third instars 5.8–12 mm. *H. schefferi* was characterized by courtesy of R. L. C. Pilgrim.

Head: Mandible broad with single large tooth except narrow with 5–6 small sharp teeth crowded together at tip in *H. schefferi* and with 6–8 teeth along margin in *Mioctenopsylla.* Anterior and posterior rows each with six setae except for seven setae in anterior row of *S. americana* and *S. tripectinata.* These *Stenoponia* have a dark pigmented plate around the egg breaker of the first instar that may be a generic character (Klein, 1964).

Thorax: Anterior row with five short setae. Posterior row with four long setae and with ventralmost short seta.

Abdominal segments 1–8: Anterior row with six short setae. Posterior row with 5–7 long setae and with ventralmost short seta; in *H. talpae,* second seta 1/2–3/4 length of first and third setae on A1–5; in *H. schefferi,* third seta or third and fourth setae as short as second seta in *H. talpae. Abdominal Segment* 9: Anterior row with five short setae. Posterior row with 6–7 long setae and with ventralmost short seta, two long setae immediately dorsad shorter than other long setae.

Anal Segment: With three ventrolateral setae. Anal comb with double row, one seta in anterior row and 4–7 setae in posterior row except for *Mioctenopsylla* with two setae in anterior row. Strut setae, 5–9.

Comments: *Hystrichopsylla* has 14 species, five being N. American, but just two of 15 *Stenoponia* are N. American. There are only two species in the N. American genus *Mioctenopsylla,* and although the adults are regarded as ceratophyllids (Ceratophyllinae), the larvae resemble these Hystrichopsyllidae except that *Mioctenopsylla* shares with Ceratophyllinae the narrow mandible and the two setae in the anterior row of the anal comb. Larval classification supports adult classification in that *Hystrichopsylla* and *Stenoponia* should be separated from the rest of the hystrichopsyllids but there are not enough differences in larval characters to suggest that these two genera should belong to more than one subfamily. Kirjakova (1971) places the larvae of *Hystrichopsylla* and *Stenoponia* in the family Ctenophthalmidae (=Ctenophthalminae, Hystrichopsyllidae).

Anomiopsyllinae: Anomiopsyllini (Hystrichopsyllidae)

Figures 34.4, 36.15, 36.16

Relationships and Diagnosis: The two ventrolateral setae on the anal segment (fig. 36.15) distinguish the Anomiopsyllini from all other larvae. In the posterior row on A1–5, the second seta is 1/2–3/4 the length of the first and third setae (fig. 36.16, A) as for first instars of the Phalacropsyllini (Neopsyllinae, Hystrichopsyllidae), *Opisocrostis tuberculatus cynomuris* (Ceratophyllinae, Ceratophyllidae) and the third instar of *Hystrichopsylla talpae* (Hystrichopsyllinae, Hystrichopsyllidae). In the posterior row on A8 and A9 of the Anomiopsyllini, there are respectively five and six setae, characters shared with the Ceratophyllinae (fig. 36.21).

Biology and Ecology: Adult *Anomiopsyllus falsicalifornicus falsicalifornicus* C. Fox and *A. f. congruens* Stewart are parasites of wood rats, *Neotoma,* in California, and *A. hiemalis hiemalis* Eads and Menzies is found on *N. micropus* in Kansas, Oklahoma and Texas. Barnes (1963, unpubl. thesis) reared *A. falsicalifornicus* on caged wood rats, *N. fuscipes* and *N. lepida,* using natural nest material. The viscous eggs are laid in the very cup of the wood rat nest and the feeding larvae are found within this inner nest where the humidity is higher; prepupae and cocoons are found in the lower humidity of the outer nest. He showed that the adult rests in the cocoon during the low summer humidity; when fall rains wet the host's nest fleas emerge if mechanically disturbed by a potential host.

Description: The following description is based on *A. falsicalifornicus* and *A. hiemalis.* Length of first instar of *A. hiemalis* 1.6–2 mm and of third instar 2.6–3.1 mm.

Head: Narrow mandible with five teeth. Anterior and posterior rows each with six setae.

Thorax: Anterior row with five short setae. Posterior row with four long setae and with ventralmost short seta.

Abdominal segments 1–7: Anterior row with six short setae, first seta longest. Posterior row with five long setae and with ventralmost short seta; second seta 1/2–3/4 length of first and third setae on A1–5. *Abdominal segment* 8: Anterior row with six short setae, first seta longest. Posterior row with five long setae, two ventral setae shortest. *Abdominal segment* 9: Anterior row with five short setae, first seta longest. Posterior row with six long setae, three ventral setae shortest.

Anal Segment: With two ventrolateral setae. Anal comb with double row of setae, 2–3 in anterior row and 5–8 in posterior row. Strut setae, 5–10.

Comments: *Anomiopsyllus* has 13 species, seven restricted to western United States and six to Mexico. Although adult *Anomiopsyllus* are hystrichopsyllids, the larvae represent a distinct group between the hystrichopsyllids and ceratophyllids.

Neopsyllinae: Phalacropsyllini (Hystrichopsyllidae)

Figures 36.4, 36.13, 36.20

Relationships and Diagnosis: The anterior row of short setae on A1–5 with the second to fifth setae in a straight line (fig. 36.20, C) distinguishes the Phalacropsyllini from all other larval groups. In the posterior row on A1–5 of the first instar, the second seta is 1/2–3/4 the length of the first and third setae as for *Opisocrostis tuberculatus cynomuris* (Ceratophyllinae, Ceratophyllidae), the third instar of *Hystrichopsylla talpae* (Hystrichopsyllinae, Hystrichopsyllidae) and all instars of the Anomiopsyllini (Anomiopsyllinae, Hystrichopsyllidae) (fig. 36.16, A). In the posterior row on A8 and A9 of the Phalacropsyllini, there are respectively five and six setae, characters shared with the Ceratophyllinae (fig. 36.21). The anal comb has a double row of setae with just one seta in the anterior row (fig. 36.13), a character shared with the Leptopsyllini (Leptopsyllinae, Leptopsyllidae).

Biology and Ecology: The adult *Catallagia charlottensis* (Baker) is a parasite of *Peromyscus* and *Microtus* in British Columbia, Washington and Oregon. *Glaucomys volans* of eastern United States is the preferred host of *Epitedia faceta* (Rothschild) which appears to have two generations, with adults emerging in the fall and early winter (Benton, in prep.). The adult *E. cavernicola* Traub parasitizes *Neotoma floridana* in eastern United States and *E. wenmanni* (Rothschild) adult infests a wide range of N. American hosts but is probably a true parasite of *Peromyscus* mice.

Description: The following description is based on *C. charlottensis, E. cavernicola, E. faceta* and *E. wenmanni.* Length of first instar 1.5–2.7 mm and of third instar 3.6–5 mm. *C. charlottensis* and *E. cavernicola* were characterized by courtesy of R. L. C. Pilgrim.

Head: Narrow mandible with 4–6 (usually five) teeth. Anterior and posterior rows each with six setae.

Thorax: Anterior row with five short setae. Posterior row with four long setae and with ventralmost short seta.

Abdominal segments 1–7: Anterior row with six short setae, first seta longest and second to fifth setae in straight line. Posterior row with five long setae and with ventralmost short seta. *Abdominal segment* 8: Anterior row with six short setae, first seta longest and second to fifth setae in straight line. Posterior row with five long setae, two ventral setae shortest. *Abdominal segment* 9: Anterior row with five short setae, first seta longest. Posterior row with six long setae, three ventral setae shortest.

Anal Segment: With three ventrolateral setae. Anal comb with double row, one small seta in anterior row and 8–10 setae in posterior row. Strut setae, 6–12.

Comments: *Catallagia* has 13 species of which 10 are found in N. America but all seven species of *Epitedia* are confined to N. America. Although adult Phalacropsyllini are hystrichopsyllids, the larvae represent a distinct group between the hystrichopsyllids and leptopsyllids.

Selected Bibliography

Hystrichopsyllidae
Ctenophthalminae: Ctenophthalmini
Cotton 1963, 1970.
Kirjakova 1964.
Lewis 1974a.
Rhadinopsyllinae: Rhadinopsyllini
Bartkowska 1972.
Cotton 1970.
Kirjakova 1971.
Lewis 1974a.
Hystrichopsyllinae and Stenoponiinae
Cotton 1970.
Kirjakova 1968b, 1971.
Klein 1964.
Lewis 1974a, 1975.
Oudemans 1913.
Anomiopsyllinae: Anomiopsyllini
Barnes 1963 (unpubl. thesis).
Lewis 1974a.
Neopsyllinae: Neopsyllini
Lewis 1974a.

ISCHNOPSYLLIDAE: Ischnopsyllinae

Figures 36.5, 36.14, 36.17

Relationships and Diagnosis: In the posterior row on A1–9, there are six setae (fig. 36.17). This chaetotaxy distinguishes these larvae from all others except *Paratyphloceras oregonensis* (Rhadinopsyllini, Rhadinopsyllinae, Hystrichopsyllidae) and some leptopsyllids (Leptopsyllinae, Leptopsyllini) which have an anal comb with a double row of setae with just one seta in the anterior row (fig. 36.13); this row in the Ischnopsyllinae has two or more setae (fig. 36.14). Rhopalopsyllids and some hystrichopsyllids (Hystrichopsyllinae and Stenoponiinae) agree in setal number on A1–8 but have 7–8 setae in the posterior row on A9 (fig. 36.19) and an anal comb like the Leptopsyllini (fig. 36.13). In the anterior row on A9, the Ischnopsyllinae have four short setae (fig. 36.17), a character shared with the Leptopsyllini (fig. 36.18).

Biology and Ecology: Adult ischnopsyllids are, with rare accidental exception, exclusively parasites of bats. According to Tracy (1978, unpubl. thesis), eggs of the N. American *Myodopsylla insignis* (Rothschild) are laid in the pelage of the bat, *Myotis lucifugus,* and either fall to the guano or are removed when the bat grooms. Eggs were observed in the guano in early May and continued to hatch until mid-July. Larvae were found from late May to late July and adults from early July until early September when infestation levels increased on bats preparing to leave nursery roosts for hibernacula. Fleas remaining in the nursery roosts after bats depart apparently do not survive and those that fail to emerge during the summer do not emerge the following season as all intact pupal cases found in the spring were dehydrated. However, fleas on the host remain active during winter hibernation and are transported back to nursery colonies in the spring where mating and egg laying commence again. In contrast, Smith (1981, unpubl. thesis) thinks that *M. insignis* overwinters in the cocoon as large numbers of fleas appeared immediately on return of bats to the nursery roost from Virginia caves where few fleas were found on hibernating bats. He also thinks that newly-emerged fleas find their hosts by crawling up walls and lower edges of the roof. Instead, A. H. Benton (personal communication) thinks that fleas get on young bats that fall to the guano before they are rescued by adults. In addition, he thinks that some adult fleas overwinter in crevices in rafters and emerge in the spring to feed and lay eggs.

Description: The following description is based on *M. insignis* and the European *Nycteridopsylla eusarca* Dampf. Length of first instars 1–2.1 mm and of third instars 3.2–4.2 mm.

Head: Narrow mandible with 6–8 teeth. Anterior and posterior rows each with six setae except for seven setae in posterior row of *N. eusarca.*

Thorax: Anterior row with five short setae except for four short setae in *N. eusarca.* Posterior row with four long setae and with ventralmost short seta which is absent in *N. eusarca.*

Abdominal segments 1–8: Anterior row with six short setae. Posterior row with six long setae, ventralmost seta shortest. *Abdominal segment* 9: Anterior row with four short setae. Posterior row with six long setae, two ventral setae shortest.

Anal Segment: With three ventrolateral setae. Anal comb with double row of setae, 2–3 in anterior row and 7–8 in posterior row. Strut setae, 7–13.

Comments: Five of 10 *Myodopsylla* are N. American species, the rest being mainly Neotropical. The 15 members of *Nycteridopsylla* are restricted to the Palaearctic Region with two exceptions which infest N. American bats. Larval classification suggests that the ischnopsyllids belong between the hystrichopsyllids and leptopsyllids which agrees with the arbitrary arrangement of Lewis (1974b).

Selected Bibliography

Hůrka 1969.
Lewis 1974b.
Smith 1981 (unpubl. thesis).
Tracy 1978 (unpubl. thesis).

LEPTOPSYLLIDAE, Leptopsyllinae: Leptopsyllini

Figures 36.4, 36.5, 36.13, 36.18, 36.21

Relationships and Diagnosis: In the posterior row on A1–5, the second seta is at most 1/4 the length of the first and third setae (fig. 36.18, B). Included here is *Conorhinopsylla stanfordi* Stewart. This chaetotaxy distinguishes these larvae from all others except *Trichopsylloides oregonensis* (Rhadinopsyllini, Rhadinopsyllinae, Hystrichopsyllidae) which has a broad, curved mandible with large distal tooth and 5–7 minute blunt teeth (fig. 36.2) and the first instar of *Hystrichopsylla talpae* (Hystrichopsyllinae, Hystrichopsyllidae) which has seven setae in the posterior row on A9 (fig. 36.19). In the first instars of the Phalacropsyllini (Neopsyllinae, Hystrichopsyllidae) and *Opisocrostis tuberculatus cynomuris* (Ceratophyllinae, Ceratophyllidae) as well as the third instar of *H. talpae* and all instars of the Anomiopsyllini (Anomiopsyllinae, Hystrichopsyllidae), the second seta in the posterior row on A1–5 is 1/2–3/4 the length of the first and third setae (fig. 36.16, A). In the posterior row on A8 and A9, the Leptopsyllini have respectively five and six setae (fig. 36.21), characters shared with the Ceratophyllinae, or six setae in both segments (fig. 36.18), characters shared with *Paratyphloceras oregonensis* (Rhadinopsyllini, Rhadinopsyllinae, Hystrichopsyllidae) and the ischnopsyllids (Ischnopsyllinae). Except for *C. stanfordi* with five short setae in the anterior row on A9 (fig. 36.21), the Leptopsyllini have four short setae in this row, a character shared with the Ischnopsyllinae (fig. 36.17). The anal comb has a double row of setae with just one seta in the anterior row (fig. 36.13), a character shared with rhopalopsyllids and some hystrichopsyllids (Rhadinopsyllinae, Hystrichopsyllinae, Stenoponiinae and Neopsyllinae).

Biology and Ecology: The adult European mouse flea, *Leptopsylla segnis* (Schönherr), is typically a parasite of the house mouse, *Mus musculus,* but is found also on house rats. *L. segnis* attacks man reluctantly and is a weak vector of plague. The British *Peromyscopsylla spectabilis* (Rothschild) adult is a parasite of the red-backed mouse, *Clethrionomys glareolus.* The adult *Conorhinopsylla stanfordi* of eastern N. America infests tree squirrels and flying squirrels, *Glaucomys,* which are preferred hosts. *C. stanfordi* seems to have one generation, with adults emerging in the fall and early winter (Benton, in prep.).

Description: The following description is based on *C. stanfordi, L. segnis* and *P. spectabilis.* The western N. American *P. selenis* (Rothschild) is very similar to *P. spectabilis* (Pilgrim, in prep.). Length of third instar of *L. segnis* 3.1–6.5 mm. Eggs of *C. stanfordi* are unique in that the surface is reticulated which distinguishes them from all other known flea eggs (A. H. Benton, personal communication).

Head: Narrow mandible with 5–9 teeth. In *L. segnis,* Kirjakova (1968b) rarely found 11 teeth. Anterior and posterior rows each with six setae.

Thorax: Anterior row with five short setae. Posterior row with four long setae and with ventralmost short seta.

Abdominal segments 1–7: Anterior row with six short setae. Posterior row with ventralmost short seta and with five long setae except second seta in A1–5 at most 1/4 length of first and third setae. *Abdominal segment* 8: Anterior row with six short setae. Posterior row with five long setae, two ventral long setae shortest. Often there is a ventralmost short seta making a total of six setae in posterior row (Elbel, 1952). *Abdominal segment* 9: Anterior row with four short setae except for five short setae in *C. stanfordi.* Posterior row with six long setae, two ventral setae shortest.

Anal Segment: With three ventrolateral setae. Anal comb with double row, one small seta in anterior row and 7–10 setae in posterior row. Strut setae, 5–9.

Comments: *Leptopsylla* has 15 species which are distributed through the Ethiopian and Palaearctic Regions except that *L. segnis* has a cosmopolitan distribution due to human transport from its apparent origin in western Asia. *Peromyscopsylla* has 18 species of which eight are found in N. America. There are just two species in the N. American genus *Conorhinopsylla* and although the adults are hystrichopsyllids (Anomiopsyllinae, Anomiopsyllini), the larvae resemble the Leptopsyllini except for the five short setae in the anterior row of abdominal segment 9 which *C. stanfordi* shares with the ceratophyllids (Ceratophyllinae). Larval classification suggests that the leptopsyllids belong between the ischnopsyllids and the ceratophyllids which agrees with the arbitrary arrangement of Lewis (1974b). Kirjakova (1971) places *L. segnis* larvae in the family Ctenophthalmidae (= Ctenophthalminae, Hystrichopsyllidae).

Selected Bibliography

Bacot and Ridewood 1914.
Cotton 1970.
Elbel 1952.
Kirjakova 1968b, 1971.
Lewis 1974a, 1974b.
Wang 1956.

CERATOPHYLLIDAE: Ceratophyllinae

Figures 36.1, 36.4, 36.5, 36.14, 36.21

Relationships and Diagnosis: Larvae are distinguished by the following combination of characters: a narrow mandible with at least five teeth (figs. 36.4, 36.5), three ventrolateral setae and an anal comb having a double row of setae with at least two setae in the anterior row (fig. 36.14). In the posterior row on A1–5 of the first instar of *Opisocrostis tuberculatus cynomuris* Jellison, the second seta is 1/2–3/4 the length of the first and third setae as for the Phalacropsyllini (Neopsyllinae, Hystrichopsyllidae), the third instar of *Hystrichopsylla talpae* (Hystrichopsyllinae, Hystrichopsyllidae) and all instars of the Anomiopsyllini (Anomiopsyllinae, Hystrichopsyllidae) (fig. 36.16, A). In the posterior row on A8 and A9, there are respectively five and six setae (fig. 36.21), characters shared with some leptopsyllids (Leptopsyllinae, Leptopsyllini) and some hystrichopsyllids (Ctenophthalminae, Rhadinopsyllinae, Anomiopsyllinae and Neopsyllinae).

Biology and Ecology: Adults of species of *Ceratophyllus* are almost exclusively bird parasites, but the remaining Ceratophyllinae primarily infest rodents. Larvae of species studied here whose adults are known to attack humans are: the nearly cosmopolitan European chicken flea, *Ceratophyllus gallinae* (Schrank); the western chicken flea, *C. niger* C. Fox of N. America; the ground squirrel flea, *Diamanus montanus* (Baker) of western N. America; the N. American *Monopsyllus wagneri* (Baker) which is usually found on the deer mouse, *Peromyscus maniculatus;* the northern rat flea, *Nosopsyllus fasciatus* (Bosc), which has been associated with the transmission of bubonic plague and replaces the Oriental rat flea in the cooler Temperate Zone of Europe and N. America; *Orchopeas howardi howardi* (Baker), a N. American squirrel flea that was imported to England. On *Glaucomys volans,* flying squirrel, *O. howardi* adults dominate throughout the warmer months with at least four generations per year, but the N. American *Opisodasys pseudarctomys* (Baker) adults show a huge peak in early spring and another in late fall with probably 3–4 generations per year (Benton, in prep.). According to Humphries (1968), *C. gallinae* takes about a month in the summer to develop from egg to adult which is about the time it takes for the bird host to lay and incubate eggs and rear a brood. Since the nest is not generally used for a second brood, the flea population is separated from its host when the fledglings leave. The adult fleas overwinter in the cocoon; tactile stimuli and a rise in temperature initiate emergence when the birds return. Benton *et al.* (1979) removed 1,470 *C. niger* larvae, many with bright red blood-colored digestive tracts, from a dead chick and they suggested that the larvae swarmed over the chick in response to the presence of blood. They watched larvae of *O. pseudarctomys* suck from a fiber soaked with the blood of its flying squirrel host. One larva bit into the body of a smaller larva and sucked its body fluids. Molyneux (1967) observed that larval *N. fasciatus* actively pursue and seize adult fleas from whom fecal blood is sucked rapidly as it passes out of the anus; water and rat urine are ingested similarly. Krampitz (1978) stated that although the mandibles of flea larvae are suited for rasping on debris, the well-developed musculature of the pharynx suggests that the sucking method of feeding is usual. He observed *Xenopsylla cheopis* suck blood as rapidly and easily as *N. fasciatus.* Adult hosts of other larval species studied here are: the Russian and N. American *C. styx riparius* Jordan and Rothschild which is found on the bank swallow, *Riparia riparia;* the 9 subspecies of *Malaraeus penicilliger* which are primarily parasites of red-backed mice, *Clethrionomys,* from Great Britain through Europe, Alaska and northwestern N. America; *Megabothris clantoni clantoni* Hubbard which infests the sagebrush vole, *Lagurus curtatus,* in Washington; *Opisocrostis tuberculatus cynomuris* of western N. America where it infests prairie dogs, *Cynomys;* the N. American *Orchopeas leucopus* (Baker) which occurs mainly on the deer mouse, *P. maniculatus;* the 8 subspecies of the N. American *O. sexdentatus* which are all specific parasites of various *Neotoma,* wood rats; *Oropsylla arctomys arctomys* (Baker), the range of which coincides with that of its woodchuck host, *Marmota monax,* in N. America; *Thrassis acamantis howelli* (Jordan) of Oregon where it is the only principal flea of *M. flaviventris,* the yellow-bellied marmot; *T. bacchi gladiolis* (Jordan) which infests the white-tailed antelope squirrel, *Ammospermophilus leucurus,* in western Utah.

Description: The following description is based on *Ceratophyllus gallinae, C. niger, C. styx riparius, Diamanus montanus, Malaraeus penicilliger mustelae* (Dale), *Megabothris clantoni clantoni, Monopsyllus wagneri, Nosopsyllus fasciatus, Opisocrostis tuberculatus cynomuris, Opisodasys pseudarctomys, Orchopeas howardi howardi, O. leucopus, O. sexdentatus sexdentatus* (Baker), *Oropsylla arctomys arctomys, Thrassis acamantis howelli* and *T. bacchi gladiolis.* Length of first instars 1.2–2.7 mm and of third instars 3.3–6.0 mm.

Head: Narrow mandible with 5–10 teeth. Anterior and posterior rows each with 5–6 setae. For the anterior row of *O. howardi,* Sikes (1930) showed seven setae but R. L. C. Pilgrim (personal communication) found only six setae.

Thorax: Anterior row with five short setae. Posterior row with four long setae and with ventralmost short seta.

Abdominal segments 1–7: Anterior row with six short setae. Posterior row with five long setae and with ventralmost short seta. *Abdominal segment* 8: Anterior row with six short setae. Posterior row with five long setae, two ventral setae shortest. *Abdominal segment* 9: Anterior row with five short setae. Posterior row with six long setae, two ventral setae shortest.

Anal Segment: With three ventrolateral setae. Anal comb with double row of setae, 2–6 in anterior row and 5–10 in posterior row. Strut setae, 5–10.

Comments: Most of the Ceratophyllinae are Holarctic or are restricted to the Nearctic or Palaearctic Regions. The 11 genera studied here contain a total of 186 species of which 48% are found in N. America. Larval classification suggests that the ceratophyllids belong after the leptopsyllids which agrees with the arbitrary arrangement of Lewis (1975).

Selected Bibliography

Bacot and Ridewood 1914.
Benton et al. 1979.
Cotton 1970.
Elbel 1951, 1952.
Harwood and James 1979.
Humphries 1968.
Kirjakova 1965.
Krampitz 1978.
Lewis 1975.
Molyneux 1967.
Poole and Underhill 1953.
Sikes 1930.
Stark et al. 1976.

BIBLIOGRAPHY

Bacot, A. W. 1911. Flea eggs. Trans. R. Ent. Soc. London 5:767, VI–VII.

Bacot, A. W. 1914. A study of the bionomics of the common rat fleas and other species associated with human habitations, with special reference to the influence of temperature and humidity at various periods of the life history of the insect. J. Hyg. 13 (Plague Suppl. III):447–654.

Bacot, A. W. and W. G. Ridewood. 1914. Observations on the larvae of fleas. Parasitology 7:157–175.

Barnes, A. M. 1963. A revision of the genus *Anomiopsyllus* Baker, 1904 (Siphonaptera: Hystrichopsyllidae) with studies on the biology of *Anomiopsyllus falsicalifornicus*. Ph.D. dissertation, Univ. Calif., 247 p.

Barnes, A. M. and F. J. Radovsky. 1969. A new *Tunga* (Siphonaptera) from the Nearctic Region with description of all stages. J. Med. Ent. 6:19–36.

Bartkowska, K. 1972. Morphology of the larva of *Rhadinopsylla* (*Actenophthalmus*) *integella* Jordan and Rothschild (Siphonaptera: Hystrichopsyllidae). Polskie Pismo Ent. 42:535–543 (translated from Polish to English by J. Müller, Christchurch, New Zealand, and provided by courtesy of R. L. C. Pilgrim).

Benton, A. H., M. Surman and W. L. Krinsky. 1979. Observations on the feeding habits of some larval fleas (Siphonaptera). J. Parasitol. 65:671–672.

Cotton, M. J. 1963. The larva of *Ctenophthalmus nobilis* (Rothschild) (Siphonaptera). Proc. R. Ent. Soc. London (A) 38:153–158.

Cotton, M. J. 1970. The comparative morphology of some species of flea larvae (Siphonaptera) associated with nests of small mammals. Ent. Gaz. 21:191–204.

Elbel, R. E. 1951. Comparative studies on the larvae of certain species of fleas (Siphonaptera). J. Parasitol. 37:119–128.

Elbel, R. E. 1952. Comparative morphology of some rat flea larvae (Siphonaptera). J. Parasitol. 38:230–238.

Freeman, R. B. and H. Madsen. 1949. A parasitic flea larva. Nature 164:187–188.

Haas, G. E. and N. Wilson. 1967. *Pulex simulans* and *P. irritans* on dogs in Hawaii. (Siphonaptera: Pulicidae). J. Med. Ent. 4:25–30.

Harwood, R. F. and M. T. James. 1979. Fleas. p. 319–341 *in* Entomology in human and animal health. Seventh Edition. MacMillan Publ. Co. New York, VI + 548 p.

Henderson, J. R. 1928. A note on some external characters of the larva of *Xenopsylla cheopis*. Parasitology 20:115–118.

Hicks, E. P. 1930. The early stages of the jigger, *Tunga penetrans*. Ann. Trop. Med. Parasitol. 24:575–586.

Hirst, L. F. 1923. On the spread of plague in the East Indies. Trans. R. Soc. Trop. Med. Hyg. 17:101–127.

Humphries, D. A. 1968. The host-finding behaviour of the hen flea, *Ceratophyllus gallinae* (Schrank) (Siphonaptera). Parasitology 58:403–414.

Hůrka, L. 1969. Die Larve des Fledermausflohes *Nycteridopsylla eusarca* Dampf (Aphaniptera: Ischnopsyllidae). Acta Ent. Bohemoslovaca 66:317–320.

Karandikar, K. R. and D. M. Munshi. 1952. Life history and bionomics of the cat flea, *Ctenocephalides felis* Bouché. J. Bombay Nat. Hist. Soc. 49:169–177.

Kirjakova, A. N. 1961. Larvae of the fleas of the family Pulicidae. I. External morphology of the larva of a cat flea, *Ctenocephalides felis* Bouché, 1835. Parasit. Sborn., Acad. Sci. USSR 20:306–323 (translated from Russian to English by R. L. C. Pilgrim, University of Canterbury, Christchurch, New Zealand).

Kirjakova, A. N. 1964. Flea larvae of the family Ctenophthalmidae. 4. Zool. Zh. 43:572–580 (translated from Russian to English by R. L. C. Pilgrim, University of Canterbury, Christchurch, New Zealand).

Kirjakova, A. N. 1965. Larvae of fleas (Aphaniptera) of the family Ceratophyllidae. 3. Ent. Rev. Engl. Transl. 44:218–224.

Kirjakova, A. N. 1968a. External morphology of larvae of fleas of the family Pulicidae (Aphaniptera) Part 6. Parazitologiya 2:548–558 (translated from Russian to English by R. L. C. Pilgrim, University of Canterbury, Christchurch, New Zealand).

Kirjakova, A. N. 1968b. Comparative morphology of larvae of certain genera of fleas (Aphaniptera). Ent. Rev. Engl. Transl. 47:41–46.

Kirjakova, A. N. 1971. The larvae of fleas (Aphaniptera), their morphology and systematics. XIII Int. Congr. Ent. Moscow (1968) 1:258 (translated from Russian to English by R. L. C. Pilgrim, University of Canterbury, Christchurch, New Zealand).

Klein, J. M. 1964. Contribution à l'étude morphologique externe des larves de puces. Les larves de *Xenopsylla buxtoni* Jordan, 1949, *Nosopsyllus iranus iranus* Wagner et Argyropulo, 1934, et *Stenoponia tripectinata irakana* Jordan, 1958 (Siphonaptera). Bull. Soc. Ent. France 69:174–196.

Krampitz, H. E. 1978. Investigations into the blood uptake of larval fleas. Proc. IV Int. Congr. Parasitol. (Warszawa): 216–217.

Lavoipierre, M. M. J., F. J. Radovsky and P. D. Budwiser. 1979. The feeding process of a tungid flea, *Tunga monositus* (Siphonaptera: Tungidae), and its relationship to the host inflammatory and repair response. J. Med. Ent. 15:187–217.

Lewis, R. E. Notes on the geographical distribution and host preference in the order Siphonaptera.
1972. Part 1. Pulicidae. J. Med. Ent. 9:511–520.
1973. Part 2. Rhopalopsyllidae, Malacopsyllidae and Vermipsyllidae. J. Med. Ent. 10:255–260.
1974a. Part 3. Hystrichopsyllidae. J. Med. Ent. 11:147–167.
1974b. Part 5. Ancistropsyllidae, Chimaeropsyllidae, Ischnopsyllidae, Leptopsyllidae and Macropsyllidae. J. Med. Ent. 11:525–540.
1975. Part 6. Ceratophyllidae. J. Med. Ent. 11:658–676.

Mitzmain, M. B. 1910. General observations on the bionomics of the rodent and human fleas. Publ. Health Bull. 38:3–34.

Molyneux, D. H. 1967. Feeding behaviour of the larval rat flea, *Nosopsyllus fasciatus* (Bosc). Nature 215:779.

Oudemans, A. C. 1913. Suctoriologisches aus Maulwurfsnestern. Tijdschr. Ent. 56:238–280.

Pausch, R. D. and G. Fraenkel. 1966. The nutrition of the larva of the Oriental rat flea, *Xenopsylla cheopis* (Rothschild). Physiol. Zool. 39:202–222.

Poole, V. V. and R. A. Underhill. 1953. Biology and life history of *Megabothris clantoni clantoni* (Siphonaptera: Dolichopsyllidae). Walla Walla Coll. Publ. Dept. Biol. Sci. Biol. Stn. 9:1–19.

Riek, E. F. 1970. Lower Cretaceous fleas. Nature 227:746–747.

Rothschild, M. 1975. Recent advances in our knowledge of the order Siphonaptera. Ann. Rev. Ent. 20:241–259.

Rothschild, M. and B. Ford. 1972. Breeding cycle of the flea *Cediopsylla simplex* is controlled by breeding cycle of host. Sci. 178:625–626.

Sharif, M. 1937. On the life history and the biology of the rat flea, *Nosopsyllus fasciatus* (Bosc). Parasitology 29:225–238.

Sikes, E. K. 1930. Larvae of *Ceratophyllus wickhami* and other species of fleas. Parasitology 22:242–259.

Smit, F. G. A. M. 1972. On some adaptive structures in Siphonaptera. Folia Parasit. (Praha) 19:5–17.

Smith, S. A. 1981. Studies on the morphology and biology of the bat flea *Myodopsylla insignis* (Rothschild) (Siphonaptera: Ischnopsyllidae). Masters thesis, Ohio State Univ., 129 p.

Srivastava, A. P. and S. L. Perti. 1971. Studies on the life history and behaviour of rat fleas. Labdev. J. Sci. Tech. 9–B:121–125.

Stark, H. E., E. G. Campos and R. E. Elbel. 1976. Description of the third-instar larva of *Monopsyllus wagneri* (Baker) (Siphonaptera: Ceratophyllidae). J. Med. Ent. 13:107–111.

Suter, P. R. 1964. Biology of *Echidnophaga gallinacea* (Westwood) and comparison of its behavior with that of other species of fleas. Acta Trop. 21:193–238 (translated from German to English in part by CDC Library, Atlanta, Georgia and provided by courtesy of H. E. Stark and in part by Miriam Rothschild, Ashton Wold, Peterborough, England).

Tracy, M. L. A. 1978. Ecology and life history of *Myodopsylla insignis*. Masters thesis, Boston Univ., 68 p.

Traub, R. 1980. The zoogeography and evolution of some fleas, lice and mammals, pp. 93–172 *in* Traub, R. and H. Starcke (eds.) Fleas. Proc. Int. Conf. Fleas. A. A. Balkema, Rotterdam, Netherlands, and Salem, NH, 480 p.

Wang, D. C. 1956. Comparative morphology of some common flea larvae (Siphonaptera). Acta Ent. Sin. 6:311–321 (translated from Chinese to English by staff of the Bishop Museum, Honolulu, Hawaii and provided by courtesy of Nixon Wilson).

Order Diptera[1]

37

B. A. Foote, Coordinator
Kent State University

INTRODUCTION

H. J. Teskey, *Biosystematics Research Centre, Ottawa*

Diptera larvae are extremely variable morphologically. Head structures range from a well-developed exposed capsule with mouthparts adapted for biting and chewing, to variously reduced structures partially or completely retracted within the thorax, and with the mouthparts altered for piercing and rasping. Rarely the head skeleton is essentially absent. The segments of the larval body may be variously fused or subdivided, conspicuously swollen in whole or in part, or cylindrical or compressed. Some or all segments may bear filamentous or tuberculous outgrowths of various kinds. Spiracles may be present on one or more of the body segments or completely absent.

Variation is so great that no individual or concise combination of characters exists by which all Diptera larvae can be distinguished from all other endopterygote insect larvae. The only character common to all Diptera larvae, but also found in representatives of several other orders, is the absence of jointed thoracic legs. This feature, coupled with the fact that the majority of free-living Diptera larvae are rather slender with active directional movement, serves to distinguish most of them. Most legless larvae of other orders are rather swollen, with movements that appear slow and undirected. Also, in Diptera larvae, spiracles may be present or completely absent on one or more segments. If more than one pair of abdominal spiracles is present, the posterior pair is larger than the others, and with few exceptions, faces posteriorly. In addition, the posterior pair is usually bordered by accessory lobes or situated in a tubular projection. In other orders the posterior abdominal spiracles are normally located laterally on the segments and without bordering lobes, and are similar in size to the other spiracles.

Diptera larvae are as variable in their living habits as in their appearance. Possibly as many as half of the North American species are aquatic, especially those belonging to the Nematocera and orthorrhaphous Brachycera. Most of the other larvae are confined to relatively moist terrestrial habitats, and a few are endoparasitic. In some cases the differentiation between aquatic and terrestrial habitats may be impossible. At the other extreme, very few Diptera larvae survive rather dry conditions.

1. Many of the figures used in this chapter are taken from the *Manual of Nearctic Diptera,* Volumes 1 and 2 (McAlpine et al. 1981, 1987), and are reproduced by permission of the Minister of Supply and Services, Canada. Those figures used in the Key to Families are acknowledged by an MD.

Aquatic larvae are found in streams and rivers of all sizes and rates of flow and depth, in lakes, ponds, stagnant pools, puddles, freshwater swamps and marshes and marine intertidal zones; in short, almost any place where water is present for at least a few weeks. The water may be fresh and clean or polluted, or brackish, acidic or alkaline, and clear or turbid. In the terminology of Merritt and Cummins (1984) some larvae are planktonic by floating or swimming in open water such as the Chaoboridae and Culicidae, where they feed as detritivores on suspended or sedimentary deposits of decomposing organic matter or as predators of living organisms. Some larvae, such as Blephariceridae and Simuliidae, cling to rocks or vegetation in the stream, some grazing on algae and associated materials on rock surfaces, while others filter suspended food particles from the water. The majority of aquatic larvae are burrowing detritivores and carnivores in bottom sediments, or herbivores in aquatic plant stems, leaves or roots.

Terrestrial larvae are to be found in equally diverse but basically very similar habitats. Many larvae are leaf, stem or root miners, some being responsible for conspicuous gall formations. Decaying plant and animal matter provides home and food for larvae of a large assemblage of species. Such habitats also contain a variety of Diptera whose larvae are predacious or parasitic on other invertebrates and vertebrates, including humans. An excellent summary of the larval feeding habits of the Muscomorphan Diptera is found in Ferrar (1987).

The basic number of instars is 4–9 for the lower Diptera (usually four), with reduction to three for most higher flies. The rate of larval development is highly variable, ranging from a few days for those maggots which are dependent on the short-term resource of a decaying carcass, to some species that live in cold, wet habitats and can take two years to complete development.

Some useful publications that provide broad biological information include Clausen, 1940 (entomophagous groups); Felt, 1940 (galls); Séguy, 1950 (general); Hennig, 1948, 1950, 1952 (general); Oldroyd, 1964 (general); Cole (1969) (western North America); Pennak (1972), (aquatic); Merritt and Cummins, 1984 (aquatic families and genera); and McAlpine, et al., 1981 (Vol. 1) and 1987 (Vol. 2) (all families).

Morphology

The following is a summary of the chapter on morphology and terminology of larvae that was prepared for the "Manual of Nearctic Diptera", Vol. 1 (McAlpine et al., 1981), and which should be consulted for a more comprehensive treatment. It has been written with the aim of providing the reader with a basic knowledge of larval morphology sufficient to appreciate the evolution of the head structure of Diptera larvae and to interpret other body features mentioned in the larval key and text. Most literature references are not given in the following text, but they are available in the "Manual of the Nearctic Diptera", are indexed at the end of this section, and included in the bibliography for the Diptera. However, several of these must be mentioned here for their special importance to the study of larval Diptera. Hennig's publications (1948, 1950, 1952) remain as the most indispensable general source books on Diptera larvae. Krivosheina (1969) and Ferrar (1987) are also valuable general reference texts. Anthon (1943b), Cook (1949) and Ludwig (1949) provide very comprehensive interpretations of the morphology of the various types of head structures. Keilin (1944), Hinton (1947) and Whitten (1955) are indispensable for information on the larval respiratory systems. A series of papers by Roberts (1969a and b, 1970, 1971a) treats the functional morphology of the mouthparts of various Diptera.

Head

The great diversity in head structure can be systematized according to certain morphological trends that correspond closely with the phylogeny of the Diptera. Three broad categories of head structure have been distinguished based on the degree of reduction and the amount of retraction within the thorax of the head capsule, and on the structure and plane of movement of the mandibles. A well-developed, fully exposed head capsule with mandibles usually bearing teeth and operating in a horizontal or oblique plane is termed **eucephalic** and typifies the larvae of most Nematocera. A **hemicephalic** head capsule is more or less reduced or incomplete posteriorly and partially retracted within the thorax, with sickle-shaped mandibles operating in a vertical plane. Such a head capsule is mainly found among the orthorrhaphous Brachycera. Further reduction and loss of sclerotization of the external parts of the head associated with its almost complete retraction within the thorax, and coupled with the further development of an internal tentoropharyngeal skeleton produces the so-called **acephalic** condition, the term used to describe the characteristically shaped cephalopharyngeal skeleton that typifies larvae of the Muscomorpha (=cyclorrhaphous Brachycera).

Although these terms are useful in defining general trends they are not mutually exclusive. Exceptions to their application include the Tipulidae, a nematocerous family with a head capsule that is more or less extensively reduced posteriorly and partially retracted within the thorax (figs. 37.1–37.6). The tipulid mandibles vary from being stout and toothed, adapted for chewing, to slender and sickle-shaped for piercing. However, they are nearly always opposed to each other in a horizontal plane. Although the head capsule of some Cecidomyiidae (figs. 37.7, 37.8) may be classed as hemicephalic or acephalic because of its retraction within the thorax and apparent reduction posteriorly as suggested by the posterolateral projections from the head capsule, the family is undoubtedly referable to the Nematocera. The larva of another nematocerous family, the Synneuridae, essentially lacks any vestige of a head capsule and is the only truly acephalic example (fig. 37.9). The mandibles of larvae of Trichoceridae, Anisopodidae and Psychodidae operate in an oblique or nearly vertical direction but the fully developed exserted head capsule proclaims their nematoceran affiliation. Despite such exceptions these fundamental differences clearly support the division of the order into the three major groups. A typical example of a eucephalic head capsule is that of *Bibio* (Bibionidae). Figures 37.10a,b show the major external morphological features. Most of these structures are known in various forms in all representatives of the Nematocera. The greatest departure from this typical example, apart from the above mentioned features, is the development in several families of a sclerotized toothed plate or a pair of plates ventrally in the region of the labium called the hypostoma. These are prominent in the larvae of Chironomidae (fig. 37.11), Simuliidae (fig. 37.12), some Tipulidae (fig. 37.13), and Psychodidae. The hypostoma appears to be used in scraping the substrate to dislodge food materials.

The antennae of Diptera larvae have one to six segments, but usually no more than three. They are commonly rather short. However, those of Blephariceridae (fig. 37.14) and Deuterophlebiidae (fig. 37.15) are more elongate and in the latter the terminal segment is biramous. Antennae of Chaoboridae are prehensile, with apical spines, and are used in capturing prey (figs. 37.16, 37.17).

The hemicephalic head capsule is typified by the Stratiomyidae (fig. 37.18), Tabanidae (figs. 37.19a,b) and Asilidae (figs. 37.20a,b). Here the attachment line of the prothoracic integument to the head capsule is well anterior of the posterior sclerotized portions of the head capsule (fig. 37.18). The invaginated portion of the capsule is always more or less reduced ventrally and sometimes reduced to one or a pair of slender dorsal metacephalic rods that project into the thorax. In larvae of the Therevidae and Scenopinidae the slender retracted metacephalic rod (figs. 37.21a,b) may not be evident without dissection, and except for the vertical orientation of the mandibles the head looks typically eucephalic. The process of retraction has progressed still further in the Empididae and Dolichopodidae (figs. 37.22, 37.23a,b) so that little of the head capsule is exposed anteriorly. The head capsule of parasitic larvae of the families Acroceridae, Nemestrinidae and Bombyliidae is reduced almost to the point of resembling the so-called acephalic condition of the Muscomorpha.

Characteristic also of the hemicephalic head capsule is the growth of the tentorial skeleton. Such growth is a major reason for the rotation of the mandibles to a vertical position. The epicondyle, or upper anterior point of articulation of the mandible, is carried ventrally on the developing tentorial phragma, thus bringing the epicondyle more nearly parallel with the outer hypocondyle.

In conjunction with the development of these tentorial phragmata is the growth from their posterior ventral margin of a pair of posterior tentorial arms (figs. 37.19a, 37.20a, 37.23b, 37.24). The tentorial arms substitute for the portions of the head capsule that are lost as points of attachment for muscles from the mouthparts and the pharynx.

The mandibles of most orthorrhaphous Brachycera are bisegmented. There is also a strong tendency in this group for fusion of the mandibles and maxillae or the loss of the mandibles. This is best shown in the Stratiomyidae where the respective structures are not clearly differentiated (fig. 37.25).

The acephalic head skeleton is the culmination of the process of retraction of the head capsule within the thorax and the corresponding loss of a sclerotized external skeleton along with the compensatory growth of an internal tentorial skeleton that becomes intimately linked with the pharynx. This has resulted in the cephalopharyngeal skeleton of larvae of the Muscomorpha (=cyclorrhaphous Brachycera). The typical structure of the cephalopharyngeal skeleton is shown in Figs. 37.26a–c. Although there is great variation in the conformation of the various sclerites among larvae of this large portion of the Diptera and between each of the three instars, the general theme remains the same for all. The presence of two sensory organs on a pair of membranous lobes of the cephalic segment must be homologous with the antennae and maxillary palpi, the only major sensory structures on the eucephalic and hemicephalic head capsules. Therefore, it follows that the cephalic segment corresponds to the anterior portion of the head capsule that has lost sclerotization.

The pharyngeal filter (figs. 37.26b,c) is a structure found in diverse taxa of Muscomorpha that strains microorganisms and particulate food from the "liquid" medium inhabited by the larva. Its presence or absence is a character used near the end of the key to families; dissection may be necessary to observe it, although it is frequently visible through the ventral integument.

Body

Although most Diptera larvae are slender, there is great diversity in body shape. The slender body forms may be quite cylindrical, as in many Tipulidae, Bibionidae, Trichoceridae and Anisopodidae (figs. 37.1, 37.27–37.29). Larvae of Cecidomyidae, Tabanidae and some Muscomorpha have a fusiform body that tapers rather gradually to a point at both ends (figs. 37.30–37.32). Larvae of some Ceratopogonidae, Chironomidae, Therevidae and Scenopinidae have an elongate serpentine body (figs. 37.33, 37.34). Some larvae, especially of many Muscomorpha, are markedly more narrowed anteriorly (figs. 37.35, 37.36). Others are dorsoventrally flattened as are many Stratiomyidae, the Lonchopteridae, Platypezidae and *Fannia* (Muscidae) (figs. 37.37–37.40). Larvae of many Syrphidae and most parasitic Diptera are very stout (figs. 37.41–37.43). Parts of the body may be conspicuously swollen as is the thorax of Chaoboridae and Culicidae (fig. 37.44) and the posterior part of the abdomen of Simuliidae (fig. 37.45). The larva of *Microdon* (Syrphidae) is hemispherical (fig. 37.46).

Segmentation

The number of body segments is most commonly 12 in the Nematocera and 11 in the Brachycera, of which the anterior three are thoracic and the remainder abdominal. Any variance from these numbers is a decrease in the number of abdominal segments in the Nematocera and an increase for the Brachycera. Thus, several families of Nematocera have only eight abdominal segments. The Blephariceridae have only seven abdominal segments, and only seven total segments since the first segment is a fusion of the head, thorax and first abdominal segment (fig. 37.14). Fusion of the three thoracic segments is an important feature of the Chaoboridae, the Culicidae (fig. 37.44) and the Simuliidae (fig. 37.45). Conversely, subdivision of segments is a feature of several taxa. A narrow intercalary pseudosegment is inserted in front of the prothorax and each of the abdominal segments of Anisopodidae (fig. 37.29). The three thoracic segments and first abdominal segment of larvae of the subfamily Psychodinae (Psychodidae) are commonly divided into two subdivisions, with the following six abdominal segments each divided into three subdivisions (fig. 37.47). Larvae of Therevidae and Scenopinidae are unique in having 20 very similar segmental divisions (fig. 37.34). Only slight size and shape differences and the typical location of the posterior spiracles on the eighth abdominal segment indicate that the first six abdominal segments are divided once and that segments 8 and possibly 9 comprise the terminal five divisions. Similar difficulties occur in interpreting segmental divisions of some Bibionidae (figs. 37.48, 37.49).

Cuticle

The integument of most Diptera larvae is non-pigmented, at most weakly sclerotized, flexible and elastic. The thickness varies widely, dependent for the most part on the habitat occupied. Larvae living exclusively in concealed damp or aquatic habitats and subject to few abrasive forces usually have a thin cuticle, whereas such larvae as Asilidae, living and moving in relatively dry soil, have a tough, leathery cuticle that protects them from abrasion and water loss. Some sclerotization of the integument may be present, e.g. the thorax of some Xylophagidae (fig. 37.50), abdominal segments of some Psychodidae (fig. 37.47) and all tergites of Lonchopteridae (fig. 37.43). Larvae of the Stratiomyiidae (fig. 37.37) and Xylomyidae have a distinctive hardened, armored integument, a result of the deposition in the integument of calcium carbonate. Larvae may be variously colored, sometimes due to surface sclerotization, but more often the result of coloring of the hemolymph.

Cuticular outgrowths of the body surface of Diptera larvae, apart from those associated with locomotion, are common. These range from minute fine microtrichia or spicules to scale-like projections, to simple or greatly modified hairs or setae, to strong spines and fleshy processes.

Apart from the microtrichia that densely cover the integument of some larvae (figs. 37.51–37.53) other cuticular outgrowths usually have a distinctive symmetrical arrangement, the pattern generally quite similar on each of the first seven abdominal segments but differing from the meso- and metathorax whose pattern, in turn, differs from the prothorax and eighth and possibly ninth segments. Such symmetrically arranged integumental hairs or setae may be extremely small and visible only under high magnification.

Locomotory Structures

Diptera larvae lack segmented thoracic limbs. Common substitutes are different types and sizes of projections on the anterior and posterior margins of 1 or more body segments, usually bearing setae or spines. These appendages are basically of two types, prolegs and creeping welts. A third quite different type of structure called a suction disc is found in larvae of Blephariceridae (Fig. 37.14) and some Psychodinae.

Prolegs are typically round or oval, ventral paired fleshy tubercles of one or two segments, and usually found on the prothorax and terminal segment or one or more of the intermediate abdominal segments. They bear one or more curved spinules near their apices. Such spinules are varied in form, sometimes very small and showing little organization, or they may be very strong, hooked crochets arranged in one or more partially or completely encircling rows at the apex of the proleg (figs. 37.54–37.58).

Creeping welts are transverse swollen ridges on the ventral and sometimes dorsal intersegmental region of usually the first seven abdominal segments (figs. 37.59, 37.60); sometimes the thoracic segments may be involved (fig. 37.61). Creeping welts normally bear spinules arranged in transverse rows. The spinules in the anterior rows of any creeping welt may be inclined anteriorly to assist in backward movement.

Other body parts may assist in locomotion. The vertically oriented mandibles of brachycerous larvae provide an anchoring point against which contraction of the body results in forward movement. Transverse rows of ambulatory combs are present on the venter of abdominal segments five to seven in larvae of Dixidae (fig. 37.62). Preanal and postanal ridges of many brachycerous larvae appear to function in the same way as creeping welts and sometimes bear spinules similar to those on the creeping welts (figs. 37.63–37.65).

Respiratory System

The respiratory system includes the internal system of tracheae and the external spiracles, the general plan of which has been well studied by Keilin (1944) and Whitten (1955). This system and especially the individual characteristics and location of functional spiracles has great significance in larval systematics. The basic number of spiracles on Diptera larvae is 10 pairs, although this number is present only in Bibionidae and the European genus *Pachyneura* (Pachyneuridae). These paired spiracles are located on the prothorax, metathorax, and each of eight abdominal segments. The spiracles have often been named for the segment on which they are located. However, the spiracles on the prothorax are possibly the mesothoracic spiracles that have migrated forward (Hinton 1947). Or as suggested by Keilin (1944) all spiracles may have originated intersegmentally and the spiracle between the prothorax and mesothorax moved forward to the prothorax and all other spiracles were displaced to the segment behind. Thus, it is more accurate to refer to the spiracles on the prothorax as the anterior spiracles. The metathoracic spiracles are sometimes called the posterior thoracic spiracles. Similarly the eighth pair of abdominal spiracles, although usually located on this segment, are situated on the apparent ninth segment in larvae of some Bibionidae, presumably as a result of backward migration. Therefore, this terminal eighth pair of spiracles is best referred to as the posterior spiracles. The intermediate abdominal spiracles are designated by the number of the segment on which they occur.

Various degrees of reduction in the number of spiracles has taken place among Diptera larvae and a convenient terminology has been devised in referring to these spiracular arrangements as diagrammatically represented in fig. 37.66. All of the various arrangements are represented among the families of Nematocera and orthorrhaphous Brachycera, but only the amphipneustic and metapneustic systems are present in larvae of the Muscomorpha.

The structure of the spiracles is diverse, not only among species but between larval instars and between spiracles on the same larva; the anterior spiracles usually differ from the abdominal spiracles, and the posterior spiracles differ from other abdominal spiracles in holopneustic and peripneustic systems. Spiracles are of three types, based on the manner in which the internal tracheae are withdrawn through the newly developed spiracles at the time of molting, as described by Keilin (1944), but for practical purposes they can be divided into symmetrical and asymmetrical spiracles. The former comprise a relatively uniform series of small spiracular openings surrounding one or more central ecdysial scars (figs. 37.67–37.71), or a simple central spiracular opening surrounded by a rather uniform sclerotized collar. The ecdysial scar is the place where the trachea of the previous larval instar was withdrawn during molting, and which subsequently was closed and sealed, usually as a conspicuous spot. In asymmetrical spiracles normally three or more spiracular openings are dominantly to one side of the ecdysial scar (figs. 37.72, 37.73). The asymmetrical spiracle is much the more common in Diptera. Only a few groups, mainly the lower Nematocera, have symmetrical spiracles.

The configuration of the spiracles, including the number and shape of the spiracular openings and their location in relation to the ecdysial scar, is an important identification feature. Such configurations differ between spiracles on various segments of the body and between spiracles of different larval instars. In many higher Diptera the number of spiracular openings is two for the first and second instars, and three for the third. The ecdysial scar may not be very evident. This is usually the case in the anterior thoracic spiracle of larvae of the Muscomorpha.

These anterior spiracles of the Muscomorpha often resemble a fan or tuft of papillae of various number and length (figs. 37.74–37.77); in addition, they are usually quite different from the posterior spiracles of the same specimen, and from the anterior spiracles of non-muscomorphan Diptera larvae.

Aquatic larvae and/or pupae may have spiracles modified for piercing air-containing plant tissues or for the intake of atmospheric air. These range from the short, sharp, barbed siphons of some plant-piercing culicids (fig. 37.78), to the more intermediate length siphons of other culicids (fig. 37.79), to the very long respiratory siphons of rat-tailed maggots (Syrphidae: *Eristalis,* fig. 37.80) that may be extended to several times the length of the body.

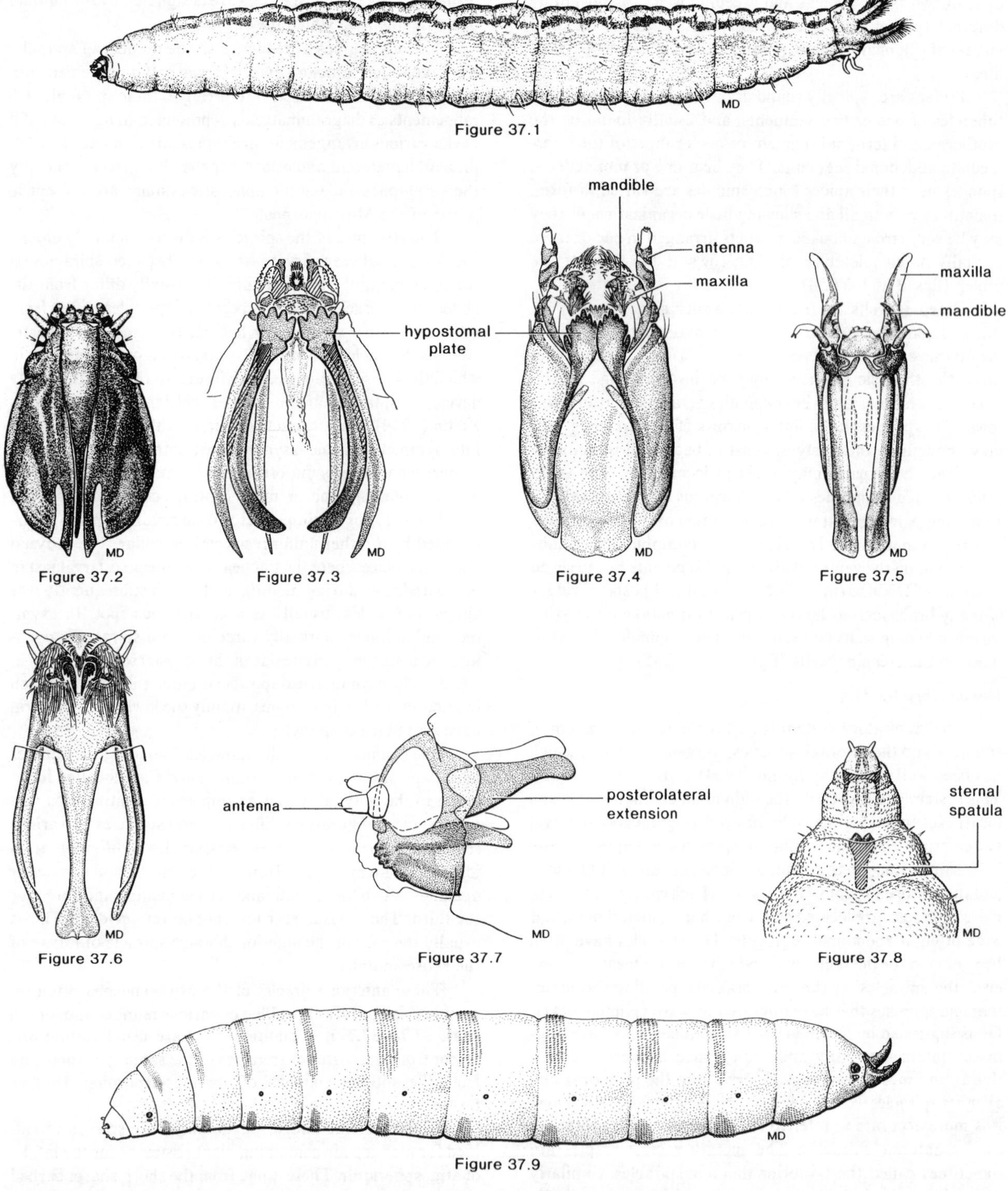

Figures 37.1–37.6. Tipulidae (Nematocera) with partially retracted or reduced head capsules: **37.1.** Larva, *Prionocera* sp.; **37.2.** *Prionocera dimidiata* (Loew), dorsal; **37.3.** *Molophilus* sp., ventral; **37.4.** *Pseudolimnophila inornata* (Osten Sacken), ventral; **37.5.** *Limnophila* sp., dorsal; **37.6.** *Pilaria recondita* (Osten Sacken), ventral (from Manual of Nearctic Diptera, Vol. 1).

Figures 37.7–37.8. Reduced and retracted head capsules of Cecidomyiidae (Nematocera): **37.7.** *Cecidomyia* sp., head, lateral; **37.8.** larva, *Dasineura* sp., head, pro- and mesothorax, ventral (from Manual of Nearctic Diptera, Vol. 1).

Figure 37.9. Synneuridae. Larva, *Synneuron decipiens* Hutson, lacking any vestige of a head capsule (from Manual of Nearctic Diptera, Vol. 1).

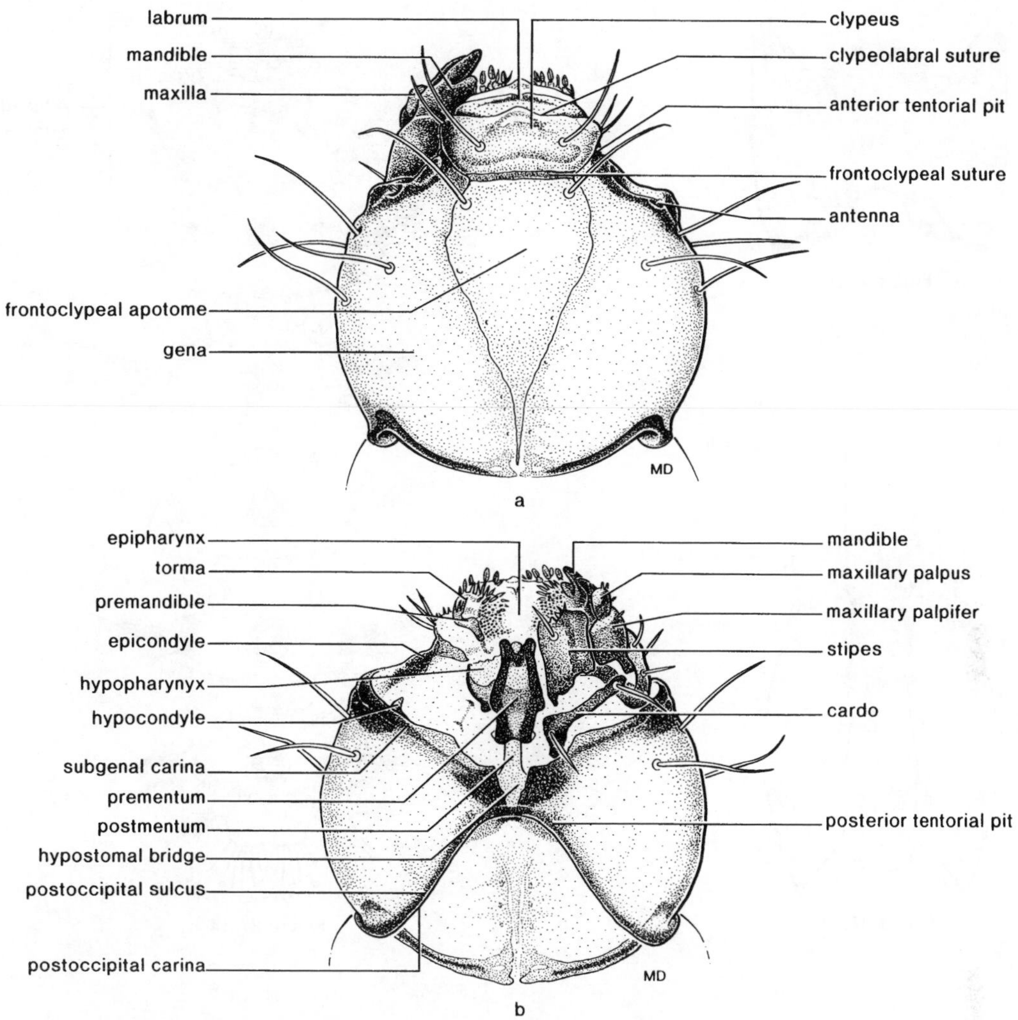

labrum
mandible
maxilla
clypeus
clypeolabral suture
anterior tentorial pit
frontoclypeal suture
antenna
frontoclypeal apotome
gena
MD
a
epipharynx
torma
premandible
epicondyle
hypopharynyx
hypocondyle
subgenal carina
prementum
postmentum
hypostomal bridge
postoccipital sulcus
postoccipital carina
mandible
maxillary palpus
maxillary palpifer
stipes
cardo
posterior tentorial pit
MD
b

Figure 37.10

epipharyngeal bar
torma
mandible
premandible
hypostoma
paralabial plate
MD

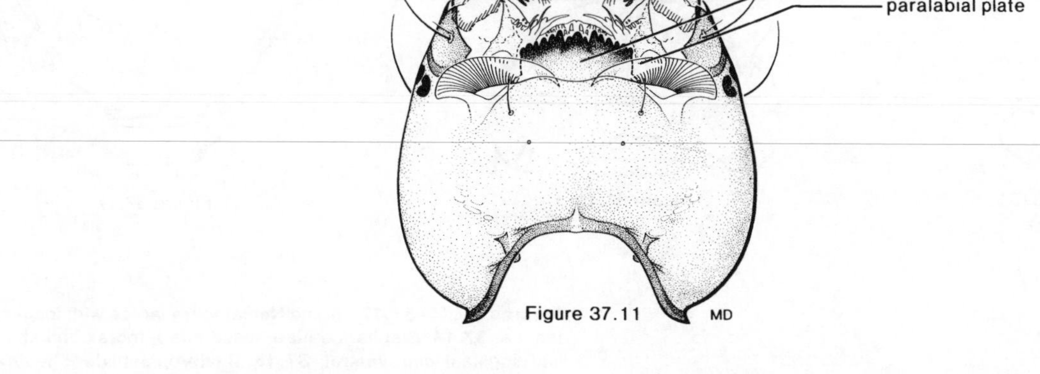

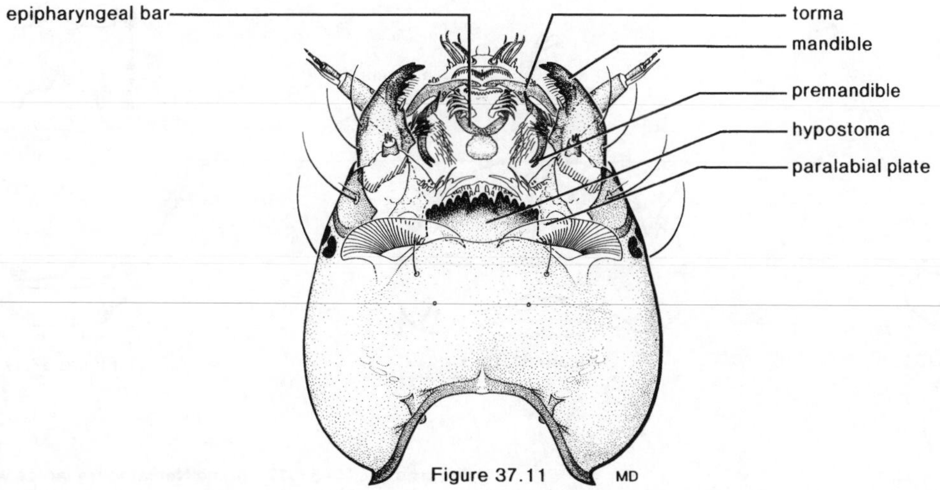

Figure 37.11

Figure 37.10a,b. Bibionidae. Head capsule structures of *Bibio* sp. **a.** dorsal; **b.** ventral, with right mandible and maxilla removed (from Manual of Nearctic Diptera, Vol. 1).

Figure 37.11. Head capsule, ventral view, with sclerotized, toothed, hypostoma: **37.11.** Chironomidae, *Chironomus* sp. (from Manual of Nearctic Diptera, Vol. 1).

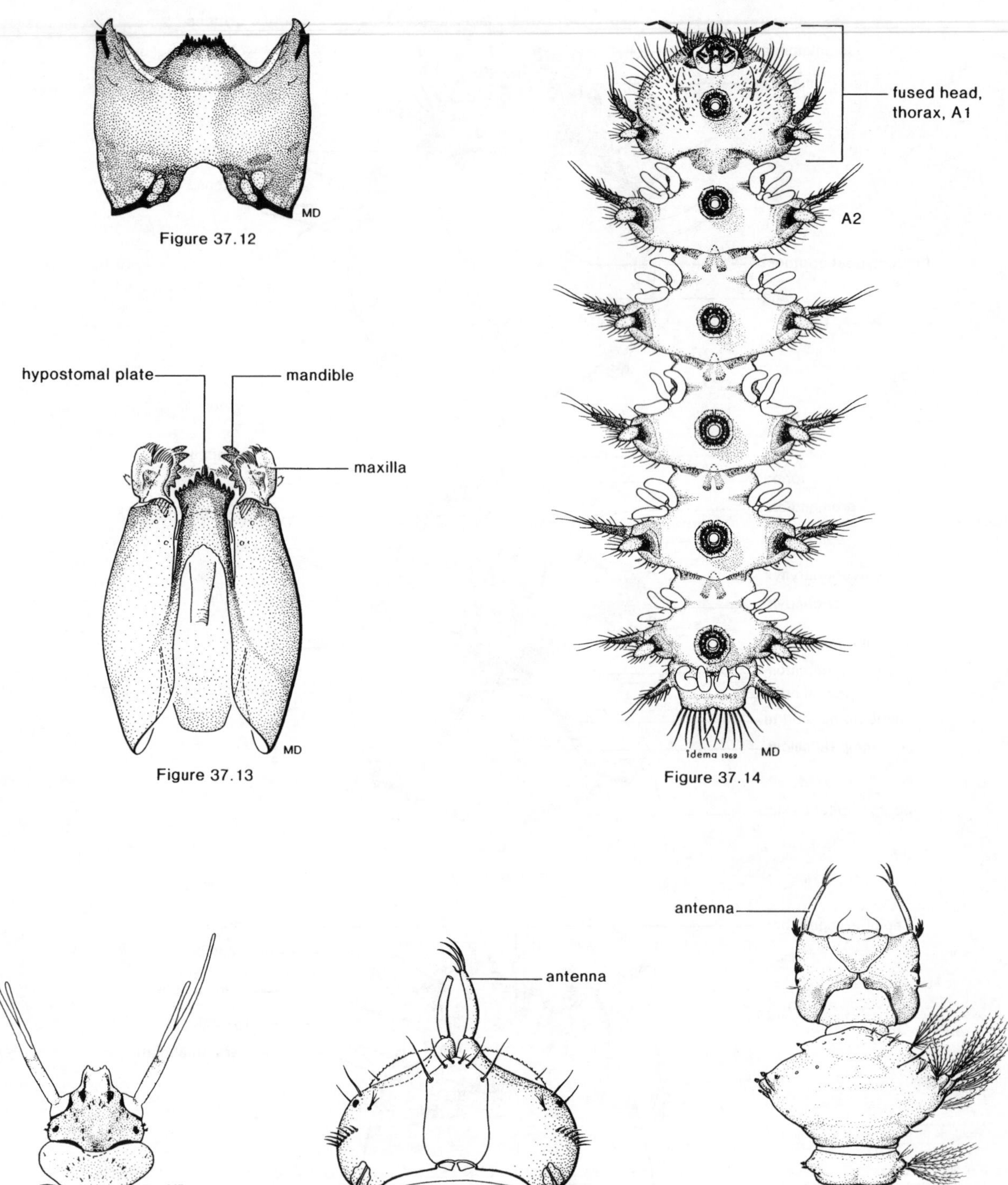

Figure 37.12

Figure 37.13

Figure 37.14

Figure 37.15

Figure 37.16

Figure 37.17

Figure 37.12–37.13. Head capsules, ventral view, with sclerotized, toothed, hypostoma: **37.12.** Simuliidae, *Simulium vittatum* Zetterstedt; **37.13.** Tipulidae, *Limonia* sp. (from Manual of Nearctic Diptera, Vol. 1).

Figures 37.14–37.17. Some Nematocera larvae with longer antennae: **37.14.** Blephariceridae, fused head, thorax and abdominal segment one, ventral; **37.15.** Deuterophlebiidae, head and prothorax; **37.16.** Chaoboridae, *Corethrella brakeleyi* (Coquillett), showing prehensile antenna; **37.17** Chaoboridae. *Eucorethra underwoodi* Underwood, showing prehensile antennae (from Manual of Nearctic Diptera, Vol. 1).

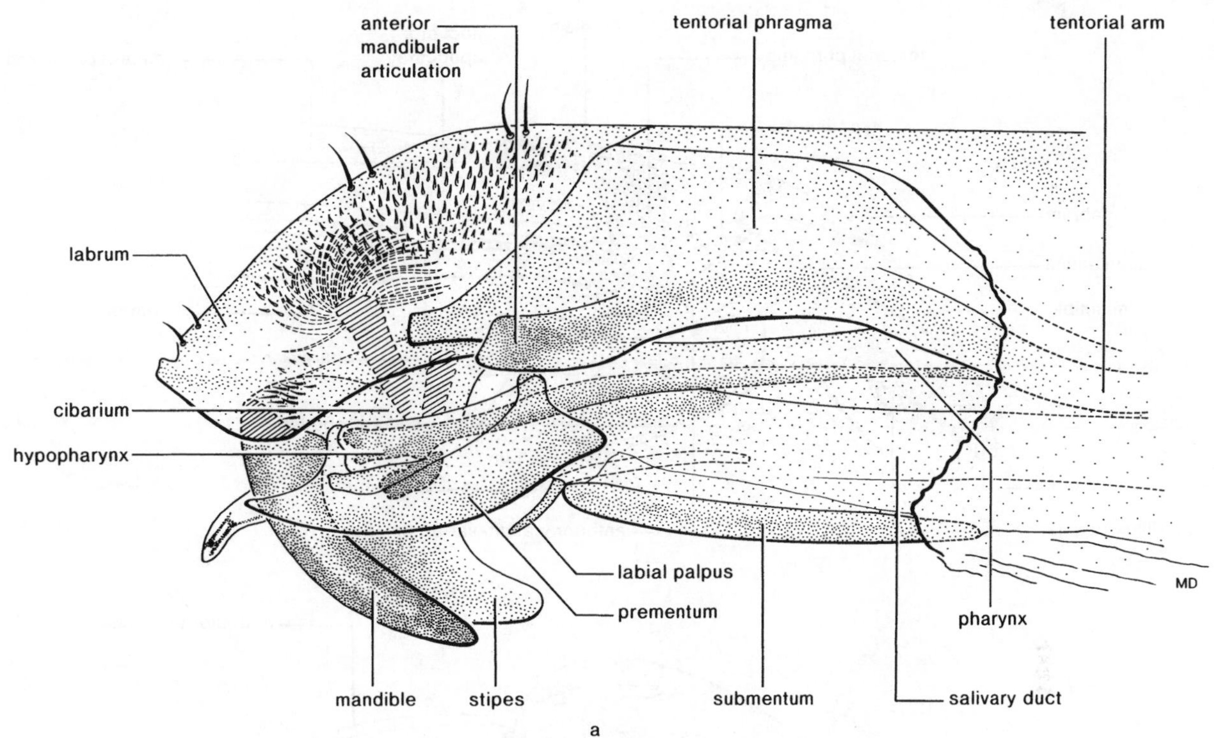

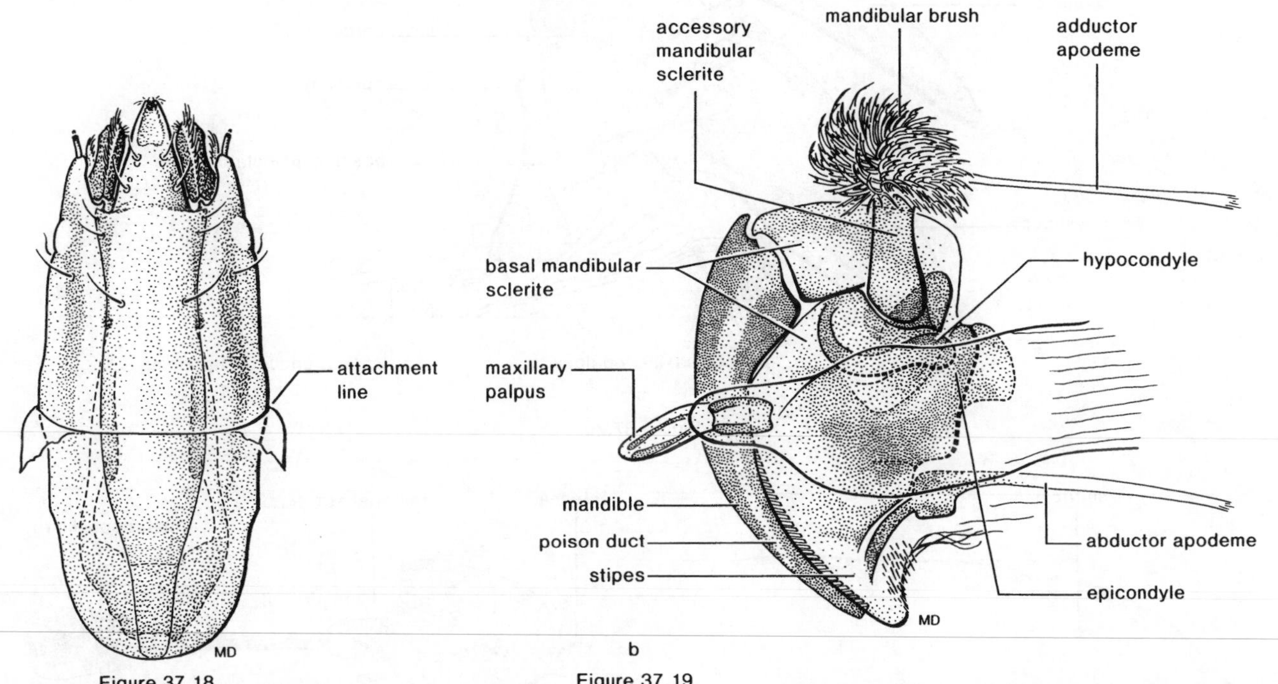

Figure 37.18

Figure 37.19

Figures 37.18–37.19. Hemicephalic head capsules: **37.18.** Stratiomyidae. *Odontomyia cincta* Olivier, dorsal; **37.19a, b.** Tabanidae. *Tabanus marginalis* Fabricius. **a.** lateral view of anterior portion with left mandibular-maxillary complex and portion of head capsule removed; **b.** left mandibular-maxillary complex, lateral view (from Manual of Nearctic Diptera, Vol. 1).

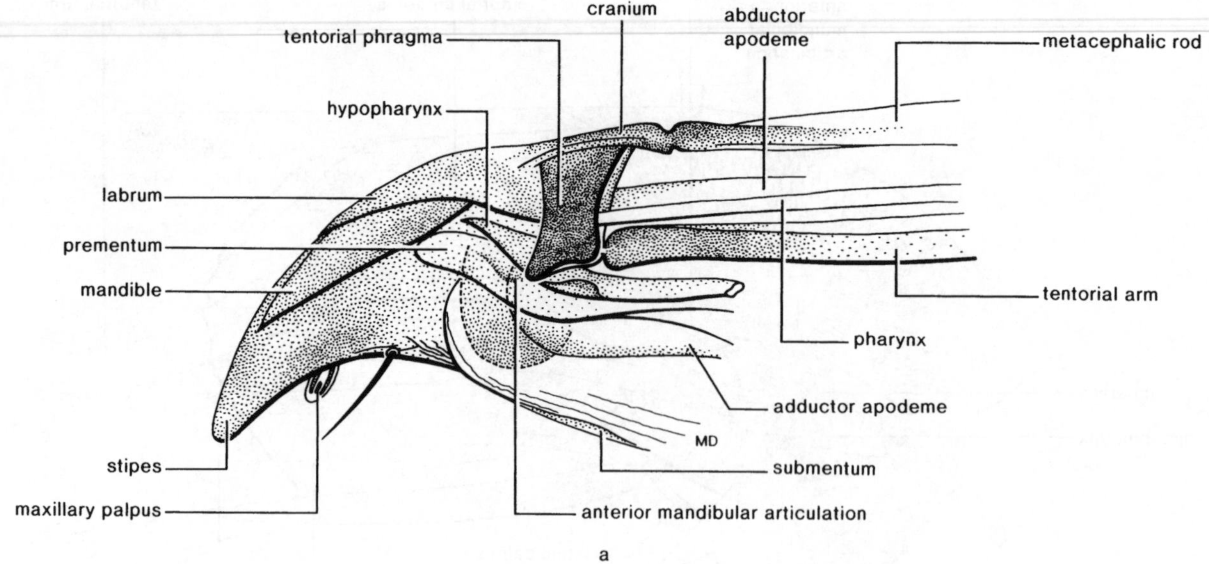

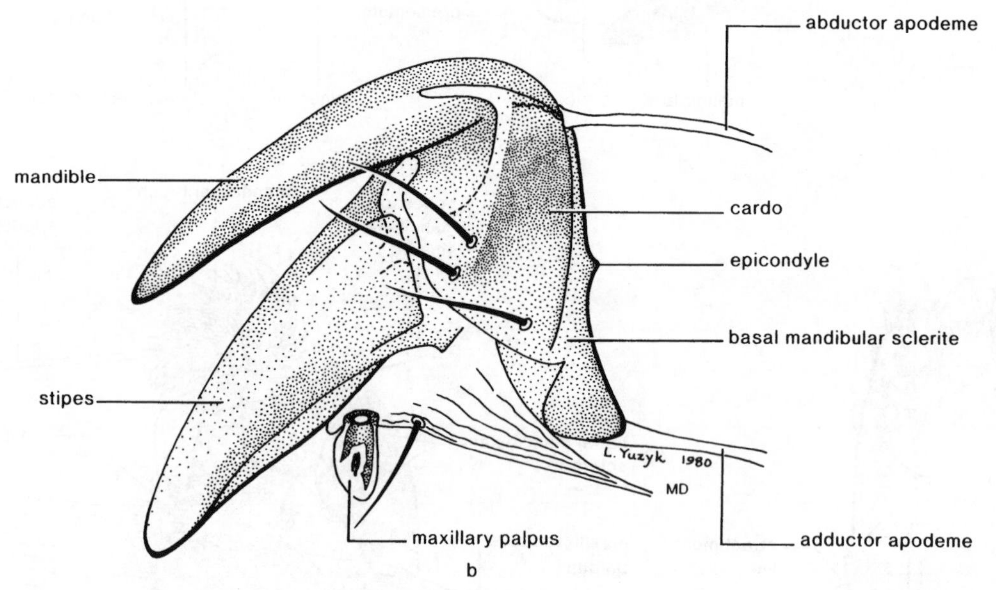

Figure 37.20

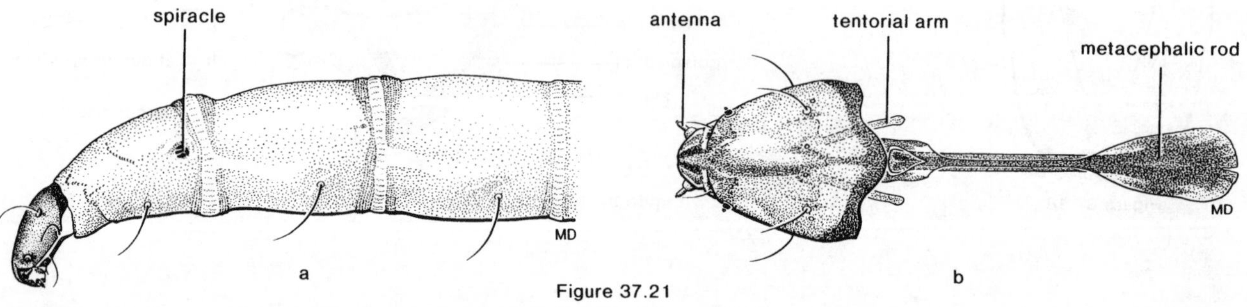

Figure 37.21

Figures 37.20. Hemicephalic head capsules: **37.20a, b.** Asilidae. *Promachus* sp. **a.** lateral view of anterior portion with left mandibular-maxillary complex and portion of head capsule removed; **b.** left mandibular-maxillary complex, lateral view (from Manual of Nearctic Diptera, Vol. 1).

Figure 37.21a,b. Therevidae. *Thereva fucata* Loew. **a.** head and thorax, lateral; **b.** head, dorsal (from Manual of Nearctic Diptera, Vol. 1).

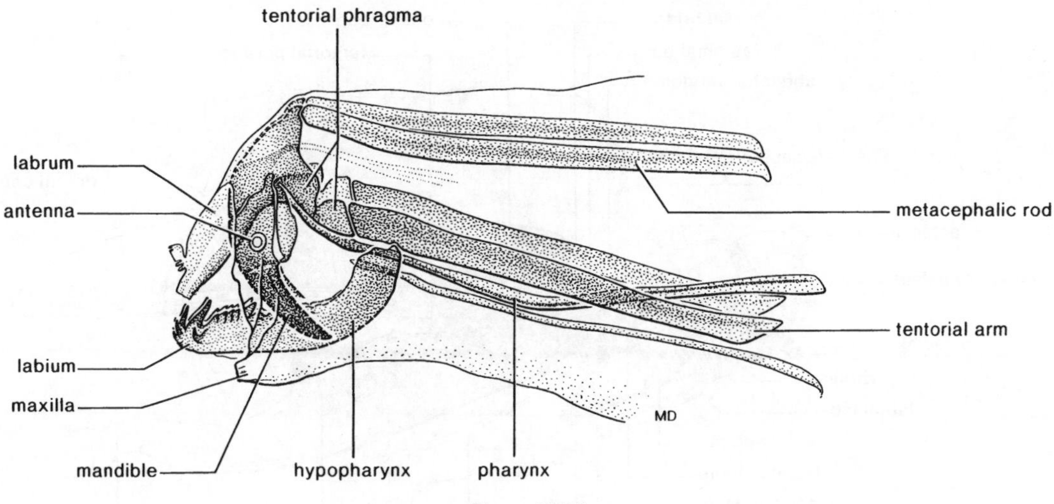

Figure 37.22

pharyngeal sclerite

tentorial arm

maxilla

mandible

labrum

maxillary palpus

antenna

MD

metacephalic rod

a

MD

b

Figure 37.23

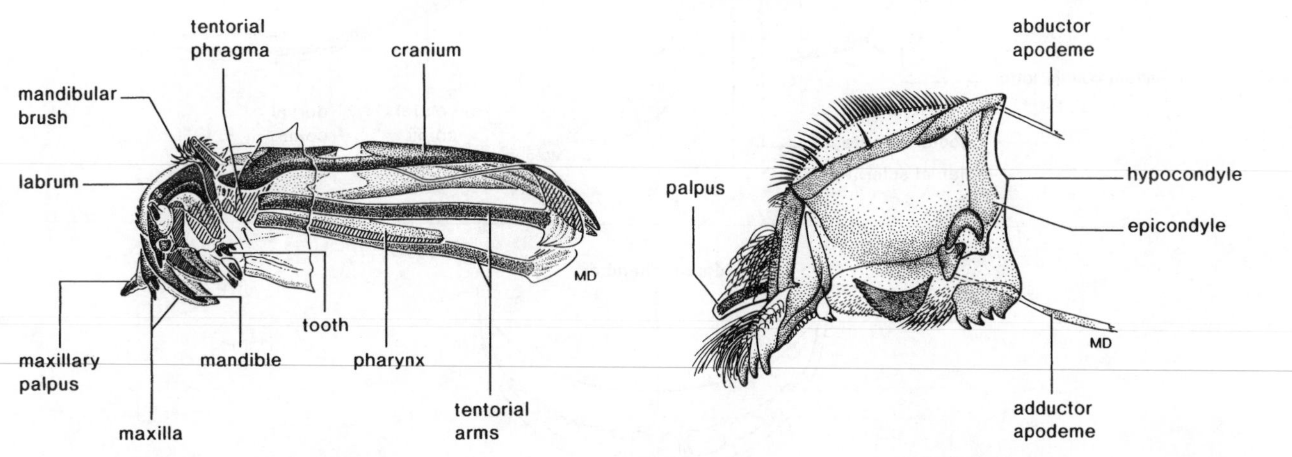

Figure 37.24

Figure 37.25

Figures 37.22–37.23a,b. Retracted and reduced heads: **37.22.** Empididae, Empidinae, larval head skeleton, lateral; **37.23a, b.** Dolichopodidae, *Medetera* sp., **a.** larva, lateral, and **b.** head skeleton, dorsal (from Manual of Nearctic Diptera, Vol. 1).

Figure 37.24. Rhagionidae, *Rhagio* sp., head capsule, lateral, showing posterior tentorial arms (from Manual of Nearctic Diptera, Vol. 1).

Figure 37.25. Stratiomyidae, *Odontomyia* sp. showing left, undifferentiated mandibular-maxillary complex (from Manual of Nearctic Diptera, Vol. 1).

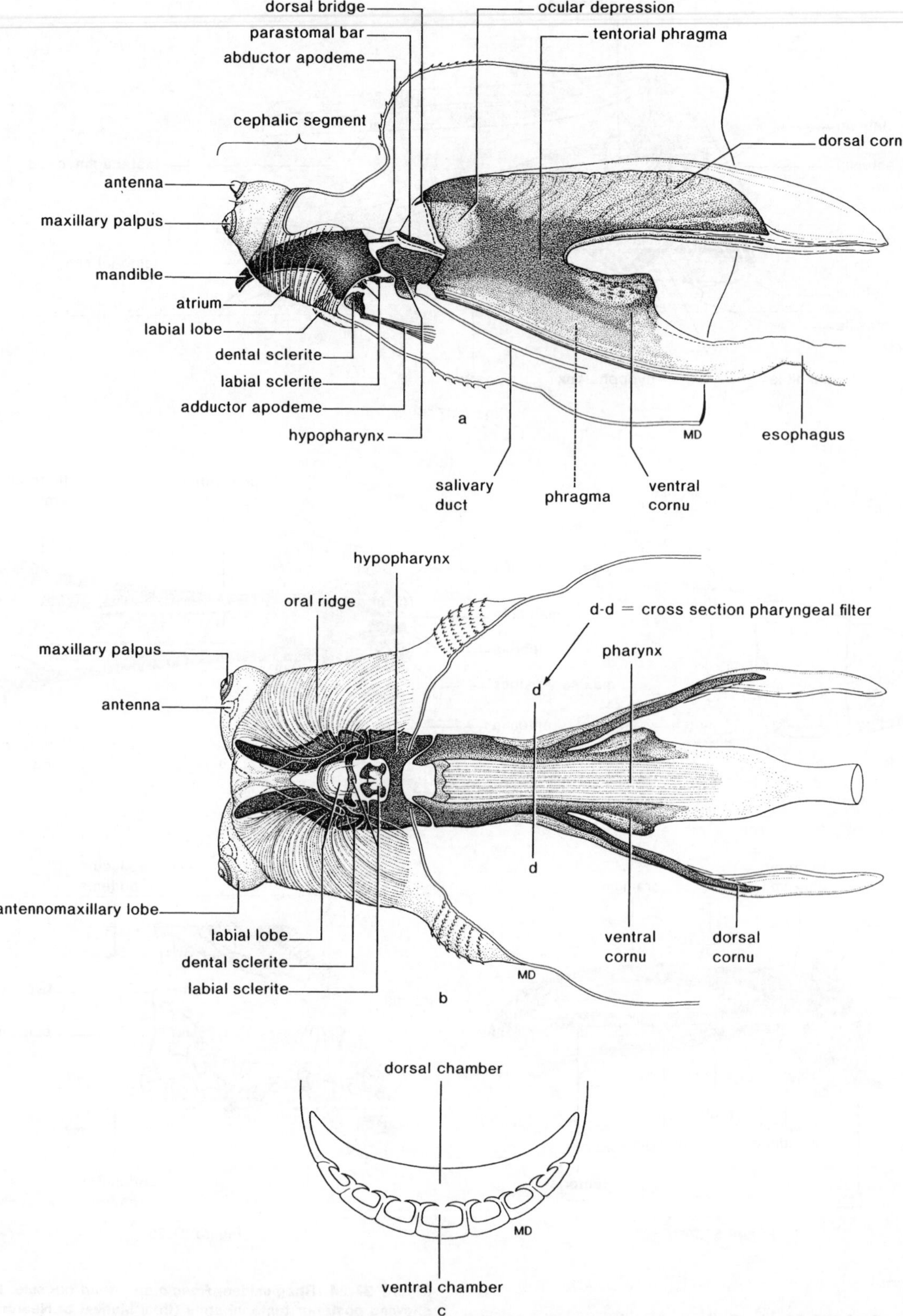

Figure 37.26a-c. Calliphoridae, cephalic segment and cephalopharyngeal skeleton of *Phormia regina* (Meigen). **a.** lateral, with left antennomaxillary lobe removed; **b.** ventral, with posterior portion of integument removed; **c.** diagrammatic x-section of pharyngeal filter (from Manual of Nearctic Diptera, Vol. 1).

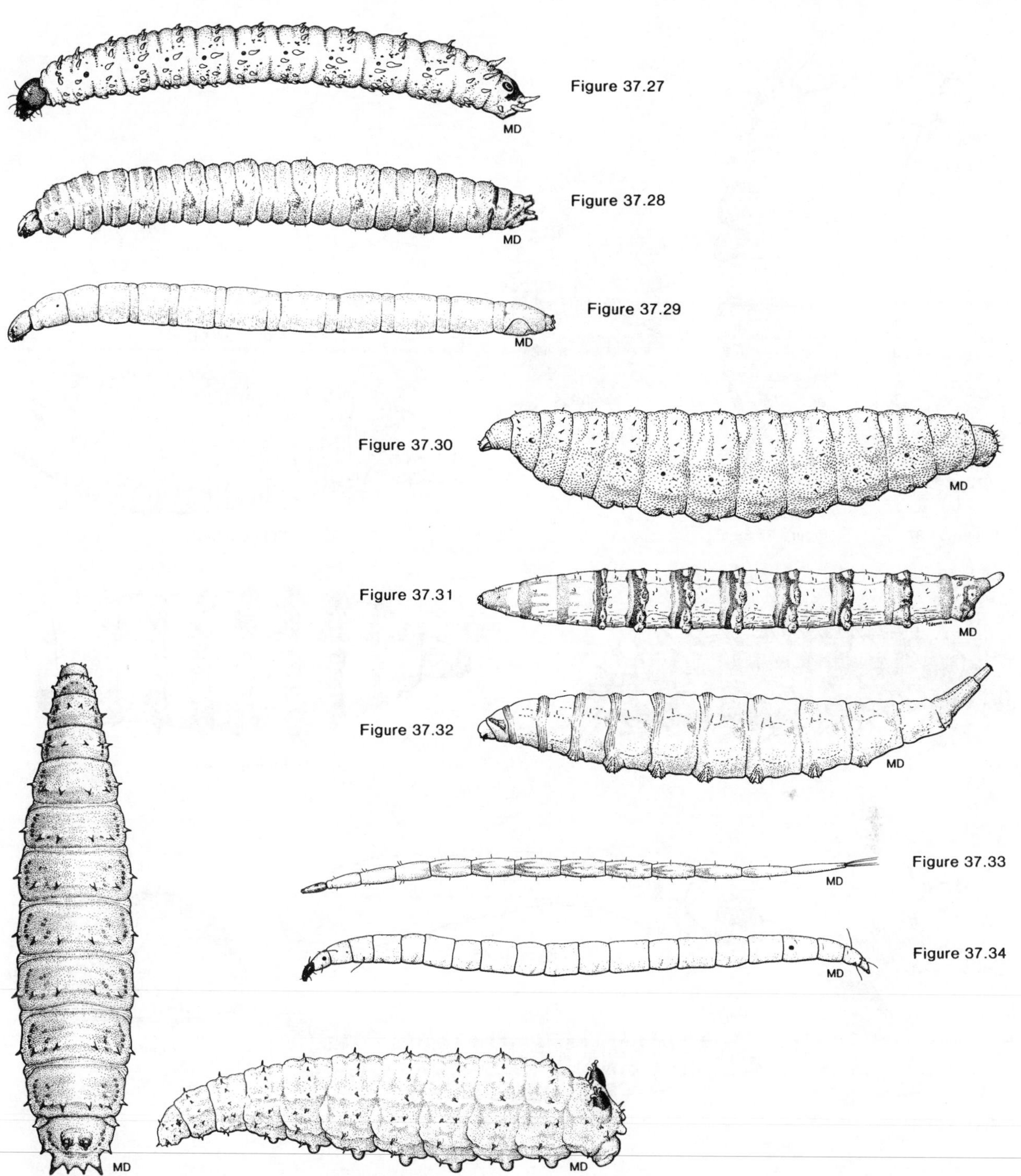

Figures 37.27–37.29 Larvae with slender, cylindrical body forms: **37.27.** Bibionidae, *Bibio* sp. lateral; **37.28.** Trichoceridae, *Trichocera* sp., lateral; **37.29.** Anisopodidae, *Sylvicola punctatus* (Fabricius), lateral (from Manual of Nearctic Diptera, Vol. 1).

Figures 37.30–37.32. Larvae with fusiform bodies: **37.30.** Cecidomyiidae, *Dasineura* sp., lateral; **37.31.** Tabanidae, *Tabanus reinwardtii* Wiedemann, lateral; **37.32.** Canacidae, *Canace macateei* Malloch, lateral (from Manual of Nearctic Diptera, Vol. 1).

Figures 37.33–37.34. Larvae with elongate, serpentine bodies: **37.33.** Ceratopogonidae, *Bezzia* sp. dorsal; **37.34.** Scenopinidae, *Scenopinus* sp., lateral (from Manual of Nearctic Diptera, Vol. 1).

Figures 37.35–37.36. Larvae that are markedly narrowed anteriorly: **37.35.** Phoridae, *Megaselia* sp., dorsal; **37.36.** Chamaemyiidae, *Leucopus simplex* Loew, lateral (from Manual of Nearctic Diptera, Vol. 1).

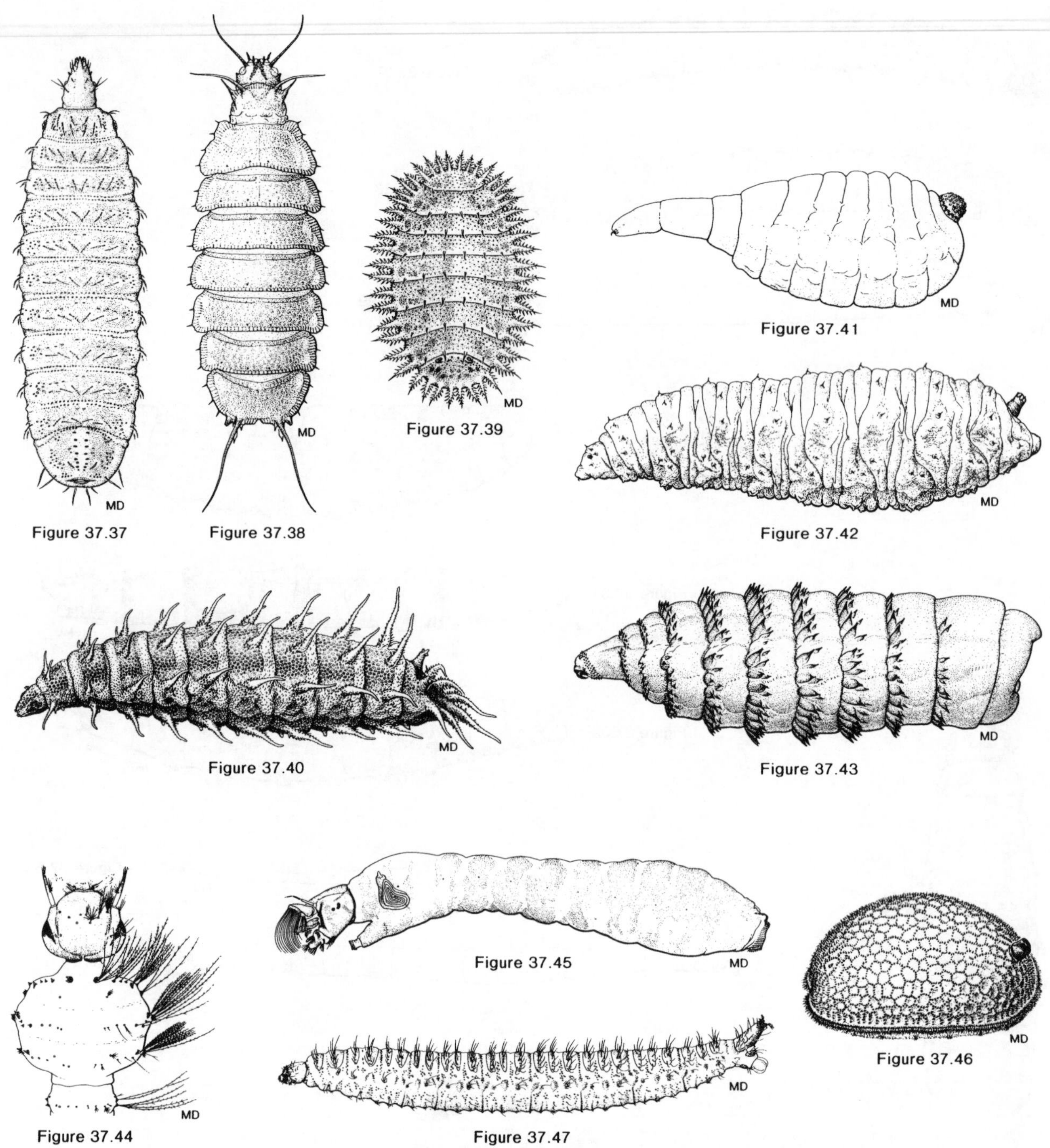

Figures 37.37–37.40. Larvae that are dorso-ventrally flattened: **37.37.** Stratiomyidae, *Berkshiria albistylum* Johnson, dorsal; **37.38.** Lonchopteridae, *Lonchoptera* sp., dorsal; **37.39.** Platypezidae, *Callomyia gilloglyorum* Kessel, dorsal; **37.40.** Muscidae, *Fannia canicularis* (L.), dorsolateral (from Manual of Nearctic Diptera, Vol. 1).

Figures 37.41–37.43. Larvae that are very stout: **37.41.** Conopidae, *Physocephala bimarginipennis* Karsch, lateral; **37.42.** Syrphidae, *Syrphus knabi* Shannon, lateral; **37.43.** Oestridae, *Gasterophilus intestinalis* (De Geer), lateral (from Manual of Nearctic Diptera, Vol. 1).

Figure 37.44. Enlarged thorax of Culicidae. *Culiseta incidens* (Thomson), head, thorax and first abdominal segment (from Manual of Nearctic Diptera, Vol. 1).

Figure 37.45. Enlarged posterior of abdomen of Simuliidae. *Simulium venustum* Say (from Manual of Nearctic Diptera, Vol. 1).

Figure 37.46. Hemispherical body of *Microdon* sp. (Syrphidae) (from Manual of Nearctic Diptera, Vol. 1).

Figure 37.47. Psychodidae, Psychodinae. *Pericoma* sp., lateral, showing subdivisions of thoracic and abdominal segments (from Manual of Nearctic Diptera, Vol. 1).

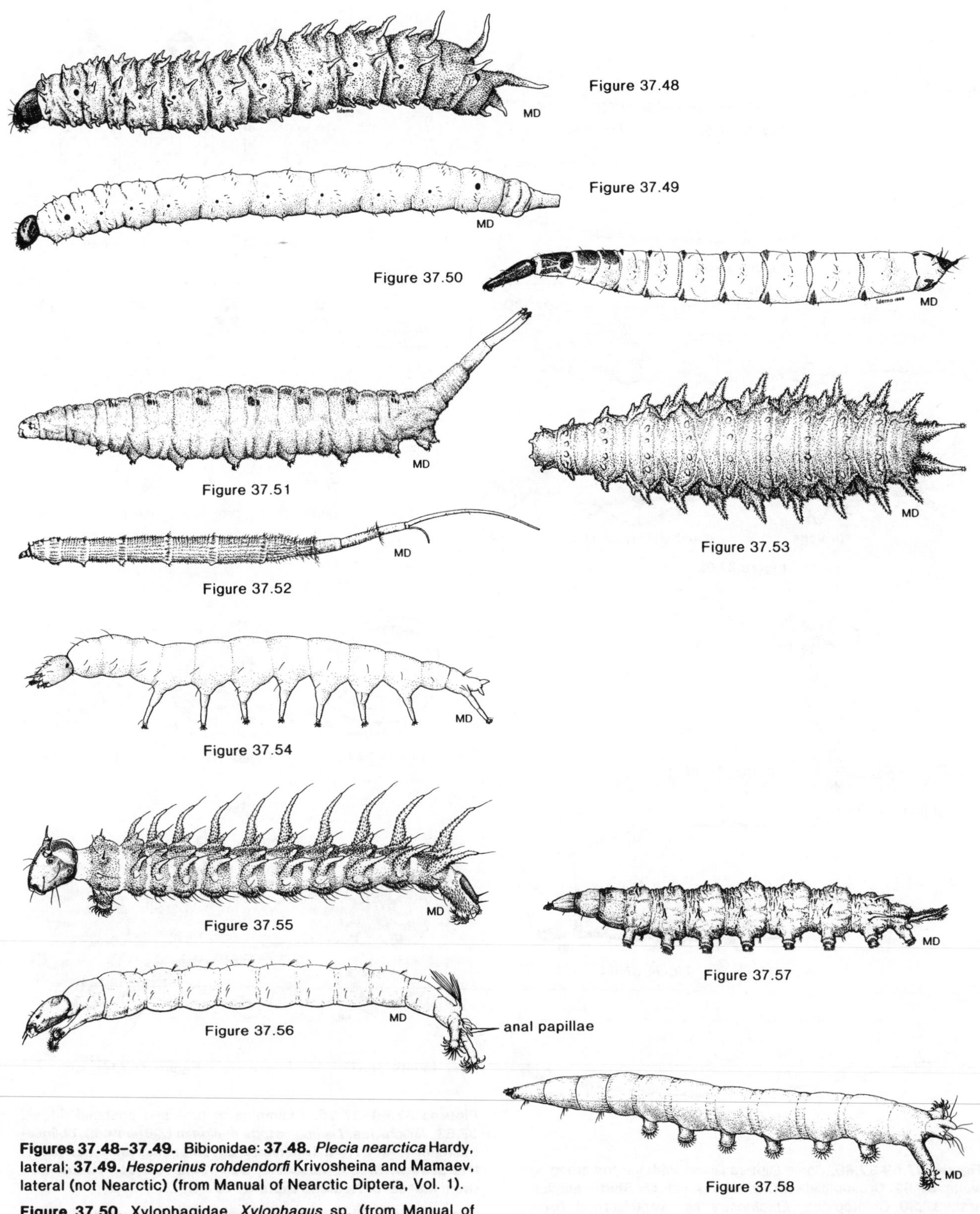

Figures 37.48–37.49. Bibionidae: **37.48.** *Plecia nearctica* Hardy, lateral; **37.49.** *Hesperinus rohdendorfi* Krivosheina and Mamaev, lateral (not Nearctic) (from Manual of Nearctic Diptera, Vol. 1).

Figure 37.50. Xylophagidae. *Xylophagus* sp. (from Manual of Nearctic Diptera, Vol. 1).

Figures 37.51–37.53. Larvae with dense integumental microtrichia: **37.51.** Ephydridae, *Ephydra* sp., lateral; **37.52.** Ptychopteridae, *Ptychoptera* sp., lateral; **37.53.** Periscelididae, *Periscelis annulata* (Fallén), dorsal (from Manual of Nearctic Diptera, Vol. 1).

Figures 37.54–37.58. Some Diptera larvae with strong crochets: **37.54.** Nymphomyiidae, *Palaeodipteron* walkeri Ide, lateral; **37.55.** Ceratopogonidae, *Atrichopogon polydactylus* Nielson, lateral; **37.56.** Chironomidae, *Ablabesmyia* sp., lateral; **37.57.** Athericidae, *Atherix* sp., lateral; **37.58.** Empididae, *Heterodromia* sp., lateral (from Manual of Nearctic Diptera, Vol. 1)

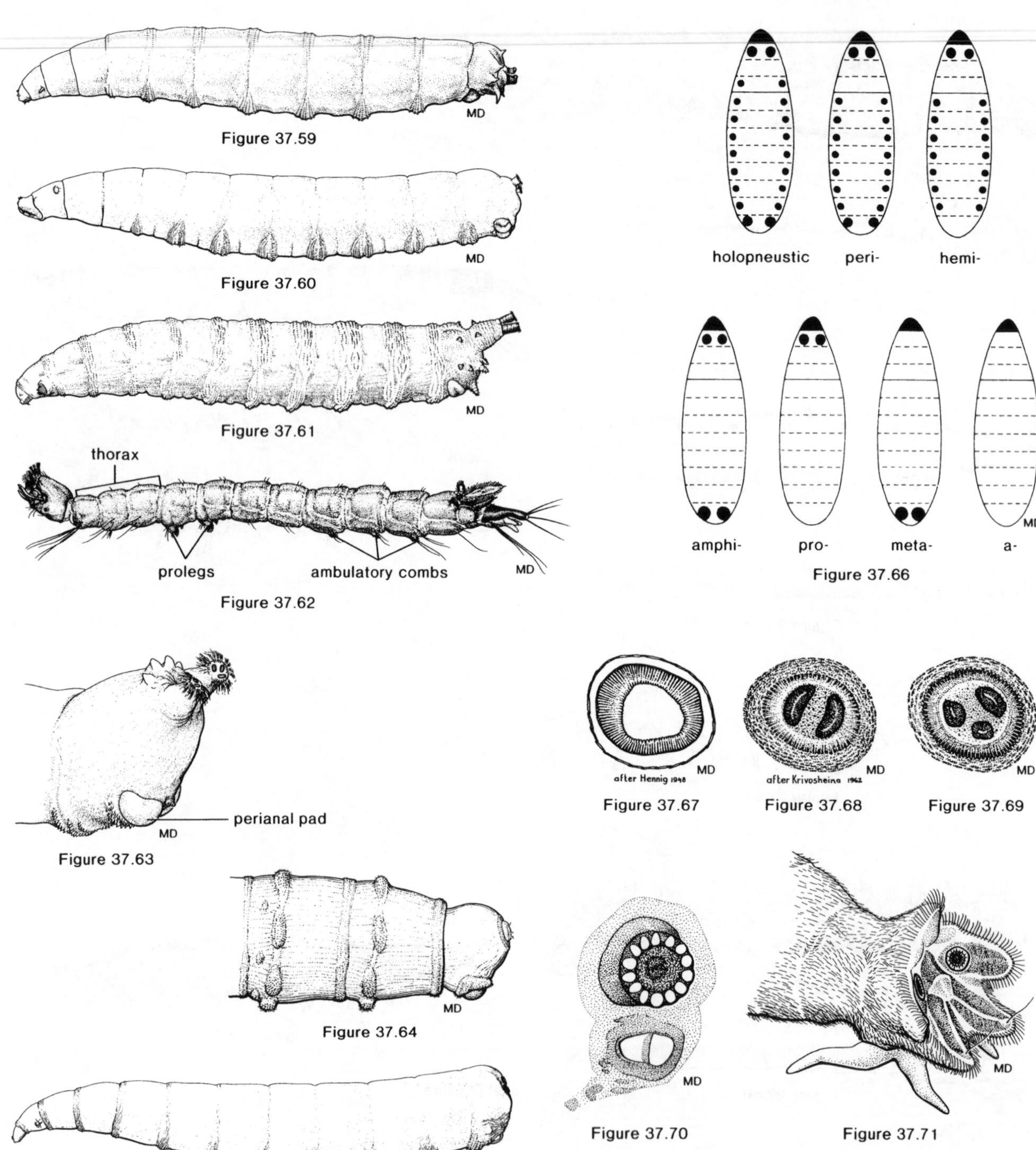

Figures 37.59-37.60. Some Diptera larvae with ventral creeping welts: **37.59.** Drosophilidae, *Chymomyza aldrichi* Sturtevant, lateral; **37.60.** Chloropidae, *Elachiptera* sp., ventrolateral (from Manual of Nearctic Diptera, Vol. 1).

Figure 37.61. A larva with some thoracic creeping welts. Drosophilidae. *Drosophila melanogaster* Meigen, lateral (from Manual of Nearctic Diptera, Vol. 1).

Figure 37.62. Dixidae. Transverse ambulatory combs on larva of *Dixa* sp., lateral (from Manual of Nearctic Diptera, Vol. 1).

Figures 37.63–37.65. Examples of pre- and postanal ridges: **37.63.** Milichiidae, *Desmometopa m-nigrum* (Zetterstedt), oblique-posterior view; **37.64.** Tabanidae, *Leucotabanus annulatus* (Say), lateral view; **37.65.** Muscidae, *Dendrophaonia* sp., lateral view (from Manual of Nearctic Diptera, Vol. 1).

Figure 37.66. Diagrammatic representation of spiracular systems in Diptera larvae (from Manual of Nearctic Diptera, Vol. 1).

Figures 37.67–37.71. Symmetrical spiracles of Diptera larvae: **37.67–37.69.** Bibionidae, **37.67.** *Penthetria* sp.; **37.68.** *Bibio* sp.; **37.69.** *Dilophus* sp.; **37.70.** Synneuridae, *Synneuron decipiens* Hutson; **37.71.** Tipulidae, *Gonomyia* sp., posteriolateral view of terminal segments (from Manual of Nearctic Diptera, Vol. 1).

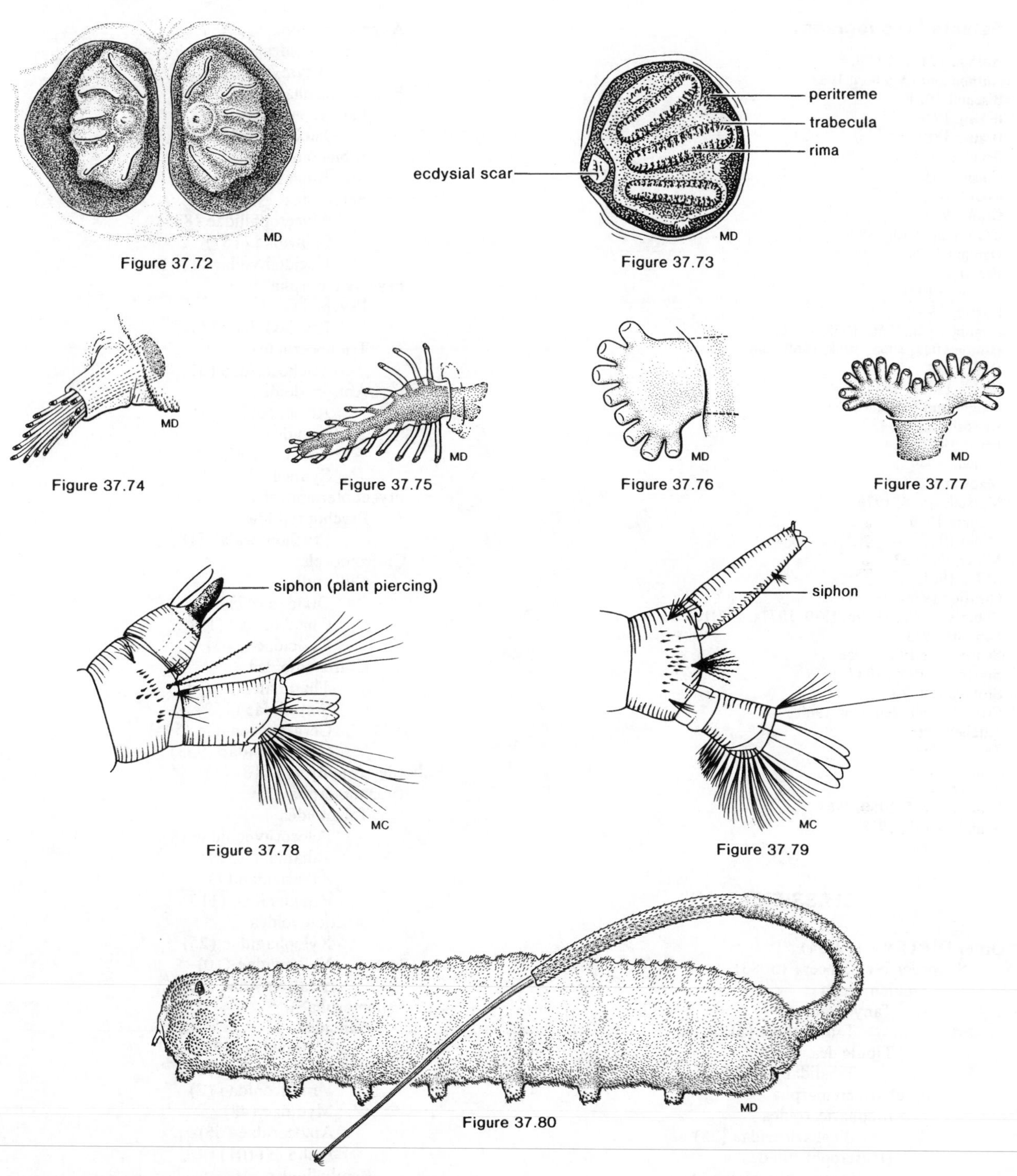

Figures 37.72–37.73. Asymmetrical spiracles of Diptera larvae: **32.72.** Tachinidae, *Zaira* sp., posterior spiracles; **37.73.** Calliphoridae, *Cynomyopsis cadaverina* (Robineau-Desvoidy), right posterior spiracle (from Manual of Nearctic Diptera, Vol. 1).

Figures 37.74–37.77. Anterior thoracic spiracles of Muscomorpha larvae: **37.74.** Drosophilidae, *Drosophila melanogaster* Meigen; **37.75.** Sepsidae, *Sepsis neocynipsea* Melander and Spuler; **37.76.** Piophilidae, *Lasiopiophila pilosa* (Staeger); **37.77.** Scathophagidae, *Scathophaga stercoraria* (L.) (from Manual of Nearctic Diptera, Vol. 1).

Figures 37.78–37.79. Culicidae. Respiratory siphons: **37.78.** *Mansonia* sp., adapted for piercing aquatic plants. **37.79.** A common type for surface respiration (from Merritt and Cummins 1984).

Figure 37.80. Syrphidae. *Eristalis tenax* (L.) larva, a species with a very long extensible respiratory siphon (from Manual of Nearctic Diptera, Vol. 1).

Selected Bibliography

Anthon 1943a, 1943b.
Anthon and Lyneborg 1968.
Bischoff 1924.
Bolwig 1946.
Brauer 1883.
Brindle 1952.
Chiswell 1955.
Cook 1944, 1949.
Craig 1967.
Crampton 1930b.
Damant 1924.
Ferrar 1987.
Harnisch 1954.
Hartley 1963.
Hennig 1948, 1950, 1952, 1973.
Hinton 1947, 1955, 1958, 1960, 1963.
Hora 1933.
Keilen 1912b, 1944.
Knight and Laffoon 1971.
Krivosheina 1969.
Lewis 1949.
Ludwig 1949.
Madwar 1937.
Matsuda 1965, 1976.
Meijere 1916.
Melin 1923.
Miller, D. 1932.
Milne 1961.
Olsufjev 1936.
Roberts 1969a, 1969b, 1970, 1971a, 1971b.
Schmitz 1938.
Schremmer 1951, 1956.
Snodgrass 1935, 1947, 1959.
Stoffolano 1970.
Strenzke and Neumann 1960.
Tatchell 1960.
Teskey 1969.
Tonnoir 1933.
Traxler 1977.
Whitten 1955, 1960, 1963.
Wigglesworth, 1938.

CLASSIFICATION[2]

Order **DIPTERA** (19424)
- Suborder Nematocera (6083)
  - Tipulomorpha
    - Tanyderoidea
      - Tanyderidae (4)
    - Tipuloidea
      - Tipulidae (1630)
  - Blephariceromorpha
    - Blephariceroidea
      - Blephariceridae (25)
    - Deuterophlebioidea
      - Deuterophlebiidae (4)
    - Nymphomyioidea
      - Nymphomyiidae (1)
  - Axymyiomorpha
    - Axymyioidea
      - Axymyiidae (2)
  - Bibionomorpha
    - Pachyneuroidea
      - Pachyneuridae (1)
    - Bibionoidea
      - Bibionidae (81)
    - Sciaroidea
      - Mycetophilidae (835)
      - Sciaridae (157)
      - Cecidomyiidae (1141)
  - Psychodomorpha
    - Psychodoidea
      - Psychodidae (117)
    - Trichoceroidea
      - Trichoceridae (31)
    - Anisopodoidea
      - Anisopodidae (9)
    - Scatopsoidea
      - Scatopsidae (79)
      - Synneuridae (3)
  - Ptychopteromorpha
    - Ptychopteroidea
      - Ptychopteridae (18)
  - Culicomorpha
    - Culicoidea
      - Dixidae (47)
      - Chaoboridae (19)
      - Culicidae (165)
    - Chironomoidea
      - Thaumaleidae (7)
      - Simuliidae (159)
      - Ceratopogonidae (595)
      - Chironomidae (953)
- Suborder Brachycera (13341)
  - Tabanomorpha
    - Tabanoidea
      - Pelecorhynchidae (8)
      - Tabanidae (329)
      - Athericidae (4)
      - Rhagionidae (115)
    - Stratiomyoidea
      - Xylophagidae (25)
      - Xylomyidae (10)
      - Stratiomyidae (259)
  - Asilomorpha
    - Asiloidea
      - Therevidae (139)
      - Scenopinidae (128)
      - Vermileonidae (2)
      - Mydidae (48)
      - Apioceridae (66)
      - Asilidae (1011)
    - Bombylioidea
      - Acroceridae (61)
      - Nemestrinidae (7)
      - Bombyliidae (904)
      - Hilarimorphidae (27)
    - Empidoidea
      - Empididae (748)
      - Dolichopodidae (1293)

2. After McAlpine et al. 1981. Manual of Nearctic Diptera. The numbers of species per family in parentheses are from a more recent 1988 compilation by North American dipterists, courtesy of F. C. Thompson, Systematic Entomology Laboratory, ARS, USDA. Many of these totals are somewhat higher than those in the text, most of which came from McAlpine et al. 1981 and 1987.

Muscomorpha (8157)
- Aschiza (1397)
  - Lonchopteroidea
    - Lonchopteridae (4)
  - Platypezoidea
    - Platypezidae (76)
    - Phoridae (393)
  - Syrphoidea
    - Syrphidae (807)
    - Pipunculidae (117)
- Schizophora-Acalyptratae (3475)
  - Conopoidea
    - Conopidae (68)
  - Nerioidea
    - Cypselosomatidae (2)
    - Micropezidae (36)
    - Neriidae (2)
  - Diopsoidea
    - Tanypezidae (2)
    - Strongylophthalmyiidae (1)
    - Psilidae (32)
    - Diopsidae (2)
  - Tephritoidea
    - Lonchaeidae (126)
    - Otitidae (134)
    - Platystomatidae (41)
    - Pyrgotidae (8)
    - Tephritidae (297)
    - Richardiidae (10)
    - Pallopteridae (9)
    - Piophilidae (40)
  - Opomyzoidea
    - Clusiidae (37)
    - Acartophthalmidae (2)
    - Odiniidae (11)
    - Agromyzidae (732)
    - Opomyzidae (12)
    - Anthomyzidae (10)
    - Aulacigastridae (5)
    - Periscelididae (3)
    - Asteiidae (19)
    - Milichiidae (44)
    - Carnidae (15)
    - Braulidae (1)
  - Sciomyzoidea
    - Coelopidae (5)
    - Dryomyzidae (11)
    - Sciomyzidae (188)
    - Ropalomeridae (1)
    - Sepsidae (34)
  - Lauxanioidea
    - Lauxaniidae (157)
    - Chamaemyiidae (57)
  - Sphaeroceroidea
    - Heleomyzidae (124)
    - Trixoscelididae (27)
    - Chyromyidae (9)
    - Rhinotoridae (1)
    - Sphaeroceridae (199)
  - Ephydroidea
    - Curtonotidae (1)
    - Drosophilidae (196)
    - Diastatidae (8)
    - Camillidae (1)
    - Ephydridae (440)
    - Chloropidae (282)
    - Cryptochetidae (1)
    - Tethinidae (26)
    - Canacidae (6)
- Schizophora-Calyptratae (3285)
  - Muscoidea
    - Scathophagidae (148)
    - Anthomyiidae (532)
    - Muscidae (725)
  - Oestroidea
    - Calliphoridae (85)
    - Oestridae (52)
    - Sarcophagidae (358)
    - Rhinophoridae (2)
    - Tachinidae (1340)
  - Hippoboscoidea
    - Hippoboscidae (30)
    - Nycteribiidae (6)
    - Streblidae (7)

DIPTERA

KEY TO FAMILIES OF LARVAE[3,4,5]

H. J. Teskey, Biosystematics Research Centre, Ottawa

With some modifications by B. A. Foote *Kent State University,* and F. W. Stehr, *Michigan State University*

Identification to family is easiest for the Nematocera and orthorrhaphous Brachycera since larvae of approximately 10% of the species representing 40% of the genera have been described, and each family (or its natural subdivisions) is relatively homogeneous. In the Muscomorpha larvae of fewer than 5% of the Nearctic species representing perhaps 10–15% of the genera have been described in enough detail for diagnosis. Therefore, in many families of the Muscomorpha, including several large ones, the known larvae may be atypical of the family. In some families of the Muscomorpha the variation is so great that they are keyed out in more than one place.

The posterior spiracles are frequently closely associated and are sometimes even borne on a single base, so they are usually referred to in the plural. The anterior spiracles are widely separated and are always referred to in the singular.

Larvae of the following families have not been described: Acartophthalmidae, Asteiidae, Camillidae, Chyromyidae, Diastatidae, Hilarimorphidae, Rhinotoridae, Richardiidae, Ropalomeridae, Strongylophthalmyiidae, Tethinidae, and Trixoscelididae.

—IMPORTANT

Key Figures: Some figures are duplicated in the key. They are usually placed in numerical sequence (e.g. 8, 23, 96, 150, etc.) on the page or facing page where they are used. If the figure is not present, it will be on the previous (←——) or following (——→) two pages. Check the bottom corners for its location.

1. Sclerotized head or skeletal elements not detectable; respiratory system peripneustic (fig. 1). Terminal abdominal segment with large, upwardly curved, horn-like tubercles (fig. 2). In decaying wood. (If the head and skeletal elements appear unsclerotized and the description for Synneuridae in the text does not fit (ie., the respiratory system is NOT peripneustic), yet there are dorsally directed posterior spiracular "thorns", check Clusiidae, p. 817, or continue with couplet 2) .... NEMATOCERA (part) ***Synneuridae*** p. 750

1′. Head capsule or internal cephalopharyngeal skeleton usually conspicuous, respiratory system usually not peripneustic, but if so, then terminal abdominal segment lacking large, curved, horn-like tubercles 2

2(1′). Larva usually less than 5 mm in length, often yellow to reddish; small postcephalic segment present between prothorax and head capsule plus 3 thoracic and abdominal segments (fig. 3). Head capsule very small, lightly sclerotized, rounded or slightly longer than broad; mouthparts poorly differentiated; antenna 2- or 3-segmented; paired cephalic bars projecting posteriorly from lateral margins (fig. 4). Venter of prothorax usually with various shaped sternal spatula (fig. 4). Respiratory system peripneustic or apneustic. Commonly inhabiting plant galls but also free-living in moist habitats .... NEMATOCERA (part) ***Cecidomyiidae*** p. 742

2′. Size and color various; postcephalic segment usually not distinguishable. Head capsule or internal cephalopharyngeal skeleton with well-differentiated mouthparts; paired cephalic bars lacking. Venter of prothorax lacking sternal spatula. Respiratory system variable 3

3. Modified from the key by H. J. Teskey in the Manual of Nearctic Diptera, 1981.

4. Illustrations selected by H. J. Teskey, with some modifications and additions by F. W. Stehr.

5. Those illustrations marked with an **MD** are from the Manual of Nearctic Diptera, Vol. 1, 1981, reproduced by permission of the Minister of Supply and Services Canada; those marked with an Ⓟ are reproduced with permission from Peterson, A. 1951, Larvae of Insects, Vol. 2; those marked with **MC** are from Merritt, R. and Cummins, K., 1984, Introduction to the Aquatic Insects of North America, reproduced by permission of Kendall/Hunt Publishing Company, Dubuque, Iowa. Modifications of some figures have been made.

3(2′). Mandibles normally opposed, moving against one another in horizontal or oblique plane, and usually with 2 or more apical teeth, rarely hook-like or sickle-shaped. Head capsule usually complete and permanently exserted (eucephalic), but if partially retracted within thorax (figs. 5, 6, 7) and incomplete as result of excisions in capsule posteriorly (Tipulidae. *See* couplet 4, figs. 12, 13), then tentorial arms lacking (if in doubt, open thorax so rear of capsule is visible) .... most NEMATOCERA 4

3′. Mandibles moving parallel to one another in vertical plane, usually hook-like or sickle-shaped (figs. 8, 9), with or without secondary apical teeth. Head capsule usually reduced posteriorly and partially or almost entirely retracted into thorax (hemicephalic) or replaced by internal cephalopharyngeal skeleton (fig. 10); if appearing complete and permanently exserted, then with slender, metacephalic rod extending into prothorax (fig. 11) (if in doubt, open thorax so rear of capsule is visible) .... BRACHYCERA 28

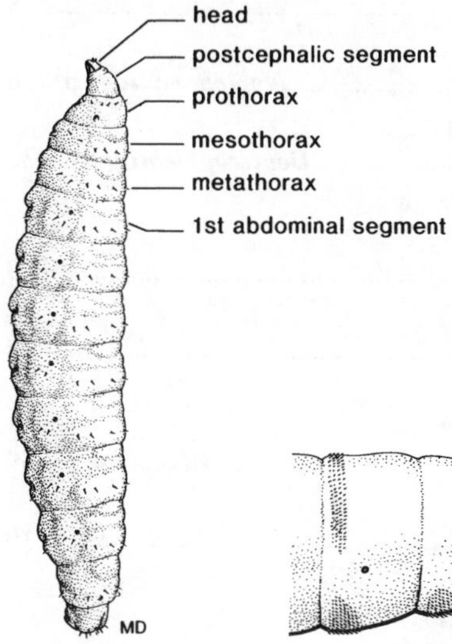

head
postcephalic segment
prothorax
mesothorax
metathorax
1st abdominal segment
MD

Figure 3

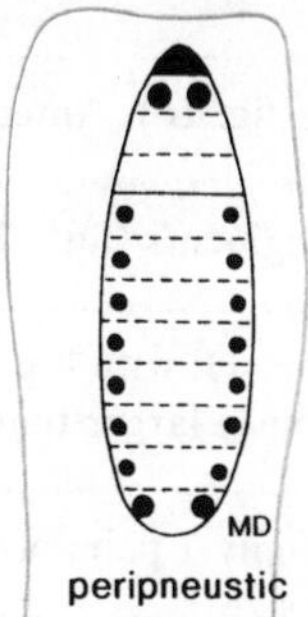

MD
peripneustic

Figure 1

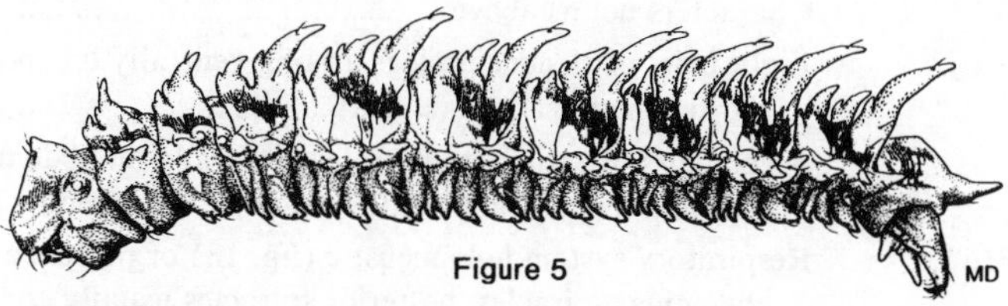

MD

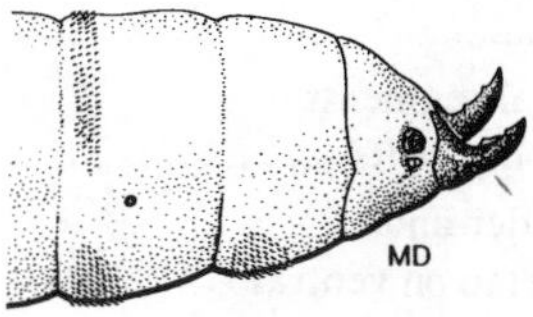

Figure 2

MD

Figure 5

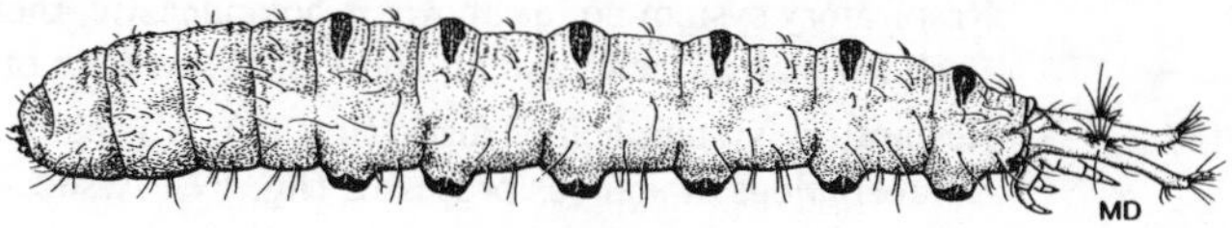

MD

Figure 6

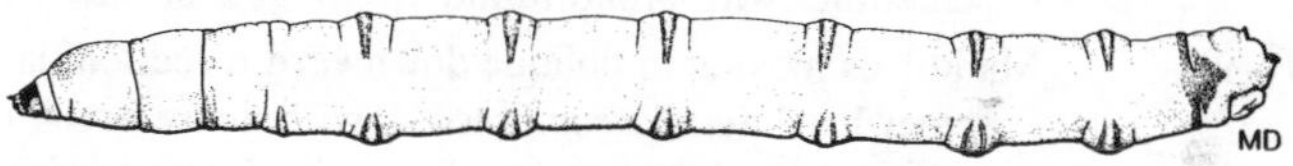

MD

Figure 7

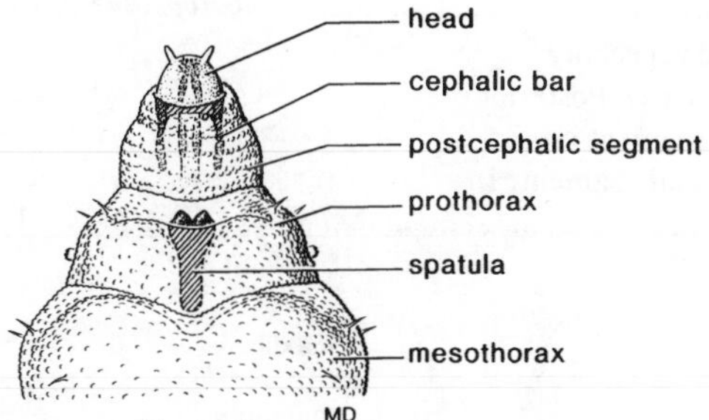

head
cephalic bar
postcephalic segment
prothorax
spatula
mesothorax
MD

Figure 4

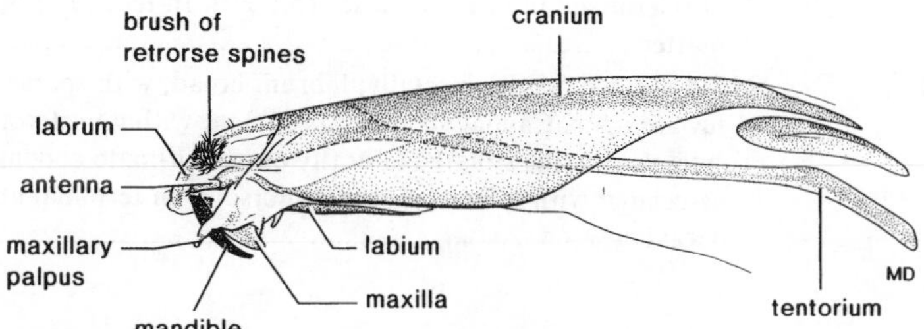

brush of retrorse spines
cranium
labrum
antenna
maxillary palpus
labium
maxilla
mandible
tentorium
MD

Figure 8

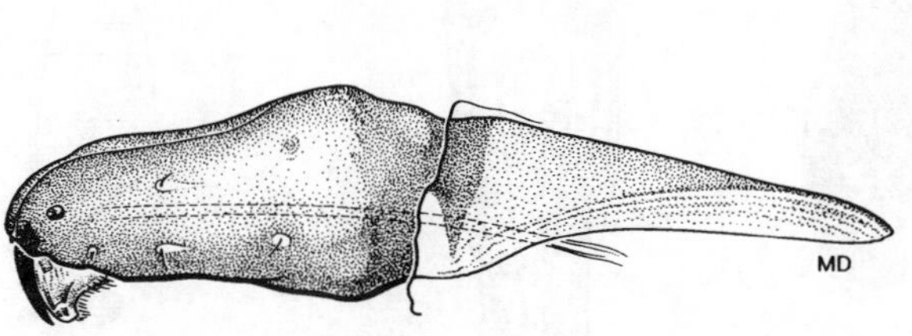

Figure 9

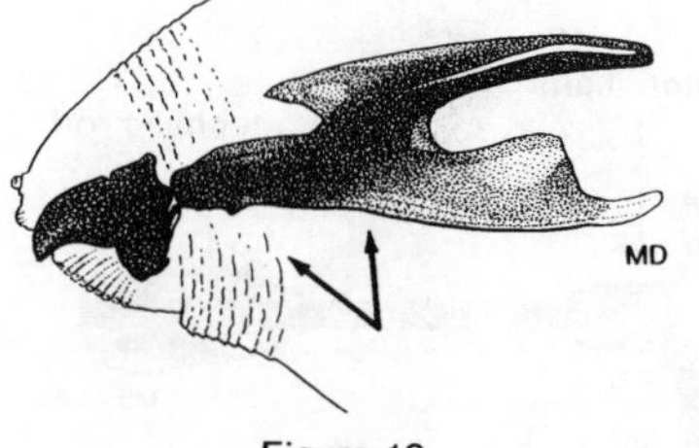

Figure 10

Figs. 11, 12, 13 ⟶

4(3). Head capsule longitudinally excised dorsally to varying degrees (figs. 12, 13), in extreme cases reduced to several slender rods (fig. 13); capsule capable of partial or complete retraction within thorax (figs. 5, 6, 7). Respiratory system usually metapneustic (fig. 14), rarely apneustic. Larvae occurring mostly in wet earth or decaying wood, occasionally in streams .... ***Tipulidae*** p. 731

4′. Head capsule complete and incapable of retraction within thorax. Respiratory system usually not metapneustic 5

5(4′). Head, thorax, and first abdominal segment fused to form compound body division (fig. 15); suctorial discs ventrally on compound division and on each of next 5 segments. Attached to objects in stream beds .... ***Blephariceridae*** p. 736

5′. Head, thorax, and first abdominal segment distinctly separated; suctorial discs present or absent 6

6(5′). A pair of prolegs, each bearing apical crochets, on each of 7 or 8 abdominal segments (figs. 16, 17). Clinging to stones in streams 7

6′. Characters not as above 8

7(6). Eight pairs of slender prolegs arising ventrally on abdomen (fig. 16). Antenna shorter than head .... ***Nymphomyiidae*** p. 736

7′. Seven pairs of rather stout prolegs laterally on abdominal segments (fig. 17). Antenna longer than head, branching near middle .... ***Deuterophlebiidae*** p. 736

8(6′). Respiratory system holopneustic (fig. 18) or peripneustic (fig. 1), with 8 pairs of abdominal spiracles; posterior spiracles usually conspicuously larger than preceding 7 pairs 9

8′. Respiratory system not as above; if holopneustic, then with only 7 pairs of abdominal spiracles and all abdominal spiracles of equal size 12

9(8). Respiratory system holopneustic (fig. 18). All segments usually bearing tuberculous or spinous processes (fig. 19). Associated with plant roots and decaying organic matter in soil .... ***Bibionidae*** p. 739

9′. Respiratory system peripneustic (fig. 1). Only caudal abdominal segments sometimes with broad tumid swellings associated with creeping welts 10

10(9′). Mandibles moving in oblique downward direction; labrum slender and somewhat laterally compressed, with dense brush of short setae on ventral apex and epipharyngeal surface. Caudal abdominal segment with pair of dorsally sclerotized lobes or broad, sclerotized shelf behind anus and ventral to posteriorly directed spiracles. Posterior spiracles either sessile or at apices of sclerotized tubular processes (fig. 20). In feces and decaying organic matter .... ***Scatopsidae*** p. 749

10′. Mandibles moving horizontally; labrum broad, with sparse setae especially toward apex. Caudal abdominal segment without sclerotized areas. Posterior spiracles sessile, situated laterally on penultimate abdominal segment or associated with spinous processes dorsally on terminal abdominal segment. In decaying wood 11

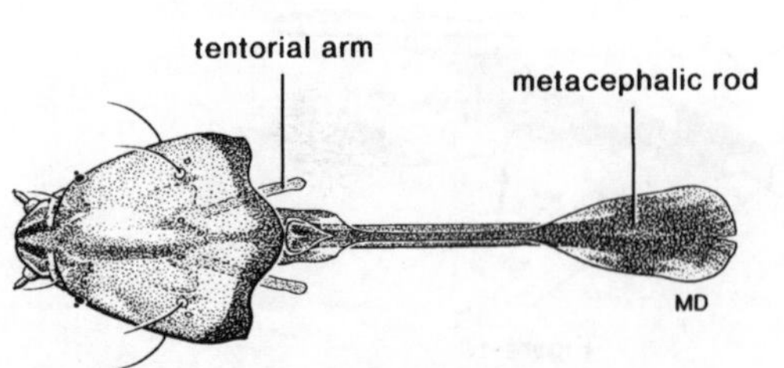

Figure 11

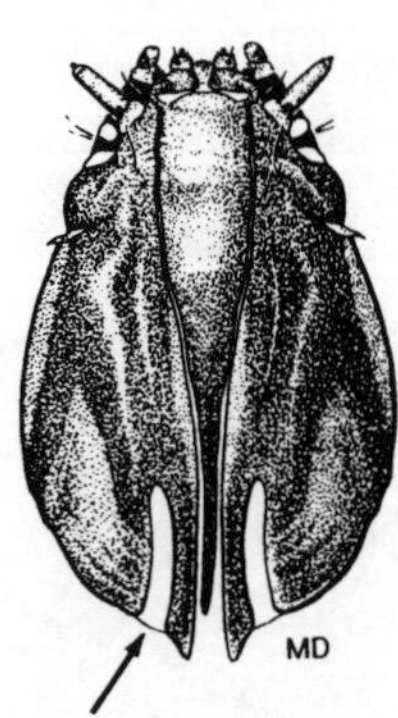

Figure 12

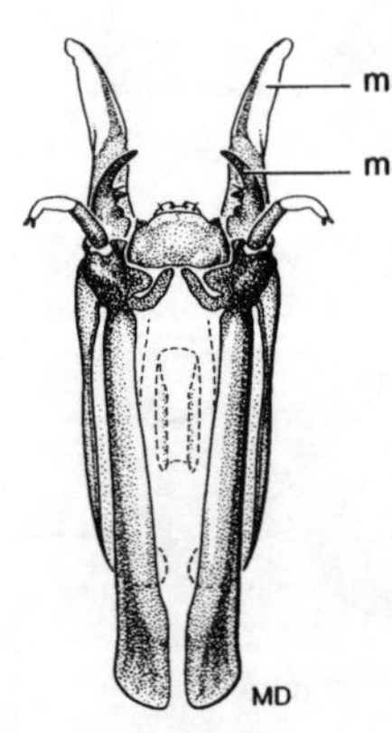

Figure 13

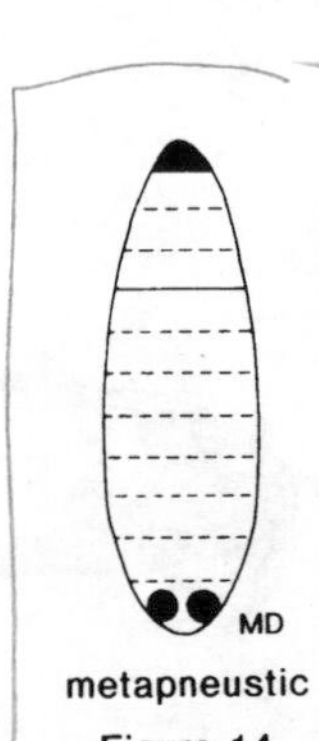

Figure 14

← Figs. 1, 5, 6, 7

11(10′). Body hairs conspicuous, longest hairs approximately half width of segment (fig. 21) .... ***Pachyneuridae*** p. 738

11′. Body hairs inconspicuous and much shorter than half width of segment .... (part) **(Ditomyiinae)** ***Mycetophilidae*** p. 741

12(8′). Respiratory system hemipneustic (fig. 22), rarely propneustic (fig. 23) or apneustic. Mandibie or maxilla or both flattened or lamellate, with inner margin serrate (fig. 25). In fungi 13

12′. Respiratory system amphipneustic (fig. 24), metapneustic (fig. 14), or apneustic. Mandible and maxilla not flattened, inner margin not serrate 14

13(12). Posterior tentorial bridge complete or nearly so (fig. 25) (bridge usually visible beneath integument within occipital cavity in preserved specimens without special treatment). Abdominal creeping welts lacking sclerotized spicules .... ***Sciaridae*** p. 742

13′. Posterior tentorial bridge absent (fig. 26), or if bridge partially formed, abdominal creeping welts with sclerotized spicules (fig. 27) .... (part) ***Mycetophilidae*** p. 741

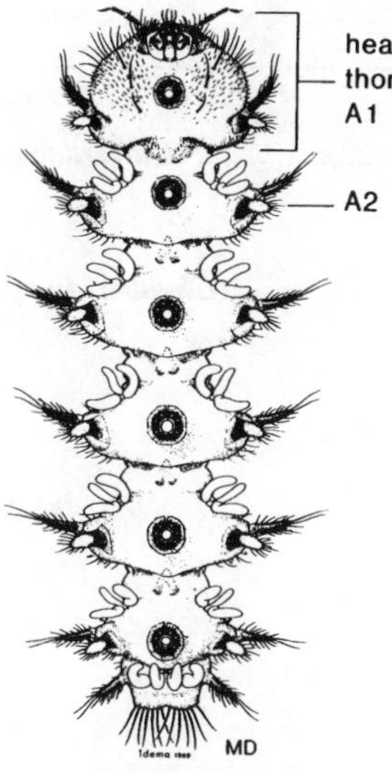

Figure 15

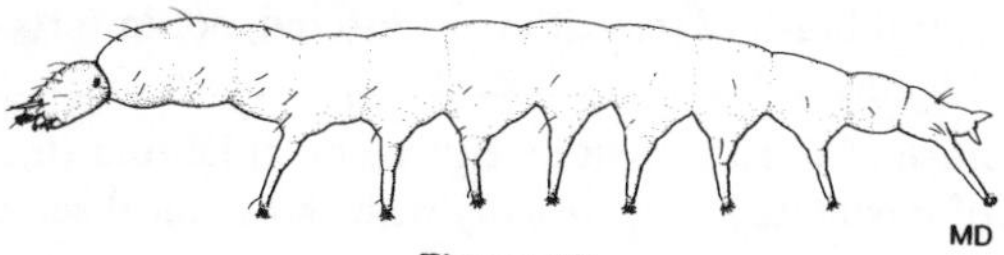

Figure 16

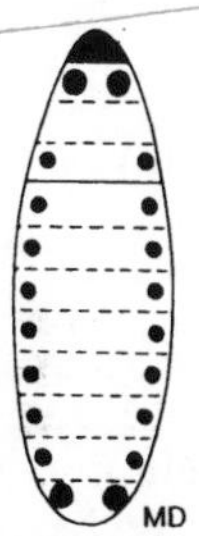

Figure 17

holopneustic
Figure 18

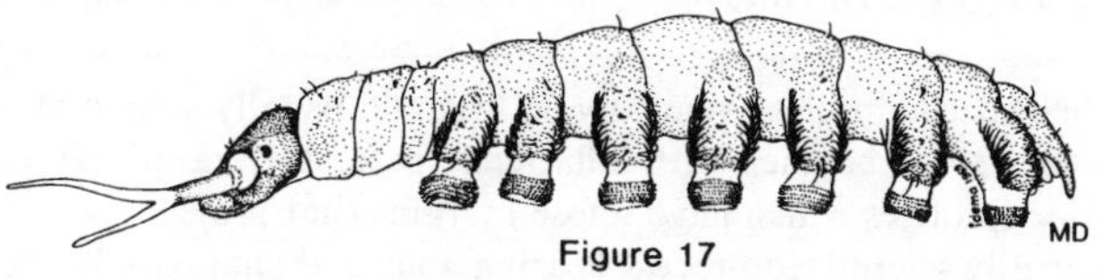

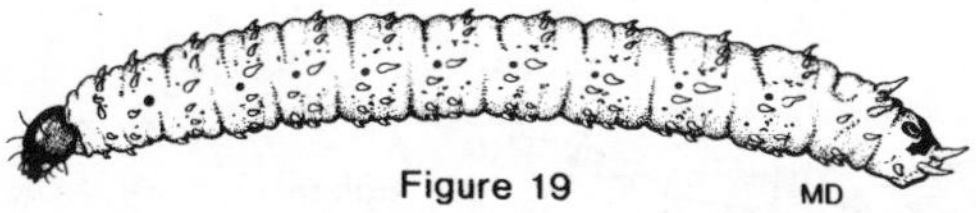

Figure 19

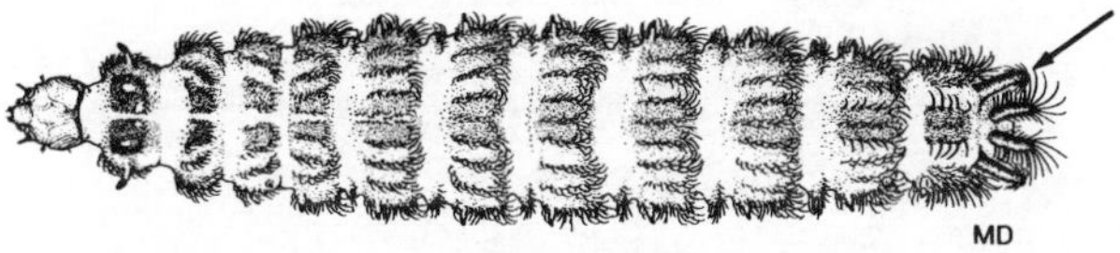

Figure 20

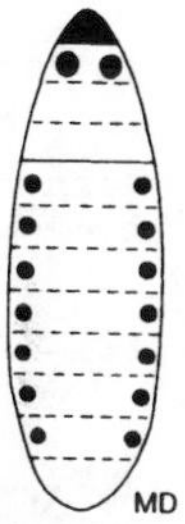

hemipneustic
Figure 22

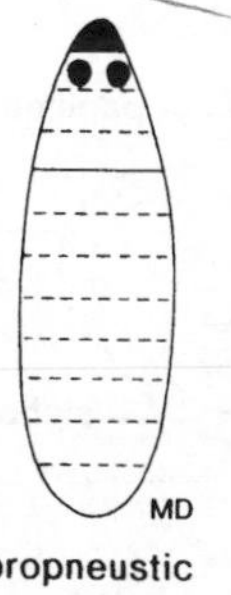

propneustic
Figure 23

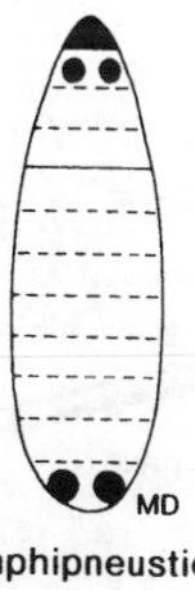

amphipneustic
Figure 24

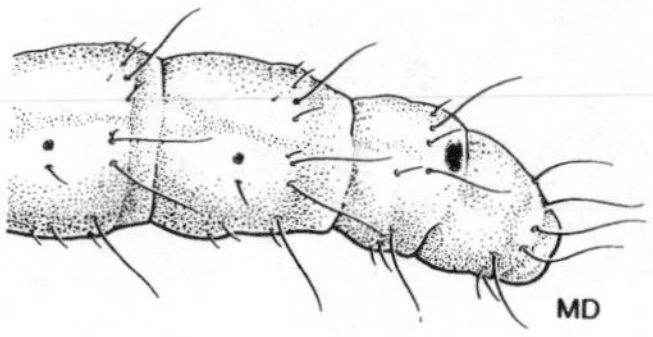

Figure 21

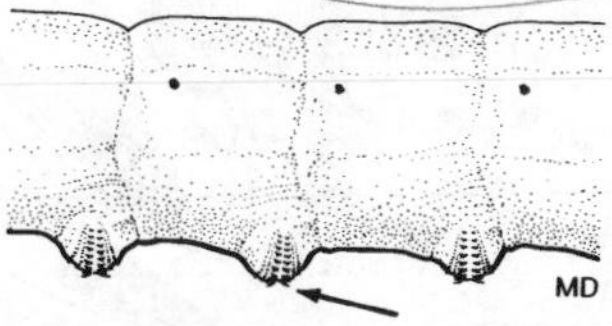

Figure 27

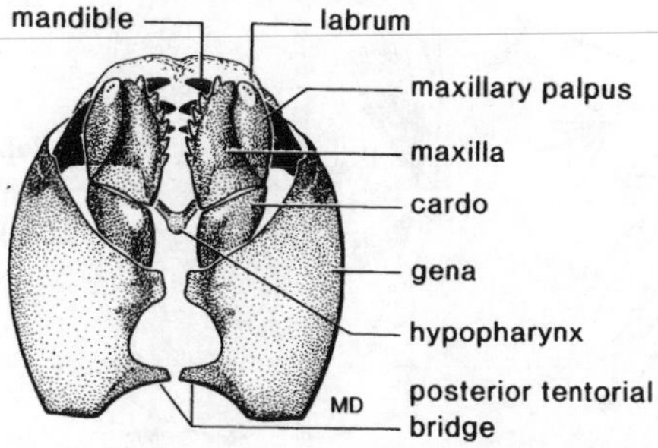

Figure 25

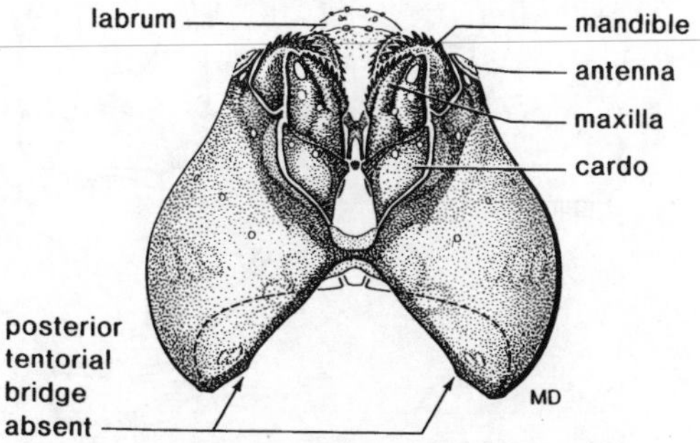

Figure 26

14(12′). Posterior spiracles always present, borne at end of slender respiratory siphon that is nearly or fully as long as body when completely extended 15

14′. Posterior spiracles borne on no more than very short siphon that is much shorter than body, or spiracles absent 16

15(14). Integument smooth, shiny, and white. Respiratory siphon not retractile. Anal papillae large and pinnately branched (fig. 28). In wet decaying logs .... ***Axymyiidae*** p. 738

15′. Integument with numerous hairs, warts, or tubercles. Respiratory siphon retractile. Anal papillae small and unbranched (fig. 29). In saturated soil having high organic content .... ***Ptychopteridae*** p. 751

16(14′). Thoracic segments fused, forming round or somewhat flattened compound segment that is wider than any abdominal segment (figs. 30, 31). Terminal abdominal segment with preanal fan-like ventral brush of setae (fig. 32). Body segments often with tufts of long setae laterally 17

16′. Thoracic segments usually individually distinguishable, about equal to or narrower than width of abdominal segments. Terminal abdominal segment without ventral brush of setae. Setae scattered, not in tufts on thoracic and abdominal segments 18

17(16). Prominent brush of setae present on either side of labrum (figs. 33, 34). Antenna of moderate length, usually with short apical setae .... ***Culicidae*** p. 754

17′. Labral setae absent or few in number and not divided into 2 groups on either side of labrum (fig. 31). Antenna sometimes prehensile, with long apical setae .... ***Chaoboridae*** p. 753

18(16′). Pair of crochet-bearing prolegs ventrally on first and usually second abdominal segments (fig. 35). Abdomen with 2 flattened posterolateral processes behind the posterior spiracles which have setose margins that project above the conical, dorsally sclerotized process bearing anus and anal papillae ventrally (fig. 35) .... ***Dixidae*** p. 751

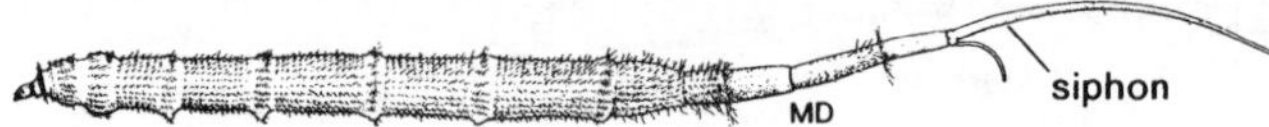

Figure 29

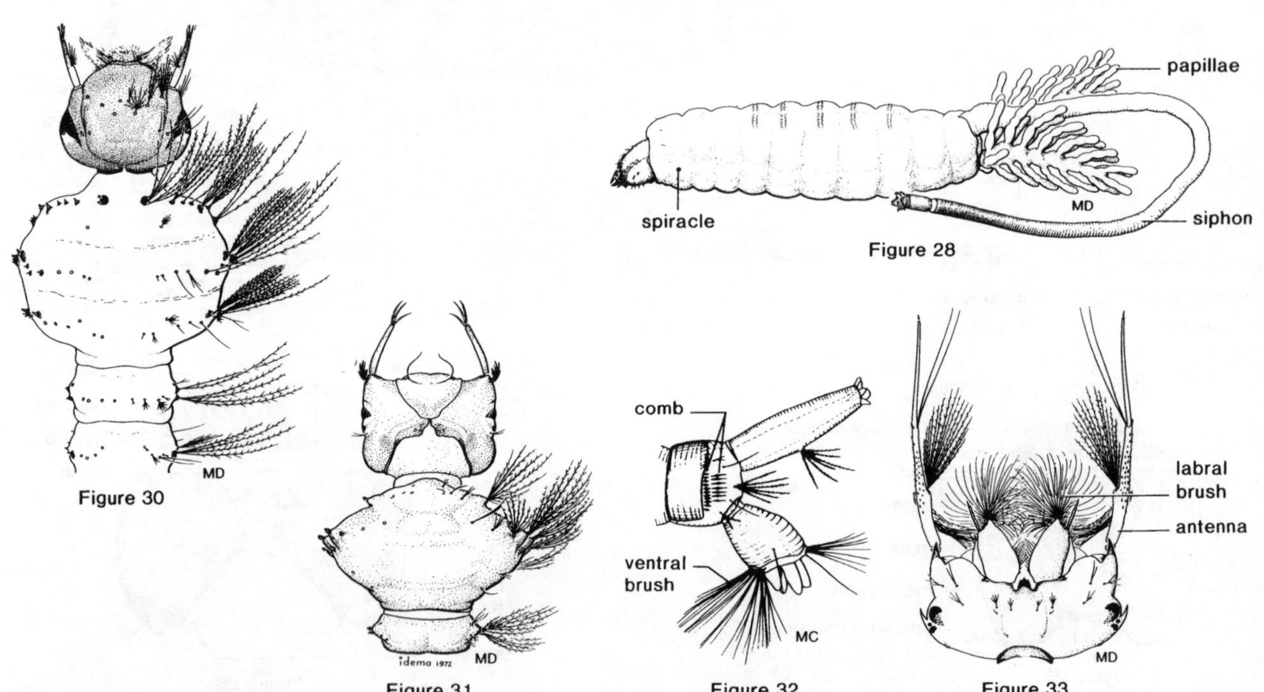

Figure 28

Figure 30

Figure 31

Figure 32

Figure 33

← Fig. 24

18′. Prolegs lacking on abdominal segments 1 and 2. Abdomen without posterior flattened, fringed postspiracular processes; abdomen lacking conical, dorsally sclerotized anal process 19

19(18′). Prothorax with 1 proleg or pair of prolegs ventrally 20

19′. Prothorax without prolegs 23

20(19). Head capsule usually with pair of conspicuous labral fans dorsolaterally. Abdomen enlarged distally; terminal segment ending in ring or circlet of numerous radiating rows of minute hooked spines. Attached to substrate in flowing water (fig. 36) .... ***Simuliidae*** p. 757

20′. Head capsule lacking labral fans. Abdomen not conspicuously swollen distally; terminal segment without radiating row of hooked spines posteriorly, but sometimes with 1 or 2 crochet-bearing anal prolegs 21

21(20′). Respiratory system amphipneustic (fig. 24); each anterior spiracle on short stalk (fig. 37); posterior spiracles opening into transverse cleft between finger-like processes on abdominal segment 8. Prothoracic and terminal prolegs unpaired. On rocks washed by film of water .... ***Thaumaleidae*** p. 755

21′. Respiratory system apneustic. Prothoracic or terminal prolegs usually paired, distinction may appear only as slight separation of apical spines or crochets 22

22(21′). Body segments with prominent tubercles or setae or both (figs. 38, 39) .... (part) **(Forcipomyiinae)** ***Ceratopogonidae*** p. 758

22′. Body segments, except sometimes caudal one, lacking prominent tubercles and setae .... ***Chironomidae*** p. 762

23(19′). Abdominal segment 8 with pair of long, filamentous processes arising laterally behind spiracles; abdominal segment 9 with similar paired processes arising dorsolaterally and also from near apex of 2 elongate cylindrical posteroventrally projecting prolegs (fig. 40). In saturated sandy gravel along streams .... ***Tanyderidae*** p. 730

23′. Distal abdominal segments without long, filamentous processes, without prolegs or with only single terminal proleg 24

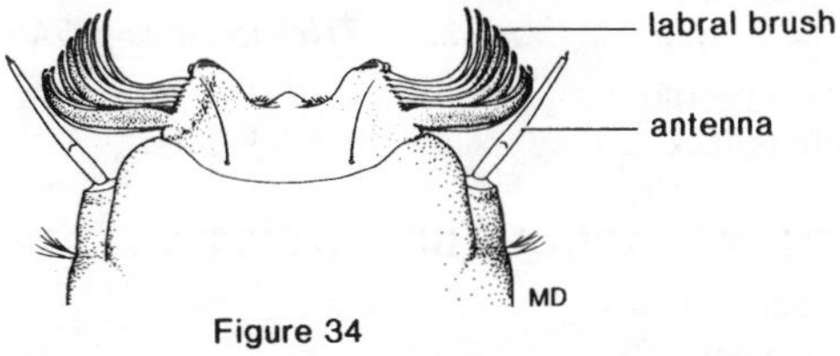

Figure 34

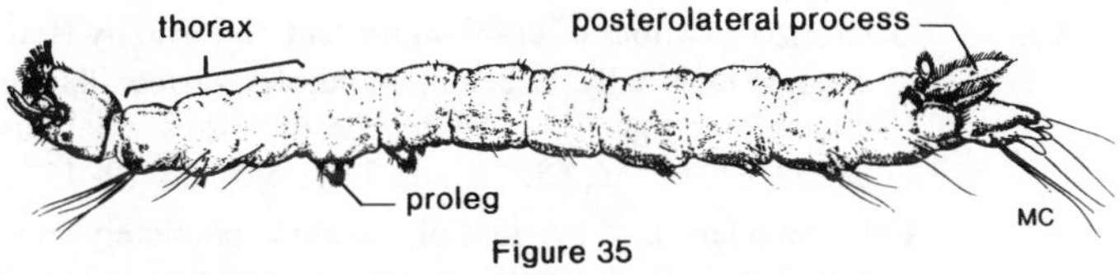

Figure 35

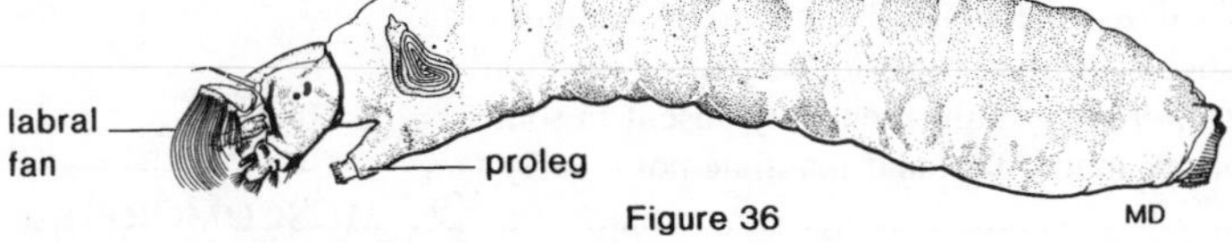

Figure 36

Figure 40

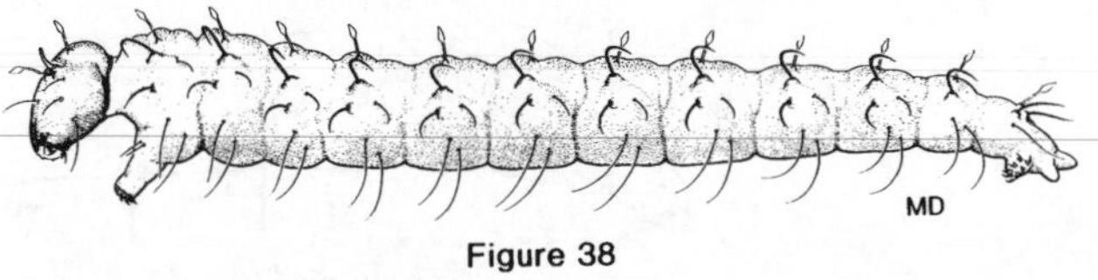

Figure 38

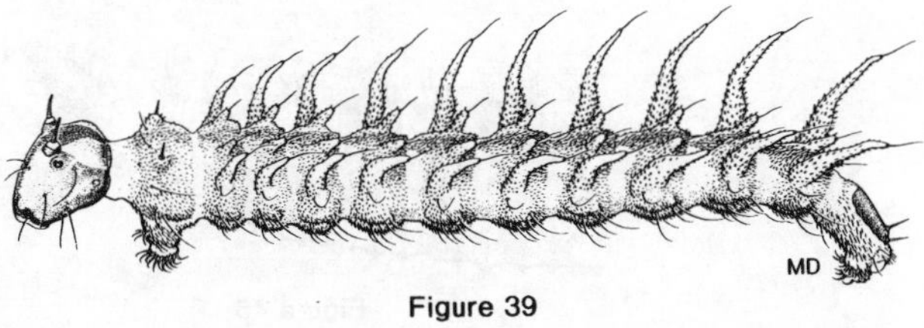

Figure 39

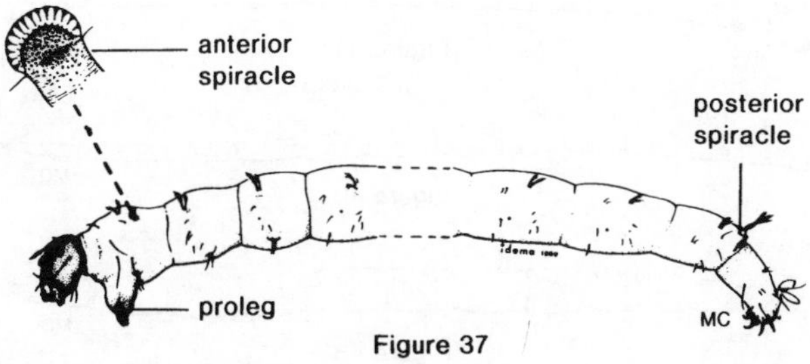

Figure 37

24(23′). Respiratory system apneustic. Larva slender, with uniform segments; integument smooth; long setae only on terminal abdominal segment (figs. 41, 42, 43) .... (part) ***Ceratopogonidae*** p. 758

24′. Respiratory system amphipneustic (fig. 24) or metapneustic (fig. 14). Larva usually somewhat wrinkled, with segments secondarily divided; distinctive setation or sclerotized plaques present on most segments 25

25(24′). Secondary segmentation of prothorax and abdominal segments evident, each of these segments with a distinct narrow annulus anteriorly (fig. 44). Posterior spiracles sessile on caudal abdominal segment and placed either laterally or apically, with apically positioned spiracles surrounded by 5 small lobes. In decaying organic material .... ***Anisopodidae*** p. 748

25′. Secondary segmentation either not apparent or usually with thoracic and first abdominal segments subdivided into 2 sections and remaining abdominal segments subdivided into 3 sections. Posterior spiracles either mounted on respiratory siphon or sessile, but if sessile and apical in position then surrounded by only 4 lobes 26

26(25′). Posterior spiracles and pair of fan-like setal brushes either borne dorsally at apical margin of sclerotized plate on caudal abdominal segment or at apex of short respiratory siphon projecting posterodorsally from caudal segment (fig. 45). Sclerotized plaque or plaques dorsally on 1 or more secondary segmental divisions. In aquatic or semiaquatic habitats or in decaying organic material .... (part) **(Psychodinae)** ***Psychodidae*** p. 746

26′. Posterior spiracles not borne on respiratory siphon. Sclerotized plaques absent dorsally 27

27(26′). Posterior spiracles situated laterally on caudal or penultimate abdominal segment. Setae on integument either prominent and systematically arranged with clavate form evident and with some very long setae on dorsum of terminal segment (**Phlebotominae**) (fig. 46), or setae short and unmodified, or setae absent (fig. 47) .... (part) **(Trichomyiinae)** ***Psychodidae*** p. 746

27′. Posterior spiracles situated on apex of terminal abdominal segment and surrounded by 4 fleshy lobes (fig. 48). Setae variable. In decaying vegetable matter .... ***Trichoceridae*** p. 746

28(3′). Sclerotized portions of cranium present and usually (but not always) partially exposed externally. Labrum, mandibles, or maxillae recognizable (check closely); see figs. 8, 11, 54, 55, 60, 69; also see habitus figs. 51, 53, 56, 57, 58, 59, 61, 62, 66, 68, 72, and 74 in couplets 29–45 .... ORTHORRHAPHOUS BRACHYCERA 29

28′. External sclerotized portions of cranium completely lacking; only membranous pseudocephalic segment anterior to prothorax remaining, this segment normally with 2 pairs of papilla-like projections, thought to be vestiges of antenna and palpi; characteristically shaped cephalopharyngeal skeleton retracted completely within prothorax (or almost entirely absent in some usually parasitic species). Labrum, mandibles, and maxillae not clearly definable .... MUSCOMORPHA 46

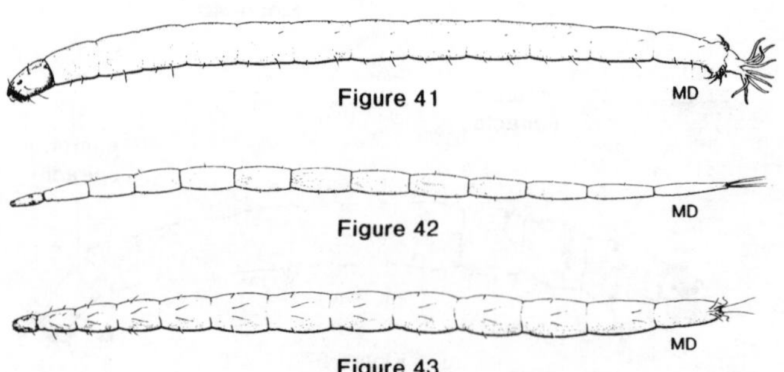

Figure 41

Figure 42

Figure 43

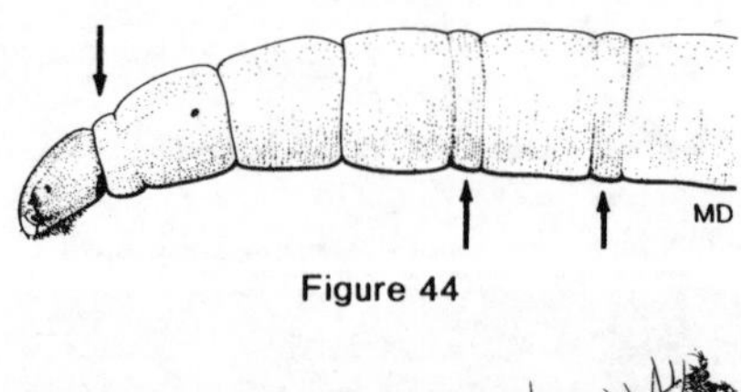

Figure 44

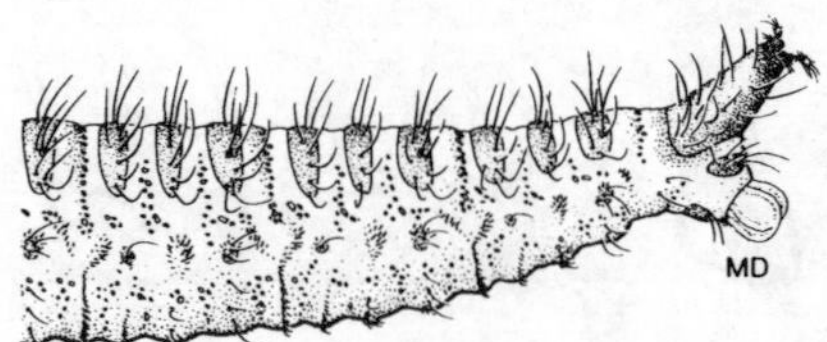

Figure 45

29(28). Large sclerotized plates present dorsally on 1 or more thoracic segments or on caudal abdominal segment or on both (figs. 49, 50); sclerotized plate on abdominal segment obliquely truncate, with projecting processes and bearing posterior spiracles. Exposed portion of head capsule darkly sclerotized and conical, mouthparts very small, projecting from tip of cone. In soil or under bark of decaying trees .... ***Xylophagidae*** p. 768

29′. Sclerotized plates usually lacking on thoracic and caudal abdominal segments, if plate occurs on caudal abdominal segment it does not bear spiracles. Exposed portion of head capsule not conical, mouthparts relatively large 30

30(29′). Body dorsoventrally depressed. Integument hardened by small roundish or hexagonal calcareous plates giving shagreened pattern to body surface (fig. 51). Head capsule always partially exposed, capable of only slight independent movement 31

30′. Body form various. Integument not hardened by calcium deposits, sometimes tough and leathery. Head capsule capable of much independent movement 32

31(30). Prothoracic and mesothoracic segments lacking shagreening on part of dorsal surface. Anus bordered anteriorly by transverse row of strong posteriorly directed teeth (fig. 52) .... ***Xylomyidae*** p. 770

31′. Prothoracic and mesothoracic segments with shagreen pattern entire dorsally. Anus not bordered anteriorly by row of teeth .... ***Stratiomyidae*** p. 770

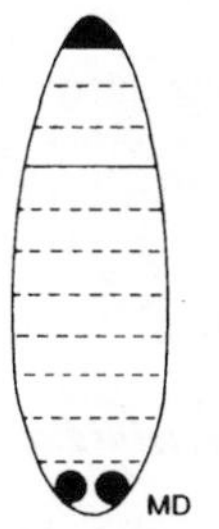

metapneustic
Figure 14

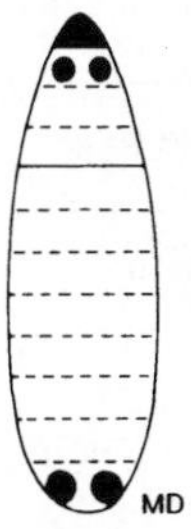

amphipneustic
Figure 24

Figure 47

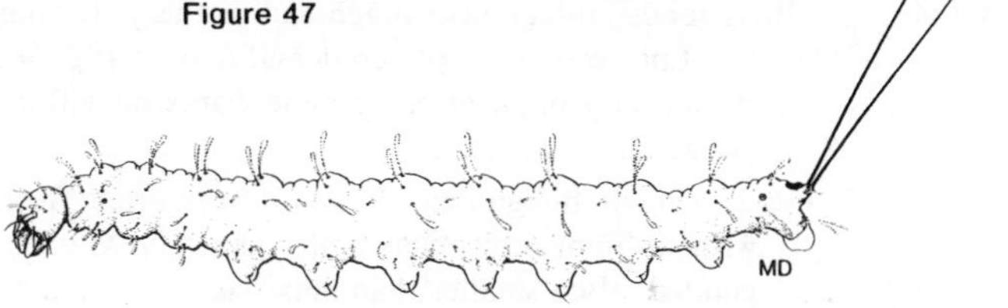

Figure 46

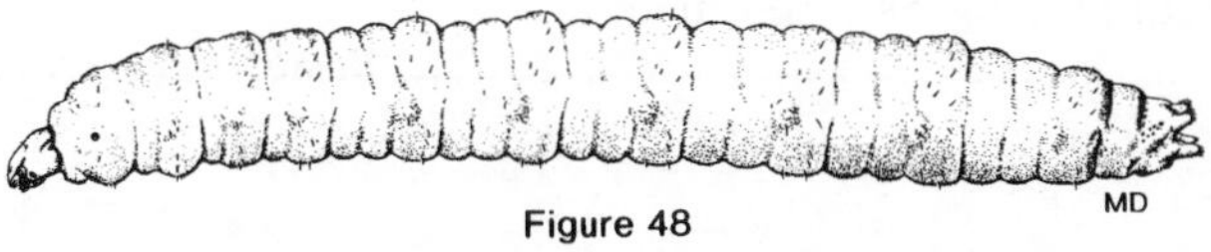

Figure 48

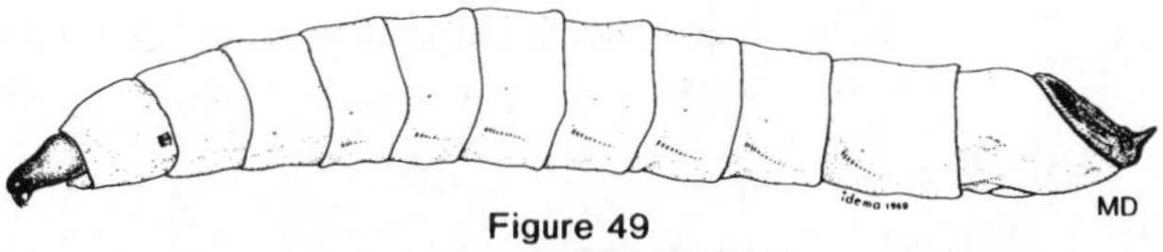

Figure 49

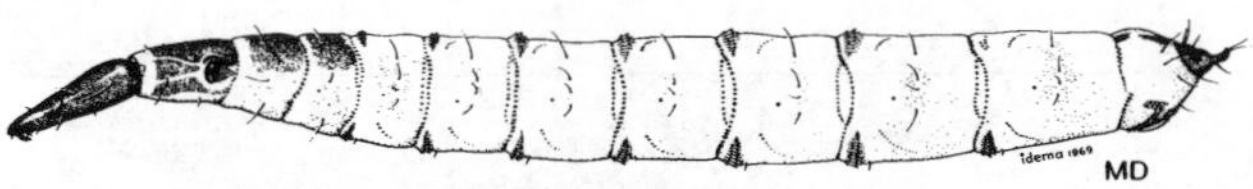

Figure 50

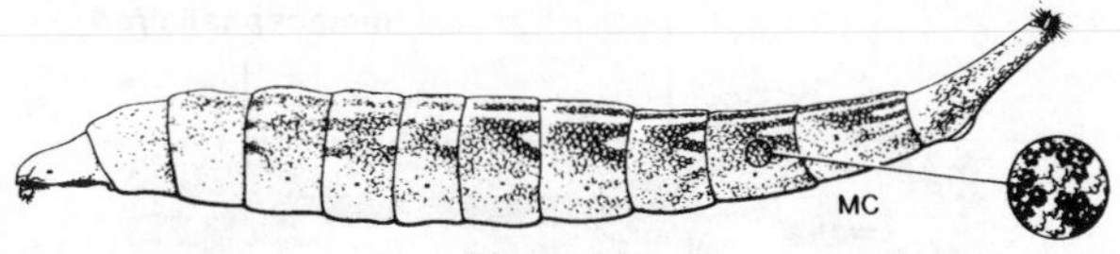

Figure 51

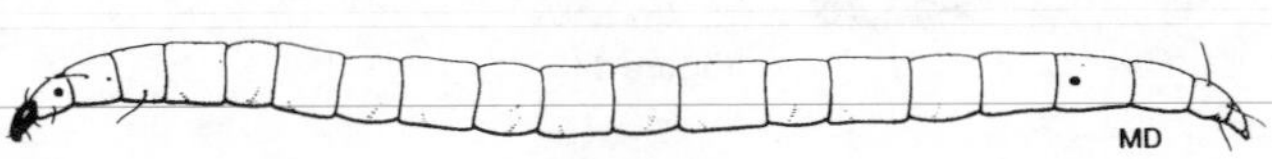

Figure 53

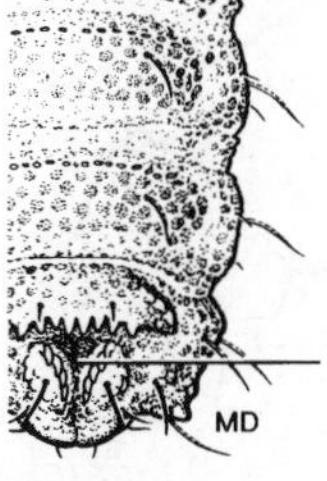

Figure 52

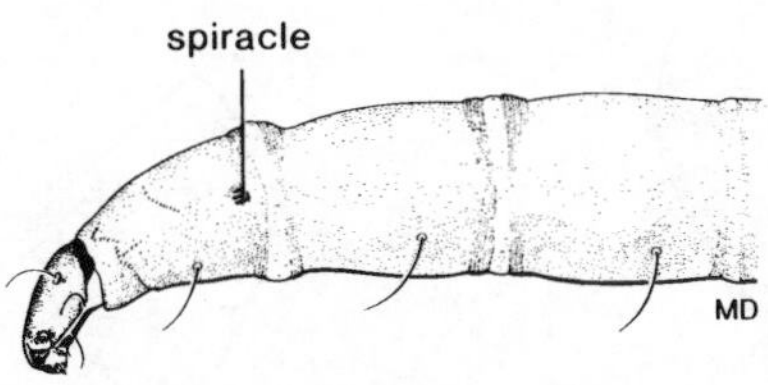

Figure 54

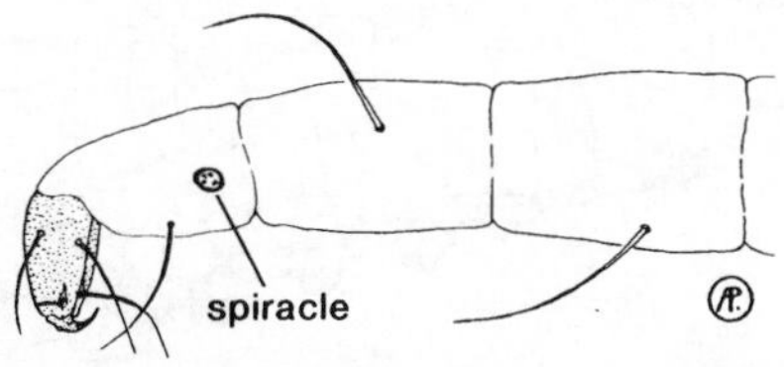

Figure 55

32(30′). Body long and slender, eel-like, apparently composed of 20 segments (fig. 53). Posterior spiracles situated laterally on fourth segment from caudal end of body. Head capsule (fig. 11) seemingly complete and permanently exserted, articulated posteriorly with slender or spatulate metacephalic rod lying within thorax 33

32′. Body not eel-like, composed of no more than 12 apparent segments. Posterior spiracles on ultimate or penultimate abdominal segment. Head capsule more or less reduced, especially posteroventrally, and partially retracted within thorax, with or without single broad or nonspatulate metacephalic rod lying within thorax, occasionally with 2 such rods 34

33(32). Metacephalic rod expanded apically, spatulate (fig. 11). Antenna minute and peg-like (fig. 11). Setae on each side of thoracic segments shorter than diameter of segments and situated ventrolaterally (fig. 54). Predacious in soil and decaying wood ..... ***Therevidae*** p. 773

33′. Metacephalic rod slender throughout. Antenna long and filamentous. Setae on each side of thoracic segments at least as long as diameter of segments, mesothoracic setae situated higher on segment than are prothoracic and metathoracic setae (fig. 55). Predacious on insects in homes, stored foods, and wood ..... ***Scenopinidae*** p. 774

34(32′). Larva parasitic within the body of other Arthropoda. Body plump and grub-like (figs. 56, 57, 58). Head usually small, almost completely retracted within thorax, only mandibles or maxillae and at least vestige of labrum visible externally 35

34′. Larva free-living. Body usually elongate and slender. Portions of head capsule and mouthparts visible externally 37

35(34). Body robust, integument tough and leathery. Terminal abdominal segment with blunt projections on posterodorsal margin (fig. 56). Maxillae large and shovel-shaped; mandibles absent. Parasitic within grasshoppers and beetle larvae ..... ***Nemestrinidae*** p. 781

35′. Body whitish, integument thin and transparent. Terminal abdominal segment without blunt projections posterodorsally. Mandibles present, slender and pointed, often smaller than maxillae 36

36(35′). Body pear-shaped, with abdomen enlarged (fig. 57). Parasitic in bodies of spiders ..... ***Acroceridae*** p. 780

36′. Body somewhat crescent-shaped, tapering toward both ends (fig. 58). Parasitic on insects ..... ***Bombyliidae*** p. 782

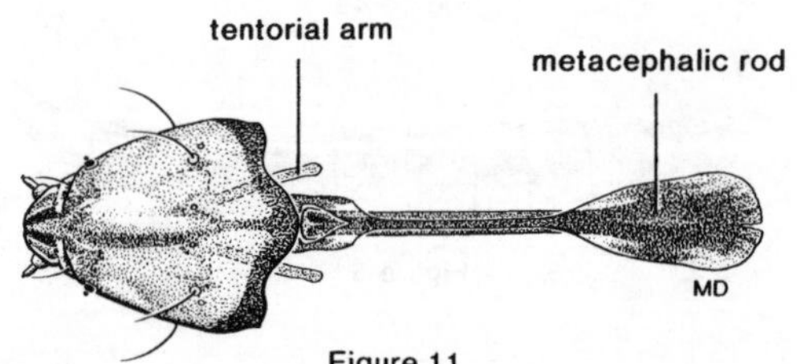

Figure 11

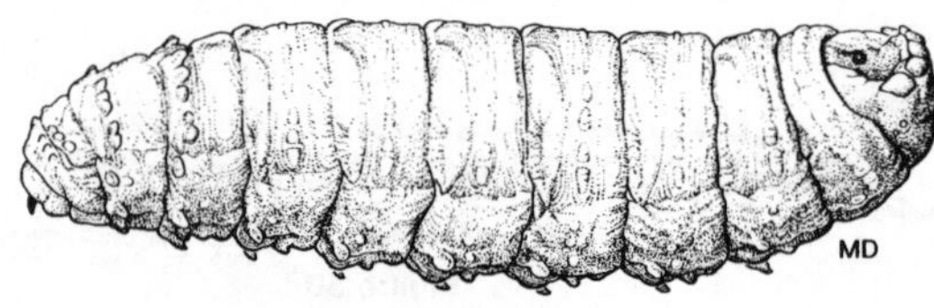

Figure 56

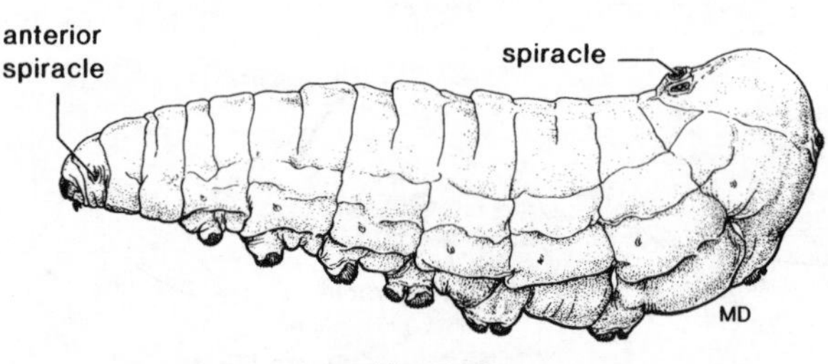

Figure 57

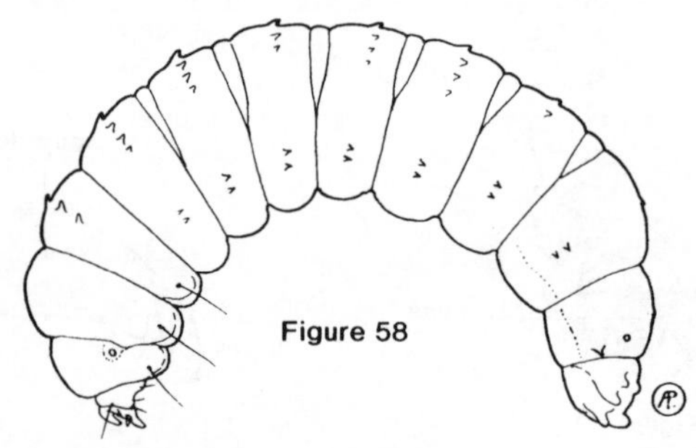
Figure 58

← Figs. 14, 24, 53, 54, 55

37(34′). Body with caudal segments distinctly enlarged (fig. 59); integument wrinkled and warty. Abdominal segment 1 with ventral proleg. Abdominal segment 7 with row of spine-like tubercles along posterodorsal margin. Posterior spiracles dorsal on abdominal segment 8. Predacious, in funnel-shaped pitfall traps in sandy soil .... ***Vermileonidae*** p. 775

37′. Body usually cylindrical; integument smooth. If present, prolegs situated on several segments. Abdominal segment 7 without dorsal row of tubercles. Posterior spiracles, if present, either situated near apex of terminal segment or placed dorsolaterally on penultimate or ultimate segment 38

38(37′). Brush of retrorse spines situated above base of each mandible (fig. 8). Portion of cranium lying within thorax continuous with anterior exposed portion without apparent break, although desclerotization may suggest bilateral division. Tentorial arms solidly connected with tentorial phragmata, except in *Rhagio* 39

38′. No brush of spines associated with mandibles. Portion of cranium (metacephalic rod or rods) lying within thorax (fig. 60) separated from anterior exposed portion by clear seam allowing independent movement in both portions. Tentorial arms flexibly attached to tentorial phragmata 42

39(38). Respiratory system apneustic. Living in beds of streams 40

39′. Respiratory system metapneustic (fig. 14) or amphipneustic (fig. 24) 41

40(39). Each abdominal segment with ventral pair of crochet-bearing prolegs (fig. 61). Slender tubercles laterally and dorsolaterally on abdominal segments 1–7 .... ***Athericidae*** p. 766

40′. Abdominal segments lacking prolegs and tubercles. Integument smooth, shiny; segmentation bead-like (fig. 62). .... ***Pelecorhynchidae*** p. 764

41(39′). Posterior spiracles either lying within fissures on either side of pair of abutting vertically linear bars (fig. 63) or borne on retractable, laterally compressed spine (figs 64, 65). Tracheal trunks closely approximated within siphon and caudal segment (fig. 65). Terminal segment without lobes or tubercles. Several or all of 7 anterior abdominal segments with *encircling* row of projections that sometimes bear apical spicules (fig. 66) and serve as prolegs. Submentum present .... ***Tabanidae*** p. 764

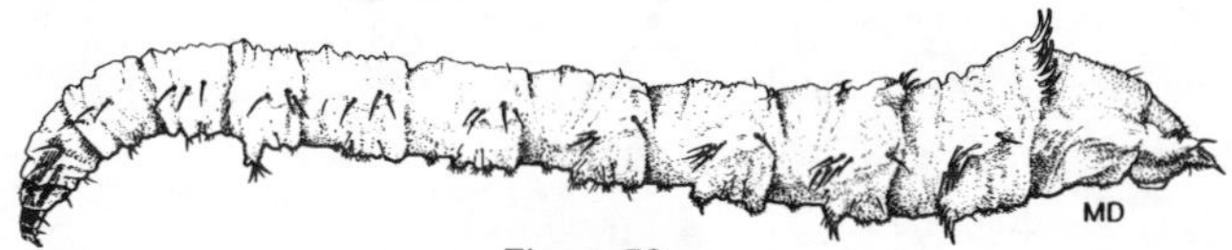

Figure 59

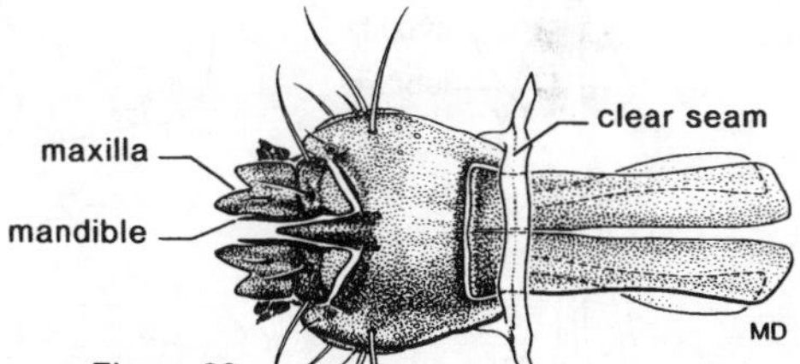

Figure 60

Figure 61

Figure 62

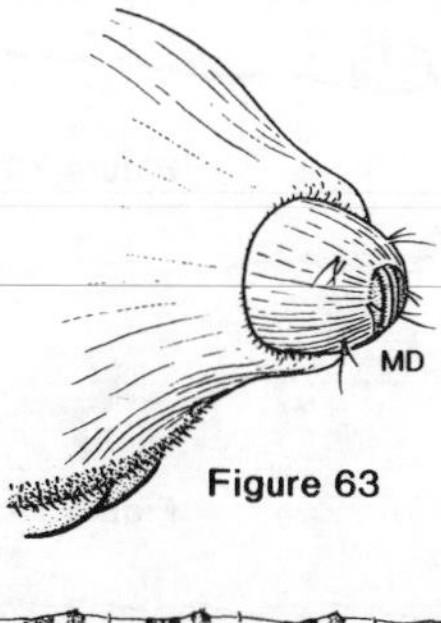

Figure 63

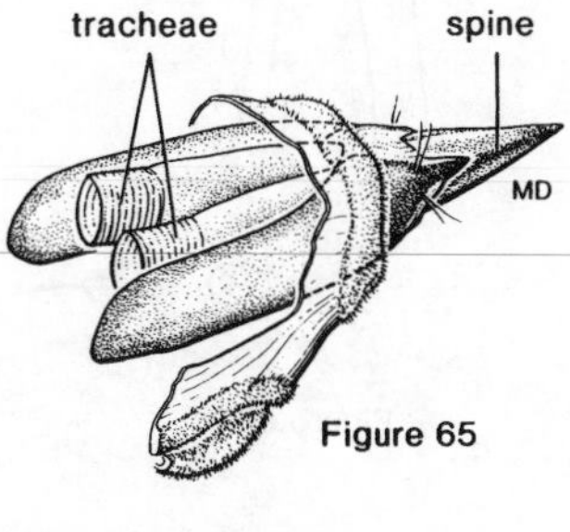

Figure 65

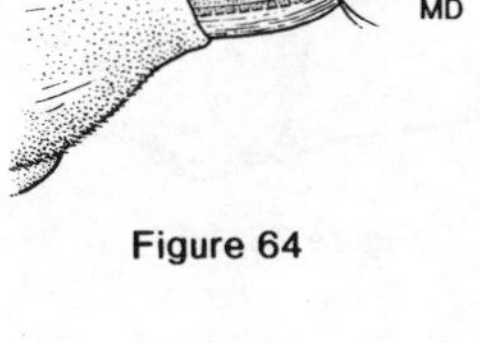

Figure 64

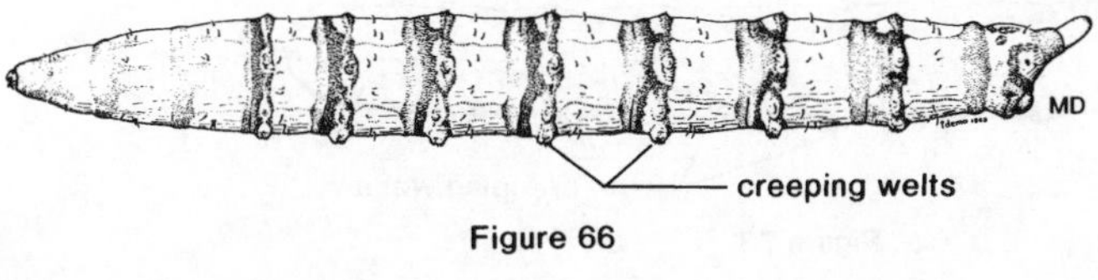

Figure 66

Fig. 8, p. 709

41′. Posterior spiracular openings exposed; each spiracle circular or oval. Tracheal trunks distinctly separated caudally. Terminal segment deeply cleft posteriorly to form 2 or 4 lobes (fig. 67) or bearing pair of sclerotized horn-like processes dorsally and pair of rounded lobes ventrally; posterior spiracles on caudal face of dorsal lobes. First 7 abdominal segments with *ventral* creeping welts (fig. 68). Submentum absent .... ***Rhagionidae*** p. 767

42(38′). Head largely membranous (fig. 69), with single narrow or broader metacephalic rod that is sometimes split almost to base. Sclerotized submentum present ventrally on head capsule. Maxillae large and heavily sclerotized, more prominent than slender mandibles. **Nine** abdominal segments. Respiratory system functionally amphipneustic, although remnants of spiracles forming holopneustic system usually visible; posterior spiracles situated laterally on abdominal segment 8. Larva usually longer than 15 mm at maturity 43

42′. Head skeletonized, with 2 slender metacephalic rods and 2 tentorial arms particularly prominent; no submentum; maxillae sometimes seemingly absent, never heavily sclerotized or more prominent than mandibles. **Eight** abdominal segments; posterior spiracles, if present, located caudally on last segment. Respiratory system amphipneustic, metapneustic or apneustic. Larva usually less than 15 mm at maturity 45

43(42). Maxillae laterally compressed, tending to cup mandibles (fig. 69), similar in length to mandibles; maxillary palpus apical (fig. 69). In logs or soil, predacious .... ***Mydidae*** p. 775

43′. Maxillae more or less dorsoventrally compressed, often toothed apically and concave ventrally to form digging structures, usually much longer than mandibles; maxillary palpus lateral (fig. 60) 44

44(43′). Abdominal segment 8 about twice as long as wide; posterior spiracles lateral near anterior margin of abdominal segment 8 .... ***Apioceridae*** p. 776

44′. Abdominal segment 8 no longer than half its diameter (fig. 70). Posterior spiracles situated dorsolaterally in distal half of segment 8 .... ***Asilidae*** p. 778

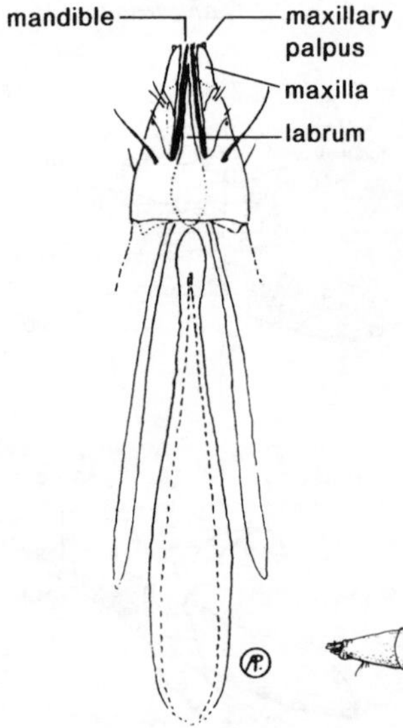

Figure 69

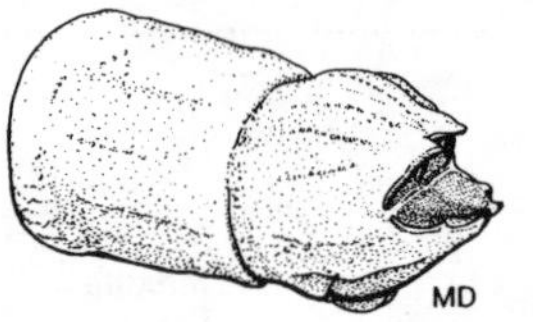

Figure 67

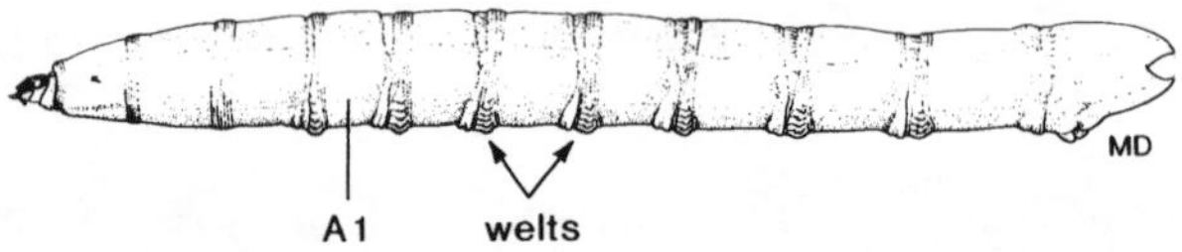

Figure 68

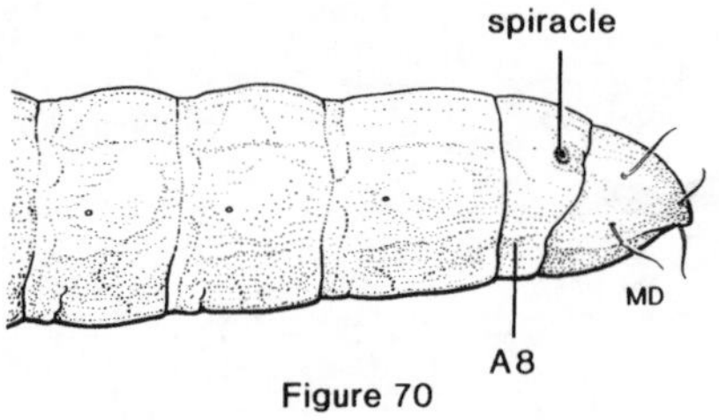

Figure 70

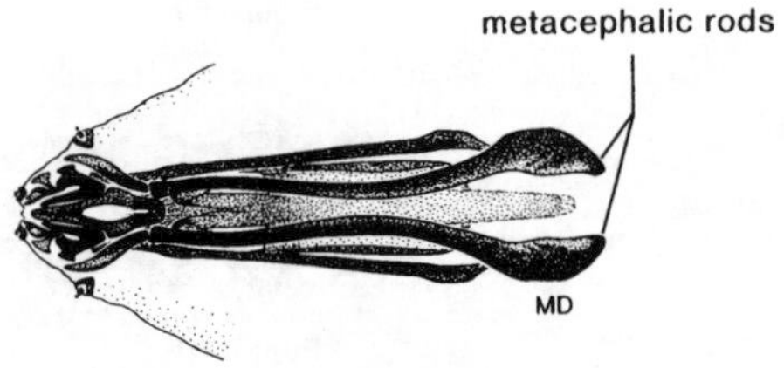

Figure 71

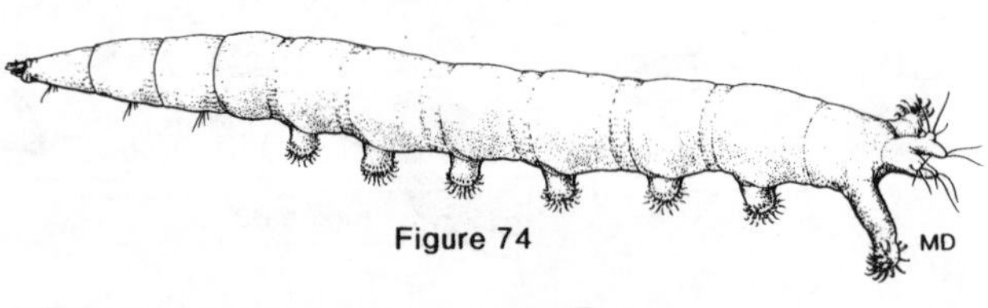

Figure 74

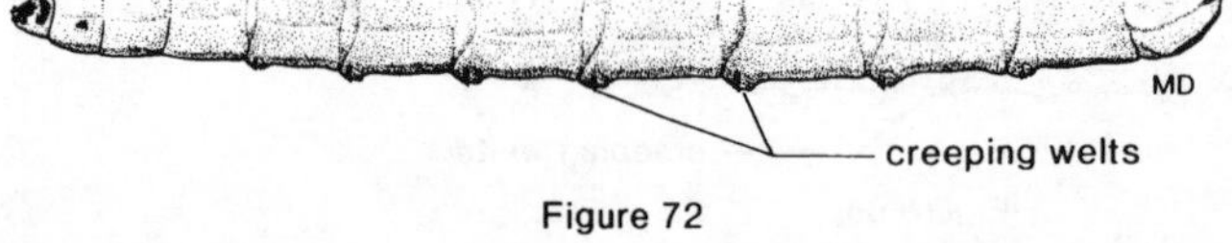

Figure 72

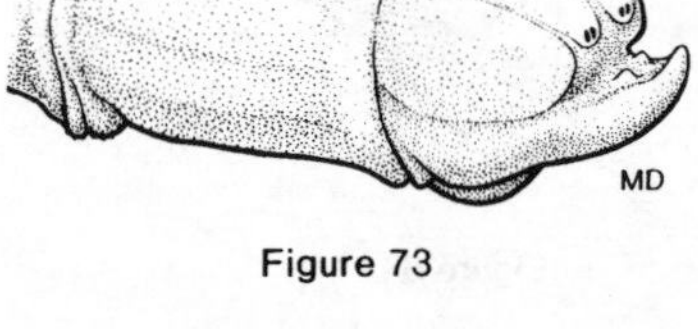

Figure 73

← Fig. 60

45(42′). Metacephalic rods moderately expanded or spatulate apically (fig. 71). Terminal abdominal segment either evenly rounded (in plant-mining species) or with 4 (rarely 2 ventral) primary lobes surrounding posterior spiracles; 1 pair of abdominal prolegs (in *Systenus*) and either 6 or 7 abdominal creeping welts (fig. 72) .... ***Dolichopodidae*** p. 786

45′. Metacephalic rods evenly slender throughout. Terminal abdominal segment either bearing single median protuberance below posterior spiracles (fig. 73) or if more than 1 terminal lobe present, then respiratory system often apneustic and with 7 or 8 pairs of crochet bearing prolegs (fig. 74) .... ***Empididae*** p. 784

46(28′). Posterior spiracles on a common, distinctive, sclerotized plate (fig. 75). Parasitic within bodies of Homoptera .... ***Pipunculidae*** p. 795

46′. Posterior spiracles not on a common sclerotized plate of form shown in fig. 75 (spiracles sometimes hidden in a pit) 47

47(46′). Anterior spiracles close together on dorsum of prothorax (fig. 76). Mandibles with longitudinal axis at oblique or right angle to remainder of cephalopharyngeal skeleton, each mandible usually bearing 2 or more pairs of equal-sized, anteriorly directed teeth. Phytophagous; mostly leaf miners, some stem miners .... ***Agromyzidae*** p. 819

47′. Anterior spiracles arising on lateral or dorsolateral surface of prothorax. Mandibles usually on same plane as remainder of cephalopharyngeal skeleton, each either bearing fewer than 2 pairs of teeth or bearing 2 or more pairs of unequally sized teeth 48

48(47′). Larva up to 2 mm long, oval to globular in shape. Two pairs of posterior spiracles present, the posterior pair sometimes united into 1 plate; spiracles on each side usually visibly joined by slender convoluted branches of felt chamber. No cephalopharyngeal skeleton. Ectoparasitic on bats 49

48′. Larva variable in length and shape. No more than 1 pair of posterior spiracles. Cephalopharyngeal skeleton usually present. Not associated with bats 50

49(48). Posterior spiracles composed of simple circular pore-like spiracular openings .... ***Nycteribiidae*** p. 878

49′. Posterior spiracles oval, crescent-shaped, or with numerous spiracular openings placed circularly on margin, or otherwise modified .... ***Streblidae*** p. 879

50(48′). Posterior spiracles projecting above body on structures ranging from short prominence (fig. 77) to very long and retractile tube (fig. 78); spiracular plates united along median margin (fig. 79). Body bearing dense pubescence or spicules or tubercles (fig. 80) .... ***Syrphidae*** p. 791

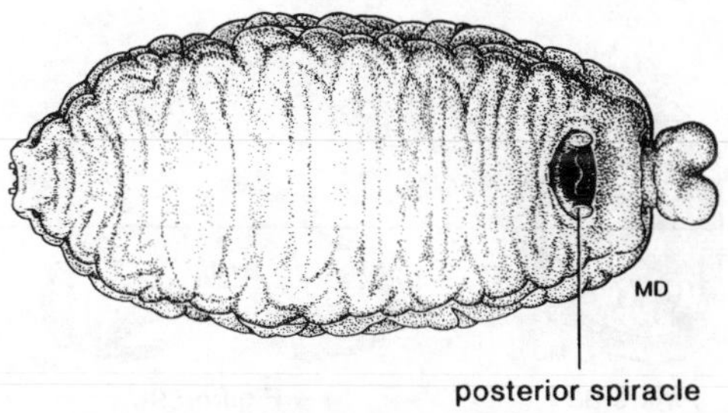

posterior spiracle

Figure 75

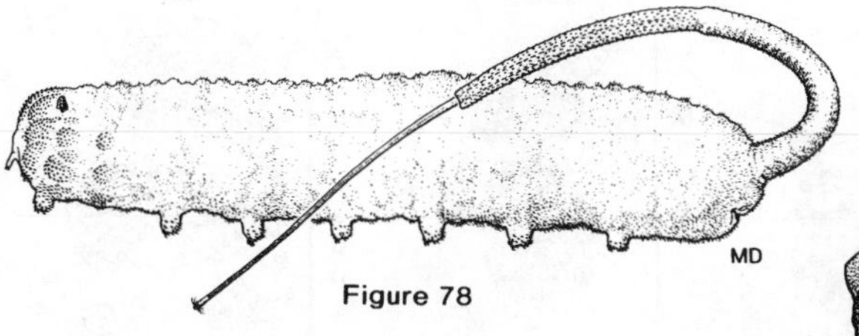

Figure 78

Figure 79

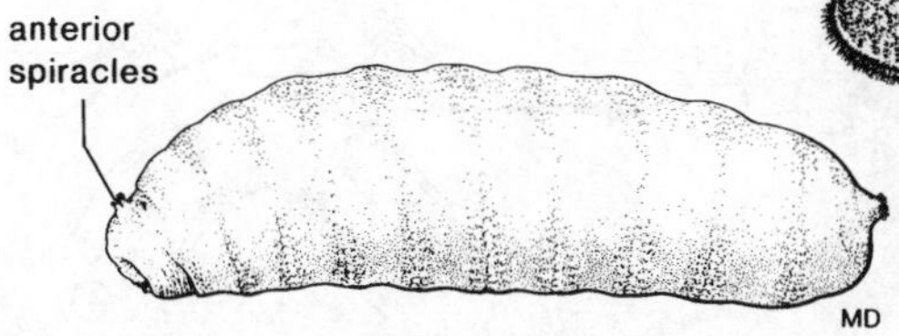

anterior spiracles

Figure 76

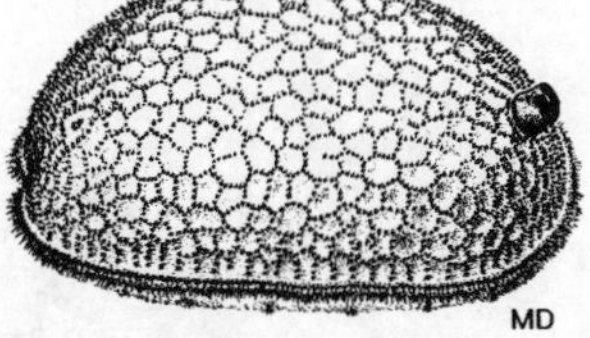

Figure 77

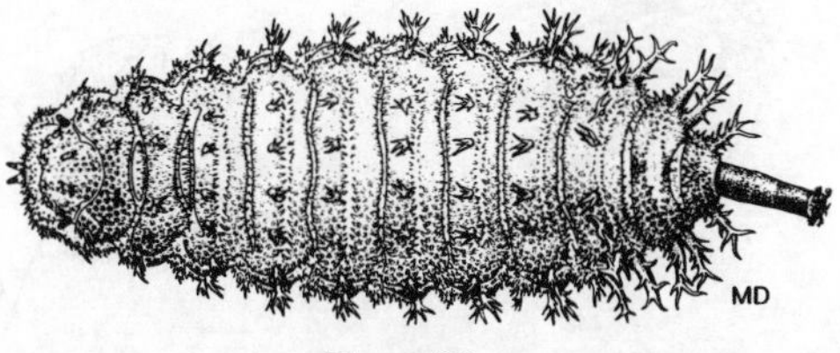

Figure 80

50′. Posterior spiracles sessile or elevated above surface of caudal abdominal segment; spiracular plates normally well-separated, but if appearing fused, then body lacking dense pubescence, prominent spicules, or tubercles 51

51(50′). Each posterior spiracle with numerous roundish, oval, or short slit-like spiracular openings (figs. 81–87); openings either randomly arranged or located along margin of spiracular plate or associated with intricately convoluted coral-like or serpentine bands; spiracles not thorn-like. Body usually highly wrinkled, or otherwise rather swollen and roundish to pear-shaped 52

51′. Each posterior spiracle with 3 isolated oval or slit-like relatively large and sometimes sinuous spiracular openings (figs. 88–92) (rarely with 4 to 6 such openings or sometimes thornlike) (fig. 93). Body usually rather slender and subcylindrical or flattened 59

52(51). Larva deposited as smooth, generally featureless oval to round prepupa having darkly sclerotized spiracular plate that often covers posterior end of body (fig. 94), some species bear integumentary setae. Ectoparasitic on birds and mammals .... ***Hippoboscidae*** p. 878

52′. Larva not as above 53

53(52′). Spiracular openings oval, arrayed in circle on margin of spiracular plate. Parasitic within bodies of grasshoppers .... (part)(*Acridomyia*) ***Anthomyiidae*** p. 853

53′. Spiracular openings distributed rather evenly over spiracular plate 54

54(53′). Posterior spiracular plates kidney-shaped, each consisting of series of curvilinear bands, each with 8–14 yellowish to orange clusters of round or oval to short bar-like spiracular openings, and with uppermost cluster extended into short spine (fig. 81). Parasitic within bodies of Scarabaeidae .... ***Pyrgotidae*** p. 809

54′. Posterior spiracular plates not as above 55

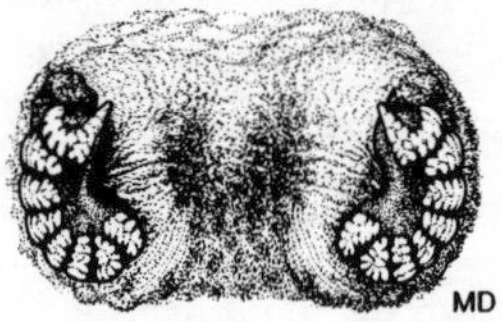

Figure 81

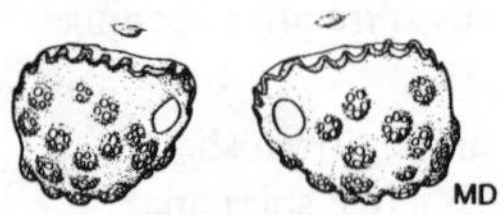

Figure 82

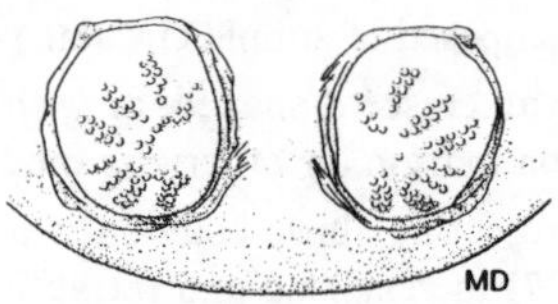

Figure 83

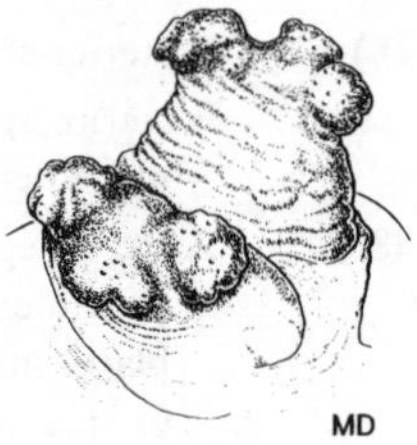

Figure 84

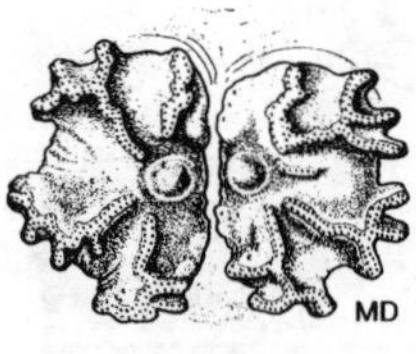

Figure 85

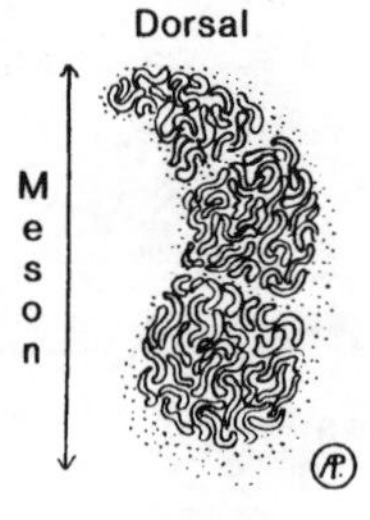

Figure 86

Figure 87

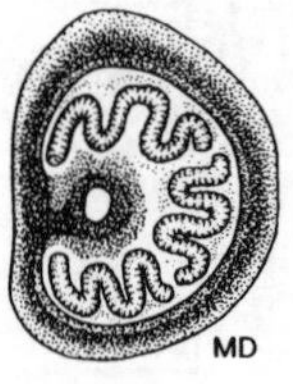

Figure 88

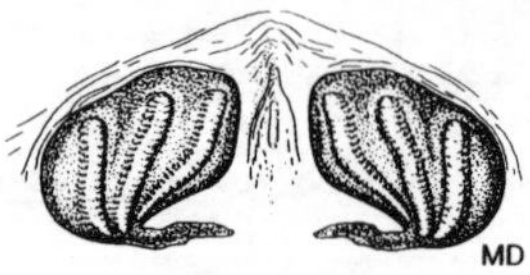

Figure 89

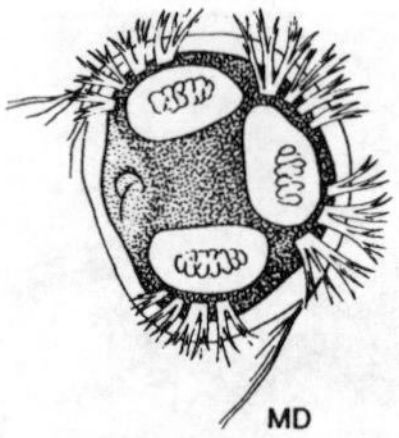

Figure 90

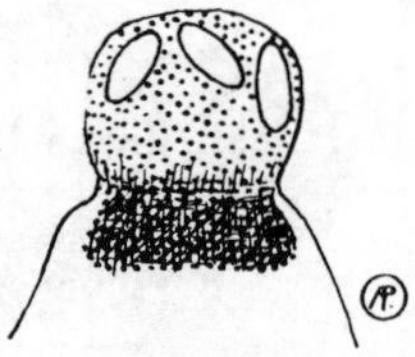

Figure 91

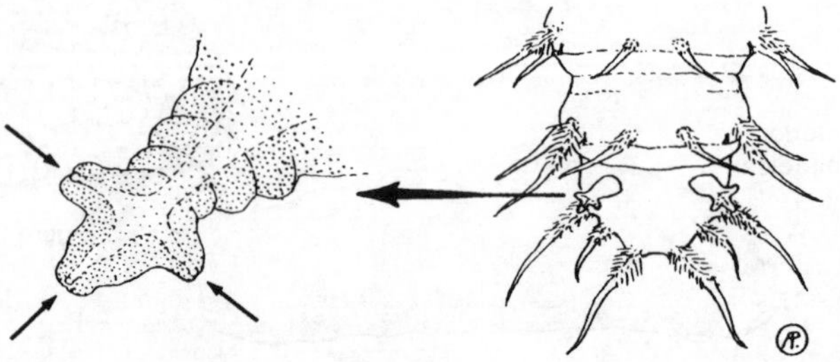

Figure 92

55(54′). Posterior spiracular plates dome-shaped, either with circular wart-like protuberances each bearing several pale spiracular openings (fig. 82) or with linear clusters of pores radiating from ecdysial scar. Parasitic on bees and wasps .... ***Conopidae*** p. 796

55′. Posterior spiracular plates not dome-shaped and without wart-like protuberances. Parasitic on other arthropods or mammals 56

56(55′). Posterior spiracles each with numerous openings elevated on coral-like sculpturing of spiracular plate; spiracular plate usually more or less 3-parted (figs. 84, 85). Parasitic on various insects and centipedes .... (part) ***Tachinidae*** p. 875

56′. Posterior spiracles not as described above. Endoparasitic in mammals 57

57(56′). Posterior spiracles each with many short serpentine lines resembling a maze, lines obviously arranged to form 3 groups on each spiracular plate (fig. 86). Integument largely covered with scale-like spines or spines (fig. 95) .... (part) **(Cuterebrinae)** ***Oestridae*** p. 867

57′. Posterior spiracles each with oval to round spiracular openings arranged to form more than 3 groups (fig. 87). Integument with spines restricted to margins of segment or to ventral surface; spines frequently not scale-like 58

58(57′). Posterior spiracles placed on dorsal surface of transverse cleft in terminal abdominal segment (fig. 96), spiracles frequently concealed within cleft when opposing surfaces are brought together .... (part) **(Oestrinae)** ***Oestridae*** p. 867

58′. Posterior spiracles not placed within cleft but on evenly rounded terminal extremity of body (fig. 97) .... (part) **(Hypodermatinae)** ***Oestridae*** p. 867

59(51′). Body stout, blunt posteriorly and strongly tapered anteriorly, with 1 or 2 rows of stout spines partially or entirely encircling most segments (fig. 98). Posterior spiracles frequently concealed within cavity on terminal abdominal segment; spiracular plate with 3 parallel vertically oriented bands (fig. 99). Endoparasitic in horses .... (part) **(Gasterophilinae)** ***Oestridae*** p. 867

59′. Characteristics not as above 60

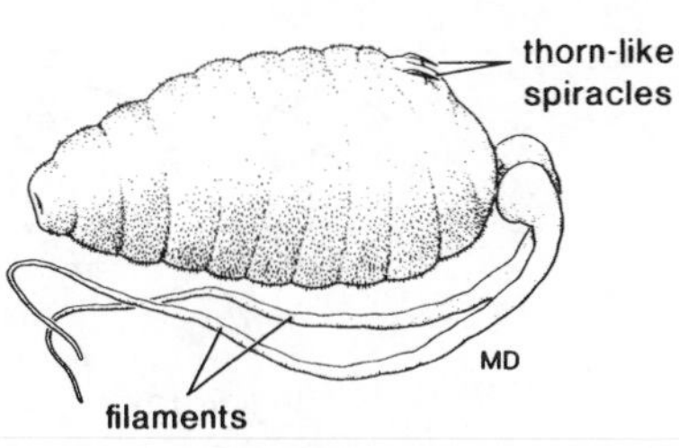

Figure 93

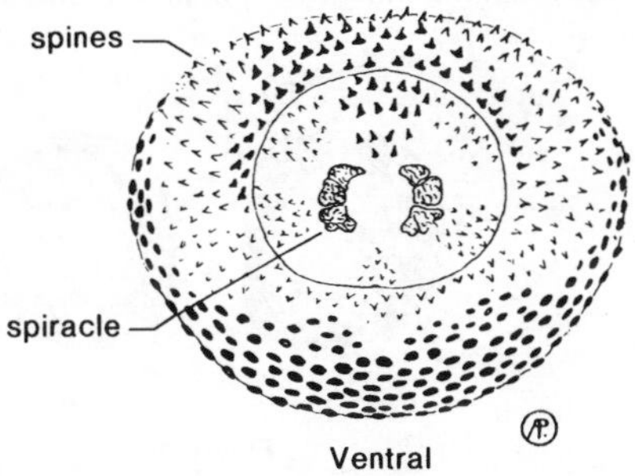

Ventral
Figure 95

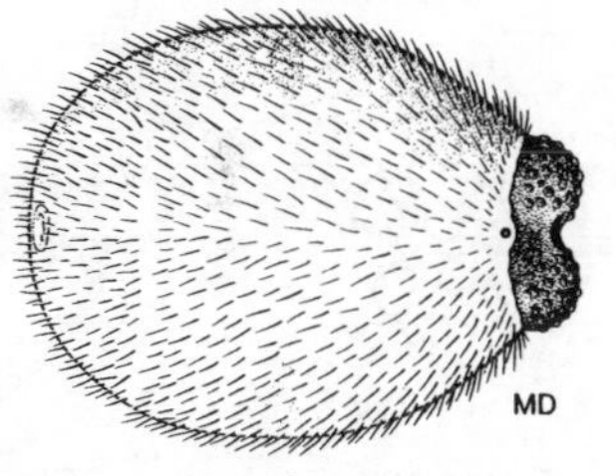

Figure 94

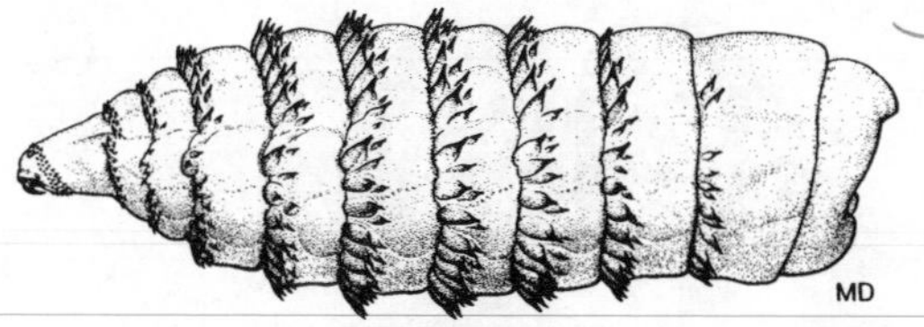

Figure 98

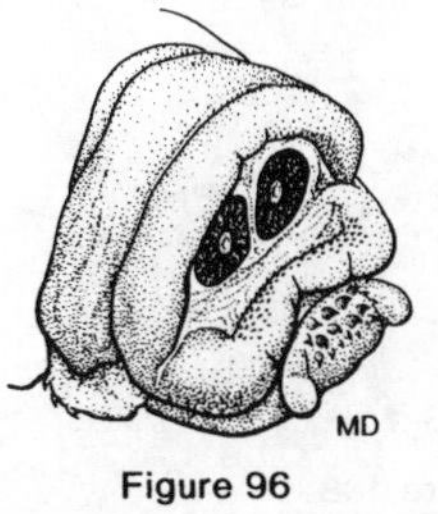

Figure 96

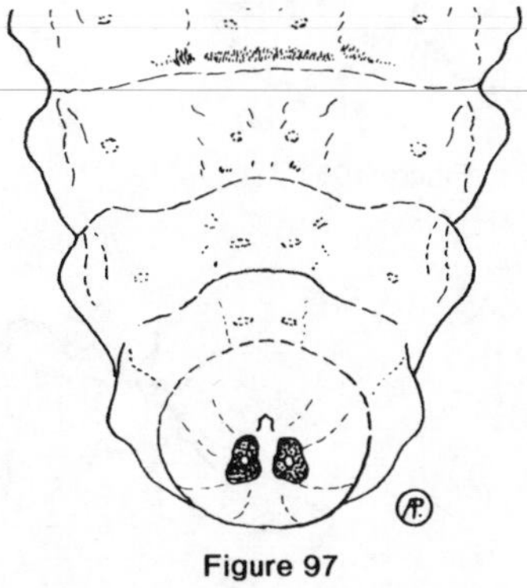

Figure 97

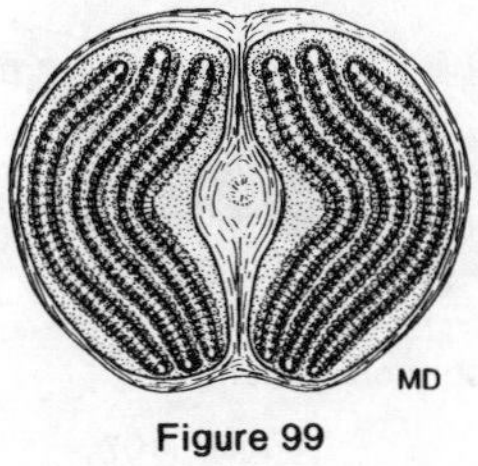

Figure 99

60(59′). Posterior spiracles on short telescopic respiratory tube that is not forked terminally; spiracles separated only by slight depression. Restricted to coastal habitats .... ***Canacidae*** p. 849

60′. Posterior spiracles either not on telescopic respiratory tube, or on telescopic tube that is conspicuously forked terminally 61

61(60′). Terminal abdominal segment with pair of slender filaments at least as long as body (fig. 93). Posterior spiracles thorn-like, situated dorsally (fig. 93). Known only from California where larvae parasitize scale insects .... ***Cryptochetidae*** p. 848

61′. Terminal abdominal segment lacking long filaments. Posterior spiracles not as above 62

62(61′). Body cylindrical, tapered anteriorly, bluntly rounded posteriorly; terminal segment often with abundant minute spicules but lacking papillae or tubercles on any segment. Anterior spiracles occasionally absent, when present simple, with 1 to many randomly arranged papillate openings at apex of short stalk (fig. 100) or nearly sessile on surface of prothorax (fig. 101). Posterior spiracles often heavily sclerotized, usually at least slightly elevated above segment; each spiracle with 3 to several short to long spiracular openings that are serpentine (fig. 102), bowed, or variously bent (figs. 103, 104), but openings occasionally straight and often following distinct ridges on spiracular plate (fig. 105); spiracular openings arranged more or less radially around ecdysial scar but almost never in predominately vertical axis. Cephalopharyngeal skeleton strong and heavy, lacking parastomal bars (fig. 106). Parasitic within insects and isopods .... ***Rhinophoridae*** p. 873
(part) ***Tachinidae*** p. 875

62′. Differing from above. If spiracular openings of posterior spiracles are serpentine, then cephalopharyngeal skeleton has accessory oral sclerites (fig. 107) 63

63(62′). Body flattened dorsoventrally, tergal plates on all segments with thin striated lateral margins. Long filaments on first 2 thoracic segments and terminal abdominal segment (fig. 108) .... ***Lonchopteridae*** p. 787

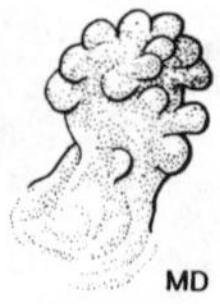

MD

Figure 100

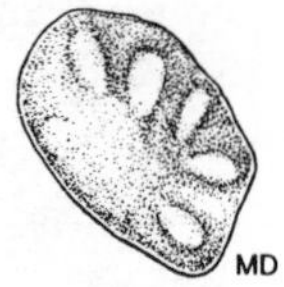

MD

Figure 101

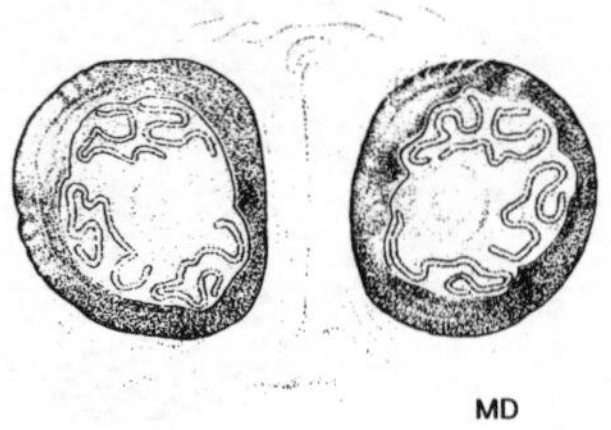

MD

Figure 102

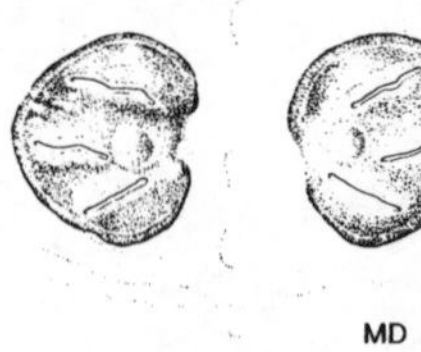

MD

Figure 103

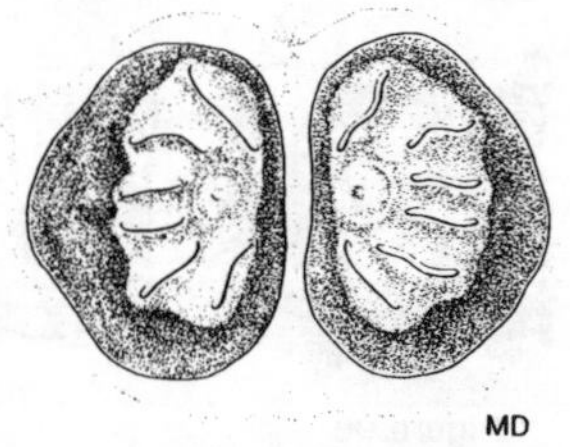

MD

Figure 104

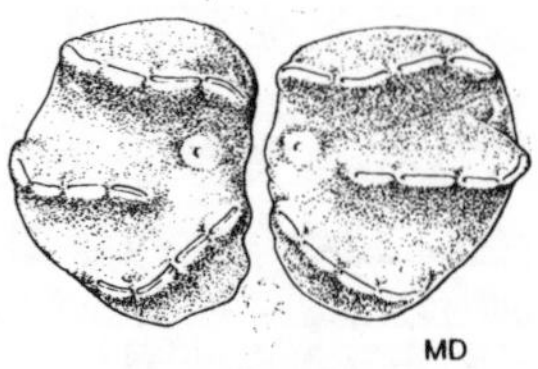

MD

Figure 105

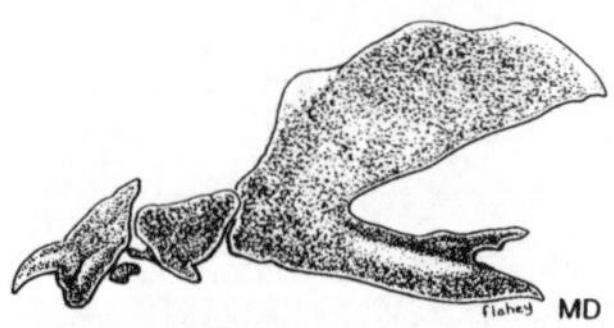

MD

Figure 106

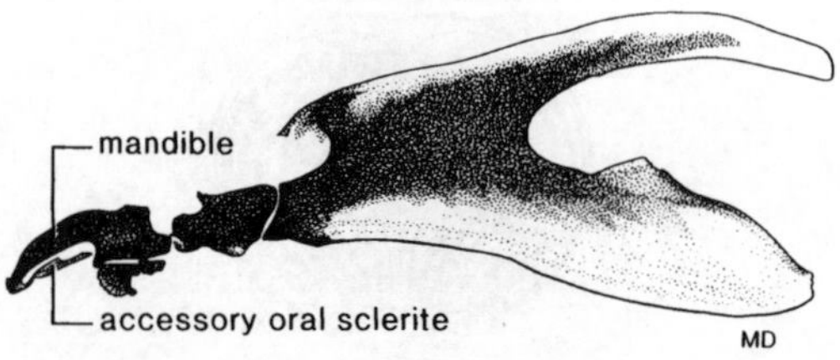

mandible

accessory oral sclerite

MD

Figure 107

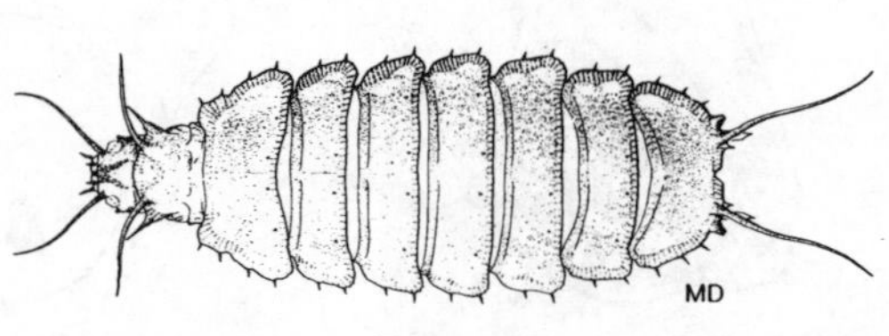

MD

Figure 108

←— Fig. 93

63′. Body often not flattened dorsoventrally, but if so, then lacking thin striated lateral margins on tergal plates of all segments. Long filamentous processes, if present, not restricted to first 2 thoracic segments and terminal abdominal segment 64

64(63′). Anterior spiracles simple, each with 1 to several sessile spiracular openings placed peripherally at apex of short tubular or conical projection (fig. 109). Body often somewhat dorsoventrally flattened. All body segments usually bearing several systematically arranged spicules or tubercles, usually with those situated laterally most prominent. Tentoropharyngeal and hypopharyngeal sclerites finely constructed and fused to each other (fig. 110); hypopharyngeal sclerite usually continuous anteriorly with single or multi-toothed median labial sclerite, or with paired mandibles, or with both structures 65

64′. Anterior spiracles either lacking or, if present, bearing 2 or more short papillae, or bearing long filaments arising on apex of spiracular stalk (fig. 111). Body not as above. Tentoropharyngeal and hypopharyngeal sclerites often more strongly constructed than above, and distinctly separated (fig. 112); hypopharyngeal sclerite fused to hook-like labial sclerite only in the first instar of some species 66

65(64). Each posterior spiracle on short, conical, apically sclerotized spiracular support; with 4 spiracular openings arranged radially around ecdysial scar (fig. 113) .... ***Platypezidae*** p. 788

65′. Posterior spiracles variously supported, with spiracular openings arranged in 2 pairs placed one behind the other (fig. 114) .... ***Phoridae*** p. 790

66(64′). First 4 segments and terminal abdominal segment with encircling rows of small strobiliform tubercles (fig. 115). Respiratory system metapneustic; posterior spiracles sessile. Tentoropharyngeal and hypopharyngeal sclerites fused to each other. Mining walls of bee combs .... ***Braulidae*** p. 826

66′. If tubercular processes present on thoracic segments, then tubercles also present on most abdominal segments. Respiratory system usually amphipneustic, with posterior spiracles elevated. Tentoropharyngeal and hypopharyngeal sclerites usually separate 67

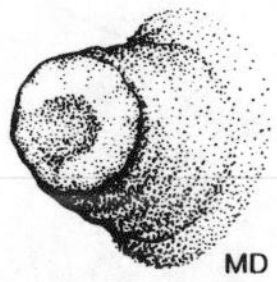

Figure 109

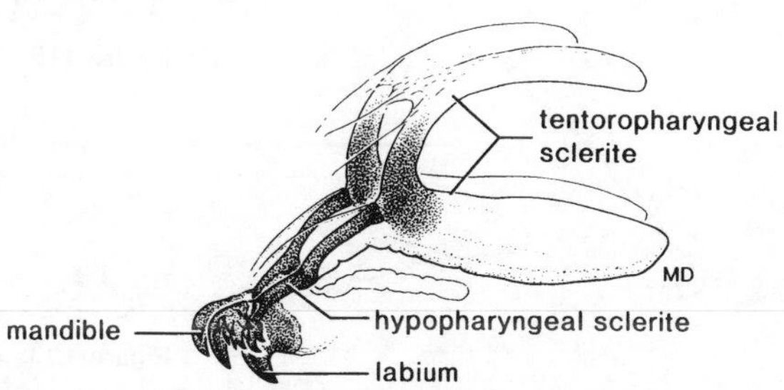

Figure 110

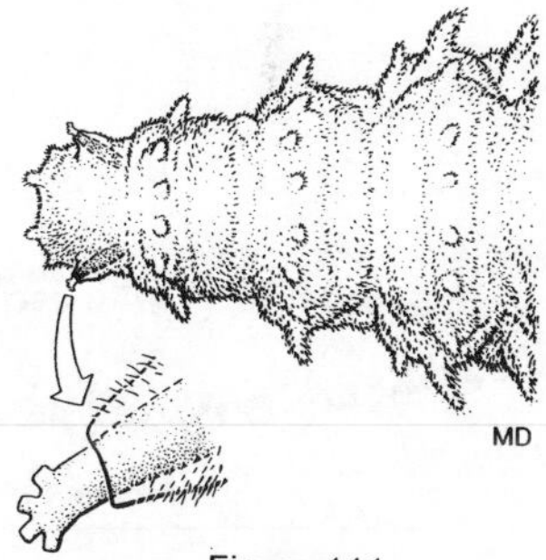

Figure 111

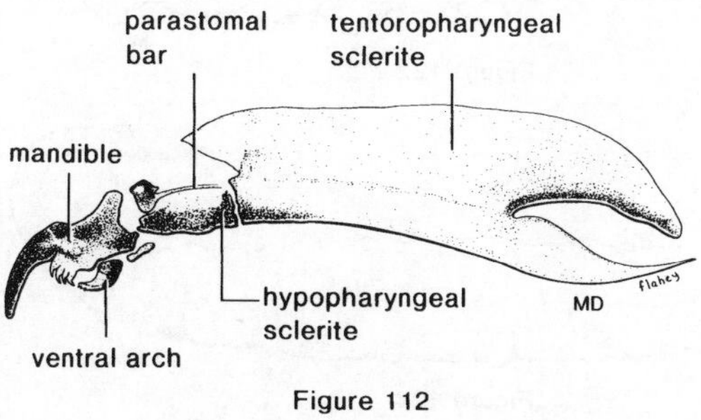

Figure 112

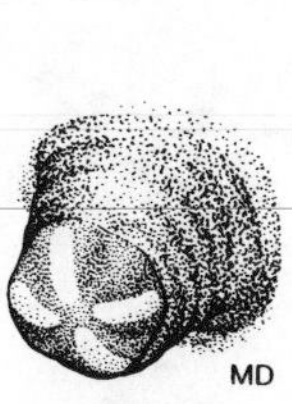

Figure 113

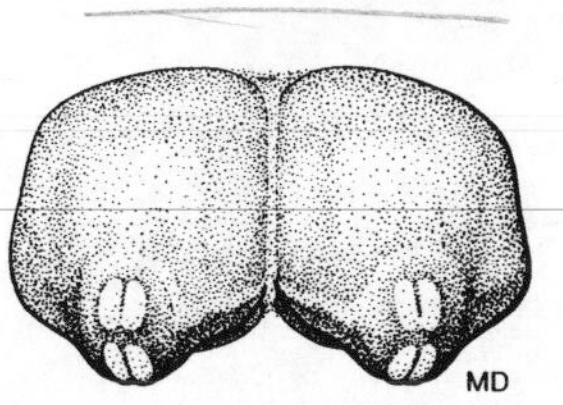

Figure 114

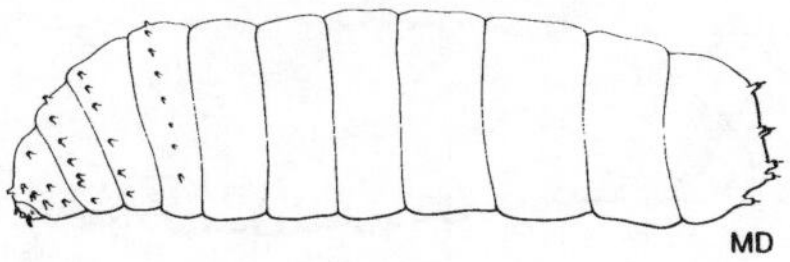

Figure 115

67(66′). Spiculate or setiferous tubercles present on several body segments preceding terminal one (fig. 116) 68

67′. Tubercles lacking, or situated only on terminal abdominal segment 70

68(67). Tubercles present only on abdominal segments. Body cylindrical (fig. 117) .... (part) ***Drosophilidae*** p. 841
(part) ***Ephydridae*** p. 843

68′ Tubercles present on both thoracic and abdominal segments. Body dorsoventrally flattened 69

69(68′). Posterior spiracles each on short, non-spiculate tuberculate process situated dorsally near anterior margin of last abdominal segment (fig. 116); process terminating in 3 lobes, each lobe bearing spiracular opening. Body tubercles pinnately setiferous .... (part) (*Fannia*) ***Muscidae*** p. 857

69′. Posterior spiracles each on long slender spiculate tuberculate process arising caudally on terminal abdominal segment (fig. 118). Other spiculate tubercles on body differing from spiracular tubercle only in being longer .... ***Periscelididae*** p. 823

70(67′). One or more body segments *densely clothed* with minute setulae or spicules (figs. 117–124), or caudal abdominal segment elongated to form respiratory tube (fig. 119); or terminal abdominal segment bearing distinctive array of 1 or more pairs of symmetrically placed papillae or tubercles (fig. 125) that are usually distinctive, but sometimes more reduced (as in fig. 126) 71

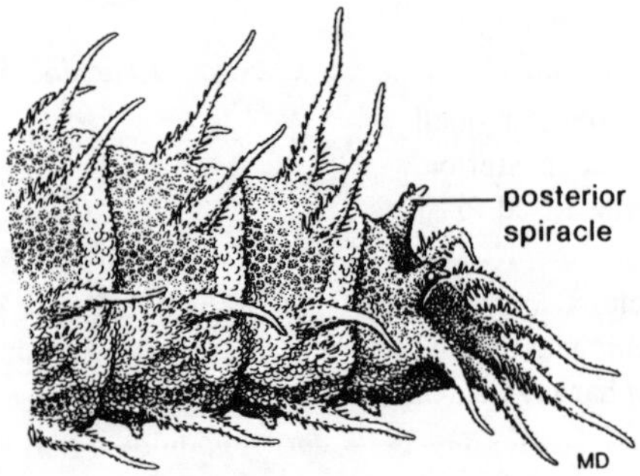

Figure 116

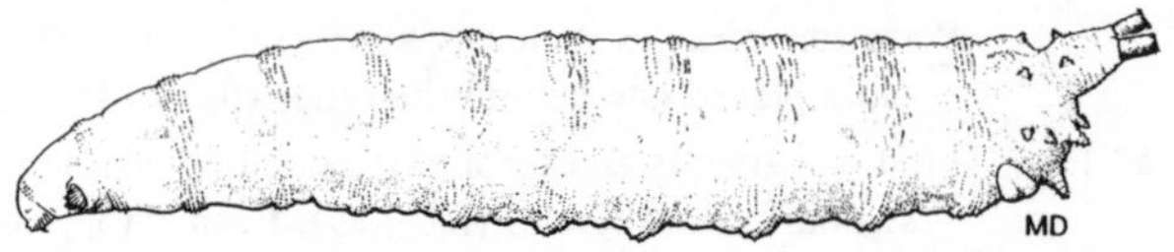

Figure 117

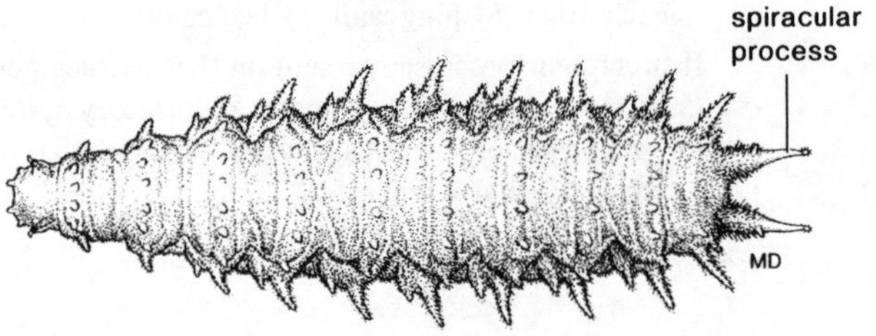

Figure 118

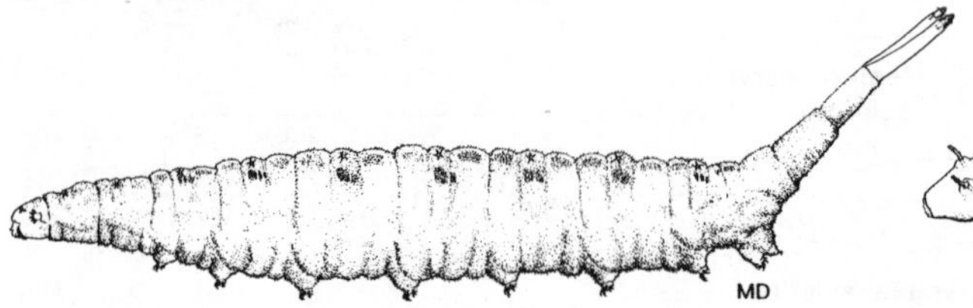

Figure 119

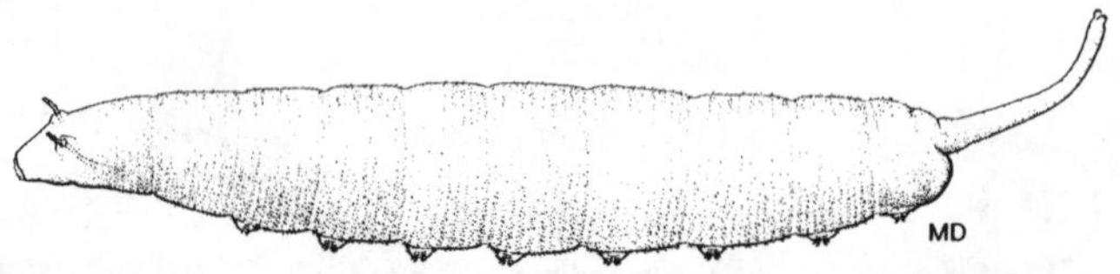

Figure 120

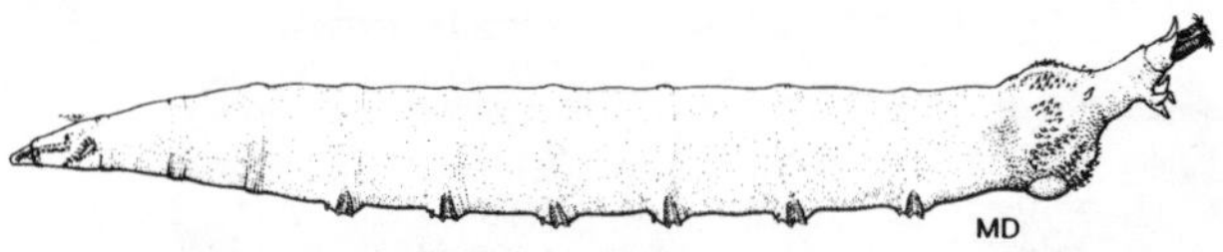

Figure 121

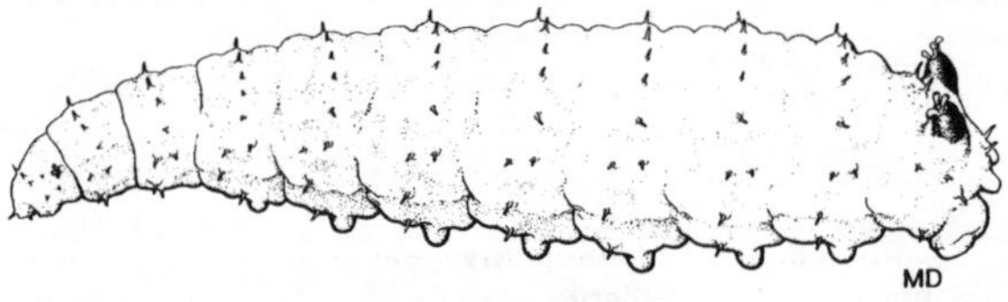

Figure 122

Figure 123

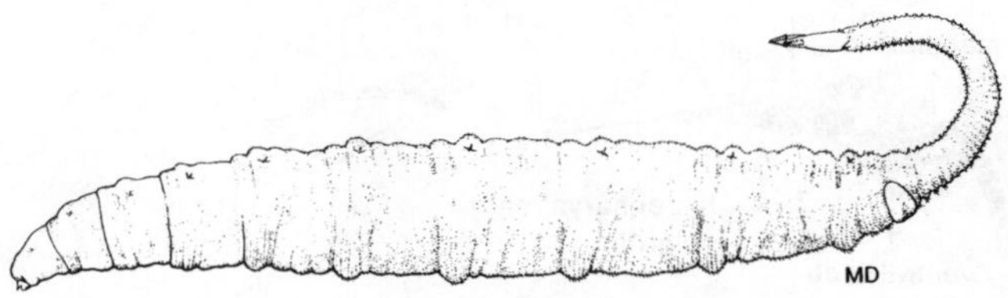

Figure 124

← Fig. 112

70′. Body segments lacking abundant setulae, spicules, papillae, or tubercles, generally featureless except for spicules on creeping welts; welts occasionally encircling anterior margins of a few segments 86

71(70). Cephalopharyngeal skeleton with ventral arch below base of mandibles (figs. 112, 127). Larva a predator or parasitoid on freshwater, shoreline, and terrestrial molluscs or their eggs .... ***Sciomyzidae*** p. 828

71′. Cephalopharyngeal skeleton lacking ventral arch below mandibles 72

72(71′). Terminal abdominal segment more or less tapered posteriorly because of close proximity or union of conical spiracular prominences (figs. 121, 128), or basal union attenuated to form respiratory tube (fig. 120); shorter spiracular prominences when not forming respiratory tube usually each with tubercle located dorsally or dorsolaterally at base (fig. 121). Integument of one or more segments extensively covered with spicules or setulae 73

72′. Terminal abdominal segment rather truncate; posterior spiracular prominences sometimes elongate, but distinctly separate and lacking tubercles on the base of each prominence (fig. 129). Spicules or setae usually present only on anterior or posterior margins of segments (fig. 129) 79

73(72). Anterior spiracle with basal stalk terminating in many long filamentous processes (fig. 130), spiracle retractile into body .... (part) (**Drosophilinae**) ***Drosophilidae*** p. 841

73′. Anterior spiracle absent or having different form than above, but if in form of elongate retractile stalk, then bearing short lateral papillae near apex of stalk (fig. 131) 74

74(73′). Posterior spiracles elongated to form hollow spine (figs. 124, 132) (*Scaptomyza* of Drosophilidae is similar, keyed in previous couplet), attached to rootlets of wetland plants .... (part) (*Notiphila*) ***Ephydridae*** p. 843

74′. Posterior spiracles not spine-like 75

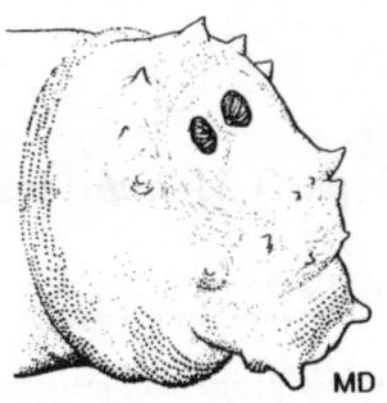

Figure 125

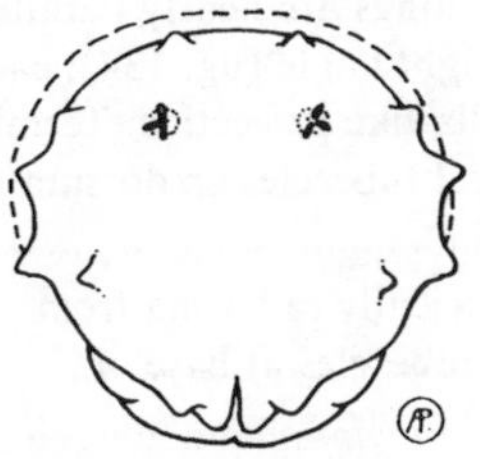

Figure 126

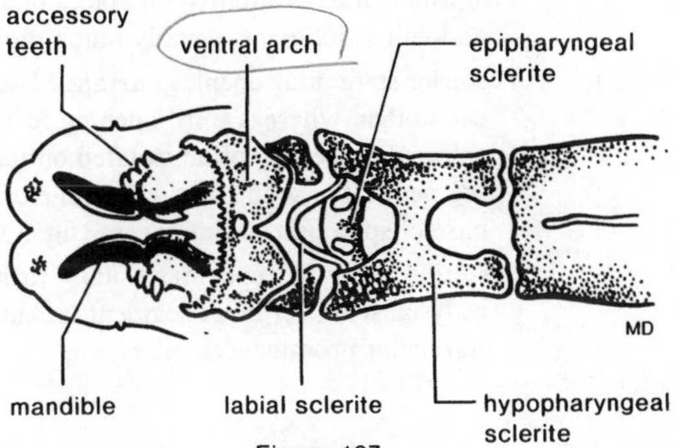

Figure 127

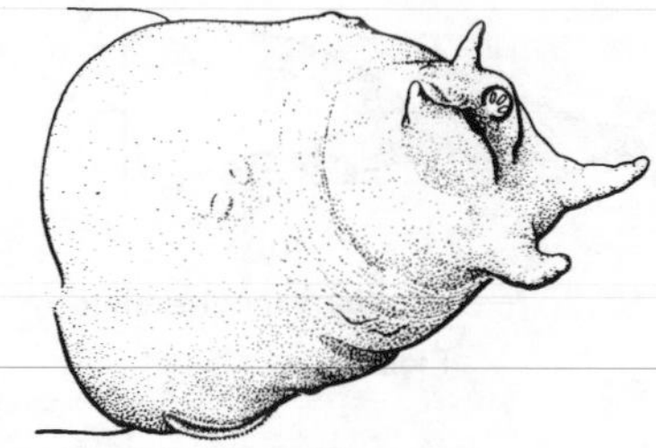

Figure 128

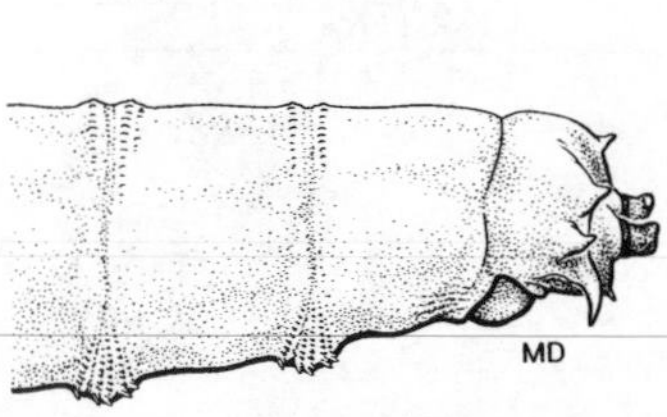

Figure 129

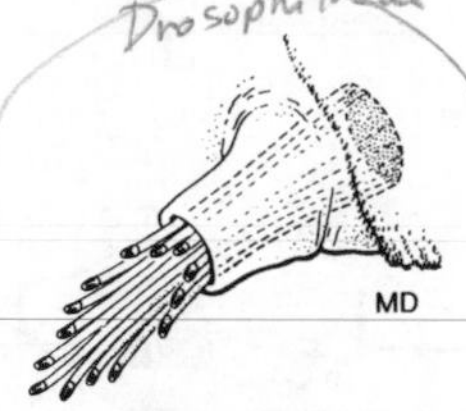

Figure 130

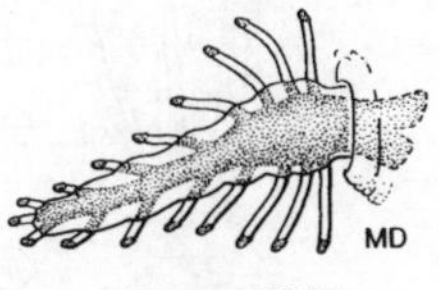

Figure 131

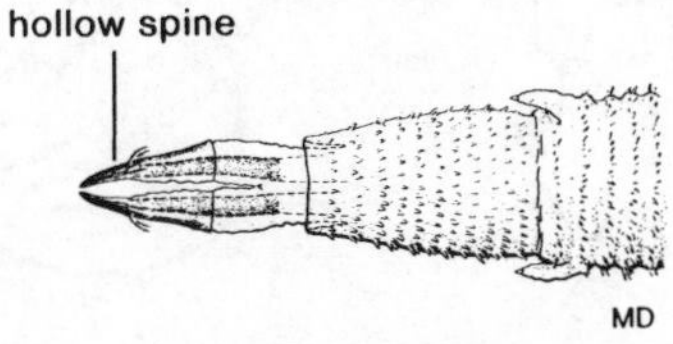

Figure 132

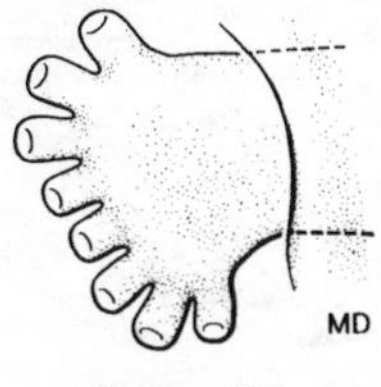

Figure 133

75(74′). Terminal abdominal segment attenuated posteriorly to form more or less elongate respiratory tube (figs. 119, 120), tube lacking tubercles dorsally and usually capable of some retraction. Spicules or setae entirely covering all abdominal segments 76

75′. Terminal abdominal segment tapered, either with very short respiratory tube (fig. 121) or with closely placed or basally fused conical spiracular prominences (fig. 128); spiracular prominences each with tubercle dorsally at or near base. Spicules or setulae not covering all surfaces of every abdominal segment 77

76(75). Posterior respiratory tube elongate, with very short and unsclerotized terminal fork, length of each branch about equal to its diameter (fig. 120) .... ***Aulacigastridae*** p. 822

76′. Posterior respiratory tube of variable length, but with terminal fork usually apically sclerotized, length of each branch greater than its diameter (fig. 119) .... (part) ***Ephydridae*** p. 843

77(75′). Spicules and pubescence extensively covering terminal abdominal segment only (fig. 121). Posterior spiracles usually with well-developed spiracular setae; each anterior spiracle with papillae projecting on either side of more or less elongate central axis (fig. 131) .... ***Sepsidae*** p. 832

77′. Spicules present either only at segmental margins of terminal abdominal segment or extensively covering other segments besides the terminal one. Posterior spiracles with spiracular setae inconspicuous or absent; each anterior spiracle with papillae projecting fan-like (fig. 133) (except in some Lauxaniidae) 78

78(77′). Posterior spiracles situated on median sloping faces of spiracular prominences (fig. 128) and appearing capable of retraction on one another. Segments immaculate except for tubercles on terminal segment and spicules on anterior ventral creeping welts of abdominal segments (fig. 128) .... ***Piophilidae*** p. 816

78′. Posterior spiracles situated on apices of spiracular prominences. Spicules on abdominal segments usually much more extensive than described above .... ***Lauxaniidae*** p. 833

79(72′). Posterior spiracular openings arranged so that 2 openings are nearly parallel to each other, whereas third opening forms nearly right angle (fig. 134); each spiracular opening often isolated on its own papilla-like projection. Terminal segment often with transverse ridge of 3 or 4 small tubercles on dorsum near base of spiracular prominences (fig. 134) .... (part) ***Milichiidae*** p. 824

79′. Posterior spiracular openings usually rather symmetrically radiating from ecdysial scar. Terminal segment lacking ridge of tubercles at base of spiracular prominences 80

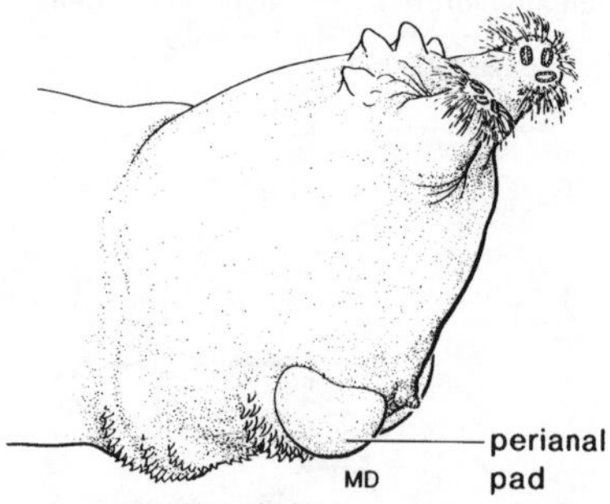

Figure 134

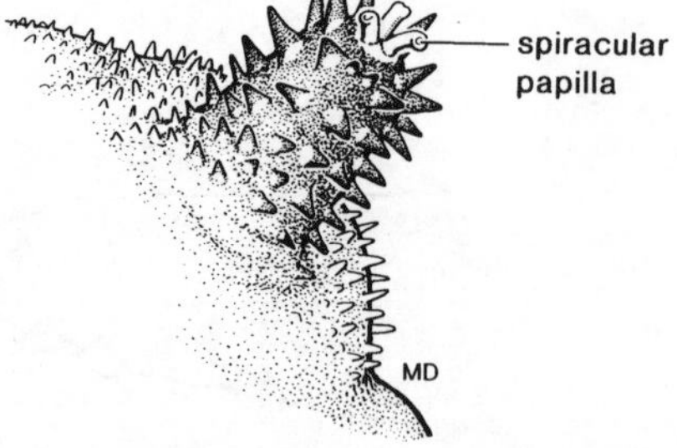

Figure 135

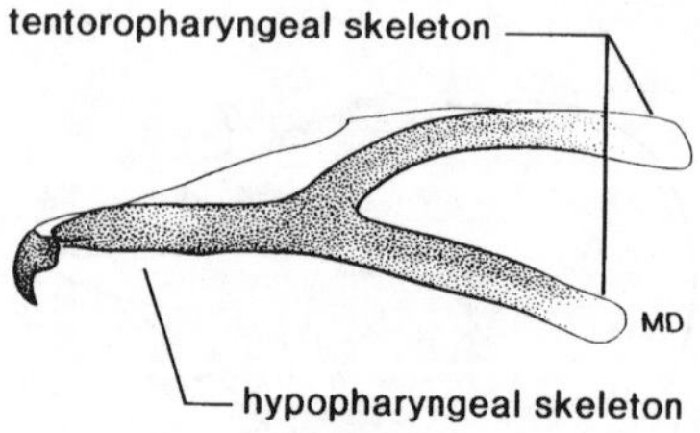

Figure 136

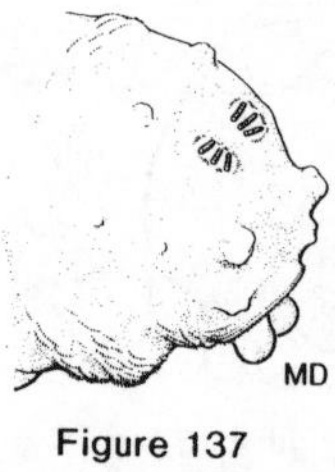

Figure 137

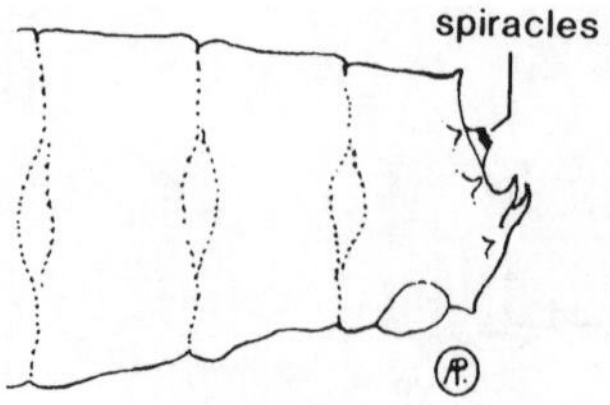

Figure 138

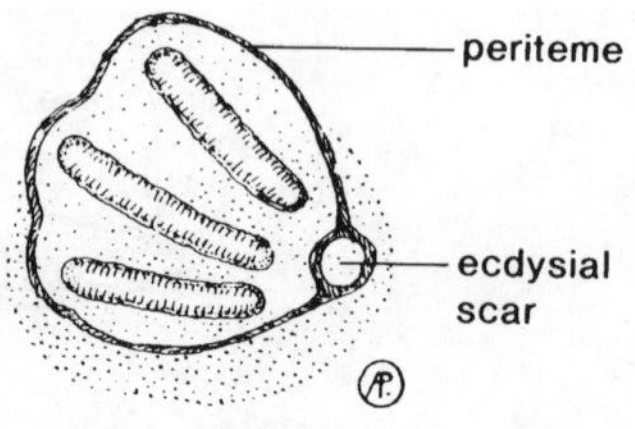

Figure 139

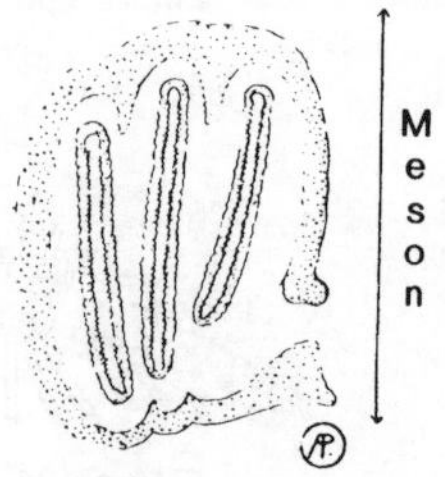

Figure 140

← Figs. 119–133

80(79′). Integument of all segments clothed with fine pubescence or spicules 81

80′. Integument of at least part of each thoracic segment free from pubescence or spicules 82

81(80). Each posterior spiracular opening on papilla-like projection from spiracular plate (fig. 135). Cephalopharyngeal skeleton with hypopharyngeal and tentoropharyngeal sclerites fused (fig. 136). Predators and parasitoids of aphids, adelgids, and coccids .... ***Chamaemyiidae*** **p. 835**

81′. Posterior spiracular openings sessile on surface of terminal segment. Hypopharyngeal and tentoropharyngeal sclerites separated .... **(part) (Steganinae) *Drosophilidae* p. 841**

82(80′). Posterior spiracles nearly or quite sessile on surface of anal segment (figs. 125, 141) and lacking a sclerotized peritreme, or with spiracular openings slit-like and with all slits oriented in a predominantly vertical (fig. 89) or median direction (fig. 139) 83

82′. Posterior spiracles distinctly elevated above plane of terminal segment (fig. 138) and longitudinal axis of one or more spiracular openings oriented dorsally or dorsomedially (fig. 160) 85

83(82). Posterior spiracles lacking sclerotized peritremes (fig. 137), with ecdysial scar situated medially or mediodorsally to spiracular openings .... ***Tephritidae*** (p. 809)

83′. Posterior spiracles with distinctly sclerotized peritremes (fig. 139), ecdysial scar, or point where spiracular openings converge, situated ventrally or ventromedially (fig. 139) 84

84(83′). Spiracular openings oriented more or less vertically (fig. 140); posterior spiracles frequently within deep spiracular cavity on terminal segment (fig. 141); ecdysial scar usually not visible; peritreme not completely encircling each spiracular plate (fig. 140) .... ***Sarcophagidae*** **p. 871**

84′. Spiracular openings obliquely or horizontally oriented (fig. 139); posterior spiracles at surface of terminal abdominal segment (fig. 125); ecdysial scar present; peritreme completely encircling each spiracular plate (fig. 139) .... ***Calliphoridae*** **p. 862**

85(82′). Anterior spiracle 2-branched, with papillae present along each diverging arm (fig. 142). Cephalopharyngeal skeleton without parastomal bars (as in fig. 10) .... **(part) *Scathophagidae* p. 850**

85′. Anterior spiracle fan-shaped (fig. 143) or tree-like (fig. 144), or parastomal bars present in cephalopharyngeal skeleton (as in fig. 112) or both features present .... **(part) *Heleomyzidae* p. 837**
(part) *Sphaeroceridae* p. 839
Curtonotidae **p. 840**
(part) (Steganinae) *Drosophilidae* p. 841
(part) *Anthomyiidae* p. 853

Figure 89

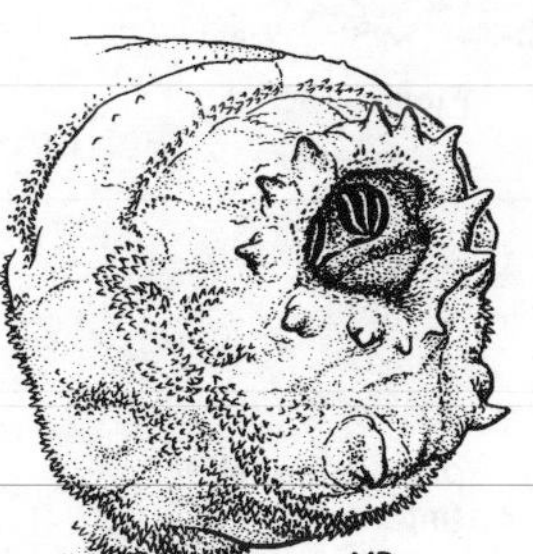

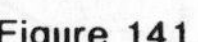

Figure 141

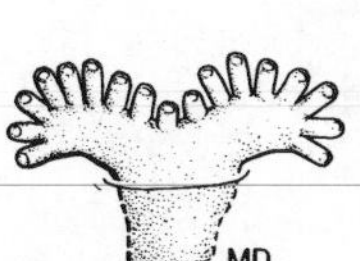

Figure 142

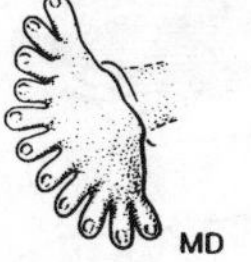

Figure 143

Figure 144

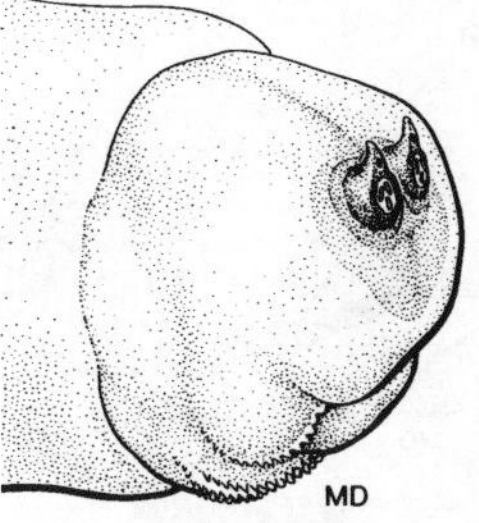

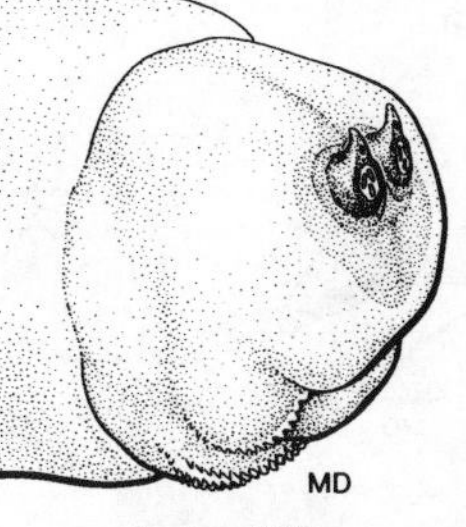

Figure 145

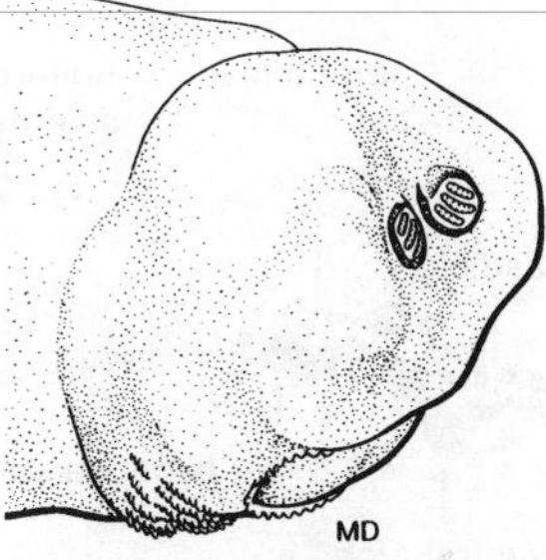

Figure 146

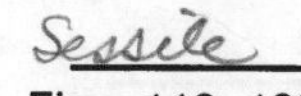

Figs. 112, 160 ⟶

86(70′). Posterior spiracles sessile on terminal segment (figs. 137, 146) 87

86′. Posterior spiracles distinctly elevated above surface of terminal segment (fig. 145) 88

87(86). Spiracular peritremes unpigmented (fig. 137) .... (part) ***Tephritidae*** p. 809

87′. Spiracular peritremes usually distinctly pigmented (fig. 146) .... (part) ***Otitidae*** p. 806
(part) ***Chloropidae*** p. 847

88(86′). Spiracular setae abundant, forming *continuous* fringe around border of each posterior spiracular plate (fig. 147); spiracular openings band- or slit-like and distinctly bowed or bent. Living in decaying seaweed .... ***Coelopidae*** p. 826

88′. Spiracular setae, if visible, not forming continuous fringe, but arranged to form 3 or 4 groups of setae (fig. 148) or scattered as individual branched setae (fig. 149); spiracular openings not as above 89

89(88′). Posterior spiracular plate either with 1 or more distinct spines (fig. 148) or lobes arising from dorsal or lateral surface or with sharp ridge along dorsal margin 90

89′. Posterior spiracular plates without lobes, spines, or ridges along dorsal margin 96

90(89). Cephalopharyngeal skeleton greatly reduced, with all portions except small mandibles unpigmented. Living in rotten wood .... ***Clusiidae*** p. 817

90′. Cephalopharyngeal skeleton of normal size, usually darkly pigmented 91

91(90′). Posterior spiracular openings short and oval, lying nearly at right angles to one another (fig. 148) 92

91′. Posterior spiracular openings radiating from ecdysial scar at distinctly less than right angles, or irregularly or peripherally located (fig. 150) 94

92(91). Tentoropharyngeal and hypopharyngeal sclerites fused (as in fig. 136); pharyngeal filter lacking. Living in roots, stems or galls of plants .... ***Psilidae*** p. 801

92′. Tentoropharyngeal and hypopharyngeal sclerites separated (as in fig. 112); pharyngeal filter present (fig. 151) 93

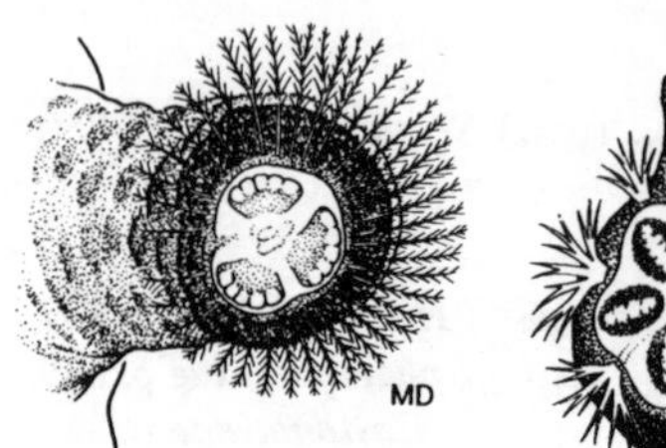

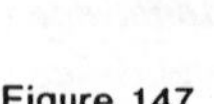

Figure 147

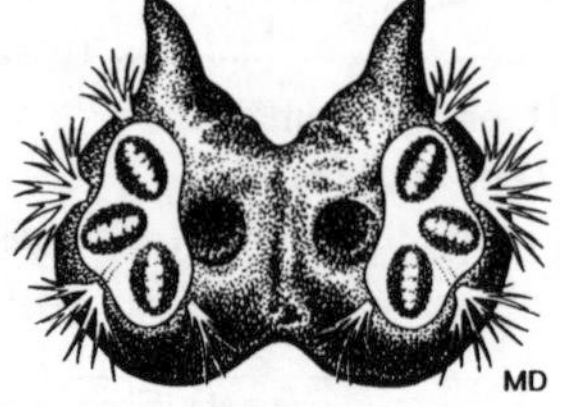

Figure 148

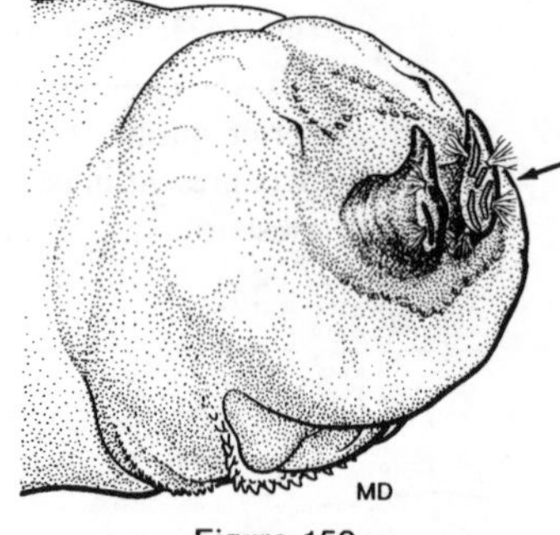

Figure 150

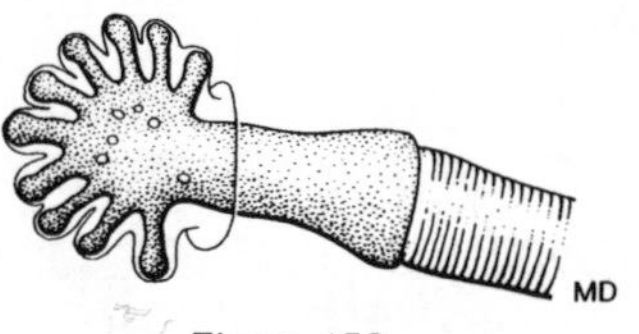

Figure 153

Figure 149

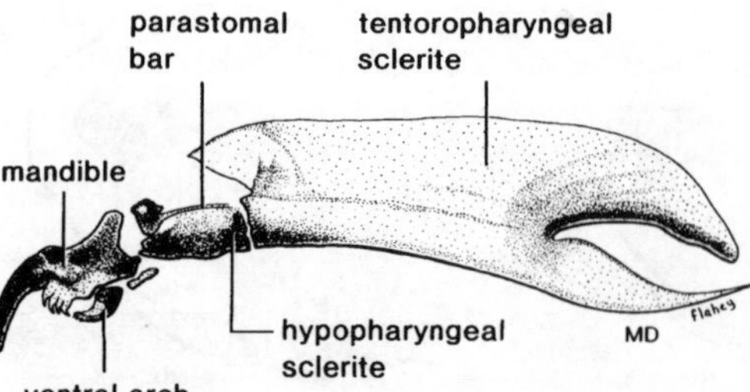

Figure 112

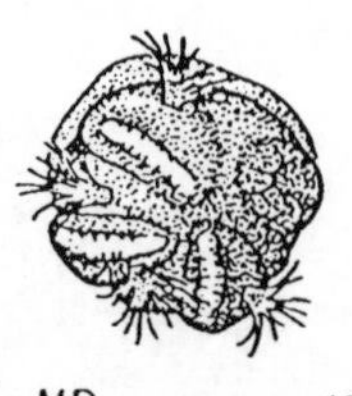

Figure 160

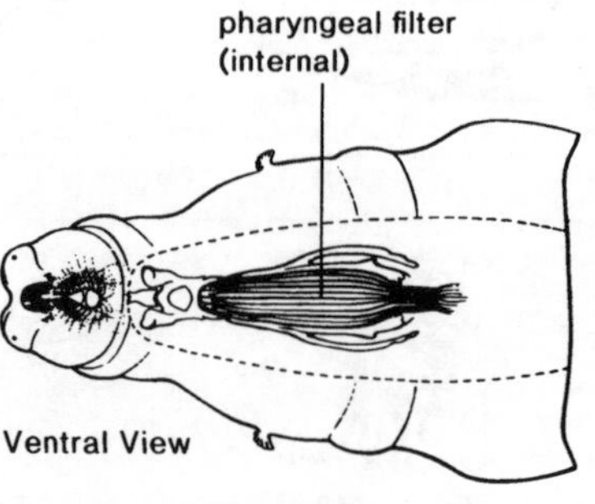

Figure 151

←— Figs. 136, 137, 142, 144–146

93(92′). Posterior spiracle with distinct dorsal spine (fig. 152). Living in damaged or decaying plant material, under bark, or in pine cones .... (part) ***Lonchaeidae*** p. 804

93′. Posterior spiracles lacking dorsal spine. Living in decaying seaweed .... (part) (*Helcomyza*) ***Dryomyzidae*** p. 828

94(91′). Pharyngeal filter present (fig. 151). Living in decaying plant material .... ***Micropezidae*** p. 797

94′. Pharyngeal filter absent 95

95(94′). Body surface smooth, glistening white, lacking tubercles, spines, and spicules. Living in roots of sugar beet .... (part) (*Tetanops*) ***Otitidae*** p. 806

95′. Body surface somewhat wrinkled; margins of terminal segment bearing tubercles. Living in stems of water lilies .... (part) (*Hydromyza*) ***Scathophagidae*** p. 850

96(89′). Anterior spiracle 2-branched (as in fig. 142). Living in decaying plant material .... (part) (*Tanypeza*) ***Tanypezidae*** p. 801

96′. Anterior spiracle unbranched 97

97(96′). Anterior spiracle with papillae either arranged semicircularly around distal part (figs. 153, 154) or projecting laterally along slender basal part (figs. 144, 155) 98

97′. Anterior spiracle with papillae arranged otherwise 99

98(97). Posterior spiracles usually on cylindrical supports (fig. 156) that are 2 or 3 times longer than diameter of each spiracular plate; 3 short, oval spiracular openings located more or less around margin of plate (fig. 157), or asymmetrically positioned in relation to both margins of spiracular plate or ecdysial scar. Pharynx with pharyngeal filter. Living in decaying material .... (part) ***Sphaeroceridae*** p. 839

98′. Posterior spiracles barely elevated to height of less than 1 diameter of spiracular plate (as in fig. 152); spiracular openings radiating from ecdysial scar at approximately 90 degrees to nearest neighbor (as in fig. 148). Pharynx lacking filter. Living in stems of grasses .... ***Opomyzidae*** p. 821

99(97′). Cephalopharyngeal skeleton with accessory oral sclerites (fig. 107), or posterior spiracles either with serpentine spiracular openings (fig. 88) or with openings distinctly angled (fig. 158) or both features present .... (part) ***Muscidae*** p. 857

99′. Neither of these characteristics evident 100

100(99′). Posterior spiracle with thorn-like projection along dorsal margin (fig. 159). Living in or near root nodules of leguminous plants .... (*Rivellia*) ***Platystomatidae*** p. 808

100′. Posterior spiracle lacking thorn-like projection dorsally, or if present, not living in root nodules 101

101(100′). Pharyngeal filter lacking. Living as stem and bud feeders of plants or as predators of beetle larvae under bark .... ***Pallopteridae*** p. 815

101′. Pharyngeal filter usually present (fig. 151). Living in decaying organic matter or as blood suckers of birds 102

102(101′). Living as blood suckers of birds .... ***Carnidae*** p. 825

102′. Not known to suck blood of birds 103

103(102′). Posterior spiracle bearing dorsal thorn-like projection (fig. 152) .... (part) ***Lonchaeidae*** p. 804
(part) ***Otitidae*** p. 806

103′. Posterior spiracle lacking dorsal thorn-like projection 104

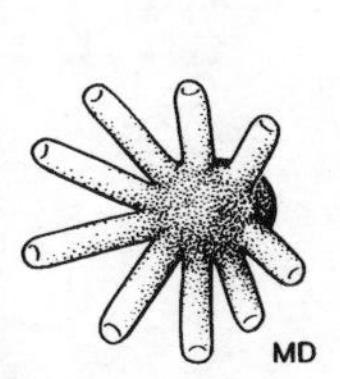

Figure 154

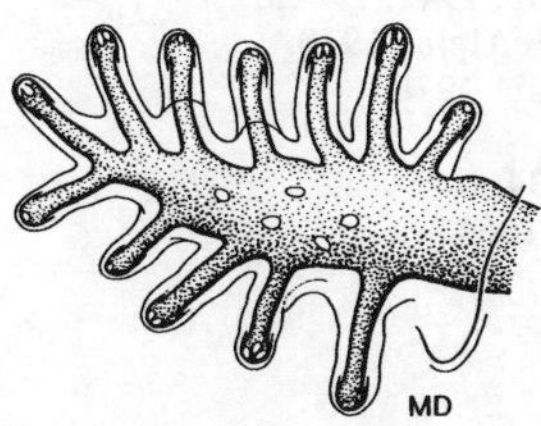

Figure 155

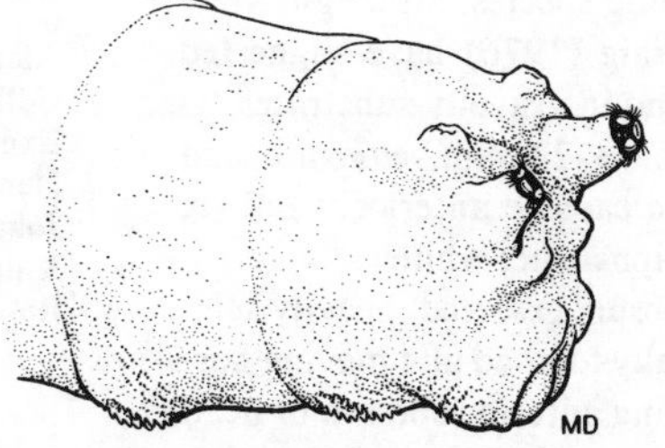

Figure 156

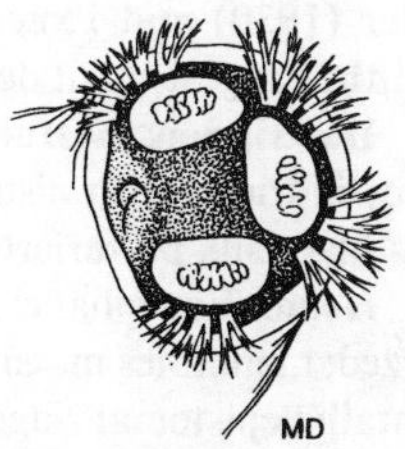

Figure 157

Figs. 88, 107, 152, 158, 159 ⟶

104(103′). Living phytophagously in stems of grasses and sedges .... ***Anthomyzidae*** p. 822
104′. Either not associated with stems of grasses and sedges, or if so, feeding saprophagously as secondary invaders of damaged tissue 105
105(104′). Living within burrows of wood-boring insects in recently felled logs .... ***Odiniidae*** p. 818
105′. Not living in burrows of wood-boring insects. Compare figures, diagnoses, descriptions and biology for: .... ***Cypselosomatidae*** p. 797
Diopsidae p. 803
(part) (*Dryomyza*) ***Dryomyzidae*** p. 828
(part) ***Chloropidae*** p. 847
(part) ***Heleomyzidae*** p. 837
(part) ***Milichiidae*** p. 824
(part) ***Otitidae*** p. 806

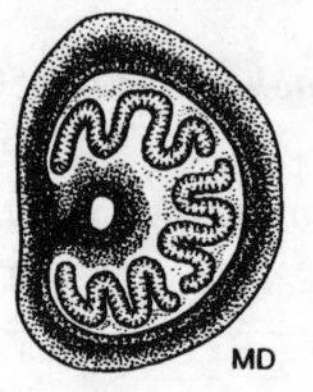

Figure 88

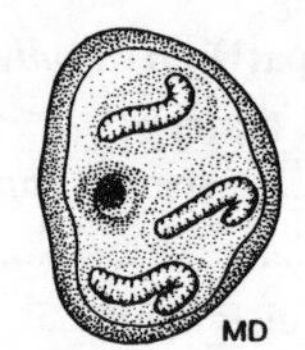

Figure 158

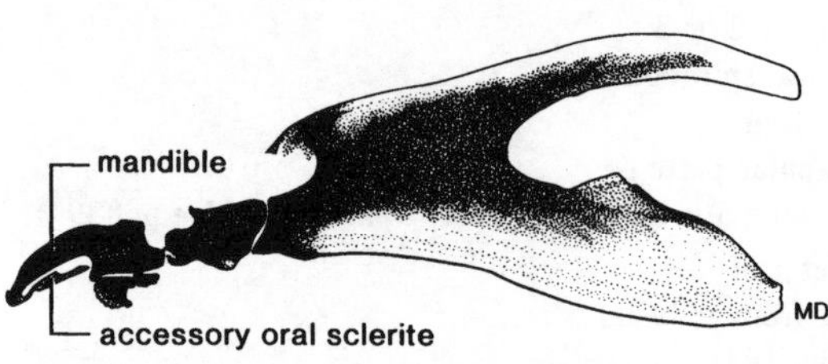

Figure 107

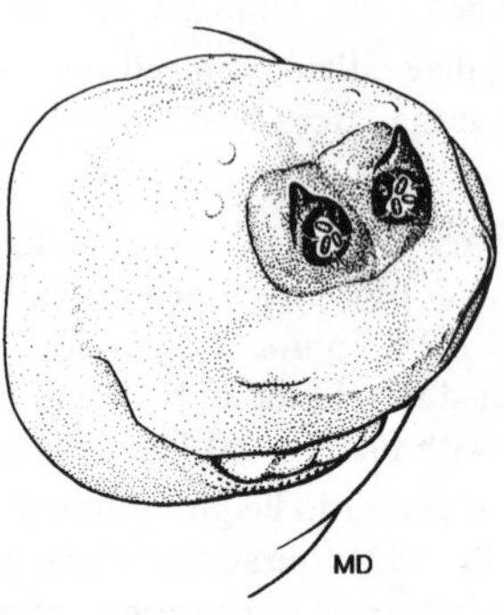

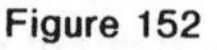
Figure 152

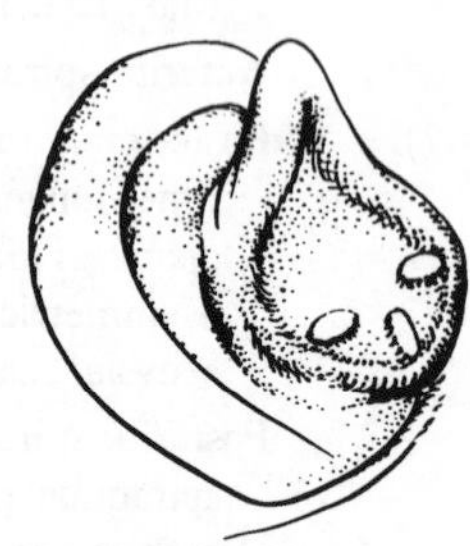
Figure 159

SUBORDER NEMATOCERA

TANYDERIDAE (TANYDEROIDEA)

B. A. Foote, *Kent State University*

The Tanyderids

Figure 37.81

Relationship and Diagnosis: According to McAlpine et al. (1981), the family is so distinct morphologically that it should be placed in a separate superfamily, the Tanyderoidea. This superfamily, in turn, is included in the infraorder Tipulomorpha along with the Tipuloidea. Alexander (1981a) has recently reviewed the family.

The worm-like larvae of Tanyderidae are distinctive due to the six long filaments and pair of prolegs on the caudal segment.

Biology and Ecology: Larvae have been found burrowing in sandy soil along large rivers. Little is known of the feeding habits of the four Nearctic species, although Alexander (1930) and Exner and Craig (1976) have suggested that they ingest plant debris found in gravelly substrates.

Description: Mature larvae 18–22 mm, elongate and subcylindrical, with distinct head capsule anteriorly and six long filaments posteriorly; amphipneustic; white.

Head: Eucephalic; head capsule exserted, heavily sclerotized; mandibles massive, apically toothed and moving horizontally; epistomal ridge bordering anterior dorsum of head with four lobes; antenna short, two-segmented; hypostomium on underside of head with four blunt teeth.

Thorax: Prothorax subdivided to form two rings; anterior spiracles posterolateral on prothorax, spiracles circular with centrally located ecdysial scar, no apparent spiracular openings; remaining segments not subdivided, all segments with scattered hairs.

Abdomen: Segments, except caudal one, very similar, bearing scattered bristles; no prolegs or creeping welts on segments preceding caudal one; last segment bearing three pairs of filaments, four simple, finger-like anal papillae, and a pair of crochet bearing prolegs, each proleg bearing long filament near apex; posterior spiracles placed laterally on penultimate segment.

Comments: According to Alexander (1981a), the world fauna consists of 37 species scattered among 11 genera. Two genera, *Protoplasa* (1 sp.) and *Protanyderus* (3 spp.) occur in America north of Mexico. Three of the species are western, with only *P. fitchii* Osten Sacken occurring in the eastern states and provinces.

Selected Bibliography

Alexander 1930, 1942, 1981a.
Colless and McAlpine 1970.
Exner and Craig 1976.
Hennig 1950.
Johannsen 1934.
Knight 1963, 1964.
Rose 1963.

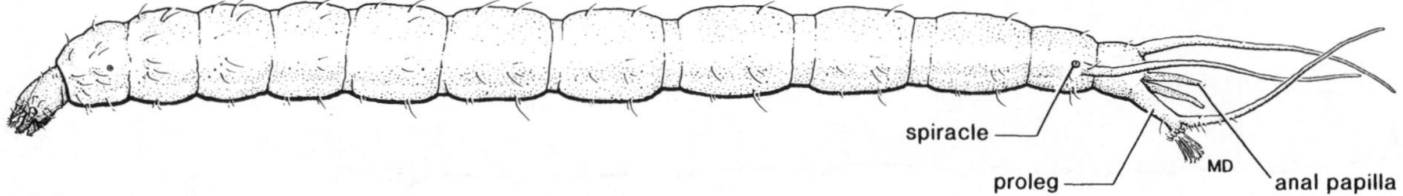

Figure 37.81. Tanyderidae. Lateral of *Protoplasa fitchii* Osten Sacken. Found in sandy soil at stream margins (from Manual of Nearctic Diptera, Vol. 1).

TIPULIDAE (TIPULOIDEA)

B. A. Foote, *Kent State University*

Crane Flies

Figures 37.82–37.98

Relationships and Diagnosis: Long recognized as being one of the most primitive families of Diptera, the Tipulidae are widely considered as being closely related to such nematocerous families as the Tanyderidae, Blephariceridae, and Deuterophlebiidae. Recently, McAlpine, et al. (1981) have placed it as the sole member of the Tipuloidea. Earlier workers, in contrast, believed the crane flies were particularly close to such families as the Anisopodidae and Trichoceridae. Alexander and Byers (1981) have summarized our knowledge of the crane flies.

Larvae can be generally recognized by their elongate, cylindrical shape, the posteriorly incised head capsule (figs. 37.83a, 37.84, 37.85 and 37.89) that is largely embedded in the thorax and may be reduced to several slender rods (fig. 37.85), and by the metapneustic respiratory system (rarely apneustic).

Biology and Ecology: There is no question concerning the evolutionary success of the tipulids as reflected in the extremely wide geographic and habitat distribution of the family. Species are recorded from all of the world's biogeographic regions and from a great diversity of environments, including marine and saline habitats. Probably the greatest number of species is associated with moist temperate regions, particularly those having dense growths of vegetation near streams, marshes, and lakes. However, there are surprising numbers of species in more arid habitats, and some taxa have been recorded from grasslands and sagebrush deserts.

Larvae are amazingly diverse in their utilization of food resources and in their selection of habitats. They have been found abundantly in freshwater, including rapidly flowing streams, marshes, and lake margins; intertidal zones; moist shoreline muds and sands; steep and vertical rock faces over which a thin film of water flows; cushions of moistened mosses and liverworts; dry to saturated wood; organic rich soils; accumulations of decaying plant material; burrows of mammals and bird nests; and in grassland and rangeland soils. Larvae are mostly saprophagous, utilizing decaying organic matter, but many species have become more specialized. Thus, some are herbivores, others are leaf miners, while still others have opted for a predacious way of life.

Description: Mature larvae 3–60 mm, usually elongate and cylindrical, with head capsule incised posteriorly and retractile into the thorax; usually metapneustic, rarely apneustic; integument bare, clothed with short pubescence, or bearing tubercles or fleshy projections; white, gray, green, or nearly black.

Head: Well sclerotized anteriorly but deeply incised posteriorly, capsule reduced to a few narrow rods in some species; antennae usually one-segmented; mandibles opposed and moving horizontally, varying from heavy and massive in many species to very slim and sickle-like in certain predacious forms; labium reduced to anteriorly toothed mentum.

Thorax: Segments very similar, last two may bear ventral creeping welts; no functional anterior spiracles.

Abdomen: Segments commonly bearing ventral (also dorsal in *Antocha*) creeping welts; integument smooth and bare, to wrinkled and leathery, occasionally bearing fleshy projections, commonly covered by fine pubescence; caudal segment usually bearing two conspicuous circular spiracles, spiracles in a few species elongated and pointed (*Erioptera*); spiracular disc usually surrounded by fleshy projections of varying number, shape, and length; anal papillae of varying shape usually present.

Comments: This is the largest family of Diptera, with well over 14,000 species described in the world. Some 1525 species in 64 genera have been recorded from America north of Mexico (Alexander and Byers 1981). It is estimated that less than 10% of the Nearctic larvae are known, and no larval descriptions are available for species belonging to 11 genera. None of the native North American species is considered to have much economic significance, although a few species can be destructive to pasture, sod grasses and rangeland plants. The European crane fly, *Tipula paludosa* Meigen can be a serious pest of grain, turf, vegetable and flower gardens in Newfoundland (Morris, 1986). It is also established on Cape Breton Island, Nova Scotia, near Vancouver, B.C., and in western Washington state and Oregon where it can be a problem in pastures and lawns (Tashiro 1987). It has caused severe damage to a wide variety of crops in Europe.

Selected Bibliography

Alexander 1920, 1922, 1942, 1967.
Alexander and Byers 1981.
Brindle 1957, 1958a, 1958b, 1958c, 1958d, 1958e, 1959a, 1959b, 1959c, 1959d, 1960a, 1960b, 1967.
Brodo 1967.
Byers 1958, 1961a, 1961b, 1975, 1984.
Chiswell 1956.

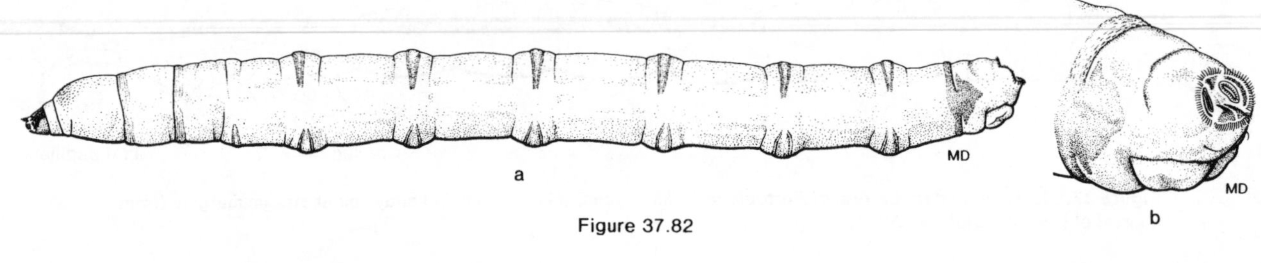

Figure 37.82

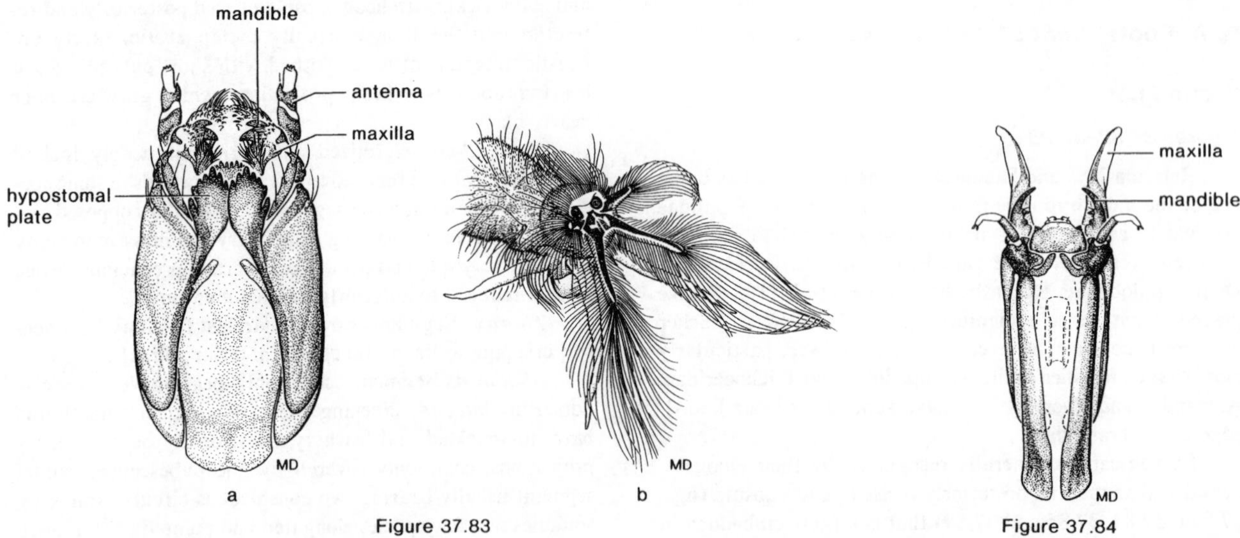

Figure 37.83

Figure 37.84

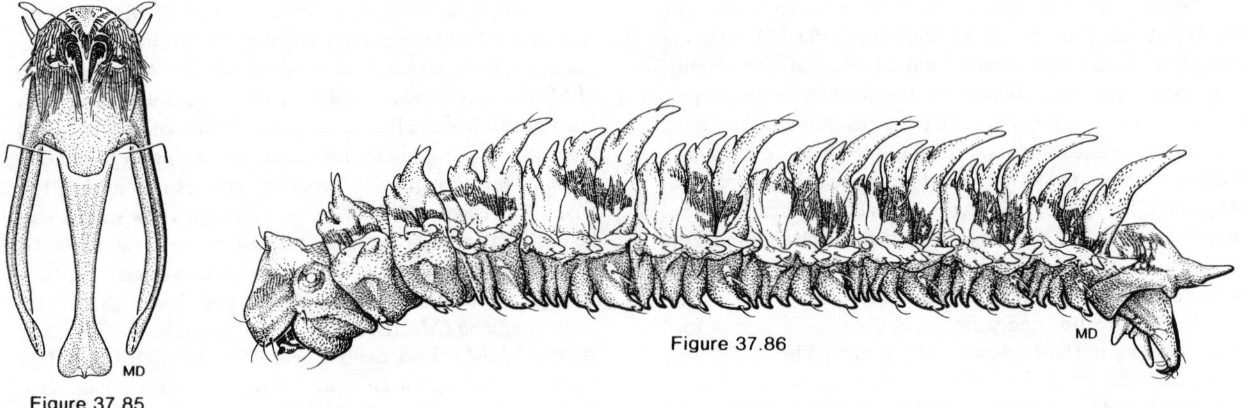

Figure 37.85

Figure 37.86

Figure 37.82a,b. Tipulidae, Limoniinae. **a.** Lateral of *Limonia* sp.; **b.** posterior spiracular disc. The saprophagous larvae are encountered in organically enriched soil (from Manual of Nearctic Diptera, Vol. 1).

Figure 37.83a,b. Tipulidae. *Pseudolimnophila inornata* (Osten Sacken). **a.** head capsule, ventral; **b.** posterior spiracular disc. The predacious larvae commonly occur in wet soils (from Manual of Nearctic Diptera, Vol. 1).

Figure 37.84. Tipulidae. Dorsal of head capsule of *Limnophila* sp., a predator that occurs in wet soils (from Manual of Nearctic Diptera, Vol. 1).

Figure 37.85. Tipulidae. Dorsal of head capsule of *Pilaria recondita* (Osten Sacken), a large predator that occurs in seeps along margins of streams, and in marshy soils (from Manual of Nearctic Diptera, Vol. 1).

Figure 37.86. Tipulidae. Lateral of *Liogma nodicornis* (Osten Sacken), a species commonly found in cushions of terrestrial mosses (from Manual of Nearctic Diptera, Vol. 1).

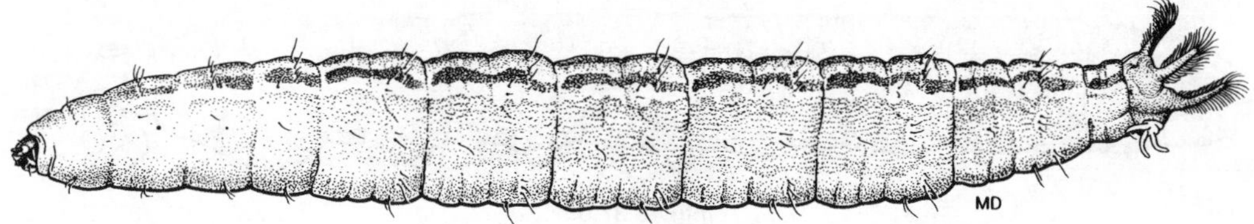

Figure 37.88

Figure 37.87

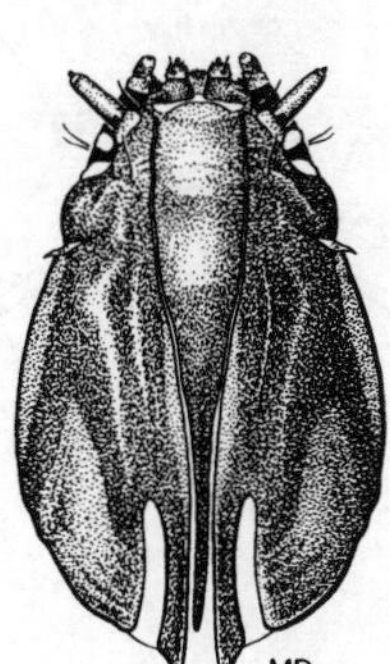

Figure 37.89

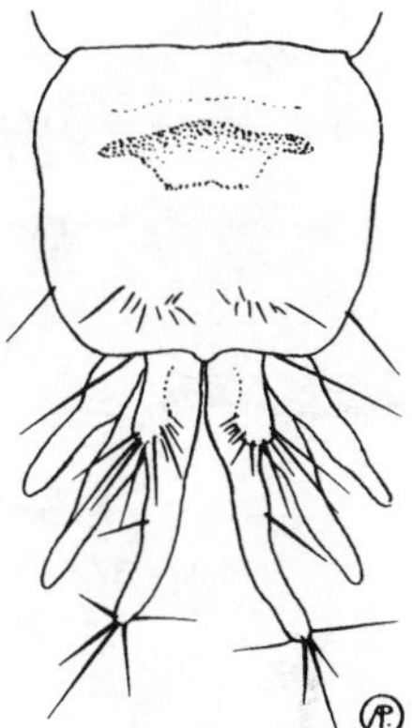

Figure 37.92

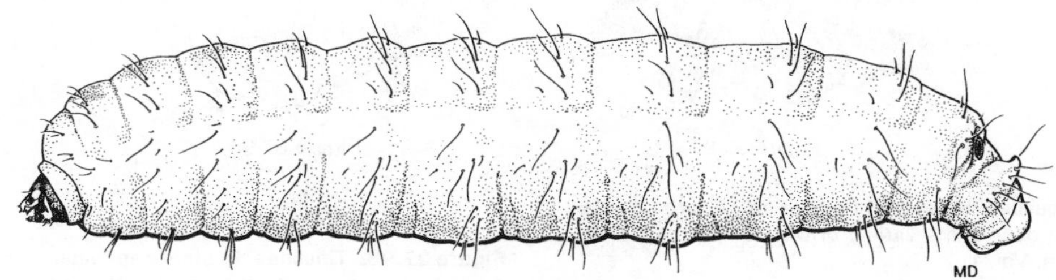

Figure 37.90

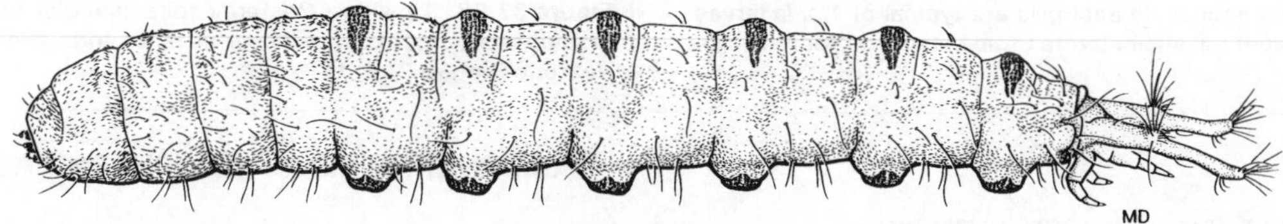

Figure 37.91

Figure 37.87. Tipulidae. Posterior spiracular disc of *Dolichopeza* sp. Larvae feed on terrestrial mosses (from Manual of Nearctic Diptera, Vol. 1).

Figure 37.88. Tipulidae. Lateral of *Prionocera* sp. Larvae are common in organic muds of shallow pools and marshes (from Manual of Nearctic Diptera, Vol. 1).

Figure 37.89. Tipulidae. Dorsal view of head capsule of *Prionocera dimidiata* (Loew). Found in aquatic detritus (from Manual of Nearctic Diptera, Vol. 1).

Figure 37.90. Tipulidae. Lateral of *Ctenophora dorsalis* Walker, a heavy-bodied larva that occurs in newly fallen logs (from Manual of Nearctic Diptera, Vol. 1).

Figure 37.91. Tipulidae. Dorsolateral of *Antocha* sp. A relatively common species in riffles of rapidly flowing streams (from Manual of Nearctic Diptera, Vol. 1).

Figure 37.92. Tipulidae. Dorsal view of caudal segment of *Antocha saxicola* Osten Sacken (from Peterson 1951).

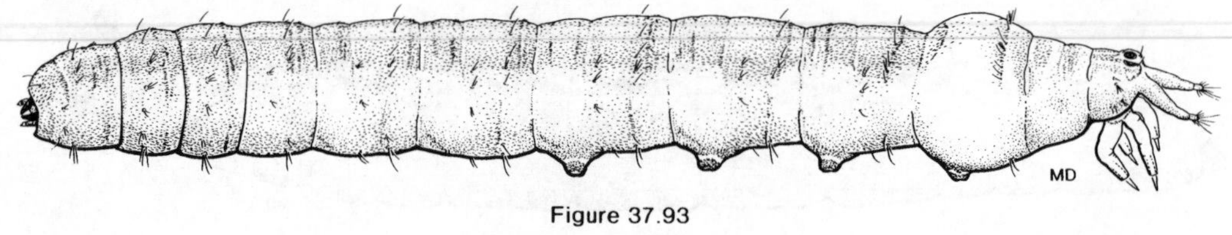

Figure 37.93

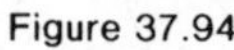

Figure 37.94

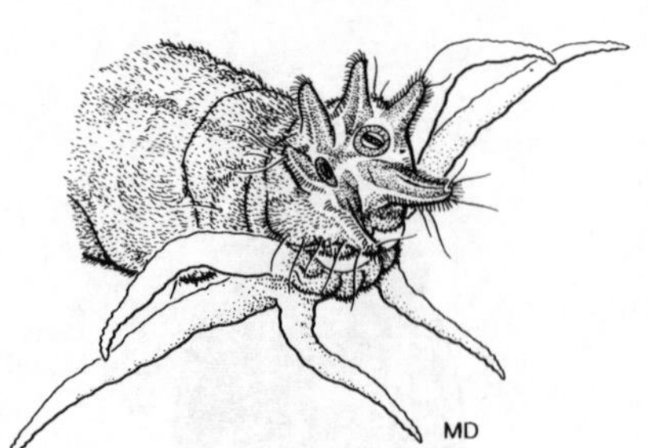

Figure 37.95

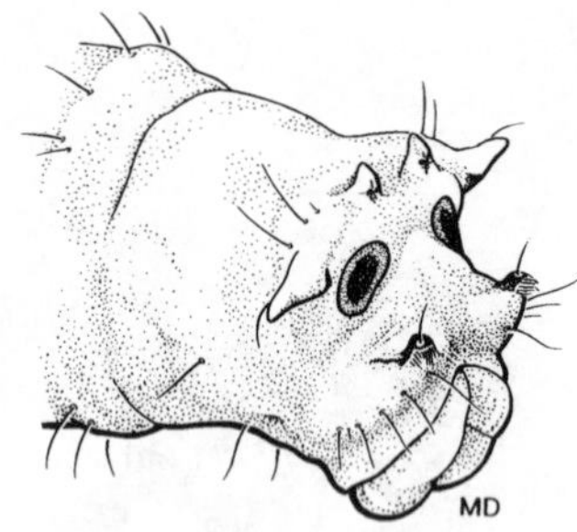

Figure 37.96

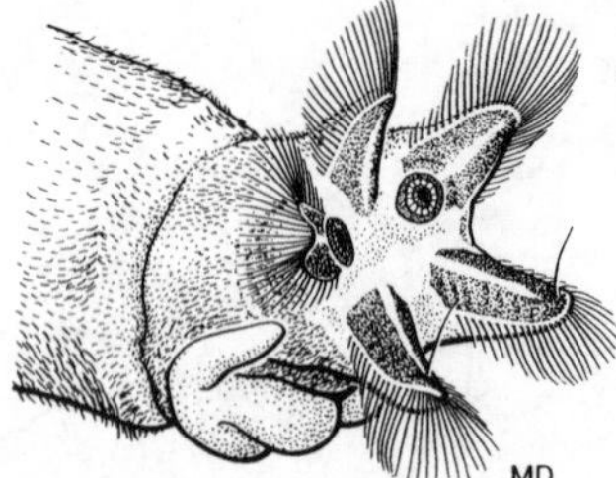

Figure 37.97

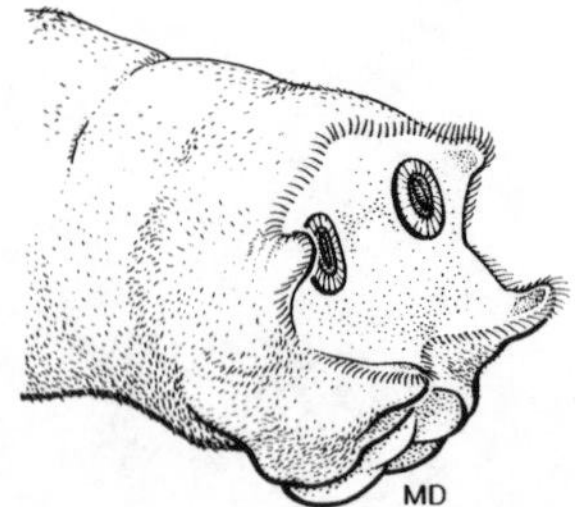

Figure 37.98

Figure 37.93. Tipulidae. Lateral of *Pedicia* sp., a large predacious species that occurs in a variety of wet soils (from Manual of Nearctic Diptera, Vol. 1).

Figure 37.94. Tipulidae. Posterior spiracular disc of *Tipula trivittata* Say, a common larva under bark of decaying logs (from Manual of Nearctic Diptera, Vol. 1).

Figure 37.95. Tipulidae. Posterior spiracular disc of *Tipula strepens* Loew. The noticeable anal gills are typical of *Tipula* larvae that occur in more aquatic habitats (from Manual of Nearctic Diptera, Vol. 1).

Foote 1963.
Hennig 1950.
Hynes 1958, 1963, 1965, 1968, 1969a, 1969b, 1969c.
Peterson, A. 1951.
Pritchard and Hall 1971.
Rogers 1926a, 1926b, 1927a, 1927b, 1927c, 1928, 1930, 1932, 1933a, 1933b, 1942, 1949.
Rogers and Byers 1956.
Saunders 1928.
Theowald 1957, 1967.
Tokunaga 1930.

Figure 37.96. Tipulidae. Posterior spiracular disc of *Ctenophora angustipennis* Loew (from Manual of Nearctic Diptera, Vol. 1).

Figure 37.97. Tipulidae. Posterior spiracular disc of *Ormosia* sp. Larvae are usually found in moist soils of wooded habitats (from Manual of Nearctic Diptera, Vol. 1).

Figure 37.98. Tipulidae. Posterior spiracular disc of *Epiphragma fascipennis* (Say). Larvae occur in fallen logs (from Manual of Nearctic Diptera, Vol. 1).

BLEPHARICERIDAE (BLEPHARICEROIDEA)

B. A. Foote, *Kent State University*

Net-Winged Midges

Figures 37.99–37.101

Relationships and Diagnosis: The Blephariceridae is presently considered to be the sole Nearctic representative of the Blephariceroidea of the infraorder Blephariceromorpha which also includes the Deuterophlebioidea and Nymphomyioidea (McAlpine et al. 1981). Alexander (1963) and Hogue (1981) have summarized our knowledge of the midges.

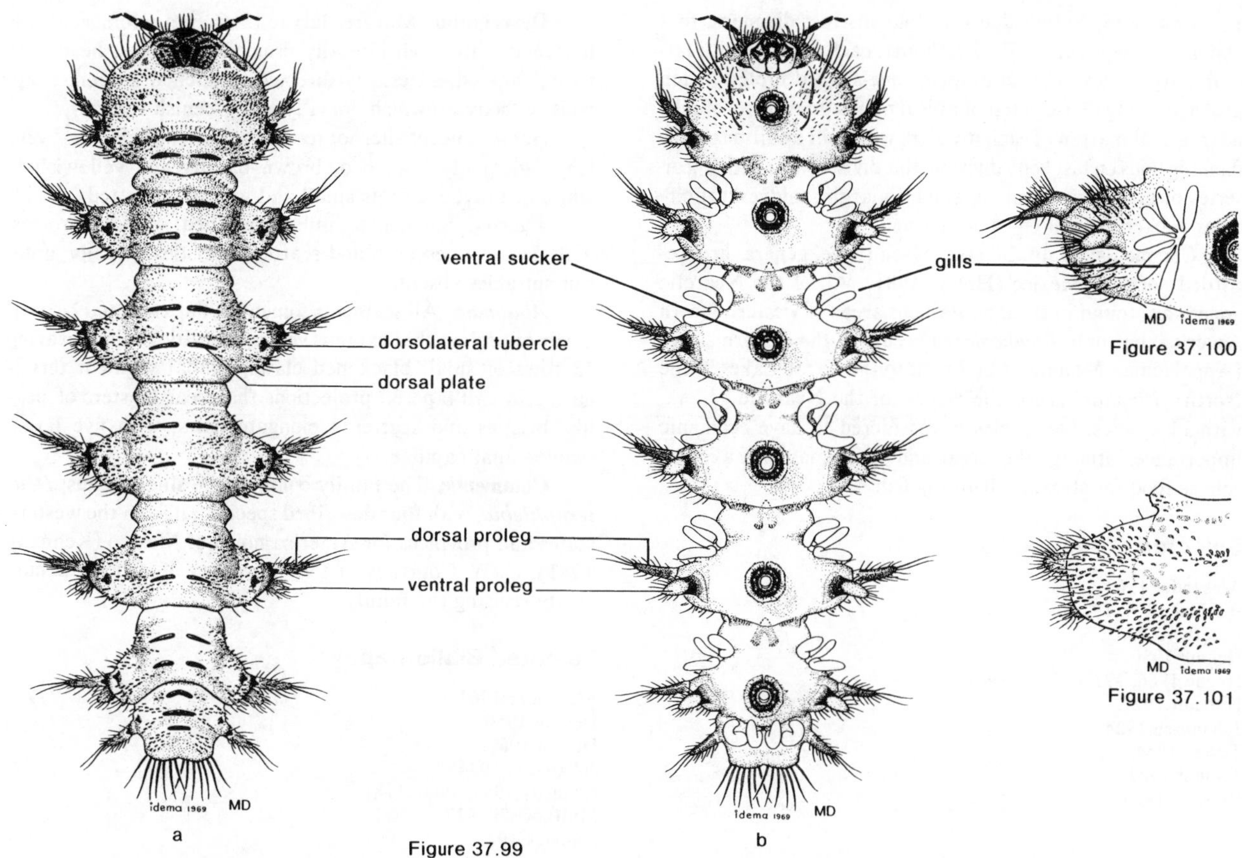

Figure 37.99a,b. Blephariceridae. *Philorus californicus* Hogue. **a.** dorsal; **b.** ventral. Larvae occur on rocks in fast-flowing mountain streams of the West (from Manual of Nearctic Diptera, Vol. 1).

Figure 37.100. Blephariceridae. Ventral of first abdominal division of *Bibliocephala grandis* Osten Sacken. Larvae occur on rocks in riffle areas of streams throughout the more mountainous areas of western North America (from Manual of Nearctic Diptera, Vol. 1).

Figure 37.101. Blephariceridae. Ventral of first abdominal segment of *Blepharicera tenuipes* (Walker). Larvae of this *Blepharicera* are found on riffle rocks in the more mountainous areas of North America (from Manual of Nearctic Diptera, Vol. 1).

The stream-inhabiting larvae are easily recognized by the complete head capsule that is fused with the thorax, the six distinct body divisions, and the six suctorial discs along the ventral side (fig. 37.99b).

Biology and Ecology: Adults are restricted to the vicinity of rapidly flowing streams and waterfalls. Females of species possessing mandibles are known to feed on other soft-bodied insects, but the feeding habits of the non-mandibulate species are unknown. Adults are thought to live only one to two weeks, with flight being largely restricted to the daylight hours. Eggs are firmly attached in small clusters to the surface of rocks that have been exposed by a dropping water level. The flattened larvae have ventral suckers that maintain close contact with the substrate, allowing them to avoid being washed downstream by the swift current. Larvae subsist largely on diatoms and perhaps other microorganisms that compose the epilithon on the surface of riffle rocks. Pupae are also firmly attached to the rock surface.

Description: Mature larvae 5–12 mm, subcylindrical, flattened on ventral side; body subdivided, with deep lateral constrictions and lobes, circular sucker along midventral line on each body division, tufts of gills on each division behind cephalic one, tufts of 3–5 branches; body dark dorsally, with integument bearing numerous, variously arranged spicules, sclerotized plates and projections, whitish ventrally.

Head: Fused with thorax and first abdominal segment to form cephalic division of body; head capsule with large opening ventrally that encloses small mouthparts; mandibles and maxillae well developed and adapted for grazing diatoms and algae from surface of rocks; antennae short to somewhat elongated, composed of two or three segments; circular sucker present midventrally.

Thorax: Forming part of cephalic division; lateral margins downturned to form cone-shaped prolegs, apex of each proleg with dense patch of fine pubescence; gill tufts present near each lateral margin in front of prolegs; circular sucker present midventrally.

Abdomen: Subdivided into four distinct divisions, first abdominal segment fused with thorax; each division with lateral margins downturned to form cone-shaped prolegs, each proleg bearing dense patch of apical pubescence; tufts of gills near lateral margin of each division, with four additional gills flanking anus on last body division, this division also with linear series of large bristles along caudal margin; midventral circular sucker present on each division.

Comments: Only 24 species and five genera are recorded north of Mexico (Hogue 1981). All of the Nearctic genera are found in the mountainous areas of western North America, but only *Blepharicera* occurs in the eastern states (Appalachian Mountains and west to the Great Lakes in the North). *Blepharicera* is the largest of the Nearctic genera, with 11 species. No species is considered to have economic importance, although the larvae and pupae may play a minor role as food for stream-inhabiting fish.

Selected Bibliography

Alexander 1963.
Bischoff 1924, 1928.
Feuerborn 1932.
Hennig 1950.
Hogue 1966, 1973b, 1978, 1981.
Hubault 1927.
Johannsen 1934.
Teskey 1984.
Tonnoir 1923.
Vimmer 1925.

DEUTEROPHLEBIIDAE (DEUTEROPHLEBIOIDEA)

B. A. Foote, *Kent State University*

Mountain Midges

Figures 37.102a,b

Relationships and Diagnosis: Thought to be related to the Blephariceridae and Nymphomyiidae, the Deuterophlebiidae are placed in the infraorder Blephariceromorpha and is the sole Nearctic family of the Deuterophlebioidea (McAlpine et al. 1981). Kennedy (1981) has summarized the current knowledge.

Larvae are easily recognized by their small size (up to 6 mm), eucephalic condition, apneustic respiratory system, and flattened body that bears seven pairs of ventrolateral crochet-bearing prolegs (figs. 37.102a,b). The head bears a pair of very long, branched antennae.

Biology and Ecology: Adults of the Nearctic species have been found only near rapidly flowing, well-aerated mountain streams of western North America. Adults fly early in the morning and live only a few hours during which mating and oviposition take place. Larvae occur in fast flowing water (usually exceeding 75 cm/s) on the upper surface of relatively smooth rocks that contain cracks and depressions. They seemingly prefer shallow areas in the stream and have been rarely encountered in water more than 30 cm deep. Larval feeding habits apparently are unknown. Pupae are found in the same habitats occupied by the larvae, although they are often formed in slight depressions in the rock surface.

Description: Mature larvae 3–6 mm, dorsoventrally flattened with seven laterally projecting prolegs; head with two elongated, anteriorly directed, branched antennae; apneustic; body brownish dorsally and greenish ventrally.

Head: Eucephalic, not retractile; flattened antennae very long, unequally branched, brown basally and yellowish to white distally; eye spots small and widely separated.

Thorax: Segments similar, flattened, without prolegs, with dense pubescence and scattered bristles dorsally; anterior spiracles absent.

Abdomen: All segments somewhat flattened and bearing a pair of prolegs that extends ventrolaterally; prolegs bearing 13 rows of small blackened claws; caudal segment terminating in two tapered projections that bear clusters of peglike bristles and scattered elongate hairs, with five large, swollen anal papillae.

Comments: The family contains the single genus, *Deuterophlebia,* with four described species, all from the western states and provinces for America north of Mexico (Kennedy 1981). G. W. Courtney of the University of Alberta is currently revising the family.

Selected Bibliography

Alexander 1963.
Hennig 1950.
Hinton 1962.
Johannsen 1934.
Kennedy 1958, 1960, 1981.
Muttkowski 1927.
Pennack 1945.

NYMPHOMYIIDAE (NYMPHOMYIOIDEA)

B. A. Foote, *Kent State University*

The Nymphomyiids

Figures 37.103a,b

Relationships and Diagnosis: The taxonomic placement of the Nymphomyiidae has been debated repeatedly by a variety of investigators. McAlpine, et al. (1981), recognizing the distinctness of the family, placed it in its own superfamily, the Nymphomyioidea, a member of the infraorder Blephariceromorpha. Kevan and Cutten (1981) have reviewed the knowledge of this small family.

The tiny, whitish, stream-inhabiting larvae (figs. 37.103a,b) are easily recognized by their elongate, cylindrical body, nine pairs of elongate prolegs, and exserted head capsule. The antennae project forward, and the prolegs bear a crown of spines apically.

Biology and Ecology: As far as known, larvae and pupae are restricted to streams in which the temperature remains relatively low throughout the year. Larvae have been found in a variety of aquatic mosses in riffles and on the surfaces of diatom-covered rocks. The elongate, ventrally directed prolegs allow larvae to cling to a substrate in rapidly flowing currents. Larvae are believed to feed primarily on algae, microscopic plants, and detritus. Adults have atrophied mouthparts and do not feed. Shortly after emerging, adults

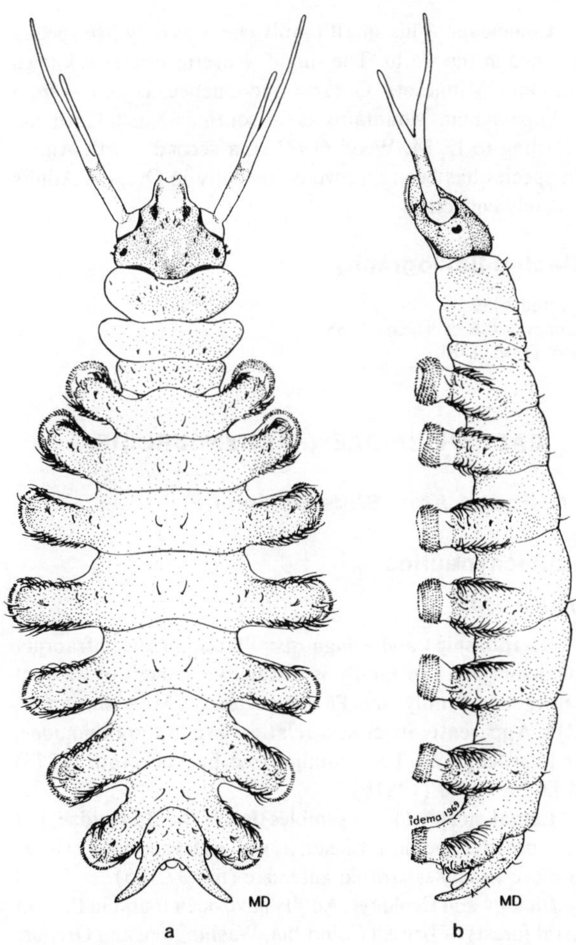

Figure 37.102a,b. Deuterophlebiidae. *Deuterophlebia nielsoni* Kennedy. **a.** dorsal; **b.** lateral. Found in swift mountain streams (from Manual of Nearctic Diptera, Vol. 1).

break off their wings at a line of weakness. It is not known whether the Nearctic species can actually fly. Wingless, mating pairs of *Paleodipteron walkeri* Ide have been observed underwater clinging to the substrate with their enlarged tarsal claws. Eggs occurring in batches of up to 50 remain attached to the body of the female after her death.

Description: Mature larvae 1–3 mm, elongate and cylindrical, with eight pairs of ventrally projecting prolegs; apneustic; white.

Head: Eucephalic, exserted, brownish; antenna elongate, three-segmented, apparently biramous and projecting forward; labrum conical, bearing numerous setae; mandible short and stout; maxilla somewhat saw-like, with two heavy spines projecting medially; labium flat, broad, and toothed along apical margin, bearing broad paralabial plates laterally.

Thorax: Segments similar, lacking prolegs, no anterior spiracles; integument with scattered bristles.

Abdomen: First seven segments bearing pair of slender, elongate, ventrally projecting prolegs with apical crowns of curved spines; penultimate segment lacking prolegs; caudal segment with pair of apparently two-segmented, tapering prolegs that terminate in a pair of pectinate plates and two

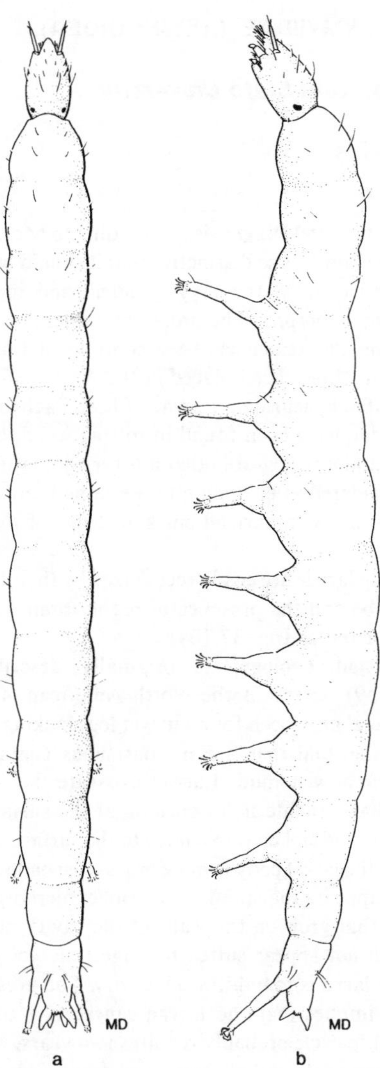

Figure 37.103a,b. Nymphomyiidae. *Palaeodipteron walkeri* Ide. **a.** dorsal; **b.** lateral. The very small, distinctive larvae occur in mats of aquatic mosses or on the surface of rocks in fast-flowing streams (from Manual of Nearctic Diptera, Vol. 1).

hook-like structures, apex of segment with four anal papillae; all segments bearing scattered bristles.

Comments: This is a small but widely distributed family containing only three genera (Kevan and Cutten 1981). The genus *Nymphomyia* occurs in Japan and the Soviet Far East, *Felicitomyia* is known only from the Himalayas, and *Palaeopteron* is Nearctic. In America north of Mexico, the single species, *P. walkeri,* has been collected in a few eastern provinces and states.

Selected Bibliography

Adler et al. 1985.
Cutten and Kevan 1970.
Ide 1965.
Kevan and Cutten 1975, 1981.
Merritt and Schlinger 1984.
Mingo and Gibbs 1976.
Teskey 1984.
Tokunaga 1935.

AXYMYIIDAE (AXYMYIOIDEA)

B. A. Foote, *Kent State University*

The Axymyiids

Figure 37.104

Relationships and Diagnosis: According to McAlpine et al. (1981), this family is so distinctive that it should be placed in its own superfamily, the Axymyioidea, and in the infraorder Axymyiomorpha. The proper placement of the species comprising this taxon has been controversial and they have, at various times, been placed in the Anisopodidae, Bibionidae, and Pachyneuridae. Larvae of both Pachyneuridae and Axymyiidae have been found in rotting wood, but they are so different morphologically that it is very doubtful if they should be considered as belonging to the same family. D. M. Wood (1981a) has summarized our knowledge of this small family.

The plump larvae are easily recognized by their elongate respiratory tube and the presence of highly branched, conspicuous anal papillae (fig. 37.104).

Biology and Ecology: As originally described by Krogstad (1959), larvae of the North American *Axymyia furcata* McAtee, have been found in wet logs lacking bark or a moss covering, and that are in continuous contact with standing water or wet mud. Larvae excavate flask-shaped cavities that have minute holes opening at the surface. The elongated breathing siphon is extended to the surface through the hole while the larval body is more or less suspended within the cavity. It appears that the larvae utilize microorganisms and/or fungi that grow on the walls of the cavity, since the mouthparts do not appear suited for ingesting wood. Typically only one larva is found in each cavity. The presence of anal papillae implies that the larvae can survive total immersion. The life cycle probably requires two years, with pupation occurring in the larval cavity, and adults emerging in late April and early May.

Description: Mature larvae 3–8 mm, body plump, with partially invaginated head capsule, an elongate terminal breathing tube, and highly branched anal papillae posteriorly; amphipneustic; white.

Head: Partially invaginated into prothorax with only exposed portion heavily sclerotized; antenna merely a sensory pit on cranium near base of each mandible; labrum small and conical, bearing setae ventrally; mandible conical, apex strongly sclerotized, bearing two teeth; maxilla greatly reduced, palpus two-segmented; labium small and finger-like; cranium with rows of bristles.

Thorax: Segments similar, bearing scattered bristles; anterior spiracles circular, lateral on prothorax.

Abdomen: First five segments with two inconspicuous rows of small spinules dorsally and ventrally; caudal segment with a greatly elongated respiratory tube; basal part of tube white and flexible, distal part sclerotized and with ring-like thickenings, apex crowned with five spine-tipped tubercles, two dorsolateral tubercles bearing small, black spiracles; two highly branched, elongate anal papillae, each papilla with three rows of branches.

Comments: This small family contains only five species described in the world. The single Nearctic species is known from Ohio, Minnesota, Ontario, and Quebec, as well as from the Appalachian Mountains as far south as North Carolina. According to D. M. Wood (1981a), a second North American species has been discovered recently in Oregon. Adults are rarely collected.

Selected Bibliography

Krogstad 1959.
Mamaev and Krivosheina 1966.
Wood, D. M. 1981a.

PACHYNEURIDAE (PACHYNEUROIDEA)

B. A. Foote, *Kent State University*

The Pachyneurids

Figure 37.105

Relationships and Diagnosis: Placed in the infraorder Bibionomorpha, the family is distinctive enough to warrant its own superfamily, the Pachyneuroidea (McAlpine et al. 1981). Apparently its closest relatives are in the Bibionidae. Our knowledge has been summarized by Vockeroth (1974) and D. M. Wood (1981b).

Larval morphology resembles that of the Bibionidae, but with a broader and flatter head, as well as larger body bristles and more widely separated antennae (fig. 37.105).

Biology and Ecology: Adults have been found in the wet coastal forests of British Columbia, Washington, and Oregon. Emergence occurs during late winter, and eggs are placed on decaying wood of red alder (*Alnus rubra*). Larvae burrow just under loosened bark or in superficial layers of rotting wood. Larval development apparently requires about a year, with pupae being formed in the larval burrows.

Description: Mature larvae 5–8 mm, elongate, with distinct head capsule and strong creeping welts on thorax; integument bearing numerous long bristles; peripneustic; white.

Head: Eucephalic, not retractile into prothorax; very broad and flattened dorsally; labrum short and somewhat bilobed; antennae widely separated, two-segmented with distal segment minute; mandible heavy, broad at base and constricted at midlength, ending in three or four teeth; maxilla with median lobe, palpus small and two-segmented; labium greatly reduced; head capsule bearing scattered bristles.

Thorax: Pro- and mesothorax with lightly sclerotized areas dorsally, all three segments with strong creeping welts ventrally; segments bearing numerous elongate bristles; anterior spiracles large, posterolaterally on prothorax, spiracles subcircular to oval and with distinct peritreme.

Abdomen: Segments very similar, bearing numerous long bristles; first seven segments bearing small, laterally placed spiracles that are about one-third the size of prothoracic spiracles; last segment with much larger, dorsolateral spiracles; segments without creeping welts.

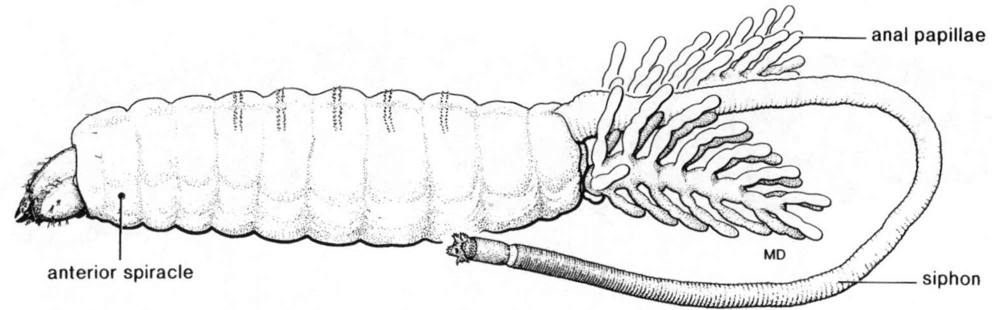

Figure 37.104. Axymyiidae. Lateral of *Axymyia furcata* McAtee. Larvae occur in cavities of rotting logs that are saturated with water (from Manual of Nearctic Diptera, Vol. 1).

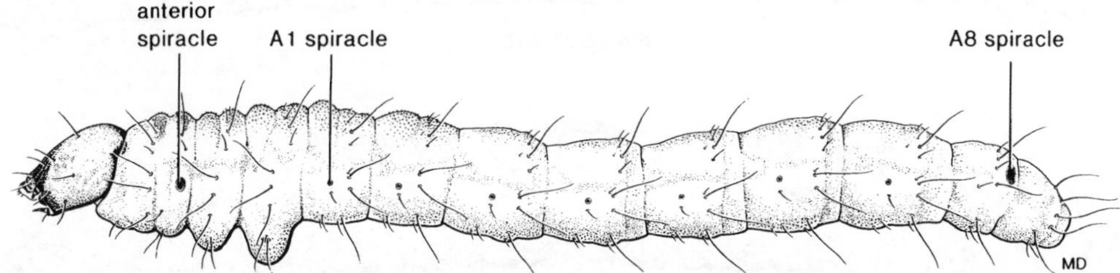

Figure 37.105. Pachyneuridae. Lateral of *Cramptonomyia spenceri* Alexander. Larvae are found in decaying alder logs in the wet coastal forests of the Pacific Northwest (from Manual of Nearctic Diptera, Vol. 1).

Comments: This is another small family, with only four species recorded for the world (Wood 1981b). Species are known from Europe, Asia, Japan, eastern Siberia, and western North America. The single described Nearctic species is *Cramptonomyia spenceri* Alexander from the western states.

Selected Bibliography

Krivosheina and Mamaev 1970.
Vockeroth 1974.
Wood, D. M. 1981b.

BIBIONIDAE (BIBIONOIDEA)

B. A. Foote, *Kent State University*

March Flies

Figures 37.106–37.111

Relationships and Diagnosis: The Bibionidae have long been recognized as a distinct taxon; this is reflected in the arrangement of McAlpine et al. (1981) who placed them as the sole member of the Bibionoidea, a group included in the infraorder Bibionomorpha along with the Pachyneuroidea and Sciaroidea. Our knowledge of this family has been summarized by Hardy (1981).

Larvae can be recognized by their pigmented, partially retractile head capsule, holopneustic respiratory system, and the tuberculous or spinose processes on the integument (figs. 37.106–37.109).

Biology and Ecology: Adults of North American species usually emerge during the spring and early summer months (giving rise to the common name, "March flies"). They frequently emerge in massive numbers and form swarms consisting of males whose swirling flight patterns attract females. Adults seemingly are short lived, surviving only a few days during which they may feed on pollen and nectar. Mated females excavate burrows in moist, friable soil in which egg masses of 200–300 eggs are laid. The larvae are mostly scavengers on decaying organic matter and are particularly abundant in moist soils containing high concentrations of decaying leaf litter. Larvae of some species apparently are more phytophagous, as they feed on roots of living plants. Although the life histories of the Nearctic species are poorly known, it is suspected that most of the life is spent in the larval stage and that overwintering occurs as larvae in soil.

Description: Mature larvae 6–25 mm, subcylindrical, slightly flattened dorsally with distinct retractile head capsule; integument usually roughened or with numerous fleshy projections, may also have spinules; little or no development of creeping welts; holopneustic; white to brownish.

Head: Eucephalic, with well sclerotized and distinctly pigmented, retractile capsule that is slightly longer than broad; bearing several scattered bristles; antenna small and setose; mandible heavily sclerotized, bearing distinct teeth apically; maxilla short and relatively broad, bearing several spine-like bristles, palpi ending in similar array of heavy bristles; labium variable, usually bilobed anteriorly.

Thorax: Spiracles lacking on mesothorax, those of prothorax large, while spiracles of metathorax are much smaller; spiracles dark, circular, with distinct peritremes; segments commonly with rows of fleshy projections and spinules.

Abdomen: All segments except penultimate one with lateral spiracles, with caudal pair largest; frequently with rows of fleshy projections but without prolegs or ventral creeping

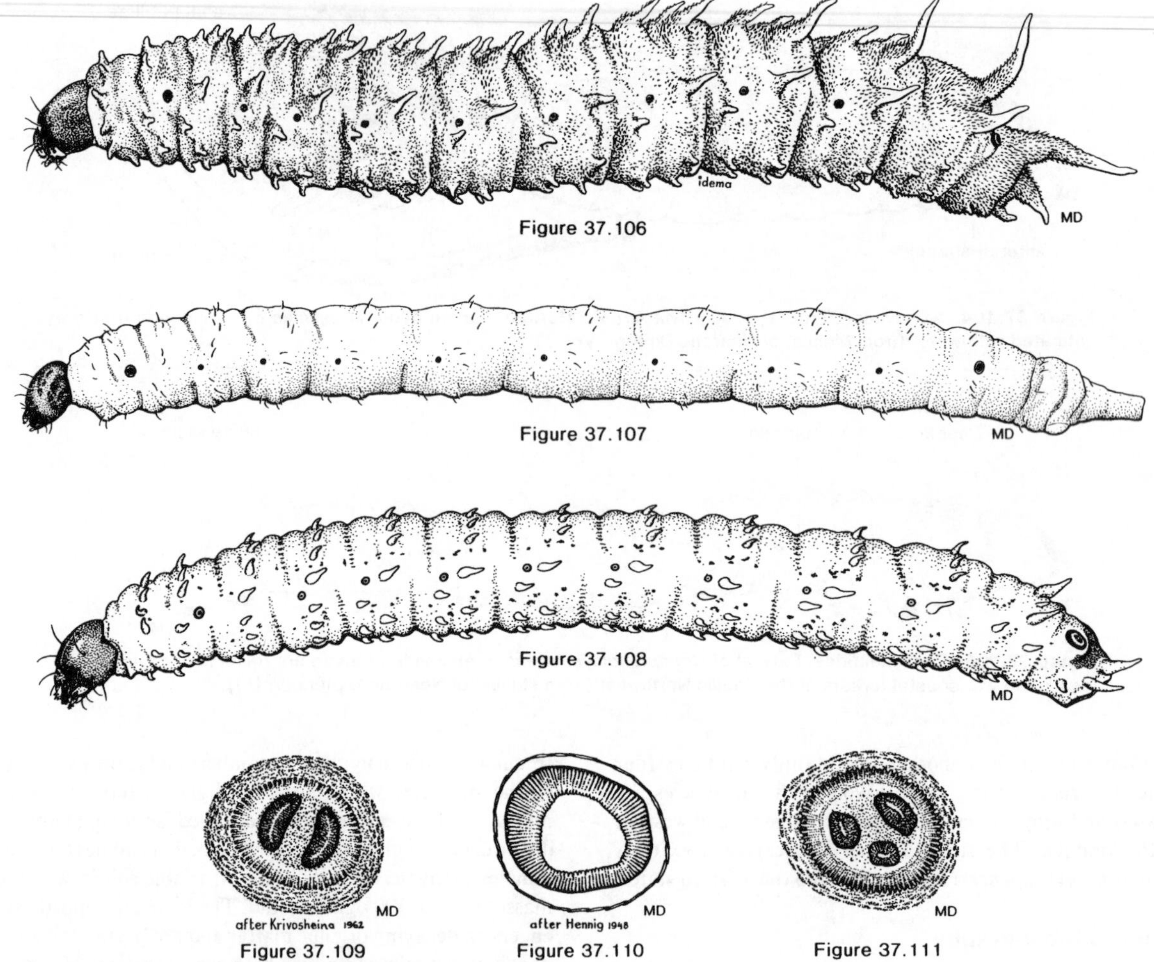

Figure 37.106. Bibionidae. Lateral of *Plecia nearctica* Hardy. The saprophagous larvae are usually encountered in soils containing high concentrations of organic matter. The adults are the famed "love bugs" of Florida (from Manual of Nearctic Diptera, Vol. 1).

Figure 37.107. Bibionidae. Lateral of *Hesperinus rohdendorfi* Krivosheina. This species does not occur in North America, but a closely related species, *H. brevifrons* Walker, has been recorded from the Rocky Mountains. The larval stages are thought to inhabit humus-rich soils (from Manual of Nearctic Diptera, Vol. 1).

Figure 37.108. Bibionidae. *Bibio* sp., lateral. The saprophagous larvae are frequently abundant in organically enriched soils, particularly in woodlands (from Manual of Nearctic Diptera, Vol. 1).

Figures 37.109–37.111. Bibionidae. Posterior spiracles: **37.109.** *Bibio* sp.; **37.110.** *Penthetria* sp.; **37.111.** *Dilophus* sp. (from Manual of Nearctic Diptera, Vol. 1).

welts; caudal segment with pair of large posterior spiracles on dorsum; spiracles circular with distinct peritremes and one to three openings.

Comments: The family contains about 700 described species in the world, and is best developed in the tropics (Hardy 1981). The fauna north of Mexico consists of 78 species and six genera in three subfamilies. Because larvae of several of the Nearctic species feed on roots and rootlets, bibionids are of some significance economically as pests of cereals, vegetables, forage crops, and pastureland plants. The lovebug, *Plecia nearctica* Hardy, is famous throughout the southeastern states for its massive mating swarms that can quickly darken the windshields of motorists.

Selected Bibliography

Brauns 1954.
Brindle 1962a.
Callahan and Denmark 1973.
Hardy 1945, 1958, 1981.
Hennig 1948.
Keilin 1919c.
Krivosheina 1962.
Malloch 1917.
Morris 1921, 1922.
Needham 1902.
Strickland 1916.
Thornhill 1976.

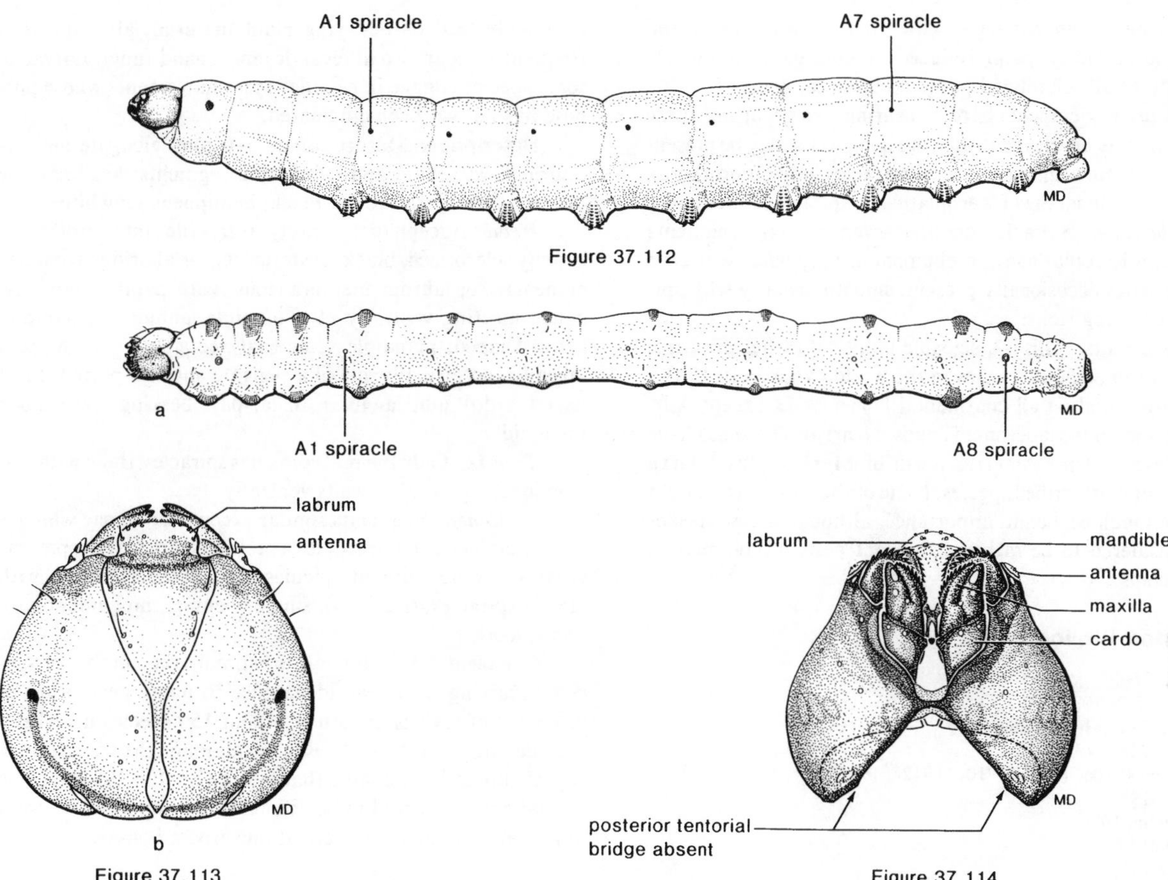

Figure 37.112. Mycetophilidae. Lateral of *Mycetophila* sp. Larvae of this genus frequently are abundant in fleshy and woody fungi (from Manual of Nearctic Diptera, Vol. 1).

Figure 37.113a,b. Mycetophilidae. *Symmerus coqulus* Garrett. **a.** lateral; **b.** head capsule, dorsal (from Manual of Nearctic Diptera, Vol. 1).

Figure 37.114. Mycetophilidae. *Mycetophila fisherae* (Laffoon). Ventral of head capsule (from Manual of Nearctic Diptera, Vol. 1).

MYCETOPHILIDAE (SCIAROIDEA)

B. A. Foote, *Kent State University*

Fungus Gnats

Figures 37.112–37.114

Relationships and Diagnosis: Presently included in the infraorder Bibionomorpha, the families closest to the Mycetophilidae are the Sciaridae and Cecidomyiidae (McAlpine, et al. 1981). Vockeroth (1981) has summarized our knowledge of this family.

Larvae usually can be recognized by their small, elongate, white bodies bearing a dark brown or black exserted head capsule. The body hairs are inconspicuous and the integument appears smooth. The respiratory system is peripneustic or hemipneustic. The larvae are distinguished from those of Sciaridae by the absence of a posterior bridge in the head capsule (fig. 37.114).

Biology and Ecology: Adult mycetophilids are most commonly encountered in moist, woodland habitats containing large amounts of decaying organic matter. They are relatively inactive during the daylight hours, and most adult activity probably takes place during times of reduced light intensity. Larvae of most North American species feed on fleshy or woody fungi but can also be found under bark of dead and dying trees and in bird nests. Some species have more specialized trophic habits. Thus, at least one species of *Boletina* feeds on liverworts, while larvae of some species of the Keroplatinae spin webs in which they capture small arthropods. Larvae of many species spin a cocoon for pupation.

Description: Mature larvae 3–13 mm, elongate to very elongate (nearly worm-like), cylindrical to strongly flattened dorsoventrally; hemipneustic or peripneustic, with 8–9 pairs of laterally placed spiracles, or functionally apneustic (Keroplatinae); creeping welts present or absent; integument bare and subshining, occasionally bearing small and inconspicuous bristles; white, with dark head capsule.

Head: Eucephalic, poorly or not retractile into prothorax, shining brown to black, bearing few short bristles; antennae very short, one-segmented, unsclerotized, or more elongate and three-segmented (*Bolitophila*); labrum fleshy; mandible broad, toothed along inner margin; maxilla composed of inner blade-like lobe and outer oval lobe, inner lobe toothed along inner margin, outer lobe membraneous and papillose centrally, palpi reduced or enlarged; labium reduced to small sclerotized plate; no tentorial arms.

Thorax: Segments bare or bearing two groups of three or four minute bristles ventrally; creeping welts may be present on one or more segments; lateral spiracles only on prothorax (no spiracles in larvae of Keroplatinae), spiracles subcircular.

Abdomen: Spiracles on first seven or eight segments (absent in Keroplatinae); integument usually bare, scattered short bristles occasionally present, smooth, usually with spinulose creeping welts.

Comments: This is a sizeable family of rather primitive flies, with some 3000 species in the world (Vockeroth 1981). They are found on all continental land masses except Antarctica and on most oceanic islands. Nearly 600 species have been described for America north of Mexico, plus a large number of undescribed species. None of the species is thought to have much economic importance, although a few species are considered to be rather significant pests of commercial mushrooms.

Selected Bibliography

Buxton 1960.
Gagné 1975b, 1981a.
Hackman and Meinwander 1979.
Hennig 1948, 1973.
Johannsen 1910a, 1910b, 1910c, 1912a, 1912b.
Keilin 1919b.
Krivosheina 1980.
Laffoon 1957.
Madwar 1917.
Mansbridge 1933.
Matile 1970.
Munroe 1974.
Plachter 1979a, 1979b, 1979c.
Vockeroth 1981.

SCIARIDAE (SCIAROIDEA)

B. A. Foote, *Kent State University*

The Sciarids, Darkwinged Fungus Gnats

Figures 37.115–37.116

Relationships and Diagnosis: Sciarids are closely related to the Mycetophilidae and Cecidomyiidae of the infraorder Bibionomorpha (McAlpine, et al. 1981). Many authors consider them to be merely a subfamily of the Mycetophilidae. Steffan (1981) has reviewed our knowledge of these flies.

Larvae closely resemble those of Mycetophilidae but differ in possessing a posterior tentorial bridge (fig. 37.115b). They can also be distinguished by the absence of spicules in the creeping welts (spicules present in Mycetophilidae).

Biology and Ecology: Sciarids are commonly found in the same kind of habitats as Mycetophilidae, being particularly abundant in forested areas having dense shade and accumulations of decaying organic matter. Several species have been encountered in somewhat unusual habitats such as caves, birds' nests, and animal burrows. Larvae are scavengers and commonly feed on decaying plant material, although they frequently occur also in fecal droppings and fungi. Larvae of some species engage in peculiar mass movements whose purpose has not yet been elucidated.

Description: Mature larvae 3–6 mm; elongate and cylindrical, with bare and subshining segments; head capsule distinct, weakly retractile, black; hemipneustic; white.

Head: Eucephalic, weakly retractile into prothorax, heavily sclerotized, black, posterior tentorial bridge complete or nearly so; labrum membraneous, with papillae dorsally; genae meeting ventrally at two points; antennae small and arrow-shaped; mandible somewhat rectangular, with teeth apically and medially; maxilla elongate and two-parted, basal part (cardo) subtriangular, distal part bearing teeth along inner side.

Thorax: Only the prothorax has spiracles, these with two openings; no creeping welts ventrally.

Abdomen: Segments similar except caudal one which is somewhat lobed and resembles a proleg; creeping welts present ventrally; welts without spicules; first seven segments with lateral spiracles, these with single opening; integument bare and smooth.

Comments: It is difficult to estimate the number of species occurring in the world because so many genera are in dire need of revisionary study. About 150 species have been recorded from the Nearctic Region, although the true number is probably at least double that figure (Steffan 1981). A few species are considered to be economically important as pests in greenhouses and commercial mushroom houses.

Selected Bibliography

Helsel and Wicklow 1979.
Hennig 1948.
Johannsen 1912b.
Steffan 1966, 1973, 1981.
White 1950.

CECIDOMYIIDAE (SCIAROIDEA)

B. A. Foote, *Kent State University*

Gall Midges

Figures 37.117–37.125

Relationships and Diagnosis: The family is closely related to the Sciaridae and Mycetophilidae. All three families are now included in the infraorder Bibionomorpha (McAlpine et al. 1981). Gagné (1981b) reviewed the family, and his comprehensive book (1989) contains keys to plant damage and a key to genera of larvae.

Most larvae are easily recognized by the presence of a sclerotized, apically forked spatula on the ventral surface of the prothorax (fig. 37.117b). The head is tiny and cone-shaped.

Biology and Ecology: The larval feeding habits are highly diverse, with different species using distinctly different food sources. A number of species, particularly in the subfamilies Lestremiinae and Porricondylinae, feed on fungi and can be found in decaying vegetation, rotting wood, and

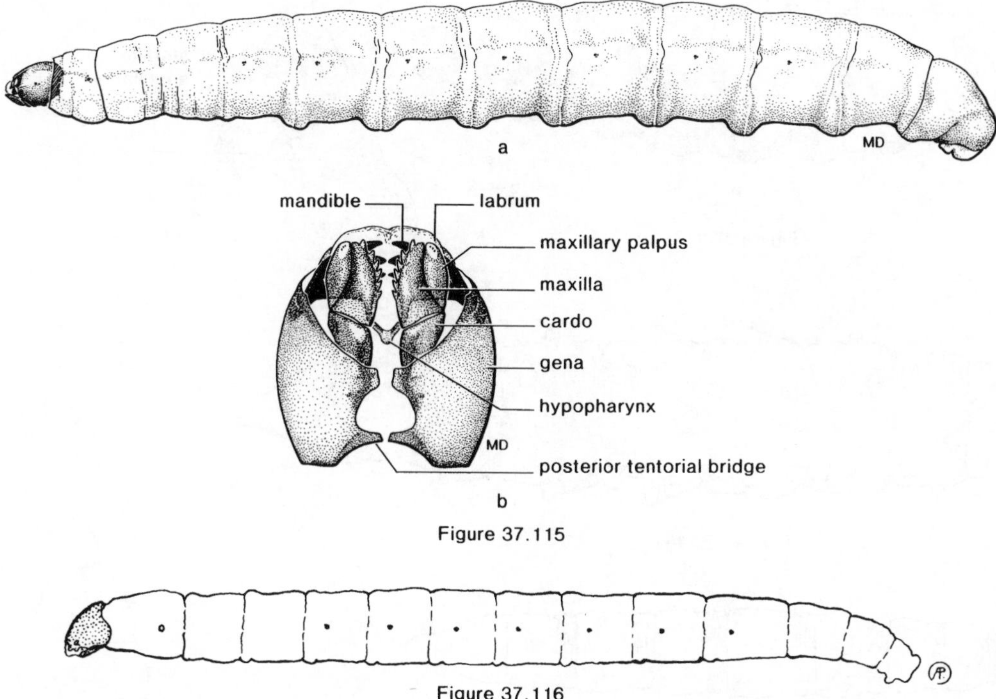

Figure 37.115a,b. Sciaridae. **a.** lateral of *Corynoptera* sp.; **b.** head capsule, ventral (from Manual of Nearctic Diptera, Vol. 1).

Figure 37.116. Sciaridae. *Sciara* sp. lateral; Larvae of this genus can be encountered in fleshy fungi (from Peterson 1951).

mushrooms. In contrast, many species of Cecidomyiinae are gall makers, utilizing a vast diversity of higher plants, with galls being formed on leaves, stems, roots, and in flower heads. Several species are predacious as larvae, feeding on such invertebrates as mites, other insect larvae, and aphids. Finally, a few species are internal parasitoids of aphids and psyllids. Adults of many species form mating swarms.

Description: Mature larvae 2–7 mm, elongate, cylindrical to somewhat flattened, with very small cone-shaped head; postcephalic segment preceding prothorax with spicules and setae; peripneustic; integument with systematically arranged setae or fleshy papillae; white, yellowish, or reddish.

Head: Very small and cone-shaped, and retractile into prothorax, with two projections (metacephalic rods) posteriorly; mouthparts reduced and modified for imbibing liquids; mandible minute; antennae two- or three-segmented, prominent and elongate.

Thorax: Prothorax of mature larvae nearly always with elongate, apically forked, sclerotized sternal spatula; integument may have rows of setae or papillae; anterior spiracles usually present laterally on prothorax; no spiracles on other thoracic segments.

Abdomen: All segments except caudal one similar, may have rows of setae or fleshy papillae and some species with clusters of spicules; each segment with lateral spiracles; caudal segment with anus terminal (Lestrimiinae and some Porricondylinae) or ventral (Cecidomyiinae).

Comments: This is a very large family. Some 3000 species have been described in the world, with America north of Mexico having about 1100 species scattered among some 160 genera (Gagné 1981b). Some species are considered to be economically important and are covered for the world by Barnes (1946–56). Examples are the gall makers (such as the alfalfa gall midge (*Asphondylia websteri* Felt)) that reduce the attractiveness or photosynthetic ability of agricultural and horticultural plants; phytophagous species that invade foliage such as the clover leaf midge, (*Dasineura trifolii* (Loew)); species that are predators of seeds, such as the clover seed midge (*Dasineura leguminicola* (Lintner)); and predators such as *Aphidoletes aphidimyza* (Rondani), a Holarctic aphid predator that can be very effective in both the field and in greenhouses. The most destructive pest is undoubtedly the Hessian fly, *Myetiola destructor* (Say) (figs. 37.122a,b) which can cause severe damage to susceptible winter wheat varieties.

Selected Bibliography

Barnes 1946–1956.
Felt 1940.
Foote, R. H. 1956.
Gagné 1969, 1975a, 1975b, 1978, 1981b, 1989.
Hawkins et al. 1986.
Hennig 1948.
Housewaert and Brewer 1972.
Jones et al. 1983.
Judd 1952.
Kieffer 1891, 1895a, 1895b, 1900.
Kim 1967.
Mamaev and Krivosheina 1965.
Mohn 1955.
Nijveldt 1969.
Osgood and Gagné 1978.
Pritchard 1947, 1948, 1951, 1960.
Roskam 1977.
Wilson 1966, 1968.
Wyatt 1967.

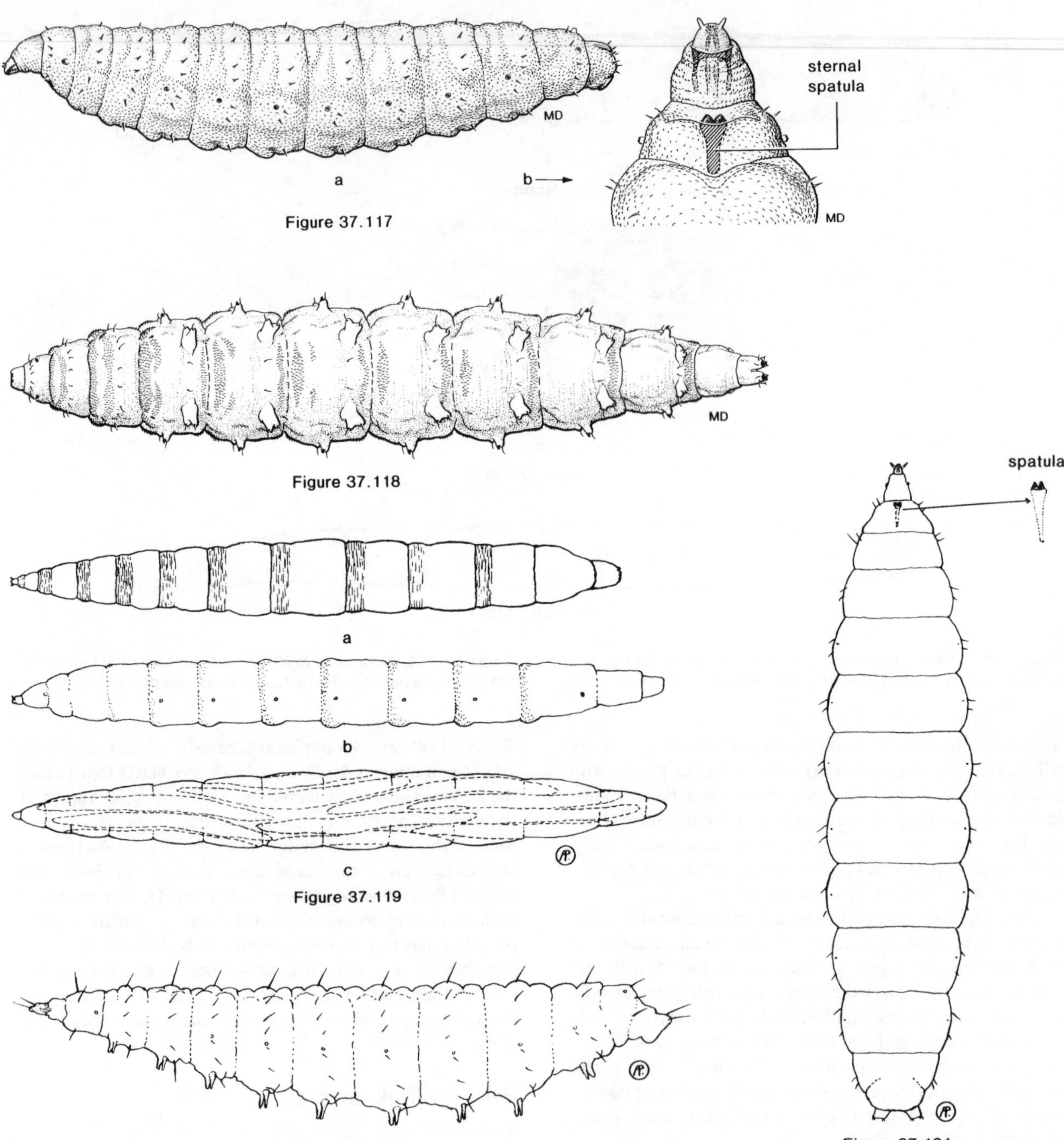

Figure 37.117a,b. Cecidomyiidae. *Dasineura* sp. **a.** lateral; **b.** ventral of anterior segments, showing the sternal spatula that typifies most larvae of Cecidomyiidae. *Dasineura leguminicola* (Lintner), the clover seed midge, is frequently abundant in red clover heads, and may feed on other clovers to a lesser extent (from Manual of Nearctic Diptera, Vol. 1).

Figure 37.118. Cecidomyiidae. Dorsal of *Cecidomyia candidipes* Foote. Feeds on white pine resin oozings, where it stimulates additional bleeding of resin (from Manual of Nearctic Diptera, Vol. 1).

Figure 37.119a-c. Cecidomyiidae. *Miastor* sp. **a.** immature larva, dorsal; **b.** mature larva, lateral; **c.** dorsal, showing living larvae within body cavity of a mature larva. The spindle-shaped larvae commonly occur in massive aggregations under the loosened, moist bark of recently fallen trees. Larvae of some species of this genus are paedogenetic (from Peterson 1951).

Figure 37.120. Cecidomyiidae. Lateral of *Lestodiplosis grassator* (Fyles). The 3 mm long larvae occur within *Phylloxera* galls on leaves of wild and domesticated grape. They are predators of the gall makers. (from Peterson 1951).

Figure 37.121. Cecidomyiidae. *Aphidoletes* sp., ventral, a pink to orange aphid predator (from Peterson 1951).

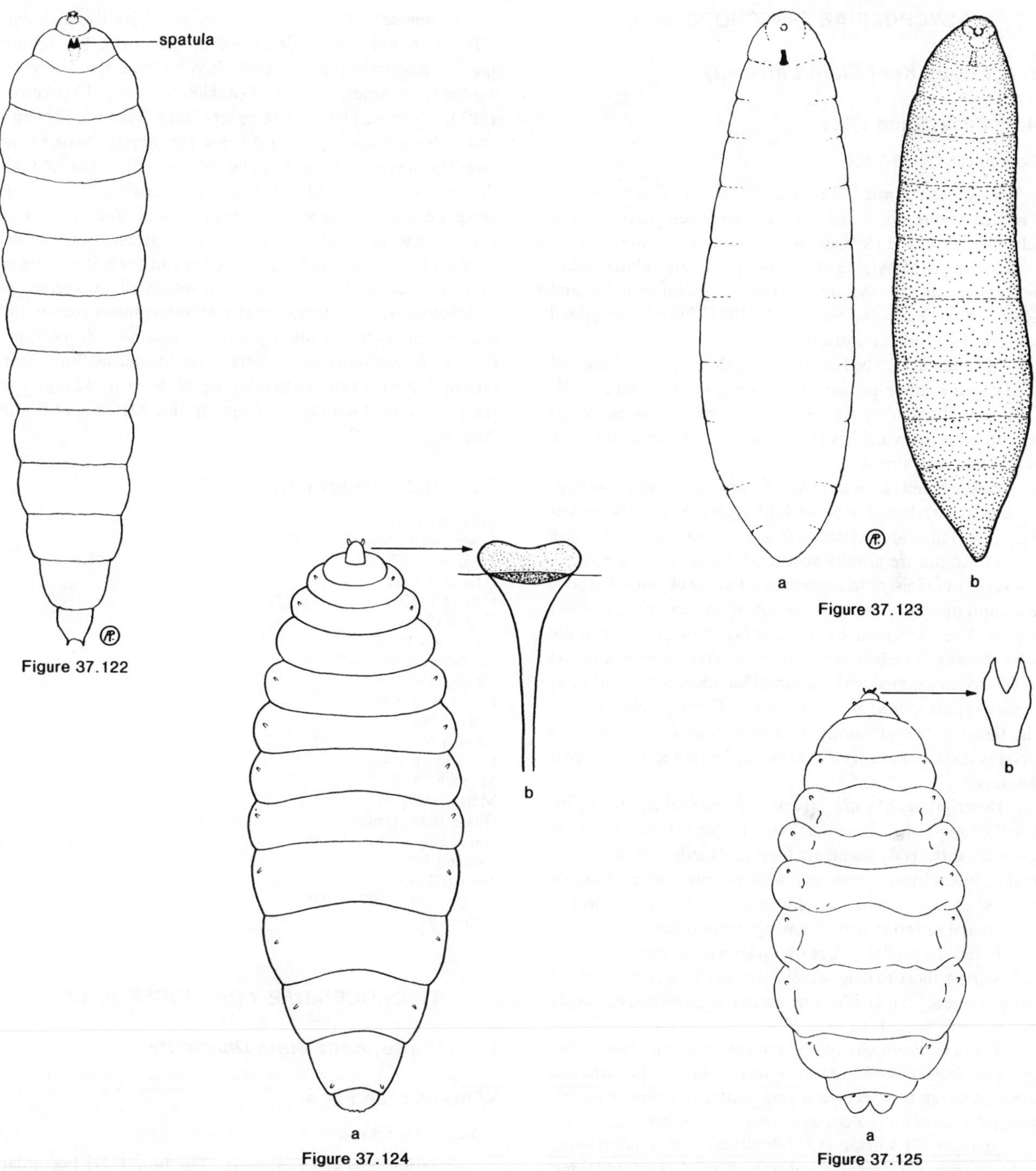

Figure 37.122

Figure 37.123

Figure 37.124

Figure 37.125

Figure 37.122. Cecidomyiidae. Ventral of *Monarthropalpus buxi* (Laboulbene), a leaf miner of boxwood foliage (from Peterson 1951).

Figure 37.123a,b. Cecidomyiidae. *Myetiola destructor* (Say), the Hessian fly, a significant pest of winter wheat from the Midwest eastward. Also feeds on barley, rye and other Poaceae. **a.** mature larva, ventral; **b.** the brown, overwintering puparium ("flaxseed") stage. Larvae feed at the base of the plant between the leaf and the stem. Two or more generations per year, depending on latitude (from Peterson 1951).

Figure 37.124a,b. Cecidomyiidae. *Contarinia negundifolia* Felt. The boxelder leaf gall midge. **a.** mature larva, dorsal; **b.** sternal spatula. Forms elongate, irregular, subglobose galls on the major veins of boxelder leaves in spring. One generation per year; overwinters in the soil (from Wilson 1966, courtesy U.S. Forest Service).

Figure 37.125a,b. Cecidomyiidae. *Rhabdophaga strobiloides* (Osten Sacken) the pine cone willow gall midge. **a.** mature larva, dorsal; **b.** sternal spatula. Forms cone-shaped galls at short tips of *Salix errocephala* and some other *Salix* species in spring. One generation per year; overwinters in gall (from Wilson 1968, courtesy U.S. Forest Service).

PSYCHODIDAE (PSYCHODOIDEA)

B. A. Foote, *Kent State University*

Moth Flies, Sand Flies

Figures 37.126–37.129

Relationships and Diagnosis: The family Psychodidae is a rather unique taxon and, as such, has been placed as the sole member of the Psychodoidea within the infraorder Psychodomorpha (McAlpine et al. 1981). Closely related families appear to be the Trichoceridae, Anisopodidae, and Scatopsidae. Quate and Vockeroth (1981) have summarized our knowledge of the moth flies.

Most larvae can be recognized by their distinct, exserted head capsule, amphipneustic respiratory system, and secondarily divided segments. Larvae of many species possess dorsal sclerotized plaques and have the posterior spiracles at the apex of a respiratory siphon.

Biology and Ecology: Adults are commonly encountered in moist, shaded habitats (although they are also rather frequently found around drains in homes, restrooms and locker rooms). Adults are usually nocturnal. Most larvae are found in a variety of moist or wet substrates. Larvae of several genera are saprophagous and can be found in decaying organic matter. A few species live in or near fast-flowing water, while some larvae of Phlebotominae develop in relatively dry desert soils often associated with mammal burrows. Adults of most species appear to feed on sugary materials or decaying matter, but those of the subfamily Phlebotominae (sand flies) are blood suckers and are usually associated with reptiles or small mammals.

Description: Mature larvae 3–6 mm, elongate, cylindrical or somewhat flattened, caudal segment may form respiratory tube; body segments frequently subdivided to form annuli; integument commonly with plaques, short hairs or bristles, or nearly bare; amphipneustic; white to brownish. Body shape differing widely among subfamilies.

Head: Eucephalic, heavily sclerotized, non-retractile, and bearing short to long setae; antennae very short, two- or three-segmented; mandible one- or two-segmented, variously shaped, with or without teeth.

Thorax: Prothorax with anterior spiracle borne laterally on short tubercle, segment may be subdivided into two annuli, bearing bristles or elongate, flattened projections, occasionally bare; no prolegs or obvious creeping welts.

Abdomen: Elongate and cylindrical in Bruchomyiinae, with first seven abdominal segments each subdivided into three annuli, and with conspicuous creeping welts ventrally; hairs on integument brush-like, and integument covered with short heavy spines. Larvae of Trichomyiinae cylindrical, segments not annulated, and integument with only few simple and very small setae. Larvae of Psychodinae cylindrical, tapering posteriorly, and ending in distinct breathing tube that bears a pair of fan-like brushes; integument bearing numerous long bristles and dorsal sclerotized plates; no ventral creeping welts; each segment subdivided into two or three annuli.

Comments: This is a relatively small family, with only a few hundred species described in the world. Ninety-one species, mostly in the subfamily Psychodinae, have been recorded from America north of Mexico (Quate and Vockeroth 1981). These are placed in 16 genera, with *Psychoda* (21 spp.) and *Telmatoscopus* (17 spp.) being the largest. None of the Nearctic species is thought to be economically important, although a few species of *Psychoda* are occasionally numerous around drains. A few species are typically found in trickling beds of sewage treatment plants. In contrast, some of the tropical blood-sucking species are very important vectors of human diseases. Sand-fly fever, transmitted by species of *Phlebotomus,* is endemic in the Mediterranean region and occasionally explodes into significant epidemics. Several species of *Phlebotomus* are vectors of leishmaniasis (kala-azar, Oriental sore), a protozoan infection of the skin and digestive tract in tropical countries of the Orient, Africa, and South America.

Selected Bibliography

Efflatoun 1921.
Feuerborn 1922a, 1923, 1927.
Fullaway 1907.
Haseman 1907.
Hennig 1950.
Hogue 1973a.
Johannsen 1934.
Keilin and Tate 1937.
Kellogg 1901.
Kloter et al. 1977.
Lewis 1973.
Lloyd 1943.
Lloyd et al. 1940.
Malloch 1917.
Muttkowski 1915.
Quate 1955, 1960.
Quate and Vockeroth 1981.
Satchell 1953.
Tonnoir 1933.
Vaillant 1963, 1971–1976.
Welch 1912.

TRICHOCERIDAE (TRICHOCEROIDEA)

B. A. Foote, *Kent State University*

Winter Crane Flies

Figures 37.130a,b

Relationships and Diagnosis: The family Trichoceridae is generally considered to be related to the Psychodidae, Anisopodidae, and Scatopsidae, although McAlpine et al. (1981) consider it distinct enough to warrant placement as the sole member of the Trichoceroidea. Alexander (1981b) has reviewed the family.

Larvae can be recognized by their eucephalic condition, annulated segments covered with fine pubescence, and the four short lobes surrounding the posterior spiracles (figs. 37.130a,b).

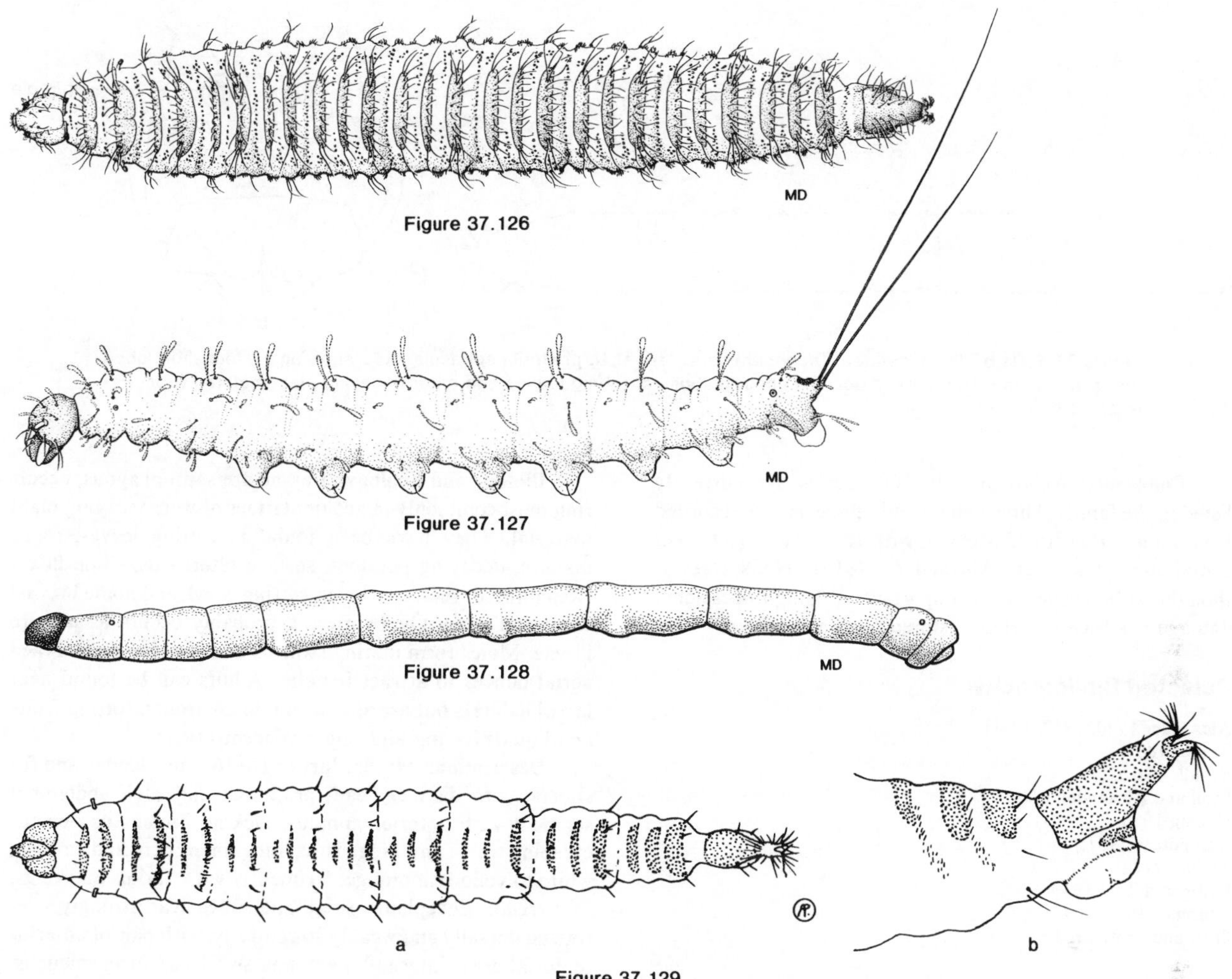

Figure 37.126. Psychodidae. Dorsal of *Pericoma* sp. Larvae occur on surface of rocks in riffles of streams (from Manual of Nearctic Diptera, Vol. 1).

Figure 37.127. Psychodidae. Lateral of *Lutzomyia vexator* Coquillett. Larvae of this species occur in moist sandy soils. The adult female is a notorious biter (from Manual of Nearctic Diptera, Vol. 1).

Figure 37.128. Psychodidae. Lateral of *Trichomyia urbica* Haliday (not Nearctic). Larvae may be encounterd in rotting wood (from Manual of Nearctic Diptera, Vol. 1).

Figure 37.129a,b. Psychodidae. *Psychoda* sp. **a.** dorsal; **b.** posterior breathing siphon, lateral. Larvae can be abundant in accumulations of decaying plant material in semi-aquatic habitats (from Peterson 1951).

Biology and Ecology: The common name, winter crane flies, reflects the fact that adults commonly emerge during the spring and fall months. Male swarms typically occur on sunny days. Larvae are commonly encountered in accumulations of decaying leaves, vegetable material, rotting fungi, and in burrows of rodents where they are scavengers.

Description: Mature larvae 6–12 mm, elongate and cylindrical, segments subdivided to form annuli, without prolegs or noticeable creeping welts; integument covered with fine pubescence and scattered bristles; amphipneustic; whitish to dark gray or brown.

Head: Eucephalic, with exserted, well-sclerotized head capsule that bears few scattered bristles; labrum somewhat conical; antenna short and papilla-like; mandible two-parted, with stout base and narrower, toothed distal part which possesses hair tufts; maxilla fleshy, two-parted; labium small and fleshy.

Thorax: Segments subdivided to form two annuli; prothorax bearing anterior spiracles laterally; spiracle circular, with several peripheral openings surrounding central ecdysial scar.

Abdomen: All but caudal segment subdivided to form three annuli; covered with fine pubescence and bearing scattered, short bristles; caudal segment ending in four short lobes, lobes fringed with short setae and weakly sclerotized on inner faces; posterior spiracles borne near base on inner side of dorsal lobes, spiracles circular with dark peritremes and central ecdysial scar.

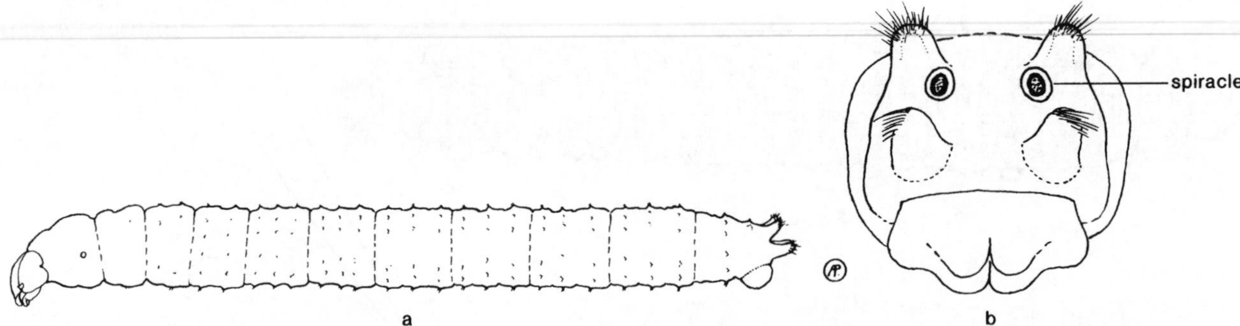

Figure 37.130a,b. Trichoceridae. *Trichocera* sp. **a.** lateral; **b.** posterior spiracular disc, showing the four short lobes that surround the disc. Larvae occur in accumulations of rotting leaves and other decaying plant material (from Peterson 1951).

Comments: Approximately 100 species are currently listed in the family. Three genera and 29 species are recorded from America north of Mexico, with the genus *Trichocera* containing 27 species (Alexander 1981b). No species is thought to be economically important, although occasional damage has been reported to stored roots and tubers.

Selected Bibliography

Alexander 1920, 1942, 1981b.
Brauer 1883.
Brindle 1962d.
Dahl and Alexander 1976.
Hennig 1950.
Karandikar 1931.
Keilin 1912a.
Keilin and Tate 1940.
Laurence 1956.
Pratt and Pratt 1984.

ANISOPODIDAE (ANISOPODOIDEA)

B. A. Foote, *Kent State University*

The Anisopodids

Figures 37.131–37.132

Relationships and Diagnosis: The correct phylogenetic placement of the Anisopodidae has puzzled workers for decades. At the present time, it is the sole Nearctic family in the Anisopodoidea (McAlpine et al. 1981). Even the species placed in the family have been questioned, and some of the taxa currently placed there have been placed previously in such families as Tipulidae, Bibionidae, and Mycetophilidae. Several different family names have been applied to some of the species groups, including Phryneidae, Rhyphidae, and Sylvicolidae, among others. B. V. Peterson (1981a) has summarized our knowledge of the family.

Larvae can be recognized by the secondary divisions of the prothoracic and abdominal segments (fig. 37.131a), by the sessile posterior spiracles, amphipneustic respiratory system, and the small, subconical head capsule.

Biology and Ecology: Larvae are saprophagous, occurring most commonly in accumulations of wet, decaying plant material. They have been found in rotting leaves, roots, manure, decaying potatoes, sewage filter beds, slime fluxes from injured trees, tree holes, rotting wood, and home brewed cider and wine. Overwintering probably occurs as mature larvae. Males form mating swarms and perform stereotyped aerial dances to attract females. Adults can be found near larval habitats but are also encountered around rotting fruits and liquids having high sugar concentrations.

Description: Mature larvae 10–16 mm, slender and fusiform, with distinct head; prothorax and each abdominal segment with anterior annulus, lacking prolegs or obvious creeping welts; amphipneustic; integument bare and smooth; white to yellowish orange, frequently with darker markings.

Head: Eucephalic, small and subconical, strongly sclerotized dorsally and weakly so ventrally, with pair of anterior tentorial arms internally; antenna small and inconspicuous, with small sensory papillae; labrum largely membranous, with fleshy protuberances bearing numerous fine setae; mandible variously sclerotized, with two parts, the apical part strongly sclerotized and bearing several strong teeth, frequently with brush of setae; maxilla soft and fleshy, lobed and bearing variety of setae and papillae, palpus short and nearly transparent; labium well developed, occasionally soft and fleshy.

Thorax: Segments longer than broad, prothorax with anterior annulus; all segments with two ventral sensory organs bearing long and short setae; anterior spiracles borne posterolaterally on prothorax, spiracle small and with 3–19 openings.

Abdomen: All segments with anterior annulus, caudal segment with additional annuli and either ending in short conical process or in five variously sized lobes; anus ventral and frequently surrounded by broad perianal pad; posterior spiracles either borne on posterior face of caudal segment, with spiracles crescent-shaped or oval and with 12 to 23 openings, or borne on side of caudal segment above the anus, with each spiracle crescent-shaped and bearing 25 openings.

Comments: This is a relatively primitive family having worldwide distribution. Six genera and some 100 species have been described, with three genera and nine species recorded

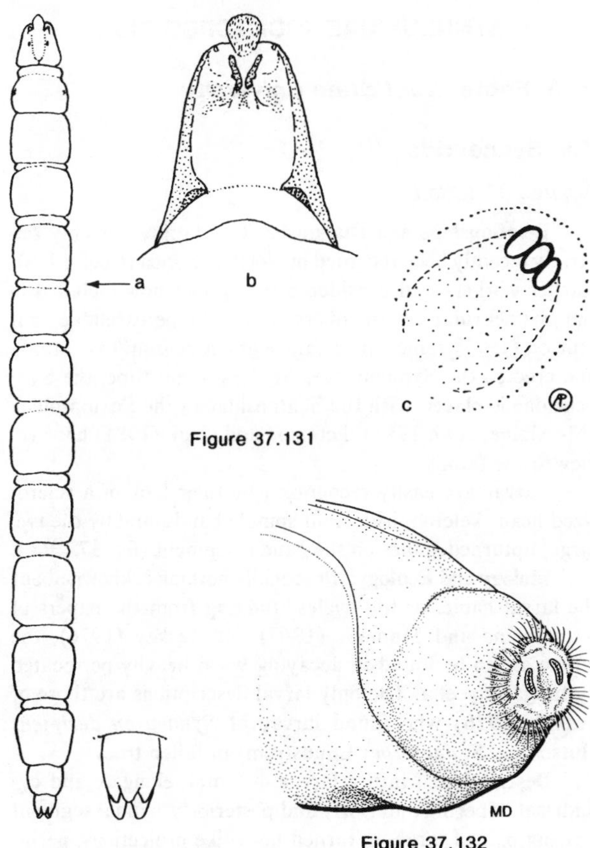

Figure 37.131a,b,c. Anisopodidae. *Sylvicola alternatus* (Say). **a.** dorsal, with an enlargement of the caudal segment; **b.** ventral of head capsule; **c.** prothoracic spiracles. Larvae can be found in decaying fruits (from Peterson 1951).

Figure 37.132. Anisopodidae. Posterolateral of spiracular disc of *Mycetobia divergens* Walker. Larvae have been found in rotting, highly organic plant debris occurring in tree holes (from Manual of Nearctic Diptera, Vol. 1).

for America north of Mexico (B. V. Peterson 1981a). According to K. G. V. Smith and Taylor (1966), larvae of *Anisopus* can cause intestinal or urinogenital myiasis. Edwards (1928) reported that anisopodid larvae had invaded partly exposed, moldy liver that had been preserved in 10% formalin.

Selected Bibliography

Alexander 1942.
Brindle 1962d.
Edwards 1928.
Hennig 1948.
Keilin 1919b.
Keilin and Tate 1940.
Lindner 1930.
Malloch 1917.
Peterson, B. V. 1981a.
Pratt and Pratt 1980.
Smith, K. G. V. and Taylor 1966.

SCATOPSIDAE (SCATOPSOIDEA)

B. A. Foote, *Kent State University*

The Scatopsids

Figures 37.133–37.135

Relationships and Diagnosis: Long considered to be rather closely related to the march flies (Bibionidae), the scatopsids have recently been shifted to a position closer to the Synneuridae and thus are no longer thought to be phylogenetically close to the Bibionidae. The two families are grouped in the Scatopsoidea, a member of the infraorder Psychodomorpha by McAlpine et al., (1981). Cook (1981a) has recently reviewed the family.

Larvae are not easily characterized due to the variation occurring within the family. In general, they can be recognized by the exserted complete head capsule, fusiform, depressed body, and the two, tube-like projections or a broad sclerotized shelf below the posterior spiracles (fig. 37.133).

Biology and Ecology: Relatively little has been published on the life cycles and larval feeding habits, and the larvae of only four or five species have been described (Cook 1981a). As far as known, they are saprophagous, feeding on a variety of decaying organic matter, near which the adults are usually found. Larvae have been reported as infesting rotting bulbs and roots, wastes from canneries and wineries, and fecal material. At least one species is known to feed on decaying tissues under the bark of dead and dying trees.

Description: Mature larvae 3–6 mm, dorsoventrally flattened, with distinct head; no prolegs, but creeping welts may be present; integument with numerous bristles and shorter hairs; peripneustic; grayish brown.

Head: Eucephalic, non-retractile, brownish and well sclerotized, bearing a few scattered bristles; antenna elongate, three-segmented; labrum slender and somewhat compressed, with dense brush of short setae apicoventrally.

Thorax: Anterior spiracles lateral on prothorax, on short or long projection; dorsal and lateral surfaces of all segments with clusters of short setae making distinct patches.

Abdomen: Segments dorsoventrally flattened, bearing clusters of short setae dorsally and laterally; lateral spiracles small, projecting or sessile; posterior spiracles larger, commonly on distinct projections arising from posterior surface of caudal segment; some species with sessile spiracles; caudal segment with pair of partially sclerotized lobes or broad sclerotized shelf below spiracles.

Comments: Seventy-three species, 18 genera, and four subfamilies occur in America north of Mexico (Cook 1981a). Larger genera are *Psectrosciara, Rhexoza,* and *Swammerdamella,* each containing 10 species. No species is economically important.

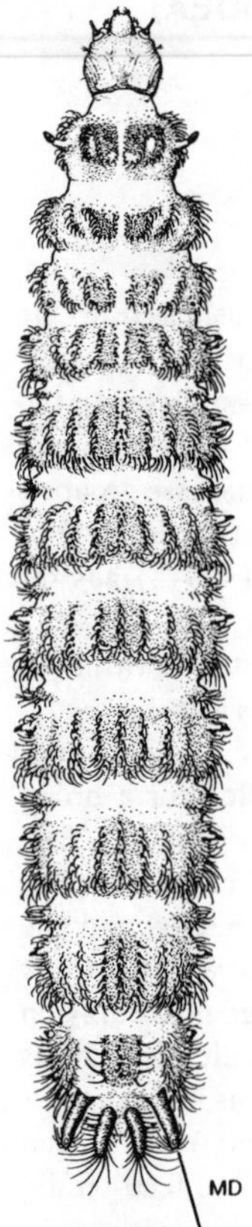

spiracle
MD
spiracle
spiracle

Figure 37.133 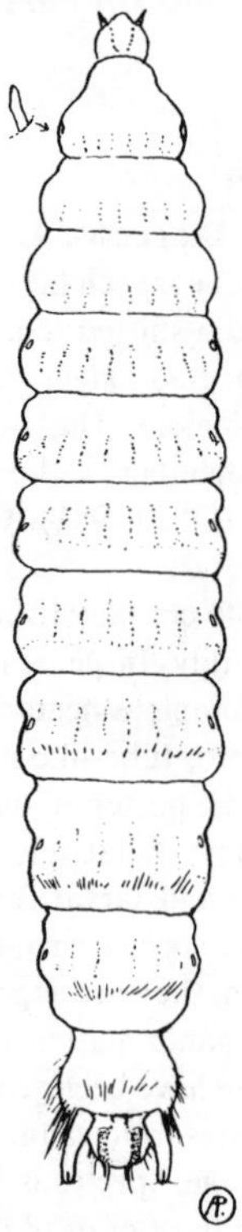 Figure 37.134 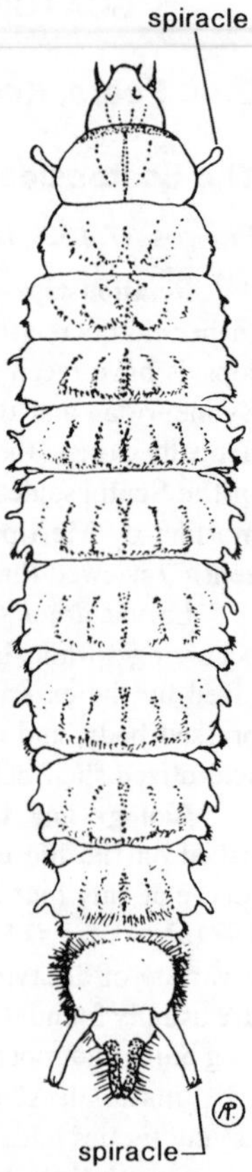 Figure 37.135

Figure 37.133. Scatopsidae. Dorsal of *Rhexoza* sp. Larvae occur under loosened bark of decaying deciduous trees (from Manual of Nearctic Diptera, Vol. 1).

Figure 37.134. Scatopsidae. Dorsal of *Rhegmoclema* sp. Larvae can be found in rotting vegetation and under bark of decaying trees (from Peterson 1951).

Figure 37.135. Scatopsidae. Dorsal of *Scatopse* sp. Larvae frequently are abundant in accumulations of moist, decaying vegetation or fruit (from Peterson 1951).

Selected Bibliography

Bovien 1935.
Cook 1955, 1956a, 1957, 1958, 1963, 1965, 1981a.
Duda 1928.
Hennig 1948.
Laurence 1953.
Lyall 1929.
Malloch 1917.
Meade and Cook 1961.
Melander 1916.
Morris 1918.

SYNNEURIDAE (SCATOPSOIDEA)

B. A. Foote, *Kent State University*

The Synneurids

Figures 37.136a,b

Relationships and Diagnosis: This family has only relatively recently been recorded in North America (Cook 1963). Earlier workers had considered the species now included in the Synneuridae as members of the Hyperoscelidae, but Hutson (1977) removed certain segregates from that family and erected the Synneuridae. At the present time, the Synneuridae is placed with the Scatopsidae in the Scatopsoidea (McAlpine, et al. 1981). Peterson and Cook (1981) have reviewed the family.

Larvae are easily recognized by their lack of a sclerotized head skeleton, iridescent spinule bands, and by the two large, upturned spines on the caudal segment (fig. 37.136a).

Biology and Ecology: Practically nothing is known about the larval habits or life cycles. Judging from the papers of Krivosheina and Mamaev (1967) and Teskey (1976), the larvae are to be found in decaying wood heavily permeated by fungal mycelia. The only larval descriptions are those of Teskey (1976) who found larvae of *Synneuron decipiens* Hutson (as *S. annulipes* Lundstrom) in fallen trees.

Description: Mature larvae 4–6 mm, elongate and cylindrical, tapering anteriorly and posteriorly, caudal segment bearing pair of large, upturned horn-like projections; peripneustic; integument of most segments dorsally and ventrally with transverse bands of minute pubescence that appear iridescent.

Head: Small and membranous, without sclerotized skeleton; antenna short and two-segmented.

Thorax: Meso- and metathorax with patches of iridescence dorsally and ventrally; anterior spiracles lateral on prothorax; each spiracle enlarged, circular, and with several suboval openings around margin; integument below spiracle somewhat thickened and sclerotized to form subcircular area.

Abdomen: First seven segments with dorsal and ventral bands of iridescence, eighth segment with only the ventral band; spiracles lateral, small and circular; caudal segment bearing pair of strong, upturned, sclerotized horns posteriorly; posterior spiracles enlarged and resembling those of prothorax.

Comments: This is a very small family, with only four species and two genera described in the world. The genera *Exiliscelis,* with one species, and *Synneuron,* with one species, have been recorded from America north of Mexico (Peterson and Cook 1981). Both species are western.

Selected Bibliography

Cook 1963.
Hutson 1977.
Krivosheina and Mamaev 1967.
Mamaev and Krivosheina 1969.
Peterson and Cook 1981.
Teskey 1976.

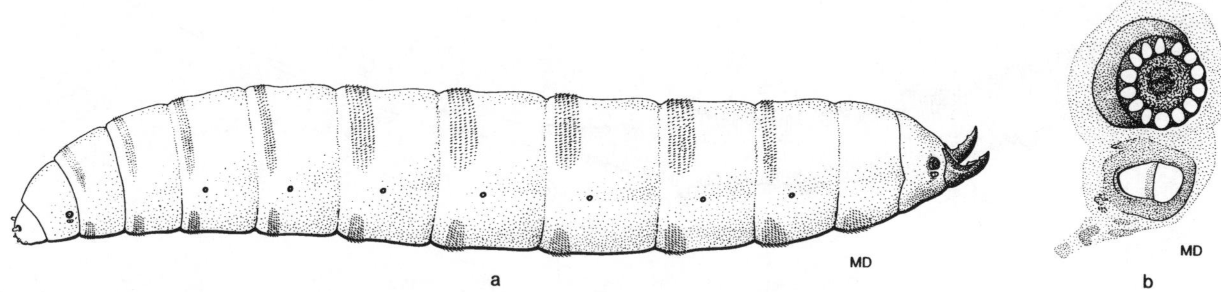

Figure 37.136a,b. Synneuridae. *Synneuron decipiens* Hutson. **a.** larva, lateral; **b.** caudal spiracle. Found in decaying wood (from Manual of Nearctic Diptera, Vol. 1).

PTYCHOPTERIDAE (PTYCHOPTEROIDEA)

B. A. Foote, *Kent State University*

Phantom Crane Flies

Figures 37.137, 37.138

Relationships and Diagnosis: McAlpine, et al. (1981) believe the family is sufficiently distinct to warrant its own superfamily and infraorder, the Ptychopteromorpha. Alexander (1981c) has reviewed the family.

Larvae possess a complete, exserted head capsule, a long and retractile breathing tube, and an integument that bears numerous hairs, warts, or tubercles (fig. 37.137). The anal papillae are small and unbranched.

Biology and Ecology: Larvae are found in marshy habitats containing large quantities of decaying organic matter. They appear to be saprophagous, ingesting finely divided particulate matter. Adults of the common, marsh-inhabiting *Bittacomorpha clavipes* (Fabricius), the phantom crane fly, have black and white banded legs and expanded, air-filled first tarsomeres that enable them to drift with air currents.

Description: Mature larvae 10–25 mm, slender and cylindrical; metapneustic, with posterior end elongated to form distinct, retractile breathing tube; integument covered with fine pubescence, some species with numerous hair-bearing papillae, warts, or tubercles; white, yellow, rusty red, brown, or black.

Head: Eucephalic, complete, small and well sclerotized; antenna apparently two-segmented; labrum bearing hair brushes laterally; mandible appearing two-segmented, well developed, with one to three outer teeth.

Thorax: Segments similar, lacking prolegs; integument with pubescence, long hairs, or papillae, no anterior spiracles.

Abdomen: First three segments with inconspicuous to well developed paired prolegs ventrally, prolegs each bearing single hook-like crochet; caudal segment greatly elongated and retractile, bearing posterior spiracles apically; anal papillae slender and unbranched.

Comments: This is a relatively small family of somewhat over 60 described species. Three genera, *Bittacomorpha, Bittacomorphella,* and *Ptychoptera,* and 16 species are known to occur in America north of Mexico (Alexander 1981c). No species has economic significance.

Selected Bibliography

Alexander 1920, 1927, 1942, 1967, 1981c.
Anthon 1943b.
Brindle 1962b.
Hansen 1979.
Hart 1898.
Hennig 1948.
Hodkinson 1973.
Johannsen 1934.
Malloch 1917.
Peus 1958.
Teskey 1984.

DIXIDAE (CULICOIDEA)

B. A. Foote, *Kent State University*

The Dixid Midges

Figures 37.139–37.141

Relationships and Diagnosis: McAlpine, et al. (1981) have placed the Dixidae in the Culicoidea along with the Culicidae and Chaoboridae, although there is still some dispute concerning this placement. Earlier authors tended to include the dixid midges within the Culicidae or Chironomidae. Peters (1981) has summarized our knowledge.

Dixid larvae (fig. 37.139a–c) are recognizable by the large exserted head capsule, ventral prolegs on the first or first and second abdominal segments, and possession of two flattened posterolateral processes behind the posterior spiracles.

Biology and Ecology: Surprisingly little is known about the biology and ecology of dixids, despite their abundance and widespread occurrence in freshwater habitats. Adults are commonly found near the larval habitat. Larvae are most commonly encountered in shallow water supporting lush growths of aquatic macrophytes. They can be found in marshes, along marshy borders of lakes, shallow ponds, bogs, and other bodies of slow-flowing or relatively stagnant water. They are associated with the surface film and frequently assume a distinctive U-shape when resting, with only the head and posterior spiracles immersed in water while the middle of the body protrudes above the surface. Larval food consists of microorganisms and finely divided detrital particles that are swept toward the mouth by a pair of labral brushes. Pupae

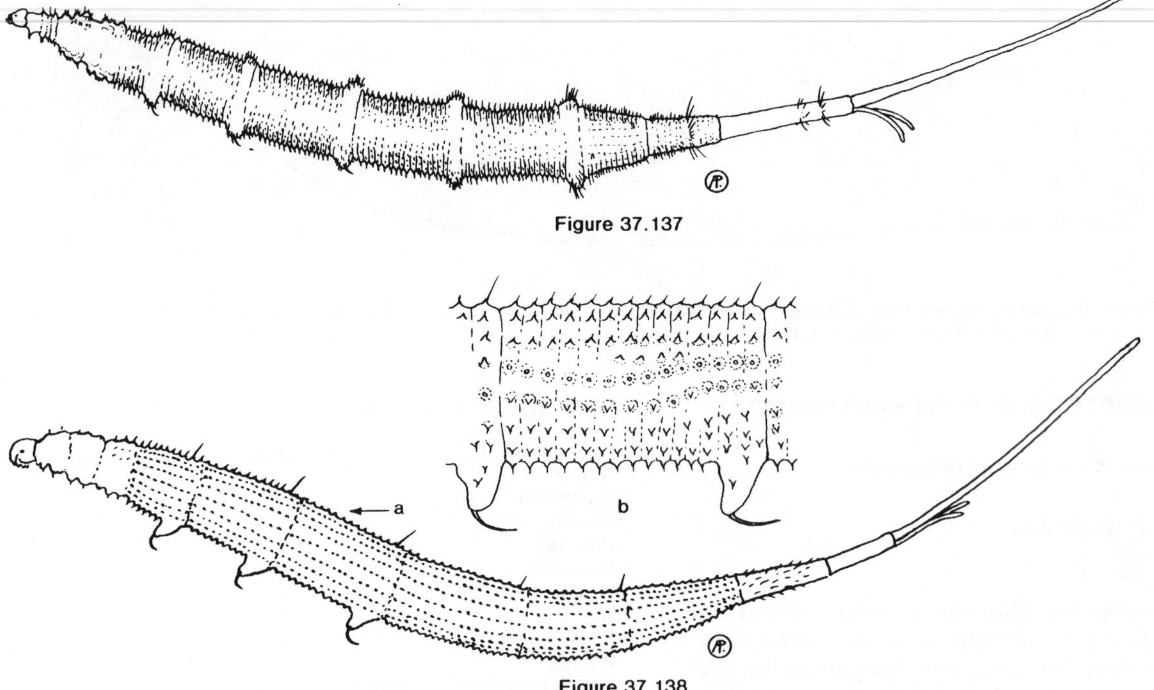

Figure 37.137. Ptychopteridae. Lateral of *Ptychoptera* sp. Larvae found in the bottom of the shallow ponds and marshes where they feed on organic sediments and decaying vegetation. The posterior end is telescopic and can be extended to the surface film for breathing (from Peterson 1951).

Figure 37.138a,b. Ptychopteridae. *Bittacomorpha clavipes* (Fabricius). **a.** lateral; **b.** lateral of third abdominal segment, showing tuberculate nature of integument as well as ventral prolegs. Found in shallow pools and marshes containing high concentrations of decaying organic matter (from Peterson 1951).

are formed on or in wet soil near the water's edge. Eggs are deposited in water by the female penetrating the surface film with the tip of her abdomen. Newly laid eggs are enclosed in a gelatinous matrix.

Description: Mature larvae 5–25 mm, elongate and cylindrical, with distinct exserted head and one or two anterior abdominal prolegs, some abdominal segments bearing ambulatory combs ventrally; caudal segment complex; metapneustic; integument of anterior abdominal segments with or without plumose hairs dorsally; brownish.

Head: Eucephalic, large and complete, exserted; ocular area large and circular; antenna elongate, one-segmented, arising from tubercle and somewhat upturned; labrum subtriangular, with brush comprised of numerous hairs; mandible somewhat rectangular, with two articulations, distal outer angle with stout seta, inner surface with teeth distally, small peg-like structures proximally, and hair comb on incisor area; maxilla with elongate, triangular endite, palpus one-segmented and elongate, and resembling antenna; labium subtriangular, complex, inner surface bearing peg-like structures and short spine-like setae.

Thorax: Segments distinct, sparsely haired dorsally but with clusters of elongate bristles ventrally, lacking spiracles; metathorax largest.

Abdomen: First seven segments subequal, pair of crochet-bearing prolegs ventrally on first one or two segments; segments 5–7 with ambulatory comb on ventral surface; anterior segments may have dense patch of complex plumose hairs dorsally; posterior spiracles borne dorsally on eighth abdominal segment, spiracle with hair-fringed postspiracular process, ninth segment with unpaired median sclerite, pair of posterolateral processes, and pair of anterolateral plates; posterolateral processes elongate, fringed with row of hydrofuge hairs; anterolateral plates subrectangular and with pecten of short spine-like setae or of small lobes bearing setae; caudal segment partially or entirely encircled by heavily sclerotized saddle that is extended posteriorly to form postanal process, latter with two or three pairs of very long bristles apically; anus and eversible anal papillae situated apically, anus bordered by anal comb consisting of compound spines or spine-like setae.

Comments: This is a small family in the Nearctic Region, with only three genera and 42 species recorded from America north of Mexico (Peters 1981). The genus *Dixa* includes 22 species; *Dixella*, 19; and *Meringodixa*, 1. None is considered to have economic significance.

Selected Bibliography

Disney 1975.
Edwards 1932.
Hennig 1950.
Hubert 1953.
Johannsen 1903, 1934.
Malloch 1917.
Nowell 1951, 1963.
Peters 1981.
Peters and Cook 1966.
Smith, F. K. 1928.
Teskey 1984.

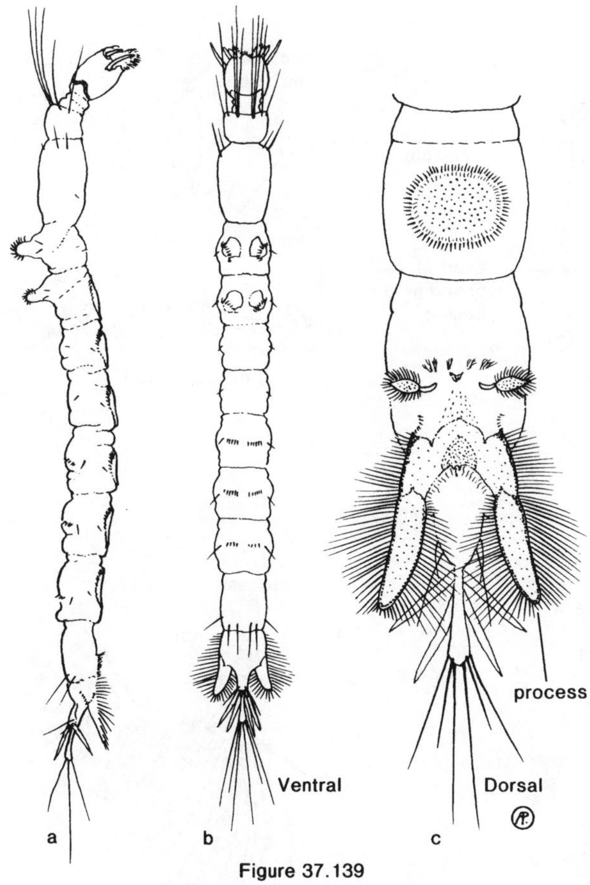

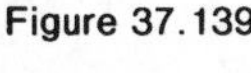

Figure 37.139

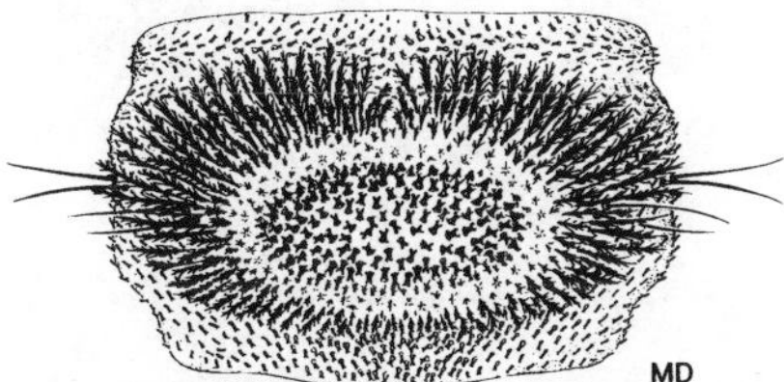

Figure 37.140

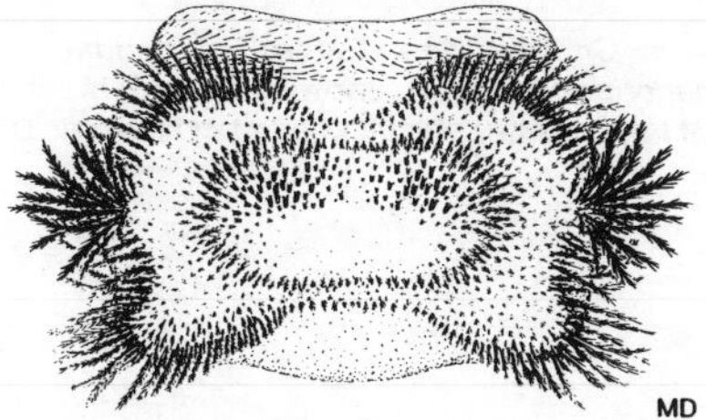

Figure 37.141

Figure 37.139a,b,c. Dixidae. *Dixa* sp. **a.** lateral; **b.** ventral; **c.** caudal end, dorsal. Larval dixids usually assume a u-shape at the surface in quiet water along the margins of streams, ponds and marshes (from Peterson 1951).

Figure 37.140. Dixidae. Dorsal surface of abdominal segment of *Dixa* sp. (from Manual of Nearctic Diptera, Vol. 1).

Figure 37.141. Dixidae. Dorsal surface of abdominal segment of *Meringodixa chalonensis* (Nowell). (from Manual of Nearctic Diptera, Vol. 1).

CHAOBORIDAE (CULICOIDEA)

B. A. Foote, *Kent State University*

The Chaoborids, Phantom Midges

Figures 37.142–37.145

Relationships and Diagnosis: Long believed to be closely related to the Culicidae and frequently considered to be a group within that family, the Chaoboridae are presently included with the Culicidae and Dixidae in the Culicoidea of the infraorder Culicomorpha (McAlpine, et al. 1981). Cook (1981b) recently reviewed the family.

Larvae are recognized by having prehensile antennae, a rather broad head, and a distinctly enlarged thorax composed of fused segments (fig. 37.142). Some species have distinct thoracic air sacs, helping to give them a translucent appearance and the name "phantom midge". Adults are found near the larval habitat.

Biology and Ecology: Larvae are aquatic, occurring primarily in standing bodies of water. They are predacious, feeding rather unselectively on a diversity of aquatic organisms, and may play some role in structuring freshwater communities. The larvae are well known for their vertical movements in the water column over 24 hour cycles (K. G. Wood, 1956).

Description: Mature larvae 6–12 mm, elongate, basically cylindrical but with thoracic segments fused and enlarged; caudal segment with preanal fan-like brush of setae ventrally; metapneustic; integument bearing numerous bristles and clusters of setae; white, with some species nearly transparent.

Head: Eucephalic, complete and non-retractile; antennae prehensile, either close together on head or widely separated, with 3–5 long, blade-like setae apically; frontoclypeal apotome enlarged, bearing variable number of modified setae near apex; labrum well developed, lacking brushes; mandible well sclerotized, bearing spines; maxilla reduced but bearing variously modified setae; labium reduced, with greatly modified setae.

Thorax: Segments fused to form enlarged compound segment, with numerous simple and branched bristles; without spiracles; some species with conspicuous air sacs visible through integument.

Abdomen: Comprised of eight obvious segments, with reduced ninth and tenth segments; anterior segments all similar, but eighth segment frequently bearing short or long breathing tube (absent in larvae of *Chaoborus*); caudal segment with four long or short papillae and several setae and setal clusters; segment also bearing fan-like cluster of pectinate setae ventrally except in species of subfamily Corethrellinae; anus often bordered by row of hooked or blade-like spines.

Comments: This is a small family of mosquito-like flies that contains only 19 species and five genera in the Nearctic Region (Cook 1981b). The largest genus is *Chaoborus* with 10 species. No species is considered to be economically important.

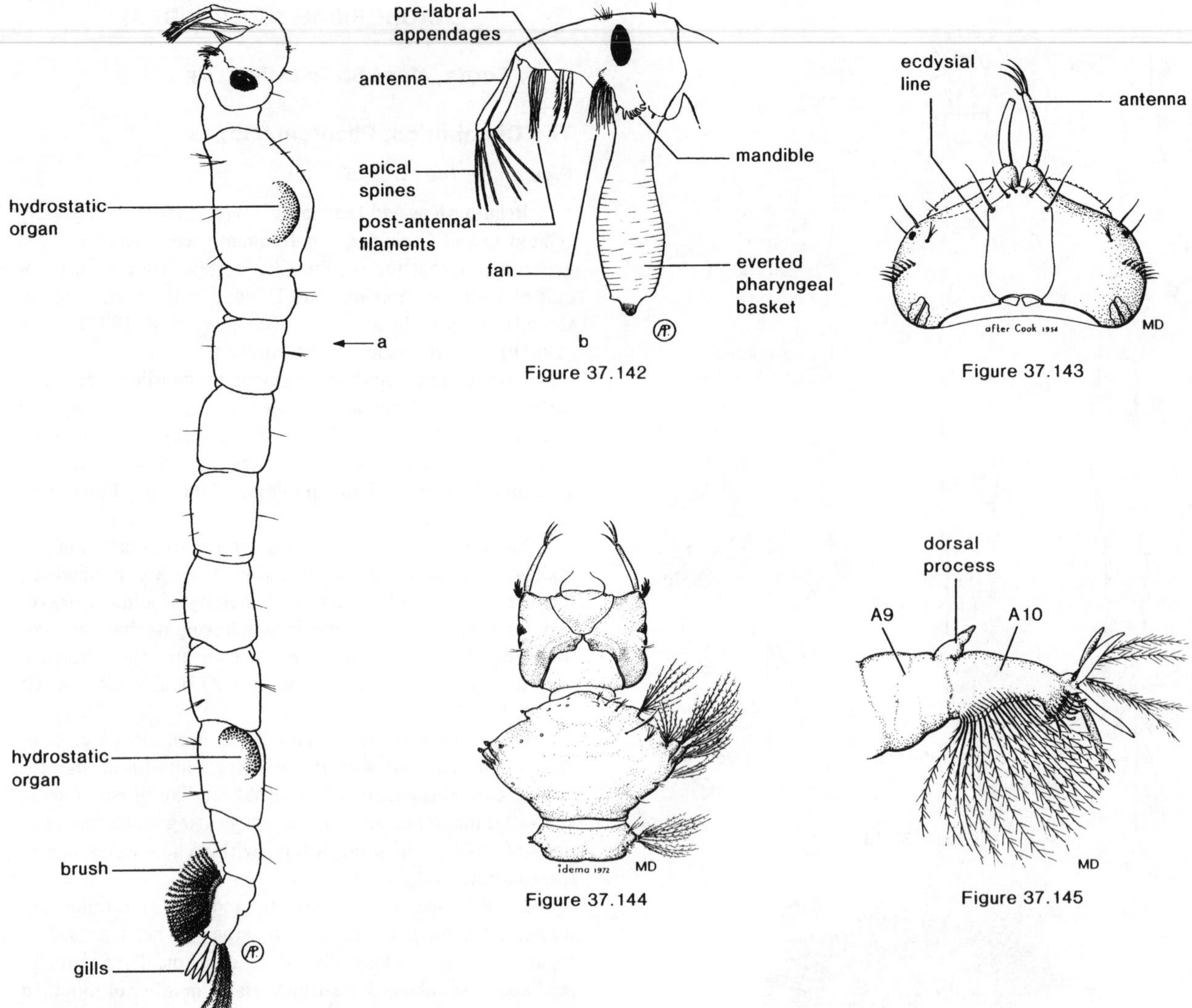

Figure 37.142a,b. Chaoboridae. *Chaoborus* sp. **a.** lateral; **b.** head capsule, lateral, showing everted pharyngeal basket. The highly predacious larvae have been collected in woodland pools, shallow ponds, marshes, tree holes, and drainage ditches (from Peterson 1951).

Figure 37.143. Chaoboridae. *Corethrella brakeleyi* (Coquillett). Head capsule, dorsal, showing the prehensile antennae. The predacious larvae occur in standing water habitats (from Manual of Nearctic Diptera, Vol. 1).

Figure 37.144. Chaoboridae. Dorsal of anterior segments of *Eucorethra underwoodi* Underwood (from Manual of Nearctic Diptera, Vol. 1).

Figure 37.145. Chaoboridae. Lateral of caudal segments of *Chaoborus punctipennis* (Say), showing distinct brush on the ventral surface of last segment (from Manual of Nearctic Diptera, Vol. 1).

Selected Bibliography

Borkent 1979.
Colless 1977.
Cook 1956b, 1981b.
Edwards 1932.
Eggleton 1932.
Hennig 1950.
Johannsen 1903, 1934.
Juday 1921.
Lake 1960, 1969.
Miller 1941.
Roth and Parma 1970.
Saether 1970, 1972.
Smith, J. B. 1902.
Stadtmann-Averfeldt 1923.
Stone 1968.
Teskey 1984.
Wood, K. G. 1956.

CULICIDAE (CULICOIDEA)

B. A. Foote, *Kent State University*

Mosquitoes

Figures 37.146–37.151

Relationships and Diagnosis: Long recognized as being closely related to the Chaoboridae, the Culicidae are now included with that family and the Dixidae in the Culicoidea of the infraorder Culicomorpha (McAlpine, et al. 1981). Stone (1981) has recently reviewed the family.

Mosquito larvae (fig. 37.148) are easily recognized by their elongate bodies, broadened thorax, labral brushes, ventral fan-like brush on the caudal segment, and respiratory siphon (absent in *Anopheles,* fig. 37.146).

Biology and Ecology: Females of most species are blood sucking, and attack a vast array of warm-blooded hosts, including humans. As such, they are important vectors of blood-inhabiting diseases of humans and domesticated animals. Larvae and the active pupae are found in still water, where they rise to the surface to breathe. They occur in open marshes, along the marshy borders of lakes, in woodland pools, in a variety of temporary habitats such as cans, bottles, bird baths, old tires, and in tree holes. A few species are associated with unusual aquatic habitats such as the small pools formed in pitcher plants and epiphytes. Eggs are deposited singly or in clusters called rafts. They may be laid in water or in areas that are subsequently flooded. Larvae of most species are filter feeders, utilizing a variety of small organic particles removed from the water by mouth brushes. Larvae of a few species are predators of other mosquito larvae and thus are considered to be beneficial.

Description: Mature larvae 4–18 mm, elongate and cylindrical to slightly flattened, head distinct and bearing labral brushes; thoracic segments typically swollen and bearing tufts of setae; caudal segment usually bearing elongate respiratory siphon (absent in *Anopheles*) dorsoposteriorly and anal papillae ventrally; metapneustic; integument commonly bearing setal tufts, scattered bristles, spinules, or scales; white to pale brownish, occasionally yellowish to green.

Head: Eucephalic, exserted and usually prognathous, commonly bearing tufts of bristles at various locations on dorsal side; antenna elongate, one-segmented, and usually bearing brushes; labrum with conspicuous brush composed of many fine hairs (reduced to stout curved rods in some species); labium sclerotized, frequently with apical teeth.

Thorax: Segments poorly defined, enlarged and fused to form swollen region bearing tufts of bristles laterally and dorsally; lacking anterior spiracles and prolegs.

Abdomen: Segments similar, well defined; integument commonly with setal tufts, scattered bristles, spinules, or scales on dorsal and lateral surfaces; eighth segment usually with distinct respiratory siphon (absent in *Anopheles*); caudal segment with sclerotized saddle dorsally or encircling sclerotized ring and long dorsal hairs, usually with conspicuous ventral brush; with two or four membranous anal papillae.

Comments: The Culicidae is a fairly large family that is widely distributed, with the center of distribution being the tropics. Some 150 species in 13 genera have been recorded from America north of Mexico (Stone 1981). A few of these are important vectors of human diseases, including species of the genus *Anopheles* that transmit malaria, *Aedes* that carry yellow fever, and different species of *Aedes, Culex,* and *Culiseta* that transmit various types of encephalitides. Malaria is now listed by the World Health Organization as the number one human disease.

Selected Bibliography

Barr 1958.
Belkin and Hogue 1959.
Breland 1949.
Carpenter and LaCasse 1955.
Darsie 1949, 1951.
Darsie and Ward 1981.
Dodge 1964.
Hennig 1950.
Jenkins and Carpenter 1946.
Johannsen 1934.
King et al. 1960.
Matheson 1929.
Nielsen and Rees 1961.
Newson 1984.
Steffan and Evenhuis 1981.
Stojanovich 1960, 1961.
Stone 1981.
Vargas 1940.
Wesenberg-Lund 1943.

THAUMALEIDAE (CHIRONOMOIDEA)

B. A. Foote, ***Kent State University***

The Thaumaleids

Figure 37.152

Relationships and Diagnosis: Long considered to be rather closely related to the Chironomidae and Simuliidae, the Thaumaleidae have been placed recently in the Chironomoidea along with the two above-named families and the Ceratopogonidae (McAlpine, et al. 1981). Stone and Peterson (1981) have summarized our knowledge.

Larvae possess unpaired prolegs on the prothorax and caudal segment, an exserted head capsule, and sclerotized dorsal plates on the thoracic and abdominal segments.

Biology and Ecology: Larvae have only been found on vertical rock surfaces wetted by a thin flow of water (Stone and Peterson 1981). The rarely encountered adults are found in the vicinity of the larval habitat. The larval food apparently consists largely of diatoms.

Description: Mature larvae 2–4 mm, elongate and cylindrical, with conspicuous prolegs on prothorax and last abdominal segment, and with distinct, exserted head capsule; integument of last two thoracic and all abdominal segments bearing scattered or clustered stout setae dorsolaterally and laterally; amphipneustic; white, with dorsum of abdominal segments sclerotized.

Head: Eucephalic, non-retractile, bearing scattered bristles; antenna reduced; hypostoma elongate and tapering, upturned to give truncated appearance to head, bearing teeth apically.

Thorax: Segments sclerotized dorsally, each bearing a few stout setae dorsolaterally and scattered bristles laterally and ventrally; prothorax with single conspicuous, crochet-bearing proleg ventrally; anterior spiracles dorsolateral on prothorax, each on a short, tube-like structure.

Abdomen: Segments sclerotized dorsally, each bearing stout setae laterally and scattered bristles elsewhere; eighth segment with transverse respiratory opening near hind margin,

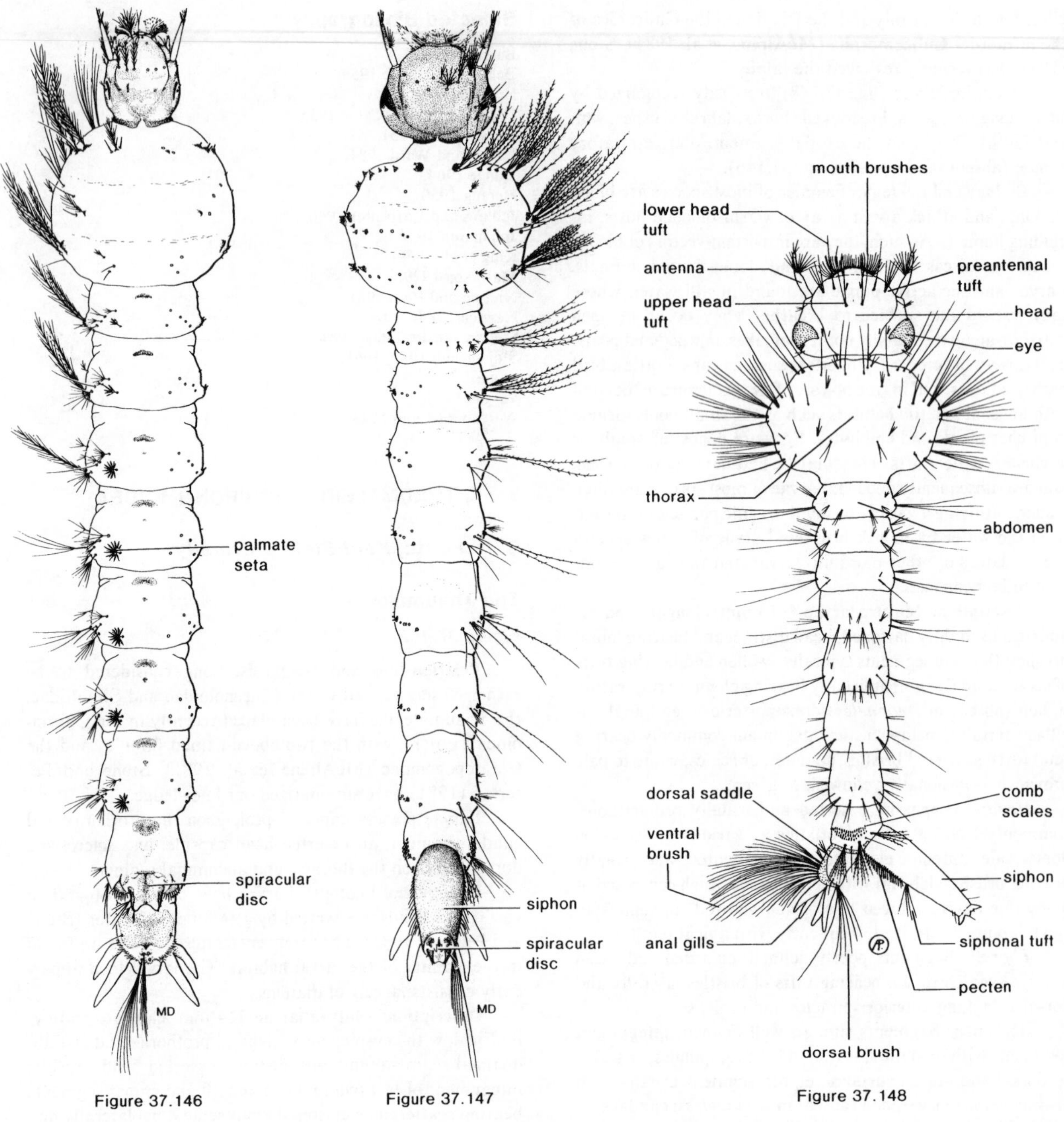

Figure 37.146

Figure 37.147

Figure 37.148

Figure 37.146. Culicidae. Dorsal of *Anopheles earlei* Vargas. Adults of various species of *Anopheles* serve as the primary vectors of malaria in tropical regions (from Manual of Nearctic Diptera, Vol. 1).

Figure 37.147. Culicidae. Dorsal of *Culiseta incidens* (Thomson) (from Manual of Nearctic Diptera, Vol. 1).

Figure 37.148. Culicidae. Dorsal of *Aedes stimulans* (Walker) (from Peterson 1951).

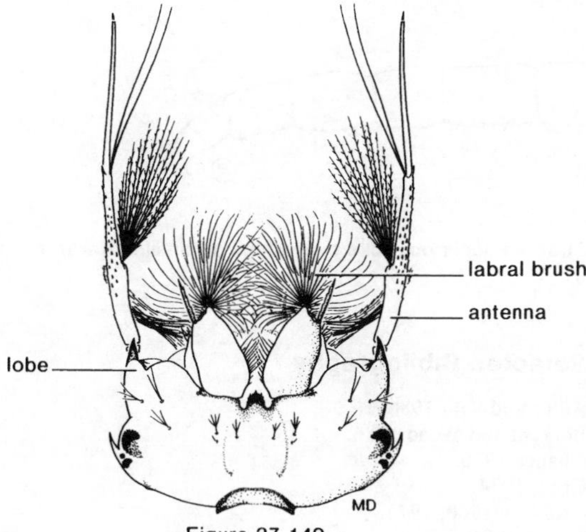

Figure 37.149

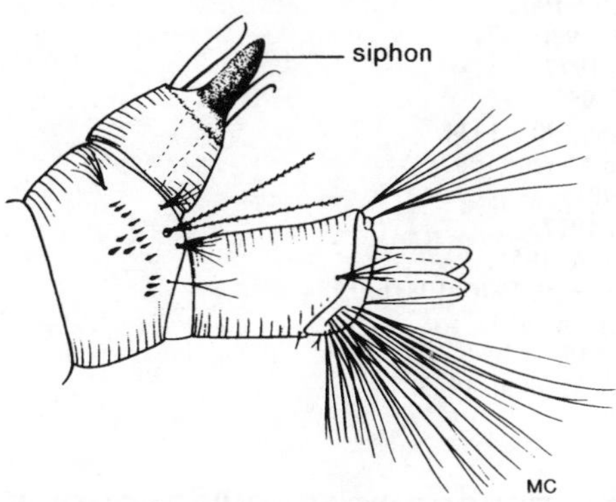

Figure 37.150

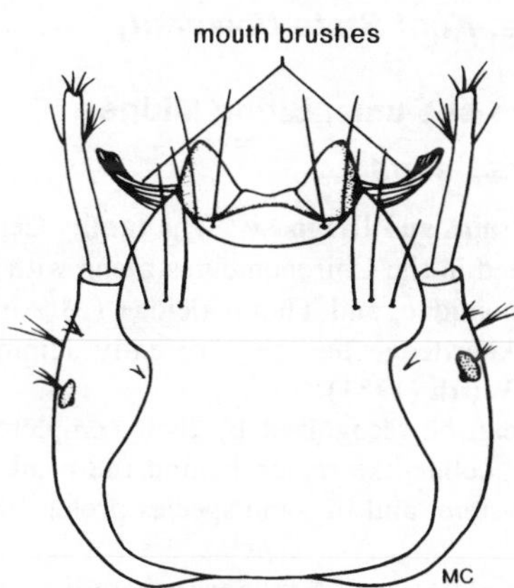

Figure 37.151

Figure 37.149. Culicidae. *Mansonia titillans* (Walker). Head capsule, dorsal (from Manual of Nearctic Diptera, Vol. 1).

Figure 37.150. Culicidae. *Mansonia* sp. Caudal segments, lateral, showing the pointed posterior siphon that can be inserted into intercellular air spaces in roots of wetland macrophytes (from Merritt and Cummins 1984).

Figure 37.151. Culicidae. Dorsal head capsule of *Toxorhynchites* sp. showing typical mouth brushes. Larvae are predators of other mosquito larvae living in tree holes (from Merritt and Cummins 1984).

opening bordered by several long, dark bristles; segment ending in distinct, hook-bearing proleg.

Comments: About 121 species in eight genera are known (Theischinger 1986). The American fauna north of Mexico consists of seven species in two genera, nearly all of which are western (Stone and Peterson 1981).

Selected Bibliography

Bischoff 1922.
Dyar and Shannon 1924.
Hennig 1950.
Leathers 1922.
Saunders 1923.
Stone 1964.
Stone and Peterson 1981.
Theischinger 1986.
Thienemann 1909.
Vaillant 1959.

SIMULIIDAE (CHIRONOMOIDEA)

B. A. Foote, *Kent State University*

Black Flies, Buffalo Gnats

Figures 37.153a–h

Relationships and Diagnosis: Although there have been disputes and disagreements concerning the correct phylogenetic affinities of the Simuliidae, there is general agreement today that they should be placed close to the Ceratopogonidae and Chironomidae. These three families plus the Thaumaleidae are presently included in the Chironomoidea of the infraorder Culicomorpha (McAlpine et al. 1981). B. V. Peterson (1981b) has recently summarized our knowledge of black flies.

Simuliid larvae are easily recognized by the complete, exserted head capsule that bears a pair of conspicuous labral fans (fig. 37.153a,b) (occasionally absent), the single prothoracic proleg, and the posteriorly swollen abdomen. The caudal segment bears numerous rows of minute, hooked spines on two modified prolegs.

Biology and Ecology: The larval and pupal stages of all black flies are aquatic, showing a strong preference for flowing water. They occur in a fair diversity of lotic habitats, ranging from small, spring-fed trickles to very slow-moving large rivers; a few are even hyporheic. The commonest feeding habit involves the filtering of small, water-borne particles via the pair of labral fans. A few species lack fans and graze on the organic film covering the substrate. Pupae are formed within cases attached to the substrate, with adults emerging under water in a bubble of air that carries them to the surface. There is commonly one generation a year for north temperate species, with overwintering as eggs or larvae. Adult emergence is concentrated in the late spring and early summer, an event well known in more northern regions and at higher elevations in more southerly ones. Females of some species are voracious blood feeders and constitute one of the most serious pests of humans and wild and domestic animals, not only because of

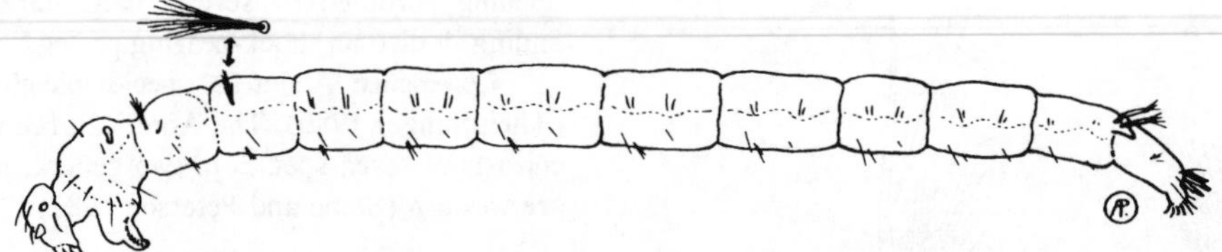

Figure 37.152. Thaumaleidae. Lateral view of *Thaumalea* sp. Larvae occur on rocks over which a thin film of water is flowing, feeding on diatoms and algae (from Peterson 1951).

their bloodsucking and subsequent reactions to the bites, but also because many species serve as vectors of human and animal diseases. Females of a few species don't suck blood, but acquire nutrients for energy from nectar.

Description: Mature larvae 5–15 mm, nearly cylindrical but with abdomen distinctly swollen posteriorly, with single proleg on venter of prothorax; apneustic; integument bare or with variety of spinules, scales, or setae; light brown to blackish brown.

Head: Eucephalic, exserted, large and distinct; anterolateral corners of frontoclypeal apotome in most species bearing a conspicuous pair of labral fans, the stalk of each fan composed of short basal and long distal segment, and the fan composed of three subunits; antenna 3-segmented, variously pigmented; hypostoma strongly sclerotized, commonly subtriangular with anterior margin bearing series of teeth; teeth often arranged into central cluster and two lateral clusters; mandible heavily sclerotized, rectangular, and bearing 3–5 large teeth apically as well as shorter teeth elsewhere; mandible also with series of simple to complex bristles; maxilla consisting of large median lobe and smaller laterobasal lobe, median lobe bearing five groups of bristles or brushes; labrum and hypopharynx apparently fused into single sclerite that bears variety of setae and sensillae.

Thorax: Segments poorly differentiated; prothorax bearing single 2-segmented ventral proleg with an apical ring of minute spinules; no anterior spiracles.

Abdomen: Basal segments rather slender but distal segments distinctly swollen; some larvae possess subconical tubercles dorsolaterally on first five segments; caudal segment bearing apical ring of many rows of minute, blackened spinules; rectum with three extrusible, simple or compound anal papillae.

Comments: The family contains some 500 species and over 60 genera in the world, with about 150 species in 11 genera recorded for America north of Mexico (B. V. Peterson 1981b). Black flies are pests for two reasons. One, of course, is their habit of sucking blood, a capability that literally makes some stream-side areas uninhabitable during the biting season. The second and more significant reason is that several species are vectors of human and animal diseases. Perhaps the most notorious is onchocerciasis, river blindness, a frightening ailment caused by the filarial worm *Onchocera volvulus* Leuckart that is widespread in suitable areas of tropical Africa, Central America, and northern South America. The medical significance of various species of Simuliidae is covered by Crosskey (1973).

Selected Bibliography

Adler and Kim 1986.
Borkent and Wood 1986.
Chance 1970.
Craig 1974.
Crosskey 1960, 1973.
Currie 1986.
Davis 1974.
Dumbleton 1962.
Edwards 1920.
Fredeen 1977.
Hennig 1950.
Johannsen 1903, 1934.
Kim and Merritt 1987.
Laird 1981.
Malloch 1917.
Peterson, A. 1951.
Peterson, B. V. 1970, 1981b, 1984.
Sommerman 1953.
Twinn 1936.
Wood, D. M. et al. 1963.

CERATOPOGONIDAE (CHIRONOMOIDEA)

B. A. Foote, *Kent State University*

Punkies, No-see-ums, Biting Midges

Figures 37.154–37.161

Relationships and Diagnosis: The family Ceratopogonidae is included in the Chironomoidea along with the Simuliidae, Chironomidae, and Thaumaleidae (McAlpine, et al. 1981). Our knowledge has been recently summarized by Downes and Wirth (1981).

Larvae can be recognized by their complete, exserted head capsule, collar-like region behind the head, apneustic respiratory system, and in some species prolegs on the prothorax.

Biology and Ecology: Adults are generally found in moist environments in or near the larval habitat. However, a few species, such as those that live within the decaying tissues of cacti, occur in very arid surroundings. Several taxa are abundant in somewhat unusual habitats such as alkaline and saline marshes and pools, coastal beaches and salt marshes, and in heavily fertilized farmland. A great majority of the species are crepuscular, although a few forms occur in bright sun. Females of most species require a protein-rich meal before eggs can be matured. Many taxa are blood suckers of mammals and birds, but a few seemingly prefer reptilian or amphibian blood. Other species feed on fluids commonly obtained

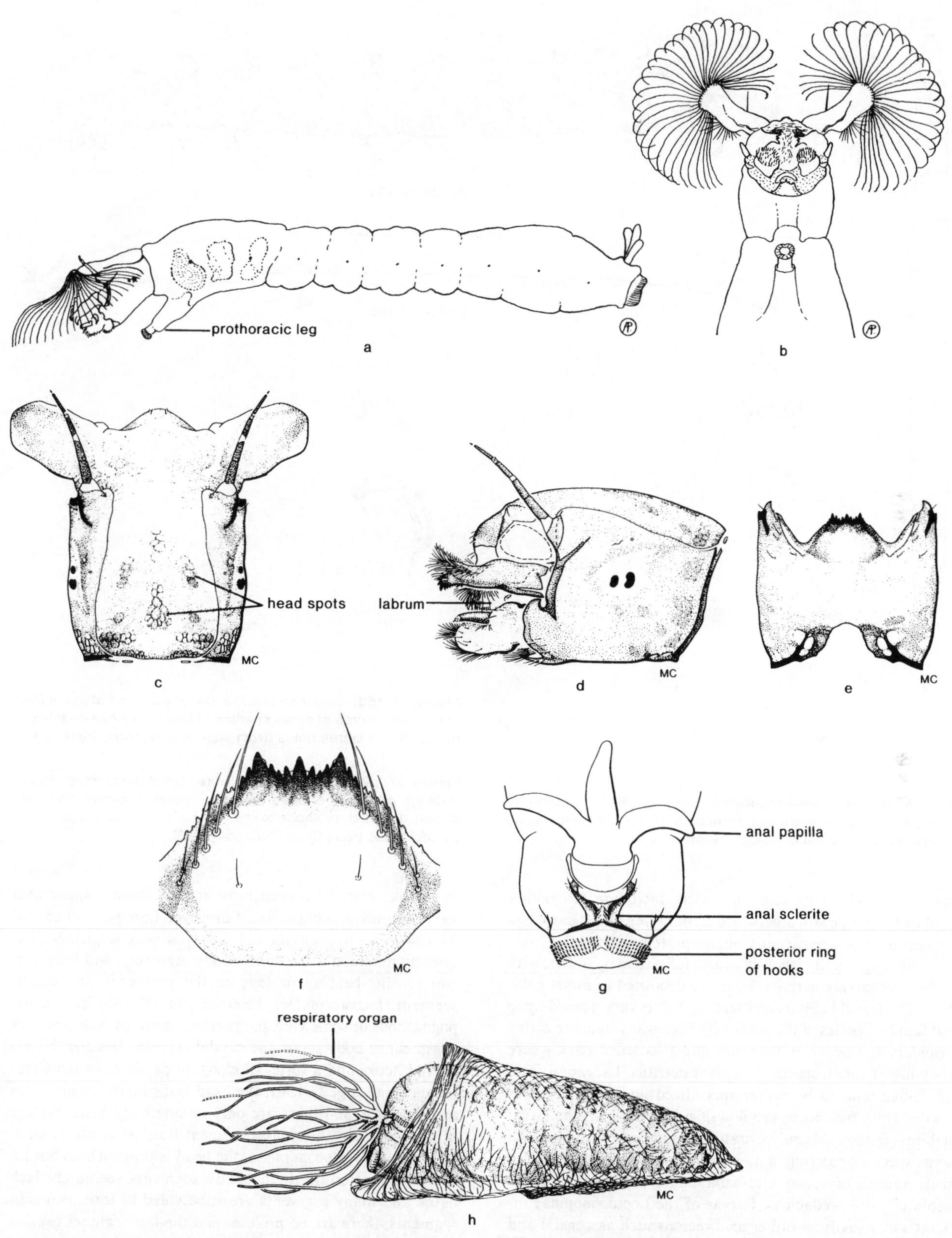

Figure 37.153a-h. Simuliidae. *Simulium vittatum* Zetterstedt. **a.** lateral; **b.** ventral view of anterior showing labral fans; **c.** head capsule, dorsal; **d.** head capsule, lateral; **e.** head capsule, ventral, appendages removed; **f.** ventral view of anterior portion of head capsule showing hypostomal teeth; **g.** dorsal view of tip of abdomen; **h.** pupa, lateral (figs. a, b from Peterson 1951, c-h from Merritt and Cummins 1984).

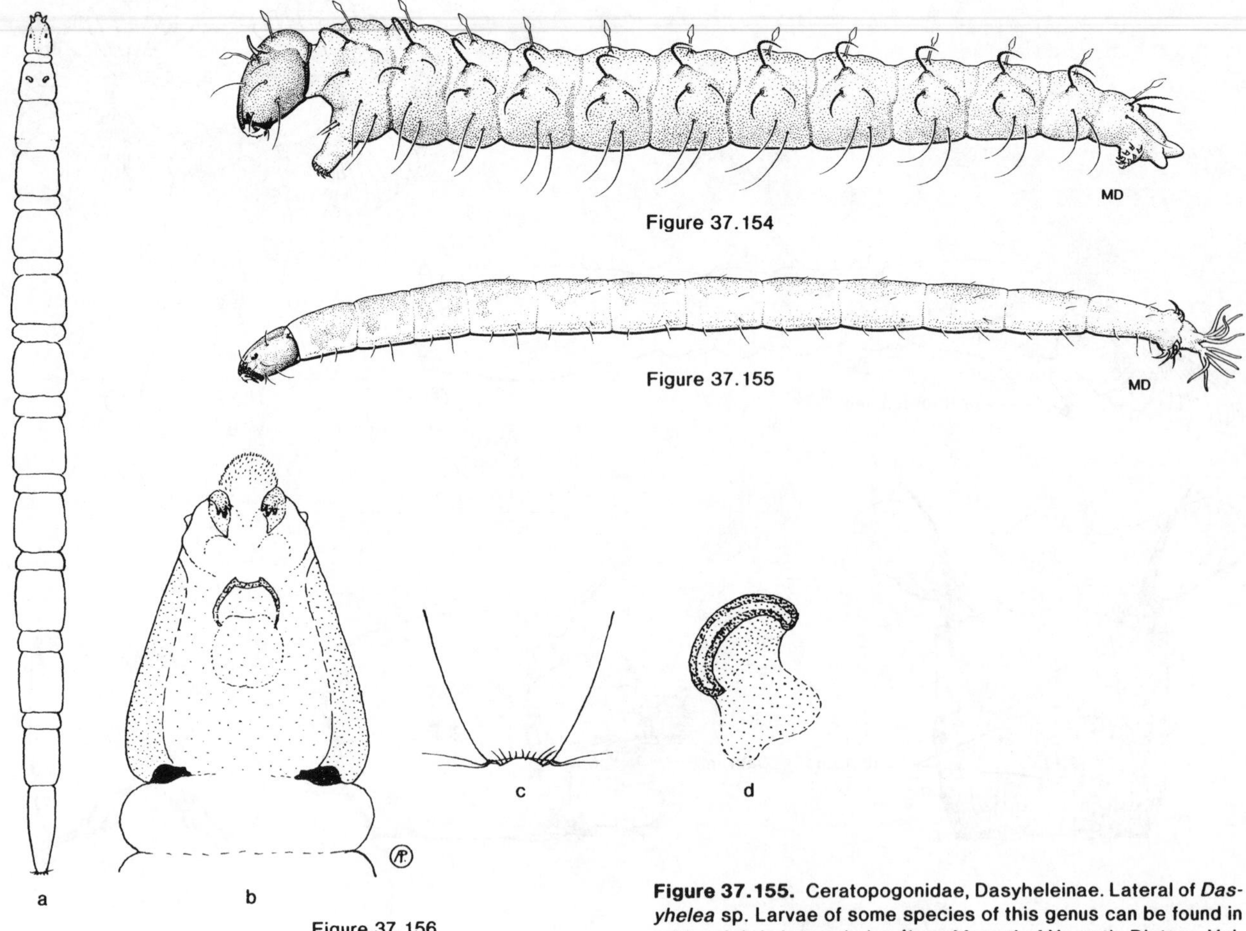

Figure 37.154. Ceratopogonidae, Forcipomyiinae. Lateral of *Forcipomyia* sp. Larvae occur under bark, in leaf litter, and among mosses (from Manual of Nearctic Diptera, Vol. 1).

Figure 37.155. Ceratopogonidae, Dasyheleinae. Lateral of *Dasyhelea* sp. Larvae of some species of this genus can be found in rotting debris in tree holes (from Manual of Nearctic Diptera, Vol. 1).

Figure 37.156a-d. Ceratopogonidae, Ceratopogoninae. *Palpomyia* sp. **a.** dorsal; **b.** head capsule, ventral; **c.** caudal tip of abdomen, dorsal; **d.** prothoracic spiracle. Larvae occur in slime fluxes on deciduous trees (from Peterson 1951).

by piercing the wing veins of insects, including dragonflies and moths, while still others feed on decaying insect carcasses or nectar. A few apparently obtain protein from pollen.

Mating usually involves swarm formation by males, with mating occurring aerially. Eggs are deposited in moist habitats. The larval habitats and feeding habits vary according to subfamily. Species of the subfamily Forcipomyiinae are active crawlers and occur in moss and under loosened bark where they ingest fungi, spores, and plant detritus. Larvae of Dasyheleinae tend to be rather specialized and have been recorded from tree holes, sap flows, and rock pools. The larval habitats in the subfamily Ceratopogoninae are variable, with some species occurring in moist soil while others are found in truly aquatic or subaquatic habitats. Several species of this subfamily are predacious. Larvae of the Leptoconopinae inhabit soil, usually in rather arid regions such as coastal and inland beaches.

Description: Mature larvae 2–10 mm, elongate and frequently eel-like, with distinct exserted head capsule and short collar-like region between head and prothorax; appearance varying among subfamilies. Larvae of Forcipomyiinae (fig. 37.154) have hypognathous heads, possess conspicuous, frequently modified setae on the body segments, and have ventral crochet-bearing prolegs on the prothorax and caudal segment. Larvae of Dasyheleinae (fig. 37.155) have a hypognathous or somewhat prognathous head capsule and lack conspicuous body setae; the caudal segment has sizeable recurved hooks and a series of elongate papillae. In the Ceratopogoninae (fig. 37.156), the head is distinctly prognathous and elongate, and there are only minute body hairs, prolegs are lacking, and the caudal segment bears elongate bristles. In larvae of Leptoconopinae, the head is prognathous but incompletely sclerotized, the body segments seemingly lack hairs, and many segments are subdivided to form two subsegments; there are no prolegs, and the last segment bears a few elongate bristles. Apneustic or propneustic.

Head: Eucephalic, strongly sclerotized and exserted; prognathous or hypognathous; mandibles well developed,

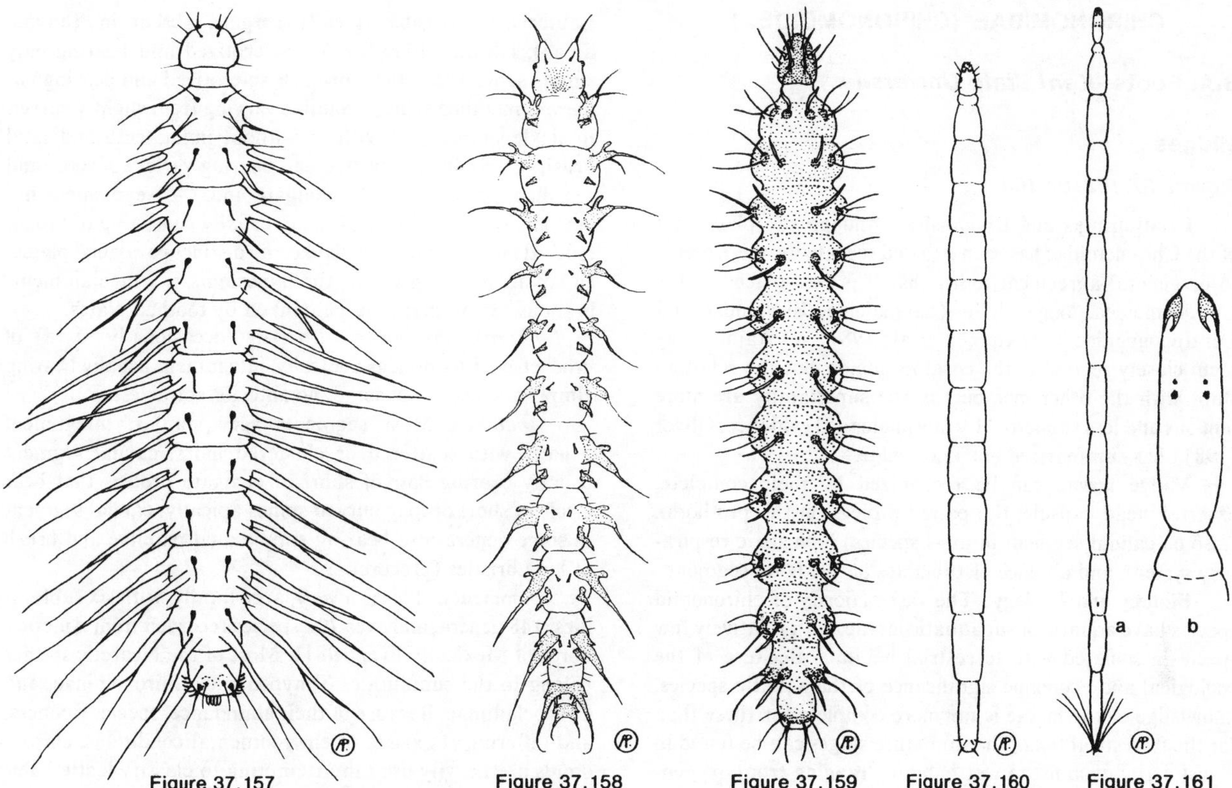

Figure 37.157 Figure 37.158 Figure 37.159 Figure 37.160 Figure 37.161

Figure 37.157. Ceratopogonidae, Forcipomyiinae. Dorsal of *Atrichopogon peregrinus* Johannsen. Larvae can be found in mats of floating algae in shallow ponds and marshes (from Peterson 1951).

Figure 37.158. Ceratopogonidae, Forcipomyiinae. Dorsal of *Atrichopogon levis* Coquillett. Larvae have been found in mats of algae growing on wet soil (from Peterson 1951).

Figure 37.159. Ceratopogonidae, Forcipomyiinae. *Forcipomyia brevipennis* (Macquart), dorsal. Found in damp situations under boards, bark, dung pats, etc. (from Peterson 1951).

Figure 37.160. Ceratopogonidae, Ceratopogoninae. *Culicoides* sp., dorsal. Found in wet, decomposing plant material (from Peterson 1951).

Figure 37.161a,b. Ceratopogonidae, Ceratopogoninae. *Bezzia glabra* (Coquillett). **a.** larva, dorsal; **b.** head, dorsal. Found in algae-covered ponds (from Peterson 1951).

toothed, not opposed to each other; conspicuous pharyngeal structure present internally, consisting of two diverging arms and a series of food-filtering combs.

Thorax: Anterior spiracles on prothorax in some species, but lacking in most. Prolegs present or absent, according to subfamily.

Abdomen: Segments may have numerous projections (*Atrichopogon*), bear modified (*Forcipomyia*), or simple setae, or appear nearly smooth and bare. Caudal segment usually bears elongate bristles and may have crochet-bearing prolegs.

Comments: Four subfamilies and some 60 genera are recognized in the world. Approximately 380 species plus a horde of undescribed ones are known for America north of Mexico (Downes and Wirth 1981). Because of their biting habits alone, the family is economically important. In addition, several species in the tropics are important vectors of human and animal diseases. These include a few filarial worms, blood parasites, and certain viruses.

Selected Bibliography

Atchley 1970.
Blanton and Wirth 1979.
Boesel 1973.
Boesel and Snyder 1944.
Downes 1978.
Downes and Wirth 1981.
Hennig 1950.
Johannsen 1937b, 1952.
Malloch 1915, 1917.
Mayer 1934.
Saunders 1956.
Wirth 1952, 1956.
Wirth and Hubert 1960.

CHIRONOMIDAE (CHIRONOMOIDEA)

B.A. Foote, *Kent State University*

Midges

Figures 37.162–37.166

Relationships and Diagnosis: Although the placement of the Chironomidae has been debated repeatedly, there seems to be general agreement today that it is best placed in the Chironomoidea along with the Thaumaleidae, Simuliidae, and Ceratopogonidae (McAlpine, et al. 1981). Certainly they seem closely related to the ceratopogonids, but the relationships with the other members of the superfamily are more tenuous and less supported by morphological evidence. Oliver (1981) has summarized our knowledge.

Midge larvae can be recognized by their complete, exserted head capsule, the paired prolegs on the prothorax (also on caudal segment in most species), apneustic respiratory system, and absence of tubercles on the body segment.

Biology and Ecology: The vast majority of chironomid species have aquatic or subaquatic larvae, with relatively few species associated with terrestrial habitats. Because of the ecological and economic significance of the aquatic species, knowledge of the larvae is far more complete for these than for the terrestrial taxa. The immature stages can be found in a vast array of moist to wet habitats, ranging from oxygen-poor sediments of deep lakes to the saturated soils of marshes, from riffles in fast-flowing streams to reservoirs on large rivers, from inland bodies of fresh or saline or highly mineralized waters to coastal marine environments. Larvae of most species, exclusive of the free-living taxa in the subfamilies Tanypodinae and Podonominae, construct cases composed of fine substrate particles bound together by a salivary secretion. Most larvae are microphagous, feeding on finely divided detritus, algal cells, and small animals. Larvae of the Tanypodinae and a few species in other subfamilies are predacious. Different species have more or less restricted periods during which adult emergence occurs. Adults frequently form mating swarms, particularly during the evening hours, with the swarms consisting almost entirely of males. Mating is initiated when a female flies into the swarm but may be completed on a nearby substrate. A few species copulate on the ground without forming swarms. Adults are frequently assumed not to feed, but a few species have been observed to ingest honeydew.

Description: Mature larvae 1.5–30 mm, elongate and cylindrical, slender with well-developed, exserted, prognathous head; commonly with pair of ventral prolegs on prothorax and on last abdominal segment; apneustic; integument usually appearing bare or with scattered small setae; whitish, yellowish, brownish, greenish, pinkish, or reddish.

Head: Eucephalic, well sclerotized, non-retractile, usually smooth but occasionally rugose or patterned and bearing simple to plumose bristles; eyespots usually noticeable on sides of head; antenna four (Tanypodinae) or five, six, or seven segmented, either retractile within cranium (Tanypodinae) or non-retractile, sometimes greatly reduced so segmentation becomes obscure; second and third segments usually with Lauterborn's organs apically (organs absent in Tanypodinae); labrum either weakly sclerotized and bearing only simple setae or sensillae, or more sclerotized and bearing numerous modified setae; mandible varying from slightly curved to sickle-shaped and with or without inner teeth and setal brushes; maxilla two-parted and bearing variety of setae and sensillae; palpus short to elongate and 1–6 segmented; hypostoma on ventral side of head, usually consisting of broad, toothed plate that may be flanked by distinct paralabial plates. In larvae of Tanypodinae, the hypostoma is a median membranous region that may be flanked by toothed plates.

Thorax: Prothorax with pair (occasionally fused) of crochet-bearing prolegs ventrally, integument usually bearing simple and scattered setae; no anterior spiracles.

Abdomen: Most segments very similar; integument usually with scattered or clustered hairs; caudal segment usually bearing pair of short or elongate prolegs that bear circle of short or long curved spines apically; caudal segment in some genera also bearing simple anal papillae and brush of long bristles (procercus).

Comments: This is a very large family, with six subfamilies, 148 genera, and over 700 species recorded from America north of Mexico (Oliver 1981). Most of the Nearctic species belong to the subfamilies Tanypodinae, Chironominae, and Orthocladiinae. Because of their abundance, species richness, and differing responses to environmental conditions, chironomids are heavily used in attempting to classify lentic habitats, particularly lakes. More recently, they have become useful indicators of water quality in flowing water systems. Additionally, numerous species are important food for fish. Mating swarms can be so abundant in certain lakeside areas that they become pestiferous.

Selected Bibliography

Arntfield 1977.
Baker 1979.
Beck 1976.
Beckett and Lewis 1982.
Berg 1950b.
Bode 1983.
Boesel 1974.
Brundin 1948, 1949, 1956.
Bryce and Hobart 1972.
Coffman and Ferrington 1984.
Curry 1958, 1965.
Davies 1976.
Downes 1969, 1974.
Fittkau 1962.
Hamilton et al. 1969.
Hansen and Cook 1976.
Hennig 1950.
Johannsen 1905, 1937a, 1937b.
Johannsen and Townes 1952.
Malloch 1915, 1917.
Mason 1973.
McCauley 1974.
Meier and Torres 1978.
Oliver 1968, 1971, 1981.
Pinder 1986.
Roback 1963, 1966a, 1966b, 1969, 1974, 1976, 1977a, 1977b, 1981.
Saether 1975.
Saunders 1928.
Schlee 1968.

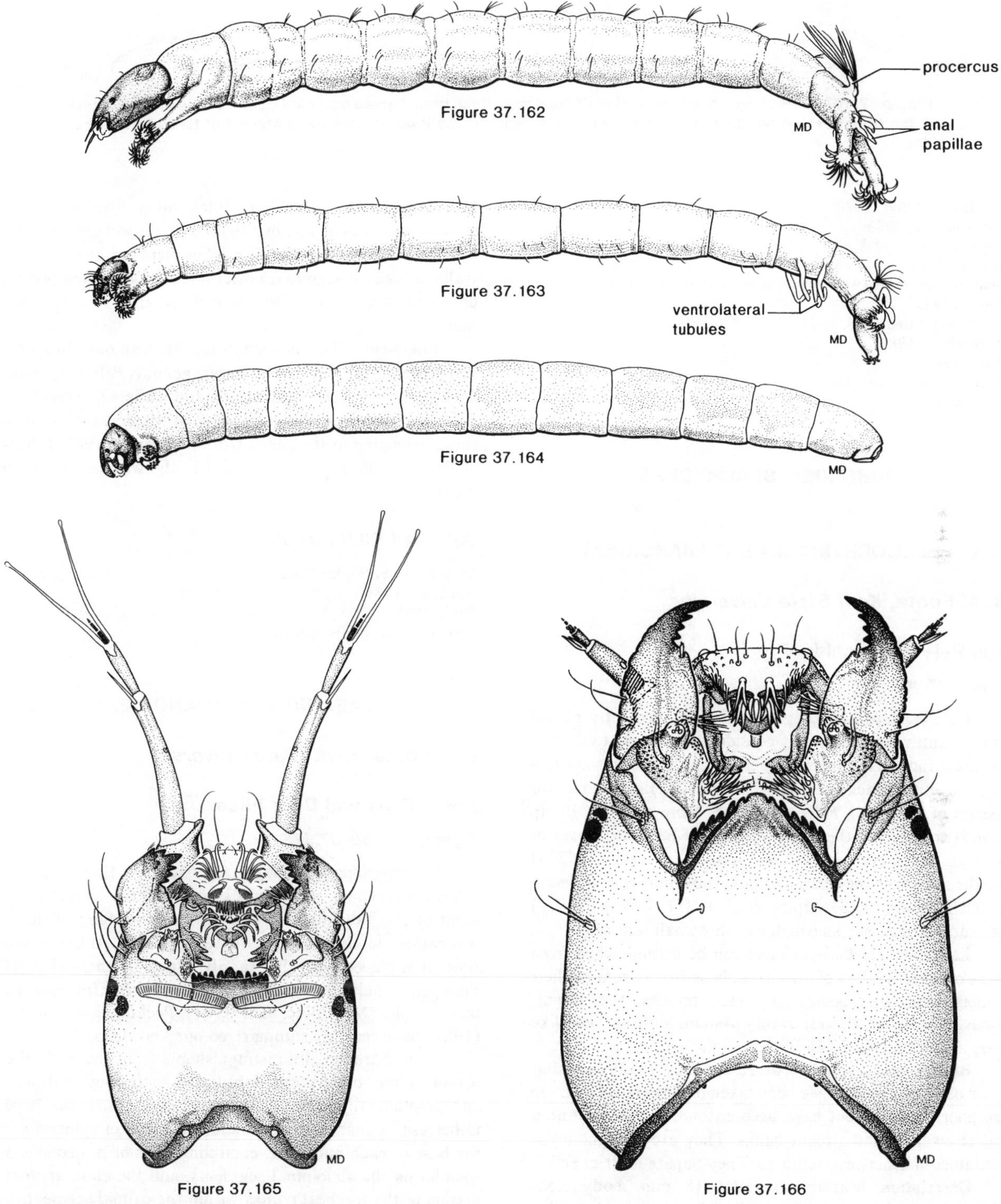

Figure 37.162. Chironomidae. *Ablabsemyia* sp., lateral. The larvae are predators in diverse aquatic habitats (from Manual of Nearctic Diptera, Vol. 1).

Figure 37.163. Chironomidae. *Chironomus* sp., lateral. Burrowers in aquatic bottom sediments (from Manual of Nearctic Diptera, Vol. 1).

Figure 37.164. Chironomidae. *Pseudosmittia* sp., lateral. Found in lotic habitats (from Manual of Nearctic Diptera, Vol. 1).

Figure 37.165. Chironomidae. Head capsule of *Micropsectra* sp., ventral. Found on bottom sediments and detritus (from Manual of Nearctic Diptera, Vol. 1).

Figure 37.166. Chironomidae. Head capsule of *Cricotopus* sp., ventral. Found on vegetation, detritus, and in sediments (from Manual of Nearctic Diptera, Vol. 1).

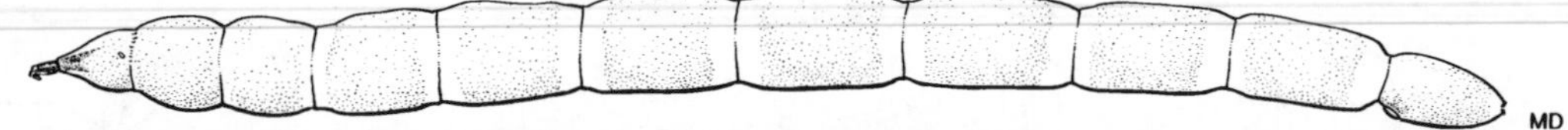

Figure 37.167. Pelecorhynchidae. Lateral of *Glutops rossi* Pechuman. Larvae occur in saturated soils along streams in the Northwest, where they are believed to prey on soft-bodied invertebrates (from Manual of Nearctic Diptera, Vol. 1).

Simpson and Bode 1980.
Simpson et al. 1983.
Soponis 1977.
Sublette 1964.
Thienemann 1937, 1944, 1954.
Townes 1945.
Wesenberg-Lund 1943.
Wiederholm 1983.
Wirth 1949.
Wirth and Sublette 1970.
Wootton 1972.

SUBORDER BRACHYCERA

PELECORHYNCHIDAE (TABANOIDEA)

B. A. Foote, *Kent State University*

The Pelecorhynchids

Figure 37.167

Relationships and Diagnosis: Species presently placed in this family were originally considered to belong to the Tabanidae, and the family name Pelecorhynchidae was not recognized until Mackerras and Fuller (1942) erected it for species of the genus *Pelecorhynchus.* More recently, Philip (1965) transferred the monotypic genus *Bequaertomyia* of western North America to this family, and Teskey (1970a) did the same with the genus *Glutops.* The family is now placed in the Tabanoidea (McAlpine et al. 1981). Teskey (1981a) has summarized our knowledge of this small family.

Larvae of Pelecorhynchidae can be distinguished from those of other families of orthorrhaphous Brachycera by their smooth, wormlike bodies that lack prolegs, the amphipneustic or apneustic respiratory system, and the bead-like body segments (fig. 37.167).

Biology and Ecology: Adults are rarely collected, but those of a few species have been taken on flowers. The larvae are poorly known, but have been encountered in saturated soil of swamps and stream banks. They are thought to be predators of other invertebrates. They pupate in drier soil.

Description: Mature larvae 10–18 mm, body elongate and cylindrical, tapering at both ends, with 11 similar non-telescoping, bead-like segments; respiratory system apneustic (*Glutops*) or amphipneustic (*Pelecorhynchus*); translucent white.

Head: Capsule capable of complete retraction into thorax, similar in structure to Tabanidae and Rhagionidae, but differing from those families in having stout spines laterally on labrum and apically on maxilla.

Thorax and Abdomen: Integument firm and thick, smooth and glossy, lacking striations, projections, or other surface irregularities; swollen lobes flanking anal slit ventrally on caudal segment; posterior spiracles, when present, on upper surface of transverse cleft at apex of caudal segment.

Comments: This is a small family, with only 46 species recognized in the world. The largest genus is *Pelecorhynchus,* containing 34 species in Australia and Chile. Only eight species are known from America north of Mexico, with seven of these belonging to the genus *Glutops* (Teskey 1981a). Most of the Nearctic species are found in the western states and provinces.

Selected Bibliography

Mackerras and Fuller 1942.
Nagatomi 1975.
Philip 1965.
Teskey 1970a, 1970b, 1981a.

TABANIDAE (TABANOIDEA)

B A. Foote, *Kent State University*

Horse Flies and Deer Flies

Figures 37.168–37.175

Relationships and Diagnosis: Although there has been some disagreement in the past concerning the correct placement of the Tabanidae, it was generally recognized that it was rather closely related to the Rhagionidae. At the present time, it is placed with that family and the Athericidae and Pelecorhynchidae in the Tabanoidea of the infraorder Tabanomorpha (McAlpine, et al. 1981). Pechuman and Teskey (1981) have recently summarized our knowledge.

Larvae are usually distinguishable from those of other orthorraphous brachyceran families by their tough, subshiny, and longitudinally striated integument, and cylindrical shape. Other useful characteristics include a brush of spines above the base of each mandible, encircling rows of projections or spicules on the abdominal segments, and the close approximation of the tracheal trunks within the caudal segment.

Biology and Ecology: Adult females of most Tabanidae are blood sucking, fulfilling their protein needs by ingesting blood of warm-blooded animals. A few species are reported to utilize blood of cold-blooded vertebrates, and several species do not use blood at all, but are thought to obtain their energy from nectar.

Larvae of most species occur in moist to saturated soils or in aquatic habitats, but some larvae have been encountered in relatively dry terrestrial sites. Larvae of most species are

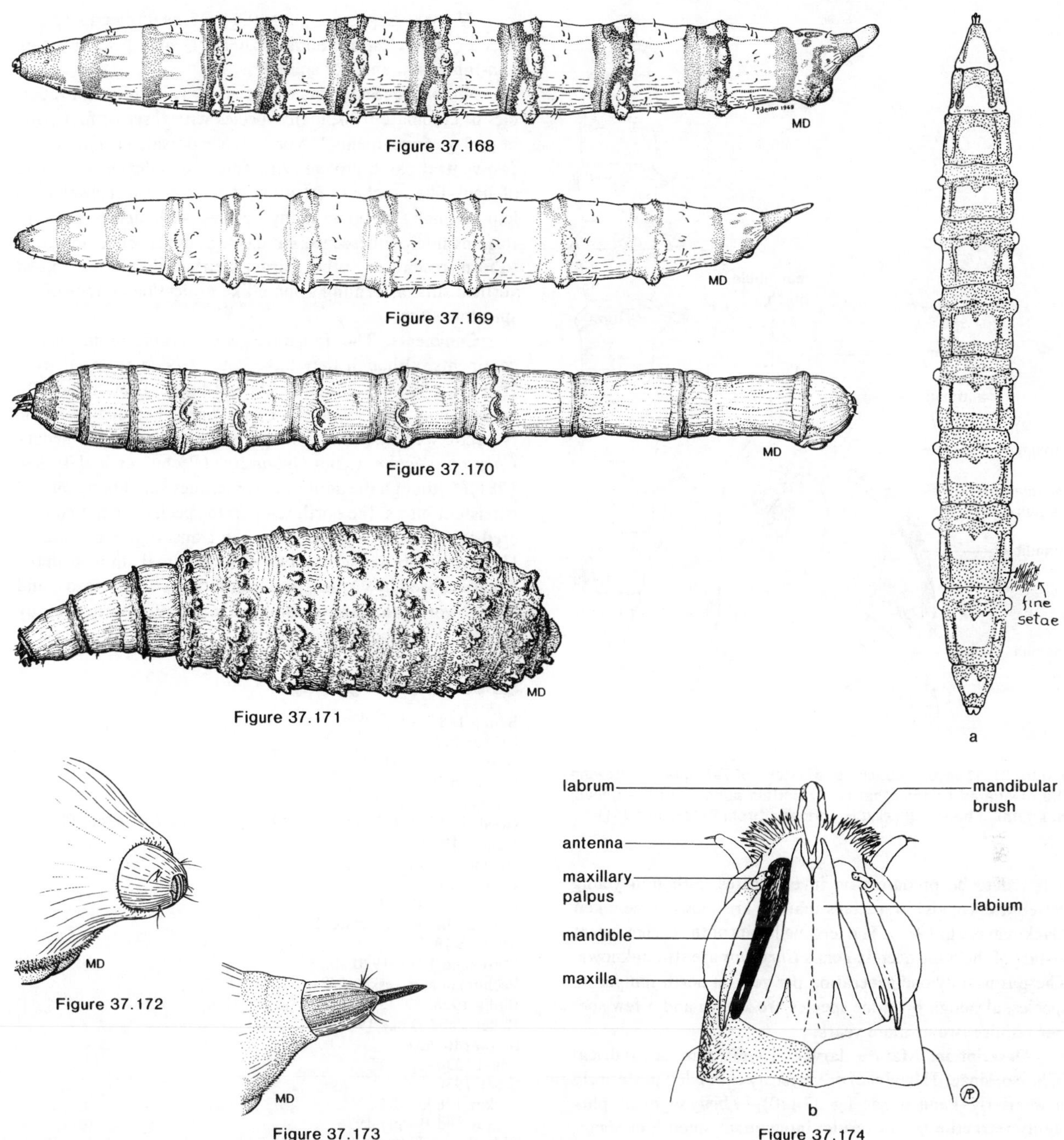

Figure 37.168. Tabanidae. *Tabanus reinwardtii* Wiedemann, lateral, a common species in the eastern states and provinces (from Manual of Nearctic Diptera, Vol. 1).

Figure 37.169. Tabanidae. *Chrysops furcatus* Walker, lateral. Larvae of this widespread species occur in wet soils (from Manual of Nearctic Diptera, Vol. 1).

Figure 37.170. Tabanidae. *Esenbeckia delta* (Hine), lateral. This species is restricted to the southwestern states (from Manual of Nearctic Diptera, Vol. 1).

Figure 37.171. Tabanidae. *Goniops chrysocoma* (Osten Sacken), lateral. The strangely swollen larvae can be found in woodland leaf litter in the eastern states (from Manual of Nearctic Diptera, Vol. 1).

Figure 37.172. Tabanidae. Posterior end of *Tabanus marginalis* Fabricius showing the slit-like spiracles (from Manual of Nearctic Diptera, Vol. 1).

Figure 37.173. Tabanidae. Posterior end of *Chrysops cincticornis* Walker showing the spine-like spiracles that can be inserted into air spaces in roots of aquatic macrophytes (from Manual of Nearctic Diptera, Vol. 1).

Figure 37.174a,b. Tabanidae. **a.** dorsal of *Tabanus atratus* Fabricius, the black horse fly; **b.** ventral of head, showing mouthparts. The very large larvae can be found in streams where they prey on soft-bodied invertebrates. The adult females are notorious biters of man and livestock (from Peterson 1951).

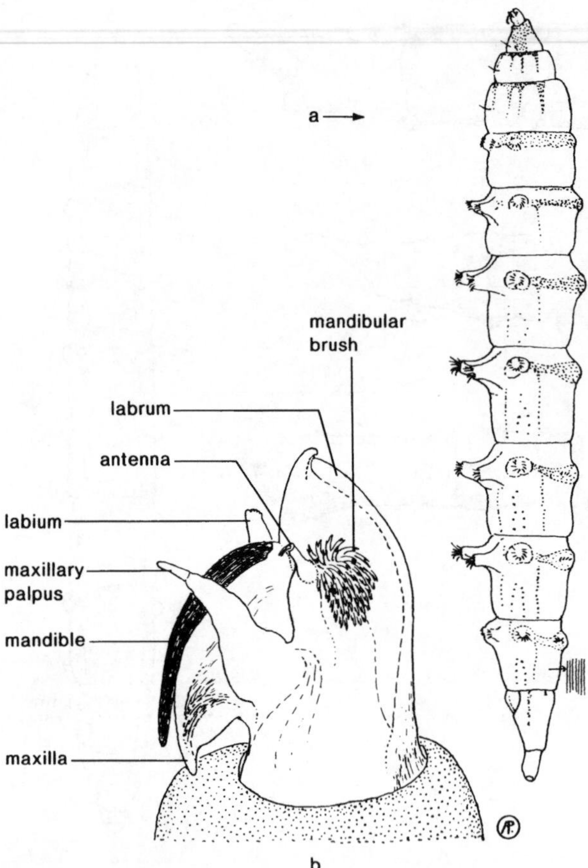

Figure 37.175a,b. Tabanidae. **a.** lateral of *Tabanus* sp. showing the elongated prolegs that typify certain species of the genus; **b.** lateral of head, showing mouthparts (from Peterson 1951).

believed to be predators on invertebrates, although young vertebrates (toads) in special situations are known to be taken (Jackman et al., 1983). The feeding habits of the earlier larval instars of the huge deer fly genus *Chrysops* are still unknown. There is usually one generation per year in north temperate species, although bivoltine species are known and a few species require two or more years.

Description: Mature larvae 15–60 mm, cylindrical (Chrysopsinae, Tabaninae) or variously expanded posteriorly or anteriorly (Pangoniinae, fig. 37.170), 11 body segments plus highly retractile head capsule; integument smooth or somewhat wrinkled or tuberculate, bare or somewhat setose, commonly with longitudinal striations; metapneustic; white, yellowish, or greenish, some species with distinct pigmented pattern.

Head: Hemicephalic, elongate and somewhat cylindrical; cranium somewhat tubular; tentorial arms slender, attached to inner wall of cranium anteriorly and free posteriorly; no metacephalic rod; labrum well developed, downturned; mandibles slender and curved, apically pointed and bearing a ventral canal for injection of paralytic poison, and possessing subdorsal brush of spines; maxillae and labium forming floor of head capsule anteriorly, maxillary palpi elongate; antennae elongate, apparently three-segmented.

Thorax: Segments generally lacking projections, prolegs, or tubercles; prothorax posteriorly may bear vestigial, non-functional anterior spiracles.

Abdomen: Segments encircled by rings of small tubercles or crenulate frills, many species with three or four pairs of prolegs on segments 1–5 or 1–7, one dorsal, one lateral and two ventral pairs; prolegs with apical spinules or with rows of hook-like crochets; integument with fine to conspicuous longitudinal striations; many species with dense, pattern-forming micropubescence; caudal segment short or elongated to form respiratory tube, spiracles either closely juxtaposed vertical slits or forming a spine-like projection at apex of respiratory tube.

Comments: The Tabanidae is a sizeable family on all continents, although they have not yet been reported from Hawaii, Greenland, and Iceland. Three subfamilies, 26 genera, and over 350 species have been recorded from the Nearctic Region, with most species belonging to the genera *Chrysops, Tabanus,* and *Hybomitra* (Pechuman and Teskey 1981). Although the adults are sometimes very abundant and persistent biters, the north temperate species are not considered to be overly significant in the transmission of disease. However, there are several human and animal ailments in the tropics caused by viruses, bacteria, rickettsiae, protozoa, and filarial worms that are transmitted to the vertebrate hosts by a variety of blood-sucking tabanids.

Selected Bibliography

Beling 1882.
Burger 1977.
Burger et al. 1981.
Cameron 1926.
Dukes et al. 1974.
Goodwin 1972, 1973a, 1973b, 1973c, 1974, 1976a, 1976b, 1976c.
Hennig 1950.
Johannsen 1935.
Lane 1975.
Leclercq 1960, 1966.
Magnarelli and Anderson 1978.
Malloch 1917.
Marchand 1917, 1920.
Pechuman and Teskey 1981.
Philip 1928, 1931.
Roberts and Dicke 1964.
Schwardt 1931.
Segal 1936.
Stone 1930.
Teskey 1969, 1984.
Teskey and Burger 1976.
Tidwell and Tidwell 1973.

ATHERICIDAE (TABANOIDEA)

B. A. Foote, *Kent State University*

Aquatic Snipe Flies

Figure 37.176

Relationships and Diagnosis: Species of this family were earlier included in the Rhagionidae until 1973 when Stuckenberg erected the family to accommodate several genera.

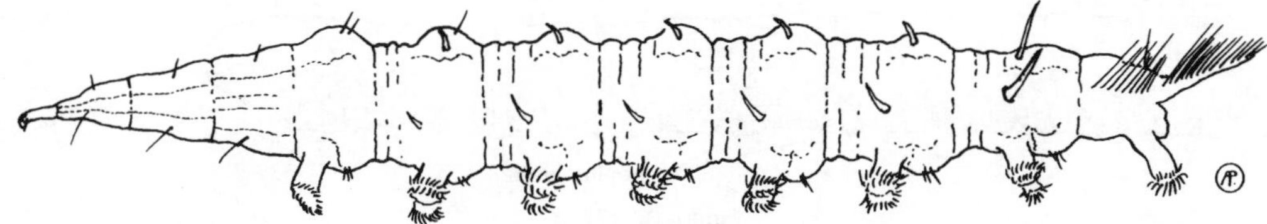

Figure 37.176. Athericidae. *Atherix variegata* Walker, lateral. The predacious larvae are abundant in riffle areas of streams (from Peterson 1951).

The family is obviously closely related to the Rhagionidae and Tabanidae, and all three are now placed in the Tabanoidea along with the Pelecorhynchidae (McAlpine et al. 1981). Webb (1981a) has summarized our knowledge of the family.

Larvae are distinguished by their crochet-bearing ventral prolegs, lateral slender tubercles, and two longer caudal tubercles. They are apneustic and occur in relatively fast-flowing streams.

Biology and Ecology: Adults of the Nearctic species are frequently found on bridges and vegetation bordering or overhanging streams. Adults of most species apparently feed on honeydew, but those of *Suragina* are blood suckers. Females of species of *Atherix* congregate in sizeable masses on overhanging streamside vegetation where they oviposit. Hatching larvae fall into the stream where they adopt a predacious mode of life in riffle areas. There is some evidence that they feed preferentially on larvae of Ephemeroptera and Chironomidae. Pupae are formed in moist soil along the edges of streams. The immature stages of the Nearctic species of *Suragina* are unknown.

Description: Mature larvae 10–17 mm, elongate and cylindrical, tapering anteriorly, with ventral prolegs; amphipneustic; integument wrinkled, tuberculate, and setose on abdominal segments; dark gray to brown.

Head: Very small, resembling that of Tabanidae, retractile into thorax; hemicephalic, no metacephalic rods but tentorial arms present; mandibles elongate and slender, subparallel; maxillary palpi elongate, 3-segmented; antennae elongate, 1-segmented, bearing four slender sense organs apically and numerous curved setae basally.

Thorax: Segments cone-shaped, each increasing in width and length caudally; lacking anterior spiracles; integument relatively bare and smooth; without prolegs.

Abdomen: Segments 1–7 with pairs of pointed tubercles subdorsally and laterally, and with paired prolegs ventrally; each proleg bearing semicircles of curved hooks apically and subapically; caudal segment with single proleg, and bearing two elongate, setose projections and the single posterior spiracle dorsomedially; integument wrinkled and setose.

Comments: This is a small family, with two Nearctic genera, *Atherix* and *Suragina*, each containing three species (Webb 1981a). Species of *Atherix* are widely distributed, but the genus *Suragina* is restricted to southwestern Texas and Mexico.

Selected Bibliography

Greene 1926.
Hennig 1952.
Johannsen 1935.
Malloch 1917.
Nagatomi 1958, 1960, 1961a, 1961b, 1962, 1977.
Neveu 1977.
Séguy 1940.
Stuckenberg 1973.
Thomas 1974a, 1974b, 1975.
Webb 1977, 1981a.

RHAGIONIDAE (TABANOIDEA)

B. A. Foote, *Kent State University*

Snipe Flies

Figures 37.177, 37.178

Relationships and Diagnosis: The generic composition of the Rhagionidae has exercised several generations of systematists, and arguments still wax and wane. The disputes have arisen because of the rather heterogenous nature of the taxa commonly included in the family. Species belonging to the genera *Atherix* and *Suragina*, with aquatic larvae, have been removed from the Rhagionidae and placed in the family Athericidae (Stuckenberg, 1973). Similarly, such genera as *Hilarimorpha, Glutops, Dialysis,* and *Vermileo* have been removed recently and placed in other families.

The correct phylogenetic placement of rhagionids has also elicited considerable debate over the years. McAlpine, et al. (1981), following a number of earlier workers, consider it to be a member of the Tabanoidea along with the Pelecorhynchidae, Tabanidae, and Athericidae. James and Turner (1981) have recently summarized our knowledge of snipe flies.

Larvae of the known species can be distinguished from those of other Brachycera in that the posterior spiracles are partially hidden by two or four lobes (fig. 37.178). Other useful characters include the cylindrical body, ventral creeping welts, amphipneustic (anterior spiracles probably nonfunctional) or metapneustic respiratory system, and the occurrence of larvae in damp, organic-rich soils.

Biology and Ecology: The larval ecology is rather poorly known, and morphological descriptions of the immature stages are scant or non-existent for most species. As far as known, larvae are predators, probably of other insect larvae, and are most frequently encountered in rotting wood, damp soils, and leaf litter.

Adults of *Symphoromyia* are blood suckers, usually attacking mammals, while those of other genera are believed to be predators of insects. Adults are usually encountered in relatively moist woodlands.

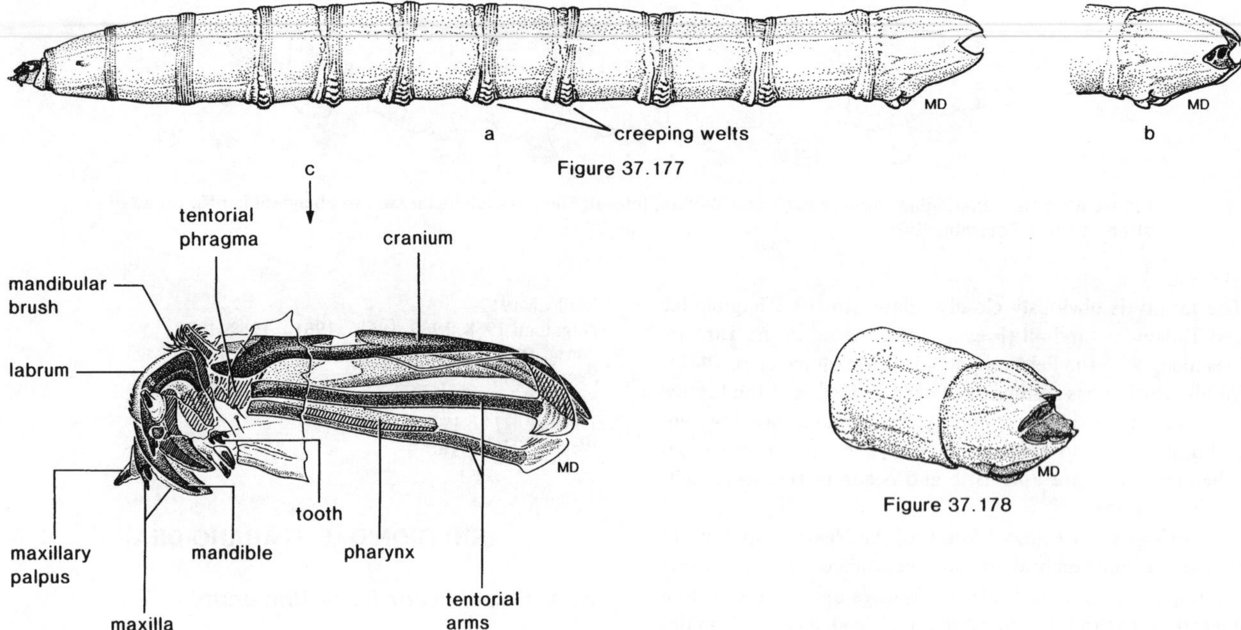

Figure 37.177a-c. Rhagionidae. *Rhagio* sp. **a.** lateral view; **b.** lateral view of posterior end showing concavity in which the spiracles are located; **c.** lateral view of head capsule and mouthparts. Found in damp, organic soils (from Manual of Nearctic Diptera, Vol. 1).

Figure 37.178. Rhagionidae. Posterior view of caudal end of *Symphoromyia* sp. The larvae occur in wet to relatively dry soils where they prey on other insects. Adults of this mostly western genus can inflict painful bites. Found in wooded areas (from Manual of Nearctic Diptera, Vol. 1).

Description: Mature larvae 6–15 mm, elongate, cylindrical, and somewhat maggot-like, tapering anteriorly, more or less truncate and variously lobed posteriorly; integument smooth, frequently with hairs and bristles; amphipneustic or metapneustic; abdomen with ventral creeping welts; white.

Head: Capsule slender, mostly retracted within thorax; open posteroventrally, with tentorial phragmata and tentorial arms, but without metacephalic rods; mandibles slender and sickle-like, articulated to a mandibular brush of posteriorly curved spines dorsally, and the maxillae laterally; antenna peg-like, apparently two-segmented.

Thorax: Anterior spiracles probably non-functional; segments encircled at segmental borders by bands of spinules; metathorax with ventral creeping welt along posterior border.

Abdomen: Segments with ventral creeping welts at segmental borders and encircled by spinule bands; caudal segment variously lobed posteriorly, commonly with two dorsal and two ventral lobes; posterior spiracles in concavity formed by posterior lobes; anterior spiracles questionably functional; posterior spiracles functional, sessile on posterior face of caudal segment. Caudal segment with variously shaped lobes or tubercles surrounding spiracles; no respiratory tube.

Comments: This is a relatively small family, consisting of some 400 species in the world. Nine genera and 108 species have been recorded for America north of Mexico (James and Turner 1981). The two largest genera are *Chrysopilus,* with 32 species, and *Symphoromyia,* with 29 species. Although certain species of *Symphoromyia* in the western states can be pestiferous due to their biting habits, none is considered to be medically or economically important. No species is known to serve as a disease vector.

Selected Bibliography

Banks 1912.
Beling 1875.
Chillcott 1961b, 1963, 1965.
Coquillett 1883.
Greene 1926.
Hardy 1949b.
Hardy and McGuire 1974.
Hennig 1952.
James and Turner 1981.
Leonard 1930.
Sommerman 1962.
Turner 1974.
Vimmer 1925.

XYLOPHAGIDAE (STRATIOMYOIDEA)

B. A. Foote, *Kent State University*

The Xylophagids

Figures 37.179–37.181

Relationships and Diagnosis: The family Xylophagidae has long puzzled systematists with respect to its proper alignment among the array of brachyceran families. McAlpine, et al. (1981) consider the family to be a member of the

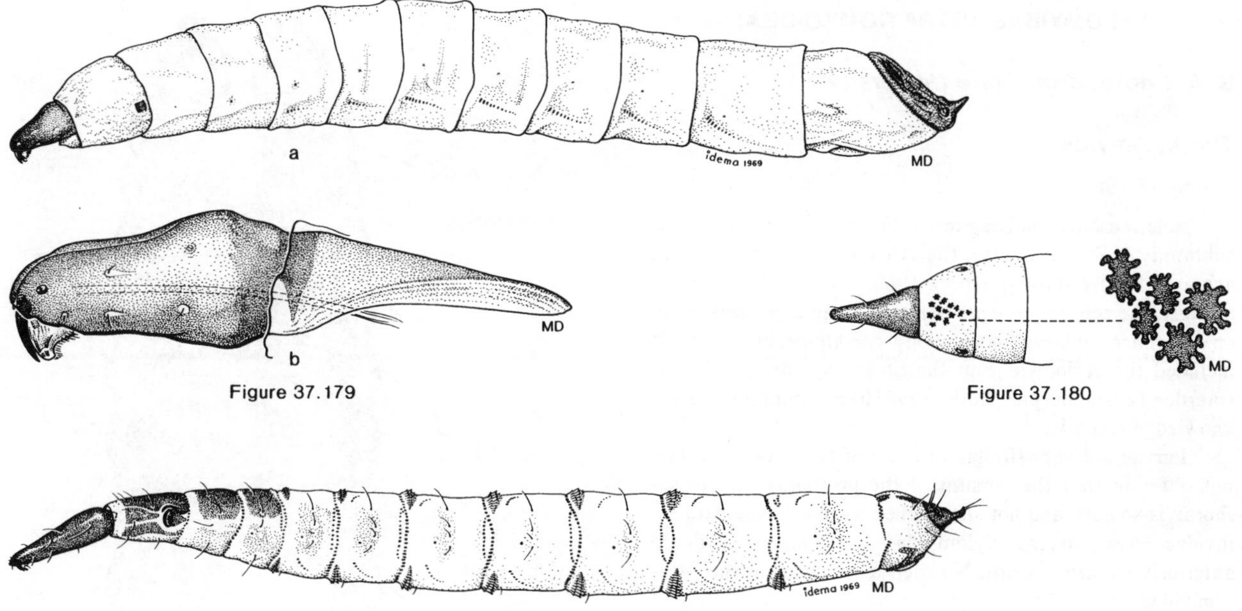

Figure 37.179a,b. Xylophagidae. **a.** *Coenomyia ferruginea* (Scopoli), lateral; **b.** lateral view of head capsule. Found in decaying wood. (from Manual of Nearctic Diptera, Vol. 1).

Figure 37.180. Xylophagidae. Dorsal view of anterior end of *Rachicerus nitidus* Johnson, with enlargement of sclerotized scales (from Manual of Nearctic Diptera, Vol. 1).

Figure 37.181. Xylophagidae. Lateral of *Xylophagus* sp., a common predator under bark of logs (from Manual of Nearctic Diptera, Vol. 1).

Stratiomyoidea, along with the Xylomyidae and Stratiomyidae. Leonard (1930), in his revision of the North American Rhagionidae, included the xylophagid species in the Rhagionidae. Hennig (1967) recognized several segregates within the "Rhagionidae" and erected the Xylomyidae, Rachiceridae, Coenomyiidae, Xylophagidae, and Rhagionidae. Since then, additional entities have been lifted to family level (Hilarimorphidae, James 1965, Webb 1974; Athericidae, Stuckenberg 1973; Vermileonidae, Nagatomi 1977; and Pelecorhynchidae, Philip *in* Stone 1965, Teskey 1970a, 1981a). As now constituted, the Xylophagidae includes only five genera in the Nearctic Region. James (1981a) has reviewed our knowledge of the family. Larvae are distinctive among the brachyceran families in possessing an array of sclerotized plates dorsally on one or more of the thoracic segments or on the dorsum of the caudal segment. In addition, the head capsule is heavily sclerotized and conical or slender and tapering.

Biology and Ecology: Adults are relatively rarely collected but are usually encountered in moist deciduous woodlands. The few larvae that have been collected were found under bark and in decaying wood. They are believed to be predacious on other insect larvae.

Description: Mature larvae 15–20 mm, elongate, subcylindrical to somewhat depressed, strongly tapered anteriorly, with distinct pointed head capsule; caudal segment with conspicuous sclerotized plate dorsally; spiracles amphipneustic, but with prothoracic spiracle darkened in some species, and with vestiges of other spiracles along the body.

Head: Conical to pointed, exserted and non-retractile, brown and heavily sclerotized; cranium solid, well developed, with two metacephalic rods and two tentorial arms extending posteriorly; mandibles and maxillae largely fused.

Thorax: With or without dorsal sclerotized plates; anterior spiracles arising posterolaterally on prothorax; spiracles subcircular with distinct peritremes; vestiges of metathoracic spiracles may be present.

Abdomen: Segments similar except caudal one; no prolegs but usually with roughened, spinulose bands on anterior borders of segments; caudal segment bearing sclerotized plate dorsally and terminating in pair of fingerlike projections; posterior spiracles surrounded by sclerotized plate, spiracles subcircular and slightly projecting.

Comments: The family is small with only five genera and 27 species recorded from America north of Mexico (James 1981a). The larger genera are *Xylophagus* (10 species) and *Dialysis* (9 species). Larval descriptions are available for species of *Coenomyia, Dialysis, Rachicerus,* and *Xylophagus,* but larvae of *Arthropeas* remain unknown.

Selected Bibliography

Beling 1875.
Brauer 1883.
Felt 1912.
Greene 1926.
Hennig 1952, 1967.
James 1965, 1981a.
Krivosheina and Mamaev 1966.
Malloch 1917.
Nagatomi 1977.
Perris 1870.
Webb 1974, 1978, 1979, 1983a, 1983b, 1983c.

XYLOMYIDAE (STRATIOMYOIDEA)

B. A. Foote, *Kent State University*

The Xylomyids

Figure 37.182

Relationships and Diagnosis: Earlier considered to be a subfamily of Stratiomyidae, the xylomyid flies recently have acquired family status (Hennig 1967, James 1981b). Interestingly, the separation from Stratiomyidae is based largely on differences in larval morphology. McAlpine, et al. (1981) included the Xylomyidae in the Stratiomyoidea of the infraorder Tabanomorpha. James (1981b) has summarized our knowledge recently.

Larvae are very similar to those of the Stratiomyidae, but differ in that the dorsum of the prothorax and mesothorax is smooth and not shagreened as it is in the Stratiomyidae. Also, in the Xylomyidae the anus is bordered anteriorly by strong teeth. No such teeth occur in the Stratiomyidae.

Biology and Ecology: Adults are rare but usually occur in relatively moist deciduous woodlands. Larvae have been encountered under bark of fallen logs where they are thought to be scavengers on decaying organic matter or possibly predacious on insect larvae.

Description: Mature larvae 8–12 mm, body elongate-oval; flattened; functionally amphipneustic; abdominal segments with integument appearing shagreened due to calcium carbonate plates; gray to nearly brown.

Head: Hemicephalic, composed of conspicuous, largely exserted head capsule; cranium well-developed, broad and heavily sclerotized; tentorial arms present, free from pharynx; maxillae with mouth brushes; mandibles reduced or absent; antennae short, 3-segmented.

Thorax: Dorsum of first two segments smooth, without shagreened pattern; anterior spiracles nearly sessile on lateral surface of prothorax, consisting of two subparallel linear openings.

Abdomen: Dorsum with distinct shagreened pattern; caudal segment with conspicuous row of posteriorly directed teeth in front of anus; posterior spiracles subcircular, located within transverse cleft at apex of caudal segment, lacking spiracular hairs; vestigial spiracles occur laterally on the abdominal segments.

Comments: This is a very small family, with only two genera (*Solva, Xylomya*) and 12 species recorded from America north of Mexico (James 1981b). Larvae of *Xylomya* remain unknown in America, but are known for some Palearctic species.

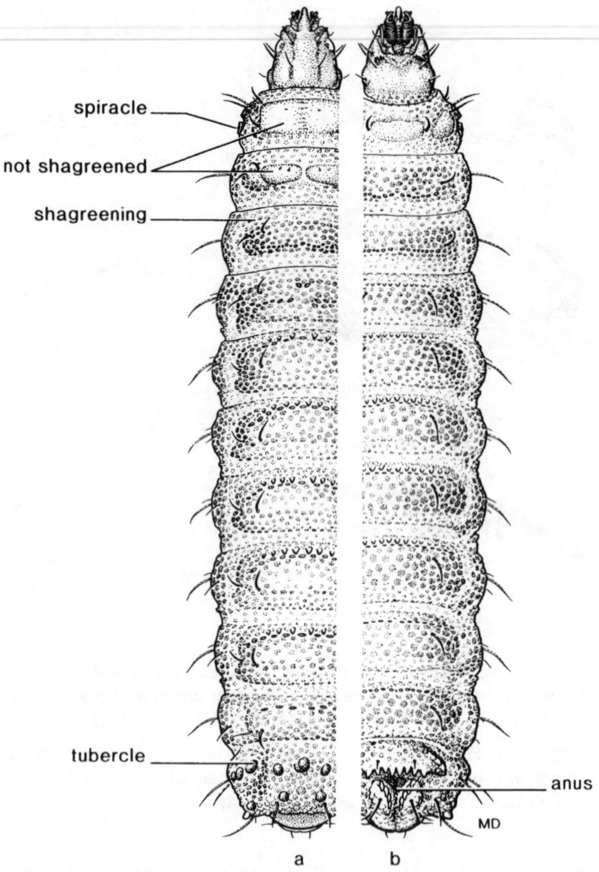

Figure 37.182a,b. Xylomyidae. *Solva pallipes* (Loew). **a.** dorsal; **b.** ventral. Found under bark (from Manual of Nearctic Diptera, Vol. 1).

Selected Bibliography

Berg 1952.
Brauer 1883.
Greene 1926.
Hennig 1952.
James 1981b.
Leonard 1930.
Malloch 1917.
McFadden 1967.
Nagatomi 1977.
Séguy 1926.
Steyskal 1947.
Teskey 1976.
Townsend 1893b.
Vimmer 1925.
Webb 1984.

STRATIOMYIDAE (STRATIOMYOIDEA)

B. A. Foote, *Kent State University*

Soldier Flies

Figures 37.183–37.190

Relationships and Diagnosis: Long considered to be closely related to the Xylophagidae, the Stratiomyidae now forms the core taxon of the Stratiomyoidea, along with the Xylophagidae and Xylomyidae (McAlpine, et al. 1981). Earlier workers tended to include the xylomyids in the Stratiomyidae, but larval structure is fairly different, thus supporting the separation into two families. McFadden (1967) and James (1981c) have summarized and reviewed our knowledge of stratiomyid larvae.

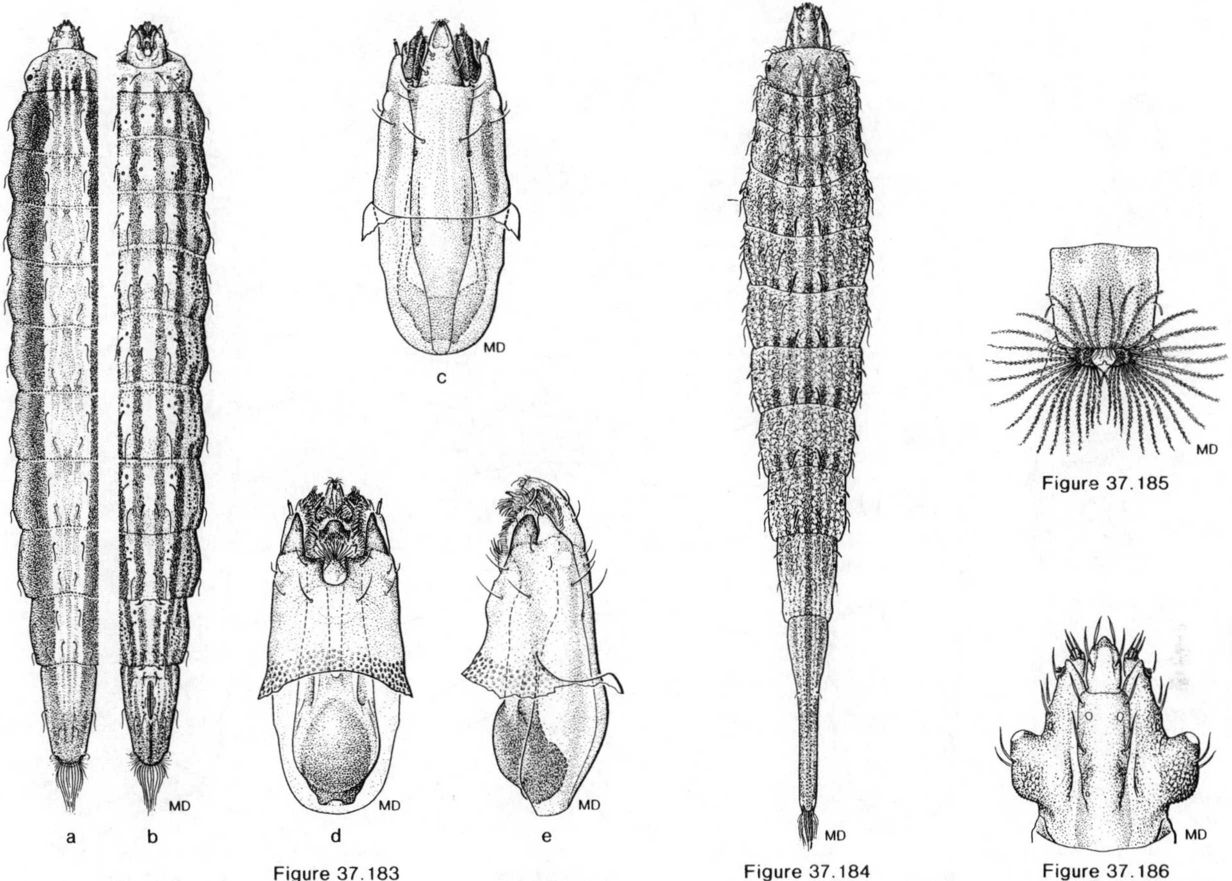

Figure 37.183a-e. Stratiomyidae. *Odontomyia cincta* Olivier. **a.** dorsal; **b.** ventral; **c,d,e.** head capsule, dorsal, ventral, lateral, respectively. Larvae are common at the water surface in ponds and marshes (from Manual of Nearctic Diptera, Vol. 1).

Figure 37.184. Stratiomyidae. *Stratiomys norma* Wiedemann, dorsal. Larvae occur at the water surface in ponds and marshes (from Manual of Nearctic Diptera, Vol. 1).

Figure 37.185. Stratiomyidae. Dorsal view of posterior end of *Myxosargus nigricornis* Greene, showing expanded spiracular hydrofuge hairs (from Manual of Nearctic Diptera, Vol. 1).

Figure 37.186. Stratiomyidae. Head of *Sargus cuprarius* (L.), dorsal, showing conspicuous ocular prominences (from Manual of Nearctic Diptera, Vol. 1).

Stratiomyid larvae are easily distinguished from those of other brachyceran families by the calcareous deposits in the integument that give a shagreened appearance to the body. Stratiomyid larvae have shagreening on the entire dorsum of the prothorax and mesothorax, in contrast to xylomyid larvae which have non-shagreened areas (fig. 37.182a).

Biology and Ecology: The often colorful adults are commonly swept from vegetation or flowers in marshy habitats. They feed mostly on nectar and pollen as well as honeydew.

The distinctive larvae of the subfamily Stratiomyinae are aquatic and usually can be found floating at the water surface of marshes, lake margins and shallow ponds. They also occur abundantly in highly organic mud along streams and ponds. Several species have larvae that are highly tolerant of saline waters or the thermal regime of hot springs. The larvae are filter feeders, utilizing a variety of small organic particles, including detritus and algal cells. Larvae of Pachygastrinae can be found under bark of fallen trees where they feed on the microorganism-laden decaying cambial tissues. A few larvae of Chiromyzinae feed on the roots of grasses. Several species in the Southwest feed as larvae within rot pockets of cactus. Larvae of *Hermetia illucens* (L.) feed in a wide variety of decaying organic matter.

Description: Mature larvae 5–35 mm, elongate and flattened, with distinct exserted head capsule; head and body bearing a variety of bristles; integument shagreened due to deposits of calcium carbonate; functionally amphipneustic, but with vestiges of lateral spiracles; white, greenish, dark gray to nearly black.

Head: Slender, elongate, non-retractile or partially retractile, ocular prominences noticeable laterally; no free pharyngeal skeleton internally; mandibles and maxillae nearly fused; antennae short and peg-like or slender and elongate.

Thorax: Distinctly broader than head; bearing short to long bristles dorsally and laterally; shagreened; anterior spiracles conspicuous on lateral surface of prothorax, vestiges of non-functional spiracles may be present on metathorax.

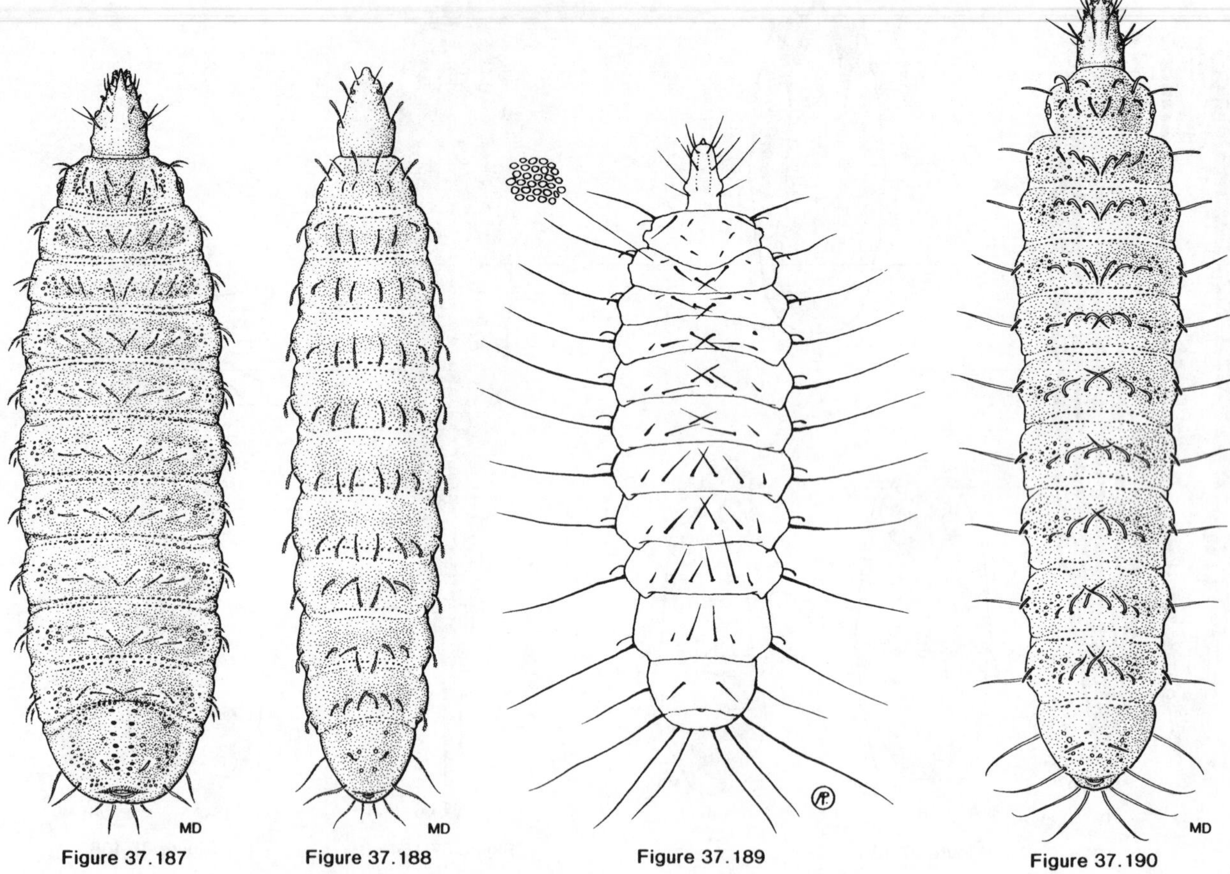

Figure 37.187. Stratiomyidae. *Berkshiria albistylum* Johnson, dorsal, a common species whose larvae can be found under bark of recently fallen trees (from Manual of Nearctic Diptera, Vol. 1).

Figure 37.188. Stratiomyidae. *Pachygaster pulchra* Loew, dorsal. Found under bark of fallen trees (from Manual of Nearctic Diptera, Vol. 1).

Figure 37.189. Stratiomyidae. *Pachygaster* sp., dorsal (from Peterson 1951).

Figure 37.190. Stratiomyidae. *Zabrachia polita* Coquillett, dorsal (from Manual of Nearctic Diptera, Vol. 1).

Abdomen: Shape varying from nearly parallel-sided and elongate-oval to distinctly tapering posteriorly with a terminal respiratory tube; segments shagreened and frequently patterned, bearing short to long bristles; vestiges of nonfunctional spiracles may be present on segments preceding caudal one; posterior spiracles located within transverse cleft at apex of caudal segment, cleft fringed by hydrofuge hairs in aquatic larvae; no prolegs on any segment.

Comments: This is a relatively large family, with over 2000 species known in the world. Forty-two genera and 254 species are recorded from the Nearctic Region (James 1981c). Larvae of a species of Chiromyzinae introduced from Australia have been reported to damage lawns in California, and a few Australian species of that subfamily are pests of sugarcane. Larvae of *Hermetia illucens* have been reported to cause human enteric myiasis.

Selected Bibliography

Berg 1952.
Bischoff 1925.
Brindle 1965b.
Brues 1928, 1932.
Cook 1950, 1953.
Faucheux 1977, 1978.
Greene 1917.
Hanson 1958.
Hart 1898.
Hennig 1952.
Hinton 1953.
James 1935, 1936a, 1936b, 1939, 1941, 1942, 1943, 1960, 1965, 1966, 1967, 1981c.
James and McFadden 1969.
James and Steyskal 1952.
James and Wirth 1967.
Johannsen 1922, 1935.
King 1916.
Kraft and Cook 1961.
Krogstad 1974.
Kuster 1934.
Lenz 1923, 1926.
Malloch 1917.
May 1961.
McFadden 1967, 1971, 1972.
Needham and Betten 1901.
Roberts 1969b.
Rozkošný 1982, 1983.

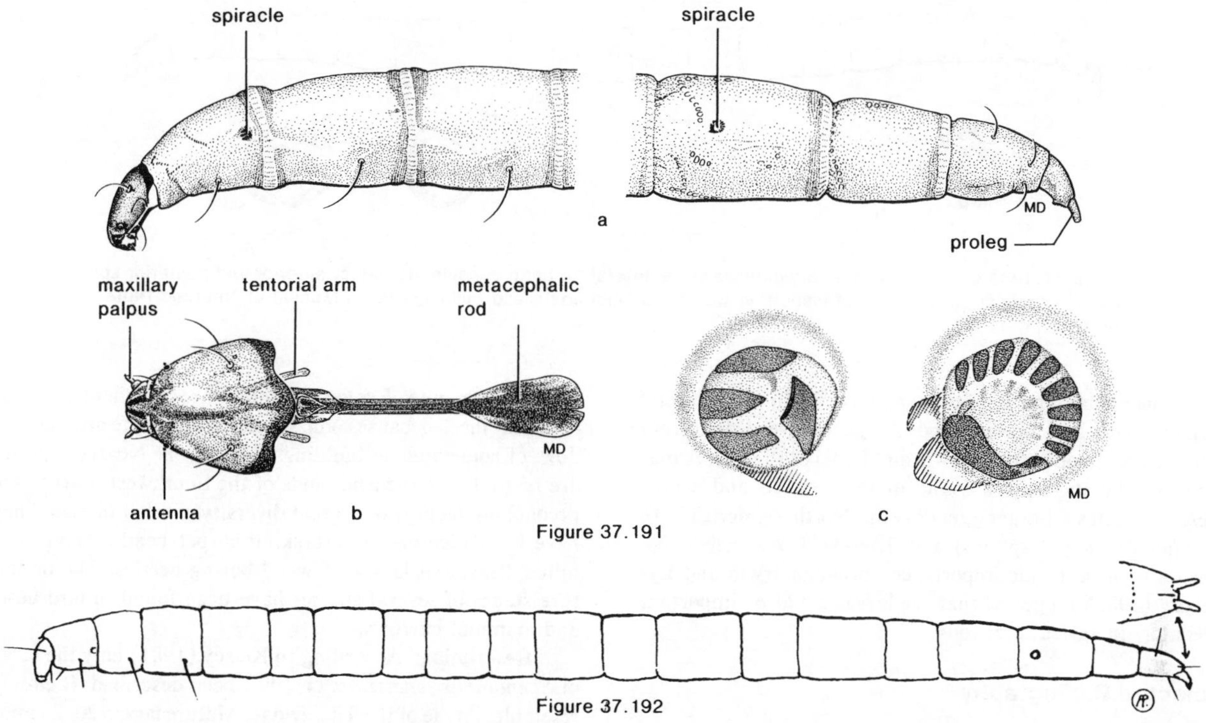

Figure 37.191a-c. Therevidae. *Thereva fucata* Loew. **a.** lateral; **b.** dorsal view of head capsule; **c.** anterior and posterior spiracles. Larvae can be found in soil in the western states (from Manual of Nearctic Diptera, Vol. 1).

Figure 37.192. Therevidae. Lateral of *Psilocephala* sp., showing the secondary divisions of the body. Formed in soil and heavily decayed (crumbly or powdery) wood (from Peterson 1951).

Séguy 1926.
Teskey 1976.
Vaillant and Delhom 1956.
Vimmer 1925.
Wilcochs and Oliver 1976.
Woodley 1981.

THEREVIDAE (ASILOIDEA)

B. A. Foote, *Kent State University*

Stiletto Flies

Figures 37.191–37.192

Relationships and Diagnosis: McAlpine et al. (1981) included the Therevidae in the Asiloidea, a group that also contains the Asilidae, Apioceridae, Mydidae, Vermileonidae, and Scenopinidae. Irwin and Lyneborg (1981) have recently summarized our knowledge.

Larvae resemble those of Scenopinidae in having a long, slender, eel-like body of 19 apparent segments, and a distinctly extended head capsule. They differ from scenopinid larvae in having the metacephalic rod broadened apically, possessing short, peg-like antennae, and having the three pairs of thoracic setae projecting at the same angle (fig. 37.191a).

Biology and Ecology: Adults are most commonly encountered in the arid lands of the western states, although they are rarely abundant. Eggs are laid in soil or surface litter, and the larvae are most often found in sandy or sandy-loam soils. Larvae of a few species occur under bark of fallen trees, in rotten fungi, and in fruit. They are voracious predators, feeding relatively non-selectively on a broad variety of soft-bodied arthropods as well as earthworms. There is some evidence that they prefer larvae of Coleoptera. Most therevids are univoltine, and many have a long larval period of up to 2 years. Adults are consumers of nectar and other plant products and seemingly prefer open, sunny habitats.

Description: Mature larvae 5–30 mm, elongate and slender, cylindrical, very similar to Scenopinidae, tapering anteriorly toward head and apex of abdomen; apparently with 20 segments; amphipneustic; white.

Head: Small, deeply pigmented, exserted head capsule present; metacephalic rod extending posteriorly into prothorax and conspicuously expanded distally; antennae short and peg-like.

Thorax: Segments similar, cylindrical, each bearing pair of long bristles ventrolaterally; anterior spiracles posterolaterally on prothorax, spiracles subcircular, with two or three openings.

Abdomen: Segments similar, subdivided to form pseudosegments; caudal segment highly tapered and ending in two retractile lobes (prolegs), bearing three pairs of long bristles; posterior spiracles apparently on antepenultimate segment, spiracles subcircular, apparently with eight spiracular openings.

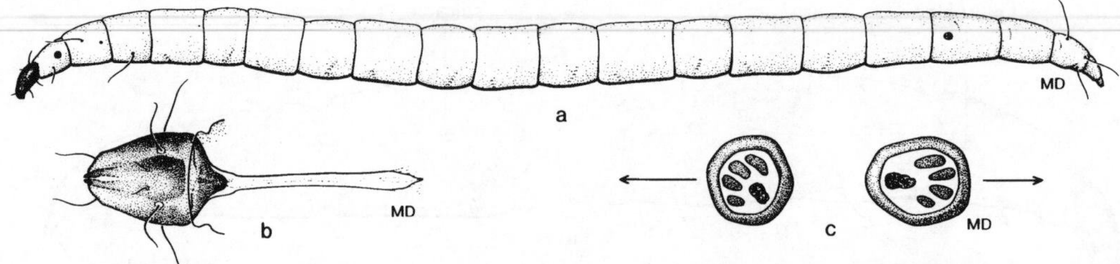

Figure 37.193a-c. Scenopinidae. *Scenopinus* sp. **a.** lateral; **b.** head capsule, dorsal; **c.** anterior and posterior spiracle. Predacious on a variety of insects in wood, mammal nests and buildings (from Manual of Nearctic Diptera, Vol. 1).

Comments: This is a fairly small family in the Nearctic Region, with 129 species and 29 genera recorded from America north of Mexico (Irwin and Lyneborg 1981). A majority of the species are found in the western and southwestern states. Larger genera in North America are *Ozodiceromya* (34 species) and *Thereva* (29 species). No species is of economic importance, although Irwin and Lyneborg (1981) suggested that the larvae could be important predators in agricultural soils.

Selected Bibliography

Beling 1875.
Brauer 1883.
Cole 1923b.
Cole (with collaboration of Schlinger) 1969.
Collinge 1909.
English 1950.
Felt 1912.
Hennig 1952.
Irwin 1972, 1976, 1977.
Irwin and Lyneborg 1980, 1981.
Malloch 1915, 1917.
Vimmer 1925.

SCENOPINIDAE (ASILOIDEA)

B. A. Foote, *Kent State University*

Window Flies

Figure 37.193

Relationships and Diagnosis: McAlpine, et al. (1981) place the Scenopinidae in the Asiloidea of the infraorder Asilomorpha, a superfamily that includes the Therevidae, Apioceridae, and Asilidae. The Scenopinidae probably are rather close phylogenetically to the Therevidae. Kelsey (1981) has summarized our knowledge of this poorly known family.

Larvae closely resemble those of Therevidae in being eel-like with 19 apparent body segments and a distinct head capsule, but differ in having a parallel-sided metacephalic rod, a much longer and thinner antenna, and for the known larvae, in having the mesothoracic setae more dorsal and projecting in a different direction than the pro- and metathoracic setae (fig. 37.193)

Biology and Ecology: Adults are frequently called "window flies" because some species accumulate around windows of homes and outbuildings. Most of the Nearctic species are restricted to the arid lands of the Southwest. Larvae are predacious, feeding on a great diversity of other insects. They have been recorded as attacking carpet beetle larvae, termites, fleas, and larvae of wood-boring beetles. The immature stages of several species have been found in bird nests and mammal burrows.

Description: According to Kelsey (1981) only the larva of *Scenopinus fenestralis* (L.) has been described. It closely resembles larvae of the Therevidae. Mature larvae 20–25 mm, elongate; slender and somewhat eel-like, with 20 apparent segments; amphipneustic, with posterior spiracles on penultimate segment; white except for dark brown or black head capsule.

Head: Distinct, conical, not retractile into prothorax, and bearing several long and conspicuous setae; capsule compact and articulating with thorax by long, parallel-sided metacephalic rod; antennae long and filamentous.

Thorax: All segments similar, each one bearing a pair of long setae; anterior spiracles posterolateral on prothorax, spiracle circular, with four oval openings.

Abdomen: Segments subdivided so that abdomen seemingly consists of more than eight segments, caudal segment distinctly tapered and bearing a few long setae; circular posterior spiracles located posterolaterally on penultimate segment (apparently the fourth segment from caudal end); spiracles with distinct peritreme and four oval openings.

Comments: This is a small family consisting of about 300 species in the world. Some 136 species and five genera have been recorded from the Nearctic Region, with most of them occurring in the southwestern states (Kelsey 1981). The largest genus is *Scenopinus*, with 48 species. No species is currently believed to be of economic significance, although the predacious larvae may play a minor role in reducing populations of such home-infesting pests as carpet beetles and termites.

Selected Bibliography

Hennig 1952.
Kelsey 1981.
Malloch 1917.
Perris 1870.
Séguy 1926.
Thompson et al. 1970.

Figure 37.194. Vermileonidae. *Vermileo comstocki* Wheeler, lateral. A predator that hides at the bottom of sand or dust pits similar to those of the Myrmeleontidae (Neuroptera) (from Manual of Nearctic Diptera, Vol. 1).

VERMILEONIDAE (ASILOIDEA)

B. A. Foote, *Kent State University*

Worm Lions

Figure 37.194

Relationships and Diagnosis: Commonly included in the Rhagionidae, the four genera now comprising the Vermileonidae were segregated from the snipe flies by Nagatomi (1977). The family was placed in the infraorder Asilomorpha and Asiloidea by McAlpine *et al.* (1981), a group that also includes the Asilidae, Apioceridae, and Therevidae, among others. Teskey (1981b) has reviewed our knowledge of this family.

Larvae are distinctive in being enlarged posteriorly, with the penultimate segment bearing numerous recurved bristles and the caudal segment ending in four lobes. The integument possesses numerous microtrichia.

Biology and Ecology: Larvae are called worm lions and form pitfall traps for capturing prey. The larval habits are remarkably similar to those of the ant lions of the neuropteran family Myrmeleontidae, since the larvae lie in wait at the bottom of the pit until prey falls into it. The larva seizes the victim with its mandibles and pulls it into the sand where the soft parts are consumed. Empty prey cadavers are subsequently thrown out of the pit. Adults of *Vermileo* feed on nectar obtained from inflorescences of Apiaceae (=Umbelliferae).

Description: Mature larvae 12–15 mm, body slender, distinctly broader posteriorly, apparently 11-segmented, integument covered by abundant microtrichia, transversely wrinkled on anterior segments; amphipneustic; white.

Head: Head capsule capable of being retracted almost completely into prothorax; retracted portion of head covered dorsally by sclerotized cranium, but open ventrally; exserted portion largely membranous; antennae 1-segmented; two lateral tentorial arms and median pharynx sclerotized, slender, and extending posteriorly nearly to end of cranium; labrum distinctly tapering anteriorly; mandibles slender, somewhat curved, and serrate apically.

Thorax: Appearing annulated due to transverse wrinkling of integument, bearing few large bristles that arise from laterally placed papillae; anterior spiracles posterolaterally placed on prothorax, spiracles with curved row of five or six short, linear openings.

Abdomen: Anterior segments appearing annulated; first segment bearing a bristle-tipped, unpaired, eversible proleg ventrally; all segments with elongate bristles laterally and with shorter, more numerous bristles ventrally, penultimate segment with row of heavy, recurved bristles posterodorsally; caudal segment apically with transverse row of four triangular lobes, the lobes fringed with short setae; posterior spiracles located on dorsal surface of caudal segment, each with a curved row of 10–12 slit-like openings.

Comments: This is a small family of only 27 species in the world distributed among four genera. Only the genus *Vermileo,* with four species, occurs in the Nearctic Region (Teskey 1981b). The four North American species are found primarily in the southwestern states and Mexico, extending northward into Oregon. They can be commonly collected in the dust on stream banks beneath bridge abutments.

Selected Bibliography

Brauer 1883.
Engel 1929.
Greene 1926.
Hemmingsen 1977.
Hennig 1952.
Leonard 1930.
Marchal 1897.
Nagatomi 1977.
Pechuman 1938.
Séguy 1926.
Teskey 1981b.
Yang 1979.

MYDIDAE (ASILOIDEA)

B. A. Foote, *Kent State University*

The Mydid Flies

Figures 37.195a–c

Relationships and Diagnosis: Although considerable controversy has erupted occasionally concerning the correct taxonomic placement of the Mydidae, it has been largely agreed in recent years that it is rather closely related to such families as the Asilidae and Apioceridae. McAlpine, et al. (1981) have included the Mydidae in the Asiloidea, a group containing six families in the infraorder Asilomorpha. Wilcox (1981) has summarized our knowledge of this family for North America.

The large, robust terrestrial larvae can be recognized by the single club-shaped metacephalic rod and the laterally compressed maxillae that tend to cup the mandibles.

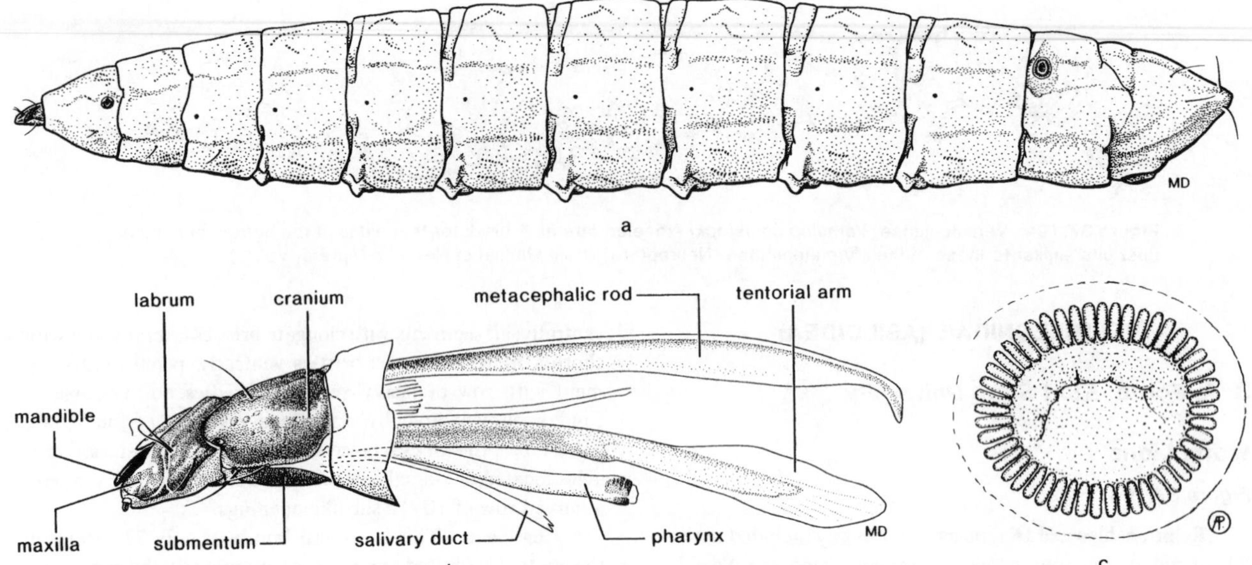

Figure 37.195a-c. Mydidae. *Mydas clavatus* Drury. **a.** lateral; **b.** lateral view of head capsule; **c.** posterior spiracle (a,b from Manual of Nearctic Diptera, Vol. 1; c from Peterson 1951).

Biology and Ecology: Very few larvae have been reared, so it is difficult to generalize about the basic feeding habits. As far as known, the larvae are predators of beetle larvae, particularly those of wood- or sod-inhabiting Scarabaeidae. There is also some question as to the adult feeding habits, with several authors suggesting that they are flower visitors. However, others suggest that adults of at least a few species are predators of other insects.

Description: Mature larvae up to 50 mm; robust and elongate, nearly cylindrical, narrowly pointed anteriorly, broader and more rounded posteriorly; amphipneustic but with vestigial spiracles along length of body; body with 12 segments plus head; cuticle smooth, white.

Head: Capsule conical, relatively small but heavily sclerotized, with scattered setae, partially retractile within prothorax. Club-shaped metacephalic rod and paired tentorial arms articulating with cranium and extending posteriorly into thorax; mandible slender, rather straight, sharply pointed, overlaid by maxilla; maxilla compressed laterally, strongly tapering anteriorly, with a small, single-segmented palpus.

Thorax: Anterior spiracle posterolaterally on prothorax, circular, biforous, and with numerous marginal openings.

Abdomen: Segments 1–7 each with four slender proleglike ridges across anteroventral margin; posterior margin of caudal segment also with ridge that is somewhat pointed posteromedially; segments 1–3 each with pair of long slender bristles ventrally; caudal segment with pair of similar bristles ventrally near apex plus a more lateral pair subapically; large rounded spiracles laterally; penultimate segment with each spiracle having a series of peripheral openings.

Comments: Although this is a relatively small family of heavy bodied flies, with only some 340 described species for the world, it is widely distributed throughout the world in dry, warm regions. Seven genera and 40 species have been recorded in America north of Mexico, with most of the species belonging to the genera *Mydas* and *Neomydas* (Wilcox 1981).

Selected Bibliography

Brauer 1883.
Genung 1959.
Greene 1917.
Hennig 1952.
Johnson 1926.
Malloch 1917.
Papavero and Wilcox 1974.
Séguy 1926.
Steyskal 1956.
Teskey 1976.
Wilcox 1981.
Wilcox and Papavero 1971.
Zikan 1942, 1944.

APIOCERIDAE (ASILOIDEA)

B. A. Foote, *Kent State University*

The Apiocerids

Figures 37.196a–c

Relationships and Diagnosis: Currently placed in the Asiloidea of the infraorder Asilomorpha (McAlpine et al. 1981), the Apioceridae have had a rather confused taxonomic history. Species have been placed by various earlier workers in such families as the Asilidae, Mydidae, and Therevidae. The family was finally characterized and unified by the work of Cazier (1941). B. V. Peterson (1981c) has recently summarized our knowledge of the family.

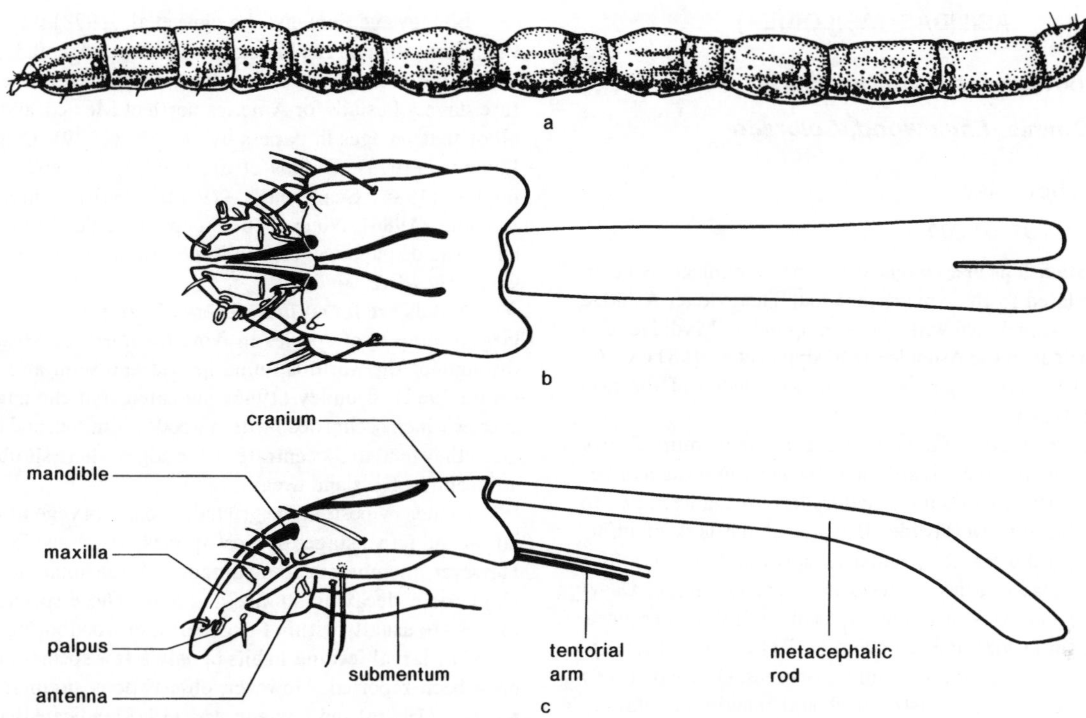

Figure 37.196a-c. Apioceridae. *Apiocera maritima* Hardy. **a.** lateral; **b,c.** head and mouthparts, dorsally and laterally. The white predacious larvae were found buried in the sand near the highwater mark at Mystery Beach, Queensland, Australia (redrawn from English 1947).

No mature larvae of the Nearctic species are known, but the known larvae are asiloid in appearance and head structure (figs. 37.196a–c).

Biology and Ecology: Practically nothing is known of the natural history and larval feeding habits. Only a single paper (Cazier 1963) includes biological observations on a Nearctic species. The North American species are usually found in the arid lands of the western and southwestern states, particularly in areas of sparse vegetation. Adults commonly rest on the apical branches of vegetation or upon the ground in somewhat shaded situations. Lavigne (1975) has suggested that adults are scavengers of decaying animal material. Gravid females insert their eggs into loose soil or sand. Larvae are known to prey on other insect larvae and possibly earthworms.

Description: Larvae are known only for *Apiocera maritima* Hardy from Australia (English 1947), *A. painteri* Cazier from Arizona (Cazier 1963, first instar only) and a species of *Tongamya* from South Africa (Irwin and Stuckenberg 1972).

Mature larvae of *Apiocera maritima* reach 51 mm; body long, slender, and white.

Head: Well-developed and elongate, retractile for about half its length into prothorax; mandible slender but heavily sclerotized, fitting into groove on dorsal side of maxilla. Mouthparts resembling those of the Asilidae.

Thorax: Segments 1 and 2 longer than 3; prothorax with thickened collar anteriorly that bears small tubercles; each segment longitudinally with pair of short setae ventrally in a furrow. Anterior spiracles situated near posterior margin of prothoracic segment; metathoracic spiracles much smaller.

Abdomen: First five segments somewhat bead-like, broad anteriorly and tapering posteriorly; segments 6–8 more cylindrical; segment 9 short, apex sharply upturned, bearing anus on ventral surface, and with four pairs of setae with posterior pair longest. Ventral surface of segments 2–6 with trace of paired processes; all segments with longitudinal striations, and with pair of lateral furrows that become deeper anteriorly; segments also with two semicircular furrows laterally near posterior margin, these cross each lateral furrow to form small raised area. Spiracles placed between lateral furrows; posterior spiracle large, lying near anterior margin of segment 8; smaller spiracles also occur on abdominal segments 1–7.

Comments: This is a small family of some 120 species and five genera in the world. Only two genera (*Apiocera* and *Rhaphiomidas*) and 29 species occur in America north of Mexido (B. V. Peterson 1981c). Twenty of these species belong to *Apiocera*. The family is thought to be in evolutionary decline and has no known economic importance.

Selected Bibliography

Cazier 1941, 1963, 1982.
English 1947.
Hennig 1952.
Hogue 1967.
Irwin and Stuckenberg 1972.
Lavigne 1975.
Peterson, B. V. 1981c.

ASILIDAE (ASILOIDEA)

B. A. Foote, *Kent State University*

Steve Dennis, *Englewood, Colorado*

The Robber Flies

Figures 37.197–37.200

Relationships and Diagnosis: Long recognized as being closely related to the Apioceridae and Therevidae, Asilidae are now also included with the Scenopinidae, Mydidae, and Vermileonidae in the Asiloidea (McAlpine, et al. 1981). G. C. Wood (1981) recently reviewed our knowledge of the taxonomic relationships.

The systematics of the family are becoming more clearly defined, however, there is still debate concerning the number of subfamilies. Various investigators have recognized two to eight subfamilies worldwide. Because of the lack of information on asilid larvae in America north of Mexico, a four subfamily classification is necessary here (Asilinae, Dasypogoninae, Leptogastrinae and Laphriinae). It should be noted that Martin (1968) and Papavero (1973) consider the Leptogastrinae to be a separate family; whereas, Oldroyd (1969) believes that the Leptogastrinae should remain a subfamily of Asilidae where it is placed in the following discussion.

The elongate, cylindrical, mature larvae have a mostly bare integument, taper at each end, and possess a single metacephalic rod (occasionally deeply forked posteriorly) that extends into the prothorax (fig. 39.197b). The mandibles lack a brush of retrorse spines. Based on an examination of asilid larvae and available literature, a key to the four subfamilies in America north of Mexico is presented below. As Knutson (1972b) indicated, only about 2% of the approximately 5,000 world species of asilids are known in any immature stage, so the key is tentative.

Biology and Ecology: Lavigne et al. (1978) summarized the information on the biology and ecology of Asilidae. Since then, limited information has been presented on the immature stages of asilids for America north of Mexico, and almost all of that on eggs in papers by Dennis, (1979); Dennis and Lavigne, (1979); Dennis et al., (1986); Lavigne and Bullington, (1984); Scarbrough, (1978b, 1981b); Schreiber and Lavigne, (1986). None of these papers included information on larvae or pupae, except in some instances to speculate on where they live.

Asilids are found in a variety of habitats. However, the largest number of species in America north of Mexico and throughout the world is found in arid and semi-arid regions (Hull, 1962). Bromley (1946) indicated that the number of species which occurs deep within woods is limited, and in these areas they tend to concentrate on the edges where shrubs occur adjacent to grassland areas.

Asilids oviposit in the ground, or on or in vegetation. The larval and pupal stages of most species are passed in the soil. However, members of the subfamily Laphriinae develop in plant roots, decaying stumps and logs. Those species found in logs are usually within the galleries of woodboring insects.

The larval feeding habits of only a few species of asilids have been reported. However, observations summarized by Knutson (1972b) and Lavigne et al. (1978) indicate that many larvae are predacious on the immature stages of other insects.

Description: Mature larvae 8–35 mm. elongate, cylindrical to somewhat dorsoventrally flattened, often tapering at each end. Shiny, white to yellowish, often with clear or translucent areas laterally. Head distinct, body comprised of 3 thoracic and 8 abdominal segments. Segments sometimes with fine longitudinal streaks or striae, or furrowed with rows of minute pores. Integument mostly bare, except for a pair of setae or bristles on each thoracic segment and setae on the last abdominal segment. Repiratory system functionally amphipneustic; vestigial spiracles may occur on abdominal segments 1–7.

KEY TO SUBFAMILIES OF ASILIDAE IN AMERICA NORTH OF MEXICO BASED ON THE LARVAE

1. Mandibles reduced or absent; maxillae widely separated apically, broadened basally; dorsal surface on outer side of maxilla with lateral incision in which lies the maxillary palpus; last abdominal segment with terminal bud-shaped formation when viewed dorsally .... **Leptogastrinae**

   Mandibles present; maxillae close together or touching apically, not broadened basally; dorsal surface or outer side of maxilla with or without lateral incision in which lies the maxillary palpus; last abdominal segment terminally rounded or with spine-like processes, not distinctly bud-shaped 2

2. Maxillae narrow without lateral incision and maxillary palpus ventrally placed; last abdominal segment may have a horizontal ridge when viewed from side .... **Asilinae**

   Maxillae broad with lateral incision on dorsal surface in which lies the maxillary palpus; last abdominal segment without horizontal ridge when viewed from side 3

3. Maxillae toothed apically; abdominal segments 1–6 with 3 or 4 pairs of tubercles (dorsal, lateral, ventral and ventrolateral) on each segment; last abdominal segment with terminal sclerotized area or plate .... **Laphriinae**

   Maxillae not toothed apically; abdominal segments 1–6 with one pair of ventral tubercles on each segment; last abdominal segment without terminal sclerotized area or plate .... **Dasypogoninae**

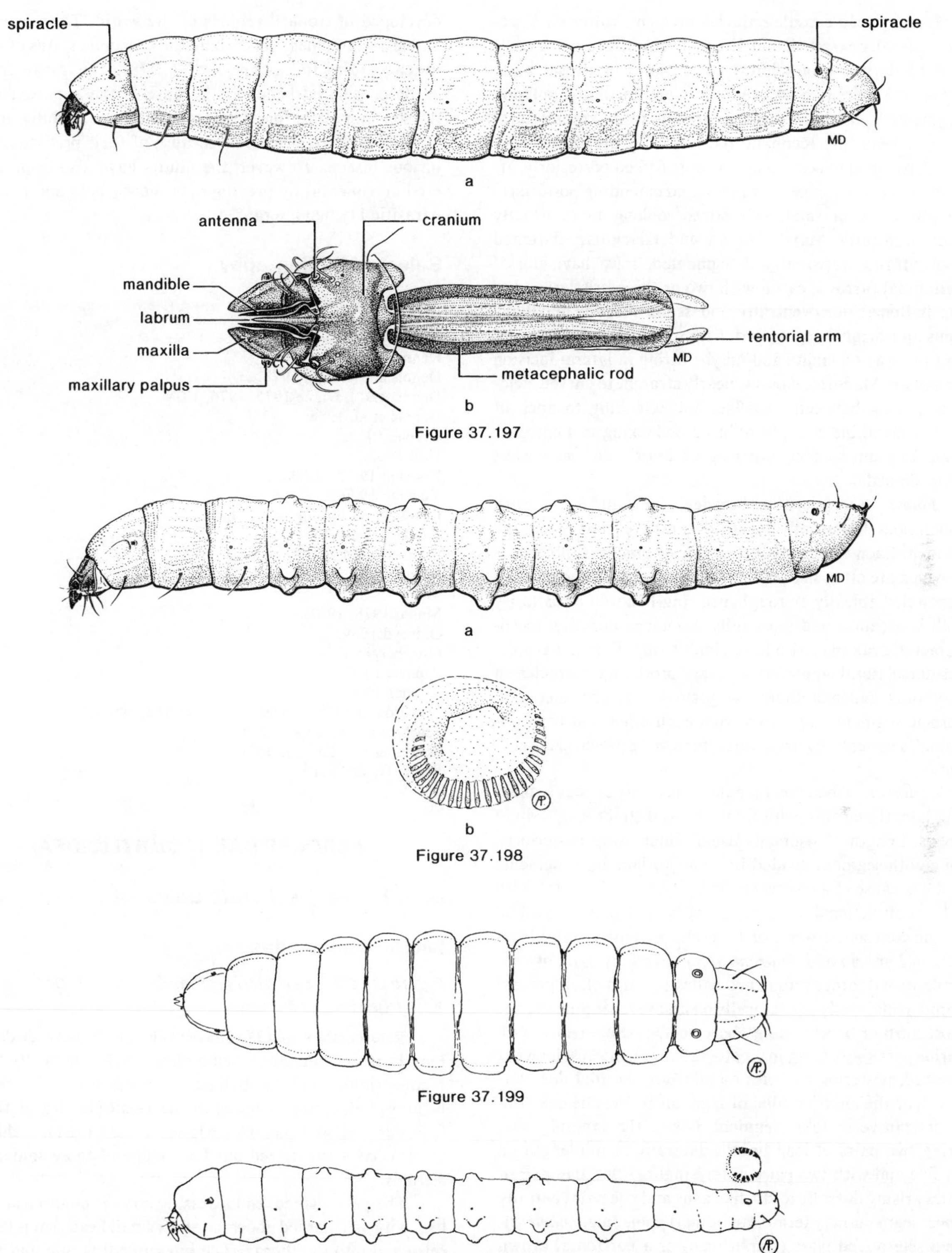

Figure 37.197a,b. Asilidae. *Promachus* sp. **a.** lateral; **b.** dorsal view of head capsule (from Manual of Nearctic Diptera, Vol. 1).

Figure 37.198a,b. Asilidae. *Laphria* sp. **a.** dorsal; **b.** posterior spiracle. Larvae are found in stumps and fallen trees (a from Manual of Nearctic Diptera, Vol. 1; b from Peterson 1951).

Figure 37.199. Asilidae. *Promachus bastardii* Macquart, dorsal. (from Peterson 1951).

Figure 37.200. Asilidae. Probably *Andrenosoma* sp., lateral (from Peterson 1951).

Head: Head capsule reduced, much narrower than prothorax, usually exserted, and directed ventrally; capsule composed of dark brown sclerotized cranium or dorsal plate bearing 1–3 pairs of bristles laterally and medially; antenna 1-segmented, small to minute, arising from anterolateral corner of head; metacephalic rod relatively broad and sclerotized, rounded to occasionally deeply forked posteriorly, attached to posterior edge of cranium and extending posteriorly into prothorax; a small sclerotized, oblong plate usually present ventrally. Maxilla broad and triangular, flattened dorsoventrally, apparently 2-segmented; may have dorsal longitudinal furrows; cardo with two or three bristles; endite long, flattened dorsoventrally and scoop-shaped; maxillary palpus appearing 2-segmented, with terminal buds, attached about midway on endite and maybe fitting in lateral incision of maxillae. Mandible narrow, nearly straight to curved, sclerotized, lying between maxillae, not extending to apex of maxilla; mandible strongly reduced or lacking in Leptogastrinae. Labrum slender, tapering anteriorly and somewhat wedge-shaped.

Thorax: Occasionally broader than abdomen, with swollen sides, but usually not strongly differentiated from the abdomen. Each segment bearing pair of bristles ventrolaterally which are often smallest on the prothorax. Prothorax may be encircled apically by roughened ring-like welt or callosity which is broadest and most fully developed dorsally; meso- and metathorax may also have similar ring. Functional spiracles dorsolateral on posterior part of prothorax, spiracles on mesothorax and metathorax very small and nonfunctional. Thoracic segments separated from each other and from abdominal segments by transverse furrow between each segment.

Abdomen: Abdominal segments also may appear swollen along lateral margins, with faint or less distinct longitudinal furrows. Length of segments usually increasing posteriorly, with eighth segment divided by some authors into segments 8 and 9 because of a transverse fold. Segments 1–7 each with small, nonfunctional spiracle laterally and pair of oval to rhombic contractile warts or tubercles anteroventrally (segments in *Laphria* and *Andrenosoma* have four pairs of contractile warts); other roughened callosities also often present anterodorsolaterally and laterally on first seven segments, but almost flush or barely rising above surface. Segment 8 with anterior part short (long in Leptogastrinae), often somewhat depressed; posterior spiracles circular and located dorsolaterally near the anterior edge of segment before the constriction or transverse fold. Segment 8 with the tapering part bearing two pairs of long bristles laterally at midlength on each side and with two pairs of terminal bristles, one pair of bristles arising dorsally to slit-like anus and one pair ventrally to anus; segment may terminate in short spine (e.g. *Dioctria*), have a sclerotized plate (Laphriinae) or a horizontal brown or whitish ridge (Asilinae).

Comments: This is a notably large family, with some 87 genera and subgenera and 983 species recorded from America north of Mexico (G. C. Wood, 1981). The larger Nearctic genera are *Efferia, Asilus, Laphria, Stenopogon, Cyrtopogon,* and *Leptogaster.* Most of the species occur in the southwestern states including Texas, and the family is best developed in tropical regions of the world. There is considerable disagreement concerning the economic status of robber flies according to Lavigne et al. (1978). The larvae are generally considered beneficial because of the apparent predation on immature forms of economic pests. Adults are also often considered beneficial because of their predation on injurious insects. However, the adults have also been considered detrimental by preying upon honey bees and beneficial parasitic Hymenoptera.

Selected Bibliography

Adamovic 1963a, 1963b.
Brindle 1962c, 1968, 1969.
Bromley 1946.
Dennis 1979.
Dennis and Gowen 1978.
Dennis and Lavigne 1975, 1976, 1979.
Dennis et al. 1986.
Greene 1917.
Hull 1962.
Knutson 1972b, 1976.
Kurkina 1979.
Lavigne and Bullington 1984.
Lavigne et al. 1978.
Malloch 1915, 1917.
Martin 1968.
Melin 1923.
Musso 1978, 1981.
Oldroyd 1969.
Osterberger 1930.
Papavero 1973.
Ritcher 1940.
Scarbrough 1978a, 1978b, 1981a, 1981b, 1981c.
Scarbrough and Sipes 1973.
Schreiber and Lavigne 1986.
Wood, G. C. 1981.

ACROCERIDAE (BOMBYLIOIDEA)

B. A. Foote, *Kent State University*

Micro-Headed Flies

Figure 37.201. See also first instar, fig. 3.130c, Vol. 1 (p. 45)

Relationships and Diagnosis: The famiy Acroceridae has long been considered as being close phylogenetically to the Nemestrinidae and Bombyliidae. These families, plus the Hilarimorphidae, are included in the Bombylioidea of the infraorder Asilomorpha (McAlpine, et al. 1981). Schlinger (1981) has summarized our knowledge of these spider parasitoids.

The pear-shaped, endoparasitic larva is plump and grublike, with an enlarged abdomen and a small head that is largely retracted into the thorax. The integument is thin and transparent. The posterior spiracles are located dorsally near the base of the caudal segment.

Biology and Ecology: Adults are generally rare and infrequently collected. A few species can be encountered on flowers, but apparently many species do not visit such sites and may even have reduced mouthparts. Several species are fast flying, but others are rather sluggish. Larvae of all reared

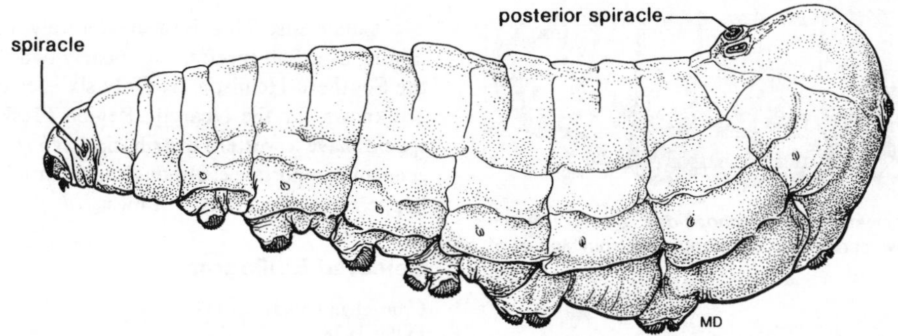

Figure 37.201. Acroceridae. *Exetasis eickstedtae* Schlinger, lateral; an internal parasitoid of spiders (from Manual of Nearctic Diptera, Vol. 1).

species are internal parasitoids of true spiders. Eggs frequently are deposited in large numbers in the vicinity of potential hosts. They may be placed on dead branches and twigs, grass stems, tree trunks, or on the ground. Each species has its own specific oviposition site. Eggs hatch into active planidia larvae which seek out the host spider. Once a larva finds a host, it burrows through the integument, usually at an intersegmental area on a leg or on the abdomen closely behind the cephalothorax. It then attaches itself to the book lungs of the host, and breathes outside air. Larvae of the subfamily Acrocerinae typically go into diapause and remain inactive for several months or even years. Once diapause is terminated, the remaining larval life is relatively short and can be completed within two or three days. The mature larva abandons its host to form a pupa in the spider's web, with emergence occurring some ten days later. The life cycle of species of Panopinae may be quite different, with the length of time spent as a second instar larva being greatly extended, and the last instar feeding for four or five days. Evidence suggests that host specificity rarely occurs, although there is a general relationship between genera of Acroceridae and families of spiders. Thus, species of the more primitive subfamily Panopinae are restricted to spiders belonging to the suborder Orthognatha, whereas species of the more derived subfamily Acrocerinae are restricted to the more advanced spider suborder Labidognatha.

Description: Hypermetamorphic, with three larval stages. First instar a planidium, 0.25–1 mm long and 0.05–0.15 mm wide. Second instar poorly known, white, without apparent sclerotized parts, and with poorly defined head, a small pharyngeal plate, a large unsegmented thoracic area, and apparently six abdominal segments. Posterior spiracles large, located in middle of rounded caudal segment or placed on sclerotized spiracular plate.

Mature larvae 3–35 mm, distinctly tapering anteriorly and swollen posteriorly; three thoracic segments, and eight or nine abdominal segments; body peripneustic, with anterior and posterior spiracles large. Mouthparts of *Exetasis eickstedtae* Schlinger consisting of an articulated mandible, a condylic pleurostomal ridge, a labrum, and a pharyngeal plate. Body with a pair of large, dorsolateral, eye-like anterior spiracles, a larger pair of posterior spiracles placed anteriorly on tergite 8, and up to seven pairs of small lateral spiracles. Each abdominal sternite with 1–3 bands of minute setae placed on small plates.

Head: Small, retractile into prothorax, consisting of dorsal cranium, ventral pharyngeal plate, and relatively small mandible anteriorly; antennae apparently absent.

Thorax: All segments similar; anterior spiracles located laterally on prothorax, large, and subcircular; apparently no spiracles on last two thoracic segments.

Abdomen: Consisting of eight or nine similar segments, the more posterior segments somewhat swollen, and all segments with ventral platelets that bear 1–3 bands of minute setae; posterior spiracles enlarged and subcircular, located on anterodorsum of segment 8; spiracles on remaining segments reduced.

Comments: The family contains about 500 described species distributed among 50 genera. There are 59 species described for the Nearctic Region, although there are at least another 20 undescribed species (Schlinger 1981). The largest North American genus is *Acrocera,* with 16 species. No species is considered to have economic significance.

Selected Bibliography

Baerg 1958.
Bechtes and Schlinger 1957.
Cole 1919.
Coyle 1971.
Emerton 1890.
Hennig 1952.
Jenks 1940.
King 1916.
Konig 1895.
Lamore 1960.
Malloch 1915.
Sabrosky 1948.
Schlinger 1952, 1960a, 1960b, 1972, 1981.
Séguy 1926.

NEMESTRINIDAE (BOMBYLIOIDEA)

B. A. Foote, *Kent State University*

The Nemestrinids

Figure 37.202. See also first instar, fig. 3.130b, Vol. 1 (p. 45)

Relationships and Diagnosis: Long assumed to be rather closely related to the Bombyliidae and Acroceridae, the Nemestrinidae are now included with those families and the Hilarimorphidae in the Bombylioidea of the infraorder

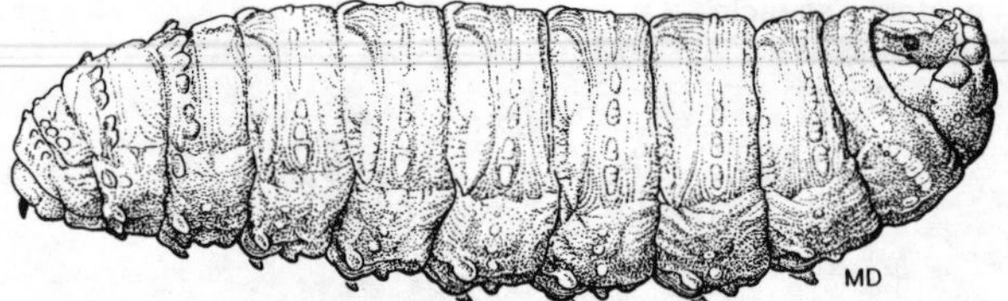

Figure 37.202. Nemestrinidae. *Trichopsidea clausa* (Osten Sacken). Lateral view, parasitic in grasshoppers (from Manual of Nearctic Diptera, Vol. 1).

Asilomorpha (McAlpine, et al. 1981). Teskey (1981c) has summarized our meager knowledge of this family.

The heavy bodied, small headed larvae are endoparasitic within the bodies of grasshoppers and beetles. The integument is tough and leathery. The body segments have wart-like projections, and the caudal segment has the postero-dorsal margin scalloped by blunt projections. The maxillae are large and somewhat shovel-shaped.

Biology and Ecology: The biology of the North American species is poorly known, and only three species of three different genera have been studied with respect to life cycles and larval feeding habits. If those studies are representative of the natural history of the family, it is apparent that hypermetamorphosis occurs, with the larvae being internal parasitoids of other insects. Grasshoppers are known to be the hosts of species belonging to two genera, while larvae of scarabaeid beetles serve as hosts of a third genus. Individual females deposit between 4000 and 5000 eggs in habitats containing large populations of potential hosts. The newly hatched larvae are minute planidia which invade the body of the host in a species specific manner. Larvae may penetrate the host's body via the intersegmental cuticle on the thorax or abdomen or via a spiracle and tracheae. In all cases they end up in the hemocoel. Once in place, larvae develop a breathing tube which extends from the parasitoid's posterior to the point of entry on the host's integument. Larvae feed on the fat body and ovarian tissue of the host, causing sterility even if it survives the parasitism. Mature larvae leave their hosts during the fall, enter the soil, and overwinter. There is a single generation per year. Adults of one species have non-functional mouthparts, while other species are known to feed on nectar and pollen.

Description: Mature larvae 5–10 mm, rather heavy bodied and cylindrical; integument leathery, roughened by arrays of wart-like tubercles, these most prominent as a transverse ridge behind posterior spiracles; metapneustic; white.

Head: Cranium broader anteriorly and posteriorly but constricted at mid-length; tentorial arms somewhat enlarged apically; no metacephalic rods; labrum small and narrow; mandibles absent; maxillae broad, spatulate and projecting forward; antennae short and peg-like.

Thorax: Segments encircled by rows of wart-like tubercles; no anterior spiracles.

Abdomen: Segments encircled by rows of wart-like projections, caudal segment with conspicuous scalloped ridge between spiracles; posterior spiracles on dorsal surface of caudal segment, each consisting of 18–26 pore-like openings surrounding central ecdysial scar.

Comments: This is a moderately sized family, containing over 250 species, with nearly two-thirds occurring in the Southern Hemisphere. Only six species in three genera are known for the Nearctic Region (Teskey, 1981c). Most species have a western distribution. None is of demonstrated economic importance, although substantial numbers of grasshoppers are parasitized at times.

Selected Bibliography

Crouzel and Salavin 1943.
Fuller 1938.
Handlirsch 1882, 1883.
Hennig 1952.
Leonide 1964.
Prescott 1955, 1960.
Séguy 1926.
Spencer 1958.
Teskey 1981c.

BOMBYLIIDAE (BOMBYLIOIDEA)

B. A. Foote, *Kent State University*

Bee Flies

Figures 37.203–37.205. See also first instar, fig. 3.130a, Vol. 1 (p. 45)

Relationships and Diagnosis: According to McAlpine, et al. (1981), the Bombyliidae form the core group of the superfamily Bombylioidea, which also includes the Acroceridae, Nemestrinidae, and Hilarimorphidae. Hall (1981) has recently summarized our knowledge of bee flies.

Most of the known larvae are recognizable by the crescentic body (fig. 37.203a) that is tapered anteriorly and posteriorly; a few are less curved. The bare integument is whitish and transparent. Mandibles are slender and pointed, and the elongate maxillae each bear a club-shaped palpus (fig. 37.203b).

Biology and Ecology: Surprisingly little is known of the natural history of this sizeable and widely distributed family. The few published studies indicate that the larvae are parasitoids of larvae and/or pupae of other insects, particularly Lepidoptera, Hymenoptera, Coleoptera, Neuroptera, and Diptera. Larvae of some species are known to prey on the egg pods of short-horned grasshoppers (Acrididae). Adults are commonly seen hovering in open, sunny areas or resting on flowers.

Description: Larvae are known for only a few species. The first instars are slender planidia. Mature larvae 9–22 mm, usually crescent-shaped, tapering at both ends; amphipneustic; white.

Head: Capsule capable of being invaginated into thorax; head consisting of broad cranium dorsally and two relatively slender tentoria and a single pharynx ventrally, tentoria typically broader distally; labrum large, broad posteriorly and distinctly narrowing anteriorly; mandibles slender, somewhat sickle-shaped; maxillae usually larger than mandibles, each bearing bristle-tipped, 1-segmented, club-shaped palpus; antennae 1-segmented, short and flattened; head bearing scattered bristles.

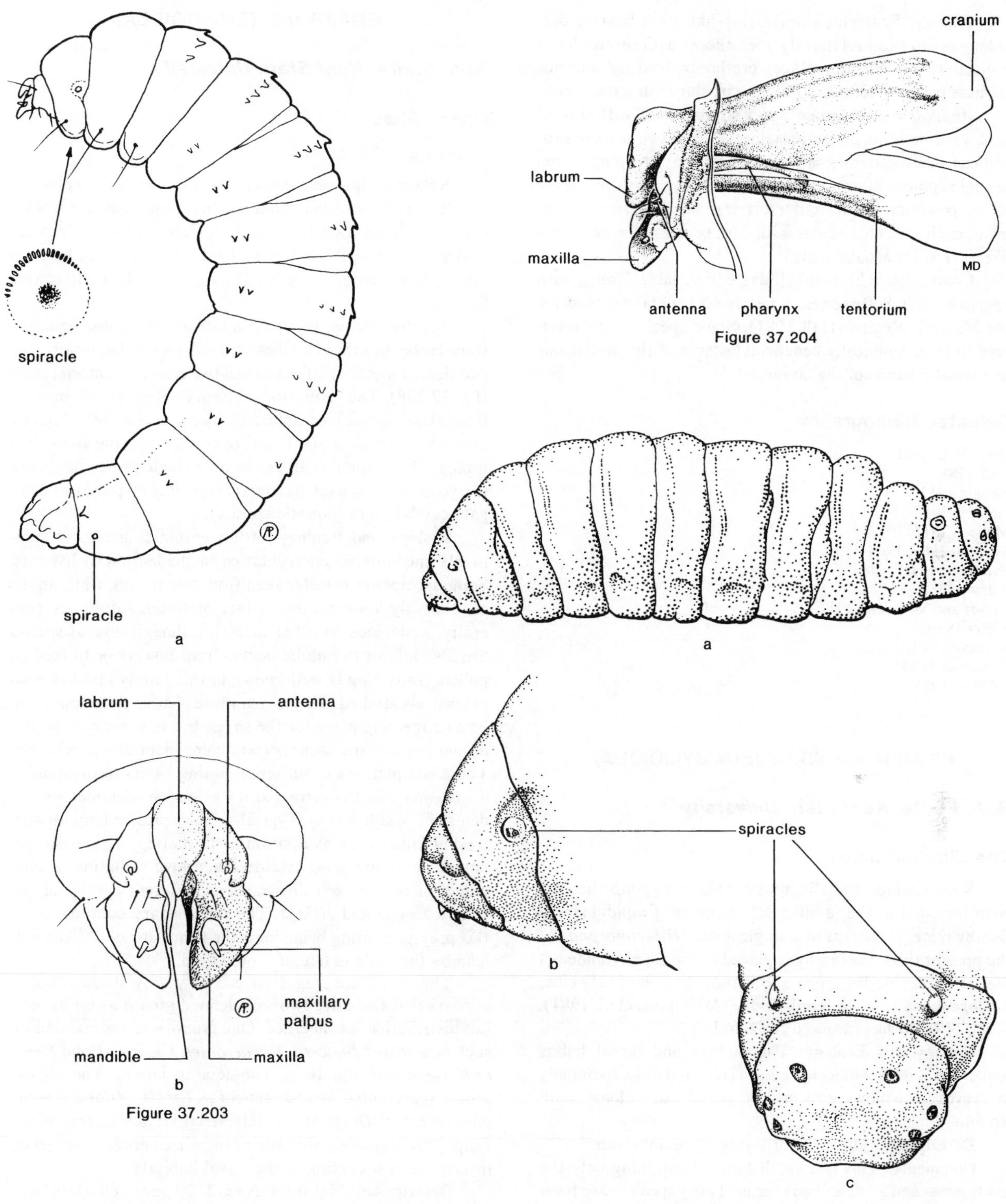

Figure 37.203a,b. Bombyliidae. Undetermined species. **a.** Lateral; **b.** anterior view of head capsule. Found in soil infested with scarabaeid larvae (from Peterson 1951).

Figure 37.204. Bombyliidae. *Sparnopolius* sp., lateral view of head capsule (from Manual of Nearctic Diptera, Vol. 1).

Figure 37.205a-c. Bombyliidae. *Villa lateralis* (Say). **a.** larva, lateral; **b.** anterior end; **c.** posterior end, dorsoposterior view. Reared from *Acrobasis vacinii* Riley (Pyralidae), the cranberry fruitworm, Allegan Co., Michigan by Doug Murray. Determination by L. V. Knutson, Syst. Ent. Lab., ARS, USDA. Drawings by P. Carrington.

Thorax: Segments similar, ring-like, each bearing pair of long bristles ventrolaterally; mesothorax with several blunt projections on dorsal surface; prothorax bearing anterior spiracle posterolaterally; spiracles circular with dark center.

Abdomen: All segments very similar, most with several backward-pointing projections dorsally, and two pairs of projections ventrolaterally; no obvious ventral creeping welts; caudal segment bluntly rounded to tapering, with few short lobes; posterior spiracles dorsolateral on penultimate segment, each spiracle circular with dark center and ring of slit-like dark spots around margin.

Comments: The Bombyliidae is a sizeable family, with approximately 800 species in nearly 65 genera recorded for the Nearctic Region (Hall 1981). Some species are considered to be economically beneficial because of the predacious or parasitic habits of the larvae.

Selected Bibliography

Berg, V. L. 1940.
Hall 1981.
Hennig 1952.
Hull 1973.
Malloch 1915, 1917.
Marston 1963, 1964, 1971.
Painter and Hall 1960.
Palmer 1982.
Parker and Wakeland 1957.
Séguy 1926.
Shelford 1913.
Townsend 1893a.
Vimmer 1925.

HILARIMORPHIDAE (BOMBYLIOIDEA)

B. A. Foote, *Kent State University*

The Hilarimorphids

Relationships and Diagnosis: This is a group that has been included in the families Rhagionidae, Empididae and Bombyliidae. It consists of a single genus, *Hilarimorpha.* At the present time, the family is placed in the Bombylioidea of the infraorder Asilomorpha and is thought to be fairly close phylogenetically to the Bombyliidae (McAlpine, et al. 1981). Webb (1981b) has reviewed the family.

Biology and Ecology: The biology and larval habits remain unknown. Adults have been collected most commonly in groves of willows growing on gravel bars along small streams.

Description: The immature stages are unknown.

Comments: This is a small family containing only the single genus and 33 described species. Thirty species have been recorded from the northern states and Canada (Webb 1981b).

Selected Bibliography

Webb 1974, 1975, 1981b.

EMPIDIDAE (EMPIDOIDEA)

B. A. Foote, *Kent State University*

Dance Flies

Figures 37.206–37.210

Relationships and Diagnosis: Obviously morphologically close to the Dolichopodidae, the Empididae are now included with that family in the Empidoidea of the infraorder Asilomorpha (McAlpine, et al. 1981). Steyskal and Knutson (1981) have summarized our knowledge of these interesting flies.

The free-living, predacious larvae can be distinguished from larvae of other families of brachyceran Diptera by the two slender metacephalic rods and two elongate tentorial arms (fig. 37.208). The respiratory system is either amphipneustic (terrestrial larvae) or apneustic (aquatic larvae). Aquatic larvae have seven or eight pairs of crochet-bearing abdominal prolegs. The caudal segment bears a single projection below the posterior spiracles (terrestrial species) or possesses elongate caudal lobes (aquatic species).

Biology and Ecology: Adult empidids are most commonly encountered on vegetation in shaded, moist habitats. Some species are readily taken from tree trunks, while adults of others fly low over the surface of water. Adults are generally predacious on other insects, although several species are also known to imbibe nectar from flowers or to feed on pollen. Swarming is well known in this family and has been extensively studied. The primary role of swarming appears to be as a meeting place for the sexes, but in some species predation has become an important component of the behavior. There is a plethora of different mating habits in this family, a situation that has attracted the attention of numerous students of insect behavior. Typically, males present females with freshly killed prey as a stimulus to mating. This basic behavior has undergone extensive adaptive radiation, a phenomenon that is well illustrated in different species of the genera *Empis* and *Hilara.* The evolutionary culmination of this prey-presenting behavior is the formation of a silken balloon by the male in lieu of a prey item.

The immature stages have been poorly studied, but it appears that there are two well defined groups as far as habitat distribution is concerned. One group, well represented by such genera as *Clinocera, Roederoides, Chelifera,* and *Hemerodromia* has aquatic or sub-aquatic larvae. The second group, represented by such genera as *Empis, Hilara, Rhamphomyia,* and *Drapetis* has terrestrial larvae. Larvae of all Empididae seemingly are rather generalized predators of other invertebrates occurring in the larval habitats.

Description: Mature larvae 3–20 mm, elongate and slender; terrestrial larvae amphipneustic, aquatic larvae usually apneustic; caudal segment terminating in single ventral median lobe in terrestrial larvae and in elongate lobes or small tubercles in aquatic larvae; abdomen with distinct prolegs in aquatic larvae but lacking prolegs in terrestrial species; white.

Head: Cephalic segment slender, retractile into prothorax, consisting of paired, slender metacephalic rods dorsally and pair of elongate tentorial arms and slender pharynx

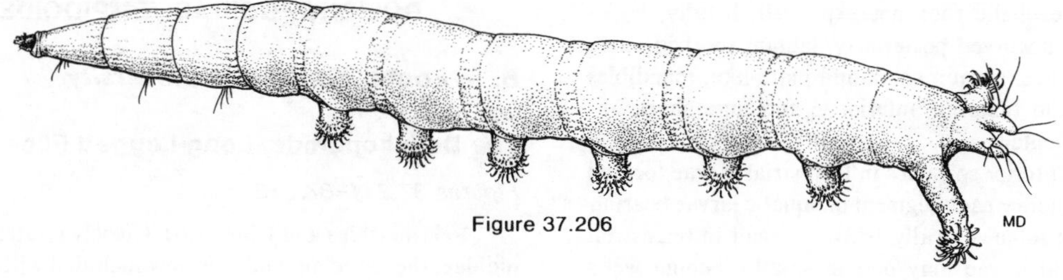

Figure 37.206

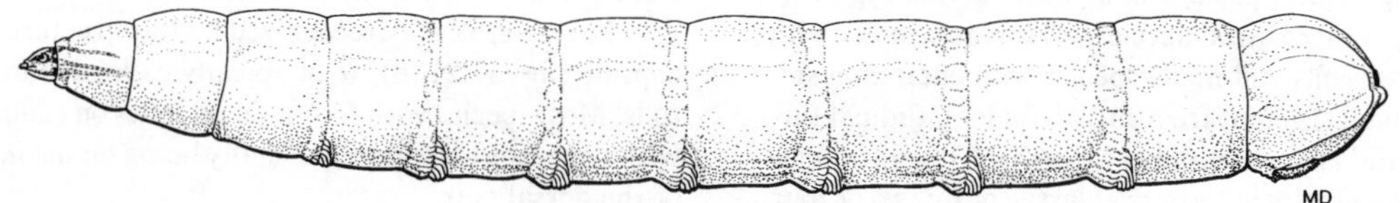

Figure 37.207

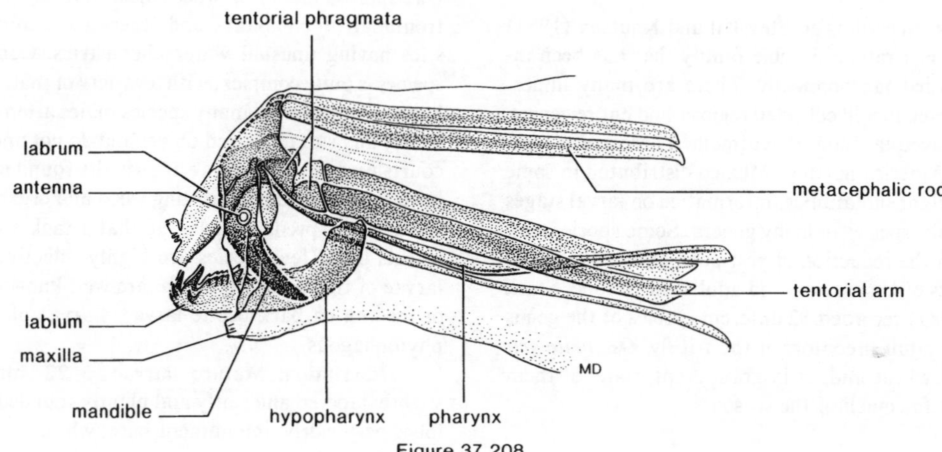

Figure 37.208

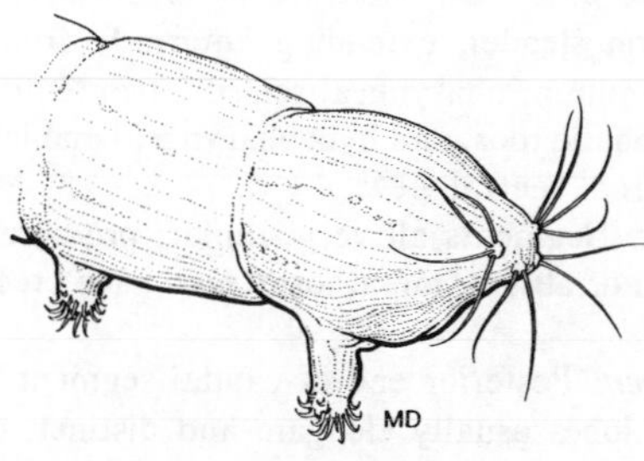

Figure 37.209

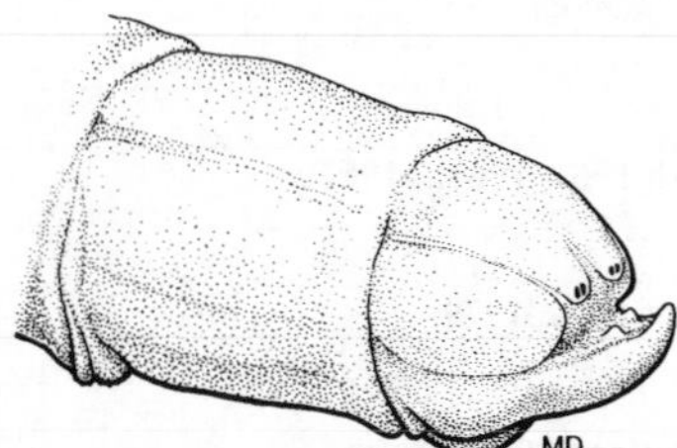

Figure 37.210

Figure 37.206. Empididae. Dorsolateral of *Hemerodromia* sp. Larvae can be found in riffle areas in streams (from Manual of Nearctic Diptera, Vol. 1).

Figure 37.207. Empididae. Lateral of *Rhamphomyia* sp. Larvae are found in soil (from Manual of Nearctic Diptera, Vol. 1).

Figure 37.208. Empididae. Mouthparts of an undetermined species of Empidinae (from Manual of Nearctic Diptera, Vol. 1).

Figure 37.209. Empididae. Posterior segments of *Chelifera* sp. (from Manual of Nearctic Diptera, Vol. 1).

Figure 37.210. Empididae. Posterior segments of *Phyllodromia* sp. (from Manual of Nearctic Diptera, Vol. 1).

ventrally; metacephalic rods not expanded distally; hypopharynx heavy, upcurved posteriorly; labium toothed along apicodorsal surface; labrum small and plate-like; mandibles broad basally and tapering anteriorly; antennae short, apparently 1-segmented.

Thorax: Anterior spiracles in terrestrial larvae located laterally on prothorax; each segment of aquatic larvae bearing a cluster of long setae ventrally, bristles absent in terrestrial larvae; terrestrial larvae may possess small creeping welts ventrally on last two thoracic segments.

Abdomen: Last seven or eight segments with distinct crochet-bearing ventral prolegs in aquatic larvae; segments without prolegs in terrestrial species; all larvae usually with ventral creeping welts and many species with encircling spinule bands; caudal segment frequently with longitudinal furrows in terrestrial larvae, bordered by undivided ventral lobe below posterior spiracles in terrestrial larvae or by two or four bristle-tipped lobes or short tubercles in aquatic species; posterior spiracles present in terrestrial larvae, but usually absent in aquatic forms; spiracles, if present, with or without spiracular hairs.

Comments: According to Steyskal and Knutson (1981) the Empididae is a rather sizeable family that has been inadequately studied taxonomically. There are many undescribed species, even in well collected regions, and entire genera have never received taxonomic treatment. There are at least 750 species in America north of Mexico distributed in some 62 genera and eight subfamilies. Information on larval stages is non-existent for species in many genera. Some species may be important in the reduction of pest populations due to the predatory habits of both larvae and adults. In Europe, Stark and Wetzel (1987) recorded 12 different species of the genus *Platypalpus* as adult predators of the frit fly, *Oscinella frit* (L.), and other wheat midges in grain crops, some of them being abundant for much of the season.

Selected Bibliography

Beling 1882.
Chillcott 1959, 1961a, 1962.
Chvala 1976.
Collin 1961.
Crane 1961.
Downes 1969.
Hennig 1952.
Hobby and Smith 1961, 1962a, 1962b.
Johannsen 1935.
Kessel 1955.
Knutson and Flint 1971, 1979.
Malloch 1917.
Melander 1947.
Needham and Betten 1901.
Newkirk 1970.
Stark and Wetzel 1987.
Steyskal and Knutson 1981.
Wilder 1983.

DOLICHOPODIDAE (EMPIDOIDEA)

B. A. Foote, *Kent State University*

The Dolichopodids, Long-Legged Flies

Figures 37.211–37.212

Relationships and Diagnosis: Closely related to the Empididae, the Dolichopodidae is now included with that family in the Empidoidea of the infraorder Asilomorpha (McAlpine, et al. 1981). Our knowledge has been summarized by Robinson and Vockeroth (1981).

Larvae are easily recognized by the structure of the head capsule (fig. 37.212b), with apically expanded metacephalic rods. Most species with four pointed lobes on caudal segment, with the posterior spiracles usually being on the inner surface of the dorsal pair.

Biology and Ecology: Most of the studied species are predacious, both as larvae and adults. Adults usually occur in moist habitats such as the margins of streams, marshes, lake shores, and damp woodlands. Several species are known from maritime habitats, and others are recorded from inland sites having unusual water chemistries. Courtship in some species is quite complex, with a variety of mating dances being known. However, in many species males attempt to mate with nearly any suitably sized object and do not undergo extensive courtship displays. Larvae are usually found in moist soil, but have also been taken in rotting wood and organic debris. Most species have predacious larvae that attack a variety of small insects but a few species are highly selective. For example, larvae of the genus *Medetera* are well known for their habit of feeding on bark beetle larvae. Larvae of *Thrypticus* are phytophagous.

Description: Mature larvae 6–22 mm, cylindrical, slightly tapered anteriorly and bluntly rounded with 4 pointed lobes posteriorly; integument bare, white.

Head: Cephalic segment usually short, unsclerotized externally, bearing four sensory lobes anteriorly; anterior pair of lobes considered to be maxillary palpi; lateral pair of lobes with peg-like antennae. Mouthparts sclerotized, brown to black; labrum slender, extending anteriorly from anterior vestige of cranium; pharyngeal sclerite, two tentorial arms, two metacephalic rods, and hypopharyngeal and labial sclerites visible from ventral view.

Thorax: Segments all very similar; prothoracic spiracles vestigial or absent; metathorax may have creeping welt ventrally.

Abdomen: Posterior end of caudal segment with 4 or more lobes, lobes usually elongate and distinct, but sometimes very short with caudal segment appearing rounded; each dorsal lobe usually bearing posterior spiracle and fringed with branched setae. Abdominal segments 1–7 with pair of ventral creeping welts on anterior margin, these occasionally encircling the segment; in *Systenus* welts of segment 1 are very large, resembling prolegs.

Comments: This is a large family of approximately 6000 described species in the world. Over 50 genera and some 1000 species have been recorded in America north of Mexico, with

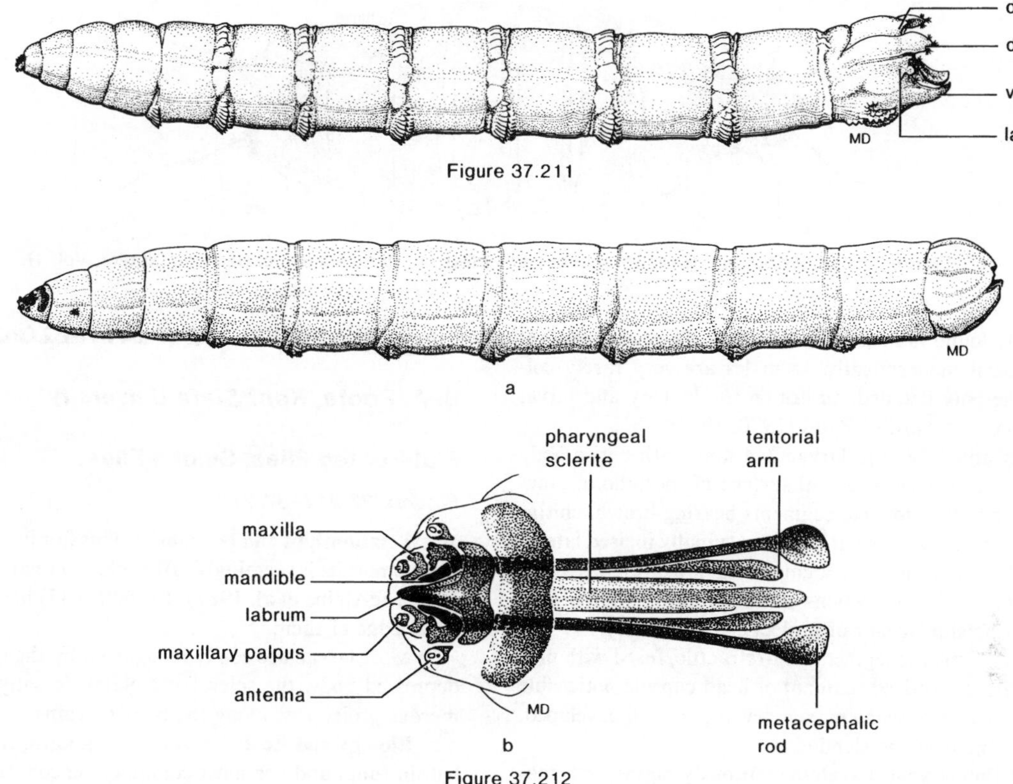

Figure 37.211. Dolichopodidae. Lateral of *Rhaphium campestre* Curran (from Manual of Nearctic Diptera, Vol. 1).

Figure 37.212a,b. Dolichopodidae. *Medetera* sp. **a.** lateral; **b.** dorsal view of mouthparts (from Manual of Nearctic Diptera, Vol. 1).

about one third of the species belonging to the genus *Dolichopus* (Robinson and Vockeroth 1981). The only species considered to be of economic importance belong to the genus *Medetera,* whose larvae attack colonies of bark beetle larvae (Scolytidae) in commercially important timber.

Selected Bibliography

Beaver 1966.
Beling 1875, 1882.
Corpus 1986.
DeLeon 1935.
Dyte 1959.
Greene 1922, 1923a, 1954.
Hennig 1952.
Hubault 1925.
Johannsen 1935.
Johannsen and Crosby 1913.
Malloch 1917.
Marchand 1918.
McGhehey and Nagel 1966.
Robinson 1964.
Robinson and Vockeroth 1981.
Roubaud 1903.
Saunders 1928.
Smith, M. E. 1952.
Teskey 1984.
Vaillant 1948, 1949, 1950.

INFRAORDER MUSCOMORPHA
DIVISION ASCHIZA

LONCHOPTERIDAE (LONCHOPTEROIDEA)

B. A. Foote, *Kent State University*

Spear-Winged Flies

Figure 37.213

Relationships and Diagnosis: The lonchopterids belong to the Division Aschiza of the suborder Brachycera, a group that includes the Phoridae, Platypezidae, Pipunculidae, and Syrphidae. It is the sole Nearctic family of the Lonchopteroidea (McAlpine, et al. 1981). Peterson (1987a) has summarized our knowledge of the family.

Larvae are strongly depressed dorsoventrally and bear conspicuous brownish sclerotized plates dorsally. The first two thoracic segments and the caudal segment bear elongate, tapering filaments that project out from the body.

Biology and Ecology: Larvae can be found in leaf litter where they feed saprophagously on decaying plant material, playing a minor role in the processing of fallen leaves. Adults

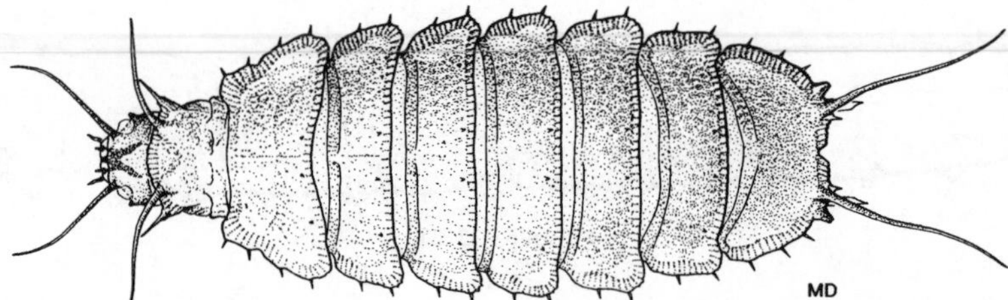

Figure 37.213. Lonchopteridae. Dorsal of *Lonchoptera* sp. (from Manual of Nearctic Diptera, Vol. 1).

are usually found in moist, shaded areas. They probably reproduce parthenogenetically, as males are very rarely collected. The only detailed studies on the biology and larval development are those of Baud (1970, 1973).

Description: Mature larvae 2–4 mm, with entire body dorsoventrally flattened; dorsal surface of mesothorax, metathorax and all abdominal segments bearing brown, chitinized plates having somewhat paler, marginally incised lateral borders. First two thoracic segments each bearing a pair of long, filament-like projections anterolaterally; caudal segment bearing similar pair of projections posteriorly.

Head: Cephalic segment non-retractile, fused with prothorax; dark sclerotized remnant of head capsule noticeable on dorsal surface of segment; sensory organs well developed, with antennae long and slender.

Cephalopharyngeal skeleton variously pigmented; tentoropharyngeal and hypopharyngeal sclerites fused; pharyngeal filter present; mandibles sickle-shaped.

Thorax: Prothorax bearing pair of long filaments anterolaterally; anterior spiracles arising anteriorly on segment in dorsolateral position; mesothorax with darkened plate dorsally and one long and two short pairs of filaments; metathorax seemingly fused with first abdominal segment, bearing distinct tergal plate.

Abdomen: All segments except caudal one similar, each bearing a darkened tergal plate; caudal segment bearing pair of long filaments, each filament subtended at base by pair of short, sharp projections.

Comments: This small family contains only 4 Nearctic species, all in the genus *Lonchoptera*. No species is known to be economically important.

Selected Bibliography

Baud 1970, 1973.
Curran 1934.
Ferrar 1987.
Hennig 1952.
Meijere 1900.
Peterson, B. V. 1987a.
Whitten 1956.

PLATYPEZIDAE (PLATYPEZOIDEA)

B. A. Foote, *Kent State University*

Flat-Footed Flies, Smoke Flies

Figures 37.214–37.218

Relationships and Diagnosis: This family of the Aschiza Muscomorpha is seemingly rather closely related to the Phoridae (McAlpine et al. 1981). Kessel (1987) has reviewed our knowledge of them.

Most larvae are easily recognized by the dorsoventrally depressed body, the sclerotized plates dorsally, and the numerous projections along the body margins.

Biology and Ecology: As far as known, the larvae feed within fungi and are most commonly encountered in mushrooms in forested habitats. Adults are usually found on low vegetation in such habitats. Males of several species form sizeable swarms at various heights above ground. A few species have a peculiar affinity for smoke and occasionally form conspicuous swarms above woodland fires.

Description: Mature larvae 2–6 mm; body in most genera dorsoventrally depressed and oval to broadly oval in outline; in a few genera the body is more cylindrical and tapering anteriorly; margins of body usually with unbranched or variously branched processes; head frequently invaginated into prothorax which is commonly turned under so the body seemingly consists of only 10 segments (only 9 in *Callomyia*); white to dark brown, many species with dark chitinized plates dorsally.

Head: Cephalic segment highly retractile, bearing short 2-segmented antennae anteriorly and palp-like sensory organs anteroventrally.

Cephalopharyngeal skeleton weakly to strongly pigmented; tentoropharyngeal and hypopharyngeal sclerites and mandibles all fused so that skeleton appears to be a single sclerite; dorsal cornua usually shorter than ventral cornua, apparently lacking pharyngeal filter; mandibles usually short, commonly with 1 to several accessory teeth.

Thorax: Prothorax usually with long pair of marginal processes projecting anterolaterally; processes unbranched and filamentous or highly branched and feather-like (*Callomyia,* fig. 37.214); processes lacking in *Agathomyia,* and *Callomyia* with two pairs; anterior spiracles simple and tubular,

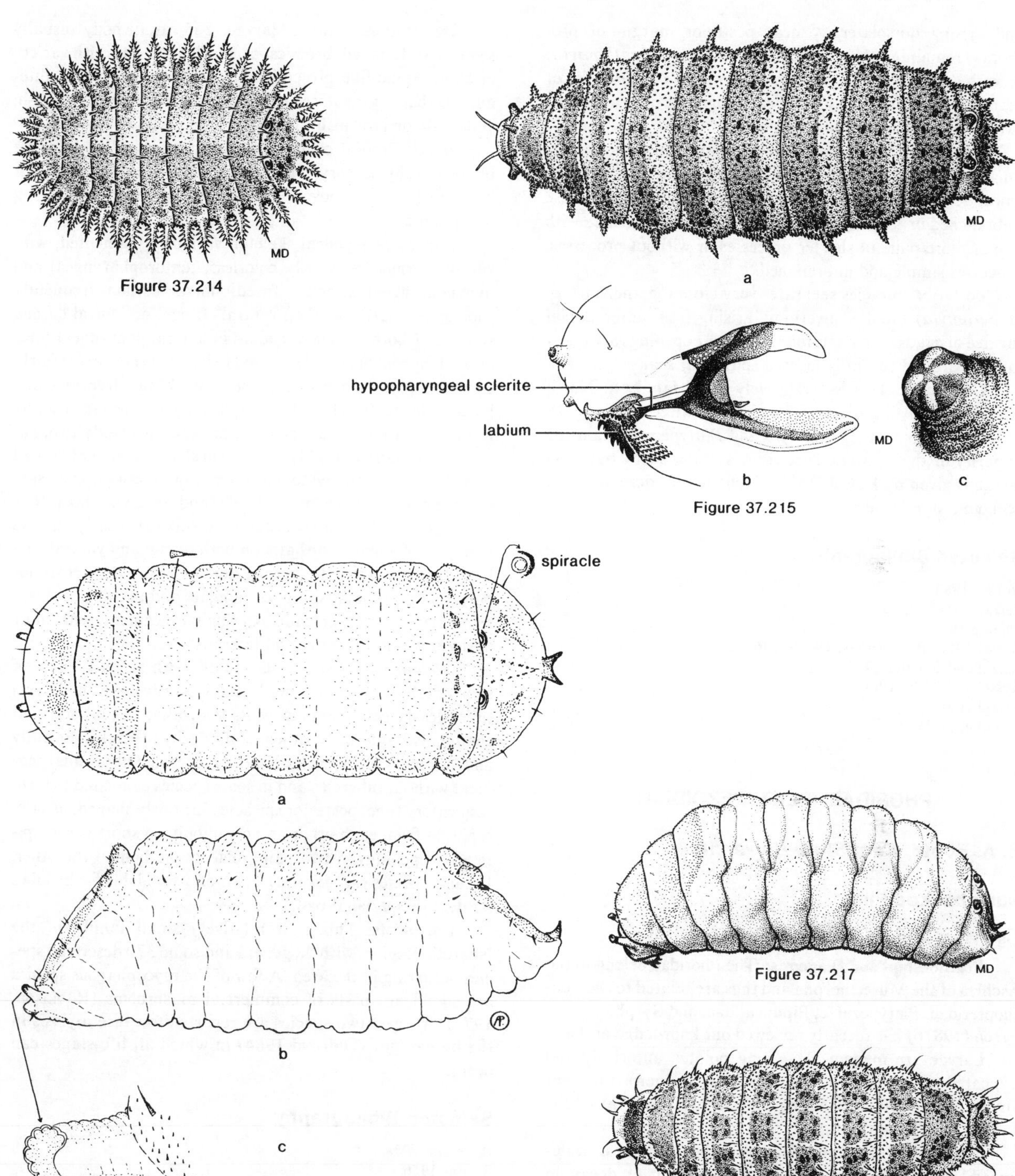

Figure 37.214. Platypezidae. Dorsal of *Callomyia gilloglyorum* Kessel (from Manual of Nearctic Diptera, Vol. 1).

Figure 37.215a-c. Platypezidae. **a.** dorsal of *Platypeza* sp.; **b.** cephalopharyngeal skeleton; **c.** posteriordorsal view of posterior spiracle (from Manual of Nearctic Diptera, Vol. 1).

Figure 37.216a-c. Platypezidae. *Clythia* sp. **a.** dorsal; **b.** lateral; **c.** anterior spiracle (from Peterson 1951).

Figure 37.217 Platypezidae. *Polyporivora polypori* (Willard), lateral (from Manual of Nearctic Diptera, Vol. 1).

Figure 37.218. Platypezidae. *Bertamyia notata* (Loew), dorsal (from Manual of Nearctic Diptera, Vol. 1).

and arising dorsolaterally near posterior margin of prothorax; remaining thoracic segments without (*Agathomyia*) or with one pair (most genera), or two pairs (*Callomyia*) of marginal processes; dorsum of each segment may also bear two or more pairs of shorter processes.

Abdomen: Segments all very similar, each without marginal processes (*Melanderomyia, Agathomyia*), or with one (most genera), or two (*Callomyia*) pairs of processes, these unbranched or feather-like (*Callomyia*); dorsal surface with two or more pairs of shorter processes or without processes; processes simple and unbranched.

Posterior spiracles separate (very close together in *Melanderomyia*) tubular, diverging, arising from anterodorsal surface of caudal segment, four spiracular openings arranged radially around centrally located spiracular scar.

Comments: This is a relatively small family in North America consisting of some 77 species scattered among 18 genera. The genera *Agathomyia* and *Platypeza* contain the majority of the Nearctic species. A key to genera based on larvae is given by Kessel (1987). None is considered to have economic significance.

Selected Bibliography

Brauer 1883.
Ferrar 1987.
Hennig 1952.
Kessel 1960, 1961, 1963a, 1963b, 1965, 1987.
Kessel and Buegler 1972.
Kessel and Kirby 1968.
Kessel et al. 1971.
Kessel et al. 1973.

PHORIDAE (PLATYPEZOIDEA)

B. A. Foote, *Kent State University*

Humpbacked Flies

Figures 37.219, 37.220a,b, 37.220y,z

Relationships and Diagnosis: The Phoridae belong to the Aschiza of the Muscomorpha and thus are related to the Lonchopteridae, Platypezidae, Pipunculidae, and Syrphidae. Peterson (1987b) has recently reviewed our knowledge of them.

Larvae are maggot-like, being pointed anteriorly and truncated posteriorly. The small size and numerous rows of tiny spine-like processes on the body segments are fairly distinctive for the family.

Biology and Ecology: The primitive and most widespread larval feeding habit is that of scavenging on decaying plant and animal material. From this ancestral condition, however, has arisen a diversity of trophic specializations. Some larvae feed within fungi, particularly mushrooms. Others are internal parasitoids of insects and other invertebrates, with a few tropical species being known as ant-decapitating flies. A few species are commensal in colonies of ants, bees, wasps, and termites. Adults are often common near the larval habitats.

Description: Mature larvae 2–8 mm; body usually somewhat flattened, broadest at mid-length, and with variety of short, spine-like projections laterally and dorsally; integument bare, granulose, spinulose, or shagreened; whitish, yellowish, or brownish.

Head: Preoral cavity posteriorly nearly filled by large bilobed fleshy sensory organ, usually bordered by simple or branching oral grooves and ridges; antenna appearing 3-segmented.

Cephalopharyngeal skeleton variously pigmented, with ventral cornua frequently colorless; tentoropharyngeal and hypopharyngeal sclerites fused, dorsal cornua frequently shorter and narrower than ventral cornua, or ventral cornua shorter, or both cornua subequal in length; pharyngeal filter present; hypopharyngeal sclerite H-shaped, very short or fairly elongate; parastomal bars present or absent, free apically; labial sclerite fused with hypopharyngeal sclerite, usually large and with distinct decurved hook, occasionally two, anteriorly. Mandibles highly variable in shape, commonly broad and plate-like with few to many accessory teeth; in some species the mandibles are more slender and somewhat hook-like.

Thorax: Anterior border of prothorax nearly always with row of spinules or hairs on both dorsal and ventral surfaces; anterior spiracles arising dorsolaterally near posterior border of prothorax; spiracles elongate and tubular with few spiracular openings apically. All thoracic segments with short fleshy tubercles dorsally and laterally.

Abdomen: All segments usually with at least 1 row of short, fleshy tubercles dorsally and laterally, with the lateral tubercles larger; creeping welts ventrally; caudal segment commonly bordered by at least four fleshy tubercles ventrally and two pairs of tubercles laterally; occasionally caudal segment without tubercles and in some species elongated to form respiratory tube; posterior spiracles variously shaped, usually separate from each other and commonly on short bases; spiracular openings in two pairs, placed one behind the other. In some species (*Megaselia*) a pair of distendable fleshy lobes borders the perianal pad.

Comments: This is a relatively small family in the Nearctic Region, with 48 genera and some 370 described species occurring in the area. A few of the mycophagous species are significant pests of commercial mushrooms (Robinson, 1977). At least two species are reported to cause myiasis in the human gut (Oldroyd 1964) in which all life stages can occur.

Selected Bibliography

Baumann 1978.
Brauns 1950.
Ferrar 1987.
Hennig 1952.
Keilin 1911.
Johannsen 1935.
Oldroyd 1964.
Pergande 1901.
Peterson, B. V. 1987b.
Robinson 1971, 1975, 1977.
Schmitz 1935, 1938, 1941.

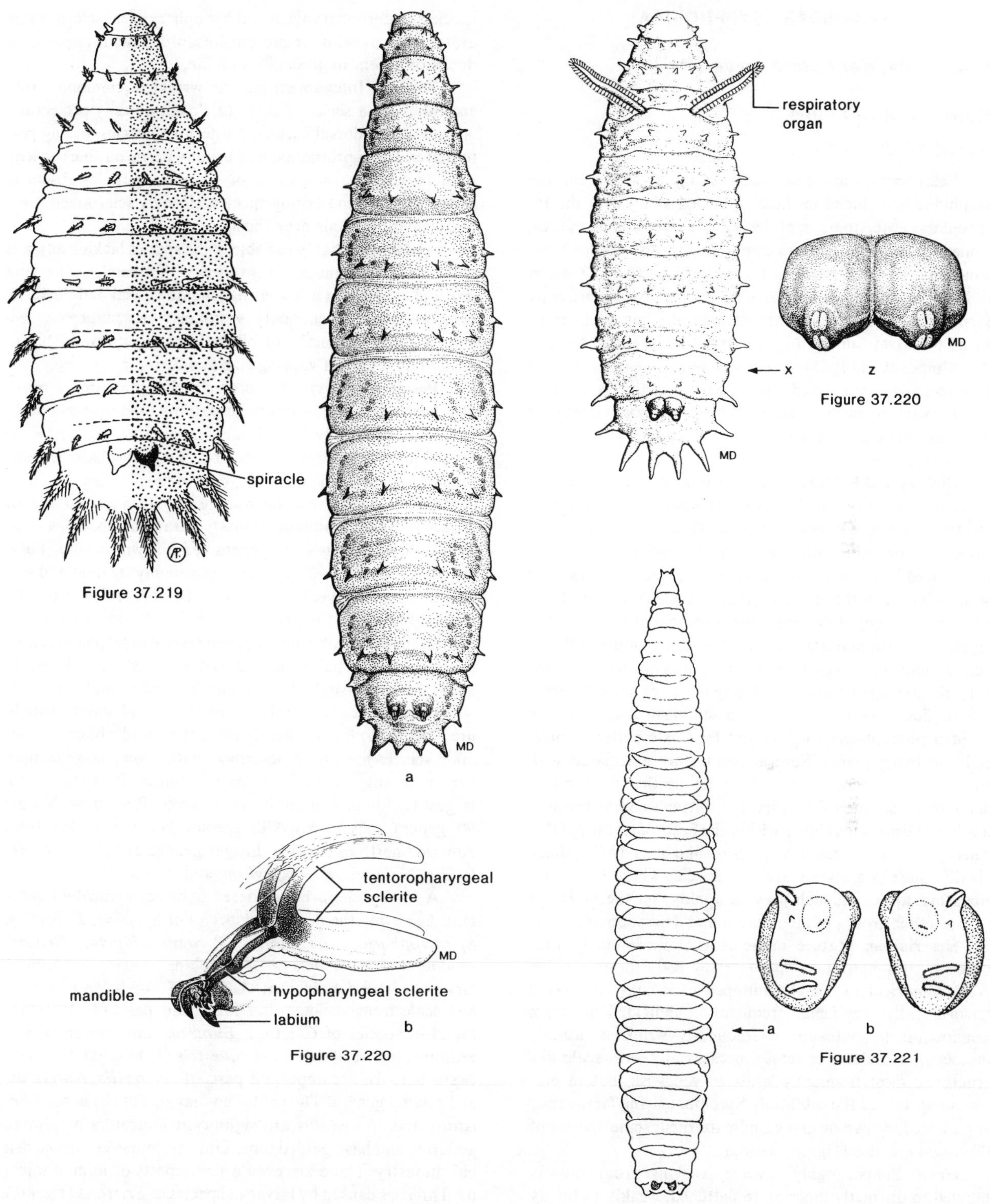

Figure 37.219. Phoridae. *Megaselia* sp. (probably *agaraci* Lintner, dorsal). From commercial mushrooms. (from Peterson 1951).

Figure 37.220a,b. Phoridae. *Megaselia* sp. **a.** dorsal; **b.** cephalopharyngeal skeleton (from Manual of Nearctic Diptera, Vol. 1).

Figure 37.220x,z. Phoridae. *Dohrniphora cornuta* (Bigot). **x.** dorsal; **z.** posterior spiracles (from Manual of Nearctic Diptera, Vol. 1).

Figure 37.221a,b. Syrphidae. *Mesogramma polita* Say. **a.** dorsal; **b.** posterior spiracular plates (from Heiss 1938).

SYRPHIDAE (SYRPHOIDEA)

B. A. Foote, *Kent State University*

Flower or Hover Flies

Figures 37.221–37.231

Relationships and Diagnosis: At the present time, the Syrphidae is included in the Syrphoidea along with the Pipunculidae (McAlpine, et al. 1981). The Syrphoidea, in turn, is one of three superfamilies comprising the infraorder Muscomorpha, Division Aschiza. Earlier authors (e.g., Stone, et al. 1965) tended to include the Conopidae in the superfamily Syrphoidea, but more recently that family has been transferred to its own superfamily and removed from the Aschiza (McAlpine, et al. 1981). Vockeroth and Thompson (1987) have recently summarized our knowledge of the Syrphidae.

Larvae are easily recognized by having their posterior spiracles united along the midline or at least very closely approximated.

Biology and Ecology: Adults are the common flower or hover flies, usually found feeding on flowers or hovering to and fro. One of the largest families of Diptera, the Syrphidae possess an amazingly diverse array of larval feeding habits and related larval structure. Perhaps the most common and basic behavior is that of saprophagy in which larvae utilize a vast assortment of decomposing organic matter. Decaying vegetation, fecal material, sewage sludge, and organically enriched muds and stagnant water are all consumed by a variety of scavenging species belonging to two or more different subfamilies. Larvae of a different assemblage of species have adopted phytophagous habits and feed externally or internally on living plants. Several species are associated with higher fungi, whereas others have larvae that feed within fungi- or yeast-laden slime fluxes or rotting wood. A few species have become inquilines within the nests of termites (*Psilota*), wasps and bees, (*Volucella*), and ants (*Microdon*). Finally, there is a diverse array of species whose larvae are predators, attacking such prey as aphids and the larvae of various other groups of Homoptera and Thysanoptera.

Description: Mature larvae 6–30 mm; commonly muscidiform, although many species show body forms ranging from cylindrical to dorsally humped to strongly depressed dorsoventrally; segments frequently subdivided to form pseudosegments; integument frequently wrinkled, usually pubescent, spinulose, or setose, occasionally with scale-like structures; most frequently white to yellowish, but in predacious species of the subfamily Syrphinae living larvae may be pink, red, brown or green and patterned; some species of *Microdon* are also brightly colored.

Head: Shape highly diverse, varying from bluntly rounded to distinctly tapering to flattened, weakly to highly retractile; antennae varying from short and fleshy to elongate and 2–3 segmented; facial mask highly variable, with oral grooves and ridges in saprophagous species, these reduced or lacking in predacious and some phytophagous taxa.

Cephalopharyngeal skeleton weakly pigmented, usually rather poorly developed except in the more saprophagous species; tentoropharyngeal and hypopharyngeal sclerites separate; pharyngeal filter present in saprophagous species, reduced or absent in predacious larvae.

Thorax: Integument bare to wrinkled, commonly tuberculate with a series of short to elongate fleshy projections on lateral and dorsal surfaces; anterior spiracles arising posterolaterally on prothorax, and usually borne on short to long tubular bases; openings oval or circular; spiracles absent in *Microdon;* in wood-boring species, strongly sclerotized rake-like structures occur near the spiracles.

Abdomen: Highly variable, some species lacking any sort of fleshy protuberances, many others with tubercles of varying size and shape, and a few with elongate lateral and dorsolateral projections; commonly wrinkled with numerous secondary folds; perianal pad bilobed, usually (always?) with fleshy rectal gills of varying size and form.

Posterior spiracles fused or closely approximate medially, borne on a long, retractile breathing tube in aquatic species (e.g., subfamily Eristalinae), on a short, non-retractile tube in many terrestrial species, or on very short bases in some predacious species (in *Chrysogaster* the spiracles are fused along midline to form a sharp spine which is inserted into air spaces in roots of aquatic plants); spiracular openings commonly three, but in some genera (e.g., *Temnostoma, Volucella*) as many as 9–25; openings commonly straight and slit-like, occasionally weakly to strongly curved, and in some species highly serpentine; spiracular hairs absent in some terrestrial species, weakly developed in several saprophagous taxa living in moist habitats, and conspicuously developed in aquatic forms, usually four in number and branching.

Comments: The family is very large and widely distributed, being recorded in nearly all of the world's biomes. Over 5000 species have been described, with a very large component occurring in the New World tropics. It is the fourth largest family of Diptera in the Nearctic Region, with some 90 genera and nearly 900 species being recorded from America north of Mexico. Larger genera are *Cheilosa, Eristalis, Melangyna, Copestylum,* and *Xylota.*

A few species are considered to be economically important. Certainly the numerous species of *Syrphus, Eupeodes, Sphaerophoria, Ocyptamus, Paragus, Pipiza, Neocnemodon,* and other genera of Syrphinae having predacious larvae are important biocontrol agents of crop-infesting aphids and scale insects. Somewhat important pests of plants are found in species of *Cheilosa, Eumerus,* and *Merodon.* For example, larvae of *Merodon equestris* (Fabricius), the Narcissus bulb fly, are important pests of *Amaryllis, Narcissus,* and related genera. The rat-tailed maggots of the genera *Eristalis* and *Helophilus* are significant indicators of sewage pollution and have gained some fame as "mousies" in the fish bait industry. There are even a few reports of human intestinal myiasis caused by larvae of species of *Eristalis* (Hermes and James 1961).

Selected Bibliography

Andries 1912.
Balachowsky and Mesnil 1936.
Benestad Hagvar 1974.
Bhatia 1939.

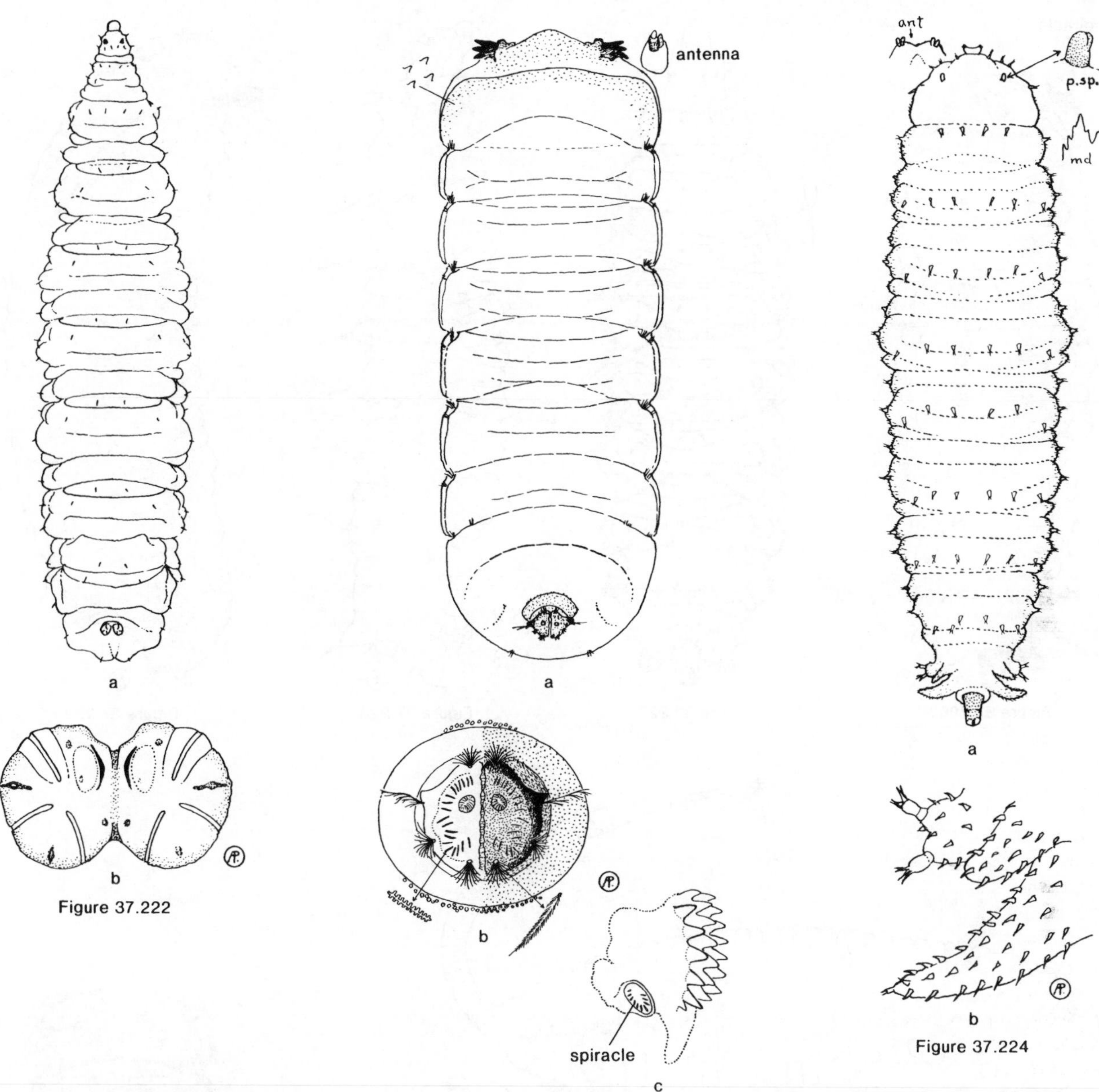

MD

Figure 37.225

Figure 37.222a,b. Syrphidae. *Syrphus vittafrons* Shannon. **a.** dorsal; **b.** posterior spiracular plates. Larvae are predators of aphids (from Peterson 1951).

Figure 37.223a-c. Syrphidae. *Temnostoma* sp. (near *balyras* (Walker)). **a.** dorsal; **b.** posterior spiracular plates; **c.** spiracular rake. Found in logs. (from Peterson 1951).

Figure 37.224a,b. Syrphidae. *Eumerus* sp. (probably *tuberculatus* Rondani, the lesser bulb fly). **a.** dorsal; **b.** tubercles bordering one side of posterior breathing tube. Larvae attack the bulbs of many spring flowers and some root crops (from Peterson 1951).

Figure 37.225. Syrphidae. *Microdon* sp., dorsal. Larvae are frequently found in carpenter ant nests (from Manual of Nearctic Diptera, Vol. 1).

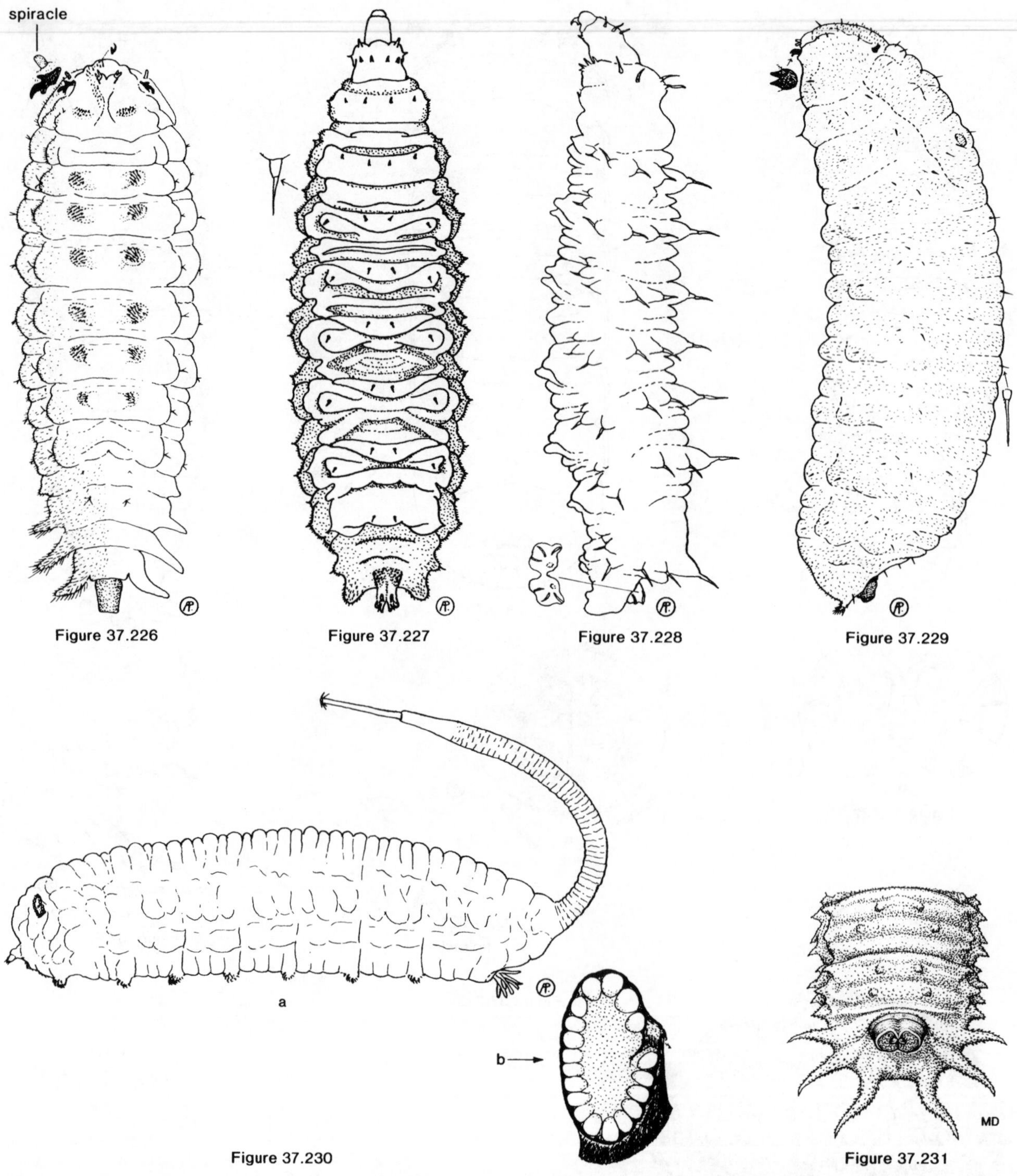

Figure 37.226. Syrphidae. *Xylota* sp., dorsal. Found under bark of decaying logs (from Peterson 1951).

Figure 37.227. Syrphidae. Dorsal of *Pipiza femoralis* Loew. Larvae occur in galls on elm trees where they prey on the gall-making aphids (from Peterson 1951).

Figure 37.228. Syrphidae. Lateral of *Baccha clavata* (Fabricius). Larvae are predators of aphids and scale insects (from Peterson 1951).

Figure 37.229. Syrphidae. Lateral of *Merodon equestris* (Fabricius), the narcissus bulb fly. Larvae are important pests of narcissus, and also attack bulbs of tulip and hyacinth (from Peterson 1951).

Figure 37.230a,b. Syrphidae. *Eristalis* sp. **a.** lateral; **b.** left anterior spiracle. Larvae are found in highly organic pools, sewage, etc. (from Peterson 1951).

Figure 37.231. Syrphidae. Dorsal of posterior segments of *Volucella bombylans* (L.) (from Manual of Nearctic Diptera, Vol. 1).

Brauer 1883.
Carrera et al. 1924.
Chandler 1968.
Coe 1939, 1942.
Cole 1923a.
Delucchi and Pschorn-Walcher 1954.
Douchette et al. 1942.
Duffield 1981.
Dusek 1962.
Dusek and Krislek 1967.
Ferrar 1987.
Fluke 1929, 1931.
Greene 1923b, 1923c.
Hartley 1958, 1963.
Heiss 1938.
Hennig 1952.
Hermes and James 1961.
Hodson 1927, 1932a, 1932b.
Ibrahim 1975.
Johannsen 1935.
Jones, C. R. 1922.
Kanervo 1942.
Knutson 1971, 1973b.
Lavallee and Wallace, 1974.
Lopez and Bonaric 1977.
Maier, 1978.
Malloch 1919.
Meijere 1916.
Metcalf 1911a, 1911b, 1912a, 1912b, 1913a, 1913b, 1916, 1917.
Novak et al. 1977.
Okuno 1970.
Roberts 1970a.
Schneider 1969.
Scott, E. J. 1939.
Stone et al. 1965.
Thompson, F. C. 1981.
Vimmer 1925.
Vockeroth 1969.
Vockeroth and Thompson 1987.
Wallace and Lavallee 1973.
Walsh and Riley 1869.
Watson 1926.
Wheeler, W. M. 1901, 1908, 1924.
Wilcox 1926, 1927.

PIPUNCULIDAE (SYRPHOIDEA)

B. A. Foote, *Kent State University*

Bigheaded Flies

Figure 37.232

Relationships and Diagnosis: The bigheaded flies are placed in the Syrphoidea of the aschizous Muscomorpha and are thought to be rather closely related to the Syrphidae (McAlpine et al. 1981). Hardy (1987) has recently reviewed our knowledge.

Larvae are endoparasitic and thus have few surface characteristics. The body is smooth and generally shows poor segmentation. Perhaps the most distinctive characteristic is the occurrence of the posterior spiracles on a single deeply pigmented sclerotized plate.

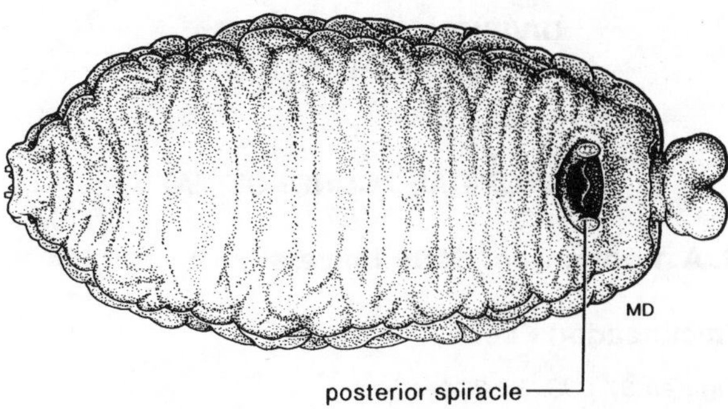

Figure 37.232. Pipunculidae. Dorsal of *Pipunculus* sp. (from Manual of Nearctic Diptera, Vol. 1).

Biology and Ecology: As far as known, the larvae are all endoparasitoids of Homoptera, particularly Cicadellidae (leafhoppers), Fulgoridae (planthoppers), and Cercopidae (froghoppers). Mature larvae leave the host's body to pupate in the soil. Adults are relatively uncommon.

Description: Mature larvae 2–4 mm; body short and swollen, without surface features except for numerous indentations dividing the body into a series of rings; in some species lateral projections occur.

Head: Cephalopharyngeal skeleton frequently vestigial, with only sickle-shaped mandibles noticeable; mandibles may have numerous, small accessory teeth.

Thorax: Anterior spiracles sessile on surface of prothorax, with four or five finger-like papillae radiating out from common center.

Abdomen: Posterior surface of caudal segment covered with a deeply or weakly pigmented plate that surrounds the posterior spiracles; each spiracle with three (four?) short-oval spiracular openings arranged in a slightly curved line along edge of spiracular plate.

Comments: There are some 400 species in the world, with a little over 100 species in 7 genera recorded from America north of Mexico. Pipunculids as parasitoids are of considerable importance in the natural control of Homoptera and they have been used in biocontrol (Hardy, 1987).

Selected Bibliography

Ferrar 1987.
Hardy 1954, 1964, 1971, 1987.
Hennig 1952.
Hough 1899.
Keilin and Thompson 1915.
Koizumi 1959.
Linnane and Osgood 1977.
Meijere 1916.
Whittaker 1969.

DIVISION SCHIZOPHORA
SECTION ACALYPTRATAE

CONOPIDAE (CONOPOIDEA)

B. A. Foote, *Kent State University*

Thickheaded Flies

Figures 37.233, 37.234

Relationships and Diagnosis: Although previously considered to be a member of the Aschiza and thus relatively close to the Syrphidae, Pipunculidae, Phoridae, Platypezidae, and Lonchopteridae, conopids have recently been transferred to the schizophorous Acalyptratae. They form a separate superfamily, Conopoidea, and apparently are not very closely related to any other family of Nearctic Diptera (McAlpine et al. 1981). Smith and Peterson (1987) have recently summarized our knowledge.

The surface of the larvae is relatively featureless, apparently reflecting their endoparasitic habits. The body is broad posteriorly and tapered anteriorly. The most distinctive characteristic is the dome-like shape of the posterior spiracular plates which bear a large number of small protuberances or have linear clusters of pores. The cephalopharygeal skeleton is reduced and lacks a hypopharyngeal sclerite.

Biology and Ecology: Larvae of all Conopidae are endoparasitic in adult wasps and bees, except for those in the subfamily Stylogasterinae which are parasitic on cockroaches and calyptrate Diptera that may be fleeing from army ants. Eggs usually are laid on the body of the host. Adults are most commonly found on flowers.

Description: Mature larvae 5–10 mm; body very swollen, particularly in posterior half, strongly narrowed and tapered anteriorly; integument frequently wrinkled, devoid of spinules and hairs.

Head: Cephalic segment weakly retractile, Cephalopharyngeal skeleton highly modified, lacking hypopharyngeal sclerite; tentoropharyngeal sclerite very broad, weakly divided into very short and apically rounded dorsal cornua and somewhat longer, apically tapering ventral cornua; no pharyngeal filtering mechanism. Mandibles with sickle-shaped hook and subrectangular basal part, no accessory teeth; dental sclerites long and slender.

Thorax: Most species metapneustic, without anterior spiracles.

Abdomen: Posterior spiracles separate, each borne on dome-shaped plate and consisting of numerous (4 up to 500) spiracular openings, the latter either scattered or arranged in lines on spiracular plate.

Comments: This is a relatively small family with 45 genera and some 800 described species. Approximately 66 species and 9 genera have been recorded in America north of Mexico. Sizeable genera are *Physoconops, Zodion,* and *Myopa.* A species of *Zodion* apparently parasitizes honey bees and thus may be a minor pest to beekeepers. Keys to genera based on puparia and eggs are given by Smith and Peterson (1987).

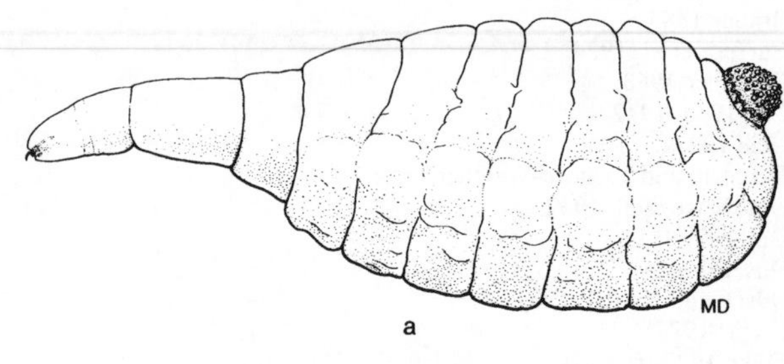

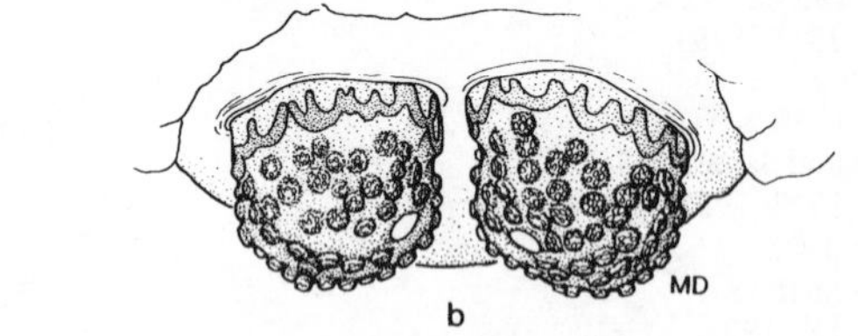

Figure 37.233

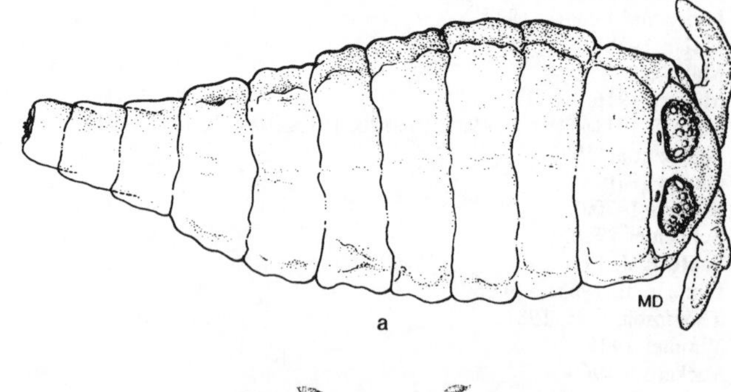

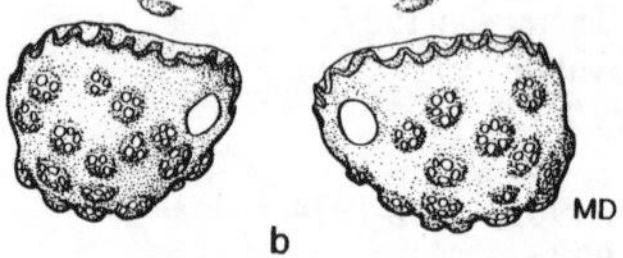

Figure 37.234

Figure 37.233a,b. Conopidae. *Physocephala bimarginipennis* Karsch. **a.** lateral; **b.** posterior spiracles (from Manual of Nearctic Diptera, Vol. 1).

Figure 37.234a,b. Conopidae. **a.** Larva of *Zodion* sp.; **b.** posterior spiracles (from Manual of Nearctic Diptera, Vol. 2).

Selected Bibliography

Aldrich 1930.
Camras 1957a, 1957b.
Ferrar 1987.
Hennig 1952.
Smith and Peterson, B. V. 1987.
Townsend 1935.
Vimmer 1925.

CYPSELOSOMATIDAE (NERIOIDEA)

B. A. Foote, *Kent State University*

The Cypselosomatids

Figure 37.235

Relationships and Diagnosis: A member of the Nerioidea, the cypselosomatids are considered to be closely related to the Neriidae and Micropezidae (McAlpine, et al. 1981). McAlpine (1987a) has recently reviewed our knowledge.

Larvae of Cypselosomatidae closely resemble those of Micropezidae and Neriidae except that they lack a distinct dorsobasal lobe on the tentoropharyngeal sclerite.

Biology and Ecology: Nothing is known of the larval biology or feeding habits of the two Nearctic species. Larvae of an Australian species, *Cypselosoma australis* McAlpine, were found in a cave where they were feeding on bat dung. Adults of the Nearctic genus *Latheticomyia* were collected in banana bait traps in mountainous areas of southern Arizona.

Description: Larvae of the Nearctic species are unknown, and the following description is based on that given for *C. australis* by D. K. McAlpine (1966).

Mature larva about 6 mm; muscidiform, tapering anteriorly and somewhat truncate posteriorly; integument bare and shining, without encircling bands of spinules.

Head: Cephalic segment retractile; facial mask with numerous transverse ridges bearing long, tooth-like serrations around preoral cavity.

Cephalopharyngeal skeleton with only mandibles and hypopharyngeal sclerite pigmented; tentoropharyngeal and hypopharyngeal sclerites separate, dorsal cornua narrower than ventral cornua, latter with slight convexity dorsobasally but without distinct lobe; pharyngeal filter present; hypopharyngeal sclerite apparently H-shaped, parastomal bars present but not connected anteriorly; apparently no epipharyngeal sclerite. Mandibles with sickle-shaped hook, no accessory teeth, apparently no dental sclerites.

Thorax: Anterior spiracles arising posterolaterally on prothorax, spiracles consisting of 8–10 finger-like papillae that arise directly on body surface of segment, without a basal stalk; apparently no ventral creeping welts.

Abdomen: All segments except caudal one similar, each bearing creeping welt ventrally, each welt composed of several rows of spines, although welt on first segment is reduced to two rows; perianal pad bilobed, surrounded laterally and posteriorly by 2–3 rows of broad spines.

Posterior spiracles borne on hemispherical prominences; spiracular plates with four oval openings distributed around margin of plate; four conspicuous, dichotomously branched spiracular hairs.

Comments: This is a small family of fewer than 15 species in the world. It is largely tropical, with only *Latheticomyia lineata* Wheeler and *L. tricolor* Wheeler being recorded in America north of Mexico (Arizona and Utah).

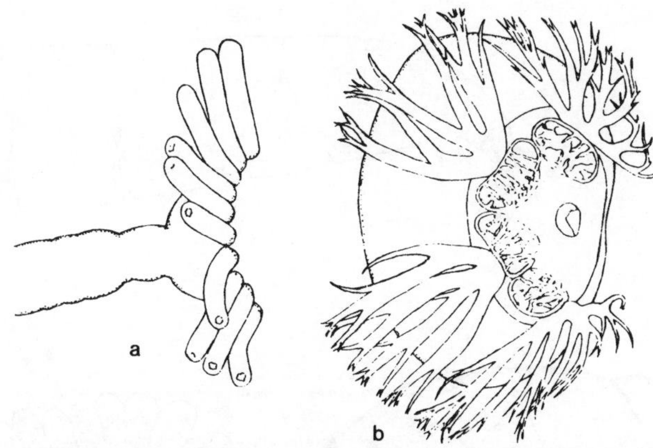

Figure 37.235a,b. Cypselosomatidae. *Cypselosoma australis* McAlpine. **a.** right anterior spiracle; **b.** left posterior spiracle. This is an Australian species. Larvae of the two Nearctic species remain unknown (from D. K. McAlpine 1966).

Selected Bibliography

Ferrar 1987.
McAlpine, D. K. 1966.
McAlpine, J. F. 1987a.
Wheeler, M. R. 1956.

MICROPEZIDAE (NERIOIDEA)

B. A. Foote, *Kent State University*

Stilt-legged Flies

Figures 37.236–37.239

Relationships and Diagnosis: The micropezids appear to be closely related to the acalyptrate families Cypselosomatidae and Neriidae, and all three are now placed in the Nerioidea (McAlpine, et al. 1981). Steyskal (1987a) has recently summarized our knowledge.

Micropezid larvae possess the basic morphological features repeatedly seen in saprophagous muscoid Diptera. They are maggot-like, relatively long and slender, and have rather featureless surfaces except for pale to slightly darkened spinule bands. Creeping welts are variously developed along the ventral surface. The facial mask has a distinctive array of food channels bordering and leading into the preoral cavity. The anterior spiracles are fan-shaped and have a variable number of marginal papillae. The posterior spiracular plates are separate and variously shaped. Mature larvae have three or four circular, oval, or serpentine spiracular openings and four clusters of spiracular hairs. The mouthparts show no special modifications, and the tentoropharyngeal sclerite has ridges forming a filtering apparatus along its ventral surface.

Biology and Ecology: As far as known, the larvae feed saprophagously within decaying plant material with the exception of two species that bore in ginger roots (Steyskal, 1964). Larvae of *Cnodacophora, Compsobata,* and *Taeniaptera* have been reared in the laboratory on decaying lettuce,

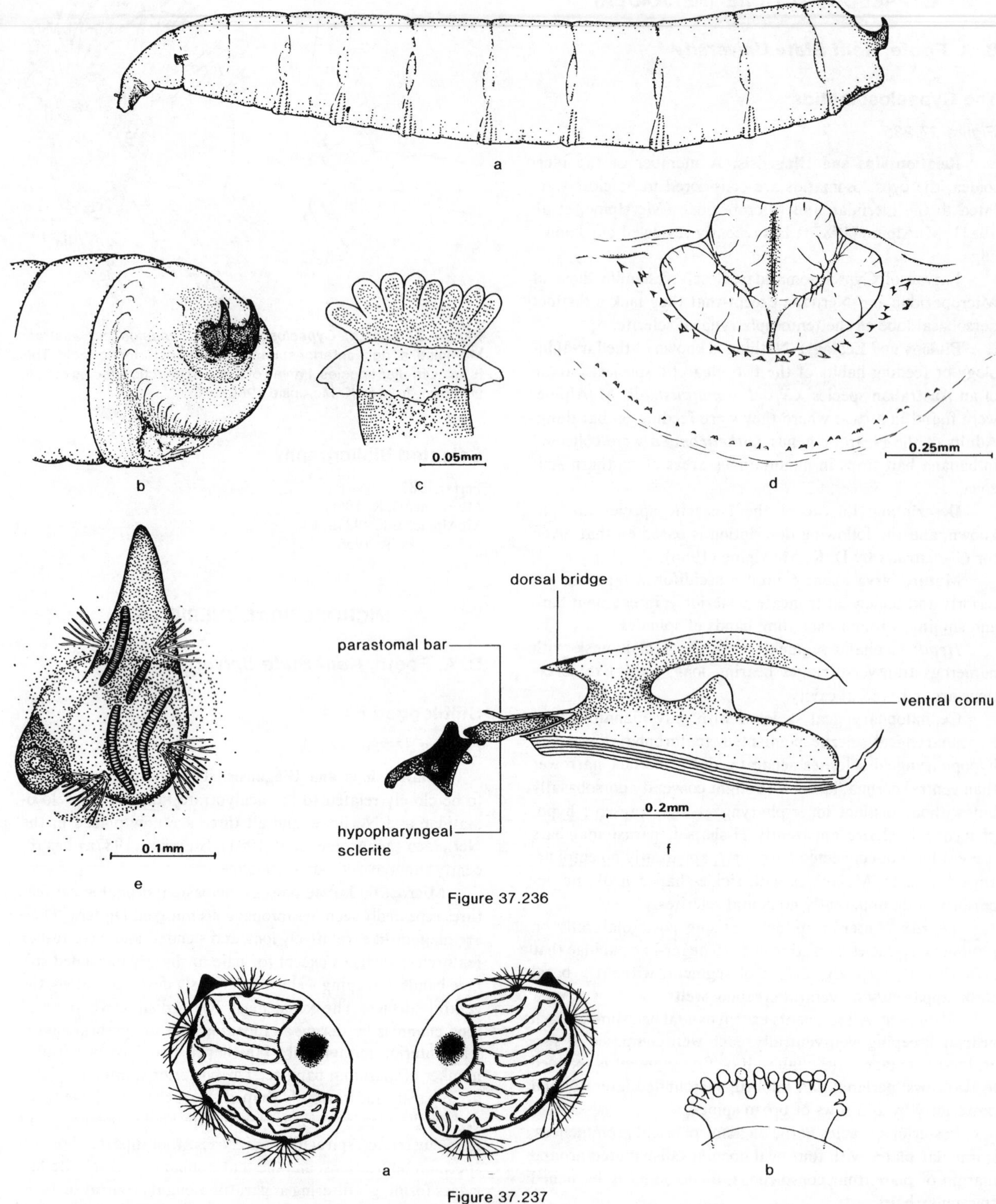

Figure 37.236a-f. Micropezidae. *Compsobata univitta* (Walker). **a.** larva, lateral; **b.** caudal end; **c.** anterior spiracle; **d.** anal pad, ventral; **e.** right spiracular plate; **f.** cephalopharyngeal skeleton. Found beneath decaying vegetation near a stream (from Teskey 1972).

Figure 37.237a,b. Micropezidae. *Rainieria antennaepes* (Say). **a.** posterior spiracles; **b.** anterior spiracle. Larvae occur in rotting accumulations of woody debris in tree holes and crotches (from Steyskal 1964).

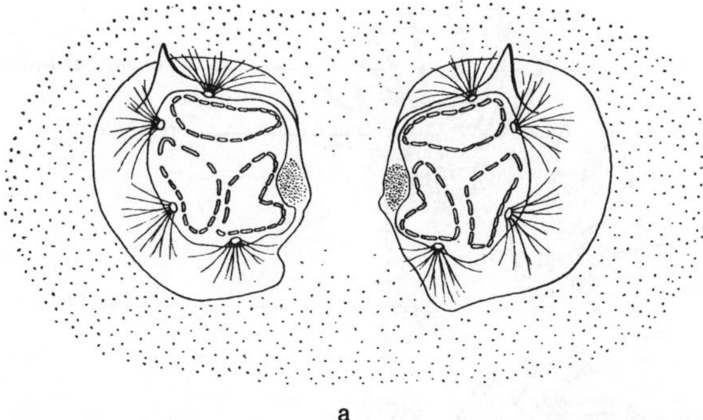
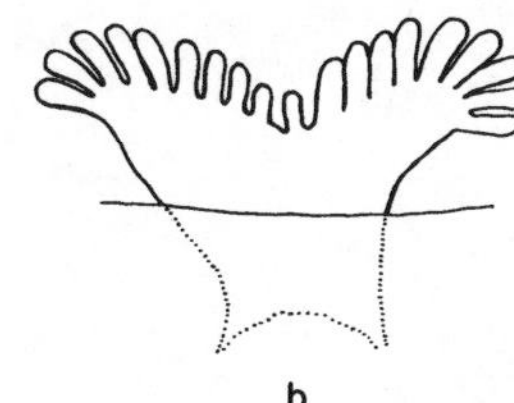

Figure 37.238

Figure 37.238a,b. Micropezidae. *Mimegralla coeruleifrons* (Macquart). **a.** posterior spiracles; **b.** anterior spiracle (from Steyskal 1964).

Figure 37.239

Figure 32.239. Micropezidae. *Taeniptera lasciva* (Fabricius). Posterior spiracles (from Steyskal 1964).

and those of *Calobatina* and *Rainieria* have been found in rot pockets of trees containing high bacteria, fungi and yeast populations. Adults are most commonly encountered in marshes and moist to mesic woodlands.

Description: Mature larvae 6–10 mm, muscidiform, relatively elongate and slender; integument mostly smooth, some species with a few pale spinule bands encircling anterior margins of thoracic and abdominal segments; generally whitish.

Head: Cephalic segment retractile, antennomaxillary organs well developed, antennae appearing 2- or 3-segmented; facial mask with noticeable, mostly unbranching oral ridges and grooves leading into preoral cavity. In *Compsobata* two short rows of rasping spinules occur on either side of preoral cavity.

Cephalopharyngeal skeleton deeply or weakly pigmented, with tentoropharyngeal and hypopharyngeal sclerites separate; dorsal cornua connected by dorsal bridge anteriorly, ventral cornua with dorsobasal lobe and commonly with apical windows; pharyngeal filter present; hypopharyngeal sclerite basically H-shaped, parastomal bars present and free apically, no obvious epipharyngeal sclerite. Mandibles usually darkly pigmented, hook with (*Compsobata*) or without accessory teeth, basal part commonly with small window.

Thorax: Anterior spiracles arising laterally or dorsolaterally on posterior half of prothorax, spiracles usually fan-shaped, with 7–20 papillae; in some species spiracles appearing somewhat bilobed.

Abdomen: Segments uniform in appearance, with well developed creeping welts ventrally, some species with pale to somewhat darkened spinule bands dorsally and laterally; caudal segment somewhat rounded, without fleshy protuberances, perianal pad bilobed and usually with scattered or clustered spinules near posterior border.

Posterior spiracular disc lacking marginal tubercles; posterior spiracles distinctly separated, usually elevated on sclerotized bases; each spiracular plate with 3–4 variously oriented spiracular openings that are narrowly elongate (*Compsobata*), oval, or mildly serpentine (*Rainieria*); plate with (*Compsobata*) or without (*Rainieria*) dorsally directed marginal spur; four dichotomously branched spiracular hairs around margin of spiracular plate.

Comments: This is a relatively small family of some 600 species that is most diverse in the tropics. Only about 30 species and 11 genera have been recorded from America north of Mexico. The largest Nearctic genus is *Micropeza* with 15 species. No species is considered to be economically important in North America.

Selected Bibliography

Berg 1947.
Bohart and Gressitt 1951.
Brindle 1965a.
Ferrar 1987.
Fischer 1932a.
Greene 1929.
Hennig 1952.
Merritt and James 1973.
Müller 1957.
Peterson, A. 1951.
Sabrosky 1942.
Séguy 1940.
Steyskal 1964, 1987a.
Teskey 1972, 1976.
Wallace 1970.

NERIIDAE (NERIOIDEA)

B. A. Foote, *Kent State University*

Cactus Flies

Figures 37.240a-d

Relationships and Diagnosis: Formerly considered to be a subfamily of Micropezidae (Hennig 1952), the neriids are now given family status. There seems little doubt that the family is close to the Micropezidae (Hennig 1958, Griffiths 1972), and McAlpine et al. (1981) included it in the Nerioidea along with the Micropezidae and Cypselosomatidae. Steyskal (1987b) has recently summarized our knowledge.

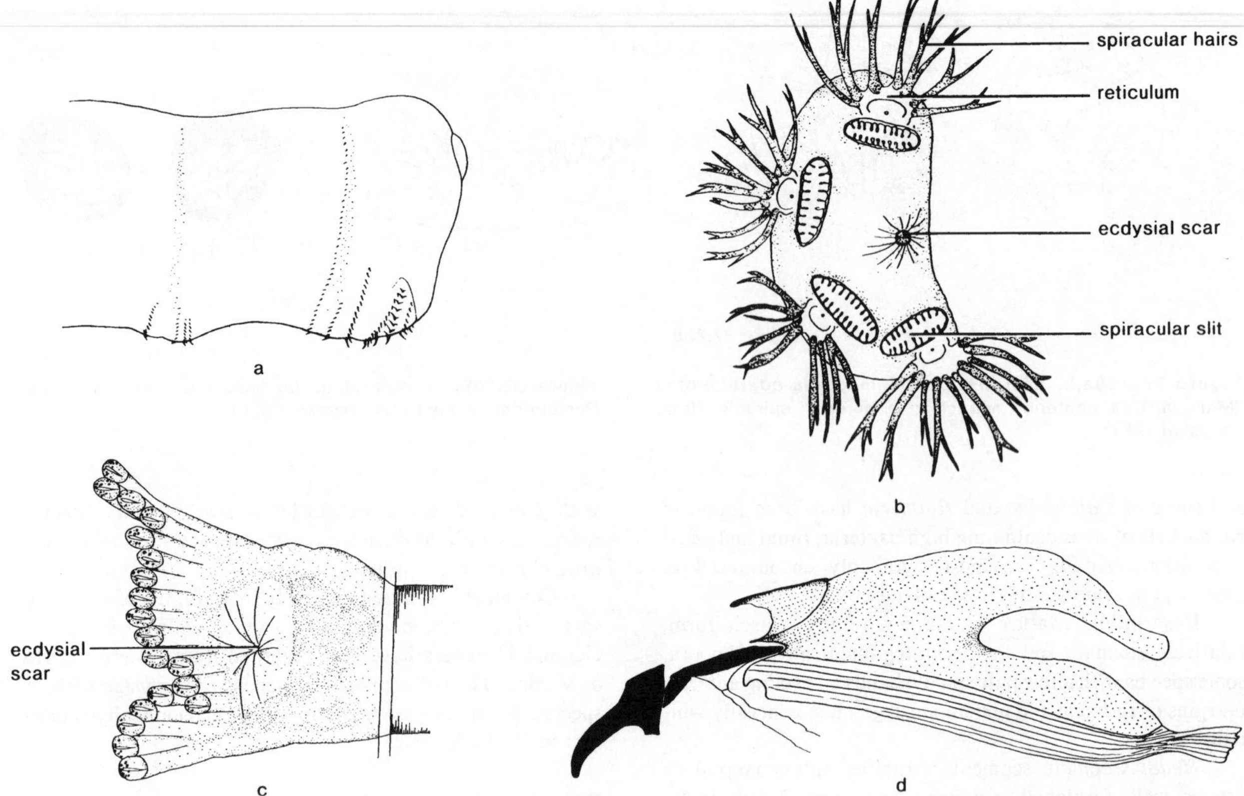

Figure 37.240a-d. Neriidae. *Odontoloxozus longicornis* (Coquillett). **a.** posterior end, lateral; **b.** posterior spiracle; **c.** anterior spiracle; **d.** cephalopharyngeal skeleton. The larvae feed in decaying cactus pads (a, d from Steyskal 1965; b, c from Olsen and Ryckman 1963).

Unfortunately, the larvae of Neriidae lack distinctive features that permit their separation from other related families of saprophagous Acalyptratae. A fairly useful character is the presence of four oval spiracular openings that are arranged around the periphery of each posterior spiracular plate (fig. 37.240b). Larvae of *Odontoloxozus* are unusually large, measuring up to nearly 13 mm, and are commonly encountered only in rot pockets in arborescent cacti (e.g., saguaro) in the desert Southwest.

Biology and Ecology: This is another family having saprophagous larvae. Adults of *Odontoloxozus longicornis* (Coquillett) are frequently abundant on columnar cactus in which larvae develop within rot pockets. Larvae of this species have also been found in rotting roots of papaya. Larval habits of the sole Nearctic species of *Oncopsia* are unknown.

Description: Mature larvae 9–12 mm, somewhat muscidiform, slender and cylindrical; integument smooth, without dorsal bands of spinules; whitish to yellowish.

Head: Cephalic segment retractile, antennomaxillary sense organs well developed; antennae seemingly 3-segmented; facial mask with numerous oral ridges and grooves leading into preoral cavity.

Cephalopharyngeal skeleton with mandibles and hypopharyngeal sclerite deeply pigmented, tentoropharyngeal sclerite light brown; tentoropharyngeal and hypopharyngeal sclerites separate, dorsal cornua connected anteriorly by dorsal bridge, ventral cornua without noticeable dorsobasal lobe or windows; pharyngeal filter present; hypopharyngeal sclerite H-shaped, apparently no epipharyngeal or labial sclerites; parastomal bars short, free apically. Mandibles rather slender, hook slightly decurved and lacking accessory teeth, basal part without windows, no dental or accessory sclerites.

Thorax: First two segments lacking ventral creeping welts, metathorax with welt at posterior segmental border; anterior spiracles arising dorsolaterally on prothorax, fan-shaped with 17–19 finger-like marginal papillae.

Abdomen: Segments very similar except that caudal segment is about twice length of penultimate segment; each segment with distinct creeping welt ventrally near border of segment, the welts consisting of rows of blackened spinules; welt on caudal segment extending dorsally approximately two-thirds height of segment; caudal segment rounded, lacking fleshy protuberances; perianal pad bilobed, surrounded by four irregular rows of blackened spinules.

Posterior spiracular disc lacking marginal tubercles; posterior spiracles nearly touching along midline, nearly sessile on surface of spiracular disc; each spiracular plate with four oval spiracular openings arranged around border of plate; four rather pale and dichotomously branched spiracular hairs.

Comments: The family is most diverse in the tropics and only *Odontoloxozus longicornis* and *Oncopsia flavifrons* (Bigot) occur in the southwestern states. Neither is considered to have economic importance.

Selected Bibliography

Berg 1947.
Cresson 1938.
Ferrar 1987.
Olsen and Ryckman 1963.
Ryckman and Olsen 1963.
Steyskal 1965, 1987b.

TANYPEZIDAE (DIOPSOIDEA)

B. A. Foote, *Kent State University*

The Tanypezids

Figures 37.241a–f

Relationships and Diagnosis: This family belongs to the Diopsoidea along with the Strongylophthalmyiidae, Psilidae, and Diopsidae (McAlpine, et al. 1981). Steyskal (1987c) recently summarized our knowledge.

Larvae have the body form typical of saprophagous acalyptrate Diptera and closely resemble micropezid larvae. Perhaps the most distinctive structure that separates them from micropezid and most otitid larvae is the bicornuate nature of the anterior spiracles (fig. 37.241b)

Biology and Ecology: Little is known of the biology. The single reared species, *Tanypeza longimana* Fallén, has saprophagous larvae that were reared in the laboratory on decaying plant materials (Foote 1970). Adults are rarely collected but occur in deciduous woodlands.

Description: Mature larvae 5–7 mm, muscidiform, tapering anteriorly and somewhat rounded posteriorly; integument smooth, without spinule bands dorsally or laterally; whitish.

Head: Cephalic segment retractile, with well developed antennomaxillary organs; antennae short, appearing 2-segmented; facial mask with numerous oral ridges and grooves leading into oral cavity, grooves unbranched or poorly branched; area immediately in front of oral cavity roughened and reticulated; no postoral spine band.

Cephalopharyngeal skeleton deeply pigmented; tentoropharyngeal and hypopharyngeal sclerites separate; ventral cornua each with basodorsal lobe, dorsal cornua connected anteriorly by dorsal bridge; pharyngeal filter present; hypopharyngeal sclerite H-shaped, labial sclerite broadly V-shaped. Mandibles with hook slender and lacking accessory teeth, basal part without windows; dental sclerite slender and curved.

Thorax: Segments without prolegs; metathorax with ventral creeping welt; anterior spiracles arising laterally on prothorax, bicornuate, with tubelike basal part and deeply bifurcate distal part, each distal branch slender and bearing 6–9 finger-like papillae.

Abdomen: Segments similar and subcylindrical, caudal segment about twice as long as penultimate segment, smooth and rounded distally; perianal pad bilobed, with each lobe broad; a short row of spinules anterior to the pad.

Posterior spiracular disc lacking marginal tubercles; posterior spiracles distinctly separated, elevated above surface of disc on short, deeply pigmented, tube-like structures; each spiracular plate pale yellow to light brown, spiracular openings oval, arranged in T-shape around ecdysial scar; three dichotomously branched spiracular hairs arising between outer ends of spiracular openings.

Comments: The tanypezids are best developed in the tropics, and only two species of the genus *Tanypeza* have been recorded in the eastern and midwestern states. No species anywhere is of economic significance.

Selected Bibliography

Ferrar 1987.
Foote 1970.
Hennig 1937, 1958.
Steyskal 1987c.

STRONGYLOPHTHALMYIIDAE (DIOPSOIDEA)

B. A. Foote, *Kent State University*

The Strongylophthalmyiids

Relationships and Diagnosis: Presently included in the Diopsoidea (McAlpine, et al. 1981), the Strongylophthalmyiidae were earlier considered to be a subfamily of Psilidae. Hennig (1958) suggested that the species warranted family rank and felt that the family had affinities with the Tanypezidae. Steyskal (1987d) recently summarized our knowledge of them.

Biology and Ecology: Nothing is known of the biology or larval feeding habits of any species of this family. The sole North American species occurs in wooded habitats. Adults of a Palaearctic species have been found on aspen logs in Finland.

Description: No larval descriptions have been published.

Comments: This is a very small family, consisting of one genus and approximately 26 species largely restricted to southern Asia. Only one species, *Strongylophthalmyia angustipennis* Melander, has been recorded from America north of Mexico.

Selected Bibliography

Ferrar 1987.
Hennig 1958.
Steyskal 1987d.

PSILIDAE (DIOPSOIDEA)

B. A. Foote, *Kent State University*

Rust Flies

Figures 37.242a–d, 37.242z

Relationships and Diagnosis: This is another member of the Diopsoidea, and is thought to be closely related to the Diopsidae (McAlpine et al. 1981). Griffiths (1972), however,

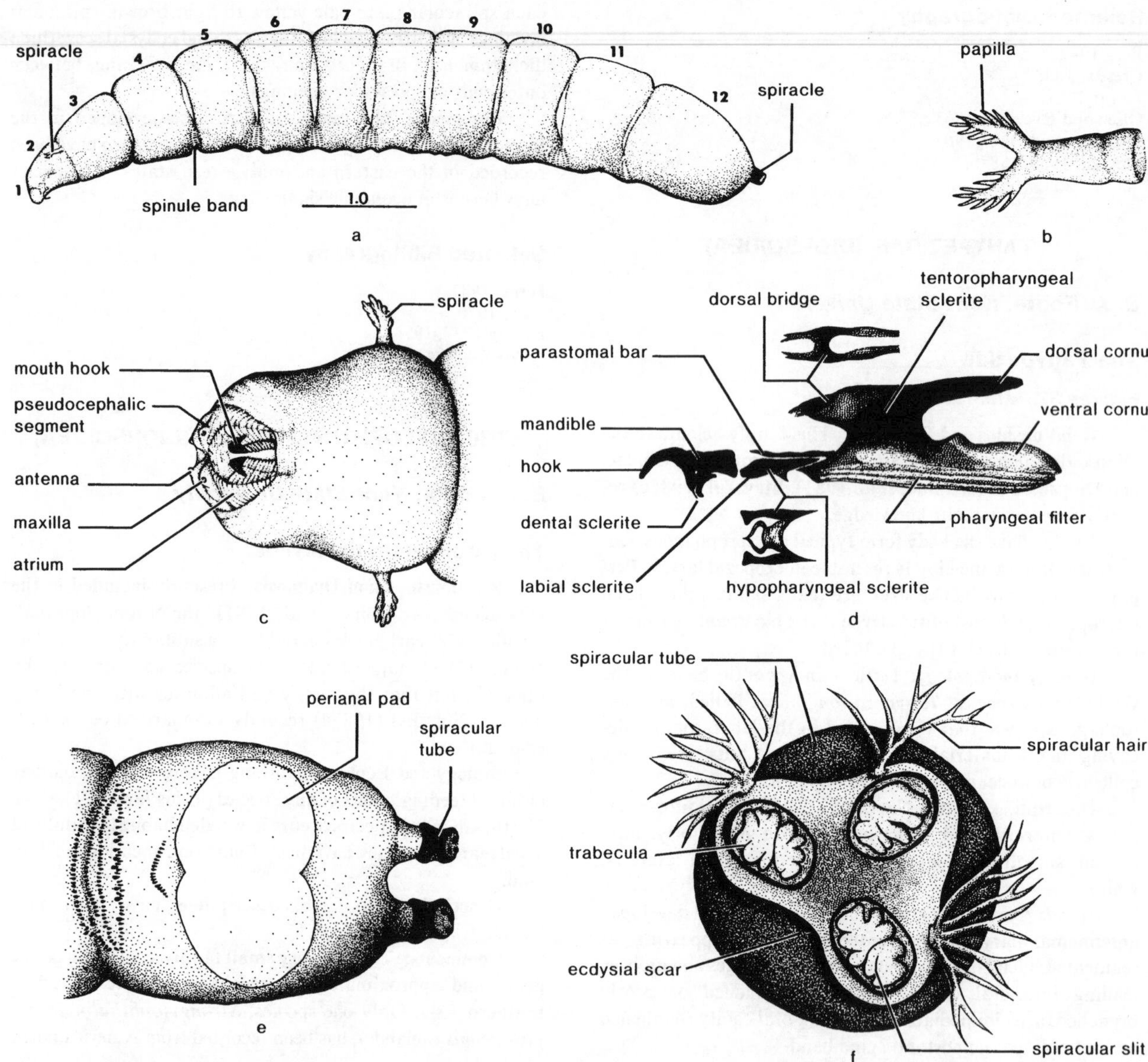

Figure 37.241a-f. Tanypezidae. *Tanypeza luteipennis* Knab and Shannon. **a.** lateral; **b.** anterior spiracle; **c.** ventral view of anterior end; **d.** cephalopharyngeal skeleton; **e.** ventral view of posterior end showing bilobed perianal pad; **f.** posterior spiracular plate (from Foote 1970. Reprinted with permission of the Entomological Society of America).

placed the Psilidae in the superfamily Nothyboidea, considering it to be most closely related to the Periscelididae. Steyskal (1987e) recently summarized our knowledge of the family.

Larvae are poorly known and not easily distinguished from those of some genera of Lonchaeidae, Platystomatidae, and Otitidae. Fairly distinctive is the presence of a thorn-like spur arising from the dorsal side of each conspicuously elevated posterior spiracle (fig. 37.243). The spiracular openings are elongate-oval and radiate out from the ecdysial scar. Larvae are elongate and possess a smooth-surfaced integument. The cephalopharyngeal skeleton is well sclerotized and has the hypopharyngeal sclerite fused with the tentoropharyngeal sclerite. A pharyngeal filter apparently is absent.

Biology and Ecology: As far as known, larvae are phytophagous, feeding within roots and stems of a considerable variety of herbaceous angiosperms. A few species develop under the bark of living trees, gaining access through wounds. Adults are usually encountered in moist or wooded habitats and are relatively uncommon.

Description: Fully grown larvae 7–10 mm, somewhat muscidiform, slender and cylindrical; integument mostly smooth, with weak spinule bands dorsally along borders of segments; near white to yellowish.

Head: Cephalic segment retractile, with very inconspicuous sensory organs; facial mask with somewhat weakened oral ridges on each antennomaxillary lobe.

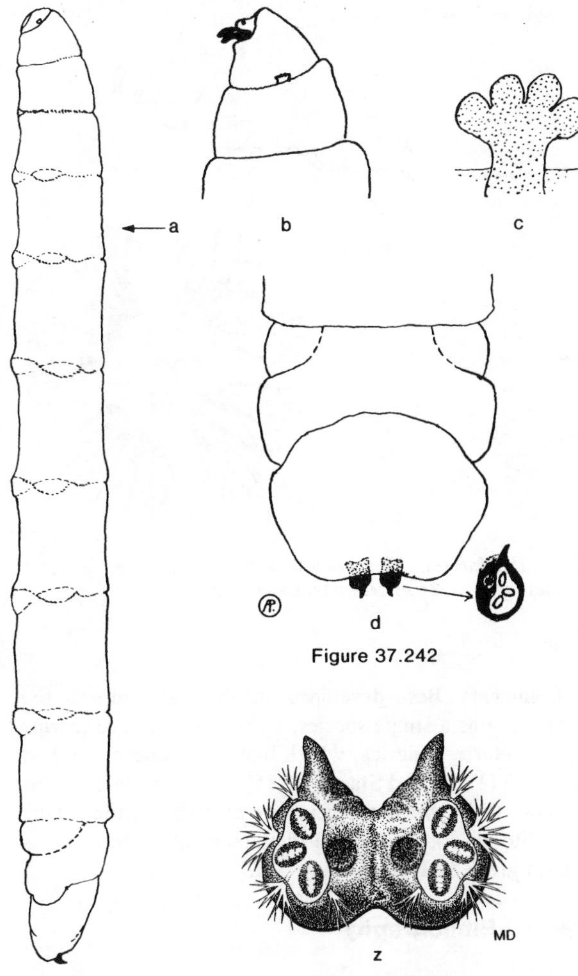

Figure 37.242a-d. Psilidae. *Psila rosae* Fabricius, the carrot rust fly. **a.** lateral; **b.** anterior segments; **c.** anterior spiracle; **d.** dorsal view of posterior segments. Larvae attack roots and underground stems of vegetables such as carrots, celery, parsnip, and parsley (from Peterson 1951).

Figure 37.242z. Psilidae. Posterior spiracular plates of *Loxocera cylindrica* Say. Stem borers in monocots (from Manual of Nearctic Diptera, Vol. 1).

Cephalopharyngeal skeleton deeply pigmented, with tentoropharyngeal and hypopharyngeal sclerites fused; cornua of tentoropharyngeal sclerite lacking obvious windows; no pharyngeal filter. Mandibles heavy, lacking accessory sclerites, no windows in basal part.

Thorax: Segments without prolegs or conspicuous creeping welts ventrally; anterior spiracles arising laterally on prothorax, small and pale, fan-shaped, with 5–14 marginal papillae.

Abdomen: Segments cylindrical, subequal in diameter except for slightly enlarged caudal segment; each segment ventrally bearing creeping welt composed of pale spinules; caudal segment rounded, without fleshly protuberances; perianal pad bilobed.

Posterior spiracular disc lacking marginal tubercles; posterior spiracles nearly contiguous or actually touching along mid-line, and distinctly elevated on deeply pigmented bases; each spiracular plate with conspicuous dorsally-projecting spine on upper margin; spiracular openings small, oval, and arranged fan-like around ecdysial scar; no spiracular hairs.

Comments: This is a relatively small family containing 34 species and 3 genera in North America. The only species known to have economic significance is *Psila rosae* (Fabricius), the carrot rust fly, a European introduction that can be quite damaging to carrots, celery and related plants.

Selected Bibliography

Ashby and Wright 1946.
Capelle 1953.
Chittenden 1902.
Ferrar 1987.
Hennig 1952.
Meijere 1941, 1947.
Pechuman 1943.
Staeler 1972.
Steyskal 1987e.
Vimmer 1925.
Vos-de Wilde 1935.

DIOPSIDAE (DIOPSOIDEA)

Terry A. H. Stasny, *West Virginia University*

Stalkeyed Flies

Figures 37.243a–d

Relationship and Diagnosis: This family of the Diopsoidea may be fairly closely related to the Psilidae (McAlpine et al. 1981). Peterson (1987c) recently reviewed our knowledge.

Larvae are saprophagous and are not distinguishable from those of several other families of acalyptrate Diptera having similar habits (e.g., Otitidae, Heleomyzidae). A fairly diagnostic structure is the occurrence of two tube-like structures bearing the posterior spiracles. They can be separated from larvae of Psilidae and many Lonchaeidae by the absence of a thorn-like projection arising from the dorsal side of the posterior spiracular plate.

Biology and Ecology: Immature stages of the single Nearctic species have not been encountered in the field, but Lavigne (1962) reared larvae in the laboratory on rotting vegetation, and Stasny (1985) reared them on decaying dandelion flowers. Adults are solitary in summer, but overwinter by the hundreds in aggregations, usually in rock crevices on stream banks (Flint 1956, Stasny 1985).

Description: Mature larvae 5–8 mm, muscidiform, tapering anteriorly and posteriorly; integument mostly bare but with irregular rows of spinules encircling most segments and with well developed creeping welts ventrally; white.

Head: Cephalic segment retractile, bearing small sensory organs anteriorly and anteroventrally; antennae short,

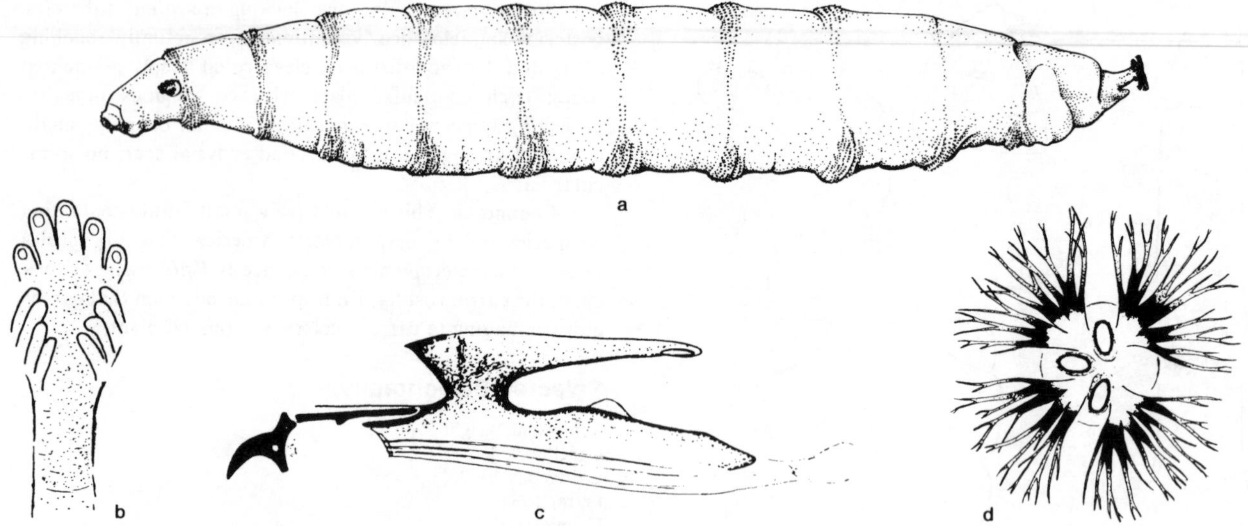

Figure 37.243a-d. Diopsidae. *Sphyracephala brevicornis* (Say). **a.** lateral; **b.** extended anterior spiracle; **c.** cephalopharyngeal skeleton, 3rd instar; **d.** posterior spiracular plate (fig. 37.343a drawn by Lana Tackett; figs. b, c, d from Stasny 1985).

apparently 2-segmented; facial mask with rows of comb-like plates bordering preoral cavity; postoral spine band apparently absent.

Cephalopharyngeal skeleton deeply pigmented; tentoropharyngeal and hypopharyngeal sclerites separate; dorsal cornua narrow, connected anteriorly by fenestrated dorsal bridge, and with narrow window posteriorly; ventral cornua broad, with dorsobasal lobe; pharyngeal filter present; hypopharyngeal sclerite very narrow when viewed laterally and H-shaped when viewed dorsally; crossbar very narrow; parastomal bars elongate and free apically; epipharyngeal and labial sclerites undeveloped. Mandibles deeply pigmented, hook somewhat sickle-shaped and lacking accessory teeth, basal part narrow and with ventral projection, and one or two small windows; dental sclerites separate, very narrow and slightly curved in lateral view; no accessory oral sclerites.

Thorax: Metathorax with ventral creeping welts near segmental border and encircled by band of pale spinules; anterior spiracles arising posterolaterally on prothorax; spiracles with short tubular base and expanded palmate distal part, with 11–14 finger-like papillae along lateral and distal borders.

Abdomen: All segments except caudal one very similar, each encircled by weak band of spinules and bearing ventral creeping welts; perianal pad bilobed, narrow and strap-like, extending onto lateral surface of caudal segment, lacking a spinule patch near posterior border, and not subtended by anal lobes; anal slit lightly sclerotized.

Posterior spiracles borne at apices of separate tubular bases, each spiracular projection subtended ventrally by short tubercle; spiracular plates with poorly developed peritremes; three elongate-oval spiracular openings that are nearly perpendicular to each other; ecdysial scar not apparent; four evenly spaced, 3–4 branched spiracular hairs that are pigmented in the middle, resulting in a dark circle (fig. 37.243d).

Comments: Best developed in the Paleotropics, the family contains a single species, *Sphyracephala brevicornis* (Say), in North America, whose biology has been reported by Lavigne (1962) and Stasny (1985). Excellent descriptions of several of the African species are available in the papers of Schillito (1940, 1960), and the biology of several species has been given by Descamps (1957).

Selected Bibliography

Descamps 1957.
Ferrar 1987.
Flint 1956.
Hennig 1952.
Lavigne 1962.
Peterson, B. V. 1987c.
Schillito 1940, 1960.
Stasny 1985.

LONCHAEIDAE (TEPHRITOIDEA)

B. A. Foote, *Kent State University*

The Lonchaeid Flies

Figures 37.244, 37.245

Relationships and Diagnosis: The placement of the Lonchaeidae within the acalyptrate Diptera has been a matter of dispute for some time. Hennig (1952) suggested that the family was close to the Chamaemyiidae. Stone et al. (1965) considered it to be a member of the superfamily Palloptero-idea and thus relatively close to the Piophilidae and Pallopteridae. Griffiths (1972) has the most radical placement in that he recognizes a new superfamily, Lonchaeoidea, for the Lonchaeidae and Cryptochetidae. The *Manual of Nearctic Diptera* (McAlpine, et al. 1981) takes a more conservative

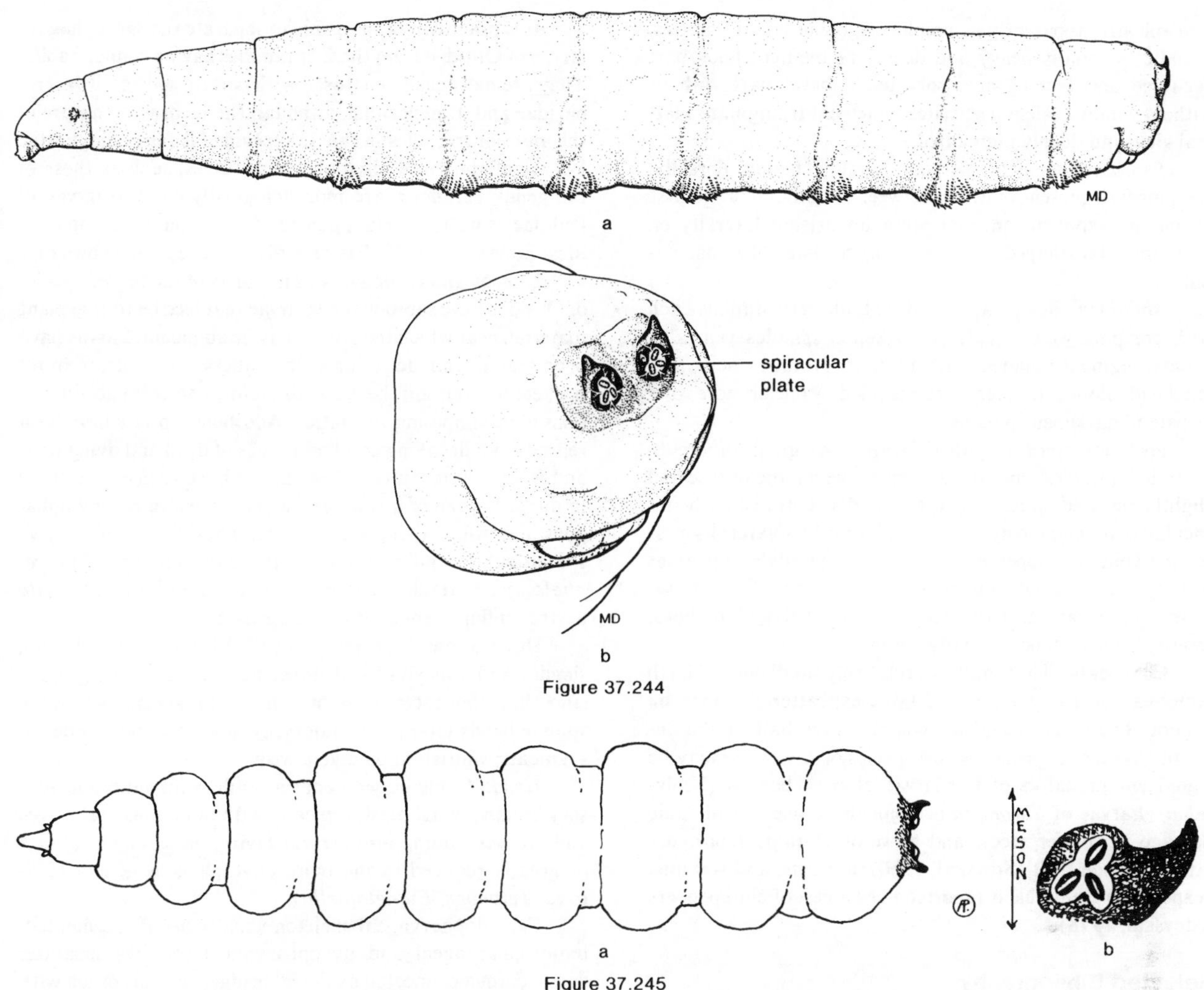

Figure 37.244a,b. Lonchaeidae. *Lonchaea corticis* Taylor. **a.** lateral; **b.** posterolateral view of posterior end (from Manual of Nearctic Diptera, Vol. 1).

Figure 37.245a,b. Lonchaeidae. *Lonchaea winnemanae* Malloch. **a.** dorsal; **b.** posterior spiracular plate. Larvae occur under bark of recently fallen deciduous trees (from Peterson 1951).

approach and places it in the Tephritoidea close to the Otitidae. McAlpine (1987b) recently summarized our knowledge.

Larvae have characteristics typical of the saprophagous Muscomorpha and thus are frequently indistinguishable from those of such families as the Otitidae, Pallopteridae, Milichiidae, Dryomyzidae, and Heleomyzidae. The body is elongate, lacks distinctive features, and is smooth surfaced except for rows of generally pale spinules that form creeping welts along the ventral side. The posterior spiracular disc lacks bordering tubercles, the spiracular plates are somewhat elevated above the disc surface, and many species have a thorn-like projection dorsally. The cephalopharyngeal skeleton is typical of saprophagous species and shows no peculiarities. Filtering ridges are present in the floor of the pharyngeal sclerite.

Biology and Ecology: Lonchaeid larvae are basically saprophagous and have been reported to develop within manure, decaying cambial tissues, rotting plant remains, and damaged fruits. A few species of the genus *Lonchaea* are thought to be predators of bark beetle larvae. Larvae of *Earomyia* are phytophagous and attack seeds within cones of coniferous trees. A few species of *Dasiops* have been reported to attack fruits of apricot and passion flower. Adults are occasionally collected near larval habitats.

Description: Mature larva 5–9 mm, muscidiform, elongate and slender; integument mostly smooth without spinule bands or pubescence; whitish to pale yellowish.

Head: Cephalic segment retractile, with well developed antennomaxillary sense organs, antennae elongate, appearing 2-segmented; facial mask with numerous unbranched oral ridges and grooves leading into preoral cavity.

Cephalopharyngeal skeleton variously pigmented, with tentoropharyngeal sclerite frequently largely pale; tentoropharyngeal and hypopharyngeal sclerites separate, dorsal cornua connected anteriorly by dorsal bridge, ventral cornua with weak dorsobasal lobe or lacking lobe, usually without windows apically; pharyngeal filter present; hypopharyngeal sclerite basically H-shaped in ventral view; parastomal bars free apically; apparently no epipharyngeal or labial sclerites. Mandibles usually heavy and deeply pigmented, hook part decurved and lacking accessory teeth; basal part with or without windows; dental sclerites somewhat triangular in lateral view and deeply pigmented.

Thorax: Segments without spinule bands dorsally; creeping welt present ventrally between metathoracic and first abdominal segment; anterior spiracles arising laterally on prothorax, fan-shaped, with 5–12 finger-like marginal papillae.

Abdomen: Segments cylindrical, all very similar, each with creeping welt of pale to darkened spinules ventrally. Caudal segment rounded, without fleshy protuberances; perianal pad bilobed, frequently surrounded by one or more rows of pale to darkened spinules.

Posterior spiracular disc lacking marginal tubercles; posterior spiracles separate, arising above middle of disc and slightly elevated on tubular, pigmented structures; each spiracular plate commonly with dorsally or dorsolaterally projecting spine on upper margin; 3 oval spiracular openings usually radiating out from ecdysial scar to give T-shape appearance; spiracular hairs pale to darkened and dichotomously branched, occasionally absent.

Comments: The family is a relatively small one in North America, consisting of some 120 species scattered among six genera. The genus *Lonchaea* contains over half of the described Nearctic species. A few species of *Lonchaea* may be significant predators of the larvae of bark beetles (Scolytidae). Larvae of *Earomyia* occasionally cause considerable damage to conifer seeds, and those of *Dasiops* reportedly attack passion fruit (Steyskal, 1980), apricots, and walnuts. A species in Columbia is reported to be a pest of chili peppers (Steyskal, 1978).

Selected Bibliography

Engel 1916.
Ferrar 1987.
Harman and Wallace 1971.
Hennig 1952.
Keifer 1930.
McAlpine, J. F. 1956, 1961, 1987b.
McAlpine and Steyskal 1982.
McAlpine and Morge 1970.
Morge 1959, 1962.
Quayle 1929.
Smith, K. G. V. 1957.
Stegmaier 1973.
Steyskal 1978, 1980.
Vimmer 1925.

OTITIDAE (TEPHRITOIDEA)

B. A. Foote, *Kent State University*

Picture-Winged Flies

Figures 37.246–37.248

Relationships and Diagnosis: The otitids have long been considered to be closely related to the Platystomatidae, Richardiidae, Tephritidae and other acalyptrate families that currently compose the Tephritoidea (McAlpine, et al. 1981).

It is practically impossible to separate the saprophagous larvae of Otitidae from those of families having similar habits. They are most easily confused with certain species of Micropezidae and Tanypezidae, although the bicornuate nature of the anterior spiracles in the Tanypezidae is fairly diagnostic. Larvae of the more phytophagous species, such as those of the genus *Tetanops,* are morphologically close to larvae of Psilidae as well as certain genera of other plant-feeding families. Steyskal (1987f) has recently reviewed our knowledge.

Biology and Ecology: The majority of the reared species of Otitidae have saprophagous larvae that feed on rotting plant material, near which the adults may be abundant. Larvae have been reared from decaying bulbs, tubers, and roots, from rot pockets in cacti and herbaceous stems, and from accumulations of decomposing vegetation. Additional species have been reared from decaying cambial tissues of dead and dying trees and from rotting fruits. A few species have been encountered in dung. Larvae of a few genera are known to be phytophagous and utilize living plant tissue. Thus, larvae of *Tritoxa* spp. attack the bulbs of onions, those of *Tetanops* (*Eurycephalomyia*) attack sugar beets, and those of *Eumetopiella* destroy inflorescences of certain grasses.

Description: Mature larvae 4–13 mm, muscidiform, slender and subcylindrical, tapering anteriorly; integument smooth, without setae or pubescence; some species with weak spinule bands laterally and dorsally along anterior border of segments; whitish to pale yellowish.

Head: Cephalic segment retractile, with noticeable sensory organs; facial mask typically with numerous oral ridges and grooves leading into preoral cavity; these may be absent or greatly reduced in the more phytophagous genera (*Tritoxa, Tetanops, Eumetopiella*).

Cephalopharyngeal skeleton usually deeply pigmented, tentoropharyngeal and hypopharyngeal sclerites separate; dorsal cornua connected by dorsal bridge, ventral cornua with distinct dorsobasal lobe and usually with distal windows; pharyngeal fiter present; hypopharyngeal sclerite H-shaped, with single transverse bar; parastomal bars free apically; apparently no epipharyngeal sclerite; labial sclerite forming V-shaped structure. Mandibles heavy, usually deeply pigmented; hook pale or pigmented, with one to several accessory teeth ventrally; basal part commonly with small to larger window; dental sclerite narrow and elongate or thicker and somewhat triangular in lateral view.

Thorax: Segments similar, metathorax with ventral creeping welt; anterior spiracles arising posterolaterally on prothorax; spiracles usually fan-shaped, with 8–15 papillae along distal border.

Abdomen: Segments all very similar, subcylindrical, each segment bearing creeping welt ventrally near anterior border, in some species spinule bands encircle the entire segment; caudal segment rounded to truncate, without fleshy protuberances; perianal pad bilobed, commonly with spinule patch near posterior border.

Posterior spiracular disc lacking marginal tubercles; posterior spiracles distinctly separated, usually weakly to distinctly elevated on pale to deeply pigmented bases; in some genera (e.g. *Pseudotephritis*) spiracles sessile on disc; each spiracular plate with pigmented peritreme and usually without

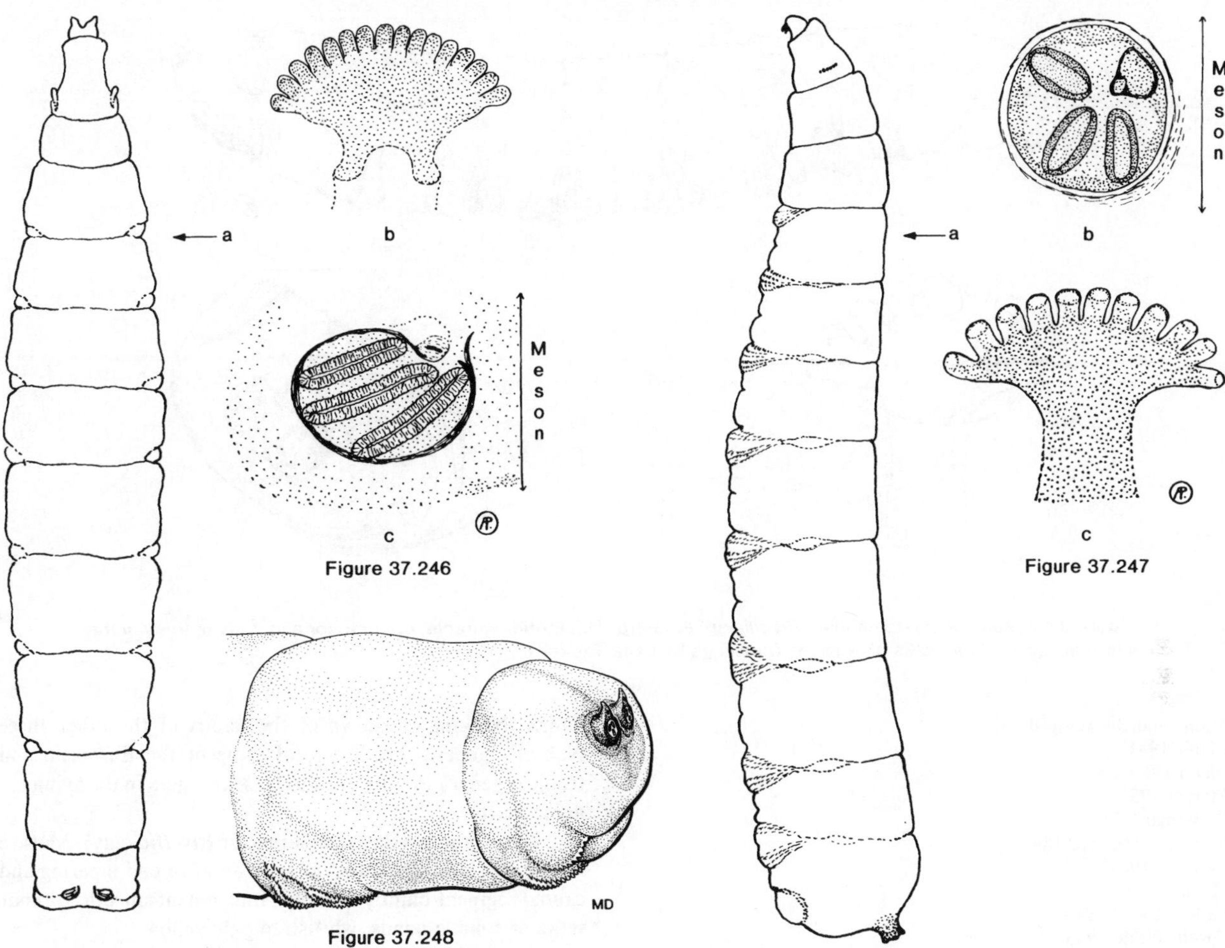

Figure 37.246

Figure 37.247

Figure 37.248

Figure 37.246a-c. Otitidae. *Pseudotephritis* sp. **a.** dorsal; **b.** anterior spiracle; **c.** posterior spiracular plate. Larvae can be found beneath the slightly loosened bark of recently fallen trees (from Peterson 1951).

Figure 37.247a-c. Otitidae. *Tritoxa flexa* (Wiedemann). **a.** lateral; **b.** posterior spiracular plate; **c.** anterior spiracle. The larvae feed on bulbs of onions (from Peterson 1951).

Figure 37.248. Otitidae. Oblique posterior view of caudal segments of *Tetanops myopaeformis* (Röder), the sugarbeet root maggot (from Manual of Nearctic Diptera, Vol. 1).

dorsal spine or spur (spur present in *Tetanops*); three narrowly oval spiracular openings radiating out from ecdysial scar; four dichotomously branched spiracular hairs usually present, each hair arising near outer border of spiracular opening, hairs absent or greatly reduced in phytophagous species.

Comments: This is a moderately large family of approximately 500 species in the world. Some 130 species and 42 genera are recorded from America north of Mexico. The largest genus is *Euxesta* with 31 species. There are very few species that have any economic importance. *Tritoxa flexa* (Wiedemann) (figs. 37.247a–c) is considered to be a minor pest of commercial onions. *Tetanops myopaeformis* (Roder), the sugarbeet root maggot (fig. 37.248), is a fairly common pest in the western states. A few species of *Euxesta* are relatively unimportant pests of corn in tropical areas.

Selected Bibliography

Allen and Foote 1967, 1975.
App 1938.
Banks 1912.
Barber 1939.
Blanton 1938.
Bohart and Gressitt 1951.
Brues 1902.
Chittenden 1911, 1927.
Drake and Decker 1932.
Ferrar 1987.
Foote 1976a.
Gojmerac 1956.
Greene 1917.
Harper 1962.
Hawley 1922.
Hennig 1952.
Hunter et al. 1913.
Hutchinson 1916.
Lobanov 1964, 1972.

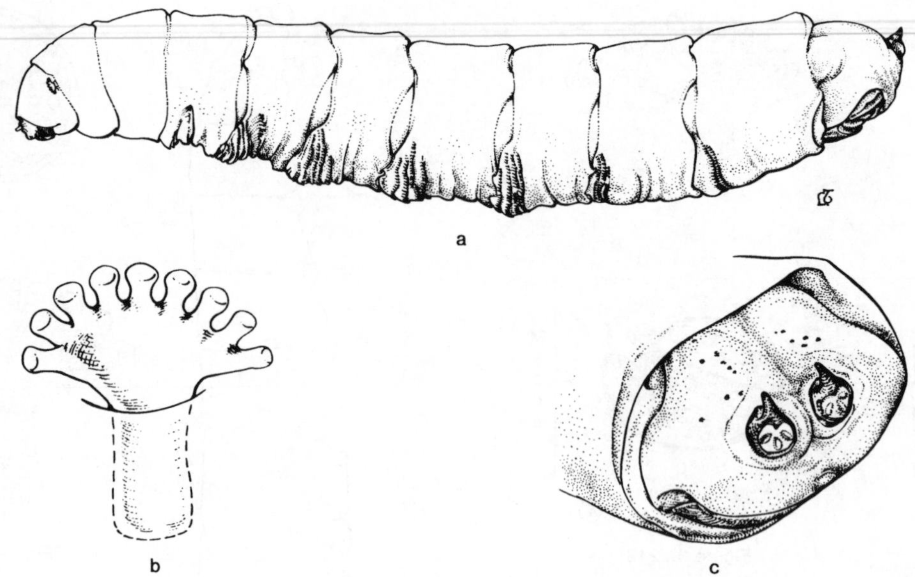

Figure 37.249a-c. Platystomatidae. *Rivellia* sp. **a.** lateral; **b.** anterior spiracle; **c.** posterior end. Larvae feed on the nitrogen-fixing root nodules of legumes (drawings by Lana Tackett).

Mahrt and Blickenstaff 1979.
Manis 1941.
Mann 1969.
Merrill 1951.
Pechuman 1937.
Riley and Howard 1894.
Steyskal 1987f.
Teskey 1976.
Valley et al. 1969.
Vos-de Wilde 1935.
Wheeler, A. G. 1973.

PLATYSTOMATIDAE (TEPHRITOIDEA)

B. A. Foote, *Kent State University*

The Platystomatid Flies

Figures 37.249a-c

Relationships and Diagnosis: This mostly tropical family is usually placed in the Tephritoidea along with such families as the Otitidae and Tephritidae (McAlpine et al. 1981).

This is another of a number of acalyptrate families that are difficult to characterize and distinguish from other families having saprophagous larvae. The presence of a spur on the upper surface of each posterior spiracle coupled with spiracular openings that are oriented at right angles to each other are fairly distinctive for larvae of the largest Nearctic genus, *Rivellia*. Occurrence of larvae in soil below leguminous plants is also useful in recognizing larvae of this genus. Steyskal (1987g) recently summarized our knowledge.

Biology and Ecology: Very little is known about the biology of the adults or the larval feeding habits. Recently, larvae of several North American species of *Rivellia* have been reported to attack the nitrogen-fixing root nodules of legumes, and it is probable that other species of the genus have similar habits. Nothing is known of the habits of the other three Nearctic genera, but larvae of some of the European and tropical genera are believed to be scavengers in decaying organic matter.

Description: (Based on larvae of *Rivellia* only). Mature larvae 4–9 mm, muscidiform, with anterior end tapering and caudal segment bluntly rounded; integument smooth, without setae or spinule bands; whitish to pale yellow.

Head: Cephalic segment retractile, with well developed sensory organs, antennae elongate and appearing 2-segmented; facial mask with several parallel, unbranched oral ridges and grooves leading into preoral cavity, which is apparently divided into two lateral parts by a median rod.

Cephalopharyngeal skeleton mostly deeply pigmented, with tentoropharyngeal and hypopharyngeal sclerites separate; dorsal cornua connected anteriorly by fenestrated dorsal bridge, ventral cornua with dorsobasal lobe and with windows distally; pharyngeal filter present. Hypopharyngeal sclerite H-shaped, parastomal bars free apically, apparently no epipharyngeal or labial sclerites. Mandibles deeply pigmented, hook weakly decurved, without accessory teeth along ventral margin; basal part without window; dental sclerite triangular in lateral view.

Thorax: Segments cylindrical and without spinule bands; metathorax ventrally with weakly developed creeping welt of pale spinules near posterior border of segment; anterior spiracles arising posterolaterally on prothorax, fan-shaped, and with 5–12 papillae arranged in single row along distal margin.

Abdomen: Segments cylindrical, very similar to each other, without dorsal spinule bands; each segment ventrally with weak creeping welt composed of pale blunt spinules; caudal segment bluntly rounded, without fleshy protuberances; perianal pad nearly circular, bilobed, appearing reticulated, lacking spinule patches in vicinity of pad.

Posterior spiracular disc lacking marginal tubercles; posterior spiracles arising on upper half of disc, distinctly separated, and somewhat elevated on deeply pigmented bases; each spiracular plate with dorsally or dorsolaterally projecting spur on upper surface, three oval to slit-like spiracular openings radiating out from ecdysial scar; openings nearly at right angles to each other; no spiracular hairs or greatly abbreviated.

Comments: This family is best developed in the tropics, particularly Africa. Nearly 1000 species are known, with four genera and somewhat over 40 species recorded from America north of Mexico. The largest genus in the Nearctic Region is *Rivellia* with 30 species. The only species having economic significance belong to the genus *Rivellia*, whose larvae consume the root nodules of agriculturally important legumes (e.g. soybeans).

Selected Bibliography

Bhattacharjee 1977.
Bibro and Foote 1986.
Diatloff 1965.
Eastman 1980.
Eastman and Wuensche 1977.
Ferrar 1987.
Foote 1985.
Hendel 1914.
Hennig 1952.
Koizumi 1957.
McAlpine 1973.
Namba 1956.
Newsom et al. 1978.
Seeger and Maldague 1960.
Steyskal 1961a, 1987g.

PYRGOTIDAE (TEPHRITOIDEA)

B. A. Foote, *Kent State University*

The Pyrgotids

Figure 37.250

Relationships and Diagnosis: This family is usually included in the Tephritoidea and is considered to be relatively closely related to the Platystomatidae, Otitidae, and Tephritidae (McAlpine et al. 1981). Steyskal (1987h) recently reviewed our knowledge.

The large, grub-shaped larva is distinctive in possessing a deep pit on the caudal segment in which are located the posterior spiracles.

Biology and Ecology: As far as known, the larvae are endoparasitoids of adult scarabaeid beetles. Adults are mostly nocturnal and are most commonly collected around lights, although at least one N. American species, *Pyrgotella chagnoni* (Johnson), is diurnal. Females oviposit through the thin cuticle on the dorsal surface of the beetle's abdomen during the time when the host is flying and the elytra are uplifted.

Description: Fully grown larvae 8–15 mm, body very robust, resembling a scarabaeid larva; slightly curved; integument smooth and shining, lacking hairs; white.

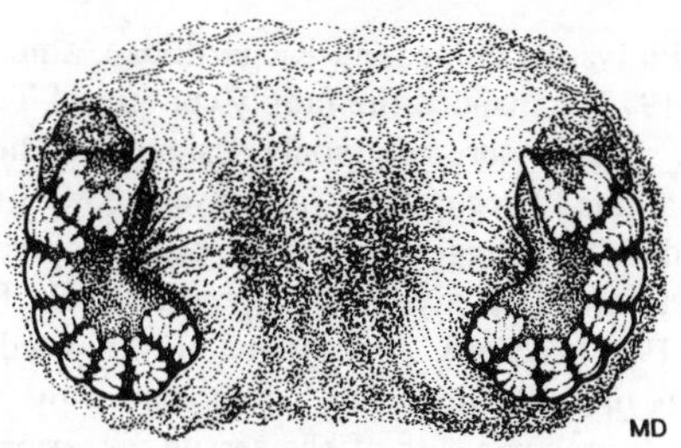

Figure 37.250. Pyrgotidae. Posterior spiracles of *Sphecomyiella valida* (Harris) (from Manual of Nearctic Diptera, Vol. 1).

Head: Cephalic segment somewhat retractile, with conspicuous sensory organs; facial mask without oral grooves or ridges.

Cephalopharyngeal skeleton relatively reduced, pigmented, with tentoropharyngeal and hypopharyngeal sclerites fused; dorsal cornua unpigmented distally, broader than ventral cornua, not connected anteriorly by dorsal bridge; no pharyngeal filter. Hypopharyngeal sclerite slender, not H-shaped; parastomal bars apparently absent. Mandibles rather slender, hook sickle-shaped, without accessory teeth; basal part with truncate ventral projection, without windows; no dental sclerites.

Thorax: Segments without prolegs or noticeable creeping welts; anterior spiracles arising laterally on prothorax, fan-like, with over 30 marginal papillae.

Abdomen: Segments swollen, without ventral welts or spinules. Posterior spiracular disc bordered by marginal tubercles, and with deep pit in which spiracles are located; posterior spiracles (fig. 37.250) distinctly separated and very large; each spiracular plate kidney-shaped, with small, dorsally directed spur arising on upper margin; spiracular openings small, bar-like, circular or oval, arranged in 6–8 slightly curved bands over surface of spiracular plate.

Comments: The family is worldwide, containing somewhat over 300 species. It is best developed in the Paleotropics. Only five genera and eight species are known from America north of Mexico. There are no economically important species, although the larvae may have some value as natural enemies of agriculturally significant Scarabaeidae.

Selected Bibliography

Clausen et al. 1933.
Ferrar 1987.
Hennig 1952.
Steyskal 1987h.

TEPHRITIDAE (TEPHRITOIDEA)

B. A. Foote, *Kent State University*

True Fruit Flies

Figures 37.251–37.266

Relationships and Diagnosis: This economically important family is the largest family of the Tephritoidea, a complex of eight North American families that includes the

Otitidae, Platystomatidae, and Richardiidae, among others. Griffiths (1972) expanded the family concept of Tephritidae to include species more commonly assigned to the families Platystomatidae, Pyrgotidae, Tachiniscidae, Otitidae, and Pallopteridae. This suggestion has not been adopted by other systematists of the acalyptrate Diptera. R. H. Foote and Steyskal (1987) recently summarized our knowledge.

Larvae of Tephritidae can be recognized by parallel or subparallel spiracular slits of the sessile posterior spiracles that also usually lack a peritreme (fig. 37.257a). The posterior spiracular disc in most species possesses an array of short, fleshy marginal tubercles.

Biology and Ecology: The larvae of nearly all species are phytophagous, infesting a broad spectrum of plant parts of a fair diversity of angiosperm families. Four major groups of larval feeding habits are commonly recognized. 1. Fruit feeders that are most frequently encountered in fleshy fruits. Genera occurring in North America that have this habit include *Anastrepha, Rhagoletis, Chetostoma, Oedecarina, Paraterellia, Rhagoletotrypeta, Zonosemata, Myoleja, Epochra* and *Toxotrypana*. 2. Consumers of seeds and other tissues in the flower heads of Asteraceae. Common Nearctic genera included here are *Gymnocarena, Acinia, Jamesomyia, Euarestoides, Procecidocharoides, Xanthaciura, Paracantha, Icterica, Euaresta, Paroxyna, Dioxyna, Trupanea, Tephritis, Neotephritis, Orellia, Neaspilota,* and *Tomoplagia*. 3. Miners within leaves and stems would include species of *Euleia, Trypeta* and *Strauzia*. 4. Gall formers living within various plant tissues include species of *Urophora, Stenopa, Procecidochares, Eutreta, Eurosta,* and *Aciurina*. In addition, there are a few species in parts of the world with very diverse hosts, including *Dirioxa pornia* (Walker) which attacks damaged and fallen fruits in Australia, the Australian *Termitorioxa termitoxena* (Bezzi) which lives in termite galleries in trees, and a group of genera in China and Southeast Asia whose larvae attack bamboo. The larval morphology commonly reflects the feeding habits. Adults are commonly collected on or near the host plants or in traps of various kinds.

Description: Mature larvae 3–16 mm, muscidiform, especially in fruit-eating and mining taxa, plump and cylindrical in seed-eating and gall-making species; integument bare to spinulose, various body segments with encircling rows of small papillae in seed predators, less so in gall formers and fruit eaters; white to yellowish; several species of seed predators have posterior spiracular disc blackened and sclerotized.

Head: Cephalic segment highly retractile into prothorax, bearing sensory organs anteriorly and anteroventrally, antennae short to elongate and appearing 2- or 3-segmented; maxillary sensory papillae highly variable in size, shape, and structure; facial mask usually with series of oral ridges and grooves leading into preoral cavity, with ridges unbranching or variously divided; some species of leaf miners have rasping teeth on mask.

Cephalopharyngeal skeleton usually deeply pigmented; tentoropharyngeal and hypopharyngeal sclerites separate; dorsal and ventral cornua frequently equally broad and variously pigmented; dorsal cornua usually connected anteriorly by fenestrated dorsal bridge; ventral cornua usually with window apically and without or with weak dorsobasal lobe; pharyngeal filter present in fruit-infesting species but absent in larvae using seeds, forming galls, or mining leaves; hypopharyngeal sclerite short and broad, basically H-shaped in ventral view; parastomal bars absent, weakly developed, or partially fused to hypopharyngeal sclerite; epipharyngeal and labial sclerites absent. Mandibles usually deeply pigmented, varying in shape from rather narrow and elongate in frugivorous species to heavy and broad in seed-eating taxa; mandible hook slender and decurved or heavy and little decurved; accessory teeth present or absent, teeth well developed in leaf-mining larvae.

Thorax: Anterior spiracles arising posterolaterally on prothorax, nearly sessile or on very short tubular bases, and frequently fan-shaped with 3–40 short to finger-like papillae along distal margin; distal part in some species slightly to strongly bicornuate.

Abdomen: Segments in many species with ventral creeping welts and lateral pads; perianal pad bilobed, commonly rather narrow, occasionally subtended by fleshy lobes.

Posterior spiracular disc usually with short to somewhat elongate fleshy tubercles around margin of disc; spiracles usually sessile on surface of disc or occasionally on very short unpigmented bases; peritremes absent or unpigmented; three straight, elongate, or oval spiracular openings that may be nearly parallel or diverging from ecdysial scar; four spiracular hairs that are well developed and branching in fruit-feeding forms, but highly reduced or even absent in seed predators and gall makers; hairs reduced to small, thorn-like structures in some seed feeders.

Comments: The Tephritidae is a large family with more than 4000 known species. Some 53 genera and over 260 species have been recorded in America north of Mexico. Because many species oviposit in fleshy fruits, there are numerous economically important species. Undoubtedly, the most important fruit pest is the Mediterranean fruit fly, *Ceratitis capitata* (Wiedemann), which is periodically introduced into the citrus-growing areas of the United States. Larvae of this highly destructive species can develop in well over 200 different fruits and vegetables. Another very injurious pest is *Anastrepha ludens* (Loew), the Mexican fruit fly, which invades commercially grown fruits in southern Texas and Central America. The papaya fruit fly, *Toxotrypana curvicauda* Gerstaeker, is occasionally destructive to papayas and mangos in Texas and Florida. Many species of the widespread genus *Rhagoletis* are important pests of cherries, apples, walnuts, blueberries, and roses, with the apple maggot, *Rhagoletis pomonella* (Walsh), being a major pest of apples. *Euleia fratria* (Loew), the parsnip leaf miner, is of considerable economic importance in the United States.

In contrast to the destructive species, certain tephritids are being investigated as potential biological control agents of introduced and native weeds. Thus species of *Paracantha,* because of the seed-predating habits of the larvae, may be potentially useful in controlling economically important species of thistle.

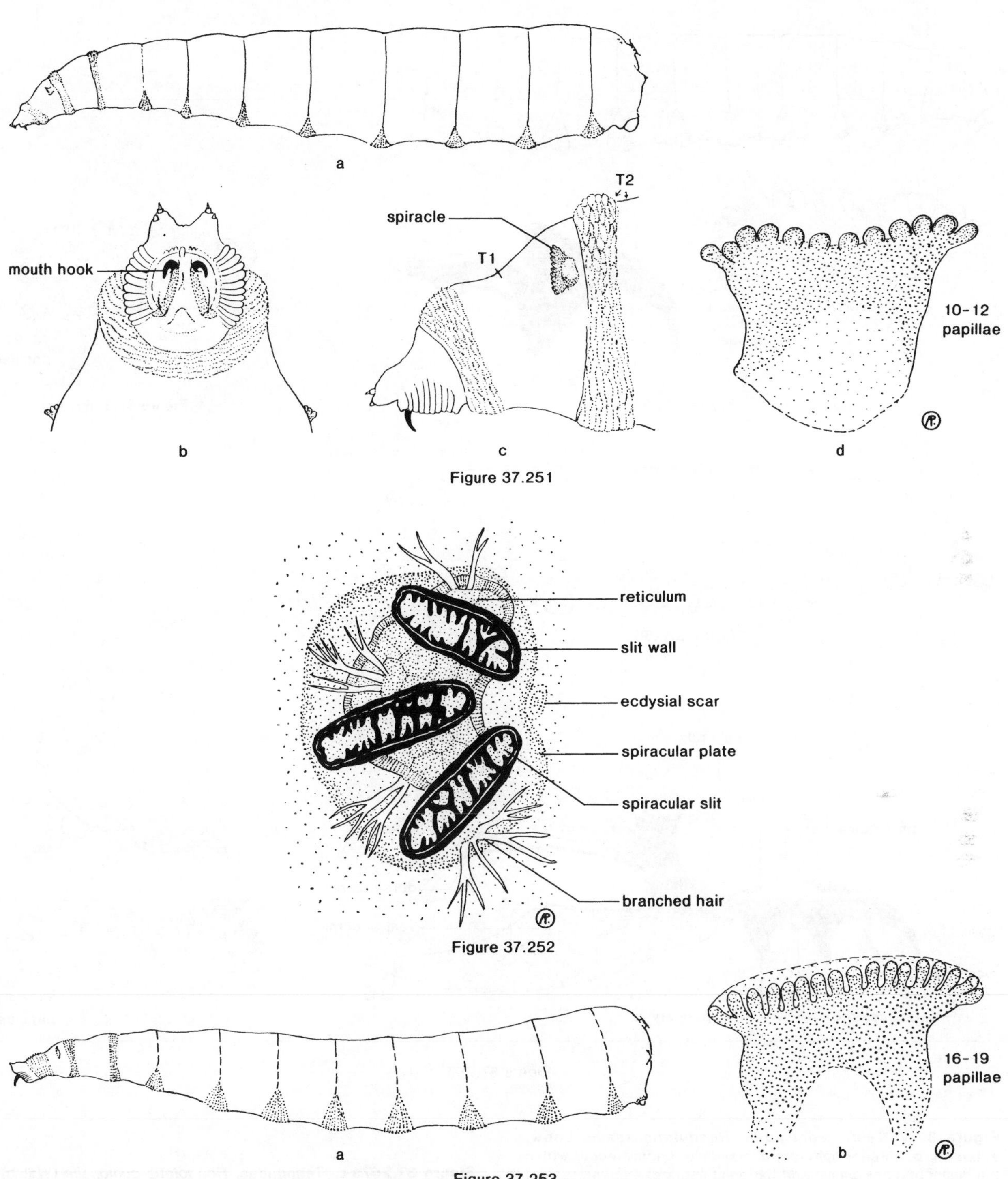

Figure 37.251

Figure 37.252

Figure 37.253

Figure 37.251a-d. Tephritidae. *Dacus dorsalis* Hendel, the Oriental fruit fly. **a.** lateral; **b.** ventral view of anterior end; **c.** lateral view of head and prothorax; **d.** anterior spiracle. Larvae can be very destructive to a wide variety of fruits. It has been recorded occasionally in California but has not yet become established in North America (from Peterson 1951).

Figure 37.252. Tephritidae. A representative left caudal spiracle (from Phillips 1946).

Figure 37.253a,b. Tephritidae. **a.** lateral, and **b.** anterior spiracle of *Dacus cucurbitae* Coquillett, the melon fly, a widely distributed species in Southeast Asia and Hawaii where its larvae have been recorded from over 80 plant species. It has not become established in the United States, although it has been recorded in California (from Peterson 1951).

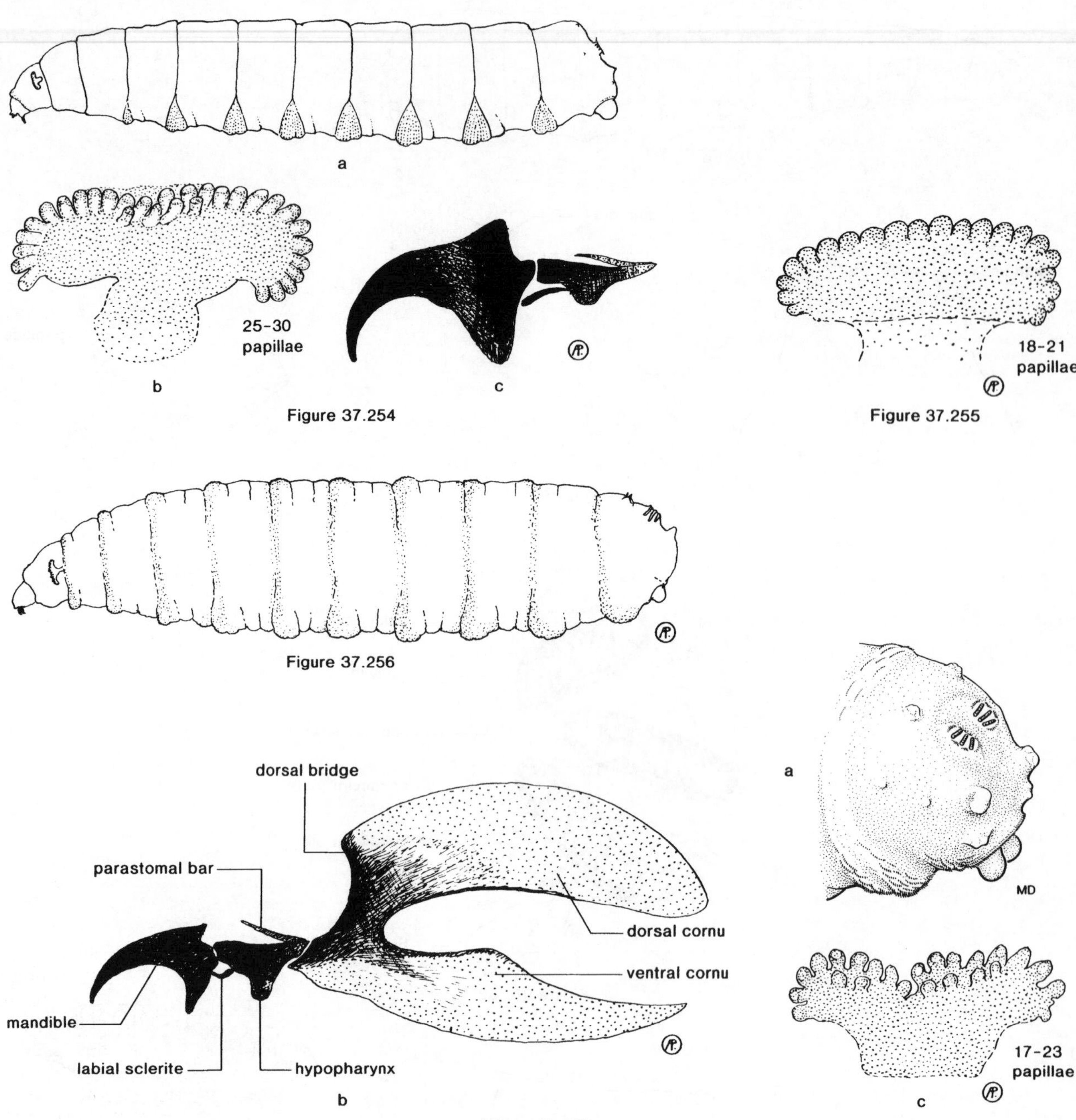

Figure 37.254a-c. Tephritidae. *Rhagoletis suavis* Loew. **a.** lateral; **b.** anterior spiracle; **c.** mandible. Larvae occur within the husks of black walnut and butternut in the eastern states, as well as in other species of *Juglans* and varities of Persian walnut (from Peterson 1951).

Figure 37.255. Tephritidae. Anterior spiracle of *Rhagoletis completa* Cresson, the walnut husk fly. It attacks black walnut and other species of *Juglans,* and can be quite destructive to varieties of Persian walnut in the Southwest and California (from Peterson 1951).

Figure 37.256. Tephritidae. *Rhagoletis mendax* Curran, the blueberry maggot. The larva is currently indistinguishable from *R. pomonella,* the apple maggot. A pest in commercial blueberries (from Peterson 1951).

Figure 37.257a-c. Tephritidae. *Rhagoletis pomonella* (Walsh), the apple maggot. **a.** Oblique posterior view of caudal segment; **b.** lateral view of cephalopharyngeal skeleton; **c.** anterior spiracle. The apple maggot is an important pest of apples in the eastern half of the United States. It is also established in parts of the West from northern California through western Oregon and southwestern Washington, where it attacks apple and hawthornes (*Crataegus* spp.). It is also abundant in Utah where it attacks cherries and hawthornes, but not apple, and in Colorado where it attacks only hawthornes (G. L. Bush, personal communication). Additional hosts include pear, plum and rose hips (fig. 37.257a from Manual of Nearctic Diptera, Vol. 1; figs. 37.257b, c from Peterson 1951).

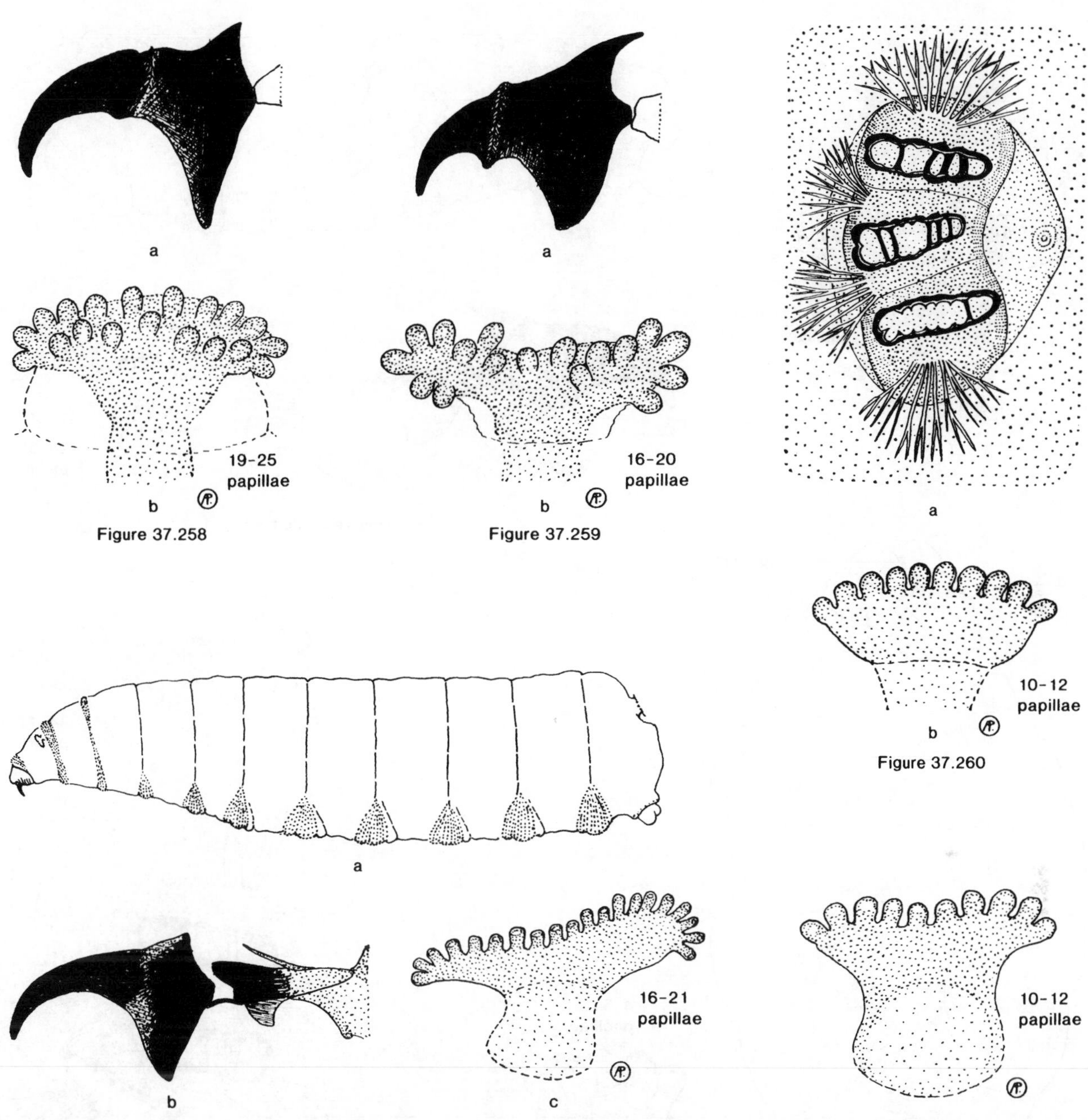

Figure 37.258a,b. Tephritidae. *Rhagoletis cingulata* (Loew) the cherry fruit fly or cherry maggot. **a.** mandible; **b.** anterior spiracle. The cherry maggot is an important enemy of cultivated cherries in the Eastern States. *Rhagoletis indifferens* Curran, the western cherry fruit fly, is an important pest of sweet cherries from Utah to the Pacific Northwest (from Peterson 1951).

Figure 37.259a,b. Tephritidae. *Rhagoletis fausta* (Osten Sacken), the black cherry fruit fly. **a.** mandible; **b.** anterior spiracle. It is occasionally damaging to commercial sour cherries both East and West. Its native host is pin cherry, *Prunus pensylvanica* (from Peterson 1951).

Figure 37.260a,b. Tephritidae. *Dacus oleae* (Gmelin), the olive fruit fly. **a.** posterior spiracle; **b.** anterior spiracle. Larvae infest fruits of wild and cultivated olives. It does not occur in North America (from Peterson 1951).

Figure 37.261a-c. Tephritidae. *Anastrepha ludens* (Loew), the Mexican fruit fly. **a.** lateral; **b.** mandible; **c.** anterior spiracle. The Mexican fruit fly is an important pest of citrus, mango, avocado and other fleshy fruits. It has been recorded from Southern California and in the Rio Grande Valley of Texas (from Peterson 1951).

Figure 37.262. Tephritidae. Anterior spiracle of *Ceratitis capitata* (Wiedemann), the Mediterranean fruit fly. The "Med fly" is probably the most important fruit-infesting species of Tephritidae in the world. It has periodically been found in the citrus groves of California and Florida, but has so far been successfully eradicated following each introduction (from Peterson 1951).

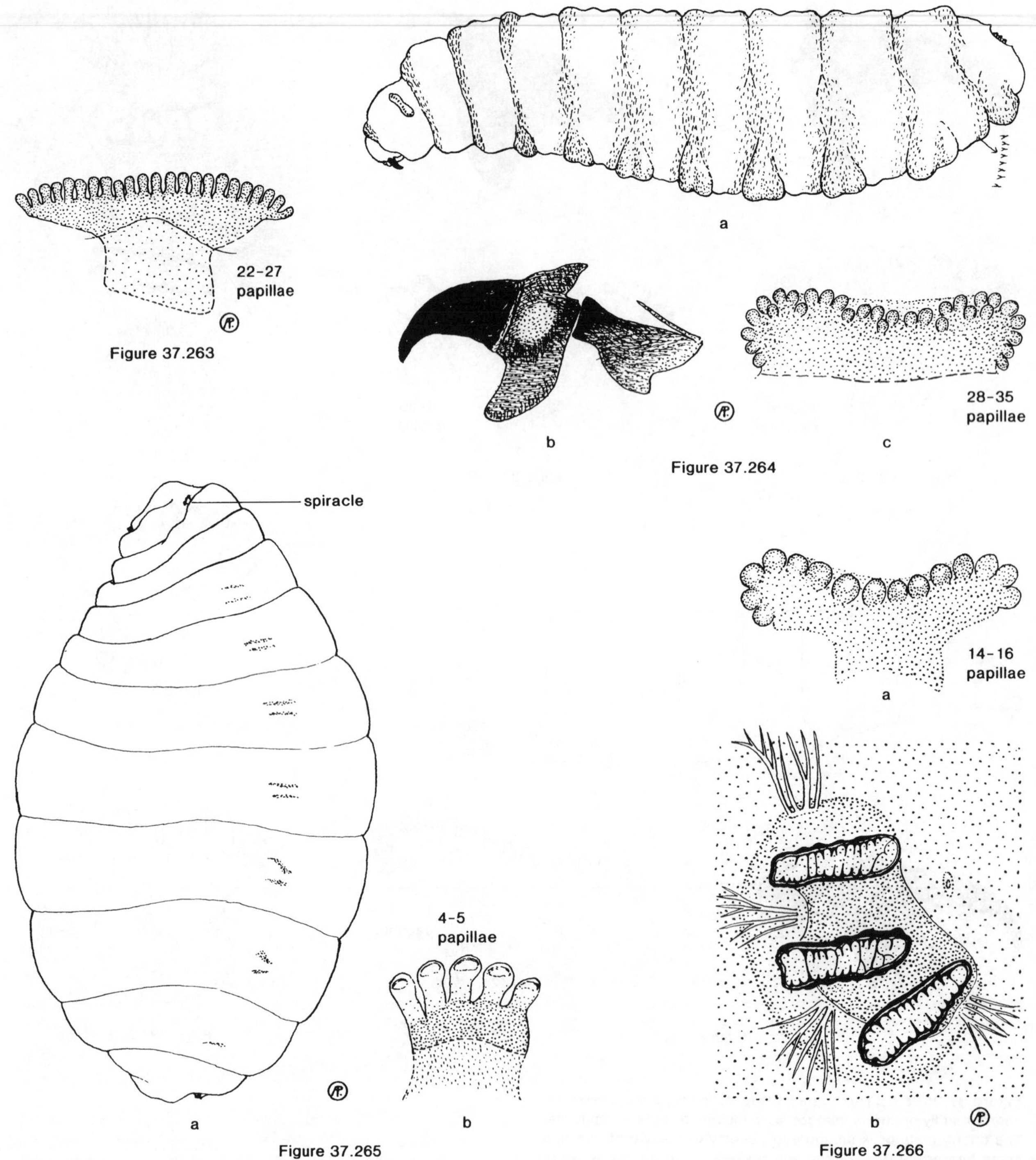

Figure 37.263. Tephritidae. Anterior spiracle of *Toxotrypana curvicauda* Gerstaecker, the papaya fruit fly, whose larvae prefer the seeds and do not attack the pulp before it is ripe (from Peterson 1951).

Figure 37.264a-c. Tephritidae. *Zonosemata electa* (Say), the pepper maggot. **a.** lateral; **b.** mandible; **c.** anterior spiracle. Larvae attack cultivated peppers, egg plant, horse nettle, *Solanum carolinense,* and are occasionally important pests in commercial peppers (from Peterson 1951).

Figure 37.265a,b. Tephritidae. *Eurosta solidaginis* Fitch. **a.** lateral; **b.** anterior spiracle. Larvae form conspicuous, rather globose galls on stems of goldenrod, *Solidago* (from Peterson 1951).

Figure 37.266a,b. Tephritidae. *Epochra canadensis* (Loew), the currant fruit fly. **a.** anterior spiracle; **b.** posterior spiracle. Infests the berries of cultivated and wild gooseberries and currants, *Ribes* spp., throughout northern North America (from Peterson 1951).

Selected Bibliography

Balduff 1959.
Banks 1912.
Benjamin 1934.
Blanc and Foote 1961.
Boyce 1934.
Bush 1965, 1966.
Butt 1937.
Cavender and Goeden 1982.
Christenson and Foote 1960.
Coquillett 1895.
Daecke 1910.
Efflatoun 1927.
Ferrar 1987.
Fischer 1932b.
Foote, B. A. 1965, 1967.
Foote, R. H. and Steyskal 1987.
Frick 1971.
Frost 1924.
Gibson and Kerby 1978.
Greene 1929.
Gurney 1912.
Hardy 1949a.
Harris 1980.
Hennig 1952.
Illingworth 1912.
Kamali and Schulz 1974.
Keilin and Tate 1943.
Lamp and McCarty 1982.
McFadden and Foote 1960.
Novak and Foote 1968, 1975.
Novak et al. 1967.
Persson 1963.
Peterson, A. 1923, 1951.
Phillips 1923, 1946.
Piper 1976.
Quayle 1929.
Reissig and Smith 1978.
Ries 1934.
Schwitzgebel and Wilbur 1943.
Silverman and Goeden 1980.
Snodgrass 1924.
Stegmaier 1967b, 1967c, 1968a, 1968b, 1968c.
Steyskal 1973, 1975.
Stoltzfus 1974, 1977, 1978.
Stoltzfus and Foote 1965.
Tauber and Tauber 1967, 1968.
Tauber and Toschi 1965a, 1965b.
Uhler 1951.
Varley 1937b, 1947.
Vimmer 1925, 1931.
Wangberg 1978, 1980.
Wasbauer 1972.

RICHARDIIDAE (TEPHRITOIDEA)

B. A. Foote, *Kent State University*

The Richardiids

Relationships and Diagnosis: The Richardiidae is commonly placed in the Tephritoidea, and is considered to be closely related to the Otitidae, Platystomatidae, and Tephritidae (McAlpine, et al. 1981). Steyskal (1987i) has recently summarized our knowledge.

Biology and Ecology: Very little is known of the biology of any species. It appears that the larvae may be saprophagous in decaying plant material, but unpublished data (Steyskal, personal communication) indicate they may damage growing plant tissue. Adults of some species have been captured in fruit-fly traps.

Description: No larval descriptions have been published, but Steyskal (1958) has described the cephalopharyngeal skeleton of *Epiplatea hondurana* Steyskal as having a short and narrow dorsal cornua, broad, but thin and pale ventral cornua, hypostomal sclerite H-shaped, and mandible with two blunt teeth. He also figures the puparium, showing the three slits of each caudal spiracle to be very close together, nearly evenly parallel to each other, and perpendicular to the vertical plane.

Comments: This is primarily a family of the Neotropics, and only six genera and 10 species are known from America north of Mexico. The largest Nearctic genus is *Sepsisoma* with three species. Most of the Nearctic species are southern and are rarely collected.

Selected Bibliography

Ferrar 1987.
Griffiths 1972.
Hennig 1952.
Steyskal 1958, 1961b, 1987i.

PALLOPTERIDAE (TEPHRITOIDEA)

B. A. Foote, *Kent State University*

The Pallopterids

Relationships and Diagnosis: The correct placement of this family has proved to be difficult. Hennig (1952) placed it in the Pallopteroidea, whereas Griffiths (1972) considered *Palloptera* to be a genus of the Tephritidae. More recently McAlpine et al. (1981) placed it in the Tephritoidea near the Piophilidae. McAlpine (1987c) recently reviewed our knowledge.

Larvae resemble those of Lonchaeidae in that the posterior spiracles bear a dorsal spur. The two families cannot be reliably separated in the larval stage.

Biology and Ecology: The life histories and larval feeding habits of the Nearctic species are unknown. In Europe, larvae have been found in flower buds and stems of Asteraceae and Apiaceae (=Umbelliferae) and are probably phytophagous. Larvae of other species have been encountered under bark of coniferous and deciduous trees where they seemingly prey on larvae of Cerambycidae and Scolytidae. Adults of the Nearctic species are usually collected in moist to mesic woodlands.

Description: No description of the larvae of any Nearctic species has been published, and it is necessary to rely on papers dealing with a few of the European species. No good figures have been published.

Fully grown larvae 4–8 mm, muscidiform, somewhat cylindrical, anterior end tapering; integument mostly smooth, lacking spinule bands; white.

Head: Cephalic segment retractile, bearing well developed sensory organs; antennae apparently 2-segmented; facial mask apparently lacking oral ridges and grooves.

Cephalopharyngeal skeleton deeply pigmented; tentoropharyngeal and hypopharyngeal sclerites separate; dorsal cornua connected anteriorly by dorsal bridge; ventral cornua with small dorsobasal lobe and usually with apical window; pharyngeal filter present; hypopharyngeal sclerite H-shaped in ventral view; parastomal bars apparently free apically; no obvious epipharyngeal or labial sclerites. Mandibles deeply pigmented, hook decurved and lacking accessory teeth, basal part commonly with windows; dental sclerite somewhat triangular in shape in lateral view.

Thorax: Segments lacking prolegs and creeping welts; anterior spiracles arising posterolaterally on prothorax, spiracles fan-shaped and with 6–7 short papillae along distal margin.

Abdomen: Segments cylindrical, all very similar, without spinule bands or ventral creeping welts; caudal segment bluntly rounded, without fleshy protuberances; perianal pad bilobed.

Posterior spiracular disc lacking marginal tubercles; posterior spiracles distinctly separated, spiracles borne on short, tubular structures; each spiracular plate with 3 oval, radially arranged spiracular openings, and with dorsally directed spur arising from upper surface of peritreme; spiracular hairs absent.

Comments: This is a small family of some 50 species worldwide. There is only one genus, *Palloptera*, and nine species in America north of Mexico. If larvae of the Nearctic species are predacious, then some species will be natural enemies of wood-boring beetles.

Selected Bibliography

Balachowsky and Mesnil 1936.
Ferrar 1987.
Hennig 1952.
McAlpine 1987c.
Meijere 1944.
Morge 1956, 1967.
Perris 1870.
Teskey 1976.

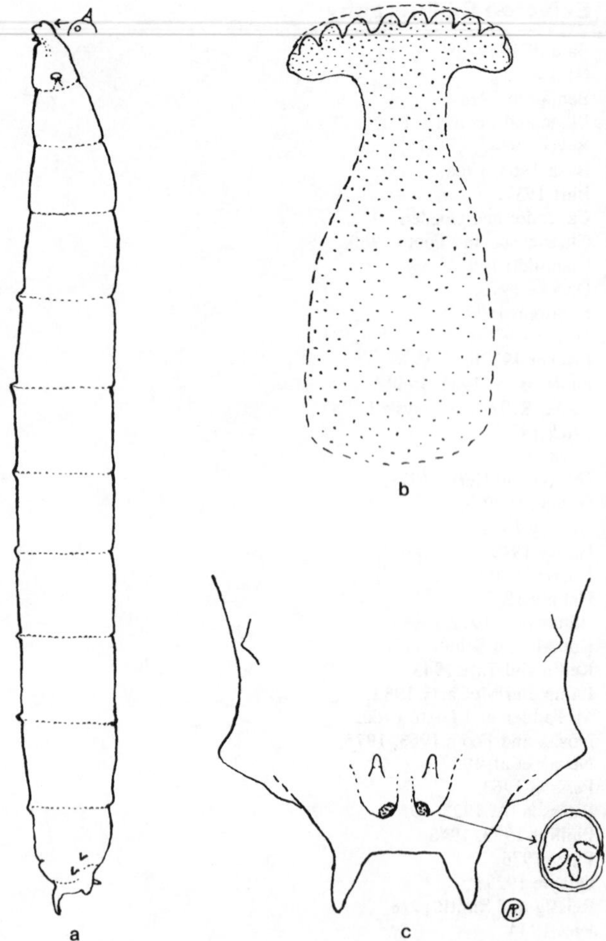

Figure 37.267a-c. Piophilidae. *Piophila casei* (L.) the cheese skipper. **a.** lateral; **b.** anterior spiracle; **c.** dorsal view of posterior end. The larva can flip through the air for several inches, and is a serious pest in the food industry, especially cheese and meat producers (from Peterson 1951).

PIOPHILIDAE (TEPHRITOIDEA)

B. A. Foote, *Kent State University*

Skipper Flies

Figures 37.267a–c

Relationships and Diagnosis: This is another family whose taxonomic placement has had a varied history. Historically, it has been commonly placed near the Lonchaeidae or Sepsidae. Hennig (1952), however, pointed out that the larvae were quite unlike those of the Sepsidae and later (1958) suggested that the piophilids should be placed in the Pallopteroidea along with the Pallopteridae and Lonchaeidae. More recently, McAlpine et al. (1981) have included it in the Tephritoidea and implied that the family is particularly close to the Pallopteridae. The previously recognized families Thyreophoridae and Neottiophilidae are now included in the Piophilidae (J. F. McAlpine, 1977). McAlpine (1987d) recently summarized our knowledge.

Larvae of Piophilidae are fairly distinctive in that the posterior spiracles are borne on the inner side of short, triangular or tube-like structures arising on the somewhat elongated caudal segment (fig. 37.267c). The two spiracles may be closely appressed to each other in living larvae, concealing the spiracular plates. In addition, the posterior spiracular disc bears 2–4 pairs of fleshy tubercles.

Biology and Ecology: Larvae of the large genus *Piophila* are scavengers that feed on decaying carcasses, preserved meat, feces, fungi, and cheese. Because the larvae are notable jumpers they are commonly called "skippers." Larvae of *Neottiophilum* and possibly *Actenoptera* are ectoparasitic blood-suckers of nestling birds, and those of *Mycetaulus* occur in rotting fungi. Larvae of other genera feed on carrion, especially carcasses in an advanced state of decay. Adults are found near the larval habitats.

Description: Mature larvae 5–12 mm; muscidiform, slender; integument smooth, without hairs, but segments with encircling banks of spinules; ventral surface of most segments with creeping welts; white to pale yellow.

Head: Cephalic segment retractile, bearing well-developed sensory organs anteriorly; facial mask with numerous oral ridges and grooves leading into preoral cavity.

Cephalopharyngeal skeleton deeply pigmented, tentoropharyngeal and hypopharyngeal sclerites separate; dorsal and ventral cornua equal in size; pharyngeal filter present. Hypopharyngeal sclerite H-shaped, parastomal bars slender and not connected apically; labial sclerite well developed. Mandibles heavy; hook decurved and lacking accessory teeth, basal part with or without small window; no dental sclerite.

Thorax: Segments without prolegs but with ventral creeping welt on border between metathorax and first abdominal segment; anterior spiracles arising posterolaterally on prothorax, each spiracle somewhat fan-shaped with 4–15 finger-like papillae along distal margin.

Abdomen: Segments cylindrical, all subequal in diameter and length, bearing creeping welts near anterior segmental borders and encircling bands of somewhat darkened spinules; caudal segment distinctive, bearing 2–4 pairs of fleshy, posteriorly directed protuberances, ventral pair longest, slender and tapering, other pairs shorter; perianal pad bilobed.

Posterior spiracular disc with posterior spiracles sessile on inner side of distinct subtriangular swellings that can be brought into juxtaposition with each other, hiding spiracular plates; plates pale with darker peritremes; no dorsal spur; 3 narrowly oval spiracular openings, radially arranged around ecdysial scar; spiracular hairs apparently absent.

Comments: This is a relatively small family of only some 70 species in the world. Fourteen genera and some 37 species are recorded from America north of Mexico (McAlpine 1977). Larvae of *Piophila casei* (L.) are known as "cheese skippers" and have been repeatedly encountered in cheeses and preserved meats. They have also caused human intestinal myiasis.

Selected Bibliography

Banks 1912.
Cockburn et al. 1975.
Ferrar 1987.
Freidberg 1981.
Hennig 1952, 1958.
James 1947.
McAlpine, J. F. 1977, 1987d.
Vimmer 1925.
Wandolleck 1898.
Zuska and Lastovka 1965.

CLUSIIDAE (OPOMYZOIDEA)

B. A. Foote, *Kent State University*

The Clusiids

Figure 37.268

Relationships and Diagnosis: Some confusion has existed historically as to the proper placement of the Clusiidae. Hennig (1952) suggested that the family was relatively close to the Micropezidae, but later (1958) placed it in his group 7, "families with unclear relationships," although he placed it next to the Agromyzidae and Acartophthalmidae. Griffiths (1972) included the Clusiidae in his prefamily Agromyzoinea, whereas McAlpine, et al. (1981) placed it in the Opomyzoidea, a large group of 12 families including the Acartophthalmidae, Odiniidae, Agromyzidae, Opomyzidae, and Milichiidae. Soós (1987) recently reviewed our knowledge.

The most distinctive feature of the larvae is the unpigmented cephalopharyngeal skeleton. This can usually be observed by focusing on the region of the skeleton through the ventral surface of the larva.

Biology and Ecology: Adults are most commonly encountered in mesic woodlands containing rotting logs. Larvae are found under bark and in decaying wood of fallen trees, logs, and stumps. Larval feeding habits are poorly known, but it is possible that they are mostly ingesting decomposer microorganisms such as bacteria, fungi, and yeasts. Larvae can leap by grasping hooklets on the caudal segments with their mouthhooks and then suddenly releasing their hold. Eggs are probably laid in cracks and crannies of rotting trees. Puparia are yellowish and are found in decaying wood or under loosened bark.

Description: Mature larvae 6–10 mm; muscidiform; integument smooth and devoid of hairs and spinules; whitish.

Head: Cephalic segment retractile, bearing inconspicuous sensory organs apically and apicoventrally; facial mask with weakly developed oral grooves leading into preoral cavity.

Cephalopharyngeal skeleton usually unpigmented or with only mouthhooks pigmented; sclerites vestigial or nearly so in some species; tentoropharyngeal and hypopharyngeal sclerites separate, not distinctive. Mandibles usually pigmented but rather poorly developed.

Thorax: Thoracic segments without prolegs or obvious ventral creeping welts; anterior spiracles arising posterolaterally on prothorax, spiracles fan-shaped with 4–6 marginal papillae.

Abdomen: Segments subequal in size, without prolegs but with weakly developed creeping welts ventrally along segmental borders; caudal segment rounded, without fleshy projections; a band of slightly darkened spinules may nearly encircle segment; perianal pad transversely elongated, tapering to blunt point laterally.

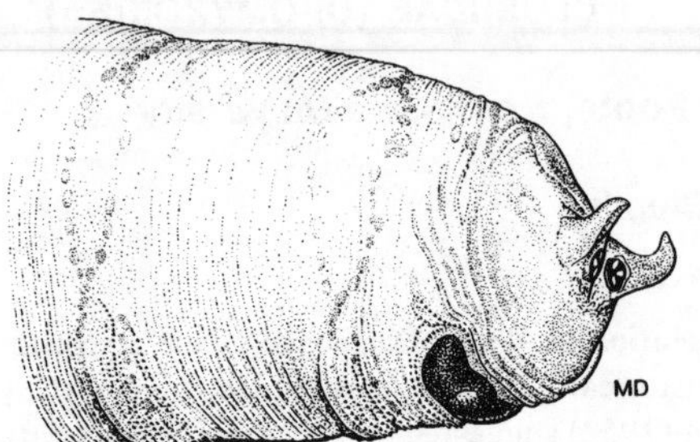

Figure 37.268. Clusiidae. Oblique posterior view of caudal segments of *Clusia* sp. (from Manual of Nearctic Diptera, Vol. 1).

Posterior spiracular disc without marginal tubercles; posterior spiracles narrowly separated, arising near middle of spiracular disc, weakly to strongly pigmented; each spiracular plate with distinct thorn directed dorsally, three spiracular openings at base of each thorn, openings narrowly oval and radiating out from ecdysial scar; no spiracular hairs.

Comments: This is a small family of some 150 species in the world. Five genera and nearly 40 species are recorded for America north of Mexico. No species are considered to have economic significance. Many of the Nearctic species are found in the eastern deciduous forest biome.

Selected Bibliography

Ferrar 1987.
Griffiths 1972.
Hennig 1952, 1958.
Malloch 1918.
Perris 1870.
Séguy 1950.
Soós, Á. 1987.

ACARTOPHTHALMIDAE (OPOMYZOIDEA)

B. A. Foote, *Kent State University*

The Acartophthalmids

Relationships and Diagnosis: This family consists of a single genus that was previously included in the Clusiidae but was elevated to family status by Hennig (1958). McAlpine et al. (1981) placed it in the Opomyzoidea close to the Clusiidae, whereas Griffiths (1972) believed that it should be included in the prefamily Tephritoinea and the Chloropidae family-group. McAlpine (1987e) recently reviewed our knowledge.

Adults are rarely collected, although one Nearctic species, *Acartophthalmus nigrinus* (Zetterstedt), is transcontinental. They have been found around carrion and rotting fungi. The larval feeding habits and morphology are unknown.

Description: No larval descriptions have been published.

Comments: This is a very small family consisting of a single genus, *Acartophthalmus,* and two species in the Nearctic Region.

Selected Bibliography

Ferrar 1987.
Griffiths 1972.
Hennig 1958.
McAlpine 1987e.

ODINIIDAE (OPOMYZOIDEA)

B. A. Foote, *Kent State University*

The Odiniids

Figures 37.269a–f

Relationships and Diagnosis: The family Odiniidae has had a rather checkered history with respect to its phylogenetic placement. Hendel (1920) and other earlier workers included the taxa currently placed in the Odiniidae within the Agromyzidae. Later, Hendel (1922) recognized the odiniids as a distinct family which he included in the Milichioidea along with the Agromyzidae, Carnidae, and Milichiidae. Hennig (1958) was unsure of the correct relationships, but suggested that it may be relatively close to the Clusiidae and Agromyzidae. Griffiths (1972) included it in his prefamily Tephritoinea and claimed that it was not particularly close to the Agromyzidae. Finally, McAlpine, et al. (1981) placed the Odiniidae in the Opomyzoidea along with 11 other families of acalyptrate Diptera. They implied that the odiniids were particularly close to the Clusiidae, Acartophthalmidae, and Agromyzidae. McAlpine (1987f) recently summarized our knowledge.

Larvae are fairly distinctive because of their small size (< 6 mm), small posterior spiracles, elongated anterior spiracles, and two fleshy lobes near the perianal pad. Their occurrence in burrows of wood-boring insects, particularly beetles, is also useful in recognizing the family.

Biology and Ecology: Adult Odiniidae are most commonly encountered in forested areas around sap flows and slime fluxes, on rotting fungi, and near decaying stumps and logs. Larvae have been found in burrows of wood-boring beetles and moths, and are believed to be scavengers of insect frass and other kinds of decaying organic matter. There is some evidence that the larvae can attack larvae of other insects.

Description: Mature larvae 3–6 mm; muscidiform, tapering anteriorly and bluntly rounded posteriorly; integument smooth, without hairs or encircling bands of spinules; creeping welts ventrally on abdominal segments; white.

Head: Cephalic segment retractile, with well developed sensory organs anteriorly and anteroventrally; antennae appearing 2-segmented; facial mask with oral grooves leading into preoral cavity.

Cephalopharyngeal skeleton variously pigmented; tentoropharyngeal and hypopharyngeal sclerites separate; dorsal cornua connected anteriorly by fenestrated dorsal bridge; pharyngeal filter present; hypopharyngeal sclerite H-shaped; parastomal bars not connected anteriorly; apparently no epipharyngeal sclerite. Mandibles basically sickle-shaped, hook without accessory teeth.

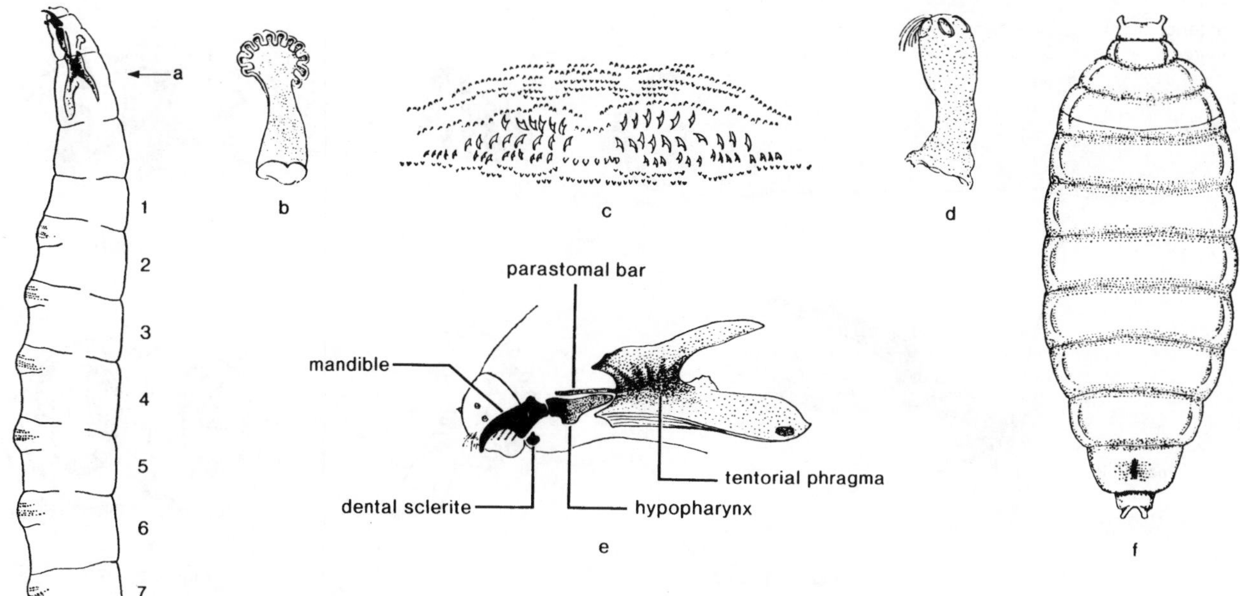

Figure 37.269a-f. Odiniidae. *Neoalticomerus seamansi* Shewell. **a.** lateral; **b.** anterior spiracle; **c.** creeping welt of fifth abdominal segment; **d.** posterior spiracle; **e.** cephalopharyngeal skeleton; **f.** puparium. Larvae of this species from eastern Canada were found in the galleries of the weevil *Cryptorhynchus lapathi* (L.) in balsam poplar (from Shewell 1960).

Thorax: All segments apparently lacking ventral creeping welts; anterior spiracles arising posterolaterally on prothorax, spiracles slender, elongate, with about six short papillae scattered around margin of somewhat broadened distal part.

Abdomen: All segments except caudal one very similar, each bearing creeping welt ventrally near posterior border; caudal segment with two short fleshy lobes near perianal pad; pad bilobed.

Posterior spiracles very small, borne on apices of short tubular structures; each spiracular plate with 3 diverging, oval spiracular openings; at least 1 branching spiracular hair present on each plate.

Comments: This is a small family of about 70 described species worldwide. Three genera and 11 species have been recorded from America north of Mexico, with most of the species occurring in the eastern states and provinces.

Selected Bibliography

Ferrar 1987.
Griffiths 1972.
Hendel 1920, 1922.
Hennig 1952, 1958.
McAlpine 1987f.
Shewell 1960.
Teskey 1976.
Vos-de Wilde 1935.
Yang 1984.
Zubkov and Kovalev 1975.

AGROMYZIDAE (OPOMYZOIDEA)

B. A. Foote, *Kent State University*

Leafminer Flies

Figures 37.270a-d, 37.270x, 37.270z

Relationships and Diagnosis: Hennig (1958) was uncertain as to the correct phylogenetic placement of the Agromyzidae within the acalyptrate Diptera and included it in his group 7 (families of uncertain relationships). He recognized the family as being monophyletic on the basis of 7 apomorphic traits possessed by the adults. Later, Hennig (1971) recognized the Agromyzoidea for the Agromyzidae and Odiniidae. Griffiths (1972) suggested a close relationship of the Clusiidae with the Agromyzidae and placed both families in his prefamily Agromyzoinea. McAlpine, et al. (1981) did not recognize the Agromyzoidea but instead placed the Agromyzidae and Clusiidae in the superfamily Opomyzoidea, a large group of 12 families. Spencer (1987) has recently summarized our knowledge.

Larvae are easily recognized by the position of the anterior spiracles on the dorsal side of the prothorax on either side of the mid-line (fig. 37.270a).

Biology and Ecology: Adults are small to minute, rarely exceeding 5 mm in length, are commonly black or black and yellow, and are frequently abundant near the larval host plants. The larvae are almost always associated with plants, with most species feeding as leaf miners, although there are also stem- and root-mining taxa as well as seed-feeding species. Most species are monophagous or oligophagous, and polyphagy is

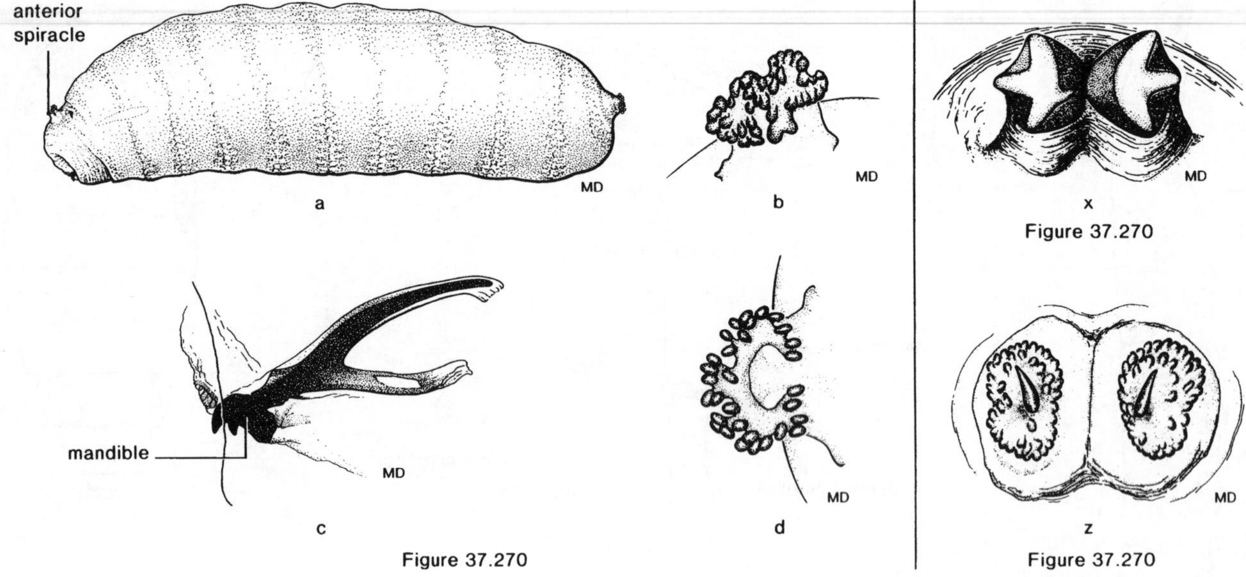

Figure 37.270a-d. Agromyzidae. *Phytomyza chelonei* Spencer. **a.** lateral; **b.** anterior spiracle; **c.** cephalopharyngeal skeleton; **d.** posterior spiracle. Larvae attack the seeds of turtlehead, *Chelone glabra* (from Manual of Nearctic Diptera, Vol. 1).

Figure 37.270x. Agromyzidae. *Agromyza albipennis* Meigen, posterior spiracles; a leaf miner (from Manual of Nearctic Diptera, Vol. 2).

Figure 37.270z. Agromyzidae. *Melanagromyza* sp., posterior spiracles. Larvae feed in stems and pods of assorted plants (from Manual of Nearctic Diptera, Vol. 2).

unusual, but it does occur in some species, e.g. *Liriomyza trifolii* (Burgess), *L. huidobrensis* (Blanchard) and *L. sativae* Blanchard, three serious leaf miners. Herbaceous and woody plants are attacked, with the vast majority being flowering plants. There are several generations a year in most species. Pupation in many species occurs in the mine, but in others larvae abandon the mine to form puparia in soil. Usually each mine contains a single larva.

Description: Fully grown larvae 4–6 mm; muscidiform, body usually somewhat depressed dorsoventrally (body long and slender in larvae feeding on cambial tissues under bark of tree branches); integument smooth, without noticeable hairs or spinules except for pale to darkened bands of spinules encircling segments and forming ventral creeping welts; whitish to yellowish.

Head: Cephalic segment retractile, with sensory organs noticeable apically and apicoventrally; facial mask without or with only weakly developed oral grooves leading into preoral cavity, usually with a number of short or pedunculate sensory papillae on each antennomaxillary lobe.

Cephalopharyngeal skeletal structure unique (fig. 37.270b) and not found in any other family of Muscomorpha. Pigmentation variable, usually with cornua less pigmented than remainder of skeleton; tentoropharyngeal and hypopharyngeal sclerites fused or very poorly separated; dorsal cornua narrow to broad, with or without noticeable window, connected anteriorly by dorsal bridge; ventral cornua commonly shorter than dorsal ones, without or with small window and lacking dorsobasal lobe; parastomal bars present or seemingly fused with hypopharyngeal sclerite; apparently no epipharyngeal or labial sclerites. Mandibles oriented nearly at right angle to end of hypopharyngeal sclerite, hook usually with one to several accessory teeth; basal parts connected by sclerotized bridge.

Thorax: Anterior spiracles dorsal on prothorax on either side of midline; spiracular shape highly variable, ranging from fan-shaped or semi-circular to bicornuate; marginal papillae varying from fewer than 8 to over 50, papillar shape ranging from short and broad to elongate and finger-like.

Abdomen: Segments except for caudal one very similar, in many species encircled completely or partially by spinule bands at segmental borders; integument of all segments usually possessing inconspicuous to noticeable sensory papillae; perianal pad commonly bilobed and bordered by one to several rows of spinules.

Posterior spiracles usually borne at apices of short or elongate lobes that are not fused basally; shape highly diverse, in some species (*Agromyza*) only three oval spiracular openings occur, but in a few taxa there are many openings (certain *Melanagromyza*), while in many other species the spiracles may be variously branched (*Melanagromyza*); sometimes with each branch bearing numerous finger-like papillae (*Ophiomyia*).

Comments: Some 27 genera and 700 species are recorded from America north of Mexico (Spencer and Steyskal, 1986). There are numerous economically important species such as *Agromyza frontella* (Rondani), the alfalfa blotch leaf miner; *Liriomyza brassicae* (Riley), a serpentine leaf miner of crucifers; *Liriomyza sativae*, the vegetable leaf miner; *Phytomyza aquilegiana* Frost, the columbine leaf miner, *Phytomyza ilicis* Curtis and *P. ilicicola* Loew, holly leaf miners; and *P. syngenesiae* (Hardy), the chrysanthemum leaf miner. Singh and Ipe (1973) contains much information on immatures.

Selected Bibliography

Balachowsky and Mesnil 1936.
Beri 1973.
Claassen 1918.
Cohen 1936.
Ferrar 1987.
Frick 1956.
Frost 1924.
Guppy 1981.
Hendrickson and Keller 1983.
Hennig 1952, 1958, 1971.
Hering 1924, 1932, 1954, 1955, 1957a.
Malloch 1915.
Meijere 1940a, 1955.
Miall and Taylor 1907.
Nowakowski 1962.
Peterson, A. 1951.
Singh and Ipe 1973.
Smulyan 1914.
Spencer 1969, 1987.
Spencer and Steyskal 1986.
Stegmaier 1967a.
Vimmer 1925, 1931.
Weaver and Dorsey 1967.

OPOMYZIDAE (OPOMYZOIDEA)

B. A. Foote, *Kent State University*

The Opomyzids

Figures 37.271, 37.272

Relationships and Diagnosis: The Opomyzidae has been treated variously by earlier workers, and was commonly included in the Anthomyzidae. Hennig (1958) recognized it as distinct from the Anthomyzidae but was unsure of its placement. Griffiths (1972) placed it within his prefamily Anthomyzoinea along with such families as Heleomyzidae, Anthomyzidae, Asteiidae, and Sphaeroceridae. Currently it and 11 other families are in the Opomyzoidea (McAlpine et al. 1981). Vockeroth (1987a) recently reviewed our knowledge.

The fused tentoropharyngeal and hypopharyngeal sclerites, absence of a pharyngeal filter, and a complicated set of reticulations around the preoral cavity are fairly distinctive. The elongated anterior spiracles (figs. 37.271, 37.272) separate some larvae from those of most species of Chloropidae.

Biology and Ecology: This is another of the numerous acalyptrate families whose biology and larval feeding habits are poorly known. Adults of *Geomyza* and *Opomyza* are usually found in moist, grassy habitats where their larvae feed within the stems of various grasses.

Description: Mature larvae 4–8 mm, muscidiform, cylindrical, slender and elongate; integument bare, without hairs; spinules present only in ventral creeping welts; whitish to pale yellow.

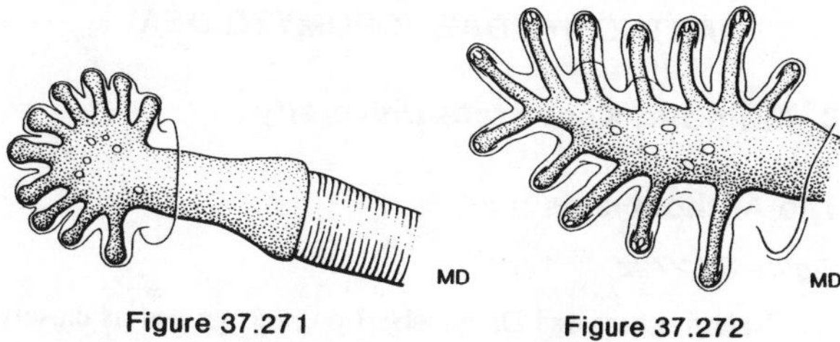

Figure 37.271 Figure 37.272

Figure 37.271. Opomyzidae. Anterior spiracle of *Opomyza petrei* Mesnil (from Manual of Nearctic Diptera, Vol. 1).

Figure 37.272. Opomyzidae. Anterior spiracle of *Geomyza balachowskyi* Mesnil (from Manual of Nearctic Diptera, Vol. 1).

Head: Cephalic segment somewhat retractile and weakly bilobed, bearing sensory organs anteriorly; antennae apical, somewhat bulbous and appearing 2-segmented; facial mask with complicated reticulate pattern on each antennomaxillary lobe.

Cephalopharyngeal skeleton deeply pigmented, with tentoropharyngeal and hypopharyngeal sclerites fused; ventral cornua apparently lacking apical windows; no pharyngeal filter; parastomal bars short; epipharyngeal sclerite absent. Mandibles deeply pigmented, hook with few accessory teeth, basal part elongated.

Thorax: Ventral creeping welt only on metathorax; anterior spiracles arising posterolaterally on prothorax, elongated (*Geomyza*) or fan-shaped (*Opomyza*), with 6–15 finger-like papillae scattered along lateral and apical margins.

Abdomen: Segments all very similar, cylindrical, each with weakly to strongly developed creeping welt ventrally; welts composed of 3–5 rows of pale to blackened spinules; no spinule bands encircling segments; caudal segment somewhat longer, without fleshy protuberances; perianal pad bilobed.

Posterior spiracular disc lacking marginal tubercles; posterior spiracles slightly elevated, arising close to each other or distinctly separated; each spiracular plate with three oval spiracular openings diverging from ecdysial scar and four dichotomously branching spiracular hairs.

Comments: This is a small family of fewer than 100 species. Three genera and 13 species, mostly in the genus *Geomyza*, have been recorded from America north of Mexico. Most of the Nearctic species occur in the western states and provinces. Because of their phytophagous feeding habits, larvae of a few species such as *Opomyza germinationis* (L.) and *O. petrei* Mesnil are fairly significant pests of cereal crops.

Selected Bibliography

Balachowsky and Mesnil 1935.
Ferrar 1987.
Hennig 1952, 1958.
Latteur 1974.
Mesnil 1934.
Nye 1958.
Thomas 1933.
Vockeroth 1961, 1987a.

ANTHOMYZIDAE (OPOMYZOIDEA)

B. A. Foote, *Kent State University*

The Anthomyzids

Figure 37.272z

Relationships and Diagnosis: Long recognized as closely related to the opomyzids, the Anthomyzidae is currently placed in the Opomyzoidea (McAlpine et al. 1981). This is a complex of 12 Nearctic families including the Clusiidae, Agromyzidae, Asteiidae, Milichiidae, and Braulidae. Vockeroth (1987b) recently summarized our knowledge.

This is another acalyptrate family whose larvae are impossible to separate with any confidence from those of certain other families having saprophagous or partially phytophagous larvae. Perhaps the elongate nature of the anterior spiracles (fig. 37.272z) with papillae arising only on one side, coupled with the small body size are the most distinctive characteristics.

Biology and Ecology: As with many of the smaller, more obscure families of acalyptrate Diptera, practically nothing is known about the life histories or larval feeding habits of the Anthomyzidae. The few published papers suggest that the larvae are phytophagous or possibly saprophagous within the stems or leaves of such herbaceous wetland plants as grasses, rushes, and cattails where the adults are sometimes common. The presence of a filtering mechanism in the floor of the pharynx suggests that the larvae are basically scavengers and thus are best considered as being secondary invaders of plant tissues that have been damaged by truly phytophagous larvae of other families.

Description: No larval stage of any Nearctic species has been described, and the following description is based on studies of European species. Mature larva 3–6 mm, muscidiform, rather slender, tapering anteriorly and bluntly rounded posteriorly; integument mostly bare except for weakly developed ventral creeping welts.

Head: Cephalic segment retractile, bearing well-developed sensory organs anteriorly and anteroventrally; facial mask with complicated network of oral grooves and ridges around preoral cavity.

Cephalopharyngeal skeleton weakly to deeply pigmented; tentoropharyngeal and hypopharyngeal sclerites separated; pharyngeal filter present; hypopharyngeal sclerite H-shaped; parastomal bars present and not connected apically; mandibles with sickle-shaped hook that may have accessory teeth along ventral side.

Thorax: Anterior spiracles arising posterolaterally on prothorax; spiracles greatly elongate, with 5–6 finger-like papillae arising along one side of spiracular axis.

Abdomen: Segments all very similar, each with weakly or more strongly developed creeping welt ventrally near segmental border.

Posterior spiracles borne on apices of widely separated, short, fleshy lobes, with three oval spiracular openings and four branching spiracular hairs.

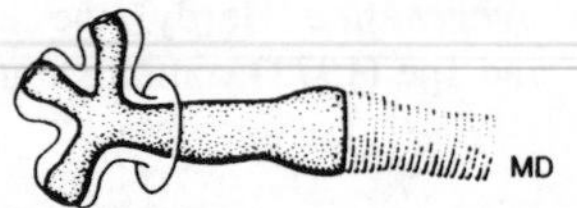

Figure 37.272z. Anthomyzidae. *Anthomyza sabulosa* Haliday. Anterior spiracle. Feeds in grass stems (from Manual of Nearctic Diptera, Vol. 2).

Comments: This is another of the numerous families of higher Diptera whose taxonomy has been largely ignored. Probably there are fewer than 200 species in the world. Three genera and 13 species are currently recognized in the Nearctic fauna, although most of the genera badly need revision. No species is considered to have economic significance.

Selected Bibliography

Anderson 1976.
Balachowskv and Mesnil 1935.
Ferrar 1987.
Hennig 1952.
Meijere 1944.
Sturtevant 1954.
Vockeroth 1987b.

AULACIGASTRIDAE (OPOMYZOIDEA)

B. A. Foote, *Kent State University*

The Aulacigasterids

Figure 37.273

Relationships and Diagnosis: Commonly placed near the Drosophilidae by earlier workers, the Aulacigastridae was considered by Hennig (1958) to be closer to the Chyromyidae and Anthomyzidae. This relationship was accepted by McAlpine et al. (1981) who placed it in their Opomyzoidea. In contrast, Griffiths (1972) felt that it belonged in the prefamily Anthomyzoinea along with such families as the Heleomyzidae, Anthomyzidae, Asteiidae, and Sphaeroceridae. Hennig (1969) expanded the family concept to include *Cyamops,* a genus previously placed in the Anthomyzidae. Teskey (1987a) recently summarized our knowledge.

Larvae are fairly distinct in possessing a telescopic respiratory tube that bears the posterior spiracles apically. The greatly elongated, filament-like anterior spiracles are unique for this family. In addition, they possess paired prolegs on several of the body segments.

Biology and Ecology: Larvae of *Aulacigaster leucopeza* (Meigen) are recorded as breeding within slime fluxes emanating from wounds on trunks of deciduous trees where the adults are found.

Description: Mature larvae 5–7 mm, basically muscidiform but somewhat flattened and with caudal segment elongated to form respiratory tube; ventral surface with paired prolegs on abdominal segments; integument covered with numerous backwardly directed spinules; whitish to somewhat darkened.

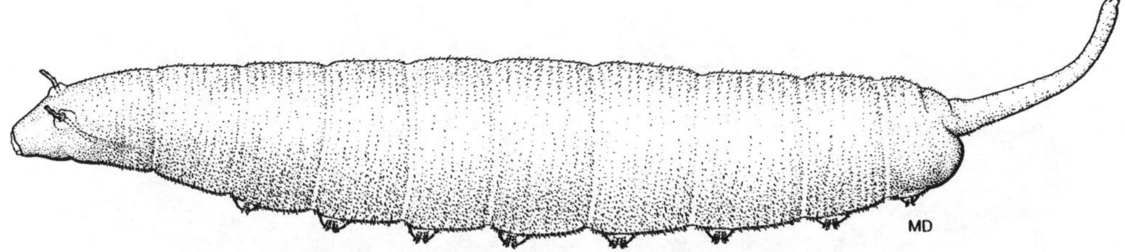

Figure 37.273. Aulacigastridae. Lateral of *Aulacigaster leucopeza* (Meigen). Larvae can be found in slime fluxes on deciduous and coniferous trees (from Manual of Nearctic Diptera, Vol. 1).

Head: Cephalic segment somewhat retractile, antennomaxillary organs well developed; antennae apparently 2-segmented; facial mask with numerous rows of spinules anterior to and lateral of preoral cavity; oral ridges and grooves present.

Cephalopharyngeal skeleton variously pigmented, with tentoropharyngeal and hypopharyngeal sclerites apparently fused; dorsal cornua narrower and shorter than ventral cornua, connected anteriorly by fenestrated dorsal bridge; ventral cornua distally unpigmented, lacking dorsobasal lobe; floor of tentoropharyngeal sclerite with filtering mechanism; hypopharyngeal sclerite somewhat elongate and H-shaped in ventral view; parastomal bars slender and free apically; apparently no epipharyngeal sclerite. Mandibles relatively narrow, consisting of elongate basal part and somewhat broader hook with 1 accessory tooth; a pair of narrow dental sclerites below basal part.

Thorax: Ventral surface lacking prolegs; anterior spiracles arising posterolaterally on thorax, each spiracle greatly elongate and filament-like, bearing 30–35 finger-like papillae.

Abdomen: Segments rather uniform except for caudal segment which is elongated to form telescopic respiratory tube; each segment bearing prolegs ventrally, each proleg divided medially so that each segment seemingly bears two spinose projections; perianal pad transversely elongated and extended up sides of caudal segment.

Posterior spiracles borne at apex of respiratory tube; spiracular plates nearly touching along midline, with three oval spiracular openings arranged around periphery of each spiracular plate; no spiracular hairs.

Comments: A very small family, the Aulacigastridae was long considered to consist of only the genus *Aulacigaster* and a single Holarctic species. Hennig (1969) transferred the genus *Cyamops* to this family. Only *Aulacigaster leucopeza* (Meigen) and *Cyamops*, with three species, are recorded from the Nearctic Region. No species is known to be economically significant.

Selected Bibliography

Davis and Zack 1978.
Ferrar 1987.
Griffiths 1972.
Hennig 1952, 1958, 1969.
Malloch and McAtee 1924.
Sabrosky 1958.
Teskey 1976, 1987a.

PERISCELIDIDAE (OPOMYZOIDEA)

B. A. Foote, *Kent State University*

The Periscelidids

Figure 37.274

Relationships and Diagnosis: The phylogenetic relationships of this rarely collected family are poorly understood, and its taxonomic placement has varied. It was frequently considered to be close to the Drosophilidae and Ephydridae, but others felt it was related to the Lauxaniidae (Hennig 1958), Psilidae or Chamaemyiidae. McAlpine et al. (1981) have included it in the 12 families of the Opomyzoidea, implying that it is relatively close to the Aulacigastridae and Asteiidae. McAlpine (1987g) recently summarized our knowledge.

Larvae are easily recognized by their small size (<5 mm), somewhat depressed body, fleshy projections laterally, and two distinctly separated, diverging respiratory tubes bearing the posterior spiracles. Their occurrence in fermenting sap flows of deciduous trees is also distinctive.

Biology and Ecology: Members of this family are rare and infrequently collected. They are widely distributed in North America and adults are usually encountered in wooded habitats around sap flows emanating from wounds in tree trunks. Larvae of *Periscelis wheeleri* (Sturtevant) have been found in fermenting sap issuing from an oak tree.

Description: No larval stage of any Nearctic species has been described, and the following information is taken from a description of a European species (Heeger 1852). Mature larvae less than 5 mm, muscidiform but somewhat depressed dorsoventrally, tapering anteriorly and broadening posteriorly; most segments have two pairs of lateral projections plus smaller tubercles dorsally; integument bearing numerous darkened spinules and longer hairs.

Head: Cephalic segment poorly retractile, usually partially obscured by prothorax, bearing antennae apically and circular sensory plates apicoventrally; facial mask with series of oral grooves and ridges around preoral cavity.

Cephalopharyngeal skeleton pigmented, with tentoropharyngeal and hypopharyngeal sclerites separated; pharyngeal filter present on floor of tentoropharyngeal sclerite.

Thorax: Segments each with fairly conspicuous pair of fleshy tubercles laterally and reduced pair dorsolaterally; without prolegs; anterior spiracles borne at tip of elongated tube-like structures, each spiracle with four blunt papillae along distal margin.

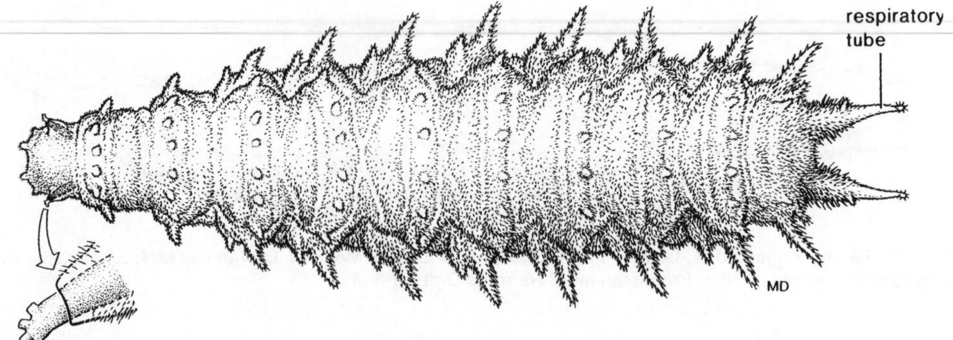

Figure 37.274. Periscelididae. Dorsal of *Periscelis annulata* (Fallén) (from Manual of Nearctic Diptera, Vol. 1).

Abdomen: Segments all similar and subequal in size; each with 3 pairs of fleshy tubercles laterally and dorsolaterally and row of four low tubercles across dorsum; caudal segment with only two pairs of lateral and one pair of dorsal tubercles; segment terminating in two elongate projections bearing posterior spiracles apically.

Posterior spiracles small, each arising at tip of elongate, tapering tube-like structure; three small and oval spiracular openings; numerous spiracular hairs arising around border of spiracular plate.

Comments: This is a very small family consisting of only the genus *Periscelis* with three species in America north of Mexico.

Selected Bibliography

Duda 1934.
Ferrar 1987.
Hennig 1952, 1958.
McAlpine 1987g.
Sturtevant 1954.

ASTEIIDAE (OPOMYZOIDEA)

B. A. Foote, *Kent State University*

The Asteiid Flies

Relationships and Diagnosis: As for many other families of Muscomorpha, the correct phylogenetic placement of the Asteiidae has puzzled a variety of dipterists over the years. Both Hendel (1922) and Malloch (1927) considered them to be very close to the Drosophilidae, a position denied by Hennig (1958) who felt that the family should be placed close to the Anthomyzidae and Opomyzidae. Later, Hennig (1971) included the Asteiidae in the Anthomyzoidea along with the Periscelididae, Aulacigastridae, Acartophthalmidae, Clusiidae, Anthomyzidae, Opomyzidae, and Chyromyidae. Recently, McAlpine et al. (1981) placed the Asteiidae in the Opomyzoidea with 11 other Nearctic families. They felt the asteiids were particularly close to the Periscelididae and Milichiidae. Sabrosky (1987a) recently reviewed our knowledge.

Biology and Ecology: Nothing is known of the larval habitats or feeding behavior. Adults are only occasionally collected, and larvae are completely unknown. Species have been taken in mesic to wet open or forested habitats, and an adult of an undescribed species was swept from an alga-encrusted coastal boulder exposed at low tide.

Description: No larvae have been described.

Comments: This is a very small, obscure family of acalyptrate flies that is rarely collected. Six genera and 20 species have been recorded for America north of Mexico, although there are several undescribed species in various collections.

Selected Bibliography

Ferrar 1987.
Hendel 1922.
Hennig 1958, 1971.
Malloch 1927.
Sabrosky 1957b, 1987a.

MILICHIIDAE (OPOMYZOIDEA)

B. A. Foote, *Kent State University*

The Milichiids

Figures 37.275a,b

Relationships and Diagnosis: Frequently considered by earlier workers to be related to such families as the Drosophilidae, Ephydridae, and Tethinidae, the milichiids have more recently been placed in an array of families that includes the Sphaeroceridae, Braulidae, Tethinidae, and Canacidae (Hennig 1958). Most recently McAlpine et al. (1981) placed it in the complex of 12 families comprising the Opomyzoidea and implied that it was particularly close to the Asteiidae and Carnidae. In contrast, Griffiths (1972) included the Milichiidae in his prefamily Tephritoinea near the Carnidae and the Chloropidae in the family-group Chloropidae. Sabrosky (1987b) recently summarized our knowledge.

Larvae are poorly differentiated from those of other families of saprophagous Muscomorpha. The elongate body coupled with the fleshy tubercles above the posterior spiracles (fig. 37.275a) are fairly distinctive.

Biology and Ecology: Adults of most species are small, dark, and commonly overlooked, although they can be quite abundant at certain times in some habitats. Although the

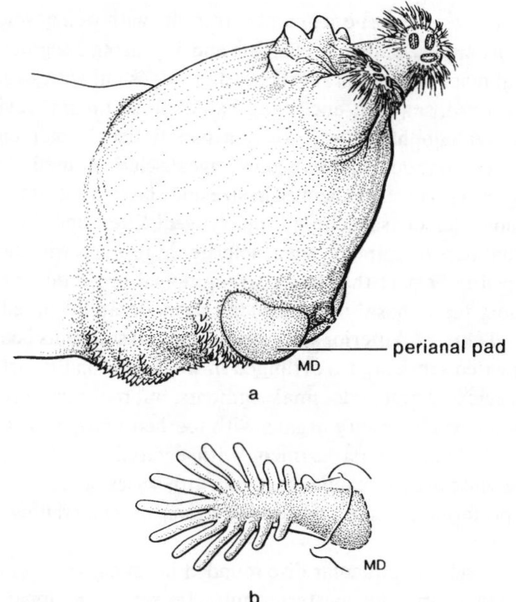

Figure 37.275a,b. Milichiidae. *Desmometopa m-nigrum* (Zetterstedt). **a.** oblique posterior view of caudal segment; **b.** anterior spiracle (from Manual of Nearctic Diptera, Vol. 1).

larval habits are poorly known, it appears that they are basically saprophagous. Larvae have been encountered in rotting plant tissue, under bark of dead trees, in fecal deposits, carrion, and bird nests. A few species have been found in leaf detritus of leaf-cutting ants. Adults of some species feed on exudates of prey captured by various predacious insects. The milichiids apparently ride on the backs of predators.

Description: Mature larvae 5–9 mm, basically muscidiform but with a very slender, elongate body; integument mostly bare except for spinules comprising ventral creeping welts; whitish.

Head: Cephalic segment weakly retractile, bearing well-developed sensory organs anteriorly and anteroventrally; facial mask with numerous oral ridges and grooves leading into preoral cavity.

Cephalopharyngeal skeleton mostly deeply pigmented; tentoropharyngeal and hypopharyngeal sclerites separate; dorsal cornua connected anteriorly by dorsal bridge; pharyngeal filter present; mandibles with sickle-shaped hook that usually lacks accessory teeth.

Thorax: Anterior spiracles arising posterolaterally on prothorax; spiracles elongate and slender, with 10–16 marginal finger-like papillae; ventral surface of metathorax with creeping welt.

Abdomen: Segments long and slender, bearing creeping welts ventrally near segmental borders; caudal segment somewhat longer; perianal pad bilobed and may be subtended by one or more fleshy tubercles.

Posterior spiracles borne at apices of basally separate fleshy structures; dorsum of caudal segment frequently with row of 3–4 fleshy tubercles immediately in front of spiracles; spiracular plate with weakly pigmented or no peritreme; spiracular openings oval, in some species (*Desmometopa*) two openings are parallel with the third opening at a distinct angle from other two; in other species all three openings radiate out from ecdysial scar; four profusely branched spiracular hairs.

Comments: This is a relatively small family of probably fewer than 200 species. Twelve genera and about 40 species are recorded for America north of Mexico. None of the species is considered to be economically important.

Selected Bibliography

Bohart and Gressitt 1951.
Engel 1930.
Ferrar 1987.
Griffiths 1972.
Hennig 1952, 1958.
Melander 1913.
Sabrosky 1958, 1987b.

CARNIDAE (OPOMYZOIDEA)

B. A. Foote, *Kent State University*

The Carnids

Relationships and Diagnosis: Commonly considered earlier as a subfamily of Milichiidae (Hennig 1958), the few species of this taxon are now generally segregated into a separate family. McAlpine, et al. (1981) included the carnids in the large Opomyzoidea and considered the family to be particularly close to the Milichiidae and Braulidae. Sabrosky (1987c) recently summarized our knowledge.

Larvae not easily distinguished from those of other saprophagous families. However, their small size (<5 mm) and extreme slenderness is fairly distinctive.

Biology and Ecology: Larvae of *Carnus hemapterus* Nitzsch, a widely distributed Holarctic species, occur in bird nests where they feed on organic debris. Interestingly, the adults have become ectoparasitic and live among the feathers of nestling birds. There is some debate as to whether the wingless (they lose them) adults suck blood or feed on moist skin secretions. Larvae of certain species of *Meoneura* have been found in dung.

Description: Mature larvae 3–5 mm, muscidiform, but strikingly slender; integument generally bare except for ventral creeping welts; white.

Head: Cephalic segment retractile, bearing well-developed sensory organs anteriorly and anteroventrally; facial mask with oral grooves and ridges around preoral cavity.

Cephalopharyngeal skeleton usually deeply pigmented; tentoropharyngeal and hypopharyngeal sclerites separate; dorsal cornua apparently connected by dorsal bridge; hypopharyngeal sclerite H-shaped; parastomal bars present but apparently not connected apically. Mandibles with narrow, sickle-shaped hook and broad basal part; apparently no dental sclerites.

Thorax: Anterior spiracles arising posterolaterally on prothorax; each spiracle with very short basal piece and a broadened distal piece with 3–5 finger-like marginal papillae; metathorax with weakly developed creeping welt ventrally.

Abdomen: Segments all very similar, with creeping welts ventrally. Posterior spiracles arising separately on short bases; each spiracular plate with three finger-like projections, each bearing a spiracular opening; apparently no spiracular hairs.

Comments: This is a very small family of fewer than 30 species in the world. Three genera and 16 species have been recorded from America north of Mexico. The largest genus is *Meoneura* with 13 species. No species is considered to be economically significant.

Selected Bibliography

Bequaert 1942.
Engel 1930.
Ferrar 1987.
Hennig 1937, 1952, 1958.
Sabrosky 1987c.
Séguy 1950.

BRAULIDAE (OPOMYZOIDEA)

B. A. Foote, *Kent State University*

Bee Lice

Figure 37.276

Relationships and Diagnosis: The highly modified body of adult Braulidae, which superficially resembles the Hippoboscidae, has presented great difficulties in determining the correct phylogenetic placement. It has been placed by different workers at various times close to the Chamaemyiidae, Drosophilidae, Milichiidae, and Sphaeroceridae. Hennig (1958) placed it in his Milichioidea along with the Sphaeroceridae, Tethinidae, Milichiidae, and Canacidae. More recently, McAlpine et al. (1981) included the Braulidae in the Opomyzoidea, relatively close to the Carnidae. Grimaldi and Underwood (1986) provide a review of the family, and Peterson (1987d) provides a summary.

Larvae are distinctive in having a sclerotized band of cuticle encircling the pseudocephalic segment, by being metapneustic, and bearing elongate, noticeable sensory organs on the first four and last body segments (fig. 37.276). They are only encountered in honey bees colonies.

Biology and Ecology: Adults and larvae of the single North American species, *Braula coeca* Nitzsch, live with honey bees. Adults usually are attached to the bodies of bees and steal honey and pollen from the cells of the honeycomb. Larvae feed on wax while boring from one cell to another, apparently causing little damage to the honeycombs. The pupa is unusual for the higher Diptera in that it is formed within the unmodified last larval cuticle and does not resemble the puparium found in the other higher Diptera.

Description: Mature larvae 2–4 mm, basically musciform, tapering anteriorly and broadly rounded posteriorly; integument bare, without hairs or spinules but appearing somewhat granular under high magnification; first four segments and caudal segment bearing numerous elongate, blunt-tipped sensory organs; whitish.

Head: Cephalic segment retractile, with well developed sensory organs, antennae elongate and appearing 2-segmented; facial mask without oral grooves and ridges but with strongly sclerotized, segment-encircling band behind preoral cavity.

Cephalopharyngeal skeleton mostly deeply pigmented; tentoropharyngeal and hypopharyngeal sclerites fused, dorsal and ventral cornua subequal in size and shape, ventral cornua without dorsobasal lobe, no pharyngeal filter, and no parastomal bars or epipharyngeal sclerite. Mandibles with sickle-shaped hook part that has a strong dorsal projection, no accessory teeth, basal part heavier and moderately curved.

Thorax: Anterior spiracles absent; all segments bearing elongated sense organs arranged in single row on the last two thoracic and first abdominal segments, but more scattered on prothorax; all sensory organs with toothed margins distally.

Abdomen: First segment with encircling row of elongate sense organs; caudal segment with series of sense organs on posterior surface; no sense organs on intervening segments.

Posterior spiracular disc rounded in caudal view, without marginal tubercles; posterior spiracles sessile on upper surface of caudal segment; three short-oval, radially arranged spiracular openings; 3–4 short spiracular hairs.

Comments: This is a very small family consisting of two genera and seven species. A tiny, wingless species, *Braula coeca* Nitzsch, has been introduced repeatedly into the United States. It is not considered to be a serious pest in apiaries.

Selected Bibliography

Ferrar 1987.
Grimaldi and Underwood 1986.
Hassanein and Abd El-Salam 1962.
Hennig 1938, 1952.
Imms 1942.
Peterson, B. V. 1987d.
Skaife 1921.
Smith, I. B. 1984.

COELOPIDAE (SCIOMYZOIDEA)

B. A. Foote, *Kent State University*

Seaweed Flies

Figures 37.277a–c

Relationships and Diagnosis: There has been relative agreement on the phylogenetic placement of the Coelopidae. Hennig (1958), Griffiths (1972), and McAlpine et al. (1981) all placed it with the Dryomyzidae, Sciomyzidae, Sepsidae, and Ropalomeridae in the Sciomyzoidea. Vockeroth (1987c) has recently summarized our knowledge.

Larvae are easily distinguished from larvae of other families of Muscomorpha by the circular fringe of spiracular hairs surrounding the posterior spiracular plates. Additionally, they are rather unique in having greatly expanded mouthhooks that bear a number of sharply pointed teeth apically. Finally, they are only found in shoreline accumulations of seaweed and kelp.

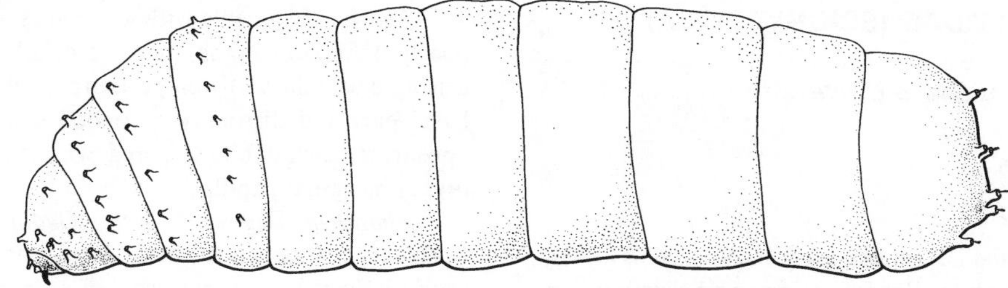

Figure 37.276. Braulidae. Lateral of *Braula coeca* Nitzsch (from Manual of Nearctic Diptera, Vol. 1).

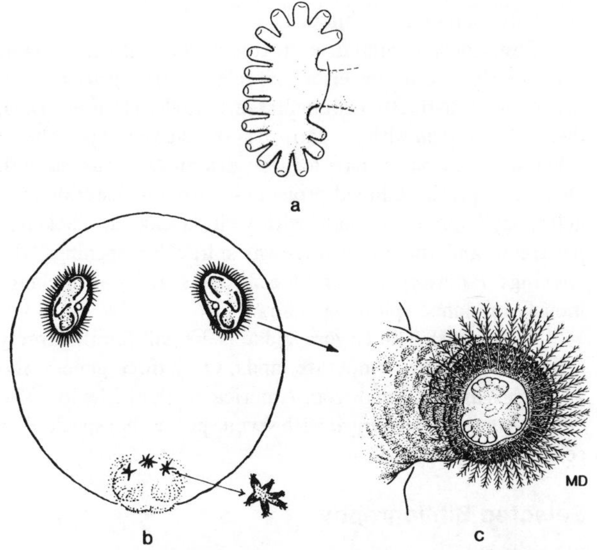

Figure 37.277a-c. Coelopidae. *Coelopa frigida* (Fabricius). **a.** anterior spiracle; **b.** posterior view of caudal segment; **c.** posterior spiracle (figs. a, c from Manual of Nearctic Diptera, Vol. 2; fig. b from Peterson 1951).

Biology and Ecology: Coelopidae are restricted to coastal areas, and adults are particularly abundant around piles of decaying seaweed. Larvae are saprophagous on decaying mats of seaweed and kelp that accumulate at the strand line.

Description: Mature larvae 9–14 mm, muscidiform, slender and cylindrical, tapering anteriorly and bluntly rounded posteriorly; integument smooth, without hairs or spinules except in ventral creeping welts; white.

Head: Cephalic segment bilobed and retractile, bearing noticeable sensory organs apically and apicoventrally; antennae apparently 2-segmented; facial mask with several rows of rather sharply pointed spinules in front of and lateral to the preoral opening; each antennomaxillary lobe with few oral ridges and grooves.

Cephalopharyngeal skeleton usually deeply pigmented, with tentoropharyngeal and hypopharyngeal sclerites separate; cornua apparently lacking windows and dorsobasal lobes; pharyngeal filter present; hypopharyngeal sclerite somewhat H-shaped in ventral view; parastomal bars slender and not connected apically; mandibles deeply pigmented, hook greatly expanded and bearing numerous sharply pointed teeth along distal margin; basal part small and without window.

Thorax: Segments without prolegs but with ventral creeping welt of darkened, pointed spinules at border of metathoracic segment; no spinule band encircling any segment; anterior spiracles arising posterolaterally on prothorax, somewhat bifurcate apically, with each branch bearing 6–10 finger-like papillae along outer border.

Abdomen: Segments all very similar, subcylindrical, caudal segment somewhat longer; each segment bearing ventral creeping welt along segmental border; caudal segment smoothly rounded in posterior view; perianal pad bilobed, with distinct spinule patch posterior to anal slit.

Posterior spiracular disc lacking marginal tubercles; posterior spiracles borne on short, tube-like, rather deeply pigmented bases; spiracles distinctly separate and arising on upper third of spiracular disc; each spiracular plate somewhat kidney-shaped, peritremes deeply pigmented, narrow, curved, and slit-like; three spiracular openings; spiracular hairs numerous, short plumose or branched, and forming distinct circular fringe around peritremes.

Comments: This is a small family of rather dark-bodied flies. Only two genera and five species have been recorded from America north of Mexico. Different species are found on each coast, with *Coelopa frigida* (Fabricius) being the only species occurring in New England. Four other species are found from Alaska to Mexico. None is considered to be of economic importance, although adults occasionally are annoying on bathing beaches, and larvae have a role in recycling nutrients in wrack accumulations. Adults and larvae are eagerly eaten by a variety of shore birds.

Selected Bibliography

Backlund 1945a.
Burnet and Thompson 1960.
Dobson 1974.
Egglishaw 1960b.
Ferrar 1987.
Griffiths 1972.
Hennig 1952, 1958.
Kompfner 1974.
Poinar 1977.
Roubaud 1901.
Scott, H. 1920
Vockeroth 1987c.

DRYOMYZIDAE (SCIOMYZOIDEA)

B. A. Foote, *Kent State University*

The Dryomyzids

Figures 37.278, 37.279

Relationships and Diagnosis: Long recognized as closely related to the Coelopidae, Ropalomeridae, and Sciomyzidae, the Dryomyzidae are now included in the five families comprising the Sciomyzoidea (McAlpine et al. 1981). Steyskal (1987j) recently reviewed our knowledge.

Larvae of *Oedoparena* are fairly distinctive in being somewhat depressed dorsoventrally and having fleshy lateral projections along the body. Their occurrence in barnacles is unique for the higher Diptera. The larval stages of *Dryomyza* and *Helcomyza* are less easily characterized, as they possess the body form of many groups of saprophagous Muscomorpha. The six pairs of variously developed tubercles around the margin of the posterior spiracular disc is fairly distinctive for *Dryomyza*.

Biology and Ecology: With respect to habitat distribution, two distinct groups can be recognized. *Dryomyza* are commonly found in moist woodland habitats, particularly those dominated by deciduous trees. In contrast, *Helcomyza* and *Oedoparena* are restricted to wet coastal areas. Larvae of *Dryomyza* probably are consumers of decaying animal matter, as reported by Barnes (1984), although Portschinshy (1910) found them in human feces, and Brauer (1883) stated that larvae of this genus live in mushrooms. *Helcomyza* larvae feed on decaying seaweed. In contrast to the species having scavenging habits, larvae of *Oedoparena glauca* (Coquillett) are predators of barnacles occurring in the rocky intertidal areas of the West Coast.

Description: Mature larvae 4–10 mm, muscidiform in *Dryomyza* and *Helcomyza,* somewhat depressed with lateral fleshy projections in *Oedoparena*; integument bearing numerous pale to darkened spinules that form rather distinct darkened rows dorsally in *Oedoparena*; whitish to yellowish.

Head: Cephalic segment somewhat retractile, with conspicuous sensory organs anteriorly and ventrally; antennae 2-segmented; facial mask with numerous oral grooves and ridges leading into preoral cavity; area behind preoral cavity with numerous posteriorly-directed spinules forming postoral spinule band.

Cephalopharyngeal skeleton deeply pigmented; tentoropharyngeal and hypopharyngeal sclerites separate; dorsal cornua connected by fenestrated dorsal bridge; ventral cornua with small (*Oedoparena*) or large (*Dryomyza*) dorsobasal lobe and usually with window posteriorly; pharyngeal filter present (*Oedoparena*?); hypopharyngeal sclerite H-shaped (*Dryomyza*) or more elongate (*Oedoparena*), parastomal bars elongate and free apically; epipharyngeal sclerite present (*Dryomyza*) or absent (*Oedoparena*), somewhat elongate-oval with small marginal indentations. Mandibles black, hook sickle-shaped, with (*Oedoparena*) or without accessory teeth (*Dryomyza*), basal part with or without small window, nearly square in outline in *Dryomyza* and with long posterior projection in *Oedoparena;* dental sclerites narrow and elongate in ventral view, somewhat triangular in lateral view.

Thorax: Segments without prolegs, metathorax bearing poorly defined creeping welt posteriorly; anterior spiracles arising posterolaterally on prothorax, with elongate, tubular, basal part and distinctly expanded distal part which may appear somewhat bilobed, and with 14–24 (40 in *Helcomyza*) marginal papillae.

Abdomen: Segments with (*Oedoparena*) or without (*Dryomyza, Helcomyza*) fleshy lateral projections; if present, each segment bearing one pair of shorter ventrolateral and one pair of longer lateral projections; caudal segment truncate (*Dryomyza*) or slightly elongate with lateral projections (*Oedoparena*), each segment with poorly defined creeping welt ventrally; perianal pad bilobed.

Posterior spiracular disc either subcircular in caudal view, with nearly sessile posterior spiracles (*Dryomyza*) or tappering, with distinctly pedunculated spiracles (*Oedoparena*); disc in *Dryomyza* with six pairs of spinule-covered peripheral tubercles, ventromost pair longest; disc in *Oedoparena* with elongate, apically bilobed projection between posterior spiracles; each spiracular plate with well developed, blackened peritreme and three elongate-oval spiracular openings, the openings radiating out from ecdysial scar; four dichotomously branched spiracular hairs.

Comments: The Dryomyzidae is a small family largely restricted to north temperate lands. Only three genera and 11 species are recorded from America north of Mexico. The largest genus is *Dryomyza* with seven species. No species has economic importance.

Selected Bibliography

Barnes 1984.
Brauer 1883.
Burger et al. 1980.
Egglishaw 1960a.
Ferrar 1987.
Hennig 1952.
Mathis and Steyskal 1980.
Schlinger 1975.
Steyskal 1957, 1987j.

SCIOMYZIDAE (SCIOMYZOIDEA)

B. A. Foote, *Kent State University*

Marsh Flies, Snail-killing Flies

Figures 37.280–37.284

Relationships and Diagnosis: Hennig (1958) suggested that the Sciomyzidae formed the central taxon of the Sciomyzoidea, which also included the Helcomyzidae, Rhopalomeridae, Coelopidae, Dryomyzidae, and Sepsidae. He felt that the sciomyzids were particularly close to the Sepsidae. Griffiths (1972) and McAlpine, et al. (1981) basically agreed with this grouping. Knutson (1987) has recently summarized our knowledge.

The most distinctive morphological characteristic of sciomyzid larvae is an anteriorly toothed sclerite, the ventral arch, found below the basal part of the mandibles (fig. 37.284b). No other family of Muscomorpha possesses a similar sclerite.

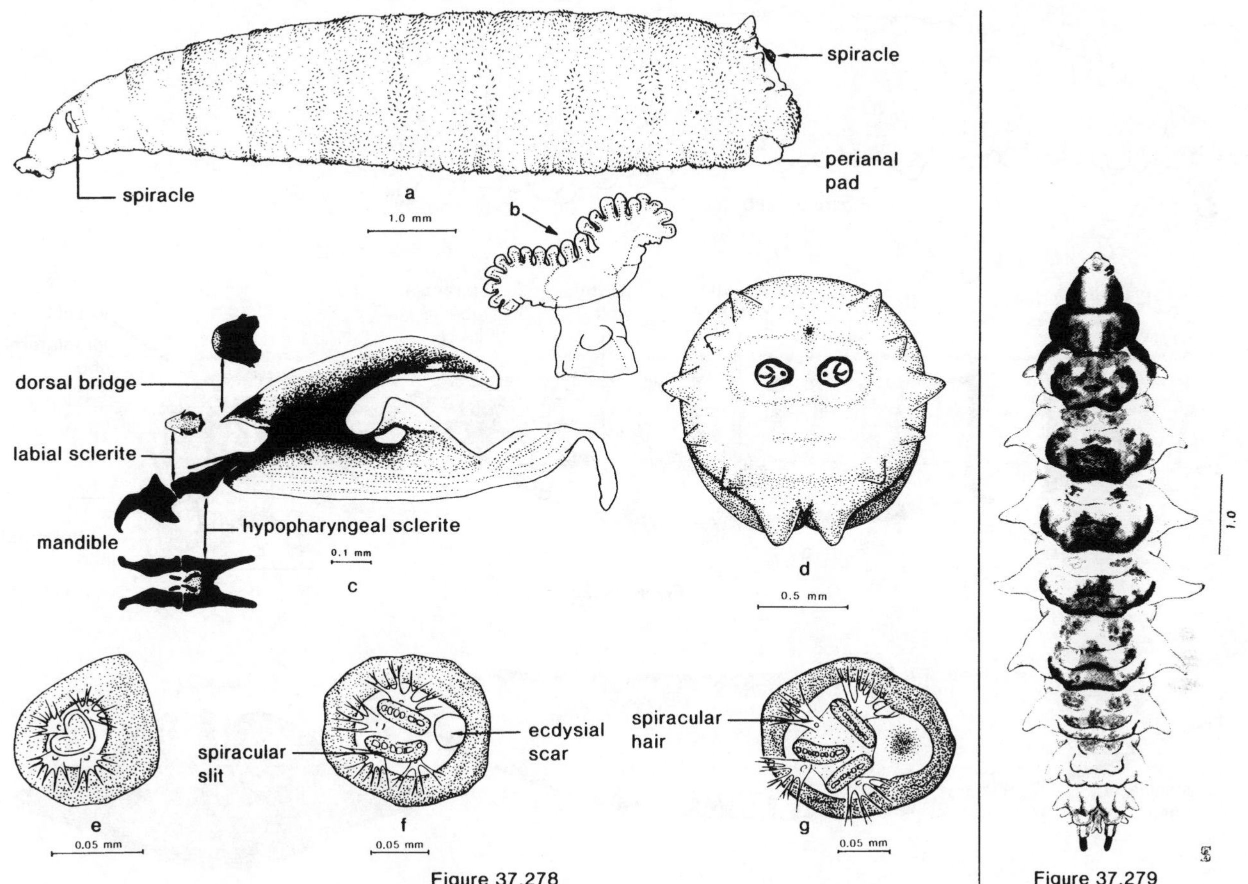

Figure 37.278a-g. Dryomyzidae. *Dryomyza anilis* Fallén. **a.** lateral; **b.** anterior spiracle; **c.** cephalopharyngeal skeleton; **d.** posterior view of caudal segment; **e, f, g.** Left caudal spiracle of 1st, 2nd and 3rd instars showing one, two and three spiracular slits respectively. Larvae are found in rotting fungi, carrion and excrement (from Barnes 1984).

Figure 37.279. Dryomyzidae. *Oedoparena glauca* (Coquillett), dorsal. Larvae are predators of barnacles along the West Coast (from Burger et al. 1980).

Biology and Ecology: The species of Sciomyzidae are united biologically by the larval habit of preying on species of Mollusca. Most of the prey species belong to the class Gastropoda, although larvae of *Renocera* feed on fingernail clams of the class Pelecypoda. Within the gastropods, a wide variety of aquatic and terrestrial species are attacked by sciomyzid larvae. The most commonly utilized prey are aquatic pulmonate snails that are consumed by numerous species of *Dictya, Elgiva, Hedria, Sepedomerus, Sepedon,* and *Tetanocera.* Larvae of many species of *Atrichomelina, Colobaea, Pherbellia, Pteromicra,* and *Sciomyza* attack aquatic snails that have been stranded by dropping water levels. A few species have acquired more terrestrial habits and are parasitoids of land snails (*Pherbellia, Oidematops, Pteromicra, Euthycera, Tetanocera*) or slugs (*Euthycera, Limnia, Tetanocera*). Larvae of *Antichaeta* are specialized for the consumption of snail eggs. Adult sciomyzids are most commonly encountered in marshy habitats and along shorelines, although species having larvae that feed on land snails can be found in moist, mesic, and even relatively arid terrestrial microhabitats. Eggs of most species are merely deposited on vegetation in habitats containing the larval prey, but a few species of *Pherbellia, Pteromicra,* and *Sciomyza* place their eggs directly on the shell of the host snail.

Description: Mature larvae 3–15 mm, muscidiform, body dorsoventrally flattened in some species, subcylindrical in most; integument smooth (Sciomyzini) to wrinkled or tuberculate (Tetanocerini); dorsal and lateral surfaces largely bare in most species of the tribe Sciomyzini (*Pherbellia, Pteromicra, Sciomyza*), but pubescent, setose, or spinulose in many species of Tetanocerini (*Dictya, Elgiva, Sepedon,* aquatic *Tetanocera*); species of Sciomyzini with ventral creeping welts and several species with encircling bands of spinules; larvae of Sciomyzini white, white to gray to nearly black in the Tetanocerini, and numerous larvae of latter tribe with dorsal pattern or appearing shagreened; living larvae of *Elgiva* green.

Head: Cephalic segment highly retractile, bearing well-developed sensory organs anteriorly and anteroventrally; antennae varying from short and fleshy to narrow and elongate, apparently 2-segmented; facial mask without oral grooves and

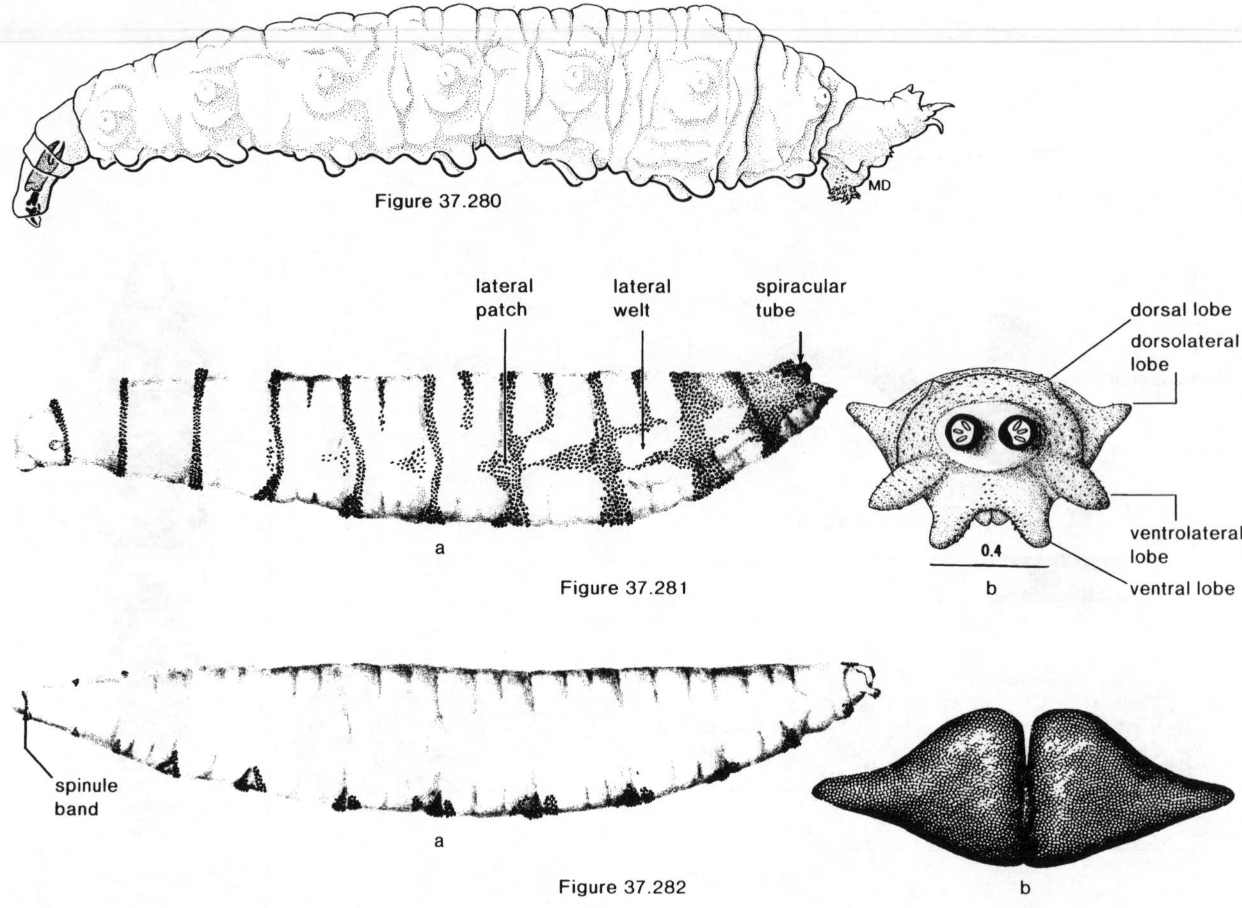

Figure 37.280. Sciomyzidae. *Sepedon* sp., lateral. Larvae prey on fresh water pulmonate snails occurring at the surface film of marshes (from Manual of Nearctic Diptera, Vol. 1).

Figures 37.281a,b. Sciomyzidae. *Pherbellia parallela* (Walker). **a.** lateral; **b.** posterior view of caudal segment. Larvae prey on stranded aquatic pulmonate snails (from Bratt et al. 1969).

Figure 37.282a,b. Sciomyzidae. *Pherbellia schoenherri maculata* (Cresson). **a.** lateral; **b.** perianal pad. Larvae prey on snails of the family Succineidae (from Bratt et al. 1969).

ridges surrounding preoral cavity; in some species preoral cavity bordered by clustered or scattered spinules; postoral spine band usually present.

Cephalopharyngeal skeleton usually deeply pigmented; tentoropharyngeal and hypopharyngeal sclerites separate; dorsal cornua commonly as broad as or broader than ventral cornua, connected anteriorly by dorsal bridge in Sciomyzini but not connected in most species of Tetanocerini; ventral cornua narrow to broad, without noticeable dorsobasal lobe but with window posteriorly; pharyngeal filter absent; hypopharyngeal sclerite usually short and broad and H-shaped in ventral view; parastomal bars well-developed and usually fused distally with epipharyngeal sclerite, the latter sclerite variously shaped but commonly transversely elongate and containing 2 or more small windows; labial sclerite between lateral arms of hypopharyngeal sclerite U- or V-shaped in ventral view. Mandibles usually deeply pigmented, hook typically sickle-shaped and decurved, basal part subrectangular, bearing two or more teeth in nearly all larvae of Tetanocerini but without teeth in species of Sciomyzini, basal part subtended by ventral arch, this sclerite typically bilobed along posterior border and toothed along anterior border; last two segments usually with vental creeping welts, and in some species encircling spinule bands.

Thorax: Last two segments in Sciomyzini usually with ventral creeping welts and, in some species, with encircling spinule bands; no obvious creeping welts or encircling bands of spinules in most species of Tetanocerini; anterior spiracles arising posterolaterally on prothorax; spiracles typically with short basal piece and somewhat expanded distal part, the latter commonly fan-shaped with 4–32 short or elongate papillae along apical margin.

Abdomen: All segments except caudal one very similar; larvae of Sciomyzini with ventral creeping welts and, in some species, with encircling spinule bands; larvae of Tetanocerini usually without creeping welts and spinule bands although integument may be spinulose, setose, or shagreened; species of *Antichaeta* with dense coat of nearly transparent spinules; perianal pad bilobed, variously shaped.

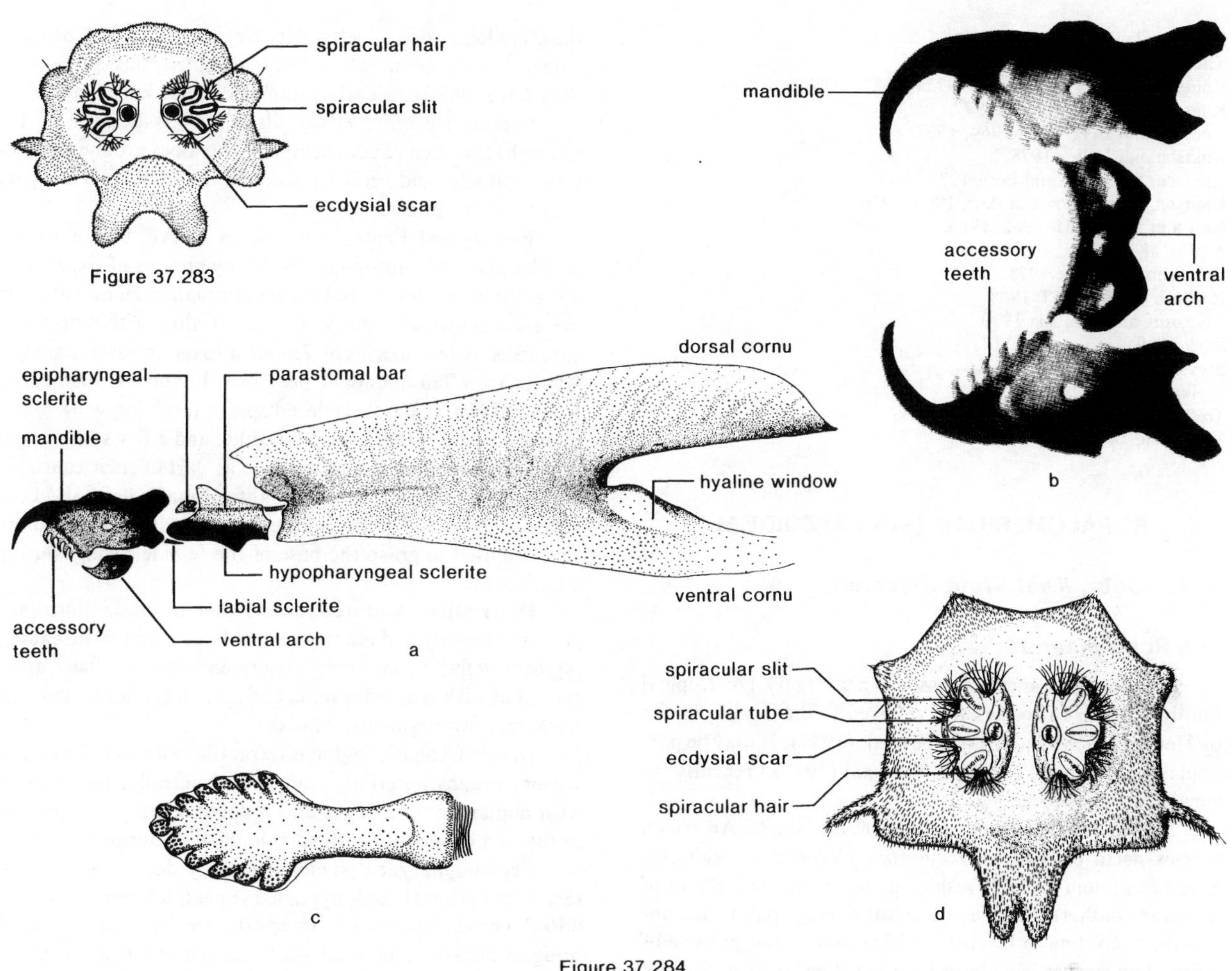

Figure 37.283. Sciomyzidae. Posterior view of caudal segment of *Dictya oxybeles* Steyskal. Larvae are predators of small snails in salt marshes along the East Coast (from Valley and Berg 1977).

Figure 37.284a-d. Sciomyzidae. *Elgiva solicita* (Harris). **a.** Cephalopharyngeal skeleton; **b.** mandibles and ventral arch; **c.** anterior spiracle; **d.** posterior view of caudal segment. Larvae prey on aquatic pulmonate snails (from Knutson and Berg 1964).

Posterior spiracular disc with 2–5 pairs of short or elongate fleshy lobes around margin, with the more ventral lobes usually being better developed; posterior spiracles arising separately near center of disc, sessile or borne on short, tubular bases; peritremes usually deeply pigmented; spiracular plate with three oval spiracular openings radiating from ecdysial scar; interspiracular bristles or hairs present, dichotomously branched in aquatic species of Tetanocerini, absent in Sciomyzini, and greatly reduced or absent in terrestrial Tetanocerini.

Comments: This is a moderate sized family of acalyptrate flies, with 65 genera and somewhat over 600 species recorded for the world. Some 175 species and 21 genera are known to occur in America north of Mexico. Larger genera include *Pherbellia, Dictya, Sepedon,* and *Tetanocera.* About two-thirds of the Nearctic species belong to the Tetanocerini.

The family has generated considerable interest as a potentially significant group in the biological control of snails of medical and veterinary importance. For example, a species of *Sepedon* has been introduced into Hawaii in an effort to control a snail that serves as the intermediate host of a cattle liver fluke. Other efforts have explored the possible role of predacious sciomyzid larvae in the control of various aquatic snails that serve as intermediate hosts of blood flukes (*Schistosoma* spp.) that cause schistosomiasis in humans.

Selected Bibliography

Barnes 1979a, 1979b, 1980a, 1980b.
Beaver, O. 1972a, 1972b.
Berg 1953, 1961, 1964a, 1964b, 1973.
Berg and Knutson 1978.
Berg et al. 1982.
Bratt et al. 1969.
Chock et al. 1961.
Davis et al. 1961.
Eckblad 1973.
Ferrar 1987.
Fisher and Orth 1964, 1983.
Foote 1959, 1971, 1973, 1976b, 1977.
Foote and Knutson 1970.

Foote et al. 1960.
Kaczynski et al. 1969.
Knutson 1966, 1970a, 1970b, 1972a, 1973a, 1973b, 1987.
Knutson and Abercrombie 1977.
Knutson and Berg 1963, 1964, 1967.
Knutson and Valley 1978.
Knutson, Rozkošný and Berg 1975.
Knutson, Stephenson and Berg 1965, 1970.
Neff and Berg 1961, 1962, 1966.
Neff et al. 1967.
Robinson and Foote 1978.
Rozkošný 1965, 1967, 1968.
Rozkošný and Knutson 1970.
Sack 1939.
Steyskal et al. 1978.
Trelka and Berg 1977.
Trelka and Foote 1970.
Valley and Berg 1977.

ROPALOMERIDAE (SCIOMYZOIDEA)

B. A. Foote, *Kent State University*

The Ropalomerids

Relationships and Diagnosis: This largely Neotropical family of some 15 species was referred to the Sciomyzoidea by Hennig (1958) and McAlpine et al. (1981). It may be particularly close to the Sepsidae. Steyskal (1987k) recently reviewed our knowledge.

Biology and Ecology: Adults of the single American species north of Mexico, *Rhytidops floridensis* (Aldrich), have been found around fresh exudates of various species of palms in southern Florida. Its larval feeding habits are unknown. According to Fischer (1932c), larvae and puparia of a Brazilian species have been found in stems of bananas and palms.

Description: No larvae of this family have been described.

Comments: This very small family is not known to contain any economically important species.

Selected Bibliography

Ferrar 1987.
Fischer 1932c.
Hennig 1952, 1958.
Steyskal 1987k.

SEPSIDAE (SCIOMYZOIDEA)

B. A. Foote, *Kent State University*

Black Scavenger Flies

Figures 37.285–37.286

Relationships and Diagnosis: The correct taxonomic placement of the Sepsidae has presented problems. Earlier workers tended to place it near the Micropezidae, although a few investigators considered it to be closely related to an array of families centered around the Drosophilidae. Hennig (1958) included the Sepsidae in the Sciomyzoidea along with the Ropalomeridae, Coelopidae, Dryomyzidae, and Sciomyzidae. McAlpine et al. (1981) accepted this placement. Steyskal (1987 l) recently summarized our knowledge.

Larvae are quite easily distinguished because of the somewhat swollen caudal segment, the lobes around the posterior spiracles and perianal pad, and the frequently elongate nature of the anterior spiracles.

Biology and Ecology: As far as known, larvae are coprophagous and saprophagous. Many species of *Sepsis* and *Themira* have been reared from mammalian feces, but there are also confirmed records of larvae feeding within decaying carcasses. A few species of *Themira* breed in sewage sludge. Adults are often abundant near larval habitats, where they frequently carry on courtship displays involving wing movements that are rather species specific, and a few species emit a pleasant-smelling sex pheromone as part of their courtship and mating behavior. Mating flights of males have also been reported. Males of many species have modified fore femora that are used to grasp the base of the female's wings during copulation.

Description: Mature larvae 5–9 mm, muscidiform, tapering anteriorly and bluntly rounded posteriorly with caudal segment definitely swollen; integument relatively bare anteriorly but with numerous hairs and spinules caudally that encircle last few segments; whitish.

Head: Cephalic segment retractile, with well developed sensory organs anteriorly and anteroventrally; facial mask with numerous oral ridges and grooves leading into preoral cavity and commonly with spinule band posterior to cavity.

Cephalopharyneal skeleton usually deeply pigmented; tentoropharyngeal and hypopharyngeal sclerites separate; dorsal cornua connected anteriorly by fenestrated dorsal bridge; ventral cornua usually broader and with distal window; pharyngeal filter present; hypopharyngeal sclerite H-shaped, parastomal bars present but not connected apically; apparently no epipharyngeal or labial sclerites. Mandibles with slender sickle-shaped hook that lacks accessory teeth, basal part commonly with window.

Thorax: Anterior spiracles arising posterolaterally on prothorax, spiracles either greatly elongated with more than 10 marginal papillae or less than twice as long as broad and with fewer than 10 papillae; papillae long and somewhat finger-like; ventral surface of metathorax with well developed creeping welt.

Abdomen: All segments with ventral creeping welts and commonly with hairs and spinules which are particularly noticeable on the last two segments; caudal segment swollen and noticeably larger than preceding segments; perianal pad bilobed, subtended by fleshy lobes laterally and by small spinule patch posteriorly.

Posterior spiracles borne at apices of distinctly elongated, somewhat tube-like structures that nearly touch each other basally; caudal segment with fleshy lobe on each side of perianal pad, a pair of rather small lobes laterally, a pair of ventrolateral lobes before spiracles, and a pair of dorsolateral lobes just in front of spiracles; spiracular plate either with three narrow and arcuate spiracular openings or with openings more oval; openings variously arranged on plate but not oriented radially; spiracular hairs usually present.

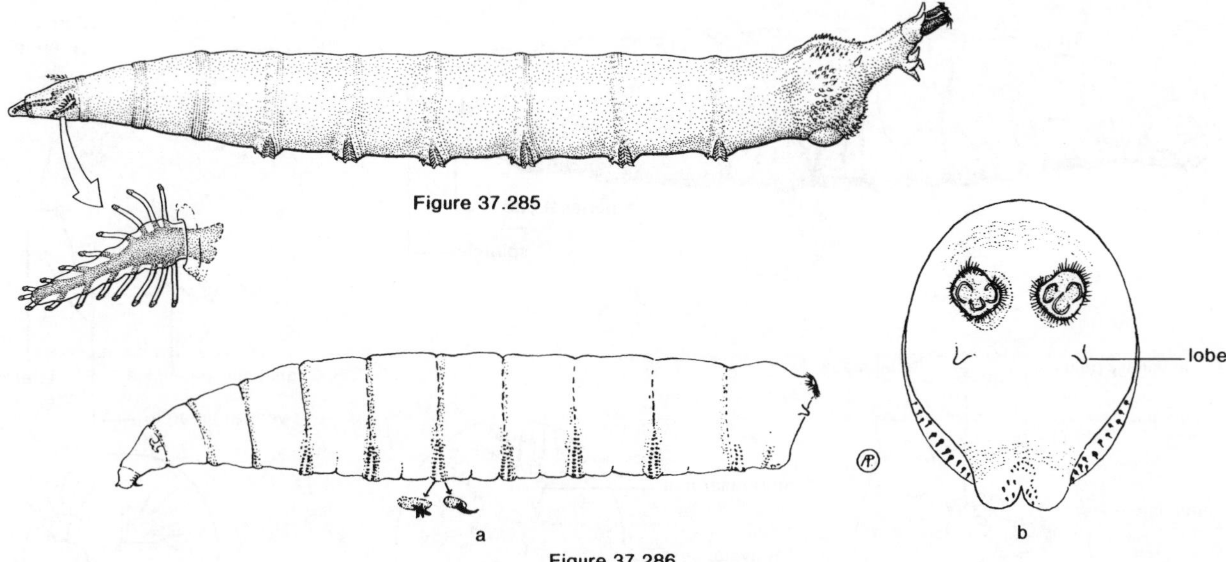

Figure 37.285. Sepsidae. Lateral of *Saltella spondylii* (Schrank). Larvae can be abundant in dung (from Manual of Nearctic Diptera, Vol. 1).

Figure 37.286a,b. Sepsidae. *Orygma luctuosa* Meigen. **a.** lateral; **b.** posterior view of caudal segment. Larvae feed on decaying kelp and seaweed at the strand line along the Northeastern Coast (from Peterson 1951).

Comments: A relatively small family of shining brown or black flies, it contains about 200 species in the world. Eight genera and some 35 species occur in America north of Mexico. The largest genera are *Sepsis* and *Themira,* each with about 12 species. Many of the species are widely distributed, with several being transcontinental. Although the family is generally considered to be unimportant economically, a few species are occasionally minor pests. *Themira putris* (L.), for example, is an abundant fly around sewage sludge deposits, and swarms of adults of other species can be annoying upon occasion. The larvae undoubtedly play important ecological roles in the decomposition of organic wastes, particularly dung, and in the recycling of nutrients.

Selected Bibliography

Ferrar 1987.
Hafez 1939a, 1939b.
Hennig 1952, 1958.
Laurence 1954.
Mangan 1976, 1977.
Merritt 1976.
Mohr 1943.
Peterson, A. 1951
Roubaud 1901.
Steyskal 1943, 1946, 1987 l.
Valiela 1970.

LAUXANIIDAE (LAUXANIOIDEA)

B. A. Foote, *Kent State University*

The Lauxaniids

Figures 37.287–37.290

Relationships and Diagnosis: At the present time, the family is placed in the Lauxanioidea along with the Chamaemyiidae (McAlpine et al. 1981). Hennig (1952) felt that the lauxaniids showed relationships with the Drosophilidae or possibly the Sciomyzidae. Later, he (Hennig, 1958) placed them in the Lauxanioidea along with the Celyphidae and Chamaemyiidae. Griffiths (1972) also placed them near the Chamaemyiidae. Shewell (1987a) recently summarized our knowledge.

It has proved impossible to distinguish all larvae of Lauxaniidae from those of other acalyptrate families having saprophagous habits. Fairly useful distinguishing features include the relatively small size (less than 7.5 mm), frequently depressed body form, spinulose integument, facial mask with rows of teeth-like structures (fig. 37.287b), and a pair of short to elongate lobes subtending the perianal pad.

Biology and Ecology: Largely confined to woodland habitats, many species have saprophagous larvae that feed most commonly on decaying mesophyll tissue of fallen leaves. Larvae have also been encountered in bird nests, mammal burrows, and rotting wood. A few larvae have been found in the stems and leaves of clover, in the ovaries of violets, and in pads of *Opuntia* cactus. A summary of the larval feeding habits is available in Miller and Foote (1975).

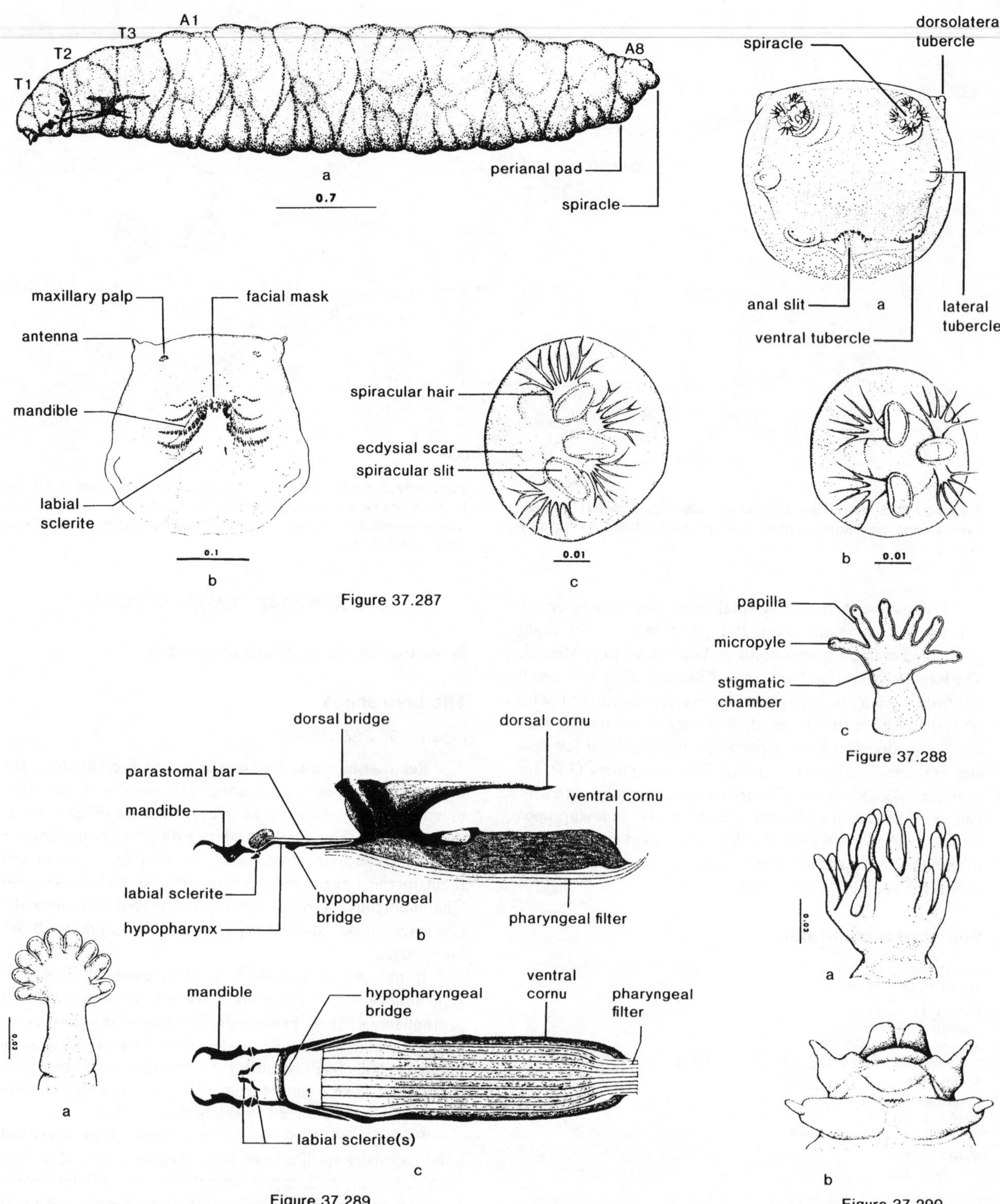

Figure 37.287a-c. Lauxaniidae. *Homoneura americana* (Wiedemann). **a.** lateral; **b.** ventral view of anterior segment; **c.** posterior spiracular plate (from Miller & Foote 1976).

Figure 37-288a-c. Lauxaniidae. *Camptoprosopella confusa* Shewell. **a.** posterior view of caudal segment; **b.** posterior spiracular plate; **c.** anterior spiracle (from Miller and Foote 1976).

Figure 37.289a-c. Lauxaniidae. *Lyciella browni* (Curran). **a.** anterior spiracle; **b.** lateral view of cephalopharyngeal skeleton; **c.** ventral view of cephalopharyngeal skeleton (from Miller and Foote 1976).

Figure 37.290a,b. Lauxaniidae. *Minettia lyraformis* Shewell. **a.** anterior spiracle; **b.** ventral view of posterior segments showing elongate lobes (from Miller and Foote 1976).

Description: Mature larvae 3–7 mm; muscidiform, frequently somewhat depressed dorsoventrally; integument usually bearing numerous spinules and short hairs, these commonly arranged in rows or distinctive patterns on more anterior segments and more scattered on posterior segments; creeping welts present ventrally; integument white to slightly darkened due to surface pubescence.

Head: Cephalic segment somewhat retractile, bearing well developed sensory organs anteriorly and anteroventrally; antennae appearing 2-segmented and somewhat elongate; facial mask with few to several rows of sharp teeth-like structures anterior and lateral to preoral cavity, and numerous oral grooves and ridges lateral to opening; some species with postoral spine band on ventral surface of segment.

Cephalopharyngeal skeleton usually deeply pigmented, tentoropharyngeal and hypopharyngeal sclerites separate; dorsal cornua slender, without windows apically but connected anteriorly by fenestrated dorsal bridge; ventral cornua with distinct dorsobasal lobe and commonly with variously shaped windows apically; hypopharyngeal sclerite H-shaped, parastomal bars slender and not connected apically; epipharyngeal sclerite present, variously shaped but commonly broad with posterior rami and usually fenestrated; epipharyngeal sclerite and parastomal bars not fused; labial sclerites commonly V-shaped or occurring as separate rods, lying between anterior arms of hypopharyngeal sclerite. Mandibles deeply pigmented, hook part usually strongly decurved and with or without accessory teeth, basal part subrectangular and with or without small window; dental sclerites appearing triangular in lateral view.

Thorax: Segments commonly well covered with numerous spinules and hairs arranged in rows; ventral creeping welt on metathorax; anterior spiracles arising posterolaterally on prothorax, highly variable in shape, but commonly stalked with 6–20 short or elongate papillae along distal margin which may be in single, double, or several rows.

Abdomen: Segments all similar except for caudal one, usually covered with numerous spinules and hairs that become more irregularly distributed on more posterior segments; all segments with ventral creeping welts; perianal pad bilobed, commonly subtended laterally by pair of short to elongate, variously shaped fleshly lobes and by spinule patch posteriorly.

Posterior spiracular disc usually subcircular but sometimes somewhat transverse, bearing 2–3 pairs of tubercles, one pair dorsolateral, another lateral, and third pair ventrolateral in position; posterior spiracles usually on short tubular bases, with distinct peritremes and three oval spiracular openings that usually radiate out from ecdysial scar; four short and branched spiracular hairs.

Comments: This is a relatively small family in the Nearctic Region, with some 150 species and 20 genera being recorded from America north of Mexico. It is considerably more diverse in the tropics. No species is considered to be economically significant.

Selected Bibliography

Becker 1895.
Bouché 1847.
Edwards 1925.
Ferrar 1987.
Griffiths 1972.
Hendel 1908.
Hennig 1952, 1958.
Malloch 1927.
Malloch and McAtee 1924.
Marchal 1897.
McAtee 1927.
McDonald et al. 1979.
Meijere 1909, 1910.
Melander 1913.
Miller, R. M. 1977a, 1977b, 1977c.
Miller and Foote 1975, 1976.
Oldroyd 1964.
Shewell 1938, 1987a.
Vimmer 1925.

CHAMAEMYIIDAE (LAUXANIOIDEA)

B. A. Foote, *Kent State University*

The Chamaemyiids

Figures 37.291–37.293

Relationships and Diagnosis: Considerable disagreement exists as to the proper placement of the Chamaemyiidae. Commonly it was included in or placed near the Agromyzidae by earlier workers. Malloch (1933) and others, however, felt that the family was more closely related to the Lauxaniidae. Hennig (1958), following Hendel (1916), included the Chamaemyiidae in the Lauxanioidea, a placement accepted by McAlpine et al. (1981) and Griffiths (1972). This superfamily includes only the Lauxaniidae and Chamaemyiidae in the Nearctic fauna. McAlpine (1987h) recently reviewed our knowledge.

Most larvae are characterized by the heavily spinulose and papillose nature of the integument, conspicuously elongate respiratory tubes, and the three papillae bearing the spiracular openings apically (fig. 37.293). Additionally, the tentoropharyngeal and hypopharyngeal sclerites are fused and there are no parastomal bars (fig. 37.291b).

Biology and Ecology: These small, usually largely gray, and obscure flies have larvae that prey on scale insects, mealy bugs, and aphids. The habits of *Chamaemyia* and *Leucopis* have been investigated fairly extensively, but relatively little is known about the larval feeding habits and natural history of the other Nearctic genera. Larvae of a few species inhabit galls formed by coccids or aphids. The sole Nearctic species of *Paraleucopis* has been reared from bird nests, but its larval food remains unknown, and McAlpine (1987h) believes it is more closely allied to the Asteiidae.

Description: Mature larvae 2–5 mm, muscidiform, tapering anteriorly and rather blunt posteriorly; integument covered with abundant spinules and hairs, each segment usually bearing one or two rows of slender, peg-like papillae; whitish to yellowish.

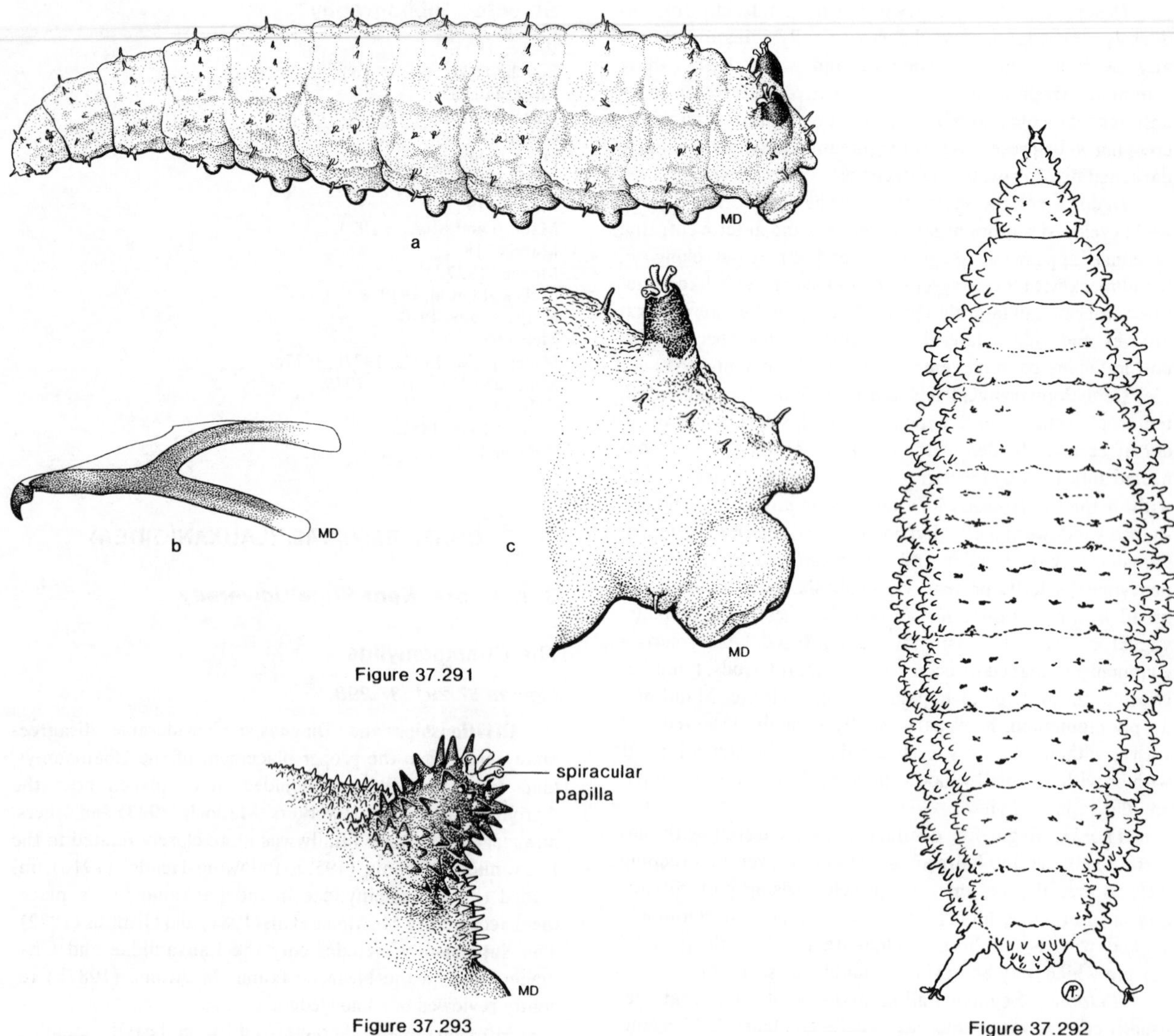

Figure 37.291a-c. Chamaemyiidae. *Leucopis simplex* Loew. **a.** lateral; **b.** lateral view of cephalopharyngeal skeleton; **c.** details of terminal abdominal segment (from Manual of Nearctic Diptera, Vol. 2).

Figure 37.292. Chamaemyiidae. Dorsal of *Leucopis* sp., a predator on aphids (from Peterson 1951).

Figure 37.293. Chamaemyiidae. Posterior spiracle of *Leucopis* sp. (from Manual of Nearctic Diptera, Vol. 2).

Head: Cephalic segment retractile, bearing sensory organs anteriorly and anteroventrally; antennae elongate and appearing 2-segmented; facial mask apparently without oral grooves and ridges around preoral opening; typically, a narrow sclerotized band occurs on dorsal and lateral surfaces of segment.

Cephalopharyngeal skeleton strongly pigmented, rather narrow; tentoropharyngeal and hypopharyngeal sclerites fused; hypopharyngeal sclerite with dorsal wings; no parastomal bars, epipharyngeal, or labial sclerites; mandibles with heavy, somewhat rectangular basal part and slender, decurved hook that lacks accessory teeth.

Thorax: Anterior spiracles arising posterolaterally on prothorax; spiracles with short basal piece and distinctly expanded distal part bearing 3–6 elongate, finger-like papillae along outer margin; integument with few to numerous spinules; no noticeable ventral creeping welts.

Abdomen: Segments all similar, each subdivided by two secondary folds so that segment seemingly consists of three rings; integument usually covered with abundant spinules and rows of papilla-like projections (integument nearly bare in *Chamaemyia*); ventral surface with well developed creeping welts that in some species resemble short paired prolegs; perianal pad circular, bilobed, frequently bordered by one to several rows of spinules (in some species of *Leucopis* small clusters of 2–4 spinules may occur).

Posterior spiracles borne on apices of elongate, widely separated, diverging respiratory tubes that usually bear many spinules and papillae; spiracular opening at apex of three slender, diverging papillae.

Comments: This is a small family that badly needs taxonomic attention, as a large number of undescribed species are known. It is best developed in the Holarctic Region. Seven genera and some 40 species are recorded for America north of Mexico. Because of the predatory habits of the larvae, a few species are important natural enemies of economically important Homoptera. Thus, Brown and Clark (1956a, 1956b, 1957) reported that certain chamaemyiid species, particularly *Neoleucopis obscura* (Haliday), were fairly significant predators of the balsam wooly aphid, *Adelges picea* (Ratzburg), in eastern Canada.

Selected Bibliography

Babaeo and Tanasiychuk 1971.
Barber 1985.
Blanchard 1964.
Brown and Clark 1956a, 1956b, 1957.
Clark and Brown 1957, 1962.
Cogan 1978.
Cumming 1959.
Delucchi and Pschorn-Walcher 1954.
Ferrar 1987.
Hendel 1916.
Hennig 1952.
Keilin 1915.
Malloch 1933, 1940.
McAlpine, J. F. 1960a, 1963, 1971, 1978, 1987h.
Mitchell and Maksymov 1977.
Mitchell and Wright 1967.
Perris 1870.
Sabrosky 1957a.
Singh Pruthi and Bhatia 1938.
Sluss and Foote 1970, 1971, 1973.
Smith, R. L. 1981.
Stevenson 1967.
Steyskal 1971a, 1972b.
Tanasiychuk 1974.
Tanasiychuk et al. 1976.
Tracewski 1983.
Tragardh 1931.
Wilson 1938.

HELEOMYZIDAE (SPHAEROCEROIDEA)

B. A. Foote, *Kent State University*

The Heleomyzids

Figures 37.294a–f

Relationships and Diagnosis: Placed with Trixoscelididae, Chyromyidae, Rhinotoridae, and Sphaeroceridae in the Sphaeroceroidea by McAlpine et al. (1981), the correct phylogenetic placement of the Heleomyzidae puzzled earlier generations of dipterists. Hennig (1952) felt that the family was closely related to the Dryomyzidae, Piophilidae, and Pallopteridae. In 1958, however, he was less sure of its correct placement and included it in his group 7—families of uncertain relationships. Griffiths (1972) included it in the prefamily Anthomyzoinea along with such families as the Anthomyzidae, Trixoscelididae and Sphaeroceridae. Gill and Peterson (1987) have summarized our knowledge.

It is impossible at the present time to distinguish larvae of Heleomyzidae from those of other families (e.g. Lonchaeidae) having saprophagous habits. Fairly distinctive is the presence of variously sized and distributed tubercles or papillae on the posterior surface of the spiracular disc.

Biology and Ecology: Larvae of the subfamily Heleomyzinae apparently are saprophagous and have been found in a variety of decaying organic matter. They have been reared from bird nests, mammal burrows, and bat caves where they feed on fecal material and rotting debris. Other species have been bred from small carcasses and accumulations of rotting plant material. In contrast, larvae of the Suillinae are usually found in mushrooms. Adults are found near the larval habitats.

Description: Mature larvae 3–13 mm, muscidiform, rather long and slender, broader toward posterior end; creeping welts on ventral surface; integument nearly bare, with few scattered spinules except for those in creeping welts; white.

Head: Cephalic segment retractile, bearing well-developed sensory organs anteriorly and anteroventrally; antennae elongate, appearing 2-segmented; facial mask with numerous, occasionally branched oral grooves and ridges leading into preoral cavity.

Cephalopharyngeal skeleton mostly deeply pigmented; tentoropharyngeal and hypopharyngeal sclerites separate; dorsal cornua narrower than ventral cornua, commonly with apical window and connected anteriorly by fenestrated dorsal bridge; ventral cornua may be less pigmented, usually with dorsobasal lobe and commonly with apical window; pharyngeal filter present; hypopharyngeal sclerite H-shaped, parastomal bars elongate but not connected apically; no epipharyngeal sclerite; labial sclerite somewhat V-shaped, between anterior arms of hypopharyngeal sclerite. Mandibles with slender, decurved hook that possesses or lacks accessory teeth, basal part subrectangular and usually with 1 or 2 small windows; lateral dental sclerites appearing triangular in lateral view, the more median dental sclerite elongate and between basal parts of mandibles.

Thorax: Anterior spiracles arising posterolaterally on prothorax, variously shaped, commonly broad with short base and 5–18 distal papillae, or more elongate and distally branched with 10–15 finger-like papillae scattered around margin.

Abdomen: All segments except caudal one quite similar, each with ventral creeping welt; some species with spinules on dorsal and lateral surfaces of more distal segments; perianal pad bilobed, usually subtended laterally by fleshy lobes and posteriorly by spinule patch.

Posterior spiracular disc subcircular with posterior spiracles arising separately on upper half or third of disc; posterior surface of disc usually with short tubercles or papillae; spiracles typically borne on short cylindrical bases; each spiracular plate with pigmented peritreme (in some species peritreme is less conspicuous as its margins are indented), three oval spiracular openings radiating out from ecdysial scar, with openings nearly at right angles to each other; in some species four weakly to profusely branched spiracular hairs.

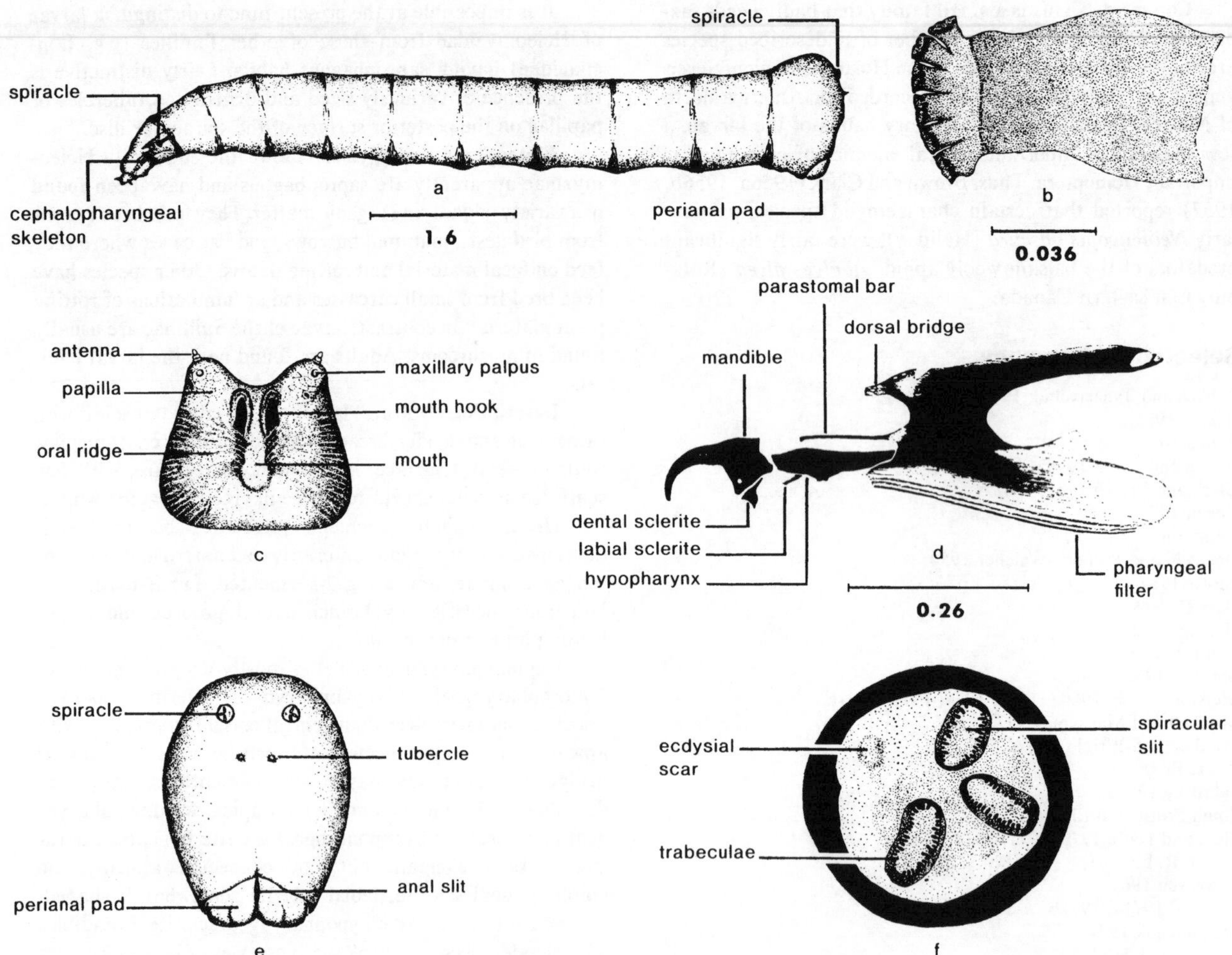

Figure 37.294a-f. Heleomyzidae. *Pseudoleria crassata* Garrett. **a.** lateral; **b.** anterior spiracle; **c.** ventral view of anterior end; **d.** lateral view of cephalopharyngeal skeleton; **e.** posterior view of caudal segment; **f.** posterior spiracular plate (from Garnett and Foote 1967. Reprinted with permission of the Entomological Society of America).

Comments: Containing fewer than 300 species, the family is best developed in the Holarctic Region. Some 23 genera and 113 species are recorded from America north of Mexico. None is considered to be of economic significance.

Selected Bibliography

Banta 1907.
Barnes, J. K. 1980a.
Buxton 1960.
Edwards 1925.
Ferrar 1987.
Garnett and Foote 1966.
Gill 1962.
Gill and Peterson 1987.
Griffiths 1972.
Hennig 1952, 1958.
Mathis 1973.
McAlpine, D. K. and Kent, 1982.
Perris 1870.
Séguy 1950.
Vimmer 1911, 1925.

TRIXOSCELIDIDAE (SPHAEROCEROIDEA)

B. A. Foote, *Kent State University*

The Trixoscelidids

Relationships and Diagnosis: Long considered to be an entity within the Heleomyzidae or the Chyromyidae, the trixoscelidids have relatively recently acquired family status. Hennig (1958) placed them next to the Heleomyzidae in his group 7, families whose phylogenetic position was unclear. Griffiths (1972) included them in his prefamily Anthomyzoinea along with such families as the Heleomyzidae, Asteiidae, Opomyzidae, and Sphaeroceridae. Recently, McAlpine et al. (1981) included them in the array of five Nearctic families comprising the Sphaeroceroidea, being closely related to the Heleomyzidae, Chyromyidae, Rhinotoridae, and Sphaeroceridae. Teskey (1987b) recently summarized our knowledge.

Biology and Ecology: The natural history and larval feeding habits are practically unknown. Adults are rarely collected, but have been taken most commonly in arid, grassy habitats. Adults of *Neossos marylandicus* Malloch have been reared from puparia occurring in bird nests, but the larval food was not determined.

Description: No larvae have been described.

Comments: This is a small family largely restricted to the Nearctic Region. Three genera and about 30 species are recorded for America north of Mexico, with *Trixoscelis* containing most of the species. Many Nearctic species are southern.

Selected Bibliography

Ferrar 1987.
Griffiths 1972.
Hennig 1958.
Melander 1952.
Teskey 1987b.

CHYROMYIDAE (SPHAEROCEROIDEA)

B. A. Foote, *Kent State University*

The Chyromyids

Relationships and Diagnosis: Earlier workers, following Hendel (1916), tended to place the Chyromyidae near the Anthomyzidae and Opomyzidae. However, other investigators considered the family as being close to the Clusiidae or Heleomyzidae. Recently, McAlpine et al. (1981) have placed it near the Trixoscelididae in the Sphaeroceroidea. McAlpine (1987i) has summarized our knowledge.

Biology and Ecology: This is another of the numerous small and obscure families of acalyptrate Diptera whose natural history is practically unknown. Adults of a few species have been reared from bird nests and mammal burrows or rotting wood. Adults have been collected rather frequently on windows or swept from vegetation.

Description: No larval stages have been described.

Comments: Only three genera and about a dozen species are recorded from America north of Mexico.

Selected Bibliography

Ferrar 1987.
Hendel 1916.
McAlpine, J. F. 1987i.
Wheeler 1961.

RHINOTORIDAE (SPHAEROCEROIDEA)

B. A. Foote, *Kent State University*

The Rhinotorids

Relationships and Diagnosis: Earlier included in the Rhopalomeridae, the rhinotorids are now given family rank and are considered to be rather closely related to the Chyromyidae and Sphaeroceridae. McAlpine et al. (1981) included the family in the Sphaeroceroidea. McAlpine (1987j) summarized our knowledge.

Biology and Ecology: Adults of the sole Nearctic species were collected in banana bait traps. A few of the tropical species have been collected from tree trunks. No life history data have been published for any species of the family.

Description: No larvae have been described.

Comments: Largely confined to the New World Tropics, a single species, *Neorhinotora diversa* (Giglio-Tos), has been recorded from Mexico, Arizona and New Mexico.

Selected Bibliography

Ferrar 1987.
McAlpine, J. F. 1987j.
Wheeler 1954.

SPHAEROCERIDAE (SPHAEROCEROIDEA)

B. A. Foote, *Kent State University*

The Sphaerocerids, Small Dung Flies

Figures 37.295, 37.296

Relationships and Diagnosis: McAlpine et al. (1981) placed the Sphaeroceridae in the Sphaeroceroidea along with the Heleomyzidae, Trixoscelididae, Chyromyidae, and Rhinotoridae. In contrast, Hennig (1958) included it in the Milichioidea, and Griffiths (1972) placed it in his prefamily Anthomyzoinea, suggesting that it was close to the families Opomyzidae, Chyromyidae, and Aulacigastridae. Marshall and Richards (1987) recently summarized our knowledge.

It is seemingly impossible to separate all larvae of this family from those belonging to other families of saprophagous Muscomorpha. Fairly distinctive features in many species are the elongate nature of the anterior spiracles (figs. 37.295, 37.296a) (also found in many Sepsidae), the comblike structures anterior to the preoral cavity (also found in certain Sepsidae and many Ephydridae), and the very slender dorsal cornua.

Biology and Ecology: The small, dark-bodied adults are most commonly encountered in moist to wet habitats that contain accumulations of decaying organic matter. Larvae are saprophagous and can be quite abundant in such microhabitats as mats of decaying seaweed, piles of rotting plant materials, fecal deposits, and sewage enriched mud. Adults of a few species hang onto the bodies of dung beetles and are thought to oviposit on the balls of dung collected by *Scarabaeus* and other genera. Several species have been reared from mammal burrows, ant mounds, and rotting fungi. Many species occurring in such wind-swept habitats as mountain tops and islands are apterous or brachypterous.

Description: Mature larvae 4–7 mm, muscidiform, tapering anteriorly and broader posteriorly, caudal segment commonly truncate or bluntly rounded, but in some species distal part elongated to form short breathing tube; integument mostly bare, ventral creeping welts present; white.

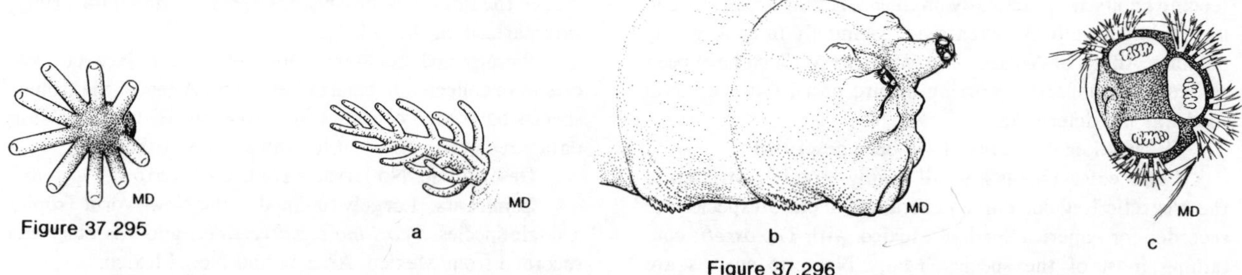

Figure 37.295. Sphaeroceridae. Anterior spiracle of *Thoracochaeta zosterae* (Haliday). Larvae feed as scavengers within stranded kelp and seaweed (from Manual of Nearctic Diptera, Vol. 1).

Figure 37.296a-c. Sphaeroceridae. *Leptocera* sp. **a.** anterior spiracle; **b.** oblique view of posterior segments; **c.** posterior spiracular plate (from Manual of Nearctic Diptera, Vol. 1).

Head: Cephalic segment retractile, bearing well developed sensory organs anteriorly and anteroventrally; antennae appearing 2-segmented; facial mask with many relatively unbranched oral grooves and ridges leading into preoral cavity, mask anterior to cavity usually with series of comb-like structures composed of backward pointing spinules.

Cephalopharyngeal skeleton usually deeply pigmented, tentoropharyngeal and hypopharyngeal sclerites separate; dorsal cornua connected anteriorly by fenestrated dorsal bridge and usually narrower than ventral cornua, the latter commonly with apical window; pharyngeal filter present; hypopharyngeal sclerite H-shaped; parastomal bars present and elongate but not connected anteriorly; apparently no epipharyngeal or labial sclerites. Mandibles with weakly decurved hook that lacks accessory teeth, and broader basal part that may have small window.

Thorax: Ventral surface of metathorax with creeping welt near posterior border; anterior spiracles arising posterolaterally on prothorax; spiracles commonly somewhat elongate and with 8–14 finger-like marginal papillae; in some species spiracles broad and with 3–12 short papillae along distal margin.

Abdomen: Segments all very similar except for caudal one, each with ventral creeping welt near segmental border; caudal segment usually bluntly rounded with posterior spiracles arising separately on short tubular structures, but in some species distal part of segment elongated to form short breathing tube; perianal pad bilobed, apparently lacking spinule patch posteriorly.

Posterior spiracular disc usually subcircular and without marginal tubercles; posterior spiracles borne on short tubelike structures, peritremes pigmented; three short-oval spiracular openings, commonly radiating out from ecdysial scar; in some species openings subparallel to each other, and in a few species they parallel the peritreme; 3–4 dichotomously branched spiracular hairs; in a few species the hairs reduced to short, thorn-like structures.

Comments: The family is found worldwide and contains over 1000 species. Some 20 genera and about 125 species are described from America north of Mexico, with many species undescribed. Although the larvae have roles in the processing of organic matter and in food chains, no species is considered to have economic significance. A few species are minor pests in tanks of sewage plants.

Selected Bibliography

Backlund 1945b.
Banta 1907.
Ferrar 1987.
Fredeen and Glen 1970.
Freeden and Taylor 1964.
Goddard 1938.
Griffiths 1972.
Guibe 1939.
Hackman 1969.
Hafez 1939b.
Hammer 1941.
Hennig 1952, 1958.
Kompfner 1974.
Laurence 1955.
Marshall, S. A. 1982a, 1982b.
Marshall and Richards 1987.
Okeley 1974.
Richards 1930, 1968.
Vogler 1900b.

CURTONOTIDAE (EPHYDROIDEA)

B. A. Foote, *Kent State University*

The Curtonotids

Relationships and Diagnosis: The family Curtonotidae was placed in the Drosophiloidea by Hennig (1958), a placement subsequently accepted by Griffiths (1972). McAlpine et al. (1981) modified the drosophiloid concept slightly and now refer to it as the Ephydroidea. McAlpine (1987k) reviewed our knowledge.

Biology and Ecology: The larval feeding habits and microhabitats are largely unknown, although Greathead (1958) found larvae of an African species feeding as scavengers in egg pods of a desert grasshopper belonging to the genus *Schistocerca.* Adults of the single North American species are rare, but are most commonly encountered while sweeping vegetation in moist habitats.

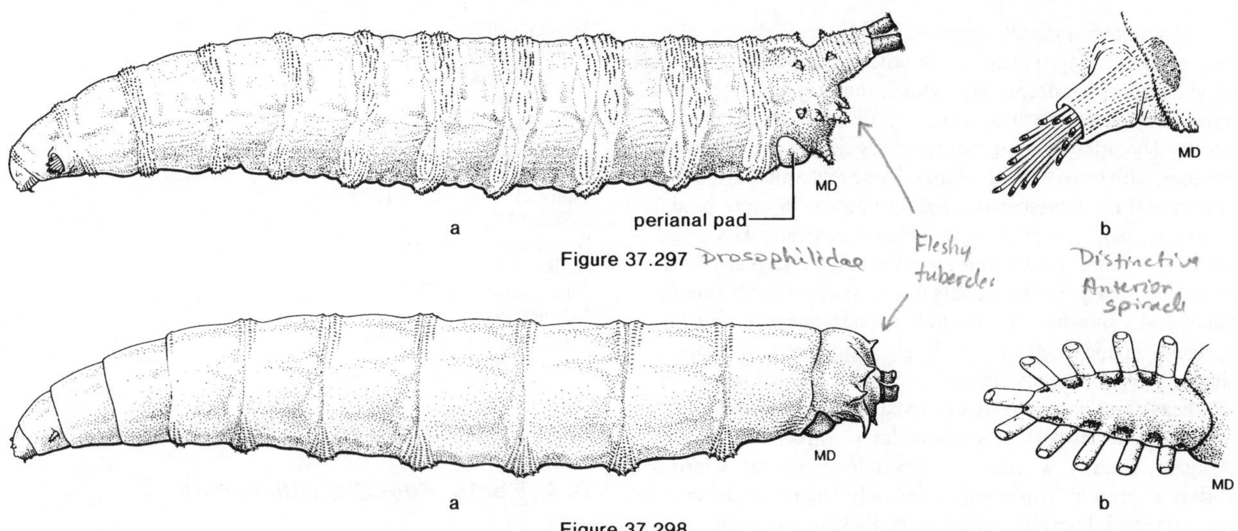

Figure 37.297a,b. Drosophilidae. *Drosophila melanogaster* Meigen. **a.** lateral; **b.** left anterior spiracle. The larvae feed in fermenting and decaying fruits and vegetables (from Manual of Nearctic Diptera, Vol. 2).

Figure 37.298a,b. Drosophilidae. *Chymomyza* sp. **a.** lateral; **b.** anterior spiracle (from Manual of Nearctic Diptera, Vol. 2).

Description: No larvae of Nearctic species have been described.

Comments: Best developed in tropical Africa and America, only one species, *Curtonotum helvum* (Loew), has been recorded from eastern North America.

Selected Bibliography

Ferrar 1987.
Greathead 1958.
Griffiths 1972.
Hennig 1952, 1958.
McAlpine, J. F. 1987k.

DROSOPHILIDAE (EPHYDROIDEA)

B. A. Foote, *Kent State University*

Pomace Flies, Small Fruit Flies, Vinegar Flies

Figures 37.297, 37.298

Relationships and Diagnosis: At the present time the family is included in the Ephydroidea along with such families as Diastatidae, Ephydridae, Chloropidae, Tethinidae, and Canacidae (McAlpine et al. 1981). It has long been considered to be closely related to the Ephydridae (Hennig 1952, 1958), but placing it relatively close to the Chloropidae is a relatively recent decision that requires further attention. Wheeler (1987) recently summarized our knowledge.

The varying morphology found in the larval stages of different species makes it difficult, if not impossible, to characterize the family as a whole. Fairly distinctive for many species is the structure of the anterior spiracle. This consists of an elongate, tubular basal piece that terminates in a varying number of filamentous projections (fig. 37.297b). Also useful for most species of the family is the tapered nature of the caudal segment and the occurrence of fleshy tubercles on the posterior spiracular disc (fig. 37.297a).

Biology and Ecology: The vast majority of the Nearctic species are woodland-inhabiting flies whose larvae are saprophagous, feeding on a great variety of decaying plant materials that are undergoing fermentation. Probably the primary energy source for the larvae is the yeast flora that is so abundant in the larval microhabitats. Larvae have been found in rotting fruits, cider presses, wine vats, vinegar factories, decaying leaves, slime fluxes on trees, fermenting cambial tissue under bark of fallen trees, rot pockets in cactus, decaying flowers, mushrooms and bracket fungi, all of which are very attractive to the abundant adults. There have been several radiations into more specialized habits. Thus larvae of some species are associated with land crabs, others prey on or parasitize mealybugs and other small Homoptera, and still other species prey on spider eggs. Larvae of the scaptomyzine drosophilids are largely leaf miners.

Description: Mature larvae 3–9 mm, muscidiform in most species, several species dorsoventrally depressed, and a few species with ventral prolegs; others may have dorsal and lateral fleshy projections; body tapering anteriorly and somewhat truncate to pointed posteriorly; integument bare to somewhat spinulose, usually with ventral creeping welts of backwardly pointing spinules; with or without encircling spinule bands on body segments; white to yellowish.

Head: Cephalic segment retractile, bearing sensory organs anteriorly and anteroventrally; antennae short to elongate, apparently 2-segmented; facial mask of saprophagous species with numerous branching oral grooves and ridges leading into preoral cavity, mask of predacious and more parasitic species usually with few or no grooves or ridges.

Cephalopharyngeal skeleton highly variable, in most species resembling that found in other saprophagous larvae of Muscomorpha; deeply pigmented, tentoropharyngeal and hypopharyngeal sclerites separate; dorsal cornua usually narrow (broader in some parasitic species); ventral cornua broader, with or without dorsobasal lobe and apical windows; pharyngeal filter present (vestigial to absent in more predacious species); hypopharyngeal sclerite basically H-shaped, parastomal bars slender and rod-like; epipharyngeal sclerite absent or weakly developed; labial sclerites commonly present below and somewhat anterior to hypopharyngeal sclerite and variously shaped. Mandibles deeply pigmented, variable in shape, hook commonly slender and sickle shaped, occasionally heavy and only slightly decurved, with or without accessory teeth, basal part subrectangular to rather narrow and elongate, usually without small windows; dental sclerites absent or present, if present, commonly triangular in shape when viewed laterally; apparently lacking accessory oral sclerites.

Thorax: Anterior spiracles arising posterolaterally on prothorax; spiracles in *Drosophila* commonly consisting of basal stalk terminating in several to many long, retractile filamentous processes; in other genera anterior spiracles elongate with marginal papillae (*Stegana*) or fan-shaped with apical papillae (*Leucophenga*).

Abdomen: Segments of many species with ventral creeping welts and encircling bands of spinules; perianal pad bilobed, variously shaped, subtended by fleshy lobes in a few species.

Posterior spiracular disc usually bearing marginal tubercles that vary in size, location, and number; posterior spiracles borne on sclerotized tube-like projections that lie close to each other (distinctly separated in species of *Leucophenga*); spiracular plates usually circular to oval with well developed peritremes; spiracles sharply pointed in some leaf-mining species of *Scaptomyza*; three oval and diverging spiracular openings; spiracular hairs present or absent.

Comments: This is a sizeable family of rather small and inconspicuous flies that are most abundant in moist shaded habitats. Over 2500 species have been described. About 180 species and 17 genera are known to occur in America north of Mexico. None of the species is considered to represent a major economic problem, although several of the cosmopolitan species such as *S. busckii* Coquillett, *D. melanogaster* Meigen, and *D. simulans* Sturtevant can be annoying pests in homes and in fruit processing plants.

Selected Bibliography

Ashburner and Wright 1978.
Ashburner et al. 1981, 1982, 1983a, 1983b.
Carson 1967.
Carson and Stalker 1951.
Chittenden 1902.
Duda 1935.
Fellows and Heed 1972.
Ferrar 1987.
Fogelman 1981.
Frost 1924.
Hackman 1955, 1959.
Heed 1957, 1968, 1971.
Hennig 1952, 1958.
Jaenicke 1978.
Kearney 1983.
Lindsay 1958.
Okada 1963.
Patterson and Stone 1952.
Pipkin, et al. 1966.
Shorrocks 1977.
Shorrocks and Woods 1973.
Spieth 1952, 1974.
Spieth and Heed 1972.
Stalker 1945.
Throckmorton 1975.
Wagner 1944.
Wheeler, M. R. 1957, 1987.

DIASTATIDAE (EPHYDROIDEA)

B. A. Foote, *Kent State University*

The Diastatids

Relationships and Diagnosis: Generally considered to be closely related to the Drosophilidae and Ephydridae (Stone, et al. 1965), it has been included in the Ephydroidea by McAlpine et al. (1981). McAlpine (1987 l) reviewed our knowledge.

Biology and Ecology: Very little is known about the life history or larval feeding habits of any species except that adults are usually collected in moist habitats. According to Hennig (1952), larvae of a European species of *Campichoeta* probably live in leaf litter. Nothing is known of the larval feeding habits of the six Nearctic species.

Description: No larval descriptions are available.

Comments: This is a small family of fewer than 30 species worldwide. Two genera, *Diastata* and *Campichoeta,* and six species are recorded for America north of Mexico.

Selected Bibliography

Duda 1934.
Ferrar 1987.
Hennig 1952.
McAlpine, J. F. 1962, 1987 l.
Melander 1913.
Séguy 1950.
Stone et al. 1965.

CAMILLIDAE (EPHYDROIDEA)

B. A. Foote, *Kent State University*

The Camillids

Relationships and Diagnosis: Commonly included in Drosophilidae by many earlier workers, it is considered to be relatively close to the Drosophilidae and Ephydridae of the Ephydroidea (McAlpine et al. 1981). McAlpine (1987m) has reviewed our knowledge.

Biology and Ecology: Nothing is known of the larval habits of the single North American species, but adults of some of the European species have been reared from nests of rodents.

Description: No larvae have been described.

Comments: The family consists of two genera and 13 species, all recorded from the Northern Hemisphere. A single species, *Camilla glabra* (Fallén) is known from eastern Canada. No species has any known economic significance.

Selected Bibliography

Duda 1934.
Ferrar 1987.
McAlpine, J. F. 1960b, 1987m.

EPHYDRIDAE (EPHYDROIDEA)

B. A. Foote, *Kent State University*

Shore Flies, Brine Flies

Figures 37.299–37.305

Relationships and Diagnosis: Long considered to be closely related to the Drosophilidae, the Ephydridae forms the central taxon of the Ephydroidea, a complex of nine Nearctic families that also includes the Diastatidae, Curtonotidae, Drosophilidae, Camillidae, Chloropidae, Cryptochetidae, Tethinidae, and Canacidae (McAlpine, et al. 1981). Hennig (1958) had a somewhat more restricted view of the Drosophiloidea (=Ephydroidea), suggesting that it consisted only of the families Curtonotidae, Drosophilidae, Diastatidae, Camillidae, and Ephydridae. He felt that the sister group of the Ephydridae was the Camillidae. Griffiths (1972) basically agreed with the arrangement presented by McAlpine et al. (1981), except that he excluded the Tethinidae and Canacidae from the superfamily containing the Ephydridae. Wirth, Mathis and Vockeroth (1987) recently reviewed our knowledge.

The tremendous variety of larval feeding habits seemingly is correlated with the broad spectrum of larval structure. This morphological diversity has made it very difficult to characterize ephydrid larvae as a unit. However, one fairly useful biological characteristic is the occurrence of larvae in wetland habitats. The elongate, retractile respiratory tube (fig. 37.304a) in many species of the Ephydrinae and Hydrelliinae is also helpful. Unfortunately, there is an array of other species, especially in the Psilopinae and Parydrinae, that lack such a tube and have sessile posterior spiracles or a pair of short, fleshy structures that bear the spiracles (fig. 37.299a). Another fairly distinctive feature seen in many microphagous larvae is the rows of comb-like structures that occur on the facial mask (fig. 37.300c). Although ventral prolegs occur in most species of Ephydrinae, such structures are absent in larvae of the remaining subfamilies. In many species, the integument bears a thin to heavy array of spinules, scales or hairs that may give a patterned appearance to the dorsum (fig. 37.300a). Finally, the labial and epipharyngeal sclerites are well developed in most species, as is a pharyngeal filter.

Biology and Ecology: Typically encountered in wetlands, the Ephydridae show wide latitude in their ability to adapt to a variety of habitats. Many species occur in freshwater marshes, but others are coastal or flourish in highly saline and alkaline inland wetlands. They are frequently encountered in thermal springs, and one species is regularly recorded from oil pools. Larvae of most species appear to be filter feeders, utilizing a vast array of microorganisms. However, in contrast to larvae of Drosophilidae, which largely consume yeasts and bacteria, the microphagous larvae of Ephydridae mainly utilize a variety of algae. Many species, especially those of the Ephydrinae, are generalists and ingest unicellular and filamentous algae of several different taxa. Other species are somewhat more oligophagous in their use, while still others are highly specialized and seemingly restricted to only a few algal taxa. For example, larvae of species of the tribe Hyadinini appear to be restricted to the consumption of blue-green algae (Cyanobacteria), whereas larvae of *Parydra* spp. preferentially feed on diatoms. Larvae of other species are not consumers of algae but can ingest and digest a variety of heterotrophic microorganisms that are associated with decaying organic matter occurring in wetland habitats. A good example of this mode of life is seen in the genus *Notiphila* whose larvae occur in organic-rich sediments and which have spiracular spines that enable them to penetrate the rootlets of hydrophytes to acquire oxygen. Still other species have larvae that are leaf miners (*Hydrellia, Psilopa*). Those of *Ochthera* prey on larvae of other insects, particularly Chironomidae. Another specialization is seen in the genus *Trimerina* whose larvae attack eggs of marsh-inhabiting spiders.

Description: Mature larvae 2–14 mm, commonly muscidiform, many species bearing posterior spiracles at apex of retractile breathing tube, and most species of Ephydrinae with ventral prolegs; integument bare, setose, spinulose, or bearing scales in different species; many species of Psilopiinae tuberculate, with short lateral and dorsal projections; white, dark gray to brown, or with distinct dorsal pattern.

Head: Cephalic segment usually retractile, bearing short to elongate, 2-segmented antennae anteriorly and circular sensory plate anteroventrally; facial mask highly diverse, many species of Ephydrinae, Parydrinae, and Notiphilinae with a series of comb-like structures anterior and lateral to preoral cavity; in many Psilopinae the mask bearing few to several unbranched or branching oral grooves and ridges around preoral opening; spine band may border opening posteriorly.

Cephalopharyngeal skeleton fairly diverse, usually deeply pigmented on all or most sclerites; tentoropharyngeal and hypopharyngeal sclerites separate in most species but fused in leaf mining larvae of *Hydrellia;* dorsal cornua usually narrower than ventral cornua and connected anteriorly by fenestrated dorsal bridge; ventral cornua usually broad and with or without dorsobasal lobe and apical windows; in many species of Ephydrinae a small lobe arises along anterior edge of tentoropharyngeal sclerite; pharyngeal filter usually present but absent in species of *Hydrellia* and *Ochthera*; hypopharyngeal sclerite H-shaped in ventral view, seemingly connected by 2 crossbars; parastomal bars usually present (absent in *Hydrellia*), occasionally fused to hypopharyngeal sclerite, free apically or fused with epipharyngeal sclerite; epipharyngeal sclerite usually present, lightly to deeply pigmented and variously shaped, usually somewhat oval or triangular with indented margins; labial sclerite usually present (absent in *Hydrellia*), narrow and bar-like, seemingly forming anterior bridge of hypopharyngeal sclerite. Mandibles usually deeply pigmented, hook commonly sickle-shaped and decurved, occasionally nearly straight, with or without accessory teeth, basal part commonly subrectangular, occasionally narrow and elongate, with or without window; dental sclerite commonly present, triangular to elongate in lateral view; accessory oral sclerites usually absent.

Thorax: At least last two segments commonly with ventral creeping welts and encircling bands of spinules at segmental borders; anterior spiracles arising posterolaterally on prothorax, commonly with tubular base and expanded distal part that bears few to many short or finger-like papillae; in some genera (e.g. *Lytogaster*) spiracles bifurcate, and in species of *Hydrellia* anterior spiracles are absent.

Abdomen: Segments commonly with creeping welts and encircling bands of spinules, dense to sparse setation, and/or spinules; in some species segments tuberculate or with short, fleshy projections; caudal segment elongated to form retractile respiratory tube bearing posterior spiracles at apex in some species; in others caudal segment short and spiracles sessile or elevated on short projections; perianal pad bilobed and variously shaped, frequently bordered posteriorly by spinule patch and in a few species subtended by fleshy projections.

Posterior spiracles variously located and of diverse shape; in many species of Ephydrinae and Hydrelliinae and some Parydrinae (*Parydra*) spiracles at apex of elongate, commonly bifurcate breathing tube; in other species spiracles separate and nearly sessile on upper half of spiracular disc; spiracles usually with well-developed peritreme, 3–4 slit-like, oval, or circular spiracular openings arranged variously on surface of spiracular plate (in *Hydrellia* and *Notiphila* spiracles elongate and spine-like); dichotomously branched spiracular hairs usually present around margin of spiracular plate between spiracular openings.

Comments: About half the size of the Drosophilidae, the Ephydridae contains approximately 1200 species worldwide. Sixty-eight genera and some 425 species are known to occur in America north of Mexico. No species is considered to be overly significant economically, although *Hydrellia griseola* (Fallén) is a leaf-mining pest of rice, and *Psilopa leucostoma* (Meigen) is occasionally damaging to the leaves of sugar beets. Recently, larvae and adults of *Scatella stagnalis* Fallén have invaded greenhouses in the southeastern states, and a species of *Ephydra* in California is considered to be a potential problem in the culture of *Spirulina,* a blue-green alga which is being used as a food supplement. In Utah, *Ephydra* spp., especially *E. cincerea* Jones and *E. riparia* Fallén (commonly called "brine flies"), can reach annoying numbers on beaches bordering Great Salt Lake. Undoubtedly, the frequently abundant larvae and adults are significant components of the diet of freshwater marsh and coastal animals.

Selected Bibliography

Aldrich 1912.
Becker 1926.
Berg 1950a.
Bergenstamm 1864.
Beyer 1939.
Blair and Foote 1984.
Bohart and Gressitt 1951.
Bokermann 1957.
Bolwig 1940.
Brauns 1939.
Brock et al. 1969.
Busacca and Foote 1978.
Cheng and Lewin 1974.
Collins 1975, 1977, 1980.
Connell and Scheiring 1982.
Crawford 1912.
Cresson 1942.
Dahl 1959.
Deonier 1965, 1971, 1972, 1974, 1979.
Deonier and Regensburg 1978.
Deonier et al. 1979.
Disney 1970.
Eastin and Foote 1971.
Ferrar 1987.
Foote, B. A. 1977, 1979, 1981a, 1981b, 1982, 1983, 1984.
Foote and Eastin 1974.
Frost 1924.
Gercke 1882.
Grigarick 1959.
Hennig 1943, 1952, 1958.
Houlihan 1969.
Johannsen 1935.
Kuenzel and Wiegert 1974.
Landis et al. 1967.
Malloch 1915.
Mathis 1979a, 1979b, 1982b.
Mathis and Simpson 1981.
Meijere 1940b, 1947.
Müller 1922.
Nemenz 1960.
Norrbom 1983.
Oldroyd 1964.
Oliveira 1958.
Packard 1871.
Ping 1921.
Runyan and Deonier 1979.
Scheiring and Foote 1973.
Scotland 1934, 1939.
Sen, P. 1931.
Simpson 1975, 1976a, 1976b, 1979.
Steinly and Runyan 1979.
Sturtevant and Wheeler 1954.
Teskey 1984.
Thorpe 1930.
Torelli 1922.
Tragardh 1903.
Tuxen 1936.
Varley 1937a.
Vimmer 1925.
Vogler 1900b.
Way et al. 1983.
Wheeler, A. G. 1982.
Williams 1938.
Wirth 1971a, 1971b.
Wirth et al. 1987.
Wirth and Stone 1956.
Zack 1983a, 1983b.
Zack and Foote 1978.

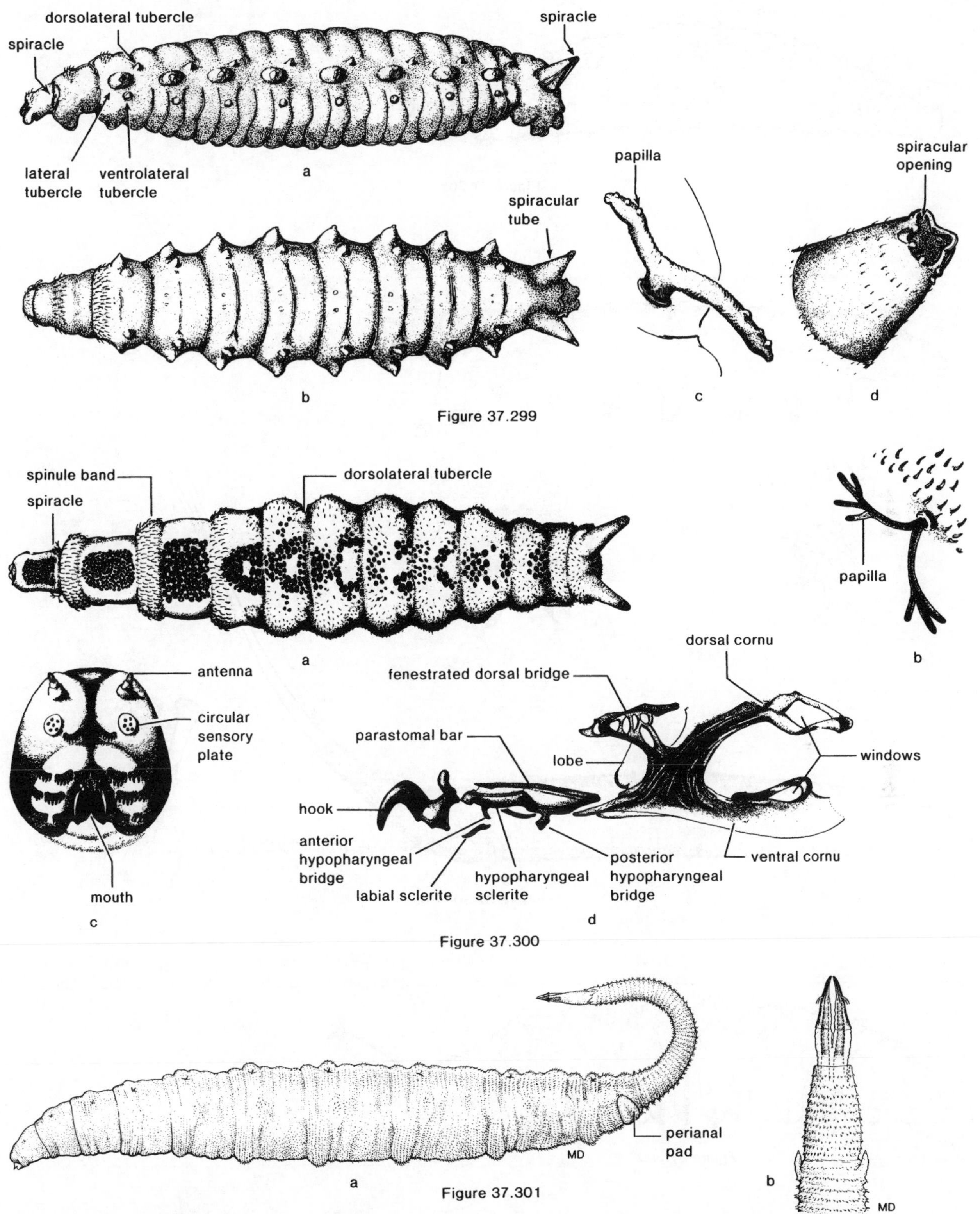

Figure 37.299a-d. Ephydridae. *Nostima approximata* Sturtevant and Wheeler. **a.** lateral; **b.** dorsal; **c.** anterior spiracle; **d.** posterior spiracle (from Foote 1983).

Figure 37.300a-d. Ephydridae. *Pelina truncatula* Loew. **a.** dorsal; **b.** anterior spiracle; **c.** ventral view of anterior end; **d.** lateral view of cephalopharyngeal skeleton (from Foote 1981b).

Figure 37.301a,b. Ephydridae. *Notiphila* sp. **a.** lateral; **b.** dorsal view of breathing tube showing sharpened spiracles that can be inserted into roots of aquatic plants (from Manual of Nearctic Diptera, Vol. 1).

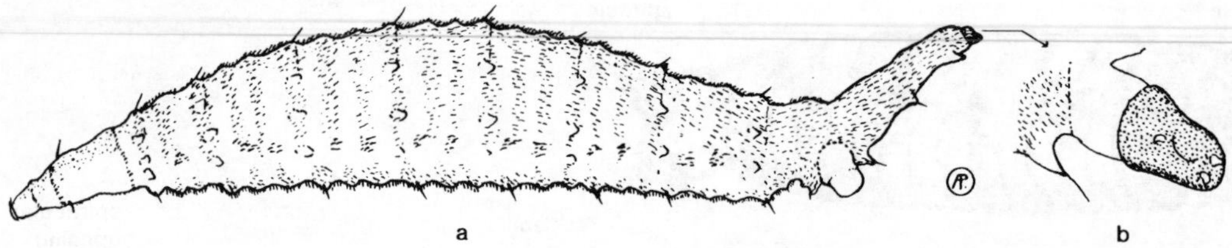

a

b

Figure 37.302

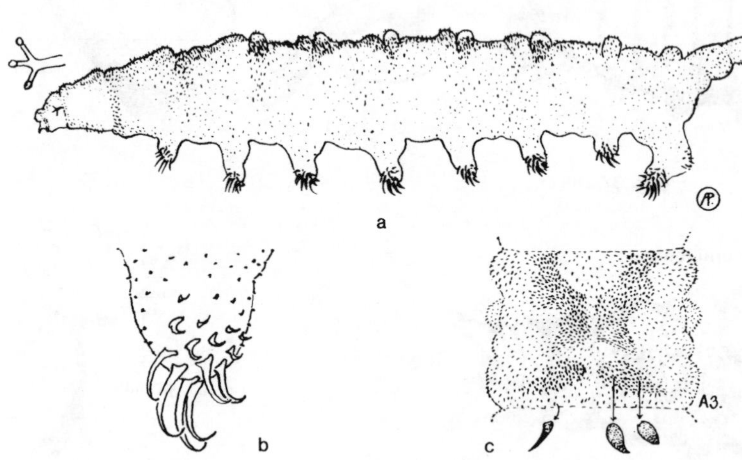

a

b

c

A3.

spiracular cap

Figure 37.303

perianal pad

a

b

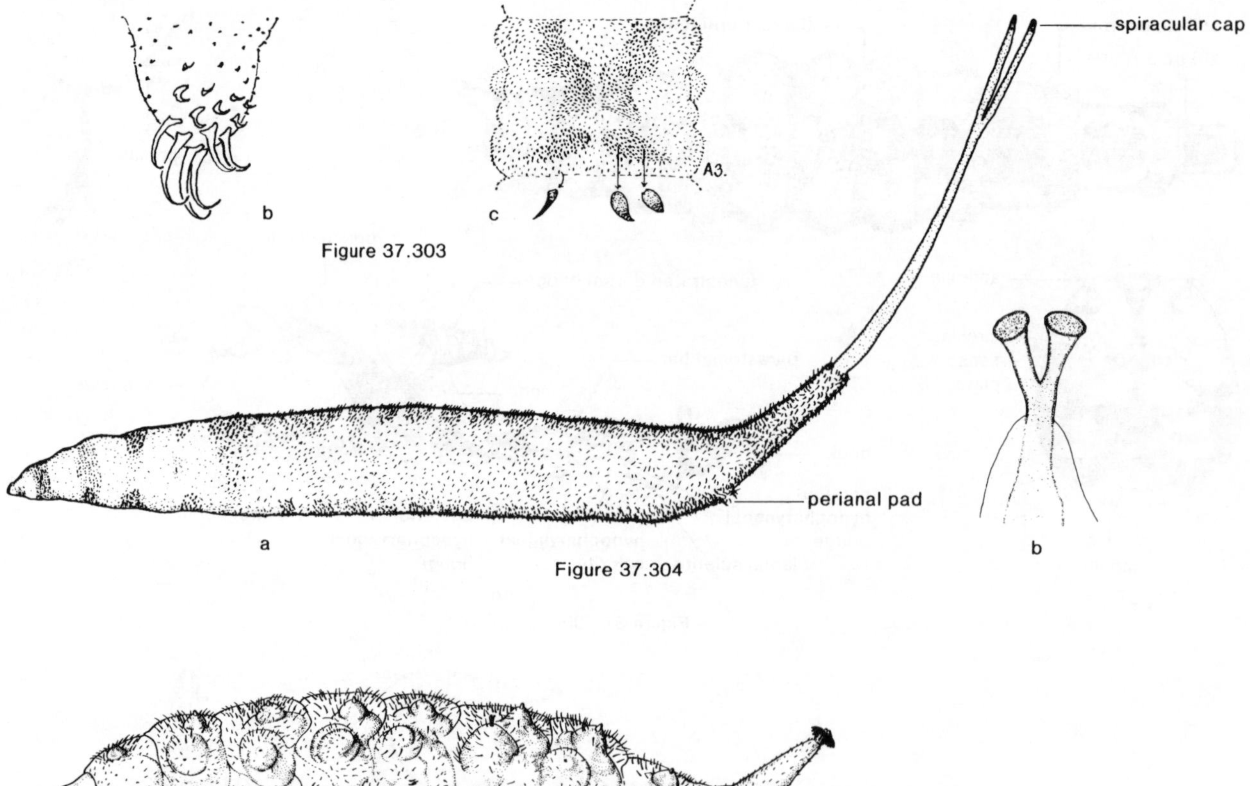

Figure 37.304

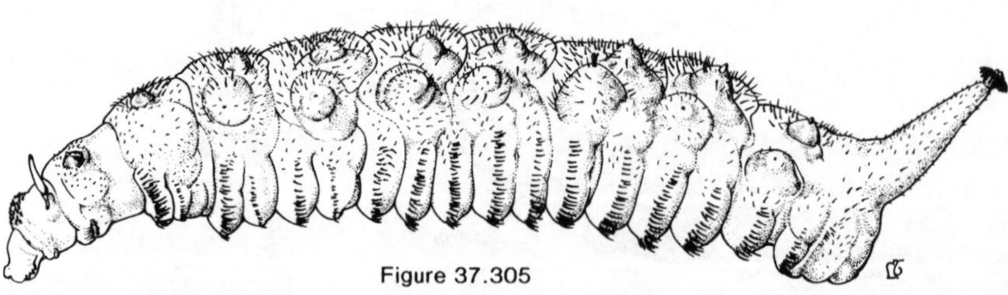

Figure 37.305

Figure 37.302. Ephydridae. Lateral of *Brachydeutera argentata* Walker (from Peterson 1951).

Figure 37.303. Ephydridae. *Ephydra riparia* Fallén. **a.** lateral; **b.** lateral view of proleg on fourth abdominal segment; **c.** dorsal view of third abdominal segment. Larvae occur in floating mats of algae in shallow ponds and marshes (from Peterson 1951).

Figure 37.304a,b. Ephydridae. *Paracoenia bisetosa* (Coquillett). **a.** lateral; **b.** anterior spiracle (from Zack 1983b. Reprinted with permission of the Entomological Society of America).

Figure 37.305. Ephydridae. *Lytogaster excavata* (Sturdevant and Wheeler), lateral (drawing by Lana Tackett).

CHLOROPIDAE (EPHYDROIDEA)

B. A. Foote, *Kent State University*

Frit Flies

Figures 37.306, 37.307

Relationships and Diagnosis: The proper placement of the Chloropidae in a phylogenetic scheme of the higher Diptera has been long debated. Hennig (1952) included it in a group centered on the family Drosophilidae, but later (1958) became undecided as to its affinities and recognized it as a separate entity within the Acalyptratae. Griffiths (1972) suggested that it was related to the Carnidae and Milichiidae and recognized it as a member of the Chloropidae family-group in his prefamily Tephritoinea. Recently, McAlpine et al. (1981) placed the Chloropidae close to the Ephydridae and Tethinidae in the large superfamily Ephydroidea. Sabrosky (1987d) recently summarized our knowledge.

Larvae are difficult to characterize because of the diversity of morphology found in species having different feeding habits. In general, they are long and cylindrical, have fan-shaped anterior spiracles with 4–9 papillae, possess ventral creeping welts, and have three short-oval openings in each posterior spiracular plate. The plates are usually borne on parallel cylindrical stigmatophores.

Biology and Ecology: Larvae possess a variety of feeding habits. A fair number are phytophagous, but many species are saprophagous. Many of the latter are best considered as being secondary invaders of plant tissue that has been damaged by other insects having more phytophagous habits. Several of the phytophagous species attack stems of grasses and cereals and a few of these are economically important. A few species are gall formers, others have larvae that attack root aphids, and still other species prey on the eggs of spiders, mantids, grasshoppers, and moths. Adults are most commonly collected in open sunny habitats dominated by grasses. They are abundant in herbaceous wetlands, and several species occur in coastal salt marshes.

Description: Mature larvae 5–10 mm, muscidiform, frequently long and slender but some species with thicker and shorter bodies; creeping welts usually present, rarely with segment-encircling spinule bands; white to yellowish.

Head: Cephalic segment somewhat retractile, bearing well-developed sensory organs anteriorly and anteroventrally; facial mask commonly with complicated network of grooves and ridges around preoral cavity, in other species the mask has only a few simple grooves and ridges, and in some species no obvious grooves can be discerned.

Cephalopharyngeal skeleton variously pigmented, frequently less pigmented on cornua; tentoropharyngeal and hypopharyngeal sclerites either distinctly separate (saprophagous species) or appearing somewhat fused (phytophagous species); dorsal cornua commonly connected by dorsal bridge anteriorly; ventral cornua with or without window distally and apparently lacking dorsobasal lobe; pharyngeal filter usually present, but absent in some truly phytophagous species; hypopharyngeal sclerite basically H-shaped; parastomal bars relatively short and free apically. Mandibles with broad basal part and a more slender, decurved hook that may have one to several accessory teeth.

Thorax: Metathorax may have ventral creeping welt; anterior spiracles arising posterolaterally on prothorax, each spiracle nearly sessile, commonly with only 4–9 marginal papillae (more than 15 in some species).

Abdomen: Segments all very similar except caudal one which is somewhat larger; each segment usually with ventral creeping welt near segmental border, and a few species with encircling band of spinules; perianal pad bilobed, not bordered posteromedially by spinule patch or rows of spinules.

Spiracular disc without marginal tubercles; posterior spiracles usually borne on apices of short tubular processes, these separate at base; each spiracular plate commonly subcircular, with blackened peritreme; three short-oval spiracular openings oriented at nearly right angles to each other; in some species (*Thaumatomyia*) spiracles lack peritremes and end in three papilla-like structures, while in at least some species of *Chlorops* the spiracles open on the inner side of two fleshy lobes; four dichotomously branched spiracular hairs usually present, but reduced or even absent in a few species.

Comments: This is fairly sizeable family of small black or yellow and black flies. About 1200 species have been described, with 55 genera and some 300 species recorded from America north of Mexico. Because of their phytophagous habits, a few species are considered as important agricultural pests, the most important North American one being the wheat stem maggot, *Meromyza americana* Fitch (fig. 37.307). The well-known frit fly, *Oscinella frit* (L.), is widely distributed in the Holarctic region where it is a destructive pest of cereals. Eye gnats of the genera *Hippelates* and *Siphunculina* are highly annoying in the tropics and probably carry conjunctivitis and certain other eye diseases.

Selected Bibliography

Balachowsky and Mesnil 1936.
Ferrar 1987.
Griffiths 1972.
Hall 1932b.
Hennig 1943, 1952, 1958.
Hermes and Burgess 1930.
Hewitt, T. R. 1914.
Johannsen 1935.
Jones, F. M. 1916.
Nye 1958.
Oettingen 1935.
Sabrosky 1987d.
Teskey et al. 1976.
Valley et al. 1969.
Vimmer 1925.
Wendt 1968.

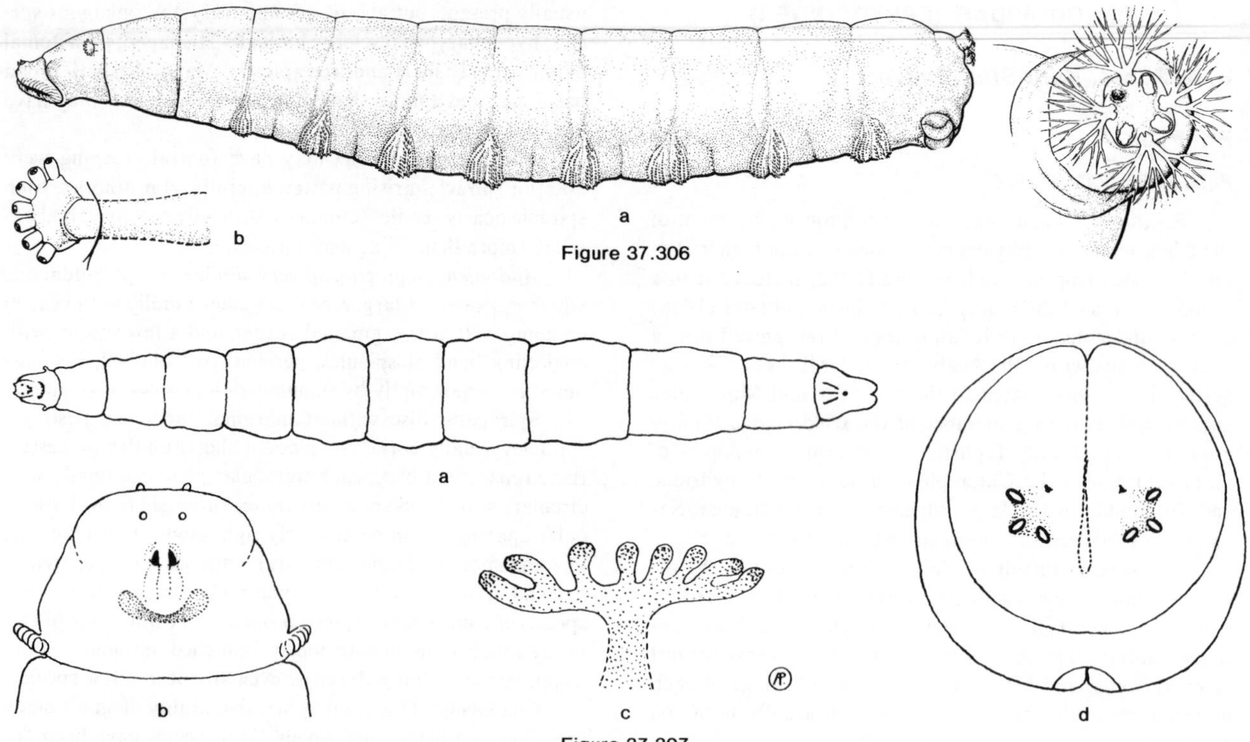

Figure 37.306a-c. Chloropidae. *Elachiptera* sp. **a.** lateral; **b.** anterior spiracle; **c.** posterior spiracular plate (from Manual of Nearctic Diptera, Vol. 1).

Figure 37.307a-d. Chloropidae. *Meromyza americana* Fitch, the wheat stem maggot. **a.** dorsal; **b.** ventral view of anterior end; **c.** anterior spiracle; **d.** caudal view of posterior end. Larvae are pests of wheat, rye, barley, oats, timothy, and several wild grasses, where they tunnel in the stems, the summer generation causing the heads to die and turn conspicuously white (from Peterson 1951).

CRYPTOCHETIDAE (EPHYDROIDEA)

B. A. Foote, *Kent State University*

The Cryptochetids

Figure 37.308. See also first instar, fig. 3.130h, Vol. 1 (p. 46)

Relationships and Diagnosis: The family Cryptochetidae has resided in various taxonomic groupings over the years. Hennig (1937) once considered it to be related to the Chamaemyiidae, Milichiidae, and Drosophilidae but later (1958) placed it in his group 7, families of uncertain relationships, suggesting that it was close to the Chamaemyiidae, Lonchaeidae, Milichiidae, and Drosophilidae. Griffiths (1972) placed it with the Lonchaeidae in his Lonchaeoidea. Recently, McAlpine et al. (1981) have considered it to be within the Ephydroidea. McAlpine (1987n) recently reviewed our knowledge.

Larvae are easily recognized by the very long filaments arising from the caudal segment and by the thorn-like posterior spiracles on the dorsal side of the last abdominal segment.

Biology and Ecology: Larvae are endoparasitoids of scale insects of the family Margarodidae. Eggs are placed into the hemocoel of the host where the newly hatched sac-like larva absorbs blood through its integument. Older larvae possess a functional digestive tract and are capable of ingesting blood, tissues, and the fat body of the host. Over a dozen larvae have been found within the body of a single scale insect. Pupation occurs within the host's body.

Description: Mature larvae 3–4 mm, basically musciform but with greatly elongated tracheal tubes projecting posteriorly from caudal segment; integument rather densely covered with spinules that may form distinct encircling spinule bands; whitish to darkened.

Head: Cephalic segment poorly retractile, with rather inconspicuous sensory organs; facial mask apparently lacking oral grooves and ridges around preoral cavity.

Cephalopharyngeal skeleton deeply pigmented; tentoropharyngeal and hypopharyngeal sclerites fused; dorsal cornua narrow, connected by dorsal bridge, and projecting dorsally nearly at right angle to broader ventral cornua; pharyngeal filter absent. Mandibles rather slender, hook without accessory teeth.

Thorax: Anterior spiracles arising in distinct depressions posterolaterally on prothorax, each spiracle slender and with 3–12 very long finger-like apical papillae.

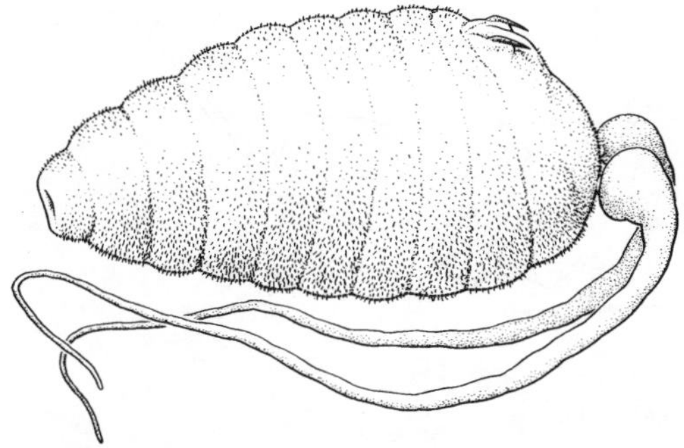

Figure 37.308. Cryptochetidae. Lateral view of *Cryptochetum yokohama* (Kuwana). Larvae of this oriental taxon closely resemble those of *C. iceryae* (Williston), a species that has been introduced into California from Australia to control the cottony cushion scale (from Manual of Nearctic Diptera, Vol. 1).

Abdomen: Rather swollen, without prolegs or obvious creeping welts ventrally; caudal segment bluntly rounded and bearing two greatly elongated tracheal tubes posteroventrally, with each filamentous tube somewhat swollen near base; posterior spiracles thorn-like, arising on dorsal side of caudal segment; each spiracle may be 3-parted, with each part possessing an oval spiracular opening.

Comments: This is a small family of some 20 species and one genus that is best developed in tropical Africa. A single species, *Cryptochetum iceryae* (Williston), was introduced into California from Australia as a biocontrol agent of the cottonycushion scale, *Icerya purchasi* Maskell.

Selected Bibliography

Ferrar 1987.
Hennig 1937, 1952, 1958.
McAlpine, J. F. 1987n.
Thorpe 1930, 1931, 1934, 1941.

TETHINIDAE (EPHYDROIDEA)

B. A. Foote, *Kent State University*

The Tethinids

Relationships and Diagnosis: Once considered to be relatively closely related to the Milichiidae (Hennig, 1958), the tethinids have been treated variously by more recent workers. Griffiths (1972) placed it close to the Odiniidae and Chloropidae in his prefamily Tephritoinea, whereas McAlpine et al. (1981) considered it to be a member of the Ephydroidea, relatively close to the Canacidae. Vockeroth (1987d) reviewed our knowledge.

Biology and Ecology: Very little is known of the biology. Adults of *Neopelomyia* and *Tethinia* are seemingly restricted to coastal habitats, but adults of other genera have been collected at inland sites. The larval feeding habits of Nearctic genera are unknown, although the habits of adults worldwide suggest that the larvae live in soil, algal masses, seaweed or bird dung in various parts of the world (Vockeroth 1987d).

Description: No Nearctic larvae have been described.

Comments: There are about 100 described species in the world, with 4 genera and 24 species recorded for America north of Mexico.

Selected Bibliography

Ferrar 1987.
Foster, 1976.
Griffiths 1972.
Hennig 1952, 1958.
Melander 1952.
Vockeroth 1987d.

CANACIDAE (EPHYDROIDEA)

B. A. Foote, *Kent State University*

Beach Flies

Figures 37.309a,b

Relationships and Diagnosis: The canacids were commonly thought to be a part of the Ephydridae by earlier workers (Hennig 1952), but are now considered to be a distinct family that is rather distant from the ephydrids. Hennig (1958) placed the Canacidae in the Milichioidea along with the Tethinidae, Milichiidae, Braulidae, and Sphaeroceridae, suggesting that it was particularly close to the Tethinidae. In contrast, Griffiths (1972) expressed uncertainty as to the correct phylogenetic relationships of the Canacidae and did not attempt to place it within any superfamily. Recently, McAlpine et al. (1981) placed it in the Ephydroidea, implying that it was particularly close to the Tethinidae. Wirth (1987) recently summarized our knowledge.

Larvae are relatively easily distinguished by the elongate anterior spiracles and the telescopic respiratory tube that bears posterior spiracles that are nearly approximate. A telescopic tube is found elsewhere only in certain species of Syrphidae, Aulacigastridae, and Ephydridae.

Biology and Ecology: Most species occur in intertidal coastal habitats where adults occur on rocky, sandy, and muddy substrates. They also occur in coastal salt marshes. Larvae are thought to feed on algae and are occasionally abundant in blue-green and green algal mats. The larvae of a freshwater species in Hawaii feed on algae growing on rocks in swift-flowing mountain streams.

Description: Mature larvae 4–7 mm, muscidiform, tapering anteriorly and posteriorly, with caudal segment forming telescopic respiratory tube; integument in some species covered with brownish spinules; in other species (e.g., *Canacea macateei* Malloch) spinules largely restricted to encircling spinule bands or to ventral creeping welts; white.

Head: Cephalic segment retractile, with well developed sensory organs anteriorly and anteroventrally; antennae appearing 2-segmented; facial mask with numerous oral grooves and ridges leading into preoral cavity.

Cephalopharyngeal skeleton mostly deeply pigmented, with tentoropharyngeal and hypopharyngeal sclerites separate; dorsal cornua distinctly narrower than ventral cornua,

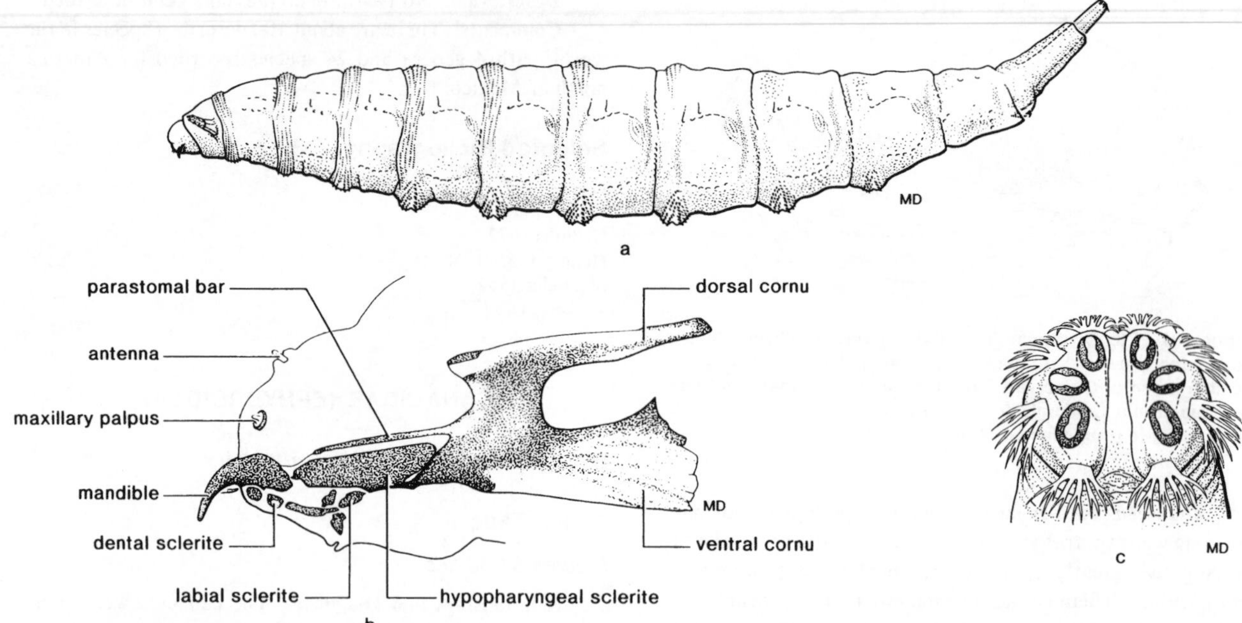

Figure 37.309a-c. Canacidae. *Canacea macateei* Malloch. **a.** lateral; **b.** cephalopharyngeal skeleton; **c.** posterior spiracles (from Manual of Nearctic Diptera, Vol. 2).

connected anteriorly by dorsal bridge; ventral cornua broad and truncate apically, commonly with dorsobasal lobe; floor of tentoropharyngeal sclerite with pharyngeal filter; hypopharyngeal sclerite basically H-shaped, with single connecting bridge; parastomal bars elongate and connected anteriorly by fenestrated epipharyngeal sclerite; four pairs of variously shaped labial sclerites below anterior end of hypopharyngeal sclerite. Mandibles relatively narrow, hook weakly decurved and lacking accessory teeth, subtended ventrally by V-shaped accessory oral sclerite, basal part apparently lacking windows; two pairs of subrectangular dental sclerites below basal part.

Thorax: Meso- and metathorax with encircling spinule bands that form creeping welts ventrally; anterior spiracles arising posterolaterally on prothorax, elongate, tubular, and retractile, with 2–6 sessile or somewhat lengthened papillae along margins of distal part.

Abdomen: Segments generally uniform and similar, covered by spinules or each segment encircled by spinule band at segmental boundaries; creeping welts noticeable; caudal segment 2-parted, with distal part elongated to form respiratory tube; perianal pad somewhat bilobed, frequently bordered by spinule patch posteromedially.

Posterior spiracles borne on apex of telescopic respiratory tube, and narrowly separated by shallow declivity; each spiracular plate with three oval spiracular openings, each opening oriented at nearly right angle to its nearest neighbor; four relatively short but profusely branched spiracular hairs around margin of each spiracular plate.

Comments: The Canacidae is a relatively small family containing about 100 species and 11 genera worldwide. Three genera, *Canacea, Canaceoides, Nocticanace,* and some 5 species are recorded from America north of Mexico. Larvae play a role in the recycling of nutrients in coastal areas.

Selected Bibliography

Ferrar 1987.
Griffiths 1972.
Hennig 1952, 1958.
Mathis 1982a.
Mathis and Freidberg 1983.
Robles and Cubit 1981.
Teskey and Valiela 1977.
Williams 1938.
Wirth 1951, 1987.

SECTION CALYPTRATAE

SCATHOPHAGIDAE (MUSCOIDEA)

Gregory A. Dahlem, *Michigan State University*

The Scathophagids

Figures 37.310–37.315

Relationships and Diagnosis: The accepted systematic position of this group has changed several times, but they are placed closest to the Anthomyiidae in McAlpine et al. (1987).

Information may be found in the literature under Cordyluridae, Scathophagidae, Anthomyiidae, or Muscidae, depending on the author's preference and date of publication. As a separate family, its closest affiliations are with the Anthomyiidae. Scathophagidae is placed with Anthomyiidae and Muscidae in the Muscoidea by McAlpine et al. (1981). Vockeroth (1987e) has recently reviewed our knowledge.

At the present time, no characters are available which will separate all larvae of Scathophagidae from those of the Anthomyiidae. The distinctly bicornuate anterior spiracles (figs. 37.310e, 37.311b) are of general usefulness, but there are many exceptions among both anthomyiids and scathophagids. In most instances, rearing to adults will be necessary for an accurate identification.

Biology and Ecology: Scathophagids have relatively diverse feeding habits as immatures. Larvae may be saprophagous, coprophagous, phytophagous, or predaceous. Due to the ambiguous nature of larval morphological characters used to separate this family, larval behavior and host associations may prove to be much more useful for identification.

Many species feed on monocotyledonous plants. *Norellisoma spinimanum* (Fallén) is a stem miner in a variety of docks, *Rumex* spp. (Disney, 1976). *Norellia spinipes* (Meigen), an Old World species, is a leaf miner in daffodils, *Narcissus* spp. (Ciampolini, 1957). Larvae of the large genus *Cordilura* feed as stem miners in several sedges, *Carex* spp., and rushes, *Scirpus* spp. *C. praeusta* Loew has been raised from the culms of the rush *Juncus effusus* (Wallace and Neff, 1971). Some species of *Nanna* are pests of rye and grasses of the genus *Phleum* in the Palearctic region (King *et al.*, 1935). The records of *Chaetosa punctipes* (Meigen) from grasses, given by Balachowsky and Mesnil (1935), were based on a mistaken identification of a species of *Nanna*.

Parallelomma spp. and *Delina* spp. are known to be leaf miners in many species of Orchidaceae and Liliaceae (Meijere, 1940a; Séguy, 1934) in the Old World. *Hexamitocera tricincta* (Loew) is a leaf miner of false Solomon's-seal, *Smilacina racemosa* (Frost, 1932). *Neochirosia atrifrons* (Coquillett) larvae produce extensive blotch mines as they feed on Indian poke, *Viratrum viride* (Neff, 1970).

Records from dicotyledonous plants are not as common. *Gimnomera cerea* (Coquillett) feeds upon the ovules and capsules of lousewort, *Pedicularis canadensis*, and *G. incisurata* Malloch has been recorded to feed on the flower parts of three species of beardtongue, *Penstemon* spp. (Neff, 1968). *Hydromyza confluens* Loew feeds on submerged petioles and floating leaves of the yellow water-lily, *Nymphaea americana* (Hickman, 1935).

Of the other genera, *Orthacheta hirtipes* Johnson is a facultative predator on *Cordilura* larvae, but they can survive and mature on the decaying plant material and bacteria inside of the stem mines alone (Neff and Wallace, 1969b). *Spaziphora* spp. are known to be free living larvae in aquatic habitats, feeding on algae, fungi, and perhaps detritus or other larvae in their surroundings (Graham, 1939). The majority of *Scatophaga* larvae appear to be coprophagous in a variety of animal dung. *S. stercoraria* (L.) is a very common sheep and cow dung feeder (Cotterell, 1920). Some species of *Scatophaga* have been reared from rotting seaweed in the Palearctic region (Backlund, 1945b).

Description: Larvae are white to cream white and may turn pale yellowish when preserved. Mature larvae 4–15 mm in length, most ranging from 6–8 mm. Most larvae nearly cylindrical (fig. 37.310a) but some more conical (fig. 37.311a). Mature larvae amphipneustic.

Head: Cephalic segment retractile and bearing well-developed sensory organs anteriorly and anteroventrally. Antennae appearing 2-segmented. Cephalopharyngeal skeleton deeply pigmented.

Thorax and Abdomen: With or without encircling spinule bands, but usually with some ventral creeping welts. Anterior spiracles arising posterolaterally on prothorax, usually bicornuate and T-, Y-, or U-shaped with variable numbers of papillae.

Posterior spiracular disc usually with 4–6 pairs of marginal tubercles and two postanal tubercles. Posterior spiracles distinctly elevated above the plane of the terminal segment on short tubes. Spiracular plates with 3 slits and, usually, 4-branched spiracular hairs. They often have a distinctive peritreme.

Comments: Thirty-nine genera are recognized from America north of Mexico (Vockeroth, 1987e), with descriptions of larvae available for 13. None of the species are considered to be of economic importance.

Selected Bibliography

Backlund 1945b.
Balachowsky and Mesnil 1935.
Ciampolini 1957.
Cotterell 1920.
Disney 1976.
Ferrar 1987.
Frost 1932.
Graham 1939.
Hennig 1952.
Hickman 1935.
James 1955b.
King et al. 1935.
Meijere 1940a.
Neff 1968, 1970.
Neff and Wallace 1969a, 1969b.
Séguy 1934.
Vockeroth 1987e.
Wallace and Neff 1971.

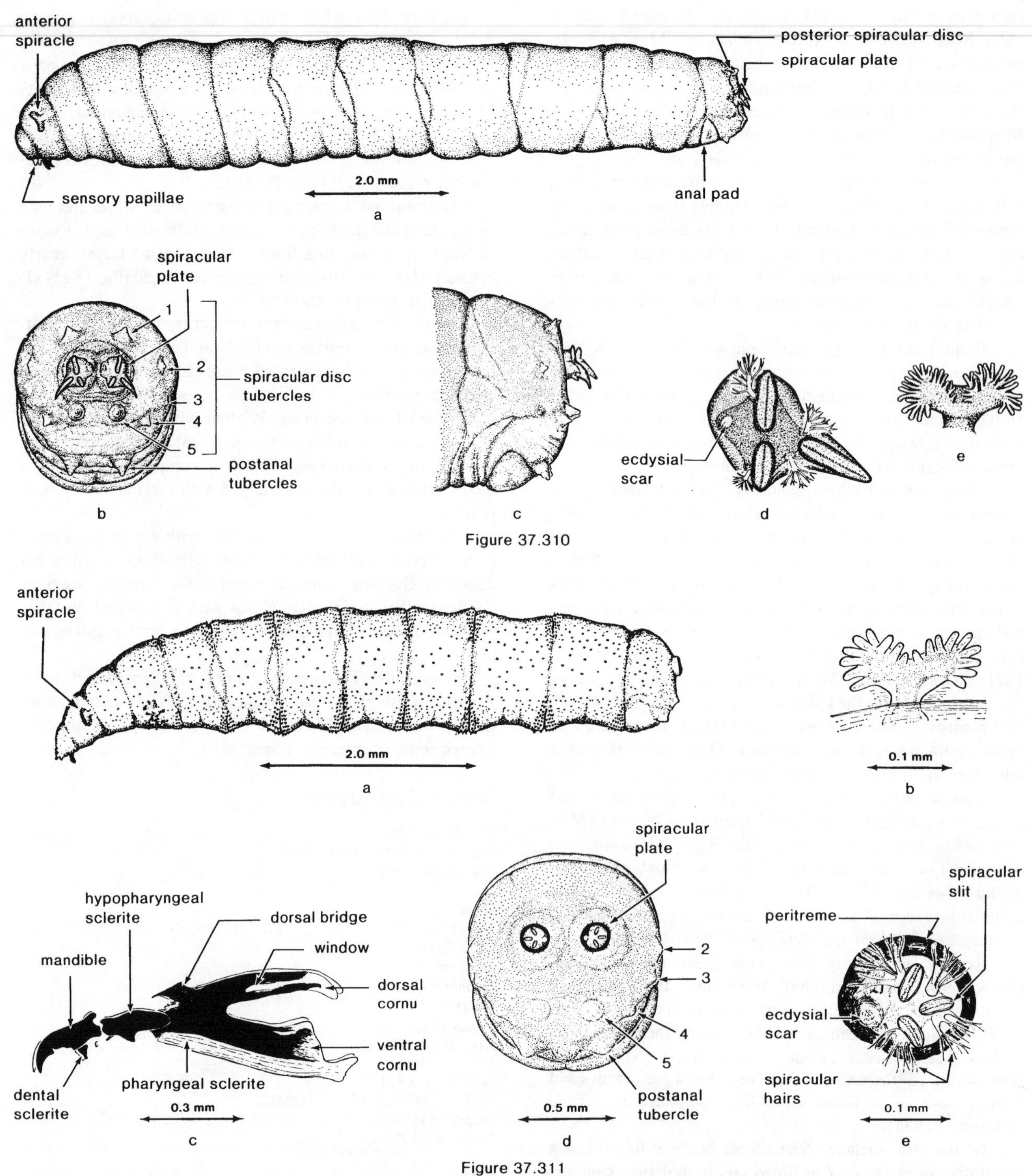

Figure 37.310a-e. Scathophagidae. *Cordilura deceptiva* Malloch. **a.** lateral; **b.** posterior of caudal segment, Nos. 1-5 are spiracular disc tubercles; **c.** lateral of caudal segment; **d.** posterior spiracular plate; **e.** anterior spiracle. Larvae mine the culms of the sedge, *Carex stricta.* (from Neff and Wallace 1969a. Reprinted with the permission of the Entomological Society of America).

Figure 37.311a-e. Scathophagidae. *Orthacheta hirtipes* Johnson. **a.** lateral; **b.** anterior spiracle; **c.** cephalopharyngeal skeleton; **d.** posterior of caudal segment, Nos. 2-5 are tubercles; **e.** posterior spiracular plate. First instar larvae bore down the center of damaged *Carex* culms infested with *Cordilura* (Scathophagidae) larvae; 2nd and 3rd instars may prey on *Cordilura* larvae and/or feed on decomposing plant material and microorganisms (from Neff and Wallace 1969b. Reprinted with permission of the Entomological Society of America).

Figure 37.312

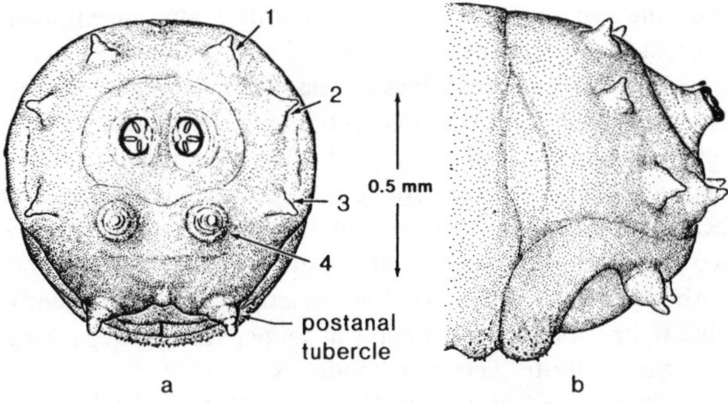

Figure 37.313

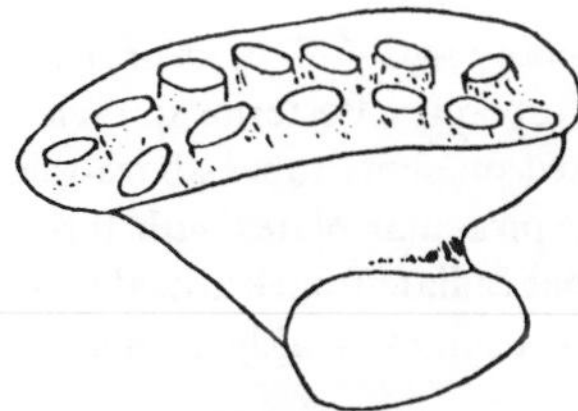

Figure 37.314

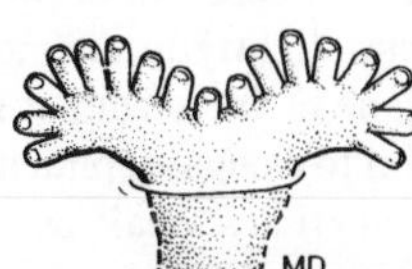

Figure 37.315

Figure 37.312a-c. Scathophagidae. *Gimnomera incisurata* Malloch. **a.** posterior of caudal segment, Nos. 1-5 are tubercles; **b.** lateral of caudal segment; **c.** posteriot spiracular plate. Larvae feed on the flower parts of beard-tongue, *Penstemon* spp. (Scrophulariaceae) (from Neff 1968).

Figure 37.313a,b. Scathophagidae. *Neochirosia atrifrons* (Coquillett). **a.** posterior of caudal segment, Nos. 1-4 are tubercles; **b.** lateral of caudal segment. Larvae form blotch mines in the leaves of *Veratrum viride* (Liliaceae) (from Neff 1970).

Figure 37.314. Scathopagidae. *Hydromyza confluens* Loew. Anterior spiracle. Larvae feed on petioles and floating leaves of yellow waterlily, *Nymphaea americana* (from Hickman 1935).

Figure 37.315. Scathophagigae. *Scathophaga stercoraria* (L.). Anterior spiracle. Larvae are frequently abundant in cattle droppings (from Manual of Nearctic Diptera, Vol. 1).

ANTHOMYIIDAE (MUSCOIDEA)

Gregory A. Dahlem, *Michigan State University*

F. C. Thompson, *Systematic Entomology Laboratory, A.R.S., U.S.D.A.*

The Anthomyiids

Figures 37.316-37.319 and Delia *Key figs. A-P*

Relationships and Diagnosis: This family is closely related to the Scathophagidae and Muscidae. Information may be found in the literature under both Anthomyiidae and Muscidae. Currently, Anthomyiidae is considered to be a separate family in the Muscoidea by McAlpine, *et al.* (1981). Huckett (1987) has recently summarized our knowledge, and Griffiths (1982–86) is revising the family.

At the present time, there are no characters known that will separate all larvae of Anthomyiidae from the related family Scathophagidae or from several more distantly related families. The larvae show a great deal of structural variation, but, in general, they are conic-cylindrical (peg-shaped) with fan-like anterior spiracles and with marginal tubercles on the posterior spiracular disc. Rearing to adults is suggested for accurate identification, especially of non-pest species.

Biology and Ecology: Anthomyiid larvae have diverse feeding habits as a group and frequently within a single species. Larvae may be saprophagous, coprophagous, phytophagous, or predacious. Larval behavior and host associations can be very important supplementary information in their identification. Most economically important *Delia* species (the root maggots) attack a much greater diversity of plant species than the common name indicates.

Many of the important species feed on the roots of cruciferous plants and other vegetables. *Delia floralis* (Fallén) (=*crucifera*), the turnip maggot, and *D. plannipalpis* (Stein) appear to be exclusively root feeders, the former feeding on turnip, cabbage, and cauliflower while the latter also feeds on radishes (Brooks 1951). *D. radicum* (L.), the cabbage maggot, usually feeds on the roots of cabbage, turnip, radish, and cauliflower, but may also feed in the growing point of turnip and cauliflower before the head is formed, in the inflorescence of cauliflower, and in the midrib and larger leaf veins of cabbage and its relatives (K. M. Smith 1927). *D. florilega* (Zetterstedt) (=*trichodactyla*) and *D. platura* (Meigen) (=*cilicrura*), the seedcorn maggot (which is only an occasional problem on seedling corn and other large seeds), feed on the roots of such cruciferous plants as turnip, Brussels sprouts, and cabbage that have already been damaged by other Diptera larvae, but they may also attack the cotyledons and shoots of beans, peas and other vegetables (Brooks 1951; Miles 1950). *Delia antiqua* (Meigen), the onion maggot, is an important pest of onions; larvae may be found feeding in the bulbs of young onions, in maturing bulbs, and on bulbs in storage. Damaged and decomposing onions are particularly attractive. *Delia echinata* (Séguy) is a spinach midrib and stem miner (Miles 1953). *Delia brunnescens* (Zetterstedt), the carnation maggot, and *D. cardui* (Meigen) are miners in stems and shoots of carnations (Séguy 1932). *D. coarctata* (Fallén) attacks the stems of wheat, rye, barley and other grasses (Keilin 1917a), and is now established in northeastern North America (McAlpine and Slight 1981). *Botanophila fugax* (Meigen) also feeds on roots of plants such as turnip and Brussels sprouts that have been previously damaged by other larvae (Brooks 1951).

The important genus *Pegomya* has been subdivided in recent years. Griffiths (1982, following Hennig 1973) divided *Pegomya* into two subgenera, with the subgenus *Pegomya* consisting of just leaf miners. The larvae of these can be easily distinguished by the numerous (at least 3) teeth on the mouthhooks (fig. 37.318a). The subgenus *Pegomya* was further divided into four sections: 1) the *bicolor* section which are miners on Centrospermae and Polygonaceae; 2) the *hyoscyami* section with the *hyoscyami* subsection on Caryophyllaceae, Chaenopodiaceae, Solanaceae, Polemoniaceae and *Lupinus* (Fabaceae), including the *hyoscyami* superspecies (the beet and spinach leaf miners) which form blotch mines in leaves of spinach and a wide variety of other plants (Cameron 1914; Frost 1924; Michelson 1980), and the *genupuncta* subsection on Asteraceae; 3) the *minuta* section with host data for one species only (*cognata* Stein on *Salicornia*); and 4) the *dorsimaculata* section with host data for only one species (*wygodzinskyi* Alburquerque on *Amaranthus*). The other subgenus, *Phoraea*, is also divided into four sections: 1) the *rubivora* section which are mainly stemborers on Rosaceae and *Equisetum;* 2) the *geniculata* and 3) the *flavoscutella* sections which feed on mushrooms (Wallace 1971); and 4) the *holmgreni* section for which no life-history data is available.

One genus, *Chirosia*, is specialized to feed only on ferns (Pterophyta; Aderkas and Peterson 1987). *Egle muscaria* (F.) and *E. parva* (Robineau-Desvoidy) larvae are noted as feeding on willow catkins (Gäbler 1933; Séguy 1923b). *Strobilomya anthracina* (Czerny) larvae tunnel in the rachis of the cones of spruce trees (Tripp 1954; Kangas and Leskinen 1944; Michelson 1988).

Many anthomyiids are coprophagous (*Emmesomyia, Hylemyia, Pegoplata, Calythea, Adia, Paregle*) and some specialize on fecal material in burrows or nests of rodents (*Eutrichota*) or gopher tortoises (*Eutrichota gopheri* Johnson). *Paregle audacula* (Harris) has been reared from horse dung (Hewitt 1907) as well as dog and human dung (Coffey, 1966). *Adia cinerella* (Fallén) has been reared from swine, cow, horse and sheep dung (Coffey, 1966). *Hylemya alcathoe* (Walker) and *Calythea micropteryx* (Thomas) larvae feed in cow dung (Coffey 1966; Merritt and Anderson 1977).

Among the other habitats of anthomyiid larvae, *Anthomyia pluvialis* (L.) and *A. procellaris* Rondani feed on decaying organic matter in bird nests (Keilin 1924a); *Lasiomma octoguttata* (Zetterstedt) has been reared from bird nests, skua pellets and dead snails (R. A. Beaver 1969); *Anthomyia confusana* Michelsen has been reared from dead snails (R. A. Beaver 1969). *Fucellia costalis* Stein and *F. rufitibia* Stein feed in wrack on ocean beaches (Kompfner, 1974); larvae of *Acridomyia* are parasitoids on grasshoppers (Acrididae); and larvae of *Leucophora* and *Eustalomyia* are known to live as inquilines or parasitoids in the nests of solitary bees and wasps (Huckett 1965c).

Description: Larvae of anthomyiids are usually white to off-white, occasionally yellow. Mature larvae 4–12 mm in length, most ranging from 7–9 mm. Most larvae peg-shaped, tapering anteriorly and bluntly rounded posteriorly, but some more cylindrical. Mature larvae amphipneustic.

Head: Cephalic segment retractile and bearing well-developed sensory organs anteriorly and anteroventrally. Antennae appearing 2-segmented, often with basal sclerotization.

Cephalopharyngeal skeleton usually deeply pigmented; tentoropharyngeal and hypopharyngeal sclerites separate or fused (many phytophagous species); dorsal cornua usually narrower than ventral cornua and commonly with narrow, elongate window posteriorly; connected anteriorly by fenestrated dorsal bridge; ventral cornua broader, with dorsobasal lobe and small to large window posteriorly; pharyngeal filter present in saprophagous species, reduced or absent in many phytophagous and predacious species; hypopharyngeal sclerite basically H-shaped; parastomal bars frequently absent; epipharyngeal sclerite absent; labial sclerite commonly narrow and rod-like or absent. Mandibles deeply pigmented and heavy, hook weakly decurved and with one or more accessory teeth, base subrectangular, with posteroventral projection and with or without small window; dental sclerites frequently absent, or if present, triangular in shape; apparently lacking accessory sclerites between mandibles.

Thorax and Abdomen: With or without encircling spinule bands, but usually with ventral creeping welts. Anterior spiracles arising posterolaterally on prothorax, usually fan-shaped and bearing 4–40 marginal papillae.

Posterior spiracular disc usually with 6–7 pairs of marginal tubercles and two postanal tubercles. Posterior spiracles very slightly to distinctly elevated on short tubes above the plane of the terminal segment. Spiracular plates with three oval to elongate spiracular slits that radiate from ecdysial scar (sometimes weak or absent). Peritremes weakly to deeply pigmented.

Comments: Twenty-three genera, with over 400 species, are recognized from America north of Mexico (Griffiths, *in litt.*). Larval descriptions are available for eight of these genera.

Several species are considered to be major agricultural pests. These include the spinach/beet leafminer, *Pegomya hyoscyami;* the onion maggot, *Delia antiqua;* the cabbage maggot, *D. radicum;* and the seedcorn maggot, *D. platura.* The common last instar *Delia* root maggots can be identified in the following key; a short synopsis of each species follows.

KEY TO THE COMMON LAST INSTAR *DELIA* ROOT MAGGOTS[1]

(Figure A is an inverted posterior view showing the tubercle relationships and names.)

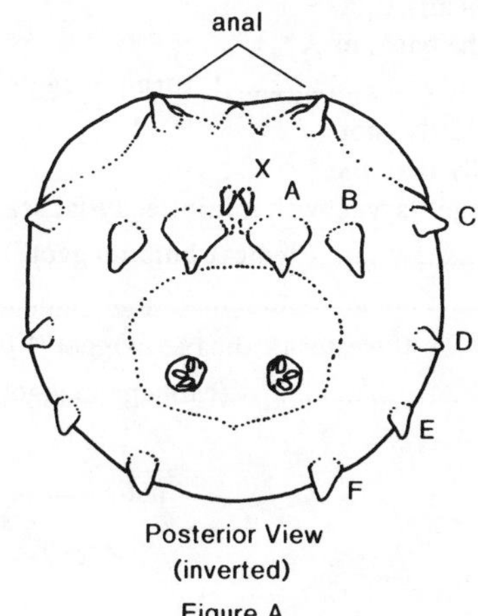

Posterior View
(inverted)
Figure A

1. Assorted other maggots belonging to different families and genera may also be found, but usually not abundantly (*see* Brooks 1951 for a few of them). Drawings C, E, K, N, O redrawn from Brooks (1949 and 1951) by Peter Carrington.

1. Posterior spiracular plate with a distinct ecdysial scar (fig. B); window in dorsal cornu of tentoropharyngeal sclerite long, reaching almost to base (fig. C) 4

1′. Posterior spiracular plate without a distinct ecdysial scar (fig. D); window in dorsal cornu of tentoropharyngeal sclerite short, extending about to middle (fig. E) 2

2. Anterior spiracle with 6–8 papillae (fig. F) 3

2′. Anterior spiracle with 10–14 papillae (fig. G); tubercles A and B contiguous basally, but widely separated apically (figs. H, I and J) .... (most) (onion maggot) *antiqua*

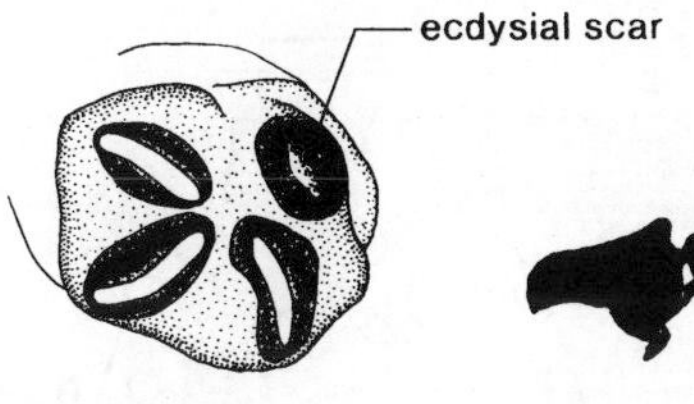

ecdysial scar

Left
radicum
Figure B

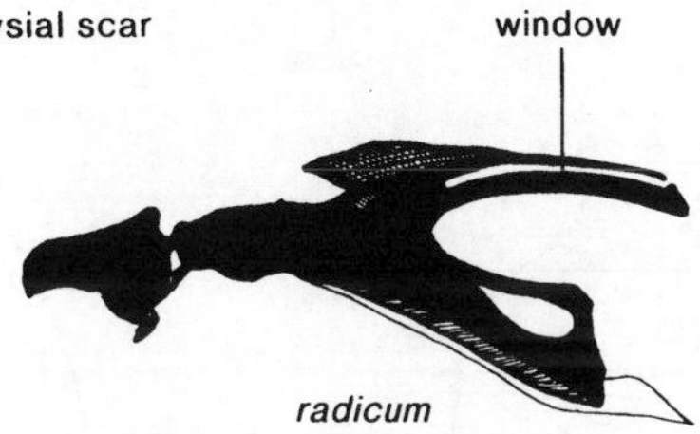

window

radicum
Figure C

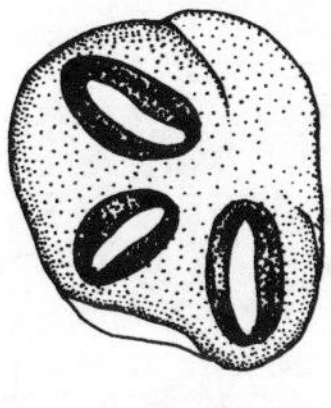

Left
antiqua
Figure D

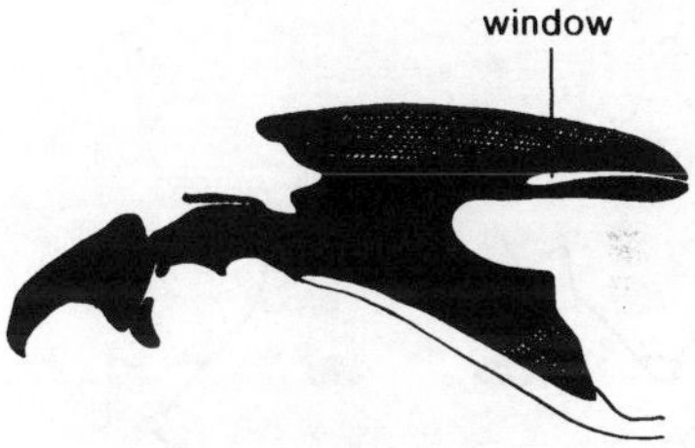

window

antiqua
Figure E

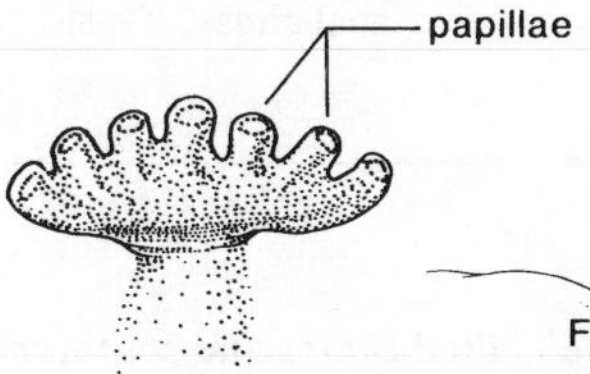

papillae

platura
Figure F

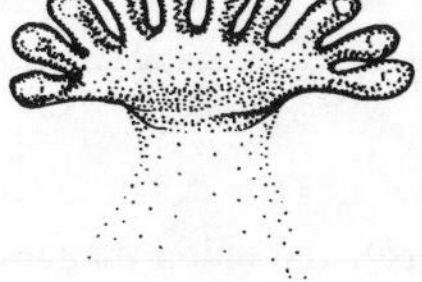

antiqua
Figure G

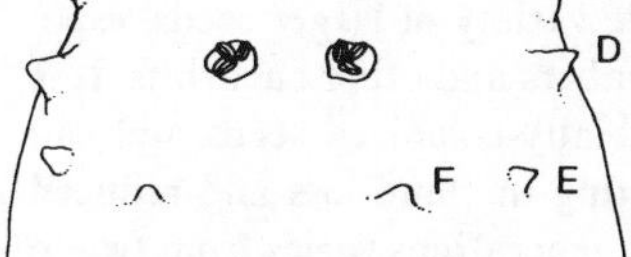

X A B C D F E

antiqua
Dorso-posterior view
Figure H

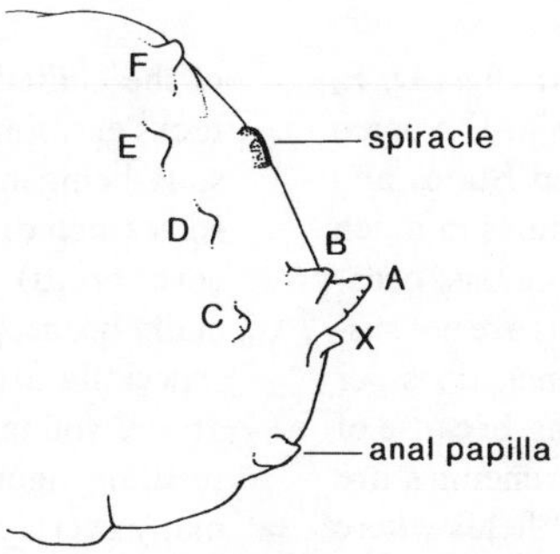

F E D B A C X spiracle anal papilla

antiqua
lateral view
Figure I

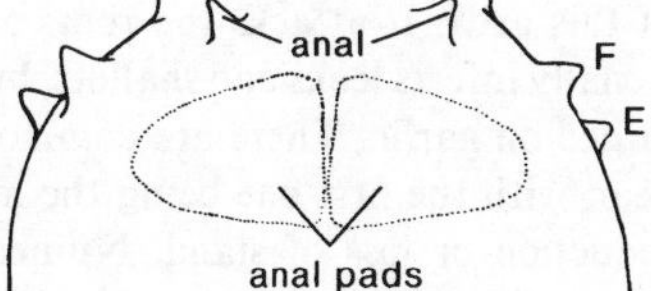

X A B anal F E anal pads

antiqua
Ventral view
Figure J

3. Tubercles A and B long, contiguous basally, with A distinctly broader than B (fig. K) .... *florilega*

3′. Tubercles A and B short, separate basally, subequal in breadth (fig. L) .... (seedcorn maggot) *platura*

4. Tubercles A and B not equal, widely separated, with A notched apically (figs. M, P) 5

4′. Tubercles A and B subequal, contiguous basally; tubercle A simple apically (not notched); tubercles X present, **the bases arising nearly in line with the bases of A and B (figs. N and O)** (*see* third choice) .... ("turnip maggot") *floralis*

4″. Tubercles A and B subequal, contiguous basally; tubercle A simple apically (not notched); tubercles X present, **the bases clearly arising more ventrally than the bases of A and B (figs. H and J)** .... (some onion maggot larvae with a faint ecdysial scar) *antiqua*

5. Tubercle A with a shallow apical notch (fig. P) .... ("wheat bulb maggot") *coarctata*

5′. Tubercle A with a deeper apical notch (fig. M) 6

6. Tubercle A distinctly separated to the base .... ("western cabbage maggot") *planipalpis*

6′. Tubercle A not distinctly separated to the base (fig. M) .... (cabbage maggot) *radicum*

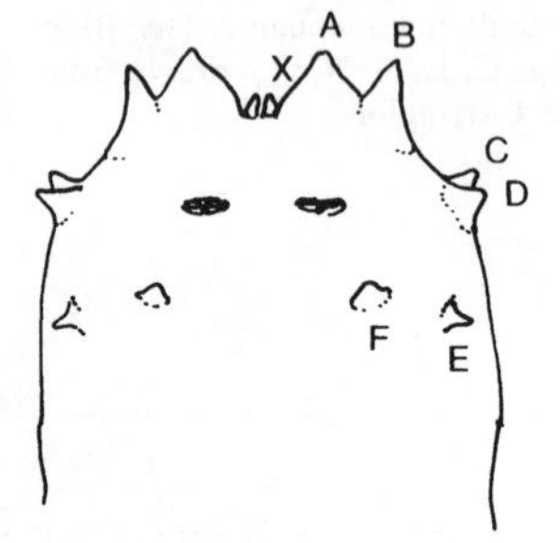

florilega
Dorso-posterior view
Figure K

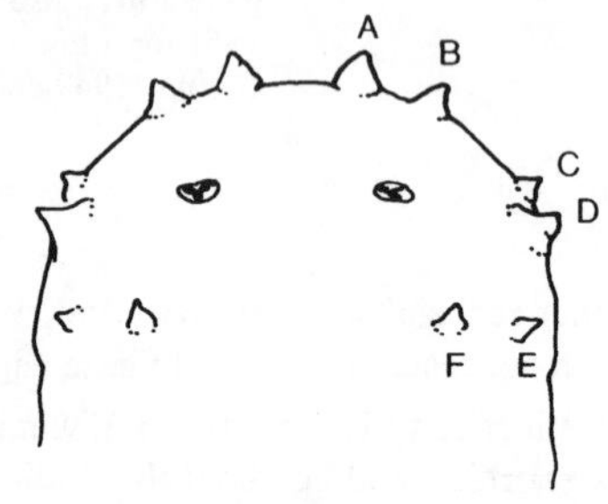

platura
Dorso-posterior view
Figure L

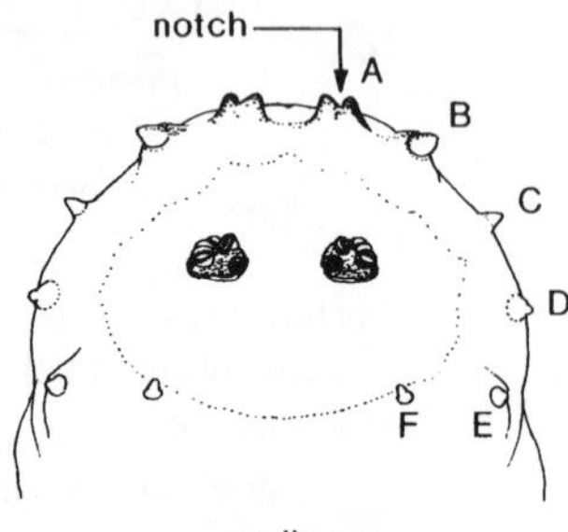

radicum
Dorso-posterior view
Figure M

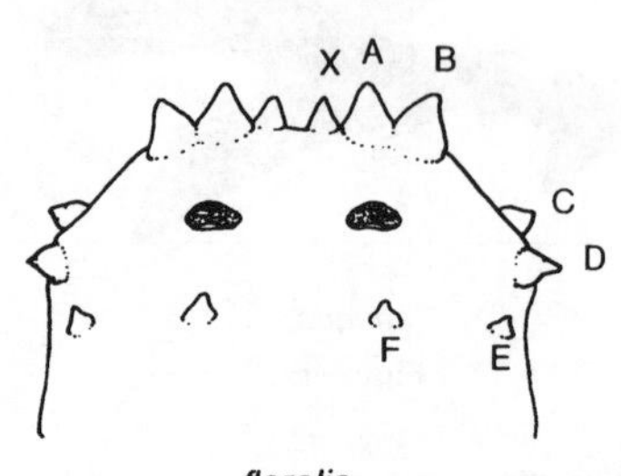

floralis
Dorso-posterior view
Figure N

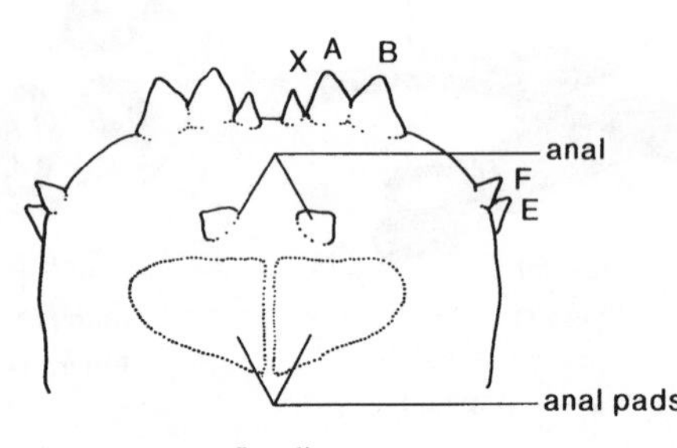

floralis
Ventral view
Figure O

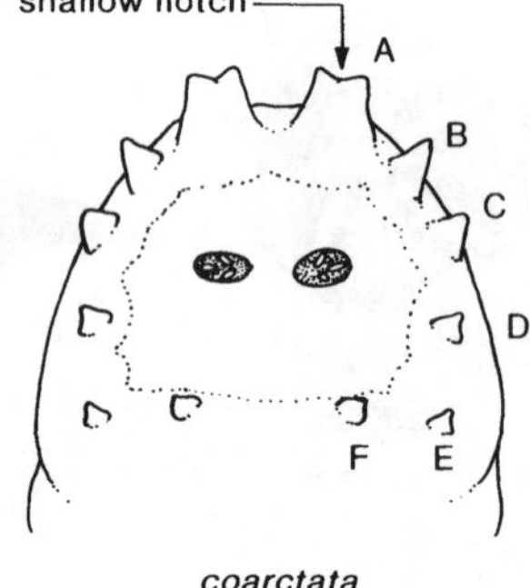

coarctata
Dorso-posterior view
Figure P

Delia antiqua **(Meigen), the onion maggot:** (figs. D, E, G–J). Believed to be native to northern Eurasia, the onion maggot ranges throughout the northern United States and southern Canada, where it is a major pest of onions in much of this area. It attacks the roots and bulbs of onions, occasionally infests leeks and shallots, but populations are not sustained on garlic. There are commonly three generations per year, with the first one being the most damaging because of reduction or loss of stand. Numerous natural enemies are known, but they are uncommon in commercial fields where insecticide treatments prevail.

Delia platura **(Meigen), the seedcorn maggot** (figs. F, L). Believed to be native to the Holarctic region, the seedcorn maggot is nearly cosmopolitan, and occurs throughout much of the United States and southern Canada. It commonly attacks germinating seeds and feeds on other organic matter in soils, being most commonly a problem in high organic content soils (including those containing manure and plowed down cover crops). It attacks a wide variety of larger seeds, especially beans, peas, corn, cucumbers and other cucurbits. It is especially attracted to microbially-colonized seeds and can attack young seedlings, resulting in stand loss and reduced seedling vigor. The number of generations varies from two to many overlapping generations per year.

Delia florilega **(Zetterstedt)** (fig. K). Holarctic in distribution like *D. platura,* and possibly of Eurasian origin, this species is similar to the seedcorn maggot in being a secondary invader that is primarily attracted by high organic matter or microbial growth. It is found from Alaska to eastern Canada, south to New Mexico in the West and New York in the East. The maggots attack a variety of larger seeds of the same crops as *D. platura*.

***Delia radicum* (L.), the cabbage maggot** (figs. B, C, M, 37.319a–d). Also believed to be native to northern Eurasia, the cabbage maggot has a distribution similar to that of the onion maggot across the northern United States and southern Canada, where it can cause significant injury to crucifers. This includes damage to the edible roots of radish, horseradish, turnip, and rutabaga, and reduced vigor, yield or stand loss to crops such as cabbage, broccoli, Brussels sprouts, cauliflower, Chinese cabbage and kale. Numerous wild mustards and relatives are also believed to be hosts, possibly providing the reservoir of non-resistant flies, since resistance to insecticides is not known in this species to date. There are commonly three generations per year, and an array of natural enemies similar to those of the onion maggot is known from insecticide-free areas.

***Delia planipalpis* (Stein), the "western cabbage maggot".** Native to western North America, this Holarctic species currently is found from the northern Great Plains westward in America and in mountainous regions as far south as California and Colorado. It attacks radish and all of the *Brassica* varieties (cabbage, cauliflower, etc.) but is of minor importance compared with the cabbage maggot, *D. radicum* (L.). It has one generation in the north and two or more farther south.

***Delia floralis* (Fallén), the turnip maggot** (figs. N, O). Native to Eurasia, the turnip maggot has a more northern distribution than the cabbage maggot, *D. radicum*, and the "western cabbage maggot," *D. planipalpis*, and has not been recorded from the lower 48 states, although its hosts are similar. It is the most important pest of these crops from Alaska across the Canadian Prairie Provinces. There is a single generation per year in North America, but more than one generation is known from parts of Europe.

***Delia coarctata* (Fallén), the "wheat bulb maggot"** (fig. P). Native to northern Eurasia, where it is a serious pest of winter wheat, the wheat bulb fly is known from Quebec, the Maritime Provinces, and Maine (McAlpine and Slight 1981). The chief wild host in North America is couch grass, *Agropyron repens* (L.), but to date it has not been reported as attacking commercial grains here. In Europe the maggots attack the roots of fall-sown grains in the spring, notably winter wheat, barley and rye, after hatching from the overwintering eggs. In contrast to our other agriculturally important species of *Delia*, there is only a single generation per year.

It should be noted, especially in the economically important species, that the species names and genera used by authors have not been very stable. For this reason, it is suggested that literature searches on these species be conducted with many combinations of species names, subgenera and genera. For example, Brooks (1951) provided a key for mature larvae of Diptera attacking the roots of cruciferous plants. In this book three of the five species names of *Delia* he used are now considered synonyms. The currently valid names for six common *Delia* species covered by Brooks in 1949 and 1951 follow:

Current Valid Name	*Brooks 1949*	*Brooks 1951*
antiqua (Meigen)	*antiqua* (Meigen)	*antiqua* (Meigen)
floralis (Fallén)	*crucifera* (Huckett)	*crucifera* (Huckett)
florilega (Zetterstedt)	*floralis* (Fallén)	*trichodactyla* (Rondani)
planipalpis (Stein)	*planipalpis* (Stein)	*planipalpis* (Stein)
platura (Meigen)	*cana* (Macquart)	*cilicrura* (Rondani)
radicum (Linnaeus)	*brassicae* (Bouché)	*brassicae* (Bouché)

Selected Bibliography

Aderkas and Peterson 1987.
Beaver, R. A. 1969.
Brooks 1949, 1951.
Cameron 1914.
Coffey 1966.
Eyer 1922.
Ferrar 1987.
Frost 1924.
Gäbler 1933.
Griffiths 1982, 1983, 1984, 1986.
Hennig 1952.
Hewitt 1907.
Huckett 1965c, 1987.
Kangas and Leskinen 1944.
Keilin 1917a, 1924a.
Kompfner 1974.
McAlpine and Slight 1981.
Merritt and Anderson 1977.
Michelson 1980, 1988.
Miles 1950, 1953.
Séguy 1923b, 1932.
Smith, K. M. 1927.
Stone et al. 1965.
Tripp 1954.
Wallace 1971.

MUSCIDAE (MUSCOIDEA)

Gregory A. Dahlem, *Michigan State University*

The Muscids

Figures 37.320–37.328

Relationships and Diagnosis: Muscids are closely related to the Anthomyiidae. Until the 1960s, the name "Anthomyiidae" was commonly used to encompass the entire Muscoidea, as currently delineated. The classification has changed and the name "Muscidae" has been adopted for part of this broad group of flies. Consequently, information may be found in the literature under both family names. At the present time, Muscidae is a separate family in the Muscoidea, along with Anthomyiidae and Scatophagidae (McAlpine *et al.* 1981). Huckett and Vockeroth (1987) have recently summarized our knowledge.

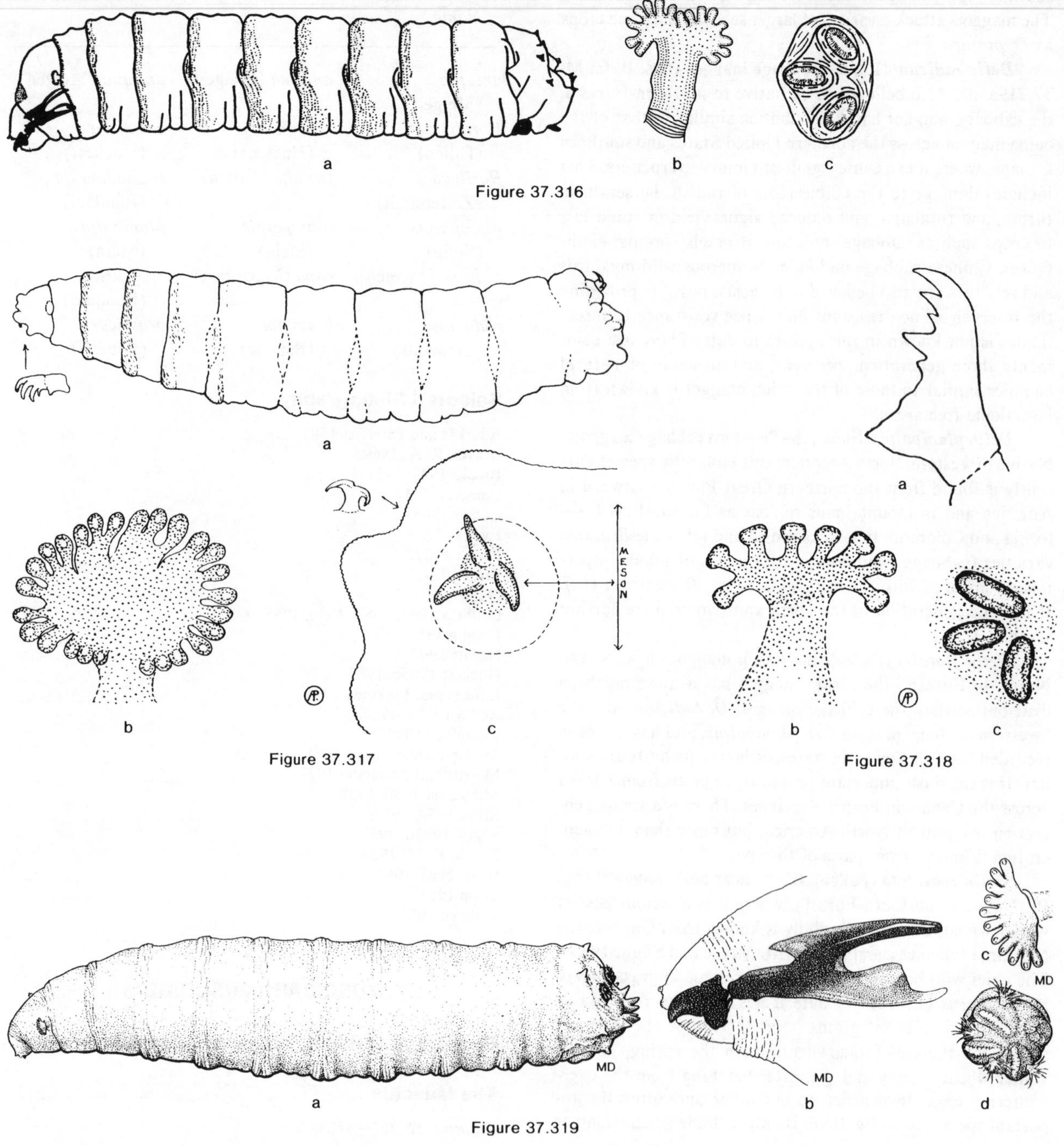

Figure 37.316a-c. Anthomyiidae. *Strobilomya anthracina* Czerny. **a.** third instar larva; **b.** anterior spiracle; **c.** posterior spiracle. Larvae tunnel in the new cones of white spruce, *Picea glauca*, in May and June, destroying the seeds (from Tripp 1954).

Figure 37.317a-c. Anthomyiidae. *Pegomya* sp. **a.** third instar larva; **b.** anterior spiracle; **c.** elevated posterior spiracle, with curved projections bearing spiracular slits. Larvae are leafminers in pale dock, *Rumex altissimus* (from Peterson 1951).

Figure 37.318a-c. Anthomyiidae. *Pegomya hyoscyami* (Panzer), the spinach/beet leafminer. **a.** mandible; **b.** anterior spiracle; **c.** posterior spiracle, which is relatively small and slightly elevated. Larvae are miners in the leaves of spinach, beets, Swiss chard, sugar beets and assorted weeds (from Peterson 1951).

Figure 37.319a-d. Anthomyiidae. *Delia radicum* (L.), the cabbage maggot. **a.** mature larva; **b.** lateral view of cephalopharyngeal skeleton; **c.** anterior spiracle; **d.** posterior spiracle. The cabbage maggot is a pest of roots of many Brassicaceae, especially cabbage and related varieties (from Manual of Nearctic Diptera, Vol. 1).

Within the Muscidae there has been disagreement in recent times over the placement of the subfamilies Fanniinae and Eginiinae. The former is often given family status by current workers, and the Eginiinae, which are Old World internal parasitoids of Diplopoda, have larvae that are quite different from the rest of the Muscidae. Future studies may reveal closer affinities with the Oestroidea (Skidmore 1985).

Generally, muscid larvae (exclusive of the Fanniinae) may be distinguished from related families by their peg-like shape, tapering anteriorly and bluntly rounded posteriorly, greatly reduced or absent spiculation on much of the body, and lack of enlarged tubercles on the spiracular disc. The Fanniinae are quite different in that the body is dorsoventrally flattened and possesses many filament-like projections (fig. 37.322a).

Biology and Ecology: Larvae of Muscidae are found in diverse habitats. They may be saprophagous, coprophagous, predacious, or rarely phytophagous. Many mature larvae whose feeding behaviors have been studied are obligate or facultative predators on other insects (especially Diptera larvae) in the medium in which they are found.

Dung is a common medium inhabited by a rich variety of muscid larvae. *Muscina levida* (Harris) has been reared from cow dung and *M. stabulans* (Fallén) from chicken dung. All known species of *Azelia, Mesembrina, Morellia,* and many species of *Hydrotaea, Ophyra, Potamia, Polietes* (*Pseudophaonia*), *Hebecnema, Mydaea, Pseudolimnophora,* and *Brontaea,* as well as the common species *Graphomya maculata* (Scopoli), *Eudasyphora cyanicolor* (Zetterstedt), *Neomyia cornicina* (Fabricius) (=*Orthellia caesarion*), *Haematobia irritans* (L.), *Stomoxys calcitrans* (L.), *Musca domestica* L. and *M. autumnalis* De Geer have all been reared from dung of a variety of livestock. *Haematobosca alcis* (Snow) has only been reared from moose dung. Many other muscids are occasionally reared from dung.

Carrion is another common medium exploited by muscid larvae. *Muscina levida* (Harris) and *Hydrotaea armipes* Fallén have been reared from dead snails. *Muscina pascuorum* (Meigen) has been reared from caterpillars and *Synthesiomyia nudiseta* (Wulp) has been reared from dead locusts, but is more commonly found in large vertebrate carrion. *Muscina prolapsa* (Harris), *Antherigona* (*Acritochaeta*) *orientalis* Schiner, several species of *Ophyra,* and *Hydrotaea dentipes* (F.) are predacious in large vertebrate carrion. Several other muscids are occasionally reared from carrion, including many species of Fanniinae.

Some species are known to feed on living vertebrates. *Muscina levida* and *M. stabulans* may cause myiasis in nestling birds. Species of *Philornis* are subcutaneous haematophages on nestling birds. *Hydrotaea armipes, H. basdeni* Collin, *Ophyra campensis* (Wiedemann), *Potamia littoralis* Robineau-Desvoidy have all been reared from various bird nests, but their food resource is unclear. Several species of *Fannia* have been implicated in human intestinal, vesicular, and aural myiasis.

Compost, rotting fruit, leaf litter, sickly parts of trees or herbaceous plants and humous soil are favorable, although often secondary, media for many muscid larvae. Fungi are also a productive medium for a variety of muscids, with *Alloeostylus diaphanus* (Wiedemann), most *Mydeae* spp., *Fannia* spp., and *Platycoenosia mikii* Strobl having been reared from a variety of fungi.

Lispoides aequifrons (Stein) and most *Spilogona* and *Limnophora* are predators found in aquatic or semiaquatic environments. Some species are found in association with running water and some with ponds or boggy pools.

Before pupariation, many muscids produce cocoons. This is common in the subfamilies Reinwardtiinae, Mydaeinae and the brontaeine Phaoniinae, and also occurs in scattered species in other subfamilies. The cocoon is formed from debris which is usually agglutinated by a frothy oral exudate.

Description: Mature larvae 6–20 mm, conical, tapering anteriorly and bluntly rounded posteriorly, greatly reduced or absent spiculation on much of the body and lack of enlarged tubercles on the spiracular disc. The anal region is rarely visible in the dorsal view. The Fanniinae are quite different in that the body is dorsoventrally flattened and possesses many filament-like projections. Aquatic larvae may possess abdominal prolegs and sometimes long paired caudal processes. Mature larvae are amphipneustic.

Head: Cephalic segment retractile, bearing sensory organs anteriorly and anteroventrally. Antennae appearing 2-segmented. Spine band commonly occurring on ventral surface behind preoral cavity.

Cephalopharyngeal skeleton deeply pigmented, usually heavy; tentoropharyngeal and hypopharyngeal sclerites separate, dorsal cornua usually narrower than ventral cornua and lacking apical windows; ventral cornua with dorsobasal lobe (fig. 37.320b) and commonly with apical windows; hypopharyngeal sclerite H-shaped, parastomal bars usually lacking; epipharyngeal and labial sclerites absent. Mandibles well developed and heavy (some species with only one mandible or with one mandible reduced in size); hook nearly straight to sickle-shaped and decurved, without accessory teeth; basal part somewhat quadrate and may have small windows; dental sclerite usually elongate and with ventral projection anteriorly when viewed laterally; rod-like accessory mandibular sclerites present (fig. 37.320b) (predacious species) or absent (saprophagous species) below mandibles.

Thorax and Abdomen: With or without encircling spinule bands, but often with ventral creeping welts. Anterior spiracles arising posterolaterally on prothorax; variable in shape.

Posterior spiracular disc usually without tubercles and postanal tubercles, although short anal papillae are often present. Posterior spiracles usually elevated above the plane of the terminal segment on short tubes which are often darkly pigmented. Spiracular plates with three spiracular slits which range in shape from straight to serpentine, with many variations in between.

Comments: Nearly 50 genera and more than 700 species are recognized in America north of Mexico (Huckett and Vockeroth 1987). Skidmore (1985) provides an excellent summary of the world literature on the biology and immature stages of Muscidae and provides many new descriptions of larvae and puparia. This work describes in detail some 450

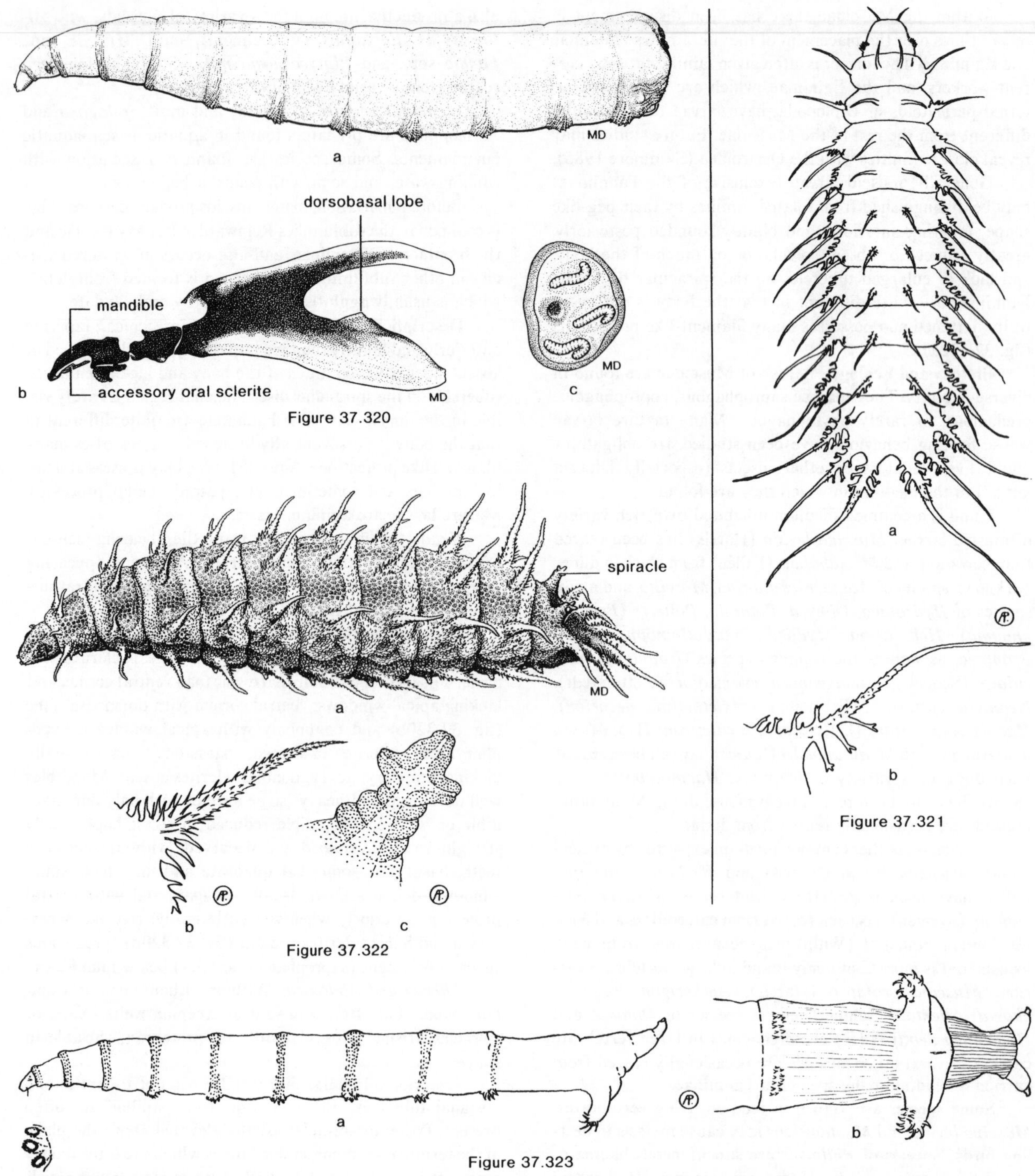

Figure 37.320a-c. Muscidae. *Potamia* sp. **a.** larva; **b.** cephalopharyngeal skeleton; **c.** right posterior spiracle. Larvae are facultative predators in a great variety of habitats (from Manual of Nearctic Diptera, Vol. 1).

Figure 37.321a,b. Muscidae. *Fannia scalaris* (Fabricius), the latrine fly. **a.** larva, dorsal; **b.** enlarged caudal filament. Larvae feed in human and vertebrate excrement and other decaying materials (from Peterson 1951).

Figure 37.322a-c. Muscidae. *Fannia canicularis* (L.), the little house fly. **a.** larva, lateral; **b.** enlarged filament; **c.** posterior spiracular stalk. Larvae feed in wet, decaying leaf litter and other rotting plant material (fig. a from Manual of Nearctic Diptera, Vol. 1; figs. b, c from Peterson 1951).

Figure 37.323a,b. Muscidae. *Lispoides aequifrons* (Stein). **a.** lateral; **b.** ventral of caudal segments. Larvae live in mosses and algae in rapidly flowing streams where they prey on insect larvae (from Peterson 1951).

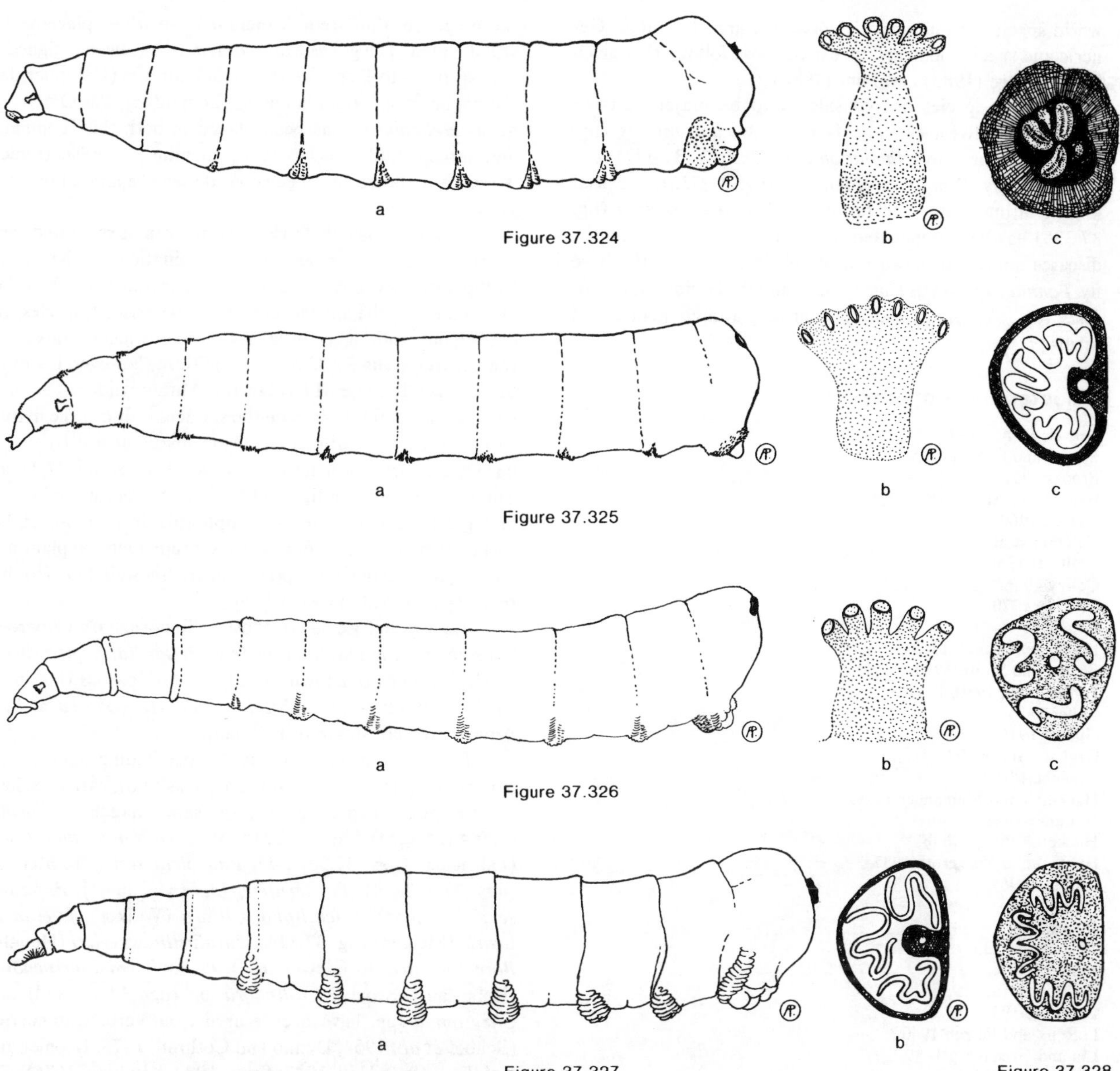

Figure 37.324a-c. Muscidae. *Muscina* sp. (probably *assimilis* (Fallén)). **a.** lateral; **b.** anterior spiracle; **c.** left posterior spiracle. Larvae have two mandibles and develop in injured or decaying plant matter. *Muscina stabulans* (Fallén), the false stable fly, has similar habits (from Peterson 1951).

Figure 37.325a-c. Muscidae. *Musca domestica* L., the house fly. **a.** lateral; **b.** anterior spiracle, with 6 or 7 orange papillae; **c.** left posterior spiracle, with 3 sinuous spiracular slits, a strong peritreme, and the ecdysial scar located toward the median. Spiracles separated by less than a spiracular width. Larvae feed in excrement, especially fresh horse manure, garbage, and all sorts of decaying plant materials (figs. a, b from Peterson 1951; fig. c from U.S. Public Health Service (CDC) Pictorial Keys (to) Arthropods of Public Health Importance, 1962).

Figure 37.326a-c. Muscidae. *Stomoxys calcitrans* (L.), the stable fly or so-called "biting house fly" (both sexes bite). **a.** larva; **b.** anterior spiracle, with 5 or 6 yellow papillae; **c.** left posterior spiracle, usually quite dark, with 3 S-shaped spiracular slits, an indistinct peritreme; and the ecdysial scar centrally located. Spiracles separated by more than a spiracular width. Larvae feed in accumulations of decaying plant materials and to a lesser extent manures (figs. a, b from Peterson 1951; fig. c from U.S. Public Health Service (CDC) Pictorial Keys (to) Arthropods of Public Health Importance, 1962).

Figure 37.327a,b. Muscidae. *Haematobia irritans* (L.), the horn fly; both sexes are blood-sucking pests of livestock, principally cattle. **a.** larva; **b.** left posterior spiracle, with 3 irregular sinuous spiracular slits, a thin peritreme, and the ecdysial scar toward the median. Larvae feed in very fresh cattle manure (fig. a from Peterson 1951; fig. b from U.S. Public Health Service (CDC) Pictorial Keys (to) Arthropods of Public Health Importance, 1962).

Figure 37.328. Muscidae. *Musca autumnalis* De Geer, the face fly, whose non-biting adults cluster on the faces of cattle and lap up secretions. Left posterior spiracle; usually quite dark, with 3 strongly sinuous spiracular slits, little or no peritreme, and the ecdysial scar toward the median. Spiracles separated by less than one spiracular width. Larvae feed primarily in cattle manure (from U.S. Public Health Service (CDC) Pictorial Keys (to) Arthropods of Public Health Importance, 1962).

world species of muscids, 275 of which are illustrated. Generic and species nomenclature used above follows that given by Skidmore (1985) and Pont (1986).

Several species are considered to be major pests of humans and livestock. The face fly, *M. autumnalis* (fig. 37.328), the horn fly, *Haematobia irritans* (fig. 37.327), and the stable fly, *Stomoxys calcitrans* (fig. 37.326), are well known biting pests. The house fly, *Musca domestica* (fig. 37.325) has been implicated in the spread of several human diseases and is well known as a household pest. The little house fly, *Fannia canicularis* (fig. 37.322) and the latrine fly, *F. scalaris* (fig. 37.321), are often annoying around homes and farms.

Selected Bibliography

Beaver, R. A. 1969.
Bohart and Gressitt 1951.
Brooks 1951.
Burger and Anderson 1974.
Buxton 1960.
Calhoun et al. 1956.
Chillcott 1960.
Coffey 1966.
Conway 1970.
Disney 1972.
Dodge and Aitken 1968.
Easton and Smith 1970.
Ferrar 1979, 1980, 1987.
Fuller 1934.
Gilbert 1919.
Graham-Smith 1916.
Greenberg 1971.
Hackman and Meinander 1979.
Hammer 1941.
Huckett 1965a, 1965b.
Huckett and Vockeroth 1987.
Ishijima 1967.
Iwasa 1980.
James 1947.
Johannsen 1935.
Keilin 1917b.
Keilin and Tate 1930.
Laurence 1954.
LeRoux and Perron 1960.
Liu and Greenberg 1989.
Marchand 1923.
Merritt and Anderson 1977.
Niblett 1955.
Pont 1986.
Skidmore 1985.
Stork 1936.
Thomson, M. and Hammer 1936.
Thomson, R. C. M. 1937.
Wallace 1971.
Wallace and Snoddy 1969.

CALLIPHORIDAE (OESTROIDEA)

Gregory A. Dahlem, *Michigan State University*

The Blow Flies

Figures 37.329–37.337

Relationships and Diagnosis: Calliphorids are closely related to the Sarcophagidae and Rhinophoridae. The main body of flies included in the family has remained fairly constant over the years, although this cannot be said for some of the subgroups that are now included, but whose placement is still debated. The genus *Melanomya* s.l. has been assigned at various times to each of four oestroid families (Calliphoridae, Rhinophoridae, Sarcophagidae, Tachinidae). The Old World genus *Helicobosca* has been placed in both the Tachinidae and Sarcophagidae, as have the Ameniinae. The Neotropical Mesembrinellinae have been considered a separate family by various authors.

Larvae are usually fairly large and can be separated from other calyptrate families by a combination of characters. Calliphorid larvae can usually be separated from the other Oestroidea by the slit pattern of the posterior spiracles, radiating out from the ecdysial scar which is incorporated into the peritreme (fig 37.329c). This pattern, however, is similar to that seen in some Muscidae and Anthomyiidae. The presence of marginal tubercles on the spiracular disc of the mature calliphorid larva will nearly always distinguish it from similar muscid larvae which usually lack tubercles (fig. 37.325a). The most common calliphorid food of vertebrate carrion will biologically separate most calliphorids from most anthomyiids (which are phytophagous, saprophagous on plant material, coprophagous or predacious). Shewell (1987b) has recently reviewed our knowledge.

Biology and Ecology: Most calliphorids are oviparous, but some taxa are larviparous (e.g. *Bellardia, Eggisops,* and *Onesia*). Some are known to be macrolarviparous (maturing one larva at a time), (e.g. *Euphumosia, Helicobosca,* and the Ameniinae and Mesembrinellinae).

The larvae have relatively diverse feeding habits, most being necrophagous, but a few are parasitoids, parasites, haematophagous, coprophagous, or saprophagous. *Phormia regina* (Meigen) (figs. 37.331a–e), *Cochliomyia macellaria* (Fabricius) (figs. 37.330a–f), *Phaenicia sericata* (Meigen) (figs. 37.333a–d), *P. coeruleiviridis* (Macquart), *P. pallescens* (Shannon), *Eucalliphora lilaea* (Walker), *Lucilia illustris* (Meigen) (fig. 37.334), *Paralucilia wheeleri* (Hough), *Boreellus atriceps* (Zetterstedt), *Protophormia terraenovae* (Robineau-Desvoidy), *Calliphora* spp. (figs. 37.332 a–d), and *Chrysomya* spp. have been reared from vertebrate carrion (Schoof *et al.*, 1954; Denno and Cothran, 1975; Deonier and Knipling, 1940; Hall, 1947; Prins, 1982; Kitching, 1976). *Cynomya cadaverina* (Robineau-Desvoidy) (figs. 37.336 a, b) is noted to use cured meats as a larval food. Many other species have been reared in laboratory conditions on decaying meat (Hall, 1947), and many of these have been reared from household garbage (Schoof *et al.*, 1954). This group of species has also occasionally been reared from excrement, but this does not seem to be the primary larval food for any of the Nearctic species. Occasionally, some may be reared from rotting fruit, vegetable or other plant materials.

Several groups of Calliphoridae are parasitoids or scavengers in land snails. In the Old World, *Helicobosca muscaria* (Meigen) and *H. palpalis* (Robineau-Desvoidy) have been recorded from snails in the genera *Thebia, Cepaea, Arianta,* and *Helix* (Rognes, 1986). *Amenia leonina* (Fabricius) of the Ameniinae has been reared from snails (Crosskey, 1965). Many, and possibly all, of the species of *Melanomya* and relatives are parasitoids of land snails (Downes, 1986).

The common cluster fly, *Pollenia rudis* (Fabricius) (figs. 37.337a–e), is a parasitoid of various species of earthworms

(Hall, 1947; Richards and Morrison, 1973; Yahake & Georfe, 1972). Different strains (or perhaps sister species) of this species seem to have marked preferences for different species of earthworms and have different larval behavior in the host. Old World members of *Onesia* have also been noted as earthworm parasitoids (Hall, 1947).

Stomorhina lunata (Fabricius) larvae are predacious upon grasshopper egg masses (Hall, 1947).

Several calliphorids are known to be parasites of vertebrates. *Bufolucilia bufonivora* (Moniez) and *B. silvarum* (Meigen) have been reported as parasites of toads and frogs and seem to base their attack in the region of the eyes (Hall, 1947; Strihnasah, 1980). Members of the genus *Protocalliphora* (figs. 37.335a,b) are obligatory, bloodsucking parasites upon nestling birds (Hall, 1947), and the larvae of *Trypocalliphora braueri* (Hendel) are obligate subcutaneous parasites on nestling and juvenile birds (Rognes, 1984).

The larvae of the screwworm, *Cochliomyia hominivorax* (Coquerel) (figs. 37.329a–f), are found in nature only in wounds of mammals and never in decaying meats (Hall, 1947). *C. macellaria,* the secondary screwworm (figs. 37.330a–f), and several other carrion associated larvae are known for their secondary invasion of wounds in livestock, humans, and other mammals (Hall, 1947; James, 1947). In the Old World, several members of the genus *Chrysomya* have been implicated in myiasis of humans, livestock and other mammals (James, 1947).

Description: Mature larvae 6–20 mm, muscidiform, tapering anteriorly and truncate or broadly rounded posteriorly; integument usually with encircling bands of spinules along anterior segmental border on last two thoracic and most abdominal segments; ventral creeping welts present; white.

Head: Cephalic segment retractile, bearing well-developed sensory organs anteriorly and anteroventrally; antennae appearing 2-segmented; facial mask with numerous simple or branched oral grooves and ridges leading into preoral cavity.

Cephalopharyngeal skeleton usually deeply pigmented; tentoropharyngeal and hypopharyngeal sclerites separate; dorsal cornua commonly longer than ventral cornua and with or without elongate window distally, connected anteriorly by dorsal bridge; ventral cornua broad, usually with distinct dorsal lobe and oval to elongate window dorsoposteriorly, pharyngeal filter usually present (absent in some predacious species); hypopharyngeal sclerite H-shaped; parastomal bars present and free apically; apparently no epipharyngeal sclerite; labial sclerite variously shaped and commonly in form of a V. Mandibles deeply pigmented, hook decurved and commonly sickle-shaped, usually without accessory teeth, basal part subrectangular and commonly with small window; dental sclerite usually triangular in lateral view; accessory oral sclerites present in few species.

Thorax: Anterior spiracles arising posterolaterally on prothorax; at least the last two segments with encircling band of spinules near anterior margin as well as ventral creeping welts; anterior spiracles simple, commonly fan-shaped with 8–15 short papillae along distal margin.

Abdomen: All segments except caudal one very similar, usually encircled by spinule bands near anterior margins and with creeping welts ventrally; perianal pad bilobed.

Posterior spiracular disc commonly with 6–7 pairs of marginal tubercles; spiracles arising on upper half of disc, usually borne on short pigmented bases; each spiracular plate with deeply pigmented, broken or unbroken peritreme, three oval spiracular openings obliquely inclined and radiating out from ecdysial scar; scar incorporated into peritreme; spiracular hairs usually absent.

Comments: This is a fairly large family worldwide, but poorly represented in the New World, and with only about 75 species and 24 genera recorded for America north of Mexico. Larger genera are *Calliphora* and *Protocalliphora,* each with 10 species.

Several species have economic significance because of the tendencies of their larvae to invade healthy tissues surrounding superficial wounds in humans and domestic animals. The screwworm, *Cochliomyia hominivorax* (Coquerel) (figs. 37.329a–f) has been responsible for massive losses of cattle in Texas and elsewhere in subtropical and tropical America. This species is slowly coming under control via use of the sterile male technique. Occasionally, eggs of *Phaenicia sericata* (Meigen) are deposited in cutaneous wounds, and the developing larvae (figs. 37.333a–d) attack necrotic tissue. In earlier times, larvae of this species were actually used medically to clean up wounds and aid the healing process. Today, unwanted myiasis due to this species occasionally appears in hospital patients. There are reports of larvae of some species of this genus moving from damaged tissues in a purulent wound to attack healthy flesh. In Australia, sheep blow flies of the genera *Calliphora* and *Lucilia* are of considerable concern to the livestock industry due to the myiasis-causing tendencies of the larvae. Another significant medical role of calliphorid flies rests on their documented ability to serve as vectors of microorganisms that cause human enteric diseases. Adults of the cluster fly, *Pollenia rudis* (Fabricius), frequently invade heated buildings during the fall months where they can become a nuisance.

Selected Bibliography

Coutant 1914.
Crosskey 1965.
Denno and Cothran 1975.
Deonier and Knipling 1940.
Downes 1986.
Ferrar 1987.
Greenberg 1971, 1973.
Greenberg and Szyska 1984.
Hall 1947.
James 1947.
Kitching 1976.
Knipling 1936b, 1939.
Laake et al. 1936.
Liu and Greenberg 1989.
Prins 1982.
Reiter and Wollenek 1982.
Richards and Morrison 1973.
Roberts 1971a.
Rognes 1984, 1986.
Sabrosky 1956.
Schoof et al. 1954.
Shewell 1987b.
Smith, K. G. V. 1986.
Strihnasah 1980.
Yahake and Georfe 1972.
Zumpt 1965.

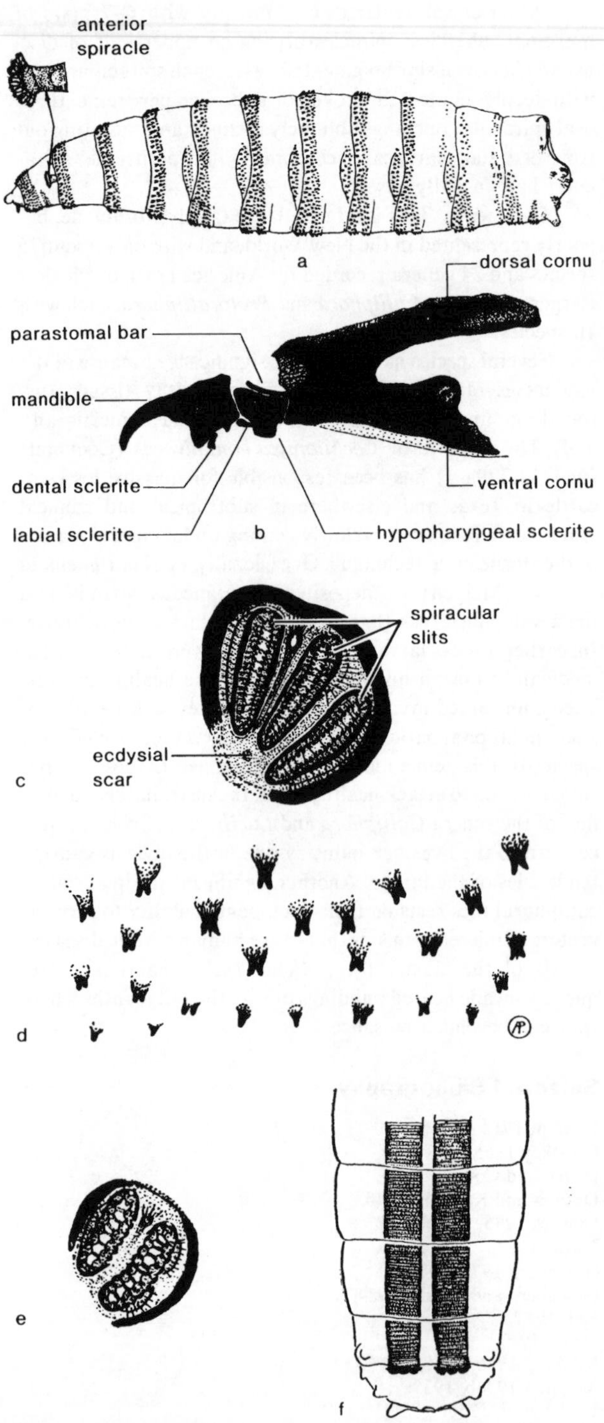

Figure 37.329a-f. Calliphoridae. *Cochliomyia hominivorax* (Coquerel), the screwworm. Last instar, except e. **a.** lateral; **b.** cephalopharyngeal skeleton; **c.** right posterior spiracle; **d.** spinules on mid-dorsum of fourth abdominal segment; **e.** 2nd instar right posterior spiracle; **f.** pigmented tracheal trunk. Larvae are distinguished from *C. macellaria,* the secondary screwworm (figs. 37.330a-f), by the pigmented tracheal trunks. Larvae feed only on living tissue of wild and domestic animals, particularly cattle, in the Southwest, although they have been greatly reduced by the sterile male release program in the U.S. and Mexico (a, b, c, e from Hall 1947; d from Peterson 1951; f from James 1947).

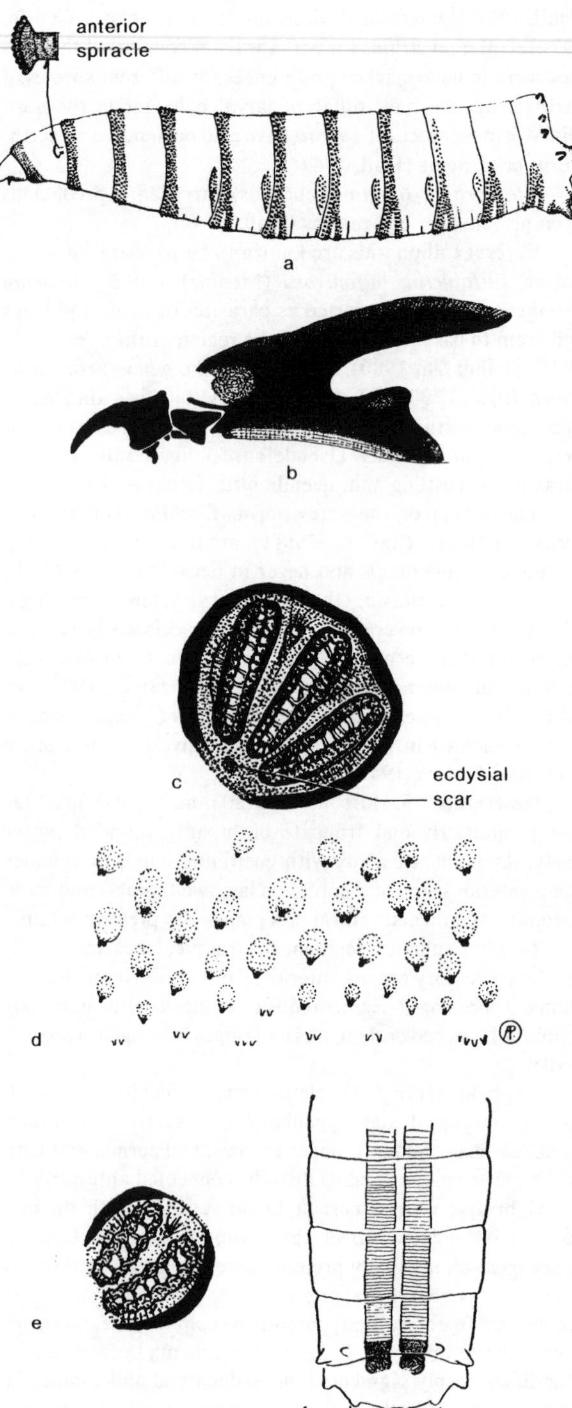

Figure 37.330a-f. Calliphoridae. *Cochliomyia macellaria* (Fabricius), the secondary screwworm. Last instar larva except e. **a.** lateral; **b.** cephalopharyngeal skeleton; **c.** posterior spiracle; **d.** mid-dorsum of fourth abdominal segment; **e.** 2nd instar right posterior spiracle; **f.** unpigmented tracheal trunk. Larvae cause secondary myiasis in wounds on wild and domestic animals, especially after injury by *C. hominivorax,* the screwworm (figs. 37.329a-f), but will also develop in carcasses (a, b, c, e, from Hall 1947; d from Peterson 1951; f from James 1947).

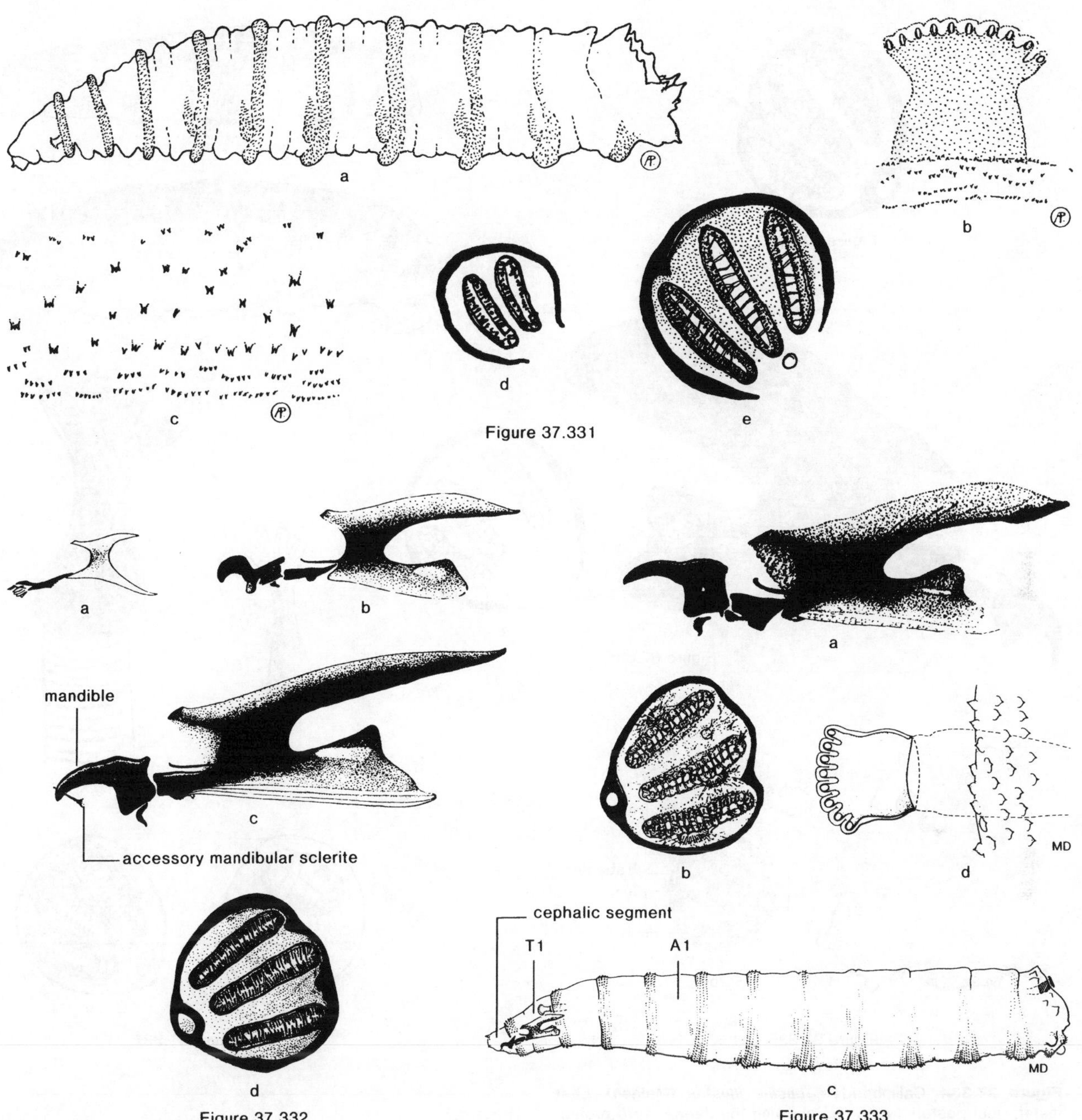

Figure 37.331a-e. Calliphoridae. *Phormia regina* (Meigen), the black blow fly. Last instar (except d) larval structures. **a.** larva; **b.** anterior spiracle; **c.** spinules on mid-dorsum of fourth abdominal segment; **d.** posterior spiracle, 2nd instar; **e.** posterior spiracle, 3rd instar. Larvae feed primarily in carcasses, but also in garbage; may cause secondary myiasis (a, b, c from Peterson 1951; d, e from Hall 1947).

Figure 37.332a-d. Calliphoridae. *Calliphora vicina* Robineau-Desvoidy. Adults are black and blue (abdomen). **a, b, c.** First, second, third instar cephalopharyngeal skeletons, respectively (note accessory mandibular sclerite in **c.** which is lacking in the genera *Phaenicia* (fig. 37.333a) and *Lucilia*.); **d.** right posterior spiracle. Worldwide in temperate regions and throughout most of North America. Most common in spring and fall. Larvae primarily in carrion, although known to cause secondary myiasis in humans and other animals (James 1947) (from Hall 1947).

Figure 37.333a-d. Calliphoridae. *Phaenicia sericata* (Meigen) The adult is a common metallic-green blow fly. Last instar larval structures (except c). **a.** cephalopharyngeal skeleton; **b.** right posterior spiracle (compare with *L. illustris*, fig. 37.334); **c.** second instar larva; **d.** anterior spiracle third instar. Widespread in North America. Common in carrion along with *L. illustris* and others; sometimes in garbage and excrement. One of the important sheep strike species (primary or secondary) in parts of the world; also found in foul wounds and was used for wound cleaning before the advent of antibiotics (a, b from Hall 1947; c, d from Manual of Nearctic Diptera, Vol. 2).

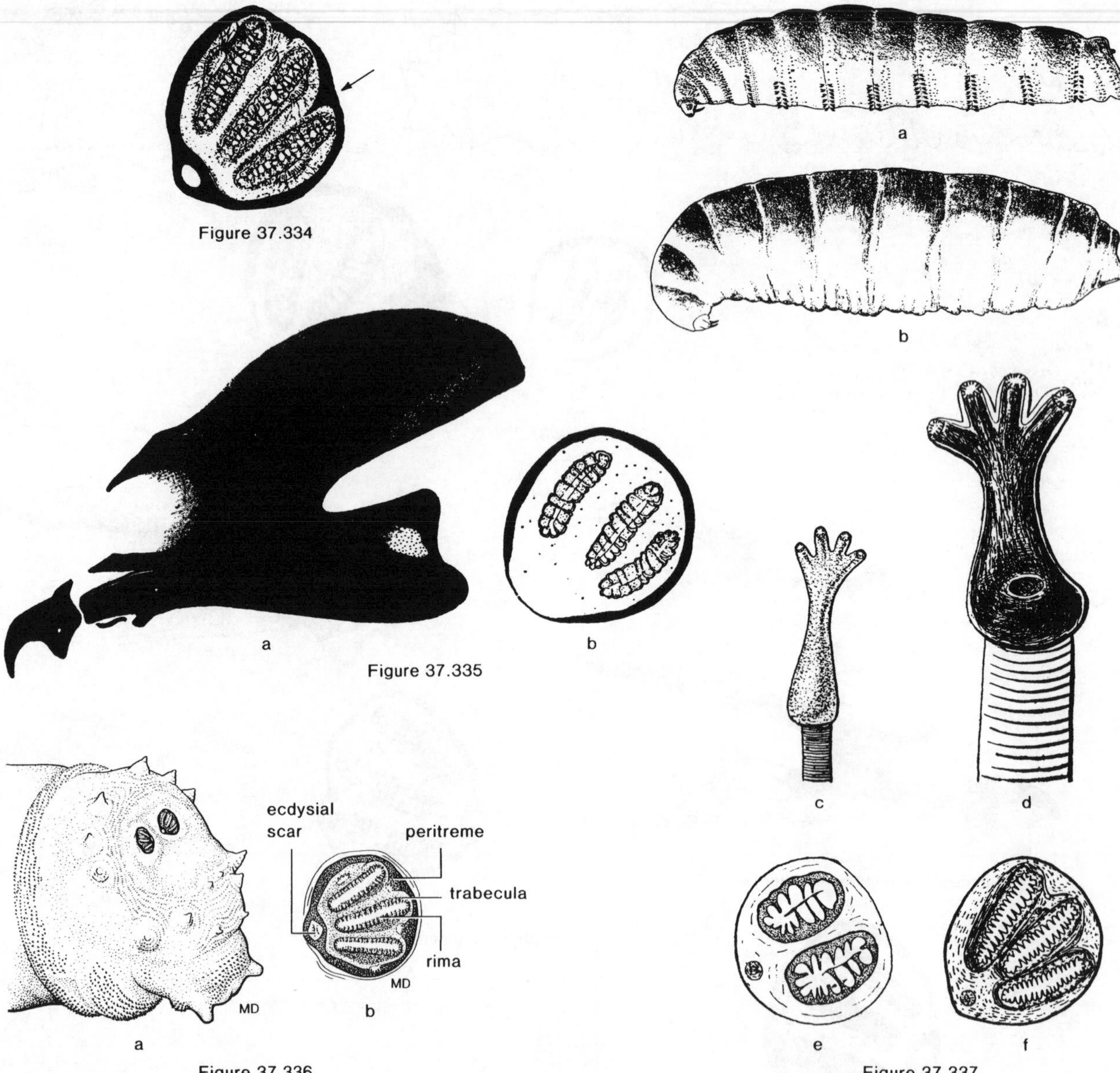

Figure 37.334. Calliphoridae. *Lucilia illustris* (Meigen). Last instar, right posterior spiracle, showing the deep inward projection of the peritreme between the median and outer spiracular slits (shallower in *Phaenicia sericata* (fig. 37.333b), but other *Phaenicia* spp. may have deep projections). Both *Lucilia* and *Phaenicia* lack the accessory mandibular sclerite found in *Calliphora* (fig 37.332c). Holarctic. The adult is a common bluish-green blowfly. An abundant species from Mexico to southern Canada, the larvae primarily feeding in carrion and occasionally in garbage and excrement (from Hall 1947).

Figure 37.335a,b. Calliphoridae. *Protocalliphora avium* (Shannon and Dobroscky). Adults are metallic blue-black. Last instar larval structures. **a.** cephalopharyngeal skeleton; **b.** right posterior spiracle. Larvae are found in the nests of crows, magpies, and passerine birds, where they are intermittent blood feeders. Jellison and Philip (1933) found them in all crow and magpie nests examined in the Bitterroot Valley of Montana, where they appeared to cause no significant injury to the nestlings, in contrast to reports of injury and even death to passerine birds (from Hall 1947).

Figure 37.336a,b. Calliphoridae. *Cynomya cadaverina* (Robineau-Desvoidy). Last instar larval structures. **a.** oblique posterior view of caudal segment of third instar larva; **b.** right posterior spiracle. Throughout the U.S. and Canada, the larvae feeding on carrion and cured meat. The adult is a blue blowfly that is most abundant in spring and fall (from Manual of Nearctic Diptera, Vol. 1).

Figure 37.337a-f. Calliphoridae. *Pollenia rudis* (Fabricius), the cluster fly. **a.** first instar larva; **b.** last instar larva; **c.** second instar anterior spiracle; **d.** last instar anterior spiracle; **e.** second instar posterior spiracle; **f.** last instar anterior spiracle. Holarctic, more common in the north. The larvae parasitize many species of earthworms. Adults black and gray, with dense crumbly yellow thoracic pile; they are a common overwintering nuisance in loosely-constructed buildings (a-e from Keilin 1915; f from Hall 1947).

OESTRIDAE (OESTROIDEA)

B. A. Foote, Kent State University

F. W. Stehr, Michigan State University

Bot Flies, Warble Flies

Figures 37.338–37.345

Relationships and Diagnosis: The taxonomic placement of the species currently included in the family Oestridae has had a long and confusing history. Earlier workers tended to subordinate the group as a subfamily or lesser entity within other families, while other workers segregated certain clusters of genera and gave them family status (e.g., the Gasterophilidae, Cuterebridae and Hypodermatidae). McAlpine et al. (1981) recognized them as a distinct family within the calyptrate Oestroidea, a taxon that also includes the Calliphoridae, Sarcophagidae, Rhinophoridae, and Tachinidae. D. M. Wood (1987a) provides an excellent summary of oestrid relationships and biology on a worldwide basis.

Larvae are easily recognized by their large swollen bodies that bear many heavy spines that may encircle the segments (fig. 37.338a), by the absence or reduced nature of the anterior spiracles, and by their endoparasitic habit in mammals.

Biology and Ecology: This is a family of heavy-bodied, bee-like flies whose buzzing oviposition behavior of some species may irritate and excite their hosts, and whose larvae are endoparasites. Females of Hypodermatinae, such as the northern cattle grub, *Hypoderma bovis* (L.), lay eggs on hairs of the legs from whence larvae penetrate the skin and eventually burrow through the tissues to the spinal canal. They later migrate to the subcutaneous tissues of the back where they form warbles under the skin and cut a respiratory opening for the caudal spiracles. Mature larvae abandon the host to form puparia in the soil in the spring and early summer.

In contrast to the Hypodermatinae, females of the Oestrinae (e.g., *Oestrus ovis* L., the sheep bot fly) are larviparous and deposit their larvae directly into the nasal cavities of the host. Larvae then move into the frontal sinuses where they attach to the mucous membrane. When mature, larvae leave the host and form puparia in the soil. Somewhat similar habits are found in species of the Cephenemyiinae whose larvae develop within the nasal cavities of members of the deer family Cervidae.

Catts (1982) has reviewed the biology of the subfamily Cuterebrinae which contains the rabbit and rodent bot flies, the human bot fly, howler monkey bots and a few others about which practically nothing is known.

The human bot fly, *Dermatobia hominis* (Linnaeus Jr.) (fig. 37.342) occurs from Taumalipas in Mexico south to northern Argentina in tropical and subtropical areas, where it commonly infests humans, cattle, dogs and many other wild and domestic mammals, creating a purulent and sometimes painful warble. The life cycle is unusual in that females capture and lay their eggs on other blood-sucking flies; when the carrier fly rests on the host, the eggs hatch, the larvae drop to the host skin, burrow in, and develop to maturity at that site, with no migration to other parts of the host body as many other oestrids do.

Species of the genus *Cuterebra* infest rabbits and rodents. The larvae of some species are among the largest bots (greater than 25 mm in length) and they infest the smallest hosts (small rodents). Sabrosky (1986) has recently monographed the North American *Cuterebra*, but the larvae are poorly known.

Four species of Gasterophilinae, belonging to the genus *Gasterophilus*, have been introduced from the Old World and infest the alimentary tract of horses, donkeys and mules. Most horses are infected to some extent (some heavily) and larvae and puparia can be collected from beneath *fresh* droppings in the spring. The three commonest species can be separated by the following key (adapted from James 1947).

Description: Mature larvae 15–30 mm, cylindrical, swollen and heavy, anterior end frequently narrower and tapering; integument commonly with numerous heavy, frequently thickened, backward projecting spinules that may form distinct bands around body or be evenly distributed over the body; white to yellowish.

Head: Frequently greatly reduced, with vestigial sensory organs; in some species the cephalic segment is recognizable and well-developed sensory organs are present.

Cephalopharyngeal skeleton highly reduced (Hypodermatinae) or quite well developed (other subfamilies); mandibles strong and commonly sickle-shaped.

Thorax: Anterior spiracles lacking or vestigial; segments anteriorly with encircling spinule bands or with spinule bands largely restricted to dorsal and ventral surfaces.

Abdomen: All or nearly all segments with spinule bands that may entirely encircle body or be ventral and dorsal in position.

Posterior spiracles quite diverse; in species of Cuterebrinae, the spiracles bear complex serpentine lines that are arranged into three groups in each spiracular plate (fig. 37.344b); in Oestrinae, the spiracles are placed in a transverse cleft of the caudal segment that can be closed; in Hypodermatinae, the spiracles occur on the posterior surface of caudal segment and are C-shaped with numerous spiracular openings (figs. 37.340b and 37.341b); in Gasterophilinae, the spiracles are oriented so that the spiracular openings of each plate are arranged in three parallel lines (fig. 37.338b).

Comments: This is a small but important family, with only 41 species and 6 genera recorded in America north of Mexico. Because of their parasitic habits, several species are important pests of domestic animals, particularly cattle, goats, sheep and horses. In addition to the human bot fly, *D. hominis*, several species have been reported as accidental in humans, where they can cause injury but do not mature.

KEY TO MATURE LARVAE OF THE THREE MOST COMMON SPECIES OF *GASTEROPHILUS**

1. Spines on anterior margins of segments arranged in a single row (fig. A) .... the throat bot, *G. nasalis* (L.)
   Spines on the anterior margins of segments arranged in a double row, those of the front row being the more strongly developed 2

2. Spines small, tapering to a fine point; spines lacking on at least the middle half of dorsum of segment 10 and on entire dorsum of segment 11 (fig. B) .... the nose bot, *G. haemorrhoidalis* (L.)
   Spines larger and stronger, blunt at the apices; only 1 or 2 pairs of spines lacking on dorsum of segment 10; segment 11 with 1 to 5 spines above lateral line on each side (fig. C) .... the horse bot, *G. intestinalis* De Geer

*Figures A, B, C from Peterson 1951

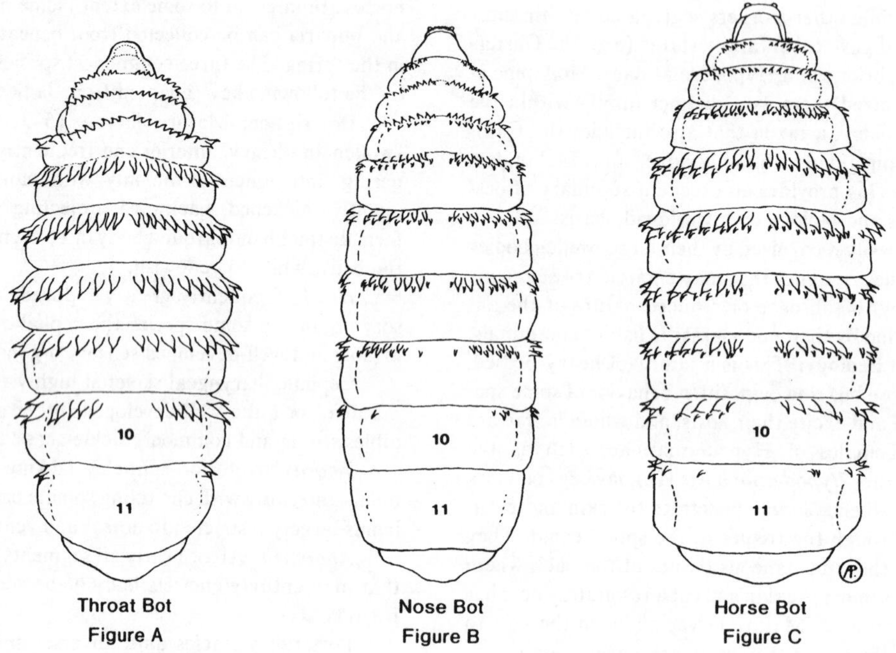

Throat Bot
Figure A

Nose Bot
Figure B

Horse Bot
Figure C

Selected Bibliography

Baird 1972, 1975.
Bennett 1955, 1972.
Bergmann 1917.
Blanchard 1892, 1894.
Brauer 1863, 1883.
Capelle 1966, 1970.
Carpenter and Pollard 1918.
Catts 1965, 1982.
Cawan 1943.
Ferrar 1987.
Hennig 1952.
Hensley 1976.
James 1947.
Knipling and Brody 1940.
Patton 1921, 1923.
Roubaud 1914.
Sabrosky 1986.
Townsend 1892.
Ullrich 1936a, 1936b, 1939.
Vimmer 1925.
Wecker 1962.
Wood, D. M. 1987a.
Zumpt 1965.

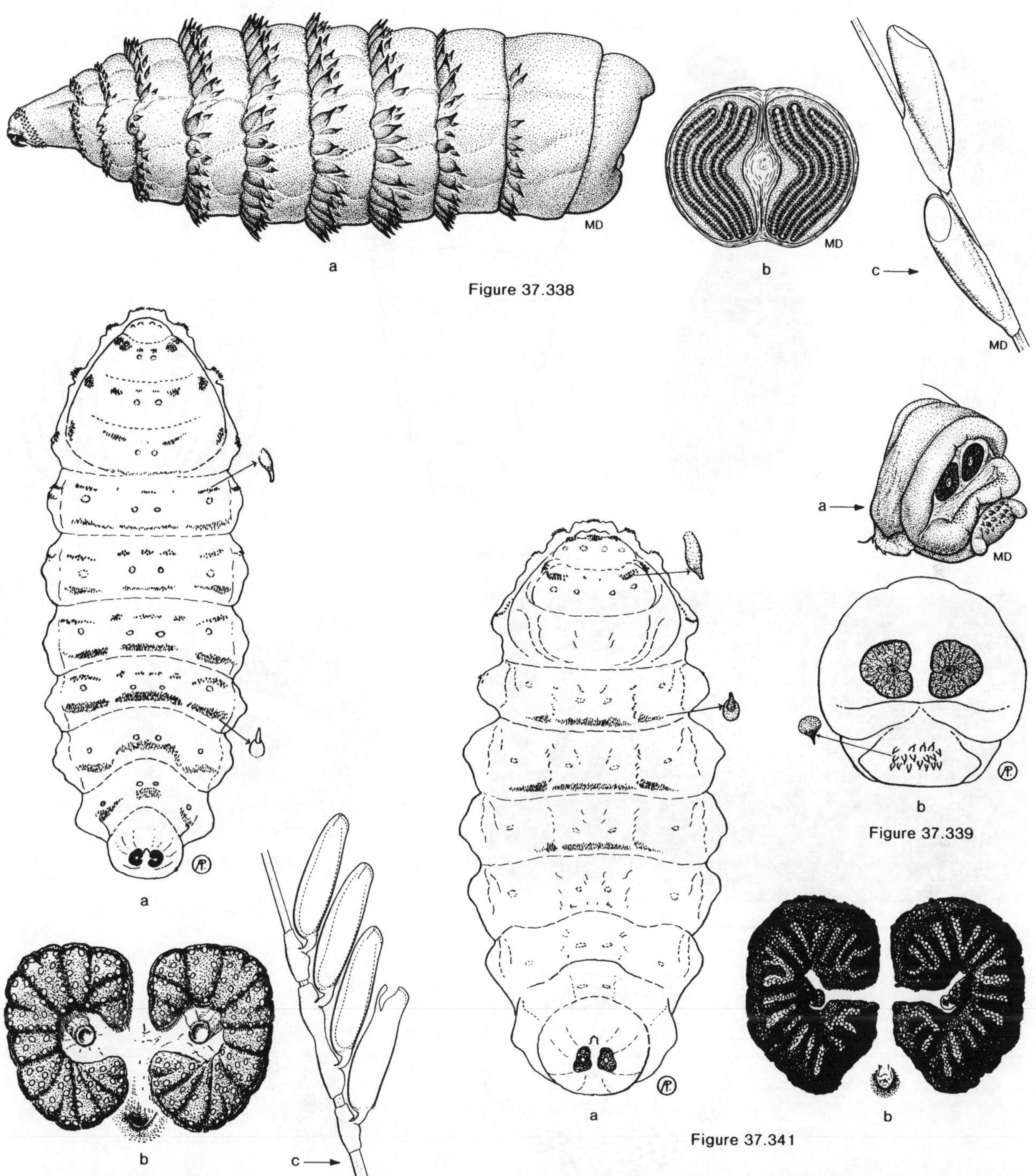

Figure 37.338a-c. Oestridae, Gasterophilinae. *Gasterophilis intestinalis* (De Geer). **a.** lateral; **b.** posterior spiracles; **c.** eggs laid on hair. Larvae of the horse bot fly can be found attached to the inner wall of the stomach, duodenum or rectum of horses, donkeys and mules (from Manual of Nearctic Diptera, Vols. 1 and 2).

Figure 37.339a,b. Oestridae, Oestrinae. *Oestrus ovis* L., the sheep bot. **a.** latero-posterior view of caudal segment; **b.** caudal view showing spiracles. Larvae develop within the nasal passages of sheep and goats (a from Manual of Nearctic Diptera, Vol. 1; b from Peterson 1951).

Figure 37.340a-c. Oestridae, Hypodermatinae. *Hypoderma lineatum* (Villers), the common cattle grub. **a.** dorsal of mature larva; **b.** posterior spiracles; **c.** eggs attached to hair. Larvae of the common cattle grub (and northern cattle grub) cause conspicuous bumps, ("warbles") on the backs of cattle and bison, and occasionally on goats, sheep, dogs, and other mammals from Canada to Mexico (a from Peterson 1951; b from James 1947).

Figure 37.341a,b. Oestridae, Hypodermatinae. *Hypoderma bovis* (L.) the northern cattle grub. **a.** dorsal of mature larva; **b.** posterior spiracle. Larvae also develop within conspicuous warbles similar to those of *H. lineatum* on the backs of cattle in Canada and the northern two-thirds of the United States (a from Peterson 1951; b from James 1947).

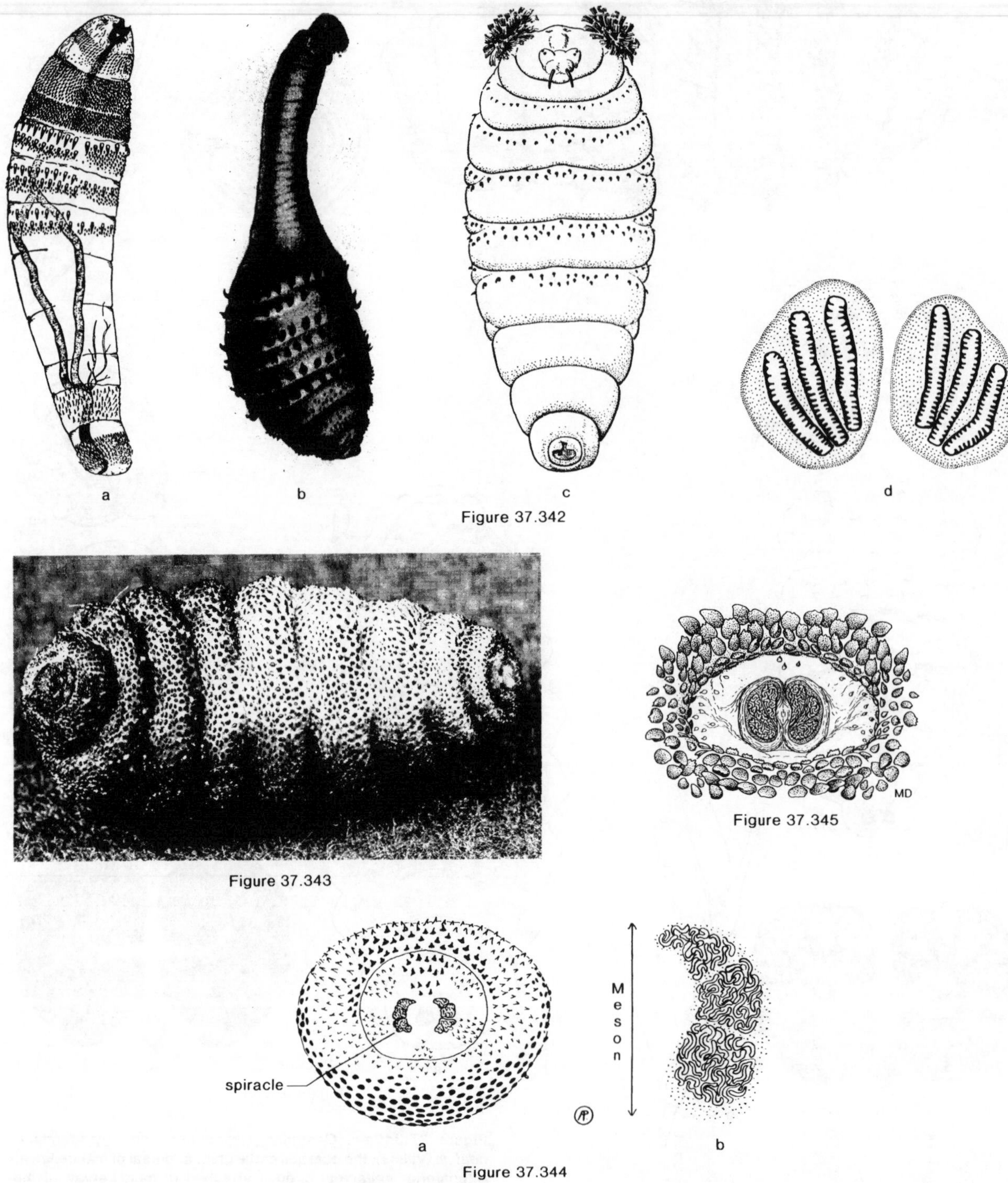

Figure 37.342a-d. Oestridae, Cuterebrinae. *Dermatobia hominis* (Linnaeus Jr.), the human bot. **a.** first instar larva; **b.** second instar larva; **c.** mature larva, ventral; **d.** caudal spiracles, last instar. The larvae are dermal parasites of humans, cattle, dogs and other mammals. Neotropical, northward to Taumalipas, Mexico (a, b, c from James 1947; d drawn by Peter Carrington).

Figure 37.343. Oestridae, Cuterebrinae. Mature larva of *Cuterebra jellisoni* Curran, a parasite of rabbits in the West (photo courtesy of R. D. Akre, Washington State University).

Figure 37.344a,b. Oestridae, Cuterebrinae. *Cuterebra* sp. **a.** caudal view of mature larva; **b.** right caudal spiracle. From neck of cat or rabbit (from Peterson 1951).

Figure 37.345. Oestridae, Cuterebrinae. Posterior spiracles of *Cuterebra emasculator* Fitch, a parasite of the eastern chipmunk, *Tamias striatus* (L.) and the gray squirrel, *Sciurus carolinensis* Gmelin (from Manual of Nearctic Diptera, Vol. 1).

SARCOPHAGIDAE (OESTROIDEA)

Gregory A. Dahlem, *Michigan State University*

Flesh Flies

Figures 37.346–37.350

Relationships and Diagnosis: Sarcophagids are closely related to the Calliphoridae and Tachinidae. In the past, some authors have considered them to be a subfamily of the Calliphoridae and some have placed the Miltograminae in the Tachinidae. Synonymous names include Stephanostomatidae and Metopiidae (in part). Sarcophagidae is currently placed in the Oestroidea, along with Tachinidae, Calliphoridae, Rhinophoridae, and Oestridae (McAlpine *et al.*, 1981).

Mature larvae are distinguished from those of other calyptrate families by most species having three nearly vertical, parallel posterior spiracular slits arising from a ventral ecdysial scar (which is frequently indistinct (fig. 37.347b) or absent (fig. 37.346c)) and by the spiracular slits usually not pointing toward the opening in the peritreme (fig. 37.346c). Sarcophagid larvae often have deeply recessed posterior spiracles (fig. 37.348b). Shewell (1987c) has recently reviewed our knowledge.

Biology and Ecology: All sarcophagids retain their eggs within a bipouched uterus and deposit first instar larvae on the larval medium or host. First instar larvae can often be obtained from pinned females. As larvae, sarcophagids have the most diverse feeding habits of all families of calyptrate flies. Many are parasitoids of other arthropods, while others are coprophagous, necrophagous, predacious, or saprophagous. The taxonomy of the adults is difficult and errors in their identification may explain some atypical rearing records in the literature.

In North America, at least 10 species of *Ravinia* have been reared from mammalian dung (Coffey, 1966; Dodge, 1956; Knipling, 1936a; Turner *et al.*, 1968). First instar larvae of *R. lherminieri* have been shown to be facultative predators of other Diptera larvae in dung (Pickens, 1981). In addition to these *Ravinia* species, *Oxysarcodexia ventricosa* (van der Wulp) and *Sarcophaga cruentata* Meigen (= *haemorrhoidalis* aut. nec Fallén) are commonly reared from dung (Aldrich, 1916; Sanders and Dobson, 1966).

In contrast to their common name, very few flesh fly larvae exploit vertebrate carrion as a food resource. *Sarcophaga bullata* Aldrich, *S. argyrostoma* Robineau-Desvoidy, *S. cooleyi* Parker, and *Blaesoxipha plinthopyga* (Wiedemann) are examples of relatively common Nearctic necrophages found in vertebrate carrion (Denno and Cothran, 1975; Graenicher, 1931). Many species, however, have been reared on liver, hamburger, and other decaying meat, but this gives little information on the true larval food resource in the field.

Several sarcophagids are reported as parasitic in, and saprophytic on, snails (R. A. Beaver, 1977; Sanjean, 1957; Lopes, 1940). Some may parasitize earthworms (Lopes, 1942), some have been reared from scorpions (Townsend, 1893c), others have been reared from millipedes (Aldrich, 1916), and several *Sarcophaga* species have been reared from spider egg sacks (Cantrell, 1981; Auten, 1925).

A few species, such as the Nearctic *Wohlfahrtia vigil* (Walker) (fig. 37.349a–d), cause cuticular myiasis in young mammals, including humans (James, 1947; Walker, 1937; Eschle and DeFoliart, 1965). *Sarcophaga citellivora* Shewell has been reared from Columbian ground squirrels in western North America (Shewell, 1950). *Anolisimyia blakeae* Dodge has been reared from subcutaneous lesions in an anole lizard (Blake, 1955), *Cistudinomyia cistudinis* (Aldrich) is a parasite of box and gopher turtles (Knipling, 1937), and *Notochaeta bufonivora* Lopes and Vogelsang parasitizes toads (Lopes and Vogelsang, 1953). *Eumacronychia nigricornis* Allen has been reared from lizard eggs (Mullen et al., 1984), *Metoposarcophaga importuna* (Walker) has been bred from terrapin eggs (Aldrich, 1916), and *Eumacronychia sternalis* Allen, which feeds on sea turtle eggs, may be the most destructive predator upon sea turtles with the exception of humans (Lopes, 1982a).

Many sarcophagids are parasitoids of, and saprophytic on a wide variety of insects. In the Orthoptera, they are well known parasitoids of grasshoppers (Smith and Finlayson, 1950; Middlekauff, 1959) and other orthopterans, such as Mormon crickets (LaRivers, 1944), rhaphidophorids (Arnaud, 1954), and mantids (Breland, 1942). A few rearing records are reported from Hemiptera (Osborn, 1919). Adult cicadas (mostly males) serve as hosts for the larvae of a few species (Soper *et al.*, 1976) where throngs of male cicadas attract them. They have also been reared from adult *Corydalus cornutus* (L.) (Megaloptera) (Aldrich, 1916). Many Coleoptera are parasitized (van Emden, 1950), and in the Diptera, several species of horse flies (Tabanidae) are hosts (Thompson, 1978). Sarcophagids are frequently reported as parasitoids of Lepidoptera (Bibby, 1942; Lejeune and Silver, 1961; Hodson, 1939; Coppel, 1960; Malo and Willis, 1961; Davis, 1960), and they are reported as parasitoids or predators of Hymenoptera such as bumble bees (Frison, 1926), honey bees (Ronna, 1936) and wasps (Dambach and Good, 1943; Nelson, 1968). A wide variety have been noted as inquilines or cleptoparasites, feeding on the stores of a variety of bees and wasps (Pape, 1987; Cross *et al.*, 1975; Peckham *et al.*, 1973; Newcomer, 1930; James, 1955a; Allen, 1926).

A few sarcophagid larvae are aquatic in the cups of pitcher plants. These include members of the genus *Fletcherimyia* and some species of *Sarcophaga* (Fish, 1976; Aldrich, 1916). These species feed on insects which drown in the water.

Description: Mature larvae usually white or off-white, 5–25 mm in length and peg-shaped, tapering anteriorly and bluntly rounded to truncate posteriorly. Posterior surface of caudal segment usually with a distinct cavity which contains the posterior spiracles. Posterior spiracular slits usually parallel and vertical. Caudal segment often with enlarged tubercles around spiracular disc. Mature larvae amphipneustic.

Head: Cephalic segment usually retractile, bearing sensory organs anteriorly and anteroventrally. Antennae appearing 2-segmented. Spinule band frequently on ventral surface behind preoral cavity.

Cephalopharyngeal skeleton (fig. 37.348a) deeply pigmented; tentoropharyngeal and hypopharyngeal sclerites separate; dorsal cornua as broad as or broader than ventral

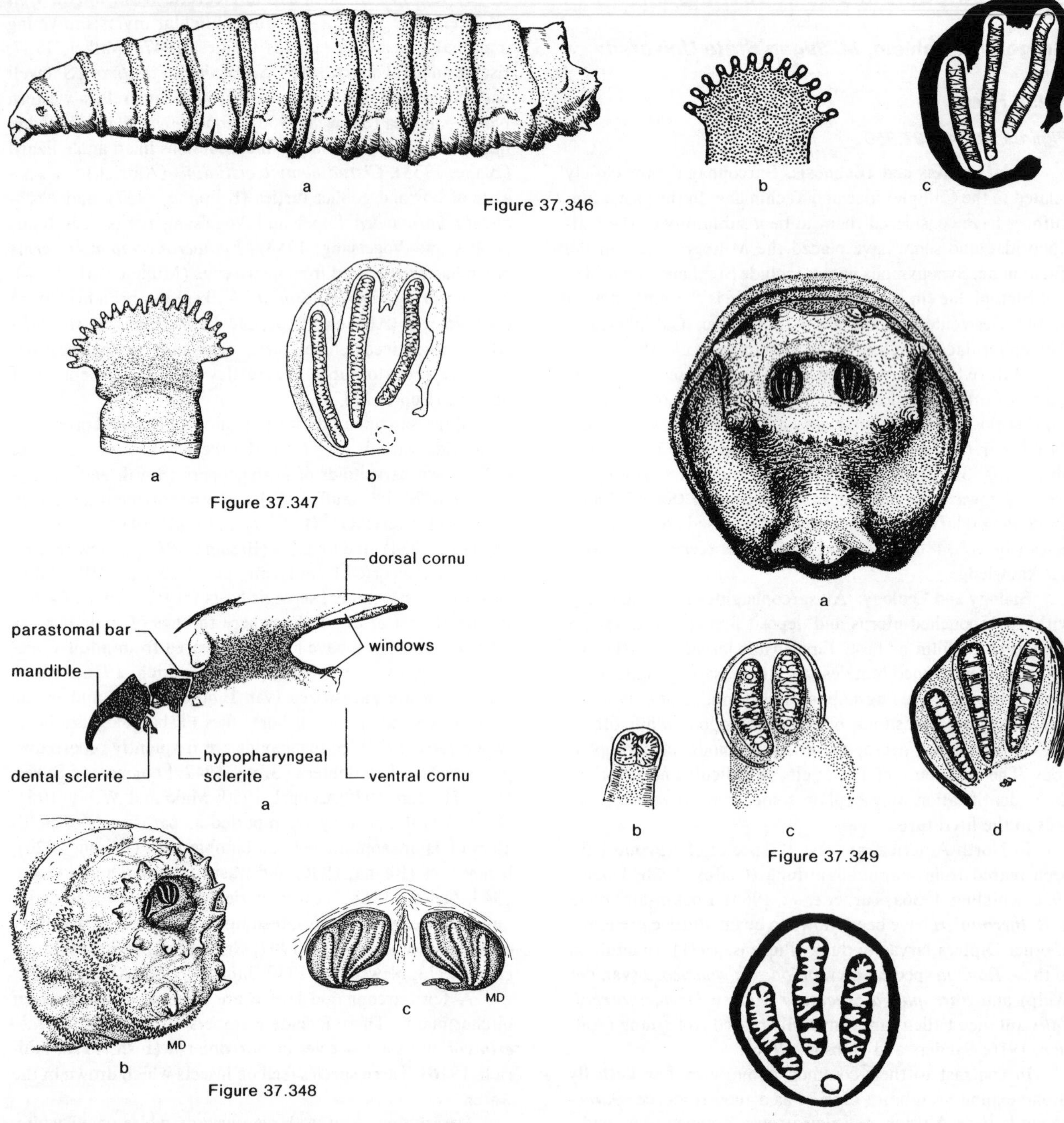

Figure 37.346a-c. Sarcophagidae, Sarcophaginae. *Sarcophaga crassipalpis* Macquart, of Eurasian origin. **a.** mature larva; **b.** anterior spiracle; **c.** left posterior spiracle, with incomplete peritreme and no ecdysial scar. Commonly found in garbage in urban areas, it also is a secondary invader of wounds in humans and other animals (from James 1947).

Figure 37.347a,b. Sarcophagidae, Sarcophaginae. *Sarcophaga bullata* Parker. **a.** anterior spiracle; **b.** posterior spiracle with incomplete peritreme and weak (or absent) ecdysial scar. Sometimes a secondary invader of the wounds of animals and occasionally humans, the larvae normally feed in carrion (from Sanjean 1957).

Figure 37.348a,b. Sarcophagidae, Sarcophaginae. *Ravinia querula* (Walker). **a.** cephalopharyngeal skeleton; **b.** oblique posterior view of caudal segment; **c.** posterior spiracles. Larvae develop in bovine dung (from Manual of Nearctic Diptera, Vol. 1).

Figure 37.349a-d. Sarcophagidae, Miltogramminae. *Wohlfahrtia vigil* (Walker), a primary parasite of young, wild and domestic mammals, including humans (especially infants). The penetration of tender unbroken skin is normal. Carrion is not a normal larviposition site. **a.** caudal view of mature larva; **b-d.** left posterior spiracles, first to third instars respectively (from James 1947).

Figure 37.350. Sarcophagidae, Miltogramminae. Left posterior spiracle of *Brachyocoma sarophagina* (Townsend), a species with a complete peritreme and a distinct ecdysial scar. Larvae are predacious on the brood of bumble bees (*Bombus* spp.) (from L. H. Townsend 1936).

cornua, connected anteriorly by fenestrated dorsal bridge and with narrow elongate window apically; ventral cornua with dorsobasal lobe, although lobe may be near apex of cornua in many species and with small window below or posterior to dorsobasal lobe; pharyngeal filter usually present; hypopharyngeal sclerite relatively short, H-shaped; parastomal bars short and free apically; epipharyngeal sclerite apparently undeveloped; labial sclerite either V-shaped or occurring as separate rods below and between anterior arms of hypopharyngeal sclerite. Mandibles deeply pigmented, hook usually sickle-like and commonly with accessory teeth, basal part subrectangular and with or without small window; dental sclerites one or two, appearing triangular or rod-like in lateral view; apparently no accessory oral sclerites between mandibles.

Thorax and Abdomen: Prothorax usually partially encircled by a spinule band; the rest of the thoracic and abdominal segments usually with well-defined encircling spinule bands and creeping welts. Spinules variable in shape and size. Anterior spiracles arising posterolaterally on prothorax and usually fan-shaped, with 10–30 short to elongate papillae.

Posterior spiracular disc with or without enlarged tubercles and postanal tubercles. Posterior spiracles usually sunken in caudal pit. Three elongate spiracular slits, usually oriented vertically, with the openings parallel or nearly so to each other. Ecdysial scar reduced or absent.

Comments: This is a very large and diverse family with over 40 genera and 300 species recorded for America north of Mexico (Downes *in* Stone *et al.*, 1965) with many species remaining undescribed. Tribal and generic nomenclature for the Sarcophagidae is currently in a state of flux.

Only a few species are considered to be economically important. The larvae of *Wohlfahrtia* species are often causative agents of cuticular myiasis in young mammals, including humans, and *Sarcophaga cruentata* has been implicated on multiple occasions in intestinal myiasis in humans. *Sarcophaga aldrichi* and *Agria affinis* are important parasitoids of a wide variety of pest Lepidoptera.

Selected Bibliography

Aldrich 1916.
Allen 1926.
Arnaud 1954.
Auten 1925.
Beaver, R. A. 1977.
Bibby 1942.
Blake 1955.
Breland 1942.
Cantrell 1981.
Coffey 1966.
Coppel 1960.
Cross et al. 1975.
Dambach and Good 1943.
Davis 1960.
Denno and Cothran 1975.
Dodge 1956.
Downes 1955.
Emden 1950.
Eschle and DeFoliart 1965.
Ferrar 1987.
Fish 1976.
Frison 1926.
Graenicher 1931, 1935.
Greene 1925.
Hall 1932a.
Hodson 1939.
James 1947, 1955a.
Kano and Sato 1951.
Knipling 1936a, 1937.
LaRivers, 1944.
Lejeune and Silver 1961.
Lopes 1940, 1942, 1982a, 1982b.
Lopes and Vogelsang 1953.
Malo and Willis 1961.
Middlekauf 1959.
Mullen et al. 1984.
Nelson 1968.
Newcomer 1930.
Newhouse et al. 1955.
Osborn 1919.
Pape 1987.
Peckham et al. 1973.
Pickens 1981.
Roback 1954.
Ronna 1936.
Sanders and Dobson 1966.
Sanjean 1957.
Shewell 1950, 1987c.
Smith and Finlayson 1950.
Soper et al. 1976.
Thompson, P. H. 1978.
Townsend, C. H. T. 1893c, 1936b.
Turner et al. 1968.
Walker 1920, 1937.

RHINOPHORIDAE (OESTROIDEA)

Gregory A. Dahlem, *Michigan State University*

The Rhinophorids

Figures 37.351–37.352

Relationships and Diagnosis: This family is closely related to the Tachinidae and Calliphoridae and has been placed as a subfamily of the Tachinidae, Calliphoridae, and Sarcophagidae. At present, most authors give it family rank, acknowledging the uniqueness and lack of evidence for a closer relation to any of the other oestroid families. McAlpine, *et al.* (1981) have placed this family in the Oestroidea along with the Calliphoridae, Oestridae, Sarcophagidae, and Tachinidae. D. M. Wood (1987b) has recently summarized our knowledge.

Although the family is small, an unambiguous demarcation of rhinophorid larvae has not yet been achieved. However, the larval habit as endoparasitoids of woodlice (Isopoda) appears to be unique among Diptera.

Biology and Ecology: The larvae of most, if not all, species are normally endoparasitoids of terrestrial isopods (Thompson, 1934; Bedding, 1973). Specific host records for the Palearctic species are given by Herting (1961) with supplements by Kugler (1978). In addition, there are several reports of rhinophorids parasitizing insects and being reared from spider egg cocoons (Pape, 1986). It is probable that rhinophorids occasionally parasitize arthropods other than isopods.

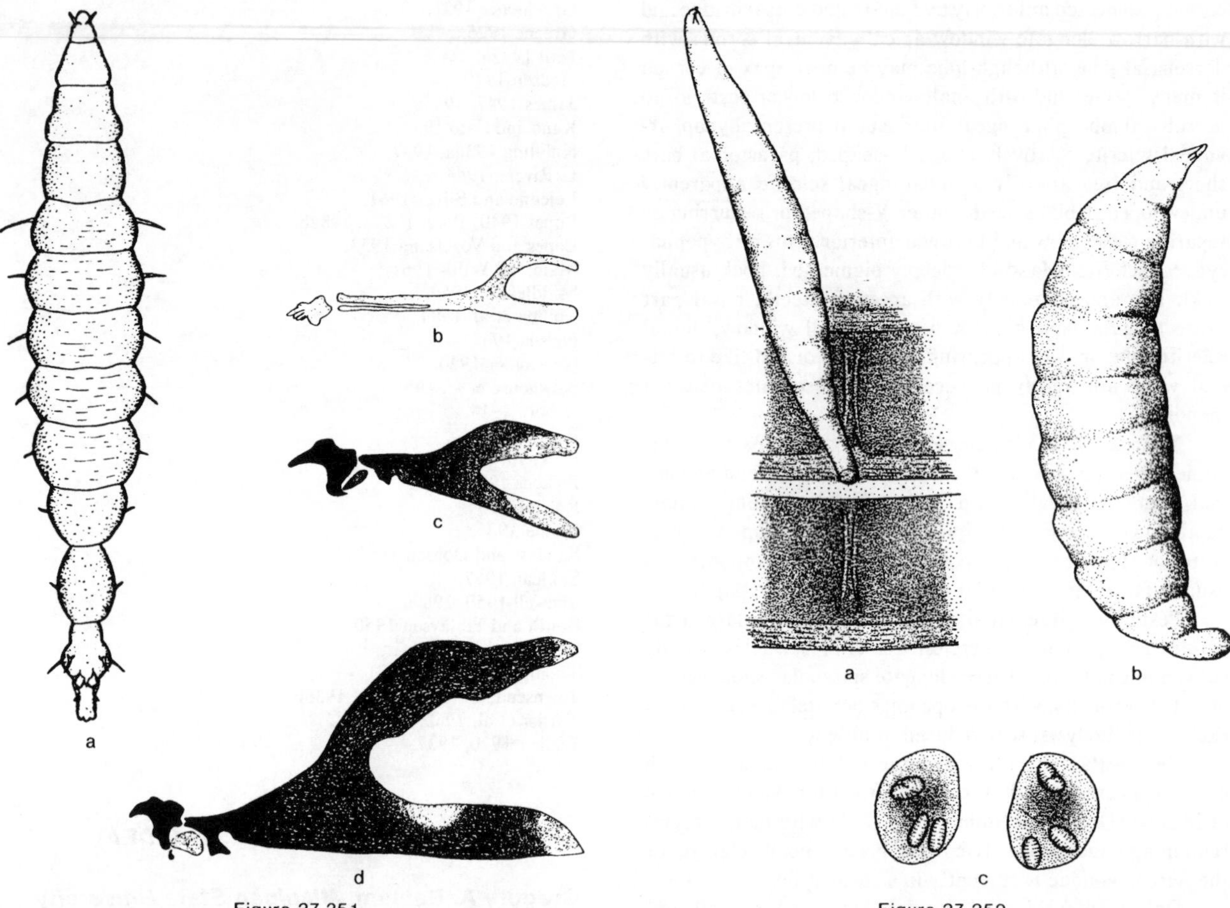

Figure 37.351

Figure 37.352

Figure 37.351a-d. Rhinophoridae. *Melanophora roralis* (L.), a parasitoid of isopods probably introduced from Europe. **a.** first instar larva; **b,c,d.** first, second, and third instar cephalopharyngeal skeletons, respectively (from Bedding 1973).

Figure 37.352a-c. Rhinophoridae. *Chaetostevenia maculata* (Fallén), a European species. **a.** second instar larva internally attached to host integument of an isopod; **b.** third instar larva; **c.** posterior spiracles (from W. R. Thompson 1934).

Description: First instar larvae (fig. 37.351a) have many more distinctive features than do the later instars (figs. 37.352a,b) and have been more thoroughly investigated (Thompson, 1934; Bedding, 1973). Mature larvae 5–10 mm, usually pale white. Most larvae spindle-shaped, tapering anteriorly and posteriorly. Mature larvae amphipneustic or metapneustic.

Head: Cephalic segment retractile and bearing sensory organs anteriorly and anteroventrally. Antennae appearing 2-segmented. Facial mask apparently without oral grooves and ridges.

Cephalopharyngeal skeleton deeply pigmented; tentoropharyngeal and hypopharyngeal sclerites separate or fused (*Melanophora roralis* (L.), figs. 37.351b,c,d); dorsal cornua slender and usually without apical window, connected anteriorly by dorsal bridge; ventral cornua broader and usually with dorsal lobe and window posteriorly; pharyngeal filter apparently absent; hypopharyngeal sclerite short and H-shaped; parastomal bars present and slender, not connected apically. Mandibles deeply pigmented; hook decurved, some species with accessory teeth, basal part subrectangular and usually without window; dental sclerite subtriangular in lateral view; no accessory oral sclerites.

Thorax and Abdomen: With or without encircling spinule bands, but usually with some ventral creeping welts. Cuticle with or without minute transparent spines. Anterior spiracles more or less reduced, arising posterolaterally on prothorax, usually with 6–15 irregularly distributed papillae apically.

Posterior spiracular disc without tubercles. Posterior spiracles distinctly elevated above the plane of the terminal segment on short tubes. Second instar larvae often have very long tubes connecting to the exterior of the isopod. Posterior spiracular plates with 3 slits.

Comments: Rhinophoridae is a small family consisting of about 100 Old World species. Only two species are known from America north of Mexico, *Melanophora roralis* (L.) (figs. 37.351a–d) and *Phyto discrepans* (Pandelle). Both of these may be immigrants, with *M. roralis* by far the most common. No species has economic significance.

Selected Bibliography

Bedding 1973.
Crosskey 1977.
Ferrar 1987.
Herting 1961.
Kugler 1978.
Pape 1986.
Thompson, W. R. 1934.
Wood, D. M. 1987b.

TACHINIDAE (OESTROIDEA)

B. A. Foote, *Kent State University*

Gregory A. Dahlem, *Michigan State University*

The Tachinids

Figures 37.353–37.362

Relationships and Diagnosis: This family is closely related to the Rhinophoridae and Sarcophagidae. Until the 1960s, the name Larvaevoridae was commonly used in the literature to encompass the family. Other synonyms for the Tachinidae include Phasiidae, Gymnosomatidae, and Dexiidae. In the past, some authors have placed the Rhinophoridae and the sarcophagid subfamily Miltogramminae in the Tachinidae. McAlpine, *et al.* (1981) have placed the tachinids in the Oestroidea along with the Calliphoridae, Sarcophagidae, Oestridae, and Rhinophoridae. Wood (1987c) has recently summarized our knowledge.

At the present time, no characters are available which will separate all larvae of Tachinidae. Mature larvae are generally conic-cylindrical, with more or less prominent anterior spiracles situated laterally, and with sessile or slightly elevated posterior spiracles with 3 (or sometimes more) straight to serpentine slits which often radiate out from the conspicuous ecdysial scar. Almost all larvae of Tachinidae are endoparasitoids of insects. Rearing to adults is suggested for accurate identification.

Biology and Ecology: The larvae of most species are endoparasitoids of a wide variety of terrestrial insects. Tachinids employ several different strategies for delivering their offspring to an acceptable host. One of these strategies is the deposition on vegetation of many microtype eggs which hatch upon ingestion by a suitable host. Another strategy is the retention of macrotype eggs by the female for deposition on or near a suitable host. Many of these species have eggs that hatch quickly and the first instar larvae penetrate the host's integument or enter a suitable body opening. A few species have eggs that hatch after developing on the host for a few days. Some tachinids appear to be larviparous, but these species lay eggs which hatch immediately upon deposition. First instar larvae of these species can be obtained from many pinned females (Ravlin 1977).

First and second instar larvae usually lie within the haemocoele of the host and obtain oxygen by a direct connection with the host's integument or tracheal system. The feeding by these larvae usually causes little harm to the host, but third instar larvae feed much more extensively on the host's body tissues and this usually results in the death of the host. Pupariation usually occurs in the soil, but sometimes may occur in the host itself.

Some species are very host specific while others have been reared from a wide variety of hosts. For example, *Compsilura concinnata* (Meigen) has been reared from approximately 200 different host species in 26 families and 3 orders of insects (Arnaud, 1978). Lepidoptera larvae are the most common hosts of tachinids, although larvae and adults of grasshoppers and stink bugs, scarab and leaf beetle larvae, and sawfly larvae are also common hosts. Other hosts include crane fly and horse fly larvae, earwigs, centipedes, cockroaches, walking sticks, and a wide variety of Hemiptera, Coleoptera, and Hymenoptera immatures and adults. Arnaud (1978) provides an excellent review of host-parasite information on North American tachinid species, and Wood (1987c) provides an excellent review of biology and behavior.

Description: Mature larvae usually white to yellowish, 6–25 mm in length and conic-cylindrical, tapering anteriorly and bluntly rounded to truncate posteriorly. Integument bare or bearing variously distributed spinules and hairs. Segmental borders sometimes with encircling bands of spinules. Mature larvae amphipneustic. First instar larvae often much more distinctive than those of later instars.

Head: Cephalic segment retractile, bearing sensory organs anteriorly and anteroventrally; antennae reduced or well-developed and apparently 2-segmented; facial mask may have oral ridges and grooves leading into preoral cavity; distinct spinule band may border posterior edge of preoral cavity.

Cephalopharyngeal skeleton usually strong and heavy, deeply pigmented (fig. 37.356b); tentoropharyngeal and hypopharyngeal sclerites separate or fused (relatively rare); dorsal cornua as broad as or broader than the ventral cornua and sometimes connected anteriorly by fenestrated dorsal bridge, commonly without windows; ventral cornu without or with greatly reduced dorsobasal lobe and may have variously shaped apical windows; pharyngeal filter lacking; hypopharyngeal sclerite short and heavy, basically H-shaped in ventral view; parastomal bars and epipharyngeal sclerite absent; labial sclerites, if present, short and rod-like. Mandibles usually heavy and deeply pigmented; hook part nearly straight or somewhat decurved and lacking accessory teeth; basal part subrectangular to somewhat elongate and usually lacking windows; dental sclerite, if present, somewhat triangular in lateral view; accessory oral sclerites absent.

Thorax and Abdomen: Meso- and metathorax and abdomen may possess creeping welts ventrally and may be encircled by spinule bands. Anterior spiracles may or may not be well developed. If present, they arise posterolaterally on prothorax and may be sessile (fig. 37.356c), or at the apex of short tubular projections with papillae-like structures present (fig. 37.358), or with spiracular openings clustered and appearing like a bunch of grapes (fig. 37.357b).

Perianal pad bilobed, variously shaped, but usually not subtended by fleshy lobes or bordered posteriorly by spinule patch. Posterior spiracles highly diverse in size, shape, and pattern. Spiracles usually borne on upper half of spiracular disc on short, rather broad projections. Spiracular openings may be numerous (fig. 37.359) and randomly scattered over

Tachinidae

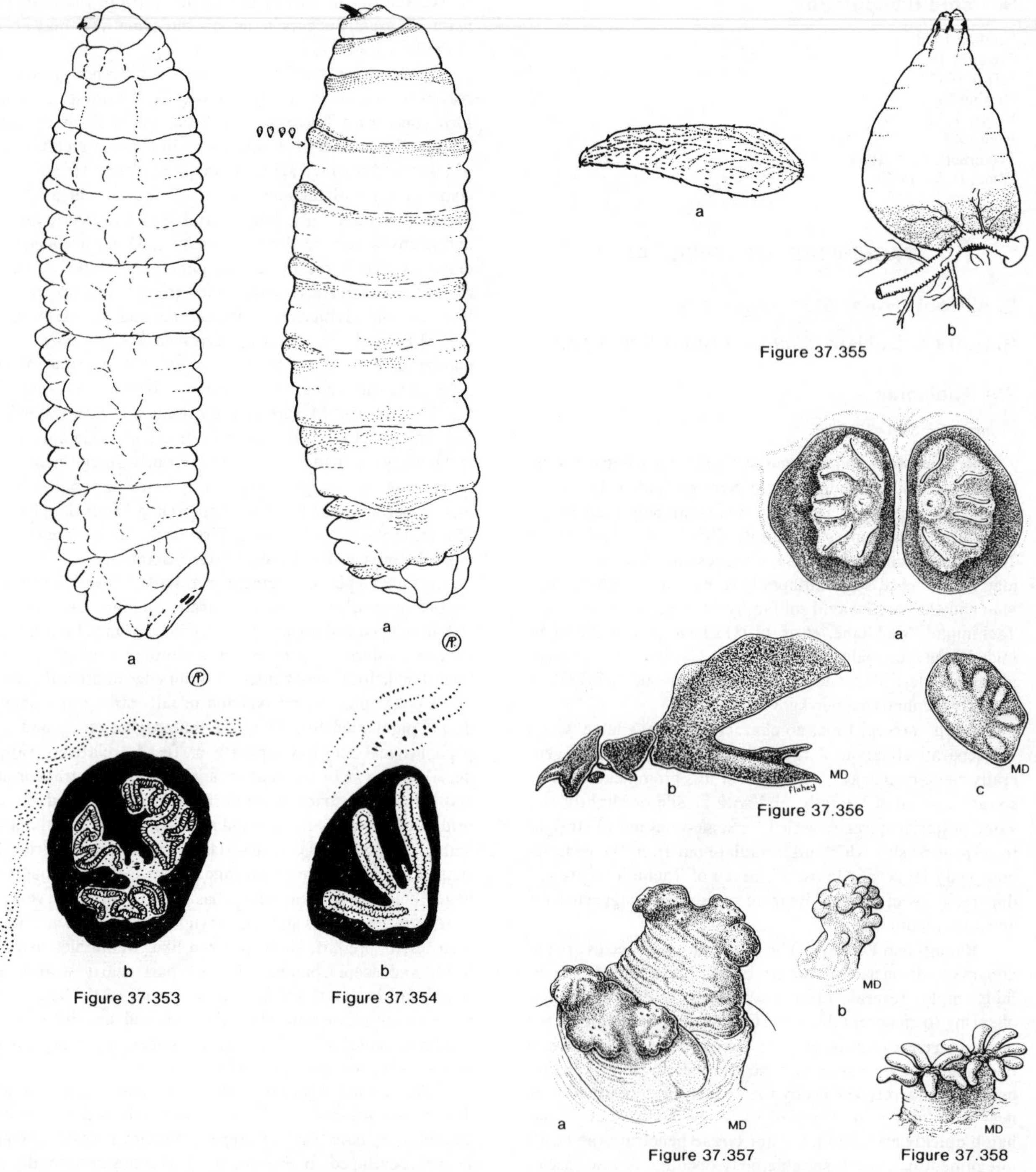

Figure 37.353a,b. Tachinidae. *Blepharipa scutellata* (Robineau-Desvoidy). **a.** lateral; **b.** posterior spiracle. Larvae are endoparasitoids of assorted macrolepidoptera (from Peterson 1951).

Figure 37.354a,b. Tachinidae. *Compsilura concinnata* (Meigen). **a.** lateral; **b.** posterior spiracle. This is an extremely polyphagous species, having been recorded from at least 200 hosts, and is commonly recovered from gypsy moth larvae (from Peterson 1951).

Figure 37.355a,b. Tachinidae. *Campogaster exigua* (Meigen), a European parasitoid of adult *Sitona* and other weevils. **a.** first instar larva; **b.** last instar larva attached to trachea (from Berry and Parker 1950).

Figure 37.356a-c. Tachinidae. *Zaira* sp. **a.** posterior spiracles; **b.** lateral view of cephalopharyngeal skeleton; **c.** anterior spiracle. Larvae are parasitoids of adult carabids and tenebrionids (from Manual of Nearctic Diptera, Vol. 1).

Figure 37.357a,b. Tachinidae. *Ceracia dentata* (Coquillett). **a.** posterior spiracles; **b.** anterior spiracle. Larvae parasitize Orthoptera, especially *Melanoplus* grasshoppers (Acrididae) (from Manual of Nearctic Diptera, Vol. 1).

Figure 37.358. Tachinidae. *Cleonice* sp., anterior spiracle. Larvae are parasitoids of chrysomelid beetle larvae (from Manual of Nearctic Diptera, Vol. 1).

Tachinidae

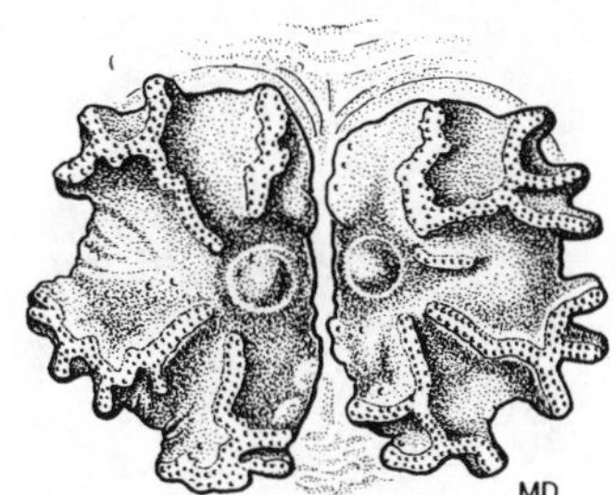

Figure 37.359

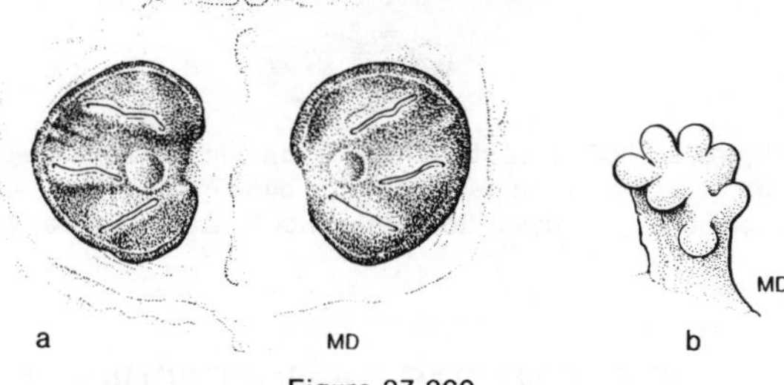

Figure 37.360

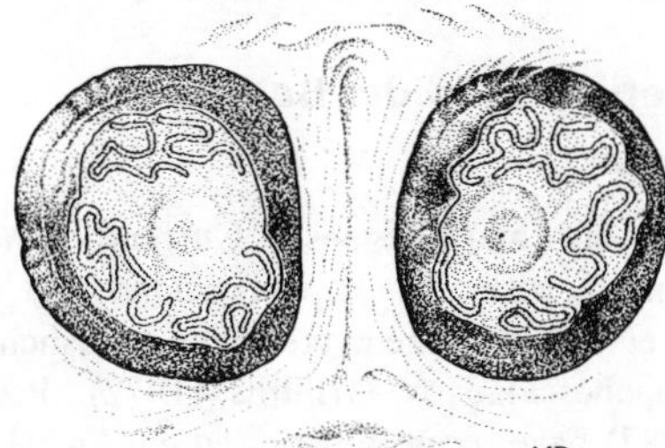

Figure 37.361

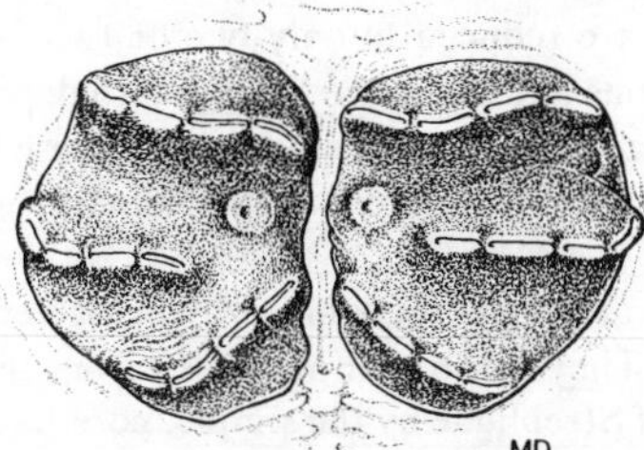

Figure 37.362

Figure 37.359. Tachinidae. *Blepharomyia* sp., caudal spiracles. Larvae parasitize Lepidoptera larvae (from Manual of Nearctic Diptera, Vol. 1).

Figure 37.360a,b. Tachinidae. *Lypha fumipennis* Brooks. **a.** posterior spiracles; **b.** anterior spiracle. Larvae are parasitoids of Lepidoptera larvae (from Manual of Nearctic Diptera, Vol. 1).

Figure 37.361. Tachinidae. *Lespesia callosamiae* Beneway, caudal spiracles. Larvae parasitize the larvae of *Callosamia promethia* (Drury) (Saturniidae) (from Manual of Nearctic Diptera, Vol. 1).

Figure 37.362 Tachinidae. *Uramya halisidotae* (Townsend), posterior spiracles. Larvae are parasitoids of *Halisidota* spp. (Arctiidae) (from Manual of Nearctic Diptera, Vol. 1).

spiracular plate which may appear tripartite (fig. 37.357a), or may be few in number and with slit-like (figs. 37.354b, 37.360a), serpentine (fig. 37.361) or oval openings which often radiate out from the ecdysial scar which may be faint. Peritreme may be poorly distinguished to well developed and deeply pigmented.

Comments: This is a very large and diverse family with over 400 genera and 1200 species recorded from the Nearctic region (Wood 1987c) with many species remaining undescribed. Tribal and generic nomenclature is currently in a state of flux.

Several species are considered to be beneficial because the parasitic larvae can have a significant impact on populations of defoliating caterpillars. Other species are parasitoids of a wide range of agricultural pests and may be significant biocontrol agents under appropriate conditions. An excellent summary of the economic value of Tachinidae as biocontrol agents of pest species is given by W. R. Thompson 1943–1958).

In the Palearctic and Oriental regions, some species are considered to be important pests of the silk industry. Datta and Mukherjee (1978) give an account of the life history of *Tricholyga bombycis* Beck, a serious parasitoid of the commercial silkworm, *Bombyx mori* L., in eastern India.

Selected Bibliography

Arnaud 1978.
Baldwin and Coppel 1949.
Beard 1940.
Berry and Parker 1950.
Bloesler 1914.
Clausen 1940.
Coppel and Maw 1954.
Datta and Mukherjee 1978.
Elsey and Rabb 1970.
Emden 1959.
Farinets 1980.
Ferrar 1987.
Finlayson 1960.
Greene 1921.
Hafez 1953a, 1953b, 1953c.
Hays 1958.
Hennig 1952.
Landis and Howard 1940.
Loudon and Attia 1981.
Maw and Coppel 1953.
Neff and Eisner 1960.
O'Hara 1982, 1988.
Ravlin 1977.
Ravlin and Stehr 1984.
Rennie and Sutherland 1920.
Rikhter and Farinets 1983.
Sabrosky 1978a, 1981.
Thompson, W. R. 1923, 1924, 1926, 1943–1958, 1954, 1963.
Wallace and Franklin 1970.
Wishart 1946.
Wood, D. M. 1987c.
Zuska 1962.

HIPPOBOSCIDAE (HIPPOBOSCOIDEA)

B. A. Foote, *Kent State University*

Louse Flies

Figure 37.363

Relationships and Diagnosis: This family, along with the Nycteribiidae and Streblidae, has long been classified under the term "Pupipara" because of their habit of depositing fully mature larvae that quickly form puparia. Thus, all three families are grouped together in the Hippoboscoidea (McAlpine et al. 1981). Maa and Peterson (1987) have recently summarized our knowledge.

As in all species of the superfamily Hippoboscoidea, larvae of the Hippoboscidae are swollen and featureless. They can be distinguished from larvae of Nycteribiidae and Streblidae by the large, blackened spiracular plate that covers nearly the entire posterior end of the body.

Biology and Ecology: As the common name "louse flies" implies, hippoboscids are ectoparasites of warm-blooded animals. Adults are strongly flattened, possess stylet-like mouthparts, and may be winged or wingless. Some species shed their wings shortly after finding a host. Adults of both sexes suck blood but seemingly cause little discomfort to their hosts. They mainly attack birds, but domestic livestock as well as wild mammals, particularly deer, are attacked. Most are quite host specific, but the winged species frequently use a broad variety of hosts. It is believed that the mammal-infesting species represent a relatively recent evolutionary development, as it appears that the family is basically a group of bird parasites (149 species). Females deposit a fat, whitish to yellowish, immobile fully mature larva which quickly forms a puparium.

Description: Mature larvae (really a prepupa) less than 3 mm. Body very swollen and featureless. Posterior end of body covered with blacked spiracular plate.

Comments: A small but widely distributed family, the Hippoboscidae contains nearly 150 species. Thirteen genera and about 30 species are recorded for America north of Mexico. The only species considered to be of economic importance is the "sheep ked," *Melophagus ovinus* (L.), a flightless species from Eurasia that can occasionally become abundant on domestic sheep. Adults of a few species, such as the "deer ked," have been known occasionally to bite humans.

Selected Bibliography

Bequaert 1954.
Carpenter 1901.
Ferrar 1987.
Ferris 1923.
Ferris and Cole 1922.
Guimaraes 1944.
Hennig 1952.
Kim 1985.
Maa and Peterson 1987.
Marshall, A. G. 1981.
Peterson, A. 1951.
Pratt, H. S. 1893.
Roberts, J. I. 1927.

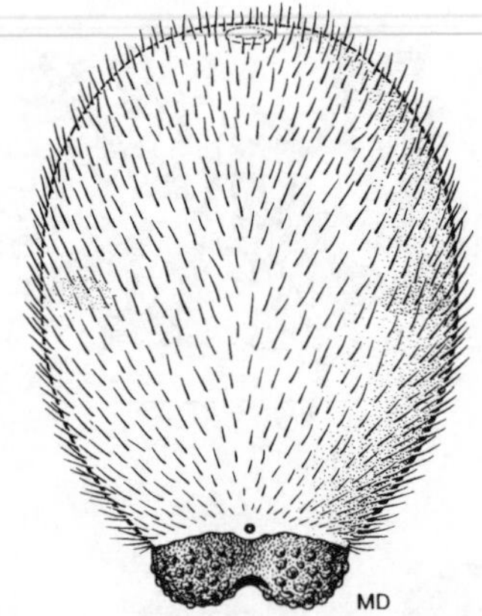

Figure 37.363. Hippoboscidae. Puparium of *Olfersia spinifera* Leach, a common parasite of frigate birds over tropical seas, and occasionally pelicans (from Manual of Nearctic Diptera, Vol. 2).

NYCTERIBIIDAE (HIPPOBOSCOIDEA)

B. A. Foote, *Kent State University*

The Nycteribiids, Spider Bat Flies

Figure 37.364

Relationships and Diagnosis: Currently placed in the Hippoboscoidea with the Hippoboscidae and Streblidae (McAlpine et al. 1981), the nycteribiids were included in the family Hippoboscidae by Griffiths (1972). Peterson and Wenzel (1987) have recently reviewed our knowledge.

Biology and Ecology: This family has become highly specialized as blood-sucking ectoparasites of bats. Adults are small and wingless, and strongly resemble spiders. The compound eyes are reduced to only one or two facets. Females are pupiparous and leave the bat hosts to deposit soft, white mature larvae on walls of the bat roost. These larvae quickly pupate, harden and darken. Adults emerge within several days.

Larvae are easily recognized by being very swollen and featureless. They can be distinguished from the closely similar larvae of Streblidae by the simple, pore-like nature of the posterior spiracles (fig. 37.364).

Description: Mature larvae less than 2 mm in length, very swollen and featureless. Posterior spiracles consisting of a simple, pore-like opening.

Comments: This is a small family that contains fewer than 260 species in the world. Most species are found in the Old World Tropics. There is only one genus in America north of Mexico, *Basilia,* with seven species that are found in bat caves in the western and southern states.

Selected Bibliography

Ferrar 1987.
Ferris 1923, 1930.
Griffiths 1972.
Hennig 1952.
Kim 1985.
Marshall, A. G. 1981.

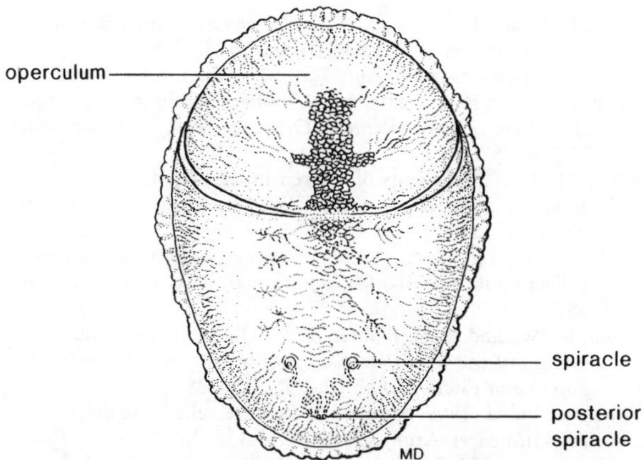

Figure 37.364. Nycteribiidae. Dorsal of puparium of *Basilia corynorhini* (Ferris), a parasite of vespertilionid bats (from Manual of Nearctic Diptera, Vol. 2).

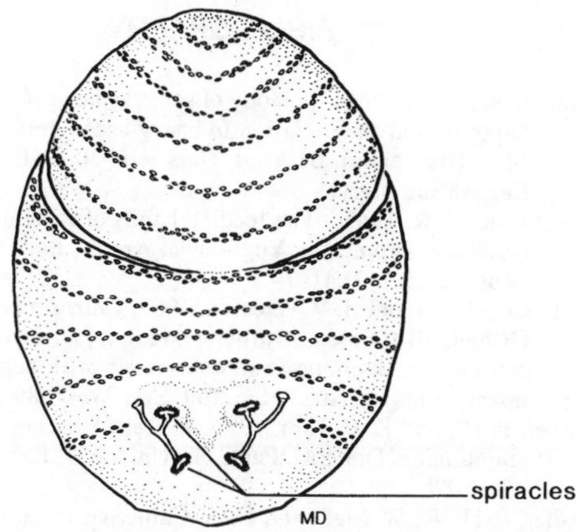

Figure 37.365. Streblidae. Dorsal of puparium of *Trichobius caecus* Edwards, a common parasite of cave-roosting bats (from Manual of Nearctic Diptera, Vol. 2).

Peterson, B. V. 1960, 1971.
Peterson and Wenzel 1987.
Rodhain and Bequaert 1915.
Ryberg 1939.
Speiser 1901.

STREBLIDAE (HIPPOBOSCOIDEA)

B. A. Foote, *Kent State University*

Bat Flies

Figure 37.365

Relationships and Diagnosis: This family has long been considered to be closely related to the Nycteribiidae and Hippoboscidae, and all three are now placed in the Hippoboscoidea (McAlpine et al. 1981). Wenzel and Peterson (1987) have recently reviewed our knowledge.

Larvae are highly distinctive in being very swollen, structureless, and less than 2 mm long. They are distinguished from larvae of Nycteribiidae by having the posterior spiracles (fig. 37.365) oval, crescent-shaped, or composed of numerous openings (in Nycteribiidae, each posterior spiracle consists of a simple pore-like opening (fig. 37.364)).

Biology and Ecology: This family, like the Nycteribiidae, consists of species that are obligate ectoparasites of bats, although some species in the Old World burrow into the skin and are thus truly endoparasitic. Adults may possess well developed wings, be brachypterous or wingless. They may be somewhat flattened and have flattened tarsi that possess strong claws, or they can be laterally compressed and resemble fleas. Females deposit mature larvae on the walls of the bat roost. The larvae quickly form puparia which may become so abundant that the cave walls become encrusted. Newly emerged adults easily find hosts and adopt blood-sucking habits.

Description: Mature larvae under 2 mm, body very swollen and structureless. Posterior spiracles oval, crescent-shaped, or with numerous spiracular openings scattered around margin of spiracular plate.

Comments: Consisting of somewhat more than 200 species, the family is particularly diverse in the New World Tropics. It thus shows a distribution pattern that contrasts with the other family of bat parasites, the Nycteribiidae, which is best developed in the tropics of the Old World. Only three genera and some nine species occur in America north of Mexico. Most species are found in the southern or western states. No species is economically important.

Selected Bibliography

Ferrar 1987.
Ferris 1923.
Guimaraes 1944.
Hennig 1952.
Jobling 1949.
Kim 1985.
Marshall, A. G. 1981.
Speiser 1900.
Wenzel and Peterson 1987.
Wenzel et al. 1966.

BIBLIOGRAPHY

Adamovic, Z. R. 1963a. Ecology of some asilid species (Asilidae, Diptera) and their relation to honey bees (*Apis mellifera* L.). Mus. Hist. Natur. Beograd, Hors serie 30:1–102. (In Serbian, English summary).

Adamovic, Z. R. 1963b. The feeding habits of some asilid species (Asilidae, Diptera) in Yugoslavia. Archiv. Biol. Nauka, Beograd 15(1–2):41–74.

Aderkas, P. von and B. V. Peterson. 1987. *Chirosia betuleti* (Ringdahl) (Diptera: Anthomyiidae), a gall-former on the ostrich fern, *Matteuccia struthiopteris*, with notes on other insect-fern associates. Proc. Ent. Soc. Wash. 89:532–547.

Adler, P. H., and K. C. Kim. 1986. The black flies of Pennsylvania (Simuliidae, Diptera). Penn. St. Univ. Agr. Exp. Sta. Bull. 856:1–88.

Adler, P. H., R. W. Light and E. A. Cameron. 1985. Habitat characteristics of *Paleodipteron walkeri* (Diptera: Nymphomyiidae). Ent. News 96:211–213.

D'Aguilar, J. 1943. Contribution à l'étude des larves de *Chlorops* Meig.. Bull. Soc. Entomol. France. 1943:153–155.

Akre, R. D., G. Alpert, and T. Alpert. 1973. Life cycle and behavior of *Microdon cothurnatus* in Washington (Diptera: Syrphidae). J. Kans. Ent. Soc. 46:327–338.

Aldrich, J. M. 1912. The biology of some western species of the dipterous genus *Ephydra*. J. N. Y. Ent. Soc. 20:77–99.

Aldrich, J. M. 1916. *Sarcophaga* and allies. Thomas Say Foundation, Volume 1, Ent. Soc. Amer. pp. 1–302.

Aldrich, J. M. 1927. The dipterous parasites of the migratory locust of tropical America, *Schistocera paranesis* Burmeister. J. Econ. Ent. 20:588–593.

Aldrich, J. M. 1930. American two-winged flies of the genus *Stylogaster* Macquart. Proc. U.S. Nat. Mus. 78(9):1–27.

Alexander, C. P. 1920. The crane-flies of New York. Part II. Biology and phylogeny. Mem. Cornell Univ. Agr. Exp. Sta. 38:699–1133.

Alexander, C. P. 1922. The biology of the North American crane-flies (Tipulidae, Diptera). VI The genus *Cladura* Osten Sacken. Pomona J. Ent. Zool. 14:1–7.

Alexander, C. P. 1927. Family Ptychopteridae (Diptera). Fascicle 188, pp. 1–12. *In* Wytsman, P., ed., Genera insectorum. Bruxelles.

Alexander, C. P. 1930. Observations on the dipterous family Tanyderidae. Proc. Linn. Soc. N.S.W. 55:221–230.

Alexander, C. P. 1942. Guide to the insects of Connecticut. Part VI. The Diptera or true flies of Connecticut. First fascicle (part). Families Tanyderidae, Ptychopteridae, Trichoceridae, Anisopodidae, and Tipulidae. Bull. Conn. St. Geol. and Nat. Hist. Surv. 64:183–486.

Alexander, C. P. 1963. Guide to the Insects of Connecticut. Part VI. The Diptera or true flies of Connecticut. Eighth fascicle. Family Blepharoceridae. Bull. Conn. St. Geol. Nat. Hist. Surv. 93:39–71.

Alexander, C. P. 1967. The crane flies of California. Bull. Calif. Insect Surv. 8:1–269.

Alexander, C. P. 1981a. Tanyderidae, pp. 149–151. *In* McAlpine, J. F., et al., eds., Manual of Nearctic Diptera. Vol. 1. Res. Br. Agr. Can. Monogr. 27.

Alexander, C. P. 1981b. Trichoceridae, pp. 301–304. *In* McAlpine, J. F., et al., eds., Manual of Nearctic Diptera. Vol. 1. Res. Br. Agr. Can. Monogr. 27.

Alexander, C. P. 1981c. Ptychopteridae, pp. 325–328. *In* McAlpine, J. F., et al., eds., Manual of Nearctic Diptera. Vol. 1. Res. Br. Agr. Can. Monogr. 27.

Alexander, C. P., and G. W. Byers. 1981. Tipulidae, pp. 153–190. *In* McAlpine, J. F., et al., eds., Manual of Nearctic Diptera. Vol. 1. Res. Br. Agr. Can. Monogr. 27.

Alexander, C. P., and J. T. Lloyd. 1914. The biology of the North American crane-flies I. The genus *Eriocera* Macquart. Pomona J. Ent. Zool. 6:12–23.

Allen, E. J., and B. A. Foote. 1967. Biology and immature stages of three species of Otitidae (Diptera) which have saprophagous larvae. Ann. Ent. Soc. Amer. 60:826–836.

Allen, E. J., and B. A. Foote. 1975. Biology and immature stages of *Tritoxa incurva* (Diptera: Otitidae). Proc. Ent. Soc. Wash. 77:246–257.

Allen, H. W. 1925. Biology of the red-tailed tachina-fly, *Winthemia quadripustulata* Fabr. Miss. Agr. Exp. Sta. Bull. No. 12:3–32.

Allen, H. W. 1926. North American species of two-winged flies belonging to the tribe Miltogrammini. Proc. U.S. Nat. Mus. 68:1–106.

Allen, M. W., and R. H. Painter. 1937. Observations on the biology of the wheat-stem maggot in Kansas, *Meromyza americana* Fitch. J. Agr. Res. 55:215–238.

Allessandrini, G. 1909. Studi ed esperienze sulle larve della *Piophila casei*. Arch. Parasitol. 13:337–389.

Anderson, H. 1976. Revision of the *Anthomyza* species of northwest Europe (Diptera: Anthomyzidae) 1. The *gracilis* group. Ent. Scand. 7:41–52.

Andries, M. 1912. Zur Systematik, Biologie und Entwicklung von *Microdon* Meigen. Zeitschr. Wiss. Zool. 103:300–361.

Anthon, H. 1943a. Zum Kopfbau der primitivsten bisher bekannten Dipteren-Larve: *Olbiogaster* sp. (Rhyphidae). Ent. Meddr. 23:303–320.

Anthon, H. 1943b. Der Kopfbau der Larven einiger nematoceren Dipterenfamilien: Rhyphidae, Trichoceridae, Psychodidae und Ptychopteridae. Spolia Zool. Mus. Haun. 3:1–61.

Anthon, H., and L. Lyneborg. 1968. The cuticular morphology of the larval head capsule in Blephariceridae (Diptera). Spolia zool. Mus. Haun. 27:7–56.

App, B. A. 1938. *Euxesta stigmatias* Loew, an otitid fly infesting ear corn in Puerto Rico. J. Agr. Univ. Puerto Rico 22:181–188.

Arnaud, P. H., Jr. 1954. *Hilarella hilarella* (Zetterstedt) (Diptera: Sarcophagidae) parasitic upon a rhaphidophorid (Orthoptera: Gryllacrididae). Can. Ent. 86:135–136.

Arnaud, P. H. 1968. The American species of the genus *Trypetisoma* Malloch (Diptera, Lauxaniidae). Wasmann J. Biol. 26:107–119.

Arnaud, P. H. 1978. A host-parasite catalog of North American Tachinidae (Diptera). U.S. Dept. Agr. Misc. Publ. 1319, 860 p.

Arntfield, P. W. 1977. Systematics and biology of the genus *Chasmatonotus* Loew (Diptera: Chironomidae: Orthocladiinae) from North America. Unpublished Ph.D. Thesis. McGill University, Montreal.

Arthur, A. P., and H. C. Coppel. 1953. Studies on dipterous parasites of the spruce budworm, *Choristoneura fumiferana* (Clem.) (Lepidoptera: Tortricidae). I. *Sarcophaga aldrichi* Park. (Diptera: Sarcophagidae). Can. J. Zool. 31:374–391.

Ashburner, M., and T. R. F. Wright, eds., 1978. The Genetics and Biology of *Drosophila*. Vol. 2b. Academic Press, N.Y. 601p.

Ashburner, M., H. L. Carson, and J. N. Thompson, Jr., eds., 1981. The Genetics and Biology of *Drosophila*. Vol. 3a. Academic Press, N.Y. 429p.

Ashburner, M., H. L. Carson, and J. N. Thompson, Jr., eds., 1982. The Genetics and Biology of *Drosophila*. Vol. 3b. Academic Press, N.Y. 428p.

Ashburner, M., H. L. Carson, and J. N. Thompson, Jr., eds., 1983a. The Genetics and Biology of *Drosophila* Vol. 3c. Academic Press, N.Y. 428p.

Ashburner, M., H. L. Carson, and J. N. Thompson, Jr., eds., 1983b. The Genetics and Biology of *Drosophila*. Academic Press, N.Y. 382p.

Ashby, D. G. and D. W. Wright. 1946. The immature stages of the carrot fly. Trans. Roy. Ent. Soc. London 97:355–379.

Atchley, W. R. 1970. A biosystematic study of the subgenus *Selfia* of *Culicoides* (Diptera: Ceratopogonidae). Kans. Univ. Sci. Bull. 49:181–336.

Auten, M. 1925. Insects associated with spider nests. Ann. Ent. Soc. Amer. 18:240–250.

Babaeo, T., and V. N. Tanasiychuk. 1971. *Leucopis (Leucopomyia) alticeps* (Diptera, Chamaemyiidae). Its biology and food relations. Zool. Zh.10:1520–1529.

Babb, Z. F. 1902. Notes on *Rhagoletis suavis* Loew, with a description of the larva and puparium (Trupaneidae). Ent. News 13:242.

Backlund, H. O. 1945a. Larva and pupa of *Heterochila buccata* Fall. compared with the supposed larva of *Helcomyza ustulata* Curtis (Dipt. Dryomyzidae). Zund, K. fysiogr. Sallsk., Lund 15:1–5.

Backlund, H. O. 1945b. Wrack fauna of Sweden and Finland. Opusc. Ent. Suppl. 5.

Baer, W. 1921. Die Tachinen als Schmarotzer der schädlichen Insekten. Ihre Lebenweise, wirtschlaftliche Bedeutung und systmatische Kennzeichnung II. Zeitschr. Angew. Ent. 6:349–423.

Baerg, W. 1958. The Tarantula. Univ. Kans. Press, Lawrence.

Baird, C. R. 1972. Development of *Cuterebra ruficrus* (Diptera: Cuterebridae) in six species of rabbits and rodents with a morphological comparison of *C. ruficrus* and *C. jellisoni* third instars. J. Med. Ent. 9:81–85.

Baird, C. R. 1975. Larval development of the rodent fly, *Cuterebra tenebrosa,* in bushy-tail wood rats and its relationship to pupal diapause. Can. J. Zool. 53:1788–1798.

Baker, A. S. 1979. Food preferences of Tanypodinae larvae (Diptera: Chironomidae). Hydrobiol. 62:283.

Balachowsky, A., and L. Mesnil. 1935, 1936. Les insectes nuisibles aux plantes cultivées, I, Paris 1935 and II, Paris 1936.

Balduff, W. V. 1959. Obligatory and facultative insects in rose hips, their recognition and bionomics. III. Biol. Monogr. 26:1–194.

Baldwin, W. F. and Coppel, H. C. 1949. The biology of *Phorocera hamata* A. & W., a tachinid parasite of sawflies. Can. Ent. 81:237–245.

Banks, N. 1912. The structure of certain dipterous larvae with particular reference to those in human foods. U.S. Dept. Agr. Bur. Ent. Tech. Bull. 22:1–44.

Banta, A. M. 1907. The fauna of Mayfields Cave. Publ. 67, Carnegie Inst. 114:45.

Barber, G. W. 1939. Injury to sweet corn by *Euxesta stigmatias* Loew in southern Florida. J. Econ. Ent. 32:879–880.

Barber, K. 1985. A taxonomic revision of the genus *Pseudodinia* Coquillett (Diptera: Chamaemyiidae). Proc. Ent. Soc. Ont. 116:105–167.

Barnes, H. F. 1946–1956. Gall Midges of Economic Importance. Vols. 1–7. Crosby Lockwood and Son Ltd., London.

Barnes, J. K. 1979a. Biology of the New Zealand genus *Neolimnia* (Diptera: Sciomyzidae). N. Z. J. Zool. 6:561–576.

Barnes, J. K. 1979b. Bionomics of the New Zealand genera *Neolimnia* and *Eulimnia* (Diptera: Sciomyzidae). J. N.Y. Ent. Soc. 86:277–278.

Barnes, J. K. 1980a. Biology and immature stages of *Helosciomyza subalpina* (Diptera: Helosciomyzidae), an ant-killing fly from New Zealand. N. Z. J. Zool. 7:221–229.

Barnes, J. K. 1980b. Taxonomy of the New Zealand genus *Eulimnia,* and biology and immature stages of *E. philpotti* (Diptera: Sciomyzidae). N. Z. J. Zool. 7:91–103.

Barnes, J. K. 1984. Biology and immature stages of *Dryomyza anilis* Fallen (Diptera: Dryomyzidae). Proc. Ent. Soc. Wash. 86:43–52.

Barr, A. R. 1958. The mosquitoes of Minnesota (Diptera: Culicidae: Culicinae). Bull. Minn. Agr. Exp. Sta. 228:1–154.

Basden, E. B. 1961. Notes on the Camillidae (Diptera) in Stroble's collection and on the biology of *Camilla.* Nat. Ent. 41:124–129.

Baud, F. 1970. Le développement post-embryonnaire de deux diptères musidorides: *Musidora furcata* Fall. et *M. lutea* Panz. Revue suisse Zool. 77:647–650.

Baud, F. 1973. Biologie et cytologie de cinq espèces du genre *Lonchoptera* Meig. (Dipt.) dont l'une est parthénogénétique et les autres bisexuées, avec quelques remarques d'ordre taxonomique. Rev. suisse Zool. 80:473–515.

Baumann, E. 1978. Rennfliegen (Diptera: Phoridae) als Blutenbesucher. Kritische Sichtung der Literatur. Flora Bd. 167:301–314.

Beard, R. L. 1940. The biology of *Anasa tristis* DeGeer, with particular reference to the tachinid parasite, *Trichopoda pennipes* Fabr. Conn. Agr. Exp. Sta. Bull. 440:595–679.

Beaver, O. 1972a. Notes on the biology of some British sciomyzid flies (Diptera: Sciomyzidae). 1. Tribe Sciomyzini. Entomol. 105:139–145.

Beaver, O. 1972b. Notes on the biology of some British sciomyzid flies (Diptera: Sciomyzidae). II. Tribe Tetanocerini. Entomol. 105:284–299.

Beaver, R. A. 1966. The biology and immature stages of two species of *Medetera* (Diptera: Dolichopodidae) associated with the bark beetle *Scolytus scolytus* (F.). Proc. Roy. Ent. Soc. (A) 41:145–154.

Beaver, R. A. 1969. Anthomyiid and muscid flies bred from snails. Ent. Mon. Mag. 105:25–26.

Beaver, R. A. 1972. Ecological studies on Diptera breeding in dead snails. I. Biology of the species found on *Cepaea nemoralis* (L.). Entomologist 105:41–52.

Beaver, R. A. 1977. Non-equilibrium "island" communities: Diptera breeding in dead snails. J. Animal Ecol. 46:783–798.

Bechtes, R. C., and E. I. Schlinger. 1957. Biological observations on *Ectemnius* with particular reference to their *Ogcodes* prey (Hymenoptera: Sphecidae.-Diptera: Acroceridae). Ent. News 48:225–232.

Beck, W. M. 1976. Biology of the larval chironomids. Fla. St. Dept. Environ. Reg. Tech. Ser. 2:1–58.

Beck, W. M. 1977. Environmental requirements and pollution tolerance of common freshwater Chironomidae. U.S. E.P.A. EPA-600; 4–77–024:1–261.

Becker, T. 1895. Dipterologische Studien. II. Sapromyzidae. Berliner Ent. Zeitschr. 40:171–264.

Becker, T. 1926. Ephydridae. Lfg. 10, pp. 1–48; Lfg. 11, pp. 49–115. *In* Lindner, E., ed., Die Fliegen der palaearktischen Region. Vol. 6, pt. 1. Schweizerbart, Stuttgart.

Beckett, D. C., and P. A. Lewis. 1982. An efficient procedure for slide mounting of larval chironomids. Trans. Amer. Microsc. Soc. 101:96–99.

Bedding, R. A. 1973. The immature stages of the Rhinophorinae (Diptera: Calliphoridae) that parasitise British Woodlice. Trans. Roy. Ent. Soc. London 125:27–44.

Beingolea, O. 1957. Notes on a chamaemyiid (Diptera) predator of the eggs of *Orthezia insignis* Douglas (Homoptera) in Peru. Bull. Brooklyn Ent. Soc. 52:118–121.

Beling, T. 1875. Beitrag zur Metamorphose zweiflügeliger Insekten. Arch. Naturg. 41:31–57.

Beling, T. 1882. Beitrag zur Metamorphose zweiflügeliger Insekten aus den Familien Tabanidae, Leptidae, Asilidae, Empididae, Dolichopodidae und Syrphidae. Arch. Naturg. 44:187–240.

Beling, T. 1886a. Beitrag zur Metamorphose der Zweiflügler-Gattung *Sciara* Meig. Wien. Ent. Ztg. 5:11–14, 71–74, 93–96, 129.

Beling, T. 1886b. Dritter Beitrag zur Naturgeschichte (Metamorphose) verschiedener Arten aus der Familie der Tipuliden. Verh. Zool.-Bot. Ges. Wien 36:171–214.

Beling, T. 1888. Beitrag zur Metamorphose einiger zweiflügeliger Insekten aus den Familien Tabanidae, Empidae und Syrphidae. Verh. Zool.-Bot. Ges. Wien 38:1–4.

Belkin, J. N., and C. L. Hogue. 1959. A review of the crabhole mosquitoes of the genus *Deinocerites* (Diptera, Culicidae). Univ. Calif. Publ. Ent. 14:411–458.

Benestad Hagvar, E. 1974. Effectiveness of larvae of *Syrphus ribesii* and *S. corollae* (Diptera: Syrphidae) as predators on *Myzus persicae* (Homoptera: Aphididae). Entomophaga 19:123.

Benjamin, F. H. 1934. Descriptions of some native trypetid flies with notes on their habits. U.S. Dept. Agr. Tech. Bull. 401.

Bennett, G. F. 1955. Studies on *Cuterebra emasculator* Fitch 1856 (Diptera: Cuterebridae) and a discussion of the status of the genus *Cephenemyia* Ltr. 1818. Can. J. Zool. 33:75–98.

Bennett, G. F. 1972. Further studies on the chipmunk warble, *Cuterebra emasculator* Fitch (Diptera: Cuterebridae). Can. J. Zool. 50:861–864.

Bequaert, J. 1942. *Carnus hemapterus* Nitzsch, an ectoparasitic fly of birds, new to America (Diptera). Bull. Brooklyn Ent. Soc. 37:140–149.

Bequaert, J. 1954. The Hippoboscidae or louse flies of mammals and birds. Part II. Taxonomy, evolution and revision of American genera and species. Ent. Amer. 34(N. S.):1–232.

Berg, C. O. 1947. Biology and metamorphosis of some Solomon Islands Diptera. Part I. Micropezidae and Neriidae. Occ. Pap. Mus. Zool. Univ. Michigan 503:1–14.

Berg, C. O. 1950a. *Hydrellia* and some other acalyptrate Diptera reared from *Potamogeton*. Ann. Ent. Soc. Amer. 43:374–398.

Berg, C. O. 1950b. Biology of certain Chironomidae reared from *Potamogeton*. Ecol. Monogr. 20:83–101.

Berg, C. O. 1952. Biology and metamorphosis of some Solomon Islands Diptera. Part. II: *Solva bergi* James (Erinnidae), with a comparison of related species. Pan-Pacif. Ent. 28:203–215.

Berg, C. O. 1953. Sciomyzid larvae (Diptera) that feed on snails. J. Parasitol. 39:630–636.

Berg, C. O. 1961. Biology of snail-killing Sciomyzidae (Diptera) of North America and Europe. Verh. XI. Intern. Kongr. Ent. (Wien 1960) 1 (1): 197–202.

Berg, C. O. 1964a. Snail control in trematode diseases: the possible value of sciomyzid larvae, snail-killing Diptera, pp. 259–309. *In* Dawes, B., ed., Advances in Parasitology. 2. Acad. Press. London.

Berg, C. O. 1964b. Snail-killing sciomyzid flies: biology of the aquatic species. Verh. Intern. Verein. Limnol. 15:926–932.

Berg, C. O. 1973. Biological control of snail-borne diseases: a review. Exp. Parasitol. 33:318–330.

Berg, C. O., B. A. Foote, L. Knutson, J. K. Barnes, S. L. Arnold, and K. Valley. 1982. Adaptive differences in phenology in sciomyzid flies, pp. 15–36. *In* Mathis, W. N., and F. C. Thompson, eds., Recent Advances in Dipteran Systematics: Commemorative Volume in Honor of Curtis W. Sabrosky. Mem. Ent. Soc. Wash. 10.

Berg, C. O., and L. Knutson. 1978. Biology and systematics of the Sciomyzidae. Annu. Rev. Entomol. 23:239–258.

Berg, K. 1937. Contributions to the biology of *Corethra* Meigen. Biol. Meddel. Kgl. Dansk Vidensk. Selsk. 13:1–101.

Berg, V. L. 1940. The external morphology of the immature stages of the bee fly, *Systoechus vulgaris* Loew, a predator of grasshopper egg pods. Can. Entomol. 72:169–178.

Bergenstamm, J. von. 1864. Über die Metamorphose von *Discomyza incurva* Fall. Verh. Zool.-Bot. Ges. Wien 14:713–716.

Bergmann, A. M. 1917. Om renens oestrider. Entomol. Tidskr. 38:1–32, 113–146.

Beri, S. K. 1973. Comparative morphological studies on the spiracles of larval Agromyzidae (Diptera). J. Nat. Hist. 7:481–491.

Berry, P. A. and Parker, H. L. 1950. Notes on parasites of *Sitona* in Europe, with especial reference to *Campogaster exigua* (Meig.) (Diptera, Larvaevoridae). Proc. Ent. Soc. Wash. 52:251–258.

Beyer, A. 1939. Morphologische, ökologische und physiologische Studien an den Larven der Fliegen: *Ephydra riparia* Fallén, *E. micans* Haliday und *Cania fumosa* Stenhammar. Kieler Meeresforschungen 3:265–320.

Bhatia, M. L. 1939. Biology, morphology and anatomy of aphidophagous syrphid larvae. Parasitol. 31:78–129.

Bhattacharjee, N. S. 1977. Preliminary studies on the effect of some pesticides on soybean nodulation. Pesticides 70:447–449.

Bibby, F. F. 1942. Some parasites of *Heliothis armigera* (Hbn.) in Texas. J. Econ. Ent. 35:943–944.

Bibro, C. M., and B. A. Foote. 1986. Larval description of *Rivellia pallida* (Diptera: Platystomatidae), a consumer of the nitrogen-fixing root nodules of hog-peanut, *Amphicarpa bracteata* (Leguminosae). Proc. Ent. Soc. Wash. 88:578–584.

Bischoff, F. C., Laake, E. W., Bundrett, H. M. and R. V. Wells. 1926. The cattle grubs or ox warbles, their biologies and suggestions for control. U.S. Dept. Agr. Bull. 1369.

Bischoff, W. 1922. Über die Kopfbildung der Dipterenlarven. Einleitung und I. Teil: Die Köpfe der Oligoneuralarven. Arch. Naturg. 88:1–51.

Bischoff, W. 1923. Die Biologie der Blepharoceriden wahrend der Entwicklung und als Imago. Verh. Intern. Ver. Limnol. 1:222–234.

Bischoff, W. 1924. Die Segmentierung der Blepharoceridenlarven und Puppen und die ökologische Begründung der Larvenphylogenie. Zool. Anz. 60:231–251.

Bischoff, W. 1925. Über die Kopfbildung der Dipterenlarven. III. Teil. Die Köpfe der Orthorrhapha-Brachycera Larven. Arch. Naturg. 90 A 8:1–105.

Bischoff, W. 1928. Die Ökologie der palaearktischen Blepharoceridae. Ergebn. Fortschr. Zool. 7:209–278.

Bjelski, B. J. 1917. On the biology of *Opomyza florum*, a pest of winter cereals. J. Appl. Ent. 1:47–76.

Blair, J. M., and B. A. Foote 1984. Resource partitioning in five sympatric species of *Scatella* (Diptera: Ephydridae). Environ. Ent. 13:1336–1339.

Blake, D. H. 1955. Note on the rearing of *Anolisimyia blakeae*, a sarcophagid fly from the American chameleon, *Anolis carolinensis* Voight (Diptera: Sarcophagidae). Proc. Ent. Soc. Wash. 57:187–188.

Blanc, F. L., and R. H. Foote. 1961. A new genus and five new species of California Tephritidae (Diptera). Pan-Pacif. Ent. 37:73–83.

Blanchard, E. E. 1964. Nuevos predatores de la familia Chamaemyiidae con informacion sobre otras especies Argentinas (dipt.). Rev. Invest. Agropec. Buenos Aires (5) 1:133–150, 7 figs.

Blanchard, R. 1892. Sur les Oestrides américains dont la larve vit dans la peau de l'homme. Annls. Soc. ent. France 109–154.

Blanchard, R. 1894. Contributions à l'étude des Diptères parasites. (2 è série). Annls. Soc. ent. France 142–160.

Blanton, F. S. 1938. Some dipterous insects reared from *Narcissus* bulbs. J. Econ. Ent. 31:113–116.

Blanton, F. S., and W. W. Wirth. 1979. The sand flies (*Culicoides*) of Florida (Diptera: Ceratopogonidae). Arthropods of Florida and Neighboring Land Areas 10:1–204.

Bloesler, W. 1914. Notes on the life history and anatomy of *Siphona plusiae* Coq. Ann. Ent. Soc. Amer. 7:301–306.

Bode, R. W. 1983. Larvae of North American *Eukiefferiella* and *Tvetenia* (Diptera: Chironomidae). Bull. N.Y. St. Mus. 452:1–40.

Boesel, M. W. 1973. The genus *Atrichopogon* (Diptera, Ceratopogonidae) in Ohio and neighboring states. Ohio J. Sci. 73:202–215.

Boesel, M. W. 1974. Observations on the Coelotanypodini of the Northeastern States with keys to the known stages. (Diptera: Chironomidae: Tanypodinae). J. Kans. Ent. Soc. 47:417.

Boesel, M. W., and E. G. Snyder. 1944. Observations on the early stages and life history of the grass punky, *Atrichopogon levis* Coq. Ann. Ent. Soc. Amer. 37:37–46.

Bohart, G. E., and J. L. Gressitt. 1951. Filth inhabiting flies of Guam. Bernice P. Bishop Museum Bull. 204:1–152.

Bokermann, W. C. A. 1957. Frog eggs parasitized by dipterous larvae. Herpetol. 13:231–232.

Bolwig, N. 1940. The description of *Scatophila unicornis* Czerny, 1900 (Ephydridae, Diptera). Proc. Roy. Entomol. Soc. London (B) 9:129–136.

Bolwig, N. 1946. Senses and sense organs of the anterior end of the housefly larva. Vidensk. Meddr. Dansk Naturh. Foren. 109:81–217.

Borg, A. 1959. Investigations on the biology and control of timothy grass flies, *Amaurosoma armillatum* Zett. and *A. flavipes* Fall. (Dipt. Cordyluridae). Swedish St. Plant Protect. Inst. Contr. 11:301–372.

Borkent, A. 1979. Systematics and bionomics of the species of the subgenus *Schadonophasma* Dyar and Shannon (*Chaoborus*, Chaoboridae, Diptera). Quaest. Ent. 15:122–255.

Borkent, A., and D. M. Wood. 1986. The first and second larval instars and the egg of *Parasimulium stonei* Peterson (Diptera: Simuliidae). Proc. Ent. Soc. Wash. 88:287–296.

Bouché, P. F. 1847. Beiträge zur Kenntnis der Insekten-Larven. Stettin. Entomol. Ztg. 8:142–146.

Bovien, P. 1935. The larval stages of *Scatopse*. Vidensk. Meddel. Naturhist. Foren. Kobenhavn 99:33–43.

Boyce, A. M. 1934. Bionomics of the walnut husk fly, *Rhagoletis cingulata*. Hilgardia 8:361–579.

Bratt, A. D., L. V. Knutson, B. A. Foote, and C. O. Berg. 1969. Biology of *Pherbellia* (Diptera: Sciomyzidae). Mem. Cornell Agr. Exp. Sta. 404:1–247.

Brauer, F. 1863. Monographie der Oestriden. Wien. C. Ueberreuter, 291p.

Brauer, F. 1883. Die Zweiflügler des Kaiserlichen Museums zu Wien III. Systematische Studien auf Grundlage der Dipterenlarven nebst einer Zusammenstellung von Beispielen aus der Literatur über dieselben und Beschreibung neuer Formen. Denkschr. Akad. Wiss. Wien Math.-Nat. 47:1–100.

Brauns, A. 1939. Zur Biologie der Meeresstrandfliege *Scatella subguttata* Meig. (Familie Ephydridae; Diptera). Zool. Anz. 126:273–285.

Brauns, A. 1950. Zur Kenntnis der Schadinsekten der Champignonkulturen. Nachr. Blatt Planzenschutzdienst, Braunschweig 2:153–156.

Brauns, A. 1954. Terricole Dipterenlarven. Musterschmidt Wissenschaftliche Verlag, Gottingen.

Breland, O. P. 1942. Dipterous parasites of adult mantids (Mantidae: Sarcophagidae). Proc. Ent. Soc. Wash. 44:19–22.

Breland, O. P. 1949. The biology and immature stages of the mosquito *Megarhinus septentrionalis* Dyar and Knab. Ann. Ent. Soc. Amer. 42:38–47.

Brindle, A. 1957. The ecological significance of the anal papillae of *Tipula* larvae (Dipt., Tipulidae). Ent. Mon. Mag. 93:202.

Brindle, A. 1958a. Notes on the larvae of the British Tipulinae. (Dipt., Tipulidae). Part 1. Ent. Mon. Mag. 94:230–232.

Brindle, A. 1958b. Notes on the larvae of the British Tipulinae. (Dipt., Tipulidae). Part 2. Ent. Mon. Mag. 94:241–244.

Brindle, A. 1958c. Notes on the larvae of the British Tipulinae. (Dipt.. Tipulidae). Part 3. Ent. Mon. Mag. 94:272–274.

Brindle, A. 1958d. Notes on the identification of *Tipula* larvae (Diptera-Tipulidae). Part 1. Ent. Gaz. 9:45–52.

Brindle, A. 1958e. A field key for the identification of *Tipula* larvae (Dipt.: Tripulidae). Ent. Gaz. 9:165–182.

Brindle, A. 1959a. Notes on the larvae of the British Tipulinae. (Dipt., Tipulidae). Part 4. Ent. Mon. Mag. 95:36–37.

Brindle, A. 1959b. Notes on the larvae of the British Tipulinae. (Dipt., Tipulidae). Part 5. Ent. Mon. Mag. 95:64–65.

Brindle, A. 1959c. Notes on the larvae of the British Tipulinae. (Dipt., Tipulidae). Part 6. Ent. Mon. Mag. 95:176–177.

Brindle, A. 1959d. Notes on the larvae of the British Tipulinae. (Dipt., Tipulidae). Part 7. Ent. Mon. Mag. 95:204–205.

Brindle, A. 1960a. The larvae and pupae of the British Tipulinae (Diptera: Tipulidae). Trans. Soc. Brit. Entomol. 14:63–114.

Brindle, A. 1960b. The larvae of the British Hexatomini (Dipt., Tipulidae). Ent. Gaz. 11:207–224.

Brindle, A. 1962a. Taxonomic notes on the larvae of British Diptera. 6. The family Bibionidae. Entomol. 95:22–26.

Brindle, A. 1962b. Taxonomic notes on the larvae of British Diptera. 9. The family Ptychopteridae. Entomologist 95:212–216.

Brindle, A. 1962c. Taxonomic notes on the larvae of British Diptera, 10. The Asilidae. Ent. 95:241–247.

Brindle, A. 1962d. Taxonomic notes on the larvae of British Diptera. 11. Trichoceridae and Anisopodidae. Entomol. 95:212–216, 285–288.

Brindle, A. 1965a. Taxonomic notes on the larvae of British Diptera. No. 19. The Micropezidae (Tylidae). Entomol. 98:83–86.

Brindle, A. 1965b. Taxonomic notes on the larvae of British Diptera. No. 23. The Geosarginae (Stratiomyidae). Entomol. 98:208–216.

Brindle, A. 1967. The larvae and pupae of the British Cylindrotominae and Limoniinae (Diptera, Tipulidae). Trans. Soc. Brit. Ent. 17:151–216.

Brindle, A. 1968. Taxonomic notes on the larvae of British Diptera. No. 25, The larvae and pupa of *Disetria rufipes* (DeGeer) (Asilidae). Entomol. 101:213–216.

Brindle, A. 1969. No. 26, The presumed larvae of *Disetria oelandica* (L.) (Asilidae). Entomol. 102:3–6.

Brittain, W. H. 1927. The cabbage maggot and its control in Nova Scotia. Nova Scotia Dept. Nat. Res. Bull. 11:1–48.

Brock, M. L., R. G. Wiegert, and T. D. Brock. 1969. Feeding by *Paracoenia* and *Ephydra* (Diptera: Ephydridae) on the microorganisms of hot springs. Ecol. 50:192–200.

Brodo, F. 1967. A review of the subfamily Cylindrotominae in North America (Diptera: Tipulidae). Kans. Univ. Sci. Bull. 47:71–115.

Bromley, S. W. 1946. Guide to the Insects of Connecticut. Part VI. The Diptera or True Flies of Connecticut. Third Fascicle. Family Asilidae. Bull. St. Conn. Geol. Nat. Hist. Surv. 69:1–48.

Brooks, A. R. 1949. The identification of the commoner root maggots of garden crops in Canada. 12 pages, 6 plates, mimeograph. Systematic Entomology, Dominion Entomology Lab., Saskatoon, Sask.

Brooks, A. R. 1951. Identification of the root maggots (Diptera: Anthomyiidae) attacking cruciferous garden crops in Canada, with notes on biology and control. Can. Ent. 83:109–120.

Brown, N. R., and R. C. Clark. 1956a. Studies of predators of the balsam woolly aphid, *Adelges piceae* (Ratz.) (Homoptera: Adelgidae). I. Field identification of *Neoleucopis obscura* (Hal.), *Leucopina americana* (Mall.) and *Cremifamia nigrocellulata* Cz. (Diptera, Chamaemyiidae). Can. Ent. 88:272–279.

Brown, N. R., and R. C. Clark. 1956b. Studies of predators of the balsam woolly aphid, *Adelges piceae* (Ratz.) (Homoptera: Adelgidae). II. An annotated list of the predators associated with the balsam woolly aphid in eastern Canada. Can. Ent. 88:678–683.

Brown, N. R., and R. C. Clark. 1957. Studies of the balsam woolly apid, *Adelges piceae* (Ratz.) (Homoptera: Adelgidae). IV. *Neoleucopis obscura* (Hal.) (Diptera: Chamaemyiidae), an introduced predator in eastern Canada. Can. Ent. 89:533–546.

Brues, C. T. 1902. Notes on the larvae of some Texan Diptera. Psyche 9:351–354.

Brues, C. T. 1928. Studies of the fauna of hot springs in the western United States and the biology of thermophilous animals. Proc. Amer. Acad. Arts Sci. 63:139–228.

Brues, C. T. 1932. Further studies on the fauna of North American hot springs. Proc. Amer. Acad. Arts Sci. 67:185–303.

Brundin, L. 1948. Über die Metamorphose der Sectio Tanytarsariae connectentes. Ark. Zool. 41 A, 2:1–22.

Brundin, L. 1949. Chironomiden und andere Bodentiere der sudschwedischen Urgebirgsseen. Rep. Inst. Freshw. Res. Drottningholm 30:1–914.

Brundin, L. 1956. Zur Systematik der Orthocladiinae (Dipt. Chironomidae). Rep. Inst. Freshw. Res. Drottningholm 37:1–185.

Bryce, D., and A. Hobart. 1972. The biology and identification of the larvae of the Chironomidae (Diptera). Ent. Gaz. 23:175–217.

Burger, J. F. 1977. The biosystematics of immature Arizona Tabanidae (Diptera). Trans. Amer. Ent. Soc. 103:145–258.

Burger, J. F. and Anderson, J. R. 1974. Taxonomy and life history of the moose fly, *Haematobosca alcis*, and its association with the moose, *Alces alces shirasi* in Yellowstone National Park. Ann. Ent. Soc. Amer. 67:204–214.

Burger, J. F., J. R. Anderson, and M. F. Knudsen. 1980. The habits and life history of *Oedoparena olauca* (Diptera: Dryomyzidae), a predator of barnacles. Proc. Ent. Soc. Wash. 82:360–377.

Burger, J. F., D. J. Lake, and M. L. McKay. 1981. The larval habitats and rearing of some common *Chrysops* species (Diptera: Tabanidae) in New Hampshire. Proc. Ent. Soc. Wash. 83:373–389.

Burnet, B., and H. Thompson. 1960. Laboratory culture of *Coelopa frigida* (Fabricius) (Diptera: Coelopidae). Proc. Roy. Ent. Soc. London (A) 35:85–89.

Busacca, J. D., and B. A. Foote. 1978. Biology and immature stages of two species of *Notiphila*, with notes on other shore flies occurring in cattail marshes (Diptera: Ephydridae). Ann. Ent. Soc. Amer. 71:457–466.

Bush, G. L. 1965. The genus *Zonosemata*, with notes on the cytology of two species (Diptera-Tephritidae). Psyche 72: 307–323.

Bush, G. L. 1966. The taxonomy, cytology, and evolution of the genus *Rhagoletis* in North America (Diptera, Tephritidae). Bull. Mus. Comp. Zool. 134:431–562.

Butt, F. H. 1937. The posterior stigmatic apparatus of trypetid larvae. Ann. Ent. Soc. Amer. 30:487–491.

Buxton, P. A. 1960. British Diptera associated with fungi. III. Flies of all families reared from about 150 species of fungi. Ent. Mon. Mag. 96:61–94.

Byers, G. W. 1958. Species recognition in immature craneflies (Diptera: Tipulidae). X Intern. Congr. Ent. (Montreal, 1956) 1:131–136.

Byers, G. W. 1961a. Biology and classification of *Chionea* (Diptera: Tipulidae). XI Intern. Congr. Ent. (Vienna, 1960) 1:188–191.

Byers, G. W. 1961b. The cranefly genus *Dolichopeza* in North America. Kans. Univ. Sci. Bull. 42:665–924.

Byers, G. W. 1975. Larva and pupa of *Idiognophomyia enniki* Alexander (Diptera: Tipulidae). Pan-Pacif. Ent. 50:282–287.

Byers, G. W. 1984. Tipulidae, pp. 491–514. *In* Merritt, R. W. and K. W. Cummins, eds., An Introduction to the Aquatic Insects of North America. 2nd ed. Kendall Hunt, Dubuque, Iowa.

Calhoun, E. L., H. R. Dodge, and R. W. Fay. 1956. Description and rearing of various stages of *Dendrophaonia scabra* (G. T.) (Diptera: Muscidae). Ann. Ent. Soc. Amer. 49:49–54.

Callahan, P. S., and H. A. Denmark. 1973. Attraction of the "lovebug," *Plecia nearctica* (Diptera: Bibionidae) to UV irradiated automobile exhaust fumes. Fla. Ent. 56:113–119.

Cameron, A. E. 1914. A contribution to the knowledge of the belladonna leafminer, *Pegomyia hyoscyami* Panz., its life history and biology. Ann. Appl. Biol. 1:43–76.

Cameron, A. E. 1918. Life history of the leaf-eating cranefly, *Cylindrotoma splendens* Doane. Ann. Ent. Soc. Amer. 11:67–89.

Cameron, W. E. 1926. Bionomics of the Tabanidae of the Canadian prairie. Bull. Entomol. Res. 17:1–42.

Camras, S. 1957a. A review of the New World *Physocephala* (Diptera: Conopidae). Ann. Ent. Soc. Amer. 50:213–218.

Camras, S. 1957b. The Conopid flies of California (Diptera). Bull. Calif. Insect Surv. 6:19–49.

Cantrell, B. K. 1981. Redescription of *Parasarcophaga reposita* (Diptera: Sarcophagidae). Australian Ent. Mag. 8:29–35.

Capelle, K. J. 1953. A revision of the genus *Loxocera* in North America with a study of geographical variation in *L. cylinderica* (Diptera, Psilidae). Ann. Ent. Soc. Amer. 46:99–114.

Capelle, K. J. 1966. The occurrence of *Oestrus ovus* L. (Diptera: Oestridae) in the bighorn sheep from Wyoming and Montana. Parasitol. 52:618–621.

Capelle, K. J. 1970. Studies of the life history and development of *Cuterebra polita* (Diptera: Cuterebridae) in four species of rodents. J. Med. Ent. 7:320–327.

Carpenter, G. H. 1901. The puparium of the grouse-fly. Irish Nat. 10:221–226.

Carpenter, G., and F. Pollard. 1918. The presence of lateral spiracles in the larva of *Hypoderma*. Proc. Roy. Irish Acad. 34:73–84.

Carpenter, S. J., and W. J. LaCasse. 1955. Mosquitoes of North America. Univ. Calif. Press, Berkeley. vii + 360p.

Carrera, M., Souza Lopes, H. De, and J. Lane. 1947. Contribuicao ao conhecimento dos Microdontinae neotropicos e descricao de duas novas especies de *Nausigaster* Williston. Rev. Brasil. Biol. 7:471–486.

Carson, H. L. 1967. The association between *Drosophila carcinophila* Wheeler and its host, the land crab *Gecarcinus ruricola* (L.). Amer. Midl. Nat. 78:324–343.

Carson, H. L., and H. D. Stalker. 1951. Natural breeding sites for some wild species of *Drosophila* in the eastern United States. Ecol. 32:317–330.

Catts, E. P. 1965. Host-parasite interrelationships in rodent bot fly infections. Trans. N. Amer. Wildl. Nat. Res. Conf. 30:184–196.

Catts, E. P. 1982. Biology of New World bot flies: Cuterebridae. Annu. Rev. Ent. 27:313–38.

Cavender, G. L. and R. D. Goeden. 1982. Life history of *Trupanea bisetosa* (Diptera: Tephritidae) on wild sunflower in southern California. Ann. Ent. Soc. Amer. 75:400–406.

Cazier, M. A. 1941. A generic review of the family Apioceratidae with a revision of the North American species (Diptera-Brachycera). Amer. Midl. Nat. 25:589–631.

Cazier, M. A. 1963. The description and bionomics of a new species of *Apiocera*, with notes on other species (Diptera: Apioceridae). Wasmann J. Biol. 21:205–234.

Cazier, M. A. 1982. A revision of the North American flies belonging to the genus *Apiocera* (Diptera, Apioceridae). Bull. Amer. Mus. Nat. Hist. 171:287–467.

Center for Disease Control. 1962. Pictorial keys (to) arthropods of public health importance. U.S. Public Health Service, Atlanta, GA., 192 pp.

Chance, M. M. 1970. The functional morphology of the mouthparts of blackfly larvae (Diptera: Simuliidae). Quaest. Ent. 6:245–284.

Chandler, A. E. F. 1968. A preliminary key of the eggs of some of the commoner aphidophagous Syrphidae (Diptera) occurring in Britain. Trans Roy. Ent. Soc. London 120:199–218.

Chassagnard, M. T. and L. Tsacas. 1974. Morphology of the head and mouth parts of imagoes of *Chrysopilus auratus* F. and of *Vermileo vermileo* G. (Diptera: Rhagionidae) (In French). Intern. J. Insect Morphol. Embryol. 3:13–32.

Cheng, L., and R. A. Lewin. 1974. Fluidization as a feeding mechanism in beach flies. Nature. 250:167–168.

Chillcott, J. G. 1959. Studies on the genus *Rhamphomyia* Meigen: A revision of the Nearctic species of the *basalis* group of the subgenus *Pararhamphomyia* Frey (Diptera: Empididae). Can. Ent. 91:257–275.

Chillcott, J. G. 1960. A revision of the Nearctic species of Fanniinae (Diptera: Muscidae). Can. Ent. (Suppl. 14):5–295.

Chillcott, J. G. 1961a. A revision of the genus *Roederioides* Coquillet (Diptera: Empididae). Can. Ent. 93(6):419–428.

Chillcott, J. G. 1961b. The genus *Bolbomyia* Loew (Diptera; Rhagionidae). Can. Ent. 93:632–636.

Chillcott, J. G. 1962. A revision of the *Platypalpus juvenis* complex in North America (Diptera: Empididae). Can. Ent. 94:113–143.

Chillcott, J. G. 1963. A new genus of Rhagionidae (Diptera) with notes and descriptions of *Bolbomyia* Loew. Can. Ent. 95:1185–1190.

Chillcott, J. G. 1965. A revision of the eastern Nearctic species of *Rhagio* Fabricius (Diptera: Rhagionidae). Can. Ent. 97:785–795.

Chiswell. J. R. 1955. On the last instar larva of *Tipula livida* van der Wulp (Diptera, Tipulidae) with notes on the frontoclypeal region of larval Tipulinae and caterpillars. Proc. R. ent. Soc. Lond., Ser. A, 30:127–136.

Chiswell, J. R. 1956. A taxonomic account of the last instar larvae of some British Tipulinae (Diptera:Tipulidae). Trans. Roy. Ent. Soc. London 108:409–484.

Chittenden, F. H. 1902. Some insects injurious to vegetable crops. Bull. U.S. Div. Ent. (N.S.) No. 33.

Chittenden, F. H. 1911. Maggots affecting yams in the south. U.S. Dept. Agr. Bur. Ent. Bull. 82(7):90.

Chittenden, F. H. 1927. *Tritoxa flexa* Wied., the black onion fly. Can. Ent. 59:1–4.

Chock, Q. C., C. J. Davis, and M. Chong. 1961. *Sepedon macropus* (Diptera: Sciomyzidae) introduced into Hawaii as a control for the liver fluke snail, *Lymnaea ollula*. J. Econ. Ent. 54:1–4.

Christenson, L. D., and R. H. Foote. 1960. Biology of fruit flies. Annu. Rev. Ent. 5:171–192.

Chu, I-Wu, and R. C. Axtell. 1971. Fine structure of the dorsal organ of the house fly larva, *Musca domestica* L. Zeitschr. Zellforsch. 117:17–34.

Chvala, M. 1976. Swarming, mating and feeding habits in Empididae (Diptera) and their significance in evolution of the family. Acta Ent. Bohemoslovaca 73:363–366.

Ciampolini, M. 1957. Reperti sulla *Norellia spinipes* Meig. (Diptera, Cordyluridae). Redia 42:259–272.

Clark, R. C., and N. R. Brown. 1957. Studies of predators of the balsam wooly aphid *Adelges piceae* (Ratz.) (Homoptera; Adelgidae). III. Field identification and some notes on the biology of *Neoleucopis pinicola* Mall. (Diptera: Chamaemyiidae). Can. Ent. 89:404–420.

Clark, R. C., and N. R. Brown. 1962. Studies of predators of the balsam wooly aphid *Adelges piceae* (Ratz.) Homoptera: Adelgidae). XI: *Cremifania nigrocellulata* Cz. (Diptera: Chamaemyiidae), an introduced predator in eastern Canada. Can. Ent. 94:1171–1175.

Claassen, P. W. 1918. Observations on the life history and biology of *Agromyza laterella* Zetterstedt. Ann. Ent. Soc. Amer. 11:9.

Clausen, C. P. 1940. Entomophagous Insects. McGraw-Hill, N.Y. 688 p.

Clausen, C. P., H. A. Jaynes, and T. R. Gardner. 1933. Further investigations of the parasites of *Popillia japonica* in the Far East. U.S. Dept. Agr. Tech. Bull. 366, 58 pp.

Cockburn, A., R. A. Barraco, T. A. Reyman, and W. H. Peck. 1975. Autopsy on an Egyptian mummy. Science 187: 1155–1160.

Coe, R. L. 1939. *Callicera yerburyi* Verrall, a synonym of *C. rufa* Schummel: further details of its life-history, with a description of the puparium. Entomol. 72:228–231.

Coe, R. L. 1942. *Rhingia campestris* Meigen: An account of its life-history and description of the early stages. Ent. Mon. Mag. 78:121–130.

Coe, R. L. 1966. Cyclorrapha. Diptera. Pipunculidae. Handbook for identification of British insects. 10(2c), 83 pp.

Coffey, M. D. 1966. Studies on the association of flies (Diptera) with dung in southeastern Washington. Ann. Ent. Soc. Amer. 59:207–218.

Coffman, W. P., and L. C. Ferrington, Jr. 1984. Chironomidae, pp. 551–652. *In* Merritt, R. W. and K. W. Cummins, eds., An Introduction to the Aquatic Insects of North America. 2nd ed. Kendall Hunt, Dubuque, Iowa.

Cogan, B. H. 1978. A revision of *Acrometopia* Schiner and closely related genera. Beitr. Ent. 28:223–250.

Cohen, M. 1936. The biology of the chrysanthemum leaf-miner, *Phytomyza atricornis* Mg. (Agromyzidae). Ann. Appl. Biol. 23:612–632.

Cole, F. R. 1919. The dipterous family Cyrtidae in North America. Trans. Amer. Ent. Soc. 45:1–79.

Cole, F. R. 1923a. Notes on the early stages of the syrphid genus *Microdon*. Pomona J. Ent. Zool. 15:19–20.

Cole, F. R. 1923b. A revision of the North American two-winged flies of the family Therevidae. Proc. U.S. Nat. Mus. 62:1–140.

Cole, F. R. (with collaboration of E. I. Schlinger) 1969. The Flies of Western North America. Univ. Calif. Press, Berkeley, xi+693 p.

Colless, D. H. 1977. A possibly unique feeding mechanism in a dipterous larva (Diptera: Culicidae: Chaoborinae). J. Aust. Ent. Soc. 16:335–340.

Colless, D. H., and D. K. McAlpine. 1970. Diptera (flies), p. 656–740. *In* The Insects of Australia. Melbourne Univ. Press, Carlton, Victoria.

Collin, J. E. 1961. British flies, vol. VI. Empididae pt. 1: Tachydrominae, pt. 2: Hybotinae, Empidinae (except *Hilara*), pt. III: Empidinae (*Hilara* only), Hemerodrominae. Cambridge Univ. Press. viii+782 p.

Collinge, W. E. 1909. Observations on the life-history and habits of *Thereva nobilitata* Fabr., and other species. J. Econ. Biol. 4:14–18.

Collins, N. C. 1975. Population biology of a brine fly (Diptera: Ephydridae) in the presence of abundant algal food. Ecol. 56:1139–1148.

Collins, N. C. 1977. Mechanisms determining the relative abundance of brine flies (Diptera: Ephydridae) in Yellowstone thermal spring effluents. Can. Ent. 109:415–422.

Collins, N. C. 1980. Developmental responses to food limitation as indicators of environmental conditions for *Ephydra cinerea* Jones (Diptera). Ecol. 61:650–661.

Connell, T. D., and J. F. Scheiring. 1982. Demography of the shore fly, *Scatella picea* (Walker) (Diptera: Ephydridae). Environ. Ent. 11:611–617.

Conway, J. A. 1970. *Ophyra capensis* Wiedemann (Dipt., Muscidae)—a new ecological niche for this species in Britain? Ent. Mon. Mag. 106:18.

Cook, E. F. 1944. The morphology of the larval heads of certain Culicidae (Diptera). Microentomology 9:38–68.

Cook, E. F. 1949. The evolution of the head in the larvae of the Diptera. Microentomology 14:1–57.

Cook, E. F. 1953. On the early stages of *Neopachygaster maculicornis* (Hine) and *Berkshiria aldrichi* (Malloch) (Diptera: Stratiomyidae). Ann. Ent. Soc. Amer. 46:293–299.

Cook, E. F. 1955. A contribution toward a monograph of the family Scatopsidae (Diptera). Part I, a revision of the genus *Rhegmoclema* Enderlein (= *Aldrovandiella* Enderlein) with particular reference to the North American species. Part II, The genera *Rhegmoclemina* Enderlein, *Parascatopse* n.g., and a new species of *Rhegmoclema*. Ann. Ent. Soc. Amer. 48:240–251, 351–364.

Cook, E. F. 1956a. A contribution toward a monograph of the Scatopsidae (Diptera). Part III. The genus *Rhexoza* Enderlein. Part IV, The genus *Swammerdamella* Enderlein. Part V, The genus *Colobostema* Enderlein. Ann. Ent. Soc. Amer. 49:1–12, 15–29, 325–332.

Cook, E. F. 1956b. The Nearctic Chaoborinae (Diptera: Culicidae). Bull. Minn. Agr. Exp. Sta. 218:1–102.

Cook, E. F. 1957. A contribution toward a monograph of the Scatopsidae (Diptera). Part VI. The genera *Scatopse* Geoffroy and *Holoplagia* Enderlein. Ann. Ent. Soc. Amer. 49:593–611.

Cook, E. F. 1958. A contribution toward a monograph of the Scatopsidae (Diptera). Part VII. The genus *Psectrosciara* Kieffer. Ann. Ent. Soc. Amer. 51:587–595.

Cook, E. F. 1963. Guide to the Insects of Connecticut. Part VI. The Diptera or True Flies of Connecticut. Eighth Fascicle. Scatopsidae and Hyperoscelidae. Bull. Conn. St. Geol. Nat. Hist. Surv. 93:1–37.

Cook, E. F. 1965. A contribution toward a monograph of the Scatopsidae (Diptera). Part VIII. The genus *Anapausis*. Part IX. The genera *Aspistes* and *Arthria*. Ann. Ent. Soc. Amer. 58:7–18, 713–721.

Cook, E. F. 1981a. Scatopsidae, pp. 313–319. *In* McAlpine J. F. et al., eds., Manual of Nearctic Diptera. Vol. 1. Res. Br. Agr. Can. Monogr. 27.

Cook, E. F. 1981b. Chaoboridae, pp. 335–339. *In* McAlpine et al., (eds). Manual of Nearctic Diptera. Vol. 1. Res. Br. Agr. Can. Monogr. 27.

Coppel, H. C. 1960. Key to adults of dipterous parasites of spruce budworm, *Choristoneura fumiferana* (Clem.) (Lepidoptera: Tortricidae). Ann. Ent. Soc. Amer. 53:94–97.

Coppel, H. C. and Maw, M. G. 1954. Studies on dipterous parasites of the spruce budworm, *Choristoneura fumiferana* (Clem.) (Lepidoptera: Tortricidae). III. *Ceromasia auricaudata* Tns. (Diptera: Tachinidae). Can. J. Zool. 32:145–156.

Coquillett, D. W. 1883. On the early stages of the dipterous fly, *Chrysopila foeda* Loew. Can. Ent. 15:112–113.

Coquillett, D. W. 1895. Two dipterous leaf-miners on garden vegetables (*Acidia fratria* Lw.). Insect Life VII: 383.

Corpus, L. D. 1986. Immature stages of *Liancalus similis* (Diptera: Dolichopodidae). J. Kans. Ent. Soc. 59:635–640.

Cotterell, G. S. 1920. The life-history and habits of the yellow dung fly (*Scatophaga stercoraria*), a possible blow fly check. Proc. Zool. Soc. London 1920:629–647.

Courtney, G. W. 1986. Discovery of the immature stages of *Parasimulium crosskeyi* Peterson (Diptera: Simuliidae), with a discussion of a unique blackfly habitat. Proc. Ent. Soc. Wash. 88:280–286.

Coutant, A. F. 1914. The habits, life history, and structure of a blood-sucking muscid larva. Parasitol. 1:135–150.

Cowan, J. M. 1943. Notes on the life history and morphology of *Cephenomyia jellisoni* Townsend and *Lipoptena depressa* Say, two dipterous parasites of the Columbian black-tailed deer. Can. Res. 21:171–187.

Coyle, F. A. 1971. Systematics and natural history of the mygalomorph spider genus *Antrodiaetus* and related genera. Bull. Mus. Comp. Zool. 141:269–402.

Craig, D. A. 1967. The eggs and embryology of some New Zealand Blephariceridae (Diptera, Nematocera) with reference to the embryology of other Nematocera. Trans. R. Soc. N.Z. 8 (18):191–206.

Craig, D. A. 1974. The labrum and cephalic fans of larval Simuliidae (Diptera: Nematocera). Can. J. Zool. 52:133–159.

Crampton, G. C. 1930a. Some anatomical details of the pupa of the archaic tanyderid dipteran *Protoplasa fitchii* O.S. Proc. Ent. Soc. Wash. 32:83–98.

Crampton, G. C. 1930b. A comparison of the more important structural details of the larva of the archaic tanyderid dipteran *Protoplasa fitchii* with other Holometabola from the standpoint of phylogeny. Bull. Brooklyn Ent. Soc. 25:239–258.

Crampton, G. C. 1942. Guide to the Insects of Connecticut. Part VI. The Diptera or True Flies of Connecticut. First Fascicle. The external morphology of Diptera. Bull. Conn. St. Geol. Nat. Hist. Surv. 64:10–165.

Crane, A. E. 1961. A study of the habits of *Ramphomyia scutellaris* Coquillet (Diptera: Empididae). Wasmann J. Biol. 19:247–263.

Crawford, D. O. 1912. The petroleum fly in California, *Psilopa petrolei* Coq. Pomona College J. Ent. Zool. 4:687–697.

Cresson, E. T., Jr. 1938. The Neriidae and Micropezidae of America north of Mexico. Trans. Amer. Ent. Soc. 64:293–366.

Cresson, E. T., Jr. 1942. Synopsis of North American Ephydridae (Diptera). Trans. Amer. Ent. Soc. 68:101–128.

Cross, E. A., Stith, M. G. and Bauman, T. R. 1975. Bionomics of the organ-pipe mud-dauber, *Trypoxylon politum* (Hymenoptera: Sphecoidea). Ann. Ent. Soc. Amer. 68:901–916.

Crosskey, R. W. 1960. A taxonomic study of the larvae of West African Simuliidae (Diptera: Nematocera) with comments on the morphology of the larval blackfly head. Bull. Br. Mus. (Nat. Hist). Ent. 10:1–74.

Crosskey, R. W. 1965. A systematic revision of the Ameniinae (Diptera: Calliphoridae). Bull. Brit. Mus. (Nat. Hist.) Ent. 16:33–140.

Crosskey, R. W. 1973. Simuliidae (black-flies, German: Kriebelmucken), pp. 109–153. *In* Smith, K. G. V., ed., Insects and other Arthropods of Medical Importance. Brit. Mus. (Nat. Hist.), London.

Crosskey, R. W. 1977. A review of the Rhinophoridae (Diptera) and a revision of the Afrotropical species. Bull. Brit. Mus. (Nat. Hist.). Entomol. 36:1–66.

Crouzel, I. S. de, and R. G. Salavin. 1943. Contribucion al estudio de las *Neorhynchocephalus argentinos* (Diptera: Nemestrinidae). Ann. Soc. Cient. Argentina 136:145–177.

Cumming, M. E. P. 1959. The biology of *Adelges cooleyi* (Gill.) (Homoptera: Phylloxeridae). Can. Ent. 91:601–617.

Curran, C. H. 1934. The North American Lonchopteridae (Diptera). Amer. Mus. Novit. 696:1–7.

Currie, D. C. 1986. An annotated list and keys to the immature black flies of Alberta (Diptera: Simuliidae). Mem Ent. Soc. Can. No. 134, 90 pp.

Curry, L. L. 1958. Larvae and pupae of the species of *Cryptochironomus* (Diptera) in Michigan. Limn. Oceanogr. 3:427–442.

Curry, L. L. 1965. A survey of environmental requirements for the midge, pp. 127–141. *In* Biological Problems in Water Pollution. U.S. Pub. Health Serv. Publ. 999-WP-25. Cincinnati.

Cutten, F. E. A., and D. K. McE. Kevan. 1970. The Nymphomyiidae (Diptera) with special reference to *Palaeodipteron walkeri* Ide and its larva in Quebec, and a description of a new genus and species from India. Can. J. Zool. 48:1–24.

Daecke, E. 1910. Trypetid galls and *Eurosta elsa* n. sp. Ent. News 21:341–343.

Dahl, C. and C. P. Alexander. 1976. A world catalogue of Trichoceridae Kertesz. 1902 (Diptera). Ent. Scand. 7:7–18.

Dahl, R. G. 1959. Studies on Scandinavian Ephydridae (Diptera, Brachycera). Opusc. Ent. Suppl. 15:1–224.

Damant, G. C. C. 1924. The adjustment of the buoyancy of the larva of *Corethra plumicornis*. J. Physiol. 59:345–356.

Dambach, C. A. and Good, E. 1943. Life history and habits of the cicada killer in Ohio. Ohio Jour. Sci. 43:32–41.

Darsie, R. F. 1949. Pupae of the anopheline mosquitoes of the northeastern United States (Diptera, Culicidae). Rev. Ent. 20:509–530.

Darsie, R. F. 1951. Pupae of the culicine mosquitoes of the northeastern United States (Diptera, Culicidae, Culicini). Mem. Cornell Univ. Agr. Exp. Sta. 304:1–67.

Darsie, R. F. and R. A. Ward. 1981. Identification and geographical distribution of mosquitoes of North America north of Mexico. Mosq. Syst. Suppl. 1:1–313.

Datta, R. K. and Mukherjee, P. K. 1978. Life history of *Tricholyga bombycis* (Diptera: Tachinidae), a parasite of *Bombyx mori* (Lepidoptera: Bombicidae). Ann. Ent. Soc. Amer. 71:767–770.

Davies, B. R. 1976. The dispersal of Chironomidae larvae: a review. J. Ent. Soc. S. Africa. 39:39–62.

Davis, C. J., Q. C. Chock, and M. Chong. 1961. Introduction of the liver fluke snail predator, *Sciomyza dorsata* (Sciomyzidae, Diptera), in Hawaii. Proc. Hawaii. Ent. Soc. 17:395–397.

Davis, E. J., III, and W. J. Turner. 1978. Biology, distribution and abundance of flesh flies (Diptera: Sarcophagidae) of the Wallowa-Whitman National Forest in northeastern Oregon. Melanderia 30:111–159.

Davis, E. J., and R. S. Zack. 1978. New host records and notes on the dipterous family Aulacigastridae. Pan-Pacif. Ent. 54:129–130.

Davis, L. 1974. Evolution of larval head-fans in Simuliidae (Diptera) as inferred from the structure and biology of *Crozetia crozetensis* (Womersley) compared with other genera. Zool. J. Linn. Soc. 55:193–224.

Davis, R. 1960. Parasites of the elm spanworm, *Ennomos subsignarius* (Hbn.), in Georgia (Lepidoptera: Geometridae). Proc. Ent. Soc. Wash. 62:247–248.

DeLeon, D. 1935. A study of *Medetera aldrichi* Wh. (Diptera: Dolichopodidae), a predator of the mountain pine beetle (*Dendroctonus monticolae* Hopk). Ent. Amer. 15:59–91.

Del Guercio, G. 1893. La mosca del Giaggiolo. Bull. Soc. Ent. Ital. 24:321–330.

Delucchi, V., and H. Pschorn-Walcher. 1954. *Cremifania nigrocellulata* Cz., ein Räuber an *Dreyfusia (Adelges) piceae* Ratz. Zeitschr. Angew. Ent. 36:84–107.

Dennis, D. S. 1979. Ethology of *Holcocephala fusca* in Virginia (Diptera: Asilidae). Proc. Ent. Soc. Wash. 81:366–378.

Dennis, D. S. and J. A. Gowen. 1978. A "nocturnal" foraging record for *Diogmites neoternatus* (Diptera: Asilidae). Proc. Ent. Soc. Wash. 80:313–314.

Dennis, D. S., and R. J. Lavigne. 1975. Comparative behavior of Wyoming robber flies. II. (Diptera: Asilidae). Univ. Wyo. Agr. Exp. Sta. Sci. Monogr. 30:1–68.

Dennis, D. S. and R. J. Lavigne. 1976. Description and notes on the pupae and pupal cases of ten species of Wyoming robber flies (Diptera: Asilidae). Proc. Ent. Soc. Wash. 78:277–303.

Dennis, D. S. and R. J. Lavigne. 1979. Ethology of *Machimus callidus* with incidental observations on *Machimus occidentalis* in Wyoming (Diptera: Asilidae). Pan-Pac. Ent. 55:208–221.

Dennis, D. S., R. J. Lavigne and S. W. Bullington. 1986. Ethology of *Efferia cressoni* with a review of the comparative ethology of the genus (Diptera: Asilidae). Proc. Ent. Soc. Wash. 88:42–55.

Denno, R. F. and Cothran, W. R. 1975. Niche relationships of a guild of necrophagous flies. Ann. Ent. Soc. Amer. 68:741–754.

Deonier, D. L. 1965. Ecological observations on Iowa shore flies (Diptera, Ephydridae). Proc. Ia. Acad. Sci. 71:496–510.

Deonier, D. L. 1971. A systematic and ecological study of Nearctic *Hydrellia* (Diptera: Ephydridae). Smiths. Cont. Zool. 68:1–147.

Deonier, D. L. 1972. Observations on mating, ovipositon, and food habits of certain shore flies (Diptera: Ephydridae). Ohio J. Sci. 72:22–29.

Deonier, D. L. 1974. Biology and descriptions of immature stages of the shore fly *Scatophila iowana* (Diptera: Ephydridae). Iowa St. J. Res. 49:17–22.

Deonier, D. L. 1979. Introduction—A prospectus of research in Ephydridae, pp. 1–19. *In* Deonier, D. L., ed., First symposium on the systematics and ecology of Ephydridae (Diptera). N. Amer. Benthol. Soc.

Deonier, D. L., and J. T. Regensburg. 1978. Biology and immature stages of *Parydra quadrituberculata* (Diptera: Ephydridae). Ann. Ent. Soc. Amer. 71:341–353.

Deonier, D. L., Mathis, W. N., and T. Regensburg. 1979. Natural history and life cycle stages of *Notiphila carinata* (Diptera: Ephydridae). Proc. Biol. Soc. Wash. 91:798–814.

Deonier, O. C. and E. F. Knipling. 1940. The biology of *Compsomyiops wheeleri* Hough and description of the larva. Ann. Ent. Soc. Amer. 33:578–582.

Descamps, 1957. Recherches morphologiques et biologiques sur les Diopsidae du Nord-Caméroun. Bull. scient. Minist. outre Mer. France 7:1–154.

Diatloff, A. 1965. Larvae of *Rivellia* attacking the root nodules of *Glycine javanica* L. J. Ent. Soc. Queensland 4:86.

Disney, R. H. L. 1970. A note on *Discomyza similis* Lamb (Dipt., Ephydridae) and other flies reared from dead snails in Cameroon. Ent. Mon. Mag. 105:250–251.

Disney, R. H. L. 1972. Some flies associated with dog dung in an English city. Ent. Mon. Mag. 108:93–94.

Disney, R. H. L. 1975. A key to the larvae, pupae, and adults of the British Dixidae (Diptera). Freshwater Biol. Assoc. Sci. Publ. 31:4–78.

Disney, R. H. L. 1976. The pre-adult stages of *Norellisoma spinimanum* (Fallén) (Dipt., Cordyluridae) and a parasitoid (Hym., Pteromalidae) of the same. Ent. Gaz. 27:263–267.

Dobson, T. 1974. Studies on the biology of the kelp-fly *Coelopa* in Great Britain. J. Nat. Hist. 8:155–177.

Dodge, H. R. 1956. New North American Sarcophagidae, with some new synonymy (Diptera). Ann. Ent. Soc. Amer. 49:182–190.

Dodge, H. R. 1963. A new *Philornis* with coprophagous larva, and some related species (Diptera: Muscidae). J. Kans. Ent. Soc. 36:239–247.

Dodge, H. R. 1964. Larval chaetotaxy and notes on the biology of *Toxorhynchites rutilis septentrionalis*. Ann. Ent. Soc. Amer. 57:46–53.

Dodge, H. R. and Aitken, T. H. G. 1968. *Philornis* flies from Trinidad (Diptera: Muscidae). J. Kansas Ent. Soc. 41:134–154.

Douchette, C. F., R. Latta, C. H. Martin, R. Schope, and P. M. Eide. 1942. Biology of the narcissus bulb fly in the Pacific Northwest. Tech. Bull. U.S. Dept. Agric. 809:1–66.

Downes, J. A. 1969. The swarming and mating flight of Diptera. Annu. Rev. Ent. 14:271–298.

Downes, J. A. 1974. The feeding habits of adult Chironomidae. Ent. Tidskr. 95:84–90.

Downes, J. A. 1978. Feeding and mating in the insectivorous Ceratopogoninae (Diptera). Mem. Ent. Soc. Can. 104:1–62.

Downes, J. A., and W. W. Wirth. 1981. Ceratopogonidae, pp. 393–421. *In* McAlpine, J. F., et al., eds., Manual of Nearctic Diptera. Vol. 1. Res. Br. Agr. Can. Monogr. 27.

Downes, W. L. 1955. Notes on the morphology and classification of the Sarcophagidae and other calyptrates. Proc. Iowa Acad. Sci. 62:514–538.

Downes, W. L., Jr. 1986. The Nearctic *Melanomya* and relatives (Diptera: Calliphoridae). A problem in calyptrate classification. Bull. N.Y. State Mus. No. 460, 35 pp.

Downes, W. L., Jr. and G. A. Dahlem. 1987. Keys to the evolution of Diptera: role of Homoptera. Env. Ent. 16:847–854.

Drake, C. J., and G. C. Decker. 1932. A scavenger fly, *Chrysomyza demandata* Fabr., breeding in corn silage. Ent. News 43:29–30.

Duda, O. 1928. Scatopsidae. Lfg. 26, Bd. II(1), pp. 1–62. *In* Lindner, E., ed., Die Fliegen der palaearktischen Region. Schweizerbart, Stuttgart.

Duda, O. 1934. Camillidae. Lfg. 81, Bd.VI(1), pp. 1–7. *In* Lindner, E. ed., Die Fliegen der palaearktischen Region. Schweizerbart, Stuttgart.

Duda, O. 1935. Drosophilidae. Lfg. 84, pp.1–64; Lfg. 86, pp. 65–118, Bd. VI(1). *In* E. Lindner, ed., Die Fliegen der palaearktischen Region. Schweizerbart, Stuttgart.

Duffield, R. M. 1981. Biology of *Microdon fusipennis* (Diptera: Syrphidae) with interpretations of the reproductive strategies of *Microdon* species found north of Mexico. Proc. Ent. Soc. Wash., 83:716–724.

Dukes, J., T. D. Edwards, and R. C. Axtell. 1974. Distribution of larval Tabanidae (Diptera) in a *Spartina alterniflora* salt marsh. J. Med. Ent. 11:79–83.

Dumbleton, L. J. 1962. Taxonomic characters in the pre-adult stages of Simuliidae (Diptera). N. A. J. Sci. 5:496–506.

Dusek, J. 1962. Beitrag zur Kenntnis von Larven der Gattung *Cheilosia* Meigen (Diptera, Syrphidae). Acta Soc. Ent. Czech. 59:68–73.

Dusek, J., and J. Krislek. 1967. Zur Kenntnis der Schwelfliegenlarven (Diptera: Syrphidae) in den Gallen der Pappelblattläuse (Homoptera: Pemphigidae). Zeitschr. Angew. Ent. 60:124–136.

Dyar, H. G., and R. C. Shannon. 1924. The American species of Thaumaleidae (Orphnephilidae) (Diptera). J. Wash. Acad. Sci. 14:432–434.

Dyte, C. E. 1959. Some interesting habits of larval Dolichopodidae (Diptera). Ent. Mon. Mag. 105:139–143.

Eastin, W. C., and B. A. Foote. 1971. Biology and immature stages of *Dichaeta caudata* (Diptera: Ephydridae). Ann. Ent. Soc. Amer. 64:271–279.

Eastman, C. E. 1980. Sampling phytophagous underground soybean arthropods, pp. 327–354. *In* Kogan, ed., Sampling Methods in Soybean Entomology. Springer. Verlag. N.Y.

Eastman, C. E., and A. L. Wuensche. 1977. A new insect damaging nodules of soybean: *Rivellia quadrifasciata* (Macquart). J. Ga. Ent. Soc. 12:190–199.

Easton, A. M. and Smith, K. G. V. 1970. The entomology of the cadaver. Medicine, Sci. Law: 208–215.

Eckblad, J. W. 1973. Experimental predation studies of malacophagous larvae of *Sepedon fuscipennis* (Diptera: Sciomyzidae) and aquatic snails. Exp. Parasitol. 33:331–342.

Edwards, F. W. 1920. On the British species of *Simultium*.-II. The early stages; with corrections and additions to part I. Bull. Ent. Res. 11:211–246.

Edwards, F. W. 1925. Insects inhabiting bird's nests. Trans. Herts. Nat. Hist. Soc. 18:132–133.

Edwards, F. W. 1928. Diptera. Fam. Protorhyphidae, Anisopodidae, Pachyneuridae, Trichoceridae. Fascicle 190, 40 pp. *In* Wytsman, P., ed., Genera insectorum, Bruxelles.

Edwards, F. W. 1932. Diptera. Fam. Culicidae. Fascicle 194, 258 pp. *In* Wytsman, P., ed., Genera insectorum, Bruxelles.

Efflatoun, H. C. 1921. The life-history of *Telmatoscopus meridionalis* Eaton. Bull. Soc. Ent. Egypte 5:22–34.

Efflatoun, H. C. 1927. On the morphology of some Egyptian Trypaneid larvae, with descriptions of some hitherto unknown forms. Bull. Soc. Ent. Egypte 11:17–50.

Eggleton, F. E. 1932. Limnetic distribution and migration of *Corethra* larvae in two Michigan lakes. Pap. Mich. Acad. Sci. 15:361–388.

Egglishaw, H. J. 1960a. The life-history of *Helcomyza* Curt. (Dipt. Dryomyzidae). Ent. Mon. Mag. 96:39–42.

Egglishaw, H. J. 1960b. Studies on the family Coelopidae (Diptera). Trans. Roy. Ent. Soc. London 112:109–140.

Ellisor, L. D. 1934. Notes on the biology and control of *Neosciara ocellaris* Comst. Iowa St. Coll. Sci. 9:25–36.

Elsey, K. D., and Rabb, R. L. 1970. Biology of *Voria ruralis* (Diptera: Tachinidae). Ann. Ent. Soc. Amer. 63:216–222.

Emden, F. I. 1950. Dipterous parasites of Coleoptera. Ent. Mon. Mag. 86:182–206.

Emden, F. I. van. 1959. Evolution of Tachinidae and their parasitism (Diptera). XV Intern. Congr. Zool. (London), 1958:664–666.

Emerton, J. H. 1890. An internal dipterous parasite of spiders. Psyche 5:404.

Engel, E. O. 1916. Beiträge zur Kenntnis einiger Dipterenlarven. Mitt. Münchn. Ent. Ges. 7:68–76.

Engel, E. O. 1919. Dipteren, die nicht Pupiparen sind, als Vogelparasiten. Zeitschr. Wiss. Ins. Biol. 15:249–258.

Engel, E. O. 1929. Notes on two larvae of South African Diptera belonging to the families Leptidae and Asilidae. Trans. Roy. Soc. S. Afr. 18:147–162.

Engel, E. O. 1930. Fliegenmaden im Schnupftabak. Zeitschr. Angew. Ent. 17:184–188.

English, K. M. J. 1947. Notes on the morphology and biology of *Apiocera maritima* Hardy (Diptera, Apioceridae). Proc. Linn. Soc. N. S. Wales 71:296–302.

English, K. M. J. 1950. Notes on the morphology and biology of *Anabarrhynchus fasciatus* Macq. and other Australian Therevidae. Proc. Linn. Soc. N. S. Wales 75:345–359.

Eschle, J. L. and DeFoliart, G. R. 1965. Rearing and biology of *Wohlfahrtia vigil* (Diptera: Sarcophagidae). Ann. Ent. Soc. Amer. 58:849–855.

Exner, K., and D. A. Craig. 1976. Larvae of Alberta Tanyderidae (Diptera: Nematocera). Quaest. Ent. 12:219–237.

Eyer, J. R. 1922. The bionomics and control of the onion maggot. Pennsylvania Agric. Exper. Sta. Bull. 171:1–16.

Farinets, S. T. 1980. First-instar larvae of Transcarpathian tachinids (Diptera: Tachinidae). II. Ent. Rev. 59:163–177.

Faucheux, M. J. 1977. Contribution à l'étude de la respiration et de la nutrition chez quelques insectes aquatiques des marais salants: *Stratiomyia, Nemotelus, Ephydra, Aedes* (Diptères). *Berosus* (Coléoptère *Sigara* (Hétéroptère); Importance des soies hydrofuges et des appareils filtrants. Bull. Soc. Sci. nat. Quest. France 75:56–68.

Faucheux, M. J. 1978. Contribution to the study of the biology of the larvae of *Stratiomyia longicornis* (Diptera: Stratiomyiidae) feeding and respiration. Annls. Soc. ent. France 14:49–72.

Fellows, D. P., and W. B. Heed. 1972. Factors affecting host plant selection in desert adapted cactiphilic *Drosophilia*. Ecol. 53:850–858.

Felt, E. P. 1912. 27th rept. of the state entomologist. Bull. N.Y. St. Mus. 155, p. 121.

Felt, E. P. 1940. Plant Galls and Gall Makers. Comstock Publ. Co. Inc., Ithaca, N.Y. viii+364 p.

Ferrar, P. 1979. The immature stages of dung-frequenting muscoid flies in Australia with notes on the species, and keys to larvae and puparia. Australian Zool., Suppl. Ser. 73:1–106.

Ferrar, P. 1980. Cocoon formation by Muscidae (Diptera). J. Australian Ent. Soc. 19:171–174.

Ferrar, P. 1987. A guide to the breeding habits and immature stages of Diptera Cyclorrhapha. E. J. Brill, Leiden/ Scandinavian Science Press, Copenhagen. Pt. 1 (text):1–478; Pt. 2 (figs.):479–907.

Ferris, G. F. 1923. Observations on the larvae of some Diptera pupipara, with description of a new species of Hippoboscidae. Parasitol. 15:54–58.

Ferris, G. F. 1930. The puparium of *Basilia corynorhini* Ferris. Ent. News 41:295–297.

Ferris, G. F., and F. R. Cole. 1922. A contribution to the knowledge of the Hippoboscidae. Parasitol. 14:178–205.

Feuerborn, H. G. 1922a. Der sexuelle Reizapparat (Schmuck-, Duft- und Berührungsorgane) der Psychodiden nach biologischen und physiologischen Gesichtspunkten untersucht. Arch. Naturgesch. 88 (A–4):1–137.

Feuerborn, H. 1922b. Das Hypopygium inversum und circumversum der Dipteren. Zool. Anz. 55:189–212.

Feuerborn, H. 1923. Die Larven der Psychodiden oder Schmetterlingsmücken. Ein Beitrag zur Ökologie des Feuchten. Verh. Intern. Ver. Theor. Angew. Limnol. 1922:181–213.

Feuerborn, H. 1927. Über Chaetotaxis und Typus der Larve und Puppe von *Psychoda*. Zool. Anz. 70:167–184.

Feuerborn, H. 1932. Die Psychodide *Maruina indica* sp. n. und ihre Beziehung zu den Blepharoceriden. Arch. Hydrobiol. Suppl. 11:55–128.

Finlayson, T. 1960. Taxonomy of cocoons and puparia, and their contents, of Canadian parasites of *Diprion hercyniae* (Htg.) (Hymenoptera: Diprionidae). Can. Ent. 92:922–941.

Fischer, C. R. 1932a. Contribucao para o conhecimento da metamorphose e posicao systematica da familia Tylidae. Rev. Ent. 2:15–24.

Fischer, C. R. 1932b. Nota taxonomica e biologica sobre *Anastrepha grandis* Macq. Rev. Entomol. 2:302–310.

Fischer, C. R. 1932c. Um genero e duas especies novas de Rhopalomeridae do Brasil, e o pupario de *Willistoniella pleoropunctata* Wied. Rev. Entomol. 2:441–450.

Fish, D. 1976. Insect-plant relationships of the insectivorous pitcher plant *Sarracenia minor*. Fla. Ent. 59:199–203.

Fisher, T. W., and R. E. Orth. 1964. Biology and immature stages of *Antichaeta testacea* Melander (Diptera: Sciomyzidae). Hilgardia 36:1–29.

Fisher, T. W. and R. E. Orth. 1983. The marsh flies of California (Diptera: Sciomyzidae). Bull. Calif. Insect Surv. 24:1–117.

Fittkau, E. J. 1962. Die Tanypodinae (Diptera: Chironomidae). Abh. Larvalsyst. Insekten 6:1–453.

Fletcher, B. S. 1987. Biology of dacine fruit flies. Ann. Rev. Ent. 32:115–144.

Flint, O. S., Jr. 1956. Hibernation of the diopsid fly, *Sphyrocephala brevicornis*. Bull. Brooklyn Ent. Soc. 51:44.

Fluke, C. L. 1929. The known predaceous and parasitic enemies of the pea- *Aphis* in North America. Wisc. Agr. Exp. Sta. Res. Bull. 93:1–47.

Fluke, C. L. 1931. Notes on certain syrphus flies related to *Xanthogramma* with descriptions of two new species. Trans. Wisc. Acad. Sci. Arts and Letters 26:289–309.

Fogelman, J. C. 1981. Utilization of food resources by *Drosophila* larvae. Genetics 97:36–37.

Foote, B. A. 1959. Biology and life history of the snail-killing flies belonging to the genus *Sciomyza* Fallén (Diptera: Sciomyzidae). Ann. Ent. Soc. Amer. 52:31–43.

Foote, B. A. 1963. Observations on the biology of *Tipula footeana* Alexander. Bull. Brooklyn Ent. Soc. 68:145–150.

Foote, B. A. 1965. Biology and immature stages of Eastern ragweed flies (Tephritidae). Proc. North Central Br. Ent. Soc. Amer. 20:105–106.

Foote, B. A. 1967. Biology and immature stages of fruit flies: the genus *Icterica* (Diptera: Tephritidae). Ann. Ent. Soc. Amer. 60:1295–1305.

Foote, B. A. 1970. The larvae of *Tanypeza longimana* (Diptera: Tanypezidae). Ann. Ent. Soc. Amer. 63:1.

Foote, B. A. 1971. Biology of *Hedria mixta* (Diptera: Sciomyzidae). Ann. Ent. Soc. Amer. 64:931–941.

Foote, B. A. 1973. Biology of *Pherbellia prefixa* (Diptera: Sciomyzidae), a parasitoid-predator of the operculate snail *Valvata sincera*. Proc. Ent. Soc. Wash. 75:141–149.

Foote, B. A. 1976a. Biology of pictured-wing flies (Diptera: Otitidae), pp. 51–59. *In* Barr, W. F., ed., Anniv. Publ. Dept. Ent. Univ. Idaho. Moscow.

Foote, B. A. 1976b. Biology and larval feeding habits of three species of *Renocera* (Diptera: Sciomyzidae) that prey on fingernail clams (Mollusca: Sphaeriidae). Ann. Ent. Soc. Amer. 69:121–133.

Foote, B. A. 1977. Biology of *Oidematops ferrugineus* (Diptera: Sciomyzidae), a parasitoid of the land snail *Stenotrema hirsutum* (Mollusca: Polygyridae). Proc. Ent. Soc. Wash. 79:609–619.

Foote, B. A. 1978. Utilization of blue-green algae by larvae of shore flies (Diptera: Ephydridae). Environ. Ent. 6:812–814.

Foote, B. A. 1979. Utilization of algae by larvae of shore flies, pp. 61–71. *In* D. L. Deonier, ed., First symposium on the systematics and ecology of Ephydridae (Diptera). No. Amer. Benthol. Soc., 147 pp.

Foote, B. A. 1981a. Biology and immature stages of *Lytogaster excavata*, a grazer of blue-green algae (Diptera: Ephydridae). Proc. Ent. Soc. Wash. 83:394–415.

Foote, B. A. 1981b. Biology and immature stages of *Pelina truncatula*, a consumer of blue-green algae (Diptera: Ephydridae). Proc. Ent. Soc. Wash. 83:607–619.

Foote, B. A. 1982. Biology and immature stages of *Setacera atrovirens*, a grazer of floating algal mats (Diptera: Ephydridae). Proc. Ent. Soc. Wash. 84:828–844.

Foote, B. A. 1983. Biology and immature stages of *Nostima approximata*, a grazer of the blue-green algal genus *Oscillatoria* (Diptera: Ephydridae). Proc. Ent. Soc. Wash. 85:472–484.

Foote, B. A. 1984. Biology of *Trimerina madizans* (Diptera: Ephydridae), a predator of spider eggs. Proc. Ent. Soc. Wash. 86:486–492.

Foote, B. A. 1985. Biology of *Rivellia pallida* (Diptera: Platystomatidae), a consumer of the nitrogen-fixing root nodules of *Amphicarpa bracteata* (Leguminosae). J. Kans. Ent. Soc. 58:27–35.

Foote, B. A., and W. C. Eastin. 1974. Biology and immature stages of *Discocerina obscurella* (Diptera: Ephydridae). Proc. Ent. Soc. Wash. 76:401–408.

Foote, B. A., and L. V. Knutson. 1970. Clam-killing fly larvae. Nature 226:466.

Foote, B. A., S. E. Neff, and C. O. Berg. 1960. Biology and immature stages of *Atrichomelina pubera* (Diptera: Sciomyzidae). Ann. Ent. Soc. Amer. 53:192–199.

Foote, R. H. 1956. Entomology. Gall midges associated with cones of western forest trees (Dipt.: Itonididae). J. Wash. Acad. Sci. 46:48–57.

Foote, R. H. and G. C. Steyskal. 1987. Tephritidae. pp. 817–831. *In* McAlpine, J. F. et al. (eds.). Manual of Nearctic Diptera, Vol. 2. Res. Br. Agr. Can. Monogr. 28.

Foster, G. A. 1976. Notes on the phylogeny of the Nearctic Tethinidae and a review of the genus *Neopelomyia* Hendel and the *Tethnia milichioides* group (Diptera). Proc. Ent. Soc. Wash. 78:336–352.

Fredeen, F. J. H. 1977. A review of the economic importance of black flies (Simuliidae) in Canada. Quaest. Ent. 17:219–229.

Fredeen, F. J. H., and G. S. Glen. 1970. The survival and development of *Leptocera caenosa* (Diptera: Sphaeroceridae) in laboratory cultures. Can. Ent. 102:164–171.

Fredeen, F. J. H. and M. E. Taylor. 1964. Borborids (Diptera: Sphaeroceridae) infesting sewage disposal tanks, with notes on the life cycle, behavior and control of *Leptocera* (*Leptocera*) *caenosa* (Rondani). Can. Ent. 96:801–808.

Freeman, J. V. 1987. Immature stages of *Tabanus conteriminatus* and keys to larvae and pupae of common Tabanidae from United States East Coast salt marches. Ann. Ent. Soc. Amer. 80:613–623.

Freidberg, A. 1981. Taxonomy, natural history and immature stages of the bone skipper, *Centrophlebomyia furcata* (Fabricius) (Diptera: Piophilidae, Thyreophorina). Ent. Scand. 12:320–326.

Frick, K. E. 1956. Revision of the *Calcomyza* species of North America (Agromyzidae). Ann. Ent. Soc. Amer. 46:284–300.

Frick, K. E. 1971. The biology of *Trypeta angustigena* Foote in central coastal California—host plants and notes (Diptera: Tephritidae). J. Wash. Acad. Sci. 61:20–24.

Frison, T. H. 1926. Contributions to the knowledge of the interrelation of the bumblebees of Illinois with their animate environment. Ann. Ent. Soc. Amer. 19:203–234.

Frost, S. 1924. A study of the leaf-mining Diptera of North America. Mem. Cornell Univ. Agr. Exp. Sta. 78:1–228.

Frost, S. W. 1932. *Cordylura tricincta* Loew, a leaf-miner on *Smilacina racemosa* (L.) Desf. Ent. News 43:75–77.

Fullaway, D. T. 1907. Immature stages of a psychodid fly. Ent. News 18:386–389.

Fuller, M. E. 1934. The insect inhabitants of carrion: a study in animal ecology. Bull. Coun. Scient. Ind. Res. Melb. 82:1–62.

Fuller, M. E. 1938. Notes on *Triclopsidea oestracea* and *Cyrtomorpha flaviscutellaris*, two dipterous enemies of grasshoppers. Proc. Linn. Soc. N. S. Wales 63:95–104.

Gäbler, H. 1933. Schäden an den weiblichen Weidenkätzchen durch *Egle* (*Hylemyia*) *parva* Rob.-Desv. (Dipt.). Mitt. Deutsche Ent. Ges. 9:27–33.

Gagné, R. J. 1969. A review of the genus *Walchomyia* including a new species reared from *Cupressus* galls in California (Diptera: Cecidomyiidae). Pan-Pacif. Ent. 45:16–19.

Gagné, R. J. 1975a. A revision of the Nearctic Stomatosematini (Diptera: Cecidomyiidae: Cecidomyiinae). Ann. Ent. Soc. Amer. 68:86–90.

Gagné, R. J. 1975b. A revision of the Nearctic species of the genus *Phronia* (Diptera: Mycetophilidae). Trans. Amer. Ent. Soc. 101:227–318.

Gagné, R. J. 1975c. A review of the Nearctic genera of Oligotrophidi with piercing ovipositors (Diptera: Cecidomyiidae). Ent. News 86:5–12.

Gagné, R. J. 1978. A systematic analysis of the pine pitch midges, *Cecidomyia* spp. (Diptera: Cecidomyiidae). U.S. Dept. Agr. Tech. Bull. 1575:1–18.

Gagné, R. J. 1981a. A monograph of *Trichonta* with a model for the distribution of Holarctic Mycetophilidae (Diptera). U.S. Dept. Agr. Tech. Bull. 1638:1–64.

Gagné, R. J. 1981b. Cecidomyiidae, pp. 257–292. *In* McAlpine, J. F., et al., eds., Manual of Nearctic Diptera. Vol. 1. Res. Br. Agr. Can. Monogr. 27.

Gagné, R. J. 1989. The plant-feeding gall midges of North America. Cornell University Press. Ithaca, New York. 370 p.

Gagné, R. J., and B. A. Hawkins. 1983. Biosystematics of the Lasiopterini (Diptera: Cecidomyiidae: Cecidomyiinae) associated with *Atriplex* spp. (Chenopodiaceae) in southern California. Ann. Ent. Soc. Amer. 76:379–383.

Garnett, W. B., R. D. Akre, and G. Sehlke. 1985. Cocoon mimicry and predation by myrmecophilous Diptera (Diptera: Syrphidae). Fla. Ent. 68:615–621.

Garnett, W. B., and B. A. Foote. 1966. Notes on the biology of certain heleomyzid flies of eastern North America (Diptera: Heleomyzidae). J. Kans. Ent. Soc. 39:552–555.

Gelhaus, J. K. 1986. Larvae of the cranefly genus *Tipula* in North America (Diptera: Tipulidae). Univ. Kans. Sci. Bull. 53:121–22.

Genung, W. G. 1959. Biological and ecological observations on *Mydas maculiventris* Westwood (Diptera: Mydaidae) as a predator of white grubs. Fla. Ent. 42:35–37.

Gercke, G. 1879. Über die Metamorphose der *Hydromyza livens* Fall. Verh. Ver. Naturwiss. Unterh. Hamburg 4:229–234.

Gercke, G. 1882. Über die Metamorphose einiger Dipteren. Verh. Ver. Naturwiss. Unterh. Hamburg 5:68–78.

Gibson, K. E., and W. H. Kerby. 1978. Seasonal life history of the walnut husk fly and husk maggot in Missouri. Environ. Ent. 7:81–87.

Gilbert, P. A. 1919. A dipterous parasite of nestling birds. Emu. 19:48–49.

Gill, G. 1962. The heleomyzid flies of America north of Mexico (Diptera: Heleomyzidae). Proc. U.S. Nat. Mus. 113:495–603.

Gill, G. D. and B. V. Peterson. 1987. Heleomyzidae, pp. 973–980. *In* McAlpine, J. F., et al., (eds.). Manual of Nearctic Diptera, Vol. 2. Res. Br. Agr. Can. Monogr. 28.

Goddard, W. H. 1938. The description of the puparia of fourteen British species of Sphaeroceridae. Trans. Soc. Brit. Ent. 5:236–258.

Goetghebuer, M. 1924. Note sur la biologie et la morphologie de *Liponeura belgica*. Annls. Biol. Lacustre. 13:107–118.

Gojmerac, W. L. 1956. Description of the sugar beet root maggot, *Tetanops myopaeformis* (von Roder), with observations on reproductive capacity. Ent. News 67:203–210.

Goodwin, J. T. 1972. Immature stages of some eastern Nearctic Tabanidae (Diptera). I. Introduction and the genus *Chrysops* Meigen. J. Georgia Ent. Soc. 7:98–109.

Goodwin, J. T. 1973a. Immature stages of some eastern Nearctic Tabanidae (Diptera). II. Genera of the tribe Diachlorini. J. Georgia Ent. Soc. 8:5–11.

Goodwin, J. T. 1973b. Immature stages of some eastern Nearctic Tabanidae (Diptera). III. The genus *Tabanus* Linnaeus. J. Georgia Ent. Soc. 8:82–99.

Goodwin, J. T. 1973c. Immature stages of some eastern Nearctic Tabanidae (Diptera). IV. The genus *Merycomyia*. J. Tenn. Acad. Sci. 48:115–118.

Goodwin, J. T. 1974. Immature stages of some eastern Nearctic Tabanidae (Diptera). V. *Stenotabanus* (*Aegialomyia*) *magnicallus* (Stone). J. Tenn. Acad. Sci. 49:14–15.

Goodwin, J. T. 1976a. Notes on some "rare" eastern Nearctic Tabanidae (Diptera); state records and host-parasite relationships for other species. Fla. Ent. 59:63–66.

Goodwin, J. T. 1976b. Immature stages of eastern Nearctic Tabanidae (Diptera): VI. Additional species of *Chrysops* Meigen. Fla. Ent. 59:343–351.

Goodwin, J. T. 1976c. Immature stages of some eastern Nearctic Tabanidae (Diptera): VII. *Haematopota* Meigen and *Whitneyomyia* Bequaert plus other Tabanini. Fla. Ent. 59:369–390.

Graenicher, S. 1931. Some observations on the biology of the Sarcophaginae (Diptera: Sarcophagidae). Ent. News 42:227–230.

Graenicher, S. 1935. Some biological notes on *Sarcophaga bullata* Park. (Diptera: Sarcophagidae). Ent. News 46:193–196.

Graham, J. F. 1939. The external features of the early stages of *Spathiophora hydromyzina* (Fall). Proc. Roy. Ent. Soc. London (B) 8:157–162.

Graham-Smith, G. S. 1916. Observations on the habits and parasites of some common flies. Parasitology 8:440–544.

Greathead, D. J. 1958. Notes on the larva and life history of *Curtonotum cuthbertsoni* Duda (Dipt., Drosophilidae), a fly associated with the desert locust *Schistocerca gregaria* (Forskal). Ent. Mon. Mag. 94:36–37.

Greenberg, B. 1971. Flies and Disease. Vol. 1. Princeton Univ. Press, Princeton, NJ. 856 p.

Greenberg, B. 1973. Flies and Disease. Vol. 2. Princeton Univ. Press, Princeton, NJ. 447 p.

Greenberg, B., and M. L. Szyska, 1984. Immature stages and biology of fifteen species of Peruvian Calliphoridae (Diptera). Ann. Ent. Soc. Amer. 77:488–517.

Greene, C. T. 1917. A contribution to the biology of North American Diptera. Proc. Ent. Soc. Wash. 19:146–161.

Greene, C. T. 1921. An illustrated synopsis of the puparia of 100 Muscoid flies. Proc. U.S. Nat. Mus. 60:1–39.

Greene, C. T. 1922. Synopsis of the N. American flies of the genus *Tachytrechus*. Proc. U.S. Nat. Mus. 60:1–21.

Greene, C. T. 1923a. The immature stages of *Hydrophorus agalina* Wheeler (Dolichopodidae). Proc. Ent. Soc. Wash. 25:66–69.

Greene, C. T. 1923b. A contribution to the biology of North American Diptera. Proc. Ent. Soc. Wash. 25:82–91.

Greene, C. T. 1923c. The larva and pupa of *Microdon megalogaster* Snow. Proc. Ent. Soc. Wash. 25:140–141.

Greene, C. T. 1925. The puparia and larvae of sarcophagid flies. Proc. U.S. Nat. Mus. 66:1–26.

Greene, C. T. 1926. Descriptions of larvae and pupae of two-winged flies belonging to the family Leptidae. Proc. U.S. Nat. Mus. 70:1–20.

Greene, C. T. 1929. Characters of the larvae and pupae of certain fruit flies. J. Agr. Res. 38:489–504.

Greene, C. T. 1954. Larva and pupa of *Thrypticus fraterculus* (Wheeler) with new original notes on the habits of the family Dolichopodidae. Ent. News. 65:89–92.

Griffiths, G. C. D. 1972. The Phylogenetic Classification of Diptera Cyclorrhapha, with Special Reference to the Structure of the Male Postabdomen. W. Junk, The Hague. 340 p.

Griffiths, G. C. D. 1980. Flies of the Nearctic region. Vol. 1. Handbook. Part 1. History of Nearctic Dipterology. E. Schweizerbart'sche Verlangsbuchhandlung. Stuttgart.

Griffiths, G. C. D. 1982–1986. Flies of the Nearctic Region. Anthomyiidae. Vol. 8. Part 2. pp. 1–160 (1982); pp. 161–288 (1983); pp. 289–408 (1984); pp. 409–600 (1984); pp. 601–728 (1986); E. Schweizerbart'sche Verlangsbuchhandlung. Stuttgart.

Grigarick, A. A. 1959. Bionomics of the rice leaf miner, *Hydrellia griseola* (Fallén) in California (Diptera: Ephydridae). Hilgardia 29:1–80.

Grimaldi, D., and B. A. Underwood. 1986. *Megabraula*, a new genus for two new species of Braulidae (Diptera), and a discussion of braulid evolution. Syst. Ent. 11:427–438.

Guberlet, J. E. and Hobson, H. H. 1940. A fly maggot attacking young birds, with an observation of its life history. Murrelet 21:65–68.

Guibe, J. 1939. Contribution à l'étude d'une espèce: *Apterina pedestris* Meigen. Thèses, Fac. Sci. Caen. No. 29, 112 pp.

Guimaraes, L. R. 1944. Sobre os primeros estadios de alguns Dipteros pupiparos. Papeis avulsos Dept. Zool., Sao Paulo 6:181–192.

Guppy, J. C. 1981. Bionomics of the alfalfa blotch leaf-miner, *Agromyza frontella* (Diptera: Agromyzidae) in eastern Ontario. Can. Ent. 113:593–600.

Gurney, W. B. 1912. Fruit-flies and other insects attacking cultivated and wild fruits in New South Wales. Agr. Gaz. N. S. Wales 23:75–80.

Hackman, W. 1955. On the genera *Scaptomyza* Hardy and *Parascaptomyza* Duda (Diptera, Drosophilidae). Notulae Ent. 35:74–91.

Hackman, W. 1959. On the genus *Scaptomyza* Hardy (Diptera, Drosophilidae) with descriptions of new species from various parts of the world. Acta Zool. Fenn. 97:1–73.

Hackman, W. 1969. A review of the zoogeography and classification of the Sphaeroceridae (Borboridae, Diptera). Notulae Ent. 49:193–210.

Hackman, W., and M. Meinander. 1979. Diptera feeding as larvae on macrofungi in Finland. Ann. Zool. Fenn. 16:50–83.

Hafez, M. 1939a. The life-history of *Sepsis impunctata* Macq. Bull. Soc. Fouad I. Ent. 23:326–332.

Hafez, M. 1939b. Ecological and biological observations on some coprophagous Sepsidae (Diptera). Proc. Roy. Ent. Soc. London (A) 23:99–104.

Hafez, M. 1940. A study of the morphology and life-history of *Sarcophaga falculata* Pand. 24:183–212.

Hafez, M. 1949. Biology of Borboridae. Proc. Roy. Ent. Soc. London. 24:1–5.

Hafez, M. 1953a. Studies on *Tachina larvarum* L. (Diptera, Tachinidae). I. Preliminary notes. Bull. Soc. Fouad. 1 Ent. 37:255–266.

Hafez, M. 1953b. Studies on *Tachina larvarum* L. (Diptera, Tachinidae). II. Morphology of the adult and of its early stages. Bull. Soc. Fouad. 1. Ent. 37:267–304.

Hafez, M. 1953c. Studies on *Tachina larvarum* L. (Diptera, Tachinidae). III. Biology and life-history. Bull. Soc. Fouad. 1. Ent. 37:305–335.

Hall, D. G. 1932a. Biology of *Sarothromyia femoralis* var. *simplex* Aldrich. (Diptera: calliphoridae). Ann. Ent. Soc. Amer. 25:641–647.

Hall, D. G. 1932b. Some studies on the breeding media, development and stages of the eyegnat *Hippelates pusio.* Amer. J. Hyg. 16:854–864.

Hall, D. G. 1947. The blowflies of North America. Thomas Say Foundation, Volume 4. 477 pp.

Hall, J. C. 1981. Bombyliidae, pp. 589–602. *In* McAlpine, J. F., et al., eds., Manual of Nearctic Diptera. Vol. 1. Res. Br. Agr. Can. Monogr. 27.

Hallock, H. C. 1929. Notes on methods of rearing Sarcophaginae and the biology of *Sarcophaga latisterna* Park. Ann. Ent. Soc. Amer. 22:246–250.

Hamilton, A. L., O. A. Saether, and D. R. Oliver. 1969. A classification of the Nearctic Chironomidae. Tech. Rep. Fish. Res. Bd. Can. 124:1–42.

Hammer, O. 1941. Biological and ecological investigations on flies associated with pasturing cattle and their excrement. Dansk Naturh. For. Kobenhavn, Vidensk. Meddel. 105:141–393.

Handlirsch, A. 1882. Die Metamorphose und Lebensweise von *Hirmoneura obscura* Meig., einem Vertreter der Dipteren-Familie Nemestrinidae. Wien. Ent. Ztg. 1:224–228.

Handlirsch, A. 1883. Beiträge zur Biologie der Dipteren. Verh. Zool.-Bot. Ges. Wien 33:243–246.

Handlirsch, A. 1883. Die Metamorphose und Lebensweise von *Hirmoneura obscura* Meig., einem Vertreter der Dipteren-Familie Nemestrinidae I und II. Wien. Ent. Ztg. 2:11–15.

Hansen, B. 1979. Life cycle and growth of two species of *Ptychoptera* (Diptera, Nematocera) in a Danish brook. Ent. Meddel. 47:33–38.

Hansen, D. C., and E. F. Cook. 1976. The systematics and morphology of the Nearctic species of *Diamesa* Meigen, 1835 (Diptera: Chironomidae). Mem. Amer. Ent. Soc. 30:1–203.

Hanson, W. J. 1958. A revision of the subgenus *Melanonemotelus* of America north of Mexico (Diptera: Stratiomyidae). Univ. Kans. Sci. Bull. 38:1351–1391.

Hardy, D. E. 1945. Revision of Nearctic Bibionidae including Neotropical *Plecia* and *Penthetria* (Diptera). Univ. Kans. Sci. Bull. 30:367–546.

Hardy, D. E. 1949a. Studies in Hawaiian fruit flies. Proc. Ent. Soc. Wash. 51:181–205.

Hardy, D. E. 1949b. The North American *Chrysopilus* (Rhagionidae-Diptera). Amer. Midl. Nat. 41:143–167.

Hardy, D. E. 1954. Studies in New World Dorilaidae (Pipunculidae: Diptera). J. Kans. Ent. Soc. 27:121–127.

Hardy, D. E. 1958. Guide to the Insects of Connecticut. Part VI. The Diptera or True Flies of Connecticut. Sixth Fascicle. Family Bibionidae. Bull. Conn. St. Geol. Nat. Hist. Surv. 87:5–45.

Hardy, D. E. 1964. Diptera: Brachycera. II–Cyclorrhapha. Insects of Hawaii. Vol. 11, 458 pp.

Hardy, D. E. 1971. Pipunculidae (Diptera) parasites on rice leafhoppers in the Oriental region. J. Hawaii. Ent. Soc. 21:79–91.

Hardy, D. E. 1981. Bibionidae. pp. 217–222 *in* McAlpine, J. F. et al., (eds.). Manual Nearctic Diptera. Vol. 1. Res. Br. Agr. Can. Monogr. 27.

Hardy, D. E. 1987. Pipunculidae. pp. 745–748. *In* McAlpine, J. F., et al. (eds.). Manual of Nearctic Diptera. Vol. 2. Res. Br. Agr. Can. Monogr. 28.

Hardy, D. E., and J. V. McGuire. 1974. The Nearctic *Ptiolina* (Rhagionidae-Diptera). J. Kans. Ent. Soc. 20:1–15.

Harman, D. M., and J. B. Wallace. 1971. Description of the immature stages of *Lonchaea corticis,* with notes on its role as a predator of the white pine weevil, *Pissodes strobi.* Ann. Ent. Soc. Amer. 64:1221–1226.

Harnisch, O. 1954. Die physiologische Bedeutung der präanalen Tubuli der Larve von *Chironomus thummi.* Zool. Anz. 153:204–211.

Harper, A. M. 1962. Life history of the sugar-beet root maggot *Tetanops myopaeformis* (Roder) (Diptera: Otitidae) in southern Alberta. Can. Ent. 94:1334–1340.

Harris, P. 1980. Establishment of *Urophora affinis* Frfld. and *U. quadrifasciatus* (Meig.) (Diptera: Tephritidae) in Canada for the biological control of diffuse and spotted knapweed. Zeitschr. Angew. Ent. 89:504–514.

Hart, C. A. 1898. On the entomology of the Illinois River and adjacent waters. Ill. St. Lab. Nat. Hist. Bull. 4:149–273.

Hartley, J. C. 1958. The root piercing spiracles of the larva of *Chrysogaster hirtella* (Diptera: Syrphidae). Proc. Roy. Ent. Soc. London (A) 33:81–87.

Hartley, J. C. 1963. The cephalopharyngeal apparatus of syrphid larvae and its relationship to other Diptera. Proc. Zool. Soc. London 141:261–280.

Haseman, L. 1907. A monograph of the North American Psychodidae, including ten new species and an aquatic psychodid from Florida. Trans. Amer. Ent. Soc. 33:299–333.

Hassanein, M. H., and A. L. Abd El-Salam. 1962. Biological studies on the bee louse, *Braula coeca* Nitzsch. Bull. Soc. Ent. Egypte 46:87–95.

Hawkins, B. A., R. D. Goeden and R. J. Gagné. 1986. Ecology and taxonomy of the *Asphondylia* spp. (Diptera: Cecidomyiidae) forming galls on *Atriplex* spp. (Chenopodiaceae) in Southern California. Entomography 4:55–107.

Hawley, I. M. 1922. The sugar-beet root maggot, a new pest of sugar beets. J. Econ. Ent. 15:388–391.

Hawley, I. M. 1932. Insects and other animal pests injurious to field beans in New York. Mem. Cornell Univ. Agr. Exp. Sta. 55:949–1037.

Hays, K. L. 1958. Descriptions of the larva and adult female of *Phorostoma tabanivora* (Hall) (Diptera: Larvaevoridae) and notes on the biology of the species. Ann. Ent. Soc. Amer. 51:552–553.

Heed, W. B. 1957. VI. Ecological and distributional notes on the Drosophilidae (Diptera) of El Salvador. Gen. of *Drosophila* 5721:62–78.

Heed, W. B. 1968. Ecology of the Hawaiian Drosophilidae. Stud. Gen. 4:387–419.

Heed, W. B. 1971. Host plant specificity and speciation in Hawaiian *Drosophila.* Taxon 20:115–121.

Heeger, E. 1852. Beiträge zur Naturgeschichte der Insecten. Fitz.-Ber. Akad. wiss. Wien. Math.-nat. kl. 9, 774–781.

Heiss, E. M. 1938. A classification of the larvae and puparia of the Syrphidae of Illinois exclusive of aquatic forms. Univ. Ill. Bull. 36:1–142.

Helsel, E. D., and D. T. Wicklow. 1979. Decomposition of rabbit feces; role of the sciarid fly *Lycoriella mali* (Diptera: Sciaridae) in energy transformations. Can. Ent. 111:213–218.

Hemmingsen, A. M. 1977. Omerlovestudier. Ent. Meddel. 45:167–188.

Hendel, F. 1908. Diptera. Fam. Muscaridae, Subfam. Lauxaninae. Fascicle 68, 66 pp. *In* Wytsman, P., ed., Genera insectorum, Bruxelles.

Hendel, F. 1914. Diptera. Fam. Muscaridae, subfam. Platystomatidae, 179 p. *In* Wytsman, P., ed., Genera insectorum 157. Bruxelles.

Hendel, F. 1916. Beiträge zur Systematik der Acalyptraten Musciden. Ent. Mitt. 5:294–299.

Hendel, F. 1920. Die paläarktischen Agromyziden (Dipt.) (Prodromus einer Monographie). Arch. Naturgesch. (1918) A, 84 (7):109–174.

Hendel, F. 1922. Die paläarktischen Muscidae acalyptratae Girsch = Haplostomata Frey nach ihren Familien und Gattungen I. Die Familien. Konowia 1:145–160.

Hendrickson, R. M., and M. A. Keller. 1983. Observations on the biology of *Liriomyza trifoliearum* (Diptera: Agromyzidae). Proc. Ent. Soc. Wash. 85:806–810.

Hennig, W. 1937. Milichiidae et Carnidae, Lfg. 115, 91 p. *In* Lindner, E., ed., Die Fliegen der palaearktischen Region. E. Schweizerbart, Stuttgart.

Hennig, W. 1938. Braulidae, Lfg. 60c. 14 p. *In* Lindner, E., ed., Die Fliegen der palaearktischen Region. Bd. 6(1). E. Schweizerbart, Stuttgart.

Hennig, W. 1943. Übersicht über die bisher bekannten Metamorphosestadien der Ephydriden, mit Neubeschreibungen nach dem Material der Deutschen Limnologischen Sunda-Expedition. Arb. Morph. Tax. Ent. 10:105–138.

Hennig, W. 1948. Die Larvenformen der Dipteren. 1. Teil. Akad.-Verlag, Berlin. 185 p.

Hennig, W. 1950. Die Larvenformen der Dipteren. 2. Teil. Akad.-Verlag. Berlin. 458 p.

Hennig, W. 1952. Die Larvenformen der Dipteren. 3. Teil. Akad.-Verlag, Berlin. 628 p.

Hennig, W. 1958. Die familien der Diptera Schizophora und ihre phylogenetischen Verwandtschaftbeziehungen. Beitr. Z. Ent. 8:505–688.

Hennig, W. 1967. Die sogenannten "niederen Brachycera" im Baltischen Bernstein (Diptera: Fam. Xylophagidae, Xylomyidae, Rhagionidae, Tabanidae). Stuttg. Beitr. Naturkunde 174:1–51.

Hennig, W. 1969. Neue Übersicht über die aus dem Baltischen Bernstein bekannten Acalyptratae. Stuttg. Beitr. Naturkunde 209:1–42.

Hennig, W. 1971. Neue Untersuchungen über die Familien der Diptera Schizophora (Diptera: Cyclorrhapha). Stuttg. Beitr. Naturkunde 226:1–76.

Hennig, W. 1973. Ordnung Diptera (Zweiflügler). Handb. Zool. 4(2) 2/31 (Lfg. 20):1–337.

Hensley, M. S. 1976. Prevalence of cuterebrid parasitism among woodmice in Virginia. J. Wildl. Dis. 12:172–179.

Hering, E. M. 1924. Minenstudien IV. Zeitschr. Morphol. Ökol. Tiere 2:217–250.

Hering, E. M. 1932. Minenstudien II. Zeitschr. Wiss. Ins. Biol. 26:157–182.

Hering, E. M. 1951. Biology of the Leaf Miners. W. Junk, The Hague. iv+420 p.

Hering, E. M. 1954. Die Larven der Agromyziden I. Tydschr. Ent. 97:115–136.

Hering, E. M. 1955. Die Larven der Agromyziden II. Tydschr. Ent. 98:257–281.

Hering, E. M. 1957a. Die Larven der Agromyziden III. Tydschr. Ent. 100:73–94.

Hering, E. M. 1957b. Bestimmungstabellen der Blattminen von Europa einschliesslich des Mittelmeerbeckens und der Kanarischen Inseln. W. Junk, The Hague.

Herms, W. B., and R. W. Burgess. 1930. A description of the immature stages of *Hippelates pusio* Loew and a brief account of its life history. J. Econ. Ent. 23:600–603.

Herms, W. B., and M. T. James. 1961. Medical Entomology. The Macmillan Co., NY. 616 p.

Herting, B. 1961. Rhinophorinae. pp. 1–36 *in* Lindner, E., ed., Die Fliegen der palaearktischen Region, 64e Lfr. 216, Stuttgart.

Hewitt, C. G. 1907. On the life history of the root maggot, *Anthomyia radicum* Meigen. J. Econ. Biol. 2:56–63.

Hewitt, C. G. 1908. The structure, development, and bionomics of the house fly, *Musca domestica* Linn., Part II. The breeding habits, development, and the anatomy of the larva. Quart. J. Microscop. Sci. 52:495–545.

Hewitt, C. G. 1912. *Fannia canicularis* L. and *Fannia scalaris* Fabr. Parasitol. 5:161–174.

Hewitt, T. R. 1914. The larva and puparium of the frit fly. Sci. Proc. Roy. Soc. Dublin 14:313–318.

Hickman, C. P. 1935. External features of the larva of *Hydromyza confluens*. Proc. Indiana Acad. Sci. 44:212–216.

Hinton, H. E. 1947. On the reduction of functional spiracles in the aquatic larvae of the Holometabola, with notes on the moulting process of spiracles. Trans. R. ent. Soc. Lond. 98:449–473.

Hinton, H. E. 1953. Some adaptations of insects to environments that are alternately dry and flooded, with some notes on the habits of the Stratiomyiidae. Trans. Brit. Soc. Ent. 11:209–227.

Hinton, H. E. 1955. On the structure, function and distribution of the prolegs of the Panorpoidea, with criticism of the Berlese-Imms theory. Trans. R. ent. Soc. Lond. 106:455–540.

Hinton, H. E. 1958. On the nature and metamorphosis of the colour pattern of *Thaumalea* (Diptera, Thaumaleidae). J. Insect Physiol. 2:249–260.

Hinton, H. E. 1960. The ways in which insects change colour. Sci. Prog., Lond. 48 (190):341–350.

Hinton, H. E. 1962. The structure and function of the spiracular gills of *Deuterophlebia* (Deuterophlebiidae) in relation to those of other Diptera. Proc. Zool. Soc. London 138:11–122.

Hinton, H. E. 1963. The ventral ecdysial lines of the head of endopterygote larvae. Trans. R. ent. Soc. Lond. 115:39–61.

Hobby, B. M., and K. G. V. Smith. 1961. The immature stages of *Rhamphomyia* (*Megacyttarus*) *anomalipennis* Mg. (Dipt., Empididae). Ent. Mon. Mag. 97:138–139.

Hobby, B. M. and K. G. V. Smith. 1962a. The bionomics of *Empis tessellata* F. (Dipt., Empididae). Ent. Mon. Mag. 98:2–10.

Hobby, B. M. and K. G. V. Smith. 1962b. The larva of the viviparous fly *Ocydromia glabricula* (Fln.) (Dipt., Empididae). Ent. Mon. Mag. 98:49–50.

Hodkinson, I. D. 1973. The immature stages of *Ptychoptera lenis lenis* (Diptera: Ptychopteridae) with notes on their biology. Can. Ent. 105:1091–1099.

Hodson, A. C. 1939. *Sarcophaga aldrichi* Parker as a parasite of *Malacosoma disstria* Hbn. J. Econ. Ent. 32:396–401.

Hodson, W. E. H. 1927. The bionomics of the lesser bulb flies, *Eumerus strigatus* Fln. and *Eumerus tuberculatus* Rond., in southwest England. Bull. Ent. Res. 17:373–384.

Hodson, W. E. H. 1932a. A comparison of the larvae of *Eumerus strigatus* Fln. and *Eumerus tuberculatus* Rond. Bull. Ent. Res. 23:247–249.

Hodson, W. E. H. 1932b. The large narcissus fly, *Merodon equestris* Fab. Bull. Ent. Res. 23:429–448.

Hopping, G. A. 1947. Notes on the seasonal development of *Medetera aldrichii* (Diptera, Dolichop.) as a predator of the Douglas fir bark beetle. Can. Ent. 79:150–153.

Hogue, C. L. 1966. The California species of *Philorus:* taxonomy, early stages and descriptions of two new species (Diptera: Blephariceridae). Contr. Sci. 99:1–22.

Hogue, C. L. 1967. The pupa of *Rhaphiomidas terminatus* Cazier (Diptera: Apioceridae). Bull. S. Calif. Acad. Sci. 66:49–53.

Hogue, C. L. 1973a. A taxonomic review of the genus *Maruina* (Diptera: Psychodidae). Sci. Bull. Nat. Hist. Mus. Los Angeles Co. 17:1–69.

Hogue, C. L. 1973b. The net-winged midges or Blephariceridae of California. Bull. Calif. Insect Surv. 15:1–83.

Hogue, C. L. 1978. The net-winged midges of eastern North America, with notes on new taxonomic characters in the family Blephariceridae (Diptera). Contr. Sci. 291:1–41.

Hogue, C. L. 1981. Blephariceridae, pp. 191–197. *In* McAlpine, J. F. et al., eds., Manual of Nearctic Diptera. Vol. 1. Res. Br. Agr. Can. Monogr. 27.

Hora, S. L. 1933. Remarks on Tonnoir's theory of the evolution of the ventral suckers of dipterous larvae. Rec. Indian Mus. 35:283–286.

Hough, G. 1899. Studies in Diptera Cyclorrhapha 1. The Pipunculidae of the U.S. Proc. Boston Soc. Nat. Hist. 29:77–86.

Houlihan, D. F. 1969. The structure and behavior of *Notiphila riparia* and *Erioptera squalida* (Dipt.). J. Zool. Lond. 159:249–267.

House, H. L. 1954. Nutritional studies with *Pseudosarcophaga affinis* (Fallén), a dipterous parasite of the spruce budworm, *Choristoneura fumiferana* (Clemens). I. a chemically defined medium and aseptic-culture technique. Can. J. Zool. 32:331–341.

Houser, J. S. 1912. The gooseberry gall midge or bud deformer (*Rhopalomyia grossulariae* Felt). J. Econ. Ent. 5:180–184.

Housewart, M. W. and J. W. Brewer. 1972. Biology of a pinyon spindle gall midge (Diptera: Cecidomyiidae). Ann. Ent. Soc. Amer. 65:331–336.

Hubault, E. 1925. Contribution à la biologie du genre *Medeterus* Fischer. Annls. Sci. nat. Zool. (10) 8:133–141.

Hubault, E. 1927. Contribution à l'étude des invertébrés torrenticoles. Bull. biol. France Belg., Suppl. IX:1–388 (165–187, 297–312)

Hubert, A. A. 1953. The biology of *Paradixa californica*. Pan-Pacif. Ent. 29:181–190.

Huckett, H. C. 1965a. The Muscidae of northern Canada, Alaska and Greenland (Diptera). Mem. Entomol. Soc. Can. 42:369 pp.

Huckett, H. C. 1965b. Muscidae. *In:* Stone, *et al.* eds. A catalog of the Diptera of America north of Mexico. Handbook No. 276. Agr. Res. Ser. U.S. Dept of Agric.

Huckett, H. C. 1965c. Subfamily Fucellinae and Anthomyiinae. *In:* Stone, *et al.* eds. A catalog of the Diptera of America north of Mexico. Agricultural Handbook No. 276. Agr. Res. Ser. U.S. Dept of Agr.

Huckett, H. C. 1987. Anthomyiidae, pp. 1099–1114. *In* McAlpine, J. F., et al., eds., Manual of Nearctic Diptera. Vol. 2. Res. Br. Agr. Can. Monogr. 28.

Huckett, H. C., and J. R. Vockeroth. 1987. Muscidae, pp. 1115–1131. *In* McAlpine, J. F. et al., eds. Manual of Nearctic Diptera, Vol. 2. Res. Br. Agr. Can. Monogr. 28.

Hull, F. M. 1962. Robber flies of the world. The genera of the family Asilidae. U.S. Natl. Mus. Bull. 224. Pts. 1 & 2. U.S. Govt. Printing Office, 907 pp.

Hull, F. M. 1973. Bee flies of the World. The genera of the family Bombyliidae. Smithsonian Inst. Press. Washington. 697 pp.

Hunter, W. D., F. C. Pratt, and J. D. Mitchell. 1913. The principal cactus insects of the United States. Bull. (n. ser.) Bur. Ent. USDA. 113:1–71.

Hutchinson, R. H. 1916. Notes on the larvae of *Euxesta notata* Wied. Proc. Ent. Soc. Wash. 18:171–175.

Hutson, A. M. 1977. A revision of the families Synneuridae and Canthyloscelidae (Diptera). Bull. Brit. Mus. Nat. Hist. (Ent.) 35(3):65–100.

Hynes, C. D. 1958. A description of the immature stages of *Limnophila* (*Eutonia*) *marchandi* Alex. Proc. Ent. Soc. Wash. 60:9–14.

Hynes, C. D. 1963. Description of the immature stages of *Cryptolabis magnistyla* Alexander (Diptera: Tipulidae). Pan-Pacif. Ent. 39:255–260.

Hynes, C. D. 1965. The immature stages of the genus *Lipsothrix* in the western United States. Pan-Pacif. Ent. 41:165–172.

Hynes, C. D. 1968. The immature stages of *Hesperoconopa dolichophallus* (Alex.) (Diptera: Tipulidae). Pan-Pacif. Ent. 44:324–327.

Hynes, C. D. 1969a. The immature stages of *Arctoconopa carbonipes* (Alex.) (Diptera: Tipulidae). Pan-Pacif. Ent. 45:1–3.

Hynes, C. D. 1969b. The immature stages of *Gonomyodes tacoma* Alex. Pan-Pacif. Ent. 45:116–119.

Hynes, C. D. 1969c. The immature stages of the genus *Rhabdomastix* (Diptera: Tipulidae). Pan-Pacif. Ent. 45:229–237.

Ibrahim, I. A. 1975. The occurrence of paedogenesis in *Eristalis* larvae (Diptera: Syrphidae). J. Med. Ent. 12:268.

Ide, F. P. 1965. A fly of the archaic family Nymphomyiidae (Diptera) from North America. Can. Ent. 97:496–507.

Illingworth, J. T. 1912. A study of the biology of the apple maggot, *Rhagoletis pomonella*. Bull. Cornell Univ. Exp. Sta. 324:129–137.

Imms, A. D. 1942. On *Braula coeca* Nitsch and its affinities. Parasitol. 34:88–100.

Irwin, M. E. 1972. Diagnosis and habitat preferences of the immature stages of three South African species of the *Xestomyza*-group (Diptera: Therevidae). Ann. Natal Mus. 21:377–389.

Irwin, M. E. 1976. Morphology of the terminalia and the known ovipositing behavior in female Therevidae (Diptera: Asiloidea) with an account of correlated adaptations and comments on phylogenetic relationships. Ann. Natal Mus. 22:913–935.

Irwin, M. E. 1977. Two new genera and four new species of the *Pherocera*-group from western North America, with observations on habitats and behavior (Diptera: Therevidae: Phycinae). Proc. Ent. Soc. Wash. 79:422–451.

Irwin, M. E., and L. Lyneborg. 1980. A revision of the Nearctic genera of Therevidae. Bull. Ill. Nat. Hist. Survey. 32:191–277.

Irwin, M. E., and L. Lyneborg. 1981. Therevidae, pp. 513–523. *In* McAlpine, J. F., et al., eds., Manual of Nearctic Diptera. Vol. 1. Res. Br. Agr. Can. Monogr. 27.

Irwin, M. E., and B. R. Stuckenberg. 1972. A description of the female, egg and first-instar larva of *Tongamya miranda*, with notes on oviposition and the habitat of the species (Diptera: Apioceridae). Ann. Natal Mus. 21:439–453.

Ishijima, H. 1967. Revision of the third stage larvae of the synanthropic flies of Japan (Diptera; Anthomyiidae, Muscidae, Calliphoridae and Sarcophagidae). Jap. Sanit. Zool. 18:47–100.

Iwasa, M. 1980. Descriptions of three species of muscid larvae and eggs newly recorded from cowdung in Japan (Diptera: Muscidae). Jap. Sanit. Zool. 31:53–56.

Jackman, R. S., S. Nowicki, D. J. Aneshansley and T. Eisner. 1983. Predatory capture of toads by fly larvae. Science 222:515–516.

Jaenike, J. 1978. Host selection by mycophagous *Drosophila*. Ecol. 59:1286–1288.

James, M. T. 1935. The genus *Hermetia* in the U.S. (Diptera, Stratiomyidae). Bull. Brooklyn Ent. Soc. 30:165–170.

James, M. T. 1936a. A review of the Nearctic Geosarginae (Diptera, Stratiomyidae). Can. Ent. 67:267–275.

James, M. T. 1936b. The genus *Odontomyia* in America north of Mexico (Diptera, Stratiomyiidae). Ann. Ent. Soc. Amer. 29:517–550.

James, M. T. 1939. A review of the Nearctic Beridinae (Diptera, Stratiomyiidae). Ann. Ent. Soc. Amer. 32:543–548.

James, M. T. 1941. A preliminary study of the new world Myxosargini (Diptera, Stratiomyidae). Lloydia, Menasha 4:300–309.

James, M. T. 1942. A review of the Myxosargini (Diptera, Stratiomyiidae). Pan-Pacif. Ent. 18:49–60.

James, M. T. 1943. A revision of the Nearctic species of *Adoxomyia* (Diptera: Stratiomyidae) Proc. Ent. Soc. Wash. 45:163–171.

James, M. T. 1947. The flies that cause myiasis in man. U.S. Dept. Agric. Misc. Publ. 631, 175 pp.

James, M. T. 1955a. A new sarcophagid parasite of *Nomia* bees. Proc. Ent. Soc. Wash. 57:283–285.

James, M. T. 1955b. The genus *Cordilura* in America north of Mexico (Scatophagidae). Ann. Ent. Soc. Amer. 48:84–100.

James, M. T. 1960. The soldier flies or Stratiomyidae of California. Bull. Calif. Insect Surv. 6:79–122.

James, M. T. 1965. Contributions to our knowledge of the Nearctic Pachygasterinae (Diptera: Stratiomyidae) Ann. Ent. Soc. Amer. 58:902–908.

James, M. T. 1966. A new genus of pachygasterine Stratiomyidae reared from cactus (Diptera). J. Kans. Ent. Soc. 39:109–112.

James, M. T. 1967. The *Hermetia comstocki* group (Diptera: Stratiomyidae). Pan-Pacif. Ent. 43:61–64.

James, M. T. 1981a. Xylophagidae, pp. 489–492. *In* McAlpine, J. F., et al., eds., Manual of Nearctic Diptera. Vol. 1. Res. Br. Agr. Can. Monogr. 27.

James, M. T. 1981b. Xylomyidae, pp. 493–495. *In* McAlpine, J. F., et al., eds., Manual of Nearctic Diptera. Vol. 1. Res. Br. Agr. Can. Monogr. 27.

James, M. T. 1981c. Stratiomyidae, pp. 497–511. *In* McAlpine, J. F., et al., eds., Manual of Nearctic Diptera. Vol. 1. Res. Br. Agr. Can. Monogr. 27.

James, M. T., and F. X. Gassner. 1974. The immature stages of the fox maggot (*Wohlfahrtia opaca* Coq.) Parasitol. 34:241–244.

James, M. T., and M. W. McFadden. 1969. The genus *Adoxomyia* in America north of Mexico (Diptera: Stratiomyidae). J. Kans. Ent. Soc. 42:260–276.

James, M. T., and G. C. Steyskal. 1952. A review of the Nearctic Stratiomyini (Diptera, Stratiomyidae) Ann. Ent. Soc. Amer. 45:385–412.

James, M. T., and W. J. Turner. 1981. Rhagionidae, pp. 483–488. *In* McAlpine, J. F., et al., eds., Manual of Nearctic Diptera. Vol. 1. Res. Br. Agr. Can. Monogr. 27.

James, M. T., and W. W. Wirth. 1967. The species of *Hermetia* of the *aurata* group (Diptera: Stratiomyidae). Proc. U.S. Nat. Mus. 123:1–19.

Jenkins, D. W., and S. J. Carpenter. 1946. Ecology of the tree-hole breeding mosquitoes of Nearctic North America. Ecol. Monogr. 16:31–47.

Jenks, G. E. 1940. The spider's "uninvited" fly brings doom. Nat. History 45:157–161.

Jobling, B. 1949. A revision of the species of the genus *Aspidoptera* Coquillett, with some notes on the larva and the puparium of *A. clovisi*, and a new synonym. Proc. Roy. Ent. Soc. London (B) 18:135–144.

Johannsen, O. A. 1903. Aquatic nematocerous Diptera. I. N.Y. St. Mus. Bull. 68:328–441.

Johannsen, O. A. 1905. Aquatic nematocerous Diptera II. N.Y. St. Mus. Bull. 86:76–331.

Johannsen, O. A. 1910a. Insect notes for 1909. Maine Agr. Exp. Sta. Bull. 177:21–44.

Johannsen, O. A. 1910b. The fungus gnats of North America. The Mycetophilidae of North America. Part I. Maine Agr. Exp. Sta. Bull. (2) 172:209–276.

Johannsen, O. A. 1910c. The fungus gnats of North America. The Mycetophilidae of North America. Part II. Maine Agr. Exp. Sta. Bull. (2) 180:125–192.

Johannsen, O. A. 1912a. The fungus gnats of North America. The Mycetophilidae of North America. Part III. Maine Agr. Exp. Sta. Bull. (1911) (2) 196:249–328.

Johannsen, O. A. 1912b. The fungus gnats of North America. The Mycetophilidae of North America. Part IV. Maine Agr. Exp. Sta. Bull. (2) 200:57–146.

Johannsen, O. A. 1921a. A seed potato maggot (*Hylemyia trichodactyla* Rondani). J. Econ. Ent. 14:503–504.

Johannsen, O. A. 1921b. The first instar of *Wohlfahrtia vigil* Walker. J. Parasitol. 7:154–155.

Johannsen, O. A. 1922. Stratiomyiid larvae and pupae of the northeastern states. J. N.Y. Ent. Soc. 30:141–153.

Johannsen, O. A. 1934. Aquatic Diptera I. Nematocera, exclusive of Chironomidae and Ceratopogonidae. Mem. Cornell Univ. Agr. Exp. Sta. 164:1–71.

Johannsen, O. A. 1935. Aquatic Diptera II. Orthorrhapha-Brachycera and Cyclorrhapha. Mem. Cornell Univ. Agr. Expt. Sta. 177:1–62.

Johannsen, O. A. 1937a. Aquatic Diptera III. Chironomidae: subfamily Tanypodinae, Diamesinae, and Orthocladinae. Mem. Cornell Univ. Agr. Exp. Sta. 205:1–84.

Johannsen, O. A. 1937b. Aquatic Diptera IV. Chironomidae: Subfamily Chironominae s. l. and Supplementary Notes to Pt. I, II and III of Aquatic Diptera. Mem. Cornell Univ. Agr. Exp. Sta. 210:1–52.

Johannsen, O. A. 1952. Guide to the Insects of Connecticut. Part VI. The Diptera or True Flies of Connecticut. Fifth Fascicle. Heleidae (Certatopogonidae). Bull Conn. St. Geol. Nat. Hist. Surv. 80:3–26.

Johannsen, O. A., and C. R. Crosby. 1913. The life history of *Thrypticus muhlenbergiae* sp. nov. Psyche 20:164–166.

Johannsen, O. A., and H. K. Townes. 1952. Guide to the Insects of Connecticut. Part VI. The Diptera or True Flies of Connecticut. Fifth Fascicle. Tendipedidae. Bull. Conn. St. Geol. Nat. Hist. Surv. 80:27–147.

Johnson, C. W. 1926. A revision of some of the North American species of Mydaidae. Proc. Boston Soc. Nat. Hist. (1925–1928) 38:131–145.

Johnson, T. H., and M. J. Bancroft. 1919. The life histories of *Musca australis* and *M. vetustissima*. Proc. Roy. Soc. Queensland 31:181–203.

Jones, C. R. 1922. A contribution to our knowledge of the Syrphidae of Colorado. Coll. Agr. Exper. Sta. Bull. 269:1–72.

Jones, F. M. 1916. Two insect associates of the California pitcher-plant, *Darlingtonia californica*. Ent. News 27:385–392.

Jones, F. M. 1920. Another pitcher-plant insect (Diptera: Sciarinae). Ent. News 31:91–94.

Jones, R. G., R. J. Gagné, and W. F. Barr. 1983. Biology and taxonomy of the *Rhopalomyia* gall midges of *Artemisia tridentata* Nuttall (Compositae) in Idaho. Contr. Amer. Ent. Soc. Inst. 21:1–90.

Juday, C. 1921. Observations on the larvae of *Corethra punctipennis* Say. Biol. Bull. 40:271–286.

Judd, W. W. 1952. Diptera and Hymenoptera reared from pine cone willow galls caused by *Rhabdophaga strobiloides* (Diptera: Itonididae). Ent. Soc. Ontario 83rd Rep.: 34–42.

Judd, W. W. 1954. Insects collected from birds nests at London, Ontario. Can. Fld. Nat. 68:122–123.

Kaczynski, V. W., J. Zuska, and C. O. Berg. 1969. Taxonomy, immature stages, and bionomics of the South American genera *Perilimnia* and *Shannonia* (Diptera: Sciomyzidae). Ann. Ent. Soc. Amer. 62:572–592.

Kamali, K. and J. T. Schulz. 1974. Biology and ecology of *Gymnocarena diffusa* (Diptera: Tephritidae) on sunflower in North Dakota. Ann. Ent. Soc. Amer. 67:695–699.

Kamal, A. S. 1958. Comparative study of thirteen species of sarcosaprophagous Calliphoridae and Sarcophagidae (Diptera). I. Bionomics. Ann. Ent. Soc. Amer. 51:261–270.

Kamal, A. S. 1959. Comparative studies of thirteen species of sarcosaprophagous Calliphoridae and Sarcophagidae (Diptera). II. Digestive enzymology. Ann. Ent. Soc. Amer. 52:167–173.

Kanervo, V. 1942. Die Unterscheidung der Larven und Puppen von *Eumerus tuberculatus* Rond. und *E. strigatus* Fall. Ann. Ent. Fenn. 8:227–233.

Kangas, E. and Leskinen, K. 1944. *Pegohylemyia anthracina* Czerny als Zapfenschädling an der Fichte. Ann. Ent. Fenn. 9:195–205.

Kano, R. & Sato, K. 1951. Notes on the flies of medical importance in Japan. (Part II). The larvae of *Sarcophaga* known in Japan. Japan. Exp. Med. 21:115–131.

Karandikar, K. R. 1931. The early stages and bionomics of *Trichocera maculipennis* (Meigen) (Diptera, Tipulidae). Trans. Roy. Ent. Soc. London. 79:249–262.

Kearney, J. 1983. Selection and utilization of natural substrates as breeding sites by woodland *Drosophila* spp. Ent. Exp. Appl. 33:63–70.

Keifer, H. H. 1930. Synopsis of the dipterous larvae found in California fruits. Mon. Bull. Dept. Agr. Calif. 19:574–581.

Keilin, D. 1911. Recherches sur la morphologie larvaire des Diptères du genre *Phora*. Bull. Sci. France Belg. (7) 44:27–88.

Keilin, D. 1912a. Recherches sur les Diptères du genre *Trichocera*. Bull. Sci. France Belg. (7) 46:172–190.

Keilin, D. 1912b. Structure du pharynx en fonction du régime chez les larves de diptères cyclorrhaphes. C. r. hebd. Séanc. Acad. Sci. Paris 155:1548–1550.

Keilin, D. 1914. Les formes adaptives des larves des Anthomyiides: les Anthomyiides à larves carnivores. Bull. Soc. ent. France 1914:496–501.

Keilin, D. 1915/1916. Recherches sur les larves de Diptères Cyclorrhaphes. Bull. Sci. France Belg. 49:15–198.

Keilin, D. 1916. Sur la viviparité chez les Diptères et sur les larves de Diptères vivipares. Arch. Zool. Exp. Gen. 55:393–415.

Keilin, D. 1917a. On the supposed first stage larva of *Leptohylemyia coarctata* Fall. Bull. Ent. Res. 8:121–123.

Keilin, D. 1917b. Recherches sur les Anthomyiides à larves carnivores. Parasitol. 9:325–450.

Keilin, D. 1919b. On the structures of the larvae and the systematic position of the genera *Mycetobia*, Mg. *Ditomyia*, Winn., and *Symmerus*, Walk. (Diptera, Nematocera). Ann. Mag. Nat. Hist. (9) 3:33–42.

Keilin, D. 1919c. On the alimentary canal and its appendages in the larvae of Scatopsidae and Bibionidae, with some remarks on the parasites of these larvae. Ent. Mon. Mag. 5:92–96.

Keilin, D. 1924a. On the life history of *Anthomyia procellaris* Rond. and *Anthomyia pluvialis* L. inhabiting the nests of birds. Parasitol. 16:150–159.

Keilin, D. 1924b. Sur la position primitive des stigmates chez les insectes et sur le sort des stigmates thoraciques. Bull. Soc. ent. France 1924:125–128.

Keilin, D. 1924c. On a case of intestinal myiasis in man produced by the larvae of a sarcophagine fly. Parasitol. 16:318–320.

Keilin, D. 1944. Respiratory systems and respiratory adaptations in larvae and pupae of Diptera. Parasitology 36:1–66.

Keilin, D., and P. Tate. 1930. On certain semicarnivorous anthomyiid larvae. Parasitol. 22:168–181.

Keilin, D., and P. Tate. 1937. A comparative account of the larvae of *Trichomyia urbica* Curtis, *Psychoda albipennis* Zett. and *Phlebotomus argentipes* Ann. & Brunn. Parasitol. 29:247–258.

Keilin, D., and P. Tate. 1940. The early stages of the families Trichoceridae and Anisopodidae (Rhyphidae) (Diptera: Nematocera). Trans. Roy. Ent. Soc. London 90:39–62.

Keilin, D., and P. Tate. 1943. The larval stages of the celery fly (*Acidia heraclei* L.) and of the braconid *Adelura apii* (Curtis) with notes upon an associated parasitic yeast-like fungus. Parasitol. 35:27–36.

Keilin, D., and H. Thompson. 1915. Sur le cycle évolutif des Pipunculides, parasites intracoelomiques des Typhlocybes. C. r. Séane Soc. Biol., Paris 67:9–12.

Kellogg, V. L. 1901. An aquatic Psychodid. Ent. News 12:46–49.

Kelsey, L. P. 1981. Scenopinidae, pp. 525–528. *In* McAlpine, J. F., et al., eds., Manual of Nearctic Diptera. Vol. 1. Res. Br. Agr. Can. Monogr. 27.

Kennedy, H. D. 1958. Biology and life history of a new species of mountain midge, *Deuterophlebia nielsoni*, from eastern California (Diptera: Deuterophlebiidae). Trans. Amer. Microscop. Soc. 7:201–228.

Kennedy, H. D. 1960. *Deuterophlebia inyoensis*, a new species of mountain midge from the alpine zone of the Sierra Nevada range, California (Diptera: Deuterophlebiidae). Trans. Amer. Microscop. Soc. 79:191–210.

Kennedy, H. D. 1981. Deuterophlebiidae, pp. 199–202. *In* McAlpine, J. F., et al., eds., Manual of Nearctic Diptera. Vol. 1. Res. Br. Agr. Can. Monogr. 27.

Kepner, W. A. 1912. The larva of *Sarcophaga*, a parasite of *Cistudo carolina* and the histology of its respiratory apparatus. Biol. Bull. 22:163–73.

Kessel, E. L. 1955. The mating activities of balloon flies. Syst. Zool. 4:97–104.

Kessel, E. L. 1960. The life cycle of *Clythia agarici* (Willard) (Diptera: Platypezidae). Wasmann J. Biol. 18:263–270.

Kessel, E. L. 1961. The fungus hosts of *Melanderomyia* (Diptera: Platypezidae). Wasmann J. Biol. 19:291–294.

Kessel, E. L. 1963a. The host fungus and distribution of *Clythia coraxa* (Diptera: Platypezidae). Wasmann J. Biol. 21:79–86.

Kessel, E. L. 1963b. The rediscovery of *Colatarsa calceata* (Snow) (Diptera: Platypezidae). Wasmann J. Biol. 21:235.

Kessel, E. L. 1965. *Microsania* as prey for *Hormopeza* (Diptera: Platypezidae and Empididae). Wasmann J. Biol. 23:225–226.

Kessel, E. L. 1987. Platypezidae, pp. 681–688. *In* McAlpine, J. F., et al., eds., Manual of Nearctic Diptera. Vol. 2. Res. Br. Agr. Can. Monogr. 28.

Kessel, E. L., and M. E. Buegler. 1972. A review of the genus *Callomyia* in North America, with the description of a new species (Diptera: Platypezidae). Wasmann J. Biol. 30:241–278.

Kessel, E. L., M. E. Buegler, and P. M. Keyes. 1973. A survey of the known larvae and puparia of Platypezidae, with a key to ten genera based on immature stages (Diptera). Wash. J. Biol. 31:233–261.

Kessel, E. L., and M. F. Kirby. 1968. *Grossoseta*, a new genus related to *Platypezina* Wahlorem, with notes on the distribution of the species of these genera in North America (Diptera: Platypezidae). Wasmann J. Biol. 26:17–31.

Kessel, E. L., E. R. Royak, and D. Gunther. 1971. Oviposition, incubation, and larval development and diapause in *Calotarsa insignis* Aldrich (Diptera: Platypezidae). Wasmann J. Biol. 29:17–27.

Kevan, D. K. McE., and F. E. A. Cutten. 1975. Canadian Nymphomyiidae (Diptera). Can. J. Zool. 53:853–866.

Kevan, D. K. McE., and F. E. A. Cutten. 1981. Nymphomyiidae, pp. 203–207. *In* McAlpine, J. F., et al., eds., Manual of Nearctic Diptera. Vol. 1. Res. Br. Agr. Can. Monogr. 27.

Kieffer, J. J. 1891. Zur Kenntnis der Weidengallmücken. Berlin. Ent. Zeit. 36:241–258.

Kieffer, J. J. 1895a. Beobachtungen über die Larven der Cecidomyinen. Wien Ent. Zeit. 14:1–16.

Kieffer, J. J. 1895b. Über Papillen bei Gallmückenlarven. Wien Ent. Zeit. 14:117–126.

Kieffer. J. J. 1899. Beiträge zur Biologie und Morphologie der Dipteren-Larven. III. Zeitschr. Ent. 4:353–354, 372–374.

Kieffer, J. J. 1900. Beiträge zur Biologie und Morphologie der Dipteren-larven. III. Zeitschr. Ent. 5:131–133, 241–42.

Kim, K. C. 1967. The North American species of the genus *Anarete* (Diptera: Cecidomyiidae). Ann. Ent. Soc. Amer. 60:521–530.

Kim, K. C., ed. 1985. Coevolution of Parasitic Arthropods and Mammals. Wiley Interscience, NY. 800 p.

Kim, K. C., and R. W. Merritt, eds. 1987. Black Flies: Ecology, Population Management, and Annotated World List. Penn. State Press, Univ. Park. 440 pp.

King, J. L. 1916. Observations on the life history of *Pterodontia flavipes* Gray (Diptera). Ann. Ent. Soc. Amer. 9:309–321.

King, L. A. L., A. Meikle, and A. Broadfoot. 1935. Observations on the timothy grass fly, *Amaurosoma armillatum* Zett. Ann. Appl. Biol. 22:267–278.

King, W. V., G. H. Bradley, C. N. Smith, and W. C. McDuffy. 1960. A handbook of the mosquitoes of the southeastern United States. U.S. Dept. Agr. Handb. 173:1–188.

Kitching, R. L. 1976. The immature stages of the Old World screw-worm fly, *Chrysomya bezziana* Villeneuve, with comparative notes on other Australasian species of *Chrysomya* (Diptera, Calliphoridae). Bull. Ent. Res. 66:195–203.

Kloter, K. O., L. R. Penner, and W. J. Widmer. 1977. Interactions between the larvae of *Psychoda alternata* and *Dohrniphora cornuta* in a trickling filter sewage bed, with descriptions of the immature stages of the latter. Ann Ent. Soc. Amer. 70:775–781.

Knight, A. W. 1963. Description of the tanyderid larva *Protanyderus margarita* Alexander from Colorado. Bull. Brooklyn Ent. Soc. 58:99–102.

Knight, A. W. 1964. Description of the tanyderid pupa *Protanyderus margarita* Alexander from Colorado. Ent. News 75:237–241.

Knight, K. L., and J. L. Laffoon. 1971. A mosquito taxonomic glossary VIII. The larval chaetotaxy. Mosquito Syst. Newsl. 3:160–194.

Knipling, E. F. 1936a. A comparative study of the first instar larvae of the genus *Sarcophaga*, with notes on the biology. J. Parasitol. 22:417–454.

Knipling, E. F. 1936b. Some specific taxonomic characters of common *Lucilia* larvae. Iowa State Coll. J. Sci. 10:275–293.

Knipling, E. F. 1937. The biology of *Sarcophaga cistudinis* Aldrich, a species of Sarcophagidae parasitic on turtles and tortoises. Proc. Ent. Soc. Wash. 39:91–101.

Knipling, E. F. 1939. A key for blowfly larvae concerned in wound and cutaneous myiasis. Ann. Ent. Soc. Amer. 32:376–383.

Knipling, E. F. and A. L. Brody. 1940. Some taxonomic characters of cuterebrine larvae, with larval descriptions of two species from Georgia. J. Parasitol. 26:33–43.

Knutson, L. V. 1966. Biology and immature stages of malacophagous flies: *Antichaeta analis, A. atriseta, A. brevipennis,* and *A. obliviosa* (Diptera: Sciomyzidae). Trans. Amer. Ent. Soc. 92:67–101.

Knutson, L. V. 1970a. Biology and immature stages of *Tetanura pallidiventris,* a parasitoid of terrestrial snails (Diptera: Sciomyzidae). Ent. Scand. 1:81–89.

Knutson, L. V. 1970b. Biology of snail-killing flies in Sweden (Diptera: Sciomyzidae). Ent. Scand. 1:307–314.

Knutson, L. V. 1971. Puparia of *Salpingogaster conopida* and *S. texana,* with notes on prey (Diptera: Syrphidae). Ent. News 82:29–38.

Knutson, L. V. 1972a. Description of the female of *Pherbecta limenitis* Steyskal (Diptera: Sciomyzidae), with notes on biology, immature stages, and distribution. Ent. News 83:15–21.

Knutson, L. V. 1972b. Pupa of *Neomochtherus angustipennis* (Hine), with notes on feeding habits of robber flies and a review of publications on morphology of immature stages (Diptera: Asilidae). Proc. Biol. Soc. Wash. 85:163–178.

Knutson, L. V. 1973a. Biology and immature stages of *Coremacera marginata* F., a predator of terrestrial snails (Diptera: Sciomyzidae). Ent. Scand. 4:123–133.

Knutson, L. V. 1973b. Taxonomic revision of the aphid killing flies of the genus *Sphaerophoria* in the western hemisphere (Syrphidae). Misc. Publ. Ent. Soc. Amer. 9:1–50.

Knutson, L. V. 1976. Key to subfamilies of robber flies based on pupal cases, with a description of the pupal case of *Doryclus distendens* (Asilidae: Megapodinae). Proc. Biol. Soc. Wash. 88:509–514.

Knutson, L. V. 1987. Sciomyzidae. pp. 927–940 *In* McAlpine, J. F. et al., eds., Manual of Nearctic Diptera. Vol. 2. Res. Br. Agr. Can. Monogr. 28.

Knutson, L., and J. Abercrombie. 1977. Biology of *Antichaeta melanosoma* (Diptera: Sciomyzidae), with notes on parasitoid Braconidae and Ichneumonidae (Hymenoptera). Proc. Ent. Soc. Wash. 79:111–125.

Knutson, L. V., and C. O. Berg. 1963. Biology and immature stages of a snail-killing fly, *Hydromya dorsalis* (Fabricius) (Diptera: Sciomyzidae). Proc. Roy. Ent. Soc. London (A) 38 (4–6):45–58.

Knutson, L. V., and C. O. Berg. 1964. Biology and immature stages of snail-killing flies: the genus *Elgiva* (Diptera: Sciomyzidae). Ann. Ent. Soc. Amer. 57:173–192.

Knutson, L. V., and C. O. Berg. 1967. Biology and immature stages of malacophagous Diptera of the genus *Knutsonia* Verbeke (Sciomyzidae). Bull. Inst. Roy. Sci. Nat. Belg. 43 (7):1–60.

Knutson, L. V., and O. S. Flint, Jr. 1971. Pupae of Empidae in pupal cocoons of Rhyacophilidae and Glossosomatidae. Proc. Ent. Soc. Wash. 73:314–320.

Knutson, L. V. and O. S. Flint, Jr. 1979. Do dance flies feed on caddisflies?—further evidence (Diptera: Empididae; Trichoptera). Proc. Ent. Soc. Wash. 81:32–33.

Knutson, L. V., R. Rozkosny, and C. O. Berg. 1975. Biology and immature stages of *Pherbina* and *Psacadina* (Diptera: Sciomyzidae). Acta. Sci. Nat. Acad. Sci. Bohem.-Brno. 9 (1):1–38.

Knutson, L. V., J. W. Stephenson, and C. O. Berg. 1965. Biology of a slug-killing fly, *Tetanocera elata* (Diptera: Sciomyzidae). Proc. Malacol. Soc. London 36:213–220.

Knutson, L. V., J. W. Stephenson, and C. O. Berg. 1970. Biosystematic studies of *Salticella fasciata* (Meigen), a snail-killing fly (Diptera: Sciomyzidae). Trans. Roy. Ent. Soc. London 122 (3):81–100.

Knutson, L., and K. Valley. 1978. Biology of a neotropical snail-killing fly, *Sepedonea isthmi* (Diptera: Sciomyzidae). Proc. Ent. Soc. Wash. 80:197–209.

Koizumi, K. 1957. Notes on dipterous pests of economic plants in Japan. Botyu-Kagaki 22:223–227.

Koizumi, K. 1959. On four dorilaid parasites of the green rice leafhopper *Nephotettix cincticeps* Uhler (Diptera). Okayama Univ. Fac. Agr. Sci. Rpt. 13:37–45.

Kompfner, H. 1974. Larvae and pupae of some wrack dipterans on a California beach (Diptera: Coelopidae, Anthomyiidae, Sphaeroceridae). Pan-Pacif. Ent. 50:44–52.

König, A. 1895. Über die Larve von *Ogcodes*. Verh. Zool.-Bot. Ges. Wien. 44:163–66.

Kraft, K. J., and E. J. Cook. 1961. A revision of the Pachygasterinae (Diptera, Stratiomyidae) of America north of Mexico. Misc. Publ. Ent. Soc. Amer. 3:1–24.

Krivosheina, N. P. 1962. The European larvae of the Bibionidae (Diptera, Nematocera), with keys to several species (In Russian). Pedobiol. 1:210–227.

Krivosheina, N. P. 1969. Ontogeny and evolution of the Diptera (in Russian). Nauka, Moscow.

Krivosheina, N. P. 1980. Larvae of the family Ditomyiidae (Diptera, Nematocera) of the fauna of the USSR. Zool. Zh. 59:546–557.

Krivosheina, N. P., and B. M. Mamaev. 1966. Die Larven der europaischen Arten der Gattung *Xylophagus* Meigen (Diptera: Xylophagidae). Beitr. Ent. 16:275–283.

Krivosheina, N. P., and B. M. Mamaev. 1967. Classification key to the larvae of arboricolous dipteran insects (In Russian). Nauka, Moscow.

Krivosheina, N. P., and B. M. Mamaev. 1970. Morphology, ecology and phylogenetic relationships of the Cramptonomyiidae (Diptera, Nematocera), a family new to the fauna of the USSR [In Russian]. Ent. Obozr. 49:886–898.

Krogstad, B. O. 1959. Some aspects of the ecology of *Axymyia furcata* McAtee (Diptera: Sylvicolidae). Proc. Minn. Acad. Sci. 27:175–177.

Krogstad, B. O. 1974. Aquatic stages of *Stratiomys normula unilimbata* Loew. (Diptera: Stratiomyiidae). J. Minn. Acad. Sci. 38:86–88.

Kuenzel, W. J., and R. G. Wiegert. 1974. Energetics of a spotted sandpiper feeding on brine fly larvae, *Paracoenia turbida,* (Diptera: Ephydridae) in a thermal spring community. Wilson Bull. 85:473–476.

Kugler, J. 1978. The Rhinophoridae (Diptera) of Israel. Israel Jour. Entomol. 12:65–106.

Kurir, A. 1965. Zur Biologie zweier aphidophager Schwebfliegen (Diptera: Syrphidae). *Heringia heringi* Zetterstedt und *Pipiza festiva* Meigen in der Gallen der Spaten Blattstieldrehgallen Pappelblattlaus (*Pemphigus spirothecae* Passerini) auf der Pyrmidempappel (*Papulua nigra* var *pyrmidalis* Spach). Zeitschr. Angew. Ent. 52:61–83.

Kurkina, L. A. 1979. The biology of *Machimus annulipes* (Diptera, Asilidae). Ent. Rev. 58:26–30.

Kuroda, M. 1958. Studies on the spiracles and cephalopharyngeal sclerites of the larvae of the agromyzid flies. (Reports I, II). Kontyu 26:142–147, 148–152.

Kuroda, M. 1960. Studies on the spiracles and cephalopharyngeal sclerites of the larvae of the agromyzid flies (Reports III, IV). Kontyu 28:48–53, 172–176.

Kuroda, M. 1961. Studies on the spiracles and cephalopharyngeal sclerites of the larvae of the Agromyzid flies (Rept. V). Kontyu 29:65–71.

Kuster, K. C. 1934. A study of the general biology, morphology of the respiratory system, and respiration of certain aquatic *Stratiomyia* and *Odontomyia* larvae (Diptera). Pap. Mich. Acad. Sci. Arts Lett. 19:605–658.

Laake, E. W. 1921. Distinguishing characters of the larval stages of ox-warbles *Hypoderma bovis* and *Hypoderma lineatum,* with descriptions of a new larval stage (Oestridae). J. Agr. Res. 21:439–457.

Laake, E. W., E. C. Cushing, and H. E. Parish. 1936. Biology of the primary screw worm fly *Cochliomyia americana* and a comparison of its stages with those of *C. macellaria.* U.S. Dept. Agr. Tech. Bull. 500, 24 pp.

Laffoon, J. L. 1957. A revision of the Nearctic species of *Fungivora* (Diptera, Mycetophilidae). Iowa St. Coll. J. Sci. (1956) 31:141–340.

Laird, M. 1981. Blackflies. Acad. Press, NY. xii+399 p.

Lake, R. W. 1960. Observations on the biology of *Eucorethra underwoodi* Underwood in Passaic County, New Jersey, and Bristol County, Massachusetts. Mosquito News 20:171–174.

Lake, R. W. 1969. Life history, habitat, and taxonomic characters of the larva of *Mochlonyx fuliginosus.* Ent. News 80:34–37.

Lamore, D. H. 1960. Cases of parasitism of the basilica spider, *Allepeira lemniscata* (Walckenaer), by the dipteran endoparasite, *Ogcodes dispar* (Macquart) (Araneida: Argiopidae and Diptera: Acroceridae). Proc. Ent. Soc. Wash. 62:65–85.

Lamp, W. D., and M. K. McCarty. 1982. Biology and predispersal seed predators of the Platte thistle, *Cirsium canescens.* J. Kans. Ent. Soc. 55:305–316.

Landis, B. J. & Howard, N. F. 1940. *Paradexodes epilachnae,* a tachinid parasite of the Mexican bean beetle. U.S. Dept. Agr. Tech. Bull. 721, 31 pp.

Landis, B. J., R. L. Wallis, and R. D. Redmond. 1967. *Psilopa leucostoma,* a new leaf miner of sugar beets in the United States. J. Econ. Ent. 60:115–118.

Lane, R. S. 1975. Immatures of some Tabanidae (Diptera) from Mendocino County, Calif. Ann. Ent. Soc. Amer. 68:803–819.

LaRivers, I. 1944. A summary of the Mormon cricket (*Anabrus simplex*) (Tettigoniidae: Orthoptera). Ent. News 55:71–77, 97–102.

Latteur, G. 1974. The cereal grass fly *Opomyza florum* (Diptera: Brachycera: Opomyzidae). Rev. Agr. 27:861–869.

Laurence, B. R. 1953. The larva of *Ectaetia* (Diptera: Scatopsidae). Entomol. Mon. Mag. 89:204–205.

Laurence, B. R. 1954. The larval inhabitants of cow pats. J. Anim. Ecol. 23:234–260.

Laurence, B. R. 1955. The ecology of some British Sphaeroceridae (Borboridae, Diptera). J. Anim. Ecol. 24:187–199.

Laurence, B. R. 1956. On the life history of *Trichocera saltator* (Harris) (Diptera, Trichoceridae). Proc. Zool. Soc. London 126:235–243.

Lavallee, A. G., J. B. Wallace. 1974. Immature stages of Milesiinae (Syrphidae) II. *Sphegina keeniana* and *Chrysogaster nitida.* J. Georgia Ent. Soc. 9:8–15.

Lavallee, A. G. and J. B. Wallace. 1974. Immature stages of Milesiinae (Syrphidae). II. *Sphegina keeniana* and *Chrysogaster nitida.* J. Georgia Ent. Soc. 9:8–15.

Lavigne, R. J. 1962. Immature stages of the stalk-eyed fly, *Sphyracephala brevicornis* (Say) (Diptera: Diopsidae), with observations on its biology. Bull. Brooklyn Ent. Soc. 107:5–14.

Lavigne, R. J. 1963. Notes on the behavior of *Stenopogon coyote* Bromley with a description of the eggs (Diptera: Asilidae). Pan-Pacif. Ent. 39:103–107.

Lavigne, R. J. 1975. Redescription of *Apiocera clavator* with notes on its behavior (Diptera: Apioceridae). Ann. Ent. Soc. Amer. 68:673–676.

Lavigne, R. J., and S. W. Bullington. 1984. Ethology of *Laphria fernaldi* (Back) (Diptera: Asilidae) in southeast Wyoming. Proc. Ent. Soc. Wash. 86:326–336.

Lavigne, R. J., D. S. Dennis, and J. A. Gowen. 1978. Asilid literature update 1956–1976, including a brief review of robber fly biology. Sci. Monogr. Univ. Wyo. Agr. Exp. Sta. 36:1–134.

Leathers, A. L. 1922. Ecological study of aquatic midges and some related insects with special reference to feeding habits. Bull. Bur. Fish. 38:1–61.

Leclercq, M. 1960. Révision systématique et biogéographique des Tabanidae (Diptera) paléarctiques. Vol. 1. Pangoniinae et Chrysopinae. Mém. Inst. r. Sci. nat. Belg. (2) 63:1–77.

Leclercq, M. 1966. Revision systematique et biogéographique des Tabanidae (Diptera) paléarctiques. Vol. 2. Tabaninae. Mém. Inst. r. Sci. nat. Belg. (2) 80:1–237.

Lejeune, R. R. and Silver, G. T. 1961. Parasites and hyperparasites of the satin moth, *Stilpnotia salicis* Linnaeus, (Lymantriidae) in British Columbia. Can. Ent. 93:456–467.

Lenz, F. 1923. Stratiomyiiden aus Quellen. Ein Beitrag zur Metamorphose der Stratiomyiiden. Arch. Naturgesch. 89, A 2, pp. 39–62.

Lenz, F. 1926. Stratiomyiiden-Larven aus dem Salzwasser. Mitt. Geogr. Ges. Lübeck 31(2):170–175.

Leonard, M. D. 1930. A revision of the dipterous family Rhagionidae (Leptidae) in the United States and Canada. Mem. Amer. Ent. Soc. 7:1–181.

Léonide, J. C. 1964. Contribution à l'étude biologique du *Neorhynchocephalus tauscher* (Fisch.) (Diptera: Nemestrinidae) et commentaires sur la biologie imaginale des némestrinides. Bull. Soc. zool. France 89:210–218.

LeRoux, E. J., and J. P. Perron. 1960. Descriptions of immature stages of *Coenosia tiorina* (F.) (Diptera: Anthomyiidae), with notes on hibernation of larvae and predation by adults. Can. Ent. 92:284–296.

Lewis, D. J. 1949. Tracheal gills in some African culicine mosquito larvae. Proc. R. ent. Soc. Lond., Ser. A, 24:60–66.

Lewis, D. J. 1973. Phlebotomidae and Psychodidae (sand-flies and moth-flies), pp. 155–179. *In* Smith, K. G. V., ed., Insects and other Arthropods of Medical Importance. Brit. Mus. (Nat. Hist.), London.

Lindner, E. 1930. Phryneidae (Anisopodidae, Rhyphidae). Lfg. 50, Bd. 2(1), pp. 1–10. *In* Lindner, E., ed. Die Fliegen der palaearktischen Region. Schweizerbart, Stuttgart.

Lindsay, S. L. 1958. Food preferences of *Drosophila* larvae. Amer. Nat. 92:279–285.

Linnane, J. P., and E. A. Osgood. 1977. *Verallia virginica* (Diptera: Pipunculidae) reared from the Saratoga spittle-bug in Maine. Proc. Ent. Soc. Wash. 79:622–623.

Linsley, E. G. 1960. Ethology of some bee-and-wasp-killing robber flies of southeastern Arizona and western New Mexico (Diptera: Asilidae). Univ. Calif. Publ. Ent. 16:357–381.

Liu, D. and B. Greenberg. 1989. Immature stages of some flies of forensic importance. Ann. Ent. Soc. Amer. 82:80–93.

Lloyd, L. L. 1943. Materials for a study in animal competition. The fauna of the sewage bacteria beds. Part II. Ann. Appl. Biol. 30:47–60.

Lloyd, L. L., J. S. Graham, and T. B. Reynoldson. 1940. Materials for a study in animal competition. The fauna of the sewage bacteria beds. Ann. Appl. Biol. 27:122–150.

Lobanov, A. M. 1964. Data on ecology and morphology of preimaginal phases of *Ceroxys urticae* L. (Diptera, Otitidae). Ent. Obozr. 43:67–70.

Lobanov, A. M. 1972. Morphology of full grown larva of *Tetanops sintenisi* (Diptera, Otitidae). Zool. Z. H. 51:146–149.

Lopes, H. S. 1940. Contribuicao ao conhecimento do genero *Udamopyga* Hall e de outro sarcophagideos que vivem im moluscos no Brasil (Diptera). Rev. Ent. 11:924–954.

Lopes, H. S. 1942. *Notochaeta aldrichi* n. sp., parasita de Oligochaeta no Brasil (Diptera, Sarcophagidae). Rev. Brasil. de Biol. 2:361–364.

Lopes, H. S. 1943. Contribuicao ao conhecimento das larvas dos Sarcophagidae com especial referencia ao esqueleto cefalico. Mem. Inst. Oswaldo Cruz 38:127–163.

Lopes, H. S. 1946. Contribuicao ao conhecimento das especies do genero *Notochaeta* Aldrich, 1916. Mem. Inst. Oswaldo Cruz 42:503–550.

Lopes, H. S. 1966. Sobre *Malacophagomyia* g. n. (Diptera, Sarcophagidae) cuyas larvas vivem em cadaveres de Gastropoda (Mollusca). Rev. Brasil Biol. 26:315–321.

Lopes, H. S. 1982a. On *Eumacronychia sternalis* Allen (Diptera: Sarcophagidae) with larvae living on eggs and hatchlings of the East Pacific Green Turtle. Rev. Brasil. de Biol. 42:425–429.

Lopes, H. S. 1982b. The importance of the mandible and clypeal arch of the first instar larvae in the classification of the Sarcophagidae (Diptera). Rev. Bras. Ent. 26:293–326.

Lopes, H. S. and Vogelsang, E. G. 1953. *Notochaeta bufonivora* n. sp., parasita de *Bufo granulosus* Spix em Venezuela (Diptera, Sarcophagidae). Ann. Acad. Brasil. Cienc. 25:139–143.

Lopez, A., and J. C. Bonaric. 1977. Notes sur une nymphe myrmécophile du genre *Microdon* (Diptera, Syrphidae): éthologie et structure tégumentaire. Annls. Soc. ent. France (n.s.) 13:131–137.

Loudon, B. J. and Attia, F. I. 1981. The immature stages of *Alophora lepidofera* (Malloch) (Diptera: Tachinidae), a native parasite of Lygaeidae (Hemiptera) in Australia. Australian Ent. Mag. 7:61–67.

Ludwig, C. E. 1949. Embryology and morphology of the larval head of *Calliphora erythrocephala* Meigen. Microentomology 14:75–111.

Lyall, E. 1929. The larva and pupa of *Scatopse fuscipes* Mg. and comparison of the known species of scatopsid larvae. Ann. Appl Biol. 16:630–642.

Maa, T. C. and B. V. Peterson. 1987. Hippoboscidae, pp. 1271–1281. *In* McAlpine, J. F. et al., eds. Manual of Nearctic Diptera, Vol. 2. Res. Br. Agr. Can. Monogr. 28.

Mackerras, M. J. 1933. Observations on the life histories, nutritional requirements and fecundity of blow-flesh flies. Bull. Ent. Res. 24:353–362.

Mackerras, M. J., and M. E. Fuller. 1942. The genus *Pelecorhynchus* (Diptera, Tabanoidea). Proc. Linn. Soc. N. S. Wales 67:9–76.

Madwar, S. 1934. Biology and morphology of *Bolitophila hybrida* Meig. Bull. Soc. Roy. Ent. Egypte (1933) 17:126–135.

Madwar, S. 1937. Biology and morphology of the immature stages of Mycetophilidae (Diptera, Nematocera). Phil. Trans. R. Soc., Ser. B, 227:1–110.

Magnarelli, L. A., and J. F. Anderson. 1978. Distribution and development of immature salt marsh Tabanidae (Diptera). J. Med. Ent. 14:573–578.

Mahrt, G. G., and C. C. Blickenstaff. 1979. Host plants of the root maggot, *Tetanops myopaeformis*. Ann. Ent. Soc. Amer. 72:627–631.

Maier, C. T. 1978. The immature stages and biology of *Mallota posticata* (Fabricius) (Diptera: Syrphidae). Proc. Ent. Soc. Wash. 80:424–40.

Malloch, J. R. 1915. Some additional records of Chironomidae for Illinois and other notes on other Illinois Diptera. Ill. State Lab. Nat. Hist. Bull. 11 (1915–18):305–364.

Malloch, J. R. 1917. A preliminary classification of Diptera, exclusive of pupipara, based upon larval and pupal characters, with keys to imagines in certain families. Part 1. Ill. St. Lab. Nat. Hist. Bull. 12:161–410.

Malloch, J. R. 1918. A revision of the dipterous family Clusiodidae (Heteroneuridae). Proc. Ent. Soc. Wash. 20:2–8.

Malloch, J. R. 1919. The larval habitat of *Chalcomyia aerea* Loew (Diptera, Syrphidae). Ent. News 30:25.

Malloch, J. R. 1927. Descriptions and figures of the puparia of *Minettia ordinaria* and *Caliope flaviceps*. Proc. Ent. Soc. Wash. 29:184.

Malloch, J. R. 1933. Acalyptrata, pp. 177–391. *In* British Museum (Natural History), Diptera of Patagonia and South Chile. 6.

Malloch, J. R. 1940. The North American genera of the dipterous subfamily Chamaemyiinae. Ann. Mag. Nat. Hist. Ser. 11, 6:265–274.

Malloch, J. R., and W. L. McAtee. 1924. Keys to flies of the families Lonchaeidae, Pallopteridae, and Sapromyzidae of the eastern U. S., with a list of the species of the District of Columbia region. Proc. U.S. Nat. Mus. 65:1–26.

Malo, F., and Willis, E. R. 1961. Life history and biological control of *Caligo eurilochus*, a pest of banana. J. Econ. Ent. 54:530–536.

Mamaev, B. M., and N. P. Krivosheina. 1965. Larvae of gall midges. Diptera. Cecidomyiidae (in Russian). Akad. Nauk USSR, Moscow.

Mamaev, B. M., and N. P. Krivosheina. 1966. New data on the taxonomy and biology of the family Axymyiidae (Diptera). Ent. Rev. 45:93–99.

Mamaev, B. M., and N. P. Krivosheina. 1969. New data on the morphology and ecology of the Hyperoscelidae (Diptera, Nematocera). Ent. Rev. 48:594–599.

Mangan, R. L. 1976. *Themira athabasca* n. sp. (Diptera: Sepsidae) with a revised key to North American *Themira* and notes on the sexual morphology of sympatric species. Ann. Ent. Soc. Amer. 69:1024–28.

Mangan, R. L. 1977. A key and selected notes for the identification of larvae of Sepsidae (Diptera) from the temperate regions of North America. Proc. Ent. Soc. Wash. 79:338–342.

Mangrum, J. F. 1942. The parasitic fly, *Zelia vertebrata* Say. Ann. Ent. Soc. Amer. 35:73–75.

Manis, H. C. 1941. Bionomics and morphology of the black onion fly, *Tritoxa flexa*. Iowa St. Coll. J. Sci. 16:96–98.

Mann, J. 1969. Cactus-feeding insects and mites. Bull. U.S. Nat. Mus. 256:1–158.

Mansbridge, G. H. 1933. On the biology of some Ceroplatinae and Macrocerinae (Diptera, Mycetophilidae), with an appendix on the chemical nature of the web fluid in larvae of Ceroplatinae by H. W. Buston. Trans. Roy. Ent. Soc. London 81:75–92.

Marchal, P. 1897. Notes d'entomologie biologiques sur une excursion en Algérie et en Tunisie. Mém. Soc. zool. France 10:5–25.

Marchand, W. 1917. Notes on the early stages of *Chrysops*. J. N. Y. Ent. Soc. 25:149–163.

Marchand, W. 1918. The larval stages of *Argyra albicans* Loew. Ent. News 29:216–220.

Marchand, W. 1920. The early stages of Tabanidae (horse flies). Monogr. Rockefeller Inst. Med. Res. 13:1–203.

Marchand, W. 1923. The larval stages of *Limnophora discreta* Stein. Bull. Brooklyn Ent. Soc. 18:58–62.

Marshall, A. G. 1981. The Ecology of Ectoparasitic Insects. Acad. Press, NY. 459 p.

Marshall, S. A. 1982a. A revision of *Halidayina* Duda (Diptera: Sphaeroceridae). Can. Ent. 114:841–847.

Marshall, S. A. 1982b. A revision of the Nearctic *Leptocera* (subgenus *Thoracochaeta*) (Diptera: Sphaeroceridae). Can. Ent. 114:63–78.

Marshall, S. A., and M. Eymann. 1981. Micro-organisms as food for the onion maggot, *Delia* (*Hylemya*) *antiqua* (Diptera: Anthomyiidae). Proc. Ent. Soc. Ontario 112:1–5.

Marshall, S. A., and O. W. Richards. 1987. Sphaeroceridae, pp. 993–1106. *In* McAlpine, J. F. et al. (eds.). Manual of Nearctic Diptera, Vol. 2. Res. Br. Agr. Can. Monogr. 28.

Marston, N. 1963. A revision of the Nearctic species of the *albofasciatus* group of the genus *Anthrax* Scopoli (Diptera: Bombyliidae). Kans. Agr. Exp. Sta. Bull. 127:3–79.

Marston, N. 1964. The biology of *Anthrax limatulus fur* (Osten Sacken), with a key to and descriptions of pupae of some species in the *Anthrax albofasciatus* and *trimaculatus* groups (Diptera: Bombyliidae). J. Kans. Ent. Soc. 37:89–105.

Marston, N. 1971. Taxonomic study of the known pupae of the genus *Anthrax* (Diptera: Bombyliidae) in North and South America. Smiths. Contr. Zool. 100:1–18.

Martin, C. H. 1968. The new family Leptogastridae (the grass flies) compared with the Asilidae (robber flies) (Diptera). J. Kans. Ent. Soc. 41:60–100.

Mason, A. C. 1922. Biology of the Papaya fruit fly, *Toxoptrypanea curvicauda*, in Florida (Trupaneidae). U.S. Dept. Agr. Bull. 1081.
Mason, W. T., Jr. 1973. An introduction to the identification of Chironomid larvae. MERC/EPA. 90 pp. Cincinnati.
Matheson, R. 1929. A Handbook of the Mosquitoes of North America. Thomas and Co., Springfield, Ill. 268 p.
Mathis, W. N. 1973. A review of the genus *Borboropsis*. Pan-Pacif. Ent. 49:373–377.
Mathis, W. N. 1979a. Studies of Notiphilinae (Diptera: Ephydridae), I: Revision of the Nearctic species of *Notiphila* Fallén, excluding the *caudata* group. Smiths. Contr. Zool. 287:1–108.
Mathis, W. N. 1979b. Studies of Ephydrinae (Diptera: Ephydridae), II: Phylogeny, classification, and zoogeography of Nearctic *Lamproscatella* Hendel. Smiths. Contr. Zool. 295:1–41.
Mathis, W. N. 1982a. Studies of Canacidae (Diptera). I: Suprageneric revision of the family, with revisions of the new tribe Dynomiellini and new genus *Isocanace*. Smithsonian Contr. Zool. No. 347, 29 pp.
Mathis, W. N. 1982b. Studies of Ephydrinae (Diptera: Ephydridae), VII: Revision of the genus *Setacera* Cresson. Smiths. Contr. Zool. 350:1–57.
Mathis, W. N., and A. Freidberg. 1983. New beach flies of the genus *Xanthocanace* Hendel with a review of the species from the western Palaearctic (Diptera: Canaceidae). Mem. Ent. Soc. Wash. 10:97–104.
Mathis, W. N., and K. W. Simpson. 1981. Studies of Ephydrinae (Diptera: Ephydridae), V: The genera *Cirrula* Cresson and *Dimecoenia* Cresson in North America. Smiths. Contr. Zool. 329:1–51.
Mathis, W. N. and G. C. Steyskal. 1980. A revision of the genus *Oedoparena* Curran (Diptera: Dryomyzidae: Dryomyzinae). Proc. Ent. Soc. Wash. 82:349–359.
Matile, L. 1970. Les diptères cavernicoles. Ann. Spéléol. 25:179–222.
Matsuda, R. 1965. Morphology and evolution of the insect head. Mem. Amer. Ent. Soc. 4:1–334.
Matsuda, R. 1976. Morphology and evolution of the insect abdomen. Pergamon Press, New York, Oxford, Toronto.
Maw, M. G. and Coppel, H. C. 1953. Studies on dipterous parasites of the spruce budworm, *Choristoneura fumiferana* (Clem.) (Lepidoptera: Tortricidae). II. *Phryxe pecosensis* (Tns.) (Diptera: Tachinidae). Can. J. Zool. 31:392–403.
May, B. M. 1961. The occurrence in New Zealand and the life history of the soldier fly *Hermetia illucens* (L.) (Diptera: Stratiomyiidae). New Z. J. Sci. 4:55–65.
Mayer, K. 1934. Die Metamorphose der Ceratopogonidae. Ein Beitrag zur Morphologie, Systematik, Ökologie und Biologie der Jugendstadien dieser Dipterenfamilie. Arch. Naturg. 3:205–288.
Mazza, S. and M. E. Jorg. 1939. *Cochliomyia hominivorax americana* C. & P., estudie de sus larvas y consideraciones sobre miasis. Univ. Buenos Aires Mis. Estud. Pat. Reg. Argent. Publ. 41:3–46.
McAlpine, D. K. 1960. A review of the Australian species of Clusiidae (Diptera, Acalyptrata). Rec. Australian Mus. 25:63–94.
McAlpine, D. K. 1966. Description and biology of an Australian species of Cypselosomatidae (Diptera), with a discussion of family relationships. Aust. J. Zool. 14:673–685.
McAlpine, D. K. 1971. Status and synonymy of the genus *Craspedochaeta* Czerny (Diptera: Clusiidae). J. Australian. Ent. Soc. 10:121–122.
McAlpine, D. K. 1973. The Australian Platystomatidae (Diptera: Schizophora) with a revision of five genera. Australian Mus. Mem. 15:1–256.
McAlpine, D. K. 1978. Description and biology of a new genus of flies related to *Anthoclusia* and representing a new family (Diptera, Schizophora, Neurochaetidae). Ann. Natal Mus. 23:273–295.
McAlpine, D. K., and D. S. Kent. 1982. Systematics of *Tapeigaster* (Diptera: Heleomyzidae) with notes on biology and larval morphology. Proc. Linn. Soc. N. S. Wales. 106:33–58.
McAlpine, J. F. 1956. Cone-infesting lonchaeids of the genus *Earomyia* Zett., with descriptions of five new species from western North America (Diptera: Lonchaeidae). Can. Ent. 88:178–196.
McAlpine, J. F. 1960a. A new species of *Leucopis* (*Leucopella*) from Chile and a key to the world genera and subgenera of Chamaemyiidae (Diptera). Can. Ent. 92:51–58.
McAlpine, J. F. 1960b. First record of the family Camillidae in the New World (Diptera). Can. Ent. 92:954–956.
McAlpine, J. F. 1961. A new species of *Dasiops* (Diptera: Lonchaeidae) injurious to apricots. Can. Ent. 93:539–544.
McAlpine, J. F. 1962. A revision of the genus *Campichoeta* Macquart (Diptera: Diastatidae). Can. Ent. 94:1–10.
McAlpine, J. F. 1963. Relationships of *Cremifania* Czerny (Diptera: Chamaemyiidae) and description of a new species. Can. Ent. 95:239–253.
McAlpine, J. F. 1971. A revision of the subgenus *Neoleucopis* (Diptera: Chamaemyiidae). Can. Ent. 103:1851–1874.
McAlpine, J. F. 1976. Systematic position of the genus *Omomyia* Coquillett and its transference to the Richardiidae (Diptera). Can. Ent. 108:849–854.
McAlpine, J. F. 1977. A revised classification of the Piophilidae, including "Neottiophilidae" and "Thyreophoridae" (Diptera: Schizophora). Mem. Ent. Soc. Can. 103:1–66.
McAlpine, J. F. 1978. A new dipterous predator of balsam woolly aphid from Europe and Canada (Diptera: Chamaemyiidae). Ent. Ger. 4:349–355.
McAlpine, J. F. 1987a. Cypselosomatidae, pp. 757–760. *In* McAlpine, J. F., et al., eds., Manual of Nearctic Diptera. Vol. 2. Res. Br. Agr. Can. Monogr. 28.
McAlpine, J. F. 1987b. Lonchaeidae, pp. 791–797. *In* McAlpine, J. F., et al., eds., Manual of Nearctic Diptera. Vol. 2. Res. Br. Agr. Can. Monogr. 28.
McAlpine, J. F. 1987c. Pallopteridae, pp. 839–843. *In* McAlpine, J. F., et al., eds., Manual of Nearctic Diptera. Vol. 2. Res. Br. Agr. Can. Monogr. 28.
McAlpine, J. F. 1987d. Piophilidae, pp. 845–852. *In* McAlpine, J. F., et al., eds., Manual of Nearctic Diptera. Vol. 2. Res. Br. Agr. Can. Monogr. 28.
McAlpine, J. F. 1987e. Acartophthalmidae, pp. 859–861. *In* McAlpine, J. F., et al., eds., Manual of Nearctic Diptera. Vol. 2. Res. Br. Agr. Can. Monogr. 28.
McAlpine, J. F. 1987f. Odiniidae, pp. 863–867. *In* McAlpine, J. F., et al., eds., Manual of Nearctic Diptera. Vol. 2. Res. Br. Agr. Can. Monogr. 28.
McAlpine, J. F. 1987g. Periscelididae. pp. 895–898. *In* McAlpine, J. F., et al., eds., Manual of Nearctic Diptera. Vol. 2. Res. Br. Agr. Can. Monogr. 28.
McAlpine, J. F. 1987h. Chamaemyiidae, pp. 965–971. *In* McAlpine, J. F., et al., eds., Manual of Nearctic Diptera. Vol. 2. Res. Br. Agr. Can. Monogr. 28.
McAlpine, J. F. 1987i. Chyromyidae, pp. 985–988. *In* McAlpine, J. F., et al., eds., Manual of Nearctic Diptera. Vol. 2. Res. Br. Agr. Can. Monogr. 28.
McAlpine, J. F. 1987j. Rhinotoridae, pp. 989–992. *In* McAlpine, J. F., et al., eds., Manual of Nearctic Diptera. Vol. 2. Res. Br. Agr. Can. Monogr. 28.
McAlpine, J. F. 1987k. Curtonotidae, pp. 1007–1010. *In* McAlpine, J. F., et al., eds., Manual of Nearctic Diptera. Vol. 2. Res. Br. Agr. Can. Monogr. 28.
McAlpine, J. F. 1987l. Diastatidae, pp. 1019–1022. *In* McAlpine, J. F., et al., eds., Manual of Nearctic Diptera. Vol. 2. Res. Br. Agr. Can. Monogr. 28.
McAlpine, J. F. 1987m. Camillidae, pp. 1023–1025. *In* McAlpine, J. F., et al., eds., Manual of Nearctic Diptera. Vol. 2. Res. Br. Agr. Can. Monogr. 28.
McAlpine, J. F. 1987n. Cryptochetidae, pp. 1069–1072. *In* McAlpine, J. F., et al., eds., Manual of Nearctic Diptera. Vol. 2. Res. Br. Agr. Can. Monogr. 28.

McAlpine J. F., and G. Morge. 1970. The identity, distribution, and biology of *Lonchaea zetterstedti* with notes on related species (Diptera: Lonchaeidae). Can. Ent. 102:1559–66.

McAlpine, J. F., B. V. Peterson, G. E. Shewell, H. J. Teskey, J. R. Vockeroth and D. M. Wood, Coordinators. 1981. Manual of Nearctic Diptera. Vol. 1. Res. Br. Agr. Can. Monogr. 27, pp. 1–674.

McAlpine, J. F. (Editor), B. V. Peterson, G. E. Shewell, H. J. Teskey, J. R. Vockeroth and D. M. Wood, (Coordinators). 1987. Manual of Nearctic Diptera, Vol. 2. Res. Br. Agr. Can. Monogr. 28, pp. 675–1332.

McAlpine, J. F., B. V. Peterson, G. E. Shewell, H. J. Teskey, J. R. Vockeroth, and D. M. Wood. 1981. Introduction, pp. 1–7. *In* McAlpine, J. F., et al., eds., Manual of Nearctic Diptera. Vol. 1. Res. Br. Agr. Can. Monogr. 27.

McAlpine, J. F., and C. Slight. 1981. The wheat bulb fly, *Delia coarctata,* in North America (Diptera: Anthomyiidae). Can. Ent. 113:615–21.

McAlpine, J. F., and G. C. Steyskal. 1982. A revision of *Neosilba* with a key to the world genera of Lonchaeidae (Diptera). Can. Ent. 114:105–138.

McAtee, W. L. 1911. Facts in the life-history of *Goniops chrysocoma.* Proc. Ent. Soc. Wash. 13:21–29.

McAtee, 1927. Bird nests as insect and arachnid hibernacula. Proc. Ent. Soc. Wash. 29:180–184.

McCauley, V. J. E. 1974. Instar differentiation in larval Chironomidae (Diptera). Can. Ent. 106:179–200.

McDonald, J. F., W. B. Heed, and M. Miranda. 1979. The larval nutrition of *Minettia flaveola* and *Phaonia parviceps* and its significance to the Hawaiian leaf-breeding *Drosophila* (Diptera: Lauxaniidae, Muscidae, Drosophilidae). Pan-Pacif. Ent. 50:78–82.

McFadden, M. W. 1967. Soldier fly larvae in America north of Mexico. Proc. U.S. Nat. Mus. 121:1–72.

McFadden, M. W. 1971. Two new species of *Ptecticus* with a key to species occurring in America north of Mexico. Pan-Pacif. Ent. 47:94–100.

McFadden, M. W. 1972. The soldier flies of Canada and Alaska (Diptera: Stratiomyiidae). 1. Beridinae, Sarginae, and Clitellarinae. Can. Ent. 104:531–562.

McFadden, M. W., and R. H. Foote. 1960. The genus *Orellia* in America north of Mexico. Proc. Ent. Soc. Wash. 62:253–261.

McGhehey, J. H., and W. P. Nagel. 1966. A technique for rearing larvae of *Medetera aldrichi* (Diptera: Dolichopodidae). Ann. Ent. Soc. Amer. 59:290–292.

Meade, A. B., and E. F. Cook. 1961. Notes on the biology of *Scatopse fuscipes* Meigen. Ent. News 72:13–18.

Meier, P. G., and H. C. Torres. 1978. A modified method for rearing midges (Diptera: Chironomidae). Great Lakes Ent. 11:89.

Meijere, J. C. H. de. 1895. Über zusammengesetzte Stigmen bei Dipteren-Larven nebst einem Beitrag zur Metamorphose von *Hydromyza livens*. Tijdschr. Ent. 38:65–100.

Meijere, J. C. H. de. 1900. Über die Larve von *Lonchoptera*. Zool. Jahrb. Syst. 14:87–132.

Meijere, J. C. H. de. 1909. Zur Kenntnis der Metamorphose der Lauxaniinae. Zeitschr. Wiss. Ins. Biol. 5:152–155.

Meijere, J. C. H. de. 1910. Nepenthes-Tiere. Ann. Jard. Bot. Buitenzorg (2) Suppl. 3:917–940.

Meijere, J. C. H. de. 1911. Über in Farnen parasitierende Hymenopteren- und Dipterenlarven. Tijdschr. Ent. 54:80–127.

Meijere, J. C. H. de. 1916. Beiträge zur Kenntnis der Dipterenlarven und -puppen. Zool. Jahrb. Syst. 40:177–322.

Meijere, J. C. H. de. 1940a. Über die Larven der in Orchideen minierenden Dipteren. Tijd. Ent. 83:122–127.

Meijere, J. C. H. de. 1940b. *Hydromyza livens* Fall. en *Notiphila brunnipes* Rob.-Desv. twee Dipteren, wier levenswijze verband houdt met *Nymphaea alba* L. Ent. Ber. 10:220–222.

Meijere, J. C. H. de. 1941. Puparien van *Loxocera* in Stengels van *Juncus*. Ent. Ber. 10:286–287.

Meijere, J. C. H. de. 1944. Over de Metamorphose van *Metopis leucocephala* Rossi, *Cacoxenus indagator* Low., *Palloptera saltuum* L., *Paranthomyza nitida* Meig. en *Hydrellia nigripes* Zett. Tijdschr. Ent. 86:57–61.

Meijere, J. C. H. de. 1947. Over eenige Dipterenlarven, waaronder een galmug, die mijngangen maakt, en twee Dipteren, die Gallen op Paddenstoelen veroorzaken. Tijdschr. Ent. 88:49–62.

Meijere, J. C. H. de. 1955. Die Larven der Agromyziden. Tijdschr. Ent. 98:1–27.

Melander, A. L. 1913. A synopsis of the dipterous groups Agromyzinae, Milichiinae, Ochthiphilinae and Geomyzinae. J. N. Y. Ent. Soc. 21:219–273, 285–300.

Melander, A. L. 1916. The dipterous family Scatopsidae. Bull. Wash. Agr. Exp. Sta. 130:1–21.

Melander, A. L. 1947. Synopsis of the Hemerodromiinae (Diptera, Empididae). J. N. Y. Ent. Soc. 55:237–273.

Melander, A. L. 1952. The American species of Trixoscelidae. J. N. Y. Ent. Soc. 60:37–52.

Melander, A. L., and N. G. Argo. A revision of the two-winged flies of the family Clusiidae. Proc. U.S. Nat. Mus. 64:1–51.

Melin, D. 1923. Contributions to the knowledge of the biology, metamorphosis and distribution of the Swedish asilids in relation to the whole family of asilids. Zool. Bidr. Upps. 8:1–317.

Merrill, L. S., Jr. 1951. Diptera reared from Michigan onions growing from seed. J. Econ. Ent. 44:1015.

Merritt, R. W. 1976. A review of the food habits of the insect fauna inhabiting cattle droppings in north central California. Pan. Pacif. Ent. 52:13–22.

Merritt, R. W. and Anderson, J. R. 1977. The effects of different pasture and rangeland ecosystems on the annual dynamics of insects in cattle droppings. Hilgardia 45:31–71.

Merritt, R. W., and M. T. James. 1973. The Micropezidae of California (Diptera). Bull. Calif. Insect Surv. 14:1–29.

Merritt, R. W. and K. W. Cummins, eds. 1984. An Introduction to the Aquatic Insects of North America. 2nd ed. Kendall/Hunt Publ. Co., Dubuque, Ia. 722 pp.

Merritt, R. W., and E. I. Schlinger. 1984. Aquatic Diptera. Part Two. Adults of Aquatic Diptera, pp. 259–283. *In* Merritt, R. W., and K. W. Cummins, eds., An Introduction to the Aquatic Insects of North America. Kendall/Hunt Publ. Co., Dubuque, Iowa.

Mesnil, L. 1934. A propos de deux Diptères nouveaux de la famille des Opomyzidae. Rev. franc. Ent. 1:191–207.

Metcalf, C. L. 1911a. A preliminary report on the life histories of two species of Syrphidae. Ohio Nat. 11:337–344.

Metcalf, C. L. 1911b. Life histories of Syrphidae II. Ohio Nat. 12:397–404.

Metcalf, C. L. 1912a. Life histories of Syrphidae III. Ohio Nat. 12:477–488.

Metcalf, C. L. 1912b. Life histories of Syrphidae IV. Ohio Nat. 12:533–541.

Metcalf, C. L. 1913a. Life histories of Syrphidae V. Ohio Nat. 13:81–95.

Metcalf, C. L. 1913b. The Syrphidae of Ohio. Ohio State Univ. Bull. 17, 31, pp. 1–123.

Metcalf, C. L. 1916. Syrphidae of Maine. Maine Agr. Exp. Sta. Bull. 253:193–264.

Metcalf, C. L. 1917. Syrphidae of Maine, Second Report. Maine Agr. Exp. Sta. Bull. 263:153–176.

Miall, L. E., and Taylor, T. H. 1907. The structure and the life-history of the holly fly. Trans. Ent. Soc. London 1907:259–283.

Michelsen, V. 1980. A revision of the beet leaf-miner complex, *Pegomyia hyoscyami* s. l. (Diptera: Anthomyiidae) Ent. Scand. 11:297–309.

Michelsen, V. 1988. A world revision of *Strobilomyia* gen. n.: the anthomyiid seed pests of conifers (Diptera: Anthomyiidae). Syst. Ent. 13:271–314.

Middlekauf, W. W. 1959. Some biological observations on *Sarcophaga falciformis,* a parasite of grasshoppers (Diptera: Sarcophagidae). Ann. Ent. Soc. Amer. 52:724–728.

Middlekauf, W. W., and R. S. Lane. 1980. Adult and immature Tabanidae (Diptera) of California. Bull. Calif. Insect Surv. 22:1–99.

Miles, M. 1950. Studies of British anthomyiid flies. I. Biology and habits of the bean seed flies, *Chortophila cilicrura* (Rond.) and *C. trichodactyla* (Rond.). Bull. Ent. Res. 41:343–354.

Miles, M. 1953. Studies of British anthomyiid flies. IV. Biology of the spinach stem fly, *Hylemyia echinata* (Séguy). Bull. Ent. Res. 44:591–597.

Miller, D. 1932. The buccopharyngeal mechanism of a blow fly larva (*Calliphora quadrimaculata* Swed.). Parasitology 24:491–499.

Miller, R. B. 1941. Some observations on *Chaoborus punctipennis* Say (Diptera, Culicidae). Can. Ent. 73:37–39.

Miller, R. M. 1977a. Ecology of Lauxaniidae (Diptera: Acalyptratae). I. Old and new rearing records with biological notes and discussion. Ann. Natal. Mus. 23:215–238.

Miller, R. M. 1977b. Taxonomy and biology of the Nearctic species of *Homoneura* (Diptera: Lauxaniidae). I. Subgenera *Mallochomyza* and *Tarsohomoneura*. Iowa St. J. Res. 52:147–176.

Miller, R. M. 1977c. Taxonomy and biology of the Nearctic species of *Homoneura* (Diptera: Lauxaniidae). II. Subgenus *Homoneura*. Iowa St. J. Res. 52:177–252.

Miller, R. M., and B. A. Foote. 1975. Biology and immature stages of eight species of Lauxaniidae (Diptera). I. Biological observations. Proc. Ent. Soc. Wash. 77:308–328.

Miller, R. M., and B. A. Foote. 1976. Biology and immature stages of eight species of Lauxaniidae (Diptera). II. Descriptions of immature stages and discussion of larval feeding habits and morphology. Proc. Ent. Soc. Wash. 78:16–37.

Milne, D. L. 1961. The function of the sternal spatula in gall midges. Proc. R. ent. Soc. Lond., Ser. A, 36:126–131.

Mingo, T. M., and K. E. Gibbs. 1976. A record of *Palaeodipteron walkeri* Ide (Diptera: Nymphomyiidae) from Maine: a species and family new to the United States. Ent. News 87:184–185.

Mitchell, R. G., and J. K. Maksymov. 1977. Observations of predation on spruce gall aphids within the gall. Entomophaga 22:179–186.

Mitchell, R. G., and K. H. Wright. 1967. Foreign predator introductions for control of the balsam wooly aphid in the Pacific Northwest. J. Econ. Ent. 60:140–147.

Mohn, E. 1955. Beiträge zur Systematik der Larven der Itonididae (=Cecidomyiidae, Diptera). 1. Teil: Porricondylinae und Itonidinae Mitteleuropas. Zoologica 105 (1,2):1–247.

Mohr, C. O. 1943. Cattle droppings as ecological units. Ecol. Monogr. 13:275–298.

Morge, G. 1956. Über Morphologie und Lebensweise der bisher unbekannten Larven von *Palloptera usta* Meigen, *Palloptera ustulata* Fallén und *Stegana coleoptrata* Scopoli. Beitr. Ent. 6:124–137.

Morge, G. 1959. Monographie der palaearktischen Lonchaeidae (Diptera). Beitr. Ent. 9:323–371, 909–945.

Morge, G. 1962. Monographie der palaearktischen Lonchaeidae (Diptera). (contd.). Beitr. Ent. 12:381–434.

Morge, G. 1967. Die Lonchaeidae und Pallopteridae Österreichs und der angrenzenden Gebiete. 2. Teil: Die Pallopteridae. Naturk. Jb. Stadt Linz 1967:141–213

Morris, H. M. 1918. The larval and pupal stages of *Scatopse notata* L. Ann. Appl. Biol. 5:102–108.

Morris, H. M. 1921. The larval and pupal stages of the Bibionidae. Bull. Ent. Res. 12:221–232.

Morris, H. M. 1922. The larval and pupal stages of the Bibionidae-Part III. Bull. Ent. Res. 13(2):189–195.

Morris, R. F. 1986. Notes on an unusual habitat for overwintering European crane fly larvae (Diptera: Tipulidae) in Newfoundland. Can. Ent. 118:1205–7.

Muir, F. 1912. Two new species of *Ascodipteron*. Bull. Mus. Comp. Zool. 54:353–366.

Muirhead-Thompson, R. C. 1937. Observations on the biology and larvae of the Anthomyiidae. Parasitol. 29:273–358.

Mullen, G. R., Trauth, S. E. and Sellers, J. C. 1984. Association of a miltogrammine fly, *Eumacronychia nigricornis* Allen (Diptera: Sarcophagidae), with the brood burrows of *Sceloporus undulatus* (Latreille) (Reptilia: Lacertillia). J. Georgia Ent. Soc. 19:1–6.

Müller, G. W. 1922. Insektenlarven an Wurzeln von Wasserpflanzen. Mitt. Naturwiss. Verein Neuvorpommern Rugen. 48/49:30–50.

Müller, H. 1957. Leguminosenknollen als Nahrungsquelle heimischer Micropezidae-(Tylidae) Larven. Beitr. Ent. 7:247–267.

Munroe, D. D. 1974. The systematics, phylogeny and zoogeography of *Symmerus* Walker and *Australosymmerus* Freeman (Diptera: Mycetophilidae: Ditomyiinae). Mem. Ent. Soc. Can. 92:1–183.

Musso, J. J. 1978. Recherches sur le développement, la nutrition et l'écologie des Asilidae (Diptera-Brachycera). Thesis, PhD, Universite de Droit, d'Economie et des Sciences d'Aix-Marseille France. 312 pp.

Musso, J. J. 1981. Morphology and development of the immature stages of some robber flies (Diptera: Brachycera: Asilidae). Ent. Generalis 7:89–104.

Muttkowski, R. A. 1915. New insect life histories I. Bull. Wisconsin Nat. Hist. Soc. 13:109–122.

Muttkowski, R. A. 1927. A new and unusual insect record for North America. Bull. Brooklyn Ent. Soc. 22:245–249.

Nagatomi, A. 1958. Life history of the Japanese aquatic snipe flies (Diptera: Rhagionidae). Hyogo Univ. Agr. Sci. Rep. Ser. Agr. Biol. 3(2):113–140.

Nagatomi, A. 1960. Studies on the aquatic snipe flies of Japan. Part II. Descriptions of the eggs (Diptera, Rhagionidae). Mushi 34:1–3.

Nagatomi, A. 1961a. Studies in the aquatic snipe flies of Japan. Part III. Descriptions of the larvae (Diptera: Rhagionidae). Mushi 35:11–27.

Nagatomi, A. 1961b. Studies in the aquatic snipe flies of Japan. Part IV. Descriptions of the pupae (Diptera, Rhagionidae). Mushi 35:29–38.

Nagatomi, A. 1962. Studies in the aquatic snipe flies of Japan. Part V. Biological notes (Diptera, Rhagionidae). Mushi 36:103–150.

Nagatomi, A. 1975. Definition of Coenomyiidae (Diptera). III. Genera excluded from the family. Proc. Japan Acad. 51:462–466.

Nagatomi, A. 1977. Classification of lower Brachycera (Diptera). J. Nat. Hist. 11:321–335.

Nagel, W. P., and T. O. Fitzgerald. 1975. *Medetera aldrichii* larval feeding behavior and prey consumption (Diptera: Dolichopodidae). Entomophaga 20:121–127.

Namba, R. 1956. A revision of the flies of the genus *Rivellia* (Otitidae, Diptera) of America north of Mexico. Proc. U.S. Nat. Mus. 106:21–84.

Needham, J. G. 1902. A remarkable occurrence of the fly *Bibio fraternus* Loew. Amer. Nat. 36:181.

Needham, J. G. 1908. Notes on the aquatic insects of Walnut Lake. Michigan St. Bd. Geol. Surv. Ann. Rep. 1907, pp. 252–271.

Needham, J. G., and C. Betten. 1901. Aquatic insects in the Adirondacks. N.Y. State Mus. Bull. 47:383–612.

Neff, S. E. 1968. Observations on the immature stages of *Gimnomera cerea* and *G. incisurata* (Diptera: Anthomyiidae, Scatophaginae). Can. Ent. 100:74–83.

Neff, S. E. 1970. Observations on the life cycle and immature stages of *Neochirosia atrifrons* (Diptera: Scatophagidae). Can. Ent. 102:1088–1093.

Neff, S. E., and C. O. Berg. 1961. Observations on the immature stages of *Protodictya hondurana* (Diptera: Sciomyzidae). Bull. Brooklyn Ent. Soc. 56:46–56.

Neff, S. E., and C. O. Berg. 1962. Biology and immature stages of *Hoplodictya spinicornis* and *H. setosa* (Diptera: Sciomyzidae). Trans. Amer. Ent. Soc. 88:77–93.

Neff, S. E., and C. O. Berg. 1966. Biology and immature stages of malacophagous Diptera of the genus *Sepedon* (Sciomyzidae). Va. Agr. Exp. Sta. Bull. 566:1–113.

Neff, S. E., and T. Eisner. 1960. Note on two tachinid parasites of the walking stick, *Anisomorpha buprestoides* (Stoll). Bull. Brooklyn Ent. Soc. 55:101–103.

Neff, S. E., L. V. Knutson, and C. O. Berg. 1967. Biology of snail-killing flies from Africa and southern Spain (Sciomyzidae: *Sepedon*). Parasitol. 57:487–505.

Neff, S. E. and J. B. Wallace. 1969a. Observations on the immature stages of *Cordilura* (*Achaetella*) *deceptiva* and *C.* (*A.*) *varipes*. Ann. Ent. Soc. Amer. 62:775–785.

Neff, S. E., and J. B. Wallace. 1969b. Biology and description of immature stages of *Orthacheta hiritipes*, a predator of *Cordilura* spp. Ann. Ent. Soc. Amer. 62:785–790.

Nelson, J. M. 1968. Parasites and symbionts of nests of *Polistes* wasps. Ann. Ent. Soc. Amer. 61:1528–1539.

Nemenz, H. 1960. On the osmotic regulation of the larvae of *Ephydra cinerea* Jones (Dipt.). J. Insect Physiol. 4:38–44.

Neveu, A. 1977. Ecologie des larves d'Athericidae (Diptera: Athericidae) dans un ruisseau Pyrénées-atlantiques. II. Production, comparison of different calculation methods. Ann. Hydrobiol. 8:45–66.

Newcomer, E. J. 1930. Notes on the habits of a digger wasp and its inquiline flies. Ann. Ent. Soc. Amer. 23:552–563.

Newhouse, V. F., Walker, D. W. and James, M. T. 1955. The immature stages of *Sarcophaga cooleyi*, *S. bullata*, and *S. shermani* (Diptera: Sarcophagidae). J. Wash. Acad. Sci. 45:15–20.

Newkirk, M. R. 1970. Biology of the longtailed dance fly, *Rhamphomyia longicauda* (Diptera: Empididae); a new look at swarming. Ann. Ent. Soc. Amer. 63:1407–1412.

Newsom, L. D., E. P. Dunigan, C. E. Eastman, R. L. Hutchison, and R. M. McPherson. 1978. Insect injury reduces nitrogen fixation in soybeans. La. Agr. 21:15–16.

Newson, H. D. 1984. Culicidae, pp. 515–533. *In* Merritt, R. W. and K. W. Cummins, eds., An Introduction to the Aquatic Insects of North America. 2nd. Kendall/Hunt Publ. Co., Dubuque, Ia.

Newstead, R. 1906. On the life history of *Stomoxys calcitrans* Linn. J. Econ. Biol. 1:157–166.

Niblett, M. 1955. Some Diptera bred from fungi. Ent. Rec. J. Var. 67:151–152.

Nielsen, J. C. 1909. Jagttagelser over entoparasitiske Muscidelarver hos Arthropoder. Ent. Meddel. 4:1–27.

Nielsen, J. C. 1911. Undersogelser over entoparasitiske Muscidelarver hos Arthropoder I. Vidensk. Meddel. Dansk Naturh. For. 63:1–26.

Nielsen, J. C. 1916. Om *Gymnopeza*-Arternes Biologi. Vidensk. Meddel. Dansk Naturh. For. 67:133–136.

Nielsen, L. T., and D. M. Rees. 1961. An identification guide to the mosquitoes of Utah. Univ. Utah Biol. Ser. 12 (3):1–63.

Nijveldt, W. 1969. Gall midges of economic importance. Vol. 8. Gall midges-miscellaneous. Crosby, Lockwood & Son Ltd., London.

Norrbom, A. L. 1983. Four acalyptrate Diptera reared from dead horseshoe crabs. Ent. News 94:117–121.

Norrbom, A. L., and K. C. Kim. 1985. Taxonomy and phylogenetic relationships of *Copromyza* Fallén (*s.s.*) (Diptera: Sphaeroceridae). Ann. Ent. Soc. Amer. 78:331–347.

Novak, J. A., and B. A. Foote. 1968. Biology and immature stages of fruit flies: *Paroxyna albiceps* (Diptera: Tephritidae). J. Kans. Ent. Soc. 41:108–119.

Novak, J. A., and B. A. Foote. 1975. Biology and immature stages of fruit flies: The genus *Stenopa* (Diptera: Tephritidae). J. Kans. Ent. Soc. 48:42–52.

Novak, J. A., W. B. Stoltzfus, E. J. Allen, and B. A. Foote. 1967. New host records for North American fruit flies (Diptera: Tephritidae). Proc. Ent. Soc. Wash. 69:146–148.

Novak, J. A., R. D. Akre, and W. B. Garnett. 1977. Keys to adults and puparia of five species of *Microdon* (Diptera: Syrphidae) from eastern Washington and northern Idaho, with descriptions of a new species. Can. Ent. 109:663–668.

Nowakowski, J. T. 1962. Introduction to a systematic revision of the family Agromyzidae (Diptera) with remarks on host plant selection by these flies. Ann. Zool. (Polska Akad. Nauk, Inst. Zool.) 20:67–183.

Nowell, W. R. 1951. The dipterous family Dixidae in western North America (Insecta: Diptera). Microentomol. 16:187–270.

Nowell, W. R. 1963. Guide to the insects of Connecticut. Part VI. The Diptera or true flies of Connecticut. Eighth Fascicle. Dixidae. Bull. Conn. St. Geol. Nat. Hist. Surv. 93:85–111.

Nye, I. W. B. 1958. The external morphology of some dipterous larvae living in the Gramininae of Britain. Trans. Roy. Ent. Soc. London 110:411–487.

Ode, P. E., and J. G. Matthysse. 1967. Bionomics of the face fly, *Musca autumnalis* DeGeer. Mem. Cornell Univ. Agr. Exp. Sta. 402:1–91.

Oettingen, H. von. 1935. Die wichtigsten grasbewohnenden Fliegenlarven Nord-Deutschlands. Nachr. Schadl. Bek. 10:62–70.

O'Hara, J. E. 1982. Classification, phylogeny, and zoogeography of the North American species of *Siphona* (Diptera: Tachinidae). Quaest. Ent. 18:261–380.

O'Hara, J. E. 1988. Survey of first instars of the Siphonini (Diptera: Tachinidae). Ent. Scand. 18:367–382.

Okada, T. 1963. Caenogenetic differentiation of mouth hooks in drosophilid larvae. Evol. 17:84–98.

Okely, E. F. 1974. Description of puparia of 23 British species of Sphaeroceridae (Diptera, Acalyptratae). Trans. Roy. Ent. Soc. London 126:41–56.

Okuno, T. 1970. Immature stages of two species of the genus *Volucella* Geoffrey (Diptera, Syrphidae). Kontyu 38:268–270.

Oldroyd, H. 1964. The natural history of flies. Weidenfeld and Nicolson, London, 324 pp.

Oldroyd, H. 1969. The family Leptogastridae (Diptera). Proc. Roy. Ent. Soc. London (B) 38:27–31.

Oliveira Albuquerque, D. de 1947. Contribuicao ao conhecimento de *Charadrella malacophaoa* Lopes. 1938. Rev. Ent. 18:101–112.

Oliveira, S. J. de. 1958. Contribuicao para o conhecimento do genero *Dimecoenia* Cresson. 1916. IV. Descricao da larva e do pupario do *Dimecoenia* grumanni Oliveira, 1954 (Diptera, Ephydridae). Rev. Brasil. Biol. 18:167–169.

Oliver, D. R. 1968. Adaptations of arctic Chironomidae. Ann. Zool. Fenn. 5:111–118.

Oliver, D. R. 1971. Life history of the Chironomidae. Annu. Rev. Ent. 16:211–230.

Oliver, D. R. 1981. Chironomidae, pp. 423–458. *In* McAlpine, J. F., et al., eds., Manual of Nearctic Diptera. Vol. 1. Res. Br. Agr. Can. Monogr. 27.

Olsen, L. E., and R. E. Ryckman. 1963. Studies on *Odontoloxozus longicornis* (Diptera: Neriidae). Part I. Life history and descriptions of immature stages. Ann. Ent. Soc. Amer. 56:454–469.

Olsufjev, N. G. 1936. The microscopic anatomy of the head and alimentary tract of *Tabanus* larvae (in Russian). Parazit. Sb. 6:247–278.

Osborn, E. 1919. Report of the assistant entomologist. Fla. Agric. Exp. Sta. Rept. 1918–1919. 60–65.

Osgood, E. A., and R. J. Gagné. 1978. Biology and taxonomy of two gall midges (Diptera: Cecidomyiidae) found in galls on balsam fir needles with description of a new species of *Paradiplosis*. Ann. Ent. Soc. Amer. 71:85–91.

Osterberger, B. A. 1930. *Erax interruptus* Macq. as a predator. J. Econ. Ent. 23:709–711.

Packard, A. S. 1871. On insects inhabiting salt water. Amer. J. Sci. Arts 1:100–110.

Painter, R. H., and J. C. Hall. 1960. A monograph of the genus *Poecilanthrax* (Diptera: Bombyliidae). Tech. Bull. Kans. Exp. Sta. 106:1–132.

Palmer, M. K. 1982. Biology and behavior of two species of *Anthrax* (Diptera: Bombyliidae), parasitoids of the larvae of tiger beetles (Coleoptera: Cicindelidae). Ann. Ent. Soc. Amer. 75: 61–70.

Papavero, N. 1973. Studies of Asilidae (Diptera) systematics and evolution. I. A preliminary classification in subfamilies. Arq. Zool., São Paulo 23:217–274.

Papavero, N. 1977. The World Oestridae (Diptera). Mammals and Continental Drift. W. Junk, The Hague. viii+240 p.

Papavero, N. and J. Wilcox. 1974. Studies of Mydidae (Diptera) systematics and evolution. I. A preliminary classification of subfamilies with the descriptions of two new genera from the Oriental and Australian regions. II. Classification of the Mydinae, with description of a new genus and a revision of *Ceriomydas* Williston. Archos Zool. Est. São Paulo 25:1–60.

Pape, T. 1986. A phylogenetic analysis of the woodlouse-flies (Diptera, Rhinophoridae). Tijd. Entomol. 129:15–34.

Pape, T. 1987. Revision of Neotropical *Metopia* Meigen (Diptera: Sarcophagidae). Syst. Entomol. 12:81–101.

Parker, J. R. 1918. The life history and habits of *Chloropisca glabra* Meig., a predacious oscinid (Chloropidae). J. Econ. Ent. 11:368.

Parker, J. R., and C. Wakeland. 1957. Grasshopper egg pods destroyed by larvae of bee flies, blister beetles, and ground beetles. U.S. Dept. Agr. Tech. Bull. 1165:1–29.

Parker, L. B. 1934. Notes on the life history and biology of *Centeter unicolor* Aldrich. J. Econ. Ent. 27:486–491.

Parrella, M. P. 1987. Biology of *Liriomyza.* Annu. Rev. Ent. 32:201–24.

Patterson, J. T., and W. S. Stone. 1952. Evolution in the genus *Drosophila.* N.Y. 610 pp.

Patton, W. W. 1921. Notes on the myiasis-producing Diptera of man and animals. Bull. Ent. Res. 12:239–261.

Patton, W. W. 1923. Identification of larvae of the Diptera which cause nasopharyngeal and aureal myiasis in man. J. Laryng. Otol. 38:18.

Pechuman, L. L. 1937. An annotated list of insects found in the bark and wood of *Ulmus americanus* L. in New York State. Bull. Brooklyn Ent. Soc. 32:8–21.

Pechuman, L. L. 1938. A synopsis of the New World species of *Vermileo* (Diptera-Rhagionidae). Bull. Brooklyn Ent. Soc. 33:84–89.

Pechuman, L. L. 1943. Notes on the biology of *Chyliza notata* (Diptera, Psilidae). Bull. Brook. Ent. Soc. 38:97.

Pechuman, L. L., and H. J. Teskey. 1981. Tabanidae, pp. 463–478. *In* McAlpine, J. F., et al., eds., Manual of Nearctic Diptera. Vol. 1. Res. Br. Agr. Can. Mongr. 27.

Peckham, D. J., Kurczewski, F. E. and Peckham, D. B. 1973. Nesting behavior of Nearctic species of *Oxybelus* (Hymenoptera: Sphecidae). Ann. Ent. Soc. Amer. 66:647–661.

Pelt, A. F. van, and S. A. van Pelt. 1972. *Microdon* (Diptera: Syrphidae) in nests of *Monomorium* (Hymenoptera: Formicidae) in Texas. Ann. Ent. Soc. Amer. 65:977–978.

Pennack, R. W. 1945. Notes on mountain midges (Deuterophlebiidae) with a description of the immature stages of a new species from Colorado. Amer. Mus. Novitates 1276:1–10.

Pennak, R. W. 1972. Fresh-water invertebrates of the United States. Wiley Interscience, N.Y., 803 pp.

Pergande, T. 1901. The ant-decapitating fly. Proc. Ent. Soc. Wash. 4:497–501.

Perris, E. 1870. Histoire des insectes du pin maritime. Diptères. Ann. Soc. ent. France 4:135–232, 321–366.

Persson, P. I. 1963. Studies on the biology and larval morphology of some Trypetidae (Dipt.) Opusc. Ent. 28:33–69.

Peters, T. M. 1981. Dixidae, pp. 329–333. *In* McAlpine, J. F., et al., eds., Manual of Nearctic Diptera. Vol. 1. Res. Br. Agr. Can. Monogr. 27.

Peters, T. M., and E. F. Cook. 1966. The Nearctic Dixidae (Diptera). Misc. Publ. Ent. Soc. Amer. 5:231–278.

Peterson, A. 1923. The pepper maggot, a new pest of peppers and eggplants. *Spilographa electa* Say (Trypetidae). Bull. N.J. Agr. Exp. Stn. 373:4–23.

Peterson, A. 1951. Larvae of Insects. Part II. Published by the author. Columbus Oh. 416 p.

Peterson, B. V. 1960. New distribution and host records for bat flies, and a key to the North American species of *Basilia* Ribeiro (Diptera: Nycteribiidae). Proc. Ent. Soc. Ont. 90:30–37.

Peterson, B. V. 1970. The *Prosimulium* of Canada and Alaska (Diptera: Simuliidae). Mem. Ent. Soc. Can. 69:1–216.

Peterson, B. V. 1971. Notes on the bat flies of Costa Rica (Diptera: Nycteribiidae). Contr. Sci. 212:1–8.

Peterson. B. V. 1977. A synopsis of the genus *Parasimulium* Malloch (Diptera: Simuliidae), with descriptions of one new subgenus and two new species. Proc. Ent. Soc. Wash. 79:96–106.

Peterson, B. V. 1981a. Anisopodidae, pp. 305–312. *In* McAlpine, J. F., et al., eds. Manual of Nearctic Diptera. Vol. 1. Res. Br. Agr. Can. Monogr. 27.

Peterson, B. V. 1981b. Simuliidae, pp. 335–391. *In* McAlpine, J. F., et al., eds., Manual of Nearctic Diptera. Vol. 1. Res. Br. Agr. Can. Monogr. 27.

Peterson, B. V. 1981c. Apioceridae, pp. 541–548. *In* McAlpine, J. F., et al., eds., Manual of Nearctic Diptera. Vol. 1. Res. Br. Agr. Can. Monogr. 27.

Peterson. B. V. 1984. Family Simuliidae, pp. 534–550. *In* Merritt, R. W. and K. W. Cummins, eds., An Introduction to the Aquatic Insects of North America. Kendall/Hunt, Dubuque, Ia.

Peterson, B. V. 1987a. Lonchopteridae, pp. 675–680. *In* McAlpine, J. F., et al., eds., Manual of Nearctic Diptera. Vol. 2. Res. Br. Agr. Can. Monogr. 28.

Peterson, B. V. 1987b. Phoridae, pp. 689–712. *In* McAlpine, J. F., et al., eds., Manual of Nearctic Diptera. Vol. 2. Res. Br. Agr. Can. Monogr. 28.

Peterson, B. V. 1987c. Diopsidae, pp. 785–789. *In* McAlpine, J. F., et al., eds., Manual of Nearctic Diptera. Vol. 2. Res. Br. Agr. Can. Monogr. 28.

Peterson, B. V. 1987d. Braulidae, pp. 913–918. *In* McAlpine, J. F., et al., eds., Manual of Nearctic Diptera. Vol. 2. Res. Br. Agr. Can. Monogr. 28.

Peterson, B. V., and E. F. Cook. 1981. Synneuridae, pp. 321–324. *In* McAlpine, J. F., et al., eds., Manual of Nearctic Diptera. Vol. 2. Res. Br. Agr. Can. Monogr. 27.

Peterson, B. V., and R. L. Wenzel. 1987. Nycteribiidae, pp. 1283–1291. *In* Manual of Nearctic Diptera. Vol. 2. Res. Br. Agr. Can. Monogr. 28.

Peus, F. 1958. Liriopeidae. Lfg. 200, pp. 10–44. *In* Lindner, E., ed., Die Fliegen der palaearktischen Region. Schweizerbart, Stuttgart.

Philip, C. B. 1928. Methods of collecting, and rearing the immature stages of Tabanidae. J. Parasitol. 14:243.

Philip, C. B. 1931. The Tabanidae (horseflies) of Minnesota, with special reference to their biologies and taxonomy. Minn. Agr. Exp. Sta. Tech. Bull. 80:1–132.

Philip, C. B. 1965. Family Pelecorhynchidae, p. 319. *In* Stone, A. et al., eds., A Catalog of the Diptera of America north of Mexico. U.S.D.A. Agr. Handb. 276.

Phillips, V. T. 1923. A revision of the Trypetidae of northeastern America. J. N.Y. Ent. Soc. 31:119–56.

Phillips, V. T. 1946. The biology and identification of trypetid larvae. Mem. Amer. Ent. Soc. 12:1–161.

Pickens, L. G. 1981. The life history and predatory efficiency of *Ravinia lherminieri* (Diptera: Sarcophagidae) on the face fly (Diptera: Muscidae). Can. Ent. 113:523–526.

Pictorial keys (to) arthropods of public health importance. 1962. Center for Disease Control, U.S. Public Health Service, Atlanta, GA., 192 pp.

Pinder, L. C. V. 1986. Biology of freshwater Chironomidae. Annu. Rev. Ent. 31:1–23.

Ping, C. 1921. The biology of *Ephydra subopaca* Loew. Mem. Cornell Univ. Agr. Exp. Sta. 49:555–616.

Piper, G.L. 1976. Bionomics of *Euarestoides acutangulus* (Diptera: Tephritidae). Ann. Ent. Soc. Amer. 69:381–386.

Pipkin, S. B., R. L. Rodrigues, and J. Leon. 1966. Plant host specificity among flower feeding Neotropical *Drosophila* (Diptera: Drosophilidae). Amer. Nat. 100:135–155.

Plachter, H. 1979a. Zur Kenntnis der Praeimaginalstadien der Pilzmücken (Diptera, Mycetophiloidea). Teil I: Gespinstbau. Zool. Jahrb. (Abt. Anat. Ontog. Tiere) 101:168–266.

Plachter, H. 1979b. Zur Kenntnis der Praeimaginalstadien der Pilzmücken (Diptera, Mycetophiloidea). Teil II: Eidonomie der Larven. Zool. Jahrb. (Abt. Anat. Ontog. Tiere) 101:271–392.

Plachter, H. 1979c. Zur Kenntnis der Praeimaginalstadien der Pilzmücken (Diptera, Mycetophiloidea). Teil III: Die Puppen. Zool. Jahrb. (Abt. Anat. Ontog. Tiere) 101:427–455.

Poinar, G. O. 1977. Observations on the kelp fly, *Coelopa vanduzeei* Cresson in southern California (Diptera: Coelopidae). Pan-Pacif. Ent. 53:81–86.

Pont, A. C. 1986. Muscidae. *In:* Soos, A. and Papp, L. eds. Catalogue of Palaearctic Diptera. Akademiai Kiado. Budapest.

Portschinsky, J. 1910. Recherches biologiques sur le *Stomoxys calcitraus* L. et biologie comparée des mouches coprophages. Work in Applied Entomology. Chem. Bur. Entomol. Sci. Comm. Central Bd. Land. Administr. Agr. 8, 8:63 pp.

Pratt, G. K., and H. D. Pratt. 1980. Notes on Nearctic *Sylvicola* (Diptera: Anisopodidae). Proc. Ent. Soc. Wash. 82:86–98.

Pratt, H. D., and G. K. Pratt. 1984. The winter crane flies of the eastern United States. Proc. Ent. Soc. Wash. 86:249–265.

Pratt, H. S. 1893. Beiträge zur Kenntnis der Pupiparen (Die Larve von *Melophagus ovinus*). Archiv Naturgesch. 59:151–200.

Prebble, M. L. 1935. *Actia diffidens* Curran, a parasite of *Peronea variana* (Fernald) in Cape Breton, Nova Scotia. Can. J. Res. 12:216–227.

Prescott, H. W. 1955. *Neorhynchocephalus sackenii* and *Trichopsidea clausa,* nemestrinid parasites of grasshoppers. Ann. Ent. Soc. Amer. 48:392–402.

Prescott, H. W. 1960. Suppression of grasshoppers by nemestrinid parasites (Diptera). Ann. Ent. Soc. Amer. 53:513–521.

Prins, A. J. 1982. Morphological and biological notes on six South African blow-flies (Diptera, Calliphoridae) and their immature stages. Ann. S. Afr. Mus. 90:201–217.

Pritchard, A. E. 1947. The North American gall midges of the tribe Micromyini, Itonididae (Cecidomyiidae) Diptera. Ent. Amer. 27:1–87.

Pritchard, A. E. 1948. The North American gall midges of the tribe Catotrichini and Catochini [Diptera: Itonididae (Cecidomyiidae)]. Ann. Ent. Soc. Amer. 40:662–671.

Pritchard, A. E. 1951. The North American gall midges of the tribe Lestremiini, Itonididae (Cecidomyiidae) Diptera. Univ. Calif. Publ. Ent. 8:239–275.

Pritchard, A. E. 1960. A new classification of the paedogenetic gall midges formerly assigned to the subfamily Heteropezinae (Diptera: Cecidomyiidae). Ann. Ent. Soc. Amer. 53:305–316.

Pritchard, G. 1983. Biology of Tipulidae. Annu. Rev. Ent. 28:1–22.

Pritchard, G., and H. A. Hall. 1971. An introduction to the biology of craneflies in a series of abandoned beaver ponds, with an account of the life cycle of *Tipula sacra* Alexander (Diptera: Tipulidae). Can. J. Zool. 49:467–482.

Quate, L. W. 1955. A revision of the Psychodidae (Diptera) in America north of Mexico. Univ. Calif. Publ. Ent. 10:103–273.

Quate, L. W. 1960. Guide to the Insects of Connecticut. Part VI. The Diptera or True Flies of Connecticut. Seventh Fascicle. Psychodidae. Bull. Conn. St. Geol. Nat. Hist. Surv. 92:1–48.

Quate, L. W., and J. R. Vockeroth. 1981. Psychodidae, pp. 293–300. *In* McAlpine, J. F., et al., eds., Manual of Nearctic Diptera. Vol. 1. Res. Br. Agr. Can. Monogr. 27.

Quayle, H. J. 1929. The Mediterranean and other fruit flies. Circ. Calif. Agr. Exp. Sta. 315, 19 p.

Quist, J. A., and M. T. James. 1973. The genus *Euparyphus* in America north of Mexico with a key to the new world genera and subgenera of Oxycerini (Diptera: Stratiomyidae). Wash. St. Ent. Soc. 11:1–26.

Ravlin, F. W. 1977. Retrieval of first instar tachinid larvae from pinned specimens. Proc. Ent. Soc. Wash. 79:518–520.

Ravlin, F. W. and Stehr, F. W. 1984. Revision of the genus *Archytas* (Diptera: Tachinidae) for America north of Mexico. Misc. Publ. Ent. Soc. Amer. No. 58, 59 pp.

Rees, N. E. 1973. Arthropod and nematode parasites, parasitoids and predators of Acrididae in America north of Mexico. U.S. Dept. Agr. Tech. Bull. 1460, 288 p.

Reissig, W. H., and D. C. Smith. 1978. Bionomics of *Rhagoletis pomonella* in *Crataegus*. Ann. Ent. Soc. Amer. 71:155–159.

Reiter, C., and G. Wollenek. 1982. Remarks on the morphology of maggots of forensically important flies. Z. Rechtsmed. 89:197–206.

Rennie, J., and C. H. Sutherland. 1920. On the life history of *Bucentes* (*Siphona*) *geniculata,* parasite of *Tipula paludosa* and other species. Parasitol. 12:199–211.

Richards, O. W. 1930. The British species of Sphaeroceridae (Borboridae). Proc. Roy. Ent. Soc. London 1930:261–345.

Richards, O. W. 1968. Sphaerocerid flies associating with doryline ants, collected by Dr. D. H. Kistner. Trans. Roy. Ent. Soc. London 120 (7):183–198.

Richards, P. G., amd F. O. Morrison. 1972. A summary of published information on the cluster fly, *Pollenia rudis* (Fabricius) (Diptera: Calliphoridae). Phytoprotection 53:103–111.

Richards, P. G., and F. O. Morrison. 1973. *Pollenia rudis* (Diptera: Calliphoridae) reared on *Allolobophora chlorotica* (Annelida: Lumbricidae). Phytoprotection 54:1–8.

Ries, D. F. 1934. Biological study of the walnut husk fly (*Rhagoletis suavis* Loew). Pap. Mich. Acad. Sci. 20:717–723.

Rikhter, V. A. and Farinets, S. I. 1983. The first instar larvae of tachinids of the subfamily Dexiinae (Diptera, Tachinidae) of the U.S.S.R. Ent. Rev. 62:142–165.

Riley, C. V. 1873. Descriptions and natural history of two insects which brave the dangers of *Sarracenia variolaris*. Trans. Acad. Sci. St. Louis 3:235–242.

Riley, C. V., and L. O. Howard. 1894. Larval food of *Euxesta notata* Wied. Insect Life 6:270.

Ristich, S. S. 1956. The host relationship of a miltogrammid fly *Senotainia trilineata* (VDW). Ohio J. Sci. 56:271–274.

Ritcher, P. O. 1940. Kentucky white grubs. Ky. Agr. Expt. Sta. Bull. 401:73–157 (Ky. Agr. Expt. Sta. Ann. Rpt. 53:73–157).

Roback, S. S. 1954. The evolution and taxonomy of the Sarcophaginae (Diptera: Sarcophagidae). Ill. Biol. Monogr. 23:1–181.

Roback, S. S. 1963. The genus *Xenochironomus* (Diptera: Tendipedidae) Kieffer, taxonomy and immature stages. Trans. Amer. Ent. Soc. 88:235–245.

Roback, S. S. 1966a. The immature stages of *Stictochironomus annulicius* (Townes) (Diptera, Tendipedidae). Ent. News 27:169–173.

Roback, S. S. 1966b. A new *Procladius* species with description of the immature stages (Diptera, Chironomidae). Ent. News 77:177–184.

Roback, S. S. 1969. The immature stages of the genus *Tanypus* Meigen. Trans. Amer. Ent. Soc. 94:407–428.

Roback, S. S. 1974. Insects (Arthropoda: Insecta), pp. 313–376. *In* Hart, C. W. and S. L. H. Fuller, eds., Pollution Ecology of Freshwater Invertebrates. Academic Press, N.Y.

Roback, S. S. 1976. The immature chironomids of the eastern United States: I. Introduction and Tanypodinae-Coleotanypodini. Proc. Acad. Nat. Sci. Phila. 127:147–201.

Roback, S. S. 1977a. The immature chironomids of the eastern United States: II. Tanypodinae: Tanypodini. Proc. Acad. Nat. Sci. Phila. 128:55–88.

Roback, S. S. 1977b. The immature chironomids of the eastern United States: III. Tanypodinae - Anatopyniini, Macropelopiini, and Natarsiini. Proc. Acad. Nat. Sci. Phila. 129:151.

Roback, S. S. 1981. The immature chironomids of the eastern USA: V. Pentaneurini Thienemannimyia group. Proc. Acad. Nat. Sci. Phila. 133:73–128.

Roberts, J. I. 1927. The anatomy and morphology of *H. equina.* Ann. Trop. Med. Parasitol. 21:11–24.

Roberts, M. J. 1969a. Pharyngeal grinding mills in some dipteran larvae. Entomologist 102:279–284.

Roberts, M. J. 1969b. Structure of the mouthparts of the larvae of the flies *Rhagio* and *Sargus* in relation to feeding habits. J. Zool. London. 159:381–398.

Roberts, M. J. 1970. The structure of the mouthparts of syrphid larvae (Diptera) in relation to feeding habits. Acta Zool. 51:43–65.

Roberts, M. J. 1971a. The structure of the mouthparts of some calypterate dipteran larvae in relation to their feeding habits. Acta. Zool., Stockh. 52:171–188.

Roberts, M. J. 1971b. The eye and its relationship to the cranial nervous system in the larva of the brachyceran *Rhagio scolopaceus* (L.) (Diptera, Rhagionidae). Proc. R. ent. Soc. Lond., Ser. A, 45(4–6):45–50.

Roberts, R. H., and R. J. Dicke. 1964. The biology and taxonomy of some immature Nearctic Tabanidae (Diptera). Ann. Ent. Soc. Amer. 57:31–40.

Robinson, H. 1964. A synopsis of the Dolichopodidae (Diptera) of the southeastern United States and adjacent regions. Misc. Publ. Ent. Soc. Amer. 4:105–192.

Robinson, H., and J. R. Vockeroth. 1981. Dolichopodidae, pp. 625–639. *In* McAlpine, J. F., et al., eds., Manual of Nearctic Diptera. Vol. 1. Res. Br. Agr. Can. Monogr. 27.

Robinson, W. H. 1971. Old and new biologies of *Megaselia* species (Diptera, Phoridae). Studia Ent. 14:321–348.

Robinson, W. H. 1975. *Megaselia* (M.) *scalaris* (Diptera: Phoridae) associated with laboratory cockroach colonies. Proc. Ent. Soc. Wash. 77:384–390.

Robinson, W. H. 1977. Phoridae (Diptera) associated with cultivated mushrooms in eastern North America. Proc. Ent. Soc. Wash. 79:452–462.

Robinson, W. H., and B. A. Foote. 1978. Biology and immature stages of *Antichaeta borealis* (Diptera: Sciomyzidae), a predator of snail eggs. Proc. Ent. Soc. Wash. 80:388–396.

Robles, C. D., and J. Cubit. 1981. Influence of biotic factors in an upper intertidal community: effects of grazing Diptera larvae on algae. Ecol. 62: 1536–1547.

Rodhain, J., and J. Bequaert. 1915. Observations sur la biologie de *Cyclopodia greeffii* Karsch, Nycteribiide parasite d'une chauve-souris congolaise. Bull. Soc. zool. France 40:248–262.

Rogers, C. E. 1977. Hosts and parasitoids of the Cecidomyiidae (Diptera) in the rolling plains of Texas. J. Kans. Ent. Soc. 50:179–186.

Rogers, J. S. 1926a. Some notes on the feeding habits of adult crane-flies. Fla. Ent. 10:5–7.

Rogers, J. S. 1926b. Notes on the biology and immature stages of *Gonomyia* (*Leiponeura*) *pleuralis* (Will.)-Tipulidae, Diptera. Fla. Ent. 10:33–38.

Rogers, J. S. 1927a. Notes on the biology of *Atarba picticornis* Osten Sacken-Tipulidae-Diptera. Fla. Ent. 10:49–54.

Rogers, J. S. 1927b. Notes on the life history, distribution and ecology of *Diotrepha mirabilis* Osten Sacken. Ann. Ent. Soc. Amer. 20:23–36.

Rogers, J. S. 1927c. Notes on the biology and immature stages of *Geranomyia* (Tipulidae, Diptera). I. *Geranomyia rostrata* (Say). Fla. Ent. 11:17–26.

Rogers, J. S. 1928. Notes on the biology of *Gnophomyia luctuosa* Osten Sacken with descriptions of the immature stages. Ann. Ent. Soc. Amer. 21:398–406.

Rogers, J. S. 1930. The summer crane-fly fauna of the Cumberland plateau in Tennessee. Occ. Pap. Mus. Zool. Univ. Mich. 215:1–50.

Rogers, J. S. 1932. On the biology of *Limonia* (*Dicranomyia*) *floridana* (Osten Sacken). Fla. Ent. 15:65–70.

Rogers, J. S. 1933a. The ecological distribution of the crane-flies of northern Florida. Ecol. Monogr. 3 (1):1–74.

Rogers, J. S. 1933b. Contributions toward a knowledge of the natural history and immature stages of the crane-flies. I. The genus *Polymera* Wiedemann. Occ. Pap. Mus. Zool. Univ. Mich. 268:1–13.

Rogers, J. S. 1942. The crane-flies (Tipulidae) of the George Reserve, Michigan. Misc. Publ. Mus. Zool. Univ. Mich. 53:1–128.

Rogers, J. S. 1949. The life history of *Meogistocera longipennis* (Macquart) (Tipulidae, Diptera), a member of the neuston fauna. Occ. Pap. Mus. Zool. Univ. Mich. 521:1–17.

Rogers, J. S., and G. W. Byers. 1956. The ecological distribution, life history, and immature stages of *Lipsothrix sylvia* (Diptera: Tipulidae). Occ. Pap. Mus. Zool. Univ. Mich. 572:1–14.

Rognes, K. 1984. Revision of the bird-parasitic blowfly genus *Trypocalliphora* Peus, 1960 (Diptera: Calliphoridae). Ent. Scand. 15:371–382.

Rognes, K. 1986. The systematic position of the genus *Helicobosca* Bezzi with a discussion of the monophyly of the calyptrate families Calliphoridae, Rhinophoridae, Sarcophagidae and Tachinidae (Diptera). Ent. Scand. 17:75–92.

Ronna, A. 1936. Observancoes biologicas sobre dois dipteros parasitas de *Apis mellifica* L. (Dipt. Phoridae, Sarcophagidae). Rev. de Ent. 6:1–9.

Root, F. M. 1923. Notes on the larval characters in the genus *Sarcophaga.* J. Parasitol. 9:227–229.

Rose, J. H. 1963. Supposed larva of *Protanyderus vipio* (Osten Sacken) discovered in California (Diptera: Tanyderidae). Pan-Pacif. Ent. 39:272–275.

Roskam, J. C. 1977. Biosystematics of insects living in female birch catkins. I. Gall midges of the genus *Semudobia* Kieffer (Diptera, Cecidomyiidae). Tijdschr. Ent. 120:153–197.

Roth, J. C., and S. Parma. 1970. A *Chaoborus* bibliography. Bull. Ent. Soc. Amer. 16:100–110.

Roubaud, E. 1901. Sur deux types de Diptères fucicoles, *Orygma luctuosa* Meig., *Coelopa pilipes* Hal. Bull. Soc. philom. (9) III:77–80.

Roubaud, E. 1903. Sur des larves marines de Dolichopodes atribuées au genre *Aphrosilus.* Bull. Mus. Hist. nat. Paris 9:338–340.

Roubaud, E. 1913. Recherches sur les Auchméromyies. Bull. scient. France Belg. 47:105–202.

Roubaud, E. 1914. Oestrides gastricoles et cavicoles de l'Afrique occidentale francaise. Bull. Soc. Path. exot. 7:212–215.

Roubaud, E. 1915. Les Muscides à larves piqueuses et suceuses de sang. C. r. Séanc. Soc. Biol., Paris 78:92–97.

Roy, D. N. 1937. On the function of the pharyngeal ridges in the larvae of *Calliphora erythrocephala.* Parasitol. 29:143–149.

Rozkošný, R. 1965. Neue Metamorphosestadien mancher *Tetanocera* Arten (Diptera: Sciomyzidae). Zool. Listy 14:367–371.

Rozkošný, R. 1967. Zur Morphologie und Biologie der Metamorphosestadien mitteleuropäischer Sciomyziden (Diptera). Acta Acad. Sci. Nat. Brno 1 (4):117–160.

Rozkošný, R. 1968. Malacophagie als Lebensweise von Dipterenlarven in Mitteleuropa. Abh. Ber. Naturkundemus. Goerlitz. 44 (2):165–170.

Rozkošný, R. 1982. Series Entomologica, Vol. 21. A biosystematic study of the European Stratiomyidae (Diptera). Vol. 1. Introduction, Beridinae, Sarginae, and Stratiomyinae. W. Junk, The Hague, viii+40 p.

Rozkošný, R. 1983. Series Entomologica, Vol. 25. A biosystematic study of the European Stratiomyidae (Diptera). Vol. 2. W. Junk, The Hague. 436 pp.

Rozkošný, R., and L. V. Knutson. 1970. Taxonomy, biology, and immature stages of Palearctic *Pteromicra,* snail-killing Diptera (Sciomyzidae). Ann. Ent. Soc. Amer. 63:1434–1459.

Runyan, J. T., and D. T. Deonier. 1979. A comparative study of *Pseudohecamede* and *Allotrichoma* (Diptera: Ephydridae), pp. 123–137. *In* Deonier, D. L., ed., First symposium on the systematics and ecology of Ephydridae (Diptera). N. Amer. Benthol. Soc., 147 pp.

Ryberg, O. 1939. Beiträge zur Kenntnis der Fortpflanzungsbiologie und Metamorphose der Fledermausfliegen Nycteribiidae. Verh. VII. Intern. Kongr. Ent. 11:1285–1299.

Ryckman, R. E., and L. E. Olsen. 1963. Studies on *Odontoloxozus longicornis* (Diptera: Neriidae) II. Distribution and ecology. Ann. Ent. Soc. Amer. 56:470–472.

Sabrosky, C. W. 1942. An unusual rearing of *Rainieria brunneipes* (Cresson). Ent. News 53:283–285.

Sabrosky, C. W. 1948. A further contribution to the classification of the North American spider parasites of the family Acroceratidae (Diptera). Am. Midl. Nat. 39:382–430.

Sabrosky, C. W. 1953. Taxonomy and host relations of the tribe Ormiini in the Western Hemisphere (Diptera: Larvaevoridae). Proc. Ent. Soc. Wash. 55:167–183.

Sabrosky, C. W. 1956. The utilization of morphological, ecological, and life history evidence in the classification of *Protocalliphora* (Diptera: Calliphoridae). Proc. 10th Intern. Congr. Ent. 1:163–164.

Sabrosky, C. W. 1957a. A new genus and two new species of *Chamaemyiidae* (Diptera) feeding on *Orthezia* scale insects. Bull. Brooklyn Ent. Soc. 52:114–117.

Sabrosky, C. W. 1957b. Synopsis of the new world species of the dipterous family Asteiidae. Ann. Ent. Soc. Amer. 50:43–61.

Sabrosky, C. W. 1958. New species and notes on North American acalyptrate Diptera. Ent. News 69:169–176.

Sabrosky, C. W. 1965. Hosts of the tachinid tribe Eutherini (Diptera). Proc. Ent. Soc. Wash. 67:61.

Sabrosky, C. W. 1978a. Tachinid parasites of *Heliothis* in the Western Hemisphere (Diptera: Lepidoptera). Proc. Ent. Soc. Wash. 80:37–42.

Sabrosky, C. W. 1978b. A third set of corrections to "A catalog of the Diptera of America north of Mexico". Ent. Soc. Amer., Bull. 24:143–144.

Sabrosky, C. W. 1981. A partial revision of the genus *Eucelatoria* (Diptera, Tachinidae), including important parasites of *Heliothis*. U.S. Dept. Agr. Tech. Bull. 1635:1–18.

Sabrosky, C. W. 1986. North American species of *Cuterebra*, the rabbit and rodent bot flies (Diptera: Cuterebridae). Thomas Say Foundation, Ent. Soc. Amer., 240 pp.

Sabrosky, C. W. 1987a. Asteiidae, pp. 899–902. *In* McAlpine, J. F., et al., eds., Manual of Nearctic Diptera. Vol. 2. Res. Br. Agr. Can. Monogr. 28.

Sabrosky, C. W. 1987b. Milichiidae, pp. 903–908. *In* McAlpine, J. F., et al., eds., Manual of Nearctic Diptera. Vol. 2. Res. Br. Agr. Can. Monogr. 28.

Sabrosky, C. W. 1987c. Carnidae, pp. 909–912. *In* McAlpine, J. F., et al., eds., Manual of Nearctic Diptera. Vol. 2. Res. Br. Agr. Can. Monogr. 28.

Sabrosky, C. W. 1987d. Chloropidae, pp. 1049–1067. *In* McAlpine, J. F., et al., eds., Manual of Nearctic Diptera. Vol. 2. Res. Br. Agr. Can. Monogr. 28.

Sabrosky, C. W., and G. C. Steyskal. 1974. The genus *Sobarocephala* (Diptera: Clusiidae) in America north of Mexico. Ann. Ent. Soc. Amer. 67:371–385.

Sack, P. 1939. Sciomyzidae. Lfg. 125, Parts 1, 2, 3. Bd. 5, 87 p. *In* Lindner, E., ed., Die Fliegen der palaearktischen Region. Schweizerbart, Stuttgart.

Saether, O. A. 1970. Nearctic and Palaearctic *Chaoborus* (Diptera: Chaoboridae). Bull. Fish. Res. Bd. Can. 174:1–57.

Saether, O. A. 1972. Chaoboridae. Das Zooplankton der Binnengewässer. 1. Teil. Binnengew. 26:257–280.

Saether, O. A. 1975. Two new species of *Heterotanytarsus* Sparck with keys to Nearctic and Palaearctic males and pupae of the genus (Diptera: Chironomidae). J. Fish. Res. Bd. Can. 32: 259–270.

Sanders, D. P. and Dobson, R. C. 1966. The insect complex associated with bovine manure in Indiana. Ann. Ent. Soc. Amer. 59:955–959.

Sanjean, J. 1957. Taxonomic studies of *Sarcophaga* larvae of New York, with notes on the adults. Cornell Univ. Agr. Exp. Sta. Mem. 349:1–115.

Satchell, G. H. 1953. On the early stages of *Bruchomvia argentina* Alexander (Diptera: Psychodidae). Proc. Roy. Ent. Soc. Lond. (A) 28:1–12.

Saunders, L. G. 1923. On the larva, pupa and systematic position of *Orphnephila testacea* Macq. Ann. Mag. Nat Hist. 11:631–640.

Saunders, L. G. 1925. On the life history, morphologie and systematic position of *Apelma* Kieff. and *Thridomyia* n.g. Parasitol. 17:252–271.

Saunders, L. G. 1928. Some marine insects of the Pacific coast of Canada. Ann. Ent. Soc. Amer. 21:521–545.

Saunders, L. G. 1956. Revision of the genus *Forcipomyia* based on characters of all stages (Diptera: Ceratopogonidae). Can. J. Zol. 34:657–705.

Scaramuzza, L. C. 1930. Preliminary report on a study of the biology of *Lixophaga diatraeae* Tns. J. Econ. Ent. 23:999–1004.

Scarbrough, A. G. 1978a. Ethology of *Cerotainia albipilosa* Curran (Diptera: Asilidae) in Maryland: predatory behavior. Proc. Ent. Soc. Wash. 80: 113–127.

Scarbrough, A. G. 1978b. Ethology of *Cerotainia albipilosa* Curran (Diptera: Asilidae) in Maryland: Courtship, mating and oviposition. Proc. Ent. Soc. Wash. 80:179–190.

Scarbrough, A. G. 1981a. Ethology of *Eudioctria tibialis* Banks (Diptera: Asilidae) in Maryland: prey, predator behavior, and enemies. Proc. Ent. Soc. Wash. 83:258–268.

Scarbrough, A. G. 1981b. Ethology of *Eudioctria tibialis* Banks (Diptera: Asilidae) in Maryland: Reproductive behavior. Proc. Ent. Soc. Wash. 83:432–443.

Scarbrough, A. G. 1981c. Ethology of *Eudioctria tibialis* Banks (Diptera: Asilidae) in Maryland: Seasonal distribution, abundance, diurnal movements and behaviors. Proc. Ent. Soc. Wash. 83:245–257.

Scarbrough, A. G., and G. Sipes. 1973. The biology of *Leptogaster flavipes* Loew in Maryland (Diptera: Asilidae). Proc. Ent. Soc. Wash. 75:441–448.

Schaefer, C. W. 1979. Feeding habits and hosts of calyptrate flies (Diptera, Brachycera, Cyclorrhapha). Ent. Generalis 5:193.

Scheiring, J. F., and B. A. Foote. 1973. Habitat distribution of the shore flies of northeastern Ohio (Diptera: Ephydridae). Ohio J. Sci. 73:152–164.

Schillito, J. F. 1940. Studies on Diopsidae. Novitates Zool. 42:147–163.

Schillito, J. F. 1960. A bibliography of the Diopsidae. J. Soc. Biblio. Nat. Hist. 3:337–350.

Schlee, D. 1968. Phylogenic studies on Chironomidae (Dipt.) with special reference to the *Corynoneura* group. Ann. Zool. Fenn. 5:130–138.

Schlinger, E. I. 1952. The emergence, feeding habits and host of *Opsebius diligens* Osten Sacken (Diptera: Acroceridae). Pan-Pacif. Ent. 28:7–12.

Schlinger, E. I. 1960a. A revision of the genus *Ogcodes* Latreille with particular reference to species of the Western Hemisphere. Proc. U.S. Nat. Mus. 11:227–336.

Schlinger, E. I. 1960b. A review of the genus *Eulonchus* Gerstaecker. Part I. The species of the *smaragdinus* group (Diptera: Acroceridae). Ann. Ent. Soc. Amer. 53:416–422.

Schlinger, E. I. 1972. A new Brazilian panopine species, *Exetasis eickstedtae* reared from the therophasid spider, *Lasiodora klugi* (Koch) with a description of its immature larval stages (Diptera: Acroceridae). Papeis Avulsos Zool. 26:73–82.

Schlinger, E. I. 1975. Diptera, pp. 436–446. *In* Smith, R. I. and J. T. Carlton, eds., Light's Manual: Intertidal Invertebrates of the Central California Coast. Univ. Calif. Press, Berkeley.

Schlinger, E. I. 1981. Acroceridae, pp. 575–584. *In* McAlpine, J. F., et al., eds., Manual of Nearctic Diptera. Vol. 1. Res. Br. Agr. Can. Monogr. 27.

Schmitz, H. 1935. Ein neuer Beitrag zur Kenntnis der *Paraspiniphora* - Arten. Broteria (Ser. Sci. Nat.) 4 (31):155–173.

Schmitz, H. 1938. Phoridae. Lfg. 123, pages 1–64 *in* E. Lindner, ed. Die Fliegen der palaearktischen Region. Schweizerbart, Stuttgart.

Schmitz, H. 1941. Über die Larve von *Chaetopleurophora ovoidialis* M. Natuurhist. Maandbl. 30:63–66.

Schneider, F. 1969. Bionomics and physiology of aphidophagous Syrphidae (Dipt.). Annu. Rev. Ent. 14:103–124.

Schoof, H. F., Mail, G. A. and Savage, E. P. 1954. Fly production sources in urban communities. J. Econ. Ent. 47:245–253.

Schreiber, E. T., and R. J. Lavigne. 1986. Ethology of *Asilus gilvipes* (Hine) (Diptera: Asilidae) associated with small mammal burrows in southeastern Wyoming. Proc. Ent. Soc. Wash. 88:711–719.

Schremmer, F. 1951. Die Mundteile der Brachycerenlarven und der Kopfbau der Larven von *Stratiomys chamaeleon* L. Öst. Zool. Z. 3:326–397.

Schremmer, F. 1956. Funktionsmorphologische Studien an Dipterenlarven. Verh. dt. zool. Ges. 1956:301–305.

Schumacher, F. 1919. *Leucopis nigricornis* Eggers, eine in Schild- und Blattlausen parasitierende Fliege. Zeitschr. Wiss. Insektenbiol. 14:304–306.

Schwardt, H. H. 1931. The biology of *Tabanus lineola* Fabr. Ann. Ent. Soc. Amer. 24:409–416.

Schwitzgebel, R. B., and D. A. Wilbur. 1943. Diptera associated with ironweed, *Vernonia interior* Small, in Kansas. J. Kans. Ent. Soc. 16:4–13.

Scotland, M. B. 1934. The animals of the *Lemna* association. Ecol. 15:290–294.

Scotland, M. B. 1939. The *Lemna* fly and some of its parasites. Ann. Ent. Soc. Amer. 32:713–718.

Scott, E. J. 1939. An account of the developmental stages of some aphidophagous Syrphidae and their parasites. Ann. Appl. Biol. 26:509–532.

Scott, H. 1920. Notes on the parasitic staphylinid *Aleochara algarum* Fauvel, and its hosts, the phycodromid flies. Ent. Mon. Mag. 56:148–157.

Seeger, J. R., and M. E. Maldague. 1960. Infestation de nodules de légumineuses en région équatoriale par des larves de *Rivellia* sp. (Dipt.). Parasitica 16:75–89.

Segal, G. 1936. Synopsis of the Tabanidae of New York, their biology and taxonomy I. The genus *Chrysops* Meigen. J. N.Y. Ent. Soc. 44:51–78, 125–154.

Séguy, E. 1923. Note sur les larves des *Muscina stabulans* et *assimilis* (Diptères). Bull. Mus. Hist. nat. Paris 29:443–445.

Séguy, E. 1923b. Diptères Anthomyiides. *In* Faune de France. 6:i–xi, 1–393.

Séguy, E. 1926. Diptera (Brachycera), *In* Faune de France 13:1–308.

Séguy E. 1928. Caractères particuliers du *Rhynchoestrus weissii.* Encycl., ent. (B) 11, Dipt. 4:97–100.

Séguy, E. 1932. Études sur les Anthomyiides. 6è note. Notes biologiques et taxonomiques sur les mouches de l'oeillet. Encycl. ent. (B) II, Dipt. 6:71–81.

Séguy, E. 1934. Diptères, Muscidae acalypterae et Scatophagidae. Faune de France, 28. Paris.

Séguy, E. 1940. Croisière du Bougainville aux iles australes françaises. IV. Diptères. Mém. Mus. Hist. nat. Paris (N. S.) 14:203–217.

Séguy, E. 1941a. Études sur les mouches parasites. II. Calliphorines, Sarcophagines et Rhinophorines de l'Europe occidentale et méridionale. Tome II. Encycl. ent. (A) 21:1–436.

Séguy, E. 1941b. Étude biologique et systématique des Sarcophagines myiasigènes du genre *Wohlfahrtia.* Annls. Parasitol. hum. comp. 18:220–232.

Séguy, E., 1950. La biologie des Diptères. Encycl. ent., 26:609 pp.

Séguy, E., and L. Pauchet. 1929. Étude sur le *Chylizosoma paridis.* Bull. Soc. linn. Nord France (1929) 418:47–63.

Seifert, R. P., and F. Hammet Seif. 1976. Natural history of insects living in inflorescenses of two species of *Heliconia.* J. N. Y. Ent. Soc. 84:233–242.

Sen, P. 1931. Preliminary observations on the early stages of *Scatella stagnalis* Fal. J. Mar. Biol. Assoc. (N.S.) 17:847–851.

Sen, P. 1938a. On the structure (anatomical histological) of the full-grown Larva of *Rhabdophaga saliciperda* Dufour. Zool. Jahrb. Anat. 65:37–62.

Sen, P. 1938b. On the biology and morphology of *Rhabdophaga saliciperda* Dufour, a common pest of willows (*Salix fragilis*). Zool. Jahrb. Anat. 65:1–36.

Sen, S. K. 1921. Life histories of Indian insects. Diptera: *Sphyracephala hearseiana* Westw. India Agr. Dept. Mem. Ent. Ser. 7:33–38.

Shaw, F. R., and E. B. Fisher. 1952. Guide to the Insects of Connecticut. Part VI. The Diptera or True Flies of Connecticut. Fifth Fascicle: Midges and gnats (part). Family Fungivoridae (=Mycetophilidae). Bull. Conn. St. Geol. Nat. Hist. Surv. 80:177–250.

Shelford, V. E. 1913. The life history of a bee-fly (*Spogostylum anale* Say) parasite of the larva of a tiger beetle (*Cicindela scutellaris* Say var. *lecontei* Halid.). Ann. Ent. Soc. Amer. 6:213–225.

Shewell, G. E. 1938. The Lauxaniidae (Diptera) of southern Quebec and adjacent regions. Can. Ent. 70:102–118, 133–142.

Shewell, G. E. 1950. A new species of *Sarcophaga* reared from the Columbian ground squirrel (Diptera: Sarcophagidae). Can. Ent. 82:245–246.

Shewell, G. E. 1960. Notes on the family Odiniidae with a key to the genera and descriptions of new species (Diptera). Can. Ent. 92:625–633.

Shewell, G. E. 1987a. Lauxaniidae, pp. 951–964. *In* McAlpine, J. F., et al., eds., Manual of Nearctic Diptera. Vol. 2. Res. Br. Agr. Can. Monogr. 28.

Shewell, G. E. 1987b. Calliphoridae, pp. 1133–1145. *In* McAlpine, J. F., et al., eds. Manual of Nearctic Diptera, Vol. 2. Res. Br. Agr. Can. Monogr. 28.

Shewell, G. E. 1987c. Sarcophagidae, pp. 1147–1186. *In* McAlpine, J. F., et al., eds. Manual of Nearctic Diptera, Vol. 2. Res. Br. Agr. Can. Monogr. 28.

Shorrocks, B. 1977. An ecological classification of European *Drosophila* species. Oecolog. 26:335–345.

Shorrocks, B., and A. M. Woods. 1973. A preliminary note on the fungus feeding species of *Drosophila.* J. Nat. Hist. 7:551–556.

Siddons, L. D. and D. N. Roy. 1940. The early stage of *Musca inferior* Stein. Ind. J. Med. Res. 27:819–822.

Siddons, L. D. and D. N. Roy. 1942. On the life-history of *Synthesiomyia nudiseta* van der Wulp, a myiasis-producing fly. Parasitol. 34:239–245.

Silverman, J., and R. D. Goeden. 1980. Life history of a fruit fly, *Procecidochares* sp., on the ragweed, *Ambrosia dumosa* (Gray) Paine, in southern California (Diptera: Tephritidae). Pan-Pacif. Ent. 56:283–288.

Simpson, K. W. 1975. Biology and immature stages of three species of Nearctic *Ochthera* (Diptera: Ephydridae). Proc. Ent. Soc. Wash. 77:129–155.

Simpson, K. W. 1976a. The mature larvae and puparia of *Ephydra* (*Halephydra*) *cinerea* Jones and *Ephydra* (*Hydrapyrus*) *hians* Say (Diptera: Ephydridae). Proc. Ent. Soc. Wash. 78:263–269.

Simpson, K. W. 1976b. Shore flies and brine flies (Diptera: Ephydridae), pp. 465–495. *In* Cheng, L., ed., Marine Insects. North Holland Publ. Co.

Simpson, K. W. 1979. Evolution of life histories in the Ephydrini, pp. 99–109. *In* Deonier, D. L., ed., First Symposium on the Systematics and Ecology of Ephydridae. N. Amer. Benthol. Soc.

Simpson, K. W., and R. W. Bode. 1980. Common larvae of Chironomidae (Diptera) from New York State streams and rivers. Bull. N.Y. St. Mus. 439:1–105.

Simpson, K. W., R. W. Bode, and P. Albu. 1983. Keys for the genus *Cricotopus* adapted from "Revision der Gattung *Cricotopus* van der Wulp und ihrer Verwandten (Diptera, Chironomidae)" by M. Hirvenoja. Bull. N.Y. St. Mus. 450:1–133.

Singh Pruthi, H., and H. L. Bhatia. 1938. Biology and general morphology of *Leucopis griseola* Fall., an important predator of *Aphis gossypii* and some other aphid pests of crops. Ind. J. Agr. 8:735–740.

Singh, S. and I. M. Ipe. 1973. The Agromyzidae from India. Mem. School Entomol. St. John's College, Agra. No. 1:(1–6), 286 pp.

Skaife, S. H. 1921. On *Braula coeca*, Nitzsch, a dipterous parasite of the honey bee. Trans. Roy. Soc. S. Africa 10: 41–48.

Skidmore, P. 1985. The biology of the Muscidae of the world. Series Entomologica. Vol. 29. Junk Publ. Boston.

Sluss, T. P., and B. A. Foote. 1970. Parthenogenesis in Chamaemyiidae (Diptera). Ann. Ent. Soc. Amer. 63:615.

Sluss, T. P., and B. A. Foote. 1971. Biology and immature stages of *Leucopis verticalis* (Diptera: Chamaemyiidae). Can. Ent. 103:1427–1434.

Sluss, T. P., and B. A. Foote. 1973. Biology and immature stages of *Chamaemyia polystigma* (Diptera: Chamaemyiidae). Can. Ent. 105:1443–1452.

Smart, J. 1956. A Handbook for the Identification of Insects of Medical Importance. 2nd ed. London (B. M.). xi+363 p.

Smith, C. N. 1933. Notes on life history and molting process of *Sarcophaga securifera* Villeneuve. Proc. Ent. Soc. Wash. 35:159–164.

Smith, F. K. 1928. Larval characters of the genus *Dixa*. J. N.Y. Ent. Soc. 36: 263–284.

Smith, I. B., Jr. 1984. Distribution of the bee louse *Braula coeca* Nitzsch in honey bee colonies and its preferences among workers, queens, and drones. J. Apic. Res. 23:171–176.

Smith, J. B. 1902. Notes on the early stages of *Corethra brakeleyi* Coq. Canad. Ent. 34:139–140.

Smith, K. G. V., 1957. Notes on the immature stages of four British species of *Lonchaea* Fln. (Diptera, Lonchaeidae). Ent. Mon. Mag. 92:402–405.

Smith, K. G. V. 1986. A manual of forensic entomology. Cornell Univ. Press, Ithaca, N.Y. 205 p.

Smith, K. G. V. and B. V. Peterson. 1987. Conopidae, pp. 749–756. *In* Manual of Nearctic Diptera. Vol. 2. Res. Br. Agr. Can. Monogr. 28.

Smith, K. G. V., and E. Taylor. 1966. *Anisopus* larvae (Diptera) in cases of intestinal and urino-genital myiasis. Nature 210:852.

Smith, K. M. 1922. A study of the life-history of the onion fly (*Hylemyia antiqua*). Ann. Appl. Biol. 9:177–183.

Smith, K. M. 1927. A study of *Hylemyia* (*Chortophila*) *brassicae* Bouché, the cabbage root fly and its parasites. Ann. Appl. Biol. 14:312–330.

Smith, M. E. 1952. Immature stages of the marine fly, *Hypocharassus pruinosus* Wh., with a review of the biology of immature Dolichopodidae. Amer. Midl. Nat. 48:421–432.

Smith, R. L. 1981. The trouble with "bobos", *Paraleucopis mexicana* Steyskal, at Kino Bay, Sonora, Mexico (Diptera: Chamaemyiidae). Proc. Ent. Soc. Wash. 83:406–412.

Smith, R. W., and T. U. Finlayson. 1950. Larvae of dipterous parasites of nymphal and adult grasshoppers. Canad. J. Res. (D) 28:81–117.

Smulyan, M. T. 1914. The marguerite fly (*Phytomyza chrysanthemi* Kow.). Mass. Agr. Exp. Sta. Bull. 157, 52 p.

Snodgrass, R. E. 1924. Anatomy and metamorphosis of the apple maggot, *Rhagoletis pomonella* Walsh. J. Agr. Res. 28, 1, pp. 1–36.

Snodgrass, R. E. 1935. Principles of insect morphology. McGraw Hill Book Co., New York, N.Y.

Snodgrass, R. E. 1947. The insect cranium and the "Epicranial suture". Smithson. Misc. Coll. 107 (7):1–52.

Snodgrass, R. E. 1959. The anatomical life of the mosquito. Smithson. Misc. Coll. 139(8):1–87.

Snyder, F. M. 1954. A review of the Nearctic *Lispe* Latreille (Diptera, Muscidae). Amer. Mus. Novitates 1675:1–38.

Sommerman, K. M. 1953. Identification of Alaskan black fly larvae (Diptera, Simuliidae). Proc. Ent. Soc. Wash. 55:258–273.

Sommerman, K. M. 1962. Alaskan snipe fly immatures and their habitat (Rhagionidae: *Symphoromyia*). Mosquito News 22:116–123.

Soós, Á. 1961. A new subspecies of *Czernyola basalis* Czerny, 1903 (Diptera: Clusiidae). Folia Ent. Hungarica (S. N.) 14:399–402.

Soós, Á. 1987. Clusiidae, pp. 853–857. *In*. McAlpine, J. F., et al., (eds.). Manual of Nearctic Diptera, Vol. 2. Res. Br. Agr. Can. Monogr. 28.

Soper, R. S., G. E. Shewell, and D. Tyrrell. 1976. *Colcondamyia auditrix* nov. sp. (Diptera: Sarcophagidae), a parasite which is attracted by the mating song of its host, *Okanagana rimosa* (Homoptera: Cicadidae). Can. Ent. 108:61–68.

Soponis, A. R. 1977. A revision of the Nearctic species of *Orthocladius* (*Orthocladius*) van der Wulp (Diptera: Chironomidae). Mem. Ent. Soc. Can. 102:1–187.

Speiser, P. 1900. Über die Art der Fortpflanzung bei Strebliden, nebst synonymischen Bemerkungen. Zool. Anz. 23:153–154.

Speiser, P. 1901. Über die Nycteribiiden, Fledermausparasiten aus der Gruppe der pupiparen Dipteren. Archiv Naturgesch. 67:11–78.

Spencer, G. J. 1958. On the Nemestrinidae of British Columbia dry range lands. 10th Intern. Congr. Ent. (Montréal, 1956) 4:503–509.

Spencer, K. A. 1969. The Agromyzidae of Canada and Alaska. Mem. Ent. Soc. Can. 64:1–311.

Spencer, K. A. 1987. Agromyzidae, pp. 869–879. *In* McAlpine, J. F., et al., eds., Manual of Nearctic Diptera. Vol. 2. Res. Br. Agr. Can. Monogr. 28.

Spencer, K. A., and G. C. Steyskal. 1986. Manual of the Agromyzidae (Diptera) of the United States. U.S. Dept. Agr. Hanbk. 638. vi+478 p.

Spieth, H. T. 1952. Mating behavior within the genus *Drosophila*. Bull. Amer. Mus. Nat. Hist. 99:395–474.

Spieth, H. T. 1974. Courtship behavior in *Drosophila*. Annu. Rev. Ent. 19:385–405.

Spieth, H. T., and W. B. Heed. 1972. Experimental systematics and ecology of *Drosophila*. Annu. Rev. Ecol. Syst. 3:269–287.

Stadtmann-Averfeldt, H. 1923. Beiträge zur Kenntnis der Stechmückenlarven. Deutsche Entomol. Zeitschr. 1923, pp. 105–152.

Staeler, E. 1972. The orientation and host plant selection of the carrot fly *Psila rosae* F. (Diptera: Psilidae). Zeitschr. Angew. Ent. 4:425–438.

Stalker, H. D. 1945. On the biology and genetics of *Scaptomyza graminum* Fallén (Diptera, Drosophilidae). Genetics 30:266–279.

Stalker, H. D. 1956. A case of polyploidy in Diptera. Genetics 42:194–199.

Stark, Von A. and T. Wetzel. 1987. Fliegen der Gattung *Platypalpus* (Diptera: Empididae)-bisher wenig beachtete Prädatoren im Getreidebestand. Zeits. angewandte Ent. 103:1–14.

Stasny, T. A. H. 1985. Biology, behavior and life cycle of *Sphyracephala brevicornis* Say (Diptera: Diopsidae). Unpublished thesis, West Virginia University, Morgantown, WV 194 pp.

Steffan, W. A. 1966. A generic revision of the family Sciaridae (Diptera) of America north of Mexico. Univ. Calif. Publ. Ent. 44:1–77.

Steffan, W. A. 1973. Polymorphism in *Plastosciara perniciosa* (Diptera: Sciaridae). Science 182:1265–1266.

Steffan, W. A. 1981. Sciaridae, pp. 247–255. *In* McAlpine, J. F., et al., eds., Manual of Nearctic Diptera. Vol. 1. Res. Br. Agr. Can. Monogr. 27.

Steffan, W. A., and N. L. Evenhuis. 1981. Biology of *Toxorhynchites*. Annu. Rev. Ent. 26:159–181.

Stegmaier, C. E., Jr. 1967a. Notes on the biology and distribution of Florida leaf-mining flies of the genus *Phytobia* Lioy, subgenus *Calycomyza* Hendel (Diptera: Agromyzidae). Fla. Ent. 50:13–26.

Stegmaier, C. E., Jr. 1967b. *Pluchea odorata*, a new host record for *Acinia picturata* (Diptera, Tephritidae). Fla. Ent. 50:53–55.

Stegmaier, C. E., Jr. 1967c. Notes on a seed-feeding Tephritidae, *Paracantha forficula* (Diptera) in Florida. Fla. Ent. 50:157–160.

Stegmaier, C. E., Jr. 1968a. Host plant records of *Dyseuaresta mexicana* (Diptera: Tephritidae) with notes on its life history in Florida. Fla. Ent. 51:20–21.

Stegmaier, C. E., Jr. 1968b. *Erigeron*, a host plant genus of Tephritidae (Diptera). Fla. Ent. 51:46–50.

Stegmaier, C. E., Jr. 1968c. Notes on the biology of *Trupanea actinobola* (Diptera: Tephritidae). Fla. Ent. 51:95–99.

Stegmaier, C. E., Jr. 1973. *Dasiops passifloris* (Diptera: Lonchaeidae), a pest of wild passion fruit in South Florida. Fla. Ent. 56:8–10.

Steinley, B. A., and J. T. Runyan. 1979. The life history of *Leptopsilopa atrimana*, pp. 139–147. *In* Deonier, D. L., ed., First Symposium on the Systematics and Ecology of Ephydridae (Diptera). N. Amer. Benthol. Soc.

Stevenson, A. B. 1967. *Leucopis simplex* (Diptera: Chamaemyiidae) and other species occurring in galls of *Phylloxera vitifoliae* (Homoptera: Phylloxeridae) in Ontario. Can. Ent. 99:815–820.

Steyskal, G. C. 1943. Old World Sepsidae in North America, with a key to the American genera (Diptera). Pan.-Pacif. Ent. 19:93–95.

Steyskal, G. C. 1946. *Themira nigricornis* Meigen in North America, with a revised key to the Nearctic species of *Themira* (Diptera: Sepsidae). Ent. News 57:93–95.

Steyskal, G. C. 1947. A revision of the Nearctic species of *Xylomyia* and *Solva* (Diptera, Erinnidae). Pap. Mich. Acad. Sci. (1945) 31:181–190.

Steyskal, G. C. 1956. The eastern species of *Nemomydas* Curran (Diptera: Mydaidae). Occ. Pap. Mus. Zool. Univ. Mich. 573:1–5.

Steyskal, G. C. 1957. A revision of the family Dryomyzidae (Diptera, Acalyptratae). Mich. Acad. of Sci. Arts Lett. 42:55–68.

Steyskal, G. C. 1958. Notes on the Richardiidae, with a review of the species known to occur in the United States (Diptera, Acalyptratae). Ann. Ent. Soc. Amer. 51:302–310.

Steyskal, G. C. 1961a. The genera of Platystomatidae and Otitidae known to occur in America north of Mexico (Diptera, Acalyptratae). Ann. Ent. Soc. Amer. 54:401–410.

Steyskal, G. C. 1961b. Two new species of *Sepsisoma* from Kansas (Diptera: Richardiidae). J. Kans. Ent. Soc. 34:83–85.

Steyskal, G. C. 1964. Larvae of Micropezidae (Diptera), including two species that bore in ginger roots. Ann. Ent. Soc. Amer. 57:292–296.

Steyskal, G. C. 1965. The third larval instar and puparium of *Odontoloxozus longicornis* (Diptera: Neriidae). Ann. Ent. Soc. Amer. 58:936–937.

Steyskal, G. C. 1971a. The genus *Paraleucopis* Malloch (Diptera: Chamaemyiidae), with one new species. Ent. News 82:1–4.

Steyskal, G. C. 1971b. Notes on some species of the genus *Copromyza*, subgenus *Borborilus*. J. Kans. Ent. Soc. 44:476–479.

Steyskal, G. C. 1972a. Notes on the genera *Homalomitra* Borgmeier, *Pyenopota* Bezzi and *Sphinalomyia* Borgmeier (Diptera: Sphaeroceridae). Proc. Ent. Soc. Wash. 73:376–378.

Steyskal, G. C. 1972b. The genus *Acrometopia* Schiner in North America (Diptera: Chamaemyiidae). Proc. Ent. Soc. Wash. 74:302.

Steyskal, G. C. 1973. Distinguishing character of the walnut husk maggots of the genus *Rhagoletis* (Diptera-Tephritidae). U.S. Dept. Agr. Coop. Econ. Ins. Rept. 23:522.

Steyskal, G. C. 1975. Recognition characters for larvae of the genus *Zonosemata* (Diptera: Tephritidae). U.S. Dept. Agr. Coop. Econ. Ins. Rept. 25:231–232.

Steyskal, G. C. 1978. A new pest of Chili peppers in Columbia (Diptera: Lonchaeidae). U.S. Dept. Agr. Coop. Plt. Pest Report. 3:72.

Steyskal, G. C. 1980. Two-winged flies of the genus *Dasiops* (Diptera: Lonchaeidae) attacking flowers or fruit of *Passiflora* (passion fruit, granadilla, curuva, etc.) Proc. Ent. Soc. Wash. 82:166–170.

Steyskal, G. C. 1987a. Micropezidae, pp. 761–767. *In* McAlpine, J. F., et al., eds., Manual of Nearctic Diptera. Vol. 2. Res. Br. Agr. Can. Monogr. 28.

Steyskal, G. C. 1987b. Neriidae, pp. 769–771. *In* McAlpine, J. F., et al., eds., Manual of Nearctic Diptera. Vol. 2. Res. Br. Agr. Can. Monogr. 28.

Steyskal, G. C. 1987c. Tanypezidae, pp. 773–776. *In* McAlpine, J. F., et al., eds., Manual of Nearctic Diptera. Vol. 2. Res. Br. Agr. Can. Monogr. 28.

Steyskal, G. C. 1987d. Strongylophthalmyiidae, pp. 777–779. *In* McAlpine, J. F., et al., eds., Manual of Nearctic Diptera. Vol. 2. Res. Br. Agr. Can. Monogr. 28.

Steyskal, G. C. 1987e. Psilidae, pp. 781–784. *In* McAlpine, J. F., et al., eds., Manual of Nearctic Diptera. Vol. 2. Res. Br. Agr. Can. Monogr. 28.

Steyskal, G. C. 1987f. Otitidae, pp. 799–808. *In* McAlpine, J. F., et al., eds., Manual of Nearctic Diptera. Vol. 2. Res. Br. Agr. Can. Monogr. 28.

Steyskal, G. C. 1987g. Platystomatidae, pp. 809–812. *In* McAlpine, J. F., et al., eds., Manual of Nearctic Diptera. Vol. 2. Res. Br. Agr. Can. Monogr. 28.

Steyskal, G. C. 1987h. Pyrgotidae, pp. 813–816. *In* McAlpine, J. F., et al., eds., Manual of Nearctic Diptera. Vol. 2. Res. Br. Agr. Can. Monogr. 28.

Steyskal, G. C. 1987i. Richardiidae, pp. 833–837. *In* McAlpine, J. F., et al., eds., Manual of Nearctic Diptera. Vol. 2. Res. Br. Agr. Can. Monogr. 28.

Steyskal, G. C. 1987j. Dryomyzidae, pp. 923–926. *In* McAlpine, J. F., et al., eds., Manual of Nearctic Diptera. Vol. 2. Res. Br. Agı. Can. Monogr. 28.

Steyskal, G. C. 1987k. Ropalomeridae, pp. 941–944. *In* McAlpine, J. F., et al., eds., Manual of Nearctic Diptera. Vol. 2. Res. Br. Agr. Can. Monogr. 28.

Steyskal, G. C. 1987l. Sepsidae, pp. 945–950. *In* McAlpine, J. F., et al., eds., Manual of Nearctic Diptera. Vol. 2. Res. Br. Agr. Can. Monogr. 28.

Steyskal, G. C., T. W. Fisher, L. V. Knutson, and R. E. Orth. 1978. Taxonomy of North American flies of the genus *Limnia* (Diptera: Sciomyzidae). Univ. Calif. Publ. Ent. 83:1–48.

Steyskal, G. C., and L. V. Knutson. 1981. Empididae, pp. 607–624. *In* McAlpine, J. F., et al., eds., Manual of Nearctic Diptera. Vol. 1. Res. Br. Agr. Can. Monogr. 27.

Stoffolano, J. G., Jr. 1970. The anal organ of *Musca autumnalis*, *M. domestica*, and *Orthellia caesarion* (Diptera: Muscidae). Ann. Ent. Soc. Amer. 63:1647–1654.

Stojanovich, C. J. 1960. Illustrated key to common mosquitoes of southeastern United States. Atlanta, Ga. Center for Disease Control, U.S. Pub. Health Ser.

Stojanovich, C. J. 1961. Illustrated key to common mosquitoes of northeastern United States. Atlanta, Ga. Center for Disease Control, U.S. Publ. Health Ser.

Stoltzfus, W. B. 1974. Biology and larval description of *Procecidocharoides penelope* (Osten Sacken) (Diptera: Tephritidae). J. Wash. Acad. Sci. 64:12–14.

Stoltzfus, W. B. 1977. The taxonomy and biology of *Eutreta* (Diptera: Tephritidae). Iowa St. J. Res. 51:369–438.

Stoltzfus, W. B. 1978. Life history and description of the immature stages of *Jamesomyia geminata* (Diptera: Tephritidae). Proc. Ent. Soc. Wash. 80:87–90.

Stoltzfus, W. B., and B. A. Foote. 1965. The use of froth masses in courtship of *Eutreta* (Diptera: Tephritidae). Proc. Ent. Soc. Wash. 67:263–264.

Stone, A. 1930. The bionomics of some Tabanidae. Ann. Ent. Soc. Amer. 23:261–304.

Stone, A. 1964. Guide to the insects of Connecticut. Part VI. The Diptera or true flies of Connecticut. Ninth Fascicle. Simuliidae and Thaumaleidae. Bull. Conn. St. Geol. Nat. Hist. Surv. 97:1–126.

Stone, A., et al., eds. 1965. A catalog of Diptera of America north of Mexico. Handbook No. 276. Agr. Res. Ser., U.S. Dept. Agr. 1696 pp.

Stone, A. 1968. The genus *Corethrella* in the United States (Diptera: Chaoboridae). Fla. Ent. 51:183–186.

Stone, A. 1981. Culicidae, pp. 341–350. *In* McAlpine, J. F., et al., eds., Manual of Nearctic Diptera. Vol. 1. Res. Br. Agr. Can. Monogr. 27.

Stone, A., and B. V. Peterson. 1981. Thaumaleidae, pp. 351–353. *In* McAlpine, J. F., et al., eds., Manual of Nearctic Diptera. Vol. 1. Res. Br. Agr. Can. Monogr. 27.

Stork, M. N. 1936. A contribution to the knowledge of the puparia of Anthomyiidae. Tijdschr. Ent. 79:94–168.

Strenzke, K., and D. Neumann. 1960. Die Variabilität der abdominalen Körperanhänge aquatischer Chironomidenlarven in Abhängigkeit von der Ionen-zusammensetzung des Mediums. Biol. Zbl. 79:199–225.

Strickland, E. H. 1916. The march fly (*Bibio abbreviatus*) in grain fields and as a pest of celery. Agr. Gaz. Can. 3:600–603.

Strihnasah, H. 1980. Mortality in a population of *Bufo bufo* resulting from the fly *Lucilia bufonivora.* Decol. 45:285–286.

Stubbs, A., and P. Chandler (eds.). 1978. A Dipterist's handbook. The Amateur Entomologist, Vol. 15, 255 pp.

Stuckenberg, B. R. 1973. The Athericidae, a new family in the lower Brachycera (Diptera). Ann. Natal Mus. 21:649–673.

Sturtevant, A. H. 1954. Nearctic flies of the family Periscelidae (Diptera) and certain Anthomyzidae referred to the family. Proc. U.S. Nat. Mus. 103:551–561.

Sturtevant, A. H., and M. R. Wheeler. 1954. Synopses of Nearctic Ephydridae (Diptera). Trans. Amer. Ent. Soc. 79:151–261.

Sublette, J. E. 1964. Chironomidae (Diptera) of Louisiana I. Systematics and immature stages of some lentic chironomids of west-central Louisiana. Tulane Stud. Zool. 11:109–150.

Tamaki, G., B. J. Landis, and J. E. Turner. 1974. *Psilopa leucostoma:* the role of *Atriplex patula* var. *hastata* in the early seasonal establishment on sugarbeets in the Northwest. Environ. Ent. 4:31–32.

Tanasiychuk, V. N. 1974. Morphology of eggs of aphid flies of the family Chamaemyiidae (Diptera). (In Russian). Ent. Obozr. 53:304–308.

Tanasiychuk, V. N., G. Remaudiere, and F. Leclant. 1976. Dipt. Chamaemyiidae predators of aphids and coccids in France. Ann. Soc. Ent. France. 12:691–698.

Tashiro, H. 1987. Turfgrass Insects of the United States and Canada. Cornell Univ. Press. Ithaca, N.Y., 391 pp.

Tatchell, R. J. 1960. A comparative account of the tracheal system of the horse bot-fly, *Gasterophilus intestinalis* (De Geer) and some other dipterous larvae. Parasitology 50:481–496.

Tate, P. 1935. The larva of *Phaonia mirabilis* Ringdahl, predatory on mosquito-larvae. Parasitol. 27:556–560.

Tauber, M. J., and C. A. Tauber. 1967. Reproductive behavior and biology of the gall former *Aciurina ferruginea* Doane (Diptera: Tephritidae). Can. J. Zool. 45:907–913.

Tauber, M. J., and C. A. Tauber. 1968. Biology of the gall former *Procecidochares stonei* on a composite. Ann. Ent. Soc. Amer. 61:553–554.

Tauber, M. J., and C. A. Toschi. 1965a. Bionomics of *Euleia fratria* (Loew) (Diptera: Tephritidae). Can. J. Zool. 43:369–379.

Tauber, M. J., and C. A. Toschi. 1965b. Life history and mating behavior of *Tephritis stigmatica* (Coquillett) (Diptera: Tephritidae). Pan-Pacif. Ent. 41:73–79.

Tawfik, M. F. S. 1966. The life-history of *Leucopis puncticornis aphidivora* Rond. (Diptera: Octphilidae). Bull. Soc. Ent. Egypte 49:47–57.

Teskey, H. J. 1969. Larvae and pupae of some eastern North American Tabanidae (Diptera). Mem. Ent. Soc. Can. 63:1–147.

Teskey, H. J. 1970a. The immature stages and phyletic position of *Glutops rossi* (Diptera: Pelecorhynchidae). Can. Ent. 102:1130–1135.

Teskey, H. J. 1970b. A review of the genus *Glutops* (Diptera: Pelecorhynchidae), with description of four new species. Can. Ent. 102:1171–1179.

Teskey, H. J. 1972. The mature larva and pupa of *Compsobata univitta* (Diptera: Micropezidae). Can. Ent. 104:295–298.

Teskey, H. J. 1976. Diptera larvae associated with trees in North America. Mem. Ent. Soc. Can. 100:1–53.

Teskey, H. J. 1981a. Pelecorhynchidae, pp. 459–461. *In* McAlpine, J. F., et al., eds., Manual of Nearctic Diptera. Vol. 1. Res. Br. Agr. Can. Monogr. 27.

Teskey, H. J. 1981b. Vermileonidae, pp. 529–532. *In* McAlpine, J. F., et al., eds., Manual of Nearctic Diptera. Vol. 1. Res. Br. Agr. Can. Monogr. 27.

Teskey, H. J. 1981c. Nemestrinidae, pp. 585–588. *In* McAlpine, J. F., et al., eds., Manual of Nearctic Diptera. Vol. 1. Res. Br. Agr. Can. Monogr. 27.

Teskey, H. J. 1984. Aquatic Diptera, Part One. Larvae of aquatic Diptera, pp. 448–466. *In* Merritt, R. W., and K. W. Cummins, eds., An Introduction to the Aquatic Insects of North America. Kendall/Hunt Publ. Co., Dubuque, Ia.

Teskey, H. J. 1987a. Aulacigastridae, pp. 891–894. *In* McAlpine, J. F., et al., eds., Manual of Nearctic Diptera. Vol. 2. Res. Br. Agr. Can. Monogr. 28.

Teskey, H. J. 1987b. Trixoscelididae, pp. 981–984. *In* McAlpine, J. F., et al., eds., Manual of Nearctic Diptera. Vol. 2. Res. Br. Agr. Can. Monogr. 28.

Teskey, H. J., and J. F. Burger. 1976. Further larvae and pupae of eastern North American Tabanidae (Diptera). Can. Ent. 108: 1085–1096.

Teskey, H. J., J. M. Clarke, and C. R. Elliott. 1976. *Hylemya extremitata* (Diptera: Anthomyiidae) and species of Chloropidae associated with injury to bromegrass, with descriptions of larvae. Can. Ent. 108:185–192.

Teskey, H. J. and I. Valiela. 1977. The mature larva and puparium of *Canace macateei* (Diptera: Canaceidae). Can. Ent. 109:545–548.

Theischinger, G. 1986. Australian Bombyliidae (Insecta: Diptera). Records Australian Mus. 38:291–317.

Theowald, B. 1957. Die Entwicklungsstadien der Tipuliden (Diptera, Nematocera) insbesondere der West-Palaearktischen Arten. Tijdschr. Ent. 100:195–308.

Theowald, B. 1967. Familie Tipulidae (Diptera, Nematocera) Larven und Puppen. Bestimmungsbücher zur Bodenfaune Europas. Lfg. 7. Akad.-Verlag.

Thienemann, A. 1909, 1911. *Orphnephila testacea* Macq. Ein Beitrag zur Kenntnis der Fauna hygropetrica. Ann. biol. lacustre 4:53–87.

Thienemann, A. 1937. Chironomiden-Metamorphosen XV. Mitt. Ent. Ges. Halle 15:22–36.

Thienemann, A. 1944. Bestimmungstabellen für die bis jetzt bekannten Larven und Puppen der Orthocladiinen. Arch. Hydrobiol. 39:551–664.

Thienemann, A. 1954. *Chironomus.* Leben, Verbreitung und wirtschäftliche Bedeutung der Chironomiden. Binnengew. 20:1–834.

Thomas, A. G. B. 1974a. Poorly known torrential Diptera. I. Larvae and adults of Athericidae (Brachycera: Orthorrhapha) from the south of France. Ann. Limnol. 10:55–84.

Thomas, A. G. B. 1974b. Diptères torrenticoles peu connus: 2. Les Athericidae (nymphes) du sud de la France (Brachycera, Orthorrhapha). Ann. Limnol. 10:121–130.

Thomas, A. G. B. 1975. Poorly known torrential Diptera: III. Athericidae of the south of France (qualitative aspects of the food of the larvae) (Brachycera, Orthorrhapha). Ann. Limnol. 11:169–188.

Thomas, F. 1933. On the bionomics and structure of some dipterous larvae infesting cereals and grasses I. *Opomyza florum* Fabr. Ann. Appl. Biol. 20:707–721.

Thomas, F. 1938. On the bionomics and structure of some dipterous larvae infesting cereals and grasses III. *Geomyza (Balioptera) tripunctata* Fall. Ann. Appl. Biol. 25:181–196.

Thompson. F. C. 1981. Revisionary notes on Nearctic *Microdon* flies (Diptera: Syrphidae). Proc. Ent. Soc. Wash. 83:725–758.

Thompson, J. H., L. V. Knutson, and O. S. Culp. 1970. Larva of *Scenopinus* sp. (Diptera: Scenopinidae) causing human urogenital myiasis. Mayo Clinic Proc. 45:597–601.

Thompson, P. H. 1978. Parasitism of adult *Tabanus subsimilis subsimilis* Bellardi (Diptera: Tabanidae) by a miltogrammine sarcophagid (Diptera: Sarcophagidae). Proc. Ent. Soc. Wash. 80:69–74.

Thompson, W. R. 1915. Sur les caractères anatomiques et éthologiques des Tachinaires du genre *Plagia* Meig. Parasitol. 78:671–674.

Thompson, W. R. 1920a. Note sur *Rhacodineura antiqua* Fall., Tachinaire parasite des Forficules. Bull. Soc. ent. France 1920, pp. 199–201.

Thompson, W. R. 1920b. Recherches sur les Dipteres parasites des larves des Sarcophagidae. Bull. Biol. France Belg. 54:313–463.

Thompson, W. R. 1920c. Sur les Diptères parasites des Isopodes terrestres. Comptes Rend. Soc. Biol. Paris 83:450–451.

Thompson, W. R. 1921. Contribution à la connaissance des formes larvaires des Sarcophagides. I. *Engyzops pecchiolii* Rond. Compt. Rend. Soc. Biol. Paris 1921, pp. 27–31.

Thompson, W. R. 1923. Recherches sur les Diptères parasites. Les larves primaires des Tachinidae du groupe des Echinomyiinae. Annls. Épiphyties 9:137–201.

Thompson, W. R. 1924. Les larves primaires des Tachinaires à oeufs microtypes. Annls. Parasitol. hum. comp. 2:185–201, 279–306.

Thompson, W. R. 1926. Recherches sur les larves des Tachinaires *Sturmia, Winthemia, Carcelia* et *Exorista*. Ann. Parasitol. Hum. Comp. 4:111–125, 207–227.

Thompson, W. R. 1934. The tachinid parasites of woodlice. Parasitol. 26:378–448.

Thompson, W. R. 1943–1958. A catalogue of the parasites and predators of insect pests. Commonwealth Inst. Biol. Control, Ottawa. 16 vols.

Thompson. W. R. 1954. *Hyalomyodes triangulifera* Loew (Diptera, Tachinidae). Canad. Entomol. 86:137–144.

Thompson, W. R. 1963. The tachinids of Trinidad. II. Echinomyiines, Dexiines, and allies. Can. J. Zool. 41:335–576.

Thomson, A. J. 1975. The biology of *Pollenia rudis,* the cluster fly (Diptera: Calliphoridae). IV. A preliminary model of the cluster fly life cycle, and some possible methods of biological control. Can. Ent. 107:855–864.

Thomson, A. J., and D. M. Davies. 1973a. The biology of *Pollenia rudis,* the cluster fly (Diptera: Calliphoridae): I. Host location by first-instar larvae. Can. Ent. 105:335–341.

Thomson, A. J. and D. M. Davies. 1973b. The biology of *Pollenia rudis,* the cluster fly (Diptera: Calliphoridae): II. Larval feeding behavior and host specificity. Can. Ent. 105:985–990.

Thomson, A. J., and D. M. Davies. 1974. The biology of *Pollenia rudis,* the cluster fly (Diptera: Calliphoridae): III. The effect of soil conditions on the host-parasite relationship. Can. Ent. 106:107–112.

Thomson, M. 1935. A comparative study of the development of the Stomoxydinae (especially *Haematobia stimulans* Meigen) with remarks on other coprophagous Muscidae. Proc. Zool. Soc. London 1935, pp. 531–550.

Thomson, M. 1938. Stuefluen og Stikfluen. Kopenhagen.

Thomson, M. and Hammer, O. 1936. The breeding media of some common flies. Bull. Ent. Res. 27:559–587.

Thomson, R. C. M. 1937. Observations on the biology of Anthomyiidae. Parasitol. 29:273–358.

Thornhill, R. C. 1976. Reproductive behavior of the lovebug, *Plecia nearctica* (Diptera: Bibionidae). Ann. Ent. Soc. Amer. 69:843–847.

Thorpe, W. H. 1930. The biology of the petroleum fly (*Psilopa petrolei*). Trans. Ent. Soc. London 78:331–344.

Thorpe, W. H. 1931. The biology, post-embryonic development and economic importance of *Cryptochaetum iceryae* (Diptera, Agromyzidae) parasitic on *Icerya purchasi* (Coccidae, Monophlebini). Proc. Zool. Soc. London 1931, pp. 929–971.

Thorpe, W. H. 1934. The biology and development of *Cryptochaetum grandicorne* (Diptera), an internal parasite of *Guerinia serratulae* (Coccidae). Quart. J. Microsc. Sci. (N.S.) 77:273–304.

Thorpe, W. H. 1941. The biology of *Cryptochaetum* (Diptera) and *Eupelmus* (Hymenoptera) parasites of *Aspidoproctus* (Coccidae) in East Africa. Parasitol. 33:149–168.

Throckmorton, L. H. 1975. The phylogeny, ecology, and geography of *Drosophila*, pp. 421–469. *In* King, R. C., ed., Handbook of Genetics, III.

Tidwell, Mac. A., and M. A. Tidwell. 1973. Larvae and pupae of five eastern North American *Tabanus* species (Diptera: Tabanidae). Ann. Ent. Soc. Am. 66:390–398.

Tokunaga, M. 1930. The morphological and biological studies on a new marine crane fly from Japan. Mem. Coll. Agr. Kyoto Imp. Univ. 10:1–127.

Tokunaga, M. 1935. On the pupae of the nymphomyiid fly (Diptera). Mushi 8:44–52.

Tolg, F. 1910. *Billaea pectinata* Mg. (*Sirostoma latum* Egg.) als Parasit von Cetoniden- und Cerambyciden-Larven. Metamorphose und äussere Morphologie der Larve. Zeitschr. Wiss. Ins. Biol. 6:208–211, 278–283, 331–336, 387–395, 426–430.

Tolg, F. 1913. Biologie und Morphologie einiger in Nonnenraupen schmarotzender Fliegenlarven. Zentralbl. Bakt. Parasitol., Inf. Krankh. 37:392–412.

Tonnoir, A. 1923. Australian Blephariceridae II. Larvae and pupae. Australian Zool. 3:47–59.

Tonnoir, A. 1933. Descriptions of remarkable Indian Psychodidae and their early stages, with a theory of the evolution of the ventral suckers of dipterous larvae. Rec. Indian Mus. 35:59–75.

Torelli, B. 1922. La *Notiphila chamaeleon* Becker sua larva rinvenute nel laghetto degli Astroni. Suppl. Ann. Mus. Zool. Napoli (N.S.), Fauna Astroni 9:1–6.

Townes, H. K., Jr. 1945. The Nearctic species of Tendipedini [Diptera, Tendipedidae (=Chironomidae)]. Amer. Midl. Nat. 34:1–206.

Townsend, C. H. T. 1892. Descriptions of Oestrid larvae taken from the jack rabbit and cotton-tail. Psyche 6:298–300.

Townsend, C. H. T. 1893a. Description of the pupa of *Toxophora virgata* O.S. Psyche 6:455–457.

Townsend, C. H. T. 1893b. The puparium and pupa of *Subula pallipes* Loew. Ent. News 4:163–165.

Townsend, C. H. T. 1893c. Dipterous parasites in their relation to economic entomology. Insect Life 6:201–204.

Townsend, C. H. T. 1934. Manual of myiology in twelve parts. Pt. 1: Development and structure, 280 pp. São Paulo.

Townsend, C. H. T. 1935. Pt. 2: Muscoid classification and habits, 296 pp., São Paulo.

Townsend, C. H. T. 1936a. Pt. 3: Oestroid classification and habits (Gymnosomatidae to Tachinidae), 255 pp. São Paulo.

Townsend, C. H. T. 1936b. Pt. 4: Oestroid classification and habits (Dexiidae to Exoristidae), 303 pp. São Paulo.

Townsend, C. H. T. 1937. Pt. 5: Muscoid generic diagnoses and data (Glossinini to Agriini), 234 pp. São Paulo.

Townsend, C. H. T. 1938a. Pt. 6: Muscoid generic diagnoses and data (Stephanostomatini to Moriniini), 309 pp. São Paulo.

Townsend, C. H. T. 1938b. Pt. 7: Oestroid generic diagnoses and data (Gymnostomatini to Senostomatini, 434 pp. São Paulo.

Townsend, C. H. T. 1939a. Pt. 8: Oestroid generic diagnoses and data (Microtropesini to Voriini), 408 pp. São Paulo.

Townsend, C. H. T. 1939b. Pt. 9: Oestroid generic diagnoses and data (Thelairini to Clythoini), 270 pp. São Paulo.

Townsend, C. H. T. 1940. Pt. 10: Oestroid generic diagnoses and data (Anacamptomyiini to Frontinini), 335 pp. São Paulo.

Townsend, C. H. T. 1941. Pt. 11: Oestroid generic diagnoses and data (Goniini to Trypherini), 342 pp. São Paulo.

Townsend, C. H. T. 1942. Pt. 12: General consideration of the Oestromuscaria, 365 pp. São Paulo.

Townsend, C. H. T. 1935. The mature larva and puparium of *Physocephala sagittaria* Say. Psyche 42:142–148.

Townsend, L. H. 1936. The mature larva and puparium of *Brachycoma sarcophagina* (Townsend) (Diptera: Metopiidae). Proc. Ent. Soc. Wash. 38:92–98.

Tracewski, K. T. 1983. Description of the immature stages of *Leucopis* sp. nr. *albipuncta* (Diptera: Chamaemyiidae) and their role as predators of the apple aphid, *Aphis pomi* (Homoptera: Aphididae). Can. Ent. 115:735–742.

Tragardh, J. 1903. Zur Anatomie und Entwicklungsphysiologie der Larve *Ephydra riparia*. Ark. Zool. 1:1–42.

Tragardh, J. 1931. Zwei forstentomologisch wichtige Fliegen. Zeitschr. Angew. Ent. 18:672–690.

Traxler, F. E. 1977. Developmental anatomy of the cephalopharyngeal apparatus of the first and second instars of *Lucilia sericata* Meigen larvae (Diptera: Calliphoridae). J. N.Y. Ent. Soc. 85:2–17.

Trelka, D. G., and C. O. Berg. 1977. Behavorial studies of the slug-killing larvae of two species of *Tetanocera* (Diptera: Sciomyzidae). Proc. Ent. Soc. Wash. 79:475–486.

Trelka, D. G., and B. A. Foote. 1970. Biology of slug-killing *Tetanocera* (Diptera: Sciomyzidae). Ann. Ent. Soc. Amer. 63:877–895.

Tripp, H. A. 1954. The instars of a maggot (*Pegohylemyia*) inhabiting white spruce cones. Can. Ent. 86:185–189.

Turner, E. C., Burton, R. P. and Gerhardt, R. R. 1968. Natural parasitism of dung-breeding Diptera: a comparison between native hosts and an introduced host, the face fly. J. Econ. Ent. 61:1012–15.

Turner, W. J. 1974. A revision of the genus *Symphoromyia* Frauenfeld (Diptera: Rhagionidae). 1. Introduction. Subgenera and species-groups. Review of biology. Can. Ent. 106:851–858.

Tuxen, S. L. 1936. Die Arten der Gattung *Scatella* in heissen Quellen. Opuscula Entomol. 1:105–111.

Twinn, C. R. 1936. The black flies of eastern Canada. Can. J. Res. 14:97–150.

Uhler, L. D. 1951. Biology and ecology of the goldenrod gall fly *Eurosta solidaginis* (Fitch). Mem. Cornell Univ. Agr. Exp. Sta. 300:1–51.

Ullrich, H. 1936a. Über die Dasselfliege des Elches, *Hypoderma alcis* spec. nov. Deutsche Tierarztl. Wochenschr. 44:577–579.

Ullrich, H. 1936b. Untersuchungen über die Biologie der Rachenbremse (Genus *Cephenomyia* Latreille) über die pathogenen Einflüsse der Rachenbremsenlarven auf ihre Wirtstiere und über Bekämpfungsmöglichkeiten der Rachenbremsenplage. Inaugural-Dissertation Berlin.

Ullrich, H. 1939. Zur Biologie der Rachenbremsen unseres einheimischen Wildes, Genus *Cephenomyia* Latreille und Genus *Pharyngomyia* Schiner. Verhandl. VII. Intern. Kongr. Entomol. (Berlin) 3:2149–2162.

Vaillant, F. 1948. Les premiers stades de *Liancalus virens* Scop. (Dolichopodidae). Bull. Soc. zoo. France 73:118–130.

Vaillant, F. 1949. Les premiers stades de *Tachytrechus notatus* Stann. et de *Syntormon zelleri* Lw. (Dolichopodidae). Bull. Soc. Zoo. France 74:122–126.

Vaillant, F. 1950. Les premiers stades de *Dolichopus griseipennis* Stann. (Dolichopodidae). Bull. Soc. Zoo. France 75:80–84.

Vaillant, F. 1959. The Thaumaleidae (Diptera) of the Appalachian Mountains. J. N.Y. Ent. Soc. 67:31–37.

Vaillant, F. 1963. Les *Maruina* d'Amérique du Nord (Dipt. Psychodidae). Bull. Soc. ent. France 68:71–91.

Vaillant, F. 1971–1976. Psychodidae-Psychodinae. Bd. 3 Lfg. 9d, pp. 1–206. *In* Lindner, E., ed., Die Fliegen der palaearktischen Region. Schweizerbart, Stuttgart.

Vaillant, F., and M. Delhom. 1956. Les formes adaptives de l'appareil bucco-pharyngien chez les larves de Stratiomyiidae (Diptera). Bull. Soc. Hist. nat. Afr. N. 47:217–250.

Valiela, I. 1970. The arthropod fauna of bovine dung in central New York and sources on its natural history. J. N.Y. Ent. Soc. 77:210–220.

Valley, K. R., and C. O. Berg. 1977. Biology and immature stages of snail-killing Diptera of the genus *Dictya* (Sciomyzidae). Search Agr. (Geneva, N.Y.) 7 (2):1–44.

Valley, K. R., J. A. Novak, and B. A. Foote. 1969. Biology and immature stages of *Eumetopiella rufipes*. Ann. Ent. Soc. Amer. 62:227–234.

Van Emden-*see* Emden

Vargas, L. 1940. Clave para identificar las larvas de *Anopheles* mexicanos. Ciencia 1:66–68.

Varley, G. C. 1937a. Aquatic insect larvae which obtain oxygen from the roots of plants. Proc. Roy. Ent. Soc. London (A) 12:55–60.

Varley, G. C. 1937b. The life-history of some trypetid flies, with descriptions of the early stages. Proc. Roy. Ent. Soc. London (A) 12:109–122.

Varley, G. C. 1947. The natural control of population balance in the knapweed gall-fly (*Urophora jaceana*). J. Anim. Ecol. 16:139–187.

Varssière, P. 1926. Contribution à l'étude biologique et systématique des Coccidae. Ann. Epiphyties 12:197–382.

Verbeke, J. 1975. Contribution à l'étude des diptères malacophages. 2. Données nouvelles sur la taxonomie et la répartition géographique des Sciomyzidae paléarctiques. Bull. Inst. roy. Sci. nat. Belg. 40:1–27.

Vimmer, A. 1911. Beiträge zur Kenntnis der cyclorrhaphen Dipterenpuppen. Acta Soc. Entomol. Bohem. 8:51–55.

Vimmer, A. 1912. Über den Hypopharynx einiger Dipterenlarven aus der Unterordnung Orthorrhapha. Soc. Ent. 27:103–105, 110–112.

Vimmer, A. 1925. Larvy a kukly dvoukridleho hmyzu Stredoevropskeho Nakladem Ceske. Prague. 348 p.

Vimmer, A. 1931. Über die Larven kleiner Fliegen, welche in der Tschechoslovak. Republik durch Ausnagen von Hyponomen den Pflanzen schaden. Arch. Prirod. Vyzkum Cech. 18:1–159.

Vockeroth, J. R. 1961. The North America species of the family Opomyzidae (Diptera: Acalyptratae). Can. Ent. 93:503–522.

Vockeroth, J. R. 1965. Scatophaginae, *in* Stone *et al.* A Catalog of the Diptera of America North of Mexico. Agric. Handbook No. 276. Agr. Res. Ser. U.S. Dept. of Agric.

Vockeroth, J. R. 1969. A revision of the genera of the Syrphini (Diptera: Syrphidae). Mem. Ent. Soc. Can. 62:1–176.

Vockeroth, J. R. 1974. Notes on the biology of *Cramptonomyia spenceri* Alexander (Diptera: Cramptonomyiidae). J. Ent. Soc. Brit. Col. 71:38–42.

Vockeroth, J. R. 1981. Mycetophilidae, pp. 223–246. *In* McAlpine, J. F., et al. eds. Manual of Nearctic Diptera. Vol. 1. Res. Br. Agr. Can. Monogr. 27.

Vockeroth, J. R. 1987a. Opomyzidae, pp. 881–885. *In* McAlpine, J. F., et al., eds., Manual of Nearctic Diptera. Vol. 2. Res. Br. Agr. Can. Monogr. 28.

Vockeroth, J. R. 1987b. Anthomyzidae, pp. 887–890. *In* McAlpine, J. F., et al., eds., Manual of Nearctic Diptera. Vol. 2. Res. Br. Agr. Can. Monogr. 28.

Vockeroth, J. R. 1987c. Coelopidae, pp. 919–922. *In* McAlpine, J. F., et al., eds., Manual of Nearctic Diptera. Vol. 2. Res. Br. Agr. Can. Monogr. 28.

Vockeroth, J. R. 1987d. Tethinidae, pp. 1073–1078. *In* McAlpine, J. F., et al., eds., Manual of Nearctic Diptera. Vol. 2. Res. Br. Agr. Can. Monogr. 28.

Vockeroth, J. R. 1987e. Scathophagidae, pp. 1085–1097. *In* McAlpine, J. F., et al., eds., Manual of Nearctic Diptera. Vol. 2. Res. Br. Agr. Can. Monogr. 28.

Vockeroth, J. R., and F. C. Thompson. 1987. Syrphidae, pp. 713–743. *In* McAlpine, J. F., et al., eds., Manual of Nearctic Diptera. Vol. 2. Res. Br. Agr. Can. Monogr. 28.

Vogler, C. H. 1900a. Beiträge zur Metamorphose der *Teichomyza fusca*. Ill. Zeitschr. Ent. 5:1–4, 17–20, 33–36.

Vogler, C. H. 1900b. Weitere Beiträge zur Kenntnis von Dipterenlarven. Ill. Zeitschr. Ent. 5:273–276, 289–292.

Vos-de Wilde, B. 1935. Contributions à l étude des larves des Cyclorrhaphes, plus spécialement des larves d'Anthomyides. Dissertation, Amsterdam.

Wagner, R. P. 1944. The nutrition of *Drosophila mulleri* and *D. aldrichi*. Growth of the larvae on a cactus extract and the microorganisms found in cactus. Univ. Texas Publ. 4445:104–128.

Walker, E. M. 1920. *Wohlfahrtia vigil* (Walker) as a human parasite (Diptera-Sarcophagidae). J. Parasitol. 7:1–7.

Walker, E. M. 1937. The larval stages of *Wohlfahrtia vigil* (Walker). J. Parasitol. 23:163–174.

Wallace, J. B. 1970. The mature larva and pupa of *Calobatina geometroides* (Cresson) (Diptera: Micropezidae). Ent. News 80:317–321.

Wallace, J. B. 1971. Immature stages of some muscoid calyptrate Diptera inhabiting mushrooms in the southern Appalachians. J. Georgia Ent. Soc. 6:218–229.

Wallace, J. B., and R. T. Franklin. 1970. *Eutheresia interrupta* (Diptera: Tachinidae), a parasite of the cerambycid *Neacanthocinus obsoletus*. J. Georgia Ent. Soc. 5:25–30.

Wallace, J. B., and A. G. Lavallee. 1973. Immature stages of Milesiinae (Syrphidae) I. *Cheilosia pallipes* and *Volucella apicalis*. J. Georgia Ent. Soc. 8:187–194.

Wallace, J. B. and Neff, S. E. 1971. Biology and immature stages of the genus *Cordilura* (Diptera: Scatophagidae) in the eastern United States. Ann. Ent. Soc. Amer. 64:1310–1330.

Wallace, J. B. and Snoddy, E. L. 1969. Notes on *Dendrophaonia querceti* with descriptions of the larva and pupa (Diptera, Muscidae). J. Georgia Ent. Soc. 4:1–4.

Walsh, B. D., and C. V. Riley. 1869. The root-louse syrphus fly (*Pipiza radicum* n. sp.). Amer. Ent. 1:83–84.

Walton, W. R. 1908. Notes on the egg and larva of *Goniops chrysocoma* U.S. Ent. News 19:464–465.

Wandolleck, B. 1898. Die Fühler der cyclorrhaphen Dipterenlarven. Zool. Anz. 21: 283–294.

Wangberg, J. K. 1978. Biology of gall formers of the genus *Valentibulla* (Diptera: Tephritidae) on rabbitbrush in Idaho. J. Kans. Ent. Soc. 51:472–483.

Wangberg, J. K. 1980. Comparative biology of gall-formers in the genus *Procecidochares* (Diptera: Tephritidae) on rabbitbrush in Idaho. J. Kans. Ent. Soc. 53:401–420.

Ward, R. D., and P. A. Ready. 1975. Chorionic sculpturing in some sandfly eggs (Diptera, Psychodidae). J. Ent. (A) (Physiol., Behavior) 50:127–134.

Wasbauer, W. S. 1972. An annotated host catalog of the fruit flies of America north of Mexico. (Diptera: Tephritidae). Bur. Ent. Calif. Dept. Agr. Occ. Pap. 19:1–172.

Watson, J. R. 1926. Citrus insects and their control. Bull. Fla. Agr. Exp. Sta. 183:367–368.

Way, M. O., F. T. Turner, and J. K. Clark. 1983. The rice leaf miner, *Hydrellia griseola*, a potential pest of rice in Texas. Southwest Ent. 8:186–189.

Weaver, J. E., and C. K. Dorsey. 1967. Larval mine characteristics of five species of leaf-mining insects in black locust, *Robinia pseudoacacia*. Ann. Ent. Soc. Amer. 60:172–186.

Webb, D. W. 1974. A revision of the genus *Hilarimorpha* (Diptera: Hilarimorphidae). J. Kans. Ent. Soc. 47:172–222.

Webb, D. W. 1975. New species of *Hilarimorpha* (Diptera: Hilarimorphidae). Ent. News 86:80–84.

Webb, D. W. 1977. The Nearctic Athericidae (Insecta: Diptera). J. Kans. Ent. Soc. 50:473–495.

Webb, D. W. 1978. A revision of the Nearctic genus *Dialysis* (Diptera: Rhagionidae) J. Kans. Ent. Soc. 51:405–431.

Webb, D. W. 1979. A revision of the Nearctic species of *Xylophagus* (Diptera, Xylophagidae). J. Kans. Ent. Soc. 52:489–523.

Webb, D. W. 1981a. Athericidae, pp. 479–482. *In* McAlpine, J. F., et al., eds., Manual of Nearctic Diptera. Vol. 1. Res. Br. Agr. Can. Monogr. 27.

Webb, D. W. 1981b. Hilarimorphidae, pp. 603–605. *In* McAlpine J. F., et al., eds. Manual of Nearctic Diptera. Vol. 1. Res. Br. Agr. Can. Monogr. 27.

Webb, D. W. 1983a. The genus *Coenomyia* (Diptera: Coenomyiidae) in the Nearctic region and notes on generic placement. Proc. Ent. Soc. Wash. 85:653–664.

Webb, D. W. 1983b. A revision of the Nearctic species of *Arthropeas* (Diptera: Coenomyiidae). Proc. Ent. Soc. Wash. 85:737–747.

Webb, D. W. 1983c. A new genus of Nearctic coenomyiid (Diptera: Coenomyiidae). Proc. Ent. Soc. Wash. 85:822–825.

Webb, D. W. 1983d. A revision of the Nearctic species of *Rachicerus* (Diptera: Rachiceridae). J. Kans. Ent. Soc. 56:298–315.

Webb, D. W. 1984. A revision of the Nearctic species of the family Solvidae (Insecta: Diptera). Trans. Amer. Ent. Soc. 110:245–293.

Webb, D. W., and E. A. Lisowski. 1983. The immature stages of *Dialysis fasciventris* (Loew) (Diptera: Coenomyiidae). Proc. Ent. Soc. Wash. 85:691–697.

Webb, J. L., and R. W. Wells. 1924. Horse-flies: biologies and relation to western agriculture. U.S. Dept. Agr. Bull. 1218:1–36.

Wecker, S. C. 1962. The effects of bot fly parasitism on a local population of the white-footed mouse. Ecol. 43:561–565.

Weiss, H. B. 1919. Notes on the early stages and larval locomotion of *Leia bivittata* Say. Psyche 26:80–82.

Welch, P. S. 1912. Observations on the life-history of a new species of *Psychoda*. Ann. Ent. Soc. Amer. 5:411–418.

Welch, P. S. 1917. Further studies on *Hydromyza confluens* Loew. Ann. Ent. Soc. Amer. 10:35–46.

Wendt, H. 1968. Faunistisch-ökologische Untersuchungen an Halmfliegen der Berliner Umgebung. Deutsche Ent. Zeitschr. 15:49–105.

Wenzel, R. L., and B. V. Peterson, 1987. Streblidae, pp. 1293–1301. *In* McAlpine, J. F. et al., eds. Manual of Nearctic Diptera, Vol. 2. Res. Br. Agr. Can. Mongr. 28.

Wenzel, R. L., V. J. Tipton, and A. Kiewlicz. 1966. The streblid bat flies of Panama, pp. 405–675. *In* Wenzel, R. L., and V. J. Tipton. The ectoparasites of Panama. Field Mus. Nat. Hist., Chicago.

Wesenberg-Lund, C. 1943. Biologie der Süsswasserinsekten. J. Springer, Berlin. 682 pp.

West, L. S. 1951. The Housefly, its Natural History, Medical Importance and Control. Comstock Publ. Co., Ithaca, N.Y. xvi+584 pp.

Wheeler, A. G., Jr. 1973. Studies on the arthropod fauna of alfalfa IV. Species associated with the crown. Can. Ent. 105:353–366.

Wheeler, A. G., Jr. 1982. *Clanoneurum americanum* (Diptera: Ephydridae), a leafminer of the littoral chenopod *Suaeda linearis*. Proc. Ent. Soc. Wash. 84:297–300.

Wheeler, M. R. 1954. *Rhinotora diversa* (Giglio-Tos from the southwestern United States (Diptera: Rhinotoridae). Wasmann J. Biol. 12:35–39.

Wheeler, M. R. 1956. *Latheticomyia*, a new genus of acalyptrate flies of uncertain family relationship. Proc. U.S. Nat. Mus. 106:305–314.

Wheeler, M. R. 1957. VII. Taxonomic and distribution studies of Nearctic and Neotropical Drosophilidae, pp. 79–114. *In* J. T. Patterson, Studies in the Genetics of *Drosophila*. Texas Univ. Publ. 5721:1–316.

Wheeler, M. R. 1961. New species of southwestern acalyptrate Diptera. Southw. Nat. 6:86–91.

Wheeler, M. R. 1987. Drosophilidae, pp. 1011–1018. *In* McAlpine, J. F., et al., eds., Manual of Nearctic Diptera. Vol. 2. Res. Br. Agr. Can. Monogr. 28.

Wheeler, W. M. 1901. *Microdon* larvae in *Pseudomyrma* nests. Psyche 9:222–224.

Wheeler, W. M. 1908. Studies on myrmecophiles III. *Microdon*. J. N.Y. Ent. Soc. 16:202–213.

Wheeler, W. M. 1924. Two extra-ordinary larval myrmecophiles from Panama. Proc. Nat. Sci. U.S.A. 10:237–244.

Wheeler, W. M. 1930. Demons of the Dust. W. W. Norton, New York.

White, M. J. D. 1950. Cytological studies of gall midges (Cecidomyiidae). Univ. Texas Publ. 5007:1–80.

Whittaker, J. B. 1969. The biology of Pipunculidae (Diptera) parasitising some British Cercopidae. Proc. Roy. Ent. Soc. London (A) 44:17–24.

Whittaker, J. D., Jr., and D. A. Esterla. 1974. Batflies (Streblidae and Nycteribiidae) in the eastern U.S., and a nycteribiid record from Saskatchewan. Ent. News 85:221–223.

Whitten, J. M. 1955. A comparative morphological study of the tracheal system in larval Diptera. Part 1. Q.J. Micros. Sci. 96:257–278.

Whitten, J. M. 1956. The tracheal system of the larva of *Lonchoptera lutea* Panzer (Diptera: Lonchopteridae). Proc. Roy. Ent. Soc. London 31(7–9):105–108.

Whitten, J. M. 1960. The tracheal system as a systematic character in larval Diptera. Syst. Zool. 8:130–139.

Whitten, J. M. 1963. The tracheal pattern and body segmentation in the blephariccerid larva. Proc. R. Ent. Soc. Lond., Ser. A, 38:39–44.

Weiderholm, T., ed. 1983. Chironomidae of the Holarctic Region. Keys and Diagnoses. Part 1. Larvae. Entomol. Scand. Suppl. 19. 457 pp.

Wigglesworth, V. B. 1938. The regulation of osmotic pressure and chloride concentration in the haemolymph of mosquito larvae. J. Exp. Biol. 15:235–237.

Wilcochs, C. R., and E. H. A. Oliver. 1976. Life cycle of Australian soldier fly, *Inopus rubriceps* (Diptera-Stratiomyiidae) in New Zealand. N. Z. J. Zool. 3:115–126.

Wilcox, J. 1926. The lesser bulb fly, *Eumerus strigatus* Fallen, in Oregon. J. Econ. Ent. 19:762–772.

Wilcox, J. 1927. Observations on the life history, habits and control of the narcissus bulb fly, *Merodon equestris* Fabr. and the lesser bulb fly, *Eumerus strigatus* Fall. in Oregon. 19th Bienn. Rept. Ore. St. Bd. Hort., pp. 149–158.

Wilcox, J. 1981. Mydidae, pp. 533–540. *In* McAlpine, J. F., et al., eds., Manual of Nearctic Diptera. Vol. 1. Res. Br. Agr. Can. Monogr. 27.

Wilcox, J., and N. Papavero. 1971. The American genera of Mydidae (Diptera), with the description of three new genera and two new species. Archos Zool. Est. S. Paulo 21:41–119.

Wilder, D. D. 1974. A revision of the genus *Syneches* Walker (Diptera: Empididae) for North America and the Antilles. Contrib. Amer. Ent. Inst. (Ann Arbor). 10(5):1–30.

Wilder, D. D. 1983. Two new Nearctic Clinocerine genera (Diptera: Empidiae). Mem. Ent. Soc. Wash. 10:166–178.

Wilkes, A., H. C. Coppel, and W. G. Mathers. 1948. Notes on the insect parasites of the spruce budworm, *Choristoneura fumiferana* (Clemens) in British Columbia. Can. Ent. 80:138–155.

Williams, F. X. 1938. Biological studies in Hawaiian waterloving insects, part III. Diptera or flies. A. Ephydridae and Anthomyiidae. Proc. Hawaii. Ent. Soc. 10:85–119.

Wilson, F. 1938. Notes on the insect enemies of *Chermes* with particular reference to *Pineus pini* Koch, and *P. strobi* Hartig. Bull. Ent. Res 29:373–389.

Wilson, L. F. 1966. Life history, habits, and damage of the boxelder leaf gall midge, *Contarina negundifolia* Felt (Diptera: Cecidomyiidae) in Michigan. Can. Ent. 98:777–84.

Wilson, L. F. 1968. Life history and habits of the pine cone willow gall midge, *Rhabdophaga strobiloides* (Diptera: Cecidomyiidae) in Michigan. Can. Ent. 100:430–433.

Wirth, W. W. 1949. A revision of the clunionine midges with descriptions of a new genus and four new species. Univ. Calif. Publ. Ent. 8:151–182.

Wirth, W. W. 1951. A revision of the dipterous family Canaceidae. Bernice P. Bishop Mus. Occ. Papers 20:245–275.

Wirth, W. W. 1952. The Heleidae of California. Univ. Calif. Publ. Ent. 9:95–266.

Wirth, W. W. 1956. New species and records of biting midges ectoparasitic on insects (Diptera; Heleidae). Ann. Ent. Soc. Amer. 49:356–364.

Wirth, W. W. 1971a. *Platygymnopa*, a new genus of Ephydridae reared from decaying snails in North America (Diptera). Can. Ent. 103:266–270.

Wirth, W. W. 1971b. The brine flies of the genus *Ephydra* in North America (Diptera: Ephydridae). Ann. Ent. Soc. Amer. 64:359–377.

Wirth, W. W. 1987. Canacidae, pp. 1079–1083. *In* McAlpine, J. F. et al., eds. Manual of Nearctic Diptera, Vol. 2. Res. Br. Agr. Can. Monogr. 28.

Wirth, W. W., W. N. Mathis and J. R. Vockeroth. 1987. Ephydridae, pp. 1027–1047. *In* McAlpine, J. F. et al., eds. Manual of Nearctic Diptera. Vol. 2. Res. Br. Agr. Can. Mongr. 28.

Wirth, W. W., and A. A. Hubert. 1960. Ceratopogonidae (Diptera) reared from cacti, with a review of the *copiosus* group of *Culicoides*. Ann. Ent. Soc. Amer. 53:639–658.

Wirth, W. W., and A. Stone. 1956. Aquatic Diptera, pp. 372–482. *In* Usinger, R. L., ed., Aquatic Insects of California, with Keys to North American Genera and California species. Univ. Calif. Press, Berkeley.

Wirth, W. W., and J. E. Sublette. 1970. A review of the Podonominae of North America with descriptions of three new species of *Trichotanypus* (Dipt.: Chir.). J. Kans. Ent. Soc. 43:335–354.

Wishart, G. 1946. *Aplomyia caesar* (Aldrich), a tachinid parasite of the European corn borer. Can. Ent. 77:157–167.

Wood, A. H. 1933. Notes on some dipterous parasites of *Schistocerca* and *Locusta* in the Sudan. Bull. Ent. Res. 24:521–530.

Wood, D. M. 1981a. Axymyiidae, pp. 209–212. *In* McAlpine, J. F., et al., eds., Manual of Nearctic Diptera. Vol. 1. Res. Br. Agr. Can. Monogr. 27.

Wood, D. M. 1981b. Pachyneuridae, pp. 213–216. *In* McAlpine, J. F., et al., eds., Manual of Nearctic Diptera. Vol. 1. Res. Br Agr. Can. Monogr. 27.

Wood, D. M. 1987a. Oestridae, pp. 1147–1158. *In* McAlpine, J. F. et al., eds. Manual of Nearctic Diptera, Vol. 2. Res. Br. Agr. Can. Monogr. 28.

Wood, D. M. 1987b. Rhinophoridae, pp. 1187–1191. *In* McAlpine, J. F., et al., eds., Manual of Nearctic Diptera. Vol. 2. Res. Br Agr. Can. Monogr. 28.

Wood, D. M. 1987c. Tachinidae, pp. 1193–1269. *In* McAlpine, J. F. et al., eds. Manual of Nearctic Diptera, Vol. 2. Res. Br. Agr. Can. Monogr. 28.

Wood, D. M., B. V. Peterson, D. M. Davies, and H. Gyorkos. 1963. The black flies (Diptera: Simuliidae) of Ontario. II. Larval identification, with descriptions and illustrations. Proc. Ent. Soc. Ont. 93:99–129.

Wood, G. C. 1981. Asilidae, pp. 549–573. *In* McAlpine, J. F., et al., eds., Manual of Nearctic Diptera. Vol. 1. Res. Br. Agr. Can. Monogr. 27.

Wood, K. G. 1956. Ecology of *Chaoborus* (Diptera: Culicidae) in an Ontario lake. Ecol. 37:639–643.

Woodley, N. 1981. A revision of the Nearctic Beridinae (Diptera: Stratiomyidae). Bull. Mus. Comp. Zool. 149(6):319–369.

Wootton, R. J. 1972. The evolution of insects in fresh water ecosystems, pp. 69–82. *In* Clark, R. B., and R. J. Wootton, eds., Essays in Hydrobiology. Univ. Exeter Press, England.

Worthley, H. N. 1924. The biology of *Trichopoda pennipes* Fab. a parasite of the common squash bug. Psyche 31:7–16, 57–77.

Wyatt, I. J. 1967. Pupal paedogenesis in the Cecidomyiidae (Diptera). 3. A reclassification of the Heteropezini. Trans. Roy. Ent. Soc. London 119:71–98.

Yahake, E. and J. A. Georfe. 1972. Rearing and immature stages of the cluster fly *Pollenia rudis* (Diptera: Calliphoridae) in Ontario. Can. Ent. 104:567–576.

Yang, C. K. 1979. A new genus and species of wormlion from China (Diptera:Rhagionidae). Entomotaxonomia 1:83–89. (in Chinese with English summary).

Yang, Z.-Q. 1984. Notes on the larva and puparium of *Odinia xanthocera* Collin (Diptera, Odiniidae). Ann. Ent. Fenn. 50:93–94.

Zack, R. S., Jr. 1983a. Biology and immature stages of *Setacera needhami* Johannsen (Diptera: Ephydridae). Proc. Ent. Soc. Wash. 85:10–25.

Zack, R. S., Jr. 1983b. Biology and immature stages of *Paracoenia bisetosa* (Coquillett) (Diptera: Ephydridae) Ann. Ent. Soc. Amer. 76:487–497.

Zack, R. S., Jr., and B. A. Foote. 1978. Utilization of algal monocultures by larvae of *Scatella stagnalis*. Environ. Ent. 7:509–511.

Zikan, J. F. 1942. Algo sobre a simbiose de *Mydas* com *Atta*. Rodriguesia VI (15):1–7.

Zikan, J. F. 1944. Novas observacoes sobre a biologia de *Mydas* (Diptera) e sua relacao com os formigueiros de sauva. Boln Minist. Agr. Rio de Janeiro 33:43–55.

Zubkov, G. A., and Kovalev, V. G. 1975. New data on the development of flies of the family Odiniidae (Diptera) Nauchnye Dokl. Vyssh. Skk. Biol. Nauki 14–19.

Zumpt, F. 1965. Myiasis in Man and Animals in the Old World. Butterworth. London. 267 p.

Zuska, J. 1962. The first instar larvae of the genus *Trixa* Meigen and remarks on the systematics and nomenclature of this genus (Diptera: Larvaevoridae). Cas. C. Spol. Ent. 59:80–86.

Zuska, J., and P. Lastovka. 1965. A review of the Czechoslovak species of the family Piophilidae with special reference to their importance to food industry (Diptera, Acalyptrata). Acta Ent. Bohemoslovaca 62:141–157.

Glossary

Abductor Muscle. A muscle, usually small, which is attached to the outer edge of the mandibular base and which moves the apex of the mandible away from the midline.

Acanthoparia (-ae). Part of the epipharynx in Scarabaeoidea; spiny marginal part of paria, bearing bristles (Coleoptera).

Accessory Air Tube. *See* Accessory Chamber.

Accessory Chamber. One of the small side chambers opening into the atrium of an annular-uniforous, annular-biforous, or annular-multiforous spiracle (Coleoptera).

Accessory Condyle. *See* Accessory Ventral Process.

Accessory Process. *See* Sensorium.

Accessory Ventral Condyle. *See* Accessory Ventral Process.

Accessory Ventral Process (condyle). A sclerotized process located on the ventral surface of the manidubular base just laterad of the mola and near the insertion of the adductor muscle (=second condyle) (Coleoptera).

Acephalic. Lacking a distinct head capsule, typical of Muscomorpha that have a cephalopharyngeal skeleton (Diptera).

Acetabulum. The cavity into which another structure fits (=fossa, depression).

Acrocercus (-i). One of the paired processes in the respiratory chamber of some hydrophilid larvae; located posterior to mesocerci.

Acroparia (-ae). Part of the epipharynx in Scarabaeoidea; anterior part of paria, bearing bristles (Coleoptera).

Acuminate. Tapering to a long point.

Adductor Muscle. A large muscle, which is attached to the mesal part of the mandibular base or to the adductor apodeme, and which moves the apex of the mandible toward the midline (bringing the two mandibles into apposition) (or downward in many Diptera).

Adecticous Pupa. A pupa without functional mandibles.

Adfrontal Area. An oblique sclerite on each side of the frons, usually extending from the base of the antennae to the epicranial suture where they meet, or to the epicranial notch if they do not meet (Lepidoptera).

Adhesive Seta. Tarsungular seta expanded at the apex (Coleoptera).

Adnasale (-alia). One of the lateral lobes of the nasale in carabid larvae (Coleoptera).

Aerenchymatous. Cells or tissues of aquatic plants that contain air.

Aereolate. A network-like pattern (Hemiptera)

Alar Area. *See* Spiracular Area and Spiracular Sclerite.

Ambrosia Beetles. Beetles which infect wood with *Endomyces* and *Ceratocystis* fungi, on which their larvae feed; members of the families Lymexylidae and Platypodidae, and certain Scolytidae, such as *Xyleborus* species (Coleoptera).

Ambulacral. Locomotory (referring to various locomotory protuberances, ambulatory warts, prolegs, etc.).

Ambulacral Wart. *See* Ampulla.

Ambulatory Wart (ambulatorial wart) *See* Ampulla.

Amphipneustic Respiratory System. One in which only one pair of thoracic spiracles and one or two posterior pairs of abdominal spiracles are open and functional.

Amphitoky. A type of parthenogensis in which both males and females are produced.

Ampulla (-ae). A fleshy protuberance located dorsally or ventrally on the abdomen and sometimes dorsally on the thorax, and used in locomotion (Lepidoptera, Coleoptera, and others).

Anal Cleft. A space or crack leading from the anal opercula to the posterior body margin (Coccidae).

Anal Fringe. Irregular posterior body margin made up of supra-anal plates and anal ring setae; Homoptera:Tachardiidae.

Anal Gills. A group of gills associated with the abdominal apex, usually consisting of from 3 to 5 tufts or papillae.

Anal Lobes. Any and various protrusions near the anus. Posterior protrusion of derma at apex of body; one on each side (Coccoidea).

Anal Opercula. One of two plates that makes up the operculum, a shieldlike structure that covers the anal opening (Coccoidea).

Anal Pad. Glabrous area surrounded by a fine cuticular line and located on one of the lateral anal lobes in Lucanidae (Coleoptera).

Anal Papilla (-ae). One of two to many membranous lobes projecting from the anal region, which may be respiratory or osmoregulatory in function; they differ from typical anal gills in being neither filamentous nor arranged in tufts.

Anal Plate. Caterpillars and other larvae. The dorsal shieldlike covering of the last abdominal segment (-suranal plate, anal shield; epiproct). Sclerotized structure associated with anal apparatus that does not contain pores and is not in the form of a ring (Coccoidea).

Anal Ring. Sclerotized structure surrounding orifice of anal opening, frequently containing pores and setae (Coccoidea).

Anal Slit. Anal opening which is narrow and either vertical or transverse.

Anal Tubes. Eversible, tubular organs arising from the anal region and armed with fine asperities; *see* Holdfast Organ (Coleoptera).

Annular Spiracle. One which has a single, circular or oval opening and no accessory chambers.

Annular-Biforous Spiracle. One which has two accessory chambers attached to the atrium.

Annular-Multiforous Spiracle. One which has three to many accessory chambers attached to the atrium.

Annular-Uniforous Spiracle. One which has one accessory chamber attached to the atrium.

Annulate (annulated). Ringed, but not truly segmented.

Annuliform Spiracle. *See* Annular Spiracle.

Antecylpeus. The anterior and usually more lightly sclerotized part of the clypeus, to which the labrum is attached.

Antennal Fossa (-ae). A cavity or depression in which the antenna is located.

Antennal Insertion. *See* Antennal Fossa.

Antepenultimate. Second from the last.

Anterad. Toward the front.

Apicad. Toward the apex or tip.

Apical Lobes Of Maxilla. Lobes attached to the stipes; *see* Galea, Lacinia, Mala.

Apneustic Respiratory System. One in which there are no functional spiracles.

Apodeme. A plate-like, cuticular invagination; visible on the surface as a suture or fold.

Apodous Larva. Legless.

Apomorphy--phic,-phous. The relatively derived state (more recent) of a sequence of homologous characters. *Compare* Plesiomorphy.

Apophysis (-ses). A narrow, rod-like cuticular invagination; usually visible on the surface as a pit.

Apotorma (-ae). Process extending forward from torma between pternotorma and mesal end of torma (or base of epitorma) (Coleoptera).

Arched Plate. Narrow, sclerotized bar in form of half circle, surrounding anterior margin of anal apparatus, associated with anal plates (Coccoidea).

Armature. Setae, spines or other sclerotized processes on the cuticle; sometimes restricted to unarticulated processes, as opposed to vestiture, which refers to setae, hairs, etc.

Arrhenotoky. Facultative parthenogenesis in which fertilized eggs give rise to females and unfertilized eggs give rise to males.

Articulating Area. *See* Postcoxale, Precoxale (Coleoptera).

Articulation. A connection or joint between two structures or sections of cuticle; may be ball and socket joint, a suture, a fold, or a conjunctiva (lightly sclerotized or membranous area between 2 more heavily sclerotized areas).

Asperate. Bearing asperities.

Asperities. Small, tooth-like, spine-like, or peg-like structures, frequently in rows or patches (microspines, microtrichiae, spinules, spinulae).

Atrium. A chamber just inside the opening of a spiracle; sometimes used as a synonym of respiratory chamber in Coleoptera (=felt chamber of Diptera larvae).

Auxiliary Setae. Setae in characteristic positions on dorsum and venter of Thysanoptera larvae of both instars but which are always very small and which have little use in identification.

Barbula (-ae). Tuft or patch of hairs or short bristles at sides of abdomen near anal region in Scarabaeoidea (Coleoptera).

Basimaxillary Membrane. *See* Maxillary Articulating Area.

Basisternum. On a thoracic segment, that area of the sternum anterior to the sternacostal suture or the pits formed by the paired sternal apophyses; when the pits or suture is absent, or on the abdomen, the anterior part of the sternum.

Beak. The protruding mouthparts of a piercing-sucking or sucking insect.

Bicameral Spiracle. *See* Annular-Biforous Spiracle.

Biforous Spiracle. A spiracle with two adjacent openings separated by a median partition.

Bilabiate Spiracle. A type of annular spiracle with an elongate-oval or elliptical opening between two lip-like structures.

Blood Channel. An internal duct or external groove in the mandible or combined mandible/maxilla of certain predacious larvae (especially Neuroptera and some Coleoptera).

Blood Gill. A gill lacking tracheae, so that oxygen passes directly into the haemocoel.

Bow. A small, curved sclerite just ventrad of the nates in anobiid larvae (Coleoptera).

Brachial Plate. Sclerotized area laterad of spiracle, containing pores (Coccoidea).

Breastbone. A sclerotized process (spatula) on the venter of the thorax caudad of the mouth of cecidomyiid larvae (Diptera).

Breathing Pocket. *See* Respiratory Chamber.

Bristle. A stout seta or hair.

Brood Chamber. A pouch formed by part of the body of the adult female in which the eggs are laid and protected (Coccoidea).

Buccal Cavity. Mouth or oral cavity.

Buccula (-ae). Elevated plates or ridges on the underside of the head, on each side of the labium of Hemiptera.

Bulla. The area of the cuticle adjacent to and partly or almost completely enclosed by the reniform sieve plate of a cribriform spiracle and containing the ecdysial scar.

ca. Circa—approximately.

Calli The paired raised areas on the anterior third of the pronotum of some Hemiptera.

Campodeiform Larva. An elongate larva, with well-developed thoracic legs and prognathous head, usually active and predacious.

Campus (-i). Bare region of the 10th or combined 9th and 10th segments of scarabaeoid larvae, located in front of or laterad of raster.

Carabiform Larva. *See* Campodeiform Larva.

Caraboid Larva. The long-legged, first instar larva of *Micromalthus* (Coleoptera).

Cardo (-ines). The basal segment of a maxilla between the head and stipes.

Carina (-inae). A ridge or keel, not necessarily high or acute.

Carinate. With a carina or carinae, keeled.

Cauda. The modified 9th abdominal tergum, sometimes termed a tail (Aphididae, Homoptera).

Caudad. Toward the rear.

Cephalad. Toward the head or anterior end.

Cephalic Segment. The anterior membranous portion of the muscomorphan head which bears the antennal and maxillary sensory papillae anteriorly. (Diptera).

Cephalopharyngeal Skeleton. The sclerotized, largely internal structures of the "head" of Muscomorpha, comprised of three main structures, the mandibles, hypopharyngeal sclerite, and tentoropharyngeal sclerite (Diptera).

Cerambycoid Larva. The legless, feeding stage larva of *Micromalthus* (Coleoptera).

Cerarii (sing., cerarius). A cluster of wax pores and at least 2 associated setae, normally located around perimeter of body and serving as foundation of waxy filaments (Coccoidea).

Cercus (-ci). An appendage (usually paired and segmented) of abdominal segment 10. Sometimes misapplied to appendages of abdominal segment 9 in Coleoptera larvae; *See* Urogomphus.

Cervical. Pertaining to the neck or cervix.

Cervical Gland. Ventral gland on the prothorax of some caterpillars.

Cervical Groove. In carabid larvae, a groove on the basal part of each parietale running from the dorsal to the lateral part (Coleoptera).

Cervical Membrane. The membrane between the head and thorax.

Cervical Triangle. *See* Epicranial Notch.

Cervicosternum. A sclerite in front of the prothoracic basisternum in some Tenebrionoidea, which represents a cervical sclerotization (Coleoptera).

Cervix. Membranous region between the head and prothorax (neck).

Chaetoparia (-ae). Part of the epipharynx in Scarabaeoidea; inner part of paria covered with bristles (Coleoptera).

Chaetotaxy. The arrangement and nomenclature of setae (*see* Setal Map).

Chalaza (-ae). Pimple-like projection of the body wall bearing a seta.

Chelate. Pincer-like, forming opposing claws.

Chelicera (-ae) One of the anterior pair of appendages in arachnids.

Cheloniform Larva. Larva which is more or less circular or ovoid and strongly flattened with concealed head, resembling a turtle.

Chewing Mouthparts. With opposable, non-sucking mouthparts.

Chitin. A nitrogenous polysaccharide occurring in the cuticle of arthropods.

Chitinized, Chitinous. Consisting of or containing chitin; formerly used to mean sclerotized.

Chorion. Outer shell of an insect egg.

Chromatophores. Large, stellate, subepidermal cells in tubuliferous Thysanoptera containing pigment granules and responsible, in part, for body color.

Chrysalis (chrysalid). The pupa of a butterfly.

Cibarial Plates. Paired series of plates bearing fringes of microtrichia and located just posterior to the epipharynx (Coleoptera).

Cibarium. The area of the mouth cavity just behind the epipharynx and above the hypopharynx.

Ciliate. Fringed or lined with fine hairs.

Circulus. Circular or rectangular area located medially on venter, posterior of hind pair of legs (Coccoidea).

Circumanal Ring. In the Psyllidae (Homoptera), the ring of pores surrounding the anus.

Clavate. Clublike, or enlarged toward the tip; e.g., clavate antennae.

Clavate Larva. One with an enlarged thoracic region, as in Buprestidae, Eucnemidae, and Cerambycidae (Coleoptera).

Claw. A sharp, curved structure (or one of two such structures) at the apex of the larval leg, representing the pretarsus or combined tarsus and pretarsus; *see* Pretarsus, Tarsungulus, Ungulus.

Cleavage Line (=ecdysial line, epicranial suture). The Y-shaped line of weakness on the head capsule, along which the integument splits at time of molting.

Cleft. Split or forked.

Cleptoparasite. One that feeds on the food stored for the host larvae.

Clithrum (-a). Part of epipharynx in Scarabaeoidea; paired, short sclerome in anterior part of margin of epipharynx, separating corypha and paria (Coleoptera).

Clubbed. With the distal part enlarged; e.g., clubbed antennae.

Clypeolabral Suture. Suture separating clypeus and labrum.

Clypeus. The sclerite between the frons and labrum.

Coadaptation. Two or more structures forming or functioning as a single unit due to their proximity, connation, or fusion.

Coarctate Larva. The third phase of hypermetamorphic larval development in Meloidae, generally equivalent to instar six, in which the larva is strongly sclerotized, has rudimentary appendages, and is immobile.

Coarctate Pupa. A pupa enclosed in a hardened case formed by the last larval skin (Diptera).

Cocoon. A silken structure spun by the larva in which the pupa is formed.

Colleterial Gland. A glandular structure accessory to the oviduct, secreting the viscous material used to cement eggs together or produce an oötheca.

Collum. Neck-like constriction at the base of the head capsule.

Comb-Hair. A seta or hair with several, parallel processes forming a comb at the end.

Comminution. The process of breaking down larger particles to smaller ones by cutting, shearing, crushing, or grinding them.

Condyle. A knoblike process forming part of an articulation by fitting into a depression (e.g., mandibular condyle).

Conjunctiva (-ae). A membranous area between two sclerites.

Connate. Immovably joined but separated by a suture or line.

Contiguous. Touching.

Convergent. Becoming closer together at one end.

Cordate. Heart-shaped; with the corners rounded.

Corneous. Sclerotized or horny.

Cornicle (=siphunculus). Paired tubular structures located dorsally on the 5th or 6th abdominal segment; length and shape varying from mere rings to elongate tubes (Aphididae); a dorso-lateral tubular structure on the posterior part of the abdomen. Alarm pheromones are expelled here.

Corniculum. A small process similar to a chalaza, but without a seta.

Corniculus (-i). *See* Urogomphus.

Cornu (ua). The dorsal or ventral rearward-projecting arm(s) of the U-shaped tentoropharyngeal sclerite (figs. 37.26a,b) (Diptera).

Coronal Suture (=epicranial stem). The median stem of the epicranial suture.

Corporotentorium. Transverse bridge connecting two halves of a tentorium; located posteriorly between the two metatentoria and the beginning of the two pretentoria.

Corypha (-ae). Part of the epipharynx in Scarabaeoidea; unpaired anterior region of epipharynx between clithra, bearing a few setae and often merged with acropariae into common apical region when clithra are absent (Coleoptera).

Costipes. The first maxillary palp segment in Hydrophiloidea; thought by some to be a distal division of the stipes (Coleoptera).

Coxal Lobe. A laterosternal area on the abdomen; *see* Laterosternite.

Coxites. Paired lateral plates of abdominal sterna of Microcoryphia and Thysanura.

Cranium. The sclerotized part of the head capsule.

Crawler. Active first instar of a scale insect.

Creeping Welt. Transverse, swollen ridge(s), usually on the anterior, ventral margins of abdominal segments 1–7 (occasionally dorsally or on the thorax), and usually bearing spinules (Diptera).

Cremaster. The terminal spined or hooked process of a pupa, used for attachment or movement.

Crenulate. Wavy or scalloped.

Crepis (-ides). Part of the epipharynx in Scarabaeoidea; a curved median crossbar forming part of the haptolachus (Coleoptera).

Cribriform Spiracle. A spiracle whose opening is covered by a perforated sieve-like (cribriform) plate, especially among Scarabaeoidea.

Cristate. Crested.

Crochets. Sclerotized, hooklike structures, usually arranged in rows or circles on the prolegs of Lepidoptera larvae (and a few others).

Cruciform Pore. Small oval structure with cross-shaped, central orifice, usually located on venter near body margin of thorax and head (Coccoidea).

Ctenidium (-ia). A comb-like row of setae (bristles).

Cultriform. Shaped like a pruning knife.

Curculionoid Larva. The legless, male-producing larva of *Micromalthus* (Coleoptera).

Cursorial. With legs for running.

Cuticle. The integument or body covering.

Cyphosomatic. With the dorsum curved or humped and the venter flat, head visible (common in Chrysomelidae).

Cyst Stage. Intermediate instar that is shaped like a pearl, and lacks legs and has reduced antennae (Coccoidea: Margarodidae).

Deciduous. Falling off at maturity or at certain periods; tending to break off due to a line of weakness at the base.

Decticous. A pupa with articulated mandibles; they can be used by the pharate adult.

Decumbent. Bent downward.

Decurved. Curved downward.

Dehiscence, Line Of. A suture on the dorsum of an insect's body along which the cuticle splits during ecdysis to permit exit of the insect.

Dehiscent. Capable of being easily broken off, like the terminal process on the abdomen of some Scraptiidae (Coleoptera).

Dentate. Toothed.

Denticle. Small tooth (on inner margin of claw (Coccoidea)).

Denticulate. With minute toothlike projections or edges; with little teeth or notches.

Dermal Papillae. Numerous small projections on surface of body.

Deutonymph. Third instar of a mite.

Dexiotorma (-ae). The right torma in Scarabaeoidea; usually slender and transverse (Coleoptera).

Diapause. A period of arrested development and reduced metabolic rate, during which growth, differentiation, and metamorphosis cease; a period of dormancy not immediately referable to adverse environmental conditions.

Dicondylous. With two condyles, or two processes articulating the mandible to the head capsule.

Digitiform, Digitate. Finger-like.

Digitiform Sensillum (-a). A flattened sensillum, often one of a group, located in a shallow cavity near the base of the last maxillary or labial palp segment; often adpressed against the surface and difficult to see.

Digits. Finger-like lobes, especially of the prothoracic spiracles of maggots (Diptera).

Dilated. Widened.

Dimorphic. Occurring in two distinctive forms.

Disc. The central portion of a structure; also a sclerotized plate.

Distad. On an appendage, toward the tip or point farthest away from the body.

DISTAL. Toward the tip or end; farthest from the body.

DISTICARDO. That part of the cardo laterad of a dividing ridge.

DISTISTIPES. Distal or apical portion of a divided stipes.

DIURNAL. Active during daylight.

DORSAD. Toward the top or back.

DORSAL ACETABULUM. A cavity on the dorsal side of the mandibular base which articulates with a condyle (dorsal mandibular articulation) on the cranium.

DORSAL CARINA (OF MANDIBLE). Carina extending basally from the dorsalmost apical tooth of the mandible (Coleoptera).

DORSAL FLANGE. Enlarged portion of dorsolateral angle of each lateral arm of the labial sclerite (Hymenoptera: Apocrita mouthparts).

DORSAL LINE. A longitudinal line along the dorsomeson.

DORSAL PLATE. Plate or disc-like area, often asperate, on the dorsal surface of the prothorax in larvae of Buprestidae (Coleoptera).

DORSAL SENSORY SPOTS. Flattened sensoria on the apical antennal segments of some Scarabaeoidea (Coleoptera).

DORSOLATERAL SUTURE. A weak line or groove just below the spiracular area and above the laterotergite on the abdomen (Coleoptera).

DORSOMESAL. At the top and near the midline.

DORSOMESON. The middle of the back or top.

DORSOPLEURAL FOLD. The major suture fold or line separating the dorsum from pleuron on both thorax and abdomen; when the pleuron is absent on the abdomen, synonymous with tergosternal fold.

DORSOPLEURAL LINE. The line of separation between the dorsum and pleuron, often marked by a fold or groove. (*See* DORSOLATERAL SUTURE and DORSOPLEURAL FOLD).

DORSOPLEURAL LOBE. *See* LATEROTERGITE.

DORSUM. The entire upper surface of the body; the upper surface of any trunk segment.

ECDYSIAL LINE. A line of weakness, often incorrectly termed a "suture", which splits during ecdysis.

ECDYSIAL SCAR. Scar located near the paired openings of a biforous spiracle or the sieve-plate of a cribriform spiracle, which represents the opening of a channel through which the spiracular lining is withdrawn during ecdysis of the elateroid type; in Scarabaeoidea usually located on bulla (Coleoptera). In the posterior spiracles of Diptera larvae, a central or marginal remnant of the previous instar's tracheal opening located in the spiracular plate.

ECDYSIAL SUTURE. Where the head capsule usually splits during molting; laterad of the adfrontal areas in Lepidoptera larvae (*see* ECDYSIAL LINE).

ECDYSIS (-SES). Molting, shedding the cuticle.

ECLOSION. Hatching from the egg; also emergence of the imago or adult insect from the pupa.

ECTOGNATHOUS. With mouthparts exserted or exposed.

ECTOPARASITE. A parasite that lives on the outside of its host.

ECTOPARASITOID. A parasitoid that lives on the outside of its host.

EGG BURSTER. A tooth, spine, or ridge on the head, thorax, or abdomen of a late embryo and used in rupturing the egg shell when hatching; often retained by first instar larvae.

8-SHAPED PORE. Flat, oval structure composed of two adjacent circles giving the appearance of an "8" (Coccoidea).

8-SHAPED TUBULAR DUCT. Cylindrical structure with a central septum giving the appearance of an "8" (Coccoidea).

ELATERIFORM LARVA. *See* WIREWORM.

ELATEROID ECDYSIAL PROCESS. A type of molting process in which the spiracular opening is blocked (biforous or cribriform type) and the spiracular lining must be shed through a side chamber, the opening of which is later visible as a scar (*see* ECDYSIAL SCAR) (Coleoptera).

EMARGINATE. With a dent or notch in the margin.

EMPODIUM (-DIA). A pad or bristlelike structure between the claws at the tip of the last tarsal segment.

ENDOCARINA (-AE). An internal ridge which is visible on the surface as a heavy line; usually occurring on the dorsal part of the cranium and sometimes contiguous with parts of the epicranial suture.

ENDOGNATHOUS. Mandibles and maxillae almost entirely enclosed by fusion of lateral margins of labium with cranial folds, e.g., Protura.

ENDOGNATHOUS MOUTHPARTS. Those which are stylet-like and enclosed within the buccal cavity.

ENDOPARASITE. A parasite that lives inside its host.

ENDOPARASITOID. A parasitoid that lives inside its host.

ENDOPTERYGOTE. With the wings developing internally; holometabolous insects.

ENDOSKELETON. A variety of internal sclerotized structures; in the cranium, consisting of the tentorium, endocarinae, hypopharyngeal bracon, hypopharyngeal suspensorium, gular ridges, etc.

ENLARGED SETAE. Robust setae that are spinelike in appearance (Coccoidea).

ENSIFORM. Resembling the shape of a sword.

ENTAL. Toward the buccal cavity.

ENTOGNATHOUS. *See* ENDOGNATHOUS.

ENTOMOPHAGOUS. Feeding on insects.

EPICRANIAL HALF. *See* EPICRANIAL PLATE.

EPICRANIAL NOTCH (= CERVICAL TRIANGLE, VERTICAL TRIANGLE). The V-shaped dorsomedial area delimited laterally by the cranial halves.

EPICRANIAL PLATE. One of the paired regions of the cranium behind the frontal arms and on either side of the epicranial stem (if present).

EPICRANIAL STEM. The median stem of the epicranial suture.

EPICRANIAL SUTURE. Inverted, Y-shaped, V-shaped, U-shaped, or lyriform suture (sometimes erroneously termed the ecdysial line or cleavage line); which separates epicranial halves on dorsum of head and whose arms extend anteriorly (ventrally) on either side of frons.

EPICRANIUM. The cranium above or behind the frons.

EPIMERON. That area of thoracic pleuron immediately adjacent to the coxa and posterior to pleural suture.

EPIPHARYNGEAL ROD. *See* LABRAL ROD.

Epipharynx. Inner or ental surface of labrum.

Epipleural Area. The area immediately above the dorsopleural fold; *see* Epipleurum, Laterotergite (Coleoptera).

Epipleural Lobe. *See* Laterotergite, Epipleurum.

Epipleurite. *See* Laterotergite.

Epipleurum. The lateral region of the tergum; *see* Laterotergite.

Epiproct. Dorsal lobe of terminal abdominal segment; *see* Suranal Plate.

Episternum. That region on the thoracic pleuron immediately adjacent to the coxa and anterior to the pleural suture.

Epistoma. Upper part of the frame surrounding the opening of the buccal cavity, continuous with pleurostoma on each side and hypostoma below; area just behind or above the labrum; used by various authors as a synonym of anteclypeus, clypeus, postclypeus, or anterior margin of frons.

Epistomal Suture (=frontoclypeal suture). The suture separating the frons from the clypeus.

Epitorma (-ae). Rod extending anteriorly from the mesal end of the laetotorma in Scarabaeoidea.

Epizygum (-a). Part of the epipharynx in Scarabaeoidea; elongate plate or bar extending from zygum toward clithrum on right side (Coleoptera).

Eruciform Larva. Caterpillar-like; a larva with a well-developed head, thoracic legs, and abdominal prolegs.

Eucephalous (eucephalic). With a well-developed head. In Diptera larvae, with the mandibles operating horizontally or obliquely.

Eusternum. The combined basisternum and sternellum; sometimes used as a synonym of basisternum.

Eversible. Capable of being everted or projected outward.

Ex. From or out of. (e.g., Ex pupa = reared from the pupa).

Exarate Pupa. A type in which the legs and wings are free from the body.

Exopterygote. The wings developing externally—winged insects with hemimetabolous metamorphosis.

Explanate. Spread out, flattened, especially at an edge.

Exserted. Projecting from the body; protruding.

Extraoral Digestion. Digestion of food outside the body cavity, accomplished by the injection of enzymes into the surrounding medium or a particular food item.

Exuvia (-ae). The cast skin. In Latin "exuviae" means clothes, booty, spoils of war. There was (is) no singular, but "exuvia" is the correct derived singular. "Exuvium" is not a correct singular form.

Falciform, Falcate. Sickle-shaped.

False Legs. Without true segments. *See* Prolegs.

Fastigium (-ia). The extreme point or front of the vertex of the head.

Felt Chamber. *See* atrium.

Filament. A long, slender structure.

File. *See* Stridulitrum.

Filiform. Hairlike or threadlike; slender and of equal diameter; with segments of equal diameter.

Filter Feeding. The extraction of small food particles from a liquid medium.

Fimbriate. Fringed with setae or fine hairs; sometimes referring to a dense fringe of modified setae.

First Grub. The second phase of hypermetamorphic larval development in Meloidae, generally incorporating instars 2–5, in which the larva is grublike (scarabaeiform) in appearance.

Flabellate. With fanlike lobes or processes (e.g., flabellate antennae).

Flagellate. Whiplike.

Flagellomere. A segment or division of the flagellum or antenna beyond its two basal segments of pedicel and scape.

Flagellum. That part of the antenna distad of the second segment.

Floricolous. Living in or on flowers.

Foliacious. Leaf-like, resembling a leaf.

Fossorial. Structured for digging.

Frass. The pelletlike excrement of insects, especially from caterpillars and wood-borers.

Frayed Seta. A seta whose apex is expanded and appears frayed.

Frons. Median sclerite on face of head delimited above by epicranial arms and below by clypeal suture; sometimes termed frontal area.

Front. *See* Frons.

Frontal Area, Frontal Region. *See* Frons.

Frontal Arms. *See* Frontal Sutures.

Frontal Sutures. The two branches of the epicranial suture on either side of the frons.

Frontoclypeal Apotome. The middorsal sclerite of the head capsule (between the genae) (Diptera).

Frontoclypeal Region. Area occupied by the frons and clypeus.

Frontoclypeal Suture. The suture separating the frons from the clypeus.

Fungivorous. Feeding on or in fungi.

Furca. A forked structure, frequently internal in the thorax of winged insects.

Furcate. Forked.

Fused. Immovably united, without a suture to mark the point of union.

Fusiform. Widest at middle and narrowed at both ends.

Galea (-ae). The outer (lateral) lobe at the apex of the maxilla.

Gena (-ae). The part of the head posterior to the mandibles; the cheek.

Geniculate. Elbowed.

Gibbose. Bearing 1 or more swellings or protuberances.

Gill. A variously formed respiratory structure in aquatic immature insects, through which they obtain dissolved oxygen.

Gill Tuft. Group of filamentous gills joined to a common stem.

Gin Trap. Protective, pinching device formed between sclerotized edges of two adjacent abdominal segments in Coleoptera pupae.

Glabrous. Smooth, without setae.

Glandular Seta. A seta which is hollow and filled with a poison or other defensive chemical; a seta bearing numerous fine projections (*see* Pubescent Seta).

GLOBOSE. Approximately spherical.

GLOSSA (-AE). The inner pair of lobes at the tip of the labium between the paraglossae.

GRESSORIAL. Having legs fitted for walking.

GRUB. A general term for a soft-bodied, relatively inactive larva usually found in a protected location. C-shaped larva living in soil; *see* SCARABAEOID LARVA (Coleoptera).

GULA. Sclerite on the venter of the head between the labium (postlabium or postmentum) and the neck; area lying between the gular sutures.

GULAMENTUM. The combined gula and submentum (or postmentum).

GULAR AREA, GULAR REGION. The area on the venter of the head immediately behind the labium; *see* HYPOSTOMA, VENTRAL HEAD CLOSURE.

GULAR SUTURE. Usually one of a pair of sutures extending from the posterior pits to the occipital foramen; sometimes paired ecdysial lines on the venter of the head, anterior to, or not associated with, the tentorial pits; occasionally a median suture on the venter of the head.

GYMNOPARIA (-AE). Part of the epipharynx in Scarabaeoidea; naked part of paria between acanthoparia and chaetoparia and behind acroparia (Coleoptera).

HAPTOLACHUS (-HI). Part of the epipharynx in Scarabaeoidea; posteromesal region of epipharynx behind pedium; composed of the nesia, the crepis, and a number of sensilla (Coleoptera).

HAPTOMERUM (-RI). Part of the epipharynx in Scarabaeoidea; anteromesal region of epipharynx, in front of pedium and behind corypha, or behind apical region consisting of united acropariae and corypha; composed of zygum, sensilla, and series of heli (Coleoptera).

HASTISETA. Spear-headed setae arising from the tergites of some dermestid larvae.

HAUSTELLATE (HAUSTATE). Mouthparts for sucking or piercing-sucking.

HELUS (-LI). Part of the epipharynx in Scarabaeoidea; coarse, fixed spine without cup located in haptomerum (Coleoptera).

HEMICEPHALIC. Head capsule reduced or incomplete posteriorly, partially retracted into the thorax, with mandibles operating vertically.

HEMIMETABOLOUS. A type of metamorphosis in which the form of the immature gradually approaches that of the adult through successive instars; also, metamorphosis in which the immatures (naiads) are aquatic (Odonata, Plecoptera and Ephemeroptera) and do not resemble the adults as closely as terrestrial groups do (Comstock, 1918).

HEMIPNEUSTIC RESPIRATORY SYSTEM. One with spiracles on the thorax and abdominal segments 1–7.

HIBERNACULUM. A protective retreat made out of silk or other material, in which a larva hides or hibernates; an overwintering retreat or shelter, usually applied to early instars.

HOLDFAST ORGAN. A structure which serves to secure a part of the body to the substrate, like the anal tubes of some beetle larvae.

HOLOMETABOLOUS. With complete metamorphosis (egg, larva, pupa, adult).

HOLOPNEUSTIC. All the spiracles open and functional.

HONEYDEW. Sweet, sticky liquid discharged from the anus by some Homoptera.

HOOK. The anterior, sickle-shaped portion of the mandible (=mouthhook) (Diptera).

HORN. A stiff, pointed cuticular process.

HOST. The organism in or on which a parasitoid or parasite lives; the plant on which an insect feeds.

HUMERAL ANGLES. The posterolateral angles of the pronotum.

HYALINE. Translucent, but not necessarily flexible or membranous.

HYGROPETRIC. Rocks with a thin layer of water trickling over them, such as a roadside seepage.

HYPERMETAMORPHOSIS. Larval development involving passage through 2 or more distinct phases, as in Meloidae; a type of complete metamorphosis in which a larva passes through two or more very different appearing instars.

HYPERPARASITOID. A parasitoid whose host is another parasitoid.

HYPOGNATHOUS. With the head vertical and mouthparts directed ventrad (*compare* PROGNATHOUS).

HYPOPHARYNGEAL BRACON. A transverse brace extending from the hypopharynx to a point on each side of the head near the dorsal mandibular articulation (Coleoptera).

HYPOPHARYNGEAL COMBS. Comb-like structures on the hypopharynx of helodid larvae, which remove collected particles from the maxillae (Coleoptera).

HYPOPHARYNGEAL SCLEROME. A heavily sclerotized bar or tooth-like structure on the hypopharynx (Coleoptera).

HYPOPHARYNGEAL SUSPENSORIUM. One or more pairs of sclerites extending dorsally from the hypopharynx on either side of the cibarium.

HYPOPHARYNX. The median mouthpart structure anterior (in a hypognathous head) to the labium (dorsal in a prognathous head) that may bear taxonomically useful structures.

HYPORHEIC. Subterranean zone of water-filled areas between the rocks beneath the stream bottom.

HYPOSTOMA. The ventral portion of the mouth frame or peristoma, extending from the base of the labium to the pleurostoma on each side; also used for the general area behind the insertions of the maxillae and labium.

HYPOSTOMAL CAVITY. In larvae with retracted ventral mouthparts, the excavation in which these mouthparts are housed; *see* HYPOSTOMA.

HYPOSTOMAL MARGIN. *See* HYPOSTOMA.

HYPOSTOMAL RIDGE. *See* HYPOSTOMA.

HYPOSTSOMAL RODS. A pair of cuticular thickenings, seen as dark lines, extending posteriorly from the vicinity of each maxillary articulation (base of cardo).

HYPOSTOMAL SPUR. A sclerite extending from each hypostoma across the stipes; this is an abbreviation of Short's (1952) term, "sclerotic spur of the hypostoma" (Hymenoptera: parasitica).

IMAGO. An adult.

IMBRICATED. Overlapping like shingles.

Incisor Area. Most of the mesal edge of the mandible, extending from just below the apex to the basal or molar region.

Incisor Edge. *See* Incisor Area.

Inner Lobe. *See* Lacinia.

Inquiline. An animal that lives in the home or nest of a different species.

Instar. The stage between molts.

Integument. The outer body wall.

Intersternite (=poststernellum). A sclerite located between the posterior edge of the prothoracic or mesothoracic venter and the anterior edge of the following segment.

Interstitial. Referring to the spaces between objects, like leaves and sticks (leaf litter), sand grains, or soil particles. An interstitial organism is of such dimensions that it can move freely within the spaces; thus in substrates composed of fine particles, only minute organisms can be called interstitial.

Invaginated Tubular Duct. Cylindrical structure with apex in form of a cup (Coccoidea).

Juga (jugum). In Hemiptera, the two lateral lobes of the head, one on each side of the tylus.

Juxtacardo (-dines). A separate part of the cardo extending from the cardo proper toward the submentum (Coleoptera).

Juxtastipes (-pites). A separate part of the stipes extending from the stipes proper toward the mentum; located basally beneath the maxillary articulating area (Coleoptera).

Keeled. With a distinct ridge (usually ventrally).

Labial Palp. A pair of small, segmented sensory structures arising on the distolateral portions of the labium.

Labial Plate. A single sclerite representing the fused prementum and postmentum.

Labial Sclerite. A sclerotized ring, usually broken dorsally, and occasionally ventrally, which represents the marginal sclerotization of the prelabium (Hymenoptera: Parasitica).

Labium. The lower lip, basically consisting of submentum, mentum, prementum, ligula and labial palps; in most insects with piercing-sucking mouthparts, the thickened, elongate segmented mouthpart which houses the stylets.

Labral Rod. One of a pair of paramedial, longitudinal sclerites in the labrum-epipharynx, which extends posteriorly from near the apex or in the middle (Coleoptera).

Labrum. The upper lip; sclerite attached to the anterior edge of the clypeus or frontoclypeal region.

Labrum-Epipharynx. The combined labrum and epipharynx.

Lacinia (-ae). The inner (mesal) lobe at the apex of the maxilla, adjacent to the galea.

Lacinia Mandibulae. *See* Prostheca.

Lacinia Mobilus. A small plate-like or blade-shaped appendage near the base of the mandible (Coleoptera).

Laetotorma (-ae). The left torma in Scarabaeoidea; transverse and often provided with various accessory structures, such as the pternotorma, epitorma, or apotorma (Coleoptera).

Lamella (-ae). A thin plate or leaf-like process.

Lamellate. Thin and plate-like; composed of lamellae.

Lamina (-ae). *See* Lamella.

Laminate. *See* Lamellate.

Lanceolate. Spear-shaped.

Larva. The stage between the egg and pupa of those insects having complete metamorphosis; the stage between the egg and adult of those insects not having complete metamorphosis (also known as nymphs); also, the 6-legged first instar of Acarina.

Larva I. The second instar of Protura; the first of two active, feeding larval instars in Thysanoptera.

Larva II. The third instar of Protura; the second of two active, feeding larval instars in Thysanoptera, usually lasting as long as or longer than the adult.

Larviform. Shaped like a larva.

Laterad. Toward the side, away from the midline.

Lateral Air Tube. *See* Accessory Chamber.

Laterosternal Fold. *See* Laterosternite.

Laterosternite. One of two or more sclerites lying on either side of an abdominal sternum, but below the pleuroventral line.

Laterotergite. One of two or more sclerites lying on either side of a thoracic or abdominal tergum, but above the dorsopleural line.

Leafminer. An insect that feeds and usually lives between the upper and lower surfaces of a leaf, often forming distinctive patterns.

Lentic. Inhabiting still waters, such as lakes, ponds, or swamps.

Lepismatoid Larva. *See* Campodeiform Larva.

Ligula. A single process at the apex of the labium which represents the fused glossae.

Ligular Sclerome. A heavily sclerotized ligula, as in Micromalthidae and Callirhipidae (Coleoptera).

Limpet-Like. Resembling a limpet; with ovoid, flattened or dorsally arched body closely attached to a substrate.

Lingula. Tonguelike process within the vasiform orifice under the operculum (Aleyrodidae)

Lotic. Inhabiting running waters, such as streams or rivers.

Lucifugous. Light-shunning.

Luminous. Producing visible light.

Lyriform, Lyrate, Lyre-Shaped. Shaped like a lyre; applied to an epicranial suture with the frontal arms sinuate, that is, bowed outwardly and then inwardly again, before curving outwardly toward the antennal insertions.

Macrophagy. Feeding on relatively large food particles, which must be broken up before ingestion.

Maculate. Marked (spotted) with pigmented areas of varying shape.

Maggot. Most larvae of higher Diptera; legless, "headless", and peg-like. A wormlike, legless larva with little or no head capsule.

Mala (-ae), Mala Maxillae. A single lobe at the apex of the maxilla, which usually represents the fused galea and lacinia, but may also be the galea alone.

Malar Sclerome. A sclerotization in the cleft at the apex of the mala in some Tenebrionoidea (Coleoptera).

Mandibular Articulations. The paired dorsal condyles and ventral acetabula at the ends of the pleurostomata, with which the mandibles articulate.

Mandibular Fossa. *See* Dorsal Acetabulum.

Margined. With a sharp edge.

Matte Patch. One of a series of ovoid patches of minute asperities on the terga and sterna of some eucnemid larvae (Coleoptera).

Maxilla (-ae). The paired mouthparts posterior (ventral) to the mandibles.

Maxillary Articulating Area. A membranous or lightly sclerotized, usually triangular area lying between the cardo, base of the stipes, and the base of the mentum or submentum (Coleoptera).

Maxillary Palp. A small, segmented sensory structure arising from the maxilla.

Maxillary Palpifer. *See* Palpifer.

Maxillula (-ae). A lobe, sometimes subdivided, on either side of the hypopharynx.

Meconium. The liquid waste products voided by the adult after emergence from the pupa.

Median. In the middle.

Mediosternal Fold. *See* Basisternum.

Mediosternite. The sternum or major mesal plate of a longitudinally divided abdominal sternum.

Medius. Jeannel's term for the leg segment following the femur in the leg of Adephaga (Coleoptera); *see* Tibia.

Mentum. That part of the labium lying between the submentum and prementum; sometimes used for the proximal sclerite of the labium (postmentum) when there are only two labial sclerites.

Mesocercus. The mesal and usually largest of the paired processes on the floor of the respiratory chamber in Hydrophilidae (Coleoptera).

Meson. An imaginary vertical middle plane of the body, dividing the body into equal halves.

Mesonotum. The dorsal part of the mesothorax.

Mesopleuron. The lateral part of the mesothorax.

Mesosternum. The ventral part of the mesothorax.

Mesotergum. Tergum or main dorsal plate of the mesothorax.

Metacephalic Rod. An elongate, sclerotized sclerite articulating dorsally to the head and extending into the prothorax (Diptera).

Metamorphosis. Change in form during development.

Metapleuron. The lateral part of the metathorax.

Metapneustic Respiratory System. One with only the last pair of spiracles functional (open).

Metasternum. The ventral part of the metathorax.

Metatarsus. The first (basal) tarsal segment.

Metatergum. Tergum or main dorsal plate of the metathorax.

Microphagy. Feeding on small to minute particles which must be concentrated and gathered into the mouth cavity.

Micropyle. The minute opening(s) in the insect egg chorion through which sperm enter.

Microspines (=aculeae, spinules). Minute spines on the body, usually visible only under magnification.

Microtrichia (-ae). Minute, ciliate processes of the cuticle; of characteristic pattern on antennae and body of larval Thysanoptera; minute spines or asperities, often not visible under lower magnifications.

Microtubular Duct. Small tubular structure, usually with 2 sclerotized bars and without an apical cup (Coccoidea).

Mola (-ae). The processing area at the inner (mesal) edge of the mandibular base.

Molar Area. The inner (mesal) edge of the mandibular base, even when this does not bear a distinct mola.

Molar Process. Sclerotized, rounded process at floor of mouth, visible when mandibles open; found in Dolerinae (Tenthredinidae).

Moniliform. With round segments like a string of beads.

Monocondylous. With one condyle articulating mandible to head capsule.

Monophagous. Feeding on only one species.

Mouthcone. The hypognathous or opisthognathous rostrum of Thysanoptera. Consists of the labrum in front, the maxillary stipites on either side and the labium behind, and conceals three feeding stylets: two maxillary and the left mandibular.

Mouthhook(s). The parallel claw-like mouthpart(s) of the larvae of higher Diptera.

Multiannulate. Consisting of numerous ring-like segments.

Multiforous (spiracle). One with three or more openings within or adjacent to the peritreme.

Multilocular Pore. Circular structure with more than 5 divisions (Coccoidea).

Muricate. With sharp, rigid points.

Muscidiform. Shaped like the peglike larvae of higher Diptera.

Mycetophagous, Mycophagous. Feeding on or in fungi.

Myiasis. A disease of animals caused by invading maggots.

Nasale(-s). The fused labrum and clypeus, or labrum, clypeus and frons; an anterior, median projection from the frons (Coleoptera).

Natatorial. Adapted for swimming; usually applied to the legs of aquatic larvae, which are lined with swimming hairs.

Nates. A structure on the 10th segment ventrad of the anus in anobiid larvae, consisting of a pair of oval lobes on either side of a longitudinal groove (Coleoptera).

Navicular. Shaped like a boat, i.e., tapered at both ends.

Neoteny. Sexual maturity in a non-adult stage.

Nesium (-ia). Part of the epipharynx in Scarabaeoidea; one of two sclerites in the haptolachus, anterior to the crepis and between the inner ends of the tormae (Coleoptera).

Nodate. With node(s) or enlargement(s).

Node. Knotlike swelling.

Nodose. Knotty.

Noninvaginated Tubular Duct. Cylindrical structure without cup-shaped apex, usually with single sclerotized band (Coccoidea).

Notum (-ta). The tergum or major plate of a thoracic dorsum.

Nuchal. Of the neck.

Nuchal Constriction. A neck-like constriction near the base of the head capsule.

Nymph. The larva of a hemimetabolous insect; the 8-legged second stage of Acarina.

Obtect Pupa. One in which the appendages are tightly appressed to the body (most Lepidoptera, for example).

Obtuse. Blunt, not pointed or acute.

Occipital Foramen. The main opening of the head capsule, leading to the cervical region or prothorax.

Ocellar Spots. Weakly pigmented, oval areas on submedial region of head (Coccoidea).

Ocellus (-li). The simple eyes of adult insects and hemimetabolous immatures. (*See* Stemma).

Ocularium. Elevated or pigmented area bearing or surrounding the stemma or simple eye.

Oligopod. An active larva with well-developed thoracic legs.

Ommatidium (-ia). A single section of a compound eye; the eye elements of Mecoptera larvae.

Oncylus (-li). One of two or more sclerites on the hypopharynx in Scarabaeoidea (Coleopetera).

Onisciform Larva. A flattened, ovate larva, resembling some terrestrial isopods.

Opercular Hooks. Paired hooks attached to the operculum in some aquatic beetle larvae.

Operculum. A circular or ovoid lid formed from the 9th tergum or 10th sternum in some beetle larvae. In the Aleyrodidae (Homoptera), the structure or lid over the lingula and within the vasiform orifice.

Opisthognathous Head. With the mouthparts directed downward and backward (as in Blattodea and some Homoptera).

Opisthosoma. The posterior region of the body, the part behind the legs (Arachnida).

Opposable. Capable of being opposed or meeting one another, as in the mandibles of chewing insects.

Oral Filter. A filter formed from dense patches of setae on the epipharynx, hypopharynx, mandibular bases, and ental maxillary surfaces.

Orthosomatic. With surfaces of body subparallel; a straight body.

Ostiole. Dermal invaginations on dorsosubmarginal areas of seventh abdominal segment and prothorax (Coccoidea).

Outer Lobe Of Maxilla. *See* Galea.

Ovipara. Sexual female that mates with a male and produces 1 or more eggs that overwinter (Aphidoidea).

Oviparous. Laying eggs.

Ovisac. Waxy secretion produced by the adult female that encloses the eggs; often encloses part or all of the adult female also (Coccoidea).

Ovoviviparous. Producing living young by the hatching of the ovum while still within the female.

Paedogenesis. Reproduction by an immature stage.

Paedogenetic. Having immature stages which are capable of reproduction.

Palidium (-ia). Part of the raster in Scarabaeoidea; a group of pali forming 1 or more rows.

Palmate Mandible. One which has a concave mesal edge and 3 or more apical teeth which are more or less perpendicular to the plane of movement.

Palp(s). A paired, segmented, sensory structure arising from the maxilla or labium.

Palpifer. A lobe on the outer edge of the maxillary stipes which bears the palp. *Compare* Palpiger.

Palpiger. A paired lobe on either side of the labial prementum which bears the palp; sometimes applied to the palp-bearing process on either maxilla or labium. *Compare* Palpifer.

Palpus (-pi). *See* Palp.

Palus (-li). A straight pointed spine forming part of the palidium (Coleoptera).

Papilla (-ae). A soft minute bump or projection.

Paracercus (-ci). A paired lateral process on the floor of the respiratory chamber in larvae of some Hydrophilidae (Coleoptera).

Paradorsal Area. Adjacent to or at the side of the dorsal area.

Paraglossa (-ae). One of a pair of lobes at the tip of the labium, lateral to the glossae.

Paragula (-ae), Paragular Plate. A sclerite on either side of the gula in some Cleroidea.

Paramedian. On either side of, but relatively close to the midline.

Parascolus (-li). A spinose projection of the body wall not more than three times as long as wide and bearing a few short branches, each with a seta at its distal end.

Parasite. An animal that lives in or on its host, at least during a part of its life cycle, feeding on it, but usually not killing it (*compare* Parasitoid).

Parasitic. Living as a parasite or parasitoid.

Parasitoid. An animal that feeds in or on a *single* host, eventually killing it (*compare* Parasite, Predator).

Paratergum (paratergite). *See* Laterotergite.

Parta (-ae). Part of epipharynx in Scarabaeoidea; lateral paired region extending from clithrum, epizygum, and haptomerum posteriorly to tormae, divided into acanthoparia, gymnoparia, chaetoparia, and acroparia, and bearing plegmatium, proplegmatium, and phobae (Coleoptera).

Parietal (areas). Behind and lateral to the frontal arms on either side of the epicranial stem.

Parietal Bands. Dorsolateral paired streaks on front of head of some Hymenoptera, slightly differentiated and often pigmented.

Parietal Plate. *See* Epicranial Plate.

Parietals. The lateral areas of the cranium, between the frontal and occipital areas.

Paronychial Appendix. *See* Pulvillus.

Pars Stridens. *See* Stridulitrum.

Parthenogenesis. Reproduction without fertilization.

Pectinate. With branches like a comb.

Pedal Lobe. Paired, fleshy, non-segmented lobe on thorax of legless larva, which is thought to be a leg rudiment.

Pedicel. The second segment of an antenna.

Pedipalps. The second pair of appendages of an arachnid.

Pedium (-ia). Part of epipharynx in Scarabaeoidea; central bare region extending between haptomerum and haptolachus and limited laterally by pariae (Coleoptera).

Pedunculate. On a slender stalk.

Pedunculate Seta. A structure on the mala of some cleroid larvae, which consists of a seta (or sometimes more than one seta) borne on a projecting process or tubercle.

Penicillum (-li), Penicillus (-li). A brush or pencil of hairs, especially at the base of the mandible.

Penultimate. Next to last.

Perforate Mandible. One which has an internal channel extending from the base to the apex.

Perianal Pad. A variably sized and shaped area bordering the anal cleft (Diptera).

Peripneustic Respiratory System. One in which there are 9 pairs of functional spiracles, one on the thorax and 8 on the abdomen.

Peristoma. The mouth frame, comprised of the epistomal ridge, pleurostomata, and hypostomata.

Peritreme. A sclerotic ring surrounding the outer spiracular opening.

Petiolate. Attached with a narrow stalk or petiole.

Pharate. The condition in which the new (next) stage is enclosed by the cuticle of the present one; e.g., a "pharate" adult is often visible beneath the pupal skin in endopterygotes and may be capable of locomotion in some.

Pharyngeal Filter. The modified pharynx of some nonpredacious, filter-feeding or saprophagous Diptera larvae (some Nematocera and Muscomorpha, and a few Stratiomyidae) that strains suspended food particles from ingested liquids before the excess fluid is regurgitated through the mouth.

Pharyngeal Skeleton. In higher Diptera the conspicuous sclerotized structures articulating with the mouthhook(s).

Pharynx. The area behind the cibarium; an enlargement of the anterior part of the oesophagus.

Phoba (-ae). Part of epipharynx in Scarabaeoidea; fringe of densely set, hair-like, often forked projections located posteriorly at inner edge of paria (Coleoptera).

Phoresy. A relationship (usually temporary) in which an individual of one species is transported by another species, but does not feed on it or otherwise injure it.

Phoretic. Exhibiting phoresy (q.v.).

Physogastric. With an enlarged or swollen abdomen.

Phytophagous. Feeding on plants.

Piceous. Very black.

Pilose. Covered with soft hair.

Pinaculum (-a). A small, flat, or slightly elevated chitinized area bearing seta(e) (especially Lepidoptera larvae).

Planidiform (larva). First active stage of certain parasitoids (Diptera and Hymenoptera).

Planidium (-ia). A type of first-instar larva that undergoes hypermetamorphosis, a larva that is legless and somewhat flattened (some Diptera and Hymenoptera).

Plantula. A lobe of the tarsus, a climbing cushion.

Plastron. A permanent physical gill consisting of a gas layer of constant volume, held in place by a hydrofuge mesh that prevents the entry of water under pressure, and a large air-water interface.

Plastron Plate. Finely perforated plate almost completely surrounding each thoracic spiracle and all abdominal spiracles except the 8th pair, in the larva of *Araeopidius* (Ptilodactylidae) (Coleoptera).

Plate. A larger sclerotized area of the body (*see* Anal or Suranal Plate, *compare* Pinaculum).

Platyform Larva. A circular, flattened, or disc-like larva.

Plectrum. The movable part of a stridulatory organ.

Plegma (-ata). Part of epipharynx in Scarabaeoidea; fold pertaining to plegmatium and proplegmatium (Coleoptera).

Plegmatium (-ia). Part of epipharynx in Scarabaeoidea; lateral, paired space with plicate, somewhat sclerotized surface; bordered by marginal spines of acanthoparia, with 1 plegma inside of each spine (Coleoptera).

Pleisiomorphy, -phic, -phous. The relatively primitive state (older) of a sequence of homologous characters. *Compare* Apomorphy.

Pleural Apophysis. The internal, bar-like ridge beneath the pleural suture.

Pleural Fold. *See* Dorsolateral Suture.

Pleural Lobe. *See* Epipleurum.

Pleurite. A pleural sclerite.

Pleuron (-ra). The lateral area of a segment; the area immediately adjacent to the coxa and pleural suture on each side of the thorax; the area lying ventral to the dorsopleural line and dorsal to the pleuroventral line on the abdomen.

Pleurostoma (-ata). The lateral portion of the peristoma or mouth frame, lying between the dorsal and ventral mandibular articulations. The lateral extension of the epistoma on either side of the cephalic structure separating the clypeus from the front of the face and usually continuous with the hypostoma (Hymenoptera: parasitica).

Plica (-ae). An integumental fold, especially on the dorsum in soft-bodied larvae.

Plumose. With many small branches; featherlike.

Polyembryony. An egg developing into more than 1 embryo.

Polymorphic. Having many forms.

Polyphagous. Feeding on several species.

Polypod Larva. One with legs on most segments.

Postclypeus. The major and more posterior region of a divided clypeus; the caudal (dorsal) portion of a transversely divided clypeus.

Postcornus. Sclerotized, horny process at apex of 10th abdominal tergum of many boring sawfly larvae and a few others.

Postcoxal Bridge. A crescent-shaped sclerite or area behind the procoxae in some tenebrionid larvae, probably consisting of the sternellum fused with the postcoxale on each side; also used as synonym of postcoxale.

Postcoxale (-alia). The area immediately posterior to the epimeron and extending posteromesally around the coxa; often not separated from the epimeron.

Postdorsal Fold. *See* Scutum.

Posterad. Toward the rear (caudad).

Posterior Spiracular Plate. The caudal plate bearing the spiracles.

Postgena (-ae). The region immediately behind the gena.

Postmentum. The proximal sclerite of the labium; sometimes divided into a submentum and mentum.

Postnotal Fold. *See* Scutum.

Postscutellum. A fold at the posterior end of an abdominal tergum in some scarabaeoid larvae (Coleoptera).

Postspiracular. Caudad of the spiracles.

Poststernellum. *See* Intersternite.

Postventrite. A sclerite behind the main sternal plate or ventrite on the abdomen of some carabid larvae (Coleoptera).

Preapical. Just before the apex.

Precoxale (-alia). The area immediately anterior to the episternum and extending anteromesally around the coxa; often not separated from the episternum.

Predator. An animal that kills its hosts (prey), requiring more than 1 to complete its life cycle (*compare* Parasite, Parasitoid).

Preeusternum. Part of the basisternum in some tenebrionoid larvae (Coleoptera).

Pregomphi. Paired processes immediately in front of the urogomphi in some beetle families, like Nitidulidae (Coleoptera).

Pregula. The area immediately anterior to a forked gular suture, lying between it and the labium.

Prehensile. Adapted for wrapping around.

Prehypopharynx (-ynges). A lobe or columnar structure lying between the hypopharynx and the ligula (Coleoptera).

Prelabial Sclerite. A sclerite below the orifice of the silk press within the labial sclerite; it may be attached to the dorsal arms and/or ventral portion of the labial sclerite or may be free (Hymenoptera: parasitica).

Prementum. The distal part of the labium, bearing the palps, glossae (and sometimes paraglossae), or the ligula.

Premolar Tooth. A sclerotized tooth at the distal end of the mola.

Prenotal Fold. *See* Prescutum.

Prepupa. The non-feeding portion of the last instar preceding the pupa in which the larva is shorter, thicker, and relatively inactive. The waxy-white third instar flea larva that has expelled all contents from the alimentary canal; the prepupal period is post-defecation to pupal molt.

Prescutum. The anteriormost subdivision of a thoracic or abdominal tergum.

Presternum. The anterior part of the prothoracic basisternum, often set off as a distinct sclerite in many beetle larvae (*see* Intersternite).

Pretarsus. The apical segment of a 6-segmented larval leg, consisting of a single claw or ungulus, or paired claws.

Preventrite. A sclerite in front of the main sternal plate or ventrite on the abdomen of some carabid larvae (Coleoptera).

Primary Sensorium (-ia) (=Primary Rhinaria). Plate-like sense organs, one located subapically on the penultimate antennal segment and one on the last segment just proximal of the unguis (Aphididae).

Primary Setae. Those setae with definite locations and names, and found on all instars.

Proboscis. Tubelike or beaklike mouthparts.

Procercus. A paired process on each side of the 8th tergum, just above the respiratory chamber in larvae of some Hydrophilidae (Coleoptera).

Prodorsal Fold. *See* Prescutum.

Produced. Extended or projecting.

Prognathous Head. One which is horizontal, so that the mouthparts are directed forward.

Prolarva. In Cantharidae, a newly hatched larva, soft-bodied, with legs incompletely developed; usually remaining in congregations (Coleoptera).

Prolegs. Fleshy, unjointed abdominal legs with or without crochets; also termed larvapods or false legs.

Proplegmatium (-ia). Part of epipharynx in Scarabaeoidea; paired space with pliate surface inside and usually somewhat in front of plegmatium (Coleoptera).

Propneustic. With only the first (anterior) pair of spiracles functional.

Propupa. The first of two or three quiescent instars in Thysanoptera; with or without wingpads, and without functional mouthparts.

Prosoma. The anterior region of the body—the fused head and thorax, bearing the legs, usually the cephalothorax (Arachnida).

Prosternum. Sternum or main ventral plate of the prothorax.

Prostheca. A hyaline or partly sclerotized, rigid or flexible, simple or complex lobe or process, or a group of hairs or specialized setae, arising from the mesal surface of the mandible just distad of the mola (Coleoptera).

Protergum. Tergum or main dorsal plate of the prothorax.

Protonymph. A second instar mite.

Protracted. Extended forward; with the head not withdrawn into the prothorax. Mouthparts—arising near the anterior margin of a prognathous head (Coleoptera).

Protracted Head. One which is not withdrawn into the prothorax.

Protracted Ventral Mouthparts. Those which are attached to the head capsule at about the same level as the ventral mandibular articulations, so that the hypostomata are more or less transverse.

Protuberance. Any projection with or without setae.

Proventriculus. Modified portion of the foregut between the oesophagus or crop and the midgut.

Proxicardo. That part of the cardo mesad of the dividing ridge.

Proximad. Toward the point of attachment.

Proximal. Nearest to the point of attachment.

Proxistipes. Proximal or basal portion of a divided stipes.

Pseudocerarii. A pair of closely associated spines on the body margin near the anal opening (Coccoidea).

Pseudocercus. *See* Urogomphus.

Pseudomola. A modified area on the mesal edge of the mandible, which is subapical or subbasal in position and which is analagous to, but not homologous with, a true basal mola.

Pseudopod (-podia). A soft, footlike appendage (common on many Diptera larvae).

Pseudopupa. *See* Coarctate Larva.

Pternotorma (-ae). Curving, stout process at outer end of laetotorma, and sometimes dexiotorma in Scarabaeoidea (Coleoptera).

Pterygote. Winged.

Pubescent. With soft, short, fine hairs.

Pubescent Seta. A seta covered with minute, fixed processes and appearing "hairy".

Pulvilliform. Lobelike or padlike; shaped like a pulvillus; e.g., pulvilliform empodium.

Pulvillus (-li). Lobe or pad beneath the base of a tarsal claw or tarsungulus.

Puncture. Depression or pit.

Pupa. The relatively inactive, transformation stage between the larva and adult of holometabolous insects. The second (or second and third) quiescent instar(s) in Thysanoptera, usually with wingpads but without functional mouthparts. In the Aleyrodidae (Homoptera), the name given to the 4th instar, the stage usually used in identification.

Pupa, Complete. This form is present in the higher Ditrysia (Lepidoptera) and has 2 or fewer movable abdominal segments in both males and females, no cocooncutter on the head, and usually no caudally directed spines on the abdomen. These pupae almost never protrude from the cocoon when the adult emerges (*compare* Pupa, Incomplete).

Pupa I. The first of two wing-bearing quiescent instars in tubuliferous Thysanoptera.

Pupa II. The second of two wing-bearing quiescent instars in tubuliferous Thysanoptera.

Pupa, Incomplete. This form is typical of the more primitive Lepidoptera with obtect pupae, and is characterized by three or more freely movable abdominal segments in the male and two or more in the female, usually with a ridge or some other kind of cocooncutter on the head, and nearly always with some caudally directed spines on the dorsum of the abdomen. They often protrude from the pupal cell when the adult emerges (*compare* Pupa, Complete).

Pupal Horns. Hornlike respiratory processes of some Diptera pupae (Culicidae, Syrphidae).

Puparium (-ia). The hardened, barrel-like last larval skin in which the pupa is formed (higher Diptera).

Pupiform Larva. In the Phylloxeridae (Homoptera), the sessile, non-feeding, pupalike immatures of the sexuales.

Pustules. Microscopic, blisterlike plaques borne by the thoracic and abdominal cuticle of larval Thysanoptera in the families Heterothripidae and Thripidae.

Pygidial Lobes. Sclerotized dermal protrusions located on the posterior margin of the pygidium (Coccoidea).

Pygidium (-ia). The 9th tergum, especially when this is sclerotized; also a sclerotized area, sometimes a concave disc, which is set off from the rest of the 9th tergum (Coleoptera).

Pygopod. A foot-like structure or one of a pair of such structures formed from the 10th sternum or from the entire 10th segment.

Pygopodium (-ia). *See* Pygopod.

Pyriform. Pear-shaped.

Quadrilocular Pore. Flat, circular structure that is divided into 4 parts (Coccoidea).

Quiescent Instars. The two or three non-moving, nonfeeding instars of Thysanoptera. *See* Propupa, Pupa, Pupa I, Pupa II.

Quinquelocular Pores. Flat, circular structures that are divided into 5 parts (Coccoidea).

Ramose (setae). Setae with branches, usually arising from the base.

Raptorial. Adapted for seizing prey (usually the forelegs).

Raster. Complex of bare areas, hairs, and spines on ventral surface of last abdominal segment, in front of anus, in Scarabaeoidea; comprised of septula, palidium, teges, tegillum (in some groups), and campus (Coleoptera).

Recurved. Curved backward.

Reniform. Kidney-shaped.

Respiratory Chamber. In most larvae of Hydrophilidae, a chamber formed from the 8th or 9th terga and enclosing a pair of large spiracles (Coleoptera).

Respiratory Plate. *See* Sieve Plate.

Respiratory Siphon. A breathing tube of aquatic larvae.

Respiratory Spine. The sharply pointed caudal spiracles of some Diptera larvae.

Respiratory Trumpet. Prothoracic respiratory protuberance, usually paired, of the pupa of some aquatic Diptera.

Retinaculum. A heavily sclerotized tooth or projection on the mesal or incisor edge of the mandible; usually found in mandibles without a basal mola (Coleoptera).

Retracted Head. One which is withdrawn into the prothorax. Mouthparts—arising near the rear of a prognathous head.

Retracted Ventral Mouthparts. Those which are attached posteriorly to the ventral mandibular articulations, so that the hypostomata are more or less parallel or obliquely longitudinal.

Retractile. Capable of being drawn in.

Retrorse. Bent backward or downward.

Rima (-ae). A marginal supporting sclerotization of the longer posterior spiracular openings of Diptera larvae.

Riparian. Living adjacent to water, usually a stream.

Rostrum. The snout (adult Coleoptera); the beak (Hemiptera); and mouthparts (Homoptera).

Rudimentary. Reduced, barely developed.

Rugose. Roughly wrinkled.

Rugosites. Small or minute elevations that are close together.

Saccoid. Swollen; sac-like.

Salivary Opening (slit, or lips). Opening of the silk or salivary glands on the extremity of the labium.

Saltatorial. Adapted for leaping or jumping.

Saprophagous. Feeding on dead or decaying materials.

Scale. A highly modified seta, which is somewhat expanded and usually flattened. The outer covering (test) of many Coccoidea.

Scansorial Wart. *See* Ampulla.

Scarabaeiform, Scarabaeoid. U- or C-shaped, with a distinct head and thoracic legs, and resembling a white grub of the Scarabaeidae.

Scissorial Area. *See* Incisor Area.

Scissorial Tooth. Large, sclerotized process on the mesal or incisor edge of the mandible of some grub-like larvae (Dascillidae and Scarabaeoidea); like a retinaculum but usually blunter, more obliquely oriented, and accompanied by a basal mola.

Sclerome. Any heavily sclerotized structure which is not plate-like.

Sclerotized. Hardened and tanned, so that it is yellow to black in color.

Scolus (-li). A branched projection of the body wall with each branch bearing a stout seta at its distal end.

Scrobe. A groove or depression in the body for reception of an appendage.

Scutellum. The third fold or sclerite of a transversely divided mesothoracic, metathoracic, or abdominal tergum.

Scutum (=postdorsal or postnotal fold). The second fold or sclerite of a transversely divided mesothoracic, metathoracic, or abdominal tergum.

Second Condyle. *See* Accessory Ventral Process.

Second Grub. The fourth phase of hypermetamorphic larval development in Meloidae, in which, following the coarctate phase, the larva again becomes grublike.

Secondary Chamber. *See* Accessory Chamber.

Secondary Sensoria (=Secondary Rhinaria). Plate-like sensory organs that may be present on antennal segments 3, 4, 5, and/or 6 of adult Aphididae.

Secondary Setae. Those setae with indefinite locations and variable in number, and usually not present on the first instar.

Seminivorous. Feeding upon seeds of grasses.

Semivoltine. Taking two years for one generation.

Sense Cone. A minute sensory cone or peg. Slender, semitransparent sensilla borne by the antennae of Thysanoptera; they are probably thin-walled multiporous chemoreceptors and in larvae have a common number and position throughout the order, though differing between taxa in shape and relative length.

Sensilla Styloconicum. A cone-like or peg-like sense organ.

Sensillum (-la). A sense organ.

Sensoria. Fleshy, peglike sensory organs.

Sensoria On Trochanter. Two small porelike structures usually arranged linearly on each surface of trochanter (Coccoidea).

Sensorial Appendage. *See* Sensorium.

Sensorium (-ia). A hyaline, sensory vesicle on the preapical antennal segment (=Supplemental Process or Supplemental Joint).

Sensory Appendix. *See* Sensorium.

Sentus (-ii). An elongate, spinose, non-branching projection of the body wall which bears a few short setae directly on its trunk (Coleoptera).

Septula (-ae). Part of raster in Scarabaeoidea; bare area between a transverse, single-rowed palidium and lower anal lip, or between 2 longitudinal or oblique rows of pali (Coleoptera).

Serial. Referring to the arrangement of the bases of the crochets.

Serrate. Sawlike.

Sessile. Attached, incapable of moving from place to place.

Seta (-ae). A hair-like or bristle-like projection of the body wall which is articulated in a socket.

Setai Map. A flat, diagrammatic drawing of the arrangement of the setae on one side of a larva (usually the left with the head to the left).

Setiferous, Setaceous, Setose. Bearing setae.

Sexupara. Female that produces both sexual males and females (Aphidoidea).

Shagreened. Roughened with numerous minute bumps.

Shield. A sclerotized plate covering part of the dorsum of a segment.

Sieve Plate. The perforated plate covering the opening of a cribriform spiracle.

Sigmoid. S-shaped.

Silk Press. Salivary opening associated with silk glands.

Simple Eyes. *See* Stemmata and Ocelli.

Simple Pore. Flat, round structure, without divisions (Coccoidea).

Sinuous (sinuate). Curved in and out; winding.

Sinus. A cavity or depression.

Siphon. A tubular external structure, usually a breathing tube.

Spatula (=breastbone). Ventral, variously shaped prothoracic sclerite characteristic of many Cedidomyiidae (Diptera).

Spatulate (spathulate). Shaped like a spatula, spoon-shaped; flat, rounded and broad at tip, narrowed at the base.

Spermatozooid Ducts. Elongate ducts that narrow from broad apex to slender dermal orifice (Coccoidea).

Spiciseta. A barbed seta found on larvae of Dermestidae (Coleoptera).

Spicule. A minute, pointed spine or process.

Spindlelike. Pointed at both ends and swollen in the middle (fusiform).

Spine. A fixed, multicellular, thorn-like process arising from the cuticle; *compare* Spur and Seta.

Spinneret. A structure from which silk is spun, usually located on the labium of larvae, and often protruding as lips or elongate process(es).

Spinula (-ae). Minute spines (microspines) (*See* Spicule).

Spinule. *See* Spinula and Spicule.

Spinulose. Bearing little spines or spinules.

Spiracular Area. Area round the spiracle when this is not a distinct sclerite.

Spiracular Atrium. Sclerotized area surrounding orifice of tracheal openings on thorax (Coccoidea).

Spiracular Chamber. *See* Atrium, Respiratory Chamber.

Spiracular Closing Apparatus. A cuticular band which surrounds the base of the spiracular atrium at the point of attachment of the trachea; it is attached to a musculated apodeme, and functions to close the spiracle.

Spiracular Disc. In amphipneustic Diptera larvae, the flattened posterior or posterior-dorsal area bearing the spiracles.

Spiracular Furrow. Depressed area on derm leading from spiracular atrium to body margin, usually lined with wax pores (Coccoidea).

Spiracular Gill. A gill associated with a spiracle, formed from either an extension of the spiracular opening or a plastron around the opening.

Spiracular Line. The line coinciding with or near or between the spiracles.

Spiracular Lobes. Lobes immediately below or immediately posterior to the lobe bearing the spiracle.

Spiracular Plate. In the posterior spiracles of Diptera larvae, the area usually defined by the peritreme (sometimes weak or absent), bearing the ecdysial scar and spiracular openings.

Spiracular Pore Band. Row of wax pores leading from spiracular opening to body margin (Coccoidea).

Spiracular Sclerite. A laterotergite containing the spiracle (Coleoptera).

Spiracular Setae. Setae on body margin adjacent to spiracle that usually are differentiated from other marginal setae (Coccoidea).

Spiracular Tube. A tubular structure containing a spiracle, with the opening at its apex.

Spiracles, Anterior and Posterior (Diptera). Anterior: the apparent prothoracic spiracle; Posterior: the most caudal abdominal spiracle, usually on the 8th segment, but apparently on the 9th in a few Bibionidae.

Spur. A movable spine.

Spurious. False.

Squama (-ae). *See* Scale.

Stadium (-ia). The time period between molts.

Stellate. Star-shaped.

Stem Mother (=Fundatrix). Female that develops from the overwintering egg (Aphidoidea).

Stemapoda. Elongated anal prolegs of some notodontid larvae.

Stemma (-ata). A simple eye of holometabolous larvae. *See* Ocellus.

Sternal. Pertaining to the sternum.

Sternal Apophysis (-ses). One of a pair of apophyses on the thoracic sterna of various insects. *See* Apophysis.

Sternal Pit. One of a pair of pits marking the invagination of a sternal apophysis.

Sternal Plate. *See* Sternum.

Sternellum. That part of the thoracic eusternum posterior to the paired sternal pits, sternacostal suture, or in the absence of either, an arbitrary line joining the centers of the two coxal cavities.

Sternite. One of two or more sclerotized plates on the venter of a thoracic or abdominal segment.

Sternum (-a). The main sclerite or area of the venter of a segment.

Stigmatic Atrium. *See* Respiratory Chamber.

Stigmatic Scar. *See* Ecdysial Scar.

Stilettiform. Shaped like a stiletto or style.

Stipes (-pites). The part of a maxilla distad to the cardo, bearing the palp, galea, and lacinia.

Stipital Sclerite. A sclerotized bar extending from the cardo along the posterior edge of the maxilla on either side of the cephalic structure; homologous with the sclerotized part of the stipes in primitive insects (Hymenoptera; parasitica).

Stipites Labii. *See* Prementum.

Stria (-ae). A fine, impressed line or groove, usually longitudinal.

Striate. With grooves (and ridges).

Stridulate. To make a noise by rubbing 2 structures together.

Stridulatory Area. A stridulitrum (located on the dorsal surface of the mandibular base in some Scarabaeoidea (Coleoptera)).

Stridulatory File. *See* Stridulitrum.

Stridulatory Organ. A structure producing sound by moving one sclerotized part, the plectrum, over another, the stridulitrum.

Stridulatory Plate. *See* Stridulitrum.

Stridulatory Teeth. Teeth located near the base of the dorsal (ental) surface of the stipes in many scarabaeoid larvae (Coleoptera).

Stridulitrum (-tra). The fixed part of a stridulatory organ; usually a finely striate plate or file, sometimes a patch of teeth or tubercles.

Struma (-ae). A mound-like projection of the body wall bearing a few chalazae.

Strut Setae. Small setae at the base of the anal struts on the tenth abdominal segment of flea larvae.

Stygobiontic. Adapted for and living in underground water (streams, aquifers, caves, etc.)

Stylate. Stylelike.

Stylet. A long and narrow, needle-like structure in piercing-sucking mouthparts; usually a modified maxilla or mandible.

Styliform. Stylet-like.

Subanal. Below the anus.

Subcortical. Living under bark, usually in the rotten cambial tissue of dead trees.

Subcuticular Pigment. Pigment granules of the epidermis. In all instars of thysanopterans, and responsible, in part, for body color; because it varies intra-specifically, its use in identification is limited.

Subdorsal Line. A line below the dorsal or addorsal line (if present).

Subequal. Nearly equal.

Submarginal Stria. *See* Proplegmatium.

Submentum. The basal part of a three-part labium, attached to, and sometimes fused with, the gula.

Submolar Lobe. A fleshy, setose lobe or hyaline area at the basal edge of the mandibular mola.

Subparallel. Nearly parallel.

Subprimary Seta. Those setae having a definite position, but appearing in the second instar (not present in the first instar); otherwise similar to primary setae.

Subscutum. A lateral area between the scutum and scutellum of an abdominal tergum.

Subspiracular Line. A line below the spiracles.

Substrate. Any substance, like soil or wood, upon or in which an organism lives.

Subulate. Awl-shaped; linear or bulbous at the base and attenuated apically.

Subventral Line. A line just above the bases of the prolegs.

Sucker. *See* Sucking Disc.

Sucking Disc. A suction device formed from the anal region in some larvae (Coleoptera). In Diptera, ventral abdominal discs that maintain Blephariceridae and *Maruina* larvae (Psychodidae) in swift water.

Sulcus (-i). A groove or furrow.

Superlingua. *See* Maxillula.

Supplemental Process. *See* Sensorium.

Supplementary Joint. *See* Sensorium.

Supra-Anal Plate. Sclerotized area laterad of anal opening but not forming an operculum (Coccoidea). A plate dorsad to the anus.

Supraspiracular Line. Above the spiracles.

Suranal (=supra-anal). Above the anus.

Suranal Plate (=anal plate, anal shield). The area on the dorsum of the last abdominal segment, frequently rather heavily sclerotized.

Suranal Process (=postcornu). A sclerotized projection on the meson of the suranal plate.

Surpedal Lobes. Lobes dorsal to prolegs and ventral to spiracular lobes.

Symplesiomorphy, -phic. The sharing by 2 or more taxa of the more primitive of a homologous pair or sequence of character states. *Compare* Synapomorphy, and *see* Plesiomorphy and Apomorphy.

Synapomorphy, -phic, -phous. The sharing by two or more taxa of the more recent (apomorphic) of a sequence of homologus characters. *See* Apomorphy and Pleisiomorphy.

Tactile Papilla. *See* Sensorium.

Talus (-li). The heavily sclerotized area on the head capsule around the base of each dorsal mandibular articulation.

Tanning. *See* Sclerotized.

Tarsungulus (-li). Fused tarsus and tarsal claw in many larval Coleoptera.

Tarsus. The fifth segment of the thoracic leg which usually is 1–5 segmented and usually bears one or two claws. If holometabolous larvae have tarsal segments, most have only one segment.

Teges (-gites). Part of raster in Scarabaeoidea; continuous patch of hooked or straight, short to minute setae (Coleoptera).

Tegillum (-la). Part of raster in Scarabaeoidea; paired patch of hooked or straight setae on each side of paired and well-developed palidia (Coleoptera).

Teneral. Refers to pale, soft-bodied, recently molted immatures or adults.

Tentorial Bridge. *See* Corporotentorium.

Tentorium (-ia). The major portion of the cephalic endoskeleton, consisting usually of paired, subparallel, posterior metatentoria and anterior pretentoria, a corporotentorium joining the two, and paired, dorsally projecting supratentoria.

Tergal. Pertaining to the tergum.

Tergal Plate. *See* Tergite.

Tergite. One of two or more sclerotized plates on the dorsum of a thoracic or abdominal tergum.

Tergosternal Fold. The junction of an abdominal tergum and sternum, when the abdominal pleuron is absent.

Terminal Filament. Long and slender apical prolongation of last abdominal segment.

Test. Waxy structure that covers insect body (Coccoidea).

Thamnophilous. Living in thickets or dense shrubbery.

Thelytoky (thelytokous). Female-producing parthenogenesis (Hymenoptera).

Thigmotactic. Contact-loving.

Thysanuriform Larva. *See* Campodeiform Larva.

Tibia. The 4th segment of the thoracic leg.

Tibiotarsus. Term applied to the tibia of Polyphaga (Coleoptera), when it is assumed that it represents a fusion of the tibia and tarsus.

Torma (-ae). One of a pair of dark scleromes originating at each end of the clypeolabral suture and extending mesally and/or posteriorly, sometimes asymmetrical, or complex with accessory structures.

Tormogen Cell. An epidermal cell which forms the setal membrane or socket (ring base).

Trabecula (-ae). Strengthening cross ridges or serrations in the posterior spiracular rimae of Diptera larvae.

Tracheal Gill. A gill containing numerous tracheae or tracheoles.

Tracheated. Supplied with tracheae.

Tracheole. A small trachea.

Transverse Fold. *See* Sternellum.

Trichobothria. Sensory hairs, especially very long ones.

Tricuspidate. Having three cusps or teeth.

Trident. Having 3 teeth, processes or points.

Trifid. Divided into three parts.

Trilocular Pores. Flat, round or triangular structures on derm that are divided into three parts (Coccoidea).

Triquetral. Triangular in section, with three flat sides.

Triungulin. The active hypermetamorphic first instar or primary larva of Meloidae, so named because of paired setae on the tarsungulus, which suggest the presence of "3 claws". The term has also been applied to primary larvae of Rhipiphoridae and Strepsiptera, although the term triungulinid is preferable for these other groups that do not have "3 claws". *See* also Planidium.

Triungulinid. *See* Triungulin.

Trochanter. The 2nd segment of a thoracic leg.

Trochantin. A very narrow sclerite sometimes located at the base of the coxa.

Truncate. Squared off at the end.

Tubercle. A small mound-like projection of the body wall, frequently bearing a seta; in most groups of beetles, any small projection of the cuticle, sometimes longer than wide, and often bearing a seta at the apex; in Coccinellidae, synonymous with verruca.

Tuft. A group of setae.

Tylus (-li). Part of epipharynx in Scarabaeoidea; sclerome covering or partly covering fused elements of epizygum, corypha and haptomerum (Coleoptera); in Hemiptera, the single, median lobe on the anterior margin of the head, margined by the lateral juga.

Tympanum (-na). A vibrating membrane; an eardrum.

Ultimate Antennal Segment, Basal Portion. In the Aphididae (Homoptera), the proximal part of the terminal segment, including the primary sensorium.

Uncus (-ci). A tooth-like or hook-like process, or occasionally a setiferous process at the inner edge of the malar apex or at the apex of the lacinia in various beetle larvae.

UNDULATE SPIRACLE. A type of biforous spiracle in which the paired openings are undulate and located at the end of a spiracular tube; found only in Chelonariidae (Coleoptera).

UNGUIS (=PROCESSUS TERMINALIS). Slender terminal part of the last antennal segment distal to the primary sensorium (Aphididae).

UNGULUS (-LI). The 6th, claw-like segment, or one of two claws, in the thoracic leg of most Adephaga and Archostemata (Coleoptera).

UNICAMERAL SPIRACLE. *See* ANNULAR-BIFOROUS SPIRACLE.

UNIFOROUS SPIRACLE. A spiracle having one opening; *see* ANNULAR SPIRACLE.

UNIVOLTINE. Having one generation per year.

UROGOMPHAL PLATE. A sclerotized, hinged plate formed from the 9th tergum, which articulates with the posterior edge of tergum 8.

UROGOMPHUS (-HI). A sclerotized, paired, dorsal process, which projects from the posterior edge of the 9th tergum and may be either fixed or articulated at the base and segmented or not (Coleoptera and others). *Compare* CERCUS.

URTICATING HAIRS. Hairs or setae connected to poison glands that cause irritation; barbed hairs also may cause irritation without poison.

VASIFORM ORIFICE. The anal structure that contains the lingula and operculum (Aleyrodidae).

VENTER. The entire under surface of the body.

VENTRAL CONDYLE. The rounded process on the ventral edge of the mandible which articulates with the head.

VENTRAL HEAD CLOSURE. The area just behind the labium, especially when it is sclerotized. *See* GULAR REGION, and HYPOSTOMA (Coleoptera).

VENTRAL PROLEGS. All prolegs on any abdominal segment except the last, which are the anal prolegs.

VENTROLATERAL SETAE. A few large setae anterior to the anal struts on the tenth abdominal segment of flea larvae.

VERMIFORM. Worm-like.

VERRUCA (VERRUCAE). A somewhat elevated area of the cuticle, bearing setae pointing in many directions like a pincushion (Lepidoptera); for Coleoptera, *See* TUBERCLE.

VERRICULE. A dense tuft of upright, parallel setae.

VERTEX. The "top" of the head.

VERTICAL FURROW. Furrow on each side of head near dorsal aspect.

VERTICAL TRIANGLE. *See* EPICRANIAL NOTCH.

VESTIBULUM (VESTIBULA). An external genital cavity formed above the seventh abdominal sternum, when the latter extends beyond the eighth.

VESTIGIAL. Weakly developed or degenerate; only a remnant left.

VESTITURE. The general surface covering such as setae, scales, hairs.

VITTA (VITTAE). A broad stripe.

VITTATE. Striped.

VIVIPAROUS. Bearing living young.

WIREWORM. A larva that is slender, heavily sclerotized, bears thoracic legs, lacks prolegs, and is nearly naked; a phytophagous click beetle larva.

XYLOPHAGOUS. Feeding on wood.

XYLOPHILOUS. Loving or attracted to wood.

ZYGUM (-GA). Part of epipharynx in Scarabaeoidea; sclerome forming anterior margin of haptomerum (Coleoptera).

Host Plant and Non-animal Substrate Index

Animals and most products of animal origin are in the main index. Very general substrates such as hardwoods, softwoods, leaves, bark, herbs, flowers, roots, wood, logs, detritus, weeds, fruits, seeds, vegetables, etc., are not entered.

Index

Corrections, Additions and Modifications for Volume 1

COVER and two inside title pages: Add: "Volume 1".

Credits: Line 3: Replace: "B. McDaniel, How to Know the Mites and Ticks. Copyright 1979" with "B. J. Kaston, How to Know the Spiders, Copyright 1978".

pages 2, 4, 6, 10, 19, 27 and in the "Index": change "Raphidiodea" to "Raphidioptera".

page 97: Right column of "Classification:" Drop the first "s" from Cordulesgastroidea and Cordulesgastridae.

page 290: In figure 26.1, switch the labels "C1" and "C2."

page 309: In figure 25, label the lower, right seta "S2".

page 314: In figure 53, change SD to SD1.

page 325: In couplet 115, "Plutellidae" should be in italics.

page 326: In couplet 125′, add "(rarely biordinal)".

page 330: In couplet 148, insert "or touching" after "spaced" and at the end add "; examine the bases carefully".

In couplet 150′, change "—a middorsal tuft" to "—a long middorsal tuft or verricule"

page 333: In line 7 change "neutral" to neural.

page 338: In couplet 204′ add "(rarely biordinal)" after "mesoseries".

page 340: Change couplet 222′ as follows. Delete "(part) ***Cosmopterigidae*** (p. 391)", and substitute "222a". Add couplet 222a.

222a(222). A1–8 with setae L1 and L2 (below the spiracle) arranged in a vertical row (one above the other); spiracles large, with wide peritremes; spinules conspicuous on the integument .... (*Petrova* spp.) ***Tortricidae*** (p. 419)

222a′. A1–8 with setae L1 and L2 (below the spiracle) arranged in a horizontal row (one in front of the other); spiracles smaller, with narrow peritremes; spinules not conspicuous .... (part) ***Cosmopterigidae*** (p. 391)

page 348: Line 7 first column, change "L2" to "L3".

page 349: Change in legends the following: Change *Hepialis* to *Hepialus* (3 places); **Figure 26.31t,** change "sp." to "*californicus* Boisduval"; and change "f.g.1" to "f.g.l". **Figure 26.31u-x,** Change "*quadriguttatus* (Grote)?" to "*argenteomaculatus* (Harris)"; change "f.g.1" to "f.g.l"; and change "n,o,p and q" to "u,v,w, and x"

page 381: In figure 26.52b change "SV2" to SV3, and "SV3" to SV2.

page 407: In "Comments", change "fig. 26.98" to fig. 26.99, and change "fig. 26.99" to fig. 26.98.

page 421: Couplet 3—delete "(p. 421)".

page 447: Add "26.160" before "a" for figure in upper right corner.

page 451: In legend 26.167 change "a,b" to a–c.

page 456: In "Comments" for Limacodidae—*Phobetron* (not *Phoebetron*).

page 458: Figure 26.182 caption, *Euclea* (not *Eudea*).

page 504: For figure 26.282, write "head" at the right end.

page 513: In figure 26.300 change "vellda" to velleda.

page 598: In figure 27.1 shorten the line pointing to the maxilla by 3 mm (it currently points to the hidden mandible)

page 729: In the second column, eliminate line 7, "*ochrodesma*".

page 730: In the first column, add:
Anabasis, 486
ochrodesma, 486

page 740: Delete "258" after "Hydrophilidae".

page 745: In the second column, replace *Ochrodesma, Acrobasis* with *ochrodesma, Anabasis* (do not capitalize "*ochrodesma*").